Course Principles of Biology I and II
Course Number **BIO 111 and 112**
LINCOLN LAND COMMUNITY COLLEGE

Alaina's
DO
NOT
THROW AWAY
:)

create™

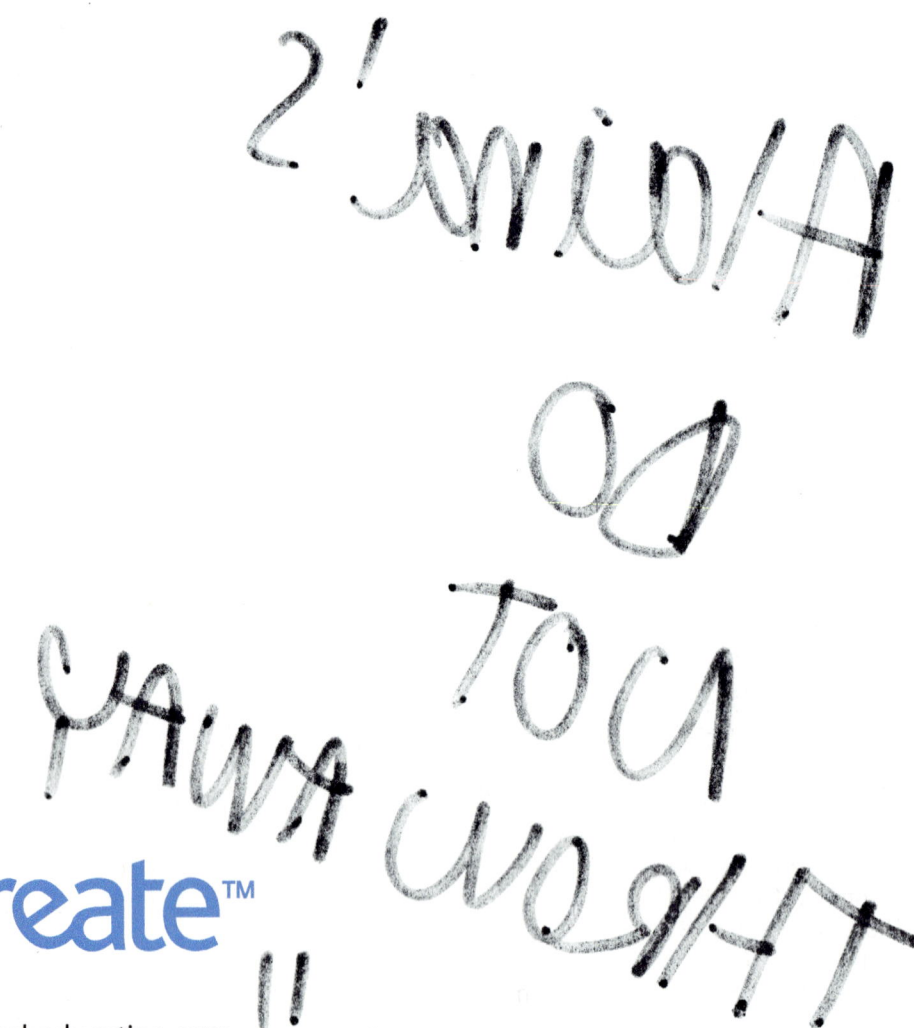

http://create.mheducation.com

Copyright 2017 by McGraw-Hill Education. All rights reserved. Printed in the United States of America. Except as permitted under the United States Copyright Act of 1976, no part of this publication may be reproduced or distributed in any form or by any means, or stored in a database or retrieval system, without prior written permission of the publisher.

This McGraw-Hill Create text may include materials submitted to McGraw-Hill for publication by the instructor of this course. The instructor is solely responsible for the editorial content of such materials. Instructors retain copyright of these additional materials.

ISBN-10: 1307088546   ISBN-13: 9781307088540

Contents

i. About the Authors 2
ii. Committed to Excellence 3
iii. Preparing Students for the Future 7
iv. Using Connect® and Biology, 11th edition 10

The Molecular Basis of Life 11

1. The Science of Biology 13
2. The Nature of Molecules and the Properties of Water 29
3. The Chemical Building Blocks of Life 45

Biology of the Cell 71

4. Cell Structure 73
5. Membranes 102
6. Energy and Metabolism 121
7. How Cells Harvest Energy 136
8. Photosynthesis 161
9. Cell Communication 182
10. How Cells Divide 200

Genetic and Molecular Biology 221

11. Sexual Reproduction and Meiosis 223
12. Patterns of Inheritance 237
13. Chromosomes, Mapping, and the Meiosis–Inheritance Connection 255
14. DNA: The Genetic Material 272
15. Genes and How They Work 294
16. Control of Gene Expression 320
17. Biotechnology 343
18. Genomics 369
19. Cellular Mechanisms of Development 391

Evolution 415

20. Genes Within Populations 417
21. The Evidence for Evolution 439
22. The Origin of Species 458
23. Systematics, Phylogenies, and Comparative Biology 478

24. Genome Evolution 498
25. Evolution of Development 515

Diversity of Life on Earth 531

26. The Origin and Diversity of Life 533
27. Viruses 547
28. Prokaryotes 564
29. Protists 585
30. Seedless Plants 608
31. Seed Plants 622
32. Fungi 639
33. Animal Diversity and the Evolution of Body Plans 658
34. Protostomes 680
35. Deuterostomes 712

Ecology and Behavior 753

36. Behavioral Biology 754
37. Ecology of Individuals and Populations 784
38. Community Ecology 807
39. Dynamics of Ecosystems 829
40. The Biosphere 852
41. Conservation Biology 879
  A. Appendix: Answer Key 903
  B. Glossary 943
  C. Credits 967
  D. Index 972

Online Supplements 1005

    Connect Online Access for Biology, 11th Edition 1006
    Access Code for the Online Supplements 1007

Credits

i. About the Authors: *Chapter from Biology, 11th Edition by Raven, Johnson, Mason, Losos, Singer, 2017* 2
ii. Committed to Excellence: *Chapter from Biology, 11th Edition by Raven, Johnson, Mason, Losos, Singer, 2017* 3
iii. Preparing Students for the Future: *Chapter from Biology, 11th Edition by Raven, Johnson, Mason, Losos, Singer, 2017* 7
iv. Using Connect® and Biology, 11th edition: *Chapter from Biology, 11th Edition by Raven, Johnson, Mason, Losos, Singer, 2017* 10

The Molecular Basis of Life 11

1. The Science of Biology: *Chapter 1 from Biology, 11th Edition by Raven, Johnson, Mason, Losos, Singer, 2017* 13
2. The Nature of Molecules and the Properties of Water: *Chapter 2 from Biology, 11th Edition by Raven, Johnson, Mason, Losos, Singer, 2017* 29
3. The Chemical Building Blocks of Life: *Chapter 3 from Biology, 11th Edition by Raven, Johnson, Mason, Losos, Singer, 2017* 45

Biology of the Cell 71

4. Cell Structure: *Chapter 4 from Biology, 11th Edition by Raven, Johnson, Mason, Losos, Singer, 2017* 73
5. Membranes: *Chapter 5 from Biology, 11th Edition by Raven, Johnson, Mason, Losos, Singer, 2017* 102
6. Energy and Metabolism: *Chapter 6 from Biology, 11th Edition by Raven, Johnson, Mason, Losos, Singer, 2017* 121
7. How Cells Harvest Energy: *Chapter 7 from Biology, 11th Edition by Raven, Johnson, Mason, Losos, Singer, 2017* 136
8. Photosynthesis: *Chapter 8 from Biology, 11th Edition by Raven, Johnson, Mason, Losos, Singer, 2017* 161
9. Cell Communication: *Chapter 9 from Biology, 11th Edition by Raven, Johnson, Mason, Losos, Singer, 2017* 182
10. How Cells Divide: *Chapter 10 from Biology, 11th Edition by Raven, Johnson, Mason, Losos, Singer, 2017* 200

Genetic and Molecular Biology 221

11. Sexual Reproduction and Meiosis: *Chapter 11 from Biology, 11th Edition by Raven, Johnson, Mason, Losos, Singer, 2017* 223
12. Patterns of Inheritance: *Chapter 12 from Biology, 11th Edition by Raven, Johnson, Mason, Losos, Singer, 2017* 237
13. Chromosomes, Mapping, and the Meiosis–Inheritance Connection: *Chapter 13 from Biology, 11th Edition by Raven, Johnson, Mason, Losos, Singer, 2017* 255

14. DNA: The Genetic Material: *Chapter 14 from Biology, 11th Edition by Raven, Johnson, Mason, Losos, Singer, 2017* 272
15. Genes and How They Work: *Chapter 15 from Biology, 11th Edition by Raven, Johnson, Mason, Losos, Singer, 2017* 294
16. Control of Gene Expression: *Chapter 16 from Biology, 11th Edition by Raven, Johnson, Mason, Losos, Singer, 2017* 320
17. Biotechnology: *Chapter 17 from Biology, 11th Edition by Raven, Johnson, Mason, Losos, Singer, 2017* 343
18. Genomics: *Chapter 18 from Biology, 11th Edition by Raven, Johnson, Mason, Losos, Singer, 2017* 369
19. Cellular Mechanisms of Development: *Chapter 19 from Biology, 11th Edition by Raven, Johnson, Mason, Losos, Singer, 2017* 391

Evolution 415

20. Genes Within Populations: *Chapter 20 from Biology, 11th Edition by Raven, Johnson, Mason, Losos, Singer, 2017* 417
21. The Evidence for Evolution: *Chapter 21 from Biology, 11th Edition by Raven, Johnson, Mason, Losos, Singer, 2017* 439
22. The Origin of Species: *Chapter 22 from Biology, 11th Edition by Raven, Johnson, Mason, Losos, Singer, 2017* 458
23. Systematics, Phylogenies, and Comparative Biology: *Chapter 23 from Biology, 11th Edition by Raven, Johnson, Mason, Losos, Singer, 2017* 478
24. Genome Evolution: *Chapter 24 from Biology, 11th Edition by Raven, Johnson, Mason, Losos, Singer, 2017* 498
25. Evolution of Development: *Chapter 25 from Biology, 11th Edition by Raven, Johnson, Mason, Losos, Singer, 2017* 515

Diversity of Life on Earth 531

26. The Origin and Diversity of Life: *Chapter 26 from Biology, 11th Edition by Raven, Johnson, Mason, Losos, Singer, 2017* 533
27. Viruses: *Chapter 27 from Biology, 11th Edition by Raven, Johnson, Mason, Losos, Singer, 2017* 547
28. Prokaryotes: *Chapter 28 from Biology, 11th Edition by Raven, Johnson, Mason, Losos, Singer, 2017* 564
29. Protists: *Chapter 29 from Biology, 11th Edition by Raven, Johnson, Mason, Losos, Singer, 2017* 585
30. Seedless Plants: *Chapter 30 from Biology, 11th Edition by Raven, Johnson, Mason, Losos, Singer, 2017* 608
31. Seed Plants: *Chapter 31 from Biology, 11th Edition by Raven, Johnson, Mason, Losos, Singer, 2017* 622
32. Fungi: *Chapter 32 from Biology, 11th Edition by Raven, Johnson, Mason, Losos, Singer, 2017* 639
33. Animal Diversity and the Evolution of Body Plans: *Chapter 33 from Biology, 11th Edition by Raven, Johnson, Mason, Losos, Singer, 2017* 658
34. Protostomes: *Chapter 34 from Biology, 11th Edition by Raven, Johnson, Mason, Losos, Singer, 2017* 680
35. Deuterostomes: *Chapter 35 from Biology, 11th Edition by Raven, Johnson, Mason, Losos, Singer, 2017* 712

Ecology and Behavior 753

36. Behavioral Biology: *Chapter 54 from Biology, 11th Edition by Raven, Johnson, Mason, Losos, Singer, 2017* 754
37. Ecology of Individuals and Populations: *Chapter 55 from Biology, 11th Edition by Raven, Johnson, Mason, Losos, Singer, 2017* 784
38. Community Ecology: *Chapter 56 from Biology, 11th Edition by Raven, Johnson, Mason, Losos, Singer, 2017* 807
39. Dynamics of Ecosystems: *Chapter 57 from Biology, 11th Edition by Raven, Johnson, Mason, Losos, Singer, 2017* 829
40. The Biosphere: *Chapter 58 from Biology, 11th Edition by Raven, Johnson, Mason, Losos, Singer, 2017* 852

41. Conservation Biology: *Chapter 59 from Biology, 11th Edition by Raven, Johnson, Mason, Losos, Singer, 2017* 879
A. Appendix: Answer Key: *Chapter A from Biology, 11th Edition by Raven, Johnson, Mason, Losos, Singer, 2017* 903
B. Glossary: *Chapter from Biology, 11th Edition by Raven, Johnson, Mason, Losos, Singer, 2017* 943
C. Credits: *Chapter from Biology, 11th Edition by Raven, Johnson, Mason, Losos, Singer, 2017* 967
D. Index: *Chapter from Biology, 11th Edition by Raven, Johnson, Mason, Losos, Singer, 2017* 972

Online Supplements 1005

Connect Online Access for Biology, 11th Edition: *Media by Raven, Johnson, Mason, Losos, Singer* 1006

Access Code for the Online Supplements 1007

About the Authors

Pictured left to right: Susan Rundell Singer, Jonathan Losos, Kenneth Mason

Kenneth Mason is currently associated with the University of Iowa, Department of Biology. His academic positions, as a teacher and researcher, include the faculty of the University of Kansas, where he designed and established the genetics lab, and taught and published on the genetics of pigmentation in amphibians. At Purdue University, he successfully developed and grew large introductory biology courses and collaborated with other faculty in an innovative biology, chemistry, and physics course supported by the National Science Foundation. At the University of Iowa, where his wife served as president of the university, he taught introductory biology and human genetics for eight years. His honor society memberships include Phi Sigma, Alpha Lambda Delta, and, by vote of Purdue pharmacy students, Phi Eta Sigma Freshman Honors Society.

Jonathan Losos is the Monique and Philip Lehner Professor for the Study of Latin America in the Department of Organismic and Evolutionary Biology and curator of herpetology at the Museum of Comparative Zoology at Harvard University. Losos's research has focused on studying patterns of adaptive radiation and evolutionary diversification in lizards. He is the recipient of several awards, including the prestigious Theodosius Dobzhansky and David Starr Jordan Prizes, the Edward Osborne Wilson Naturalist Award, and the Daniel Giraud Elliot Medal from the National Academy of Sciences. Losos has published more than 150 scientific articles.

Susan Rundell Singer is the Laurence McKinley Gould Professor of Natural Sciences in the Department of Biology at Carleton College in Northfield, Minnesota, where she has taught introductory biology, plant biology, genetics, and plant development for nearly 30 years. Her research focuses on the development and evolution of flowering plants and genomics learning. Singer has authored numerous scientific publications on plant development and coauthored education reports including *Vision and Change* and "America's Lab Report." A fellow of the American Association for the Advancement of Science (AAAS), she has received the American Society of Plant Biology's Excellence in Teaching Award and the Botanical Society's Bessey Award, served on the National Academies Board on Science Education, and chaired several National Research Council study committees including the committee that produced *Discipline-Based Education Research*.

Lead Digital Author

Ian Quitadamo is a Professor with a dual appointment in Biological Sciences and Science Education at Central Washington University in Ellensburg, WA. He teaches introductory and majors biology courses and cell biology, genetics, and biotechnology as well as science teaching methods courses for future science teachers and interdisciplinary content courses in alternative energy and sustainability. Dr. Quitadamo was educated at Washington State University and holds a bachelor's degree in biology, master's degree in genetics and cell biology, and an interdisciplinary Ph.D. in science, education, and technology. Previously a researcher of tumor angiogenesis, he now investigates the behavioral and neurocognitive basis of critical thinking and has published numerous studies of factors that improve student critical-thinking performance. He has received the Crystal Apple award for teaching excellence and led multiple initiatives in critical thinking and assessment. He is active nationally in helping transform university faculty practices. He is a coauthor of *Biology*, 11th ed., by Mader and Windelspecht (2013), and is the lead digital author for *Biology*, 3rd and 4th ed., by Brooker (2014 and 2017), *Biology*, 10th ed., by Raven (2014), *Understanding Biology* by Mason (2015), and *Principles of Biology* by Brooker (2015), all published by McGraw-Hill. For fun, Dr. Quitadamo practices Kyokushin full contact karate and is a 5th degree blackbelt.

Committed to Excellence

With the new 11th edition, Raven and Johnson's *Biology* continues the momentum built over the last three editions. This edition provides an unmatched comprehensive text fully integrated with a continually evolving, state-of-the-art digital environment. We have used this digital environment in the revision of *Biology*. The McGraw-Hill SmartBook© for the 10th edition provided data on student responses, and thus identify material that students find difficult. This "heat-mapping" technology is unique in the industry, and allows us to direct editing to difficult areas, or problem areas for students. The text continues to be a leader with an organization that emphasizes important biological concepts, while keeping the student engaged with learning outcomes that allow assessment of progress in understanding these concepts. An inquiry-based approach with robust, adaptive tools for discovery and assessment in both text and digital resources provides the intellectual challenge needed to promote student critical thinking and ensure academic success. A major strength of both text and digital resources is assessment across multiple levels of Bloom's taxonomy that develops critical-thinking and problem-solving skills in addition to comprehensive factual knowledge. McGraw-Hill's Connect® platform offers a powerful suite of online tools that are linked to the text and now include new quantitative assessment tools. The adaptive learning system helps students learn faster, study efficiently, and retain more knowledge of key concepts.

The 11th edition continues our tradition of providing the student with clear learning paths that emphasize data analysis and quantitative reasoning. Additional embedded eBook resources link to asides that delve more deeply into quantitative aspects.

As a team, we continually strive to improve the text by integrating the latest cognitive and best practices research with methods that are known to positively affect learning. We have multiple features that are focused on scientific inquiry, including an increased quantitative emphasis in the Scientific Thinking figures. We continue to use the concise, accessible, and engaging writing style of past editions while maintaining the clear emphasis on evolution and scientific inquiry that have made this a leading textbook of choice for majors biology students. Our emphasis on evolution combined with integrated cell and molecular biology and genomics offers our readers a student-friendly text that is modern and well balanced.

The 11th edition continues to employ the aesthetically stunning art program that the Raven and Johnson *Biology* text is known for. Complex topics are represented clearly and succinctly, helping students to build the mental models needed to understanding biology.

Insights into the diversity of life that are provided by molecular tools have led to a continued updating of these topics in the 11th edition. The diversity unit reflects the most current research on eukaryotic phylogenies, blending molecular, morphological, and development viewpoints. The biotechnology and genomics chapters have been completely revised to reflect changes in these fast-moving areas of modern biology. These are just a few examples of the many changes in the 11th edition of *Biology* that provide students with scientifically accurate context, historical perspective, and relevant supporting details essential to a modern understanding of life science.

As the pace of scientific discovery continues to provide new insights into the foundation of life on Earth, our author team will continue to use every means possible to ensure students are as prepared as possible to engage in biological topics. Our goal now, as it has always been, is to ensure student success. To that end, we approached this revision differently. To help guide our revision for this 11th edition, we were able to incorporate student usage data and input, derived from thousands of our SmartBook® users. SmartBook "heat maps" provided a quick visual snapshot of chapter usage data and the relative difficulty students experienced in mastering the content. With these data, we were able to hone not only our text content but also the SmartBook probes.

- If the data indicated that the subject was more difficult than other parts of the chapter, as evidenced by a high proportion of students responding incorrectly to the probes, we revised or reorganized the content to be as clear and illustrative as possible.
- In other cases, if one or more of the SmartBook probes for a section was not as clear as it might be or did not appropriately reflect the content, we edited the probe, rather than the text.

Below is an example of one of the heat maps from Chapter 8. The color-coding in highlighted sections indicate the various levels of difficulty students experienced in learning the material; topics highlighted in red being the most challenging for students.

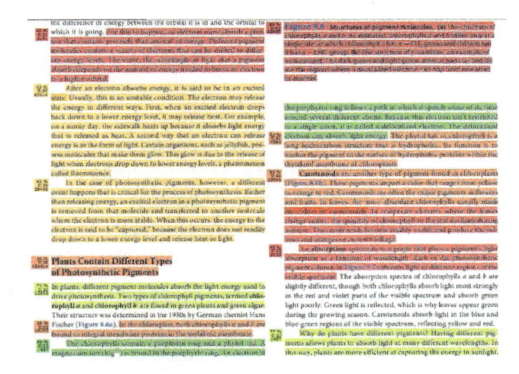

We're excited about the 11th edition of this quality textbook providing a learning path for a new generation of students. All of us have extensive experience teaching undergraduate biology, and we've used this knowledge as a guide in producing a text that is up to date, beautifully illustrated, and pedagogically sound for the student. We've also worked to provide clear explicit learning outcomes, and more closely integrate the text with its media support materials to provide instructors with an excellent complement to their teaching.

Ken Mason, Jonathan Losos, Susan Rundell Singer

Cutting Edge Science
Changes to the 11th Edition

Part I: The Molecular Basis of Life

The revision for the 11th edition started with the end users—the students. As described earlier, the authors analyzed SmartBook student usage data collected over the life of the 10th edition. The data from SmartBook revealed content areas where students struggled and based on that information, the authors revised the text to improve clarity. All chapters were evaluated using the SmartBook heat-mapping data.

Other content-specific changes incude:

Part II: Biology of the Cell

Chapter 5—New information on how phospholipid composition differs in different membranes and how this can affect function was added. The chapter was reorganized and a new figure was added to highlight this material.

Chapter 6—Figures 6.4 and 6.5 were revised for clarity and accuracy.

Chapter 9—A new evolutionary aside on the Ras superfamily of small GTPases was added.

Part III: Genetic and Molecular Biology

The overall organization of this section remains the same. We have retained the split of transmission genetics into two chapters as it has proved successful for students.

Content changes in the molecular genetics portion of this section continue to update material that is the most rapidly changing in the entire book. We also continue to refine the idea that RNA plays a much greater role now than appreciated in the past.

Chapter 14—Extensive editing for clarity was done based on heat-map data.

Chapter 15—Extensive updating was done, including a rewritten section on eukaryotic transcription that emphasizes a more modern perspective. Extensive editing throughout the chapter for clarity was done based on heat-map data.

Chapter 16—Extensive updating was done, including a rewritten section on eukaryotic chromatin structure to emphasize a more modern view, taking new high-throughput data into account.

Chapter 17—The chapter was significantly revised to reflect advances in molecular biology techniques such as quantitative reverse-transcription PCR, the CRISPR/Cas9 and related gene-editing technologies, and genetic engineering approaches. Additional modifications include the revision or addition of relevant, engaging applications of biotechnology such as wastewater treatment, biofuel production, and disease detection and treatment.

Chapter 18—Changes to this chapter focus on updating content related to recent advances in sequencing technologies and include the addition of new comparative discussions on the human, wheat, and cancer genome projects. Also, a new analytical commentary on the ENCODE project helps students think critically about recent findings in the genomics field. The addition of a section on the applications of genomics helps students appreciate the social relevance of an often abstract subject area.

Part IV: Evolution

Chapter 20—The section on genetic variation was substantially revised to include recent genomic surveys quantifying the extent of genetic variation across the genome in humans. The text was clarified to say that "dominant" has no connotation of selective superiority. In addition, revisions were made to the information on the current cost in human lives of evolutionary change in microbes evolving resistance to antibiotics.

Chapter 21—The coverage of Darwin's finches and the peppered moth stories were revised to incorporate new information. A section on dating of fossils using radioactive decay was added.

Chapter 22—New additions include examples of geographic variation in the black rat snake incorporating recent phylogenetic analyses of DNA data; reproductive character displacement involving *Phlox* plants in Texas; and adaptive radiation in plants using the Hawaiian *Lobelia*. The phylogeny of Darwin's finches was modified based on current research. New data appear on the evolution of developmental regulation of beak shape in Darwin's finches, along with a new section on mass extinction.

Chapter 23—A new section was added on the relationship between phylogenetics and taxonomic classifications. The figures illustrating how phylogenetics works and how character evolution is interpreted on an evolutionary tree were revised.

New data are included on the evolution of saber-teeth in mammals.

Chapter 24—Updates include new data on comparative genomics. Consideration of primate genomes is expanded, including human and Neanderthal.

Chapter 25—The introduction was rewritten. The cichlid and stickleback examples were updated with new information.

Part V: Diversity of Life on Earth

Chapter 26—The information on geological dating and taxonomic classification was moved to other chapters where the discussions were more relevant.

Chapter 27—This chapter was extensively revised and updated, including a new section on giant viruses and material on the recent Ebola outbreak in Africa.

Chapter 28—Extensive updates include a new section on CRISPR systems, which provide adaptive immunity in bacteria.

Chapter 29—Updated discussions appear on microfossils, malaria vaccines, micronucleus, and *Chlamydomonas genome*. Changes in section headings more clearly describe section contents. Numerous figures were updated to reflect changes in the text.

Chapters 30—The introduction has been modified to provide an overview of land plant evolution. Major trends in the evolution of land plants are now emphasized. The discussion on the effects of mutations on diploid versus haploid bodies has been clarified. The difference between animal and plant life cycles has been emphasized. Throughout the chapter, distinctions between sporophyte and gametophyte generations are clearly described. The significance of hornworts in land plant evolution is described. Section headings were changed to more clearly describe section contents.

Chapter 31—The reduction in the complexity of the gametophyte generation in the evolution of land plants is emphasized. Distinctions between gamete and gametophyte, male and female gametophyte, zygote and embryo, and gymnosperms and angiosperms have been clarified. A discussion of the hypothesis for the rapid expansion of the world's biomes by the angiosperms has been added. The development of the female gametophyte has been described in more detail. The significance of double fertilization has been described.

Chapter 32—The development of hyphae during the evolution of fungi is described. The significance of above-ground spore dispersal structures is emphasized. The characteristics of each fungus group are clearly and concisely described. The significance of fungi in rumen biology has been described. Section headings were changed to more clearly describe section contents. Numerous figures were updated to reflect changes in the text.

Chapters 33–35—These chapters have been streamlined, eliminating extraneous information that was outside the scope of the main topics in the chapters. Throughout, changes were made in the species used as illustrative examples. Information on number of species in different taxa was updated.

Chapter 34—The information on medical infection rates to various invertebrate groups was updated.

Chapter 35—The phylogeny of chordates was updated. A discussion of evolution of tortoises and new information on the sensory abilities of the platypus were added. The phylogeny of primates was updated. New information was added on the genome of Neanderthals and understanding of the evolution of modern humans.

Part VI: Plant Form and Function

Throughout the plant chapters, corrections have been made so that $2n$ and n refer to the sporophyte and gametophyte generations, respectively, and x refers to the number of sets of chromosomes.

Chapter 36—The anatomical positions of components of plant tissues are more clearly presented. Structural differences between angiosperms and gymnosperms are emphasized. Distinctions between similar structures in different tissues—for example, pits in xylem versus pores in phloem—are made.

Chapter 37—The significance of water potential gradients in water transport is clarified and emphasized. The association between anaerobic conditions and poor root growth is described.

Chapter 38—The mechanism of closing in the Venus flytrap leaf has been updated.

Chapter 41—The concept of alternation of generations has been clarified and emphasized.

Part VII: Animal Form and Function

Chapter 46—A new illustration of hinge joints was added.

Chapter 47—The section on pancreas function was revised.

Chapter 48—Information on cutaneous respiration in turtles was added. The discussion of gas exchange in the capillaries was revised.

Chapter 49—The illustration and explanation of components of blood cells were revised, along with the explanation of how blood clotting works.

Chapter 52—New information on facultative parthenogenesis in vertebrates was added. Information on birth control was updated.

Part VIII: Ecology and Behavior

Chapter 54—Information on the social behaviors and brains of prairie and montane voles was updated. The discussion of orientation and migration, plus the section on evolution of mate choice in frogs were revised. Extraneous examples were eliminated to streamline the chapter.

Chapter 55—The information on human population growth and population demographics for several countries was updated using current statistics.

Chapter 56— Extraneous material was removed to streamline the chapter.

Chapter 57—Figures and explanations of trophic cascades, how effects move from one level of the food web to the next, and the discussions of trophic levels and island biogeography were revised. New ideas are presented on why the tropics are so biologically rich.

Chapter 58—Up-to-date information appears on global warming and global ozone levels, with an illustration of how the Earth revolves around the sun. A section was added on new human diseases that come from animals (zoonotic diseases).

Chapter 59—Information on human population growth in biodiversity hot spots, human health toll of West Nile Virus, and the recovery of the peregrine falcon was updated.

A Note From the Authors

A revision of this scope relies on the talents and efforts of many people working behind the scenes and we have benefited greatly from their assistance.

Beatrice Sussman was the copyeditor for this edition. She has labored many hours and always improves the clarity and consistency of the text. She has made significant contributions to the quality of the final product.

We were fortunate to work again with MPS to update the art program and improve the layout of the pages. Our close collaboration resulted in a text that is pedagogically effective as well as more beautiful than any other biology text on the market.

We have the continued support of an excellent team at McGraw-Hill. Justin Wyatt, preceded by Rebecca Olson, the brand managers for *Biology* have been steady leaders during a time of change. Lead Product Developer Liz Sievers, provided support in so many ways it would be impossible to name them all. April Southwood, content project manager, and David Hash, designer, ensured our text was on time and elegantly designed. Patrick Reidy, executive marketing manager, is always a sounding board for more than just marketing, and many more people behind the scenes have all contributed to the success of our text. This includes the digital team, whom we owe a great deal for their efforts to continue improving our Connect assessment tools.

Throughout this edition we have had the support of spouses and children, who have seen less of us than they might have liked because of the pressures of getting this revision completed. They have adapted to the many hours this book draws us away from them, and, even more than us, looked forward to its completion.

In the end, the people we owe the most are the generations of students who have used the many editions of this text. They have taught us at least as much as we have taught them, and their questions and suggestions continue to improve the text and supplementary materials.

Finally, we need to thank instructors from across the country who are continually sharing their knowledge and experience with us through market feedback and symposia. The feedback we received shaped this edition. All of these people took time to share their ideas and opinions to help us build a better edition of *Biology* for the next generation of introductory biology students, and they have our heartfelt thanks.

Preparing Students for the Future

Developing Critical Thinking with the Help of . . .

Detailed Feedback in Connect®

Learning is a process of iterative development, of making mistakes, reflecting, and adjusting over time. The question and test banks in Connect® for *Biology*, 11th edition, are more than direct assessments; they are self-contained learning experiences that systematically build student learning over time.

For many students, choosing the right answer is not necessarily based on applying content correctly; it is more a matter of increasing their statistical odds of guessing. A major fault with this approach is students don't learn how to process the questions correctly, mostly because they are repeating and reinforcing their mistakes rather than reflecting and learning from them. To help students develop problem-solving skills, all higher level Blooms questions in Connect are supported with hints, to help students focus on important information for answering the questions, and detailed feedback that walks students through the problem-solving process, using Socratic questions in a decision-tree-style framework to scaffold learning, where each step models and reinforces the learning process.

The feedback for each higher level Blooms question (Apply, Analyze, Evaluate) follows a similar process: Clarify Question, Gather Content, Choose Answer, Reflect on Process.

Unpacking the Concepts

We've taken problem solving a step further. In each chapter, three to five higher level Blooms questions in the question and test banks are broken out by the steps of the detailed feedback. Rather than leaving it up to the student to work through the detailed feedback, a second version of the question is presented in a stepwise format. Following the problem-solving steps, students need to answer questions about earlier steps, such as "What is the key concept addressed by the question?" before proceeding to answer the question. A professor can choose which version of the question to include in the assignment based on the problem-solving skills of the students.

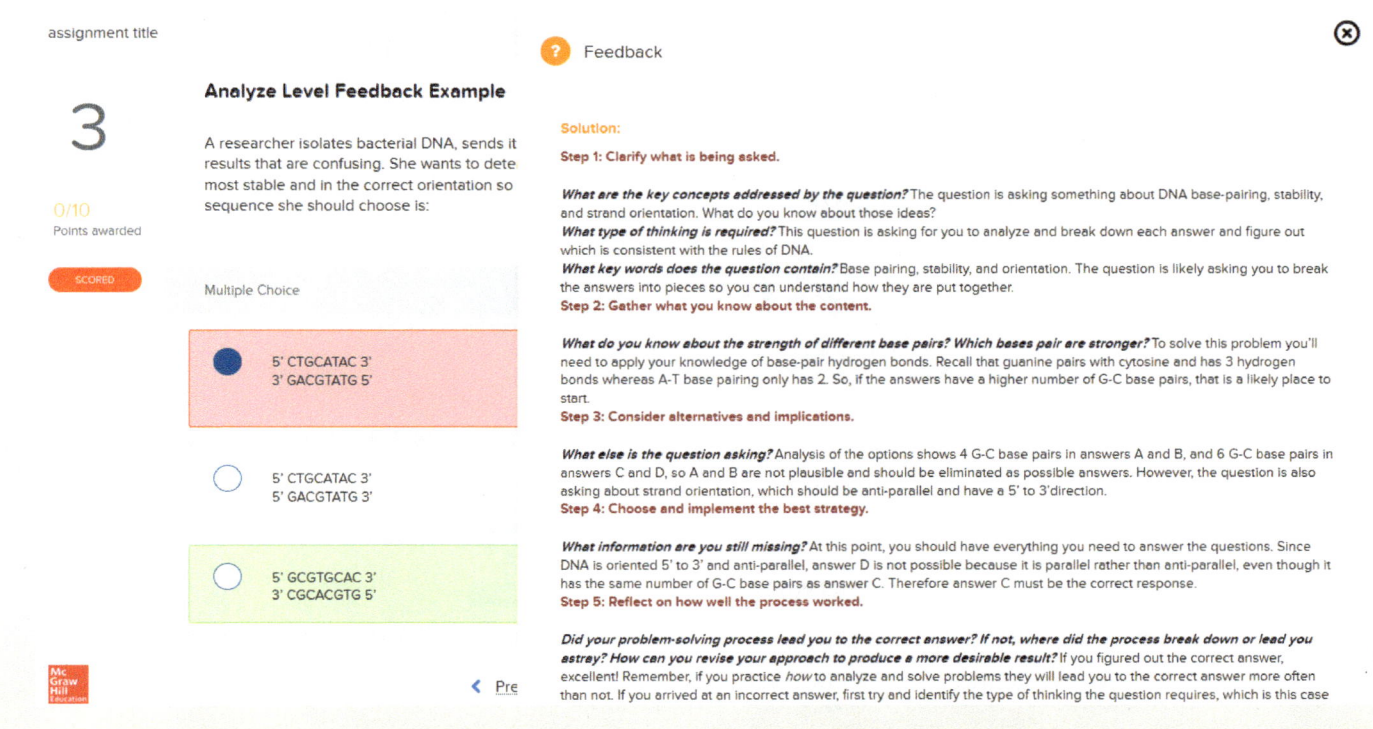

---

assignment title

3

0/10
Points awarded

SCORED

Analyze Level Feedback Example

A researcher isolates bacterial DNA, sends it results that are confusing. She wants to dete most stable and in the correct orientation so sequence she should choose is:

Multiple Choice

- ● 5' CTGCATAC 3'
  3' GACGTATG 5'

- ○ 5' CTGCATAC 3'
  5' GACGTATG 3'

- ○ 5' GCGTGCAC 3'
  3' CGCACGTG 5'

< Pre

? Feedback

Solution:
Step 1: Clarify what is being asked.

What are the key concepts addressed by the question? The question is asking something about DNA base-pairing, stability, and strand orientation. What do you know about those ideas?
What type of thinking is required? This question is asking for you to analyze and break down each answer and figure out which is consistent with the rules of DNA.
What key words does the question contain? Base pairing, stability, and orientation. The question is likely asking you to break the answers into pieces so you can understand how they are put together.
Step 2: Gather what you know about the content.

What do you know about the strength of different base pairs? Which bases pair are stronger? To solve this problem you'll need to apply your knowledge of base-pair hydrogen bonds. Recall that guanine pairs with cytosine and has 3 hydrogen bonds whereas A-T base pairing only has 2. So, if the answers have a higher number of G-C base pairs, that is a likely place to start.
Step 3: Consider alternatives and implications.

What else is the question asking? Analysis of the options shows 4 G-C base pairs in answers A and B, and 6 G-C base pairs in answers C and D, so A and B are not plausible and should be eliminated as possible answers. However, the question is also asking about strand orientation, which should be anti-parallel and have a 5' to 3' direction.
Step 4: Choose and implement the best strategy.

What information are you still missing? At this point, you should have everything you need to answer the questions. Since DNA is oriented 5' to 3' and anti-parallel, answer D is not possible because it is parallel rather than anti-parallel, even though it has the same number of G-C base pairs as answer C. Therefore answer C must be the correct response.
Step 5: Reflect on how well the process worked.

Did your problem-solving process lead you to the correct answer? If not, where did the process break down or lead you astray? How can you revise your approach to produce a more desirable result? If you figured out the correct answer, excellent! Remember, if you practice *how* to analyze and solve problems they will lead you to the correct answer more often than not. If you arrived at an incorrect answer, first try and identify the type of thinking the question requires, which is this case

Strengthen Problem Solving Skills and Key Concept Development with Connect®

SmartBook with Learning Resources

To help students understand key concepts, SmartBook® for *Biology*, 11th edition, is enhanced with Learning Resources. Based on student usage data, derived from thousands of SmartBook users of the tenth edition, concepts that proved more challenging for students are supported with Learning Resources to enhance the textbook presentation. Learning Resources, such as animations or tutorials, are indicated in SmartBook adjacent to the textbook content. If a student is struggling with a concept based on his/her performance on the SmartBook questions, the student is given an option to review the Learning Resource or the student can click on the Learning Resources at any time.

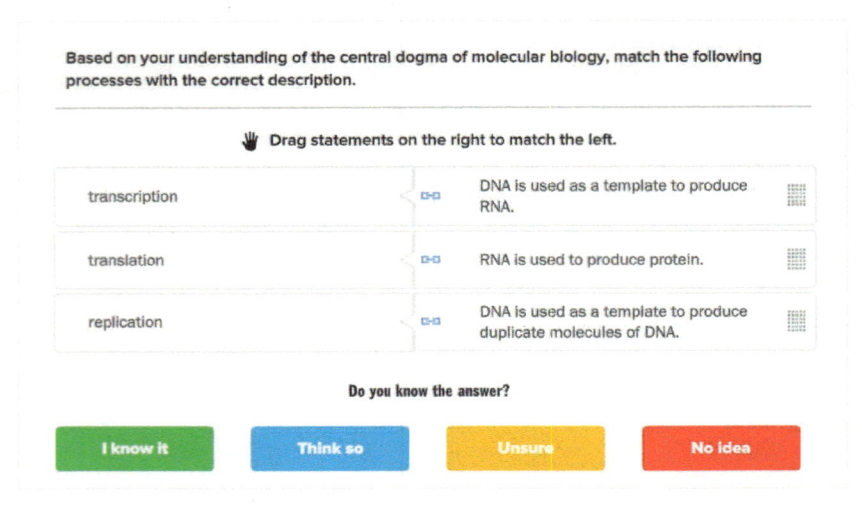

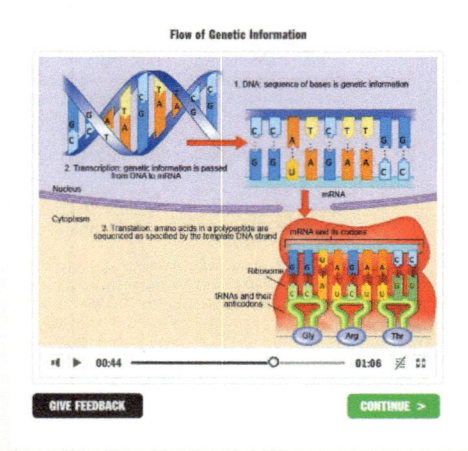

Scientific Thinking Art

Key illustrations in every chapter highlight how the frontiers of knowledge are pushed forward by a combination of hypothesis and experimentation. These figures begin with a hypothesis, then show how it makes explicit predictions, tests these by experiment and finally demonstrates what conclusions can be drawn, and where this leads. Scientific Thinking figures provide a consistent framework to guide the student in the logic of scientific inquiry. Each illustration concludes with open-ended questions to promote scientific inquiry.

SCIENTIFIC THINKING

Hypothesis: *The plasma membrane is fluid, not rigid.*
Prediction: *If the membrane is fluid, membrane proteins may diffuse laterally.*
Test: *Fuse mouse and human cells, then observe the distribution of membrane proteins over time by labeling specific mouse and human proteins.*

Result: *Over time, hybrid cells show increasingly intermixed proteins.*
Conclusion: *At least some membrane proteins can diffuse laterally in the membrane.*
Further Experiments: *Can you think of any other explanation for these observations? What if newly synthesized proteins were inserted into the membrane during the experiment? How could you use this basic experimental design to rule out this or other possible explanations?*

Figure 5.5 Test of membrane fluidity.

Data Analysis Questions

It's not enough that students learn concepts and memorize scientific facts, a biologist needs to analyze data and apply that knowledge. Data Analysis questions inserted throughout the text challenge students to analyze data and Interpret experimental results, which shows a deeper level of understanding.

Supporting Material Provided Online

Evolutionary Asides are inserted at relevant places in the book. The student links to this online content through the Evolutionary Aside icon found in the eBook. Evolutionary Asides provide additional examples or discussions of evolutionary topics related to the textual discussion.

Quantitative Asides are inserted at relevant places in the book. The student links to this online content through the Quantitative Aside icon found in the eBook. Quantitative Asides provide additional examples or expanded discussions of a quantitative aspect of the topic under discussion.

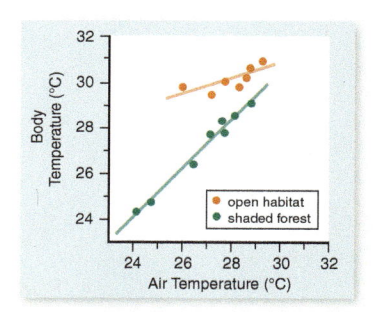

Figure 55.3 Behavioral adaptation. In open habitats, the Puerto Rican crested lizard, *Anolis cristatellus*, maintains a relatively constant temperature by seeking out and basking in patches of sunlight; as a result, it can maintain a relatively high temperature even when the air is cool. In contrast, in shaded forests, this behavior is not possible, and the lizard's body temperature conforms to that of its surroundings.

Inquiry question When given the opportunity, lizards regulate their body temperature to maintain a temperature optimal for physiological functioning. Would lizards in open habitats exhibit different escape behaviors from lizards in shaded forest?

Data analysis Can the slope of the line tell us something about the behavior of the lizard?

Inquiry Questions

Questions that challenge students to think about and engage in what they are reading at a more sophisticated level.

The Molecular Basis of Life

Part I The Molecular Basis of Life

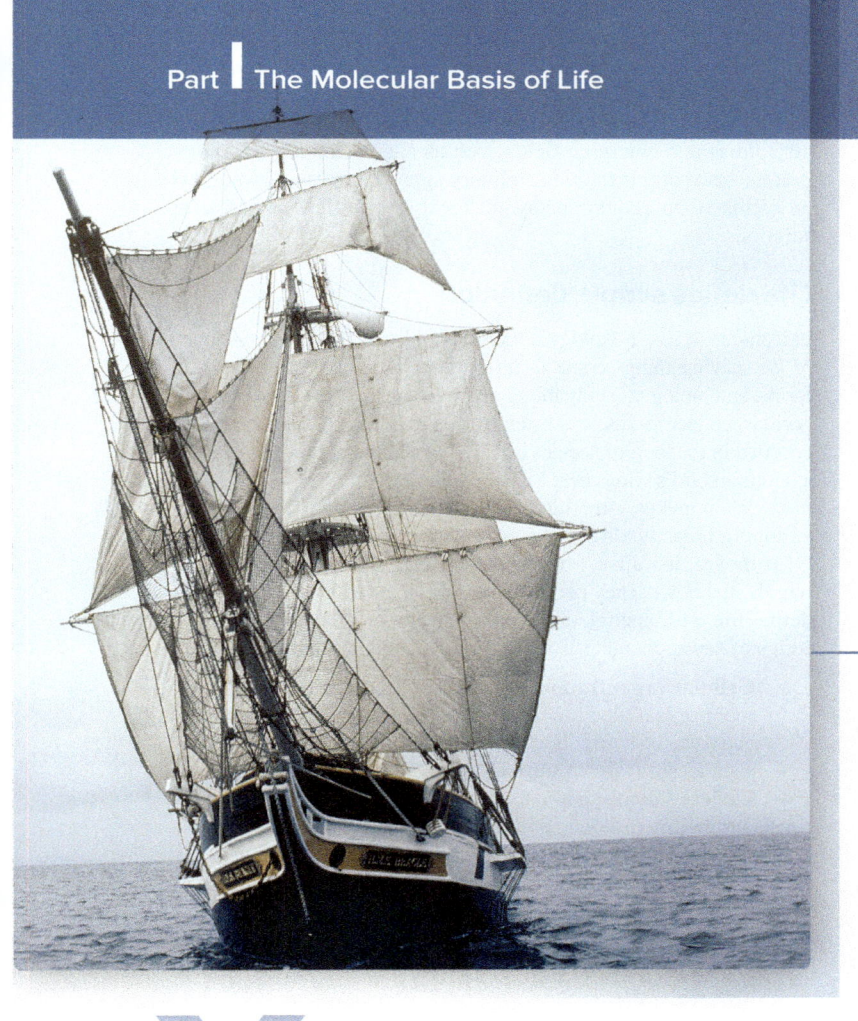

CHAPTER 1

The Science of Biology

Chapter Contents

1.1 The Science of Life
1.2 The Nature of Science
1.3 An Example of Scientific Inquiry: Darwin and Evolution
1.4 Unifying Themes in Biology

Introduction

You are about to embark on a journey—a journey of discovery about the nature of life. More than 180 years ago, a young English naturalist named Charles Darwin set sail on a similar journey on board H.M.S. *Beagle*; a replica of this ship is pictured here. What Darwin learned on his five-year voyage led directly to his development of the theory of evolution by natural selection, a theory that has become the core of the science of biology. Darwin's voyage seems a fitting place to begin our exploration of biology—the scientific study of living organisms and how they have evolved. Before we begin, however, let's take a moment to think about what biology is and why it's important.

1.1 The Science of Life

Learning Outcomes

1. Compare biology to other natural sciences.
2. Describe the characteristics of living systems.
3. Characterize the hierarchical organization of living systems.

This is the most exciting time to be studying biology in the history of the field. The amount of information available about the natural world has exploded in the last 42 years since the construction of the first recombinant DNA molecule. We are now in a position to ask and answer questions that previously were only dreamed of.

The 21st century began with the completion of the sequence of the human genome. The largest single project in the history of biology took about 20 years. Yet less than 15 years later, we can sequence an entire genome in a matter of days. This flood of sequence data and genomic analysis are altering the landscape of biology. These and other discoveries are also moving into the clinic as never before with new tools for diagnostics and treatment. With robotics, advanced imaging, and analytical techniques, we have tools available that were formerly the stuff of science fiction.

In this text, we attempt to draw a contemporary picture of the science of biology, as well as provide some history and experimental perspective on this exciting time in the discipline. In this introductory chapter, we examine the nature of biology and the foundations of science in general to put into context the information presented in the rest of the text.

Biology unifies much of natural science

The study of biology is a point of convergence for the information and tools from all of the natural sciences. Biological systems are the most complex chemical systems on Earth, and their many functions are both determined and constrained by the principles of chemistry and physics. Put another way, no new laws of nature can be gleaned from the study of biology—but that study does illuminate and illustrate the workings of those natural laws.

The intricate chemical workings of cells can be understood using the tools and principles of chemistry. And every level of biological organization is governed by the nature of energy transactions first studied by thermodynamics. Biological systems do not represent any new forms of matter, and yet they are the most complex organization of matter known. The complexity of living systems is made possible by a constant source of energy—the Sun. The conversion of this radiant energy into organic molecules by photosynthesis is one of the most beautiful and complex reactions known in chemistry and physics.

The way we do science is changing to grapple with increasingly difficult modern problems. Science is becoming more interdisciplinary, combining the expertise from a variety of traditional disciplines and emerging fields such as nanotechnology. Biology is at the heart of this multidisciplinary approach because biological problems often require many different approaches to arrive at solutions.

Life defies simple definition

In its broadest sense, biology is the study of living things—*the science of life*. Living things come in an astounding variety of shapes and forms, and biologists study life in many different ways. They live with gorillas, collect fossils, and listen to whales. They read the messages encoded in the long molecules of heredity and count how many times a hummingbird's wings beat each second.

What makes something "alive"? Anyone could deduce that a galloping horse is alive and a car is not, but why? We cannot say, "If it moves, it's alive," because a car can move, and gelatin can wiggle in a bowl. They certainly are not alive. Although we cannot define life with a single simple sentence, we can come up with a series of seven characteristics shared by living systems:

- **Cellular organization.** All organisms consist of one or more cells. Often too tiny to see, cells carry out the basic activities of living. Each cell is bounded by a membrane that separates it from its surroundings.
- **Ordered complexity.** All living things are both complex and highly ordered. Your body is composed of many different kinds of cells, each containing many complex molecular structures. Many nonliving things may also be

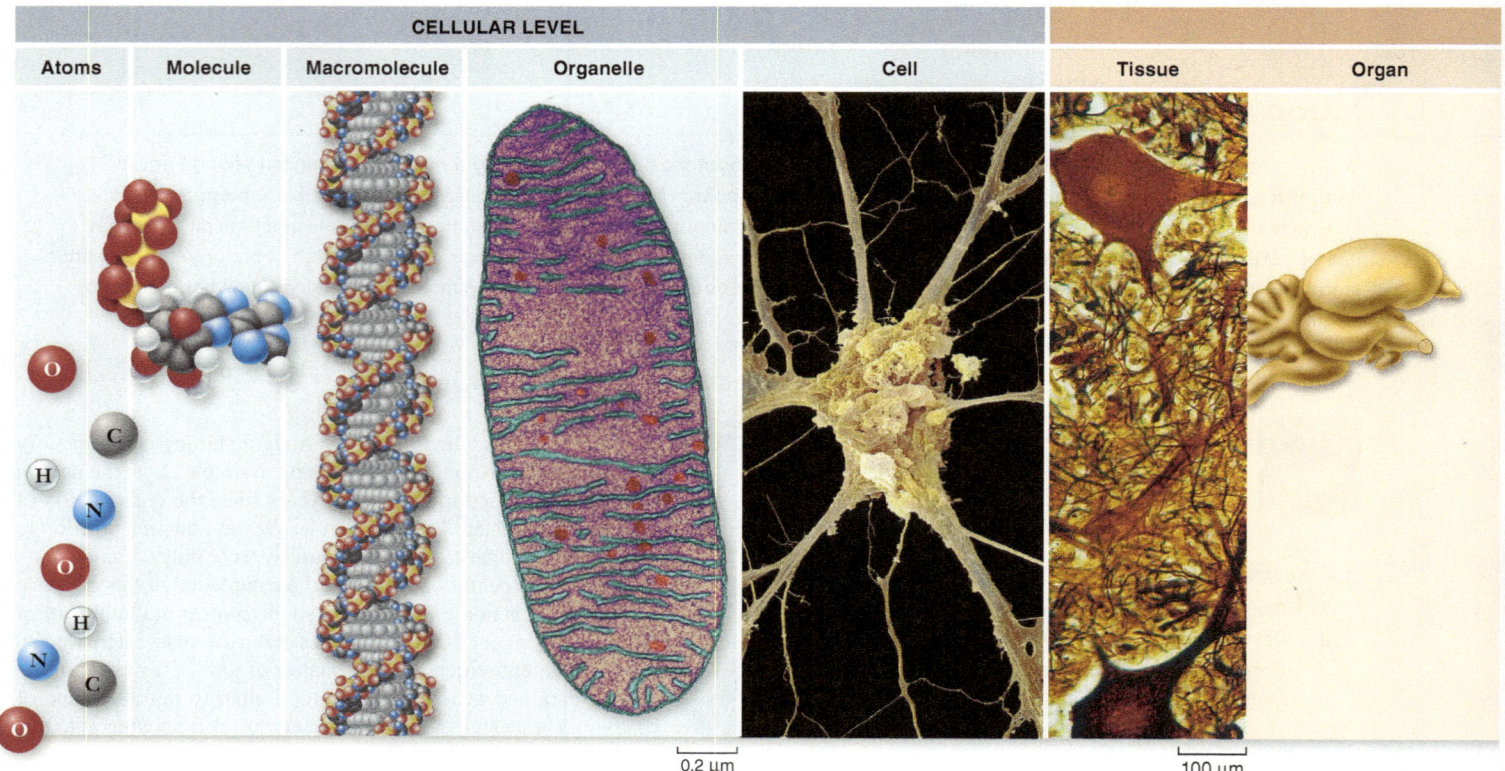

CELLULAR LEVEL

| Atoms | Molecule | Macromolecule | Organelle | Cell | Tissue | Organ |

complex, but they do not exhibit this degree of ordered complexity.
- **Sensitivity.** All organisms respond to stimuli. Plants grow toward a source of light, and the pupils of your eyes dilate when you walk into a dark room.
- **Growth, development, and reproduction.** All organisms are capable of growing and reproducing, and they all possess hereditary molecules that are passed to their offspring, ensuring that the offspring are of the same species.
- **Energy utilization.** All organisms take in energy and use it to perform many kinds of work. Every muscle in your body is powered with energy you obtain from your diet.
- **Homeostasis.** All organisms maintain relatively constant internal conditions that are different from their environment, a process called **homeostasis.** For example, your body temperature remains stable despite changes in outside temperatures.
- **Evolutionary adaptation.** All organisms interact with other organisms and the nonliving environment in ways that influence their survival, and as a consequence, organisms evolve adaptations to their environments.

Living systems show hierarchical organization

The organization of the biological world is hierarchical—that is, each level builds on the level below it:

1. **The cellular level.** At the cellular level (figure 1.1), **atoms,** the fundamental elements of matter, are joined together into clusters called **molecules.** Complex biological molecules are assembled into tiny structures called **organelles** within membrane-bounded units we call **cells.** The cell is the basic unit of life. Many independent organisms are composed only of single cells. Bacteria are single cells, for example. All animals and plants, as well as most fungi and algae, are multicellular—composed of more than one cell.

2. **The organismal level.** Cells in complex multicellular organisms exhibit three levels of organization. The most basic level is that of **tissues,** which are groups of similar cells that act as a functional unit. Tissues, in turn, are grouped into **organs**—body structures composed of several different tissues that act as a structural and functional unit. Your brain is an organ composed of nerve cells and a variety of associated tissues that form protective coverings and contribute blood. At the third level of organization, organs are grouped into **organ systems.** The nervous system, for example, consists of sensory organs, the brain and spinal cord, and neurons that convey signals.

Figure 1.1 Hierarchical organization of living systems. Life forms a hierarchy of organization from atoms to complex multicellular organisms. Atoms are joined together to form molecules, which are assembled into more complex structures such as organelles. These in turn form subsystems that provide different functions. Cells can be organized into tissues, then into organs and organ systems such as the goose's nervous system pictured. This organization then extends beyond individual organisms to populations, communities, ecosystems, and finally the biosphere.

ORGANISMAL LEVEL		POPULATIONAL LEVEL				
Organ system	Organism	Population	Species	Community	Ecosystem	Biosphere

3. **The populational level.** Individual organisms can be categorized into several hierarchical levels within the living world. The most basic of these is the **population**—a group of organisms of the same species living in the same place. All populations of a particular kind of organism together form a **species,** its members similar in appearance and able to interbreed. At a higher level of biological organization, a **biological community** consists of all the populations of different species living together in one place.
4. **The ecosystem level.** At the highest tier of biological organization, populations of organisms interact with each other and their physical environment. Together populations and their environment constitute an ecological system, or **ecosystem.** For example, the biological community of a mountain meadow interacts with the soil, water, and atmosphere of a mountain ecosystem in many important ways.
5. **The biosphere.** The entire planet can be thought of as an ecosystem that we call the biosphere.

As you move up this hierarchy, the many interactions occurring at lower levels can produce novel properties. These so-called **emergent properties** may not be predictable. Examining individual cells, for example, gives little hint about the whole animal. Many weather phenomena, such as hurricanes, are actually emergent properties of many interacting meteorological variables. It is because the living world exhibits many emergent properties that it is difficult to define "life."

The previous descriptions of the common features and organization of living systems begins to get at the nature of what it is to be alive. The rest of this book illustrates and expands on these basic ideas to try to provide a more complete account of living systems.

> ### Learning Outcomes Review 1.1
> Biology as a science brings together other natural sciences, such as chemistry and physics, to study living systems. Life does not have a simple definition, but living systems share a number of properties that together describe life. Living systems can be organized hierarchically, from the cellular level to the entire biosphere, with emergent properties that may exceed the sum of the parts.
>
> ■ *Can you study biology without studying other sciences?*

1.2 The Nature of Science

Learning Outcomes
1. *Compare the different types of reasoning used by biologists.*
2. *Demonstrate how to formulate and test a hypothesis.*

Much like life itself, the nature of science defies simple description. For many years scientists have written about the "scientific method" as though there is a single way of doing science. This oversimplification has contributed to confusion on the part of nonscientists about the nature of science.

At its core, science is concerned with developing an increasingly accurate understanding of the world around us using observation and reasoning. To begin with, we assume that natural forces acting now have always acted, that the fundamental nature of the universe has not changed since its inception, and that it is not changing now. A number of complementary approaches allow understanding of natural phenomena—there is no one "scientific method."

Scientists also attempt to be as objective as possible in the interpretation of the data and observations they have collected. Because scientists themselves are human, this is not completely possible, but because science is a collective endeavor subject to scrutiny, it is self-correcting. One person's results are verified by others, and if the results cannot be repeated, they are rejected.

Much of science is descriptive

The classic vision of the scientific method is that observations lead to hypotheses that in turn make experimentally testable predictions. In this way, we dispassionately evaluate new ideas to arrive at an increasingly accurate view of nature. We discuss this way of doing science later in this section but it is important to understand that much of science is purely descriptive: In order to understand anything, the first step is to describe it completely. Much of biology is concerned with arriving at an increasingly accurate description of nature.

The study of biodiversity is an example of descriptive science that has implications for other aspects of biology in addition to societal implications. Efforts are currently under way to classify all life on Earth. This ambitious project is purely descriptive, but it will lead to a much greater understanding of biodiversity as well as the effect our species has on biodiversity.

One of the most important accomplishments of molecular biology at the dawn of the 21st century was the completion of the sequence of the human genome. Many new hypotheses about human biology will be generated by this knowledge, and many experiments will be needed to test these hypotheses, but the determination of the sequence itself was descriptive science.

Science uses both deductive and inductive reasoning

The study of logic recognizes two opposite ways of arriving at logical conclusions: deductive and inductive reasoning. Science makes use of both of these methods, although induction is the primary way of reasoning in hypothesis-driven science.

Deductive reasoning

Deductive reasoning applies general principles to predict specific results. More than 2200 years ago, the Greek scientist Eratosthenes used Euclidean geometry and deductive reasoning to accurately estimate the circumference of the Earth (figure 1.2). Deductive reasoning is the reasoning of mathematics and philosophy, and it is used to test the validity of general ideas in all branches of

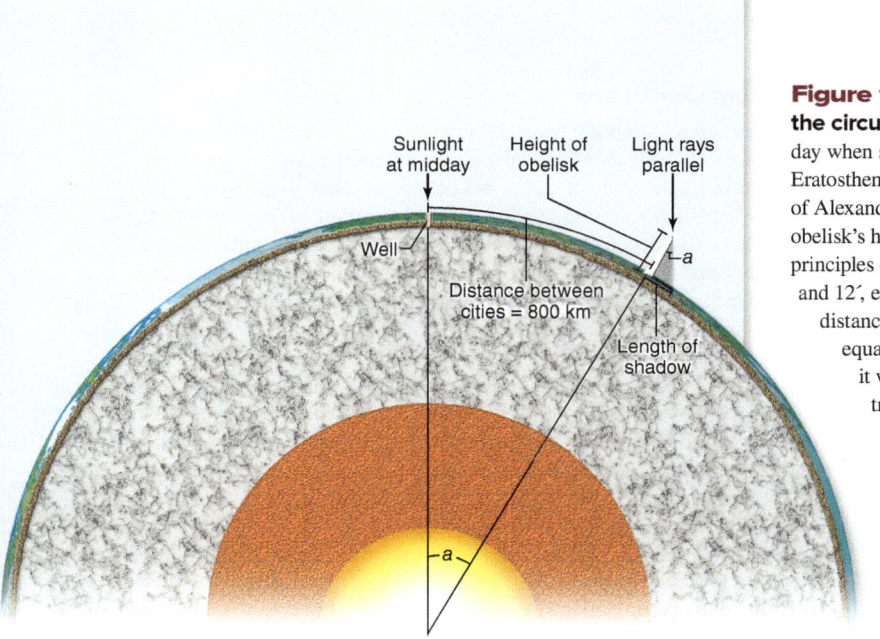

Figure 1.2 Deductive reasoning: How Eratosthenes estimated the circumference of the Earth using deductive reasoning. *1.* On a day when sunlight shone straight down a deep well at Syene in Egypt, Eratosthenes measured the length of the shadow cast by a tall obelisk in the city of Alexandria, about 800 kilometers (km) away. *2.* The shadow's length and the obelisk's height formed two sides of a triangle. Using the recently developed principles of Euclidean geometry, Eratosthenes calculated the angle, *a,* to be 7° and 12´, exactly 1/50 of a circle (360°). *3.* If angle *a* is 1/50 of a circle, then the distance between the obelisk (in Alexandria) and the well (in Syene) must be equal to 1/50 the circumference of the Earth. *4.* Eratosthenes had heard that it was a 50-day camel trip from Alexandria to Syene. Assuming a camel travels about 18.5 km per day, he estimated the distance between obelisk and well as 925 km (using different units of measure, of course). *5.* Eratosthenes thus deduced the circumference of the Earth to be 50 × 925 = 46,250 km. Modern measurements put the distance from the well to the obelisk at just over 800 km. Using this distance Eratosthenes's value would have been 50 × 800 = 40,000 km. The actual circumference is 40,075 km.

knowledge. For example, if all mammals by definition have hair, and you find an animal that does not have hair, then you may conclude that this animal is not a mammal. A biologist uses deductive reasoning to infer the species of a specimen from its characteristics.

Inductive reasoning

In **inductive reasoning,** the logic flows in the opposite direction, from the specific to the general. Inductive reasoning uses specific observations to construct general scientific principles. For example, if poodles have hair, and terriers have hair, and every dog that you observe has hair, then you may conclude that all dogs have hair. Inductive reasoning leads to generalizations that can then be tested. Inductive reasoning first became important to science in the 1600s in Europe, when Francis Bacon, Isaac Newton, and others began to use the results of particular experiments to infer general principles about how the world operates.

An example from modern biology is the role of homeobox genes in development. Studies in the fruit fly, *Drosophila melanogaster,* identified genes that could cause dramatic changes in developmental fate, such as a leg appearing in the place of an antenna. These genes have since been found in essentially all multicellular animals analyzed. This led to the general idea that homeobox genes control developmental fate in animals.

Hypothesis-driven science makes and tests predictions

Scientists establish which general principles are true from among the many that might be true through the process of systematically testing alternative proposals. If these proposals prove inconsistent with experimental observations, they are rejected as untrue. Figure 1.3 illustrates the process.

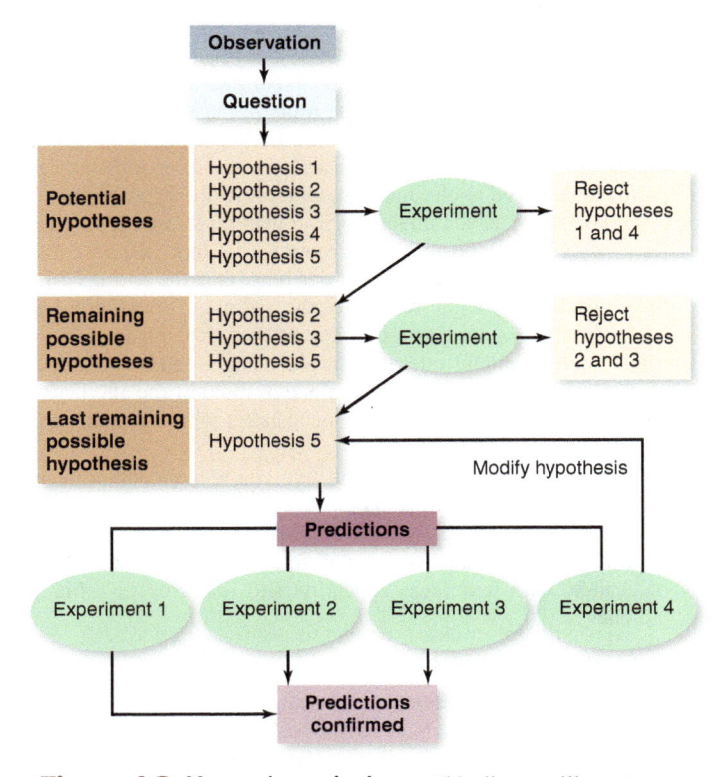

Figure 1.3 How science is done. This diagram illustrates how scientific investigations proceed. First, scientists make observations that raise a particular question. They develop a number of potential explanations (hypotheses) to answer the question. Next, they carry out experiments in an attempt to eliminate one or more of these hypotheses. Then, predictions are made based on the remaining hypotheses, and further experiments are carried out to test these predictions. The process can also be iterative. As experimental results are performed, the information can be used to modify the original hypothesis to fit each new observation.

After making careful observations, scientists construct a **hypothesis**, which is a suggested explanation that accounts for those observations. A hypothesis is a proposition that might be true. Those hypotheses that have not yet been disproved are retained. They are useful because they fit the known facts, but they are always subject to future rejection if, in the light of new information, they are found to be incorrect.

This is usually an ongoing process with a hypothesis changing and being refined with new data. For instance, geneticists George Beadle and Edward Tatum studied the nature of genetic information to arrive at their "one-gene/one-enzyme" hypothesis (see chapter 15). This hypothesis states that a gene represents the genetic information necessary to make a single enzyme. As investigators learned more about the molecular nature of genetic information, the hypothesis was refined to "one-gene/one-polypeptide" because enzymes can be made up of more than one polypeptide. With still more information about the nature of genetic information, other investigators found that a single gene can specify more than one polypeptide, and the hypothesis was refined again.

Testing hypotheses

We call the test of a hypothesis an **experiment.** Suppose you enter a dark room. To understand why it is dark, you propose several hypotheses. The first might be, "There is no light in the room because the light switch is turned off." An alternative hypothesis might be, "There is no light in the room because the lightbulb is burned out." And yet another hypothesis might be, "I am going blind." To evaluate these hypotheses, you would conduct an experiment designed to eliminate one or more of the hypotheses.

For example, you might test your hypotheses by flipping the light switch. If you do so and the room is still dark, you have disproved the first hypothesis: Something other than the setting of the light switch must be the reason for the darkness. Note that a test such as this does not prove that any of the other hypotheses are true; it merely demonstrates that the one being tested is not. A successful experiment is one in which one or more of the alternative hypotheses is demonstrated to be inconsistent with the results and is thus rejected.

As you proceed through this text, you will encounter many hypotheses that have withstood the test of experiment. Many will continue to do so; others will be revised as new observations are made by biologists. Biology, like all science, is in a constant state of change, with new ideas appearing and replacing or refining old ones.

Establishing controls

Often scientists are interested in learning about processes that are influenced by many factors, or **variables.** To evaluate alternative hypotheses about one variable, all other variables must be kept constant. This is done by carrying out two experiments in parallel: a test experiment and a control experiment. In the **test experiment,** one variable is altered in a known way to test a particular hypothesis. In the **control experiment,** that variable is left unaltered. In all other respects the two experiments are identical, so any difference in the outcomes of the two experiments must result from the influence of the variable that was changed.

Much of the challenge of experimental science lies in designing control experiments that isolate a particular variable from other factors that might influence a process.

Using predictions

A successful scientific hypothesis needs to be not only valid but also useful—it needs to tell us something we want to know. A hypothesis is most useful when it makes predictions because those predictions provide a way to test the validity of the hypothesis. If an experiment produces results inconsistent with the predictions, the hypothesis must be rejected or modified. In contrast, if the predictions are supported by experimental testing, the hypothesis is supported. The more experimentally supported predictions a hypothesis makes, the more valid the hypothesis is.

As an example, in the early history of microbiology it was known that nutrient broth left sitting exposed to air becomes contaminated. Two hypotheses were proposed to explain this observation: spontaneous generation and the germ hypothesis. Spontaneous generation held that there was an inherent property in organic molecules that could lead to the spontaneous generation of life. The germ hypothesis proposed that preexisting microorganisms that were present in the air could contaminate the nutrient broth.

These competing hypotheses were tested by a number of experiments that involved filtering air and boiling the broth to kill any contaminating germs. The definitive experiment was performed by Louis Pasteur, who constructed flasks with curved necks that could be exposed to air, but that would trap any contaminating germs. When such flasks were boiled to sterilize them, they remained sterile, but if the curved neck was broken off, they became contaminated (figure 1.4).

SCIENTIFIC THINKING

Question: *What is the source of contamination that occurs in a flask of nutrient broth left exposed to the air?*
Germ Hypothesis: *Preexisting microorganisms present in the air contaminate nutrient broth.*
Prediction: *Sterilized broth will remain sterile if microorganisms are prevented from entering flask.*
Spontaneous Generation Hypothesis: *Living organisms will spontaneously generate from nonliving organic molecules in broth.*
Prediction: *Organisms will spontaneously generate from organic molecules in broth after sterilization.*
Test: *Use swan-necked flasks to prevent entry of microorganisms. To ensure that broth can still support life, break swan-neck after sterilization.*

Flask is sterilized by boiling the broth.   Unbroken flask remains sterile.   Broken flask becomes contaminated after exposure to germ-laden air.

Result: *No growth occurs in sterile swan-necked flasks. When the neck is broken off, and the broth is exposed to air, growth occurs.*
Conclusion: *Growth in broth is of preexisting microorganisms.*

Figure 1.4 Experiment to test spontaneous generation versus germ hypothesis.

This result was predicted by the germ hypothesis—that when the sterile flask is exposed to air, airborne germs are deposited in the broth and grow. The spontaneous generation hypothesis predicted no difference in results with exposure to air. This experiment disproved the hypothesis of spontaneous generation and supported the hypothesis of airborne germs under the conditions tested.

Reductionism breaks larger systems into their component parts

Scientists use the philosophical approach of **reductionism** to understand a complex system by reducing it to its working parts. Reductionism has been the general approach of biochemistry, which has been enormously successful at unraveling the complexity of cellular metabolism by concentrating on individual pathways and specific enzymes. By analyzing all of the pathways and their components, scientists now have an overall picture of the metabolism of cells.

Reductionism has limits when applied to living systems, however—one of which is that enzymes do not always behave exactly the same in isolation as they do in their normal cellular context. A larger problem is that the complex interworking of many interconnected functions leads to emergent properties that cannot be predicted based on the workings of the parts. For example, ribosomes are the cellular factories that synthesize proteins, but this function could not be predicted based on analysis of the individual proteins and RNA that make up the structure. On a higher level, understanding the physiology of a single Canada goose would not lead to predictions about flocking behavior. The emerging field of systems biology uses mathematical and computational models to deal with the whole as well as understanding the interacting parts.

Biologists construct models to explain living systems

Biologists construct models in many different ways for a variety of uses. Geneticists construct models of interacting networks of proteins that control gene expression, often even drawing cartoon figures to represent that which we cannot see. Population biologists build models of how evolutionary change occurs. Cell biologists build models of signal transduction pathways and the events leading from an external signal to internal events. Structural biologists build actual models of the structure of proteins and macromolecular complexes in cells.

Models provide a way to organize how we think about a problem. Models can also get us closer to the larger picture and away from the extreme reductionist approach. The working parts are provided by the reductionist analysis, but the model shows how they fit together. Often these models suggest other experiments that can be performed to refine or test the model.

As researchers gain more knowledge about the actual flow of molecules in living systems, more sophisticated kinetic models can be used to apply information about isolated enzymes to their cellular context. In systems biology, this modeling is being applied on a large scale to regulatory networks during development, and even to modeling an entire bacterial cell.

The nature of scientific theories

Scientists use the word **theory** in two main ways. The first meaning of theory is a proposed explanation for some natural phenomenon, often based on some general principle. Thus, we speak of the principle first proposed by Newton as the "theory of gravity." Such theories often bring together concepts that were previously thought to be unrelated.

The second meaning of theory is the body of interconnected concepts, supported by scientific reasoning and experimental evidence, that explains the facts in some area of study. Such a theory provides an indispensable framework for organizing a body of knowledge. For example, quantum theory in physics brings together a set of ideas about the nature of the universe, explains experimental facts, and serves as a guide to further questions and experiments.

To a scientist, theories are the solid ground of science, expressing ideas of which we are most certain. In contrast, to the general public, the word theory usually implies the opposite—a *lack* of knowledge, or a guess. Not surprisingly, this difference often results in confusion. In this text, theory will always be used in its scientific sense, in reference to an accepted general principle or body of knowledge.

Some critics outside of science attempt to discredit evolution by saying it is "just a theory." The hypothesis that evolution has occurred, however, is an accepted scientific fact—it is supported by overwhelming evidence. Modern evolutionary theory is a complex body of ideas, the importance of which spreads far beyond explaining evolution. Its ramifications permeate all areas of biology, and it provides the conceptual framework that unifies biology as a science. Again, the key is how well a hypothesis fits the observations. Evolutionary theory fits the observations very well.

Research can be basic or applied

In the past it was fashionable to speak of the "scientific method" as consisting of an orderly sequence of logical, either–or steps. Each step would reject one of two mutually incompatible alternatives, as though trial-and-error testing would inevitably lead a researcher through the maze of uncertainty to the ultimate scientific answer. If this were the case, a computer would make a good scientist. But science is not done this way.

As the British philosopher Karl Popper has pointed out, successful scientists without exception design their experiments with a pretty fair idea of how the results are going to come out. They have what Popper calls an "imaginative preconception" of what the truth might be. Because insight and imagination play such a large role in scientific progress, some scientists are better at science than others—just as Bruce Springsteen stands out among songwriters or Claude Monet stands out among Impressionist painters.

Some scientists perform *basic research*, which is intended to extend the boundaries of what we know. These individuals typically work at universities, and their research is usually supported by grants from various agencies and foundations.

The information generated by basic research contributes to the growing body of scientific knowledge, and it provides the scientific foundation utilized by *applied research*. Scientists who conduct applied research are often employed in some kind of industry. Their work may involve the manufacture of food additives, the creation of new drugs, or the testing of environmental quality.

Research results are written up and submitted for publication in scientific journals, where the experiments and conclusions are reviewed by other scientists. This process of careful evaluation, called *peer review,* lies at the heart of modern science. It helps to ensure that faulty research or false claims are not given the authority of scientific fact. It also provides other scientists with a starting point for testing the reproducibility of experimental results. Results that cannot be reproduced are not taken seriously for long.

Learning Outcomes Review 1.2

Much of science is descriptive, amassing observations to gain an accurate view. Both deductive reasoning and inductive reasoning are used in science. Scientific hypotheses are suggested explanations for observed phenomena. Hypotheses need to make predictions that can be tested by controlled experiments. Theories are coherent explanations of observed data, but they may be modified by new information.

- *How does a scientific theory differ from a hypothesis?*

1.3 An Example of Scientific Inquiry: Darwin and Evolution

Learning Outcomes

1. Examine Darwin's theory of evolution by natural selection as a scientific theory.
2. Describe the evidence that supports the theory of evolution.

Darwin's theory of evolution explains and describes how organisms on Earth have changed over time and acquired a diversity of new forms. This famous theory provides a good example of how a scientist develops a hypothesis and how a scientific theory grows and wins acceptance.

Charles Robert Darwin (1809–1882; figure 1.5) was an English naturalist who, after 30 years of study and observation, wrote one of the most famous and influential books of all time. This book, *On the Origin of Species by Means of Natural Selection,* created a sensation when it was published, and the ideas Darwin expressed in it have played a central role in the development of human thought ever since.

The idea of evolution existed prior to Darwin

In Darwin's time, most people believed that the different kinds of organisms and their individual structures resulted from direct actions of a Creator (many people still believe this). Species were

Figure 1.5 Charles Darwin. This newly rediscovered photograph taken in 1881, the year before Darwin died, appears to be the last ever taken of the great biologist.

thought to have been specially created and to be unchangeable over the course of time.

In contrast to these ideas, a number of earlier naturalists and philosophers had presented the view that living things must have changed during the history of life on Earth. That is, **evolution** has occurred, and living things are now different from how they began. Darwin's contribution was a concept he called *natural selection,* which he proposed as a coherent, logical explanation for this process, and he brought his ideas to wide public attention.

Darwin observed differences in related organisms

The story of Darwin and his theory begins in 1831, when he was 22 years old. He was part of a five-year navigational mapping expedition around the coasts of South America (figure 1.6), aboard H.M.S. *Beagle*. During this long voyage, Darwin had the chance to study a wide variety of plants and animals on continents and islands and in distant seas. Darwin observed a number of phenomena that were of central importance to his reaching his ultimate conclusion.

Repeatedly, Darwin saw that the characteristics of similar species varied somewhat from place to place. These geographical patterns suggested to him that lineages change gradually as species migrate from one area to another. On the Galápagos Islands, 960 km (600 miles) off the coast of Ecuador, Darwin encountered a variety of different finches on the various islands. The 14 species, although related, differed slightly in appearance, particularly in their beaks (figure 1.7).

Darwin thought it was reasonable to assume that all these birds had descended from a common ancestor arriving from the South American mainland several million years ago. Eating different foods on different islands, the finches' beaks had changed during their descent—"descent with modification," or evolution. (These finches are discussed in more detail in chapters 21 and 22.)

8 part | The Molecular Basis of Life

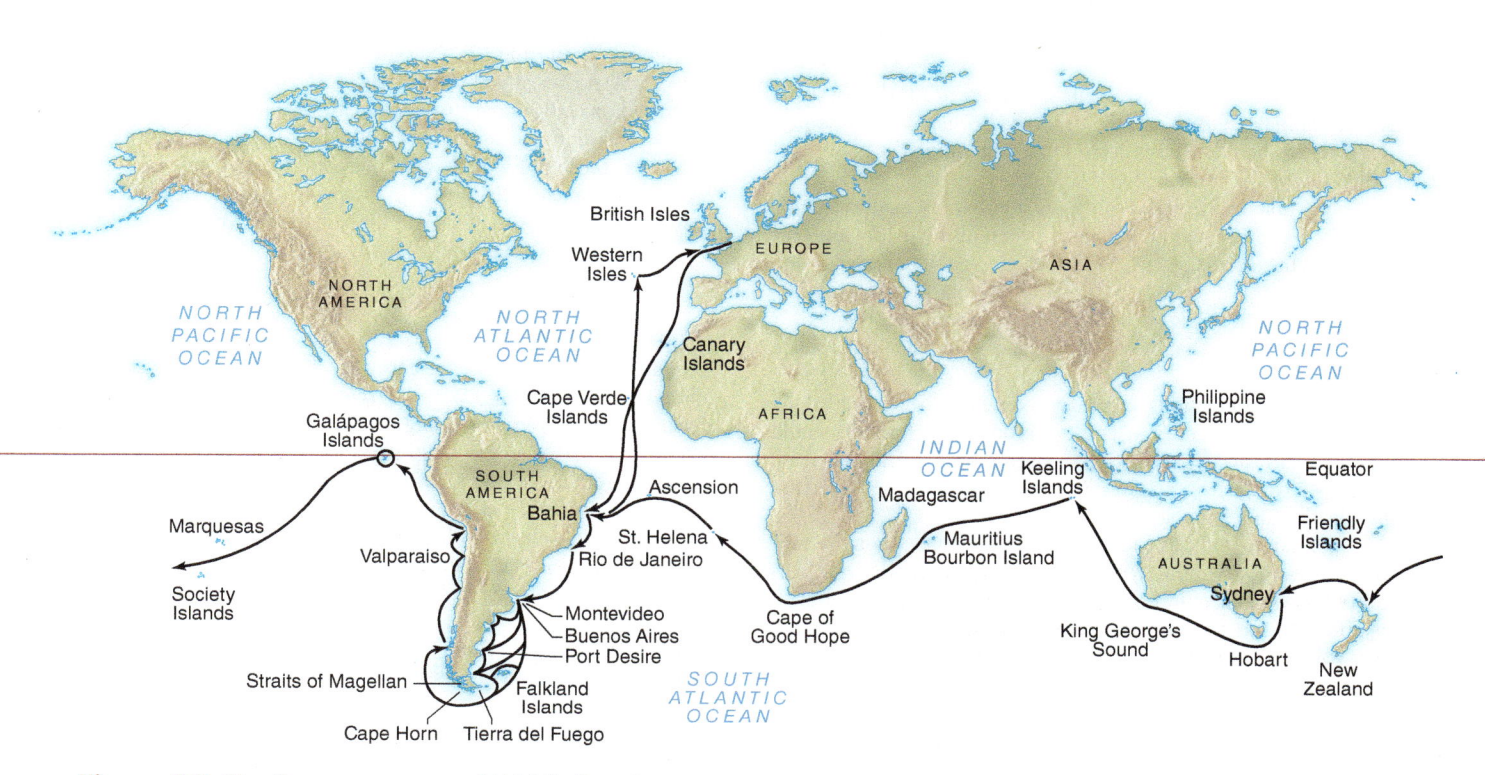

Figure 1.6 The five-year voyage of H.M.S. Beagle. Most of the time was spent exploring the coasts and coastal islands of South America, such as the Galápagos Islands. Darwin's studies of the animals of the Galápagos Islands played a key role in his eventual development of the concept of evolution by means of natural selection.

In a more general sense, Darwin was struck by the fact that the plants and animals on these relatively young volcanic islands resembled those on the nearby coast of South America. If each one of these plants and animals had been created independently and simply placed on the Galápagos Islands, why didn't they resemble the plants and animals of islands with similar climates—such as those off the coast of Africa, for example? Why did they resemble those of the adjacent South American coast instead?

Darwin proposed natural selection as a mechanism for evolution

It is one thing to observe the results of evolution, but quite another to understand how it happens. Darwin's great achievement lies in his ability to move beyond all the individual observations to formulate the hypothesis that evolution occurs because of natural selection.

Figure 1.7 Three Galápagos finches and what they eat. On the Galápagos Islands, Darwin observed 14 different species of finches differing mainly in their beaks and feeding habits. These three finches eat very different food items, and Darwin surmised that the different shapes of their bills represented evolutionary adaptations that improved their ability to eat the foods available in their specific habitats.

Darwin and Malthus

Of key importance to the development of Darwin's insight was his study of Thomas Malthus's *An Essay on the Principle of Population* (1798). In this book, Malthus stated that populations of plants and animals (including humans) tend to increase geometrically, while humans are able to increase their food supply only arithmetically. Put another way, population increases by a multiplying factor—for example, in the series 2, 6, 18, 54, the starting number is multiplied by 3. Food supply increases by an additive factor—for example, the series 2, 4, 6, 8 adds 2 to each starting number. Figure 1.8 shows the difference that these two types of relationships produce over time.

Because populations increase geometrically, virtually any kind of animal or plant, if it could reproduce unchecked, would cover the entire surface of the world surprisingly quickly. Instead, populations of species remain fairly constant year after year, because death limits population numbers.

Sparked by Malthus's ideas, Darwin saw that although every organism has the potential to produce more offspring than can survive, only a limited number actually do survive and produce further offspring. Combining this observation with what he had seen on the voyage of the *Beagle*, as well as with his own experiences in breeding domestic animals, Darwin made an important association: Individuals possessing physical, behavioral, or other attributes that give them an advantage in their environment are more likely to survive and reproduce than those with less advantageous traits. By surviving, these individuals gain the opportunity to pass on their favorable characteristics to their offspring. As the frequency of these characteristics increases in the population, the nature of the population as a whole will gradually change. Darwin called this process *selection*.

Natural selection

Darwin was thoroughly familiar with variation in domesticated animals, and he began *On the Origin of Species* with a detailed discussion of pigeon breeding. He knew that animal breeders selected certain varieties of pigeons and other animals, such as dogs, to produce certain characteristics, a process Darwin called **artificial selection.**

Artificial selection often produces a great variation in traits. Domestic pigeon breeds, for example, show much greater variety than all of the wild species found throughout the world. Darwin thought that this type of change could occur in nature, too. Surely if pigeon breeders could foster variation by artificial selection, nature could do the same—a process Darwin called **natural selection.**

Darwin drafts his argument

Darwin drafted the overall argument for evolution by natural selection in a preliminary manuscript in 1842. After showing the manuscript to a few of his closest scientific friends, however, Darwin put it in a drawer, and for 16 years turned to other research. No one knows for sure why Darwin did not publish his initial manuscript—it is very thorough and outlines his ideas in detail.

The stimulus that finally brought Darwin's hypothesis into print was an essay he received in 1858. A young English naturalist named Alfred Russel Wallace (1823–1913) sent the essay to Darwin from Indonesia; it concisely set forth the hypothesis of evolution by means of natural selection, a hypothesis Wallace had developed independently of Darwin. After receiving Wallace's essay, friends of Darwin arranged for a joint presentation of their ideas at a seminar in London. Darwin then completed his own book, expanding the 1842 manuscript he had written so long ago, and submitted it for publication.

The predictions of natural selection have been tested

More than 130 years have elapsed since Darwin's death in 1882. During this period, the evidence supporting his theory has grown progressively stronger. We briefly explore some of this evidence here; in chapter 21, we will return to the theory of evolution by natural selection and examine the evidence in more detail.

The fossil record

Darwin predicted that the fossil record would yield intermediate links between the great groups of organisms—for example, between fishes and the amphibians thought to have arisen from them, and between reptiles and birds. Furthermore, natural selection predicts the relative positions in time of such transitional forms. We now know the fossil record to a degree that was unthinkable in the

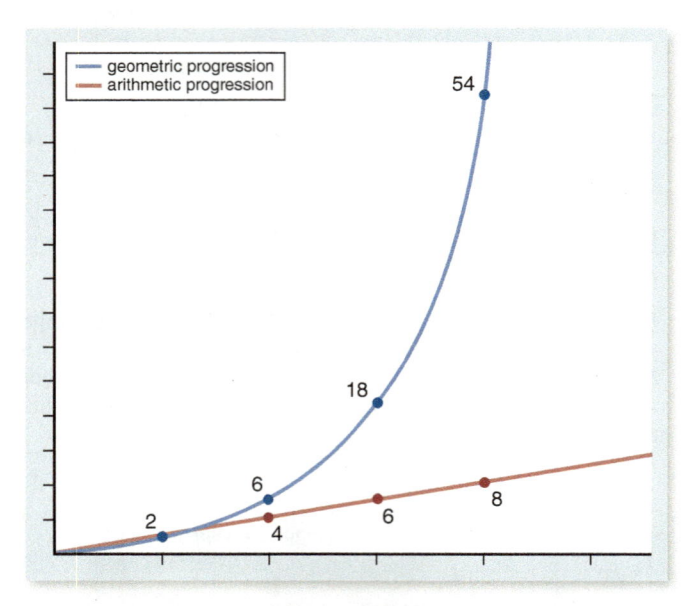

Figure 1.8 Geometric and arithmetic progressions. A geometric progression increases by a constant factor (for example, the curve shown increases ×3 for each step), whereas an arithmetic progression increases by a constant difference (for example, the line shown increases +2 for each step). Malthus contended that the human growth curve was geometric, but the human food production curve was only arithmetic.

Data analysis What is the effect of reducing the constant factor for a geometric progression? How would this change the curve in the figure?

Inquiry question Might this effect be achieved with humans? How?

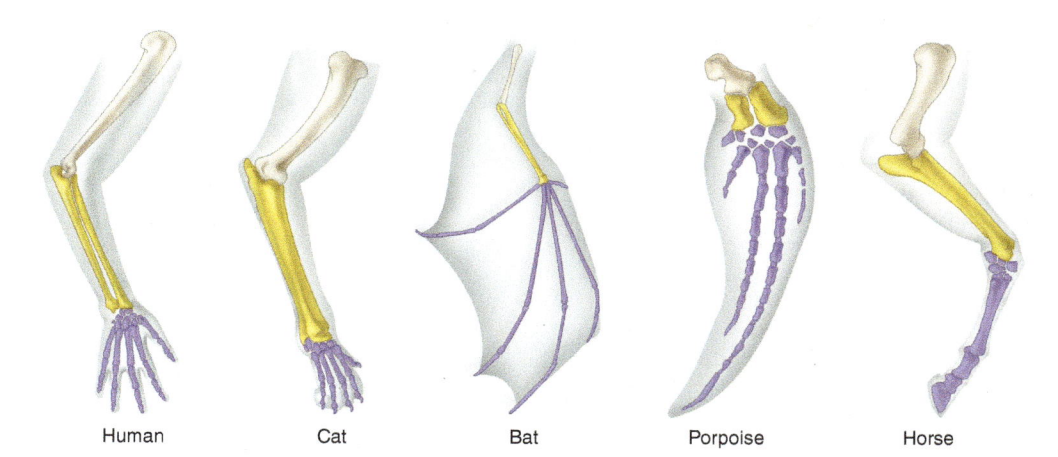

Figure 1.9 Homology among vertebrate limbs. The forelimbs of these five vertebrates show the ways in which the relative proportions of the forelimb bones have changed in relation to the particular way of life of each organism.

19th century, and although truly "intermediate" organisms are hard to determine, paleontologists have found what appear to be transitional forms and found them at the predicted positions in time.

Recent discoveries of microscopic fossils have extended the known history of life on Earth back to about 3.5 billion years ago (BYA). The discovery of other fossils has supported Darwin's predictions and has shed light on how organisms have, over this enormous time span, evolved from the simple to the complex. For vertebrate animals especially, the fossil record is rich and exhibits a graded series of changes in form, with the evolutionary sequence visible for all to see.

The age of the Earth

Darwin's theory predicted the Earth must be very old, but some physicists argued that the Earth was only a few thousand years old. This bothered Darwin, because the evolution of all living things from some single original ancestor would have required a great deal more time. Using evidence obtained by studying the rates of radioactive decay, we now know that the physicists of Darwin's time were very wrong: The Earth was formed about 4.5 BYA.

The mechanism of heredity

Darwin received some of his sharpest criticism in the area of heredity. At that time, no one had any concept of genes or how heredity works, so it was not possible for Darwin to explain completely how evolution occurs.

Even though Gregor Mendel was performing his experiments with pea plants in Brünn, Austria (now Brno, the Czech Republic), during roughly the same period, genetics was established as a science only at the start of the 20th century. When scientists began to understand the laws of inheritance (discussed in chapters 12 and 13), this problem with Darwin's theory vanished.

Comparative anatomy

Comparative studies of animals have provided strong evidence for Darwin's theory. In many different types of vertebrates, for example, the same bones are present, indicating their evolutionary past. Thus, the forelimbs shown in figure 1.9 are all constructed from the same basic array of bones, modified for different purposes.

These bones are said to be **homologous** in the different vertebrates—that is, they have the same evolutionary origin, but they now differ in structure and function. They are contrasted with **analogous** structures, such as the wings of birds and butterflies, which have similar function but different evolutionary origins.

Molecular evidence

Evolutionary patterns are also revealed at the molecular level. By comparing the genomes (that is, the sequences of all the genes) of different groups of animals or plants, we can more precisely specify the degree of relationship among the groups. A series of evolutionary changes over time should involve a continual accumulation of genetic changes in the DNA.

This difference can be seen clearly in the protein hemoglobin (figure 1.10). Rhesus monkeys, which like humans are primates, have fewer differences from humans in the 146-amino-acid

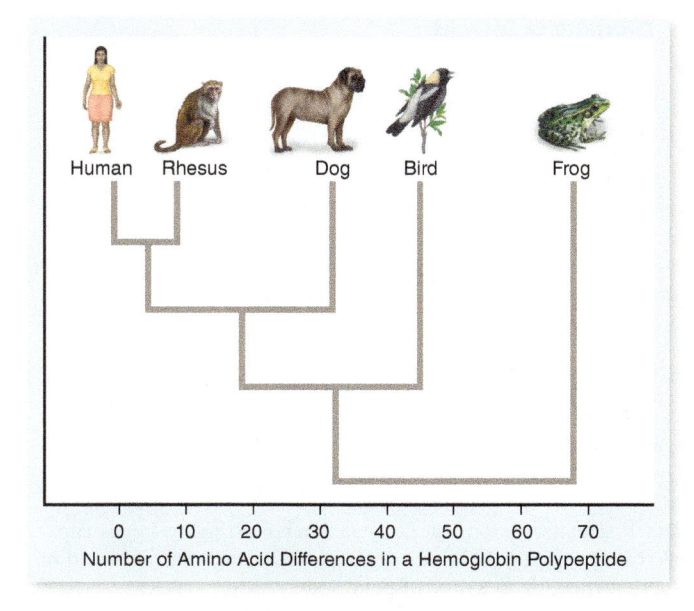

Figure 1.10 Molecules reflect evolutionary patterns. Vertebrates that are more distantly related to humans have a greater number of amino acid differences in the hemoglobin polypeptide.

Inquiry question Where do you imagine a snake might fall on the graph? Why?

hemoglobin β chain than do more distantly related mammals, such as dogs. Nonmammalian vertebrates, such as birds and frogs, differ even more.

The sequences of some genes, such as the ones specifying the hemoglobin proteins, have been determined in many organisms, and the entire time course of their evolution can be laid out with confidence by tracing the origins of particular nucleotide changes in the gene sequence. The pattern of descent obtained is called a **phylogenetic tree.** It represents the evolutionary history of the gene, its "family tree." Molecular phylogenetic trees agree well with those derived from the fossil record, which is strong direct evidence of evolution. The pattern of accumulating DNA changes represents, in a real sense, the footprints of evolutionary history.

Learning Outcomes Review 1.3
Darwin observed differences in related organisms and proposed the hypothesis of evolution by natural selection to explain these differences. The predictions generated by natural selection have been tested and continue to be tested by analysis of the fossil record, genetics, comparative anatomy, and even the DNA of living organisms.

- *Does Darwin's theory of evolution by natural selection explain the origin of life?*

1.4 Unifying Themes in Biology

Learning Outcomes
1. Discuss the unifying themes in biology.
2. Contrast living and nonliving systems.

The study of biology encompasses a large number of different subdisciplines, ranging from biochemistry to ecology. In all of these, however, unifying themes can be identified. Among these are cell theory, the molecular basis of inheritance, the relationship between structure and function, evolution, and the emergence of novel properties.

Living systems are organized into cells

As was stated at the beginning of this chapter, all organisms are composed of cells, life's basic units (figure 1.11). Cells were discovered by Robert Hooke in England in 1665, using one of the first microscopes, one that magnified 30 times. Not long after that, the Dutch scientist Anton van Leeuwenhoek used microscopes capable of magnifying 300 times and discovered an amazing world of single-celled life in a drop of pond water.

In 1839, the German biologists Matthias Schleiden and Theodor Schwann, summarizing a large number of observations by themselves and others, concluded that all living organisms consist of cells. Their conclusion has come to be known as the **cell theory.** Later, biologists added the idea that all cells come from preexisting cells. The cell theory, one of the basic ideas in biology, is the foundation for understanding the reproduction and growth of all organisms.

a.
60 μm

b.
500 μm

Figure 1.11 Cellular basis of life. All organisms are composed of cells. Some organisms, including the protists, shown in part *(a)* are single-celled. Others, such as the plant shown in cross section in part *(b)* consist of many cells.

The molecular basis of inheritance explains the continuity of life

Even the simplest cell is incredibly complex—more intricate than any computer. The information that specifies what a cell is like—its detailed plan—is encoded in **deoxyribonucleic acid (DNA),** a long, cablelike molecule. Each DNA molecule is formed from two long chains of building blocks, called nucleotides, wound around each other (see chapter 14). Four different nucleotides are found in DNA, and the sequence in which they occur encodes the cell's information. Specific sequences of several hundred to many thousand nucleotides make up a **gene,** a discrete unit of information.

The continuity of life from one generation to the next—heredity—depends on the faithful copying of a cell's DNA into daughter cells. The entire set of DNA instructions that specifies a cell is called its *genome*. The sequence of the human genome, 3 billion nucleotides long, was decoded in rough draft form in 2001, a triumph of scientific investigation.

The relationship between structure and function underlies living systems

One of the unifying themes of molecular biology is the relationship between structure and function. Function in molecules, and larger macromolecular complexes, is dependent on their structure.

Although this observation may seem trivial, it has far-reaching implications. We study the structure of molecules and macromolecular complexes to learn about their function. When we know the function of a particular structure, we can infer the function of similar structures found in different contexts, such as in different organisms.

Biologists study both aspects, looking for the relationships between structure and function. On the one hand, this allows similar structures to be used to infer possible similar functions. On the other hand, this knowledge also gives clues as to what kinds of structures may be involved in a process if we know about the functionality.

For example, suppose that we know the structure of a human cell's surface receptor for insulin, the hormone that controls uptake of glucose. We then find a similar molecule in the membrane of a cell from a different species—perhaps even a very different organism, such as a worm. We might conclude that this membrane molecule acts as a receptor for an insulin-like molecule produced by the worm. In this way, we might be able to discern the evolutionary relationship between glucose uptake in worms and in humans.

The diversity of life arises by evolutionary change

The unity of life that we see in certain key characteristics shared by many related life-forms contrasts with the incredible diversity of living things in the varied environments of Earth. The underlying unity of biochemistry and genetics argues that all life has evolved from the same origin event. The diversity of life arises by evolutionary change leading to the present biodiversity we see.

Biologists divide life's great diversity into three great groups, called domains: Bacteria, Archaea, and Eukarya (figure 1.12). The domains Bacteria and Archaea are composed of single-celled organisms *(prokaryotes)* with little internal structure, and the domain Eukarya is made up of organisms *(eukaryotes)* composed of a complex, organized cell or multiple complex cells.

Within Eukarya are four main groups called kingdoms (figure 1.12). Kingdom Protista consists of all the unicellular eukaryotes except yeasts (which are fungi), as well as the multicellular algae. Because of the great diversity among the protists, many biologists feel kingdom Protista should be split into several kingdoms.

Kingdom Plantae consists of organisms that have cell walls of cellulose and obtain energy by photosynthesis. Organisms in

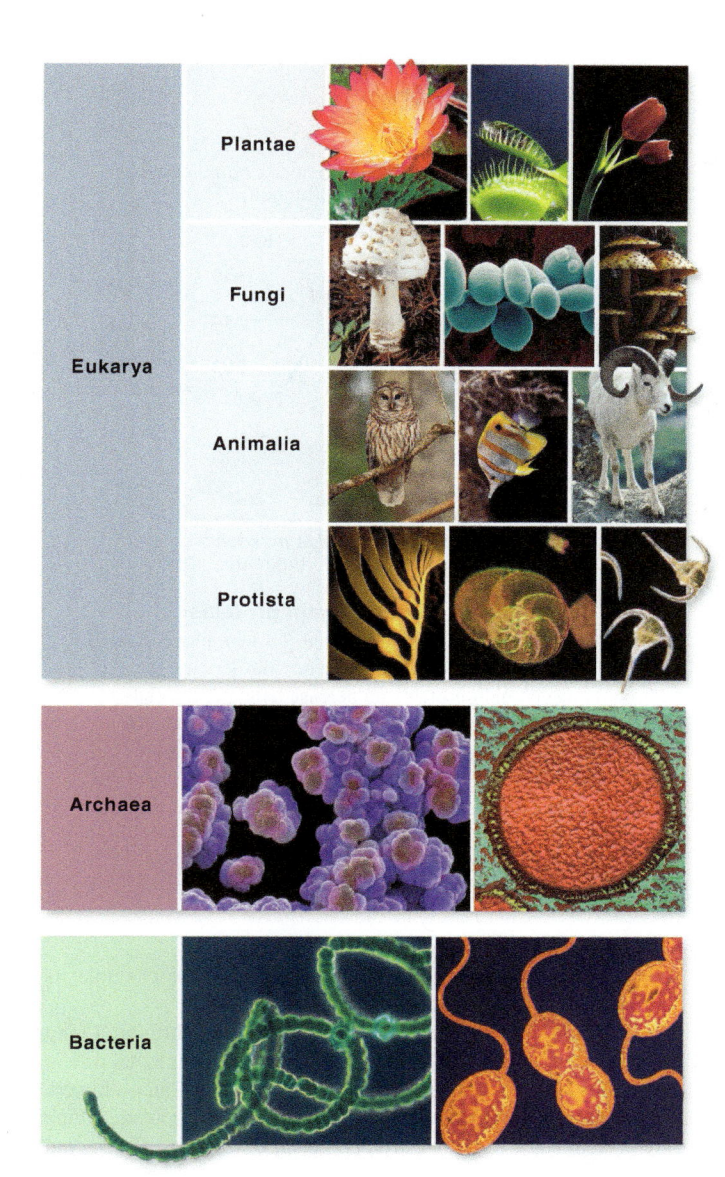

Figure 1.12 The diversity of life. Biologists categorize all living things into three overarching groups called domains: Bacteria, Archaea, and Eukarya. Domain Eukarya is composed of four kingdoms: Plantae, Fungi, Animalia, and Protista.

the kingdom Fungi have cell walls of chitin and obtain energy by secreting digestive enzymes and then absorbing the products they release from the external environment. Kingdom Animalia contains organisms that lack cell walls and obtain energy by first ingesting other organisms and then digesting them internally.

Evolutionary conservation explains the unity of living systems

Biologists agree that all organisms alive today have descended from some simple cellular creature that arose about 3.5 BYA. Some of the characteristics of that earliest organism have been preserved.

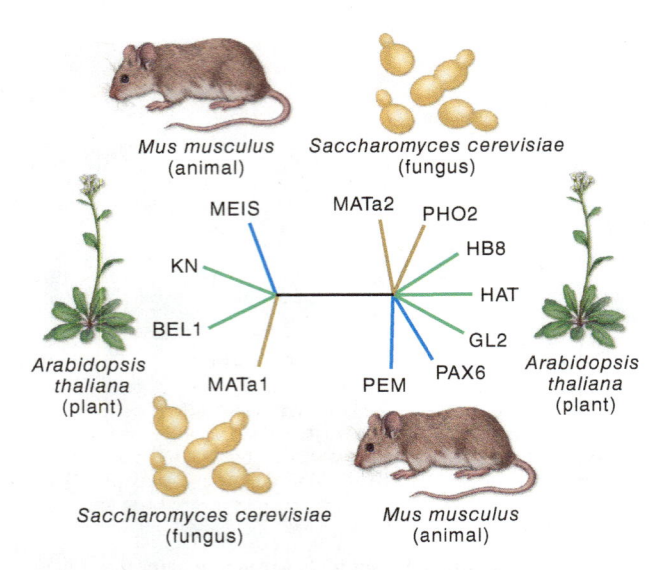

Figure 1.13 Tree of homeodomain proteins.
Homeodomain proteins are found in fungi *(brown)*, plants *(green)*, and animals *(blue)*. Based on their sequence similarities, these 11 different homeodomain proteins (uppercase letters at the ends of branches) fall into two groups, with representatives from each kingdom in each group. That means, for example, the mouse homeodomain protein PAX6 is more closely related to fungal and flowering plant proteins, such as PHO2 and GL2, than it is to the mouse protein MEIS.

The storage of hereditary information in DNA, for example, is common to all living things.

Evolutionary conservation of characteristics through a long line of descent usually reflects that they have a fundamental role in the biology of the organism—one not easily changed once adopted. A good example is provided by the homeodomain proteins, which play critical roles in early development in eukaryotes. Conserved characteristics can be seen in approximately 1850 homeodomain proteins, distributed among three different kingdoms of organisms (figure 1.13). The homeodomain proteins are powerful developmental tools that evolved early, and for which no better alternative has arisen.

Cells are information-processing systems

One way to think about cells is as highly complex nanomachines that process information. The information stored in DNA is used to direct the synthesis of cellular components, and the particular set of components can differ from cell to cell. The way that proteins fold in space is a form of information that is three-dimensional, and interesting properties emerge from the interaction of these shapes in macromolecular complexes. The control of gene expression allows differentiation of cell types in time and space, leading to changes over developmental time into different tissue types—even though all cells in an organism carry the same genetic information.

Cells also process information that they receive about the environment. Cells sense their environment through proteins in their membranes, and this information is transmitted across the membrane to elaborate signal-transduction chemical pathways that can change the functioning of a cell.

This ability of cells to sense and respond to their environment is critical to the function of tissues and organs in multicellular organisms. A multicellular organism can regulate its internal environment, maintaining constant temperature, pH, and concentrations of vital ions. This homeostasis is possible because of elaborate signaling networks that coordinate the activities of different cells in different tissues.

Living systems exist in a nonequilibrium state

A key feature of living systems is that they are open systems that function far from thermodynamic equilibrium. This has a number of implications for their behavior. A constant supply of energy is necessary to maintain a stable nonequilibrium state. Consider the state of the nucleic acids, and proteins in all of your cells: At equilibrium they are not polymers, they would all be hydrolyzed to monomer nucleotides and amino acids. Second, nonequilibrium systems exhibit self-organizing properties not seen in equilibrium systems.

These self-organizing properties of living systems show up at different levels of the hierarchical organization. At the cellular level, macromolecular complexes such as the spindle necessary for chromosome separation can self-organize. At the population level, a flock of birds, a school of fish, or the bacteria in a biofilm are all also self-organizing. This kind of interacting behavior of individual units leads to emergent properties that are not predictable from the nature of the units themselves.

Emergent properties are properties of collections of molecules, cells, individuals, that are distinct from the categorical properties that can be described by such statistics as mean and standard deviation. The mathematics necessary to describe these kind of interacting systems is nonlinear dynamics. The emerging field of systems biology is beginning to model biological systems in this way. The kinds of feedback and feedforward loops that exist between molecules in cells, or neurons in a nervous system, lead to emergent behaviors like human consciousness.

Learning Outcomes Review 1.4

Biology is a broad and complex field, but we can identify unifying themes in this complexity. Cells are the basic unit of life, and they are information-processing machines. The structures of molecules, macromolecular complexes, cells, and even higher levels of organization are related to their functions. The diversity of life can be classified and organized based on similar features; biologists identify three large domains that encompass six kingdoms. Living organisms are able to use energy to construct complex molecules from simple ones, and are thus not in a state of thermodynamic equilibrium.

■ *How do viruses fit into our definitions of living systems?*

Chapter Review

1.1 The Science of Life

Biology unifies much of natural science.
The study of biological systems is interdisciplinary because solutions require many different approaches to solve a problem.

Life defies simple definition.
Although life is difficult to define, living systems have seven characteristics in common. They are composed of one or more cells; are complex and highly ordered; can respond to stimuli; can grow, reproduce, and transmit genetic information to their offspring; need energy to accomplish work; can maintain relatively constant internal conditions (homeostasis); and are capable of evolutionary adaptation to the environment.

Living systems show hierarchical organization.
The hierarchical organization of living systems progresses from atoms to the biosphere. At each higher level, emergent properties arise that are greater than the sum of the parts.

1.2 The Nature of Science

At its core, science is concerned with understanding the nature of the world by using observation and reasoning.

Much of science is descriptive.
Science is concerned with developing an increasingly accurate description of nature through observation and experimentation.

Science uses both deductive and inductive reasoning.
Deductive reasoning applies general principles to predict specific results. Inductive reasoning uses specific observations to construct general scientific principles.

Hypothesis-driven science makes and tests predictions.
Hypotheses are based on observations, and generate testable predictions. Experiments involve a test where a variable is manipulated, and a control where the variable is not manipulated. If the predictions cannot be verified the hypothesis is rejected.

Reductionism breaks larger systems into their component parts.
Reductionism attempts to understand a complex system by breaking it down into its component parts. It is limited because parts may act differently when isolated from the larger system.

Biologists construct models to explain living systems.
A model provides a way of organizing our thinking about a problem; models may also suggest experimental approaches.

The nature of scientific theories.
Scientists use the word *theory* in two main ways: as a proposed explanation for some natural phenomenon and as a body of concepts that explains facts in an area of study.

Research can be basic or applied.
Basic research extends the boundaries of what we know; applied research seeks to use scientific findings in practical areas such as agriculture, medicine, and industry.

1.3 An Example of Scientific Inquiry: Darwin and Evolution

Darwin's theory of evolution shows how a scientist develops a hypothesis and sets forth evidence, as well as how a scientific theory grows and gains acceptance.

The idea of evolution existed prior to Darwin.
A number of naturalists and philosophers had suggested living things had changed during Earth's history. Darwin's contribution was the concept of natural selection as a mechanism for evolutionary change.

Darwin observed differences in related organisms.
During the voyage of the H.M.S. *Beagle,* Darwin had an opportunity to observe worldwide patterns of diversity.

Darwin proposed natural selection as a mechanism for evolution.
Darwin noted that species produce many more offspring than will survive and reproduce. He observed that traits can be changed by artificial selection. Darwin proposed that individuals possessing traits that increase survival and reproductive success become more numerous in populations over time. Darwin called this descent with modification (natural selection). Alfred Russel Wallace independently came to the same conclusions.

The predictions of natural selection have been tested.
Natural selection has been tested using data from many fields. Among these are the fossil record; the age of the Earth, determined by rates of radioactive decay to be 4.5 billion years; genetic experiments showing that traits can be inherited as discrete units; comparative anatomy and the study of homologous structures; and molecular data that provide evidence for changes in DNA and proteins over time.

Taken together, these findings strongly support evolution by natural selection. No data to conclusively disprove evolution have been found.

1.4 Unifying Themes in Biology

Living systems are organized into cells.
The cell is the basic unit of life and is the foundation for understanding growth and reproduction in all organisms.

The molecular basis of inheritance explains the continuity of life.
Hereditary information, encoded in genes found in the DNA molecule, is passed on from one generation to the next.

The relationship between structure and function underlies living systems.
The function of macromolecules and their complexes is dictated by and dependent on their structure. Similarity of structure and function from one life-form to another may indicate an evolutionary relationship.

The diversity of life arises by evolutionary change.
Living organisms appear to have had a common origin from which a diversity of life arose by evolutionary change. They can be grouped into three domains comprising six kingdoms based on their differences.

Evolutionary conservation explains the unity of living systems.
The underlying similarities in biochemistry and genetics support the contention that all life evolved from a single source.

Cells are information-processing systems.
Cells can sense and respond to environmental changes through proteins located on their cell membranes. Differential expression of stored genetic information is the basis for different cell types.

Living systems exist in a nonequilibrium state.
Organisms are open systems that need a constant supply of energy to maintain their stable nonequilibrium state. Living things are able to self-organize, creating levels of complexity that may exhibit emergent properties.

Review Questions

UNDERSTAND

1. Which of the following is NOT a property of life?
   a. Energy utilization
   b. Movement
   c. Order
   d. Homeostasis

2. The process of inductive reasoning involves
   a. the use of general principles to predict a specific result.
   b. the generation of specific predictions based on a belief system.
   c. the use of specific observations to develop general principles.
   d. the use of general principles to support a hypothesis.

3. A hypothesis in biology is best described as
   a. a possible explanation of an observation.
   b. an observation that supports a theory.
   c. a general principle that explains some aspect of life.
   d. an unchanging statement that correctly predicts some aspect of life.

4. A scientific theory is
   a. a guess about how things work in the world.
   b. a statement of how the world works that is supported by experimental data.
   c. a belief held by many scientists.
   d. Both a and c are correct.

5. The cell theory states that
   a. cells are small.
   b. cells are highly organized.
   c. there is only one basic type of cell.
   d. all living things are made up of cells.

6. The molecule DNA is important to biological systems because
   a. it can be replicated.
   b. it encodes the information for making a new individual.
   c. it forms a complex, double-helical structure.
   d. nucleotides form genes.

7. The organization of living systems is
   a. linear with cells at one end and the biosphere at the other.
   b. circular with cells in the center.
   c. hierarchical with cells at the base, and the biosphere at the top.
   d. chaotic and beyond description.

8. The idea of evolution
   a. was original to Darwin.
   b. was original to Wallace.
   c. predated Darwin and Wallace.
   d. Both a and b are correct.

APPLY

1. What is the significance of Pasteur's experiment to test the germ hypothesis?
   a. It proved that heat can sterilize a broth.
   b. It demonstrated that cells can arise spontaneously.
   c. It demonstrated that some cells are germs.
   d. It demonstrated that cells can only arise from other cells.

2. Which of the following is NOT an example of reductionism?
   a. Analysis of an isolated enzyme's function in an experimental assay
   b. Investigation of the effect of a hormone on cell growth in a Petri dish
   c. Observation of the change in gene expression in response to specific stimulus
   d. An evaluation of the overall behavior of a cell

3. How is the process of natural selection different from that of artificial selection?
   a. Natural selection produces more variation.
   b. Natural selection makes an individual better adapted.
   c. Artificial selection is a result of human intervention.
   d. Artificial selection results in better adaptations.

4. If you found a fossil for a modern organism next to the fossil of a dinosaur, this would
   a. argue against evolution by natural selection.
   b. have no bearing on evolution by natural selection.
   c. indicate that dinosaurs may still exist.
   d. Both b and c are correct.

5. The theory of evolution by natural selection is a good example of how science proceeds because
   a. it rationalizes a large body of observations.
   b. it makes predictions that have been tested by a variety of approaches.
   c. it represents Darwin's belief of how life has changed over time.
   d. Both b and c are correct.

6. In which domain of life would you find only single-celled organisms?
   a. Eukarya
   b. Bacteria
   c. Archaea
   d. Both b and c are correct.

7. Evolutionary conservation occurs when a characteristic is
   a. important to the life of the organism.
   b. not influenced by evolution.
   c. no longer functionally important.
   d. found in more primitive organisms.

SYNTHESIZE

1. Exobiology is the study of life on other planets. In recent years, scientists have sent various spacecraft out into the galaxy in search for extraterrestrial life. Assuming that all life shares common properties, what should exobiologists be looking for as they explore other worlds?

2. The classic experiment by Pasteur (figure 1.4) tested the hypothesis that cells arise from other cells. In this experiment cell growth was measured following sterilization of broth in a swan-necked flask or in a flask with a broken neck.
   a. Which variables were kept the same in these two experiments?
   b. How does the shape of the flask affect the experiment?
   c. Predict the outcome of each experiment based on the two hypotheses.
   d. Some bacteria (germs) are capable of producing heat-resistant spores that protect the cell and allow it to continue to grow after the environment cools. How would the outcome of this experiment have been affected if spore-forming bacteria were present in the broth?

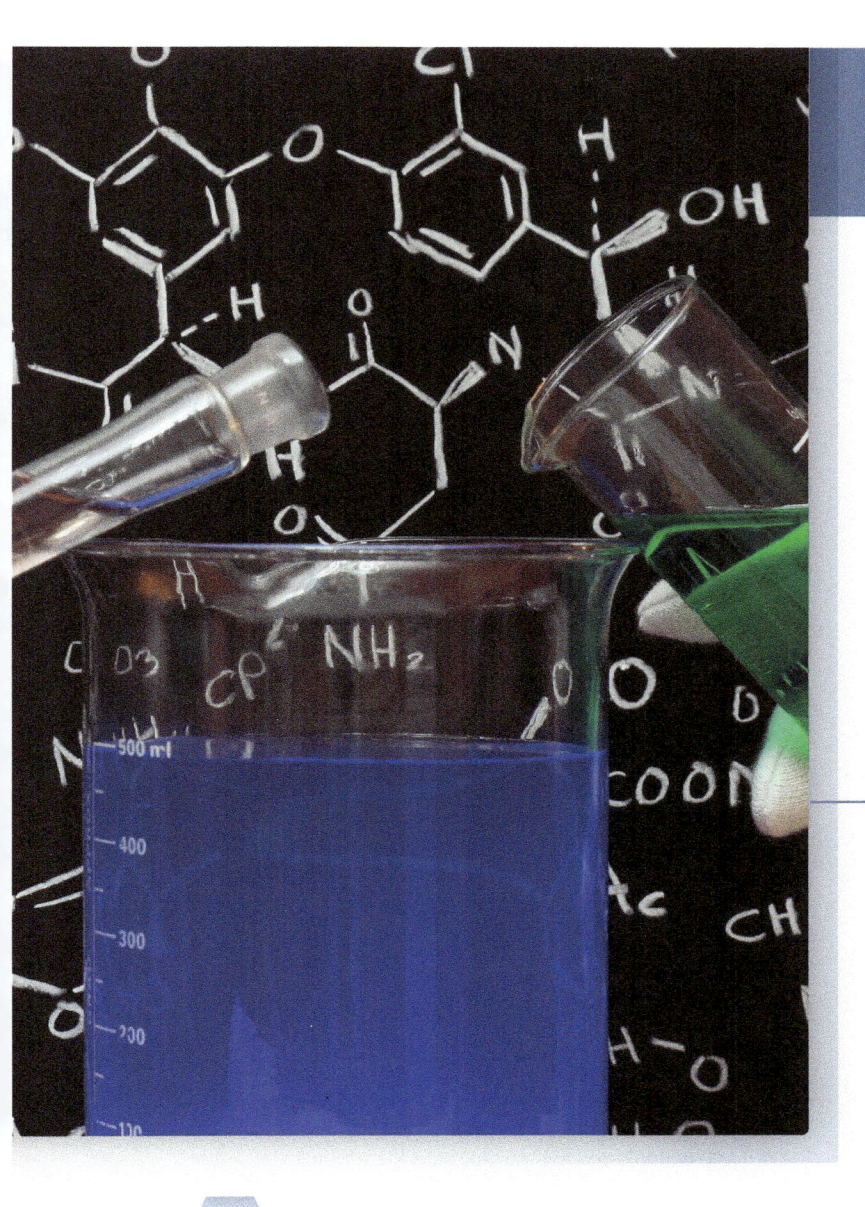

CHAPTER 2

The Nature of Molecules and the Properties of Water

Chapter Contents

2.1 The Nature of Atoms
2.2 Elements Found in Living Systems
2.3 The Nature of Chemical Bonds
2.4 Water: A Vital Compound
2.5 Properties of Water
2.6 Acids and Bases

Introduction

About 12.5 billion years ago (BYA), an enormous explosion probably signaled the beginning of the universe. This explosion started a process of star building and planetary formation that eventually led to the formation of Earth, about 4.5 BYA. Around 3.5 BYA, life began on Earth and started to diversify. To understand the nature of life on Earth, we first need to understand the nature of the matter that forms the building blocks of all life.

The earliest speculations about the world around us included this most basic question, "What is it made of?" The ancient Greeks recognized that larger things may be built of smaller parts. This concept was formed into a solid experimental scientific idea in the early 20th century, when physicists began trying to break atoms apart. From those humble beginnings to the huge particle accelerators used by the modern physicists of today, the picture of the atomic world emerges as fundamentally different from the tangible, macroscopic world around us.

To understand how living systems are assembled, we must first understand a little about atomic structure, about how atoms can be linked together by chemical bonds to make molecules, and about the ways in which these small molecules are joined together to make larger molecules, until finally we arrive at the structures of cells and then of organisms. Our study of life on Earth therefore begins with physics and chemistry. For many of you, this chapter will be a review of material encountered in other courses.

2.1 The Nature of Atoms

Learning Outcomes
1. Define an element based on its composition.
2. Describe the relationship between atomic structure and chemical properties.
3. Explain where electrons are found in an atom.

Any substance in the universe that has mass and occupies space is defined as *matter*. All matter is composed of extremely small particles called **atoms**. Because of their size, atoms are difficult to study. Not until early in the 20th century did scientists carry out the first experiments revealing the physical nature of atoms (figure 2.1).

SCIENTIFIC THINKING

Hypothesis: *Atoms are composed of diffuse positive charge with embedded negative charge (electrons).*

Prediction: *If alpha (α) particles, which are helium nuclei, are shot at a thin foil of gold, the α particles will not be deflected much by the diffuse positive charge or by the light electrons.*

Test: *α Particles are shot at a thin sheet of gold foil surrounded by a detector screen, which shows flashes of light when hit by the particles.*

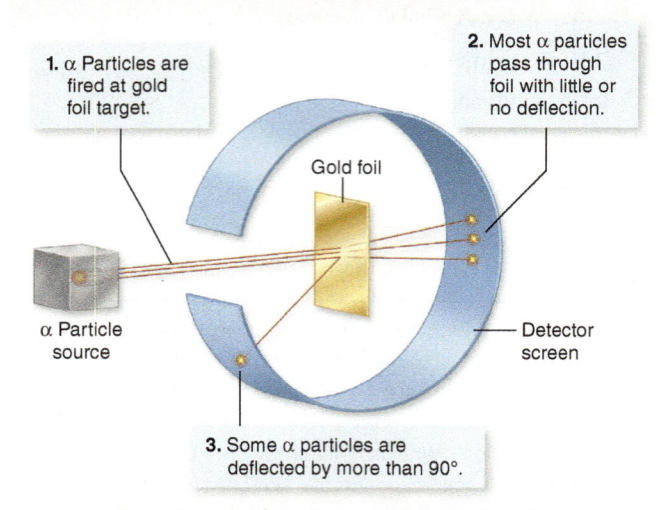

1. α Particles are fired at gold foil target.
2. Most α particles pass through foil with little or no deflection.
3. Some α particles are deflected by more than 90°.

Result: *Most particles are not deflected at all, but a small percentage of particles are deflected at angles of 90° or more.*
Conclusion: *The hypothesis is not supported. The large deflections observed led to a view of the atom as composed of a very small central region containing positive charge (the nucleus) surrounded by electrons.*
Further Experiments: *How does the Bohr atom with its quantized energy for electrons extend this model?*

Figure 2.1 Rutherford scattering experiment. Large-angle scattering of α particles led Rutherford to propose the existence of the nucleus.

Atomic structure includes a central nucleus and orbiting electrons

Objects as small as atoms can be "seen" only indirectly, by using complex technology such as tunneling microscopy (figure 2.2). We now know a great deal about the complexities of atomic structure, but the simple view put forth in 1913 by the Danish physicist Niels Bohr provides a good starting point for understanding atomic theory. Bohr proposed that every atom possesses an orbiting cloud of tiny subatomic particles called *electrons* whizzing around a core, like the planets of a miniature solar system. At the center of each atom is a small, very dense nucleus formed of two other kinds of subatomic particles: *protons* and *neutrons* (figure 2.3).

Atomic number

Different atoms are defined by the number of protons, a quantity called the *atomic number*. Atoms with the same atomic number (that is, the same number of protons) have the same chemical properties and are said to belong to the same element. Formally speaking, an *element* is any substance that cannot be broken down to any other substance by ordinary chemical means.

Within the nucleus, the cluster of protons and neutrons is held together by a force that works only over short, subatomic distances. Each proton carries a positive (+) charge, and each neutron has no charge. Each electron carries a negative (−) charge. Typically, an atom has one electron for each proton and is thus electrically neutral. The chemical behavior of an atom is due to the number and configuration of electrons, as we will see later in this section.

Atomic mass

The terms *mass* and *weight* are often used interchangeably, but they have slightly different meanings. *Mass* refers to the amount of a substance, but *weight* refers to the force gravity exerts on a substance. An object has the same mass whether it is on the Earth or the Moon, but its weight will be greater on the Earth because the Earth's gravitational force is greater than the Moon's. The *atomic mass* of an atom is equal to the sum of the masses of its protons and neutrons. Atoms that occur naturally on Earth contain from 1 to 92 protons and up to 146 neutrons.

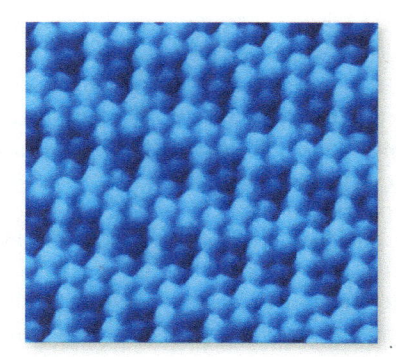

Figure 2.2 Scanning-tunneling microscope image. The scanning-tunneling microscope is a nonoptical way of imaging that allows atoms to be visualized. This image shows a lattice of oxygen atoms *(dark blue)* on a rhodium crystal *(light blue)*.

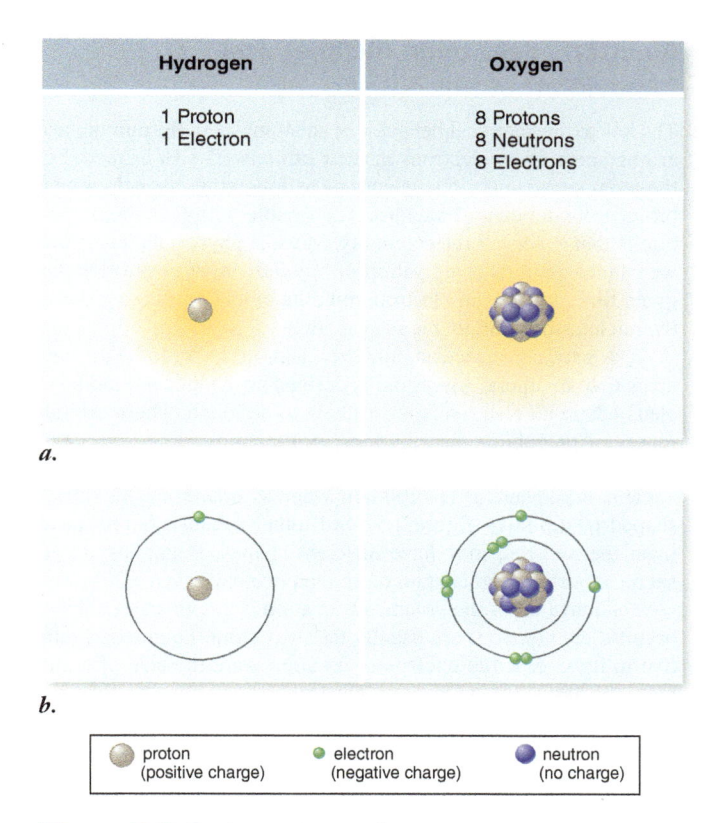

Figure 2.3 Basic structure of atoms. All atoms have a nucleus consisting of protons and neutrons, except hydrogen, the smallest atom, which usually has only one proton and no neutrons in its nucleus. Oxygen typically has eight protons and eight neutrons in its nucleus. In the simple "Bohr model" of atoms pictured here, electrons spin around the nucleus at a relatively far distance. *a.* Atoms are depicted as a nucleus with a cloud of electrons (not shown to scale). *b.* The electrons are shown in discrete energy levels. These are described in greater detail in the text.

The mass of atoms and subatomic particles is measured in units called *daltons*. To give you an idea of just how small these units are, note that it takes 602 million million billion (6.02×10^{23}) daltons to make 1 gram (g). A proton weighs approximately 1 dalton (actually 1.007 daltons), as does a neutron (1.009 daltons). In contrast, electrons weigh only 1/1840 of a dalton, so they contribute almost nothing to the overall mass of an atom.

Electrons

The positive charges in the nucleus of an atom are neutralized, or counterbalanced, by negatively charged electrons, which are located in regions called **orbitals** that lie at varying distances around the nucleus. Atoms with the same number of protons and electrons are electrically neutral—that is, they have no net charge, and are therefore called *neutral atoms*.

Electrons are maintained in their orbitals by their attraction to the positively charged nucleus. Sometimes other forces overcome this attraction, and an atom loses one or more electrons. In other cases, atoms gain additional electrons. Atoms in which the number of electrons does not equal the number of protons are known as *ions,* and they are charged particles. An atom having more protons than electrons has a net positive charge and is called a **cation.** For example, an atom of sodium (Na) that has lost one electron becomes a sodium ion (Na$^+$), with a charge of +1. An atom having fewer protons than electrons carries a net negative charge and is called an **anion.** A chlorine atom (Cl) that has gained one electron becomes a chloride ion (Cl$^-$), with a charge of −1.

Isotopes

Although all atoms of an element have the same number of protons, they may not all have the same number of neutrons. Atoms of a single element that possess different numbers of neutrons are called **isotopes** of that element.

Most elements in nature exist as mixtures of different isotopes. Carbon (C), for example, has three isotopes, all containing six protons (figure 2.4). Over 99% of the carbon found in nature exists as an isotope that also contains six neutrons. Because the total mass of this isotope is 12 daltons (6 from protons plus 6 from neutrons), it is referred to as carbon-12 and is symbolized ^{12}C. Most of the rest of the naturally occurring carbon is carbon-13, an isotope with seven neutrons. The rarest carbon isotope is carbon-14, with eight neutrons. Unlike the other two isotopes, carbon-14 is unstable: This means that its nucleus tends to break up into elements with lower atomic numbers. This nuclear breakup, which emits a significant amount of energy, is called *radioactive decay,* and isotopes that decay in this fashion are **radioactive isotopes.**

Some radioactive isotopes are more unstable than others, and therefore they decay more readily. For any given isotope, however, the rate of decay is constant. The decay time is usually

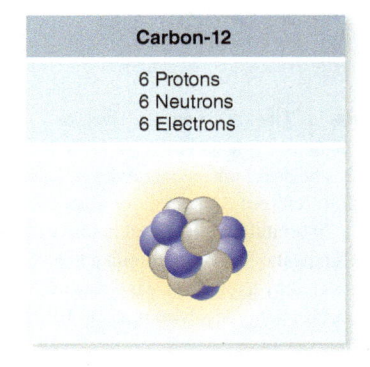

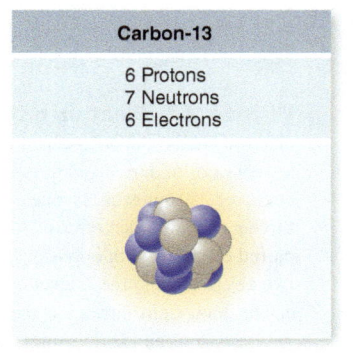

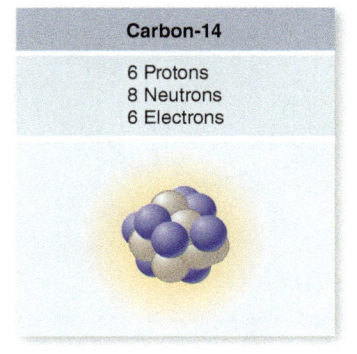

Figure 2.4 The three most abundant isotopes of carbon. Isotopes of a particular element have different numbers of neutrons.

expressed as the *half-life*, the time it takes for one-half of the atoms in a sample to decay. Carbon-14, for example, often used in the carbon dating of fossils and other materials, has a half-life of 5730 years. A sample of carbon containing 1 g of carbon-14 today would contain 0.5 g of carbon-14 after 5730 years, 0.25 g 11,460 years from now, 0.125 g 17,190 years from now, and so on. By determining the ratios of the different isotopes of carbon and other elements in biological samples and in rocks, scientists are able to accurately determine when these materials formed.

Radioactivity has many useful applications in modern biology. Radioactive isotopes are one way to label, or "tag," a specific molecule and then follow its progress, either in a chemical reaction or in living cells and tissue. The downside, however, is that the energetic subatomic particles emitted by radioactive substances have the potential to severely damage living cells, producing genetic mutations and, at high doses, cell death. Consequently, exposure to radiation is carefully controlled and regulated. Scientists who work with radioactivity follow strict handling protocols and wear radiation-sensitive badges to monitor their exposure over time to help ensure a safe level of exposure.

Electrons determine the chemical behavior of atoms

The key to the chemical behavior of an atom lies in the number and arrangement of its electrons in their orbitals. The Bohr model of the atom shows individual electrons as following distinct circular orbits around a central nucleus. The trouble with this simple picture is that it doesn't reflect reality. Modern physics indicates that we cannot pinpoint the position of any individual electron at any given time. In fact, an electron could be anywhere, from close to the nucleus to infinitely far away from it.

A particular electron, however, is more likely to be in some areas than in others. An orbital is defined as the area around a nucleus where an electron is most likely to be found. These orbitals represent probability distributions for electrons—that is, regions more likely to contain an electron. Some electron orbitals near the nucleus are spherical (*s* orbitals), whereas others are dumbbell-shaped (*p* orbitals) (figure 2.5). Still other orbitals, farther away from the nucleus, may have different shapes. Regardless of its shape, no orbital can contain more than two electrons.

Almost all of the volume of an atom is empty space. This is because the electrons are usually far away from the nucleus, relative to its size. If the nucleus of an atom were the size of a golf ball, the orbit of the nearest electron would be a mile away. Consequently, the nuclei of two atoms never come close enough in nature to interact with each other. It is for this reason that an atom's electrons, not its protons or neutrons, determine its chemical behavior, and it also explains why the isotopes of an element, all of which have the same arrangement of electrons, behave the same way chemically.

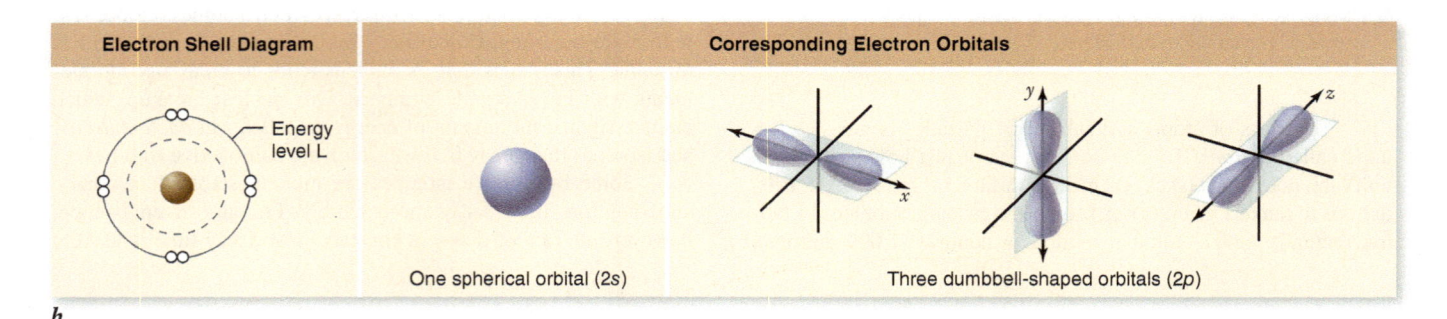

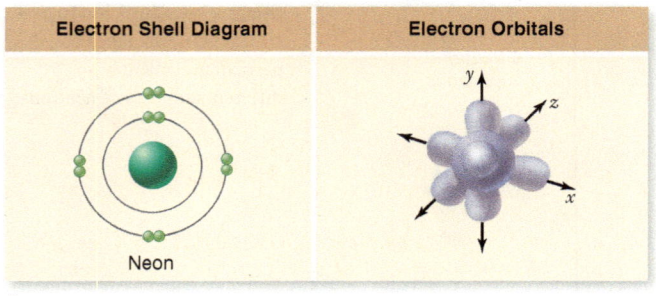

Figure 2.5 Electron orbitals. *a.* The lowest energy level, or electron shell—the one nearest the nucleus—is level K. It is occupied by a single *s* orbital, referred to as 1*s*. *b.* The next highest energy level, L, is occupied by four orbitals: one *s* orbital (referred to as the 2*s* orbital) and three *p* orbitals (each referred to as a 2*p* orbital). Each orbital holds two paired electrons with opposite spin. Thus, the K level is populated by two electrons, and the L level is populated by a total of eight electrons. *c.* The neon atom shown has the L and K energy levels completely filled with electrons and is thus unreactive.

Atoms contain discrete energy levels

Because electrons are attracted to the positively charged nucleus, it takes work to keep them in their orbitals, just as it takes work to hold a grapefruit in your hand against the pull of gravity. The formal definition of energy is the ability to do work.

The grapefruit held above the ground is said to possess *potential energy* because of its position. If you release it, the grapefruit falls, and its potential energy is reduced. On the other hand, if you carried the grapefruit to the top of a building, you would increase its potential energy. Electrons also have a potential energy that is related to their position. To oppose the attraction of the nucleus and move the electron to a more distant orbital requires an input of energy, which results in an electron with greater potential energy. The chlorophyll that makes plants green captures energy from light during photosynthesis in this way. As you'll see in chapter 8—light energy excites electrons in the chlorophyll molecule. Moving an electron closer to the nucleus has the opposite effect: Energy is released, usually as radiant energy (heat or light), and the electron ends up with less potential energy (figure 2.6).

One of the initially surprising aspects of atomic structure is that electrons within the atom have discrete **energy levels.** These discrete levels correspond to quanta (singular, quantum), which means specific amount of energy. To use the grapefruit analogy again, it is as though a grapefruit could only be raised to particular floors of a building. Every atom exhibits a ladder of potential energy values, a discrete set of orbitals at particular energetic "distances" from the nucleus.

Because the amount of energy an electron possesses is related to its distance from the nucleus, electrons that are the same distance from the nucleus have the same energy, even if they occupy different orbitals. Such electrons are said to occupy the same energy level. The energy levels are denoted with letters K, L, M, and so on (figure 2.6). Be careful not to confuse energy levels, which are drawn as rings to indicate an electron's *energy,* with orbitals, which have a variety of three-dimensional shapes and indicate an electron's most likely *location.* Electron orbitals are arranged so that as they are filled, this fills each energy level in successive order. This filling of orbitals and energy levels is what is responsible for the chemical reactivity of elements.

During some chemical reactions, electrons are transferred from one atom to another. In such reactions, the loss of an electron is called **oxidation,** and the gain of an electron is called *reduction.*

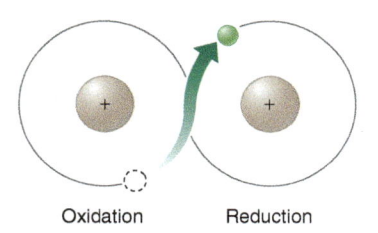

Notice that when an electron is transferred in this way, it keeps its energy of position. In organisms, chemical energy is stored in high-energy electrons that are transferred from one atom to another in reactions involving oxidation and reduction (described in chapter 7). When the processes of oxidation and reduction are coupled, which often happens, one atom or molecule is oxidized, while another is reduced in the same reaction. We call these combinations *redox reactions.*

Learning Outcomes Review 2.1

An atom consists of a nucleus of protons and neutrons surrounded by a cloud of electrons. For each atom, the number of protons is the atomic number; atoms with the same atomic number constitute an element. Atoms of a single element that have different numbers of neutrons are called isotopes. Electrons, which determine the chemical behavior of an element, are located about a nucleus in orbitals representing discrete energy levels. No orbital can contain more than two electrons, but each energy level consists of multiple orbitals, and thus contains many electrons with the same energy.

- *If the number of protons exceeds the number of neutrons, is the charge on the atom positive or negative?*
- *If the number of protons exceeds electrons?*

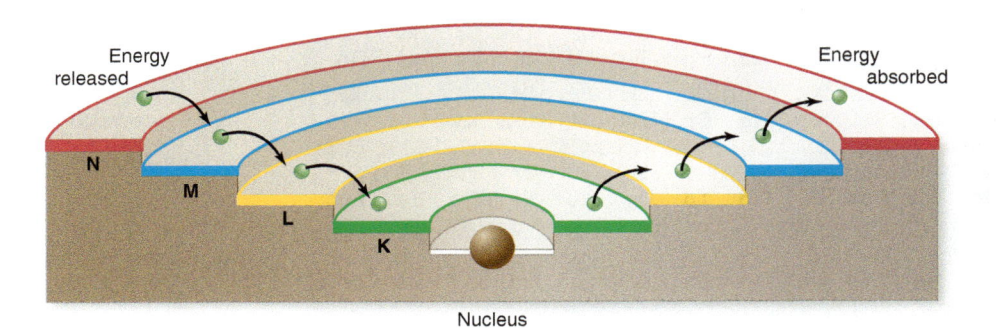

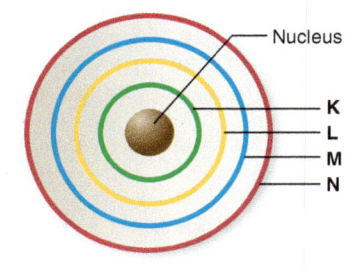

Figure 2.6 Atomic energy levels. Electrons have energy of position. When an atom absorbs energy, an electron moves to a higher energy level, farther from the nucleus. When an electron falls to lower energy levels, closer to the nucleus, energy is released. The first two energy levels are the same as shown in figure 2.5.

2.2 Elements Found in Living Systems

Learning Outcomes
1. Relate atomic structure to the periodic table of the elements.
2. List the important elements found in living systems.

Ninety elements occur naturally, each with a different number of protons and a different arrangement of electrons. When the 19th-century Russian chemist Dmitri Mendeleev arranged the known elements in a table according to their atomic number, he discovered one of the great generalizations of science: The elements exhibit a pattern of chemical properties that repeats itself in groups of eight. This periodically repeating pattern lent the table its name: the periodic table of elements (figure 2.7).

The periodic table displays elements according to atomic number and properties

The eight-element periodicity that Mendeleev found is based on the interactions of the electrons in the outermost energy level of the different elements. These electrons are called **valence electrons**, and their interactions are the basis for the elements' differing chemical properties. For most of the atoms important to life, the outermost energy level can contain no more than eight electrons; the chemical behavior of an element reflects how many of the eight positions are filled. Elements possessing all eight electrons in their outer energy level (two for helium) are *inert*, or nonreactive. These elements, which include helium (He), neon (Ne), argon (Ar), and so on, are called the *noble gases*. In sharp contrast, elements with seven electrons (one fewer than the maximum number of eight) in their outer energy level, such as fluorine (F), chlorine (Cl), and bromine (Br), are highly reactive. They tend to gain the extra electron needed to fill the energy level. Elements with only one electron in their outer energy level, such as lithium (Li), sodium (Na), and potassium (K), are also very reactive, but they tend to lose the single electron in their outer level.

Mendeleev's periodic table leads to a useful generalization, the **octet rule**, or *rule of eight* (Latin *octo*, "eight"): Atoms tend to establish completely full outer energy levels. For the main group elements of the periodic table, the rule of eight is accomplished by one filled *s* orbital and three filled *p* orbitals (figure 2.8). The exception to this is He, in the first row, which needs only two electrons to fill the 1*s* orbital. Most chemical behavior of biological interest can be predicted quite accurately from this simple rule, combined with the tendency of atoms to balance positive and negative charges. For instance, you read earlier that sodium ion (Na$^+$) has lost an electron, and chloride ion (Cl$^-$) has gained an electron. In the section 2.3, we describe how these ions react to form table salt.

Of the 90 naturally occurring elements on Earth, only 12 (C, H, O, N, P, S, Na, K, Ca, Mg, Fe, Cl) are found in living systems in more than trace amounts (0.01% or higher). These elements all have atomic numbers less than 21, and thus, have low atomic masses. Of these 12, the first 4 elements (carbon, hydrogen, oxygen, and nitrogen) constitute 96.3% of the weight of your body. The majority of molecules that make up your body (other than water) are compounds of carbon, which we call *organic* compounds.

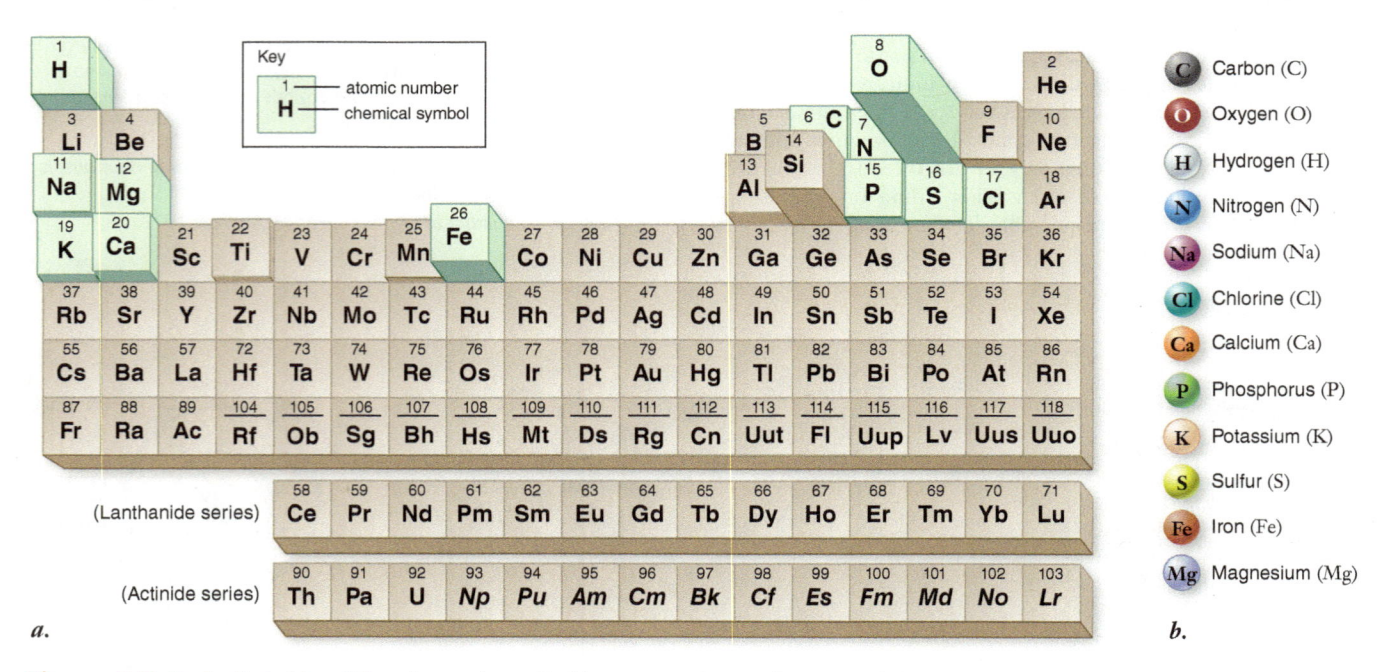

Figure 2.7 Periodic table of the elements. *a.* In this representation, the frequency of elements that occur in the Earth's crust is indicated by the height of the block. Elements shaded in green are found in living systems in more than trace amounts. *b.* Common elements found in living systems are shown in colors that will be used throughout the text.

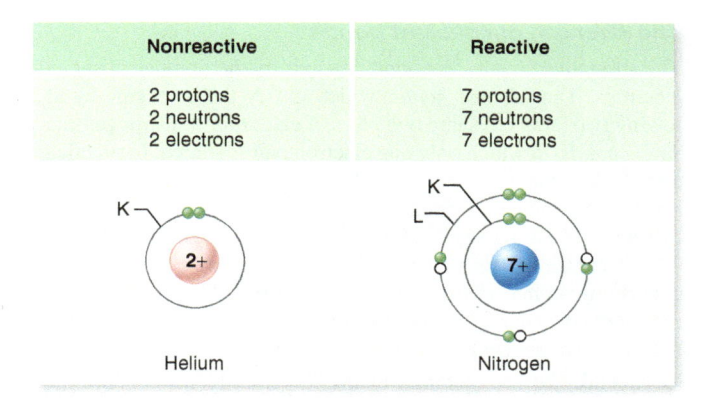

Figure 2.8 Electron energy levels for helium and nitrogen. Green balls represent electrons, blue ball represents the nucleus with number of protons indicated by number of (+) charges. Note that the helium atom has a filled K shell and is thus unreactive, whereas the nitrogen atom has five electrons in the L shell, three of which are unpaired, making it reactive.

These organic compounds contain primarily these four elements (CHON), explaining their prevalence in living systems. Some trace elements, such as zinc (Zn) and iodine (I), play crucial roles in living processes even though they are present in tiny amounts. Iodine deficiency, for example, can lead to enlargement of the thyroid gland, causing a bulge at the neck called a goiter.

Learning Outcomes Review 2.2

The periodic table shows the elements in terms of atomic number and repeating chemical properties. Only 12 elements are found in significant amounts in living organisms: C, H, O, N, P, S, Na, K, Ca, Mg, Fe, and Cl.

- Why are the noble gases more stable than other elements in the periodic table?

2.3 The Nature of Chemical Bonds

Learning Outcomes

1. Predict which elements are likely to form ions.
2. Explain how molecules are formed from atoms joined by covalent bonds.
3. Contrast polar and nonpolar covalent bonds.

A group of atoms held together by energy in a stable association is called a *molecule*. When a molecule contains atoms of more than one element, it is called a *compound*. The atoms in a molecule are joined by *chemical bonds;* these bonds can result when atoms with opposite charges attract each other (ionic bonds), when two atoms share one or more pairs of electrons (covalent bonds), or when atoms interact in other ways (table 2.1). We will start by examining *ionic bonds,* which form when atoms with opposite electrical charges (ions) attract.

TABLE 2.1 Bonds and Interactions

Name	Basis of Interaction	Strength
Covalent bond	Sharing of electron pairs	Strong
Ionic bond	Attraction of opposite charges	↕
Hydrogen bond	Sharing of H atom	
Hydrophobic interaction	Forcing of hydrophobic portions of molecules together in presence of polar substances	
van der Waals attraction	Weak attractions between atoms due to oppositely polarized electron clouds	Weak

Ionic bonds form crystals

Common table salt, the molecule sodium chloride (NaCl), is a lattice of ions in which the atoms are held together by ionic bonds (figure 2.9). Sodium has 11 electrons: 2 in the inner energy

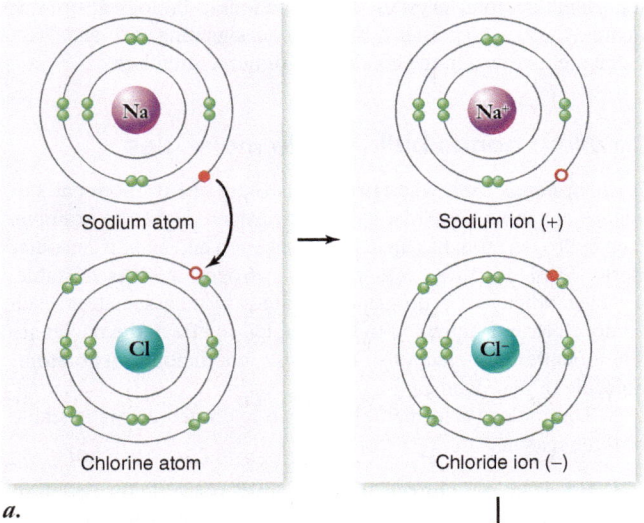

Figure 2.9 The formation of ionic bonds by sodium chloride.
a. When a sodium atom donates an electron to a chlorine atom, the sodium atom is oxidized and the chlorine atom reduced. This produces a positively charged sodium ion, and a negatively charged chloride ion. *b.* The electrostatic attraction of oppositely charged ions leads to the formation of a lattice of Na⁺ and Cl⁻.

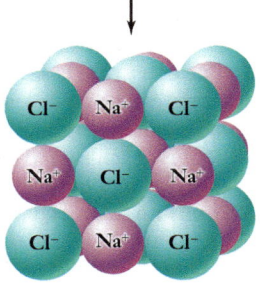

b. NaCl crystal

chapter 2 The Nature of Molecules and the Properties of Water 23

level (K), 8 in the next level (L), and 1 in the outer (valence) level (M). The single, unpaired valence electron has a strong tendency to join with another unpaired electron in another atom. A stable configuration can be achieved if the valence electron is lost to another atom that also has an unpaired electron. The loss of this electron results in the formation of a positively charged sodium ion, Na$^+$.

The chlorine atom has 17 electrons: 2 in the K level, 8 in the L level, and 7 in the M level. As you can see in the figure, one of the orbitals in the outer energy level has an unpaired electron (red circle). The addition of another electron fills that level and causes a negatively charged chloride ion, Cl$^-$, to form.

When placed together, metallic sodium and gaseous chlorine react swiftly and explosively, as the sodium atoms are oxidized, donating electrons to chlorine atoms, reducing them, and forming Na$^+$ and Cl$^-$ ions. Because opposite charges attract, the Na$^+$ and Cl$^-$ remain associated in an *ionic compound,* NaCl, which is electrically neutral. The electrical attractive force holding NaCl together, however, is not directed specifically between individual Na$^+$ and Cl$^-$ ions, and no individual sodium chloride molecules form. Instead, the force exists between any one ion and *all* neighboring ions of the opposite charge. The ions aggregate in a crystal matrix with a precise geometry. Such aggregations are what we know as salt crystals. If a salt such as NaCl is placed in water, the electrical attraction of the water molecules disrupts the forces holding the ions in their crystal matrix, causing the salt to dissolve into a roughly equal mixture of free Na$^+$ and Cl$^-$ ions.

Because living systems always include water, ions are more important than ionic crystals. Important ions in biological systems include Ca^{2+}, which is involved in cell signaling, K$^+$ and Na$^+$, which are involved in the conduction of nerve impulses.

Covalent bonds build stable molecules

Covalent bonds form when two atoms share one or more pairs of valence electrons. Consider gaseous hydrogen (H$_2$) as an example. Each hydrogen atom has an unpaired electron and an unfilled outer energy level; for these reasons, the hydrogen atom is unstable. However, when two hydrogen atoms are in close association, each atom's electron is attracted to both nuclei. In effect, the nuclei are able to share their electrons. The result is a diatomic (two-atom) molecule of hydrogen gas.

The molecule formed by the two hydrogen atoms is stable for three reasons:

1. **It has no net charge.** The diatomic molecule formed as a result of this sharing of electrons is not charged because it still contains two protons and two electrons.
2. **The octet rule is satisfied.** Each of the two hydrogen atoms can be considered to have two orbiting electrons in its outer energy level. This state satisfies the octet rule, because each shared electron orbits both nuclei and is included in the outer energy level of both atoms.
3. **It has no unpaired electrons.** The bond between the two atoms also pairs the two free electrons.

Unlike ionic bonds, covalent bonds are formed between two individual atoms, giving rise to true, discrete molecules.

The strength of covalent bonds

The strength of a covalent bond depends on the number of shared electrons. Thus *double bonds,* which satisfy the octet rule by allowing two atoms to share two pairs of electrons, are stronger than *single bonds,* in which only one electron pair is shared. In practical terms, more energy is required to break a double bond than a single bond. The strongest covalent bonds are *triple bonds,* such as those that link the two nitrogen atoms of nitrogen gas molecules (N$_2$).

Covalent bonds are represented in chemical formulas as lines connecting atomic symbols. Each line between two bonded atoms represents the sharing of one pair of electrons. The *structural formulas* of hydrogen gas and oxygen gas are H—H and O=O, respectively, and their *molecular formulas* are H$_2$ and O$_2$. The structural formula for N$_2$ is N≡N.

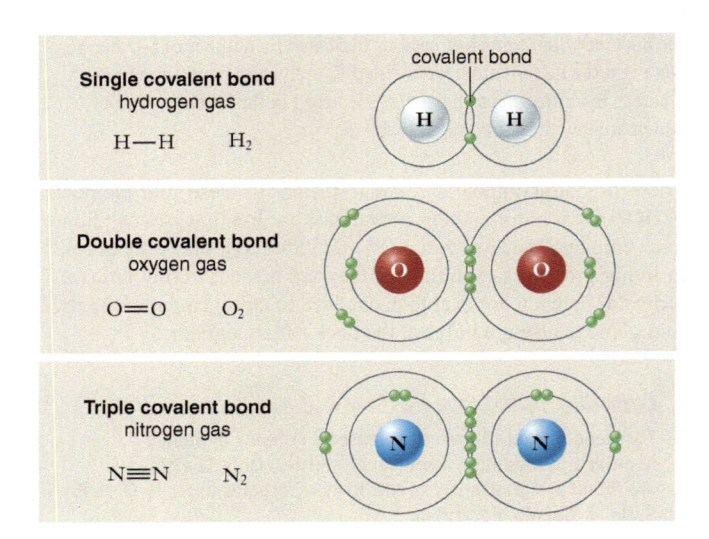

Molecules with several covalent bonds

A vast number of biological compounds are composed of more than two atoms. An atom that requires two, three, or four additional electrons to fill its outer energy level completely may acquire them by sharing its electrons with two or more other atoms.

For example, the carbon atom (C) contains six electrons, four of which are in its outer energy level and are unpaired. To satisfy the octet rule, a carbon atom must form four covalent bonds. Because four covalent bonds may form in many ways, carbon atoms are found in many different kinds of molecules. CO$_2$ (carbon dioxide), CH$_4$ (methane), and C$_2$H$_5$OH (ethanol) are just a few examples.

Polar and nonpolar covalent bonds

Atoms differ in their affinity for electrons, a property called **electronegativity.** In general, electronegativity increases left to right across a row of the periodic table and decreases down the column. Thus the elements in the upper-right corner have the highest electronegativity.

For bonds between identical atoms, for example, between two hydrogen or two oxygen atoms, the affinity for electrons is obviously the same, and the electrons are equally shared. Such bonds are termed **nonpolar.** The resulting compounds (H_2 or O_2) are also referred to as nonpolar.

For atoms that differ greatly in electronegativity, electrons are not shared equally. The shared electrons are more likely to be closer to the atom with greater electronegativity, and less likely to be near the atom of lower electronegativity. In this case, although the molecule is still electrically neutral (same number of protons as electrons), the distribution of charge is not uniform. This unequal distribution results in regions of partial negative charge near the more electronegative atom, and regions of partial positive charge near the less electronegative atom. Such bonds are termed **polar covalent bonds,** and the molecules polar molecules. When drawing polar molecules, these partial charges are usually symbolized by the lowercase Greek letter delta (δ). The partial charge seen in a polar covalent bond is relatively small—far less than the unit charge of an ion. For biological molecules, we can predict polarity of bonds by knowing the relative electronegativity of a small number of important atoms (table 2.2). Notice that although C and H differ slightly in electronegativity, this small difference is negligible, and C—H bonds are considered nonpolar.

Because of its importance in the chemistry of water, we will explore the nature of polar and nonpolar molecules in the section 2.4. Water (H_2O) is a polar molecule with electrons more concentrated around the oxygen atom.

Chemical reactions alter bonds

The formation and breaking of chemical bonds, which is the essence of chemistry, is termed a *chemical reaction*. All chemical reactions involve the shifting of atoms from one molecule or ionic compound to another, without any change in the number or identity of the atoms. For convenience, we refer to the original molecules before the reaction starts as *reactants,* and the molecules resulting from the chemical reaction as *products.* For example:

$$6H_2O + 6CO_2 \longrightarrow C_6H_{12}O_6 + 6O_2$$
$$\text{reactants} \longrightarrow \text{products}$$

You may recognize this reaction as a simplified form of the photosynthesis reaction, in which water and carbon dioxide are combined to produce glucose and oxygen. Most animal life ultimately depends on this reaction, which takes place in plants. (Photosynthetic reactions will be discussed in detail in chapter 8.)

TABLE 2.2	Relative Electronegativities of Some Important Atoms
Atom	**Electronegativity**
O	3.5
N	3.0
C	2.5
H	2.1

The extent to which chemical reactions occur is influenced by three important factors:

1. **Temperature.** Heating the reactants increases the rate of a reaction because the reactants collide with one another more often. (Care must be taken that the temperature is not so high that it destroys the molecules.)
2. **Concentration of reactants and products.** Reactions proceed more quickly when more reactants are available, allowing more frequent collisions. An accumulation of products typically slows the reaction and, in reversible reactions, may speed the reaction in the reverse direction.
3. **Catalysts.** A catalyst is a substance that increases the rate of a reaction. It doesn't alter the reaction's equilibrium between reactants and products, but it does shorten the time needed to reach equilibrium, often dramatically. In living systems, proteins called enzymes catalyze almost every chemical reaction.

Many reactions in nature are reversible. This means that the products may themselves be reactants, allowing the reaction to proceed in reverse. We can write the preceding reaction in the reverse order:

$$C_6H_{12}O_6 + 6O_2 \longrightarrow 6H_2O + 6CO_2$$
$$\text{reactants} \longrightarrow \text{products}$$

This reaction is a simplified version of the oxidation of glucose by cellular respiration, in which glucose is broken down into water and carbon dioxide in the presence of oxygen. Virtually all organisms carry out forms of glucose oxidation; details are covered later, in chapter 7.

Learning Outcomes Review 2.3

An ionic bond is an attraction between ions of opposite charge in an ionic compound. A covalent bond is formed when two atoms share one or more pairs of electrons. Complex biological compounds are formed in large part by atoms that can form one or more covalent bonds: C, H, O, and N. A polar covalent bond is formed by unequal sharing of electrons. Nonpolar bonds exhibit equal sharing of electrons.

■ *How is a polar covalent bond different from an ionic bond?*

2.4 Water: A Vital Compound

Learning Outcomes

1. Relate how the structure of water leads to hydrogen bonds.
2. Describe water's cohesive and adhesive properties.

Of all the common molecules, only water exists as a liquid at the relatively low temperatures that prevail on the Earth's surface. Three-fourths of the Earth is covered by liquid water (figure 2.10).

a. Solid   *b.* Liquid   *c.* Gas

Figure 2.10 Water takes many forms. *a.* When water cools below 0°C, it forms beautiful crystals, familiar to us as snow and ice. *b.* Ice turns to liquid when the temperature is above 0°C. *c.* Liquid water becomes steam when the temperature rises above 100°C, as seen in this hot spring at Yellowstone National Park.

When life was beginning, water provided a medium in which other molecules could move around and interact, without being held in place by strong covalent or ionic bonds. Life evolved in water for 2 billion years before spreading to land. And even today, life is inextricably tied to water. About two-thirds of any organism's body is composed of water, and all organisms require a water-rich environment, either inside or outside it, for growth and reproduction. It is no accident that tropical rain forests are bursting with life, whereas dry deserts appear almost lifeless except when water becomes temporarily plentiful, such as after a rainstorm.

Water's structure facilitates hydrogen bonding

Water has a simple molecular structure, consisting of an oxygen atom bound to two hydrogen atoms by two single covalent bonds (figure 2.11). The resulting molecule is stable: It satisfies the octet rule, has no unpaired electrons, and carries no net electrical charge. The electronegativity of O is much greater than that of H (see table 2.2), and so the bonds between these atoms are highly polar. *The polarity of water underlies water's chemistry and the chemistry of life.*

The single most outstanding chemical property of water is its ability to form weak chemical associations, called **hydrogen bonds.** These bonds form between the partially negative O atoms and the partially positive H atoms of two water molecules. Although these bonds have only 5–10% of the strength of covalent bonds, they are important to DNA and protein structure, and thus responsible for much of the chemical organization of living systems.

If we consider the shape of a water molecule, we see that its two covalent bonds have a partial charge at each end: δ^- at the oxygen end and δ^+ at the hydrogen end. The most stable arrangement of these charges is a *tetrahedron (a pyramid with a triangle as its base),* in which the two negative and two positive charges are approximately equidistant from one another. The oxygen atom lies at the center of the tetrahedron, the hydrogen atoms occupy two of the apexes (corners), and the partial negative charges occupy the other two apexes (figure 2.11*b*). The bond angle between the two covalent oxygen–hydrogen bonds is 104.5°. This value is slightly less than the bond angle of a regular tetrahedron, which would be 109.5°. In water, the partial negative charges occupy more space than the partial positive regions, so the oxygen–hydrogen bond angle is slightly compressed.

Water molecules are cohesive

The polarity of water allows water molecules to be attracted to one another—that is, water is *cohesive.* The oxygen end of each water molecule, which is δ^-, is attracted to the hydrogen end, which is δ^+, of other molecules. The attraction produces hydrogen bonds among water molecules (figure 2.12). Each hydrogen bond is individually very weak and transient, lasting on average only a hundred-billionth (10^{-11}) of a second. The cumulative effects of large numbers of these bonds, however, can be enormous. Water forms an abundance of hydrogen bonds, which are responsible for many of its important physical properties (table 2.3).

Water's cohesion is responsible for its being a liquid, not a gas, at moderate temperatures. The cohesion of liquid water is also responsible for its **surface tension.** Small insects can walk on water (figure 2.13) because at the air–water interface, all the surface water molecules are hydrogen-bonded to molecules below them.

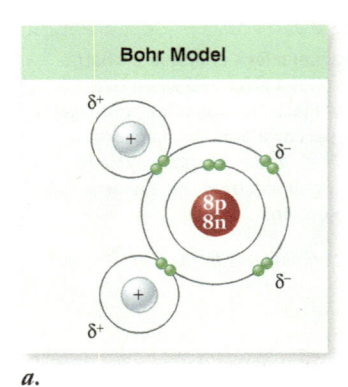

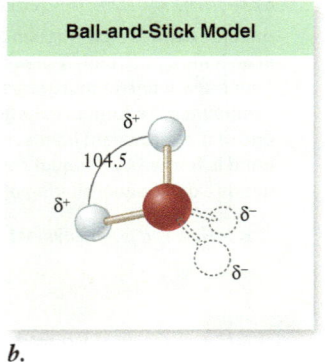

Figure 2.11 Water has a simple molecular structure. *a.* Each water molecule is composed of one oxygen atom and two hydrogen atoms. The oxygen atom shares one electron with each hydrogen atom. *b.* The greater electronegativity of the oxygen atom makes the water molecule polar: Water carries two partial negative charges (δ^-) near the oxygen atom and two partial positive charges (δ^+), one on each hydrogen atom. *c.* Space-filling model shows what the molecule would look like if it were visible.

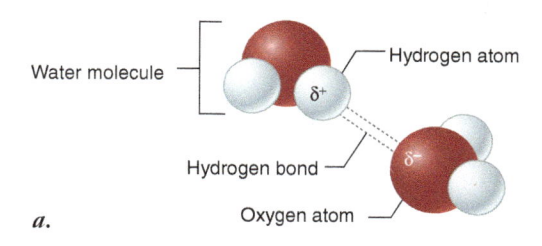

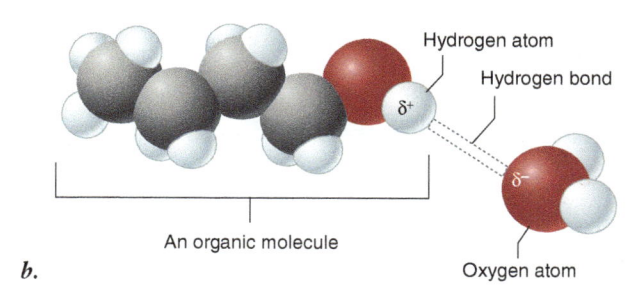

Figure 2.12 Structure of a hydrogen bond. *a.* Hydrogen bond between two water molecules. *b.* Hydrogen bond between an organic molecule (*n*-butanol) and water. H in *n*-butanol forms a hydrogen bond with oxygen in water. This kind of hydrogen bond is possible any time H is bound to a more electronegative atom (see table 2.2).

Figure 2.13 Cohesion. Some insects, such as this water strider, literally walk on water. Because the surface tension of the water is greater than the force of one foot, the strider glides atop the surface of the water rather than sinking. The high surface tension of water is due to hydrogen bonding between water molecules.

Water molecules are adhesive

The polarity of water causes it to be attracted to other polar molecules as well. This attraction for other polar substances is called *adhesion*. Water adheres to any substance with which it can form hydrogen bonds. This property explains why substances containing polar molecules get "wet" when they are immersed in water, but those that are composed of nonpolar molecules (such as oils) do not.

The attraction of water to substances that have electrical charges on their surface is responsible for capillary action. If a glass tube with a narrow diameter is lowered into a beaker of water, the water will rise in the tube above the level of the water in the beaker, because the adhesion of water to the glass surface, drawing it upward, is stronger than the force of gravity, pulling it downward. The narrower the tube, the greater the electrostatic forces between the water and the glass, and the higher the water rises (figure 2.14).

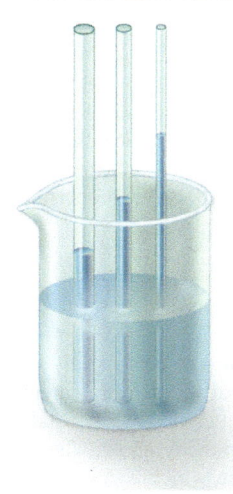

Figure 2.14 Adhesion. Capillary action causes the water within a narrow tube to rise above the surrounding water level; the adhesion of the water to the glass surface, which draws water upward, is stronger than the force of gravity, which tends to pull it down. The narrower the tube, the greater the surface area available for adhesion for a given volume of water, and the higher the water rises in the tube.

TABLE 2.3	The Properties of Water	
Property	**Explanation**	**Example of Benefit to Life**
Cohesion	Hydrogen bonds hold water molecules together.	Leaves pull water upward from the roots; seeds swell and germinate.
High specific heat	Hydrogen bonds absorb heat when they break and release heat when they form, minimizing temperature changes.	Water stabilizes the temperature of organisms and the environment.
High heat of vaporization	Many hydrogen bonds must be broken for water to evaporate.	Evaporation of water cools body surfaces.
Lower density of ice	Water molecules in an ice crystal are spaced relatively far apart because of hydrogen bonding.	Because ice is less dense than water, lakes do not freeze solid, allowing fish and other life in lakes to survive the winter.
Solubility	Polar water molecules are attracted to ions and polar compounds, making these compounds soluble.	Many kinds of molecules can move freely in cells, permitting a diverse array of chemical reactions.

Learning Outcomes Review 2.4

Because of its polar covalent bonds, water can form hydrogen bonds with itself and with other polar molecules. Hydrogen bonding is responsible for water's cohesion, the force that holds water molecules together, and its adhesion, which is its ability to "stick" to other polar molecules. Capillary action results from both of these properties.

- If water were made of C and H instead of H and O, would it still be cohesive and adhesive?

2.5 Properties of Water

Learning Outcomes

1. Illustrate how hydrogen bonding affects the properties of water.
2. Explain the relevance of water's unusual properties for living systems.
3. Identify the dissociation products of water.

Water moderates temperature through two properties: its high specific heat and its high heat of vaporization. Water also has the unusual property of being less dense in its solid form, ice, than as a liquid. Water acts as a solvent for polar molecules and exerts an organizing effect on nonpolar molecules. All these properties result from its polar nature.

Water's high specific heat helps maintain temperature

The temperature of any substance is a measure of how rapidly its individual molecules are moving. In the case of water, a large input of thermal energy is required to break the many hydrogen bonds that keep individual water molecules from moving about. Therefore, water is said to have a high **specific heat,** which is defined as the amount of heat 1 g of a substance must absorb or lose to change its temperature by 1 degree Celsius (°C). Specific heat measures the extent to which a substance resists changing its temperature when it absorbs or loses heat. Because polar substances tend to form hydrogen bonds, the more polar it is, the higher is its specific heat. The specific heat of water (1 calorie/g/°C) is twice that of most carbon compounds and nine times that of iron. Only ammonia, which is more polar than water and forms very strong hydrogen bonds, has a higher specific heat than water (1.23 cal/g/°C). Still, only 20% of the hydrogen bonds are broken as water heats from 0° to 100°C.

Because of its high specific heat, water heats up more slowly than almost any other compound and holds its temperature longer. Because organisms have a high water content, water's high specific heat allows them to maintain a relatively constant internal temperature. The heat generated by the chemical reactions inside cells would destroy the cells if not for the absorption of this heat by the water within them.

Water's high heat of vaporization facilitates cooling

The **heat of vaporization** is defined as the amount of energy required to change 1 g of a substance from a liquid to a gas. A considerable amount of heat energy (586 cal) is required to accomplish this change in water. As water changes from a liquid to a gas it requires energy (in the form of heat) to break its many hydrogen bonds. The evaporation of water from a surface cools that surface. Many organisms dispose of excess body heat by evaporative cooling, for example, through sweating in humans and many other vertebrates.

Solid water is less dense than liquid water

At low temperatures, water molecules are locked into a crystal-like lattice of hydrogen bonds, forming solid ice (see figure 2.10a). Interestingly, ice is less dense than liquid water because the hydrogen bonds in ice space the water molecules relatively far apart. This unusual feature enables icebergs to float. If water did not have this property, nearly all bodies of water would be ice, with only the shallow surface melting every year. The buoyancy of ice is important ecologically because it means bodies of water freeze from the top down and not the bottom up. Because ice floats on the surface of lakes in the winter and the water beneath the ice remains liquid, fish and other animals keep from freezing.

Polar molecules and ions are soluble in water

Water molecules gather closely around any substance that bears an electrical charge, whether that substance carries a full charge (ion) or a charge separation (polar molecule). For example, sucrose (table sugar) is composed of molecules that contain polar hydroxyl (OH) groups. A sugar crystal dissolves rapidly in water because water molecules can form hydrogen bonds with individual hydroxyl groups of the sucrose molecules. Therefore, sucrose is said to be *soluble* in water. Water is termed the *solvent,* and sugar is called the *solute.* Every time a sucrose molecule dissociates, or breaks away, from a solid sugar crystal, water molecules surround it in a cloud, forming a *hydration shell* that prevents it from associating with other sucrose molecules. Hydration shells also form around ions such as Na^+ and Cl^- (figure 2.15).

Water organizes nonpolar molecules

Water molecules always tend to form the maximum possible number of hydrogen bonds. When nonpolar molecules such as oils, which do not form hydrogen bonds, are placed in water, the water molecules act to exclude them. The nonpolar molecules aggregate, or clump together, thus minimizing their disruption of the hydrogen bonding of water. In effect, they shrink from contact with water, and for this reason they are referred to as **hydrophobic** (Greek *hydros,* "water," and *phobos,* "fearing"). In contrast, polar molecules, which readily form hydrogen bonds with water, are said to be **hydrophilic** ("water-loving").

The tendency of nonpolar molecules to aggregate in water is known as **hydrophobic exclusion.** By forcing the hydrophobic portions of molecules together, water causes these molecules to assume particular shapes. This property can also affect the structure of

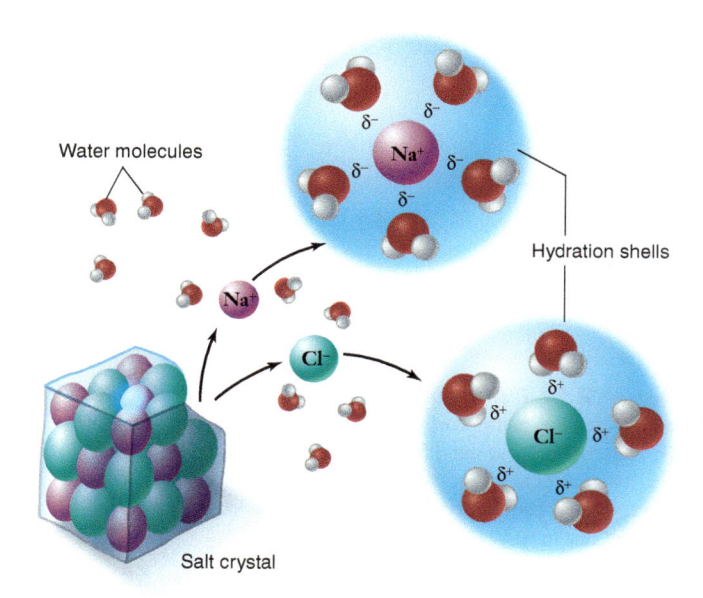

Figure 2.15 Why salt dissolves in water. When a crystal of table salt dissolves in water, individual Na+ and Cl− ions break away from the salt lattice and become surrounded by water molecules. Water molecules orient around Na+ so that their partial negative poles face toward the positive Na+; water molecules surrounding Cl− orient in the opposite way, with their partial positive poles facing the negative Cl−. Surrounded by hydration shells, Na+ and Cl− never reenter the salt lattice.

proteins, DNA, and biological membranes. In fact, the interaction of nonpolar molecules and water is critical to living systems.

Water can form ions

The covalent bonds of a water molecule sometimes break spontaneously. In pure water at 25°C, only 1 out of every 550 million water molecules undergoes this process. When it happens, a proton (hydrogen atom nucleus) dissociates from the molecule. Because the dissociated proton lacks the negatively charged electron it was sharing, its positive charge is no longer counterbalanced, and it becomes a hydrogen ion, H+. The rest of the dissociated water molecule, which has retained the shared electron from the covalent bond, is negatively charged and forms a hydroxide ion, OH−. This process of spontaneous ion formation is called *ionization*:

$$H_2O \longrightarrow OH^- + H^+$$
water   hydroxide ion   hydrogen ion (proton)

At 25°C, 1 liter (L) of water contains one ten-millionth (or 10^{-7}) mole of H+ ions. A **mole** (mol) is defined as the weight of a substance in grams that corresponds to the atomic masses of all of the atoms in a molecule of that substance. In the case of H+, the atomic mass is 1, and a mole of H+ ions would weigh 1 g. One mole of any substance always contains 6.02×10^{23} molecules of the substance. Therefore, the **molar concentration** of hydrogen ions in pure water, represented as [H+], is 10^{-7} mol/L. (In reality, the H+ usually associates with another water molecule to form a hydronium ion, H_3O^+.)

Learning Outcomes Review 2.5

Water has a high specific heat so it does not change temperature rapidly, which helps living systems maintain a near-constant temperature. Water's high heat of vaporization allows cooling by evaporation. Solid water is less dense than liquid water because the hydrogen bonds space the molecules farther apart. Polar molecules are soluble in a water solution, but water tends to exclude nonpolar molecules. Water dissociates to form H+ and OH−.

■ *How does the fact that ice floats affect life in a lake?*

2.6 Acids and Bases

Learning Outcomes
1. Define acids, bases, and the pH scale.
2. Relate changes in pH to changes in [H+].

The concentration of hydrogen ions, and concurrently of hydroxide ions, in a solution is described by the terms *acidity* and *basicity,* respectively. Pure water, having an [H+] of 10^{-7} mol/L, is considered to be neutral—that is, neither acidic nor basic. Recall that for every H+ ion formed when water dissociates, an OH− ion is also formed, meaning that the dissociation of water produces H+ and OH− in equal amounts.

The pH scale measures hydrogen ion concentration

The *pH scale* (figure 2.16) is a more convenient way to express the hydrogen ion concentration of a solution. This scale defines *pH,* which stands for "power of hydrogen," as the negative logarithm of the hydrogen ion concentration in the solution:

$$pH = -\log [H^+]$$

Because the logarithm of the hydrogen ion concentration is simply the exponent of the molar concentration of H+, the pH equals the exponent times −1. For water, therefore, an [H+] of 10^{-7} mol/L corresponds to a pH value of 7. This is the neutral point—a balance between H+ and OH−—on the pH scale. This balance occurs because the dissociation of water produces equal amounts of H+ and OH−.

Note that, because the pH scale is *logarithmic,* a difference of 1 on the scale represents a 10-fold change in [H+]. A solution with a pH of 4 therefore has 10 times the [H+] of a solution with a pH of 5 and 100 times the [H+] of a solution with a pH of 6.

Acids

Any substance that dissociates in water to increase the [H+] (and lower the pH) is called an **acid.** The stronger an acid is, the more hydrogen ions it produces and the lower its pH. For example, hydrochloric acid (HCl), which is abundant in your stomach, ionizes completely in water. A dilution of 10^{-1} mol/L of HCl dissociates to form 10^{-1} mol/L of H+, giving the solution a pH of 1. The pH of champagne, which bubbles because of the carbonic acid dissolved in it, is about 2.

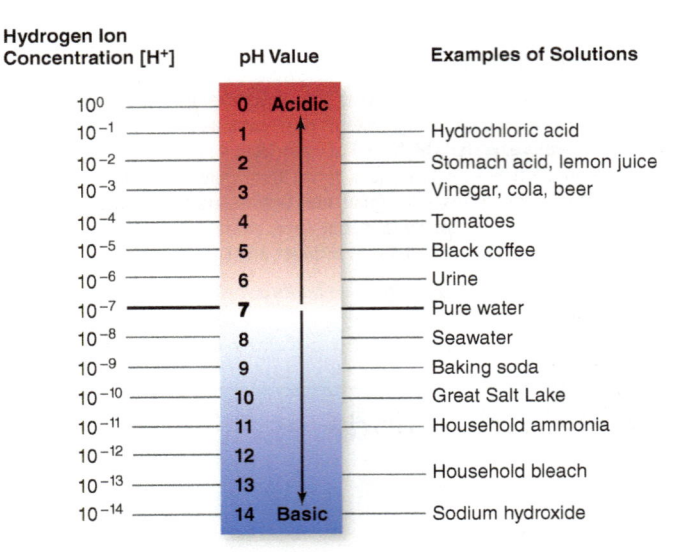

Figure 2.16 The pH scale. The pH value of a solution indicates its concentration of hydrogen ions. Solutions with a pH less than 7 are acidic, whereas those with a pH greater than 7 are basic. The scale is logarithmic, which means that a pH change of 1 represents a 10-fold change in the concentration of hydrogen ions. Thus, lemon juice is 100 times more acidic than tomato juice, and seawater is 10 times more basic than pure water, which has a pH of 7.

Bases

A substance that combines with H^+ when dissolved in water, and thus lowers the $[H^+]$, is called a **base.** Therefore, basic (or alkaline) solutions have pH values above 7. Very strong bases, such as sodium hydroxide (NaOH), have pH values of 12 or more. Many common cleaning substances, such as ammonia and bleach, accomplish their action because of their high pH.

Buffers help stabilize pH

The pH inside almost all living cells, and in the fluid surrounding cells in multicellular organisms, is fairly close to neutral, 7. Most of the enzymes in living systems are extremely sensitive to pH. Often even a small change in pH will alter their shape, thereby disrupting their activities. For this reason, it is important that a cell maintain a constant pH level.

But the chemical reactions of life constantly produce acids and bases within cells. Furthermore, many animals eat substances that are acidic or basic. Cola drinks, for example, are moderately strong (although dilute) acidic solutions. Despite such variations in the concentrations of H^+ and OH^-, the pH of an organism is kept at a relatively constant level by buffers (figure 2.17).

A **buffer** is a substance that resists changes in pH. Buffers act by releasing hydrogen ions when a base is added and absorbing hydrogen ions when acid is added, with the overall effect of keeping $[H^+]$ relatively constant.

Within organisms, most buffers consist of pairs of substances, one an acid and the other a base. The key buffer in human blood is an acid–base pair consisting of carbonic acid (acid) and bicarbonate (base). These two substances interact in a pair of reversible reactions. First, carbon dioxide (CO_2) and H_2O join to form carbonic acid (H_2CO_3), which in a second reaction dissociates to yield bicarbonate ion (HCO_3^-) and H^+.

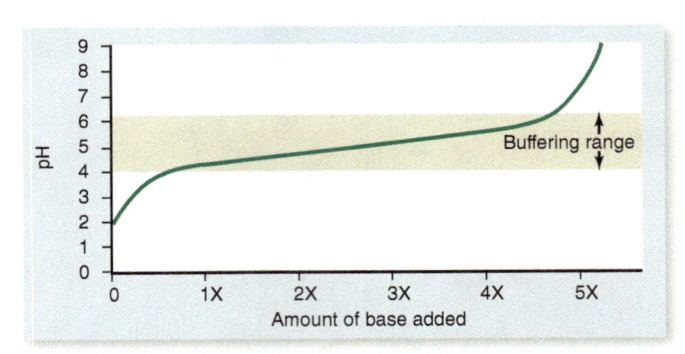

Figure 2.17 Buffers minimize changes in pH. Adding a base to a solution neutralizes some of the acid present, and so raises the pH. Thus, as the curve moves to the right, reflecting more and more base, it also rises to higher pH values. A buffer makes the curve rise or fall very slowly over a portion of the pH scale, called the "buffering range" of that buffer.

Data analysis If we call each step on the x-axis one volume of base, how many volumes of base must be added to change the pH from 4 to 6?

If some acid or other substance adds H^+ to the blood, the HCO_3^- acts as a base and removes the excess H^+ by forming H_2CO_3. Similarly, if a basic substance removes H^+ from the blood, H_2CO_3 dissociates, releasing more H^+ into the blood. The forward and reverse reactions that interconvert H_2CO_3 and HCO_3^- thus stabilize the blood's pH:

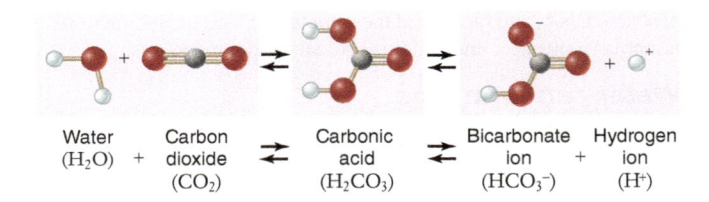

Water (H_2O) + Carbon dioxide (CO_2) ⇌ Carbonic acid (H_2CO_3) ⇌ Bicarbonate ion (HCO_3^-) + Hydrogen ion (H^+)

The reaction of carbon dioxide and water to form carbonic acid is a crucial one because it permits carbon, essential to life, to enter water from the air. The Earth's oceans are rich in carbon because of the reaction of carbon dioxide with water.

In a condition called blood acidosis, human blood, which normally has a pH of about 7.4, drops to a pH of about 7.1. This condition is fatal if not treated immediately. The reverse condition, blood alkalosis, involves an increase in blood pH of a similar magnitude and is just as serious.

Learning Outcomes Review 2.6

Acid solutions have a high $[H^+]$, and basic solutions have a low $[H^+]$ (and therefore a high $[OH^-]$). The pH of a solution is the negative logarithm of its $[H^+]$. Low pH values indicate acids, and high pH values indicate bases. Even small changes in pH can be harmful to life. Buffer systems in organisms help to maintain pH within a narrow range.

■ *A change of 2 pH units indicates what change in $[H^+]$?*

Chapter Review

2.1 The Nature of Atoms

All matter is composed of atoms (figure 2.3).

Atomic structure includes a central nucleus and orbiting electrons.

Electrically neutral atoms have the same number of protons as electrons. Atoms that gain or lose electrons are called ions.

Elements are defined by the number of protons in the nucleus, the atomic number. Atomic mass is the sum of the mass of protons and neutrons. Isotopes are forms of a single element with different atomic mass due to different numbers of neutrons. Radioactive isotopes are unstable.

Electrons determine the chemical behavior of atoms.

The potential energy of electrons increases as distance from the nucleus increases. Electron orbitals are probability distributions. *s*-Orbitals are spherical; other orbitals have different shapes, such as the dumbbell-shaped *p*-orbitals.

Atoms contain discrete energy levels.

Energy levels correspond to quanta (singular, quantum) of energy, a "ladder" of energy levels that an electron may have.

The loss of electrons from an atom is called oxidation. The gain of electrons is called reduction. Electrons can be transferred from one atom to another in coupled redox reactions.

2.2 Elements Found in Living Systems

The periodic table displays elements according to atomic number and properties.

Atoms tend to establish completely full outer energy levels (the octet rule). Elements with filled outermost orbitals are inert.

Ninety elements occur naturally in the Earth's crust. Twelve of these elements are found in living organisms in greater than trace amounts: C, H, O, N, P, S, Na, K, Ca, Mg, Fe, and Cl.

Compounds of carbon are called organic compounds. The majority of molecules in living systems are composed of C bound to H, O, and N.

2.3 The Nature of Chemical Bonds

Molecules contain two or more atoms joined by chemical bonds. Compounds contain two or more different elements.

Ionic bonds form crystals.

Ions with opposite electrical charges form ionic bonds, such as NaCl (figure 2.9*b*).

Covalent bonds build stable molecules.

A molecule formed by a covalent bond is stable because it has no net charge, the octet rule is satisfied, and it has no unpaired electrons. Covalent bonds may be single, double, or triple, depending on the number of pairs of electrons shared. Nonpolar covalent bonds involve equal sharing of electrons between atoms. Polar covalent bonds involve unequal sharing of electrons.

Chemical reactions alter bonds.

Temperature, reactant concentration, and the presence of catalysts affect reaction rates. Most biological reactions are reversible, such as the conversion of carbon dioxide and water into carbohydrates.

2.4 Water: A Vital Compound

Water's structure facilitates hydrogen bonding.

Hydrogen bonds are weak interactions between a partially positive H in one molecule and a partially negative O in another molecule (figure 2.11).

Water molecules are cohesive.

Cohesion is the tendency of water molecules to adhere to one another due to hydrogen bonding. The cohesion of water is responsible for its surface tension.

Water molecules are adhesive.

Adhesion occurs when water molecules adhere to other polar molecules. Capillary action results from water's adhesion to the sides of narrow tubes, combined with its cohesion.

2.5 Properties of Water

Water's high specific heat helps maintain temperature.

The specific heat of water is high because it takes a considerable amount of energy to disrupt hydrogen bonds.

Water's high heat of vaporization facilitates cooling.

Breaking hydrogen bonds to turn liquid water into vapor takes a lot of energy. Many organisms lose excess heat through evaporative cooling, such as sweating.

Solid water is less dense than liquid water.

Hydrogen bonds are spaced farther apart in the solid phase of water than in the liquid phase. As a result, ice floats.

Polar molecules and ions are soluble in water.

Water's polarity makes it a good solvent for polar substances and ions. Polar molecules or portions of molecules are attracted to water (hydrophilic). Molecules that are nonpolar are repelled by water (hydrophobic). Water makes nonpolar molecules clump together.

Water organizes nonpolar molecules.

Nonpolar molecules will aggregate to avoid water. This maximizes the hydrogen bonds that water can make. This hydrophobic exclusion can affect the structure of DNA, proteins, and biological membranes.

Water can form ions.

Water dissociates into H^+ and OH^-. The concentration of H^+, shown as $[H^+]$, in pure water is 10^{-7} mol/L.

2.6 Acids and Bases (figure 2.16)

The pH scale measures hydrogen ion concentration.

pH is defined as the negative logarithm of $[H^+]$. Pure water has a pH of 7. A difference of 1 pH unit means a 10-fold change in $[H^+]$.

Acids have a greater $[H^+]$ and therefore a lower pH; bases have a lower $[H^+]$ and therefore a higher pH.

Buffers help stabilize pH.

Carbon dioxide and water react reversibly to form carbonic acid. A buffer resists changes in pH by absorbing or releasing H^+. The key buffer in the human blood is the carbonic acid/bicarbonate pair.

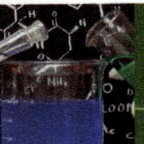

Review Questions

UNDERSTAND

1. The property that distinguishes an atom of one element (carbon, for example) from an atom of another element (oxygen, for example) is
   a. the number of electrons.
   b. the number of protons.
   c. the number of neutrons.
   d. the combined number of protons and neutrons.

2. If an atom has one valence electron—that is, a single electron in its outer energy level—it will most likely form
   a. one polar, covalent bond.
   b. two nonpolar, covalent bonds.
   c. two covalent bonds.
   d. an ionic bond.

3. An atom with a net positive charge must have more
   a. protons than neutrons.
   b. protons than electrons.
   c. electrons than neutrons.
   d. electrons than protons.

4. The isotopes carbon-12 and carbon-14 differ in
   a. the number of neutrons.
   b. the number of protons.
   c. the number of electrons.
   d. Both b and c are correct.

5. Which of the following is NOT a property of the elements most commonly found in living organisms?
   a. The elements have a low atomic mass.
   b. The elements have an atomic number less than 21.
   c. The elements possess eight electrons in their outer energy level.
   d. The elements are lacking one or more electrons from their outer energy level.

6. Ionic bonds arise from
   a. shared valence electrons.
   b. attractions between valence electrons.
   c. charge attractions between valence electrons.
   d. attractions between ions of opposite charge.

7. A solution with a high concentration of hydrogen ions
   a. is called a base.
   b. is called an acid.
   c. has a high pH.
   d. Both b and c are correct.

APPLY

1. Using the periodic table on page 22, which of the following atoms would you predict should form a positively charged ion (cation)?
   a. Fluorine (F)
   b. Neon (Ne)
   c. Potassium (K)
   d. Sulfur (S)

2. Refer to the element pictured. How many covalent bonds could this atom form?
   a. Two
   b. Three
   c. Four
   d. None

3. A molecule with polar covalent bonds would
   a. be soluble in water.
   b. not be soluble in water.
   c. contain atoms with very similar electronegativity.
   d. Both b and c are correct.

4. Hydrogen bonds are formed
   a. between any molecules that contain hydrogen.
   b. only between water molecules.
   c. when hydrogen is part of a polar bond.
   d. when two atoms of hydrogen share an electron.

5. If you shake a bottle of oil and vinegar then let it sit, it will separate into two phases because
   a. the nonpolar oil is soluble in water.
   b. water can form hydrogen bonds with the oil.
   c. polar oil is not soluble in water.
   d. nonpolar oil is not soluble in water.

6. The decay of radioactive isotopes involves changes to the nucleus of atoms. Explain how this differs from the changes in atoms that occur during chemical reactions.

SYNTHESIZE

1. Elements that form ions are important for a range of biological processes. You have learned something about the cations sodium (Na^+), calcium (Ca^{2+}), and potassium (K^+) in this chapter. Use your knowledge of the definition of a cation to identify other examples from the periodic table.

2. A popular theme in science fiction literature has been the idea of silicon-based life-forms in contrast to our carbon-based life. Evaluate the possibility of silicon-based life based on the chemical structure and potential for chemical bonding of a silicon atom.

3. Efforts by NASA to search for signs of life on Mars have focused on the search for evidence of liquid water rather than looking directly for biological organisms (living or fossilized). Use your knowledge of the influence of water on life on Earth to construct an argument justifying this approach.

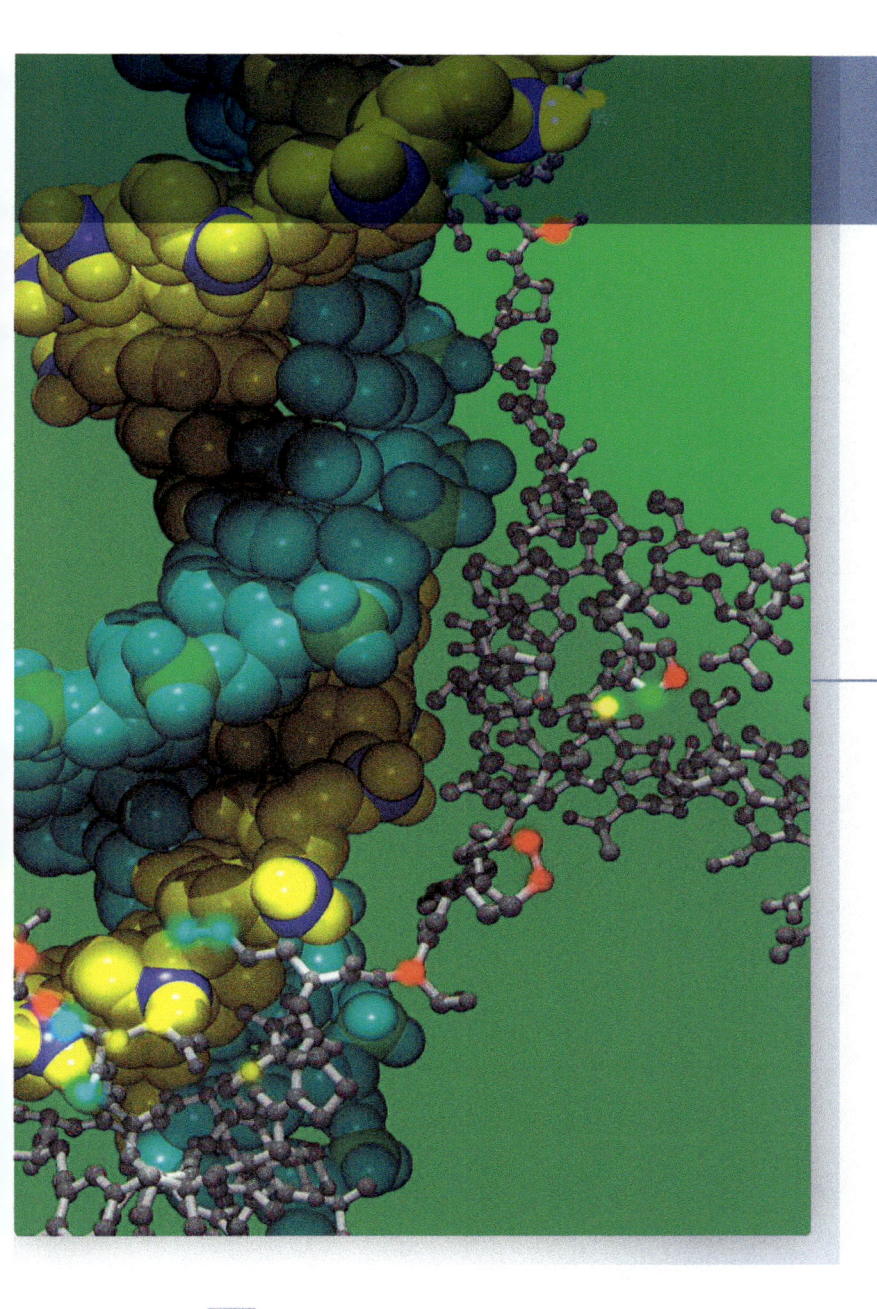

CHAPTER 3

The Chemical Building Blocks of Life

Chapter Contents

3.1 Carbon: The Framework of Biological Molecules

3.2 Carbohydrates: Energy Storage and Structural Molecules

3.3 Nucleic Acids: Information Molecules

3.4 Proteins: Molecules with Diverse Structures and Functions

3.5 Lipids: Hydrophobic Molecules

Introduction

A cup of water contains more molecules than there are stars in the sky. But many molecules are much larger than water molecules. Many thousands of distinct biological molecules are long chains made of thousands or even billions of atoms. These enormous assemblages, which are almost always synthesized by living things, are *macromolecules*. As you may know, biological macromolecules can be divided into four categories: *carbohydrates, nucleic acids, proteins,* and *lipids,* and they are the basic chemical building blocks from which all organisms are composed.

We take the existence of these classes of macromolecules for granted now, but as late as the 19th century many theories of "vital forces" were associated with living systems. One such theory held that cells contained a substance, protoplasm, that was responsible for the chemical reactions in living systems. Any disruption of cells was thought to disturb the protoplasm. Such a view makes studying the chemical reactions of cells in the lab (in vitro) impossible. The demonstration of fermentation in a cell-free system marked the beginning of modern biochemistry (figure 3.1). This approach involves studying biological molecules outside of cells to infer their role inside cells. Because these biological macromolecules all involve carbon-containing compounds, we begin with a brief summary of carbon and its chemistry.

SCIENTIFIC THINKING

Hypothesis: Chemical reactions, such as the fermentation reaction in yeast, are controlled by enzymes and do not require living cells.
Prediction: If yeast cells are broken open, these enzymes should function outside of the cell.
Test: Yeast is mixed with quartz sand and diatomaceous earth and then ground in a mortar and pestle. The resulting paste is wrapped in canvas and subjected to 400–500 atm pressure in a press. Fermentable and nonfermentable substrates are added to the resulting fluid, with fermentation being measured by the production of CO_2.

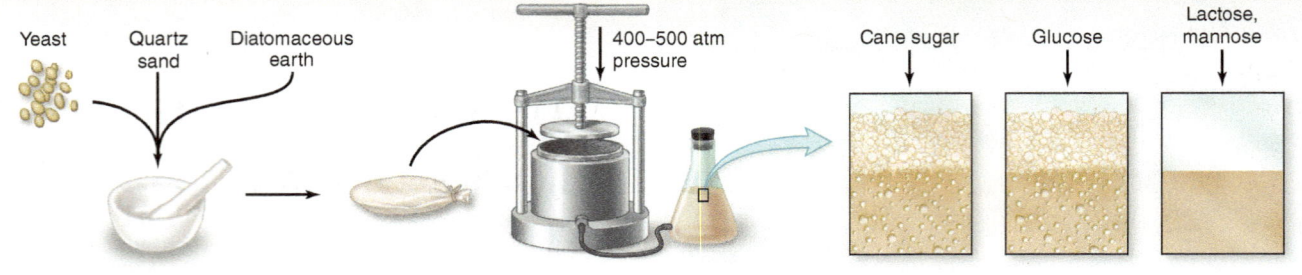

Result: When a fermentable substrate (cane sugar, glucose) is used, CO_2 is produced; when a nonfermentable substrate (lactose, mannose) is used, no CO_2 is produced. In addition, visual inspection of the fluid shows no visible yeast cells.
Conclusion: The hypothesis is supported. The fermentation reaction can occur in the absence of live yeast.
Historical Significance: Although this is not precisely the intent of the original experiment, it represents the first use of a cell-free system. Such systems allow for the study of biochemical reactions in vitro and the purification of proteins involved. We now know that the "fermentation reaction" is actually a complex series of reactions. Would such a series of reactions be your first choice for this kind of demonstration?

Figure 3.1 The demonstration of cell-free fermentation. The German chemist Eduard Buchner's (1860–1917) demonstration of fermentation by fluid produced from yeast, but not containing any live cells, both argued against the protoplasm theory and provided a method for future biochemists to examine the chemistry of life outside of cells.

3.1 Carbon: The Framework of Biological Molecules

Learning Outcomes
1. Describe the relationship between functional groups and macromolecules.
2. Recognize the different kinds of isomers.
3. List the different kinds of biological macromolecules.

In chapter 2, we reviewed the basics of atomic structure and chemical bonding. Biological systems obey all the laws of chemistry. Thus, chemistry forms the basis of living systems.

The framework of biological molecules consists predominantly of carbon atoms bonded to other carbon atoms or to atoms of oxygen, nitrogen, sulfur, phosphorus, or hydrogen. Because carbon atoms can form up to four covalent bonds, molecules containing carbon can form straight chains, branches, or even rings, balls, tubes, and coils.

Molecules consisting only of carbon and hydrogen are called *hydrocarbons*. Because carbon–hydrogen covalent bonds store considerable energy, hydrocarbons make good fuels. Gasoline, for example, is rich in hydrocarbons, and propane gas, another hydrocarbon, consists of a chain of three carbon atoms, with eight hydrogen atoms bound to it. The chemical formula for propane is C_3H_8. Its structural formula is

$$\begin{array}{c} H \quad H \quad H \\ | \quad | \quad | \\ H-C-C-C-H \\ | \quad | \quad | \\ H \quad H \quad H \end{array}$$ Propane structural formula

Theoretically speaking, the length of a chain of carbon atoms is unlimited. As described in the rest of this chapter, the four main types of biological molecules often consist of huge chains of carbon-containing compounds.

Functional groups account for differences in molecular properties

Carbon and hydrogen atoms both have very similar electronegativities. Electrons in C—C and C—H bonds are therefore evenly distributed, with no significant differences in charge over the molecular surface. For this reason, hydrocarbons are nonpolar. Most biological molecules produced by cells, however, also contain other atoms. Because these other atoms frequently have different electronegativities (see table 2.2), molecules containing them exhibit regions of partial positive or negative charge. They are polar.

These molecules can be thought of as a C—H core to which specific molecular groups, called **functional groups,** are attached. One such common functional group is —OH, called a *hydroxyl group.*

Functional groups have definite chemical properties that they retain no matter where they occur. Both the hydroxyl and carbonyl (C=O) groups, for example, are polar because of the electronegativity of the oxygen atoms (see chapter 2). Other common functional groups are the acidic carboxyl (COOH), phosphate (PO_4^-), and the basic amino (NH_2) group. Many of these functional groups can also participate in hydrogen bonding. Hydrogen bond donors and acceptors can be predicted based on their electronegativities shown in table 2.2. Figure 3.2 illustrates these biologically important functional groups and lists the macromolecules in which they are found.

Isomers have the same molecular formulas but different structures

Organic molecules having the same molecular or empirical formula can exist in different forms called **isomers.** If there are differences in the actual structure of their carbon skeleton, we call them *structural isomers.* In section 3.2, you will see that glucose and fructose are structural isomers of $C_6H_{12}O_6$. Another form of isomers, called *stereoisomers,* have the same carbon skeleton but differ in how the groups attached to this skeleton are arranged in space.

Enzymes in biological systems usually recognize only a single, specific stereoisomer. A subcategory of stereoisomers, called *enantiomers,* are actually mirror images of each other. A molecule that has mirror-image versions is called a *chiral* molecule. When carbon is bound to four different molecules, this inherent asymmetry exists (figure 3.3).

Chiral compounds are characterized by their effect on polarized light. Polarized light has a single plane, and chiral molecules rotate this plane either to the right (Latin, *dextro*) or left (Latin, *levo*). We therefore call the two chiral forms *D* for *dextrorotatory* and *L* for *levorotatory.* Living systems tend to produce only a single enantiomer of the two possible forms; for example, in most organisms we find primarily D-sugars and L-amino acids.

Functional Group	Structural Formula	Example	Found In
Hydroxyl	—OH	Ethanol	carbohydrates, proteins, nucleic acids, lipids
Carbonyl	—C(=O)—	Acetaldehyde	carbohydrates, nucleic acids
Carboxyl	—C(=O)OH	Acetic acid	proteins, lipids
Amino	—NH₂	Alanine	proteins, nucleic acids
Sulfhydryl	—S—H	Cysteine	proteins
Phosphate	—O—P(=O)(O⁻)—O⁻	Glycerol phosphate	nucleic acids
Methyl	—CH₃	Alanine	proteins

Figure 3.2 The primary functional chemical groups. These groups tend to act as units during chemical reactions and give specific chemical properties to the molecules that possess them. Amino groups, for example, make a molecule more basic, and carboxyl groups make a molecule more acidic. These functional groups are also not limited to the examples in the "Found In" column but are widely distributed in biological molecules.

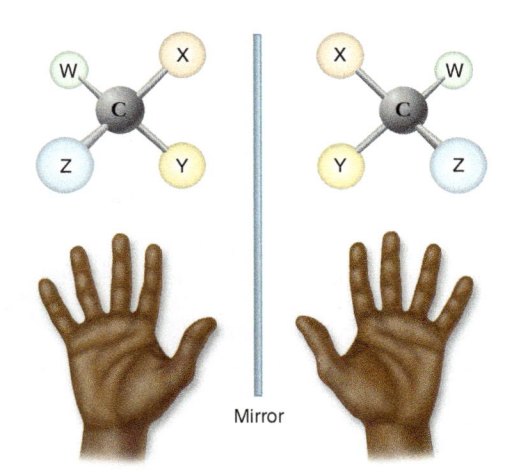

Figure 3.3 Chiral molecules. When carbon is bound to four different groups, the resulting molecule is said to be chiral (from Greek *cheir,* meaning "hand"). A chiral molecule will have stereoisomers that are mirror images. The two molecules shown have the same four groups but cannot be superimposed, much like your two hands cannot be superimposed but must be flipped to match. These types of stereoisomers are called *enantiomers.*

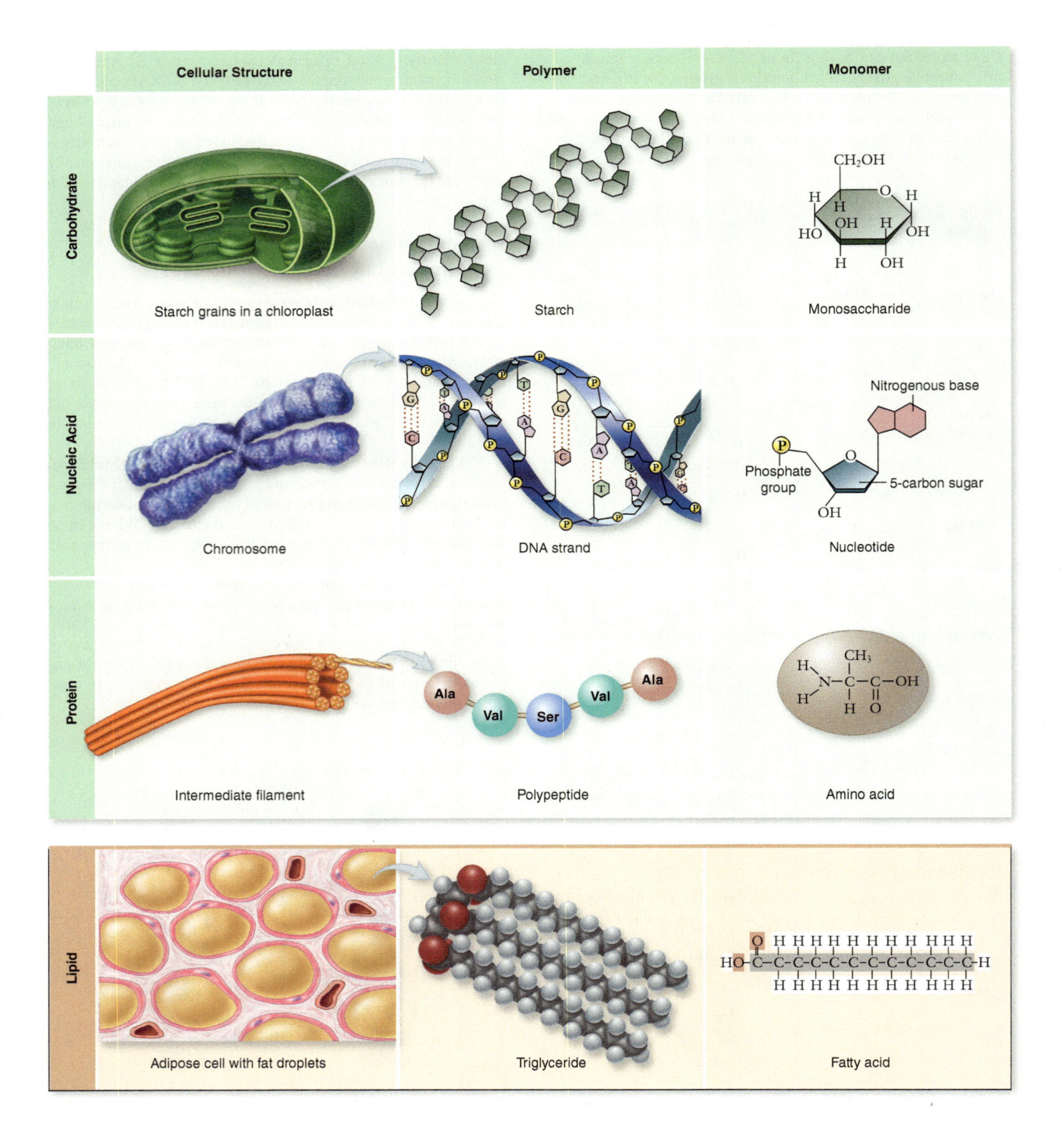

Figure 3.4 Polymer macromolecules. The four major biological macromolecules are shown. Carbohydrates, nucleic acids, and proteins all form polymers and are shown with the monomers used to make them. Lipids do not fit this simple monomer–polymer relationship. The triglyceride shown is constructed from glycerol and fatty acids. All four types of macromolecules are also shown in their cellular context.

TABLE 3.1 Macromolecules

Macromolecule	Subunit	Function	Example
CARBOHYDRATES			
Starch, glycogen	Glucose	Energy storage	Potatoes
Cellulose	Glucose	Structural support in plant cell walls	Paper; strings of celery
Chitin	Modified glucose	Structural support	Crab shells
NUCLEIC ACIDS			
DNA	Nucleotides	Encodes genes	Chromosomes
RNA	Nucleotides	Needed for gene expression	Messenger RNA
PROTEINS			
Functional	Amino acids	Catalysis; transport	Hemoglobin
Structural	Amino acids	Support	Hair; silk
LIPIDS			
Triglycerides (animal fat, oils)	Glycerol and three fatty acids	Energy storage	Butter; corn oil; soap
Phospholipids	Glycerol, two fatty acids, phosphate, and polar R groups	Cell membranes	Phosphatidylcholine
Prostaglandins	Five-carbon rings with two nonpolar tails	Chemical messengers	Prostaglandin E (PGE)
Steroids	Four fused carbon rings	Membranes; hormones	Cholesterol; estrogen
Terpenes	Long carbon chains	Pigments; structural support	Carotene; rubber

Biological macromolecules include carbohydrates, nucleic acids, proteins, and lipids

Remember that biological macromolecules are traditionally grouped into carbohydrates, nucleic acids, proteins, and lipids (table 3.1). In many cases, these macromolecules are polymers. A **polymer** is a long molecule built by linking together a large number of small, similar chemical subunits called **monomers**. They are like railroad cars coupled to form a train. The nature of a polymer is determined by the monomers used to build the polymer. Here are some examples. Complex carbohydrates such as starch are polymers composed of simple ring-shaped sugars. Nucleic acids (DNA and RNA) are polymers of nucleotides, and proteins are polymers of amino acids (figure 3.4). These long chains are built via chemical reactions termed *dehydration reactions* and are broken down by *hydrolysis reactions*. Lipids are macromolecules, but they really don't follow the monomer–polymer relationship. However, lipids are formed through dehydration reactions, which link the fatty acids to glycerol.

The dehydration reaction

Despite the differences between monomers of these major polymers, the basic chemistry of their synthesis is similar: To form a covalent bond between two monomers, an —OH group is removed from one monomer, and a hydrogen atom (H) is removed from the other (figure 3.5a). This reaction is the same for joining nucleotides when synthesizing DNA or joining glucose units together to make starch. This reaction is also used to link fatty acids to glycerol in lipids. This chemical reaction is called condensation, or a **dehydration reaction,** because the removal of —OH and —H is the same as the removal of a molecule of water (H_2O). For every subunit added to a macromolecule, one water molecule is removed. These and other biochemical reactions require that the reacting substances are held close together and that the correct chemical bonds are stressed and broken. This process of positioning and stressing, termed *catalysis,* is carried out within cells by enzymes.

The hydrolysis reaction

Cells disassemble polymers into their constituent monomers by reversing the dehydration reaction—a molecule of water is added instead of removed (figure 3.5b). In this reaction, called hydrolysis, a hydrogen atom is attached to one subunit and a hydroxyl group to the other, breaking the covalent bond joining the subunits. When you eat a potato, which contains starch (see section 3.2), your body breaks the starch down into glucose units by hydrolysis. The potato plant built the starch molecules originally by dehydration reactions.

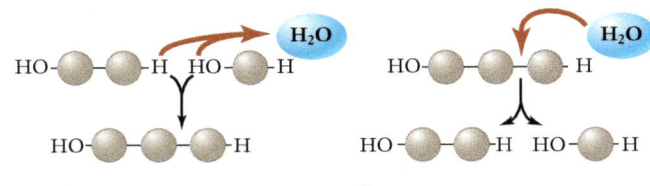

a. Dehydration reaction   *b.* Hydrolysis reaction

Figure 3.5 Making and breaking macromolecules.
a. Biological macromolecules are polymers formed by linking monomers together through dehydration reactions. This process releases a water molecule for every bond formed. *b.* Breaking the bond between subunits involves hydrolysis, which reverses the loss of a water molecule by dehydration.

Learning Outcomes Review 3.1

Functional groups account for differences in chemical properties in organic molecules. Isomers are compounds with the same empirical formula but different structures. This difference may affect biological function. Macromolecules are polymers consisting of long chains of similar subunits that are joined by dehydration reactions and are broken down by hydrolysis reactions.

- What is the relationship between dehydration and hydrolysis?

3.2 Carbohydrates: Energy Storage and Structural Molecules

Learning Outcomes

1. Describe the structure of simple sugars with three to six carbons.
2. Relate the structure of polysaccharides to their functions.

Monosaccharides are simple sugars

Carbohydrates are a loosely defined group of molecules that all contain carbon, hydrogen, and oxygen in the molar ratio 1:2:1. Their empirical formula (which lists the number of atoms in the molecule with subscripts) is $(CH_2O)_n$, where n is the number of carbon atoms. Because they contain many carbon–hydrogen (C—H) bonds, which release energy when oxidation occurs, carbohydrates are well suited for energy storage. Sugars are among the most important energy-storage molecules, and they exist in several different forms.

The simplest of the carbohydrates are the **monosaccharides** (Greek *mono*, "single," and Latin *saccharum*, "sugar"). Simple sugars contain as few as three carbon atoms, but those that play the central role in energy storage have six (figure 3.6). The empirical formula of 6-carbon sugars is:

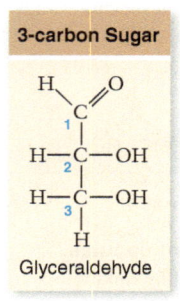

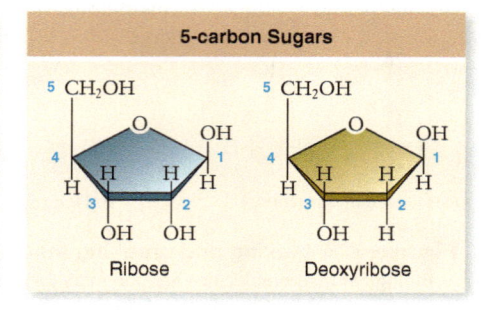

$$C_6H_{12}O_6 \quad \text{or} \quad (CH_2O)_6$$

Six-carbon sugars can exist in a straight-chain form, but dissolved in water (an aqueous environment) they almost always form rings.

The most important of the 6-carbon monosaccharides for energy storage is glucose, which you first encountered in the examples of chemical reactions in chapter 2. Glucose has seven energy-storing C—H bonds (figure 3.7). Depending on the orientation of the carbonyl group (C=O) when the ring is closed, glucose can exist in two different forms: alpha (α) or beta (β).

Sugar isomers have structural differences

Glucose is not the only sugar with the formula $C_6H_{12}O_6$. Both structural isomers and stereoisomers of this simple 6-carbon skeleton exist in nature. Fructose is a structural isomer that differs in the position of the carbonyl carbon (C=O); galactose is a stereoisomer that differs in the position of —OH and —H groups relative to the ring (figure 3.8). These differences often account for substantial functional differences between the isomers. Your taste buds can discern them: Fructose tastes much sweeter than glucose, despite the fact that both sugars have identical chemical composition. Enzymes that act on different sugars can distinguish both the structural and stereoisomers of this basic 6-carbon skeleton. The different stereoisomers of glucose are also important in the polymers that can be made using glucose as a monomer, as you will see later in this section.

Disaccharides serve as transport molecules in plants and provide nutrition in animals

Most organisms transport sugars within their bodies. In humans, the glucose that circulates in the blood does so as a simple monosaccharide. In plants and many other organisms, however, glucose is converted into a transport form before it is moved from place to place within the organism. In such a form, it is less readily metabolized during transport.

Transport forms of sugars are commonly made by linking two monosaccharides together to form a **disaccharide** (Greek *di*, "two"). Disaccharides serve as effective reservoirs of glucose because the enzymes that normally use glucose in the organism cannot break the bond linking the two monosaccharide subunits. Enzymes that can do so are typically present only in the tissue that uses glucose.

Transport forms differ depending on which monosaccharides are linked to form the disaccharide. Glucose forms transport disaccharides with itself and with many other monosaccharides,

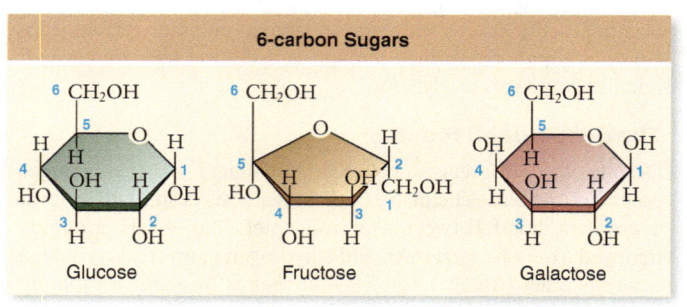

3-carbon Sugar	5-carbon Sugars	6-carbon Sugars
Glyceraldehyde	Ribose, Deoxyribose	Glucose, Fructose, Galactose

Figure 3.6 Monosaccharides. Monosaccharides, or simple sugars, can contain as few as three carbon atoms and are often used as building blocks to form larger molecules. The 5-carbon sugars ribose and deoxyribose are components of nucleic acids (see figure 3.15). The carbons are conventionally numbered (in *blue*) from the more oxidized end.

Figure 3.7 Structure of the glucose molecule. Glucose is a linear, 6-carbon molecule that forms a six-membered ring in solution. Ring closure occurs such that two forms can result: α-glucose and β-glucose. These structures differ only in the position of the —OH bound to carbon 1. The structure of the ring can be represented in many ways; shown here are the most common, with the carbons conventionally numbered so that the forms can be compared easily. The heavy lines in the ring structures represent portions of the molecule that are projecting out of the page toward you.

Figure 3.8 Isomers and stereoisomers. Glucose, fructose, and galactose are isomers with the empirical formula $C_6H_{12}O_6$. A structural isomer of glucose, such as fructose, has identical chemical groups bonded to different carbon atoms. Notice that this results in a five-membered ring in solution (see figure 3.6). A stereoisomer of glucose, such as galactose, has identical chemical groups bonded to the same carbon atoms but in different orientations (the —OH at carbon 4).

including fructose and galactose. When glucose forms a disaccharide with the structural isomer fructose, the resulting disaccharide is *sucrose,* or table sugar (figure 3.9a). Sucrose is the form most plants use to transport glucose and is the sugar that most humans and other animals eat. Sugarcane and sugar beets are rich in sucrose.

When glucose is linked to the stereoisomer galactose, the resulting disaccharide is *lactose,* or milk sugar. Many mammals supply energy to their young in the form of lactose. Adults often have greatly reduced levels of lactase, the enzyme required to cleave lactose into its two monosaccharide components, and thus they cannot metabolize lactose efficiently. This can result in lactose intolerance in humans. Most of the energy that is channeled into lactose production is therefore reserved for offspring. For this reason, lactose as an energy source is primarily for offspring in mammals.

Polysaccharides provide energy storage and structural components

Polysaccharides are longer polymers made up of monosaccharides that have been joined through dehydration reactions. **Starch,** a storage polysaccharide, consists entirely of α-glucose molecules linked in long chains. **Cellulose,** a structural polysaccharide, also consists of glucose molecules linked in chains, but these molecules are β-glucose. Because starch is built from α-glucose we call the linkages α linkages; cellulose has β linkages.

Starches and glycogen

Organisms store the metabolic energy contained in monosaccharides by converting them into disaccharides, such as *maltose* (figure 3.9b). These are then linked together into the insoluble polysaccharides called *starches.* These polysaccharides differ mainly in how the polymers branch.

The starch with the simplest structure is *amylose.* It is composed of many hundreds of α-glucose molecules linked together in long, unbranched chains. Each linkage occurs between the carbon 1 (C-1) of one glucose molecule and the C-4 of another, making them α-(1⟶4) linkages (figure 3.10a). The long chains of amylose tend to coil up in water, a property that renders amylose insoluble. Potato starch is about 20% amylose (figure 3.10b).

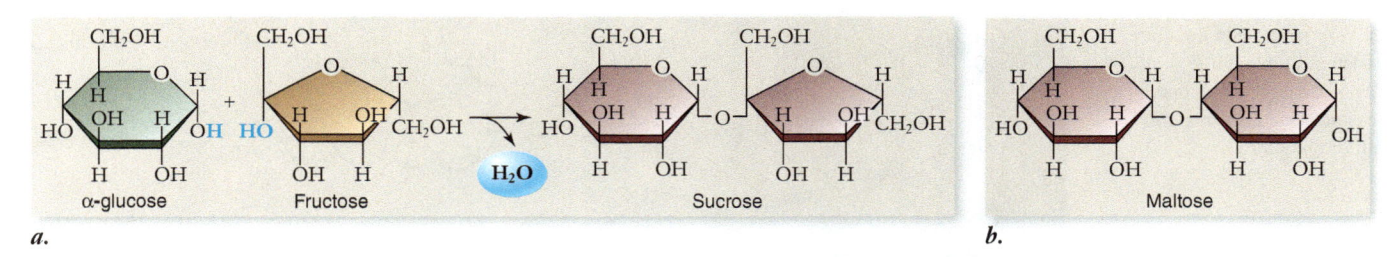

Figure 3.9 How disaccharides form. Some disaccharides are used to transport glucose from one part of an organism's body to another; one example is sucrose *(a)*, which is found in sugarcane. Other disaccharides, such as maltose *(b)*, are used in grain for storage.

chapter **3** The Chemical Building Blocks of Life

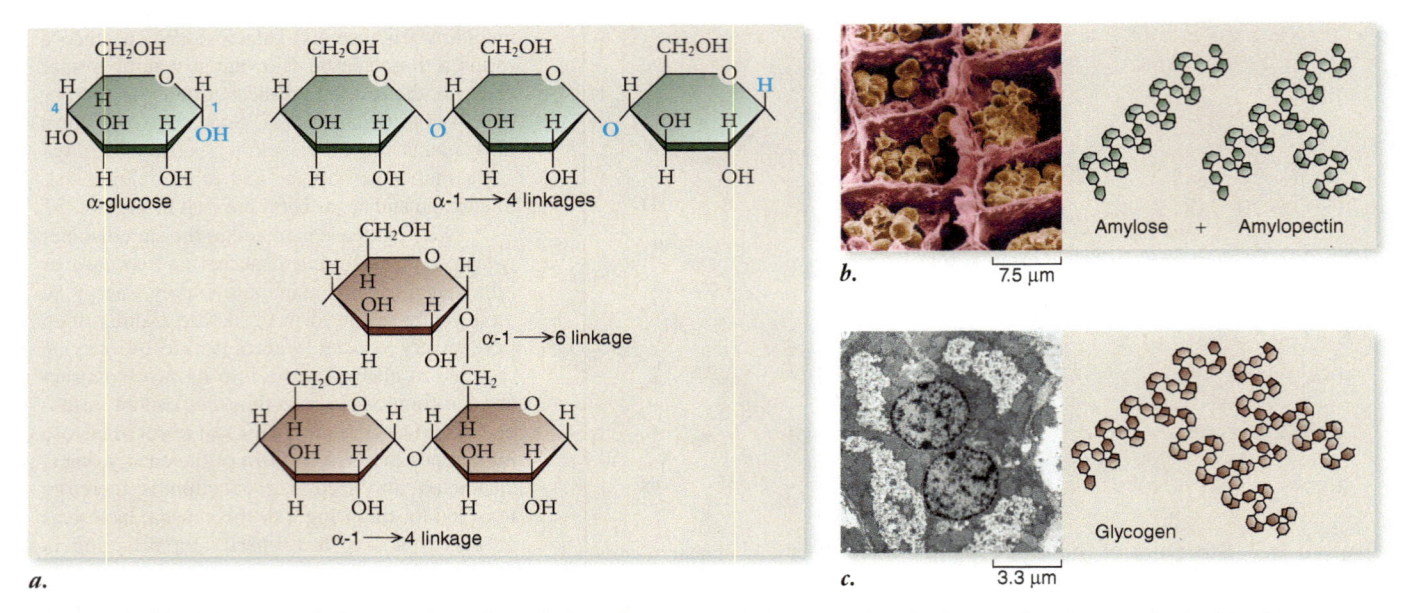

Figure 3.10 Polymers of glucose: Starch and glycogen. *a.* Starch chains consist of polymers of α-glucose subunits joined by α-(1⟶4) glycosidic linkages. These chains can be branched by forming similar α-(1⟶6) glycosidic bonds. These storage polymers then differ primarily in their degree of branching. *b.* Starch is found in plants and is composed of amylose and amylopectin, which are unbranched and branched, respectively. The branched form is insoluble and forms starch granules in plant cells. *c.* Glycogen is found in animal cells and is highly branched and also insoluble, forming glycogen granules.

Most plant starch, including the remaining 80% of potato starch, is a somewhat more complicated variant of amylose called *amylopectin*. Pectins are branched polysaccharides with the branches occurring due to bonds between the C-1 of one molecule and the C-6 of another [α-(1⟶6) linkages]. These short amylose branches consist of 20 to 30 glucose subunits (figure 3.10*b*).

The comparable molecule to starch in animals is **glycogen**. Like amylopectin, glycogen is an insoluble polysaccharide containing branched amylose chains. Glycogen has a much longer average chain length and more branches than plant starch (figure 3.10*c*).

Cellulose

Although some chains of sugars store energy, others serve as structural material for cells. For two glucose molecules to link together, the glucose subunits must be of the same form. *Cellulose* is a polymer of β-glucose (figure 3.11). The bonds between adjacent

Figure 3.11 Polymers of glucose: Cellulose. Starch chains consist of α-glucose subunits, and cellulose chains consist of β-glucose subunits. *a.* Thus the bonds between adjacent glucose molecules in cellulose are β-(1⟶4) glycosidic linkages. *b.* Cellulose is unbranched and forms long fibers. Cellulose fibers can be very strong and are quite resistant to metabolic breakdown, which is one reason wood is such a good building material.

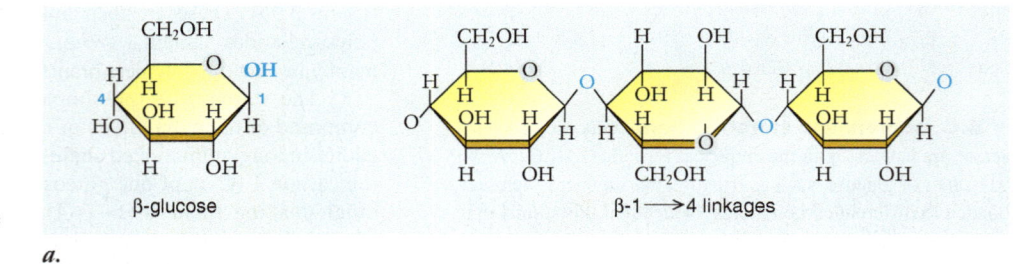

glucose molecules still exist between the C-1 of the first glucose and the C-4 of the next glucose, but these are β-(1⟶4) linkages.

The properties of a chain of glucose molecules consisting of all β-glucose are very different from those of starch. These long, unbranched β-linked chains make tough fibers. Cellulose is the chief component of plant cell walls (see figure 3.11b). It is chemically similar to amylose, with one important difference: The starch-hydrolyzing enzymes that occur in most organisms cannot break the bond between two β-glucose units because they only recognize α linkages.

Because cellulose cannot be broken down readily by most animals, it works well as a biological structural material. But some animals, such as cows, are able to utilize cellulose aided by symbiotic bacteria and protists in their digestive tracts. These organisms provide the necessary enzymes for cleaving the β-(1⟶4) linkages, thus providing access to a rich source of energy.

Chitin

Chitin, the structural material found in arthropods and many fungi, is a polymer of *N*-acetylglucosamine, a substituted version of glucose. When cross-linked by proteins, it forms a tough, resistant surface material that serves as the hard exoskeleton of insects and crustaceans (figure 3.12; see chapter 34). Few organisms are able to digest chitin, but most possess a chitinase enzyme, probably to protect against fungi.

Learning Outcomes Review 3.2

Monosaccharides have three to six or more carbon atoms typically arranged in a ring form. Disaccharides consist of two linked monosaccharides; polysaccharides are long chains of monosaccharides. Structural differences between sugar isomers can lead to functional differences. Starches are branched polymers of α-glucose used for energy storage. Cellulose in plants consists of unbranched chains of β-glucose that are not easily digested.

■ *How do the structures of starch, glycogen, and cellulose affect their function?*

Figure 3.12 Chitin. Chitin is the principal structural element in the external skeletons of many invertebrates, such as this lobster.

3.3 Nucleic Acids: Information Molecules

Learning Outcomes

1. Describe the structure of nucleotides.
2. Contrast the structures of DNA and RNA.
3. Discuss the functions of DNA and RNA.
4. Recognize other nucleotides involved in energy metabolism.

The biochemical activity of a cell depends on production of a large number of proteins, each with a specific sequence. The information necessary to produce the correct proteins is passed through generations of organisms, even though the proteins themselves are not inherited.

Nucleic acids carry information inside cells, just as disks contain the information in a computer or road maps display information needed by travelers. Two main varieties of nucleic acids are **deoxyribonucleic acid (DNA;** figure 3.13) and **ribonucleic acid (RNA).**

Genetic information is stored in DNA, and short-lived copies of this are made in the form of RNA, which is then used to direct the synthesis of proteins during the process of gene expression (as discussed in detail in chapter 15). Unique among macromolecules, nucleic acids are able to serve as templates for producing precise copies of themselves. This characteristic allows genetic information to be preserved during cell division and during the reproduction of organisms.

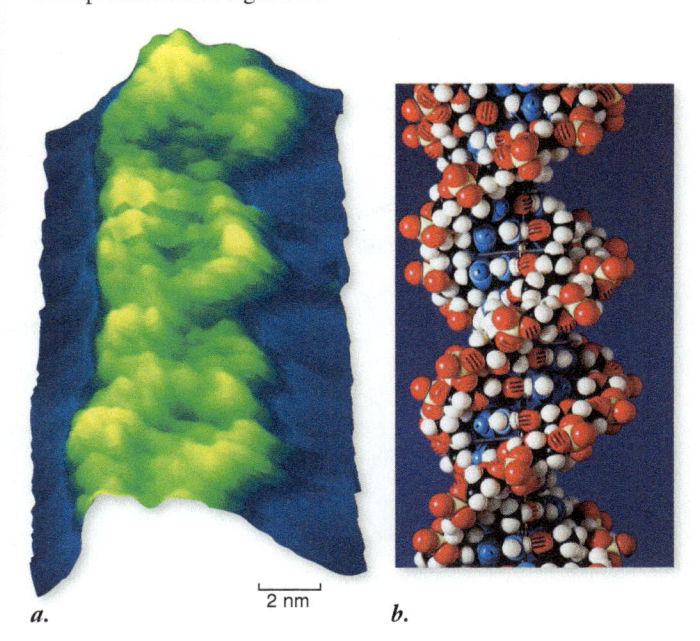

a. *b.*

Figure 3.13 Images of DNA. *a.* A scanning-tunneling micrograph of DNA (false color; 2,000,000×) showing approximately three turns of the DNA double helix. *b.* A space-filling model for comparison to the image of actual DNA in *(a)*.

The role of RNA in cells is much more complicated: RNA carries information, is part of the organelle responsible for protein synthesis, and recent work indicates it is also involved in the control of gene expression. As a carrier of information, the form of RNA called **messenger RNA (mRNA)** consists of transcribed single-stranded copies of portions of the DNA. These transcripts serve as blueprints specifying the amino acid sequences of proteins. This process will be described in detail in chapter 15.

Nucleic acids are nucleotide polymers

Nucleic acids are long polymers of repeating subunits called **nucleotides.** Each nucleotide consists of three components: a pentose, or 5-carbon sugar (ribose in RNA and deoxyribose in DNA); a phosphate ($-PO_4^-$) group; and an organic nitrogenous (nitrogen-containing) base (figure 3.14). Nucleotides can form polymers by joining the phosphate of one nucleotide to a hydroxyl group on the sugar of another nucleotide by a dehydration reaction. This forms a *phosphodiester bond* linking the two sugars through a phosphate. A **nucleic acid,** then, is simply a chain of 5-carbon sugars linked together by phosphodiester bonds with a nitrogenous base protruding from each sugar (see figure 3.15a). These chains of nucleotides, *polynucleotides,* have polarity, or different ends: a phosphate on one end and an —OH from a sugar on the other end. We conventionally refer to these ends as 5′ ("five-prime," $-PO_4^-$) and 3′ ("three-prime," —OH) taken from the carbon numbering of the sugar (figure 3.15a).

Nucleotides have five types of nitrogenous bases (figure 3.15b). Two of these are large, double-ring molecules called *purines* that are each found in both DNA and RNA; the two purines are adenine (A) and guanine (G). The other three bases are single-ring molecules called *pyrimidines* that include cytosine (C, in both DNA and RNA), thymine (T, in DNA only), and uracil (U, in RNA only).

DNA stores genetic information

Organisms use sequences of nucleotides in DNA to encode the information specifying the amino acid sequences of their proteins. This method of encoding information is very similar to the way in which sequences of letters encode information in a sentence.

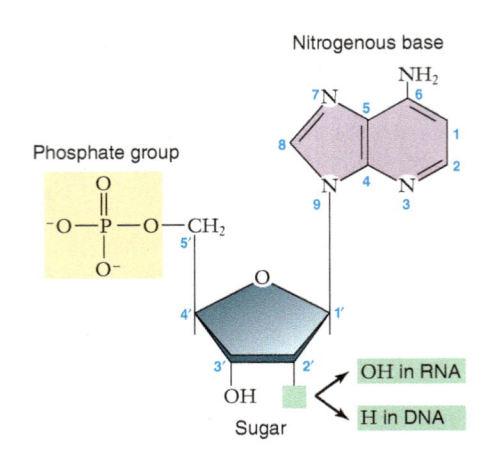

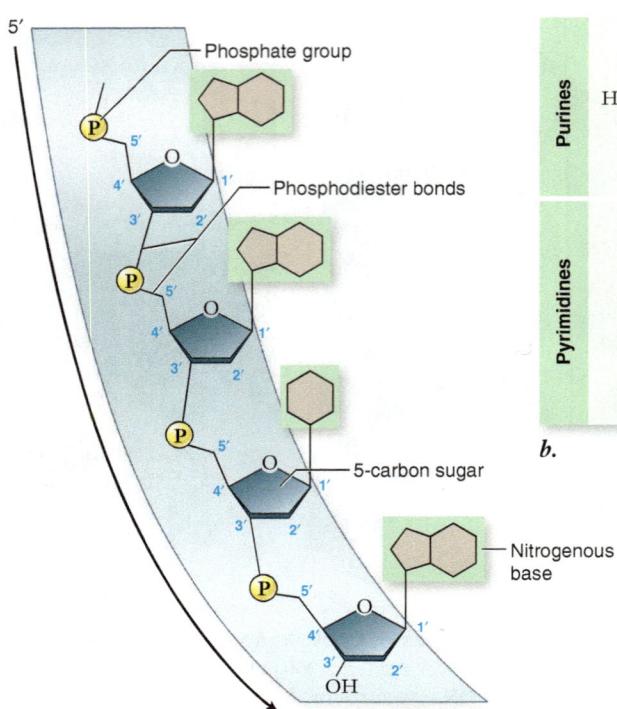

Figure 3.14 Structure of a nucleotide. The nucleotide subunits of DNA and RNA are made up of three elements: a 5-carbon sugar (ribose or deoxyribose), an organic nitrogenous base (adenine is shown here), and a phosphate group. Notice that all the numbers on the sugar are given as "primes" (1′, 2′, etc.) to distinguish them from the numbering on the rings of the bases.

Figure 3.15 The structure of a nucleic acid and the organic nitrogenous bases. *a.* In a nucleic acid, nucleotides are linked to one another via phosphodiester bonds formed between the phosphate of one nucleotide and the sugar of the next nucleotide. We call this the sugar-phosphate backbone, with the organic bases protrude from this chain. The backbone also has different ends: a 5′ phosphate end and a 3′ hydroxyl end (the blue numbers come from the numbers in the sugars). *b.* The organic nitrogenous bases can be either purines or pyrimidines. The base thymine is found in DNA. The base uracil is found in RNA.

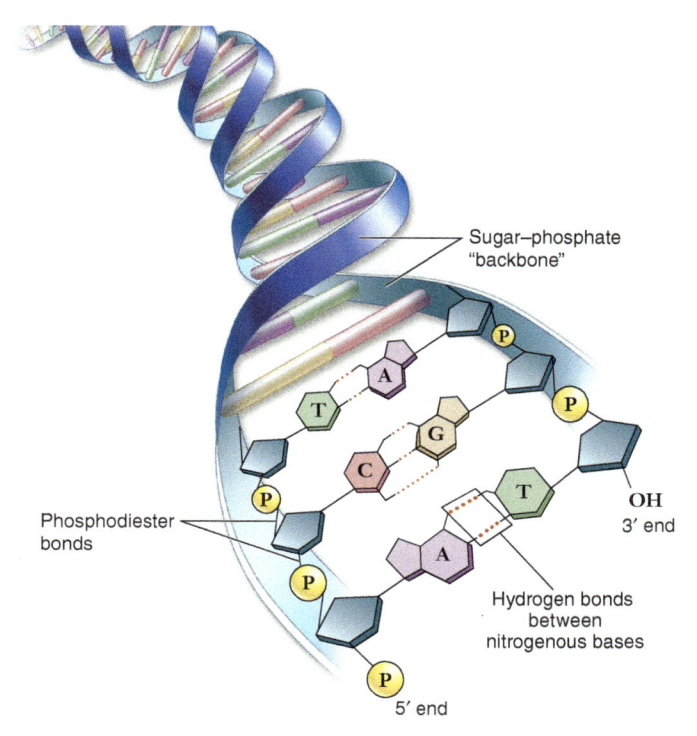

Figure 3.16 The structure of DNA. DNA consists of two polynucleotide chains running in opposite directions wrapped about a single helical axis. Hydrogen bond formation (dashed lines) between the nitrogenous bases, called base-pairing, causes the two chains of DNA to bind to each other and form a double helix.

A sentence written in English consists of a combination of the 26 different letters of the alphabet in a certain order; the code of a DNA molecule consists of different combinations of the four types of nucleotides in specific sequences, such as CGCTTACG.

DNA molecules in organisms exist as two chains wrapped about each other in a long linear molecule in eukaryotes, and a circular molecule in most prokaryotes. The two strands of a DNA polymer wind around each other like the outside and inside rails of a spiral staircase. Such a spiral shape is called a helix, and a helix composed of two chains is called a **double helix.** Each step of DNA's helical staircase is composed of a base-pair. The pair consists of a base in one chain attracted by hydrogen bonds to a base opposite it on the other chain (figure 3.16).

The base-pairing rules arise from the most stable hydrogen bonding configurations between the bases: Adenine pairs with thymine (in DNA) or with uracil (in RNA), and cytosine pairs with guanine. The bases that participate in base-pairing are said to be **complementary** to each other. Additional details of the structure of DNA and how it interacts with RNA in the production of proteins are presented in chapters 14 and 15.

In eukaryotic organisms, the DNA is further complexed with protein to form structures we call chromosomes. This actually forms a higher order structure that affects the function of DNA as it is involved in the control of gene expression (see chapter 16).

RNA has many roles in a cell

RNA is similar to DNA, but with two major chemical differences. First, RNA molecules contain ribose sugars, in which the C-2 is bonded to a hydroxyl group. (In DNA, a hydrogen atom replaces this hydroxyl group.) Second, RNA molecules use uracil in place of thymine. Uracil has a similar structure to thymine, except that one of its carbons lacks a methyl (—CH$_3$) group.

RNA is produced by transcription (copying) from DNA, and is usually single-stranded (figure 3.17). The role of RNA in cells is quite varied: it carries information in the form of **mRNA,** it is part of the ribosome, in the form of **ribosomal RNA (rRNA),** and it carries amino acids in the form of **transfer RNA (tRNA).** There has been a revolution of late in how we view RNA since it has been found to function as an enzyme, and other forms of RNA are involved in regulating gene expression (explored in more detail in chapter 16).

Other nucleotides are vital components of energy reactions

In addition to serving as subunits of DNA and RNA, nucleotide bases play other critical roles in the life of a cell. For example, adenine is a key component of the molecule **adenosine triphosphate**

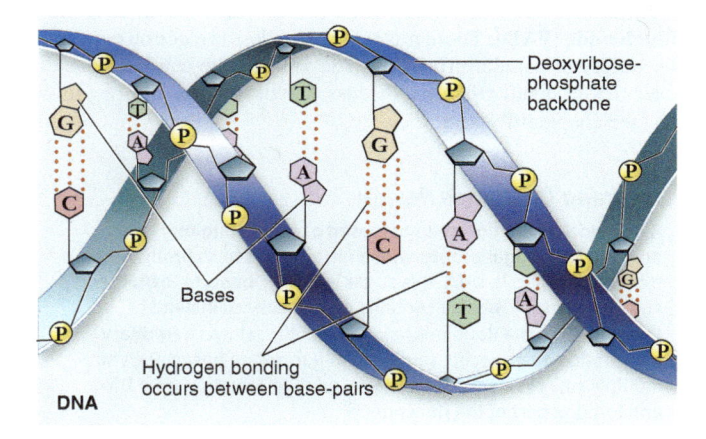

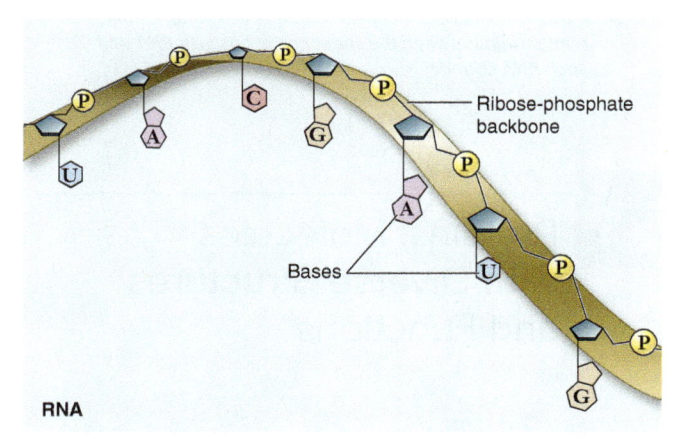

Figure 3.17 DNA versus RNA. DNA forms a double helix, uses deoxyribose as the sugar in its sugar–phosphate backbone, and uses thymine among its nitrogenous bases. RNA is usually single-stranded, uses ribose as the sugar in its sugar–phosphate backbone, and uses uracil in place of thymine.

Figure 3.18 ATP. Adenosine triphosphate (ATP) contains adenine, a 5-carbon sugar, and three phosphate groups.

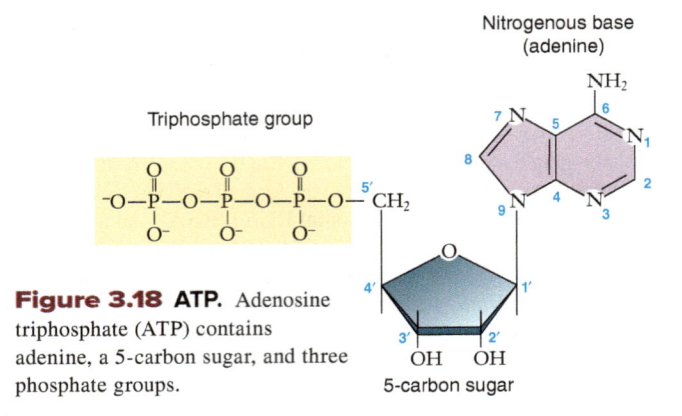

(**ATP**; figure 3.18)—the energy currency of the cell. Cells use ATP as energy in a variety of transactions, the way we use money in society. ATP is used to drive energetically unfavorable chemical reactions, to power transport across membranes, and to power the movement of cells.

Two other important nucleotide-containing molecules are **nicotinamide adenine dinucleotide (NAD$^+$)** and **flavin adenine dinucleotide (FAD)**. These molecules function as electron carriers in a variety of cellular processes. You will see the action of these molecules in detail when we discuss photosynthesis and respiration (see chapters 7 and 8).

Learning Outcomes Review 3.3

A nucleic acid is a polymer composed of alternating phosphate and 5-carbon sugar groups with a nitrogenous base protruding from each sugar. In DNA, this sugar is deoxyribose. In RNA, the sugar is ribose. RNA also contains the base uracil instead of thymine. DNA is a double-stranded helix that stores hereditary information as a specific sequence of nucleotide bases. RNA has multiple roles in a cell, including carrying information from DNA and forming part of the ribosome.

- If an RNA molecule is copied from a DNA strand, what is the relationship between the sequence of bases in RNA and each DNA strand?

3.4 Proteins: Molecules with Diverse Structures and Functions

Learning Outcomes

1. Describe the possible levels of protein structure.
2. Explain how motifs and domains contribute to protein structure.
3. Understand the relationship between amino acid sequence and their three-dimensional structure.

Proteins are the most diverse group of biological macromolecules, both chemically and functionally. Because proteins have so many different functions in cells we could not begin to list them all. We can, however, group these functions into the following seven categories. This list is a summary only, however; the function of proteins is relevant to most topics in biology:

1. **Enzyme catalysis.** Enzymes are biological catalysts that facilitate specific chemical reactions. Because of this property, the appearance of enzymes was one of the most important events in the evolution of life. Enzymes are three-dimensional globular proteins that fit snugly around the molecules they act on. This fit facilitates chemical reactions by stressing particular chemical bonds.
2. **Defense.** Other globular proteins use their shapes to "recognize" foreign microbes and cancer cells. These cell-surface receptors form the core of the body's endocrine and immune systems.
3. **Transport.** A variety of globular proteins transport small molecules and ions. The transport protein hemoglobin, for example, transports oxygen in the blood. Membrane transport proteins help move ions and molecules across the membrane.
4. **Support.** Protein fibers play structural roles. These fibers include keratin in hair, fibrin in blood clots, and collagen. The last one, collagen, forms the matrix of skin, ligaments, tendons, and bones and is the most abundant protein in a vertebrate body.
5. **Motion.** Muscles contract through the sliding motion of two kinds of protein filaments: actin and myosin. Contractile proteins also play key roles in the cell's cytoskeleton and in moving materials within cells.
6. **Regulation.** Small proteins called hormones serve as intercellular messengers in animals. Proteins also play many regulatory roles within the cell—turning on and shutting off genes during development, for example. In addition, proteins receive information, acting as cell-surface receptors.
7. **Storage.** Calcium and iron are stored in the body by binding as ions to storage proteins.

Table 3.2 summarizes these functions and includes examples of the proteins that carry them out in the human body.

Proteins are polymers of amino acids

Proteins are linear polymers made with 20 different amino acids. **Amino acids,** as their name suggests, contain an amino group (—NH$_2$) and an acidic carboxyl group (—COOH). The specific order of amino acids determines the protein's structure and function. Many scientists believe amino acids were among the first molecules formed on the early Earth. It seems highly likely that the oceans that existed early in the history of the Earth contained a wide variety of amino acids.

Amino acid structure

The generalized structure of an amino acid is shown as amino and carboxyl groups bonded to a central carbon atom, with an additional hydrogen and a functional side group indicated

TABLE 3.2	The Many Functions of Protein		
Function	**Class of Protein**	**Examples**	**Examples of Use**
Enzyme catalysis	Enzymes	Glycosidases	Cleave polysaccharides
		Proteases	Break down proteins
		Polymerases	Synthesize nucleic acids
		Kinases	Phosphorylate sugars and proteins
Defense	Immunoglobulins	Antibodies	Mark foreign proteins for elimination
	Toxins	Snake venom	Blocks nerve function
	Cell-surface antigens	MHC* proteins	"Self"-recognition
Transport	Circulating transporters	Hemoglobin	Carries O_2 and CO_2 in blood
		Myoglobin	Carries O_2 and CO_2 in muscle
		Cytochromes	Electron transport
	Membrane transporters	Sodium–potassium pump	Excitable membranes
		Proton pump	Chemiosmosis
		Glucose transporter	Transports glucose into cells
Support	Fibers	Collagen	Forms cartilage
		Keratin	Forms hair, nails
		Fibrin	Forms blood clots
Motion	Muscle	Actin	Contraction of muscle fibers
		Myosin	Contraction of muscle fibers
Regulation	Osmotic proteins	Serum albumin	Maintains osmotic concentration of blood
	Gene regulators	*lac* Repressor	Regulates transcription
	Hormones	Insulin	Controls blood glucose levels
		Vasopressin	Increases water retention by kidneys
		Oxytocin	Regulates uterine contractions and milk production
Storage	Ion-binding	Ferritin	Stores iron, especially in spleen
		Casein	Stores ions in milk
		Calmodulin	Binds calcium ions

*MHC, major histocompatibility complex.

by R. These components completely fill the bonds of the central carbon:

$$H_2N-\underset{\underset{H}{|}}{\overset{\overset{R}{|}}{C}}-COOH$$

The unique character of each amino acid is determined by the nature of the R group. Notice that unless the R group is an H atom, as in glycine, amino acids are chiral and can exist as two enantiomeric forms: D or L. In living systems, only the L-amino acids are found in proteins, and D-amino acids are rare.

The R group also determines the chemistry of amino acids. Serine, in which the R group is —CH$_2$OH, is a polar molecule. Alanine, which has —CH$_3$ as its R group, is nonpolar. The 20 common amino acids are grouped into five chemical classes, based on their R group:

1. Nonpolar amino acids, such as leucine, often have R groups that contain —CH$_2$ or —CH$_3$.
2. Polar uncharged amino acids, such as threonine, have R groups that contain oxygen (or —OH).
3. Charged amino acids, such as glutamic acid, have R groups that contain acids or bases that can ionize.
4. Aromatic amino acids, such as phenylalanine, have R groups that contain an organic (carbon) ring with alternating single and double bonds. These are also nonpolar.
5. Amino acids that have special functions have unique properties. Some examples are methionine, which is often the first amino acid in a chain of amino acids; proline, which causes kinks in chains; and cysteine, which links chains together.

Each amino acid affects the shape of a protein differently, depending on the chemical nature of its side group. For example, portions of a protein chain with numerous nonpolar amino acids tend to fold into the interior of the protein by hydrophobic exclusion.

Peptide bonds

In addition to its R group, each amino acid, when ionized, has a positive amino (NH_3^+) group at one end and a negative carboxyl (COO^-) group at the other. The amino and carboxyl groups on a pair of amino acids can undergo a dehydration reaction to form a covalent bond. The covalent bond that links two amino acids is called a **peptide bond** (figure 3.19). The two amino acids linked by such a bond are not free to rotate around the N—C linkage because the peptide bond has a partial double-bond character. This is different from the N—C and C—C bonds to the central carbon of the amino acid. This lack of rotation about the peptide bond is one factor that determines the structural character of the coils and other regular shapes formed by chains of amino acids.

A protein is composed of one or more long unbranched chains. Each chain is called a **polypeptide** and is composed of amino acids linked by peptide bonds. The terms *protein* and *polypeptide* tend to be used loosely and may be confusing. For proteins that include only a single polypeptide chain, the two terms are synonymous.

The pioneering work of Frederick Sanger in the early 1950s provided the evidence that each kind of protein has a specific amino acid sequence. Using chemical methods to remove successive amino acids and then identify them, Sanger succeeded in determining the amino acid sequence of insulin. In so doing he demonstrated clearly that this protein had a defined sequence, which was the same for all insulin molecules in the solution. Although many different amino acids occur in nature, only 20 commonly occur in proteins. Of these 20, 8 are called essential amino acids because humans cannot synthesize them and thus must get them from their diets. Figure 3.20 illustrates these 20 amino acids and their side groups.

Proteins have levels of structure

The shape of a protein determines its function. One way to study the shape of something as small as a protein is to look at it with very short wavelength energy—in other words, with X-rays. X-rays can be passed through a crystal of protein to produce a diffraction pattern. This pattern can then be analyzed by a painstaking procedure that allows the investigator to build up a three-dimensional picture of the position of each atom. The first protein to be analyzed in this way was myoglobin, and the related protein hemoglobin was analyzed soon thereafter.

As more and more proteins were studied, a general principle became evident: In every protein studied, essentially all the internal amino acids are nonpolar ones—amino acids such as leucine, valine, and phenylalanine. Water's tendency to hydrophobically exclude nonpolar molecules literally shoves the nonpolar portions of the amino acid chain into the protein's interior (figure 3.21). This tendency forces the nonpolar amino acids into close contact with one another, leaving little empty space inside. Polar and charged amino acids are restricted to the surface of the protein, except for the few that play key functional roles.

The structure of proteins is usually discussed in terms of a hierarchy of four levels: *primary, secondary, tertiary,* and *quaternary* (figure 3.22). We will examine this view and then integrate it with a more modern approach arising from our increasing knowledge of protein structure.

Primary structure: Amino acid sequence

The **primary structure** of a protein is its amino acid sequence. Because the R groups that distinguish the amino acids play no role in the peptide backbone of proteins, a protein can consist of any sequence of amino acids. Thus, because any of 20 different amino acids might appear at any position, a protein containing 100 amino acids could form any of 20^{100} different amino acid sequences (that's the same as 10^{130}, or 1 followed by 130 zeros—more than the number of atoms known in the universe). This important property of proteins permits great diversity.

Consider the protein hemoglobin, the protein your blood uses to transport oxygen. Hemoglobin is composed of two α-globin peptide chains and two β-globin peptide chains. The α-globin chains differ from the β-globin ones in the sequence of amino acids. Furthermore, any alteration in the normal sequence of either of the types of globin proteins, even by a single amino acid, can have drastic effects on how the protein functions.

Secondary structure: Hydrogen bonding patterns

The amino acid side groups are not the only portions of proteins that form hydrogen bonds. The peptide groups of the main chain can also do so. These hydrogen bonds can be with water or with other peptide groups. If the peptide groups formed too many hydrogen bonds with water, the proteins would tend to behave like a random coil and wouldn't produce the kinds of globular

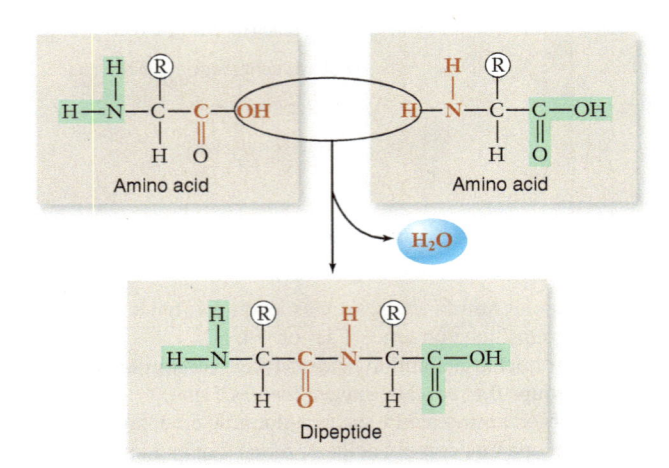

Figure 3.19 The peptide bond. A peptide bond forms when the amino end of one amino acid joins to the carboxyl end of another. Reacting amino and carboxyl groups are shown in red and nonreacting groups are highlighted in green. Notice that the resulting dipeptide still has an amino end and a carboxyl end. Because of the partial double-bond nature of peptide bonds, the resulting peptide chain cannot rotate freely around these bonds.

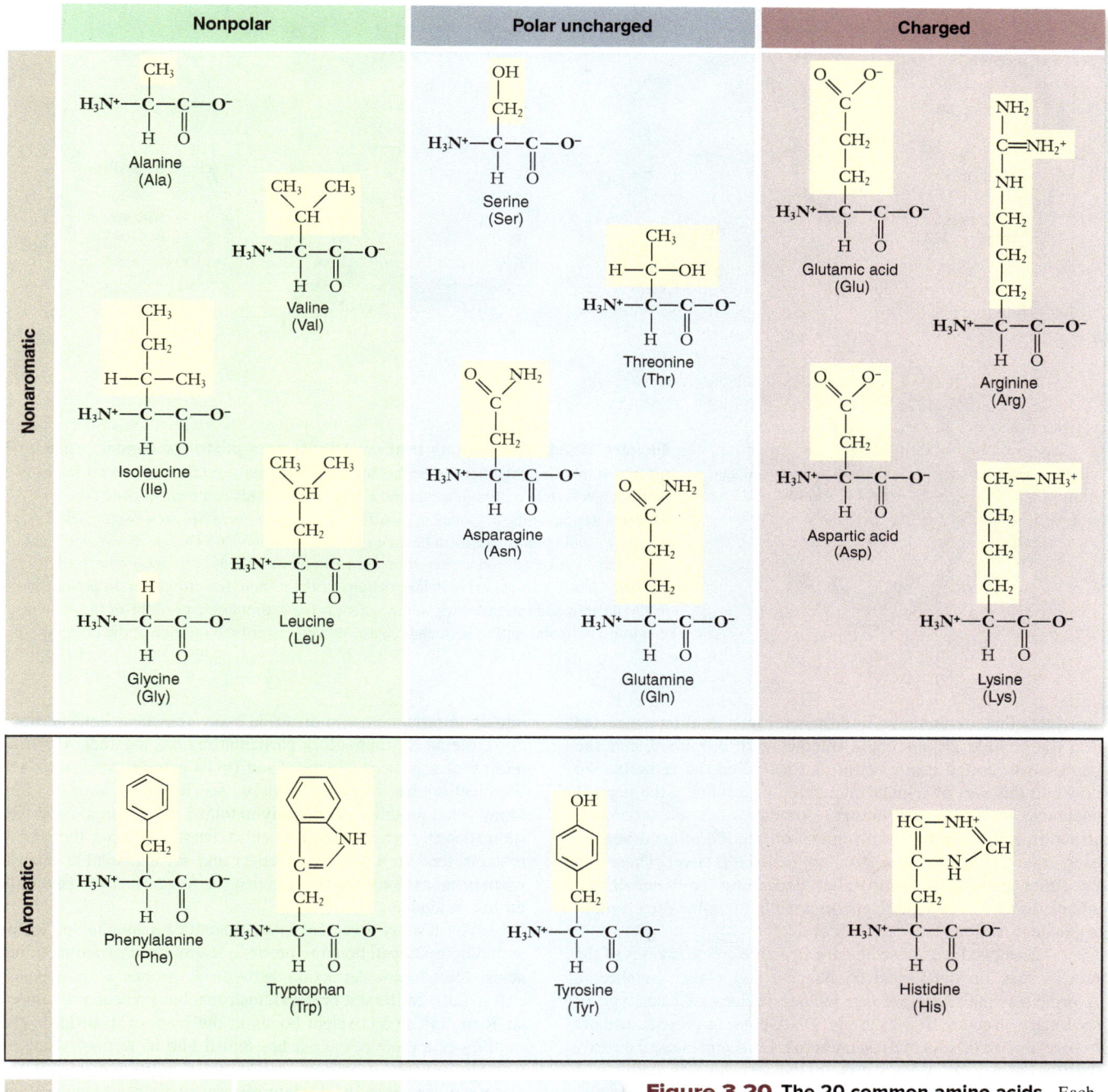

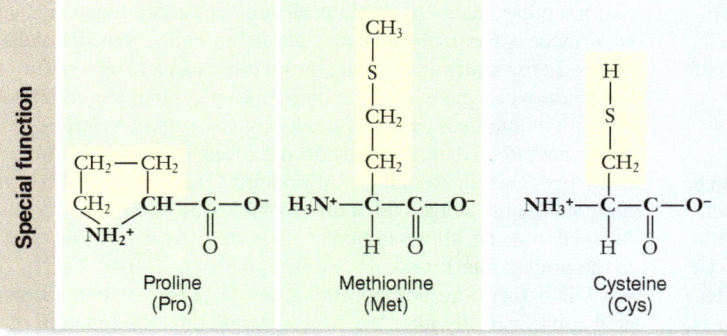

Figure 3.20 The 20 common amino acids. Each amino acid has the same chemical backbone, but differs in the side, or R, group. Seven of the amino acids are nonpolar because they have —CH$_2$ or —CH$_3$ in their R groups. Two of the seven contain ring structures with alternating double and single bonds, which classifies them also as aromatic. Another five are polar because they have oxygen or a hydroxyl group in their R groups. Five others are capable of ionizing to a charged form. The remaining three special-function amino acids have chemical properties that allow them to help form links between protein chains or kinks in proteins.

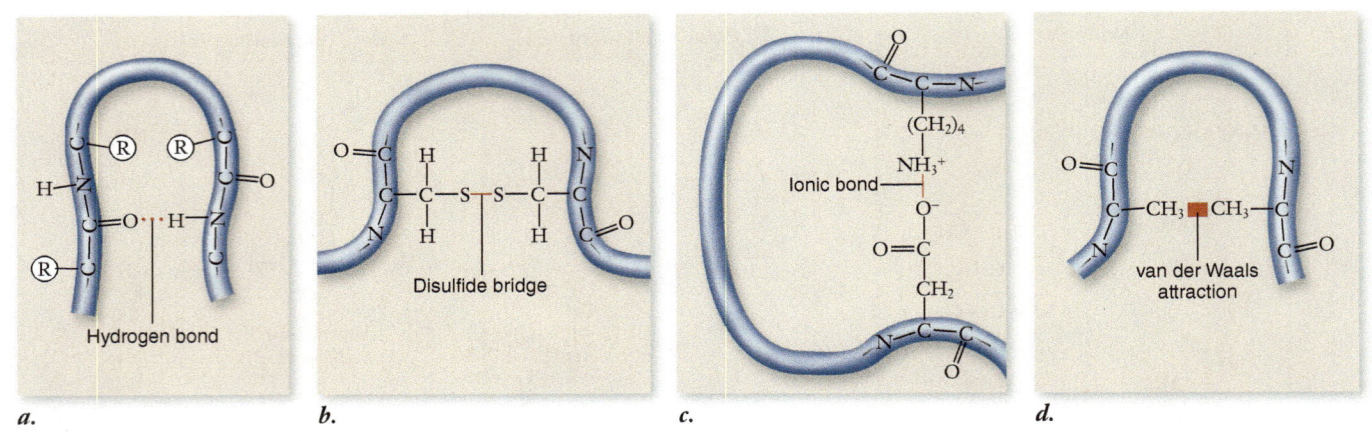

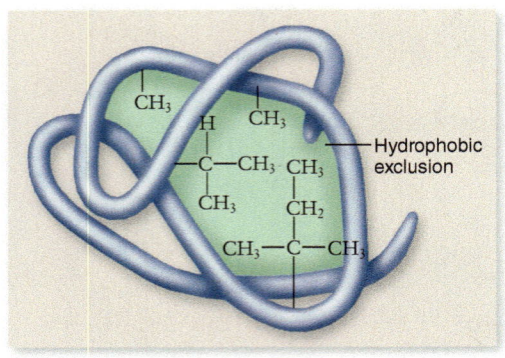

Figure 3.21 Interactions that contribute to a protein's shape. Aside from the bonds that link together the amino acids in a protein, several other weaker forces and interactions stabilize protein structure. *a.* Hydrogen bonds can form between the different amino acids. *b.* Covalent disulfide bridges can form between two cysteine side chains. *c.* Ionic bonds can form between groups with opposite charge. *d.* van der Waals attractions, which are weak attractions between atoms due to oppositely polarized electron clouds, can occur. *e.* Polar portions of the protein tend to gather on the outside of the protein and interact with water, whereas the hydrophobic portions of the protein, including nonpolar amino acid chains, are shoved toward the interior of the protein.

structures that are common in proteins. Linus Pauling suggested that the peptide groups could interact with one another if the peptide was coiled into a spiral that he called the **α helix.** We now call this sort of regular interaction of groups in the peptide backbone **secondary structure.** Another form of secondary structure can occur between regions of peptide aligned next to each other to form a planar structure called a **β sheet.** These can be either parallel or antiparallel depending on whether the adjacent sections of peptide are oriented in the same direction, or opposite direction.

These two kinds of secondary structure create regions of the protein that are cylindrical (α helices) and planar (β sheets). A protein's final structure can include regions of each type of secondary structure. For example, DNA-binding proteins usually have regions of α helix that can lay across DNA and interact directly with the bases of DNA. Porin proteins that form holes in membranes are composed of β sheets arranged to form a pore in the membrane. Finally in hemoglobin, the α- and β-globin peptide chains that make up the final molecule each have characteristic regions of secondary structure.

Tertiary structure: Folds and links

The final folded shape of a globular protein is called its **tertiary structure.** This tertiary structure contains regions that have secondary structure and determines how these are further arranged in space to produce the overall structure. A protein is initially driven into its tertiary structure by hydrophobic exclusion from water. Ionic bonds between oppositely charged R groups bring regions into close proximity, and disulfide bonds (covalent links between two cysteine R groups) lock particular regions together. The final folding of a protein is determined by its primary structure—the chemical nature of its side groups (see figures 3.21 and 3.22). Many small proteins can be fully unfolded ("denatured") and will spontaneously refold into their characteristic shape. Other larger proteins tend to associate together and form insoluble clumps when denatured, such as the film that can form when you heat milk for hot chocolate.

The tertiary structure is stabilized by a number of forces including hydrogen bonding between R groups of different amino acids, electrostatic attraction between R groups with opposite charge (also called salt bridges), hydrophobic exclusion of nonpolar R groups, and covalent bonds in the form of disulfides. The stability of a protein, once it has folded into its tertiary shape, is strongly influenced by how well its interior fits together. When two nonpolar chains in the interior are very close together, they experience a form of molecular attraction called van der Waals forces. Individually quite weak, these forces can add up to a strong attraction when many of them come into play, like the combined strength of hundreds of hooks and loops on a strip of Velcro. These forces are effective only over short distances, however. No "holes" or cavities exist in the interior of proteins. The variety of different nonpolar amino acids, with a different-sized R group with its own distinctive shape, allows nonpolar chains to fit very precisely within the protein interior.

It is therefore not surprising that changing a single amino acid can drastically alter the structure, and thus the function of a

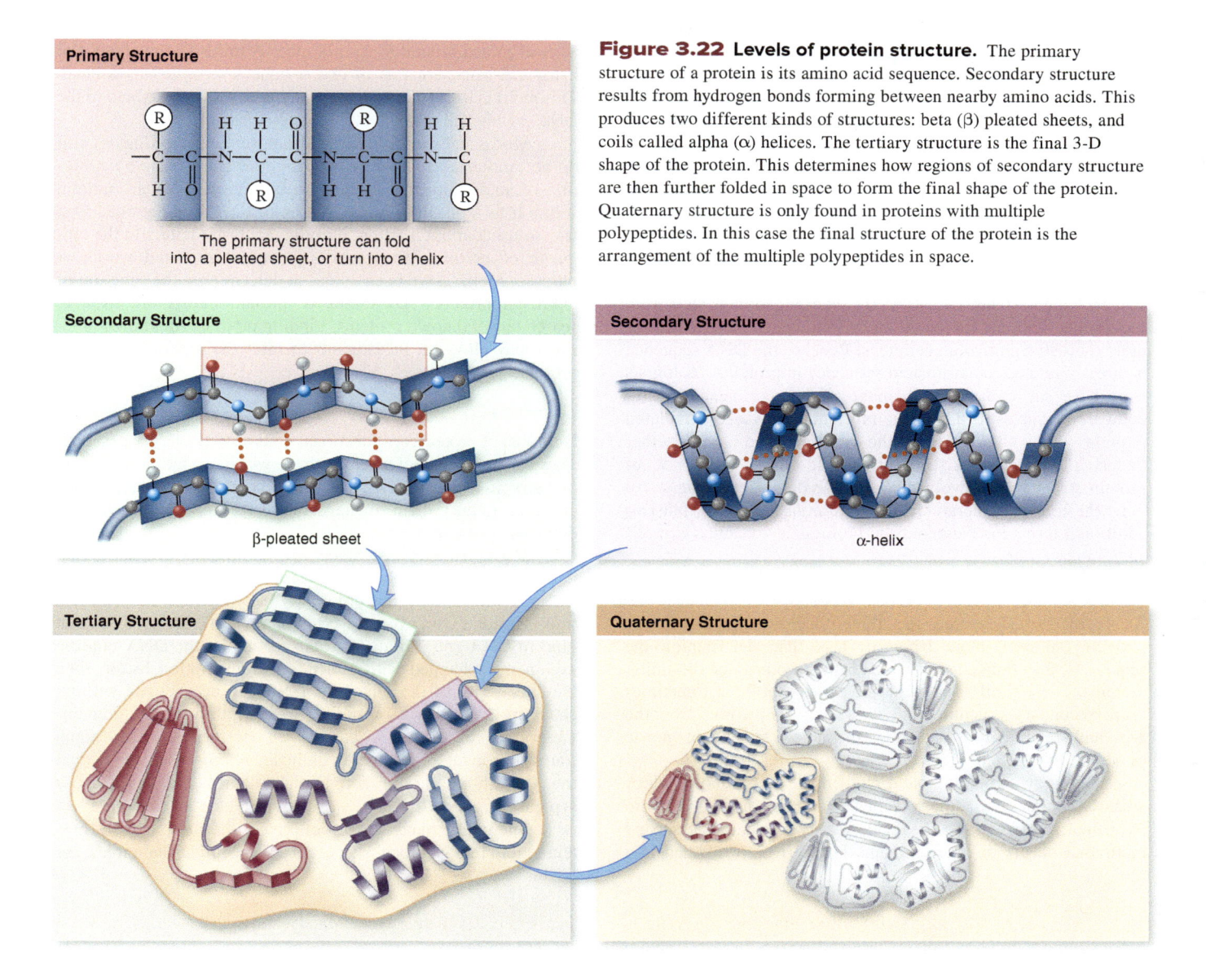

Figure 3.22 Levels of protein structure. The primary structure of a protein is its amino acid sequence. Secondary structure results from hydrogen bonds forming between nearby amino acids. This produces two different kinds of structures: beta (β) pleated sheets, and coils called alpha (α) helices. The tertiary structure is the final 3-D shape of the protein. This determines how regions of secondary structure are then further folded in space to form the final shape of the protein. Quaternary structure is only found in proteins with multiple polypeptides. In this case the final structure of the protein is the arrangement of the multiple polypeptides in space.

protein. The sickle cell version of hemoglobin (HbS), for example, is a change of a single glutamic acid for a valine in the β-globin chain. This change substitutes a charged amino acid for a nonpolar one on the surface of the protein, leading the protein to become sticky and form clumps. Another variant of hemoglobin called HbE, actually the most common in human populations, causes a change from glutamic acid to lysine at a different site in the β-globin chain. In this case the structural change is not as dramatic, but it still impairs function, resulting in blood disorders called anemia and thalassemia. More than 700 structural variants of hemoglobin are known, with up to 7% of the world's population being carriers of forms that are medically important.

Quaternary structure: Subunit arrangements

When two or more polypeptide chains associate to form a functional protein, the individual chains are referred to as subunits of the protein. The arrangement of these subunits is termed its **quaternary structure.** In proteins composed of subunits, the interfaces where the subunits touch one another are often nonpolar, and they play a key role in transmitting information between the subunits about individual subunit activities.

Remember that the protein hemoglobin is composed of two α-chain subunits and two β-chain subunits. Each α- and β-globin chain has a primary structure consisting of a specific sequence of amino acids. This then assumes a characteristic secondary structure consisting of α helices and β sheets that are then arranged into a specific tertiary structure for each α- and β-globin subunit. Lastly, these subunits are then arranged into their final quaternary structure. This is the final structure of the protein. For proteins that consist of only a single peptide chain, the enzyme lysozyme for example, the tertiary structure is the final structure of the protein.

chapter **3** *The Chemical Building Blocks of Life* **49**

Motifs and domains are structural elements of proteins

To directly determine the sequence of amino acids in a protein is a laborious task. Although the process has been automated, it remains slow and difficult.

The ability to sequence DNA changed this situation rather suddenly. Originally done manually, the Human Genome Project drove the development of automated sequencing. This increased throughput significantly, but the advent of next-generation sequencing technologies resulted in quantum increases in sequence data. Today, over 40,000 bacterial genomes have been sequenced, and almost 8000 eukaryotic genomes, including more than 80 mammalian genomes. Because the DNA sequence is directly related to amino acid sequence in proteins, biologists now have an enormous database of protein sequences to compare and analyze. This new information has also stimulated thought about the logic of the genetic code and whether underlying patterns exist in protein structure. Our view of protein structure has evolved with these data. Researchers still view the four-part hierarchical structure as important, but two additional terms have entered the biologist's vocabulary: motif and domain.

Motifs

As biologists discovered the three-dimensional structure of proteins (an even more laborious task than determining the sequence), they noticed similarities between otherwise dissimilar proteins. These similar structures are called **motifs,** or sometimes "supersecondary structure." The term *motif* is borrowed from the arts and refers to a recurring thematic element in music or design.

One very common protein motif is the β-α-β motif, which creates a fold or crease; the so-called "Rossmann fold" at the core of nucleotide-binding sites in a wide variety of proteins. A second motif that occurs in many proteins is the β barrel, which is a β sheet folded around to form a tube. A third type of motif, the helix-turn-helix, consists of two α helices separated by a bend. This motif is important because many proteins use it to bind to the DNA double helix (figure 3.23; see also chapter 16).

Motifs indicate a logic to structure that investigators still do not understand. Do they simply represent a reuse by evolution of something that already works, or are they an optimal solution to a problem, such as how to bind a nucleotide? One way to think about it is that if amino acids are letters in the language of proteins, then motifs represent repeated words or phrases. Motifs have been useful in determining the function of unknown proteins. Databases of protein motifs are used to search new unknown proteins. Finding motifs with known functions may allow an investigator to infer the function of a new protein.

Domains

Domains of proteins are functional units within a larger structure. They can be thought of as substructure within the tertiary structure of a protein (figure 3.23). To continue the metaphor: Amino acids are letters in the protein language, motifs are words or phrases, and domains are paragraphs.

Most proteins are made up of multiple domains that perform different parts of the protein's function. In many cases, these domains can be physically separated. For example, transcription factors (discussed in chapter 16) are proteins that bind to DNA and initiate its transcription. If the DNA-binding region is exchanged with a different transcription factor, then the specificity of the factor for DNA can be changed without changing its ability to stimulate transcription. Such "domain-swapping" experiments have been performed with many transcription factors, and they indicate, among other things, that the DNA-binding and activation domains are functionally separate.

These functional domains of proteins may also help the protein to fold into its proper shape. As a polypeptide chain folds, the

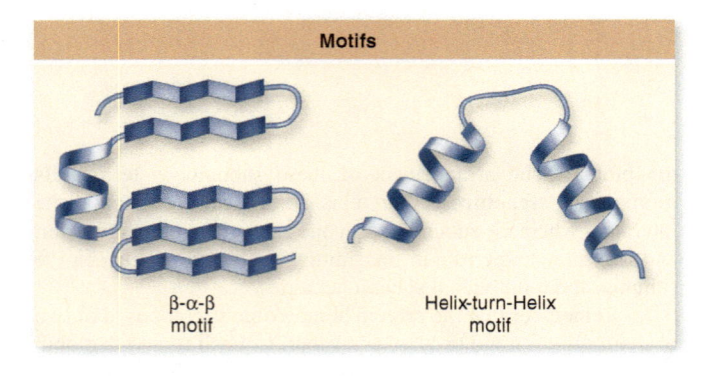

Figure 3.23 Motifs and domains. The elements of secondary structure can combine, fold, or crease to form motifs. These motifs are found in different proteins and can be used to predict function. Proteins also are made of larger domains, which are functionally distinct parts of a protein. The arrangement of these domains in space is the tertiary structure of a protein.

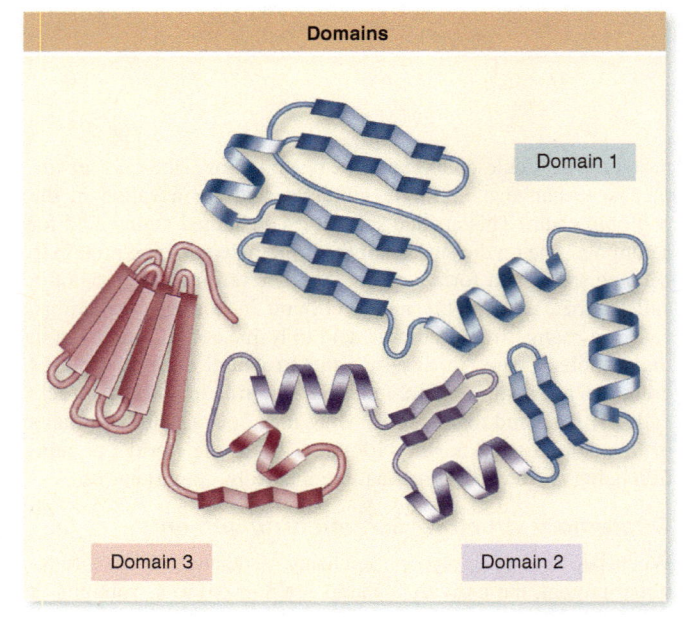

domains take their proper shape, each more or less independently of the others. This action can be demonstrated experimentally by artificially producing the fragment of a polypeptide that forms the domain in the intact protein, and showing that the fragment folds to form the same structure as it exhibits in the intact protein. A single polypeptide chain connects the domains of a protein, like a rope tied into several adjacent knots.

Domains can also correspond to the structure of the genes that encode them. Later, in chapter 15, you will see that genes in eukaryotes are often in pieces within the genome, and these pieces, called *exons*, sometimes encode the functional domains of a protein. This finding led to the idea of evolution acting by shuffling protein-encoding domains.

The process of folding relies on chaperone proteins

Originally, biochemists thought that newly made proteins fold spontaneously, randomly trying out different configurations as hydrophobic interactions with water shoved nonpolar amino acids into the protein's interior until the final structure was arrived at. We now know this view is too simple. Protein chains can fold in so many different ways that trial and error would simply take too long. In addition, as the open chain folds its way toward its final form, nonpolar "sticky" interior portions are exposed during intermediate stages. If these intermediate forms are placed in a test tube in an environment identical to that inside a cell, they stick to other, unwanted protein partners, forming a gluey mess.

How do cells avoid having their proteins clump into a mass? A vital clue came in studies of unusual mutations that prevent viruses from replicating in bacterial cells. It turns out that the virus proteins produced inside the cells could not fold properly. Further study revealed that normal cells contain **chaperone proteins,** which help other proteins to fold correctly.

Molecular biologists have now identified many proteins that act as molecular chaperones. This large class of proteins can be divided into subclasses, and representatives have been found in essentially every organism that has been examined. At least some of these proteins have been shown to be necessary for viability, illustrating their fundamental importance. Many are so-called heat shock proteins, produced in large amounts in response to elevated temperature. High temperatures cause proteins to unfold, and heat shock chaperone proteins help the cell's proteins to refold properly.

One class of these proteins, called chaperonins, has been extensively studied. In the bacterium *Escherichia coli (E. coli)*, one example is the essential protein GroE chaperonin. In mutants in which the GroE chaperonin is inactivated, fully 30% of the bacterial proteins fail to fold properly. Chaperonins associate to form a large macromolecular complex that resembles a cylindrical container. Proteins can move into the container, and the container itself can change its shape considerably (figure 3.24). Experiments have shown that an improperly folded protein can enter the chaperonin and be refolded. Although we don't know exactly how this happens, it seems to involve changes in the hydrophobicity of the interior of the chamber.

The flexibility of the structure of chaperonins is amazing. We tend to think of proteins as being fixed structures, but this is clearly not the case for chaperonins and this flexibility is necessary for their function. It also illustrates that even domains that may be very widely separated in a very large protein are still functionally connected. The folding process within a chaperonin harnesses the hydrolysis of ATP to power these changes in structure necessary for function. This entire process can occur in a cyclic manner until the appropriate structure is achieved. Cells use these chaperonins both to accomplish the original folding of some proteins and to restore the structure of incorrectly folded ones.

Improper folding of proteins can result in disease

Chaperone protein deficiencies may be implicated in certain diseases in which key proteins are improperly folded. Cystic fibrosis is a hereditary disorder in which a mutation disables a vital protein

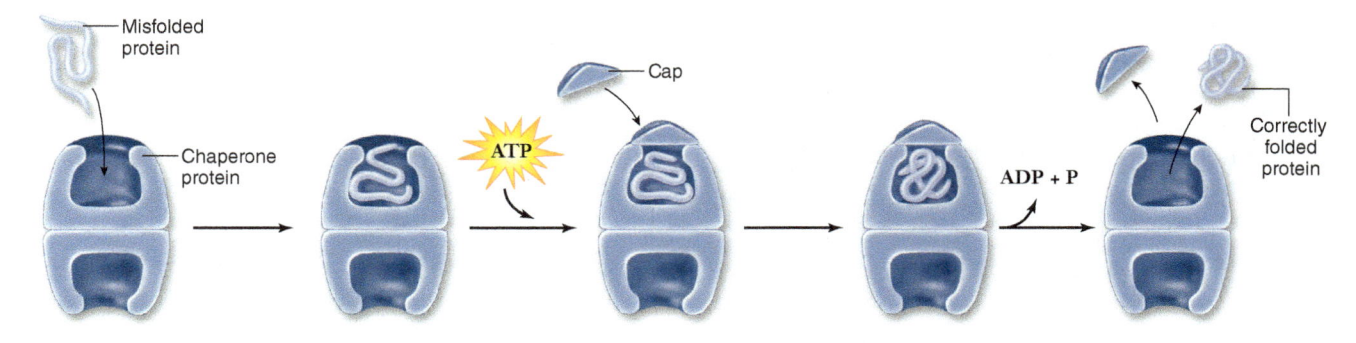

Chance for protein to refold

Figure 3.24 How one type of chaperone protein works. This barrel-shaped chaperonin is from the GroE family of chaperone proteins. It is composed of two identical rings each with seven identical subunits, each of which has three distinct domains. An incorrectly folded protein enters one chamber of the barrel, and a cap seals the chamber. Energy from the hydrolysis of ATP fuels structural alterations to the chamber, changing it from hydrophobic to hydrophilic. This change allows the protein to refold. After a short time, the protein is ejected, either folded or unfolded, and the cycle can repeat itself.

that moves ions across cell membranes. As a result, people with cystic fibrosis have thicker than normal mucus. This results in breathing problems, lung disease, and digestive difficulties, among other things. One interesting feature of the molecular analysis of this disease has been the number of different mutations found in human populations. One diverse class of mutations all result in problems with protein folding. The number of different mutations that can result in improperly folded proteins may be related to the fact that the native protein often fails to fold properly.

Denaturation inactivates proteins

If a protein's environment is altered, the protein may change its shape or even unfold completely. This process is called **denaturation** (figure 3.25). Proteins can be denatured when the pH, temperature, or ionic concentration of the surrounding solution changes.

Denatured proteins are usually biologically inactive. This action is particularly significant in the case of enzymes. Because practically every chemical reaction in a living organism is catalyzed by a specific enzyme, it is vital that a cell's enzymes work properly.

The traditional methods of food preservation, salt curing and pickling, involve denaturation of proteins. Prior to the general availability of refrigerators and freezers, the only practical way to keep microorganisms from growing in food was to keep the food in

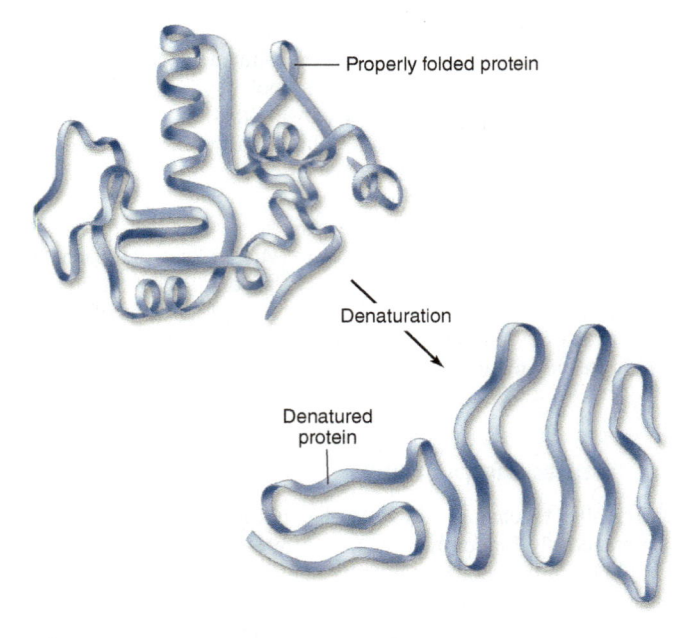

Figure 3.25 Protein denaturation. Environmental changes, such as variation in temperature or pH, can cause a protein to unfold and lose its shape. This loss of structure is called denaturation. Denatured proteins are biologically inactive.

SCIENTIFIC THINKING

Hypothesis: The 3-D structure of a protein is the thermodynamically stable structure. It depends only on the primary structure of the protein and the solution conditions.
Prediction: If a protein is denatured and allowed to renature under native conditions, it will refold into the native structure.
Test: Ribonuclease is treated with a reducing agent to break disulfide bonds and is then treated with urea to completely unfold the protein. The disulfide bonds are re-formed under nondenaturing conditions to see if the protein refolds properly.

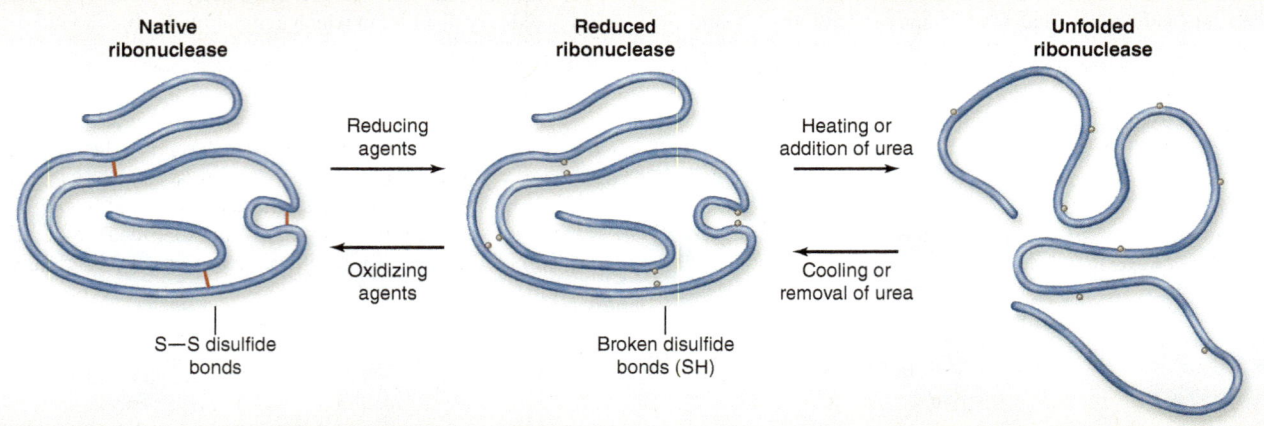

Result: Denatured ribonuclease refolds properly under nondenaturing conditions.
Conclusion: The hypothesis is supported. The information in the primary structure (amino acid sequence) is sufficient for refolding to occur. This implies that protein folding results in the thermodynamically stable structure.
Further Experiments: If the disulfide bonds were allowed to re-form under denaturing conditions, would we get the same result? How can we rule out that the protein had not been completely denatured and therefore retained some structure?

Figure 3.26 Primary structure determines tertiary structure.

a solution containing a high concentration of salt or vinegar, which denatured the enzymes of most microorganisms and prevented them from growing on the food.

Most enzymes function within a very narrow range of environmental conditions. Blood-borne enzymes that course through a human body at a pH of about 7.4 would rapidly become denatured in the highly acidic environment of the stomach. Conversely, the protein-degrading enzymes that function at a pH of 2 or less in the stomach would be denatured in the relatively basic pH of the blood. Similarly, organisms that live near oceanic hydrothermal vents have enzymes that work well at these extremes of temperature (over 100°C). They cannot survive in cooler waters, because their enzymes do not function properly at lower temperatures. Any given organism usually has a tolerance range of pH, temperature, and salt concentration. Within that range, its enzymes maintain the proper shape to carry out their biological functions.

When a protein's normal environment is reestablished after denaturation, a small protein may spontaneously refold into its natural shape, driven by the interactions between its nonpolar amino acids and water (figure 3.26). This process is termed *renaturation*, and it was first established for the enzyme ribonuclease (RNase). The renaturation of RNase led to the doctrine that primary structure determines tertiary structure. Larger proteins can rarely refold spontaneously, however, because of the complex nature of their final shape, so this simple idea needs to be qualified.

The fact that some proteins can spontaneously renature implies that tertiary structure is strongly influenced by primary structure. In an extreme example, the *E. coli* ribosome can be taken apart and put back together experimentally. Although this process requires temperature and ion concentration shifts, it indicates an amazing degree of self-assembly. That complex structures can arise by self-assembly is a key idea in the study of modern biology.

It is important to distinguish denaturation from **dissociation**. For proteins with quaternary structure, the subunits may be dissociated (separated) without losing their individual tertiary structure. For example, the four subunits of hemoglobin may dissociate into four individual molecules (two α-globins and two β-globins) without denaturation of the folded globin proteins. They readily reassume their four-subunit quaternary structure.

Learning Outcomes Review 3.4

Proteins are molecules with diverse functions. They are constructed from 20 different kinds of amino acids. Protein structure can be viewed at four levels: (1) the amino acid sequence, or primary structure; (2) coils and sheets, called secondary structure; (3) the three-dimensional shape, called tertiary structure; and (4) individual polypeptide subunits associated in a quaternary structure. Different proteins often have similar substructures called motifs and can be broken down into functional domains. Proteins have a narrow range of conditions in which they fold properly; outside that range, proteins tend to unfold (denaturation). Under some conditions, denatured proteins can refold and become functional again (renaturation).

■ How does our knowledge of protein structure help us to predict the function of unknown proteins?

3.5 Lipids: Hydrophobic Molecules

Learning Outcomes
1. Describe the structure of triglycerides.
2. Explain how fats function as energy-storage molecules.
3. Apply knowledge of the structure of phospholipids to the formation of membranes.

Lipids are a somewhat loosely defined group of molecules with one main chemical characteristic: They are insoluble in water. Storage fats such as animal fat are one kind of lipid. Oils such as those from olives, corn, and coconut are also lipids, as are waxes such as beeswax and earwax. Even some vitamins are lipids!

Lipids have a very high proportion of nonpolar carbon–hydrogen (C—H) bonds, and so long-chain lipids cannot fold up like a protein to confine their nonpolar portions away from the surrounding aqueous environment. Instead, when they are placed in water, many lipid molecules spontaneously cluster together and expose what polar (hydrophilic) groups they have to the surrounding water, while confining the nonpolar (hydrophobic) parts of the molecules together within the cluster. You may have noticed this effect when you add oil to a pan containing water, and the oil beads up into cohesive drops on the water's surface. This spontaneous assembly of lipids is of paramount importance to cells, as it underlies the structure of cellular membranes.

Fats consist of complex polymers of fatty acids attached to glycerol

Many lipids are built from a simple skeleton made up of two main kinds of molecules: fatty acids and glycerol. Fatty acids are long-chain hydrocarbons with a carboxylic acid (COOH) at one end. Glycerol is a 3-carbon polyalcohol (three —OH groups). Many lipid molecules consist of a glycerol molecule with three fatty acids attached, one to each carbon of the glycerol backbone. Because it contains three fatty acids, a fat molecule is commonly called a **triglyceride** (the more accurate chemical name is *triacylglycerol*). This basic structure is depicted in figure 3.27. The three fatty acids of a triglyceride need not be identical, and often they are very different from one another. The hydrocarbon chains of fatty acids vary in length. The most common are even-numbered chains of 14 to 20 carbons. The many C—H bonds of fats serve as a form of long-term energy storage.

If all of the internal carbon atoms in a fatty acid chain are bonded to two hydrogen atoms, we call this **saturated**, which refers to its having the maximum hydrogen atoms possible (figure 3.27). A fatty acid with double bonds between one or more pairs of successive carbon atoms will have fewer hydrogen atoms, and thus is said to be **unsaturated**. Fatty acids with one double bond are called monounsaturated, and those with more than one double bond are termed **polyunsaturated**. Most naturally occurring unsaturated fatty acids have double bonds with a *cis* configuration, where the carbon chain is on the same side before and after the double bond

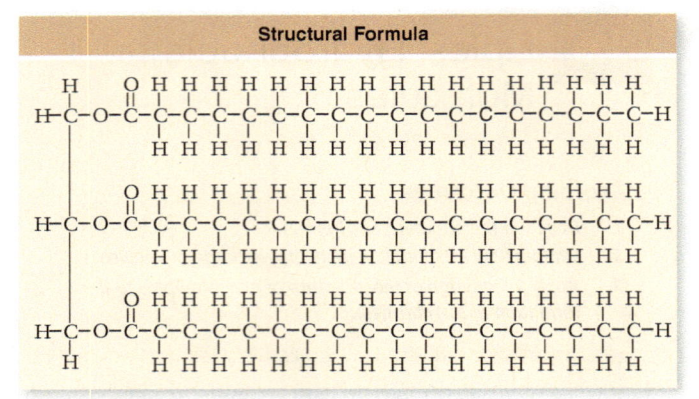

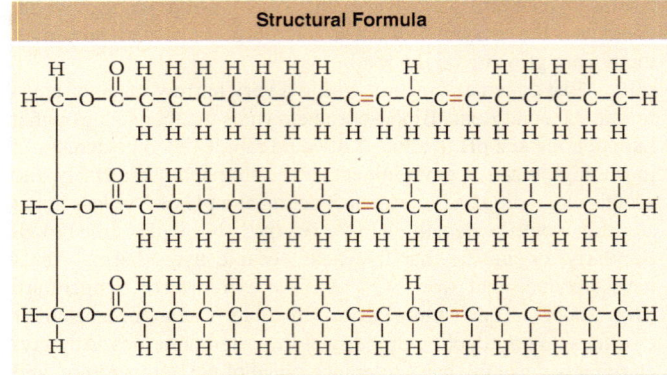

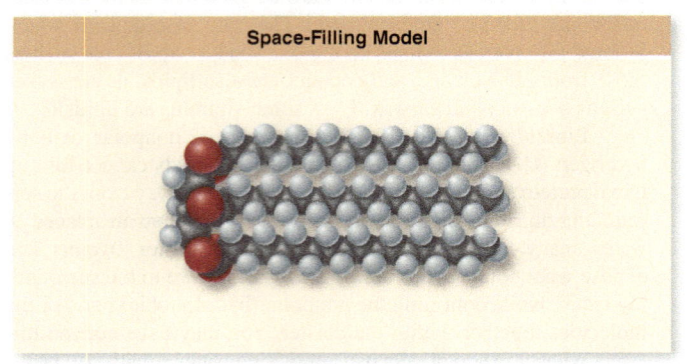

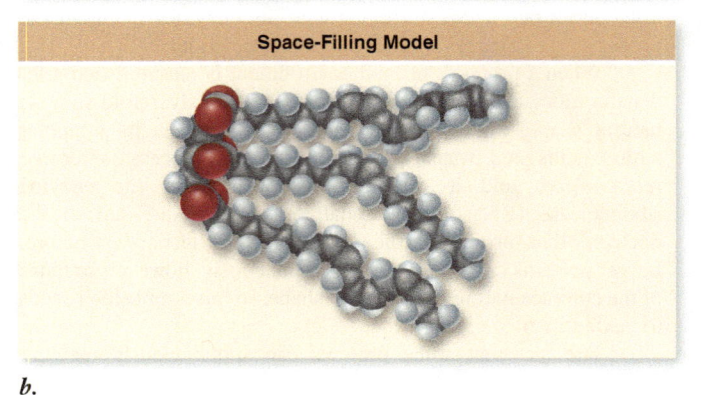

Figure 3.27 Saturated and unsaturated fats. *a.* A saturated fat is composed of triglycerides that contain three saturated fatty acids (the kind that have no double bonds). A saturated fat therefore has the maximum number of hydrogen atoms bonded to its carbon chain. Most animal fats are saturated. *b.* Unsaturated fat is composed of triglycerides that contain three unsaturated fatty acids (the kind that have one or more double bonds). These have fewer than the maximum number of hydrogen atoms bonded to the carbon chain. This example includes both a monounsaturated and two polyunsaturated fatty acids. Plant fats are typically unsaturated. The many kinks of the double bonds prevent the triglyceride from closely aligning, which makes them liquid oils at room temperature.

(double bonds in fatty acids in figure 3.27b are all *cis*). When fats are partially hydrogenated industrially, this can produce double bonds with a *trans* configuration where the carbon chain is on opposite sides before and after the double bond. These are the so-called trans fats. These have been linked to elevated levels of low-density lipoprotein (LDL) "bad cholesterol" and lowered levels of high-density lipoprotein (HDL) "good cholesterol." This condition is thought to be associated with an increased risk for coronary heart disease.

The correlation of dietary trans fats and coronary artery disease led the FDA in the United States to remove trans fats from the list of generally regarded as safe (GRAS) compounds. This is not the same as a ban, as they can be used as a dietary supplement, but along with product labeling, it has been effective in helping to remove trans fats from most common sources.

Having double bonds changes the behavior of the molecule because free rotation cannot occur about a C=C double bond as it can with a C—C single bond. This characteristic mainly affects melting point—that is, whether the fatty acid is a solid fat or a liquid oil at room temperature. Fats containing polyunsaturated fatty acids have low melting points because their fatty acid chains bend at the double bonds, preventing the fat molecules from aligning closely with one another. Most saturated fats, such as animal fat or those in butter, are solid at room temperature.

Placed in water, triglycerides spontaneously associate together, forming fat globules that can be very large relative to the size of the individual molecules. Because fats are insoluble in water, they can be deposited at specific locations within an organism, such as in vesicles of adipose tissue.

Organisms contain many other kinds of lipids besides fats (figure 3.28). *Terpenes* are long-chain lipids that are components of many biologically important pigments, such as chlorophyll and the visual pigment retinal. Rubber is also a terpene. *Steroids,* another class of lipid, are composed of four carbon rings. Most animal cell membranes contain the steroid cholesterol. Other steroids, such as testosterone and estrogen, function as hormones in multicellular animals. *Prostaglandins* are a group of about 20 lipids that are modified fatty acids, with two nonpolar "tails" attached to a 5-carbon ring. Prostaglandins act as local chemical messengers in many vertebrate tissues. Chapter 45 explores the effects of some of these complex fatty acids.

Fats are excellent energy-storage molecules

Most fats contain over 40 carbon atoms. The ratio of energy-storing C—H bonds in fats is more than twice that of carbohydrates (see section 3.2), making fats much more efficient molecules for

a. Terpene (citronellol)

b. Steroid (cholesterol)

Figure 3.28 Other kinds of lipids. *a.* Terpenes are found in biological pigments, such as chlorophyll and retinal, and *(b)* steroids play important roles in membranes and as the basis for a class of hormones involved in chemical signaling.

storing chemical energy. On average, fats yield about 9 kilocalories (kcal) of chemical energy per gram, as compared with about 4 kcal/g for carbohydrates.

Most fats produced by animals are saturated (except some fish oils), whereas most plant fats are unsaturated (see figure 3.27). The exceptions are the tropical plant oils (palm oil and coconut oil), which are saturated even though they are liquid at room temperature.

When an organism consumes excess carbohydrate, it is converted into starch, glycogen, or fats reserved for future use. The reason that many humans in developed countries gain weight as they grow older is that the amount of energy they need decreases with age, but their intake of food does not. Thus, an increasing proportion of the carbohydrates they ingest is converted into fat.

A diet heavy in fats is one of several factors thought to contribute to heart disease, particularly atherosclerosis. In atherosclerosis, sometimes referred to as "hardening of the arteries," fatty substances called plaque adhere to the lining of blood vessels, blocking the flow of blood. Fragments of a plaque can break off from a deposit and clog arteries to the brain, causing a stroke.

Phospholipids form membranes

Complex lipid molecules called **phospholipids** are among the most important molecules of the cell because they form the core of all biological membranes. An individual phospholipid can be thought of as a substituted triglyceride—that is, a triglyceride with a phosphate replacing one of the fatty acids. The basic structure of a phospholipid includes three kinds of subunits:

1. *Glycerol,* a 3-carbon alcohol, in which each carbon bears a hydroxyl group. Glycerol forms the backbone of the phospholipid molecule.
2. *Fatty acids,* long chains of —CH$_2$ groups (hydrocarbon chains) ending in a carboxyl (—COOH) group. Two fatty acids are attached to the glycerol backbone in a phospholipid molecule.
3. *A phosphate group* (—PO$_4^{2-}$) attached to one end of the glycerol. The charged phosphate group usually has a charged organic molecule linked to it, such as choline, ethanolamine, or the amino acid serine.

The phospholipid molecule can be thought of as having a polar "head" at one end (the phosphate group) and two long, very nonpolar "tails" at the other (figure 3.29). This structure is essential for how these molecules function, although it first appears

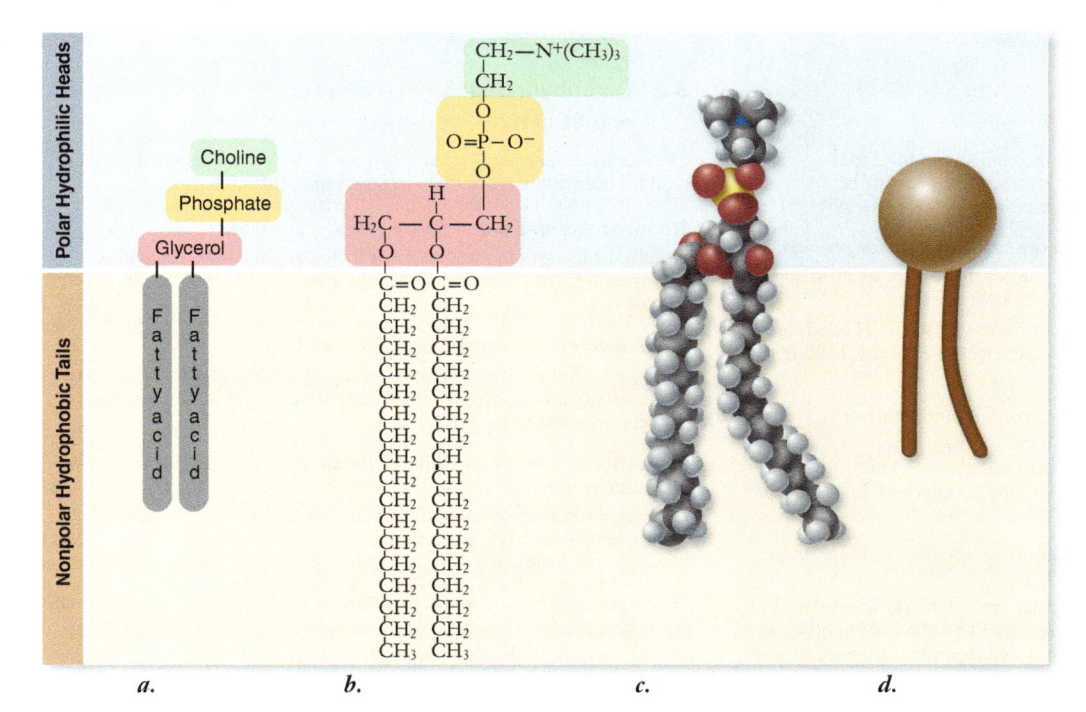

Figure 3.29 Phospholipids. The phospholipid phosphatidylcholine is shown as *(a)* a schematic, *(b)* a formula, *(c)* a space-filling model, and *(d)* an icon used in depictions of biological membranes.

a.   *b.*   *c.*   *d.*

chapter **3** The Chemical Building Blocks of Life   55

paradoxical. Why would a molecule need to be soluble in water, but also not soluble in water? The formation of a membrane shows the unique properties of such a structure.

In water, the nonpolar tails of nearby lipid molecules aggregate away from the water, forming spherical *micelles*, with the tails facing inward (figure 3.30a). This is actually how detergent molecules work to make grease soluble in water. The grease is soluble within the nonpolar interior of the micelle and the polar surface of the micelle is soluble in water. With phospholipids, a more complex structure forms in which two layers of molecules line up, with the hydrophobic tails of each layer pointing toward one another, or inward, leaving the hydrophilic heads oriented outward, forming a bilayer (figure 3.30b). Lipid bilayers are the basic framework of biological membranes, discussed in detail in chapter 5.

Learning Outcomes Review 3.5

Triglycerides are made of fatty acids linked to glycerol. Fats can contain twice as many C—H bonds as carbohydrates and thus they store energy efficiently. Because the C—H bonds in lipids are nonpolar, they are not water-soluble and aggregate together in water. Phospholipids replace one fatty acid with a hydrophilic phosphate group. This allows them to spontaneously form bilayers, which are the basis of biological membranes.

■ *Why do phospholipids form membranes while triglycerides form insoluble droplets?*

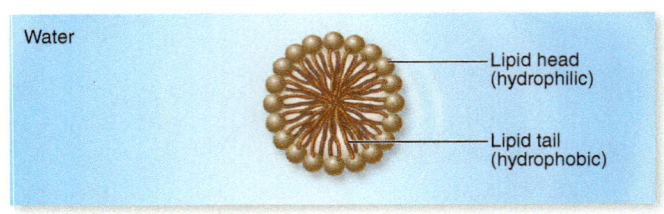

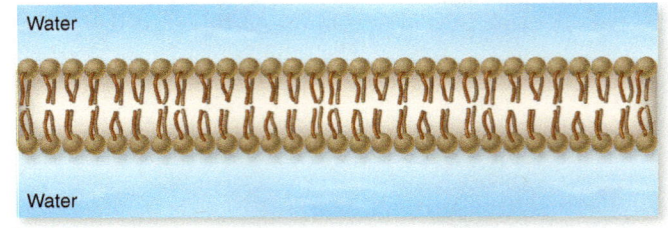

Figure 3.30 Lipids spontaneously form micelles or lipid bilayers in water. In an aqueous environment, lipid molecules orient so that their polar (hydrophilic) heads are in the polar medium, water, and their nonpolar (hydrophobic) tails are held away from the water. *a.* Droplets called micelles can form, or *(b)* phospholipid molecules can arrange themselves into two layers; in both structures, the hydrophilic heads extend outward and the hydrophobic tails inward. This second example is called a phospholipid bilayer.

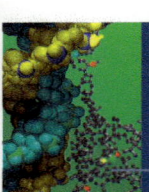

Chapter Review

3.1 Carbon: The Framework of Biological Molecules

Carbon, the backbone of all biological molecules, can form four covalent bonds and make long chains. Hydrocarbons consist of carbon and hydrogen, and their bonds store considerable energy.

Functional groups account for differences in molecular properties.
Functional groups are small molecular entities that confer specific chemical characteristics when attached to a hydrocarbon.
Carbon and hydrogen have similar electronegativity so C—H bonds are not polar. Oxygen and nitrogen have greater electronegativity, leading to polar bonds.

Isomers have the same molecular formulas but different structures.
Structural isomers are molecules with the same formula but different structures; stereoisomers differ in how groups are attached. Enantiomers are mirror-image stereoisomers.

Biological macromolecules include carbohydrates, nucleic acids, proteins, and lipids.
Most important biological macromolecules are polymers—long chains of monomer units. Biological polymers are formed by elimination of water (H and OH) from two monomers (dehydration reaction). They are broken down by adding water (hydrolysis).

3.2 Carbohydrates: Energy Storage and Structural Molecules

The empirical formula of a carbohydrate is $(CH_2O)_n$. Carbohydrates are used for energy storage and as structural molecules.

Monosaccharides are simple sugars.
Simple sugars contain three to six or more carbon atoms. Examples are glyceraldehyde (3 carbons), deoxyribose (5 carbons), and glucose (6 carbons).

Sugar isomers have structural differences.
The general formula for 6-carbon sugars is $C_6H_{12}O_6$, and many isomeric forms are possible. Living systems often have enzymes for converting isomers from one to the other.

Disaccharides serve as transport molecules in plants and provide nutrition in animals.
Plants convert glucose into the disaccharide sucrose for transport within their bodies. Female mammals produce the disaccharide lactose to nourish their young.

Polysaccharides provide energy storage and structural components.
Glucose is used to make three important polymers: glycogen (in animals), and starch and cellulose (in plants). Chitin is a related structural material found in arthropods and many fungi.

3.3 Nucleic Acids: Information Molecules

Deoxyribonucleic acid (DNA) and ribonucleic acid (RNA) are polymers composed of nucleotide monomers. Cells use nucleic acids for information storage and transfer.

Nucleic acids are nucleotide polymers.
Nucleic acids contain four different nucleotide bases. In DNA these are adenine, guanine, cytosine, and thymine. In RNA, thymine is replaced by uracil.

DNA stores genetic information.
DNA exists as a double helix held together by specific base pairs: adenine with thymine and guanine with cytosine. The nucleic acid sequence constitutes the genetic code.

RNA has many roles in a cell.
RNA is made by copying DNA. RNA carries information from DNA and forms part of the ribosome. RNA can also be an enzyme and affect gene expression.

Other nucleotides are vital components of energy reactions.
Adenosine triphosphate (ATP) provides energy in cells; NAD^+ and FAD transport electrons in cellular processes.

3.4 Proteins: Molecules with Diverse Structures and Functions

Most enzymes are proteins. Proteins also provide defense, transport, motion, and regulation, among many other roles.

Proteins are polymers of amino acids.
Amino acids are joined by peptide bonds to make polypeptides. The 20 common amino acids are characterized by R groups that determine their properties.

Proteins have levels of structure.
Protein structure is defined by the following hierarchy: primary (amino acid sequence), secondary (hydrogen bonding patterns), tertiary (three-dimensional folding), and quaternary (associations between two or more polypeptides).

Motifs and domains are structural elements of proteins.
Motifs are similar structural elements found in dissimilar proteins. They can create folds, creases, or barrel shapes. Domains are functional subunits or sites within a tertiary structure.

The process of folding relies on chaperone proteins.
Chaperone proteins assist in the folding of proteins. Heat shock proteins are an example of chaperone proteins.

Improper folding of proteins can result in disease.
Some forms of cystic fibrosis and Alzheimer disease are associated with misfolded proteins.

Denaturation inactivates proteins.
Denaturation refers to an unfolding of tertiary structure, which usually destroys function. Some denatured proteins may recover function when conditions are returned to normal. This implies that primary structure strongly influences tertiary structure.

Dissociation refers to separation of quaternary subunits with no changes to their tertiary structure.

3.5 Lipids: Hydrophobic Molecules

Lipids are insoluble in water because they have a high proportion of nonpolar C—H bonds.

Fats consist of complex polymers of fatty acids attached to glycerol.
Many lipids exist as triglycerides, three fatty acids connected to a glycerol molecule. Saturated fatty acids contain the maximum number of hydrogen atoms. Unsaturated fatty acids contain one or more double bonds between carbon atoms.

Fats are excellent energy-storage molecules.
The energy stored in the C—H bonds of fats is more than twice that of carbohydrates: 9 kcal/g compared with 4 kcal/g. For this reason, excess carbohydrate is converted to fat for storage.

Phospholipids form membranes.
Phospholipids contain two fatty acids and one phosphate attached to glycerol. In phospholipid-bilayer membranes, the phosphate heads are hydrophilic and cluster on the two faces of the membrane, and the hydrophobic tails are in the center.

Review Questions

UNDERSTAND

1. How is a polymer formed from multiple monomers?
   a. From the growth of the chain of carbon atoms
   b. By the removal of an —OH group and a hydrogen atom
   c. By the addition of an —OH group and a hydrogen atom
   d. Through hydrogen bonding

2. Why are carbohydrates important molecules for energy storage?
   a. The C—H bonds found in carbohydrates store energy.
   b. The double bonds between carbon and oxygen are very strong.
   c. The electronegativity of the oxygen atoms means that a carbohydrate is made up of many polar bonds.
   d. They can form ring structures in the aqueous environment of a cell.

3. Plant cells store energy in the form of _____, and animal cells store energy in the form of _____.
   a. fructose; glucose
   b. disaccharides; monosaccharides
   c. cellulose; chitin
   d. starch; glycogen

4. Which carbohydrate would you find as part of a molecule of RNA?
   a. Galactose
   b. Deoxyribose
   c. Ribose
   d. Glucose

5. A molecule of DNA or RNA is a polymer of
   a. monosaccharides.
   b. nucleotides.
   c. amino acids.
   d. fatty acids.

6. What makes cellulose different from starch?
   a. Starch is produced by plant cells, and cellulose is produced by animal cells.
   b. Cellulose forms long filaments, and starch is highly branched.
   c. Starch is insoluble, and cellulose is soluble.
   d. All of the choices are correct.

7. What monomers make up a protein?
   a. Monosaccharides   c. Amino acids
   b. Nucleotides   d. Fatty acids

8. A triglyceride is a form of _____ composed of _____.
   a. lipid; fatty acids and glucose
   b. lipid; fatty acids and glycerol
   c. carbohydrate; fatty acids
   d. lipid; cholesterol

APPLY

1. You can use starch or glycogen as an energy source, but not cellulose because
   a. starch and cellulose have similar structures.
   b. cellulose and glycogen have similar structures.
   c. starch and glycogen have similar structures.
   d. your body makes starch but not cellulose.

2. Which of the following is NOT a difference between DNA and RNA?
   a. Deoxyribose sugar versus ribose sugar
   b. Thymine versus uracil
   c. Double-stranded versus single-stranded
   d. Phosphodiester versus hydrogen bonds

3. Which part of an amino acid has the greatest influence on the overall structure of a protein?
   a. The (—NH$_2$) amino group
   b. The R group
   c. The (—COOH) carboxyl group
   d. Both a and c are correct.

4. A mutation that alters a single amino acid within a protein can alter
   a. the primary level of protein structure.
   b. the secondary level of protein structure.
   c. the tertiary level of protein structure.
   d. All of the choices are correct.

5. Two different proteins have the same domain in their structure. From this we can infer that they have
   a. the same primary structure.
   b. similar function.
   c. very different functions.
   d. the same primary structure but different function.

6. What aspect of triglyceride structure accounts for their insolubility in water?
   a. The COOH group of fatty acids
   b. The nonpolar C—H bonds in fatty acids
   c. The OH groups in glycerol
   d. The C═C bonds found in unsaturated fatty acids

7. The spontaneous formation of a lipid bilayer in an aqueous environment occurs because
   a. the polar head groups of the phospholipids can interact with water.
   b. the long fatty acid tails of the phospholipids can interact with water.
   c. the fatty acid tails of the phospholipids are hydrophobic.
   d. Both a and c are correct.

SYNTHESIZE

1. How do the four biological macromolecules differ from one another? How does the structure of each relate to its function?

2. Hydrogen bonds and hydrophobic interactions each play an important role in stabilizing and organizing biological macromolecules. Consider the four macromolecules discussed in this chapter. Describe how these affect the form and function of each type of macromolecule. Would a disruption in the hydrogen bonds affect form and function? Hydrophobic interactions?

3. Plants make both starch and cellulose. Would you predict that the enzymes involved in starch synthesis could also be used by the plant for cellulose synthesis? Construct an argument to explain this based on the structure and function of the enzymes and the polymers synthesized.

Biology of the Cell

Part II Biology of the Cell

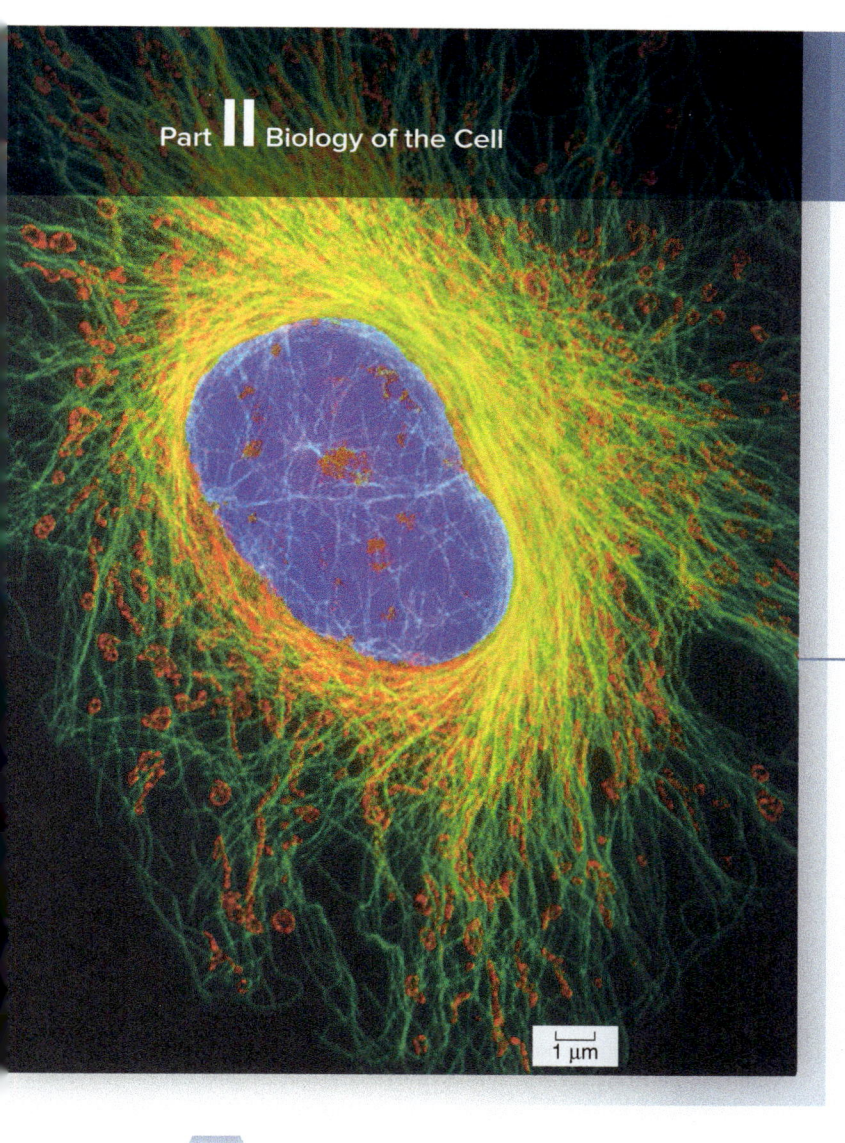

CHAPTER 4

Cell Structure

Chapter Contents

4.1 Cell Theory
4.2 Prokaryotic Cells
4.3 Eukaryotic Cells
4.4 The Endomembrane System
4.5 Mitochondria and Chloroplasts: Cellular Generators
4.6 The Cytoskeleton
4.7 Extracellular Structures and Cell Movement
4.8 Cell-to-Cell Interactions

Introduction

All organisms are composed of cells. The gossamer wing of a butterfly is a thin sheet of cells and so is the glistening outer layer of your eyes. The burger or tomato you eat is composed of cells, and its contents soon become part of your cells. Some organisms consist of a single cell too small to see with the unaided eye. Others, such as humans, are composed of many specialized cells, such as the fibroblast cell shown in the striking fluorescence micrograph on this page. Cells are so much a part of life that we cannot imagine an organism that is not cellular in nature. In this chapter, we take a close look at the internal structure of cells. In chapters 5 to 10, we will focus on cells in action—how they communicate with their environment, grow, and reproduce.

4.1 Cell Theory

Learning Outcomes

1. Discuss the cell theory.
2. Describe the factors that limit cell size.
3. Categorize structural and functional similarities in cells.

Cells are characteristically microscopic in size. Although there are exceptions, a typical eukaryotic cell is 10 to 100 micrometers (μm) (10 to 100 millionths of a meter) in diameter, although most prokaryotic cells are only 1 to 10 μm in diameter.

Because cells are so small, they were not discovered until the invention of the microscope in the 17th century. English natural philosopher Robert Hooke was the first to observe cells in 1665, naming the shapes he saw in cork *cellulae* (Latin, "small rooms"). This is known to us as *cells*. Another early microscopist, Dutch Anton van Leeuwenhoek, first observed living cells, which he termed "animalcules," or little animals. After these early efforts, a century and a half passed before biologists fully recognized the importance of cells. In 1838, German botanist Matthias Schleiden stated that all plants "are aggregates of fully individualized, independent, separate beings, namely the cells themselves." In 1839, German physiologist Theodor Schwann reported that all animal tissues also consist of individual cells. Thus, the cell theory was born.

Cell theory is the unifying foundation of cell biology

The cell theory was proposed to explain the observation that all organisms are composed of cells. It sounds simple, but it is a far-reaching statement about the organization of life.

In its modern form, the *cell theory* includes the following three principles:

1. All organisms are composed of one or more cells, and the life processes of metabolism and heredity occur within these cells.
2. Cells are the smallest living things, the basic units of organization of all organisms.
3. Cells arise only by division of a previously existing cell.

Although life likely evolved spontaneously in the environment of early Earth, biologists have concluded that no additional cells are originating spontaneously at present. Rather, life on Earth represents a continuous line of descent from those early cells.

Cell size is limited

Most cells are relatively small for reasons related to the diffusion of substances into and out of them. The rate of diffusion is affected by a number of variables, including (1) surface area available for diffusion, (2) temperature, (3) concentration gradient of diffusing substance, and (4) the distance over which diffusion must occur. As the size of a cell increases, the length of time for diffusion from the outside membrane to the interior of the cell increases as well. Larger cells need to synthesize more macromolecules, have correspondingly higher energy requirements, and produce a greater quantity of waste. Molecules used for energy and biosynthesis must be transported through the membrane. Any metabolic waste produced must be removed, also passing through the membrane. The rate at which this transport occurs depends on both the distance to the membrane and the area of membrane available. For this reason, an organism made up of many relatively small cells has an advantage over one composed of fewer, larger cells.

The advantage of small cell size is readily apparent in terms of the **surface area-to-volume ratio.** As a cell's size increases, its volume increases much more rapidly than its surface area. For a spherical cell, the surface area is proportional to the square of the radius, whereas the volume is proportional to the cube of the radius. Thus, if the radii of two cells differ by a factor of 10, the larger cell will have 10^2, or 100 times, the surface area, but 10^3, or 1000 times, the volume of the smaller cell (figure 4.1).

The cell surface provides the only opportunity for interaction with the environment, because all substances enter and exit a cell via this surface. The membrane surrounding the cell plays a key

Figure 4.1 Surface area-to-volume ratio. As a cell gets larger, its volume increases at a faster rate than its surface area. If the cell radius increases by 10 times, the surface area increases by 100 times, but the volume increases by 1000 times. A cell's surface area must be large enough to meet the metabolic needs of its volume.

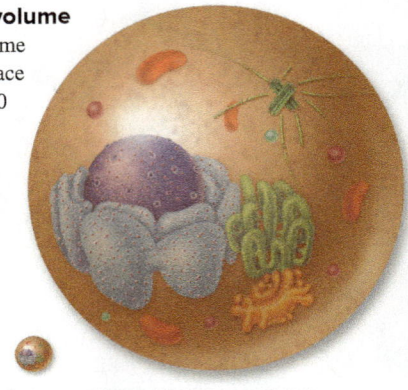

Cell radius (r)	1 unit	10 unit
Surface area ($4\pi r^2$)	12.57 unit²	1257 unit²
Volume ($\frac{4}{3}\pi r^3$)	4.189 unit³	4189 unit³
Surface Area / Volume	3	0.3

role in controlling cell function. Because small cells have more surface area per unit of volume than large ones, control over cell contents is more effective when cells are relatively small.

Although most cells are small, some quite large cells do exist. These cells have apparently overcome the surface area-to-volume problem by one or more adaptive mechanisms. For example, some cells, such as skeletal muscle cells, have more than one nucleus, allowing genetic information to be spread around a large cell. Some other large cells, such as neurons, are long and skinny, so that any given point within the cell is close to the plasma membrane. This permits diffusion between the inside and outside of the cell to still be rapid.

Microscopes allow visualization of cells and components

Other than egg cells, not many cells are visible to the naked eye (figure 4.2). Most are less than 50 μm in diameter, far smaller than the period at the end of this sentence. So, to visualize cells we need the aid of technology. The development of microscopes and their refinement over the centuries has allowed us to continually explore cells in greater detail.

The resolution problem

How do we study cells if they are too small to see? The key is to understand why we can't see them. The reason we can't see such small objects is the limited resolution of the human eye. *Resolution* is the minimum distance two points can be apart and still be distinguished as two separate points. When two objects are closer together than about 100 μm, the light reflected from each strikes the same photoreceptor cell at the rear of the eye. Only when the objects are farther than 100 μm apart can the light from each strike different cells, allowing your eye to resolve them as two distinct objects rather than one.

Types of microscopes

One way to overcome the limitations of our eyes is to increase magnification so that small objects appear larger. The first microscopists used glass lenses to magnify small cells and cause them to appear larger than the 100-μm limit imposed by the human eye. The glass lens increases focusing power. Because the glass lens makes the object appear closer, the image on the back of the eye is bigger than it would be without the lens.

Modern *light microscopes,* which operate with visible light, use two magnifying lenses (and a variety of correcting lenses) to achieve very high magnification and clarity (table 4.1). The first lens focuses the image of the object on the second lens, which magnifies it again and focuses it on the back of the eye. Microscopes that magnify in stages using several lenses are called *compound microscopes.* They can resolve structures that are separated by at least 200 nanometers (nm).

Light microscopes, even compound ones, are not powerful enough to resolve many of the structures within cells. For example, a cell membrane is only 5 nm thick. Why not just add another magnifying stage to the microscope to increase its resolving power? This doesn't work because when two objects are closer than a few hundred nanometers, the light beams reflecting from the two images start to overlap each other. The only way two light beams can get closer together and still be resolved is if their wavelengths are shorter. One way to avoid overlap is by using a beam of electrons rather than a beam of light. Electrons have a much shorter wavelength, and an *electron microscope,* employing electron beams, has 1000 times the resolving power of a light microscope.

Transmission electron microscopes, so called because the electrons used to visualize the specimens are transmitted through the material, are capable of resolving objects only 0.2 nm apart—which is only twice the diameter of a hydrogen atom!

A second kind of electron microscope, the *scanning electron microscope,* beams electrons onto the surface of the specimen. The electrons reflected back from the surface, together with other electrons that the specimen itself emits as a result of the bombardment, are amplified and transmitted to a screen, where the image can be viewed and photographed. Scanning electron microscopy yields striking three-dimensional images. This technique has improved our understanding of many biological and physical phenomena (table 4.1).

Using stains to view cell structure

Although resolution remains a physical limit, we can improve the images we see by altering the sample. Certain chemical stains increase the contrast between different cellular components. Structures within the cell absorb or exclude the stain differentially, producing contrast that aids resolution.

Stains that bind to specific types of molecules have made these techniques even more powerful. This method uses antibodies that bind, for example, to a particular protein. This process, called *immunohistochemistry,* uses antibodies generated in animals such as rabbits or mice. When these animals are injected with specific proteins, they produce antibodies that bind to the injected protein. The antibodies are then purified and chemically bonded to enzymes, to stains, or to fluorescent molecules. When cells are incubated in a solution containing the antibodies, the antibodies bind to cellular structures that contain the target molecule and can

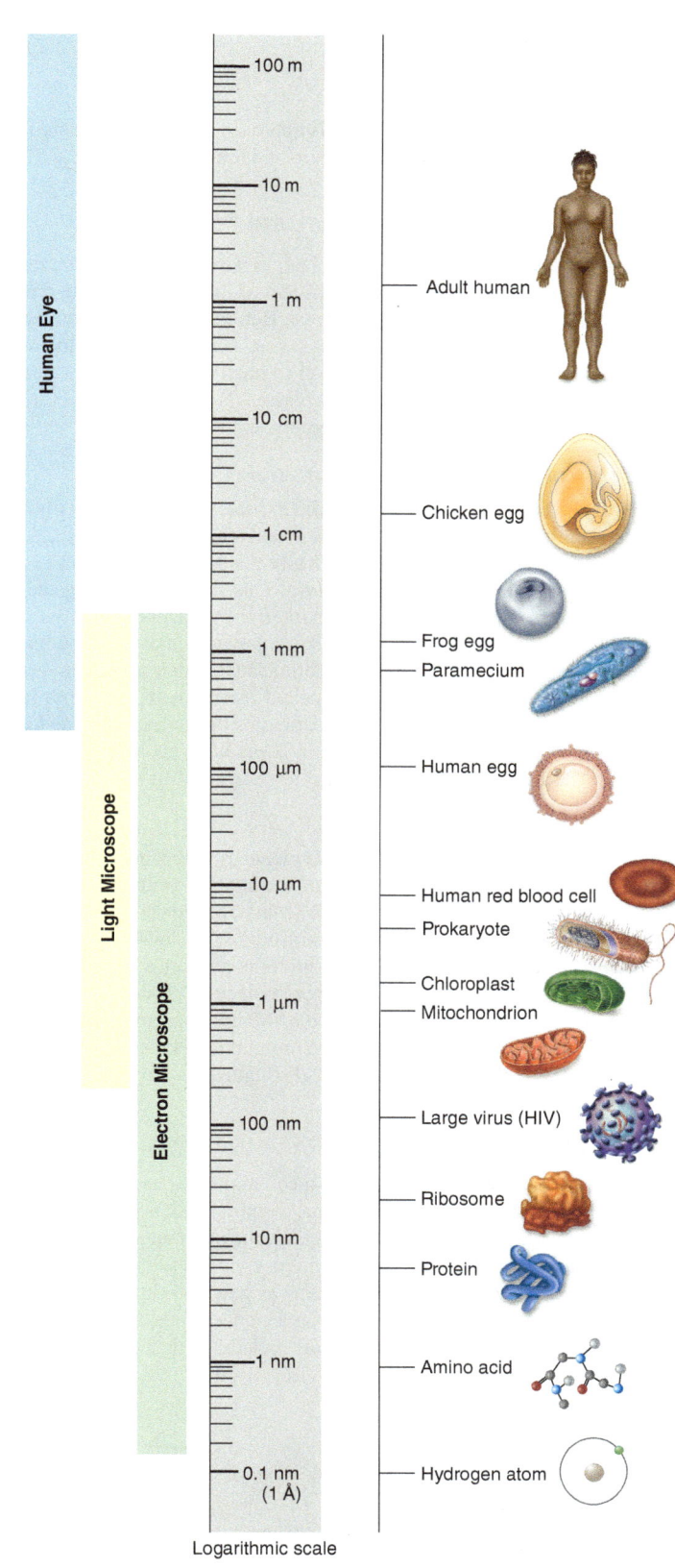

Figure 4.2 The size of cells and their contents. Except for vertebrate eggs, which can typically be seen with the unaided eye, most cells are microscopic in size. Prokaryotic cells are generally 1 to 10 μm across.
$1 \text{ m} = 10^2 \text{ cm} = 10^3 \text{ mm} = 10^6 \text{ μm} = 10^9 \text{ nm}$

TABLE 4.1	Microscopes
LIGHT MICROSCOPES	
Bright-field microscope: Light is transmitted through a specimen, giving little contrast. Staining specimens improves contrast but requires that cells be fixed (not alive), which can distort or alter components.	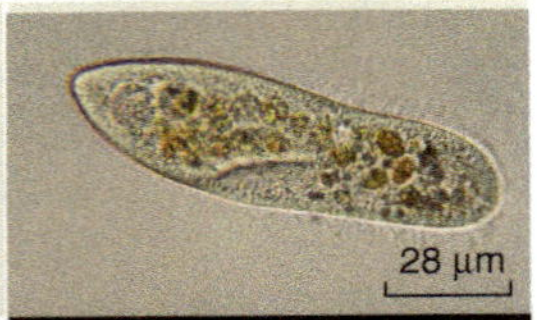 28 μm
Dark-field microscope: Light is directed at an angle toward the specimen. A condenser lens transmits only light reflected off the specimen. The field is dark, and the specimen is light against this dark background.	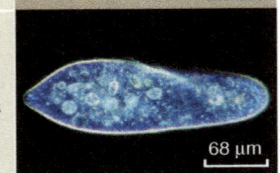 68 μm
Phase-contrast microscope: Components of the microscope bring light waves out of phase, which produces differences in contrast and brightness when the light waves recombine.	 33 μm
Differential-interference–contrast microscope: Polarized light is split into two beams that have slightly different paths through the sample. Combining these two beams produces greater contrast, especially at the edges of structures.	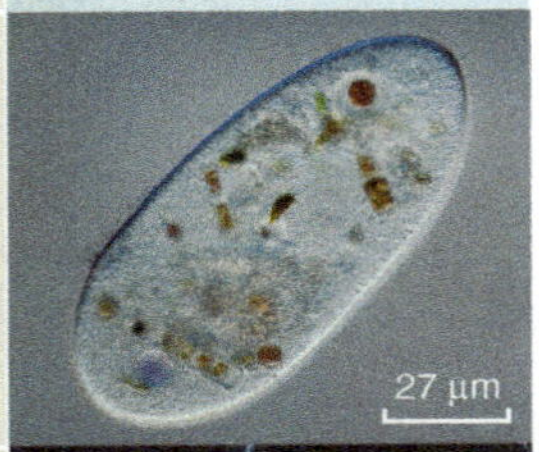 27 μm
Fluorescence microscope: Fluorescent stains absorb light at one wavelength, then emit it at another. Filters transmit only the emitted light.	10 μm
Confocal microscope: Light from a laser is focused to a point and scanned across the fluorescently stained specimen in two directions. This produces clear images of one plane of the specimen. Other planes of the specimen are excluded to prevent the blurring of the image. Multiple planes can be used to reconstruct a 3-D image.	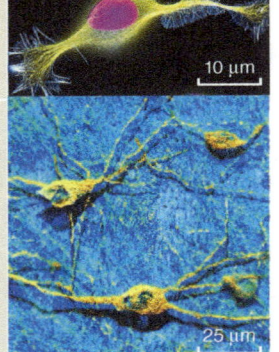 25 μm
ELECTRON MICROSCOPES	
Transmission electron microscope: A beam of electrons is passed through the specimen. Electrons that pass through are used to expose film. Areas of the specimen that scatter electrons appear dark. False coloring enhances the image.	3 μm
Scanning electron microscope: An electron beam is scanned across the surface of the specimen, and electrons are knocked off the surface. Thus, the topography of the specimen determines the contrast and the content of the image. False coloring enhances the image.	7 μm

be seen with light microscopy. This approach has been used extensively in the analysis of cell structure and function.

All cells share many structural features

The general plan of cellular organization varies between different organisms, but despite these modifications, all cells resemble one another in certain fundamental ways. Before we begin a detailed examination of cell structure, let's first summarize four major features all cells have in common: (1) a nucleoid or nucleus where genetic material is located, (2) cytoplasm, (3) *ribosomes* to synthesize proteins, and (4) a plasma membrane.

Centrally located genetic material

Every cell contains DNA, the hereditary molecule. In **prokaryotes**, the simplest organisms, most of the genetic material lies in a single circular molecule of DNA. It typically resides near the center of the cell in an area called the **nucleoid.** This area is not segregated, however, from the rest of the cell's interior by membranes.

By contrast, the DNA of eukaryotes, which are more complex organisms, is contained in the nucleus, which is surrounded by a double-membrane structure called the **nuclear envelope.** In both types of organisms, the DNA contains the genes that code for the proteins synthesized by the cell. (Details of nucleus structure are described in section 4.3.)

The cytoplasm

A semifluid matrix called the **cytoplasm** fills the interior of the cell. The cytoplasm contains all of the sugars, amino acids, and proteins the cell uses to carry out its everyday activities. Although it is an aqueous medium, cytoplasm is more like Jell-O than water due to the high concentration of proteins and other macromolecules. We call any discrete macromolecular structure in the cytoplasm specialized for a particular function an **organelle.** The part of the cytoplasm that contains organic molecules and ions in solution is called the **cytosol** to distinguish it from the larger organelles suspended in this fluid.

The plasma membrane

The **plasma membrane** encloses a cell and separates its contents from its surroundings. The plasma membrane is a phospholipid bilayer about 5 to 10 nm (5 to 10 billionths of a meter) thick, with proteins embedded in it. Viewed in cross section with the electron microscope, such membranes appear as two dark lines separated by a lighter area. This distinctive appearance arises from the tail-to-tail packing of the phospholipid molecules that make up the membrane (see chapter 5).

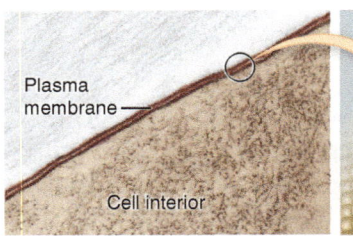

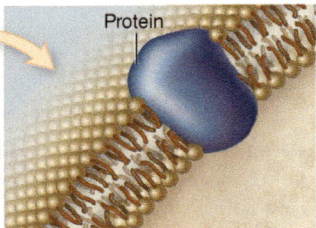

The proteins of the plasma membrane are generally responsible for a cell's ability to interact with the environment. *Transport proteins* help molecules and ions move across the plasma membrane, either from the environment to the interior of the cell or vice versa. *Receptor proteins* induce changes within the cell when they come in contact with specific molecules in the environment, such as hormones, or with molecules on the surface of neighboring cells. These molecules can function as *markers* that identify the cell as a particular type. This interaction between cell surface molecules is especially important in multicellular organisms, whose cells must be able to recognize one another as they form tissues.

We'll examine the structure and function of cell membranes more thoroughly in chapter 5.

Learning Outcomes Review 4.1

All organisms are single cells or aggregates of cells, and all cells arise from preexisting cells. Cell size is limited primarily by the efficiency of diffusion across the plasma membrane. As a cell becomes larger, its volume increases more quickly than its surface area. Past a certain point, diffusion cannot support the cell's needs. All cells are bounded by a plasma membrane and filled with cytoplasm. The genetic material is found in the central portion of the cell; and in eukaryotic cells, it is contained in a membrane-bounded nucleus.

■ Would finding life on Mars change our view of cell theory?

4.2 Prokaryotic Cells

Learning Outcomes

1. Describe the organization of prokaryotic cells.
2. Distinguish between bacterial and archaeal cell types.

When cells were visualized with microscopes, two basic cellular architectures were recognized: eukaryotic and prokaryotic. These terms refer to the presence or absence, respectively, of a membrane-bounded nucleus that contains genetic material. We have already mentioned that in addition to lacking a nucleus, prokaryotic cells do not have an internal membrane system or numerous membrane-bounded organelles.

Prokaryotic cells have relatively simple organization

Prokaryotes are the simplest organisms. Prokaryotic cells are small. They consist of cytoplasm surrounded by a plasma membrane and are encased within a rigid **cell wall**. They have no distinct interior compartments (figure 4.3). A prokaryotic cell is like a one-room cabin in which eating, sleeping, and watching TV all occur.

Prokaryotes are very important in the ecology of living organisms. Some harvest light by photosynthesis, others break down dead organisms and recycle their components. Still others cause disease or have uses in many important industrial processes. Prokaryotes have two main domains: archaea and bacteria. Chapter 28 covers prokaryotic diversity in more detail.

> **Inquiry question** What modifications would you include if you wanted to make a cell as large as possible?

Although prokaryotic cells do contain organelles like **ribosomes,** which carry out protein synthesis, most lack the membrane-bounded organelles characteristic of eukaryotic cells. It was long thought that prokaryotes also lack the elaborate cytoskeleton found in eukaryotes, but we have now found they have molecules related to both actin and tubulin, which form two of the cytoskeletal elements described in section 4.6. The strength and shape of the cell is determined by the cell wall and not these cytoskeletal elements (figure 4.3). However, cell wall structure is influenced by the cytoskeleton. For instance, the presence of actin like MreB fibers running the length of the cell lead to perpendicular cell-wall fibers that produce a rod-shaped cell. This can be seen when MreB protein is removed, cells become spherical rather than rod-shaped. During cell division, cell-wall deposition is influenced by the tubulin-like FtsZ protein (see chapter 10).

Figure 4.3 Structure of a prokaryotic cell. Generalized cell organization of a prokaryote. The nucleoid is visible as a dense central region segregated from the cytoplasm. Some prokaryotes have hairlike growths (called pili [singular, *pilus*]) on the outside of the cell.

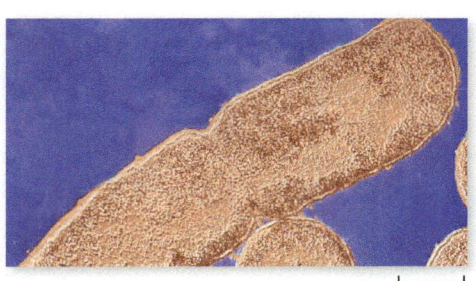

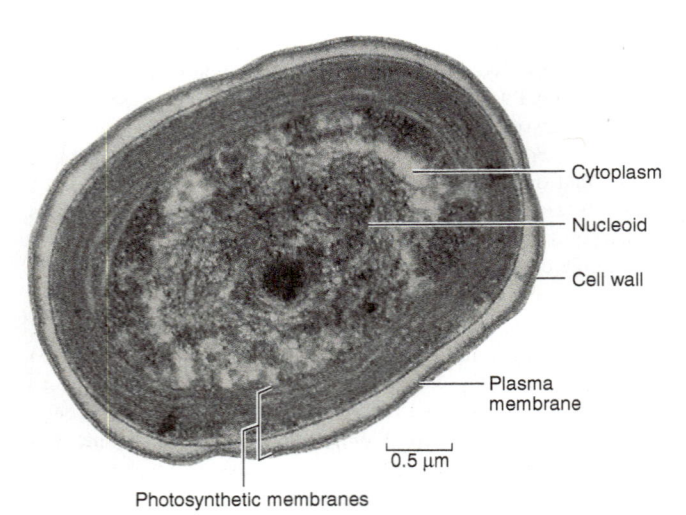

Figure 4.4 Electron micrograph of a photosynthetic bacterial cell. Extensive folded photosynthetic membranes are shown in green in this false colored electron micrograph of a *Prochloron* cell.

The plasma membrane of a prokaryotic cell carries out some of the functions organelles perform in eukaryotic cells. For example, some photosynthetic bacteria, such as the cyanobacterium *Prochloron* (figure 4.4), have an extensively folded plasma membrane, with the folds extending into the cell's interior. These membrane folds contain the bacterial pigments connected with photosynthesis. In eukaryotic plant cells, photosynthetic pigments are found in the inner membrane of the chloroplast.

Because a prokaryotic cell contains no membrane-bounded organelles, the DNA, enzymes, and other cytoplasmic constituents have access to all parts of the cell. Reactions are not compartmentalized as they are in eukaryotic cells, and the whole prokaryote operates as a single unit.

Bacterial cell walls consist of peptidoglycan

Most bacterial cells are encased by a strong **cell wall.** This cell wall is composed of *peptidoglycan,* which consists of a carbohydrate matrix (polymers of sugars) that is cross-linked by short polypeptide units. Details about the structure of this cell wall are discussed in chapter 28. Cell walls protect the cell, maintain its shape, and prevent excessive uptake or loss of water. The exception is the class Mollicutes, which includes the common genus *Mycoplasma,* which lack a cell wall. Plants, fungi, and most protists also have cell walls but with a chemical structure different from peptidoglycan.

The susceptibility of bacteria to antibiotics often depends on the structure of their cell walls. The drugs penicillin and vancomycin, for example, interfere with the ability of bacteria to cross-link the peptides in their peptidoglycan cell wall. Like removing all the nails from a wooden house, this destroys the integrity of the structural matrix, which can no longer prevent water from rushing in and swelling the cell to bursting.

Some bacteria also secrete a jellylike protective capsule of polysaccharide around the cell. Many disease-causing bacteria have such a capsule, which enables them to adhere to teeth, skin, food—or to practically any surface that can support their growth.

Archaea have unusual membrane lipids

We are still learning about the physiology and structure of archaea. Many of these organisms are difficult to culture in the laboratory, and so this group has not yet been studied in detail. More is known about their genetic makeup than about any other feature.

The cell walls of archaea are composed of various chemical compounds, including polysaccharides and proteins, and possibly even inorganic components. A common feature distinguishing archaea from bacteria is the nature of their membrane lipids. The chemical structure of archaeal lipids is distinctly different from that of lipids in bacteria and can include saturated hydrocarbons that are covalently attached to glycerol at both ends, such that their

Figure 4.5 Some prokaryotes move by rotating their flagella. *a.* The photograph shows *Vibrio cholerae,* the microbe that causes the serious disease cholera. *b.* The bacterial flagellum is a complex structure. The motor proteins, powered by a proton gradient, are anchored in the plasma membrane. Two rings are found in the cell wall. The motor proteins cause the entire structure to rotate. *c.* As the flagellum rotates it creates a spiral wave down the structure. This powers the cell forward.

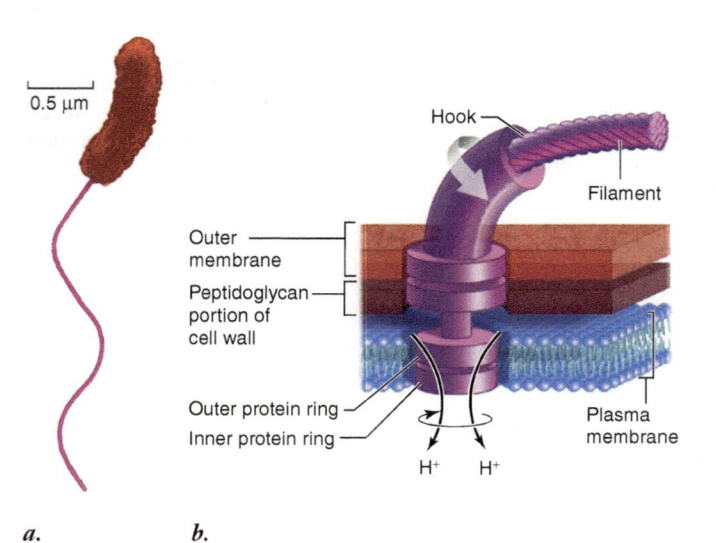

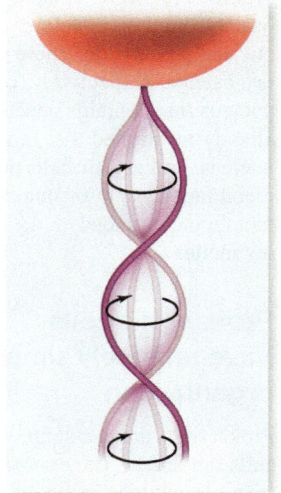

membrane is a monolayer. These features seem to confer greater thermal stability to archaeal membranes, although the trade-off seems to be an inability to alter the degree of saturation of the hydrocarbons—meaning that archaea with this characteristic cannot adapt to changing environmental temperatures.

The cellular machinery that replicates DNA and synthesized proteins in archaea is more closely related to eukaryotic systems than to bacterial systems. Even though they share a similar overall cellular architecture with prokaryotes, archaea appear to be more closely related on a molecular basis to eukaryotes.

Some prokaryotes move by means of rotating flagella

Flagella (singular, *flagellum*) are long, threadlike structures protruding from the surface of a cell that are used in locomotion. Prokaryotic flagella are protein fibers that extend out from the cell. There may be one or more per cell, or none, depending on the species. Bacteria can swim at speeds of up to 70 cell lengths per second by rotating their flagella like screws (figure 4.5). The rotary motor uses the energy stored in a gradient that transfers protons across the plasma membrane to power the movement of the flagellum. Interestingly, the same principle, in which a proton gradient powers the rotation of a molecule, is used in eukaryotic mitochondria and chloroplasts by an enzyme that synthesizes ATP (see chapters 7 and 8).

Learning Outcomes Review 4.2

Prokaryotes are small cells that lack complex interior organization. The two domains of prokaryotes are archaea and bacteria. The cell wall of bacteria is composed of peptidoglycan, which is not found in archaea. Archaea have cell walls made from a variety of polysaccharides and peptides, as well as membranes containing unusual lipids. Some bacteria move using a rotating flagellum.

- *What features do bacteria and archaea share?*

4.3 Eukaryotic Cells

Learning Outcomes

1. Compare the organization of eukaryotic and prokaryotic cells.
2. Discuss the role of the nucleus in eukaryotic cells.
3. Describe the role of ribosomes in protein synthesis.

Eukaryotic cells (figures 4.6 and 4.7) are far more complex than prokaryotic cells. The hallmark of the eukaryotic cell is compartmentalization. This is achieved through a combination of an extensive **endomembrane system** that weaves through the cell interior and by numerous *organelles*. These organelles include membrane-bounded structures that form compartments within which multiple biochemical processes can proceed simultaneously and independently.

Plant cells often have a large, membrane-bounded sac called a **central vacuole,** which stores proteins, pigments, and waste materials. Both plant and animal cells contain **vesicles**—smaller sacs that store and transport a variety of materials. Inside the nucleus, the DNA is wound tightly around proteins and packaged into compact units called **chromosomes.**

All eukaryotic cells are supported by an internal protein scaffold, the **cytoskeleton.** Although the cells of animals and some protists lack cell walls, the cells of fungi, plants, and many protists have strong cell walls composed of cellulose or chitin fibers embedded in a matrix of other polysaccharides and proteins. Through the rest of this chapter, we will examine the internal components of eukaryotic cells in more detail.

The nucleus acts as the information center

The largest and most easily seen organelle within a eukaryotic cell is the **nucleus** (Latin, "kernel" or "nut"), first described by the Scottish botanist Robert Brown in 1831. Nuclei are roughly spherical in shape, and in animal cells, they are typically located in the central region of the cell (figure 4.8*a*). In some cells, a network of fine cytoplasmic filaments seems to cradle the nucleus in this position.

The nucleus is the repository of the genetic information that enables the synthesis of nearly all proteins of a living eukaryotic cell. Most eukaryotic cells possess a single nucleus, although the cells of fungi and some other groups may have from several to many nuclei. Mammalian erythrocytes (red blood cells) lose their nuclei when they mature. Many nuclei exhibit a dark-staining zone called the **nucleolus,** which is a region where intensive synthesis of ribosomal RNA is taking place.

The nuclear envelope

The surface of the nucleus is bounded by *two* phospholipid bilayer membranes, which together make up the **nuclear envelope** (figure 4.8). The outer membrane of the nuclear envelope is continuous with the cytoplasm's interior membrane system, called the *endoplasmic reticulum* (described in section 4.4).

Scattered over the surface of the nuclear envelope are what appear as shallow depressions in the electron micrograph but are in fact structures called **nuclear pores** (figure 4.8*b, c*). These pores form 50 to 80 nm apart at locations where the two membrane layers of the nuclear envelope come together. The structure consists of a central framework with eightfold symmetry that is embedded in the nuclear envelope. This is bounded by a cytoplasmic face with eight fibers, and a nuclear face with a complex ring that forms a basket beneath the central ring. The pore allows ions and small molecules to diffuse freely between nucleoplasm and cytoplasm, while controlling the passage of proteins and RNA–protein complexes. Transport across the pore is controlled and consists mainly of the import of proteins that function in the nucleus, and the export to the cytoplasm of RNA and RNA–protein complexes formed in the nucleus.

The inner surface of the nuclear envelope is covered with a network of fibers that make up the nuclear lamina (figure 4.8*d*). This is composed of intermediate filament fibers called *nuclear lamins.*

Figure 4.6 Structure of an animal cell. In this generalized diagram of an animal cell, the plasma membrane encases the cell, which contains the cytoskeleton and various cell organelles and interior structures suspended in a semifluid matrix called the cytoplasm. Some kinds of animal cells possess fingerlike projections called microvilli. Other types of eukaryotic cells—for example, many protist cells—may possess flagella, which aid in movement, or cilia, which can have many different functions.

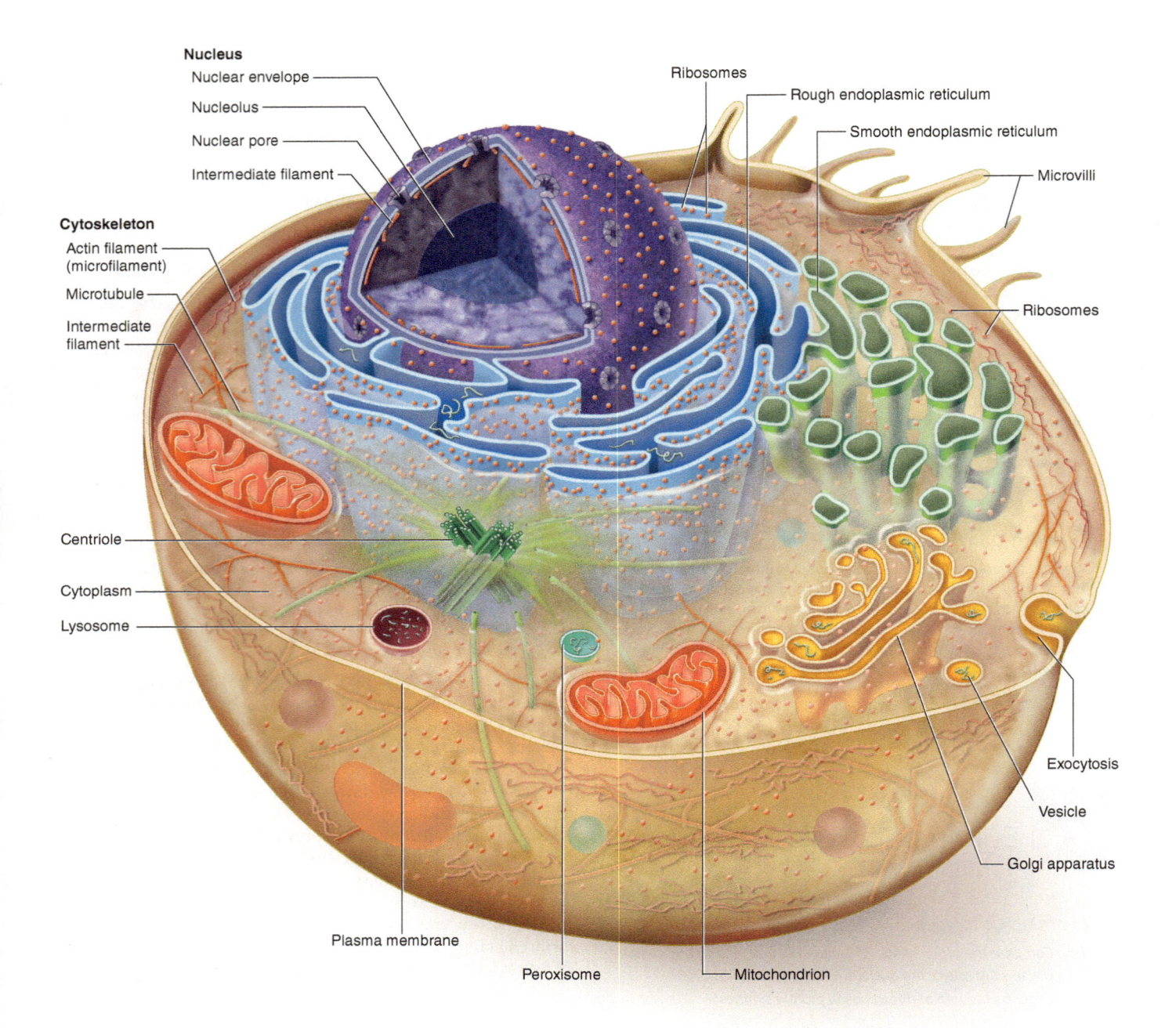

Figure 4.7 Structure of a plant cell. Most mature plant cells contain a large central vacuole, which occupies a major portion of the internal volume of the cell, and organelles called chloroplasts, within which photosynthesis takes place. The cells of plants, fungi, and some protists have cell walls, although the composition of the walls varies among the groups. Plant cells have cytoplasmic connections to one another through openings in the cell wall called plasmodesmata. Flagella occur in sperm of a few plant species, but are otherwise absent from plant and fungal cells. Centrioles are also usually absent.

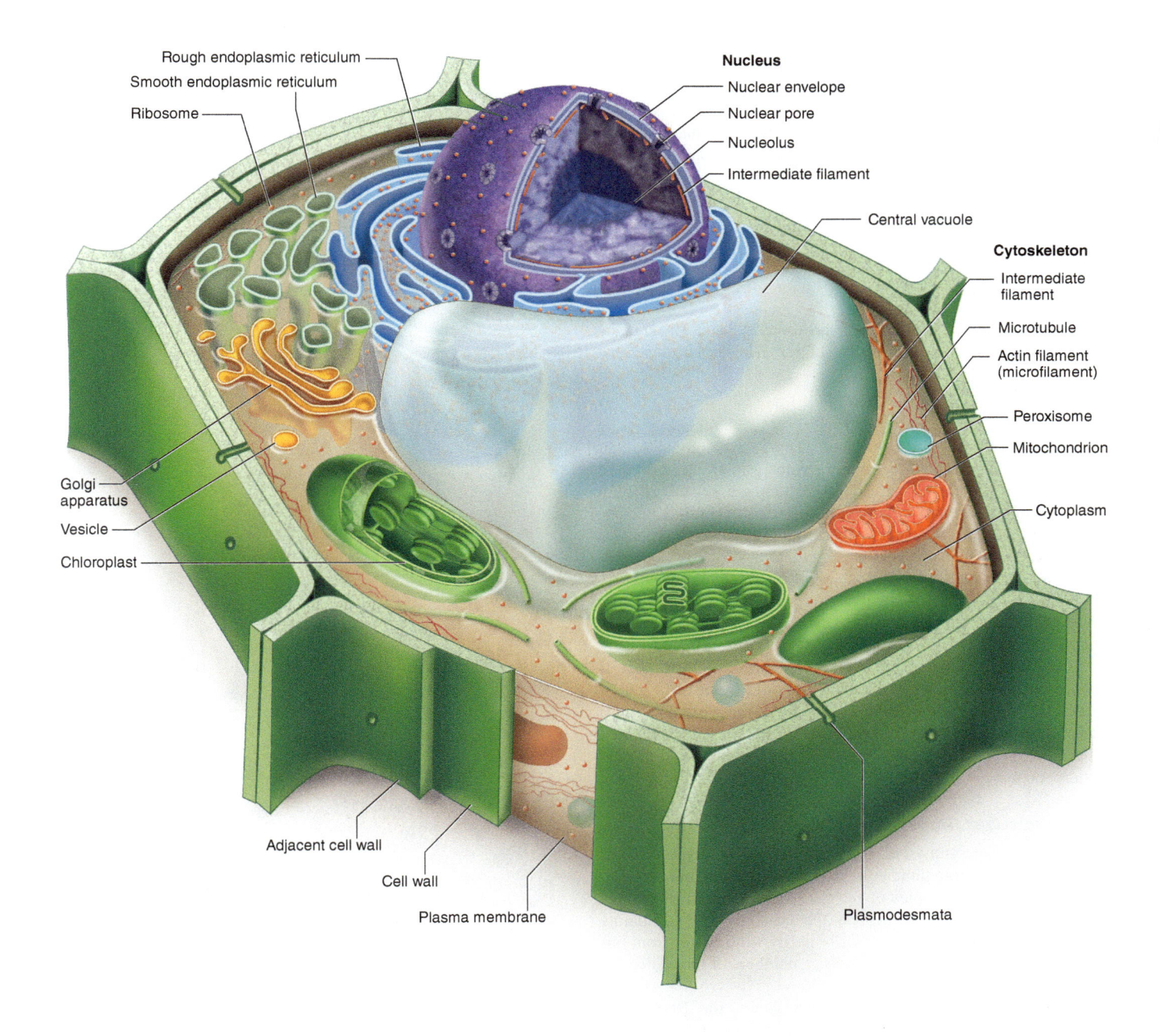

Figure 4.8 The nucleus. *a.* The nucleus is composed of a double membrane called the nuclear envelope, enclosing a fluid-filled interior containing chromatin. The individual nuclear pores extend through the two membrane layers of the envelope. The close-up of the nuclear pore shows the central hub, cytoplasmic ring with fibers, and nuclear ring with basket. *b.* A freeze-fracture electron micrograph (see figure 5.4) of a cell nucleus, showing many nuclear pores. *c.* A transmission electron micrograph of the nuclear membrane showing a single nuclear pore. The dark material within the pore is protein, which acts to control access through the pore. *d.* The nuclear lamina is visible as a dense network of fibers made of intermediate filaments. The nucleus has been colored purple in the micrographs.

This structure gives the nucleus its shape and is also involved in the deconstruction and reconstruction of the nuclear envelope that accompanies cell division.

Chromatin: DNA packaging

In both prokaryotes and eukaryotes, DNA is the molecule that stores genetic information. In eukaryotes, the DNA is divided into multiple linear chromosomes, which are organized with proteins into a complex structure called **chromatin**. It is becoming clear that the very structure of chromatin affects the function of DNA. Changes in gene expression that do not involve changes in DNA sequence, so-called epigenetic changes, involve alterations in chromatin structure (see chapter 16). Although still not fully understood, this offers an exciting new view of many old ideas.

Chromatin is usually in a more extended form that is organized in the nucleus, although we still do not fully understand this organization. When cells divide, the chromatin must be further compacted into a more highly condensed state that forms the X-shaped chromosomes visible in the light microscope.

The nucleolus: Ribosomal subunit manufacturing

Before cells can synthesize proteins in large quantity, they must first construct a large number of ribosomes to carry out this synthesis. Hundreds of copies of the genes encoding the ribosomal RNAs are clustered together on the chromosome, facilitating ribosome construction. By transcribing RNA molecules from this cluster, the cell rapidly generates large numbers of the molecules needed to produce ribosomes.

The clusters of ribosomal RNA genes, the RNAs they produce, and the ribosomal proteins all come together within the nucleus during ribosome production. These ribosomal assembly areas are easily visible within the nucleus as one or more dark-staining regions called nucleoli (singular, *nucleolus*). Nucleoli can be seen under the light microscope even when the chromosomes are uncoiled.

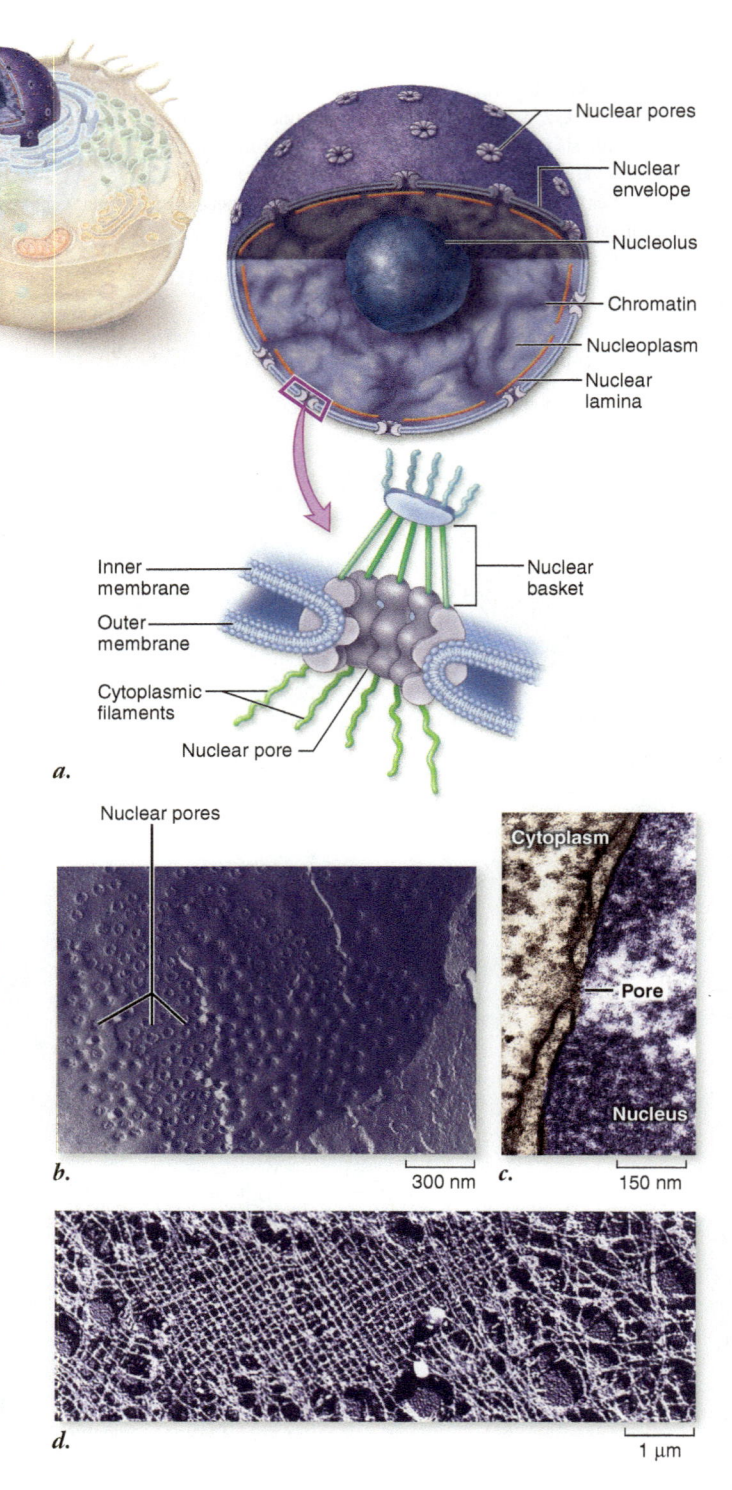

Ribosomes are the cell's protein synthesis machinery

Although the DNA in a cell's nucleus encodes the amino acid sequence of each protein in the cell, the proteins are not assembled there. A simple experiment demonstrates this: If a brief pulse of radioactive amino acid is administered to a cell, the radioactivity shows up associated with newly made protein in the cytoplasm,

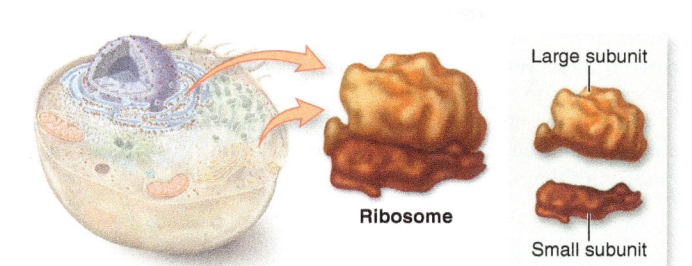

Figure 4.9 A ribosome. Ribosomes consist of a large and a small subunit composed of rRNA and protein. The individual subunits are synthesized in the nucleolus and then move through the nuclear pores to the cytoplasm, where they assemble to translate mRNA. Ribosomes serve as sites of protein synthesis.

not in the nucleus. When investigators first carried out these experiments, they found that protein synthesis is associated with large RNA–protein complexes (called ribosomes) outside the nucleus.

Ribosomes are among the most complex molecular assemblies found in cells. Each ribosome is composed of two subunits (figure 4.9), each of which is composed of a combination of RNA, called **ribosomal RNA (rRNA),** and proteins. The subunits join to form a functional ribosome only when they are actively synthesizing proteins. This complicated process requires the two other main forms of RNA: **messenger RNA (mRNA),** which carries coding information from DNA, and **transfer RNA (tRNA),** which carries amino acids. Ribosomes use the information in mRNA to direct the synthesis of a protein. This process will be described in more detail in chapter 15.

Ribosomes are found either free in the cytoplasm or associated with internal membranes, as described in section 4.4. Free ribosomes synthesize proteins that are found in the cytoplasm, nuclear proteins, mitochondrial proteins, and proteins found in other organelles not derived from the endomembrane system. Membrane-associated ribosomes synthesize membrane proteins, proteins found in the endomembrane system, and proteins destined for export from the cell.

Ribosomes can be thought of as "universal organelles" because they are found in all cell types from all three domains of life. As we build a picture of the minimal essential functions for cellular life, ribosomes will be on the short list. Life is protein-based, and ribosomes are the factories that make proteins.

Learning Outcomes Review 4.3

In contrast to prokaryotic cells, eukaryotic cells exhibit compartmentalization. Eukaryotic cells contain an endomembrane system and organelles that carry out specialized functions. The nucleus, composed of a double membrane connected to the endomembrane system, contains the cell's genetic information. Material moves between the nucleus and cytoplasm through nuclear pores. Ribosomes translate mRNA, which is transcribed from DNA in the nucleus, into polypeptides that make up proteins. Ribosomes are a universal organelle found in all known cells.

■ *Would you expect cells in different organs in complex animals to have the same structure?*

4.4 The Endomembrane System

Learning Outcomes

1. Identify the different parts of the endomembrane system.
2. Contrast the different functions of internal membranes and compartments.
3. Evaluate the importance of each step in the protein-processing pathway.

The interior of a eukaryotic cell is packed with membranes that form an elaborate internal, or endomembrane, system. This endomembrane system fills the cell, dividing it into compartments, channeling the passage of molecules through the interior of the cell, and providing surfaces for the synthesis of lipids and some proteins. The presence of these membranes in eukaryotic cells marks one of the fundamental distinctions between eukaryotes and prokaryotes.

The largest of the internal membranes is called the **endoplasmic reticulum (ER).** The ER is composed of a phospholipid bilayer embedded with proteins. The ER has functional subdivisions, described here, and forms a variety of structures from folded sheets to complex tubular networks (figure 4.10). The ER also may be connected to the cytoskeleton, which can affect

Figure 4.10 The endoplasmic reticulum. Rough ER (RER), blue in the drawing, is composed more of flattened sacs and forms a compartment throughout the cytoplasm. Ribosomes associated with the cytoplasmic face of the RER extrude newly made proteins into the interior, or lumen. The smooth ER (SER), green in the drawing, is a more tubelike structure connected to the RER. The micrograph has been colored to match the drawing.

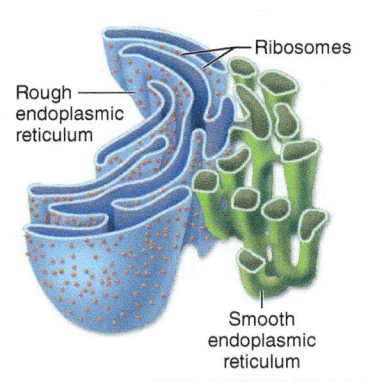

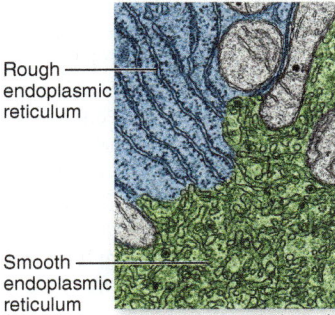

ER structure and growth. The two largest compartments in eukaryotic cells are the inner region of the ER, called the **cisternal space,** or **lumen,** and the region exterior to it, the cytosol, which is the fluid component of the cytoplasm containing dissolved organic molecules such as proteins and ions.

The rough ER is a site of protein synthesis

The **rough ER (RER)** gets its name from its pebbly surface appearance. The RER is not easily visible with a light microscope, but it can be seen using the electron microscope. It appears to be composed primarily of flattened sacs, the surfaces of which are bumpy with ribosomes (figure 4.10).

The proteins synthesized on the surface of the RER are destined to be exported from the cell, sent to lysosomes or vacuoles (described later in this section), or embedded in the plasma membrane. These proteins enter the cisternal space as a first step in the pathway that will sort proteins to their eventual destinations. This pathway also involves vesicles and the Golgi apparatus. The sequence of the protein being synthesized determines whether the ribosome will become associated with the ER or remain a cytoplasmic ribosome.

In the ER, newly synthesized proteins can be modified by the addition of short-chain carbohydrates to form **glycoproteins.** Those proteins destined for secretion are separated from other products and later packaged into vesicles that move to the Golgi for further modification and packaging for transport to other cellular locations.

The smooth ER has multiple roles

Regions of the ER with relatively few bound ribosomes are referred to as **smooth ER (SER).** The SER has a variety of structures ranging from a network of tubules, to flattened sacs, to higher order tubular arrays. The membranes of the SER contain many embedded enzymes. Enzymes anchored within the ER are involved in the synthesis of a variety of carbohydrates and lipids. Steroid hormones are synthesized in the SER as well. The majority of membrane lipids are assembled in the SER and then sent to whatever parts of the cell need membrane components. Membrane proteins in the plasma membrane and other cellular membrane are inserted by ribosomes on the RER.

An important function of the SER is to store intracellular Ca^{2+}. This keeps the cytoplasmic level low, allowing Ca^{2+} to be used as a signaling molecule. In muscle cells, for example, Ca^{2+} is used to trigger muscle contraction. In other cells, Ca^{2+} release from SER stores is involved in diverse signaling pathways.

The ratio of SER to RER is not fixed but depends on a cell's function. In multicellular animals such as ourselves, great variation exists in this ratio. Cells that carry out extensive lipid synthesis, such as those in the testes, intestine, and brain, have abundant SER. Cells that synthesize proteins that are secreted, such as antibodies, have much more extensive RER.

Another role of the SER is the modification of foreign substances to make them less toxic. In the liver, the enzymes of the SER carry out this detoxification. This action can include neutralizing substances that we have taken for a therapeutic reason, such as penicillin. Thus, relatively high doses are prescribed for some drugs to offset our body's efforts to remove them. Liver cells have extensive SER as well as enzymes that can process a variety of substances by chemically modifying them.

The Golgi apparatus sorts and packages proteins

Flattened stacks of membranes form a complex called the **Golgi body,** or **Golgi apparatus** (figure 4.11). These structures are named for Camillo Golgi, the 19th-century Italian physician who first identified them. The individual stacks of membrane are called **cisternae** (Latin, "collecting vessels"), and they vary in number within the Golgi body from 1 or a few in protists, to 20 or more in animal cells and to several hundred in plant cells. In vertebrates individual Golgi are linked to form a Golgi ribbon. They are especially abundant in glandular cells, which manufacture and secrete substances.

The Golgi apparatus functions in the collection, packaging, and distribution of molecules synthesized at one location and used at another within the cell or even outside of it. A Golgi body has a front and a back, with distinctly different membrane compositions at these opposite ends. The front, or receiving end, is called the *cis* face and is usually located near the ER. Materials arrive at the *cis* face in transport vesicles that bud off the ER and exit the

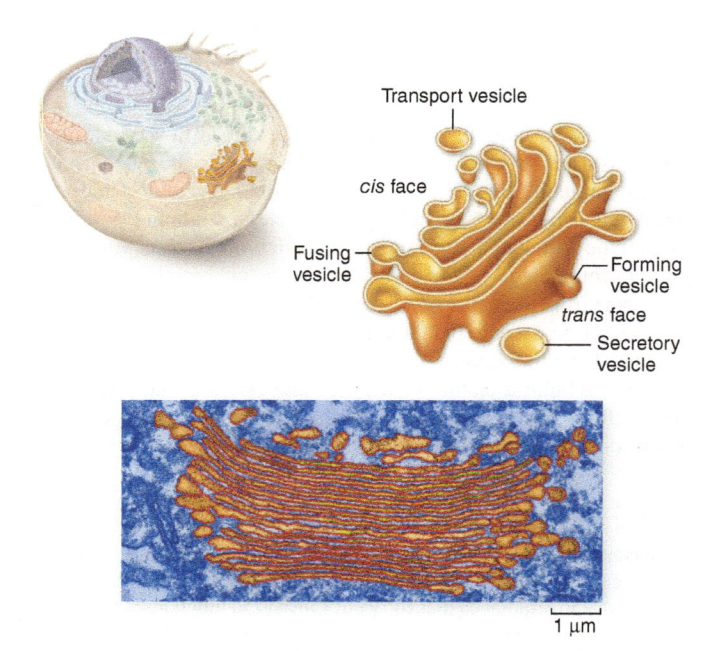

Figure 4.11 The Golgi apparatus. The Golgi apparatus is a smooth, concave, membranous structure. It receives material for processing in transport vesicles on the *cis* face and sends the material packaged in transport or secretory vesicles off the *trans* face. The substance in a vesicle could be for export out of the cell or for distribution to another region within the same cell.

trans face, where they are discharged in secretory vesicles (figure 4.12). How material transits through the Golgi has been a source of much contention. Models include maturation of the individual cisternae from *cis* to *trans*, transport between cisternae by vesicles, and direct tubular connections. Although there is probably transport of material by all of these, it now appears that the primary mechanism is cisternal maturation.

Proteins and lipids manufactured on the rough and smooth ER membranes are transported into the Golgi apparatus and modified as they pass through it. The most common alteration is the addition or modification of short sugar chains, forming glycoproteins and glycolipids. In many instances, enzymes in the Golgi apparatus modify existing glycoproteins and glycolipids made in the ER by cleaving a sugar from a chain or by modifying one or more of the sugars. These are then packaged into small, membrane-bounded vesicles that pinch off from the *trans* face of the Golgi. These vesicles then diffuse to other locations in the cell, distributing the newly synthesized molecules to their appropriate destinations.

Another function of the Golgi apparatus is the synthesis of cell-wall components. Noncellulose polysaccharides that form part of the cell wall of plants are synthesized in the Golgi apparatus and sent to the plasma membrane, where they can be added to the cellulose that is assembled on the exterior of the cell. Other polysaccharides secreted by plants are also synthesized in the Golgi apparatus.

Lysosomes contain digestive enzymes

Membrane-bounded digestive vesicles, called **lysosomes,** are also components of the endomembrane system. They arise from the Golgi apparatus. They contain high levels of degrading enzymes, which catalyze the rapid breakdown of proteins, nucleic acids, lipids, and carbohydrates. Throughout the lives of eukaryotic cells, lysosomal enzymes break down old organelles and recycle their component molecules. This makes room for newly formed organelles. For example, mitochondria are replaced in some tissues every 10 days.

The digestive enzymes in the lysosome are optimally active at acid pH. Lysosomes are activated by fusing with a food vesicle produced by *phagocytosis* (a specific type of endocytosis; see chapter 5) or by fusing with an old or worn-out organelle. The fusion event activates proton pumps in the lysosomal membrane, resulting in a lower internal pH. As the interior pH falls, the arsenal of digestive enzymes contained in the lysosome is activated. This leads to the degradation of macromolecules in the food vesicle or the destruction of the old organelle.

A number of human genetic disorders, collectively called lysosomal storage disorders, affect lysosomes. For example, the genetic abnormality called Tay–Sachs disease is caused by the loss of function of a single lysosomal enzyme (hexosaminidase). This enzyme is necessary to break down a membrane glycolipid found in nerve cells. Accumulation of glycolipid in lysosomes affects nerve cell function, leading to a variety of clinical symptoms such as seizures and muscle rigidity.

In addition to breaking down organelles and other structures within cells, lysosomes eliminate other cells that the cell has

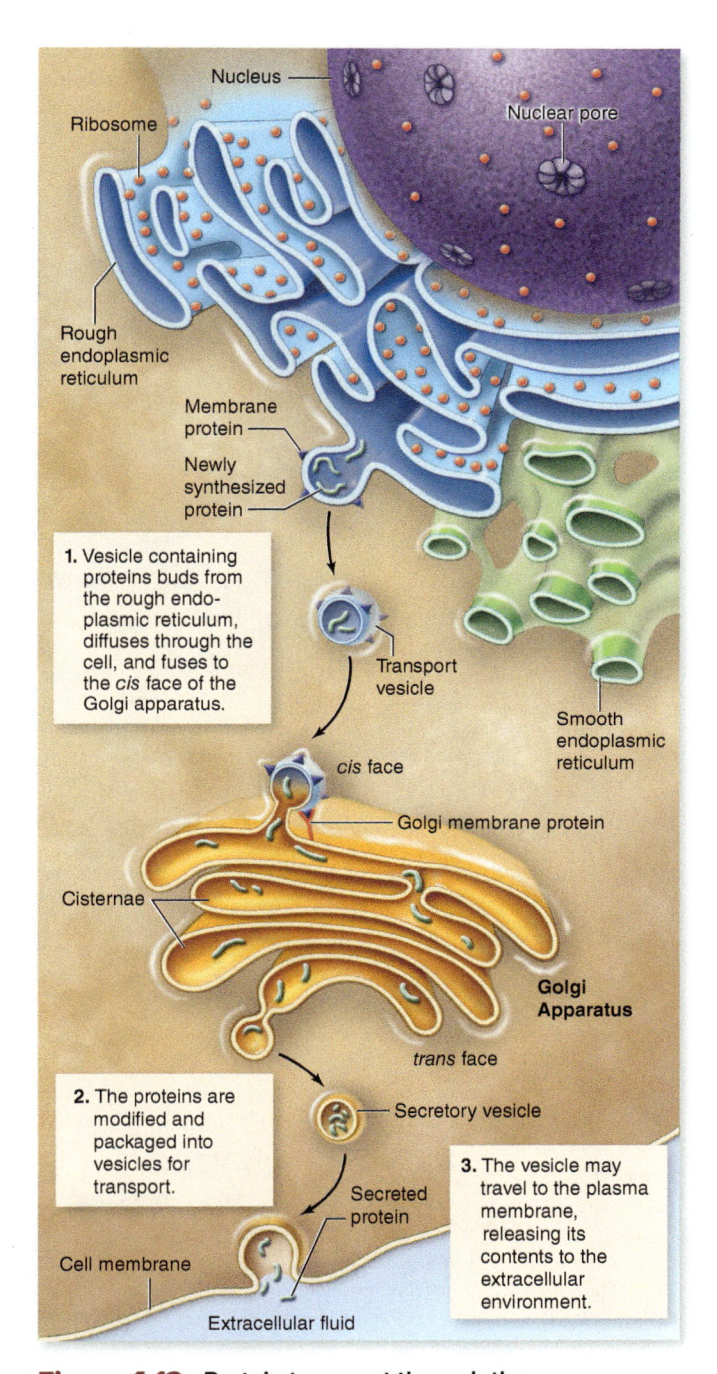

Figure 4.12 Protein transport through the endomembrane system. Proteins synthesized by ribosomes on the RER are translocated into the internal compartment of the ER. These proteins may be used at a distant location within the cell or secreted from the cell. They are transported within vesicles that bud off the RER. These transport vesicles travel to the *cis* face of the Golgi apparatus. There they can be modified and packaged into vesicles that bud off the *trans* face of the Golgi apparatus. Vesicles leaving the *trans* face transport proteins to other locations in the cell, or fuse with the plasma membrane, releasing their contents to the extracellular environment.

engulfed by phagocytosis. When a white blood cell, for example, phagocytizes a passing pathogen, lysosomes fuse with the resulting "food vesicle," releasing their enzymes into the vesicle and degrading the material within (figure 4.13).

Microbodies are a diverse category of organelles

Eukaryotic cells contain a variety of enzyme-bearing, membrane-enclosed vesicles called **microbodies**. These are found in the cells of plants, animals, fungi, and protists. The distribution of enzymes into microbodies is one of the principal ways eukaryotic cells organize their metabolism.

Peroxisomes: Peroxide utilization

An important type of microbody is the **peroxisome** (figure 4.14), which contains enzymes involved in the oxidation of fatty acids. If these oxidative enzymes were not isolated within microbodies, they would tend to short-circuit the metabolism of the cytoplasm, which often involves adding hydrogen atoms to oxygen. Because many peroxisomal proteins are synthesized by cytoplasmic ribosomes, the organelles themselves were long thought to form by the addition of lipids and proteins, leading to growth. As they grow larger, they divide to produce new peroxisomes. Although division of peroxisomes still appears to occur, it is now clear that peroxisomes can form from the fusion of ER-derived vesicles. These vesicles then import peroxisomal proteins to form a mature peroxisome. Genetic screens have isolated some 32 genes that encode proteins involved in biogenesis and maintenance of peroxisomes. The human genetic diseases called peroxisome biogenesis disorders (PBDs) can be caused by mutations in some of these genes.

Peroxisomes get their name from the hydrogen peroxide produced as a by-product of the activities of oxidative enzymes.

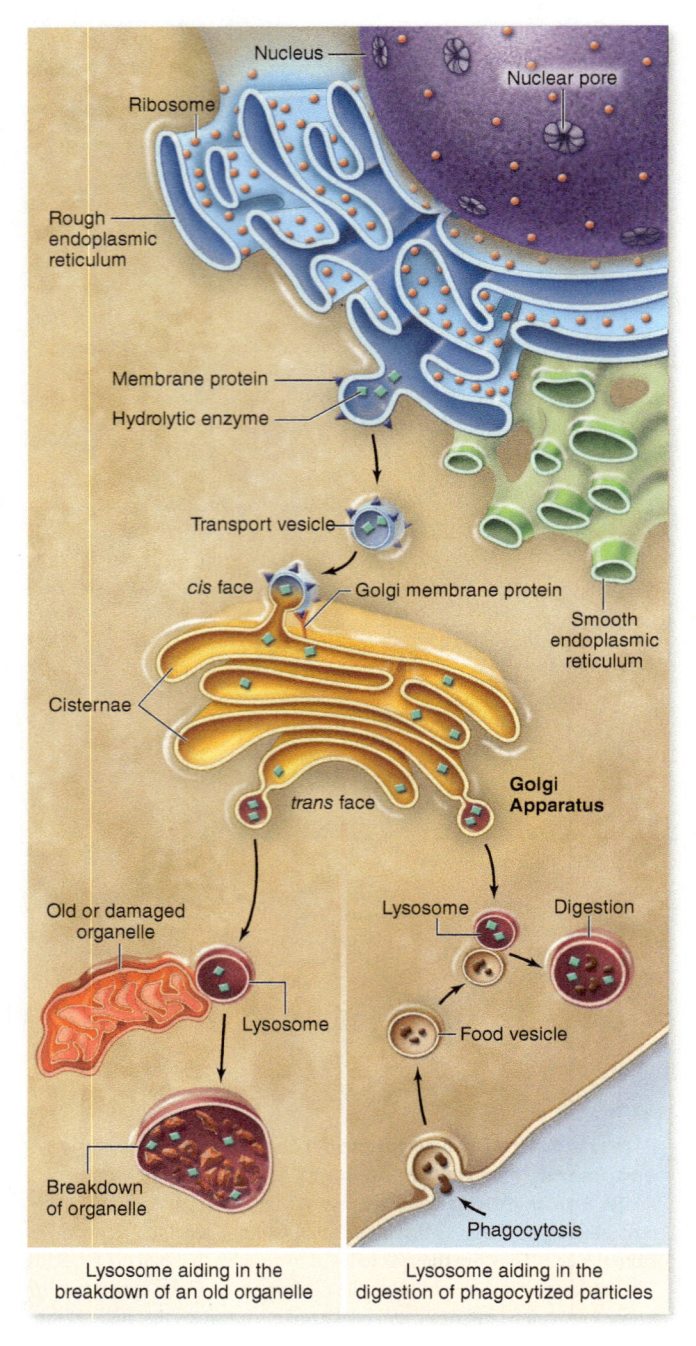

Figure 4.13 Lysosomes. Lysosomes are formed from vesicles budding off the Golgi. They contain hydrolytic enzymes that digest particles or cells taken into the cell by phagocytosis, and break down old organelles.

Figure 4.14 A peroxisome. Peroxisomes are spherical organelles that may contain a large crystal structure composed of protein. Peroxisomes contain digestive and detoxifying enzymes that produce hydrogen peroxide as a by-product. A peroxisome has been colored green in the electron micrograph.

Hydrogen peroxide is dangerous to cells because of its violent chemical reactivity. However, peroxisomes also contain the enzyme catalase, which breaks down hydrogen peroxide into its harmless constituents—water and oxygen.

Plants use vacuoles for storage and water balance

Plant cells have specialized membrane-bounded structures called **vacuoles**. The most conspicuous example is the large central vacuole seen in most plant cells (figure 4.15). In fact, *vacuole* actually means blank space, referring to its appearance in the light microscope. The membrane surrounding this vacuole is called the **tonoplast** because it contains channels for water that are used to help the cell maintain its tonicity, or osmotic balance (see osmosis in chapter 5).

For many years biologists assumed that only one type of vacuole existed and that it served multiple functions. The functions assigned to this vacuole included water balance and storage of both useful molecules (such as sugars, ions, and pigments) and waste products. The vacuole was also thought to store enzymes involved in the breakdown of macromolecules and those used in detoxifying foreign substances. Old textbooks of plant physiology referred to vacuoles as the attic of the cell for the variety of substances thought to be stored there.

Studies of tonoplast transporters and the isolation of vacuoles from a variety of cell types have led to a more complex view of vacuoles. These studies have made it clear that different vacuolar types can be found in different cells. These vacuoles are specialized, depending on the function of the cell.

The central vacuole is clearly important for a number of roles in all plant cells. The central vacuole and the water channels of the tonoplast maintain the tonicity of the cell, allowing the cell to expand and contract, depending on conditions. The central vacuole is also involved in cell growth by occupying most of the volume of the cell. Plant cells grow by expanding the vacuole, rather than by increasing cytoplasmic volume.

Vacuoles with a variety of functions are also found in some types of fungi and protists. One form is the contractile vacuole, found in some protists, which can pump water and is used to maintain water balance in the cell. Other vacuoles are used for storage or to segregate toxic materials from the rest of the cytoplasm. The number and kind of vacuoles found in a cell depends on the needs of the particular cell type.

Learning Outcomes Review 4.4

The endoplasmic reticulum (ER) is an extensive system of folded membranes that spatially organize the cell's biosynthetic activities. Smooth ER (SER) is the site of lipid and membrane synthesis and is used to store Ca^{2+}. Rough ER (RER) is covered with ribosomes and is a site of protein synthesis. Proteins from the RER are transported by vesicles to the Golgi apparatus where they are modified, packaged, and distributed to their final location. Lysosomes are vesicles that contain digestive enzymes used to degrade materials such as invaders or worn-out components. Peroxisomes carry out oxidative metabolism that generates peroxides. Vacuoles are membrane-bounded structures with roles ranging from storage to cell growth in plants. They are also found in some fungi and protists.

■ *How do ribosomes on the RER differ from cytoplasmic ribosomes?*

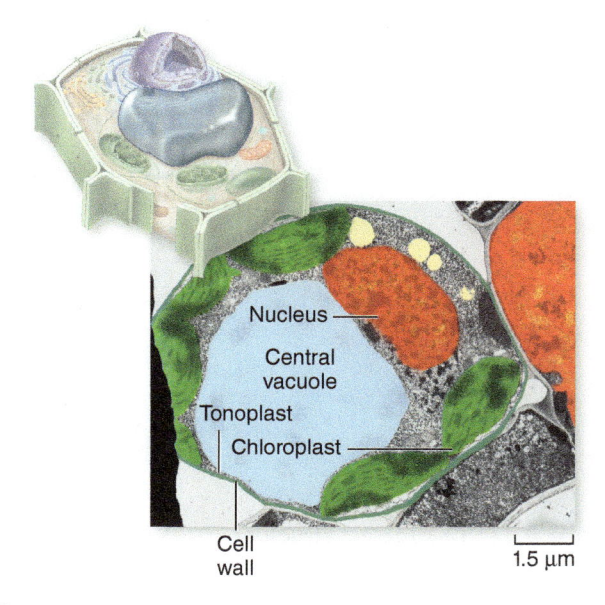

Figure 4.15 The central vacuole. A plant's central vacuole stores dissolved substances and can expand in size to increase the tonicity of a plant cell. Micrograph shown with false color.

4.5 Mitochondria and Chloroplasts: Cellular Generators

Learning Outcomes

1. Describe the structure of mitochondria and chloroplasts.
2. Compare the function of mitochondria and chloroplasts.
3. Explain the probable origin of mitochondria and chloroplasts.

Mitochondria and chloroplasts share structural and functional similarities. Structurally, they are both surrounded by a double membrane, and both contain their own DNA and protein synthesis machinery. Functionally, they are both involved in energy metabolism, as we will explore in detail in chapters 7 and 8 on energy metabolism and photosynthesis.

Chloroplasts use light to generate ATP and sugars

Plant cells and cells of other eukaryotic organisms that carry out photosynthesis typically contain from one to several hundred **chloroplasts.** Chloroplasts bestow an obvious advantage on the organisms that possess them: They can manufacture their own food. Chloroplasts contain the photosynthetic pigment chlorophyll that gives most plants their green color.

The chloroplast, like the mitochondrion, is surrounded by two membranes (figure 4.17). However, chloroplasts are larger and more complex than mitochondria. In addition to the outer and inner membranes, which lie in close association with each other, chloroplasts have closed compartments of stacked membranes called **grana** (singular, *granum*), which lie inside the inner membrane.

A chloroplast may contain a hundred or more grana, and each granum may contain from a few to several dozen disk-shaped structures called **thylakoids.** On the surface of the thylakoids are the light-capturing photosynthetic pigments, to be discussed in depth in chapter 8. Surrounding the thylakoid is a fluid matrix called the *stroma*. The enzymes used to synthesize glucose during photosynthesis are found in the stroma.

Like mitochondria, chloroplasts contain DNA, but many of the genes that specify chloroplast components are also located in the nucleus. Some of the elements used in photosynthesis,

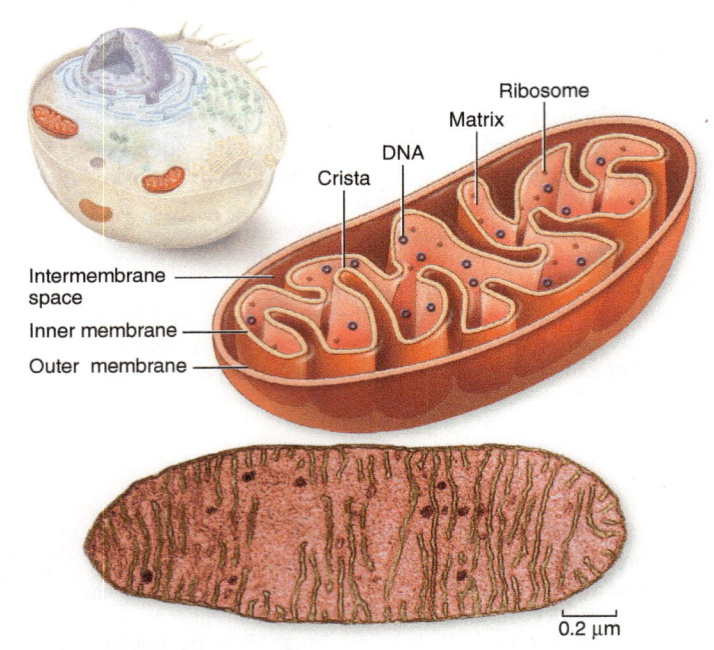

Figure 4.16 Mitochondria. The inner membrane of a mitochondrion is shaped into folds called cristae that greatly increase the surface area for oxidative metabolism. A mitochondrion in cross section and cut lengthwise is shown colored red in the micrograph.

Mitochondria metabolize sugar to generate ATP

Mitochondria (singular, *mitochondrion*) are typically tubular or sausage-shaped organelles about the size of bacteria that are found in all types of eukaryotic cells (figure 4.16). Mitochondria are bounded by two membranes: a smooth outer membrane, and an inner folded membrane with numerous contiguous layers called **cristae** (singular, *crista*).

The cristae partition the mitochondrion into two compartments: a **matrix,** lying inside the inner membrane; and an outer compartment, or **intermembrane space,** lying between the two mitochondrial membranes. On the surface of the inner membrane, and also embedded within it, are proteins that carry out oxidative metabolism, the oxygen-requiring process by which energy in macromolecules is used to produce ATP (see chapter 7).

Mitochondria have their own DNA; this DNA contains several genes that produce proteins essential to the mitochondrion's role in oxidative metabolism. Thus, the mitochondrion, in many respects, acts as a cell within a cell, containing its own genetic information specifying proteins for its unique functions. The mitochondria are not fully autonomous, however, because most of the genes that encode the enzymes used in oxidative metabolism are located in the cell nucleus.

A eukaryotic cell does not produce brand-new mitochondria each time the cell divides. Instead, the mitochondria themselves divide in two, doubling in number, and these are partitioned between the new cells. Most of the components required for mitochondrial division are encoded by genes in the nucleus and are translated into proteins by cytoplasmic ribosomes. Mitochondrial replication is, therefore, impossible without nuclear participation, and mitochondria thus cannot be grown in a cell-free culture.

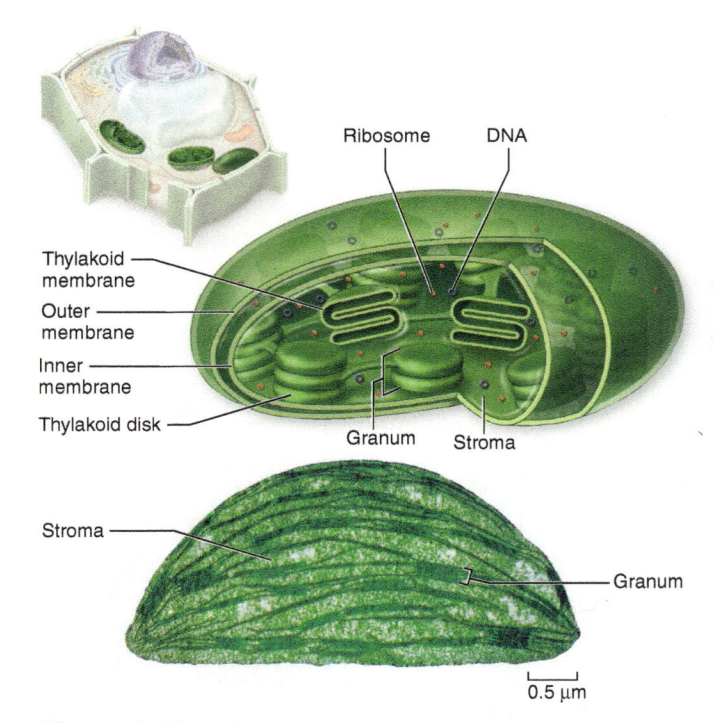

Figure 4.17 Chloroplast structure. The inner membrane of a chloroplast surrounds a membrane system of stacks of closed chlorophyll-containing vesicles called thylakoids, within which photosynthesis occurs. Thylakoids are typically stacked one on top of the other in columns called grana. The chloroplast has been colored green in the micrograph.

including the specific protein components necessary to accomplish the reaction, are synthesized entirely within the chloroplast.

Other DNA-containing organelles in plants, called *leucoplasts,* lack pigment and a complex internal structure. In root cells and some other plant cells, leucoplasts may serve as starch-storage sites. A leucoplast that stores starch (amylose) is sometimes termed an **amyloplast.** These organelles—chloroplasts, leucoplasts, and amyloplasts—are collectively called **plastids.** All plastids are produced by the division of existing plastids.

> **Inquiry question** Mitochondria and chloroplasts both generate ATP. What structural features do they share?

Mitochondria and chloroplasts arose by endosymbiosis

Symbiosis is a close relationship between organisms of different species that live together. As noted in chapter 29, the theory of **endosymbiosis** proposes that some of today's eukaryotic organelles evolved by a symbiosis arising between two cells that were each free-living. One cell, a prokaryote, was engulfed by and became part of another cell, which was the precursor of modern eukaryotes (figure 4.18).

According to the endosymbiont theory, the engulfed prokaryotes provided their hosts with certain advantages associated with their special metabolic abilities. Two key eukaryotic organelles are believed to be the descendants of these endosymbiotic prokaryotes: mitochondria, which are thought to have originated as bacteria capable of carrying out oxidative metabolism, and chloroplasts, which apparently arose from photosynthetic bacteria. This is discussed in detail in chapter 29.

Learning Outcomes Review 4.5

Mitochondria and chloroplasts have similar structures, with an outer membrane and an extensive inner membrane compartment. Both mitochondria and chloroplasts have their own DNA, but both also depend on nuclear genes for some functions. Mitochondria and chloroplasts are both involved in energy conversion: Mitochondria metabolize sugar to produce ATP, whereas chloroplasts harness light energy to produce ATP and synthesize sugars. Endosymbiosis theory proposes that both mitochondria and chloroplasts arose as prokaryotic cells were engulfed by a eukaryotic precursor.

- *Many proteins in mitochondria and chloroplasts are encoded by nuclear genes. In light of the endosymbiont hypothesis, how might this come about?*

4.6 The Cytoskeleton

Learning Outcomes

1. Contrast the structure and function of different fibers in the cytoskeleton.
2. Illustrate the role of microtubules in intracellular transport.

The cytoplasm of all eukaryotic cells is crisscrossed by a network of protein fibers that supports the shape of the cell and anchors organelles to fixed locations. This network, called the cytoskeleton, is a dynamic system, constantly assembling and disassembling. Individual fibers consist of polymers of identical protein subunits that attract one another and spontaneously assemble into long chains. Fibers disassemble in the same way, as one subunit after another breaks away from one end of the chain.

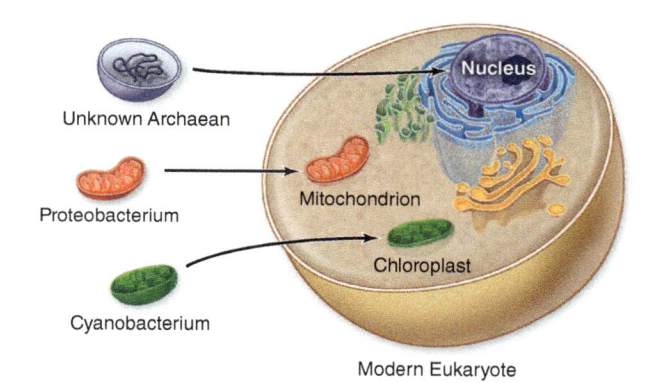

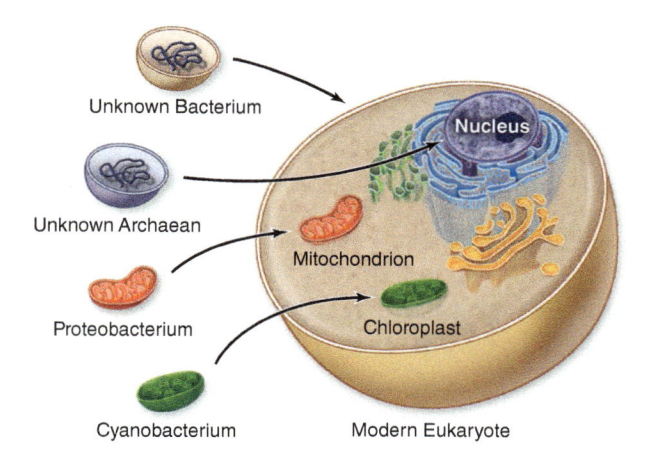

Figure 4.18 Possible origins of eukaryotic cells. Both mitochondria and chloroplasts are thought to have arisen by endosymbiosis when a free-living cell is taken up but not digested. The nature of the engulfing cell is unknown. Two possibilities are (1) the engulfing cell *(top)* is an archaean that gave rise to the nuclear genome and cytoplasmic contents; and (2) the engulfing cell *(bottom)* consists of a nucleus derived from an archaean in a bacterial cell. This could arise by a fusion event or by engulfment of the archaean by the bacterium.

Three types of fibers compose the cytoskeleton

Eukaryotic cells may contain the following three types of cytoskeletal fibers, each formed from a different kind of subunit: (1) actin filaments, sometimes called microfilaments; (2) microtubules; and (3) intermediate filaments.

Actin filaments (microfilaments)

Actin filaments are long fibers about 7 nm in diameter. Each filament is composed of two protein chains loosely twined together like two strands of pearls (figure 4.19). Each "pearl," or subunit, on the chain is the globular protein **actin.** Actin filaments exhibit polarity—that is, they have plus (+) and minus (−) ends. These designate the direction of growth of the filaments. Actin molecules spontaneously form these filaments, even in a test tube.

Cells regulate the rate of actin polymerization through other proteins that act as switches, turning on polymerization when appropriate. Actin filaments are responsible for cellular movements such as contraction, crawling, "pinching" during division, and formation of cellular extensions.

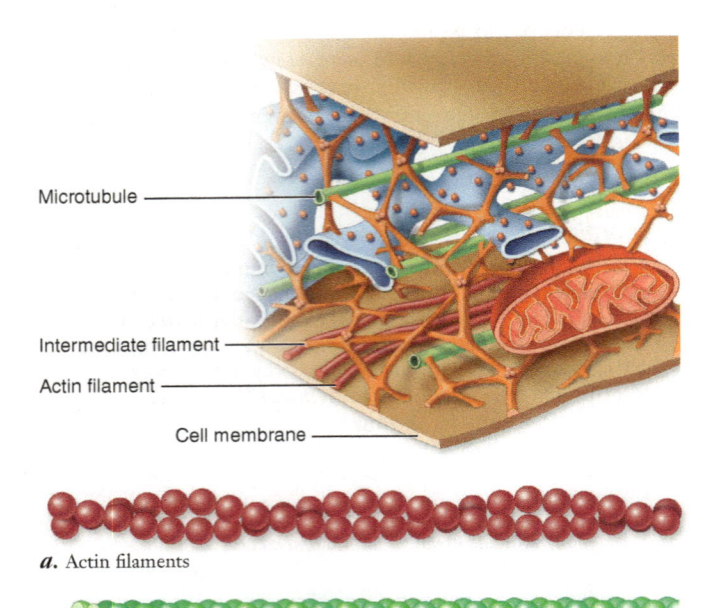

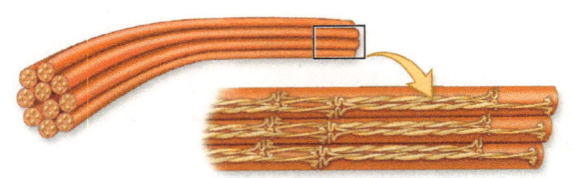

Microtubules

Microtubules, the largest of the cytoskeletal elements, are hollow tubes about 25 nm in diameter, each composed of a ring of 13 protein protofilaments (figure 4.19). Globular proteins consisting of dimers of α- and β-*tubulin* subunits polymerize to form the 13 protofilaments. The protofilaments are arrayed side by side around a central core, giving the microtubule its characteristic tube shape.

In many cells, microtubules form from nucleation centers near the center of the cell and radiate toward the periphery. They are in a constant state of flux, continually polymerizing and depolymerizing. The average half-life of a microtubule ranges from as long as 10 minutes in a nondividing animal cell to as short as 20 seconds in a dividing animal cell. The ends of the microtubule are designated as plus (+) (away from the nucleation center) or minus (−) (toward the nucleation center).

Along with facilitating cellular movement, microtubules organize the cytoplasm and are responsible for moving materials within the cell itself, as described shortly.

Intermediate filaments

The most durable element of the cytoskeleton in animal cells is a system of tough, fibrous protein molecules twined together in an overlapping arrangement (figure 4.19). These **intermediate filaments** are characteristically 8 to 10 nm in diameter—between the size of actin filaments and microtubules. Once formed, intermediate filaments are stable and usually do not break down.

Intermediate filaments constitute a mixed group of cytoskeletal fibers. The most common type, composed of protein subunits called *vimentin,* provides structural stability for many kinds of cells. *Keratin,* another class of intermediate filament, is found in epithelial cells (cells that line organs and body cavities) and associated structures such as hair and fingernails. The intermediate filaments of nerve cells are called *neurofilaments*.

Centrosomes are microtubule-organizing centers

Centrioles are barrel-shaped organelles found in the cells of animals and most protists. They occur in pairs, usually located at right angles to each other near the nuclear membranes (figure 4.20). The region surrounding the pair in almost all animal cells is referred to as a *centrosome*. Surrounding the centrioles in the centrosome is the **pericentriolar material,** which contains ring-shaped structures composed of tubulin. The pericentriolar material can nucleate the assembly of microtubules in animal cells. Structures with this function are called *microtubule-organizing centers*. The centrosome is also responsible for the reorganization of microtubules that occurs during cell division. The centrosomes of plants and fungi lack

Figure 4.19 Molecules that make up the cytoskeleton.
a. Actin filaments: Actin filaments, also called *microfilaments,* are made of two strands of the globular protein actin twisted together. They are often found in bundles or in a branching network. Actin filaments in many cells are concentrated below the plasma membrane in bundles known as stress fibers, which may have a contractile function. *b. Microtubules:* Microtubules are composed of α- and β-tubulin protein subunits arranged side by side to form a tube. Microtubules are comparatively stiff cytoskeletal elements and have many functions in the cell including intracellular transport and the separation of chromosomes during mitosis. *c. Intermediate filaments:* Intermediate filaments are composed of overlapping staggered tetramers of protein. These tetramers are then bundled into cables. This molecular arrangement allows for a ropelike structure that imparts tremendous mechanical strength to the cell.

Figure 4.20 Centrioles. Each centriole is composed of nine triplets of microtubules. Centrioles are usually not found in plant cells. In animal cells they help to organize microtubules.

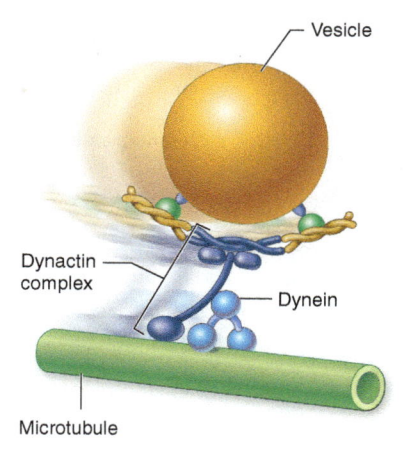

Figure 4.21 Molecular motors. Vesicles can be transported along microtubules using motor proteins that use ATP to generate force. The vesicles are attached to motor proteins by connector molecules, such as the dynactin complex shown here. The motor protein dynein moves the connected vesicle along microtubules.

centrioles, but still contain microtubule-organizing centers. You will learn more about the actions of the centrosomes when we describe the process of cell division in chapter 10.

The cytoskeleton helps move materials within cells

Actin filaments and microtubules often orchestrate their activities to affect cellular processes. For example, during cell reproduction (see chapter 10), newly replicated chromosomes move to opposite sides of a dividing cell because they are attached to shortening microtubules. Then, in animal cells, a belt of actin pinches the cell in two by contracting like a purse string.

Muscle cells also use actin filaments, which slide along filaments of the motor protein myosin when a muscle contracts. The fluttering of an eyelash, the flight of an eagle, and the awkward crawling of a baby all depend on these cytoskeletal movements within muscle cells.

Not only is the cytoskeleton responsible for the cell's shape and movement, but it also provides a scaffold that interacts with the ER and other cytoplasmic macromolecules. Enzymes involved in cell metabolism bind to actin filaments, as do ribosomes. By moving and anchoring particular enzymes near one another, the cytoskeleton helps organize the cell's activities.

In both animals and fungi, the ER is associated with both microtubules and actin filaments. In animal cells, if we visualize cellular structures with fluorescent labels, the tubules of the ER align closely with microtubules. This may be involved in the growth and distribution of the ER. In yeast, microfilaments are involved in inheritance of the ER during cell division.

Molecular motors

All eukaryotic cells must move materials from one place to another in the cytoplasm. One way cells do this is by using the channels of the endoplasmic reticulum as an intracellular highway. Material can also be moved using vesicles loaded with cargo that can move along the cytoskeleton like a railroad track. For example, in a nerve cell with an axon that may extend far from the cell body, vesicles can be moved along tracks of microtubules from the cell body to the end of the axon.

Four components are required to move material along microtubules: (1) a vesicle or organelle that is to be transported, (2) a motor protein that provides the energy-driven motion, (3) a connector molecule that connects the vesicle to the motor molecule, and (4) microtubules on which the vesicle will ride like a train on a rail (figure 4.21).

The direction a vesicle is moved depends on the type of motor protein involved and the fact that microtubules are organized with their plus ends toward the periphery of the cell. In one case, a protein called kinectin binds vesicles to the motor protein *kinesin*. Kinesin uses ATP to power its movement toward the cell periphery, dragging the vesicle with it as it travels along the microtubule toward the plus end (figure 4.22). As nature's tiniest motors, these

SCIENTIFIC THINKING

Hypothesis: Kinesin molecules can act as molecular motors and move along microtubules using energy from ATP.

Test: A microscope slide is covered with purified kinesin. Purified microtubules are added in a buffer containing ATP. The microtubules are monitored under a microscope using a video recorder to capture any movement.

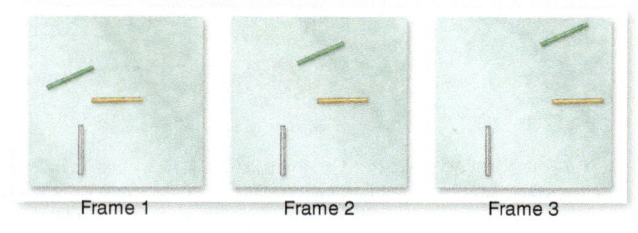

Frame 1   Frame 2   Frame 3

Result: Over time, the movement of individual microtubules can be observed in the microscope. This is shown schematically in the figure by the movement of specific microtubules shown in color.

Conclusion: Kinesin acts as a molecular motor moving along (in this case actually moving) microtubules.

Further Experiments: Are there any further controls that are not shown in this experiment? What additional conclusions could be drawn by varying the amount of kinesin sticking to the slide?

Figure 4.22 Demonstration of kinesin as molecular motor. Microtubules can be observed moving over a slide coated with kinesin.

TABLE 4.2 Eukaryotic Cell Structures and Their Functions

Structure	Description	Function
Plasma membrane	Phospholipid bilayer with embedded proteins	Regulates what passes into and out of cell; cell-to-cell recognition; connection and adhesion; cell communication
Nucleus	Structure (usually spherical) that contains chromosomes and is surrounded by double membrane	Instructions for protein synthesis and cell reproduction; contains genetic information
Chromosomes	Long threads of DNA that form a complex with protein	Contain hereditary information used to direct synthesis of proteins
Nucleolus	Site of genes for rRNA synthesis	Synthesis of rRNA and ribosome assembly
Ribosomes	Small, complex assemblies of protein and RNA, often bound to ER	Sites of protein synthesis
Endoplasmic reticulum (ER)	Network of internal membranes	Intracellular compartment forms transport vesicles; participates in lipid synthesis and synthesis of membrane or secreted proteins
Golgi apparatus	Stacks of flattened vesicles	Packages proteins for export from cell; forms secretory vesicles
Lysosomes	Vesicles derived from Golgi apparatus that contain hydrolytic digestive enzymes	Digest worn-out organelles and cell debris; digest material taken up by endocytosis
Microbodies	Vesicles that are formed from incorporation of lipids and proteins and that contain oxidative and other enzymes	Isolate particular chemical activities from rest of cell
Mitochondria	Bacteria-like elements with double membrane	"Power plants" of the cell; sites of oxidative metabolism
Chloroplasts	Bacteria-like elements with double membrane surrounding a third, thylakoid membrane containing chlorophyll, a photosynthetic pigment	Sites of photosynthesis
Cytoskeleton	Network of protein filaments	Structural support; cell movement; movement of vesicles within cells
Flagella (cilia)	Cellular extensions with 9 + 2 arrangement of pairs of microtubules	Motility or moving fluids over surfaces
Cell wall	Outer layer of cellulose or chitin; or absent	Protection; support

proteins pull the transport vesicles along the microtubular tracks. Another set of vesicle proteins, called the dynactin complex, binds vesicles to the motor protein *dynein* (figure 4.22), which directs movement in the opposite direction along microtubules toward the minus end, inward toward the cell's center. (Dynein is also involved in the movement of eukaryotic flagella, as discussed in section 4.7.) The destination of a particular transport vesicle and its content is thus determined by the nature of the linking protein embedded within the vesicle's membrane.

The major eukaryotic cell structures and their respective functions are summarized in table 4.2.

Learning Outcomes Review 4.6
The three principal fibers of the cytoskeleton are actin filaments (microfilaments), microtubules, and intermediate filaments. These fibers interact to modulate cell shape and permit cell movement. They also act to move materials within the cytoplasm. Material is also moved in large cells using vesicles and molecular motors. The motor proteins move vesicles along tracks of microtubules.

- *What advantage does the cytoskeleton give to large eukaryotic cells?*

4.7 Extracellular Structures and Cell Movement

Learning Outcomes
1. Describe how cells move.
2. Identify the different cytoskeletal elements involved in cell movement.
3. Classify the elements of extracellular matrix in animal cells.

Essentially all cell motion is tied to the movement of actin filaments, microtubules, or both. Intermediate filaments act as intracellular tendons, preventing excessive stretching of cells. Actin filaments play a major role in determining the shape of cells. Because actin filaments can form and dissolve so readily, they enable some cells to change shape quickly.

Some cells crawl

The arrangement of actin filaments within the cell cytoplasm allows cells to crawl, literally! Crawling is a significant cellular phenomenon, essential to such diverse processes as inflammation, clotting, wound healing, and the spread of cancer. White blood cells in particular exhibit this ability. Produced in the bone marrow, these cells are released into the circulatory system and then eventually crawl out of venules and into the tissues to destroy potential pathogens.

At the leading edge of a crawling cell, actin filaments rapidly polymerize, and their extension forces the edge of the cell forward. This extended region is stabilized when microtubules polymerize into the newly formed region. Overall forward movement of the cell is then achieved through the action of the protein **myosin,** which is best known for its role in muscle contraction. Myosin motors along the actin filaments contract, pulling the contents of the cell toward the newly extended front edge.

Cells crawl when these steps occur continuously, with a leading edge extending and stabilizing, and then motors contracting to pull the remaining cell contents along. Receptors on the cell surface can detect molecules outside the cell and stimulate extension in specific directions, allowing cells to move toward particular targets.

Flagella and cilia aid movement

In section 4.2, we described the structure of prokaryotic flagella. Eukaryotic cells have a completely different kind of flagellum, consisting of a circle of nine microtubule pairs surrounding two central microtubules. This arrangement is referred to as the *9 + 2 structure* (figure 4.23).

As pairs of microtubules move past each other using arms composed of the motor protein dynein, the eukaryotic flagellum *undulates,* or waves up and down, rather than rotates. When examined carefully, each flagellum proves to be an outward projection of the cell's interior, containing cytoplasm and enclosed by the plasma membrane. The microtubules of the flagellum are derived from a **basal body,** situated just below the point where the flagellum protrudes from the surface of the cell.

The flagellum's complex microtubular apparatus evolved early in the history of eukaryotes. Today the cells of many multicellular and some unicellular eukaryotes no longer possess flagella and are nonmotile. Other structures, called **cilia** (singular, *cilium*), with

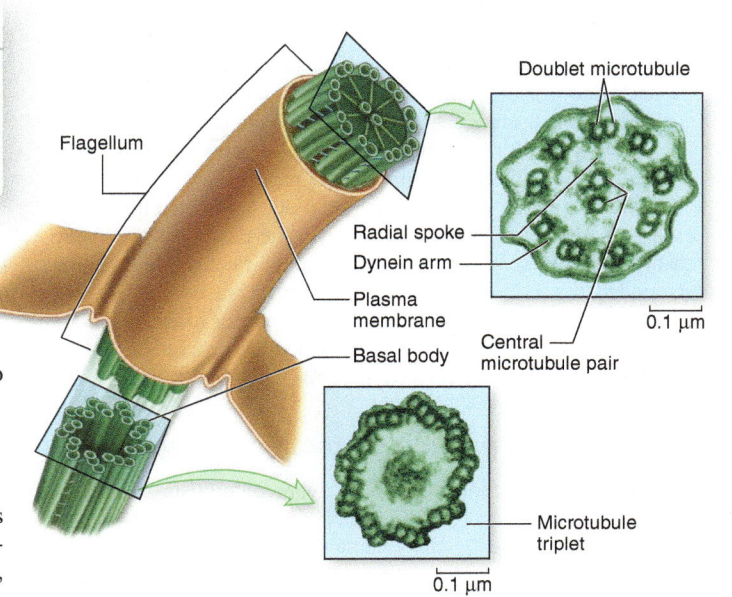

Figure 4.23 Flagella and cilia. A eukaryotic flagellum originates directly from a basal body. The flagellum has two microtubules in its core connected by radial spokes to an outer ring of nine paired microtubules with dynein arms (9 + 2 structure). The basal body consists of nine microtubule triplets connected by short protein segments. The structure of cilia is similar to that of flagella, but cilia are usually shorter.

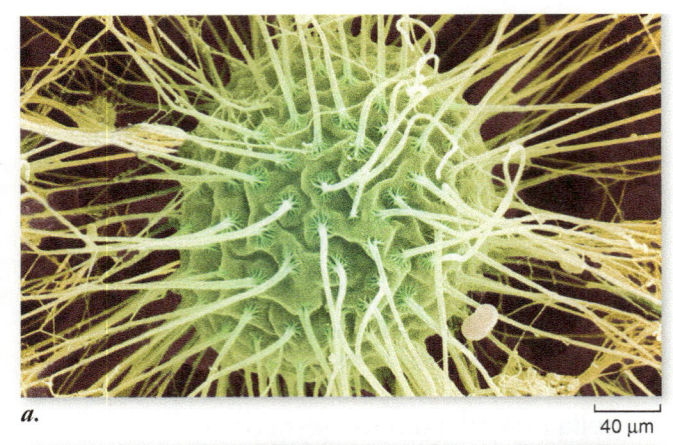

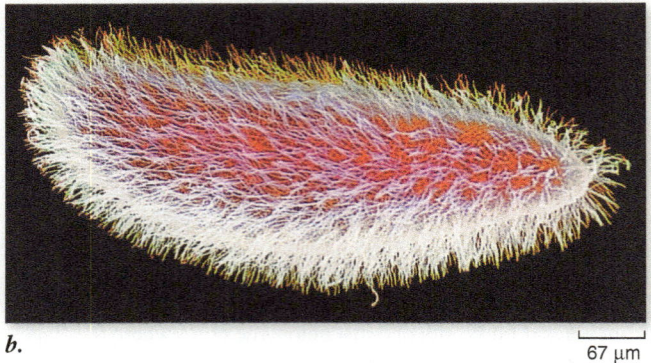

Figure 4.24 Flagella and cilia. *a.* A green alga with numerous flagella that allow it to move through the water.
b. Paramecia are covered with many cilia, which beat in unison to move the cell. The cilia can also be used to move fluid into the paramecium's mouth to ingest material.

an organization similar to the 9 + 2 arrangement of microtubules can still be found within them. Cilia are short cellular projections that are often organized in rows. They are more numerous than flagella on the cell surface, but have the same internal structure.

In many multicellular organisms, cilia carry out tasks far removed from their original function of propelling cells through water. In several kinds of vertebrate tissues, for example, the beating of rows of cilia move water over the tissue surface. The sensory cells of the vertebrate ear also contain conventional cilia surrounded by actin-based stereocilia; sound waves bend these structures and provide the initial sensory input for hearing. Thus, the 9 + 2 structure of flagella and cilia appears to be a fundamental component of eukaryotic cells (figure 4.24).

Plant cell walls provide protection and support

The cells of plants, fungi, and many types of protists have cell walls, which protect and support the cells. The cell walls of these eukaryotes are chemically and structurally different from prokaryotic cell walls. In plants and protists, the cell walls are composed of fibers of the polysaccharide cellulose, whereas in fungi, the cell walls are composed of chitin.

In plants, **primary walls** are laid down when the cell is still growing. Between the walls of adjacent cells a sticky substance, called the **middle lamella,** glues the cells together (figure 4.25). Some plant cells produce strong **secondary walls,** which are deposited inside the primary walls of fully expanded cells.

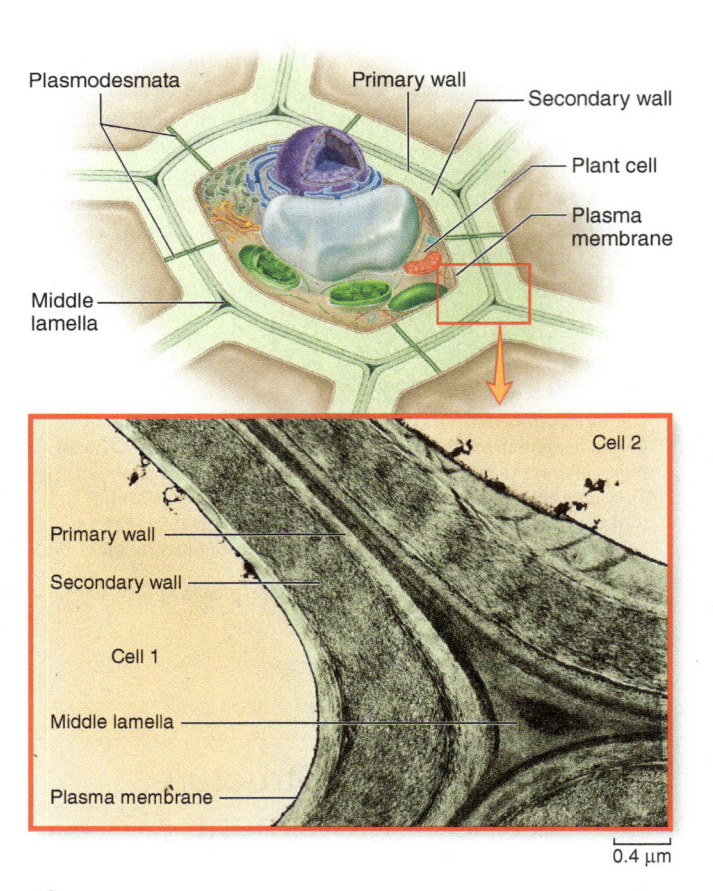

Figure 4.25 Cell walls in plants. Plant cell walls are thick, strong, and rigid. Primary cell walls are laid down when the cell is young. Thicker secondary cell walls may be added later when the cell is fully grown.

Animal cells secrete an extracellular matrix

Animal cells lack the cell walls that encase plants, fungi, and most protists. Instead, animal cells secrete an elaborate mixture of glycoproteins into the space around them, forming the *extracellular matrix (ECM)* (figure 4.26). The fibrous protein collagen, the same protein found in cartilage, tendons, and ligaments may be abundant in the ECM. Strong fibers of collagen and another fibrous protein, elastin, are embedded within a complex web of other glycoproteins, called proteoglycans, that form a protective layer over the cell surface.

> **Inquiry question** The passageways of the human trachea (the path of air flow into and out of the lungs) are known to be lined with ciliated cells. What function could these cilia perform?

The ECM of some cells is attached to the plasma membrane by a third kind of glycoprotein, *fibronectin*. Fibronectin molecules bind not only to ECM glycoproteins but also to proteins called **integrins.** Integrins are an integral part of the plasma membrane, extending into the cytoplasm, where they are attached to the

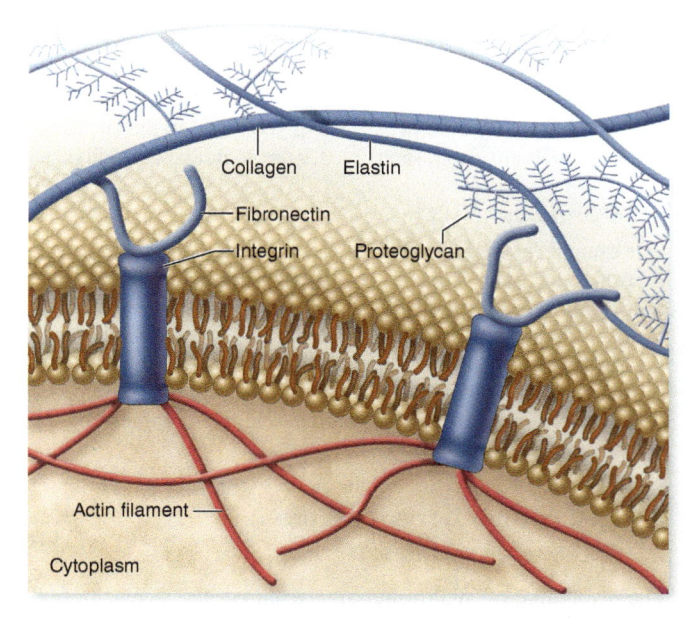

Figure 4.26 The extracellular matrix. Animal cells are surrounded by an extracellular matrix composed of various glycoproteins that give the cells support, strength, and resilience.

microfilaments and intermediate filaments of the cytoskeleton. Linking ECM and cytoskeleton, integrins allow the ECM to influence cell behavior in important ways. They can alter gene expression and cell migration patterns by a combination of mechanical and chemical signaling pathways. In this way, the ECM can help coordinate the behavior of all the cells in a particular tissue.

Table 4.3 compares and reviews the features of three types of cells.

Learning Outcomes Review 4.7

Cell movement involves proteins. These can either be internal in the case of crawling cells that use actin and myosin, or external in the case of cells powered by cilia or flagella. Eukaryotic cilia and flagella are different from prokaryotic flagella because they are composed of bundles of microtubules in a 9 + 2 array. They undulate rather than rotate.

Plant cells have a cellulose-based cell wall. Animal cells lack a cell wall. In animal cells, the cytoskeleton is linked to a web of glycoproteins called the extracellular matrix.

- *What cellular roles are performed by microtubules and microfilaments and not intermediate filaments?*

TABLE 4.3 A Comparison of Prokaryotic, Animal, and Plant Cells

	Prokaryote	Animal	Plant
EXTERIOR STRUCTURES			
Cell wall	Present (protein-polysaccharide)	Absent	Present (cellulose)
Cell membrane	Present	Present	Present
Flagella/cilia	Flagella may be present	May be present (9 + 2 structure)	Absent except in sperm of a few species (9 + 2 structure)
INTERIOR STRUCTURES			
Endoplasmic reticulum	Absent	Usually present	Usually present
Ribosomes	Present	Present	Present
Microtubules	Absent	Present	Present
Centrioles	Absent	Present	Absent
Golgi apparatus	Absent	Present	Present
Nucleus	Absent	Present	Present
Mitochondria	Absent	Present	Present
Chloroplasts	Absent	Absent	Present
Chromosomes	Single; circle of DNA	Multiple; DNA–protein complex	Multiple; DNA–protein complex
Lysosomes	Absent	Usually present	Present
Vacuoles	Absent	Absent or small	Usually a large single vacuole

4.8 Cell-to-Cell Interactions

Learning Outcomes
1. Differentiate between types of cell junctions.
2. Describe the roles of surface proteins.

A basic feature of multicellular animals is the formation of diverse kinds of *tissue*, such as skin, blood, or muscle, where cells are organized in specific ways. Cells must also be able to communicate with each other and have markers of individual identity. All of these functions—connections between cells, markers of cellular identity, and cell communication—involve membrane proteins and proteins secreted by cells. As an organism develops, the cells acquire their identities by carefully controlling the *expression* of those genes, turning on the specific set of genes that encode the functions of each cell type. Table 4.4 provides a summary of the kinds of connections seen between cells that are explored in this section.

Surface proteins give cells identity

One key set of genes functions to mark the surfaces of cells, identifying them as being of a particular type. When cells make contact, they "read" each other's cell-surface markers and react accordingly. Cells that are part of the same tissue type recognize each other, and they frequently respond by forming connections between their surfaces to better coordinate their functions.

Glycolipids

Most tissue-specific cell-surface markers are glycolipids—that is, lipids with carbohydrate heads. The glycolipids on the surface of red blood cells are also responsible for the A, B, and O blood types.

MHC proteins

One example of the function of cell-surface markers is the recognition of "self" and "nonself" cells by the immune system. This function is vital for multicellular organisms, which need to defend themselves against invading or malignant cells. The immune system of vertebrates uses a particular set of markers to distinguish self from nonself cells, encoded by genes of the *major histocompatibility complex (MHC)*. Cell recognition in the immune system is covered in chapter 51.

Cell connections mediate cell-to-cell adhesion

The evolution of multicellularity required the acquisition of molecules that can connect cells to each other. It appears that multicellularity arose independently in different lineages, but the types of connections between cells are remarkably conserved, and many of the proteins involved are ancient.

The nature of the physical connections between the cells of a tissue in large measure determines what the tissue is like. Indeed, a tissue's proper functioning often depends critically on how the individual cells are arranged within it. Just as a house cannot maintain its structure without nails and cement, so a tissue cannot maintain its characteristic architecture without the appropriate cell junctions. Cell junctions can be characterized by both their

TABLE 4.4 Cell-to-Cell Connections and Cell Identity

Type of Connection	Structure	Function	Example
Surface markers	Variable, integral proteins or glycolipids in plasma membrane	Identify the cell	MHC complexes, blood groups, antibodies
Septate junctions Tight junctions	Tightly bound, leakproof, fibrous claudin protein seal that surrounds cell	Hold cells together such that materials pass through but not between the cells	Junctions between epithelial cells in the gut
Adhesive junction (desmosome)	Variant cadherins, desmocollins, bind to intermediate filaments of cytoskeleton	Creates strong flexible connections between cells. Found in vertebrates	Epithelium
Adhesive junction (adherens junction)	Classical cadherins, bind to microfilaments of cytoskeleton	Connects cells together. Oldest form of cell junction, found in all multicellular organisms	Tissues with high mechanical stress, such as the skin
Adhesive junction (hemidesmosome, focal adhesion)	Integrin proteins bind cell to extracellular matrix	Provides attachment to a substrate	Involved in cell movement and important during development
Communicating junction (gap junction)	Six transmembrane connexon/pannexin proteins creating a pore that connects cells	Allows passage of small molecules from cell to cell in a tissue	Excitable tissue such as heart muscle
Communicating junction (plasmodesmata)	Cytoplasmic connections between gaps in adjoining plant cell walls	Communicating junction between plant cells	Plant tissues

visible structure in the microscope, and the proteins involved in the junction.

Adhesive junctions

Adhesive junctions appear to have been the first to evolve. Primitive forms can even be found in sponges, and they are found in all animal species. They mechanically attach the cytoskeleton of a cell to the cytoskeletons of other cells or to the extracellular matrix. These junctions are found in tissues subject to mechanical stress, such as muscle and skin epithelium.

Adherens junctions are based on the protein **cadherin**, which is a Ca^{2+}-dependent adhesion molecule with very wide phylogenetic distribution. Cadherin is a single-pass transmembrane protein with an extracellular domain that can interact with the extracellular domain of a cadherin in an adjacent cell to join the cells together (figure 4.27). Adherens junctions are found in animals ranging from jellyfish to vertebrates. Cadherins found in these junctions are called classical cadherins and are broken down into types I and II. When cells bearing either type I or type II cadherins are mixed, they sort into populations joined by I to I or by II to II interactions. There is some evidence for interactions between type I and type II cadherins, but they are not as strong. On the cytoplasmic side, the cadherins interact indirectly through other proteins with actin to form flexible connections between cells (figure 4.27).

Desmosomes are a cadherin-based junction unique to vertebrates. They contain the cadherins desmocollin and desmoglein, which interact with intermediate filaments of cytoskeletons instead of actin. Desmosomes join adjacent cells (figure 4.28b). These connections support tissues against mechanical stress.

Hemidesmosomes and focal adhesions connect cells to the basal lamina or other ECM. In this case the proteins that interact with the ECM are called integrins. The integrins are members of a large superfamily of cell-surface receptors that bind to a protein component of the extracellular matrix. At least 20 different integrins exist, each with a differently shaped binding domain. These junctions also connect to the cytoskeleton of cells: actin filaments at focal adhesions and intermediate filaments at hemidesmosomes.

Septate, or Tight, junctions

Septate junctions are found in both invertebrates and vertebrates and form a barrier that can seal off a sheet of cells. The proteins found at these junctions have been given different names in different systems; in *Drosophila*, the proteins include Discs Large and Neurexin. Their wide distribution indicates that they probably evolved soon after or with adherens junctions.

Tight junctions are unique to vertebrates and contain proteins called Claudins because of their ability to occlude or block substances from passing between cells. This form of junction between cells acts as a wall within the tissue, keeping molecules on one side or the other (figure 4.28a).

Creating sheets of cells. The cells that line an animal's digestive tract are organized in a sheet only one cell thick. One surface of the sheet faces the inside of the tract, and the other faces the extracellular space, where blood vessels are located. Tight junctions encircle each cell in the sheet, like a belt cinched around a person's waist. The junctions between neighboring cells are so securely attached that there is no space between them for leakage. Hence, nutrients absorbed from the food in the digestive tract must pass directly through the cells in the sheet to enter the bloodstream because they cannot pass through spaces between cells.

The tight junctions between the cells lining the digestive tract also partition the plasma membranes of these cells into separate compartments. Transport proteins in the membrane facing the inside of the tract carry nutrients from that side to the cytoplasm of the cells. Other proteins, located in the membrane on the opposite side of the cells, transport those nutrients from the cytoplasm to the extracellular fluid, where they can enter the bloodstream. Tight junctions effectively segregate the proteins on opposite sides of the sheet, preventing them from drifting within the membrane from one side of the sheet to the other. When tight junctions are experimentally disrupted, just this sort of migration occurs.

Communicating junctions

The proteins involved in the junctions previously described can be found in some single-celled organisms as well. The evolution of multicellularity also led to a new form of cellular connection: the *communicating junctions*. These junctions allow communication between cells by diffusion through small openings. Communicating junctions permit small molecules or ions to pass from one cell to the other. In animals, these direct communication channels between cells are called *gap junctions*, and in plants, *plasmodesmata*.

Gap junctions in animals. Gap junctions are found in both invertebrates and vertebrates. In invertebrates they are formed by proteins known as pannexins. In vertebrates pannexin-base gap

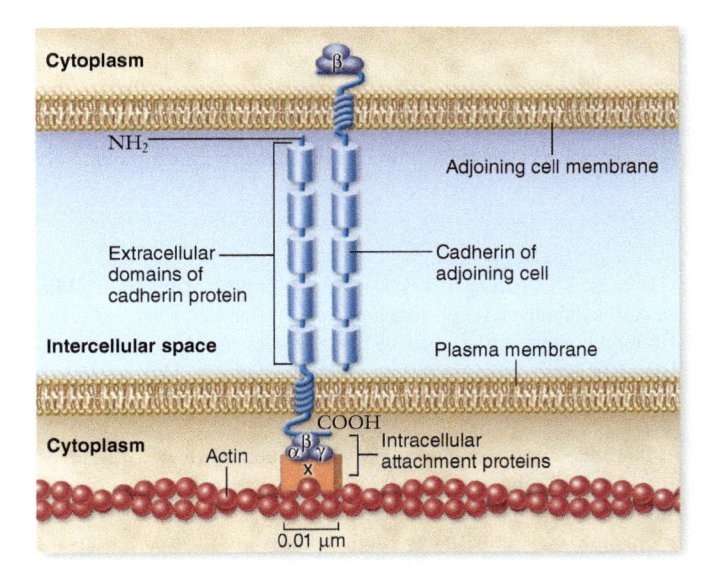

Figure 4.27 A cadherin-mediated junction. The cadherin molecule is anchored to actin in the cytoskeleton and passes through the membrane to interact with the cadherin of an adjoining cell.

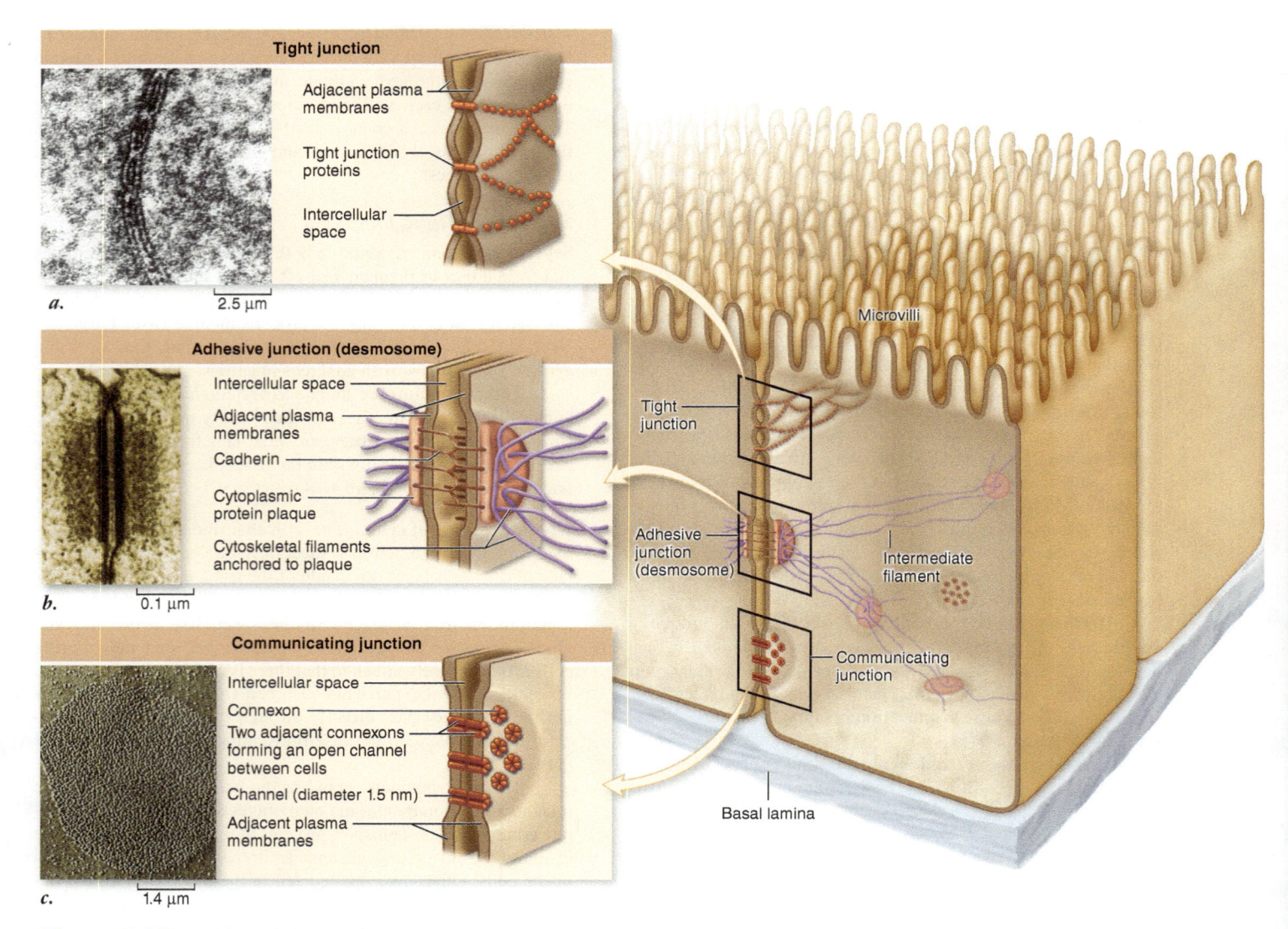

Figure 4.28 Cell junction types in animal epithelium. Here, the diagram of gut epithelial cells on the right illustrates the comparative structures and locations of common cell junctions. The detailed models on the left show the structures of the three major types of cell junctions: *(a)* tight junction; *(b)* adhesive junction—the example shown is a desmosome; *(c)* communicating junction—the example shown is a gap junction.

junctions exist, but there is an additional type based on similar proteins called connexons. In each case, a structure is formed by complexes of six identical transmembrane proteins (figure 4.28c). The proteins are arranged in a circle to create a channel through the plasma membrane that protrudes several nanometers from the cell surface. A gap junction forms when the connexons/pannexins of two cells align perfectly, creating an open channel that spans the plasma membranes of both cells.

Gap junctions provide passageways large enough to permit small substances, such as simple sugars and amino acids, to pass from one cell to the next. Yet the passages are small enough to prevent the passage of larger molecules, such as proteins.

Gap junction channels are dynamic structures that can open or close in response to a variety of factors, including Ca^{2+} and H^+ ions. This gating serves at least one important function. When a cell is damaged, its plasma membrane often becomes leaky. Ions in high concentrations outside the cell, such as Ca^{2+}, flow into the damaged cell and close its gap junction channels. This isolates the cell and prevents the damage from spreading.

Plasmodesmata in plants. In plants, cell walls separate every cell from all others. Cell–cell junctions occur only at holes or gaps in the walls, where the plasma membranes of adjacent cells can come into contact with one another. Cytoplasmic connections that form across the touching plasma membranes are called **plasmodesmata** (singular, *plasmodesma*) (figure 4.29). The majority of living cells within a higher plant are connected to their neighbors by these junctions.

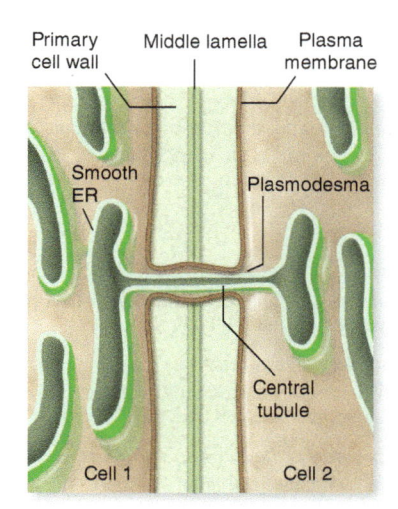

Figure 4.29 Plasmodesmata. Plant cells can communicate through specialized openings in their cell walls, called plasmodesmata, where the cytoplasm of adjoining cells are connected.

Plasmodesmata function much like gap junctions in animal cells, although their structure is more complex. Unlike gap junctions, plasmodesmata are lined with plasma membrane and contain a central tubule that connects the endoplasmic reticulum of the two cells.

Learning Outcomes Review 4.8

The evolution of multicellularity required the acquisition of cell adhesion molecules to connect cells together. Cell connections fall into three basic categories: (1) adhesive junctions provide strength and flexibility; (2) tight, or septate, junctions help to make sheets of cells that form watertight seals; and (3) communicating junctions, including gap junctions in animals and plasmodesmata in plants, allow passage of some materials between cells. Cells in multicellular organisms have distinct identity and connections. Cell identity is conferred by surface glycoproteins, which include the MHC proteins that are important in the immune system.

■ *How do cell junctions help to form tissues?*

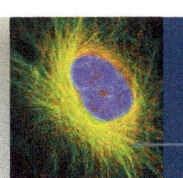

Chapter Review

4.1 Cell Theory

Cell theory is the unifying foundation of cell biology.
All organisms are composed of one or more cells. Cells arise only by division of preexisting cells.

Cell size is limited.
Cell size is constrained by the diffusion distance. As cell size increases, diffusion becomes inefficient.

Microscopes allow visualization of cells and components.
Magnification gives better resolution than is possible with the naked eye. Staining with chemicals enhances contrast of structures.

All cells share many structural features.
All cells have centrally located DNA, a semifluid cytoplasm, and an enclosing plasma membrane.

4.2 Prokaryotic Cells (figure 4.3)

Prokaryotic cells have relatively simple organization.
Prokaryotic cells contain DNA and ribosomes, but they lack a nucleus, an internal membrane system, and membrane-bounded organelles. A rigid cell wall surrounds the plasma membrane.

Bacterial cell walls consist of peptidoglycan.
Peptidoglycan is composed of carbohydrate cross-linked with short peptides.

Archaea lack peptidoglycan.
Archaeal cell walls do not contain peptidoglycan, and they have unique plasma membranes.

Some prokaryotes move by means of rotating flagella.
Prokaryotic flagella rotate because of proton transfer across the plasma membrane.

4.3 Eukaryotic Cells (figures 4.6 and 4.7)

Eukaryotic cells have a membrane-bounded nucleus, an endomembrane system, and many different organelles.

The nucleus acts as the information center.
The nucleus is surrounded by an envelope of two phospholipid bilayers; the outer layer is contiguous with the ER. Pores allow exchange of small molecules. The nucleolus is a region of the nucleoplasm where rRNA is transcribed and ribosomes are assembled.

In most prokaryotes, DNA is organized into a single circular chromosome. In eukaryotes, numerous chromosomes are present.

Ribosomes are the cell's protein synthesis machinery.
Ribosomes translate mRNA to produce polypeptides. They are found in all cell types.

4.4 The Endomembrane System

The endoplasmic reticulum (ER) creates channels and passages within the cytoplasm (figure 4.10).

The rough ER is a site of protein synthesis.
The rough ER (RER), studded with ribosomes, synthesizes and modifies proteins and manufactures membranes.

The smooth ER has multiple roles.
The smooth endoplasmic reticulum (SER) lacks ribosomes; it is involved in carbohydrate and lipid synthesis and detoxification.

The Golgi apparatus sorts and packages proteins.
The Golgi apparatus receives vesicles from the ER, modifies and packages macromolecules, and transports them (figure 4.11).

Lysosomes contain digestive enzymes.
Lysosomes break down macromolecules and recycle the components of old organelles (figure 4.13).

Microbodies are a diverse category of organelles.

Plants use vacuoles for storage and water balance.

4.5 Mitochondria and Chloroplasts: Cellular Generators

Mitochondria and chloroplasts have a double-membrane structure, contain their own DNA, and can divide independently.

Mitochondria metabolize sugar to generate ATP.
The inner membrane of mitochondria is extensively folded into layers called cristae. Proteins on the surface and in the inner membrane carry out metabolism to produce ATP (figure 4.16).

Chloroplasts use light to generate ATP and sugars.
Chloroplasts capture light energy via thylakoid membranes arranged in stacks called grana, and use it to synthesize glucose (figure 4.17).

Mitochondria and chloroplasts arose by endosymbiosis.
The endosymbiont theory proposes that mitochondria and chloroplasts were once prokaryotes engulfed by another cell.

4.6 The Cytoskeleton

The cytoskeleton consists of crisscrossed protein fibers that support the shape of the cell and anchor organelles (figure 4.19).

Three types of fibers compose the cytoskeleton.
Actin filaments, or microfilaments, are long, thin polymers involved in cellular movement. Microtubules are hollow structures that move materials within a cell. Intermediate filaments serve a wide variety of functions.

Centrosomes are microtubule-organizing centers.
Centrosomes help assemble the nuclear division apparatus of animal cells (figure 4.20).

The cytoskeleton helps move materials within cells.
Molecular motors move vesicles along microtubules, like a train on a railroad track. Kinesin and dynein are two motor proteins.

4.7 Extracellular Structures and Cell Movement

Some cells crawl.
Cell crawling occurs as actin polymerization forces the cell membrane forward, while myosin pulls the cell body forward.

Flagella and cilia aid movement.
Eukaryotic flagella have a 9 + 2 structure and arise from a basal body. Cilia are shorter and more numerous than flagella.

Plant cell walls provide protection and support.
Plants have cell walls composed of cellulose fibers. The middle lamella, between cell walls, holds adjacent cells together.

Animal cells secrete an extracellular matrix.
Glycoproteins are the main component of the extracellular matrix (ECM) of animal cells.

4.8 Cell-to-Cell Interactions (figure 4.28)

Surface proteins give cells identity.
Glycolipids and MHC proteins on cell surfaces help distinguish self from nonself.

Cell connections mediate cell-to-cell adhesion.
Cell junctions include tight junctions, adhesive junctions, and communicating junctions. In animals, gap junctions allow the passage of small molecules between cells. In plants, plasmodesmata penetrate the cell wall and connect cells.

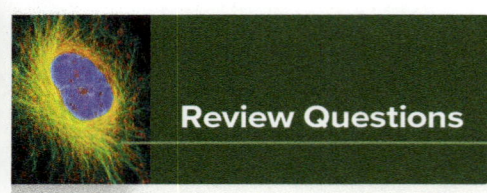

Review Questions

UNDERSTAND

1. Which of the following statements is NOT part of the cell theory?
   a. All organisms are composed of one or more cells.
   b. Cells come from other cells by division.
   c. Cells are the smallest living things.
   d. Eukaryotic cells have evolved from prokaryotic cells.

2. All cells have all of the following except
   a. plasma membrane.
   b. genetic material.
   c. cytoplasm.
   d. cell wall.

3. Eukaryotic cells are more complex than prokaryotic cells. Which of the following are found only in a eukaryotic cell?
   a. Cell wall
   b. Plasma membrane
   c. Endoplasmic reticulum
   d. Ribosomes

4. Which of the following are differences between bacteria and archaea?
   a. The molecular architecture of their cell walls
   b. The type of ribosomes found in each
   c. Archaea have an internal membrane system that bacteria lack.
   d. Both a and b are correct.

5. The cytoskeleton includes
   a. microtubules made of actin filaments.
   b. microfilaments made of tubulin.
   c. intermediate filaments made of twisted fibers of vimentin and keratin.
   d. smooth endoplasmic reticulum.

6. The smooth endoplasmic reticulum is
   a. involved in protein synthesis.
   b. a site of protein glycosylation.
   c. used to store a variety of ions.
   d. the site of lipid and membrane synthesis.

7. Plasmodesmata in plants and gap junctions in animals are functionally similar in that
   a. each is used to anchor layers of cells.
   b. they form channels between cells that allow diffusion of small molecules.
   c. they form tight junctions between cells.
   d. they are anchored to the extracellular matrix.

APPLY

1. The most important factor that limits the size of a cell is the
   a. quantity of proteins and organelles a cell can make.
   b. rate of diffusion of small molecules.
   c. surface area-to-volume ratio of the cell.
   d. amount of DNA in the cell.

2. All eukaryotic cells possess each of the following except
   a. mitochondria.
   b. cell wall.
   c. cytoskeleton.
   d. nucleus.

3. Adherens junctions, which contain cadherin, are found in all animals. Given this, which of the following predictions is most likely?
   a. Cadherins would not be found in the ancestor to all animals.
   b. Cadherins would be found in prokaryotes.
   c. Cadherins would be found in the ancestor to all animals.
   d. Cadherins would be found in vertebrates but not invertebrates.

4. Different motor proteins like kinesin and myosin are similar in that they can
   a. interact with microtubules.
   b. use energy from ATP to produce movement.
   c. interact with actin.
   d. do both a and b.

5. The protein sorting pathway involves the following organelles/compartments in order:
   a. SER, RER, transport vesicle, Golgi.
   b. RER, lysosome, Golgi.
   c. RER, transport vesicle, Golgi, final destination.
   d. Golgi, transport vesicle, RER, final destination.

6. Chloroplasts and mitochondria have many common features because both
   a. are present in plant cells.
   b. arose by endosymbiosis.
   c. function to oxidize glucose.
   d. function to produce glucose.

7. Eukaryotic cells are composed of three types of cytoskeletal filaments. How are these three filaments similar?
   a. They contribute to the shape of the cell.
   b. They are all made of the same type of protein.
   c. They are all the same size and shape.
   d. They are all equally dynamic and flexible.

SYNTHESIZE

1. The smooth endoplasmic reticulum is the site of synthesis of the phospholipids that make up all the membranes of a cell—especially the plasma membrane. Use the diagram of an animal cell (figure 4.6) to trace a pathway that would carry a phospholipid molecule from the SER to the plasma membrane. What endomembrane compartments would the phospholipids travel through? How can a phospholipid molecule move between membrane compartments?

2. Use the information provided in table 4.3 to develop a set of predictions about the properties of mitochondria and chloroplasts if these organelles were once free-living prokaryotic cells. How do your predictions match with the evidence for endosymbiosis?

3. In evolutionary theory, homologous traits are those with a similar structure and function derived from a common ancestor. Analogous traits represent adaptations to a similar environment, but from distantly related organisms. Consider the structure and function of the flagella found on eukaryotic and prokaryotic cells. Are the flagella an example of a homologous or analogous trait? Defend your answer.

4. The protist *Giardia intestinalis* is the organism associated with water-borne diarrheal diseases. *Giardia* is an unusual eukaryote because it seems to lack mitochondria. Provide two possible evolutionary scenarios for this in the context of the endosymbiotic theory.

102 Principles of Biology I and II

CHAPTER 5

Membranes

Chapter Contents

5.1 The Structure of Membranes
5.2 Phospholipids: The Membrane's Foundation
5.3 Proteins: Multifunctional Components
5.4 Passive Transport Across Membranes
5.5 Active Transport Across Membranes
5.6 Bulk Transport by Endocytosis and Exocytosis

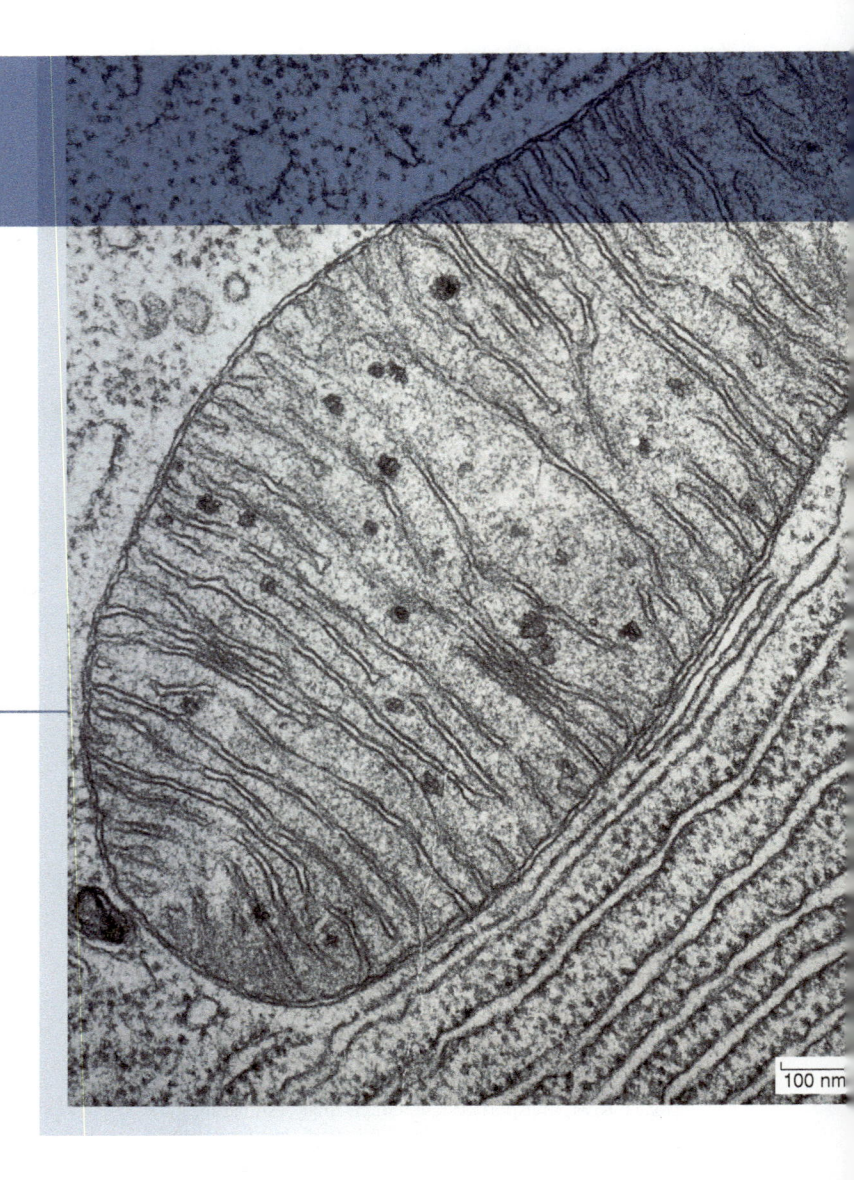

Introduction

A cell's interactions with the environment are critical, a give-and-take that never ceases. Without it, life could not exist. Living cells are encased within a lipid membrane through which few water-soluble substances can pass. The membrane also contains protein passageways that permit specific substances to move into and out of the cell and allow the cell to exchange information with its environment. Eukaryotic cells also contain internal membranes like those of the mitochondrion and endoplasmic reticulum pictured here. We call the delicate skin of lipids with embedded protein molecules that encase the cell a *plasma membrane*. This chapter examines the structure and function of this remarkable membrane.

5.1 The Structure of Membranes

Learning Outcomes

1. *Describe the components of biological membranes.*
2. *Explain the fluid mosaic model of membrane structure.*

The membranes that encase all living cells are two phospholipid sheets that are only 5 to 10 nm thick; more than 10,000 of these sheets piled on one another would just equal the thickness of this sheet of paper. Biologists established the components of membranes—not only lipids, but also proteins and other molecules—through biochemical assays, but the organization of the membrane components remained elusive.

We begin by considering the theories that have been advanced about membrane structure. We then look at the individual components of membranes more closely.

The fluid mosaic model shows proteins embedded in a fluid lipid bilayer

The lipid layer that forms the foundation of a cell's membranes is a bilayer formed of **phospholipids.** These phospholipids include primarily the glycerol phospholipids (figure 5.1), and the sphingolipids such as sphingomyelin (figure 5.2). Note that although these look superficially similar, they are built on a different carbon skeleton. For many years, biologists thought that the protein components of the cell membrane covered the inner and outer surfaces of the phospholipid bilayer like a coat of paint. An early model portrayed the membrane as a sandwich; a phospholipid bilayer between two layers of globular protein.

In 1972, S. Jonathan Singer and Garth J. Nicolson revised the model in a simple but profound way: They proposed that the globular proteins are *inserted* into the lipid bilayer, with their nonpolar segments in contact with the nonpolar interior of the bilayer and their polar portions protruding out from the membrane surface. In this model, called the *fluid mosaic model,* a mosaic of proteins floats in or on the fluid lipid bilayer like boats on a pond (figure 5.3).

We now recognize two categories of membrane proteins based on their association with the membrane. *Integral membrane proteins* are embedded in the membrane, and *peripheral proteins* are associated with the surface of the membrane.

Cellular membranes consist of four component groups

A eukaryotic cell contains many membranes. Although they are not all identical, they share the same fundamental architecture. Cell membranes are assembled from four components (table 5.1):

1. **Phospholipid bilayer.** Every cell membrane is composed of phospholipids in a bilayer. The other components of the membrane are embedded within the bilayer, which provides a flexible matrix and, at the same time, imposes a barrier to permeability. Animal cell membranes also contain cholesterol, a steroid with a polar hydroxyl group (–OH). Plant cells have other sterols, but little or no cholesterol.
2. **Transmembrane proteins.** A major component of every membrane is a collection of proteins that float in the lipid bilayer. These proteins have a variety of functions, including transport and communication across the membrane. Many integral membrane proteins are not fixed in position. They can move about, just as the phospholipid molecules do. Some membranes are crowded with proteins, but in others, the proteins are more sparsely distributed.
3. **Interior protein network.** Membranes are structurally supported by intracellular proteins that reinforce the membrane's shape. For example, a red blood cell has a characteristic biconcave shape because a scaffold made

Figure 5.1 Different views of phospholipid structure. Phospholipids are composed of glycerol *(pink)* linked to two fatty acids and a phosphate group. The phosphate group *(yellow)* can have additional molecules attached, such as the positively charged choline *(green)* shown. Phosphatidylcholine is a common component of membranes. It is shown in *(a)* with its chemical formula, *(b)* as a space-filling model, and *(c)* as the icon that is used in most of the figures in this chapter.

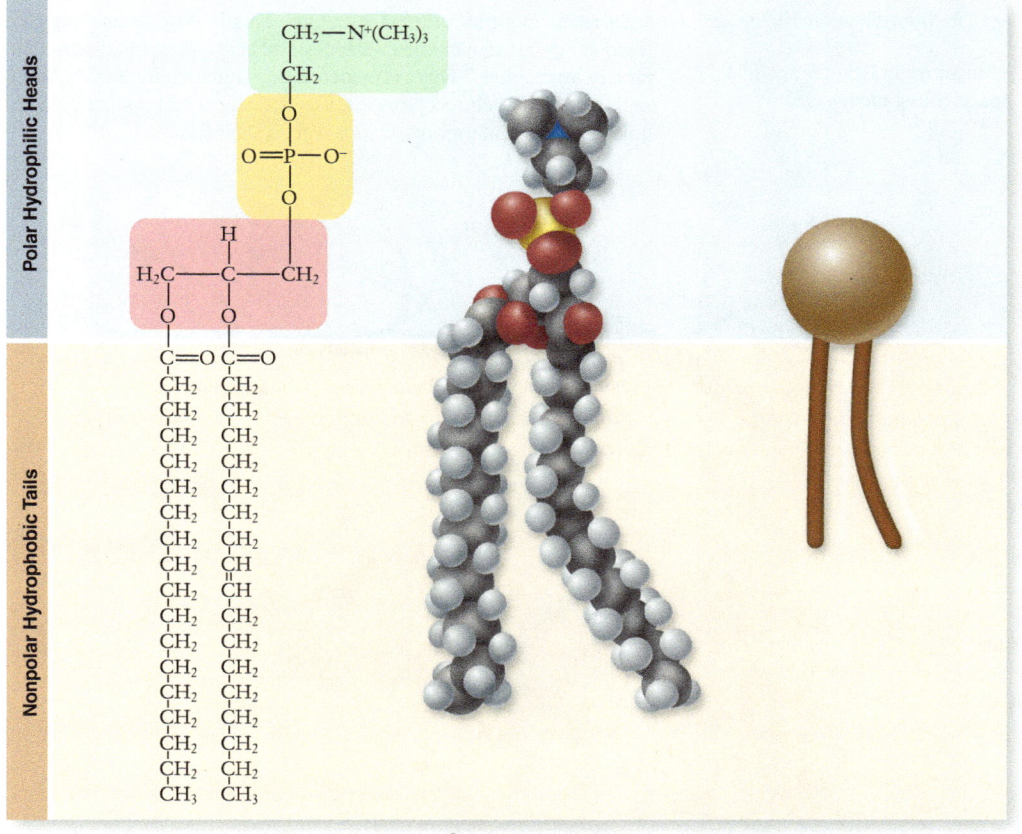

a. Formula *b.* Space-filling model *c.* Icon

of a protein called spectrin links proteins in the plasma membrane with actin filaments in the cell's cytoskeleton.

Membranes use networks of other proteins to control the lateral movements of some key membrane proteins, anchoring them to specific sites.

4. **Cell-surface markers.** As you learned in chapter 4, membrane sections assemble in the endoplasmic reticulum, transfer to the Golgi apparatus, and then are transported to the plasma membrane. The ER adds chains of sugar molecules to membrane proteins and lipids, converting them into **glycoproteins** and **glycolipids.** Different cell types exhibit different varieties of these glycoproteins and glycolipids on their surfaces, which act as cell identity markers.

Cellular membranes have an organized substructure

Originally, it was believed that because of its fluidity, the plasma membrane was uniform, with lipids and proteins free to diffuse rapidly in the plane of the membrane. However, in the last decade evidence has accumulated suggesting the plasma membrane is not homogeneous and contains microdomains with distinct lipid and protein composition. This was first observed in epithelial cells in which the lipid composition of the apical and basal membranes was shown to be distinctly different. Theoretical work also showed that lipids can exist in either a disordered or an ordered phase within a bilayer.

This led to the idea of lipid microdomains called *lipid rafts* that are heavily enriched in cholesterol and sphingolipids. These lipids appear to interact with each other, and with raft-associated proteins—together forming an ordered structure. This is now technically defined as "dynamic nanometer-sized, sterol and sphingolipid-enriched protein assemblies." There is evidence that signaling molecules, such as the B- and T-cell receptors discussed in chapter 51, associate with lipid rafts and that this association affects their function.

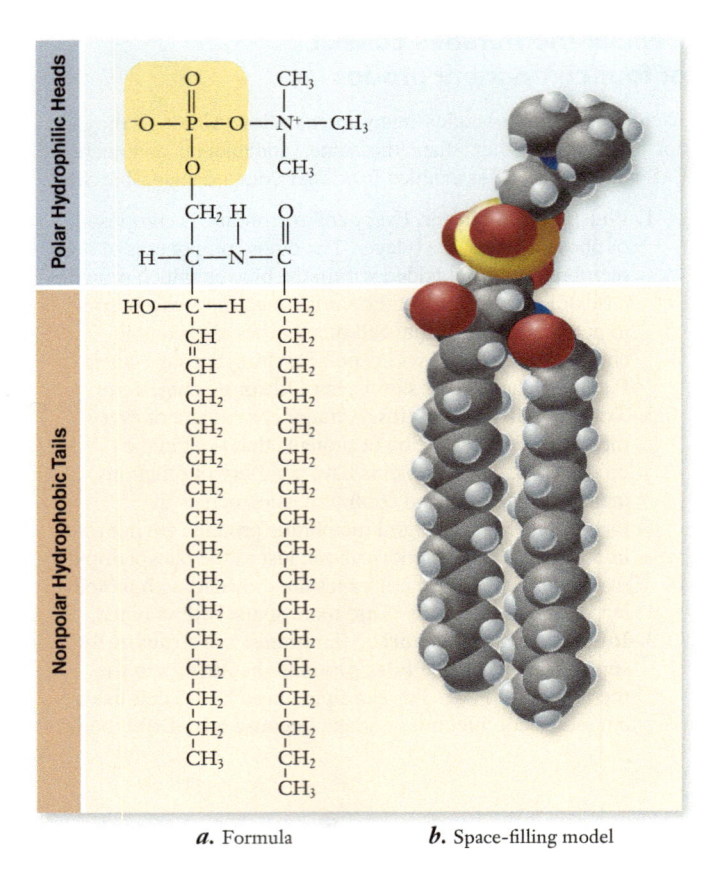

Figure 5.2 Sphingomyelin. Sphingomyelin is a sphingolipid found in animal cells. *a.* Formula. *b.* Space-filling model.

Figure 5.3 The fluid mosaic model of cell membranes. Integral proteins protrude through the plasma membrane, with nonpolar regions that tether them to the membrane's hydrophobic interior. Carbohydrate chains are often bound to the extracellular portion of these proteins, forming glycoproteins. Peripheral membrane proteins are associated with the surface of the membrane. Membrane phospholipids can be modified by the addition of carbohydrates to form glycolipids. Inside the cell, actin filaments and intermediate filaments interact with membrane proteins. Outside the cell, many animal cells have an elaborate extracellular matrix composed primarily of glycoproteins.

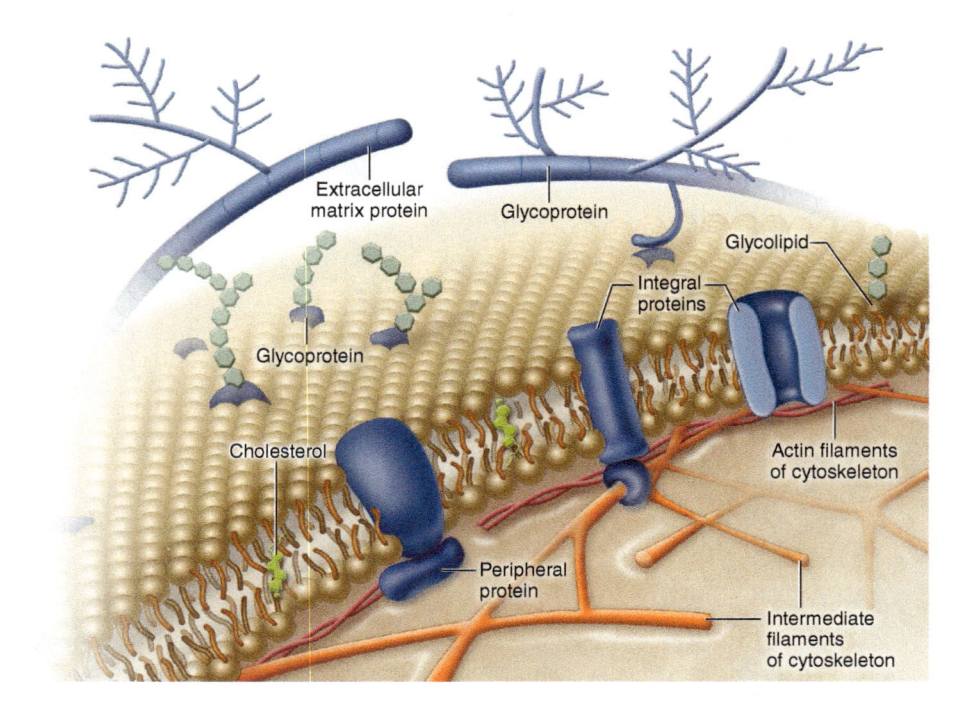

TABLE 5.1 Components of the Cell Membrane

Component	Composition	Function	How It Works	Example
Phospholipid bilayer	Phospholipid molecules	Provides permeability barrier, matrix for proteins	Excludes water-soluble molecules from nonpolar interior of bilayer and cell	Bilayer of cell is impermeable to large water-soluble molecules, such as glucose
Transmembrane proteins	Carriers	Actively or passively transport molecules across membrane	Move specific molecules through the membrane in a series of conformational changes	Glycophorin carrier for sugar transport; sodium–potassium pump
	Channels	Passively transport molecules across membrane	Create a selective tunnel that acts as a passage through membrane	Sodium and potassium channels in nerve, heart, and muscle cells
	Receptors	Transmit information into cell	Signal molecules bind to cell-surface portion of the receptor protein. This alters the portion of the receptor protein within the cell, inducing activity	Specific receptors bind peptide hormones and neurotransmitters
Interior protein network	Spectrins	Determine shape of cell	Form supporting scaffold beneath membrane, anchored to both membrane and cytoskeleton	Red blood cell
	Clathrins	Anchor certain proteins to specific sites, especially on the exterior plasma membrane in receptor-mediated endocytosis	Proteins line coated pits and facilitate binding to specific molecules	Localization of low-density lipoprotein receptor within coated pits
Cell-surface markers	Glycoproteins	"Self" recognition	Create a protein/carbohydrate chain shape characteristic of individual	Major histocompatibility complex protein recognized by immune system
	Glycolipid	Tissue recognition	Create a lipid/carbohydrate chain shape characteristic of tissue	A, B, O blood group markers

Electron microscopy has provided structural evidence

Electron microscopy allows biologists to examine the delicate, filmy structure of a cell membrane. We discussed two types of electron microscopes in chapter 4: the transmission electron microscope (TEM) and the scanning electron microscope (SEM). Both provide illuminating views of membrane structure.

When examining cell membranes with electron microscopy, specimens must be prepared for viewing. In one method of preparing a specimen, the tissue of choice is embedded in a hard epoxy matrix. The epoxy block is then cut with a microtome, a machine with a very sharp blade that makes incredibly thin, transparent "epoxy shavings" less than 1 μm thick that peel away from the block of tissue.

These shavings are placed on a grid, and a beam of electrons is directed through the grid with the TEM. At the high magnification an electron microscope provides, resolution is good enough to reveal the double layers of a membrane. False color can be added to the micrograph to enhance detail.

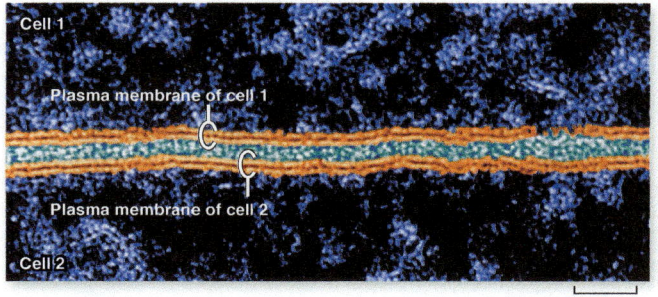

Freeze-fracturing a specimen is another way to visualize the inside of the membrane (figure 5.4). The tissue is embedded in a medium and quick frozen with liquid nitrogen. The frozen tissue is then "tapped" with a knife, causing a crack between the phospholipid layers of membranes. Proteins, carbohydrates, pits, pores, channels, or any other structure affiliated with the membrane will pull apart (whole, usually) and stick with one or the other side of the split membrane.

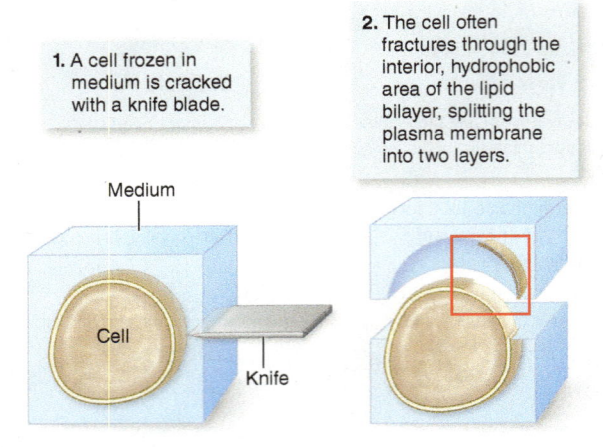

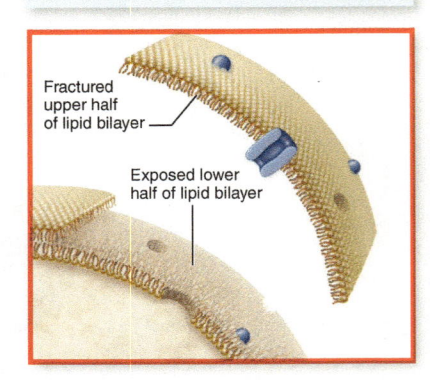

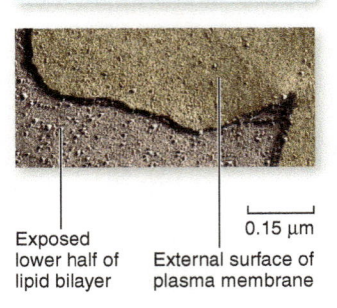

Figure 5.4 Viewing a plasma membrane with freeze-fracture microscopy.

1. A cell frozen in medium is cracked with a knife blade.
2. The cell often fractures through the interior, hydrophobic area of the lipid bilayer, splitting the plasma membrane into two layers.
3. The plasma membrane separates such that proteins and other embedded membrane structures remain within one or the other layers of the membrane.
4. The exposed membrane is coated with platinum, which forms a replica of the membrane. The underlying membrane is dissolved away, and the replica is then viewed with electron microscopy.

Next, a very thin coating of platinum is evaporated onto the fractured surface, forming a replica or "cast" of the surface. After the topography of the membrane has been preserved in the cast, the actual tissue is dissolved away, and the cast is examined with electron microscopy, creating a textured and three-dimensional view of the membrane.

Learning Outcomes Review 5.1

Cellular membranes contain four components: (1) a phospholipid bilayer, (2) transmembrane proteins, (3) an internal protein network providing structural support, and (4) cell-surface markers composed of glycoproteins and glycolipids. The fluid mosaic model of membrane structure includes both the fluid nature of the membrane and the mosaic composition of proteins floating in the phospholipid bilayer. Transmission electron microscopy (TEM) and scanning electron microscopy (SEM) have provided evidence supporting the fluid mosaic model.

- If the plasma membrane were just a phospholipid bilayer, how would this affect its function?

5.2 Phospholipids: The Membrane's Foundation

Learning Outcomes

1. List the different components of phospholipids.
2. Explain how membranes form spontaneously.
3. Describe the factors involved in membrane fluidity.

Lipidomics, the field defining the number and biological function of lipids, is revealing a significant diversity in membrane lipids. Although there are over 1000 distinct lipids identified in cells, we can organize them into only three classes: glycerol phospholipids (see figure 5.1), sphingolipids (see figure 5.2), and sterols such as cholesterol. The classical phospholipid bilayer consists of a combination of glycerol phospholipids and sphingolipids. The glycerol phospholipids are the most diverse with head groups that can have both positive and negative charge (zwitterionic), or primarily negative charge (anionic). Phospholipids can also vary in the length and composition of the fatty acid tail, with it being either saturated or *cis*-unsaturated (see section 3.5). Sphingolipids usually contain saturated hydrocarbon chains. These components are not uniformly distributed in biological membranes, and different cellular compartments have distinct membrane lipid composition, as discussed later in this section.

Phospholipids spontaneously form bilayers

Phospholipids spontaneously form bilayers because of their amphipathic structure. The polar head groups are hydrophilic, whereas the nonpolar hydrocarbon tails are hydrophobic. The two nonpolar fatty acids extend in one direction, roughly parallel to each other, and the polar phosphate group points in the other direction. To represent this structure, phospholipids are often diagrammed as a polar head with two dangling nonpolar tails, as in figure 5.1c.

What happens when a collection of phospholipid molecules is placed in water? The polar water molecules repel the long, nonpolar tails of the phospholipids while seeking partners for hydrogen bonding. Because of the polar nature of the water molecules, the nonpolar tails of the phospholipids end up packed closely together, sequestered as far as possible from water. Every phospholipid molecule is oriented with its polar head toward water and its nonpolar tails away. When two layers form with the tails facing each other, no tails ever come in contact with water. The resulting structure is the phospholipid bilayer. Phospholipid bilayers form spontaneously, driven by the tendency of water molecules to form the maximum number of hydrogen bonds.

The nonpolar interior of a lipid bilayer impedes the passage of any water-soluble substances through the bilayer, just as a layer of oil impedes the passage of a drop of water. This barrier to water-soluble substances is the key biological property of the lipid bilayer.

SCIENTIFIC THINKING

Hypothesis: *The plasma membrane is fluid, not rigid.*
Prediction: *If the membrane is fluid, membrane proteins may diffuse laterally.*
Test: *Fuse mouse and human cells, then observe the distribution of membrane proteins over time by labeling specific mouse and human proteins.*

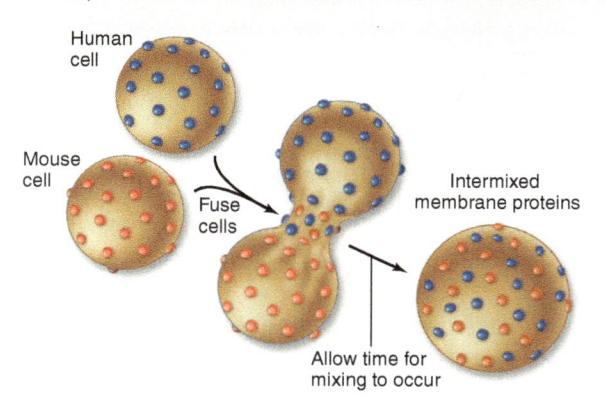

Result: *Over time, hybrid cells show increasingly intermixed proteins.*
Conclusion: *At least some membrane proteins can diffuse laterally in the membrane.*
Further Experiments: *Can you think of any other explanation for these observations? What if newly synthesized proteins were inserted into the membrane during the experiment? How could you use this basic experimental design to rule out this or other possible explanations?*

Figure 5.5 Test of membrane fluidity.

The phospholipid bilayer is fluid

A lipid bilayer is stable because water's affinity for hydrogen bonding never stops. Just as surface tension holds a soap bubble together, even though it is made of a liquid, so the hydrogen bonding of water holds a membrane together. Although water drives phospholipids into a bilayer configuration, it does not have any effect on the mobility of phospholipids and their nonlipid neighbors in the bilayer. Because phospholipids interact relatively weakly with one another, individual phospholipids and unanchored proteins are comparatively free to move about within the membrane. This can be demonstrated vividly by fusing cells and watching their proteins intermix with time (figure 5.5).

Membrane fluidity varies with lipid composition

The degree of membrane fluidity changes with the composition of the membrane itself. Much like triglycerides can be solid or liquid at room temperature, depending on their fatty acid composition, membrane fluidity can be altered by changing the membrane's lipid composition. Glycerol phospholipids that are saturated, or mono *cis*-unsaturated tend to make the membrane less fluid as they pack well. Similarly, the sphingolipids, which are usually unsaturated, also make the membrane less fluid.

Changes in the environment can have drastic effects on the membranes of single-celled organisms such as bacteria. Increasing temperature makes a membrane more fluid, and decreasing temperature makes it less fluid. Bacteria have evolved mechanisms to maintain a constant membrane fluidity despite fluctuating temperatures. Some bacteria contain enzymes called *fatty acid desaturases* that can introduce double bonds into fatty acids in membranes. Genetic studies, involving either the inactivation of these enzymes or the introduction of them into cells that normally lack them, indicate that the action of these enzymes confers cold tolerance. At colder temperatures, the double bonds introduced by fatty acid desaturase make the membrane more fluid, counteracting the environmental effect of reduced temperature.

Phospholipid composition affects membrane structure

Although most lipids are synthesized in the ER, the composition of the ER membrane, the Golgi stack, and the plasma membrane are quite distinct. Cells maintain this distinct lipid composition despite trafficking between these organelles as part of their function. These differences in lipid composition affect the structure and function of each membrane compartment (figure 5.6).

The plasma membrane has a high concentration of phosphatidylcholine and sphingomyelin, both of which have a cylindrical shape, and make a dense gel-like membrane. This membrane is given fluidity

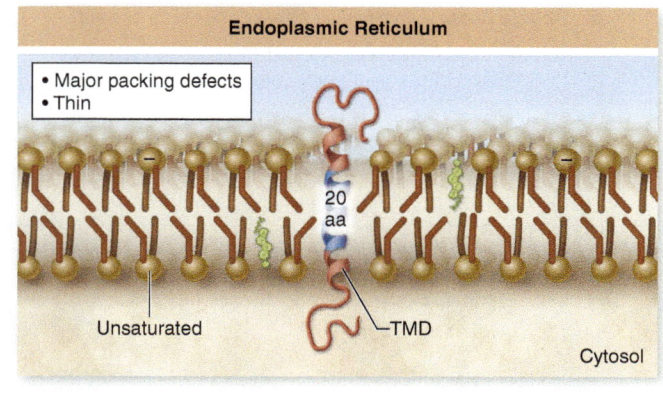

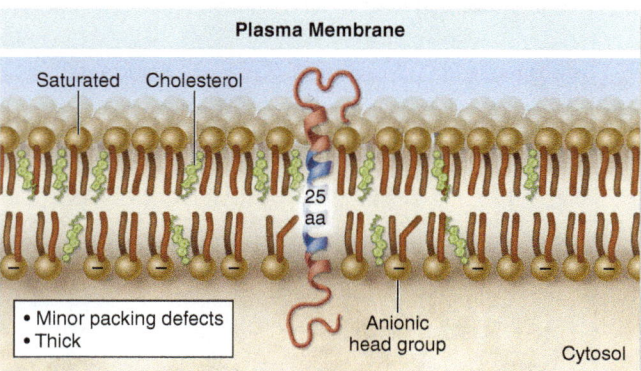

Figure 5.6 Comparing endoplasmic reticulum and plasma membrane. The different composition of membrane lipids in the ER and plasma membrane result in different membrane structures. The plasma membrane is thicker and less permeable, the ER membrane forms tubes and sheets and is more dynamic.

by the incorporation of cholesterol, which interacts with the nonpolar tails. There is also more cholesterol in the outer leaflet compared to the cytoplasmic leaflet. The unsaturated hydrocarbon chains also lead to a thicker membrane, as seen by the average length of a transmembrane domain for membrane proteins of around 25 amino acids. The overall structure of the plasma membrane is thus a fluid, but relatively rigid, membrane that forms an excellent barrier.

The ER membrane contains mainly unsaturated phospholipids that make the membrane more fluid, but also introduce curvature because they tend to form cones or inverted cones rather than cylinders. There is little to no cholesterol in the ER membrane. The ER membrane is also thinner than the plasma membrane, which is reflected in the length of around 20 amino acids for the transmembrane domains of ER proteins.

Learning Outcomes Review 5.2

Biological membranes contain glycerol phospholipids, sphingolipids, and cholesterol. In water, phospholipid molecules spontaneously form a bilayer, with phosphate groups facing out toward the water and lipid tails facing in, where they are sequestered from water. Membrane fluidity varies with composition and conditions: Unsaturated fats disturb packing of the lipid tails and make the membrane more fluid, as do higher temperatures. The phospholipid composition of different membrane compartments varies despite lipid movement between compartments.

■ *Would a phospholipid bilayer form in a nonpolar solvent?*

5.3 Proteins: Multifunctional Components

Learning Outcomes

1. Illustrate the functions of membrane proteins.
2. Illustrate how proteins can associate with the membrane.
3. Identify a transmembrane domain.

Cell membranes contain a complex assembly of proteins enmeshed in the fluid soup of phospholipid molecules. This very flexible organization permits a broad range of interactions with the environment, some directly involving membrane proteins.

Proteins and protein complexes perform key functions

Although cells interact with their environment through their plasma membranes in many ways, we will focus on six key classes of membrane protein in this chapter and in chapter 9 (figure 5.7):

1. **Transporters.** Membranes are very selective, allowing only certain solutes to enter or leave the cell, either through channels or carriers composed of proteins.

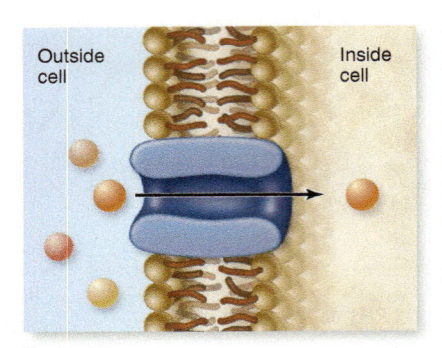

Transporter

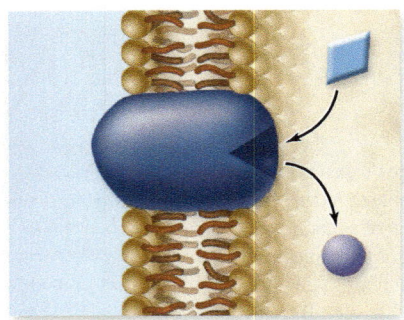

Enzyme

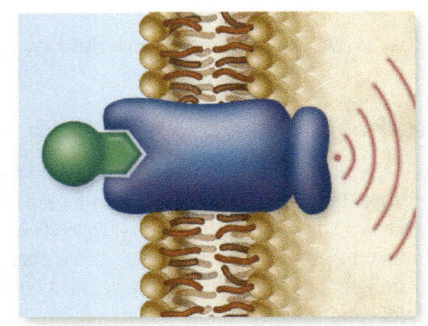

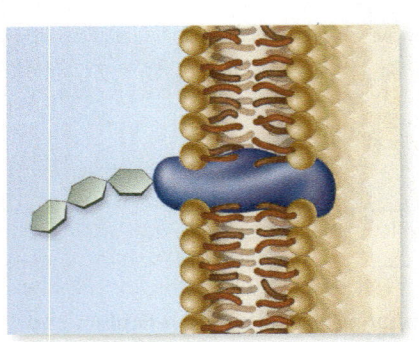

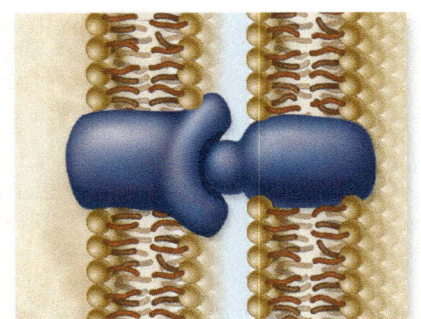

Cell-surface receptor

Cell-surface identity marker

Cell-to-cell adhesion

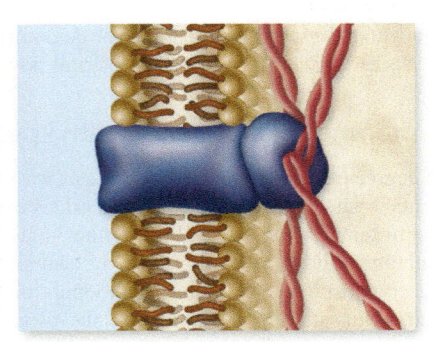

Attachment to the cytoskeleton

Figure 5.7 Functions of plasma membrane proteins. Membrane proteins act as transporters, enzymes, cell-surface receptors, and cell-surface identity markers, as well as aiding in cell-to-cell adhesion and securing the cytoskeleton.

2. **Enzymes.** Cells carry out many chemical reactions on the interior surface of the plasma membrane, using enzymes attached to the membrane.
3. **Cell-surface receptors.** Membranes are exquisitely sensitive to chemical messages, which are detected by receptor proteins on their surfaces.
4. **Cell-surface identity markers.** Membranes carry cell-surface markers that identify them to other cells. Most cell types carry their own ID tags, specific combinations of cell-surface proteins and protein complexes such as glycoproteins that are characteristic of that cell type.
5. **Cell-to-cell adhesion proteins.** Cells use specific proteins to glue themselves to one another. Some act by forming temporary interactions, and others form a more permanent bond. (See chapter 4.)
6. **Attachments to the cytoskeleton.** Surface proteins that interact with other cells are often anchored to the cytoskeleton by linking proteins.

Structural features of membrane proteins relate to function

As we've just detailed, membrane proteins can serve a variety of functions. These diverse functions arise from the diverse structures of these proteins, yet they also have common structural features related to their role as membrane proteins.

The anchoring of proteins in the bilayer

Some membrane proteins are attached to the surface of the membrane by special molecules that associate strongly with phospholipids. Like a ship tied to a floating dock, these anchored proteins are free to move about on the surface of the membrane tethered to a phospholipid. The anchoring molecules are modified lipids that have (1) nonpolar regions that insert into the internal portion of the lipid bilayer and (2) chemical bonding domains that link directly to proteins.

Inquiry question According to the fluid mosaic model, membranes are held together by hydrophobic interactions. Considering the forces that some cells may experience, why do membranes not break apart every time an animal moves?

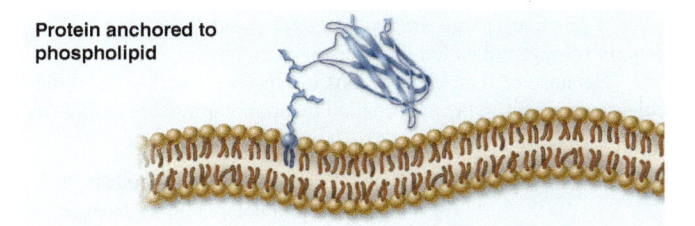

Protein anchored to phospholipid

In contrast, other proteins actually span the lipid bilayer (transmembrane proteins). The part of the protein that extends through the lipid bilayer and that is in contact with the nonpolar interior are α helices or β-pleated sheets (see chapter 3) that consist of nonpolar amino acids. Because water avoids nonpolar amino acids, these portions of the protein are held within the interior of the lipid bilayer. The polar ends protrude from both sides of the membrane. Any movement of the protein out of the membrane, in either direction, brings the nonpolar

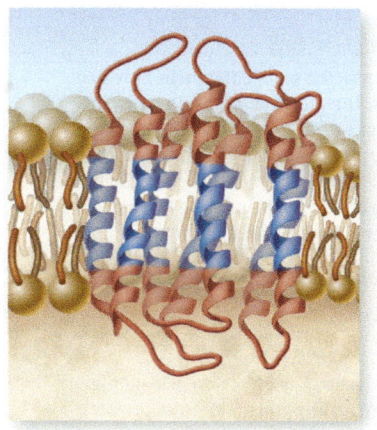

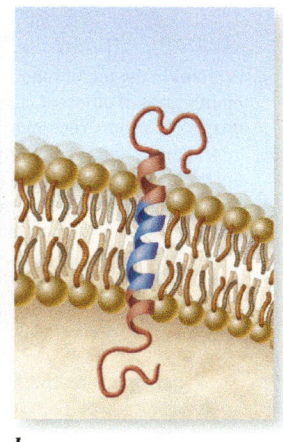

a. *b.*

Figure 5.8 Transmembrane domains. Integral membrane proteins have at least one hydrophobic transmembrane domain (shown in *blue*) to anchor them in the membrane. *a.* Receptor protein with seven transmembrane domains. *b.* Protein with single transmembrane domain.

regions of the protein into contact with water, which "shoves" the protein back into the interior. These forces prevent the transmembrane proteins from simply popping out of the membrane and floating away.

Transmembrane domains

Cell membranes contain a variety of different transmembrane proteins, which differ in the way they traverse the lipid bilayer. The primary difference lies in the number of times that the protein crosses the membrane. Each membrane-spanning region is called a **transmembrane domain.** These domains are composed of hydrophobic amino acids usually arranged into α helices (figure 5.8).

Proteins need only a single transmembrane domain to be anchored in the membrane, but they often have more than one such domain. An example of a protein with a single transmembrane domain is the linking protein that attaches the spectrin network of the cytoskeleton to the interior of the plasma membrane.

Biologists classify some types of receptors based on the number of transmembrane domains they have, such as G protein–coupled receptors with seven membrane-spanning domains (chapter 9). These receptors respond to external molecules, such as epinephrine, and initiate a cascade of events inside the cell.

Another example is bacteriorhodopsin, one of the key transmembrane proteins that carries out photosynthesis in halophilic (salt-loving) archaea. It contains seven nonpolar helical segments that traverse the membrane, forming a structure within the membrane through which protons pass during the light-driven pumping of protons.

Pores

Some transmembrane proteins have extensive nonpolar regions with secondary configurations of β-pleated sheets instead of α helices (see chapter 3). The β sheets form a characteristic motif, folding back and forth in a cylinder so the sheets arrange themselves like a pipe through the membrane. This forms a polar environment in the interior of the β sheets spanning the membrane. This so-called β *barrel*, open on both ends, is a common feature of the porin class of proteins that are found within the outer membrane of some bacteria. The openings allow molecules to pass through the membrane.

Learning Outcomes Review 5.3

Proteins in the membrane confer the main differences between membranes of different cells. Their functions include transport, enzymatic action, reception of extracellular signals, cell-to-cell interactions, and cell identity markers. Peripheral proteins can be anchored in the membrane by modified lipids. Integral membrane proteins span the membrane and have one or more hydrophobic regions, called transmembrane domains, that anchor them.

- *Why are transmembrane domains hydrophobic?*

? Inquiry question Based only on amino acid sequence, how would you recognize an integral membrane protein?

5.4 Passive Transport Across Membranes

Learning Outcomes
1. Compare simple diffusion and facilitated diffusion.
2. Differentiate between channel proteins and carrier proteins.
3. Predict the direction of water movement by osmosis.

Many substances can move in and out of the cell without the cell's having to expend energy. This type of movement is termed **passive transport.** Some ions and molecules can pass through the membrane fairly easily and do so because of a *concentration gradient*—a difference between the concentration on the inside of the membrane and that on the outside. Some substances also move in response to a gradient, but do so through specific channels formed by proteins in the membrane.

Transport can occur by simple diffusion

Molecules and ions dissolved in water are in constant random motion. This random motion causes a net movement of these substances from regions of high concentration to regions of lower concentration, a process called **diffusion** (figure 5.9).

Net movement driven by diffusion will continue until the concentration is the same in all regions. Consider what happens when you add a drop of colored ink to a bowl of water. Over time the ink becomes dispersed throughout the solution. This is due to diffusion of the ink molecules. In the context of cells, we are usually concerned with differences in concentration of molecules across the plasma membrane. We need to consider the relative concentrations both inside and outside the cell, as well as how readily a molecule can cross the membrane.

The major barrier to crossing a biological membrane is the hydrophobic interior that repels polar molecules but not nonpolar molecules. If a concentration difference exists for a nonpolar molecule, it will move across the membrane until the concentration is equal on both sides. At this point, movement in both directions still occurs, but there is no net change in either direction. This includes molecules like O_2 and nonpolar organic molecules such as steroid hormones.

The plasma membrane has limited permeability to small polar molecules and very limited permeability to larger polar molecules and ions. The movement of water, one of the most important polar molecules, is discussed separately.

Proteins allow membrane diffusion to be selective

Many important molecules required by cells cannot easily cross the plasma membrane. These molecules can still enter the cell by diffusion through specific channel proteins or carrier proteins embedded in the plasma membrane, provided there is a higher concentration of the molecule outside the cell than inside. We call this process of diffusion mediated by a membrane protein **facilitated diffusion. Channel proteins** have a hydrophilic interior that provides an aqueous channel through which polar molecules can pass when the channel is open. **Carrier proteins,** in contrast to channels, bind specifically to the molecule they assist, much like an enzyme binds to its substrate. These channels and carriers are usually selective for one type of molecule, and thus the cell membrane is said to be **selectively permeable.**

Facilitated diffusion of ions through channels

You saw in chapter 2 that atoms with an unequal number of protons and electrons have an electric charge and are called ions. Those that carry a positive charge are called *cations* and those that carry a negative charge are called *anions*.

Because of their charge, ions interact well with polar molecules such as water, but are repelled by nonpolar molecules such as

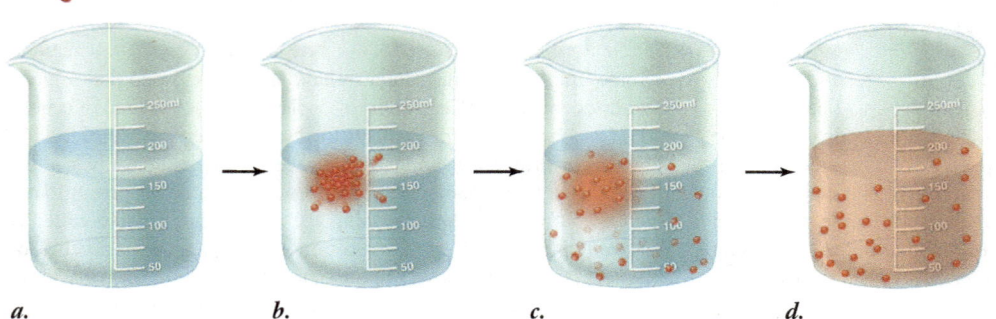

Figure 5.9 Diffusion. If a drop of colored ink is dropped into a beaker of water (*a*) its molecules dissolve (*b*) and diffuse (*c*). Eventually, diffusion results in an even distribution of ink molecules throughout the water (*d*).

a.   *b.*   *c.*   *d.*

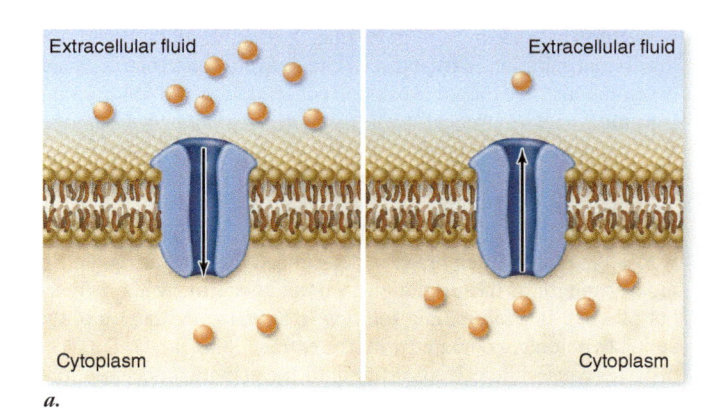

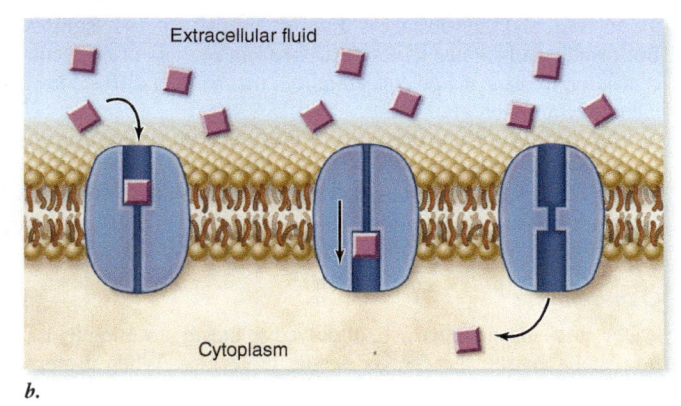

Figure 5.10 Facilitated diffusion. Diffusion can be facilitated by membrane proteins. *a.* The movement of ions through a channel is shown. On the left the concentration is higher outside the cell, so the ions move into the cell. On the right the situation is reversed. In both cases, transport continues until the concentration is equal on both sides of the membrane. At this point, ions continue to cross the membrane in both directions, but there is no net movement in either direction. *b.* Carrier proteins bind specifically to the molecules they transport. In this case, the concentration is higher outside the cell, so molecules bind to the carrier on the outside. The carrier's shape changes, allowing the molecule to cross the membrane. This is reversible, so net movement continues until the concentration is equal on both sides of the membrane.

the interior of the plasma membrane. Therefore, ions cannot move between the cytoplasm of a cell and the extracellular fluid without the assistance of membrane transport proteins.

Ion channels possess a hydrated interior that spans the membrane. Ions can diffuse through the channel in either direction, depending on their relative concentration across the membrane (figure 5.10). Some channel proteins can be opened or closed in response to a stimulus. These channels are called *gated channels,* and depending on the nature of the channel, the stimulus can be either chemical or electrical.

Three conditions determine the direction of net movement of the ions: (1) their relative concentrations on either side of the membrane, (2) the voltage difference across the membrane and for the gated channels, and (3) the state of the gate (open or closed). A voltage difference is an electrical potential difference across the membrane called a *membrane potential.* Changes in membrane potential form the basis for transmission of signals in the nervous system and some other tissues. (We discuss this topic in detail in chapter 43.) Each type of channel is specific for a particular ion, such as calcium (Ca^{2+}), sodium (Na^+), potassium (K^+), or chloride (Cl^-), or in some cases, for more than one cation or anion. Ion channels play an essential role in signaling by the nervous system.

Facilitated diffusion by carrier proteins

Carrier proteins can help transport both ions and other solutes, such as some sugars and amino acids, across the membrane. Transport through a carrier is still a form of diffusion and therefore requires a concentration difference across the membrane.

Carriers must bind to the molecule they transport, so the relationship between concentration and rate of transport differs from that due to simple diffusion. As concentration increases, transport by simple diffusion shows a linear increase in rate of transport. But when a carrier protein is involved, a concentration increase means that more of the carriers are bound to the transported molecule. At high enough concentrations all carriers will be occupied, and the rate of transport will be constant. This means that the carrier exhibits *saturation.*

This situation is somewhat like that of a stadium (the cell) where a crowd must pass through turnstiles to enter. If there are unoccupied turnstiles, you can go right through, but when all are occupied, you must wait. When ticket holders are passing through the gates at maximum speed, the rate at which they enter cannot increase, no matter how many are waiting outside.

Facilitated diffusion in red blood cells

Several examples of facilitated diffusion can be found in the plasma membrane of vertebrate red blood cells (RBCs). One RBC carrier protein, for example, transports a different molecule in each direction: chloride ion (Cl^-) in one direction and bicarbonate ion (HCO_3^-) in the opposite direction. As you will learn in chapter 48, this carrier is important in the uptake and release of carbon dioxide.

The glucose transporter is a second vital facilitated diffusion carrier in RBCs. Red blood cells keep their internal concentration of glucose low through a chemical trick: They immediately add a phosphate group to any entering glucose molecule, converting it to a highly charged glucose phosphate that can no longer bind to the glucose transporter, and therefore cannot pass back across the membrane. This maintains a steep concentration gradient for unphosphorylated glucose, favoring its entry into the cell.

The glucose transporter that assists the entry of glucose into the cell does not appear to form a channel in the membrane. Instead, this transmembrane protein appears to bind to a glucose molecule and then to flip its shape, dragging the glucose through the bilayer and releasing it on the inside of the plasma membrane. After it releases the glucose, the transporter reverts to its original shape and is then available to bind the next glucose molecule that comes along outside the cell.

Osmosis is the movement of water across membranes

The cytoplasm of a cell contains ions and molecules, such as sugars and amino acids, dissolved in water. The mixture of these substances and water is called an *aqueous solution.* Water is termed the **solvent,** and the substances dissolved in the water are **solutes.** Both water and solutes tend to diffuse from regions of high concentration to ones of low concentration; that is, they diffuse down their concentration gradients.

When two regions are separated by a membrane, what happens depends on whether the solutes can pass freely through that membrane. Most solutes, including ions and sugars, are not lipid-soluble and, therefore, are unable to cross the lipid bilayer. The concentration gradient of these solutes can lead to the movement of water.

Osmosis

Water molecules interact with dissolved solutes by forming hydration shells around the charged solute molecules. When a membrane separates two solutions with different concentrations of solutes, the concentrations of *free* water molecules on the two sides of the membrane also differ. The side with higher solute concentration has tied up more water molecules in hydration shells and thus has fewer free water molecules.

As a consequence of this difference, free water molecules move down their concentration gradient, toward the higher solute concentration. This net diffusion of water across a membrane toward a higher solute concentration is called **osmosis** (figure 5.11).

The concentration of *all* solutes in a solution determines the **osmotic concentration** of the solution. If two solutions have unequal osmotic concentrations, the solution with the higher concentration is **hypertonic** (Greek *hyper,* "more than"), and the solution with the lower concentration is **hypotonic** (Greek *hypo,* "less than"). When two solutions have the same osmotic concentration, the solutions are **isotonic** (Greek *iso,* "equal"). The terms *hyperosmotic, hypoosmotic,* and *isosmotic* are also used to describe these conditions.

A cell in any environment can be thought of as a plasma membrane separating two solutions: the cytoplasm and the extracellular fluid. The direction and extent of any diffusion of water across the plasma membrane is determined by comparing the osmotic strength of these solutions. Put another way, water diffuses out of a cell in a hypertonic solution (that is, the cytoplasm of the cell is hypotonic, compared with the extracellular fluid). This loss of water causes the cell to shrink until the osmotic concentrations of the cytoplasm and the extracellular fluid become equal.

Aquaporins: Water channels

The transport of water across the membrane is complex. Studies on artificial membranes show that water, despite its polarity, can cross the membrane, but this flow is limited. Water flow in living cells is facilitated by **aquaporins,** which are specialized channels for water.

A simple experiment demonstrates this. If an amphibian egg is placed in hypotonic spring water (the solute concentration in the cell is higher than that of the surrounding water), it does not swell. If aquaporin mRNA is then injected into the egg, the channel proteins are expressed and appear in the egg's plasma membrane. Water can now diffuse into the egg, causing it to swell.

More than 11 different kinds of aquaporins have been found in mammals. These fall into two general classes: those that are specific for only water, and those that allow other small hydrophilic molecules, such as glycerol or urea, to cross the membrane as well. This latter class explains how some membranes allow the easy passage of small hydrophilic substances.

The human genetic disease, hereditary (nephrogenic) diabetes insipidus (NDI), has been shown to be caused by a nonfunctional aquaporin protein. This disease causes the excretion of large volumes of dilute urine, illustrating the importance of aquaporins to our physiology.

Osmotic pressure

What happens to a cell in a hypotonic solution? (That is, the cell's cytoplasm is hypertonic relative to the extracellular fluid.) In this situation, water diffuses into the cell from the extracellular fluid, causing the cell to swell. The pressure of the cytoplasm pushing out against the cell membrane, or hydrostatic pressure, increases. The amount of water that enters the cell depends on the difference in solute concentration between the cell and the extracellular fluid. This is measured as **osmotic pressure,** defined as the force needed to stop osmotic flow.

If the membrane is strong enough, the cell reaches an equilibrium, at which the osmotic pressure, which tends to drive water into the cell, is exactly counterbalanced by the hydrostatic pressure, which tends to drive water back out of the cell. However, a plasma membrane by itself cannot withstand large internal pressures, and an isolated cell under such conditions would burst like an overinflated balloon (figure 5.12).

Accordingly, it is important for animal cells, which only have plasma membranes, to maintain osmotic balance. In contrast, the cells of prokaryotes, fungi, plants, and many protists are surrounded by strong cell walls, which can withstand high internal pressures without bursting.

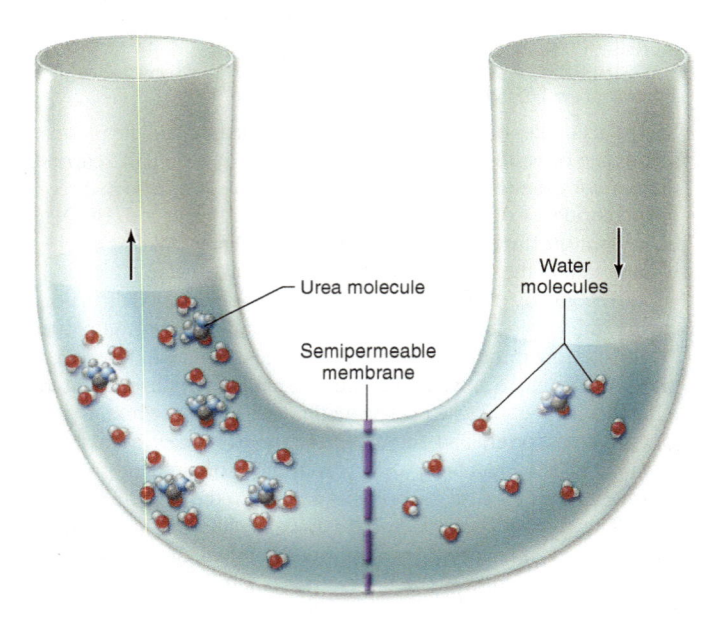

Figure 5.11 Osmosis. Concentration differences in charged or polar molecules that cannot cross a semipermeable membrane result in movement of water, which can cross the membrane. Water molecules form hydrogen bonds with charged or polar molecules creating a hydration shell around them in solution. A higher concentration of polar molecules (urea) shown on the left side of the membrane leads to water molecules gathering around each urea molecule. These water molecules are no longer free to diffuse across the membrane. The polar solute has reduced the concentration of free water molecules, creating a gradient. This causes a net movement of water by diffusion from right to left in the U-tube, raising the level on the left and lowering the level on the right.

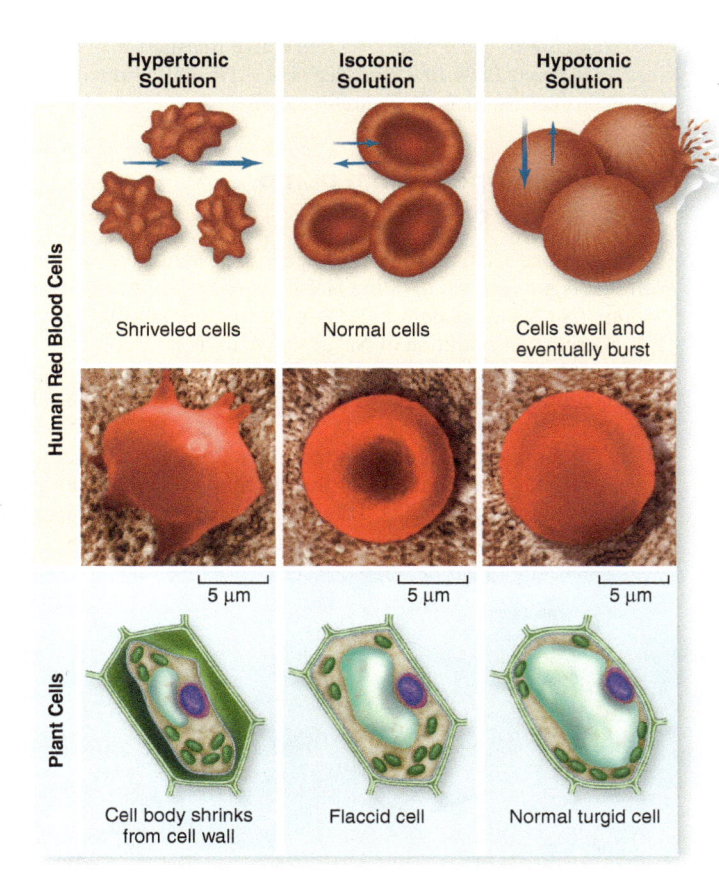

Figure 5.12 How solutes create osmotic pressure. In a hypertonic solution, water moves out of the cell, causing the cell to shrivel. In an isotonic solution, water diffuses into and out of the cell at the same rate, with no change in cell size. In a hypotonic solution, water moves into the cell. Direction and amount of water movement is shown with blue arrows *(top)*. As water enters the cell from a hypotonic solution, pressure is applied to the plasma membrane until the cell ruptures. Water enters the cell due to osmotic pressure from the higher solute concentration in the cell. Osmotic pressure is measured as the force needed to stop osmosis. The strong cell wall of plant cells can withstand the hydrostatic pressure to keep the cell from rupturing. This is not the case with animal cells.

Maintaining osmotic balance

Organisms have developed many strategies for solving the dilemma posed by being hypertonic to their environment and therefore having a steady influx of water by osmosis:

Extrusion. Some single-celled eukaryotes, such as the protist *Paramecium,* use organelles called contractile vacuoles to remove water. Each vacuole collects water from various parts of the cytoplasm and transports it to the central part of the vacuole, near the cell surface. The vacuole possesses a small pore that opens to the outside of the cell. By contracting rhythmically, the vacuole pumps out (extrudes) through this pore the water that is continuously drawn into the cell by osmotic forces.

Isosmotic Regulation. Some organisms that live in the ocean adjust their internal concentration of solutes to match that of the surrounding seawater. Because they are isosmotic with respect to their environment, no net flow of water occurs into or out of these cells.

Many terrestrial animals solve the problem in a similar way, by circulating a fluid through their bodies that bathes cells in an isotonic solution. The blood in your body, for example, contains a high concentration of the protein albumin, which elevates the solute concentration of the blood to match that of your cells' cytoplasm.

Turgor. Most plant cells are hypertonic to their immediate environment, containing a high concentration of solutes in their central vacuoles. The resulting internal hydrostatic pressure, known as **turgor pressure,** presses the plasma membrane firmly against the interior of the cell wall, making the cell rigid. Most green plants depend on turgor pressure to maintain their shape, and thus they wilt when they lack sufficient water.

Learning Outcomes Review 5.4

Passive transport involves diffusion, which requires a concentration gradient. Hydrophobic molecules can diffuse directly through the membrane (simple diffusion). Polar molecules and ions can also diffuse through the membrane, but only with the aid of a channel or carrier protein (facilitated diffusion). Channel proteins assist by forming a hydrophilic passageway through the membrane, whereas carrier proteins bind to the molecule they assist. Water passes through the membrane and through aquaporins in response to solute concentration differences inside and outside the cell. This process is called osmosis.

■ *If you require intravenous (IV) medication in the hospital, what should the concentration of solutes in the IV solution be relative to your blood cells?*

5.5 Active Transport Across Membranes

Learning Outcomes

1. Differentiate between active transport and diffusion.
2. Describe the function of the Na^+/K^+ pump.
3. Explain the energetics of coupled transport.

Diffusion, facilitated diffusion, and osmosis are passive transport processes that move materials down their concentration gradients, but cells can also actively move substances across a cell membrane *up* their concentration gradients. This process requires the expenditure of energy, typically from ATP, and is therefore called **active transport.**

Active transport uses energy to move materials against a concentration gradient

Like facilitated diffusion, active transport involves highly selective protein carriers within the membrane that bind to the transported substance, which could be an ion or a simple molecule, such as a sugar, an amino acid, or a nucleotide. These carrier proteins are called **uniporters** if they transport a single type of molecule and symporters or antiporters if they transport two different molecules together. **Symporters** transport two molecules in the same direction, and **antiporters** transport two molecules in opposite directions. These terms can also be used to describe facilitated diffusion carriers.

Active transport is one of the most important functions of any cell. It enables a cell to take up additional molecules of a substance that is already present in its cytoplasm in concentrations higher than in the extracellular fluid. Active transport also enables a cell to move substances out of its cytoplasm and into the extracellular fluid, despite higher external concentrations.

The use of energy from ATP in active transport may be direct or indirect. Let's first consider how ATP is used directly to move ions against their concentration gradients.

The sodium–potassium pump runs directly on ATP

More than one-third of all of the energy expended by an animal cell that is not actively dividing is used in the active transport of sodium (Na^+) and potassium (K^+) ions. Most animal cells have a low internal concentration of Na^+, relative to their surroundings, and a high internal concentration of K^+. They maintain these concentration differences by actively pumping Na^+ out of the cell and K^+ in.

The remarkable protein that transports these two ions across the cell membrane is known as the **sodium–potassium pump** (Na^+/K^+ pump) (figure 5.13). This carrier protein uses the energy

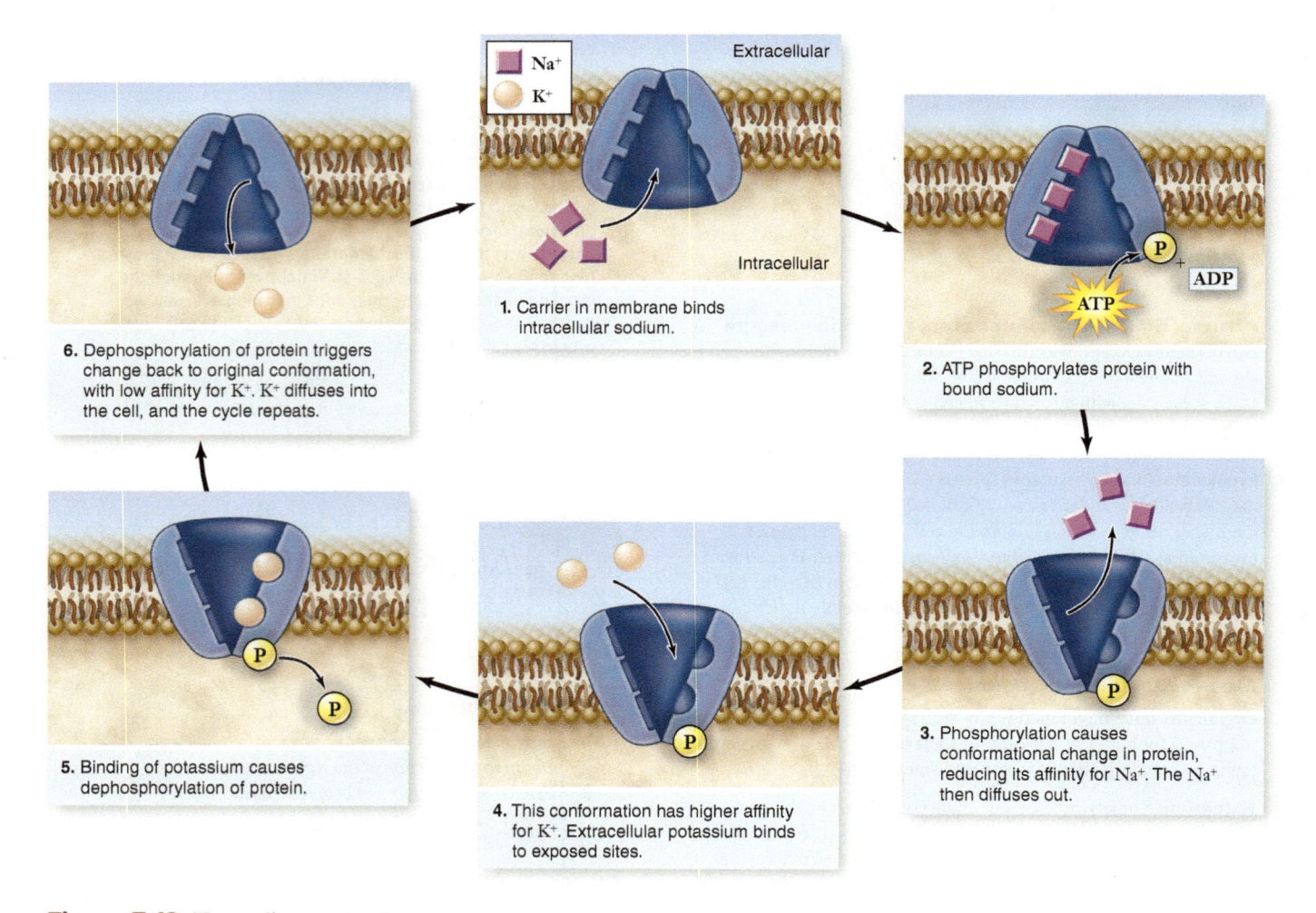

1. Carrier in membrane binds intracellular sodium.

2. ATP phosphorylates protein with bound sodium.

3. Phosphorylation causes conformational change in protein, reducing its affinity for Na^+. The Na^+ then diffuses out.

4. This conformation has higher affinity for K^+. Extracellular potassium binds to exposed sites.

5. Binding of potassium causes dephosphorylation of protein.

6. Dephosphorylation of protein triggers change back to original conformation, with low affinity for K^+. K^+ diffuses into the cell, and the cycle repeats.

Figure 5.13 The sodium–potassium pump. The protein carrier known as the sodium–potassium pump transports sodium (Na^+) and potassium (K^+) across the plasma membrane. For every three Na^+ transported out of the cell, two K^+ are transported into it. The sodium–potassium pump is fueled by ATP hydrolysis. The affinity of the pump for Na^+ and K^+ is changed by adding or removing phosphate (P), which changes the conformation of the protein.

stored in ATP to move these two ions. In this case, the energy is used to change the conformation of the carrier protein, which changes its affinity for either Na+ ions or K+ ions. This is an excellent illustration of how subtle changes in the structure of a protein affect its function.

The important characteristic of the Na+/K+ pump is that it is an active transport mechanism, transporting Na+ and K+ from areas of low concentration to areas of high concentration. This transport is the opposite of passive transport by diffusion; it is achieved only by the constant expenditure of metabolic energy. The Na+/K+ pump works through the following series of conformational changes in the transmembrane protein (summarized in figure 5.13):

Step 1. Three Na+ bind to the cytoplasmic side of the protein, causing the protein to change its conformation.

Step 2. In its new conformation, the protein binds a molecule of ATP and cleaves it into adenosine diphosphate (ADP) and phosphate (or P_i for inorganic phosphate). ADP is released, but the phosphate group is covalently linked to the protein. The protein is now phosphorylated.

Step 3. The phosphorylation of the protein induces a second conformational change in the protein. This change translocates the three Na+ across the membrane, so they now face the exterior. In this new conformation, the protein has a low affinity for Na+, and the three bound Na+ break away from the protein and diffuse into the extracellular fluid.

Step 4. The new conformation has a high affinity for K+, two of which bind to the extracellular side of the protein as soon as it is free of the Na+.

Step 5. The binding of the K+ causes another conformational change in the protein, this time resulting in the hydrolysis of the bound phosphate group.

Step 6. Freed of the phosphate group, the protein reverts to its original shape, exposing the two K+ to the cytoplasm. This conformation has a low affinity for K+, so the two bound K+ dissociate from the protein and diffuse into the interior of the cell. The original conformation has a high affinity for Na+. When these ions bind, they initiate another cycle.

In every cycle, three Na+ leave the cell and two K+ enter. The changes in protein conformation that occur during the cycle are rapid, enabling each carrier to transport as many as 300 Na+ per second. The Na+/K+ pump appears to exist in all animal cells, although cells vary widely in the number of pump proteins they contain.

Coupled transport uses ATP indirectly

Some molecules are moved against their concentration gradient by using the energy stored in a gradient of a different molecule. In this process, called *coupled transport,* the energy released as one molecule moves down its concentration gradient is captured and used to move a different molecule against its gradient. As you just saw, the energy stored in ATP molecules can be used to create a gradient of Na+ and K+ across the membrane. These gradients can then be used to power the transport of other molecules across the membrane.

As one example, let's consider the active transport of glucose across the membrane in animal cells. Glucose is such an important molecule that there are a variety of transporters for it, one of which was discussed earlier under passive transport. In a multicellular organism, intestinal epithelial cells can have a higher concentration of glucose inside the cell than outside, so these cells need to be able to transport glucose against its concentration gradient. This requires energy and a different transporter than the one involved in facilitated diffusion of glucose.

The active glucose transporter uses the Na+ gradient produced by the Na+/K+ pump as a source of energy to power the movement of glucose into the cell. In this system, both glucose and Na+ bind to the transport protein, which allows Na+ to pass into the cell down its concentration gradient, capturing the energy and using it to move glucose into the cell. In this kind of cotransport, both molecules are moving in the same direction across the membrane; therefore the transporter is a symporter (figure 5.14).

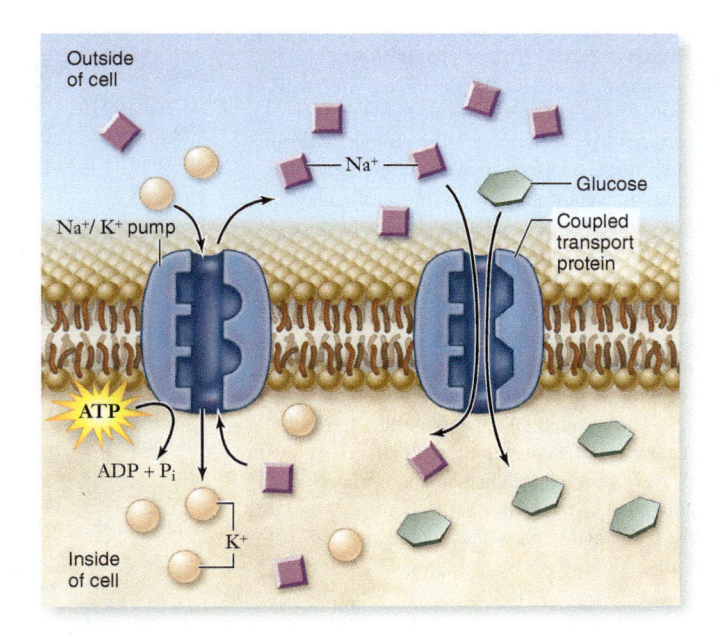

Figure 5.14 Coupled transport. A membrane protein transports Na+ into the cell, down its concentration gradient, at the same time it transports a glucose molecule into the cell. The gradient driving the Na+ entry allows sugar molecules to be transported against their concentration gradient. The Na+ gradient is maintained by the Na+/K+ pump. ADP = adenosine diphosphate; ATP = adenosine triphosphate; P_i = inorganic phosphate

In a related process, called *countertransport*, the inward movement of Na⁺ is coupled with the outward movement of another substance, such as Ca²⁺ or H⁺. As in cotransport, both Na⁺ and the other substance bind to the same transport protein, which in this case is an antiporter, as the substances bind on opposite sides of the membrane and are moved in opposite directions. In countertransport, the cell uses the energy released as Na⁺ moves down its concentration gradient into the cell to eject a substance against its concentration gradient. In both cotransport and countertransport, the potential energy in the concentration gradient of one molecule is used to transport another molecule against its concentration gradient. They differ only in the direction that the second molecule moves relative to the first.

Learning Outcomes Review 5.5

Active transport requires both a carrier protein and energy, usually in the form of ATP, to move molecules against a concentration gradient. The Na⁺/K⁺ pump uses ATP to moved Na⁺ in one direction and K⁺ in the other to create and maintain concentration differences of these ions. In coupled transport, a favorable concentration gradient of one molecule is used to move a different molecule against its gradient, such as in the transport of glucose by Na⁺.

■ *Can active transport involve a channel protein? Why or why not?*

5.6 Bulk Transport by Endocytosis and Exocytosis

Learning Outcomes

1. Distinguish between endocytosis and exocytosis.
2. Illustrate how endocytosis can be specific.

The lipid nature of cell plasma membranes raises a second problem. The substances cells require for growth are mostly large, polar molecules that cannot cross the hydrophobic barrier a lipid bilayer creates. How do these substances get into cells? Two processes are involved in this **bulk transport:** *endocytosis* and *exocytosis*.

Bulk material enters the cell in vesicles

In **endocytosis,** the plasma membrane envelops food particles and fluids. Cells use three major types of endocytosis: phagocytosis, pinocytosis, and receptor-mediated endocytosis (figure 5.15). Like active transport, these processes also require energy expenditure.

Figure 5.15 Endocytosis. Both (*a*) phagocytosis and (*b*) pinocytosis are forms of endocytosis. *c.* In receptor-mediated endocytosis, cells have pits coated with the protein clathrin that initiate endocytosis when target molecules bind to receptor proteins in the plasma membrane. Photo inserts (false color has been added to enhance distinction of structures): (*a*) A TEM of phagocytosis of a bacterium, *Rickettsia tsutsugamushi,* by a mouse peritoneal mesothelial cell. The bacterium enters the host cell by phagocytosis and replicates in the cytoplasm. (*b*) A TEM of pinocytosis in a smooth muscle cell. (*c*) A coated pit appears in the plasma membrane of a developing egg cell, covered with a layer of proteins. When an appropriate collection of molecules gathers in the coated pit, the pit deepens and will eventually seal off to form a vesicle.

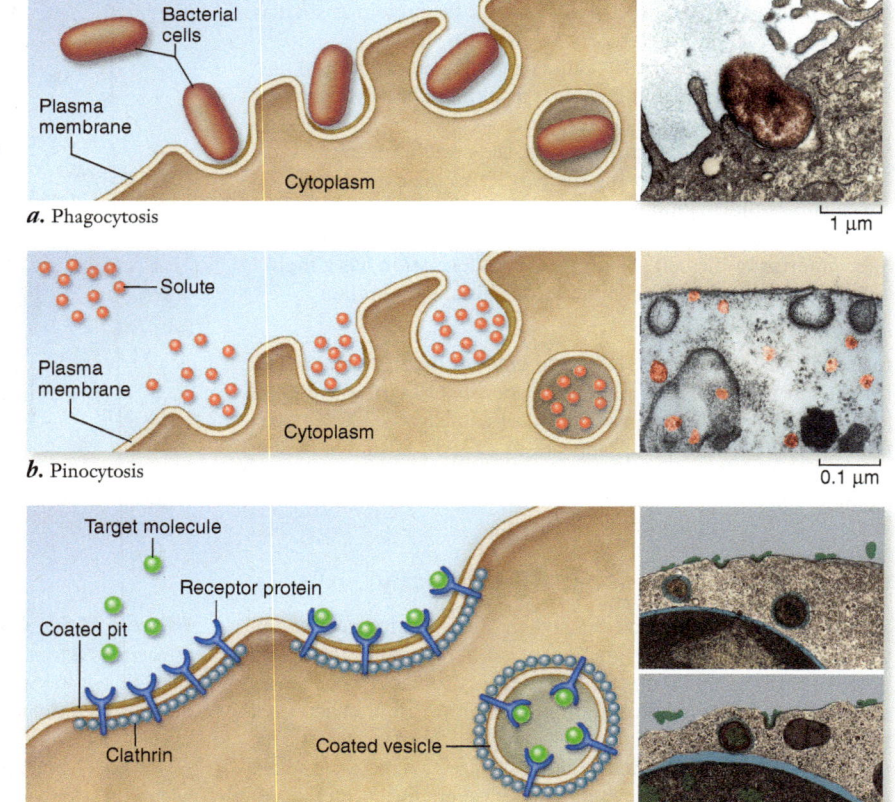

Phagocytosis and pinocytosis

If the material the cell takes in is particulate (made up of discrete particles), such as an organism or some other fragment of organic matter (figure 5.15a), the process is called **phagocytosis** (Greek *phagein,* "to eat," + *cytos,* "cell"). If the material the cell takes in is liquid (figure 5.15b), the process is called **pinocytosis** (Greek *pinein,* "to drink"). Pinocytosis is common among animal cells. Mammalian egg cells, for example, "nurse" from surrounding cells; the nearby cells secrete nutrients that the maturing egg cell takes up by pinocytosis.

Virtually all eukaryotic cells constantly carry out these kinds of endocytotic processes, trapping particles and extracellular fluid in vesicles and ingesting them. Endocytosis rates vary from one cell type to another. They can be surprisingly high; some types of white blood cells ingest up to 25% of their cell volume each hour.

Receptor-mediated endocytosis

Molecules are often transported into eukaryotic cells through **receptor-mediated endocytosis.** These molecules first bind to specific receptors in the plasma membrane—they have a conformation that fits snugly into the receptor. Different cell types contain a characteristic battery of receptor types, each for a different kind of molecule in their membranes.

The portion of the receptor molecule that lies inside the membrane is trapped in an indented pit coated on the cytoplasmic side with the protein *clathrin.* Each pit acts like a molecular mousetrap, closing over to form an internal vesicle when the right molecule enters the pit (figure 5.15c). The trigger that releases the trap is the binding of the properly fitted target molecule to the embedded receptor. When binding occurs, the cell reacts by initiating endocytosis; the process is highly specific and very fast. The vesicle is now inside the cell carrying its cargo.

One type of molecule that is taken up by receptor-mediated endocytosis is low-density lipoprotein (LDL). LDL molecules bring cholesterol into the cell where it can be incorporated into membranes. Cholesterol plays a key role in determining the stiffness of the body's membranes. In the human genetic disease familial hypercholesterolemia, the LDL receptors lack tails, so they are never fastened in the clathrin-coated pits and as a result, do not trigger vesicle formation. The cholesterol stays in the bloodstream of affected individuals, accumulating as plaques inside arteries and leading to heart attacks.

It is important to understand that endocytosis in itself does not bring substances directly into the cytoplasm of a cell. The material taken in is still separated from the cytoplasm by the membrane of the vesicle.

Material can leave the cell by exocytosis

The reverse of endocytosis is **exocytosis,** the discharge of material from vesicles at the cell surface (figure 5.16). In plant cells, exocytosis is an important means of exporting the materials needed to construct the cell wall through the plasma membrane. Among protists, contractile vacuole discharge is considered a form of exocytosis. In animal cells, exocytosis provides a mechanism for secreting many hormones, neurotransmitters, digestive enzymes, and other substances.

The mechanisms for transport across cell membranes are summarized in table 5.2.

Learning Outcomes Review 5.6

Large molecules and other bulky materials can enter a cell by endocytosis and leave the cell by exocytosis. These processes require energy. Endocytosis may be mediated by specific receptor proteins in the membrane that trigger the formation of vesicles.

- *What feature unites transport by receptor-mediated endocytosis, transport by a carrier, and catalysis by an enzyme?*

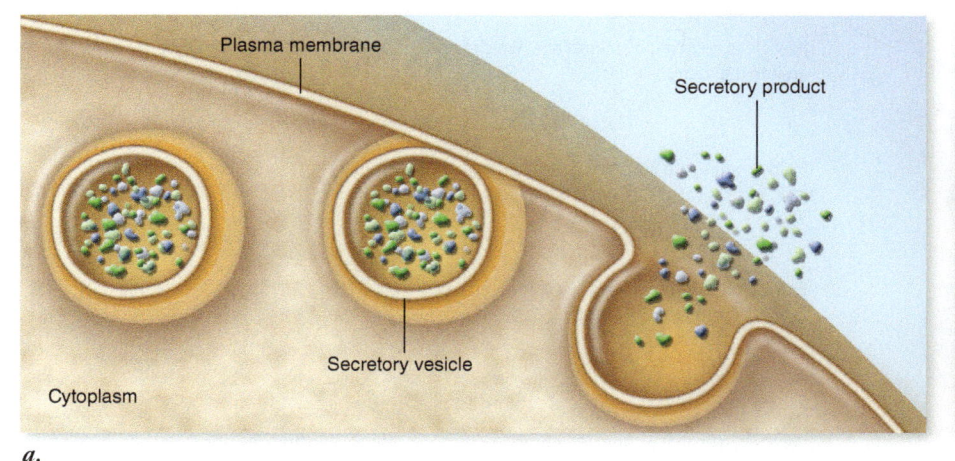

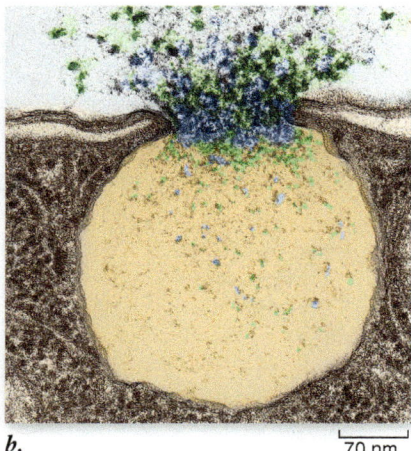

Figure 5.16 Exocytosis. *a.* Proteins and other molecules are secreted from cells in small packets called vesicles, whose membranes fuse with the plasma membrane, releasing their contents outside the cell. *b.* A false-colored transmission electron micrograph showing exocytosis.

TABLE 5.2 — Mechanisms for Transport Across Cell Membranes

Process	How It Works	Example
PASSIVE PROCESSES		
Diffusion		
Direct	Random molecular motion produces net migration of nonpolar molecules toward region of lower concentration	Movement of oxygen into cells
Facilitated Diffusion		
Protein channel	Polar molecules or ions move through a protein channel; net movement is toward region of lower concentration	Movement of ions in or out of cell
Protein carrier	Molecule binds to carrier protein in membrane and is transported across; net movement is toward region of lower concentration	Movement of glucose into cells
Osmosis		
Aquaporins	Diffusion of water across the membrane via osmosis; requires osmotic gradient	Movement of water into cells placed in a hypotonic solution
ACTIVE PROCESSES		
Active Transport		
Protein carrier		
Na^+/K^+ pump	Carrier uses energy to move a substance across a membrane against its concentration gradient	Na^+ and K^+ against their concentration gradients
Coupled transport	Molecules are transported across a membrane against their concentration gradients by the cotransport of sodium ions or protons down their concentration gradients	Coupled uptake of glucose into cells against its concentration gradient using a Na^+ gradient
Endocytosis		
Membrane vesicle		
Phagocytosis	Particle is engulfed by membrane, which folds around it and forms a vesicle	Ingestion of bacteria by white blood cells
Pinocytosis	Fluid droplets are engulfed by membrane, which forms vesicles around them	"Nursing" of human egg cells
Receptor-mediated endocytosis	Endocytosis triggered by a specific receptor, forming clathrin-coated vesicles	Cholesterol uptake
Exocytosis		
Membrane vesicle	Vesicles fuse with plasma membrane and eject contents	Secretion of mucus; release of neurotransmitters

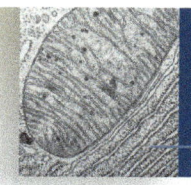

Chapter Review

5.1 The Structure of Membranes

The fluid mosaic model shows proteins embedded in a fluid lipid bilayer.

Membranes are sheets of phospholipid bilayers with associated proteins (figure 5.3). Hydrophobic regions of a membrane are oriented inward and hydrophilic regions oriented outward. In the fluid mosaic model, proteins float on or in the lipid bilayer.

Cellular membranes consist of four component groups.

In eukaryotic cells, membranes have four components: a phospholipid bilayer, transmembrane proteins (integral membrane proteins), an interior protein network, and cell-surface markers. The interior protein network is composed of cytoskeletal filaments and peripheral membrane proteins, which are associated with the membrane but are not an integral part. Membranes contain glycoproteins and glycolipids on the surface that act as cell identity markers.

Cholesterol and sphingolipid can associate to form microdomains. The two leaflets of the plasma membrane are also not identical.

Electron microscopy has provided structural evidence.

Transmission electron microscopy (TEM) and scanning electron microscopy (SEM) have confirmed the structure predicted by the fluid mosaic model.

5.2 Phospholipids: The Membrane's Foundation

Phospholipids are composed of two fatty acids and a phosphate group linked to a three-carbon glycerol molecule.

Phospholipids spontaneously form bilayers.

The phosphate group of a phospholipid is polar and hydrophilic; the fatty acids are nonpolar and hydrophobic, and they orient away from the polar head of the phospholipids. The nonpolar interior of the lipid bilayer impedes the passage of water and water-soluble substances.

The phospholipid bilayer is fluid.

Hydrogen bonding of water keeps the membrane in its bilayer configuration; however, phospholipids and unanchored proteins in the membrane are loosely associated and can diffuse laterally.

Membrane fluidity varies with lipid composition.

Membrane fluidity depends on the fatty acid composition of the membrane. Unsaturated fats tend to make the membrane more fluid because of the "kinks" of double bonds in the fatty acid tails. Temperature also affects fluidity.

Phospholipid composition affects membrane structure.

Different membrane compartments have different phospholipid composition. This affects both structure and function of different membranes, and persists despite membrane traffic.

5.3 Proteins: Multifunctional Components

Proteins and protein complexes perform key functions.

Transporters are integral membrane proteins that carry specific substances through the membrane. Enzymes often occur on the interior surface of the membrane. Cell-surface receptors respond to external chemical messages and change conditions inside the cell; cell identity markers on the surface allow recognition of the body's cells as "self." Cell-to-cell adhesion proteins glue cells together; surface proteins that interact with other cells anchor to the cytoskeleton.

Structural features of membrane proteins relate to function.

Surface proteins are attached to the surface by nonpolar regions that associate with polar regions of phospholipids. Transmembrane proteins may cross the bilayer a number of times, and each membrane-spanning region is called a transmembrane domain. Such a domain is composed of hydrophobic amino acids usually arranged in α helices. In certain proteins, β-pleated sheets in the nonpolar region form a pipelike passageway having a polar environment. An example is the porin class of proteins.

5.4 Passive Transport Across Membranes

Transport can occur by simple diffusion.

Simple diffusion is the passive movement of a substance along a chemical or electrical gradient. Biological membranes pose a barrier to hydrophilic polar molecules, while they allow hydrophobic substances to diffuse freely.

Proteins allow membrane diffusion to be selective.

Ions and large hydrophilic molecules cannot cross the phospholipid bilayer. Diffusion can still occur via channel or carrier proteins by facilitated diffusion. Channels allow the diffusion of specific ions, by forming an aqueous pore in the membrane. Carrier proteins bind to the molecules they transport, much like an enzyme. The rate of transport by a carrier is limited by the number of carriers in the membrane.

Osmosis is the movement of water across membranes.

The direction of movement due to osmosis depends on the solute concentration on either side of the membrane (figures 5.11 & 5.12). Solutions can be isotonic, hypotonic, or hypertonic. Cells in an isotonic solution are in osmotic balance; cells in a hypotonic solution will gain water; and cells in a hypertonic solution will lose water. Aquaporins are water channels that facilitate the diffusion of water.

5.5 Active Transport Across Membranes

Active transport uses energy to move materials against a concentration gradient.

Active transport uses specialized protein carriers that couple a source of energy to transport. They are classified based on the number of molecules and direction of transport. Uniporters transport a specific molecule in one direction; symporters transport two molecules in the same direction; and antiporters transport two molecules in opposite directions.

The sodium–potassium pump runs directly on ATP.

The sodium–potassium pump moves Na^+ out of the cell and K^+ into the cell against their concentration gradients using ATP. In every cycle of the pump, three Na^+ leave the cell and two K^+ enter it. This pump appears to be almost universal in animal cells.

Coupled transport uses ATP indirectly.

Coupled transport uses a concentration gradient of one molecule to move another against a gradient in the same direction. Countertransport is similar, but the two molecules move in opposite directions.

5.6 Bulk Transport by Endocytosis and Exocytosis

Bulk transport moves large quantities of substances that cannot pass through the cell membrane.

Bulk material enters the cell in vesicles.

In endocytosis, the cell membrane surrounds material and pinches off to form a vesicle. In receptor-mediated endocytosis, specific molecules bind to receptors on the cell membrane.

Material can leave the cell by exocytosis.

In exocytosis, material in a vesicle is discharged when the vesicle fuses with the membrane.

Review Questions

UNDERSTAND

1. The fluid mosaic model of the membrane describes the membrane as
   a. containing a significant quantity of water in the interior.
   b. composed of fluid phospholipids on the outside and protein on the inside.
   c. composed of protein on the outside and fluid phospholipids on the inside.
   d. made of proteins and lipids that can freely move.

2. What chemical property characterizes the interior of the phospholipid bilayer?
   a. It is hydrophobic.
   b. It is hydrophilic.
   c. It is polar.
   d. It is saturated.

3. The transmembrane domain of an integral membrane protein
   a. is composed of hydrophobic amino acids.
   b. often forms an α-helical structure.
   c. can cross the membrane multiple times.
   d. All of the choices are correct.

4. The specific function of a membrane within a cell is determined by the
   a. degree of saturation of the fatty acids within the phospholipid bilayer.
   b. location of the membrane within the cell.
   c. presence of lipid rafts and cholesterol.
   d. type and number of membrane proteins.

5. The movement of water across a membrane is dependent on
   a. the solvent concentration.
   b. the solute concentration.
   c. the presence of carrier proteins.
   d. membrane potential.

6. If a cell is in an isotonic environment, then
   a. the cell will gain water and burst.
   b. no water will move across the membrane.
   c. the cell will lose water and shrink.
   d. osmosis still occurs, but there is no net gain or loss of cell volume.

7. Which of the following is NOT a mechanism for bringing material into a cell?
   a. Exocytosis
   b. Endocytosis
   c. Pinocytosis
   d. Phagocytosis

APPLY

1. A bacterial cell that can alter the composition of saturated and unsaturated fatty acids in its membrane lipids is adapted to a cold environment. If this cell is shifted to a warmer environment, it will react by
   a. increasing the amount of cholesterol in its membrane.
   b. altering the amount of protein present in the membrane.
   c. increasing the degree of saturated fatty acids in its membrane.
   d. increasing the percentage of unsaturated fatty acids in its membrane.

2. What variable(s) influence(s) whether a nonpolar molecule can move across a membrane by passive diffusion?
   a. The structure of the phospholipids bilayer
   b. The difference in concentration of the molecule across the membrane
   c. The presence of transport proteins in the membrane
   d. All of the choices are correct.

3. Which of the following does NOT contribute to the selective permeability of a biological membrane?
   a. Specificity of the carrier proteins in the membrane
   b. Selectivity of channel proteins in the membrane
   c. Hydrophobic barrier of the phospholipid bilayer
   d. Hydrogen bond formation between water and phosphate groups

4. How are *active* transport and *coupled* transport related?
   a. They both use ATP to move molecules.
   b. Active transport establishes a concentration gradient, but coupled transport doesn't.
   c. Coupled transport uses the concentration gradient established by active transport.
   d. Active transport moves one molecule, but coupled transport moves two.

5. A cell can use the process of facilitated diffusion to
   a. concentrate a molecule such as glucose inside a cell.
   b. remove all of a toxic molecule from a cell.
   c. move ions or large polar molecules across the membrane regardless of concentration.
   d. move ions or large polar molecules from a region of high concentration to a region of low concentration.

SYNTHESIZE

1. Figure 5.5 describes a classic experiment demonstrating the ability of proteins to move within the plane of the cell's plasma membrane. The following table outlines three different experiments using the fusion of labeled mouse and human cells.

Experiment	Conditions	Temperature (°C)	Result
1	Fuse human and mouse cells	37	Intermixed membrane proteins
2	Fuse human and mouse cells in presence of ATP inhibitors	37	Intermixed membrane proteins
3	Fuse human and mouse cells	4	No intermixing of membrane proteins

What conclusions can you reach about the movement of these proteins?

2. Each compartment of the endomembrane system of a cell is connected to the plasma membrane. Create a simple diagram of a cell including the RER, Golgi apparatus, vesicle, and the plasma membrane. Starting with the RER, use two different colors to represent the inner and outer halves of the bilayer for each of these membranes. What do you observe?

3. The distribution of lipids in the ER membrane is symmetric, that is, it is the same in both leaflets of the membrane. The Golgi apparatus and plasma membrane do not have symmetric distribution of membrane lipids. What kinds of processes could achieve this outcome?

CHAPTER 6

Energy and Metabolism

Chapter Contents

6.1 The Flow of Energy in Living Systems
6.2 The Laws of Thermodynamics and Free Energy
6.3 ATP: The Energy Currency of Cells
6.4 Enzymes: Biological Catalysts
6.5 Metabolism: The Chemical Description of Cell Function

Introduction

Life can be viewed as a constant flow of energy, channeled by organisms to do the work of living. Each of the significant properties by which we define life—order, growth, reproduction, responsiveness, and internal regulation—requires a constant supply of energy. Both the lion and the giraffe need to eat to provide energy for a wide variety of cellular functions. Deprived of a source of energy, life stops. Therefore, a comprehensive study of life would be impossible without discussing bioenergetics, the analysis of how energy powers the activities of living systems. In this chapter, we focus on energy—what it is and how it changes during chemical reactions.

6.1 The Flow of Energy in Living Systems

Learning Outcomes
1. Differentiate between kinetic and potential energy.
2. Identify the source of energy for the biosphere.
3. Describe the nature of redox reactions.

Thermodynamics is the branch of chemistry concerned with energy changes. Cells are governed by the laws of physics and chemistry, so we must understand these laws in order to understand how cells function.

Energy can take many forms

Energy is defined as the capacity to do work. We think of energy as existing in two states: kinetic energy and potential energy (figure 6.1). **Kinetic energy** is the energy of motion. Moving objects perform work by causing other matter to move. **Potential energy** is stored energy. Objects that are not actively moving but have the capacity to do so possess potential energy. A boulder perched on a hilltop has gravitational potential energy. As it begins to roll downhill, some of its potential energy is converted into kinetic energy. Much of the work that living organisms carry out involves transforming potential energy into kinetic energy.

Energy can take many forms: mechanical energy, heat, sound, electric current, light, or radioactivity. Because it can exist in so many forms, energy can be measured in many ways. Heat is the most convenient way of measuring energy because all other forms of energy can be converted into heat. In fact, the term *thermodynamics* means "heat changes."

The unit of heat most commonly employed in biology is the kilocalorie (kcal). One kilocalorie is equal to 1000 calories (cal). One calorie is the heat required to raise the temperature of one gram of water one degree Celsius (°C). (You are probably more used to seeing the term *Calorie* with a capital C. This is used on food labels and is actually the same as kilocalorie.) Another energy unit, often used in physics, is the *joule;* one joule equals 0.239 cal.

The Sun provides energy for living systems

Energy flows into the biological world from the Sun. It is estimated that sunlight provides the Earth with more than 13×10^{23} calories per year, or 40 million billion calories per second! Plants, algae, and certain kinds of bacteria capture a fraction of this energy through photosynthesis.

In photosynthesis, energy absorbed from sunlight is used to combine small molecules (water and carbon dioxide) into more complex ones (sugars). This process converts carbon from an inorganic to an organic form. In the process, energy from sunlight is stored as potential energy in the covalent bonds between atoms in the sugar molecules.

Breaking the bonds between atoms requires energy. In fact, the strength of a covalent bond is measured by the amount of energy required to break it. For example, it takes 98.8 kcal to break one mole (6.023×10^{23}) of the carbon–hydrogen (C—H) bonds found in organic molecules. Fat molecules have many C—H bonds,

a. Potential energy

b. Kinetic energy

Figure 6.1 Potential and kinetic energy. *a.* Objects that have the capacity to move but are not moving have potential energy. The energy required for the girl to climb to the top of the slide is stored as potential energy. *b.* Objects that are in motion have kinetic energy. The stored potential energy is released as kinetic energy as the girl slides down.

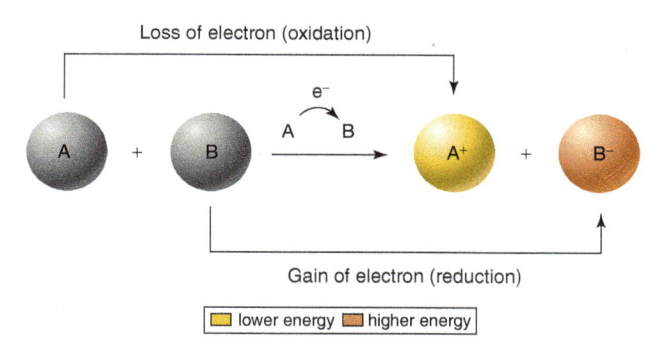

Figure 6.2 Redox reactions. Oxidation is the loss of an electron; reduction is the gain of an electron. In this example, the charges of molecules A and B appear as superscripts in each molecule. Molecule A loses energy as it loses an electron, and molecule B gains that energy as it gains an electron.

and breaking those bonds provides lots of energy. This is one reason animals store fat. The oxidation of one mole of a 16-carbon fatty acid that is completely saturated with hydrogens yields 2340 kcal.

Redox reactions transfer electrons

During a chemical reaction, the energy stored in chemical bonds may be used to make new bonds. In some of these reactions, electrons actually pass from one atom or molecule to another. An atom or molecule that loses an electron is said to be oxidized, and the process by which this occurs is called **oxidation.** The name comes from the fact that oxygen is the most common electron acceptor in biological systems. Conversely, an atom or molecule that gains an electron is said to be reduced, and the process is called *reduction*. The reduced form of a molecule has a higher level of energy than the oxidized form (figure 6.2).

Oxidation and reduction always take place together, because every electron that is lost by one atom through oxidation is gained by another atom through reduction. Therefore, chemical reactions of this sort are called **oxidation–reduction,** or **redox,** *reactions*. Oxidation–reduction reactions play a key role in the flow of energy through biological systems.

In chapters 7 and 8, you will learn the details of how organisms derive energy from the oxidation of organic compounds via respiration, as well as from the energy in sunlight via photosynthesis.

Learning Outcomes Review 6.1

Energy is defined as the capacity to do work. The two forms of energy are kinetic energy, or energy of motion, and potential energy, or stored energy. The ultimate source of energy for living systems is the Sun. Organisms derive their energy from redox reactions. In oxidation, a molecule loses an electron; in reduction, a molecule gains an electron.

■ *What energy source might ecosystems at the bottom of the ocean use?*

6.2 The Laws of Thermodynamics and Free Energy

Learning Outcomes
1. Explain the laws of thermodynamics.
2. Relate free energy changes to the outcome of chemical reactions.
3. Contrast the course of a reaction with and without an enzyme catalyst.

All activities of living organisms—growing, running, thinking, singing, reading these words—involve changes in energy. A set of two universal laws we call the laws of thermodynamics govern all energy changes in the universe, from nuclear reactions to a bird flying through the air.

The First Law states that energy cannot be created or destroyed

The **First Law of Thermodynamics** concerns the amount of energy in the universe. Energy cannot be created or destroyed; it can only change from one form to another (from potential to kinetic, for example). The total amount of energy in the universe remains constant.

The lion eating a giraffe at the beginning of this chapter is acquiring energy. Rather than creating new energy or capturing the energy in sunlight, the lion is merely transferring some of the potential energy stored in the giraffe's tissues to its own body, just as the giraffe obtained the potential energy stored in the plants it ate while it was alive.

Within any living organism, chemical potential energy stored in some molecules can be shifted to other molecules and stored in different chemical bonds. It can also be converted into other forms, such as kinetic energy, light, or electricity. During each conversion, some of the energy dissipates into the environment as **heat,** which is a measure of the random motion of molecules (and therefore a measure of one form of kinetic energy). Energy continuously flows through the biological world in one direction, with new energy from the Sun constantly entering the system to replace the energy dissipated as heat.

Heat can be harnessed to do work only when there is a heat gradient—that is, a temperature difference between two areas. Cells are too small to maintain significant internal temperature differences, so heat energy is incapable of doing the work of cells. Instead, cells must rely on chemical reactions for energy.

Although the total amount of energy in the universe remains constant, the energy available to do work decreases as more of it is progressively lost as heat.

The Second Law states that some energy is lost as disorder increases

The **Second Law of Thermodynamics** concerns the transformation of potential energy into heat, or random molecular motion during any energy transaction. It states that the disorder in the universe, more formally called **entropy,** is continuously increasing. Put simply, disorder is more likely than order. For example, it is much more likely that a column of bricks will tumble over than that a pile of bricks will arrange themselves spontaneously to form a column.

In general, energy transformations proceed spontaneously to convert matter from a more ordered, less stable form to a less ordered, but more stable form. For this reason, the second law is sometimes called "time's arrow." Looking at the photographs in figure 6.3, you could put the pictures into correct sequence using the information that time had elapsed with only natural processes occurring. Although it might be great if our rooms would straighten themselves up, we know from experience how much work it takes to do so.

The Second Law of Thermodynamics can also be stated simply as "entropy increases." When the universe formed, it held all the potential energy it will ever have. It has become progressively more disordered ever since, with every energy exchange increasing the amount of entropy.

Chemical reactions can be predicted based on changes in free energy

It takes energy to break the chemical bonds that hold the atoms in a molecule together. Heat energy, because it increases atomic motion, makes it easier for the atoms to pull apart. Both chemical bonding and heat have a significant influence on a molecule. Chemical bonding reduces disorder; heat increases it. The net effect, the amount of energy actually available to break and subsequently form other chemical bonds, is called the *free energy* of that molecule. In a more general sense, **free energy** is defined as the energy available to do work in any system.

For a molecule within a cell, where pressure and volume usually do not change, the free energy is denoted by the symbol G (for "Gibbs free energy"). G is equal to the energy contained in a molecule's chemical bonds (called **enthalpy** and designated H) together with the energy term (TS) related to the degree of disorder in the system, where S is the symbol for *entropy* and T is the absolute temperature expressed in the Kelvin scale ($K = °C + 273$):

$$G = H - TS$$

Chemical reactions break some bonds in the reactants and form new ones in the products. Consequently, reactions can produce changes in free energy. When a chemical reaction occurs under conditions of constant temperature, pressure, and volume—as do most biological reactions—the change symbolized by the Greek capital letter delta, Δ, in free energy (ΔG) is simply:

$$\Delta G = \Delta H - T\Delta S$$

The change in free energy is useful because it allows us to predict whether a particular chemical reaction is spontaneous or not. The change in free energy is calculated as energy of products minus energy of reactants, so if the products have *more* free energy than the reactants, ΔG is positive. Such reactions are not spontaneous because they require an input of energy. This can be because either the bond energy (H) is higher, or the disorder (S) in the system is lower. We call this an **endergonic** reaction. A plot of free energy over the course of reaction shows this graphically (figure 6.4). In the case of an endergonic reaction, it is "uphill" from reactants to products, and not spontaneous.

Conversely, if the products have less free energy than the reactants, ΔG is negative, and the reaction will proceed spontaneously. The negative ΔG can be because the bond energy (H) is lower, or the disorder (S) is higher, or both. These reactions release the excess free energy as heat and are called **exergonic** reactions.

Figure 6.3 Entropy in action. As time elapses, the room shown at right becomes more disorganized. Entropy has increased in this room. It takes energy to restore it to the ordered state shown at left.

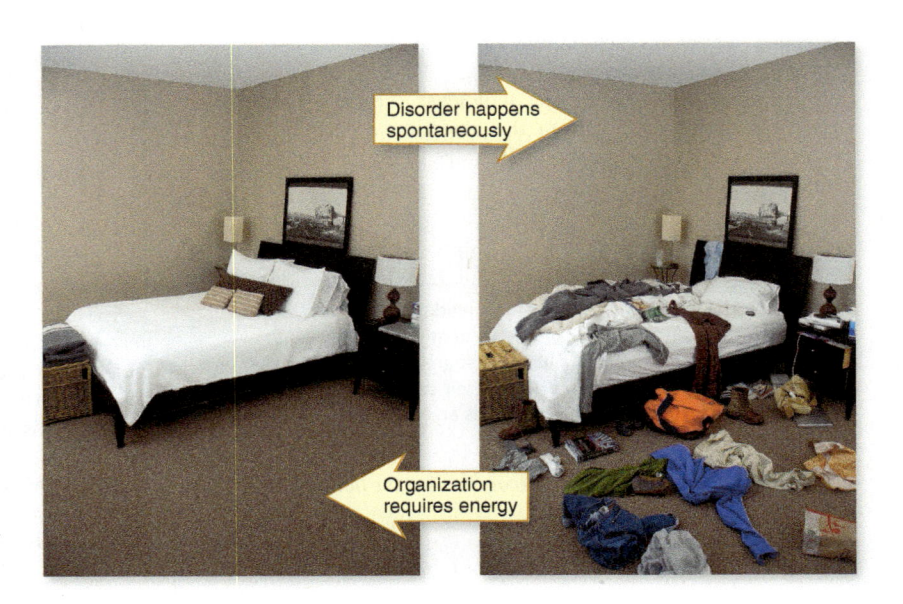

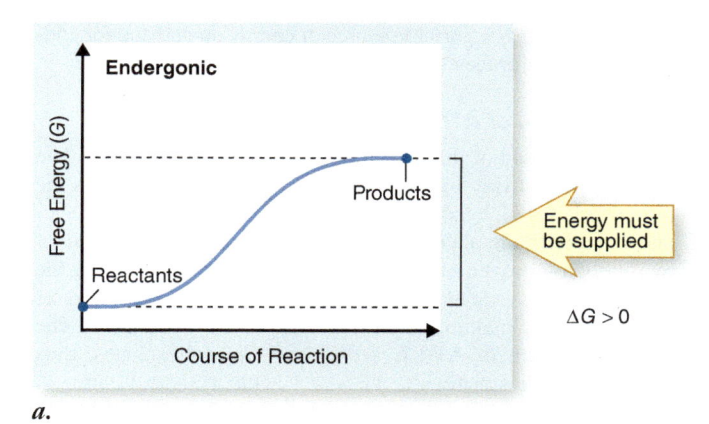

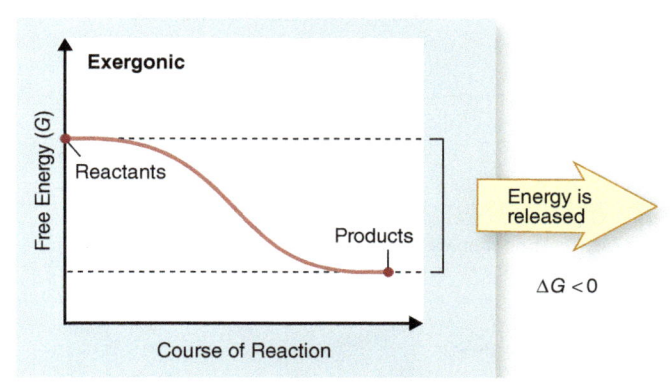

Figure 6.4 Energy in chemical reactions. *a.* In an endergonic reaction, the products of the reaction contain more energy than the reactants, and the extra energy must be supplied for the reaction to proceed. *b.* In an exergonic reaction, the products contain less energy than the reactants, and the excess energy is released.

Our plot of free energy for the reaction is now "downhill" and the reaction will be spontaneous (figure 6.4). Note that *spontaneous* does not mean the same thing as *instantaneous*. A spontaneous reaction may proceed very slowly.

Because chemical reactions are reversible, a reaction that is exergonic in the forward direction will be endergonic in the reverse direction. For each reaction, an equilibrium exists at some point between the relative amounts of reactants and products. This equilibrium has a numeric value and is called the *equilibrium constant*. This characteristic of reactions provides us with another way to think about free energy changes: An exergonic reaction has an equilibrium favoring the products, and an endergonic reaction has an equilibrium favoring the reactants.

Spontaneous chemical reactions require activation energy

If all chemical reactions that release free energy tend to occur spontaneously, why haven't all such reactions already occurred? Consider the gasoline tank of your car: The oxidation of the hydrocarbons in gasoline is an exergonic reaction, but your gas tank does not spontaneously explode. One reason is that most reactions require an input of energy to get started. In the case of your car, this input consists of the electrical sparks in the engine's cylinders, producing a controlled explosion.

Activation energy

Before new chemical bonds can form, even bonds that contain less energy, existing bonds must first be broken, and that requires energy input. The extra energy needed to destabilize existing chemical bonds and initiate a chemical reaction is called **activation energy** (figure 6.5).

The rate of an exergonic reaction depends on the activation energy required for the reaction to begin. Reactions with larger activation energies tend to proceed more slowly because fewer molecules succeed in getting over the initial energy hurdle. The rate of reactions can be increased in two ways: (1) by increasing the energy of reacting molecules or (2) by lowering activation energy. Chemists often drive important industrial reactions by increasing the energy of the reacting molecules, which is frequently accomplished simply by heating up the reactants. The other strategy is to use a catalyst to lower the activation energy.

How catalysts work

Stressing particular chemical bonds can make them easier to break. The process of influencing chemical bonds in a way that lowers the activation energy needed to initiate a reaction is called **catalysis,** and substances that accomplish this are known as *catalysts* (figure 6.5).

Catalysts exert their action by affecting an intermediate stage in a reaction—the transition state. The energy needed to reach this transition state is the activation energy. Catalysts stabilize this transition state, thus lowering activation energy.

Catalysts cannot violate the basic laws of thermodynamics; they cannot make an endergonic reaction proceed spontaneously. By reducing the activation energy, a catalyst accelerates both the forward and the reverse reactions by exactly the same amount.

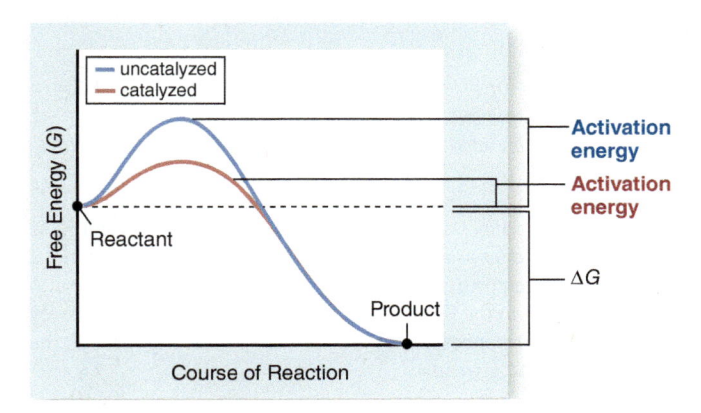

Figure 6.5 Activation energy and catalysis. Exergonic reactions do not necessarily proceed rapidly because activation energy must be supplied to destabilize existing chemical bonds. Catalysts accelerate particular reactions by lowering the amount of activation energy required to initiate the reaction. Catalysts do not alter the free-energy change produced by the reaction.

Therefore, a catalyst does not alter the proportion of reactant ultimately converted into product.

To understand this, imagine a bowling ball resting in a shallow depression on the side of a hill. Only a narrow rim of dirt below the ball prevents it from rolling down the hill. Now imagine digging away that rim of dirt. If you remove enough dirt from below the ball, it will start to roll down the hill—but removing dirt from below the ball will *never* cause the ball to roll up the hill. Removing the lip of dirt simply allows the ball to move freely; gravity determines the direction it then travels.

Similarly, the direction in which a chemical reaction proceeds is determined solely by the difference in free energy between reactants and products. Like digging away the soil below the bowling ball on the hill, catalysts reduce the energy barrier that is preventing the reaction from proceeding. Only exergonic reactions can proceed spontaneously, and catalysts cannot change that. What catalysts *can* do is make a reaction proceed much faster. In living systems, enzymes act as catalysts.

Learning Outcomes Review 6.2

The First Law of Thermodynamics states that energy cannot be created or destroyed. The Second Law states that disorder, or entropy, is increasing. Free-energy changes (ΔG) can predict whether chemical reactions take place. Reactions with a negative ΔG occur spontaneously, and those with a positive ΔG do not. Energy needed to initiate a reaction is termed activation energy. Catalysts stabilize an intermediate transition state, lowering activation energy and accelerating reactions.

- Can an enzyme make an endergonic reaction exergonic?

6.3 ATP: The Energy Currency of Cells

Learning Outcomes

1. Describe the role of ATP in short-term energy storage.
2. Distinguish which bonds in ATP are "high energy."

The chief "currency" all cells use for their energy transactions is the nucleotide *adenosine triphosphate (ATP)*. ATP powers almost every energy-requiring process in cells, from making sugars, to supplying activation energy for chemical reactions, to actively transporting substances across membranes, to moving through the environment and growing.

Cells store and release energy in the bonds of ATP

You saw in chapter 3 that nucleotides serve as the building blocks for nucleic acids, but they play other cellular roles as well. ATP is used as a building block for RNA molecules, and it also has a critical function as a portable source of energy on demand for endergonic cellular processes.

The structure of ATP

ATP is composed of three smaller components (figure 6.6). The first component is a 5-carbon sugar, ribose, which serves as the framework to which the other two subunits are attached. The second component is adenine, an organic molecule composed of two carbon–nitrogen rings. Each of the nitrogen atoms in the ring has an unshared pair of electrons and weakly attracts hydrogen ions, making adenine chemically a weak base. The third component of ATP is a chain of three phosphates, thus adenosine *tri*phosphate.

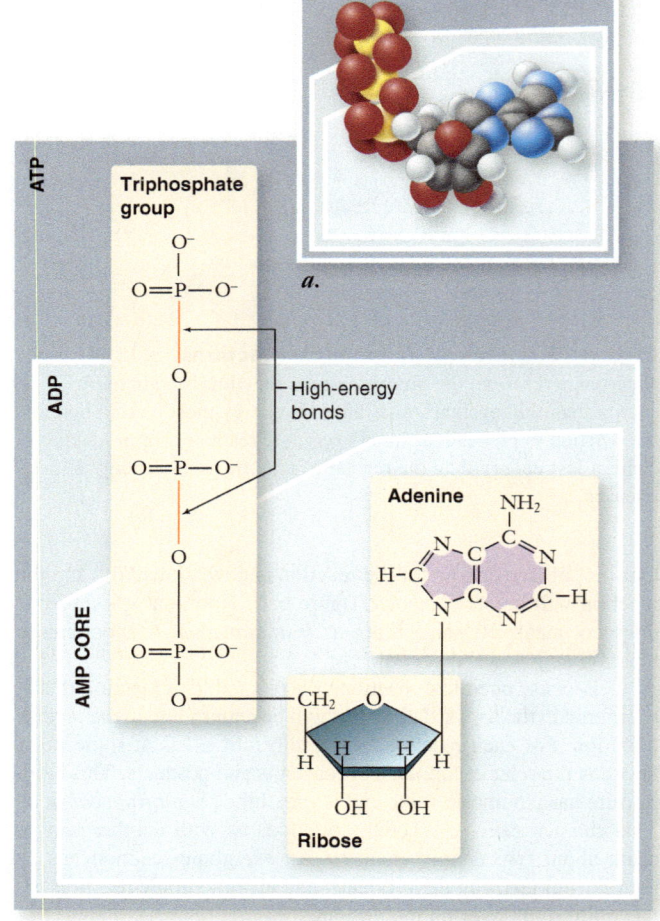

Figure 6.6 The ATP molecule. The model (*a*) and the structural diagram (*b*) both show that ATP has a core of AMP. Addition of one phosphate to AMP yields ADP, and addition of a second phosphate yields ATP. These two terminal phosphates are attached by high-energy bonds so that removing either by hydrolysis is an exergonic reaction that releases energy. ADP, adenosine diphosphate; AMP, adenosine monophosphate; ATP, adenosine triphosphate

How ATP stores energy

The key to how ATP stores energy lies in its triphosphate group. Phosphate groups are highly negatively charged, and thus they strongly repel one another. This electrostatic repulsion makes the covalent bonds joining the phosphates unstable. The molecule is often referred to as a "coiled spring," with the phosphates straining away from one another.

The unstable bonds holding the phosphates together in the ATP molecule have a low activation energy and are easily broken by hydrolysis. When they break, they can transfer a considerable amount of energy. In other words, the hydrolysis of ATP has a negative ΔG, and the energy it releases can be used to perform work.

In most reactions involving ATP, only the outermost high-energy phosphate bond is hydrolyzed, cleaving off the phosphate group on the end. When this happens, ATP becomes *adenosine diphosphate (ADP)* plus an **inorganic phosphate (P_i),** and energy equal to 7.3 kcal/mol is released under standard conditions. The liberated phosphate group usually attaches temporarily to some intermediate molecule. When that molecule is dephosphorylated, the phosphate group is released as P_i.

Both of the two terminal phosphates can be hydrolyzed to release energy, leaving *adenosine monophosphate (AMP)*, but the third phosphate is not attached by a high-energy bond. With only one phosphate group, AMP has no other phosphates to provide the electrostatic repulsion that makes the bonds holding the two terminal phosphate groups high-energy bonds.

ATP hydrolysis drives endergonic processes

Cells use ATP to drive endergonic processes. Remember that endergonic reactions do not proceed spontaneously because their products possess more free energy than their reactants; ΔG for an endergonic reaction is positive. If the ATP hydrolysis releases more energy than the other reaction consumes, then coupling the two reactions will produce an overall ΔG for the coupled reactions that is negative. That is, energy released by the hydrolysis of ATP can supply the energy needed by the endergonic reaction. Because most endergonic reactions in cells require less energy than is released by ATP hydrolysis, ATP can provide most of the energy a cell needs. ATP also powers other cellular functions such as generating force in muscles, or creating concentration gradients of important ions, both processes that require energy.

> **Data analysis** Consider the reaction:
> glutamate + $NH_3 \longrightarrow$ glutamine ($\Delta G = +3.4$ kcal/mol). If this reaction is coupled to ATP hydrolysis ($\Delta G = -7.3$ kcal/mol), what would be the overall ΔG? Would this process be endergonic or exergonic?

ATP cycles continuously

The same feature that makes ATP an effective energy donor—the instability of its phosphate bonds—make it a poor long-term energy-storage molecule. Fats and carbohydrates serve that function better.

Cells make and use ATP cyclically: Cells use exergonic reactions to provide energy to synthesize ATP from ADP + P_i; they then use the hydrolysis of ATP to provide energy to drive endergonic processes (figure 6.7). Most cells do not maintain large stockpiles of ATP. Instead, they typically have only a few seconds' supply of ATP at any given time, and they continually produce more from ADP and P_i. It is estimated that even a sedentary individual turns over an amount of ATP in one day roughly equal to his or her body weight. This statistic makes clear the importance of ATP synthesis. In chapters 7 and 8 we will explore in detail the cellular mechanisms for synthesizing ATP.

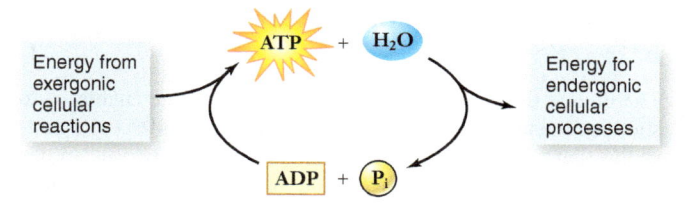

Figure 6.7 The ATP cycle. ATP is synthesized and hydrolyzed in a cyclic fashion. The synthesis of ATP from ADP + P_i is endergonic and is powered by exergonic cellular reactions. The hydrolysis of ATP to ADP + P_i is exergonic, and the energy released is used to power endergonic cellular functions such as muscle contraction. ADP, adenosine diphosphate; ATP, adenosine triphosphate; P_i, inorganic phosphate

> **Learning Outcomes Review 6.3**
>
> ATP is a nucleotide with three phosphate groups. Endergonic cellular processes can be driven by coupling to the exergonic hydrolysis of the two terminal phosphates. The bonds holding the terminal phosphate groups together are easily broken, releasing energy like a coiled spring. The cell is constantly building ATP using exergonic reactions and breaking it down to drive endergonic reactions.
>
> ■ *If the molecular weight of ATP is 507.18 g/mol, and the ΔG for hydrolysis is –7.3 kcal/mol, how much energy is released over the course of the day by a 100-kg man?*

6.4 Enzymes: Biological Catalysts

Learning Outcomes

1. *Discuss the specificity of enzymes.*
2. *Explain how enzymes bind to their substrates.*
3. *List the factors that influence the rate of enzyme-catalyzed reactions.*

The chemical reactions within living organisms are regulated by controlling the points at which catalysis takes place. Life itself, therefore, can be seen as regulated by catalysts. This is possible because the catalysts in living systems, enzymes, are also highly specific. Most enzymes are proteins, although RNA molecules can also have catalytic activity.

Enzymes lower the activation energy of reactions

The unique three-dimensional shape of an enzyme enables it to stabilize a temporary association between **substrates**—the molecules that will undergo the reaction. By bringing two substrates together in the correct orientation or by stressing particular chemical bonds of a substrate, an enzyme lowers the activation energy required for new bonds to form. The reaction thus proceeds much more quickly than it would without the enzyme. The enzyme itself is not changed or consumed in the reaction, so only a small amount of an enzyme is needed, and it can be used over and over.

As an example of how an enzyme works, let's consider the reaction of carbon dioxide and water to form carbonic acid. This important enzyme-catalyzed reaction occurs in vertebrate red blood cells:

$$CO_2 + H_2O \rightleftharpoons H_2CO_3$$
carbon dioxide — water — carbonic acid

This reaction may proceed in either direction, but because it has a large activation energy, the reaction is very slow in the absence of an enzyme: Perhaps 200 molecules of carbonic acid form in an hour in a cell in the absence of any enzyme. Reactions that proceed this slowly are of little use to a cell. Vertebrate red blood cells overcome this problem by employing an enzyme within their cytoplasm called *carbonic anhydrase* (enzyme names usually end in "-ase"). Under the same conditions, but in the presence of carbonic anhydrase, an estimated 600,000 molecules of carbonic acid form every *second!* Thus, the enzyme increases the reaction rate by more than one million times.

Thousands of different kinds of enzymes are known, each catalyzing one or a few specific chemical reactions. By facilitating particular chemical reactions, the enzymes in a cell determine the course of metabolism—the collection of all chemical reactions—in that cell.

Different types of cells contain different sets of enzymes, and this difference contributes to structural and functional variations among cell types. For example, the chemical reactions taking place within a red blood cell differ from those that occur within a nerve cell, in part because different cell types contain different arrays of enzymes.

Active sites of enzymes conform to fit the shape of substrates

Most enzymes are globular proteins with one or more pockets or clefts, called **active sites,** on their surface (figure 6.8). Substrates bind to the enzyme at these active sites, forming an **enzyme–substrate complex** (see figure 6.10). For catalysis to occur within the complex, a substrate molecule must fit precisely into an active site. When that happens, amino acid side groups of the enzyme end up very close to certain bonds of the substrate. These side groups interact chemically with the substrate, usually stressing or distorting a particular bond and consequently lowering the activation energy needed to break the bond. After the bonds of the substrates are broken, or new bonds are formed, the substrates have been converted to products. These products then dissociate from the enzyme, leaving the enzyme ready to bind its next substrate and begin the cycle again.

Proteins are not rigid. The binding of a substrate induces the enzyme to adjust its shape slightly, leading to a better *induced fit* between enzyme and substrate (figure 6.9). This interaction may

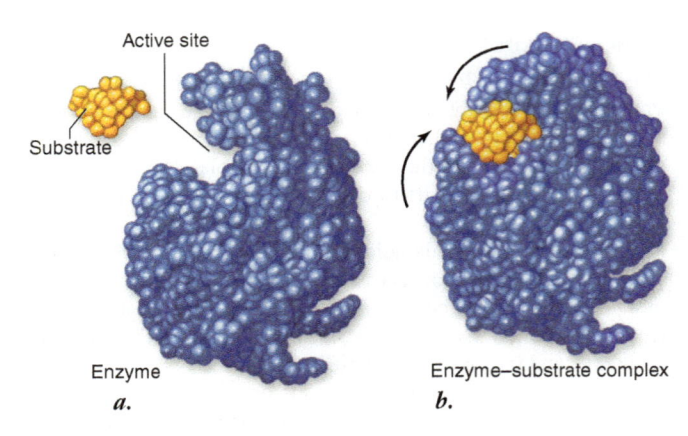

Figure 6.8 Enzyme binding its substrate. *a.* The active site of the enzyme lysozyme fits the shape of its substrate, a peptidoglycan that makes up bacterial cell walls. *b.* When the substrate, indicated in yellow, slides into the groove of the active site, the protein is induced to alter its shape slightly and bind the substrate more tightly. This alteration of the shape of the enzyme to better fit the substrate is called induced fit.

SCIENTIFIC THINKING

Hypothesis: *Protein structure is flexible, not rigid.*
Prediction: *Antibody–antigen binding can involve a change in protein structure.*
Test: *Determine crystal structure of a fragment of a specific antibody with no antigen bound, and with antigen bound for comparison.*

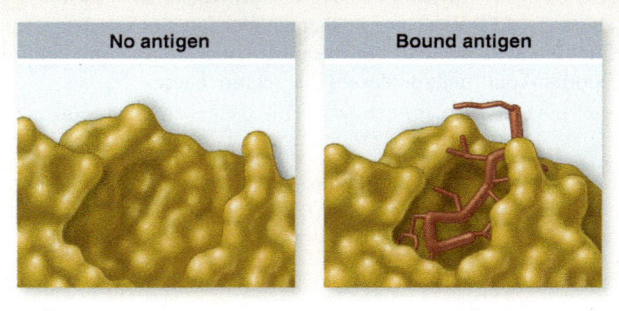

Result: *After binding, the antibody folds around the antigen forming a pocket.*
Conclusion: *In this case, binding involves an induced-fit kind of change in conformation.*
Further Experiments: *Why is this experiment easier to do with an antibody than with an enzyme? Can this experiment be done with an enzyme?*

Figure 6.9 Induced-fit binding of antibody to antigen.

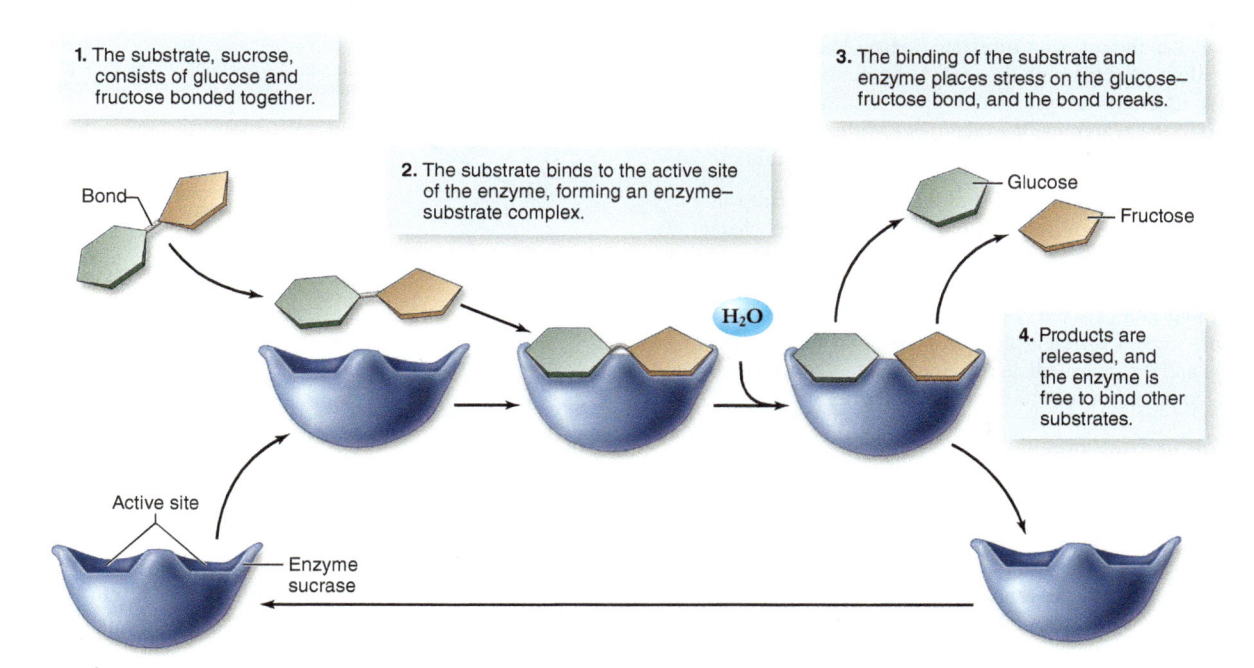

Figure 6.10 The catalytic cycle of an enzyme. Enzymes increase the speed at which chemical reactions occur, but they are not altered permanently themselves as they do so. In the reaction illustrated here, the enzyme sucrase is splitting the sugar sucrose into two simpler sugars: glucose and fructose.

also facilitate the binding of other substrates; in such cases, one substrate "activates" the enzyme to receive other substrates.

Enzymes occur in many forms

Although many enzymes are suspended in the cytoplasm of cells, not attached to any structure, other enzymes function as integral parts of cell membranes and organelles. Enzymes may also form associations called *multienzyme complexes* to carry out reaction sequences. And it is now clear that some enzymes may contain catalytic RNA rather than being only protein.

Multienzyme complexes

Often several enzymes catalyzing different steps of a sequence of reactions are associated with one another in noncovalently bonded assemblies called **multienzyme complexes.** The bacterial pyruvate dehydrogenase multienzyme complex, shown in figure 6.11, contains enzymes that carry out three sequential reactions in oxidative metabolism. Each complex has multiple copies of each of the three enzymes—60 protein subunits in all. The many subunits work together to form a molecular machine that performs multiple functions.

Multienzyme complexes offer the following significant advantages in catalytic efficiency:

1. The rate of any enzyme reaction is limited by how often the enzyme collides with its substrate. If a series of sequential reactions occurs within a multienzyme complex, the product of one reaction can be delivered to the next enzyme without releasing it to diffuse away.

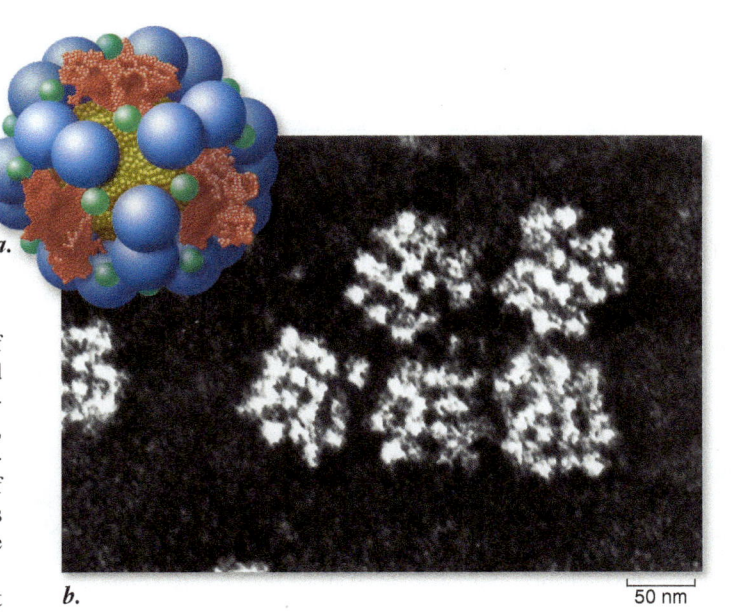

Figure 6.11 A complex enzyme: pyruvate dehydrogenase. Pyruvate dehydrogenase, which catalyzes the oxidation of pyruvate, is one of the most complex enzymes known. *a.* A model of the enzyme showing the arrangement of the 60 protein subunits. *b.* Many of the protein subunits are clearly visible in the electron micrograph.

chapter **6** *Energy and Metabolism* 115

2. Because the reacting substrate doesn't leave the complex while it goes through the series of reactions, unwanted side reactions are prevented.
3. All of the reactions that take place within the multienzyme complex can be controlled as a unit.

In addition to pyruvate dehydrogenase, which controls entry to the Krebs cycle during aerobic respiration (see chapter 7), several other key processes in the cell are catalyzed by multienzyme complexes. One well-studied system is the fatty acid synthetase complex that catalyzes the synthesis of fatty acids from two-carbon precursors. Seven different enzymes make up this multienzyme complex, and the intermediate reaction products remain associated with the complex for the entire series of reactions.

Nonprotein enzymes

The original enzymes discovered and studied were all proteins. For many years, we assumed that all enzymes were proteins until Thomas R. Cech and colleagues at the University of Colorado reported in 1981 an RNA splicing reaction that did not require protein. Around the same time, Sidney Altman and Norman Pace were studying the enzyme RNase P, and found that this enzyme was composed of both protein and RNA, and further the RNA was the catalytic molecule. Like protein enzymes, these RNA catalysts, which are loosely called "ribozymes," greatly accelerate the rate of particular biochemical reactions and show extraordinary substrate specificity.

Research has revealed at least two sorts of ribozymes. Some ribozymes have folded structures and catalyze reactions on themselves, a process called *intra*molecular catalysis. Other ribozymes act on other molecules without being changed themselves, a process called *inter*molecular catalysis.

The most striking example of the role of RNA as enzyme is emerging from recent work on the structure and function of the ribosome. For many years, it was thought that RNA was a structural framework for this vital organelle, but it is now clear that ribosomal RNA plays a key role in ribosome function. The ribosome itself is a ribozyme.

The ability of RNA, an informational molecule, to act as a catalyst has stirred great excitement because it seems to address the question, "Which came first, the protein or the nucleic acid?" It now appears likely that RNA evolved first and may have catalyzed the formation of the first proteins.

Enzyme function is sensitive to environmental factors

The rate of an enzyme-catalyzed reaction is affected by the concentrations of both the substrate and the enzyme that works on it. In addition, any chemical or physical factor that alters the enzyme's three-dimensional shape—such as temperature, pH, and the binding of regulatory molecules—can affect the enzyme's ability to catalyze the reaction.

Temperature

Increasing the temperature of an uncatalyzed reaction increases its rate because the additional heat increases random molecular movement. This motion can add stress to molecular bonds and affect the activation energy of a reaction.

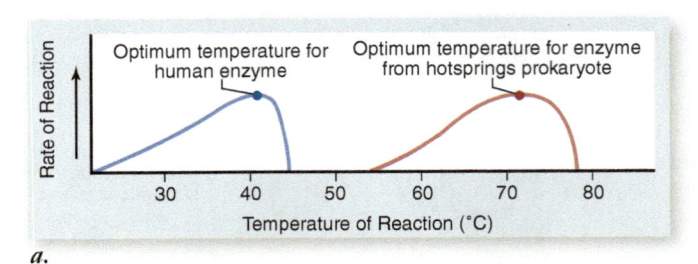

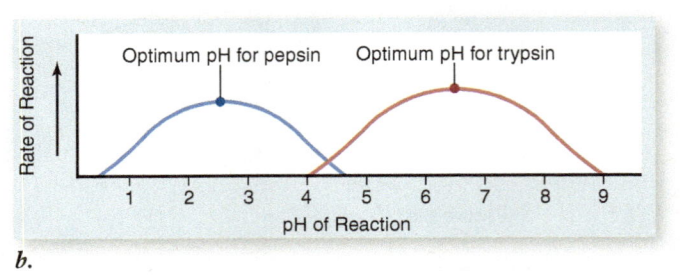

Figure 6.12 Enzyme sensitivity to the environment. The activity of an enzyme is influenced by both (*a*) temperature and (*b*) pH. Most human enzymes, such as the protein-degrading enzyme trypsin, work best at temperatures of about 40°C and within a pH range of 6 to 8. The hot springs prokaryote tolerates a higher environmental temperature and a correspondingly higher temperature optimum for enzymes. Pepsin works in the acidic environment of the stomach and has a lower optimum pH.

The rate of an enzyme-catalyzed reaction also increases with temperature, but only up to a point called the *optimum temperature* (figure 6.12*a*). Below this temperature, the hydrogen bonds and hydrophobic interactions that determine the enzyme's shape are not flexible enough to permit the induced fit that is optimum for catalysis. Above the optimum temperature, these forces are too weak to maintain the enzyme's shape against the increased random movement of the atoms in the enzyme. At higher temperatures, the enzyme denatures, as described in chapter 3.

Most human enzymes have an optimum temperature between 35°C and 40°C—a range that includes normal body temperature. Prokaryotes that live in hot springs have more stable enzymes (that is, enzymes held together more strongly), so the optimum temperature for those enzymes can be 70°C or higher. In each case the optimal temperature for the enzyme corresponds to the "normal" temperature usually encountered in the body or the environment, depending on the type of organism.

pH

Ionic interactions between oppositely charged amino acid residues, such as glutamic acid (−) and lysine (+), also hold enzymes together. These interactions are sensitive to the hydrogen ion concentration of the fluid in which the enzyme is dissolved, because changing that concentration shifts the balance between positively and negatively charged amino acid residues. For this reason, most enzymes have an *optimum pH* that usually ranges from pH 6 to 8.

Enzymes able to function in very acidic environments are proteins that maintain their three-dimensional shape even in the presence of high hydrogen ion concentrations. The enzyme pepsin, for example, digests proteins in the stomach at pH 2, a very acidic level (figure 6.12b).

Inhibitors and activators

Enzyme activity is also sensitive to the presence of specific substances that can bind to the enzyme and cause changes in its shape. Through these substances, a cell is able to regulate which of its enzymes are active and which are inactive at a particular time. This ability allows the cell to increase its efficiency and to control changes in its characteristics during development. A substance that binds to an enzyme and *decreases* its activity is called an **inhibitor.** Very often, the end product of a biochemical pathway acts as an inhibitor of an early reaction in the pathway, a process called *feedback inhibition* (discussed in section 6.5).

Enzyme inhibition occurs in two ways: **Competitive inhibitors** compete with the substrate for the same active site, occupying the active site and thus preventing substrates from binding; **noncompetitive inhibitors** bind to the enzyme in a location other than the active site, changing the shape of the enzyme and making it unable to bind to the substrate (figure 6.13).

Many enzymes can exist in either an active or inactive conformation; such enzymes are called *allosteric enzymes*. Most noncompetitive inhibitors bind to a specific portion of the enzyme called an **allosteric site.** These sites serve as chemical on/off switches; the binding of a substance to the site can switch the enzyme between its active and inactive configurations. A substance that binds to an allosteric site and reduces enzyme activity is called an **allosteric inhibitor** (figure 6.13b).

This kind of control is also used to activate enzymes. An **allosteric activator** binds to allosteric sites to keep an enzyme in its active configuration, thereby *increasing* enzyme activity.

Enzyme cofactors

Enzyme function is often assisted by additional chemical components known as **cofactors.** These can be metal ions that are often found in the active site participating directly in catalysis. For example, the metallic ion zinc is used by some enzymes, such as protein-digesting carboxypeptidase, to draw electrons away from their position in covalent bonds, making the bonds less stable and easier to break. Other metallic elements, such as molybdenum and manganese, are also used as cofactors. Like zinc, these substances are required in the diet in small amounts.

When the cofactor is a nonprotein organic molecule, it is called a **coenzyme.** Many of the small organic molecules essential in our diets that we call vitamins function as coenzymes. For example, the B vitamins B_6 and B_{12} both function as coenzymes for a number of different enzymes. Modified nucleotides are also used as coenzymes.

In numerous oxidation–reduction reactions that are catalyzed by enzymes, the electrons pass in pairs from the active site of the enzyme to a coenzyme that serves as the electron acceptor. The coenzyme then transfers the electrons to a different enzyme, which releases them (and the energy they bear) to the substrates in another reaction. Often, the electrons combine with protons (H^+) to form hydrogen atoms. In this way, coenzymes shuttle energy in the form of hydrogen atoms from one enzyme to another in a cell. The role of coenzymes and the specifics of their action will be explored in detail in chapters 7 and 8.

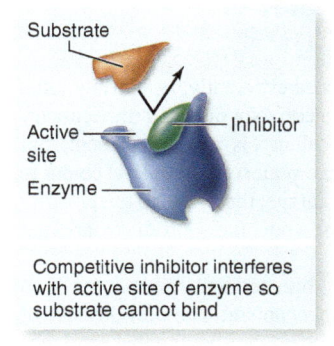

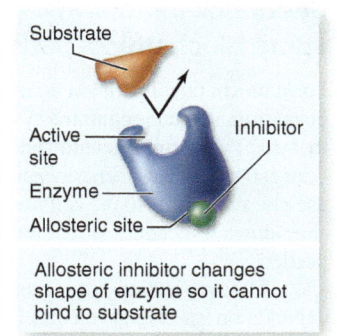

Learning Outcomes Review 6.4

Enzymes are biological catalysts that accelerate chemical reactions inside the cell. Enzymes bind to their substrates based on molecular shape, which allows them to be highly specific. Enzyme activity is affected by conditions such as temperature and pH and the presence of inhibitors or activators. Some enzymes also require an inorganic cofactor or an organic coenzyme.

- Why do proteins and RNA function as enzymes but DNA does not?

6.5 Metabolism: The Chemical Description of Cell Function

Learning Outcomes

1. Explain the kinds of reactions that make up metabolism.
2. Discuss what is meant by a metabolic pathway.
3. Recognize that metabolism is a product of evolution.

Living chemistry, the total of all chemical reactions carried out by an organism, is called **metabolism.** Those chemical reactions that expend energy to build up molecules are called *anabolic* reactions, or **anabolism.** Reactions that harvest energy by breaking down molecules are called *catabolic* reactions, or **catabolism.** A detailed discussion of metabolism is outside of the scope of this book, but this section provides an overview of key concepts of metabolism.

Figure 6.13 How enzymes can be inhibited. *a.* In competitive inhibition, the inhibitor has a shape similar to the substrate and competes for the active site of the enzyme. *b.* In noncompetitive inhibition, the inhibitor binds to the enzyme at the allosteric site, a place away from the active site, effecting a conformational change in the enzyme, making it unable to bind to its substrate.

a. Competitive inhibition — Competitive inhibitor interferes with active site of enzyme so substrate cannot bind.

b. Noncompetitive inhibition — Allosteric inhibitor changes shape of enzyme so it cannot bind to substrate.

Biochemical pathways organize chemical reactions in cells

Organisms contain thousands of different kinds of enzymes that catalyze a bewildering variety of reactions. Many of these reactions in a cell occur in sequences called **biochemical pathways.** In such pathways, the product of one reaction becomes the substrate for the next (figure 6.14). Biochemical pathways are the organizational units of metabolism—the elements an organism controls to achieve coherent metabolic activity.

Many sequential enzyme steps in biochemical pathways take place in specific compartments of the cell; for example, the steps of the Krebs cycle (see chapter 7) occur in the matrix inside mitochondria in eukaryotes. By determining where many of the enzymes that catalyze these steps are located, we can "map out" a model of metabolic processes in the cell.

Biochemical pathways may have evolved in stepwise fashion

In the earliest cells, the first biochemical processes probably involved energy-rich molecules scavenged from the environment. Most of the molecules necessary for these processes are thought to have existed independently in the "organic soup" of the early oceans.

The first catalyzed reactions were probably simple, one-step reactions that brought these molecules together in various combinations. Eventually, the energy-rich molecules became depleted in the external environment, and only organisms that had evolved some means of making those molecules from other substances could survive. Thus, a hypothetical reaction,

$$F + G \longrightarrow H$$

where two energy-rich molecules (F and G) react to produce compound H and release energy, became more complex when the supply of F in the environment ran out.

A new reaction was added in which the depleted molecule, F, is made from another molecule, E, which was also present in the environment:

$$E \longrightarrow F + G \longrightarrow H$$

When the supply of E was in turn exhausted, organisms that were able to make E from some other available precursor, D, survived. When D was depleted, those organisms in turn were replaced by ones able to synthesize D from another molecule, C:

$$C \longrightarrow D \longrightarrow E \longrightarrow F + G \longrightarrow H$$

This hypothetical biochemical pathway would have evolved slowly through time, with the final reactions in the pathway evolving first and earlier reactions evolving later.

Looking at the pathway now, we would say that the "advanced" organism, starting with compound C, is able to synthesize H by means of a series of steps. This is how the biochemical pathways within organisms are thought to have evolved—not all at once, but one step at a time, backward.

Feedback inhibition regulates some biochemical pathways

For a biochemical pathway to operate efficiently, its activity must be coordinated and regulated by the cell. Not only is it unnecessary to synthesize a compound when plenty is already present, but doing so would waste energy and raw materials that could be put to use elsewhere. It is to the cell's advantage, therefore, to temporarily shut down biochemical pathways when their products are not needed.

The regulation of simple biochemical pathways often depends on an elegant feedback mechanism: The end-product of the pathway binds to an allosteric site on the enzyme that catalyzes the first reaction in the pathway. This mode of regulation is called **feedback inhibition** (figure 6.15).

In the hypothetical pathway we just described, the enzyme catalyzing the reaction C $\longrightarrow$ D would possess an allosteric site for H, the end-product of the pathway. As the pathway churned out its product and the amount of H in the cell increased, it would become more likely that an H molecule would encounter the allosteric site on the C $\longrightarrow$ D enzyme. Binding to the allosteric site

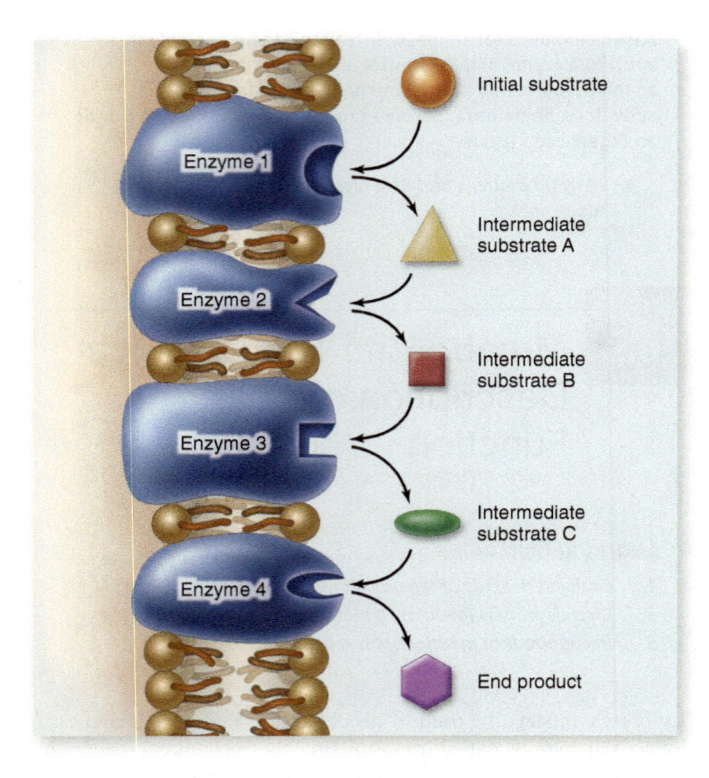

Figure 6.14 A biochemical pathway. The original substrate is acted on by enzyme 1, changing the substrate to a new intermediate, substrate A, recognized as a substrate by enzyme 2. Each enzyme in the pathway acts on the product of the previous stage. These enzymes may be either soluble or arranged in a membrane as shown.

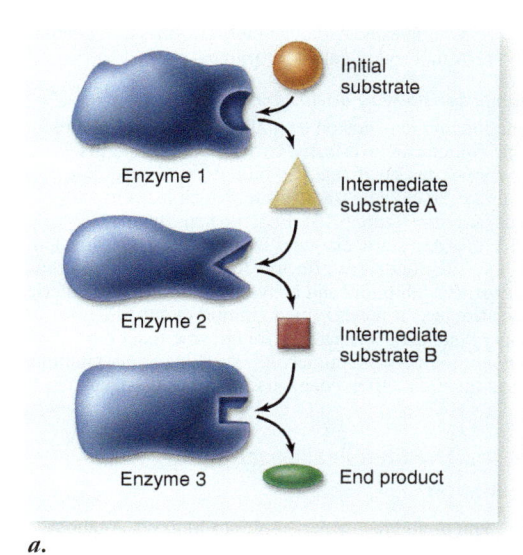

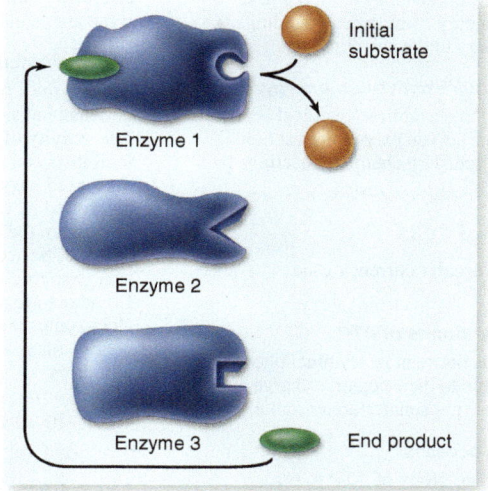

Figure 6.15 Feedback inhibition. *a.* A biochemical pathway with no feedback inhibition. *b.* A biochemical pathway in which the final end-product becomes the allosteric inhibitor for the first enzyme in the pathway. In other words, the formation of the pathway's final end-product stops the pathway. The pathway could be the synthesis of an amino acid, a nucleotide, or another important cellular molecule.

would essentially shut down the reaction C ⟶ D and in turn effectively shut down the whole pathway.

In this chapter we have reviewed the basics of energy and its transformations as carried out in living systems. Chemical bonds are the primary location of energy storage and release, and cells have developed elegant methods of making and breaking chemical bonds to create the molecules they need. Enzymes facilitate these reactions by serving as catalysts. In chapter 7 and 8 you will learn the details of the mechanisms by which organisms harvest, store, and utilize energy.

Learning Outcomes Review 6.5

Metabolism is the sum of all chemical reactions in a cell. Anabolic reactions use energy to build up molecules. Catabolic reactions release energy by breaking down molecules. In a metabolic pathway, the end-product of one reaction is the substrate for the next reaction. Evolution may have favored organisms that could use precursor molecules to synthesize a nutrient. Over time, more reactions would be linked together as novel enzymes arose by mutation.

- Is a catabolic pathway likely to be subject to feedback inhibition?

Chapter Review

6.1 The Flow of Energy in Living Systems

Thermodynamics is the study of energy changes.

Energy can take many forms.
Energy is the capacity to do work. Potential energy is stored energy, and kinetic energy is the energy of motion. Energy can take many forms: mechanical, heat, sound, electric current, light, or radioactive radiation. Energy is measured in units of heat known as kilocalories.

The Sun provides energy for living systems.
Photosynthesis stores light energy from the Sun as potential energy in the covalent bonds of sugar molecules. Breaking these bonds in living cells releases energy for use in other reactions.

Redox reactions transfer electrons.
Oxidation is a reaction involving the loss of electrons. Reduction is the gain of electrons (figure 6.2). These two reactions take place together and are therefore termed redox reactions.

6.2 The Laws of Thermodynamics and Free Energy

The First Law states that energy cannot be created or destroyed.
Virtually all activities of living organisms require energy. Energy changes form as it moves through organisms and their biochemical systems, but it is not created or destroyed.

The Second Law states that some energy is lost as disorder increases.
The disorder, or entropy, of the universe is continuously increasing. In an open system like the Earth, which is receiving energy from the Sun, this may not be the case. To increase order, however, energy must be expended. In energy conversions, some energy is always lost as heat.

Chemical reactions can be predicted based on changes in free energy.
Free energy (G) is the energy available to do work in any system. Changes in free energy (ΔG) predict the direction of reactions. Reactions with a negative ΔG are spontaneous (exergonic) reactions,

and reactions with a positive ΔG are not spontaneous (endergonic). Endergonic chemical reactions absorb energy from the surroundings, whereas exergonic reactions release energy to the surroundings.

Spontaneous chemical reactions require activation energy.
Activation energy is the energy required to destabilize chemical bonds and initiate chemical reactions (figure 6.5). Even exergonic reactions require this activation energy. Catalysts speed up chemical reactions by lowering the activation energy.

6.3 ATP: The Energy Currency of Cells

Adenosine triphosphate (ATP) is the molecular currency used for cellular energy transactions.

Cells store and release energy in the bonds of ATP.
The energy of ATP is stored in the bonds between its terminal phosphate groups. These groups repel each other due to their negative charge and therefore the covalent bonds joining these phosphates are unstable.

ATP hydrolysis drives endergonic processes.
Enzymes hydrolyze the terminal phosphate group of ATP to release energy for reactions. If ATP hydrolysis is coupled to an endergonic reaction with a positive ΔG with magnitude less than that for ATP hydrolysis, the two reactions together will be exergonic.

ATP cycles continuously.
ATP hydrolysis releases energy to drive endergonic reactions, and it is synthesized with energy from exergonic reactions (figure 6.7).

6.4 Enzymes: Biological Catalysts

Enzymes lower the activation energy of reactions.
Enzymes lower the activation energy needed to initiate a chemical reaction.

Active sites of enzymes conform to fit the shape of substrates.
Substrates bind to the active site of an enzyme. Enzymes adjust their shape to the substrate so there is a better fit (figure 6.8).

Enzymes occur in many forms.
Enzymes can be free in the cytosol or exist as components bound to membranes and organelles. Enzymes involved in a biochemical pathway can form multienzyme complexes. Although most enzymes are proteins, some are actually RNA molecules, called ribozymes.

Enzyme function is sensitive to environmental factors.
An enzyme's functionality depends on its ability to maintain its three-dimensional shape, which can be affected by temperature and pH. The activity of enzymes can be affected by inhibitors. Competitive inhibitors compete for the enzyme's active site, which leads to decreased enzyme activity (figure 6.13). Enzyme activity can be controlled by effectors. Allosteric enzymes have a second site, located away from the active site, that binds effectors to activate or inhibit the enzyme. Noncompetitive inhibitors and activators bind to the allosteric site, changing the structure of the enzyme to inhibit or activate it. Cofactors are nonorganic metals necessary for enzyme function. Coenzymes are nonprotein organic molecules, such as certain vitamins, needed for enzyme function. Often coenzymes serve as electron acceptors.

6.5 Metabolism: The Chemical Description of Cell Function

Metabolism is the sum of all biochemical reactions in a cell. Anabolic reactions require energy to build up molecules, and catabolic reactions break down molecules and release energy.

Biochemical pathways organize chemical reactions in cells.
Chemical reactions in biochemical pathways use the product of one reaction as the substrate for the next.

Biochemical pathways may have evolved in stepwise fashion.
In the primordial "soup" of the early oceans, many reactions were probably single-step reactions combining two molecules. As one of the substrate molecules was depleted, organisms having an enzyme that could synthesize the substrate would have a selective advantage. In this manner, biochemical pathways are thought to have evolved "backward" with new reactions producing limiting substrates for existing reactions.

Feedback inhibition regulates some biochemical pathways.
Biosynthetic pathways are often regulated by the end product of the pathway. Feedback inhibition occurs when the end-product of a reaction combines with an enzyme's allosteric site to shut down the enzyme's activity (figure 6.15).

Review Questions

UNDERSTAND

1. A covalent bond between two atoms represents what kind of energy?
   a. Kinetic energy
   b. Potential energy
   c. Mechanical energy
   d. Solar energy

2. During a redox reaction the molecule that gains an electron has been
   a. reduced and now has a higher energy level.
   b. oxidized and now has a lower energy level.
   c. reduced and now has a lower energy level.
   d. oxidized and now has a higher energy level.

3. An endergonic reaction has the following properties
   a. $+\Delta G$ and the reaction is spontaneous.
   b. $+\Delta G$ and the reaction is not spontaneous.
   c. $-\Delta G$ and the reaction is spontaneous.
   d. $-\Delta G$ and the reaction is not spontaneous.

4. A spontaneous reaction is one in which
   a. the reactants have a higher free energy than the products.
   b. the products have a higher free energy than the reactants.
   c. an input of energy is required.
   d. entropy is decreased.

5. What is *activation energy*?
   a. The thermal energy associated with random movements of molecules

b. The energy released through breaking chemical bonds
c. The difference in free energy between reactants and products
d. The energy required to initiate a chemical reaction

6. Which of the following is NOT a property of a catalyst?
   a. A catalyst reduces the activation energy of a reaction.
   b. A catalyst lowers the free energy of the reactants.
   c. A catalyst does not change as a result of the reaction.
   d. A catalyst works in both the forward and reverse directions of a reaction.

7. Where is the energy stored in a molecule of ATP?
   a. Within the bonds between nitrogen and carbon
   b. In the carbon-to-carbon bonds found in the ribose
   c. In the phosphorus-to-oxygen double bond
   d. In the bonds connecting the two terminal phosphate groups

APPLY

1. Cells use ATP to drive endergonic reactions because
   a. ATP is the universal catalyst.
   b. energy released by ATP hydrolysis makes ΔG for coupled reactions more negative.
   c. energy released by ATP hydrolysis makes ΔG for coupled reactions more positive.
   d. the conversion of ATP to ADP is also endergonic.

2. Which of the following statements is NOT true about enzymes?
   a. Enzymes use the three-dimensional shape of their active site to bind reactants.
   b. Enzymes lower the activation energy for a reaction.
   c. Enzymes make ΔG for a reaction more negative.
   d. Enzymes can catalyze the forward and reverse directions of a reaction.

3. ATP hydrolysis has a ΔG of −7.4 kcal/mol. Can an endergonic reaction with a ΔG of 12 kcal/mol be "driven" by ATP hydrolysis?
   a. No, the overall ΔG is still positive.
   b. Yes, the overall ΔG would now be negative.
   c. Yes, but only if an enzyme is used to lower ΔG.
   d. No, overall ΔG is now negative.

4. An online auction site offers a perpetual-motion machine. You decide not to bid on this because
   a. there is not enough energy in the universe to power this machine.
   b. the First Law says you cannot create energy.
   c. the Second Law says that energy loss due to entropy will not allow for perpetual motion.
   d. it could work, but would require a strong catalyst.

5. Enzymes have similar responses to both changes in temperature and pH. The effect of both is on the
   a. rate of movement of the substrate molecules.
   b. strength of the chemical bonds within the substrate.
   c. three-dimensional shape of the enzyme.
   d. rate of movement of the enzyme.

6. Feedback inhibition is an efficient way to control a metabolic pathway because the
   a. first enzyme in a pathway is inhibited by its own product.
   b. last enzyme in a pathway is inhibited by its own product.
   c. first enzyme in a pathway is inhibited by the end-product of the pathway.
   d. last enzyme in a pathway is inhibited by the end-product of the pathway.

SYNTHESIZE

1. Examine the graph showing the rate of reaction versus temperature for an enzyme–catalyzed reaction in a human.
   a. Describe what is happening to the enzyme at around 40°C.
   b. Explain why the line touches the x-axis at approximately 20°C and 45°C.
   c. Average body temperature for humans is 37°C. Suggest a reason why the temperature optimum of this enzyme is greater than 37°C.

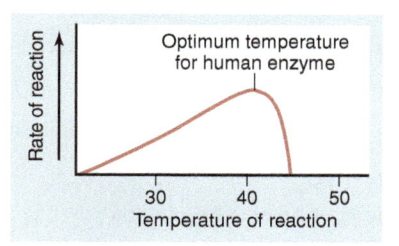

2. Phosphofructokinase functions to add a phosphate group to a molecule of fructose 6-phosphate. This enzyme functions early in glycolysis, an energy-yielding biochemical pathway discussed in chapter 7. The enzyme has an active site that binds fructose and ATP. An allosteric inhibitory site also binds ATP when cellular levels of ATP are very high.
   a. Predict the rate of the reaction if the levels of cellular ATP are low.
   b. Predict the rate of the reaction if levels of cellular ATP are very high.
   c. Describe what is happening to the enzyme when levels of ATP are very high.

CHAPTER 7

How Cells Harvest Energy

Chapter Contents

7.1 Overview of Respiration
7.2 Glycolysis: Splitting Glucose
7.3 The Oxidation of Pyruvate Produces Acetyl-CoA
7.4 The Krebs Cycle
7.5 The Electron Transport Chain and Chemiosmosis
7.6 Energy Yield of Aerobic Respiration
7.7 Regulation of Aerobic Respiration
7.8 Oxidation Without O_2
7.9 Catabolism of Proteins and Fats
7.10 Evolution of Metabolism

Introduction

Life is driven by energy. All the activities organisms carry out—the swimming of bacteria, the purring of a cat, your thinking about these words—use energy. In this chapter, we discuss the processes all cells use to derive chemical energy from organic molecules and to convert that energy to ATP. Then, in chapter 8, we will examine photosynthesis, which uses light energy to make chemical energy. We consider the conversion of chemical energy to ATP first because all organisms—both the plant, a photosynthesizer, and the caterpillar feeding on the plant, pictured in the photo—are capable of harvesting energy from chemical bonds. Energy harvest via respiration is a universal process.

7.1 Overview of Respiration

Learning Outcomes
1. Characterize oxidation–dehydrogenation reactions in biological systems.
2. Explain the role of electron carriers in energy metabolism.
3. Describe the role of ATP in biological systems.

Plants, algae, and some bacteria harvest the energy of sunlight through photosynthesis, converting radiant energy into chemical energy. These organisms, along with a few others that use chemical energy in a similar way, are called **autotrophs** ("self-feeders"). All other organisms live on the organic compounds autotrophs produce, using them as food, and are called **heterotrophs** ("fed by others"). At least 95% of the kinds of organisms on Earth—all animals and fungi, and most protists and prokaryotes—are heterotrophs. Autotrophs also extract energy from organic compounds—they just have the additional capacity to use the energy from sunlight to synthesize these compounds. The process by which energy is harvested is **cellular respiration**—the oxidation of organic compounds to extract energy from chemical bonds.

Cellular oxidations are usually also dehydrogenations

Most foods contain a variety of carbohydrates, proteins, and fats, all rich in energy-laden chemical bonds. Carbohydrates and fats, as you recall from chapter 3, possess many carbon–hydrogen (C—H) bonds, as well as carbon–oxygen (C—O) bonds.

The job of extracting energy from the complex organic mixture in most foods is tackled in stages. First, enzymes break down the large molecules into smaller ones, a process called digestion (see chapter 47). Then, other enzymes dismantle these fragments a bit at a time, harvesting energy from C—H and other chemical bonds at each stage.

The reactions that break down these molecules share a common feature: They are oxidations. Energy metabolism is therefore concerned with redox reactions, and to understand the process we must follow the fate of the electrons lost from the food molecules.

These reactions are not the simple transfer of electrons, however; they are also **dehydrogenations.** That is, the electrons lost are accompanied by protons, so that what is really lost is a hydrogen atom, not just an electron.

Cellular respiration is the complete oxidation of glucose

In chapter 6, you learned that an atom that loses electrons is said to be *oxidized,* and an atom accepting electrons is said to be *reduced.* Oxidation reactions are often coupled with reduction reactions in living systems, and these paired reactions are called *redox reactions.* Cells utilize enzyme-facilitated redox reactions to take energy from food sources and convert it to ATP.

Redox reactions

Oxidation–reduction reactions play a key role in the flow of energy through biological systems because the electrons that pass from one atom to another carry energy with them. The amount of energy an electron possesses depends on its orbital position, or energy level, around the atom's nucleus. When this electron departs from one atom and moves to another in a redox reaction, the electron's energy is transferred with it.

Figure 7.1 shows how an enzyme catalyzes a redox reaction involving an energy-rich substrate molecule, with the help of a cofactor, **nicotinamide adenosine dinucleotide (NAD$^+$)**. In this reaction, NAD$^+$ accepts a pair of electrons from the substrate, along with a proton, to form **NADH** (this process is described in more detail shortly). The oxidized product is now released from the enzyme's active site, as is NADH.

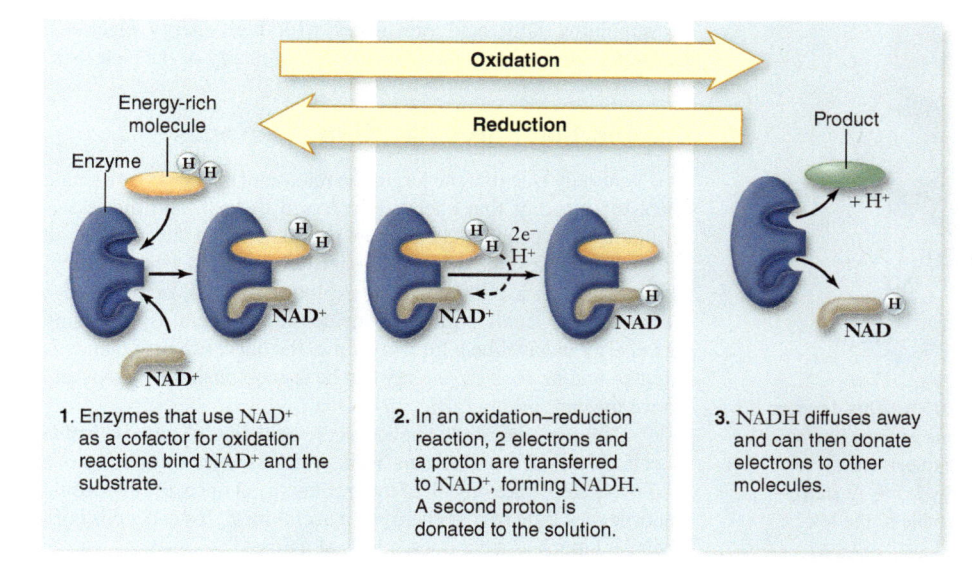

Figure 7.1 Oxidation–reduction reactions often employ cofactors. Cells use a chemical cofactor called nicotinamide adenosine dinucleotide (NAD$^+$) to carry out many oxidation–reduction reactions. Two electrons and a proton are transferred to NAD$^+$ with another proton donated to the solution. Molecules that gain electrons are said to be reduced, and ones that lose energetic electrons are said to be oxidized. NAD$^+$ oxidizes energy-rich molecules by acquiring their electrons (in the figure, this proceeds 1 ⟶ 2 ⟶ 3) and then reduces other molecules by giving the electrons to them (in the figure, this proceeds 3 ⟶ 2 ⟶ 1). NADH is the reduced form of NAD$^+$.

1. Enzymes that use NAD$^+$ as a cofactor for oxidation reactions bind NAD$^+$ and the substrate.
2. In an oxidation–reduction reaction, 2 electrons and a proton are transferred to NAD$^+$, forming NADH. A second proton is donated to the solution.
3. NADH diffuses away and can then donate electrons to other molecules.

In the overall process of cellular energy harvest dozens of redox reactions take place, and a number of molecules, including NAD⁺, act as electron acceptors. During each transfer of electrons energy is released. This energy may be captured and used to make ATP or to form other chemical bonds; the rest is lost as heat.

At the end of this process, high-energy electrons from the initial chemical bonds have lost much of their energy, and these depleted electrons are transferred to a final electron acceptor (figure 7.2). When this acceptor is oxygen, the process is called **aerobic respiration.** When the final electron acceptor is an inorganic molecule other than oxygen, the process is called **anaerobic respiration,** and when it is an organic molecule, the process is called **fermentation.**

"Burning" carbohydrates

Chemically, there is little difference between the catabolism of carbohydrates in a cell and the burning of wood in a fireplace. In both instances, the reactants are carbohydrates and oxygen, and the products are carbon dioxide, water, and energy:

$$\underset{\text{glucose}}{C_6H_{12}O_6} + \underset{\text{oxygen}}{6O_2} \longrightarrow \underset{\text{carbon dioxide}}{6CO_2} + \underset{\text{water}}{6H_2O} + \text{energy (heat and ATP)}$$

The change in free energy in this reaction is −686 kcal/mol (or −2870 kJ/mol) under standard conditions (that is, at room temperature, 1 atm pressure, and so forth). In the conditions that exist inside a cell, the energy released can be as high as −720 kcal/mol (−3012 kJ/mol) of glucose. This means that under actual cellular conditions, more energy is released than under standard conditions.

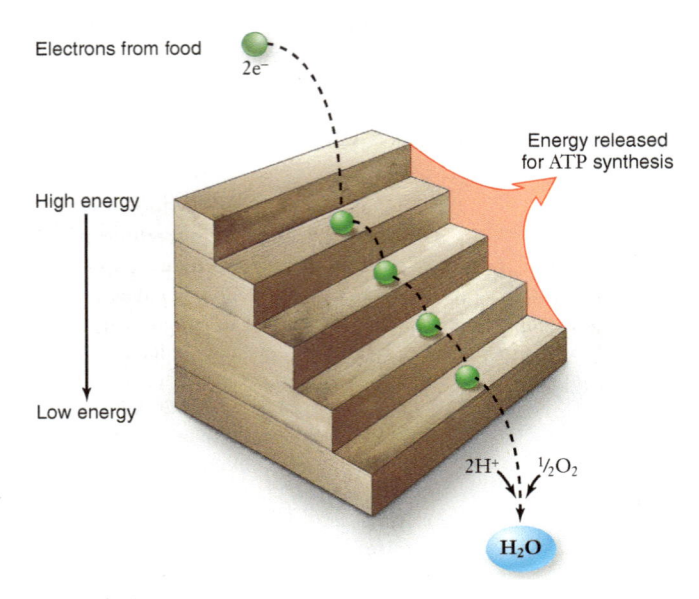

Figure 7.2 How electron transport works. This diagram shows how ATP is generated when electrons transfer from one energy level to another. Rather than releasing a single explosive burst of energy, electrons "fall" to lower and lower energy levels in steps, releasing stored energy with each fall as they tumble to the lowest (most electronegative) electron acceptor, O₂.

The same amount of energy is released whether glucose is catabolized or burned, but when it is burned, most of the energy is released as heat. Cells harvest useful energy from the catabolism of glucose by using a portion of the energy to drive the production of ATP.

Electron carriers play a critical role in energy metabolism

During respiration, glucose is oxidized to CO_2. If the electrons were given directly to O_2, the reaction would be combustion, and cells would burst into flames. Instead, as you have just seen, the cell transfers the electrons to intermediate electron carriers, then eventually to O_2.

Many forms of electron carriers are used in this process: (1) soluble carriers that move electrons from one molecule to another, (2) membrane-bound carriers that form a redox chain, and (3) carriers that move within the membrane. The common feature of all of these carriers is that they can be reversibly oxidized and reduced. Some of these carriers, such as the iron-containing cytochromes, can carry just electrons, and some carry both electrons and protons.

NAD⁺ is one of the most important electron (and proton) carriers. As shown on the left in figure 7.3, the NAD⁺ molecule is composed of two nucleotides bound together. The two nucleotides that make up NAD⁺, nicotinamide monophosphate (NMP) and adenosine monophosphate (AMP), are joined head-to-head by their phosphate groups. The two nucleotides serve different functions in the NAD⁺ molecule: AMP acts as the core, providing a shape recognized by many enzymes; NMP is the active part of the molecule, because it is readily reduced—that is, it easily accepts electrons.

When NAD⁺ acquires two electrons and a proton from the active site of an enzyme, it is reduced to NADH, shown on the right in figure 7.3. The NADH molecule now carries the two energetic electrons and can supply them to other molecules and reduce them.

This ability to supply high-energy electrons is critical to both energy metabolism and to the biosynthesis of many organic molecules, including fats and sugars. In animals, when ATP is plentiful, the reducing power of the accumulated NADH is diverted to supplying fatty acid precursors with high-energy electrons, reducing them to form fats and storing the energy of the electrons.

Respiration harvests energy in stages

It is generally true that the larger the release of energy in any single step, the more of that energy is released as heat, and the less is available to be channeled into more useful paths. In the combustion of gasoline, the same amount of energy is released whether all of the gasoline in a car's gas tank explodes at once, or burns in a series of very small explosions inside the cylinders. By releasing the energy in gasoline a little at a time, the harvesting efficiency is greater, and more of the energy can be used to push the pistons and move the car.

The same principle applies to the oxidation of glucose inside a cell. If all of the electrons were transferred to oxygen in one explosive step, releasing all of the free energy at once, the cell would recover very little of that energy in a useful form. Instead, cells burn their fuel much as a car does, a little at a time.

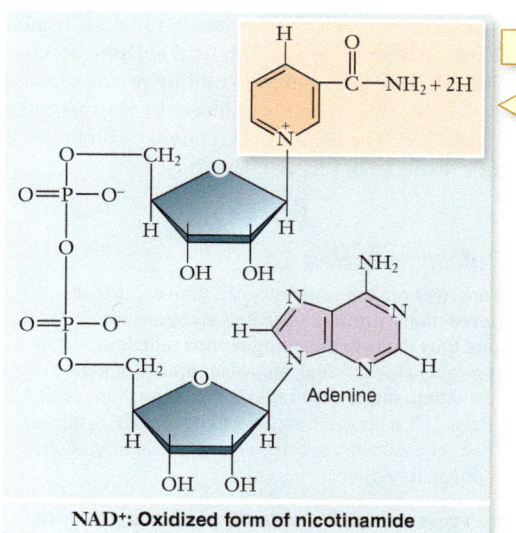

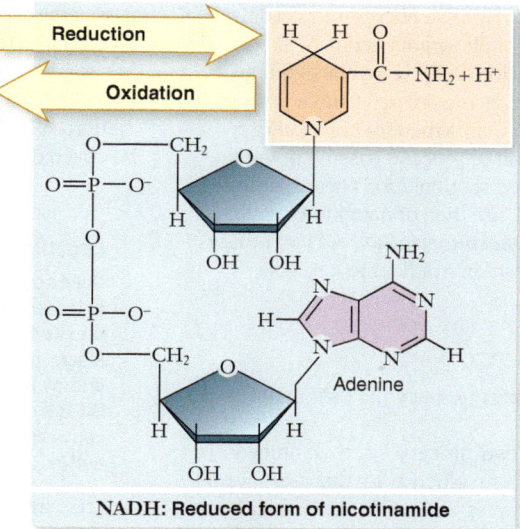

Figure 7.3 NAD⁺ and NADH. This dinucleotide serves as an "electron shuttle" during cellular respiration. NAD⁺ accepts a pair of electrons and a proton from catabolized macromolecules and is reduced to NADH.

NAD⁺: Oxidized form of nicotinamide

NADH: Reduced form of nicotinamide

The electrons in the C—H bonds of glucose are stripped off in stages in the series of enzyme-catalyzed reactions collectively referred to as glycolysis and the Krebs cycle. The electrons are removed by transferring them to NAD⁺, as described earlier, or to other electron carriers.

The energy released by all of these oxidation reactions is also not all released at once (see figure 7.2). The electrons are passed to another set of electron carriers called the **electron transport chain,** which is located in the mitochondrial inner membrane. Movement of electrons through this chain produces potential energy in the form of an electrochemical gradient. We examine this process in more detail later in section 7.5.

ATP plays a central role in metabolism

Chapter 6 introduced ATP as the energy currency of the cell. Cells use ATP to power most of those activities that require work—one of the most obvious of which is movement. Tiny fibers within muscle cells pull against one another when muscles contract. Mitochondria can move a meter or more along the narrow nerve cells that extend from your spine to your feet. Chromosomes are pulled apart by microtubules during cell division. All of these movements require the expenditure of energy by ATP hydrolysis. Cells also use ATP to drive endergonic reactions that would otherwise not occur spontaneously (see chapter 6).

How does ATP drive an endergonic reaction? The enzyme that catalyzes a particular reaction has two binding sites on its surface: one for the reactant and another for ATP. The ATP site hydrolyzes the terminal phosphate of ATP, releasing over 7 kcal ($\Delta G = -7.3$ kcal/mol) of energy. This provides the energy absorbed by the endergonic reaction. Thus endergonic reactions coupled to ATP hydrolysis become favorable.

The many steps of cellular respiration have as their ultimate goal the production of ATP. ATP synthesis is itself an endergonic reaction, which uses energy from the exergonic reactions of cellular respiration.

Cells make ATP by two fundamentally different mechanisms

The synthesis of ATP can be accomplished by two distinct mechanisms: one that involves chemical coupling with an intermediate bound to phosphate, and another that relies on an electrochemical gradient of protons for the potential energy to phosphorylate ADP.

1. In *substrate-level phosphorylation*, ATP is formed by transferring a phosphate group directly to ADP from a phosphate-bearing intermediate, or substrate (figure 7.4). During **glycolysis,** the initial breakdown of glucose (see section 7.2), the chemical bonds of glucose are shifted

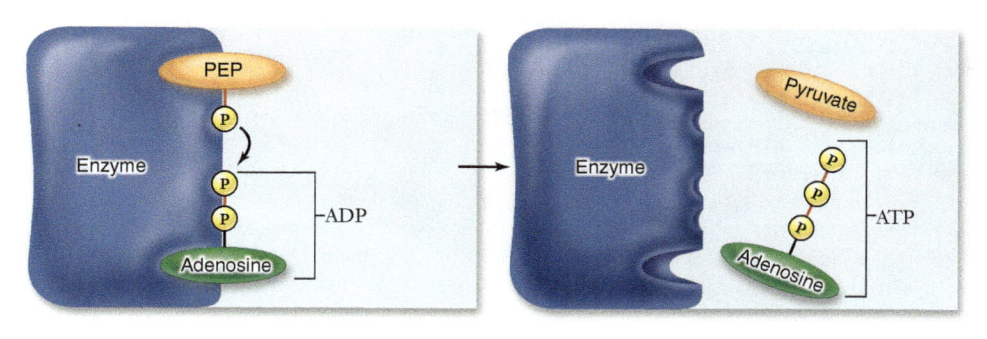

Figure 7.4 Substrate-level phosphorylation. Some molecules, such as phosphoenolpyruvate (PEP), possess a high-energy phosphate (P) bond similar to the bonds in ATP. When PEP's phosphate group is transferred enzymatically to ADP, the energy in the bond is conserved, and ATP is created.

around in reactions that provide the energy required to form ATP by substrate-level phosphorylation.

2. In **oxidative phosphorylation,** ATP is synthesized by the enzyme **ATP synthase,** using energy from a proton (H^+) gradient. This gradient is formed by high-energy electrons from the oxidation of glucose passing down an electron transport chain (see section 7.5). These electrons, with their energy depleted, are then donated to oxygen, hence the term *oxidative phosphorylation*. ATP synthase uses the energy from the proton gradient to catalyze the reaction:

$$ADP + P_i \longrightarrow ATP$$

Eukaryotes and aerobic prokaryotes produce the vast majority of their ATP this way.

In most organisms, these two processes are combined. To harvest energy to make ATP from glucose in the presence of oxygen, the cell carries out a complex series of enzyme-catalyzed reactions that remove energetic electrons via oxidation reactions. These electrons are then used in an electron transport chain that passes the electrons down a series of carriers while translocating protons into the intermembrane space. The final electron acceptor in aerobic respiration is oxygen, and the resulting proton gradient provides energy for the enzyme ATP synthase to phosphorylate ADP to ATP (figure 7.5). The details of this complex process will be covered in the remainder of this chapter.

Learning Outcomes Review 7.1

Cells acquire energy from the complete oxidation of glucose. In these redox reactions, protons as well as electrons are transferred, and thus they are dehydrogenation reactions. Electron carriers aid in the gradual, stepwise release of the energy from oxidation, rather than rapid combustion. The result is the synthesis of ATP, a portable source of energy. ATP synthesis can occur by two mechanisms: substrate-level phosphorylation and oxidative phosphorylation.

■ *Why don't cells just link the oxidation of glucose directly to cellular functions that require the energy?*

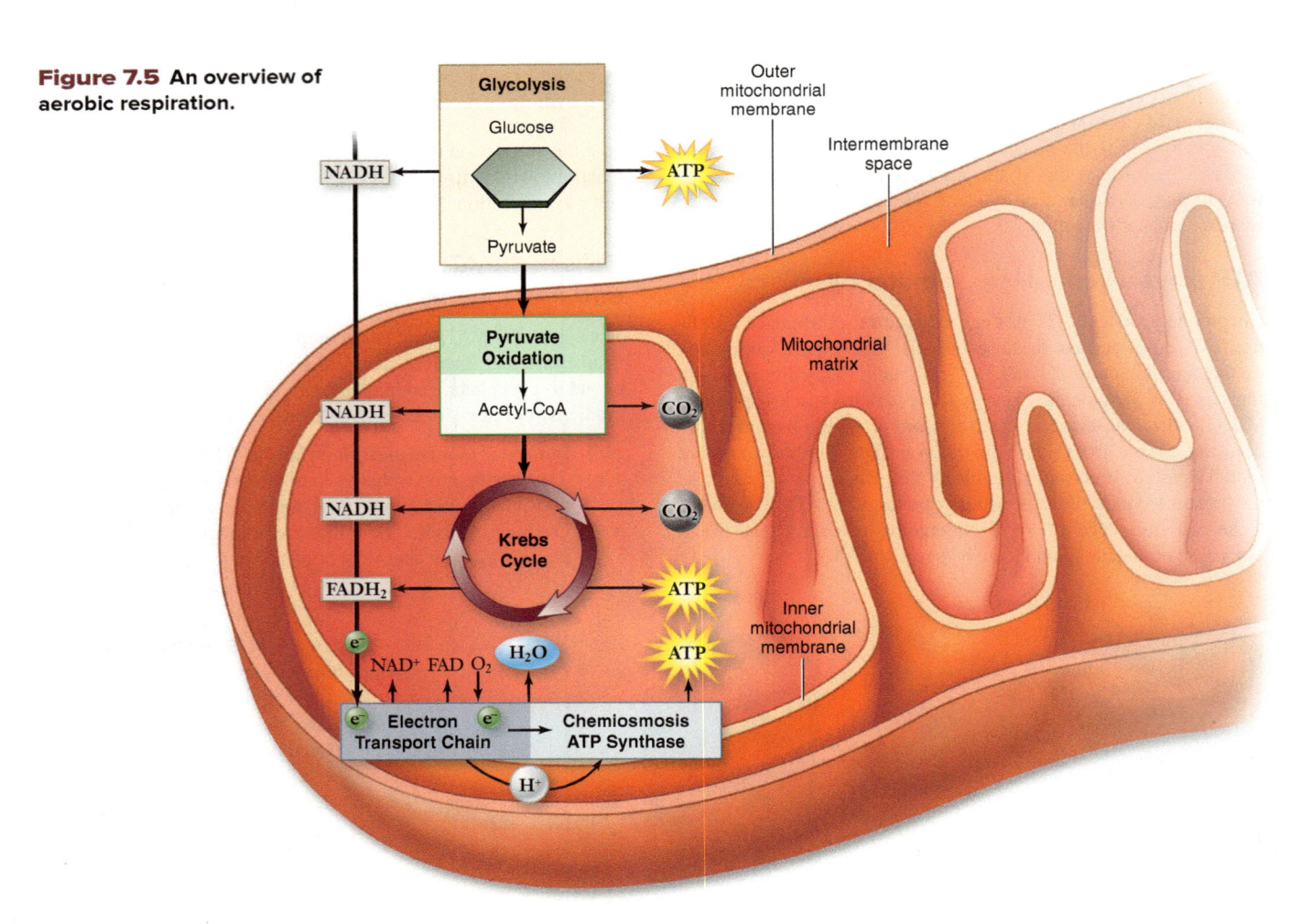

Figure 7.5 An overview of aerobic respiration.

7.2 Glycolysis: Splitting Glucose

Learning Outcomes
1. Describe the process of glycolysis.
2. Calculate the energy yield from glycolysis.
3. Distinguish between aerobic respiration and fermentation.

Glucose molecules can be dismantled in many ways, but primitive organisms evolved a glucose-catabolizing process that releases enough free energy to drive the synthesis of ATP in enzyme-coupled reactions. Glycolysis occurs in the cytoplasm and converts glucose into two 3-carbon molecules of pyruvate (figure 7.6). For each molecule of glucose that passes through this transformation, the cell nets two ATP molecules.

Glycolysis converts glucose into two pyruvate, forming two ATP and two NADH in the process

The first half of glycolysis consists of five sequential reactions that convert one molecule of glucose into two molecules of the 3-carbon compound **glyceraldehyde 3-phosphate (G3P)**. These reactions require the expenditure of ATP, so they constitute an endergonic process. In the second half of glycolysis, five more reactions convert G3P into pyruvate in an energy-yielding process that generates ATP.

Priming reactions The first three reactions "prime" glucose by changing it into a compound that can be readily cleaved into two 3-carbon phosphorylated molecules. Two of these reactions transfer a phosphate from ATP, so this step requires the cell to use two ATP molecules.

Cleavage This 6-carbon diphosphate sugar is then split into two 3-carbon monophosphate sugars. One of these is G3P, and the other is converted into G3P. The G3P then undergoes a series of reactions that eventually yields more energy than was spent priming (figure 7.7).

Oxidation and ATP formation Each G3P is oxidized, transferring two electrons (and one proton) to NAD$^+$, thus forming NADH. A molecule of P$_i$ is also added to G3P to produce 1,3-bisphosphoglycerate (BPG). The phosphate incorporated can be transferred to ADP by substrate-level phosphorylation (see figure 7.4) to allow a positive yield of ATP at the end of the process.

Another four reactions convert BPG into pyruvate. In the process, the phosphates are transferred to ADP to yield two ATP per G3P. The entire process is shown in detail in figure 7.7.

Each glucose molecule is split into two G3P molecules, so the overall reaction sequence has a net yield of two molecules of ATP, as well as two molecules of NADH and two of pyruvate:

    4 ATP (2 ATP for each of the 2 G3P molecules)
    − 2 ATP (used in the two reactions in the first step)

    2 ATP (net yield for entire process)

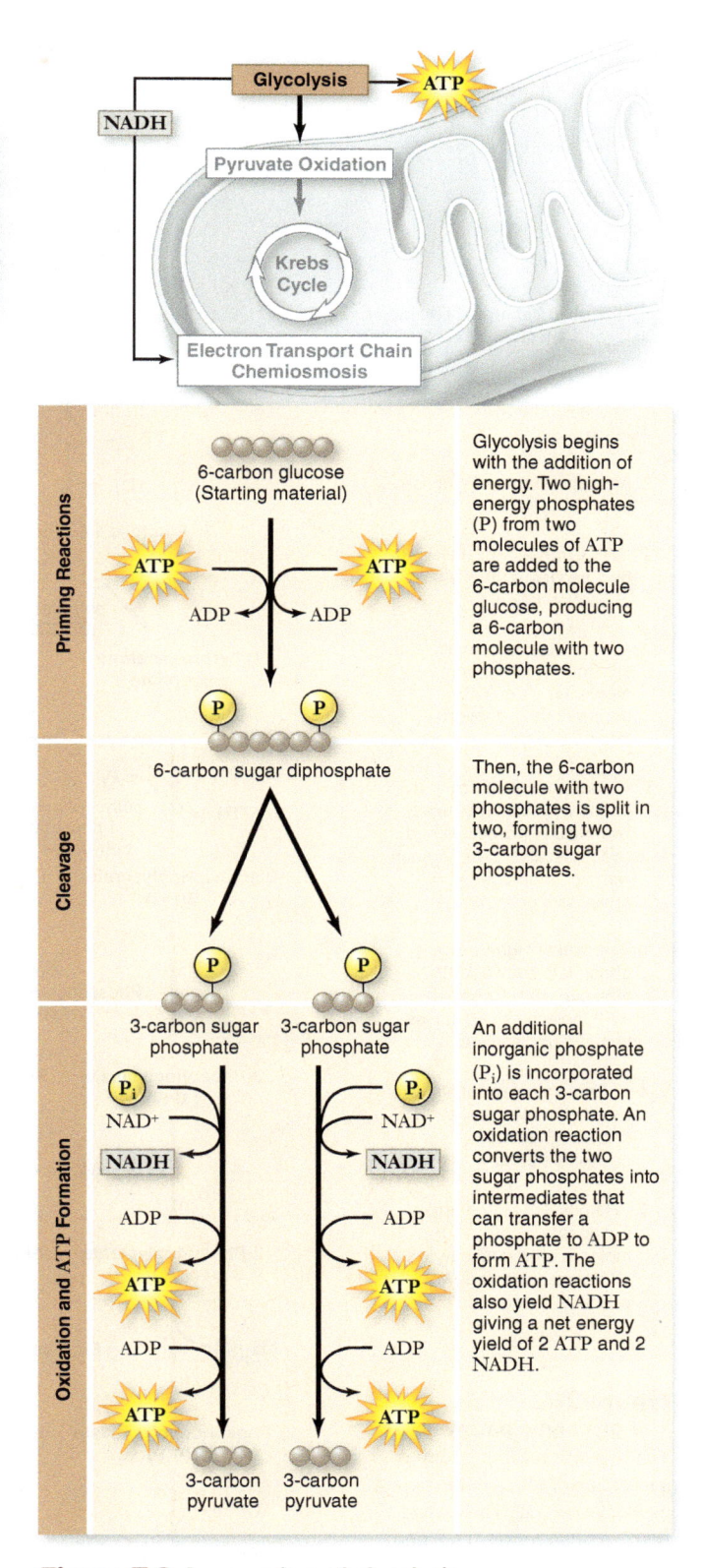

Figure 7.6 An overview of glycolysis.

Principles of Biology I and II

Glycolysis: The Reactions

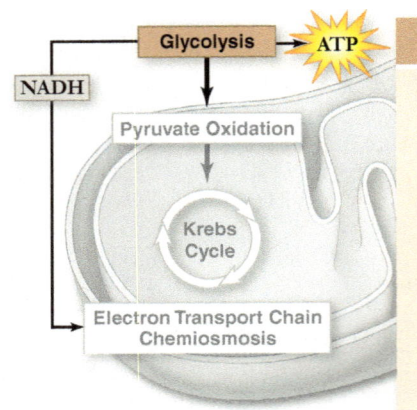

1. Phosphorylation of glucose by ATP.

2–3. Rearrangement, followed by a second ATP phosphorylation.

4–5. The 6-carbon molecule is split into two 3-carbon molecules—one G3P, another that is converted into G3P in another reaction.

6. Oxidation followed by phosphorylation produces two NADH molecules and two molecules of BPG, each with one high-energy phosphate bond.

7. Removal of high-energy phosphate by two ADP molecules produces two ATP molecules and leaves two 3PG molecules.

8–9. Removal of water yields two PEP molecules, each with a high-energy phosphate bond.

10. Removal of high-energy phosphate by two ADP molecules produces two ATP molecules and two pyruvate molecules.

Figure 7.7
The glycolytic pathway.
The first five reactions convert a molecule of glucose into two molecules of G3P. The second five reactions convert G3P into pyruvate.

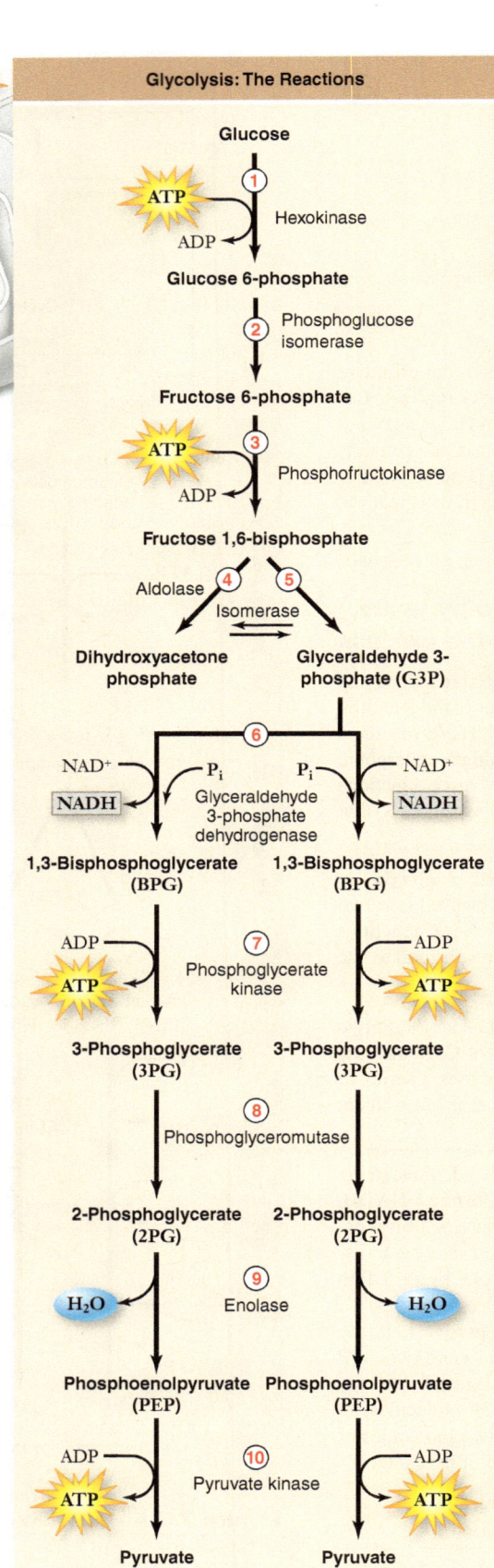

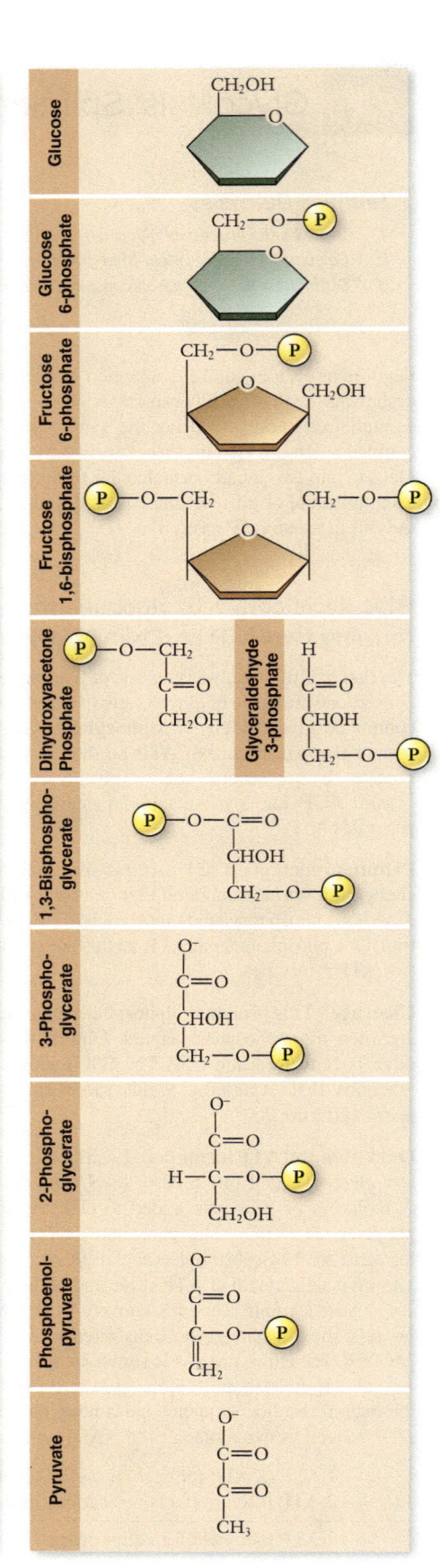

Reaction pathway:

Glucose → (1) Hexokinase, ATP → ADP → Glucose 6-phosphate → (2) Phosphoglucose isomerase → Fructose 6-phosphate → (3) Phosphofructokinase, ATP → ADP → Fructose 1,6-bisphosphate → (4) Aldolase / (5) Isomerase → Dihydroxyacetone phosphate ⇌ Glyceraldehyde 3-phosphate (G3P) → (6) Glyceraldehyde 3-phosphate dehydrogenase, NAD$^+$ + P$_i$ → NADH → 1,3-Bisphosphoglycerate (BPG) → (7) Phosphoglycerate kinase, ADP → ATP → 3-Phosphoglycerate (3PG) → (8) Phosphoglyceromutase → 2-Phosphoglycerate (2PG) → (9) Enolase, H$_2$O → Phosphoenolpyruvate (PEP) → (10) Pyruvate kinase, ADP → ATP → Pyruvate

Overview: Glycolysis (ATP, NADH) → Pyruvate Oxidation → Krebs Cycle → Electron Transport Chain / Chemiosmosis

Structures:
- Glucose
- Glucose 6-phosphate
- Fructose 6-phosphate
- Fructose 1,6-bisphosphate
- Dihydroxyacetone Phosphate
- Glyceraldehyde 3-phosphate
- 1,3-Bisphosphoglycerate
- 3-Phosphoglycerate
- 2-Phosphoglycerate
- Phosphoenolpyruvate
- Pyruvate

The hydrolysis of one molecule of ATP yields a ΔG of -7.3 kcal/mol under standard conditions. Thus cells harvest a maximum of 14.6 kcal of energy per mole of glucose from glycolysis.

A brief history of glycolysis

Although the ATP yield from glycolysis is low, it is actually quite efficient, with just under 40% of the energy released being trapped as ATP. For more than a billion years during the anaerobic first stages of life on Earth, glycolysis was the primary way heterotrophic organisms generated ATP from organic molecules.

Like many biochemical pathways, glycolysis is believed to have evolved backward—the last steps in the process being the most ancient. Thus, the second half of glycolysis, the ATP-yielding breakdown of G3P, may have been the original process. The synthesis of G3P from glucose would have appeared later, perhaps when alternative sources of G3P were depleted.

Why does glycolysis take place in modern organisms, since its energy yield in the absence of oxygen is comparatively little? There are several possible answers. First, the process is energetically efficient, and better than the alternative—no ATP. Second, evolution is an incremental process: Change occurs by improving on past successes. In catabolic metabolism, glycolysis satisfied the one essential evolutionary criterion—it was an improvement. Cells that could not carry out glycolysis were at a competitive disadvantage, and only cells capable of glycolysis survived. Later improvements in catabolic metabolism built on this framework to increase the yield of ATP as oxygen became available as an oxidizing agent. Metabolism evolved as one layer of reactions added to another. Nearly every present-day organism carries out glycolysis, as a metabolic memory of its evolutionary past.

The last section of this chapter discusses the evolution of metabolism in more detail.

NADH must be recycled to continue respiration

Consider the net reaction of the glycolytic sequence:

$$\text{glucose} + 2\text{ ADP} + 2\text{ P}_i + 2\text{ NAD}^+ \longrightarrow 2\text{ pyruvate} + 2\text{ ATP} + 2\text{ NADH} + 2\text{H}^+ + 2\text{H}_2\text{O}$$

You can see that three changes occur in glycolysis: (1) glucose is converted into two molecules of pyruvate; (2) two molecules of ADP are converted into ATP via substrate-level phosphorylation; and (3) two molecules of NAD^+ are reduced to NADH. This leaves the cell with two problems: extracting the energy that remains in the two pyruvate molecules, and regenerating NAD^+ to be able to continue glycolysis.

Recycling NADH

As long as glucose is available, a cell can continually churn out ATP by glycolysis to drive its activities. However, this process accumulates NADH and depletes the pool of NAD^+ molecules. Cells do not contain a large amount of NAD^+ so for glycolysis to continue, NADH must be recycled into NAD^+. The NADH is oxidized back to NAD^+ by reducing another molecule. Cells can do this in two ways, and which one is used depends on whether O_2 is available (figure 7.8):

1. **Aerobic respiration.** Oxygen has a high affinity for electrons, making it an excellent electron acceptor. Electrons are transferred through a series of membrane carriers, ultimately reducing oxygen and forming water. This process occurs in the mitochondria of eukaryotic cells in the presence of oxygen. Because air is rich in oxygen, this process is also referred to as *aerobic metabolism*. A significant amount of ATP is also produced.
2. **Fermentation.** When oxygen is unavailable, an organic molecule can accept electrons. The organic molecules used are quite varied and include acetaldehyde in ethanolic fermentation or pyruvate itself in lactic acid fermentation. This reaction plays an important role in the metabolism of most organisms, even those capable of aerobic respiration.

The fate of pyruvate

The fate of the pyruvate that is produced by glycolysis depends on which of these two processes takes place. The aerobic respiration path starts with the oxidation of pyruvate to produce acetyl coenzyme

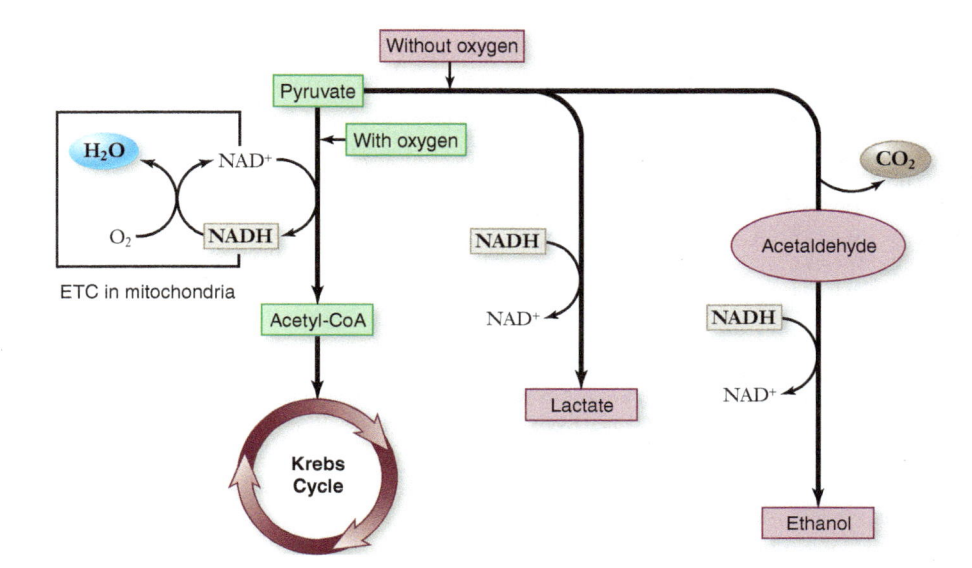

Figure 7.8 The fate of pyruvate and NADH produced by glycolysis. In the presence of oxygen, NADH is oxidized by the electron transport chain (ETC) in mitochondria using oxygen as the final electron acceptor. This regenerates NAD^+, allowing glycolysis to continue. The pyruvate produced by glycolysis is oxidized to acetyl-CoA, which enters the Krebs cycle. In the absence of oxygen, pyruvate is instead reduced, oxidizing NADH and regenerating NAD^+, thus allowing glycolysis to continue. Direct reduction of pyruvate, as in muscle cells, produces lactate. In yeast, carbon dioxide is first removed from pyruvate, producing acetaldehyde, which is then reduced to ethanol.

A (acetyl-CoA), which is then further oxidized in a series of reactions called the Krebs cycle. The fermentation path, by contrast, uses the reduction of all or part of pyruvate to oxidize NADH back to NAD^+. We examine aerobic respiration next; fermentation is described in detail in section 7.8.

Learning Outcomes Review 7.2

Glycolysis splits the 6-carbon molecule glucose into two 3-carbon molecules of pyruvate. This process uses two ATP molecules in "priming" reactions and eventually produces four molecules of ATP per glucose for a net yield of two ATP. The oxidation reactions of glycolysis require NAD^+ and produce NADH. When oxygen is abundant, NAD^+ is regenerated in the electron transport chain, using O_2 as an acceptor. When oxygen is absent, NAD^+ is regenerated in a fermentation reaction using an organic molecule as an electron receptor.

- Does glycolysis taking place in the cytoplasm argue for or against the endosymbiotic origin of mitochondria?

7.3 The Oxidation of Pyruvate Produces Acetyl-CoA

Learning Outcome

1. Diagram how the oxidation of pyruvate links glycolysis with the Krebs cycle.

In the presence of oxygen, the pyruvate produced by glycolysis can be further oxidized. In eukaryotic organisms, the extraction of additional energy from pyruvate takes place exclusively inside mitochondria. In prokaryotes, similar reactions take place in the cytoplasm and at the plasma membrane.

The cell harvests pyruvate's considerable energy in two steps. First, pyruvate is oxidized to produce a 2-carbon compound and CO_2, while reducing NAD^+ to NADH. Next, the 2-carbon compound is oxidized to CO_2 by the reactions of the Krebs cycle.

Pyruvate is oxidized in a "decarboxylation" reaction that cleaves off one of pyruvate's three carbons in the form of CO_2 (figure 7.9). The remaining 2-carbon compound, called an acetyl group, becomes bound to coenzyme A, producing *acetyl-CoA*. This oxidation is also a dehydrogenation, so a pair of electrons and one associated proton are transferred to NAD^+, reducing it to NADH, with a second proton donated to the solution.

The reaction involves three intermediate stages, and it is catalyzed within mitochondria by a *multienzyme complex*. As chapter 6 noted, a multienzyme complex organizes a series of enzymatic steps so that the chemical intermediates do not diffuse away or undergo other reactions. Within the complex, subunits pass the substrates from one enzyme to the next without releasing them. The enzyme that performs these concerted reactions is called *pyruvate dehydrogenase*, and it is one of the largest enzymes known; it contains 60 subunits! The reaction can be summarized as follows:

$$\text{pyruvate} + NAD^+ + CoA \longrightarrow \text{acetyl-CoA} + NADH + CO_2 + H^+$$

The molecule of NADH produced is used later to produce ATP. The acetyl group is fed into the Krebs cycle, with the CoA being recycled for another oxidation of pyruvate. The Krebs cycle then completes the oxidation of the original carbons from glucose.

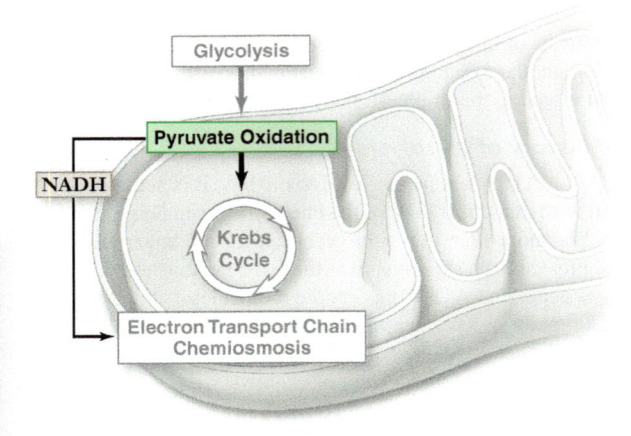

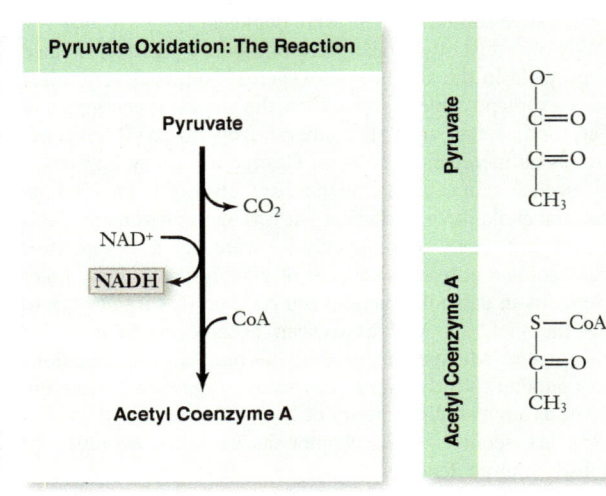

Figure 7.9 The oxidation of pyruvate. This complex reaction uses NAD^+ to accept electrons, reducing it to NADH. The product, acetyl coenzyme A (acetyl-CoA), feeds the acetyl unit into the Krebs cycle, and the CoA is recycled for another oxidation of pyruvate. NADH provides energetic electrons for the electron transport chain.

Learning Outcome Review 7.3

Pyruvate is oxidized in the mitochondria to produce acetyl-CoA and CO_2. Acetyl-CoA is the molecule that links glycolysis and the reactions of the Krebs cycle.

- What are the advantages and disadvantages of a multienzyme complex?

7.4 The Krebs Cycle

Learning Outcomes
1. Relate the nine reactions of the Krebs cycle to the flow of carbon and electrons in the cycle.
2. Diagram the oxidation reactions in the Krebs cycle.

The *Krebs cycle* allows the oxidation of 2-carbon units in the form of acetyl groups bound to CoA (acetyl-CoA). These can come from the oxidation of pyruvate, or from the oxidation of fatty acids (see section 7.9). The acetyl group is added to a 4-carbon acid, oxaloacetate. The resulting 6-carbon molecule is citric acid, thus the cycle is also called the citric acid cycle, and the TCA cycle (for tricarboxylic acid). The reactions of the Krebs cycle convert citric back to oxaloacetate, generating CO_2 and transferring electrons and protons to the electron carriers NADH and $FADH_2$. A single ATP is generated during the cycle as well, but most of the energy released is retained in the form of the electrons in NADH and $FADH_2$ that can be used by the electron transport chain to generate a *proton gradient* to drive ATP synthesis.

An overview of the Krebs cycle

The reactions of the Krebs cycle take place in the mitochondrial matrix. They take in acetyl units from acetyl-CoA, convert them into CO_2, transferring electrons and protons to NADH and $FADH_2$ (figure 7.10).

The first reaction combines the 4-carbon oxaloacetate with the acetyl group to produce the 6-carbon citrate molecule. Five more steps, which have been simplified in figure 7.10, convert citrate to a 5-carbon intermediate and then to the 4-carbon succinate. During these reactions, two NADH and one ATP are produced.

Succinate undergoes three additional reactions, also simplified in the figure, to become oxaloacetate. During these reactions, one more NADH is produced; in addition, a molecule of flavin adenine dinucleotide (FAD), another cofactor, becomes reduced to $FADH_2$.

The specifics of each reaction are described next.

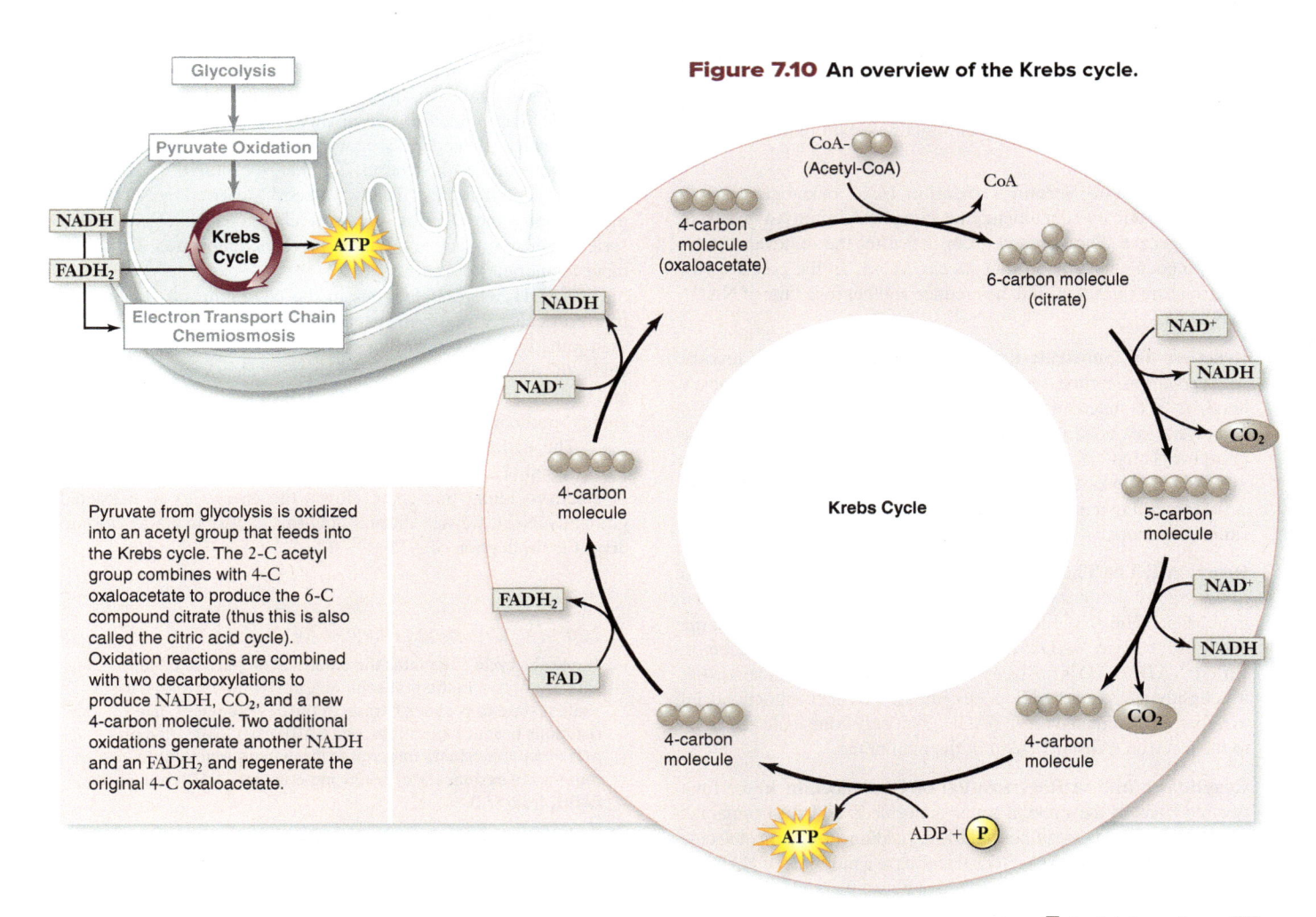

Figure 7.10 An overview of the Krebs cycle.

Pyruvate from glycolysis is oxidized into an acetyl group that feeds into the Krebs cycle. The 2-C acetyl group combines with 4-C oxaloacetate to produce the 6-C compound citrate (thus this is also called the citric acid cycle). Oxidation reactions are combined with two decarboxylations to produce NADH, CO_2, and a new 4-carbon molecule. Two additional oxidations generate another NADH and an $FADH_2$ and regenerate the original 4-C oxaloacetate.

The Krebs cycle extracts electrons and synthesizes one ATP

Figure 7.11 summarizes the sequence of the Krebs cycle reactions. A 2-carbon group from acetyl-CoA enters the cycle at the beginning, and two CO_2 molecules, one ATP, and four pairs of electrons are produced.

Reaction 1: Condensation Citrate is formed from acetyl-CoA and oxaloacetate. This condensation reaction is irreversible, committing the 2-carbon acetyl group to the Krebs cycle. The reaction is inhibited when the cell's ATP concentration is high and stimulated when it is low. The result is that when the cell possesses ample amounts of ATP, the Krebs cycle shuts down, and acetyl-CoA is channeled into fat synthesis.

Reactions 2 and 3: Isomerization Before the oxidation reactions can begin, the hydroxyl (—OH) group of citrate must be repositioned. This rearrangement is done in two steps: First, a water molecule is removed from one carbon; then water is added to a different carbon. As a result, an —H group and an —OH group change positions. The product is an isomer of citrate called *isocitrate*. This rearrangement facilitates the subsequent reactions.

Reaction 4: The First Oxidation In the first energy-yielding step of the cycle, isocitrate undergoes an oxidative decarboxylation reaction. First, isocitrate is oxidized, yielding a pair of electrons that reduce a molecule of NAD^+ to NADH. Then the oxidized intermediate is decarboxylated; the central carboxyl group splits off to form CO_2, yielding a 5-carbon molecule called α-ketoglutarate.

Reaction 5: The Second Oxidation Next, α-ketoglutarate is decarboxylated by a multienzyme complex similar to pyruvate dehydrogenase. The succinyl group left after the removal of CO_2 joins to coenzyme A, forming *succinyl-CoA*. In the process, two electrons are extracted, and they reduce another molecule of NAD^+ to NADH.

Reaction 6: Substrate-Level Phosphorylation The linkage between the 4-carbon succinyl group and CoA is a high-energy bond. In a coupled reaction similar to those that take place in glycolysis, this bond is cleaved, and the energy released drives the phosphorylation of guanosine diphosphate (GDP), forming guanosine triphosphate (GTP). GTP can transfer a phosphate to ADP converting it into ATP. The 4-carbon molecule that remains is called *succinate*.

Reaction 7: The Third Oxidation Next, succinate is oxidized to *fumarate* by an enzyme located in the inner mitochondrial membrane. The free-energy change in this reaction is not large enough to reduce NAD^+. Instead, FAD is the electron acceptor. Unlike NAD^+, FAD is not free to diffuse within the mitochondrion; it is tightly associated with its enzyme in the inner mitochondrial membrane. Its reduced form, $FADH_2$, can only contribute electrons to the electron transport chain in the membrane.

Reactions 8 and 9: Regeneration of Oxaloacetate In the final two reactions of the cycle, a water molecule is added to fumarate, forming *malate*. Malate is then oxidized, yielding a 4-carbon molecule of *oxaloacetate* and two electrons that reduce a molecule of NAD^+ to NADH. Oxaloacetate, the molecule that began the cycle, is now free to combine with another 2-carbon acetyl group from acetyl-CoA and begin the cycle again.

Glucose becomes CO_2 and potential energy

In the process of aerobic respiration, glucose is entirely consumed. The 6-carbon glucose molecule is cleaved into two 3-carbon pyruvate molecules during glycolysis. One of the carbons of each pyruvate is then lost as CO_2 in the conversion of pyruvate to acetyl-CoA. The two other carbons from acetyl-CoA are lost as CO_2 during the oxidations of the Krebs cycle.

All that is left to mark the passing of a glucose molecule into six CO_2 molecules is its energy, some of which is preserved in four ATP molecules and in the reduced state of 12 electron carriers. Ten of these carriers are NADH molecules; the other two are $FADH_2$.

Following the electrons in the reactions reveals the direction of transfer

As you examine the changes in electrical charge in the reactions that oxidize glucose, a good strategy for keeping the transfers clear is always to *follow the electrons*. For example, in glycolysis, an enzyme extracts two hydrogens—that is, two electrons and two protons—from glucose and transfers both electrons and one of the protons to NAD^+. The other proton is released as a hydrogen ion, H^+, into the surrounding solution. This transfer converts NAD^+ into NADH—that is, two negative electrons ($2e^-$) and one positive proton (H^+) are added to one positively charged NAD^+ to form NADH, which is electrically neutral.

As mentioned in section 7.1, energy captured by NADH is not harvested all at once. The two electrons carried by NADH are passed along the electron transport chain, which consists of a series of electron carriers, mostly proteins, embedded within the inner membranes of mitochondria.

NADH delivers electrons to the beginning of the electron transport chain, and oxygen captures them at the end. The oxygen then joins with hydrogen ions to form water. At each step in the chain, the electrons move to a slightly more electronegative carrier, and their positions shift slightly. Thus, the electrons move *down* an energy gradient.

The entire process of electron transfer releases a total of 53 kcal/mol (222 kJ/mol) under standard conditions. The transfer of electrons along this chain allows the energy to be extracted gradually. Next, we will discuss how this energy is put to work to drive the production of ATP.

Learning Outcomes Review 7.4

The Krebs cycle completes the oxidation of glucose begun with glycolysis. In the first segment, acetyl-CoA is added to oxaloacetate to produce citrate. In the next segment, five reactions produce succinate, two **NADH** from **NAD^+**, and one ATP. Finally, succinate undergoes three more reactions to regenerate oxaloacetate, producing one more **NADH** and one **$FADH_2$** from **FAD**.

- *What happens to the electrons removed from glucose at this point?*

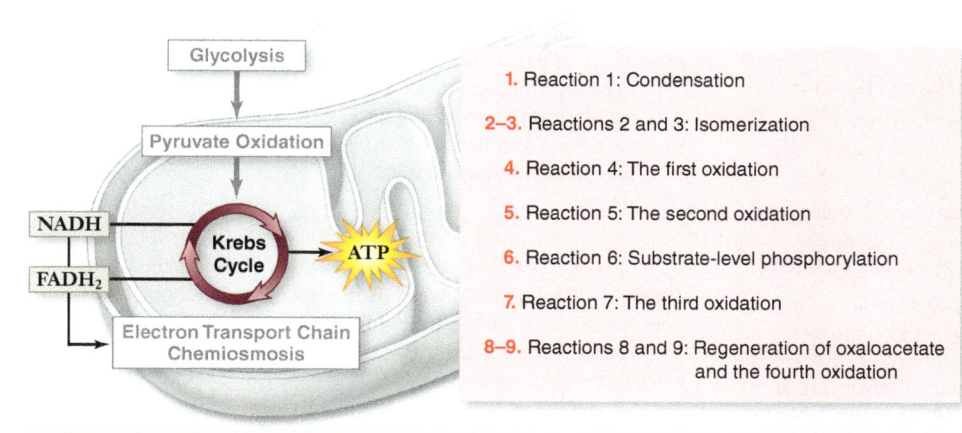

Figure 7.11 The Krebs cycle. This series of reactions takes place within the matrix of the mitochondrion. For the complete breakdown of a molecule of glucose, the two molecules of acetyl-CoA produced by glycolysis and pyruvate oxidation each have to make a trip around the Krebs cycle. Follow the different carbons through the cycle, and notice the changes that occur in the carbon skeletons of the molecules and where oxidation reactions take place as they proceed through the cycle.

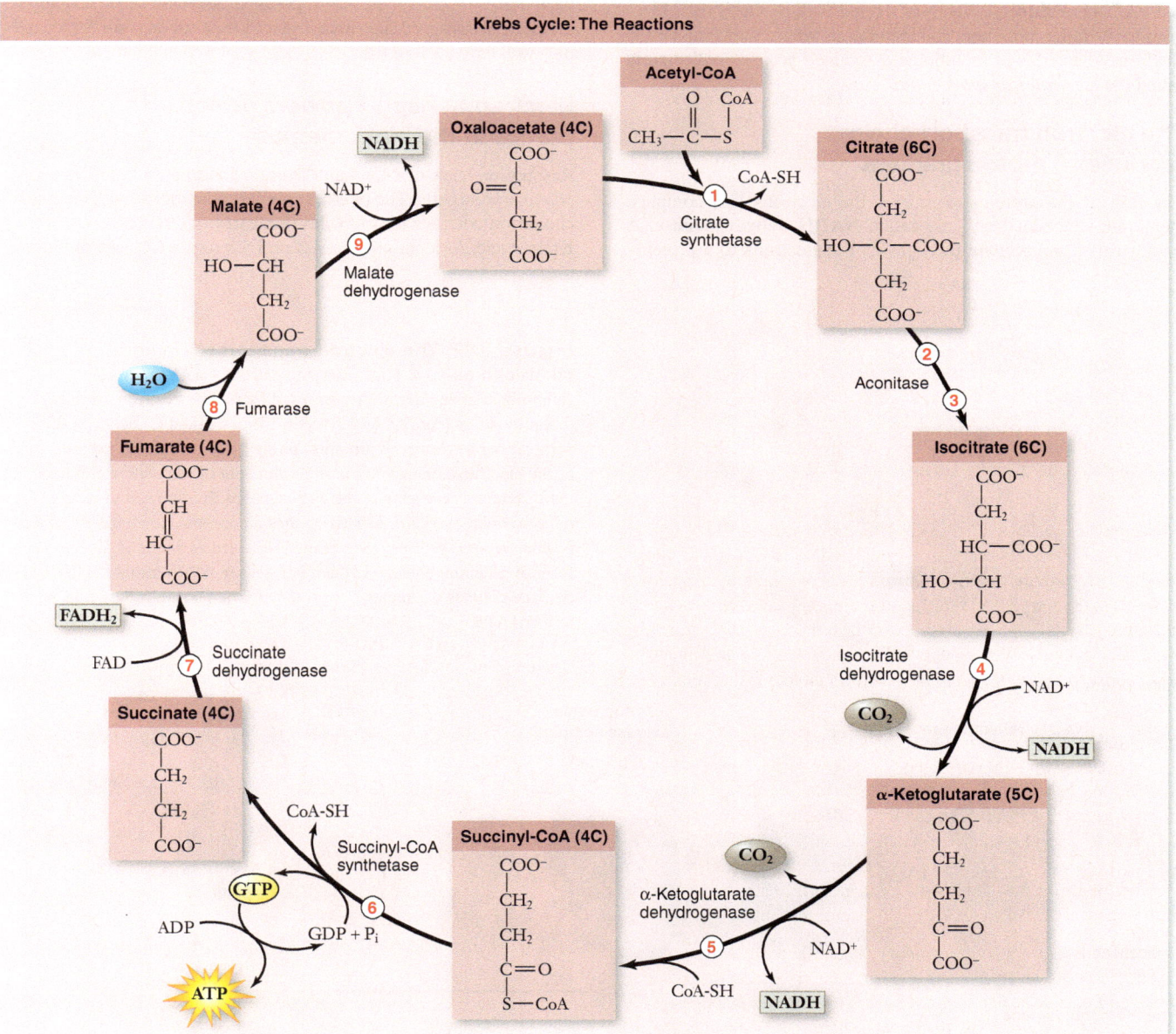

1. Reaction 1: Condensation
2–3. Reactions 2 and 3: Isomerization
4. Reaction 4: The first oxidation
5. Reaction 5: The second oxidation
6. Reaction 6: Substrate-level phosphorylation
7. Reaction 7: The third oxidation
8–9. Reactions 8 and 9: Regeneration of oxaloacetate and the fourth oxidation

Krebs Cycle: The Reactions

chapter **7** How Cells Harvest Energy 133

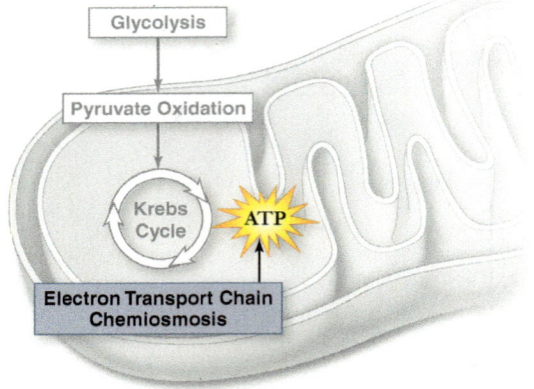

7.5 The Electron Transport Chain and Chemiosmosis

Learning Outcome
1. Describe the structure and function of the electron transport chain.
2. Diagram how the proton gradient connects electron transport with ATP synthesis.

The NADH and FADH$_2$ molecules formed during aerobic respiration each contain a pair of electrons that were gained when NAD$^+$ and FAD were reduced. The NADH and FADH$_2$ carry their electrons to the inner mitochondrial membrane, where they transfer the electrons to a series of membrane-associated proteins collectively called the *electron transport chain*.

The electron transport chain produces a proton gradient

The first of the proteins to receive the electrons is a complex, membrane-embedded enzyme called **NADH dehydrogenase**. A carrier called *ubiquinone* then passes the electrons to a protein–cytochrome complex called the *bc$_1$ complex*. Each complex in the chain operates as a proton pump, driving a proton out across the membrane into the intermembrane space (figure 7.12a).

The electrons are then carried by another carrier, *cytochrome c*, to the cytochrome oxidase complex. This complex uses four electrons to reduce a molecule of oxygen. Each oxygen then combines with two protons to form water:

$$O_2 + 4H^+ + 4e^- \longrightarrow 2H_2O$$

In contrast to NADH, which contributes its electrons to NADH dehydrogenase, FADH$_2$, which is located in the inner mitochondrial membrane, feeds its electrons to ubiquinone, which is also in the membrane. Electrons from FADH$_2$ thus "skip" the first step in the electron transport chain.

The plentiful availability of a strong electron acceptor, oxygen, is what makes oxidative respiration possible. As you'll see in chapter 8, the electron transport chain used in aerobic respiration is similar to, and may well have evolved from, the chain employed in photosynthesis.

Electron transport powers proton pumps in the inner membrane

Respiration takes place within the mitochondria present in virtually all eukaryotic cells. The internal compartment, or matrix, of a mitochondrion contains the enzymes that carry out the reactions of the Krebs cycle. As mentioned in section 7.1, protons (H$^+$) are produced

Figure 7.12 The electron transport chain and chemiosmosis. *a.* High-energy electrons harvested from catabolized molecules are transported by mobile electron carriers (ubiquinone, marked Q, and cytochrome c, marked C) between three complexes of membrane proteins. These three complexes use portions of the electrons' energy to pump protons out of the matrix and into the intermembrane space. The electrons are finally used to reduce oxygen, forming water. *b.* This creates a concentration gradient of protons across the inner membrane. This electrochemical gradient is a form of potential energy that can be used by ATP synthase. This enzyme couples the reentry of protons to the phosphorylation of ADP to form ATP.

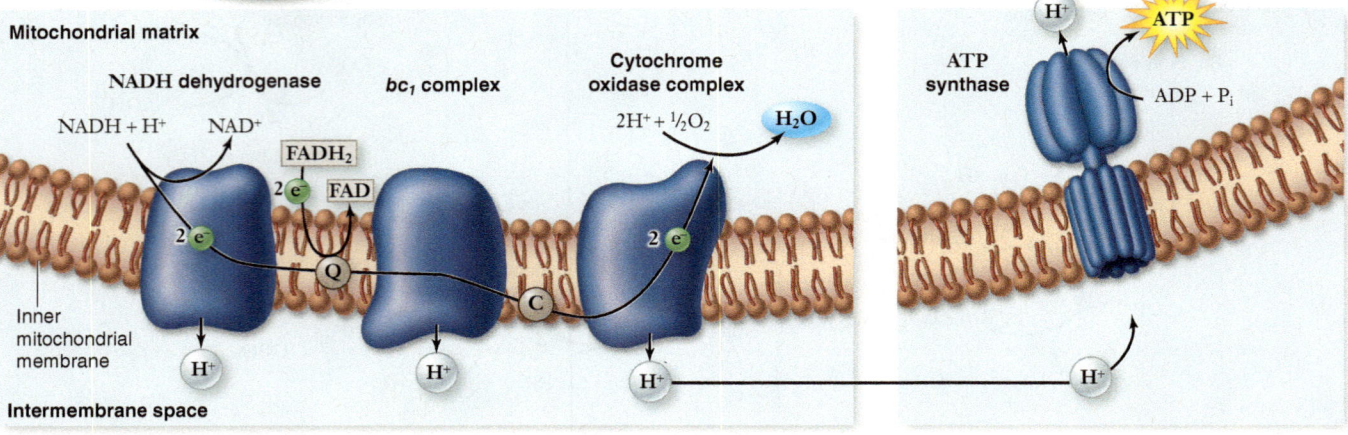

a. The electron transport chain

b. Chemiosmosis

when electrons are transferred to NAD$^+$. As the electrons harvested by oxidative respiration are passed along the electron transport chain, the energy they release transports protons out of the matrix and into the outer compartment called the intermembrane space.

Three transmembrane complexes of the electron transport chain in the inner mitochondrial membrane actually accomplish the proton transport (figure 7.12a). The flow of highly energetic electrons induces a change in the shape of pump proteins, which causes them to transport protons across the membrane. The electrons contributed by NADH activate all three of these proton pumps, whereas those contributed by FADH$_2$ activate only two because of where they enter the chain. In this way a proton gradient is formed between the intermembrane space and the matrix.

Chemiosmosis utilizes the electrochemical gradient to produce ATP

Because the mitochondrial matrix is negative compared with the intermembrane space, positively charged protons are attracted to the matrix. The higher outer concentration of protons also tends to drive protons back in by diffusion, but because membranes are relatively impermeable to ions, this process occurs only very slowly. Most of the protons that reenter the matrix instead pass through ATP synthase, an enzyme that uses the energy of the gradient to catalyze the synthesis of ATP from ADP and P$_i$. Because the chemical formation of ATP is driven by a diffusion force similar to osmosis, this process is referred to as *chemiosmosis* (figure 7.12b). The newly formed ATP is transported by facilitated diffusion to the many places in the cell where enzymes require energy to drive endergonic reactions. This chemiosmotic mechanism for the coupling of electron transport and ATP synthesis was controversial when it was proposed. Over the years, experimental evidence accumulated to support this hypothesis (figure 7.13).

The energy released by the reactions of cellular respiration ultimately drives the proton pumps that produce the proton gradient. The proton gradient provides the energy required for the synthesis of ATP. Figure 7.14 summarizes the overall process.

SCIENTIFIC THINKING

Hypothesis: ATP synthase enzyme uses a proton gradient to provide energy for phosphorylation reaction.
Prediction: The source of the proton gradient should not matter. A proton gradient formed by the light-driven pump bacteriorhodopsin should power phosphorylation in the light but not in the dark.
Test: Artificial vesicles are made with bacteriorhodopsin and ATP synthase, and ATP synthase alone. These are illuminated with light and assessed for ATP production.

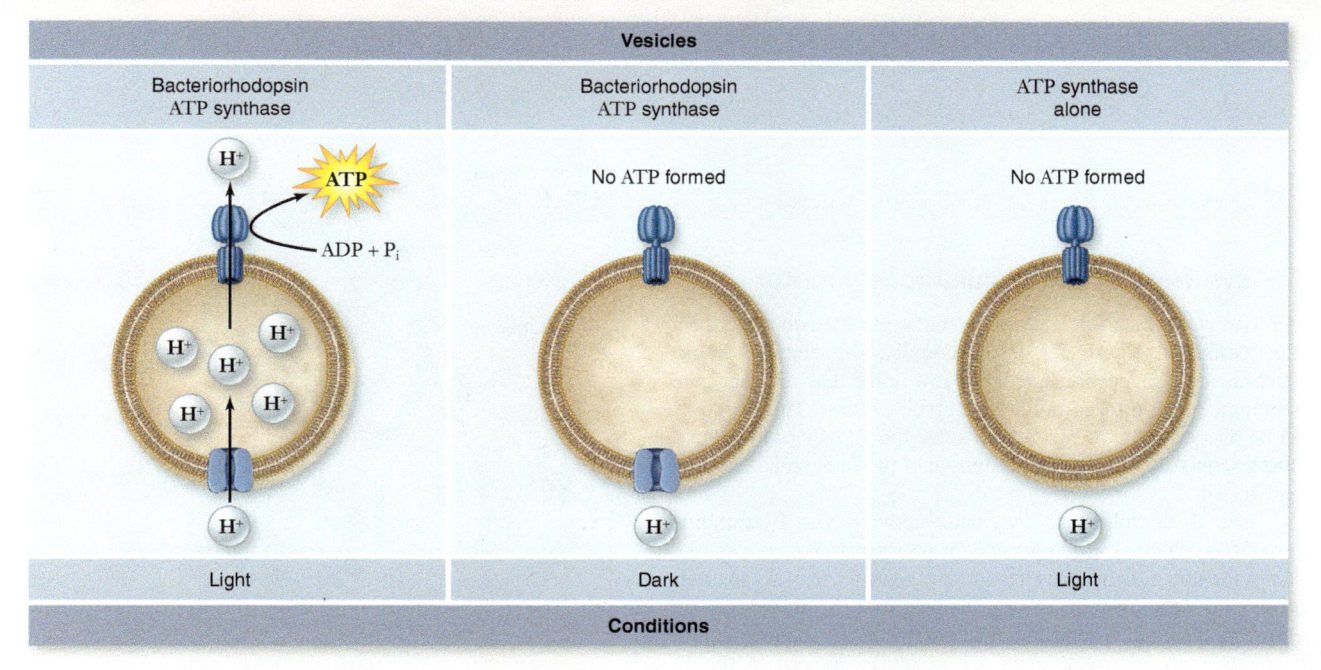

Vesicles		
Bacteriorhodopsin ATP synthase	Bacteriorhodopsin ATP synthase	ATP synthase alone
ATP formed	No ATP formed	No ATP formed
Light	Dark	Light
Conditions		

Result: The vesicle with both bacteriorhodopsin and ATP synthase can form ATP in the light but not in the dark. The vesicle with ATP synthase alone cannot form ATP in the light.
Conclusion: ATP synthase is able to utilize a proton gradient for energy to form ATP.
Further Experiments: What other controls would be appropriate for this type of experiment? Explain why this experiment is a more direct test of the chemiosmotic hypothesis than the Jagendorf acid bath experiment (see figure 8.16).

Figure 7.13 Evidence for the chemiosmotic synthesis of ATP by ATP synthase.

Figure 7.14 Aerobic respiration in the mitochondria. The entire process of aerobic respiration is shown in cellular context. Glycolysis occurs in the cytoplasm with the pyruvate and NADH produced entering the mitochondria. Here, pyruvate is oxidized and fed into the Krebs cycle to complete the oxidation process. All the energetic electrons harvested by oxidations in the overall process are transferred by NADH and $FADH_2$ to the electron transport chain. The electron transport chain uses the energy released during electron transport to pump protons across the inner membrane. This creates an electrochemical gradient that contains potential energy. The enzyme ATP synthase uses this gradient to phosphorylate ADP to form ATP.

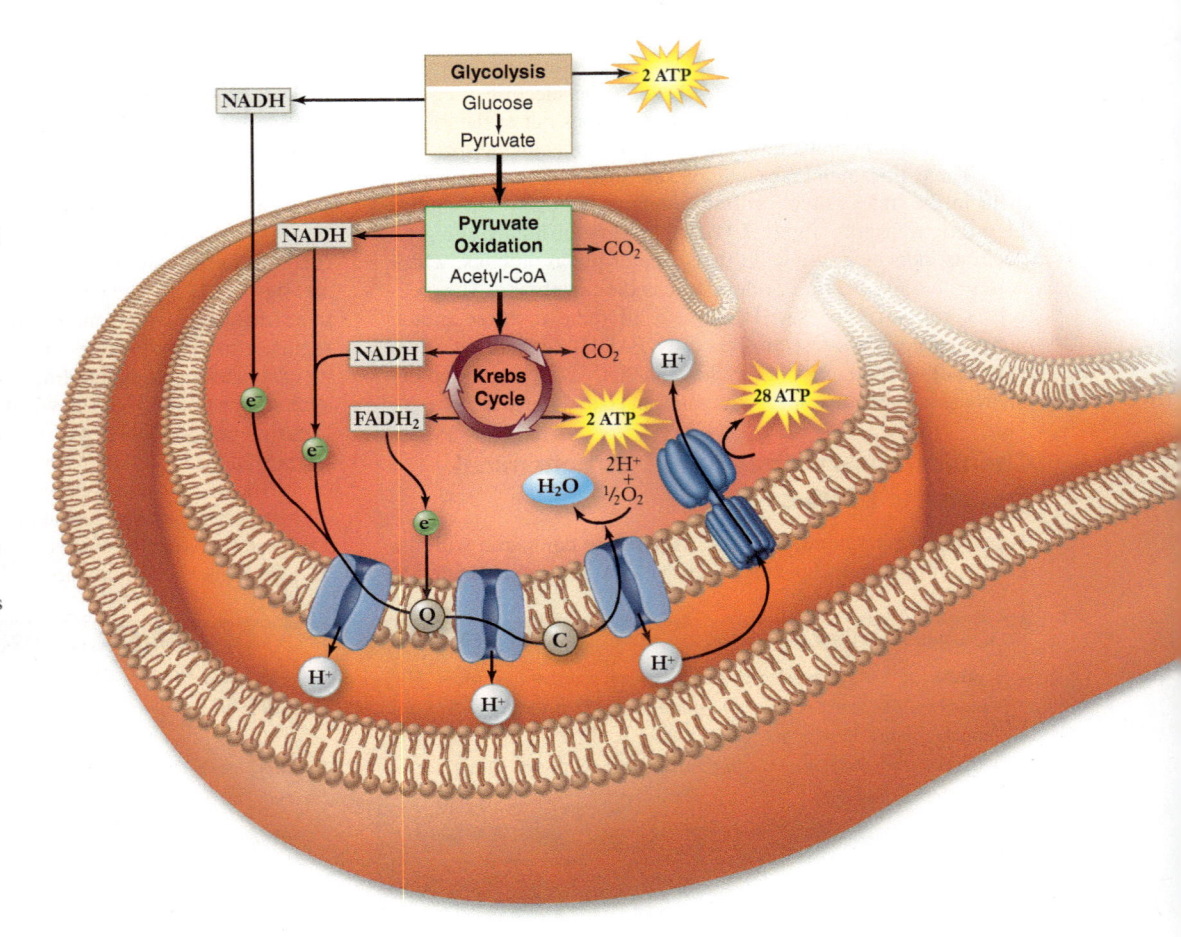

ATP synthase is a molecular rotary motor

ATP synthase uses a fascinating molecular mechanism to perform ATP synthesis (figure 7.15). Structurally, the enzyme has a membrane-bound portion and a narrow stalk that connects the membrane portion to a knoblike catalytic portion. This complex can be dissociated into two subportions: the F_0 membrane-bound complex, and the F_1 complex composed of the stalk and a knob, or head domain.

The F_1 complex has enzymatic activity. The F_0 complex contains a channel through which protons move across the membrane down their concentration gradient. As they do so, their movement causes part of the F_0 complex and the stalk to rotate relative to the knob. The mechanical energy of this rotation is used to change the conformation of the catalytic domain in the F_1 complex.

Thus, the synthesis of ATP is achieved by a tiny rotary motor, the rotation of which is driven directly by a gradient of protons. The flow of protons is like that of water in a hydroelectric power plant. Like the flow of water driven by gravity causes a turbine to rotate and generate electrical current, the proton gradient produces the energy that drives the rotation of the ATP synthase generator.

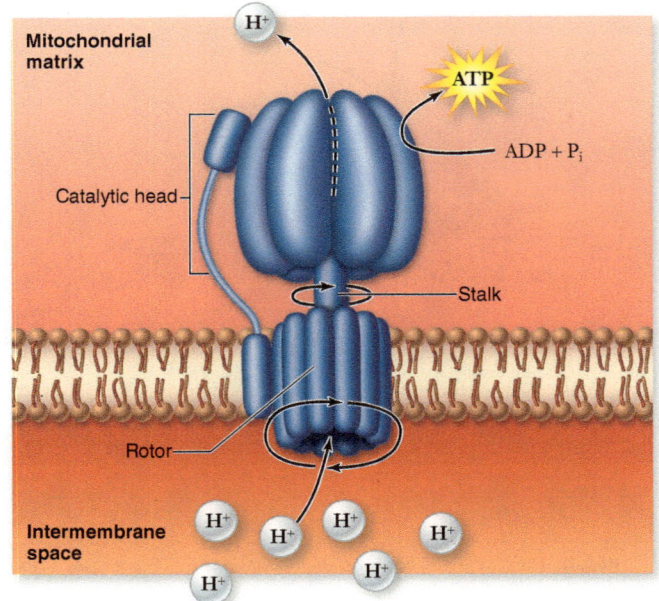

Figure 7.15 The ATP rotary engine. Protons move across the membrane down their concentration gradient. The energy released causes the rotor and stalk structures to rotate. This mechanical energy alters the conformation of the ATP synthase enzyme to catalyze the formation of ATP.

Learning Outcomes Review 7.5

The electron transport chain receives electrons from **NADH** and **FADH₂** and passes them down the chain to oxygen. The protein complexes of the electron transport chain, in the inner membrane of mitochondria, use the energy from electron transfer to pump protons across the membrane, creating an electrochemical gradient. The enzyme **ATP synthase** uses this gradient to drive the endergonic reaction of phosphorylating **ADP** to **ATP**.

- How would poking a small hole in the outer membrane affect ATP synthesis?

7.6 Energy Yield of Aerobic Respiration

Learning Outcome

1. Calculate the number of ATP molecules produced by aerobic respiration.

How much metabolic energy (in the form of ATP) does a cell gain from aerobic breakdown of glucose? This simple question has actually been a source of some controversy in biochemistry.

The theoretical yield for eukaryotes is 30 molecules of ATP per glucose molecule

The number of molecules of ATP produced by ATP synthase per molecules of glucose depends on the number of protons transported across the inner membrane, and the number of protons needed per ATP synthesized. The number of protons transported per NADH and FADH₂ is 10 and 6 H⁺, respectively. Each ATP synthesized requires 4 H⁺, leading to $10/4 = 2.5$ ATP/NADH, and $6/4 = 1.5$ ATP/FADH₂.

To finish the bookkeeping: oxidizing glucose to pyruvate via glycolysis yields 2 ATP directly, and $2 \times 2.5 = 5$ ATP from NADH. The oxidation of pyruvate to acetyl-CoA yields another $2 \times 2.5 = 5$ ATP from NADH. Lastly, the Krebs cycle produces 2 ATP directly, $6 \times 2.5 = 15$ ATP from NADH, and $2 \times 1.5 = 3$ ATP from FADH₂. Summing all of these leads to 32 ATP for respiration (figure 7.16).

This number is accurate for bacteria, but it does not hold for eukaryotes because the NADH produced in the cytoplasm by glycolysis needs to be transported into the mitochondria by active transport, which costs one ATP per NADH transported. This reduces the predicted yield for eukaryotes to 30 ATP.

Calculation of P/O ratios has changed over time

The value for the amount of ATP synthesized per O₂ molecule reduced is called the phosphate-to-oxygen ratio (P/O ratio). Both theoretical calculations, and direct measurement of this value, have been contentious issues. When theoretical calculations were first made, we lacked detailed knowledge of the respiratory chain, and the mechanism for coupling electron transport to ATP synthesis. Since redox reactions occur at three sites for NADH and two sites for FADH₂, it was assumed that three molecules of ATP were produced per NADH and two per FADH₂. We now know that assumption was overly simplistic.

Understanding that a proton gradient is the link between electron transport and ATP synthesis changed the nature of the calculations. We need to know the number of protons pumped during electron transport: 10 H⁺ per NADH, and 6 H⁺ per FADH₂. Then we need to know the number of protons needed per ATP. Since ATP synthase is a rotary motor, this calculation depends on the number of binding sites for ATP, and the number of protons required for rotation. We know that ATP synthase has three binding sites for ATP. If 12 protons are used per rotation, you get the value of 4 H⁺ per ATP

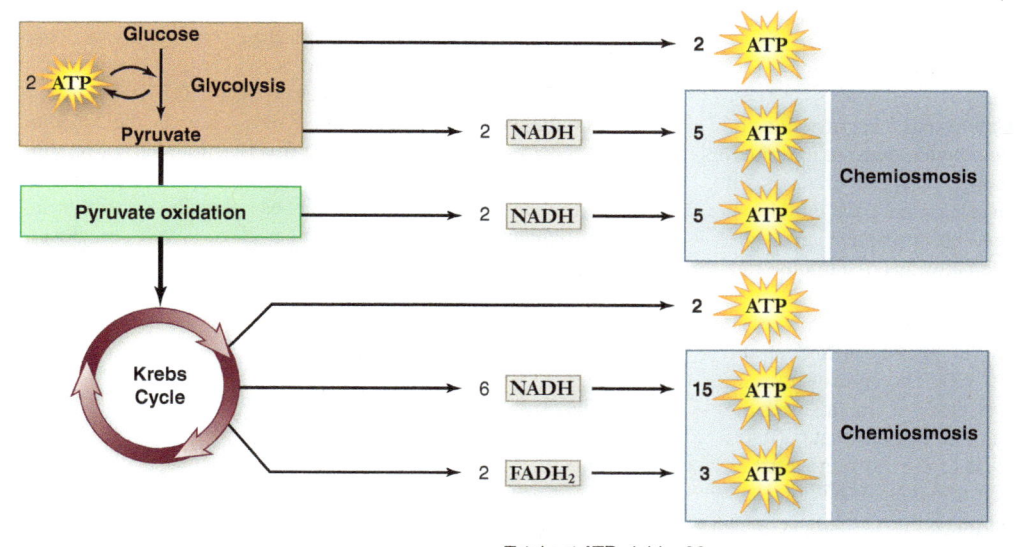

Figure 7.16 Theoretical ATP yield. The theoretical yield of ATP harvested from glucose by aerobic respiration totals 32 molecules. In eukaryotes this is reduced to 30 because it takes 1 ATP to transport each molecule of NADH that is generated by glycolysis in the cytoplasm into the mitochondria.

Total net ATP yield = 32
(30 in eukaryotes)

used in the previous calculation. Actual measurements of the P/O ratio have been problematic, but now appear to be at most 2.5.

We can also calculate how efficiently respiration captures the free energy released by the oxidation of glucose in the form of ATP. The amount of free energy released by the oxidation of glucose is 686 kcal/mol, and the free energy stored in each ATP is 7.3 kcal/mol. Therefore, a eukaryotic cell harvests about (7.3 × 30)/686 = 32% of the energy available in glucose. (By comparison, a typical car converts only about 25% of the energy in gasoline into useful energy.)

The higher energy yield of aerobic respiration was one of the key factors that fostered the evolution of heterotrophs. As this mechanism for producing ATP evolved, nonphotosynthetic organisms became more effective at using respiration to extract energy from molecules derived from other organisms. As long as some organisms captured energy by photosynthesis, others could exist solely by feeding on them.

Learning Outcome Review 7.6

Passage of electrons down the electron transport chain produces roughly 2.5 molecules of ATP per molecule of NADH (1.5 ATP per FADH$_2$). This process plus the ATP from substrate-level phosphorylation can yield a maximum of 32 ATP for the complete oxidation of glucose. NADH generated in the cytoplasm of eukaryotes yields only two ATP/NADH due to the cost of transport into the mitochondria, lowering the yield to 30 ATP.

- *How does chemiosmosis allow for noninteger numbers of ATP/NADH?*

7.7 Regulation of Aerobic Respiration

Learning Outcome

1. Understand the control points for cellular respiration.

When cells possess plentiful amounts of ATP, the key reactions of glycolysis, the Krebs cycle, and fatty acid breakdown are inhibited, slowing ATP production. The regulation of these biochemical pathways by the level of ATP is an example of feedback inhibition. Conversely, when ATP levels in the cell are low, ADP levels are high, and ADP activates enzymes in the pathways of carbohydrate catabolism to stimulate the production of more ATP.

Control of glucose catabolism occurs at two key points in the catabolic pathway, namely at a point in glycolysis and at the beginning of the Krebs cycle (figure 7.17). The control point in glycolysis is the enzyme phosphofructokinase, which catalyzes the conversion of fructose phosphate to fructose bisphosphate. This is the first reaction of glycolysis that is not readily reversible, committing the substrate to the glycolytic sequence. ATP itself is an allosteric inhibitor (see chapter 6) of phosphofructokinase, as is the Krebs cycle intermediate citrate. High levels of both ATP and citrate inhibit phosphofructokinase. Thus, under conditions when

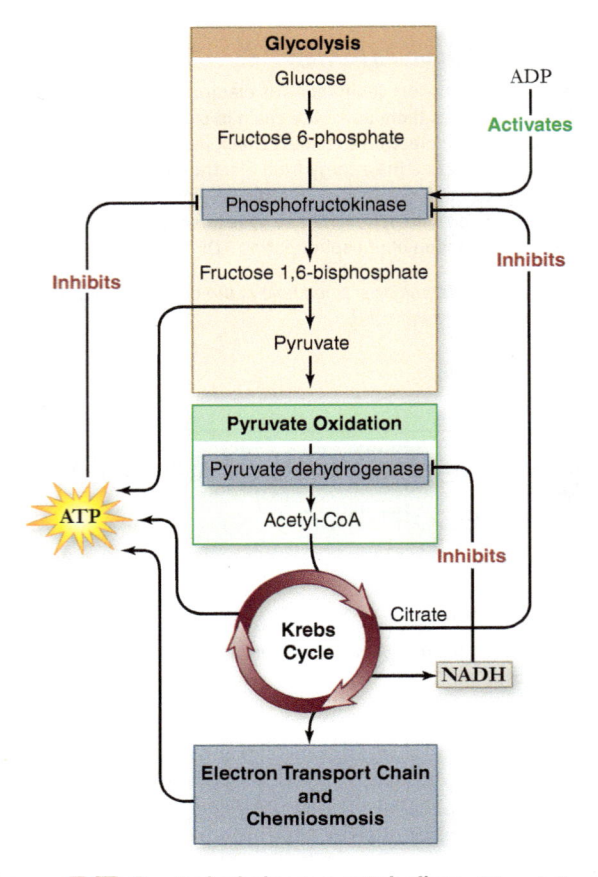

Figure 7.17 Control of glucose catabolism. The relative levels of ADP and ATP and key intermediates NADH and citrate control the catabolic pathway at two key points: the committing reactions of glycolysis and the Krebs cycle.

ATP is in excess, or when the Krebs cycle is producing citrate faster than it is being consumed, glycolysis is slowed.

The main control point in the oxidation of pyruvate occurs at the committing step in the Krebs cycle with the enzyme pyruvate dehydrogenase, which converts pyruvate to acetyl-CoA. This enzyme is inhibited by high levels of NADH, a key product of the Krebs cycle.

Another control point in the Krebs cycle is the enzyme citrate synthetase, which catalyzes the first reaction, the conversion of oxaloacetate and acetyl-CoA into citrate. High levels of ATP inhibit citrate synthetase (as well as phosphofructo-kinase, pyruvate dehydrogenase, and two other Krebs cycle enzymes), slowing down the entire catabolic pathway.

Learning Outcome Review 7.7

Respiration is controlled by levels of ATP in the cell and levels of key intermediates in the process. The control point for glycolysis is the enzyme phosphofructokinase, which is inhibited by ATP or citrate (or both). The main control point in oxidation of pyruvate is the enzyme pyruvate dehydrogenase, inhibited by NADH.

- *How does feedback inhibition ensure economic production of ATP?*

7.8 Oxidation Without O₂

Learning Outcomes
1. Compare anaerobic and aerobic respiration.
2. Distinguish between fermentation and anaerobic respiration.

In the presence of oxygen, cells can use oxygen to produce a large amount of ATP. But even when no oxygen is present to accept electrons, some organisms can still respire *anaerobically,* using inorganic molecules as final electron acceptors for an electron transport chain.

For example, many prokaryotes use sulfur, nitrate, carbon dioxide, or even inorganic metals as the final electron acceptor in place of oxygen (figure 7.18). The free energy released by using these other molecules as final electron acceptors is not as great as that using oxygen because they have a lower affinity for electrons. The amount of ATP produced is less, but the process is still respiration and not fermentation.

Methanogens use carbon dioxide

Among the heterotrophs that practice anaerobic respiration are Archaea such as thermophiles and methanogens. Methanogens use carbon dioxide (CO_2) as the electron acceptor, reducing CO_2 to CH_4 (methane). The hydrogens are derived from organic molecules produced by other organisms. Methanogens are found in diverse environments, including soil and the digestive systems of ruminants like cows.

Sulfur bacteria use sulfate

Evidence of a second anaerobic respiratory process among primitive bacteria is seen in a group of rocks about 2.7 BYA, known as the Woman River iron formation. Organic material in these rocks is enriched for the light isotope of sulfur, ^{32}S, relative to the heavier isotope, ^{34}S. No known geochemical process produces such enrichment, but biological sulfur reduction does, in a process still carried out today by certain prokaryotes.

In this sulfate respiration, the prokaryotes derive energy from the reduction of inorganic sulfates (SO_4) to hydrogen sulfide (H_2S). The hydrogen atoms are obtained from organic molecules other organisms produce. These prokaryotes thus are similar to methanogens, but they use SO_4 as the oxidizing (that is, electron-accepting) agent in place of CO_2.

The early sulfate reducers set the stage for the evolution of photosynthesis, creating an environment rich in H_2S. As discussed in chapter 8, the first form of photosynthesis obtained hydrogens from H_2S using the energy of sunlight.

Fermentation uses organic compounds as electron acceptors

In the absence of oxygen, cells that cannot utilize an alternative electron acceptor for respiration must rely exclusively on glycolysis to produce ATP. Under these conditions, the electrons generated

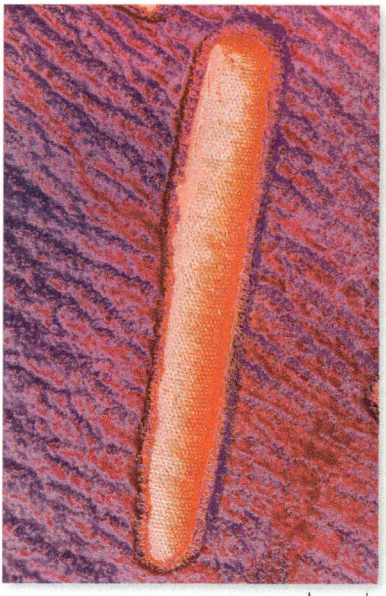

Figure 7.18 Sulfur-respiring prokaryote. *a.* The micrograph shows the archaeal species *Thermoproteus tenax*. This organism can use elemental sulfur as a final electron acceptor for anaerobic respiration. *b. Thermoproteus* is often found in sulfur-containing hot springs such as the Norris Geyser Basin in Yellowstone National Park, shown here.

by glycolysis are donated to organic molecules in a process called *fermentation*. This process recycles NAD$^+$, the electron acceptor that allows glycolysis to proceed.

Bacteria carry out more than a dozen kinds of fermentation reactions, often using pyruvate or a derivative of pyruvate to accept the electrons from NADH. Organic molecules other than pyruvate and its derivatives can be used as well; the important point is that the process regenerates NAD$^+$:

organic molecule + NADH $\longrightarrow$ reduced organic molecule + NAD$^+$

Often the reduced organic compound is an organic acid—such as acetic acid, butyric acid, propionic acid, or lactic acid—or an alcohol.

Ethanol fermentation

Eukaryotic cells are capable of only a few types of fermentation. In one type, which occurs in yeast, the molecule that accepts electrons from NADH is derived from pyruvate, the end-product of glycolysis.

Yeast enzymes remove a terminal CO$_2$ group from pyruvate through decarboxylation, producing a 2-carbon molecule called acetaldehyde. The CO$_2$ released causes bread made with yeast to rise. The acetaldehyde accepts a pair of electrons from NADH, producing NAD$^+$ and ethanol (ethyl alcohol) (figure 7.19).

This particular type of fermentation is of great interest to humans, because it is the source of the ethanol in wine and beer. Ethanol is a by-product of fermentation that is actually toxic to yeast; as it approaches a concentration of about 12%, it begins to kill the yeast. That explains why naturally fermented wine contains only about 12% ethanol.

Lactic acid fermentation

Most animal cells regenerate NAD$^+$ without decarboxylation. Muscle cells, for example, use the enzyme lactate dehydrogenase to transfer electrons from NADH back to the pyruvate that is produced by glycolysis. This reaction converts pyruvate into lactic acid and regenerates NAD$^+$ from NADH (figure 7.19). It therefore closes the metabolic circle, allowing glycolysis to continue as long as glucose is available.

Circulating blood removes excess lactate, the ionized form of lactic acid, from muscles, but when removal cannot keep pace with production, the accumulating lactic acid interferes with muscle function and contributes to muscle fatigue.

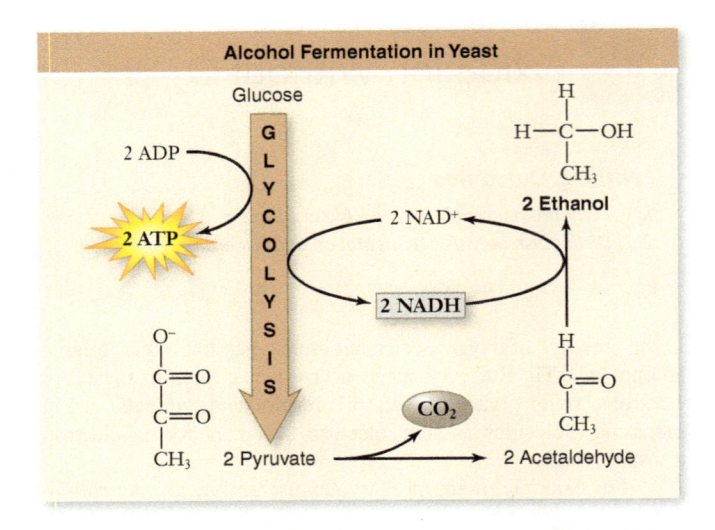

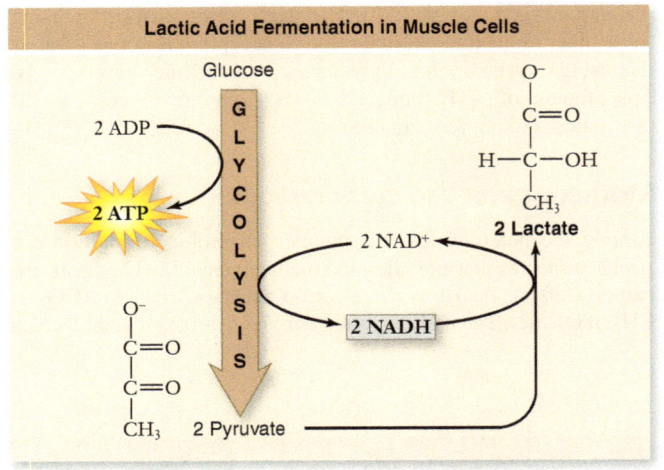

Figure 7.19 Fermentation. Yeasts carry out the conversion of pyruvate to ethanol. Muscle cells convert pyruvate into lactate, which is less toxic than ethanol. In each case, the reduction of a metabolite of glucose has oxidized NADH back to NAD$^+$ to allow glycolysis to continue under anaerobic conditions.

7.9 Catabolism of Proteins and Fats

Learning Outcomes
1. Identify the entry points for proteins and fats in energy metabolism.
2. Recognize the importance of key intermediates in metabolism.

Learning Outcomes Review 7.8

Nitrate, sulfur, and CO$_2$ are all used as terminal electron acceptors in anaerobic respiration of different organisms.
Organic molecules can also accept electrons in fermentation reactions that regenerate NAD$^+$. Fermentation reactions produce a variety of compounds, including ethanol in yeast and lactic acid in humans.

- *In what kinds of ecosystems would you expect to find anaerobic respiration?*

Thus far we have focused on the aerobic respiration of glucose, which organisms obtain from the digestion of carbohydrates or from photosynthesis. Organic molecules other than glucose,

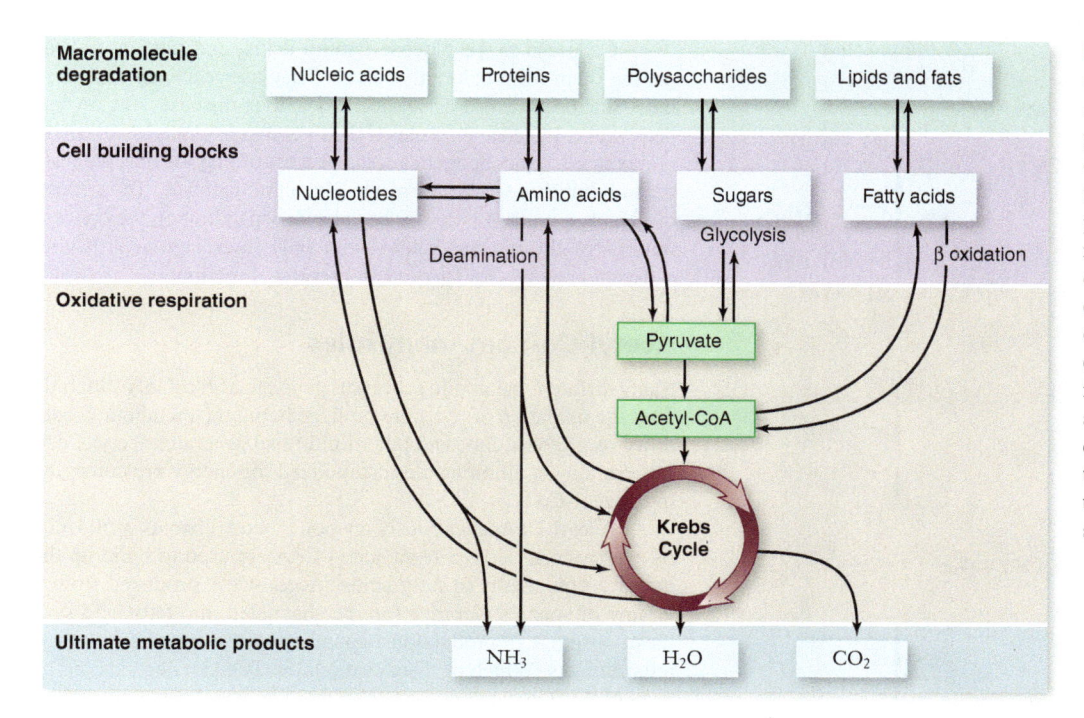

Figure 7.20 How cells extract chemical energy. All eukaryotes and many prokaryotes extract energy from organic molecules by oxidizing them. The first stage of this process, breaking down macromolecules into their constituent parts, yields little energy. The second stage, oxidative or aerobic respiration, extracts energy, primarily in the form of high-energy electrons, and produces water and carbon dioxide. Key intermediates in these energy pathways are also used for biosynthetic pathways, shown by reverse arrows.

particularly proteins and fats, are also important sources of energy (figure 7.20).

Catabolism of proteins removes amino groups

Proteins are first broken down into their individual amino acids. The nitrogen-containing side group (the amino group) is then removed from each amino acid in a process called **deamination**. A series of reactions converts the carbon chain that remains into a molecule that enters glycolysis or the Krebs cycle. For example, alanine is converted into pyruvate, glutamate into α-ketoglutarate (figure 7.21), and aspartate into oxaloacetate. The reactions of glycolysis and the Krebs cycle then extract the high-energy electrons from these molecules and put them to work making ATP.

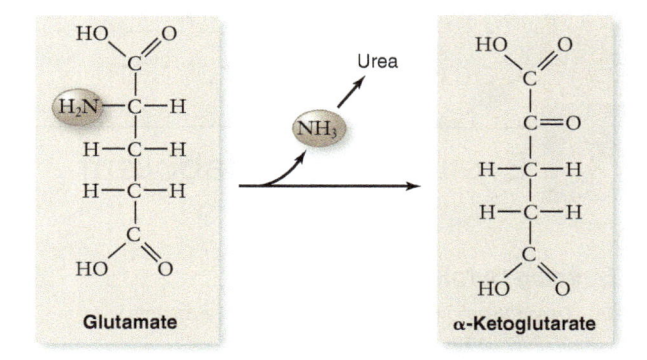

Figure 7.21 Deamination. After proteins are broken down into their amino acid constituents, the amino groups are removed from the amino acids to form molecules that participate in glycolysis and the Krebs cycle. For example, the amino acid glutamate becomes α-ketoglutarate, a Krebs cycle intermediate, when it loses its amino group.

Catabolism of fatty acids produces acetyl groups for the Krebs cycle

Fats are broken down into fatty acids plus glycerol. Long-chain fatty acids typically have an even number of carbons, and the many C—H bonds provide a rich harvest of energy. Fatty acids are oxidized in the matrix of the mitochondrion. Enzymes remove the 2-carbon acetyl groups from the end of each fatty acid until the entire fatty acid is converted into acetyl groups (figure 7.22). Each acetyl group is combined with coenzyme A to form acetyl-CoA. This process is known as **β oxidation**. This process is oxygen-dependent, which explains why aerobic exercise burns fat, but anaerobic exercise does not.

How much ATP does the catabolism of fatty acids produce? Let's compare a hypothetical 6-carbon fatty acid with the 6-carbon glucose molecule, which we've said yields about 30 molecules of ATP in a eukaryotic cell. Two rounds of β oxidation would convert the fatty acid into three molecules of acetyl-CoA. Each round requires one molecule of ATP to prime the process, but it also produces one molecule of NADH and one of FADH$_2$. These molecules together yield four molecules of ATP (assuming 2.5 ATP per NADH, and 1.5 ATP per FADH$_2$).

The oxidation of each acetyl-CoA in the Krebs cycle ultimately produces an additional 10 molecules of ATP. Overall, then, the ATP yield of a 6-carbon fatty acid is approximately: 8 (from two rounds of β oxidation) − 2 (for priming those two rounds) + 30 (from oxidizing the three acetyl-CoAs) = 36 molecules of ATP. Therefore, the respiration of a 6-carbon fatty acid yields 20% more ATP than the respiration of glucose.

Moreover, a fatty acid of that size would weigh less than two thirds as much as glucose, so a gram of fatty acid contains more than twice as many kilocalories as a gram of glucose. You can see from this fact why fat is a storage molecule for excess energy in

Figure 7.22
β oxidation.

Through a series of reactions known as β oxidation, the last two carbons in a fatty acid combine with coenzyme A to form acetyl-CoA, which enters the Krebs cycle. The fatty acid, now two carbons shorter, enters the pathway again and keeps reentering until all its carbons have been used to form acetyl-CoA molecules. Each round of β oxidation uses one molecule of ATP and generates one molecule each of FADH$_2$ and NADH.

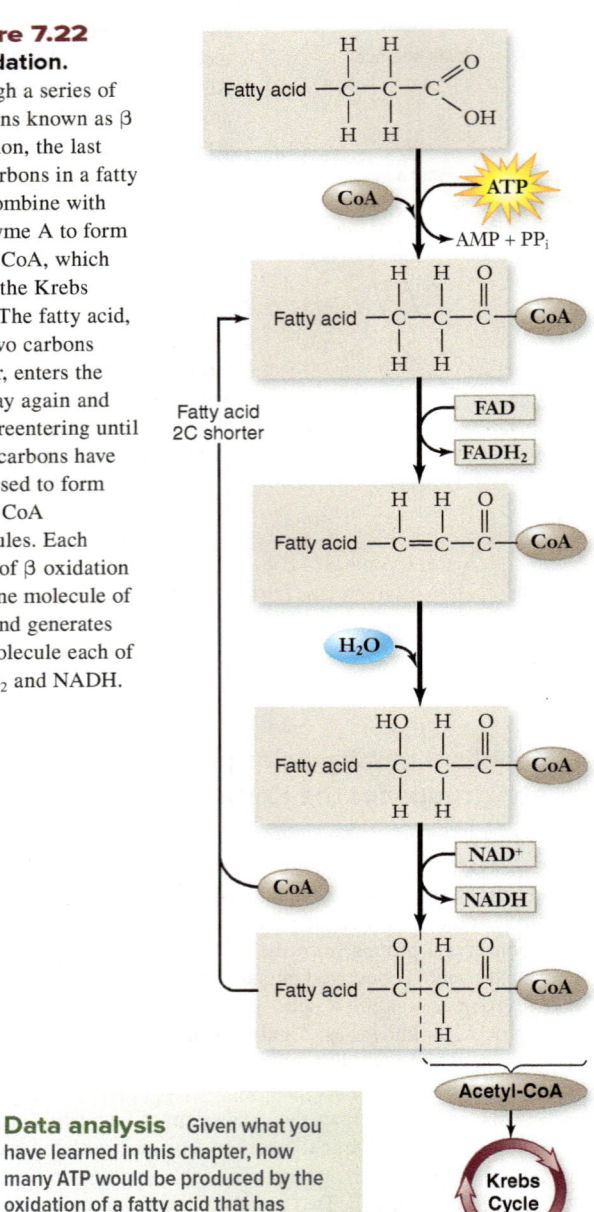

Data analysis Given what you have learned in this chapter, how many ATP would be produced by the oxidation of a fatty acid that has 16 carbons?

Cells can make glucose, amino acids, and fats, as well as getting them from external sources. They use reactions similar to those that break down these substances. In many cases, the reverse pathways even share enzymes if the free-energy changes are small. For example, gluconeogenesis, the process of making new glucose, uses all but three enzymes of the glycolytic pathway. Thus, much of glycolysis runs forward or backward, depending on the concentrations of the intermediates—with only three key steps having different enzymes for forward and reverse directions.

Acetyl-CoA has many roles

Many different metabolic processes generate acetyl-CoA. Not only does the oxidation of pyruvate produce it, but the metabolic breakdown of proteins, fats, and other lipids also generates acetyl-CoA. Indeed, almost all molecules catabolized for energy are converted into acetyl-CoA.

Acetyl-CoA has a role in anabolic metabolism as well. Units of two carbons derived from acetyl-CoA are used to build up the hydrocarbon chains in fatty acids. Acetyl-CoA produced from a variety of sources can therefore be channeled into fatty acid synthesis or into ATP production, depending on the organism's energy requirements. Which of these two options is taken depends on the level of ATP in the cell.

When ATP levels are high, the oxidative pathway is inhibited, and acetyl-CoA is channeled into fatty acid synthesis. This explains why many animals (humans included) develop fat reserves when they consume more food than their activities require. Alternatively, when ATP levels are low, the oxidative pathway is stimulated, and acetyl-CoA flows into energy-producing oxidative metabolism.

Learning Outcomes Review 7.9

Proteins can be broken into their constituent amino acids, which are then deaminated and can enter metabolism at glycolysis or different steps of the Krebs cycle. Fats can be broken into units of acetyl-CoA by β oxidation and then fed into the Krebs cycle. Many metabolic processes can be used reversibly, to either build up (anabolism) or break down (catabolism) the major biological macromolecules. Key intermediates, such as pyruvate and acetyl-CoA, connect these processes.

- Can fats be oxidized in the absence of O_2?

many types of animals. If excess energy were stored instead as carbohydrate, as it is in plants, animal bodies would have to be much bulkier.

A small number of key intermediates connect metabolic pathways

Oxidation pathways of food molecules are interrelated in that a small number of key intermediates, such as pyruvate and acetyl-CoA, link the breakdown from different starting points. These key intermediates allow the interconversion of different types of molecules, such as sugars and amino acids (see figure 7.20).

7.10 Evolution of Metabolism

Learning Outcome

1. Describe one possible hypothesis for the evolution of metabolism.

We talk about cellular respiration as a continuous series of stages, but it is important to note that these stages evolved over time, and metabolism has changed a great deal in that time. Both anabolic processes and catabolic processes evolved in concert with each other. We do not know the details of this biochemical evolution, or the order of appearance of these processes. Therefore the following timeline is based on the available geochemical evidence and represents a hypothesis rather than a strict timeline.

The earliest life-forms degraded carbon-based molecules present in the environment

The most primitive forms of life are thought to have obtained chemical energy by degrading, or breaking down, organic molecules that were abiotically produced—that is, carbon-containing molecules formed by inorganic processes on the early Earth.

The first major event in the evolution of metabolism was the origin of the ability to harness chemical bond energy. At an early stage, organisms began to store this energy in the bonds of ATP.

The evolution of glycolysis also occurred early

The second major event in the evolution of metabolism was glycolysis, the initial breakdown of glucose. As proteins evolved diverse catalytic functions, it became possible to capture a larger fraction of the chemical bond energy in organic molecules by breaking chemical bonds in a series of steps.

Glycolysis undoubtedly evolved early in the history of life on Earth, because this biochemical pathway has been retained by all living organisms. It is a chemical process that does not appear to have changed for more than 2 billion years.

Anoxygenic photosynthesis allowed the capture of light energy

The third major event in the evolution of metabolism was anoxygenic photosynthesis. Early in the history of life, a different way of generating ATP evolved in some organisms. Instead of obtaining energy for ATP synthesis by reshuffling chemical bonds, as in glycolysis, these organisms developed the ability to use light to pump protons out of their cells and to use the resulting proton gradient to power the production of ATP through chemiosmosis.

Photosynthesis evolved in the absence of oxygen and works well without it. Dissolved H_2S, present in the oceans of the early Earth beneath an atmosphere free of oxygen gas, served as a ready source of hydrogen atoms for building organic molecules. Free sulfur was produced as a by-product of this reaction.

Oxygen-forming photosynthesis used a different source of hydrogen

The substitution of H_2O for H_2S in photosynthesis was the fourth major event in the history of metabolism. Oxygen-forming photosynthesis employs H_2O rather than H_2S as a source of hydrogen atoms and their associated electrons. Because it garners its electrons from reduced oxygen rather than from reduced sulfur, it generates oxygen gas rather than free sulfur.

More than 2 BYA, small cells capable of carrying out this oxygen-forming photosynthesis, such as cyanobacteria, became the dominant forms of life on Earth. Oxygen gas began to accumulate in the atmosphere. This was the beginning of a great transition that changed conditions on Earth permanently. Our atmosphere is now 20.9% oxygen, every molecule of which is derived from an oxygen-forming photosynthetic reaction.

Nitrogen fixation provided new organic nitrogen

Nitrogen is available from dead organic matter, and from chemical reactions that generated the original organic molecules. For life to expand, a new source of nitrogen was needed. Nitrogen fixation was the fifth major step in the evolution of metabolism. Proteins and nucleic acids cannot be synthesized from the products of photosynthesis because both of these biologically critical molecules contain nitrogen. Obtaining nitrogen atoms from N_2 gas, a process called *nitrogen fixation*, requires breaking an $N\equiv N$ triple bond.

This important reaction evolved in the hydrogen-rich atmosphere of the early Earth, where no oxygen was present. Oxygen acts as a poison to nitrogen fixation, which today occurs only in oxygen-free environments or in oxygen-free compartments within certain prokaryotes.

Aerobic respiration utilized oxygen

Respiration is the sixth and final event in the history of metabolism. Aerobic respiration employs the same kind of proton pumps as photosynthesis and is thought to have evolved as a modification of the basic photosynthetic machinery.

Biologists think that the ability to carry out photosynthesis without H_2S first evolved among purple nonsulfur bacteria, which obtain their hydrogens from organic compounds instead. It was perhaps inevitable that among the descendants of these respiring photosynthetic bacteria, some would eventually do without photosynthesis entirely, subsisting only on the energy and electrons derived from the breakdown of organic molecules. The mitochondria within all eukaryotic cells are thought to be descendants of these bacteria.

The complex process of aerobic metabolism developed over geological time, as natural selection favored organisms with more efficient methods of obtaining energy from organic molecules. The process of photosynthesis, as you have seen in this concluding section, has also developed over time, and the rise of photosynthesis changed life on Earth forever. Chapter 8 explores photosynthesis in detail.

Learning Outcome Review 7.10

Major milestones in the evolution of metabolism include the evolution of pathways to extract energy from organic compounds, the pathways of photosynthesis, and those of nitrogen fixation. Photosynthesis began as an anoxygenic process that later evolved to produce free oxygen, thus allowing the evolution of aerobic metabolism.

- What evidence can you cite for this hypothesis of the evolution of metabolism?

Chapter Review

7.1 Overview of Respiration (figure 7.5)

Cellular oxidations are usually also dehydrogenations.

Cellular respiration is the complete oxidation of glucose.
Aerobic respiration uses oxygen as the final electron acceptor for redox reactions. Anaerobic respiration utilizes inorganic molecules as acceptors, and fermentation uses organic molecules.

Electron carriers play a critical role in energy metabolism.
Electron carriers can be reversibly oxidized and reduced. For example, NAD^+ is reduced to NADH by acquiring two electrons; NADH supplies these electrons to other molecules to reduce them.

Respiration harvests energy in stages.
Mitochondria of eukaryotic cells move electrons in steps via the electron transport chain to capture energy efficiently.

ATP plays a central role in metabolism.
The ultimate goal of cellular respiration is synthesis of ATP, which is used to power most of the cell's activities.

Cells make ATP by two fundamentally different mechanisms.
Substrate-level phosphorylation transfers a phosphate directly to ADP (figure 7.4). Oxidative phosphorylation generates ATP via the enzyme ATP synthase, powered by a proton gradient.

7.2 Glycolysis: Splitting Glucose (figures 7.6 & 7.7)

Glycolysis converts glucose into two pyruvate, forming two ATP and two NADH in the process.
Priming reactions add two phosphates to glucose; this is cleaved into two 3-carbon molecules of glyceraldehyde 3-phosphate (G3P). Oxidation of G3P transfers electrons to NAD^+, yielding NADH. After four more reactions, the final product is two molecules of pyruvate. Glycolysis produces 2 net ATP, 2 NADH, and 2 pyruvate.

Glycolysis is an ancient process with a low energy yield, but it can be efficient with up to 40% of available energy trapped as ATP. Glycolysis was probably the first catabolic reaction to evolve.

NADH must be recycled to continue respiration.
In the presence of oxygen, pyruvate is oxidized to acetyl-CoA, which can be oxidized by the Krebs cycle. This process leads to a large amount of ATP. In the absence of oxygen, a fermentation reaction uses all or part of pyruvate to oxidize NADH.

In the presence of oxygen, NADH passes electrons to the electron transport chain. In the absence of oxygen, NADH passes the electrons to an organic molecule such as acetaldehyde (fermentation).

7.3 The Oxidation of Pyruvate Produces Acetyl-CoA (figure 7.9)

Pyruvate is oxidized to yield 1 CO_2, 1 NADH, and 1 acetyl-CoA. Acetyl-CoA enters the Krebs cycle as 2-carbon acetyl units.

7.4 The Krebs Cycle (figures 7.10 & 7.11)

An overview of the Krebs cycle.

The Krebs cycle extracts electrons and synthesizes one ATP.
The first reaction is an irreversible condensation that produces citrate; it is inhibited when ATP is plentiful. The second and third reactions rearrange citrate to isocitrate. The fourth and fifth reactions are oxidations; in each reaction, one NAD^+ is reduced to NADH. The sixth reaction is a substrate-level phosphorylation producing GTP, and from that ATP. The seventh reaction is another oxidation that reduces FAD to $FADH_2$. Reactions eight and nine regenerate oxaloacetate, including one final oxidation that reduces NAD^+ to NADH.

Glucose becomes CO_2 and potential energy.
As a glucose molecule is broken down to CO_2, some of its energy is preserved in 4 ATP, 10 NADH, and 2 $FADH_2$.

Following the electrons in the reactions reveals the direction of transfer.

7.5 The Electron Transport Chain and Chemiosmosis (figure 7.12)

The electron transport chain produces a proton gradient.
In the inner mitochondrial membrane, NADH is oxidized to NAD^+ by NADH dehydrogenase. Electrons move through ubiquinone and the bc_1 complex to cytochrome oxidase, where they join with H^+ and O_2 to form H_2O. This results in three protons being pumped into the intermembrane space. For $FADH_2$, electrons are passed directly to ubiquinone. Thus only two protons are pumped into the intermembrane space.

Electron transport powers proton pumps in the inner membrane.

Chemiosmosis utilizes the electrochemical gradient to produce ATP.

ATP synthase is a molecular rotary motor.
Protons diffuse back into the mitochondrial matrix via the ATP synthase channel. The enzyme uses this energy to synthesize ATP (figure 7.15).

7.6 Energy Yield of Aerobic Respiration

The theoretical yield for eukaryotes is 30 molecules of ATP per glucose molecule (figure 7.16).

Calculation of P/O ratios has changed over time.

7.7 Regulation of Aerobic Respiration

Glucose catabolism is controlled by the concentration of ATP molecules and intermediates in the Krebs cycle (figure 7.17).

7.8 Oxidation Without O_2

In the absence of oxygen other final electron acceptors can be used for respiration.

Methanogens use carbon dioxide.

Sulfur bacteria use sulfate.

Fermentation uses organic compounds as electron acceptors (figure 7.19).
Fermentation is the regeneration of NAD^+ by oxidation of NADH and reduction of an organic molecule. In yeast, pyruvate is decarboxylated, then reduced to ethanol. In animals, pyruvate is reduced directly to lactate.

7.9 Catabolism of Proteins and Fats

Catabolism of proteins removes amino groups (figure 7.21).

Catabolism of fatty acids produces acetyl groups for the Krebs cycle.
Fatty acids are converted to acetyl groups by successive rounds of β oxidation (figure 7.22). These acetyl groups feed into the Krebs cycle to be oxidized and generate NADH for electron transport.

A small number of key intermediates connect metabolic pathways.

Acetyl-CoA has many roles.

With high ATP, acetyl-CoA is converted into fatty acids.

7.10 Evolution of Metabolism

Major milestones are recognized in the evolution of metabolism; the order of events is hypothetical.

The earliest life-forms degraded carbon-based molecules present in the environment.

The evolution of glycolysis also occurred early.

Anoxygenic photosynthesis allowed the capture of light energy.

Oxygen-forming photosynthesis used a different source of hydrogen.

Nitrogen fixation provided new organic nitrogen.

Aerobic respiration utilized oxygen.

Review Questions

UNDERSTAND

1. An *autotroph* is an organism that
   a. extracts energy from organic sources.
   b. converts energy from sunlight into chemical energy.
   c. relies on the energy produced by other organisms as an energy source.
   d. does both a and b.

2. Which of the following processes is (are) required for the complete oxidation of glucose?
   a. The Krebs cycle
   b. Glycolysis
   c. Pyruvate oxidation
   d. All of the choices are correct.

3. Which of the following is NOT a product of glycolysis?
   a. ATP
   b. Pyruvate
   c. CO_2
   d. NADH

4. Glycolysis produces ATP by
   a. phosphorylating organic molecules in the priming reactions.
   b. the production of glyceraldehyde 3-phosphate.
   c. substrate-level phosphorylation.
   d. the reduction of NAD^+ to NADH.

5. What is the role of NAD^+ in the process of cellular respiration?
   a. It functions as an electron carrier.
   b. It functions as an enzyme.
   c. It is the final electron acceptor for anaerobic respiration.
   d. It is a nucleotide source for the synthesis of ATP.

6. The reactions of the Krebs cycle occur in the
   a. inner membrane of the mitochondria.
   b. intermembrane space of the mitochondria.
   c. cytoplasm.
   d. matrix of the mitochondria.

7. The electrons carried by NADH and $FADH_2$ can be
   a. pumped into the intermembrane space.
   b. transferred to the ATP synthase.
   c. moved between proteins in the inner membrane of the mitochondrion.
   d. transported into the matrix of the mitochondrion.

APPLY

1. Which of the following is NOT a true statement regarding cellular respiration?
   a. Enzymes catalyze reactions that transfer electrons.
   b. Electrons have a higher potential energy at the end of the process.
   c. Carbon dioxide gas is a by-product.
   d. The process involves multiple redox reactions.

2. The direct source of energy for the ATP produced by ATP synthase comes from
   a. the electron transport chain.
   b. a proton gradient.
   c. substrate-level phosphorylation.
   d. the oxidation reactions occurring during respiration.

3. Anaerobic respiration
   a. occurs in humans in the absence of O_2.
   b. occurs in yeast and is how we make beer and wine.
   c. yields less energy than aerobic respiration because other final electron acceptors have lower affinity for electrons than O_2.
   d. yields more energy than aerobic respiration because other final electron acceptors have higher affinity for electrons than O_2.

4. What is the importance of fermentation to cellular metabolism?
   a. It generates glucose for the cell in the absence of O_2.
   b. It oxidizes NADH to NAD^+ during electron transport.
   c. It oxidizes NADH to NAD^+ in the absence of O_2.
   d. It reduces NADH to NAD^+ in the absence of O_2.

5. The link between electron transport and ATP synthesis
   a. is a high-energy intermediate like phosphoenol pyruvate.
   b. is the transfer of electrons to ATP synthase.
   c. is a proton gradient.
   d. depends on the absence of oxygen.

6. A chemical agent that makes holes in the inner membrane of the mitochondria would
   a. stop the movement of electrons down the electron transport chain.
   b. stop ATP synthesis.
   c. stop the Krebs cycle.
   d. All of the choices are correct.

7. Yeast cells that have mutations in genes that encode enzymes in glycolysis can still grow on glycerol. They are able to utilize glycerol because it
   a. enters glycolysis after the step affected by the mutation.
   b. can feed into the Krebs cycle and generate ATP via electron transport and chemiosmosis.
   c. can be utilized by fermentation.
   d. can donate electrons directly to the electron transport chain.

SYNTHESIZE

1. Use the following table to outline the relationship between the molecules and the metabolic reactions.

Molecules	Glycolysis	Cellular Respiration
Glucose		
Pyruvate		
Oxygen		
ATP		
CO$_2$		

2. Human babies and hibernating or cold-adapted animals are able to maintain body temperature (a process called *thermogenesis*) due to the presence of brown fat. Brown fat is characterized by a high concentration of mitochondria. These brown fat mitochondria have a special protein located within their inner membranes. *Thermogenin* is a protein that functions as a passive proton transporter. Propose a likely explanation for the role of brown fat in thermogenesis based on your knowledge of metabolism, transport, and the structure and function of mitochondria.

3. Recent data indicate a link between colder temperatures and weight loss. If adults retain brown fat, how could this be explained?

Biology, 11th Edition 161

CHAPTER 8

Photosynthesis

Chapter Contents

- **8.1** Overview of Photosynthesis
- **8.2** The Discovery of Photosynthetic Processes
- **8.3** Pigments
- **8.4** Photosystem Organization
- **8.5** The Light-Dependent Reactions
- **8.6** Carbon Fixation: The Calvin Cycle
- **8.7** Photorespiration

Introduction

The rich diversity of life that covers our Earth would be impossible without photosynthesis. Almost every oxygen atom in the air we breathe was once part of a water molecule, liberated by photosynthesis. All the energy released by the burning of coal, firewood, gasoline, and natural gas, and by our bodies' burning of all the food we eat—directly or indirectly—has been captured from sunlight by photosynthesis. It is vitally important, then, that we understand photosynthesis. Research may enable us to improve crop yields and land use, important goals in an increasingly crowded world. In chapter 7, we described how cells extract chemical energy from food molecules and use that energy to power their activities. In this chapter, we examine photosynthesis, the process by which organisms such as the aptly named sunflowers in the picture capture energy from sunlight and use it to build food molecules that are rich in chemical energy.

8.1 Overview of Photosynthesis

Learning Outcomes

1. *Explain the reaction for photosynthesis.*
2. *Describe the structure of the chloroplast.*

Life is powered by sunshine. The energy used by most living cells comes ultimately from the Sun and is captured by plants, algae, and bacteria through the process of photosynthesis.

The diversity of life is only possible because our planet is awash in energy streaming Earthward from the Sun. Each day, the radiant energy that reaches Earth equals the power from about 1 million Hiroshima-sized atomic bombs. Photosynthesis captures about 1% of this huge supply of energy (an amount equal to 10,000 Hiroshima bombs) and uses it to provide the energy that drives all life.

Photosynthesis combines CO₂ and H₂O, producing glucose and O₂

Photosynthesis occurs in a wide variety of organisms, and it comes in different forms. These include a form of photosynthesis that does not produce oxygen (anoxygenic) and a form that does (oxygenic). Anoxygenic photosynthesis is found in four different bacterial groups: purple bacteria, green sulfur bacteria, green nonsulfur bacteria, and heliobacteria. Oxygenic photosynthesis is found in cyanobacteria, seven groups of algae, and essentially all land plants. These two types of photosynthesis share similarities in the types of pigments they use to trap light energy, but they differ in the arrangement and action of these pigments.

In the case of plants, photosynthesis takes place primarily in the leaves. Figure 8.1 illustrates the levels of organization in a plant leaf. As you learned in chapter 4, the cells of plant leaves contain organelles called chloroplasts, which carry out the photosynthetic process. No other structure in a plant cell is able to carry out photosynthesis (figure 8.2).

Photosynthesis takes place in three stages:

1. capturing energy from sunlight;
2. using the energy to make ATP and to reduce the compound NADP⁺, an electron carrier, to NADPH; and
3. using the ATP and NADPH to power the synthesis of organic molecules from CO₂ in the air.

The first two stages require light and are commonly called the **light-dependent reactions.**

The third stage, the formation of organic molecules from CO₂, is called **carbon fixation.** This process takes place via a cyclic series of reactions. As long as ATP and NADPH are available, the carbon fixation reactions can occur either in the presence or in the absence of light, and so these reactions are also called the **light-independent reactions.**

The following simple equation summarizes the overall process of photosynthesis:

$$6CO_2 + 12H_2O + \text{light} \longrightarrow C_6H_{12}O_6 + 6H_2O + 6O_2$$

carbon dioxide + water → glucose + water + oxygen

You may notice that this equation is the reverse of the reaction for respiration. In respiration, glucose is oxidized to CO_2 using O_2 as an electron acceptor. In photosynthesis, CO_2 is reduced to glucose using electrons gained from the oxidation of water. The oxidation of H_2O and the reduction of CO_2 requires energy that is provided by light. Although this statement is an oversimplification, it provides a useful "global perspective."

In plants, photosynthesis takes place in chloroplasts

In chapter 7, you saw that a mitochondrion's complex structure of internal and external membranes contribute to its function. The same is true for the structure of the chloroplast.

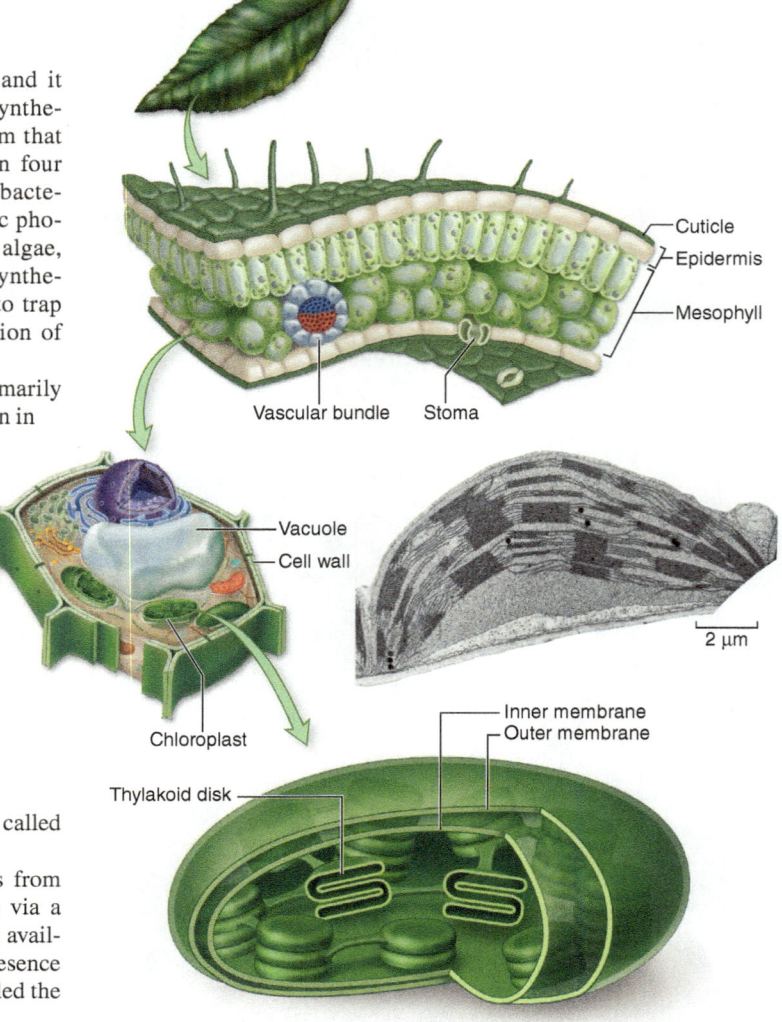

Figure 8.1 Journey into a leaf. A plant leaf possesses a thick layer of cells (the mesophyll) rich in chloroplasts. The inner membrane of the chloroplast is organized into flattened structures called thylakoid disks, which are stacked into columns called grana. The rest of the interior is filled with a semifluid substance called stroma.

The internal membrane of chloroplasts, called the *thylakoid membrane,* is a continuous phospholipid bilayer organized into flattened sacs that are found stacked on one another in columns called *grana* (singular, *granum*). The thylakoid membrane contains **chlorophyll** and other photosynthetic pigments for capturing light energy along with the machinery to make ATP. Connections between grana are termed *stroma lamella.*

Surrounding the thylakoid membrane system is a semiliquid substance called **stroma.** The stroma houses the enzymes needed to assemble organic molecules from CO_2 using energy from ATP coupled with reduction via NADPH. In the thylakoid membrane, photosynthetic pigments are clustered together to form **photosystems,** which show distinct organization within the thylakoid.

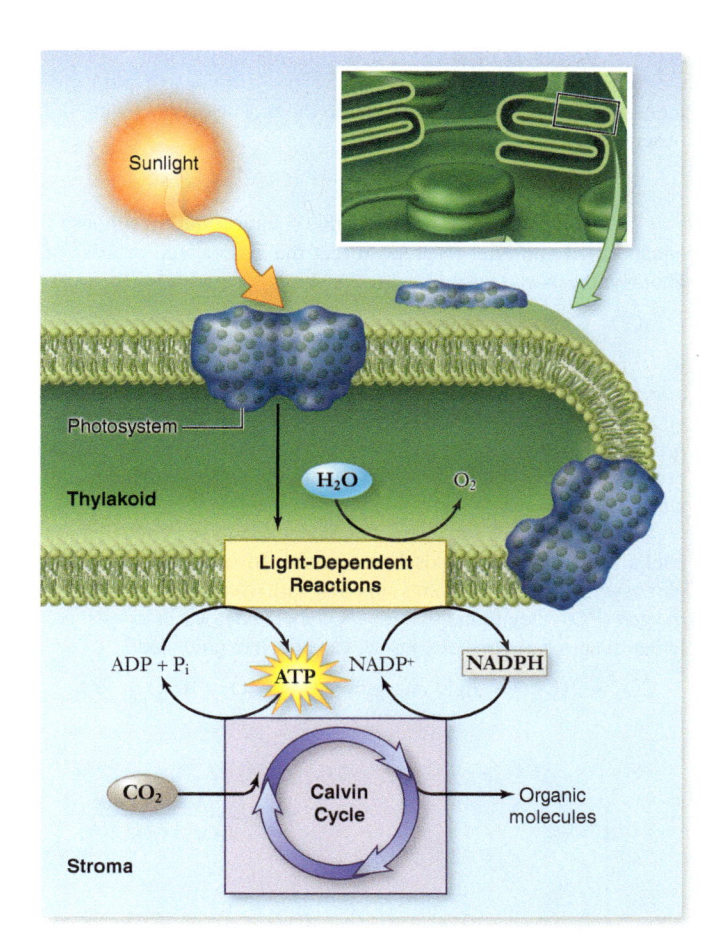

Figure 8.2 Overview of photosynthesis. In the light-dependent reactions, photosystems in the thylakoid absorb photons of light and use this energy to generate ATP and NADPH. Electrons lost from the photosystems are replaced by the oxidation of water, producing O_2 as a by-product. The ATP and NADPH produced by the light reactions is used during carbon fixation via the Calvin cycle in the stroma.

Each pigment molecule within the photosystem is capable of capturing photons, which are packets of energy. When light of a proper wavelength strikes a pigment molecule in the photosystem, the resulting excitation passes from one pigment molecule to another.

The excited electron is not transferred physically—rather, its *energy* passes from one molecule to another. The passage is similar to the transfer of kinetic energy along a row of upright dominoes. If you push the first one over, it falls against the next, and that one against the next, and so on, until all of the dominoes have fallen down.

Eventually, the energy arrives at a key chlorophyll molecule in contact with a membrane-bound protein that can accept an electron. The energy is transferred as an excited electron to that protein, which passes it on to a series of other membrane proteins that put the energy to work making ATP and NADPH. These compounds are then used to build organic molecules. The photosystem thus acts as a large antenna, gathering the light energy harvested by many individual pigment molecules.

Learning Outcomes Review 8.1

Photosynthesis consists of light-dependent reactions that require sunlight, and others that convert CO_2 into organic molecules. The overall reaction is essentially the reverse of respiration and produces O_2 as a by-product. The chloroplast's inner membrane, the thylakoid, is the site in which photosynthetic pigments are clustered, allowing passage of energy from one molecule to the next. The thylakoid membrane is organized into flattened sacs stacked in columns called grana.

■ *How is the structure of the chloroplast similar to the mitochondria?*

8.2 The Discovery of Photosynthetic Processes

Learning Outcomes

1. Describe experiments that support our understanding of photosynthesis.
2. Differentiate between the light-dependent and light-independent reactions.

The story of how we learned about photosynthesis begins over 300 years ago, and it continues to this day. It starts with curiosity about how plants manage to grow, often increasing their organic mass considerably.

Plants do not increase mass from soil and water alone

From the time of the Greeks, plants were thought to obtain their food from the soil, literally sucking it up with their roots. A Belgian doctor, Jan Baptista van Helmont (1580–1644) thought of a simple way to test this idea.

He planted a small willow tree in a pot of soil, after first weighing the tree and the soil. The tree grew in the pot for several years, during which time van Helmont added only water. At the end of five years, the tree was much larger, its weight having increased by 74.4 kg. However, the soil in the pot weighed only 57 g less than it had five years earlier. With this experiment, van Helmont demonstrated that the substance of the plant was not produced only from the soil. He incorrectly concluded, however, that the water he had been adding mainly accounted for the plant's increased biomass.

A hundred years passed before the story became clearer. The key clue was provided by the English scientist Joseph Priestly (1733–1804). On the 17th of August, 1771, Priestly put a living sprig of mint into air in which a wax candle had burnt out. On the

27th of the same month, Priestly found that another candle could be burned in this same air. Somehow, the vegetation seemed to have restored the air. Priestly found that while a mouse could not breathe candle-exhausted air, air "restored" by vegetation was not "at all inconvenient to a mouse." The key clue was that *living vegetation adds something to the air.*

How does vegetation "restore" air? Twenty-five years later, the Dutch physician Jan Ingenhousz (1730–1799) solved the puzzle. He demonstrated that air was restored only in the presence of sunlight and only by a plant's green leaves, not by its roots. He proposed that the green parts of the plant carry out a process that uses sunlight to split carbon dioxide into carbon and oxygen. He suggested that the oxygen was released as O_2 gas into the air, while the carbon atom combined with water to form carbohydrates. Other research refined his conclusions, and by the end of the 19th century, the overall reaction for photosynthesis could be written as:

$$CO_2 + H_2O + \text{light energy} \longrightarrow (CH_2O) + O_2$$

It turns out, however, that there's more to it than that. When researchers began to examine the process in more detail in the 20th century, the role of light proved to be unexpectedly complex.

Photosynthesis includes both light-dependent and light-independent reactions

At the beginning of the 20th century, the English plant physiologist F. F. Blackman (1866–1947) came to the surprising conclusion that photosynthesis is in fact a multistage process, only one portion of which uses light directly.

Blackman measured the effects of different light intensities, CO_2 concentrations, and temperatures on photosynthesis. As long as light intensity was relatively low, he found photosynthesis could be accelerated by increasing the amount of light, but not by increasing the temperature or CO_2 concentration (figure 8.3). At high light intensities, however, an increase in temperature or CO_2 concentration greatly accelerated photosynthesis.

Blackman concluded that photosynthesis consists of an initial set of what he called "light" reactions, that are largely independent of temperature but depend on light, and a second set of "dark" reactions (more properly called light-independent reactions), that seemed to be independent of light but limited by CO_2.

Do not be confused by Blackman's labels—the so-called "dark" reactions occur in the light (in fact, they require the products of the light-dependent reactions); his use of the word *dark* simply indicates that light is not *directly* involved in those reactions.

Blackman found that increased temperature increased the rate of the light-independent reactions, but only up to about 35°C. Higher temperatures caused the rate to fall off rapidly. Because many plant enzymes begin to be denatured at 35°C, Blackman concluded that enzymes must carry out the light-independent reactions.

O_2 comes from water, not from CO_2

In the 1930s, C. B. van Niel (1897–1985) working at the Hopkins Marine Station at Stanford, discovered that purple sulfur bacteria do not release oxygen during photosynthesis; instead, they convert hydrogen sulfide (H_2S) into globules of pure elemental sulfur that accumulate inside them. The process van Niel observed was:

$$CO_2 + 2H_2S + \text{light energy} \longrightarrow (CH_2O) + H_2O + 2S$$

The striking parallel between this equation and Ingenhousz's equation led van Niel to propose that the generalized process of photosynthesis can be shown as:

$$CO_2 + 2H_2A + \text{light energy} \longrightarrow (CH_2O) + H_2O + 2A$$

In this equation, the substance H_2A serves as an electron donor. In photosynthesis performed by green plants, H_2A is water, whereas in purple sulfur bacteria, H_2A is hydrogen sulfide. The product, A, comes from the splitting of H_2A. Therefore, the O_2 produced during green plant photosynthesis results from splitting water, not carbon dioxide.

When isotopes came into common use in the early 1950s, van Niel's revolutionary proposal was tested. Investigators examined photosynthesis in green plants supplied with water containing heavy oxygen (^{18}O); they found that the ^{18}O label ended up in oxygen gas rather than in carbohydrate, just as van Niel had predicted:

$$CO_2 + 2H_2{}^{18}O + \text{light energy} \longrightarrow (CH_2O) + H_2O + {}^{18}O_2$$

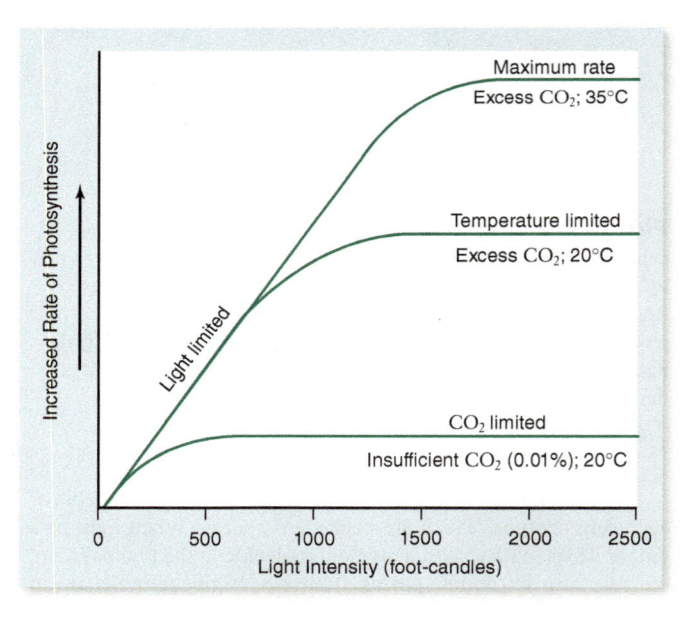

Figure 8.3 Discovery of the light-independent reactions. Blackman measured photosynthesis rates under differing light intensities, CO_2 concentrations, and temperatures. As this graph shows, light is the limiting factor at low light intensities, but temperature and CO_2 concentration are the limiting factors at higher light intensities. This implies the existence of reactions using CO_2 that involve enzymes.

Data analysis Blackman found that increasing light intensity above 2000 foot-candles did not lead to any further increase in the rate of photosynthesis. Can you suggest a hypothesis that would explain this?

In algae and green plants, the carbohydrate typically produced by photosynthesis is glucose. The complete balanced equation for photosynthesis in these organisms thus becomes:

$$6CO_2 + 12H_2O + \text{light energy} \longrightarrow C_6H_{12}O_6 + 6H_2O + 6O_2$$

ATP and NADPH from light-dependent reactions reduce CO₂ to make sugars

In his pioneering work on the light-dependent reactions, van Niel proposed that the H^+ ions and electrons generated by the splitting of water were used to convert CO_2 into organic matter in a process he called *carbon fixation*. In the 1950s, Robin Hill (1899–1991) demonstrated that van Niel was right, light energy could be harvested and used in a reduction reaction. Chloroplasts isolated from leaf cells were able to reduce a dye and release oxygen in response to light. Later experiments showed that the electrons released from water were transferred to $NADP^+$ and that illuminated chloroplasts deprived of CO_2 accumulate ATP. If CO_2 is introduced, neither ATP nor NADPH accumulate, and the CO_2 is assimilated into organic molecules.

These experiments are important for three reasons: First, they firmly demonstrate that photosynthesis in plants occurs within chloroplasts. Second, they show that the light-dependent reactions use light energy to reduce $NADP^+$ and to manufacture ATP. Third, they confirm that the ATP and NADPH from this early stage of photosynthesis are then used in the subsequent reactions to reduce carbon dioxide, forming simple sugars.

Learning Outcomes Review 8.2

Early experiments indicated that plants "restore" air to usable form—that is, produce oxygen—but only in the presence of sunlight. Further experiments showed that there are both light-dependent and independent reactions. The light-dependent reactions produce O_2 from H_2O, and generate ATP and NADPH. The light-independent reactions synthesize organic compounds through carbon fixation.

■ *Where does the carbon in your body come from?*

8.3 Pigments

Learning Outcomes

1. Discuss how pigments are important to photosynthesis.
2. Relate the absorption spectrum of a pigment to its color.

For plants to make use of the energy of sunlight, some biochemical structure must be present in chloroplasts and the thylakoids that can absorb this energy. Molecules that absorb light energy in the visible range are termed **pigments.** We are most familiar with them as dyes that impart a certain color to clothing or other materials. The color that we see is the color that is not absorbed—that is, it is reflected. To understand how plants use pigments to capture light energy, we must first review current knowledge about the nature of light.

Light is a form of energy

The wave nature of light produces an electromagnetic spectrum that differentiates light based on its wavelength (figure 8.4). We are most familiar with the visible range of this spectrum because we can actually see it, but visible light is only a small part of the entire spectrum. Visible light can be divided into its separate colors by the use of a prism, which separates light based on wavelength.

A particle of light, termed a **photon,** acts like a discrete bundle of energy. We use the wave concept of light to understand different colors of light and the particle nature of light to understand the energy transfers that occur during photosynthesis. Thus, we will refer both to wavelengths of light and to photons of light throughout the chapter.

The energy in photons

The energy content of a photon is inversely proportional to the wavelength of the light: Short-wavelength light contains photons of higher energy than long-wavelength light (figure 8.4). X-rays, which contain a great deal of energy, have very short wavelengths—much shorter than those of visible light.

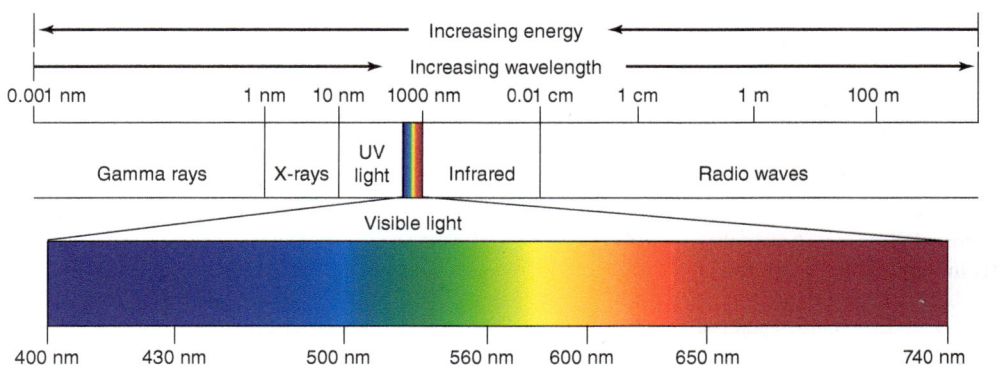

Figure 8.4 The electromagnetic spectrum. Light is a form of electromagnetic energy conveniently thought of as a wave. The shorter the wavelength of light, the greater its energy. Visible light represents only a small part of the electromagnetic spectrum between 400 and 740 nm.

A beam of light is able to remove electrons from certain molecules, creating an electrical current. This phenomenon is called the **photoelectric effect,** and it occurs when photons transfer energy to electrons. The strength of the photoelectric effect depends on the wavelength of light—that is, short wavelengths are much more effective than long ones in producing the photoelectric effect because they have more energy.

In photosynthesis, chloroplasts are acting as photoelectric devices: They absorb sunlight and transfer the excited electrons to a carrier. As we unravel the details of this process, it will become clear how this process traps energy and uses it to synthesize organic compounds.

Each pigment has a characteristic absorption spectrum

When a photon strikes a molecule with the amount of energy needed to excite an electron, then the molecule will absorb the photon raising the electron to a higher energy level. Whether the photon's energy is absorbed depends on how much energy it carries (defined by its wavelength), and also on the chemical nature of the molecule it hits.

As described in chapter 2, electrons occupy discrete energy levels in their orbits around atomic nuclei. To boost an electron into a different energy level requires just the right amount of energy, just as reaching the next rung on a ladder requires you to raise your foot just the right distance. A specific atom, therefore, can absorb only certain photons of light—namely, those that correspond to the atom's available energy levels. As a result, each molecule has a characteristic **absorption spectrum,** the range and efficiency of photons it is capable of absorbing.

As mentioned earlier in this section, pigments are good absorbers of light in the visible range. Organisms have evolved a variety of different pigments, but only two general types are used in green plant photosynthesis: chlorophylls and carotenoids. In some organisms, other molecules also absorb light energy.

Chlorophyll absorption spectra

Chlorophylls absorb photons within narrow energy ranges. Two kinds of chlorophyll in plants, chlorophyll *a* and chlorophyll *b,* preferentially absorb violet-blue and red light (figure 8.5). Neither of these pigments absorbs photons with wavelengths between about 500 and 600 nm; light of these wavelengths is reflected. When these reflected photons are subsequently absorbed by the retinal pigment in our eyes, we perceive them as green.

Chlorophyll *a* is the main photosynthetic pigment in plants and cyanobacteria and is the only pigment that can act directly to convert light energy to chemical energy. **Chlorophyll *b*,** acting as an **accessory pigment,** or secondary light-absorbing pigment, complements and adds to the light absorption of chlorophyll *a.*

Chlorophyll *b* has an absorption spectrum shifted toward the green wavelengths. Therefore, chlorophyll *b* can absorb photons that chlorophyll *a* cannot, greatly increasing the proportion of the photons in sunlight that plants can harvest. In addition, a variety of different accessory pigments are found in plants, bacteria, and algae.

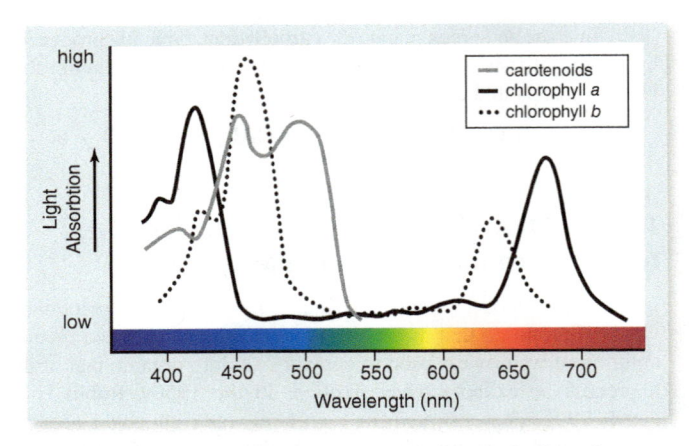

Figure 8.5 Absorption spectra for chlorophyll and carotenoids. The peaks represent wavelengths of light of sunlight absorbed by the two common forms of photosynthetic pigment, chlorophylls *a* and *b,* and the carotenoids. Chlorophylls absorb predominantly violet-blue and red light in two narrow bands of the spectrum and reflect green light in the middle of the spectrum. Carotenoids absorb mostly blue and green light and reflect orange and yellow light.

Structure of chlorophylls

Chlorophylls absorb photons by means of an excitation process analogous to the photoelectric effect. These pigments contain a complex ring structure, called a *porphyrin ring,* with alternating single and double bonds. At the center of the ring is a magnesium atom (figure 8.6).

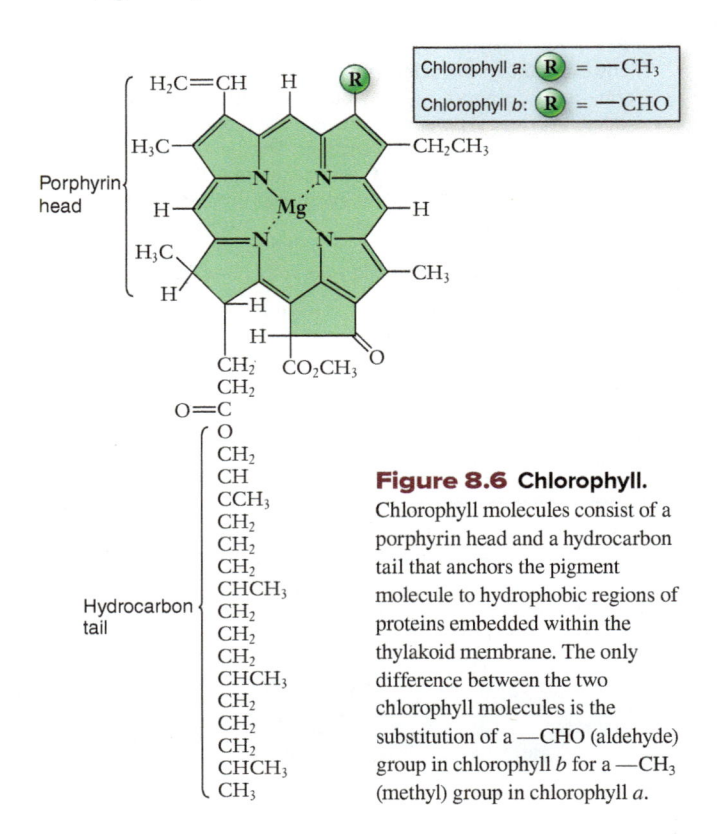

Figure 8.6 Chlorophyll. Chlorophyll molecules consist of a porphyrin head and a hydrocarbon tail that anchors the pigment molecule to hydrophobic regions of proteins embedded within the thylakoid membrane. The only difference between the two chlorophyll molecules is the substitution of a —CHO (aldehyde) group in chlorophyll *b* for a —CH₃ (methyl) group in chlorophyll *a.*

SCIENTIFIC THINKING

Hypothesis: *All wavelengths of light are equally effective in promoting photosynthesis.*

Prediction: *Illuminating plant cells with light broken into different wavelengths by a prism will produce the same amount of O_2 for all wavelengths.*

Test: *A filament of algae immobilized on a slide is illuminated by light that has passed through a prism. Motile bacteria that require O_2 for growth are added to the slide.*

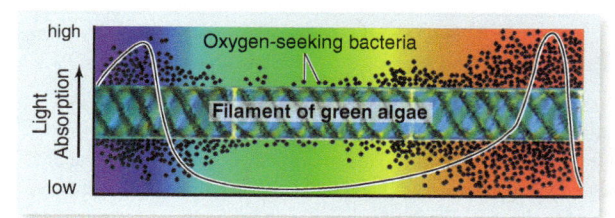

Result: *The bacteria move to regions of high O_2, or regions of most active photosynthesis. This is in the purple/blue and red regions of the spectrum.*

Conclusion: *All wavelengths are not equally effective at promoting photosynthesis. The most effective constitute the action spectrum for photosynthesis.*

Further Experiments: *How does the action spectrum relate to the various absorption spectra in figure 8.5?*

Figure 8.7 Determination of an action spectrum for photosynthesis.

Photons excite electrons in the porphyrin ring, which are then channeled away through the alternating carbon single- and double-bond system. Electrons not associated with a single atom or bond are said to be delocalized. Different side groups attached to the outside of the ring alter the absorption properties of different types of chlorophyll (figure 8.6). The precise absorption spectrum is also influenced by the association of chlorophyll with different proteins.

The **action spectrum** of photosynthesis—that is, the relative effectiveness of different wavelengths of light in promoting photosynthesis—corresponds to the absorption spectrum for chlorophylls. This is demonstrated in the experiment in figure 8.7. All plants, algae, and cyanobacteria use chlorophyll *a* as their primary pigments.

It is reasonable to ask why these photosynthetic organisms do not use a pigment like retinal (the pigment in our eyes), which has a broad absorption spectrum that covers the range of 500 to 600 nm. The most likely hypothesis involves *photoefficiency*. Although retinal absorbs a broad range of wavelengths, it does so with relatively low efficiency. Chlorophyll, in contrast, absorbs in only two narrow bands, but does so with high efficiency. Therefore, plants and most other photosynthetic organisms achieve far higher overall energy capture rates with chlorophyll than with other pigments.

Carotenoids and other accessory pigments

Carotenoids consist of carbon rings linked to chains with alternating single and double bonds. They can absorb photons with a wide range of energies, although they are not always highly efficient in transferring this energy. Carotenoids assist in photosynthesis by capturing energy from light composed of wavelengths that are not efficiently absorbed by chlorophylls (figure 8.8; see also figure 8.5).

Carotenoids also perform a valuable role in scavenging free radicals. The oxidation–reduction reactions that occur in the chloroplast can generate destructive free radicals. Carotenoids can act as general-purpose antioxidants to lessen damage. Thus carotenoids have a protective role in addition to their role as light-absorbing molecules. This protective role is not surprising, because unlike the chlorophylls, carotenoids are found in many different kinds of organisms, including members of all three domains of life.

You may have heard that eating carrots can enhance vision. If this effect is real, it is probably due to the high content of β-carotene in carrots. This carotenoid consists of two molecules of vitamin A joined together. The oxidation of vitamin A produces retinal, the pigment used in vertebrate vision.

Phycobiliproteins are accessory pigments found in cyanobacteria and some algae. These pigments contain a system of

Figure 8.8 Fall colors are produced by carotenoids and other accessory pigments. During the spring and summer, chlorophyll in leaves masks the presence of carotenoids and other accessory pigments. When cool fall temperatures cause leaves to cease manufacturing chlorophyll, the chlorophyll is no longer present to reflect green light, and the leaves reflect the orange and yellow light that carotenoids and other pigments do not absorb.

alternating double bonds similar to those found in other pigments and molecules that transfer electrons. Phycobiliproteins can be organized to form another light-harvesting complex that can absorb green light, which is reflected by chlorophyll. These complexes are probably ecologically important to cyanobacteria, helping them to exist in low-light situations in oceans. In this habitat, green light remains because red and blue light has been absorbed by green algae closer to the surface.

Learning Outcomes Review 8.3

A pigment is a molecule that can absorb light energy; its absorption spectrum shows the wavelengths at which it absorbs energy most efficiently. A pigment's color results from the wavelengths it does not absorb, which we then see. The main photosynthetic pigment is chlorophyll, which exists in several forms with slightly different absorption spectra. Many photosynthetic organisms have accessory pigments with absorption spectra different from chlorophyll; these increase light capture.

- *What is the difference between an action spectrum and an absorption spectrum?*

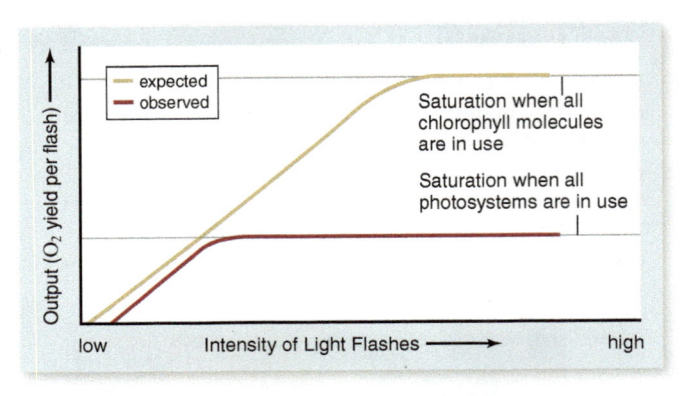

Figure 8.9 Saturation of photosynthesis. When photosynthetic saturation is achieved, further increases in intensity cause no increase in output. This saturation occurs far below the level expected for the number of individual chlorophyll molecules present. This led to the idea of organized photosystems, each containing many chlorophyll molecules. These photosystems saturate at a lower O_2 yield than that expected for the number of individual chlorophyll molecules.

Data analysis Draw the curves for photosystems that have a greater or lesser number of chlorophyll molecules than the curve shown.

8.4 Photosystem Organization

Learning Outcomes

1. Describe the nature of photosystems.
2. Contrast the function of reaction center and antenna chlorophyll molecules.

One way to study the role that pigments play in photosynthesis is to measure the correlation between the output of photosynthesis and the intensity of illumination—that is, how much photosynthesis is produced by how much light. Experiments on plants show that the output of photosynthesis increases linearly at low light intensities, but finally becomes saturated (no further increase) at high-intensity light. Saturation occurs because all of the light-absorbing capacity of the plant is in use.

Production of one O_2 molecule requires many chlorophyll molecules

Given the saturation observed with increasing light intensity, the next question is how many chlorophyll molecules have actually absorbed a photon. The question can be phrased this way: "Does saturation occur when all chlorophyll molecules have absorbed photons?" Finding an answer required being able to measure both photosynthetic output (on the basis of O_2 production) and the number of chlorophyll molecules present.

Using the unicellular algae *Chlorella,* investigators could obtain these values. Illuminating a *Chlorella* culture with pulses of light with increasing intensity should increase the yield of O_2 per pulse until the system becomes saturated. Then O_2 production can be compared with the number of chlorophyll molecules present in the culture.

The observed level of O_2 per chlorophyll molecule at saturation, however, turned out to be only one molecule of O_2 per 2500 chlorophyll molecules (figure 8.9). This result was very different from what was expected, and it led to the idea that light is absorbed not by independent pigment molecules, but rather by clusters of chlorophyll and accessory pigment molecules (photosystems). Light is absorbed by any one of hundreds of pigment molecules in a photosystem, and each pigment molecule transfers its excitation energy to a single molecule with a lower energy level than the others.

A generalized photosystem contains an antenna complex and a reaction center

In chloroplasts and all but one class of photosynthetic prokaryotes, light is captured by photosystems. Each photosystem is a network of chlorophyll *a* molecules, accessory pigments, and associated proteins held within a protein matrix on the surface of the photosynthetic membrane. Like a magnifying glass focusing light on a precise point, a photosystem channels the excitation energy gathered by any one of its pigment molecules to a specific molecule, the reaction center chlorophyll. This molecule then passes the energy out of the photosystem as excited electrons that are put to work driving the synthesis of ATP and organic molecules.

A photosystem thus consists of two closely linked components: (1) an *antenna complex* of hundreds of pigment molecules that gather photons and feed the captured light energy to the reaction center; and (2) a *reaction center* consisting of one or more chlorophyll *a* molecules in a matrix of protein, that passes excited electrons out of the photosystem.

The antenna complex

The **antenna complex** is also called a light-harvesting complex, which accurately describes its role. This light-harvesting complex captures photons from sunlight (figure 8.10) and channels them to the reaction center chlorophylls.

In chloroplasts, light-harvesting complexes consist of a web of chlorophyll molecules linked together and held tightly in the thylakoid membrane by a matrix of proteins. Varying amounts of carotenoid accessory pigments may also be present. The protein matrix holds individual pigment molecules in orientations that are optimal for energy transfer.

The excitation energy resulting from the absorption of a photon passes from one pigment molecule to an adjacent molecule on its way to the reaction center. After the transfer, the excited electron in each molecule returns to the low-energy level it had before the photon was absorbed. Consequently, it is energy, not the excited electrons themselves, that passes from one pigment molecule to the next. The antenna complex funnels the energy from many electrons to the reaction center.

The reaction center

The **reaction center** is a transmembrane protein–pigment complex. The reaction center of purple photosynthetic bacteria is simpler than the one in chloroplasts but better understood. A pair of bacteriochlorophyll *a* molecules acts as a trap for photon energy, passing an excited electron to an acceptor precisely positioned as its neighbor. Note that here in the reaction center, the excited electron itself is transferred, and not just the energy, as was the case in the pigment–pigment transfers of the antenna complex. This difference allows the energy absorbed from photons to move away from the chlorophylls, and it is the key conversion of light into chemical energy.

Figure 8.11 shows the transfer of excited electrons from the reaction center to the primary electron acceptor. By energizing an electron of the reaction center chlorophyll, light creates a strong electron donor where none existed before. The chlorophyll transfers the energized electron to the primary acceptor (a molecule of quinone), reducing the quinone and converting it to a strong electron donor. A nearby weak electron donor then passes a low-energy electron to the chlorophyll, restoring it to its original condition. The quinone transfers its electrons to another acceptor, and the process is repeated.

In plant chloroplasts, water serves as this weak electron donor. When water is oxidized in this way, oxygen is released along with two protons (H^+).

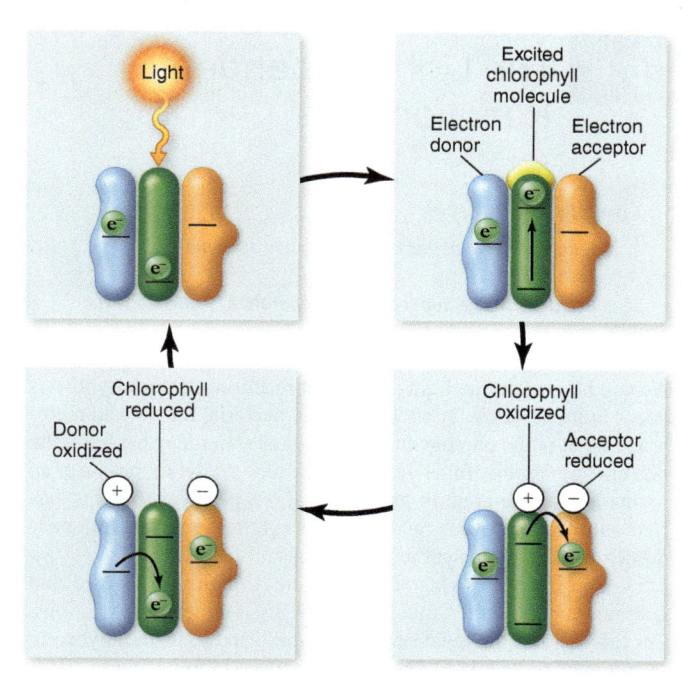

Figure 8.11 Converting light to chemical energy. When a chlorophyll in the reaction center absorbs a photon of light, an electron is excited to a higher energy level. This light-energized electron can be transferred to the primary electron acceptor, reducing it. The oxidized chlorophyll then fills its electron "hole" by oxidizing a donor molecule. The source of this donor varies with the photosystem, as discussed in the text.

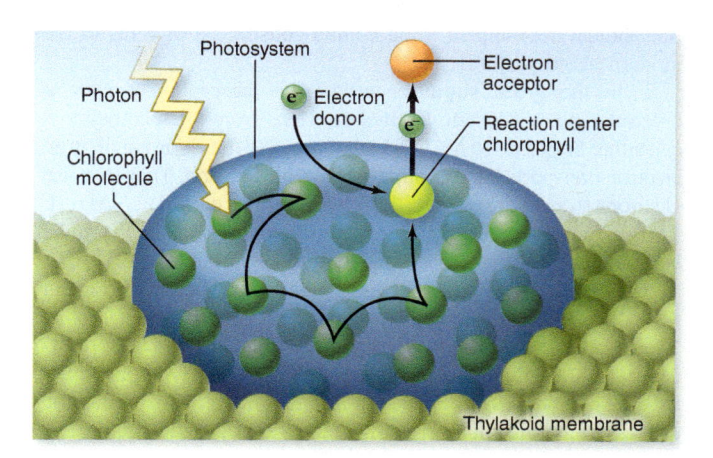

Figure 8.10 How the antenna complex works. When light of the proper wavelength strikes any pigment molecule within a photosystem, the light is absorbed by that pigment molecule. The excitation energy is then transferred from one molecule to another within the cluster of pigment molecules until it encounters the reaction center chlorophyll *a*. When excitation energy reaches the reaction center chlorophyll, electron transfer is initiated.

Learning Outcomes Review 8.4

Chlorophylls and accessory pigments are organized into photosystems found in the thylakoid membrane. The photosystem can be subdivided into an antenna complex, which is involved in light harvesting, and a reaction center, where the photochemical reactions occur. In the reaction center, an excited electron is passed to an acceptor; this transfers energy away from the chlorophylls and is key to the conversion of light into chemical energy.

■ *Why were photosystems an unexpected finding?*

8.5 The Light-Dependent Reactions

Learning Outcomes
1. Compare the function of the two photosystems in green plants.
2. Explain how the light reactions generate ATP and NADPH.

As you have seen, the light-dependent reactions of photosynthesis occur in membranes. In photosynthetic bacteria, the plasma membrane itself is the photosynthetic membrane. In many bacteria, the plasma membrane folds in on itself repeatedly to produce an increased surface area. In plants and algae, photosynthesis is carried out by chloroplasts, which are thought to be the evolutionary descendants of photosynthetic bacteria.

The internal thylakoid membrane is highly organized and contains the structures involved in the light-dependent reactions. For this reason, the reactions are also referred to as the thylakoid reactions. The thylakoid reactions take place in four stages:

1. **Primary photoevent.** A photon of light is captured by a pigment. This primary photoevent excites an electron within the pigment.
2. **Charge separation.** This excitation energy is transferred to the reaction center, which transfers an energetic electron to an acceptor molecule, initiating electron transport.
3. **Electron transport.** The excited electrons are shuttled along a series of electron carrier molecules embedded within the photosynthetic membrane. Several of them react by transporting protons across the membrane, generating a proton gradient. Eventually the electrons are used to reduce a final acceptor, NADPH.
4. **Chemiosmosis.** The protons that accumulate on one side of the membrane now flow back across the membrane through ATP synthase where chemiosmotic synthesis of ATP takes place, just as it does in aerobic respiration (see chapter 7).

These four processes make up the two stages of the light-dependent reactions mentioned at the beginning of this chapter. Steps 1 through 3 represent the stage of capturing energy from light; step 4 is the stage of producing ATP (and, as you'll see, NADPH). In the rest of this section we discuss the evolution of photosystems and the details of photosystem function in the light-dependent reactions.

Some bacteria use a single photosystem

Photosynthetic pigment arrays are thought to have evolved more than 2 BYA in bacteria similar to the purple and green bacteria alive today. In these bacteria, a single photosystem is used that generates ATP via electron transport. This process returns the electrons back to the reaction center. For this reason, it is called cyclic photophosphorylation. These systems do not produce oxygen and so are also anoxygenic.

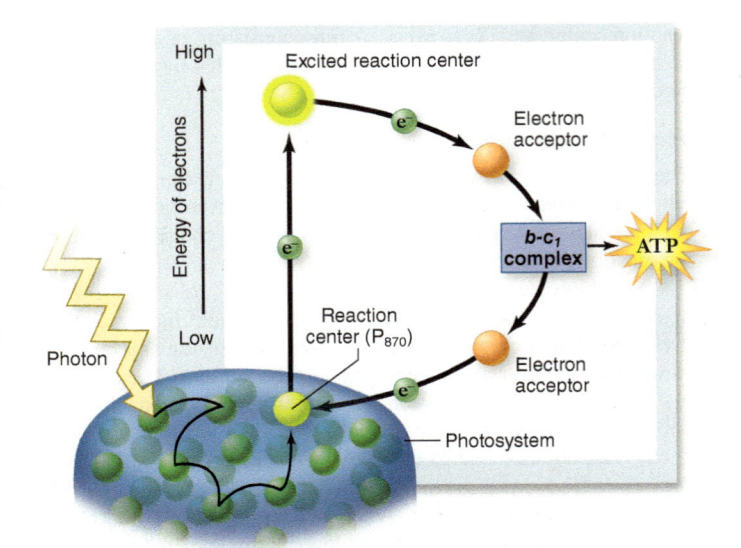

Figure 8.12 The path of an electron in purple nonsulfur bacteria. When a light-energized electron is ejected from the photosystem reaction center (P870), it returns to the photosystem via a cyclic path that produces ATP but not NADPH.

In the purple nonsulfur bacteria, peak absorption occurs at a wavelength of 870 nm (near infrared, not visible to the human eye), and thus the reaction center pigment is called P_{870}. Absorption of a photon by chlorophyll P_{870} does not raise an electron to a high enough level to be passed to NADP, so they must generate reducing power in a different way.

When the P_{870} reaction center absorbs a photon, the excited electron is passed to an electron transport chain that passes the electrons back to the reaction center, generating a proton gradient for ATP synthesis (figure 8.12). The proteins in the purple bacterial photosystem appear to be homologous to the proteins in the modern photosystem II.

In the green sulfur bacteria, peak absorption occurs at a wavelength of 840 nm. Excited electrons from this photosystem can either be passed to NADPH, or returned to the chlorophyll by an electron transport chain similar to the purple bacteria. They then use electrons from hydrogen sulfide to replace those passed to NADPH. The proteins in the green sulfur bacterial photosystem appear to be homologous to the proteins in the modern photosystem I.

Neither of these systems generates sufficient oxidizing power to oxidize H_2O. They are both anoxygenic and anaerobic. The linked photosystems of cyanobacteria and plant chloroplasts generate the oxidizing power necessary to oxidize H_2O, allowing it to serve as a source of both electrons and protons. This production of O_2 by oxygenic photosynthesis literally changed the atmosphere of the world.

Chloroplasts have two connected photosystems

In contrast to the sulfur bacteria, plants have two linked photosystems. This overcomes the limitations of cyclic photophosphorylation by providing an alternative source of electrons from the oxidation of water. The oxidation of water also generates O_2, thus oxygenic photosynthesis. The noncyclic transfer of electrons also produces NADPH, which can be used in the biosynthesis of carbohydrates.

One photosystem, called **photosystem I,** has an absorption peak of 700 nm, so its reaction center pigment is called P_{700}. This photosystem can pass electrons to NADPH similarly to the photosystem found in the sulfur bacteria discussed earlier in this section. The other photosystem, called **photosystem II,** has an absorption peak of 680 nm, so its reaction center pigment is called P_{680}. This photosystem can generate an oxidation potential high enough to oxidize water. Working together, the two photosystems carry out a noncyclic transfer of electrons that generate both ATP and NADPH.

The photosystems were named I and II in the order of their discovery, and not in the order in which they operate in the light-dependent reactions. In plants and algae, the two photosystems are specialized for different roles in the overall process of oxygenic photosynthesis. Photosystem I transfers electrons ultimately to $NADP^+$, producing NADPH. The electrons lost from photosystem I are replaced by electrons from photosystem II. Photosystem II with its high oxidation potential can oxidize water to replace the electrons transferred to photosystem I. Thus there is an overall flow of electrons from water to NADPH.

These two photosystems are connected by a complex of electron carriers called the **cytochrome/b_6-f complex** (explained shortly). This complex can use the energy from the passage of electrons to move protons across the thylakoid membrane to generate the proton gradient used by an ATP synthase enzyme.

The two photosystems work together in noncyclic photophosphorylation

Evidence for the action of two photosystems came from experiments that measured the rate of photosynthesis using two light beams of different wavelengths: one red and the other far-red. Using both beams produced a rate greater than the sum of the rates using individual beams of these wavelengths (figure 8.13). This surprising result, called the *enhancement effect,* can be explained by a mechanism involving two photosystems acting in series (that is, one after the other), one photosystem absorbs preferentially in the red, the other in the far-red.

Plants use photosystems II and I in series, first one and then the other, to produce both ATP and NADPH. This two-stage process is called **noncyclic photophosphorylation** because the path of the electrons is not a circle—the electrons ejected from the photosystems do not return to them, but rather end up in NADPH. The photosystems are replenished with electrons obtained by splitting water.

The scheme shown in figure 8.14, called a *Z diagram,* illustrates the two electron-energizing steps, one catalyzed by each photosystem. The horizontal axis shows the progress of the light reactions and the relative positions of the complexes, and the vertical axis shows relative energy levels of electrons. The electrons originate from water, which holds onto its electrons very tightly (redox potential = +820 mV), and end up in NADPH, which holds its electrons much more loosely (redox potential = −320 mV).

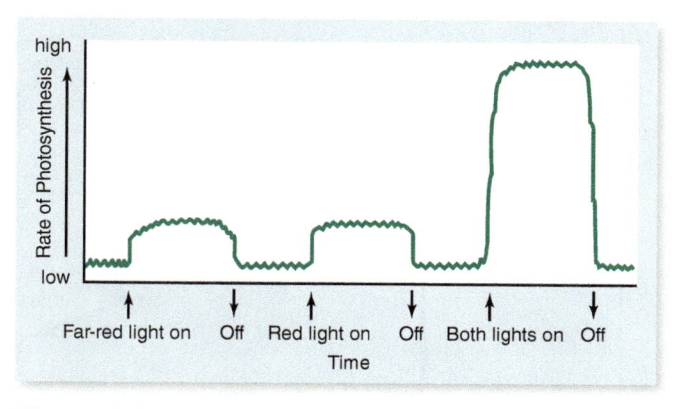

Figure 8.13 The enhancement effect. The rate of photosynthesis when red and far-red light are provided together is greater than the sum of the rates when each wavelength is provided individually. This result baffled researchers in the 1950s. Today, it provides key evidence that photosynthesis is carried out by two photochemical systems that act in series. One absorbs maximally in the far red, the other in the red portion of the spectrum.

Data analysis If "both lights on" showed a rate equal to the sum of each light, what would you conclude?

Photosystem II acts first. High-energy electrons generated by photosystem II are used to synthesize ATP and are then passed to photosystem I to drive the production of NADPH. For every pair of electrons obtained from a molecule of water, one molecule of NADPH and slightly more than one molecule of ATP are produced.

Photosystem II

The reaction center of photosystem II closely resembles the reaction center of purple bacteria. It consists of a core of 10 transmembrane protein subunits with electron transfer components and two P_{680} chlorophyll molecules arranged around this core. The light-harvesting antenna complex consists of molecules of chlorophyll *a* and accessory pigments bound to several protein chains. The reaction center of photosystem II differs from the reaction center of the purple bacteria in that it also contains four manganese atoms. These manganese atoms are essential for the oxidation of water.

Although the chemical details of the oxidation of water are not entirely clear, the outline is emerging. Four manganese atoms are bound in a cluster to reaction center proteins. Two water molecules are also bound to this cluster of manganese atoms. When the reaction center of photosystem II absorbs a photon, an electron in a P_{680} chlorophyll molecule is excited, which transfers this electron to an acceptor. The oxidized P_{680} then removes an electron from a manganese atom. The oxidized manganese atoms, with the aid of reaction center proteins, remove electrons from oxygen atoms in the two water molecules. This process requires the reaction center to absorb four photons to complete the oxidation of two water molecules, producing one O_2 in the process.

Figure 8.14 Z diagram of photosystems I and II. Two photosystems work sequentially and have different roles. Photosystem II passes energetic electrons to photosystem I via an electron transport chain. The electrons lost are replaced by oxidizing water. Photosystem I uses energetic electrons to reduce NADP⁺ to NADPH.

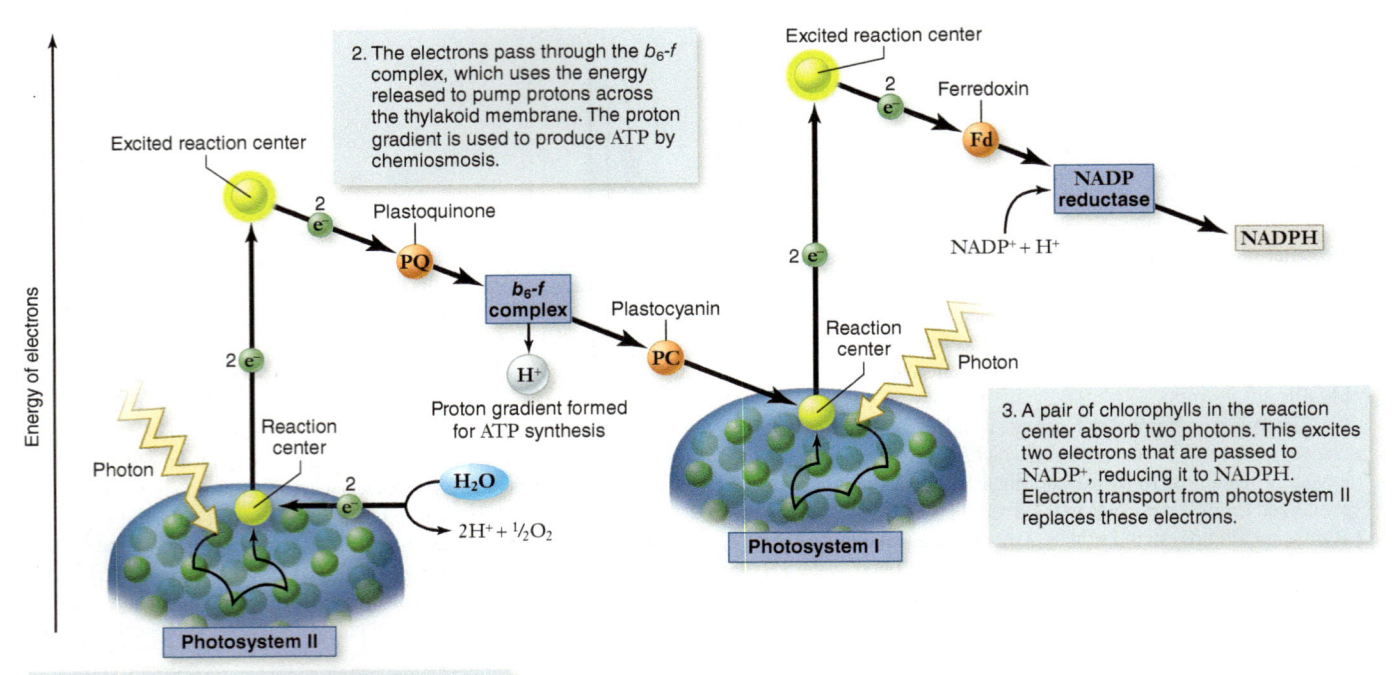

1. A pair of chlorophylls in the reaction center absorb two photons of light. This excites two electrons that are transferred to plastoquinone (PQ). Loss of electrons from the reaction center produces an oxidation potential capable of oxidizing water.

2. The electrons pass through the b_6-f complex, which uses the energy released to pump protons across the thylakoid membrane. The proton gradient is used to produce ATP by chemiosmosis.

3. A pair of chlorophylls in the reaction center absorb two photons. This excites two electrons that are passed to NADP⁺, reducing it to NADPH. Electron transport from photosystem II replaces these electrons.

The role of the b_6-f complex

The primary electron acceptor for the light-energized electrons leaving photosystem II is a quinone molecule. The reduced quinone that results from accepting a pair of electrons (*plastoquinone*) is a strong electron donor; it passes the excited electron pair to a proton pump called the b_6-f **complex** embedded within the thylakoid membrane (figure 8.15). This complex closely resembles the bc_1 complex in the respiratory electron transport chain of mitochondria, discussed in chapter 7.

Arrival of the energetic electron pair causes the b_6-f complex to pump a proton into the thylakoid space. A small, copper-containing protein called *plastocyanin* then carries the electron pair to photosystem I.

Photosystem I

The reaction center of photosystem I consists of a core transmembrane complex consisting of 12 to 14 protein subunits with two bound P_{700} chlorophyll molecules. Energy is fed to it by an antenna complex consisting of chlorophyll *a* and accessory pigment molecules.

Photosystem I accepts an electron from plastocyanin into the "hole" created by the exit of a light-energized electron. The absorption of a photon by photosystem I boosts the electron leaving the reaction center to a very high energy level. The electrons are passed to an iron–sulfur protein called *ferredoxin*. Unlike photosystem II and the bacterial photosystem, the plant photosystem I does not rely on quinones as electron acceptors.

Making NADPH

Photosystem I passes electrons to ferredoxin on the stromal side of the membrane (outside the thylakoid). The reduced ferredoxin carries an electron with very high potential. Two of them, from two molecules of reduced ferredoxin, are then donated to a molecule of NADP⁺ to form NADPH. The reaction is catalyzed by the membrane-bound enzyme *NADP reductase*.

Because the reaction occurs on the stromal side of the membrane and involves the uptake of a proton in forming NADPH, it contributes further to the proton gradient established during photosynthetic electron transport. The function of the two photosystems is summarized in figure 8.15.

ATP is generated by chemiosmosis

Protons are pumped from the stroma into the thylakoid compartment by the b_6-f complex. The splitting of water also produces added protons that contribute to the gradient. The thylakoid membrane is impermeable to protons, so this creates an electrochemical gradient that can be used to synthesize ATP.

ATP synthase

The chloroplast has ATP synthase enzymes in the thylakoid membrane that form a channel, allowing protons to cross back out into the stroma. These channels protrude like knobs on the external surface of the thylakoid membrane. As protons pass out of the thylakoid through the ATP synthase channel, ADP is phosphorylated to ATP and released into the stroma (figure 8.15). The stroma contains the enzymes that catalyze the reactions of carbon fixation—the Calvin cycle reactions.

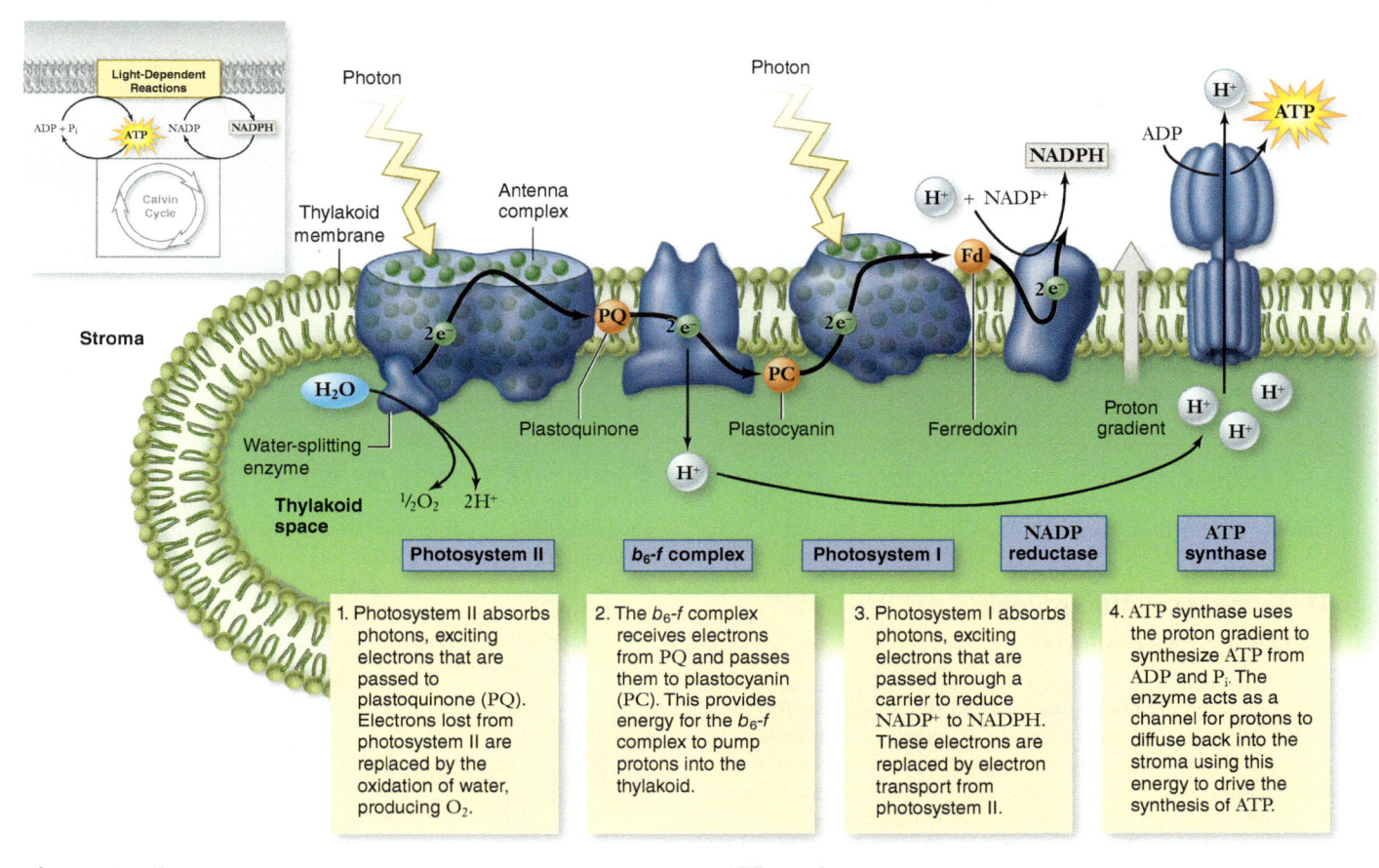

Figure 8.15 The photosynthetic electron transport system and ATP synthase. The two photosystems are arranged in the thylakoid membrane joined by an electron transport system that includes the b_6-f complex. These function together to create a proton gradient that is used by ATP synthase to synthesize ATP.

1. Photosystem II absorbs photons, exciting electrons that are passed to plastoquinone (PQ). Electrons lost from photosystem II are replaced by the oxidation of water, producing O_2.
2. The b_6-f complex receives electrons from PQ and passes them to plastocyanin (PC). This provides energy for the b_6-f complex to pump protons into the thylakoid.
3. Photosystem I absorbs photons, exciting electrons that are passed through a carrier to reduce $NADP^+$ to NADPH. These electrons are replaced by electron transport from photosystem II.
4. ATP synthase uses the proton gradient to synthesize ATP from ADP and P_i. The enzyme acts as a channel for protons to diffuse back into the stroma using this energy to drive the synthesis of ATP.

This mechanism is the same as that seen in the mitochondrial ATP synthase, and, in fact, the two enzymes are evolutionarily related. This similarity in generating a proton gradient by electron transport and ATP by chemiosmosis illustrates the similarities in structure and function in mitochondria and chloroplasts. Evidence for this chemiosmotic mechanism for photophosphorylation was actually discovered earlier (figure 8.16) and formed the background for experiments using the mitochondrial ATP synthase.

SCIENTIFIC THINKING

Figure 8.16 The Jagendorf acid bath experiment.

Hypothesis: Photophosphorylation is coupled to electron transport by a proton gradient.
Prediction: If a proton gradient can be formed artificially, then isolated chloroplasts will phosphorylate ADP in the dark.
Test: Isolated chloroplasts are incubated in acid medium, then transferred in the dark to a basic medium to create an artificial proton gradient.

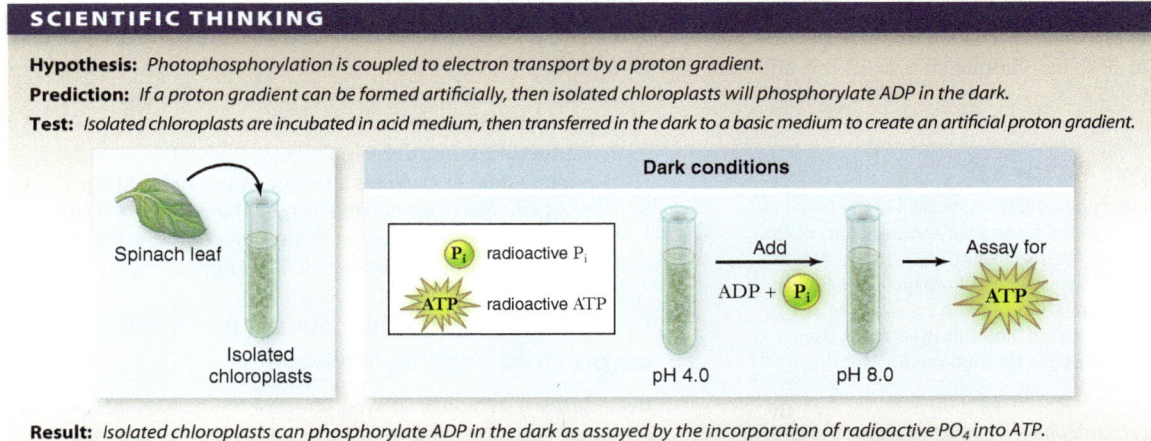

Result: Isolated chloroplasts can phosphorylate ADP in the dark as assayed by the incorporation of radioactive PO_4 into ATP.
Conclusion: The energy from electron transport in the chloroplast is coupled to the phosphorylation of ADP by a proton gradient.
Further Experiments: How would the use of agents that inhibit electron transport affect this outcome? How could such agents be used as a further test of the hypothesis?

The production of additional ATP

The passage of an electron pair from water to NADPH in noncyclic photophosphorylation generates one molecule of NADPH and slightly more than one molecule of ATP. But as you will learn in section 8.6, building organic molecules takes more energy than that—it takes 1.5 ATP molecules per NADPH molecule to fix carbon.

To produce the extra ATP, many plant species are capable of short-circuiting photosystem I, switching photosynthesis into a *cyclic photophosphorylation* mode, so that the light-excited electron leaving photosystem I is used to make ATP instead of NADPH. The energetic electrons are simply passed back to the b_6-f complex, rather than passing on to NADP$^+$. The b_6-f complex pumps protons into the thylakoid space, adding to the proton gradient that drives the chemiosmotic synthesis of ATP. The relative proportions of cyclic and noncyclic photophosphorylation in these plants determine the relative amounts of ATP and NADPH available for building organic molecules.

Thylakoid structure reveals components' locations

The four complexes responsible for the light-dependent reactions—namely photosystems I and II, cytochrome b_6-f, and ATP synthase—are not randomly arranged in the thylakoid. Researchers are beginning to image these complexes with the atomic force microscope, which can resolve nanometer scale structures, and a picture is emerging in which photosystem II is found primarily in the grana, whereas photosystem I and ATP synthase are found primarily in the stroma lamella. Photosystem I and ATP synthase may also be found in the edges of the grana that are not stacked. The cytochrome b_6-f complex is found in the borders between grana and stroma lamella. One possible model for the arrangement of the complexes is shown in figure 8.17.

The thylakoid itself is no longer thought of only as stacked disks. Some models of the thylakoid, based on electron microscopy and other imaging, depict the grana as folds of the interconnecting stroma lamella. This kind of arrangement is more similar to the folds seen in bacterial photosynthesis, and it would therefore allow for more flexibility in how the various complexes are arranged relative to one another.

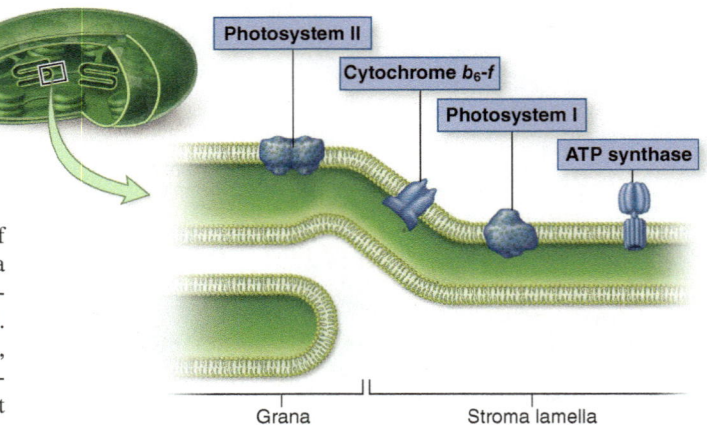

Figure 8.17 Model for the arrangement of complexes within the thylakoid. The arrangement of the two kinds of photosystems and the other complexes involved in photosynthesis is not random. Photosystem II is concentrated within grana, especially in stacked areas. Photosystem I and ATP synthase are concentrated in stroma lamella and the edges of grana. The cytochrome b_6-f complex is in the margins between grana and stroma lamella. This is one possible model for this arrangement.

8.6 Carbon Fixation: The Calvin Cycle

Learning Outcomes
1. Describe carbon fixation.
2. Demonstrate how six CO_2 molecules can be used to make one glucose.

Carbohydrates contain many C—H bonds and are highly reduced compared with CO_2. To build carbohydrates, cells use energy and a source of electrons produced by the light-dependent reactions of the thylakoids:

1. **Energy.** ATP (provided by cyclic and noncyclic photophosphorylation) drives the endergonic reactions.
2. **Reduction potential.** NADPH (provided by photosystem I) provides a source of protons and the energetic electrons needed to bind them to carbon atoms. Much of the light energy captured in photosynthesis ends up invested in the energy-rich C—H bonds of sugars.

Calvin cycle reactions convert inorganic carbon into organic molecules

Because early research showed temperature dependence, photosynthesis was predicted to involve enzyme-catalyzed reactions. These reactions form a cycle of enzyme-catalyzed steps much like the Krebs cycle of respiration. Unlike the Krebs cycle, however,

---

Learning Outcomes Review 8.5
The chloroplast has two photosystems located in the thylakoid membrane that are connected by an electron transport chain. Photosystem I passes an electron to NADPH. This electron is replaced by one from photosystem II. Photosystem II can oxidize water to replace the electron it has lost. A proton gradient is built up in the thylakoid space, and this gradient is used to generate ATP as protons pass through the ATP synthase enzyme.

■ *If the thylakoid membrane were leaky to protons, would ATP still be produced? Would NADPH?*

carbon fixation is geared toward producing new compounds, so the nature of the cycles is quite different.

The cycle of reactions that allow carbon fixation is called the **Calvin cycle,** after its discoverer, Melvin Calvin (1911–1997). Because the first intermediate of the cycle, phosphoglycerate, contains three carbon atoms, this process is also called **C₃ photosynthesis.**

The key step in this process—the event that makes the reduction of CO_2 possible—is the attachment of CO_2 to a highly specialized organic molecule. Photosynthetic cells produce this molecule by reassembling the bonds of two intermediates in glycolysis—fructose 6-phosphate and glyceraldehyde 3-phosphate (G3P)—to form the energy-rich 5-carbon sugar **ribulose 1,5-bisphosphate (RuBP).**

CO_2 reacts with RuBP to form a transient 6-carbon intermediate that immediately splits into two molecules of the 3-carbon *3-phosphoglycerate (PGA)*. This overall reaction is called the *carbon fixation reaction* because inorganic carbon (CO_2) has been incorporated into an organic form: the acid PGA. The enzyme that carries out this reaction, **ribulose bisphosphate carboxylase/oxygenase** (usually abbreviated **rubisco**) is a large, 16-subunit enzyme found in the chloroplast stroma.

Carbon is transferred through cycle intermediates, eventually producing glucose

We will consider how the Calvin cycle can produce one molecule of glucose, although this glucose is not produced directly by the cycle (figure 8.18). In a series of reactions, six molecules of CO_2 are bound to six RuBP by rubisco to produce 12 molecules of PGA

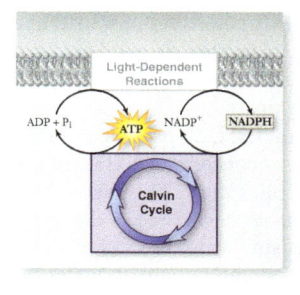

Figure 8.18 The Calvin cycle. The Calvin cycle accomplishes carbon fixation: converting inorganic carbon in the form of CO_2 into organic carbon in the form of carbohydrates. The cycle can be broken down into three phases: (1) carbon fixation, (2) reduction, and (3) regeneration of RuBP. For every six CO_2 molecules fixed by the cycle, a molecule of glucose can be synthesized from the products of the reduction reactions, G3P. The cycle uses the ATP and NADPH produced by the light reactions.

(containing 12 × 3 = 36 carbon atoms in all, 6 from CO_2 and 30 from RuBP). The 36 carbon atoms then undergo a cycle of reactions that regenerates the six molecules of RuBP used in the initial step (containing 6 × 5 = 30 carbon atoms). This leaves two molecules of *glyceraldehyde 3-phosphate (G3P)* (each with three carbon atoms) as the net gain. (You may recall G3P as also being the product of the first half of glycolysis, described in chapter 7.) These two molecules of G3P can then be used to make one molecule of glucose.

The net equation of the Calvin cycle is:

$$6CO_2 + 18\ ATP + 12\ NADPH + water \longrightarrow$$
$$2\ glyceraldehyde\ 3\text{-}phosphate + 16\ P_i + 18\ ADP + 12\ NADP^+$$

With six full turns of the cycle, six molecules of carbon dioxide enter, two molecules of G3P are produced, and six molecules of RuBP are regenerated. Thus six turns of the cycle produce two G3P that can be used to make a single glucose molecule. The six turns of the cycle also incorporated six CO_2 molecules, providing enough carbon to synthesize glucose, although the six carbon atoms do not all end up in this molecule of glucose.

Phases of the cycle

The Calvin cycle can be thought of as divided into three phases: (1) carbon fixation, (2) reduction, and (3) regeneration of RuBP. The carbon fixation reaction generates two molecules of the 3-carbon acid PGA; PGA is then reduced to G3P by reactions that are essentially a reverse of part of glycolysis; finally, the PGA is used to regenerate RuBP. Three turns around the cycle incorporate enough carbon to produce a new molecule of G3P, and six turns incorporate enough carbon to synthesize one glucose molecule.

We now know that light is required *indirectly* for different segments of the CO_2 reduction reactions. Five of the Calvin cycle enzymes—including rubisco—are light-activated—that is, they become functional or operate more efficiently in the presence of light. Light also promotes transport of required 3-carbon intermediates across chloroplast membranes. And finally, light promotes the influx of Mg^{2+} into the chloroplast stroma, which further activates the enzyme rubisco.

Output of the Calvin cycle

Glyceraldehyde 3-phosphate is a 3-carbon sugar, a key intermediate in glycolysis. Much of it is transported out of the chloroplast to the cytoplasm of the cell, where the reversal of several reactions in glycolysis allows it to be converted to fructose 6-phosphate and glucose 1-phosphate. These products can then be used to form sucrose, a major transport sugar in plants. (Sucrose, table sugar, is a disaccharide made of fructose and glucose.)

In times of intensive photosynthesis, G3P levels rise in the stroma of the chloroplast. As a consequence, some G3P in the chloroplast is converted to glucose 1-phosphate. This takes place in a set of reactions analogous to those occurring in the cytoplasm, by reversing several reactions similar to those of glycolysis. The glucose 1-phosphate is then combined into an insoluble polymer, forming long chains of starch stored as bulky starch grains in the cytoplasm. These starch grains represent stored glucose for later use.

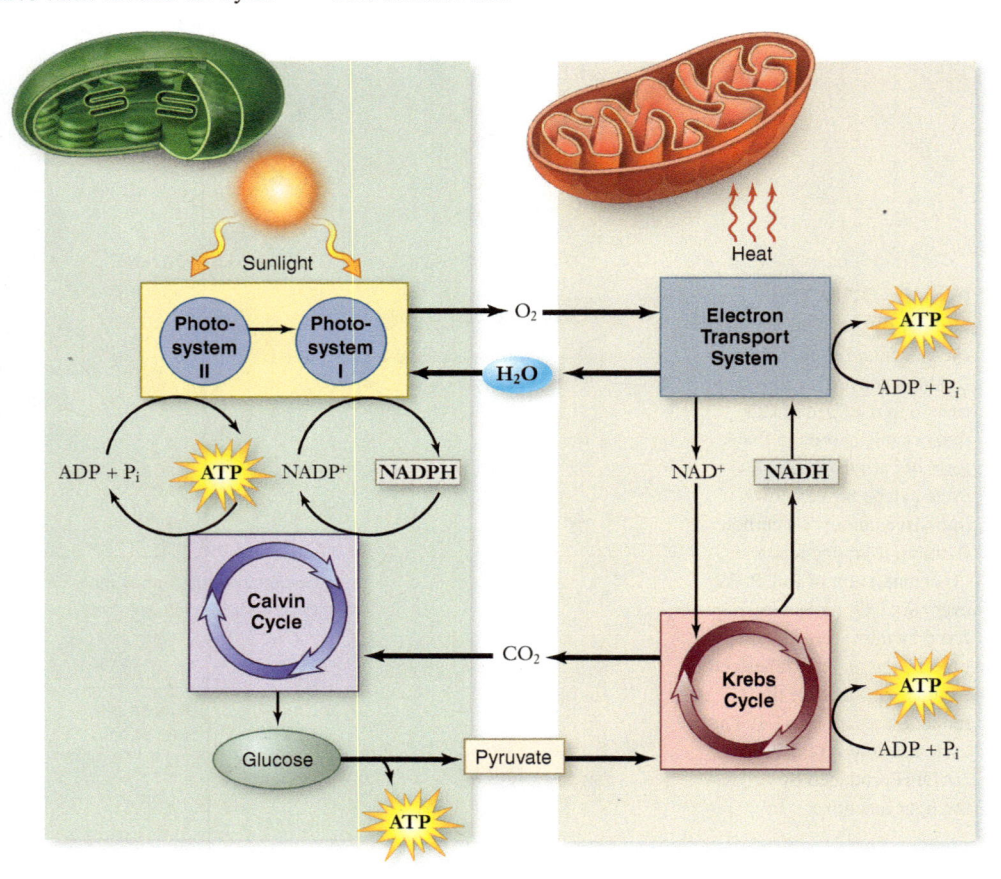

Figure 8.19 Chloroplasts and mitochondria: completing an energy cycle. Water and O_2 cycle between chloroplasts and mitochondria within a plant cell, as do glucose and CO_2. Cells with chloroplasts take in CO_2 and H_2O and produce glucose and O_2. Cells without chloroplasts, such as animal cells, take in glucose and O_2 and produce CO_2 and H_2O. This leads to global cycling of carbon through photosynthesis and respiration (see figure 57.1).

The energy cycle

The energy-capturing metabolisms of the chloroplasts studied in this chapter and the mitochondria studied in chapter 7 are intimately related (figure 8.19). Photosynthesis uses the products of respiration as starting substrates, and respiration uses the products of photosynthesis as starting substrates. The production of glucose from G3P even uses part of the ancient glycolytic pathway, run in reverse. Also, the principal proteins involved in electron transport and ATP production in plants are evolutionarily related to those in mitochondria.

Photosynthesis is but one aspect of plant biology, although it is an important one. In chapters 36 through 41, we examine plants in more detail. We have discussed photosynthesis as a part of cell biology because photosynthesis arose long before plants did, and because most organisms depend directly or indirectly on photosynthesis for the energy that powers their lives.

Learning Outcomes Review 8.6

Carbon fixation takes place in the stroma of the chloroplast, where inorganic CO_2 is incorporated into an organic molecule. The key intermediate is the 5-carbon sugar RuBP that combines with CO_2 in a reaction catalyzed by the enzyme rubisco. The cycle can be broken down into three stages: carbon fixation, reduction, and regeneration of RuBP. ATP and NADPH from the light reactions provide energy and electrons for the reduction reactions, which produce G3P. Glucose is synthesized when two molecules of G3P are combined.

- How does the Calvin cycle compare with glycolysis?

8.7 Photorespiration

Learning Outcomes

1. Distinguish between how rubisco acts to make RuBP and how it oxidizes RuBP.
2. Compare the function of carbon fixation in the C_3, C_4, and CAM pathways.

Figure 8.20 Stoma. A closed stoma in the leaf of a tobacco plant. Each stoma is formed from two guard cells whose shape changes with turgor pressure to open and close. Under dry conditions plants close their stomata to conserve water.

Evolution does not necessarily result in optimum solutions. Rather, it favors workable solutions that can be derived from features that already exist. Photosynthesis is no exception. Rubisco, the enzyme that catalyzes the key carbon-fixing reaction of photosynthesis, provides a decidedly suboptimal solution. This enzyme has a second enzymatic activity that interferes with carbon fixation, namely that of *oxidizing* RuBP. In this process, called **photorespiration**, O_2 is incorporated into RuBP, which undergoes additional reactions that actually release CO_2. Hence, photorespiration releases CO_2, essentially undoing carbon fixation.

Photorespiration reduces the yield of photosynthesis

The carboxylation and oxidation of RuBP are catalyzed at the same active site on rubisco, and CO_2 and O_2 compete with each other at this site. Under normal conditions at 25°C, the rate of the carboxylation reaction is four times that of the oxidation reaction, meaning that 20% of photosynthetically fixed carbon is lost to photorespiration.

This loss rises substantially as temperature increases, because under hot, arid conditions, specialized openings in the leaf called *stomata* (singular, *stoma*) (figure 8.20) close to conserve water. This closing also cuts off the supply of CO_2 entering the leaf and does not allow O_2 to exit (figure 8.21). As a result, the low-CO_2 and high-O_2 conditions within the leaf favor photorespiration.

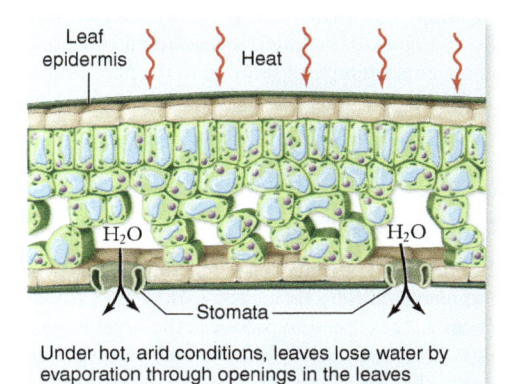

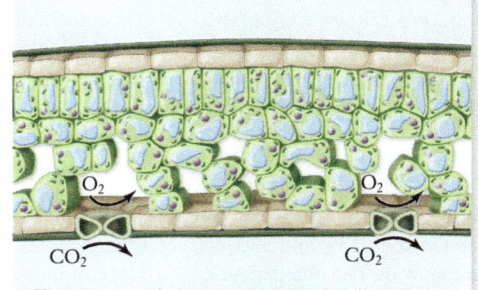

Figure 8.21 Conditions favoring photorespiration. In hot, arid environments, stomata close to conserve water, which also prevents CO_2 from entering and O_2 from exiting the leaf. The high-O_2/low-CO_2 conditions favor photorespiration.

Under hot, arid conditions, leaves lose water by evaporation through openings in the leaves called stomata.

The stomata close to conserve water but as a result, O_2 builds up inside the leaves, and CO_2 cannot enter the leaves.

Figure 8.22 Comparison of C₃ and C₄ pathways of carbon fixation. *a.* The C₃ pathway uses the Calvin cycle to fix carbon. All reactions occur in mesophyll cells using CO₂ that diffuses in through stomata. *b.* The C₄ pathway incorporates CO₂ into a 4-carbon molecule of malate in mesophyll cells. This is transported to the bundle-sheath cells where it is converted back into CO₂ and pyruvate, creating a high level of CO₂. This allows efficient carbon fixation by the Calvin cycle.

Plants that fix carbon using only C₃ photosynthesis (the Calvin cycle) are called **C₃ plants** (figure 8.22a). Other plants add CO₂ to phosphoenolpyruvate (PEP) to form a 4-carbon molecule. This reaction is catalyzed by the enzyme PEP *carboxylase*. This enzyme has two advantages over rubisco: it has a much greater affinity for CO₂ than rubisco, and it does not have oxidase activity.

The 4-carbon compound produced by PEP carboxylase undergoes further modification, only to be eventually decarboxylated. The CO₂ released by this decarboxylation is then used by rubisco in the Calvin cycle. This allows CO₂ to be pumped directly to the site of rubisco, which increases the local concentration of CO₂ relative to O₂, minimizing photorespiration. The 4-carbon compound produced by PEP carboxylase allows CO₂ to be stored in an organic form, to then be released in a different cell, or at a different time to keep the level of CO₂ high relative to O₂.

The reduction in the yield of carbohydrate as a result of photorespiration is not trivial. C₃ plants lose between 25% and 50% of their photosynthetically fixed carbon in this way. The rate depends largely on temperature. In tropical climates, especially those in which the temperature is often above 28°C, the problem is severe, and it has a major effect on tropical agriculture.

The two main groups of plants that initially capture CO₂ using PEP carboxylase differ in how they maintain high levels of CO₂ relative to O₂. In **C₄ plants** (figure 8.22b), the capture of CO₂ occurs in one cell and the decarboxylation occurs in an adjacent cell. This represents a spatial solution to the problem of photorespiration. The second group, **CAM plants,** perform both reactions in the same cell, but capture CO₂ using PEP carboxylase at night, then decarboxylate during the day. CAM stands for **crassulacean acid metabolism,** after the plant family Crassulaceae (the stonecrops, or hens-and-chicks), in which it was first discovered. This mechanism represents a temporal solution to the photorespiration problem.

C₄ plants have evolved to minimize photorespiration

The C₄ plants include corn, sugarcane, sorghum, and a number of other grasses. These plants initially fix carbon using PEP carboxylase in mesophyll cells. This reaction produces the organic acid oxaloacetate, which is converted to malate and transported to bundle-sheath cells that surround the leaf veins. Within the bundle-sheath cells, malate is decarboxylated to produce

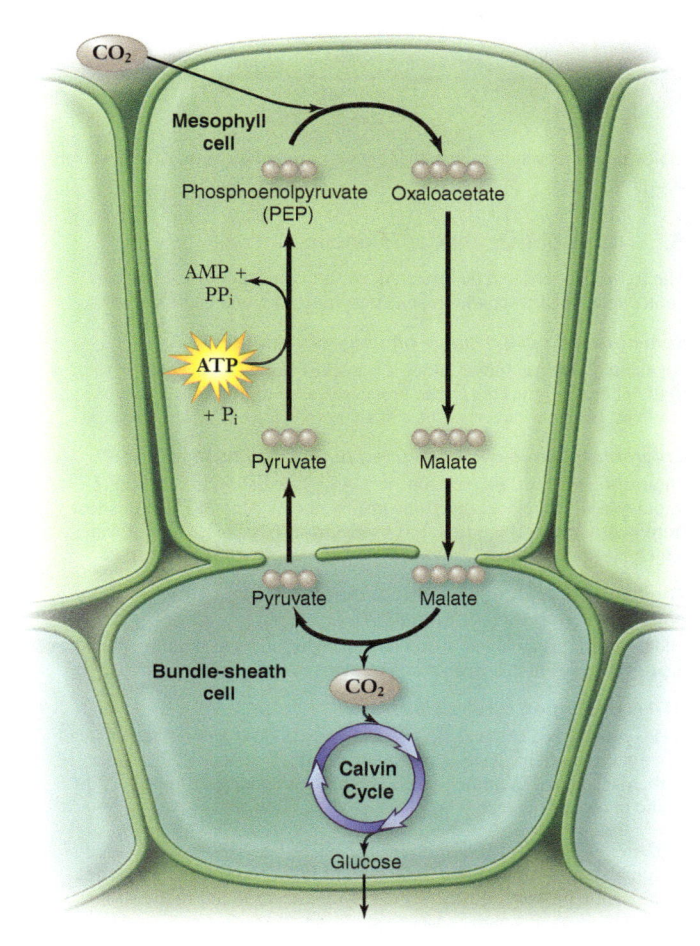

Figure 8.23 Carbon fixation in C₄ plants. This process is called the C₄ pathway because the first molecule formed, oxaloacetate, contains four carbons. The oxaloacetate is converted to malate, which moves into bundle-sheath cells where it is decarboxylated back to CO₂ and pyruvate. This produces a high level of CO₂ in the bundle-sheath cells that can be fixed by the usual C₃ Calvin cycle with little photorespiration. The pyruvate diffuses back into the mesophyll cells, where it is converted back to PEP to be used in another C₄ fixation reaction.

pyruvate and CO₂ (figure 8.23). Because the bundle-sheath cells are impermeable to CO₂, the local level of CO₂ is high and carbon fixation by rubisco and the Calvin cycle is efficient. The pyruvate produced by decarboxylation is transported back to the mesophyll cells, where it is converted back to PEP, thereby completing the cycle.

The C₄ pathway, although it overcomes the problems of photorespiration, does have a cost. The conversion of pyruvate back to PEP requires breaking two high-energy bonds in ATP. Thus each CO₂ transported into the bundle-sheath cells cost the equivalent of two ATP. To produce a single glucose, this requires 12 additional ATP compared with the Calvin cycle alone. Despite this additional cost, C₄ photosynthesis is advantageous in hot dry climates where photorespiration would remove more than half of the carbon fixed by the usual C₃ pathway alone.

Figure 8.24 Carbon fixation in CAM plants. CAM plants also use both C₄ and C₃ pathways to fix carbon and minimize photorespiration. In CAM plants, the two pathways occur in the same cell but are separated in time: The C₄ pathway is utilized to fix carbon at night, then CO₂ is released from these accumulated stores during the day to drive the C₃ pathway. This achieves the same effect of minimizing photorespiration, while also minimizing loss of water by opening stomata at night when temperatures are lower.

The Crassulacean acid pathway splits photosynthesis into night and day

A second strategy to decrease photorespiration in hot regions has been adopted by the CAM plants. These include many succulent (water-storing) plants, such as cacti, pineapples, and some members of about two dozen other plant groups.

In these plants, the stomata open during the night and close during the day (figure 8.24). This pattern of stomatal opening and closing is the reverse of that in most plants. CAM plants initially fix CO₂ using PEP carboxylase to produce oxaloacetate. The oxaloacetate is often converted into other organic acids, depending on the particular CAM plant. These organic compounds accumulate during the night and are stored in the vacuole. Then during the day, when the stomata are closed, the organic acids are decarboxylated to yield high levels of CO₂. These high levels of CO₂ drive the Calvin cycle and minimize photorespiration.

Like C₄ plants, CAM plants use both C₃ and C₄ pathways. They differ in that they use both of these pathways in the same cell: the C₄ pathway at night and the C₃ pathway during the day. In C₄ plants the two pathways occur in different cells.

Learning Outcomes Review 8.7

Rubisco can also oxidize RuBP under conditions of high O₂ and low CO₂. In plants that use only C₃ metabolism (Calvin cycle), up to 20% of fixed carbon is lost to this photorespiration. Plants adapted to hot, dry environments are capable of storing CO₂ as a 4-carbon molecule and avoiding some of this loss; they are called C₄ plants. In CAM plants, CO₂ is fixed at night into a C₄ organic compound; in the daytime, this compound is used as a source of CO₂ C₃ metabolism when stomata are closed to prevent water loss.

■ *How do C₄ plants and CAM plants differ?*

Chapter Review

8.1 Overview of Photosynthesis

Photosynthesis is the conversion of light energy into chemical energy (figure 8.2).

Photosynthesis combines CO_2 and H_2O, producing glucose and O_2.
Photosynthesis has three stages: absorbing light energy, using this energy to synthesize ATP and NADPH, and using the ATP and NADPH to convert CO_2 to organic molecules. The first two stages consist of light-dependent reactions, and the third stage of light-independent reactions.

In plants, photosynthesis takes place in chloroplasts.
Chloroplasts contain internal thylakoid membranes and a fluid matrix called stroma. The photosystems involved in energy capture are found in the thylakoid membranes, and enzymes for assembling organic molecules are in the stroma.

8.2 The Discovery of Photosynthetic Processes

Plants do not increase mass from soil and water alone.
Early investigations revealed that plants produce O_2 from carbon dioxide and water in the presence of light.

Photosynthesis includes both light-dependent and light-independent reactions.
The light-dependent reactions require light; the light-independent reactions occur in both daylight and darkness. The rate of photosynthesis depends on the amount of light, the CO_2 concentration, and temperature.

O_2 comes from water, not from CO_2.
The use of isotopes revealed the individual origins and fates of different molecules in photosynthetic reactions.

ATP and NADPH from light-dependent reactions reduce CO_2 to make sugars.
Carbon fixation requires ATP and NADPH, which are products of the light-dependent reactions. As long as these are available, CO_2 is reduced by enzymes in the stroma to form simple sugars.

8.3 Pigments

Light is a form of energy.
Light exists both as a wave and as a particle (photon). Light can remove electrons from some metals by the photoelectric effect, and in photosynthesis, chloroplasts act as photoelectric devices.

Each pigment has a characteristic absorption spectrum (figure 8.5).
Chlorophyll *a* is the only pigment that can convert light energy into chemical energy. Chlorophyll *b* is an accessory pigment that increases the harvest of photons for photosynthesis.

Carotenoids and other accessory pigments further increase a plant's ability to harvest photons.

8.4 Photosystem Organization (figure 8.10)

Production of one O_2 molecule requires many chlorophyll molecules.
Measurement of O_2 output led to the idea of photosystems—clusters of pigment molecules that channel energy to a reaction center.

A generalized photosystem contains an antenna complex and a reaction center.
A photosystem is a network of chlorophyll *a*, accessory pigments, and proteins embedded in the thylakoid membrane. Pigment molecules of the antenna complex harvest photons and feed light energy to the reaction center. The reaction center is composed of two chlorophyll *a* molecules in a protein matrix that pass an excited electron to an electron acceptor.

8.5 The Light-Dependent Reactions

The light reactions can be broken down into four processes: primary photoevent, charge separation, electron transport, and chemiosmosis.

Some bacteria use a single photosystem (figure 8.12).
An excited electron moves along a transport chain and eventually returns to the photosystem. This cyclic process is used to generate a proton gradient. In some bacteria, this can also produce NADPH.

Chloroplasts have two connected photosystems (figure 8 14).
Photosystem I transfers electrons to $NADP^+$, reducing it to NADPH. Photosystem II replaces electrons lost by photosystem I. Electrons lost from photosystem II are replaced by electrons from oxidation of water, which also produces O_2.

The two photosystems work together in noncyclic photophosphorylation (figure 8.14).
Photosystem II and photosystem I are linked by an electron transport chain; the b_6-f complex in this chain pumps protons into the thylakoid space.

ATP is generated by chemiosmosis.
ATP synthase is a channel enzyme; as protons flow through the channel down their gradient, ADP is phosphorylated producing ATP, similar to the mechanism in mitochondria. Plants can make additional ATP by cyclic photophosphorylation.

Thylakoid structure reveals components' locations.
Imaging studies suggest that photosystem II is primarily found in the grana, while photosystem I and ATP synthase are found in the stroma lamella.

8.6 Carbon Fixation: The Calvin Cycle (figure 8.18)

Calvin cycle reactions convert inorganic carbon into organic molecules.
The Calvin cycle, also known as C_3 photosynthesis, uses CO_2, ATP, and NADPH to build simple sugars.

Carbon is transferred through cycle intermediates, eventually producing glucose.
The Calvin cycle occurs in three stages: carbon fixation via the enzyme rubisco's action on RuBP and CO_2; reduction of the resulting 3-carbon PGA to G3P, generating ATP and NADPH; and regeneration of RuBP. Six turns of the cycle fix enough carbon to produce two excess G3Ps used to make one molecule of glucose.

8.7 Photorespiration

Photorespiration reduces the yield of photosynthesis.
Rubisco can catalyze the oxidation of RuBP, reversing carbon fixation. Dry, hot conditions tend to increase this reaction.

C_4 plants have evolved to minimize photorespiration.
C_4 plants fix carbon by adding CO_2 to a 3-carbon molecule, forming oxaloacetate. Carbon is fixed in one cell by the C_4 pathway, then CO_2 is released in another cell for the Calvin cycle (figure 8.23).

The Crassulacean acid pathway splits photosynthesis into night and day.
CAM plants use the C_4 pathway during the day when stomata are closed, and the Calvin cycle at night in the same cell (figure 8.24).

Review Questions

UNDERSTAND

1. The *light-dependent* reactions of photosynthesis are responsible for the production of
   a. glucose.
   b. CO_2.
   c. ATP and NADPH.
   d. H_2O.

2. Which region of a chloroplast is associated with the capture of light energy?
   a. Thylakoid membrane
   b. Outer membrane
   c. Stroma
   d. Both a and c are correct.

3. The colors of light that are most effective for photosynthesis are
   a. red, blue, and violet.
   b. green, yellow, and orange.
   c. infrared and ultraviolet.
   d. All colors of light are equally effective.

4. During noncyclic photosynthesis, photosystem I functions to _____, and photosystem II functions to _____.
   a. synthesize ATP; produce O_2
   b. reduce $NADP^+$; oxidize H_2O
   c. reduce CO_2; oxidize NADPH
   d. restore an electron to its reaction center; gain an electron from water

5. How is a reaction center pigment in a photosystem different from a pigment in the antenna complex?
   a. The reaction center pigment is a chlorophyll molecule.
   b. The antenna complex pigment can only reflect light.
   c. The reaction center pigment loses an electron when it absorbs light energy.
   d. The antenna complex pigments are not attached to proteins.

6. The ATP and NADPH from the light reactions are used
   a. in glycolysis in roots.
   b. directly in most biochemical reactions in the cell.
   c. during the reactions of the Calvin cycle to produce glucose.
   d. to synthesize chlorophyll.

7. The carbon fixation reaction converts
   a. inorganic carbon into an organic acid.
   b. CO_2 into glucose.
   c. inactive rubisco into active rubisco.
   d. an organic acid into CO_2.

8. C_4 plants initially fix carbon by
   a. the same pathway as C_3 plants, but they modify this product.
   b. incorporating CO_2 into oxaloacetate, which is converted to malate.
   c. incorporating CO_2 into citrate via the Krebs cycle.
   d. incorporating CO_2 into glucose via reverse glycolysis.

APPLY

1. The overall flow of electrons in the light reactions is from
   a. antenna pigments to the reaction center.
   b. H_2O to CO_2.
   c. photosystem I to photosystem II.
   d. H_2O to NADPH.

2. If you could measure pH within a chloroplast, where would it be lowest?
   a. In the stroma
   b. In the lumen of the thylakoid
   c. In the cytoplasm immediately outside the chloroplast
   d. In the antenna complex

3. The excited electron from photosystem I
   a. can be returned to the reaction center to generate ATP by cyclic photophosphorylation.
   b. is replaced by oxidizing H_2O.
   c. is replaced by an electron from photosystem II.
   d. Both a and c are correct.

4. If the Calvin cycle runs through six turns
   a. all of the fixed carbon will end up in the same glucose molecule.
   b. 12 carbons will be fixed by the process.
   c. enough carbon will be fixed to make one glucose, but they will not all be in the same molecule.
   d. one glucose will be converted into six CO_2.

5. Which of the following are similarities between the structure and function of mitochondria and chloroplasts?
   a. They both create internal proton gradients by electron transport.
   b. They both generate CO_2 by oxidation reactions.
   c. They both have a double membrane system.
   d. Both a and c are correct.

6. Given that the C_4 pathway gets around the problems of photorespiration, why don't all plants use it?
   a. It is a more recent process, and many plants have not had time to evolve this pathway.
   b. It requires extra enzymes that many plants lack.
   c. It requires special transport tissues that many plants lack.
   d. It also has an energetic cost.

7. If the thylakoid membrane became leaky to ions, what would you predict to be the result on the light reactions?
   a. It would stop ATP production.
   b. It would stop NADPH production.
   c. It would stop the oxidation of H_2O.
   d. All of the choices are correct.

8. The overall process of photosynthesis
   a. results in the reduction of CO_2 and the oxidation of H_2O.
   b. results in the reduction of H_2O and the oxidation of CO_2.
   c. consumes O_2 and produces CO_2.
   d. produces O_2 from CO_2.

SYNTHESIZE

1. Compare and contrast the fixation of carbon in C_3, C_4, and CAM plants.

2. Diagram the relationship between the reactants and products of photosynthesis and respiration.

3. Do plant cells need mitochondria? Explain your answer.

CHAPTER 9

Cell Communication

Chapter Contents

9.1 Overview of Cell Communication
9.2 Receptor Types
9.3 Intracellular Receptors
9.4 Signal Transduction Through Receptor Kinases
9.5 Signal Transduction Through G Protein–Coupled Receptors

Introduction

Springtime is a time of rebirth and renewal. Trees that have appeared dead produce new leaves and buds, and flowers sprout from the ground. For sufferers of seasonal allergy, this is not quite such a pleasant time. The pollen in the micrograph and other allergens produced stimulate the immune system to produce the molecule histamine and other molecules that form cellular signals. These signals cause inflammation, mucus secretion, vasodilation, and other responses that together cause the runny nose, itching, watery eyes, and other symptoms that make up the allergic reaction. We treat allergy symptoms by using drugs called antihistamines that interfere with this cellular signaling. The popular drug loratadine (better known as Claritin), for example, acts by blocking the receptor for histamine, thus preventing its action.

We will begin this chapter with a general overview of signaling, and the kinds of receptors cells use to respond to signals. Then we will look in more detail at how these different types of receptors can elicit a response from cells, and finally, how cells make connections with one another.

9.1 Overview of Cell Communication

Learning Outcomes

1. Discriminate between methods of signaling based on distance from source to reception.
2. Describe how phosphorylation can affect protein function.

Communication between cells is common in nature. Cell signaling occurs in all multicellular organisms, providing an indispensable mechanism for cells to influence one another. Effective signaling requires a signaling molecule, called a **ligand,** and a molecule to which the signal binds, called a **receptor protein.** The interaction of these two components initiates the process of *signal transduction*, which converts the information in the signal into a cellular response (figure 9.1).

The cells of multicellular organisms use a variety of molecules as signals, including but not limited to, peptides, large proteins, individual amino acids, nucleotides, and steroids and other lipids. Even dissolved gases such as NO (nitric oxide) are used as signals.

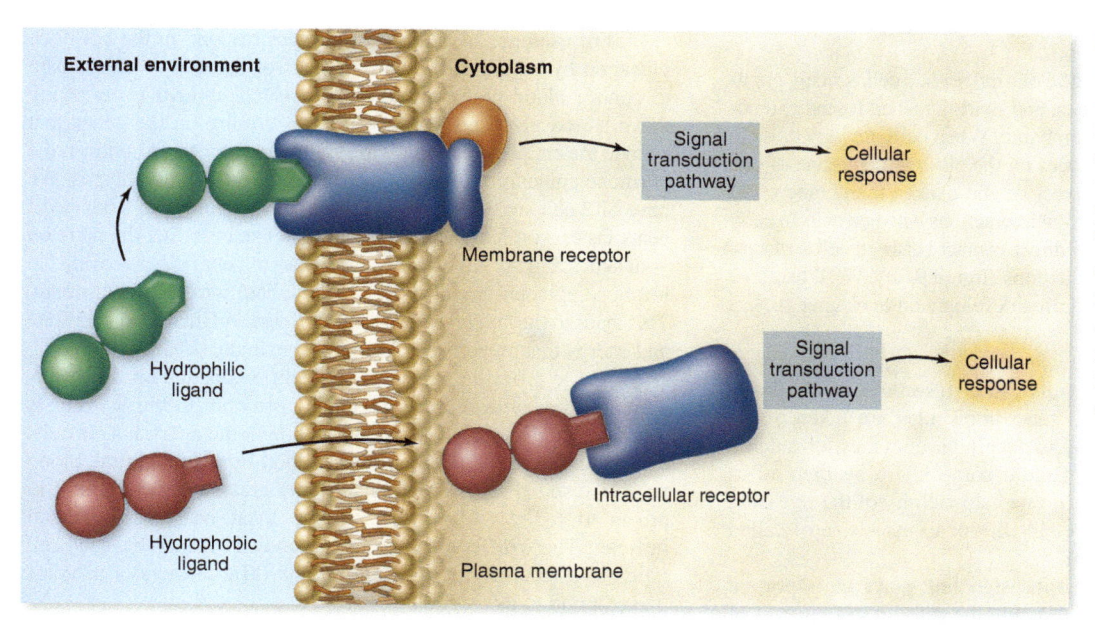

Figure 9.1 Overview of cell signaling. Cell signaling involves a signal molecule called a ligand, a receptor, and a signal transduction pathway that produces a cellular response. The location of the receptor can either be intracellular, for hydrophobic ligands that can cross the membrane, or in the plasma membrane, for hydrophilic ligands that cannot cross the membrane.

Any cell of a multicellular organism is exposed to a constant stream of signals. At any time, hundreds of different chemical signals may be present in the environment surrounding the cell. Each cell responds only to certain signals, however, and ignores the rest, like a person following the conversation of one or two individuals in a noisy, crowded room.

How does a cell "choose" which signals to respond to? The number and kind of receptor molecules determine this. When a ligand approaches a receptor protein that has a complementary shape, the two can bind, forming a complex. This binding induces a change in the receptor protein's shape, ultimately producing a response in the cell via a signal transduction pathway. In this way, a given cell responds to the signaling molecules that fit the particular set of receptor proteins it possesses and ignores those for which it lacks receptors.

Signaling is defined by the distance from source to receptor

Cells can communicate through any of four basic mechanisms, depending primarily on the distance between the signaling and responding cells (figure 9.2). These mechanisms are (1) direct contact, (2) paracrine signaling, (3) endocrine signaling, and (4) synaptic signaling.

In addition to using these four basic mechanisms, some cells actually send signals to themselves, secreting signals that bind to specific receptors on their own plasma membranes. This process, called *autocrine signaling,* is thought to play an important role in reinforcing developmental changes, and it is an important component of signaling in the immune system (see chapter 51).

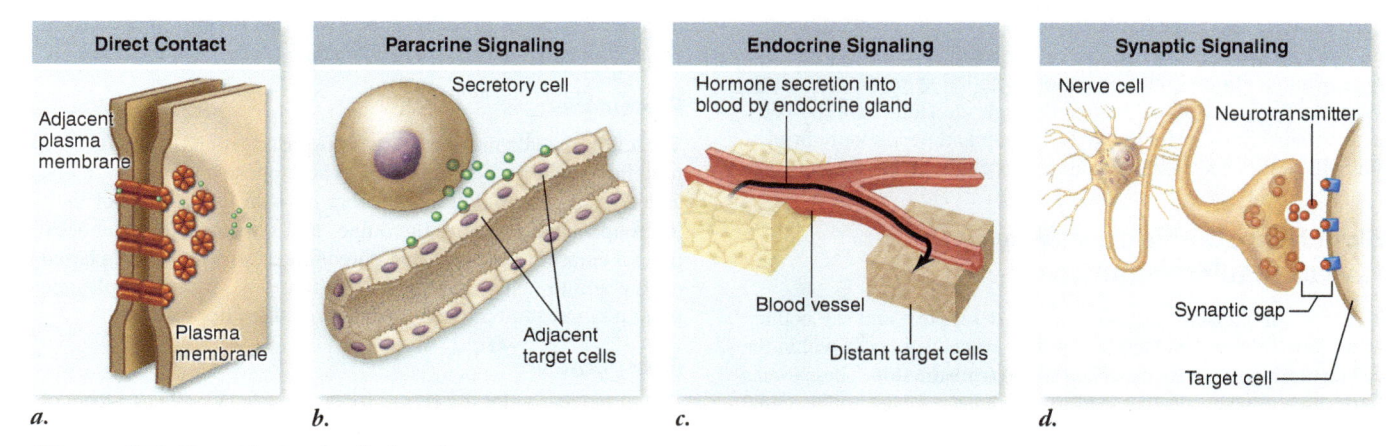

Figure 9.2 Four kinds of cell signaling. Cells communicate in several ways. *a.* Two cells in direct contact with each other may send signals across gap junctions. *b.* In paracrine signaling, secretions from one cell have an effect only on cells in the immediate area. *c.* In endocrine signaling, hormones are released into the organism's circulatory system, which carries them to the target cells. *d.* Chemical synapse signaling involves transmission of signal molecules, called neurotransmitters, from a neuron over a small synaptic gap to the target cell.

Direct contact

As you saw in chapter 5, the surface of a eukaryotic cell is richly populated with proteins, carbohydrates, and lipids attached to and extending outward from the plasma membrane. When cells are very close to one another, some of the molecules on the plasma membrane of one cell can be recognized by receptors on the plasma membrane of an adjacent cell. Many of the important interactions between cells in early development occur by means of direct contact between cell surfaces. Cells also signal through gap junctions (figure 9.2a). We'll examine contact-dependent interactions during development in chapter 19.

Paracrine signaling

Signal molecules released by cells can diffuse through the extracellular fluid to other cells. If those molecules are taken up by neighboring cells, destroyed by extracellular enzymes, or quickly removed from the extracellular fluid in some other way, their influence is restricted to cells in the immediate vicinity of the releasing cell. Signals with such short-lived, local effects are called **paracrine** signals (figure 9.2b).

Like direct contact, paracrine signaling plays an important role in early development, coordinating the activities of clusters of neighboring cells. The immune response in vertebrates also involves paracrine signaling between immune cells (see chapter 51).

Endocrine signaling

A released signal molecule that remains in the extracellular fluid may enter the organism's circulatory system and travel widely throughout the body. These longer-lived signal molecules, which may affect cells very distant from the releasing cell, are called **hormones,** and this type of intercellular communication is known as **endocrine signaling** (figure 9.2c). Chapter 45 discusses endocrine signaling in detail. Both animals and plants use this signaling mechanism extensively.

Synaptic signaling

In animals, the cells of the nervous system provide rapid communication with distant cells. Their signal molecules, **neurotransmitters,** do not travel to the distant cells through the circulatory system as hormones do. Rather, the long, fiberlike extensions of nerve cells release neurotransmitters from their tips very close to the target cells (figure 9.2d). The association of a neuron and its target cell is called a **chemical synapse,** and this type of intercellular communication is called **synaptic signaling.** Whereas paracrine signals move through the fluid between cells, neuro-transmitters cross the synaptic gap and persist only briefly. We will examine synaptic signaling more fully in chapter 43.

Signal transduction pathways lead to cellular responses

The types of signaling just outlined are descriptive and say nothing about how cells respond to signals. The events that occur within the cell on receipt of a signal are called **signal transduction.** These events form discrete pathways that lead to a cellular response to the signal received by receptors. Knowledge of these signal transduction pathways has accumulated over many years of work and indicates a high degree of complexity that explains how in some cases different cell types can have the same response to different signals, and in other cases different cell types can have a different response to the same signal.

For example, a variety of cell types respond to the hormone glucagon by mobilizing glucose as part of the body's mechanism to control blood glucose (see chapter 45). This involves breaking down stored glycogen into glucose and turning on the genes that encode the enzymes necessary to synthesize glucose. In contrast, the hormone epinephrine has diverse effects on different cell types. We have all been startled or frightened by a sudden event. Your heart beats faster, you feel more alert, and you can even feel the hairs on your skin stand up. All of this is due in part to your body releasing the hormone epinephrine (also called adrenaline) into the bloodstream. This leads to the heightened state of alertness and increased heart rate and energy that prepare us to respond to extreme situations.

These differing effects of epinephrine depend on the different cell types with receptors for this hormone. In the liver, cells are stimulated to mobilize glucose, while in the heart muscle cells contract more forcefully to increase blood flow. In addition, blood vessels respond by expanding in some areas and contracting in others to redirect blood flow to the liver, heart, and skeletal muscles. These different reactions depend on the fact that each cell type has a receptor for epinephrine, but different sets of proteins that respond to this signal.

Phosphorylation is key in control of protein function

The function of a signal transduction pathway is to change the behavior or nature of a cell. This action may require changing the composition of proteins that make up a cell or altering the activity of cellular proteins. Many proteins are inactive or nonfunctional as they are initially synthesized and require modification after synthesis for activation. In other cases, a protein may be deactivated by modification. A major source of control for protein function is the addition or removal of phosphate groups, called **phosphorylation** or **dephosphorylation,** respectively.

As you learned in chapters 7 and 8, the end result of the metabolic pathways of cellular respiration and photosynthesis was the phosphorylation of ADP to ATP. The ATP synthesized by these processes can donate phosphate groups to proteins. The phosphorylation of proteins alters their function by either turning their activity on or off. This is one way that the information from extracellular signals can result in changes in cellular activities.

Protein kinases

The class of enzyme that adds phosphate groups from ATP to proteins is called a *protein kinase*. These phosphate groups can be added to the three amino acids that have an OH as part of their R group, namely serine, threonine, and tyrosine. We categorize protein kinases as either serine–threonine or tyrosine kinases based on the amino acids they modify (figure 9.3). Most cytoplasmic protein kinases fall into the serine–threonine kinase class.

Phosphatases

Part of the reason for the versatility of phosphorylation as a form of protein modification is that it is reversible. Another class of enzymes called **phosphatases** removes phosphate groups, reversing the action of kinases (figure 9.3). Thus, a protein activated by a kinase will be deactivated by a phosphatase, and a protein deactivated by a kinase will be activated by a phosphatase.

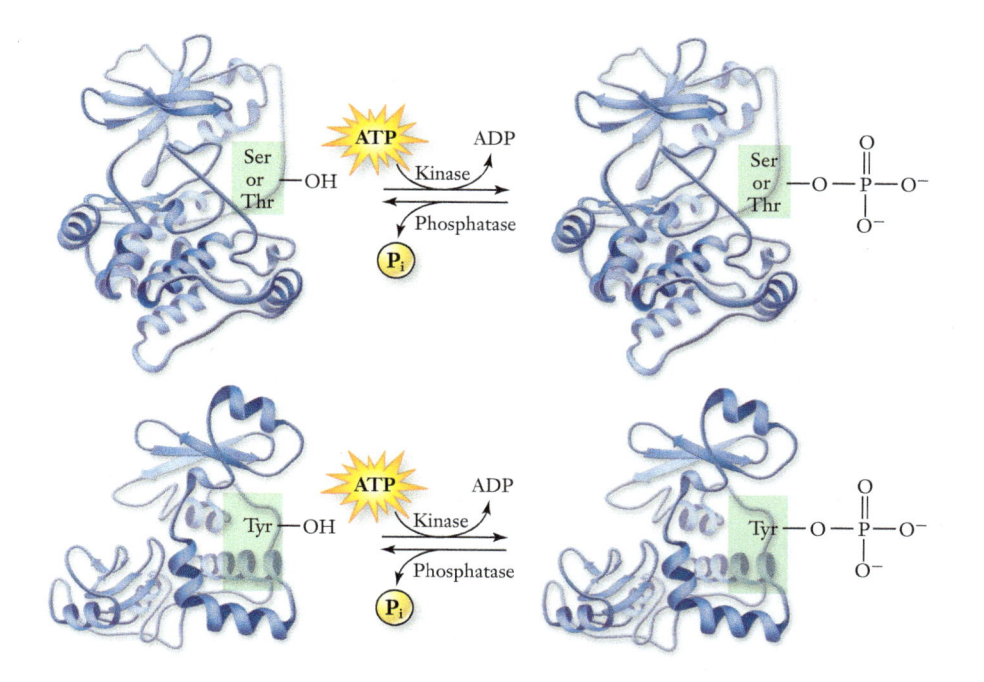

Figure 9.3 Phosphorylation of proteins. Many proteins are controlled by their phosphorylation state—that is, they are activated by phosphorylation and deactivated by dephosphorylation or the reverse. The enzymes that add phosphate groups are called kinases. These form two classes depending on the amino acid the phosphate is added to, either serine–threonine kinases or tyrosine kinases. The action of kinases is reversed by protein phosphatase enzymes.

Learning Outcomes Review 9.1

Cell communication involves chemical signals, or ligands, that bind to cellular receptors. Binding of ligand to receptor initiates signal transduction pathways that lead to a cellular response. Different cells may have the same response to one signal and the same signal can also elicit different responses in different cells. The phosphorylation–dephosphorylation of proteins is a common mechanism of controlling protein function found in signaling pathways.

- How are receptor ligand interactions similar to enzyme substrate interactions?

9.2 Receptor Types

Learning Outcome
1. Contrast the different types of receptors.

Understanding cell signaling requires first understanding the receptors themselves. For a cell to be able to respond to a specific signaling molecule, it must have a receptor that can bind to the signaling molecule. This specific binding interaction of receptor and ligand is an example of molecular recognition, where one molecule interacts with another based on their complementary shapes. This interaction causes subtle changes in the structure of the receptor, thereby activating it. This is the beginning of any signal transduction pathway.

Receptors are defined by location

Receptors can be differentiated based on their location and the chemical nature of their ligands. Intracellular receptors bind to hydrophobic ligands that can easily cross the plasma membrane to enter the cell. In contrast, cell-surface or membrane receptors bind to hydrophilic ligands outside the cell because these ligands cannot easily cross the membrane (see figure 9.1). Membrane receptors are transmembrane proteins with cytoplasmic and extracellular domains that allow them to interact with other molecules both inside and outside the cell. Table 9.1 summarizes the types of receptors and communication mechanisms discussed in this chapter.

Membrane receptors include three subclasses

When a receptor is a transmembrane protein, the ligand binds to the receptor outside of the cell and never actually crosses the plasma membrane. In this case, the receptor itself, and not the signaling molecule is responsible for information crossing the membrane. Membrane receptors can be categorized based on their structure and function.

Channel-linked receptors

Chemically gated ion channels are receptor proteins that allow the passage of ions (figure 9.4a). The receptor proteins that bind many neurotransmitters have the same basic structure. Each is a membrane protein with multiple transmembrane domains, meaning that the chain of amino acids threads back and forth across the plasma membrane several times. In the center of the protein is a pore that connects the extracellular fluid with the cytoplasm. The pore is big enough for ions to pass through, so the protein functions as an **ion channel.**

The channel is said to be chemically gated because it opens only when a chemical (the neurotransmitter) binds to it. The type of ion that flows across the membrane when a chemically gated ion channel opens depends on the shape and charge structure of the channel. Sodium, potassium, calcium, and chloride ions all have specific ion channels.

TABLE 9.1	Receptors Involved in Cell Signaling		
Receptor Type	**Structure**	**Function**	**Example**
Intracellular Receptors	No extracellular signal-binding site	Receives signals from lipid-soluble or noncharged, nonpolar small molecules	Receptors for NO, steroid hormone, vitamin D, and thyroid hormone
Cell-Surface Receptors			
Chemically gated ion channels	Multipass transmembrane protein forming a central pore	Molecular "gates" triggered chemically to open or close	Neurons
Enzymatic receptors	Single-pass transmembrane protein	Binds signal extracellularly; catalyzes response intracellularly	Phosphorylation of protein kinases
G protein–coupled receptors	Seven-pass transmembrane protein with cytoplasmic binding site for G protein	Binding of signal to receptor causes GTP to bind a G protein; G protein, with attached GTP, detaches to deliver the signal inside the cell	Peptide hormones, rod cells in the eyes

The acetylcholine receptor found in muscle cell membranes functions as an Na$^+$ channel. When the receptor binds to its ligand, the neurotransmitter acetylcholine, the channel opens allowing Na$^+$ to flow into the muscle cell. This is a critical step linking the signal from a motor neuron to muscle cell contraction (see chapter 46).

Enzymatic receptors

Many cell-surface receptors either act as enzymes or are directly linked to enzymes (figure 9.4b). When a signal molecule binds to the receptor, it activates the enzyme. In almost all cases, these enzymes are **protein kinases,** enzymes that add phosphate groups

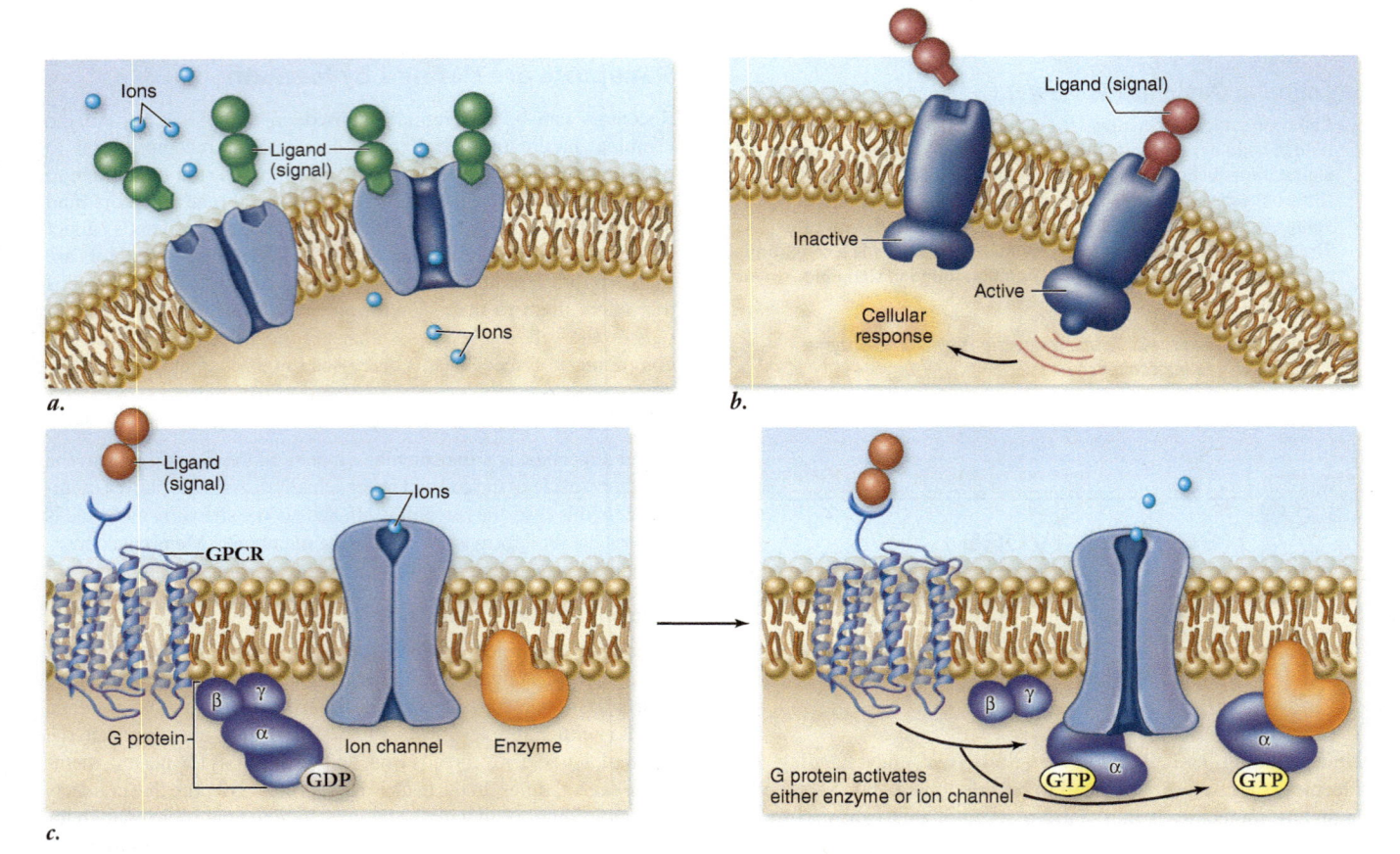

Figure 9.4 Cell-surface receptors. *a.* Chemically gated ion channels form a pore in the plasma membrane that can be opened or closed by chemical signals. They are usually selective, allowing the passage of only one type of ion. *b.* Enzymatic receptors bind to ligands on the extracellular surface. A catalytic region on their cytoplasmic portion transmits the signal across the membrane by acting as an enzyme in the cytoplasm. *c.* G protein–coupled receptors (GPCRs) bind to ligands outside the cell and to G proteins inside the cell. The G protein then activates an enzyme or ion channel, transmitting signals from the cell's surface to its interior.

to proteins. We discuss these receptors in detail in section 9.4 of this chapter.

G Protein–coupled receptors

A third class of cell-surface receptors acts indirectly on enzymes or ion channels in the plasma membrane with the aid of an assisting protein, called a **G protein.** The G protein, which is so named because it binds the nucleotide *guanosine triphosphate* (GTP), can be thought of as being inserted between the receptors and the enzyme (effector). That is, the ligand binds to the receptor, activating it, which activates the G protein, which in turn activates the effector protein (figure 9.4c). These receptors are discussed in detail in section 9.5.

Membrane receptors can generate second messengers

Some enzymatic receptors and most G protein–coupled receptors utilize other substances to relay the message within the cytoplasm. These other substances, small molecules or ions called **second messengers,** alter the behavior of cellular proteins by binding to them and changing their shape. (The original signal molecule is considered the "first messenger.") Two common second messengers are **cyclic adenosine monophosphate (cyclic AMP, or cAMP)** and calcium ions. The role of these second messengers will be explored in more detail in section 9.5.

Learning Outcome Review 9.2

Receptors may be internal (intracellular receptors) or external (membrane receptors). Membrane receptors include channel-linked receptors, enzymatic receptors, and G protein–coupled receptors. Signal transduction through membrane receptors often involves the production of a second signaling molecule, or second messenger, inside the cell.

- Would a hydrophobic molecule be expected to have an internal or membrane receptor?

9.3 Intracellular Receptors

Learning Outcomes

1. Describe the chemical nature of ligands for intracellular receptors.
2. Diagram the pathway of signal transduction through intracellular receptors.

Many cell signals are lipid-soluble or very small molecules that can readily pass through the plasma membrane of the target cell and into the cell, where they interact with an *intracellular receptor*. Some of these ligands bind to protein receptors located in the cytoplasm, others pass across the nuclear membrane as well and bind to receptors within the nucleus.

Steroid hormone receptors affect gene expression

Of all of the receptor types discussed in this chapter, the action of the steroid hormone receptors is the simplest and most direct.

Steroid hormones form a large class of compounds, including cortisol, estrogen, progesterone, and testosterone, that share a common nonpolar structure. Estrogen, progesterone, and testosterone are involved in sexual development and behavior (see chapter 52). Other steroid hormones, such as cortisol, also have varied effects depending on the target tissue, ranging from the mobilization of glucose to the inhibition of white blood cells to control inflammation. Their anti-inflammatory action is the basis of their use in medicine.

The nonpolar structure allows these hormones to cross the membrane and bind to intracellular receptors. The location of steroid hormone receptors prior to hormone binding is cytoplasmic, but their primary site of action is in the nucleus. Binding of

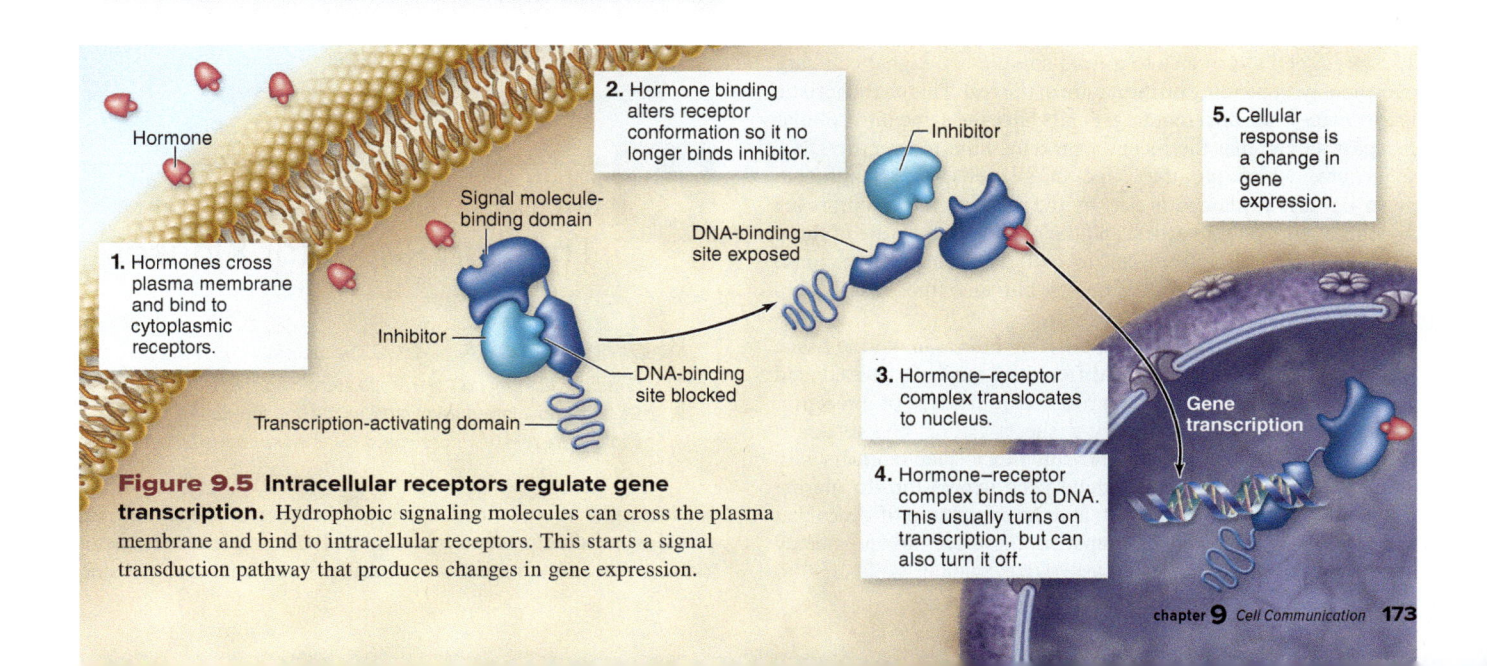

Figure 9.5 Intracellular receptors regulate gene transcription. Hydrophobic signaling molecules can cross the plasma membrane and bind to intracellular receptors. This starts a signal transduction pathway that produces changes in gene expression.

1. Hormones cross plasma membrane and bind to cytoplasmic receptors.
2. Hormone binding alters receptor conformation so it no longer binds inhibitor.
3. Hormone–receptor complex translocates to nucleus.
4. Hormone–receptor complex binds to DNA. This usually turns on transcription, but can also turn it off.
5. Cellular response is a change in gene expression.

the hormone to the receptor causes the complex to shift from the cytoplasm to the nucleus (figure 9.5). As the ligand–receptor complex makes it all the way to the nucleus of the cell, these receptors are often called **nuclear receptors.**

Steroid receptor action

The primary function of steroid hormone receptors, as well as receptors for a number of other small, lipid-soluble signal molecules such as vitamin D and thyroid hormone, is to act as regulators of gene expression (see chapter 16).

All of these receptors have similar structures; the genes that code for them appear to be the evolutionary descendants of a single ancestral gene. Because of their structural similarities, they are all part of the *nuclear receptor superfamily.*

Each of these receptors has three functional domains:

1. a hormone-binding domain,
2. a DNA-binding domain, and
3. a domain that can interact with coactivators to affect the level of gene transcription.

In its inactive state, the receptor typically cannot bind to DNA because an inhibitor protein occupies the DNA-binding site. When the signal molecule binds to the hormone-binding site, the conformation of the receptor changes, releasing the inhibitor and exposing the DNA-binding site, allowing the receptor to attach to specific nucleotide sequences on the DNA (figure 9.5). This binding activates (or, in a few instances, suppresses) particular genes, usually located adjacent to the hormone-binding sequences. In the case of cortisol, which is a glucocorticoid hormone that can increase levels of glucose in cells, a number of different genes involved in the synthesis of glucose have binding sites for the hormone receptor complex.

The lipid-soluble ligands that intracellular receptors recognize tend to persist in the blood far longer than water-soluble signals. Most water-soluble hormones break down within minutes, and neurotransmitters break down within seconds or even milliseconds. In contrast, a steroid hormone such as cortisol or estrogen persists for hours.

Specificity and the role of coactivators

The target cell's response to a lipid-soluble cell signal can vary enormously, depending on the nature of the cell. This characteristic is true even when different target cells have the same intracellular receptor. Given that the receptor proteins bind to specific DNA sequences, which are the same in all cells, this may seem puzzling. It is explained in part by the fact that the receptors act in concert with **coactivators,** and the number and nature of these molecules can differ from cell to cell. Thus, a cell's response depends on not only the receptors but also the coactivators present.

The hormone estrogen has different effects in uterine tissue than in mammary tissue. This differential response is mediated by coactivators and not by the presence or absence of a receptor in the two tissues. In mammary tissue, a critical coactivator is lacking and the hormone–receptor complex instead interacts with another protein that acts to reduce gene expression. In uterine tissue, the coactivator is present, and the expression of genes that encode proteins involved in preparing the uterus for pregnancy are turned on.

Other intracellular receptors act as enzymes

A very interesting example of a receptor acting as an enzyme is found in the receptor for nitric oxide (NO). This small gas molecule diffuses readily out of the cells where it is produced and passes directly into neighboring cells, where it binds to the enzyme guanylyl cyclase. Binding of NO activates this enzyme, enabling it to catalyze the synthesis of *cyclic guanosine monophosphate (cGMP),* an intracellular messenger molecule that produces cell-specific responses such as the relaxation of smooth muscle cells.

When the brain sends a nerve signal to relax the smooth muscle cells lining the walls of vertebrate blood vessels, acetylcholine released by the nerve cell binds to receptors on epithelial cells. This causes an increase in intracellular Ca^{2+} in the epithelial cell that stimulates nitric oxide synthase to produce NO. The NO diffuses into the smooth muscle, where it increases the level of cGMP, leading to relaxation. This relaxation allows the vessel to expand and thereby increases blood flow. This explains the use of nitroglycerin to treat the pain of angina caused by constricted blood vessels to the heart. The nitroglycerin is converted by cells to NO, which then acts to relax the blood vessels.

The drug sildenafil (better known as Viagra) also functions via this signal transduction pathway by binding to and inhibiting the enzyme cGMP phosphodiesterase, which breaks down cGMP. This keeps levels of cGMP high, thereby stimulating production of NO. The reason for Viagra's selective effect is that it binds to a form of cGMP phosphodiesterase found in cells in the penis. This allows relaxation of smooth muscle in erectile tissue, thereby increasing blood flow.

Learning Outcomes Review 9.3

Hydrophobic signaling molecules can cross the membrane and bind to intracellular receptors. The steroid hormone receptors act by directly influencing gene expression. On binding hormone, the hormone–receptor complex moves into the nucleus to turn on (or sometimes turn off) gene expression. This may also require a coactivator that functions with the hormone–receptor complex. Thus, the cell's response to a hormone depends on the presence of a receptor and coactivators as well.

- *Would these types of intracellular receptors be fast acting, or have effects of longer duration?*

9.4 Signal Transduction Through Receptor Kinases

Learning Outcomes

1. Compare the function of RTKs to steroid hormone receptors.
2. Describe how information crosses the membrane in RTKs.
3. Explain the role of kinase cascades in signal transduction.

In section 9.1, you read that protein kinases phosphorylate proteins to alter protein function and that the most common kinases act on the amino acids serine, threonine, and tyrosine.

The **receptor tyrosine kinases (RTKs)** influence the cell cycle, cell migration, cell metabolism, and cell proliferation—virtually all aspects of the cell are affected by signaling through these receptors. Alterations to the function of these receptors and their signaling pathways can lead to cancers in humans and other animals.

Some of the earliest examples of cancer-causing genes, or oncogenes, involve RTK function (discussed in chapter 10). The avian erythroblastosis virus carries an altered form of the epidermal growth factor receptor that lacks most of its extracellular domain. When this virus infects a cell the altered receptors produced are stuck in the "on" state. The continuous signaling from this receptor leads to cells that have lost the normal controls over growth.

Receptor tyrosine kinases are a large class of membrane receptors in animal cells that recognize hydrophilic ligands. Plants possess receptors with a similar overall structure and function, but they are serine–threonine kinases. These plant receptors have been named **plant receptor kinases.**

Because these receptors are performing similar functions in plant and animal cells but differ in their substrates, the duplication and divergence of each kind of receptor kinase probably occurred after the plant–animal divergence. The proliferation of these types of signaling molecules is thought to coincide with the independent evolution of multicellularity in each group.

In this section, we will concentrate on the RTK family of receptors that has been extensively studied in a variety of animal cells.

RTKs are activated by autophosphorylation

Receptor tyrosine kinases have a relatively simple structure consisting of a single transmembrane domain, an extracellular ligand-binding domain, and an intracellular kinase domain. This kinase domain contains the catalytic site of the receptor, which acts as a protein kinase that adds phosphate groups to tyrosines. On ligand binding to a specific receptor, two of these receptor–ligand complexes associate together (often referred to as dimerization) and phosphorylate each other, a process called *autophosphorylation* (figure 9.6).

The autophosphorylation event transmits across the membrane the signal that began with the binding of the ligand to the receptor. The next step, propagation of the signal in the cytoplasm, can take a variety of different forms. These forms include activation of the tyrosine kinase domain to phosphorylate other

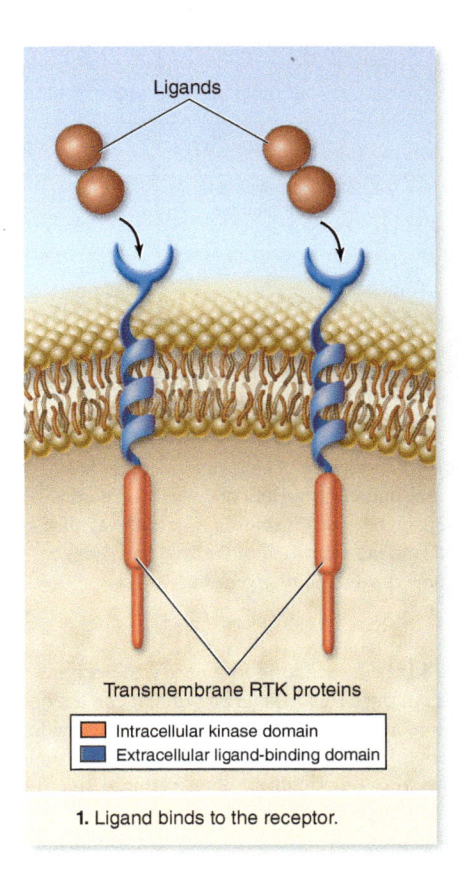

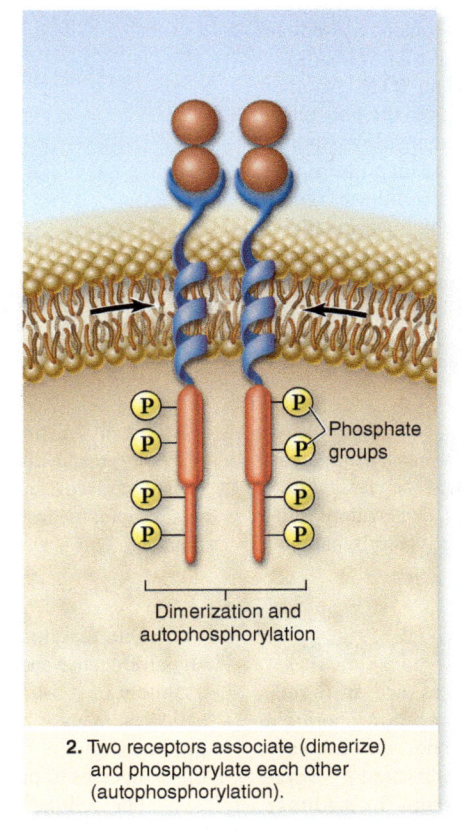

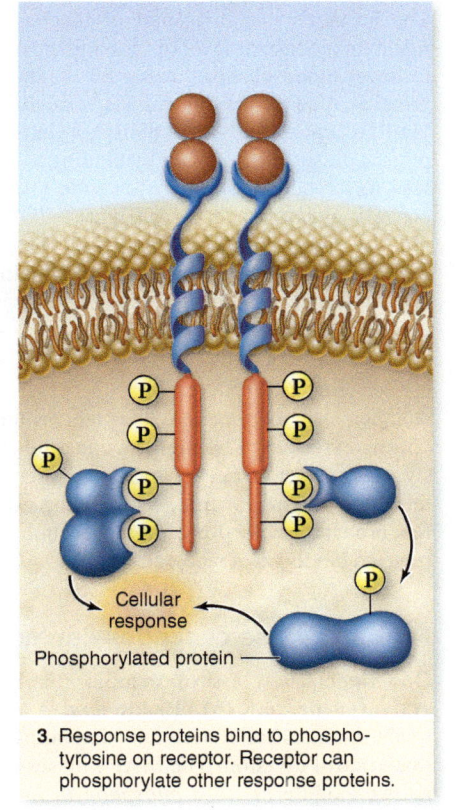

Figure 9.6 Activation of a receptor tyrosine kinase (RTK). These membrane receptors bind hormones or growth factors that are hydrophilic and cannot cross the membrane. The receptor is a transmembrane protein with an extracellular ligand-binding domain and an intracellular kinase domain. Signal transduction pathways begin with response proteins binding to phosphotyrosine on receptor, and by receptor phosphorylation of response proteins.

intracellular targets or interaction of other proteins with the phosphorylated receptor.

The cellular response after activation depends on the possible response proteins in the cell. Two different cells can have the same receptor yet a different response, depending on what response proteins are present in the cytoplasm. For example, fibroblast growth factor stimulates cell division in fibroblasts but stimulates nerve cells to differentiate rather than to divide.

Phosphotyrosine domains mediate protein–protein interactions

One way that the signal from the receptor can be propagated in the cytoplasm is via proteins that bind specifically to phosphorylated tyrosines in the receptor. When the receptor is activated, regions of the protein outside of the catalytic site are phosphorylated. This creates "docking" sites for proteins that bind specifically to phosphotyrosine. The proteins that bind to these phosphorylated tyrosines can initiate intracellular events to convert the signal from the ligand into a response (figure 9.6).

The insulin receptor

The use of docking proteins is illustrated by the insulin receptor. The hormone insulin is part of the body's control system to maintain a constant level of blood glucose. The role of insulin is to lower blood glucose, acting by binding to an RTK. Another protein called the *insulin response protein* binds to the phosphorylated receptor and is itself phosphorylated. The insulin response protein passes the signal on by binding to additional proteins that lead to the activation of the enzyme glycogen synthase, which converts glucose to glycogen (figure 9.7), thereby lowering blood glucose. Other proteins activated by the insulin receptor act to inhibit the synthesis of enzymes involved in making glucose, and to increase the number of glucose transporter proteins in the plasma membrane.

Adapter proteins

Another class of proteins, **adapter proteins,** can also bind to phosphotyrosines. These proteins themselves do not participate in signal transduction but act as a link between the receptor and proteins that initiate downstream signaling events. For example, the Ras protein discussed later in this section is activated by adapter proteins binding to a receptor.

Protein kinase cascades can amplify a signal

One important class of cytoplasmic kinases are **mitogen-activated protein (MAP) kinases.** A *mitogen* is a chemical that stimulates cell division by activating the normal pathways that control division. The MAP kinases are activated by a signaling module called a *phosphorylation cascade* or a **kinase cascade.** This module is a series of protein kinases that phosphorylate each other in succession. The final step in the cascade is the activation by phosphorylation of MAP kinase itself (figure 9.8).

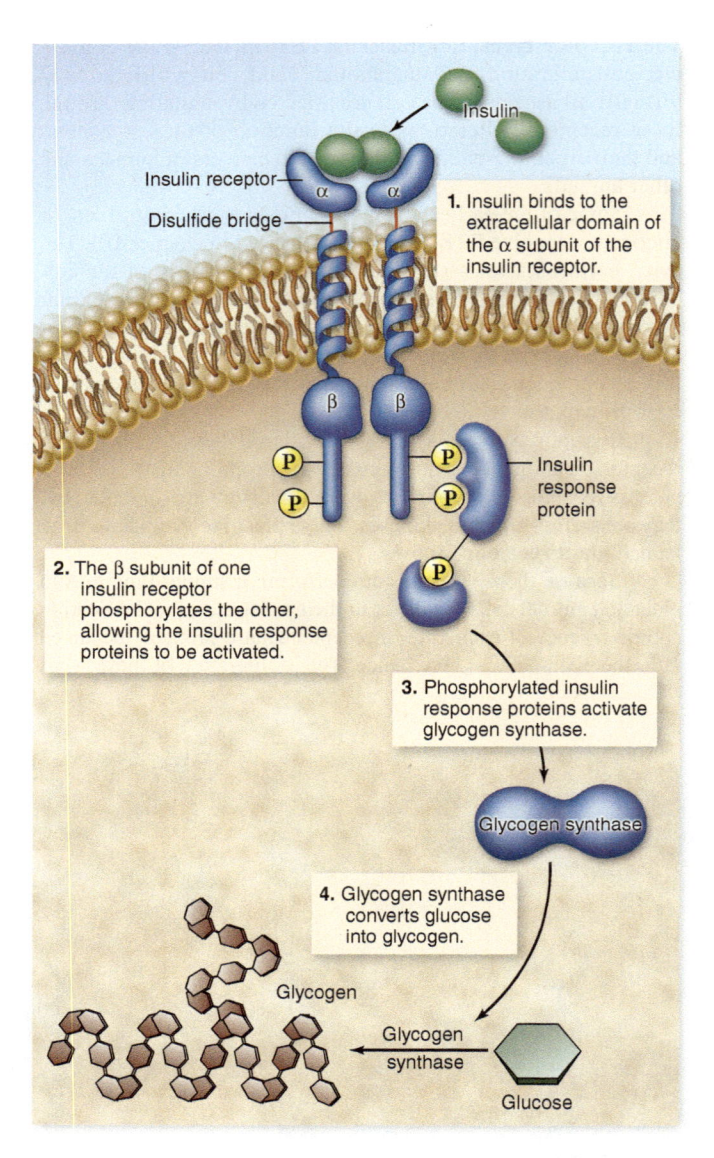

Figure 9.7 The insulin receptor. The insulin receptor is a receptor tyrosine kinase that initiates a variety of cellular responses related to glucose metabolism. One signal transduction pathway that this receptor mediates leads to the activation of the enzyme glycogen synthase. This enzyme converts glucose to glycogen.

1. Insulin binds to the extracellular domain of the α subunit of the insulin receptor.
2. The β subunit of one insulin receptor phosphorylates the other, allowing the insulin response proteins to be activated.
3. Phosphorylated insulin response proteins activate glycogen synthase.
4. Glycogen synthase converts glucose into glycogen.

One function of a kinase cascade is to amplify the original signal. Because each step in the cascade is an enzyme, it can act on a number of substrate molecules. With each enzyme in the cascade acting on many substrates this produces a large amount of the final product (figure 9.8). This allows a small number of initial signaling molecules to produce a large response.

The cellular response to this cascade in any particular cell depends on the targets of the MAP kinase, but usually involves phosphorylating transcription factors that then activate gene expression (see chapter 16). An example of this kind of signaling through growth factor receptors is provided in chapter 10 and

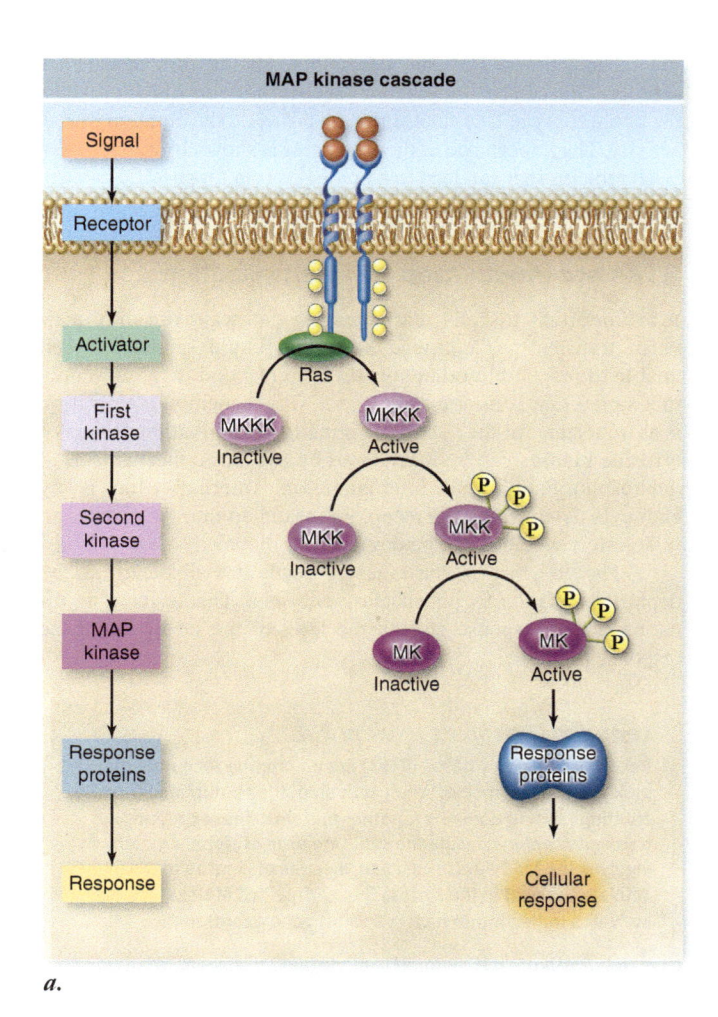

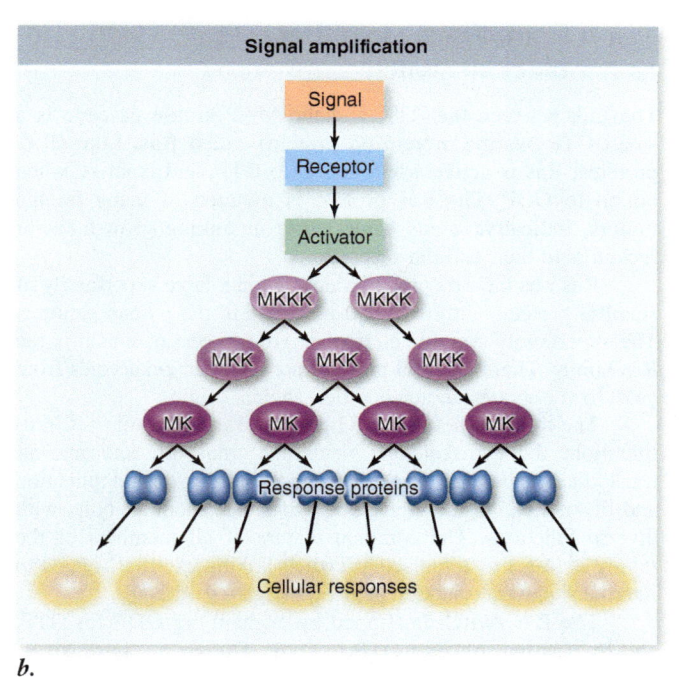

Figure 9.8 MAP kinase cascade leads to signal amplification. *a.* Phosphorylation cascade is shown as a flowchart on the left. The corresponding cellular events are shown on the right, beginning with the receptor in the plasma membrane. Each kinase is named starting with the last, the MAP kinase (MK), which is phosphorylated by a MAP kinase kinase (MKK), which is in turn phosphorylated by a MAP kinase kinase kinase (MKKK). The cascade is linked to the receptor protein by an activator protein. *b.* At each step the enzymatic action of the kinase on multiple substrates leads to amplification of the signal.

illustrates how signal transduction initiated by a growth factor can control the process of cell division through a kinase cascade.

Scaffold proteins organize kinase cascades

The proteins in a kinase cascade need to act sequentially to be effective. One way the efficiency of this process can be increased is to organize them in the cytoplasm. Proteins called *scaffold proteins* are thought to organize the components of a kinase cascade into a single protein complex, the ultimate in a signaling module. The scaffold protein binds to each individual kinase such that they are spatially organized for optimal function (figure 9.9).

The advantages of this kind of organization are many. A physically arranged sequence is clearly more efficient than one that depends on diffusion to produce the appropriate order of events. This organization also allows the segregation of signaling modules in different cytoplasmic locations.

The disadvantage of this kind of organization is that it reduces the amplification effect of the kinase cascade. Enzymes held in one place are not free to find new substrate molecules, but must rely on substrates being nearby.

The best-studied example of a scaffold protein comes from mating behavior in budding yeast. Yeast cells respond to mating pheromones with changes in cell morphology and gene expression, mediated by a protein kinase cascade. A protein called Ste5 was originally identified as a protein required for mating behavior, but no enzymatic activity could be detected for this protein. It has now been shown that this protein interacts with all of the members of the kinase cascade and acts as a scaffold protein that organizes the cascade and insulates it from other signaling pathways.

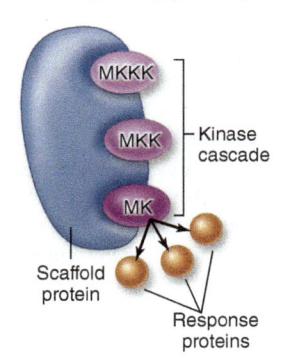

Figure 9.9 Kinase cascade can be organized by scaffold proteins. The scaffold protein binds to each kinase in the cascade, organizing them so each substrate is next to its enzyme. This organization also sequesters the kinases from other signaling pathways in the cytoplasm.

Ras is a small G protein that acts as a molecular switch

The link between the RTK and the MAP kinase cascade is a small GTP-binding protein (G protein) called **Ras.** Like all G proteins, Ras is active when bound to GTP, and inactive when bound to GDP. The Ras protein is mutated in many human tumors, indicative of its central role in linking growth factor receptors to their cellular response.

Ras was the first protein identified in a large superfamily of small G proteins with over 150 members in the human genome. The superfamily consists of five subgroups, one of which is the Ras family. These small G proteins are found in eukaryotes from yeast to vertebrates, indicating their ancient origin.

The roles of these small G proteins vary, affecting cell proliferation, the cytoskeleton, membrane transport, and nuclear transport. They are an excellent example of how gene duplication and diversification allow evolution to create modular units with diverse functions. The common feature of all members of the family is to act as a molecular switch linking external signals to internal signal transduction pathways (figure 9.10).

The Ras switch is flipped by exchanging GDP for GTP, and by Ras hydrolyzing GTP to GDP. The switch is sensitive to outside signals that regulate the activation and deactivation of Ras. Guanine nucleotide exchange factors (GEFs) activate Ras by stimulating the exchange of GDP for GTP. When a growth factor receptor is activated, it binds to an adapter protein that acts as a GEF. The activated Ras protein then activates the first kinase in the MAP kinase cascade (see figure 9.8 and chapter 10).

The action of Ras can be terminated by its intrinsic GTPase activity. This can be stimulated by a GAP protein, which provides the opportunity to fine-tune signaling based on the duration of Ras activity. The importance of these proteins is shown by mutations in GAP proteins that can lead to a predisposition for specific cancers such as neurofibromatosis.

RTKs are inactivated by internalization

It is important to cells that signaling pathways are only activated transiently. Continued activation could render the cell unable to respond to other signals or to respond inappropriately to a signal that is no longer relevant. Consequently, inactivation is as important for the control of signaling as activation. Receptor tyrosine kinases can be inactivated by two basic mechanisms—dephosphorylation and internalization. Internalization is by endocytosis, in which the receptor is taken up into the cytoplasm in a vesicle where it can be degraded or recycled.

The enzymes in the kinase cascade are all controlled by dephosphorylation by phosphatase enzymes. This leads to termination of the response at both the level of the receptor and the response proteins.

Learning Outcomes Review 9.4

Receptor tyrosine kinases (RTKs) are membrane receptors that can phosphorylate tyrosine. When activated, they autophosphorylate, creating binding domains for other proteins. These proteins transmit the signal inside the cell. One form of signaling pathway involves the MAP kinase cascade, a series of kinases that each activate the next in the series. This ends with a MAP kinase that activates transcription factors to alter gene expression.

- *Ras protein is mutated in many human cancers. What are possible reasons for this?*

9.5 Signal Transduction Through G Protein–Coupled Receptors

Learning Outcomes

1. Contrast signaling through GPCRs and RTKs.
2. Relate the function of second messengers to signal transduction pathways.

The single largest category of receptor type in animal cells is **G protein–coupled receptors (GPCRs),** so named because the receptors act by coupling with a G protein. These receptors bind diverse ligands, including ions, organic odorants, peptides, proteins, and lipids. Light-sensing receptors are also part of this family, so we could even count photons as "ligands."

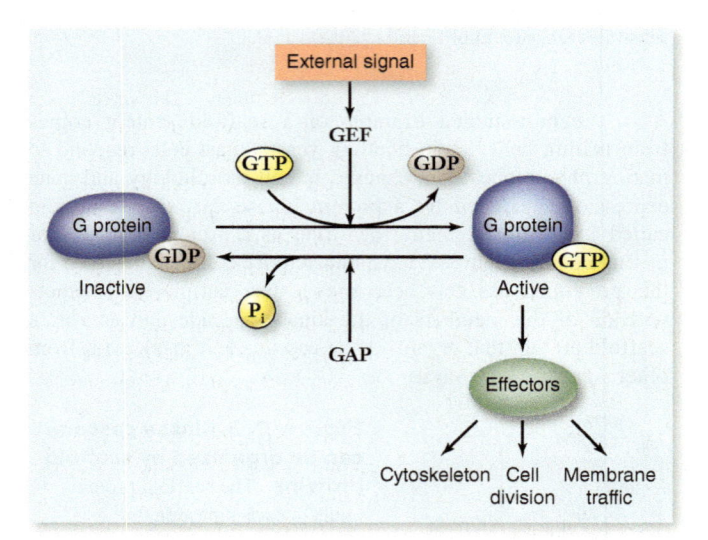

Figure 9.10 Small G proteins act as molecular switches. Small G proteins, such as Ras, link external signals to internal signal transduction pathways. External signals activate guanine nucleotide exchange proteins (GEF), which activate the G protein. The G protein can be inactivated by its weak intrinsic GTPase activity, which can be stimulated by activating proteins (GAP).

This superfamily of proteins also has a characteristic structure with seven transmembrane domains that anchor the receptors in the membrane. This arrangement of seven transmembrane domains is highly conserved and is used to search for new members in sequenced genomes. The analysis of many animal genomes indicates that GPCRs are the largest gene family in most animals. They have been found in virtually all types of eukaryotic organisms, indicating an ancient origin with duplication and divergence leading to a wide array of signaling pathways.

The latest count of genes encoding GPCRs in the human genome is 865, with about half of these encoding odorant receptors involved in the sense of taste and smell. In the mouse, over 1000 different odorant receptors are involved in the sense of smell. The family of GPCRs has been subdivided into five groups based on structure and function: Rhodopsin, Secretin, Adhesion, Glutamate, and Frizzled/Taste 2. The names refer to the first discovered member of each group; for example, Rhodopsin is the GPCR involved in light sensing in mammals. In this section, we will concentrate on the basic mechanism of activation and some of the possible signal transduction pathways.

G proteins link receptors with effector proteins

The function of the G protein in signaling by GPCRs is to provide a link between a receptor that receives signals and effector proteins that produce cellular responses. The G protein functions as a switch that is turned on by the receptor. In its "on" state, the G protein activates effector proteins to cause a cellular response.

All G proteins are active when bound to GTP and inactive when bound to GDP. The main difference between the G proteins in GPCRs and the small G proteins described in section 9.4 is that these G proteins are composed of three subunits, called α, β, and γ. As a result, they are often called *heterotrimeric G proteins*. When a ligand binds to a GPCR and activates its associated G protein, the G protein exchanges GDP for GTP and dissociates into two parts consisting of the G_α subunit bound to GTP, and the G_β and G_γ subunits together ($G_{\beta\gamma}$). The signal can then be transmitted by either the G_α or the $G_{\beta\gamma}$ components stimulating effector proteins. The hydrolysis of bound GTP to GDP by G_α causes reassociation of the heterotrimer, restoring the "off" state of the system (figure 9.11).

The effector proteins are usually enzymes. An effector protein might be a protein kinase that phosphorylates proteins to directly propagate the signal, or it may produce a second messenger to initiate a signal transduction pathway.

Effector proteins produce multiple second messengers

Often, the effector proteins activated by G proteins produce a second messenger. Two of the most common effectors are *adenylyl cyclase* and *phospholipase C*, which produce cAMP and IP_3 plus DAG, respectively.

Cyclic AMP

All animal cells studied thus far use cAMP as a second messenger (see chapter 45). When a signaling molecule binds to a GPCR that uses the enzyme **adenylyl cyclase** as an effector, a large amount of cAMP is produced within the cell

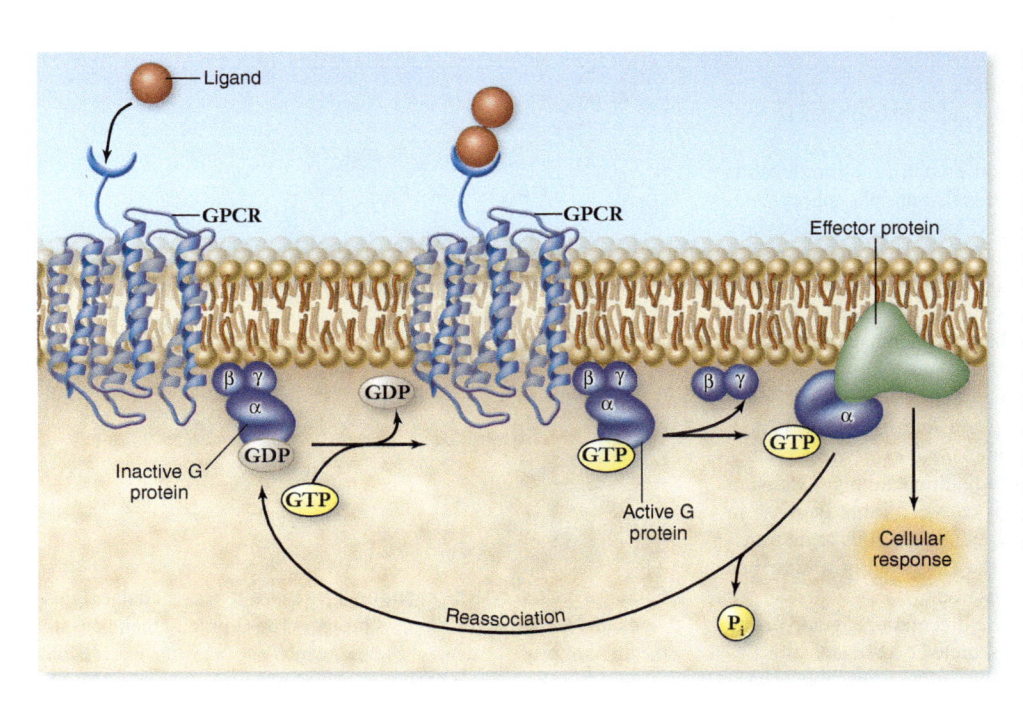

Figure 9.11 The action of G protein–coupled receptors. G protein–coupled receptors act through a heterotrimeric G protein that links the receptor to an effector protein. When ligand binds to the receptor, it activates an associated G protein, exchanging GDP for GTP. The active G protein complex dissociates into G_α and $G_{\beta\gamma}$. The G_α subunit (bound to GTP) is shown activating an effector protein. The effector protein may act directly on cellular proteins or produce a second messenger to cause a cellular response. G_α can hydrolyze GTP inactivating the system, then reassociate with $G_{\beta\gamma}$.

Figure 9.12 Production of second messengers. Second messengers are signaling molecules produced within the cell. *a.* The nucleotide ATP is converted by the enzyme adenylyl cyclase into cyclic AMP, or cAMP, and pyrophosphate (PP$_i$). *b.* The inositol phospholipid PIP$_2$ is composed of two lipids and a phosphate attached to glycerol. The phosphate is also attached to the sugar inositol. This molecule can be cleaved by the enzyme phospholipase C to produce two different second messengers: DAG, made up of the glycerol with the two lipids, and IP$_3$, inositol triphosphate.

(figure 9.12a). The cAMP then binds to and activates the enzyme protein kinase A (PKA), which adds phosphates to specific proteins in the cell (figure 9.13).

The effect of this phosphorylation on cell function depends on the identity of the cell and the proteins that are phosphorylated. In muscle cells, for example, PKA activates an enzyme necessary to break down glycogen and inhibits another enzyme necessary to synthesize glycogen. This leads to an increase in glucose available to the muscle. By contrast, in the kidney the action of PKA leads to the production of water channels that can increase the permeability of tubule cells to water.

Disruption of cAMP signaling can have a variety of effects. The symptoms of the disease cholera are due to altered cAMP levels in cells in the gut. The bacterium *Vibrio cholerae* produces a toxin that binds to a GPCR in the epithelium of the gut, causing it to be locked into an "on" state. This causes a large increase in intracellular cAMP that, in these cells, causes Cl⁻ ions to be transported out of the cell. Water follows the Cl⁻, leading to diarrhea and dehydration characteristic of the disease.

The molecule cAMP is also an extracellular signal. In the slime mold *Dictyostelium discoideum*, secreted cAMP acts as a signal for aggregation under conditions of starvation. Experiments have shown that the receptor for this signal is also a GPCR (figure 9.14).

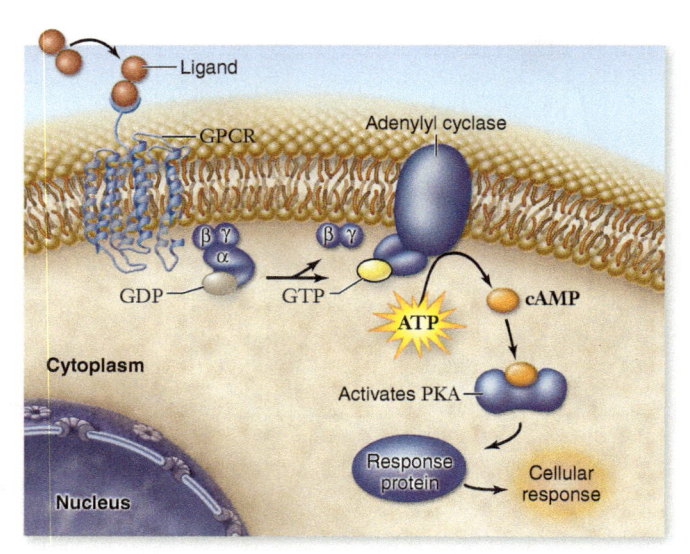

Figure 9.13 cAMP signaling pathway. Extracellular signal binds to a GPCR, activating a G protein. The G protein then activates the effector protein adenylyl cyclase, which catalyzes the conversion of ATP to cAMP. The cAMP then activates protein kinase A (PKA), which phosphorylates target proteins to cause a cellular response.

SCIENTIFIC THINKING

Question: *What is the receptor for cAMP?*

Hypothesis: *A previously identified G protein–coupled receptor is the cAMP receptor.*

Prediction: *If the function of the cAMP receptor is removed, then cells will not respond to starvation by aggregating.*

Test: *Use G protein–coupled receptor gene to direct synthesis of antisense RNA complementary to the normal mRNA. This will eliminate gene expression by the cellular copy of the G protein–coupled receptor.*

Result: *Cells transformed with the antisense construct do not aggregate normally.*

Conclusion: *Previously identified G protein–coupled receptor is the cAMP receptor, which controls the aggregation response.*

Further Experiments: *How can this kind of experiment be used to unravel other aspects of this signaling system?*

Figure 9.14 The receptor for cAMP in *D. discoideum* is a GPCR.

Inositol phosphates

A common second messenger is produced from the molecules called inositol phospholipids. These are inserted into the plasma membrane by their lipid ends and have the *inositol phosphate* portion protruding into the cytoplasm. The most common inositol phospholipid is phosphatidylinositol-4,5-bisphosphate (PIP_2). This molecule is a substrate of the effector protein phospholipase C, which cleaves PIP_2 to yield **diacylglycerol (DAG)** and **inositol-1,4,5-trisphosphate (IP_3)** (see figure 9.12b).

Both of these compounds then act as second messengers with a variety of cellular effects. DAG, like cAMP, can activate a protein kinase, in this case protein kinase C (PKC).

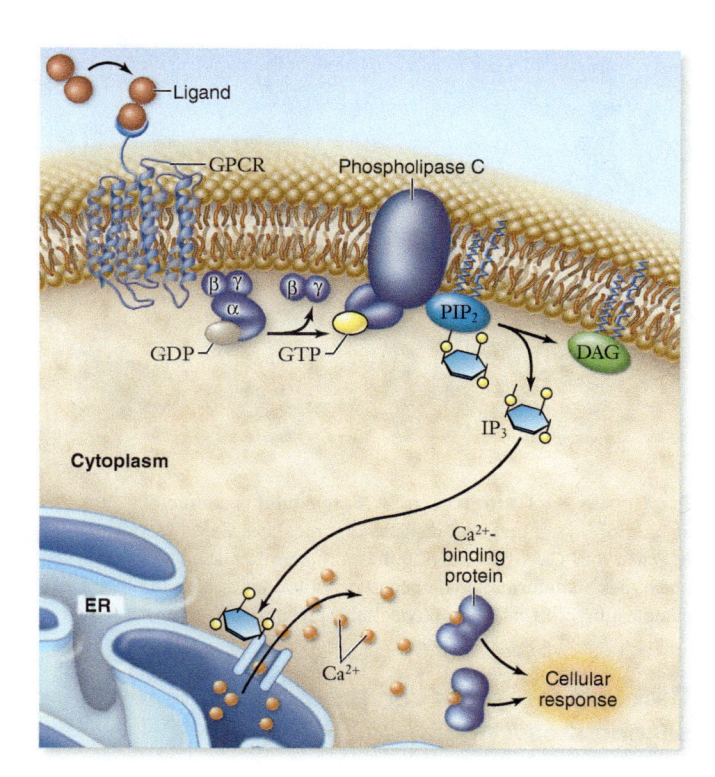

Figure 9.15 Inositol phospholipid and Ca^{2+} signaling.
Extracellular signal binds to a GPCR activating a G protein. The G protein activates the effector protein phospholipase C, which converts PIP_2 to DAG and IP_3. IP_3 is then bound to a channel-linked receptor on the endoplasmic reticulum (ER) membrane, causing the ER to release stored Ca^{2+} into the cytoplasm. The Ca^{2+} then binds to Ca^{2+}-binding proteins such as calmodulin and PKC to cause a cellular response.

Calcium

Calcium ions (Ca^{2+}) serve widely as second messengers. Ca^{2+} levels inside the cytoplasm are normally very low (less than 10^{-7} M), whereas outside the cell and in the endoplasmic reticulum, Ca^{2+} levels are quite high (about 10^{-3} M). The endoplasmic reticulum has receptor proteins that act as ion channels to release Ca^{2+}. One of the most common of these receptors can bind the second messenger IP_3 to release Ca^{2+}, linking signaling through inositol phosphates with signaling by Ca^{2+} (figure 9.15).

The result of the outflow of Ca^{2+} from the endoplasmic reticulum depends on the cell type. For example, in skeletal muscle cells Ca^{2+} stimulates muscle contraction but in endocrine cells it stimulates the secretion of hormones.

Ca^{2+} initiates some cellular responses by binding to *calmodulin*, a 148-amino-acid cytoplasmic protein that contains four binding sites for Ca^{2+} (figure 9.16). When four Ca^{2+} ions are bound to calmodulin, the calmodulin/Ca^{2+} complex is able to bind to other proteins to activate them. These proteins include protein kinases, ion channels, receptor proteins, and cyclic nucleotide phosphodiesterases. These many uses of Ca^{2+} make it one of the most versatile second messengers in cells.

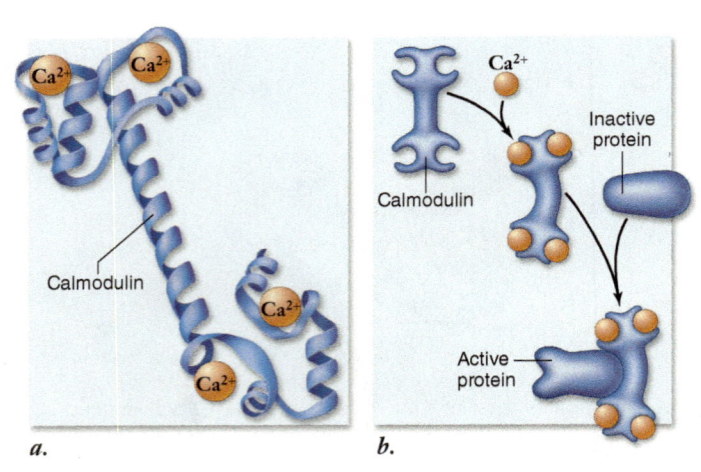

Figure 9.16 Calmodulin. *a.* Calmodulin is a protein containing 148 amino acid residues that mediates Ca²⁺ function. *b.* When four Ca²⁺ are bound to the calmodulin molecule, it undergoes a conformational change that allows it to bind to other cytoplasmic proteins and effect cellular responses.

which allows them to bind epinephrine. They differ mainly in their cytoplasmic domains, which interact with G proteins. This leads to different isoforms activating different G proteins, thereby leading to different signal transduction pathways.

Thus, in the heart, muscle cells have one isoform of the receptor that, when bound to epinephrine, activates a G protein that activates adenylyl cyclase, leading to increased cAMP. This increases the rate and force of contraction. In the intestine, smooth muscle cells have a different isoform of the receptor that, when bound to epinephrine, activates a different G protein that inhibits adenylyl cyclase, which decreases cAMP. This has the result of relaxing the muscle.

G protein–coupled receptors and receptor tyrosine kinases can activate the same pathways

Different receptor types can affect the same signaling module. For example, RTKs were shown to activate the MAP kinase cascade, but GPCRs can also activate this same cascade. Similarly,

Different receptors can produce the same second messengers

As mentioned in section 9.1, the two hormones glucagon and epinephrine can both stimulate liver cells to mobilize glucose. The reason that these different signals have the same effect is that they both act by the same signal transduction pathway to stimulate the breakdown and inhibit the synthesis of glycogen.

The binding of either hormone to its receptor activates a G protein that simulates adenylyl cyclase. The production of cAMP leads to the activation of PKA, which in turn activates another protein kinase called phosphorylase kinase. Activated phosphorylase kinase then activates glycogen phosphorylase, which cleaves off units of glucose 6-phosphate from the glycogen polymer (figure 9.17). The action of multiple kinases again leads to amplification such that a few signaling molecules result in a large number of glucose molecules being released.

At the same time, PKA also phosphorylates the enzyme glycogen synthase, but in this case it inhibits the enzyme, thus preventing the synthesis of glycogen. In addition, PKA phosphorylates other proteins that activate the expression of genes encoding the enzymes needed to synthesize glucose. This convergence of signal transduction pathways from different receptors leads to the same result—glucose is mobilized.

Receptor subtypes can lead to different effects in different cells

We also saw in section 9.1, how a single signaling molecule, epinephrine, can have different effects in different cells. One way this happens is through the existence of multiple forms of the same receptor. The receptor for epinephrine actually has nine different subtypes, or isoforms. These are encoded by different genes and are actually different receptor molecules. The sequences of these proteins are very similar, especially in the ligand-binding domain,

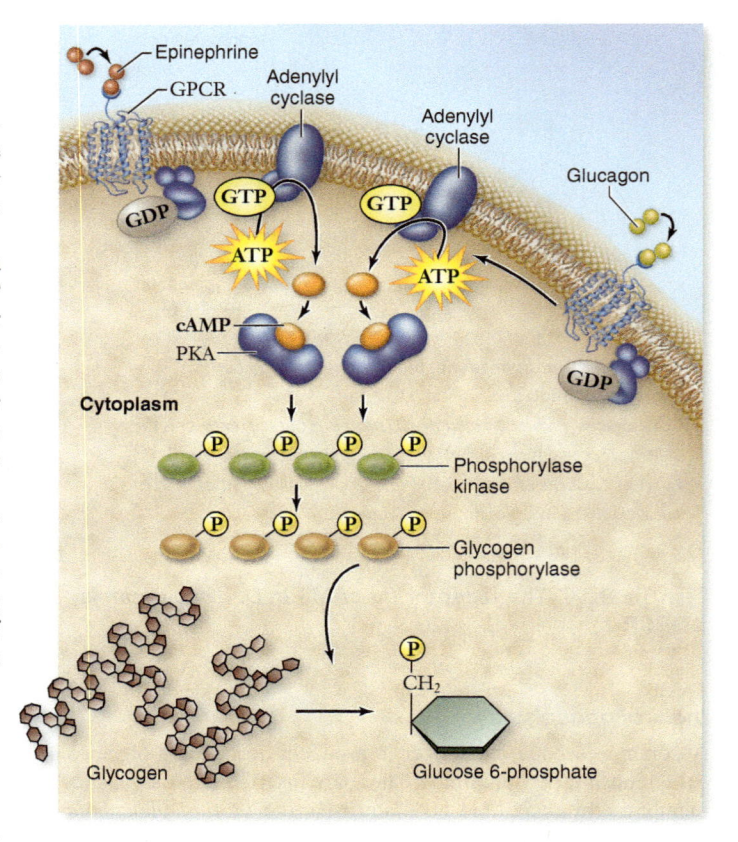

Figure 9.17 Different receptors can activate the same signaling pathway. The hormones glucagon and epinephrine both act through GPCRs. Each of these receptors acts via a G protein that activates adenylyl cyclase, producing cAMP. The activation of PKA begins a kinase cascade that leads to the breakdown of glycogen.

the activation of phospholipase C was mentioned earlier in this section in the context of GPCR signaling, but it can also be activated by RTKs.

This cross-reactivity may appear to introduce complications into cell function, but in fact it provides the cell with an incredible amount of flexibility. Cells have a large, but limited number of intracellular signaling modules, which can be turned on and off by different kinds of membrane receptors. This leads to signaling networks that interconnect possible cellular effectors with multiple incoming signals.

The Internet represents an example of a network in which many different kinds of computers are connected globally. This network can be broken down into subnetworks that are connected to the overall network. Because of the nature of the connections, when you send an e-mail message across the Internet, it can reach its destination through many different pathways. Likewise, the cell has interconnected networks of signaling pathways in which many different signals, receptors, and response proteins are interconnected. Specific pathways like the MAP kinase cascade, or signaling through second messengers like cAMP and Ca^{2+}, represent subnetworks within the global signaling network. A specific signal can activate different pathways in different cells, or different signals can activate the same pathway. We do not yet understand the cell at this level, but the field of systems biology is moving toward such global understanding of cell function.

Learning Outcomes Review 9.5

Signaling through GPCRs uses a three-part system—a receptor, a G protein, and an effector protein. G proteins are active when bound to GTP and inactive when bound to GDP. A ligand binding to the receptor activates the G protein, which then activates the effector protein. Effector proteins include adenylyl cyclase, which produces the second messenger cAMP. Another effector protein, phospholipase C, cleaves the inositol phosphates and results in the release of Ca^{2+} from the ER.

- *There are far more GPCRs than any other receptor type. What is a possible explanation for this?*

Chapter Review

9.1 Overview of Cell Communication (figure 9.1)

Cell communication requires signal molecules, called ligands, binding to specific receptor proteins producing a cellular response.

Signaling is defined by the distance from source to receptor (figure 9.2).

Direct contact—molecules on the plasma membrane of one cell contact the receptor molecules on an adjacent cell.

Paracrine signaling—short-lived signal molecules are released into the extracellular fluid and influence neighboring cells.

Endocrine signaling—long-lived hormones enter the circulatory system and are carried to target cells some distance away.

Synaptic signaling—short-lived neurotransmitters are released by neurons into the gap, called a synapse, between nerves and target cells.

Signal transduction pathways lead to cellular responses.

Intracellular events initiated by a signaling event are called signal transduction.

Phosphorylation is key in control of protein function.

Proteins can be controlled by phosphate added by kinase and removed by phosphatase enzymes.

9.2 Receptor Types (figure 9.4)

Receptors are defined by location.

Receptors are broadly defined as intracellular or cell-surface receptors (membrane receptors).

Membrane receptors are transmembrane proteins that transfer information across the membrane, but not the signal molecule.

Membrane receptors include three subclasses.

Channel-linked receptors are chemically gated ion channels that allow specific ions to pass through a central pore.

Enzymatic receptors are enzymes activated by binding a ligand; these enzymes are usually protein kinases.

G protein–coupled receptors interact with G proteins that control the function of effector proteins: enzymes or ion channels.

Membrane receptors can generate second messengers.

Some enzymatic and most G protein–coupled receptors produce second messengers, to relay messages in the cytoplasm.

9.3 Intracellular Receptors (figure 9.5)

Many cell signals are lipid-soluble and readily pass through the plasma membrane and bind to receptors in the cytoplasm or nucleus.

Steroid hormone receptors affect gene expression.

Steroid hormones bind cytoplasmic receptors, then are transported to the nucleus. Thus, they are called nuclear receptors. These can directly affect gene expression, usually activating transcription of the genes they control.

Nuclear receptors have three functional domains: hormone-binding, DNA-binding, and transcription-activating domains.

Ligand binding changes receptor shape, releasing an inhibitor occupying the DNA-binding site.

A cell's response to a lipid-soluble signal depends on the hormone–receptor complex and the other protein coactivators present.

Other intracellular receptors act as enzymes.

9.4 Signal Transduction Through Receptor Kinases

Receptor kinases in plants and animals recognize hydrophilic ligands and influence the cell cycle, cell migration, cell metabolism, and cell proliferation.

Because they are involved in growth control, alterations of receptor kinases and their signaling pathways can lead to cancer.

RTKs are activated by autophosphorylation.
The activated receptor can also phosphorylate other intracellular proteins.

Phosphotyrosine domains mediate protein–protein interactions.
Adapter proteins can bind to phosphotyrosine and act as links between the receptors and downstream signaling events.

Protein kinase cascades can amplify a signal.

Scaffold proteins organize kinase cascades.
Scaffold proteins and protein kinases form a single complex where the enzymes act sequentially and are optimally functional.
Internalized receptors are degraded or recycled.

Ras is a small G protein that acts as a molecular switch.
Small G proteins act as molecular switches linking external signals to signal transduction pathways.

RTKs are inactivated by internalization.

9.5 Signal Transduction Through G Protein–Coupled Receptors (figure 9.11)

G protein–coupled receptors function through activation of G proteins.

G proteins link receptors with effector proteins.
G proteins are active bound to GTP and inactive bound to GDP. Receptors promote exchange of GDP for GTP.
The activated G protein dissociates into two parts, G_α and $G_{\beta\gamma}$, each of which can act on effector proteins.
G_α also hydrolyzes GTP to GDP to inactivate the G protein.

Effector proteins produce multiple second messengers.
Two common effector proteins are adenylyl cyclase and phospholipase C, which produce second messengers known as cAMP, and DAG and IP_3, respectively.
Ca^{2+} is also a second messenger. Ca^{2+} release is triggered by IP_3 binding to channel-linked receptors in the ER.
Ca^{2+} can bind to a cytoplasmic protein calmodulin, which in turn activates other proteins, producing a variety of responses.

Different receptors can produce the same second messengers.
Different GPCR receptors can converge to activate the same effector enzyme and thus produce the same second messenger.

Receptor subtypes can lead to different effects in different cells.
Epinephrine causes increased contraction in heart muscle but relaxation in smooth muscle.

G protein–coupled receptors and receptor tyrosine kinases can activate the same pathways.
Both RTKs and GPCRs can activate MAP kinase cascades.

Review Questions

UNDERSTAND

1. Paracrine signaling is characterized by ligands that are
    a. produced by the cell itself.
    b. secreted by neighboring cells.
    c. present on the plasma membrane of neighboring cells.
    d. secreted by distant cells.

2. Signal transduction pathways
    a. are necessary for signals to cross the membrane.
    b. include the intracellular events stimulated by an extracellular signal.
    c. include the extracellular events stimulated by an intracellular signal.
    d. are only found in cases where the signal can cross the membrane.

3. The function of a _____ is to add phosphates to proteins, whereas a _____ functions to remove the phosphates.
    a. tyrosine; serine
    b. protein phosphatase; protein dephosphatase
    c. protein kinase; protein phosphatase
    d. receptor; ligand

4. Which of the following receptor types is NOT a membrane receptor?
    a. Channel-linked receptor
    b. Enzymatic receptor
    c. G protein–coupled receptor
    d. Steroid hormone receptors

5. How does the function of an intracellular receptor differ from that of a membrane receptor?
    a. The intracellular receptor binds a ligand.
    b. The intracellular receptor binds DNA.
    c. The intracellular receptor activates a kinase.
    d. The intracellular receptor functions as a second messenger.

6. Signaling through receptor tyrosine kinases often
    a. leads to the production of the second messenger cAMP.
    b. leads to the production of the second messenger IP_3.
    c. stimulates gene expression directly.
    d. leads to the activation of a cascade of kinase enzymes.

7. What is the function of Ras during tyrosine kinase cell signaling?
   a. It activates the opening of channel-linked receptors.
   b. It is an enzyme that synthesizes second messengers.
   c. It links the receptor protein to the MAP kinase pathway.
   d. It phosphorylates other enzymes as part of a pathway.

8. Which of the following best describes the immediate effect of ligand binding to a G protein–coupled receptor?
   a. The G protein trimer releases a GDP and binds a GTP.
   b. The G protein trimer dissociates from the receptor.
   c. The G protein trimer interacts with an effector protein.
   d. The α subunit of the G protein becomes phosphorylated.

APPLY

1. The action of steroid hormones is often longer-lived than that of peptide hormones. This is because they
   a. enter the cell and act like enzymes for a longer period of time.
   b. they turn on gene expression to produce proteins that persist in the cell.
   c. result in the production of second messengers that act directly on cellular processes.
   d. stimulate G proteins that act directly on cellular processes.

2. The ion Ca^{2+} can act as a second messenger because it is
   a. produced by the enzyme calcium synthase.
   b. normally at a high level in the cytoplasm.
   c. normally at a low level in the cytoplasm.
   d. stored in the cytoplasm.

3. Different receptors can have the same effect on a cell. One reason for this is that
   a. most receptors produce the same second messenger.
   b. different isoforms of receptors bind different ligands, but stimulate the same signaling pathway.
   c. signal transduction pathways intersect—the same pathway can be stimulated by different receptors.
   d. all receptors converge on the same signal transduction pathways.

4. In comparing small G proteins like Ras and GPCR proteins, we can say that
   a. both proteins have intrinsic GTPase activity that stops signaling.
   b. both proteins are active bound to GTP.
   c. Ras is active bound to GDP and GPCRs are active bound to GTP.
   d. both a and b are true.

5. The same signal can have different effects in different cells because there
   a. are different receptor subtypes that initiate different signal transduction pathways.
   b. may be different coactivators in different cells.
   c. may be different target proteins in different cells' signal transduction pathways.
   d. All of the choices are correct.

6. The receptors for steroid hormones and peptide hormones are fundamentally different because
   a. of the great difference in size of the molecule.
   b. peptides are one of the four major polymers and steroids are simple ringed structures.
   c. peptides are hydrophilic and steroids are hydrophobic.
   d. peptides are hydrophobic and steroids are hydrophilic.

SYNTHESIZE

1. Describe the common features found in all examples of cellular signaling discussed in this chapter. Provide examples to illustrate your answer.

2. The sheet of cells that form the gut epithelium folds into peaks called villi and valleys called crypts. The cells within the crypt region secrete a protein, Netrin-1, that becomes concentrated within the crypts. Netrin-1 is the ligand for a receptor protein that is found on the surface of all gut epithelial cells. Netrin-1 binding triggers a signal pathway that promotes cell growth. Gut epithelial cells undergo apoptosis (cell death) in the absence of Netrin-1 ligand binding.
   a. How would you characterize the type of signaling (autocrine, paracrine, endocrine) found in this system?
   b. Predict where the greatest amount of cell growth and cell death would occur in the epithelium.
   c. The loss of the Netrin-1 receptor is associated with some types of colon cancer. Suggest an explanation for the link between this signaling pathway and tumor formation.

CHAPTER 10

How Cells Divide

Chapter Contents

- 10.1 Bacterial Cell Division
- 10.2 Eukaryotic Chromosomes
- 10.3 Overview of the Eukaryotic Cell Cycle
- 10.4 Interphase: Preparation for Mitosis
- 10.5 M Phase: Chromosome Segregation and the Division of Cytoplasmic Contents
- 10.6 Control of the Cell Cycle

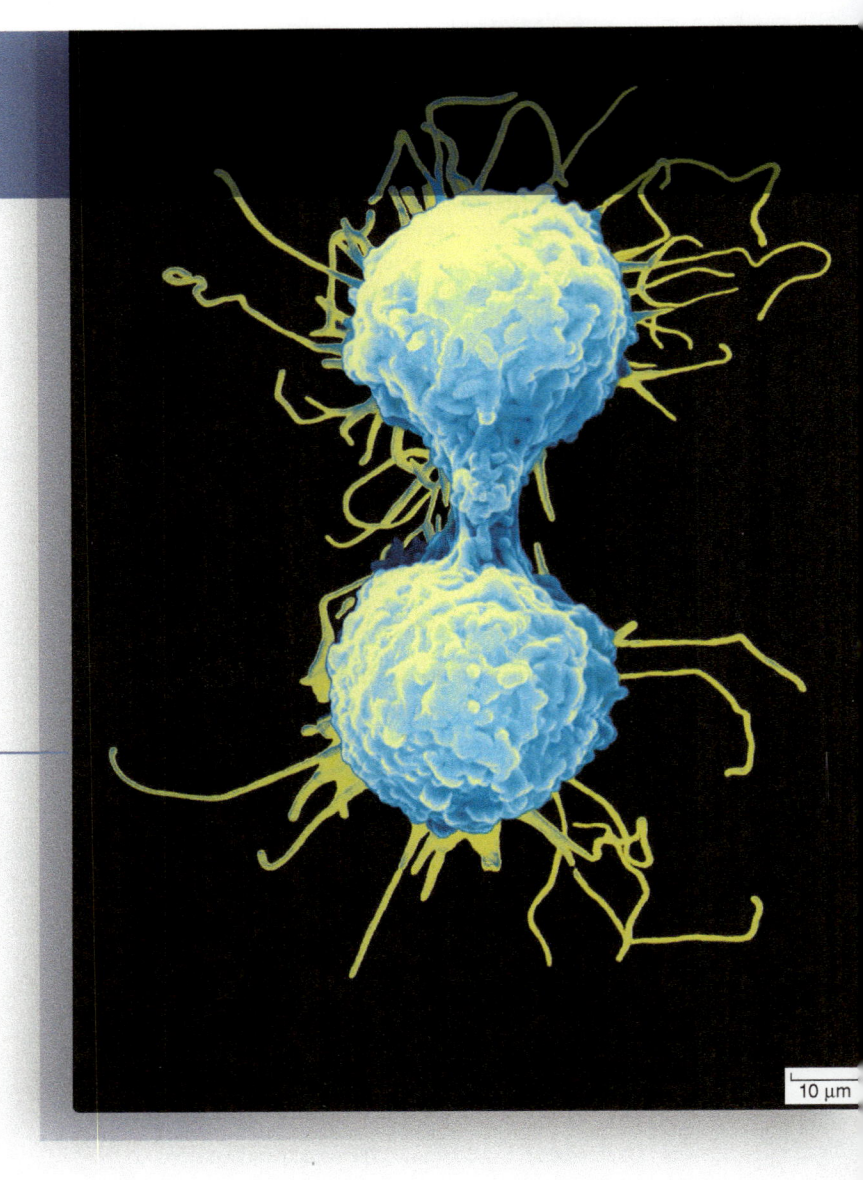

Introduction

All species of organisms—bacteria, alligators, the weeds in a lawn—grow and reproduce. From the smallest creature to the largest, all species produce offspring like themselves and pass on the hereditary information that makes them what they are. In this chapter, we examine how cells, like the white blood cell shown in the figure, divide and reproduce. Cell division is necessary for the growth of organisms, for wound healing, and to replace cells that are lost regularly, such as those in your skin and in the lining of your gut. The mechanism of cell reproduction and its biological consequences have changed significantly during the evolution of life on Earth. The process is complex in eukaryotes, involving both the replication of chromosomes and their separation into daughter cells. Much of what we are learning about the causes of cancer relates to how cells control this process, and in particular their tendency to divide, a mechanism that in broad outline remains the same in all eukaryotes.

10.1 Bacterial Cell Division

Learning Outcome
1. Describe the process of binary fission.

Bacteria divide as a way of reproducing themselves. Although bacteria exchange DNA, they do not have a sexual cycle like eukaryotes. Thus all growth in a bacterial population is due to division to produce new cells. The reproduction of bacteria is clonal—that is, each cell produced by cell division is an identical copy of the original cell.

Binary fission is a simple form of cell division

Cell division in both bacterial and eukaryotic cells produces two new cells with the same genetic information as the original. Despite the differences in these cell types, the essentials of the process are the same: duplication and segregation of genetic information into daughter cells, and division of cellular contents. We begin by looking at the simpler process, **binary fission,** which occurs in bacteria.

Most bacteria have a genome made up of a single, circular DNA molecule. In spite of its apparent simplicity, the DNA molecule of the bacterium *Escherichia coli* is actually on the order of 500 times longer than the cell itself! Thus, this "simple" structure is actually packaged very tightly to fit into the cell. Although not found in a nucleus, the DNA is located in a region called the *nucleoid* that is distinct from the cytoplasm around it.

The compaction and organization of the nucleoid involves a class of proteins called structural maintenance of chromosome, or SMC, proteins. These are ancient proteins that have diversified over evolutionary time to fulfill a variety of roles related to DNA organization in different lineages. In eukaryotes the cohesin and condensin proteins discussed throughout this chapter are SMC proteins.

During binary fission, the chromosome is replicated, and the two products are partitioned to each end of the cell prior to the actual division of the cell. One key feature of bacterial cell division is that replication and partitioning of the chromosome occur as a concerted process. In contrast, DNA replication in eukaryotic cells occurs early in division, and chromosome separation occurs much later.

Bacterial cells control chromosome separation and septum formation

Binary fission begins with the replication of the bacterial DNA at a specific site—the origin of replication (see chapter 14)—and proceeds both directions around the circular DNA to a specific site of termination (figure 10.1). The cell grows by elongation, and division occurs roughly at midcell. For many years, it was thought that newly replicated *E. coli* DNA molecules were passively segregated by attachment to and growth of the membrane as the cell elongated. Experiments that follow the movement of the origin of replication show that it is at midcell prior to replication, then the

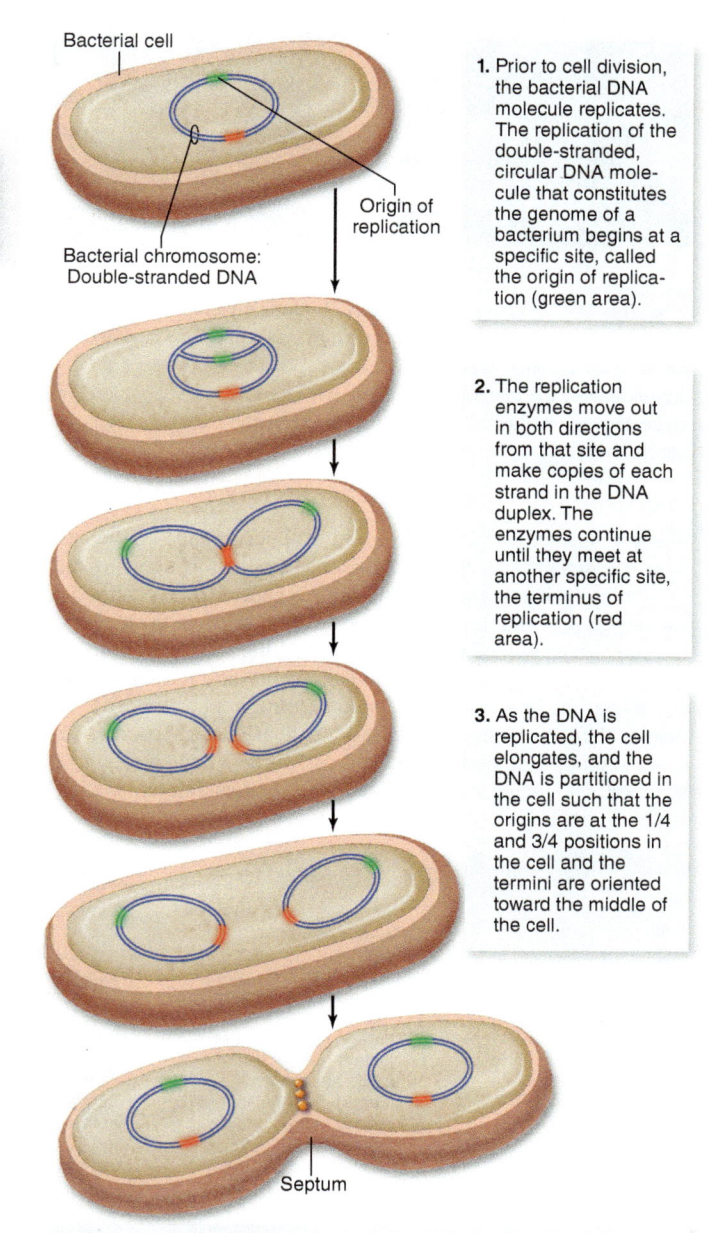

1. Prior to cell division, the bacterial DNA molecule replicates. The replication of the double-stranded, circular DNA molecule that constitutes the genome of a bacterium begins at a specific site, called the origin of replication (green area).

2. The replication enzymes move out in both directions from that site and make copies of each strand in the DNA duplex. The enzymes continue until they meet at another specific site, the terminus of replication (red area).

3. As the DNA is replicated, the cell elongates, and the DNA is partitioned in the cell such that the origins are at the 1/4 and 3/4 positions in the cell and the termini are oriented toward the middle of the cell.

4. Septation then begins, in which new membrane and cell wall material begin to grow and form a septum at approximately the midpoint of the cell. A protein molecule called FtsZ (orange dots) facilitates this process.

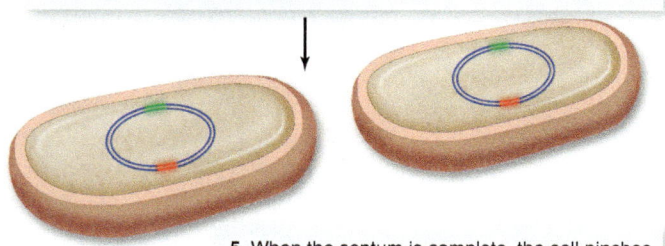

5. When the septum is complete, the cell pinches in two, and two daughter cells are formed, each containing a bacterial DNA molecule.

Figure 10.1 Binary fission.

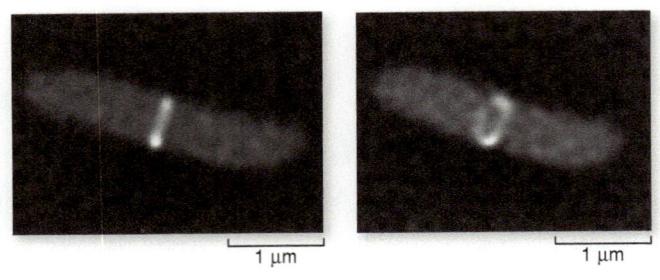

Figure 10.2 The FtsZ protein. In these dividing *E. coli* bacteria, the FtsZ protein is labeled with fluorescent dye to show its location during binary fission. The protein assembles into a ring at approximately the midpoint of the cell, where it facilitates septation and cell division. Bacteria carrying mutations in the *FtsZ* gene are unable to divide.

newly replicated origins move toward opposite ends of the cell. This movement is faster than the rate of elongation, showing that growth alone is not enough. The origins appear to be captured at the one quarter and three quarter positions relative to the length of the cell, which will be midcell of the resulting daughter cells.

Although the actual mechanism of chromosome segregation is unclear, the order of events is not. During replication, first the origin, then the rest of the newly replicated chromosomes are moved to opposite ends of the cell as two new nucleoids are assembled. The final event of replication is decatenation (untangling) of the final replication products. After replication and segregation, the midcell region is cleared of daughter nucleoids, and division occurs. The force behind chromosome segregation has been attributed to DNA replication itself, transcription, and the polymerization of actinlike molecules. At this point, no single model appears to explain the process, and it may involve more than one.

The cell's other components are partitioned by the growth of new membrane and production of the **septum** (figure 10.1). This process, termed **septation,** usually occurs at the midpoint of the cell. It begins with the formation of a ring composed of many copies of the protein FtsZ (figure 10.2). Next, accumulation of a number of other proteins occurs, including ones embedded in the membrane. This structure contracts inward radially until the cell pinches off into two new cells. The midcell location of the FtsZ ring is caused by an oscillation between the two poles of an inhibitor of FtsZ formation.

The FtsZ protein is found in most prokaryotes, including archaea. It can form filaments and rings, and three-dimensional crystals show a high degree of similarity to eukaryotic tubulin. However, its role in bacterial division is quite different from the role of tubulin in mitosis in eukaryotes.

The evolution of eukaryotic cells included much more complex genomes composed of multiple linear chromosomes housed in a membrane-bounded nucleus. These complex genomes may be possible due to the evolution of mechanisms that delay chromosome separation after replication. Although it is unclear how this ability to keep chromosomes together evolved, it does seem more closely related to binary fission than we once thought (figure 10.3).

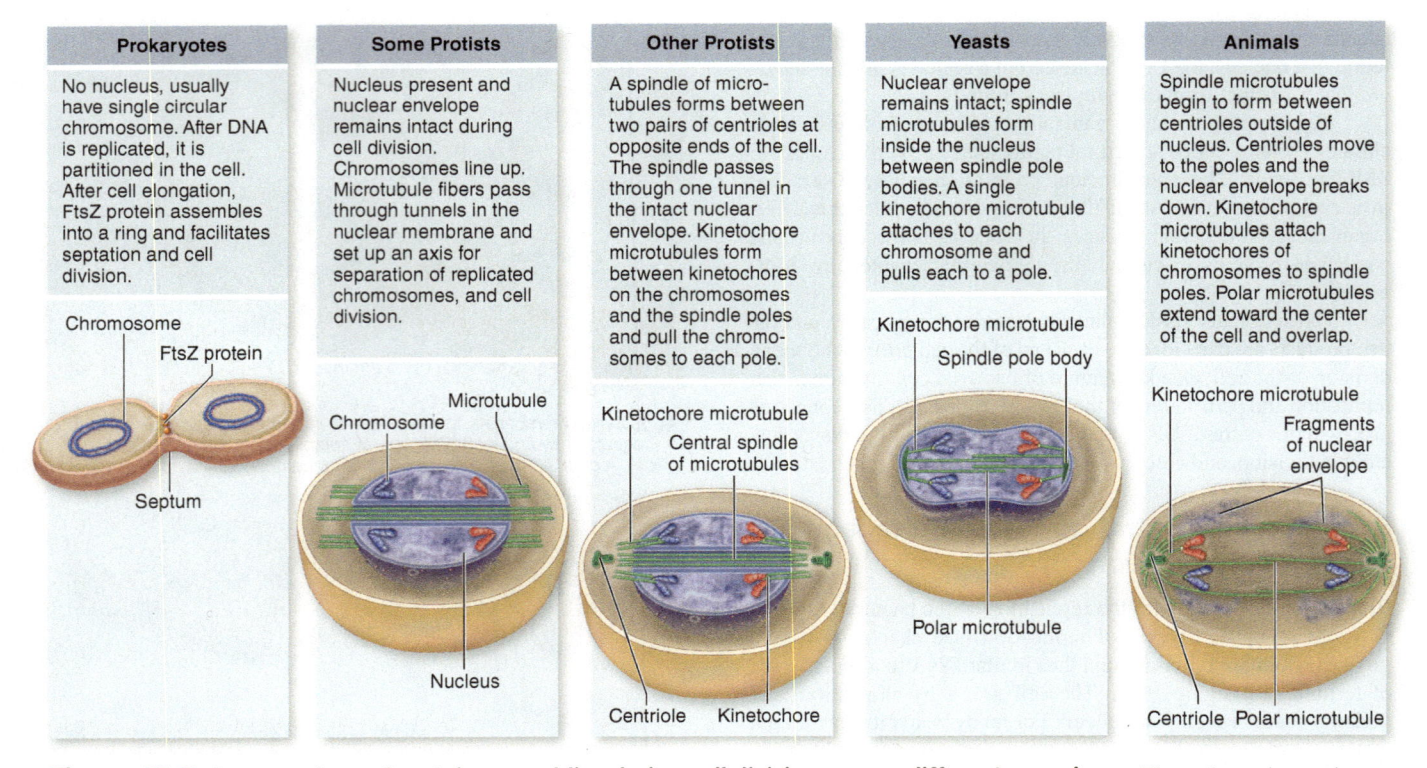

Prokaryotes	Some Protists	Other Protists	Yeasts	Animals
No nucleus, usually have single circular chromosome. After DNA is replicated, it is partitioned in the cell. After cell elongation, FtsZ protein assembles into a ring and facilitates septation and cell division.	Nucleus present and nuclear envelope remains intact during cell division. Chromosomes line up. Microtubule fibers pass through tunnels in the nuclear membrane and set up an axis for separation of replicated chromosomes, and cell division.	A spindle of microtubules forms between two pairs of centrioles at opposite ends of the cell. The spindle passes through one tunnel in the intact nuclear envelope. Kinetochore microtubules form between kinetochores on the chromosomes and the spindle poles and pull the chromosomes to each pole.	Nuclear envelope remains intact; spindle microtubules form inside the nucleus between spindle pole bodies. A single kinetochore microtubule attaches to each chromosome and pulls each to a pole.	Spindle microtubules begin to form between centrioles outside of nucleus. Centrioles move to the poles and the nuclear envelope breaks down. Kinetochore microtubules attach kinetochores of chromosomes to spindle poles. Polar microtubules extend toward the center of the cell and overlap.

Figure 10.3 A comparison of protein assemblies during cell division among different organisms. The prokaryotic protein FtsZ has a structure that is similar to that of the eukaryotic protein tubulin. Tubulin is the protein component of microtubules, which are fibers that eukaryotic cells use to construct the spindle apparatus that is used to separate chromosomes.

Learning Outcome Review 10.1

Most bacteria divide by binary fission, a form of cell division in which DNA replication and segregation occur simultaneously. This process involves active partitioning of the single bacterial chromosome and positioning of the site of septation.

- Would binary fission work as well if bacteria had many chromosomes?

10.2 Eukaryotic Chromosomes

Learning Outcomes

1. Describe the structure of eukaryotic chromosomes.
2. Distinguish between homologues and sister chromatids.
3. Contrast replicated and nonreplicated chromosomes.

TABLE 10.1 Chromosome Number in Selected Eukaryotes

Group	Total Number of Chromosomes
FUNGI	
Neurospora (haploid)	7
Saccharomyces (a yeast)	16
INSECTS	
Mosquito	6
Drosophila	8
Honeybee	diploid females 32, haploid males 16
Silkworm	56
PLANTS	
Haplopappus gracilis	2
Garden pea	14
Corn	20
Bread wheat	42
Sugarcane	80
Horsetail	216
Adder's tongue fern	1262
VERTEBRATES	
Opossum	22
Frog	26
Mouse	40
Human	46
Chimpanzee	48
Horse	64
Chicken	78
Dog	78

Chromosomes were first observed by the German embryologist Walther Flemming (1843–1905) in 1879, while he was examining the rapidly dividing cells of salamander larvae. When Flemming looked at the cells through what would now be a rather primitive light microscope, he saw minute threads within their nuclei that appeared to be dividing lengthwise. Flemming called their division **mitosis,** based on the Greek word *mitos,* meaning "thread."

Chromosome number varies among species

Since their initial discovery, chromosomes have been found in the cells of all eukaryotes examined. Their number may vary enormously from one species to another. A few kinds of organisms have only a single pair of chromosomes, whereas some ferns have more than 500 pairs (table 10.1). Most eukaryotes have between 10 and 50 chromosomes in their body cells.

Human cells each have 46 chromosomes, consisting of 23 nearly identical pairs (figure 10.4). Each of these 46 chromosomes contains hundreds or thousands of genes that play important roles in determining how a person's body develops and functions. Human embryos missing even one chromosome, a condition called *monosomy,* do not survive in most cases. Having an extra copy of any one chromosome, a condition called *trisomy,* is usually fatal except where the smallest chromosomes are involved. (You'll learn more about human chromosome abnormalities in chapter 13.)

Eukaryotic chromosomes exhibit complex structure

Researchers have learned a great deal about chromosome structure and composition in the more than 135 years since their discovery. But despite intense research, the exact structure of eukaryotic chromosomes during the cell cycle remains unclear. The structures described in this chapter represent the currently accepted model.

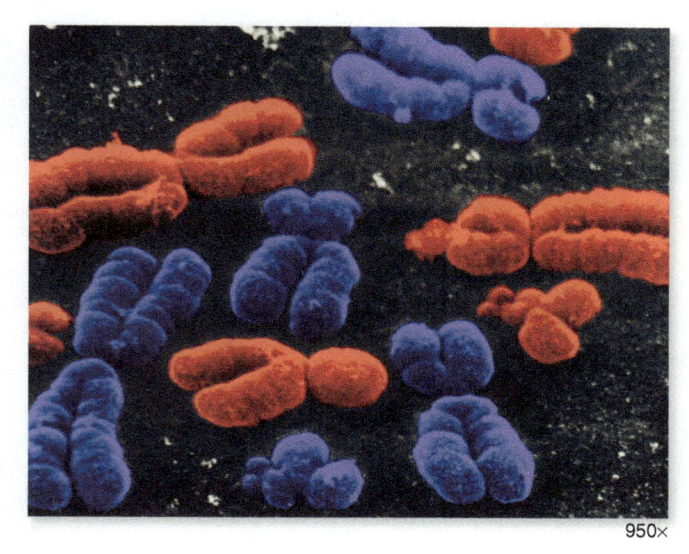

Figure 10.4 Human chromosomes. This scanning electron micrograph shows human chromosomes as they appear immediately before nuclear division. Each DNA molecule has already replicated, forming identical copies held together at a visible constriction called the centromere. False color has been added to the chromosomes.

Composition of chromatin

Chromosomes are composed of **chromatin,** a complex of DNA and protein; most chromosomes are about 40% DNA and 60% protein. A significant amount of RNA is also associated with chromosomes because chromosomes are the sites of RNA synthesis.

Each chromosome contains a single DNA molecule that runs uninterrupted through the chromosome's entire length. A typical human chromosome contains about 140 million (1.4×10^8) nucleotides in its DNA. If we think of each nucleotide as a "word," then the amount of information an average chromosome contains would fill about 280 printed books of 1000 pages each, with 500 "words" per page.

If we could lay out the strand of DNA from a single chromosome in a straight line, it would be about 5 cm (2 in.) long. Fitting such a strand into a cell nucleus is like cramming a string the length of a football field into a baseball—and that's only 1 of 46 chromosomes! In the cell, however, the DNA is compacted, allowing it to fit into a much smaller space than would otherwise be possible.

The organization of chromatin in the nondividing nucleus is not well understood, but geneticists have recognized for years that some domains of chromatin, called **heterochromatin,** are not expressed, and other domains of chromatin, called **euchromatin,** are expressed. This genetically measurable state is also related to the physical state of chromatin, although researchers are just beginning to see the details.

Chromosome structure

If we gently disrupt a eukaryotic nucleus and examine the DNA with an electron microscope, we find that it resembles a string of beads (figure 10.5). Every 200 nucleotides (nt), the DNA duplex (double strand) is coiled around a core of eight **histone proteins.** Unlike most proteins, which have an overall negative charge, histones are positively charged because of an abundance of the basic amino acids arginine and lysine. Thus, they are strongly attracted to the negatively charged phosphate groups of the DNA, and the histone cores act as "magnetic forms" that promote and guide the coiling of the DNA. The complex of DNA and histone proteins is termed a **nucleosome.**

The DNA wrapped in nucleosomes is further coiled into an even more compact structure called the *solenoid*. The precise path of this higher order folding of chromatin is still a subject of debate, but it leads to a fiber with a diameter of 30 nm and thus is called the 30-nm fiber. This 30-nm fiber is the usual state of interphase (nondividing) chromatin.

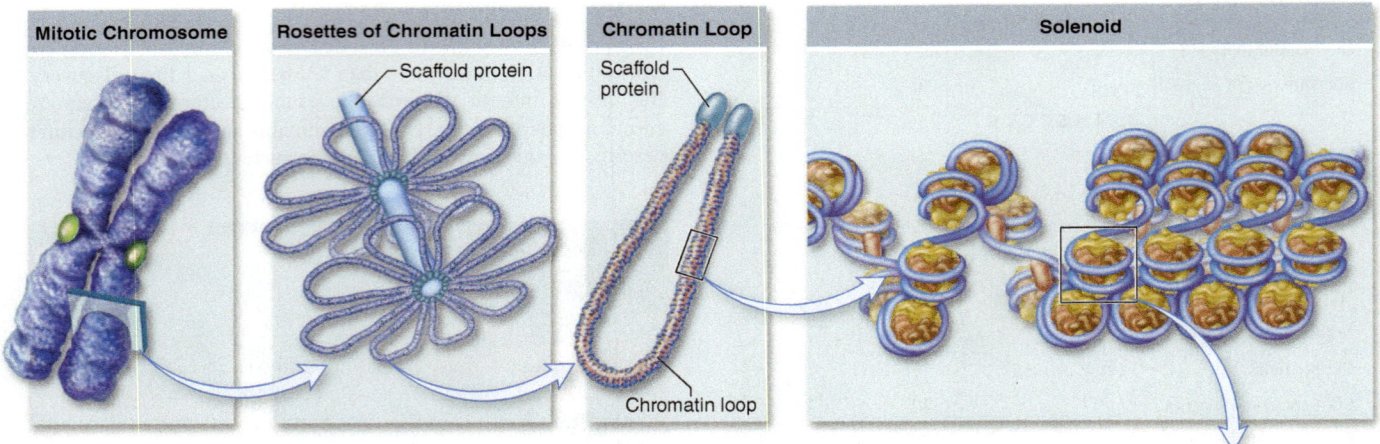

Figure 10.5 Levels of eukaryotic chromosomal organization. Each chromosome consists of a long double-stranded DNA molecule. These strands require further packaging to fit into the cell nucleus. The DNA duplex is tightly bound to and wound around proteins called histones. The DNA-wrapped histones are called nucleosomes. The nucleosomes are further coiled into the solenoid. This solenoid is then organized into looped domains. The precise organization of mitotic chromosomes is unknown, but the solenoid is further condensed around a preexisting scaffolding of proteins. The arrangement shown is one of many possibilities.

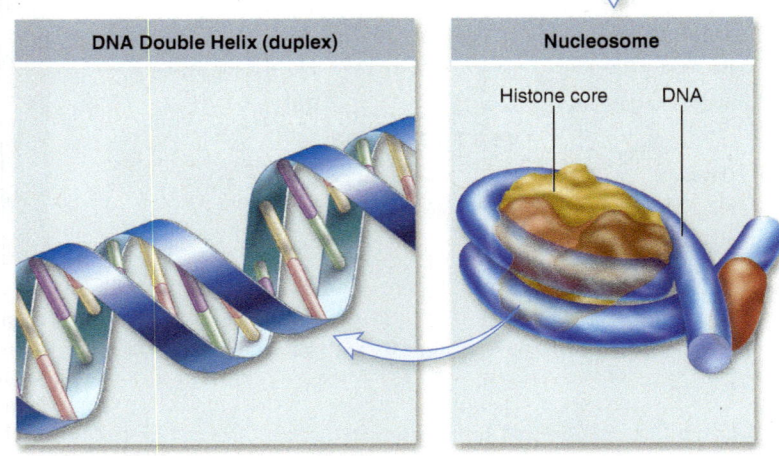

During mitosis, proteins are assembled into a scaffold that provides a framework for the final level of compaction. This gives chromosomes their familiar X-shaped structure, and facilitates separation by the mitotic machinery described later. The exact nature of the compaction is unknown, but one longstanding model involves looping of chromatin fibers from the scaffold like the fibers on a wire brush. A complex of proteins called condensin, which are evolutionarily related to bacterial SMC proteins, are necessary for compaction.

Chromosome karyotypes

Chromosomes vary in size, staining properties, the location of the centromere (a constriction found on all chromosomes, described shortly), the relative length of the two arms on either side of the centromere, and the positions of constricted regions along the arms. The particular array of chromosomes an individual organism possesses is called its **karyotype.** The karyotype in figure 10.6 shows the set of chromosomes from a normal human cell.

When defining the number of different chromosomes in a species, geneticists count the **haploid (n)** number of chromosomes. This refers to one complete set of chromosomes necessary to define an organism. For humans and many other species, the total number of chromosomes in a cell is called the **diploid (2n)**

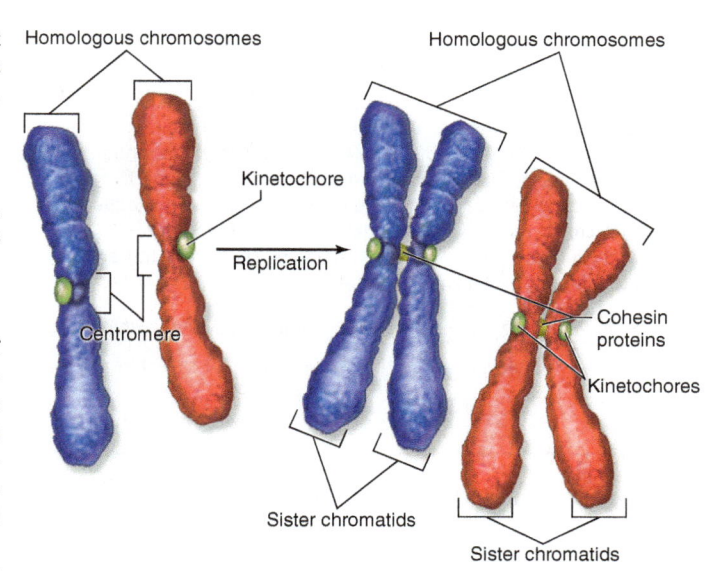

Figure 10.7 The difference between homologous chromosomes and sister chromatids. Homologous chromosomes are the maternal and paternal copies of the same chromosome—say, chromosome number 16. Sister chromatids are the two replicas of a single chromosome held together at their centromeres by cohesin proteins after DNA replication. The kinetochore (described later in section 10.4) is composed of proteins found at the centromere that attach to microtubules during mitosis.

number, which is twice the haploid number. For humans, the haploid number is 23 and the diploid number is 46. Diploid chromosomes reflect the equal genetic contribution that each parent makes to offspring. We refer to the maternal and paternal chromosomes as being **homologous,** and each one of the pair is termed a **homologue.**

Chromosome replication

Chromosomes as seen in a karyotype are only present for a brief period during cell division. Prior to replicating, each chromosome is composed of a single DNA molecule that is arranged into the 30-nm fiber described earlier in this section. After replication, each chromosome is composed of two identical DNA molecules held together by a complex of proteins called **cohesins.** As the chromosomes become more condensed and arranged about the protein scaffold, they become visible as two strands that are held together at the centromere. At this point, we still call this one chromosome, but it is composed of two **sister chromatids** (figure 10.7).

The fact that the products of replication are held together is critical to the division process. One problem that a cell must solve is how to ensure that each new cell receives a complete set of chromosomes. If we were designing a system, we might use some kind of label to identify each chromosome, much like most of us use when we duplicate files on a computer. Instead of labeling chromosomes, the cell glues replication products together at the centromere. The process of mitosis then separates all of these copies at the same time, ensuring that each daughter cell gets one copy of each chromosome.

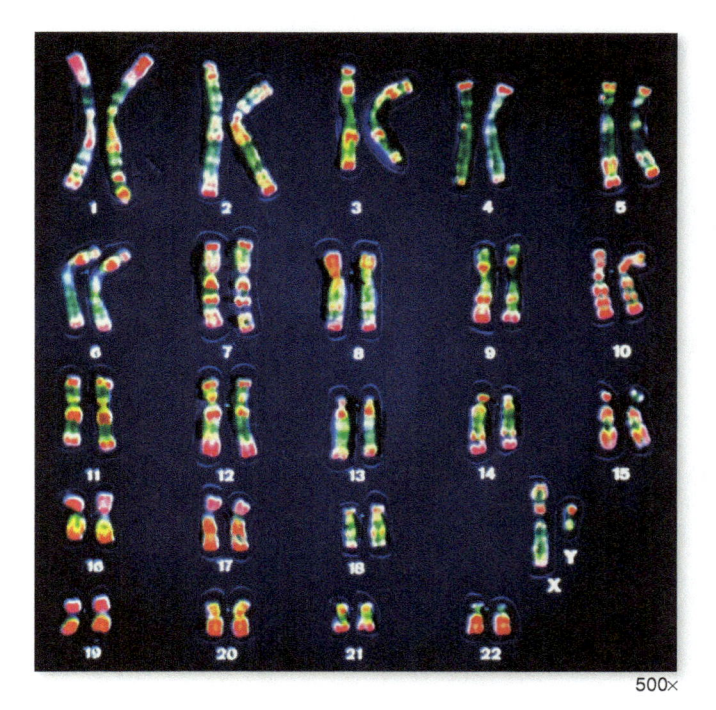

Figure 10.6 A human karyotype. The individual chromosomes that make up the 23 pairs differ widely in size and in centromere position. In this preparation of a male karyotype, the chromosomes have been specifically stained to indicate differences in their composition and to distinguish them clearly from one another. Notice that members of a chromosome pair are very similar but not identical.

Learning Outcomes Review 10.2

Eukaryotic chromosomes are complex structures that can be compacted for cell division. During interphase, DNA is coiled around proteins into a structure called a nucleosome. The string of nucleosomes is further coiled into a solenoid (30-nm fiber). Diploid cells contain a maternal and paternal copy, or homologue, for each chromosome. After chromosome replication, each homologue consists of two sister chromatids. The chromatids are held together by proteins called cohesins.

- Is chromosome number related to organismal complexity?

10.3 Overview of the Eukaryotic Cell Cycle

Learning Outcome
1. Describe the eukaryotic cell cycle.

Compared with prokaryotes, the increased size and more complex organization of eukaryotic genomes required radical changes in the partitioning of replicated genomes into daughter cells. The **cell cycle** requires the duplication of the genome, its accurate segregation, and the division of cellular contents.

The cell cycle is divided into five phases

The cell cycle is divided into phases based on the key events of genome duplication and segregation. The cell cycle is usually diagrammed as in figure 10.8:

- **G₁ (gap phase 1)** is the primary growth phase of the cell. The term *gap phase* refers to its filling the gap between cytokinesis and DNA synthesis. For most cells, this is the longest phase.
- **S (synthesis)** is the phase in which the cell synthesizes a replica of the genome.
- **G₂ (gap phase 2)** is the second growth phase, and preparation for separation of the newly replicated genome. This phase fills the gap between DNA synthesis and the beginning of mitosis. During this phase microtubules begin to reorganize to form a spindle.
  G₁, S, and G₂ together constitute **interphase**, the portion of the cell cycle between cell divisions.
- **Mitosis** is the phase of the cell cycle in which the spindle apparatus assembles, binds to the chromosomes, and moves the sister chromatids apart. Mitosis is the essential step in the separation of the two daughter genomes. It is traditionally subdivided into five stages: prophase, prometaphase, metaphase, anaphase, and telophase.
- **Cytokinesis** is the phase of the cell cycle when the cytoplasm divides, creating two daughter cells. In animal cells, the microtubule spindle helps position a contracting ring of actin that constricts like a drawstring to pinch the cell in two. In cells with a cell wall, such as plant cells, a plate forms between the dividing cells.

Mitosis and cytokinesis together are usually referred to collectively as M phase, to distinguish the dividing phase from interphase.

The duration of the cell cycle varies depending on cell type

The time it takes to complete a cell cycle varies greatly. Cells in animal embryos can complete their cell cycle in under 20 min; the shortest known animal nuclear division cycles occur in fruit fly embryos (8 min). These cells simply divide their nuclei as quickly as they can replicate their DNA, without cell growth. Half of their cycle is taken up by S, half by M, and essentially none by G₁ or G₂.

Because mature cells require time to grow, most of their cycles are much longer than those of embryonic tissue. Typically, a dividing mammalian cell completes its cell cycle in about 24 hr, but some cells, such as certain cells in the human liver, have cell cycles lasting more than a year. During the cycle, growth occurs throughout the G₁ and G₂ phases, as well as during the S phase. The M phase takes only about an hour, a small fraction of the entire cycle.

Most of the variation in the length of the cell cycle between organisms or cell types occurs in the G₁ phase. Cells often pause in G₁ before DNA replication and enter a resting state called the **G₀ phase**; cells may remain in this phase for days to years before resuming cell division. At any given time, most of the cells in an

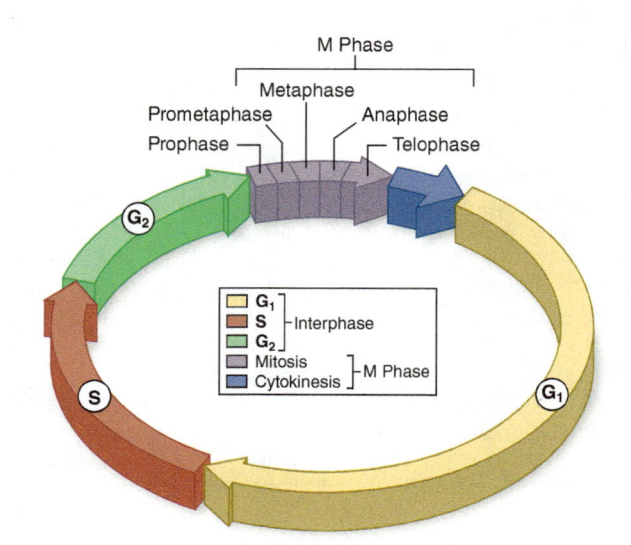

Figure 10.8 The cell cycle. The cell cycle is depicted as a circle. The first gap phase, G₁, involves growth and preparation for DNA synthesis. During S phase, a copy of the genome is synthesized. The second gap phase, G₂, prepares the cell for mitosis. During mitosis, replicated chromosomes are partitioned. Cytokinesis divides the cell into two cells with identical genomes.

animal's body are in G_0 phase. Some, such as muscle and nerve cells, remain there permanently; others, such as liver cells, can resume G_1 phase in response to factors released during injury.

Learning Outcome Review 10.3

Cell division in eukaryotes is a complex process that involves five phases: a first gap phase (G_1); a DNA synthesis phase (S); a second gap phase (G_2); mitosis (M), during which chromatids are separated; and cytokinesis in which a cell becomes two separate cells.

- When during the cycle is a cell irreversibly committed to dividing?

10.4 Interphase: Preparation for Mitosis

Learning Outcomes

1. Describe the events that take place during interphase.
2. Illustrate the connection between sister chromatids after S phase.

The events that occur during interphase—the G_1, S, and G_2 phases—are very important for the successful completion of mitosis. During G_1, cells undergo the major portion of their growth. During the S phase, each chromosome replicates to produce two sister chromatids, which remain attached to each other at the centromere. In the G_2 phase, the chromosomes coil even more tightly.

The **centromere** is a point of constriction on the chromosome containing repeated DNA sequences that bind specific proteins. These proteins make up a disklike structure called the **kinetochore.** This disk functions as an attachment site for microtubules necessary to separate the chromosomes during cell division (figure 10.9). As seen in figure 10.6, each chromosome's centromere is located at a characteristic site along the length of the chromosome.

After the S phase, the sister chromatids appear to share a common centromere, but at the molecular level the DNA of the centromere has actually already replicated, so there are two complete DNA molecules. This means that two chromatids are held together by cohesin proteins at the centromere, and each chromatid has its own set of kinetochore proteins (figure 10.10). In multicellular animals, most of the cohesins that hold sister chromatids together after replication appear to be replaced by condensin as the chromosomes are condensed. This leaves the chromosomes still attached tightly at the centromere, but loosely attached elsewhere.

The cell grows throughout interphase. The G_1 and G_2 segments of interphase are periods of active growth, during which

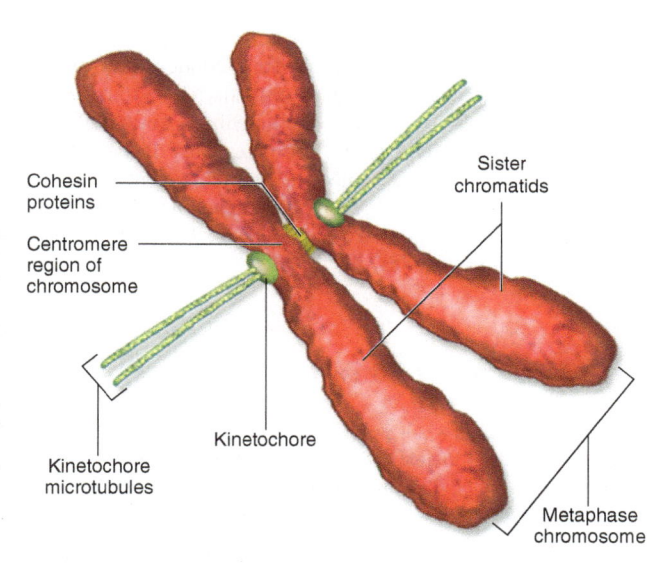

Figure 10.9 Kinetochores. Separation of sister chromatids during mitosis depends on microtubules attaching to proteins found in the kinetochore. These kinetochore proteins are assembled on the centromere of chromosomes. The centromeres of the two sister chromatids are held together by cohesin proteins.

proteins are synthesized and cell organelles are produced. The cell's DNA replicates only during the S phase of the cell cycle.

After the chromosomes have replicated in S phase, they remain fully extended and uncoiled, although cohesin proteins are associated with them at this stage. In G_2 phase, they begin the

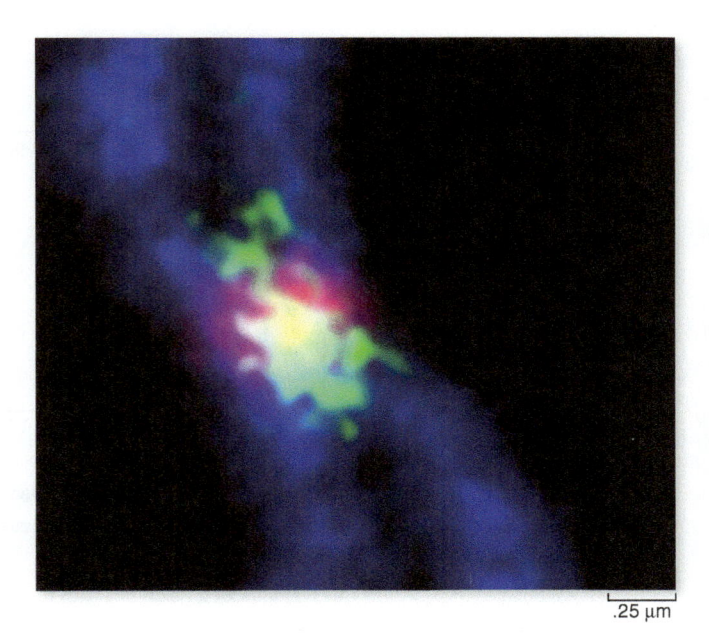

Figure 10.10 Proteins found at the centromere. In this image DNA, a human mitotic chromosome has been stained for condensin *(blue)*, cohesin *(green)* and the centromere specific histone CENP-A *(red)*. The cohesin proteins hold the centromeres of sister chromatids together.

process of condensation, coiling ever more tightly. Special *motor proteins* are involved in the rapid final condensation of the chromosomes that occurs early in mitosis. Also during G_2 phase, the cells begin to assemble the machinery they will later use to move the chromosomes to opposite poles of the cell. In animal cells, a pair of microtubule-organizing centers called *centrioles* replicate, producing one for each pole. All eukaryotic cells undertake an extensive synthesis of **tubulin**, the protein that forms microtubules.

Learning Outcomes Review 10.4

Interphase includes the G_1, S, and G_2 phases of the cell cycle. During interphase, the cell grows; replicates chromosomes, organelles, and centrioles; and synthesizes components needed for mitosis, including tubulin. Cohesin proteins hold chromatids together at the centromere of each chromosome.

- How would a mutation that deleted cohesin proteins affect cell division?

10.5 M Phase: Chromosome Segregation and the Division of Cytoplasmic Contents

Learning Outcomes

1. Describe the phases of mitosis.
2. Explain the importance of metaphase.
3. Compare cytokinesis in plants and animals.

The process of mitosis is one of the most dramatic and beautiful biological processes that we can easily observe. In our attempts to understand this process, we have divided it into discrete phases but it should always be remembered that this is a dynamic, continuous process, not a set of discrete steps. This process is shown both schematically and in micrographs in figure 10.11.

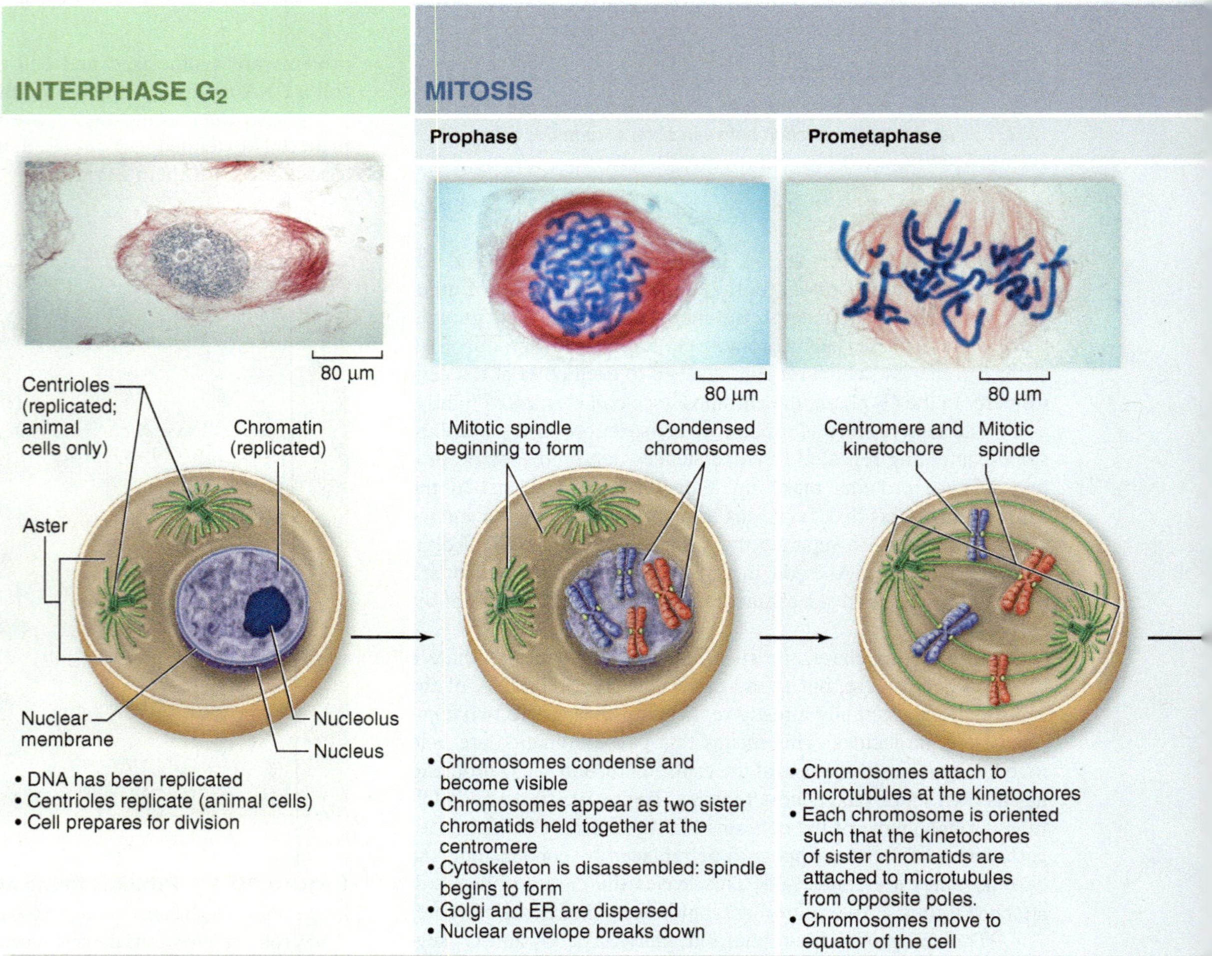

Figure 10.11 Mitosis and cytokinesis. Mitosis is conventionally divided into five stages—prophase, prometaphase, metaphase, anaphase, and telophase—which together act to separate duplicated chromosomes. This is followed by cytokinesis, which divides the cell into two separate cells. Photos depict mitosis and cytokinesis in a plant, the African blood lily (*Haemanthus katharinae*), with chromosomes stained blue and microtubules stained red. Drawings depict mitosis and cytokinesis in animal cells.

INTERPHASE G_2

80 μm

Centrioles (replicated; animal cells only)
Chromatin (replicated)
Aster
Nuclear membrane
Nucleolus
Nucleus

- DNA has been replicated
- Centrioles replicate (animal cells)
- Cell prepares for division

MITOSIS

Prophase

80 μm

Mitotic spindle beginning to form
Condensed chromosomes

- Chromosomes condense and become visible
- Chromosomes appear as two sister chromatids held together at the centromere
- Cytoskeleton is disassembled: spindle begins to form
- Golgi and ER are dispersed
- Nuclear envelope breaks down

Prometaphase

80 μm

Centromere and kinetochore
Mitotic spindle

- Chromosomes attach to microtubules at the kinetochores
- Each chromosome is oriented such that the kinetochores of sister chromatids are attached to microtubules from opposite poles.
- Chromosomes move to equator of the cell

During prophase, the mitotic apparatus forms

When the chromosome condensation initiated in G_2 phase reaches the point at which individual condensed chromosomes first become visible with the light microscope, the first stage of mitosis, **prophase,** has begun. The condensation process continues throughout prophase; consequently, chromosomes that start prophase as minute threads appear quite bulky before its conclusion. Ribosomal RNA synthesis ceases when the portion of the chromosome bearing the rRNA genes is condensed.

The spindle and centrioles

The assembly of the **spindle** apparatus that will later separate the sister chromatids occurs during prophase. The normal microtubule structure in the cell disassembled in the G_2 phase is replaced by the spindle. In animal cells, the two centriole pairs formed during G_2 phase begin to move apart early in prophase, forming between them an axis of microtubules referred to as spindle fibers. By the time the centrioles reach the opposite poles of the cell, they have established a bridge of microtubules, called the spindle apparatus, between them. In plant cells, a similar bridge of microtubular fibers forms between opposite poles of the cell, although centrioles are absent in plant cells.

In animal cell mitosis, the centrioles extend a radial array of microtubules toward the nearby plasma membrane when they reach the poles of the cell. This arrangement of microtubules is called an **aster.** Although the aster's function is not fully understood, it probably braces the centrioles against the membrane and stiffens the point of microtubular attachment during the retraction of the spindle. Plant cells, which have rigid cell walls, do not form asters.

Breakdown of the nuclear envelope

During the formation of the spindle apparatus, the nuclear envelope breaks down, and the endoplasmic reticulum reabsorbs its components. At this point, the microtubular spindle fibers extend completely across the cell, from one pole to the other. Their orientation determines the plane in which the cell will subsequently divide, through the center of the cell at right angles to the spindle apparatus.

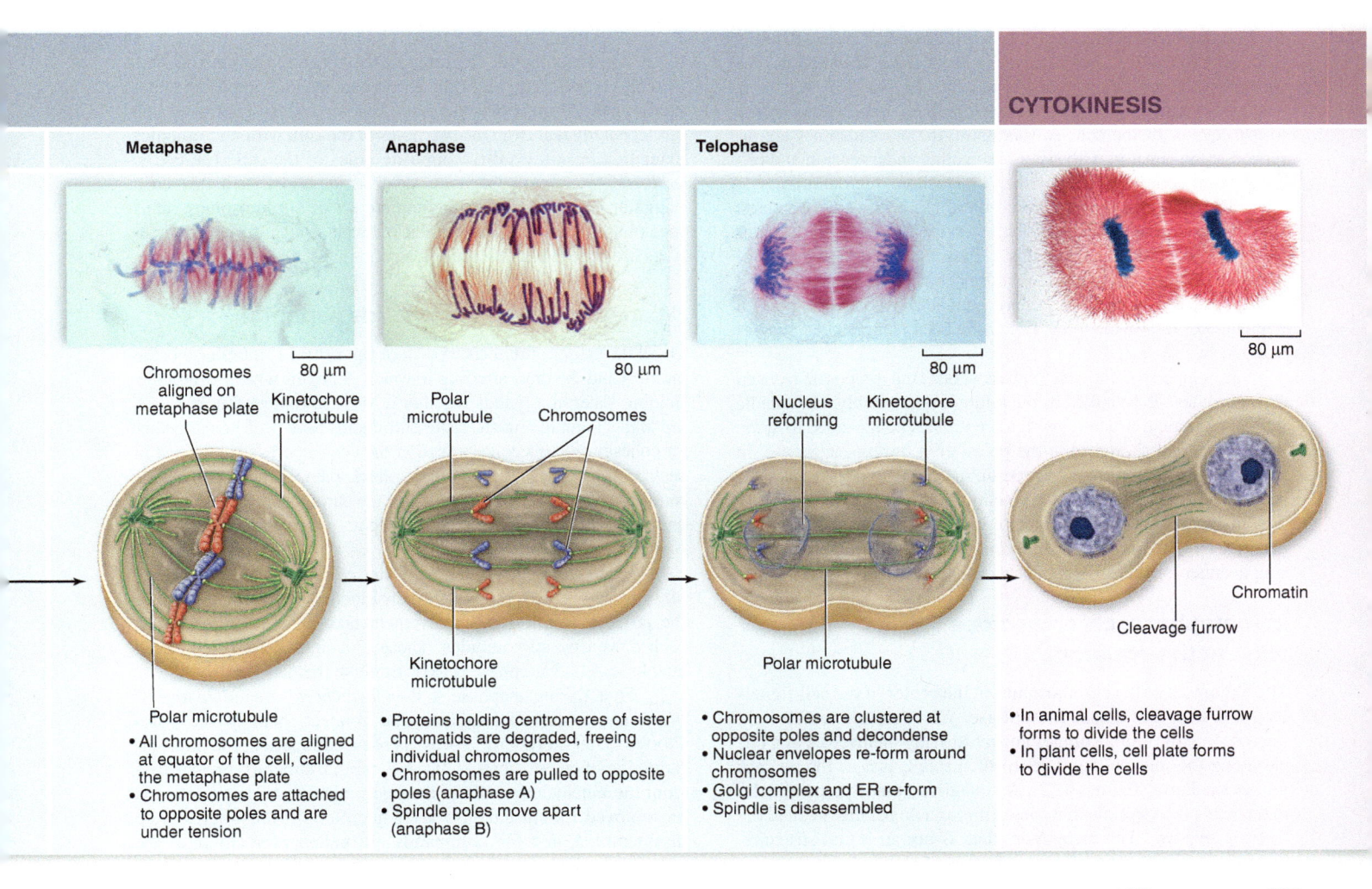

Metaphase — 80 μm
- Chromosomes aligned on metaphase plate
- Kinetochore microtubule
- Polar microtubule
- All chromosomes are aligned at equator of the cell, called the metaphase plate
- Chromosomes are attached to opposite poles and are under tension

Anaphase — 80 μm
- Polar microtubule
- Chromosomes
- Kinetochore microtubule
- Proteins holding centromeres of sister chromatids are degraded, freeing individual chromosomes
- Chromosomes are pulled to opposite poles (anaphase A)
- Spindle poles move apart (anaphase B)

Telophase — 80 μm
- Nucleus reforming
- Kinetochore microtubule
- Polar microtubule
- Chromosomes are clustered at opposite poles and decondense
- Nuclear envelopes re-form around chromosomes
- Golgi complex and ER re-form
- Spindle is disassembled

CYTOKINESIS — 80 μm
- Chromatin
- Cleavage furrow
- In animal cells, cleavage furrow forms to divide the cells
- In plant cells, cell plate forms to divide the cells

During prometaphase, chromosomes attach to the spindle

The transition from prophase to **prometaphase** occurs following the disassembly of the nuclear envelope. During prometaphase the condensed chromosomes become attached to the spindle by their kinetochores. Each chromosome possesses two kinetochores, one attached to the centromere region of each sister chromatid (see figure 10.9).

Microtubule attachment

As prometaphase continues, a second group of microtubules grow from the poles of the cell toward the centromeres. These microtubules are captured by the kinetochores on each pair of sister chromatids. This results in the kinetochores of each sister chromatid being connected to opposite poles of the spindle.

This bipolar attachment is critical to the process of mitosis; any mistakes in microtubule positioning can be disastrous. For example, the attachment of the kinetochores of both sister chromatids to the same pole leads to a failure of sister chromatid separation, and they will be pulled to the same pole ending up in the same daughter cell, with the other daughter cell missing that chromosome.

Movement of chromosomes to the cell center

Each chromosome is attached to the spindle by microtubules running from opposite poles to the kinetochores of sister chromatids. The chromosomes are being pulled simultaneously toward each pole, leading to a jerky motion that eventually pulls all of the chromosomes to the equator of the cell. At this point, the chromosomes are arranged at the equator, with sister chromatids under tension and oriented toward opposite poles by their kinetochore microtubules.

The force that moves chromosomes has been of great interest since the process of mitosis was first observed. Two basic mechanisms have been proposed to explain this: (1) assembly and disassembly of microtubules provides the force to move chromosomes, and (2) motor proteins located at the kinetochore and poles of the cell pull on microtubules to provide force. Data have been obtained that support both mechanisms.

In support of the microtubule-shortening proposal, isolated chromosomes can be pulled by microtubule disassembly. The spindle is a very dynamic structure, with microtubules being added to at the kinetochore and shortened at the poles, even during metaphase. In support of the motor protein proposal, multiple motor proteins have been identified as kinetochore proteins, and inhibition of the motor protein dynein slows chromosome separation at anaphase. Like many phenomena that we analyze in living systems, the answer is not a simple either–or choice; both mechanisms are probably at work.

In metaphase, chromosomes align at the equator

The alignment of the chromosomes in the center of the cell signals the third stage of mitosis, **metaphase.** When viewed with a light microscope, the chromosomes appear to array themselves in a circle along the inner circumference of the cell, just as the equator girdles the Earth (figure 10.12). An imaginary plane perpendicular to the axis of the spindle that passes through this circle is called the *metaphase plate*. The metaphase plate is not an actual structure, but rather an indication of the future axis of cell division.

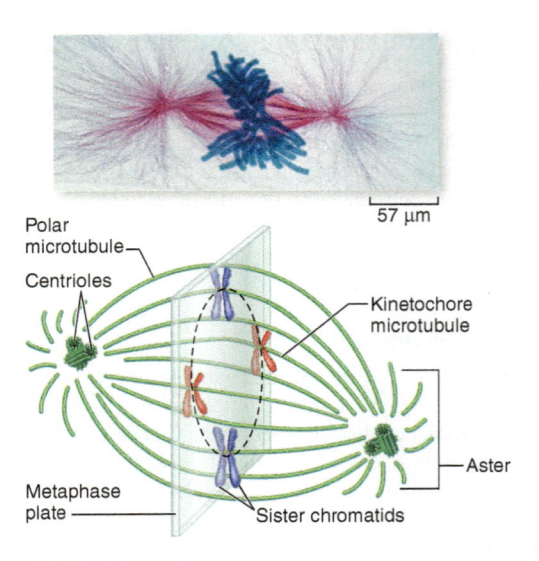

Figure 10.12 Metaphase. In metaphase, the chromosomes are arrayed at the midpoint of the cell. The imaginary plane through the equator of the cell is called the metaphase plate. As the spindle itself is a three-dimensional structure, the chromosomes are arrayed in a rough circle on the metaphase plate.

Positioned by the microtubules attached to the kinetochores of their centromeres, all of the chromosomes line up on the metaphase plate. At this point their centromeres are neatly arrayed in a circle, equidistant from the two poles of the cell, with microtubules extending back toward the opposite poles of the cell. The cell is prepared to properly separate sister chromatids, such that each daughter cell will receive a complete set of chromosomes. Thus metaphase is really a transitional phase in which all the preparations are checked before the action continues.

At anaphase, the chromatids separate

Of all the stages of mitosis, shown in figure 10.11, **anaphase** is the shortest and the most amazing to watch. It begins when the proteins holding sister chromatids together at the centromere are removed. Up to this point in mitosis, sister chromatids have been held together by cohesin proteins concentrated at the centromere, as described in section 10.2. The key event in anaphase, then, is the simultaneous removal of these proteins from all of the chromosomes. The control and details of this process are discussed in section 10.6 in the context of control of the entire cell cycle.

Freed from each other, the sister chromatids are pulled rapidly toward the poles to which their kinetochores are attached. In the process, two forms of movement take place simultaneously, each driven by microtubules. These movements are often called anaphase A and anaphase B to distinguish them.

First, during anaphase A, the *kinetochores are pulled toward the poles* as the microtubules that connect them to the poles shorten. This shortening process is not a contraction; the microtubules do not get any thicker. Instead, tubulin subunits are removed from the kinetochore ends of the microtubules. As more subunits are removed, the chromatid-bearing microtubules are progressively disassembled, and the chromatids are pulled ever closer to the poles of the cell.

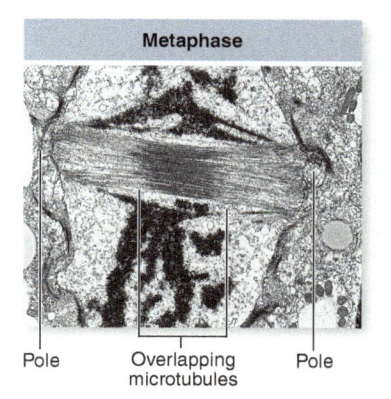

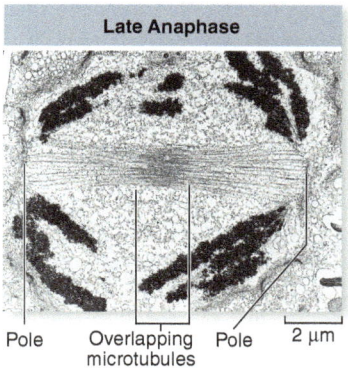

Figure 10.13 Microtubules slide past each other as the chromosomes separate. In these electron micrographs of dividing diatoms, the overlap of the microtubules lessens markedly during spindle elongation as the cell passes from metaphase to anaphase. During anaphase, B the poles move farther apart as the chromosomes move toward the poles.

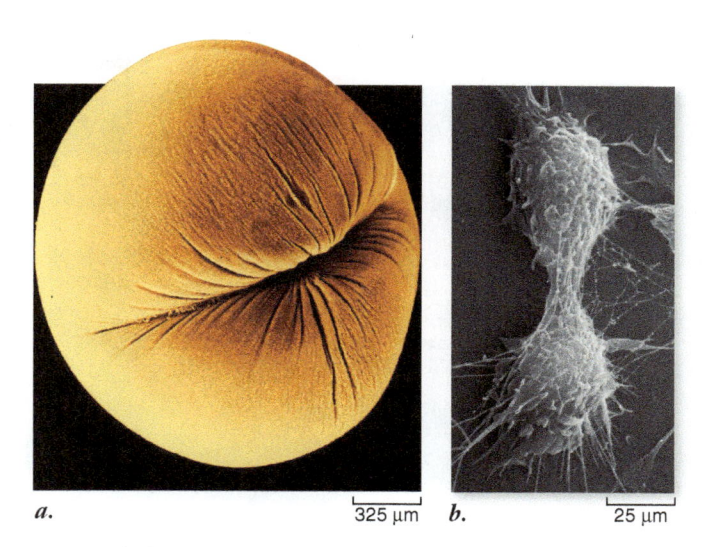

Figure 10.14 Cytokinesis in animal cells. *a.* A cleavage furrow forms around a dividing frog egg. *b.* The completion of cytokinesis in an animal cell. The two daughter cells are still joined by a thin band of cytoplasm occupied largely by microtubules.

Second, during anaphase B, the *poles move apart* as microtubular spindle fibers physically anchored to opposite poles slide past each other, away from the center of the cell (figure 10.13). Because another group of microtubules attach the chromosomes to the poles, the chromosomes move apart, too. If a flexible membrane surrounds the cell, it becomes visibly elongated.

When the sister chromatids separate in anaphase, the accurate partitioning of the replicated genome—the essential element of mitosis—is complete.

During telophase, the nucleus re-forms

In **telophase,** the spindle apparatus disassembles as the microtubules are broken down into tubulin monomers that can be used to construct the cytoskeletons of the daughter cells. A nuclear envelope forms around each set of sister chromatids, which can now be called chromosomes because they are no longer attached at the centromere. The chromosomes soon begin to uncoil into the more extended form that permits gene expression. One of the early group of genes expressed after mitosis is complete are the rRNA genes, resulting in the reappearance of the nucleolus.

Telophase can be viewed as a reversal of the process of prophase, bringing the cell back to the state of interphase. Mitosis is complete at the end of telophase. The eukaryotic cell has partitioned its replicated genome into two new nuclei positioned at opposite ends of the cell. Other cytoplasmic organelles, including mitochondria and chloroplasts (if present), were reassorted to areas that will separate and become the daughter cells.

Cell division is still not complete at the end of mitosis, however, because the division of the cell body proper has not yet begun. The phase of the cell cycle when the cell actually divides is called **cytokinesis.** It generally involves the cleavage of the cell into roughly equal halves.

In animal cells, a belt of actin pinches off the daughter cells

In animal cells and the cells of all other eukaryotes that lack cell walls, cytokinesis is achieved by means of a constricting belt of actin filaments. As these filaments slide past one another, the diameter of the belt decreases, pinching the cell and creating a **cleavage furrow** around the cell's circumference (figure 10.14*a*).

As constriction proceeds, the furrow deepens until it eventually slices all the way into the center of the cell. At this point, the cell is divided in two (figure 10.14*b*).

In plant cells, a cell plate divides the daughter cells

Plant cell walls are far too rigid to be squeezed in two by actin filaments. Instead, these cells assemble membrane components in their interior, at right angles to the spindle apparatus. This expanding membrane partition, called a **cell plate,** continues to grow outward until it reaches the interior surface of the plasma membrane and fuses with it, effectively dividing the cell in two (figure 10.15). Cellulose is then laid down on the new membranes, creating two new cell walls. The space between the daughter cells becomes impregnated with pectins and is called a *middle lamella*.

In fungi and some protists, daughter nuclei are separated during cytokinesis

In most fungi and some groups of protists, the nuclear membrane does not dissolve, and as a result, all the events of mitosis occur entirely *within* the nucleus. Only after mitosis is complete in these

Figure 10.15
Cytokinesis in plant cells. In this photomicrograph and companion drawing, a cell plate is forming between daughter nuclei. The cell plate forms from the fusion of Golgi-derived vesicles. Once the plate is complete, there will be two cells.

19,000×

Vesicles containing membrane components fusing to form cell plate

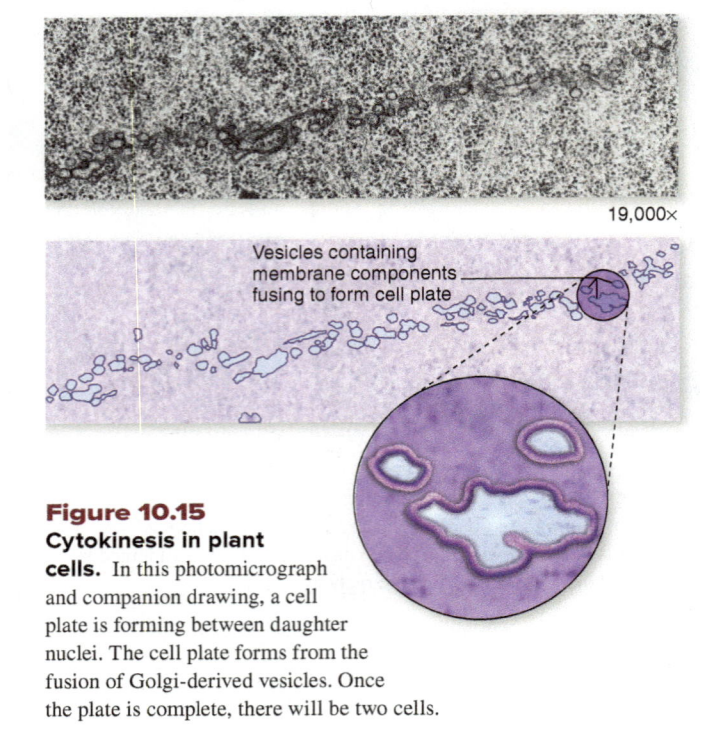

organisms does the nucleus divide into two daughter nuclei; then, during cytokinesis, one nucleus goes to each daughter cell. This separate nuclear division phase of the cell cycle does not occur in plants, animals, or most protists.

After cytokinesis in any eukaryotic cell, the two daughter cells contain all the components of a complete cell. Whereas mitosis ensures that both daughter cells contain a full complement of chromosomes, no similar mechanism ensures that organelles such as mitochondria and chloroplasts are distributed equally between the daughter cells. But as long as at least one of each organelle is present in each cell, the organelles can replicate to reach the number appropriate for that cell.

Learning Outcomes Review 10.5

Mitosis is divided into phases: prophase, prometaphase, metaphase, anaphase, and telophase. The early phases involve restructuring the cell to create the microtubule spindle that pulls chromosomes to the equator of the cell in metaphase. Chromatids for each chromosome remain attached at the centromere by cohesin proteins. Chromatids are then pulled to opposite poles during anaphase when cohesin proteins are destroyed. The nucleus is re-formed in telophase, and cytokinesis then divides the cell cytoplasm and organelles. In animal cells, actin pinches the cell in two; in plant cells, a cell plate forms in the middle of the dividing cell.

- *What would happen to a chromosome that loses cohesin protein between sister chromatids before metaphase?*

10.6 Control of the Cell Cycle

Learning Outcomes
1. Distinguish the role of checkpoints in the control of the cell cycle.
2. Characterize the role of the anaphase-promoting complex/cyclosome in mitosis.
3. Describe cancer in terms of cell-cycle control.

Our knowledge of how the cell cycle is controlled, although still incomplete, has grown enormously in the past 45 years. Our current view integrates two basic concepts. First, the cell cycle has two irreversible points: the replication of genetic material and the separation of the sister chromatids. Second, the cell cycle can be put on hold at specific points called *checkpoints*. At any of these checkpoints, the process is checked for accuracy and can be halted if there are errors. This leads to extremely high fidelity overall for the entire process. The checkpoint organization also allows the cell cycle to respond to both the internal state of the cell, including nutritional state and integrity of genetic material, and to signals from the environment, which are integrated at major checkpoints.

Research uncovered cell cycle control factors

The history of investigation into control of the cell cycle is instructive in two ways. First, it allows us to place modern observations into context; second, we can see how biologists using very different approaches often end up at the same place. The following brief history introduces three observations and then shows how they can be integrated into a single mechanism.

Discovery of MPF

Research on the activation of frog oocytes led to the discovery of a substance that was first called *maturation-promoting factor (MPF)*. Frog oocytes, which go on to become egg cells, become arrested near the end of their development at the G_2 stage before meiosis I, which is the division leading to the production of gametes (see chapter 11). They remain in this arrested state and await hormonal signaling to complete this division process.

Cytoplasm taken from a variety of actively dividing cells could prematurely induce cell division when injected into oocytes (figure 10.16). These experiments indicated the presence of a positive regulator of cell-cycle progression in the cytoplasm of dividing cells: MPF. These experiments also fit well with cell fusion experiments done with mitotic and interphase cells that also indicated a cytoplasmic positive regulator that could induce mitosis (figure 10.16).

Further studies highlighted two key aspects of MPF. First, MPF activity varied during the cell cycle: low in early G_2, rising throughout this phase, and then peaking in mitosis (figure 10.17). Second, the enzymatic activity of MPF involved the phosphorylation of proteins. This second point is not surprising given the importance of phosphorylation as a reversible switch on

SCIENTIFIC THINKING

Hypothesis: There are positive regulators of cell division.

Prediction: Frog oocytes are arrested in G_2 of meiosis I. They can be induced to mature (undergo meiosis) by progesterone treatment. If maturing oocytes contain a positive regulator of cell division, injection of cytoplasm should induce an immature oocyte to undergo meiosis.

Test: Oocytes are induced with progesterone, then cytoplasm from these maturing cells is injected into immature oocytes.

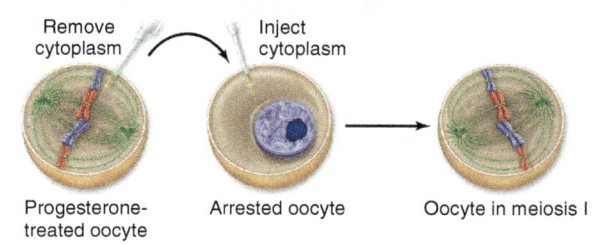

Result: Injected oocytes progress from G_2 into meiosis I.
Conclusion: The progesterone treatment causes production of a positive regulator of maturation: maturation promoting factor (MPF).

Prediction: If mitosis is driven by positive regulators, then cytoplasm from a mitotic cell should cause a G_1 cell to enter mitosis.
Test: M phase cells are fused with G_1 phase cells, then the nucleus from the G_1 phase cell is monitored microscopically.

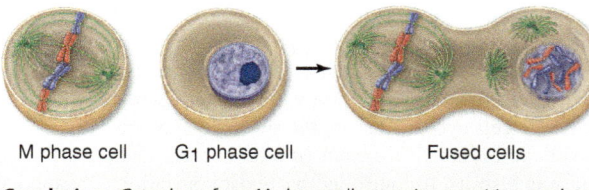

Conclusion: Cytoplasm from M phase cells contains a positive regulator that causes a cell to enter mitosis.

Further Experiments: How can both of these experiments be rationalized? What would be the next step in characterizing these factors?

Figure 10.16 Discovery of positive regulator of cell division.

the activity of proteins (see chapter 9). The first observation indicated that MPF itself was not always active, but rather was being regulated with the cell cycle, and the second showed the possible enzymatic activity of MPF.

Discovery of cyclins

Other researchers examined proteins produced during the early divisions in sea urchin embryos. They identified proteins that were produced in synchrony with the cell cycle, and named them **cyclins** (figure 10.17). These observations were extended in another marine invertebrate, the surf clam. Two forms of cyclin were found that cycled at slightly different times, reaching peaks at the G_1/S and G_2/M boundaries. Despite much effort, no identified enzymatic activity was associated with these proteins. Their hallmark was the timing of their production and not any intrinsic activity.

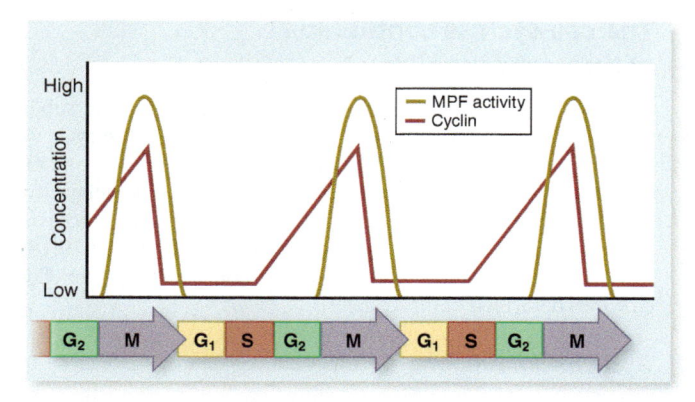

Figure 10.17 Correlation of MPF activity, amount of cyclin protein, and stages of the cell cycle. Cyclin concentration and MPF activity are shown plotted versus stage of the cell cycle. MPF activity changes in a repeating pattern through the cell cycle. This also correlates with the level of mitotic cyclin in the cell, which shows a similar pattern. The reason for this correlation is that cyclin is actually one component of MPF, the other being a cyclin-dependent kinase (Cdk). Together these act as a positive regulator of cell division.

Genetic analysis of the cell cycle

Geneticists using two different yeasts, budding yeast and fission yeast, as model systems set out to determine the genes necessary for control of the cell cycle. By isolating mutants that were halted during division, they identified genes that were necessary for cell cycle progression. These studies indicated that in yeast, there were two critical control points: the commitment to DNA synthesis, called START, as it meant committing to divide, and the commitment to mitosis. One particular gene, named *cdc2*, from fission yeast, was shown to be critical for passing both of these boundaries.

MPF is cyclin plus cdc2

All of these findings came together in an elegant fashion with the following observations. First, the protein encoded by the *cdc2* gene was shown to be a protein kinase. Second, the purification and identification of MPF showed that it was composed of both a cyclin component and a kinase component. Last, the kinase itself was the cdc2 protein!

The cdc2 protein was the first identified **cyclin-dependent kinase (Cdk)**—that is, a protein kinase enzyme that is only active when complexed with cyclin. This finding led to the renaming of MPF as *mitosis*-promoting factor, as its role was clearly more general than simply promoting the maturation of frog oocytes.

These Cdk enzymes are the key positive drivers of the cell division cycle. They are often called the engine that drives cell division. The control of the cell cycle in higher eukaryotes is much more complex than the simple single-engine cycle of yeast, but the yeast model remains a useful framework for understanding more complex regulation. The discovery of Cdks and their role in the cell cycle is an excellent example of the progressive nature of science.

The cell cycle is controlled at three checkpoints

Although we divide the cell cycle into phases and further subdivide mitosis, cells actively control only three points in the cycle we call checkpoints: G_1/S, G_2/M, and late metaphase (the spindle checkpoint). These checkpoints allow the cycle to be delayed or halted when necessary. The cell uses these checkpoints to both assess its internal state and integrate external signals (figure 10.18). Passage through these checkpoints is controlled by the Cdk enzymes described in this chapter.

G_1/S checkpoint

The **G_1/S checkpoint** is the primary point at which the cell "decides" whether or not to divide. This checkpoint is therefore the primary point at which external signals can influence events of the cycle. It is the phase during which growth factors (discussed later in this section) affect the cycle and also the phase that links cell division to cell growth and nutrition.

In yeast systems, where the majority of the genetic analysis of the cell cycle has been performed, this checkpoint is called START. In animals, it is called the restriction point (R point). In all systems, once a cell has made this irreversible commitment to replicate its genome, it has committed to divide. Damage to DNA can halt the cycle at this point, as can starvation conditions or lack of growth factors.

G_2/M checkpoint

The **G_2/M checkpoint** has received a large amount of attention because of its complexity and its importance as the stimulus for the events of mitosis. Historically, Cdks active at this checkpoint were first identified as MPFs, a term that has now evolved into **M phase-promoting factor (MPF)**.

Passage through this checkpoint represents the commitment to mitosis. This checkpoint assesses the success of DNA replication and can stall the cycle if DNA has not been accurately replicated. DNA-damaging agents result in arrest at this checkpoint as well as at the G_1/S checkpoint.

Spindle checkpoint

The **spindle checkpoint** ensures that all of the chromosomes are attached to the spindle in preparation for anaphase. The second irreversible step in the cycle is the separation of chromosomes during anaphase, and therefore it is critical that they are properly arrayed at the metaphase plate.

Cyclin-dependent kinases drive the cell cycle

The primary molecular mechanism of cell cycle control is phosphorylation, which you may recall is the addition of a phosphate group to the amino acids serine, threonine, and tyrosine in proteins (see chapter 9). The enzymes that accomplish this phosphorylation are the Cdks (figure 10.19).

The action of Cdks

The first important cell cycle kinase was identified in fission yeast and named Cdc2 (now also called Cdk1). In yeast, this Cdk can partner with different cyclins at different points in the cell cycle (figure 10.20).

Even in the simplified cycle of the yeasts, we are left with the important question of what controls the activity of the Cdks during the cycle. For many years, a common view was that cyclins drove the cell cycle—that is, the periodic synthesis and destruction of cyclins acted as a clock. It later became clear that the Cdc2 kinase is also itself controlled by phosphorylation: Phosphorylation at one site activates Cdc2, and phosphorylation at another site inactivates it (see figure 10.19). Full activation of the Cdc2 kinase requires complexing with a cyclin and the appropriate pattern of phosphorylation.

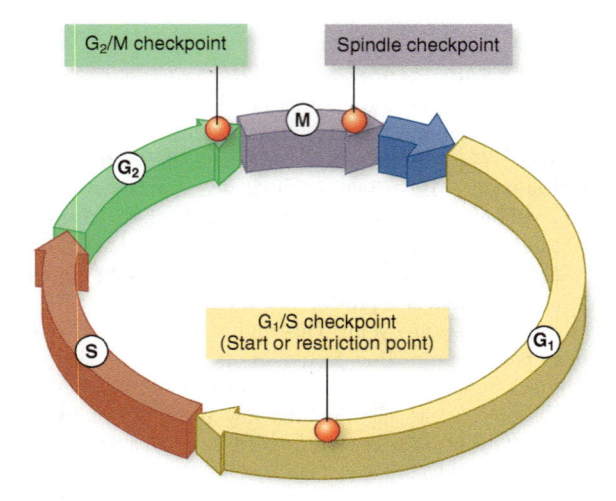

Figure 10.18 Control of the cell cycle. Cells use a centralized control system to check whether proper conditions have been achieved before passing three key checkpoints in the cell cycle.

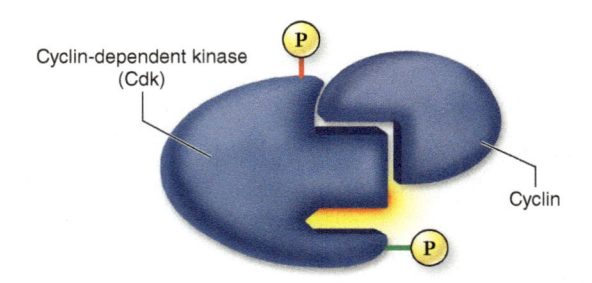

Figure 10.19 Cdk enzyme forms a complex with cyclin. Cdk is a protein kinase that activates numerous cell proteins by phosphorylating them. Cyclin is a regulatory protein required to activate Cdk. This complex is also called mitosis-promoting factor (MPF). The activity of Cdk is also controlled by the pattern of phosphorylation: Phosphorylation at one site (represented by the *red* site) inactivates the Cdk, and phosphorylation at another site (represented by the *green* site) activates the Cdk.

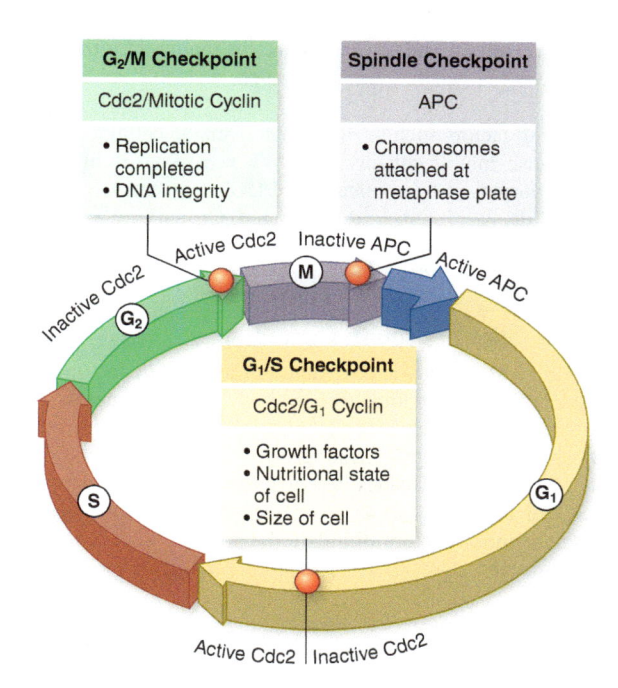

Figure 10.20 Checkpoints of the yeast cell cycle. The simplest cell cycle that has been studied in detail is the fission yeast. This is controlled by three main checkpoints and a single Cdk enzyme, called Cdc2. The Cdc2 enzyme partners with different cyclins to control the G_1/S and G_2/M checkpoints. The spindle checkpoint is controlled by the anaphase-promoting complex (APC).

As the G_1/S checkpoint is approached, the triggering signal in yeast appears to be the accumulation of G_1 cyclins. These form a complex with Cdc2 to create the active G_1/S Cdk, which phosphorylates a number of targets that bring about the increased enzyme activity for DNA replication.

The action of MPF

MPF and its role at the G_2/M checkpoint has been extensively analyzed in a number of different experimental systems. The control of MPF is sensitive to agents that disrupt or delay replication and to agents that damage DNA. It was once thought that MPF was controlled solely by the level of the M phase-specific cyclins, but it has now become clear that this is not the case.

Although M phase cyclin is necessary for MPF function, activity is controlled by inhibitory phosphorylation of the kinase component, Cdc2. The critical signal in this process is the removal of the inhibitory phosphates by a protein, phosphatase. This action forms a molecular switch based on positive feedback because the active MPF further activates its own activating phosphatase.

The checkpoint assesses the balance of the kinase that adds inhibitory phosphates with the phosphatase that removes them. Damage to DNA acts through a complex pathway that includes damage sensing and a response to tip the balance toward the inhibitory phosphorylation of MPF. Later in this section, we describe how some cancers overcome this inhibition.

The anaphase-promoting complex

The molecular details of the sensing system at the spindle checkpoint are not clear. The presence of all chromosomes at the metaphase plate and the tension on the microtubules between opposite poles are both important. The signal is transmitted through the **anaphase-promoting complex,** also called the *cyclosome (APC/C)*.

The function of the APC/C is to trigger anaphase itself. As described in section 10.5, the sister chromatids at metaphase are still held together by the protein complex cohesin. The APC does not act directly on cohesin, but rather acts by marking a protein called *securin* for destruction. The securin protein acts as an inhibitor of another protease called *separase* that is specific for one component of the cohesin complex. Once inhibition is lifted, separase destroys cohesin.

This process has been analyzed in detail in budding yeast, where it has been shown that the separase enzyme specifically degrades a component of cohesin called Scc1. This leads to the release of the sister chromatids and results in their sudden movement toward opposite poles during anaphase.

In vertebrates, most cohesin is removed from the sister chromatids during chromosome condensation, possibly with cohesin being replaced by condensin. At metaphase, the majority of the cohesin that remains on vertebrate chromatids is concentrated at the centromere (see figure 10.10). The destruction of this cohesin explains the anaphase movement of chromosomes and the apparent "division" of the centromeres.

The APC/C has two main roles in mitosis: it activates the protease that removes the cohesins holding sister chromatids together, and it is necessary for the destruction of mitotic cyclins to drive the cell out of mitosis. The APC/C complex marks proteins for destruction by the proteosome, the organelle responsible for the controlled degradation of proteins (see chapter 16). The signal to degrade a protein is the addition of a molecule called *ubiquitin*, and the APC/C acts as a ubiquitin ligase. As we learn more about the APC/C and its functions, it is clear that the control of its activity is a key regulator of the cell cycle.

In multicellular eukaryotes, many Cdks and external signals act on the cell cycle

The major difference between more complex animals and single-celled eukaryotes such as fungi and protists is twofold: First, multiple Cdks control the cycle as opposed to the single Cdk in yeasts; and second, animal cells respond to a greater variety of external signals than do yeasts, which primarily respond to signals necessary for mating.

In higher eukaryotes there are more Cdk enzymes and more cyclins that can partner with these multiple Cdks, but their basic role is the same as in the yeast cycle. A more complex cell cycle is shown in figure 10.21. These more complex controls allow the integration of more input into control of the cycle. With the evolution of more complex forms of organization (tissues, organs, and organ systems), more complex forms of cell-cycle control evolved as well.

A multicellular body's organization cannot be maintained without severely limiting cell proliferation—so that only certain cells divide, and only at appropriate times. The way cells inhibit

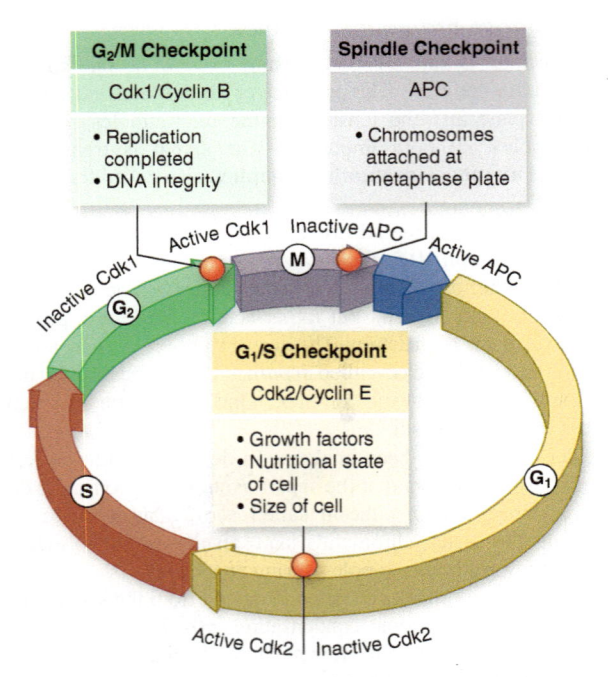

Figure 10.21 Checkpoints of the mammalian cell cycle. The more complex mammalian cell cycle is shown. This cycle is still controlled through three main checkpoints. These integrate internal and external signals to control progress through the cycle. These inputs control the state of two different Cdk–cyclin complexes and the anaphase-promoting complex (APC).

individual growth of other cells is apparent in mammalian cells growing in tissue culture: A single layer of cells expands over a culture plate until the growing border of cells comes into contact with neighboring cells, and then the cells stop dividing. If a sector of cells is cleared away, neighboring cells rapidly refill that sector and then stop dividing again on cell contact.

How are cells able to sense the density of the cell culture around them? When cells come in contact with one another, receptor proteins in the plasma membrane activate a signal transduction pathway that acts to inhibit Cdk action. This prevents entry into the cell cycle.

Growth factors and the cell cycle

Growth factors act by triggering intracellular signaling systems. Fibroblasts, for example, possess numerous receptors on their plasma membranes for one of the first growth factors to be identified, **platelet-derived growth factor (PDGF)**. The PDGF receptor is a receptor tyrosine kinase (RTK) that initiates a MAP kinase cascade to stimulate cell division (discussed in chapter 9).

PDGF was discovered when investigators found that fibroblasts would grow and divide in tissue culture only if the growth medium contained blood serum. Serum is the liquid that remains in blood after clotting; blood plasma, the liquid from which cells have been removed without clotting, would not work. The researchers hypothesized that platelets in the blood clots were releasing into the serum one or more factors required for fibroblast growth. Eventually, they isolated such a factor and named it PDGF.

Growth factors such as PDGF can override cellular controls that otherwise inhibit cell division. When a tissue is injured, a blood clot forms, and the release of PDGF triggers neighboring cells to divide, helping to heal the wound. Only a tiny amount of PDGF (approximately 10^{-10} M) is required to stimulate cell division in cells with PDGF receptors.

Characteristics of growth factors

Over 50 different proteins that function as growth factors have been isolated, and more undoubtedly exist. A specific cell surface receptor recognizes each growth factor, its binding site fitting that growth factor precisely. These growth factor receptors often initiate MAP kinase cascades in which the final kinase enters the nucleus and activates transcription factors by phosphorylation. These transcription factors stimulate the production of G_1 cyclins and the proteins that are necessary for cell-cycle progression (figure 10.22).

The cellular selectivity of a particular growth factor depends on which target cells bear its unique receptor. Some growth factors, such as PDGF and epidermal growth factor (EGF), affect a broad range of cell types, but others affect only specific types. For example, nerve growth factor (NGF) promotes the growth of certain classes of neurons, and erythropoietin triggers cell division in red blood cell precursors. Most animal cells need a combination of several different growth factors to overcome the various controls that inhibit cell division.

The G_0 phase

If cells are deprived of appropriate growth factors, they stop at the G_1 checkpoint of the cell cycle. With their growth and division arrested, they remain in this dormant G_0 phase.

The ability to enter G_0 accounts for the incredible diversity seen in the length of the cell cycle in different tissues. Epithelial cells lining the human gut divide more than twice a day, constantly renewing this lining. By contrast, liver cells divide only once every year or two, spending most of their time in the G_0 phase. Mature neurons and muscle cells usually never leave G_0.

Cancer is a failure of cell-cycle control

The unrestrained, uncontrolled growth of cells in humans leads to the disease called **cancer.** Cancer is essentially a disease of cell division—a failure of cell division control.

The *p53* gene

One of the critical players in this control system has been identified. Officially dubbed *p53,* this gene plays a key role in the G_1 checkpoint of cell division.

The gene's product, the p53 protein, monitors the integrity of DNA, checking that it is undamaged. If the p53 protein detects damaged DNA, it halts cell division and stimulates the activity of special enzymes to repair the damage. Once the DNA has been repaired, p53 allows cell division to continue. In cases

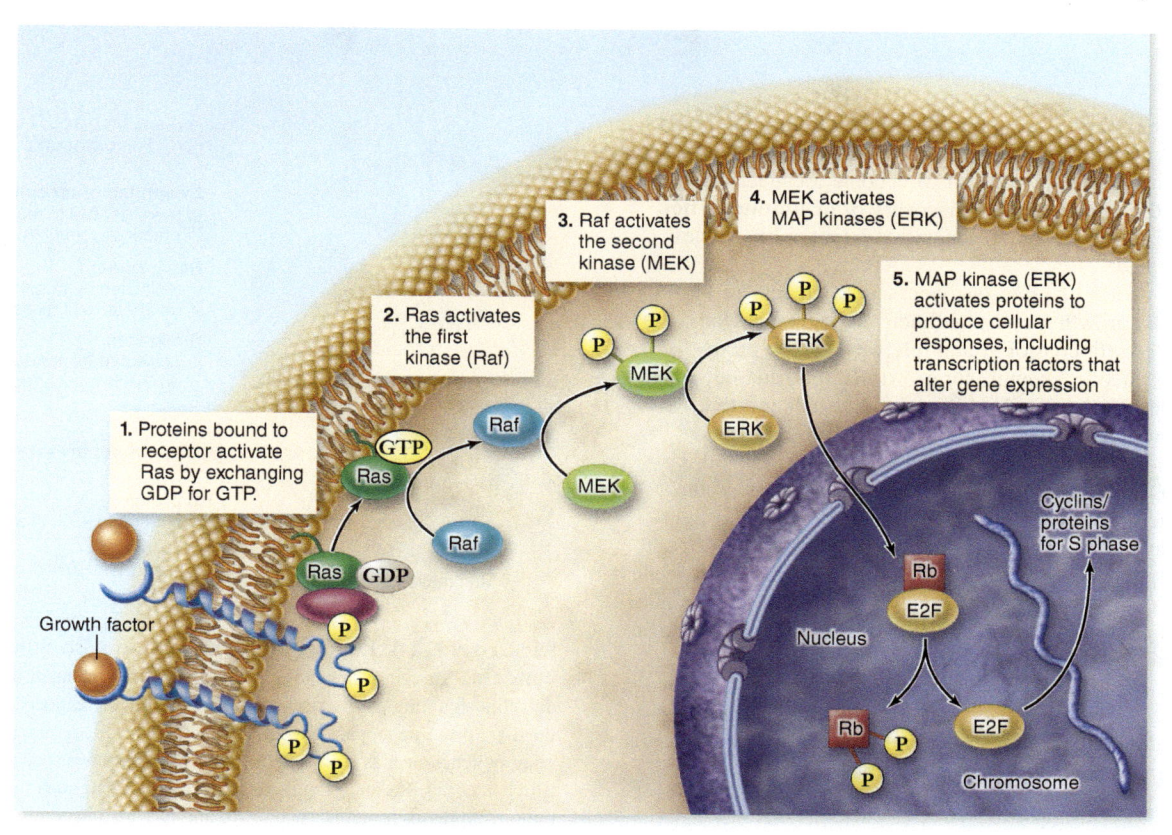

Figure 10.22 The cell proliferation-signaling pathway. Binding of a growth factor sets in motion a MAP kinase intracellular signaling pathway (described in chapter 9), which activates nuclear regulatory proteins that trigger cell division. In this example, when the nuclear retinoblastoma protein (Rb) is phosphorylated, another nuclear protein (the transcription factor E2F) is released and is then able to stimulate the production of cyclin and other proteins necessary for S phase.

where the DNA damage is irreparable, p53 then directs the cell to kill itself.

By halting division in damaged cells, the *p53* gene prevents the development of many mutated cells, and it is therefore considered a **tumor-suppressor gene** although its activities are not limited to cancer prevention. Scientists have found that *p53* is entirely absent or damaged beyond use in the majority of cancerous cells they have examined. It is precisely because *p53* is nonfunctional that cancer cells are able to repeatedly undergo cell division without being halted at the G_1 checkpoint (figure 10.23).

Proto-oncogenes

The disease we call cancer is actually many different diseases, depending on the tissue affected. The common theme in all

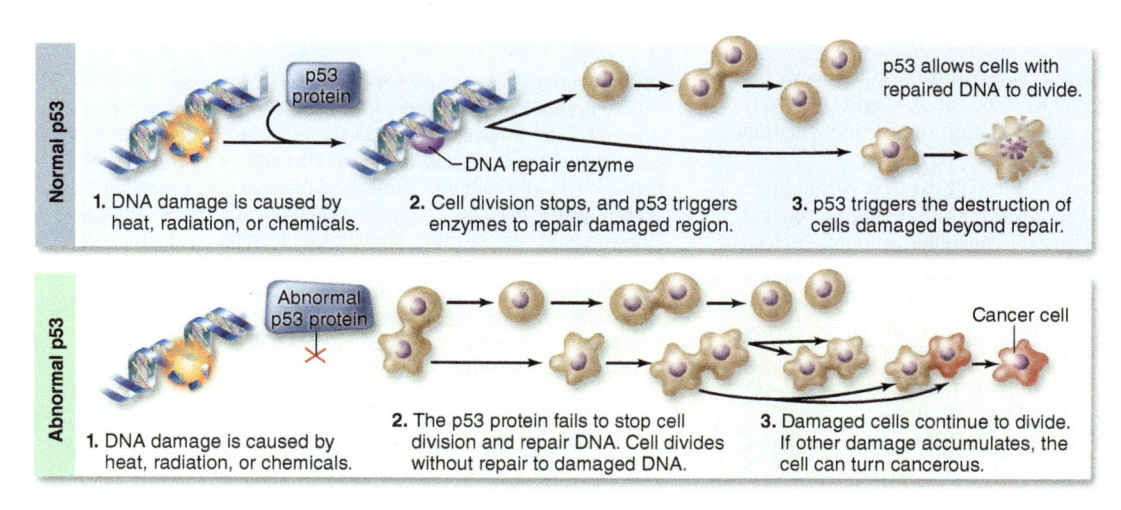

**Figure 10.23
Cell division, cancer, and p53 protein.**
Normal p53 protein monitors DNA, destroying cells that have irreparable damage to their DNA. Abnormal p53 protein fails to stop cell division and repair DNA. As damaged cells proliferate, cancer develops.

chapter **10** *How Cells Divide* 203

cases is the loss of control over the cell cycle. Research has identified numerous so-called **oncogenes,** genes that can, when introduced into a cell, cause it to become a cancer cell. This identification then led to the discovery of **proto-oncogenes,** which are normal cellular genes that become oncogenes when mutated.

The action of proto-oncogenes is often related to signaling by growth factors, and their mutation can lead to loss of growth control in multiple ways. Some proto-oncogenes encode receptors for growth factors, and others encode proteins involved in signal transduction that act after growth factor receptors. If a receptor for a growth factor becomes mutated such that it is permanently "on," the cell is no longer dependent on the presence of the growth factor for cell division. This is analogous to a light switch that is stuck on: The light will always be on. PDGF and EGF receptors both fall into the category of proto-oncogenes. Only one copy of a proto-oncogene needs to undergo this mutation for uncontrolled division to take place; thus, this change acts like a dominant mutation.

The number of proto-oncogenes identified has grown to more than 50 over the years. This line of research connects our understanding of cancer with our understanding of the molecular mechanisms governing cell-cycle control.

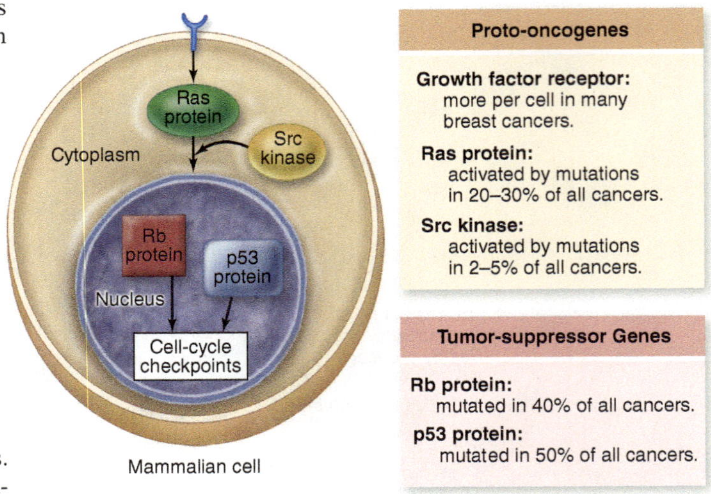

Tumor-suppressor genes

After the discovery of proto-oncogenes, a second category of genes related to cancer was identified: the tumor-suppressor genes. We mentioned earlier in this section that the *p53* gene acts as a tumor-suppressor gene, and a number of other such genes exist.

Both copies of a tumor-suppressor gene must lose function for the cancerous phenotype to develop, in contrast to the mutations in proto-oncogenes. Put another way, the proto-oncogenes act in a dominant fashion, and tumor suppressors act in a recessive fashion.

The first tumor suppressor identified was the **retinoblastoma susceptibility gene** *(Rb),* which predisposes individuals for a rare form of cancer that affects the retina of the eye. Despite the fact that a cell heterozygous for a mutant *Rb* allele is normal, it is inherited as a dominant in families. The reason is that inheriting a single mutant copy of *Rb* means the individual has only one "good" copy left, and during the hundreds of thousands of divisions that occur to produce the retina, any error that damages the remaining good copy leads to a cancerous cell. A single cancerous cell in the retina then leads to the formation of a retinoblastoma tumor.

The role of the Rb protein in the cell cycle is to integrate signals from growth factors. The Rb protein is called a "pocket protein" because it has binding pockets for other proteins. Its role is therefore to bind important regulatory proteins and prevent them from stimulating the production of the necessary cell-cycle proteins, such as cyclins or Cdks (see figure 10.21) discussed earlier in this section.

The binding of Rb to other proteins is controlled by phosphorylation: When it is dephosphorylated, it can bind a variety of regulatory proteins, but loses this capacity when phosphorylated. The action of growth factors results in the phosphorylation of Rb protein by a Cdk. This then brings us full circle, because the phosphorylation of Rb releases previously bound regulatory proteins, resulting in the production of S phase cyclins that are necessary for the cell to pass the G_1/S boundary and begin chromosome replication.

Figure 10.24 summarizes the types of genes that can cause cancer when mutated.

Proto-oncogenes

Growth factor receptor: more per cell in many breast cancers.

Ras protein: activated by mutations in 20–30% of all cancers.

Src kinase: activated by mutations in 2–5% of all cancers.

Tumor-suppressor Genes

Rb protein: mutated in 40% of all cancers.

p53 protein: mutated in 50% of all cancers.

Figure 10.24 Key proteins associated with human cancers. Mutations in genes encoding key components of the cell division–signaling pathway are responsible for many cancers. Among them are proto-oncogenes encoding growth factor receptors, protein relay switches such as Ras protein, and kinase enzymes such as Src, which act after Ras and growth factor receptors. Mutations that disrupt tumor-suppressor proteins, such as Rb and p53, also foster cancer development.

Learning Outcomes Review 10.6

Cyclin proteins are produced in synchrony with the cell cycle. These proteins complex with cyclin-dependent kinases to drive the cell cycle. Three checkpoints exist in the cell cycle: the G_1/S checkpoint, the G_2/M checkpoint, and the spindle checkpoint. The cell cycle can be halted at these checkpoints if the process is not accurate. The anaphase-promoting complex/cyclosome (APC/C) triggers anaphase by lifting inhibition on a protease that removes cohesin holding chromatids together. The loss of cell cycle control leads to cancer, which can occur by a combination of two basic mechanisms: proto-oncogenes that gain function to become oncogenes, and tumor-suppressor genes that lose function and allow cell proliferation.

■ *How can you distinguish between a tumor-suppressor gene and a proto-oncogene?*

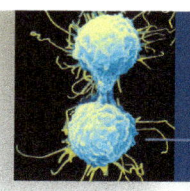

Chapter Review

10.1 Bacterial Cell Division

Binary fission is a simple form of cell division.
Prokaryotic cell division is clonal, resulting in two identical cells. Bacterial DNA replication and partitioning of the chromosome are concerted processes.

Bacterial cells control chromosome separation and septum formation.
DNA replication begins at a specific point, the origin, and proceeds bidirectionally to a specific termination site. Newly replicated chromosomes are segregated to opposite poles at the same time as they are replicated. New cells are separated by septation, which involves insertion of new cell membrane and other cellular materials at the midpoint of the cell. A ring of FtsZ and proteins embedded in the cell membrane expands radially inward, pinching the cell into two new cells.

10.2 Eukaryotic Chromosomes

Chromosome number varies among species.
The gain or loss of chromosomes is usually lethal.

Eukaryotic chromosomes exhibit complex structure.
Chromosomes are composed of chromatin, a complex of DNA, and protein. Heterochromatin is not expressed and euchromatin is expressed. The DNA of a single chromosome is a very long, double-stranded fiber. The DNA is wrapped around a core of eight histones to form a nucleosome, which can be further coiled into a 30-nm fiber in interphase cells. During mitosis, chromosomes are further condensed by arranging coiled 30-nm fibers radially around a protein scaffold.

Newly replicated chromosomes remain attached at a constricted area called a centromere, consisting of repeated DNA sequences. After replication, a chromosome consists of two sister chromatids held together at the centromere by a complex of proteins called cohesins (figure 10.7).

10.3 Overview of the Eukaryotic Cell Cycle (figure 10.8)

The cell cycle is divided into five phases.
The phases of the cell cycle are gap 1 (G_1), synthesis (S), gap 2 (G_2), mitosis, and cytokinesis (C). G_1, S, and G_2 are collectively called interphase, and mitosis and cytokinesis together are called M phase.

The duration of the cell cycle varies depending on cell type.
The length of a cell cycle varies with age, cell type, and species. Cells can exit G_1 and enter a nondividing phase called G_0; the G_0 phase can be temporary or permanent.

10.4 Interphase: Preparation for Mitosis

G_1, S, and G_2 are the three subphases of interphase. G_1 is the primary growth phase; during S phase, DNA synthesis occurs. G_2 phase occurs after S phase and before mitosis.

The centromere binds proteins assembled into a disklike structure called a kinetochore where microtubules attach during mitosis. The centromeric DNA is replicated, but the two DNA strands are held together by cohesin proteins.

10.5 M Phase: Chromosome Segregation and the Division of Cytoplasmic Contents (figure 10.11)

During prophase, the mitotic apparatus forms.
In prophase, chromosomes condense, the spindle is formed, and the nuclear envelope disintegrates. In animals cells, centriole pairs separate and migrate to opposite ends of the cell, establishing the axis of nuclear division.

During prometaphase, chromosomes attach to the spindle.

In metaphase, chromosomes align at the equator.
Chromatids of each chromosome are connected to opposite poles by kinetochore microtubules. They are held at the equator of the cell by the tension of being pulled toward opposite poles.

At anaphase, the chromatids separate.
At this point, cohesin proteins holding sister chromatids together at the centromeres are destroyed, and the chromatids are pulled to opposite poles. This movement is called anaphase A, and the movement of poles farther apart is called anaphase B.

During telophase, the nucleus re-forms.
Telophase reverses the events of prophase and prepares the cell for cytokinesis.

In animal cells, a belt of actin pinches off the daughter cells.
A contractile ring of actin under the membrane contracts during cytokinesis.

In plant cells, a cell plate divides the daughter cells.
Fusion of vesicles produces a new membrane in the middle of the cell to produce the cell plate.

In fungi and some protists, daughter nuclei are separated during cytokinesis.

10.6 Control of the Cell Cycle (figure 10.18)

Research uncovered cell-cycle control factors.
Experiments showed that there are positive regulators of mitosis, and that there are proteins produced in synchrony with the cell cycle (cyclins). The positive regulators are cyclin-dependent kinases (Cdks). Cdks are complexes of a kinase and a regulatory molecule called cyclin. They phosphorylate proteins to drive the cell cycle.

The cell cycle is controlled at three checkpoints.
Checkpoints are points at which the cell can assess the accuracy of the process and stop if needed. The G_1/S checkpoint is a commitment to divide; the G_2/M checkpoint ensures DNA integrity; and the spindle checkpoint ensures that all chromosomes are attached to spindle fibers, with bipolar orientation.

Cyclin-dependent kinases drive the cell cycle.
The cycle progresses by the action of Cdks. Yeast have only one CDK enzyme; vertebrates have more than four enzymes. During the G_1 phase, G_1 cyclin combines with Cdc2 kinase to form the Cdk that triggers entry into S phase.

The anaphase-promoting complex/cyclosome (APC/C) activates a protease that removes cohesins holding the centromeres of sister chromatids together; the result is to trigger anaphase, separating the chromatids and drawing them to opposite poles. The APC/C also triggers destruction of mitotic cyclins to exit mitosis.

In multicellular eukaryotes, many Cdks and external signals act on the cell cycle.
Growth factors, like platelet-derived growth factor (PDGF), stimulate cell division. This acts through a MAP kinase cascade that results in the production of cyclins and activation of Cdks to stimulate cell division in fibroblasts after tissue injury.

Cancer is a failure of cell-cycle control.
Mutations in proto-oncogenes have dominant, gain-of-function effects leading to cancer. Mutations in tumor-suppressor genes are recessive; loss of function of both copies leads to cancer.

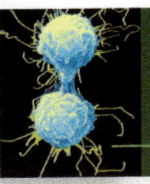

Review Questions

UNDERSTAND

1. Binary fission in prokaryotes does not require the
   a. replication of DNA.
   b. elongation of the cell.
   c. separation of daughter cells by septum formation.
   d. assembly of the nuclear envelope.

2. Chromatin is composed of
   a. RNA and protein.
   b. DNA and protein.
   c. sister chromatids.
   d. chromosomes.

3. What is a nucleosome?
   a. A region in the cell's nucleus that contains euchromatin
   b. A region of DNA wound around histone proteins
   c. A region of a chromosome made up of multiple loops of chromatin
   d. A 30-nm fiber found in chromatin

4. What is the role of cohesin proteins in cell division?
   a. They organize the DNA of the chromosomes into highly condensed structures.
   b. They hold the DNA of the sister chromatids together.
   c. They help the cell divide into two daughter cells.
   d. They connect microtubules and chromosomes.

5. The kinetochore is a structure that functions to
   a. connect the centromere to microtubules.
   b. connect centrioles to microtubules.
   c. aid in chromosome condensation.
   d. aid in chromosomes cohesion.

6. Separation of the sister chromatids occurs during
   a. prophase.
   b. prometaphase.
   c. anaphase.
   d. telophase.

7. Why is cytokinesis an important part of cell division?
   a. It is responsible for the proper separation of genetic information.
   b. It is responsible for the proper separation of the cytoplasmic contents.
   c. It triggers the movement of a cell through the cell cycle.
   d. It allows cells to halt at checkpoints.

8. What steps in the cell cycle represent irreversible commitments?
   a. The S/G_2 checkpoint
   b. The G_1/S checkpoint
   c. Anaphase
   d. Both b and c are correct.

APPLY

1. Cyclin-dependent kinases (Cdks) are regulated by
   a. the periodic destruction of cyclins.
   b. bipolar attachment of chromosomes to the spindle.
   c. DNA synthesis.
   d. Both a and b are correct.

2. The bacterial SMC proteins, eukaryotic cohesin proteins, and condensin proteins share a similar structure. Functionally they all
   a. interact with microtubules.
   b. can act as kinase enzymes.
   c. interact with DNA to compact or hold strands together.
   d. connect chromosomes to cytoskeletal elements.

3. Genetically, proto-oncogenes act in a dominant fashion. This is because
   a. there is only one copy of each proto-oncogene in the genome.
   b. they act in a gain-of-function fashion to turn on the cell cycle.
   c. they act in a loss-of-function fashion to turn off the cell cycle.
   d. they require that both genomic copies are altered to affect function.

4. The metaphase to anaphase transition involves
   a. new force being generated to pull the chromatids apart.
   b. an increase in force on sister chromatids to pull them apart.
   c. completing DNA replication of centromeres allowing chromosomes to be pulled apart.
   d. loss of cohesion between sister chromatids.

5. The main difference between bacterial cell division and eukaryotic cell division is that
   a. because bacteria only have one chromosome, they can count the number of copies in the cell.
   b. eukaryotes mark their chromosomes to identify them and bacteria do not.
   c. bacterial DNA replication and chromosome segregation are concerted processes but in eukaryotes they are separated in time.
   d. None of the above is correct.

6. In animal cells, cytokinesis is accomplished by a contractile ring containing actin. The related process in bacteria is
   a. chromosome segregation, which also appears to use an actinlike protein.
   b. septation via a ring of FtsZ protein, which is an actinlike protein.
   c. cytokinesis, which requires formation of a cell plate via vesicular fusion.
   d. septation via a ring of FtsZ protein, which is a tubulin-like protein.

SYNTHESIZE

1. Regulation of the cell cycle is very complex and involves multiple proteins. In yeast, a complex of cdc2 and a mitotic cyclin is responsible for moving the cell past the G_2/M checkpoint. The activity of the cyclin-dependent kinase cdc2 is inhibited when it is phosphorylated by the kinase, Wee-1. What would you predict would be the phenotype of a Wee-1 mutant yeast? What other genes could be altered in a Wee-1 deficient mutant strain that would make the cells act normally?

2. Review your knowledge of signaling pathways (chapter 9). Create an outline illustrating how a growth factor (ligand) can lead to the production of a cyclin protein that would trigger S phase.

3. Compare and contrast how mutations in cellular proto-oncogenes and in tumor-suppressor genes can lead to cancer cells.

Genetic and Molecular Biology

Part III Genetic and Molecular Biology

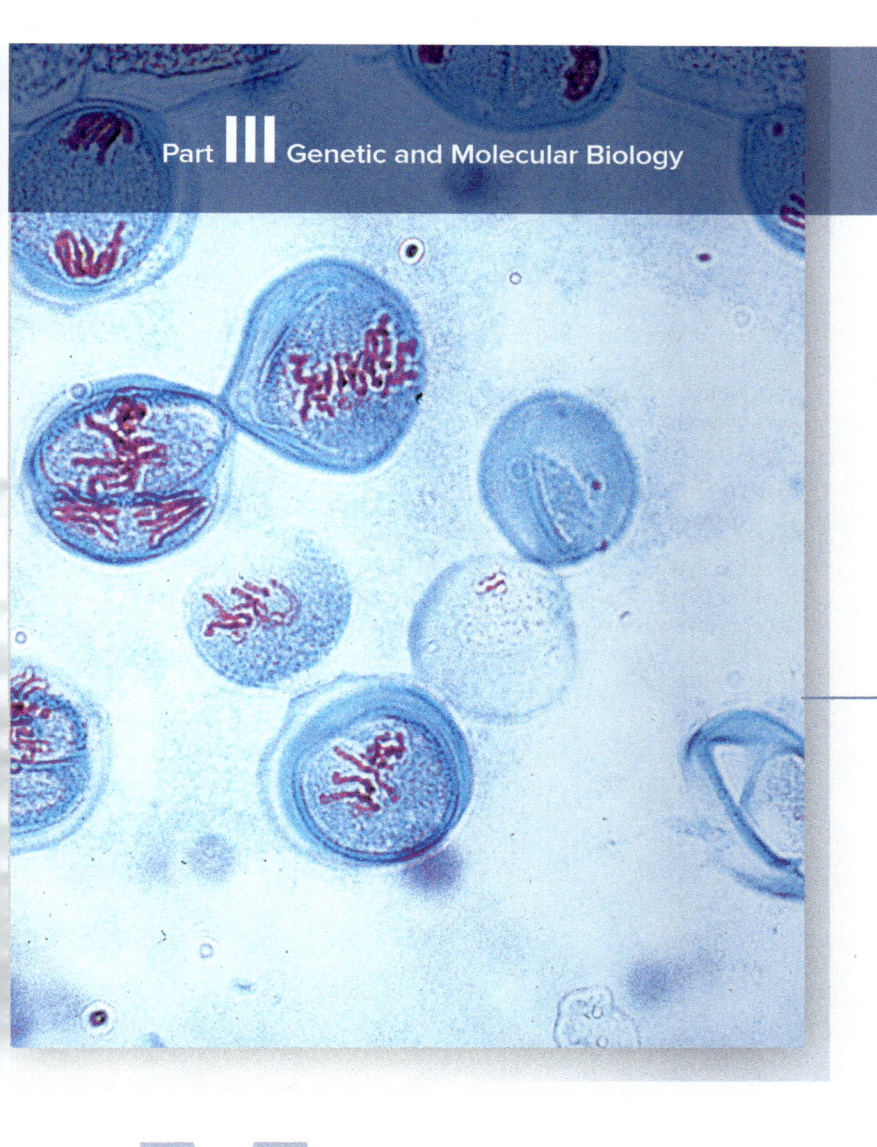

CHAPTER 11

Sexual Reproduction and Meiosis

Chapter Contents

11.1 Sexual Reproduction Requires Meiosis
11.2 Features of Meiosis
11.3 The Process of Meiosis
11.4 Summing Up: Meiosis Versus Mitosis

Introduction

Most animals and plants reproduce sexually. Gametes of opposite sex unite to form a cell that, dividing repeatedly by mitosis, eventually gives rise to an adult body with some 100 trillion cells. The gametes that form the initial cell are the products of a special form of cell division called meiosis, visible in the photo above, and the subject of this chapter. Meiosis is far more intricate than mitosis, and the details behind it are not as well understood. The basic process, however, is clear. Also clear are the profound consequences of sexual reproduction: It plays a key role in generating the tremendous genetic diversity that is the raw material of evolution.

11.1 Sexual Reproduction Requires Meiosis

Learning Outcomes

1. Characterize the function of meiosis in sexual reproduction.
2. Distinguish between germ-line and somatic cells.

The essence of sexual reproduction is the genetic contribution of two cells. This mode of reproduction imposes difficulties for sexually reproducing organisms that biologists recognized early on. We are only recently making progress on the underlying mechanism for the elaborate behavior of chromosomes during meiosis. To begin, we briefly consider the history of meiosis and its relationship to sexual reproduction.

Meiosis reduces the number of chromosomes

Only a few years after Walther Flemming's discovery of chromosomes in 1879, Belgian cytologist Edouard van Beneden was surprised to find different numbers of chromosomes in different types of cells in the roundworm *Ascaris*. Specifically, he observed that the **gametes** (eggs and sperm) each contained two chromosomes, but all of the nonreproductive cells, or **somatic cells,** of embryos and mature individuals each contained four.

From his observations, van Beneden proposed in 1883 that an egg and a sperm (gametes), each containing half the complement of chromosomes found in other cells, fuse to produce a single cell called a **zygote.** The zygote, like all of the cells ultimately derived from it, contains two copies of each chromosome. The fusion of gametes to form a new cell is called **fertilization,** or **syngamy.**

It was clear even to early investigators that gamete formation must involve some mechanism that reduces the number of chromosomes to half the number found in other cells. If it did not, the chromosome number would double with each fertilization, and after only a few generations, the number of chromosomes in each cell would become impossibly large. For example, in just 10 generations, the 46 chromosomes present in human cells would increase to over 47,000 (46×2^{10}).

The number of chromosomes does not explode in this way because of a special reduction division, **meiosis.** Meiosis occurs during gamete formation, producing cells with half the normal number of chromosomes. The subsequent fusion of two of these cells ensures a consistent chromosome number from one generation to the next.

Sexual life cycles have both haploid and diploid stages

Meiosis and fertilization together constitute a cycle of reproduction. Two sets of chromosomes are present in the somatic cells of adult individuals, making them *diploid* cells, but only one set is present in the gametes, which are thus *haploid.* Reproduction that involves this alternation of meiosis and fertilization is called **sexual reproduction.** Its outstanding characteristic is that offspring inherit chromosomes from *two* parents (figure 11.1). You, for example, inherited 23 chromosomes from your mother (maternal homologue), and 23 from your father (paternal homologue).

The life cycles of all sexually reproducing organisms follow a pattern of alternation between diploid and haploid chromosome numbers, but there is some variation in the life cycles. Many types of algae, for example, spend the majority of their life cycle in a haploid state. The zygote undergoing meiosis produces haploid cells that then undergo mitosis. Some plants and some algae alternate between a multicellular haploid phase and a multicellular diploid phase (specific examples can be found in chapters 30 and 31). In most animals, the diploid state dominates; the zygote first undergoes mitosis to produce diploid cells. Then later in the life cycle, some of these diploid cells undergo meiosis to produce haploid gametes (figure 11.2).

Germ-line cells are set aside early in animal development

In animals, the single diploid zygote undergoes mitosis to give rise to all of the cells in the adult body. The cells that will eventually undergo meiosis to produce gametes are set aside from somatic cells early in the course of development. These cells are referred to as **germ-line cells.**

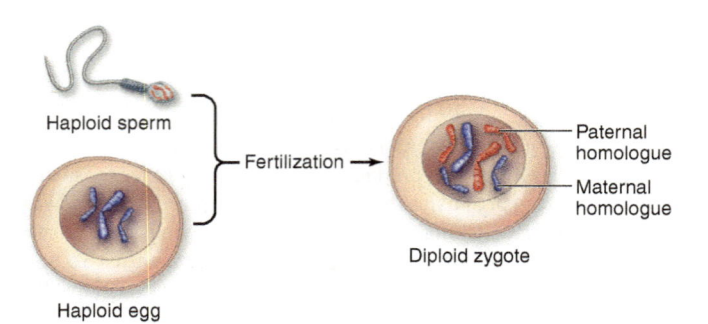

Figure 11.1 Diploid cells carry chromosomes from two parents. A diploid cell contains two versions of each chromosome, a maternal homologue contributed by the haploid egg of the mother, and a paternal homologue contributed by the haploid sperm of the father.

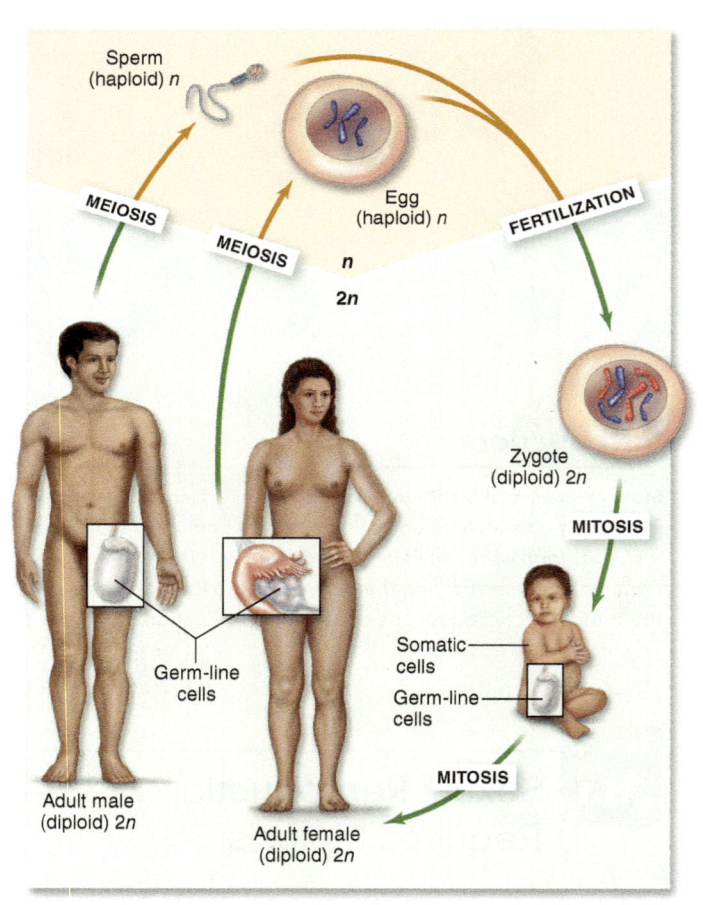

Figure 11.2 The sexual life cycle in animals. In animals, the zygote undergoes mitotic divisions and gives rise to all the cells of the adult body. Germ-line cells are set aside early in development and undergo meiosis to form the haploid gametes (eggs or sperm). The rest of the body cells are called somatic cells.

Both somatic cells and germline cells are diploid. The critical difference is that somatic cells undergo mitosis to form genetically identical, diploid daughter cells, but germ-line cells can undergo meiosis to produce haploid gametes (figure 11.2).

Learning Outcomes Review 11.1

Sexual reproduction involves the genetic contribution of two cells, each from a different individual. Meiosis produces haploid cells with half the number of chromosomes. Fertilization then unites these haploid cells to restore the diploid state of the next generation. Only germ-line cells are capable of meiosis. All other cells in the body, termed somatic cells, can undergo only mitotic division.

- Germ-line cells undergo meiosis, but how can the body maintain a constant supply of these cells?

11.2 Features of Meiosis

Learning Outcomes

1. Describe how homologous chromosomes pair during meiosis.
2. Explain why meiosis I is called the reductive division.

The mechanism of meiotic cell division varies in important details in different organisms. These variations are particularly evident in the chromosomal separation mechanisms: Those found in protists and fungi are very different from those in plants and animals, which we describe here.

Meiosis in a diploid organism consists of two rounds of division, called **meiosis I** and **meiosis II,** with each round containing prophase, metaphase, anaphase, and telophase stages. Before describing the details of this process, we first examine the features of meiosis that distinguish it from mitosis.

Homologous chromosomes pair early in meiosis

During early prophase I of meiosis, homologous chromosomes find each other and become closely associated, a process called pairing, or **synapsis** (figure 11.3a). Despite a long history of investigation, molecular details remain unclear. Biologists have used electron microscopy, data from genetic crosses, and biochemical analysis to shed light on synapsis. Thus far the results of their investigations have not been integrated into a complete picture.

The synaptonemal complex

It is clear that homologous chromosomes find their proper partners and become intimately associated during prophase I. This process includes the formation in many species of an elaborate structure called the **synaptonemal complex,** joining the paired homologues. The synaptonemal complex consists of a central element connected by filaments to two sets of lateral elements that interact with the homologues. The structure of the synaptonemal complex appears similar in all systems that have been examined, although its exact function is unclear. A representative example is shown in figure 11.3b. This results in the paired homologues, each composed of two sister chromatids, all being closely associated during this phase of meiosis. This structure is also called a *tetrad* or *bivalent*.

The exchange of genetic material between homologues

While homologues are paired during prophase I, another process unique to meiosis occurs: genetic **recombination,** or **crossing over.** This process literally allows the homologues to exchange chromosomal material. The cytological observation of this phenomenon is called crossing over, and its detection genetically is called recombination—because alleles of genes

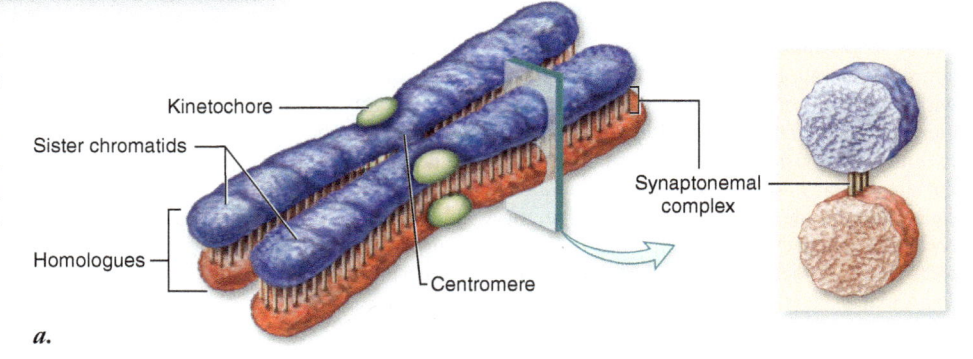

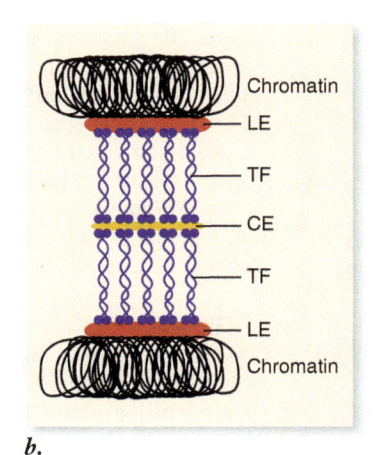

Figure 11.3 Unique features of meiosis. *a.* Homologous chromosomes pair during prophase I of meiosis. This process, called synapsis, produces homologues connected by a structure called the synaptonemal complex. The paired homologues can physically exchange parts by crossing over, or recombination. The sister chromatids of each homologue are also held together by cohesin proteins, which is not shown for clarity. *b.* A schematic drawing of the synaptonemal complex. Two lateral elements (LE) are joined to a central element (CE) by transverse filaments (TF). The chromatin of each homologue is attached to the lateral elements.

that were formerly on separate homologues can now be found on the same homologue. (Genetic recombination is covered in detail in chapter 13.)

The sites of crossing over are called **chiasmata** (singular, *chiasma*), and these sites of contact are maintained until anaphase I. The physical connection of homologues due to crossing over and the continued connection of the sister chromatids lock homologues together.

Homologue association and separation

The association between the homologues persists throughout meiosis I and dictates the behavior of the chromosomes. During metaphase I, the paired homologues move to the metaphase plate and become oriented with homologues of each pair attached to opposite poles of the spindle. By contrast, in mitosis homologues do not associate and behave independently of one another.

Then, during anaphase I, homologues are pulled to opposite poles for each pair of chromosomes. This again is in contrast to mitosis, where sister chromatids, not homologues, are pulled to opposite poles.

You can now see why the first division is termed the "reduction division"—it results in daughter cells that contain one homologue from each chromosome pair. The second meiotic division does not further reduce the number of chromosomes; it will merely separate the sister chromatids for each homologue.

Meiosis consists of two divisions with one round of DNA replication

The most obvious distinction between meiosis and mitosis is the simple observation that meiosis involves two successive divisions with no replication of genetic material between them. One way to view this is that DNA replication must be suppressed between the two meiotic divisions. Because of the behavior of chromosomes during meiosis I, the resulting cells contain one replicated copy of each chromosome. A division that acts like mitosis, without DNA replication, converts these cells into ones with a single copy of each chromosome. This is the last key to understanding meiosis: The second meiotic division is like mitosis with no chromosome duplication.

Learning Outcomes Review 11.2

Meiosis is characterized by the pairing of homologous chromosomes during prophase I. In many species, an elaborate structure called the synaptonemal complex forms between homologues. During this pairing, homologues may exchange chromosomal material at sites called chiasmata. In meiosis I, the homologues separate from each other, reducing the chromosome number to the haploid state (thus the reductive division). It is followed by a second division without replication, during which sister chromatids become separated. The result of meiosis I and II is four haploid cells.

■ *If sister chromatids separated at the first division, would meiosis still work?*

11.3 The Process of Meiosis

Learning Outcomes
1. Describe the behavior of chromosomes through both meiotic divisions.
2. Explain the importance of monopolar attachment of homologous pairs at metaphase I.
3. Differentiate between the events of anaphase I and anaphase II of meiosis.

To understand meiosis, it is necessary to carefully follow the behavior of chromosomes during each division. The first meiotic division depends on each homologous pair behaving as a unit and not as individual chromosomes, as they do in mitosis. This is accomplished by a complex set of processes that together join homologues until anaphase I, when they are separated.

Prophase I sets the stage for the reductive division

Meiotic cells have an interphase period that is similar to mitosis with G_1, S, and G_2 phases. After interphase, germ-line cells enter meiosis I. In prophase I, the DNA coils tighter, and individual chromosomes first become visible under the light microscope as a matrix of fine threads. Because the DNA has already replicated before the onset of meiosis, each of these threads actually consists of two sister chromatids joined at their centromeres. In prophase I, homologous chromosomes become closely associated in synapsis, and exchange segments by crossing over, which leaves all four chromatids associated.

Synapsis

During interphase in germ-line cells, the ends of the chromatids seem to be attached to the nuclear envelope at specific sites. The sites the homologues attach to are adjacent, so that during prophase I the members of each homologous pair of chromosomes are brought close together. Homologous pairs then align side by side, apparently guided by heterochromatin sequences, in the process of synapsis.

This association joins homologues along their entire length. The sister chromatids of each homologue are also joined by the cohesin complex in a process called *sister chromatid cohesion*. Sister chromatid cohesion also occurs in mitosis, but in meiosis the cohesin complex contains a meiosis-specific cohesin. This brings all four chromatids for each set of paired homologues into close association.

Crossing over

Along with the synaptonemal complex that forms during prophase I (figure 11.3), another kind of structure appears at the same time that recombination occurs. These are called *recombination nodules,* and they are thought to contain the enzymatic machinery necessary to break and rejoin chromatids of homologous chromosomes.

Crossing over involves a complex series of events in which DNA segments are exchanged between nonsister chromatids. Reciprocal crossovers between nonsister chromatids are controlled such that each chromosome arm usually has one or a few crossovers per meiosis, no matter what the size of the chromosome. Human chromosomes typically have two or three.

When crossing over is complete, the synaptonemal complex breaks down, and the homologous chromosomes become less tightly associated but remain attached by chiasmata. At this point, there are four chromatids for each type of chromosome (two homologous chromosomes, each of which consists of two sister chromatids).

The four chromatids are held together in two ways: (1) The two sister chromatids of each homologue, the products of DNA replication, are held together by cohesin proteins (sister chromatid cohesion); and (2) exchange of material by crossing over between homologues locks all four chromatids together.

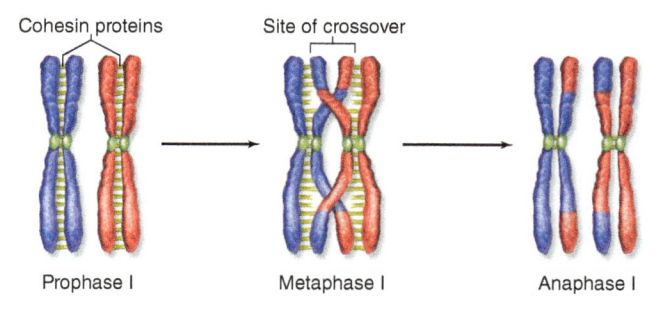

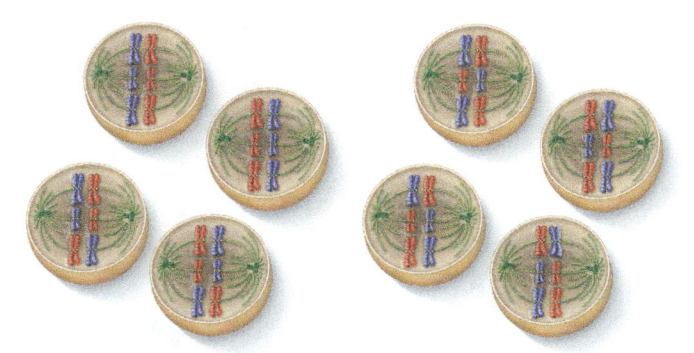

Figure 11.5 Random orientation of chromosomes on the metaphase plate. The number of possible chromosome orientations equals 2 raised to the power of the number of chromosome pairs. In this hypothetical cell with three chromosome pairs, eight (2^3) possible orientations exist. Each orientation produces gametes with different combinations of parental chromosomes.

While the elaborate behavior of chromosome pairing is taking place, other events must occur during prophase I. The nuclear envelope must be dispersed, along with the interphase structure of microtubules. The microtubules re-form into a spindle, just as in mitosis.

At metaphase I, paired homologues are aligned

Because of the events of prophase I, each of the paired homologues are locked together as a bivalent. As these bivalents capture spindle fibers, they move to the center of the cell, where they are aligned as paired homologues and not individual chromosomes.

The kinetochores of sister chromatids act as a unit to capture polar microtubules. This results in microtubules from opposite poles becoming attached to the kinetochores of *homologues,* and not to those of sister chromatids (figure 11.4).

The ability of sister centromeres to behave as a unit during meiosis I is not understood. It has been suggested, based on electron microscope data, that the centromere–kinetochore complex of sister chromatids is compacted during meiosis I, allowing them to function as a single unit.

The monopolar attachment of kinetochores of sister chromatids would be disastrous in mitosis, but it is critical to meiosis I. It produces tension on the paired homologues, pulling them to the equator of the cell. In this way, each joined pair of homologues lines up on the metaphase plate (figure 11.4).

The orientation of each pair on the spindle axis is random; either the maternal or the paternal homologue may be oriented toward a given pole (figure 11.5; see also figure 11.6).

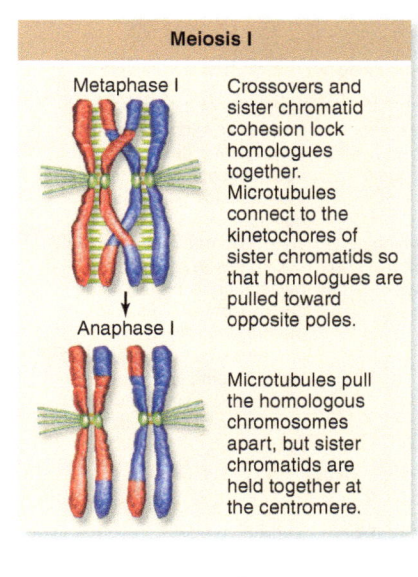

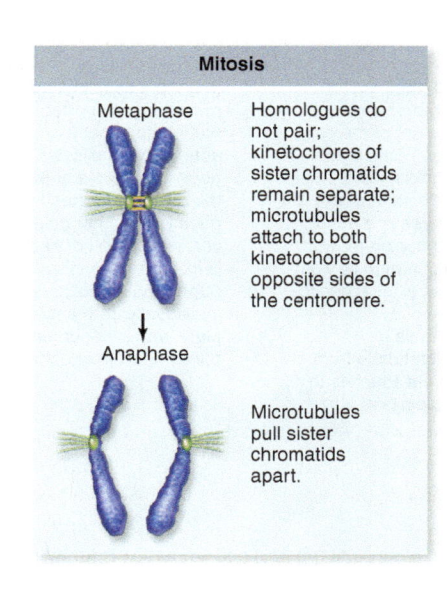

Figure 11.4 Alignment of chromosomes differs between meiosis I and mitosis. In metaphase I of meiosis I, the chiasmata and connections between sister chromatids hold homologous chromosomes together; paired kinetochores for sister chromatids of each homologue become attached to microtubules from one pole. By the end of meiosis I, connections between sister chromatid arms, but not centromeres, are broken as microtubules shorten, pulling the homologous chromosomes apart. In mitosis, microtubules from opposite poles attach to the kinetochore of each sister centromere; when the connections between sister centromeres are broken microtubules shorten, pulling the sister chromatids to opposite poles.

chapter **11** *Sexual Reproduction and Meiosis* 211

228 Principles of Biology I and II

MEIOSIS I

Prophase I

In prophase I, the chromosomes begin to condense, and the spindle of microtubules begins to form. The DNA has been replicated, and each chromosome consists of two sister chromatids attached at the centromere. The cell illustrated here has four chromosomes, or two pairs of homologues. Homologous chromosomes pair along their entire length during synapsis. Crossing over occurs, forming chiasmata, which hold homologous chromosomes together.

Labels: Chromosome (replicated), Spindle, Sister chromatids, Paired homologous chromosomes, Chiasma

Metaphase I

In metaphase I, the pairs of homologous chromosomes align at the equator of the cell. Chiasmata keep homologous pairs together and microtubules from opposite poles attach to sister kinetochores of each homologue, producing tension. A kinetochore microtubule from one pole of the cell attaches to one homologue of a chromosome, while a kinetochore microtubule from the other cell pole attaches to the other homologue of a pair.

Labels: Kinetochore microtubule, Homologue pair on metaphase plate

Anaphase I

In anaphase I, kinetochore microtubules shorten, and homologous pairs are pulled apart. One duplicated homologue goes to one pole of the cell, and the other duplicated homologue goes to the other pole. Sister chromatids do not separate. This is in contrast to mitosis, in which duplicated homologues line up individually on the metaphase plate, and sister chromatids are pulled apart in anaphase.

Labels: Sister chromatids, Homologous chromosomes

Telophase I

In telophase I, the separated homologues form a cluster at each pole of the cell, and the nuclear envelope re-forms around each daughter cell nucleus. Cytokinesis may occur. The resulting two cells have half the number of chromosomes of the original cell: In this example, each nucleus contains two chromosomes (versus four in the original cell). Each chromosome consists of two sister chromatids, but sister chromatids are not identical because crossing over has occurred.

Labels: Nonidentical sister chromatids, Chromosome, Homologous chromosomes

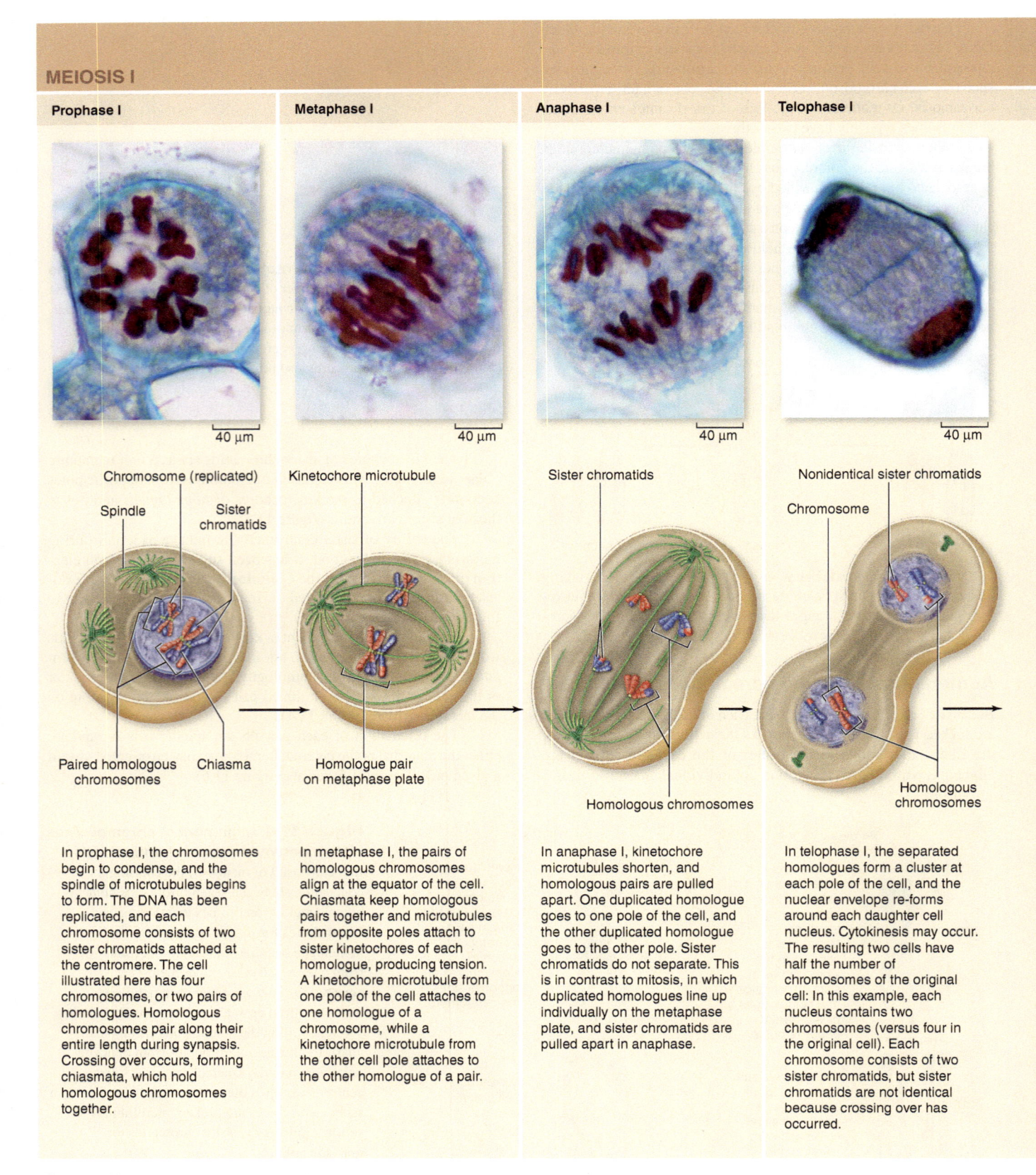

Figure 11.6 The stages of meiosis. Meiosis in plant cells (photos) and animal cells (drawings) is shown.

Biology, 11th Edition   229

MEIOSIS II

Prophase II	Metaphase II	Anaphase II	Telophase II
40 μm	40 μm	40 μm	40 μm
Spindle	Sister chromatids	Kinetochore microtubule	Nuclear membrane re-forming
Nuclear membrane breaking down	Chromosome	Sister chromatids	
Following a brief interphase, with no S phase, meiosis II begins. During prophase II, a new spindle apparatus forms in each cell, and the nuclear envelope breaks down. In some species the nuclear envelope does not re-form in telophase I, obviating the need for nuclear envelope breakdown.	In metaphase II, chromosomes consisting of sister chromatids joined at the centromere align along the metaphase plate in each cell. Now, kinetochore microtubules from opposite poles attach to kinetochores of sister chromatid, as in mitosis.	When microtubules shorten in anaphase II, sister chromatids are pulled to opposite poles of the cells, as in mitosis.	In telophase II, the nuclear membranes re-form around four different clusters of chromosomes. After cytokinesis, four haploid cells result. No two cells are alike due to the random alignment of homologous pairs at metaphase I and crossing over during prophase I.

chapter 11 *Sexual Reproduction and Meiosis*   213

During anaphase I, homologues are pulled to opposite poles

In anaphase I, the microtubules of the spindle fibers begin to shorten. As they shorten, the connections between homologues at chiasmata are broken, allowing homologues to be pulled to opposite poles.

Anaphase I comes about by the release of sister chromatid cohesion along the chromosome arms, but not at the centromeres. This release is the result of the destruction of meiosis-specific cohesin in a process analogous to anaphase in mitosis. The difference is that the destruction is inhibited at the centromeres by a mechanism that is discussed in section 11.4.

As a result of this release, the homologues are pulled apart, but not the sister chromatids. Each homologue moves to one pole, taking both sister chromatids with it. When the spindle fibers have fully contracted, each pole has a complete haploid set of chromosomes consisting of one member of each homologous pair.

Because of the random orientation of homologous chromosomes on the metaphase plate, a pole may receive either the maternal or the paternal homologue from each chromosome pair. As a result, the genes on different chromosomes assort independently; that is, meiosis I results in the **independent assortment** of maternal and paternal chromosomes into the gametes (see chapter 12).

Telophase I completes meiosis I

By the beginning of telophase I, the chromosomes have segregated into two clusters, one at each pole of the cell. Now the nuclear membrane re-forms around each daughter nucleus.

Because each chromosome within a daughter nucleus had replicated before meiosis I began, each now contains two sister chromatids attached by a common centromere. Note that *the sister chromatids are no longer identical* because of the crossing over that occurred in prophase I (figure 11.6); as you will see, this change has important implications for genetic variability.

Cytokinesis, the division of the cytoplasm and its contents, may or may not occur after telophase I. The second meiotic division, meiosis II, occurs after an interval of variable length.

Segregation can occur without crossing over

The preceding description of meiosis I relies on the observation that homologues are held together by chiasmata and by sister chromatid cohesion. This connection produces the critical behavior of chromosomes during metaphase I and anaphase I, when paired homologues move together to the metaphase plate and then move to opposite poles.

Although this connection of homologues is the rule, there are exceptions. In fruit fly *(Drosophila)* males for example, there is no recombination, and yet meiosis proceeds accurately, a process called **achiasmate segregation** ("without chiasmata"). This seems to involve an alternative mechanism for joining homologues and then allowing their segregation during anaphase I. Telomeres and other heterochromatic sequences have been implicated, but the details are not known.

Despite these exceptions, the vast majority of species that have been examined use the formation of chiasmata and sister chromatid cohesion to hold homologues together for segregation during anaphase I.

Meiosis II is like a mitotic division without DNA replication

Typically, interphase between meiosis I and meiosis II is brief and does not include an S phase: Meiosis II resembles a normal mitotic division. Prophase II, metaphase II, anaphase II, and telophase II follow in quick succession (figure 11.6):

Prophase II. At the two poles of the cell, the clusters of chromosomes enter a brief prophase II, each nuclear envelope breaking down as a new spindle forms.

Metaphase II. In metaphase II, spindle fibers from opposite poles bind to kinetochores of each sister chromatid, allowing each chromosome to migrate to the metaphase plate as a result of tension on the chromosomes from polar microtubules pulling on sister centromeres. This process is the same as metaphase during a mitotic division.

Anaphase II. The spindle fibers contract, and the cohesin complex joining the centromeres of sister chromatids is finally destroyed, allowing sister chromatids to be pulled to opposite poles. This process is essentially the same as anaphase during a mitotic division.

Telophase II. Finally, the nuclear envelope re-forms around the four sets of daughter chromosomes. Cytokinesis then follows.

The final result of this division is four cells, each containing a complete haploid set of chromosomes. The cells that contain these haploid nuclei may develop directly into gametes, as they do in animals. Alternatively, they may themselves divide mitotically, as they do in plants, fungi, and many protists, eventually producing greater numbers of gametes or, as in some plants and insects, adult individuals with varying numbers of chromosome sets.

Errors in meiosis produce aneuploid gametes

It is critical that the process of meiosis be accurate because any failure produces gametes without the correct number of chromosomes. Failure of chromosomes to move to opposite poles during either meiotic division is called *nondisjunction,* and it produces one gamete that lacks a chromosome and one that has two copies. Gametes with an improper number of chromosomes are called **aneuploid gametes.** In humans, this condition is the most common cause of spontaneous abortion. The implications of aneuploid gametes are explored in more detail in chapter 13.

Learning Outcomes Review 11.3

Sister chromatid cohesion, combined with crossing over, connects homologous chromosomes during meiosis I. The centromere of each homologue shows monopolar attachment, leading to the alignment of homologous pairs at metaphase I. Loss of cohesion on the arms but not the centromere leads to homologues moving to opposite poles during anaphase I. During anaphase II, cohesin proteins holding sister chromatids together at the centromere are removed, allowing them to move to opposite poles.

- *What would be the result of improper disjunction at anaphase I? At anaphase II?*

11.4 Summing Up: Meiosis Versus Mitosis

Learning Outcomes
1. Describe the distinct features of meiosis.
2. Describe the differences in chromatid cohesion in meiosis and mitosis.
3. Explain the importance of the suppression of replication between meiotic divisions.

The key to meiosis is understanding the differences between meiosis and mitosis. The basic machinery in both processes is the same, but the behavior of chromosomes is distinctly different during the first meiotic division (figure 11.7).

Meiosis is characterized by four distinct features:

1. Homologous pairing and crossing over joins maternal and paternal homologues during meiosis I.
2. Sister chromatids remain connected at the centromere and segregate together during anaphase I.
3. Kinetochores of sister chromatids are attached to the same pole in meiosis I and to opposite poles in mitosis.
4. DNA replication is suppressed between the two meiotic divisions.

Although the underlying molecular mechanisms are unclear, we will consider what we know of each of these features in the rest of this section.

Homologous pairing is specific to meiosis

The pairing of homologues during prophase I of meiosis is the first deviation from mitosis and sets the stage for all of the subsequent differences (figure 11.7). How homologues find each other and become aligned is one of the great mysteries of meiosis. Some cytological evidence implicates telomeres and other specific sites as being necessary for pairing, but this finding does little to clarify the essential process.

The process of sister chromatid cohesion is similar to mitosis, but involves cohesin proteins that contain meiosis-specific subunits. In yeast, the protein Rec8 replaces the mitotic Scc1 protein as part of the cohesin complex. You saw in chapter 10 that Scc1 is destroyed during anaphase of mitosis to allow sister chromatids to be pulled to opposite poles. The replacement of this critical cohesin component with a meiosis-specific version seems to be a common feature in systems analyzed to date.

Synaptonemal complex proteins have been identified in diverse species, but these proteins show little sequence conservation, but do have similarities in structure where this has been analyzed. This may explain the similarity of structures observed cytologically. The transverse elements, while showing no sequence conservation, do share the feature of coiled-coil domains that promote protein–protein interactions.

The molecular details of the recombination process that produces crossing over are complex, but many of the proteins involved have been identified. The process is initiated with the introduction of a double-strand break in one homologue. This explains the similarity in the machinery necessary for meiotic recombination and the machinery involved in the repair of double-strand breaks in DNA. Recombination probably first evolved as a repair mechanism and was later co-opted for use in disjoining chromosomes. The importance of recombination for proper disjunction is clear from the observation in many organisms that loss of function for recombination proteins also results in higher levels of meiotic nondisjunction.

Centromeres of sister chromatids remain connected through meiosis I

Meiosis I is characterized by the segregation of homologues, not sister chromatids, during anaphase. For this separation to occur, the centromeres of sister chromatids must cosegregate, or move to the same pole, during anaphase I. This means that meiosis-specific cohesin proteins must first be removed from the chromosome arms, then later from sister centromeres.

Homologues are joined by chiasmata, and sister chromatid cohesion around the site of exchange then holds homologues together. The destruction of Rec8 protein on the chromosome arms appears to be what allows homologues to be pulled apart at anaphase I.

This leaves the key distinction between meiosis and mitosis being the maintenance of sister chromatid cohesion at the centromere during all of meiosis I, but the loss of cohesion from the chromosome arms during anaphase I (figure 11.7). Some light was shed on this problem with the identification of conserved proteins, called Shugoshin (a Japanese term meaning "guardian spirit") required for cohesin protection from separase-mediated cleavage during meiosis I (figure 11.8). Mice have two Shugoshins: Sgo-1 and Sgo-2. Depletion of Sgo-2 results in early sister chromatid separation. This leaves the problem of why Sgo-2 acts only at anaphase I and not anaphase II. It has been suggested that the tension produced by anaphase II causes Sgo-2 to migrate from the centromere to the kinetochore.

MEIOSIS I

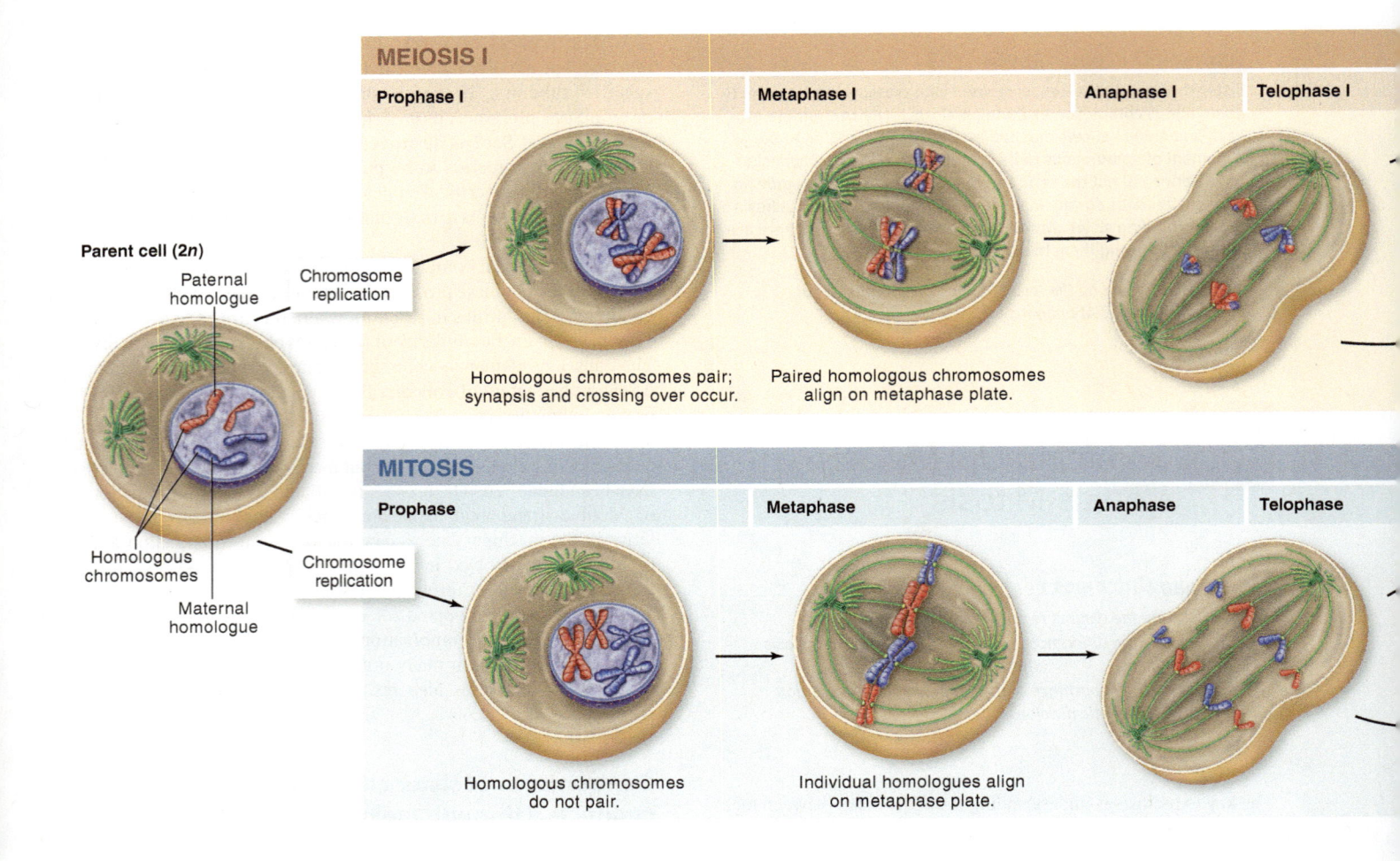

Prophase I → **Metaphase I** → **Anaphase I** → **Telophase I**

Parent cell (2n)
- Paternal homologue
- Homologous chromosomes
- Maternal homologue

Chromosome replication

Homologous chromosomes pair; synapsis and crossing over occur.

Paired homologous chromosomes align on metaphase plate.

MITOSIS

Prophase → **Metaphase** → **Anaphase** → **Telophase**

Chromosome replication

Homologous chromosomes do not pair.

Individual homologues align on metaphase plate.

Sister kinetochores are attached to the same pole during meiosis I

The cosegregation of sister centromeres requires that the kinetochores of sister chromatids are attached to the same pole during meiosis I. This attachment is in contrast to both mitosis (figure 11.7) and meiosis II, in which sister kinetochores must become attached to opposite poles.

The underlying basis of this monopolar attachment of sister kinetochores is unclear, but it seems to be based on structural differences between centromere–kinetochore complexes in meiosis I and in mitosis. Mitotic kinetochores visualized with the electron microscope appear to be recessed, making bipolar attachment more likely. Meiosis I kinetochores protrude more, making monopolar attachment easier.

It is clear that both the maintenance of sister chromatid cohesion at the centromere and monopolar attachment are required for the segregation of homologues that distinguishes meiosis I from mitosis.

Replication is suppressed between meiotic divisions

After a mitotic division, a new round of DNA replication must occur before the next division. For meiosis to succeed in halving the number of chromosomes, this replication must be suppressed between the two divisions. The detailed mechanism of suppression of replication between meiotic division is unknown. One clue is the observation that the level of one of the cyclins, cyclin B, is reduced between meiotic divisions, but is not lost completely, as it is between mitotic divisions.

During mitosis, the destruction of mitotic cyclin is necessary for a cell to enter another division cycle. The result of this maintenance of cyclin B between meiotic divisions in germ-line cells is the failure to form initiation complexes necessary for DNA replication to proceed. This failure to form initiation complexes appears to be critical to suppressing DNA replication.

MEIOSIS II

Prophase II | Metaphase II | Anaphase II | Telophase II

Homologous chromosomes separate; sister chromatids remain together.

Chromosomes align, sister chromatids separate, and four haploid cells result, each containing half the original number of homologues.

Four daughter cells (each *n*)

Two daughter cells (each 2*n*)

Sister chromatids separate, cytokinesis occurs, and two cells result, each containing the original number of homologues.

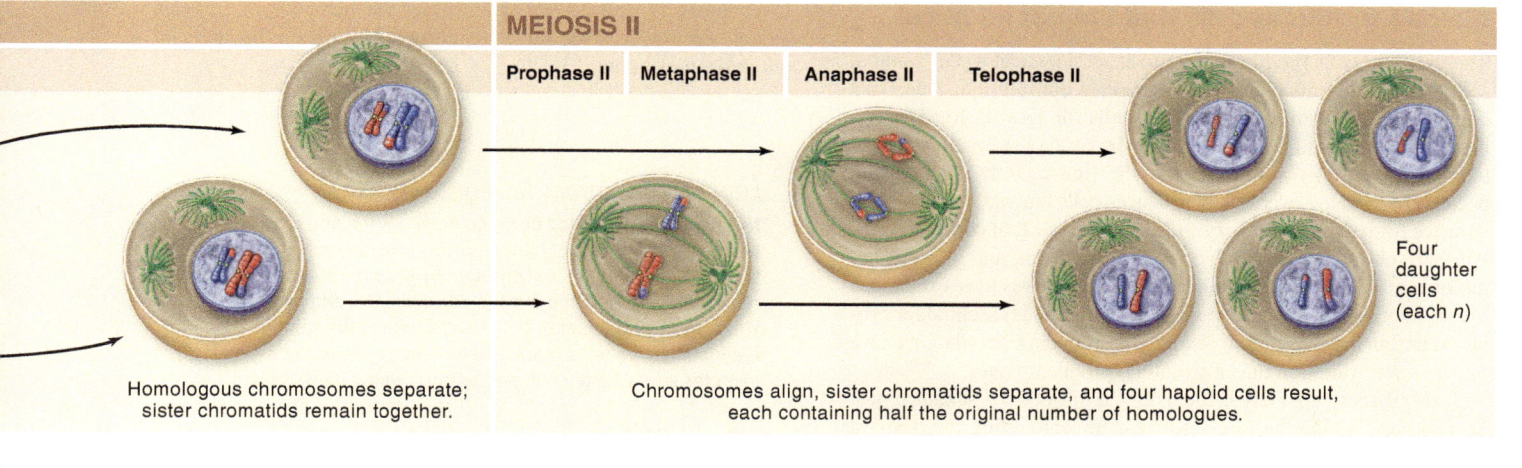

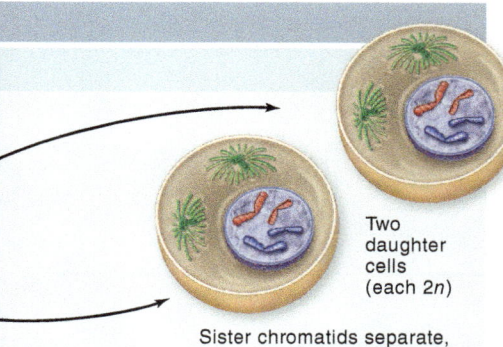

Figure 11.7 A comparison of meiosis and mitosis. Meiosis involves two nuclear divisions with no DNA replication between them. It thus produces four daughter cells, each with half the original number of chromosomes. Crossing over occurs in prophase I of meiosis. Mitosis involves a single nuclear division after DNA replication. It thus produces two daughter cells, each containing the original number of chromosomes.

? Inquiry question If the chromosomes of a mitotic cell behaved the same as chromosomes in meiosis I, would the resulting cells have the proper chromosomal constitution?

SCIENTIFIC THINKING

Question: Why are cohesin proteins at the centromeres of sister chromatids not destroyed at anaphase I of meiosis?
Hypothesis: Meiosis-specific cohesin component Rec8 is protected by another protein at centromeres.
Prediction: If Rec8 and the centromere protecting protein are both expressed in mitotic cells, chromosome separation will be prevented. This is lethal to a dividing cell.
Test: Fission yeast strain is designed to produce Rec8 instead of normal mitotic cohesin. These cells are transformed with a cDNA library that expresses all cellular proteins. Transformed cells are duplicated onto media containing dye for dead cells (allows expression of Rec8 and cDNA), and media that will result in loss of plasmid cDNA (expresses only Rec8). Cells containing cDNA for protecting protein will be dead in presence of Rec8.

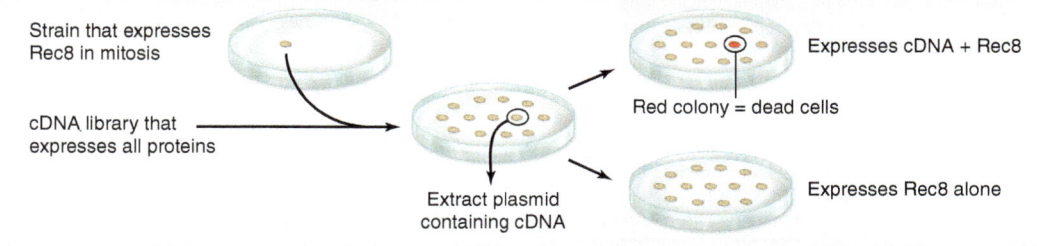

Strain that expresses Rec8 in mitosis

cDNA library that expresses all proteins

Extract plasmid containing cDNA

Expresses cDNA + Rec8
Red colony = dead cells

Expresses Rec8 alone

Result: Transformed cells that die on the plates where Rec8 is coexpressed with cDNA identify the protecting protein. When the cDNA is extracted and analyzed, the encoded protein localizes to the centromeres of meiotic cells.
Conclusion: This screen identifies a protein with Rec8 protecting activity.
Further Experiments: If the gene encoding the protecting protein is deleted from cells, what would be the expected phenotype? In mitotic cells? In meiotic cells?

Figure 11.8 Identification of meiosis-specific cohesin protector.

Meiosis produces cells that are not identical

The daughter cells produced by mitosis are identical to the parental cell, at least in terms of their chromosomal constitution. This exact copying is critical to producing new cells for growth, development, and for wound healing. Meiosis, because of the random orientation of different chromosomes at the first meiotic division and because of crossing over, rarely produces cells that are identical. The gametes from meiosis all carry an entire haploid set of chromosomes, but these chromosomes are a mixture of maternal and paternal homologues; furthermore, the homologues themselves have exchanged material by crossing over. The resulting variation is essential for evolution and is the reason that sexually reproducing populations have much greater variation than asexually reproducing ones.

Meiosis is not only critical for the process of sexual reproduction, but is also the foundation for understanding the basis of heredity. The different cells produced by meiosis form the basis for understanding the behavior of observable traits in genetic crosses.

In chapters 12 and 13 we will follow the behavior of traits in genetic crosses and see how this correlates with the behavior of chromosomes in meiosis.

Learning Outcomes Review 11.4

Meiosis is characterized by homologue pairing and crossing over; by loss of sister chromatid cohesion in the arms, but not at the centromere at the first division; by the suppression of DNA replication between the two meiotic divisions; and by sister kinetochores attachment to the same pole of the spindle. If replication were not suppressed between meiosis I and meiosis II, gametes would be diploid, and zygotes would be tetraploid.

■ What features of meiosis lead to genetic variation in the products?

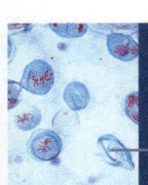

Chapter Review

11.1 Sexual Reproduction Requires Meiosis (figure 11.2)

Meiosis reduces the number of chromosomes.
Eggs and sperm are haploid (1n) cells, which contain one set of all chromosomes, and products of meiotic division.

Sexual life cycles have both haploid and diploid stages.
During fertilization, or syngamy, the fusion of two haploid gametes results in a diploid (2n) zygote, which contains two sets of chromosomes. Meiosis and fertilization constitute a reproductive cycle in sexual organisms as they alternate between diploid and haploid chromosome numbers. Somatic cells divide by mitosis and form the body of an organism.

Germ-line cells are set aside early in animal development.
Cells that eventually will form haploid gametes by meiosis are called germ-line cells. These are set aside early in development in animals.

11.2 Features of Meiosis

Homologous chromosomes pair early in meiosis.
The pairing of homologous chromosomes, called synapsis, occurs during early prophase I. Paired homologues are often joined by the synaptonemal complex (figure 11.3). During synapsis, crossing over occurs between homologous chromosomes, exchanging chromosomal material. Because the homologues are paired, they move as a unit to the metaphase plate during metaphase I. During anaphase I, homologues of each pair are pulled to opposite poles, producing two cells that each have one complete set of chromosomes.

Meiosis consists of two divisions with one round of DNA replication.
Meiosis II is like mitosis but without replication of DNA. Sister chromatids are pulled to opposite poles to yield four haploid cells.

11.3 The Process of Meiosis (figure 11.6)

Prophase I sets the stage for the reductive division.
Meiotic cells have an interphase period similar to mitosis with G_1, S, and G_2 phases. This is followed by prophase I in which homologous chromosomes align along their entire length. The sister chromatids are held together by cohesin proteins. Homologues exchange chromosomal material by crossing over, which assists in holding the homologues together during meiosis I. The nuclear envelope disperses and the spindle apparatus forms.

At metaphase I, paired homologues are aligned.
Spindle fibers attach to the kinetochores of the homologues; the kinetochores of sister chromatids behave as a single unit. Homologues of each pair become attached by kinetochore microtubules to opposite poles, and homologous pairs move to the metaphase plate as a unit. The orientation of each homologous pair on the equator is random; either the maternal or paternal homologue may be oriented toward a given pole.

During anaphase I, homologues are pulled to opposite poles.
During anaphase I the homologues of each pair are pulled to opposite poles as kinetochore microtubules shorten. Loss of sister chromatid cohesion on the arms but not at the centromeres allows homologues to separate, but sister chromatids to stay together. This is due to the loss of cohesin proteins on the arms but not at the centromere. At the end of anaphase I each pole has a complete set of haploid chromosomes, consisting of one member of each homologous pair. Because of the random orientation of homologous pairs at metaphase I, meiosis I results in the independent assortment of maternal and paternal chromosomes in gametes.

Telophase I completes meiosis I.
During telophase I the nuclear envelope re-forms around each daughter nucleus. This phase does not occur in all species. Cytokinesis may or may not occur after telophase I.

Segregation can occur without crossing over.

Although homologues are usually held together by chiasmata, some systems are able to segregate chromosomes without this.

Meiosis II is like a mitotic division without DNA replication.

A brief interphase with no DNA replication occurs after meiosis I. During meiosis II, cohesin proteins at the centromeres that hold sister chromatids together are destroyed, allowing each to migrate to opposite poles of the cell. The result of meiosis I and II is four cells, each containing haploid sets of chromosomes that are not identical. Once completed, the haploid cells may produce gametes or divide mitotically to produce even more gametes or haploid adults.

Errors in meiosis produce aneuploid gametes.

Errors occur during meiosis because of nondisjunction, the failure of chromosomes to move to opposite poles. It may result in aneuploid gametes: one gamete with no chromosome, and another gamete with two copies of a chromosome.

11.4 Summing Up: Meiosis Versus Mitosis (figure 11.7)

Four distinct features of meiosis I are not found in mitosis: Maternal and paternal homologues pair, and exchange genetic information by crossing over; the kinetochores of sister chromatids function as a unit during meiosis I, allowing sister chromatids to cosegregate during anaphase I; kinetochores of sister chromatids are connected to a single pole in meiosis I and to opposite poles in mitosis; and DNA replication is suppressed between meiosis I and meiosis II.

Homologous pairing is specific to meiosis.

How homologues find each other during meiosis is not known. The proteins of the synaptonemal complex do not seem to be conserved in different species, but there are meiosis-specific cohesin proteins. These are involved in the differential destruction of cohesins on the arms versus the centromere during meiosis I. The recombination process that occurs between paired homologues is better known. This process uses proteins involved in DNA repair and starts with a double-stranded break in DNA.

Centromeres of sister chromatids remain connected through meiosis I

Shugoshin protein protects centromeric cohesin in anaphase I, so that sister chromatids remain connected. Cohesins on the arms are not protected and are thus degraded during anaphase I, allowing homologues to move to opposite poles.

Sister kinetochores are attached to the same pole during meiosis I.

Kinetochores of sister chromatids must be attached to the same spindle fibers (monopolar attachment) to segregate together.

Replication is suppressed between meiotic divisions.

Suppression of replication may be related to the maintenance of some cyclin proteins that are degraded at the end of mitosis.

Meiosis produces cells that are not identical.

Because of the independent assortment of homologues and the process of crossing over, gametes show great variation.

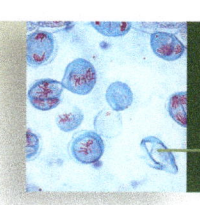

Review Questions

UNDERSTAND

1. In comparing somatic cells and gametes, somatic cells are
   a. diploid with half the number of chromosomes.
   b. haploid with half the number of chromosomes.
   c. diploid with twice the number of chromosomes.
   d. haploid with twice the number of chromosomes.

2. What are *homologous* chromosomes?
   a. The two halves of a replicated chromosome
   b. Two identical chromosomes from one parent
   c. Two genetically identical chromosomes, one from each parent
   d. Two genetically similar chromosomes, one from each parent

3. Chiasmata form
   a. between homologous chromosomes.
   b. sister chromatids.
   c. between replicated copies of the same chromosomes.
   d. sex chromosomes but not autosomes.

4. Crossing over involves each of the following with the exception of
   a. the transfer of DNA between two nonsister chromatids.
   b. the transfer of DNA between two sister chromatids.
   c. the formation of a synaptonemal complex.
   d. the alignment of homologous chromosomes.

5. During anaphase I
   a. sister chromatids separate and move to the poles.
   b. homologous chromosomes move to opposite poles.
   c. homologous chromosomes align at the middle of the cell.
   d. all the chromosomes align independently at the middle of the cell.

6. At metaphase I the kinetochores of sister chromatids are
   a. attached to microtubules from the same pole.
   b. attached to microtubules from opposite poles.
   c. held together with cohesin proteins.
   d. not attached to any microtubules.

7. What occurs during anaphase of meiosis II?
   a. The homologous chromosomes align.
   b. Sister chromatids are pulled to opposite poles.
   c. Homologous chromosomes are pulled to opposite poles.
   d. The haploid chromosomes line up.

APPLY

1. Which of the following does NOT contribute to genetic diversity?
   a. Independent assortment
   b. Recombination
   c. Metaphase of meiosis II
   d. Metaphase of meiosis I

2. How does DNA replication differ between mitosis and meiosis?
   a. DNA replication takes less time in meiosis because the cells are haploid.
   b. During meiosis, there is only one round of replication for two divisions.
   c. During mitosis, there is only one round of replication every other division.
   d. DNA replication is exactly the same in mitosis and meiosis.
3. Which of the following is NOT a distinct feature of meiosis?
   a. Pairing and exchange of genetic material between homologous chromosomes
   b. Attachment of sister kinetochores to spindle microtubules
   c. Movement of sister chromatids to the same pole
   d. Suppression of DNA replication
4. Which phase of meiosis I is most similar to the comparable phase in mitosis?
   a. Prophase I
   b. Metaphase I
   c. Anaphase I
   d. Telophase I
5. Structurally, meiotic cohesins have different components than mitotic cohesins. This leads to the following functional difference:
   a. During metaphase I, the sister kinetochores become attached to the same pole.
   b. Centromeres remain attached during anaphase I of meiosis.
   c. Centromeres remain attached through both divisions.
   d. Centromeric cohesins are destroyed at anaphase I, and cohesins along the arms are destroyed at anaphase II.
6. Mutations that affect DNA repair often also affect the accuracy of meiosis. This is because
   a. the proteins involved in the repair of double-strand breaks are also involved in crossing over.
   b. the proteins involved in DNA repair are also involved in sister chromatid cohesion.
   c. DNA repair only occurs on condensed chromosomes such as those found in meiosis.
   d. cohesin proteins are also necessary for DNA repair.

SYNTHESIZE

1. Diagram the process of meiosis for an imaginary cell with six chromosomes in a diploid cell.
   a. How many homologous pairs are present in this cell? Create a drawing that distinguishes between homologous pairs.
   b. Label each homologue to indicate whether it is maternal (M) or paternal (P).
   c. Draw a new cell showing how these chromosomes would arrange themselves during metaphase of meiosis I. Do all the maternal homologues have to line up on the same side of the cell?
   d. How would this picture differ if you were diagramming anaphase of meiosis II?
2. Mules are the offspring of the mating of a horse and a donkey. Mules are unable to reproduce. A horse has a total of 64 chromosomes, whereas donkeys have 62 chromosomes. Use your knowledge of meiosis to predict the diploid chromosome number of a mule. Propose a possible explanation for the inability of mules to reproduce.
3. Compare the processes of *independent assortment* and *crossing over*. Which process has the greatest influence on genetic diversity?
4. Aneuploid gametes are cells that contain the wrong number of chromosomes. Aneuploidy occurs as a result of *nondisjunction*, or lack of separation of the chromosomes during either phase of meiosis.
   a. At what point in meiotic cell division would nondisjunction occur?
   b. Imagine a cell had a diploid chromosome number of four. Create a diagram to illustrate the effects of nondisjunction of one pair of homologous chromosomes in meiosis I versus meiosis II.

CHAPTER 12

Patterns of Inheritance

Chapter Contents

- **12.1** The Mystery of Heredity
- **12.2** Monohybrid Crosses: The Principle of Segregation
- **12.3** Dihybrid Crosses: The Principle of Independent Assortment
- **12.4** Probability: Predicting the Results of Crosses
- **12.5** The Testcross: Revealing Unknown Genotypes
- **12.6** Extensions to Mendel

Introduction

Every living creature is a product of the long evolutionary history of life on Earth. All organisms share this history, but as far as we know, only humans wonder about the processes that led to their origin and investigate the possibilities. We are far from understanding everything about our origins, but we have learned a great deal. Like a partially completed jigsaw puzzle, the boundaries of this elaborate question have fallen into place, and much of the internal structure is becoming apparent. In this chapter, we discuss one piece of the puzzle—the enigma of heredity. Why do individuals, like the children in this picture, differ so much in appearance despite the fact that we are all members of the same species? And, why do members of a single family tend to resemble one another more than they resemble members of other families?

12.1 The Mystery of Heredity

Learning Outcomes

1. Describe explanations for inheritance prior to Mendel.
2. Explain the advantages of Mendel's experimental system.

As far back as written records go, patterns of resemblance among the members of particular families have been noted and commented on (figure 12.1), but there was no coherent model to explain these patterns. Before the 20th century, two concepts provided the basis for most thinking about heredity. The first was that heredity occurs within species. The second was that traits are transmitted directly from parents to offspring. Taken together, these ideas led to a view of inheritance as resulting from a blending of traits within fixed, unchanging species.

Figure 12.1 Heredity and family resemblance. Family resemblances are often strong—a visual manifestation of the mechanism of heredity.

Inheritance itself was viewed as traits being borne through fluid, usually identified as blood, that led to their blending in offspring. This older idea persists today in the use of the term "bloodlines" when referring to the breeding of domestic animals such as horses.

Taken together, however, these two classical assumptions led to a paradox. If no variation enters a species from outside, and if the variation within each species blends in every generation, then all members of a species should soon have the same appearance. It is clear that this does not happen—individuals within most species differ from one another, and they differ in characteristics that are transmitted from generation to generation.

Early plant biologists produced hybrids with puzzling results

The first investigator to achieve and document successful experimental **hybridizations** was Josef Kölreuter, who in 1760 cross-fertilized (or crossed, for short) different strains of tobacco and obtained fertile offspring. The hybrids differed in appearance from both parent strains. When individuals within the hybrid generation were crossed, their offspring were highly variable. Some of these offspring resembled plants of the hybrid generation (their parents), but a few resembled the original strains (their grandparents). The variation observed in second-generation offspring contradicts the theory of direct transmission. This can be seen as the beginning of modern genetics.

Over the next hundred years, other investigators elaborated on Kölreuter's work. T. A. Knight, an English landholder, in 1823 crossed two varieties of the garden pea, *Pisum sativum* (figure 12.2). One of these varieties had green seeds, and the other had yellow seeds. Both varieties were **true-breeding,** meaning that the offspring produced from self-fertilization remained uniform from one generation to the next. All of the progeny (offspring) of the cross between the two varieties had yellow seeds. Among the offspring of these hybrids, however, some plants produced yellow seeds and others, less common, produced green seeds.

Other investigators made observations similar to Knight's, namely that alternative forms of observed traits were being distributed among the offspring. A modern geneticist would say the alternative forms of each trait were **segregating** among the progeny of a mating, meaning that some offspring exhibited one form of a trait (yellow seeds), and other offspring from the same mating exhibited a different form (green seeds). This segregation of alternative forms of a trait provided the clue that led Gregor Mendel to his understanding of the nature of heredity.

Within these deceptively simple results were the makings of a scientific revolution. Nevertheless, another century passed before the process of segregation was fully appreciated.

Mendel was the first to quantify the results of his crosses

Born in 1822 to peasant parents in Austria, Gregor Mendel (figure 12.3) was educated in a monastery and went on to study science and mathematics at the University of Vienna, where he failed his examinations for a teaching certificate. He returned to the monastery and spent the rest of his life there, eventually becoming abbot. In the garden of the monastery, Mendel initiated his own series of experiments on plant hybridization. The results of these experiments would ultimately change our views of heredity irrevocably.

Practical considerations for use of the garden pea

For his experiments, Mendel chose the garden pea, the same plant Knight and others had studied. The choice was a good one for several reasons. First, many earlier investigators had produced hybrid peas by crossing different varieties, so Mendel knew that he could expect to observe segregation of traits among the offspring.

Figure 12.2 The garden pea, *Pisum sativum.* Easy to cultivate and able to produce many distinctive varieties, the garden pea was a popular experimental subject in investigations of heredity as long as a century before Gregor Mendel's experiments.

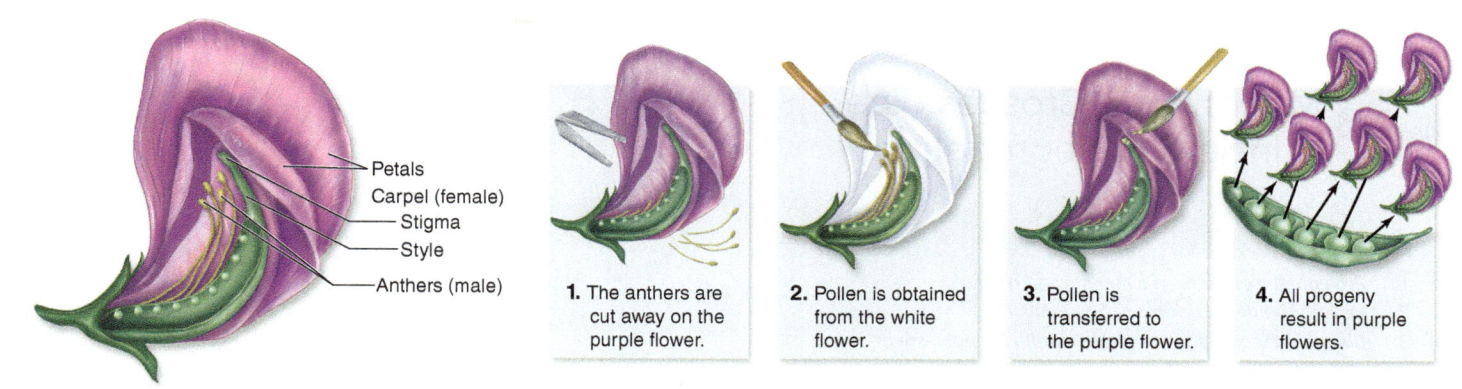

Figure 12.3 How Mendel conducted his experiments. In a pea plant flower, petals enclose both the male anther (containing pollen grains, which give rise to haploid sperm) and the female carpel (containing ovules, which give rise to haploid eggs). This ensures self-fertilization will take place unless the flower is disturbed. Mendel collected pollen from the anthers of a white flower, then placed that pollen onto the stigma of a purple flower with anthers removed. This cross-fertilization yields all hybrid seeds that give rise to purple flowers. Using pollen from a white flower to fertilize a purple flower gives the same result.

Inquiry question What confounding problems could have been seen if Mendel had chosen another plant with exposed male and female structures?

Second, a large number of pure varieties of peas were available. Mendel initially examined 34 varieties. Then, for further study, he selected lines that differed with respect to seven easily distinguishable traits, such as round versus wrinkled seeds and yellow versus green seeds, the latter a trait that Knight had studied.

Third, pea plants are small and easy to grow, and they have a relatively short generation time. A researcher can therefore conduct experiments involving numerous plants, grow several generations in a single year, and obtain results relatively quickly.

A fourth advantage of studying peas is that both the male and female sexual organs are enclosed within each pea flower (figure 12.3), and gametes produced by the male and female parts of the same flower can fuse to form viable offspring, a process termed **self-fertilization.** This self-fertilization takes place automatically within an individual flower if it is not disturbed. It is also possible to prevent self-fertilization by removing a flower's male parts before fertilization occurs, then introduce pollen from a different strain, thus performing *cross-pollination* that results in *cross-fertilization* (figure 12.3).

Mendel's experimental design

Mendel was careful to focus on only a few specific differences between the plants he was using and to ignore the countless other differences he must have seen. He also had the insight to realize that the differences he selected must be comparable. For example, he recognized that trying to study the inheritance of round seeds versus tall height would be useless.

Mendel usually conducted his experiments in three stages:

1. Mendel allowed plants of a given variety to self-cross for multiple generations to assure himself that the traits he was studying were indeed true-breeding—that is, transmitted unchanged from generation to generation.
2. Mendel then performed crosses between true-breeding varieties exhibiting alternative forms of traits. He also performed **reciprocal crosses:** using pollen from a white-flowered plant to fertilize a purple-flowered plant, then using pollen from a purple-flowered plant to fertilize a white-flowered plant.
3. Finally, Mendel permitted the hybrid offspring produced by these crosses to self-fertilize for several generations, allowing him to observe the inheritance of alternative forms of a trait. Most important, he counted the numbers of offspring exhibiting each trait in each succeeding generation.

This quantification of results is what distinguished Mendel's research from that of earlier investigators, who only noted differences in a qualitative way. Mendel's mathematical analysis of experimental results led to the inheritance model that we still use today.

Learning Outcomes Review 12.1

Prior to Mendel, concepts of inheritance did not form a consistent model. The dominant view was of blending inheritance, in which traits of parents were carried by fluid and "blended" in offspring. Plant hybridizers before Mendel, however, had already cast doubt on this model by observing characteristics in hybrids that seemed to change in second-generation offspring. Mendel's experiments with plants involved quantifying types of offspring and mathematically analyzing his observations.

- *Which was more important to Mendel's success: his approach, or his choice of experimental material?*

12.2 Monohybrid Crosses: The Principle of Segregation

Learning Outcomes

1. Evaluate the outcome of a monohybrid cross.
2. Explain Mendel's Principle of Segregation.
3. Compare the segregation of alleles with the behavior of homologues in meiosis.

A *monohybrid cross* is a cross that follows only two variations on a single trait, such as white- and purple-colored flowers. This deceptively simple kind of cross can lead to important conclusions about the nature of inheritance.

The seven characteristics, or characters, Mendel studied in his experiments possessed two variants that differed from one another in ways that were easy to recognize and score (figure 12.4). We examine in detail Mendel's crosses with flower color. His experiments with other characters were similar, and they produced similar results.

The F_1 generation exhibits only one of two traits with no blending

When Mendel crossed white-flowered and purple-flowered plants, the hybrid offspring he observed did not have flowers of intermediate color, as the hypothesis of blending inheritance would predict. Instead, in every case the flower color of the offspring resembled that of one of their parents. These offspring are customarily referred to as the **first filial generation,** or **F_1.** In a cross of white-flowered and purple-flowered plants, the F_1 offspring all had purple flowers, as other scientists had reported before Mendel.

Mendel referred to the form of each trait expressed in the F_1 plants as **dominant,** and to the alternative form that was not expressed in the F_1 plants as **recessive.** For each of the seven pairs of contrasting traits that Mendel examined, one of the pair proved to be dominant and the other recessive.

The F_2 generation exhibits a 3:1 ratio of both traits

After allowing individual F_1 plants to mature and self-fertilize, Mendel collected and planted the seeds from each plant to see what the offspring in the **second filial generation,** or **F_2,** would look like. He found that although most F_2 plants had purple flowers, some exhibited white flowers, the recessive trait. Although hidden in the F_1 generation, the recessive trait had reappeared among some F_2 individuals.

Believing the proportions of the F_2 types would provide some clue about the mechanism of heredity, Mendel counted the numbers of each type among the F_2 progeny. In the cross between the purple-flowered F_1 plants, he observed a total of 929 F_2 individuals. Of these, 705 (75.9%) had purple flowers, and 224 (24.1%) had white flowers (figure 12.4). Approximately ¼ of the F_2 individuals, therefore, exhibited the recessive form of the character.

Dominant	Recessive	F_2 Generation
1. Flower Color		
Purple	White	705 Purple: 224 White — 3.15:1
2. Seed Color		
Yellow	Green	6022 Yellow: 2001 Green — 3.01:1
3. Seed Texture		
Round	Wrinkled	5474 Round: 1850 Wrinkled — 2.96:1
4. Pod Color		
Green	Yellow	428 Green: 152 Yellow — 2.82:1
5. Pod Shape		
Inflated	Constricted	882 Inflated: 299 Constricted — 2.95:1
6. Flower Position		
Axial	Terminal	651 Axial: 207 Terminal — 3.14:1
7. Plant Height		
Tall	Short	787 Tall: 277 Short — 2.84:1

Figure 12.4 Mendel's seven traits. Mendel studied how differences among varieties of peas were inherited when the varieties were crossed. Similar experiments had been done before, but Mendel was the first to quantify the results and appreciate their significance. Results are shown for seven different monohybrid crosses. The F_1 generation is not shown in the table.

Mendel observed the same numerical result with the other six characters he examined: Of the F_2 individuals, ¾ exhibited the dominant trait, and ¼ displayed the recessive trait (figure 12.4). In other words, the dominant-to-recessive ratio among the F_2 plants was always close to 3:1.

The 3:1 ratio is actually 1:2:1

Mendel went on to examine how the F_2 plants passed traits to subsequent generations. He found that plants exhibiting the recessive trait were always true-breeding. For example, the white-flowered F_2 individuals only produced white-flowered offspring when they were self-fertilized. By contrast, ⅓ of the dominant, purple-flowered F_2 individuals (¼ of all F_2 offspring) were true-breeding, but ⅔ were not. This last class of plants produced a 3:1 ratio of dominant and recessive individuals in the third filial generation (F_3).

This result suggested that, for the entire sample, the 3:1 ratio that Mendel observed in the F_2 generation was really a disguised 1:2:1 ratio: ¼ true-breeding dominant individuals, ½ not-true-breeding dominant individuals, and ¼ true-breeding recessive individuals (figure 12.5).

> **Data analysis** In the previous set of crosses, if the purple F_1 were backcrossed to the white parent, what would be the phenotypic ratio? The genotypic ratio?

Mendel's Principle of Segregation explains monohybrid observations

From his experiments, Mendel was able to understand four things about the nature of heredity:

- The plants he crossed did not produce progeny of intermediate appearance, as a hypothesis of blending inheritance would have predicted. Instead, different plants inherited each trait intact, as a discrete characteristic.
- For each pair of alternative forms of a trait, one alternative was not expressed in the F_1 hybrids, although it reappeared in some F_2 individuals. *The trait that "disappeared" must therefore be latent (present but not expressed) in the F_1 individuals.*
- The pairs of alternative traits examined were segregated among the progeny of a particular cross, some individuals exhibiting one trait and some the other.
- These alternative traits were expressed in the F_2 generation in the ratio of ¾ dominant to ¼ recessive. This characteristic 3:1 segregation is referred to as the **Mendelian ratio** for a monohybrid cross.

Mendel's five-element model

Mendel's results can be explained with a simple model that has stood the test of time. Using more modern language than Mendel, this can be summarized as follows:

1. Parents do not transmit physiological traits directly to their offspring. Rather, they transmit discrete information for the traits, what Mendel called "factors." We now call these factors *genes*.

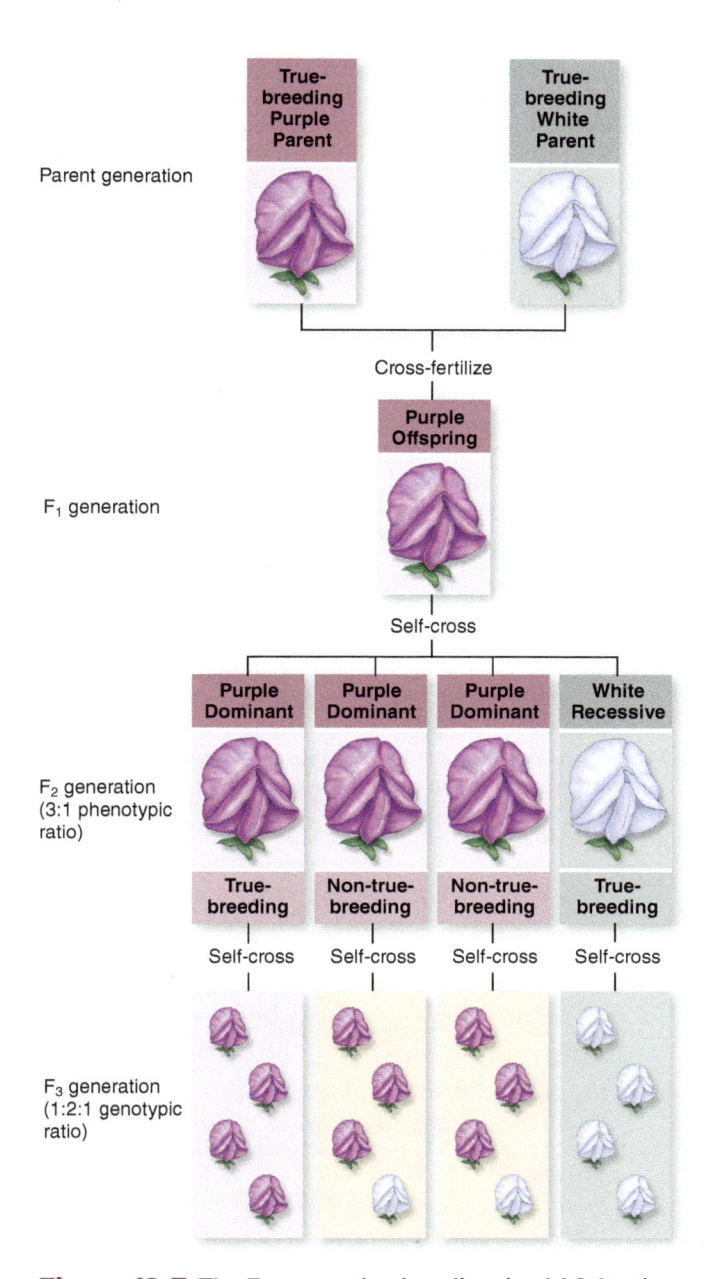

Figure 12.5 The F_2 generation is a disguised 1:2:1 ratio. By allowing the F_2 generation to self-fertilize, Mendel found from the offspring (F_3) that the ratio of F_2 plants was 1 true-breeding dominant: 2 not-true-breeding dominant: and 1 true-breeding recessive.

2. Each individual receives one copy of each gene from each parent. We now know that genes are carried on chromosomes, and each adult individual is diploid, with one set of chromosomes from each parent.
3. Not all copies of a gene are identical. The alternative forms of a gene are called **alleles.** When two haploid gametes containing the same allele fuse during fertilization, the resulting offspring is said to be **homozygous.** When the two haploid gametes contain different alleles, the resulting offspring is said to be **heterozygous.**

4. The two alleles remain discrete—they neither blend with nor alter each other. Therefore, when the individual matures and produces its own gametes, the alleles segregate randomly into these gametes.
5. The presence of a particular allele does not ensure that the trait it encodes will be expressed. In heterozygous individuals, only one allele is expressed (the dominant one), and the other allele is present but unexpressed (the recessive one).

Geneticists now refer to the total set of alleles that an individual contains as the individual's **genotype**. The physical appearance or other observable characteristics of that individual, which result from an allele's expression, is termed the individual's **phenotype**. In other words, the genotype is the blueprint, and the phenotype is the visible outcome in an individual.

This also allows us to present Mendel's ratios in more modern terms. The 3:1 ratio of dominant to recessive is the monohybrid phenotypic ratio. The 1:2:1 ratio of homozygous dominant to heterozygous to homozygous recessive is the monohybrid genotypic ratio. The genotypic ratio "collapses" into the phenotypic ratio due to the action of the dominant allele making the heterozygote appear the same as homozygous dominant.

The Principle of Segregation

Mendel's model accounts for the ratios he observed in a neat and satisfying way. His main conclusion—that alternative alleles for a character segregate from each other during gamete formation and remain distinct—has since been verified in many other organisms. It is commonly referred to as Mendel's first law of heredity, or the **Principle of Segregation**. It can be simply stated as: *The two alleles for a gene segregate during gamete formation and are rejoined at random, one from each parent, during fertilization.*

The physical basis for allele segregation is the behavior of chromosomes during meiosis. As you saw in chapter 11, homologues for each chromosome disjoin during anaphase I of meiosis. The second meiotic division then produces gametes that contain only one homologue for each chromosome.

It is a tribute to Mendel that his analysis arrived at the correct scheme, even though he had no knowledge of the cellular mechanisms of inheritance; neither chromosomes nor meiosis had yet been described.

The Punnett square allows symbolic analysis

To test his model, Mendel first expressed it in terms of a simple set of symbols. He then used the symbols to interpret his results.

Consider again Mendel's cross of purple-flowered with white-flowered plants. By convention, we assign the symbol *P* (uppercase) to the dominant allele, associated with the production of purple flowers, and the symbol *p* (lowercase) to the recessive allele, associated with the production of white flowers.

In this system, the genotype of an individual that is true-breeding for the recessive white-flowered trait would be designated *pp*. Similarly,

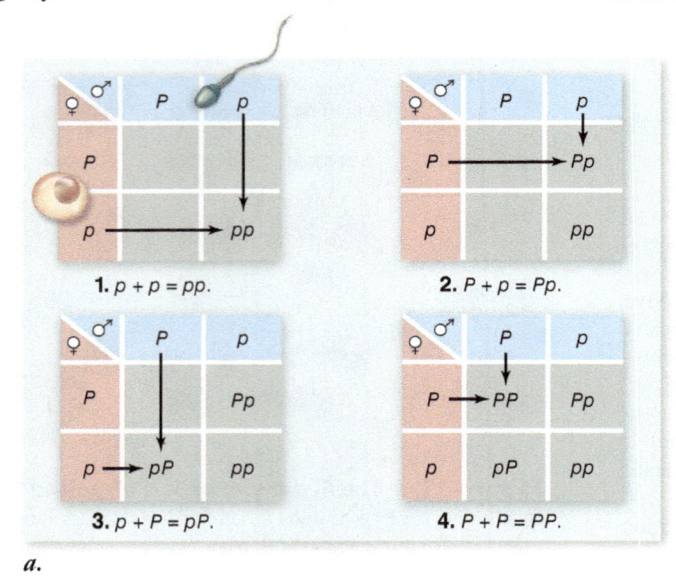

Figure 12.6 Using a Punnett square to analyze Mendel's cross.
a. To make a Punnett square, place the different female gametes along the side of a square and the different male gametes along the top. Each potential zygote is represented as the intersection of a vertical line and a horizontal line. *b.* In Mendel's cross of purple by white flowers, each parent makes only one type of gamete. The F_1 are all purple, *Pp*, heterozygotes. These F_1 offspring make two types of gametes that can be combined to produce three kinds of F_2 offspring: *PP* homozygous dominant (purple); *Pp* heterozygous (also purple); and *pp* homozygous recessive (white). The phenotypic ratio is 3 purple:1 white. The genotypic ratio is 1*PP*:2*Pp*:1*pp*.

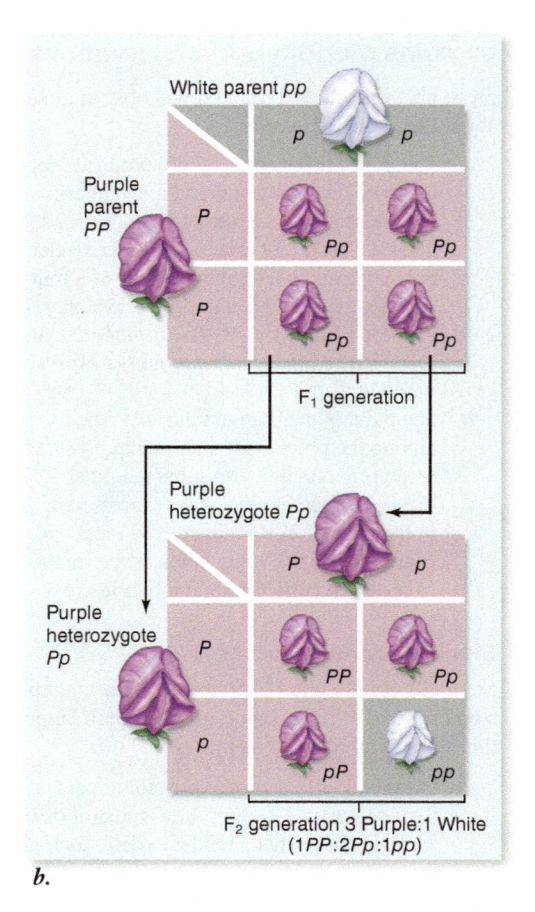

TABLE 12.1 Some Dominant and Recessive Traits in Humans

Recessive Traits	Phenotypes	Dominant Traits	Phenotypes
Albinism	Lack of melanin pigmentation	Middigital hair	Presence of hair on middle segment of fingers
Alkaptonuria	Inability to metabolize homogentisic acid	Brachydactyly	Short fingers
Red-green color blindness	Inability to distinguish red or green wavelengths of light	Huntington disease	Degeneration of nervous system, starting in middle age
Cystic fibrosis	Abnormal gland secretion, leading to liver degeneration and lung failure	Phenylthiocarbamide (PTC) sensitivity	Ability to taste PTC as bitter
Duchenne muscular dystrophy	Wasting away of muscles during childhood	Camptodactyly	Inability to straighten the little finger
Hemophilia	Inability of blood to clot properly, some clots form but the process is delayed	Hypercholesterolemia (the most common human Mendelian disorder)	Elevated levels of blood cholesterol and risk of heart attack
Sickle cell anemia	Defective hemoglobin that causes red blood cells to curve and stick together	Polydactyly	Extra fingers and toes

the genotype of a true-breeding purple-flowered individual would be designated *PP*. In contrast, a heterozygote would be designated *Pp* (dominant allele first). Using these conventions and denoting a cross between two strains with ×, we can symbolize Mendel's original purple × white cross as *PP* × *pp*.

Because a white-flowered parent (*pp*) can produce only *p* gametes, and a true-breeding purple-flowered parent (*PP*, *homozygous dominant*) can produce only *P* gametes, the union of these gametes can produce only heterozygous *Pp* offspring in the F_1 generation. Because the *P* allele is dominant, all of these F_1 individuals will have purple flowers.

When F_1 individuals are allowed to self-fertilize, the *P* and *p* alleles segregate during gamete formation to produce both *P* gametes and *p* gametes. Gametes will be randomly combined during fertilization to form F_2 individuals. The F_2 possibilities may be visualized in a simple diagram called a **Punnett square,** named after its originator, the English geneticist R. C. Punnett (figure 12.6a). Mendel's model, analyzed in terms of a Punnett square, clearly predicts that the F_2 generation should consist of ¾ purple-flowered plants and ¼ white-flowered plants, a phenotypic ratio of 3:1 (figure 12.6b).

Some human traits exhibit dominant/recessive inheritance

A number of human traits have been shown to display both dominant and recessive inheritance (table 12.1 provides a sample of these). Researchers cannot perform controlled crosses in humans the way Mendel did with pea plants; instead geneticists study crosses that have already been performed—in other words, family histories. The organized methodology we use is a **pedigree,** a consistent graphical representation of matings and offspring over multiple generations for a particular trait. The information in the pedigree may allow geneticists to deduce a model for the mode of inheritance of the trait. In analyzing these pedigrees, it is important to realize that disease-causing alleles are usually quite rare in the general population.

A dominant pedigree: Juvenile glaucoma

One of the most extensive pedigrees yet produced traced the inheritance of a form of blindness caused by a dominant allele. The disease allele causes a form of hereditary juvenile glaucoma. The disease causes degeneration of nerve fibers in the optic nerve, leading to blindness.

This pedigree followed inheritance over three centuries, following the origin back to a couple in a small town in northwestern France who died in 1495. A small portion of this pedigree is shown in figure 12.7. The dominant nature of the trait is obvious from the

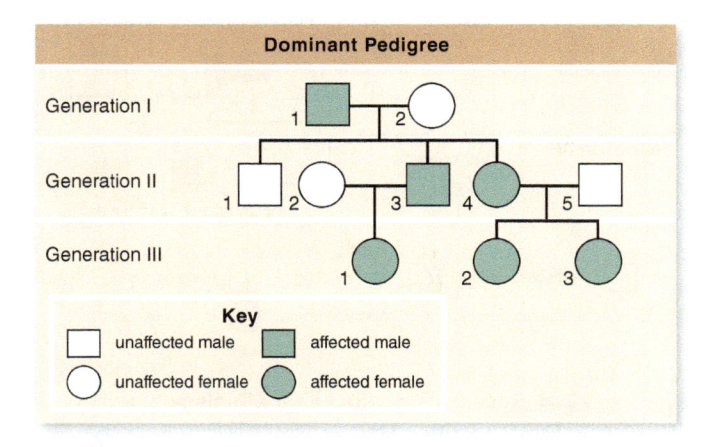

Figure 12.7 Dominant pedigree for hereditary juvenile glaucoma. Males are shown as squares and females are shown as circles. Affected individuals are shown shaded. The dominant nature of this trait can be seen in the trait appearing in every generation, a feature of dominant traits.

Data analysis If one of the affected females in the third generation married an unaffected male, could she produce unaffected offspring? If so, what are the chances of having unaffected offspring?

fact that every generation shows the trait. This is extremely unlikely for a recessive trait as it would require large numbers of unrelated individuals to be carrying the disease allele.

A recessive pedigree: Albinism

An example of inheritance of a recessive human trait is albinism, a condition in which the pigment melanin is not produced. Long thought to be due to a single gene, multiple genes are now known that lead to albinism; the common feature is the loss of pigment from hair, skin, and eyes. The loss of pigment makes albinistic individuals sensitive to the sun. The tanning effect we are all familiar with from exposure to the sun is due to increased numbers of pigment-producing cells, and increased production of pigment. This is lacking in albinistic individuals due to the lack of any pigment to begin with.

The pedigree in figure 12.8 is for a form of albinism due to a nonfunctional allele of the enzyme tyrosinase, which is required for the formation of melanin pigment. The genetic characteristics of this form of albinism are: Females and males are affected equally, most affected individuals have unaffected parents, a single affected parent usually does not have affected offspring, and affected offspring are more frequent when parents are related. Each of these features can be seen in figure 12.8, and all of this fits a recessive mode of inheritance.

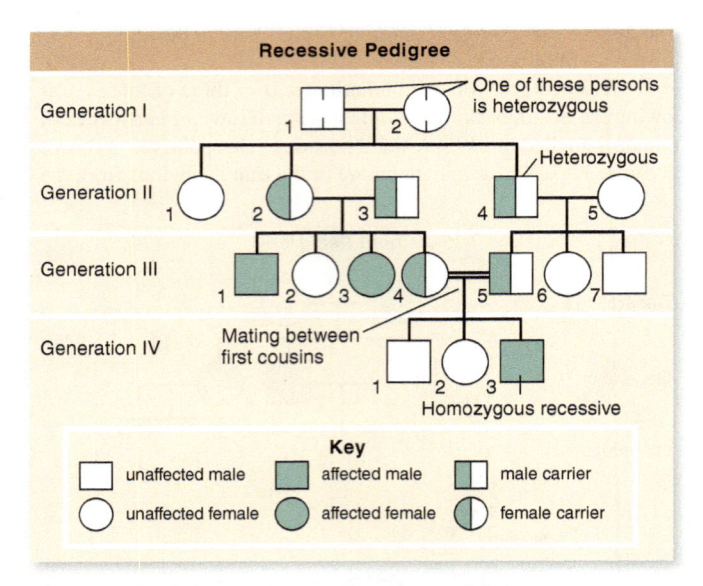

Figure 12.8 Recessive pedigree for albinism. One of the two individuals in the first generation must be heterozygous and individuals II-2 and II-4 must be heterozygous. Notice that for each affected individual, neither parent is affected, but both must be heterozygous (carriers). The double line indicates a consanguineous mating (between relatives) that, in this case, produced affected offspring.

Inquiry question From the standpoint of genetic disease, why is it never advisable for close relatives to mate and have children?

Learning Outcomes Review 12.2

Mendel's monohybrid crosses refute the idea of blending. One trait disappears in the first generation (F_1), then reappears in a predictable ratio in the next (F_2). The trait observable in the F_1 is called dominant, and the other recessive. In the F_2, the ratio of observed dominant offspring to recessive is 3:1, and this represents a ratio of 1 homozygous dominant to 2 heterozygous to 1 homozygous recessive. The Principle of Segregation states that alleles segregate into different gametes, which randomly combine at fertilization. The physical basis for segregation is the separation of homologues during anaphase I of meiosis.

- What fraction of tall F_2 plants are true-breeding?

12.3 Dihybrid Crosses: The Principle of Independent Assortment

Learning Outcomes
1. Evaluate the outcome of a dihybrid cross.
2. Explain Mendel's Principle of Independent Assortment.
3. Compare the segregation of alleles for different genes with the behavior of different homologues in meiosis.

The Principle of Segregation explains the behavior of alternative forms of a single trait in a monohybrid cross. The next step is to extend this to follow the behavior of two different traits in a single cross: a **dihybrid cross.**

With an understanding of the behavior of single traits, Mendel went on to ask if different traits behaved independently in hybrids. He first established a series of true-breeding lines of peas that differed in two of the seven characters he had studied. He then crossed contrasting pairs of the true-breeding lines to create heterozygotes. These heterozygotes are now doubly heterozygous, or dihybrid. Finally, he self-crossed the dihybrid F_1 plants to produce an F_2 generation, and counted all progeny types.

Traits in a dihybrid cross behave independently

Consider a cross involving different seed shape alleles (round, *R*, and wrinkled, *r*) and different seed color alleles (yellow, *Y*, and green, *y*). Crossing round yellow (*RR YY*) with wrinkled green (*rr yy*), produces heterozygous F_1 individuals having the same phenotype (namely round and yellow) and the same genotype (*Rr Yy*). Allowing these dihybrid F_1 individuals to self-fertilize produces an F_2 generation.

The F_2 generation exhibits four types of progeny in a 9:3:3:1 ratio

In analyzing these results, we first consider the number of possible phenotypes. We expect to see the two parental phenotypes: round yellow and wrinkled green. If the traits behave independently, then we can also expect one trait from each parent to produce plants with round green seeds and others with wrinkled yellow seeds.

Next consider what types of gametes the F_1 individuals can produce. Again, we expect the two types of gametes found in the parents: *RY* and *ry*. If the traits behave independently, then we can also expect the gametes *Ry* and *rY*. Using modern language, two genes each with two alleles can be combined four ways to produce these gametes: *RY, ry, Ry,* and *rY*.

A dihybrid Punnett square

We can then construct a Punnett square with these gametes to generate all possible progeny. This is a 4 × 4 square with 16 possible outcomes. Filling in the Punnett square produces all possible offspring (figure 12.9). From this we can see that there are 9 round yellow, 3 wrinkled yellow, 3 round green, and 1 wrinkled green. This predicts a phenotypic ratio of 9:3:3:1 for traits that behave independently.

Mendel's Principle of Independent Assortment explains dihybrid results

What did Mendel actually observe? From a total of 556 seeds from self-fertilized dihybrid plants, he observed the following results:

- 315 round yellow (signified *R__ Y__*, where the underscore indicates the presence of either allele),
- 108 round green *(R__ yy)*,
- 101 wrinkled yellow *(rr Y__)*, and
- 32 wrinkled green *(rr yy)*.

These results are very close to a 9:3:3:1 ratio. (The expected 9:3:3:1 ratio for 556 offspring is 313:104:104:35.)

The alleles of two genes appeared to behave independently of each other. Mendel referred to this phenomenon as the traits assorting independently. Note that this *independent assortment* of different genes in no way alters the segregation of individual pairs of alleles for each gene. Round versus wrinkled seeds occur in a ratio of approximately 3:1 (423:133); so do yellow versus green seeds (416:140). Mendel obtained similar results for other pairs of traits.

We call this Mendel's second law of heredity, or the **Principle of Independent Assortment.** This can also be stated simply: *In a dihybrid cross, the alleles of each gene assort independently.* A more precise statement would be stated: *The segregation of different allele pairs is independent.* This statement more closely ties independent assortment to the behavior of chromosomes during meiosis (see chapter 11). The independent alignment of different homologous chromosome pairs during metaphase I leads to the independent segregation of the different allele pairs.

Learning Outcomes Review 12.3

Mendel's analysis of dihybrid crosses revealed that the segregation of allele pairs for different traits is independent; this finding is known as Mendel's Principle of Independent Assortment. When individuals that differ in two traits are crossed, and their progeny are intercrossed, the result is four different types that occur in a ratio of 9:3:3:1, Mendel's dihybrid ratio. This occurs because of the independent behavior of different homologous pairs of chromosomes during meiosis I.

- *Which is more important in terms of explaining Mendel's laws, meiosis I or meiosis II?*

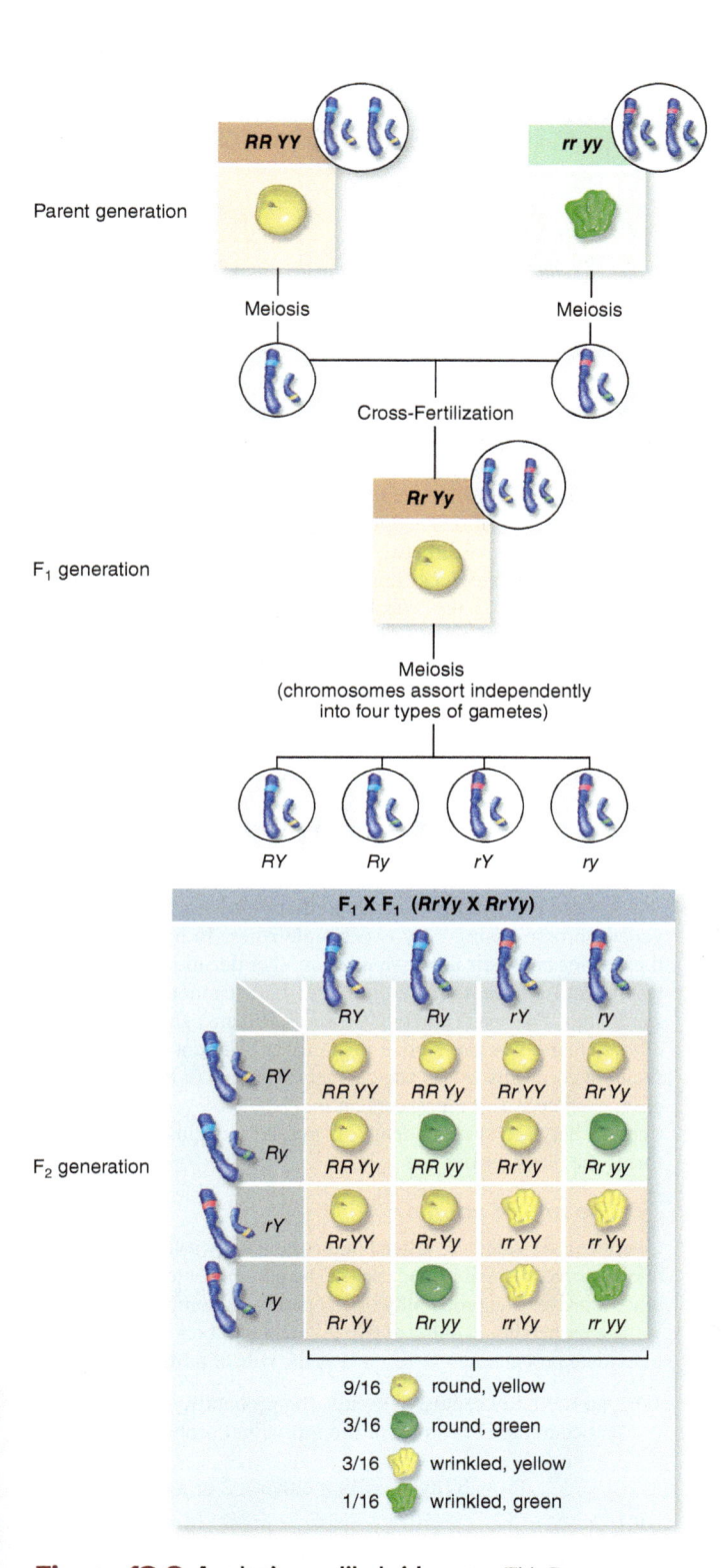

Figure 12.9 Analyzing a dihybrid cross. This Punnett square shows the results of Mendel's dihybrid cross between plants with round yellow seeds and plants with wrinkled green seeds. The ratio of the four possible combinations of phenotypes is predicted to be 9:3:3:1, the ratio that Mendel found.

12.4 Probability: Predicting the Results of Crosses

Learning Outcomes
1. Explain the rule of addition and the rule of multiplication.
2. Apply the rules of probability to genetic crosses.

Probability allows us to predict the likelihood of the outcome of random events. Because the behavior of different chromosomes during meiosis is independent, we can use probability to predict the outcome of crosses. The probability of an event that is certain to happen is equal to 1. In contrast, an event that can never happen has a probability of 0. Therefore, probabilities for all other events have fractional values, between 0 and 1. For instance, when you flip a coin, two outcomes are possible; there is only one way to get the event "heads" so the probability of heads is one divided by two, or ½. In the case of genetics, consider a pea plant heterozygous for the flower color alleles *P* and *p*. This individual can produce two types of gametes in equal numbers, again due to the behavior of chromosomes during meiosis. There is one way to get a *P* gamete, so the probability of any particular gamete carrying a *P* allele is 1 divided by 2 or ½, just like the coin toss.

Two probability rules help predict monohybrid cross results

We can use probability to make predictions about the outcome of genetic crosses using only two simple rules. Before we describe these rules and their uses, we need another definition. We say that two events are *mutually exclusive* if both cannot happen at the same time. The heads and tails of a coin flip are examples of mutually exclusive events. Notice that this is different from two consecutive coin flips where you can get two heads or two tails. In this case, each coin flip represents an *independent event*. It is the distinction between independent and mutually exclusive events that forms the basis for our two rules.

The rule of addition

Consider a six-sided die instead of a coin: for any roll of the die, only one outcome is possible, and each of the possible outcomes are mutually exclusive. The probability of any particular number coming up is ⅙. The probability of either of two different numbers is the sum of the individual probabilities, or restated as the **rule of addition**:

For two mutually exclusive events, the probability of either event occurring is the sum of the individual probabilities.

Probability of rolling either a 2 or a 6
is = ⅙ + ⅙ = 2/6 = ⅓

To apply this to our cross of heterozygous purple F_1, four mutually exclusive outcomes are possible: *PP*, *Pp*, *pP*, and *pp*. The probability of being heterozygous is the same as the probability of being either *Pp* or *pP*, or ¼ plus ¼, or ½:

Probability of F_2 heterozygote = ¼ *Pp* + ¼ *pP* = ½

In the previous example, of 379 total offspring, we would expect about 190 to be heterozygotes. (The actual number is 189.5.)

The rule of multiplication

The second rule, and by far the most useful for genetics, deals with the outcome of independent events. This is called the **product rule**, or **rule of multiplication**, and it states that the probability of two independent events both occurring is the *product* of their individual probabilities.

We can apply this to a monohybrid cross in which offspring are formed by gametes from each of two parents. For any particular outcome then, this is due to two independent events: the formation of two different gametes. Consider the purple F_1 parents from section 12.2. They are all *Pp* (heterozygotes), so the probability that a particular F_2 individual will be *pp* (homozygous recessive) is the probability of receiving a *p* gamete from the male (½) times the probability of receiving a *p* gamete from the female (½), or ¼:

Probability of *pp* homozygote = ½ *p* (male parent) × ½ *p* (female parent) = ¼ *pp*

This is actually the basis for the Punnett square that we used before. Each cell in the square was the product of the probabilities of the gametes that contribute to the cell. We then use the addition rule to sum the probabilities of the mutually exclusive events that make up each cell.

We can use the result of a probability calculation to predict the number of homozygous recessive offspring in a cross between heterozygotes. For example, out of 379 total offspring, we would expect about 95 to exhibit the homozygous recessive phenotype. (The actual calculated number is 94.75.)

Dihybrid cross probabilities are based on monohybrid cross probabilities

Probability analysis can be extended to the dihybrid case. For our purple F_1 by F_1 cross, there are four possible outcomes, three of which show the dominant phenotype. Thus the probability of any offspring showing the dominant phenotype is ¾, and the probability of any offspring showing the recessive phenotype is ¼. Now we can use this and the product rule to predict the outcome of a dihybrid cross. We will use our example of seed shape and color from section 12.3, but now analyze it using probability.

If the alleles affecting seed shape and seed color segregate independently, then the probability that a particular pair of alleles for seed shape would occur together with a particular pair of alleles for seed color is the product of the individual probabilities for each pair. For example, the probability that an individual with wrinkled green seeds (*rr yy*) would appear in the F_2 generation would be equal to the probability of obtaining wrinkled seeds (¼) times the probability of obtaining green seeds (¼), or 1/16:

Probability of *rr yy* = ¼ *rr* × ¼ *yy* = 1/16 *rr yy*

Because of independent assortment, we can think of the dihybrid cross as consisting of two independent monohybrid crosses; because these are independent events, the product rule applies. So, we can calculate the probabilities for each dihybrid phenotype:

Probability of round yellow ($R\_\_ Y\_\_$) =
$\frac{3}{4} R\_\_ \times \frac{3}{4} Y\_\_ = \frac{9}{16}$

Probability of round green ($R\_\_ yy$) =
$\frac{3}{4} R\_\_ \times \frac{1}{4} yy = \frac{3}{16}$

Probability of wrinkled yellow ($rr Y\_\_$) =
$\frac{1}{4} rr \times \frac{3}{4} Y\_\_ = \frac{3}{16}$

Probability of wrinkled green ($rr yy$) =
$\frac{1}{4} rr \times \frac{1}{4} yy = \frac{1}{16}$

The hypothesis that color and shape genes are independently sorted thus predicts that the F_2 generation will display a 9:3:3:1 phenotypic ratio. These ratios can be applied to an observed total offspring to predict the expected number in each phenotypic group. The underlying logic and the results are the same as obtained using the Punnett square.

Data analysis Purple-flowered, round, yellow peas are crossed to white-flowered, wrinkled, green peas to yield a purple-flowered, round, yellow F_1. If this F_1 is self-crossed, what proportion of progeny should be purple-flowered, round, yellow?

Learning Outcomes Review 12.4

The rule of addition states that the probability of either of two events occurring is the sum of their individual probabilities. The rule of multiplication states that the probability of two independent events both occurring is the product of their individual probabilities. These rules can be applied to genetic crosses to determine the probability of particular genotypes and phenotypes. Results can then be compared against these predictions.

- *If genes A and B assort independently, in a cross of Aa Bb by aa Bb, what is the probability of having the dominant phenotype for both genes?*

12.5 The Testcross: Revealing Unknown Genotypes

Learning Outcome

1. Interpret data from testcrosses to infer unknown genotypes.

To test his model further, Mendel devised a simple and powerful procedure called the **testcross**. In a testcross, an individual with unknown genotype is crossed with the homozygous recessive genotype—that is, the recessive parental variety. The contribution of the homozygous recessive parent can be ignored, because this parent can contribute only recessive alleles.

Consider a purple-flowered pea plant. It is impossible to tell whether such a plant is homozygous or heterozygous simply by looking at it. To learn its genotype, you can perform a testcross to a white-flowered plant. In this cross, the two possible test plant genotypes will give different results (figure 12.10):

Alternative 1: Unknown individual is homozygous dominant (*PP*)
$PP \times pp$: All offspring have purple flowers (*Pp*).

Alternative 2: Unknown individual is heterozygous (*Pp*)
$Pp \times pp$: ½ of offspring have white flowers (*pp*), and ½ have purple flowers (*Pp*).

Put simply, the appearance of the recessive phenotype in the offspring of a testcross indicates that the test individual is heterozygous for the gene in question.

For each pair of alleles Mendel investigated, he observed phenotypic F_2 ratios of 3:1 (see figure 12.4) and testcross ratios of 1:1, just as his model had predicted. Testcrosses can also be used to determine the genotype of an individual when two genes are involved. Mendel often performed testcrosses to verify the genotypes of dominant-appearing F_2 individuals.

An F_2 individual exhibiting both dominant traits ($A\_\_ B\_\_$) might have any of the following genotypes: *AABB, AaBB, AABb,* or

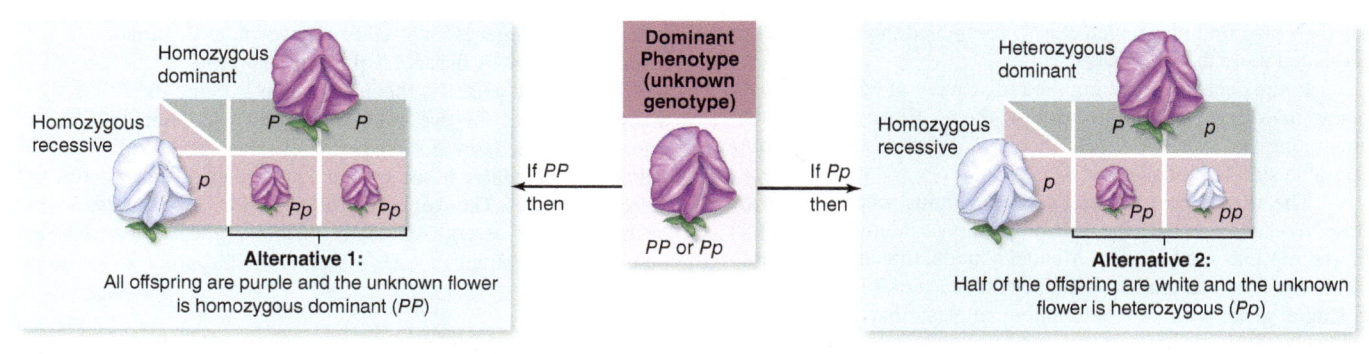

Figure 12.10 Using a testcross to determine unknown genotypes. Individuals with a dominant phenotype, such as purple flowers, can be either homozygous for the dominant allele, or heterozygous. Crossing an unknown purple plant to homozygous recessive *(white)* allows determination of its genotype. The Punnett "square" for each alternative shows only one row for the homozygous recessive white strain because they produce only *p*-bearing gametes.

AaBb. By crossing dominant-appearing F₂ individuals with homozygous recessive individuals (that is, *A__ B__* × *aabb*), Mendel was able to determine whether either or both of the traits bred true among the progeny, and so to determine the genotype of the F₂ parent.

> **Data analysis** Diagram all four possible testcrosses to determine genotypes for individuals that appear dominant for two traits.

Testcrossing is a powerful tool that simplifies genetic analysis. We will use this method of analysis in chapter 13, when we explore genetic mapping.

Learning Outcome Review 12.5

Individuals showing the dominant phenotype can be either homozygous dominant or heterozygous. Unknown genotypes can be revealed using a testcross, which is a cross to a homozygous recessive individual. Heterozygotes produce both dominant and recessive phenotypes in equal numbers as a result of the testcross.

- In a dihybrid testcross of a doubly heterozygous individual, what would be the expected phenotypic ratio?

12.6 Extensions to Mendel

Learning Outcomes

1. Describe how assumptions in Mendel's model result in oversimplification.
2. Discuss a genetic explanation for continuous variation.
3. Explain the genetic basis for observed alterations to Mendel's ratios.

Although Mendel's results did not receive much notice during his lifetime, three different investigators independently rediscovered his pioneering paper in 1900, 16 years after his death. They came across it while searching the literature in preparation for publishing their own findings, which closely resembled those Mendel had presented more than 30 years earlier.

In the decades following the rediscovery of Mendel's ideas, many investigators set out to test them. However, scientists attempting to confirm Mendel's theory often had trouble obtaining the same simple ratios he had reported.

The reason that Mendel's simple ratios were not always observed had to do with the traits that others examined. A number of assumptions are built into Mendel's model that are oversimplifications. These assumptions include that each trait is specified by a single gene with two alternative alleles; that there are no environmental effects; and that gene products act independently. The idea of dominance also hides a wealth of biochemical complexity. In this section, you'll see how Mendel's simple ideas can be extended to provide a more complete view of genetics (table 12.2).

In polygenic inheritance, more than one gene can affect a single trait

Often, the relationship between genotype and phenotype is more complicated than a single allele producing a single trait. Most phenotypes also do not reflect simple two-state cases like purple or white flowers.

Consider Mendel's crosses between tall and short pea plants. In reality, the "tall" plants actually have normal height, and the "short" plants are dwarfed by an allele at a single gene. But in most species, including humans, height varies over a continuous range, rather than having discrete values. This continuous distribution of a phenotype has a simple genetic explanation: More than one gene is at work. The mode of inheritance operating in this case is often called **polygenic inheritance.**

In reality, few phenotypes result from the action of only one gene. Instead, most characters reflect multiple additive contributions to the phenotype by several genes. When multiple genes act jointly to influence a character, such as height or weight, the character often shows a range of small differences. When these genes segregate independently, a gradation in the degree of difference can be observed when a group consisting of many individuals is examined (figure 12.11). We call this gradation **continuous variation,** and we call such traits **quantitative traits.** The greater the number of genes influencing a character, the more continuous the expected distribution of the versions of that character.

This continuous variation in traits is similar to blending different colors of paint: Combining one part red with seven parts white, for example, produces a much lighter shade of pink than does combining five parts red with three parts white. Different ratios of red to white result in a continuum of shades, ranging from pure red to pure white.

Often, variations can be grouped into categories, such as different height ranges. Plotting the numbers in each height category produces a curve called a *histogram,* such as that shown in figure 12.11. The bell-shaped histogram approximates an idealized *normal distribution,* in which the central tendency is characterized by the mean, and the spread of the curve indicates the amount of variation.

Traits once thought to be due to single genes are now known to be affected by multiple genes. Human eye color was thought to be brown dominant to blue, but this proves incorrect. Although the exact number of genes involved is not known, most variation in eye color is explained by the action of two to four genes. In the emerging field of forensic genetics, this information is being used to design assays to predict the eye color of unknown individuals. Similar efforts are aimed toward predicting the hair and skin color of unknown individuals based on our knowledge of the genetics of these phenotypes. These efforts do allow probabilistic predictions to be made about several externally visible characteristics (EVCs) based on genotyping.

In pleiotropy, a single gene can affect more than one trait

Not only can more than one gene affect a single trait, but a single gene can affect more than one trait. Considering the complexity of

TABLE 12.2	When Mendel's Laws/Results May Not Be Observed	
Genetic Occurrence	**Definition**	**Examples**
Polygenic inheritance	More than one gene can affect a single trait.	• Four genes are involved in determining eye color. • Human height
Pleiotropy	A single gene can affect more than one trait.	• A pleiotropic allele dominant for yellow fur in mice is recessive for a lethal developmental defect. • Cystic fibrosis • Sickle cell anemia
Multiple alleles for one gene	Genes may have more than two alleles.	ABO blood types in humans
Dominance not always complete	• In incomplete dominance the heterozygote is intermediate. • In codominance no single allele is dominant, and the heterozygote shows some aspect of both homozygotes.	• Japanese four o'clocks • Human blood groups
Environmental factors	Genes may be affected by the environment.	Siamese cats
Gene interaction	Products of genes can interact to alter genetic ratios.	• The production of a purple pigment in corn • Coat color in mammals

biochemical pathways and the interdependent nature of organ systems in multicellular organisms, this should be no surprise.

An allele that has more than one effect on phenotype is said to be **pleiotropic.** The pioneering French geneticist Lucien Cuenot studied yellow fur in mice, a dominant trait, and found he was unable to obtain a pure-breeding yellow strain by crossing individual yellow mice with each other. Individuals homozygous for the yellow allele died, because the yellow allele was pleiotropic: One effect was yellow coat color, but another was a lethal developmental defect.

Data analysis When Cuenot crossed yellow mice, what ratio of yellow to wild-type mice did he observe?

A pleiotropic allele may be dominant with respect to one phenotypic consequence (yellow fur) and recessive with respect to another (lethal developmental defect). Pleiotropic effects are difficult to predict, because a gene that affects one trait often performs other, unknown functions.

Pleiotropic effects are characteristic of many inherited disorders in humans, including cystic fibrosis and sickle cell anemia (discussed in chapter 13). In these disorders, multiple symptoms (phenotypes) can be traced back to a single gene defect. Cystic fibrosis patients exhibit clogged blood vessels, overly sticky mucus, salty sweat, liver and pancreas failure, and several other symptoms. It is often difficult to deduce the nature of the primary defect from the range of a gene's pleiotropic effects. As it turns out, all these symptoms of cystic fibrosis are pleiotropic effects of a single defect, a mutation in a gene that encodes a chloride ion transmembrane channel.

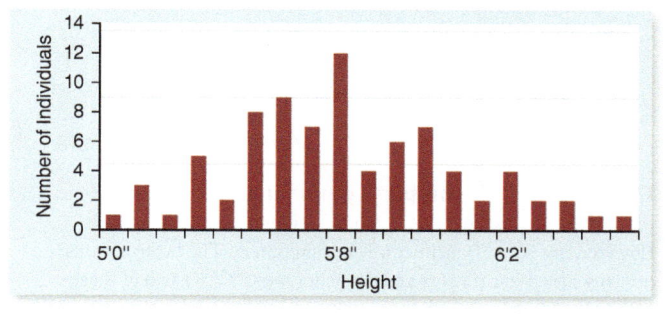

Figure 12.11 Height is a continuously varying trait.
The photo and accompanying graph show variation in height among 83 students in Dr. Hude's genetics classes from 2005–2007 at the University of Notre Dame. Because many genes contribute to height and tend to segregate independently of one another, the cumulative contribution of different combinations of alleles to height forms a *continuous* distribution of possible heights, in which the extremes are much rarer than the intermediate values. Variation can also arise due to environmental factors such as nutrition.

Genes may have more than two alleles

Mendel always looked at genes with two alternative alleles. Although any diploid individual can carry only two alleles for a gene, there may be more than two alleles in a population. The example of ABO blood types in humans, described later in this section, involves an allelic series with three alleles.

If you think of a gene as a sequence of nucleotides in a DNA molecule, then the number of possible alleles is huge because even a single nucleotide change could produce a new allele. In reality, the number of alleles possible for any gene is constrained, but usually more than two alleles exist for any gene in an outbreeding population. The dominance relationships of these alleles cannot be predicted, but can be determined by observing the phenotypes for the various heterozygous combinations.

Dominance is not always complete

Mendel's idea of dominant and recessive traits can seem hard to explain in terms of modern biochemistry. For example, if a recessive trait is caused by the loss of function of an enzyme encoded by the recessive allele, then why should a heterozygote, with only half the activity of this enzyme, have the same appearance as a homozygous dominant individual?

The answer is that enzymes usually act in pathways and not alone. These pathways, as you have seen in chapters 6 to 8, can be highly complex in terms of inputs and outputs, and they can sometimes tolerate large reductions in activity of single enzymes in the pathway without reductions in the level of the end-product. When this is the case, complete dominance will be observed; however, not all genes act in this way.

Incomplete dominance

In **incomplete dominance,** the phenotype of the heterozygote is intermediate between the two homozygotes. For example, in a cross between red- and white-flowering Japanese four o'clocks, described in figure 12.12, all the F_1 offspring have pink flowers—indicating that neither red nor white flower color was dominant. Looking only at the F_1, we might conclude that this is a case of blending inheritance. But when two of the F_1 pink flowers are crossed, they produce red-, pink-, and white-flowered plants in a 1:2:1 ratio. In this case the phenotypic ratio is the same as the genotypic ratio because all three genotypes can be distinguished.

Codominance

Most genes in a population possess several different alleles, and often no single allele is dominant; instead, each allele has its own effect, and the heterozygote shows some aspect of the phenotype of both homozygotes. The alleles are said to be **codominant.**

Codominance can be distinguished from incomplete dominance by the appearance of the heterozygote. In incomplete dominance, the heterozygote is intermediate between the two homozygotes, whereas in codominance, some aspect of both alleles is seen in the heterozygote. One of the clearest human examples is found in the human blood groups.

The different phenotypes of human blood groups are based on the response of the immune system to proteins on the surface of red blood cells. In homozygotes a single type of protein is found on the surface of cells, and in heterozygotes, two kinds of protein are found, leading to codominance.

The human ABO blood group system

The gene that determines ABO blood types encodes an enzyme that adds sugar molecules to proteins on the surface of red blood cells. These sugars act as recognition markers for the immune system (see chapter 51). The gene that encodes the enzyme, designated I, has three common alleles: I^A, whose product adds galactosamine; I^B, whose product adds galactose; and i, which codes for a protein that does not add a sugar.

SCIENTIFIC THINKING

Hypothesis: The pink F_1 observed in a cross of red and white Japanese four o'clock flowers is due to failure of dominance and is not an example of blending inheritance.

Prediction: If pink F_1 are self-crossed, they will yield progeny the same as the Mendelian monohybrid genotypic ratio. This would be 1 red: 2 pink: 1 white.

Test: Perform the cross and count progeny.

Result: When this cross is performed, the expected outcome is observed.
Conclusion: Flower color in Japanese four o'clock plants exhibits incomplete dominance.
Further Experiments: How many offspring would you need to count to be confident in the observed ratio?

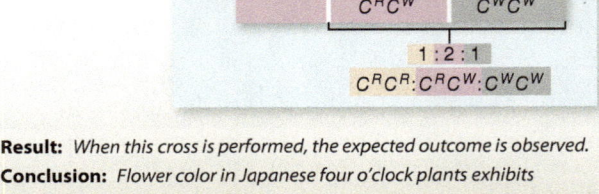

Figure 12.12 Incomplete dominance. In a cross between a red-flowered (genotype C^RC^R) Japanese four o'clock and a white-flowered one (C^WC^W), neither allele is dominant. The heterozygous progeny have pink flowers and the genotype C^RC^W. If two of these heterozygotes are crossed, the phenotypes of their progeny occur in a ratio of 1:2:1 (red:pink:white).

The three alleles of the I gene can be combined to produce six different genotypes. An individual heterozygous for the I^A and I^B alleles produces both forms of the enzyme and exhibits both galactose and galactosamine on red blood cells. Because both alleles are expressed simultaneously in heterozygotes, the I^A and I^B alleles are codominant. Both I^A and I^B are dominant over the i

Alleles	Blood Type	Sugars Exhibited	Donates and Receives
I^AI^A, I^Ai (I^A dominant to i)	A	Galactosamine	Receives A and O Donates to A and AB
I^BI^B, I^Bi (I^B dominant to i)	B	Galactose	Receives B and O Donates to B and AB
I^AI^B (codominant)	AB	Both galactose and galactosamine	Universal receiver Donates to AB
ii (i is recessive)	O	None	Receives O Universal donor

Figure 12.13 ABO blood groups illustrate both codominance and multiple alleles. There are three alleles of the I gene: I^A, I^B, and i. I^A and I^B are both dominant to i (see types A and B), but codominant to each other (see type AB). The genotypes that give rise to each blood type are shown with the associated phenotypes in terms of sugars added to surface proteins and the body's reaction after a blood transfusion.

allele, because both I^A and I^B alleles lead to sugar addition, whereas the i allele does not. The different combinations of the three alleles produce four different phenotypes (figure 12.13):

1. Type A individuals add only galactosamine. They are either I^AI^A homozygotes or I^Ai heterozygotes (two genotypes).
2. Type B individuals add only galactose. They are either I^BI^B homozygotes or I^Bi heterozygotes (two genotypes).
3. Type AB individuals add both sugars and are I^AI^B heterozygotes (one genotype).
4. Type O individuals add neither sugar and are ii homozygotes (one genotype).

These four different cell-surface phenotypes are called the **ABO blood groups.**

A person's immune system can distinguish among these four phenotypes. If a type A individual receives a transfusion of type B blood, the recipient's immune system recognizes the "foreign" antigen (galactose) and attacks the donated blood cells, causing them to clump, or agglutinate. The same thing would happen if the donated blood is type AB. However, if the donated blood is type O, no immune attack occurs, because there are no galactose antigens.

In general, any individual's immune system can tolerate a transfusion of type O blood, and so type O is termed the "universal donor." Because neither galactose nor galactosamine is foreign to type AB individuals (whose red blood cells have both sugars), those individuals may receive any type of blood, and type AB is termed the "universal recipient." Nevertheless, matching blood is preferable for any transfusion.

Phenotypes may be affected by the environment

Another assumption, implicit in Mendel's work, is that the environment does not affect the relationship between genotype and phenotype. For example, the soil in the abbey yard where Mendel performed his experiments was probably not uniform, and yet its possible effect on the expression of traits was ignored. But in reality, although the expression of genotype produces phenotype, the environment can affect this relationship.

Environmental effects are not limited to the external environment. For example, the alleles of some genes encode heat-sensitive products that are affected by differences in internal body temperature. The *ch* allele in Himalayan rabbits and Siamese cats encodes a heat-sensitive version of the enzyme tyrosinase, which as you may recall is involved in albinism (figure 12.14). The Ch version of the enzyme is inactivated at temperatures above about 33°C. At the surface of the torso and head of these animals, the temperature is above 33°C and tyrosinase is inactive, producing a whitish coat. At the extremities, such as the tips of the ears and tail, the temperature is usually below 33°C and the enzyme is active, allowing production of melanin that turns the coat in these areas a dark color.

> **Inquiry question** Many studies of identical twins separated at birth have revealed phenotypic differences in their development (height, weight, etc.). If these are identical twins, can you propose an explanation for these differences?

In epistasis, interactions of genes alter genetic ratios

The last simplistic assumption in Mendel's model is that the products of genes do not interact. But the products of genes may not act independently of one another, and the interconnected behavior of gene products can change the ratio expected by independent assortment, even if the genes are on different chromosomes that do exhibit independent assortment.

Given the interconnected nature of metabolism, it should not come as a surprise that many gene products are not independent. Genes that act in the same metabolic pathway, for example, should show some form of dependence at the level of function. In such cases, the ratio Mendel would predict is not readily observed, but it is still there in an altered form.

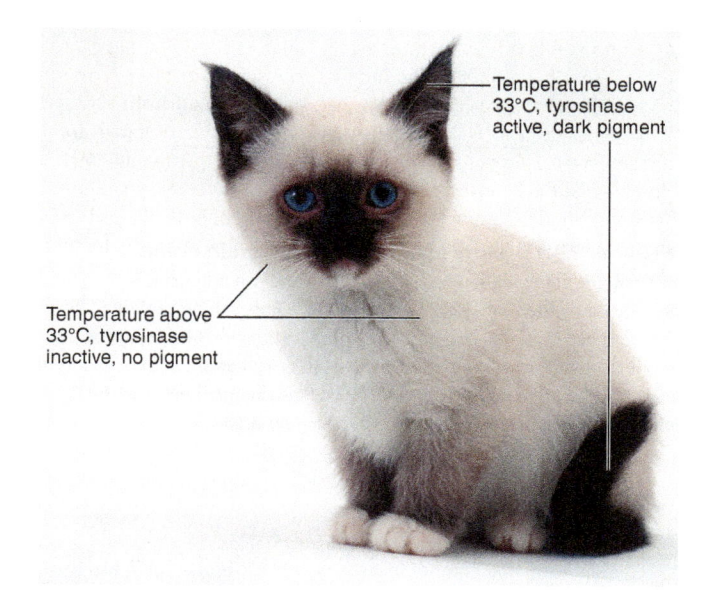

Figure 12.14 Siamese cat. The pattern of coat color is due to an allele that encodes a temperature-sensitive form of the enzyme tyrosinase.

In the tests of Mendel's ideas that followed the rediscovery of his work, scientists had trouble obtaining Mendel's simple ratios, particularly with dihybrid crosses. Sometimes, it was not possible to successfully identify each of the four phenotypic classes expected, because two or more of the classes looked alike.

An example of this comes from the analysis of particular varieties of corn, *Zea mays*. Some commercial varieties exhibit a purple pigment called anthocyanin in their seed coats, whereas others do not. In 1918, geneticist R. A. Emerson crossed two true-breeding corn varieties, each lacking anthocyanin pigment. Surprisingly, all of the F_1 plants produced purple seeds.

When two of these pigment-producing F_1 plants were crossed to produce an F_2 generation, 56% were pigment producers and 44% were not. This is clearly not what Mendel's ideas would lead us to expect. Emerson correctly deduced that two genes were involved in producing pigment, and that the second cross had thus been a dihybrid cross. According to Mendel's theory, gametes in a dihybrid cross could combine in 16 equally possible ways—so the puzzle was to figure out how these 16 combinations could occur in the two phenotypic groups of progeny. Emerson multiplied the fraction that were pigment producers (0.56) by 16 to obtain 9, and multiplied the fraction that lacked pigment (0.44) by 16 to obtain 7. Emerson therefore had a *modified ratio* of 9:7 instead of the usual 9:3:3:1 ratio (figure 12.15).

This modified ratio is easily understood by considering the function of the products encoded by these genes. When gene products act sequentially, as in a biochemical pathway, an allele expressed as a defective enzyme early in the pathway blocks the flow of material through the rest of the pathway. In this case, it is impossible to judge whether the later steps of the pathway are functioning properly. This type of gene interaction, where one gene can interfere with the expression of another, is called **epistasis**.

The pigment anthocyanin is the product of a two-step biochemical pathway:

$$\text{starting molecule (colorless)} \xrightarrow{\text{enzyme 1}} \text{intermediate (colorless)} \xrightarrow{\text{enzyme 2}} \text{anthocyanin (purple)}$$

To produce pigment, a plant must possess at least one functional copy of each enzyme's gene. The dominant alleles encode functional enzymes, and the recessive alleles encode nonfunctional enzymes. Of the 16 genotypes predicted by random assortment, 9 contain at least one dominant allele of both genes; they therefore produce purple progeny. The remaining 7 genotypes lack dominant alleles at *either or both* loci (3 + 3 + 1 = 7) and so produce colorless progeny, giving the phenotypic ratio of 9:7 that Emerson observed (figure 12.15).

You can see that although this ratio is not the expected dihybrid ratio, it is a modification of the expected ratio.

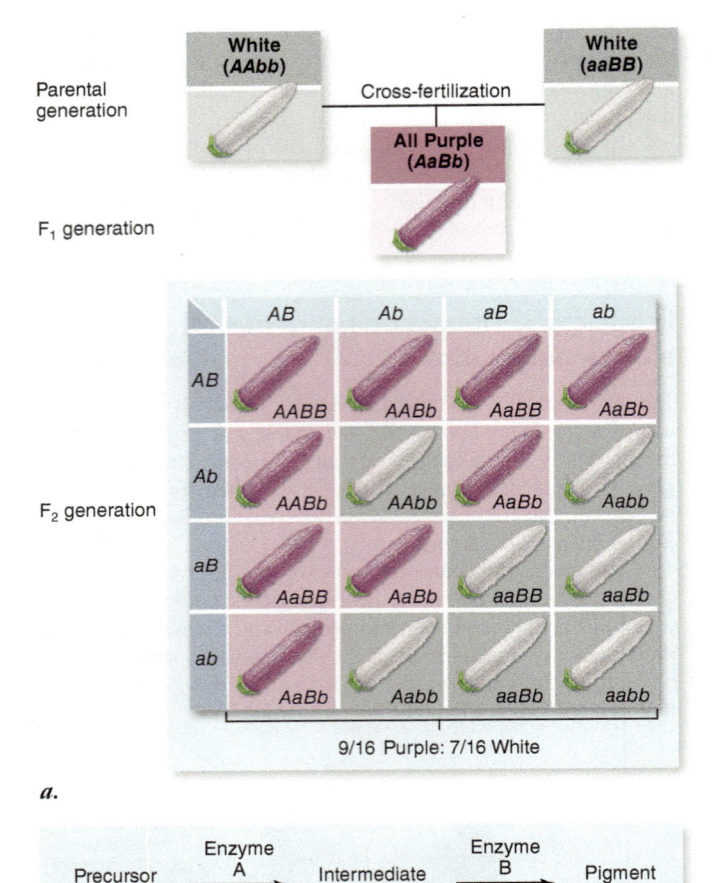

Figure 12.15 How epistasis affects grain color.
a. Crossing some white varieties of corn yields an all purple F_1. Self-crossing the F_1 yields 9 purple:7 white. This can be explained by the presence of two genes, each encoding an enzyme necessary for the production of purple pigment. Unless both enzymes are active (genotype is A_B_), no pigment is expressed. *b.* The biochemical pathway for pigment production with enzymes encoded by *A* and *B* genes.

Data analysis Mouse coat color is affected by a number of genes, including one that causes total loss of pigment (albinism) and another that leads to black/brown fur. A black mouse is crossed to another black mouse, yielding progeny with a ratio of 9 black:3 brown:4 albino. How can you explain these data if these two genes segregate independently?

Learning Outcomes Review 12.6
Mendel's model assumes that each trait is specified by one gene with only two alleles, no environmental effects alter a trait, and gene products act independently. All of these prove to be oversimplifications. Traits produced by the action of multiple genes (polygenic inheritance) have continuous variation. One gene can affect more than one trait (pleiotropy). Genes may have more than two alleles, and these may not show simple dominance. In incomplete dominance, the heterozygote is intermediate between the two homozygotes, and in codominance the heterozygote shows aspects of both homozygotes, both of which alter the monohybrid ratio. The action of genes is not always independent, which can result in modified dihybrid ratios.

- In the cross in figure 12.15, what proportion of F_2 will be white because they are homozygous recessive for one of the two genes?

Chapter Review

12.1 The Mystery of Heredity

Early plant biologists produced hybrids with puzzling results.

Plant breeders noticed that some forms of a trait can disappear in one generation only to reappear later—that is, they segregate rather than blend.

Mendel was the first to quantify the results of his crosses.

Mendel's experiments involved reciprocal crosses between true-breeding pea varieties followed by one or more generations of self-fertilization. His mathematical analysis of experimental results led to the present model of inheritance.

12.2 Monohybrid Crosses: The Principle of Segregation (figure 12.5)

The F_1 generation exhibits only one of two traits with no blending.

Mendel called the trait visible in the F_1 the dominant trait; the other he termed recessive.

The F_2 generation exhibits a 3:1 ratio of both traits.

When F_1 plants are self-fertilized, the F_2 shows a consistent ratio of 3 dominant:1 recessive. We call this 3:1 ratio the Mendelian monohybrid ratio.

The 3:1 ratio is actually 1:2:1.

Mendel then examined the F_2 and found the recessive F_2 plants always bred true, but only one out of three dominant F_2 bred true. This means the 3:1 ratio is actually 1 true-breeding dominant:2 non-true-breeding dominant:1 recessive.

Mendel's Principle of Segregation explains monohybrid observations.

Traits are determined by discrete factors we now call genes. These exist in alternative forms we call alleles. Individuals carrying two identical alleles for a gene are said to be homozygous, and individuals carrying different alleles are said to be heterozygous. The genotype is the entire set of alleles of all genes possessed by an individual. The phenotype is the individual's appearance due to these alleles.

The Principle of Segregation states that during gamete formation, the two alleles of a gene separate (segregate). Parental alleles then randomly come together to form the diploid zygote. The physical basis of segregation is the separation of homologues during anaphase of meiosis I.

The Punnett square allows symbolic analysis.

Punnett squares are formed by placing the gametes from one parent along the top of the square with the gametes from the other parent along the side. Zygotes formed from gamete combinations form the blocks of the square (figure 12.6).

Some human traits exhibit dominant/recessive inheritance.

Certain human traits have been found to have a Mendelian basis (table 12.1). Inheritance patterns in human families can be analyzed and inferred using a pedigree diagram of earlier generations.

12.3 Dihybrid Crosses: The Principle of Independent Assortment (figure 12.9)

Traits in a dihybrid cross behave independently.

If parents differing in two traits are crossed, the F_1 will be all dominant. Each F_1 parent can produce four different gametes that can be combined to produce 16 possible outcomes in the F_2. This yields a phenotypic ratio of 9:3:3:1 of the four possible phenotypes.

Mendel's Principle of Independent Assortment explains dihybrid results.

The Principle of Independent Assortment states that different traits segregate independently of one another. The physical basis of independent assortment is the independent behavior of different pairs of homologous chromosomes during meiosis I.

12.4 Probability: Predicting the Results of Crosses

Two probability rules help predict monohybrid cross results.

The rule of addition states that the probability of two independent events occurring is the sum of their individual probabilities. The rule of multiplication, or product rule, states that the probability of two independent events *both* occurring is the product of their individual probabilities.

Dihybrid cross probabilities are based on monohybrid cross probabilities.

A dihybrid cross is essentially two independent monohybrid crosses. The product rule applies and can be used to predict the cross's outcome.

12.5 The Testcross: Revealing Unknown Genotypes (figure 12.10)

In a testcross, an unknown genotype is crossed with a homozygous recessive genotype. The F_1 offspring will all be the same if the unknown genotype is homozygous dominant. The F_1 offspring will exhibit a 1:1 dominant:recessive ratio if the unknown genotype is heterozygous.

12.6 Extensions to Mendel

In polygenic inheritance, more than one gene can affect a single trait.

Many traits, such as human height, are due to multiple additive contributions by many genes, resulting in continuous variation.

In pleiotropy, a single gene can affect more than one trait.

A pleiotropic effect occurs when an allele affects more than one trait. These effects are difficult to predict.

Genes may have more than two alleles.

There may be more than two alleles of a gene in a population. Given the possible number of DNA sequences, this is not surprising.

Dominance is not always complete.

In incomplete dominance the heterozygote exhibits an intermediate phenotype; the monohybrid genotypic and phenotypic ratios are the same (figure 12.12). Codominant alleles each contribute to the phenotype of a heterozygote.

Phenotypes may be affected by the environment.

Genotype determines phenotype, but the environment will have an effect on this relationship. Environment means both external and internal factors. For example, in Siamese cats, a temperature-sensitive enzyme produces more pigment in the colder peripheral areas of the body.

In epistasis, interactions of genes alter genetic ratios.

Genes encoding enzymes that act in a single biochemical pathway are not independent. In corn, anthocyanin pigment production requires the action of two enzymes. Doubly heterozygous individuals for these enzymes yield a 9:7 ratio when self-crossed (figure 12.15).

Review Questions

UNDERSTAND

1. What property distinguished Mendel's investigation from previous studies?
   a. Mendel used true-breeding pea plants.
   b. Mendel quantified his results.
   c. Mendel examined many different traits.
   d. Mendel examined the segregation of traits.

2. The F_1 generation of the monohybrid cross purple (*PP*) × white (*pp*) flower pea plants should
   a. all have white flowers.
   b. all have a light purple or blended appearance.
   c. all have purple flowers.
   d. have ¾ purple flowers, and ¼ white flowers.

3. The F_1 plants from the previous question are allowed to self-fertilize. The phenotypic ratio for the F_2 should be
   a. all purple.
   b. 1 purple:1 white.
   c. 3 purple:1 white.
   d. 3 white:1 purple.

4. Which of the following is NOT a part of Mendel's five-element model?
   a. Traits have alternative forms (what we now call alleles).
   b. Parents transmit discrete traits to their offspring.
   c. If an allele is present it will be expressed.
   d. Traits do not blend.

5. An organism's _____ is/are determined by its _____.
   a. genotype; phenotype
   b. phenotype; genotype
   c. alleles; phenotype
   d. genes; alleles

6. Phenotypes like height in humans, which show a continuous distribution, are usually the result of
   a. an alteration of dominance for multiple alleles of a single gene.
   b. the presence of multiple alleles for a single gene.
   c. the action of one gene on multiple phenotypes.
   d. the action of multiple genes on a single phenotype.

APPLY

1. Japanese four o'clocks that are red and tall are crossed to white short ones, producing an F_1 that is pink and tall. If these genes assort independently, and the F_1 is self-crossed, what would you predict for the ratio of F_2 phenotypes?
   a. 3 red tall:1 white short
   b. 1 red tall:2 pink short:1 white short
   c. 3 pink tall:6 red tall:3 white tall:1 pink short:2 red short:1 white short
   d. 3 red tall:6 pink tall:3 white tall:1 red short:2 pink short:1 white short

2. If the two genes in the previous question showed complete linkage, what would you predict for an F_2 phenotypic ratio?
   a. 1 red tall:2 pink short:1 white short
   b. 1 red tall:2 red short:1 white short
   c. 1 pink tall:2 red tall:1 white short
   d. 1 red tall:2 pink tall:1 white short

3. What is the probability of obtaining an individual with the genotype *bb* from a cross between two individuals with the genotype *Bb*?
   a. ½
   b. ¼
   c. ⅛
   d. 0

4. In a cross of *Aa Bb cc* X *Aa Bb Cc*, what is the probability of obtaining an individual with the genotype *AA Bb Cc*?
   a. 1/16
   b. 3/16
   c. 1/64
   d. 3/64

5. When you cross true-breeding tall and short tobacco plants you get an F_1 that is intermediate in height. When this F_1 is self-crossed, it yields an F_2 with a continuous distribution of heights. What is the best explanation for these data?
   a. Height is determined by a single gene with incomplete dominance.
   b. Height is determined by a single gene with many alleles.
   c. Height is determined by the additive effects of many genes.
   d. Height is determined by epistatic genes.

6. Mendel's model assumes that each trait is determined by a single factor with alternate forms. We now know that this is too simplistic and that
   a. a single gene may affect more than one trait.
   b. a single trait may be affected by more than one gene.
   c. a single gene always affects only one trait, but traits may be affected by more than one gene.
   d. a single gene can affect more than one trait, and traits may be affected by more than one gene.

SYNTHESIZE

1. Create a Punnett square for the following crosses and use this to predict phenotypic ratio for dominant and recessive traits. Dominant alleles are indicated by uppercase letters and recessive are indicated by lowercase letters. For parts b and c, predict ratios using probability and the product rule.
   a. A monohybrid cross between individuals with the genotype *Aa* and *Aa*
   b. A dihybrid cross between two individuals with the genotype *AaBb*
   c. A dihybrid cross between individuals with the genotype *AaBb* and *aabb*

2. Explain how the events of meiosis can explain both segregation and independent assortment.

3. In mice, there is a yellow strain that when crossed yields 2 yellow:1 black. How could you explain this observation? How could you test this with crosses?

4. In mammals, a variety of genes affect coat color. One of these is a gene with mutant alleles that results in the complete loss of pigment, or albinism. Another controls the type of dark pigment with alleles that lead to black or brown colors. The albinistic trait is recessive, and black is dominant to brown. Two black mice are crossed and yield 9 black:4 albino:3 brown. How would you explain these results?

Biology, 11th Edition

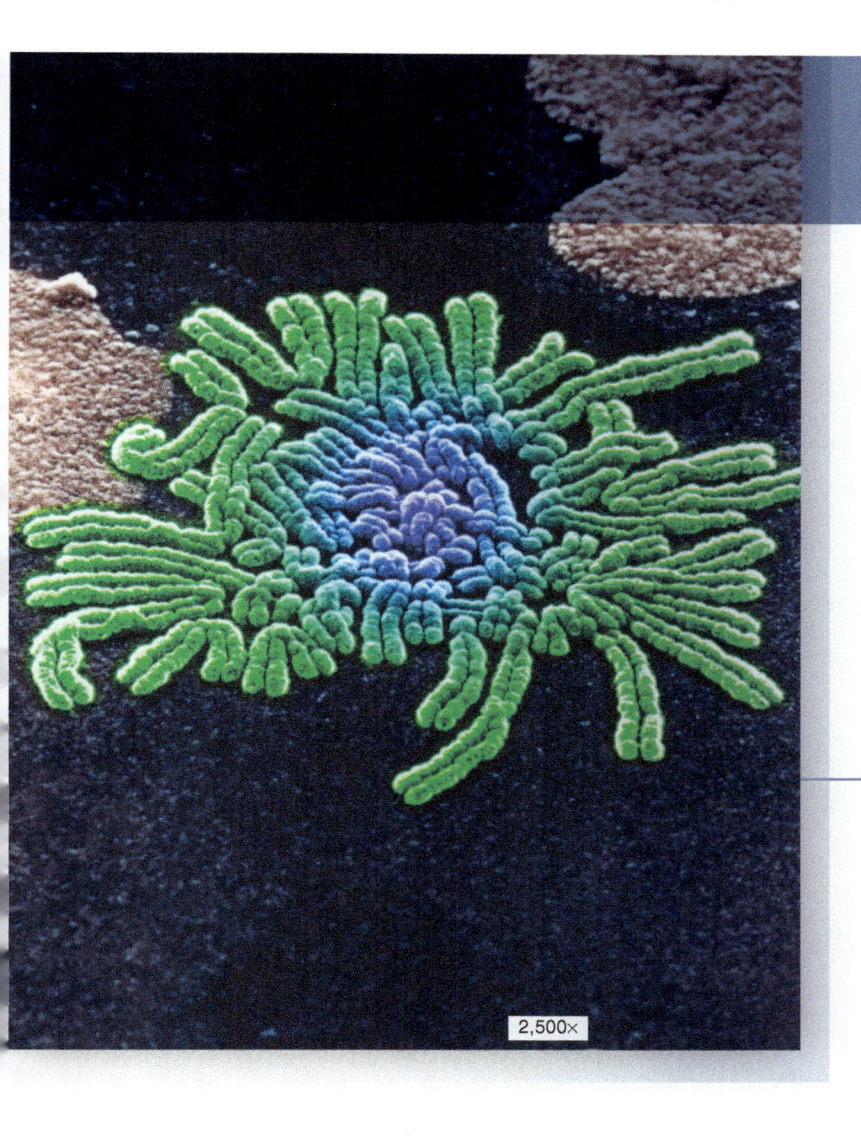

CHAPTER 13

Chromosomes, Mapping, and the Meiosis–Inheritance Connection

Chapter Contents

13.1 Sex Linkage and the Chromosomal Theory of Inheritance

13.2 Sex Chromosomes and Sex Determination

13.3 Exceptions to the Chromosomal Theory of Inheritance

13.4 Genetic Mapping

13.5 Selected Human Genetic Disorders

Introduction

Mendel's experiments opened the door to understanding inheritance, but many questions remained. In the early part of the 20th century, we did not know the nature of the factors whose behavior Mendel had described. The next step, which involved many researchers in the early part of that century, was uniting information about the behavior of chromosomes, seen in the picture, and the inheritance of traits. The basis for Mendel's principles of segregation and independent assortment lie in events that occur during meiosis.

The behavior of chromosomes during meiosis not only explains Mendel's principles, but leads to new and different approaches to the study of heredity. The ability to construct genetic maps is one of the most powerful tools of classical genetic analysis. The tools of genetic mapping developed in flies and other organisms in combination with information from the Human Genome Project now allow us to determine the location of genes and isolate those that are involved in genetic diseases.

13.1 Sex Linkage and the Chromosomal Theory of Inheritance

Learning Outcomes
1. Describe sex-linked inheritance in fruit flies.
2. Explain the evidence for genes being on chromosomes.

A central role for chromosomes in heredity was first suggested in 1900 by the German geneticist Carl Correns, in one of the papers announcing the rediscovery of Mendel's work. Soon after, observations that similar chromosomes paired with one another during meiosis led directly to the **chromosomal theory of inheritance,** first formulated by the American Walter Sutton in 1902.

Morgan correlated the inheritance of a trait with sex chromosomes

In 1910, Thomas Hunt Morgan, studying the fruit fly *Drosophila melanogaster,* discovered a mutant male fly with white eyes instead of red (figure 13.1). Morgan immediately set out to determine whether this new trait would be inherited in a Mendelian fashion. He first crossed the mutant male to a normal red-eyed female to see whether the red-eyed or white-eyed trait was dominant. All of the F_1 progeny had red eyes, so Morgan concluded that red eye color was dominant over white.

The F_1 cross

Following the experimental procedure established by Mendel, Morgan then crossed the red-eyed flies from the F_1 generation with each other. Of the 4252 F_2 progeny Morgan examined, 782 (18%) had white eyes. Although the ratio of red eyes to white eyes in the F_2 progeny was greater than 3:1, the results of the cross nevertheless provided clear evidence that eye color segregates. However, something about the outcome was strange and totally unpredicted by Mendel's theory—*all of the white-eyed F_2 flies were males* (figure 13.2)!

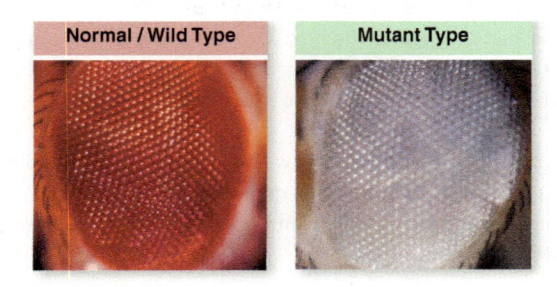

Figure 13.1 Red-eyed (wild-type) and white-eyed (mutant) *Drosophila*. Mutations are heritable alterations in genetic material. By studying the inheritance pattern of white and red alleles (located on the X chromosome), Morgan first demonstrated that genes are on chromosomes.

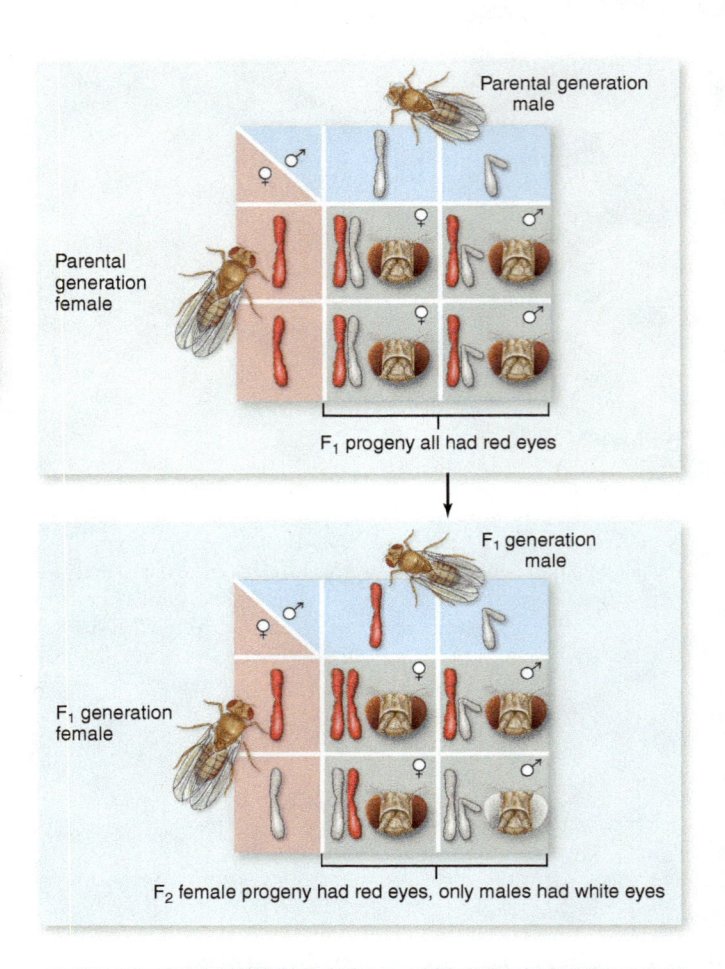

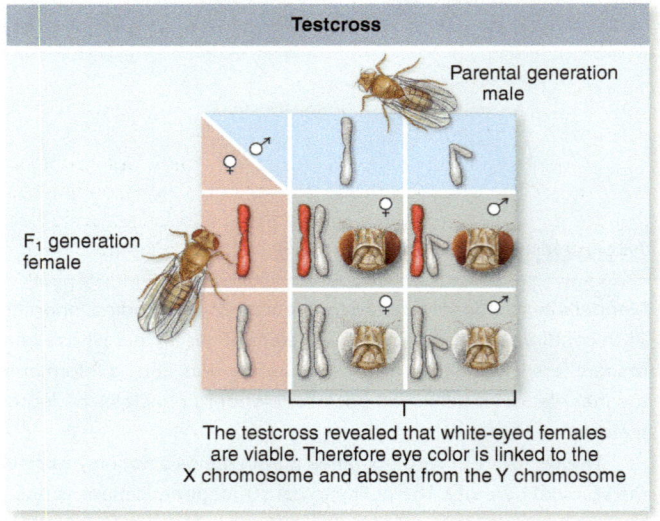

Figure 13.2 The chromosomal basis of sex linkage. White-eyed male flies are crossed to red-eyed females. The F_1 flies all have red eyes, as expected for a recessive white-eye allele. In the F_2, all of the white-eyed flies are male because the Y chromosome lacks the white-eye *(white)* gene. Inheritance of the sex chromosomes correlates with eye color, showing the *white* gene is on the X chromosome.

The testcross

Morgan sought an explanation for this result. One possibility was simply that white-eyed female flies don't exist; such individuals might not be viable for some unknown reason. To test this idea, Morgan testcrossed the female F_1 progeny with the original white-eyed male. He obtained white-eyed and red-eyed flies of both sexes in a 1:1:1:1 ratio, just as Mendel's theory had predicted (figure 13.2). Therefore, white-eyed female flies are viable. Given that white-eyed females can exist, Morgan turned to the nature of the chromosomes in males and females for an explanation.

The gene for eye color lies on the X chromosome

In *Drosophila*, the sex of an individual is determined by the number of copies it has of a particular chromosome, the **X chromosome.** Observations of *Drosophila* chromosomes revealed that female flies have two X chromosomes, but male flies have only one. In males, the single X chromosome pairs in meiosis with a dissimilar partner called the **Y chromosome.** These two chromosomes are termed **sex chromosomes** because of their association with sex.

During meiosis, a female produces only X-bearing gametes, but a male produces both X-bearing and Y-bearing gametes. When fertilization involves an X sperm, the result is an XX zygote, which develops into a female; when fertilization involves a Y sperm, the result is an XY zygote, which develops into a male.

The solution to Morgan's puzzle is that the gene causing the white-eye trait in *Drosophila* resides only on the X chromosome—it is absent from the Y chromosome. (We now know that the Y chromosome in flies carries almost no functional genes.) A trait determined by a gene on the X chromosome is said to be **sex-linked,** or X-linked, because it is associated with the sex of the individual. Knowing the white-eye trait is recessive to the red-eye trait, we can now see that Morgan's result was a natural consequence of the Mendelian segregation of chromosomes (figure 13.2).

Morgan's experiment was one of the most important in the history of genetics because it presented the first clear evidence that the genes determining Mendelian traits do indeed reside on the chromosomes, as Sutton had proposed. Mendelian traits segregate in genetic crosses because homologues separate during gamete formation.

Learning Outcomes Review 13.1

Morgan showed that the trait for white eyes in *Drosophila* segregated with the sex of offspring. X and Y chromosomes also segregate with sex. This correlates the behavior of a trait with the behavior of chromosomes. This finding supported the chromosomal theory of inheritance, which states that traits are carried on chromosomes.

- *What are the expectations for a cross of white-eyed females to red-eyed males?*

13.2 Sex Chromosomes and Sex Determination

Learning Outcomes

1. Describe the relationship between sex chromosomes and sex determination.
2. Explain the genetic consequences of dosage compensation in mammals.

The structure and number of sex chromosomes vary in different species (table 13.1). In the fruit fly, *Drosophila*, females are XX and males XY, which is also the case for humans and other mammals. However, in birds, the male has two Z chromosomes, and the female has a Z and a W chromosome. Some insects, such as grasshoppers, have no Y chromosome—females are XX and males are characterized as XO (O indicates the absence of a chromosome).

In humans, the Y chromosome generally determines maleness

In chapter 10, you learned that humans have 46 chromosomes (23 pairs). Twenty-two of these pairs are perfectly matched in both males and females and are called **autosomes.** The remaining pair are the sex chromosomes: XX in females, and XY in males.

35,000×

TABLE 13.1 Sex Determination in Some Organisms

		Female	Male
Humans, *Drosophila*		XX	XY
Birds		ZW	ZZ
Grasshoppers		XX	XO
Honeybees		Diploid	Haploid

The Y chromosome in males is highly condensed. Because few genes on the Y chromosome are expressed, recessive alleles on a male's single X chromosome have no *active* counterpart on the Y chromosome.

The "default" setting in human embryonic development leads to female development. Some of the active genes on the Y chromosome, notably the *SRY* gene, are responsible for the masculinization of genitalia and secondary sex organs, producing features associated with "maleness" in humans. Consequently, any individual with *at least one Y chromosome* is normally a male.

The exceptions to this rule actually provide support for this mechanism of sex determination. For example, translocation of part of the Y chromosome to the X chromosome can cause otherwise XX individuals to develop as male. There is also a genetic disorder that causes a failure to respond to the androgen hormones (androgen insensitivity syndrome) that causes XY individuals to develop as female. Lastly, mutations in *SRY* itself can cause XY individuals to develop as females.

This form of sex determination seen in humans is shared among mammals, but is not universal in vertebrates. Among fishes and some species of reptiles, environmental factors can cause changes in the expression of this sex-determining gene, and thus in the sex of the adult individual.

Some human genetic disorders display sex linkage

From ancient times, people have noted conditions that seem to affect males to a greater degree than females. Red-green color blindness is one well-known condition that is more common in males because the gene affected is carried on the X chromosome.

Another example is **hemophilia,** a disease that affects a single protein in a cascade of proteins involved in the formation of blood clots. Thus, in an untreated hemophiliac, even minor cuts will not stop bleeding. This form of hemophilia is caused by an X-linked recessive allele; women who are heterozygous for the allele are asymptomatic carriers, and men who receive an X chromosome with the recessive allele exhibit the disease.

The allele for hemophilia was introduced into a number of different European royal families by Queen Victoria of England. Because these families kept careful genealogical records, we have an extensive pedigree for this condition. In the five generations after Victoria, 10 of her male descendants have had hemophilia as shown in the pedigree in figure 13.3.

The Russian house of Romanov inherited this condition through Alexandra Feodorovna, a granddaughter of Queen Victoria.

Figure 13.3 The royal hemophilia pedigree. Queen Victoria, shown at the bottom center of the photo, was a carrier for hemophilia. Two of Victoria's four daughters, Alice and Beatrice, inherited the hemophilia allele from Victoria. Two of Alice's daughters are standing behind Victoria (wearing feathered boas): Princess Irene of Prussia *(right)* and Alexandra *(left)*, who would soon become czarina of Russia. Both Irene and Alexandra were also carriers of hemophilia. From the pedigree, it is clear that Alice introduced hemophilia into the Russian and Prussian royal houses, and Victoria's daughter Beatrice introduced it into the Spanish royal house. Victoria's son Leopold, himself a victim, also transmitted the disorder in a third line of descent. Half-shaded symbols represent carriers with one normal allele and one defective allele; fully shaded symbols represent affected individuals.

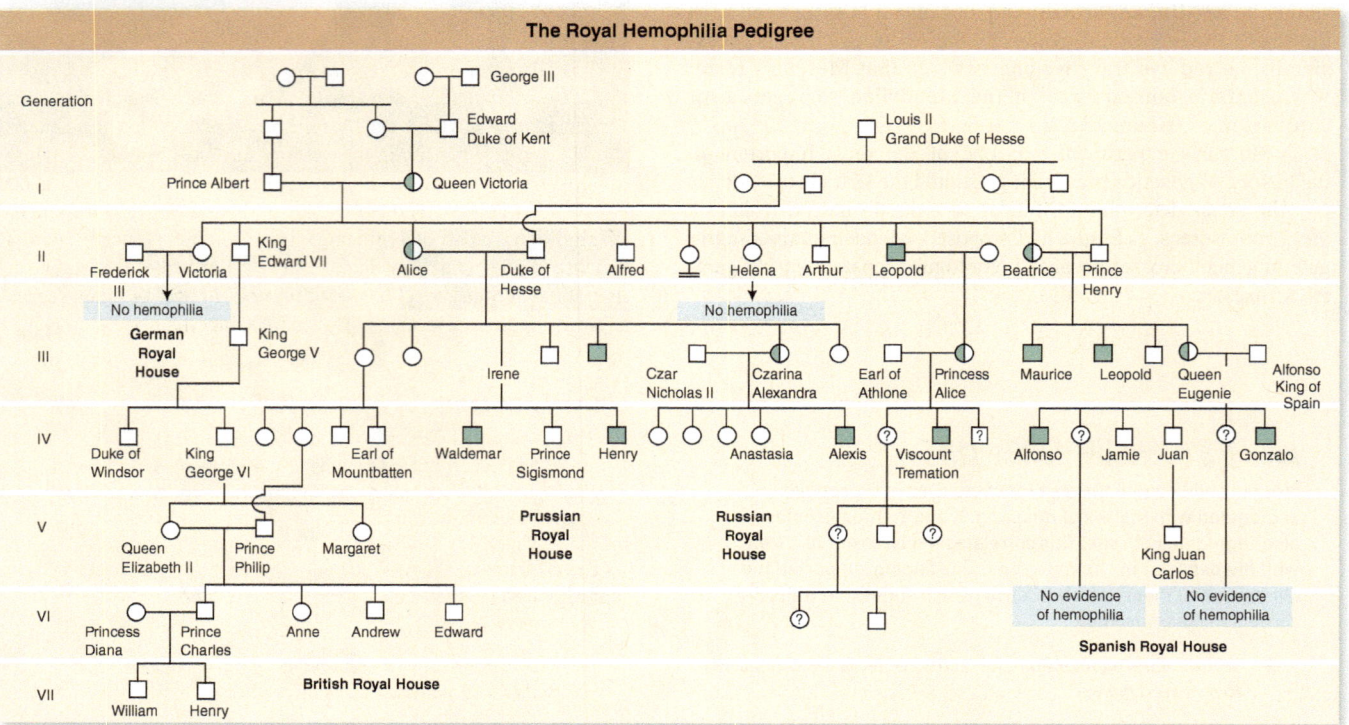

She married Czar Nicholas II, and their only son, Alexis, was afflicted with the disease. The entire family was executed during the Russian revolution. (A woman who had long claimed to be Anastasia, a surviving daughter, was shown not to be a Romanov using modern genetic techniques to test her remains.)

Ironically, this condition has not affected the current British royal family, because Victoria's son Edward, who became King Edward VII, did not receive the hemophilia allele. All of the subsequent rulers of England are his descendants.

Dosage compensation prevents doubling of sex-linked gene products

Although males have only one copy of the X chromosome and females have two, female cells do not produce twice as much of the proteins encoded by genes on the X chromosome. Instead, one of the X chromosomes in females is inactivated early in embryonic development, shortly after the embryo's sex is determined. This inactivation is an example of **dosage compensation**, which ensures an equal level of expression from the sex chromosomes despite a differing number of sex chromosomes in males and females. In *Drosophila,* by contrast, dosage compensation is achieved by increasing the level of expression on the male X chromosome.

Which X chromosome is inactivated in females varies randomly from cell to cell. If a woman is heterozygous for a sex-linked trait, some of her cells will express one allele and some the other. The inactivated X chromosome is highly condensed, making it visible as an intensely staining **Barr body,** seen below, attached to the nuclear membrane.

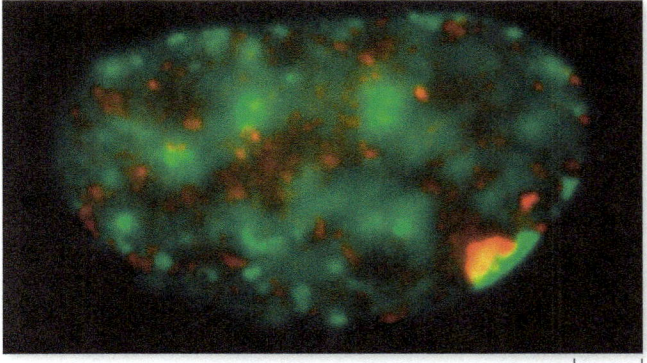

X-chromosome inactivation can lead to genetic mosaics

X-chromosome inactivation to produce dosage compensation is not unique to humans but is true of all mammals. Females that are heterozygous for X-chromosome alleles are **genetic mosaics:** Their individual cells may express different alleles, depending on which chromosome is inactivated.

One example is the calico cat, a female that has a patchy distribution of dark fur, orange fur, and white fur (figure 13.4). The dark fur and orange fur are due to heterozygosity for a gene on the X chromosome that determines pigment color. One allele results in dark fur, and another allele results in orange fur. Which of these colors is observed in any particular patch is due to inactivation of one X chromosome: If the chromosome containing the orange allele is inactivated, then the fur will be dark, and vice versa.

The patchy distribution of color, and the presence of white fur, is due to a second gene that is epistatic to the fur color gene (see chapter 12). That is, the presence of this second gene produces a patchy distribution of pigment, with some areas totally lacking pigment. In the areas that lack pigment, the effect of either fur color allele is masked. Thus, in this one animal we can see an excellent example of both epistasis and X inactivation.

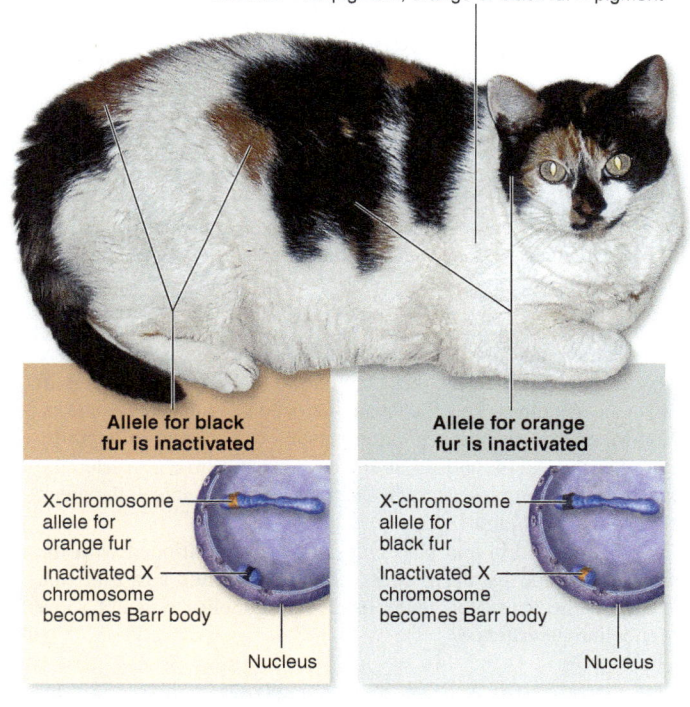

Figure 13.4 A calico cat. The cat is heterozygous for alleles of a coat color gene that produce either black fur or orange fur. This gene is on the X chromosome, so the different-colored fur is due to inactivation of one X chromosome. The patchy distribution and white color is due to a second gene that is epistatic to the coat color gene and thus masks its effects.

Learning Outcomes Review 13.2

Sex determination begins with the presence or absence of certain chromosomes termed the sex chromosomes. Additional factors may influence sex determination in different species. In humans, males are XY, and therefore they exhibit recessive traits for alleles on the X chromosome. In mammalian females, one X chromosome in each cell becomes inactivated to balance the levels of gene expression. This random inactivation can lead to genetic mosaics.

- *Would you expect an XXX individual to be viable? If so, would that individual be male or female?*

13.3 Exceptions to the Chromosomal Theory of Inheritance

Learning Outcome
1. Describe the inheritance pattern for genes contained in a chloroplast or mitochondrion DNA.

Although the chromosomal theory explains most inheritance, there are exceptions. Primarily, these are due to the presence of DNA in organelle genomes, specifically in mitochondria and chloroplasts. Non-Mendelian inheritance via organelles was studied in depth by Ruth Sager, who in the face of universal skepticism constructed the first map of chloroplast genes in *Chlamydomonas*, a unicellular green alga, in the 1960s and 1970s.

Mitochondria and chloroplasts are not partitioned with the nuclear genome by the process of meiosis. Thus any trait that is due to the action of genes in these organelles will not show Mendelian inheritance.

Mitochondrial genes are inherited from the female parent

Organelles are usually inherited from only one parent, generally the mother. When a zygote is formed, it receives an equal contribution of the nuclear genome from each parent, but it gets all of its mitochondria from the egg cell, which contains a great deal more cytoplasm (and thus organelles). As the zygote divides, these original mitochondria divide as well and are partitioned randomly.

As a result, the mitochondria in every cell of an adult organism can be traced back to the original maternal mitochondria present in the egg. This mode of uniparental (one-parent) inheritance from the mother is called **maternal inheritance**.

In humans, the disease Leber's hereditary optic neuropathy (LHON) shows maternal inheritance. The genetic basis of this disease is a mutant allele for a subunit of NADH dehydrogenase. The mutant allele reduces the efficiency of electron flow in the electron transport chain in mitochondria (see chapter 7), in turn reducing overall ATP production. Some nerve cells in the optic system are particularly sensitive to reduction in ATP production, resulting in neural degeneration.

A mother with this disease will pass it on to all of her progeny, whereas a father with the disease will not pass it on to any of his progeny. Note that unlike sex-linked inheritance, in maternal inheritance males and females are equally affected.

Chloroplast genes may also be passed on uniparentally

The inheritance pattern of chloroplasts is also usually maternal, although both paternal and biparental inheritance of chloroplasts may be observed in some species. Carl Correns first hypothesized in 1909 that chloroplasts were responsible for inheritance of variegation (mixed green and white leaves) in the plant commonly known as the four o'clock, *Mirabilis jalapa*. The offspring exhibited the phenotype of the female parent, regardless of the male's phenotype.

In Sager's work on *Chlamydomonas*, resistance to the antibiotic streptomycin was shown to be transmitted via the chloroplast DNA from only the mt^+ mating type. The mt^- mating type does not contribute chloroplast DNA to the zygote formed by fusion of mt^+ and mt^- gametes.

Learning Outcome Review 13.3

The genomes of mitochondria and chloroplasts divide independently of the nucleus. These organelles are carried in the cytoplasm of the egg cell, so any traits determined by these genomes are maternally inherited and thus do not follow Mendelian rules. In some species, however, chloroplasts may be passed on paternally or biparentally.

■ *How can you explain the lack of mt^- chloroplast DNA in* Chlamydomonas *zygotes from mt^- by mt^+ crosses?*

13.4 Genetic Mapping

Learning Outcomes
1. Describe how genes on the same chromosome will segregate.
2. Explain the relationship between frequency of recombinant progeny and map distance.
3. Calculate map distances from the frequency of recombinants in testcrosses.

We have seen that Mendelian traits are determined by genes located on chromosomes and that the independent assortment of Mendelian traits reflects the independent assortment of chromosomes in meiosis. This is fine as far as it goes, but it is still incomplete. Of Mendel's seven traits in figure 12.4, six are on different chromosomes and two are on the same chromosome, yet all show independent assortment with one another. The two on the same chromosome should not behave the same as those that are on different chromosomes. In fact, organisms generally have many more genes that assort independently than the number of chromosomes. This means that independent assortment cannot be due only to the random alignment of different chromosomes during meiosis.

Inquiry question Mendel did not examine plant height and pod shape in his dihybrid crosses. The genes for these traits are very close together on the same chromosome. How would this have changed Mendel's results?

The solution to this problem is found in an observation that was introduced in chapter 11: the crossing over of homologues during meiosis. In prophase I of meiosis, homologues appear to physically exchange material by crossing over (figure 13.5). In chapter 11, you saw how this was part of the mechanism that allows homologues, and not sister chromatids, to disjoin at anaphase I.

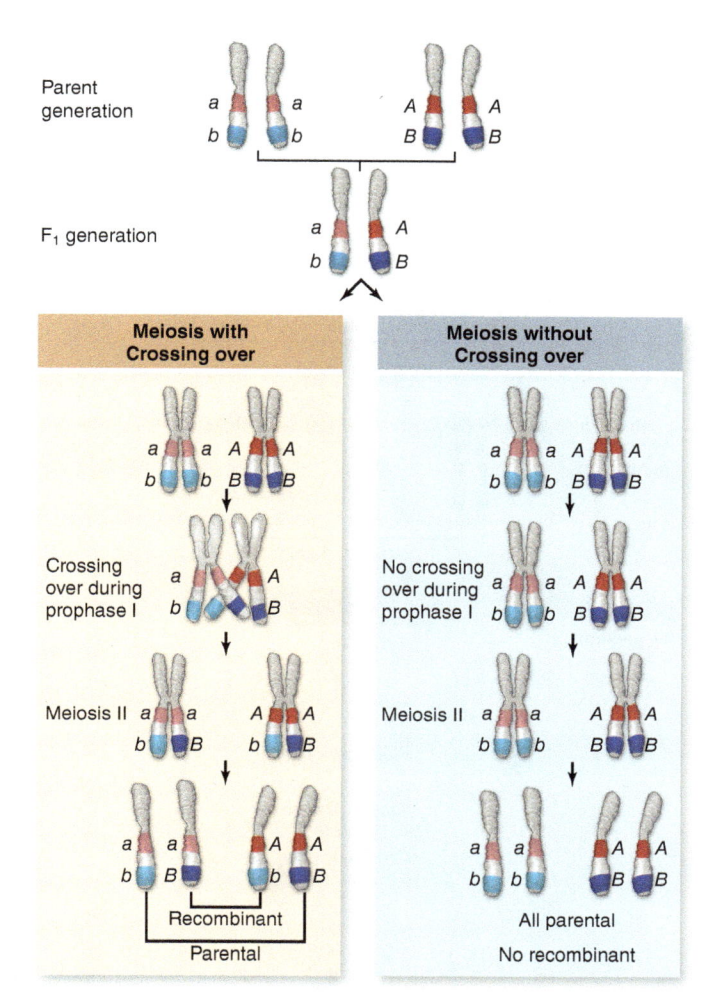

Figure 13.5 Crossing over exchanges alleles on homologues. When a crossover occurs between two loci, it leads to the production of recombinant chromosomes. If no crossover occurs, then the chromosomes will carry the parental combination of alleles.

Genetic recombination exchanges alleles on homologues

Consider a dihybrid cross performed using the Mendelian framework. Two true-breeding parents that each differ with respect to two traits are crossed, producing doubly heterozygous F_1 progeny. If the genes for the two traits are on a single chromosome, then during meiosis we would expect alleles for both loci to segregate together and produce only gametes that resemble the two parental types. But if a crossover occurs between the two loci, then each homologue would carry one allele from each parent and produce gametes that combine these parental traits (figure 13.5). We call gametes with this new combination of alleles *recombinant* gametes as they are formed by recombining the parental alleles.

The first investigator to provide evidence for this was Morgan, who studied three genes on the X chromosome of *Drosophila*. He found an excess of parental types, which he explained as due to the genes all being on the X chromosome and therefore coinherited (inherited together). He went further, suggesting that the recombinant genotypes were due to crossing over between homologues during meiosis.

Experiments performed independently by Barbara McClintock and Harriet Creighton in maize, and by Curt Stern in *Drosophila* provided evidence for this physical exchange of genetic material. The experiment done by Creighton and McClintock is detailed in figure 13.6. In this experiment, they used a chromosome with two alterations visible under a microscope: a knob on one end of the chromosome and an extension of the other end making it longer. In addition to these visible markers, this chromosome also carried a gene that determines kernel color (colored or colorless) and a gene that determines kernel texture (waxy or starchy).

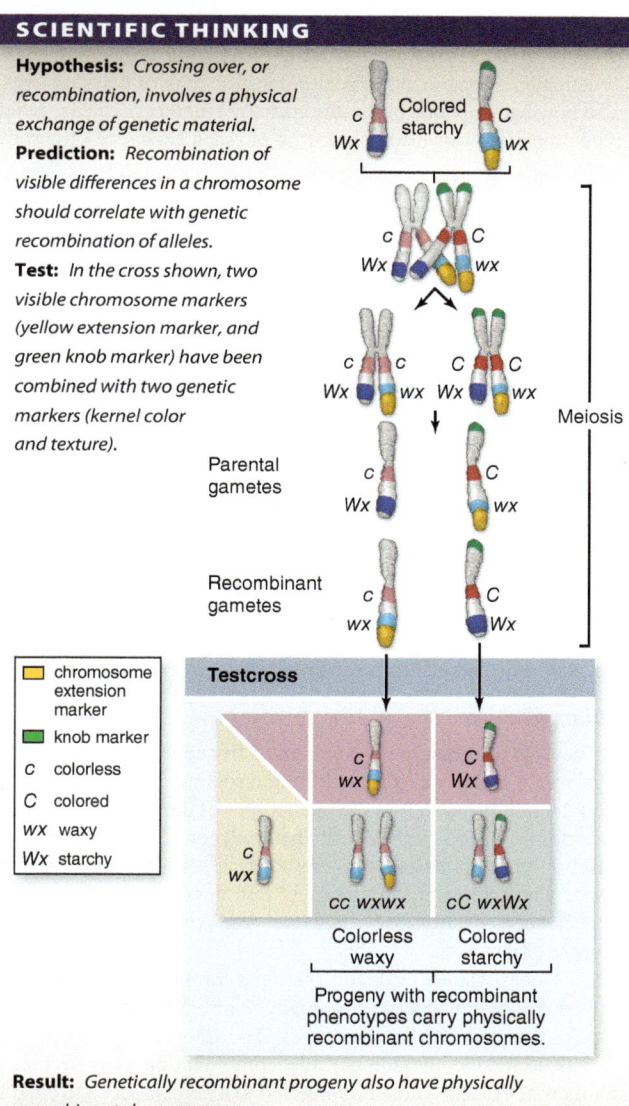

SCIENTIFIC THINKING

Hypothesis: Crossing over, or recombination, involves a physical exchange of genetic material.
Prediction: Recombination of visible differences in a chromosome should correlate with genetic recombination of alleles.
Test: In the cross shown, two visible chromosome markers (yellow extension marker, and green knob marker) have been combined with two genetic markers (kernel color and texture).

Result: Genetically recombinant progeny also have physically recombinant chromosomes.
Conclusion: A physical exchange of genetic material accompanied genetic recombination.
Further Experiments: This experiment was performed using maize. What other genetic model system would you use to test this?

Figure 13.6 The Creighton and McClintock experiment.

The long chromosome, which also had the knob, carried the dominant colored allele for kernel color *(C)* and the recessive waxy allele for kernel texture *(wx)*. Heterozygotes were constructed with this chromosome paired with a visibly normal chromosome carrying the recessive colorless allele for kernel color *(c)* and the dominant starchy allele for kernel texture *(Wx)* (figure 13.6). These plants appeared colored and starchy because they were heterozygous for both loci, and they were also heterozygous for the two visibly distinct chromosomes.

These plants, heterozygous for both chromosomal and genetic markers, were testcrossed to colorless waxy plants with normal appearing chromosomes. The progeny were analyzed for both physical recombination (using a microscope to observe chromosome appearance) and genetic recombination (by examining the phenotype of progeny). The results were striking: All of the progeny that were genetically recombinant (appear colored starchy or colorless waxy) also now had only one of the chromosomal markers. That is, genetic recombination was accompanied by physical exchange of chromosomal material.

Recombination is the basis for genetic maps

The ability to map the location of genes on chromosomes using data from genetic crosses is one of the most powerful tools of genetics. The insight that allowed this technique, like many great insights, is so simple as to seem obvious in retrospect.

Morgan had already suggested that the frequency with which a particular group of recombinant progeny appeared was a reflection of the relative location of genes on the chromosome. An undergraduate in Morgan's laboratory, Alfred Sturtevant put this observation on a quantitative basis. Sturtevant reasoned that the frequency of recombination observed in crosses could be used as a measure of genetic distance. That is, as physical distance on a chromosome increases, so does the probability of recombination (crossover) occurring between the gene loci. Using this logic, the frequency of recombinant gametes produced is a measure of their distance apart on a chromosome.

Linkage data

To be able to measure recombination frequency easily, investigators use a testcross instead of intercrossing the F_1 progeny as Mendel did. In a testcross the phenotypes of the progeny reflect the gametes produced by the doubly heterozygous F_1 individual. In the case of recombination, progeny that appear parental have not undergone crossover, and progeny that appear recombinant have experienced a crossover between the two loci in question (figure 13.5).

Genes that are close together on a chromosome are said to be linked. We define linkage genetically as an excess of parental genotypes over recombinant genotypes. The frequency of recombination is defined as the number of recombinant progeny divided by total progeny. This value is converted to a percentage, and each 1% of recombination represents one map unit. This unit has been named the centimorgan (cM) for T. H. Morgan, although it is also called simply a map unit (m.u.) as well.

Constructing maps

Constructing genetic maps then becomes a simple process of performing testcrosses with doubly heterozygous individuals and counting progeny to determine percent recombination. This is best shown with an example using a two-point cross.

Drosophila homozygous for two mutations, vestigial wings *(vg)* and black body *(b)*, are crossed to flies homozygous for the wild type, or normal alleles, of these genes *(vg⁺ b⁺)*. The doubly heterozygous F_1 progeny are then testcrossed to homozygous

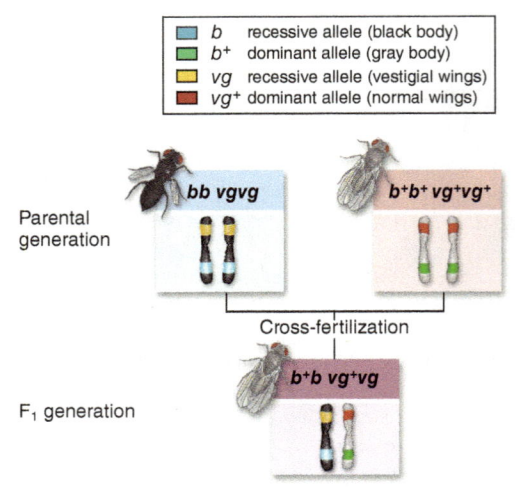

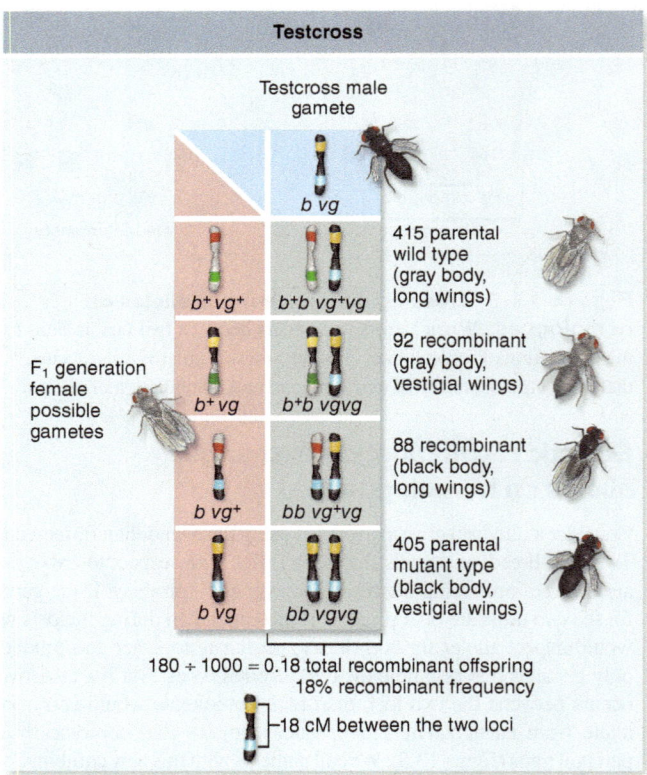

Figure 13.7 Two-point cross to map genes. Flies homozygous for long wings *(vg⁺)* and gray bodies *(b⁺)* are crossed to flies homozygous for vestigial wings *(vg)* and black bodies *(b)*. Both vestigial wings and black body are recessive to the normal (wild-type) long wings and gray body. The F_1 progeny are then testcrossed to homozygous vestigial black to produce the progeny for mapping. Data are analyzed in the text.

recessive individuals *(vg b/vg b)*, and progeny are counted (figure 13.7). The data are shown here:

vestigial wings, black body *(vg b)*	405	(parental)
long wings, gray body *(vg⁺ b⁺)*	415	(parental)
vestigial wings, gray body *(vg b⁺)*	92	(recombinant)
long wings, black body *(vg⁺ b)*	88	(recombinant)
Total Progeny	1000	

The sum of the recombinant progeny is divided by total progeny to produce the recombination frequency. In this case the recombination frequency is 92 + 88 divided by 1000, or 0.18. Converting this to a percentage yields 18 cM as the map distance between these two loci.

Multiple crossovers can yield independent assortment results

As the distance separating loci increases, the probability of recombination occurring between them during meiosis also increases. What happens when more than one recombination event occurs?

If homologues undergo two crossovers between loci, then the parental combination is restored. This leads to an underestimate of the true genetic distance because not all events can be noted. As a result, the relationship between true distance on a chromosome and the recombination frequency is not linear. It begins as a straight line, but the slope decreases; the curve levels off at a recombination frequency of 0.5 (figure 13.8).

At long distances, multiple events between loci become frequent. In this case, odd numbers of crossovers (1, 3, 5) produce recombinant gametes, and no crossover or even numbers of crossovers (0, 2, 4) produce parental gametes. At large enough distances, these frequencies are about equal, leading to the number of recombinant gametes being equal to the number of parental gametes, and the loci exhibit independent assortment! This is how Mendel could use two loci on the same chromosome and have them assort independently.

Data analysis What would Mendel have observed in a dihybrid cross if the two loci were 10 cM apart on the same chromosome? Is this likely to have led him to the idea of independent assortment?

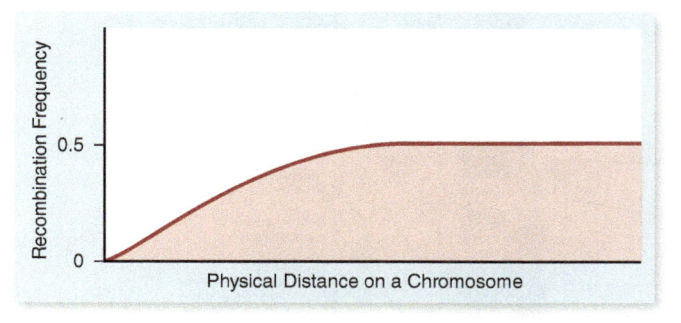

Figure 13.8 Relationship between true distance and recombination frequency. As distance on a chromosome increases, the recombinants are not all detected due to double crossovers. This leads to a curve that levels off at 0.5.

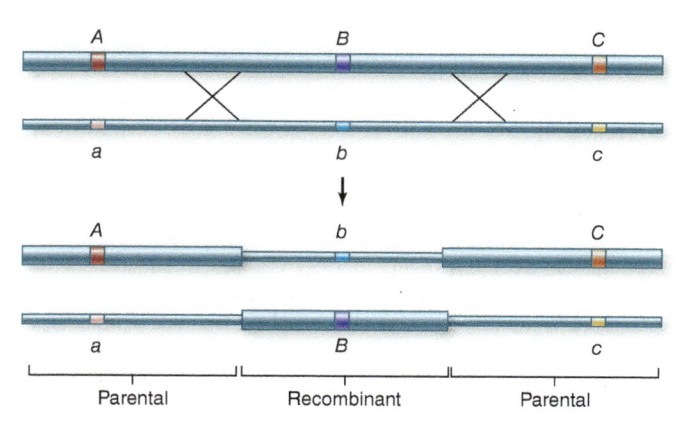

Figure 13.9 Use of a three-point cross to order genes. In a two-point cross, the outside loci appear parental for double crossovers. With the addition of a third locus, the two crossovers can still be detected because the middle locus will be recombinant. This double crossover class should be the least frequent, so whatever locus has recombinant alleles in this class must be in the middle.

Three-point crosses can be used to put genes in order

Because multiple crossovers reduce the number of observed recombinant progeny, longer map distances are not accurate. As a result, when geneticists try to construct maps from a series of two-point crosses, determining the order of genes is problematic. Using three loci instead of two, or a three-point cross, can help solve the problem.

In a three-point cross, the gene in the middle allows us to see recombination events on either side. For example, a double crossover for the two outside loci is actually a single crossover between the middle locus and each outside locus (figure 13.9).

The probability of two crossovers is equal to the product of the probability of each individual crossover, each of which is relatively low. Therefore, in any three-point cross, the class of offspring with two crossovers is the least frequent class. Analyzing these individuals to see which locus is recombinant identifies the locus that lies in the middle of the three loci in the cross (figure 13.9).

In practice, geneticists use three-point crosses to determine the order of genes, then use data from the closest two-point crosses to determine distances. Longer distances are generated by simple addition of shorter distances. This avoids using inaccurate measures from two-point crosses between distant loci.

Genetic maps can be constructed for the human genome

Human genes can be mapped, but the data must be derived from historical pedigrees, such as those of the royal families of Europe mentioned in section 13.2. The principle is the same—genetic distance is still proportional to recombination frequency—but the analysis requires the use of complex statistics and summing data from many families.

The difficulty of mapping in humans

Looking at nonhuman animals with extensive genetic maps, the majority of genetic markers have been found at loci where alleles cause morphological changes, such as variant eye color, body color, or wing morphology in flies. In humans, such alleles generally, but not always, correspond to what we consider disease states. In the early 1980s, the number of markers for the human genome numbered in the hundreds. Because the human genome is so large, however, this low number of markers would never provide dense enough coverage to use for mapping.

Another consideration is that the disease-causing alleles are those that we wish to map, but they occur at low frequencies in the population. Any one family would be highly unlikely to carry multiple disease alleles, the segregation of which would allow for mapping.

Anonymous markers

This situation changed with the development of **anonymous markers,** genetic markers that can be detected using molecular techniques, but that do not cause a detectable phenotype. The nature of these markers has evolved with technology, leading to a standardized set of markers scattered throughout the genome. These markers, which have a relatively high density, can be detected using techniques that are easy to automate. As a result of analysis, geneticists now have several thousand markers to work with, instead of hundreds, and have produced a human genetic map that would have been unthinkable 30 years ago (figure 13.10). (In chapters 17 and 18, you'll learn about some of the molecular techniques that have been developed for use with genomes.)

Single-nucleotide polymorphisms (SNPs)

The information developed from sequencing the human genome can then be used to identify and map single bases that differ between individuals. Any differences between individuals in populations are termed *polymorphisms;* polymorphisms affecting a single base of a gene locus are called **single-nucleotide polymorphisms (SNPs).** It is estimated that there are about 10 million SNPs in the human genome. These are being identified, and mapped by an international consortium, with the data being publicly available. This resource provides a powerful tool for modern genetic analysis.

The power of gene mapping applies to more than just the relatively small number of genes that show simple Mendelian inheritance. The development of a high-resolution genetic map, and the characterization of millions of SNPs, opens up the possibility of being able to characterize complex quantitative traits in humans as well.

On a more practical level, the types of molecular markers described earlier in this section are used in forensic analysis. Although not quite as rapid as some television programs would have you believe, this does allow rapid DNA testing of crime scene samples to help eliminate or confirm crime suspects and for paternity testing.

Learning Outcomes Review 13.4

Crossing over during meiosis exchanges alleles on homologues. This recombination of alleles can be used to map the location of genes. Genes that are close together are said to be linked, and exhibit an excess of parental versus recombinant types in a testcross. The frequency of recombination in testcrosses is used as a measure of genetic distance. Loci separated by large distances have multiple crossovers between them, which can lead to independent assortment.

- *If two genes assort independently, can you tell if they are on a single chromosome and far apart or on two different chromosomes?*

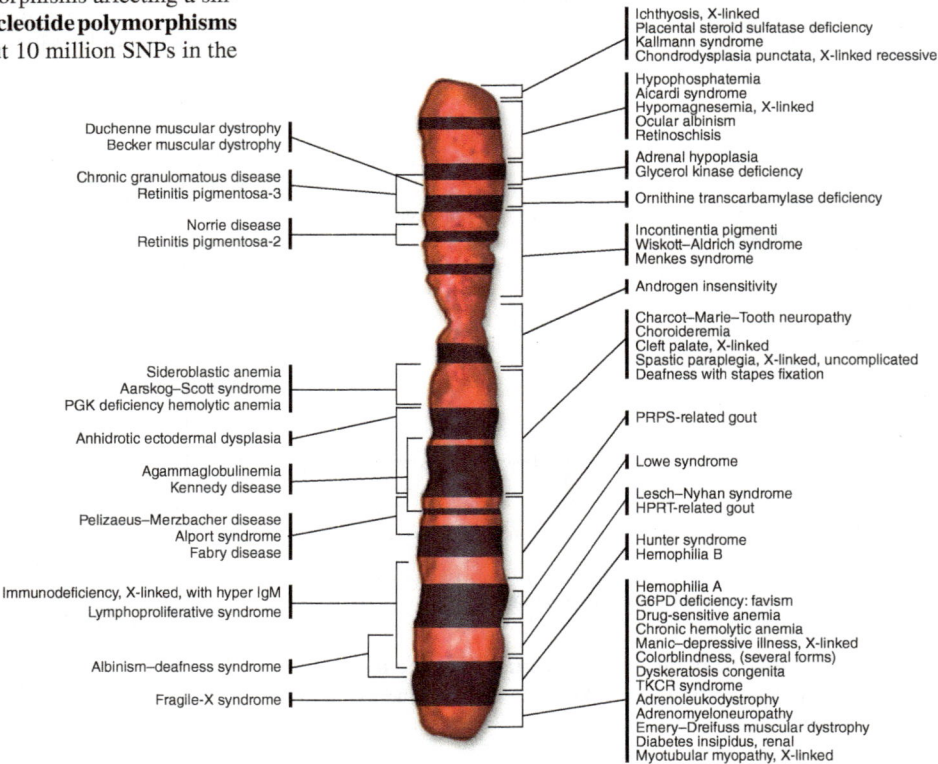

Figure 13.10 The human X-chromosome gene map. Only a partial map for the human X chromosome is presented here; a more detailed map would require a much larger figure. The black bands represent staining patterns that can be seen under the microscope, and the constriction represents the centromere. Analysis of the sequence of the X chromosome indicates 1805 genes on the X chromosome: 821 of these have mutant alleles that have been identified as affecting disease states. Of these, 59 diseases are shown that have been mapped to specific segments of the X chromosome, indicated by brackets.

13.5 Selected Human Genetic Disorders

Learning Outcomes
1. Explain how mutations can cause disease.
2. Describe the consequences of nondisjunction in humans.
3. Recognize how genomic imprinting can lead to non-Mendelian inheritance.

Diseases that run in families have been known for many years. These can be nonlife-threatening like albinism, or may result in premature death like Huntington's, which were used as examples of recessive and dominant traits in humans previously. A small sample of diseases due to alterations of alleles of a single gene is provided in table 13.2. We will discuss the nature of these genetic changes later, in chapter 15. In this section, we discuss some of the genetic disorders that have been found in human populations.

Sickle cell anemia is due to altered hemoglobin

The first human disease shown to be the result of a mutation in a protein was sickle cell anemia. It is caused by a defect in the oxygen carrier molecule, hemoglobin, that leads to impaired oxygen delivery to tissues. The defective hemoglobin molecules stick to one another, leading to stiff, rodlike structures that alter the shape of the red blood cells that carry them. These red blood cells take on a characteristic shape that led to the name "sickle cell" (figure 13.11).

Individuals homozygous for the sickle cell allele exhibit intermittent illness and reduced life span. Individuals heterozygous for the sickle cell allele are indistinguishable from normal individuals in a normal oxygen environment, although their red cells do exhibit reduced ability to carry oxygen.

The sickle cell allele is particularly prevalent in people of African descent. In some regions of Africa, up to 45% of the population is heterozygous for the trait, and 6% are homozygous.

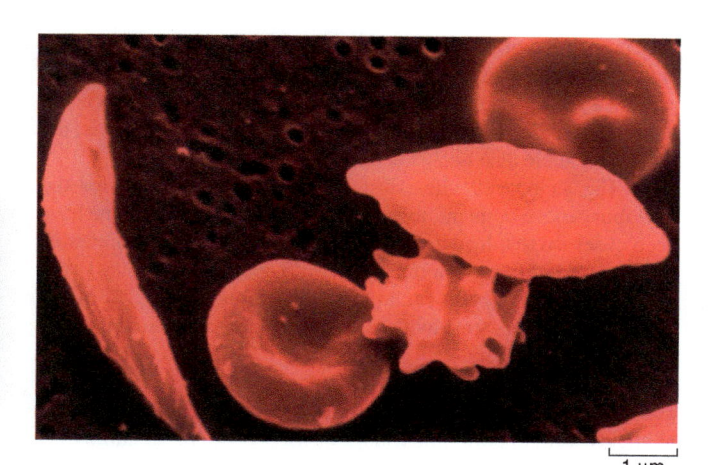

Figure 13.11 Sickle cell anemia. In individuals homozygous for the sickle cell trait, many of the red blood cells have sickled or irregular shapes, such as the cell on the far left.

TABLE 13.2 Some Important Genetic Disorders

Disorder	Symptom	Defect	Dominant/Recessive	Frequency Among Human Births
Cystic fibrosis	Mucus clogs lungs, liver, and pancreas.	Failure of chloride ion transport mechanism	Recessive	1/2500 (Caucasians)
Sickle cell anemia	Blood circulation is poor.	Abnormal hemoglobin molecules	Recessive	1/600 (African Americans)
Tay–Sachs disease	Central nervous system deteriorates in infancy.	Defective enzyme (hexosaminidase A)	Recessive	1/3500 (Ashkenazi Jews)
Phenylketonuria	Brain fails to develop in infancy; treatable with dietary restriction.	Defective enzyme (phenylalanine hydroxylase)	Recessive	1/12,000
Hemophilia	Blood fails to clot.	Defective blood-clotting factor VIII	X-linked recessive	1/10,000 (Caucasian males)
Huntington disease	Brain tissue gradually deteriorates in middle age.	Production of an inhibitor of brain cell metabolism	Dominant	1/24,000
Muscular dystrophy (Duchenne)	Muscles waste away.	Degradation of myelin coating of nerves stimulating muscles	X-linked recessive	1/3700 (males)
Hypercholesterolemia	Excessive cholesterol levels in blood lead to heart disease.	Abnormal form of cholesterol cell surface receptor	Dominant	1/500

This proportion of heterozygotes is higher than would be expected on the basis of chance alone. It turns out that heterozygosity confers a greater resistance to the blood-borne parasite that causes malaria. In regions of central Africa where malaria is endemic, the sickle cell allele also occurs at a high frequency.

The sickle cell allele is not the end of the story for the β-globin gene; a large number of other alterations of this gene have been observed that lead to anemias. In fact, for hemoglobin, which is composed of two α-globins and two β-globins, over 700 structural variants have been cataloged. It is estimated that 7% of the human population worldwide are carriers for different inherited hemoglobin disorders.

The Human Gene Mutation Database has cataloged the nature of many disease alleles, including the sickle cell allele. The majority of alleles seem to be simple changes. Almost 56% of the more than 115,000 alleles in the Human Gene Mutation Database are single-base substitutions. Another 22% are due to small insertions or deletions of less than 20 bases. This careful survey of disease-causing alterations has also revealed more large-scale changes. This is now termed *copy number variation (CNV)* and appears more important than previously thought.

Nondisjunction of chromosomes changes chromosome number

The failure of homologues or sister chromatids to separate properly during meiosis is called **nondisjunction.** This failure leads to the gain or loss of a chromosome, a condition called **aneuploidy.** The frequency of aneuploidy in humans has been estimated at 7–10% of clinically recognized conceptions.

Nondisjunction of autosomes

Humans who have lost even one copy of an autosome are called **monosomics,** and generally do not survive embryonic development. In all but a few cases, humans who have gained an extra autosome (called **trisomics**) also do not survive. Data from clinically recognized spontaneous abortions indicate levels of aneuploidy as high as 35%.

Five of the smallest human autosomes—those numbered 13, 15, 18, 21, and 22—can be present as three copies and still allow the individual to survive, at least for a time. The presence of an extra chromosome 13, 15, or 18 causes severe developmental defects, and infants with such a genetic makeup die within a few months. In contrast, individuals who have an extra copy of chromosome 21 or, more rarely, chromosome 22, usually survive to adulthood. In these people, the maturation of the skeletal system is delayed, so they generally are short and have poor muscle tone. Their mental development is also affected, and children with trisomy 21 show some degree of intellectual disability.

The developmental defect produced by trisomy 21 (figure 13.12) was first described in 1866 by J. Langdon Down; for this reason, it is called **Down syndrome.** About 1 in every 750 children exhibits Down syndrome, and the frequency is comparable in all racial groups. Similar conditions also occur in chimpanzees and other related primates.

In humans, the defect occurs when a particular small portion of chromosome 21 is present in three copies instead of two. In 97% of the cases examined, all of chromosome 21 is present in three copies. In the other 3%, a small portion of chromosome 21 containing the critical segment has been added to another chromosome by a process called *translocation* (see chapter 15); it exists along with the normal two copies of chromosome 21. This latter condition is known as *translocation Down syndrome.*

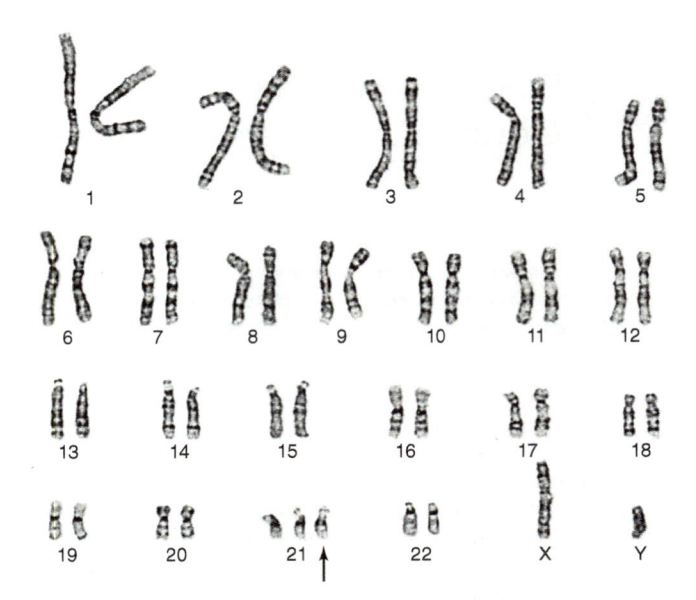

Figure 13.12 Down syndrome. As shown in this male karyotype, Down syndrome is associated with trisomy of chromosome 21 (arrow shows third copy of chromosome 21).

In mothers younger than 20 years of age, the risk of giving birth to a child with Down syndrome is about 1 in 1700; in mothers 20 to 30 years old, the risk is only about 1 in 1400. However, in mothers 30 to 35 years old, the risk rises to 1 in 750, and by age 45, the risk is as high as 1 in 16 (figure 13.13).

Interesting differences exist in the spectrum of genetic errors that arise in the male and female germ line. This can be traced to differences in gametogenesis between males and females. New mutations that affect one or a few bases occur primarily in the male germ line where stem cells are dividing mitotically for renewal throughout life. Chromosomal abnormalities arise more frequently on the female side where gametes are set aside in early development, and arrested in meiosis I with homologues paired. This is also thought to be the molecular basis of the increase in aneuploidy with maternal age.

Nondisjunction of sex chromosomes

Individuals who gain or lose a sex chromosome do not generally experience the severe developmental abnormalities caused by similar changes in autosomes. Although such individuals have somewhat abnormal features, they often reach maturity and in some cases may be fertile.

X chromosome nondisjunction. When X chromosomes fail to separate during meiosis, some of the gametes produced possess both X chromosomes, and so are XX gametes; the other gametes have no sex chromosome and are designated "O" (figure 13.14).

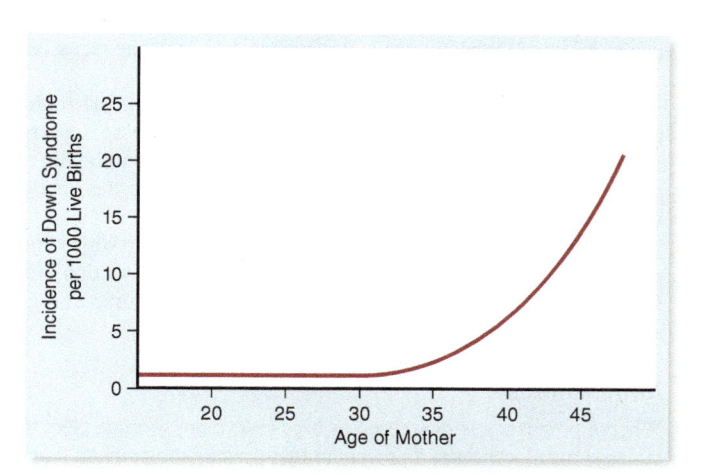

Figure 13.13 Correlation between maternal age and the incidence of Down syndrome. As women age, the chances they will bear a child with Down syndrome increase. After a woman reaches 35, the frequency of Down syndrome rises rapidly.

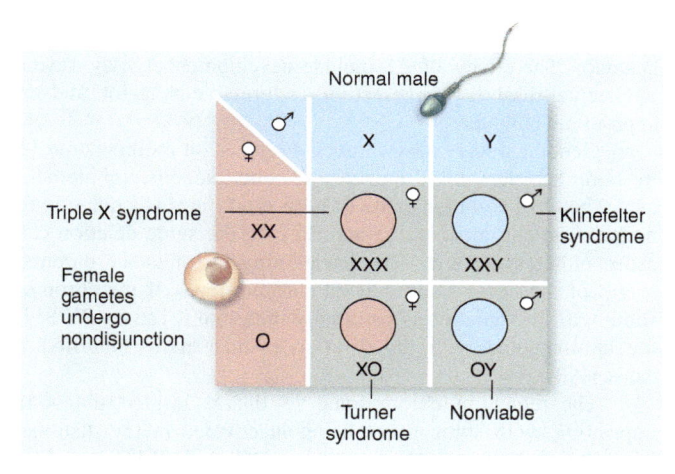

Figure 13.14 How nondisjunction can produce abnormalities in the number of sex chromosomes. When nondisjunction occurs in the production of female gametes, the gamete with two X chromosomes (XX) produces Klinefelter males (XXY) and triple X females (XXX). The gamete with no X chromosome (O) produces Turner females (XO) and nonviable OY males lacking any X chromosome.

Data analysis Over a five-year period between ages 20 and 25, the incidence of Down syndrome increases 0.1 per thousand; over a five-year period between ages 35 and 40, the incidence increases to 8.0 per thousand, 80 times as great. The period of time is the same in both instances. What has changed?

Inquiry question Can you think of two nondisjunction scenarios that would produce an XXY male?

If an XX gamete combines with an X gamete, the resulting XXX zygote develops into a female with one functional X chromosome and two Barr bodies. She may be taller in stature but is otherwise normal in appearance.

If an XX gamete instead combines with a Y gamete, the effects are more serious. The resulting XXY zygote develops into a male who has many female body characteristics and, in some cases but not all, diminished mental capacity. This condition, called *Klinefelter syndrome,* occurs in about 1 out of every 1000 male births.

If an O gamete fuses with a Y gamete, the resulting OY zygote is nonviable and fails to develop further; humans cannot survive when they lack the genes on the X chromosome. But if an O gamete fuses with an X gamete, the XO zygote develops into a sterile female of short stature, with a webbed neck and sex organs that never fully mature during puberty. The mental abilities of an XO individual are in the low-normal range. This condition, called *Turner syndrome,* occurs roughly once in every 5000 female births.

Y chromosome nondisjunction. The Y chromosome can also fail to separate in meiosis, leading to the formation of YY gametes. When these gametes combine with X gametes, the XYY zygotes develop into fertile males of normal appearance. The frequency of the XYY genotype *(Jacob syndrome)* is about 1 per 1000 newborn males.

Genomic imprinting depends on the parental origin of alleles

By the late 20th century, geneticists were confident that they understood the basic mechanisms governing inheritance. It came as quite a surprise to find that for some genes inheritance depends on the parent of origin. Even stranger, the only two groups that show this pattern are flowering plants and mammals. In **genomic imprinting,** the phenotype caused by a specific allele is exhibited when the allele comes from one parent, but not from the other.

The basis for genomic imprinting is the expression of a gene depending on passage through maternal or paternal germ lines. Some genes are inactivated in the paternal germ line and therefore are not expressed in the zygote. Other genes are inactivated in the maternal germ line, with the same result. This condition makes the zygote effectively haploid for an imprinted gene. The expression of variant alleles of imprinted genes depends on the parent of origin. Furthermore, imprinted genes seem to be concentrated in particular regions of the genome. These regions include genes that are both maternally and paternally imprinted.

Prader–Willi and Angelman syndromes

An example of genomic imprinting in humans involves the two diseases Prader–Willi syndrome (PWS) and Angelman syndrome (AS). The effects of PWS include respiratory distress, obesity, short stature, mild intellectual disability, and obsessive–compulsive behavior. The effects of AS include developmental delay, severe intellectual disability, hyperactivity, aggressive behavior, and inappropriate laughter.

Genetic studies have implicated genes on chromosome 15 for both disorders, but the pattern of inheritance is complementary. The most common cause of both syndromes is a deletion of material on chromosome 15 and, in fact, the same deletion can cause either syndrome. The determining factor is the parental origin of the normal and deleted chromosomes. If the chromosome with the deletion is paternally inherited it causes PWS; if the chromosome with the deletion is maternally inherited it causes AS.

The region of chromosome 15 that is lost is subject to imprinting, with some genes being inactivated in the maternal germ line, and others in the paternal germ line. In PWS, genes are inactivated in the maternal germ line, such that deletion or other functional loss of paternally derived alleles produces the syndrome. The opposite is true for AS syndrome: Genes are inactivated in the paternal germ line, such that loss of maternally derived alleles leads to the syndrome.

Imprinting is an example of epigenetics

Genomic imprinting is an example of the more general phenomenon: **epigenetic inheritance.** An epigenetic change is defined as a mitotically and/or meiotically stable change in gene function that does not involve a change in DNA sequence. An example from this chapter is X-chromosome inactivation where an entire chromosome is silenced, and the effect is inherited through many mitotic divisions. In some mouse and rat models, this even includes affects of maternal diet on F_2 animals.

Epigenetic mechanisms include changes in DNA methylation and histone modifications. Noncoding RNAs and nuclear organization have also been implicated. Alterations in chromatin structure and the accessibility of DNA seems to be a point of convergence for multiple epigenetic mechanisms (see chapter 16). In some well-studied cases, the pattern of imprinting that occurs in the male and female germ line is due to male- and female-specific patterns of DNA methylation and alterations to the proteins that are involved in chromosome structure.

Some genetic defects can be detected early in pregnancy

Although most genetic disorders cannot yet be cured, progress toward successful therapies is being made in many cases. In the absence of therapy, the alternative is to avoid producing children with these conditions. **Genetic counseling** is aimed at identifying parents at risk for having children with genetic defects and of assessing the genetic state of early embryos.

Pedigree analysis

One way of assessing risks is through pedigree analysis, often employed as an aid in genetic counseling. By analyzing a person's pedigree, it is sometimes possible to estimate the likelihood that the person is a carrier for certain disorders. For example, if a counseling client's family history reveals that a relative has been afflicted with a recessive genetic disorder, such as cystic fibrosis, it is possible that the client is a heterozygous carrier of the recessive allele for that disorder.

If a couple is expecting a child, and pedigree analysis indicates that both of them have a significant chance of being carriers of a deleterious recessive allele, the pregnancy is said to be high-risk. In such cases, a significant probability exists that their child will exhibit the clinical disorder.

Another class of high-risk pregnancy is that in which the mothers are older than 35. As discussed earlier in this section, the frequency of Down syndrome increases dramatically in the pregnancies of older women (see figure 13.13).

Amniocentesis

When a pregnancy is diagnosed as high-risk, many women elect to undergo **amniocentesis,** a procedure that permits the prenatal diagnosis of many genetic disorders. In the fourth month of pregnancy, a sterile hypodermic needle is inserted into the expanded uterus of the mother, removing a small sample of the amniotic fluid that bathes the fetus (figure 13.15). Within the fluid are free-floating cells derived from the fetus; once removed, these cells can be grown in cultures in the laboratory.

During amniocentesis, the position of the needle and that of the fetus are usually observed by means of *ultrasound*. The sound waves used in ultrasound are not harmful to mother or fetus, and they permit the person withdrawing the amniotic fluid to do so without damaging the fetus. In addition, ultrasound can be used to examine the fetus for signs of major abnormalities. However, about 1 out of 200 amniocentesis procedures may result in fetal death and miscarriage.

Chorionic villus sampling

Physicians have increasingly turned to a less invasive procedure for genetic screening called **chorionic villus sampling (CVS).** Using this method, the physician removes cells from the chorion, a membranous part of the placenta that nourishes the fetus (figure 13.15). This procedure can be used earlier in pregnancy (by the eighth week) and yields results much more rapidly than does amniocentesis. Risks from chorionic villus sampling are comparable to those for amniocentesis.

To test for certain genetic disorders, genetic counselors look for three characteristics in the cultures of cells obtained from amniocentesis or chorionic villus sampling. First, analysis of the karyotype can reveal aneuploidy and gross chromosomal alterations. Second, in many cases it is possible to test directly for the proper functioning of enzymes affected by genetic disorders. As examples, the lack of the enzyme responsible for breaking down phenylalanine indicates phenylketonuria (PKU); the absence of the enzyme responsible for the breakdown of gangliosides indicates Tay–Sachs disease; and so forth. Additionally, with information from the Human Genome Project, more disease alleles for genetic disorders are known. If there are a small number of alleles for a specific disease in the population, these can be identified as well.

With the changes in human genetics brought about by the Human Genome Project (see chapter 18), it is possible to design tests for many more diseases. Difficulties still exist in discerning the number and frequency of disease-causing

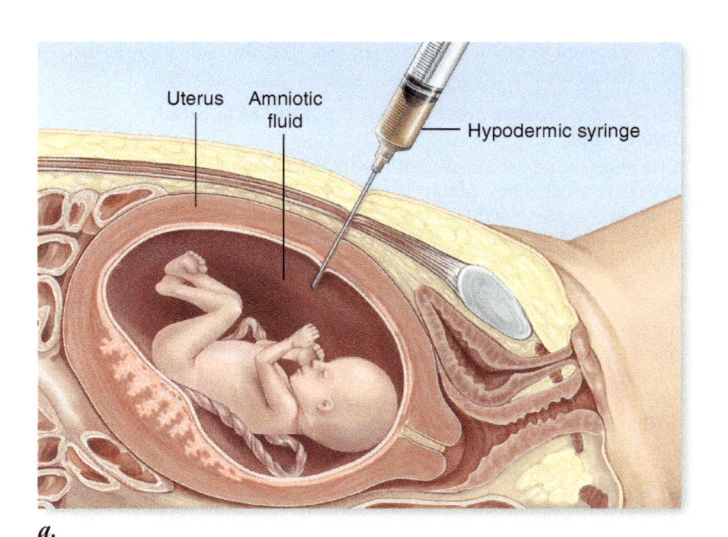

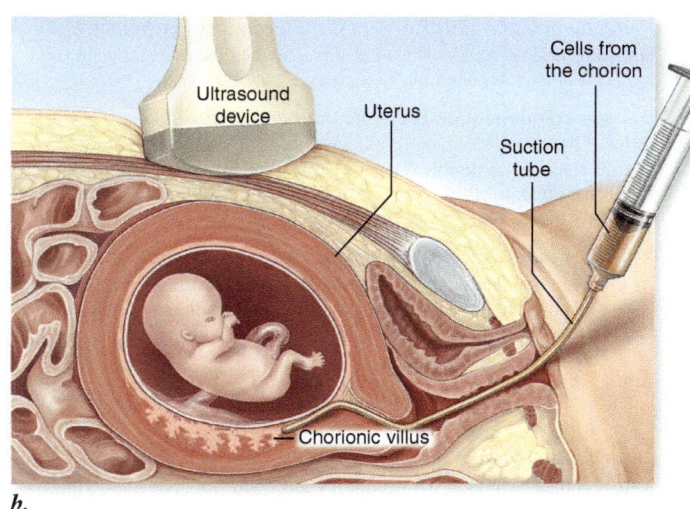

Figure 13.15 Two ways to obtain fetal cells. *a.* In amniocentesis, a needle is inserted into the amniotic cavity, and a sample of amniotic fluid containing some free cells derived from the fetus is withdrawn into a syringe. *b.* In chorionic villus sampling, cells are removed by suction with a tube inserted through the cervix. This can be done as early as the eighth to tenth week. In each case, the cells can be grown in culture, then examined for karyotypes, and used in biochemical and genetic tests.

alleles, but these problems are not insurmountable. At present, more than 200 disorders can be detected by this test. This is bound to be expanded to include alleles that do not directly lead to disease states but that predispose a person for a particular disease.

Inquiry question Based on what you read in this chapter, what reasons could a mother have to undergo CVS, considering its small but potential risks?

Learning Outcomes Review 13.5

Mutations in DNA that result in altered proteins can cause hereditary diseases. Pedigree studies and genetic testing may clarify the risk of disease. At the chromosome level, nondisjunction during meiosis can result in gametes with too few or too many chromosomes, most of which produce inviable offspring. Imprinting refers to inactivation of alleles depending on which parent the alleles come from; offspring in whom imprinting occurs appear haploid for the affected gene even though they are diploid.

- During spermatogenesis, is there any difference in outcome between first- and second-division nondisjunction?

Chapter Review

13.1 Sex Linkage and the Chromosomal Theory of Inheritance

Morgan correlated the inheritance of a trait with sex chromosomes (figure 13.2).
Morgan crossed red-eyed and white-eyed flies and found differences in inheritance based on the sex of offspring. All white-eyed offspring were males, but testcrosses showed that white-eyed females were possible, supporting the idea that the white-eye gene was on the X chromosome.

The gene for eye color lies on the X chromosome.
The inheritance of eye color in *Drosophila* segregates with the X chromosome, a phenomenon termed sex-linked inheritance.

13.2 Sex Chromosomes and Sex Determination

Sex determination in animals is usually associated with a chromosomal difference. In some animals, females have two similar sex chromosomes and males have sex chromosomes that differ. In other species, females have sex chromosomes that differ (table 13.1).

In humans, the Y chromosome generally determines maleness.
The Y chromosome is highly condensed and does not have active counterparts to most genes on the X chromosome. The *SRY* gene on the Y chromosome is responsible for the masculinization of genitalia and secondary sex organs. An XY individual can develop into a sterile female due to mutations in the *SRY* gene or the failure of the embryo to respond to androgens.

Some human genetic disorders display sex linkage (figure 13.3).
Human genetic disorders show sex linkage when the relevant gene is on the X chromosome; hemophilia is an example.

Dosage compensation prevents doubling of sex-linked gene products.
In fruit flies, males double the gene expression from their single X chromosome. In mammals, one of the X chromosomes in a female is randomly inactivated during development.

X-chromosome inactivation can lead to genetic mosaics.
In a mammalian female that is heterozygous for X-chromosome alleles, X inactivation produces a mosaic pattern, as shown in the coat color of calico cats (figure 13.4).

13.3 Exceptions to the Chromosomal Theory of Inheritance

Mitochondrial genes are inherited from the female parent.
Mitochondria have their own genomes and divide independently; they are passed to offspring in the cytoplasm of the egg cell.

Chloroplast genes may also be passed on uniparentally.
Chloroplasts also reside in the cytoplasm, have their own genomes, and divide independently. They are usually inherited maternally.

13.4 Genetic Mapping

Mendel's independent assortment is too simplistic. Genes on the same chromosome may or may not segregate independently.

Genetic recombination exchanges alleles on homologues.
Homologous chromosomes may exchange alleles by crossing over (figure 13.5). This occurs by breakage and rejoining of chromosomes as shown by crosses in which chromosomes carry both visible and genetic markers (figure 13.6).

Recombination is the basis for genetic maps.
Genes close together on a single chromosome are said to be linked. The farther apart two linked genes are, the greater the frequency of recombination. This allows genetic maps to be constructed based on recombination frequency. A map unit is expressed as the percentage of recombinant progeny.

Multiple crossovers can yield independent assortment results.
The probability of multiple crossovers increases with distance between two genes and results in an underestimate of recombination frequency. The maximum recombination frequency is 50%, the same value as for independent assortment.

Three-point crosses can be used to put genes in order (figure 13.9).
If three genes are used instead of two, data from multiple crossovers can be used to order genes. Longer map distances fail to reflect the effect of multiple crossovers and thus underestimate true distance. By evaluating intervening genes with less separation, more accurate distances can be obtained.

Genetic maps can be constructed for the human genome.
Human genetic mapping was difficult because it required multiple disease-causing alleles segregating in a family. The process has been made easier by the use of anonymous markers, identifiable molecular markers that do not cause a phenotype. Single-nucleotide polymorphisms (SNPs) can be used to detect differences between individuals for identification.

13.5 Selected Human Genetic Disorders

Sickle cell anemia is due to altered hemoglobin.
The phenotypes in sickle cell anemia can all be traced to alterations in the structure of hemoglobin that affect the shape of red blood cells. Over 700 variants of hemoglobin structure have been characterized, some of which also cause disorders.

Nondisjunction of chromosomes changes chromosome number.
Nondisjunction is the failure of homologues or sister chromatids to separate during meiosis. The result is aneuploidy: monosomy or trisomy of a chromosome in the zygote. Most aneuploidies are lethal, but some, such as trisomy 21 in humans (Down syndrome), can result in viable offspring. Sex chromosome nondisjunction can produce XX, YY, or gametes with no sex chromosomes (figure 13.14). Nondisjunction is more common on the female's side, and the frequency increases with age.

Genomic imprinting depends on the parental origin of alleles.
In genomic imprinting, the expression of a gene depends on parent of origin. This is an example of epigenetics, where a heritable change in phenotype is not due to a change to DNA. Imprinted genes appear to be inactivated by methylation. Imprinting produces a haploid phenotype.

Some genetic defects can be detected early in pregnancy.
Genetic defects in humans can be determined by pedigree analysis, amniocentesis, or chorionic villus sampling.

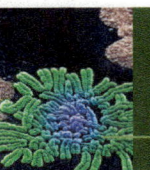

Review Questions

UNDERSTAND

1. Why is the white-eye phenotype always observed in males carrying the white-eye allele?
   a. Because the trait is dominant
   b. Because the trait is recessive
   c. Because the allele is located on the X chromosome and males only have one X
   d. Because the allele is located on the Y chromosome and only males have Y chromosomes

2. In an organism's genome, *autosomes* are
   a. the chromosomes that differ between the sexes.
   b. chromosomes that are involved in sex determination.
   c. only inherited from the mother (maternal inheritance).
   d. all of the chromosomes other than sex chromosomes.

3. What cellular process is responsible for genetic recombination?
   a. The independent alignment of homologous pairs during meiosis I
   b. Separation of the homologues in meiosis I
   c. Separation of the chromatids during meiosis II
   d. Crossing over between homologues

4. The map distance between two genes is determined by the
   a. recombination frequency.
   b. frequency of parental types.
   c. ratio of genes to length of a chromosome.
   d. ratio of parental to recombinant progeny.

5. How many map units separate two alleles if the recombination frequency is 0.07?
   a. 700 cM   b. 70 cM   c. 7 cM   d. 0.7 cM

6. How does maternal inheritance of mitochondrial genes differ from sex linkage?
   a. Mitochondrial genes do not contribute to the phenotype of an individual.
   b. Because mitochondria are inherited from the mother, only females are affected.
   c. Since mitochondria are inherited from the mother, females and males are equally affected.
   d. Mitochondrial genes must be dominant. Sex-linked traits are typically recessive.

7. Which of the following genotypes due to nondisjunction of sex chromosomes is lethal?
   a. XXX   b. XXY   c. OY   d. XO

APPLY

1. A recessive sex-linked gene in humans leads to a loss of sweat glands. A woman heterozygous for this will
   a. have no sweat glands.
   b. have normal sweat glands.
   c. have patches of skin with and without sweat glands.
   d. have an excess of sweat glands.

2. As real genetic distance increases, the distance calculated by recombination frequency becomes an
   a. overestimate due to multiple crossovers that cannot be scored.
   b. underestimate due to multiple crossovers that cannot be scored.
   c. underestimate due to multiple crossovers adding to recombination frequency.
   d. overestimate due to multiple crossovers adding to recombination frequency.

3. Down syndrome is the result of trisomy for chromosome 21. Why is this trisomy viable and trisomy for most other chromosomes is not?
   a. Chromosome 21 is a large chromosome and excess genetic material is less harmful.
   b. Chromosome 21 behaves differently in meiosis I than the other chromosomes.
   c. Chromosome 21 is a small chromosome with few genes so this does less to disrupt the genome.
   d. Chromosome 21 is less prone to nondisjunction than other chromosomes.

4. Genes that are on the same chromosome can show independent assortment
   a. when they are far enough apart for two crossovers to occur.
   b. when they are far enough apart that odd numbers of crossovers is about equal to even.
   c. only if recombination is low for that chromosome.
   d. only if the genes show genomic imprinting.

5. The A and B genes are 10 cM apart on a chromosome. If an A B/a b heterozygote is testcrossed to a b/a b, how many of each progeny class would you expect out of 100 total progeny?
   a. 25 A B, 25 a b, 25 A b, 25 a B
   b. 10 A B, 10 a b
   b. 45 A B, 45 a b
   d. 45 A B, 45 a b, 5 A b, 5 a B

6. During the process of spermatogenesis, a nondisjunction event that occurs during the second division would be
   a. worse than the first division because all four meiotic products would be aneuploid.
   b. better than the first division because only two of the four meiotic products would be aneuploid.
   c. the same outcome as the first division with all four products aneuploid.
   d. the same outcome as the first division as only two products would be aneuploid.

SYNTHESIZE

1. Color blindness is caused by a sex-linked, recessive gene. If a woman, whose father was color blind, marries a man with normal color vision, what percentage of their children will be color blind? What percentage of male children? Of female children?

2. Assume that the genes for seed color and seed shape are located on the same chromosome. A plant heterozygous for both genes is testcrossed wrinkled green with the following results:

green, wrinkled	645
green, round	36
yellow, wrinkled	29
yellow, round	590

   What were the genotypes of the parents, and how far apart are these genes?

3. A low frequency of calico cats are male (about 1/3000). How could this arise?

CHAPTER 14

DNA: The Genetic Material

Chapter Contents

14.1 The Nature of the Genetic Material
14.2 DNA Structure
14.3 Basic Characteristics of DNA Replication
14.4 Prokaryotic Replication
14.5 Eukaryotic Replication
14.6 DNA Repair

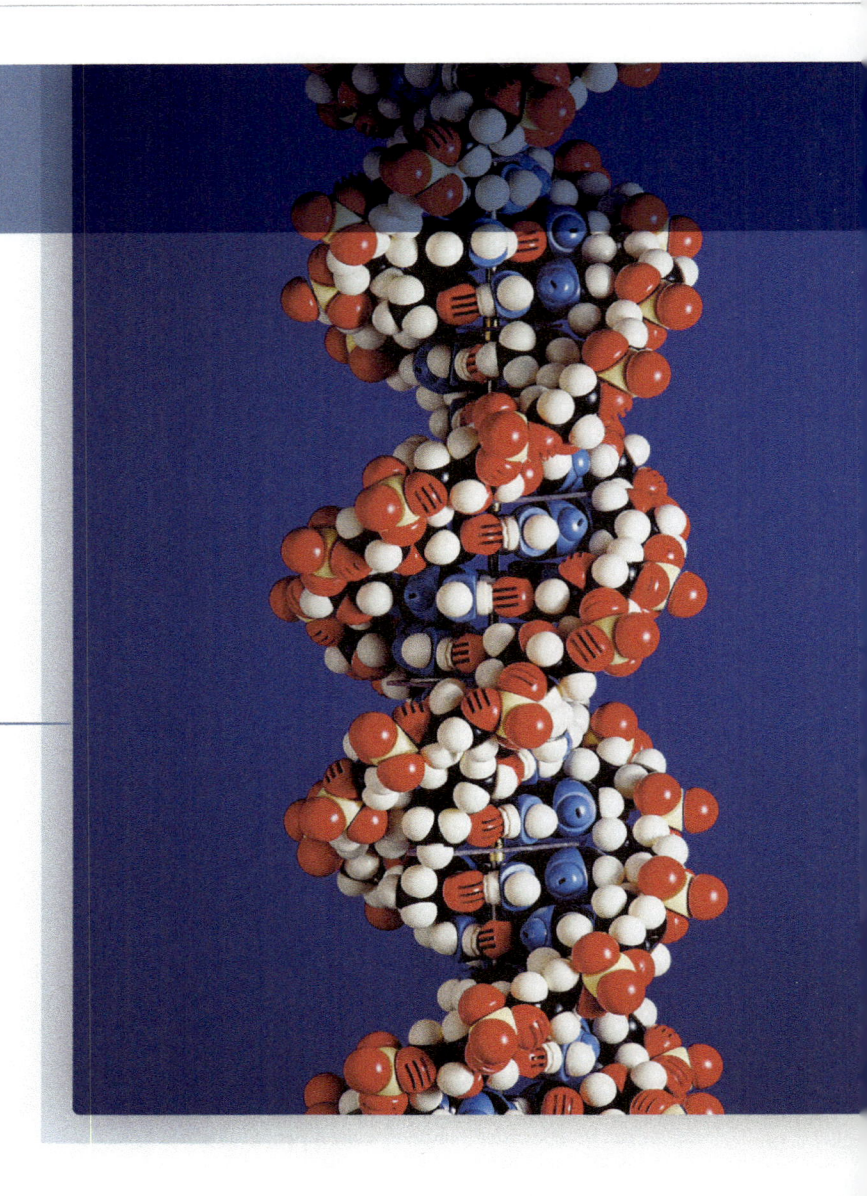

Introduction

The rediscovery of Mendel at the turn of the 20th century led to a period of rapid discovery of genetic mechanisms detailed in chapters 12 and 13. One question not directly answered for more than 50 years was perhaps the most simple: What are genes actually made of? Genes were known to be on chromosomes, but they are complex structures composed of DNA, RNA, and protein. This chapter describes the chain of experiments that led to our current understanding of DNA, modeled in the picture above, and of the molecular mechanisms of heredity. These experiments are among the most elegant in science. The elucidation of the structure of DNA was the beginning of a molecular era, whose pace is only accelerating today.

14.1 The Nature of the Genetic Material

Learning Outcomes

1. Describe the experiments of Griffith and Avery.
2. Evaluate the evidence for DNA as genetic material.

In chapters 12 and 13, you learned about the nature of inheritance and how genes, which contain the information to specify traits, are located on chromosomes. This finding led to the question of what part of the chromosome actually contains the genetic information. Specifically, biologists wondered about the chemical identity of the genetic information. They knew that chromosomes are composed primarily of both protein and DNA. Which of these organic molecules actually makes up the genes?

Starting in the late 1920s and continuing for about 30 years, a series of investigations addressed this question. DNA consists of four chemically similar nucleotides. In contrast, protein contains

20 different amino acids that are much more chemically diverse than nucleotides. These characteristics seemed initially to indicate greater informational capacity in protein than in DNA.

However, experiments began to reveal evidence in favor of DNA. We describe three of those major findings in this section.

Griffith finds that bacterial cells can be transformed

The first clue came in 1928 with the work of the British microbiologist Frederick Griffith. Griffith was trying to make a vaccine that would protect against influenza, which was thought at the time to be caused by the bacteria *Streptococcus pneumoniae*. There are two forms of this bacteria: The normal virulent form that causes pneumonia, and a mutant, nonvirulent form that does not. The normal virulent form of this bacterium is referred to as the S form because it forms smooth colonies on a culture dish. The mutant, nonvirulent form, which lacks an enzyme needed to manufacture the polysaccharide coat, is called the R form because it forms rough colonies.

Griffith performed a series of simple experiments in which mice were infected with these bacteria, then monitored for disease symptoms (figure 14.1). Mice infected with the virulent S form died from pneumonia, whereas infection with the nonvirulent R form had no effect. This result shows that the polysaccharide coat is necessary for virulence. If the virulent S form is first killed with heat, infection does not harm the mice, showing that the coat itself is not sufficient to cause disease. Lastly, infecting mice with a mixture of heat-killed S form with live R form caused pneumonia and death in the mice. This was unexpected as neither treatment alone caused disease. Furthermore, high levels of live S form bacteria were found in the lungs of the dead mice.

Somehow, the information specifying the polysaccharide coat had passed from the dead, virulent S bacteria to the live, coatless R bacteria in the mixture, permanently altering the coatless R bacteria into the virulent S variety. Griffith called this transfer of virulence from one cell to another **transformation.** Our modern interpretation is that genetic material was actually transferred between the cells.

Avery, MacLeod, and McCarty identify the transforming principle

The agent responsible for transforming *Streptococcus* went undiscovered until 1944. In a classic series of experiments, Oswald Avery and his coworkers Colin MacLeod and Maclyn McCarty identified the substance responsible for transformation in Griffith's experiment.

They first prepared the mixture of dead S *Streptococcus* and live R *Streptococcus* that Griffith had used. Then they removed as much of the protein as they could from their preparation, eventually achieving 99.98% purity. They found that despite the removal of nearly all protein, the transforming activity was not reduced.

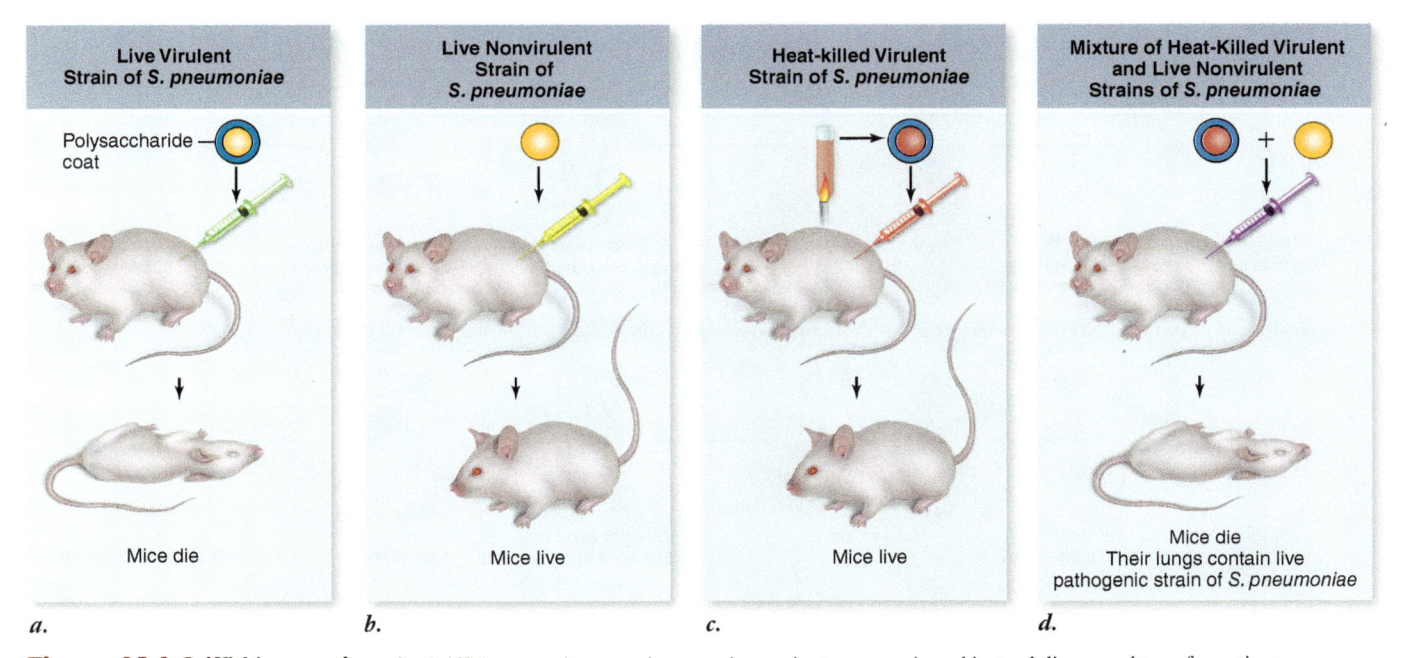

Figure 14.1 Griffith's experiment. Griffith was trying to make a vaccine against pneumonia and instead discovered transformation.
a. Injecting live virulent bacteria into mice produces pneumonia. Injection of nonvirulent bacteria (*b*) or heat-killed virulent bacteria (*c*) had no effect.
d. However, a mixture of heat-killed virulent and live nonvirulent bacteria produced pneumonia in the mice. This indicates the genetic information for virulence was transferred from dead, virulent cells to live, nonvirulent cells, transforming them from nonvirulent to virulent.

Moreover, the properties of this substance resembled those of DNA in several ways:

1. The elemental composition agreed closely with that of DNA.
2. When spun at high speeds in an ultracentrifuge, it migrated to the same level (density) as DNA.
3. Extracting lipids and proteins did not reduce transforming activity.
4. Protein-digesting enzymes did not affect transforming activity, nor did RNA-digesting enzymes.
5. DNA-digesting enzymes destroyed all transforming activity.

These experiments supported the identity of DNA as the substance transferred between cells by transformation and indicated that the genetic material, at least in this bacterial species, is DNA.

Hershey and Chase demonstrate that phage genetic material is DNA

Avery's results were not widely accepted at first because many biologists continued to believe that proteins were the genetic material. But additional evidence supporting Avery's conclusion was provided in 1952 by Alfred Hershey and Martha Chase, who experimented with viruses that infect bacteria. These viruses are called **bacteriophages,** or more simply, **phages.**

Viruses, described in more detail in chapter 27, are much simpler than cells; they generally consist of genetic material (DNA or RNA) surrounded by a protein coat. When a phage infects a bacterial cell, it first binds to the cell's outer surface and then injects its genetic information into the cell. Once inside the cell, the viral genetic information uses the cell's information-processing machinery to make viral proteins, leading to the production of thousands of new viruses. The buildup of viruses eventually causes the cell to burst, or lyse, releasing progeny phage.

The phage used by Hershey and Chase contains only DNA and protein, providing the simplest possible system to differentiate the roles of DNA and protein. If they could identify the molecule injected into the cell, they would identify the viruses' genetic material. To do this, they needed a method to introduce unique labels into both DNA and protein. Nucleotides contain phosphorus, but proteins do not, and some amino acids contain sulfur, but DNA does not. Thus, the radioactive ^{32}P isotope will

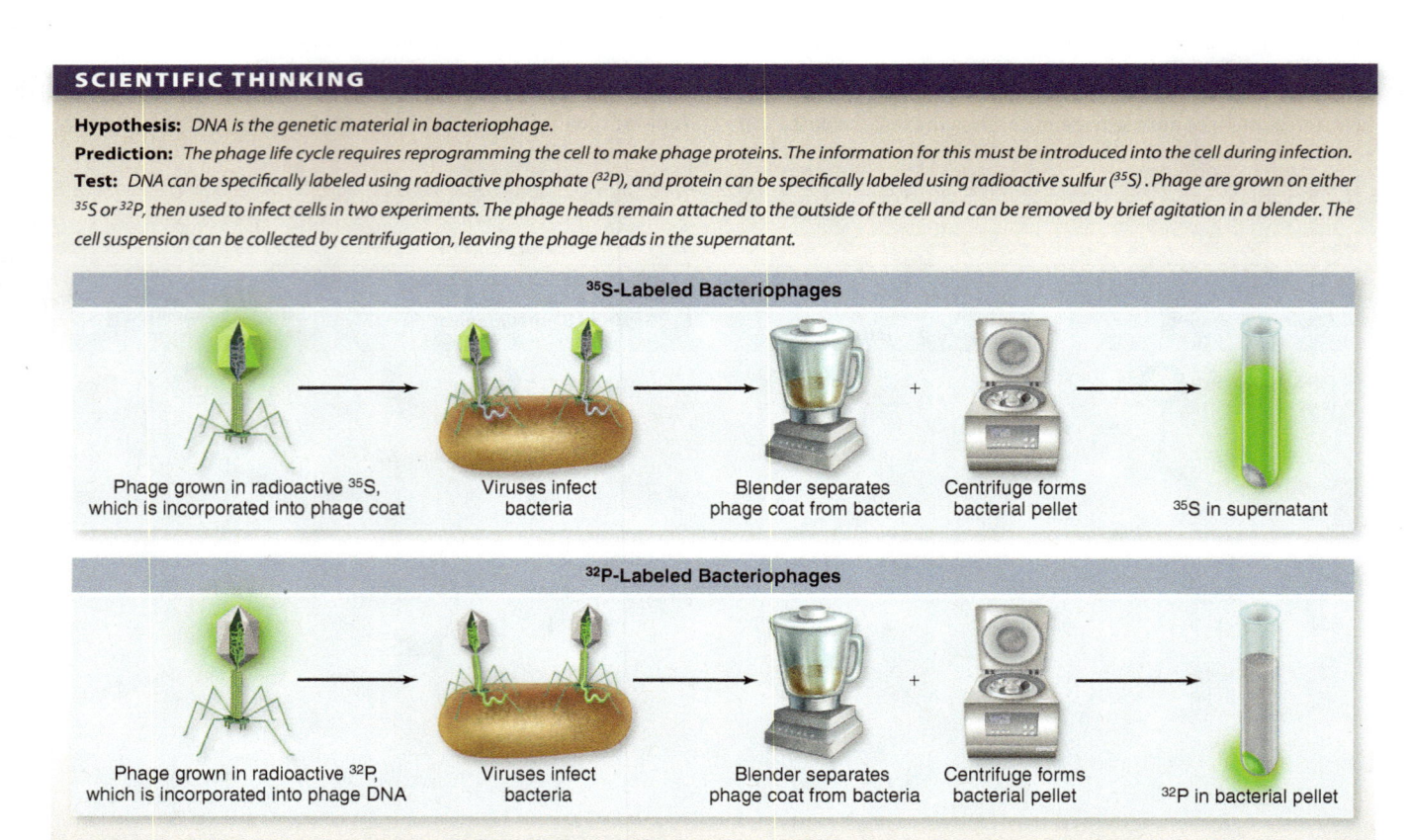

SCIENTIFIC THINKING

Hypothesis: DNA is the genetic material in bacteriophage.
Prediction: The phage life cycle requires reprogramming the cell to make phage proteins. The information for this must be introduced into the cell during infection.
Test: DNA can be specifically labeled using radioactive phosphate (^{32}P), and protein can be specifically labeled using radioactive sulfur (^{35}S). Phage are grown on either ^{35}S or ^{32}P, then used to infect cells in two experiments. The phage heads remain attached to the outside of the cell and can be removed by brief agitation in a blender. The cell suspension can be collected by centrifugation, leaving the phage heads in the supernatant.

^{35}S-Labeled Bacteriophages

Phage grown in radioactive ^{35}S, which is incorporated into phage coat → Viruses infect bacteria → Blender separates phage coat from bacteria + Centrifuge forms bacterial pellet → ^{35}S in supernatant

^{32}P-Labeled Bacteriophages

Phage grown in radioactive ^{32}P, which is incorporated into phage DNA → Viruses infect bacteria → Blender separates phage coat from bacteria + Centrifuge forms bacterial pellet → ^{32}P in bacterial pellet

Result: When the experiment is done, only ^{32}P makes it into the cell in any significant quantity.
Conclusion: Thus, DNA must be the molecule that is used to reprogram the cell.
Further Experiments: How does this experiment complement or extend the work of Avery on the identity of the transforming principle?

Figure 14.2 Hershey–Chase experiment showed DNA is genetic material for phage.

label DNA specifically, and the isotope ^{35}S will label proteins specifically. The two isotopes are easily distinguished based on the particles they emit when they decay.

Two experiments were performed (figure 14.2). In one, viruses were grown on a medium containing ^{32}P, which was incorporated into DNA; in the other, viruses were grown on medium containing ^{35}S, which was incorporated into coat proteins. Each culture of labeled viruses was then used to infect separate bacterial cultures.

After infection, the bacterial cell suspension was agitated in a blender to remove the infecting viral particles from the surfaces of the bacteria. This step ensured that only the part of the virus that had been injected into the bacterial cells—that is, the genetic material—would be detected.

Each bacterial suspension was then centrifuged to produce a pellet of cells for analysis. In the ^{32}P experiment, a large amount of radioactive phosphorus was found in the cell pellet, but in the ^{35}S experiment, very little radioactive sulfur was found in the pellet (figure 14.2). Hershey and Chase deduced that DNA, and not protein, constituted the genetic information that viruses inject into bacteria.

Learning Outcomes Review 14.1

Experiments with pneumonia-causing bacteria showed that virulence could be passed from one cell to another, a phenomenon termed transformation. When the factor responsible for transformation was purified, it was shown to be DNA. Labeling experiments with phage also indicated that the genetic material was DNA and not protein.

- Why was protein an attractive candidate for the genetic material?

14.2 DNA Structure

Learning Outcomes

1. Explain how the Watson–Crick structure rationalized the data available to them.
2. Evaluate the significance of complementarity for DNA structure and function.

A Swiss chemist, Friedrich Miescher, discovered DNA in 1869, only four years after Mendel's work was published—although it is unlikely that Miescher knew of Mendel's experiments.

Miescher extracted a white substance from the nuclei of human cells and fish sperm. The proportion of nitrogen and phosphorus in the substance was different from that found in any other known constituent of cells, which convinced Miescher that he had discovered a new biological substance. He called this substance "nuclein" because it seemed to be specifically associated with the nucleus. Because Miescher's nuclein was slightly acidic, it came to be called *nucleic acid*.

DNA's components were known, but its three-dimensional structure was a mystery

Although the three-dimensional structure of the DNA molecule was not elucidated until Watson and Crick, it was known that it contained three main components (figure 14.3):

1. a 5-carbon sugar;
2. a phosphate (PO$_4$) group; and

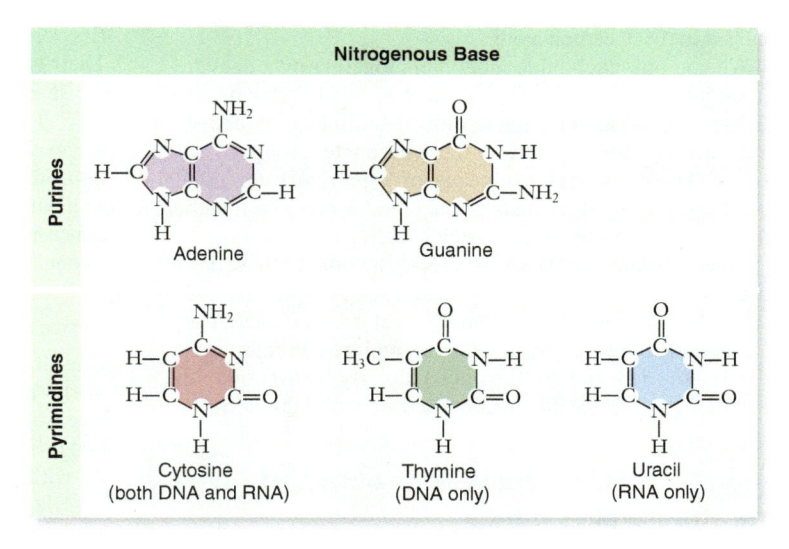

Figure 14.3 Nucleotide subunits of DNA and RNA. The nucleotide subunits of DNA and RNA are composed of three components: a 5-carbon sugar (deoxyribose in DNA and ribose in RNA); a phosphate group; and a nitrogenous base (either a purine or a pyrimidine).

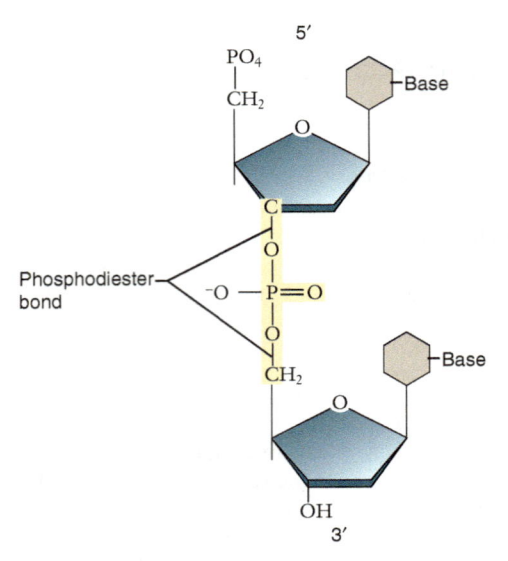

Figure 14.4 A phosphodiester bond.

3. a nitrogen-containing (nitrogenous) base. The base may be a **purine** (adenine, A, or guanine, G), a two-ringed structure; or a **pyrimidine** (thymine, T, or cytosine, C), a single-ringed structure. RNA contains the pyrimidine uracil (U) in place of thymine.

We need a way to be able to unambiguously refer to any carbon atom in a nucleotide. The convention used is that the numbers referring to carbons in the sugar include a prime symbol (') to distinguish them from the numbers that refer to atoms in the bases. The ribose sugars found in nucleic acids consist of a five-membered ring with four carbon atoms and an oxygen atom. As illustrated in figure 14.3, the carbon atoms are numbered 1' to 5', proceeding clockwise from the oxygen atom.

Under this numbering scheme, the phosphate group is attached to the 5' carbon atom of the sugar, and the base is attached to the 1' carbon atom. In addition, a free hydroxyl (—OH) group is attached to the 3' carbon atom.

We can link nucleotide monomers together by a dehydration reaction involving the 5' phosphate of one nucleotide with the 3' hydroxyl of another nucleotide. This linkage is called a **phosphodiester bond** because the phosphate group is now linked to the two sugars by means of a pair of ester bonds (figure 14.4). Many thousands of nucleotides can join together via these linkages to form long nucleic acid polymers.

Linear strands of DNA or RNA, no matter how long, almost always have a free 5' phosphate group at one end and a free 3' hydroxyl group at the other. Therefore, every DNA and RNA molecule has an intrinsic polarity, and we can refer unambiguously to each end of the molecule. By convention, the sequence of bases is usually written in the 5'-to-3' direction.

Chargaff, Franklin, and Wilkins obtained some structural evidence

To understand the model that Watson and Crick proposed, we need to review the evidence that they had available to construct their model.

Chargaff's rules

A careful study carried out by Erwin Chargaff showed that the nucleotide composition of DNA molecules varied in complex ways, depending on the source of the DNA. This strongly suggested that DNA was not a simple repeating polymer and that it might have the information-encoding properties genetic material requires. Despite DNA's complexity, however, Chargaff observed an important underlying regularity in the ratios of the bases found in native DNA: *The amount of adenine present in DNA always equals the amount of thymine, and the amount of guanine always equals the amount of cytosine.* Two important findings from this work are often called *Chargaff's rules*:

1. The proportion of A always equals that of T, and the proportion of G always equals that of C, or: A = T, and G = C.
2. The ratio of G–C to A–T varies with different species.

As mounting evidence indicated that DNA stored the hereditary information, investigators began to puzzle over how such a seemingly simple molecule could carry out such a complex coding function.

X-ray diffraction patterns of DNA

The technique of X-ray diffraction provided more direct information about the possible structure of DNA. In X-ray diffraction, crystals of a molecule are bombarded with a beam of X-rays. The rays are bent, or diffracted, by the molecules they encounter, and the diffraction pattern is recorded on photographic film. The patterns resemble the ripples created by tossing a rock into a smooth lake. When analyzed mathematically, the diffraction pattern can yield information about the three-dimensional structure of a molecule.

The problem with using this technique with DNA was that in the 1950s, it was impossible to obtain true crystals of natural DNA. However, British researcher Maurice Wilkins learned how to prepare uniformly oriented DNA fibers, and with graduate student Ray Gosling succeeded in obtaining the first crude diffraction information on natural DNA in 1950. Their early X-ray photos suggested that the DNA molecule has the shape of a helix.

The British chemist Rosalind Franklin (figure 14.5a) continued this work, perfecting the technique to obtain ever clearer "pictures" of the oriented DNA fibers. The clearest of these images (figure 14.5b) confirmed that DNA was a helix, and allowed calculation of the dimensions of the molecule, indicating a diameter of about 2 nm and a complete helical turn every 3.4 nm.

Tautomeric forms of bases

One piece of evidence important to Watson and Crick was the form of the bases themselves. Because of the alternating double and single bonds in the bases, they actually exist in equilibrium between two different forms when in solution. The different forms have to do with keto (C=O) versus enol (C—OH) groups and amino (—NH$_2$) versus imino (=NH) groups that are attached to the bases. These structural forms are called *tautomers*.

The importance of this distinction is that the two forms have very different hydrogen-bonding possibilities. The predominant

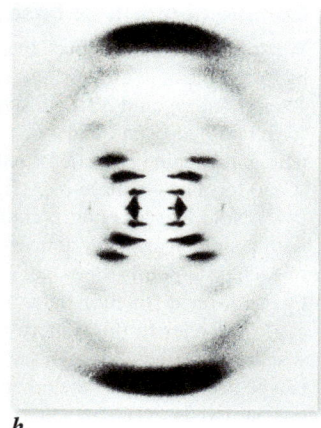

Figure 14.5 Rosalind Franklin's X-ray diffraction patterns. *a.* Rosalind Franklin. *b.* This X-ray diffraction photograph of DNA fibers, made in 1953 by Rosalind Franklin, was interpreted to show the helical structure of DNA.

forms of the bases contain the keto and amino groups (see figure 14.3), but a prominent biochemistry text of the time actually contained the opposite, and incorrect, information. Legend has it that Watson learned the correct forms while having lunch with a biochemist friend.

The Watson–Crick model fits the available evidence

Learning informally of Franklin's results before they were published in 1953, American chemist James Watson and English molecular biologist Francis Crick, two young investigators at Cambridge University, quickly worked out a likely structure for the DNA molecule (figure 14.6), which we now know was substantially correct. Watson and Crick did not perform a single experiment themselves related to DNA structure; rather, they built detailed molecular models based on the information available.

The key to the model was their understanding that each DNA molecule is actually made up of *two* chains of nucleotides that are intertwined—the double helix.

The phosphodiester backbone

The two strands of the double helix are made up of long polymers of nucleotides, and as described earlier in this section, each strand is made up of repeating sugar and phosphate units joined by phosphodiester bonds (figure 14.7). We call this the *phosphodiester backbone* of the molecule. The two strands of the backbone are then wrapped about a common axis forming a double helix (figure 14.8). The helix is often compared to a spiral staircase, in which the two strands of the double helix are the handrails on the staircase.

Figure 14.6 The DNA double helix. James Watson *(left)* and Francis Crick *(right)* deduced the structure of DNA in 1953 from Chargaff's rules, knowing the proper tautomeric forms of the bases and using Franklin's diffraction studies.

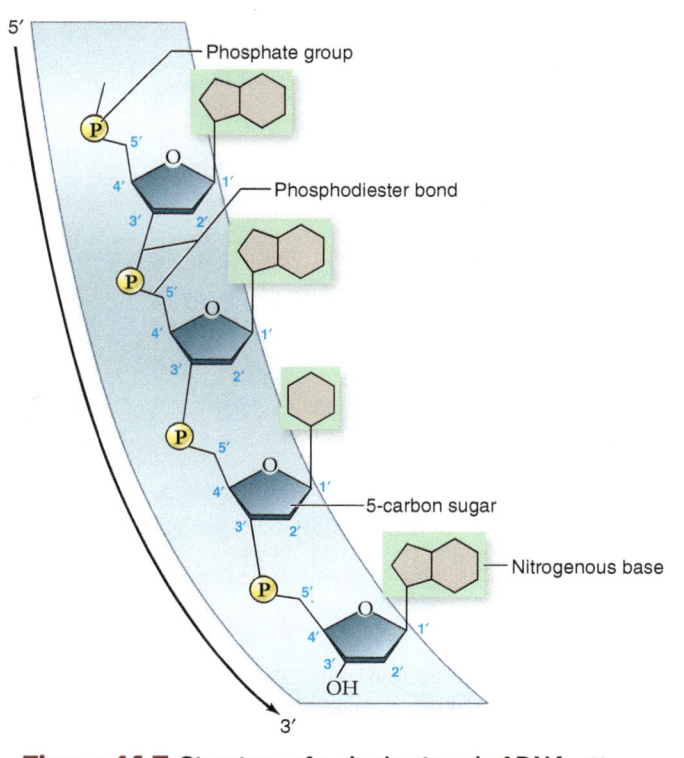

Figure 14.7 Structure of a single strand of DNA. The phosphodiester backbone is composed of alternating sugar and phosphate groups. The bases are attached to each sugar.

chapter **14** *DNA: The Genetic Material* **261**

Complementarity of bases

Watson and Crick proposed that the two strands were held together by formation of hydrogen bonds between bases on opposite strands. These bonds would result in specific **base-pairs:** Adenine (A) can form two hydrogen bonds with thymine (T) to form an A–T base-pair, and guanine (G) can form three hydrogen bonds with cytosine (C) to form a G–C base-pair (figure 14.9).

Note that this configuration also pairs a two-ringed purine with a single-ringed pyrimidine in each case, so that the diameter of each base-pair is the same. This consistent diameter is indicated by the X-ray diffraction data.

We refer to this pattern of base-pairing as *complementary*, which means that although the strands are not identical, they each can be used to specify the other by base-pairing. If the sequence of one strand is ATGC, then the complementary strand sequence must be TACG. This characteristic becomes critical for DNA replication and expression, as you will see in section 14.3.

The Watson–Crick model also explained Chargaff's results: In a double helix, adenine forms two hydrogen bonds with thymine, but it will not form hydrogen bonds properly with cytosine. Similarly, guanine forms three hydrogen bonds with cytosine, but it will not form hydrogen bonds properly with thymine. Because of this base-pairing, adenine and thymine always occur in the same proportions in any DNA molecule, as do guanine and cytosine.

Antiparallel configuration

As stated earlier in this section, a single phosphodiester strand has an inherent polarity, meaning that one end terminates in a 3′ OH

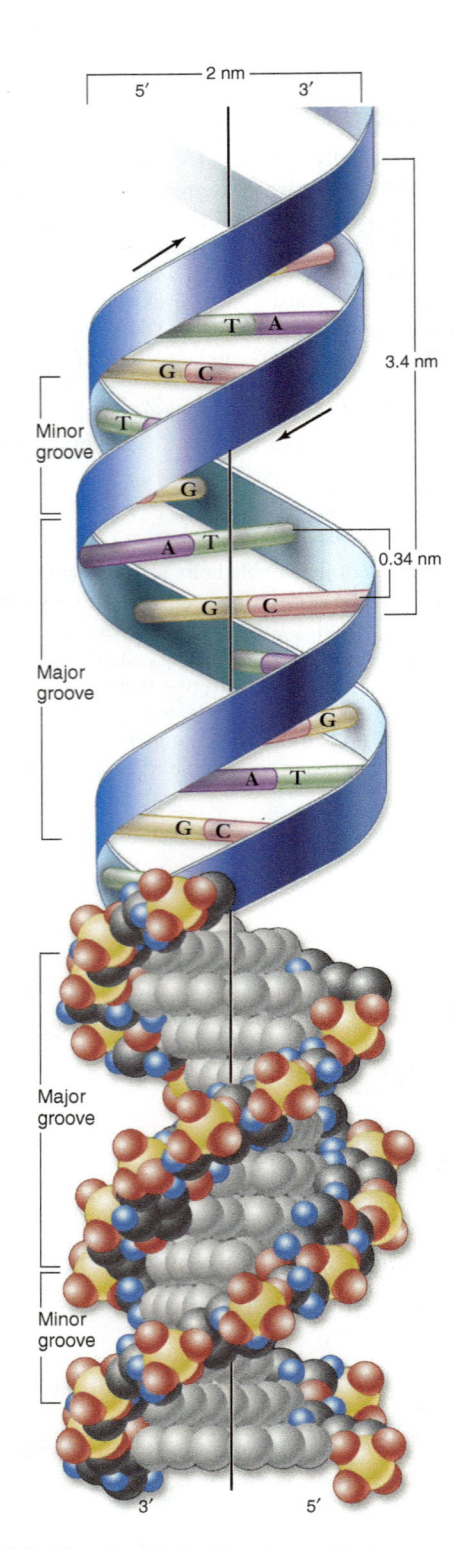

Figure 14.8 The double helix. Shown with the phosphodiester backbone as a ribbon on top and a space-filling model on the bottom. The bases protrude into the interior of the helix where they hold it together by base-pairing. The backbone forms two grooves, the larger major groove and the smaller minor groove.

Figure 14.9 Base-pairing holds strands together. The hydrogen bonds that form between A and T and between G and C are shown with dashed lines. These produce A–T and G–C base-pairs that hold the two strands together. This always pairs a purine with a pyrimidine, keeping the diameter of the double helix constant.

Data analysis Explain how the Watson–Crick model accounts for the data discussed in the text.

and the other end terminates in a 5′ PO₄. Strands are thus referred to as having either a 5′-to-3′ or a 3′-to-5′ polarity. Two strands could be put together in two ways: with the polarity the same in each (parallel) or with the polarity opposite (antiparallel). Native double-stranded DNA always has the antiparallel configuration, with one strand running 5′ to 3′ and the other running 3′ to 5′ (see figure 14.8). In addition to its complementarity, this antiparallel nature also has important implications for DNA replication.

The Watson–Crick DNA molecule

In the Watson and Crick model, a DNA molecule consists of two phosphodiester strands wrapped about a common helical axis, with the bases extending to interior of the helix. The sequence of bases on the two strands are complementary, so they form base-pairs that hold the strands together. Each strand is also polar, with a 5′ and a 3′ end, and they are oriented in the opposite (antiparallel) direction (see figures 14.8 and 14.9).

Although each individual hydrogen bond is low energy, the sum of thousands, or millions, of such bonds makes a very stable molecule. This also means that a DNA molecule can have regions where the helix can be "opened" up without affecting the stability of the entire molecule, which is critical for its function.

Although the Watson–Crick model provided a rational structural for DNA, researchers had to answer further questions about how DNA could be replicated, a crucial step in cell division, and also about how cells could repair damaged or otherwise altered DNA. We explore these questions in the rest of this chapter. (In chapter 15, we continue with the genetic code and the connection between the code and protein synthesis.)

Learning Outcomes Review 14.2

Chargaff showed that in DNA, the amount of adenine was equal to the amount of thymine, and the amount of guanosine was equal to that of cytosine. X-ray diffraction studies by Franklin and Wilkins indicated that DNA formed a helix. Watson and Crick built a model consisting of two antiparallel strands wrapped in a helix about a common axis. The two strands are held together by hydrogen bonds between the bases: Adenine pairs with thymine and guanine pairs with cytosine. The two strands are thus complementary to each other.

- *Why was information about the proper tautomeric form of the bases critical?*

14.3 Basic Characteristics of DNA Replication

Learning Outcomes

1. Illustrate the products of semiconservative replication.
2. Describe the requirements for DNA replication.

The accurate replication of DNA prior to cell division is a basic and crucial function. Research has revealed that this complex process requires the participation of a large number of cellular proteins. Before geneticists could look for these details, however, they needed to perform some groundwork on the general mechanisms.

Meselson and Stahl demonstrate the semiconservative mechanism

The Watson–Crick model immediately suggested that the basis for copying the genetic information is complementarity. One chain of the DNA molecule may have any conceivable base sequence, but this sequence completely determines the sequence of its partner in the duplex.

In replication, the sequence of parental strands must be duplicated in daughter strands. That is, one parental helix with two strands must yield two daughter helices with four strands. The two daughter molecules are then separated during the course of cell division.

Three models of DNA replication are possible (figure 14.10):

1. In a *conservative model,* both strands of the parental duplex would remain intact (conserved), and new DNA copies would consist of all-new molecules. Both daughter strands would contain all-new molecules.

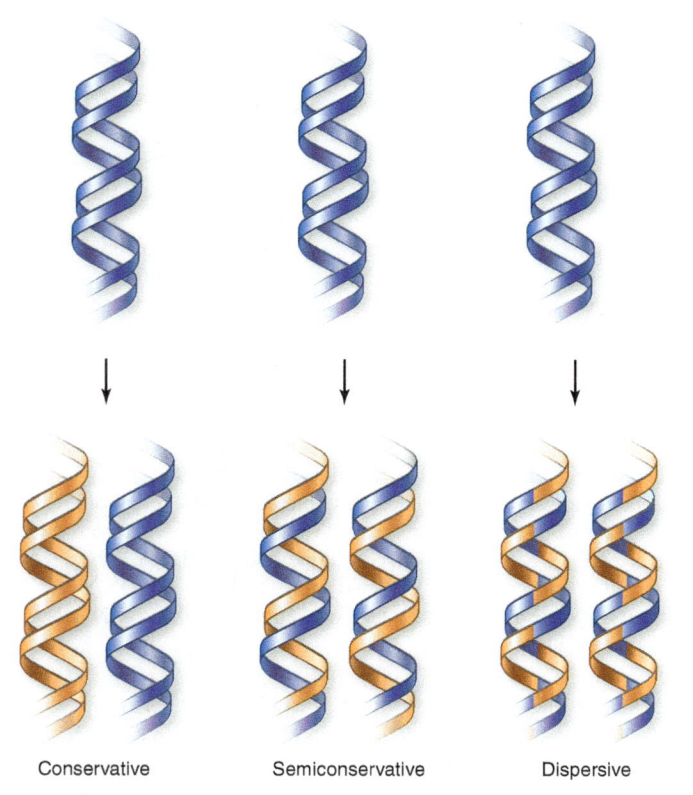

Figure 14.10 Three possible models for DNA replication. The conservative model produces one entirely new molecule and conserves the old. The semiconservative model produces two hybrid molecules of old and new strands. The dispersive model produces hybrid molecules with each strand a mixture of old and new.

2. In a **semiconservative model,** one strand of the parental duplex remains intact in daughter strands (semiconserved); a new complementary strand is built for each parental strand consisting of new molecules. Daughter strands would consist of one parental strand and one newly synthesized strand.
3. In a *dispersive model,* copies of DNA would consist of mixtures of parental and newly synthesized strands; that is, the new DNA would be dispersed throughout each strand of both daughter molecules after replication.

Notice that these three models suggest general mechanisms of replication, without specifying any molecular details of the process.

The Meselson–Stahl experiment

The three models for DNA replication were evaluated in 1958 by Matthew Meselson and Franklin Stahl. To distinguish between these models, they labeled DNA and then followed the labeled DNA through two rounds of replication (figure 14.11).

The label Meselson and Stahl used was a heavy isotope of nitrogen (^{15}N), not a radioactive label. Molecules containing ^{15}N have a greater density than those containing the common ^{14}N isotope. Ultracentrifugation can be used to separate molecules that have different densities.

Bacteria were grown in a medium containing ^{15}N, which became incorporated into the bases of the bacterial DNA. After several generations, the DNA of these bacteria was denser than that of bacteria grown in a medium containing the normally available ^{14}N. Meselson and Stahl then transferred the bacteria from the ^{15}N medium to ^{14}N medium and collected the DNA at various time intervals.

The DNA for each interval was dissolved in a solution containing a heavy salt, cesium chloride. This solution was spun at very high speeds in an ultracentrifuge. The enormous centrifugal forces caused cesium ions to migrate toward the bottom of the centrifuge tube, creating a gradient of cesium concentration, and thus of density. Each DNA strand floated or sank in the gradient until it reached the point at which its density exactly matched the density of the cesium at that location. Because ^{15}N strands are denser than ^{14}N strands, they migrated farther down the tube.

The DNA collected immediately after the transfer of bacteria to new ^{14}N medium was all of one density equal to that of ^{15}N DNA alone. However, after the bacteria completed a first round of DNA replication, the density of their DNA had decreased to a value intermediate between ^{14}N DNA alone and ^{15}N DNA. After the second round of replication, two density classes of DNA were observed: one intermediate and one equal to that of ^{14}N DNA (figure 14.11).

Interpretation of the Meselson–Stahl findings

Meselson and Stahl compared their experimental data with the results that would be predicted on the basis of the three models:

1. The conservative model was not consistent with the data because after one round of replication, two densities should have been observed: DNA strands would either be all-heavy (parental) or all-light (daughter). This model is rejected.

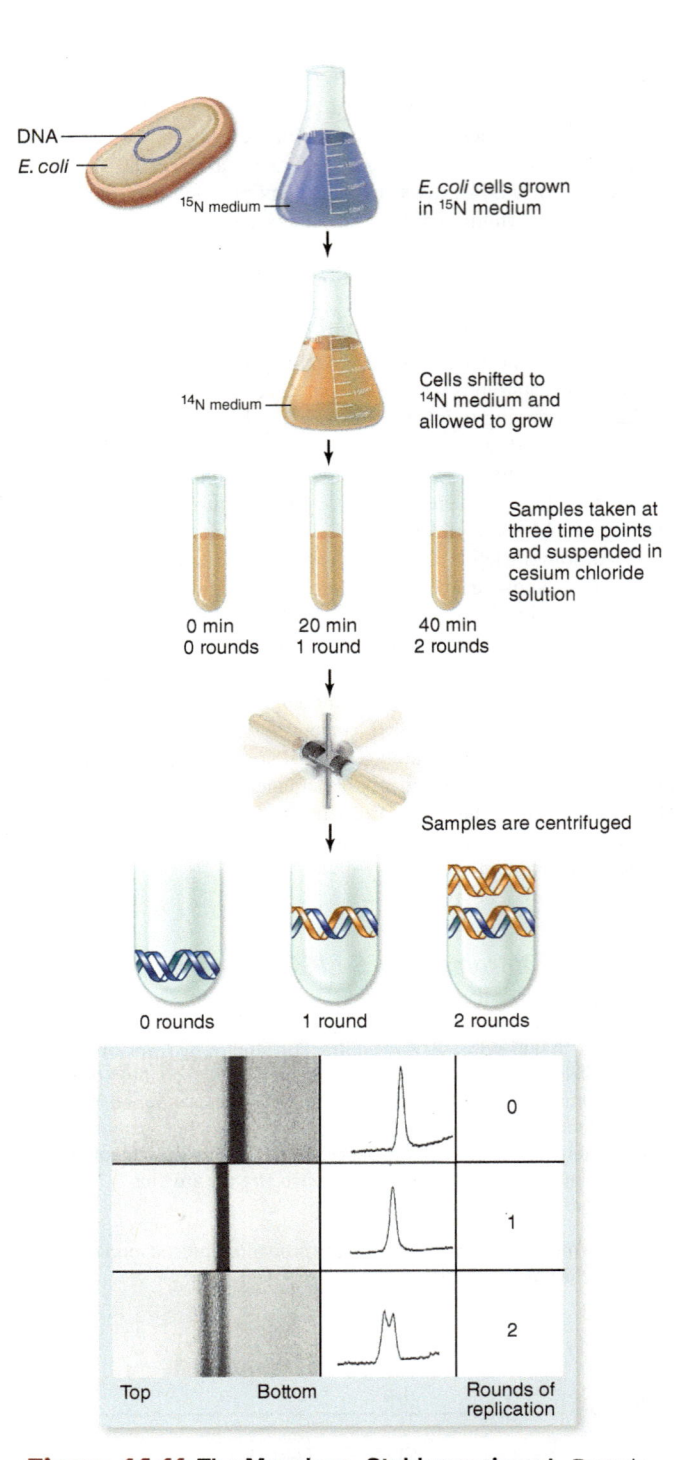

Figure 14.11 The Meselson–Stahl experiment. Bacteria grown in heavy ^{15}N medium are shifted to light ^{14}N medium and grown for two rounds of replication. Samples are taken at time points corresponding to zero, one, and two rounds of replication and centrifuged in cesium chloride to form a gradient. The actual data are shown at the bottom with the interpretation of semiconservative replication shown schematically.

Data analysis What would you predict for the products of a third round of replication?

2. The semiconservative model is consistent with all observations: After one round of replication, a single density would be predicted because all DNA molecules would have a light strand and a heavy strand. After two rounds of replication, half of the molecules would have two light strands, and half would have a light strand and a heavy strand—and so two densities would be observed. Therefore, the results support the semiconservative model.
3. The dispersive model was consistent with the data from the first round of replication, because in this model, every DNA helix would consist of strands that are mixtures of ½ light (new) and ½ heavy (old) molecules. But after two rounds of replication, the dispersive model would still yield only a single density; DNA strands would be composed of ¾ light and ¼ heavy molecules. Instead, two densities were observed. Therefore, this model is also rejected.

The basic mechanism of DNA replication is semiconservative. At the simplest level, then, DNA is replicated by opening up a DNA helix and making copies of both strands to produce two daughter helices, each consisting of one old strand and one new strand.

DNA replication requires a template, nucleotides, and a polymerase enzyme

Replication requires three things: something to copy, something to do the copying, and the building blocks to make the copy. The parental DNA molecules serve as a template, enzymes perform the actions of copying the template, and the building blocks are nucleoside triphosphates.

The process of replication can be thought of as having a beginning where the process starts; a middle where the majority of building blocks are added; and an end where the process is finished. We use the terms *initiation, elongation,* and *termination* to describe a biochemical process. Although this may seem overly simplistic, in fact, discrete functions are usually required for initiation and termination that are not necessary for elongation.

A number of enzymes work together to accomplish the task of assembling a new strand, but the enzyme that actually matches the existing DNA bases with complementary nucleotides and then links the nucleotides together to make the new strand is **DNA polymerase** (figure 14.12). All DNA polymerases that have been examined have several common features. They all add new bases to the 3′ end of existing strands. That is, they synthesize in a 5′-to-3′ direction by extending a strand base-paired to the template.

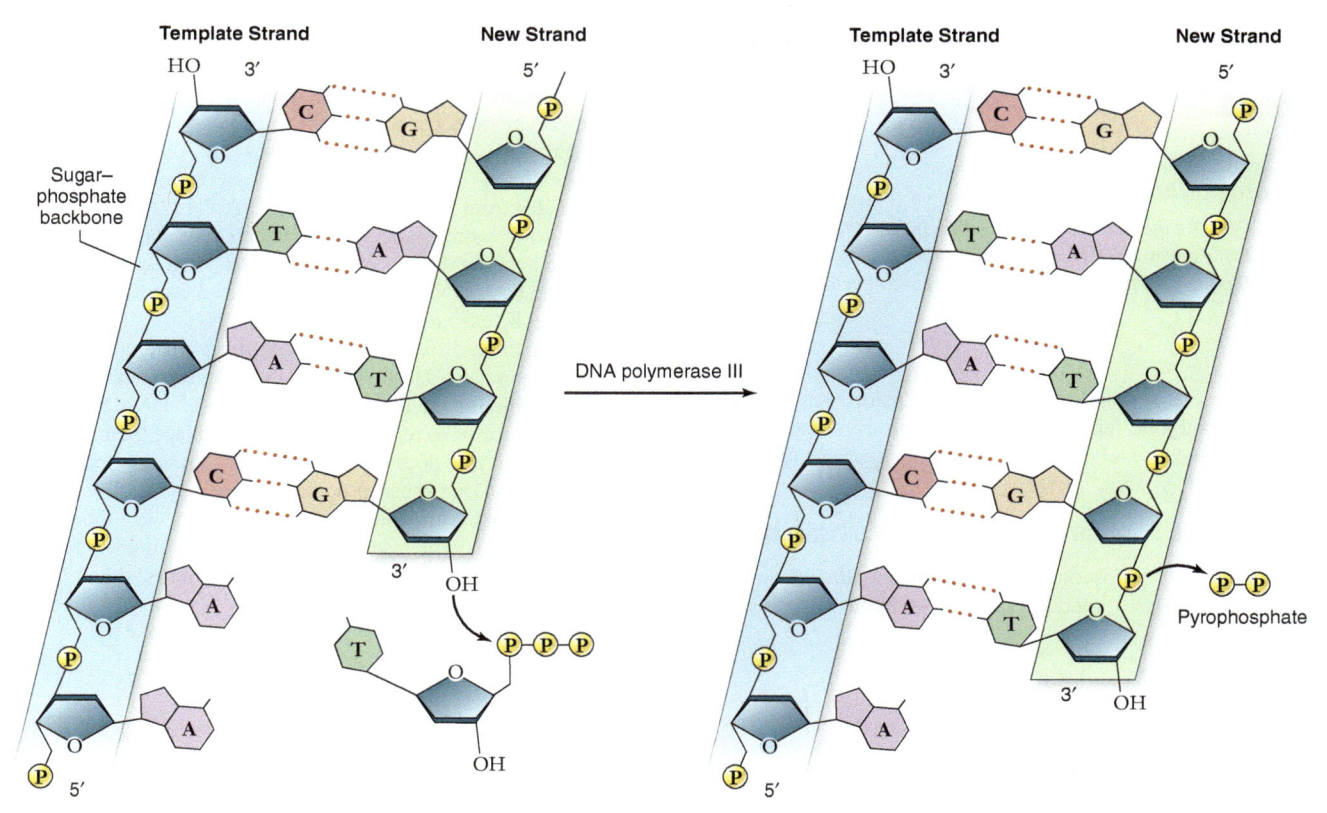

Figure 14.12 Action of DNA polymerase. DNA polymerases add nucleotides to the 3′ end of a growing chain. The nucleotide added depends on the base that is in the template strand. Each new base must be complementary to the base in the template strand. With the addition of each new nucleoside triphosphate, two of its phosphates are cleaved off as pyrophosphate.

? Inquiry question Why do you think it is important that the sugar–phosphate backbone of DNA is held together by covalent bonds, and the cross-bridges between the two strands are held together by hydrogen bonds?

All DNA polymerases also require a *primer* to begin synthesis; they cannot begin without a strand of RNA or DNA base-paired to the template. RNA polymerases do not have this requirement, so they usually synthesize the primers.

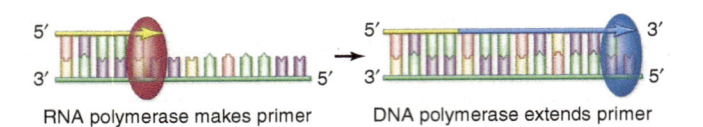

RNA polymerase makes primer → DNA polymerase extends primer

Learning Outcomes Review 14.3

Meselson and Stahl showed that the basic mechanism of replication is semiconservative: Each new DNA helix is composed of one old strand and one new strand. The process of replication requires a template to copy, nucleoside triphosphate building blocks, and the enzyme DNA polymerase. DNA polymerases synthesize DNA in a 5′-to-3′ direction from a primer, usually RNA.

- In the Meselson–Stahl experiment, what would the results be if the DNA was denatured prior to separation by ultracentrifugation?

14.4 Prokaryotic Replication

Learning Outcomes

1. Describe the functions of E. coli DNA polymerases.
2. Explain why replication is discontinuous on one strand.
3. Diagram the functions found at the replication fork.

To build up a more detailed picture of replication, we first concentrate on prokaryotic replication using *E. coli* as a model. We can then look at eukaryotic replication primarily in how it differs from the prokaryotic system.

Prokaryotic replication starts at a single origin

Replication in *E. coli* initiates at a specific site, the origin (called *oriC*), and ends at a specific site, the terminus. The sequence of *oriC* consists of repeated nucleotides that bind an initiator protein and an AT-rich sequence that can be opened easily during initiation of replication. (A–T base-pairs have only two hydrogen bonds, compared with the three hydrogen bonds in G–C base-pairs.)

After initiation, replication proceeds bidirectionally from this unique origin to the unique terminus (figure 14.13). We call the DNA controlled by an origin a **replicon**. In this case, the chromosome plus the origin forms a single replicon.

E. coli has at least three different DNA polymerases

As mentioned in section 14.3, DNA polymerase refers to a class of enzymes that use a DNA template to assemble a new complementary strand. The first DNA polymerase isolated in *E. coli* was **DNA polymerase I (Pol I)**. At first, investigators assumed this polymerase was all that was required for DNA replication. Later, a mutant was isolated with no Pol I activity, but that could still replicate its chromosome. Two additional polymerases were isolated from this strain of *E. coli*: **DNA polymerase II (Pol II)** and **DNA polymerase III (Pol III)**. As with all known DNA polymerases, all three of these enzymes synthesize polynucleotide strands only in the 5′-to-3′ direction and require a primer.

In addition to adding nucleotides to a growing DNA strand, some polymerases can also remove nucleotides, or act as a nuclease. Enzymes that act as nucleases are classified as either **endonucleases** (cut DNA internally) or **exonucleases** (remove nucleotides from the end of DNA). DNA Pol I, Pol II, and Pol III have 3′-to-5′ exonuclease activity, which serves as a proofreading function because it allows the enzyme to "back up" and remove a mispaired base. DNA Pol I also has a 5′-to-3′ exonuclease activity, which can be used to remove RNA primers.

The three different polymerases have different roles in the replication process. DNA Pol III is the main replication enzyme; it is responsible for the bulk of DNA synthesis. DNA Pol I acts on the lagging strand to remove primers and replace them with DNA. The Pol II enzyme does not appear to play a role in replication but is involved in DNA repair processes.

For many years, these three polymerases were thought to be the only DNA polymerases in *E. coli*, but several others have been identified. There are now five known polymerases, although not all are active in DNA replication.

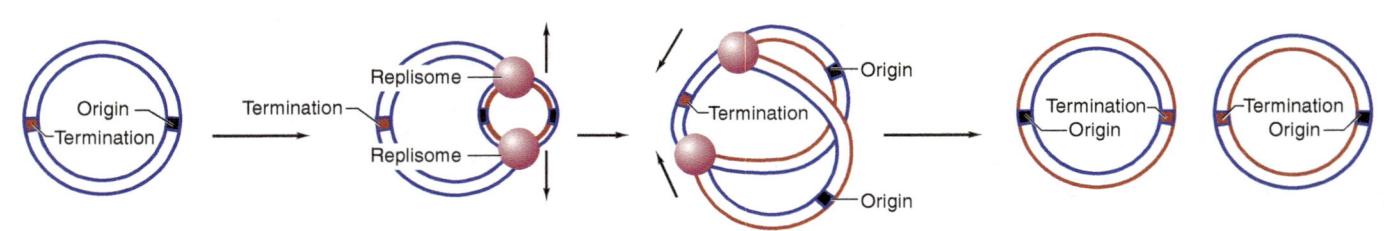

Figure 14.13 Replication is bidirectional from a unique origin. Replication initiates from a unique origin. Two separate replisomes are loaded onto the origin and initiate synthesis in the opposite directions on the chromosome. These two replisomes continue in opposite directions until they come to a unique termination site.

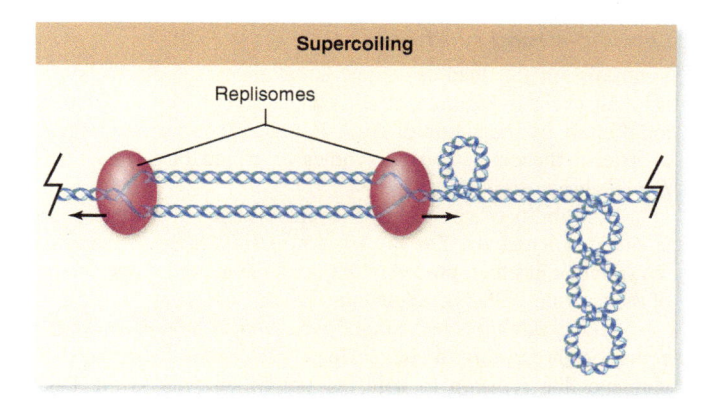

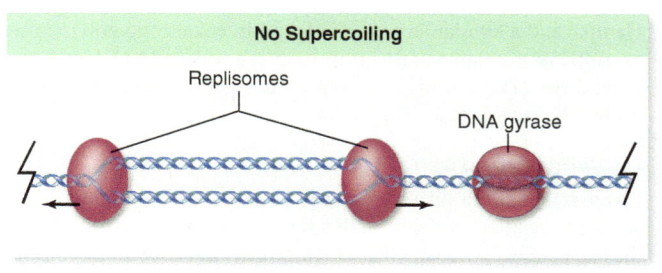

Figure 14.14 Unwinding the helix causes torsional strain. If the ends of a linear DNA molecule are constrained, as they are in the cell, unwinding the helix produces torsional strain. This can cause the double helix to further coil in space (supercoiling). The enzyme DNA gyrase can relieve supercoiling.

Unwinding DNA requires energy and causes torsional strain

Although most DNA polymerases can unwind DNA during synthesis, replication is more efficient if the helix is unwound ahead of the polymerase. An enzyme with DNA unwinding activity is called a **helicase.** This process requires energy in the form of ATP, and can progressively unwind DNA, forming single strands. These single strands are not stable because the hydrophobic bases are exposed to water. Cells solve this problem with another protein, single-strand-binding protein (SSB), that will coat exposed single strands.

The unwinding of the two strands introduces torsional strain in the DNA molecule. Imagine two rubber bands twisted together. If you now unwind the rubber bands, what happens? The rubber bands, already twisted about each other, will further coil in space. When this happens with a DNA molecule it is called **supercoiling** (figure 14.14). The branch of mathematics that studies how forms twist and coil in space is called *topology,* and therefore we describe this coiling of the double helix as the *topological state* of DNA. This state describes how the double helix itself coils in space. You have already seen an example of this coiling with DNA wrapped about histone proteins in the nucleosomes of eukaryotic chromosomes (see chapter 10).

Enzymes that can alter the topological state of DNA are called **topoisomerases.** Topoisomerase enzymes act to relieve the torsional strain caused by unwinding and to prevent this supercoiling from happening. **DNA gyrase** is the topoisomerase involved in DNA replication (figure 14.14).

Replication is semidiscontinuous

Earlier, DNA was described as being antiparallel—meaning that one strand runs in the 3'-to-5' direction, and its complementary strand runs in the 5'-to-3' direction. The antiparallel nature of DNA combined with the nature of the polymerase enzymes puts constraints on the replication process. Because polymerases can synthesize DNA in only one direction, and the two DNA strands run in opposite directions, polymerases on the two strands must be synthesizing DNA in opposite directions (figure 14.15).

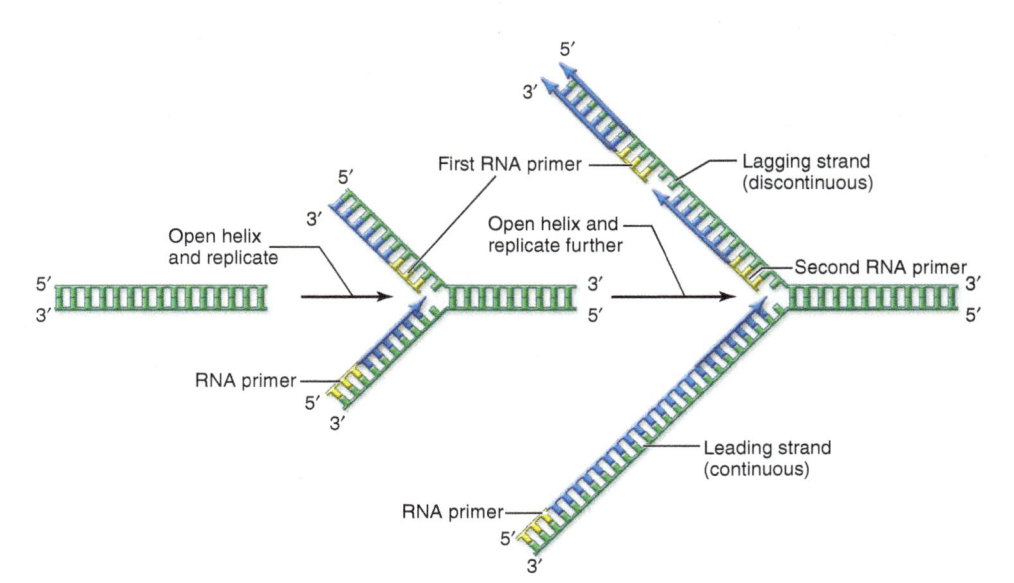

Figure 14.15 Replication is semidiscontinuous. The 5'-to-3' synthesis of the polymerase and the antiparallel nature of DNA mean that only one strand, the leading strand, can be synthesized continuously. The other strand, the lagging strand, must be made in pieces, each with its own primer.

The requirement of DNA polymerases for a primer means that on one strand primers need to be added as the helix is opened up (figure 14.15). This means that one strand can be synthesized in a continuous fashion from an initial primer, but the other strand must be synthesized in a discontinuous fashion with multiple priming events and short sections of DNA being assembled. The strand that is continuous is called the **leading strand,** and the strand that is discontinuous is the **lagging strand.** DNA fragments synthesized on the lagging strand are named **Okazaki fragments** in honor of the man who first experimentally demonstrated discontinuous synthesis. They introduce a need for even more enzymatic activity on the lagging strand, as is described next.

Synthesis occurs at the replication fork

The partial opening of a DNA helix to form two single strands has a forked appearance, and is thus called the **replication fork.** All of the enzymatic activities that we have discussed plus a few more are found at the replication fork (table 14.1). Synthesis on the leading strand and on the lagging strand proceed in different ways, however.

Priming

The primers required by DNA polymerases during replication are synthesized by the enzyme *DNA primase*. This enzyme is an RNA polymerase that synthesizes short stretches of RNA 10 to 20 bp (base-pairs) long that function as primers for DNA polymerase. Later on, the RNA primer is removed and replaced with DNA.

Leading-strand synthesis

Synthesis on the leading strand is relatively simple. A single priming event is required, and then the strand can be extended indefinitely by the action of DNA Pol III. If the enzyme remains attached to the template, it can synthesize around the entire circular *E. coli* chromosome.

The ability of a polymerase to remain attached to the template is called *processivity*. The Pol III enzyme is a large multisubunit enzyme that has high processivity due to the action of one subunit of the enzyme, called the β *subunit* (figure 14.16a).

The β subunit is made up of two identical protein chains that come together to form a circle. This circle can be loaded onto the template like a clamp to hold the Pol III enzyme to the DNA (figure 14.16b). This structure is therefore referred to as the "sliding clamp," and a similar structure is found in eukaryotic polymerases as well. For the clamp to function, it must be opened and then closed around the DNA. A multisubunit protein called the clamp loader accomplishes this task. This function is also found in eukaryotes.

Lagging-strand synthesis

Because synthesis on the lagging strand is discontinuous, more steps are required to replicate this strand. Primase is needed to synthesize primers for each Okazaki fragment, and then all these RNA primers must be removed and replaced with DNA. Finally, the fragments need to be stitched together.

The Okazaki fragments themselves are synthesized by DNA Pol III, like the leading strand, but the primers are removed and replaced with DNA by DNA Pol I. With its 5′-to-3′ exonuclease

TABLE 14.1	DNA Replication Enzymes of *E. coli*		
Protein	**Role**	**Size (kDa)**	**Molecules per Cell**
Helicase	Unwinds the double helix	300	20
Primase	Synthesizes RNA primers	60	50
Single-strand-binding protein	Stabilizes single-stranded regions	74	300
DNA gyrase	Relieves torque	400	250
DNA polymerase III	Synthesizes DNA	≈900	20
DNA polymerase I	Erases primer and fills gaps	103	300
DNA ligase	Joins the ends of DNA segments; DNA repair	74	300

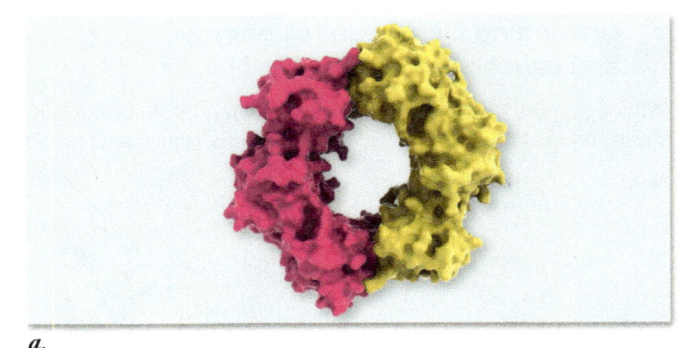

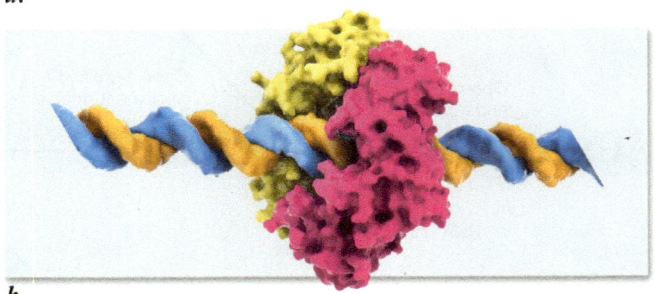

Figure 14.16 The DNA polymerase sliding clamp. *a.* The β subunit forms a ring that can encircle DNA. *b.* The β subunit is shown attached to the DNA. This forms the "sliding clamp" that keeps the polymerase attached to the template.

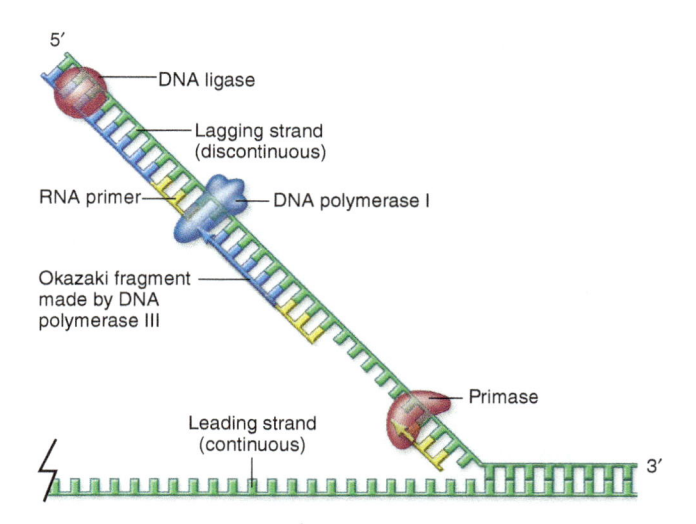

Figure 14.17 Lagging-strand synthesis. The action of primase synthesizes the primers needed by DNA polymerase III (not shown). These primers are removed by DNA polymerase I using its 5′-to-3′ exonuclease activity, then extending the previous Okazaki fragment to replace the RNA. The nick between Okazaki fragments after primer removal is sealed by DNA ligase.

activity, Pol I can remove primers and then replace them with its usual polymerase activity. The Pol I extends the Okazaki fragment "behind" it, while removing the RNA primer in "front" of it.

This leaves only the last phosphodiester bond to be formed where synthesis by Pol I ends. This is done by **DNA ligase,** which seals this "nick," acting to join the Okazaki fragments into complete strands. All of this activity on the lagging strand is summarized in figure 14.17.

> **Inquiry question** What is the role of DNA ligase? What would happen to DNA replication in a cell where this enzyme is not functional?

Termination

Termination occurs at a specific site located roughly opposite *oriC* on the circular chromosome. The last stages of replication produce two daughter molecules that are intertwined like two rings in a chain. These intertwined molecules are unlinked by the same enzyme that relieves torsional strain at the replication fork—DNA gyrase.

The replisome contains all the necessary enzymes for replication

The enzymes involved in DNA replication form a macro molecular assembly called the **replisome.** This assembly forms the "replication organelle," much as the ribosome is the organelle that synthesizes protein. The replisome has two main subcomponents: the *primosome*, and a complex of two DNA Pol III enzymes, one for each strand. The primosome consists of primase and helicase, along with a number of accessory proteins.

Despite our calling one strand the lagging strand, the two Pol III enzymes in the replisome are active on both strands simultaneously. How can the two strands be synthesized in the same direction when the strands are antiparallel? The best model involves a loop formed in the lagging strand, allowing the two polymerases to move in the same direction (figure 14.18).

The two Pol III complexes include two synthetic core subunits, each with its own β sliding clamp subunit. The entire replisome complex is held together by a number of proteins,

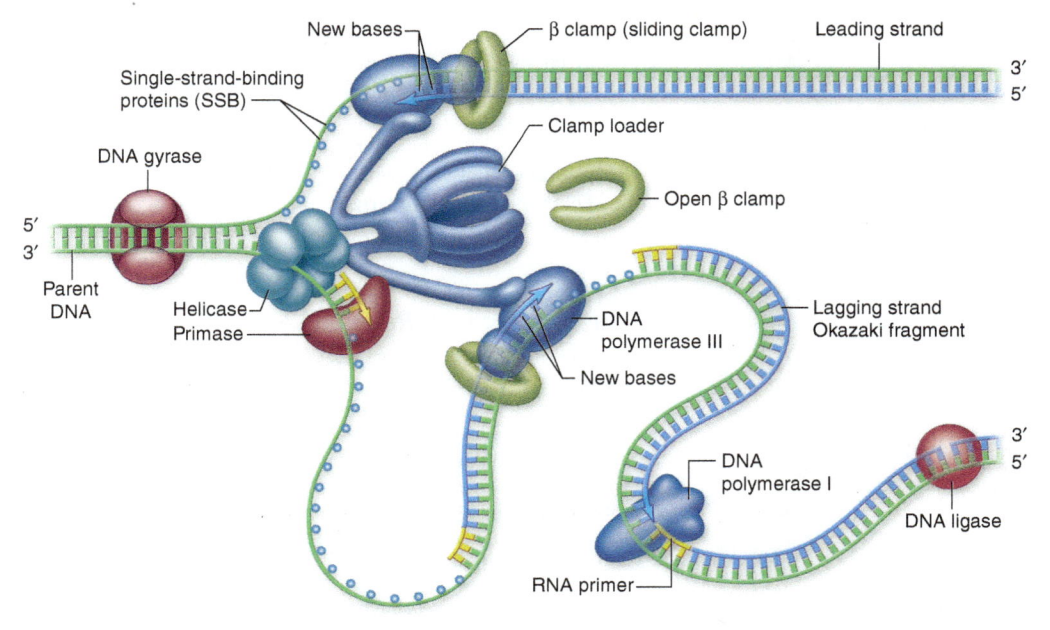

Figure 14.18 The replication fork. A model for the structure of the replication fork with two polymerase III enzymes held together by a large complex of accessory proteins. These include the "clamp loader," which loads the β subunit sliding clamp periodically on the lagging strand. The polymerase III on the lagging strand periodically releases its template and reassociates along with the β clamp. The loop in the lagging-strand template allows both polymerases to move in the same direction despite DNA being antiparallel. Primase, which makes primers for the lagging-strand fragments, and helicase are also associated with the central complex. Polymerase I removes primers and ligase joins the fragments together.

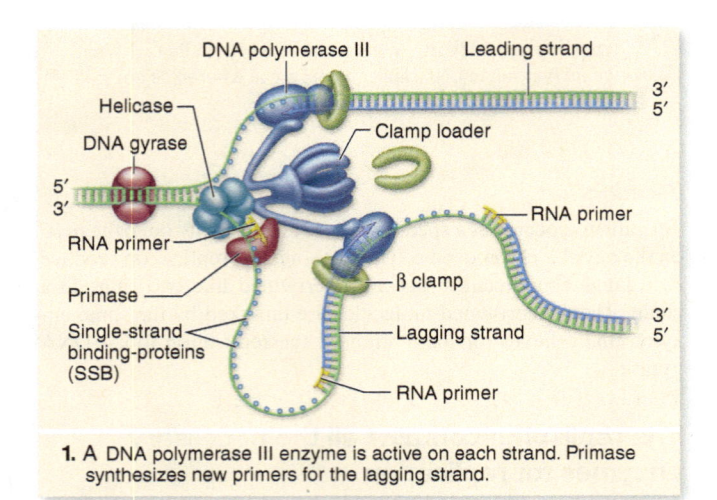

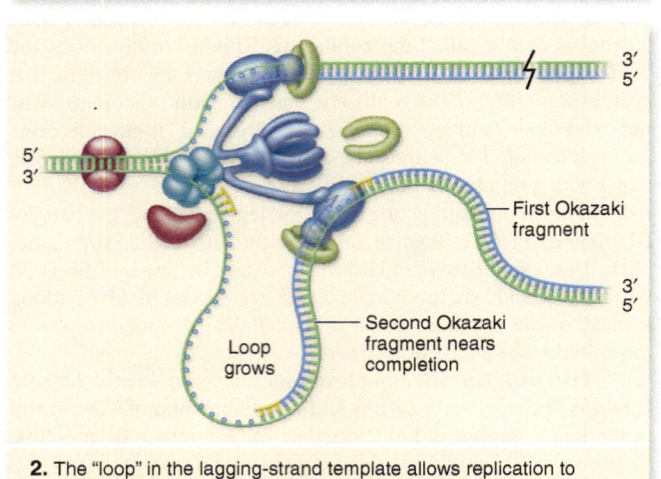

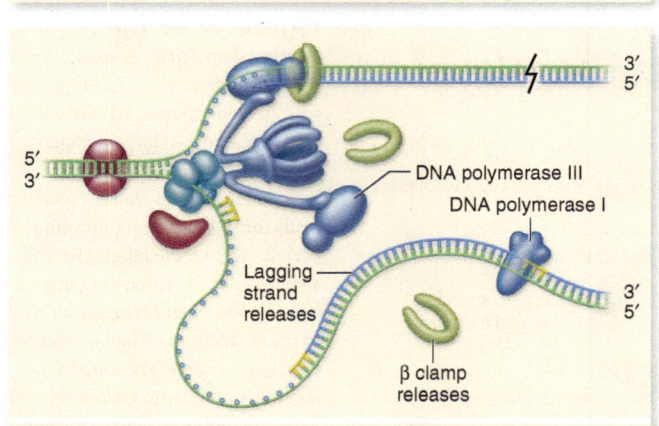

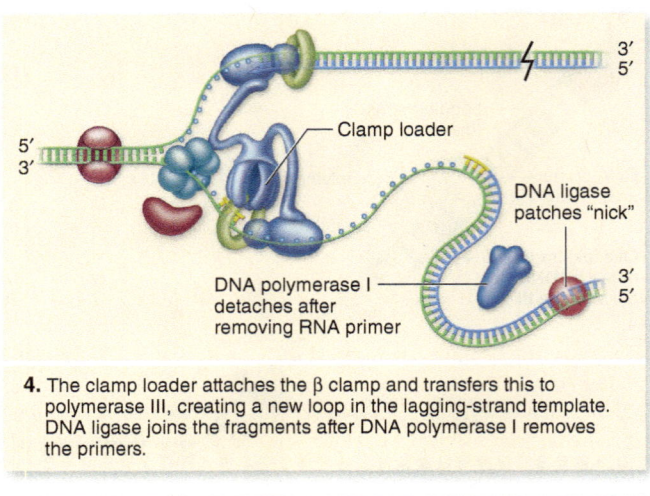

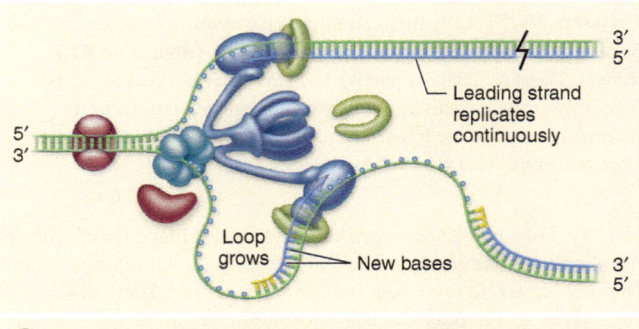

including the clamp loader. As Pol III finishes an Okazaki fragment, the clamp loader loads a β subunit onto the next fragment, and transfers the Pol III to this new β subunit (figure 14.18).

Current evidence also indicates that this replication complex is probably stationary, with the DNA strand moving through it like thread in a sewing machine, rather than the complex moving along the DNA strands. This stationary complex also pushes the newly synthesized DNA outward, which may aid in chromosome segregation. This process is summarized in figure 14.19.

Learning Outcomes Review 14.4

E. coli has three DNA polymerases: DNA Pol I, II, and III. Synthesis on one strand is discontinuous because DNA is antiparallel, and polymerases only synthesize in the 5′-to-3′ direction. Replication occurs at the replication fork, where the two strands are separated. Assembled here is a massive complex, the replisome, containing DNA polymerase III, primase, helicase, and other proteins. The lagging strand requires DNA polymerase I to remove the primers and replace them with DNA, and ligase to join Okazaki fragments.

■ *How are the nuclease functions of the different polymerases used during replication?*

Figure 14.19 DNA synthesis by the replisome. The semidiscontinuous synthesis of DNA is illustrated in stages using the model from figure 14.18.

14.5 Eukaryotic Replication

Learning Outcomes
1. Compare eukaryotic replication with prokaryotic.
2. Explain the function of telomeres.
3. Evaluate the role of telomerase in cell division.

Eukaryotic replication is complicated by two main factors: the larger amount of DNA organized into multiple chromosomes, and the linear structure of the chromosomes. This process requires new enzymatic activities only for dealing with the ends of chromosomes; otherwise the basic enzymology is the same.

Eukaryotic replication uses multiple origins

The sheer amount of DNA and how it is packaged constitute a problem for eukaryotes (figure 14.20). Eukaryotes usually have multiple chromosomes that are each larger than the *E. coli* chromosome. If only a single unique origin existed for each chromosome, the length of time necessary for replication would be prohibitive. This problem is solved by using multiple origins of replication for each chromosome, resulting in multiple *replicons*.

The origins are not as sequence-specific as *oriC*, and their recognition seems to depend on chromatin structure as well as on sequence. The number of origins used can also be adjusted during the course of development, so that early on, when cell divisions need to be rapid, more origins are activated. Each origin must be used only once per cell cycle.

The enzymology of eukaryotic replication is more complex

The replication machinery of eukaryotes is similar to that found in bacteria, but it is larger and more complex. The initiation phase of replication requires more factors to assemble both helicase and primase complexes onto the template, then load the polymerase with its sliding clamp unit.

The eukaryotic primase is interesting in that it is a complex of both an RNA polymerase and a DNA polymerase. It first makes short RNA primers, then extends these with DNA to produce the final primer. The reason for this added complexity is unclear.

The main replication polymerase itself is also a complex of two different enzymes that work together. One is called *DNA polymerase epsilon* (Pol ε) and the other *DNA polymerase delta* (Pol δ). The sliding clamp subunit that allows the enzyme complex to stay attached to the template is called PCNA (for proliferating cell nuclear antigen). This unusual name reflects the fact that PCNA was first identified as an antibody-inducing protein in proliferating (dividing) cells. The PCNA sliding clamp forms a trimer, but this structure is similar to the β subunit sliding clamp. The clamp loader is also similar to the bacterial structure. Despite the additional complexity, the action of the replisome is similar to that described in section 14.4 for *E. coli,* and the replication fork has essentially the same components.

Archaeal replication proteins are similar to eukaryotic proteins

Despite their lack of a membrane-bounded nucleus, archaeal replication proteins are more similar to eukaryotes than to bacterial. The main replication polymerase is most similar to eukaryotic Pol δ, and the sliding clamp is similar to the PCNA protein. The clamp loading complex is also more similar to eukaryotic than bacterial. The most interesting conclusion from all of these data are that all three domains of life have similar functions involved in replicating chromosomes. All three domains assemble similar protein complexes with clamp loader, sliding clamp, two polymerases, helicase, and primase at the replication fork.

Linear chromosomes have specialized ends

The specialized structures found on the ends of eukaryotic chromosomes are called **telomeres.** These structures protect

Figure 14.20 DNA of a single human chromosome. This chromosome has been relieved of most of its packaging proteins, leaving the DNA in its native form. The residual protein scaffolding appears as the dark material in the lower part of the micrograph.

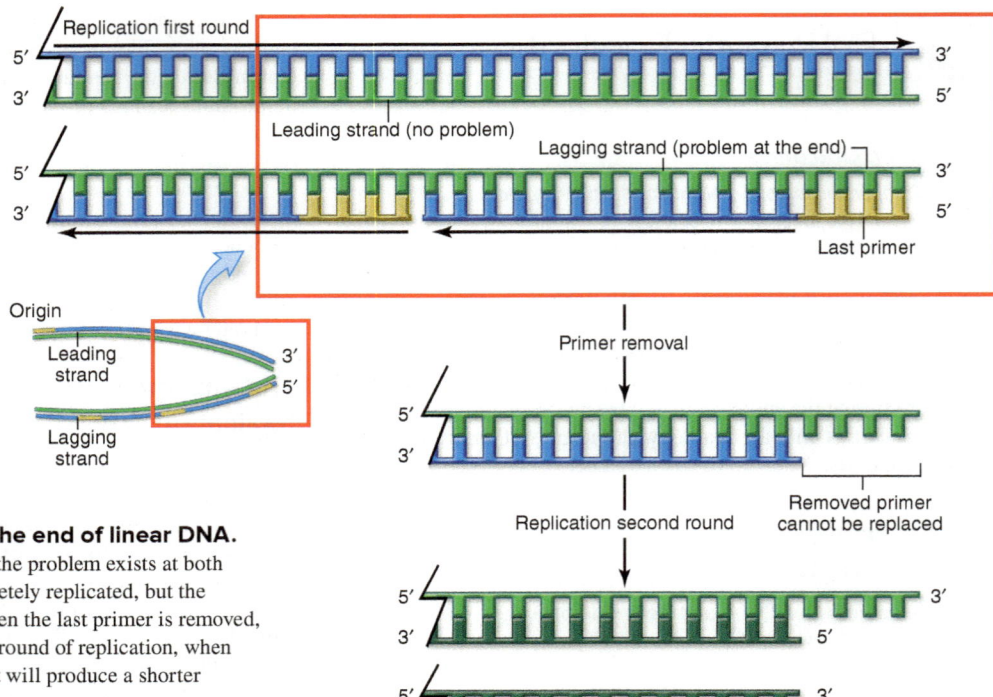

Figure 14.21 Replication of the end of linear DNA. Only one end is shown for simplicity; the problem exists at both ends. The leading strand can be completely replicated, but the lagging strand cannot be finished. When the last primer is removed, it cannot be replaced. During the next round of replication, when this shortened template is replicated, it will produce a shorter chromosome.

the ends of chromosomes from nucleases and maintain the integrity of linear chromosomes. These telomeres are composed of specific DNA sequences, but they are not made by the replication complex.

Replicating ends

The very structure of a linear chromosome causes a cell problems in replicating the ends. The directionality of polymerases, combined with their requirement for a primer, create this problem.

Consider a simple linear molecule like the one in figure 14.21. Replication of one end of each template strand is simple, namely the 5' end of the leading-strand template. When the polymerase reaches this end, synthesizing in the 5'-to-3' direction, it eventually runs out of template and is finished.

But on the other strand's end, the 3' end of the lagging strand, removal of the last primer on this end leaves a gap. This gap cannot be primed, meaning that the polymerase complex cannot finish this end properly. The result would be a gradual shortening of chromosomes with each round of cell division (figure 14.21).

The action of telomerase

When the sequence of telomeres was determined, they were found to be composed of short repeated sequences of DNA. This repeating nature is easily explained by their synthesis. They are made by an enzyme called **telomerase**, which uses an internal RNA as a template and not the DNA itself (figure 14.22).

Figure 14.22 Action of telomerase. Telomerase contains an internal RNA that the enzyme uses as a template to extend the DNA of the chromosome end. Multiple rounds of synthesis by telomerase produce repeated sequences. This single strand is completed by normal synthesis using it as a template (not shown).

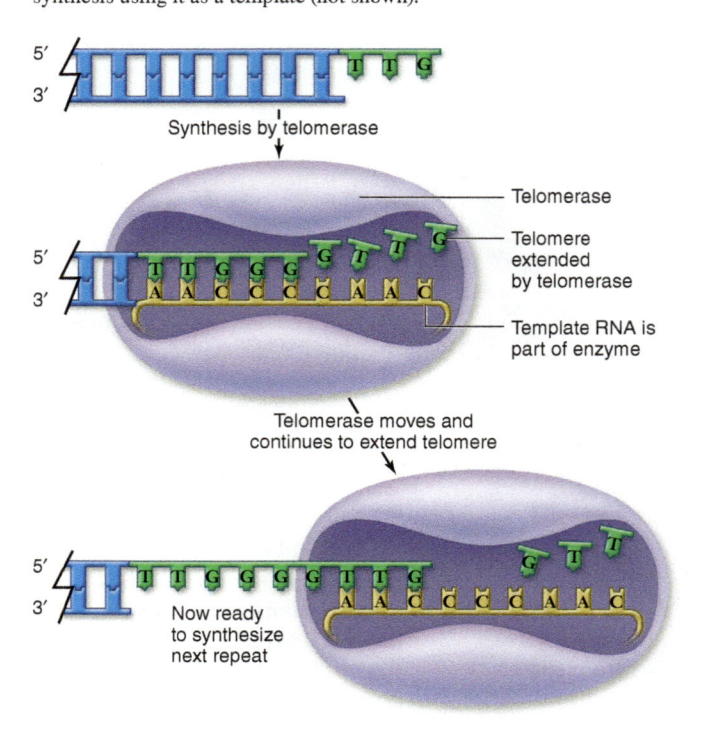

The use of the internal RNA template allows short stretches of DNA to be synthesized, composed of repeated nucleotide sequences complementary to the RNA of the enzyme. The other strand of these repeated units is synthesized by the usual action of the replication machinery copying the strand made by telomerase.

Telomerase, aging, and cancer

A gradual shortening of the ends of chromosomes occurs in the absence of telomerase activity. During embryonic and childhood development in humans, telomerase activity is high, but it is low in most somatic cells of the adult. The exceptions are cells that must divide as part of their function, such as lymphocytes. The activity of telomerase in somatic cells is kept low by preventing the expression of the gene encoding this enzyme.

Evidence for the shortening of chromosomes in the absence of telomerase was obtained by producing mice with no telomerase activity. These mice appear to be normal for up to six generations, but they show steadily decreasing telomere length that eventually leads to nonviable offspring.

This finding indicates a relationship between cell senescence (aging) and telomere length. Normal cells undergo only a specified number of divisions when grown in culture. This limit is at least partially based on telomere length.

Support for the relationship between senescence and telomere length comes from experiments in which telomerase was introduced into fibroblasts in culture. These cells have their lifespan increased relative to controls that have no added telomerase. Interestingly, these cells do not show the hallmarks of malignant cells, indicating that activation of telomerase alone does not make cells malignant.

A relationship has been found, however, between telomerase and cancer. Cancer cells do continue to divide indefinitely, and this would not be possible if their chromosomes were being continually shortened. Cancer cells generally show activation of telomerase, which allows them to maintain telomere length; but this is clearly only one aspect of conditions that allow them to escape normal growth controls.

> **? Inquiry question** How does the structure of eukaryotic genomes affect replication? Does this introduce problems that are not faced by prokaryotes?

Learning Outcomes Review 14.5

Eukaryotic replication is complicated by a large amount of DNA organized into chromosomes, and by the linear nature of chromosomes. Eukaryotes replicate a large amount of DNA in a short time by using multiple origins of replication. Linear chromosomes end in telomeres, and the length of telomeres is correlated with the ability of cells to divide. The enzyme telomerase synthesizes the telomeres. Cancer cells show activation of telomerase, which extends the ability of the cells to divide.

- *What might be the result of abnormal shortening of telomeres or a lack of telomerase activity?*

14.6 DNA Repair

Learning Outcomes

1. Explain why DNA repair is critical for cells.
2. Describe the different forms of DNA repair.

As you learned in section 14.4, many DNA polymerases have 3′-to-5′ exonuclease activity that allows "proofreading" of added bases. This action increases the accuracy of replication, but errors still occur. Without error correction mechanisms, cells would accumulate errors at an unacceptable rate, leading to high levels of deleterious or lethal mutations. A balance must exist between the introduction of new variation by mutation, and the effects of deleterious mutations on the individual.

Cells are constantly exposed to DNA-damaging agents

In addition to errors in DNA replication, cells are constantly exposed to agents that can damage DNA. These agents include radiation, such as UV light and X-rays, and chemicals in the environment. Agents that damage DNA can lead to mutations, and any agent that increases the number of mutations above background levels is called a **mutagen.**

The number of potentially mutagenic agents that organisms encounter is huge. Sunlight itself includes radiation in the UV range and is thus mutagenic. Ozone normally screens out much of the harmful UV radiation in sunlight, but some remains. The relationship between sunlight and mutations is shown clearly by the increase in skin cancer in regions of the southern hemisphere that are underneath a seasonal "ozone hole."

Organisms also may encounter mutagens in their diet in the form of either contaminants in food or natural plant products that can damage DNA. When a simple test was designed to detect mutagens, screening of possible sources indicated an amazing diversity of mutagens in the environment and in natural sources. As a result, consumer products are now screened to reduce the load of mutagens we are exposed to, but we cannot escape natural sources.

DNA repair restores damaged DNA

Cells cannot escape exposure to mutagens, but systems have evolved that enable cells to repair some damage. These DNA repair systems are vital to continued existence, whether a cell is a free-living, single-celled organism or part of a complex multicellular organism.

The importance of DNA repair is indicated by the multiplicity of repair systems that have been discovered and characterized. All cells that have been examined show multiple pathways for repairing damaged DNA and for reversing errors that occur during replication. These systems are not perfect, but they do reduce the mutational load on organisms to an acceptable level. In the rest of this section, we illustrate the action of DNA repair by concentrating on two examples drawn from these multiple repair pathways.

Repair can be either specific or nonspecific

DNA repair falls into two general categories: specific and nonspecific. Specific repair systems target a single kind of lesion in DNA and repair only that damage. Nonspecific forms of repair use a single mechanism to repair multiple kinds of lesions in DNA.

Photorepair: A specific repair mechanism

Photorepair is specific for damage that results from photochemical reactions of DNA and UV light. The most common photoproduct involves adjacent thymine bases linked together as thymine dimers (figure 14.23). These can be repaired by multiple pathways, including photorepair by an enzyme called photolyase. This enzyme binds to thymine dimers, then uses energy from visible light to cleave the dimer and restore the thymine bases (figure 14.23).

Genetic and biochemical analysis has revealed two distinct photolyase enzymes that recognize different photoproducts. These enzymes appear to be evolutionarily related to cryptochrome proteins that absorb visible light, but have no photolyase activity. In mammals, cryptochromes are part of the biochemical mechanism that links metabolic activities to daily light/dark cycles.

Photorepair is found in a wide spectrum of organisms ranging from bacteria to animals. Early reports indicated human skin was capable of photorepair, but later reports could find no activity. It now appears that placental mammals lack any photolyase, although they have cryptochromes.

Excision repair: A nonspecific repair mechanism

A common form of nonspecific repair is **excision repair**. In this pathway, a damaged region is removed, or excised, and is then replaced by DNA synthesis (figure 14.24). In *E. coli*, this action is accomplished by proteins encoded by the *uvr A*, *B*, and *C* genes.

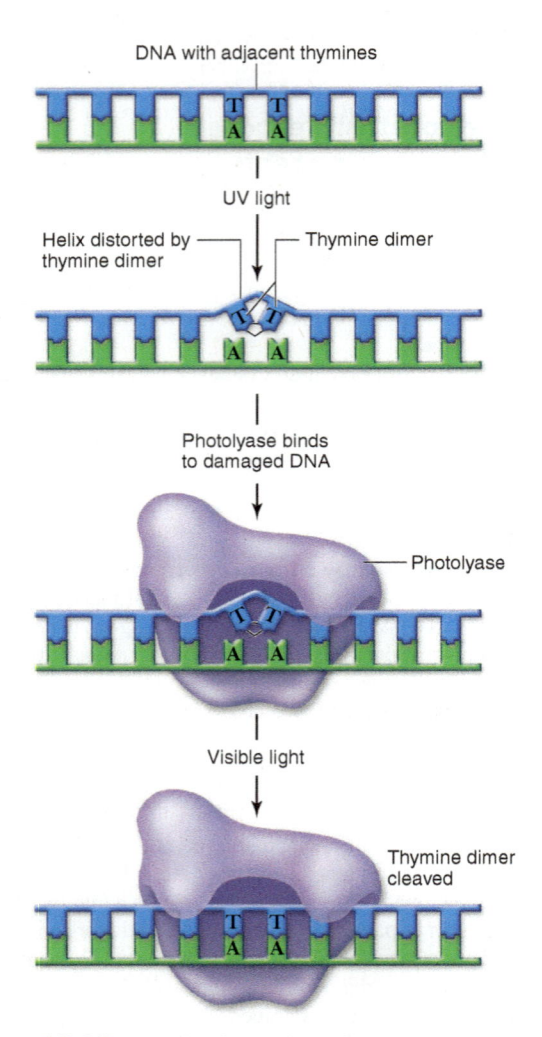

Figure 14.23 Repair of thymine dimer by photorepair.
UV light can catalyze a photochemical reaction to form a covalent bond between two adjacent thymines, thereby creating a thymine dimer. A photolyase enzyme recognizes the damage and binds to the thymine dimer. The enzyme absorbs visible light and uses the energy to cleave the thymine dimer.

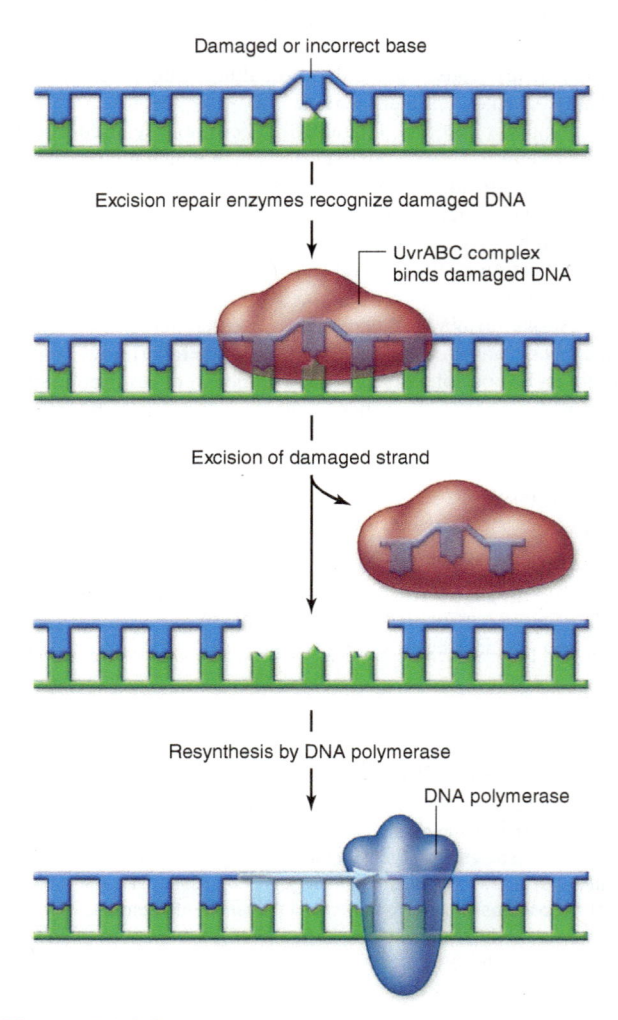

Figure 14.24 Repair of damaged DNA by excision repair. Damaged DNA is recognized by the UvrABC complex, which binds to the damaged region and removes it. Synthesis by DNA polymerase replaces the damaged region. DNA ligase finishes the process (not shown).

Although these genes were identified based on mutations that increased sensitivity of the cell to UV light (hence the "uvr" in their names), their proteins can act on damage due to other mutagens.

Excision repair follows three steps: (1) recognition of damage, (2) removal of the damaged region, and (3) resynthesis using the information on the undamaged strand as a template (figure 14.24). Recognition and excision are accomplished by the UvrABC complex. The UvrABC complex binds to damaged DNA and then cleaves a single strand on either side of the damage, removing it. In the synthesis stage, DNA Pol I or Pol II replaces the damaged DNA. This restores the original information in the damaged strand by using the information in the complementary strand.

Other repair pathways

Cells have other forms of nonspecific repair, and these fall into two categories: error-free and error-prone. It may seem strange to have an error-prone pathway, but it can be thought of as a last-ditch effort to save a cell that has been exposed to such massive damage that it has overwhelmed the error-free systems. In fact, this system in *E. coli* is part of what is called the "SOS response."

Cells can also repair damage that produces breaks in DNA. These systems use enzymes related to those involved in recombination during meiosis (see chapter 11). It is thought that recombination uses enzymes that originally evolved for DNA repair.

The number of different systems and the wide spectrum of damage that can be repaired illustrate the importance of maintaining the integrity of the genome. Accurate replication of the genome is useless if a cell cannot reverse errors that can occur during this process or repair damage due to environmental causes.

? Inquiry question Cells are constantly exposed to DNA-damaging agents, ranging from UV light to by-products of oxidative metabolism. How does the cell deal with this, and what would happen if the cell had no way of dealing with this?

Learning Outcomes Review 14.6

The ability to repair DNA is critical because of replication errors and the constant presence of damaging agents that can cause mutation. Cells have multiple repair pathways; some of these systems are specific for a single type of damage, such as photorepair that reverses thymine dimers caused by UV light. Other systems are nonspecific, such as excision repair that removes and replaces damaged regions.

■ *Could a cell survive with no form of DNA repair?*

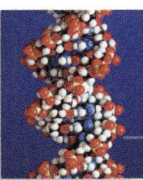

Chapter Review

14.1 The Nature of the Genetic Material

Griffith finds that bacterial cells can be transformed.
Nonvirulent *S. pneumoniae* could take up an unknown substance from a virulent strain and become virulent.

Avery, MacLeod, and McCarty identify the transforming principle.
The transforming substance could be inactivated by DNA-digesting enzymes, but not by protein-digesting enzymes.

Hershey and Chase demonstrate that phage genetic material is DNA.
Radioactive labeling showed that the infectious agent of phage is its DNA, and not its protein.

14.2 DNA Structure

DNA's components were known, but its three-dimensional structure was a mystery.
The nucleotide building blocks for DNA contain deoxyribose and the bases adenine (A), guanine (G), cytosine (C), and thymine (T). Phosphodiester bonds are formed between the 5′ phosphate of one nucleotide and the 3′ hydroxyl of another nucleotide (figure 14.4).

Chargaff, Franklin, and Wilkins obtained some structural evidence.
Chargaff found equal amounts of adenine and thymine, and of cytosine and guanine, in DNA. The bases exist primarily in keto and enol forms that exhibit hydrogen bonding. X-ray diffraction studies by Franklin and Wilkins indicated that DNA had a helical structure.

The Watson–Crick model fits the available evidence (figures 14.8 & 14.9).
DNA consists of two antiparallel polynucleotide strands wrapped about a common helical axis. These strands are held together by hydrogen bonds forming specific base-pairs (A–T and G–C). The two strands are complementary; one strand can specify the other.

14.3 Basic Characteristics of DNA Replication

Meselson and Stahl demonstrate the semiconservative mechanism (figure 14.11).
Semiconservative replication uses each strand of a DNA molecule to specify the synthesis of a new strand. Meselson and Stahl showed this by using a heavy isotope of nitrogen and separating the replication products. Replication produces two new molecules each composed of one new strand and one old strand.

DNA replication requires a template, nucleotides, and a polymerase enzyme.
All new DNA molecules are produced by DNA polymerase copying a template. All known polymerases synthesize new DNA in the 5′-to-3′ direction. These enzymes also require a primer. The building blocks used in replication are deoxynucleotide triphosphates with high-energy bonds; they do not require any additional energy.

14.4 Prokaryotic Replication

Prokaryotic replication starts at a single origin.
The *E. coli* origin has AT-rich sequences that are easily opened. The chromosome and its origin form a replicon.

E. coli has at least three different DNA polymerases.
Some DNA polymerases can also degrade DNA from one end, called exonuclease activity. Pol I, II, and III all have 3′-to-5′ exonuclease activity that can remove mispaired bases. Pol I can remove bases in the 5′-to-3′ direction, important to removing RNA primers.

Unwinding DNA requires energy and causes torsional strain.
DNA helicase uses energy from ATP to unwind DNA. The torsional strain introduced is removed by the enzyme DNA gyrase.

Replication is semidiscontinuous.
Replication is discontinuous on one strand (figure 14.15). The continuous strand is called the leading strand, and the discontinuous strand is called the lagging strand.

Synthesis occurs at the replication fork.
The partial opening of a DNA strand forms two single-stranded regions called the replication fork. At the fork, synthesis on the leading strand requires a single primer, and the polymerase stays attached to the template because of the β subunit that acts as a sliding clamp. On the lagging strand, DNA primase adds primers periodically, and DNA Pol III synthesizes the Okazaki fragments. DNA Pol I removes primer segments, and DNA ligase joins the fragments.

The replisome contains all the necessary enzymes for replication.
The replisome consists of two copies of Pol III, DNA primase, DNA helicase, and a number of accessory proteins. It moves in one direction by creating a loop in the lagging strand, allowing the antiparallel template strands to be copied in the same direction (figures 14.18 & 14.19).

14.5 Eukaryotic Replication

Eukaryotic replication uses multiple origins.
The sheer size and organization of eukaryotic chromosomes requires multiple origins of replication to be able to replicate DNA in the time available in S phase.

The enzymology of eukaryotic replication is more complex.
The eukaryotic primase synthesizes a short stretch of RNA and then switches to making DNA. This primer is extended by the main replication polymerase, which is a complex of two enzymes. The sliding clamp subunit was originally identified as protein produced by proliferating cells and is called PCNA.

Archaeal replication proteins are similar to eukaryotic proteins.
The replication proteins of archaea, including the sliding clamp, clamp loader, and DNA polymerases, are more similar to those of eukaryotes than to prokaryotes.

Linear chromosomes have specialized ends.
The ends of linear chromosomes are called telomeres. They are made by telomerase, not by the replication complex. Telomerase contains an internal RNA that acts as a template to extend the DNA of the chromosome end. Adult cells lack telomerase activity, and telomere shortening correlates with senescence.

14.6 DNA Repair

Cells are constantly exposed to DNA-damaging agents.
Errors from replication and damage induced by agents such as UV light and chemical mutagens can lead to mutations.

DNA repair restores damaged DNA.
Without repair mechanisms, cells would accumulate mutations until inviability occurred.

Repair can be either specific or nonspecific.
Photolyase enzymes use visible light to reverse UV induced damage. There are two species of enzyme specific for different kinds of damage. Mammals lack the enzyme. Excision repair is nonspecific. In prokaryotes, the uvr system can remove a damaged region of DNA.

Review Questions

UNDERSTAND

1. What was the key finding from Griffith's experiments using live and heat-killed pathogenic bacteria?
   a. Bacteria with a smooth coat could kill mice.
   b. Bacteria with a rough coat are not lethal.
   c. DNA is the genetic material.
   d. Genetic material can be transferred from dead to live bacteria.

2. Which of the following is NOT a component of DNA?
   a. The pyrimidine uracil
   b. Five-carbon sugars
   c. The purine adenine
   d. Phosphate groups

3. Chargaff studied the composition of DNA from different sources and found that
   a. the number of phosphate groups always equals the number of 5-carbon sugars.
   b. the proportions of A equal that of C and G equals T.
   c. the proportions of A equal that of T and G equals C.
   d. purines bind to pyrimidines.

4. The bonds that hold two complementary strands of DNA together are
   a. hydrogen bonds.
   b. peptide bonds.
   c. ionic bonds.
   d. phosphodiester bonds.

5. The basic mechanism of DNA replication is semiconservative with two new molecules,
   a. each with new strands.
   b. one with all new strands and one with all old strands.
   c. each with one new and one old strand.
   d. each with a mixture of old and new strands.

6. One common feature of all DNA polymerases is that they
   a. synthesize DNA in the 3′-to-5′ direction.
   b. synthesize DNA in the 5′-to-3′ direction.
   c. synthesize DNA in both directions by switching strands.
   d. do not require a primer.

7. Which of the following is NOT part of the Watson–Crick model of the structure of DNA?
   a. DNA is composed of two strands.
   b. The two DNA strands are oriented in parallel (5′-to-3′).
   c. Purines bind to pyrimidines.
   d. DNA forms a double helix.

APPLY

1. If one strand of a DNA is 5′ ATCGTTAAGCGAGTCA 3′, then the complementary strand would be:
   a. 5′ TAGCAATTCGCTCAGT 3′.
   b. 5′ ACTGAGCGAATTGCTA 3′.
   c. 5′ TGACTCGCTTAACGAT 3′.
   d. 5′ ATCGTTAAGCGAGTCA 3′.

2. Hershey and Chase used radioactive phosphorus and sulfur to
   a. label DNA and protein uniformly.
   b. differentially label DNA and protein.
   c. identify the transforming principle.
   d. Both b and c are correct.

3. The Meselson and Stahl experiment used a density label to be able to
   a. determine the directionality of DNA replication.
   b. differentially label DNA and protein.
   c. distinguish between newly replicated and old strands.
   d. distinguish between replicated DNA and RNA primers.

4. The difference in leading- versus lagging-strand synthesis is a consequence of
   a. only the physical structure of DNA.
   b. only the activity of DNA polymerase enzymes.
   c. both the physical structure of DNA and the action of polymerase enzyme.
   d. the larger size of the lagging strand.

5. If the activity of DNA ligase was removed from replication, this would have a greater effect on
   a. synthesis on the lagging strand versus the leading strand.
   b. synthesis on the leading strand versus the lagging strand.
   c. priming of DNA synthesis versus actual DNA synthesis.
   d. photorepair of DNA versus DNA replication.

6. Successful DNA synthesis requires all of the following *except*
   a. helicase.
   b. endonuclease.
   c. DNA primase.
   d. DNA ligase.

7. The synthesis of telomeres
   a. uses DNA polymerase, but without the sliding clamp.
   b. uses enzymes involved in DNA repair.
   c. requires telomerase, which does not use a template.
   d. requires telomerase, which uses an internal RNA as a template.

8. When mutations that affected DNA replication were isolated, two kinds were found. In cultures that were not synchronized (that is, not all dividing at the same time), one class put an immediate halt to replication, whereas the other put a much slower stop to the process. The first class affects functions at the replication fork like polymerase and primase. The second class affects functions necessary for
   a. elongation on the lagging but not the leading strand.
   b. elongation on the leading but not the lagging strand.
   c. initiation: cells complete replication but cannot start a new round.
   d. the sliding clamp: loss makes the polymerase slower.

SYNTHESIZE

1. The work by Griffith provided the first indication that DNA was the genetic material. Review the four experiments outlined in figure 14.1. Predict the likely outcome for the following variations on this classic research.
   a. Heat-killed pathogenic and heat-killed nonpathogenic
   b. Heat-killed pathogenic and live nonpathogenic in the presence of an enzyme that digests proteins (proteases)
   c. Heat-killed pathogenic and live nonpathogenic in the presence of an enzyme that digests DNA (endonuclease)

2. In the Meselson–Stahl experiment, a control experiment was done to show that the hybrid bands after one round of replication were in fact two complete strands, one heavy and one light. Using the same experimental setup as detailed in the text, how can this be addressed?

3. Enzyme function is critically important for the proper replication of DNA. Predict the consequence of a loss of function for each of the following enzymes.
   a. DNA gyrase
   b. DNA polymerase III
   c. DNA ligase
   d. DNA polymerase I

CHAPTER 15

Genes and How They Work

Chapter Contents

15.1 The Nature of Genes
15.2 The Genetic Code
15.3 Prokaryotic Transcription
15.4 Eukaryotic Transcription
15.5 Eukaryotic pre-mRNA Splicing
15.6 The Structure of tRNA and Ribosomes
15.7 The Process of Translation
15.8 Summarizing Gene Expression
15.9 Mutation: Altered Genes

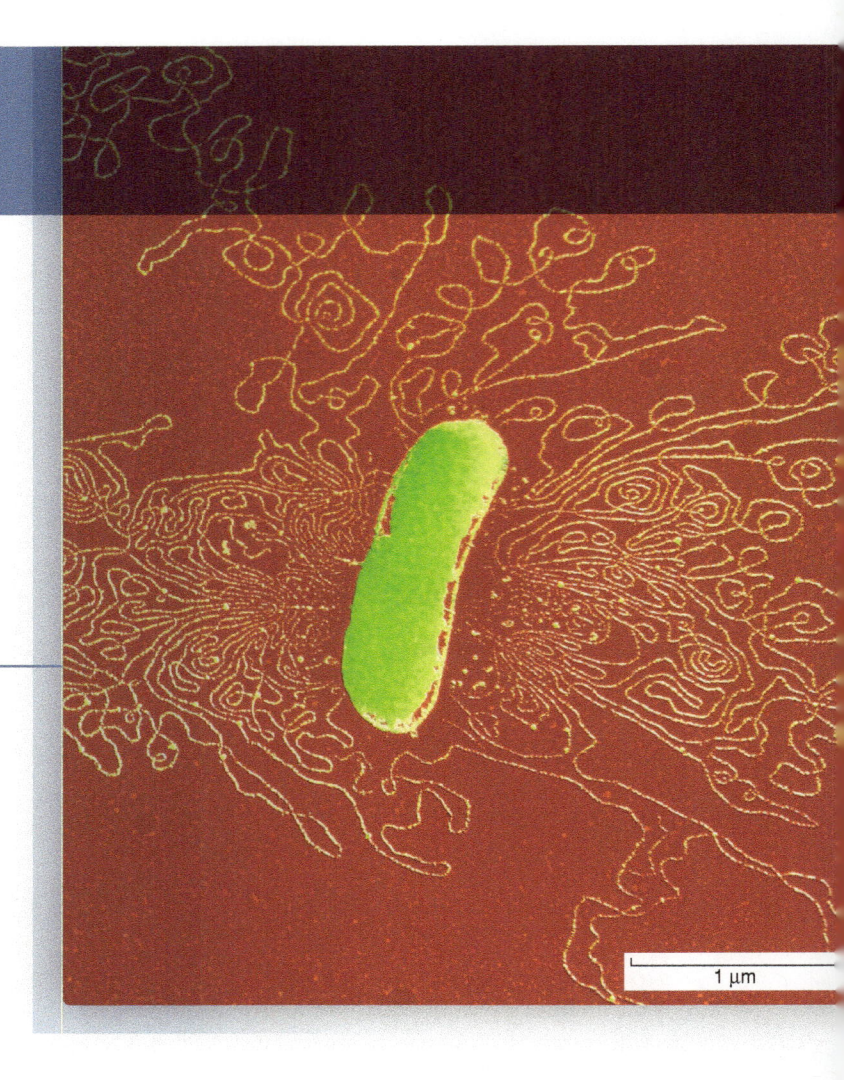

Introduction

You've seen how genes specify traits and how those traits can be followed in genetic crosses. You've also seen that the information in genes resides in the DNA molecule; the picture above shows all the DNA within the entire *E. coli* chromosome. Information in DNA is replicated by the cell and then partitioned equally during the process of cell division. The information in DNA is much like a blueprint for a building. The construction of the building uses the information in the blueprint, but requires building materials and carpenters and other skilled laborers using a variety of tools working together to actually construct the building. Similarly, the information in DNA requires nucleotide and amino acid building blocks, multiple forms of RNA, and many proteins acting in a coordinated fashion to make up the structure of a cell.

We now turn to the nature of the genes themselves and how cells extract the information in DNA in the process of gene expression. Gene expression can be thought of as the conversion of genotype into the phenotype.

15.1 The Nature of Genes

Learning Outcomes

1. Evaluate the evidence for the one-gene/one-polypeptide hypothesis.
2. Distinguish between transcription and translation.
3. List the roles played by RNA in gene expression.

We know that DNA encodes proteins, but this knowledge alone tells us little about how the information in DNA can control cellular functions. Researchers had evidence that genetic mutations affected proteins, and in particular enzymes, long before the structure and code of DNA was known. In this section, we review the evidence of the link between genes and enzymes.

Garrod concluded that inherited disorders can involve specific enzymes

In 1902, the British physician Archibald Garrod noted that certain diseases among his patients seemed to be more prevalent in particular families. By examining several generations of these families, he found that some diseases behaved as though caused by simple recessive alleles. Garrod concluded that these disorders were Mendelian traits, and that they had resulted from changes in the hereditary information in an ancestor of the affected families.

Garrod investigated the disorder in alkaptonuria in detail. Patients with alkaptonuria produce urine containing homogentisic acid (alkapton), which is rapidly oxidized in air, turning the urine black. In normal individuals, homogentisic acid is broken down into simpler substances. Garrod concluded that patients with alkaptonuria lack the enzyme necessary to catalyze this breakdown, further speculating that many other inherited diseases might also reflect enzyme deficiencies.

Beadle and Tatum showed that genes specify enzymes

Garrod was actually ahead of his time as the connection between Mendelian alleles and enzymes was not codified for almost 40 years. In 1941 the link between traits followed in crosses and proteins was demonstrated by a series of experiments by George Beadle and Edward Tatum at Stanford University. They used the then new technology of inducing mutations by X-rays to collect a series of mutants for detailed study. Their work is summarized in figure 15.1, and described next.

Neurospora crassa, *the bread mold*

Beadle and Tatum chose the bread mold *Neurospora crassa* for their experiments. This fungus can be grown readily in the laboratory on a defined medium consisting of only a carbon source (glucose), a vitamin (biotin), plus inorganic salts. This is called minimal medium because it represents the minimal requirements to support growth. Any cells that can grow on minimal medium must be able to synthesize all necessary biological molecules.

When Beadle and Tatum exposed spores to X-rays, they expected to find some mutants that would be unable to grow on minimal medium. These so-called **nutritional mutants** result from damage to genes encoding functions necessary to make important biological molecules.

Nutritional mutants

To identify nutritional mutants, Beadle and Tatum first grew cultures on rich medium, then placed subcultures of individual fungal cells onto minimal medium. This identifies any cells that had lost the ability to make compounds necessary for growth. They concentrated on the ability to synthesize the amino acid arginine, where the biosynthetic pathway was known.

To identify mutants unable to make arginine, they selected cultures that would grow only on minimal medium plus arginine. This resulted in a collection of independent mutants that were all unable to synthesize arginine, and that each could be genetically mapped to different chromosomal locations. This defined four genes they named *argE, argF, argG,* and *argH*.

One gene/one polypeptide

With these tools in hand, Beadle and Tatum were able to genetically dissect the arginine biosynthetic pathway. By supplementing

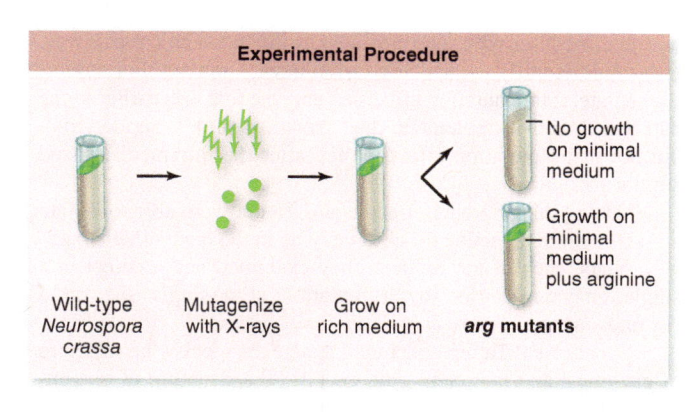

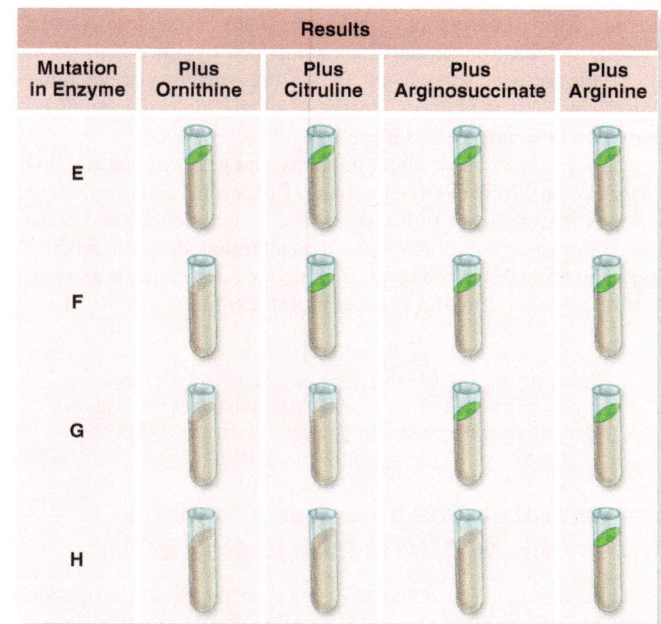

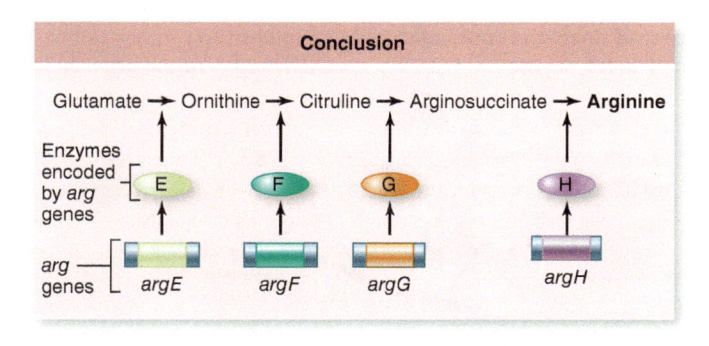

Figure 15.1 The Beadle and Tatum experiment.
Wild-type *Neurospora* were mutagenized with X-rays to produce mutants deficient in the synthesis of arginine *(top panel)*. The specific defect in each mutant was identified by growing on medium supplemented with intermediates in the biosynthetic pathway for arginine *(middle panel)*. A mutant will grow only on media supplemented with an intermediate produced after the defective enzyme in the pathway for each mutant. The enzymes in the pathway can then be correlated with genes on chromosomes *(bottom panel)*.

minimal medium with intermediates in the biochemical pathway, they could identify the specific lesion in each mutant. For any intermediate, if the mutation affects an enzyme that acts earlier in the pathway than the supplement, then growth should be supported— but not if the mutation affects a step after the intermediate used (figure 15.1).

Using this approach, Beadle and Tatum were able to isolate a mutant strain defective for each enzyme in the biosynthetic pathway. Thus, each of the mutants they examined had a defect in a single enzyme, caused by a mutation at a single site on a chromosome.

Beadle and Tatum concluded that genes specify the structure of enzymes, and that each gene encodes the structure of one enzyme (figure 15.1). They called this relationship the *one-gene/one-enzyme hypothesis*. Today, because many enzymes contain multiple polypeptide subunits, each encoded by a separate gene, the relationship is generalized to the one-gene/one-polypeptide hypothesis. This hypothesis clearly states the molecular relationship between genotype and phenotype.

As you learn more about genomes and gene expression, this clear relationship is overly simplistic. Eukaryotic genes are more complex than those of prokaryotes, and some enzymes are composed, at least in part, of RNA, itself an intermediate in the production of proteins. Nevertheless, one gene/one polypeptide is a useful starting point for thinking about gene expression.

Data analysis If you made a double mutant with *argE* and *argG*, which kind(s) of media would this strain grow on? In general, what can a double mutant tell you about the order of genes?

The central dogma describes information flow in cells as DNA to RNA to protein

The conversion of genotype to phenotype requires information stored in DNA to be converted to protein. The nature of information flow in cells was first described by Francis Crick as the **central dogma of molecular biology.** Information passes in one direction from the gene (DNA) to an RNA copy of the gene, and the RNA copy directs the sequential assembly of a chain of amino acids into a protein (figure 15.2). Stated briefly,

$$DNA \longrightarrow RNA \longrightarrow protein$$

The central dogma provides an intellectual framework that describes information flow in biological systems. We call the DNA-to-RNA step **transcription** because it produces an exact copy of the DNA, much as a legal transcription contains the exact words of a court proceeding. The RNA-to-protein step is termed **translation** because it requires translating from the nucleic acid to the protein "languages."

Since the original formulation of the central dogma, a class of viruses called **retroviruses** was discovered that can convert their RNA genome into a DNA copy, using the viral enzyme **reverse transcriptase.** This conversion violates the direction of information flow of the central dogma, and the discovery forced an updating of the possible flow of information to include this "reverse" flow from RNA to DNA.

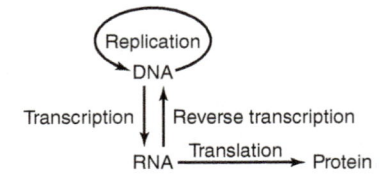

Transcription makes an RNA copy of DNA

The process of transcription produces an RNA copy of the information in DNA. That is, transcription is the DNA-directed synthesis of RNA by the enzyme RNA polymerase (figure 15.3). This process uses the principle of complementarity, described in chapter 14, to use DNA as a template to make RNA.

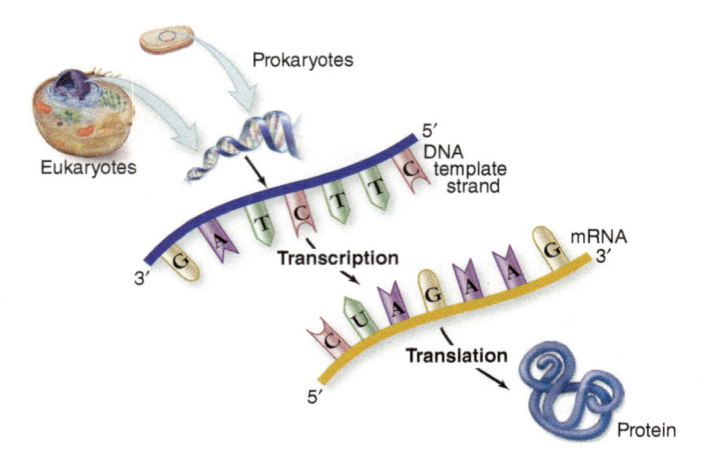

Figure 15.2 The central dogma of molecular biology. DNA is transcribed to make mRNA, which is translated to make a protein.

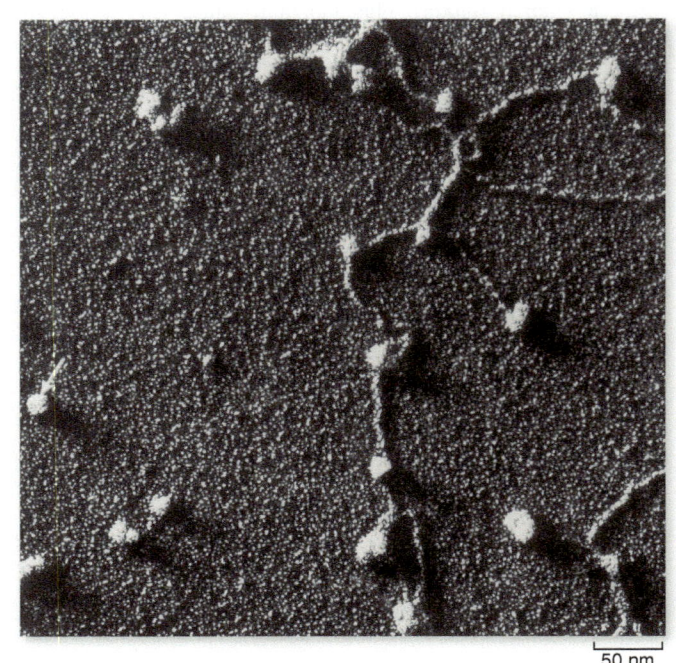

Figure 15.3 RNA polymerase. Platinum shadowed electron micrograph (freeze-fracture TEM) showing RNA polymerase molecules bound to DNA strands.

Because DNA is double-stranded and RNA is single-stranded, only one of the two DNA strands needs to be copied. We call the strand that is copied the **template strand.** The RNA transcript's sequence is complementary to the template strand. The strand of DNA not used as a template is called the **coding strand.** It has the same sequence as the RNA transcript, except that U (uracil) in the RNA is T (thymine) in the DNA-coding strand. Another naming convention for the two strands of the DNA is to call the coding strand the sense strand, as it has the same "sense" as the RNA. The template strand would then be the antisense strand:

```
Coding (sense)      5'-TCAGCCGTCAGCT-3'  ⎤
Template (antisense) 3'-AGTCGGCAGTCGA-5' ⎦ DNA
                          Transcription
                              ↓
Coding  5'-UCAGCCGUCAGCU-3'  mRNA
```

The RNA transcript used to direct the synthesis of polypeptides is termed *messenger RNA (mRNA)*. Its name reflects its use by the cell to carry the DNA message to the ribosome for processing.

> **Inquiry question** RNA polymerase has no proofreading capacity. How does this affect the error rate in transcription compared with DNA replication? Why do you think it is more important for DNA polymerase than for RNA polymerase to proofread?

Translation uses information in RNA to synthesize proteins

The process of translation is by necessity much more complex than transcription. In this case, RNA cannot be used as a direct template for a protein because there is no complementarity—that is, a sequence of amino acids cannot be aligned to an RNA template based on any kind of "chemical fit." Molecular geneticists suggested that some kind of adapter molecule must exist that can interact with both RNA and amino acids, and *transfer RNA (tRNA)* was found to fill this role. This need for an intermediary adds a level of complexity to the process that is not seen in either DNA replication or transcription of RNA.

Translation takes place on the ribosome, the cellular protein-synthesis machinery, and it requires the participation of multiple kinds of RNA and many proteins. Here we provide an outline of the processes; all are described in detail in the rest of the chapter.

RNA has multiple roles in gene expression

All RNAs are synthesized from a DNA template by transcription. Gene expression requires the participation of multiple kinds of RNA, each with different roles in the overall process. Here is a brief summary of these roles, which are described in detail in the remainder of the chapter:

Messenger RNA. Even before the details of gene expression were unraveled, geneticists recognized that there must be an intermediate form of the information in DNA that can be transported out of the eukaryotic nucleus to the cytoplasm for ribosomal processing. This hypothesis was called the "messenger hypothesis," and we retain this language in the name *messenger RNA (mRNA)*.

Ribosomal RNA. The class of RNA found in ribosomes is called **ribosomal RNA (rRNA).** There are multiple forms of rRNA, and rRNA is found in both ribosomal subunits. This rRNA is critical to the function of the ribosome.

Transfer RNA. The intermediary adapter molecule between mRNA and amino acids is **transfer RNA (tRNA).** Transfer RNA molecules have amino acids covalently attached to one end and an anticodon that can base-pair with an mRNA codon at the other. The tRNAs act to interpret information in mRNA and to help position the amino acids on the ribosome.

Small nuclear RNA. Small nuclear RNAs (snRNAs) are part of the machinery that is involved in nuclear processing of eukaryotic "pre-mRNA." We discuss this splicing reaction later in section 15.5.

SRP RNA. In eukaryotes, where some proteins are synthesized by ribosomes on the rough endoplasmic reticulum (RER), this process is mediated by the **signal recognition particle,** or **SRP,** described in section 15.7. The SRP contains both RNA and proteins.

Small RNAs. This class of RNA includes both **micro-RNA (miRNA)** and **small interfering RNA (siRNA).** These are involved in the control of gene expression discussed in chapter 16.

> ### Learning Outcomes Review 15.1
> Garrod showed that altered enzymes can cause metabolic disorders. Beadle and Tatum demonstrated that each gene encodes a unique enzyme. Genetic information flows from DNA (genes) to protein (enzymes) using messenger RNA as an intermediate. Transcription converts information in DNA into an RNA transcript, and translation converts this information into protein. RNA comes in several varieties having different functions; these include mRNA (the transcript), tRNA (the intermediary), and rRNA (in ribosomes), as well as snRNA, SNP RNA, and small RNAs (miRNA, siRNA).
>
> ■ *Why do cells need an adapter molecule like tRNA between RNA and protein?*

15.2 The Genetic Code

Learning Outcomes
1. Summarize the experiments that revealed the genetic code.
2. Describe the characteristics of the genetic code.
3. Identify the relationship between codons and amino acids.

How does a sequence of nucleotides in a DNA molecule specify the sequence of amino acids in a polypeptide? The answer to this essential question came in 1961, through an experiment led by Francis Crick and Sydney Brenner. That experiment was so elegant and the result so critical to understanding the genetic code that we describe it here in detail.

The code is read in groups of three

Crick and Brenner reasoned that the genetic code most likely consisted of a series of blocks of information called **codons,** each corresponding to an amino acid in the encoded protein. They further hypothesized that the information within one codon was probably a sequence of three nucleotides. With four DNA nucleotides (G, C, T, and A), using two in each codon can produce only 4^2, or 16, different codons—not enough to code for 20 amino acids. However, three nucleotides results in 4^3, or 64, different combinations of three, more than enough.

Do genetic messages include spaces?

In theory, the sequence of codons in a gene could be punctuated with nucleotides between the codons that are not used, like the spaces that separate the words in this sentence. Alternatively, the codons could lie immediately adjacent to each other, forming a continuous sequence of nucleotides.

If the information in the genetic message is separated by spaces, then altering any single word would not affect the entire sentence. In contrast, if all of the words are run together but read in groups of three, then any alteration that is not in groups of three would alter the entire sentence. These two ways of using information in DNA imply different methods of translating the information into protein.

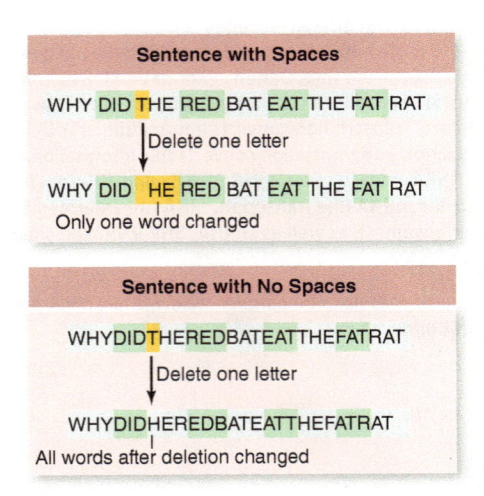

Demonstrating the lack of spaces

To choose between these alternative mechanisms, Crick and his colleagues used a chemical to create mutations that caused single-base insertions or deletions from a viral DNA molecule. They then showed that combining an insertion with a deletion restored function even though either one individually displayed loss of function. In this case, only the region between the insertion or deletion would be altered. By choosing a region of the gene that encoded a part of the protein not critical to function, this small change did not cause a change in phenotype.

When they combined a single deletion or two deletions near each other, the genetic message shifted, altering all of the amino acids after the deletion. When they made three deletions, however, the protein after the deletions was normal. They obtained the same results when they made additions to the DNA consisting of 1, 2, or 3 nt (nucleotides).

Thus, Crick and Brenner concluded that the genetic code is read in increments of three nucleotides (in other words, it is a triplet code), and that reading occurs continuously without punctuation between the 3-nt units (figure 15.4).

SCIENTIFIC THINKING

Hypothesis: The genetic code is read in groups of three bases.
Prediction: If the genetic code is read in groups of three, then deletion of one or two bases would shift the reading frame after the deletion. Deletion of three bases, however, would produce a protein with a single amino acid deleted but no change downstream.
Test: Single-base deletion mutants are collected, each of which exhibits a mutant phenotype. Three of these deletions in a single region are combined to assess the effect of deletion of three bases.

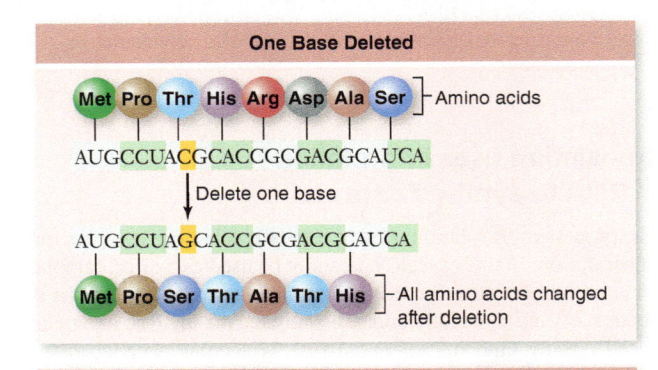

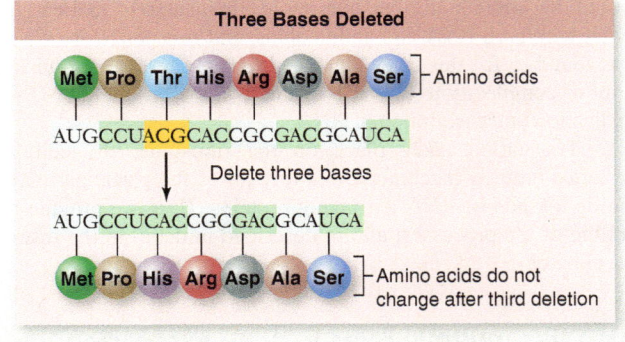

One Base Deleted

Met Pro Thr His Arg Asp Ala Ser — Amino acids

AUGCCUACGCACCGCGACGCAUCA

↓ Delete one base

AUGCCUAGCACCGCGACGCAUCA

Met Pro Ser Thr Ala Thr His — All amino acids changed after deletion

Three Bases Deleted

Met Pro Thr His Arg Asp Ala Ser — Amino acids

AUGCCUACGCACCGCGACGCAUCA

↓ Delete three bases

AUGCCUCACCGCGACGCAUCA

Met Pro His Arg Asp Ala Ser — Amino acids do not change after third deletion

Result: The combination of three deletions does not have the same drastic effect as the loss of one or two bases.
Conclusion: The genetic code is read in groups of three.
Further Experiments: If you also had mutants with one single-base additions, what would be the effect of combining a deletion and an addition?

Sentence with Spaces

WHY DID THE RED BAT EAT THE FAT RAT
↓ Delete one letter
WHY DID HE RED BAT EAT THE FAT RAT
Only one word changed

Sentence with No Spaces

WHYDIDTHEREDBATEATTHEFATRAT
↓ Delete one letter
WHYDIDHEREDBATEATTHEFATRAT
All words after deletion changed

Figure 15.4 The genetic code is triplet.

These experiments indicate the importance of the **reading frame** for the genetic message. Because there is no punctuation, the reading frame established by the first codon in the sequence determines how all subsequent codons are read. We now call the kinds of mutations that Crick and Brenner used **frameshift mutations** because they alter the reading frame of the genetic message.

Nirenberg and others deciphered the code

The determination of which of the 64 possible codons encoded each particular amino acids was one of the greatest triumphs of 20th-century biochemistry. Accomplishing this decryption depended on two related technologies: (1) cell-free biochemical systems that would support protein synthesis from a defined RNA, and (2) the ability to produce synthetic, defined RNAs that could be used in the cell-free system.

During a five-year period from 1961 to 1966, work performed primarily in American biochemist Marshall Nirenberg's laboratory led to the elucidation of the genetic code. Nirenberg's group first showed that adding the synthetic RNA molecule polyU (an RNA molecule consisting of a string of uracil nucleotides) to their cell-free systems produced the polypeptide polyphenylalanine (a string of phenylalanine amino acids). Therefore, UUU encodes phenylalanine.

Next they used enzymes to produce RNA polymers with more than one nucleotide. These polymers allowed them to infer the composition of many of the possible codons, but not the order of bases in each codon.

The researchers then were able to use enzymes to synthesize defined 3-base sequences that could be tested for binding to the protein-synthesis machinery. This so-called *triplet-binding assay* allowed them to identify 54 of the 64 possible triplets.

The organic chemist H. Gobind Khorana provided the final piece of the puzzle by using organic synthesis to produce artificial RNA molecules of defined sequence, and then examining what polypeptides they directed in cell-free systems. The combination of all of these methods allowed the determination of all 64 possible 3-nt sequences, and the full genetic code was determined (table 15.1).

The code is degenerate but specific

Some obvious features of the code jump out of table 15.1. First, 61 of the 64 possible codons are used to specify amino acids. Three codons, UAA, UGA, and UAG, are reserved for another function: they signal "stop" and are known as **stop codons.** The only other form of "punctuation" in the code is that AUG is used to signal "start" and is therefore the **start codon.** In this case the codon has a dual function because it also encodes the amino acid methionine (Met).

You can see that 61 codons are more than enough to encode 20 amino acids. That leaves lots of extra codons. One way to deal with this abundance would be to use only 20 of the 61 codons, but

TABLE 15.1 The Genetic Code

First Letter	Second Letter: U	Second Letter: C	Second Letter: A	Second Letter: G	Third Letter
U	UUU Phe Phenylalanine	UCU Ser Serine	UAU Tyr Tyrosine	UGU Cys Cysteine	U
	UUC Phe Phenylalanine	UCC Ser Serine	UAC Tyr Tyrosine	UGC Cys Cysteine	C
	UUA Leu Leucine	UCA Ser Serine	UAA "Stop"	UGA "Stop"	A
	UUG Leu Leucine	UCG Ser Serine	UAG "Stop"	UGG Trp Tryptophan	G
C	CUU Leu Leucine	CCU Pro Proline	CAU His Histidine	CGU Arg Arginine	U
	CUC Leu Leucine	CCC Pro Proline	CAC His Histidine	CGC Arg Arginine	C
	CUA Leu Leucine	CCA Pro Proline	CAA Gln Glutamine	CGA Arg Arginine	A
	CUG Leu Leucine	CCG Pro Proline	CAG Gln Glutamine	CGG Arg Arginine	G
A	AUU Ile Isoleucine	ACU Thr Threonine	AAU Asn Asparagine	AGU Ser Serine	U
	AUC Ile Isoleucine	ACC Thr Threonine	AAC Asn Asparagine	AGC Ser Serine	C
	AUA Ile Isoleucine	ACA Thr Threonine	AAA Lys Lysine	AGA Arg Arginine	A
	AUG Met Methionine; "Start"	ACG Thr Threonine	AAG Lys Lysine	AGG Arg Arginine	G
G	GUU Val Valine	GCU Ala Alanine	GAU Asp Aspartate	GGU Gly Glycine	U
	GUC Val Valine	GCC Ala Alanine	GAC Asp Aspartate	GGC Gly Glycine	C
	GUA Val Valine	GCA Ala Alanine	GAA Glu Glutamate	GGA Gly Glycine	A
	GUG Val Valine	GCG Ala Alanine	GAG Glu Glutamate	GGG Gly Glycine	G

A codon consists of three nucleotides read in the sequence shown. For example, ACU codes for threonine. The first letter, A, is in the First Letter column; the second letter, C, is in the Second Letter column; and the third letter, U, is in the Third Letter column. Each of the mRNA codons is recognized by a corresponding anticodon sequence on a tRNA molecule. Many amino acids are specified by more than one codon. For example, threonine is specified by four codons, which differ only in the third nucleotide (ACU, ACC, ACA, and ACG).

Figure 15.5 Transgenic pig. The piglet on the right is a conventional piglet. The piglet on the left was engineered to express a gene from jellyfish that encodes green fluorescent protein. The color of this piglet's nose is due to expression of this introduced gene. Such transgenic animals indicate the universal nature of the genetic code.

this is not what cells do. In reality, all 61 codons are used, making the code **degenerate,** which means that some amino acids are specified by more than one codon. The reverse, however, in which a single codon would specify more than one amino acid, is never found.

This degeneracy is not uniform. Some amino acids have only one codon, and some have up to six. In addition, the degenerate base usually occurs in position 3 of a codon, such that the first two positions are the same, and two or four of the possible nucleotides at position 3 encode the same amino acid. (The nature of protein synthesis on ribosomes explains how this codon usage works, and it is discussed in section 15.7.)

The code is practically universal, but not quite

The genetic code is the same in almost all organisms. The universality of the genetic code is among the strongest evidence that all living things share a common evolutionary heritage. Because the code is universal, genes can be transferred from one organism to another and can be successfully expressed in their new host (figure 15.5). This universality of gene expression is central to many of the advances of genetic engineering discussed in chapter 17.

In 1979, investigators began to determine the complete nucleotide sequences of the mitochondrial genomes in humans, cattle, and mice. It came as something of a shock when these investigators learned that the genetic code used by these mammalian mitochondria was not quite the same as the "universal code" that has become so familiar to biologists.

In the mitochondrial genomes, what should have been a stop codon, UGA, was instead read as the amino acid tryptophan; AUA was read as methionine rather than as isoleucine; and AGA and AGG were read as stop codons rather than as arginine. Furthermore, minor differences from the universal code have also been found in the genomes of chloroplasts and in ciliates (certain types of protists).

Thus, it appears that the genetic code is not quite universal. Some time ago, presumably after they began their endosymbiotic existence, mitochondria and chloroplasts began to read the code differently, particularly the portion associated with "stop" signals.

Inquiry question The genetic code is almost universal. Why do you think it is nearly universal?

Learning Outcomes Review 15.2

The genetic code was shown to be nucleotide base triplets with two forms of punctuation and no spaces: Three bases code for an amino acid, and the groups of three are read in order. Sixty-one codons specify amino acids, one of which also codes for "start," and three codons indicate "stop," for 64 total. Because some amino acids are specified by more than one codon, the code is termed degenerate. All codons encode only one amino acid, however.

- What would be the outcome if a codon specified more than one amino acid?

15.3 Prokaryotic Transcription

Learning Outcomes

1. Describe the transcription process in bacteria.
2. Differentiate among initiation, elongation, and termination of transcription.
3. Define the unique features of prokaryotic transcription.

We begin an examination of gene expression by describing the process of transcription in prokaryotes. The description of eukaryotic transcription in section 15.4 will concentrate on their differences from prokaryotes.

Figure 15.6 Bacterial RNA polymerase and transcription initiation. *a.* RNA polymerase has two forms: core polymerase and holoenzyme. *b.* The σ subunit of the holoenzyme recognizes promoter elements at −35 and −10 and binds to the DNA. The helix is opened at the −10 region, and transcription begins at the start site at +1.

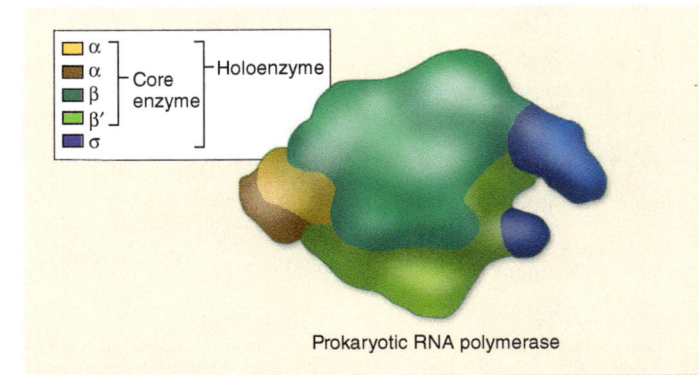

Prokaryotic RNA polymerase

a.

Prokaryotes have a single RNA polymerase

The single **RNA polymerase** of prokaryotes exists in two forms: the *core polymerase* and the *holoenzyme*. The core polymerase can synthesize RNA using a DNA template, but it cannot initiate synthesis accurately. The holoenzyme can accurately initiate synthesis.

The core polymerase is composed of four subunits: two identical α subunits, a β subunit, and a β′ subunit (figure 15.6a). The two α subunits help to hold the complex together and can bind to regulatory molecules. The active site of the enzyme is formed by the β and β′ subunits, which bind to the DNA template and the ribonucleotide triphosphate precursors.

The *holoenzyme*, which can properly initiate synthesis, is formed by the addition of a σ (sigma) subunit to the core polymerase (figure 15.6a). Its ability to recognize specific signals in DNA allows RNA polymerase to locate the beginning of genes, which is critical to its function. Note that initiation of mRNA synthesis does not require a primer, in contrast to DNA replication.

Initiation occurs at promoters

Accurate initiation of transcription requires two sites in DNA: one called a **promoter** that forms a recognition and binding site for the RNA polymerase, and the actual **start site.** The polymerase also needs a signal to end transcription, which we call a **terminator.** We then refer to the region from promoter to terminator as a **transcription unit.**

The action of the polymerase moving along the DNA can be thought of as analogous to water flowing in a stream. We can speak of sites on the DNA as being "upstream" or "downstream" of the start site. We can also use this comparison to form a simple system for numbering bases in DNA to refer to positions in the transcription unit. The first base transcribed is called **+1**, and this numbering continues downstream until the last base is transcribed. Any bases upstream of the start site receive negative numbers, starting at **−1**.

The promoter is a short sequence found upstream of the start site and is therefore not transcribed by the polymerase. Two 6-base sequences are common to bacterial promoters: One is located 35 nt upstream of the start site (−35), and the other is located 10 nt upstream of the start site (−10) (figure 15.6b). These two sites provide the promoter with asymmetry; they indicate not only the site of initiation, but also the direction of transcription.

The binding of RNA polymerase to the promoter is the first step in transcription. Promoter binding is controlled by the σ subunit of the RNA polymerase holoenzyme, which recognizes the −35 sequence in the promoter and positions the RNA polymerase at the correct start site, oriented to transcribe in the correct direction.

Once bound to the promoter, the RNA polymerase begins to unwind the DNA helix at the −10 site (figure 15.6b). The polymerase covers a region of about 75 bp but only unwinds about 12 to 14 bp.

> **Inquiry question** The prokaryotic promoter has two distinct elements that are not identical. How is this important to the initiation of transcription?

Elongation adds successive nucleotides

In prokaryotes, the transcription of the RNA chain usually starts with ATP or GTP. One of these forms the 5′ end of the chain, which grows in the 5′-to-3′ direction as ribonucleotides are added. As the RNA polymerase molecule leaves the promoter region, the σ factor is no longer required, although it may remain in association with the enzyme.

This process of leaving the promoter, called *clearance,* or *escape,* involves more than just synthesizing the first few nucleotides of the transcript and moving on, because the enzyme has made strong contacts to the DNA during initiation. It is necessary to break these contacts with the promoter region to be able to move progressively down the template. The enzyme goes through conformational changes during this clearance stage, and subsequently contacts less of the DNA than it does during the initial promoter binding.

The region containing the RNA polymerase, the DNA template, and the growing RNA transcript is called the

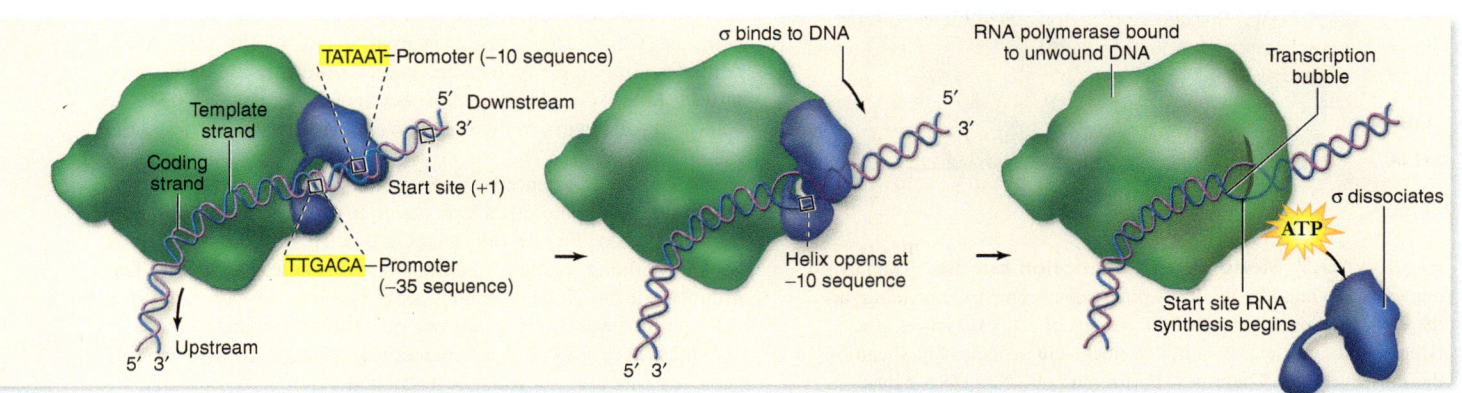

b.

transcription **bubble** because it contains a locally unwound "bubble" of DNA (figure 15.7). Within the bubble, the first 9 bases of the newly synthesized RNA strand temporarily form a helix with the template DNA strand. This stabilizes the positioning of the 3' end of the RNA so it can interact with an incoming ribonucleotide triphosphate. The enzyme itself covers about 50 bp of DNA around this transcription bubble.

The transcription bubble created by RNA polymerase moves down the bacterial DNA at a constant rate, about 50 nt/sec, with the growing RNA strand protruding from the bubble. After the transcription bubble passes, the now-transcribed DNA is rewound as it leaves the bubble.

Termination occurs at specific sites

The end of a bacterial transcription unit is marked by terminator sequences that signal "stop" to the polymerase. Reaching these sequences causes the formation of phosphodiester bonds to cease, the RNA–DNA hybrid within the transcription bubble to dissociate, the RNA polymerase to release the DNA, and the DNA within the transcription bubble to rewind.

The simplest terminators consist of a series of G–C base-pairs followed by a series of A–T base-pairs. The RNA transcript of this stop region can form a double-stranded structure in the GC region called a *hairpin*, which is followed by four or more uracil (U) ribonucleotides (figure 15.8). Formation of the

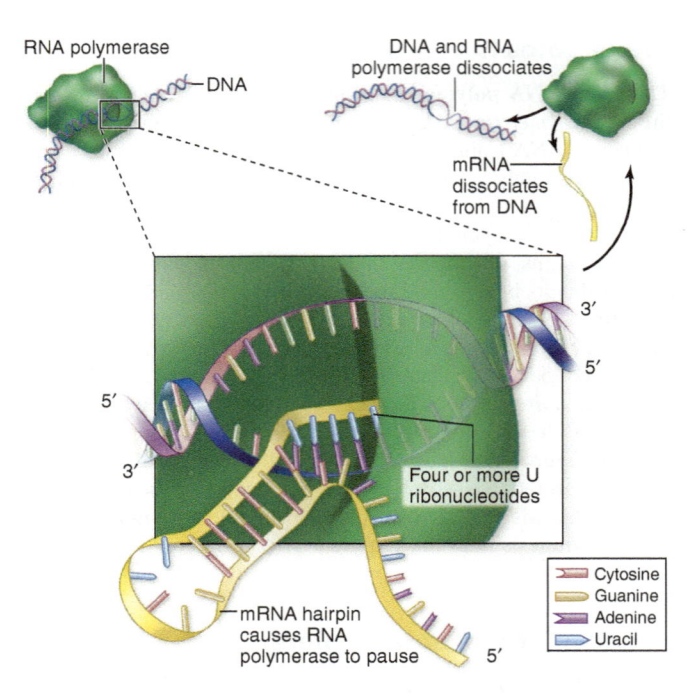

Figure 15.8 Bacterial transcription terminator. The self-complementary G–C region forms a double-stranded stem with a single-stranded loop called a hairpin. The stretch of U's forms a less stable RNA–DNA hybrid that falls off the enzyme.

hairpin causes the RNA polymerase to pause, placing it directly over the run of four uracils. The pairing of U with the DNA's A is the weakest of the four hybrid base-pairs, and it is not strong enough to hold the hybrid strands when the polymerase pauses. Instead, the RNA strand dissociates from the DNA within the transcription bubble, and transcription stops. A variety of protein factors also act at these terminators to aid in terminating transcription.

Prokaryotic transcription is coupled to translation

In prokaryotes, the mRNA produced by transcription begins to be translated before transcription is finished—that is, they are *coupled* (figure 15.9). As soon as a 5' end of the mRNA becomes available, ribosomes are loaded onto this to begin translation. (This coupling cannot occur in eukaryotes because transcription occurs in the nucleus, and translation occurs in the cytoplasm.)

Another difference between prokaryotic and eukaryotic gene expression is that the mRNA produced in prokaryotes may contain multiple genes. Prokaryotic genes are often organized such that genes encoding related functions are clustered together. This grouping of functionally related genes is referred to as an **operon.** An operon is a single transcription unit that encodes multiple enzymes necessary for a biochemical pathway. By clustering genes by function, they can be regulated together, a topic that we return to in chapter 16.

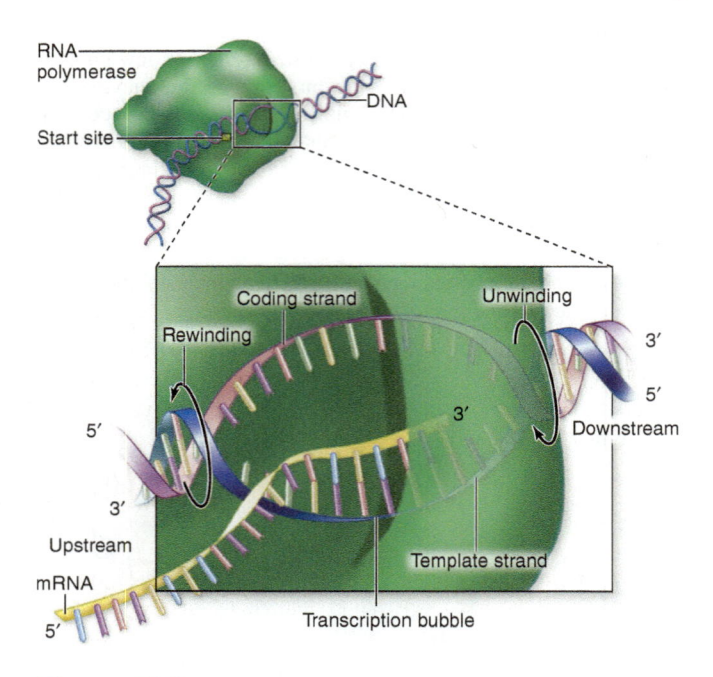

Figure 15.7 Model of a transcription bubble. The DNA duplex is unwound by the RNA polymerase complex, rewinding at the end of the bubble. One of the strands of DNA functions as a template, and nucleotide building blocks are added to the 3' end of the growing RNA. There is a short region of RNA–DNA hybrid within the bubble.

The basic mechanism of transcription by RNA polymerase is similar in all systems. In fact, the RNA polymerases of bacteria, archaea, and eukaryotes all appear to be descended from the same ancestral enzyme. However, the details of RNA production in eukaryotes differ enough to merit consideration. An obvious difference is immediately clear: Bacteria have a single RNA polymerase, while eukaryotes have three different RNA polymerases, which differ in structure and function. The enzyme RNA polymerase I transcribes rRNA, RNA polymerase II transcribes mRNA and some small nuclear RNAs, and RNA polymerase III transcribes tRNA and some other small RNAs. Together, these three enzymes accomplish all transcription in eukaryotic cells.

Each polymerase has its own promoter

The three eukaryotic RNA polymerases require different control elements in the DNA to allow each polymerase to recognize where to initiate transcription. Each polymerase recognizes a different promoter structure.

RNA polymerase I promoters

RNA polymerase I promoters at first puzzled biologists, because comparisons of rRNA genes between species showed no similarities outside the coding region. The current view is that these promoters are specific for each species, which explains the lack of conserved sequences in cross-species comparisons.

RNA polymerase II promoters

Most of the first eukaryotic genes cloned contained a sequence, TATA, upstream of the start site. The similarity of the TATA sequence to the prokaryotic −10 element led to the idea that this was the promoter. This has proved too simple and has been replaced by the idea of a "core promoter" that can be composed of a number of distinct elements, including the TATA box.

RNA polymerase III promoters

Promoters for RNA polymerase III also were a source of surprise for biologists examining the control of eukaryotic gene expression. A common technique for analyzing regulatory regions was to make successive deletions from the 5′ end of genes until enough was deleted to abolish specific transcription. In the case of tRNA genes, 5′ deletions had no effect on expression! The promoters were found within the actual gene itself. This has not proved to be the case for all polymerase III genes, but appears to be for most.

The core promoter binds factors that recruit RNA Pol II

The initiation at RNA polymerase II promoters is analogous to prokaryotic initiation, but instead of a single factor allowing promoter recognition, eukaryotes use a host of transcription factors. These factors act cooperatively to recruit RNA polymerase II to a promoter. The transcription factors interact with RNA polymerase II to form an initiation complex at the promoter (figure 15.10). We explore this complex in detail in chapter 16 when we describe the control of gene expression.

When the transcript reaches about 20 nucleotides, it is modified by the addition of GTP to the 5′ PO_4^- group, forming what is

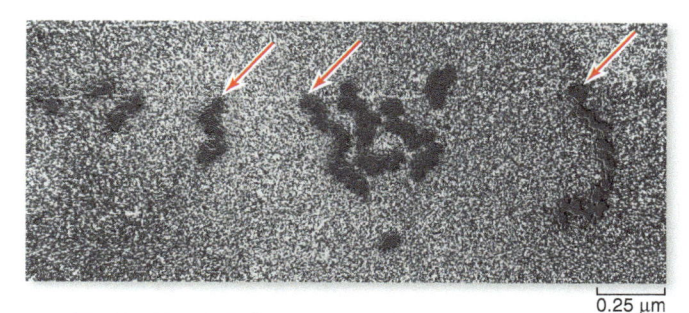

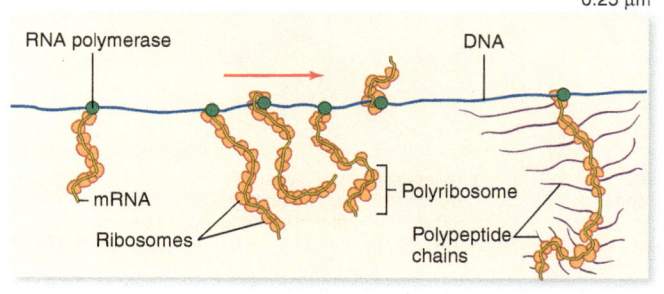

Figure 15.9 Transcription and translation are coupled in prokaryotes. In this micrograph of gene expression in *E. coli*, translation is occurring during transcription. The arrows point to RNA polymerase enzymes, and ribosomes are attached to the mRNAs extending from the polymerase. Polypeptides being synthesized by ribosomes, which are not visible in the micrograph, have been added to the last mRNA in the drawing.

Learning Outcomes Review 15.3

Transcription in bacteria is accomplished by RNA polymerase, which has two forms: a core polymerase and a holoenzyme. Initiation is accomplished by the holoenzyme form, which can accurately recognize promoter sequences. Elongation consists of RNA synthesis by the core enzyme, which adds RNA nucleotides in sequence until it reaches a terminator where synthesis stops, then the transcript is released. In prokaryotes, translation of the RNA transcript begins before transcription is finished, making the processes coupled.

- *Yeast are unicellular organisms like bacteria; would you expect them to have the same transcription/translation coupling?*

15.4 Eukaryotic Transcription

Learning Outcomes

1. *Differentiate promoters for the three eukaryotic polymerases.*
2. *Describe the processing of eukaryotic transcripts.*
3. *Explain the differences between bacterial and eukaryotic elongation of transcription.*

1. A transcription factor recognizes and binds to the TATA box sequence, which is part of the core promoter.

2. Other transcription factors are recruited, and the initiation complex begins to build.

3. Ultimately, RNA polymerase II associates with the transcription factors and the DNA, forming the initiation complex, and transcription begins.

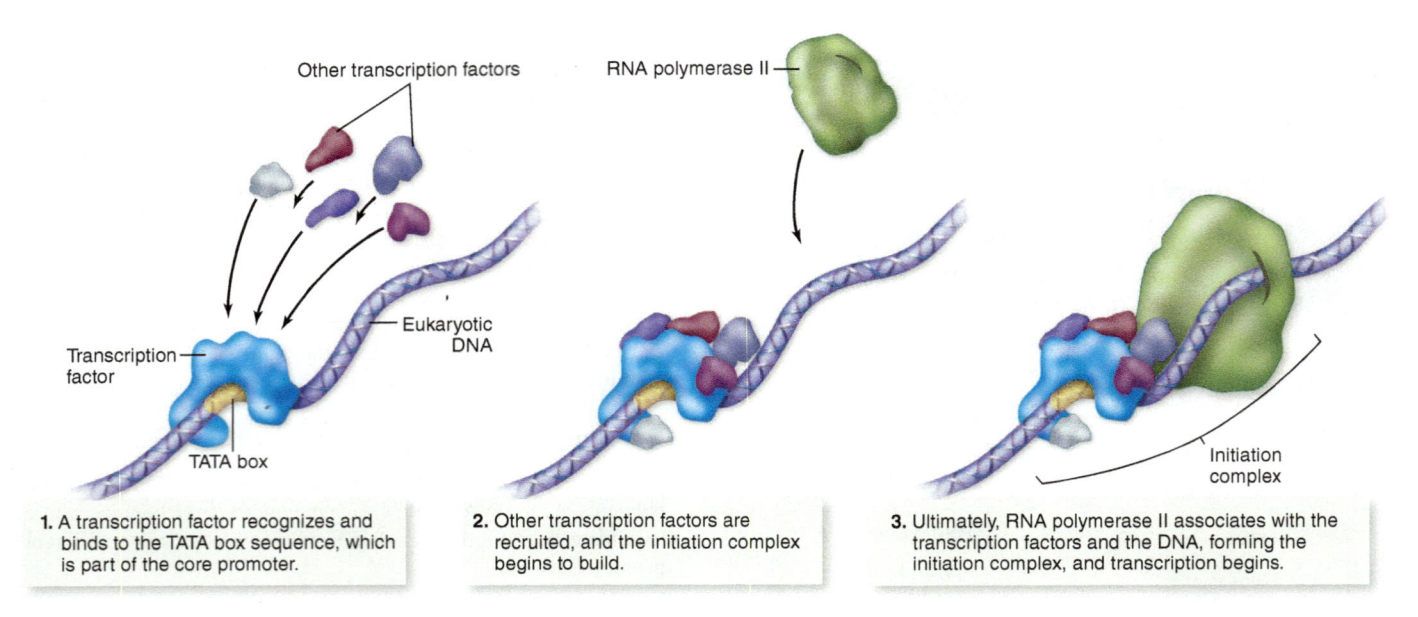

Figure 15.10 Eukaryotic initiation complex. Unlike transcription in prokaryotic cells, in which the RNA polymerase recognizes and binds to the promoter, eukaryotic transcription requires the binding of transcription factors to the promoter before RNA polymerase II binds to the DNA. The association of transcription factors and RNA polymerase II at the promoter is called the initiation complex.

known as the **5′ cap** (figure 15.11). This cap is joined to the transcript by its 5′ end; the only 5′-to-5′ bond found in nucleic acids. The G in the GTP is also modified by the addition of a methyl group, so it is often called a **methyl-G cap.** This structure is important for translation, RNA stability, and further processing.

Another event that can occur during early elongation, is that the Pol may pause. This usually occurs within 50 bp of initiation, and now appears to be a common phenomenon. Initially described for heat shock genes, where it allows the rapid induction of transcription upon heat shock, global screens of promoter occupancy indicate that a majority of genes in multicellular organisms may harbor a paused polymerase. These data are difficult to interpret as they cannot distinguish between abortive short transcripts, and a paused Pol that can restart. For the latter, a complex interplay of promoter elements, elongation factors, nucleosome occupancy, and even the capping enzyme all participate. The integration of the control of transcript initiation, and of release from promoter proximal pausing is not clear at this point.

The transcription elongation complex can recruit other factors

The transcription elongation complex contains factors that arrive after initiation. This appears to be coordinated by a structural feature of the largest subunit called the carboxy terminal domain (CTD). The CTD consists of 7 amino acids repeated many times (52 repeats in humans). The CTD contains serines that can be phosphorylated, and acts as a platform to recruit factors involved in elongation and RNA-modifying enzymes. For example, the capping enzyme just described interacts with the CTD soon after initiation.

The elongation factors that interact with the CTD include ones that are involved in the release of pausing, and in progressing through nucleosomes. Early in vitro studies indicated that RNA Pol cannot transcribe DNA that is assembled into chromatin. Clearly this is not the case in vivo, and this is due to elongation factors that interact with the Pol, and with nucleosomes to result in removal of nucleosomes and repositioning behind the polymerase. These factors include FACT (facilitates chromatin transcription), factors that modify histones, and chromatin-remodeling complexes (discussed in chapter 16).

Polyadenylation is involved in terminating mRNA transcripts

A major difference between prokaryotes and eukaryotes is that in eukaryotes, the end of the transcript is not the end of the mRNA.

Figure 15.11 Posttranscriptional modifications to 5′ and 3′ ends. Eukaryotic mRNA molecules are modified in the nucleus with the addition of a methylated GTP to the 5′ end of the transcript, called the 5′ cap, and a long chain of adenine residues to the 3′ end of the transcript, called the 3′ poly-A tail.

The eukaryotic transcript is cleaved downstream of a specific site (AAUAAA) while the polymerase is elongating. A series of 100 to 200 adenine (A) residues, called the **3′ poly-A tail,** is added after cleavage by the enzyme poly-A polymerase (figure 15.11). Poly-A polymerase also associates with the elongating polymerase via the CTD. After the cleavage reaction, an exonuclease progressively removes the downstream RNA, which causes the transcription complex to dissociate from the template. The poly-A tail appears to play a role in the stability of mRNAs by protecting them from degradation (see chapter 16). Termination of transcripts that are not polyadenylated uses an alternate mechanism requiring other factors.

Primary transcripts are modified to produce mRNAs

Eukaryotic genes may contain noncoding sequences that have to be removed to produce the final mRNA. This process, called pre-mRNA splicing, is accomplished by an organelle called the **spliceosome.** This complex topic is discussed in section 15.5. It is clear from the preceding discussion that the mRNA in eukaryotes is not the same as the primary transcript. With a modified 5′ end, and a 3′ end produced by another enzyme, and the removal of introns, the mRNA is a highly modified form of the primary transcript.

On the surface, this appears to be almost a different process from prokaryotes where transcription and translation are coupled. There is an interesting parallel in that the modifications of the transcript all occur during transcription, as translation does in prokaryotes.

> **Learning Outcomes Review 15.4**
>
> Eukaryotes have three RNA polymerases called polymerase I, II, and III. Each synthesizes a different RNA and recognizes its own promoter. The RNA polymerase I promoter is species-specific. The polymerase II core promoter consists of multiple elements. The polymerase III promoter is internal to the gene. Polymerase II is responsible for mRNA synthesis. Elongation by Pol II requires different factors, and modifying enzymes act during transcription to cap, polyadenylate and splice the transcript.
>
> ■ Does the complexity of the eukaryotic genome require three polymerases?

15.5 Eukaryotic pre-mRNA Splicing

> **Learning Outcomes**
> 1. Explain the relationship between genes and proteins in prokaryotes and eukaryotes.
> 2. Describe the splicing reaction for pre-mRNA.
> 3. Illustrate how splicing changes the nature of genes.

The first genes isolated were prokaryotic genes found in *E. coli* and its viruses. A clear picture of the nature and some of the control of gene expression emerged from these systems before any eukaryotic genes were isolated. It was assumed that although details would differ, the outline of gene expression in eukaryotes would be similar. The world of biology was in for a shock with the isolation of the first genes from eukaryotic organisms.

Eukaryotic genes may contain interruptions

Many eukaryotic genes appeared to contain sequences that were not represented in the mRNA. It is hard to exaggerate how unexpected this finding was. A basic tenet of molecular biology based on *E. coli* was that a gene was *colinear* with its protein product— that is, the sequence of bases in the gene corresponds to the sequence of bases in the mRNA, which in turn corresponds to the sequence of amino acids in the protein.

In the case of eukaryotes, genes can be interrupted by sequences that are not represented in the mRNA and the protein. The term "split genes" was used at the time, but the nomenclature that has stuck describes the unexpected nature of these sequences. We call the noncoding DNA that interrupts the sequence of the gene "intervening sequences," or **introns,** and we call the coding sequences **exons** because they are expressed (figure 15.12).

The spliceosome is the splicing organelle

It is still true that the mature eukaryotic mRNA is colinear with its protein product, but a gene that contains introns is not. Imagine

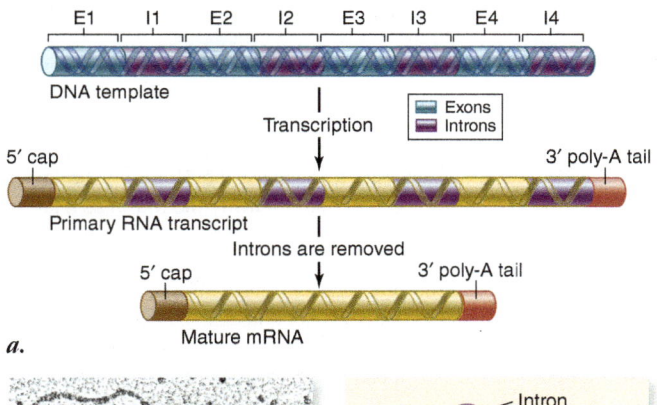

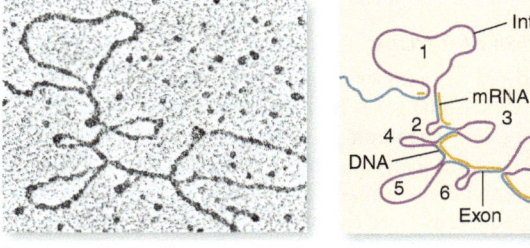

Figure 15.12 Eukaryotic genes contain introns and exons. *a.* Eukaryotic genes contain sequences that form the coding sequence called exons and intervening sequences called introns. *b.* An electron micrograph showing hybrids formed with the mRNA and the DNA of the ovalbumin gene, which has seven introns. Introns within the DNA sequence have no corresponding sequence in the mRNA and thus appear as seven loops. *c.* A schematic drawing of the micrograph.

> **Data analysis** Illustrate how the gene in figure 15.12 could encode multiple transcripts.

looking at an interstate highway from a satellite. Scattered randomly along the thread of concrete would be cars, some moving in clusters, others individually; most of the road would be bare. That is what a eukaryotic gene is like—scattered exons embedded within much longer sequences of introns.

In humans, only 1 to 1.5% of the genome is devoted to the exons that encode proteins; 24% is devoted to the noncoding introns within which these exons are embedded.

The splicing reaction

The obvious question is—How do eukaryotic cells deal with the noncoding introns? The answer is that the primary transcript is cut and put back together to produce the mature mRNA. The latter process is referred to as **pre-mRNA splicing,** and it occurs in the nucleus prior to the export of the mRNA to the cytoplasm.

The intron–exon junctions are recognized by **small nuclear ribonucleoprotein particles,** called **snRNPs** (pronounced "snurps"). The snRNPs are complexes composed of snRNA and protein. These snRNPs then cluster together with other associated proteins to form a larger complex called the **spliceosome,** which is responsible for the splicing, or removal, of the introns.

For splicing to occur accurately, the spliceosome must be able to recognize intron–exon junctions. Introns all begin with the same 2-base sequence and end with another 2-base sequence that tags them for removal. In addition, within the intron there is a conserved A nucleotide, called the *branch point,* which is important for the splicing reaction (figure 15.13).

The splicing process begins with cleavage of the 5′ end of the intron. This 5′ end becomes attached to the 2′ OH of the branch point A, forming a branched structure called a *lariat* due to its resemblance to a cowboy's lariat in a rope (figure 15.13). The 3′ end of the first exon is then used to displace the 3′ end of the intron, joining the two exons together and releasing the intron as a lariat.

The splicing process takes place during transcription. The transcript exits the enzyme near the CTD, allowing recruitment of splicing factors and the spliceosomal RNPs. There is evidence that the rate of transcription, which is not constant, can affect splice site selection. Pausing of the enzyme and the formation of secondary structure in the transcript can make splice sites inaccessible.

Distribution of introns

No rules govern the number of introns per gene or the sizes of introns and exons. Some genes have no introns; others may have 50. The sizes of exons range from a few nucleotides to 7500 nt, and the sizes of introns are equally variable. The presence of introns partly explains why so little of a eukaryotic genome is actually composed of "coding sequences" (see chapter 18 for results from the Human Genome Project).

One explanation for the existence of introns suggests that exons represent functional domains of proteins, and that the intron–exon arrangements found in genes represent the shuffling of these functional units over long periods of evolutionary time. This hypothesis, called *exon shuffling,* was proposed soon after the discovery of introns and has been the subject of much debate over the years.

The recent flood of genomic data has shed light on this issue by allowing statistical analysis of the placement of introns and on intron–exon structure. This analysis has provided support for the exon shuffling hypothesis for many genes; however, it is also clearly not universal, because all proteins do not show this kind of

1. snRNA forms base-pairs with 5′ end of intron, and at branch site.

2. snRNPs associate with other factors to form spliceosome.

3. 5′ end of intron is removed and forms bond at branch site, forming a lariat. The 3′ end of the intron is then cut.

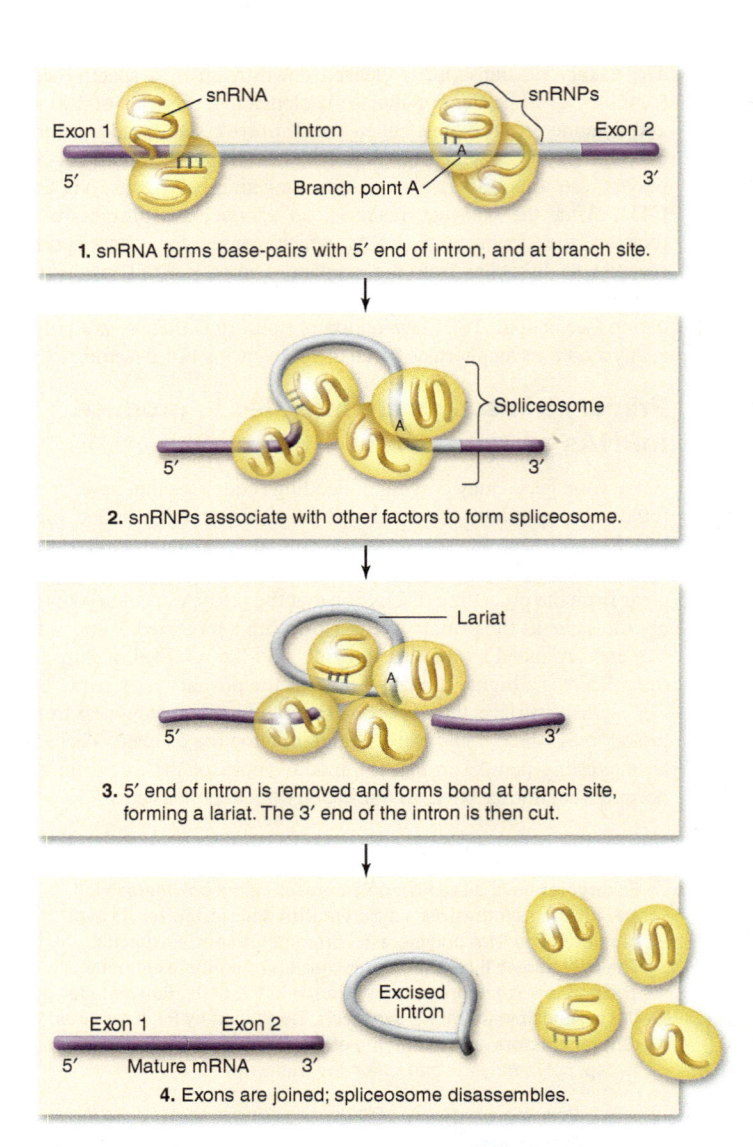

4. Exons are joined; spliceosome disassembles.

Figure 15.13 Pre-mRNA splicing by the spliceosome. Particles called snRNPs contain snRNA that interacts with the 5′ end of an intron and with a branch site internal to the intron. Several snRNPs come together with other proteins to form the spliceosome. As the intron forms a loop, the 5′ end is cut and linked to a site near the 3′ end of the intron. The intron forms a lariat that is excised, and the exons are spliced together. The spliceosome then disassembles and releases the spliced mRNA.

pattern. It is possible that introns do not have a single origin, and therefore cannot be explained by a single hypothesis.

Splicing can produce multiple transcripts from the same gene

One consequence of the splicing process is greater complexity in gene expression in eukaryotes. A single primary transcript can be spliced into different mRNAs by the inclusion of different sets of exons, a process called **alternative splicing.**

Evidence indicates that the normal pattern of splicing is important to an organism's function. Up to half of known human genetic disorders may be due to altered splicing. Mutations in the signals for splicing can introduce new splice sites or can abolish normal patterns of splicing. (In chapter 16 we consider how alternative splicing can be used to regulate gene expression.)

The availability of the human genome sequence, and high-throughput systems to sequence transcripts have led to a flood of data comparing transcripts from different tissues to the genome. One estimate, based on analysis of human and mouse transcripts, is that 95% of mammalian genes produce alternative transcripts, with an average of 4 transcripts per gene. Other estimates of the average number of transcripts per gene have produced both higher and lower numbers. Another recent study found similar overall levels of alternative splicing with as many as 25 different transcripts from individual genes. However, this same analysis indicated that most genes had a single predominant transcript. This leaves us with an incomplete picture, but further work should clarify this.

It is important to note that these analyses are global surveys using next-generation sequencing methods to analyze RNA populations from different tissues. The possible functions of the protein products of these splice variants have been investigated for only a small fraction of the potentially spliced genes. These data, however, are part of the explanation for how approximately 20,000 human genes can encode the more than 100,000 different proteins estimated to exist in human cells.

Learning Outcomes Review 15.5

In prokaryotes, genes appear to be colinear with their protein products. Eukaryotic genes, by contrast, contain exon regions, which are expressed, and intron sequences, which interrupt the exons. The introns are removed by the spliceosome in a process that leaves the exons joined together. Alternative splicing can generate different mRNAs, and thus different proteins, from the same gene. This appears to be widespread in multicellular organisms.

■ What advantages would alternative splicing confer on an organism?

15.6 The Structure of tRNA and Ribosomes

Learning Outcomes
1. Explain why the tRNA charging reaction is critical to translation.
2. Identify the tRNA-binding sites in the ribosome.

The ribosome is the key organelle in translation, but it also requires the participation of mRNA, tRNA, and a host of other factors. Critical to this process is the interaction of the ribosomes with tRNA and mRNA. To understand this, we first examine the structure of the tRNA adapter molecule and the ribosome itself.

Aminoacyl-tRNA synthetases attach amino acids to tRNA

Each amino acid must be attached to a tRNA with the correct anticodon for protein synthesis to proceed. This covalent attachment is accomplished by the action of activating enzymes called **aminoacyl-tRNA synthetases.** One of these enzymes is present for each of the 20 common amino acids.

tRNA structure

Transfer RNA is a bifunctional molecule that must be able to interact with mRNA and with amino acids. The structure of tRNAs is highly conserved in all living systems, and it can be formed into a cloverleaf type of structure based on intramolecular base-pairing that produces double-stranded regions. This primary structure is then folded in space to form an L-shaped molecule that has two functional ends: the **acceptor stem** and the **anticodon loop** (figure 15.14).

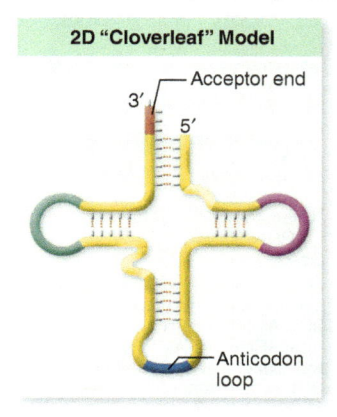

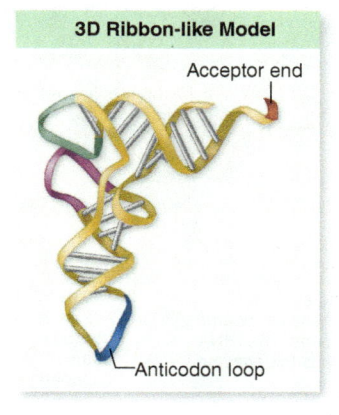

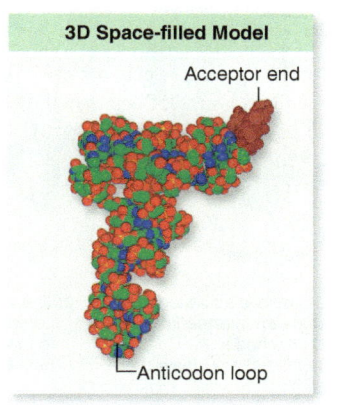

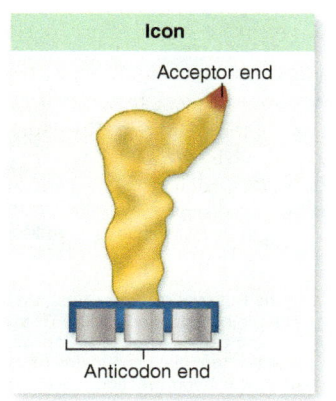

Figure 15.14 The structure of tRNA. Base-pairing within the molecule creates three stem and loop structures in a characteristic cloverleaf shape. The loop at the bottom of the cloverleaf contains the anticodon sequence, which can base-pair with codons in the mRNA. Amino acids are attached to the free, single-stranded —OH end of the acceptor stem. In its final 3-D structure, the loops of tRNA are folded into the final L-shaped structure.

The acceptor stem is the 3′ end of the molecule, which always ends in 5′ CCA 3′. The amino acid is attached to this end of the molecule. The anticodon loop is the bottom loop of the cloverleaf, and it can base-pair with codons in mRNA.

The charging reaction

The aminoacyl-tRNA synthetases must be able to recognize specific tRNA molecules as well as their corresponding amino acids. Although 61 codons code for amino acids, there are actually not 61 tRNAs in cells, although the number varies from species to species. Therefore, some aminoacyl-tRNA synthetases must be able to recognize more than one tRNA—but each recognizes only a single amino acid.

The reaction catalyzed by the enzymes is called the tRNA **charging reaction,** and the product is an amino acid joined to a tRNA, now called a *charged tRNA*. An ATP molecule provides energy for this endergonic reaction. The charged tRNA produced by the reaction is an activated intermediate that can undergo the peptide bond-forming reaction without an additional input of energy.

The charging reaction joins the acceptor stem to the carboxyl terminus of an amino acid (figure 15.15). Keeping this directionality in mind is critical to understanding the function of the ribosome, because each peptide bond will be formed between the amino group of one amino acid and the carboxyl group of another amino acid.

The correct attachment of amino acids to tRNAs is important because the ribosome does not verify this attachment. Ribosomes can only ensure that the codon–anticodon pairing is correct. In an elegant experiment, cysteine was converted chemically to alanine after the charging reaction, when the amino acid was already attached to tRNA. When this charged tRNA was used in an in vitro protein synthesis system, alanine was incorporated in the place of cysteine, showing that the ribosome cannot "proofread" the amino acids attached to tRNA.

In a very real sense, therefore, the charging reaction is the actual translation step; amino acids are incorporated into a peptide based solely on the tRNA anticodon and its interaction with the mRNA.

The ribosome has multiple tRNA-binding sites

The synthesis of any biopolymer can be broken down into initiation, elongation, and termination—you have seen this division for DNA replication as well as for transcription. In the case of translation, or protein synthesis, all three of these steps take place on the ribosome, a large macromolecular assembly consisting of rRNA and proteins. Details of the process by which the two ribosome subunits are assembled during initiation are described shortly.

For the ribosome to function it must be able to bind to at least two charged tRNAs at once so that a peptide bond can be formed between their amino acids, as described in the overview in section 15.1. The bacterial ribosome contains three binding sites, summarized in figure 15.16:

- The **P site** (peptidyl) binds to the tRNA attached to the growing peptide chain.
- The **A site** (aminoacyl) binds to the tRNA carrying the next amino acid to be added.
- The **E site** (exit) binds the tRNA that carried the previous amino acid added.

Transfer RNAs move through these sites successively during the process of elongation. Relative to the mRNA, the sites are arranged 5′ to 3′ in the order E, P, and A. The incoming charged

1. In the first step of the reaction, the amino acid is activated. The amino acid reacts with ATP to produce an intermediate with the carboxyl end of the amino acid attached to AMP. The two terminal phosphates (pyrophosphates) are cleaved from ATP in this reaction.

2. The amino acid-AMP complex remains bound to the enzyme. The tRNA next binds to the enzyme.

3. The second step of the reaction transfers the amino acid from AMP to the tRNA, producing a charged tRNA and AMP. The charged tRNA consists of a specific amino acid attached to the 3′ acceptor stem of its RNA.

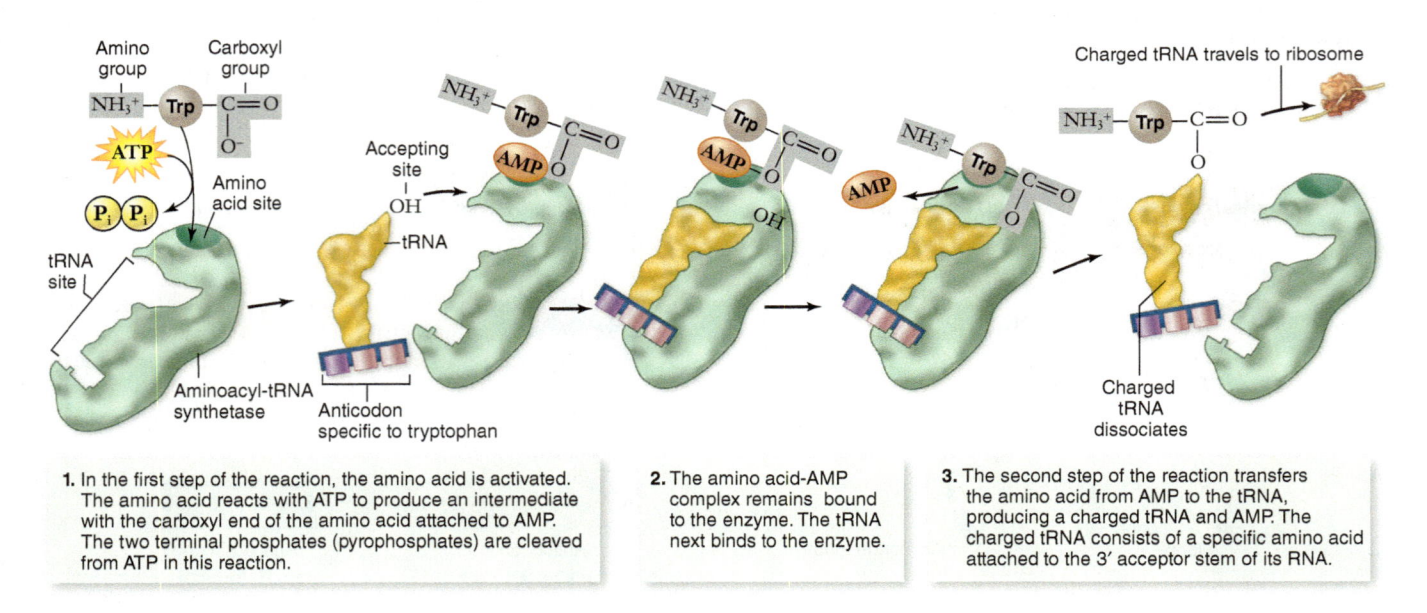

Figure 15.15 tRNA charging reaction. There are 20 different aminoacyl-tRNA synthetase enzymes each specific for one amino acid, such as tryptophan (Trp). The enzyme must also recognize and bind to the tRNA molecules with anticodons specifying that amino acid, ACC for tryptophan. The reaction uses ATP and produces an activated intermediate that will not require further energy for peptide bond formation.

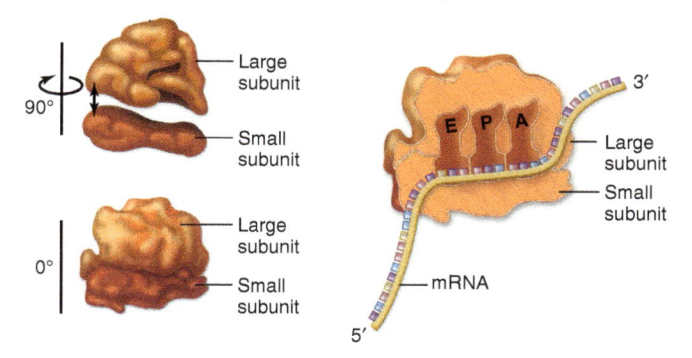

Figure 15.16 Ribosomes have two subunits. Ribosome subunits come together and apart as part of a ribosome cycle. The smaller subunit fits into a depression on the surface of the larger one. Ribosomes have three tRNA-binding sites: aminoacyl site (A), peptidyl site (P), and exit site (E).

tRNAs enter the ribosome at the A site, transit through the P site, and then leave via the E site.

The ribosome has both decoding and enzymatic functions

The two functions of the ribosome involve decoding the transcribed message and forming peptide bonds. The decoding function resides primarily in the small subunit of the ribosome. The formation of peptide bonds requires the enzyme **peptidyl transferase,** which resides in the large subunit.

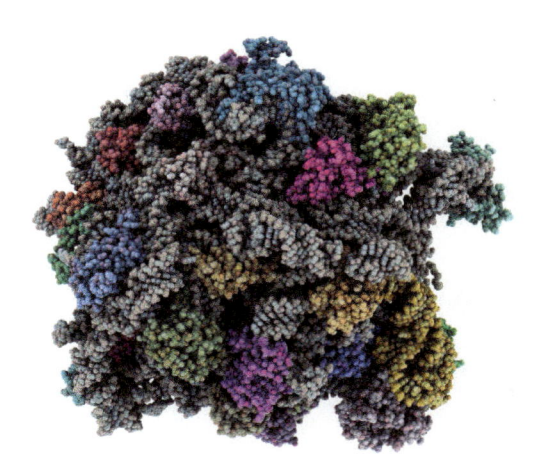

Figure 15.17 3-D structure of eukaryotic ribosome. The complete atomic structure of the yeast large ribosomal subunit is shown. The ribosomal RNA is beige, blue and pale green, all other colors are different ribosomal proteins. The faces of each ribosomal subunit are lined with rRNA such that their interaction with tRNAs, amino acids, and mRNA all involve rRNA. Proteins are absent from the active site but abundant everywhere on the surface. The proteins stabilize the structure by interacting with adjacent RNA strands.

Our view of the ribosome has changed dramatically over time. Initially, molecular biologists assumed that the proteins in the ribosome carried out its function and that the rRNA was a structural scaffold necessary to hold the proteins in the correct position. Now this view has mostly been reversed; the ribosome is seen instead as rRNAs that are held in place by proteins. The faces of the two subunits that interact with each other are lined with rRNA, and the parts of both subunits that interact with mRNA, tRNA, and amino acids are also primarily rRNA (figure 15.17). It is now thought that the peptidyl transferase activity resides in an rRNA in the large subunit.

Learning Outcomes Review 15.6

Transfer RNA has two functional regions, one that bonds with an amino acid, and the other that can base-pair with mRNA. The tRNA charging reaction joins the carboxyl end of an amino acid to the 3′ acceptor stem of its tRNA; without charged tRNAs, translation cannot take place. This reaction is catalyzed by 20 different aminoacyl-tRNA synthetases, one for each amino acid. The ribosome has three different binding sites for tRNA, one for the tRNA adding to the growing peptide chain (P site), one for the next charged tRNA (A site), and one for the previous tRNA, which is now without an amino acid (E site). The ribosome can be thought of as having both a decoding function and an enzymatic function.

- *What would be the effect on translation of a mutant tRNA that has an anticodon complementary to a stop codon?*

15.7 The Process of Translation

Learning Outcomes
1. Describe the process of translation initiation.
2. Explain the elongation cycle.
3. Compare translation on the RER and in the cytoplasm.

The process of translation is one of the most complex and energy-expensive tasks that cells perform. An overview of the process is perhaps deceptively simple: The mRNA is threaded through the ribosome, while tRNAs carrying amino acids bind to the ribosome, where they interact with mRNA by base-pairing with the mRNA's codons. The ribosome and tRNAs position the amino acids such that peptide bonds can be formed between each new amino acid and the growing polypeptide.

Initiation requires accessory factors

As mentioned in section 15.2, the start codon is AUG, which also encodes the amino acid methionine. The ribosome usually uses the first AUG it encounters in an mRNA strand to signal the start of translation.

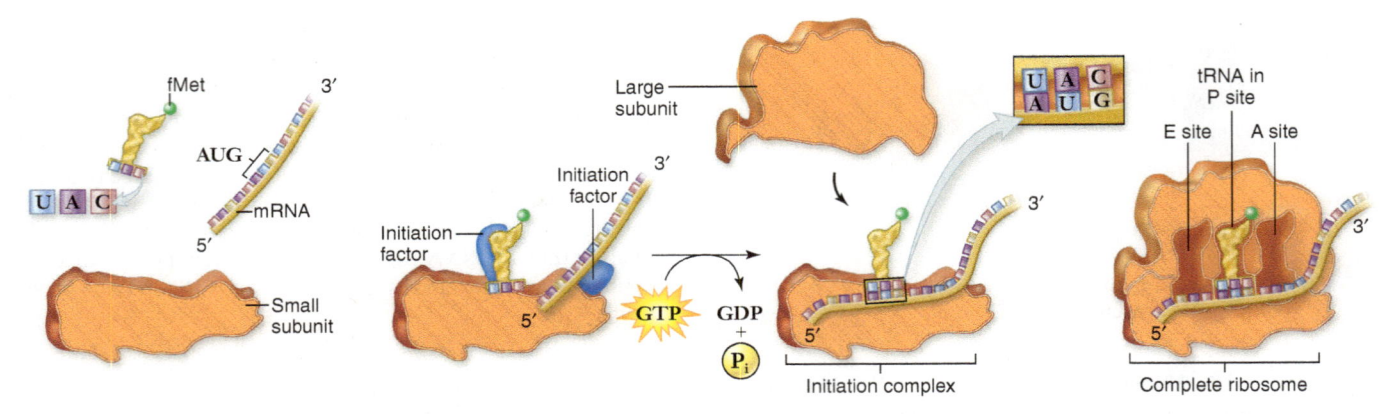

Figure 15.18 Initiation of translation. In prokaryotes, initiation factors play key roles in positioning the small ribosomal subunit, the initiator tRNA^{fMet}, and the mRNA. When the tRNA^{fMet} is positioned over the first AUG codon of the mRNA, the large ribosomal subunit binds, forming the E, P, and A sites where successive tRNA molecules bind to the ribosomes, and polypeptide synthesis begins. Ribosomal subunits are shown as a cutaway sectioned through the middle.

Prokaryotic initiation

In prokaryotes, the **initiation complex** includes a special **initiator tRNA** molecule charged with a chemically modified methionine, *N-formylmethionine*. The initiator tRNA is shown as tRNA^{fMet}. The initiation complex also includes the small ribosomal subunit and the mRNA strand (figure 15.18). The small subunit is positioned correctly on the mRNA due to a conserved sequence in the 5' end of the mRNA called the **ribosome-binding sequence (RBS)** that is complementary to the 3' end of a small subunit rRNA.

A number of initiation factors mediate this interaction of the ribosome, mRNA, and tRNA^{fMet} to form the initiation complex. These factors are involved in initiation only and are not part of the ribosome.

Once the complex of mRNA, initiator tRNA, and small ribosomal subunit is formed, the large subunit is added, and translation

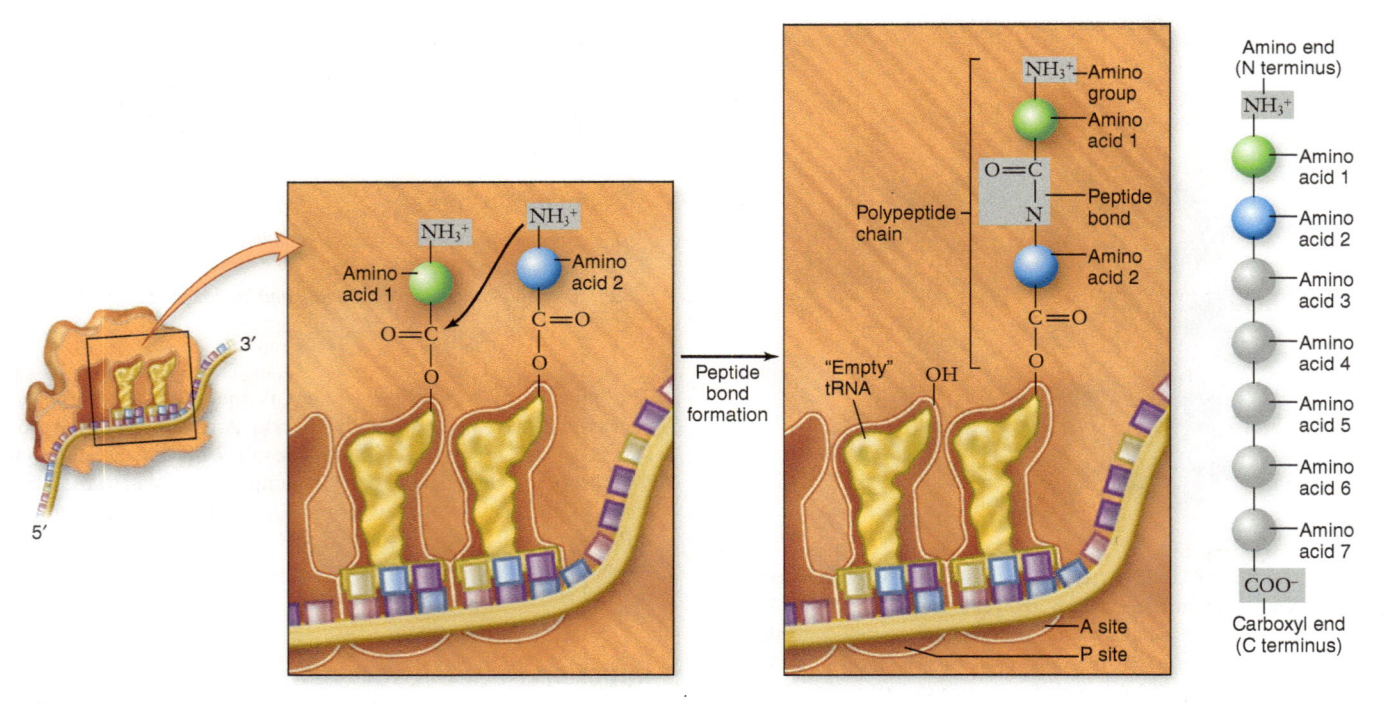

Figure 15.19 Peptide bond formation. Peptide bonds are formed between a "new" charged tRNA in the A site and the growing chain attached to the tRNA in the P site. The bond forms between the amino group of the new amino acid and the carboxyl group of the growing chain. This breaks the bond between the growing chain and its tRNA, transferring it to the A site as the new amino acid remains attached to its tRNA.

can begin. With the formation of the complete ribosome, the initiator tRNA is bound to the P site with the A site empty.

Eukaryotic initiation

Initiation in eukaryotes is similar, although it differs in two important ways. First, in eukaryotes, the initiating amino acid is methionine rather than *N*-formylmethionine. Second, the initiation complex is far more complicated than in prokaryotes, containing nine or more protein factors, many consisting of several subunits. Eukaryotic mRNAs also lack an RBS. The small subunit binds to the mRNA initially by binding to the 5′ cap of the mRNA.

Elongation adds successive amino acids

When the entire ribosome is assembled around the initiator tRNA and mRNA, the second charged tRNA can be brought to the ribosome and bind to the empty A site. This requires an **elongation factor** called **EF-Tu,** which binds to the charged tRNA and to GTP.

A peptide bond can then form between the amino acid of the initiator tRNA and the newly arrived charged tRNA in the A site. The geometry of this bond relative to the two charged tRNAs is critical to understanding the process. Remember that an amino acid is attached to a tRNA by its carboxyl terminus. The peptide bond is formed between the amino end of the incoming amino acid (in the A site) and the carboxyl end of the growing chain (in the P site) (figure 15.19).

The addition of successive amino acids is a series of events that occur in a cyclic fashion. Figure 15.20 shows the details of the elongation cycle:

1. **Matching tRNA anticodon with mRNA codon.** Each new charged tRNA comes to the ribosome bound to EF-Tu and GTP. The charged tRNA binds to the A site if its anticodon is complementary to the mRNA codon in the A site.

    After binding, GTP is hydrolyzed, and EF-Tu–GDP dissociates from the ribosome where it is recycled by another factor. This two-step binding and hydrolysis of GTP is thought to increase the accuracy of translation.

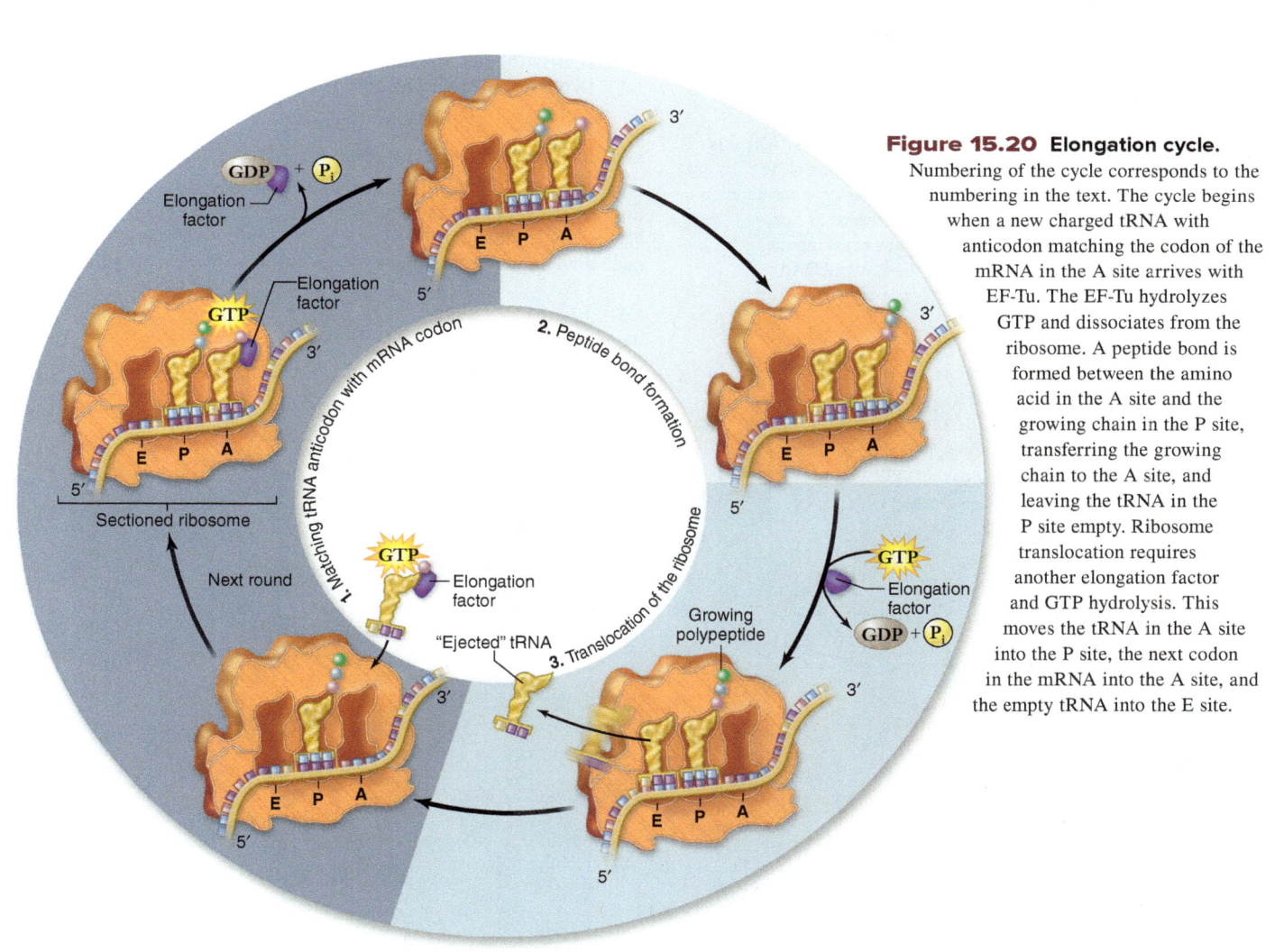

Figure 15.20 Elongation cycle. Numbering of the cycle corresponds to the numbering in the text. The cycle begins when a new charged tRNA with anticodon matching the codon of the mRNA in the A site arrives with EF-Tu. The EF-Tu hydrolyzes GTP and dissociates from the ribosome. A peptide bond is formed between the amino acid in the A site and the growing chain in the P site, transferring the growing chain to the A site, and leaving the tRNA in the P site empty. Ribosome translocation requires another elongation factor and GTP hydrolysis. This moves the tRNA in the A site into the P site, the next codon in the mRNA into the A site, and the empty tRNA into the E site.

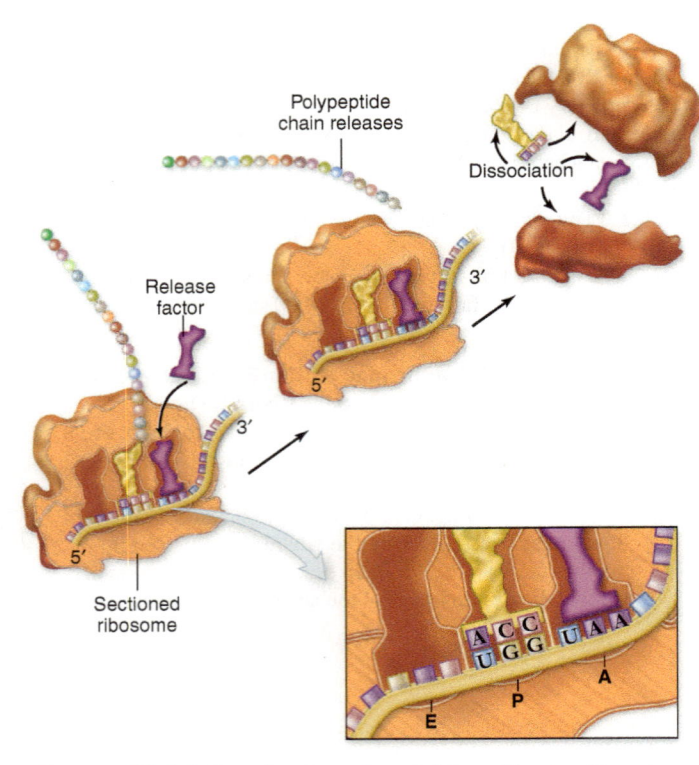

Figure 15.21 Termination of protein synthesis. There is no tRNA with an anticodon complementary to any of the three termination signal codons. When a ribosome encounters a termination codon, it stops translocating. A specific protein release factor facilitates the release of the polypeptide chain by breaking the covalent bond that links the polypeptide to the P site tRNA.

2. **Peptide bond formation.** Peptidyl transferase, located in the large subunit, catalyzes the formation of a peptide bond between the amino group of the amino acid in the A site and the carboxyl group of the growing chain. This also breaks the bond between the growing chain and the tRNA in the P site leaving it empty (no longer charged). The overall result of this is to transfer the growing chain to the tRNA in the A site.

3. **Translocation of the ribosome.** After the peptide bond has been formed, the ribosome moves relative to the mRNA and the tRNAs. The next codon in the mRNA shifts into the A site, and the tRNA with the growing chain moves to the P site. The uncharged tRNA formerly in the P site is now in the E site, and it will be ejected in the next cycle. This translocation step requires the accessory factor EF-G and the hydrolysis of another GTP.

This elongation cycle continues with each new amino acid added. The ribosome moves down the mRNA in a 5'-to-3' direction, reading successive codons. The tRNAs move through the ribosome in the opposite direction, from the A site to the P site and finally the E site, before they are ejected as empty tRNAs, which can be charged with another amino acid and then used again.

Wobble pairing

As mentioned, there are fewer tRNAs than codons. This situation is easily rationalized because the pairing between the 3' base of the codon and the 5' base of the anticodon is less stringent than normal. In some tRNAs, the presence of modified

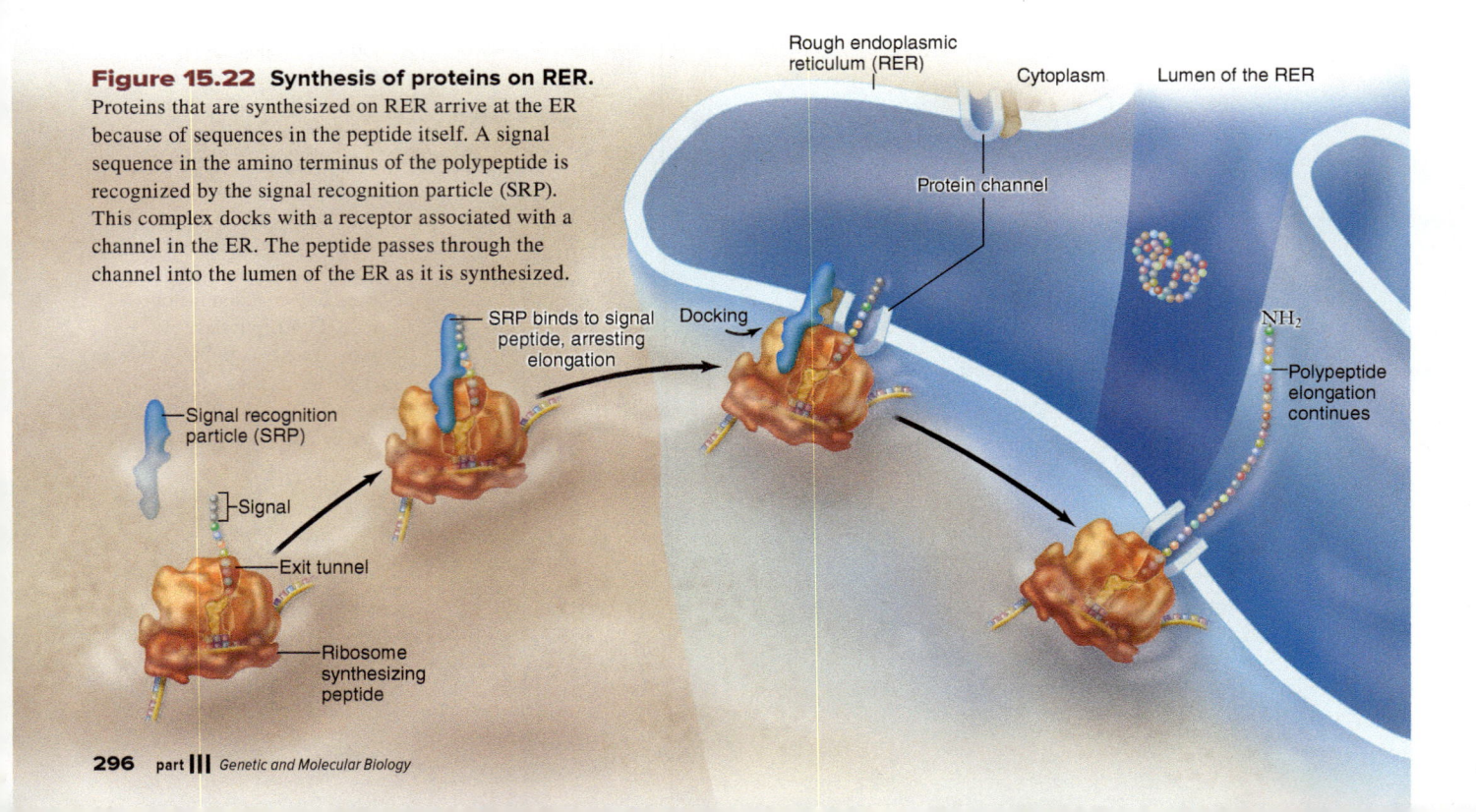

Figure 15.22 Synthesis of proteins on RER. Proteins that are synthesized on RER arrive at the ER because of sequences in the peptide itself. A signal sequence in the amino terminus of the polypeptide is recognized by the signal recognition particle (SRP). This complex docks with a receptor associated with a channel in the ER. The peptide passes through the channel into the lumen of the ER as it is synthesized.

bases with less accurate pairing in the 5′ position of the anticodon enhances this flexibility. This effect is referred to as **wobble pairing** because these tRNAs can "wobble" a bit on the mRNA, so that a single tRNA can "read" more than one codon in the mRNA.

> **? Inquiry question** How is the wobble phenomenon related to the number of tRNAs and the degeneracy of the genetic code?

Termination requires accessory factors

Elongation continues in this fashion until a chain-terminating stop codon is reached (for example, UAA in figure 15.21). These stop codons do not bind to tRNA; instead, they are recognized by release factors, proteins that release the newly made polypeptide from the ribosome.

Proteins may be targeted to the ER

In eukaryotes, translation can occur either in the cytoplasm or on the RER. Proteins that are translated on the RER are targeted there based on their own initial amino acid sequence. The ribosomes found on the RER are actively translating and are not permanently bound to the ER.

A polypeptide that starts with a short series of amino acids called a **signal sequence** is specifically recognized and bound by a cytoplasmic complex of proteins called the **signal recognition particle (SRP)**. The complex of signal sequence and SRP is in turn recognized by a receptor protein in the ER membrane. The binding of the ER receptor to the signal sequence/SRP complex holds the ribosome engaged in translation of the protein on the ER membrane, a process called *docking* (figure 15.22).

As the protein is assembled, it passes through a channel formed by the docking complex and into the interior ER compartment, the cisternal space. This is the basis for the docking metaphor—the ribosome is not actually bound to the ER itself, but with the newly synthesized protein entering the ER, the ribosome is like a boat tied to a dock with a rope.

The basic mechanism of protein translocation across membranes by the SRP and its receptor and channel complex has been conserved across all three cell types: eukaryotes, bacteria, and archaea. Given that only eukaryotic cells have an endomembrane system, this universality may seem curious; however, bacteria and archaea both export proteins through their plasma membrane, and the mechanism used is similar to the way in which eukaryotes move proteins into the cisternal space of the ER.

Once within the ER cisternal space, or lumen, the newly synthesized protein can be modified by the addition of sugars (glycosylation) and transported by vesicles to the Golgi apparatus (see chapter 4). This is the beginning of the protein-trafficking pathway that can lead to other intracellular targets, to incorporation into the plasma membrane, or to release outside of the cell itself.

> **Learning Outcomes Review 15.7**
>
> Translation initiation involves the interaction of the small ribosomal subunit with mRNA and a charged initiator tRNA. The elongation cycle involves bringing in new charged tRNAs to the ribosome's A site, forming peptide bonds between amino acids, and translocating the ribosome along the mRNA chain. The tRNAs transit through the ribosome from A to P to E sites during the process. In eukaryotes, signal sequences of a newly forming polypeptide may target it and its ribosome to be moved to the RER. Polypeptides formed on the RER enter the cisternal space rather than being released into the cytoplasm.
>
> ■ *What stages of translation require energy?*

15.8 Summarizing Gene Expression

Because of the complexity of the process of gene expression, it is worth stepping back to summarize some key points:

- The process of gene expression converts information in the genotype into the phenotype.
- A copy of the gene in the form of mRNA is produced by transcription, and the mRNA is used to direct the synthesis of a protein by translation.
- Both transcription and translation can be broken down into initiation, an elongation cycle, and termination—processes that produce their respective polymers. (The same is true for DNA replication.)
- Eukaryotic gene expression is much more complex than that of prokaryotes.

The structure of eukaryotic genes with interrupted coding sequences complicates both the process of gene expression and the nature of genetic information. It means that processing must occur between transcription and translation, and that one gene can produce multiple messages. Transcription in eukaryotes also takes place in the nucleus, whereas translation takes place in the cytoplasm. This necessitates that the mRNA be transported through nuclear pores to the cytoplasm prior to translation. The entire eukaryotic process is summarized in figure 15.23, and differences in gene expression between prokaryotes and in eukaryotes are summarized in table 15.2.

> **Learning Outcome Review 15.8**
>
> The greater complexity of eukaryotic gene expression is related to the functional organization of the cell, with DNA in the nucleus and ribosomes in the cytoplasm. The differences in gene expression between prokaryotes and eukaryotes is mainly in detail, but some differences have functional significance.

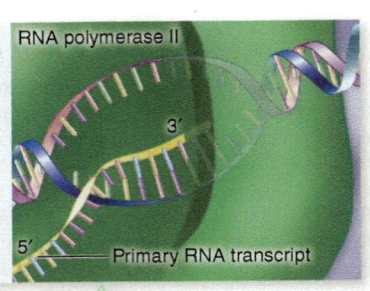

Figure 15.23 An overview of gene expression in eukaryotes.

1. RNA polymerase II in the nucleus copies one strand of the DNA to produce the primary transcript.

2. The primary transcript is processed by addition of a 5′ methyl-G cap, cleavage and polyadenylation of the 3′ end, and removal of introns. The mature mRNA is then exported through nuclear pores to the cytoplasm.

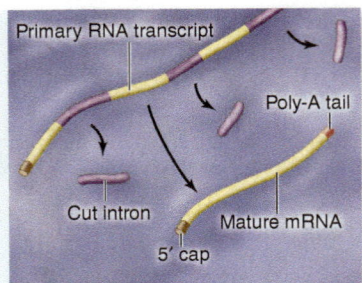

3. The 5′ cap of the mRNA associates with the small subunit of the ribosome. The initiator tRNA and large subunit are added to form an initiation complex.

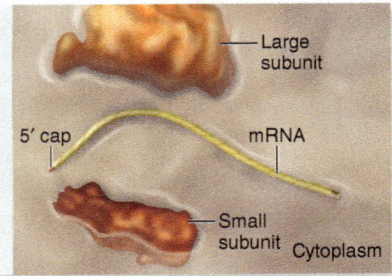

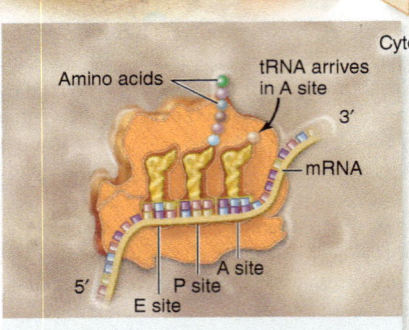

4. The ribosome cycle begins with the growing peptide attached to the tRNA in the P site. The next charged tRNA binds to the A site with its anticodon complementary to the codon in the mRNA in this site.

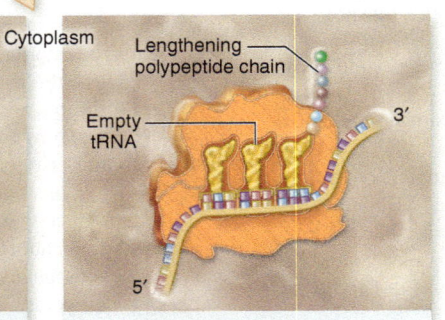

5. Peptide bonds form between the amino terminus of the next amino acid and the carboxyl terminus of the growing peptide. This transfers the growing peptide to the tRNA in the A site, leaving the tRNA in the P site empty.

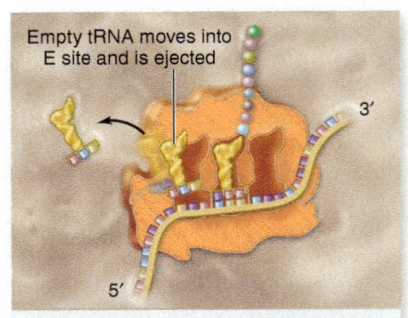

6. Ribosome translocation moves the ribosome relative to the mRNA and its bound tRNAs. This moves the growing chain into the P site, leaving the empty tRNA in the E site and the A site ready to bind the next charged tRNA.

TABLE 15.2 Differences Between Prokaryotic and Eukaryotic Gene Expression

Characteristic	Prokaryotes	Eukaryotes
Introns	No introns, although some archaeal genes possess them.	Most genes contain introns.
Number of genes in mRNA	Several genes may be transcribed into a single mRNA molecule. Often these have related functions and form an operon, which helps coordinate regulation of biochemical pathways.	Only one gene per mRNA molecule; regulation of pathways accomplished in other ways.
Site of transcription and translation	No membrane-bounded nucleus; transcription and translation are coupled.	Transcription in nucleus; mRNA is transported to the cytoplasm for translation.
Initiation of translation	Begins at AUG codon preceded by special sequence that binds the ribosome.	Begins at AUG codon preceded by the 5′ cap (methylated GTP) that binds the ribosome.
Modification of mRNA after transcription	None; translation begins before transcription is completed. Transcription and translation are coupled.	A number of modifications while the mRNA is in the nucleus: Introns are removed and exons are spliced together; a 5′ cap is added; a poly-A tail is added.

15.9 Mutation: Altered Genes

Learning Outcomes
1. Describe the effects of different point mutations.
2. Explain the nature of triplet repeat expansion.
3. List the different chromosomal mutations and their effects.

A mutation is an heritable change in the genetic material. A change as simple as altering a single base can result in an amino acid substitution that can lead to a debilitating clinical phenotype. This is illustrated for the case of sickle cell anemia in figure 15.24. In the sickle cell allele, a single A is changed to a T, resulting in a glutamic acid being replaced with a valine. The substitution of nonpolar valine causes the β chains to aggregate into polymers, which consequently alters the shape of the cells, leading to the disease state.

We now consider the spectrum of changes that can occur to DNA sequences, and their consequences. We will wrap up this discuss by considering recent data on the rate of these events in humans.

Point mutations affect a single site in the DNA

A mutation that alters a single base is termed a **point mutation,** or **single nucleotide variation (SNV).** The mutation can be either the substitution of one base for another, or the deletion or addition of a single base (or a small number of bases) (figure 15.25).

Base substitution

The substitution of one base pair for another in DNA is called a **base substitution mutation.** Because of the degenerate nature of the genetic code, base substitution may or may not alter the amino acid encoded. If the new codon from the base substitution still encodes the same amino acid, we say the mutation is *silent* (figure 15.25b). When base substitution changes an amino acid in a protein, it is also called a **missense mutation** as the "sense" of the codon produced after transcription of the mutant gene will be altered (figure 15.25c). These fall into two classes, *transitions* and *transversions*. A transition does not change the type of bases in the base pair—that is, a pyrimidine is substituted for a pyrimidine, or purine for purine. In contrast, a transversion does change the type of bases in a base pair—that is, pyrimidine to purine or the reverse. A variety of human genetic diseases, including sickle cell anemia, are caused by base substitutions.

Nonsense mutations

A special category of base substitution arises when a base is changed such that the transcribed codon is converted to a stop codon (figure 15.25d). We call these **nonsense mutations** because the mutation does not make "sense" to the translation apparatus. The stop codon results in premature termination of translation and leads to a truncated protein. How short the resulting protein is depends on where a stop codon has been introduced in the gene.

Frameshift mutations

The addition or deletion of a single base has much more profound consequences than does the substitution of one base for another. These mutations are called *frameshift mutations* because they alter the reading frame in the mRNA downstream of the mutation. This class of mutations was used by Crick and Brenner, as described in section 15.2, to infer the nature of the genetic code.

Changing the reading frame early in a gene, and thus in its mRNA transcript, means that the majority of the protein will be altered. Frameshifts also can cause premature termination of translation because 3 in 64 codons are stop codons, which represents a high probability in the sequence that has been randomized by the frameshift.

Triplet repeat expansion mutations

Given the long history of molecular genetics, and the relatively short time that molecular analysis has been possible on humans, it is surprising that a new kind of mutation was discovered in humans.

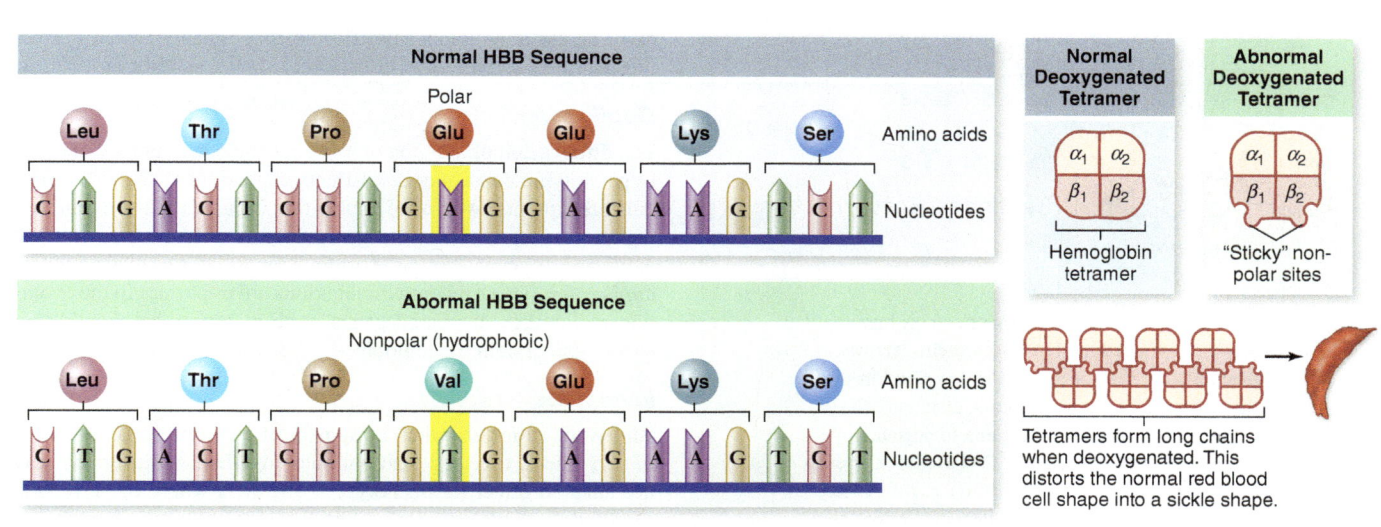

Figure 15.24 Sickle cell anemia is caused by an altered protein. Hemoglobin is composed of a tetramer of two α-globin and two β-globin chains. The sickle cell allele of the β-globin gene contains a single base change resulting in the substitution of Val for Glu. This creates a hydrophobic region on the surface of the protein that is "sticky," leading to their association into long chains that distort the shape of the red blood cells.

However, one of the first genes isolated that was associated with human disease, the gene for *Huntington disease,* provided a new kind of mutation. The gene for Huntington contains a triplet sequence of DNA that is repeated, and this repeat unit is expanded in the disease allele relative to the normal allele. Since this initial discovery, at least 20 other human genetic diseases appear to be due to this mechanism. The prevalence of this kind of mutation is unknown, but at present humans and mice are the only organisms in which they have been observed, implying that they may be limited to vertebrates, or even mammals. No such mutation has ever been found in *Drosophila,* for example.

The expansion of the triplet can occur in the coding region or in noncoding transcribed DNA. In the case of Huntington disease, the repeat unit is actually in the coding region of the gene where the triplet encodes glutamine, and expansion results in a polyglutamine region in the protein. A number of other neurodegenerative disorders also show this kind of mutation. In the case of fragile-X syndrome, an inherited form of intellectual disability, the repeat is in noncoding DNA.

Chromosomal mutations change the structure of chromosomes

Point mutations affect a single site in a chromosome, but more extensive changes can alter the structure of the chromosome itself, resulting in **chromosomal mutations.** Many human cancers are associated with chromosomal abnormalities, so these are of great clinical relevance. We briefly consider possible alterations to chromosomal structure, all of which are summarized in figure 15.26.

Deletions

A **deletion** is the loss of a portion of a chromosome. Frameshifts can be caused by one or more small deletions, but much larger regions of a chromosome may also be lost. If too much information is lost, the deletion is usually fatal to the organism.

One human syndrome that is due to a deletion is *cri-du-chat,* which is French for "cry of the cat" after the noise made by children with this syndrome. Cri-du-chat syndrome is caused by a large deletion from the short arm of chromosome 5. It usually results in early death, although many affected individuals show a normal lifespan. It has a variety of effects, including respiratory problems.

Duplications

The **duplication** of a region of a chromosome may or may not lead to phenotypic consequences. Effects depend upon the location of the "breakpoints" where the duplication occurred. If the duplicated region does not lie within a gene, there may be no effect. If the duplication occurs next to the original region, it is termed a *tandem duplication.* These tandem duplications are important in the evolution of families of related genes, such as the globin family that encode the protein hemoglobin.

Inversions

An **inversion** results when a segment of a chromosome is broken in two places, reversed, and put back together. An inversion may not have an effect on phenotype if the sites where the inversion occurs do not break within a gene. In fact, although humans all have the "same" genome, the order of genes in all individuals in a population is not precisely the same due to inversions that occur in different lineages.

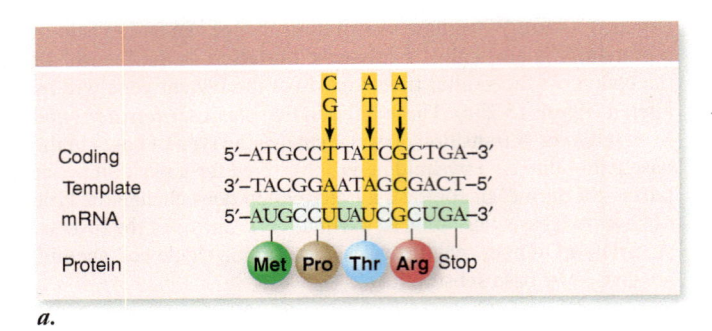

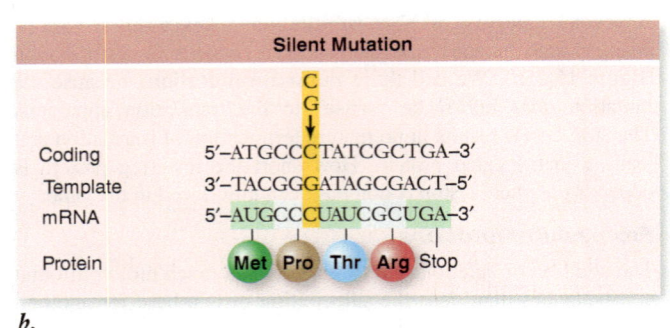

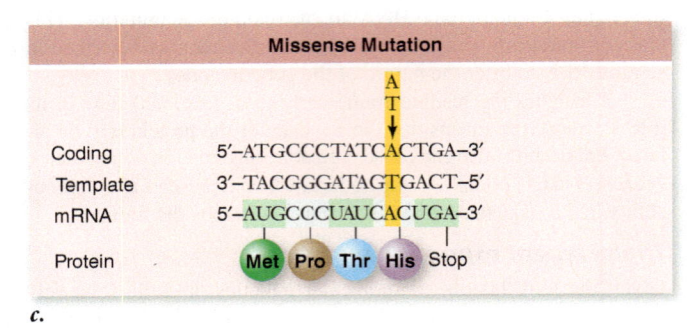

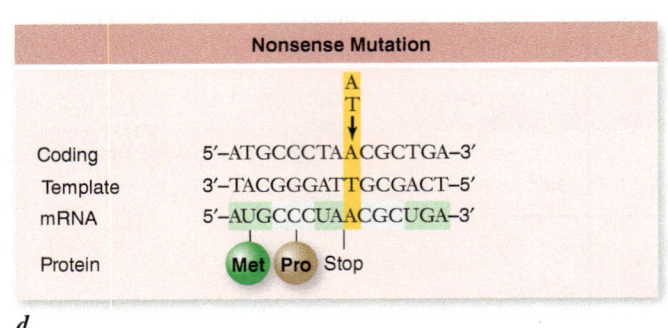

Figure 15.25 Types of mutations. *a.* A hypothetical gene is shown with encoded mRNA and protein. Arrows above the gene indicate sites of mutations described in the rest of the figure. *b.* Silent mutation. A change in the third position of a codon is often silent due to degeneracy in the genetic code. In this case T/A to C/G mutation does not change the amino acid encoded (proline). *c.* Missense mutation. The G/C to A/T mutation changes the amino acid encoded from arginine to histidine. *d.* Nonsense mutation. The T/A to A/T mutation produces a UAA stop codon in the mRNA.

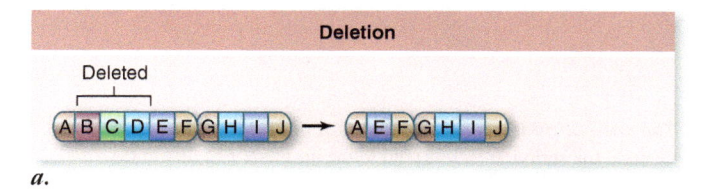

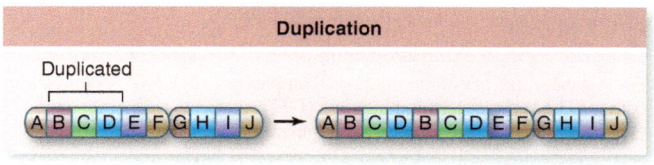

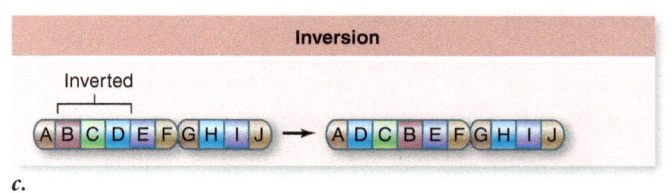

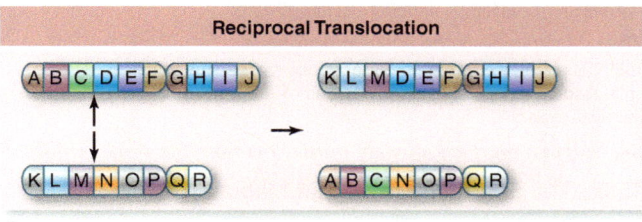

Figure 15.26 Chromosomal mutations. Larger-scale changes in chromosomes are also possible. Material can be deleted (*a*), duplicated (*b*), and inverted (*c*). Translocations occur when one chromosome is broken and becomes part of another chromosome. This often occurs where both chromosomes are broken and exchange material, an event called a reciprocal translocation (*d*).

Translocations

If a piece of one chromosome is broken off and joined to another chromosome, we call this a **translocation.** Translocations are complex because they can cause problems during meiosis, particularly when two different chromosomes try to pair with each other during meiosis I.

Translocations can also move genes from one chromosomal region to another in a manner that changes the expression of genes in the region involved. Two forms of leukemia have been shown to be associated with translocations that move oncogenes into regions of a chromosome where they are expressed inappropriately in blood cells.

Mutations are the starting point of evolution

If no changes occurred in genomes over time, then there could be no evolution. Too much change, however, is harmful to the individual with a greatly altered genome. Thus a delicate balance must exist between the amount of new variation that arises in a species and the health of individuals in the species. This topic is explored in more detail in chapter 20, when we consider evolution and population genetics.

The larger scale alteration of chromosomes has also been important in evolution, although its role is poorly understood. It is clear that gene families arise by the duplication of an ancestral gene, followed by the functional divergence of the duplicated copies. It is also clear that even among closely related species, the number and arrangements of genes on chromosomes can differ. Large-scale rearrangements may have occurred.

Human mutation rates have been directly measured

For many years, we could only estimate human mutation rates using indirect methods. The first attempt goes back to the 1930s and J.B.S. Haldane's pioneering work on hemophilia. Now with the ability to quickly and cheaply sequence entire genomes, we can directly measure human mutations rates by comparing the genomes parents and offspring.

A recent study of 78 mother, father, offspring "trios" from Iceland produced a rate of 1.2×10^{-8} SNV mutations per base-pair per generation. This corresponds to about 70 new mutations per birth. Data from this study also indicated that 76% of those new SNV mutations arose in the paternal genome, and the rate increases with the father's age. Other methods, including Haldane's original work, have found this paternal bias. This is probably a reflection of the stem cell population that gives rise to sperm accumulating mutations with age. Small insertions and deletions of less than 50 bases, called indels, occur at a rate about one-fifth to one-tenth the SNV rate. Much larger deletions or duplications, called **copy number variation (CNV),** occur at an even lower rate. This is estimated at about 1 new large CNV per 42 births. Some of the mobile genetic elements that litter our genome (see chapter 18) can still move. The rate for mobile element insertion appears to be on the order of about 1 per 20 births.

These new data will continue to shed light on human evolution, especially in comparison with analysis of other primate species. This is also relevant in terms of human disease as de novo mutations now appear to make a potentially larger contribution to human genetic disease than previously thought.

Learning Outcomes Review 15.9

Point mutations (single-base changes, additions, or deletions) include missense mutations that cause substitution of one amino acid for another, nonsense mutations that halt transcription, and frameshift mutations that throw off the correct reading of codons. Triplet repeat expansion is the abnormal duplication of a codon with each round of cell division. Mutations affecting chromosomes include deletions, duplications, inversions, and translocations.

■ *Would an inversion or duplication always be expected to have a phenotype?*

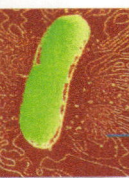

Chapter Review

15.1 The Nature of Genes

Garrod concluded that inherited disorders can involve specific enzymes.
Garrod found that alkaptonuria is due to an altered enzyme.

Beadle and Tatum showed that genes specify enzymes.
Neurospora mutants unable to synthesize arginine were found to lack specific enzymes. Beadle and Tatum advanced the "one gene/one polypeptide" hypothesis (figure 15.1).

The central dogma describes information flow in cells as DNA to RNA to protein (figure 15.2).
We call the DNA strand copied to mRNA the template (antisense) strand; the other the coding (sense) strand.

Transcription makes an RNA copy of DNA.

Translation uses information in RNA to synthesize proteins.
An adapter molecule, tRNA, is required to connect the information in mRNA into the sequence of amino acids.

RNA has multiple roles in gene expression.

15.2 The Genetic Code

The code is read in groups of three.
Crick and Brenner showed that the code is nonoverlapping and is read in groups of three. This finding established the concept of reading frame.

Nirenberg and others deciphered the code.
A codon consists of 3 nucleotides, so there are 64 possible codons. Three codons signal "stop," and one codon signals "start" and also encodes methionine. Thus 61 codons encode the 20 amino acids.

The code is degenerate but specific.
Many amino acids have more than one codon, but each codon specifies only a single amino acid.

The code is practically universal, but not quite.
In some mitochondrial and protist genomes, a stop codon is read as an amino acid; otherwise the code is universal.

15.3 Prokaryotic Transcription

Prokaryotes have a single RNA polymerase.
Prokaryotic RNA polymerase exists in two forms: core polymerase, which can synthesize mRNA; and holoenzyme, core plus σ factor, which can accurately initiate synthesis (figure 15.6).

Initiation occurs at promoters.
Initiation requires a start site and a promoter. The promoter is upstream of the start site, and binding of RNA polymerase holoenzyme to its −35 region positions the polymerase properly.

Elongation adds successive nucleotides.
Transcription proceeds in the 5′-to-3′ direction. The transcription bubble contains RNA polymerase, the locally unwound DNA template, and the growing mRNA transcript (figure 15.7).

Termination occurs at specific sites.
Terminators consist of complementary sequences that form a double-stranded hairpin loop where the polymerase pauses (figure 15.8).

Prokaryotic transcription is coupled to translation.
Translation begins while mRNAs are still being transcribed.

15.4 Eukaryotic Transcription

RNA polymerase I transcribes rRNA; polymerase II transcribes mRNA and some snRNAs; polymerase III transcribes tRNA.

Each polymerase has its own promoter.

The core promoter binds factors that recruit RNA Pol II.
General transcription factors bind to the promoter and recruit the polymerase to form the initiation complex. As the transcript reaches about 20 nucleotides, it is capped, and the polymerase may pause. Release of pausing requires elongation factors.

The transcription elongation complex can recruit other factors.
Some general factors in the initiation complex are replaced by elongation factors. The carboxy terminal domain (CTD) interacts with elongation factors and enzymes that modify the transcript.

Polyadenylation is involved in terminating mRNA transcripts.
The enzyme poly-A polymerase cleaves the transcript and adds 1 to 200 adenine (A) residues to the 3′ end.

Primary transcripts are modified to produce mRNAs.
The mRNA has a 5′ cap, a 3′ poly-A tail, and introns are removed.

15.5 Eukaryotic pre-mRNA Splicing

Eukaryotic genes may contain interruptions.
Coding DNA (an exon) is interrupted by noncoding introns. These introns are removed by splicing (figure 15.13).

The spliceosome is the splicing organelle.
snRNPs recognize intron–exon junctions and recruit spliceosomes. The spliceosome ultimately joins the 3′ end of the first exon to the 5′ end of the next exon.

Splicing can produce multiple transcripts from the same gene.

15.6 The Structure of tRNA and Ribosomes

Aminoacyl-tRNA synthetases attach amino acids to tRNA.
The tRNA charging reaction attaches the carboxyl terminus of an amino acid to the 3′ end of the correct tRNA (figure 15.15).

The ribosome has multiple tRNA-binding sites (figure 15.16).
A charged tRNA first binds to the A site, then moves to the P site where its amino acid is bonded to the peptide chain, and finally, without its amino acid, moves to the E site from which it is released.

The ribosome has both decoding and enzymatic functions.
Ribosomes hold tRNAs and mRNA in position for a ribosomal enzyme to form peptide bonds.

15.7 The Process of Translation

Initiation requires accessory factors.
In prokaryotes, initiation-complex formation is aided by the ribosome-binding sequence (RBS) of mRNA, complementary to a small subunit. Eukaryotes use the 5′ cap for the same function.

Elongation adds successive amino acids (figure 15.20).
As the ribosome moves along the mRNA, new amino acids from charged tRNAs are added to the growing peptide.

Termination requires accessory factors.
Stop codons are recognized by termination factors.

Proteins may be targeted to the ER.
In eukaryotes, proteins with a signal sequence in their amino terminus bind to the SRP, and this complex docks on the ER.

15.9 Mutation: Altered Genes

Point mutations affect a single site in the DNA.
Base substitutions exchange one base for another, and frameshift mutations involve the addition or deletion of a base. Triplet repeat expansion mutations can cause genetic diseases.

Chromosomal mutations change the structure of chromosomes.
Chromosomal mutations include additions, deletions, inversions, or translocations.

Mutations are the starting point of evolution.
Human mutation rates have been directly measured.
There are about 70 new mutations per generation.

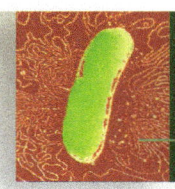

Review Questions

UNDERSTAND

1. The experiments with nutritional mutants in *Neurospora* by Beadle and Tatum provided evidence that
    a. bread mold can be grown in a lab on minimal media.
    b. X-rays can damage DNA.
    c. cells need enzymes.
    d. genes specify enzymes.

2. What is the *central dogma* of molecular biology?
    a. DNA is the genetic material.
    b. Information passes from DNA directly to protein.
    c. Information passes from DNA to RNA to protein.
    d. One gene encodes only one polypeptide.

3. In the genetic code, one codon
    a. consists of three bases.
    b. specifies a single amino acid.
    c. specifies more than one amino acid.
    d. Both a and b are correct.

4. Eukaryotic transcription differs from prokaryotic in that
    a. eukaryotes have only one RNA polymerase.
    b. eukaryotes have three RNA polymerases.
    c. prokaryotes have three RNA polymerases.
    d. Both a and c are correct.

5. An anticodon would be found on which of the following types of RNA?
    a. snRNA (small nuclear RNA)   c. tRNA (transfer RNA)
    b. mRNA (messenger RNA)   d. rRNA (ribosomal RNA)

6. RNA polymerase binds to a _____ to initiate _____.
    a. mRNA; translation
    b. promoter; transcription
    c. primer; transcription
    d. transcription factor; translation

7. During translation, the codon in mRNA is actually "read" by
    a. the A site in the ribosome.   c. the anticodon in a tRNA.
    b. the P site in the ribosome.   d. the anticodon in an amino acid.

APPLY

1. You have mutants that all affect the same biochemical pathway. If feeding an intermediate in the pathway supports growth, this tells you that the enzyme encoded by the affected gene
    a. acts after the intermediate used.
    b. acts before the intermediate used.
    c. must act to produce the intermediate.
    d. must not act to produce the intermediate.

2. The splicing process
    a. occurs in prokaryotes.
    b. joins introns together.
    c. can produce multiple mRNAs from the same transcript.
    d. only joins exons for each gene in one way.

3. The enzyme that forms peptide bonds is called peptidyl *transferase* because it transfers
    a. a new amino acid from a tRNA to the growing peptide.
    b. the growing peptide from a tRNA to the next amino acid.
    c. the peptide from one amino acid to another.
    d. the peptide from the ribosome to a charged tRNA.

4. In comparing gene expression in prokaryotes and eukaryotes
    a. eukaryotic genes can produce more than one protein.
    b. prokaryotic genes can produce more than one protein.
    c. both produce mRNAs that are colinear with the protein.
    d. Both a and c are correct.

5. The codon CCA could be mutated to produce
    a. a silent mutation.   c. a stop codon.
    b. a codon for Lys.   d. Both a and b are correct.

6. An inversion will
    a. necessarily cause a mutant phenotype.
    b. only cause a mutant phenotype if the inversion breakpoints fall within a gene.
    c. halt transcription in the inverted region because the chromosome is now backward.
    d. interfere with translation of genes in the inverted region.

7. What is the relationship between mutations and evolution?
    a. Mutations make genes better.
    b. Mutations can create new alleles.
    c. Mutations happened early in evolution, but not now.
    d. There is no relationship between evolution and genetic mutations.

SYNTHESIZE

1. A template strand of DNA has the following sequence:

    3' – CGTTACCCGAGCCGTACGATTAGG – 5'

    Use the sequence information to determine
    a. the predicted sequence of the mRNA for this gene.
    b. the predicted amino acid sequence of the protein.

2. Frameshift mutations often result in truncated proteins. Explain this observation based on the genetic code.

3. Describe how each of the following mutations will affect the final protein product (protein begins with start codon). Name the type of mutation.

    Original template strand:
    3' – CGTTACCCGAGCCGTACGATTAGG – 5'
    a. 3' – CGTTACCCGAGCCGTAACGATTAGG – 5'
    b. 3' – CGTTACCCGATCCGTACGATTAGG – 5'
    c. 3' – CGTTACCCGAGCCGTTCGATTAGG – 5'

4. There are a number of features that are unique to bacteria, and others that are unique to eukaryotes. Could any of these features offer the possibility to control gene expression in a way that is unique to either eukaryotes or bacteria?

Principles of Biology I and II

CHAPTER 16

Control of Gene Expression

Chapter Contents

16.1 Control of Gene Expression
16.2 Regulatory Proteins
16.3 Prokaryotic Regulation
16.4 Eukaryotic Regulation
16.5 Chromatin Structure and Gene Expression
16.6 Eukaryotic Posttranscriptional Regulation
16.7 Protein Degradation

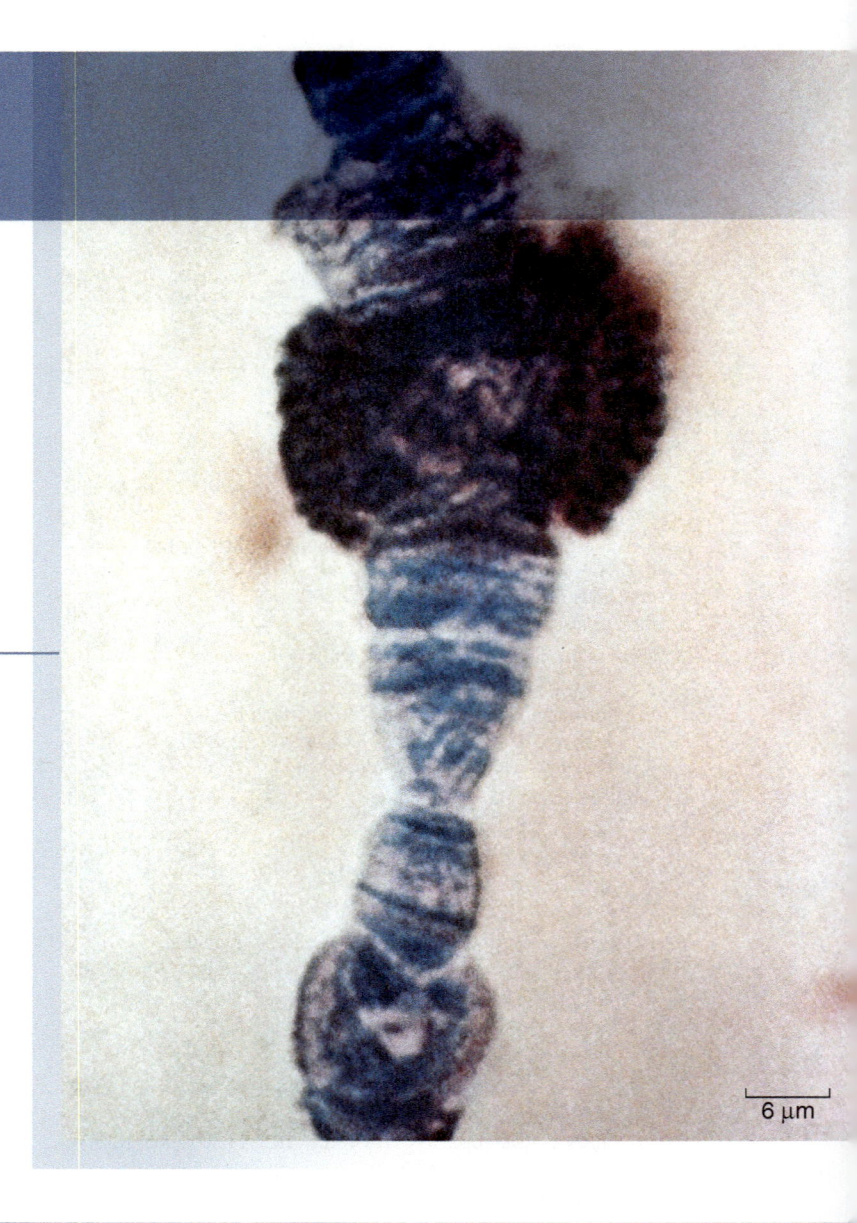

6 µm

Introduction

In a symphony, various instruments play their own parts at different times; the musical score determines which instruments play when. Similarly, in an organism, different genes are expressed at different times, with a "genetic score," written in regulatory regions of the DNA, determining which genes are active when. The picture shows the expanded "puff" of this *Drosophila* chromosome, which represents genes that are being actively expressed. Gene expression and how it is controlled is our topic in this chapter.

16.1 Control of Gene Expression

Learning Outcomes

1. Identify when gene expression is usually controlled.
2. Describe the usual action of regulatory proteins.
3. List differences between control of gene expression in prokaryotes and eukaryotes.

Control of gene expression is essential to all organisms. In prokaryotes, cells adapt to environmental changes by changing gene expression. In multicellular eukaryotes, it is critical for directing development and maintaining homeostasis. In fact, different cell types in a multicellular organism have distinct functions based on the proteins they express. All cells in a multicellular organism share a set of "housekeeping" genes that sustain the cell, but they also express a unique set of proteins that distinguish each cell type.

Control can occur at all levels of gene expression

You learned in chapter 15, that gene expression is the conversion of genotype to phenotype—the flow of information from DNA to produce functional proteins that control cellular activities. This traditional view of gene expression includes controlling the process primarily at the level of transcription initiation. Although this view remains valid, evidence is accumulating that both the extent of the genome transcribed, and the control of this transcription in multicellular organisms is even more complex than previously expected.

In this chapter, we will develop the control of initiation of transcription in some detail because of its importance, and because it is still the best-studied mechanism for the control of gene expression. With this framework in place, we will also consider how chromatin structure affects gene expression and how control can be exerted posttranscriptionally as well. The latter topic will lead us into the exciting new world of regulatory RNA molecules.

RNA polymerase is key to transcription, and it must have access to the DNA helix and must be capable of binding to the gene's promoter for transcription to begin. **Regulatory proteins** act by modulating the ability of RNA polymerase to bind to the promoter. This idea of controlling the access of RNA polymerase to a promoter is common to both prokaryotes and eukaryotes, but the details differ greatly, as you will see.

These regulatory proteins bind to specific nucleotide sequences on the DNA that are usually only 10 to 15 nt in length. (Even a large regulatory protein has a "footprint," or binding area, of only about 20 nt.) Hundreds of these regulatory sequences have been characterized, and each provides a binding site for a specific protein that is able to recognize the sequence. Binding of the protein either *blocks* transcription by getting in the way of RNA polymerase or *stimulates* transcription by facilitating the binding of RNA polymerase to the promoter.

Control strategies in prokaryotes are geared to adjust to environmental changes

Control of gene expression is accomplished very differently in prokaryotes than it is in eukaryotes. Prokaryotic cells have been shaped by evolution to grow and divide as rapidly as possible, enabling them to exploit transient resources. Proteins in prokaryotes turn over rapidly, allowing these organisms to respond quickly to changes in their external environment by changing patterns of gene expression.

In prokaryotes, the primary function of gene control is to adjust the cell's activities to its immediate environment. Changes in gene expression alter which enzymes are present in response to the quantity and type of nutrients and the amount of oxygen available. Almost all of these changes are fully reversible, allowing the cell to adjust its enzyme levels up or down in response to environment changes.

Control strategies in eukaryotes maintain homeostasis and drive development

The cells of multicellular organisms, in contrast, have been shaped by evolution to be protected from transient changes in their immediate environment. Most of them experience fairly constant conditions. Indeed, *homeostasis*—the maintenance of a constant internal environment—is a hallmark of multicellular organisms. Cells in such organisms respond to signals in their immediate environment (such as growth factors and hormones) by altering gene expression, and in doing so they participate in regulating the body as a whole.

Perhaps the most important role for controlled changes in gene expression is the development of the organism itself. This occurs by coordinated changes in gene expression that occur both over developmental time and in different tissues. Understanding this complex program has been a major goal of developmental genetics, and this work has uncovered circuits of gene expression that have been preserved over very long evolutionary time (see chapter 19). The same regulatory circuit may also be used within an organism to pattern different structures in different places at different developmental times. For example, the gene *sonic hedgehog* acts to pattern the neural tube, then later to pattern digits in developing limbs.

Unicellular eukaryotes also use different control mechanisms from those of prokaryotes. All eukaryotes have a membrane-bounded nucleus, use similar mechanisms to condense DNA into chromosomes, and have the same gene expression machinery, all of which differ from those of prokaryotes.

Learning Outcomes Review 16.1

Gene expression is usually controlled at the level of transcription initiation. Regulatory proteins bind to specific DNA sequences and affect the binding of RNA polymerase to promoters. Individual proteins may either prevent or stimulate transcription. In prokaryotes, regulation is focused on adjusting the cell's activities to the environment to ensure viability. In multicellular eukaryotes, regulation is geared to maintaining internal homeostasis, and even in unicellular forms, this control has mechanisms to deal with a bounded nucleus and multiple chromosomes.

- *Would you expect the control of gene expression in a unicellular eukaryote like yeast to be more like that of humans or E. coli?*

16.2 Regulatory Proteins

Learning Outcomes
1. Explain how proteins can interact with base pairs without unwinding the helix.
2. Describe the common features of DNA-binding motifs.

The ability of regulatory proteins to bind to *specific* DNA sequences and either block transcription or facilitate the binding of RNA polymerase is the basis for transcriptional control. To understand how cells control gene expression, it is first necessary to gain a clear picture of this molecular recognition process.

Proteins can interact with DNA through the major groove

It is not actually necessary for the DNA helix to be unwound for regulatory proteins to be able to interact with specific sequences of bases. When a protein binds to the surface of DNA, the edges of the bases are exposed in the two grooves in the molecule. The nature of the helix produces two grooves in the cylindrical structure of the helix: a larger major groove, and a smaller minor groove (see figure 14.8 for review). Within the deeper major groove, the nucleotides' hydrogen bond donors and acceptors are accessible. The pattern created by these chemical groups is unique for each of the four possible base-pair arrangements, providing a ready way for a protein nestled in the groove to read the sequence of bases (figure 16.1).

DNA-binding domains interact with specific DNA sequences

Protein–DNA recognition is critical to all DNA functions, including replication, gene expression, and the control of these activities. The number of proteins that can interact with DNA to affect gene expression is quite large, but determining the structure of a subset of these led to the discovery of a small number of protein motifs that actually bind to DNA. These motifs probably arose relatively early in evolution, and have been reused with alterations in a wide variety of different proteins. These DNA-binding motifs share the property of interacting with specific sequences of bases, usually through the major groove of the DNA helix.

DNA-binding motifs are the key structure within the DNA-binding domain of these proteins. This domain is a functionally distinct part of the protein necessary to bind to DNA in a sequence-specific manner. Regulatory proteins also need to be able to interact with the transcription apparatus, which is accomplished by a different regulatory domain.

Note that two proteins that share the same DNA-binding domain do not necessarily bind to the same DNA sequence. The similarities in the DNA-binding motifs appear in their three-dimensional structure, and not in the specific contacts that they make with DNA.

Several common DNA-binding motifs are shared by many proteins

A limited number of common DNA-binding motifs are found in a wide variety of different proteins. Four of the best known are detailed here to give the sense of how DNA-binding proteins interact with DNA.

The helix-turn-helix motif

The most common DNA-binding motif is the **helix-turn-helix**, constructed from two α-helical segments of the protein linked by a short, nonhelical segment, the "turn" (figure 16.2a). As the first motif recognized, the helix-turn-helix motif has since been identified in hundreds of DNA-binding proteins.

A close look at the structure of a helix-turn-helix motif reveals how proteins containing such motifs interact with the major groove of DNA. The helical segments of the motif interact with one another, so that they are held at roughly right angles. When this motif is pressed against DNA, one of the helical segments (called the *recognition helix*) fits snugly in the major groove of the DNA molecule, and the other butts up against the outside of the

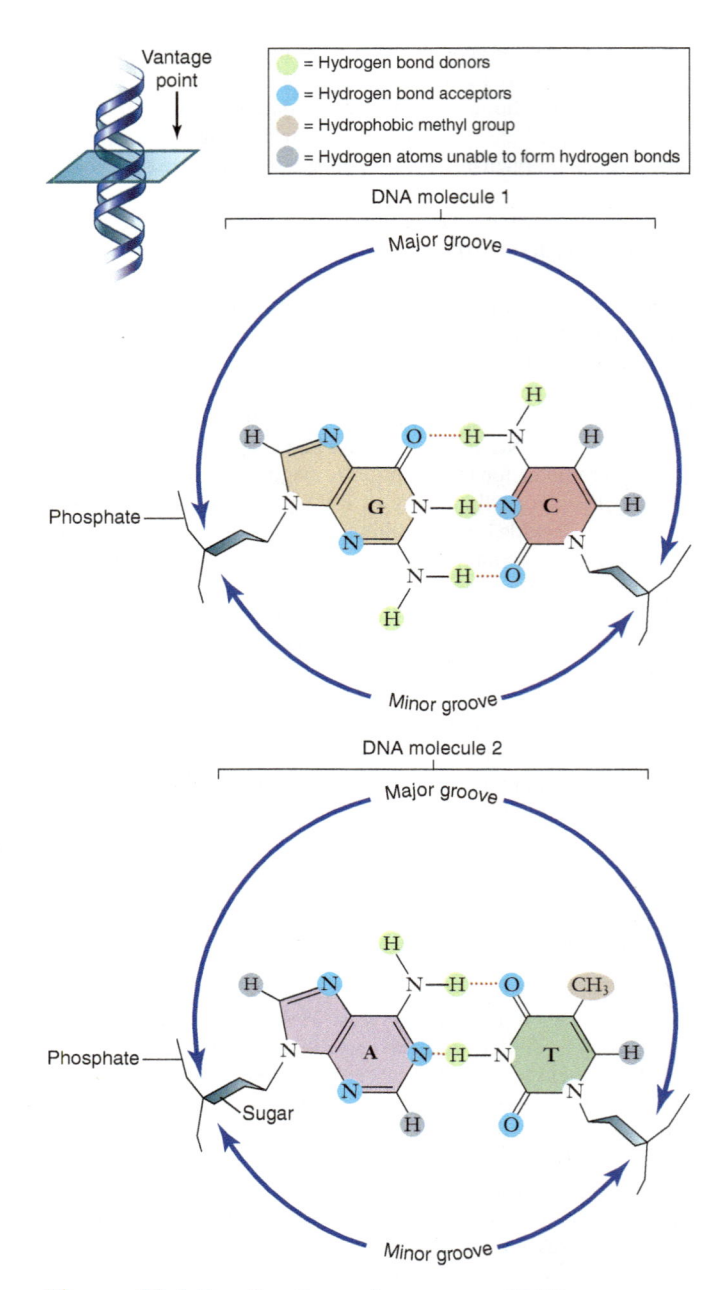

Figure 16.1 Reading the major groove of DNA. Looking down into the major groove of a DNA helix, we can see the edges of the bases protruding into the groove. Each of the four possible base-pair arrangements (two are shown here) extends a unique set of chemical groups into the groove, indicated in this diagram by differently colored circles. A regulatory protein can identify the base-pair arrangement by this characteristic signature.

DNA molecule, helping to ensure the proper positioning of the recognition helix.

Most DNA-regulatory sequences recognized by helix-turn-helix motifs occur in symmetrical pairs. Such sequences are bound by proteins containing two helix-turn-helix motifs separated by 3.4 nanometers (nm), the distance required for one turn of the DNA helix (figure 16.2a). Having *two* protein–DNA-binding sites doubles the zone of contact between protein and DNA and greatly strengthens the affinity between them.

The homeodomain motif

A special class of helix-turn-helix motif, the **homeodomain**, plays a critical role in development in a wide variety of eukaryotic organisms, including humans. A class of mutations in *Drosophila* called homeotic mutations cause one body part to be replaced by another. Analysis of the affected genes led to the discovery of the sequence of 60 amino acids that form the homeodomain. The most conserved part of the homeodomain contains a recognition helix of a helix-turn-helix motif. This was the first indication that developmental mechanisms are ancient, and shared across greater phylogenetic distance than once was thought. At the level of our genes, we are much closer to a fly than you might imagine.

The zinc finger motif

A different kind of DNA-binding motif uses one or more zinc atoms to coordinate its binding to DNA. Called **zinc fingers**, these motifs exist in several forms. In one form, a zinc atom links an α-helical segment to a β-sheet segment so that the helical segment fits into the major groove of DNA.

This sort of motif often occurs in clusters, the β-sheets spacing the helical segments so that each helix contacts the major groove. The effect is like a hand wrapped around the DNA with the fingers lying in the major groove. The more zinc fingers in the cluster, the more the protein associates with DNA.

The leucine zipper motif

In another DNA-binding motif, the name actually refers to a dimerization motif that allows different subunits of a protein to associate with the DNA. This so-called **leucine zipper** is created where a region on one subunit containing several hydrophobic amino acids (usually leucines) interacts with a similar region on the other subunit. This interaction holds the subunits together and creates a Y-shaped structure where the two arms of the Y are helical regions that each fit into the major groove of DNA but on opposite sides of the helix (figure 16.2b), holding the DNA like a pair of tongs.

Learning Outcomes Review 16.2

A DNA helix exhibits a major groove and a minor groove; regulatory proteins interact with DNA by accessing bases along the major groove. These proteins all contain DNA-binding motifs, and they often include one or two α-helical segments. These motifs form the active part of the DNA-binding domain, and another domain of the protein interacts with the transcription apparatus.

- What would be the effect of a mutation in a helix-turn-helix protein that altered the spacing of the two helices?

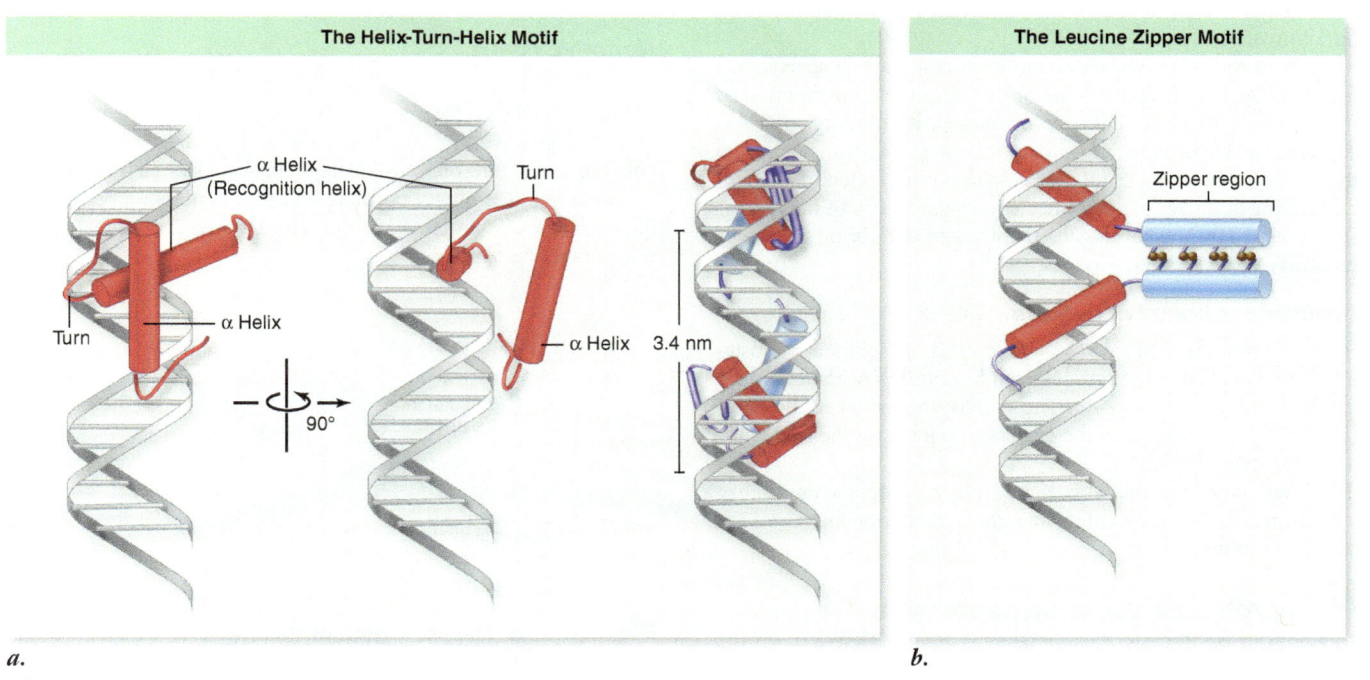

Figure 16.2 Major DNA-binding motifs. Two different DNA-binding motifs are pictured interacting with DNA. *a.* The helix-turn-helix motif binds to DNA using one α helix, the recognition helix, to interact with the major groove. The other helix positions the recognition helix. Proteins with this motif are usually dimers, with two identical subunits, each containing the DNA-binding motif. The two copies of the motif *(red)* are separated by 3.4 nm, precisely the spacing of one turn of the DNA helix. *b.* The leucine zipper acts to hold two subunits in a multisubunit protein together, thereby allowing α-helical regions to interact with DNA.

16.3 Prokaryotic Regulation

Learning Outcomes
1. Contrast control by induction and control by repression.
2. Explain control of gene expression in the *lac* operon.
3. Explain control of gene expression in the *trp* operon.

The details of regulation can be revealed by examining mechanisms used by prokaryotes to control the initiation of transcription. Prokaryotes and eukaryotes share some common themes, but they have some profound differences as well. Later, in sections 16.4 and 16.5, we discuss eukaryotic systems and concentrate on how they differ from the simpler prokaryotic systems.

Control of transcription can be either positive or negative

Control at the level of transcription initiation can be either positive or negative. **Positive control** increases the frequency of initiation, and **negative control** decreases the frequency of initiation. Each of these forms of control are mediated by regulatory proteins, but the proteins have opposite effects.

Negative control by repressors

Negative control is mediated by proteins called **repressors**. Repressors are proteins that bind to regulatory sites on DNA called **operators** to prevent or decrease the initiation of transcription. They act as a kind of roadblock to prevent the polymerase from initiating effectively.

Repressors do not act alone; each responds to specific effector molecules. Effector binding can alter the conformation of the repressor to either enhance or abolish its binding to DNA. These repressor proteins are allosteric proteins with an active site that binds DNA and a regulatory site that binds effectors. Effector binding at the regulatory site changes the ability of the repressor to bind DNA (see chapter 6 for more details on allosteric proteins).

Positive control by activators

Positive control is mediated by another class of regulatory, allosteric proteins called *activators* that can bind to DNA and stimulate the initiation of transcription. These activators enhance the binding of RNA polymerase to the promoter to increase the frequency of transcription initiation.

Activators are the logical and physical opposites of repressors. Effector molecules can either enhance or decrease activator binding.

Prokaryotes adjust gene expression in response to environmental conditions

Changes in the environments that bacteria and archaea encounter often result in changes in gene expression. In general, genes encoding proteins involved in catabolic pathways (breaking down molecules) respond oppositely from genes encoding proteins involved in anabolic pathways (building up molecules). In the discussion that follows, we describe enzymes in the catabolic pathway that transports and utilizes the sugar lactose. Later, in this section, the anabolic pathway that synthesizes the amino acid tryptophan.

As mentioned in chapter 15, prokaryotic genes are often organized into operons, multiple genes that are part of a single transcription unit having a single promoter. Genes that are involved in the same metabolic pathway are often organized in this fashion. The proteins necessary for the utilization of lactose are encoded by the ***lac* operon**, and the proteins necessary for the synthesis of tryptophan are encoded by the ***trp* operon.**

> **Inquiry question** What advantage might a bacterium get by linking into a single operon several genes, all of the products of which contribute to a single biochemical pathway?

Induction and repression

If a bacterium encounters lactose, it begins to make the enzymes necessary to utilize lactose. When lactose is not present, however, there is no need to make these proteins. Thus, we say that the synthesis of the proteins is *induced* by the presence of lactose. **Induction** therefore occurs when enzymes for a certain pathway are produced in response to a substrate.

When tryptophan is available in the environment, a bacterium will not synthesize the enzymes necessary to make tryptophan. If tryptophan ceases to be available, then the bacterium begins to make these enzymes. **Repression** occurs when bacteria capable of making biosynthetic enzymes do not produce them. In the case of both induction and repression, the bacterium is adjusting to produce the enzymes that are optimal for its immediate environment.

Negative control

Control of the initiation of transcription can be either positive or negative. On the surface, repression may appear to be negative and induction positive, but this is not the case. In fact, regulation of both the *lac* and *trp* operons are negatively controlled by repressor proteins. This seemingly opposite behavior is possible because in the two repressor proteins interact with DNA and their respective effectors in the opposite way.

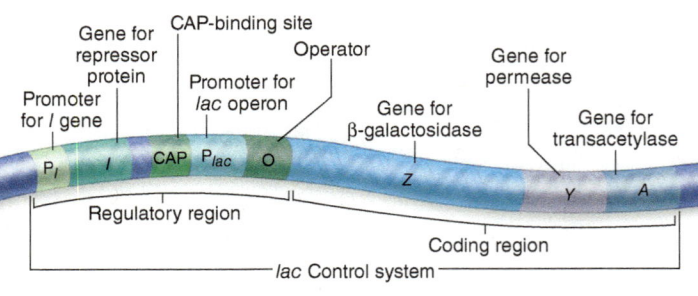

Figure 16.3 The *lac* region of the *Escherichia coli* chromosome. The *lac* operon consists of a promoter, an operator, and three genes (*lac Z, Y,* and *A*) that encode proteins required for the metabolism of lactose. In addition, there is a binding site for the catabolite activator protein (CAP), which affects RNA polymerase binding to the promoter. The *I* gene encodes the repressor protein, which can bind the operator and block transcription of the *lac* operon.

For either mechanism to work, the molecule in the environment, such as lactose or tryptophan, must produce the proper effect on the gene being regulated. In the case of *lac* induction, the presence of lactose must *prevent* a repressor protein from binding to its regulatory sequence. In the case of *trp* repression, by contrast, the presence of tryptophan must *cause* a repressor protein to bind to its regulatory sequence.

These responses are opposite because the needs of the cell are opposite in anabolic versus catabolic pathways. Each pathway is examined in detail to show how protein–DNA interactions allow the cell to respond to environmental conditions.

The *lac* operon is negatively regulated by the *lac* repressor

The control of gene expression in the *lac* operon was elucidated by the pioneering work of Jacques Monod and François Jacob.

The *lac* operon consists of the genes that encode functions necessary to utilize lactose: β-galactosidase *(lacZ)*, lactose permease *(lacY)*, and lactose transacetylase *(lacA)*, plus the regulatory regions necessary to control the expression of these genes (figure 16.3). In addition, the gene for the *lac* repressor *(lacI)* is linked to the rest of the *lac* operon and is thus considered part of the operon although it has its own promoter. The arrangement of the control regions upstream of the coding region is typical of most prokaryotic operons, although the linked repressor is not.

Action of the repressor

Initiation of transcription of the *lac* operon is controlled by the *lac* repressor. The repressor binds to the operator, which is adjacent to the promoter (figure 16.4a). This binding prevents RNA polymerase from binding to the promoter. This DNA binding is sensitive to the presence of lactose: The repressor binds DNA in the absence of lactose, but not in the presence of lactose.

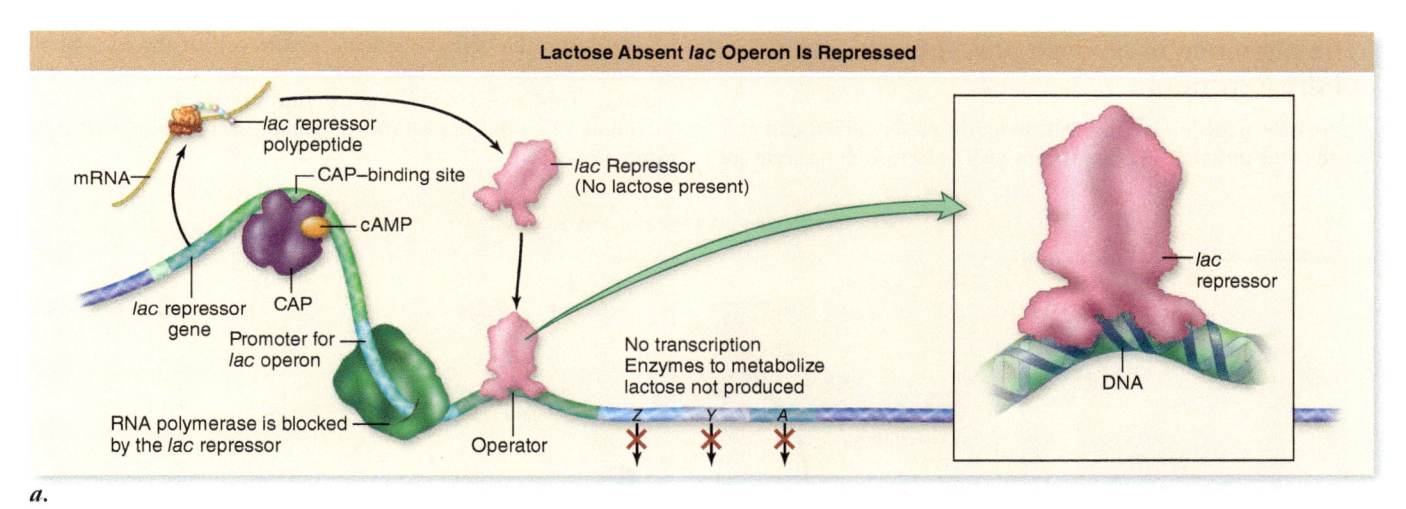

Figure 16.4 Induction of the *lac* operon. *a.* In the absence of lactose the *lac* repressor binds to DNA at the operator site, thus preventing transcription of the operon. When the repressor protein is bound to the operator site, the *lac* operon is shut down (repressed). *b.* The *lac* operon is transcribed (induced) when CAP is bound and when the repressor is not bound. Allolactose binding to the repressor alters the repressor's shape so it cannot bind to the operator site and block RNA polymerase activity.

Data analysis What would the phenotype be of a repressor mutation that prevents DNA binding? Inducer binding? How do these compare with a mutation in the operator that prevents repressor binding?

Interaction of repressor and inducer

In the absence of lactose, the *lac* repressor binds to the operator, and the operon is repressed (figure 16.4a). The effector that controls the DNA binding of the repressor is a metabolite of lactose, allolactose, which is produced when lactose is available. Allolactose binds to the repressor, altering its conformation so that it no longer can bind to the operator (figure 16.4b). The operon is now induced. Since allolactose allows induction of the operon, it is usually called the inducer.

As the level of lactose falls, allolactose concentrations decrease, making it no longer available to bind to the repressor and allowing the repressor to bind to DNA again. Thus this system of negative control by the *lac* repressor and its inducer, allolactose, allows the cell to respond to changing levels of lactose in the environment.

Even in the absence of lactose, the *lac* operon is expressed at a very low level. When lactose becomes available, it is transported into the cell and enough allolactose is produced that induction of the operon can occur.

The presence of glucose prevents induction of the *lac* operon

Glucose repression is a mechanism for the preferential use of glucose in the presence of other sugars such as lactose. If bacteria are grown in the presence of both glucose and lactose, the *lac* operon is not induced. As glucose is used up, the *lac* operon is induced, allowing lactose to be used as an energy source.

Despite the name *glucose repression*, this mechanism involves an activator protein that can stimulate transcription from multiple catabolic operons, including the *lac* operon. This activator, **catabolite activator protein (CAP),** is an allosteric protein with cAMP as an effector. This protein is also called **cAMP response protein (CRP)** because it binds cAMP, but we will use the name CAP to emphasize its role as a positive regulator. CAP alone does not bind to DNA, but binding of the effector cAMP to CAP changes its conformation such that it can bind to DNA (figure 16.5). The level of cAMP in cells is reduced in the presence of glucose so that no stimulation of transcription from CAP-responsive operons takes place.

The CAP–cAMP system was long thought to be the sole mechanism of glucose repression. But more recent research has indicated that the presence of glucose inhibits the transport of lactose into the cell. This deprives the cell of the *lac* operon inducer, allolactose, allowing the repressor to bind to the operator. This mechanism, called **inducer exclusion,** is now thought to be the main form of glucose repression of the *lac* operon.

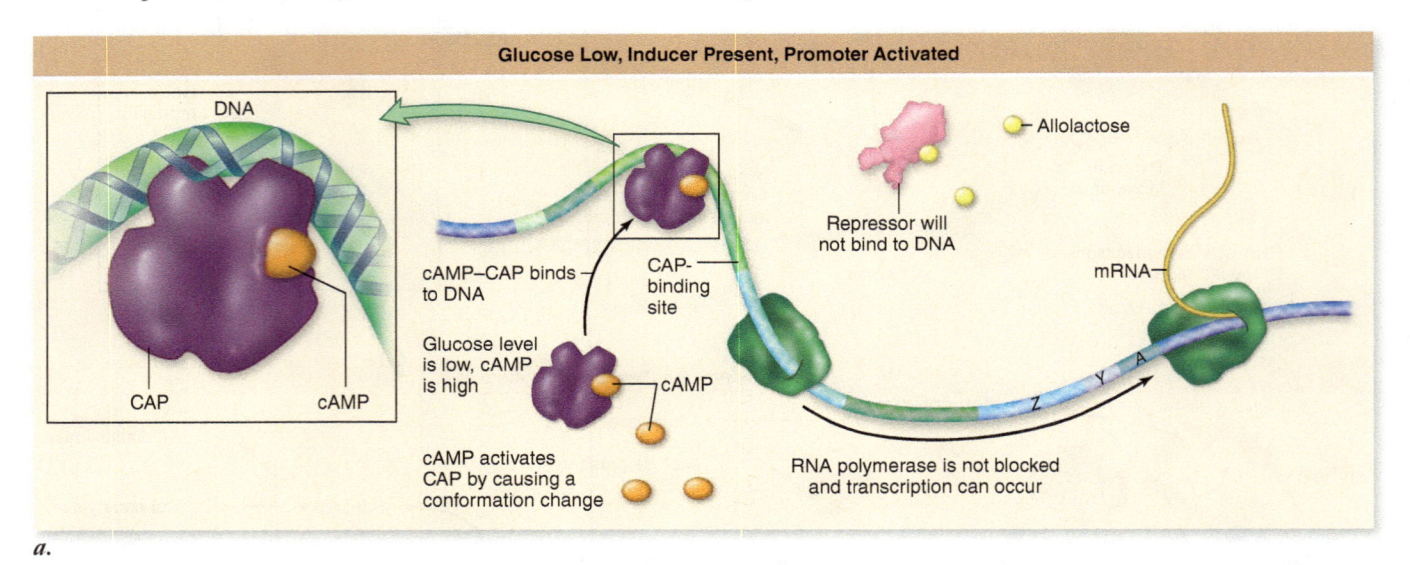

Figure 16.5 Effect of glucose on the *lac* operon. Expression of the *lac* operon is controlled by a negative regulator (repressor) and a positive regulator (CAP). The action of CAP is sensitive to glucose levels. *a.* For CAP to bind to DNA, it must bind to cAMP. When glucose levels are low, cAMP is abundant and binds to CAP. The CAP–cAMP complex causes the DNA to bend around it. This brings CAP into contact with RNA polymerase (not shown), making polymerase binding to the promoter more efficient. *b.* High glucose levels produce two effects: cAMP is scarce so CAP is unable to activate the promoter, and the transport of lactose is blocked (inducer exclusion).

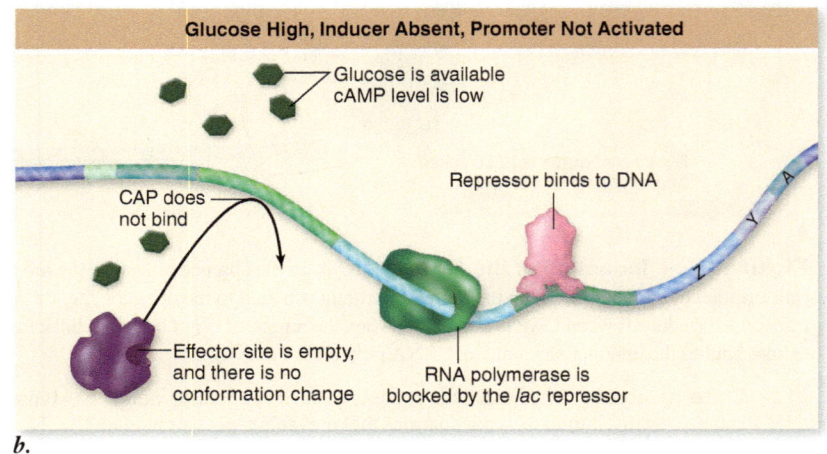

Given that inducer exclusion occurs, the role of CAP in the absence of glucose seems superfluous. But in fact, the positive control of CAP–cAMP is necessary because the promoter of the *lac* operon alone is not efficient in binding RNA polymerase. This inefficiency is overcome by the action of the positive control of the CAP–cAMP activator. Thus, the highest levels of expression occur in the absence of glucose and in the presence of lactose. In this case the presence of the activator, and absence of the repressor combine to produce the highest levels of expression (figure 16.5).

The *trp* operon is controlled by the *trp* repressor

Like the *lac* operon, the *trp* operon consists of a series of genes that encode enzymes involved in the same biochemical pathway. In the case of the *trp* operon these enzymes are necessary for synthesizing tryptophan. The regulatory region that controls transcription of these genes is located upstream of the genes. The *trp* operon is controlled by a repressor encoded by a gene located outside the *trp* operon. The *trp* operon is continuously expressed in the absence of tryptophan and is not expressed in the presence of tryptophan.

The *trp* repressor is a helix-turn-helix protein that binds to the operator site located adjacent to the *trp* promoter (figure 16.6). In the absence of tryptophan, the *trp* repressor does not bind to its operator, allowing expression of the operon, and production of the enzymes necessary to make tryptophan.

When levels of tryptophan rise, then tryptophan (the *corepressor*) binds to the repressor and alters its conformation, allowing it to bind to its operator. Binding of the

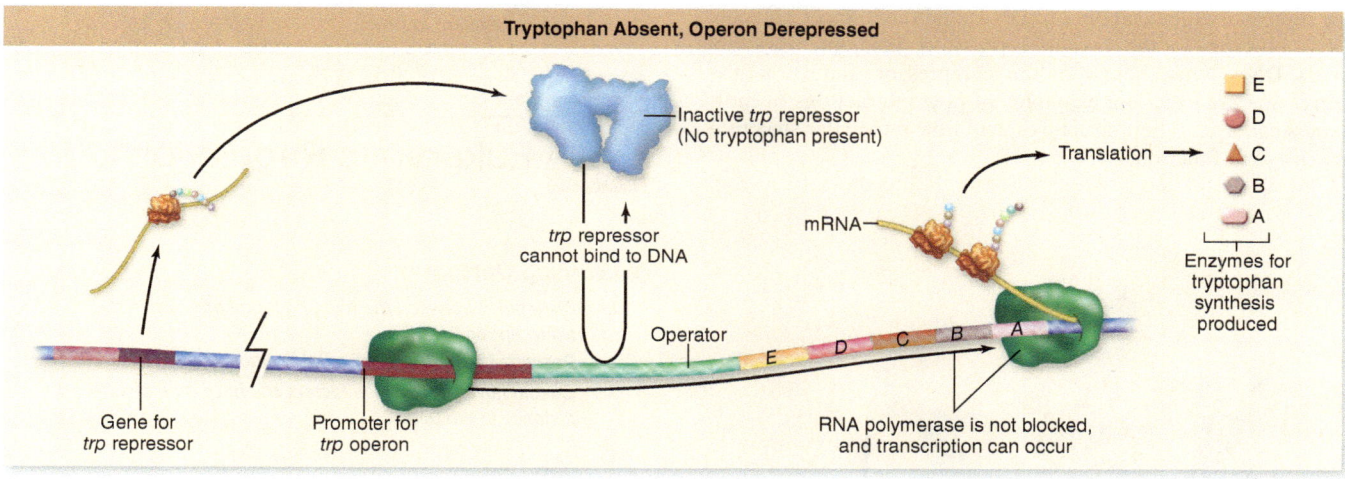

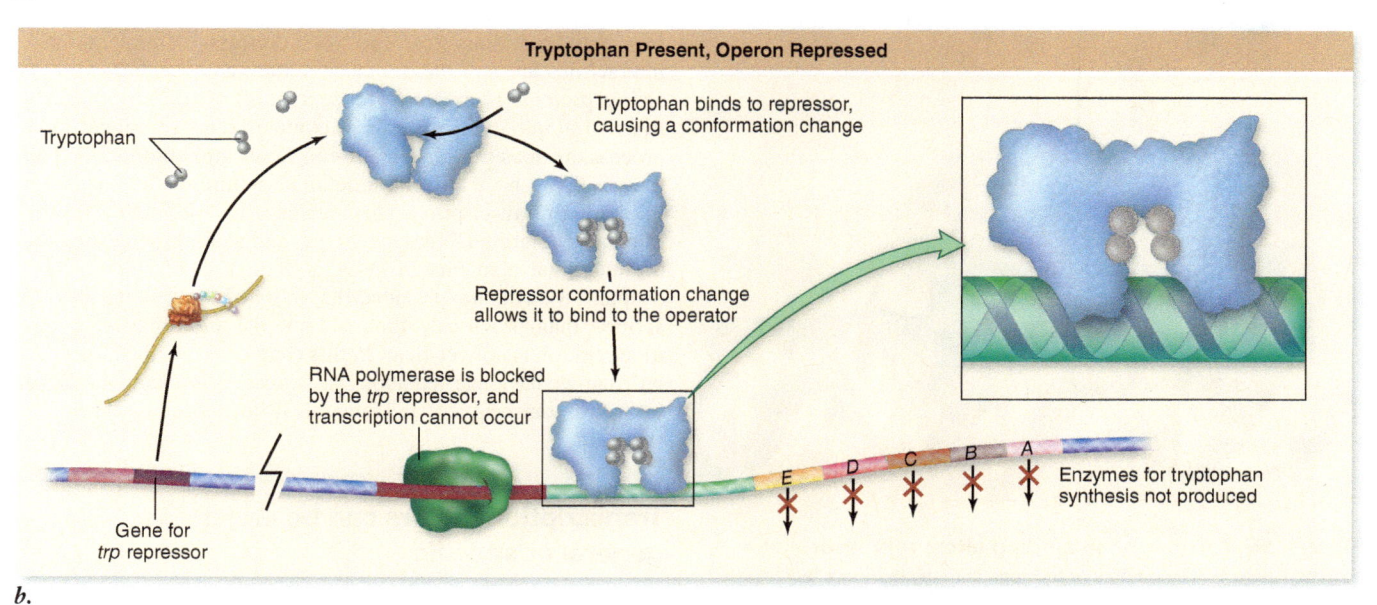

Figure 16.6 How the *trp* operon is controlled. The tryptophan operon encodes the enzymes necessary to synthesize tryptophan.
a. The tryptophan repressor alone cannot bind to DNA. The promoter is free to function, and RNA polymerase transcribes the operon.
b. When tryptophan is present, it binds to the repressor, altering its conformation so it now binds DNA. The tryptophan–repressor complex binds tightly to the operator, preventing RNA polymerase from initiating transcription.

repressor–corepressor complex to the operator prevents RNA polymerase from binding to the promoter. The actual change in repressor structure due to tryptophan binding is an alteration of the orientation of a pair of helix-turn-helix motifs that allows their recognition helices to fit into adjacent major grooves of the DNA (figure 16.7).

When tryptophan is present and bound to the repressor and this complex is bound to the operator, the operon is said to be *repressed*. As tryptophan levels fall, the repressor alone cannot bind to the operator, allowing expression of the operon. In this state, the operon is said to be **derepressed,** distinguishing this state from induction (see figure 16.6).

The key to understanding how both induction and repression can be due to negative regulation is knowledge of the behavior of repressor proteins and their effectors. In induction, the repressor alone can bind to DNA, and the inducer prevents DNA binding. In the case of repression, the repressor only binds DNA when bound to the corepressor. Induction and repression are excellent examples of how interactions of molecules can affect their structures, and how molecular structure is critical to function.

Learning Outcomes Review 16.3

Induction occurs when expression of genes in a pathway is turned on in response to a substrate; repression occurs when expression is prevented in response to a substrate. The *lac* operon is negatively controlled by a repressor protein that binds to DNA, thus preventing transcription. When lactose is present, the operon is turned on; allolactose binds to the repressor, which then no longer binds to DNA. This operon is also positively regulated by an activator protein. The *trp* operon is negatively controlled by a repressor protein that must be bound to tryptophan in order to bind to DNA. In the absence of tryptophan, the repressor cannot bind DNA, and the operon is derepressed.

■ What would be the effect on regulation of the trp operon of a mutation in the trp repressor that can still bind to trp, but no longer bind to DNA?

16.4 Eukaryotic Regulation

Learning Outcomes

1. Distinguish between the role of general and specific transcription factors.
2. Describe the formation of a Pol II initiation complex.
3. Explain how transcription factors can have an effect from a distance in the DNA.

The control of transcription in eukaryotes is much more complex than in prokaryotes. The basic concepts of protein–DNA interactions are still valid, but the nature and number of interacting proteins is greater. The separation of transcription and translation into different cellular compartments complicates the process, as does the organization of DNA into chromatin. Over the years, we have built up a detailed picture of the control of initiation of transcription. This view is still valid, but the advent of next-generation DNA sequencing, and genome-wide analysis of functional elements in DNA is changing this picture (see chapter 18). This has revealed that more of the genome is transcribed than previously thought, but the significance of this transcription is not yet clear. In this chapter, we will concentrate on what we have learned about how proteins interact with DNA to control transcription, and how this is influenced by chromatin structure.

Transcription factors can be either general or specific

In chapter 15, we introduced the concept of transcription factors. Eukaryotic transcription requires a variety of these protein factors, which fall into two categories: *general transcription*

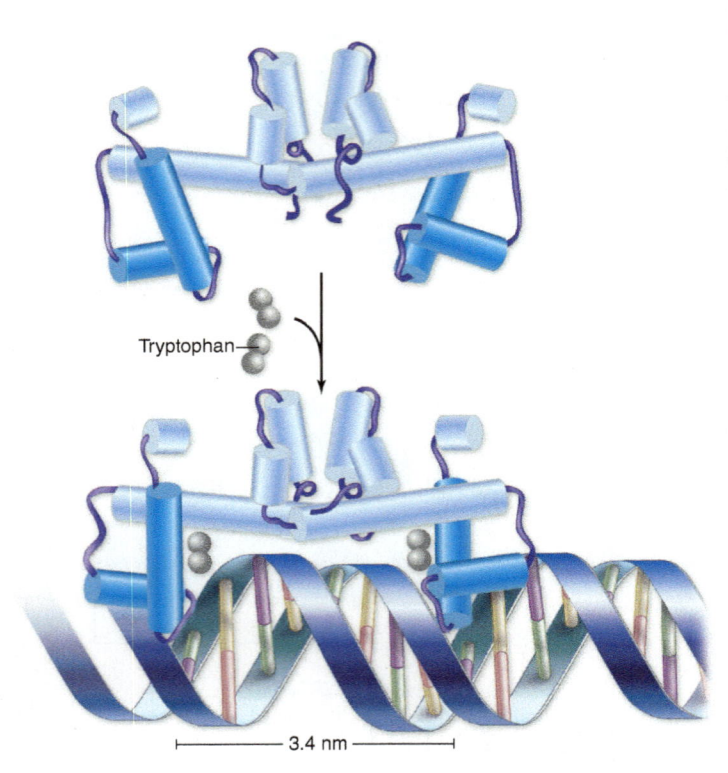

Figure 16.7 Tryptophan binding alters repressor conformation. The binding of tryptophan to the repressor increases the distance between the two recognition helices in the repressor, allowing the repressor to fit snugly into two adjacent portions of the major groove in DNA.

factors and *specific transcription factors*. General factors are necessary for the assembly of a transcription apparatus and recruitment of RNA polymerase II to a promoter. Specific factors increase the level of transcription in certain cell types or in response to signals.

General transcription factors

Transcription of RNA polymerase II templates (the majority being genes that encode protein products) requires more than just RNA polymerase II to initiate transcription. A host of **general transcription factors** are also necessary to establish productive initiation. These factors are required for transcription to occur, but they do not increase the rate above this basal rate.

General transcription factors are named with letter designations that follow the abbreviation TFII, for "transcription factor RNA polymerase II." The most important of these factors, TFIID, contains the TATA-binding protein that recognizes the TATA box sequence found in many eukaryotic promoters.

Binding of TFIID is followed by binding of TFIIE, TFIIF, TFIIA, TFIIB, and TFIIH and a host of accessory factors called *transcription-associated factors*, or TAFs. The *initiation complex* that results (figure 16.8) is clearly much more complex than the bacterial RNA polymerase holoenzyme binding to a promoter. And there is yet another level of complexity: The initiation complex, although capable of initiating synthesis at a basal level, does not achieve transcription at a high level without the participation of other, specific factors.

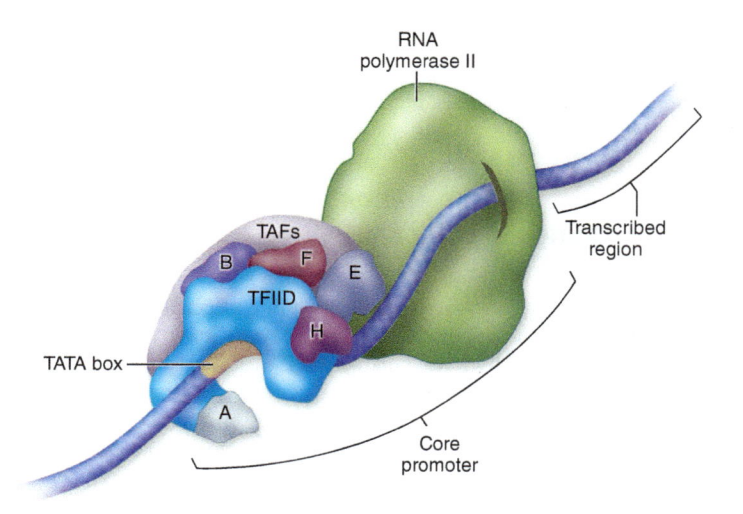

Figure 16.8 Formation of a eukaryotic initiation complex. The general transcription factor, TFIID, binds to the TATA box and is joined by the other general factors, TFIIE, TFIIF, TFIIA, TFIIB, and TFIIH. This complex is added to by a number of transcription-associated factors (TAFs) that together recruit the RNA Pol II molecule to the core promoter.

Specific transcription factors

Specific transcription factors act in a tissue- or time-dependent manner to stimulate higher levels of transcription than the basal level. The number and diversity of these factors are overwhelming. Some sense can be made of this proliferation of factors by concentrating on the DNA-binding motif, as opposed to the specific factors.

A key common theme that emerges from the study of these factors is that specific transcription factors, called *activators*, have a domain organization. Each factor consists of a DNA-binding domain and a separate activating domain that interacts with the transcription apparatus, and these domains are essentially independent in the protein. If the DNA-binding domains are "swapped" between different factors the binding specificity for the factors is switched without affecting their ability to activate transcription.

Promoters and enhancers are binding sites for transcription factors

Promoters, as mentioned in chapter 15, contain DNA-binding sites for general transcription factors. These factors then mediate the binding of RNA polymerase II to the promoter (and also the binding of RNA polymerases I and III to their specific promoters). In contrast, the holoenzyme portion of the RNA polymerase of prokaryotes can directly recognize a promoter and bind to it.

Enhancers were originally defined as DNA sequences necessary for high levels of transcription that can act independently of position or orientation. At first, this concept seemed counterintuitive, especially since molecular biologists had been conditioned by prokaryotic systems to expect control regions to be immediately upstream of the coding region. It turns out that enhancers are the binding site of the specific transcription factors. The ability of enhancers to act over large distances was at first puzzling, but investigators now think this action is accomplished by DNA bending to form a loop, positioning the enhancer closer to the promoter.

Although more important in eukaryotic systems, this looping was first demonstrated using prokaryotic DNA-binding proteins (figure 16.9). The important point is that the linear distance separating two sites on the chromosome does not have to translate to great physical distance, because the flexibility of DNA allows bending and looping. An activator bound to an enhancer can thus be brought into contact with the transcription factors bound to a distant promoter (figure 16.10).

Coactivators and mediators link transcription factors to RNA polymerase II

Other factors specifically mediate the action of transcription factors. These *coactivators* and *mediators* are also necessary for activation of transcription by the transcription factor. They act

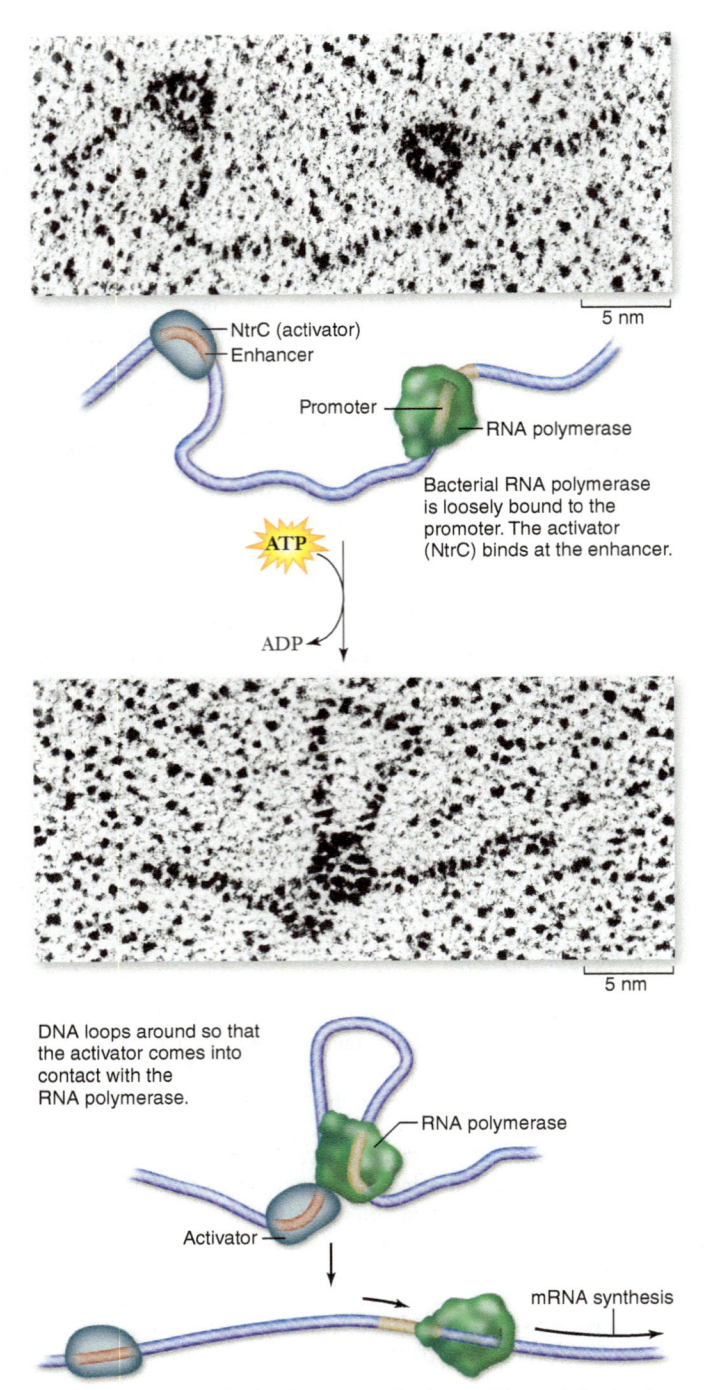

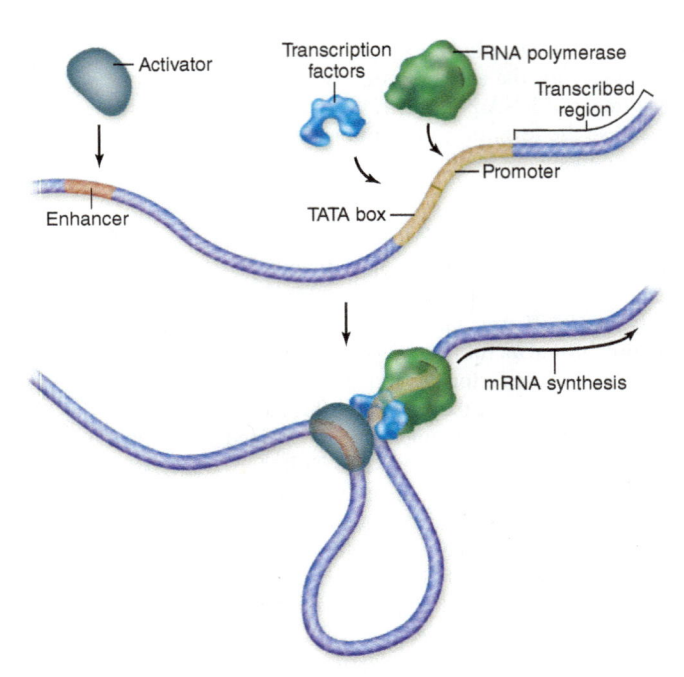

Figure 16.10 How enhancers work. The enhancer site is located far away from the gene being regulated. Binding of an activator *(gray)* to the enhancer allows the activator to interact with the transcription factors *(blue)* associated with RNA polymerase, stimulating transcription.

by binding the transcription factor and then binding to another part of the transcription apparatus. Mediators are essential to the function of some transcription factors, but not all transcription factors require them. The number of coactivators is much smaller than the number of transcription factors because the same coactivator can be used with multiple transcription factors.

The transcription complex brings things together

Although a few general principles apply to a broad range of situations, nearly every eukaryotic gene—or group of genes with coordinated regulation—represents a unique case. Virtually all genes that are transcribed by RNA polymerase II need the same suite of general factors to assemble an initiation complex, but the assembly of this complex and its ultimate level of transcription depend on specific transcription factors that in combination make up the **transcription complex** (figure 16.11).

The makeup of eukaryotic promoters, therefore, is either very simple, if we consider only what is needed for the initiation complex, or very complicated, if we consider all factors that may bind in a complex and affect transcription. This kind of combinatorial gene regulation leads to great flexibility because it can respond to the many signals a cell may receive affecting transcription, allowing integration of these signals.

Figure 16.9 DNA looping caused by proteins. When the bacterial activator NtrC binds to an enhancer, it causes the DNA to loop over to a distant site where RNA polymerase is bound, thereby activating transcription. Although such enhancers are rare in prokaryotes, they are common in eukaryotes.

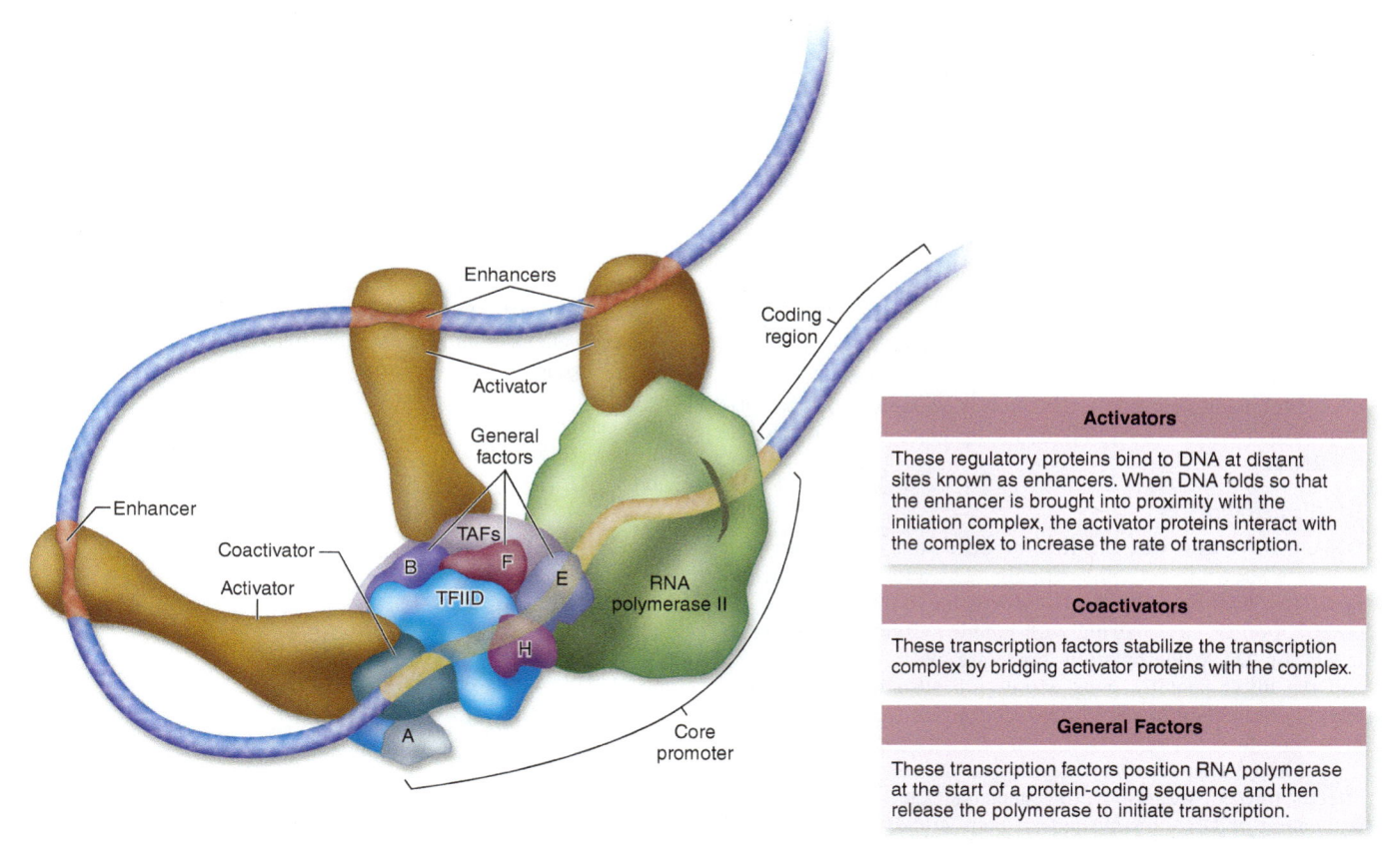

Figure 16.11 Interactions of various factors within the transcription complex. All specific transcription factors bind to enhancer sequences that may be distant from the promoter. These proteins can then interact with the initiation complex by DNA looping to bring the factors into proximity with the initiation complex. As detailed in the text, some transcription factors, called activators, can directly interact with the RNA polymerase II or the initiation complex, whereas others require additional coactivators.

Activators
These regulatory proteins bind to DNA at distant sites known as enhancers. When DNA folds so that the enhancer is brought into proximity with the initiation complex, the activator proteins interact with the complex to increase the rate of transcription.

Coactivators
These transcription factors stabilize the transcription complex by bridging activator proteins with the complex.

General Factors
These transcription factors position RNA polymerase at the start of a protein-coding sequence and then release the polymerase to initiate transcription.

Inquiry question How do eukaryotes coordinate the activation of many genes whose transcription must occur at the same time?

Learning Outcomes Review 16.4
In eukaryotes, initiation requires general transcription factors that bind to the promoter and recruit RNA polymerase II to form an initiation complex. General factors produce the basal level of transcription. Specific transcription factors, which bind to enhancer sequences, can increase the level of transcription. Enhancers can act at a distance because DNA can loop, bringing an enhancer and a promoter closer together. Additional coactivators and mediators link certain specific transcription factors to RNA polymerase II.

- *What would be the effect of a mutation that results in the loss of a general transcription factor versus the loss of a specific factor?*

16.5 Chromatin Structure and Gene Expression

Learning Outcomes
1. Describe at least two kinds of epigenetic mark.
2. Explain the function of chromatin-remodeling complexes.

In a diploid human cell, about 6×10^9 bp of DNA is wrapped into 3×10^7 nucleosomes. Further compaction forms the chromatin that fits about 2 m of DNA into a 5- to 10-μm nucleus. This obviously complicates the process of transcription. The way the cell deals with this conundrum is to selectively modulate the structure of chromatin by modifying DNA and histones, to allow the DNA transactions necessary for life. These alterations to chromatin are thought to be the basis for epigenetics—heritable changes in phenotype not due to changes in DNA sequence.

Methylation of DNA, modification of histones, and noncoding RNA all affect chromatin structure

One definition of an *epigenetic alteration* is that it must persist in the absence of the initiating stimulus, and be inherited through cell division. Alterations to chromatin structure can satisfy this definition, although the mechanisms of mitotic inheritance are not yet clear. Chromatin structure is affected by a wide variety of modifications to histones as well as DNA methylation, and even noncoding RNAs.

DNA methylation

DNA methylation was the first modification of chromosome structure shown to act epigenetically. The addition of a methyl group to cytosine creates 5-methylcytosine, but this change has no effect on its base-pairing with guanine (figure 16.12). High levels of DNA methylation correlate with inactive genes, and the allele-specific gene expression seen in genomic imprinting (see chapter 13) is mediated at least in part by DNA methylation.

Methylation satisfies the stringent definition of epigenetic effector through well-known mechanisms. The semiconservative replication of DNA methylated on both strands produces hemimethylated DNA, which becomes fully methylated by the action of a maintenance methylase. In humans, DNMT1 is the maintenance methylase, and other methylases can initiate the process, but do not need to remain active.

X-chromosome inactivation

Mammalian females inactivate one X chromosome as a form of dosage compensation to equalize X-chromosome expression in both sexes. First proposed by Mary Lyon in 1961, this continues to be an area of active research today. This is also the first example of a long noncoding RNA acting to regulate gene expression. The region of the chromosome that initiates the inactivation process contains a gene called *X-inactivation-specific transcript (Xist)*, which does not encode a protein. The process of inactivation is quite complex, but a streamlined version is that *Xist* "coats" the entire inactive X chromosome, and acts to recruit the chromatin regulator polycomb repressive complex 2 (PRC2). This leads to the modification of histones associated with inactive chromatin.

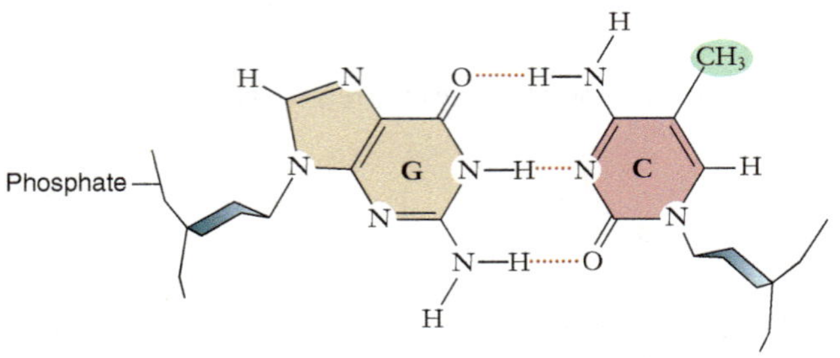

Histone modifications

The modification of histones is a complex topic as there are a large number of possible modifications, and four possible histones that can be modified. Modifications include acetylation and methylation of lysine; and phosphorylation of serine, threonine, and tyrosine. In general, acetylation, especially of H3, is correlated with active sites of transcription, both in regulatory regions and in the transcribed region of the gene itself. But methylation of the same histone (H3) can have the opposite effect, depending on the lysine methylated.

Some transcription activators alter chromatin structure

The control of eukaryotic transcription requires the presence of many different factors to activate transcription. Some activators seem to interact directly with the initiation complex or with coactivators that themselves interact with the initiation complex, as described in section 16.4. Other cases are not so clear. The emerging consensus is that some coactivators have been shown to be histone acetylases. In these cases, it appears that transcription is increased by removing higher order chromatin structure that would prevent transcription (figure 16.13). Some corepressors have been shown to be histone deacetylases as well.

These observations have led to the suggestion that a "histone code" might exist, analogous to the genetic code. This histone code is postulated to underlie the control of chromatin structure and, thus, of access of the transcription machinery to DNA.

Chromatin-remodeling complexes also change chromatin structure

The outline of how alterations to chromatin structure can regulate gene expression is beginning to emerge. A key discovery is the existence of so-called chromatin-remodeling complexes. These large complexes of proteins include enzymes that modify histones and DNA and that also change chromatin structure itself.

One class of these remodeling factors, ATP-dependent chromatin-remodeling factors, function as molecular motors that affect DNA and histones. These ATP-dependent remodeling factors use energy from ATP to alter the relationships between histones and DNA. They can catalyze four different changes in histone/DNA binding (figure 16.14): (1) sliding of nucleosome along DNA, which changes the position of a nucleosome on the DNA; (2) creation of a remodeled state where DNA is more accessible; (3) removal of nucleosomes from DNA; and (4) replacement of histones with variant histones. These functions all act to make DNA more accessible to regulatory proteins that in turn, affect gene expression.

Figure 16.12 DNA methylation. Cytosine is methylated, creating 5-methylcytosine. Because the methyl group *(green)* is positioned to the side, it does not interfere with the hydrogen bonds of a G–C base-pair, but it can be recognized by proteins.

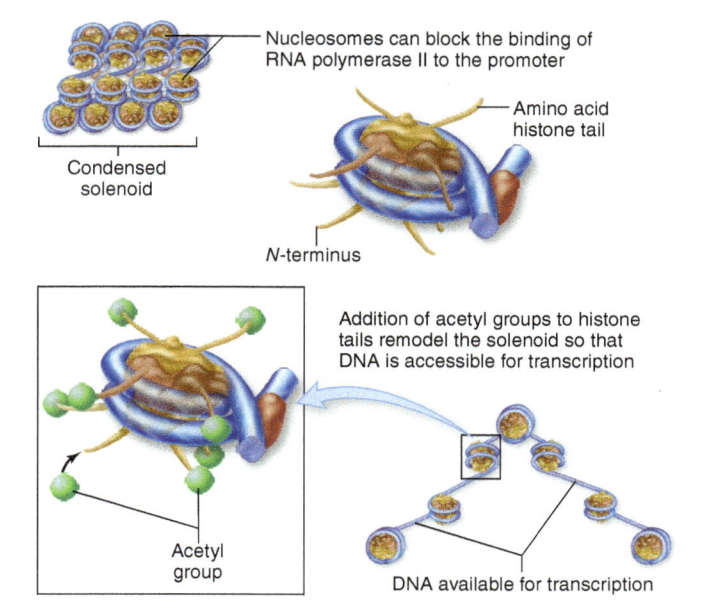

Figure 16.13 Histone modification affects chromatin structure. DNA in eukaryotes is organized first into nucleosomes and then into higher order chromatin structures. The histones that make up the nucleosome core have amino tails that protrude. These amino tails can be modified by the addition of acetyl groups. The acetylation alters the structure of chromatin, making it accessible to the transcription apparatus.

Learning Outcomes Review 16.5

Eukaryotic DNA is packaged into chromatin, adding a structural challenge to transcription. Changes in chromatin structure correlate with modification of DNA and histones, and appear to be the basis for epigenetic changes. Some transcriptional activators modify histones by acetylation. Large chromatin-remodeling complexes include enzymes that alter the structure of chromatin, making DNA more accessible to regulatory proteins.

- *Genes that are turned on in all cells are called "housekeeping" genes. Explain the idea behind this name.*

16.6 Eukaryotic Posttranscriptional Regulation

Learning Outcomes

1. Explain how small RNAs can affect gene expression.
2. Differentiate between the different kinds of posttranscriptional regulation.

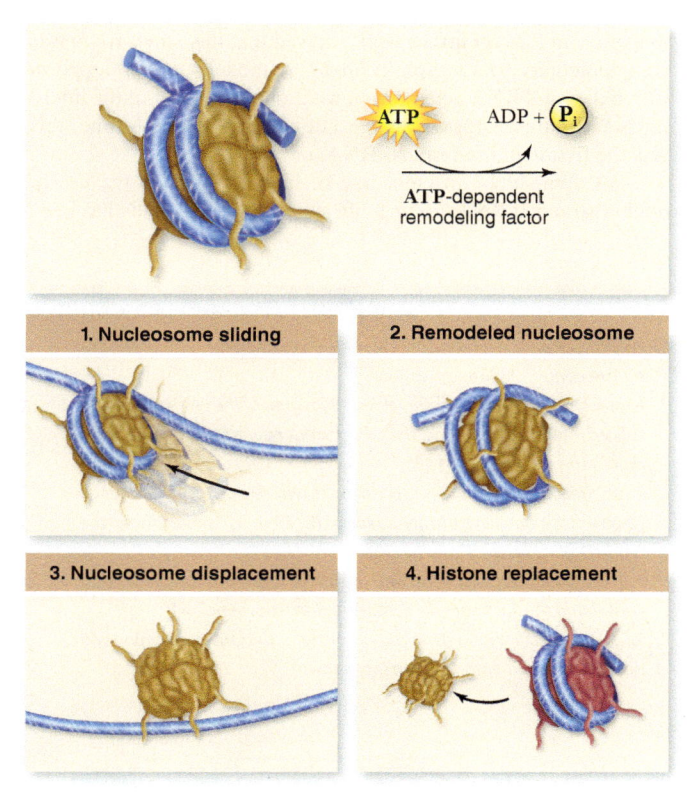

Figure 16.14 Function of ATP-dependent remodeling factors. ATP-dependent remodeling factors use the energy from ATP to alter chromatin structure. They can *(1)* slide nucleosomes along DNA to reveal binding sites for proteins; *(2)* create a remodeled state of chromatin where the DNA is more accessible; *(3)* completely remove nucleosomes from DNA; and *(4)* replace histones in nucleosomes with variant histones.

The separation of transcription in the nucleus and translation in the cytoplasm in eukaryotes provides possible points of regulation that do not exist in prokaryotes. For many years we thought of this as "alternative" forms of regulation, but it now appears that they play a much more central role than previously suspected. In this section we will consider several of these mechanisms for controlling gene expression, beginning with the exciting new area of regulation by small RNAs.

Small RNAs act after transcription to control gene expression

Developmental genetics has provided important insights into the regulation of gene expression. A striking example is the discovery of small RNAs that affect gene expression. The mutant *lin-4* was known to alter developmental timing in the worm *C. elegans*, and genetic studies had shown that *lin-4* regulated another gene, *lin-14*. When Ambros, Lee, and Feinbaum isolated the *lin-4* gene in 1992, they found it did not encode a protein product. Instead, the *lin-4* gene encoded only two small RNA molecules, one of 22 nt and the other of 61 nt. Furthermore, the 22-nt RNA was derived from the longer 61-nt RNA. Further work showed that this small RNA was complementary to a region in *lin-14*. A model was developed in which the lin-4 RNA acted as a translational repressor of the lin-14 mRNA (figure 16.15). Although not called that at the time, this was the first identified **micro-RNA, or miRNA.**

A completely different line of inquiry involved the use of double-stranded RNAs to turn off gene expression. This has been shown to act via another class of small RNA called **small interfering RNAs, or siRNAs.** These may be experimentally introduced, derived from invading viruses, or even encoded in the genome. The use of siRNA to control gene expression revealed the existence of cellular mechanisms for the control of gene expression via small RNAs.

Since its discovery, gene silencing by small RNAs has been a source of great interest for both its experimental uses, and as an explanation for posttranslational control of gene expression. Research has uncovered a wealth of new types of small RNAs, but we will confine ourselves to the two classes of miRNA and siRNA, as these are well established and illustrate the RNA silencing machinery.

miRNA genes

The discovery of the role of miRNAs in gene expression initially appeared to be confined to nematodes because the *lin-4* gene did not have any obvious homologues in other systems. Seven years later, a second gene, *let-7*, was discovered in the same pathway in *C. elegans*. The *let-7* gene also encoded a 22-nt RNA that could influence translation. In this case, homologues for *let-7* were immediately found in both *Drosophila* and humans.

As an increasing number of miRNAs were discovered in different organisms, miRNA gene discovery has turned to computer searching and high-throughput methods such as microarrays and new next-generation sequencing. A database devoted to miRNAs currently lists 1881 known human miRNA sequences.

Genes for miRNA are found in a variety of locations, including the introns of expressed genes, and they are often clustered with multiple miRNAs in a single transcription unit. They are also found in regions of the genome that were previously considered transcriptionally silent. This finding is particularly exciting because other work looking at transcription across animal genomes has found that much of what we thought was transcriptionally silent is actually not.

miRNA biogenesis and function

The production of a functional miRNA begins in the nucleus and ends in the cytoplasm with an ~22-nt RNA that functions to repress gene expression (figure 16.16). The initial transcript of a miRNA gene occurs by RNA polymerase II producing a transcript called the pri-miRNA. The region of this transcript containing the miRNA can fold back on itself and base-pair to form a stem-and-loop structure. This is cleaved in the nucleus by a nuclease called Drosha that trims the miRNA to just the stem-and-loop structure, which is now called the pre-miRNA. This pre-miRNA is exported from the nucleus through a nuclear pore bound to the protein exportin 5. Once in the cytoplasm, the pre-miRNA is further cleaved by another nuclease called Dicer to produce a short double-stranded RNA containing the miRNA. The miRNA is loaded into a complex of proteins called an **RNA-induced silencing complex, or RISC.** The RISC includes the RNA-binding protein Argonaute (Ago), which interacts with the miRNA. The complementary strand is either removed by a nuclease or is removed during the loading process.

SCIENTIFIC THINKING

Hypothesis: The region of the lin-14 gene complementary to the lin-4 miRNA controls lin-14 expression.

Prediction: If the lin-4 complementary region of the lin-14 gene is spliced into a reporter gene, then this reporter gene should show regulation similar to lin-14.

Test: Recombinant DNA is used to make two versions of a reporter gene (β-galactosidase). In transgenic worms (C. elegans), expression of the reporter gene produces a blue color.

1. The β-galactosidase gene with the lin-14 3' untranslated region containing the lin-4 complementary region (shown below)
2. The β-galactosidase gene with a control 3' untranslated region with no lin-4 complementary region (not shown)

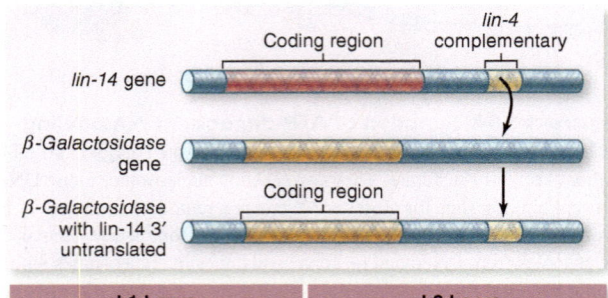

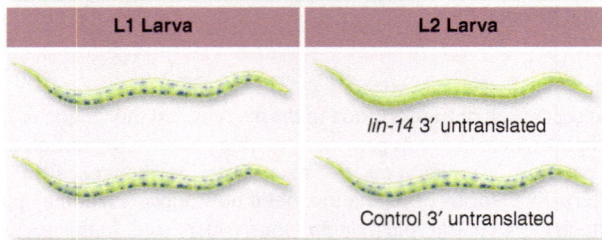

Result:
1. Transgenic worms with reporter gene plus lin-14 3' untranslated region show expression in L1 but not L2 stage larvae. This is the pattern expected for the lin-14 gene, which is controlled by lin-4.
2. Transgenic worms with reporter gene with control 3' untranslated region do not show expression pattern expected for control by lin-4.

Conclusion: The 3' untranslated region from lin-14 is sufficient to turn off gene expression in L2 larvae.

Further Experiments: What expression pattern would you predict for these constructs in a mutant that lacks lin-4 function?

Figure 16.15 Control of *lin-14* gene expression. The *lin-14* gene is controlled by the *lin-4* gene. This is mediated by a region of the 3' untranslated region of the *lin-14* mRNA that is complementary to *lin-4* miRNA.

In plants, the cleavage of the mRNA by the RISC is common and seems to be related to the more precise complementarity found between plant miRNAs and their targets than that found in animal systems.

RNA interference

Small RNA-mediated gene silencing has been known for a number of years. Some confusion arose in the nomenclature in this area because work in different systems led to a profusion of names. However, RNA interference, cosuppression, and posttranscriptional gene silencing all act through similar biochemical mechanisms. The term **RNA interference** is currently the most commonly used and involves the production of siRNAs.

The production of siRNAs is similar to that of miRNAs, except that they arise from a long piece of double-stranded RNA (figure 16.17). This can be either a very long region of self complementarity, or from two complementary RNAs. These long double-stranded RNAs are processed by Dicer to yield multiple siRNAs that are loaded into an Ago containing RISC. The siRNAs usually have near-perfect complementarity to their target mRNAs, and the result is cleavage of the mRNA by the siRNA containing RISC.

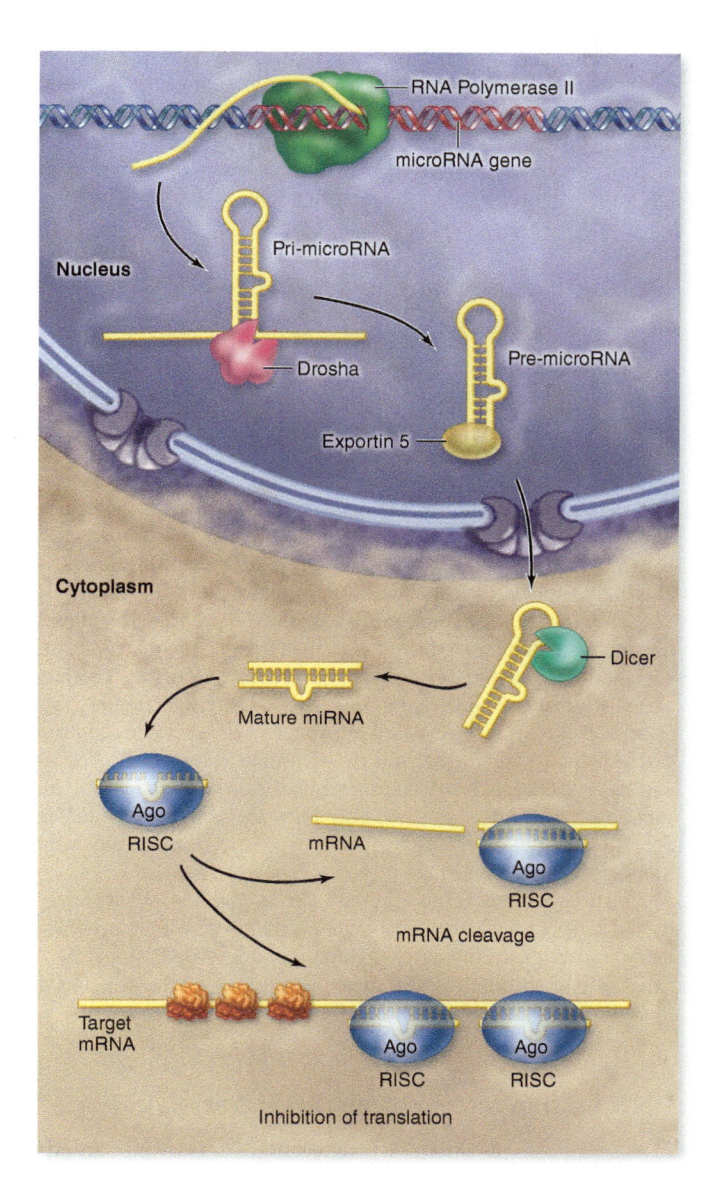

Figure 16.16 Biogenesis and function of miRNA.
Genes for miRNAs are transcribed by RNA polymerase II to produce a pri-miRNA. This is processed by the Drosha nuclease to produce the pre-miRNA, which is exported from the nucleus bound to export factor Exportin 5. Once in the cytoplasm, the pre-miRNA is processed by Dicer nuclease to produce the mature miRNA. The miRNA is loaded into a RISC, which can act to either cleave target mRNAs, or to inhibit translation of target mRNAs.

At this point, the RISC is targeted to repress the expression of other genes based on sequence complementarity to the miRNA. The complementary region is usually in the 3′ untranslated region of genes, and the result can be cleavage of the mRNA or inhibition of translation. It appears that in animals, the inhibition of translation is more common than the cleavage of the mRNA, although the precise mechanism of this inhibition is still unclear.

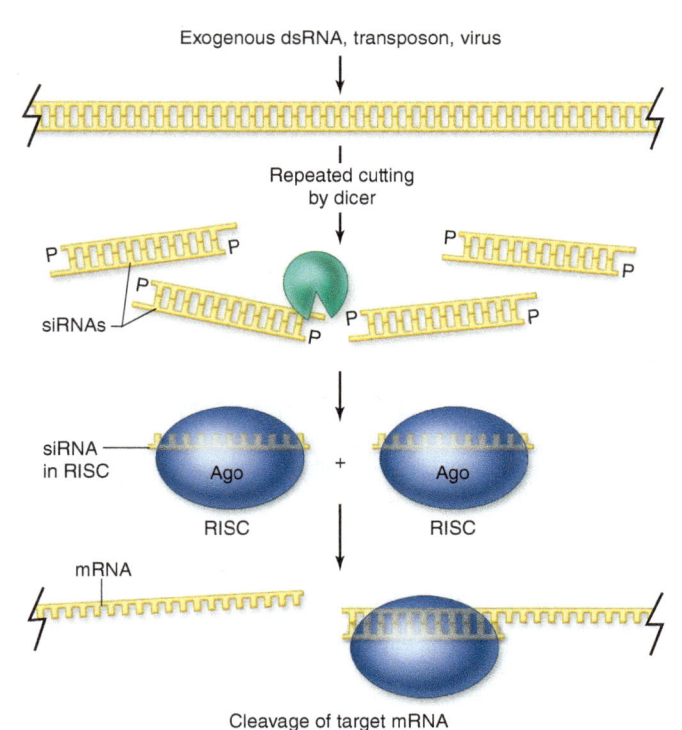

Figure 16.17 Biogenesis and function of siRNA. SiRNAs can arise from a variety of sources that all produce long double-stranded regions of RNA. The double-stranded RNA is processed by Dicer nuclease to produce a number of siRNAs that are each loaded onto their own RISC. The RISC then cleaves target mRNA.

The source of the double-stranded RNA to produce siRNAs can be either from the cell or from outside the cell. From the cell itself, genes can produce RNAs with long regions of self-complementarity that fold back to produce a substrate for Dicer in the cytoplasm. They can also arise from repeated regions of the genome that contain transposable elements. Exogenous double-stranded RNAs can be introduced experimentally or by infection with a virus.

Small RNAs may have evolved to protect the genome

The observation that viral RNA can be degraded via the RNA-silencing pathway may point toward the evolutionary origins of small RNAs. A related observation is that RNA silencing can control the action of transposons as well. In both mice and fruit flies, genetic evidence supports the involvement of the RNA interference machinery in the germ line where a specific class of small RNA appears to be involved in silencing transposons during spermatogenesis and oogenesis. Thus the origins of this mechanism may be an ancient pathway for protection of the genome from assault from both within and without. The conservation of key proteins suggests that the ancestor to all eukaryotes had some form of RNA-silencing pathway.

Distinguishing miRNAs and siRNAs

The biogenesis of both miRNA and siRNA involves cleavage by Dicer and incorporation into a RISC complex. The main thing that distinguishes these small RNAs is their targets: miRNAs tend to repress genes different from their origin, whereas endogenous siRNAs tend to repress the genes they were derived from. Additionally, siRNAs are used experimentally to turn off the expression of genes. This technique uses the cells' RNA-silencing machinery to turn off a gene by introducing a double-stranded RNA complementary to the gene.

The two classes of small RNA have other differences. When multiple species are examined, miRNAs tend to be evolutionarily conserved, but siRNAs do not. Although the biogenesis for both is similar in terms of the nucleases involved, the actual structure of the double-stranded RNAs is not the same. The transcript of miRNA genes form stem-loop structures containing the miRNA, but the double-stranded RNAs generating siRNAs may be bimolecular, or very long stem-loops. These longer double-stranded regions lead to multiple siRNAs, whereas only a single miRNA is generated from a pre-miRNA.

Small RNAs can mediate heterochromatin formation

RNA-silencing pathways have also been implicated in the formation of heterochromatin in fission yeast, plants, and *Drosophila*. In fission yeast, centromeric heterochromatin formation is driven by siRNAs produced by the action of the Dicer nuclease. This heterochromatin formation also involves modification of histone proteins and thus connects RNA interference with chromatin-remodeling complexes in this system. It is not yet clear how widespread this phenomenon is.

Plants are an interesting case in that they have a variety of small RNA species. The RNA interference pathway in plants is more complex than that in animals, with multiple forms of Dicer nuclease proteins and Argonaute RNA-binding proteins. One class of endogenous siRNA can lead to heterochromatin formation by DNA methylation and histone modification.

Alternative splicing can produce multiple proteins from one gene

The latest estimates of the frequency of alternative splicing detailed in chapter 15 emphasize its importance. However, the functional significance of these data is still not clear. Here we will consider some well characterized examples.

Alternative splicing can change the splicing events that occur during different stages of development or in different tissues. An example of developmental differences is found in *Drosophila*, in which sex determination is the result of a complex series of alternative splicing events that differ in males and females.

An excellent example of tissue-specific alternative splicing in action is found in two different human organs: the thyroid gland and the hypothalamus in chapter 45.) These two organs produce two distinct hormones: *calcitonin* and *CGRP* (calcitonin-gene-related peptide) as part of their function. Calcitonin controls calcium uptake and the balance of calcium in tissues such as bones and teeth. CGRP is involved in a number of neural and endocrine functions. Although these two hormones are used for very different physiological purposes, they are produced from the same transcript (figure 16.18).

The synthesis of one product versus another is determined by tissue-specific factors that regulate the processing of the primary transcript. In the case of calcitonin and CGRP, pre-mRNA splicing is controlled by different factors that are present in the thyroid and in the hypothalamus.

RNA editing alters mRNA after transcription

In some cases, the editing of mature mRNA transcripts can produce an altered mRNA that is not truly encoded in the genome—an unexpected possibility. RNA editing was first discovered as the insertion of uracil residues into some RNA transcripts in protozoa, and it was thought to be an anomaly.

RNA editing of a different sort has since been found in mammalian species, including humans. In this case, the editing involves chemical modification of a base to change its base-pairing properties, usually by deamination. For example, both deamination of cytosine to uracil and deamination of adenine to inosine have been observed (inosine pairs as G would during translation).

The 5-HT serotonin receptor

RNA editing has been observed in some brain receptors for opiates in humans. One of these receptors, the serotonin (5-HT) receptor, is edited at multiple sites to produce a total of 12 different isoforms of the protein.

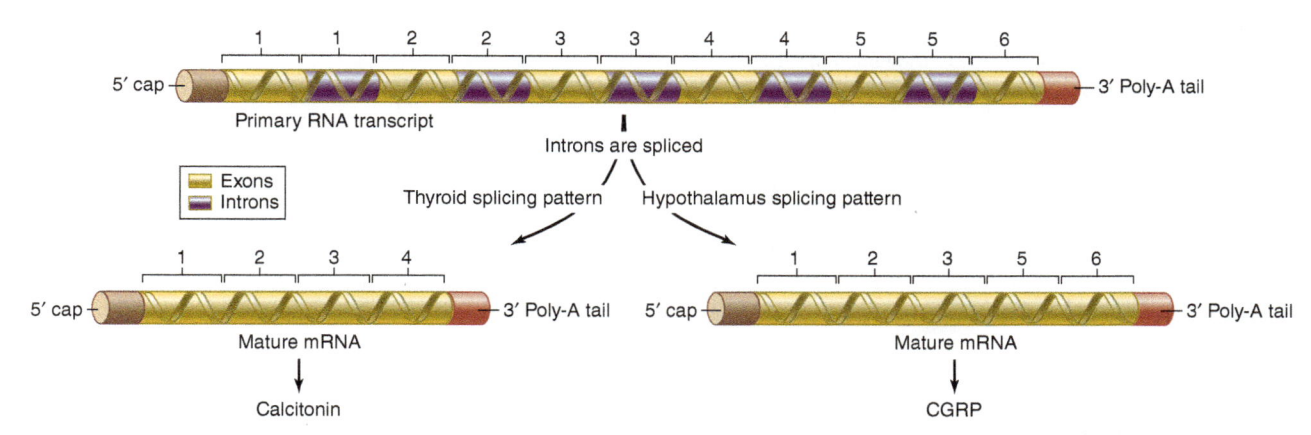

Figure 16.18 Alternative splicing. Many primary transcripts can be spliced in different ways to give rise to multiple mRNAs. In this example, in the thyroid the primary transcript is spliced to contain four exons encoding the protein calcitonin. In the hypothalamus the fourth exon, which contains the poly-A site used in the thyroid, is skipped and two additional exons are added to encode the protein calcitonin-gene-related peptide (CGRP).

It is unclear how widespread RNA editing is, but it is further evidence that the information encoded within genes is not the end of the story for protein production.

mRNA must be transported out of the nucleus for translation

Processed mRNA transcripts exit the nucleus through the nuclear pores (described in chapter 4). The passage of a transcript across the nuclear membrane is an active process that requires the transcript to be recognized by receptors lining the interior of the pores. Specific portions of the transcript, such as the poly-A tail, appear to play a role in this recognition.

There is little hard evidence that gene expression is regulated at this point, although it could be. On average, about 10% of primary transcripts consists of exons that will make up mRNA sequences, but only about 5% of the total mRNA produced as primary transcript ever reaches the cytoplasm. This observation suggests that about half of the exons in primary transcripts never leave the nucleus, but it is unclear whether the disappearance of this mRNA is selective.

Initiation of translation can be controlled

The translation of a processed mRNA transcript by ribosomes in the cytoplasm involves a complex of proteins called *translation factors*. In at least some cases, gene expression is regulated by modification of one or more of these factors. In other instances, **translation repressor proteins** shut down translation by binding to the beginning of the transcript, so that it cannot attach to the ribosome.

In humans, the production of ferritin (an iron-storing protein) is normally shut off by a translation repressor protein called aconitase. Aconitase binds to a 30-nt sequence at the beginning of the ferritin mRNA, forming a stable loop to which ribosomes cannot bind. When iron enters the cell, the binding of iron to aconitase causes the aconitase to dissociate from the ferritin mRNA, freeing the mRNA to be translated and increasing ferritin production 100-fold.

The degradation of mRNA is controlled

Another aspect that affects gene expression is the stability of mRNA transcripts in the cell cytoplasm. Unlike prokaryotic mRNA transcripts, which typically have a half-life of about 3 min, eukaryotic mRNA transcripts are very stable. For example, β-globin gene transcripts have a half-life of over 10 hr, an eternity in the fast-moving metabolic life of a cell.

The transcripts encoding regulatory proteins and growth factors, however, are usually much less stable, with half-lives of less than 1 hr. What makes these particular transcripts so unstable? In many cases, they contain specific sequences near their 3′ ends that make them targets for enzymes that degrade mRNA. A sequence of A and U nucleotides near the 3′ poly-A tail of a transcript promotes removal of the tail, which destabilizes the mRNA.

Loss of the poly-A tail leads to rapid degradation by 3′ to 5′ RNA exonucleases. Another consequence of this loss is the stimulation of decapping enzymes that remove the 5′ cap leading to degradation by 5′ to 3′ RNA exonucleases.

Other mRNA transcripts contain sequences near their 3′ ends that are recognition sites for endonucleases, which cause these transcripts to be digested quickly. The short half-lives of the mRNA transcripts of many regulatory genes are critical to the function of those genes because they enable the levels of regulatory proteins in the cell to be altered rapidly.

A review of various methods of posttranscriptional control of gene expression is provided in figure 16.19.

1. Initiation of transcription
Transcription is controlled by the frequency of initiation. This involves transcription factors that bind to promoters and enhancers.

2. RNA splicing
Gene expression can be controlled by altering the rate of splicing in eukaryotes. Alternative splicing can produce multiple mRNAs from one gene.

3. Passage through the nuclear membrane
Gene expression can be regulated by controlling access to or efficiency of transport channels.

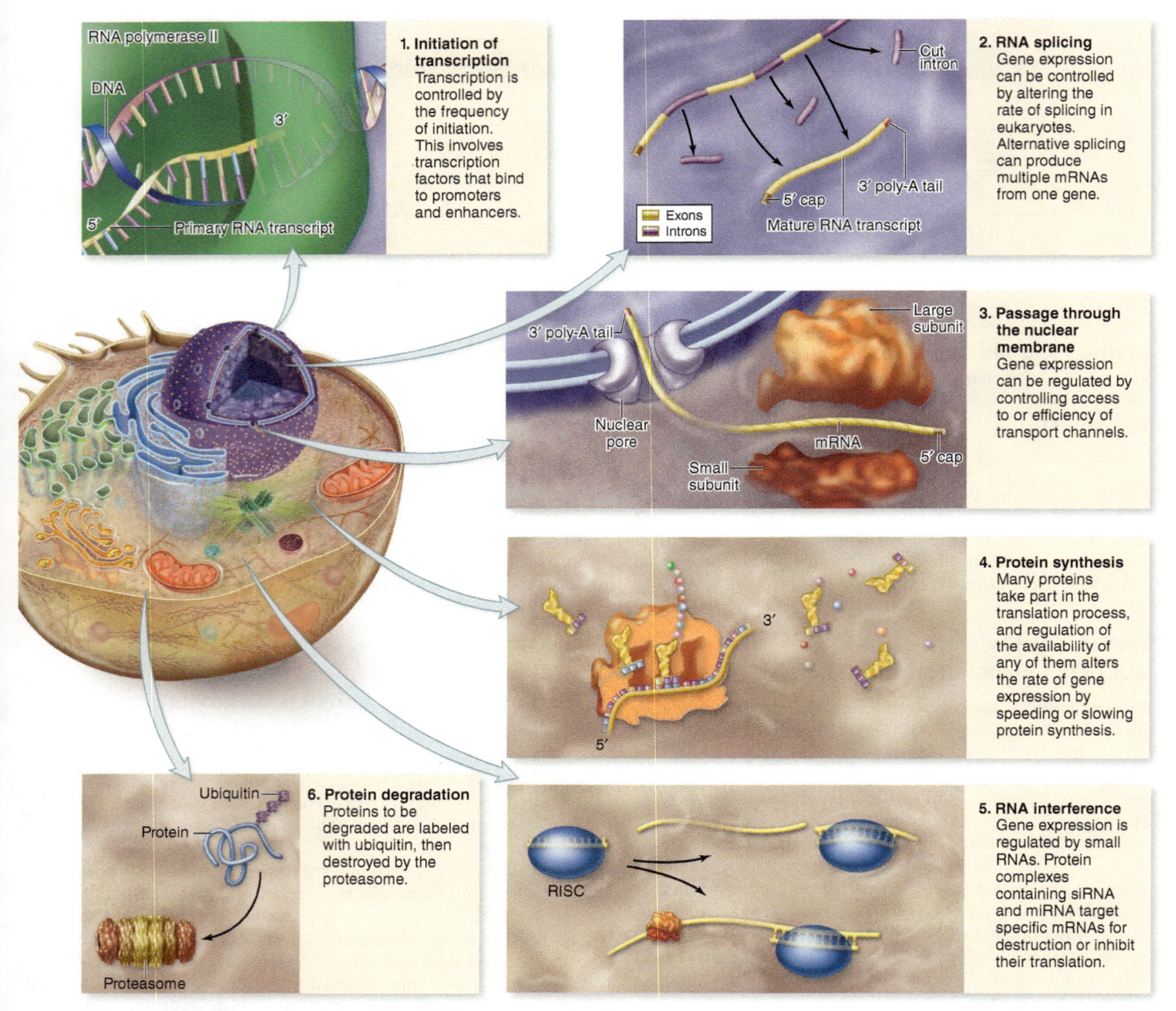

4. Protein synthesis
Many proteins take part in the translation process, and regulation of the availability of any of them alters the rate of gene expression by speeding or slowing protein synthesis.

6. Protein degradation
Proteins to be degraded are labeled with ubiquitin, then destroyed by the proteasome.

5. RNA interference
Gene expression is regulated by small RNAs. Protein complexes containing siRNA and miRNA target specific mRNAs for destruction or inhibit their translation.

Figure 16.19 Mechanisms for control of gene expression in eukaryotes.

Learning Outcomes Review 16.6
Small RNAs control gene expression by either selective degradation of mRNA, inhibition of translation, or alteration of chromatin structure. Multiple mRNAs can be formed from a single gene via alternative splicing, which can be tissue- and developmentally specific. The sequence of an mRNA transcript can also be altered by RNA editing.

■ *How could the phenomenon of RNA interference be used in drug design?*

16.7 Protein Degradation

Learning Outcomes
1. Describe the role of ubiquitin in the degradation of proteins.
2. Explain the function of the proteasome.

If all of the proteins produced by a cell during its lifetime remained in the cell, serious problems would arise. Protein labeling studies in the 1970s indicated that eukaryotic cells turn over proteins in a controlled manner. That is, proteins are continually being synthesized and degraded. Although this protein turnover is not as rapid as in prokaryotes, it indicates that a system regulating protein turnover is important.

Proteins can become altered chemically, rendering them nonfunctional; in addition, the need for any particular protein may be transient. Proteins also do not always fold correctly, or they may become improperly folded over time. These changes can lead to loss of function or other chemical behaviors, such as aggregating into insoluble complexes. In fact, a number of neurodegenerative diseases, such as Alzheimer dementia, Parkinson disease, and mad cow disease, are related to proteins that aggregate, forming characteristic plaques in brain cells. Thus, in addition to normal turnover of proteins, cells need a mechanism to get rid of old, unused, and incorrectly folded proteins.

Enzymes called **proteases** can degrade proteins by breaking peptide bonds, converting a protein into its constituent amino acids. Although there is an obvious need for these enzymes, they clearly cannot be floating around in the cytoplasm active at all times.

One way that eukaryotic cells handle such problems is to confine destructive enzymes to a specific cellular compartment. You may recall from chapter 4 that lysosomes are vesicles that contain digestive enzymes, including proteases. Lysosomes are used to remove proteins and old or nonfunctional organelles, but this system is not specific for particular proteins. Cells need another regulated pathway to remove proteins that are old or unused, but leave the rest of cellular proteins intact.

Addition of ubiquitin marks proteins for destruction

Eukaryotic cells solve this problem by marking proteins for destruction, then selectively degrading them. The mark that cells use is the attachment of a **ubiquitin** molecule. Ubiquitin, so named because it is found in essentially all eukaryotic cells (that is, it is ubiquitous), is a 76–amino-acid protein that can exist as an isolated molecule or in longer chains that are attached to other proteins.

The longer chains are added to proteins in a stepwise fashion by an enzyme called *ubiquitin ligase* (figure 16.20). This reaction requires ATP and other proteins, and it takes place in a multistep, regulated process. Proteins that have a ubiquitin chain attached are called *polyubiquitinated,* and this state is a signal to the cell to destroy this protein.

Two basic categories of proteins become ubiquitinated: those that need to be removed because they are improperly folded or nonfunctional, and those that are produced and degraded in a controlled fashion by the cell. An example of the latter are the cyclin proteins that help to drive the cell cycle (see chapter 10). When these proteins have fulfilled their role in active division of the cell, they become polyubiquitinated and are removed. In this way, a cell can control entry into cell division or maintain a nondividing state.

The proteasome degrades polyubiquitinated proteins

The cellular organelle that degrades proteins marked with ubiquitin is the **proteasome,** a large cylindrical complex that proteins enter at one end and exit the other as amino acids or peptide fragments (figure 16.21).

The proteasome complex contains a central region that has protease activity and regulatory components at each end. Although not membrane-bounded, this organelle can be thought of as a form of compartmentalization on a very small scale. By using a two-step process, first to mark proteins for destruction, then to process them through a large complex, proteins to be degraded are isolated from the rest of the cytoplasm.

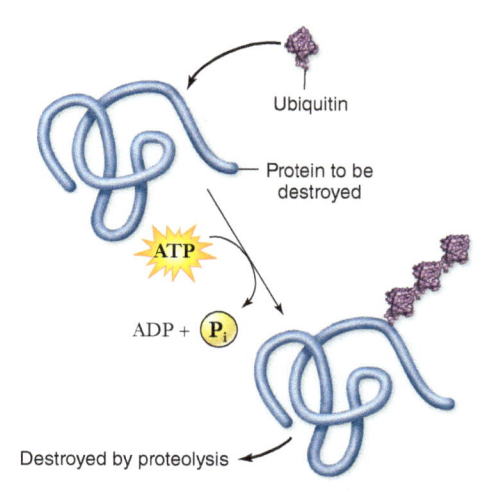

Figure 16.20 Ubiquitination of proteins. Proteins that are to be degraded are marked with ubiquitin. The enzyme ubiquitin ligase uses ATP to add ubiquitin to a protein. When a series of these have been added, the polyubiquitinated protein is destroyed.

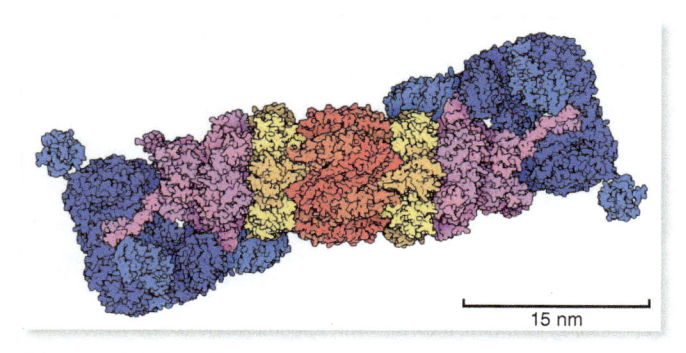

Figure 16.21 The *Drosophila* proteasome. The central complex contains the proteolytic activity, and the flanking regions act as regulators. Proteins enter one end of the cylinder and are cleaved to peptide fragments that exit the other end.

The process of ubiquitination followed by degradation by the proteasome is called the *ubiquitin–proteasome pathway*. It can be thought of as a cycle in that the ubiquitin added to proteins is not itself destroyed in the proteasome. As the proteins are degraded, the ubiquitin chain itself is simply cleaved back into ubiquitin units that can then be reused (figure 16.22).

Learning Outcomes Review 16.7
Control of protein degradation in eukaryotes involves addition of the protein ubiquitin, which marks the protein for destruction. The proteasome, a cylindrical complex with protease activity in its center, recognizes ubiquitinated proteins and breaks them down, much like a shredder destroys documents. Ubiquitin is recycled unchanged.

- If the ubiquitination process was not tightly controlled, what effect would this have on a cell?

Inquiry question What are two reasons a cell would polyubiquitinate a polypeptide?

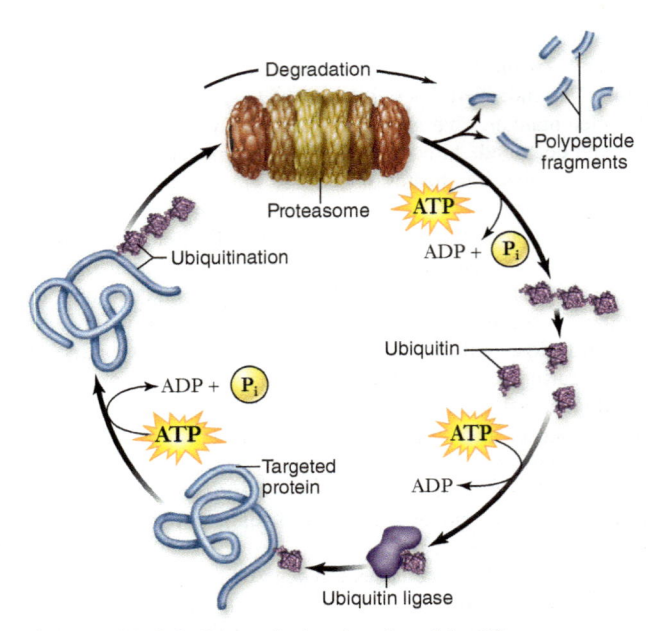

Figure 16.22 Degradation by the ubiquitin–proteasome pathway. Proteins are first ubiquitinated, then enter the proteasome to be degraded. In the proteasome, the polyubiquitin is removed and then is later "deubiquitinated" to produce single ubiquitin molecules that can be reused.

Chapter Review

16.1 Control of Gene Expression

Control can occur at all levels of gene expression.
Transcription is controlled by regulatory proteins that modulate the ability of RNA polymerase to bind to the promoter. These may either block transcription or stimulate it.

Control strategies in prokaryotes are geared to adjust to environmental changes.

Control strategies in eukaryotes maintain homeostasis and drive development.

16.2 Regulatory Proteins

Proteins can interact with DNA through the major groove.
A DNA double helix exhibits a major groove and a minor groove; bases in the major groove are accessible to regulatory proteins.

DNA-binding domains interact with specific DNA sequences.
A region of the regulatory protein that can bind to the DNA is termed a DNA-binding motif (figure 16.2).

Several common DNA-binding motifs are shared by many proteins.

Common motifs include the helix-turn-helix motif, the homeodomain motif, the zinc finger motif, and the leucine zipper.

16.3 Prokaryotic Regulation

Control of transcription can be either positive or negative.
Negative control is mediated by proteins called repressors that interfere with transcription. Positive control is mediated by a class of regulatory proteins called activators that stimulate transcription.

Prokaryotes adjust gene expression in response to environmental conditions.
The *lac* operon is induced in the presence of lactose; that is, the enzymes to utilize lactose are only produced when lactose is present. The *trp* operon is repressed; that is, the enzymes needed to produce tryptophan are turned off when tryptophan is present.

The lac operon is negatively regulated by the lac repressor.
The *lac* operon is induced when the effector (allolactose) binds to the repressor, altering its conformation such that it no longer binds DNA (figure 16.4).

The presence of glucose prevents induction of the lac operon.
Maximal expression of the *lac* operon requires positive control by catabolite activator protein (CAP) complexed with cAMP. When glucose is low, cAMP is high. Glucose repression involves both inducer

exclusion, in which lactose is prevented from entering the cell, and the control of CAP function by the level of glucose.

The trp operon is controlled by the trp repressor.
The *trp* operon is repressed when tryptophan, acting as a corepressor, binds to the repressor, altering its conformation such that it can bind to DNA and turn off the operon. This prevents expression in the presence of excess *trp*.

16.4 Eukaryotic Regulation

Transcription factors can be either general or specific.
General transcription factors are needed to assemble the transcription apparatus and recruit RNA polymerase II at the promoter. Specific factors act in a tissue- or time-dependent manner to stimulate higher rates of transcription.

Promoters and enhancers are binding sites for transcription factors.
General factors bind to the promoter to recruit RNA polymerase. Specific factors bind to enhancers, which may be distant from the promoter but can be brought closer by DNA looping.

Coactivators and mediators link transcription factors to RNA polymerase II (figure 16.11).
Some, but not all, transcription factors require a mediator. The number of coactivators is small because a single coactivator can be used with multiple transcription factors.

The transcription complex brings things together.

16.5 Chromatin Structure and Gene Expression
The packaging of DNA into nucleosomes complicates transcription. Cells can alter chromatin structure by modifying DNA or histones. This is thought to be the basis for epigenetics.

Methylation of DNA, modification of histones, and noncoding RNA all affect chromatin structure.
Methylation of DNA bases, primarily cytosine, correlates with genes that have been "turned off." X-chromosome inactivation involves a noncoding RNA and polycomb repressor complex.

Some transcription activators alter chromatin structure.
Acetylation of histones results in active regions of chromatin.

Chromatin-remodeling complexes also change chromatin structure.
Chromatin-remodeling complexes contain enzymes that move, reposition, and transfer nucleosomes.

16.6 Eukaryotic Posttranscriptional Regulation

Small RNAs act after transcription to control gene expression.
RNA interference is mediated by siRNAs formed by cleavage of double-stranded RNA by the Dicer nuclease. The siRNA is bound to a protein, Argonaute, in an RNA-induced silencing complex (RISC). The RISC can cleave mRNA or inhibit translation. Another class of small RNA, miRNA, is formed by the action of two nucleases, Drosha and Dicer, on RNA stem-and-loop structures. These also form a RISC that can either degrade mRNA or stop translation.

Small RNAs may have evolved to protect the genome.
Viral RNAs are degraded and transposons are silenced in the germ line by RNA interference. The origins of this machinery are ancient.

Small RNAs can mediate heterochromatin formation.
In fission yeast, *Drosophila*, and plants, RNA interference pathways lead to the formation of heterochromatin.

Alternative splicing can produce multiple proteins from one gene.
In response to tissue-specific factors, alternative splicing of pre-mRNA from one gene can result in multiple proteins.

RNA editing alters mRNA after transcription.

mRNA must be transported out of the nucleus for translation.

Initiation of translation can be controlled.
Translation factors may be modified to control initiation; translation repressor proteins can bind to the beginning of a transcript so that it cannot attach to the ribosome.

The degradation of mRNA is controlled.
An mRNA transcript is relatively stable, but it may carry targets for enzymes that degrade it more quickly as needed by the cell.

16.7 Protein Degradation

Addition of ubiquitin marks proteins for destruction.
In eukaryotes, proteins targeted for destruction have ubiquitin added to them as a marker.

The proteasome degrades polyubiquitinated proteins.
A cell organelle—the cylindrical proteasome—degrades ubiquitinated proteins that pass through it.

Review Questions

UNDERSTAND

1. In prokaryotes, control of gene expression usually occurs at the
   a. splicing of pre-mRNA into mature mRNA.
   b. initiation of translation.
   c. initiation of transcription.
   d. All of the choices are correct.

2. Regulatory proteins interact with DNA by
   a. unwinding the helix and changing the pattern of base-pairing.
   b. binding to the sugar–phosphate backbone of the double helix.
   c. unwinding the helix and disrupting base-pairing.
   d. binding to the major groove of the double helix and interacting with base-pairs.

3. In *E. coli*, induction in the *lac* operon and repression in the *trp* operon are both examples of
   a. negative control by a repressor.
   b. positive control by a repressor.
   c. negative control by an activator.
   d. positive control by a repressor.

4. The *lac* operon is controlled by two main proteins. These proteins
   a. both act in a negative fashion.
   b. both act in a positive fashion.
   c. act in the opposite fashion, one negative and one positive.
   d. act at the level of translation.

5. In eukaryotes, binding of RNA polymerase to a promoter requires the action of
   a. specific transcription factors.
   b. general transcription factors.
   c. repressor proteins.
   d. inducer proteins.

6. In eukaryotes, the regulation of gene expression occurs
   a. only at the level of transcription.
   b. only at the level of translation.
   c. at the level of transcription initiation, or posttranscriptionally.
   d. only posttranscriptionally.

7. In the *trp* operon, the repressor binds to DNA
   a. in the absence of *trp*.
   b. in the presence of *trp*.
   c. in either the presence or absence of *trp*.
   d. only when *trp* is needed in the cell.

APPLY

1. The *lac* repressor, the *trp* repressor and CAP are all
   a. negative regulators of transcription.
   b. positive regulators of transcription.
   c. allosteric proteins that bind to DNA and an effector.
   d. proteins that can bind DNA or other proteins.

2. Specific transcription factors in eukaryotes interact with enhancers, which may be a long distance from the promoter. These transcription factors then
   a. alter the structure of the DNA between enhancer and promoter.
   b. do not interact with the transcription apparatus.
   c. can interact with the transcription apparatus via DNA looping.
   d. can interact with the transcription apparatus by removing the intervening DNA.

3. Repression in the *trp* operon and induction in the *lac* operon are both mechanisms that
   a. would only be possible with positive regulation.
   b. allow the cell to control the level of enzymes to fit environmental conditions.
   c. would only be possible with negative regulation.
   d. cause the cell to make the enzymes from these two operons all the time.

4. Regulation by small RNAs and alternative splicing are similar in that both
   a. act after transcription.
   b. act via RNA/protein complexes.
   c. regulate the transcription machinery.
   d. Both a and b are correct.

5. Eukaryotic mRNAs differ from prokaryotic mRNAs in that they
   a. usually contain more than one gene.
   b. are colinear with the genes that encode them.
   c. are not colinear with the genes that encode them.
   d. Both a and c are correct.

6. In the cell cycle, cyclin proteins are produced in concert with the cycle. This likely involves
   a. control of initiation of transcription of cyclin genes, and ubiquitination of cyclin proteins.
   b. alternative splicing of cyclin genes to produce different cyclin proteins.
   c. RNA editing to produce the different cyclin proteins.
   d. transcription/translation coupling.

7. A mechanism of control in *E. coli* not discussed in this chapter involves pausing of ribosomes allowing a transcription terminator to form in the mRNA. In eukaryotic fission yeast, this mechanism should
   a. be common since they are unicellular.
   b. not be common since they are unicellular.
   c. not occur as transcription occurs in the nucleus and translation in the cytoplasm.
   d. not occur due to possibility of alternative splicing.

SYNTHESIZE

1. You have isolated a series of mutants affecting regulation of the *lac* operon. All of these are constitutive, that is, they express the *lac* operon all the time. You also have both mutant and wild-type alleles for each mutant in all combinations, and on F′ plasmids, which can be introduced into cells to make the cell diploid for the relevant genes. How would you use these tools to determine which mutants affect DNA binding sites on DNA, and which affect proteins that bind to DNA?

2. Examples of positive and negative control of transcription can be found in the regulation of expression of the bacterial operons *lac* and *trp*. Use these two operon systems to describe the difference between positive and negative regulation.

3. What forms of eukaryotic control of gene expression are unique to eukaryotes? Could prokaryotes use the mechanisms, or are they due to differences in these cell types?

4. The number and type of proteins found in a cell can be influenced by genetic mutation and regulation of gene expression. Discuss how these two processes differ.

CHAPTER 17

Biotechnology

0.3 µm

Chapter Contents

17.1 Recombinant DNA

17.2 Amplifying DNA Using the Polymerase Chain Reaction

17.3 Creating, Correcting, and Analyzing Genetic Variation

17.4 Constructing and Using Transgenic Organisms

17.5 Environmental Applications

17.6 Medical Applications

17.7 Agricultural Applications

Introduction

Biotechnology involves manipulating biological systems to generate useful products or to improve environmental, medical, or agricultural, processes. Biotechnology is not a new human endeavor, however: The use of commercial yeast to produce leavened bread dates to ancient Egypt and fermentation to produce wine probably predates this by several thousand years. Modern biotechnology combines recent discoveries in molecular biology and genetics with more traditional biotechnologies such as selective breeding and hybridization. Modern biotechnology would not be possible, however, without advances made in other disciplines: Progress in computer science accelerated genome sequencing, and discoveries in chemical engineering moved nanobiotechnology out of the realm of science fiction.

In this chapter, we explore modern biotechnology by first considering advances in molecular biology and genetics such as the use of plasmids and other recombinant DNA techniques. Then, we consider how biotechnology can be used to solve socially relevant problems in environmental science, medicine, and agriculture.

17.1 Recombinant DNA

Learning Outcomes

1. Describe how restriction endonucleases and ligases are used to make recombinant DNA.
2. Explain how DNA fragments can be separated with gel electrophoresis and why this is useful.
3. Describe the construction and uses of recombinant DNA libraries.

Biotechnology is not necessarily a discipline of the 21st century. For example, the initial domestication of dogs from wolves by our hunter-gatherer ancestors some 15,000 years ago, and their subsequent selective breeding has produced the 180 existing dog breeds. Later, our agriculturalist ancestors began to domesticate crops, and the development of hybrid corn in the early 20th century substantially increased yield. Recent discoveries in genetics and molecular biology have ushered in a modern era of biotechnological innovation and discovery. The ability to isolate and manipulate DNA revolutionized biotechnology, accelerated rates of discovery, and made possible novel applications of biotechnology. The construction of **recombinant DNA,** a single

DNA molecule made from two different sources, began in the mid-1970s.

In 1972, Paul Berg created the first recombinant DNA molecule when he inserted viral DNA into bacterial DNA. Building on Berg's work in 1973, Herbert Boyer and Stanley Cohen introduced genes from the toad *Xenopus laevis* into bacteria and showed that the genes could be passed from generation to generation and were expressed in the bacteria. This conclusively demonstrated that the genes from slowly reproducing animals could be replicated and expressed in much more rapidly growing bacteria. In the following three decades, the discoveries of Berg, Boyer, and Cohen would become the basis of biotechnologies from whole-genome sequencing to the production of insulin in bacteria.

Restriction endonucleases cleave DNA at specific sites

These early cloning experiments were some of the first to utilize new tools called restriction enzymes. Arising from basic research on how the host range of a bacterial virus can be restricted, these enzymes recognize specific sequences of DNA, and act as a nuclease to cleave the DNA. Originally identified by Werner Arbor studying the host restriction phenomenon, the first enzyme that cleaved DNA at its recognition site was isolated by Hamilton Smith from the bacterium *Haemophilus influenza* and thus named HindIII. Arbor, Smith, and Daniel Nathans shared the 1978 Nobel Prize in Physiology or Medicine for this work. The discovery of restriction endonucleases is important for two reasons: First, their ability to cut DNA into specific fragments, and second for their use in genome mapping (see chapter 18).

How restriction endonucleases work

There are three types of restriction enzymes, but only type II cleaves at precise locations, which enable creation of recombinant molecules. These enzymes recognize a specific DNA sequence, ranging from 4 bases to 12 bases, and cleave the DNA at a specific base within this sequence (figure 17.1). The recognition sites for most type II enzymes are palindromes. A linguistic palindrome is a word or phrase that reads the same forward and in reverse, such as the name "Hanah." The palindromic DNA sequence reads the same from 5' to 3' on one strand as it does on the complementary strand.

Given this kind of sequence, cutting the DNA at the same base on either strand produces staggered cuts that leaves "sticky ends," or overhangs. These short, unpaired sequences are the same for any DNA that is cut by this enzyme. This allows DNA molecules from different sources to be easily joined together due to the short regions of complementarity in the overhangs (figure 17.1). Although less common, some type II restriction enzymes can cut both strands in the same position, producing blunt, not sticky, ends. Blunt-cut ends can be joined with other blunt-cut ends.

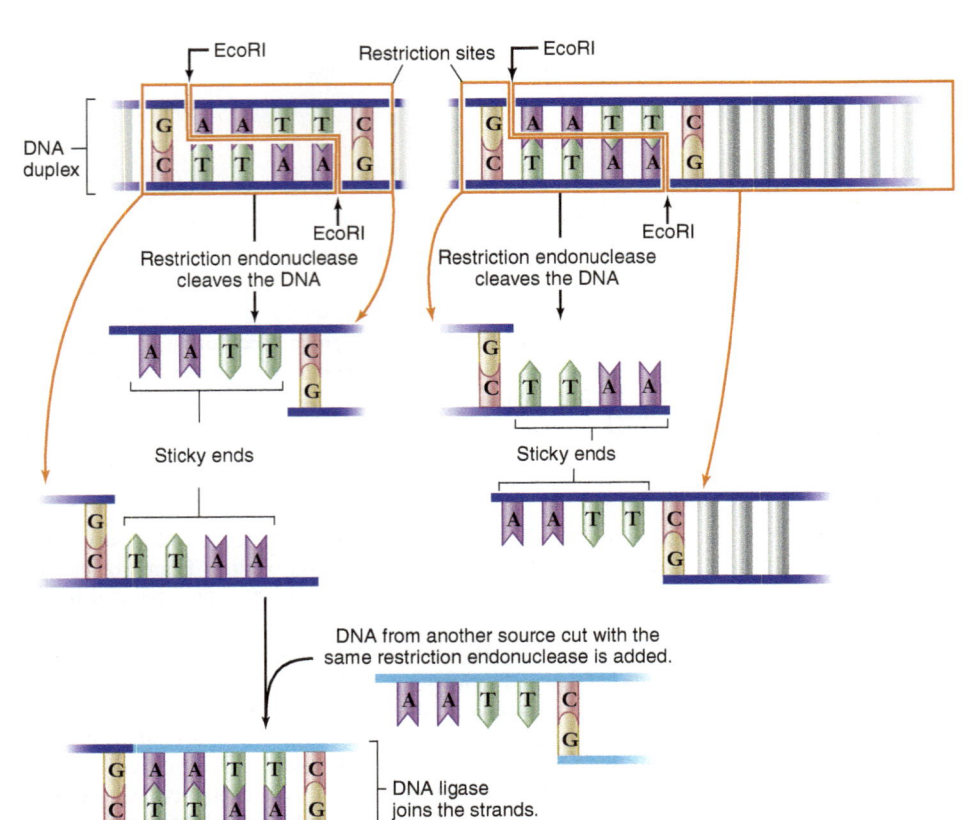

Figure 17.1 Many restriction endonucleases produce DNA fragments with sticky ends. The restriction endonuclease EcoRI always cleaves the sequence 5'–GAATTC–3' between G and A. Because the same sequence occurs on both strands, both are cut. However, the two sequences run in opposite directions on the two strands. As a result, single-stranded tails called "sticky ends" are produced that are complementary to each other. These complementary ends can then be joined to a fragment from another DNA that is cut with the same enzyme. These two molecules can then be joined by DNA ligase to produce a recombinant molecule.

Gel electrophoresis separates DNA fragments

To create a piece of recombinant DNA, two or more fragments of DNA must be joined together. Before this can happen, the fragments of DNA must be purified, or isolated. A common technique to separate different fragments of cut DNA so they can be isolated is gel electrophoresis. This technique takes advantage of the negative charge on DNA by using an electrical field to provide the force necessary to separate DNA molecules based on size.

The gel, which is made of either agarose or polyacrylamide and spread thinly on supporting material, provides a three-dimensional matrix that separates molecules based on size (figure 17.2). The gel is submerged in a buffer solution containing ions that can carry current and is subjected to an electrical field. The strong negative charges from the phosphate groups in the DNA backbone cause it to migrate toward the positive pole. The gel acts as a sieve to separate DNA molecules based on size: The larger the molecule, the slower it will move through the gel matrix. Over a given period, smaller molecules migrate farther than larger ones. The DNA in gels can be visualized using a fluorescent dye that binds to DNA. Gels can be made that are used solely to analyze the sizes of DNA fragments, or so that

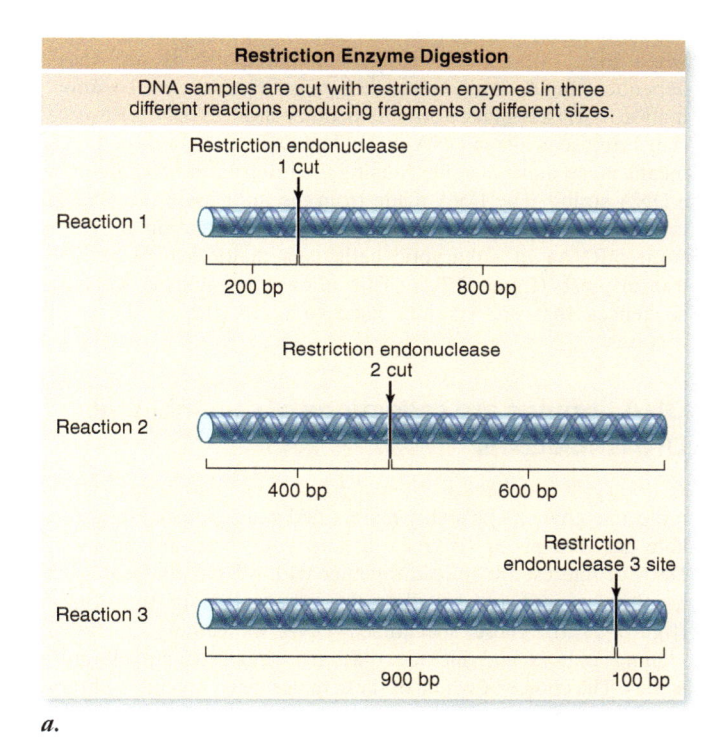

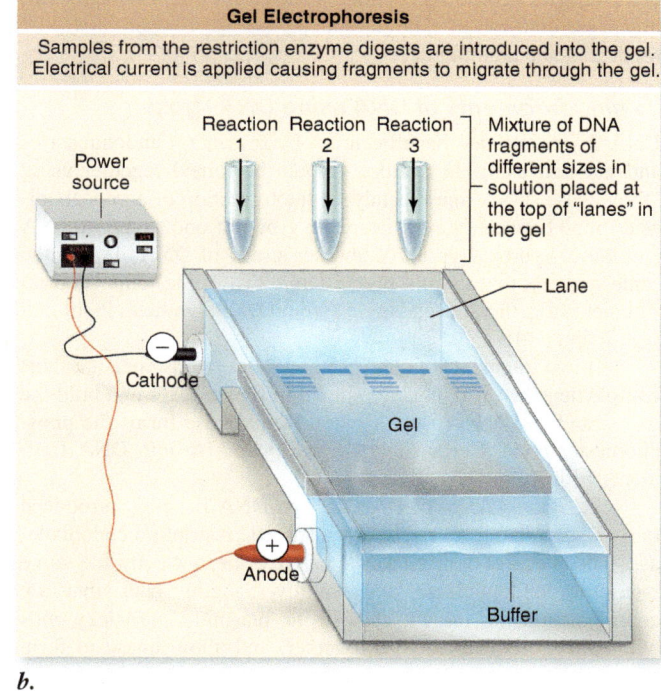

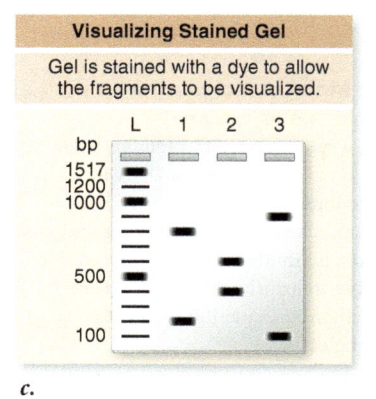

Figure 17.2 Gel electrophoresis separates DNA fragments based on size. *a.* Three restriction enzymes are used to cut DNA into specific pieces, depending on each enzyme's recognition sequence. *b.* The fragments are loaded into a gel (agarose or polyacrylamide), and an electrical current is applied. The DNA fragments migrate through the gel based on size, with larger fragments moving more slowly. *c.* This results in a pattern of fragments separated based on size, with the smaller fragments migrating farther than larger ones. A series of fragments of known sizes produces a ladder so that sizes of fragments of unkown size can be estimated (bp = base pairs; L = ladder; 1, 2, 3 = fragments from piece of DNA cut with restriction endonucleases 1, 2, and 3, respectively).

individual fragments of DNA can be cut from the gel, purified, and prepared for further use.

DNA fragments are combined to make recombinant molecules

The next step in constructing a recombinant DNA molecule is to combine pieces of DNA from different sources into a single molecule. The recombinant DNA can be stably replicated in a host organism such as the common gut bacterium, *Escherichia coli*. The most common strategy to create a stable recombinant DNA molecule is to add a fragment of DNA of interest—for example, a gene sequence to be studied—into a bacterial plasmid. The recombinant plasmid can then be introduced to *E. coli*, where it will be replicated and maintained.

Joining fragments of DNA using DNA ligase

DNA fragments that have been cut by restriction endonucleases and purified using an agarose gel can be joined together using DNA ligase. DNA ligase catalyzes the formation of a phosphodiester bond between the 5′ phosphate group of one strand of DNA and the 3′ hydroxyl group of another strand of DNA. This is the same reaction that is used by the DNA ligase that joins Okazaki fragments together during lagging strand DNA synthesis in the cell (see chapter 14).

If two fragments of cut DNA have sticky ends, then the short complementary overhangs (as shown in figure 17.1) will hold the two strands of DNA together while DNA ligase forms the phosphodiester bond. Certain DNA ligases can also join DNA fragments with blunt ends.

In figure 17.3, a piece of foreign DNA is being introduced into plasmid that has been cut by a specific restriction endonuclease. The piece of foreign DNA has also been cut by the same restriction endonuclease and has overhanging ends that are complementary to the cut ends in the plasmid. The sticky ends will hold the two molecules together, and allow ligase to form phosphodiester bonds to join them, creating a recombinant plasmid. The whole process of producing recombinant DNA molecules is often called *molecular cloning*.

Reverse transcriptase makes DNA from RNA

Although the "central dogma" of molecular biology indicates that the flow of information is from DNA to RNA to protein, there are several instances in biology when DNA can be made using the information in RNA. Reverse transcriptase, first identified in a class of viruses called retroviruses, can use RNA as a template to synthesize a DNA molecule (see chapter 27).

Retroviruses, like the human immunodeficiency virus (HIV), have RNA genomes but must convert the RNA to DNA when they infect cells. Reverse transcriptase is an RNA-dependent DNA polymerase; this means that it uses the information in RNA to make a complementary strand of DNA. Reverse transcriptase is also a DNA-dependent DNA polymerase, which means it can make new DNA using the information contained in a DNA strand. The DNA made from the information in mRNA is called **complementary DNA (cDNA)**. The creation of cDNA from mRNAs of eukaryotic cells is a major use of reverse transcriptase (figure 17.4). This allows the analysis of only sequences that are actually used to direct the synthesis of proteins.

DNA libraries are collections of recombinant DNA molecules

A library is a collection of many different recombinant DNA molecules that can be stably maintained and replicated in a suitable host organism. To create a library, different fragments of DNA of interest are spliced together with a DNA molecule that has been engineered to allow the propagation of the library (figure 17.5). These specialized DNA molecules are called *cloning vectors* and include plasmids, and artificial chromosomes. The choice of which of these to use when making a library depends on what the library will be used for and the sizes of the

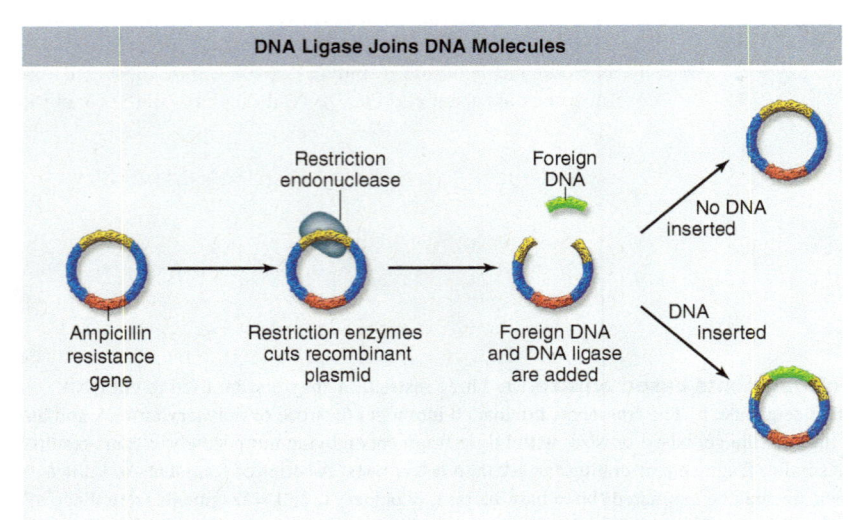

Figure 17.3 Molecular cloning with vectors. Plasmids are cut using restriction endonucleases at a specific site, and foreign DNA and DNA ligase are added. If foreign DNA is successfully incorporated into the plasmid, a recombinant plasmid is created. In a process called transformation, the plasmids can be introduced into bacteria. Spreading transformed cells on a medium containing the antibiotic ampicillin selects for plasmid-containing cells.

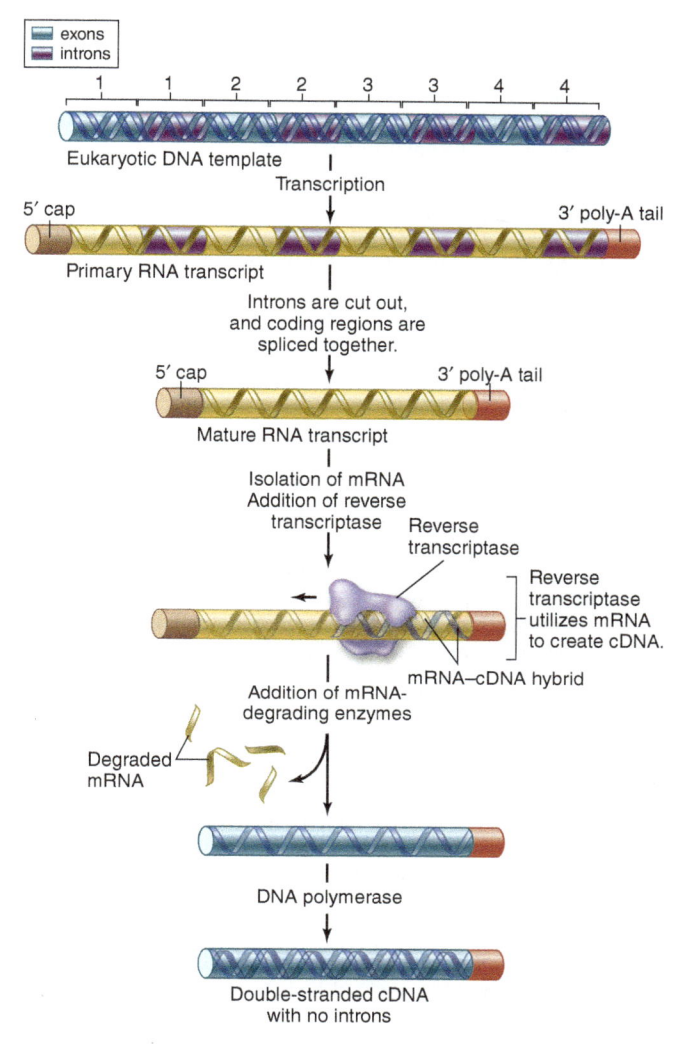

Figure 17.4 Making cDNA from mRNA using reverse transcription. A mature mRNA transcript is usually much smaller than the gene producing it due to the removal of introns. mRNA is isolated from the cytoplasm of a cell, which the enzyme reverse transcriptase uses as a template to make a DNA strand complementary to the mRNA. The newly made strand of DNA is the template for the enzyme DNA polymerase, which assembles a complementary DNA strand along it, producing cDNA—a double-stranded DNA version of the mature mRNA.

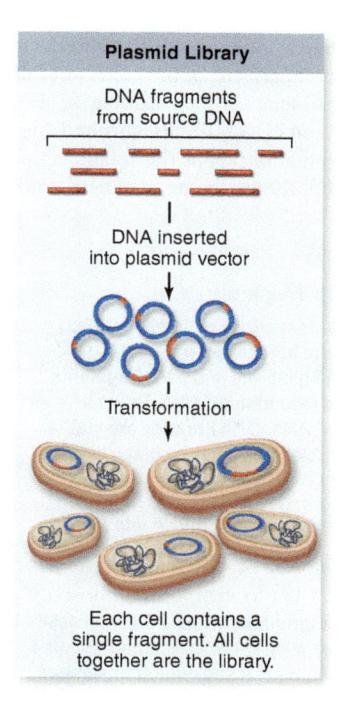

Figure 17.5 Creating a plasmid library. Fragments of DNA with the correct sticky ends can be joined with cut plasmid vectors so that each plasmid contains a different piece of DNA. Transformation of the recombinant plasmids into bacterial cells produces a library of bacteria with each bacterium containing plasmid with a unique DNA insert.

Libraries can be made that can produce proteins to be studied, for determining genome sequence and structure, or for isolating particular genes.

Constructed libraries can be stored by being introduced into an appropriate host organism (figure 17.5). Growing the host cells will then "amplify" the library and produce many copies of recombinant molecules that can be isolated or analyzed. Libraries can be introduced into bacteria or yeast by treating the cells to make them take up DNA from the environment in a process called *transformation*. The host organism for the library depends on the cloning vector used and the purpose of the library. If only small pieces of DNA are contained in the library, then bacterial plasmids work well and each bacterial cell in the library will contain an individual recombinant plasmid. When the cells in the library are grown in culture, they will replicate the cloned DNAs passively during growth. For libraries with larger pieces of DNA (such as genomic fragments), bacterial or yeast artificial chromosomes are used and the DNA library is stored in bacteria or yeast.

Reverse transcriptase can be used to make cDNA libraries

Cells from a specific developmental time point, or a specific tissue, can be used to isolate mRNA to use as a template for cDNA synthesis (see figure 17.4). The resulting cDNAs can then be used to construct a library that represents only the genes expressed at a specific time point, or in a specific tissue. Genomic libraries made from different cells or tissues of an organism all contain similar sequences, but this will not be true for cDNA libraries. A brain cDNA library will contain many brain-specific sequences not represented in a fibroblast cDNA library, which in turn will contain fibroblast-specific sequences not represented in the brain cDNA library.

pieces of DNA to be held. Cloning vectors have several important features:

1. a sequence that allows replication in a host organism (such as *E. coli* or yeast);
2. a selectable marker, such as resistance to an antibiotic; and
3. sequences that allow DNA fragments to be added (restriction endonuclease or recombination sites).

Once a library has been made, different techniques are used to identify and isolate specific recombinant molecules of interest.

Because a collection of cDNAs made from a tissue reflects the mRNA present in that tissue, comparisons of the collections of cDNAs produced from different tissues can be informative. For example, comparisons made between cDNA libraries from different cell types might provide information about the relationships between the genes being expressed, the proteome, and cell structure and function.

Learning Outcomes Review 17.1

Type II restriction endonucleases cleave DNA at specific sites. Gel electrophoresis uses an electric field to separate DNA fragments according to size. DNA ligase will join DNA fragments from different sources to produce recombinant DNA. Reverse transcription converts mRNA into cDNA. DNA libraries are stable collections of recombinant DNA molecules used for many kinds of genetic analyses.

- *Why is it important to be able to convert mRNA into DNA?*

Data analysis The human genome contains 3 billion base-pairs. If you construct a genomic library of the human genome with 500-bp fragments of DNA, what is the minimum number of clones in your library?

17.2 Amplifying DNA Using the Polymerase Chain Reaction

Learning Outcomes

1. Relate the process of DNA replication to PCR.
2. Compare and contrast PCR, RT-PCR, and quantitative RT-PCR.

The development of the polymerase chain reaction (PCR) in 1993 accelerated the construction of recombinant DNA molecules and ushered in a new era of genetic engineering. PCR mimics the processes of DNA replication to produce millions of copies of a DNA sequence (a process called amplification) without the need to insert the DNA fragment into a cloning vector and introduce it into a host cell.

A major benefit of producing DNA fragments using PCR is that the exact sequence of DNA to be amplified can be chosen. When DNA molecules are cloned into vectors based on producing DNA fragments by restriction endonuclease digestion, the DNA fragments available are limited by the location of restriction endonuclease sites.

To amplify a DNA sequence using PCR, two 20- to 25-nucleotide-long DNA sequences called *primers* are chemically synthesized. One primer is complementary to one strand of the DNA at one end of the chosen sequence, and the other primer is complementary to the opposite strand at the other end of the chosen sequence. If the primers are allowed to bind to their complementary DNA sequences, DNA polymerase can extend the two primers so that two new strands of DNA are produced similar to

DNA replication in cells. These two strands of DNA are complementary and contain the original primer binding sites. Repeating this process over and over results in an exponential increase in the number of DNA molecules. A DNA sequence up to about 10,000 base-pairs long can be amplified this way.

PCR mimics DNA replication

PCR mimics the processes of DNA replication and an individual reaction is cycled through a series of steps where each step is analogous to a step in DNA replication:

1. **Denaturation.** Heat is used to separate strands of double-stranded DNA.
2. **Annealing of primers.** Primers provide the 3′ OH required for elongation by DNA polymerase.
3. **Synthesis.** DNA polymerase makes new DNA.

New DNA molecules are synthesized in an exponential manner when these steps are cyclically repeated; a 25-cycle PCR produces 2^{25} molecules of DNA from a single target molecule! When PCR was initially developed, the steps were done manually using a mixture containing the DNA to be amplified (called the template DNA), and the primers specific for the template sequence. The mixture was heated to 95°C to separate the two strands of DNA, cooled to a temperature at which the primers could anneal to the template DNA, and then a polymerase was added to synthesize new DNA. Unfortunately, when the mixture was heated in the next cycle, the DNA polymerase was denatured and new polymerase had to be added. Because a typical PCR involves 25 to 30 cycles, this made PCR tedious and labor intensive.

PCR was made faster and less labor intensive due to two innovations. First, a thermostable DNA polymerase called **Taq polymerase** was isolated from the thermophilic bacterium, *Thermus aquaticus*. This overcame the need to add new DNA polymerase during every cycle of PCR because Taq polymerase is not denatured at the temperatures used in the denaturation step of the PCR cycle. The second innovation was the development of machines with heating blocks that can be accurately and rapidly cycled over large temperature ranges.

A typical PCR is now set up in a single tube containing the template DNA, the primers, nucleotides, and a thermostable DNA polymerase. Although Taq polymerase is still commonly used, other more accurate thermostable polymerases are also used. The tube is placed into a PCR machine that can quickly cycle between the denaturation temperature, the primer annealing temperature, and the DNA synthesis temperature (figure 17.6). Twenty-five cycles of PCR can be performed in as little as an hour. These developments have made PCR one of the most common techniques for generating specific DNA molecules for a variety of uses. Modifications of the basic PCR procedure have been developed that allow disease diagnosis, genome sequencing, and analysis of gene expression.

Reverse transcription PCR makes amplified DNA from mRNA

In addition to using a variety of types of DNA (plasmids, genomic DNA) as template, PCR can also be performed on cDNA made from mRNA. Isolated mRNA is incubated with reverse transcriptase, then the resulting cDNA is used as a template for PCR. The

Figure 17.6 The polymerase chain reaction.
The polymerase chain reaction (PCR) allows a single DNA sequence to be amplified for analysis. The process involves using short primers for DNA synthesis that flank the region to be amplified and *(1)* repeated rounds of denaturation, *(2)* annealing of primers, and *(3)* synthesis of DNA. The enzyme used for synthesis is a thermostable DNA polymerase that can work at the high temperatures needed for denaturation of template DNA. The reaction is performed in a thermocycler machine that can be programmed to change temperatures quickly and accurately. The annealing temperature used depends on the length and base composition of the primers. Details of the synthesis process have been simplified here to illustrate the amplification process. Newly synthesized strands are shown in light blue with primers in green. If there was one DNA molecule to start, then at the end of cycle one there would be two molecules, at the end of cycle two, four molecules, and at the end of cycle three, eight molecules.

combination of these two techniques is called *reverse transcription PCR (RT-PCR)*. RT-PCR is useful for three reasons. First, it allows the creation of recombinant DNA molecules containing DNA copies of only the exons of genes. Second, it allows study of the structure and function of gene products. Third, it can be used to determine relative levels of gene expression in cells and tissues.

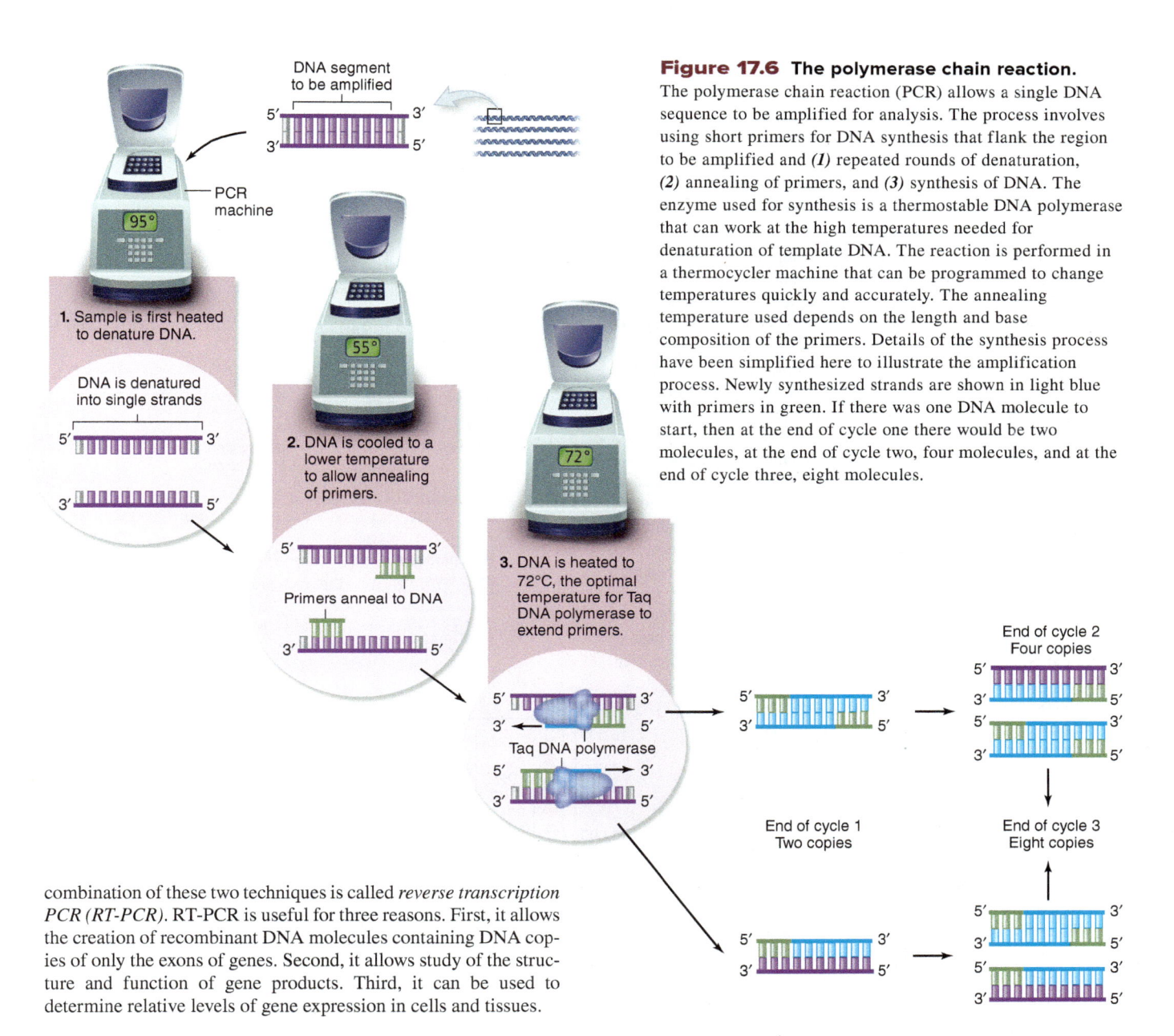

Quantitative RT-PCR can determine levels of an mRNA

Cells often respond to changes in environmental conditions, both internal and external, with changes in gene expression. Analysis of gene expression involves accurate quantification of cellular mRNA levels. One of the fastest and easiest ways to measure relative changes in gene expression is using *reverse transcription quantitative PCR (RT-qPCR)*. This involves isolating mRNA, using reverse transcriptase to convert this to cDNA, then using PCR to amplify specific cDNAs using gene-specific primers. The amount of DNA produced can be measured in real time by the PCR machine. For this reason, quantitative PCR is also known as real-time PCR. Two common techniques are used, both of which depend on the use of fluorescent dyes.

DNA quantification using fluorescent DNA-binding dyes

The simplest way to quantify the amount of DNA made during the PCR part of RT-qPCR is to add a dye to the PCR reaction that binds nonspecifically to DNA. If DNA is present, the dye will bind to the DNA and fluoresce when illuminated by a laser. After each cycle of PCR, a laser in the PCR machine illuminates the reaction and detectors measure the amount of fluorescence. As the amount of DNA increases with each PCR cycle, more dye becomes bound to the DNA, and the fluorescence increases. The data can be output to a computer to plot a graph of fluorescence vs the number of PCR cycles.

Imagine two qPCR reactions using cDNA made from the same amount of mRNA isolated from two different cell types. During the qPCR part of the experiment fluorescence might increase in one of the samples faster than in the other. In the qPCR where fluorescence appeared faster there was initially more mRNA for a specific gene because more DNA molecules were made as PCR proceeded. Computers can be used to quantify the relative differences between samples to investigate differences in gene expression under different conditions.

DNA quantification using fluorescent DNA-binding probes

A more accurate way to quantify the amount of DNA produced by a PCR reaction is to introduce a fluorescently labeled probe into the qPCR reaction. This probe consists of a 20- to 30-nucleotides-long sequence of single-stranded DNA that is complementary to a portion of the DNA being analyzed. Probes are chemically synthesized, and can be modified for different applications; for example, a fluorescent molecule can be added to one end of the probe.

In their use in qPCR, a probe complementary to a short region of the template cDNA has a fluorescent molecule added to one end of the probe. When illuminated with one wavelength of light, the fluorophore emits a different wavelength of light that can be measured by the PCR machine. A so-called quencher molecule is added to the other end of the probe. This quencher will absorb the light emitted by the fluorophore, thus preventing detection. As long as the fluorophore and quencher are attached to the same probe, they are close enough that fluorescence from the fluorophore is quenched. If the fluorophore and quencher are separated, then fluorescence can be detected.

If the probe is included in the PCR reaction, it can be used to detect the amount of DNA produced. After the denaturation step, when the reaction is cooled to the temperate at which the primers can anneal to the template, the qPCR probe also anneals to one of the strands of template between the sites where the two PCR primers anneal. As the polymerase extends the PCR primers it will run into the annealed probe on one strand. As the Taq polymerase has an exonuclease activity (see chapter 14), it can degrade the probe and complete copying the template. By degrading the probe, the fluorophore is released from the quencher allowing its fluorescence to be detected (figure 17.7a). The PCR cycles continue and with every round, there will be an increase in fluorescence that is proportional to the amount of DNA being made. The more template (cDNA made from mRNA) there is to begin with, the faster the fluorescent signal will increase (figure 17.7b).

PCR is the basis of many sequencing technologies

It took 13 years, from 1990 to 2003, to obtain a finished version of the sequence of the human genome at an estimated cost of $2.7 billion. Today, a human genome can be sequenced in as little as three days at a cost close to $1000. DNA sequencing requires that many copies of fragments of the genome be obtained. For the initial Human Genome Project, this involved cloning small fragments of genomic DNA into cloning vector that were engineered to facilitate the manual sequencing of DNA (see chapter 18). These clones then had to be grown in a suitable host, and the DNA isolated for sequencing. This process is both labor-intensive, and time-consuming.

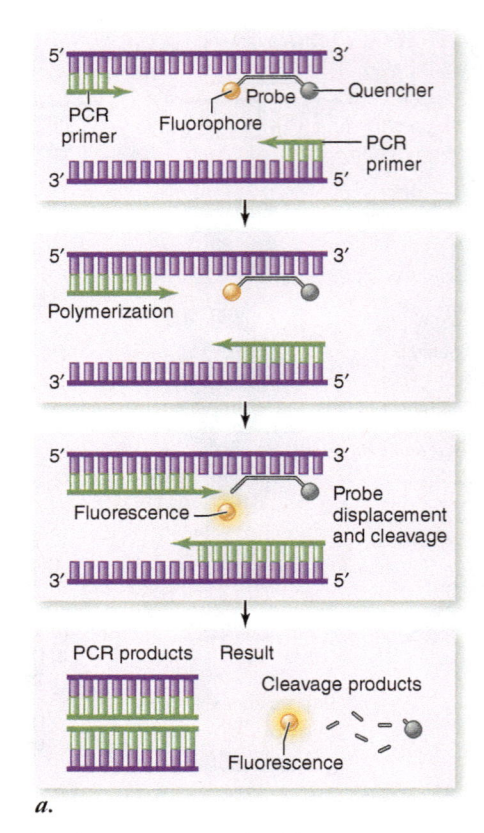

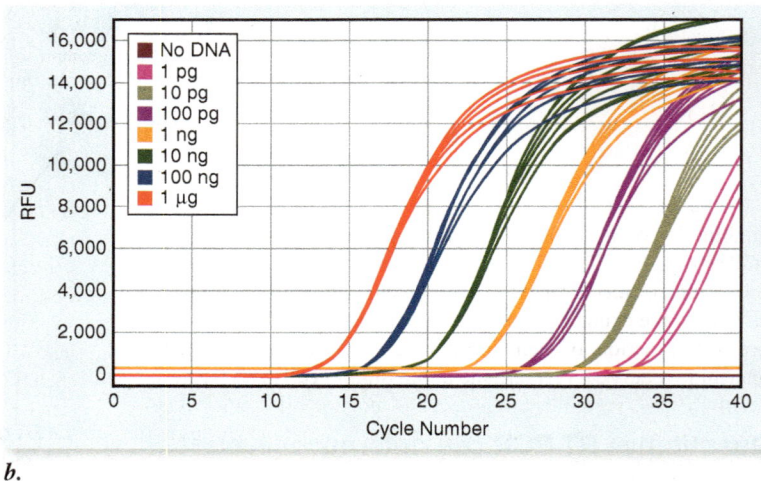

Figure 17.7 Quantification of mRNA levels by qRT-PCR. *a.* Degradation of a fluorescent probe as amplification occurs results in an increase in fluorescence. *b.* As DNA is amplified, the amount of fluorescence increases proportionally to the number of molecules of DNA present. The more DNA that is present at the start of the reaction, the more quickly fluorescence appears during the PCR (pg, picograms; ng, nanograms).

PCR overcame the need to use cloning vectors to obtain enough DNA to be sequenced. Many new sequencing technologies, called "next-generation" sequencing (NGS) use variations of PCR to produce the DNA that will be sequenced (discussed in chapter 18).

Without PCR, sequencing genomes quickly and cheaply would not be possible. The value of being able to quickly and cheaply resequence genomes is that it allows comparison between genotypes and the identification of genetic variations that could be responsible for disease, response to certain drugs, or be involved in general health.

Learning Outcomes Review 17.2

The polymerase chain reaction (PCR) is used to rapidly amplify specific DNA sequences from small amounts of starting material. RT-PCR allows DNA copies of expressed mRNAs to be amplified, and RT-qPCR allows the relative quantification of specific mRNAs. Use of PCR in next-generation sequencing has contributed to a reduction in time and cost to sequence a genome.

■ *How are PCR and DNA replication similar? How do they differ?*

Inquiry question Suppose you wanted a copy of a section of a eukaryotic genome that included the introns and exons. Would the creation of cDNA be a good way to go about this? Explain.

17.3 Creating, Correcting, and Analyzing Genetic Variation

Learning Outcomes

1. Describe how genetic information can be used to identify an unknown individual.
2. Explain how we can construct mutations in genes in vitro.
3. Describe the pros and cons of RNA interference and direct genome editing.

Individuals in sexually reproducing populations are phenotypically diverse, yet all members of a species usually have the same set of genes. Most diversity seen in populations is produced by complex interactions between genetic variation, epigenetic differences, and environmental variability. Advances in manipulating and analyzing DNA have produced new ways to study the genetic variation that contributes to phenotypic variation. Analyzing genetic variation is important because it contributes to understanding differences between individuals within a species, and between species. Recent advances in DNA technology allow us to create or repair mutant alleles as a way to investigate gene function.

Forensics uses DNA fingerprinting to identify individuals

It is sometimes desirable to be able to identify an individual based on a small amount of tissue or bodily fluids. This occurs during the investigation of crimes, and in the identification of victims after catastrophic events. Although more sensitive techniques have been developed (see chapter 18), forensics continues to use DNA fingerprinting as a primary technique. This is at least in part due to the large databases that have been assembled by law enforcement organizations.

DNA fingerprinting takes advantage of short, repeated sequences that vary among individuals. Short tandem repeats (STRs), typically 2 to 4 nt long, are not part of coding or regulatory regions of genes and mutate over generations so that the length of the repeats varies. We say that the population is polymorphic for these molecular markers. A combination of these markers can be used as DNA "fingerprints" in criminal investigations and other identification applications (figure 17.8).

This is also another example of the flexibility and utility of the PCR process. Primers are designed to flank a region known to contain an STR. These primers can then be used to amplify the STR-containing region from very small amounts of starting material. The analysis can also be done using the same technology used for automated sequencing (see chapter 18). Since 1997, 13 STRs have been established as the standard of evidence for identification in court, and approved identification kits are available commercially. The 13 STRs form the basis of a federal profiling database called CODIS (Combined DNA Index System). New DNA fingerprints can be compared with those of known individuals in CODIS.

DNA fingerprinting has also been used to exonerate wrongly convicted individuals who had been imprisoned for years before DNA analysis was widely available, as well as to identify individuals after catastrophes. After the September 11, 2001, attacks on the World Trade Centers in New York, DNA fingerprinting was the only means for identifying some of the victims. After a devastating earthquake in the Republic of Haiti in 2010, DNA fingerprinting was used to reunite children with their families.

PCR can be used to create point mutations in specific genes

Depending on its location in the genome, a change in one or a few nucleotides may or may not have an effect on phenotype. Any change to the genome that results in a phenotypic change we

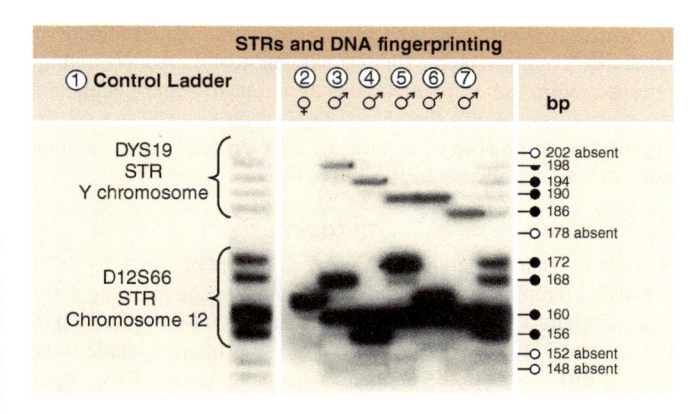

Figure 17.8 Using STRs and DNA fingerprinting to identify individuals. A Y-chromosome STR distinguishes between men and women *(top)* because it is absent in women (column 2). The STR occurs in different lengths in men, depending on the number of repeats (columns 3–7). A second STR found on chromosome 12 appears in both men and women *(bottom)*.

recognize as a mutation. For years geneticists created mutations randomly using chemicals (mutagens), but now we can create mutations in vitro. Studying mutations allows us to better understand the normal function of the altered DNA. The ultimate use of this approach is to be able to replace a wild-type gene with a mutant copy to test the function of the mutated gene.

PCR-based site-specific mutagenesis

Remember that the amplification of a region of DNA by PCR involves a template molecule, which is often a gene or cDNA copy of a gene, and a pair of primers used to initiate DNA replication of the template. If the sequence of one of the primers is made to be slightly different than the template sequence, then as DNA amplification occurs most of the DNA molecules made in the PCR will end up with a specific mutation at a specific site. In this way different mutations can be introduced into a gene or cDNA and the effects of those mutations can then be analyzed in a variety of ways, some of which we explore in this section.

PCR-based random mutagenesis

As we previously discussed, one of the polymerases used in PCR is called Taq polymerase. Unlike the polymerases used to replicate DNA in cells, Taq polymerase is not able to correct errors it makes during DNA synthesis. Taq polymerase will randomly introduce a mutation approximately every 45,000 nucleotides synthesized. Changing the chemical conditions of the PCR can further increase the mutation rate so that by performing repeated rounds of PCR, randomly mutagenized cloned genes or cDNA can be obtained. The effects of these mutations on gene product function can then be analyzed in a variety of ways.

RNA interference can reduce the level of a gene product

Although the ability to make and analyze mutants in vitro is extremely powerful, in many experimental systems it is not possible to introduce these constructs back into a living cell to assess their affects on phenotype. The technique of RNA interference (RNAi) allows researchers to reduce the amount of a gene product in living cells/organisms. This is the equivalent of a mutation that eliminates the function of a gene product. RNA interference acts to degrade or block translation of a specific mRNA in cells, which in turn reduces the level of the encoded protein. This technique takes advantage of systems in cells that have evolved to use small RNAs to control the level of gene expression posttranscriptionally (see chapter 16 for details).

Using RNAi to reduce or eliminate the production of a specific protein requires the production of a short double-stranded RNA complementary to the mRNA that encodes the protein. A variety of techniques exist to create and introduce double-stranded RNA into cells or organisms. RNAi then depends on the cellular mechanisms described in chapter 16. For any organism where we have a complete genome sequence, a library of short double-stranded RNA molecules can be created to target every gene in the genome. These libraries can be used to systematically reduce expression of individual proteins or combinations of proteins so their roles in cells and organisms can be investigated.

New technologies allow direct editing of the genome

PCR based mutagenesis allows the in vitro alteration of gene sequences and RNAi allows the reduction of gene products; however both are indirect ways to investigate gene function in living cells. Until very recently, it has not been possible to alter DNA in a living cell: to edit the genome. The discovery of several proteins that can interact with specific DNA sequences has led to techniques that allow direct, in vivo editing of organism's genome. This means it is now possible to quickly and easily change a gene's sequence, inactivate a gene, or interfere with the expression of the gene directly in the cell. By systematically modifying genes in this way, we will be better able to understand gene functions in vivo. There are two main ways to edit genes in vivo: the TALE (transcription activator-like effector) proteins and the CRISPR (clustered regularly interspaced short palindromic repeats)/Cas9 system.

TALE proteins

TALE proteins were identified due to their ability to bind DNA and their role in disease caused by the plant pathogen, *Xanthomonas*. These proteins contain multiple sets of a repeating 34-amino-acid sequence called the TALE repeat domain. The middle pair of amino acids in the domain is highly variable, whereas the surrounding amino acids are usually the same. Depending on the two amino acids present in the middle two positions, a different nucleotide will be bound by the repeat.

By combining different TALE repeat domains, different sequences of DNA can be bound by the TALE protein. If a recombinant DNA molecule is constructed that fuses DNA encoding a TALE protein to a DNA encoding a nuclease, this now encodes a novel sequence-specific nuclease. The protein produced from this recombinant DNA is called a TALE-nuclease fusion, or TALEN. A recombinant DNA molecule encoding the TALEN can be introduced into cells and the TALEN will hone in on a particular gene in the genome, and the nuclease can cut the DNA at that site. When the cell tries to repair the cut DNA, errors are sometimes made, and sequences can be lost. If this happens, the gene may become inactive. Variations on this strategy have been developed to completely replace a gene containing sequences recognized by the TALEN.

The CRISPR/Cas9 system

Much as the study of host restriction in bacteria led to the discovery of restriction endonucleases, the study of how some bacteria could become "immune" to infection by a virus led to the discovery of the CRISPR/Cas9 system. Some bacteria appear to have a kind of adaptive immunity that involves a complex system that has evolved to incorporate part of the viral genome into the bacteria's genome. This viral DNA can then be used to target future infection by a virus with the same DNA sequence. A bacterium uses the viral DNA to transcribe an RNA that is used as a "guide" by a nuclease (Cas9) to target DNA with the same sequence. As it is much simpler to design and synthesize a guide RNA than to construct a TALEN, this system appears simpler and more flexible.

Once inside a cell and associated with one another, a guide RNA and the Cas9 protein are able to seek out and associate with a gene that matches the guide RNA sequence (figure 17.9a).

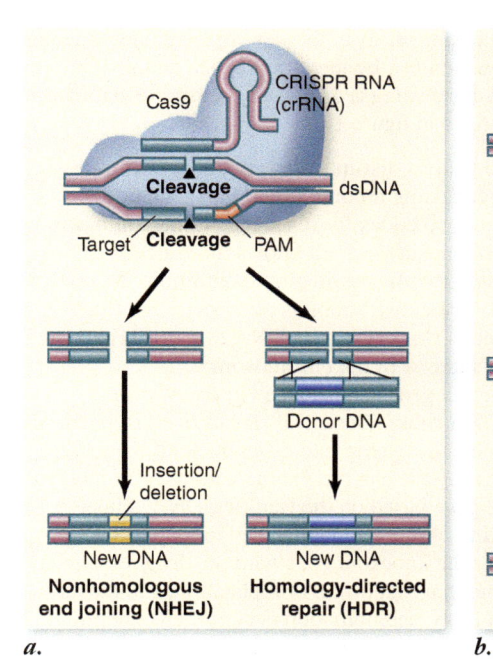

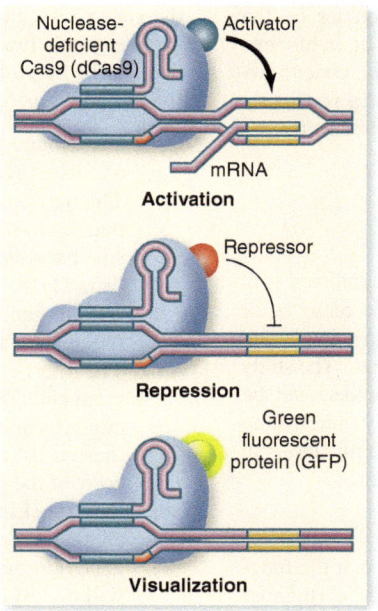

Figure 17.9 The CRISPR/Cas9 genome editing system. *a.* A sequence-specific guide RNA targets Cas9 to a target sequence where Cas9 cuts the target sequence. Repair can be inaccurate resulting in gene inactivation, or accurate in the presence of a donor DNA sequence that results in replacement of the DNA target sequence with the donor sequence. *b.* Recombinant DNA technology can be used to target Cas9 to a gene and turn it on (activation), off (inactivation), or reveal its location in the genome.

Source: New England BioLabs Inc., www.neb.com/tools-and-resources/feature-articles/crispr-cas9-and-targeted-genome-editing-a-new-era-in-molecular-biology

Cas9 is a nuclease and can cut the DNA with which it is associated. Inaccurate repair of the cut DNA can result in small deletions or insertions to the genomic sequence that can result in loss of gene function. Alternatively, if linear, double-stranded DNA is available, homology-directed repair can accurately repair the cut. By providing linear, double-stranded DNA with a mutation, changes to the sequence of the gene can be made. Alternatively, if a defective allele of a gene is targeted, a mutation can be replaced with a normal gene sequence. This technique can inactivate the genome of the human immunodeficiency virus (HIV) that is integrated into cells grown in vitro and has potential to treat certain genetic diseases.

Gene expression can be regulated using this system as well. Gene expression can be activated or repressed by constructing a fusion between Cas9 lacking nuclease activity and a transcriptional activator or repressor, respectively. It is also possible to create fusions between fluorescent proteins and nuclease-deficient Cas9 proteins (figure 17.9b). This could simplify the detection and analysis of oncogenes that are amplified in certain cancers.

Learning Outcomes Review 17.3

Individuals can be identified using DNA fingerprinting. A specific set of STR loci makes up the CODIS database. PCR can be used to create site-specific or random mutations in DNA. RNA interference can be used to reduce levels of gene products. Genome-editing techniques can be used to remove gene function and to introduce new alleles in the chromosome.

- *How does genome editing differ from creating specific mutations in vitro?*
- *Compare and contrast the TALE and Crispr/Cas9 systems of genome editing.*

17.4 Constructing and Using Transgenic Organisms

Learning Outcomes

1. Explain how the universal nature of the genetic code allows transgenesis.
2. Compare and contrast knockout, knockin, and conditional knockout mice.
3. Describe how transgenic plants are created.

Transgenic organisms contain a gene from a different species, called a *transgene*, which has been incorporated into the genome through genetic engineering. In addition to adding genes to an organism's genome, it is also possible to remove and change genes within the genome. Genetically modified organisms are organisms that have been genetically altered by techniques other than conventional breeding. The construction of genetically modified organisms has revealed the function of many genes and has applications in medicine, environmental biology, and agriculture.

The universal nature of the genetic code allows transgenesis

A human gene can be placed into a mouse genome and the cells of the mouse will express the gene. Similarly, a human gene can be placed into the genome of an *E. coli* bacterium and the bacteria will make the encoded protein. This is because the cells of organisms of different species interpret genetic information identically. The genetic code used to decipher the information in a gene is the same code used in bacteria, archaea, and eukaryotes. For example, the amino acid phenylalanine will always be coded for by the codon UUU; this is true if we look at the genetic code in humans or in *E. coli*. There are minor variations to the genetic code, however. For example, the genetic code of mitochondria is slightly different from the genetic code used to interpret nuclear genes.

Removal or addition of genes can reveal function

The techniques for making recombinant DNA can be combined with other techniques to create organisms that have genes removed or have genes added. In some cases, a gene can be removed and then replaced with a version that has been genetically altered. The study of these animals and plants allows scientists to better understand the relationships between genotype and phenotype. By introducing mutant alleles into organisms, the effects of specific mutations in disease processes can be studied in the whole organism.

"Knockout" mice

A "knockout" animal has had a gene inactivated so that the function of the gene is lost. It is possible to inactivate genes in a number of different experimental organisms. The most relevant to human genetics is the mouse, which has a rich genetic history. The effect of knocking out a gene can be assessed in the adult mouse—or if the gene is essential for survival, the developmental stage requiring the gene's function can be identified. A streamlined description of the steps in creating a knockout mouse are outlined as follows and illustrated in figure 17.10:

1. The cloned gene is disrupted by replacing part of it with a marker gene using recombinant DNA techniques. The marker gene codes for resistance to the antibiotic neomycin in bacteria, which allows mouse cells to survive when grown in a medium containing the related drug G418. The construction is done such that the marker gene is flanked by the DNA normally flanking the gene of interest in the chromosome.
2. The interrupted gene is introduced into **embryonic stem cells (ES cells)**. These cells are derived from early embryos and can develop into different adult tissues. In these cells, the gene can recombine with the chromosomal copy of the gene based on the flanking DNA. This is the same kind of recombination used to map genes (see chapter 13). The knockout gene with the drug resistance gene does not have an origin of replication, and thus it will be lost if no recombination occurs. Cells are grown in medium containing G418 to select for recombination events. (Only those containing the marker gene can grow in the presence of G418.)

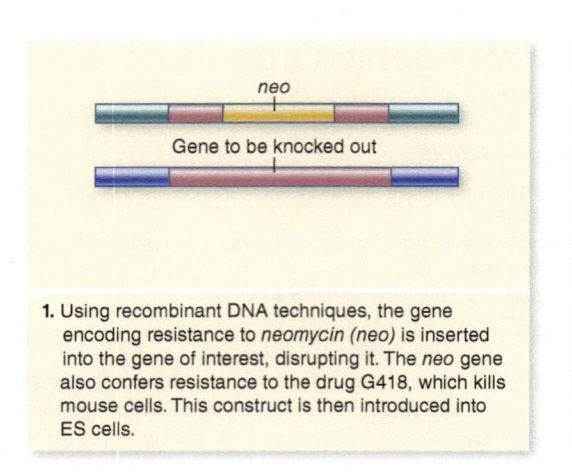

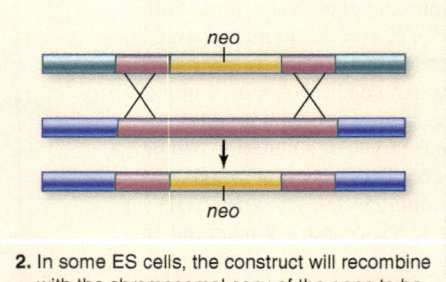

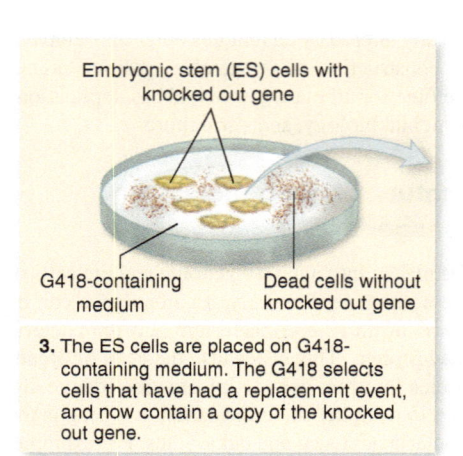

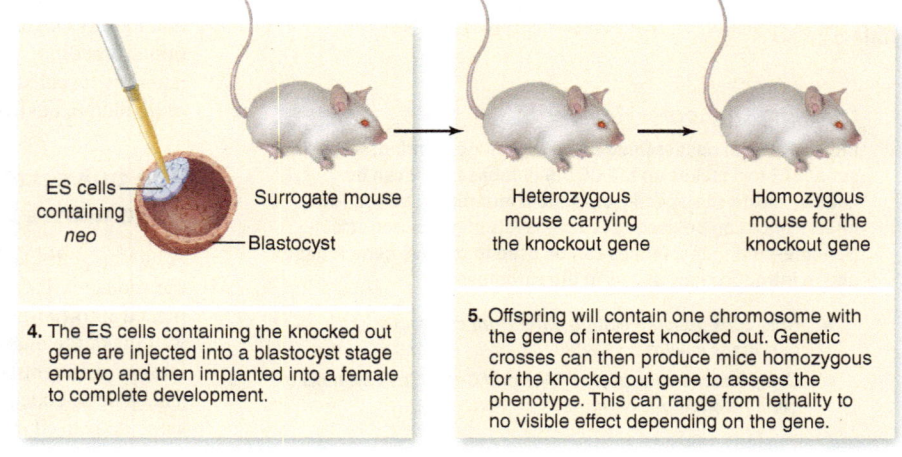

Figure 17.10 Construction of a knockout mouse. Steps in the construction of a knockout mouse. Some technical details have been omitted, but the basic concept is shown.

1. Using recombinant DNA techniques, the gene encoding resistance to *neomycin (neo)* is inserted into the gene of interest, disrupting it. The *neo* gene also confers resistance to the drug G418, which kills mouse cells. This construct is then introduced into ES cells.

2. In some ES cells, the construct will recombine with the chromosomal copy of the gene to be knocked out. This replaces the chromosomal copy with the *neo* disrupted construct. This is the equivalent to a double crossover event in a genetic cross.

3. The ES cells are placed on G418-containing medium. The G418 selects cells that have had a replacement event, and now contain a copy of the knocked out gene.

4. The ES cells containing the knocked out gene are injected into a blastocyst stage embryo and then implanted into a female to complete development.

5. Offspring will contain one chromosome with the gene of interest knocked out. Genetic crosses can then produce mice homozygous for the knocked out gene to assess the phenotype. This can range from lethality to no visible effect depending on the gene.

3. ES cells containing the knocked-out allele are injected into a mouse embryo early in its development, which is then implanted into a pseudopregnant female (a female that has been mated with a vasectomized male and as a result has a receptive uterus). Some of the cells in the pups born to this female have one of the two alleles for the gene of interest knocked-out and so these animals are chimeras. The chimeric mice are mated with a normal mouse and some of the offspring will be completely heterozygous. In some cases there may be a phenotype that can be detected and analyzed if heterozygosity affects any biological function. If the mutation is recessive and not lethal, heterozygotes can then be crossed to generate homozygous mice. The homozygous mice can be analyzed for phenotypes.

"Knockin" mice

Knockin mice have a normal allele replaced with an allele that has a specific genetic alteration. The ability to eliminate alleles provides insight into a complete loss of gene function, a so-called null allele, but does not allow analysis of alleles that alter but do not completely remove function. Specific mutations can be constructed in vitro, then introduced into the mouse to assess the affects of the alterations. This allows an investigator to use naturally occurring alleles, targeted mutations based on other information, or literally any alteration desired. Mutations can be introduced that result in complete loss of function, partial loss of function, or even gain of function of the gene product.

Conditional inactivation allows the study of essential genes

There are two major shortcomings to knocking out genes in mice. First, a gene knockout results in all cells of all tissues being affected by the inactivation. This is problematic for genes whose products affect more than one tissue. The phenotypic analysis of a knockout that affects multiple tissues can be complex and may not even be possible. The second problem has to do with developmentally important genes. A gene that is essential for early development will lead to embryonic lethality, and may leave nothing to analyze. Because of these limitations, ways to remove genes from the genome in specific tissues and at specific developmental times were developed. Conditional inactivation allows a gene to be knocked-out in cells of specific tissues or at a specific time in development.

Conditionally knocking out a gene commonly uses the Cre-Lox recombination system from the bacterial virus called P1. Cre is an enzyme that catalyzes recombination between short DNA sequences called Lox sequences. If a gene of interest is flanked by Lox sites, then in the presence of the Cre enzyme, recombination between the Lox sites will remove the intervening sequence of DNA (the gene of interest) as a circular DNA that will be degraded (figure 17.11).

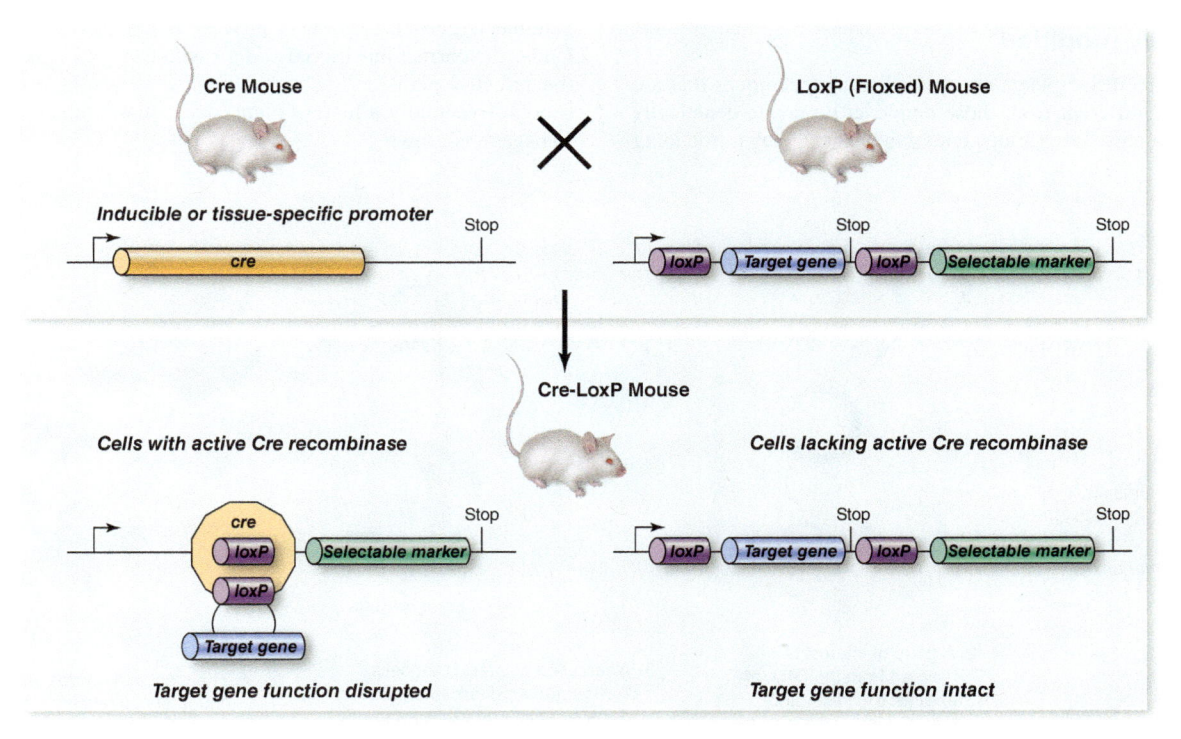

Figure 17.11 Creating conditional and tissue-specific knockout mice. A Cre mouse expressing the Cre recombinase from an inducible or tissue-specific promoter is crossed with a mouse containing a "floxed" target gene. The "floxed" mouse is created using the same techniques for making a knockin mouse. Progeny will contain both the Cre-expressing gene and the "floxed" target gene. The promoter expressing Cre is turned on in the presence of an inducing molecule, or in a particular tissue.

Tissue-specific gene knockouts

To perform a conditional inactivation using Cre-Lox, two lines of transgenic mice are constructed. The first line expresses the Cre enzyme from a promoter that is active only in the tissue of interest. In these mice, Cre will be produced only in the tissue of interest. The second line is a knockin mouse that contains the gene of interest flanked by Lox sequences. A sequence of DNA flanked by Lox sites is said to be "floxed."

Once the two lines are created, genetic crosses produce offspring in which all cells of the mouse contain the Cre recombinase and the allele to be knocked-out. However, only in the tissues that express the Cre enzyme will there be recombination to remove the "floxed" DNA. This allows the effect of loss of function for the floxed gene in that tissue to be analyzed. Other tissues in the mouse will retain the normal allele and should not contribute to any phenotype associated with loss of the gene (figure 17.11).

Developmental knockouts

To knock out a gene at a specific stage in development, a similar strategy is used as for making a tissue specific knockout. The only difference is that the Cre-expressing transgenic line will be engineered to express Cre at a specific developmental stage instead of in a specific tissue. An additional wrinkle on this strategy is to fuse Cre to a promoter that can be turned on when the mouse ingests a particular compound in its food or water. This allows analysis of developmental events after birth.

Plants require different techniques to be genetically modified

Genetically modified plants are created using techniques that are considerably different from those employed to create genetically modified animals. Transgenic plants can be created by introducing foreign DNA into plant cells by electroporation, physical bombardment, chemical treatment, and bacterial transfer. Whichever technique is used, it is not generally possible to specifically target a particular gene sequence, and integration of transgenes into plant genomes is random. Although this allows the introduction of a transgene to be expressed in the plant, it is a problem for knocking out a specific gene. It is desirable to introduce transgenes into crop species that are capable of increasing resistance to disease, frost, herbicides, and drought; also, creating crop plants with modified nutritional profiles would be beneficial to populations where nutritional deficiency is a problem.

Transforming plants with a bacterial pathogen

Transgenes can be transferred into the genomes of certain plant species from the plant pathogen *Agrobacterium tumefaciens*. This bacterium harbors a plasmid called the **Ti (tumor-inducing) plasmid.** The Ti plasmid contains genes that normally cause the formation of a plant tumor—called a gall—in tissues of infected plants. The plasmid also contains genetic sequences responsible for transferring part of the plasmid from the bacterium into plant cells. The Ti plasmid can be isolated from *Agrobacterium*, and the genes responsible for gall formation can be removed and replaced with a gene of interest. Recombinant Ti plasmids containing a gene of interest can be reintroduced into *Agrobacterium* that can then be used to infect cells isolated from plants. During the infection process the bacterium transfers part of its Ti plasmid, including the gene of interest, into the plant cell and that piece of DNA can become integrated into the plant cell's genome (figure 17.12). This process is called transformation. Under the correct nutritional and growth-stimulating conditions, the infected plant cells can be induced to produce roots and shoots. Eventually a mature plant in which all cells contain the transgene is grown.

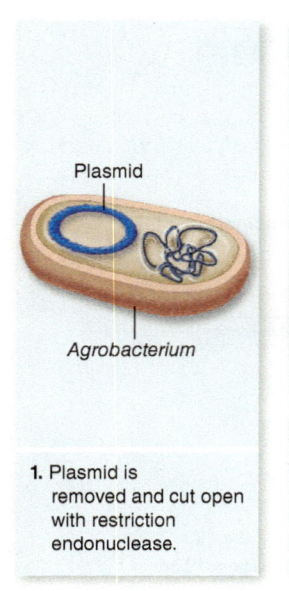

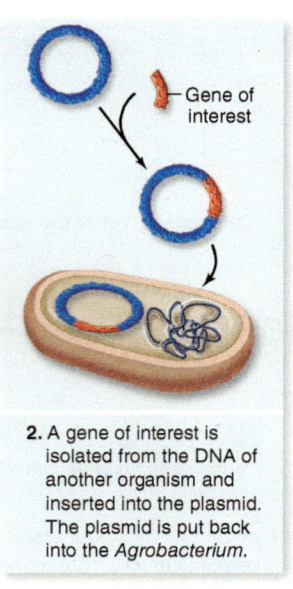

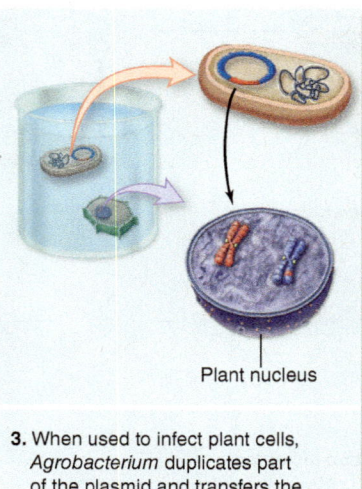

1. Plasmid is removed and cut open with restriction endonuclease.

2. A gene of interest is isolated from the DNA of another organism and inserted into the plasmid. The plasmid is put back into the *Agrobacterium*.

3. When used to infect plant cells, *Agrobacterium* duplicates part of the plasmid and transfers the new gene into a chromosome of the plant cell.

4. The plant cell divides, and each daughter cell receives the new gene. These cultured cells can be used to grow a new plant with the introduced gene.

Figure 17.12 Creating transgenic plants using *Agrobacterium* transformation. Steps in the generation of transgenic plants using *Agrobacterium tumefaciens* transformation.

Transformation via particle bombardment

Agrobacterium cannot infect all plant species, so not all plant species can be transformed using the method just described. When transformation of plant cells by *Agrobacterium* is not possible, DNA can be introduced into plant cells using physical techniques. A common approach to physically introducing recombinant DNA into plant cells is to coat gold or tungsten nanoparticles with recombinant DNA and to fire the particles at fragments of plant tissue. The DNA is carried into cells on the tiny particles and when in the cells the DNA may integrate into the genome. The plant tissue can then be grown in vitro, induced to differentiate, and eventually grown into a mature plant.

Learning Outcomes Review 17.4

The near-universal nature of the genetic code allows genes to be moved between different species. Knockout mice have specific alleles removed from the genome. Knockin mice have specific alleles replaced with an altered copy. Conditionally knocked-in alleles can be activated in specific tissues or at specific times during development. Plants can be transformed with Ti plasmid DNA from *Agrobacterium* or by nanoparticle bombardment.

■ *Under what circumstances would it be appropriate to make a conditional knockout mouse versus a normal knockout mouse, and when might you need to make a knockin mouse?*

17.5 Environmental Applications

Learning Outcomes

1. Describe the benefits of biofuel production from algae.
2. Describe the factors that limit microbial degradation of hydrocarbons in natural environments.
3. Explain the importance of microorganisms in wastewater treatment.

Environmental biotechnology is the use of biological processes to protect, repair, and reduce human impacts on the environment or to increase the sustainability of existing resources. Many applications of biotechnology in the environmental sciences focus on the recycling of waste materials such that the waste of one process can be the input to another process. Additionally, there is currently much interest in the use of biological processes in generating renewable fuel sources. Environmental biotechnology is not a new field as biological processes have been used to manage wastewater since the mid-19th century. Technical advances in chemistry, genomics, genetics, and physics have accelerated our potential to develop technologies needed to tackle human-made environmental problems such as the global energy and water availability crises.

Algae can be used to manufacture biofuels

Biofuels are fuels derived from biomass made from recently fixed carbon sources, as opposed to fuels derived from "ancient" biomass such as fossil fuels. Common sources of biomass for biofuel generation include crop plants and algae. There are two common forms of liquid biofuel: ethanol derived from the fermentation of plant carbohydrates and biodiesel produced from the chemical transformation of plant or algal lipids. There are a number of species of algae that can be used to produce biodiesel, however, all are microalgae with cell diameters in the range of 3 to 30 μm.

Biofuels produced from microalgae have several advantages over petroleum-based fossil fuels. They are renewable as long as carbon dioxide can be fixed into biomass by photosynthesis. They are more environmentally friendly than fossil fuels, given that carbon in the fuel is derived from atmospheric carbon dioxide, a greenhouse gas that contributes to climate change. The production of biodiesel from microalgae can also be combined with processes such as wastewater treatment or coupled to carbon dioxide capture systems of fossil fuel burning industrial plants.

Microalgae can be cultured in a variety of ways. The simplest culture system involves the creation of large open ponds. Although relatively cheap to construct and maintain, these ponds suffer from contamination by environmental microorganisms or algal predators. Also, they require large areas of land and cannot sustain very dense algal cultures. More expensive, but more easily controlled, are large self-contained culture systems that supply a source of water with dissolved nutrients and are combined with a light harvesting system through which the algal culture circulates (figure 17.13). These photobioreactors are more easily controlled

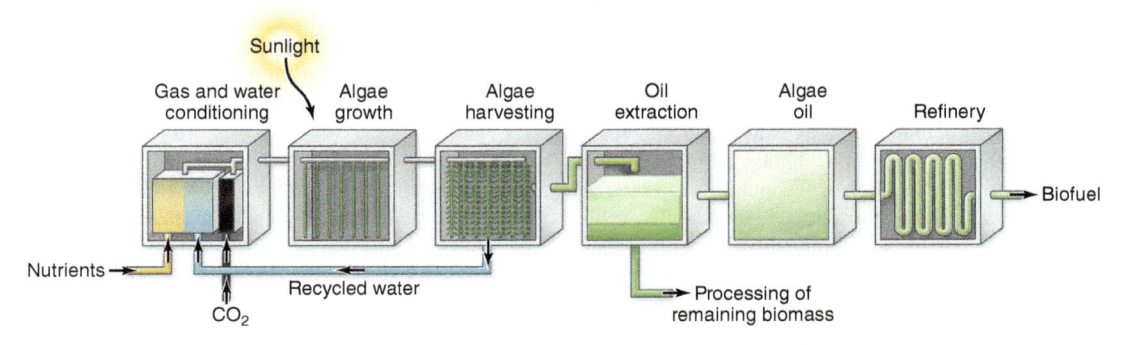

Figure 17.13 Using microalgae to make biofuel. Microalgae grown under certain conditions accumulate up to 50% of their biomass as lipids. Growth can be in ponds or, as shown here, photobioreactors. In some cases, carbon dioxide can be fed into the photobioreactors from industrial plants. Lipids harvested from the algae are processed and refined into biofuel.

environments and can be fed with wastewater. Carbon dioxide can be bubbled through the system to provide a constant source of carbon for fixation by photosynthesis. Photosynthesis produces reduced carbon that can be metabolically converted to acetyl-coA, which is the building block of fatty acids.

Some species of microalgae can also use heterotrophic nutritional strategies. These species absorb sugars from the culture medium and use those sugars for the formation of biomass for growth, or for the production of lipids. Yet other species can use both heterotrophic and photoautotrophic nutritional strategies simultaneously and these species produce some of the highest concentrations of lipid in their cells. Under conditions of stress, the algal cells can be composed of 40 to 50% triacylglycerides.

A variety of techniques can be used to harvest the lipids from the algae, and this is one of the most expensive parts of algal biodiesel production. Once the lipids have been isolated from the algae, chemical treatments separate the glycerol from the fatty acid chains in the triglyceride. Further refinement yields a biodiesel product.

> **Inquiry question** How might you genetically engineer microalgae so that they are more efficient at making the lipids used to create biofuels?

Microorganisms can degrade hydrocarbons in the environment

Our dependence on hydrocarbon-derived fuels—be they petroleum- or biofuel-based—has an inevitable and serious consequence: Those fuels can enter ecosystems and cause environmental damage. Certain hydrocarbons are persistent in terrestrial, aquatic, and marine ecosystems for decades and can accumulate on and in organisms, leading to death.

Hydrocarbons that persist in the environment can dissipate slowly by evaporation of volatile compounds, by dissolution and dispersion in watery environments, and by oxidation in the presence of light. Importantly, certain microorganisms are capable of metabolizing hydrocarbon pollutants. The degradation or metabolism of hydrocarbon pollutants by microorganisms is called *bioremediation*. The rate at which these microorganisms metabolize hydrocarbons is affected by a number of variables including temperature, physical and chemical properties of the pollutant, oxygen availability, pH, salinity, other microorganisms present, and perhaps most critically, nutrient availability.

Lab studies have shown that nutritional supplementation of certain microorganisms leads to a significant increase in the rate at which hydrocarbon pollutants can be metabolized. The addition of certain nutrients and other chemical additives to a polluted environment can increase the rate of bioremediation of hydrocarbon pollutants. Unfortunately, the increases in decontamination seen in the environment are not as great as those in the lab. An interesting possibility to increase the rate of decontamination by microorganisms is genetic engineering to optimize the pathways involved in hydrocarbon uptake and catabolism.

Bioremediation of hydrocarbons is an important part of a hydrocarbon contamination cleanup plan. Importantly, when metabolized by microorganisms, hydrocarbons end up being metabolized into biomass or catabolized into carbon dioxide, so no additional toxic compounds are released into the environment.

Wastewater treatment relies on the action of microorganisms

Another example of using microorganisms to decontaminate a natural resource is the treatment of industrial and commercial wastewater. Wastewater can consist of raw sewage containing human and other animal fecal material as well as industrial contaminants. Untreated release of this wastewater would wreak havoc on ecosystems and spread disease.

Wastewater is treated in two, or three, stages called primary, secondary, and advanced treatment. Microorganisms play critical roles in the secondary and sometimes advanced treatment stages. Primary treatment involves physical and chemical techniques to remove large and small particulate material. Secondary treatment can involve digestion by anaerobic bacteria, aerobic bacteria, or both (figure 17.14).

Aerobic treatment of wastewater

Material from a primary treatment plant is fed into large aerated tanks that are continuously agitated where oxidative breakdown of any remaining organic material occurs. The bacteria and other organisms in the aerobic tanks form aggregates called floc. Degradation of the remaining carbon containing compounds in the wastewater occurs on this floc, which is later removed in settling tanks. Wastewater is also treated on beds of crushed rock that contain bacterial communities, called biofilms. These perform the same function as the floc in agitated tanks.

Treated water from the aerobic digester can be released into the environment, or further processed to produce potable water. Advanced treatment processes are used to remove inorganic contaminants such as phosphorus. This can be done by chemical addition, or using bacterial species capable of metabolizing dissolved phosphorus. The metabolized phosphorus is accumulated into cytoplasmic granules in the bacteria, which can be harvested and disposed of.

Anaerobic treatment of wastewater effluent

Anaerobic digestion of wastewater occurs in sealed reactors called sludge digesters. In the sludge digester, anaerobic bacteria catabolize polysaccharides, lipids, and proteins remaining after primary treatment into their constituent monomers. The resulting fatty acids, amino acids, and sugars, are metabolized into products such as acetate, hydrogen gas, and carbon dioxide. Archaebacteria in the sludge digester can convert these molecules to methane, carbon dioxide, and water. The methane can be burned to heat or power the wastewater treatment plant. Some of the sludge produced by the digester is cycled back into the system to maintain the population of bacteria required for digestion.

> ### Learning Outcomes Review 17.5
> Environmental biotechnology uses biological processes to protect or repair the environment. Microalgae can be used to produce biofuel. Benefits include increased sustainability, and coupling fuel production to wastewater treatment. Nutrient supplementation of contaminated environments can stimulate hydrocarbon breakdown. Wastewater treatment processes using microorganisms occur in several stages of which aerobic and anaerobic digestion involves the use of microorganisms.
>
> ■ *Could genetic engineering improve wastewater treatment?*

Figure 17.14 Using microorganisms to treat wastewater. *a.* Organic matter in wastewater can be digested by aerobic bacteria in mixing basins. *b.* Alternatively, wastewater can be trickled over crushed rocks, and bacteria on the surface of the rocks aerobically digest the organic material in the wastewater. *c.* Organic material can also be removed from wastewater through the actions of anaerobic microorganisms. This process produces methane.

Source: Willey, J. M., Sherwood, L., Woolverton, C. J., Prescott, L. M., & Willey, J. M. (2011). Microbiology. New York: McGraw-Hill.

medicine is not a new endeavor; however, advances in genetics, genomics, and recombinant DNA technologies have accelerated the rate of discovery in the last 20 years.

Important proteins can be produced using recombinant DNA

Some diseases are due to the body's inability to produce a specific protein. We treat these diseases by providing the missing protein. Traditionally, these are isolated from animal tissues, but now we can express proteins in bacteria and other systems using recombinant DNA. An example is insulin, the first medical product produced in this way. Insulin is produced in the pancreas then transported via the bloodstream. In target tissues insulin binds a receptor to initiate a process that leads to the uptake of glucose. Type I diabetes occurs when the pancreas cannot produce sufficient insulin, ultimately resulting in a rise in blood sugar. Symptoms of high blood glucose include thirst, frequent urination, and hunger. Long-term complications include blindness, nerve damage, coma, and ultimately death. Type I diabetes can be managed with daily insulin injections.

For a long time, insulin was harvested from pigs, purified, and used to treat type I diabetes. In the 1980s, it became possible to express human insulin in *E. coli* and the yeast *S. cerevisiae*. The insulin could be purified and used in place of the pig derived protein to treat type I diabetes. In human cells, insulin is expressed as a single mRNA that is translated as a single polypeptide. The polypeptide is cleaved in the cytoplasm to produce two short chains that become linked by disulfide bonds (figure 17.15a).

If the cDNA corresponding to the coding regions of the insulin gene is expressed in *E. coli*, one long polypeptide is produced because *E. coli* cannot posttranslationally process proteins in the same

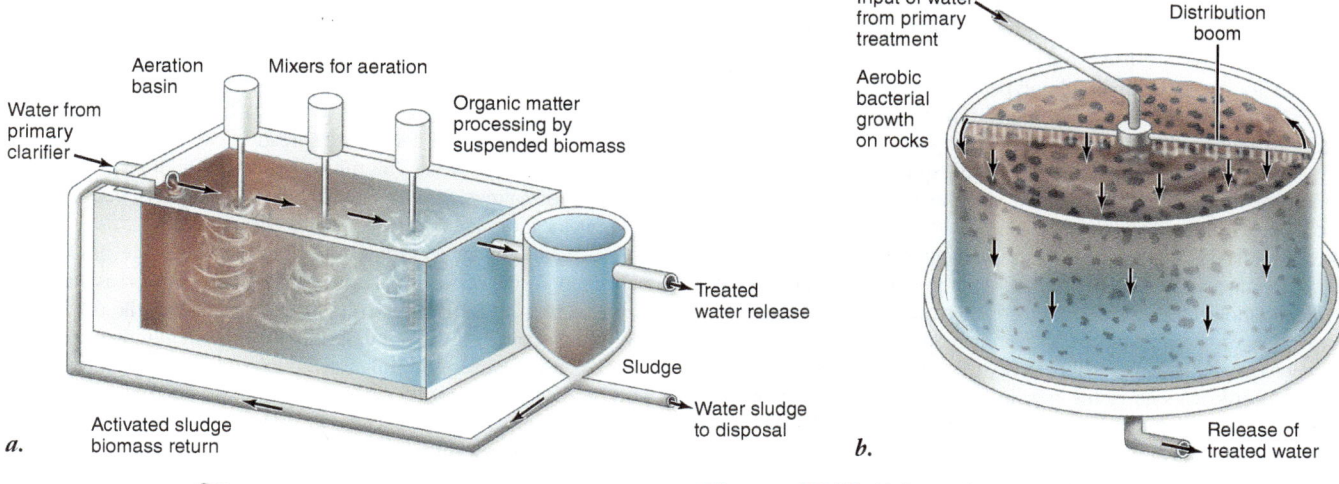

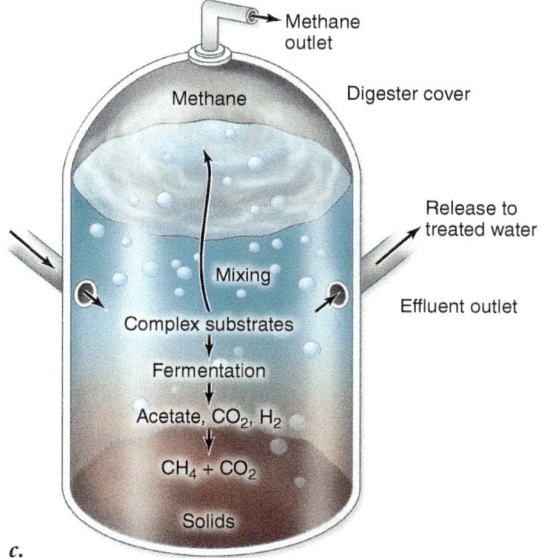

17.6 Medical Applications

Learning Outcomes

1. Describe how recombinant DNA has been used in medical treatments and diagnostics.
2. Compare and contrast FISH and gene chip technologies for diagnosing disease.
3. Describe how immunoassays can be used to diagnose viral infections.

The manipulation of biological systems has led to major advances in health care. Specifically, biotechnology has provided novel diagnostic tools, allowed the treatment of previously untreatable genetic conditions and infectious diseases, and given us ways to make novel medicines. The application of biotechnology to

way as do human cells (figure 17.15a). Rather than expressing insulin from a single cDNA corresponding to the whole insulin gene, two cDNAs are used. Each cDNA corresponds to the gene sequence for the two individual polypeptides that make up the functional insulin molecule. Each of these cDNAs is isolated and incorporated into a plasmid vector that can be expressed in *E. coli*.

Proteins from two strains of *E. coli* expressing the individual insulin polypeptides can be purified and mixed together to make a functional insulin molecule (figure 17.15b). When purified so that no other bacterial proteins contaminate the insulin, the insulin reaches very high concentrations. Unfortunately, at these high concentrations, the insulin molecules stick together and their ability to activate glucose transporters on the cell surface is severely reduced. Genetic engineering techniques, like the ones previously described, have been used to mutate specific codons in the insulin cDNA sequences so that the resulting proteins don't stick together. The consequence of this is that high concentrations of highly effective insulin can be obtained.

Fluorescent in situ hybridization can detect gross chromosomal rearrangements

Fluorescent in situ hybridization (FISH) takes advantage of the ability of DNA to be reversibly denatured and renatured. Many techniques in molecular biology use hybridization to detect a specific DNA in a complex mixture. Hybridization uses a probe, the sequence you want to find, with either a fluorescent or radioactive label. Both probe and target DNA are denatured and mixed together. When conditions are restored to allow renaturation, the probe will find any complementary sequences in the mixture and form a hybrid. In the case of FISH, the target DNA is an actual metaphase spread of chromosomes, or cells with intact chromosomes. This allows the detection of gross chromosomal abnormalities, such as large deletions, inversions, duplications and translocations (see chapter 15 for descriptions).

Aneuploidy and chromosomal abnormalities are associated with many forms of cancer. In about 25% of invasive breast cancers a gene called *HER2* has been duplicated. Instead of the usual two copies expected in a diploid cell, there can be hundreds of copies. HER2 is a receptor protein that activates signal transduction pathways leading to cell proliferation. Therefore, more copies of *HER2* can lead to uncontrolled cell growth and tumor formation. Some drugs can block cell signaling through the HER2 protein, stopping or preventing this kind of tumor from growing. However such drugs are only useful in treating breast tumors where the *HER2* gene is duplicated. Thus knowledge about whether or not *HER2* has been duplicated is vital for clinicians prescribing the drug. In this way, treatment can be personalized based on a patient's specific genetic profile.

FISH can be used to detect HER2 in cells taken from a tumor biopsy. The cells from the tumor are collected and treated to expose

a.

b.

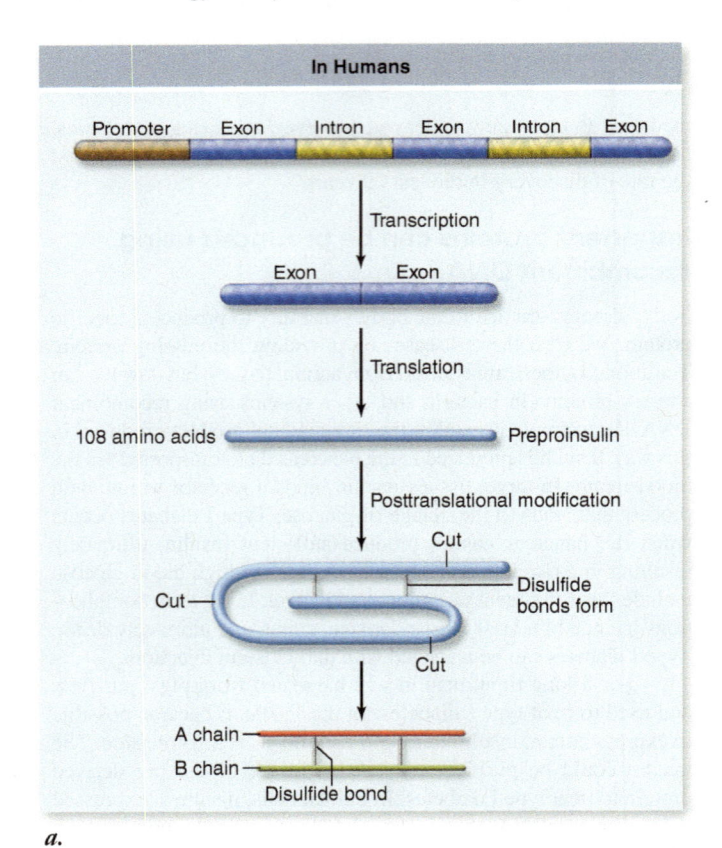

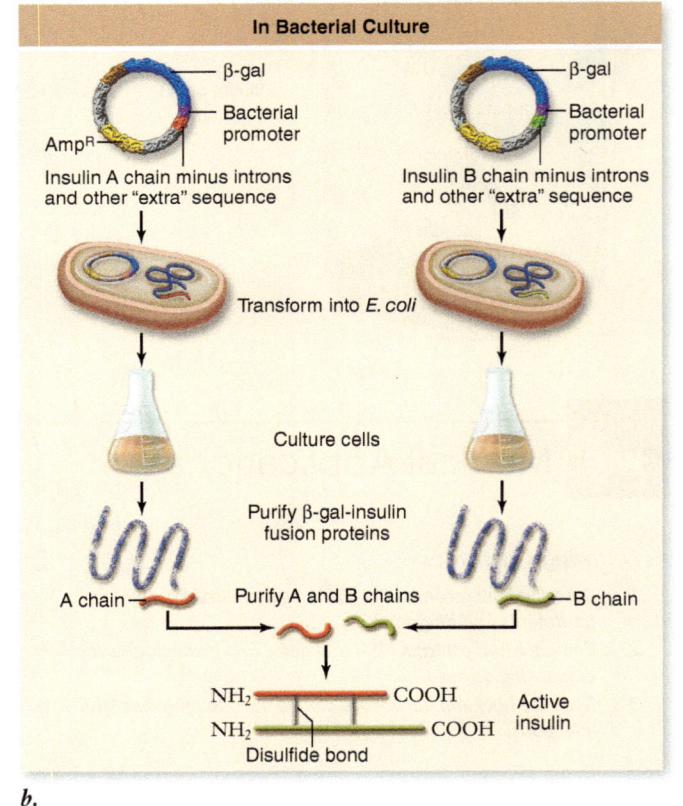

Figure 17.15 Making insulin in genetically engineered *E. coli*. *a.* In human cells one preproinsulin polypeptide is processed posttranslationally into insulin chains A and B that associate via disulfide bonds to form mature insulin. *b.* Two cDNAs corresponding to the gene sequences for insulin chain A and chain B are cloned into plasmids and introduced into different *E. coli*. Cultures of the two types of *E. coli* express either chain A or chain B. Chains A and B are purified from the different *E. coli* and when mixed, associate into active mature insulin.

the chromosomes. A fluorescent labeled HER2 probe is added, and the probe DNA strands will hybridize with complementary sequences of DNA in the chromosome (figure 17.16a). The cells on the slide can then be visualized using a microscope that illuminates samples with UV light causing the probe to fluoresce.

In normal cells, or in cells from breast cancers that do not have multiple copies of the *HER2* gene, there would be just two small dots of fluorescence seen in the microscope (figure 17.16b). If the breast cancer cells contain more than two copies of the gene, then there will be multiple, much larger dots of fluorescence (figure 17.16c). This kind of information, along with other tests, allows the personalization of medicine so that drugs are administered only when necessary.

Data analysis If you performed a FISH analysis on human cells in G_2 of the cell cycle using a probe that recognized telomeric DNA (DNA on the tips of the chromosome), how many spots would you expect to see down the microscope?

Gene chips can identify genetic markers for disease

Gene chips, also known as DNA microarrays, are collections of hundreds to thousands of different DNA sequences that are spotted onto a solid surface such as a microscope slide. The DNA sequences are usually short chemically synthesized pieces of DNA, or short sequences of DNA amplified by PCR or RT-PCR. By using different gene chips, and different experimental conditions, a variety of kinds of questions can be assessed.

A disease marker, or biomarker, is a biological molecule found under certain conditions, but not found in other conditions. For example, some molecules are only produced under disease conditions, whereas others are only found in the healthy state. Thus, biomarkers can be diagnostic of a disease state, be informative about the progress of a disease, or help monitor the response to therapy. In the case of gene chips, the biomarker is the mRNA synthesized from a particular gene. For example, individuals with certain autoimmune diseases or chronic inflammatory conditions will express different sets of genes in immune cells than will healthy individuals. Gene chips can be used to determine which genes are being expressed and the relative levels of expression for different genes.

Gene chips can also be used to diagnose genetic disorders. Some genetic diseases are caused by a problem with a single gene, whereas others are characterized by problems with many genes. Gene chips can be made that identify mutations in a single gene, or are able to detect many genetic changes that together make an individual susceptible to a particular genetic disorder. It is even possible to create a gene chip that is able to identify a specific bacterial species responsible for a particular infectious disease. In some cases, it is possible to determine the antibiotic resistance possessed by the pathogen. By knowing the specific pathogen and its antibiotic resistance status, antibiotics that have a greater chance of eliminating the pathogen and curing the disease can be administered.

Taken together, technologies such as FISH and gene chips have revolutionized the way that medicine is administered. The more that can be learned about the specific underlying causes of a disease on an individual basis, the more specific and personalized can be the health care provided. These technologies face some challenges, however. First, they require great skill to be successfully used; second, they are relatively expensive; and third, not all diseases are caused or influenced by factors specific to the individual. For example, some autoimmune and inflammatory diseases are caused by disruption of the relationship between bacteria living on and in your body and your

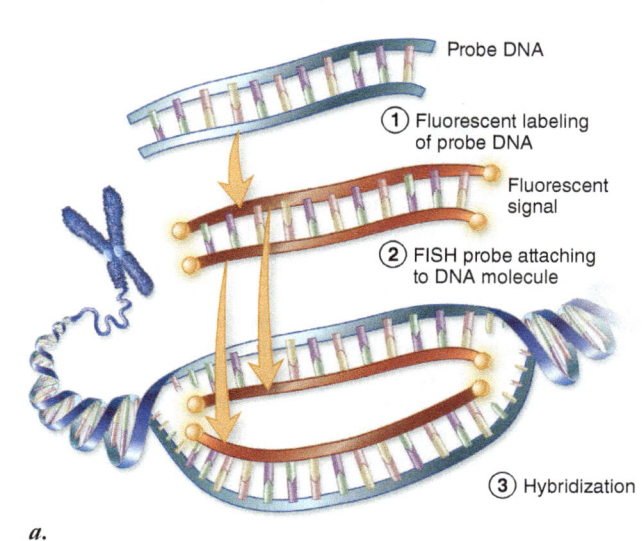

Figure 17.16 Fluorescent in situ hybridization.
a. Fluorescent probes bind to specific DNA sequences in chromosomes.
b. Only two spots of fluorescence are visible in the nuclei of the tumor cells indicating that *HER2* is not amplified. *c.* Multiple large fluorescent spots in the tumor cells indicate amplification of *HER2*.

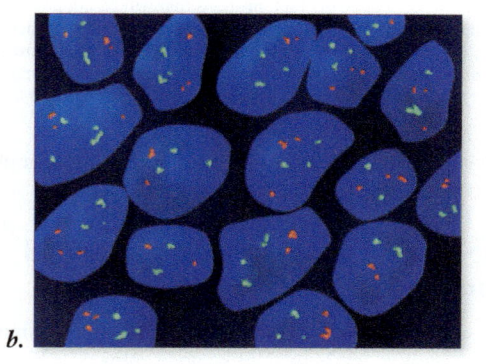

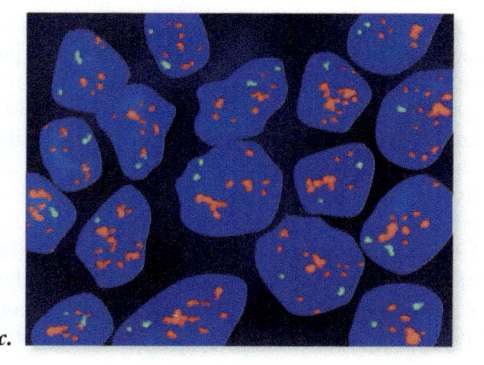

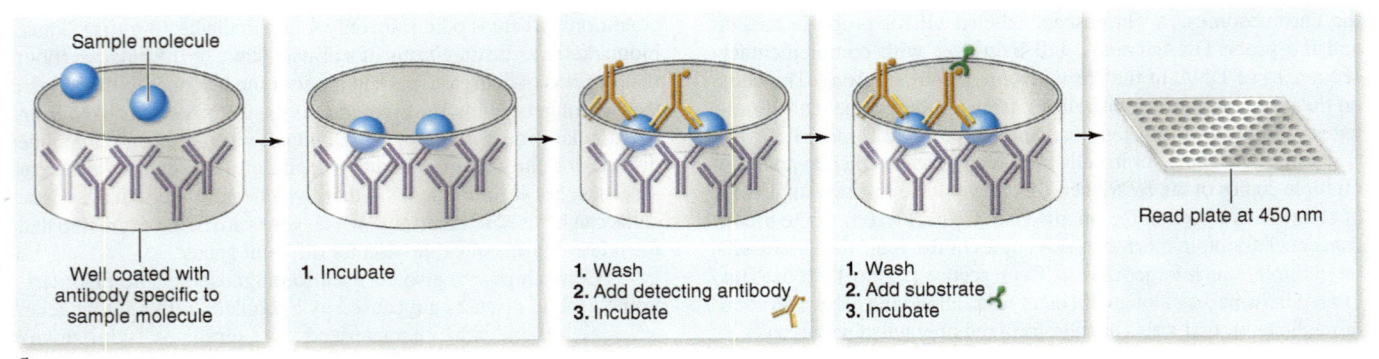

Figure 17.17 Detecting virus using an immunoassay.
a. Steps in an immunoassay to detect a virus in a sample. *b.* Modern HIV immunoassays have both viral proteins and antibodies to the virus bound to the well of the plate. Virus and antibodies against the virus in a sample will bind to the plate and can be detected using another antibody that produces a colored substance.

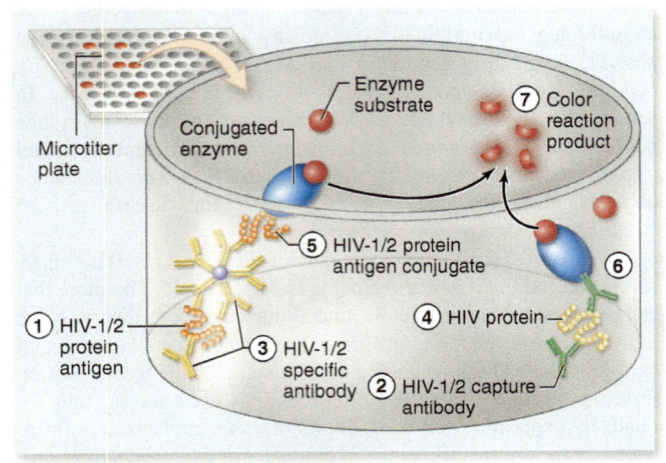

b. 4th generation HIV antibody and p24 ELISA test (1) HIV proteins and (2) HIV capture antibodies are attached to wells; (3) patient sample is added to wells that may contain antibodies to virus which can bind the HIV protein in the well or (4) HIV proteins that will be bound by the HIV capture antibodies; (5) addition of HIV protein with an attached enzyme or (6) anti-HIV protein antibody with an attached enzyme is added to the well; finally, (7) a substrate acted upon by the enzyme is added that is converted to a colored product.

amount of fluorescence or colored product made by the enzyme, is proportional to the amount of virus present.

A variation of this technology is used in the diagnosis of human immunodeficiency virus (HIV) infection (figure 17.17*b*). In HIV immunoassays, an antibody recognizing HIV and a specific protein from HIV are coated onto the well of a microtiter plate. When a sample from a patient is added to the well, the virus binds to the HIV antibody, and any antibodies that are present in the blood that have been produced in response to the infection will recognize and bind to the HIV protein in the well. After washing, a mix of antibodies recognizing HIV protein linked to an enzyme and a specific HIV protein linked to an enzyme are added to the well. After more washing, any enzymes linked to the antibodies or HIV proteins that remain bound produce an easily detectable color change.

Stem cell therapy has the potential to treat chronic diseases

Every year millions of people suffer from loss of function in tissues or organs that can only be remedied by replacement or repair of the tissue, or by long-term drug treatment. For example, the disease multiple sclerosis is caused when the immune system attacks the central nervous system. This can lead to loss of sight, speech, and mobility. Individuals so afflicted may suffer symptoms for many years before requiring permanent supportive care. The field of medicine dealing with these diseases and injuries is called *regenerative medicine*. There is growing evidence that stem cells are able to successfully repair or replaced damaged tissue.

Stem cells are populations of cells that have the potential to regenerate themselves and to produce cells that can differentiate into other cell types. **Embryonic stem cells** are pluripotent, meaning they are able to differentiate into all of the cell types found in an adult body (see chapter 19 for details). These cells are only found in the very early stages of embryonic development. After this point, stem cell populations have less potential for differentiation. Some stem cell types may only be able to become one or two differentiated types of cells. Stem cells are also found in adults and are responsible for the natural regeneration of tissue that must

body. In that case, learning about only your genetics or your physiology may not yield a complete answer about a disease.

Immunoassays can rapidly diagnose viral infections

Immunoassays are tests that allow the detection of a molecule through the use of an antibody. Unlike gene chips or FISH, immunoassays are relatively cheap and easy to use. Antibodies are protein molecules produced by most animals in response to infection by pathogens. Antibodies very specifically recognize a pathogen and target it for destruction by the immune system.

Biotechnology companies can make antibodies against a range of viruses and bacteria. These antibodies can be used in immunoassays to detect whether there is virus in a sample collected from a patient. An antibody against a specific virus can be coated into a plastic well on a microtiter plate (figure 17.17*a*). When an unknown sample is added to the well, if the virus is present, it will be bound by the antibody. Bound virus can be detected using a second antibody against the virus that has been linked to a fluorescent molecule, or an enzyme that produces a colored product from a suitable substrate. The

occur to replace lost or damaged cells. More recently, it has been possible to induce stem cells from adult cell types by using genetic engineering techniques to express specific sets of transcription factors. This reprogramming phenomenon is explored in chapter 19.

The therapeutic use of stem cells has received much hype in the popular press; however, there are only a few cases when it has been used successfully in the treatment of specific diseases or injuries. The transplant of bone marrow, which contains stem cells capable of producing numerous types of blood cell, has been used since the 1960s to treat a number of blood disorders, including various types of leukemia and anemia. This form of stem cell transplantation has also been successful in stopping progression of multiple sclerosis in some patients because it results in a complete reprogramming of the immune system that is attacking the nervous system. Although the neuronal damage is not repaired in these patients, other types of stem cell therapies have shown some promise in reducing neurological symptoms.

The majority of approved stem cell therapies use **mesenchymal stem cells.** These stem cells can differentiate into certain cell types that form connective tissues such as bone and cartilage. These therapies can promote connective tissue formation and repair, and reduce inflammation. Another area in which stem cell therapies have moved through clinical trials is for the treatment of heart attacks and the resulting tissue damage. Despite some successes, stem cell therapies face significant hurdles before they can realize the potential promised by researchers. It takes approximately 10 years and over $1 billion to bring a new medicine to the market; for stem cell therapy clinical trials, patient recruitment is difficult; regulatory agency oversight is extensive; and public perception of the use of stem cells remains biased by political and social beliefs.

Learning Outcomes Review 17.6

Recombinant DNA technology has allowed human gene products to be mass-produced. FISH can be used to detect gross chromosomal rearrangements. This can provide valuable diagnostic information and allow the personalization of treatment. Gene chips can be used to determine the patterns of expression of sets of hundreds of genes at a time. Immunoassays provide a rapid and cheap way to diagnose viral infections such as HIV. Stem cells can be transplanted to treat chronic and degenerative diseases; however, the technology faces political and social challenges.

- *Explain how genetic engineering techniques would be used to produce a specific protein product.*

17.7 Agricultural Applications

Learning Outcomes

1. Describe the benefits of creating transgenic crops.
2. Identify the most common forms of genetically modified plants.
3. Evaluate issues on each side of the transgenic crop debate.

Possibly the greatest impact of biotechnology has been in its application to agriculture. Transgenic crops are being created that resist disease, are tolerant of herbicides and drought, and have improved nutritional quality. Plants are also being used to produce pharmaceuticals, and domesticated animals are being genetically modified to produce biologically active compounds.

Herbicide-resistant crops allow for no-till planting

Tilling is the process where soils are turned over to prepare them for planting. This acts to remove weeds, which compete with crop plants for nutrients, and improve irrigation. Soils that are not tilled have greater levels of organic and inorganic nutrients that promote growth, often retain more water, and are less prone to soil erosion. There is an economic cost to tilling because it requires expensive equipment and takes time to perform. Weeds can also be removed from crop fields by the applications of herbicides, but most common herbicides are not particularly discriminating and will kill a variety of different plant species. If crops are planted without the need for tilling, farms can be more efficient—which can translate into economical gains.

Agriculturally important broadleaf plants such as soybean have been genetically engineered to be resistant to **glyphosate,** a powerful, biodegradable herbicide that kills most actively growing plants (figure 17.18). Glyphosate works by inhibiting an enzyme called 5-enolpyruvylshikimate-3-phosphate (EPSP) synthase, which plants require to make aromatic amino acids. Animals do not make aromatic amino acids; rather, they are obtained from the diet, so animals living in and around crops, or in waterways receiving crop run off, are unaffected by glyphosate. To make glyphosate-resistant plants, scientists used a Ti plasmid to insert extra copies of the EPSP synthase gene into plants. These engineered plants produce 20 times the normal level of EPSP synthase, enabling them to synthesize aromatic amino acids and grow despite glyphosate's inhibition of the enzyme (figure 17.18).

These advances are of great interest to farmers because a crop resistant to glyphosate does not have to be weeded through tilling. Because glyphosate is a broad-spectrum herbicide, farmers don't need to employ a variety of different herbicides, most of which kill only a few kinds of weeds. Furthermore, unlike many other herbicides, glyphosate breaks down readily in the environment.

Six important crop plants have been modified to be glyphosate-resistant: maize (corn), cotton, soybeans, canola, sugarbeet, and alfalfa. The use of glyphosate-resistant soy has been especially popular, accounting for 60% of the global area of transgenic crops grown worldwide. In the United States, 90% of soy currently grown is transgenic. Other commercially important crops, such as barley, rice, and wheat, have been made resistant to a range of other herbicides.

Bt crops are resistant to some insect pests

Many commercially important plants are sensitive to insect pests, and the usual defense against this has been insecticides. Over 40% of the chemical insecticides used today are targeted against boll weevils, bollworms, and other insects that eat cotton plants. Scientists have produced plants that are resistant to insect

SCIENTIFIC THINKING

Hypothesis: Petunias can acquire tolerance to the herbicide glyphosate by overexpressing EPSP synthase.
Prediction: Transgenic petunia plants with a chimeric EPSP synthase gene with strong promoter will be glyphosate tolerant.
Test:
1. Use restriction enzymes and ligase to "paste" the cauliflower mosaic virus promoter (35S) to the EPSP synthase gene and insert the construct in Ti plasmids.
2. Transform Agrobacterium with the recombinant plasmid.
3. Infect petunia cells and regenerate plants. Regenerate uninfected plants as controls.
4. Challenge plants with glyphosate.

Result: Glyphosate kills control plants, but not transgenic plants.
Conclusion: Additional EPSP synthase provides glyphosate tolerance.
Further Experiments: The transgenic plants are tolerant, but not resistant (note bleaching at shoot tip). How could you determine if additional copies of the gene would increase tolerance? Can you think of any downsides to expressing too much EPSP synthase in petunia?

Figure 17.18 Genetically engineered herbicide resistance.

pests, removing the need to use many externally applied insecticides.

The approach is to insert into crop plants genes encoding proteins harmful to the pests, but harmless to other organisms. The most commonly used protein is a toxin produced by the soil bacterium *Bacillus thuringiensis* (Bt toxin). When insects ingest Bt toxin, endogenous enzymes convert it into an insect-specific toxin, causing paralysis and death. Because these enzymes are not found in other animals, the protein is harmless to them.

Many of the same crops that have been modified for herbicide resistance have also been modified for insect resistance using the Bt toxin. The use of Bt maize is the second most common genetically modified (GM) crop, representing 14% of global area of GM crops in nine countries. The global distribution of these crops is also similar to the herbicide-resistant relatives.

Given the popularity of both of these types of crop modifications, it is not surprising that they have also been combined, so-called stacked GM crops, in both maize and cotton. Stacked crops now represent 9% of global area of GM crops.

Golden Rice shows the potential of transgenic crops

Golden Rice is a transgenic cultivar of rice, genetically modified to express β-carotene, the precursor of vitamin A. Ingested β-carotene can be converted to vitamin A in the body to alleviate the symptoms of vitamin A deficiency. The World Health Organization (WHO) estimates that vitamin A deficiency affects between 140 and 250 million preschool children worldwide, 250,000 to 500,000 of whom become blind. The deficiency is especially severe in developing countries where the major food is rice.

Golden Rice is named for its distinctive color imparted by the presence of β-carotene in the endosperm (the outer layer of rice that has been milled). Rice does not normally make β-carotene in endosperm tissue, but does produce a precursor, geranylgeranyl diphosphate (GGPP), that can be converted by three enzymes—phytoene synthase, phytoene desaturase, and lycopene β-cyclase—to β-carotene. These three genes were altered using recombinant DNA technologies to be expressed in endosperm and introduced into rice to complete the biosynthetic pathway producing β-carotene in endosperm (figure 17.19).

This case of genetic engineering is interesting for two reasons. First, it introduces a new biochemical pathway in tissue of the transgenic plants. Second, expression of β-carotene in the endosperm could not have been achieved by conventional breeding as no known rice cultivar produces these enzymes in endosperm. A second-generation rice cultivar that makes much higher levels of β-carotene has also been produced by using the gene for phytoene synthase from maize in place of the original daffodil gene.

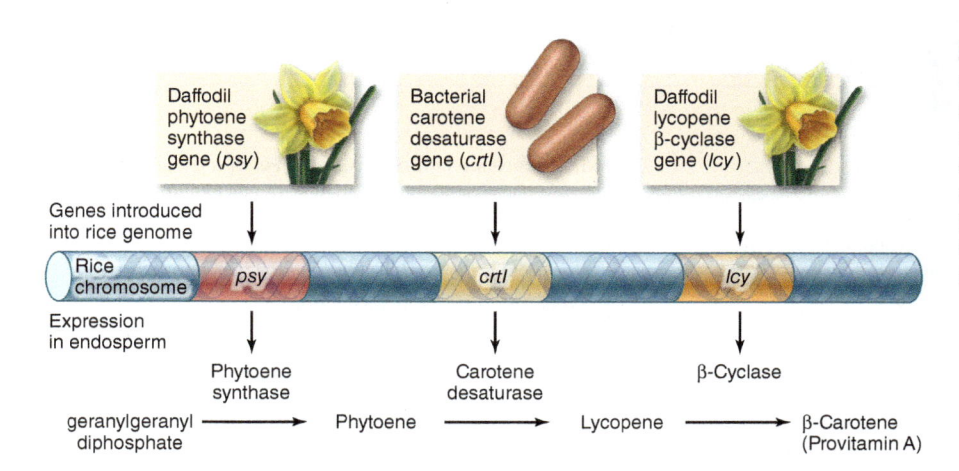

Figure 17.19 Construction of Golden Rice. Three daffodil genes were added to the rice genome to allow synthesis of β-carotene in endosperm. The source of the genes and the pathway for synthesis of β-carotene is shown. The result is Golden Rice, which contains enriched levels of β-carotene in endosperm.

Golden Rice was originally constructed in a public facility in Switzerland and made available for free with no commercial incentives. Since its creation, Golden Rice has been improved by public groups and by industry scientists, and these improved versions are also being made available without commercial strings attached.

Transgenic crops raise a number of social issues

The adoption of transgenic crops has been resisted in some places for a variety of reasons. Some people have wondered about the safety of these crops for human consumption, the likelihood of introduced genes moving into wild relatives, and the possible loss of biodiversity associated with these crops.

On the side promoting the use of transgenic crops are the multinational companies using these technologies to produce seeds for the various transgenic crops. Also on this side are groups noting the benefits of transgenic crops like Golden Rice in feeding the developing world and preventing nutritional deficiencies. On the opposing side are a variety of political and social organizations that raise concerns for the ecological and health implications of genetically modified foods. Scientists can be found on both sides of the debate.

The controversy originally centered on the safety of consuming foods containing introduced genes. In the United States, this issue has been "settled" for the crops already mentioned, and a large amount of transgenic soy and maize is consumed in this country. There remains a perceived risk of allergic reactions and longer-term effects, although there is no documentation of adverse effects on human health. Existing crops will be monitored for adverse effects, and each new modification will require regulatory approval for human consumption.

Also concerning is the potential for spread of the transgenes from a transgenic crop to a non-transgenic crop. The amount of cross-pollination between plants of the same species varies. Soybeans tend to self-pollinate with limited outcrossing. Corn, however, freely outcrosses and so genes from transgenic plants move more frequently to nontransgenic corn plants. This is a concern for organic farmers with fields close to transgenic corn fields because current regulations for organic certification exclude transgenic crops. The amount of outcrossing depends on the proximity of crops, wind speed, and direction, as well as temperature and humidity. For plants that depend on pollinators for reproduction, the range of species the pollinator visits and the distance it travels are also factors.

Gene flow can also occur between related species through hybridization. In the United States, at least 15 weedy species have hybridized with crop plants. For two of the major U.S. crops, corn and soybean, no closely related weedy species reproduce with them, so this is an unlikely mechanism for acquiring herbicide resistance. Sunflowers, however, were first domesticated in the United States, and the risk of a transgene moving into a weedy population is a concern. One study revealed that 10 to 33% of wild sunflowers had hybridized with domesticated varieties.

In the case of herbicide-resistant transgenic crop plants, the repeated application of a single herbicide over a number of years creates selective pressure on weedy species. Over time, weeds with resistance to the herbicide will increase in frequency, and the herbicide becomes ineffective. Almost 200 weed species worldwide have acquired resistance to one or more herbicides. Management practices that reduce herbicide use and vary the herbicides used can slow the appearance of herbicide resistance.

Learning Outcomes Review 17.7

To date, herbicide resistance, pathogen protection, nutritional enhancement, and drug production have been targets of agricultural genetic engineering. Controversy regarding the use of transgenic plants has centered on the potential of unforeseen effects on human health and on the environment.

- *Grasses resistant to glyphosate have been constructed, but never released as a product. What might keep a grass seed company from releasing this?*

Chapter Review

17.1 Recombinant DNA

Restriction endonucleases cleave DNA at specific sites.
DNA fragments generated by type II restriction endonucleases can be recombined with other DNA fragments cut with the same enzyme (figure 17.1).

Gel electrophoresis separates DNA fragments.
DNA fragments can be separated based on size through a gel matrix driven by an electric field (figure 17.2).

DNA fragments are combined to make recombinant molecules.
A recombinant DNA molecule consists of a hybrid made from DNA from two different sources. DNA ligase catalyzes formation of a phosphodiester bond between nucleotides to form a recombinant molecule (figure 17.3).

Reverse transcriptase makes DNA from RNA.
Reverse transcriptase uses RNA as a template to make a DNA molecule. The DNA is called cDNA (figure 17.4).

DNA libraries are collections of recombinant DNA molecules.
DNA libraries hold pieces of DNA and can be stored and replicated in a host organism (figure 17.5).

17.2 Amplifying DNA using the Polymerase Chain Reaction

PCR mimics DNA replication.
The polymerase chain reaction (PCR) amplifies a single small DNA fragment using two short primers flanking the region to be amplified. This involves cycles of heating and cooling, and synthesis with the thermostable Taq polymerase (figure 17.6).

Reverse transcription PCR makes amplified DNA from mRNA.
Reverse transcriptase can be used to make cDNA, a DNA copy of an mRNA molecule. This can be combined with PCR to amplify specific gene sequences that lack regulatory elements and introns.

Quantitative RT-PCR can determine levels of an mRNA.
Combining reverse transcription to make cDNA with PCR amplification in the presence of fluorescent DNA binding dyes or special fluorescently labeled primers allows quantitation of a specific mRNA (figure 17.7).

PCR is the basis of many sequencing technologies.
The cost of sequencing genomes has dropped significantly due to the development of novel DNA sequencing technologies. Many of these technologies rely on variations of PCR to generate material for sequencing.

17.3 Creating, Correcting, and Analyzing Genetic Variation

Forensics uses DNA fingerprinting to identify individuals.
DNA fingerprinting uses short tandem repeats (STRs) that vary in number between individuals in a population. The CODIS database utilizes data on 13 different STR loci to store DNA fingerprint data (figure 17.8).

PCR can be used to create point mutations in specific genes.
Changing nucleotides in a primer used to initiate DNA replication in PCR can be used to create site-specific mutations in a specific DNA sequence. The inherent error rate of some DNA polymerases used in PCR can also be used to create random mutations.

RNA interference can reduce the level of a gene product.
Short sequences of double-stranded RNA can be introduced into cells or some organisms to reduce the expression level of specific genes. Libraries of double-stranded RNA molecules can be analyzed to identify specific proteins involved in specific biological processes.

New technologies allow direct editing of the genome.
Genes can be edited in vivo using the TALEN enzymes and the CRISPR/Cas9 system. Both of these can be used to cleave DNA at a specific site, and allow for error prone repair to alter a gene. If a specific DNA fragment is also introduced, it can be used to direct repair and allow direct genome editing (figure 17.9).

17.4 Constructing and Using Transgenic Organisms

The universal nature of the genetic code allows transgenesis.
With a few exceptions, all organisms use the same genetic code. The universality of the genetic code means that genes from one organism can be expressed in other organisms.

Removal or addition of genes can reveal function.
In "knockout" mice, a gene is inactivated to study its function (figure 17.10). In "knockin" mice, a gene is replaced with a variant to assess the effects of specific genetic changes.

Conditional inactivation allows the study of essential genes.
Conditional inactivation allows the analysis of removing gene function from a specific tissue or at a specific developmental time point. This allows essential genes to be studied. One technology used is the Cre-Lox system (figure 17.11).

Plants require different techniques to be genetically modified.
Transgenic plants with desirable traits can be created by transformation with *Agrobacterium* and a recombinant Ti plasmid or particle bombardment (figure 17.12). Targeted gene knockouts are not currently possible.

17.5 Environmental Applications

Algae can be used to manufacture biofuels.
Lipids harvested from microalgae can be converted into biodiesel. Different species of microalgae can produce up to 50% of their biomass in the form of lipids (figure 17.13).

Microorganisms can degrade hydrocarbons in the environment.
Nutritional supplementation of aquatic, terrestrial, and marine environments can stimulate hydrocarbon degradation by some bacterial species. This can help remove toxic oil products from the environment.

Wastewater treatment relies on the action of microorganisms.
Wastewater must be decontaminated before release into the environment or reuse. The bulk of organic matter is removed from wastewater by anaerobic and/or aerobic bacteria. Further treatments can remove some inorganic contaminants (figure 17.14).

17.6 Medical Applications

Important proteins can be produced using recombinant DNA.
Recombinant DNA technology has been used to create bacteria and yeast that express human insulin that can be used to treat diabetes (figure 17.15). Recombinant DNA technology has increased the potency of this insulin.

Fluorescent in situ hybridization allows detection of gross chromosomal rearrangements.

In FISH, fluorescently labeled DNA probes hybridize to complementary DNA sequences in chromosomes. FISH can be used to identify chromosomal alterations involved in disease. Identification of amplifications of the *HER2* gene can help design personalized treatments (figure 17.16).

Gene chips can identify genetic markers for disease.

Patterns of gene expression associated with diseased or healthy states can be analyzed using gene chips. These data can potentially aid in design of clinical interventions.

Immunoassays can rapidly diagnose viral infections.

Antibodies can be used to detect the presence of specific molecules in clinical or experimental samples. Antibodies coated onto a solid surface bind specific molecules, and a second antibody linked to a fluorescent molecule or enzyme is used for detection (figure 17.17).

Stem cell therapy has the potential to treat chronic diseases.

Bone marrow transplants have been successfully used to treat some blood cancers, anemias, and some cases of multiple sclerosis. Mesenchymal stem cells can be used to repair some damaged connective tissues. Significant political, social, and logistic challenges remain.

17.7 Agricultural Applications

Herbicide-resistant crops allow for no-till planting.

Herbicide-resistant crops are widely used in the United States, reducing both the need for tilling and the associated fossil fuel consumption (figure 17.18).

Bt crops are resistant to some insect pests.

The development of insect-resistant Bt plants reduces the use of insecticides.

Golden Rice shows the potential of transgenic crops.

Golden Rice produces β-carotene, the precursor for vitamin A (figure 17.19). Dietary use of Golden Rice can reduce blindness in developing countries where rice is a staple crop.

Transgenic crops raise a number of social issues.

Use of transgenic plants raises concerns about allergic reactions, although no adverse effects have been documented. The spread of transgenes into noncultivated plants is being followed closely. To date almost 200 weed species worldwide have acquired resistance to herbicides.

Review Questions

UNDERSTAND

1. You study a gene known to be important in the early stages of heart development. Loss of the gene is also suspected to play a role in triggering lung cancer. What kind of transgenic mouse would you use to study your gene in vivo?
   a. Knockout
   b. Knockin
   c. Conditional knockout
   d. All of the choices are correct.

2. What is the basis of separation of different DNA fragments by gel electrophoresis?
   a. The positive charge on DNA
   b. The size of the DNA fragments
   c. The sequence of the fragments
   d. The presence of a dye

3. If a PCR is started using 10 pieces of template DNA, how many pieces of DNA would there be after 10 cycles?
   a. About 100
   b. About 1000
   c. About 10,000
   d. About 10^{10}

4. FISH analysis of a breast tumor biopsy for *HER2* gene reveals two spots of fluorescence in cells. What conclusion do these data support about the tumor?
   a. It is likely to contain high levels of the HER2 mRNA.
   b. It is likely to contain high levels of the HER2 protein.
   c. The *HER2* gene is amplified, and treatments targeting the HER2 protein may not be suitable.
   d. The *HER2* gene is not amplified, and so particular treatments targeting the HER2 protein may not be suitable.

5. In terms of studying gene function, what is the main benefit that genome editing has over RNAi?
   a. Genome editing can eliminate gene function; RNAi simply reduces the levels of gene products.
   b. Genome editing does not require that recombinant DNA be produced, whereas RNAi always does.
   c. Genome editing always works, but RNAi rarely works.
   d. Genome editing is done in bacteria, which are easier to manipulate than the eukaryotic cells in which RNAi is done.

6. The Ti plasmid of *Agrobacterium* usually induces tumors called galls on infected plants. When *Agrobacterium* is used to make transgenic plants, why does no gall form?
   a. The introduced transgene blocks the action of the gall-inducing genes.
   b. The gall-inducing genes are removed and replaced with the transgene.
   c. The Ti plasmid is removed from *Agrobacterium* used to make transgenic plants.
   d. The Ti plasmid is mutated by PCR so that it cannot replicate in infected plants.

APPLY

1. You set up four qRT-PCR reactions all containing the same amount of total RNA. When the reactions are complete, you see that you have 10,000 units of fluorescence after 20 cycles for sample A, after 25 cycles for sample B, after 18 cycles for sample C, and after 22 cycles for sample D. Which of the samples had the most copies of target mRNA at the start of the qRT-PCR?
   a. Sample A
   b. Sample B
   c. Sample C
   d. Sample D

2. Which of the following statements is accurate for DNA replication in your cells, but not PCR?
   a. DNA primers are required.
   b. DNA polymerase is stable at high temperatures.
   c. Ligase is essential.
   d. dNTPs are necessary.
3. What potential problems must be considered in creating a transgenic bacterium with the human insulin gene isolated from genomic DNA to produce insulin?
   a. The genetic code of bacteria is significantly different from the genetic code of humans.
   b. The bacterial cell will be unable to posttranslationally process the insulin peptide sequence.
   c. There is no way to get the bacterium to transcribe high levels of a human gene.
   d. Both a and b present problems.

SYNTHESIZE

1. Many human proteins, such as hemoglobin, are only functional as an assembly of multiple subunits. Assembly of these functional units occurs within the endoplasmic reticulum and Golgi apparatus of a eukaryotic cell. Discuss what limitations, if any, exist to the large-scale production of genetically engineered hemoglobin.
2. Amyloid beta is a proteolytic product of a protein found in the brain called amyloid precursor protein (APP). At high concentrations, amyloid beta clumps together and may be a cause of Alzheimer disease. How might you go about studying, in a mouse model, whether particular mutations maintained the normal function of APP, while preventing clumping of the amyloid beta derived from APP?

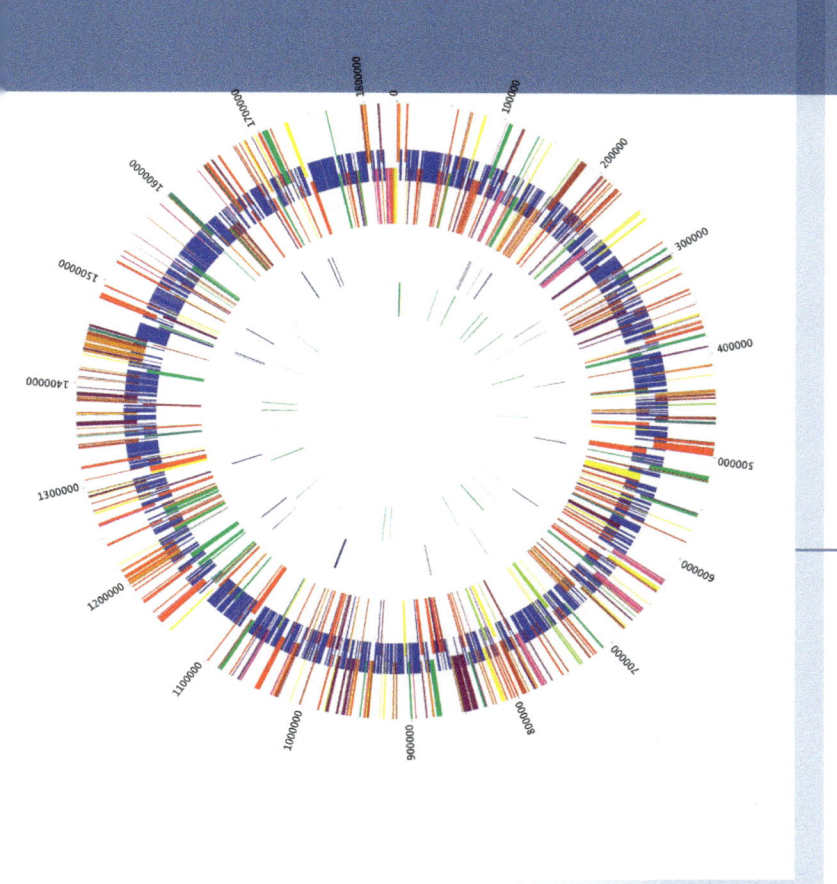

CHAPTER 18

Genomics

Chapter Contents

18.1 Mapping Genomes

18.2 Sequencing Genomes

18.3 Genome Projects

18.4 Genome Annotation and Databases

18.5 Comparative and Functional Genomics

18.6 Applications of Genomics

Introduction

The pace of discovery in biology in the last 30 years has been dramatic. Starting with the isolation of the first genes in the mid-1970s, researchers obtained the first complete genome sequence, that of the bacterium *Haemophilus influenza* (shown in the diagram above with genes of similar function similarly colored), by the mid-1990s. A complete draft sequence of the human genome had been completed by the turn of the 21st century. Put another way, science moved from cloning a single gene, to determining the sequence of a million base-pairs in 20 years, to determining the sequence of 3 billion base-pairs in five more years, to now being able to sequence 20 billion base-pairs in a few hours for as little as $1000. In chapter 17, you learned about biotechnology, and in this chapter you will see how some of those technologies have been applied to the analysis of whole genomes. Genomics integrates classical and molecular genetics with biotechnology to study the structure, function, and relatedness of genomes.

18.1 Mapping Genomes

Learning Outcomes

1. Distinguish between a genetic map and a physical map.
2. Describe the pros and cons of restriction mapping, chromosome mapping, and mapping by STS.
3. Relate the information from a genetic map to the information from a physical map.

We use maps to find locations, and depending on the nature of the location we may use different types of map. In genomics, a gene can be located to a chromosome, to a subregion of a chromosome, and finally to a precise location in the sequence of chromosomal DNA. A map produced at the DNA sequence level requires knowing the entire sequence of the genome, something that was once technologically impossible. Knowing the entire sequence of a genome is useless, however, without other information such as what parts of the genome affect particular phenotypes. This sort of information resides in genetic maps. To make an analogy from geographic maps: finding a single gene within the sequence of the human genome is like trying to find a house on a map of the world.

Genomics uses many approaches to analyze entire genomes

We construct different kinds of maps of genomes from different kinds of data, and for different kinds of analysis. Fundamentally, there are two kinds of maps: genetic maps and physical maps. Both of these types of maps use **genetic markers,** which can be any detectable differences between individuals. **Genetic maps** are abstract maps that place the relative location of genetic markers on chromosomes based on recombination frequency (see chapter 13). **Physical maps** precisely position genetic markers in the genome, with the ultimate physical map being the complete DNA sequence of a genome. Ultimately, genomics unifies many different types of maps, or views, of the genome to facilitate large-scale analysis.

Although not technically the study of the genome, the relationship between the genome and the expressed RNAs, and between the genome and the expressed proteins, can be studied. Genomic studies have shed light on evolutionary relationships, genetic susceptibility to disease, and development.

Genetic maps provide relative distances between genetic markers

Genetic maps, also called linkage maps, use the process of recombination during meiosis, which alters the linkage relationships of alleles on chromosomes (see chapter 13). These "genetic markers" can be genes, as detected by phenotypic differences, or differences in DNA sequences that can be detected by PCR or restriction endonuclease digestion, as described in chapter 17. Classical genetic mapping uses controlled crosses to determine the number of recombinant progeny for specific pairs of markers. Obviously, this is not practical in humans, so we use pre-existing data in the form of family relationships and pedigrees, and use statistical analysis to determine recombination frequencies.

Distances on a genetic map are measured in centimorgans (cM) where 1 cM corresponds to 1% recombination frequency between two loci. Two markers 1 cM apart are relatively close together as there is only a 1% chance that they will be separated during meiosis. Early human genetic maps consisted of linked genes for disease susceptibility that had been mapped to specific chromosomes, but no overall map. This did not change until the advent of molecular markers that did not cause detectable phenotypes (see chapter 13). Not only does this provide many polymorphic markers, but they can be readily integrated with molecular maps. The human genetic map is now quite dense, with a marker roughly every 1 cM.

Genetic maps are very important because they locate genetic traits on a chromosome but they also have limitations. First, the distribution of crossover, or recombination events, is not random. In fact, there are now extensive data indicating the presence of hot spots for recombination in humans, mice, and most other organisms examined. Additionally, recombination frequencies can vary between individuals, and are affected by chromatin structure. All of this means that the relationship between genetic distance (in centimorgans), and actual physical distance (in base-pairs [bp]) varies across the genome. But, with a complete genome sequence, and a dense genetic map, they can still be superimposed to provide complementary views of the genome.

Physical maps provide absolute distances between genetic markers

Unlike a genetic map, which provides the relative position of a marker in the genome, a physical map provides the absolute location of a marker. The first physical maps used the early tools of molecular biology: restriction enzymes. As technologies became more sophisticated, so too did the nature of physical maps.

The ultimate form of a physical map is the placement of many genetic markers on the complete DNA sequence of a genome. Distances between markers on a physical map are measured in base-pairs (1000 base-pairs [bp] equal 1 kilobase [kb]). Segments of DNA can be physically mapped without knowing the DNA sequence, or whether the DNA encodes any genes. In fact, many genome sequencing projects begin by constructing physical maps. There are three general types of physical maps: (1) restriction maps, constructed using restriction endonucleases; (2) chromosome maps (created using techniques discussed in chapter 17); and (3) sequence-tagged site (STS) maps.

Restriction maps

Restriction maps are generally not suitable for DNA molecules greater than about 50 kb, which means that they are only useful for some organelle and some viral genomes. The first physical maps were created by cutting DNA with single restriction enzymes and with combinations of restriction enzymes (figure 18.1). The analysis of the patterns of fragments generated is used to generate a map.

Restriction mapping can be applied to larger genomes if the DNA is first fragmented into smaller pieces. In terms of larger pieces of DNA, this process is repeated and then used to put the pieces back together, based on size and overlap, into a contiguous segment of the genome, called a **contig.**

Chromosome maps

Cytologists studying chromosomes with light microscopes found that by using different stains, they could produce reproducible patterns of bands on the chromosomes. In this way, they could identify all of the chromosomes and divide them into subregions based on banding pattern. The use of different stains allows for the construction of a cytological map of the entire genome. These maps break each chromosome into a "p arm" and a "q arm" and then break these into regions and subregions based on banding patterns. All later maps have been correlated with this low-resolution map. These large-scale physical maps are like a map of an entire country in that they encompass the whole genome but do so at low resolution.

Cytological maps are used to characterize chromosomal abnormalities associated with human diseases, such as chronic myelogenous leukemia. In this disease, a reciprocal translocation occurs between chromosome 9 and chromosome 22 (figure 18.2a), resulting in an altered form of tyrosine kinase that is always turned on, causing white blood cell proliferation and consequently, leukemia.

1. Multiple copies of a segment of DNA are cut with restriction enzymes.

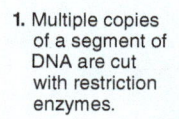

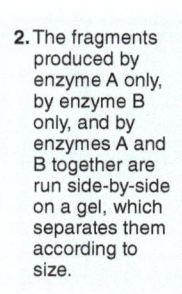

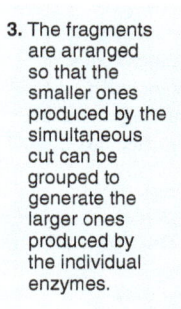

2. The fragments produced by enzyme A only, by enzyme B only, and by enzymes A and B together are run side-by-side on a gel, which separates them according to size.

3. The fragments are arranged so that the smaller ones produced by the simultaneous cut can be grouped to generate the larger ones produced by the individual enzymes.

4. A physical map is constructed.

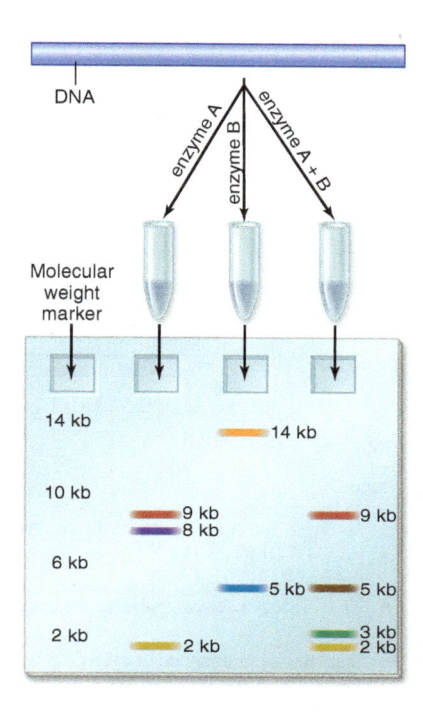

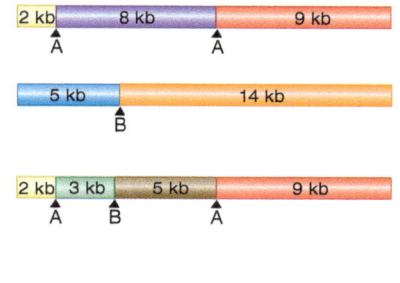

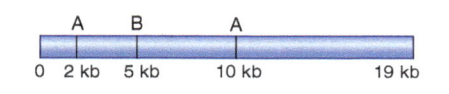

Figure 18.1 Restriction enzymes can be used to create a physical map. DNA is digested with two different restriction enzymes singly and in combination. The DNA fragments are separated based on size using gel electrophoresis. The location of sites can be deduced by comparing the sizes of fragments from the individual reactions with the combined reaction.

Fluorescent in situ hybridization, which we discussed in chapter 17, can also be used to generate chromosomal maps (figure 18.2b). Initially the level of resolution of FISH was approximately 1 Mb (megabase); this means that two points on a chromosome closer than 1 Mb could not be distinguished as separate points. This limitation made FISH only really useful for determining whether a particular DNA sequence was found on a particular chromosome. Technical advances in the 1990s allowed the use of interphase chromosomes as well as metaphase chromosomes, and this improved the resolution of FISH to 25 kb. This advance meant the technique could be used for more fine mapping.

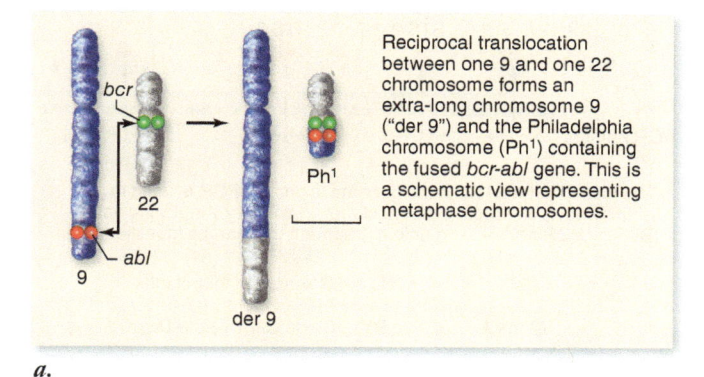

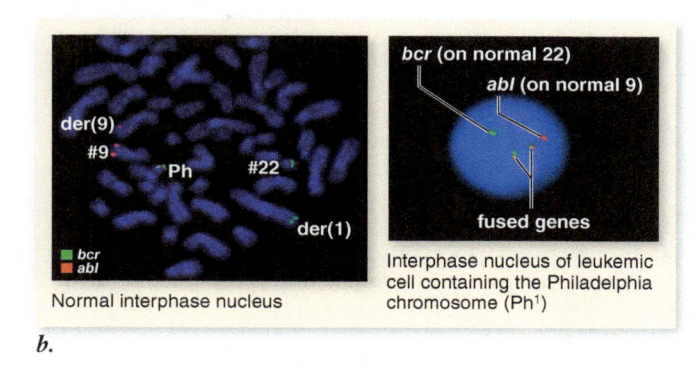

a. Reciprocal translocation between one 9 and one 22 chromosome forms an extra-long chromosome 9 ("der 9") and the Philadelphia chromosome (Ph¹) containing the fused *bcr-abl* gene. This is a schematic view representing metaphase chromosomes.

b.

Figure 18.2 Use of fluorescence in situ hybridization (FISH) to correlate cloned DNA with cytological maps.
a. Karyotype of human chromosomes showing the translocation between chromosomes 9 and 22. *b.* FISH using a *bcr* (green) and *abl* (red) probe. The yellow color indicates the fused genes (*red* plus *green* fluorescence combined). The *abl* gene and the fused *bcr-abl* gene both encode a tyrosine kinase, but the fused gene is always expressed.

? Inquiry question How can you explain the pattern of red, green, and yellow dots seen in the right panel of figure 18.2b?

Sequence-tagged site maps

Restriction mapping allows relatively high-throughput, technically easy, high-resolution mapping of relatively small DNA molecules. FISH, on the other hand, provides low-resolution mapping of large DNA molecules and is technically challenging. Sequence-tagged site (STS) mapping provides a powerful alternative that combines the best of restriction mapping with the best of mapping via FISH. STS mapping allows the rapid construction of high-resolution physical maps of large DNA molecules with few technical challenges.

An STS is a short, unique stretch of genomic DNA that can be amplified by PCR (see chapter 17). STS mapping asks whether two STSs are on the same DNA molecule. This involves randomly fragmenting DNA into overlapping fragments. If two STS markers are commonly found together on fragments of DNA, then they are relatively close together. If they are not commonly found together on fragments of DNA, then they are likely to be farther apart (figure 18.3). Much like genetic mapping is based on the frequency

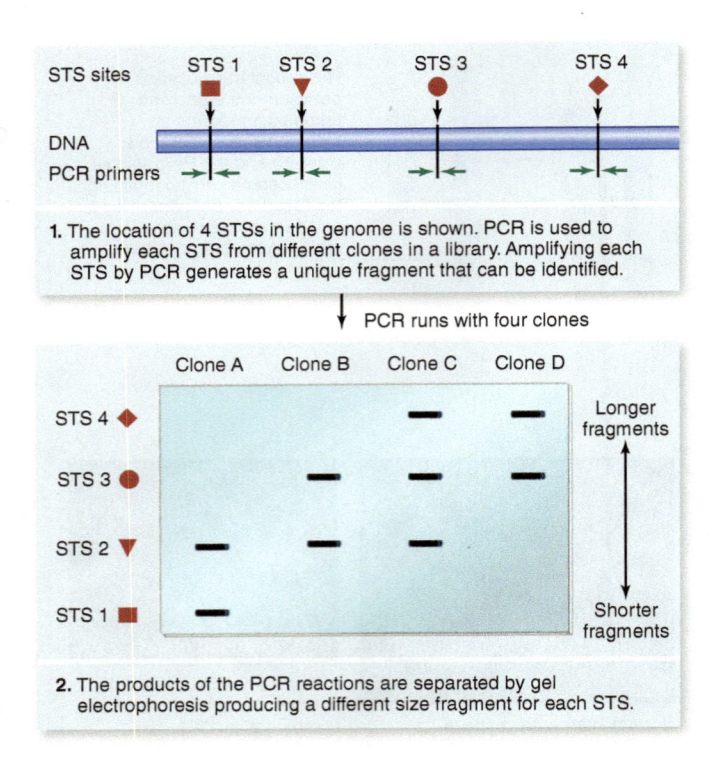

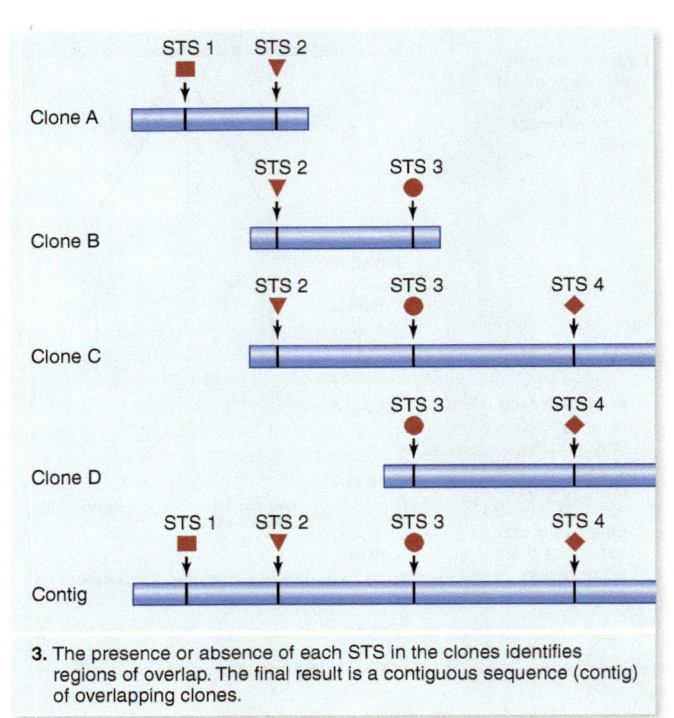

Figure 18.3 Creating a physical map with sequence-tagged sites. The presence of landmarks called sequence-tagged sites, or STSs, in the human genome made it possible to begin creating a physical map large enough in scale to provide a foundation for sequencing the entire genome. *(1)* Primers *(green arrows)* that recognize unique STSs are added to cloned DNA, followed by DNA amplification via polymerase chain reaction (PCR). *(2)* PCR products are separated based on size on a DNA gel, and the STSs contained in each clone are identified. *(3)* Cloned DNA segments are aligned based on STSs to create a contig.

Data analysis You could construct the physical map using a subset of the clones in panel 2. List the possible combinations that would allow you to construct the physical map with the fewest number of clones.

with which markers are separated by recombination, STS mapping is based on the frequency markers can be separated by breaking DNA into fragments. Markers closer together will be separated less frequently than markers that are farther apart. The farther apart two markers are, the more likely a break can occur between them.

Fragments of DNA can be pieced together using the STSs by identifying overlapping regions in fragments. Because of the high density of STSs in the human genome and the relative ease of identifying an STS in a DNA clone, investigators were able to develop physical maps on the huge scale of the 3.2 billion bp genome in the mid-1990s (figure 18.3). STSs provide a scaffold for assembling genome sequences.

Physical maps can be correlated with genetic maps

Genetic maps can identify relevant regions of the genome that affect specific phenotypes, but they tell us nothing about the nature of these "genes." The correlation of physical maps with genetic maps allows us to find the sequence of genes that have been mapped genetically.

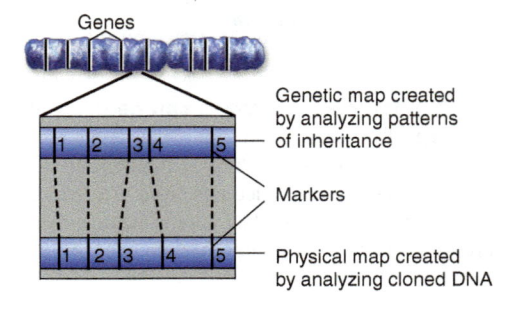

The problem in finding genes is that the resolution of genetic maps is at present not nearly as good as the resolution of the genome sequence. In the human genome, markers that are 1 cM apart may be as much as a million base-pairs apart.

Because the markers used to construct genetic maps are now primarily molecular markers, they can be easily located within a genome sequence. Similarly, any gene that has been cloned can be placed within the genome sequence and can be mapped genetically.

This provides an automatic correlation between the two maps. Genes that have been mapped genetically can, in theory, be readily located within the DNA sequence of a genome. In practice, this can be challenging as the relationship between genetic distance and physical distance varies across the genome. So 1 cM of genetic distance can actually be composed of different numbers of base-pairs in different regions of a genome.

Learning Outcomes Review 18.1

Maps of genomes can be either physical maps or genetic maps. Physical maps include cytogenetic maps of chromosome banding, or restriction maps. Genetic maps are correlated with physical maps by using DNA markers such as sequence-tagged sites (STSs) unique to each genome. While genetic maps position markers, such as genes, relative to one another, physical maps position them at a specific genetic location on a sequence of DNA. Genetic and physical maps are used to aid in the sequencing phase of genome projects.

- *How could it be that two markers on a genetic map are 10 cM apart, while two markers in a different part of the same genetic map are 12 cM apart, but both sets of markers are separated by the same amount of DNA?*

18.2 Sequencing Genomes

Learning Outcomes

1. Discriminate between dideoxy terminator sequencing and next-generation sequencing technologies.
2. Compare and contrast clone-contig and shotgun methods of genome sequencing and assembly.

The ultimate physical map is the base-pair sequence of an entire genome. Large-scale, high-throughput DNA sequencing has made whole-genome sequencing realistic for smaller labs and biotechnology companies. Conceptually, all approaches to sequencing DNA adapt DNA replication to an in vitro environment. Older approaches use modified nucleotides to stop or pause replication and limit the replication to just one strand of the DNA. The rapid advances in genomics have been enabled by newer and much faster automated sequencing technologies.

Dideoxy terminator sequencing remains important in genome sequencing

In the last 10 years there have been huge improvements to the technologies used to sequence DNA. There are still situations, however, where the original DNA sequencing methods are useful. One original method for sequencing DNA developed by Fred Sanger in the 1970s mimics DNA replication. This technique depends on the use of **dideoxynucleotides** in DNA sequencing reactions. The dideoxynucleotides act as chain terminators, and halt replication when they are incorporated. All DNA nucleotides lack a hydroxyl group at the 2′ carbon of the sugar, but dideoxynucleotides also have no 3′ OH group. Recall from chapter 14 that each additional base is added to the growing 3′ end of a strand of DNA by a reaction with the OH group of the previous nucleotide to the triphosphate of the next nucleotide (see figure 14.12). A dideoxynucleotide can be incorporated, but then cannot be added to, as it lacks the 3′ OH, and thus terminates the growing chain.

Dideoxy terminator sequencing was originally performed manually where the experimenter would perform four separate reactions, each with a single dideoxynucleotide and the four deoxynucleotides. When a dideoxynucleotide is incorporated, it will terminate synthesis at a specific base. Thus, each reaction produces a set of nested fragments that each end in a specific base, and the four reactions together represent fragments that terminate at each possible base. The fragments were separated by high-resolution gel electrophoresis, which can separate fragments differing in length by a single base. This produces a ladder of fragments that allow the sequence to be read from bottom (smallest fragment) to top (longest fragments).

To automate this, each dideoxynucleoide is labeled with a fluorescent dye, and all four are added to a single DNA synthesis reaction. As before, the random incorporation of dideoxynucleotides produces a set of fragments that end in specific bases, but in this case, each base is represented by a color. The dye-labeled nested fragments are separated in a capillary tube by electrophoresis and as each molecule passes a laser, it fluoresces. The nucleotide terminating the DNA strand is interpreted based on the color of the fluorescence and automatically reported to a computer (figure 18.4). Although this approach is still labor intensive, automation allows thousands of sequencing reactions to be performed in parallel. This means it is possible to sequence nearly 900 kb of DNA in a single day. In some ways, this was the first step in creating the massively parallel sequencing platforms that characterized the next-generation of sequencing technologies.

Next-generation sequencing uses massively parallel technologies to increase throughput

Next-generation sequencing (NGS) technologies first appeared 10 or so years ago. Since then, there have been huge improvements in the speed of sequencing, in the cost per sequenced base, and in the length of sequence read per sequencing reaction. The result is a staggering increase in the number of genomes sequenced and resequenced in the past 10 years. In the future, this abundance of data

library by conventional cloning (see chapter 17). Second, instead of thousands of reactions being prepared and analyzed simultaneously, up to millions of sequencing reactions can be performed simultaneously; it is this feature of the technology that has led to the term "massively parallel." Third, the time-consuming electrophoresis previously required is not needed. Instead, sequencing reactions occur simultaneously in solution and are read directly by the sequencing equipment (figure 18.5). Because of the increase in

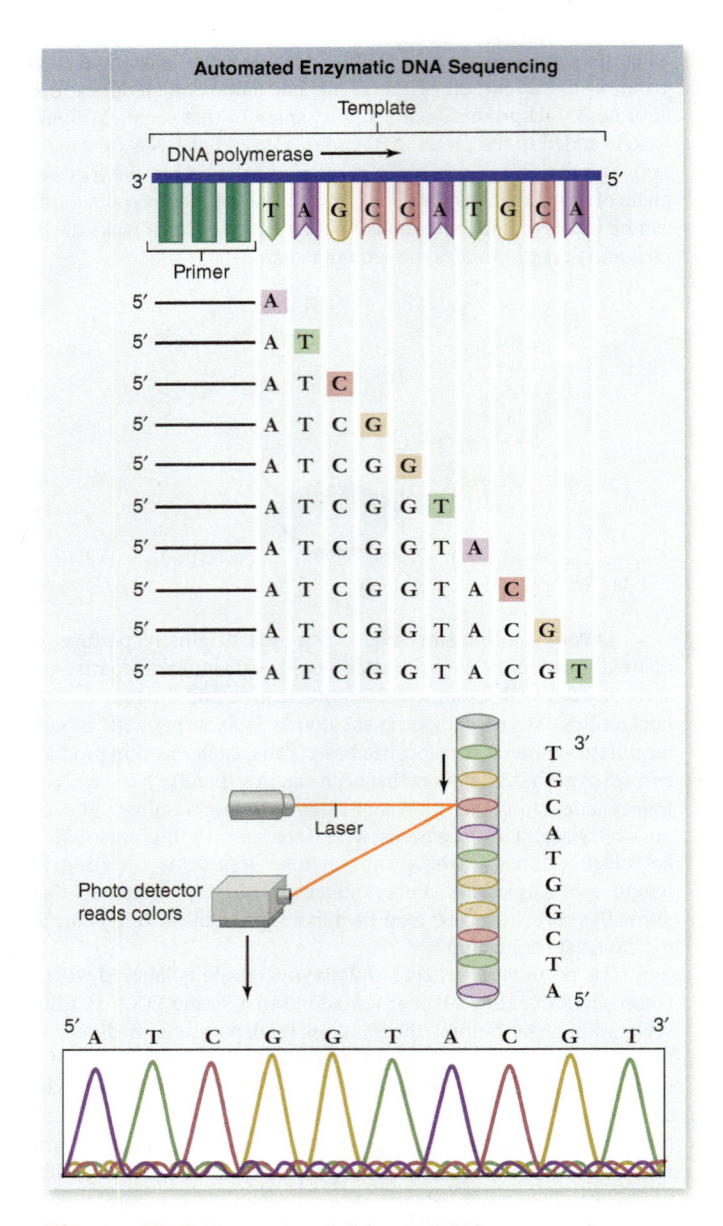

Figure 18.4 Automated dideoxy DNA sequencing. The sequence to be determined is shown at the top as a template strand for DNA polymerase with a primer attached. In automated sequencing, each ddNTP is labeled with a different color fluorescent dye, which allows the reaction to be done in a single tube. The fragments generated by the reactions are shown. When these are electrophoresed in a capillary tube, a laser at the bottom of the tube excites the dyes, and each will emit a different color that is detected by a photodetector.

could be translated into discoveries that reveal novel evolutionary relationships, improve the personalization of health care, and improve diagnostic medicine.

There are numerous NGS technologies in the market, but all have some common features. First, unlike dideoxy sequencing, DNA can be sequenced without a need to first construct a genomic

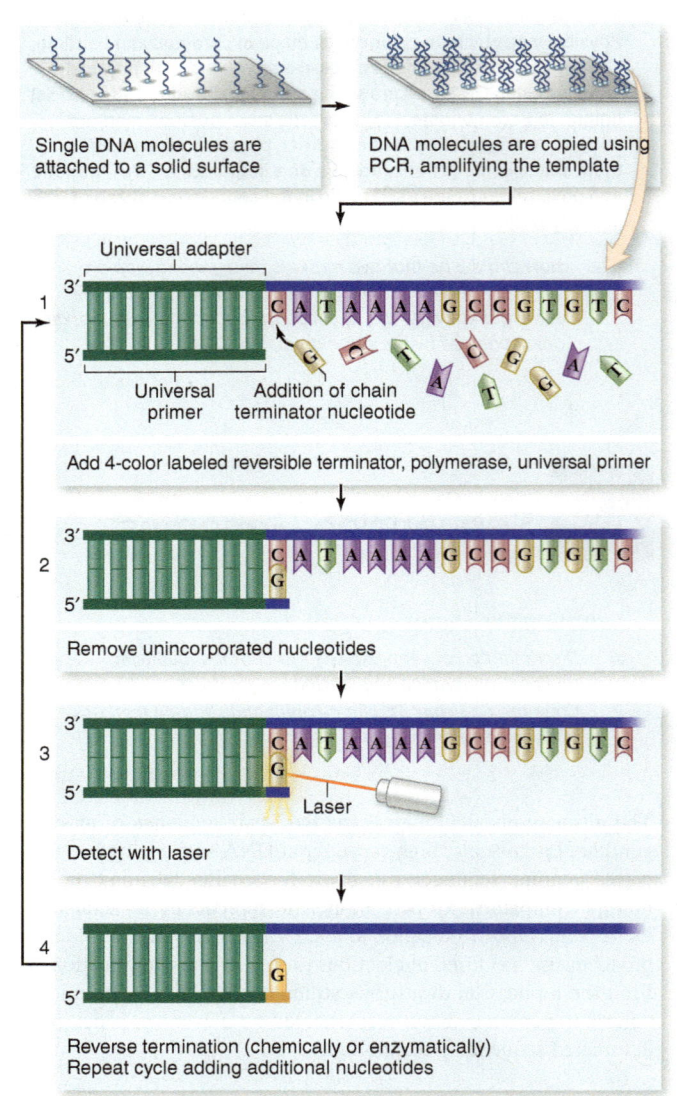

Figure 18.5 Illumina next-generation DNA sequencing. In Illumina next-generation sequencing, the template DNA molecules are attached to a solid surface and amplified by PCR to create islands of DNA. Four fluorescently labeled terminator nucleotides are then added, and after one is incorporated into new DNA, synthesis stops. The color of the incorporated nucleotide is read, the nucleotide is chemically altered so that another nucleotide can be added. This process of single-nucleotide addition is repeated to read the sequence of the DNA in each island.

sequencing speed and the reduction in cost, some bacterial genomes can be sequenced in as little as a few hours!

Although sequencing speed and cost has dropped dramatically, a drawback of some of the technologies is the amount of DNA sequence—called *read length*—produced per sequencing reaction. Read length varies significantly between the different technologies and ranges from 35 bases up to a rare example of 20,000 bases. Given the variability in the read length, different technologies are suitable for different applications; some will be good for genome sequencing, whereas others will be more appropriate for diagnostic medicine. Most of the common technologies produce read lengths in the range of 50 to 1000 bp. Shorter read lengths greatly complicate the assembly of the individual pieces of sequence into a finished genome. Because of this problem, novel algorithms for genome assembly have been created.

In 2014, it was reported that a human genome had been sequenced for $1000. This represents a cost reduction of nearly 10,000-fold relative to the initial cost of sequencing a human-sized genome approximately 10 years ago.

Sequenced genome fragments are assembled into complete sequences

The most complex task in a large-scale sequencing project is the assembly of individual pieces of sequence information into a complete genome. A genome is first fragmented into overlapping pieces that are sequenced, but then must be reassembled by matching short overlapping regions. The shorter the sequenced fragments, and the more repetitive DNA present in the genome, the harder it is to piece together the puzzle. Two strategies can be employed: (1) assemble portions of a chromosome first, and then figure out how the bigger pieces fit together (**clone-contig assembly**); or (2) try to assemble all the pieces at once, instead of in a stepwise fashion (**shotgun assembly**).

Shotgun assembly

A benefit of shotgun assembly is that it does not rely on any genetic or physical maps. This means that no prior information about the sequenced genome is required. The first genome ever sequenced, the genome of the bacterium *Haemophilus influenzae*, was sequenced and then assembled using the shotgun approach. The genome of the bacterium was broken into fragments approximately 1.5 to 2 kb in length which were cloned into suitable vectors (see chapter 17). The ends of the fragments were sequenced and then, based on overlaps in the sequence, the pieces of DNA sequence could be stitched back together to form the whole 1.8-Mb genome (figure 18.6a). This strategy rarely gives a completely assembled genome sequence. A variety of other approaches are used to close the gaps that become obvious only when the sequences have been assembled.

Clone-contig assembly

Clone-contig assembly methods (figure 18.6b) are used to assemble the pieces of sequenced DNA from more complex genomes such as those of eukaryotes. Unlike shotgun assembly, it requires a physical map to build the completely reassembled genomic sequence. A genome is fragmented into larger fragments—approximately 1 to 1.5 Mb. Then, those fragments are mapped into a larger contiguous segment of DNA, called a **contig**, using genetic markers such as STSs or restriction endonuclease sites. The fragments of DNA 1 to 1.5 Mb (megabases) in size can be sequenced and assembled using the shotgun approach (figure 18.6a)

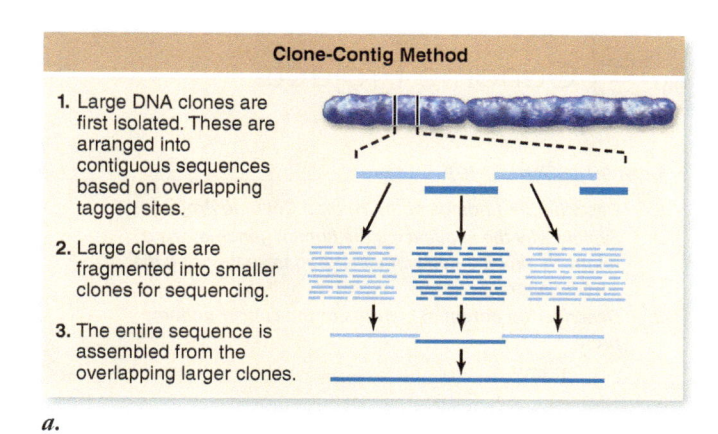

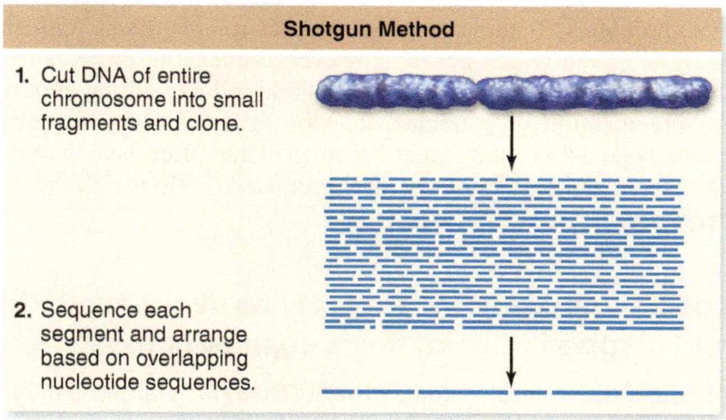

Figure 18.6 Comparison of sequencing and genome assembly methods. *a.* The clone-contig method uses large clones assembled into overlapping regions by STSs. Once assembled, these can be fragmented into smaller clones for sequencing. *b.* In the shotgun method, the entire genome is fragmented into small clones and sequenced. Computer algorithms assemble the final DNA sequence based on overlapping nucleotide sequences.

Learning Outcomes Review 18.2

Because of the enormous size of some genomes, sequencing requires the use of automated sequencers running many samples in parallel. Two approaches have been developed for whole-genome sequencing and assembly: one that uses clones already aligned by physical mapping (clone-contig sequencing and assembly), and one that involves sequencing random clones and using a computer to assemble the final sequence (shotgun sequencing and assembly). In either case, significant computing power is necessary to assemble a final sequence.

- What are the pros and cons of the different approaches to whole-genome sequencing and assembly?

18.3 Genome Projects

Learning Outcomes
1. Describe the findings of the Human Genome Project.
2. Explain why the sequencing the human genome was a relatively simple endeavor compared to sequencing the wheat genome.
3. Describe the potential benefits of the cancer genome project.

Automated-sequencing technology has produced huge amounts of sequence data. This has allowed researchers studying complex problems to move beyond approaches restricted to the analysis of individual genes. However, sequencing projects in themselves are descriptive endeavors that tell us nothing about the organization of genomes, let alone the function of gene products and how they may be interrelated. Research using data from genome projects has produced both insight and perplexity.

The Human Genome Project has sequenced and mapped most of the human genome

Although the Human Genome Project (HGP) officially began in 1990, its roots can be traced to the mid-1980s. Creation of genetic and physical maps of the human genome were the major endeavors that ultimately led to the actual sequencing and assembly of the genome. By 1995, a physical map covering 94% of the human genome, with markers spaced on average every 199 kb had been created. Then, in 1996 a genetic map was published that placed markers approximately every 1.6 cM. Since then, the number of markers and the resolution of the maps has increased significantly. At the same time, advances in sequencing automation reduced the cost of sequencing to approximately 40 cents per base. Ultimately, this laid the groundwork for the clone-contig and shotgun approaches used to sequence and assemble the human genome.

In 1998, Craig Venter, founder of The Institute for Genomic Research, declared that his private, for-profit company would compete directly with the HGP to sequence the human genome. Venter claimed that his shotgun sequencing and assembly strategy would be faster and cheaper than the clone-contig approach of the HGP. Although possibly true, Venter's company, now called Celera Genomics, also used the publicly available physical maps of the HGP. It remains unclear if Venter's approach would have succeeded for a large complex genome. It clearly was successful for the much smaller and simpler genome of *Haemophilus influenzae*. In June 2000, a draft sequence of the human genome was jointly announced by the publicly funded HGP and Venter's privately funded company, although several publications acknowledged that many gaps and errors existed. As much of the missing sequence data was unlikely to cover genes, there was widespread consensus that this was a fully functional, if incomplete, version of the human genome.

In April 2003, the Human Genome Sequencing Consortium announced that sequencing of the human genome was complete. The difference between the draft sequence and the finished sequence is that the finished sequence has fewer sequencing errors, fewer gaps, and more coverage than the draft version. This finished genome took 13 years to complete and cost the U.S. taxpayers approximately $2.7 billion based on the value of the U.S. dollar in 1991; today, certain very expensive, very high-throughput sequencers can sequence a human genome for close to $1000 in as little as 2 to 3 hours.

For many years, geneticists estimated the number of genes in the human genome to be around 100,000. One of the findings from the Human Genome Project was that the number of genes in the human genome is about 20,000. This represents only about 1.5 times as many genes as the fruit fly and nearly half as many genes as rice (figure 18.7). Humans, mice, and the puffer fish have approximately the same number of genes; clearly, the complexity of an organism is not a simple function of the number of genes in its genome, nor of the size of its genome.

Data analysis If the human genome contains approximately 3 billion base-pairs, 20,000 genes, and only 1% codes for genes, what is the approximate length of a gene?

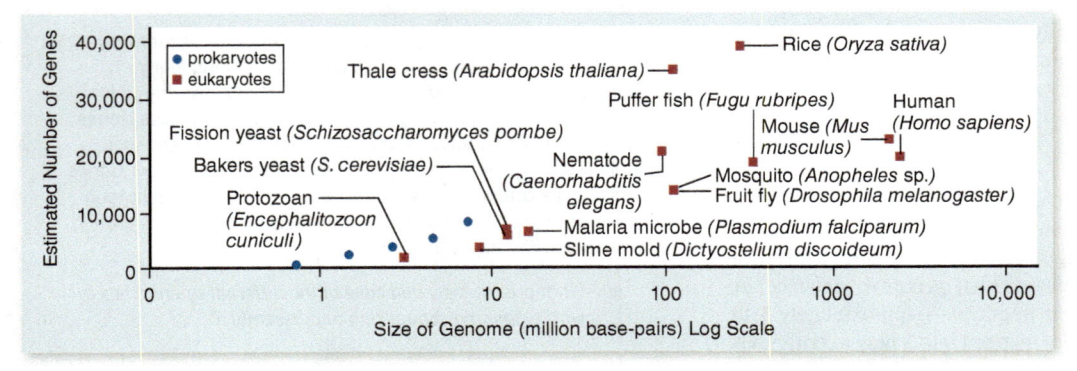

Figure 18.7 Size and complexity of genomes. In general, eukaryotic genomes are larger and have more genes than prokaryotic genomes, although the size of the organism is not the determining factor. The mouse genome is larger than the human genome and the rice genome is larger than both human and mouse genomes.

The Wheat Genome Project illustrates the difficulties of assembling complex genomes

Wheat, *Triticum aestivum*, is used to feed about 30% of the world's population. Better understanding wheat genomics can inform breeding practices to increase yield and improve drought and disease resistance. It can also provide insight into the evolution of wheat species. Given expanding populations, dwindling water resources, and global climate change, the application of scientific knowledge to the improvement of crop plants will be critical in the establishment of secure global food resources.

A version of the wheat genome was released in 2012; however, this version covered 70% of only the gene containing regions and did not map specific genes to specific chromosomes. In 2014, researchers released a draft genome sequence that identified 124,201 gene loci and placed 75,000 of them along chromosomes. As part of this, the whole sequence of the largest wheat chromosome was sequenced, annotated, and assembled to form a structure that is seven times larger than the entire genome of the model plant species *Arabidopsis thaliana*.

The wheat genome is an allohexaploid (see chapter 24) produced by two hybridization events. This means that somatic cells in wheat contain three distinct genomes designated A, B, and D, with each consisting of approximately 5.5 billion bp of DNA, arranged into 21 chromosomes (7 per ancestral genome). There are three challenges facing attempts to sequence and assemble the wheat genome: (1) its 16.5-Gb (gigabase) size, (2) its composition of 80% repetitive DNA, and (3) its polyploid nature.

The polyploid nature of the wheat genome means that gene duplication, redundancy in sequences, and repetitive DNA are common. The fastest genome sequencing and assembly approach, the shotgun approach, cannot be used on this kind of complex genome. With so much repetitive DNA and gene duplication, it is difficult to assemble the pieces of sequenced DNA into the correct chromosomes. This means that the clone-contig approach needs to be used on individual chromosomes; this approach takes significantly more time and money. It is hoped that the remaining 20 wheat chromosomes can be sequenced and assembled by 2017.

Cancer genome projects seek the genetic causes of cancer

Cancer kills approximately 1500 people a day in the United States. As average life span increases, this rate will likely continue to increase. Many years of study have revealed the outline of the genetic basis of cancer: More than one somatic mutation is necessary to convert a normal cell to a cancer cell—it is a progressive series of mutations. We have also identified two categories of genes involved: oncogenes that can cause cancer with gain-of-function mutations, or inappropriate activation, and tumor-suppressor genes that lead to cancer with loss-of-function mutations. These can also be thought of as positive, "accelerator" type mutations, and negative, "brake" type mutations (see chapter 10).

Genomics has helped to fill out this simple framework by comparing tumor genomes with the genomes of matched normal tissues. At this point, more than 5000 tumors and matched normal-tissue samples, representing more than 20 tumor types, have been sequenced. This has led to the identification of genes found associated with one or more tumor types. Mutations that are found in a tumor genome are divided into "driver mutations," which affect progress toward cancer, and "passenger mutations," which accumulate but do not lead to cancer. More than 200 different genes have been identified as potential drivers, including both oncogenes and tumor suppressor genes, and this is not yet a complete catalog. These affect signal transduction pathways, DNA repair pathways, the control of gene expression, chromatin structure, and even metabolism. Some genes are mutated in multiple tumor types, and different tumors may have from a few to hundreds of mutations.

This avalanche of information has both clarified, and provided new problems for analysis. There is no simple model where mutations in a set of genes will always lead to a specific type of cancer, but there are regularities. It is also clear that even specific types of tumors have heterogeneity. In the long run, characterizing each type of tumor will help in both diagnosis, and in designing treatments based on the genotype of the patient and the mutations in each patient's tumor.

Learning Outcomes Review 18.3

The Human Genome Project cost $2.7 billion and took 13 years to complete. By comparison, the much larger and complex wheat genome could be sequenced and assembled at a fraction of that cost and in half as much time. The sequencing and assembly of large, complex, highly repetitive genomes is technologically challenging. Cancer genomics hopes to identify sets of mutations that could indicate predisposition to cancer, predict disease progression, or anticipate responsiveness to particular treatments.

- *Explain why complex genomes, such as the wheat genome, are technically harder to sequence than relatively simple genomes.*

18.4 Genome Annotation and Databases

Learning Outcomes

1. Explain why we need to annotate genomes.
2. Compare and contrast the types of DNA found in genomes.
3. Evaluate the conclusions of the ENCODE project.

The genome sequencing and assembly stages of a genome project produce vast amounts of data. Add to this data the additional data generated from subsequent analyses of genome sequences and structures, and an immediate problem of how to organize and store the data becomes apparent. The sequence data and the information produced by analyses of the sequence data are stored in a variety of searchable, often publicly available, databases.

Although laborious to produce, the genome sequence itself is of little intrinsic interest. More important is the number and kind of genes present in the genome, and how these genes contribute to phenotype. The assignment of this kind of information to a genome sequence is called *genome annotation*.

Genome annotation assigns function to DNA sequences in genomes

An annotation is a record attached to a DNA sequence held in a database that identifies the DNA sequence as a gene. The record can contain information about the gene, its structure and function, and the product it ultimately produces (RNA or protein). The annotation may also include evidence about how the assignment of gene function was made. The overarching goal is to attempt to add biologically relevant information to the DNA sequences in the genome.

Genome annotation is largely an automated process that is checked as needed by experts. The annotation process often begins by using computer algorithms to look for regions of DNA that are likely to contain a gene. The algorithms may look for sequences that contain regulatory regions such as promoters, or they may search for a start codon followed by codons that encode amino acids, and eventually a stop codon. Sequences such as this are called **open reading frames (ORFs)**. It is also possible to search for splicing signals to predict intron/exon structure, and consequently predict the sequence and structure of an RNA that might be produced, and thus, the amino acid sequence coded for by the putative gene.

> **Inquiry question** Given the sequence of a genome, how might you predict the number of proteins produced by a genome? Why might your estimate be inaccurate?

Inferring function across species: the BLAST algorithm

Once a potential gene or gene product has been identified, it is possible to infer function by finding a similar sequence in other species. The tool that makes this possible is a search algorithm called the Basic Local Alignment Search Tool (BLAST). BLAST searching can also be automated to speed up the annotation process. Using a networked computer, potential gene or gene product sequences are submitted to the BLAST server and a search is conducted for similar sequences already deposited into the sequence databases. Significant sequence similarity with other genes or gene products in a different species could suggest conservation of function.

Using computer programs to search for genes, to compare genomes, and to assemble genomes are only a few of the new genomics approaches called **bioinformatics**.

Expressed sequence tags identify genes that are transcribed

Another way to identify potential gene sequences in genomes is to isolate total mRNA produced in different tissues and convert this to cDNA for sequencing, and then map the cDNA onto the genomic sequence (figure 18.8). This process can be simplified by just sequencing one or both ends of as many cDNAs as possible. This can be automated, and these short sections of cDNA have been named **expressed sequence tags (ESTs)**. An EST is another form of STS that can be included in physical maps. ESTs can also be used in BLAST searches; sequence similarity with ESTs from another species may provide insight into the function of the encoded protein.

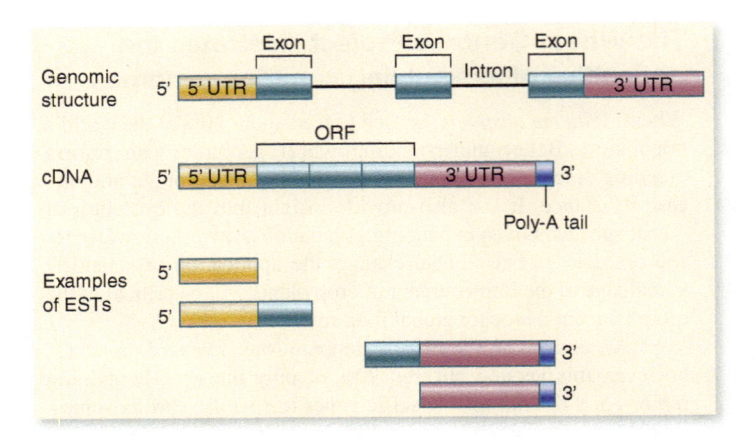

Figure 18.8 ESTs can be used to map expressed genes onto the genome. Sequencing the end of ESTs produces sequence data that can be aligned with predicted cDNA structure and mapped onto corresponding genomic structures. By sequencing many ESTs, complete genomic structures can be determined.

ESTs have been used to identify 87,000 cDNAs in different human tissues. About 80% of these cDNAs were previously unknown. The estimated 20,000 genes of the human genome can result in these 87,000 different cDNAs because of *alternative splicing* (see chapter 15).

Genomes contain both coding and noncoding sequences

We call DNA sequences that are used to produce a protein, coding sequences (CDS). This leaves all the sequences that do not result in the synthesis of polypeptide as noncoding sequences. Genome annotation identifies the distribution of coding sequences in the genome, and subsequent experimentation can reveal the function of those genes and their products. In some cases, functions of noncoding DNA is known; for example, genes that produce functional RNA molecules such as tRNAs, rRNAs, and miRNAs are functionally well characterized. On the other hand, there is much noncoding DNA that is less well characterized.

Noncoding DNA in eukaryotes

A notable characteristic of eukaryotic genomes is the amount of noncoding DNA they possess. The Human Genome Project revealed a particularly startling picture: Each of your cells contains nearly 2 m of DNA, but only about 2.5 cm of that DNA is coding sequence! Nearly 99% of the DNA in your cells is noncoding DNA.

Genes coding for proteins are scattered in clumps amongst the noncoding DNA in the human genome. Seven major types of noncoding DNA have been described in the human genome (table 18.1 shows the composition of the human genome, including noncoding DNA):

Noncoding DNA within genes. As discussed in chapter 15, a human gene is made up of numerous pieces of coding DNA (exons) interspersed with lengths of noncoding DNA

TABLE 18.1	Classes of DNA Sequences Found in the Human Genome
Class	**Description**
Protein-encoding genes	Translated portions of the approximately 20,000 genes scattered about the chromosomes
Introns	Noncoding DNA that constitutes the great majority of each human gene
Segmental duplications	Regions of the genome that have been duplicated
Pseudogenes (inactive genes)	Sequence that has characteristics of a gene but is not a functional gene
Structural DNA	Constitutive heterochromatin, localized near centromeres and telomeres
Simple sequence repeats	Stuttering repeats of a few nucleotides such as CGG, repeated thousands of times
Transposable elements	21%: Long interspersed elements (LINEs), which are active transposons 13%: Short interspersed elements (SINEs), which are active transposons 8%: Retrotransposons, which contain long terminal repeats (LTRs) at each end 3%: DNA transposon fossils
Noncoding RNA	RNAs that do not encode a protein but have regulatory functions, many of which are currently unknown

(introns). Introns comprise about 24% of the human genome, whereas the exons comprise less than 1.5%.

Structural DNA. Some regions of the chromosomes remain highly condensed, tightly coiled, and untranscribed. These regions, called *constitutive heterochromatin,* tend to be localized around the centromere or near the ends of chromosomes, at telomeres.

Simple sequence repeats. Scattered within the genome are **simple sequence repeats (SSRs).** An SSR is a one- to six-nucleotide sequence such as CA or CGG, repeated thousands of times. SSRs can arise from DNA replication errors, and their number can change due to errors in homologous recombination. SSRs make up about 3% of the human genome.

Segmental duplications. These are blocks of genomic sequences of 10,000 to 300,000 bp that have duplicated and moved either within a chromosome, or to a nonhomologous chromosome.

Pseudogenes. These inactive genes may have lost function because of mutation.

Transposable elements. Forty-five percent of the human genome consists of DNA sequences that can move from one place in the genome to another. Some of these sequences code for proteins that allow their movement, but many do not. Because of the significance of these elements, they are described in this section.

microRNA genes. Hidden within the noncoding DNA is a mechanism for controlling gene expression. Specifically, DNA was once considered "junk" has been shown to encode microRNAs (miRNAs), which are processed after transcription to lengths of 21 to 23 nt, but are never translated. About 10,000 unique miRNAs have been identified that are complementary to one or more mature mRNAs. These miRNAs regulate some of the complex developmental processes in eukaryotes by downregulating translation.

Long, noncoding RNA. In addition to the many small RNAs such as microRNAs that are not translated into protein but serve a regulatory role, tens of thousands of longer noncoding RNAs likely regulate gene expression. This recently discovered, hidden world of regulatory networks reveals a new level of complexity in the precise control of gene expression. Long noncoding RNAs have important roles in physiology and development. Only approximately 200 long, noncoding RNAs have been functionally characterized and the biological functions, if any, of the others remain unclear.

Transposable elements: Mobile DNA

Transposable elements, also called *transposons* and *mobile genetic elements,* are sequences of DNA able to move from one location in the genome to another. Transposable elements move around in different ways. In some cases, the transposon is duplicated, and the duplicated DNA moves to a new place in the genome. When this happens, the number of copies of the transposon increases. Other types of transposons are excised without duplication and insert themselves elsewhere in the genome. The role of transposons in genome evolution is discussed in chapter 24.

Human chromosomes contain four types of transposable elements. About 21% of the genome consists of **long interspersed elements (LINEs).** These ancient and very successful elements are about 6000 bp long, and they encode the enzymes needed for transposition. LINEs encode a reverse transcriptase enzyme that can make a cDNA copy of the transcribed LINE RNA. The result is a double-stranded DNA fragment that can reinsert into the genome rather than undergo translation into a protein. Since these elements use an RNA intermediate, they are called retrotransposons.

Short interspersed elements (SINEs) are similar to LINEs, but they do not encode the transposition enzymes and so cannot move without using the transposition machinery of LINEs. Nested within the genome's LINEs are over half a million copies of a

SINE element called Alu (named for a restriction enzyme that cuts within the sequence). The Alu SINE is 300 bp and represents 10% of the human genome. An Alu SINE can use the enzymes of the LINE of which it is a part to jump to a new chromosome location. Alu sequences can also jump into coding DNA sequences, causing mutations.

Two other sorts of transposable elements are also found in the human genome. First, about 8% of the human genome is made from a type of retrotransposon called a **long terminal repeat (LTR)**. Although the transposition mechanism is a bit different from that of LINEs, LTRs also use reverse transcriptase to create double-stranded copies of itself that can integrate into the cell's genome. Second, dead transposons occupy approximately 3% of the genome; these are transposons inactivated by mutation that can no longer move.

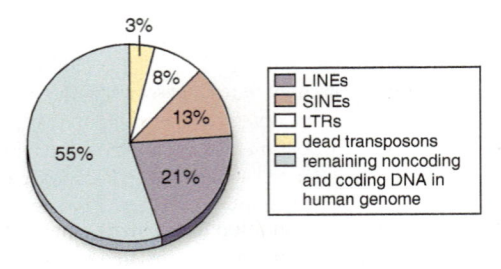

Inquiry question How could you show whether a sequence of DNA was functional or not?

The ENCODE project seeks to identify all functional elements in the human genome

The presence of noncoding DNA in genomes has been known for many years, but the biological importance of this DNA remained unclear. Since no function was assigned, this noncoding DNA was often referred to as "junk DNA." Extensive analyses of the noncoding DNA have lead some to suggest that the term "junk DNA" should be reevaluated. It may be best to reframe the DNA of no known function as "nonfunctional DNA."

The Encyclopedia of DNA Elements (ENCODE) project is a collaborative effort to identify all functional elements in the human genome. The primary conclusion of their work was that 80% of the DNA sequences in the human genome is functional. Of that 80%, approximately 62% is defined as transcriptionally active, although the majority of the transcribed, or potentially transcribed, RNAs have no known function. There has been much debate in the scientific community regarding what these numbers actually mean in terms of biological function, and the ENCODE consortium has revised its estimate of functionality to somewhere between 20 and 80%.

ENCODE's definition of function

ENCODE's definition of a functional element includes any DNA sequence that results in protein production, that is transcribed, or that has a distinct and reproducible biochemical signature (figure 18.9). A biochemical signature appears when a protein binds to a DNA sequence, when chromatin bends to form long-range interactions, or when a DNA sequence exhibits a particular

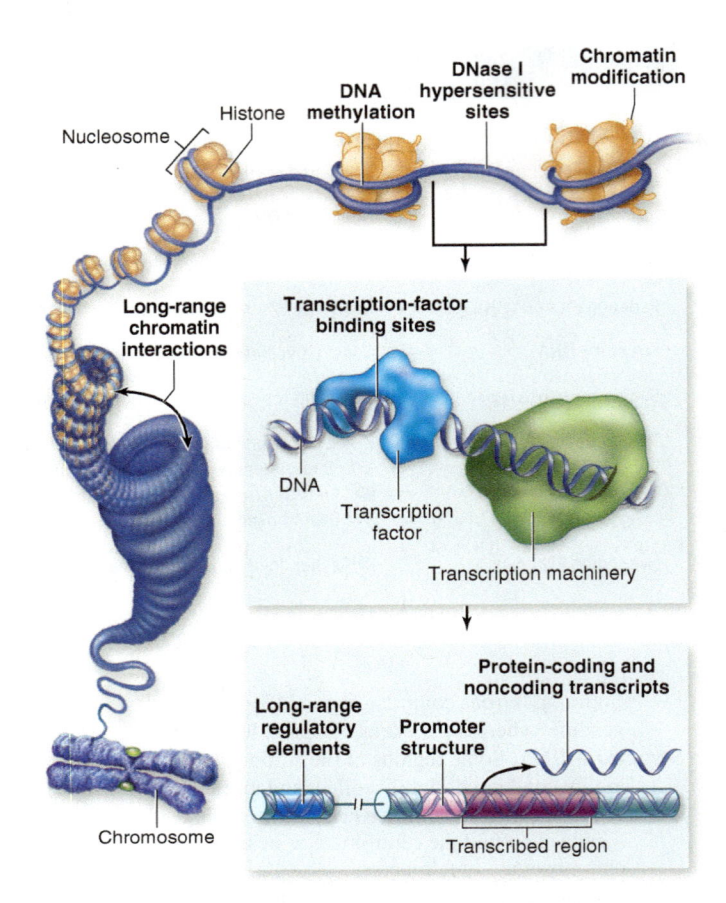

Figure 18.9 Functional DNA elements defined by ENCODE. ENCODE uses a biochemical signature definition of function. Biochemical signatures include methylation patterns of DNA sequences, modifications of histones in chromatin, sensitivity to DNase I, which suggests regions of transcriptional activity, sites of transcription factor binding, sequences identified as promoters, production of coding or noncoding transcripts, long-range regulatory elements such as enhancers, and long-range chromatin interactions.

pattern of histone modification. Given that there is likely significant difference between these functionalities in different cell types, ENCODE derived their 80% value by testing the genomes of 144 different cultured human cell lines for functional elements.

Some criticism has been directed at ENCODE for their use of cultured cell lines and stem cell lines. These cell lines may not be reflective of the situation in the genome of cells of an organism and may be more transcriptionally active than cells in tissues. Critics also claim that there are likely to be significant differences between functional elements in different developmental stages of an organism and under different environmental conditions.

Biological function

The functional definition used by ENCODE is not necessarily synonymous with biological function. A sequence of DNA that is transcribed, and therefore biochemically functional according to

ENCODE, may not produce an RNA with a function necessary for the organism. Also, just because one or several transcribed RNAs have a demonstrated biological effect, does not mean that all transcribed RNAs have biological effects. A natural extension of the definition of functional element could include any DNA sequence that is replicated; given such a biochemical definition, 100% of the genome would be functional as 100% of the genome is replicated during cell division.

Selected-effect function

ENCODEs conclusions have also drawn fire from some evolutionary biologists. Most evolutionary biologists, and many biologists in general, subscribe to a selected-effect definition of function. The selected-effect definition of function requires that the function of a characteristic is the one selected for by purifying selection. For example, the selected function of the mammalian lung is gas exchange. Selective pressures have shaped that function. The rising and falling of the chest during breathing is caused by the inflation and deflations of the lungs, but that rising and falling action has not been selected for as the lung's function.

If a genetic sequence has a particular function, this will eventually be destroyed by the accumulation of random mutations over time, unless this is counteracted by natural selection. The only way that a functional sequence is protected from being destroyed is if purifying selection acts to resist the accumulation of mutations that degrade the function. Many biologists would argue that only DNA sequences under purifying selection should be considered functional. Such sequences are more likely to be conserved in a specific lineage and are also more likely to be conserved in related species. Based on these criteria, comparative genomics studies suggest that as little as 5 to 15% of the genome is functional.

Clearly, there are some conflicting perspectives on the topic of the functionality of DNA sequences in the human genome, and ENCODE's calculation of 80% functionality remains debatable. This being said, the technologies and means of data collection advanced by the ENCODE effort significantly advance our potential for understanding functions of DNA sequences in genomes. Similarly, ENCODE's data regarding the biochemical functions of noncoding DNA provide an impressive map that can be used by others to explore the biological relevance of the identified functional elements in the human genome.

Learning Outcomes Review 18.4

Once a genome has been sequenced, assembled, and deposited into an appropriate database, functional elements of the genome are annotated with useful information. Functional elements can be identified in a variety of ways including by inference using the BLAST program. Only a fraction of the DNA in a genome contains information to make proteins; the remaining noncoding DNA may have function. Because of ambiguity in the definition of "function," conclusions of what percentage of a genome is composed of functional DNA require critical evaluation.

- *What explanation could you suggest, based on principles of natural selection, for the many repeated transposable elements in the human genome?*

18.5 Comparative and Functional Genomics

Learning Outcomes

1. Explain the value of using comparative genomics to learn about the properties of genomes.
2. Explain how functional genomics is used to learn about the functions of genomes.
3. Describe the relationships between the genome, the transcriptome, and the proteome.

Comparative genomics uses information from one genome to learn about a second genome. For example, the known function of a gene in one organism's genome can be used to hypothesize about the function of a similar gene in a related organism's genome. It turns out that 60% of the genes involved in triggering human cancer are conserved in the fruit fly genome. So, the functions of those genes can be analyzed in the fruit fly, which is significantly easier than analyzing them in humans. Comparative genomics also reveals information about the relatedness of organisms, about how different organisms perform similar biological functions, about what makes different species different, and even about the minimum number of genes it would take to build a functional cell—an engineering feat of interest to synthetic biologists discussed in section 18.6.

Functional genomics is an extension of genomics that investigates the relationship between genotype and phenotype. The phenotype of an organism is determined by the pattern of expression of genes and by the interaction of the organism with the environment. For example, by comparing the gene expression pattern in a healthy cell with the gene expression pattern in a diseased cell, it is possible to learn which genes might be involved in the disease process.

Comparative genomics reveals conserved regions in genomes

One of the striking lessons learned from the sequence of the human genome is how similar humans are to other organisms. More than half of the genes of *Drosophila* have human counterparts. Among mammals, the similarities are even greater. Humans have only 300 genes that have no counterpart in the mouse genome. The use of comparative genomics to ask evolutionary questions is also a field of great promise. The comparison of the many prokaryotic genomes already sequenced indicates a greater degree of lateral gene transfer than was previously suspected.

Comparative genomics based on synteny

Comparative genomics allows for a large-scale approach to comparing genomes by taking advantage of synteny. *Synteny* refers to the conserved arrangements of segments of DNA in related genomes. Physical mapping techniques can be used to look for synteny in genomes that have not been sequenced. Comparisons with the sequenced, syntenous segment in another species can provide information about the unsequenced genome.

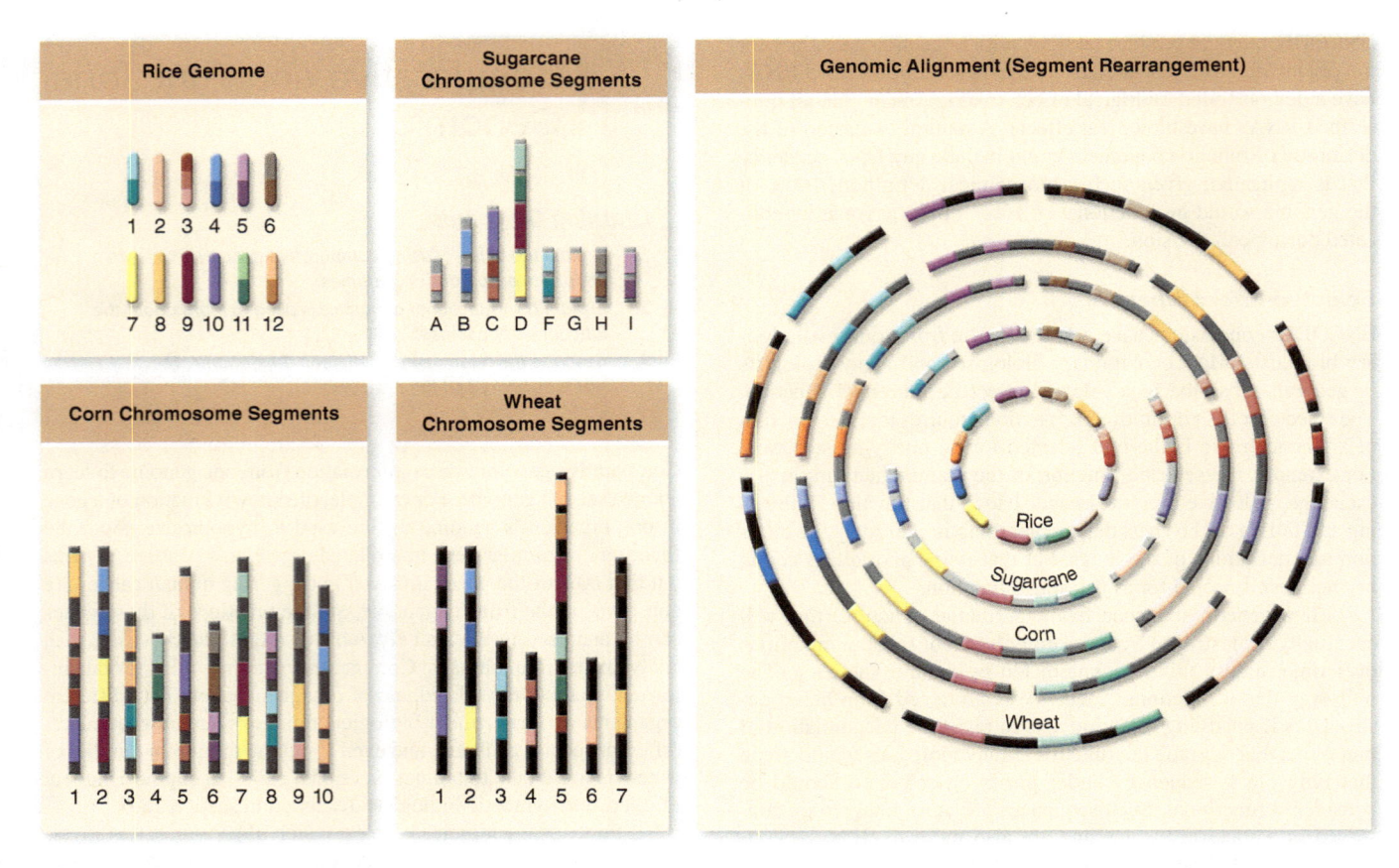

Figure 18.10 Grain genomes are rearrangements of similar chromosome segments. Shades of the same color represent pieces of DNA that are conserved among the different species but have been rearranged. By splitting the individual chromosomes of major grass species into segments and rearranging the segments, researchers have found that the genome components of rice, sugarcane, corn, and wheat are highly conserved. This implies that the order of the segments in the ancestral grass genome has been rearranged by recombination as the grasses have evolved.

To illustrate this, consider rice and its grain relatives, corn, barley, and wheat. Only rice and corn genomes have been fully sequenced. Even though these plants diverged more than 50 million years ago (MYA), the chromosomes of rice, corn, wheat, and other grass crops show extensive synteny (figure 18.10). In a genomic sense, "rice is wheat" and the information gleaned from the physical and genetic maps of corn and rice genomes can be used to speed up the ongoing sequencing of the wheat genome.

By understanding the rice and corn genomes at the level of its DNA sequence, identification and isolation of genes from grains with larger genomes should be much easier. DNA sequence analysis of cereal grains could be valuable for identifying genes associated with disease resistance, crop yield, nutritional quality, and growth capacity.

Functional genomics reveals gene function at the genome level

To fully understand how a genome contributes to the structure and function of an organism, we need to characterize the products of the genome—the RNA molecules and the proteins. This information is essential to understanding cell biology, physiology, development, and evolution. Functional genomics allows connections to be drawn between the genotype of the organism and the phenotype of the organism.

Functional genomics uses a range of high-throughput techniques to learn about the products produced from a genome, when those products are produced, and how they change during development or in response to the environment. Functional genomics can be broken into three separate, but related, categories: (1) the study of the all the RNA molecules produced by a genome (the transcriptome); (2) the study of the proteins produced from the genome (the proteome); and (3) the study of the interactions and products of interactions made between proteins. We consider proteomics later in this section, but first we summarize several of the technologies used to study the transcriptome.

DNA microarrays

DNA microarrays were created for the analysis of gene expression at the whole-genome level. The arrays allow questions to be asked about how gene expression patterns change during development, in response to environmental stimuli, and even during the development of disease. This kind of analyses requires accurate annotation of a genome and using this knowledge to construct an array containing all of the coding sequences (figure 18.11).

SCIENTIFIC THINKING

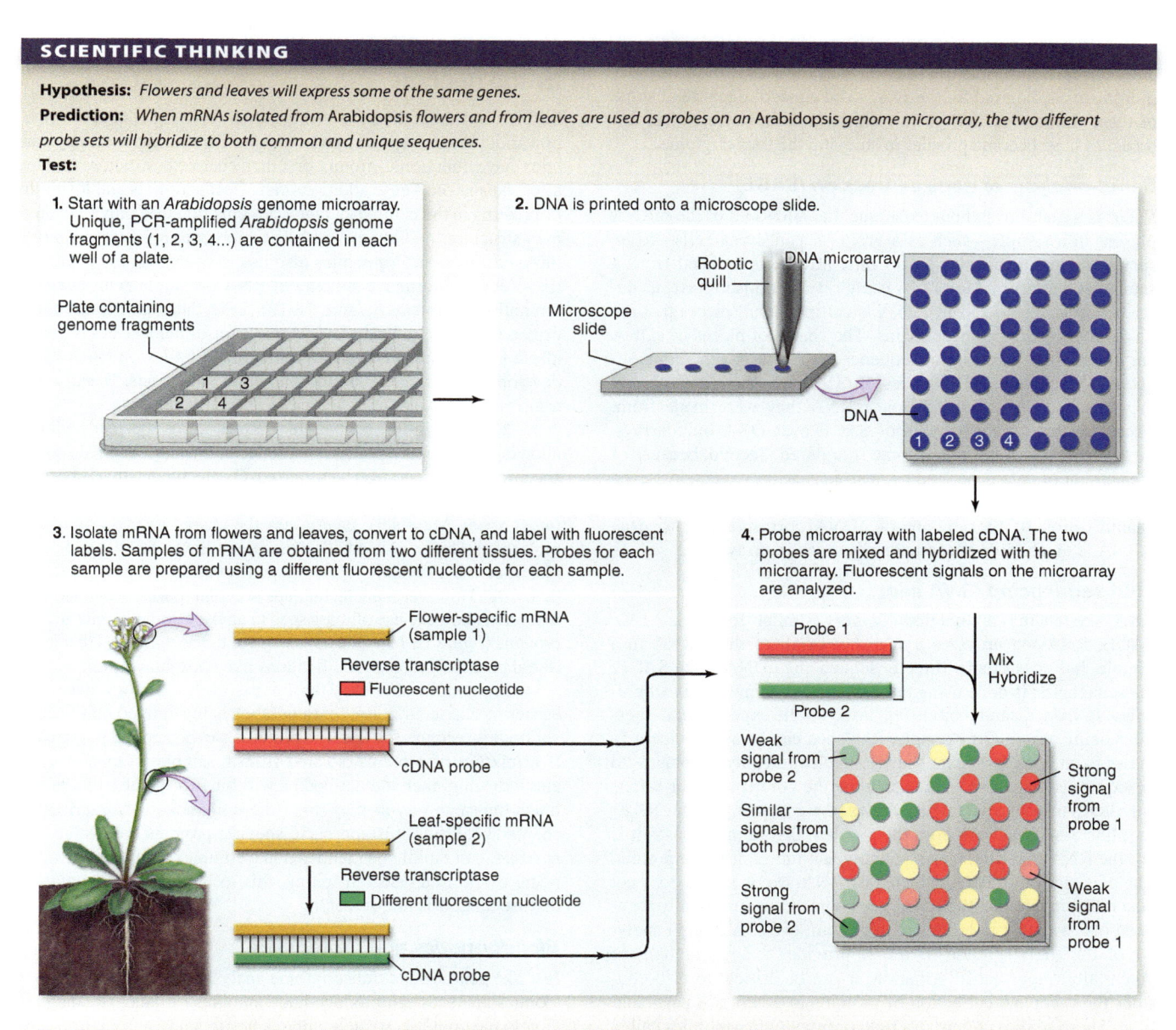

Hypothesis: Flowers and leaves will express some of the same genes.
Prediction: When mRNAs isolated from *Arabidopsis* flowers and from leaves are used as probes on an *Arabidopsis* genome microarray, the two different probe sets will hybridize to both common and unique sequences.
Test:

1. Start with an *Arabidopsis* genome microarray. Unique, PCR-amplified *Arabidopsis* genome fragments (1, 2, 3, 4...) are contained in each well of a plate.

2. DNA is printed onto a microscope slide.

3. Isolate mRNA from flowers and leaves, convert to cDNA, and label with fluorescent labels. Samples of mRNA are obtained from two different tissues. Probes for each sample are prepared using a different fluorescent nucleotide for each sample.

4. Probe microarray with labeled cDNA. The two probes are mixed and hybridized with the microarray. Fluorescent signals on the microarray are analyzed.

Result: Yellow spots represent sequences that hybridized to cDNA from both flowers and leaves. Red spots represent genes expressed only in flowers. Green spots represent genes expressed only in leaves.
Conclusion: Some *Arabidopsis* genes are expressed in both flowers and leaves, but there are genes expressed in flowers but not leaves and leaves but not flowers.
Further Experiments: How could you use microarrays to determine whether the genes expressed in both flowers and leaves are housekeeping genes or are unique to flowers and leaves?

Figure 18.11 Microarrays.

To prepare a microarray, fragments of DNA are applied onto a glass or silicone slide by a robot at indexed locations to produce an array of DNA dots. Each DNA fragment corresponds to specific gene sequences such that the entire array represents all possible coding transcripts, with unique locations in the array. The use of a microarray is shown in figure 18.11. Chips can be used in hybridization experiments with labeled mRNA from different sources. This gives a high-level view of genes that are active and inactive in a variety of different conditions or states. Arrays can also be prepared so that the spotted DNA contains variations of the DNA sequence so that only transcripts from genes with single-nucleotide polymorphisms are detected.

Microarrays have some limitations, however. The sequence of a genome must be known to prepare a microarray with appropriate sequences. Also, if a microarray is being prepared to look for rare polymorphic transcripts in a sample, then knowledge of those polymorphisms is required for the array to be created. Because of some of these limitations, and given technological advances, other approaches have become popular in studying the transcriptome.

Serial-analysis of gene expression (SAGE)

SAGE is a high-throughput technique that allows all of the mRNA present in a sample—such as a group of cancerous cells—to be surveyed simultaneously. To do this, mRNA is isolated from a sample and converted to cDNA using reverse transcriptase, as described in chapter 17. The cDNA is cut into small pieces that are linked together into longer chains. The chains of pieces of cDNA are cloned into vectors and sequenced using next-generation sequencing techniques. The pieces of cDNA are short enough that they can be used to identify the mRNA they were made from. There are three main benefits of SAGE over DNA microarrays. First, no knowledge of the genome is required. Second, because of the amount of data generated by next-generation sequencing, even the most rare mRNA molecules can be identified. Third, the quantification of the amount of cDNA corresponding to each mRNA is much more accurate than with microarray data.

RNA sequencing (RNA-seq)

RNA sequencing is functionally very similar to SAGE. Like SAGE, RNA-seq provides a snapshot of all of the mRNA in a sample, but does this by directly sequencing cDNA. Like SAGE, the sequencing is done using next-generation sequencing technologies. RNA-seq can be modified based on the experimental questions being asked. For example, RNA-seq can be used to identify intron-exon boundaries, map single-nucleotide polymorphisms to unsequenced genomes, and determine the contribution of different alleles to phenotype when alleles of a gene are present. RNA-seq, like SAGE, suffers from the problem that in many cases it is not the RNA that produces a phenotype—rather, it is the protein. The presence of an RNA detected by RNA-seq or SAGE does not take into account that the RNA may be posttranscriptionally regulated in a way that interferes with protein levels. Although analysis of the transcriptome of a cell provides a large amount of potentially important information, it can be critical to also consider the complete collection of proteins in a cell at a particular time. These studies comprise a field of functional genomics called *proteomics*.

Proteomics catalogs proteins encoded by the genome

Information about the proteome is harder to interpret than information about the genome or transcriptome. This is because many proteins have functional groups or other chemical structures added after translation. These additions can affect activities and locations of proteins in the cell. Also, many proteins function only in quaternary structures and interact with one another to regulate cell structure and function. Proteomics also has to overcome the challenge that—due to alternative splicing of RNA—a single gene can code for multiple proteins (figure 18.12). Like the transcriptome, the proteome is highly dynamic. The sequence of nucleotides is generally constant, but the transcriptome varies based on tissue type, developmental stage, and environmental conditions. Because the transcriptome is dynamic, the proteome is dynamic.

Many high-throughput techniques have been developed allowing study of the proteome. The technique used depends on the specific research question being asked. One technique that is becoming more common to analyze the proteome is *mass spectroscopy (mass spec)*. Mass spec determines the charge to mass ratio of a molecule, which with small molecules allows for unambiguous identification. For many years, this was not used with proteins due to their size. However, new techniques combine purification and ionization to allow the use of mass spec to analyze complex mixtures of proteins (figure 18.13). One advantage of this technique is that it can reveal posttranslational modifications like phosphorylation.

Proteomics also uses the microarray technology described earlier in this section. Rather than DNA being spotted on the chip, antibodies recognizing specific proteins can be applied to the chip. If a mixture of proteins isolated from a sample is applied to an antibody chip, then the antibodies will bind to specific proteins in much the same way as they do in the immunoassays described in chapter 17. Using this approach, specific proteins, known as *biomarkers*, can rapidly be identified in a complex sample. As well as being useful in a research setting, this technique has potential for disease diagnosis and screening.

Bioinformatics and proteomics

Just like genome and transcriptome analysis, high-throughput proteomics analyses can produce huge amounts of data. The data are largely meaningless without approaches to study and interpret the data. At the interface between computer science and biology lies

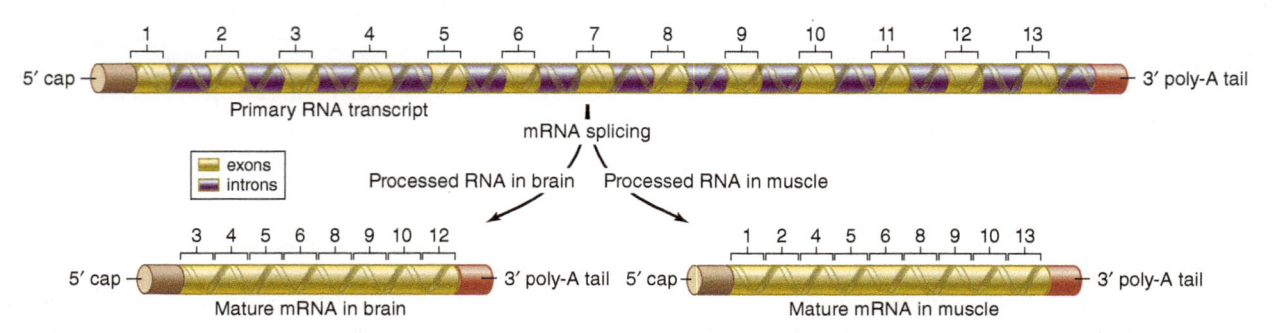

Figure 18.12 Alternative splicing can result in the production of different mRNAs from the same coding sequence.
In some cells, exons can be excised along with neighboring introns, resulting in different proteins. Alternative splicing explains why approximately 20,000 human genes can code for three to four times as many proteins.

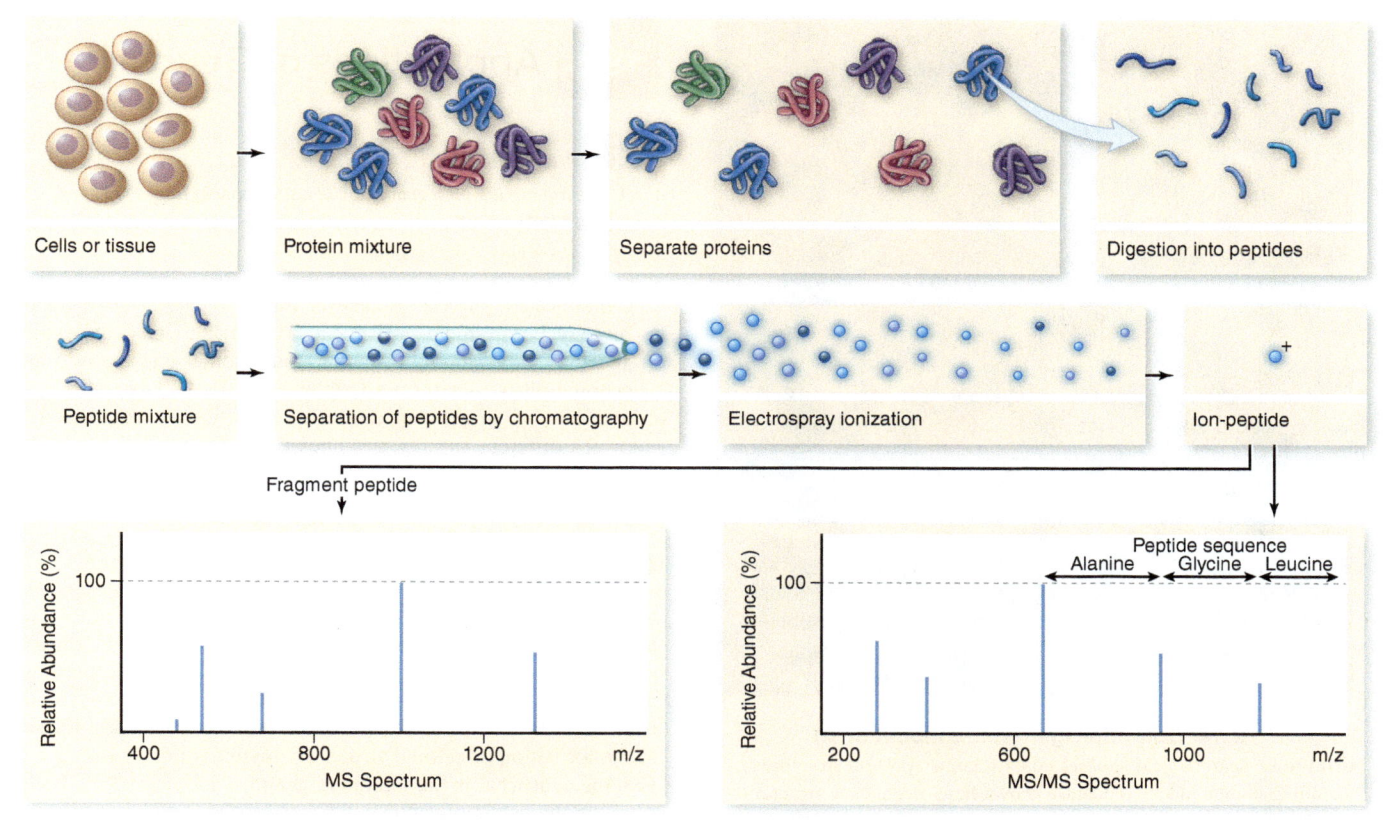

Figure 18.13 Mass spectroscopy can be used to analyze proteins. Proteins can be isolated from cells or tissues and partially separated using gels or biochemical techniques. Partially purified or individual proteins are digested into small peptides by proteases and ionized. The charged peptides can be separated by mass in a spectrometer. Computer algorithms can match the peptide fragments with predicted digestion patterns to tentatively identify proteins. Alternatively, individual peptides can be further fragmented and amino acid composition can be determined by mass spectrometry. (MS spectrum, abundance of an ion as a function of mass to charge ratio of the ion; MS/MS spectrum, abundance of an ion as a function of mass to charge ratio of the ion when the ions are the product of tandem ionization events; m/z, the mass of an ion divided by the charge of an ion commonly referred to as the mass-to-charge ratio)

the discipline of bioinformatics. Bioinformatics is the application of computer programming, mathematics, and modeling to the analysis of large sets of biological data. The application of bioinformatics to proteomics allows the rapid identification of proteins discovered by high-throughput profiling experiments, the annotation of genomes using information from mass spectroscopy and protein chips, and the prediction of protein structure and function.

> **Inquiry question** How might you annotate a gene's function in a genome database if provided with the sequence of amino acids in a protein that was obtained from a proteomics experiment?

Predicting protein structure and function

The use of new computer methods to quickly identify and characterize large numbers of proteins is a distinguishing feature between traditional protein biochemistry and proteomics. As with genomics, the challenge is one of scale.

Ideally, we would like to use a gene sequence to infer the structure and function of the encoded protein. At present, we can only do this by comparing a new protein sequence with the sequences of already identified proteins. Inferences about structure and function are sometimes possible due to the fact that similar sequences of amino acids will produce a similar structure. However, it is also true that similar structures can be produced from very different sequences of amino acids. In these cases, similar functions can only be inferred if structural information is available.

Inference of function from structure is an even more challenging task. Proteins with similar sequences and structures often have similar functions, but there are many examples where this is not the case. There are many similar functions performed by proteins with significantly different structures. It is helpful if we have information about the evolutionary history of a protein. We can be more confident in our inferences about shared structures and functions if we are comparing proteins derived from a common ancestor that have not diverged in function.

Until recently, it has been very difficult to predict the variety of ways in which chains of amino acids could fold into secondary and tertiary structures (an example of predicted protein structure is shown in figure 18.14). One type of computer algorithm attempts to predict protein structure by first looking for stretches of amino acids

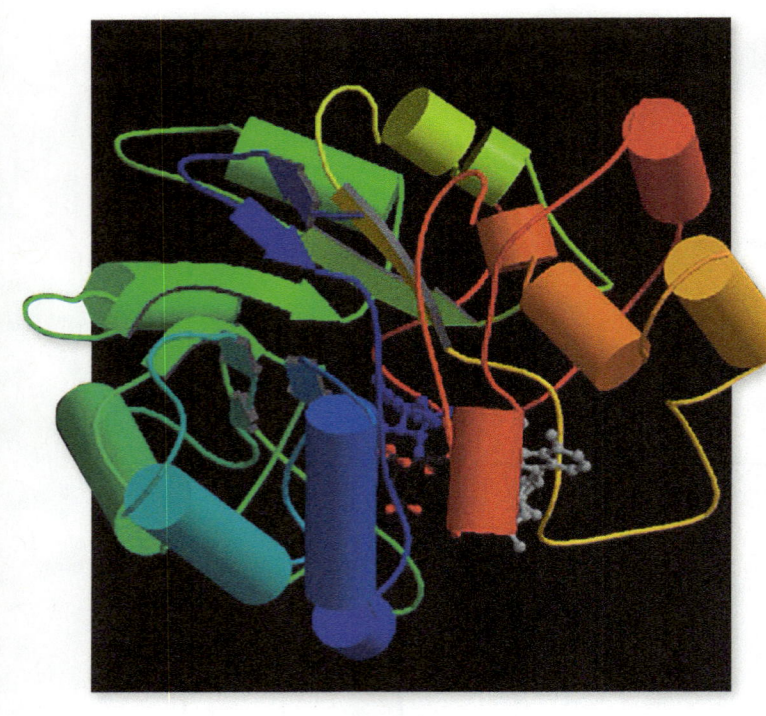

Figure 18.14 Computer-generated model of an enzyme. Searchable databases contain known protein structures, including human aldose reductase shown here. Secondary structural motifs are shown in different colors.

that have characteristics of a protein domain. Then, the algorithm takes short fragments from the protein domain and compares them to known folds in proteins whose structures are known. A computer program then uses these data to build a model of how all the fragments might be arranged to produce specific structures. When the same structure appears independently from different simulations, there is a higher probability that the structure is correct.

? Inquiry question What is the relationship between genome, transcriptome, and proteome?

Learning Outcomes Review 18.5

Comparisons of different genomes allow geneticists to infer structural, functional, and evolutionary relationships between genes and proteins, as well as relationships between species. Functional genomics provides tools to begin to understand the functions of the genes in a genome. DNA microarrays, SAGE, and RNA-seq can be used to learn about the transcriptome, but there are pros and cons to the different approaches. Proteomics involves analyses of the proteins produced by a genome under specific conditions. Although it is possible to make some predictions about protein structure and function, the analyses are complex and require powerful computers.

■ *Why is establishment of a species' transcriptome an important step in studying its proteome?*

18.6 Applications of Genomics

Learning Outcome
1. Evaluate the ethical and social issues arising from the use of genomics data.

Genomics can be used to answer fundamental questions about biology. For example, we can better understand what makes one species fundamentally different from another and the role the genome plays in determining phenotype by understanding the functions of a genome, and the evolutionary relationships between different species' genomes. With the application of the information we now have about genome structure and function comes a number of social and ethical concerns.

Genome data were used to create the first "synthetic" cell

Life is characterized by a set of properties emerging from the assembly of atoms into molecules, molecules into macromolecules, and macromolecules into cellular structures. The pathways controlling the synthesis of those molecules and the processes leading to the formation of cellular structures are under the control of the genome. But what is the minimum genome size required to build an autonomously replicating cell?

We can get some idea of the smallest minimum genome required to sustain life by finding the smallest genome in a living organism. The smallest viral genome is 5386 nt, but many scientists do not consider viruses to be alive as they are unable to replicate independently. The genome of the semiautonomous mitochondrion is about 16,600 nt long; however, the mitochondrion is unable to replicate independently of its cellular "host." The smallest free-living bacterial genome currently known is that of *Mycoplasma genitalium* at about 500,000 nt. This implies that the minimum viable genome for a living cell is somewhere between 16,600 and 500,000 nt long.

In 2010, scientists at the J. Craig Venter Institute announced that they had designed a *Mycoplasma mycoides* genome using a computer, chemically synthesized the DNA, assembled the DNA into a chromosome inside yeast cells, and transferred the chromosome into the bacterium *Mycoplasma capricolum*. The genome of the recipient bacterium was destroyed in the process and the new synthetic genome was able to control the growth of the new bacterial cell, named *Mycoplasma mycoides* JCVI-syn1.0 (figure 18.15). This approach may allow experimental verification of a minimal genome.

Genomics can help to identify and treat disease

The genomics revolution has yielded millions of new genes to be investigated. The potential of genomics to improve human health is enormous. Mutations in a single gene can explain a limited number of hereditary diseases. However, genomics potentially allows us to look at any trait that is influenced by genetics.

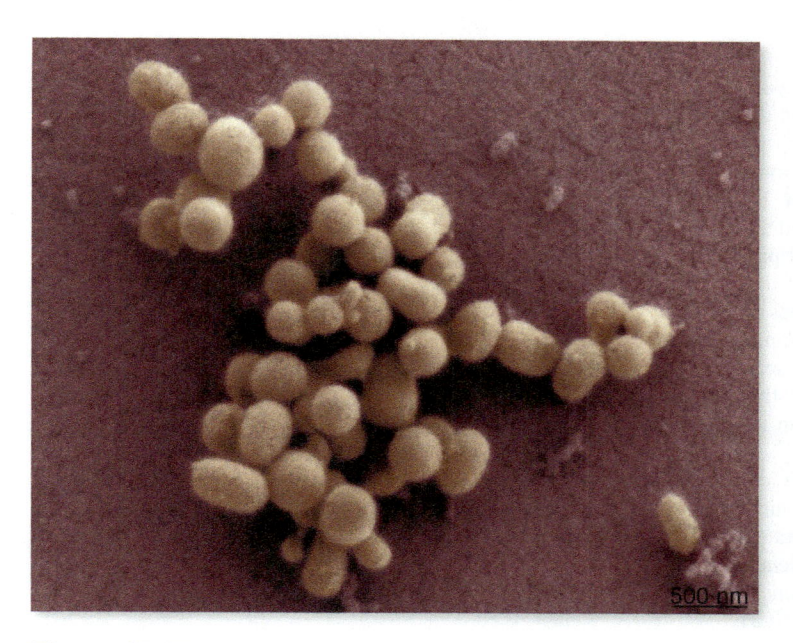

Figure 18.15 Scanning electron micrograph of synthetic *Mycoplasma mycoides* JCVI-syn1.0 bacteria.

Although proteomics will likely lead to new pharmaceuticals, the immediate effect of genomics is being seen in diagnostics. Both improved technology and gene discovery are enhancing the diagnosis of genetic abnormalities. Diagnostics are also being used to identify individuals. For example, short tandem repeats (STRs), discovered through genomic research, were among the forensic diagnostic tools used to identify remains of victims of the September 11th, 2001, terrorist attack on the World Trade Center in New York City.

Following the September 11th attacks in the United States, there was increased concern about biological weapons. Five people died and 17 more were infected with anthrax after envelopes containing spores were sent through the U.S. mail. Genome sequencing allowed identification of the specific source of the bacteria used. A difference of only 10 bp between strains allowed the FBI to trace the source to a single vial of anthrax used in a vaccine research program at U.S. Army Medical Research Institute for Infectious Diseases. In the end, no one was formally charged by the FBI in these attacks, although a researcher under suspicion committed suicide, and another was exonerated. In the wake of this incident, the Centers for Disease Control and Prevention (CDC) has ranked bacteria and viruses that are likely targets for bioterrorism (table 18.2).

Our knowledge of the human genome has also informed our search for the genetic basis of complex diseases. Studies referred to as genome-wide association studies (GWAS), look for association between genetic markers, like SNPs, and traits or disease states. These studies are ongoing for traits like height—and for virtually every disease that has been shown to have a genetic component. Although a large number of genes have been identified this way, they have yet to identify all of the genetic variability we have previously measured.

Genomics raises social and ethical issues

Genomics provides opportunities for understanding the relationship between our genotype and phenotype, as well as great potential for better diagnosing and treating some diseases. With these opportunities come social and ethical responsibilities regarding the use of the information. There must be careful consideration of the ethical and social consequences regarding ownership of genetic information and its uses, the potential misuses of the information, and even whether all genomics experiments should be performed.

Rebuilding the 1918 influenza virus

In May of 2005, a series of papers describing the sequencing and reassembly of the 1918 Spanish influenza virus were published. The 1918 influenza virus killed between 20 and 50 million people. During the 1990s, several research groups including some at the Centers for Disease Control (CDC) in Atlanta and the Armed Forces Institute of Pathology in Washington, D.C., found fragments of the viral genome, sequenced them, and reconstructed the entire virus. The virus was found to have a similar mortality rate as the original virus when tested in other mammals such as mice. It has obviously not been tested in humans.

These experiments raised serious concerns for some scientists, politicians, public interest groups, and the public in general. On the one hand, being able to study the genome of the virus, and the properties of the virus itself, may lead to more effective vaccines and help prevent future pandemics. Critics of these experiments claim that the vague potential benefits of recreating the virus are not worth the risk should an unfortunate or negligent scientist accidentally or deliberately release the virus. Many people also expressed concerns about the information being used by bioterrorists to create a weaponized form of the virus.

After extensive reviews by the scientific community, careful scrutiny by the CDC, and oversight from the National Science Advisory Board for Biosecurity, it was decided that the benefits of making the work publicly available outweighed the risks. Of course, such decisions are subjective, but by being well informed about genomics, it is possible to make a more reasoned decision about such a controversial topic.

TABLE 18.2 High-Priority Pathogens for Genomic Research

Pathogen	Disease	Genome*
Variola major	Smallpox	Complete
Bacillus anthracis	Anthrax	Complete
Yersinia pestis	Plague	Complete
Clostridium botulinum	Botulism	Complete
Francisella tularensis	Tularemia	Complete
Filoviruses	Ebola and Marburg hemorrhagic fever	Both are complete
Arenaviruses	Lassa fever and Argentine hemorrhagic fever	Both are complete

*These viruses and bacteria have multiple strains. "Complete" indicates that at least one has been sequenced. For example, the Florida strain of anthrax was the first to be sequenced.

Owning and patenting genes

You may think that you are the owner of the information in your genes, but until 2013 it appeared that someone else could own the rights to the sequences of your genes. In the 1990s, a company called Myriad Genetics patented two genes that were known to be involved in the development of some forms of breast cancer. By patenting the genes *BRCA1* and *BRCA2*, the company secured exclusive rights to use the sequences to test for the presence of particular variants of the genes. This formed the basis of a test that provided information regarding predisposition to certain types of breast cancer. Given the increased risks of breast cancer in women carrying particular variants of *BRCA1* and *BRCA2*, and the seriousness of a potential cancer diagnosis, many people believed that not being able to secure a second diagnostic opinion was an infringement of civil liberties.

After a lengthy battle through federal courts driven largely by the American Civil Liberties Union (ACLU), a case against Myriad Genetics arrived at the Supreme Court of the United States. In a unanimous ruling, the Court ruled that because Myriad Genetics had not invented the genes, they could not patent the sequences of the genes. In their ruling, the Court noted that certain derived sequences—cDNA sequences, for example—and their uses are patentable because they have been created.

> **Inquiry question** What arguments could be used to support an application for the patent of a synthetic gene that increases the yield of lipids in algae used for biofuel production?

> **Learning Outcomes Review 18.6**
> Genome sequencing, annotation, and comparative and functional genomics led to the discovery of genes involved in disease. Some of those genes can be used to better diagnose disease or to provide more accurate prognoses. Genomics can also be used to learn about disease outbreaks and epidemiology. Genomic knowledge carries social and ethical responsibilities. By consultation with bioethicists, policy and various interest groups, regulatory agencies can become better informed about experimentation and appropriate data use.
>
> ■ *Does the cell created by replacement of its genome with a sequence constructed by humans represent a new form of life?*

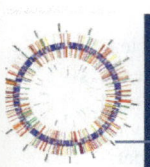

Chapter Review

18.1 Mapping Genomes

Genomics uses many approaches to analyze entire genomes.
Genomics uses genetic and physical maps to assign genetic landmarks, called markers, to positions in the genome.

Genetic maps provide relative distances between genetic markers.
Linkage mapping is used to determine the relative positions of genetic markers in a genome. Linkage mapping relies on the separation of markers during recombination; the farther apart the markers, the more likely they will be separated.

Physical maps provide absolute distances between genetic markers.
Restriction maps, chromosome maps, and STS maps can be generated to assign markers to an actual location in the sequence of a genome (figures 18.1–18.3). Physical mapping requires some knowledge about the sequence of DNA in the genome.

Physical maps can be correlated with genetic maps.
Physical and genetic maps can be correlated. Any gene that can be cloned can be placed within the genome sequence and mapped. However, absolute correspondence of distances cannot be accomplished.

18.2 Sequencing Genomes

Dideoxy terminator sequencing remains important in genome sequencing.
Although sequencing technologies have advanced hugely in recent years, there are still cases when dideoxy terminator sequencing is used. This relies on altered nucleotide chemistry to terminate DNA synthesis with a fluorescent nucleotide (figure 18.4).

Next-generation sequencing uses massively parallel technologies to increase throughput.
The cost and time to sequence whole genomes has reduced about 10,000-fold since the start of the Human Genome Project. Next-generation sequencers eliminate the need to directly clone pieces of DNA and analyze isolated DNA in thousands of simultaneous reactions to produce massive amounts of data (figure 18.5).

Sequenced genome fragments are assembled into complete sequences.
Shotgun sequencing and assembly methods recreate the DNA sequence of a genome from very short pieces of slightly overlapping DNA. Clone-contig sequencing and assembly methods uses larger pieces of DNA and a tiling approach to build larger and larger overlapping fragments until a whole genome is reassembled (figure 18.6).

18.3 Genome Projects

The Human Genome Project has sequenced and mapped most of the human genome.
A competitive race to sequence and assemble the human genome resulted in rapid advances in technology and completion of the project ahead of schedule. The human genome contains significantly fewer genes than predicted (figure 18.7).

The Wheat Genome Project illustrates the difficulties of assembling complex genomes.
Large and highly repetitive genomes are more expensive, and take longer, to sequence and assemble. Complex genome structures require the use of the clone-contig approach.

Cancer genome projects seek the genetic causes of cancer.
By identifying similarities and differences in the genomes and epigenomes of the same cancers, it may be possible to improve

diagnostics, predict relapse of disease, and more appropriately administer medications.

18.4 Genome Annotation and Databases

Genome annotation assigns function to DNA sequences in genomes.
Finding open-reading frames in DNA sequences identifies genes. Search tools that look for sequences related to a suspected gene can sometimes be used to infer gene function. The identification of an EST can support the annotation process by providing additional information (figure 18.8).

Genomes contain both coding and noncoding sequences.
Protein-encoding DNA includes single-copy genes, segmental duplications, multigene families, and tandem clusters. Noncoding DNA in eukaryotes makes up about 99% of DNA. Approximately 45% of the human genome is composed of mobile transposable elements, including LINEs, SINEs, and LTRs (table 18.1).

The ENCODE project seeks to identify all functional elements in the human genome.
The ENCODE consortium predicts that anywhere from 20 to 80% of the human genome is functional. Critics of their approach question the utility of a definition of function that does not integrate evolutionary constraints on genome function (figure 18.9).

18.5 Comparative and Functional Genomics

Comparative genomics reveals conserved regions in genomes.
More than half of the genes of *Drosophila* have human counterparts. The biggest difference between our genome and the chimpanzee genome is in transposable elements.

Functional genomics reveals gene function at the genome level.
Functional genomics uses high-throughput experimental approaches and bioinformatics to analyze gene function and gene products. DNA microarrays allow the expression of all of the genes in a cell to be monitored at once (figure 18.11). SAGE and RNA-seq provide more comprehensive ways to study the transcriptome.

Proteomics catalogs proteins encoded by the genome.
Proteomics characterizes all of the proteins produced by a cell. Proteomics is complicated due to the dynamic nature of the proteome, posttranslational modifications, and the use of alternative splicing to produce variant proteins from the same gene (figures 18.12–18.14).

18.6 Applications of Genomics

Genome data were used to create the first "synthetic" cell.
Designing a genome in a computer, chemically synthesizing the DNA of the genome, assembling the genome in yeast, and adding the genome to a bacterium created an autonomously replicating cell (figure 18.15).

Genomics can help to identify and treat disease.
The identification of certain genes by genome projects means that some diseases can be more easily and reliably diagnosed. In the future, fully personalized medicine may mean that individuals with particular genotypes receive tailored treatments.

Genomics raises social and ethical issues.
Social groups and bioethicists carefully monitor experiments such as those involving the creation of a synthetic cell, and the reconstruction of dangerous viruses. Individual genes cannot be patented, but synthetic products such as cDNA corresponding to the gene can be.

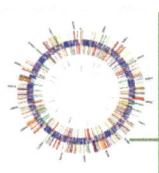

Review Questions

UNDERSTAND

1. A genetic map provides
   a. the sequence of the DNA in a genome.
   b. the relative position of genes on chromosomes.
   c. the location of sites of restriction enzyme cleavage in a known sequence of DNA.
   d. the banding pattern of a chromosome.

2. What is an STS?
   a. A unique sequence within the DNA that can be used for mapping
   b. A repeated sequence within the DNA that can be used for mapping
   c. An upstream element that allows for mapping of the 3′ region of a gene
   d. A sequence of DNA that overlaps with other sequenced fragments in clone-contig genome assembly

3. Approximately how many genes are there in the human genome?
   a. 2500
   b. 10,000
   c. 20,000
   d. 100,000

4. An open reading frame (ORF) is distinguished by the presence of
   a. a stop codon.
   b. a start codon.
   c. a sequence of DNA long enough to encode a protein.
   d. All of the choices are correct.

5. What is a BLAST search?
   a. A mechanism for aligning consensus regions during whole-genome sequencing
   b. A search for similar gene sequences from other species
   c. A method of screening a DNA library
   d. A method for identifying ORFs

6. Why is repetitive DNA a problem in shotgun sequencing and genome assembly?
   a. The repetitive DNA cannot easily be sequenced.
   b. There is a lack of uniqueness between sequenced fragments.
   c. The repetitive DNA makes the genome too GC-rich.
   d. It becomes too expensive and too time-intensive.

7. What is a proteome?
   a. The collection of all genes encoding proteins
   b. The collection of all proteins encoded by the genome
   c. The collection of all proteins present in a cell
   d. The amino acid sequence of a protein

8. Why can the transcriptome not be used to predict the proteome with complete accuracy?
   a. It cannot be sequenced like the genome can be.
   b. The transcriptome is too dynamic to be used to make predictions.
   c. Not all genes are transcribed.
   d. Many transcripts are alternatively spliced to produce different proteins.

APPLY

1. If genomics found that the same mutation was present in all cancer patients that failed to respond to a particular treatment, what might be concluded?
   a. That the mutation was responsible for the failure to respond to treatment
   b. That the mutation would be a good diagnostic tool
   c. That children of those patients have an increased risk for the same cancer
   d. That the patients all have the same type of cancer

2. ENCODE define function in a biochemical way. Which of the following genomics approaches could increase the accuracy of their estimate of genome functionality in humans?
   a. Functional genomics
   b. Comparative genomics
   c. Genome sequencing and assembly
   d. Choices a, b, and c would increase the accuracy of their estimate.

3. If I synthesized a protein by chemically linking amino acids together and added a few extra amino acids that made the protein more stable and therefore better at treating a disease, what argument would support a patent for the protein?
   a. The protein is beneficial to humans and so I should be paid for my discovery.
   b. The protein is synthetic and different from the natural form.
   c. The protein was synthesized chemically.
   d. The protein is expensive to make and a patent would help me offset my costs of production.

4. The genome of the frog *Xenopus tropicalis* is tetraploid. What challenges might this present for genome sequencing and assembly?
   a. There may be duplicated pieces of genome that make genome assembly difficult.
   b. There may be LINEs that make sequencing difficult.
   c. Genetic and physical maps cannot be made.
   d. Both b and c are correct.

5. What information can be obtained from a DNA microarray?
   a. The sequence of a particular gene
   b. The presence of genes within a specific tissue
   c. The pattern of gene expression
   d. Differences between genomes

6. Which of the following is true regarding microarray technology and cancer?
   a. A DNA microarray can determine the type of cancer.
   b. A DNA microarray can measure the response of a cancer to therapy.
   c. A DNA microarray can be used to predict whether the cancer will move from one place to another in the body.
   d. All of the choices are correct.

7. Comparative proteomics involves comparing the proteomes from two different cells or tissues from different conditions. In some cancer cells the Retinoblastoma protein often appears to have a slightly higher molecular mass than in noncancerous cells even though the proteins have identical amino acid sequences. What might explain this difference?
   a. Posttranslational modification
   b. Alternative splicing
   c. Association with different proteins
   d. Mutation of the gene encoding the protein

SYNTHESIZE

1. You are in the early stages of a genome-sequencing project. You have isolated a number of clones from a bacterial artificial chromosome (BAC) library and mapped the inserts in these clones using STSs. Use the STSs shown to align the clones into a contiguous sequence of the genome (a contig).

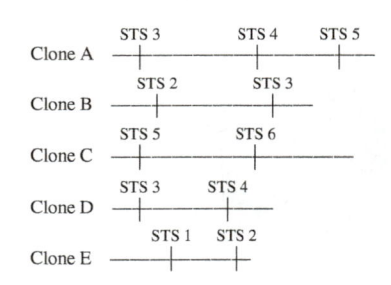

Clone	STSs
Clone A	STS 3, STS 4, STS 5
Clone B	STS 2, STS 3
Clone C	STS 5, STS 6
Clone D	STS 3, STS 4
Clone E	STS 1, STS 2

2. Genomic research can be used to determine if an outbreak of an infectious disease is natural or "intentional." Explain what a genomic researcher would be looking for in a suspected intentional outbreak of a disease like anthrax.

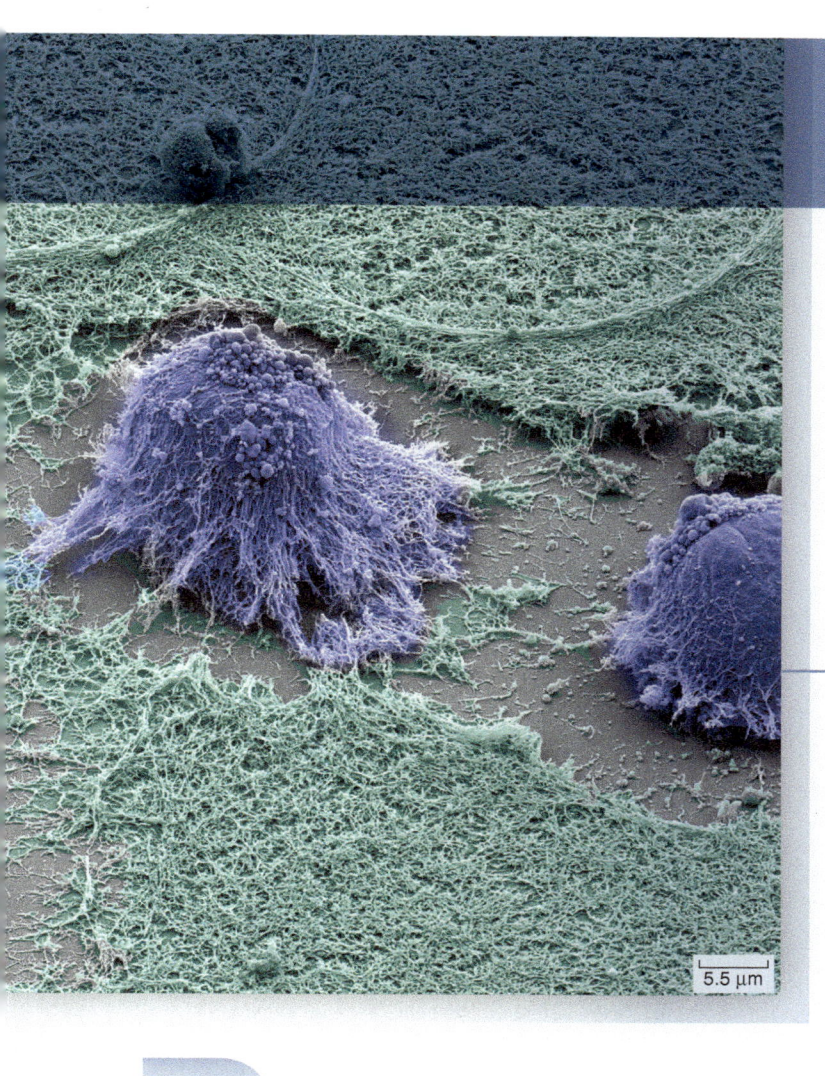

CHAPTER 19

Cellular Mechanisms of Development

Chapter Contents

19.1 The Process of Development
19.2 Cell Division
19.3 Cell Differentiation
19.4 Nuclear Reprogramming
19.5 Pattern Formation
19.6 Morphogenesis

Introduction

Recent work with different kinds of stem cells, like those pictured, have captured the hopes and imagination of the public. For thousands of years, humans have wondered how organisms arise, grow, change, and mature. We are now in an era when long-standing questions may be answered, and new possibilities for regenerative medicine seem on the horizon.

We have explored gene expression from the perspective of individual cells, examining the diverse mechanisms cells employ to control the transcription of particular genes. Now we broaden our perspective and look at the unique challenge posed by the development of a single cell, the fertilized egg, into a multicellular organism. In the course of this developmental journey, a pattern of gene expression takes place that causes particular lines of cells to proceed along different paths, spinning an incredibly complex web of cause and effect. Yet, for all its complexity, this developmental program works with impressive precision. In this chapter, we explore the mechanisms of development at the cellular and molecular level.

19.1 The Process of Development

Development can be defined as the process of systematic, gene-directed changes through which an organism forms the successive stages of its life cycle. Development is a continuum, and explorations of development can be focused on any point along this continuum. The study of development plays a central role in unifying the understanding of both the similarities and diversity of life on Earth.

We can divide the overall process of development into four subprocesses:

- **Cell division.** A developing plant or animal begins as a fertilized egg, or zygote, that must undergo cell division to produce the new individual. In all cases early development involves extensive cell division, but in many cases it does not include much growth as the egg cell itself is quite large.
- **Differentiation.** As cells divide, orchestrated changes in gene expression result in differences between cells that ultimately result in cell specialization. In differentiated cells,

certain genes are expressed at particular times, but other genes may not be expressed at all.

- **Pattern formation.** Cells in a developing embryo must become oriented to the body plan of the organism the embryo will become. Pattern formation involves cells' abilities to detect positional information that guides their ultimate fate.
- **Morphogenesis.** As development proceeds, the form of the body—its organs and anatomical features—is generated. Morphogenesis may involve cell death, cell division, cell migration, changes in cell shape and differentiation.

Despite the overt differences between groups of plants and animals, most multicellular organisms develop according to molecular mechanisms that are fundamentally very similar. This observation suggests that these mechanisms evolved very early in the history of multicellular life.

19.2 Cell Division

Learning Outcomes
1. Characterize the role of cell division in early development.
2. Describe the use of C. elegans to track cell lineages.
3. Distinguish differences in cell division between animals and plants.

When a frog tadpole hatches out of its protective coats, it is roughly the same overall mass as the fertilized egg from which it came. Instead of being made up of just one cell, however, the tadpole consists of about a million cells, which are organized into tissues and organs with different functions. Thus, the very first process that must occur during embryogenesis is cell division.

Immediately following fertilization, the diploid zygote undergoes a period of rapid mitotic divisions that ultimately result in an early embryo comprised of dozens to thousands of diploid cells. In animal embryos, the timing and number of these divisions are species-specific and are controlled by a set of molecules that we examined in chapter 10: the *cyclins* and *cyclin-dependent kinases (Cdks)*. These molecules exert control over checkpoints in the cycle of mitosis.

Development begins with cell division

In animal embryos, the period of rapid cell division following fertilization is called **cleavage**. During cleavage, the enormous mass of the zygote is subdivided into a larger and larger number of smaller and smaller cells, called **blastomeres** (figure 19.1). Hence, cleavage is not accompanied by any increase in the overall size of the embryo. The G_1 and G_2 phases of the cell cycle, during which a cell increases its mass and size, are extremely shortened or eliminated during cleavage (figure 19.2).

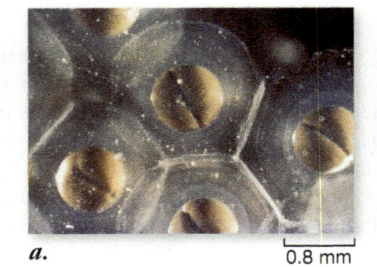

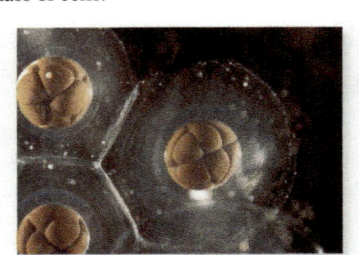

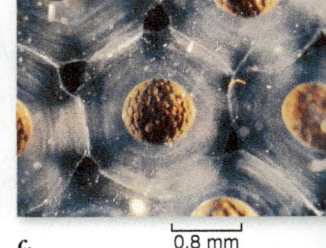

Because of the absence of the two gap/growth phases, the rapid rate of mitotic divisions during cleavage is never again approached in the lifetime of any animal. For example, zebrafish blastomeres divide once every several minutes during cleavage, to create an embryo with a thousand cells in just under 3 hr! In contrast, cycling adult human intestinal epithelial cells divide on average only once every 19 hr. A comparison of the different patterns of cleavage can be found in chapter 53.

When external sources of nutrients become available—for example, during larval feeding stages or after implantation of mammalian embryos—daughter cells can increase in size following cytokinesis, and an overall increase in the size of the organism occurs as more cells are produced.

Every cell division is known in the development of *C. elegans*

One of the most completely described models of development is the tiny nematode *Caenorhabditis elegans*. Only about 1 mm long, the adult worm consists of 959 somatic cells.

Because *C. elegans* is transparent, individual cells can be followed as they divide. By observing them, researchers have learned how each of the cells that make up the adult worm is derived from the fertilized egg. As shown on the lineage map in figure 19.3a, the egg divides into two cells, and these daughter cells continue to divide. Each horizontal line on the map represents one round of cell division. The length of each vertical line represents the time between cell divisions, and the end of each vertical line represents one fully differentiated cell. In figure 19.3b, the major organs of the worm are color-coded to match the colors of the corresponding groups of cells on the lineage map.

Some of these differentiated cells, such as some cells that generate the worm's external cuticle, are "born" after only 8 rounds of cell division; other cuticle cells require as many as 14 rounds. The cells that make up the worm's pharynx, or feeding organ, are born after 9 to 11 rounds of division, whereas cells in the gonads require up to 17 divisions.

Exactly 302 nerve cells are destined for the worm's nervous system. Exactly 131 cells are programmed to die, mostly within minutes of their "birth." The fate of each cell is the same in every *C. elegans* individual, except for the cells that will become eggs and sperm.

Figure 19.1 Cleavage divisions in a frog embryo. *a.* The first cleavage division divides the egg into two large blastomeres. *b.* After two more divisions, four small blastomeres sit on top of four large blastomeres, each of which continues to divide to produce (*c*) a compact mass of cells.

a. 0.8 mm   *b.* 0.8 mm   *c.* 0.8 mm

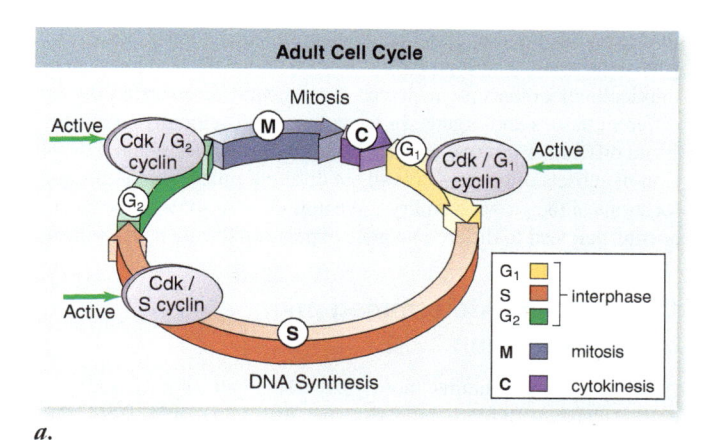

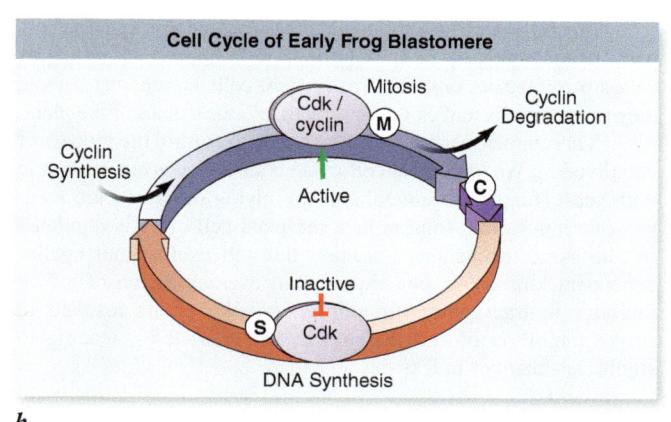

Figure 19.2 Cell cycle of adult cell and embryonic cell. In contrast to the cell cycle of adult somatic cells (*a*), the dividing cells of early frog embryos lack G_1 and G_2 stages (*b*), enabling the cleavage stage nuclei to rapidly cycle between DNA synthesis and mitosis. Large stores of cyclin mRNA are present in the unfertilized egg. Periodic degradation of cyclin proteins correlates with exiting from mitosis. Cyclin degradation and Cdk inactivation allow the cell to complete mitosis and initiate the next round of DNA synthesis.

Plant growth occurs in specific areas called meristems

A major difference between animals and plants is that most animals are mobile, at least in some phase of their life cycles, and therefore they can move away from unfavorable circumstances. Plants, in contrast, are anchored in position and must simply endure whatever environment they experience. Plants compensate for this restriction by allowing development to accommodate local circumstances.

Instead of creating a body in which every part is specified to have a fixed size and location, a plant assembles its body throughout its life span from a few types of modules, such as leaves, roots, branch nodes, and flowers. Each module has a rigidly controlled structure and organization, but how the modules are utilized is quite flexible—they can be adjusted to environmental conditions.

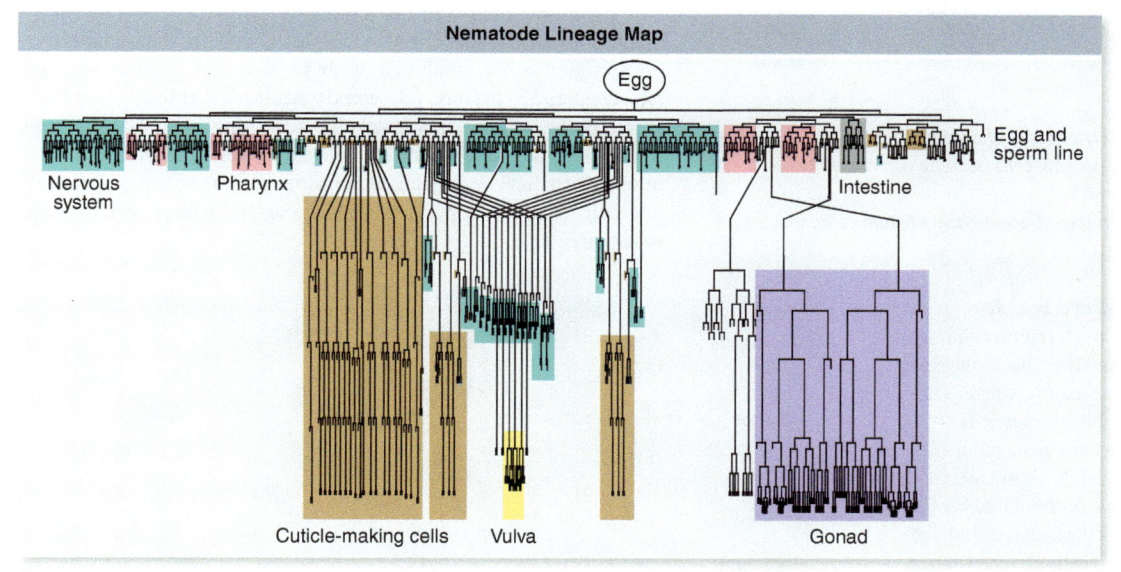

Figure 19.3 Studying embryonic cell division and development in the nematode. Development in *C. elegans* has been mapped out such that the fate of each cell from the single egg cell has been determined. *a.* The lineage map shows the number of cell divisions from the egg, and the color coding links their placement in (*b*) the adult organism.

chapter **19** *Cellular Mechanisms of Development* 377

Plants develop by building their bodies outward, creating new parts from groups of stem cells that are contained in structures called **meristems.** As meristematic stem cells continually divide, they produce cells that can differentiate into the tissues of the plant.

This simple scheme indicates a need to control the process of cell division. We know that cell-cycle control genes are present in both yeast (fungi) and animal cells, implying that these are a eukaryotic innovation—and in fact, the plant cell cycle is regulated by the same mechanisms, namely through cyclins and cyclin-dependent kinases. In one experiment, overexpression of a Cdk inhibitor in transgenic *Arabidopsis thaliana* plants resulted in strong inhibition of cell division in leaf meristems, leading to significant changes in leaf size and shape.

Learning Outcomes Review 19.2
In animal embryos, a series of rapid cell divisions that skip the G_1 and G_2 phases convert the fertilized egg into many cells with no change in size. In the nematode *C. elegans,* every cell division leading to the adult form is known, and this pattern is invariant, allowing biologists to trace development in a cell-by-cell fashion. In plants, growth is restricted to specific areas called meristems, where undifferentiated stem cells are retained.

- How are early cell divisions in an embryo different from in an adult organism?

19.3 Cell Differentiation

Learning Outcomes
1. Describe the progressive nature of determination.
2. Illustrate with examples how cells become committed to developmental pathways.
3. Differentiate between the different types of stem cells.

In chapter 16, we examined the mechanisms that control eukaryotic gene expression. These processes are critical for the development of multicellular organisms, in which life functions are carried out by different tissues and organs. In the course of development, cells become different from one another because of the differential expression of subsets of genes—not only at different times, but in different locations of the growing embryo. We now explore some of the mechanisms that lead to differential gene expression during development.

Cells become determined prior to differentiation

A human body contains more than 210 types of differentiated cells. These differentiated cells are distinguishable from one another by the particular proteins that they synthesize, their morphologies, and their specific functions. A molecular decision to become a particular type of differentiated cell occurs prior to any overt changes in the cell. This molecular decision-making process is called **cell determination,** and it commits a cell to a particular developmental pathway.

Tracking determination
Determination is often not visible in the cell and can only be detected experimentally. The standard experiment to test whether a cell or group of cells is determined is to move the donor cell(s) to a different location in a host (recipient) embryo. If the cells of the transplant develop into the same type of cell as they would have if left undisturbed, then they are judged to be already determined (figure 19.4).

The time course of determination can be assessed by transplantation experiments. For example, a cell in the prospective brain region of an amphibian embryo at the early gastrula stage (see chapter 53) has not yet been determined; if transplanted elsewhere in the embryo, it will develop according to the site of transplant. By the late gastrula stage, however, additional cell interactions have occurred, determination has taken place, and the cell will develop as neural tissue no matter where it is transplanted.

Figure 19.4 The standard test for determination. The gray ovals represent embryos at early stages of development. The cells to the right normally develop into head structures, whereas the cells to the left usually form tail structures. If prospective tail cells from an early embryo are transplanted to the opposite end of a host embryo, they develop according to their new position into head structures. These cells are not determined. At later stages of development, the tail cells are determined since they now develop into tail structures after transplantation into the opposite end of a host embryo!

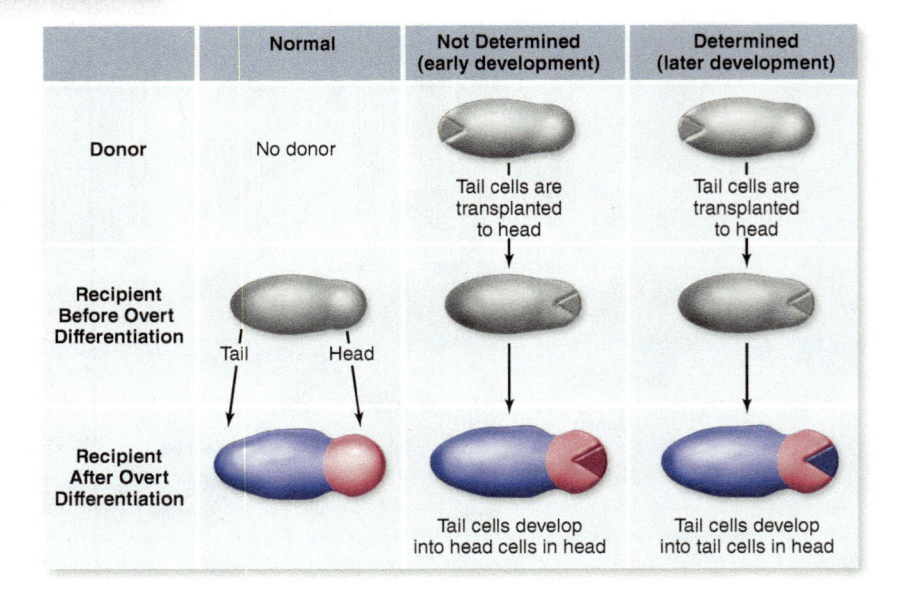

Determination often takes place in stages, with a cell first becoming partially committed, acquiring positional labels that reflect its location in the embryo. These labels can have a great influence on how the pattern of the body subsequently develops. In a chicken embryo, tissue at the base of the leg bud normally gives rise to the thigh. If this tissue is transplanted to the tip of the identical-looking wing bud, which would normally give rise to the wing tip, the transplanted tissue will develop into a toe rather than a thigh. The tissue has already been determined as leg, but it is not yet committed to being a particular part of the leg. Therefore, it can be influenced by the positional signaling at the tip of the wing bud to form a tip (but in this case, a tip of leg).

The molecular basis of determination

Cells initiate developmental changes by using transcription factors to change patterns of gene expression. When genes encoding these transcription factors are activated, one of their effects is to reinforce their own activation. This reinforcement makes the developmental switch deterministic, initiating a chain of events that leads down a particular developmental pathway.

Cells in which a set of regulatory genes have been activated may not actually undergo differentiation until some time later, when other factors interact with the regulatory protein and cause it to activate still other genes. Nevertheless, once the initial "switch" is thrown, the cell is fully committed to its future developmental path.

Cells become committed to follow a particular developmental pathway in one of two ways:

1. via the differential inheritance of cytoplasmic determinants, which are maternally produced and deposited into the egg during oogenesis; or
2. via cell–cell interactions.

The first situation can be likened to a person's social status being determined by who his or her parents are and what he or she has inherited. In the second situation, the person's social standing is determined by interactions with his or her neighbors. Clearly both can be powerful factors in the development and maturation of that individual.

Determination can be due to cytoplasmic determinants

Many invertebrate embryos provide good visual examples of cell determination through the differential inheritance of cytoplasmic determinants. Tunicates are marine invertebrates (see chapter 35), and most adults have simple, saclike bodies that are attached to the underlying substratum. Tunicates are placed in the phylum Chordata, however, due to the characteristics of their swimming, tadpolelike larval stage, which has a dorsal nerve cord and notochord (figure 19.5a). The muscles that move the tail develop on either side of the notochord.

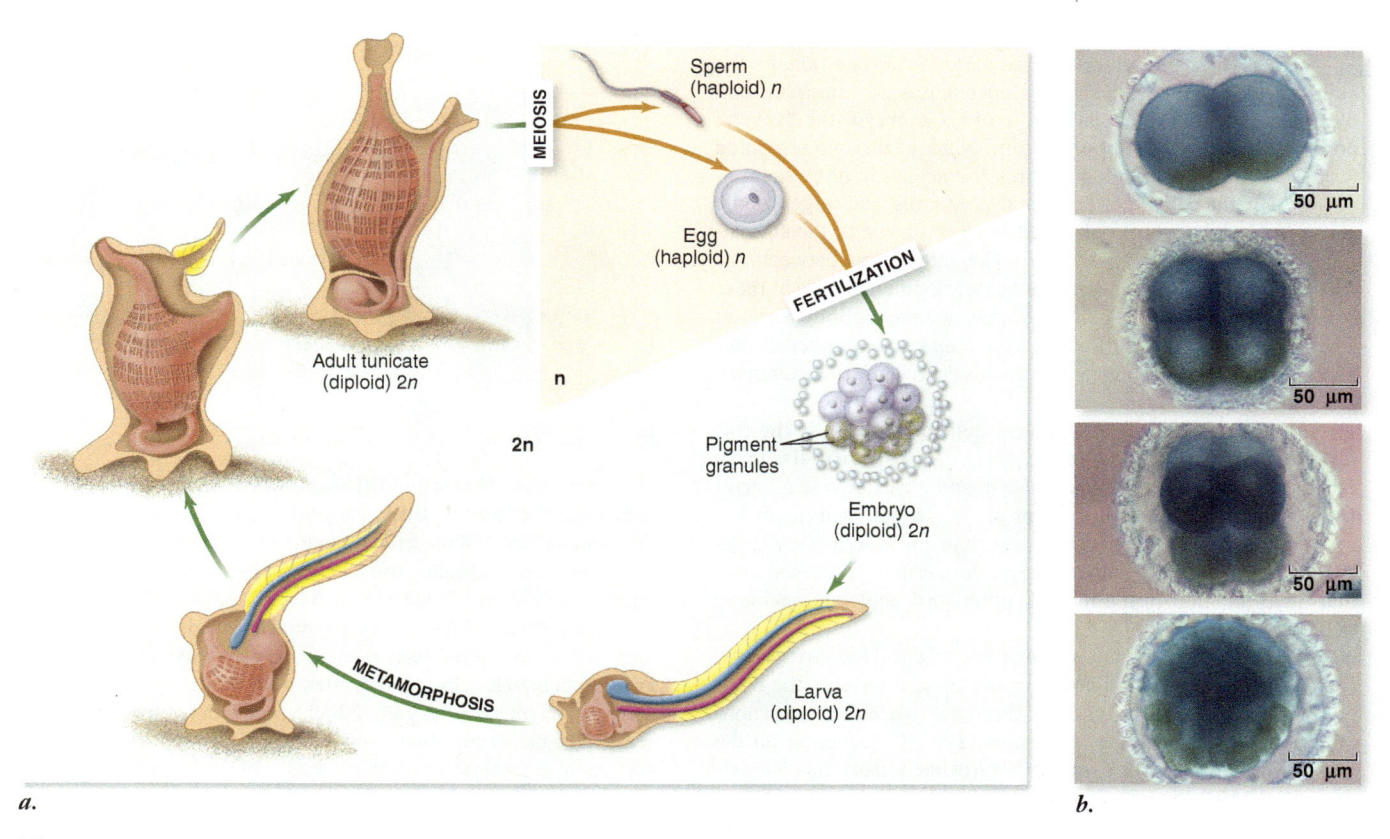

Figure 19.5 Muscle determinants in tunicates. *a.* The life cycle of a solitary tunicate. Muscle cells that move the tail of the swimming tadpole are arranged on either side of the notochord and nerve cord. The tail is lost during metamorphosis into the sedentary adult. *b.* The egg of the tunicate *Styela* contains bright yellow-colored pigment granules. These become asymmetrically localized in the egg following fertilization, and cells that inherit the yellow-colored granules during cleavage will become the larval muscle cells. Embryos at the 2-cell, 4-cell, 8-cell, and 64-cell stages are shown. The tadpole tail will grow out from the lower region of the embryo in the bottom panel.

In many tunicate species, yellow-colored pigment granules become asymmetrically localized in the egg following fertilization and subsequently segregate to the tail muscle cell progenitors during cleavage (figure 19.5b). When these pigment granules are shifted experimentally into other cells that normally do not develop into muscle, their fate is changed and they become muscle cells. Thus, the molecules that flip the switch for muscle development appear to be associated with the pigment granules.

The next step is to determine the identity of the molecules involved. Experiments indicate that the female parent provides the egg with mRNA encoded by the *macho-1* gene. The elimination of *macho-1* function leads to a loss of tail muscle in the tadpole, and the misexpression of *macho-1* mRNA leads to the formation of additional (ectopic) muscle cells from nonmuscle lineage cells. The *macho-1* gene product has been shown to be a transcription factor that can activate the expression of several muscle-specific genes.

Induction can lead to cell differentiation

In chapter 9, we examined a variety of ways by which cells communicate with one another. We can demonstrate the importance of cell–cell interactions in development by separating the cells of an early frog embryo and allowing them to develop independently.

Under these conditions, blastomeres from one pole of the embryo (the "animal pole") develop features of ectoderm, and blastomeres from the opposite pole of the embryo (the "vegetal pole") develop features of endoderm. None of the two separated groups of cells ever develop features characteristic of mesoderm, the third main cell type. If animal-pole cells and vegetal-pole cells are placed next to each other, however, some of the animal-pole cells develop as mesoderm. The interaction between the two cell types triggers a switch in the developmental path of these cells. This change in cell fate due to interaction with an adjacent cell is called **induction**. Signaling molecules act to alter gene expression in the target cells, in this case, some of the animal-pole cells.

Another example of inductive cell interactions is the formation of the notochord and mesenchyme, a specific tissue, in tunicate embryos. Muscle, notochord, and mesenchyme all arise from mesodermal cells that form at the vegetal margin of the 32-cell stage embryo. These prospective mesodermal cells receive signals from the underlying endodermal precursor cells that lead to the formation of notochord and mesenchyme (figure 19.6).

The chemical signal is a member of the *fibroblast growth factor (FGF)* family of signaling molecules. It induces the overlying marginal zone cells to differentiate into either notochord (anterior) or mesenchyme (posterior). The FGF receptor on the marginal zone cells is a receptor tyrosine kinase that signals through a MAP kinase cascade to activate a transcription factor that turns on gene expression resulting in differentiation (figure 19.7).

This example is also a case of two cells responding differently to the same signal. The presence or absence of the *macho-1* muscle determinant controls this difference in cell fate. In the presence of *macho-1*, cells differentiate into mesenchyme; in its

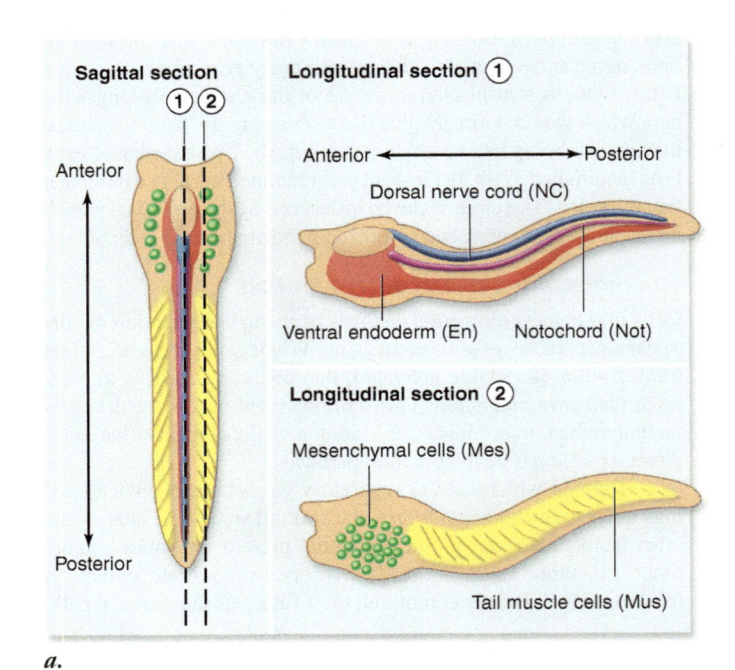

Figure 19.6 Inductive interactions contribute to cell fate specification in tunicate embryos. *a.* Internal structures of a tunicate larva. To the left is a sagittal section through the larva with dotted lines indicating two longitudinal sections. Section 1, through the midline of a tadpole, shows the dorsal nerve cord (NC), the underlying notochord (Not) and the ventral endoderm cells (En). Section 2, a more lateral section, shows the mesenchymal cells (Mes) and the tail muscle cells (Mus). *b.* View of the 32-cell stage looking up at the endoderm precursor cells. Fibroblast growth factor (FGF) secreted by these cells is indicated with light-green arrows. Only the surfaces of the marginal cells that directly border the endoderm precursor cells bind FGF signal molecules. Note that the posterior vegetal blastomeres also contain the *macho-1* determinants *(red and white stripes)*. *c.* Cell fates have been fixed by the 64-cell stage. Colors are as in *(a)*. Cells on the anterior margin of the endoderm precursor cells become notochord and nerve cord, respectively, whereas cells that border the posterior margin of the endoderm cells become mesenchyme and muscle cells, respectively.

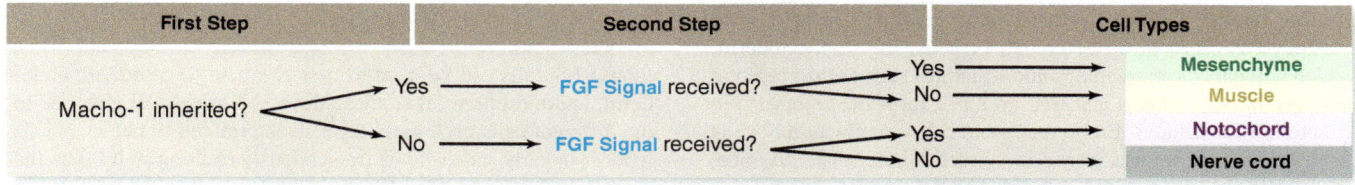

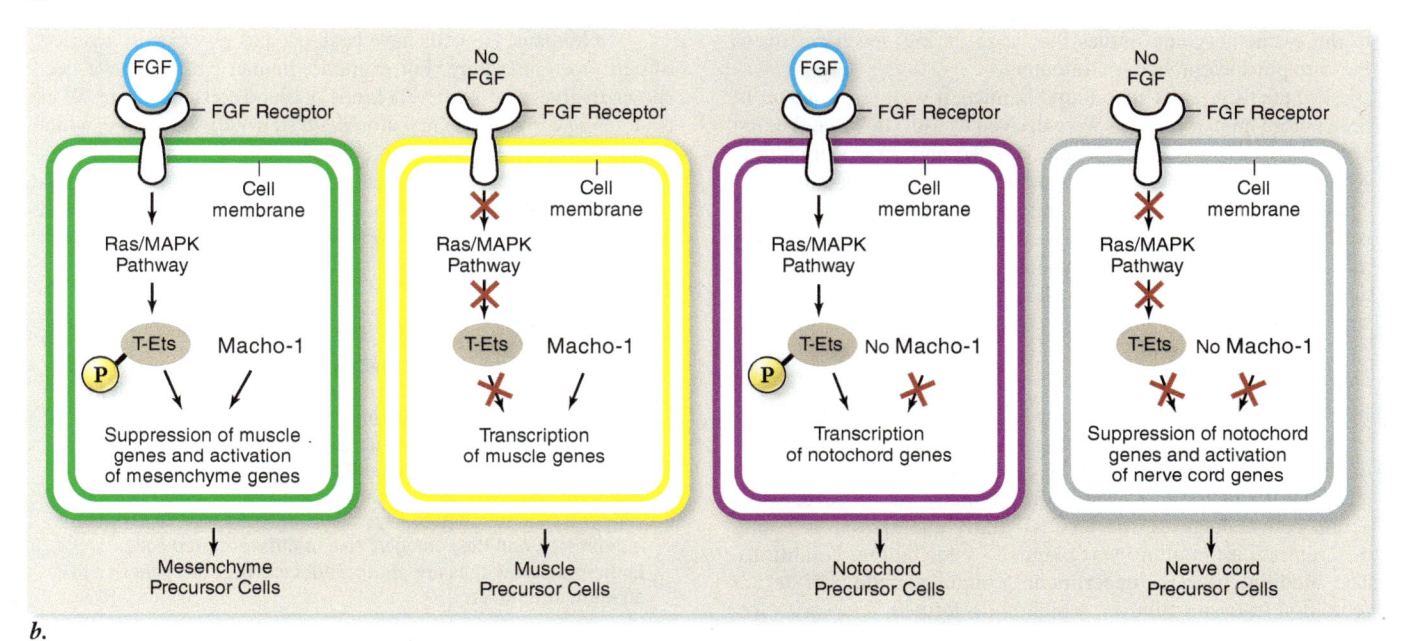

Figure 19.7 Model for cell fate specification by Macho-1 muscle determinant and FGF signaling. *a.* Two-step model of cell fate specification in vegetal marginal cells of the tunicate embryo. The first step is inheritance (or not) of muscle *macho-1* mRNA. The second step is FGF signaling from the underlying endoderm precursor cells. *b.* Posterior vegetal margin cells inherit *macho-1* mRNA. Signaling by FGF activates a Ras/MAP kinase pathway that produces the transcription factor T-Ets. Macho-1 protein and T-Ets suppress muscle-specific genes and turn on mesenchyme specific genes *(green cells)*. In cells with Macho-1 that do not receive the FGF signal, Macho-1 alone turns on muscle-specific cells *(yellow cells)*. Anterior vegetal margin cells do not inherit *macho-1* mRNA. If these cells receive the FGF signal, T-Ets turns on notochord-specific genes *(purple cells)*. In cells that lack Macho-1 and FGF, notochord-specific genes are suppressed and nerve cord-specific genes are activated *(gray cells)*.

Data analysis What type of cells would develop if you injected embryos with a reagent that blocked the FGF receptor, thus preventing its signaling? What about with a reagent that turned on the FGF receptor, thereby causing it to be always on?

absence, cells differentiate into notochord. Thus, the combination of *macho-1* and FGF signaling leads to four different cell types (figure 19.7).

Stem cells can divide and produce cells that differentiate

It is important, both during development, and even in the adult animal, to have cells set aside that can divide but are not determined for only a single cell fate. We call cells that are capable of continued division but that can also give rise to differentiated cells, **stem cells.** These cells can be characterized based on the degree to which they have become determined. At one extreme, we call a cell that can give rise to any tissue in an organism **totipotent.** In mammals, the only cells that can give rise to both the embryo and the extraembryonic membranes are the zygote and early blastomeres from the first few cell divisions. Cells that can give rise to all of the cells in the organism's body are called **pluripotent.** A stem cell that can give rise to a limited number of cell types, such as the cells that give rise to the different blood cell types, are called **multipotent.** Then at the other extreme, **unipotent** stem cells give rise to only a single cell type, such as the cells that give rise to sperm cells in males.

Embryonic stem cells are pluripotent cells derived from embryos

Embryonic stem cells (ES cells) are a form of pluripotent stem cells isolated in the laboratory. These cells are made from mammalian embryos at the blastocyst stage of development. The blastocyst consists of an outer ball of cells, the trophectoderm, which will become extraembryonic structures such as the placenta, and the inner cell mass that will go on to form the embryo (see chapter 53 for details). Embryonic stem cells are essentially cells from the inner cell mass grown in culture (figure 19.8). In mice, these cells have been shown to be able to develop into any type of cell in the tissues of the adult. However, these cells cannot give rise to the extraembryonic tissues that arise during development, so they are pluripotent, but not totipotent.

Once these cells were found in mice, it was only a matter of time before human ES cells were derived as well. In 1998, the first human ES cells (hES cells) were isolated and grown in culture. Although there are differences between human and mouse ES cells, there are also substantial similarities. These embryonic stem cells hold great promise for regenerative medicine based on their potential to produce any cell type as described here. These cells have also been the source of much controversy and ethical discussion due to their embryonic origin.

Differentiation in culture

In addition to their possible therapeutic uses, ES cells offer a way to study the differentiation process in culture. The manipulation of these cells by additions to the culture media will allow us to tease out the factors involved in differentiation at the level of the actual cell undergoing the process. Early attempts at assessing differentiation in culture was plagued by the culture conditions. The medium in early experiments contained fetal calf serum (common in tissue culture), which is ill-defined, and varies lot-to-lot. More recently, more defined culture conditions have been found that allow greater reproducibility in controlling differentiation in culture.

Using more defined media, ES cells have been used to recapitulate in culture the early events in mouse development. Thus mouse ES cells can be used to first give rise to ectoderm, endoderm, and mesoderm, then these three cell types will give rise to the different cells each germ layer is determined to become. This work is in early stages but is tremendously exciting as it offers the promise of understanding the molecular cues that are involved in the stepwise determination of different cell types.

In humans, ES cells have been used to give rise to a variety of cell types in culture. For example, human ES cells have been shown to give rise to different kinds of blood cells in culture. Work is under way to produce hematopoietic stem cells in culture, which could be used to replace such cells in patients with diseases that affect blood cells. Human ES cells have also been used to produce cardiomyocytes in culture. These cells could be used to replace damaged heart tissue after heart attacks.

Learning Outcomes Review 19.3

Cell differentiation is preceded by determination, where the cell becomes committed to a developmental pathway, but has not yet differentiated. Differential inheritance of cytoplasmic factors can cause determination and differentiation, as can interactions between neighboring cells (induction). Inductive changes are mediated by signaling molecules that trigger transduction pathways. Stem cells are able to divide indefinitely, and they can give rise to differentiated cells. Embryonic stem cells are pluripotent cells that can give rise to all adult structures.

- How could you distinguish whether a cell becomes determined by induction or because of cytoplasmic factors?

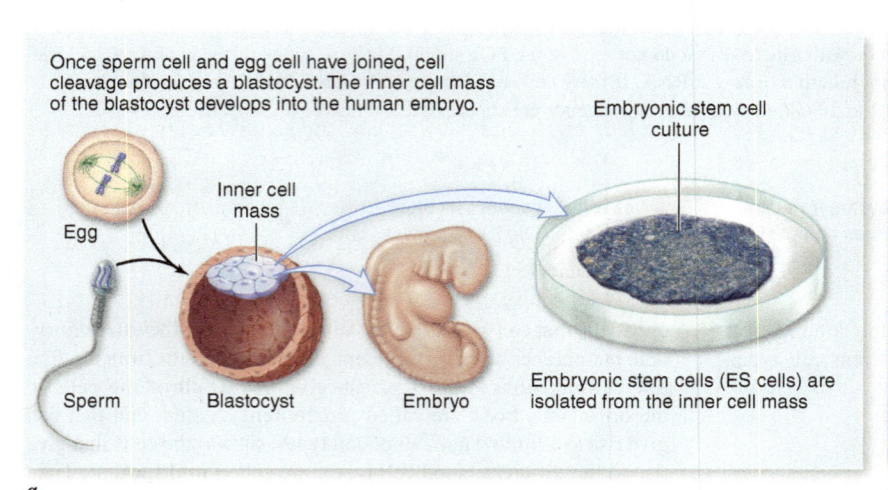

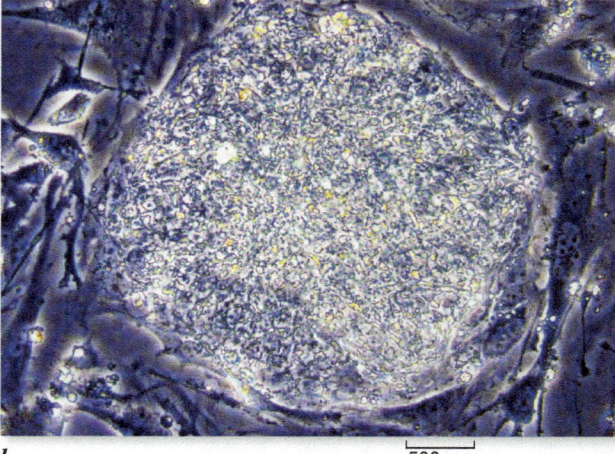

Figure 19.8 Isolation of embryonic stem cells. *a.* Early cell divisions lead to the blastocyst stage that consists of an outer layer and an inner cell mass, which will go on to form the embryo. Embryonic stem cells (ES cells) can be isolated from this stage by disrupting the embryo and plating the cells. Stem cells removed from a six-day blastocyst can be established in culture and maintained indefinitely in an undifferentiated state. *b.* Human embryonic stem cells. This mass in the photograph is a colony of undifferentiated human embryonic stem cells being studied in the developmental biologist James Thomson's research lab at the University of Wisconsin–Madison.

19.4 Nuclear Reprogramming

Learning Outcomes
1. Contrast different methods of nuclear reprogramming.
2. Differentiate between reproductive and therapeutic cloning.

The process of going from a single-celled zygote to a complex multicellular animal does not involve any irreversible changes to the DNA. This implies that the changes that underlie determination and differentiation are **epigenetic** processes. An epigenetic change does not change the DNA, but is stable through cell division. These types of changes include DNA methylation, and the modification of histone proteins, which can affect chromatin structure and gene expression (see chapter 16). During the normal course of development, this epigenetic program changes from differentiated adult cells, to germ cells, to the zygote itself.

Reversal of determination has allowed cloning

The experimental reprogramming of nuclei has a long and interesting history. This is of interest to understand the basic process, and also raises the prospect of creating patient-specific populations of specific cell types to replace cells lost to disease or trauma. This has led to a fascinating path with many twists and turns that continues today.

Experiments carried out in the 1950s showed that single cells from fully differentiated tissue of an adult plant could develop into entire, mature plants. The cells of an early cleavage stage mammalian embryo are also totipotent. When mammalian embryos naturally split in two, identical twins result. If individual blastomeres are separated from one another, any one of them can produce a completely normal individual. In fact, this type of procedure has been used to produce sets of four or eight identical offspring in the commercial breeding of particularly valuable lines of cattle.

Early research in amphibians

An early question in developmental biology was whether the production of differentiated cells during development involved irreversible changes to cells. Experiments carried out in the 1950s by Robert Briggs and Thomas King, and by John Gurdon in the 1960s and 1970s showed nuclei could be transplanted between cells. Using very fine pipettes (hollow glass tubes), these researchers sucked the nucleus out of a frog or toad egg and replaced the egg nucleus with a nucleus sucked out of a body cell taken from another individual.

The conclusions from these experiments are somewhat contradictory. On the one hand, cells do not appear to undergo any truly irreversible changes, such as loss of genes. On the other hand, the more differentiated the cell type, the less successful the nucleus in directing development when transplanted. This led to the concept of *nuclear reprogramming,* that is, a nucleus from a differentiated cell undergoes epigenetic changes that must be reversed to allow the nucleus to direct development. The early work on amphibians showed that tadpoles' intestinal cell nuclei could be reprogrammed to produce viable adult frogs. These animals not only can be considered clones, but they show that tadpole nuclei can be completely reprogrammed. However, nuclei from adult differentiated cells could only be reprogrammed to produce tadpoles, but not viable, fertile adults. Thus this work showed that adult nuclei have remarkable developmental potential, but cannot be reprogrammed to be totipotent.

Early research in mammals

Given the work done in amphibians, much effort was put into nuclear transfer in mammals, primarily mice and cattle. Not only did this not result in reproducible production of cloned animals, but this work led to the discovery of imprinting through the production of embryos with only maternal or paternal input (see chapter 13 for more information on imprinting). These embryos never developed, and showed different kinds of defects depending on whether the maternal or paternal genome was the sole contributor.

Successful nuclear transplant in mammals

These results stood until a sheep was cloned using the nucleus from a cell of an early embryo in 1984. The key to this success was in picking a donor cell very early in development. This exciting result was soon replicated by others in a host of other organisms, including pigs and monkeys. Only early embryo cells seemed to work, however.

Geneticists at the Roslin Institute in Scotland reasoned that the egg and donated nucleus would need to be at the same stage of the cell cycle for successful development. To test this idea, they performed the following procedure (figure 19.9):

1. They removed differentiated mammary cells from the udder of a six-year-old sheep. The cells were grown in tissue culture, and then the concentration of serum nutrients was substantially reduced for five days, causing them to pause at the beginning of the cell cycle.
2. In parallel preparation, eggs obtained from a ewe were enucleated.
3. Mammary cells and egg cells were surgically combined in a process called **somatic cell nuclear transfer (SCNT)** in January of 1996. Mammary cells and eggs were fused to introduce the mammary nucleus into egg.
4. Twenty-nine of 277 fused couplets developed into embryos, which were then placed into the reproductive tracts of surrogate mothers.
5. A little over five months later, on July 5, 1996, one sheep gave birth to a lamb named Dolly, the first clone generated from a fully differentiated animal cell.

Dolly matured into an adult ewe, and she was able to reproduce the old-fashioned way, producing six lambs. Thus, Dolly

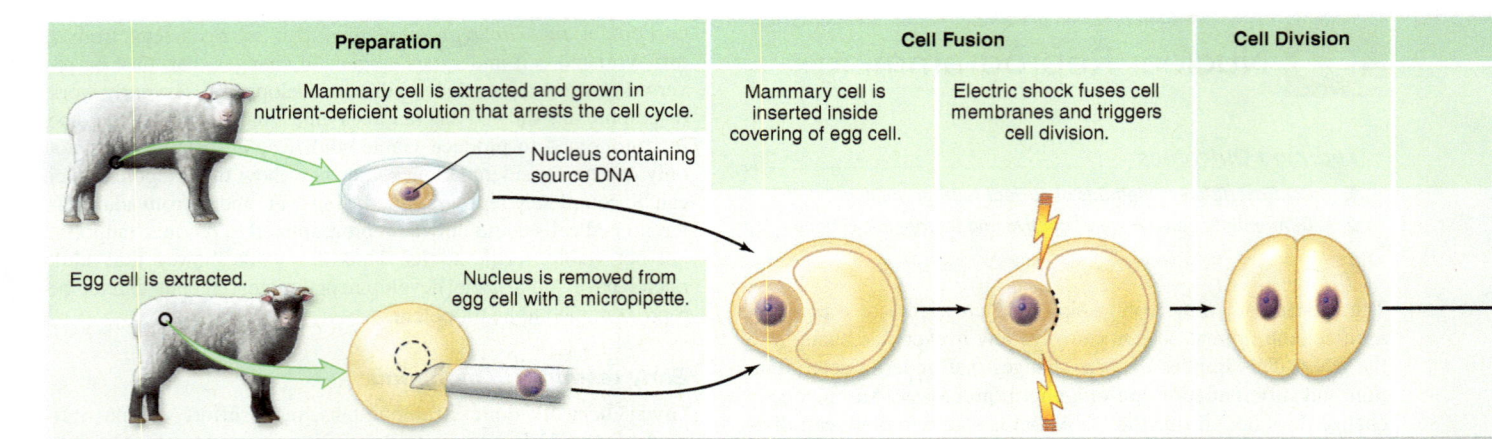

Figure 19.9 Proof that determination in animals is reversible. Scientists combined a nucleus from an adult mammary cell with an enucleated egg cell to successfully clone a sheep, named Dolly, who grew to be a normal adult and bore healthy offspring. This experiment, the first successful cloning of an adult animal, shows that a differentiated adult cell can be used to drive all of development.

Data analysis The sheep used for the donor nucleus had a different pattern of pigmentation than the donor egg. Why is this important, and which animal should Dolly resemble?

established beyond all dispute that determination in animals is reversible—that with the right techniques, the nucleus of a fully differentiated cell *can* be reprogrammed to be totipotent.

Reproductive cloning has inherent problems

The term **reproductive cloning** refers to the process just described, in which scientists use SCNT to create an animal that is genetically identical to another animal. Since the announcement of Dolly's birth in 1997, scientists have successfully cloned one or more cats, dogs, rabbits, rats, mice, cattle, goats, pigs, and mules. All of these procedures used some form of adult cell.

Low success rate and age-associated diseases

The efficiency in all reproductive cloning is quite low—only 3 to 5% of adult nuclei transferred to donor eggs result in live births. In addition, many clones that are born usually die soon thereafter of liver failure or infections. Many become oversized, a condition known as *large offspring syndrome (LOS)*. In 2003, three of four cloned piglets developed to adulthood, but all three suddenly died of heart failure at less than six months of age.

Dolly herself was euthanized at the relatively young age of six. Although she was put down because of virally induced lung cancer, she had been diagnosed with advanced-stage arthritis a year earlier. Thus, one difficulty in using genetic engineering and cloning to improve livestock is production of enough healthy animals.

Lack of imprinting

One reason for these problems lies in a phenomenon discussed in chapter 13: *genomic imprinting*. Imprinted genes are expressed differently depending on parental origin—that is, they are turned off in either egg or sperm, and this "setting" continues through development into the adult. Normal mammalian development depends on precise genomic imprinting.

During normal development, there are a variety of epigenetic programming and reprogramming events that occur. The cloning protocol attempts to sidestep this and return the nucleus of a differentiated cell directly back to the state of a zygote. This involves changes in the state of the chromatin, and probably its organization as well. Until we understand all that this process involves, it is difficult to pinpoint the precise reasons for the low efficiency.

Nuclear reprogramming has been accomplished by use of defined factors

Stimulated by the discovery of ES cells and success in the reproductive cloning of mammals, work turned to finding ways to reprogram adult cells to pluripotency without the use of embryos (figure 19.10). One approach was to fuse an ES cell to a differentiated cell. These fusion experiments showed that the nucleus of the differentiated cell could be reprogrammed by exposure to ES cell cytoplasm. Of course, the resulting cells are tetraploid (four copies of the genome), which limits their experimental and practical utility. Another line of research showed that primordial germ cells explanted into culture can give rise to cells that act similar to ES cells after extended time in culture.

All of these different lines of inquiry showed that reprogramming of somatic nuclei was possible. Investigations into the characteristics of pluripotency identified a set of transcription factors that were active in ES cells. Then in 2006 Shinya Yamanaka and his coworkers showed that introducing genes that encode four

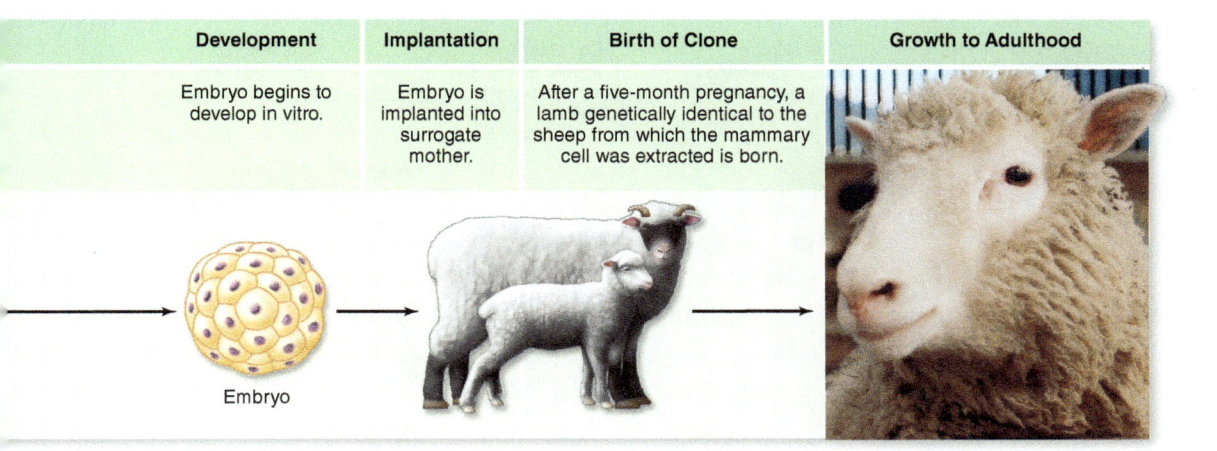

Development	Implantation	Birth of Clone	Growth to Adulthood
Embryo begins to develop in vitro.	Embryo is implanted into surrogate mother.	After a five-month pregnancy, a lamb genetically identical to the sheep from which the mammary cell was extracted is born.	

of these transcription factors, —Oct4, Sox2, c-Myc, and Klf4— could reprogram fibroblast cells in culture. Following introduction of the transcription factors genes, cells were selected that express a target gene regulated by Oct4 and Sox2, and these cells appeared to be pluripotent. These were named induced pluripotent stem cells, or iPS cells.

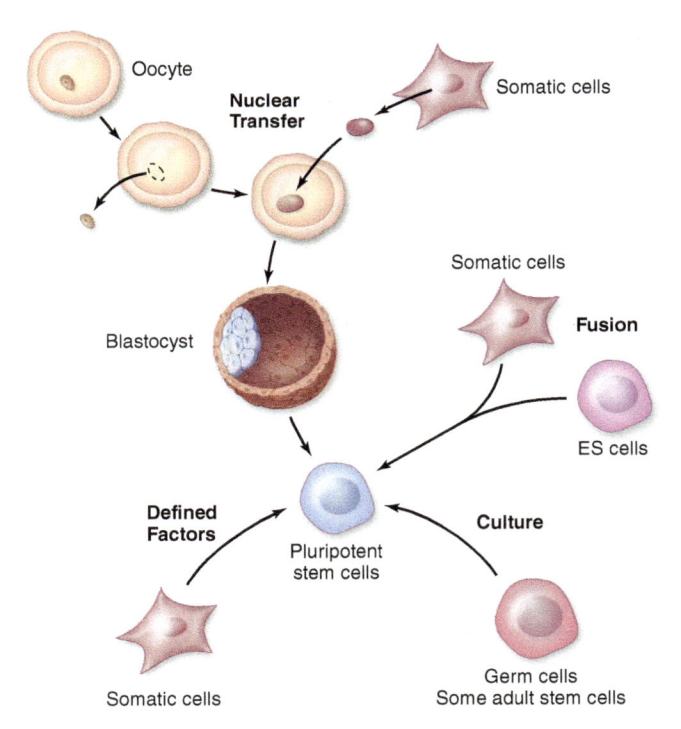

Figure 19.10 Methods to reprogram adult cell nuclei. Cells taken from adult organisms can be reprogrammed to pluripotent cells in a number of different ways. Nuclei from somatic cells can be transplanted into oocytes as during cloning. Somatic cells can be fused to ES cells created by some other means. Germ cells, and some adult stem cells, after prolonged culture appear to be reprogrammed. Recent work has shown that somatic cells in culture can be reprogrammed by introduction of defined factors.

The protocol has been refined by selection for another target gene known to be critical to the pluripotent state: *Nanog*. These *Nanog*-expressing iPS cells appear to be similar to ES cells in terms of developmental potential, as well as gene expression pattern. There is some indication that their chromatin structure, and thus their epigenetic state, may not be the same as that of ES cells.

It is worth asking what this work has taught us about the pluripotent state and the differentiated state. It is becoming clear that reprogramming is a multistep process. When this is done in culture, only a subset of cells make each step, thus explaining why the entire process is inefficient. Starting from a fibroblast, cells first change shape, becoming more spherical, and divide more rapidly. They then reverse part of their developmental program, becoming more like epithelial cells, a so-called mesenchyme-to-epithelial transition.

Lastly, the stable expression of the core pluripotency regulatory factors Oct4, Sox2, and Nanog is established. The pluripotent state is maintained by a combination of transcription factors and chromatin structure (epigenetic changes). Whole-genome analysis of epigenetic markers has revealed changes in methylation and acetylation of histones, but the significance of specific modifications is not clear. In addition, iPS cells do not appear to have the same epigenetic signature as ES cells. The significance of this is also not clear.

This technology has now been used to construct ES cells from patients with the inherited neurological disorder spinal muscular atrophy. These ES cells differentiate in culture into motor neurons that show the phenotype of the disease. The ability to derive disease-specific stem cells is an incredible advance for research on such diseases. This will allow us to study the cells affected by genetic diseases and to screen for possible therapeutics.

Pluripotent cells offer potential of tissue replacement

Pluripotent cell types themselves have potential for therapeutic applications. One way to solve the problem of graft rejection, such as in skin grafts in severe burn cases, is to produce patient-specific lines of ES cells. The first method to accomplish this was called **therapeutic cloning** and it uses the same SCNT procedure that created Dolly to assemble an embryo. The nucleus is removed from a skin cell and inserted into an egg whose nucleus has already been removed. The egg with its skin cell nucleus is allowed to form a blastocyst-stage embryo. This artificial embryo is then used to derive ES cells for transfer to injured tissue (figure 19.11).

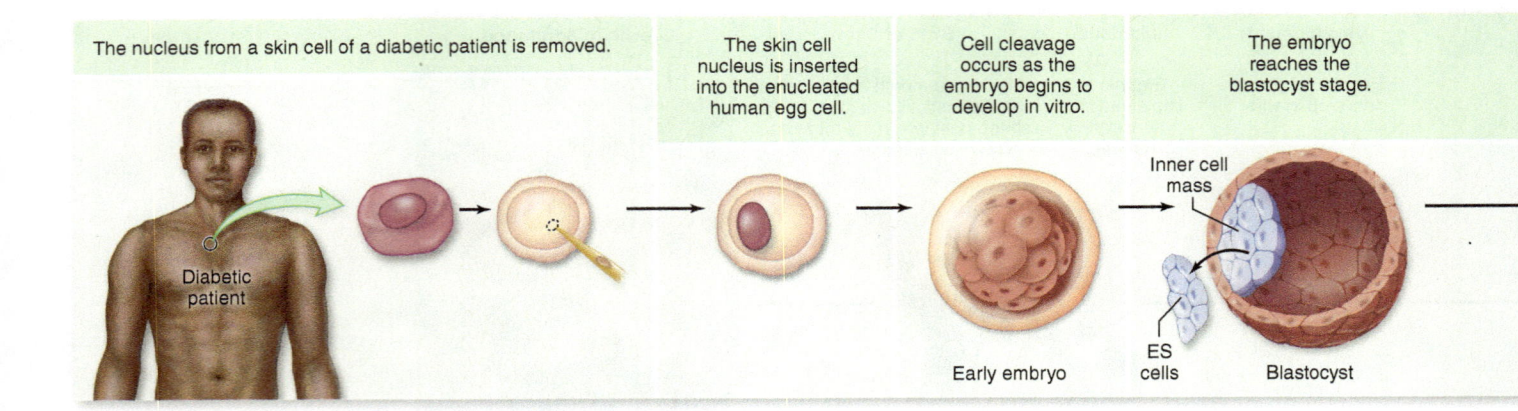

Therapeutic cloning successfully addresses one key problem that must be solved before stem cells can be used to repair human tissues damaged by heart attack, nerve injury, diabetes, or Parkinson disease—the problem of immune acceptance. Since stem cells are cloned from a person's own tissues, they pass the immune system's "self" identity check, and the body readily accepts them. A clinical trial is under way in Japan to use iPS cells to treat blindness. The patients suffer from age-related macular degeneration, the most common cause of blindness in the elderly. This is the first clinical trial using this technology, and early reports suggest some improvement of vision. We are a long way from tissue engineering, but this is a first step in that direction.

Learning Outcomes Review 19.4

Cloning has long been practiced in plants. In animals, cells from early-stage embryos are also totipotent, but attempts to use adult nuclei for cloning led to mixed results. The nucleus of a differentiated cell requires reprogramming to be totipotent. This appears to be necessary at least in part because of genomic imprinting. Nuclei may be reprogrammed by fusion with an embryonic stem cell, which produces a tetraploid cell, or through the introduction of four important transcription factors. That reprogramming is possible was shown by reproductive cloning via somatic cell nuclear transfer (SCNT). In therapeutic cloning, the goal is to produce replacement tissue using a patient's own cells.

- *What changes must occur to produce a totipotent cell from a differentiated nucleus?*

19.5 Pattern Formation

Learning Outcomes

1. Describe A/P axis formation in *Drosophila*.
2. Describe D/V axis formation in *Drosophila*.
3. Explain the importance of homeobox-containing genes in development.

For cells in multicellular organisms to differentiate into appropriate cell types, they must gain information about their relative locations in the body. All multicellular organisms seem to use positional information to determine the basic pattern of body compartments and, thus, the overall architecture of the adult body. This positional information then leads to intrinsic changes in gene activity, so that cells ultimately adopt a fate appropriate for their location.

Pattern formation is an unfolding process. In the later stages, it may involve morphogenesis of organs, but during the earliest events of development, the basic body plan is laid down, along with the establishment of the anterior–posterior (A/P, head-to-tail) axis and the dorsal–ventral (D/V, back-to-front) axis. Thus, pattern formation can be considered the process of taking a radially symmetrical cell and imposing two perpendicular axes to define the basic body plan, which in this way becomes bilaterally symmetrical. Developmental biologists use the term **polarity** to refer to the acquisition of axial differences in developing structures.

The fruit fly *Drosophila melanogaster* is the best understood animal in terms of the genetic control of early patterning. We will concentrate on the *Drosophila* system here, and later in chapter 53 we will examine axis formation in vertebrates in the context of their overall development.

A hierarchy of gene expression that begins with maternally expressed genes controls the development of *Drosophila*. To understand the details of these gene interactions, we first need to briefly review the stages of *Drosophila* development.

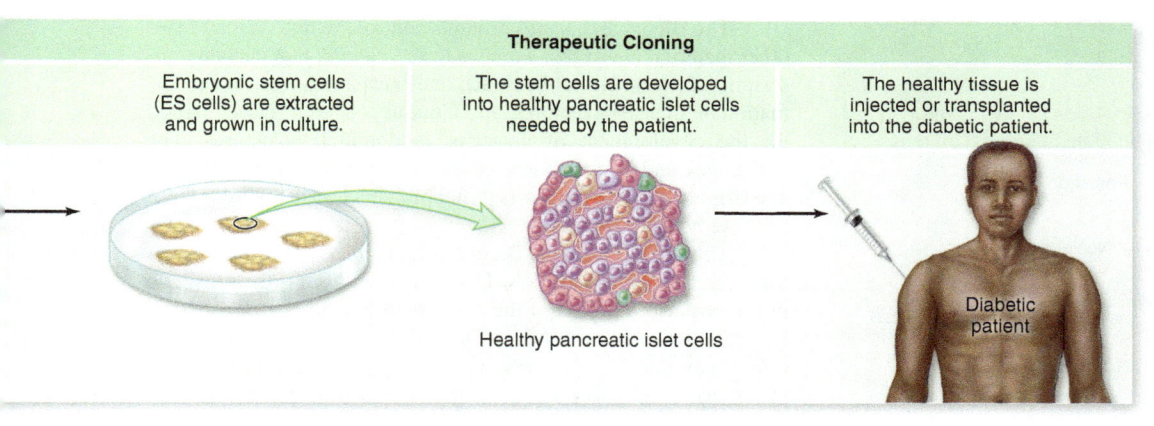

Figure 19.11 How human embryos might be used for therapeutic cloning. In therapeutic cloning, after initial stages to reproductive cloning, the embryo is broken apart and its embryonic stem cells are extracted. These are grown in culture and used to replace the diseased tissue of the individual who provided the DNA. This is useful only if the disease in question is not genetic, as the stem cells are genetically identical to the patient.

Drosophila embryogenesis produces a segmented larva

Drosophila and many other insects produce two different kinds of bodies during their development: the first, a tubular eating machine called a **larva,** and the second, an adult flying sex machine with legs and wings. The passage from one body form to the other, called **metamorphosis,** involves a radical shift in development (figure 19.12). In this chapter, we concentrate on the process of going from a fertilized egg to a larva, which is termed *embryogenesis.*

Prefertilization maternal contribution

The development of an insect like *Drosophila* begins before fertilization, with the construction of the egg. Specialized *nurse cells* that help the egg grow move some of their own maternally encoded mRNAs into the maturing oocyte (figure 19.12a).

Following fertilization, the maternal mRNAs are transcribed into proteins, which initiate a cascade of sequential gene activations. Embryonic nuclei do not begin to function (that is, to direct new transcription of genes) until approximately 10 nuclear divisions have occurred. Therefore, the action of maternal, rather than zygotic, genes determines the initial course of *Drosophila* development.

Postfertilization events

After fertilization, 12 rounds of nuclear division without cytokinesis produce about 4000 nuclei, all within a single cytoplasm. All of the nuclei within this **syncytial blastoderm** (figure 19.12b) can freely communicate with one another, but nuclei located in different sectors of the egg encounter different maternal products.

Once the nuclei have spaced themselves evenly along the surface of the blastoderm, membranes grow between them to form the **cellular blastoderm.** Embryonic folding and primary tissue development soon follow, in a process fundamentally similar to that seen in vertebrate development. Within a day of fertilization, embryogenesis creates a segmented, tubular body—which is destined to hatch out of the protective coats of the egg as a larva.

Morphogen gradients form the basic body axes in *Drosophila*

Pattern formation in the early *Drosophila* embryo requires positional information encoded in labels that can be read by cells. The unraveling of this puzzle, work that earned the 1995 Nobel Prize for researchers Christiane Nüsslein-Volhard and Eric Wieschaus, is summarized in figure 19.13. We now know that two different genetic pathways control the establishment of A/P and D/V polarity in *Drosophila*.

Anterior–posterior axis

Formation of the A/P axis begins during maturation of the oocyte and is based on opposing gradients of two different proteins: **Bicoid** and **Nanos.** These protein gradients are established by an interesting mechanism.

Nurse cells in the ovary secrete maternally produced *bicoid* and *nanos* mRNAs into the maturing oocyte where they are differentially transported along microtubules to opposite poles of the oocyte (figure 19.14a). This differential transport comes about due to the use of different motor proteins to move the two mRNAs. The *bicoid* mRNA then becomes anchored in the cytoplasm at the end of the oocyte closest to the nurse cells, and this end will develop into the anterior end of the embryo. *Nanos* mRNA becomes anchored to the opposite end of the oocyte, which will become the posterior end of the embryo. Thus, by the end of oogenesis, the *bicoid* and *nanos* mRNAs are already set to function as cytoplasmic determinants in the fertilized egg (figure 19.14b).

Following fertilization, translation of the anchored mRNA and diffusion of the proteins away from their respective sites of synthesis create opposing gradients of each protein: Highest levels of Bicoid protein are at the anterior pole of the embryo (figure 19.15a), and highest levels of the Nanos protein are at the posterior pole. Concentration gradients of soluble molecules can specify different cell fates along an axis, and proteins that act in this way, like Bicoid and Nanos, are called **morphogens.** The importance of these morphogens can be seen by the effects of loss-of-function mutants: Loss of Bicoid produces an embryo with only posterior sides, and loss of Nanos protein produces an embryo with only anterior sides.

The Bicoid and Nanos proteins control the translation of two other maternal messages, *hunchback* and *caudal*, that encode transcription factors. **Hunchback** activates genes required for the formation of anterior structures, and **Caudal** activates genes required for the development of posterior (abdominal) structures. The *hunchback* and *caudal* mRNAs are evenly distributed across the egg (figure 19.15b), so how is it that proteins translated from these mRNAs become localized?

The answer is that Bicoid protein binds to and inhibits translation of *caudal* mRNA. Therefore, *caudal* is only translated in the posterior regions of the egg where Bicoid is absent. Similarly, Nanos protein binds to and prevents translation of the *hunchback* mRNA. As a result, *hunchback* is only translated in the anterior regions of the egg (figure 19.15c). Thus, shortly after fertilization, four protein gradients exist in the embryo: anterior–posterior gradients of Bicoid and Hunchback proteins, and posterior–anterior gradients of Nanos and Caudal proteins (figure 19.15c).

Dorsal–ventral axis

The dorsal–ventral axis in *Drosophila* is established by actions of the *dorsal* gene product. Once again the process begins in the ovary, when maternal transcripts of the *dorsal* gene are put into the oocyte. However, unlike *bicoid* or *nanos*, the *dorsal* mRNA does not become asymmetrically localized. Instead, a series of steps are required for Dorsal to carry out its function.

First, the oocyte nucleus, which is located to one side of the oocyte, synthesizes *gurken* mRNA. The *gurken* mRNA then accumulates in a crescent between the nucleus and the membrane on that side of the oocyte (figure 19.16a). This will be the future dorsal side of the embryo.

The Gurken protein is a soluble cell-signaling molecule, and when it is translated and released from the oocyte, it binds to receptors in the membranes of the overlying follicle cells (figure 19.16b). These cells then differentiate into a dorsal morphology. Meanwhile, no Gurken signal is released from the other side of the oocyte, and the follicle cells on that side of the oocyte adopt a ventral fate.

Following fertilization, a signaling molecule is differentially activated on the ventral surface of the embryo in a complex sequence of steps. This signaling molecule then binds to a membrane receptor in the ventral cells of the embryo and activates a signal transduction pathway in those cells. Activation of this pathway results in the selected transport of the Dorsal protein (which is everywhere) into ventral nuclei, forming a gradient along the D/V axis. The Dorsal protein levels are highest in the nuclei of ventral cells (figure 19.16c).

(Note that many *Drosophila* genes are named for the mutant phenotype that results from a loss of function in that gene. A lack of *dorsal* function produces dorsalized embryos with no ventral structures.)

The Dorsal protein is a transcription factor, and once it is transported into nuclei, it activates genes required for the proper development of ventral structures, simultaneously repressing genes that specify dorsal structures. Hence, the product of the *dorsal* gene ultimately directs the development of ventral structures.

Although profoundly different mechanisms are involved, the unifying factor controlling the establishment of both A/P and D/V polarity in *Drosophila* is that *bicoid*, *nanos*, *gurken*, and *dorsal* are

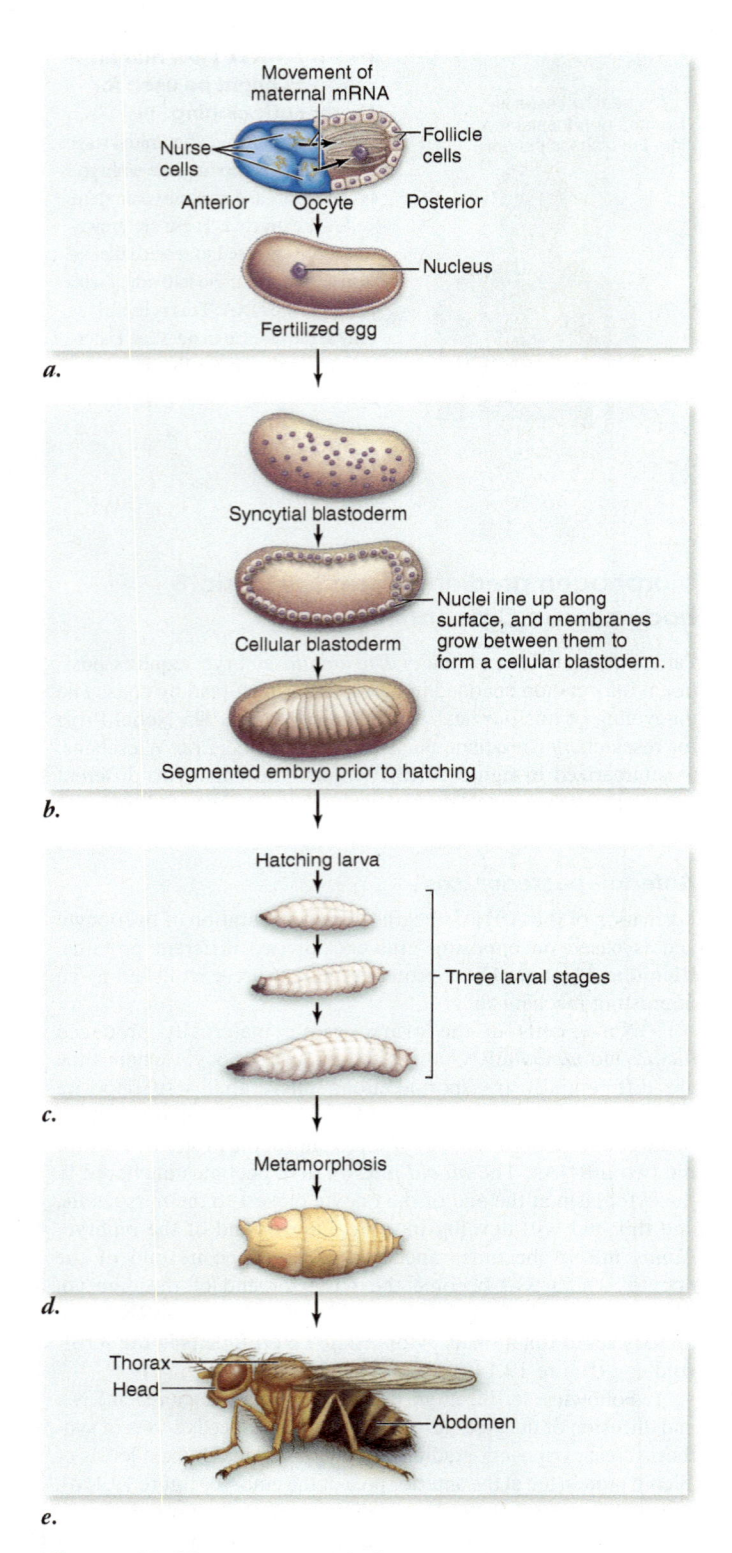

Figure 19.12 The path of fruit fly development. Major stages in the development of *Drosophila melanogaster* include formation of the (*a*) egg, (*b*) syncytial and cellular blastoderm, (*c*) larval instars, (*d*) pupa and metamorphosis into (*e*) a sexually mature adult.

Establishing the Polarity of the Embryo

Fertilization of the egg triggers the production of Bicoid protein from maternal RNA in the egg. The Bicoid protein diffuses through the egg, forming a gradient. This gradient determines the polarity of the embryo, with the head and thorax developing in the zone of high concentration (*green* fluorescent dye in antibodies that bind bicoid protein allows visualization of the gradient).

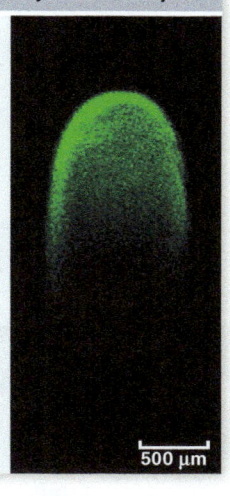

500 μm

Setting the Stage for Segmentation

About 2½ hours after fertilization, Bicoid protein turns on a series of brief signals from so-called gap genes. The gap proteins act to divide the embryo into large blocks. In this photo, fluorescent dyes in antibodies that bind to the gap proteins Krüppel (*orange*) and Hunchback (*green*) make the blocks visible; the region of overlap is yellow.

500 μm

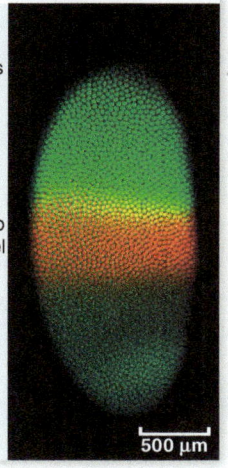

Figure 19.13 Body organization in an early *Drosophila* embryo. In these fluorescent microscope images by 1995 Nobel laureate Christiane Nüsslein-Volhard and Sean Carroll, we watch a *Drosophila* egg pass through the early stages of development, in which the basic segmentation pattern of the embryo is established. The proteins in the photographs were made visible by binding fluorescent antibodies to each specific protein.

Laying Down the Fundamental Regions

About 0.5 hr later, the gap genes switch on the "pair-rule" genes, which are each expressed in seven stripes. This is shown for the pair-rule gene *hairy*. Some pair-rule genes are only required for even-numbered segments while others are only required for odd numbered segments.

500 μm

Forming the Segments

The final stage of segmentation occurs when a "segment-polarity" gene called *engrailed* divides each of the seven regions into anterior and posterior compartments of the future segments. This occurs a little after the formation of the cellular blastoderm (see figure 19.12). The curved appearance of the embryo at this stage is because of a phenomenon called germ band extension that causes the embryo to fold over itself.

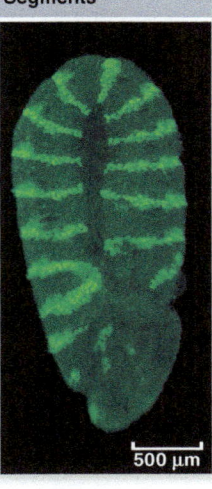

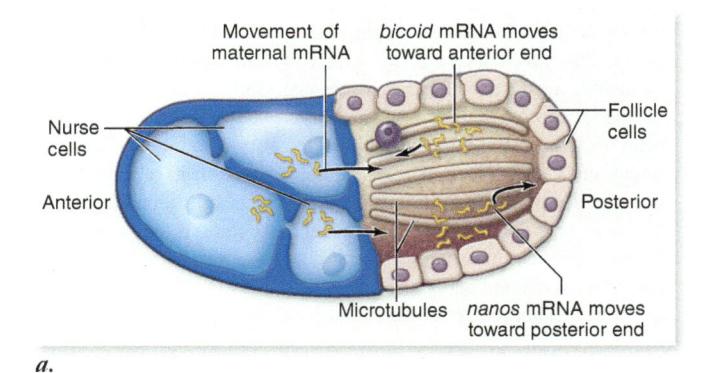

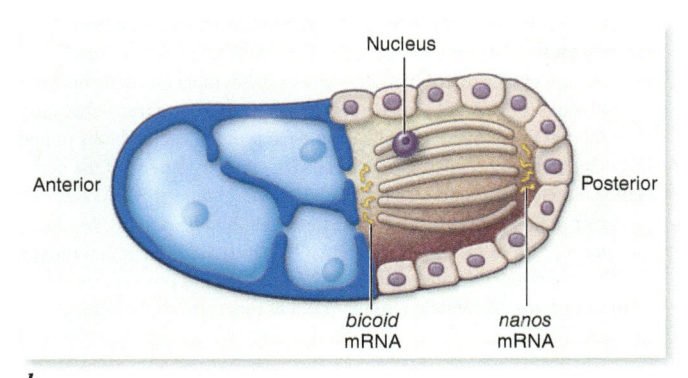

500 μm

Figure 19.14 Specifying the A/P axis in *Drosophila* embryos I. *a.* In the ovary, nurse cells secrete maternal mRNAs into the cytoplasm of the oocyte. Clusters of microtubules direct oocyte growth and maturation. Motor proteins travel along the microtubules transporting molecules in two directions. *Bicoid* mRNAs are transported toward the anterior pole of the oocyte, *nanos* mRNA is transported toward the posterior pole of the oocyte. *b.* A mature oocyte, showing localization of *bicoid* mRNAs to the anterior pole and *nanos* mRNAs to the posterior pole.

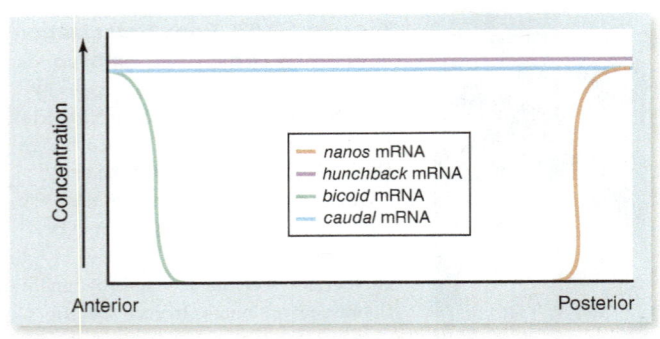

a. Oocyte mRNAs

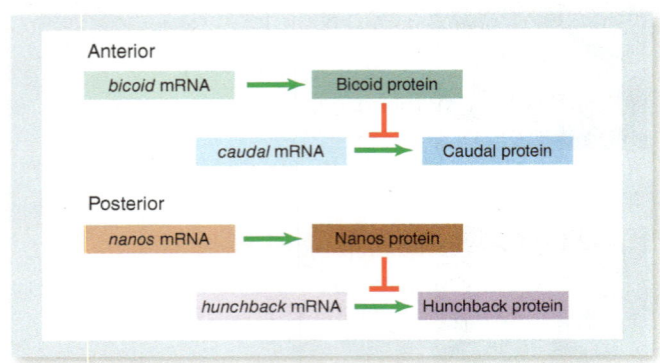

b. After fertilization

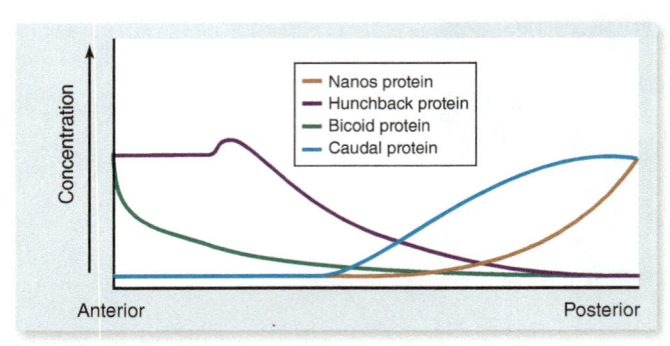

c. Early cleavage embryo proteins

Figure 19.15 Specifying the A/P axis in *Drosophila* embryos II. *a.* Unlike *bicoid* and *nanos*, *hunchback* and *caudal* mRNAs are evenly distributed throughout the cytoplasm of the oocyte. *b.* Following fertilization, *bicoid* and *nanos* mRNAs are translated into protein, making opposing gradients of each protein. Bicoid binds to and represses translation of *caudal* mRNAs (in anterior regions of the egg). Nanos binds to and represses translation of *hunchback* mRNAs (in posterior regions of the egg). *c.* Translation of *hunchback* mRNAs in anterior regions of the egg will create a Hunchback gradient that mirrors the Bicoid gradient. Translation of *caudal* mRNAs in posterior regions of the embryo will create a Caudal gradient that mirrors the Nanos gradient.

all maternally expressed genes. The polarity of the future embryo in both instances is therefore laid down in the oocyte using information coming from the maternal genome.

The preceding discussion simplifies events, but the outline is clear: Polarity is established by the creation of morphogen

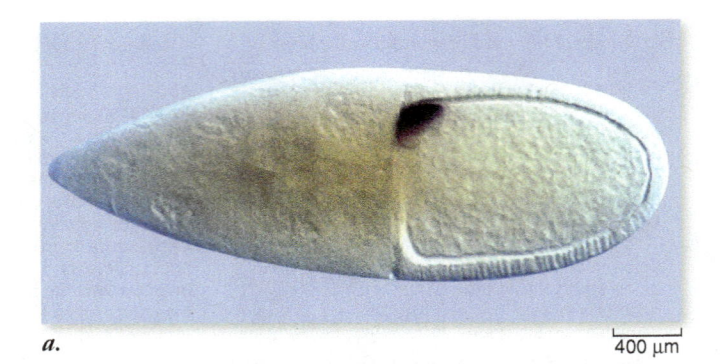

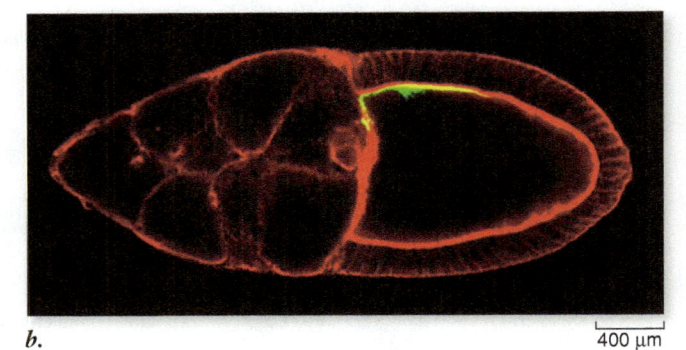

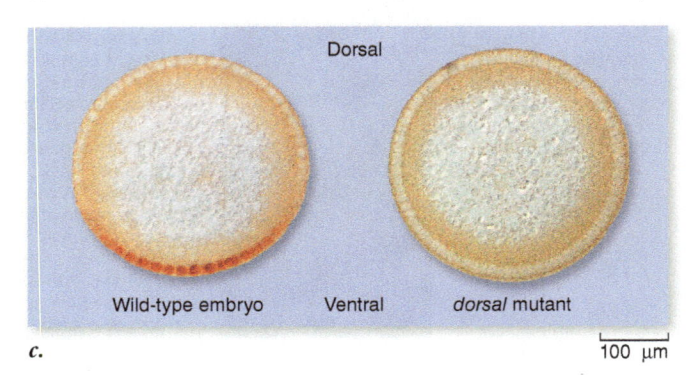

Figure 19.16 Specifying the D/V axis in *Drosophila* embryos. *a.* The *gurken* mRNA *(dark stain)* is concentrated between the oocyte nucleus (not visible) and the dorsal, anterior surface of the oocyte. *b.* In a more mature oocyte, Gurken protein *(yellow stain)* is secreted from the dorsal anterior surface of the oocyte, forming a gradient along the dorsal surface of the egg. Gurken then binds to membrane receptors in the overlying follicle cells. Double staining for actin *(red)* shows the cell boundaries of the oocyte, nurse cells, and follicle cells. *c.* For these images, cellular blastoderm stage embryos were cut in cross section to visualize the nuclei of cells around the perimeter of the embryos. Dorsal protein *(dark stain)* is localized in nuclei on the ventral surface of the blastoderm in a wild-type embryo *(left)*. The *dorsal* mutant on the right will not form ventral structures, and Dorsal is not present in ventral nuclei of this embryo.

gradients in the embryo based on maternal information in the egg. These gradients then drive the expression of the zygotic genes that will actually pattern the embryo. This reliance on a hierarchy of regulatory genes is a unifying theme for all of development.

The body plan is produced by sequential activation of genes

Let us now return to the process of pattern formation in *Drosophila* along the A/P axis. Determination of structures is accomplished by the sequential activation of three classes of **segmentation genes.** These genes create the hallmark segmented body plan of a fly, which consists of three fused head segments, three thoracic segments, and eight abdominal segments (see figure 19.12e).

To begin, Bicoid protein exerts its profound effect on the organization of the embryo by activating the translation and transcription of *hunchback* mRNA (which is the first mRNA to be transcribed after fertilization). *Hunchback* is a member of a group of nine genes called the **gap genes.** These genes map out the initial subdivision of the embryo along the A/P axis (see figure 19.13).

All of the gap genes encode transcription factors, which, in turn, regulate the expression of eight or more **pair-rule genes.** Each of the pair-rule genes, such as *hairy,* produces seven distinct bands of protein, which appear as stripes when visualized with fluorescent reagents (see figure 19.13). These bands subdivide the broad gap regions and establish boundaries that divide the embryo into seven zones. When mutated, each of the pair-rule genes alters every other body segment.

All of the pair-rule genes also encode transcription factors, and they, in turn, regulate the expression of each other and of a group of nine or more **segment polarity genes.** The segment polarity genes are each expressed in 14 distinct bands of cells, which subdivide each of the seven zones specified by the pair-rule genes (see figure 19.13). The *engrailed* gene, for example, divides each of the seven zones established by *hairy* into anterior and posterior compartments. The segment polarity genes encode proteins that function in cell–cell signaling pathways. Thus, they function in inductive events—which occur *after* the syncytial blastoderm is divided into cells—to fix the anterior and posterior fates of cells within each segment.

In summary, within 3 hr after fertilization, a highly orchestrated cascade of segmentation gene activity transforms the broad gradients of the early embryo into a periodic, segmented structure with A/P and D/V polarity. The activation of the segmentation genes depends on the free diffusion of maternally encoded morphogens, which is only possible within the syncytial blastoderm of the early *Drosophila* embryo.

Figure 19.17 Mutations in homeotic genes. Three separate mutations in the bithorax complex caused this fruit fly to develop an additional second thoracic segment, with accompanying wings.

Segment identity arises from the action of homeotic genes

With the basic body plan laid down, the next step is to give identity to the segments of the embryo. A highly interesting class of *Drosophila* mutants has provided the starting point for understanding the creation of segment identity.

In these mutants, a particular segment seems to have changed its identity—that is, it has characteristics of a different segment. In wild-type flies, a pair of legs emerges from each of the three thoracic segments, but only the second thoracic segment has wings. Mutations in the *Ultrabithorax* gene cause a fly to grow an extra pair of wings, as though it has two second thoracic segments (figure 19.17). Even more bizarre are mutations in *Antennapedia,* which cause legs to grow out of the head in place of antennae!

Thus, mutations in these genes lead to the appearance of perfectly normal body parts in inappropriate places. Such mutants are termed *homeotic mutants* because the transformed body part looks similar (homeotic) to another. The genes in which such mutants occur are therefore called **homeotic genes.**

Homeotic gene complexes

In the early 1950s, geneticist and Nobel laureate Edward Lewis discovered that several homeotic genes, including *Ultrabithorax,* map together on the third chromosome of *Drosophila* in a tight cluster called the **bithorax complex.** Mutations in these genes all affect body parts of the thoracic and abdominal segments, and Lewis concluded that the genes of the bithorax complex control the development of body parts in the rear half of the thorax and all of the abdomen.

Interestingly, the order of the genes in the bithorax complex mirrors the order of the body parts they control, as though the genes are activated serially. Genes at the beginning of the cluster switch on development of the thorax; those in the middle control the anterior part of the abdomen; and those at the end affect the posterior tip of the abdomen.

A second cluster of homeotic genes, the **Antennapedia complex,** was discovered in 1980 by Thomas Kaufman. The Antennapedia complex governs the anterior end of the fly, and the order of genes in this complex also corresponds to the order of segments they control (figure 19.18a).

The homeobox

An interesting relationship was discovered after the genes of the bithorax and Antennapedia complexes were cloned and sequenced. These genes all contain a conserved sequence of 180 nucleotides that codes for a 60-amino-acid, DNA-binding domain. Because this domain was found in all of the homeotic genes, it was named the *homeodomain,* and the DNA that encodes it is called the homeobox. Thus, the term ***Hox*** **gene** now refers to a homeobox-containing gene that specifies the identity of a body part. These genes function as transcription factors that bind DNA using their homeobox domain.

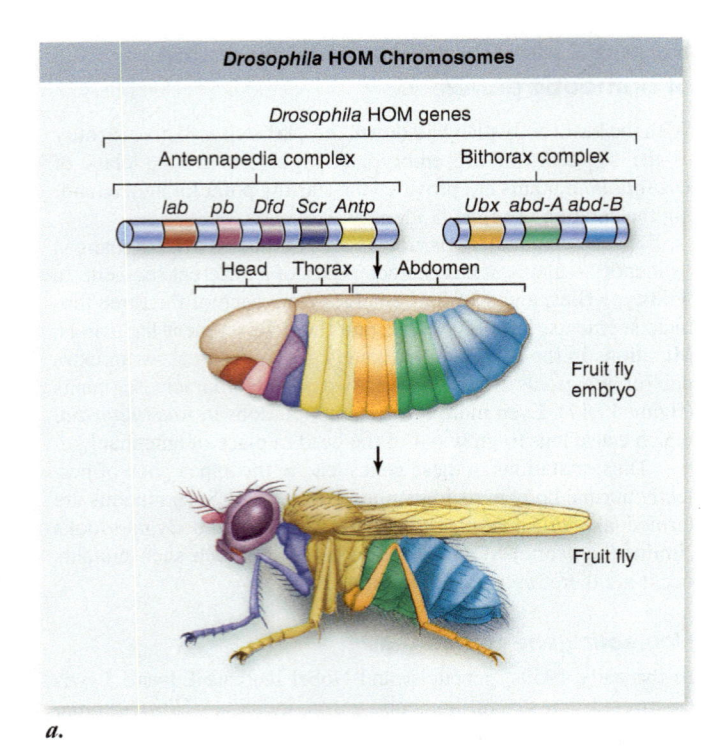

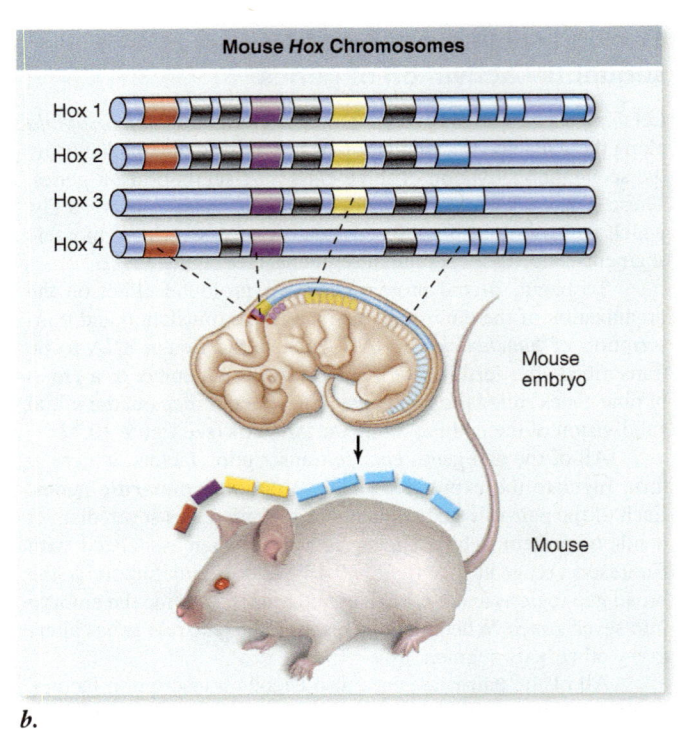

Figure 19.18 A comparison of homeotic gene clusters in the fruit fly *Drosophila melanogaster* and the mouse *Mus musculus*. *a. Drosophila* homeotic genes. Called the homeotic gene complex, or HOM complex, the genes are grouped into two clusters: the Antennapedia complex (anterior) and the bithorax complex (posterior). *b.* The *Drosophila* HOM genes and the mouse *Hox* genes are related genes that control the regional differentiation of body parts in both animals. These genes are located on a single chromosome in the fly and on four separate chromosomes in mammals. In this illustration, the genes are color-coded to match the parts of the body along the A/P axis in which they are expressed. Note that the order of the genes along the chromosome(s) is mirrored by their pattern of expression in the embryo and in structures in the adult fly.

Clearly, the homeobox distinguishes portions of the genome that are devoted to pattern formation. How the *Hox* genes do this is the subject of much current research. Scientists believe that the ultimate targets of *Hox* gene function must be genes that control cell behaviors associated with organ morphogenesis.

Evolution of homeobox-containing genes

A large amount of research has been devoted to analyzing the clustered complexes of *Hox* genes in other organisms. These investigations have led to a fairly coherent view of homeotic gene evolution.

It is now clear that the *Drosophila* bithorax and Antennapedia complexes represent two parts of a single cluster of genes. In vertebrates, there are four copies of *Hox* gene clusters. As in *Drosophila*, the spatial domains of *Hox* gene expression correlate with the order of the genes on the chromosome (figure 19.18b). The existence of four *Hox* clusters in vertebrates is viewed by many as evidence that two duplication events of the entire genome have occurred in the vertebrate lineage.

This idea raises the issue of when the original cluster arose. To answer this question, researchers have turned to more primitive organisms, such as *Amphioxus* (now called *Branchiostoma*), a lancelet chordate (see chapter 35). The finding of only one cluster of *Hox* genes in *Amphioxus* implies that indeed there have been two duplications in the vertebrate lineage, at least of the *Hox* cluster. Given the single cluster in arthropods, this finding implies that the common ancestor to all animals with bilateral symmetry had a single *Hox* cluster as well.

The next logical step is to look at even more-primitive animals: the radially symmetrical cnidarians such as *Hydra* (see chapter 34). Thus far, *Hox* genes have been found in a number of cnidarian species, and sequence analyses suggest that cnidarian *Hox* genes are also arranged into clusters. Thus, the appearance of the ancestral *Hox* cluster likely preceded the divergence between radial and bilateral symmetries in animal evolution.

Pattern formation in plants is also under genetic control

The evolutionary split between plant and animal cell lineages occurred about 1.6 BYA, before the appearance of multicellular organisms with defined body plans. The implication is that multicellularity evolved independently in plants and animals. Because of the activity of meristems, additional modules can be added to plant bodies throughout their lifetimes. In addition, plant flowers and roots have a radial organization, in contrast to the bilateral symmetry of most animals. We may therefore expect that the genetic control of pattern formation in plants is fundamentally different from that of animals.

Although plants have homeobox-containing genes, they are not organized into complexes of *Hox* genes similar to the ones that determine regional identity of developing structures in animals. Instead, the predominant homeotic gene family in plants appears to be the **MADS-box** genes.

MADS-box genes are a family of transcriptional regulators found in most eukaryotic organisms, including plants, animals, and fungi. The MADS-box is a conserved DNA-binding and dimerization domain, named after the first five genes to be discovered with this domain. Only a small number of *MADS*-box genes are found in animals, where their functions include the control of cell proliferation and tissue-specific gene expression in postmitotic muscle cells. They do not appear to play a role in the patterning of animal embryos.

In contrast, the number and functional diversity of *MADS*-box genes increased considerably during the evolution of land plants, and there are more than 100 *MADS*-box genes in the *Arabidopsis* genome. In flowering plants, the *MADS*-box genes dominate the control of development, regulating such processes as the transition from vegetative to reproductive growth, root development, and floral organ identity.

Although distinct from genes in the *Hox* clusters of animals, plant *Hox* genes encode transcription factors that have important developmental functions. One such example is the family of *knottedlike homeobox (knox)* genes, which are important regulators of shoot apical meristem development in both seed-bearing and non-seed-bearing plants. Mutations that affect expression of *knox* genes produce changes in leaf and petal shape, suggesting that these genes play an important role in generating leaf form.

Learning Outcomes Review 19.5

Pattern formation in animals involves the coordinated expression of a hierarchy of genes. Gradients of morphogens in *Drosophila* specify A/P and D/V axes, then lead to sequential activation of segmentation genes. Bicoid and Nanos protein gradients determine the A/P axis. The protein Dorsal determines the D/V axis, but activation requires a series of steps beginning with the oocyte's Gurken protein. The action of homeotic genes provide segment identity. These genes, which include a DNA-binding homeodomain sequence, are called *Hox* genes (for *homeobox* genes), and they are organized into clusters. Plants use a different set of developmental control genes called *MADS*-box genes.

- *Why would you expect homeotic genes to be conserved across species evolution?*

19.6 Morphogenesis

Learning Outcomes

1. Discuss the importance of cell shape changes and cell migration in development.
2. Explain how cell death can contribute to morphogenesis.
3. Describe the role of the extracellular matrix in cell migration.

At the end of cleavage, the *Drosophila* embryo still has a relatively simple structure: It comprises several thousand identical-looking cells, which are present in a single layer surrounding a central yolky region. The next step in embryonic development is **morphogenesis**—the generation of ordered form and structure.

Morphogenesis is the product of changes in cell structure and cell behavior. Animals regulate the following processes to achieve morphogenesis:

- the number, timing, and orientation of cell divisions;
- cell growth and expansion;
- changes in cell shape;
- cell migration; and
- cell death.

Plant and animal cells are fundamentally different in that animal cells have flexible surfaces and can move, but plant cells are immotile and encased within stiff cellulose walls. Each cell in a plant is fixed into position when it is created. Thus, animal cells use cell migration extensively during development, whereas plants use the other four mechanisms but lack cell migration. We consider the morphogenetic changes in animals here, and plant morphogenesis is detailed in chapter 41.

Cell division during development may result in unequal cytokinesis

The orientation of the mitotic spindle determines the plane of cell division in eukaryotic cells. The coordinated function of microtubules and their motor proteins determines the respective position of the mitotic spindle within a cell (see chapter 10). If the spindle is centrally located in the dividing cell, two equal-sized daughter cells will result. If the spindle is off to one side, one large daughter cell and one small daughter cell will result.

The great diversity of cleavage patterns in animal embryos is determined by differences in spindle placement. In many cases, the fate of a cell is determined by its relative placement in the embryo during cleavage. For example, in preimplantation mammalian embryos, cells on the outside of the embryo usually differentiate into trophectoderm cells, which form only extraembryonic structures later in development (for example, a part of the placenta). In contrast, the embryo proper is derived from the inner cell mass, cells which, as the name implies, are in the interior of the embryo.

Cells change shape and size as morphogenesis proceeds

In animals, cell differentiation is often accompanied by profound changes in cell size and shape. For example, the large nerve cells that connect your spinal cord to the muscles in your big toe develop long processes called *axons* that span this entire distance. The cytoplasm of an axon contains microtubules, which are used for motor-driven transport of materials along the length of the axon.

As another example, muscle cells begin as *myoblasts*, undifferentiated muscle precursor cells. They eventually undergo conversion into the large, multinucleated *muscle fibers* that make up mammalian skeletal muscles. These changes begin with the expression of the

MyoD1 gene, which encodes a transcription factor that binds to the promoters of muscle-determining genes to initiate these changes.

Programmed cell death is a necessary part of development

Not every cell produced during development is destined to survive. For example, human embryos have webbed fingers and toes at an early stage of development. The cells that make up the webbing die in the normal course of morphogenesis. As another example, vertebrate embryos produce a very large number of neurons, ensuring that enough neurons are available to make the necessary synaptic connections, but over half of these neurons never make connections and die in an orderly way as the nervous system develops.

Unlike accidental cell deaths due to injury, these cell deaths are planned—and indeed required—for proper development and morphogenesis. Cells that die due to injury typically swell and burst, releasing their contents into the extracellular fluid. This form of cell death is called **necrosis.** In contrast, cells programmed to die shrivel and shrink in a process called **apoptosis,** which means "falling away," and their remains are taken up by surrounding cells.

Genetic control of apoptosis

Apoptosis occurs when a "death program" is activated. All animal cells appear to possess such programs. In *C. elegans,* the same 131 cells always die during development in a predictable and reproducible pattern.

Work on *C. elegans* showed that three genes are central to this process. Two (*ced-3* and *ced-4*) activate the death program itself; if either is mutant, those 131 cells do not die, and go on instead to form nervous tissue and other tissue. The third gene (*ced-9*) represses the death program encoded by the other two: All 1090 cells of the *C. elegans* embryo die in *ced-9* mutants. In *ced-9/ced-3* double mutants, all 1090 cells live, which suggests that *ced-9* inhibits cell death by functioning prior to *ced-3* in the apoptotic pathway (figure 19.19a).

The mechanism of apoptosis appears to have been highly conserved during the course of animal evolution. In human nerve cells, the *Apaf1* gene is similar to *ced-4* of *C. elegans* and activates the cell death program, and the human *bcl-2* gene acts similarly to *ced-9* to repress apoptosis. If a copy of the human *bcl-2* gene is transferred into a nematode with a defective *ced-9* gene, *bcl-2* suppresses the cell death program of *ced-3* and *ced-4*.

The mechanism of apoptosis

The product of the *C. elegans ced-4* gene is a protease that activates the product of the *ced-3* gene, which is also a protease. The human *Apaf1* gene is actually named for its role: *A*poptotic *p*rotease *a*ctivating *f*actor. It activates two proteases called caspases that have a role similar to the Ced-3 protease in *C. elegans* (figure 19.19b). When the final proteases are activated, they chew up proteins in important cellular structures such as the cytoskeleton and the nuclear lamina, leading to cell fragmentation.

The role of Ced-9/Bcl-2 is to inhibit this program. Specifically, it inhibits the activating protease, preventing the activation of the destructive proteases. The entire process is thus controlled by an inhibitor of the death program.

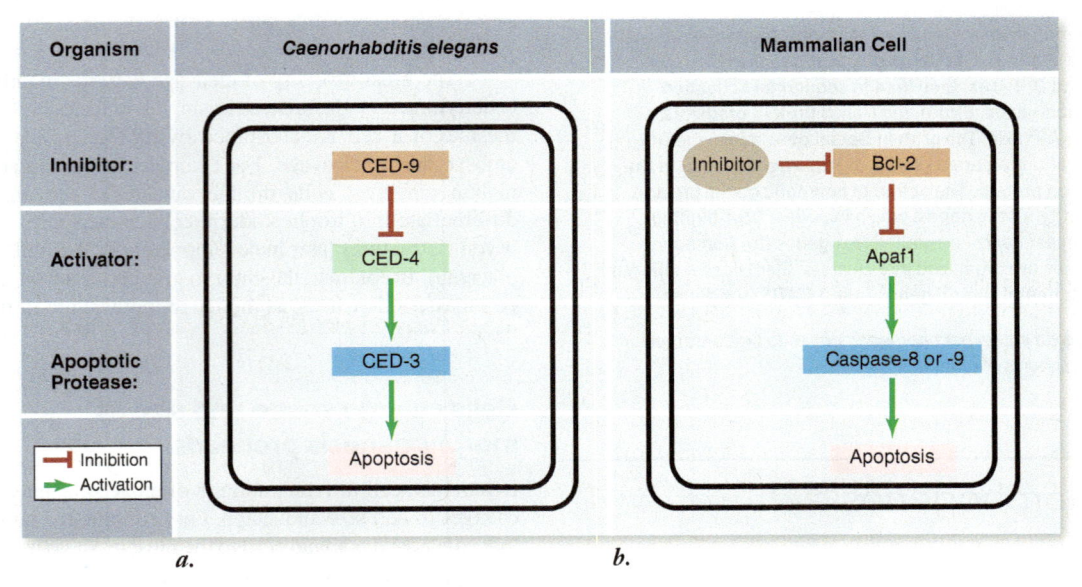

Figure 19.19 Programmed cell death pathway. Apoptosis, or programmed cell death, is necessary for the normal development of all animals. *a.* In the developing nematode, for example, two genes, *ced-3* and *ced-4*, code for proteins that cause the programmed cell death of 131 specific cells. In the other (surviving) cells of the developing nematode, the product of a third gene, *ced-9*, represses the death program encoded by *ced-3* and *ced-4*. *b.* The mammalian homologues of the apoptotic genes in *C. elegans* are *bcl-2* (*ced-9* homologue), *Apaf1* (*ced-4* homologue), and *caspase-8* or *-9* (*ced-3* homologues). In the absence of any cell survival factor, Bcl-2 is inhibited and apoptosis occurs. In the presence of nerve growth factor (NGF) and NGF receptor binding, Bcl-2 is activated, thereby inhibiting apoptosis.

Both internal and external signals control the state of the Ced-9/Bcl-2 inhibitor. For example, in the human nervous system, neurons have a cytoplasmic inhibitor of Bcl-2 that allows the death program to proceed (see figure 19.19b). In the presence of nerve growth factor, a signal transduction pathway leads to the cytoplasmic inhibitor being inactivated, allowing Bcl-2 to inhibit apoptosis and the nerve cell to survive.

Cell migration gets the right cells to the right places

The migration of cells is important during many stages of animal development. The movement of cells involves both adhesion and the loss of adhesion. Adhesion is necessary for cells to get "traction," but cells that are initially attached to others must lose this adhesion to be able to leave a site.

Cell movement also involves cell-to-substrate interactions, and the extracellular matrix may control the extent or route of cell migration. The central paradigm of morphogenetic cell movements in animals is a change in cell adhesiveness, which is mediated by changes in the composition of macromolecules in the plasma membranes of cells or in the extracellular matrix. Cell-to-cell interactions are often mediated through cadherins, but cell-to-substrate interactions often involve integrin-to-extracellular-matrix (ECM) interactions.

Cadherins

Cadherins are a large gene family, with over 80 members identified in humans. In the genomes of *Drosophila, C. elegans,* and humans, the cadherins can be sorted into several subfamilies that exist in all three genomes.

The cadherin proteins are all transmembrane proteins that share a common motif, the *cadherin domain,* a 110-amino-acid domain in the extracellular portion of the protein that mediates Ca^{2+}-dependent binding between like cadherins (homophilic binding).

Experiments in which cells are allowed to sort in vitro illustrate the function of cadherins. Cells with the same cadherins adhere specifically to one another, while not adhering to other cells with different cadherins. If cell populations with different cadherins are dispersed and then allowed to reaggregate, they sort into two populations of cells based on the nature of the cadherins on their surface.

An example of the action of cadherins can be seen in the development of the vertebrate nervous system. All surface ectoderm cells of the embryo express E-cadherin. The formation of the nervous system begins when a central strip of cells on the dorsal surface of the embryo turns off E-cadherin expression and turns on N-cadherin expression. In the process of **neurulation,** the formation of the neural tube (see chapter 53), the central strip of N-cadherin-expressing cells folds up to form the tube. The neural tube pinches off from the overlying cells, which continue to express E-cadherin. The surface cells outside the tube differentiate into the epidermis of the skin, whereas the neural tube develops into the brain and spinal cord of the embryo.

Integrins

In some tissues, such as connective tissue, much of the volume of the tissue is taken up by the spaces *between* cells. These spaces are filled with a network of molecules secreted by surrounding cells, termed a *matrix*. In connective tissue such as cartilage, long polysaccharide chains are covalently linked to proteins (proteoglycans), within which are embedded strands of fibrous protein (collagen, elastin, and fibronectin). Migrating cells traverse this matrix by binding to it with cell surface proteins called integrins.

Integrins are attached to actin filaments of the cytoskeleton and protrude out from the cell surface in pairs, like two hands. The "hands" grasp a specific component of the matrix, such as collagen or fibronectin, thus linking the cytoskeleton to the fibers of the matrix. In addition to providing an anchor, this binding can initiate changes within the cell, alter the growth of the cytoskeleton, and activate gene expression and the production of new proteins.

The process of **gastrulation** (described in detail in chapter 53), during which the hollow ball of animal embryonic cells folds in on itself to form a multilayered structure, depends on fibronectin–integrin interactions. For example, injection of antibodies against either fibronectin or integrins into salamander embryos blocks binding of cells to fibronectin in the ECM and inhibits gastrulation. The result is like a huge traffic jam following a major accident on a freeway: Cells (cars) keep coming, but they get backed up since they cannot get beyond the area of inhibition (accident site) (figure 19.20). Similarly, a targeted

---

SCIENTIFIC THINKING

Hypothesis: Fibronectin is required for cell migration during gastrulation.

Prediction: Blocking fibronectin with antifibronectin antibodies before gastrulation should prevent cell movement.

Test: Staged salamander embryos were injected either with antifibronectin antibody, or with preimmune serum as a control, prior to gastrulation. Cell movements were then monitored photographically.

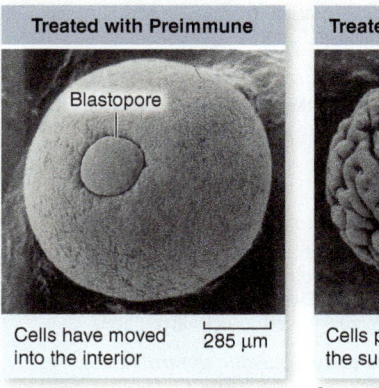

Treated with Preimmune — Blastopore — Cells have moved into the interior — 285 μm

a.

Treated with Antifibronectin — Blastopore — Cells pile up on the surface — 285 μm

b.

Result: The experimental embryos injected with antifibronectin antibody show extremely aberrant gastrulation where cells pile up and do not enter the interior of the embryo. Control embryos gastrulate normally.

Conclusion: Fibronectin is required for cells to migrate into the interior of the embryo during gastrulation.

Further Experiments: How can this same system be used to analyze the role of fibronectin in other early morphogenetic events?

Figure 19.20 Fibronectin is necessary for cell migration during gastrulation.

knockout of the fibronectin gene in mice resulted in gross defects in the migration, proliferation, and differentiation of embryonic mesoderm cells.

Thus, cell migration is largely a matter of changing patterns of cell adhesion. As a migrating cell travels, it continually extends projections that probe the nature of its environment. Tugged this way and that by different tentative attachments, the cell literally feels its way toward its ultimate target site.

Learning Outcomes Review 19.6

Morphogenesis is the generation of ordered form and structure. This process proceeds along with cell differentiation. The primary mechanisms of morphogenesis are cell shape change and cell migration. Apoptosis is programmed cell death that is a necessary part of morphogenesis. Cell migration in animals involves alternating changes in adhesion brought about by cadherins and integrins.

■ Why is cell death important to morphogenesis?

Chapter Review

19.1 The Process of Development

Development is the sequence of systematic, gene-directed changes throughout a life cycle. The four subprocesses of development are cell division, cell differentiation, pattern formation, and morphogenesis.

19.2 Cell Division

Development begins with cell division.

In animals, cleavage stage divisions divide the fertilized egg into numerous smaller cells called blastomeres. During cleavage the G_1 and G_2 phases of the cell cycle are shortened or eliminated (figure 19.2).

Every cell division is known in the development of C. elegans.

The lineage of 959 adult somatic *Caenorhabditis elegans* cells is invariant. Knowledge of the differentiation sequence and outcome allows study of developmental mechanisms.

Plant growth occurs in specific areas called meristems.

Plant growth continues throughout the life span from meristematic stem cells that can divide and differentiate into any plant tissue.

19.3 Cell Differentiation

Cells become determined prior to differentiation.

The process of determination commits a cell to a particular developmental pathway prior to its differentiation. This is not visible but can be tracked experimentally. Determination is due to differential inheritance of cytoplasmic factors or cell-to-cell interactions.

Determination can be due to cytoplasmic determinants.

In tunicates, determination of tail muscle cells depends on the presence of mRNA for the Macho-1 transcription factor, which is deposited in the egg cytoplasm during gamete formation.

Induction can lead to cell differentiation.

Induction occurs when one cell type produces signal molecules that induce gene expression in neighboring target cells.

In frogs, cells from animal and vegetal poles do not develop into mesoderm when isolated. In tunicates, signaling by the growth factor FGF induces mesoderm development.

Stem cells can divide and produce cells that differentiate.

Stem cells replace themselves by division and produce cells that differentiate. Totipotent stem cells can give rise to any cell type including extraembryonic tissues; pluripotent cells can give rise to all cells of an organism; and multipotent stem cells can give rise to many kinds of cells.

Embryonic stem cells are pluripotent cells derived from embryos.

Embryonic stem cells are derived from the inner cell mass of the blastocyst (figure 19.8). They can differentiate into any adult tissue in a mouse.

19.4 Nuclear Reprogramming

Reversal of determination has allowed cloning.

Cells undergo no irreversible changes during development. However, transplanted nuclei from older donors are less able to direct complete development. The cloning of the sheep Dolly showed that the nucleus of an adult cell can be reprogrammed to be totipotent (figure 19.9).

Reproductive cloning has inherent problems.

Reproductive cloning has a low success rate, and clones often develop age-associated diseases. Cloning is inefficient because of the difficulty of reprogramming epigenetic modifications.

Nuclear reprogramming has been accomplished by use of defined factors.

Adult cells can be converted into pluripotent cells by introduction of four genes for transcription factors. These induced pluripotent cells appear to be similar to ES cells.

Pluripotent cells offer potential of tissue replacement

The use of cells cloned from a patient's cells to replace damaged tissue could avoid the problem of transplant rejection. Clinical trials are underway to treat a form of blindness using iPS cells.

19.5 Pattern Formation

Drosophila embryogenesis produces a segmented larva.

The maternal contribution of mRNA along with the postfertilization events of cellular blastoderm formation produce a segmented embryo.

Morphogen gradients form the basic body axes in Drosophila.

Pattern formation produces two perpendicular axes in a bilaterally symmetrical organism. Positional information leads to changes in gene activity so cells adopt a fate appropriate for their location.

Formation of the anterior/posterior (A/P) axis is based on opposing gradients of morphogens, Bicoid and Nanos, synthesized from maternal mRNA (figures 19.14, & 19.15).

The dorsal/ventral (D/V) axis is established by a gradient of the Dorsal transcription factor. Successive action of transcription factors divides the embryo into segments.

The body plan is produced by sequential activation of genes.

Segment identity arises from the action of homeotic genes.

Homeotic genes, called *Hox* genes because they contain a DNA sequence called the homeobox, give identity to embryo segments.

Hox genes are found in four clusters in vertebrates.

Pattern formation in plants is also under genetic control.

Plants have *MADS*-box genes that control the transition from vegetative to reproductive growth, root development, and floral organ identity.

19.6 Morphogenesis

Cell division during development may result in unequal cytokinesis.

Cells change shape and size as morphogenesis proceeds.

Depending on the orientation of the mitotic spindle, cells of equal or different sizes can arise. Morphogenesis involves changes in cell shape and size and cell migration.

Programmed cell death is a necessary part of development.

Apoptosis, the programmed death of cells, removes structures once they are no longer needed (figure 19.19).

Cell migration gets the right cells to the right places.

The migration of cells requires both adhesion and loss of adhesion between cells and their substrate.

Cell-to-cell interactions are often mediated by cadherin proteins, whereas cell-to-substrate interactions may involve integrin-to-extracellular-matrix interactions.

Integrins bind to fibers found in the extracellular matrix. This action can alter the cytoskeleton and activate gene expression.

Review Questions

UNDERSTAND

1. During development, cells become
   a. differentiated before they become determined.
   b. determined before they become differentiated.
   c. determined by the loss of genetic material.
   d. differentiated by the loss of genetic material.

2. Determination can occur by
   a. the action of cytoplasmic determinants.
   b. induction by other cells.
   c. the loss of chromosomes during cell division.
   d. Both a and b are correct.

3. The rapid divisions that occur early in development are made possible by shortening
   a. M phase.
   b. S phase.
   c. G_1 and G_2 phases.
   d. All of the choices are correct.

4. A pluripotent cell is one that can
   a. give rise to every cell type in an organisms body.
   b. produce an indefinite supply of a single cell type.
   c. produce a limited amount of a specific cell type.
   d. produce multiple cell types.

5. Plant meristems
   a. are only present during development.
   b. contain stem cells.
   c. undergo meiosis.
   d. All of the choices are correct.

6. Pattern formation involves cells determining their position in the embryo. One mechanism that can accomplish this is
   a. the loss of genetic material.
   b. alterations of chromosome structure.
   c. gradients of morphogens.
   d. changes in the cell cycle.

7. The process of nuclear reprogramming
   a. is a normal part of pattern formation.
   b. reverses the changes that occur during differentiation.
   c. requires the introduction of new DNA.
   d. is not possible with mammalian cells.

APPLY

1. What is the common theme in cell determination by induction or cytoplasmic determinants?
   a. The activation of transcription factors
   b. The activation of cell division
   c. A change in gene expression
   d. Both a and c are correct.

2. The process of reproductive cloning
   a. shows that nuclear reprogramming is possible.
   b. is very efficient in mammals.
   c. always produces adult animals that are identical to the donor.
   d. Both a and b are correct.

3. Production of anterior–posterior and dorsal–ventral axes in the fruit fly *Drosophila*
   a. both use gradients of mRNA.
   b. are conceptually similar but mechanistically different.
   c. use the exact same mechanisms.
   d. both use gradients of protein.

4. For pattern formation to occur, the cells in the developing embryo must
   a. "know" their position in the embryo.
   b. be determined during the earliest divisions.
   c. differentiate as they are "born."
   d. must all be reprogrammed after each cell division.

5. The genes that encode the morphogen gradients in *Drosophila* were all identified in mutant screens. A mutation that removes the gradient necessary for the A/P morphogen gradient would be expected to
   a. affect the larvae but not the adult.
   b. affect the adult but not the larvae.
   c. be lethal and lead to an abnormal embryo.
   d. produce replacement of one adult structure with another.

6. What would be the likely result of a mutation of the *bcl-2* gene on the level of apoptosis?
   a. No change
   b. A decrease in apoptosis
   c. An increase in apoptosis
   d. An initial decrease, followed by an increase in apoptosis

7. *MADS*-box, and *Hox* genes are
   a. found only in plants and animals, respectively.
   b. found only in animals and plants, respectively.
   c. have similar roles in development in plants and animals, respectively.
   d. have similar roles in development in animals and plants, respectively.

SYNTHESIZE

1. The fate map for *C. elegans* (refer to figure 19.3) diagrams development of a multicellular organism from a single cell. Use this fate map to determine the number of cell divisions required to establish the population of cells that will become (a) the nervous system and (b) the gonads.

2. Carefully examine the *C. elegans* fate map in figure 19.3. Notice that some of the branchpoints (daughter cells) do *not* go on to produce more cells. What is the cellular mechanism underlying this pattern?

3. You have generated a set of mutant embryonic mouse cells. Predict the developmental consequences for each of the following mutations.
   a. Knockout mutation for N-cadherin
   b. Knockout mutation for integrin
   c. Deletion of the cytoplasmic domain of integrin

4. Assume you have the factors in hand necessary to reprogram an adult cell, and the factors necessary to induce differentiation to any cell type. How could these be used to replace a specific damaged tissue in a human patient?

Evolution

Part **IV** Evolution

CHAPTER **20**

Genes Within Populations

Chapter Contents

- 20.1 Genetic Variation and Evolution
- 20.2 Changes in Allele Frequency
- 20.3 Five Agents of Evolutionary Change
- 20.4 Quantifying Natural Selection
- 20.5 Natural Selection's Role in Maintaining Variation
- 20.6 Selection Acting on Traits Affected by Multiple Genes
- 20.7 Experimental Studies of Natural Selection
- 20.8 Interactions Among Evolutionary Forces
- 20.9 The Limits of Selection

Introduction

No other human is exactly like you (unless you have an identical twin). Often the particular characteristics of an individual have an important bearing on its survival, on its chances to reproduce, and on the success of its offspring. Evolution is driven by such factors, as different alleles rise and fall in populations. These deceptively simple matters lie at the core of evolutionary biology, which is the topic of this chapter and chapters 21 through 25.

20.1 Genetic Variation and Evolution

Learning Outcomes

1. *Define* evolution *and* population genetics.
2. *Explain the difference between evolution by natural selection and the inheritance of acquired characteristics.*

The term *genetic variation* refers to the different alleles of genes found within individuals of a population. Natural populations contain a wealth of such variation. In this chapter, we explore genetic variation in natural populations and consider the evolutionary forces that cause allele frequencies in natural populations to change.

The word *evolution* is widely used in the natural and social sciences. It refers to how an entity—be it a social system, a gas, or a planet—changes through time. Although development of the modern concept of evolution in biology can be traced to Darwin's landmark work, *On the Origin of Species,* the first five editions of his book never actually used the term. Rather, Darwin used the phrase "descent with modification."

Although many more complicated definitions have been proposed, Darwin's words probably best capture the essence of biological evolution: Through time, species accumulate differences; as a result, descendants differ from their ancestors. In this way, new species arise from existing ones.

Many processes can lead to evolutionary change

You have already learned about the development of Darwin's ideas in chapter 1. Darwin was not the first to propose a theory of evolution. Rather, he followed a long line of earlier philosophers and naturalists who deduced that the many kinds of organisms around us were produced by a process of evolution.

Unlike his predecessors, however, Darwin proposed natural selection as the mechanism of evolution. Natural selection produces evolutionary change when some individuals in a population possess certain inherited characteristics and then produce more surviving offspring than individuals lacking these characteristics. As a result, the population gradually comes to include more and more individuals with the advantageous characteristics. In this way, the population evolves and becomes better adapted to its local circumstances.

A rival theory, championed by the prominent biologist Jean-Baptiste Lamarck, was that evolution occurred by the **inheritance of acquired characteristics.** According to Lamarck, changes that individuals acquired during their lives were passed on to their offspring. For example, Lamarck proposed that ancestral giraffes with short necks tended to stretch their necks to feed on tree leaves, and this extension of the neck was passed on to subsequent generations, leading to the long-necked giraffe (figure 20.1, left). In Darwin's theory, by contrast, the variation is not created by experience, but is the result of preexisting genetic differences among individuals (figure 20.1, right).

One way to monitor how populations change through time is to look at changes in the frequencies of alleles of a gene from one generation to the next. Natural selection, by favoring individuals with certain alleles, can lead to change in such *allele frequencies,* but it is not the only process that can do so. Allele frequencies can also change when mutations occur repeatedly, changing one allele to another, and when migrants bring alleles into a population. In addition, when populations are small, the frequencies of alleles can change randomly as the result of chance events. Often, natural selection overwhelms the effects of these other processes, but as you will see in this is not always the case.

Evolution can result from any process that causes a change in the genetic composition of a population. We cannot talk about evolution, therefore, without also considering **population genetics,** the study of the properties of genes in populations.

Populations contain ample genetic variation

Biologists have always wanted to know how much genetic variation exists in natural populations. The ability to ask this question has been limited by the techniques available to analyze variation at different levels: proteins, genes, and now genomes. The story of genetic variation in natural populations is told using increasingly sophisticated tools for detecting differences.

Initial approaches to understanding variation examined the most obvious differences—the morphological. In many cases, such phenotypic variation is the result of underlying genetic differences. In such cases, natural populations usually show substantial genetic variation, as figure 20.2 illustrates.

Other examples are closer to home. For example, the human blood groups are the result of genetic differences among individuals. Chemical analysis has revealed the existence of more than 30 blood

LAMARCK'S THEORY:
Acquired variation is passed on to descendants.

Proposed ancestor of giraffes has characteristics of modern-day okapi. → Stretching → The giraffe ancestor lengthened its neck by stretching to reach tree leaves its.

→ Reproduction →

The longer neck that resulted from stretching is passed on to offspring

DARWIN'S THEORY:
Natural selection or genetically based variation leads to evolutionary change.

Some individuals born happen to have longer necks due to genetic differences.

→ Reproduction →

Individuals pass on their traits to next generation.

Natural selection → Reproduction →

Over many generations, longer-necked individuals are more successful, perhaps because they can feed on taller trees, and pass the long-neck trait on to their offspring.

Figure 20.1 Two ideas of how giraffes might have evolved long necks.

Figure 20.2 Polymorphic variation. This natural population of lupines, *Lupinus,* exhibits considerable variation in flower color. Individual differences are inherited and passed on to offspring.

group genes in humans, in addition to the ABO locus. At least one third of these genes are routinely found in several alternative allelic forms in human populations. In addition to these, more than 45 variable genes encode other proteins in human blood cells and plasma that are not considered blood groups. In short, the considerable variation that exists in blood cells and plasma indicate that many genetically variable genes are present in this one system alone.

The first approach to directly assay genetic variation within populations was the use of electrophoresis to separate proteins produced by alternative alleles of enzyme-encoding genes (see chapter 17). As DNA analysis tools were developed in the laboratory, they were adapted for use with samples collected from natural populations. This led to, in order of increasing detail, examining restriction fragment length polymorphisms (RFLPs, see chapter 17), sequencing specific genes, and, most recently, sequencing entire genomes.

One of the most useful tools, both for genetic mapping and for the analysis of population-level variation, has been single-nucleotide polymorphisms (SNPs). These are defined as single-base differences between individuals that exist in the population at more than 1% (see chapter 24). An initial large-scale international effort identified several million SNPs in the human genome.

Such variation is also being assayed in many other species of interest. The results for most species are similar to those in humans: The closer you look, the more variation you see. One reason for the appeal of SNPs is that we have been able to automate the analysis of multiple samples. This approach is being applied to many species of scientific and economic interest. The U.S. storehouse for such information, the National Center for Biotechnology Information (NCBI), now has a database with SNPs found in over 100 species, revealing the near ubiquity of genetic variation and providing tools for assessing patterns in human and natural populations.

As genomes are sequenced more thoroughly, more and more variation is being found. Sequencing the complete genome of individuals, rather than just parts of it, is becoming easier and more affordable; consequently researchers are now beginning to compare the extent of genetic variation within populations throughout the entire genome.

Not surprisingly, this effort is most advanced for human populations. The 1000 Genomes Project established the goal of initially sequencing the genome of 1000 people in a preliminary phase, with the ultimate goal of including 2500 people from five major areas of the world. This project is still in its early stages, but already extensive genetic variation has been documented. Overall, the project has detected among all individuals 14 million SNPs, 1.3 million short sections in which DNA has been inserted or deleted, and 20,000 genes in which allelic differences among individuals cause major changes in the proteins the gene produces. In addition, each individual was found, on average, to have 250 to 300 alleles that render a gene inoperative, as well as 50 to 100 alleles known to be associated with inherited disorders. Moreover, some parts of the genome harbor much more variation than other parts. The populations examined differed relatively little in genetic variation, but some differences occur in a few genes, some of which appear to be related to adaptation to different environments.

Studies of genomic variation in natural populations are just beginning and are of lesser scope, although this will no doubt change in the years to come. One study of wild populations of *Drosophila melanogaster* sequenced the entire genomes of individuals in two populations, one from North Carolina (37 individuals) and the other from Africa (6 individuals). The researchers found extensive variation in both *D. melanogaster* populations, although interestingly, some regions of the genome were more variable than others—just as with humans—and patterns of variation differed between the populations. Undoubtedly, the number of species for which many individuals are sequenced will skyrocket in the next few years, substantially enhancing our understanding of population variation.

Learning Outcomes Review 20.1

Evolution can be described as descent with modification. Natural selection occurs when individuals carrying certain alleles leave more offspring than those without the alleles. Natural populations generally contain considerable amounts of genetic variation. Population genetics studies this variability through statistical analyses.

- *Why is genetic variation in a population necessary for evolution to occur?*

20.2 Changes in Allele Frequency

Learning Outcomes

1. *Explain the Hardy–Weinberg principle.*
2. *Describe the characteristics of a population that is in Hardy–Weinberg equilibrium.*
3. *Demonstrate how the operation of evolutionary processes can be detected.*

Genetic variation within natural populations was a puzzle to Darwin and his contemporaries in the mid-1800s. Although Mendel performed his experiments during this same time period, his work was largely unknown. Selection, scientists then thought, should always favor an optimal form, and so tend to eliminate variation. Moreover, the theory of *blending inheritance*—in which offspring were expected to be phenotypically intermediate relative to their parents—was widely accepted. If blending inheritance were correct, then the effect of any new genetic variant would quickly be diluted to the point of disappearance in subsequent generations. For example, if a mutation made an individual 4 inches taller, its offspring, the result of a mating with an individual without that mutation, would be only 2 inches taller, and in the next generation's offspring would be only 1 inch taller, and so on, until the effect completely disappeared.

The Hardy–Weinberg principle allows prediction of genotype frequencies

Following the rediscovery of Mendel's research, two people in 1908 solved the puzzle of why genetic variation persists—Godfrey H. Hardy, an English mathematician, and Wilhelm Weinberg, a German physician. These workers were initially confused about why, after many generations, a population didn't come to be composed solely of individuals with the dominant phenotype. The conclusion they independently came to was that the original proportions of the genotypes in a population will remain constant from generation to generation, as long as the following assumptions are met:

1. No mutation takes place.
2. No genes are transferred to or from other sources (no immigration or emigration takes place).
3. Mating is random (individuals do not choose mates based on their phenotype or genotype).
4. The population size is very large.
5. No selection occurs.

Because the genotypes' proportions do not change, they are said to be in **Hardy–Weinberg equilibrium.**

The Hardy–Weinberg equation with two alleles: A binomial expansion

In algebraic terms, the Hardy–Weinberg principle is written as an equation. Consider a population of 100 cats in which 84 are black and 16 are white. The frequencies of the two phenotypes would be 0.84 (or 84%) black and 0.16 (or 16%) white. Based on these phenotypic frequencies, can we deduce the underlying frequency of genotypes?

If we assume that the white cats are homozygous recessive for an allele we designate as b, and the black cats are either homozygous dominant BB or heterozygous Bb, we can calculate the frequencies of the two alleles in the population from the proportion of black and white individuals, assuming that the population is in Hardy–Weinberg equilibrium.

Let the letter p designate the frequency of the B allele and the letter q the frequency of the alternative allele. We call these **gene frequencies.** Because there are only two alleles, p plus q must always equal 1 (that is, the total population). In addition, we know that the sum of the three genotype frequencies must also equal 1. The probability that an individual will have two B alleles is simply the probability that each of its alleles is a B. The probability of two events happening independently is the product of the probability of each event; in this case, the probability that the individual received a B allele from its father is equal to the frequency of the allele in the population, p, and the probability the individual received a B allele from its mother is also p, so the probability that both happened is $p * p = p^2$ (figure 20.3). By the same reasoning, the probability that an individual will have two b alleles is q^2.

What about the probability that an individual will be a heterozygote? There are two ways this could happen: The individual could receive a B from its father and a b from its mother, or vice versa. The probability of the first case is $p * q$ and the probability of

	Generation One		
Phenotypes	84%	16%	
Genotypes	BB	Bb	bb
Frequency of genotype in population	0.36	0.48	0.16
Frequency of gametes	0.36 + 0.24 = **0.60B**	0.24 + 0.16 = **0.40b**	

Generation Two

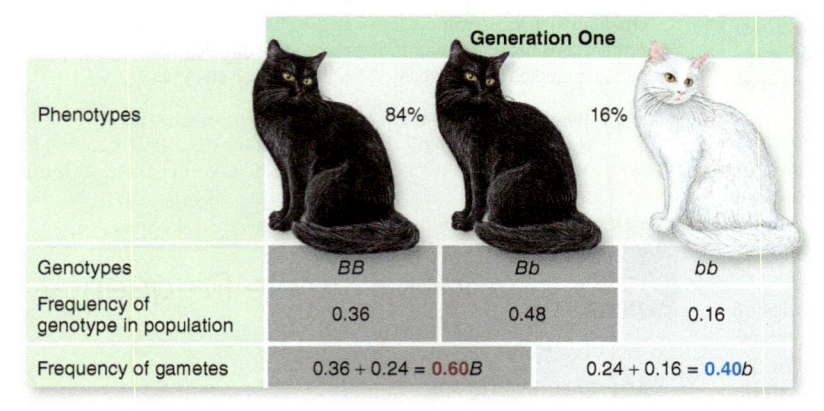

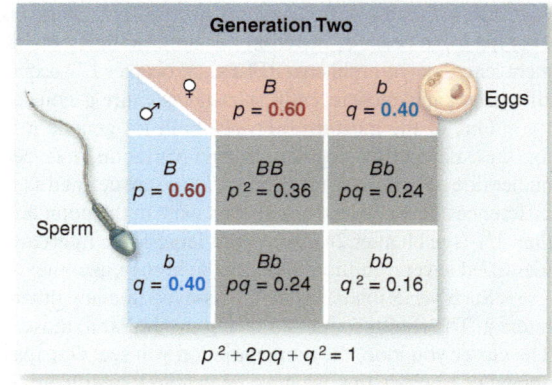

♂ \ ♀	B $p = 0.60$	b $q = 0.40$	Eggs
B $p = 0.60$	BB $p^2 = 0.36$	Bb $pq = 0.24$	
b $q = 0.40$	Bb $pq = 0.24$	bb $q^2 = 0.16$	

$p^2 + 2pq + q^2 = 1$

Figure 20.3 The Hardy–Weinberg equilibrium. In the absence of factors that alter them, the frequencies of gametes, genotypes, and phenotypes remain constant generation after generation.

Data analysis If all white cats died, what proportion of the kittens in the next generation would be white?

the second case is $q * p$. Because the result in either case is that the individual is a heterozygote, the probability of that outcome is the sum of the two probabilities, or $2pq$.

So, to summarize, if a population is in Hardy–Weinberg equilibrium with allele frequencies of p and q, then the probability that an individual will have each of the three possible genotypes is $p^2 + 2pq + q^2$. You may recognize this as the *binomial expansion:*

$$(p + q)^2 = p^2 + 2pq + q^2$$

Finally, we may use these probabilities to predict the distribution of genotypes in the population, again assuming that the population is in Hardy–Weinberg equilibrium. If the probability that any individual is a heterozygote is $2pq$, then we would expect the proportion of heterozygous individuals in the population to be $2pq$; similarly, the frequency of *BB* and *bb* homozygotes would be expected to be p^2 and q^2.

Let us return to our example. Remember that 16% of the cats are white. If white is a recessive trait, then this means that such individuals must have the genotype *bb*. If the frequency of this genotype is $q^2 = 0.16$ (the frequency of white cats), then q (the frequency of the *b* allele) = 0.4 (0.4 is the square root of 0.16). Because $p + q = 1$, therefore, p, the frequency of allele *B*, would be $1.0 - 0.4 = 0.6$ (remember, the frequencies must add up to 1). We can now easily calculate the expected **genotype frequencies:** Homozygous dominant *BB* cats would make up the p^2 group, and the value of $p^2 = (0.6)^2 = 0.36$, or 36 homozygous dominant *BB* individuals in a population of 100 cats. The heterozygous cats have the *Bb* genotype and would have the frequency corresponding to $2pq$, or $(2 * 0.6 * 0.4) = 0.48$, or 48 heterozygous *Bb* individuals.

Using the Hardy–Weinberg equation to predict frequencies in subsequent generations

The Hardy–Weinberg equation is another way of expressing the Punnett square described in chapter 12, with two alleles assigned frequencies, p and q. Figure 20.3 allows you to trace genetic reassortment during sexual reproduction and see how it affects the frequencies of the *B* and *b* alleles during the next generation.

In constructing this diagram, we have assumed that the union of sperm and egg in these cats is random, so that all combinations of *b* and *B* alleles occur. The alleles are therefore mixed randomly and are represented in the next generation in proportion to their original occurrence. Each individual egg or sperm in each generation has a 0.6 chance of receiving a *B* allele ($p = 0.6$) and a 0.4 chance of receiving a *b* allele ($q = 0.4$).

In the next generation, therefore, the chance of combining two *B* alleles is p^2, or 0.36 (that is, 0.6 * 0.6), and approximately 36% of the individuals in the population will continue to have the *BB* genotype. The frequency of *bb* individuals is q^2 (0.4 * 0.4) and so will continue to be about 16%, and the frequency of *Bb* individuals will be $2pq$ (2 * 0.6 * 0.4), or on average, 48%.

Phenotypically, if the population size remains at 100 cats, we would still see approximately 84 black individuals (with either *BB* or *Bb* genotypes) and 16 white individuals (with the *bb* genotype). Allele, genotype, and phenotype frequencies have remained unchanged from one generation to the next, despite the reshuffling of genes that occurs during meiosis and sexual reproduction.

One important point to remember is that the terms "dominant" and "recessive" refer only to the phenotype of the heterozygote and in no way imply that one allele is selectively superior to another. For this reason, in Hardy–Weinberg equilibrium, when selection is not operating, allele frequencies would not be expected to change regardless of whether one trait is dominant over another.

Hardy–Weinberg predictions can be applied to data to find evidence of evolutionary processes

The lesson from the example of black and white cats is that if all five of the assumptions listed earlier in this section hold true, the allele and genotype frequencies will not change from one generation to the next. But in reality, most populations in nature will not fit all five assumptions. The primary utility of this method is to determine whether some evolutionary process or processes are operating in a population and, if so, to suggest hypotheses about what they may be.

Suppose, for example, that the observed frequencies of the *BB*, *bb*, and *Bb* genotypes in a different population of cats were 0.6, 0.2, and 0.2, respectively. We can calculate the allele frequencies for *B* as follows: 60% (0.6) of the cats have two *B* alleles, 20% have one, and 20% have none. This means that the average number of *B* alleles per cat is 1.4 [$(0.6 \times 2) + (0.2 \times 1) + (0.2 \times 0) = 1.4$]. Because each cat has two alleles for this gene, the frequency is $1.4/2.0 = 0.7$. Similarly, you should be able to calculate that the frequency of the *b* allele = 0.3.

If the population were in Hardy–Weinberg equilibrium, then, according to the equation earlier in this section, the frequency of the *BB* genotype would be $0.7^2 = 0.49$, lower than it really is. Similarly, you can calculate that there are fewer heterozygotes and more *bb* homozygotes than expected; then clearly, the population is not in Hardy–Weinberg equilibrium.

What could cause such an excess of homozygotes and deficit of heterozygotes? A number of possibilities exist, including (1) natural selection favoring homozygotes over heterozygotes, (2) individuals choosing to mate with genetically similar individuals (because *BB* * *BB* and *bb* * *bb* matings always produce homozygous offspring, but only half of *Bb* * *Bb* produce heterozygous offspring, such mating patterns would lead to an excess of homozygotes), or (3) an influx of homozygous individuals from outside populations (or conversely, emigration of heterozygotes to other populations). By detecting a lack of Hardy–Weinberg equilibrium, we can generate potential hypotheses that we can then investigate directly.

The operation of evolutionary processes can be detected in a second way. If all of the Hardy–Weinberg assumptions are met, then allele frequencies will stay the same from one generation to the next. Changes in allele frequencies between generations would indicate that one of the assumptions is not met.

Suppose, for example, that the frequency of *b* was 0.53 in one generation and 0.61 in the next. Again, there are a number of possible explanations: For example, (1) selection favoring individuals with *b* over *B*, (2) immigration of *b* into the population or emigration of *B* out of the population, or (3) high rates of mutation that more commonly occur from *B* to *b* than vice versa. Another possibility is that the population is a small one, and that the change represents the random fluctuations that result because, simply by chance, some individuals pass on more of their genes than others. We will discuss how each of these processes is studied in the rest of the chapter.

Learning Outcomes Review 20.2

The Hardy–Weinberg principle states that in a large population with no selection and random mating, the proportion of alleles does not change through the generations. Finding that a population is not in Hardy–Weinberg equilibrium indicates that one or more evolutionary agents are operating.

- If you know the genotype frequencies in a population, how can you determine whether the population is in Hardy–Weinberg equilibrium?
- What would you conclude if you found a population not in Hardy–Weinberg equilibrium? What would be your next step?

20.3 Five Agents of Evolutionary Change

Learning Outcomes

1. Define the five processes that can cause evolutionary change.
2. Explain how these processes can cause populations to deviate from Hardy–Weinberg Equilibrium

The five assumptions of the Hardy–Weinberg principle also indicate the five agents that can lead to evolutionary change in populations. They are mutation, gene flow, nonrandom mating, genetic drift in small populations, and the pressures of natural selection. Any one of these may bring about changes in allele or genotype proportions.

Mutation changes alleles

Mutation from one allele to another can obviously change the proportions of particular alleles in a population. Mutation rates are generally so low that they have little effect on the Hardy–Weinberg proportions of common alleles. A typical gene mutates about once per 100,000 cell divisions. Because this rate is so low, other evolutionary processes are usually more important in determining how allele frequencies change.

Nonetheless, mutation is the ultimate source of genetic variation and thus makes evolution possible (figure 20.4a). It is important to remember, however, that the likelihood of a particular mutation occurring is not affected by natural selection; that is, mutations do not occur more frequently in situations in which they would be favored by natural selection.

Gene flow occurs when alleles move between populations

Gene flow is the movement of alleles from one population to another. It can be a powerful agent of change. Sometimes gene flow is obvious, as when an animal physically moves from one place to another. If the characteristics of the newly arrived individual differ from those of the animals already there, and if the newcomer is adapted well enough to the new area to survive and mate successfully, the genetic composition of the receiving population may be altered.

Other important kinds of gene flow are not as obvious. These subtler movements include the drifting of gametes or the immature stages of plants or marine animals from one place to another (figure 20.4b). Pollen, the male gamete of flowering plants, is often carried great distances by insects and other animals that visit flowers. Seeds may also blow in the wind or be carried by animals to new populations far from their place of origin. In addition, gene flow may also result from the mating of individuals belonging to adjacent populations.

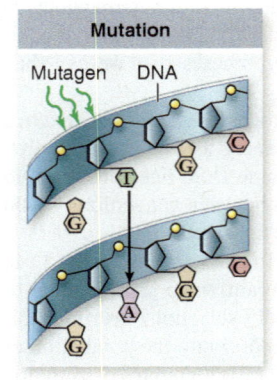

Mutation	Gene Flow	Nonrandom Mating	Genetic Drift	Selection
a. The ultimate source of variation. Individual mutations occur so rarely that mutation alone usually does not change allele frequency much.	b. A very potent agent of change. Individuals or gametes move from one population to another.	c. Inbreeding is the most common form. It does not alter allele frequency but reduces the proportion of heterozygotes.	d. Statistical accidents. The random fluctuation in allele frequencies increases as population size decreases.	e. The only agent that produces *adaptive* evolutionary changes.

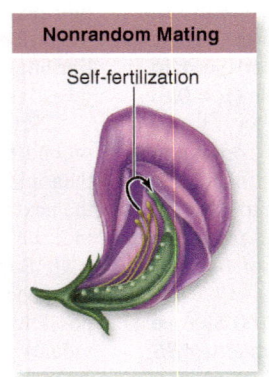

Figure 20.4 Five agents of evolutionary change. *a.* Mutation. *b.* Gene flow, *c.* Nonrandom mating. *d.* Genetic drift. *e.* Selection.

Consider two populations initially different in allele frequencies: In population 1, $p = 0.2$ and $q = 0.8$; in population 2, $p = 0.8$ and $q = 0.2$. Gene flow brings alleles into a population in proportion to their frequency in the source population. In this case, it would disproportionately bring in alleles that are rare in the recipient population because those alleles are common in the source. Thus, allele frequencies will change from generation to generation, and the populations will not be in Hardy–Weinberg equilibrium. Only when allele frequencies reach 0.5 for both alleles in both populations will equilibrium be attained. This example also indicates that gene flow tends to homogenize allele frequencies among populations.

Nonrandom mating shifts genotype frequencies

Individuals with certain genotypes sometimes mate with one another more commonly than would be expected on a random basis, a phenomenon known as *nonrandom mating* (figure 20.4c). **Assortative mating,** in which phenotypically similar individuals mate, is a type of nonrandom mating that causes the frequencies of particular genotypes to differ greatly from those predicted by the Hardy–Weinberg principle.

Assortative mating does not change the frequency of the individual alleles because it does not change the reproductive success of individuals, but rather changes with whom they mate. Because phenotypically similar individuals are likely to be genetically similar and thus are also more likely to produce offspring with two copies of the same allele, assortative mating will increase the proportion of homozygotes in the next generation. This is why populations of self-fertilizing plants consist primarily of homozygous individuals.

Inbreeding results when closely related individuals mate with each other. Because relatives are genetically similar, inbreeding will have the same result as assortative mating and increase the proportion of homozygotes. In fact, because relatives tend to be phenotypically similar, assortative mating can lead to inbreeding.

By contrast, **disassortative mating,** in which phenotypically different individuals mate, produces an excess of heterozygotes.

Genetic drift may alter allele frequencies in small populations

In small populations, frequencies of particular alleles may change drastically by chance alone. Such changes in allele frequencies occur randomly, as if the frequencies were drifting from their values. These changes are thus known as **genetic drift** (figure 20.4d). For this reason, a population must be large to be in Hardy–Weinberg equilibrium: The smaller the population, the greater change in allele frequencies from generation to generation as a result of genetic drift.

If the gametes of only a few individuals form the next generation, the alleles they carry may by chance not be representative of the parent population from which they were drawn, as illustrated in figure 20.5. In this example, a small number of individuals are removed from a bottle. By chance, most of the individuals removed are green, so the new population has a much higher population of green individuals than the parent generation had.

Suppose, for example, that four individuals form the next generation, and that by chance they are two *Bb* heterozygotes and two *BB* homozygotes—that is, the allele frequencies in the next generation would be $p = 0.75$ and $q = 0.25$. In fact, if you were to replicate this experiment 1000 times, each time randomly drawing four individuals from the parental population, then in about 8 of the 1000 experiments, one of the two alleles would be missing entirely.

This result leads to an important conclusion: Genetic drift can lead to the loss of alleles in isolated populations. Alleles that initially are uncommon are particularly vulnerable (figure 20.5).

A set of small populations that are isolated from one another may come to differ strongly as a result of genetic drift, even if the forces of natural selection are the same for both. Because of genetic drift, sometimes harmful alleles may increase in frequency in small populations, despite selective disadvantage, and favorable alleles may be lost even though they are selectively advantageous.

Although genetic drift occurs in any population, it is particularly likely in populations that were founded by a few individuals or in which the population was reduced to a very small number at some time in the past. Such founder effects or bottlenecks are particularly clear examples of genetic drift.

The founder effect

Sometimes one or a few individuals disperse and become the founders of a new, isolated population at some distance from their place of origin. These pioneers are not likely to carry all the alleles present in the source population. Thus, some alleles may be lost from the new population, and others may change drastically in frequency. In some cases, previously rare alleles in the source population may be a significant fraction of the new population's genetic endowment. This phenomenon is called the **founder effect.**

Founder effects are not rare in nature. Many self-pollinating plants start new populations from a single seed. Founder effects have been particularly important in the evolution of organisms on distant oceanic islands, such as the Hawaiian and Galápagos Islands. Most of the organisms in such areas probably derive from one or a few initial founders. Although rare, such events are occasionally observed, such as when a mass of vegetation carrying

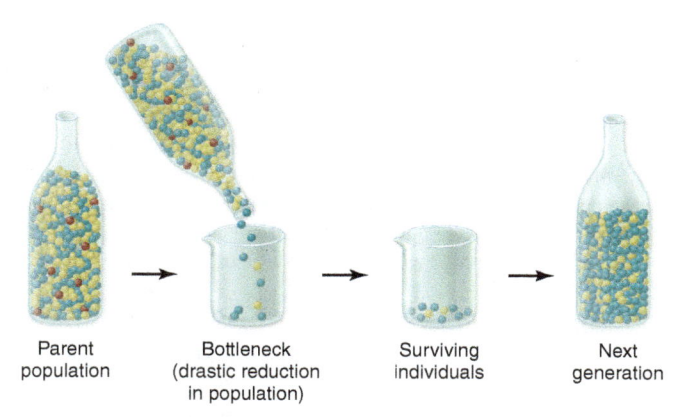

Figure 20.5 Genetic drift: a bottleneck effect. The parent population contains roughly equal numbers of green and yellow individuals and a small number of red individuals. By chance, the few remaining individuals that contribute to the next generation are mostly green. The bottleneck occurs because so few individuals form the next generation, as might happen after an epidemic or a catastrophic storm.

? Inquiry question Why are rare alleles particularly likely to be lost in a population bottleneck?

several iguanas washed up on the shore of the Caribbean island of Anguilla in 1996, leading to the establishment of a population that still occurs there to this day.

In a similar way, isolated human populations begun by relatively few individuals are often dominated by genetic features characteristic of their founders. Amish populations in the United States, for example, have unusually high frequencies of a number of conditions, such as polydactylism (the presence of a sixth finger).

The bottleneck effect

Even if organisms do not move from place to place, occasionally their populations may be drastically reduced in size. This may result from flooding, drought, epidemic disease, and other natural forces, or from changes in the environment. The few surviving individuals may constitute a random genetic sample of the original population (unless some individuals survive specifically because of their genetic makeup). The resulting alterations and loss of genetic variability have been termed the **bottleneck effect.**

The genetic variation of some living species appears to be severely depleted, probably as the result of a bottleneck effect in the past. For example, the northern elephant seal, which breeds on the western coast of North America and nearby islands, was nearly hunted to extinction in the 19th century and was reduced to a single population containing perhaps no more than 20 individuals on the island of Guadalupe off the coast of Baja, California (figure 20.6). As a result of this bottleneck, the species has lost almost all of its genetic variation, even though the seal populations have rebounded and now number in the tens of thousands and breed in locations as far north as near San Francisco.

Any time a population becomes drastically reduced in numbers, such as in endangered species, the bottleneck effect is a potential problem. Even if population size rebounds, the lack of variability may mean that the species remains vulnerable to extinction—a topic to which we will return in chapter 59.

Selection favors some genotypes over others

As Darwin pointed out, some individuals leave behind more progeny than others, and the rate at which they do so is affected by phenotype and behavior. We describe the results of this process as **selection** (see figure 20.4e). In *artificial selection,* a breeder selects for the desired characteristics. In *natural selection,* environmental conditions determine which individuals in a population produce the most offspring.

Evolution by natural selection occurs when the following conditions are met:

1. **Phenotypic variation must exist among individuals in a population.** Natural selection works by favoring individuals with some traits over individuals with alternative traits. If no variation exists, natural selection cannot operate.
2. **Variation among individuals must result in differences in the number of offspring surviving in the next generation.** This is natural selection. Because of their phenotype or behavior, some individuals are more successful than others in producing offspring. Although many traits are phenotypically variable, individuals exhibiting variation do not always differ in survival and reproductive success.
3. **Phenotypic variation must have a genetic basis.** For natural selection (see item 2) to result in evolutionary change, the selected differences must have a genetic basis. Not all variation has a genetic basis—even genetically identical individuals may be phenotypically quite distinctive if they grow up in different environments. Such environmental effects are common in nature. In many turtles, for example, individuals that hatch from eggs laid in moist soil are heavier, with longer and wider shells, than individuals from nests in drier areas.

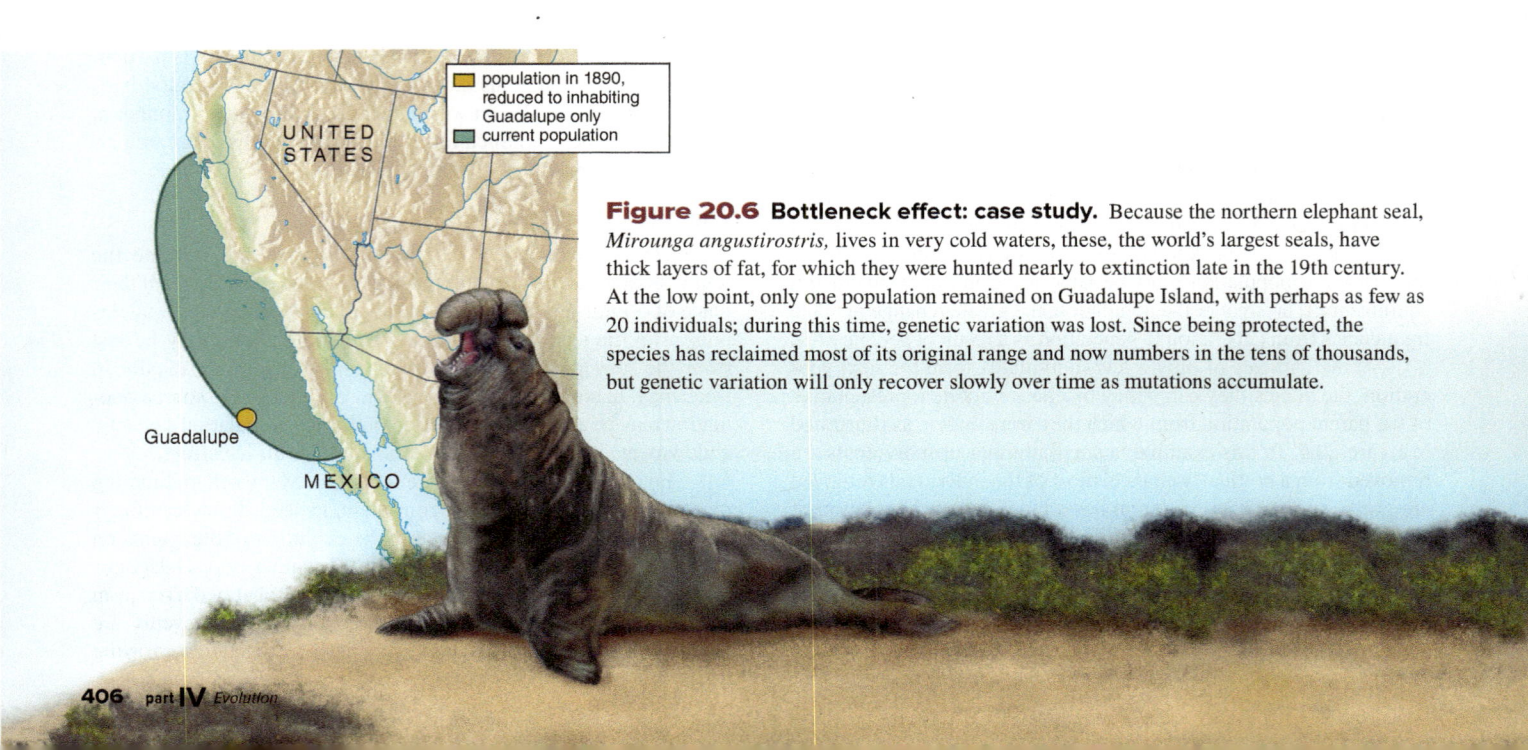

Figure 20.6 Bottleneck effect: case study. Because the northern elephant seal, *Mirounga angustirostris,* lives in very cold waters, these, the world's largest seals, have thick layers of fat, for which they were hunted nearly to extinction late in the 19th century. At the low point, only one population remained on Guadalupe Island, with perhaps as few as 20 individuals; during this time, genetic variation was lost. Since being protected, the species has reclaimed most of its original range and now numbers in the tens of thousands, but genetic variation will only recover slowly over time as mutations accumulate.

When phenotypically different individuals do not differ genetically, then differences in the number of their offspring will not alter the genetic composition of the population in the next generation, and thus, no evolutionary change will have occurred.

It is important to remember that natural selection and evolution are not the same—the two concepts often are incorrectly equated. Natural selection is a process, whereas evolution is the historical record, or outcome, of change through time. Natural selection (the process) can lead to evolution (the outcome), but natural selection is only one of several processes that can result in evolutionary change. Moreover, natural selection can occur without producing evolutionary change; only if variation is genetically based will natural selection lead to evolution.

Selection to avoid predators

The result of evolution driven by natural selection is that populations become better adapted to their environment. Many of the most dramatic documented instances of adaptation involve genetic changes that decrease the probability of capture by a predator. The caterpillar larvae of the common sulphur butterfly *Colias eurytheme* usually exhibit a pale green color, providing excellent camouflage against the alfalfa plants on which they feed. An alternative bright yellow color morph is reduced to very low frequency because this color renders the larvae highly visible on the food plant, making it easier for bird predators to see them (see figure 20.4e).

One of the most dramatic examples of background matching involves ancient lava flows in the deserts of the American Southwest. In these areas, the black rock formations produced when the lava cooled contrast starkly with the surrounding bright glare of the desert sand. Populations of many species of animals occurring on these rocks—including lizards, rodents, and a variety of insects—are dark in color, whereas sand-dwelling populations in surrounding areas are much lighter (figure 20.7).

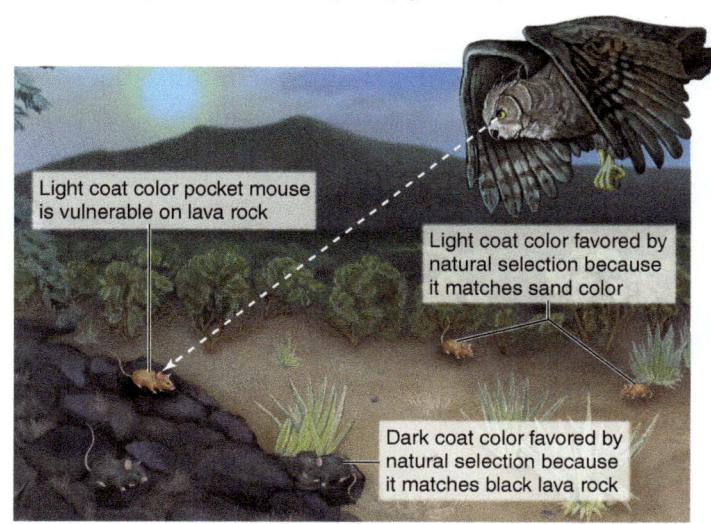

Figure 20.7 Pocket mice from the Tularosa Basin of New Mexico whose color matches their background. Black lava formations are surrounded by desert, and selection favors coat color in pocket mice that matches their surroundings. Genetic studies indicate that the differences in coat color are the result of small differences in the DNA of alleles of a single gene.

Predation is the likely cause for these differences in color. Laboratory studies have confirmed that predatory birds such as owls are adept at picking out individuals occurring on backgrounds to which they are not adapted.

Selection to match climatic conditions

Many studies of selection have focused on genes encoding enzymes, because in such cases the investigator can directly assess the consequences to the organism of changes in the frequency of alternative enzyme alleles.

Often investigators find that enzyme allele frequencies vary with latitude, so that one allele is more common in northern populations, but is progressively less common at more southern locations. A superb example is seen in studies of a fish, the mummichog *(Fundulus heteroclitus)*, which ranges along the eastern coast of North America. In this fish, geographic variation occurs in allele frequencies for the gene that produces the enzyme lactate dehydrogenase, which catalyzes the conversion of pyruvate to lactate (see section 7.8).

Biochemical studies show that the enzymes formed by these alleles function differently at different temperatures, thus explaining their geographic distributions. The form of the enzyme more frequent in the north is a better catalyst at low temperatures than is the enzyme from the south. Moreover, studies indicate that at low temperatures, individuals with the northern allele swim faster, and presumably survive better, than individuals with the alternative allele.

Selection for pesticide and microbial resistance

A particularly clear example of selection in natural populations is provided by studies of pesticide resistance in insects. The widespread use of insecticides has led to the rapid evolution of resistance in more than 500 pest species. The cost of this evolution, in terms of crop losses and increased pesticide use, has been estimated at $3 to $8 billion per year.

In the housefly, the resistance allele at the *pen* gene decreases the uptake of insecticide, whereas alleles at the *kdr* and *dld-r* genes decrease the number of target sites, thus decreasing the binding ability of the insecticide (figure 20.8). Other alleles enhance the ability of the insects' enzymes to identify and detoxify insecticide molecules.

Single genes are also responsible for resistance in other organisms. For example, Norway rats are normally susceptible to the pesticide warfarin, which diminishes the clotting ability of the rat's blood and leads to fatal hemorrhaging. However, a resistance allele at a single gene reduces the ability of warfarin to bind to its target enzyme and thus renders it ineffective.

Selection imposed by humans has also led to the evolution of resistance to antibiotics in many disease-causing pathogens. For example, *Staphylococcus aureus*, which causes staph infections, was initially treated by penicillin. However, within four years of mass-production of the drug, evolutionary change in *S. aureus* modified an enzyme so that it would attack penicillin and render it inactive. Since that time, several other drugs have been developed to attack the microbe, and each time resistance has evolved. As a result, staph infections have re-emerged as a major health threat. The speed by which adaptation occurs may be surprising, but remember that microbes have enormous population sizes. Even

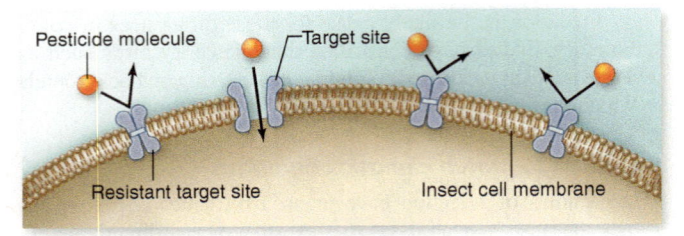

a. Insect cells with resistance allele at *pen* gene: decreased uptake of the pesticide.

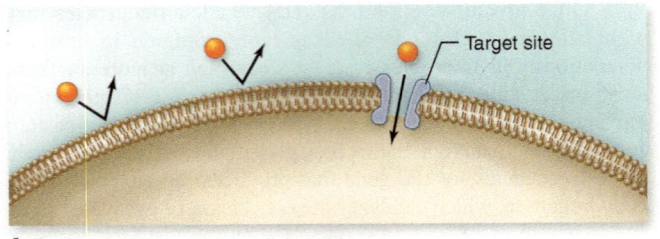

b. Insect cells with resistance allele at *kdr* gene: decreased number of target sites for the pesticide.

Figure 20.8 Selection for pesticide resistance. Resistance alleles at genes such as *pen* and *kdr* allow insects to be more resistant to pesticides. Insects that possess these resistance alleles have become more common through selection.

though mutation rates are low, the vast size of these populations guarantees a steady supply of new mutations for natural selection to utilize. The effect of such evolution on human health is enormous. In the United States alone, 2 million people each year become ill due to antibiotic-resistant bacteria, and 23,000 die from such infections.

Learning Outcomes Review 20.3

Five factors can bring about deviation from the predicted Hardy–Weinberg genotype frequencies. Of these, only selection regularly produces adaptive evolutionary change, but the genetic constitution of populations, and thus the course of evolution, can also be affected by mutation, gene flow, nonrandom mating, and genetic drift.

- *How do each of these processes cause populations to vary from Hardy–Weinberg equilibrium?*

20.4 Quantifying Natural Selection

Learning Outcomes

1. Define evolutionary fitness.
2. Explain the different components of fitness.
3. Demonstrate how the success of different phenotypes can be compared by calculating their relative fitness.

Selection occurs when individuals with one phenotype leave more surviving offspring in the next generation than individuals with an alternative phenotype. Evolutionary biologists quantify reproductive success as **fitness,** the number of surviving offspring left in the next generation.

Fitness is a relative concept; the most fit phenotype is simply the one that produces, on average, the greatest number of offspring.

A phenotype with greater fitness usually increases in frequency

Suppose, for example, that in a population of toads, two phenotypes exist: green and brown. Suppose, further, that green toads leave, on average, 4.0 offspring in the next generation, but brown toads leave only 2.5. By custom, the most fit phenotype is assigned a fitness value of 1.0, and other phenotypes are expressed as relative proportions. In this case, the fitness of the green phenotype would be 4.0/4.0 = 1.000, and the fitness of the brown phenotype would be 2.5/4.0 = 0.625. The difference in fitness would therefore be 1.000 − 0.625 = 0.375. A difference in fitness of 0.375 is quite large; natural selection in this case strongly favors the green phenotype.

If differences in color have a genetic basis, then we would expect evolutionary change to occur; the frequency of green toads should be substantially greater in the next generation. Further, if the fitness of two phenotypes remained unchanged, we would expect alleles for the brown phenotype eventually to disappear from the population.

> **Inquiry question** Why might the frequency of green toads not increase in the next generation, even if color differences have a genetic basis?

Fitness may consist of many components

Although selection is often characterized as "survival of the fittest," differences in survival are only one component of fitness.

Even if no differences in survival occur, selection may operate if some individuals are more successful than others in attracting mates. In many territorial animal species, for example, large males mate with many females, and small males rarely get to mate. Selection with respect to mating success is termed *sexual selection;* we describe this topic more fully in the discussion of behavioral biology in chapter 54.

In addition, the number of offspring produced per mating is also important. Large female frogs and fish lay more eggs than do smaller females, and thus they may leave more offspring in the next generation.

Fitness is therefore a combination of survival, mating success, and number of offspring per mating. Selection favors phenotypes with the greatest fitness, but predicting fitness from a single component can be tricky because traits favored for one component of fitness may be at a disadvantage for others. As an example, in water striders, larger females lay more eggs per day (figure 20.9). Thus, natural selection at this stage favors large size. However, larger females also die at a younger age and thus have fewer

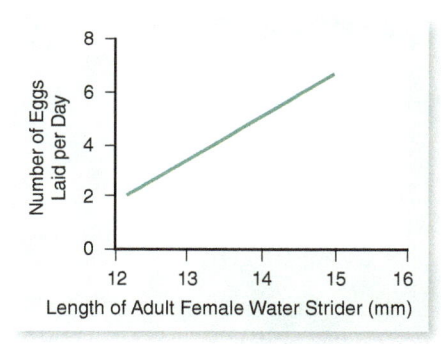

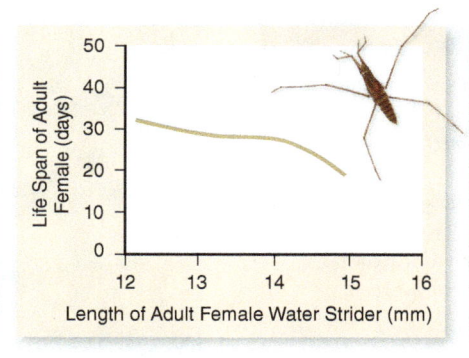

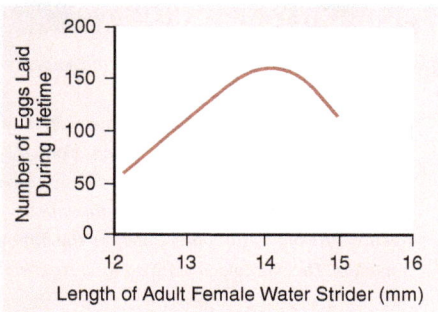

Figure 20.9 Body size and egg-laying in water striders. Larger female water striders lay more eggs per day *(left panel)*, but also survive for a shorter period of time *(center panel)*. As a result, intermediate-sized females produce the most offspring over the course of their entire lives and thus have the highest fitness *(right panel)*.

Inquiry question What evolutionary change in body size might you expect? If the number of eggs laid per day was not affected by body size, would your prediction change?

Data analysis Assuming that the values on the x-axis represent the final body size of different animals and that the y-axis represents egg-laying rate and survival once they reach that size, how many eggs would you expect a 12-mm water strider to lay? And a 15-mm strider?

opportunities to reproduce than smaller females. Overall, the two opposing directions of selection cancel each other out, and the intermediate-sized females leave the most offspring in the next generation.

Learning Outcomes Review 20.4

Fitness is defined by an organism's reproductive success relative to other members of its population. This success is determined by how long it survives, how often it mates, and how many offspring it produces per mating. Relative fitness assigns numerical values to different phenotypes relative to the most fit phenotype.

- *Is one of these factors always the most important in determining reproductive success? Explain.*

20.5 Natural Selection's Role in Maintaining Variation

Learning Outcomes

1. Define frequency-dependent selection, oscillating selection, and heterozygote advantage.
2. Explain how these processes affect the amount of genetic variation in a population.

Natural selection has been discussed as a process that removes variation from a population by favoring one allele over others at a gene locus. However, in some circumstances, selection can do exactly the opposite and actually maintain population variation.

Frequency-dependent selection may favor either rare or common phenotypes

In some circumstances, the fitness of a phenotype depends on its frequency within the population, a phenomenon termed **frequency-dependent selection.** This type of selection favors certain phenotypes depending on how commonly or uncommonly they occur.

Negative frequency-dependent selection

In negative frequency-dependent selection, rare phenotypes are favored by selection—thus, fitness has a negative relationship with phenotype frequencing, hence the term *negative frequency-dependent*. Assuming a genetic basis for phenotypic variation, such selection will have the effect of making rare alleles more common, thus maintaining variation.

Negative frequency-dependent selection can occur for many reasons. For example, it is well known that animals or people searching for something form a "search image." That is, they become particularly adept at picking out certain objects. Consequently, predators may form a search image for common prey phenotypes. Rare forms may thus be preyed upon less frequently.

An example is fish predation on an insect, the water boatman, which occurs in three different colors. Experiments indicate that each of the color types is preyed upon disproportionately when it is the most common one; fish eat more of the common-colored insects than would occur by chance alone (figure 20.10).

Another cause of negative frequency dependence is resource competition. If genotypes differ in their resource requirements, as occurs in many plants, then the rarer genotype will have fewer competitors. When the different resource types are equally abundant,

SCIENTIFIC THINKING

Question: Does negative frequency-dependent selection maintain variation in a population?

Hypothesis: Fish may disproportionately capture water boatmen (a type of aquatic insect) with the most common color.

Experiment: Place predatory fish in different aquaria with the different frequencies of the color types in each aquarium.

Result: Fish prey disproportionately on the common color in each aquarium. The rare color in each aquarium generally survives best.

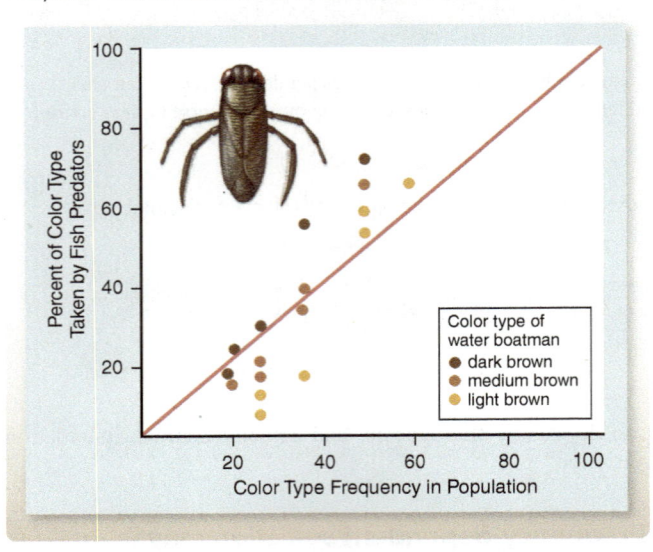

Figure 20.10 Frequency-dependent selection.

Data analysis How can this experiment distinguish between frequency-dependent and directional selection?

Figure 20.11 Positive frequency-dependent selection. In some cases, rare individuals stand out from the rest and draw the attention of predators; thus, in these cases, common phenotypes have the advantage (positive frequency-dependent selection).

the rarer genotype will be at an advantage relative to the more common genotype.

Positive frequency-dependent selection

Positive frequency-dependent selection has the opposite effect; by favoring common forms, it tends to eliminate variation from a population. For example, predators don't always select common individuals. In some cases, "oddballs" stand out from the rest and attract attention (figure 20.11).

The strength of selection should change through time as a result of frequency-dependent selection. In negative frequency-dependent selection, rare genotypes should become increasingly common, and their selective advantage will decrease correspondingly. Conversely, in positive frequency dependence, the rarer a genotype becomes, the greater the chance it will be selected against.

In oscillating selection, the favored phenotype changes as the environment changes

In some cases, selection favors one phenotype at one time and another phenotype at another time, a phenomenon called **oscillating selection.** If selection repeatedly oscillates in this fashion, the effect will be to maintain genetic variation in the population.

One example, discussed in chapter 21, concerns the medium ground finch of the Galápagos Islands. In times of drought, the supply of small, soft seeds is depleted, but there are still enough large seeds around. Consequently, birds with big bills are favored. However, when wet conditions return, the ensuing abundance of small seeds favors birds with smaller bills.

Oscillating selection and frequency-dependent selection are similar because in both cases the form of selection changes through time. But it is important to recognize that they are not the same: In oscillating selection, the fitness of a phenotype does not depend on its frequency; rather, environmental changes lead to the oscillation in selection. In contrast, in frequency-dependent selection, it is the change in frequencies themselves that leads to the changes in fitness of the different phenotypes.

In some cases, heterozygotes may exhibit greater fitness than homozygotes

If heterozygotes are favored over homozygotes, then natural selection actually tends to maintain variation in the population. This **heterozygote advantage** favors individuals with copies of both alleles, and thus works to maintain both alleles in the population. Some evolutionary biologists believe that heterozygote advantage is pervasive and can explain the high levels of genetic variation observed in natural populations. Others, however, believe that it is relatively rare.

The best documented example of heterozygote advantage is sickle cell anemia, a hereditary disease affecting hemoglobin in humans. Individuals with sickle cell anemia exhibit symptoms of severe anemia and abnormal red blood cells that are irregular in shape, with a great number of long, sickle-shaped cells

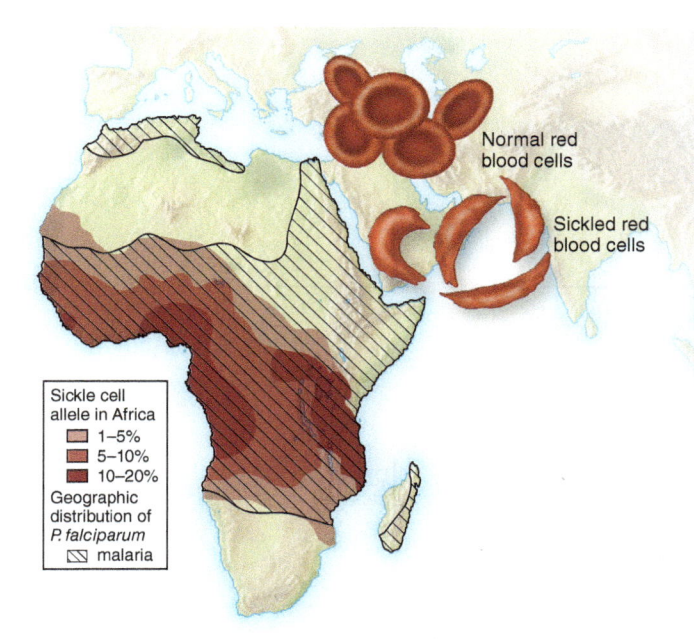

Figure 20.12 Frequency of sickle cell allele and distribution of *Plasmodium falciparum* malaria. The red blood cells of people homozygous for the sickle cell allele collapse into sickled shapes when the oxygen level in the blood is low. The distribution of the sickle cell allele in Africa coincides closely with that of *P. falciparum* malaria.

(figure 20.12). Chapter 13 discusses why the sickle cell mutation (*S*) causes red blood cells to sickle.

The average incidence of the *S* allele in central African populations is about 0.12, far higher than that found among African Americans. From the Hardy–Weinberg principle, you can calculate that 1 in 5 central African individuals is heterozygous at the *S* allele, and 1 in 100 is homozygous and develops the fatal form of the disorder. People who are homozygous for the sickle cell allele almost never reproduce because they usually die before they reach reproductive age.

Why, then, is the *S* allele not eliminated from the central African population by selection rather than being maintained at such high levels? As it turns out, one of the leading causes of illness and death in central Africa, especially among young children, is malaria. People who are heterozygous for the sickle cell allele (and thus do not suffer from sickle cell anemia) are much less susceptible to malaria. The reason is that when the parasite that causes malaria, *Plasmodium falciparum,* enters a red blood cell, it causes extremely low oxygen tension in the cell, which leads to sickling in cells of individuals either homozygous or heterozygous for the sickle cell allele (but not in individuals that do not have the sickle cell allele). Such cells are quickly filtered out of the bloodstream by the spleen, thus eliminating the parasite (The spleen's filtering effect is what leads to anemia in persons homozygous for the sickle cell allele because large numbers of red blood cells become sickle-shaped and are removed; in the case of heterozygotes, only those cells containing the *Plasmodium* parasite sickle, whereas the remaining cells are not affected, and thus anemia does not occur). Thus, homozygotes with the sickle cell allele all die of anemia, homozygotes for the non-sickle cell allele are vulnerable to malaria, and heterozygotes have the highest fitness.

Consequently, even though most homozygous recessive individuals die at a young age, the sickle cell allele is maintained at high levels in these populations because it is associated with resistance to malaria in heterozygotes and also, for reasons not yet fully understood, with increased fertility in female heterozygotes. Figure 20.12 shows the overlap between regions where sickle cell anemia is found and where malaria is prevalent.

For people living in areas where malaria is common, having the sickle cell allele in the heterozygous condition has adaptive value (see figure 20.12). Among African Americans, however, many of whose ancestors have lived for many generations in a country where malaria is now essentially absent, the environment does not place a premium on resistance to malaria. Consequently, no adaptive value counterbalances the ill effects of the disease; in this nonmalarial environment, selection is acting to eliminate the *S* allele. Only 1 in 375 African Americans develops sickle cell anemia, far fewer than in central Africa.

Learning Outcomes Review 20.5

Selection can maintain variation within populations in a number of ways. Negative frequency-dependent selection tends to favor rare phenotypes. Oscillating selection favors different phenotypes at different times. In some cases, heterozygotes have a selective advantage that may act to retain alleles that are deleterious in the homozygous state.

- How would genetic variation in a population change if heterozygotes had the lowest fitness?
- Explain the difference between negative frequency-dependent and oscillating selection. Over long periods of time, which is more likely to maintain variation in a population?

20.6 Selection Acting on Traits Affected by Multiple Genes

Learning Outcomes

1. Define and contrast disruptive, directional, and stabilizing selection.
2. Explain the evolutionary outcome of each of these types of selection.

In nature, many traits—perhaps most—are affected by more than one gene. The interactions between genes are typically complex, as you saw in chapter 12. For example, alleles of many different genes play a role in determining human height (see figure 12.11). In such cases, selection operates on all the genes, influencing most strongly those that make the greatest contribution to the phenotype. Such traits exhibit a continuous distribution of phenotypes, often conforming to a normal ("bell") curve, rather than several discretely different phenotypes (such as "red" and "black").

Disruptive selection removes intermediates

In some situations, selection acts to eliminate intermediate types, a phenomenon called **disruptive selection** (figure 20.13*a*). A clear

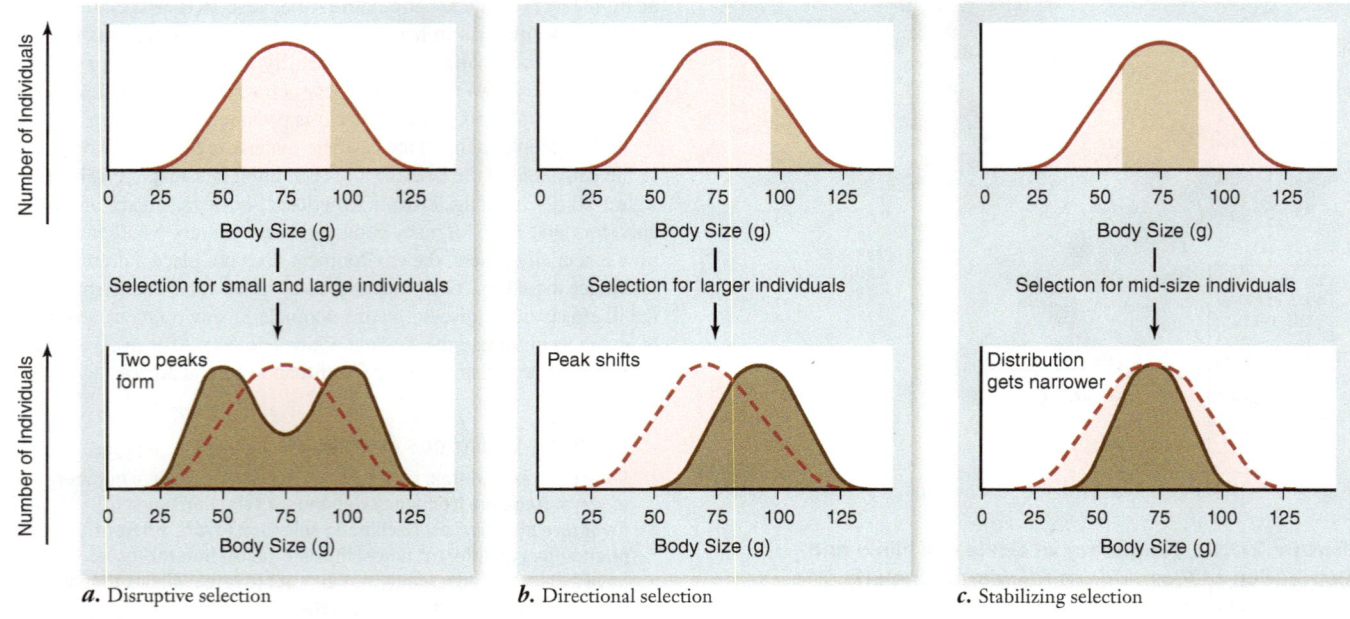

Figure 20.13 Three kinds of selection. The top panels show the populations before selection has occurred (under the solid red line). Within the population, those favored by selection are shown in light brown. The bottom panels indicate what the populations would look like in the next generation. The dashed red lines are the distribution of the original population and the solid, dark brown lines are the true distribution of the population in the next generation. *a.* In disruptive selection, individuals in the middle of the range of phenotypes of a certain trait are selected against, and the extreme forms of the trait are favored. *b.* In directional selection, individuals concentrated toward one extreme of the array of phenotypes are favored. *c.* In stabilizing selection, individuals with midrange phenotypes are favored, with selection acting against both ends of the range of phenotypes.

example is the different beak sizes of the African black-bellied seedcracker finch (figure 20.14). Populations of these birds contain individuals with large and small beaks, but very few individuals with intermediate-sized beaks.

As their name implies, these birds feed on seeds, and the available seeds fall into two size categories: large and small. Only large-beaked birds can open the tough shells of large seeds, whereas birds with the smaller beaks are more adept at handling small seeds. Birds with intermediate-sized beaks are at a disadvantage with both seed types—they are unable to open large seeds and too clumsy to efficiently process small seeds. Consequently, selection acts to eliminate the intermediate phenotypes, in effect partitioning (or "disrupting") the population into two phenotypically distinct groups.

Directional selection eliminates phenotypes on one end of a range

When selection acts to eliminate one extreme from an array of phenotypes, the genes promoting this extreme become less frequent in the population and may eventually disappear. This form of selection is called **directional selection** (see figure 20.13*b*). Thus, in the *Drosophila* population illustrated in figure 20.15, eliminating flies that move toward light causes the population over time to contain fewer individuals with alleles promoting such behavior. If you were to pick an individual at random from a later generation of flies, there is a smaller chance that the fly would spontaneously move toward light

SCIENTIFIC THINKING

Question: Does disruptive selection promote differences in beak size in the African black-bellied seedcracker finches (*Pyrenestes ostrinus*)?

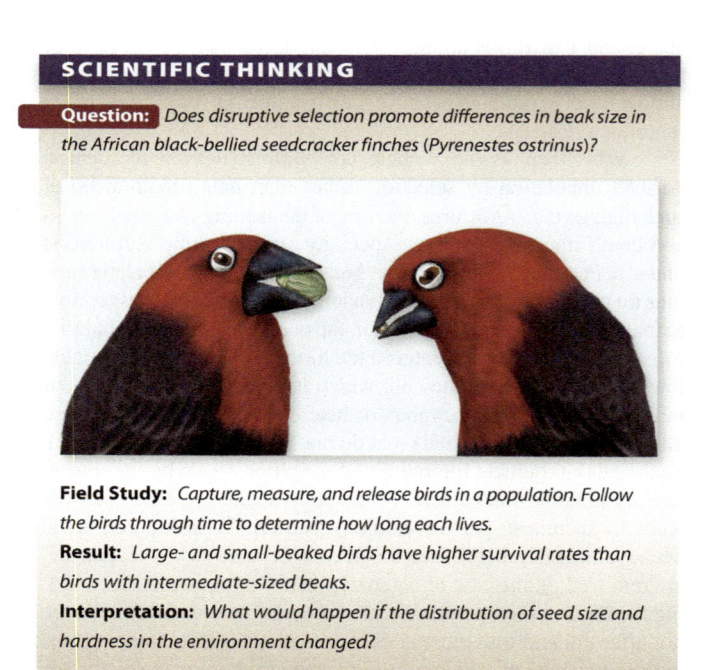

Field Study: Capture, measure, and release birds in a population. Follow the birds through time to determine how long each lives.
Result: Large- and small-beaked birds have higher survival rates than birds with intermediate-sized beaks.
Interpretation: What would happen if the distribution of seed size and hardness in the environment changed?

Figure 20.14 Disruptive selection for large and small beaks. Differences in beak size in the black-bellied seedcracker finch of west Africa are the result of disruptive selection.

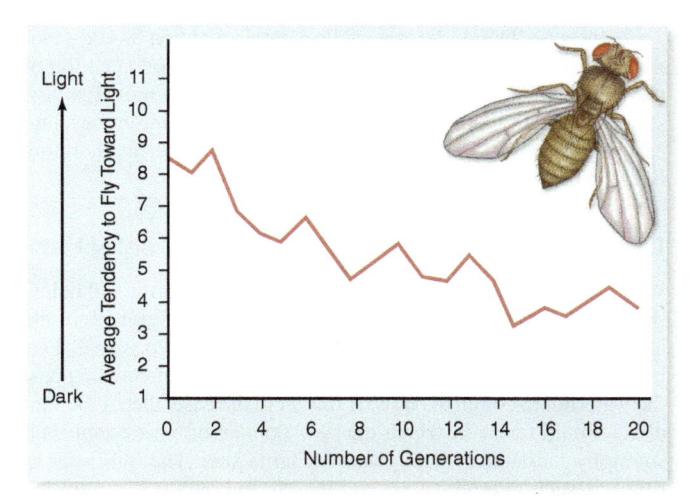

Figure 20.15 Directional selection for negative phototropism in *Drosophila*. Flies that moved toward light were discarded, and only flies that moved away from light were used as parents for the next generation. This procedure was repeated for 20 generations, producing substantial evolutionary change.

> **Inquiry question** What would happen if after 20 generations, experimenters started keeping flies that moved toward the light and discarded the others?

than if you had selected a fly from the original population. Artificial selection has changed the population in the direction of being less attracted to light. Directional selection often occurs in nature when the environment changes, favoring traits at one end of the phenotypic distribution.

Stabilizing selection favors individuals with intermediate phenotypes

When selection acts to eliminate both extremes from an array of phenotypes, the result is to increase the frequency of the already common intermediate type. This form of selection is called **stabilizing selection** (see figure 20.13c). In effect, selection is operating to prevent change away from this middle range of values. Selection does not change the most common phenotype of the population, but rather makes it even more common by eliminating extremes. Many examples are known. In humans, infants with intermediate weight at birth have the highest survival rate (figure 20.16). In ducks and chickens, eggs of intermediate weight have the highest hatching success.

Learning Outcomes Review 20.6

In disruptive selection, intermediate forms of a trait diminish; in stabilizing selection, intermediates increase, whereas in disruptive selection they decrease. Directional selection shifts frequencies toward one end or the other and may eventually eliminate alleles entirely.

- *How does directional selection differ from frequency-dependent selection?*

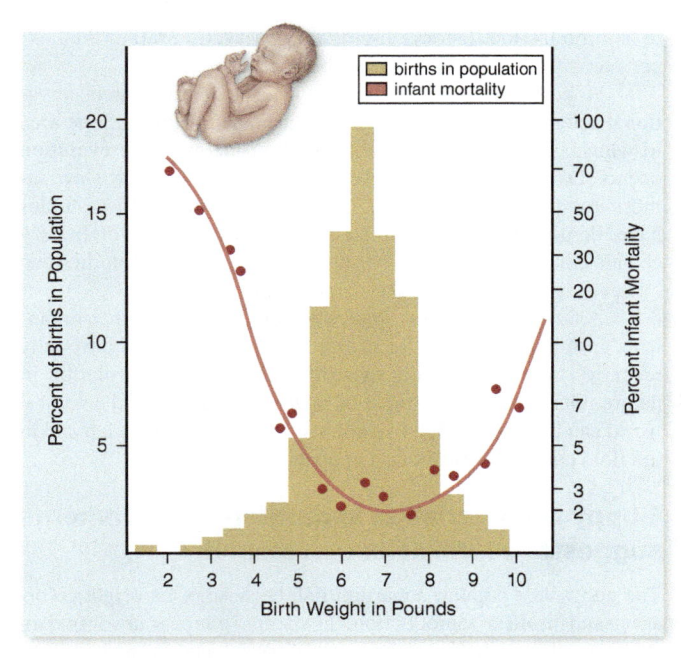

Figure 20.16 Stabilizing selection for birth weight in humans. The death rate among babies (red curve; right *y*-axis) is lowest at an intermediate birth weight; both smaller and larger babies have a greater tendency to die than those around the most frequent weight (tan area; left *y*-axis) of between 7 and 8 pounds. Recent medical advances have reduced mortality rates for small and large babies.

> **Inquiry question** As improved medical technology leads to decreased infant mortality rates, how would you expect the distribution of birth weights in the population to change?

> **Data analysis** Sketch what a graph would look like that showed the absolute number of infant mortality deaths versus birth weight.

20.7 Experimental Studies of Natural Selection

Learning Outcome

1. Explain how experiments can be used to test evolutionary hypotheses.

To study evolution, biologists have traditionally investigated what has happened in the past, sometimes many millions of years ago. To learn about dinosaurs, a paleontologist looks at dinosaur fossils. To study human evolution, an anthropologist looks at human fossils and, increasingly, examines the "family tree" of mutations that have accumulated in human DNA over millions of years. In this traditional approach, evolutionary biology is similar to astronomy and history, relying on observation rather than experimentation to examine ideas about past events.

Nonetheless, evolutionary biology is not entirely an observational science. Darwin was right about many things, but one area in which he was mistaken concerns the pace at which evolution occurs. Darwin thought that evolution occurred at a very slow, almost imperceptible pace. But in recent years many case studies have demonstrated that in some circumstances, evolutionary change can occur rapidly. Consequently, experimental studies can be devised to test evolutionary hypotheses.

Although laboratory studies on fruit flies and other organisms have been common for nearly a century, scientists have only recently started conducting experimental studies of evolution in nature. One excellent example of how observations of the natural world can be combined with rigorous experiments in the lab and in the field concerns research on guppies.

Guppy color variation in different environments suggests natural selection at work

The guppy is a popular aquarium fish because of its bright coloration and prolific reproduction. In nature, guppies are found in small streams in northeastern South America and in many mountain streams on the nearby island of Trinidad. One interesting feature of several of the streams is that they have waterfalls. Amazingly, guppies and some other fish are capable of colonizing portions of the stream above the waterfall.

The killifish is a particularly good colonizer; apparently on rainy nights, it will wriggle out of the stream and move through the damp leaf litter. Guppies are not so proficient, but they are good at swimming upstream. During flood seasons, rivers sometimes overflow their banks, creating secondary channels that move through the forest. On these occasions, guppies may be able to swim upstream in the secondary channels and invade the pools above waterfalls.

By contrast, some species are not capable of such dispersal and thus are only found in streams below the first waterfall. One species whose distribution is restricted by waterfalls is the pike cichlid, a voracious predator that feeds on other fish, including guppies.

Because of these barriers to dispersal, guppies can be found in two very different environments. In pools just below the waterfalls, predation by the pike cichlid is a substantial risk, and rates of survival are relatively low. But in similar pools just above the waterfall, the only predator present is the killifish, which rarely preys on guppies.

Guppy populations above and below waterfalls exhibit many differences. In the high-predation pools, guppies exhibit drab coloration. Moreover, they tend to reproduce at a younger age and attain relatively smaller adult sizes. Male fish above the waterfall, in contrast, are colorful (figure 20.17), mature later, and grow to larger sizes.

These differences suggest the operation of natural selection. In the low-predation environment, males display gaudy colors and spots that they use to court females. Moreover, larger males are most successful at holding territories and mating with females, and larger females lay more eggs. Thus, in the absence of predators, larger and more colorful fish may have produced more offspring, leading to the evolution of those traits.

In pools below the waterfall, natural selection would favor different traits. Colorful males are likely to attract the attention of the pike cichlid, and high predation rates mean that most fish live short lives. Individuals that are more drab and shunt energy into early reproduction, rather than growth to a larger size, are therefore likely to be favored by natural selection.

Experimentation reveals the agent of selection

Although the differences between guppies living above and below the waterfalls suggest evolutionary responses to differences in the strength of predation, alternative explanations are possible. Perhaps, for example, only very large fish are capable of crawling past the waterfall to colonize pools. If this were the case, then a founder effect would occur in which the new population was established solely by individuals with genes for large size. The only way to rule out such alternative possibilities is to conduct a controlled experiment.

The laboratory experiment

The first experiments were conducted in large pools in laboratory greenhouses. At the start of the experiment, a group of 2000

Figure 20.17 The evolution of protective coloration in guppies. In pools below waterfalls where predation is high, male guppies are drab in color. In the absence of the highly predatory pike cichlid in pools above waterfalls, male guppies are much more colorful and attractive to females. The killifish is also a predator, but it only rarely eats guppies. The evolution of these differences in guppies can be experimentally tested.

guppies was divided equally among 10 large pools. Six months later, pike cichlids were added to four of the pools and killifish to another four, with the remaining two pools left to serve as "no-predation" controls.

Fourteen months later (which corresponds to 10 guppy generations), the scientists compared the populations. The guppies in the killifish and control pools were indistinguishable—brightly colored and large. In contrast, the guppies in the pike cichlid pools were smaller and drab in coloration (figure 20.18).

These results established that predation can lead to rapid evolutionary change, but do these laboratory experiments reflect what occurs in nature?

SCIENTIFIC THINKING

Question: Does the presence of predators affect the evolution of guppy color?

Hypothesis: Predation on the most colorful individuals will cause a population to become increasingly dull through time. Conversely, in populations with few or no predators, increased color will evolve.

Experiment: Establish laboratory populations of guppies in large pools with or without predators.

Result: The populations with predators evolved to have fewer spots, while the populations in pools without predators evolved more spots.

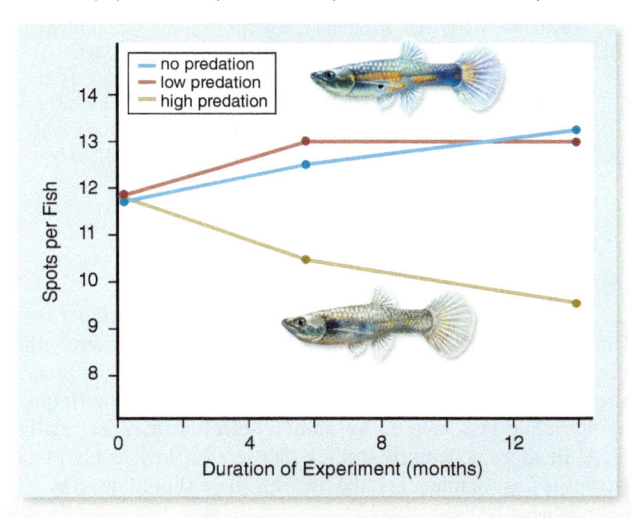

Interpretation: Why does color increase in the absence of predators? How would you test your hypothesis?

Figure 20.18 Evolutionary change in spot number.
Guppy populations raised for 10 generations in low-predation or no-predation environments in laboratory greenhouses evolved a greater number of spots, whereas selection in more dangerous environments, such as the pools with the highly predatory pike cichlid, led to less conspicuous fish. The same results are seen in field experiments conducted in pools above and below waterfalls.

? Inquiry question How do these results depend on the manner by which the guppy predators locate their prey?

The field experiment

To find out whether the laboratory results were an accurate reflection of natural processes, the scientists located two streams that had guppies in pools below a waterfall, but not above it. As in other Trinidadian streams, the pike cichlid was present in the lower pools, but only the killifish was found above the waterfalls.

The scientists then transplanted guppies to the upper pools and returned at several-year intervals to monitor the populations. Despite originating from populations in which predation levels were high, the transplanted populations rapidly evolved the traits characteristic of low-predation guppies: they matured late, attained greater size, and had brighter colors. The control populations in the lower pools, by contrast, continued to be drab and to mature early and at a smaller size. Laboratory analysis confirmed that the variations between the populations were the result of genetic differences.

These results demonstrate that substantial evolutionary change can occur in less than 12 years. More generally, these studies indicate how scientists can formulate hypotheses about how evolution occurs and then test these hypotheses in natural conditions. The results give strong support to the theory of evolution by natural selection.

Learning Outcome Review 20.7

Although much of evolutionary theory is derived from observation, experiments are sometimes possible in natural settings. Studies have revealed that traits can shift in populations in a relatively short time. The data obtained from evolutionary experiments can be used to refine theoretical assumptions.

■ *What experiments could you design to test other examples of natural selection, such as the evolution of pesticide resistance or background color matching?*

20.8 Interactions Among Evolutionary Forces

Learning Outcomes

1. Discuss how evolutionary processes can work simultaneously, but in opposing ways.
2. Evaluate what determines the evolutionary outcome when multiple processes are operating simultaneously.

The amount of genetic variation in a population may be determined by the relative strength of different evolutionary processes. Sometimes these processes act together, and in other cases they work in opposition.

Mutation and genetic drift may counter selection

In theory, if allele B mutates to allele b at a high enough rate, allele b could be maintained in the population, even if natural selection strongly favored allele *B*. In nature, however, mutation rates are rarely high enough to counter the effects of natural selection.

The effect of natural selection also may be countered by genetic drift. Both of these processes may act to remove variation from a population. But selection is a nonrandom process that operates to increase the representation of alleles that enhance survival and reproductive success, whereas genetic drift is a random process in which any allele may increase. Thus, in some cases, drift may lead to a decrease in the frequency of an allele that is favored by selection. In some extreme cases, drift may even lead to the loss of a favored allele from a population.

Remember, however, that the magnitude of drift is inversely related to population size; consequently, natural selection is expected to overwhelm drift, except when populations are very small.

Gene flow may promote or constrain evolutionary change

Gene flow can be either a constructive or a constraining force. On one hand, gene flow can spread a beneficial mutation that arises in one population to other populations. On the other hand, gene flow can impede adaptation within a population by the continual flow of inferior alleles from other populations.

Consider two populations of a species that live in different environments. In this situation, natural selection might favor different alleles—*B* and *b*—in the two populations. In the absence of other evolutionary processes such as gene flow, the frequency of *B* would be expected to reach 100% in one population and 0% in the other. However, if gene flow occurred between the two populations, then the less favored allele would continually be reintroduced into each population. As a result, the frequency of the alleles in the populations would reflect a balance between the rate at which gene flow brings the inferior allele into a population, and the rate at which natural selection removes it.

A classic example of gene flow opposing natural selection occurs on abandoned mine sites in Great Britain. Although mining activities ceased hundreds of years ago, the concentration of metal ions in the soil is still much greater than in surrounding areas. Large concentrations of heavy metals are generally toxic to plants, but alleles at certain genes confer the ability to grow on soils high in heavy metals. The ability to tolerate heavy metals comes at a price, however; individuals with the resistance allele exhibit lower growth rates on nonpolluted soil. Consequently, we would expect the resistance allele to occur with a frequency of 100% on mine sites and 0% elsewhere.

Heavy-metal tolerance has been studied intensively in the slender bent grass *Agrostis tenuis*, in which the resistance allele occurs at intermediate levels in many areas (figure 20.19). The explanation relates to the reproductive system of this grass, in which pollen, the floral equivalent of sperm, is dispersed by the wind. As a result, pollen grains—and the alleles they carry—can move great distances, leading to levels of gene flow between mine sites and unpolluted areas high enough to counteract the effects of natural selection.

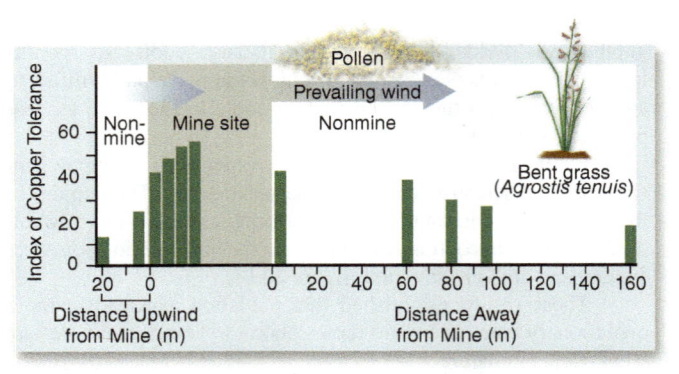

Figure 20.19 Degree of copper tolerance in grass plants on and near ancient mine sites. Individuals with tolerant alleles have decreased growth rates on unpolluted soil. Thus, we would expect copper tolerance to be 100% on mine sites and 0% on nonmine sites. However, prevailing winds blow pollen containing nontolerant alleles onto the mine site and tolerant alleles beyond the site's borders. The amount of pollen received decreases with distance, which explains the changes in levels of tolerance. The index of copper tolerance is calculated as the growth rate of a plant on soil with high concentrations of copper relative to growth rate on soils with low levels of copper; the higher the index, the more tolerant the plant is of heavy metal pollution.

Data analysis Examine the index of copper tolerance on nonmine areas. What does it suggest about the factor responsible? In particular, how does the index change as a function of distance from the mine on the right-hand side? And how does the relationship between distance and index value differ on the two sides of the mine? What process might be responsible for such patterns?

In general, the extent to which gene flow can hinder the effects of natural selection should depend on the relative strengths of the two processes. In species in which gene flow is generally strong, such as in birds and wind-pollinated plants, the frequency of the allele less favored by natural selection may be relatively high. In more sedentary species that exhibit low levels of gene flow, such as salamanders, the favored allele should occur at a frequency near 100%.

Learning Outcomes Review 20.8

Allele frequencies sometimes reflect a balance between opposing processes. Gene flow, for example, may increase some alleles while natural selection decreases them. Where several processes are involved, observed frequencies depend on the relative strength of the processes.

- *Under what circumstances might evolutionary processes operate in the same direction, and what would be the outcome?*

20.9 The Limits of Selection

Learning Outcomes
1. Define pleiotropy and epistasis.
2. Explain how these phenomena may affect the evolutionary response to selective pressure.

Although selection is the most powerful of the principal agents of genetic change, there are limits to what it can accomplish. These limits result from multiple phenotypic effects of alleles, lack of genetic variation upon which selection can act, and interactions between genes.

Genes have multiple effects

Alleles often affect multiple aspects of a phenotype (the phenomenon of *pleiotropy;* see chapter 12). These multiple effects tend to set limits on how much a phenotype can be altered.

For example, selecting for large clutch size in chickens eventually leads to eggs with thinner shells that break more easily. For this reason, we could never produce chickens that lay eggs twice as large as the best layers do now. Likewise, we cannot produce gigantic cattle that yield twice as much meat as our leading breeds, or corn with an ear at the base of every leaf, instead of just at the bases of a few leaves.

Evolution requires genetic variation

Genetic variation is a prerequisite for evolutionary change; without it, natural selection cannot produce evolution.

Over 80% of the gene pool of the thoroughbred horses racing today goes back to 31 ancestors from the late 18th century. Despite intense directional selection on thoroughbreds, their performance times have not improved for more than 50 years (figure 20.20). Decades of intense selection presumably have removed variation from the population at a rate greater than mutation can replenish it, such that little genetic variation now remains, and evolutionary change is not possible.

In some cases, phenotypic variation for a trait may never have had a genetic basis. The compound eyes of insects are made up of hundreds of visual units, termed ommatidia (described in chapter 34). In some individuals, the left eye contains more ommatidia than the right. In other individuals, the right eye contains more than the left (figure 20.21). However, despite intense selection experiments in the laboratory, scientists have never been able to produce a line of fruit flies that consistently has more ommatidia in the left eye than in the right.

The reason is that separate genes do not exist for the left and right eyes. Rather, the same genes affect both eyes, and differences in the number of ommatidia result from differences that occur as the eyes are formed in the development process. Thus, despite the existence of phenotypic variation, no underlying genetic variation is available for selection to favor.

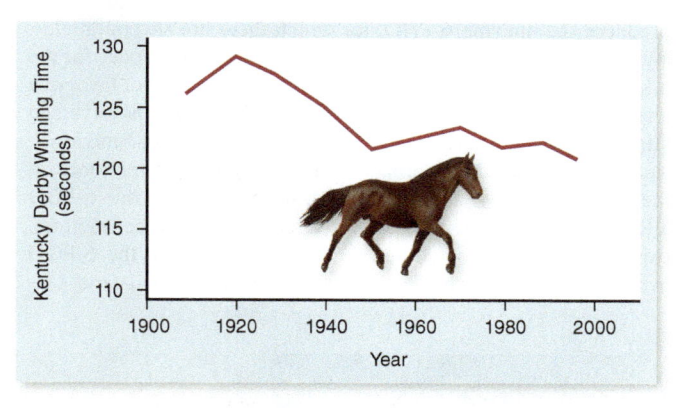

Figure 20.20 **Selection for increased speed in racehorses is no longer effective.** Kentucky Derby winning speeds have not improved significantly since 1950.

? Inquiry question What might explain the lack of change in winning speeds?

Gene interactions affect fitness of alleles

As discussed in chapter 12, *epistasis* is the phenomenon in which an allele for one gene may have different effects, depending on alleles present at other genes. Because of epistasis, the selective advantage of an allele at one gene may vary from one genotype to another. If a population is polymorphic for a second gene, then selection on the first gene may be constrained because different alleles are favored in different individuals of the same population.

Studies on bacteria illustrate how selection on alleles for one gene can depend on which alleles are present at other genes. In *E. coli,* two biochemical pathways exist to break down gluconate, each using enzymes produced by different genes. One gene

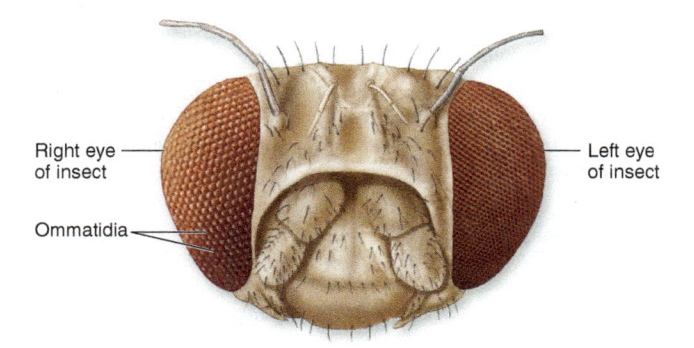

Figure 20.21 **Phenotypic variation in insect ommatidia.** In some individuals, the number of ommatidia in the left eye is greater than the number in the right.

produces the enzyme 6-PGD, for which there are several alleles. When the common allele for the second gene, which codes for the other biochemical pathway, is present, selection does not favor one allele over another at the 6-PGD gene. In some *E. coli*, however, an alternative allele at the second gene occurs that is not functional. The bacteria with this alternative allele are forced to rely only on the 6-PGD pathway, and in this case, selection favors one 6-PGD allele over another. Thus, epistatic interactions exist between the two genes, and the outcome of natural selection on the 6-PGD gene depends on which alleles are present at the second gene.

Learning Outcomes Review 20.9

In pleiotropy, a single gene affects multiple traits; in epistasis, interaction between alleles of different genes affects a single trait. Both these conditions can constrain the effects of natural selection.

- How can epistasis and pleiotropy constrain the evolutionary response to natural selection?

Chapter Review

20.1 Genetic Variation and Evolution

Many processes can lead to evolutionary change.
Darwin proposed that evolution of species occurs by the process of natural selection. Other processes can also lead to evolutionary change.

Populations contain ample genetic variation.
For a population to be able to evolve, it must contain genetic variation. DNA testing shows that natural populations generally have substantial variation.

20.2 Changes in Allele Frequency (figure 20.3)

The Hardy–Weinberg principle allows prediction of genotype frequencies.
Hardy–Weinberg equilibrium exists when observed genotype frequencies match the prediction from calculated frequencies. It occurs only when evolutionary processes are not acting to shift the distribution of alleles or genotypes in the population.

Hardy–Weinberg predictions can be applied to data to find evidence of evolutionary processes.
If genotype frequencies are not in Hardy–Weinberg equilibrium, then evolutionary processes must be at work.

20.3 Five Agents of Evolutionary Change (figure 20.4)

Mutation changes alleles.
Mutations are the ultimate source of genetic variation. Because mutation rates are low, mutation usually is not responsible for deviations from Hardy–Weinberg equilibrium.

Gene flow occurs when alleles move between populations.
Gene flow is the migration of new alleles into a population. It can introduce genetic variation and can homogenize allele frequencies between populations.

Nonrandom mating shifts genotype frequencies.
Assortative mating, in which similar individuals tend to mate, increases homozygosity; disassortative mating increases the frequency of heterozygotes.

Genetic drift may alter allele frequencies in small populations.
Genetic drift refers to random shifts in allele frequency. Its effects may be severe in small populations.

Selection favors some genotypes over others.
For evolution by natural selection to occur, genetic variation must exist, it must result in differential reproductive success, and it must be inheritable.

20.4 Quantifying Natural Selection

A phenotype with greater fitness usually increases in frequency.
Fitness is defined as the reproductive success of an individual. Relative fitness refers to the success of one genotype relative to others in a population. Usually, the genotype with highest relative fitness increases in frequency in the next generation.

Fitness may consist of many components.
Reproductive success is determined by how long an individual survives, how often it mates, and how many offspring it has per reproductive event.

20.5 Natural Selection's Role in Maintaining Variation

Frequency-dependent selection may favor either rare or common phenotypes.
Negative frequency-dependent selection favors rare phenotypes and maintains variation within a population. Positive frequency-dependent selection favors the common phenotype and leads to decreased variation.

In oscillating selection, the favored phenotype changes as the environment changes.
If environmental change is cyclical, selection would favor first one phenotype, then another, maintaining variation.

In some cases, heterozygotes may exhibit greater fitness than homozygotes.
Heterozygote advantage favors individuals with both alleles.

20.6 Selection Acting on Traits Affected by Multiple Genes (figure 20.12)

Disruptive selection removes intermediates.
When intermediate phenotypes are at a disadvantage, a population may exhibit a bimodal trait distribution.

Directional selection eliminates phenotypes at one end of a range.
Directional selection tends to shift the mean value of the population toward the favored end of the distribution.

Stabilizing selection favors individuals with intermediate phenotypes.
Stabilizing selection eliminates both extremes and increases the frequency of an intermediate type. The population may have the same mean value, but with decreased variation.

20.7 Experimental Studies of Natural Selection

The hypothesis that natural selection leads to evolutionary change can be tested experimentally.

Guppy color variation in different environments suggests natural selection at work.

Experimentation reveals the agent of selection.

Guppies in natural populations subject to different predators were shown to undergo color change over generations.

20.8 Interactions Among Evolutionary Forces

Mutation and genetic drift may counter selection.

In theory, a high rate of mutation could oppose natural selection, but this rarely happens. Genetic drift also can work counter to natural selection.

Gene flow may promote or constrain evolutionary change.

Gene flow can spread a beneficial mutation to other populations, but it can also impede adaptation due to influx of alleles with low fitness in a population's environment.

20.9 The Limits of Selection

Genes have multiple effects.

Pleiotropic genes, which have multiple effects, set limits on how much a phenotype can be altered. Even if one affected trait is favored, other affected traits may not be.

Evolution requires genetic variation.

Intense selection pressure may remove genetic variation.

Gene interactions affect fitness of alleles.

In epistasis, fitness of one allele may vary depending on the genotype of a second gene.

Review Questions

UNDERSTAND

1. Assortative mating
   a. affects genotype frequencies expected under Hardy–Weinberg equilibrium.
   b. affects allele frequencies expected under Hardy–Weinberg equilibrium.
   c. has no effect on the genotypic frequencies expected under Hardy–Weinberg equilibrium because it does not affect the relative proportion of alleles in a population.
   d. increases the frequency of heterozygous individuals above Hardy–Weinberg expectations.

2. When the environment changes from year to year and different phenotypes have different fitness in different environments
   a. natural selection will operate in a frequency-dependent manner.
   b. the effect of natural selection may oscillate from year to year, favoring alternative phenotypes in different years.
   c. genetic variation is not required to get evolutionary change by natural selection.
   d. None of the choices is correct.

3. Many factors can limit the ability of natural selection to cause evolutionary change, including
   a. a conflict between reproduction and survival as seen in Trinidadian guppies.
   b. lack of genetic variation.
   c. pleiotropy.
   d. All of the choices are correct.

4. Stabilizing selection differs from directional selection because
   a. in the former, phenotypic variation is reduced but the average phenotype stays the same, whereas in the latter both the variation and the mean phenotype change.
   b. the former requires genetic variation, but the latter does not.
   c. intermediate phenotypes are favored in directional selection.
   d. None of the choices is correct.

5. Founder effects and bottlenecks are
   a. expected only in large populations.
   b. mechanisms that increase genetic variation in a population.
   c. two different modes of natural selection.
   d. forms of genetic drift.

6. *Relative fitness*
   a. refers to the survival rate of one phenotype compared to that of another.
   b. is the physical condition of an individual's siblings and cousins.
   c. refers to the reproductive success of a phenotype.
   d. None of the choices is correct.

7. For natural selection to result in evolutionary change
   a. variation must exist in a population.
   b. reproductive success of different phenotypes must differ.
   c. variation must be inherited from one generation to the next.
   d. All of the choices are correct.

APPLY

1. In a population of red (dominant allele) or white flowers in Hardy–Weinberg equilibrium, the frequency of red flowers is 91%. What is the frequency of the red allele?
   a. 9%
   b. 30%
   c. 91%
   d. 70%

2. Genetic drift and natural selection can both lead to rapid rates of evolution. However,
   a. genetic drift works fastest in large populations.
   b. only drift leads to adaptation.
   c. natural selection requires genetic drift to produce new variation in populations.
   d. both processes of evolution can be slowed by gene flow.

3. Suppose that the relationship between birth weight and infant mortality, instead of being at a minimum at intermediate sizes, changed such that babies born at 5 or 10 pounds had the highest survival, with a valley in between such that 7.5-pound babies had

low survival rate. How would you expect the distribution of birth weights to change over time?

a. It would not change.
b. The distribution would shift to the right.
c. The distribution would become bimodal, with two peaks and the mean value unchanged.
d. The distribution would become bimodal, with two peaks and the mean value shifted to the right.

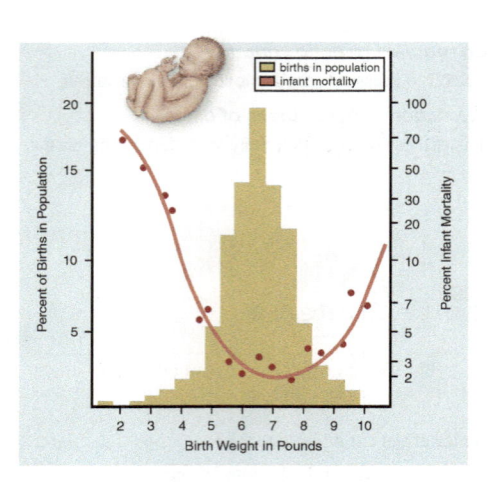

SYNTHESIZE

1. In Trinidadian guppies a combination of elegant laboratory and field experiments builds a very compelling case for predator-induced evolutionary changes in color and life history traits. It is still possible, although not likely, that there are other differences between the sites above and below the falls aside from whether predators are present. What additional studies could strengthen the interpretation of the results?

2. On large, black lava flows in the deserts of the southwestern United States, populations of many types of animals are composed primarily of black individuals. By contrast, on small lava flows, populations often have a relatively high proportion of light-colored individuals. How can you explain this difference?

3. Based on a consideration of how strong artificial selection has helped eliminate genetic variation for speed in thoroughbred horses, we are left with the question of why, for many traits like speed (continuous traits), there is usually abundant genetic variation. This is true even for traits we know are under strong selection. Where does genetic variation ultimately come from, and how does the rate of production compare with the strength of natural selection? What other mechanisms can maintain and increase genetic variation in natural populations?

CHAPTER 21

The Evidence for Evolution

Chapter Contents

21.1 The Beaks of Darwin's Finches: Evidence of Natural Selection

21.2 Peppered Moths and Industrial Melanism: More Evidence of Selection

21.3 Artificial Selection: Human-Initiated Change

21.4 Fossil Evidence of Evolution

21.5 Anatomical Evidence for Evolution

21.6 Convergent Evolution and the Biogeographical Record

21.7 Darwin's Critics

Introduction

As we discussed in chapter 1, when Darwin proposed his revolutionary theory of evolution by natural selection, little actual evidence existed to bolster his case. Instead, Darwin relied on observations of the natural world, logic, and results obtained by breeders working with domestic animals. Since his day, however, the evidence for Darwin's theory has become overwhelming.

The case is built upon two pillars: First, evidence that natural selection can produce evolutionary change, and second, evidence from the fossil record that evolution has occurred. The whale skeleton pictured here is the Vogtle whale *(Georgiacetus vogtlensis)*, which is the oldest whale fossil found in North America (40 million years old). The pelvic bones and hindlimbs reveal a link between land mammals and whales. In addition to evidence from studies of natural selection and fossils, information from many different areas of biology—fields as different as anatomy, molecular biology, and biogeography—is only interpretable scientifically as being the outcome of evolution.

21.1 The Beaks of Darwin's Finches: Evidence of Natural Selection

Learning Outcomes
1. Describe how the species of Darwin's finches have adapted to feed in different ways.
2. Explain how climatic variation drives evolutionary change in the medium ground finch.

As you learned in chapter 20, a variety of processes can produce evolutionary change. Most evolutionary biologists, however, agree with Darwin's thinking that natural selection is the primary process responsible for evolution. Although we cannot travel back through time, modern-day evidence allows us to test hypotheses about how evolution proceeds and confirms the power of natural selection as an agent of evolutionary change. This evidence comes from both the field and the laboratory and from both natural and human-altered situations.

Darwin's finches are a classic example of evolution by natural selection. When he visited the Galápagos Islands off the coast of Ecuador in 1835, Darwin collected 31 specimens of finches from three islands. Darwin, not an expert on birds, had trouble identifying the specimens, believing by examining their beaks that his collection contained wrens, "gross-beaks," and blackbirds, all birds with which he was familiar from his native England.

Upon Darwin's return home, ornithologist John Gould informed Darwin that his collection was in fact a closely related group of distinct species, all similar to one another except for their beaks. In all, 14 species are now recognized.

Galápagos finches exhibit variation related to food gathering

The diversity of Darwin's finches is illustrated in figure 21.1. The ground finches feed on seeds that they crush in their powerful beaks; species with smaller and narrower beaks, such as the warbler finch, eat insects. Other species include fruit and bud eaters, and species that feed on cactus fruits and the insects they attract; some populations of the sharp-beaked ground finch even include "vampires" that sometimes creep up on seabirds and use their sharp beaks to pierce the seabirds' skin and drink their blood. Perhaps most remarkable are the tool users, woodpecker finches that pick up a twig, cactus spine, or leaf stalk, trim it into shape with their beaks, and then poke it into dead branches to pry out grubs.

The correspondence between the beaks of the finch species and their food source suggested to Darwin that natural selection had shaped them. In *The Voyage of the Beagle,* Darwin wrote, "Seeing this gradation and diversity of structure in one small, intimately related group of birds, one might really fancy that from an original paucity of birds in this archipelago, one species has been taken and modified for different ends."

Woodpecker finch (*Cactospiza pallida*)

Large ground finch (*Geospiza magnirostris*)

Cactus finch (*Geospiza scandens*)

Warbler finch (*Certhidea olivacea*)

Vegetarian tree finch (*Platyspiza crassirostris*)

Figure 21.1 Darwin's finches. These species show differences in beaks and feeding habits among Darwin's finches. This diversity arose when an ancestral finch colonized the islands and diversified into habitats lacking other types of small birds. The beaks of several species resemble those of different families of birds on the mainland. For example, the warbler finch has a beak very similar to warblers, to which it is not closely related.

Modern research has verified Darwin's selection hypothesis

Darwin's observations suggest that differences among species in beak size and shape have evolved as the species adapted to use different food resources, but can this hypothesis be tested? In chapter 20, you read that the theory of evolution by natural selection requires that three conditions be met:

1. Phenotypic variation must exist in the population.
2. This variation must lead to differences among individuals in lifetime reproductive success.
3. Phenotypic variation among individuals must be genetically transmissible to the next generation.

The key to successfully testing Darwin's proposal proved to be patience. For more than 40 years, starting in 1973, Peter and Rosemary Grant of Princeton University and their students have studied the medium ground finch on a tiny island in the center of the Galápagos called Daphne Major. These finches feed preferentially on small, tender seeds, produced in abundance by plants in wet years. The birds resort to larger, drier seeds, which are harder to crush, only when small seeds become depleted during long periods of dry weather, when plants produce few seeds.

The Grants quantified beak shape among the medium ground finches of Daphne Major by carefully measuring beak depth (height of beak, from top to bottom, at its base) on individual birds. Measuring many birds every year, they were able to assemble for the first time a detailed portrait of evolution in action. The Grants found that not only did a great deal of variation in beak depth exist among members of the population, but the average beak depth changed from one year to the next in a predictable fashion.

During droughts, plants produced few seeds, and all available small seeds were quickly eaten, leaving large seeds as the major remaining source of food. As a result, birds with shorter and deeper, more powerful beaks survived better, because they were better able to break open these large seeds. Consequently, the birds evolved to have blunter beaks in the next generation. Then, when normal rains returned and small seeds recovered in abundance, average beak depth and length of the population returned to their original size (figure 21.2a).

Conversely, in particularly wet years, plants flourished, producing an abundance of small seeds; as a result, birds with long and shallow beaks were favored, and beaks became pointier.

Could these changes in beak dimension be evidence of evolution by natural selection? If so, the variation among individuals in beak size must be genetically based. An alternative possibility might be that the changes in beak depth do not reflect changes in gene frequencies, but rather are simply a response to diet—for example, perhaps crushing large seeds causes a growing bird to develop a larger beak.

To rule out this possibility, the Grants tested for a genetic basis to beak size variation by measuring the relationship of parent beak size to offspring beak size, examining many broods over several years. The depth of the beak was very similar between parents and offspring regardless of environmental conditions (figure 21.2b), suggesting that the differences among individuals in beak size reflect genetic differences, and therefore that the year-to-year changes in average beak depth represent evolutionary change resulting from natural selection.

a.

b.

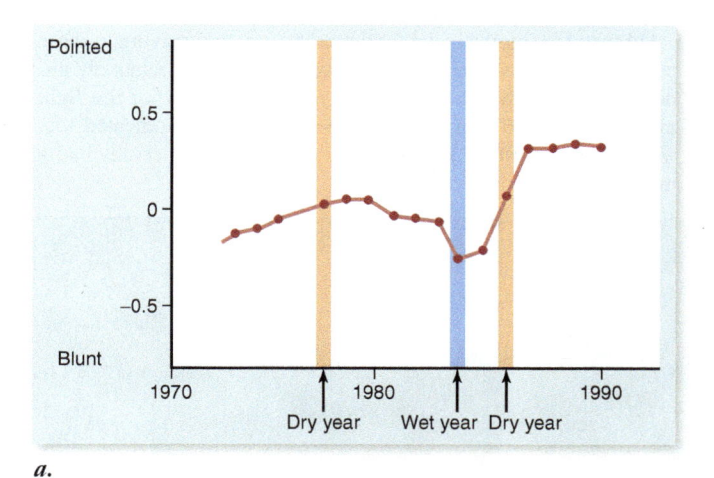

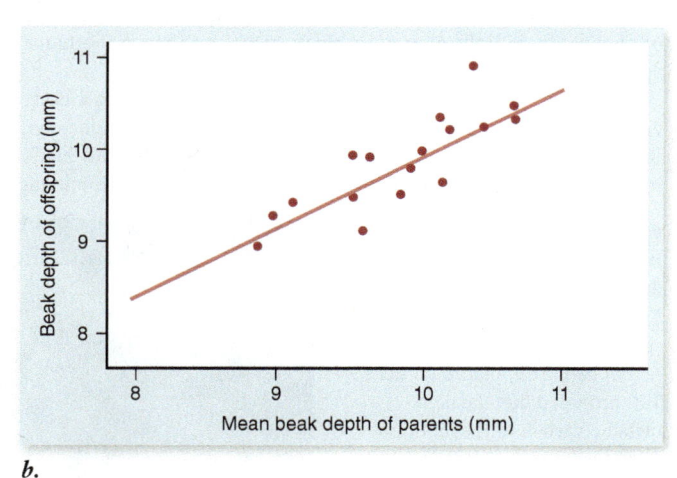

Figure 21.2 Evidence that natural selection alters beak shape in the medium ground finch, *Geospiza fortis,*. *a.* In dry years, when only large, tough seeds are available, the mean beak depth increases. (The y-axis measures beak shape relative to the average beak shape across all years.) In wet years, when many small seeds are available, mean beak depth decreases. *b.* Beak depth is inherited from parents to offspring.

Inquiry question What would the relationship in figure 21.2b look like if beak shape were not determined genetically, but rather by an environmental factor, such as what a nestling bird ate during its growth period?

Data analysis Suppose that a male with a beak depth of 10 mm mated with a female with a beak depth of 8 mm. What would the expected beak depth of the offspring be? Would it matter if the female's beak was 10 mm and the male's 6 mm?

Learning Outcomes Review 21.1

Among Darwin's finches, natural selection has been responsible for changes in the shape of the beak corresponding to characteristics of the available food supply. Because beak morphology is a genetically transmissible trait, a beak better suited to the distribution of available seed types would become more common in subsequent generations.

- Suppose that the act of eating hard seeds caused birds to develop bigger beaks. Would this lead to an evolutionary increase in beak size after a drought?

21.2 Peppered Moths and Industrial Melanism: More Evidence of Selection

Learning Outcomes

1. Explain the relationship between pollution and color evolution in peppered moths.
2. Distinguish between demonstrating that evolution has occurred and understanding the mechanism that caused it.

When the environment changes, natural selection often may favor a trait that previously wasn't favored. One classic example concerns the peppered moth, *Biston betularia*. Adults come in a range of shades, from light gray with black speckling (hence the name "peppered" moth) to jet black (melanic).

Extensive genetic analysis has shown that the moth's body color is a genetic trait that reflects different alleles of a single gene. Recent molecular genetic studies have demonstrated that all-black individuals are the descendants of a single mutation. This dominant allele was present but very rare in populations before 1850. From that time on, dark individuals increased in frequency in moth populations near industrialized centers until they made up almost 100% of these populations.

Biologists soon noticed that in industrialized regions where the dark moths were common, the tree trunks were darkened almost black by the soot of pollution, which also killed many of the light-colored lichens on tree trunks.

Light-colored moths decreased in polluted areas

Why did dark moths gain a survival advantage around 1850? In 1896, a British amateur moth collector named J. W. Tutt proposed what became the most commonly accepted hypothesis explaining the decline of the light-colored moths. He suggested that peppered forms were more visible to predators on sooty trees that have lost their lichens. Consequently, birds ate the peppered moths resting on the trunks of trees during the day. The black forms, in contrast, had an advantage because they were camouflaged (figure 21.3).

Although Tutt initially had no evidence, British ecologist Bernard Kettlewell tested the hypothesis in the 1950s by releasing equal numbers of dark and light individuals into two sets of woods: one near heavily polluted Birmingham, and the other in unpolluted Dorset. If Tutt was correct, then dark moths should have survived better in the Birmingham woods and light moths in the Dorset woods. Kettlewell then set up lights in the woods to attract moths to traps to see how many of both kinds of moths survived. To identify the moths he had released, he had marked the released moths with a dot of paint on the underside of their wings, where birds could not see it.

In the polluted area near Birmingham, Kettlewell recaptured only 19% of the light moths, but 40% of the dark ones. This indicated that dark moths had a far better chance of surviving in these polluted woods, where tree trunks were dark. In the relatively unpolluted Dorset woods, Kettlewell recovered 12.5% of the light moths but only 6% of the dark ones. This result indicated that where the tree trunks were still light-colored, light moths had a much better chance of survival.

Figure 21.3 Tutt's hypothesis explaining industrial melanism. These photographs show preserved specimens of the peppered moth, *Biston betularia*, placed on trees. Tutt proposed that the dark melanic variant of the moth is more visible to predators on unpolluted trees *(left)*, whereas the light "peppered" moth is more visible to predators on bark blackened by industrial pollution *(right)*.

Kettlewell later solidified his argument by placing moths on trees and filming birds looking for food. Sometimes the birds actually passed right over a moth that was the same color as its background.

Recently, an enormous six-year study involving the release of nearly 5000 moths confirmed Kettlewell's findings. Conducted in an unpolluted forest, the study found that dark-colored moths disappeared at a rate 10% higher than light-colored moths. In addition, direct observations of 250 feeding events revealed that dark moths were captured by birds substantially more often than light-colored moths.

When environmental conditions reverse, so does selection pressure

In industrialized areas throughout Eurasia and North America, dozens of other species of moths have evolved in the same way as the peppered moth. The term **industrial melanism** refers to the phenomenon in which darker individuals come to predominate over lighter ones.

In Great Britain, the air pollution that promoted industrial melanism began to reverse following enactment of the Clean Air Act in 1956. Beginning in 1959, the *Biston* population at Caldy Common outside Liverpool has been sampled each year. The frequency of the melanic (dark) form has dropped from a high of 93% in 1959 to less than 5% in 2002 (figure 21.4).

The drop is consistent with a 15% selective disadvantage acting against moths with the dominant melanic allele; it correlates well with a significant drop in air pollution, particularly with a lowering of the levels of sulfur dioxide and suspended particulates, both of which act to darken trees.

Interestingly, the same reversal of melanism occurred in the United States. Of 576 peppered moths collected at a field station near Detroit from 1959 to 1961, 515 were melanic, a frequency of 89%. The American Clean Air Act, passed in 1963, led to significant reductions in air pollution. Resampled in 2001, the frequency of melanics at the Detroit field station also had dropped to less than 5%; a similar decreasing trend was also observed in Pennsylvania (figure 21.4). The moth populations in Liverpool, Detroit, and Pensylvania, all part of the same natural experiment, exhibit strong evidence for natural selection.

The agent of selection may be difficult to pin down

Although the evidence for natural selection in the case of the peppered moth is strong, Tutt's hypothesis that predation on color-mismatched moths is the underlying mechanism of selection is currently being reevaluated. Researchers have noted that the recent selection against melanism does not appear to correlate with changes in the abundance of light-colored tree lichens.

At Caldy Common, the light form of the peppered moth began to increase in frequency long before lichens began to reappear on the trees. At the Detroit field station, the lichens never changed significantly as the dark moths first became dominant and then declined over a 30-year period. In fact, investigators have not been able to find peppered moths on Detroit trees at all, whether covered with lichens or not. Some evidence suggests the moths rest on leaves in the treetops during the day, but no one is sure. Could poisoning by pollution rather than predation by birds be the agent of natural selection on the moths? Perhaps—but to date, only selection resulting from bird predation has been demonstrated.

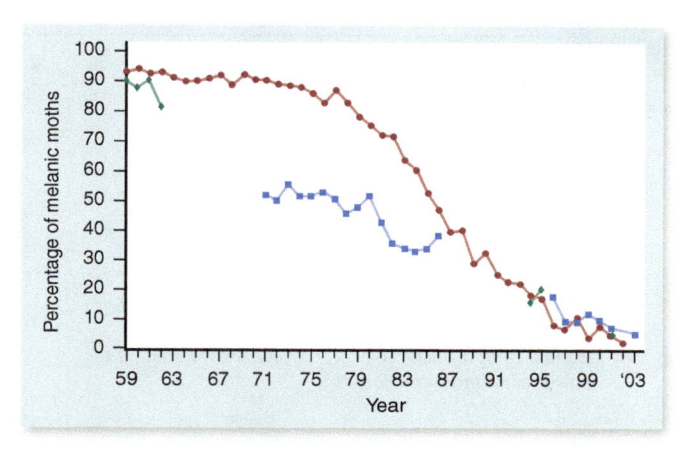

Figure 21.4 Selection against melanism. The red circles indicate the frequency of melanic *Biston betularia* moths at Caldy Common in Great Britain. Green diamonds indicate frequencies of melanic *B. betularia* in Michigan, and the blue squares indicate corresponding frequencies in Pennsylvania.

> **Inquiry question** What can you conclude from the fact that the frequency of melanic moths decreased to the same degree in the two locations?

Researchers supporting the bird predation hypothesis point out that a bird's ability to detect moths may depend less on the presence or absence of lichens, and more on other ways in which the environment is darkened by industrial pollution. Pollution tends to cover all objects in the environment with a fine layer of particulate dust, which tends to decrease how much light surfaces reflect. In addition, pollution has a particularly severe effect on birch trees, which are light in color. Both effects would tend to make the environment darker, and thus would favor darker moths by protecting them from predation by birds.

Despite this uncertainty over the agent of selection, the overall pattern is clear. Kettlewell's experiments established indisputably that selection favors dark moths in polluted habitats and light moths in pristine areas. The increase and subsequent decrease in the frequency of melanic moths, correlated with levels of pollution independently on two continents, demonstrates clearly that this selection drives evolutionary change.

The current reconsideration of the agent of natural selection illustrates well the way in which scientific progress is achieved: Hypotheses, such as Tutt's, are put forth and then tested. If rejected, new hypotheses are formulated, and the process begins anew.

Learning Outcomes Review 21.2

Natural selection has favored the dark form of the peppered moth in areas subject to severe air pollution, perhaps because on darkened trees they are less easily seen by moth-eating birds. As pollution has abated, selection has in turn shifted to favor the light form. Although selection is clearly occurring, further research is required to understand whether predation by birds is the agent of selection.

- *How would you test the idea that predation by birds is the agent of selection on moth coloration?*

21.3 Artificial Selection: Human-Initiated Change

Learning Outcomes
1. Contrast the processes of artificial and natural selection.
2. Explain what artificial selection demonstrates about the power of natural selection.

Humans have imposed selection upon plants and animals since the dawn of civilization. Just as in natural selection, such **artificial selection** operates by favoring individuals with certain phenotypic traits, allowing them to reproduce and pass their genes on to the next generation. Assuming that phenotypic differences are genetically determined, this directional selection should lead to evolutionary change, and indeed it has.

Artificial selection, imposed in laboratory experiments, agriculture, and the domestication process, has produced substantial change in almost every case in which it has been applied. This success is strong proof that selection is an effective evolutionary process.

Experimental selection produces changes in populations

With the rise of genetics as a field of science in the 1920s and 1930s, researchers began conducting experiments to test the hypothesis that selection can produce evolutionary change. A favorite subject was the laboratory fruit fly, *Drosophila melanogaster*. Geneticists have imposed selection on just about every conceivable aspect of the fruit fly—including body size, eye color, growth rate, life span, and exploratory behavior—with a consistent result: Selection for a trait leads to strong and predictable evolutionary response.

In one classic experiment, scientists selected for fruit flies with many bristles (stiff, hairlike structures) on their abdomens. At the start of the experiment, the average number of bristles was 9.5. Each generation, scientists picked out the 20% of the population with the greatest number of bristles and allowed them to reproduce, thus establishing the next generation. After 86 generations of this directional selection, the average number of bristles had quadrupled, to nearly 40! In another experiment, fruit flies in one population were selected for high numbers of bristles, while fruit flies in the other cage were selected for low numbers of bristles. Within 35 generations, the populations did not overlap at all in range of variation (figure 21.5).

Similar experiments have been conducted on a wide variety of other laboratory organisms. For example, by selecting for rats that were resistant to tooth decay, in less than 20 generations scientists were able to increase the average time for onset of decay from barely over 100 days to greater than 500 days.

SCIENTIFIC THINKING

Question: Can artificial selection lead to substantial evolutionary change?
Hypothesis: Strong directional selection will quickly lead to a large shift in the mean value of the population.
Experiment: In one population, every generation pick out the 20% of the population with the most bristles and allow them to reproduce to form the next generation. In the other population, do the same with the 20% with the fewest number of bristles.

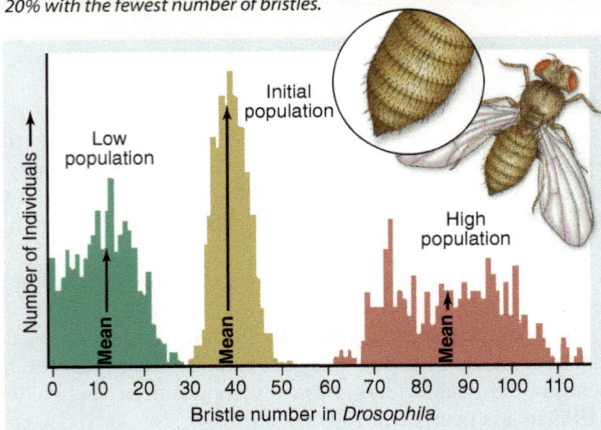

Result: After 35 generations, mean number of bristles has changed substantially in both populations.
Interpretation: Note that at the end of the experiment, the range of variation lies outside the range seen in the initial population. Selection can move a population beyond its original range because mutation and recombination continuously introduce new variation into populations.

Figure 21.5 Artificial selection can lead to rapid and substantial evolutionary change.

? Inquiry question What would happen if, within a population, both small and large individuals were allowed to breed, but middle-sized ones were not?

Agricultural selection has led to extensive modification of crops and livestock

Familiar livestock, such as cattle and pigs, and crops, such as corn and strawberries, are greatly different from their wild ancestors (figure 21.6). These differences have resulted from generations of human selection for desirable traits, such as greater milk production and larger corn ear size.

An experiment with corn demonstrates the ability of artificial selection to rapidly produce major change in crop plants. In 1896, agricultural scientists began selecting for the oil content of corn kernels, which initially was 4.5%. Just as in the fruit fly experiments, the top 20% of all individuals were allowed to reproduce. By 1986, at which time 90 generations had passed, average oil content of the corn kernels had increased approximately 450%.

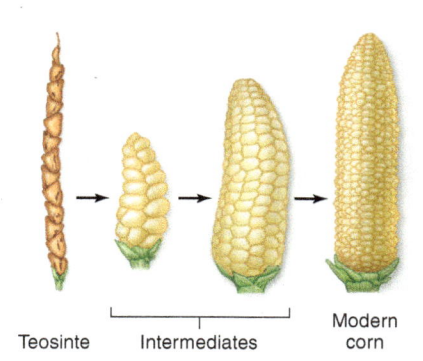

Figure 21.6 Corn looks very different from its ancestor. Teosinte, which can be found today in a remote part of Mexico, is very similar to the ancestor of modern corn. Artificial selection has transformed it into the form we know today.

Domesticated breeds have arisen from artificial selection

Human-imposed selection has produced a great variety of breeds of cats, dogs (figure 21.7), pigeons, and other domestic animals. In some cases, breeds have been developed for particular purposes. Greyhound dogs, for example, resulted from selection for maximal running ability, resulting in an animal with long legs, a long tail for balance, an arched back to increase stride length, and great muscle mass. By contrast, the odd proportions of the ungainly dachshund resulted from selection for dogs that could enter narrow holes in pursuit of badgers. In other cases, varieties have been selected primarily for their appearance, such as the many colorful breeds of pigeons or cats.

Domestication also has led to unintentional selection for some traits. In recent years, as part of an attempt to domesticate the silver fox, Russian scientists have chosen the most docile animals in each generation and allowed them to reproduce. Within 40 years, most foxes were exceptionally tame, not only allowing themselves to be petted, but also whimpering to get attention and sniffing and licking their caretakers (figure 21.8). In many respects, they had become no different from domestic dogs.

It was not only their behavior that changed, however. These foxes also began to exhibit other traits seen in some dog breeds, such as different color patterns, floppy ears, curled tails, and shorter legs and tails. Presumably, the genes responsible for docile behavior either affect these traits as well or are closely linked to the genes for these other traits (the phenomena of pleiotropy and linkage, which are discussed in chapters 12 and 13).

Can selection produce major evolutionary changes?

Given that we can observe the results of selection operating over a relatively short time, most scientists think that natural selection is the process responsible for the evolutionary changes documented in the fossil record. Some critics of evolution accept that selection can lead to changes within a species, but contend that such changes are relatively minor in scope and not equivalent to the substantial changes documented in the fossil record. In other words, it is one thing to change the number of bristles on a fruit fly or the size of an ear of corn, and quite another to produce an entirely new species.

This argument does not fully appreciate the extent of change produced by artificial selection. Consider, for example, the existing breeds of dogs, all of which have been produced since wolves were first domesticated, perhaps 10,000 years ago. If the various dog breeds did not exist and a paleontologist found fossils of animals similar to dachshunds, greyhounds, mastiffs, and chihuahuas, there is no question that they would be considered different species. Indeed, the differences in size and shape exhibited by these breeds are greater than those between members of different genera in the family Canidae—such as coyotes, jackals, foxes, and wolves—which have been evolving separately for 5 to 10 million years. Consequently, the claim that artificial selection produces only minor changes is clearly incorrect. If selection operating over a period of only 10,000 years can produce such substantial differences, it should be powerful enough, over the course of many millions of years, to produce the diversity of life we see around us today.

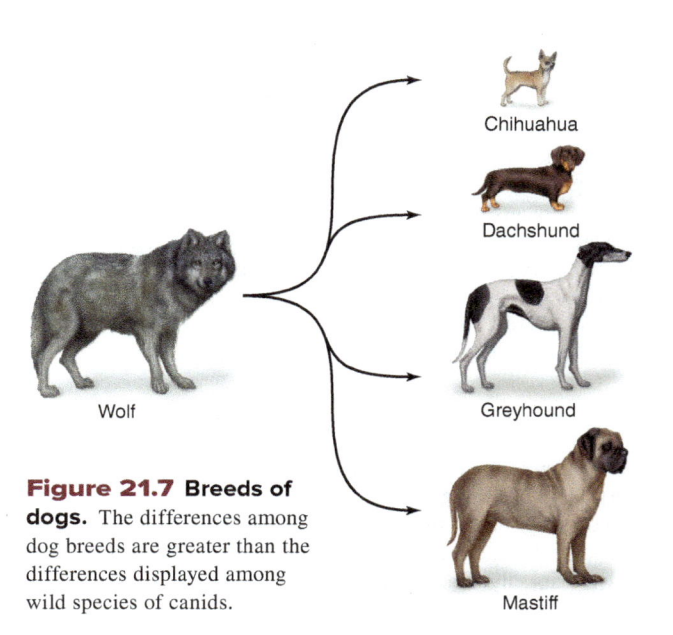

Figure 21.8 Domesticated foxes. After 40 years of selectively breeding the tamest individuals, artificial selection has produced silver foxes that are not only as friendly as domestic dogs, but also exhibit many physical traits seen in dog breeds.

Figure 21.7 Breeds of dogs. The differences among dog breeds are greater than the differences displayed among wild species of canids.

Learning Outcomes Review 21.3

In artificial selection, humans choose which plants or animals to mate in an attempt to conserve desirable traits. Rapid and substantial results can be obtained over a very short time, often in a few generations. From this we can see that natural selection is capable of producing major evolutionary change.

- In what circumstances might artificial selection fail to produce a desired change?

21.4 Fossil Evidence of Evolution

Learning Outcomes
1. Describe how fossils are formed.
2. Explain the importance of the discovery of transitional fossils.
3. Name the evolutionary trends revealed by the study of horse evolution.

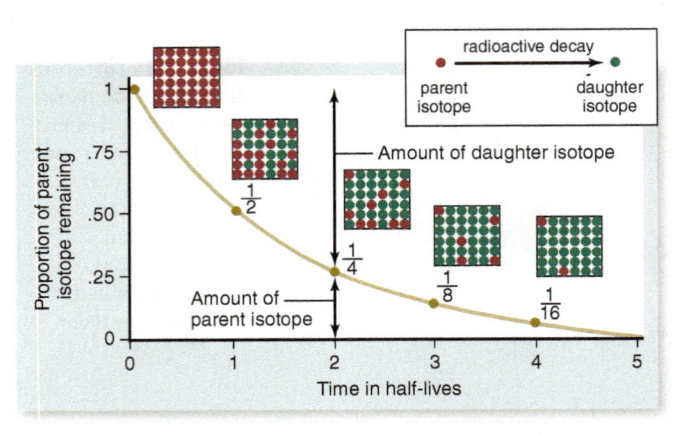

Figure 21.9 Isotopic decay. Isotopes decay at a known rate, called their half-life. After 1 half-life, one-half of the original amount of parent isotope has transformed into a daughter isotope. After each successive half-life, one-half of the remaining amount of the parent isotope is transformed.

The most direct evidence that evolution has occurred is found in the fossil record. Today we have a far more complete understanding of this record than was available in Darwin's time.

Fossils are the preserved remains of once-living organisms. They include specimens preserved in amber, Siberian permafrost, and dry caves, as well as the more common fossils preserved as rocks.

Rock fossils are created when three events occur. First, the organism must become buried in sediment; then, the calcium in bone or other hard tissue must mineralize; and finally, the surrounding sediment must eventually harden to form rock.

The process of fossilization occurs only rarely. Usually, animal or plant remains decay or are scavenged before the process can begin. In addition, many fossils occur in rocks that are inaccessible to scientists. When they do become available, they are often destroyed by erosion and other natural processes before they can be collected. As a result, only a very small fraction of the species that have ever existed (estimated by some to be as many as 500 million) are known from fossils. Nonetheless, the fossils that have been discovered are sufficient to provide detailed information on the course of evolution through time.

The age of fossils can be estimated

By dating the rocks in which fossils occur, we can get an accurate idea of how old the fossils are. In Darwin's day, rocks were dated by their position with respect to one another *(relative dating)*; rocks in lower strata are generally older because young rocks form on top of older ones. Knowing the relative positions of sedimentary rocks and the rates of erosion of different kinds of sedimentary rocks in different environments, geologists of the 19th century derived a fairly accurate idea of the relative ages of rocks.

Today, geologists can determine the absolute age of rocks using isotopic dating. At the time a rock forms, some elements exist as different isotopes. Over time the less stable isotope is converted into the other isotope and the ratio of the two forms changes.

For example, potassium is one of the most common atoms in organisms. All potassium (K) atoms have the same number of protons, but the isotopes of K vary in the number of neutrons they have. ^{40}K has 19 protons and 21 neutrons and is less stable than ^{39}K or ^{41}K. ^{40}K is converted (decays) over time and forms ^{40}Ar (argon). The ^{40}K half-life is 1.25 billion years. That is, it takes 1.25 billion years for the amount of ^{40}K to decrease by 50%. The long half-life makes it useful for dating ancient fossils by determining the ratio of ^{40}K to ^{40}Ar (figure 21.9).

For events that occurred more recently, radiocarbon dating can be used. Carbon in the form of atmospheric CO_2, with a mix of isotopes ^{14}C and ^{12}C, is incorporated into plants via photosynthesis. The relative amount of ^{14}C to ^{12}C decreases with a half-life of about 5700 years. Other isotopes can be used for more intermediate dates.

Fossils present a history of evolutionary change

When fossils are arrayed according to their age, from oldest to youngest, they often provide evidence of successive evolutionary change. At the largest scale, the fossil record documents the course of life through time, from the origin of first prokaryotic and then eukaryotic organisms, through the evolution of fishes, the rise of land-dwelling organisms, the reign of the dinosaurs, and on to the origin of humans. In addition, the fossil record shows the waxing and waning of biological diversity through time, such as the periodic mass extinctions that have reduced the number of living species. These topics are discussed at greater length in chapter 26.

Fossils document evolutionary transitions

Given the low likelihood of fossil preservation and recovery, it is not surprising that there are gaps in the fossil record. Nonetheless, intermediate forms are often available to illustrate how the major transitions in life occurred.

Undoubtedly the most famous of these is the oldest known bird, *Archaeopteryx* (meaning "ancient feather") which lived around 165 million years ago (MYA) (figure 21.10). This species is

Figure 21.10 Fossil of *Archaeopteryx*, the first bird. The remarkable preservation of this specimen reveals soft parts usually not preserved in fossils; the presence of feathers makes clear that *Archaeopteryx* was a bird, despite the presence of many dinosaurian traits. To see a picture of what this animal may have looked like in life, see figure 35.29. Fossil evidence documenting the evolutionary descent of birds from dinosaurs is considered in greater detail in chapter 23 (see figure 23.12).

clearly intermediate between birds and dinosaurs. Its feathers, similar in many respects to those of birds today, clearly reveal that it is a bird. Nonetheless, in many other respects—for example, possession of teeth, a bony tail, and other anatomical characteristics—it is indistinguishable from some carnivorous dinosaurs. Indeed, it is so similar to these dinosaurs that several specimens lacking preserved feathers were misidentified as dinosaurs and lay in the wrong natural history museum cabinet for several decades before the mistake was discovered!

Archaeopteryx reveals a pattern commonly seen in intermediate fossils—rather than being intermediate in every trait, such fossils usually exhibit some traits like their ancestors and others like their descendants. In other words, traits evolve at different rates and different times; expecting an intermediate form to be intermediate in every trait would not be correct.

The first *Archaeopteryx* fossil was discovered in 1859, the year Darwin published *On the Origin of Species*. Since then, paleontologists have continued to fill in the gaps in the fossil record. Today, the fossil record is far more complete, particularly among the vertebrates; fossils have been found linking all the major groups.

Recent years have seen spectacular discoveries, closing some of the major remaining gaps in our understanding of vertebrate evolution. For example, a four-legged aquatic mammal was discovered only recently that provides important insights concerning the evolution of whales and dolphins from land-dwelling, hoofed ancestors (figure 21.11). Similarly, a fossil snake with legs has shed light on the evolution of snakes, which are descended from lizards that gradually became more and more elongated with the simultaneous reduction and eventual disappearance of the limbs. In chapter 35, we discuss the most recent such discovery, *Tiktaalik*, a species that bridged the gap between fish and the first amphibians (see figure 35.17).

On a finer scale, evolutionary change within some types of animals is known in exceptional detail. For example, about 200 MYA, oysters underwent a change from small, curved shells to larger, flatter ones, with progressively flatter fossils seen in the fossil record over a period of 12 million years. A host of other examples illustrate similar records of successive change. The demonstration of this successive change is one of the strongest lines of evidence that evolution has occurred.

The evolution of horses is a prime example of evidence from fossils

One of the most studied cases in the fossil record concerns the evolution of horses. Modern-day members of the family Equidae include horses, zebras, donkeys, and asses, all of which are large, long-legged, fast-running animals adapted to living on open grasslands. These species, all classified in the genus *Equus*, are the last living descendants of a long lineage that has produced 34 genera since its origin in the Eocene period, approximately 55 MYA. Examination of these fossils has provided a particularly

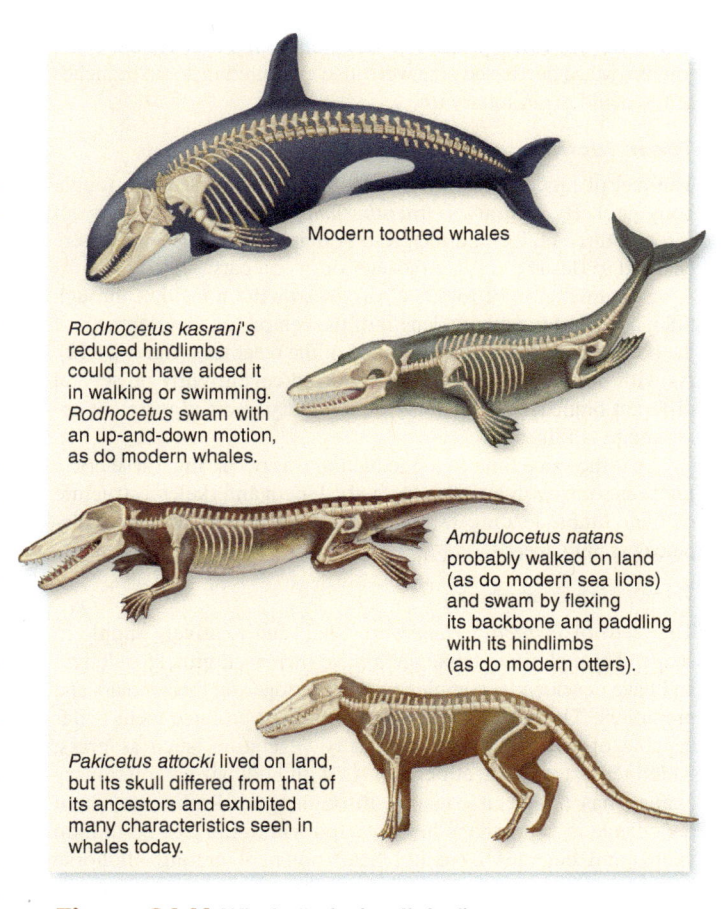

Figure 21.11 Whale "missing links." The discoveries of *Ambulocetus*, *Rodhocetus*, and *Pakicetus* have filled in the gaps between whales and their hoofed mammal ancestors. The features of *Pakicetus* illustrate that intermediate forms are not intermediate in all characteristics; rather, some traits evolve before others. In the case of the evolution of whales, changes occurred in the skull prior to evolutionary modification of the limbs. All three fossil forms occurred in the Eocene period, 45–55 MYA.

well-documented case of how evolution has proceeded through adaptation to changing environments.

The first horse

The earliest known members of the horse family, species in the genus *Hyracotherium*, didn't look much like modern-day horses at all. Small, with short legs and broad feet, these species occurred in wooded habitats, where they probably browsed on leaves and herbs and escaped predators by dodging through openings in the forest vegetation. The evolutionary path from these diminutive creatures to the workhorses of today has involved changes in a variety of traits, including size, toe reduction, and tooth size and shape (figure 21.12).

Changes in size

The first species of horses were as big as a large house cat or a medium-sized dog. By contrast, modern equids can weigh more than 500 kg. Examination of the fossil record reveals that horses changed little in size for their first 30 million years, but since then, a number of different lineages have exhibited rapid and substantial increases. However, evolution has not been unidirectional and trends toward decreased size were also exhibited in some branches of the equid evolutionary tree.

Toe reduction

The feet of modern horses have a single toe enclosed in a tough, bony hoof. By contrast, *Hyracotherium* had four toes on its front feet and three on its hind feet. Rather than hooves, these toes were encased in fleshy pads like those of dogs and cats.

Examination of fossils clearly shows the transition through time: a general increase in length of the central toe, development of the bony hoof, and reduction and loss of the other toes (figure 21.12). As with body size, these trends occurred concurrently on several different branches of the horse evolutionary tree and were not exhibited by all lineages.

At the same time as toe reduction was occurring, these horse lineages were evolving changes in the length and skeletal structure of their limbs, leading to animals capable of running long distances at high speeds.

Tooth size and shape

The teeth of *Hyracotherium* were small and relatively simple in shape. Through time, horse teeth have increased greatly in length and have developed a complex pattern of ridges on their molars and premolars. The effect of these changes is to produce teeth better capable of chewing tough and gritty vegetation, such as grass, which tends to wear teeth down. As with body size, evolutionary change has not been constant through time. Rather, much of the change in tooth shape has occurred within the past 20 million years, and changes have not been constant among all horse lineages.

All of these changes may be understood as adaptations to changing global climates. In particular, during the late Miocene and early Oligocene epochs (approximately 20 to 25 MYA), grasslands became widespread in North America, where much of horse evolution occurred. As horses shifted from forests to grasslands, high-speed locomotion probably became more important to escape predators. By contrast, the greater flexibility provided by multiple toes and shorter limbs, which was advantageous for ducking

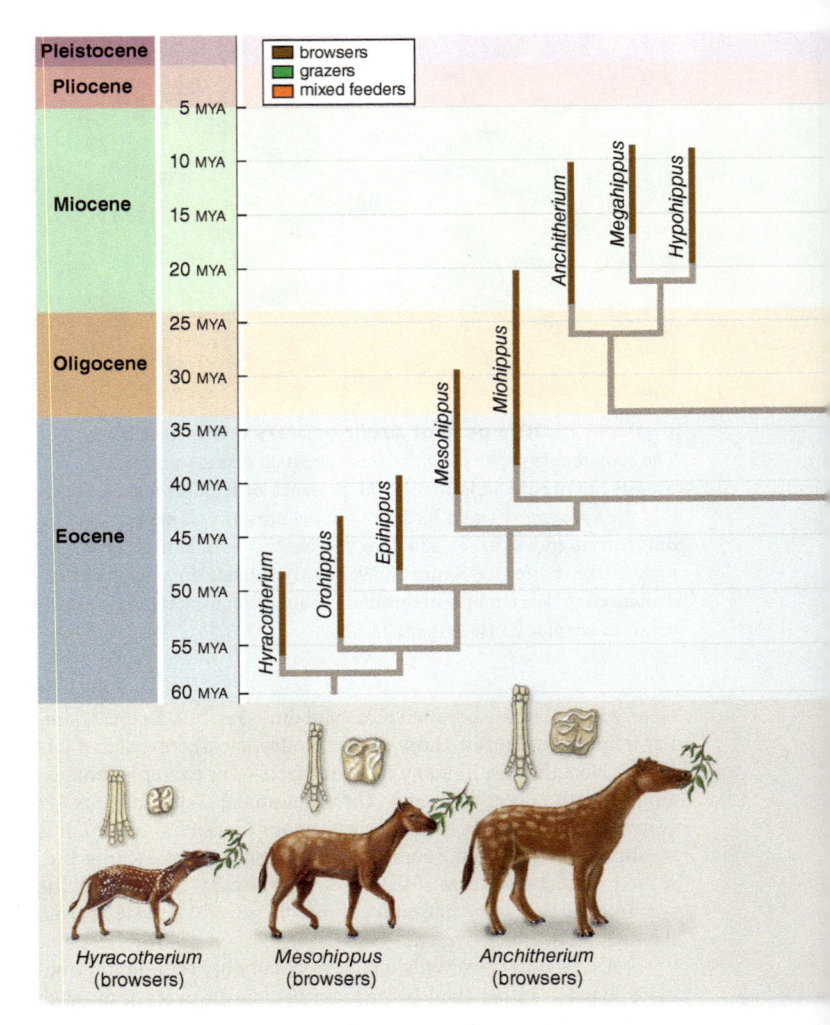

Figure 21.12 Evolutionary change in body size of horses. Lines indicate evolutionary relationships of the horse family. Horse evolution is more like a bush than a single-trunk tree; diversity was much greater in the past than it is today. In general, there has been a trend toward larger size, more complex molar teeth, and fewer toes, but this trend has exceptions. For example, a relatively recent form, *Nannippus*, evolved in the opposite direction, toward decreased size.

Inquiry question Why might the evolutionary line leading to *Nannippus* have experienced an evolutionary decrease in body size?

through complex forest vegetation, was no longer beneficial. At the same time, horses were eating grasses and other vegetation that contained more grit and other hard substances, thus favoring teeth better suited for withstanding such materials.

Evolutionary trends

For many years, horse evolution was held up as an example of constant evolutionary change through time. Some even saw in the record of horse evolution evidence for a progressive, guiding force, consistently pushing evolution in a single direction, toward longer limbs, fewer toes, and larger and more complex teeth. We now know that such views are misguided, and that the course of evolutionary change over millions of years is rarely so simple.

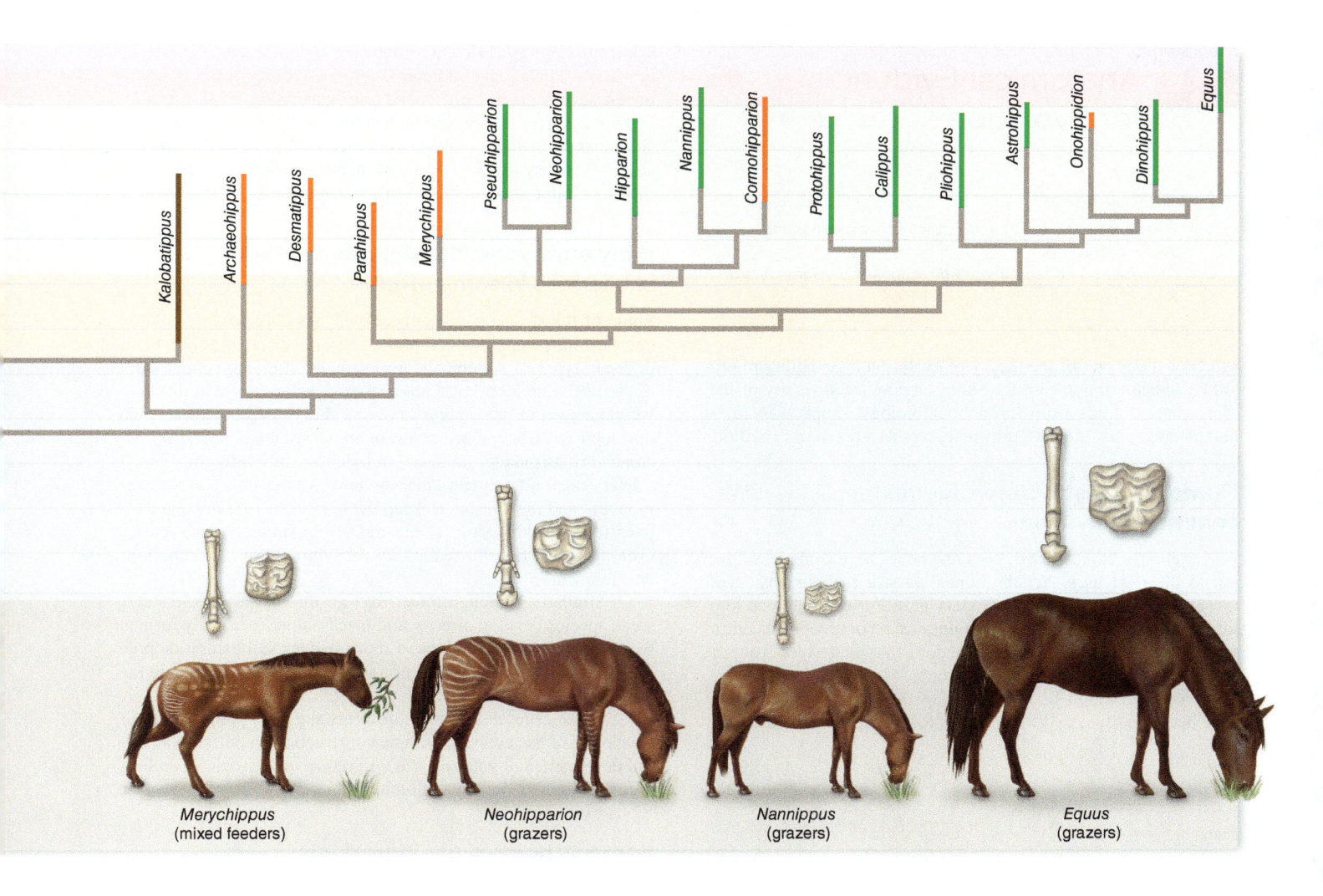

Rather, the fossils demonstrate that even though overall trends have been evident in a variety of characteristics, evolutionary change has been far from constant and uniform through time. Instead, rates of evolution have varied widely, with long periods of little observable change and some periods of great change. Moreover, when changes happen, they often occur simultaneously in different lineages of the horse evolutionary tree.

Finally, even when a trend exists, exceptions, such as the evolutionary decrease in body size exhibited by some lineages, are not uncommon. These patterns are usually discovered for any group of plants and animals for which we have an extensive fossil record, as you will see when we discuss human evolution in chapter 35.

Horse diversity

One reason that horse evolution was originally conceived of as linear through time may be that modern horse diversity is relatively limited. For this reason it is easy to mentally picture a straight line from *Hyracotherium* to modern-day *Equus*. But today's limited horse diversity—only one surviving genus—is unusual. In fact, at the peak of horse diversity in the Miocene epoch, 13 genera of horses could be found in North America alone. These species differed in body size and in a wide variety of other characteristics. Presumably, they lived in different habitats and exhibited different dietary preferences. Had this diversity existed to modern times, early evolutionary biologists would likely have had a different outlook on horse evolution.

Learning Outcomes Review 21.4

Fossils form when an organism is preserved in a matrix such as amber, permafrost, or rock. They can be used to construct a record of evolutionary transitions over long periods of time, which allows us to understand how major changes in evolution occur. The extensive fossil record for horses provides a detailed view of evolutionary diversification of this group, although trends are not constant and uniform and may include exceptions.

■ *Why might rates and direction of evolutionary change vary through time?*

chapter 21 *The Evidence for Evolution* **431**

21.5 Anatomical Evidence for Evolution

Learning Outcomes
1. Explain the evolutionary significance of homologous and vestigial structures.
2. Describe how patterns of early development provide evidence for evolution.

Much of the power of the theory of evolution is its ability to provide a sensible framework for understanding the diversity of life. Many observations from throughout biology simply cannot be understood in any meaningful way except as a result of evolution.

Homologous structures suggest common derivation

As vertebrates have evolved, the same bones have sometimes been put to different uses. Yet the bones are still recognizable, their presence betraying their evolutionary past. For example, the forelimbs of vertebrates are all **homologous structures**—structures with different appearances and functions that all derived from the same body part in a common ancestor.

You can see in figure 21.13 how the bones of the forelimb have been modified in different ways for different mammals. Why should these very different structures be composed of the same bones—a single upper forearm bone, a pair of lower forearm bones, several small carpals, and one or more digits? If evolution had not occurred, this would indeed be a riddle. But when we consider that all of these animals are descended from a common ancestor, it is easy to understand that natural selection has modified the same initial starting blocks to serve very different purposes.

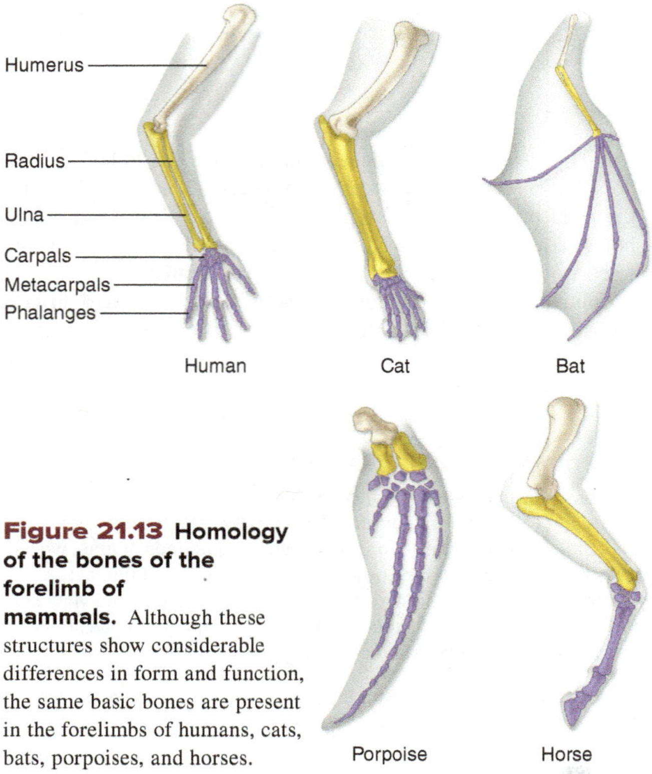

Early embryonic development shows similarities in some groups

Some of the strongest anatomical evidence supporting evolution comes from comparisons of how organisms develop. Embryos of different types of vertebrates, for example, often are similar early on, but become more different as they develop. Early in their development vertebrate embryos possess pharyngeal pouches, which develop into different structures. In humans, for example, they become various glands and ducts; in fish, they turn into gill slits. At a later stage, all primate embryos have a long tail, but whereas monkeys and most primates keep the tail, all we and our ape relatives retain is the coccyx at the end of our spine. Human fetuses even possess a fine fur (called *lanugo*) during the fifth month of development.

Similarly, although most frogs go through a tadpole stage, some species develop directly and hatch out as little, fully formed frogs. However, the embryos of these species still exhibit tadpole features, such as the presence of a tail, which disappear before the froglet hatches (figure 21.14).

These relict developmental forms suggest strongly that our development has evolved, with new instructions modifying ancestral developmental patterns. We will return to the topic of embryonic development and evolution in chapter 25.

Some structures are imperfectly suited to their use

Because natural selection can only work on the variation present in a population, it should not be surprising that some organisms do not appear perfectly adapted to their environments. For example, most animals with long necks have many neck vertebrae for enhanced flexibility: Geese have up to 25, and plesiosaurs, the long-necked reptiles that patrolled the seas during the age of dinosaurs, had as many as 76. By contrast, giraffes have only 7 very long neck vertebrae. Why haven't they evolved more, like other long-necked animals? It turns out that almost all mammals have only 7 neck vertebrae. Because mammal species have no variation in vertebra number among individuals in a population, natural selection has nothing to work with, and thus there is no way for selection to lead to increased numbers of vertebrae (why other types of animals are more variable is a question for which we currently don't have an answer). In the absence of variation in vertebra number, selection led to an evolutionary increase in vertebra size to produce the long neck of the giraffe.

An excellent example of an imperfect design is the eye of vertebrate animals, in which the photoreceptors face backward, toward the wall of the eye (figure 21.15a). As a result, the nerve fibers extend not backward, toward the brain, but forward into the eye chamber, where they slightly obstruct light. Moreover,

Figure 21.13 Homology of the bones of the forelimb of mammals. Although these structures show considerable differences in form and function, the same basic bones are present in the forelimbs of humans, cats, bats, porpoises, and horses.

(Human, Cat, Bat, Porpoise, Horse — labeled: Humerus, Radius, Ulna, Carpals, Metacarpals, Phalanges)

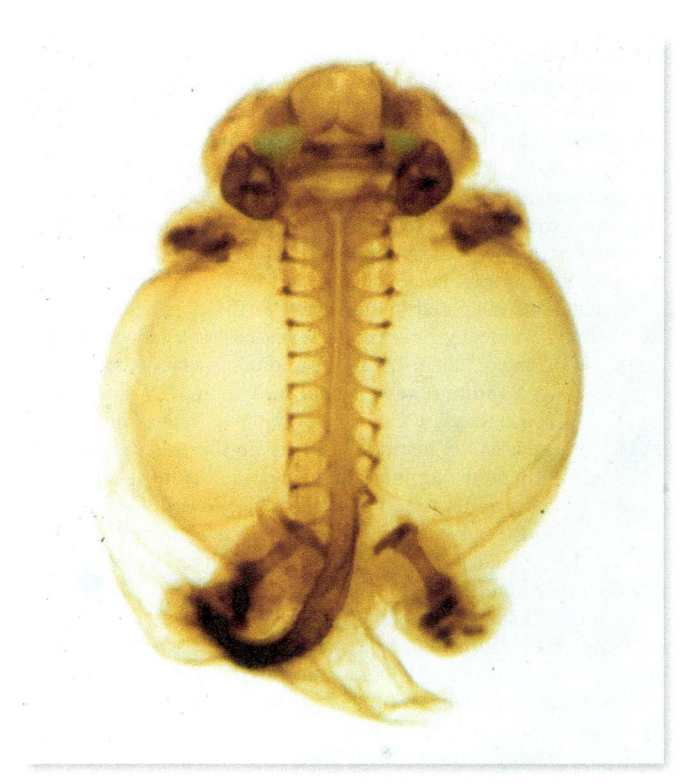

Figure 21.14 Developmental features reflect evolutionary ancestry. Some species of frogs have lost the tadpole stage. Nonetheless, tadpole features first appear and then disappear during development in the egg.

these fibers bundle together to form the optic nerve, which exits through a hole at the back of the eye, creating a blind spot.

By contrast, the eye of mollusks—such as squid and octopuses—are more optimally designed: The photoreceptors face forward, and the nerve fibers exit at the back, neither obstructing light nor creating a blind spot (figure 21.15b).

Such examples illustrate that natural selection is like a tinkerer, working with whatever material is available to craft a workable solution, rather than like an engineer, who can design and build the best possible structure for a given task. Workable, but imperfect, structures such as the vertebrate eye are an expected outcome of evolution by natural selection.

Vestigial structures can be explained as holdovers from the past

Many organisms possess **vestigial structures** that have no apparent function, but resemble structures their ancestors possessed. Humans, for example, possess a complete set of muscles for wiggling their ears, just like many other mammals do. Although these muscles allow other mammals to move their ears to pinpoint sounds such as the movements or growl of a predator, they have little purpose in humans other than amusement.

As other examples, boa constrictors have hip bones and rudimentary hind legs. Manatees (a type of aquatic mammal often referred to as "sea cows") have fingernails on their fins, which evolved from legs. Blind cave fish, which never see the light of day, have small, nonfunctional eyes. Figure 21.16 illustrates the skeleton of a baleen whale, which contains pelvic bones, as other mammal skeletons do, even though such bones serve no known function in the whale.

The human vermiform appendix is apparently vestigial; it represents the degenerate terminal part of the cecum, the blind pouch or sac in which the large intestine begins. In other mammals, such as mice, the cecum is the largest part of the large intestine and functions in storage—usually of bulk cellulose in herbivores. Although some functions have been suggested, it is difficult to assign any current function to the human vermiform appendix. In many respects, it can be a dangerous organ: appendicitis, which results from infection of the appendix, can be fatal.

Vestigial traits can also be seen in the genomes of many organisms. For example, the icefish (figure 21.17) is a bizarre-looking, nearly see-through fish that lives in the frigid waters of the Antarctic. The icefish's transparency results not only from a lack of pigment in its body structures, but also from the near invisibility of its blood.

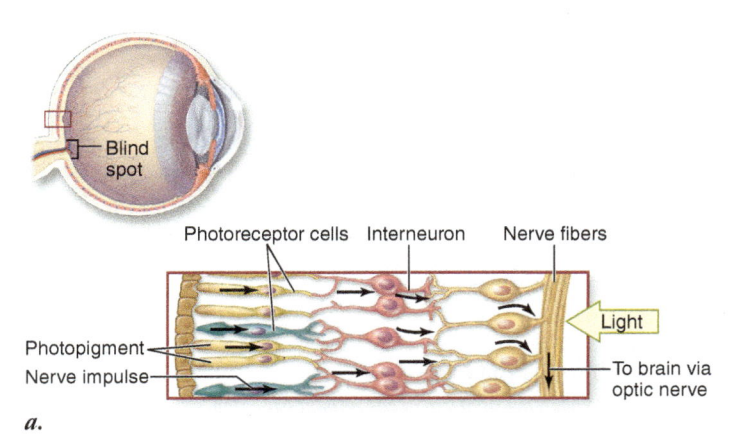

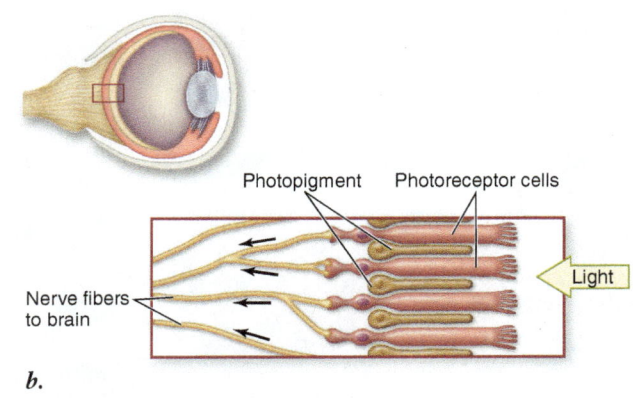

Figure 21.15 The eyes of vertebrates and mollusks. *a.* Photoreceptors of vertebrates point backward, whereas (*b*) those of mollusks face forward. As a result, vertebrate nerve fibers pass in front of the photoreceptor—and where they bundle together and exit the eye, a blind spot is created. Mollusks' eyes have neither of these problems.

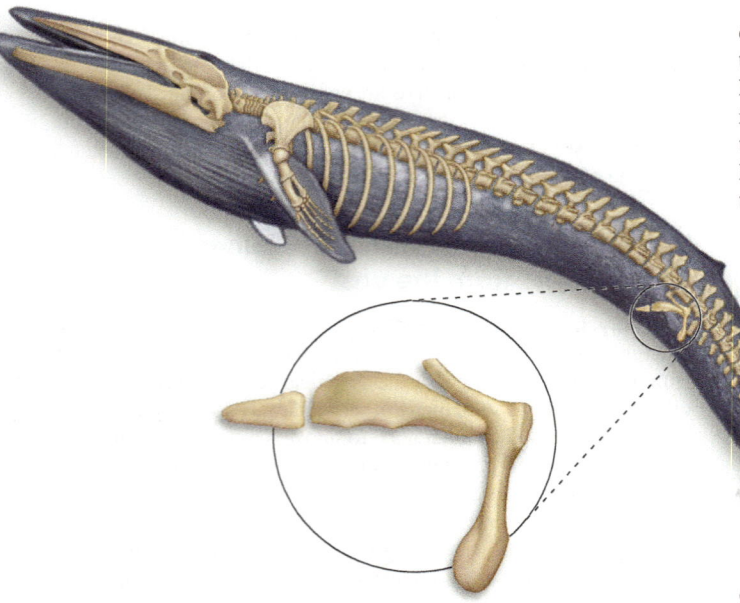

Figure 21.16 Vestigial structures. The skeleton of a whale reveals the presence of pelvic bones. These bones resemble those of other mammals, but are only weakly developed in the whale and have no apparent function.

Our blood is red due to the presence of red blood cells, which contain hemoglobin, the molecule that transports oxygen from the lungs to the tissues (see chapter 49). However, oxygen concentration in water increases as temperature decreases. The waters of the Antarctic, which are about 0°C, contain so much oxygen that the fish do not need special molecules to carry oxygen. The result is that these fish do not have hemoglobin, and consequently their blood is colorless. Nonetheless, when the DNA of icefish was examined, scientists discovered that they have the same gene that produces hemoglobin as that found in other vertebrates. However, the icefish hemoglobin gene has a variety of mutations that render it nonfunctional, and thus the icefish does not produce hemoglobin. The presence of this inoperative version of the hemoglobin gene in icefish means that its ancestors had hemoglobin; however, once the icefish's progenitors occupied the cold waters of the Antarctic and lost the need for hemoglobin, mutations that would prevent the production of hemoglobin and thus would be filtered out by natural selection in other species were able to persist in the population. Just by chance, some of these mutations increased in frequency in the population through time, eventually becoming established in all individuals and knocking out the fish's ability to produce hemoglobin.

Pseudogenes, sometimes called *fossil genes* because they are traces of previously functioning genes, such as the hemoglobin gene in the icefish, are actually quite common in the genomes of most organisms and are discussed in chapter 24: When a trait disappears, the gene does not just vanish from the genome; rather, some mutation renders it inactive, and once that occurs, other mutations can accumulate.

It is difficult to understand vestigial structures such as these as anything other than evolutionary relicts, holdovers from the past. However, the existence of vestigial structures argues strongly for the common ancestry of the members of the groups that share them, regardless of how different those groups have subsequently become.

All of these anatomical lines of evidence—homology, development, and imperfect and vestigial structures—are readily understandable as a result of descent with modification, that is, evolution.

Learning Outcomes Review 21.5

Comparisons of the anatomy of different living animals often reveal evidence of shared ancestry. In cases of homology, the same organ has evolved to carry out different functions. In other cases, an organ is still present, usually in diminished form, even though it has lost its function altogether; such an organ or structure is termed vestigial.

- How might homologous and vestigial structures be explained other than as a result of evolutionary descent with modification?

21.6 Convergent Evolution and the Biogeographical Record

Learning Outcomes

1. Explain the principle of convergent evolution.
2. Demonstrate how the biogeographical distribution of plant and animal species on islands provides evidence of evolutionary diversification.

Figure 21.17 Photo of an icefish. This nearly transparent fish is found in the frigid waters of the Antarctic.

Biogeography, the study of the geographic distribution of species, reveals that different geographical areas sometimes exhibit groups of plants and animals of strikingly similar appearance, even though the organisms may be only distantly related.

It is difficult to explain so many similarities as the result of coincidence. Instead, natural selection appears to have favored parallel evolutionary adaptations in similar environments. Because selection in these instances has tended to favor changes that made the two groups more alike, their phenotypes have converged. This form of evolutionary change is referred to as **convergent evolution.**

Marsupials and placentals demonstrate convergence

In the best-known case of convergent evolution, two major groups of mammals—marsupials and placentals—have evolved in very similar ways in different parts of the world. Marsupials are a group in which the young are born in a very immature condition and held in a pouch until they are ready to emerge into the outside world. In placentals, by contrast, offspring are not born until they can safely survive in the external environment (with varying degrees of parental care).

Australia separated from the other continents more than 70 MYA; at that time, both marsupials and placental mammals had evolved, but in different places. In particular, only marsupials occurred in Australia. As a result of this separation, the only placental mammals in Australia today are bats and a few colonizing rodents (which arrived relatively recently), and the continent is dominated by marsupials.

What are the Australian marsupials like? To an astonishing degree, they resemble the placental mammals living today on the other continents (figure 21.18). The similarity between some individual members of these two sets of mammals argues strongly that they are the result of convergent evolution, similar forms having evolved in different, isolated areas because of similar selective pressures in similar environments.

Convergent evolution is a widespread phenomenon

When species interact with the environment in similar ways, they often are exposed to similar selective pressures, and they therefore frequently develop the same evolutionary adaptations. Consider, for example, fast-moving marine predators (figure 21.19). The hydrodynamics of moving through water require a streamlined body shape to minimize friction. It is no coincidence that dolphins, sharks, and tuna—among the fastest of marine species—have all evolved to have the same basic shape. We can infer as well that ichthyosaurs—marine reptiles that lived during the Age of the Dinosaurs—exhibited a similar lifestyle.

Island trees exhibit a similar phenomenon. Most islands are covered by trees (or were until the arrival of humans). Careful inspection of these trees, however, reveals that they are not closely related to the trees with which we are familiar. Although they have all the characteristics of trees, such as being tall and having a tough outer covering, in many cases island trees are members of plant families that elsewhere exist only as flowers, shrubs, or other small bushes. For example, on many islands, the native trees are members of the sunflower family.

Why do these plants evolve into trees on islands? Probably because seeds from trees rarely make it to isolated islands. As a result, those species that do manage to colonize distant islands face an empty ecological landscape upon arrival. In the absence of other treelike plants, natural selection often would favor individual plants that could capture the most sunlight for photosynthesis, and the result is the evolution of similar treelike forms on islands throughout the world.

Niche	Burrower	Anteater	Nocturnal Insectivore	Climber	Glider	Stalking Predator	Chasing Predator
Placental Mammals	Mole	Lesser anteater	Grasshopper mouse	Ring-tailed lemur	Flying squirrel	Ocelot	Wolf
Australian Marsupials	Marsupial mole	Numbat	Marsupial mouse	Spotted cuscus	Flying phalanger	Tasmanian quoll	Thylacine

Figure 21.18 Convergent evolution. Many marsupial species in Australia resemble placental mammals occupying similar ecological niches elsewhere in the rest of the world. Marsupials evolved in isolation after Australia separated from other continents.

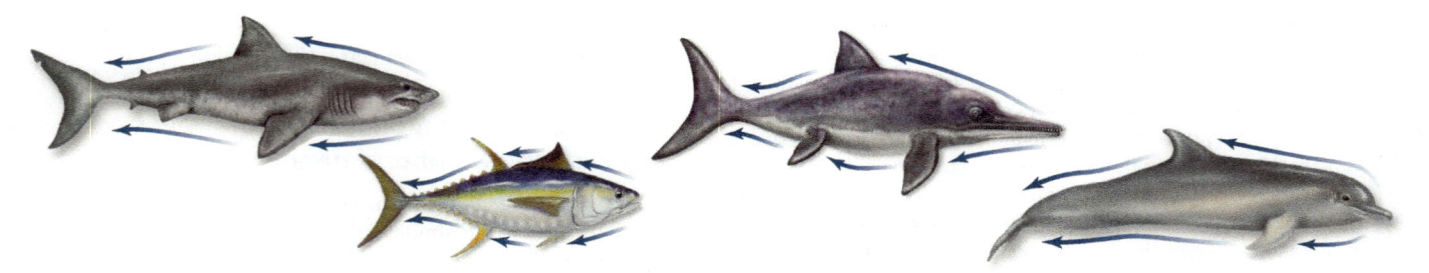

Figure 21.19 Convergence among fast-swimming predators. Fast movement through water requires a streamlined body form, which has evolved numerous times.

Convergent evolution is even seen in humans. People in most populations stop producing lactase, the enzyme that digests milk, some time in childhood. However, individuals in African and European populations that raise cattle produce lactase throughout their lives. DNA analysis indicates that the retention of lactase production into adulthood is the result of different mutations in Africa and Europe, which indicates that the populations have independently acquired this adaptation.

Biogeographical studies provide further evidence of evolution

Darwin made several important observations during his voyage around the world. He noted that islands often are missing plants and animals common on continents, such as frogs and land mammals. Accidental human introductions have proved that these species can survive if they are released on islands, so lack of suitable habitat is not the cause. In addition, those species that are present on islands often have diverged from their continental relatives and sometimes—as with Darwin's finches and the island trees just discussed—occupy ecological niches used by other species on continents. Lastly, island species usually are more closely related to species on nearby continents, even though the environment on continents and nearby islands often is not very similar.

Darwin deduced the explanation for these phenomena. Many islands have never been connected to continental areas. The species that occur there arrived by dispersing across the water. Some species, those that can fly, float, or swim, are more likely to get to the island than others. Some, like frogs, are particularly vulnerable to dehydration in saltwater and have almost no chance of island colonization.

The absence of some types of plants and animals provides opportunity to those that do arrive; as a result, colonizers, which usually come from nearby areas, often evolve into many species exhibiting great ecological and morphological diversity. This phenomenon, termed adaptive radiation, is discussed in chapter 22.

Learning Outcomes Review 21.6

Convergence is the evolution of similar forms in different lineages when exposed to similar selective pressures. The biogeographical distribution of species often reflects the outcome of evolutionary diversification with closely related species in nearby areas.

■ *Why does convergent evolution occur and why might species occupying similar environments in different localities sometimes not exhibit it?*

21.7 Darwin's Critics

Learning Outcomes

1. Characterize the criticisms of evolutionary theory and list counterarguments that can be made.
2. Distinguish between hypothesis and theory in scientific usage.

In the century and a half since he proposed it, Darwin's theory of evolution by natural selection has become nearly universally accepted by biologists, but has been a source of controversy among some members of the general public. Here we discuss seven principle objections that critics raise to the teaching of evolution as biological fact, along with some answers that scientists present in response:

1. **Evolution is not solidly demonstrated.** "Evolution is just a theory," Darwin's critics point out, as though *theory* meant a lack of knowledge, or some kind of guess.

    Scientists, however, use the word *theory* in a very different sense than the general public does. They use *theory* only for those ideas that are strongly supported by many lines of evidence, like the theories of gravity and evolution. It is important to recognize that both evolution and gravitation are "just theories." The term *theory* is not the equivalent of a "hunch" or a "notion," or even an initial hypothesis. Rather, it is a well-tested phenomenon that rationalizes the data available.

2. **There are no fossil intermediates.** "No one ever saw a fin on the way to becoming a leg," critics claim, pointing to the many gaps in the fossil record in Darwin's day.

    Since that time, however, many fossil intermediates in vertebrate evolution have indeed been found. A clear line of fossils now traces the transition between hoofed mammals and whales, between reptiles and mammals, between dinosaurs and birds, and between apes and humans. The fossil evidence of evolution from one major type to another is compelling.

3. **The intelligent design argument.** "The organs of living creatures are too complex for a random process to have produced—the existence of a clock is evidence of the existence of a clockmaker."

    Evolution by natural selection is not a random process. Quite the contrary, by favoring those variations that lead to the highest reproductive fitness, natural selection is a nonrandom process that can construct highly complex organs by incrementally improving them from one generation to the next.

For example, the intermediates in the evolution of the mammalian ear can be seen in fossils, and many intermediate "eyes" are known in various invertebrates. These intermediate forms arose because they have value—being able to detect light slightly is better than not being able to detect it at all. Complex structures such as eyes evolved as a progression of slight improvements. Moreover, inefficiencies of certain designs, such as the vertebrate eye and the existence of vestigial structures, do not support the idea of an intelligent designer.

4. **Evolution violates the Second Law of Thermodynamics.** "A jumble of soda cans doesn't by itself jump neatly into a stack—things become more disorganized due to random events, not more organized."

Biologists point out that this argument ignores what the second law really says: Disorder increases in a closed system, which the Earth most certainly is not. Energy continually enters the biosphere from the Sun, fueling life and all the processes that organize it.

5. **Proteins are too improbable.** "Hemoglobin has 141 amino acids. The probability that the first one would be leucine is 1/20, and that all 141 would be the ones they are by chance is $(1/20)^{141}$, an impossibly rare event."

This argument illustrates a lack of understanding of probability and statistics—probability cannot be used to argue backward. The probability that a student in a classroom has a particular birthdate is 1/365; arguing this way, the probability that everyone in a class of 50 would have the birthdates that they do is $(1/365)^{50}$, and yet there the class sits, all with their actual birthdates.

6. **Natural selection cannot account for major changes in evolution.** "No scientist has come up with an experiment in which fish evolve into frogs and leap away from predators."

Can we extrapolate from our understanding that natural selection produces relatively small changes that are observable in populations *within* species to explain the major differences observed *between* species? Most biologists who have studied the problem think so. The differences between breeds produced by artificial selection—such as chihuahuas, mastiffs, and greyhounds—are more distinctive than the differences between some wild species, and laboratory selection experiments sometimes create forms that cannot interbreed and thus would in nature be considered different species. Thus, production of radically different forms has indeed been observed, repeatedly. These changes usually take millions of years, and they are seen clearly in the fossil record.

7. **The irreducible complexity argument.** Because each part of a complex cellular mechanism such as blood clotting is essential to the overall process, the intricate machinery of the cell cannot be explained by evolution from simpler stages. Without all of the parts functioning, the system wouldn't not work. Consequently, evolution couldn't have built the structure because intermediate stages, with not all parts in place, would not function and thus would not be favored by natural selection.

What's wrong with this argument is that each part of a complex molecular machine evolves as part of the whole system. Natural selection can act on a complex system because at every stage of its evolution, the system functions. Parts that improved function are added. Subsequently, other parts may be modified or even lost, so that parts that were not essential when they first evolved become essential. In this way, an "irreducible complex" structure can evolve by natural selection. The same process works at the molecular level.

For example, snake venom initially evolved as enzymes to increase the ability of snakes to digest large prey items, which were captured by biting the prey and then constricting them with coils. Subsequently, the digestive enzymes evolved to become increasingly lethal. Rattlesnakes kill large prey by injecting them with venom, letting them go, and then tracking them down and eating them after they die. To do so, they have evolved extremely toxic venom, highly modified syringe-like front teeth, and many other characteristics. Take away the fangs or the venom and the rattlesnakes can't feed—what initially evolved as nonessential parts are now indispensable; irreducible complexity has evolved by natural selection.

The mammalian blood clotting system similarly has evolved from much simpler systems. The core clotting system evolved at the dawn of the vertebrates more than 500 MYA, and it is found today in primitive fishes such as lampreys. One hundred million years later, as vertebrates continued to evolve, proteins were added to the clotting system, making it sensitive to substances released from damaged tissues. Fifty million years later, a third component was added, triggering clotting by contact with the jagged surfaces produced by injury. At each stage, as the clotting system evolved to become more complex, its overall performance came to depend on the added elements. Thus, blood clotting has become "irreducibly complex" as the result of Darwinian evolution.

Statements that various structures could not have been built by natural selection have repeatedly been made over the past 150 years. In many cases, after detailed scientific study, the likely path by which such structures have evolved has been discovered.

Learning Outcomes Review 21.7

Darwin's theory of evolution is controversial to some in the general public. Objections are often based on a misunderstanding of the theory. In scientific usage, a hypothesis is an educated guess, whereas a theory is an explanation that fits available evidence and has withstood rigorous testing.

- *Suppose someone suggests that humans originally came from Mars. Would this be a hypothesis or a theory, and how could it be tested?*

Chapter Review

21.1 The Beaks of Darwin's Finches: Evidence of Natural Selection

Galápagos finches exhibit variation related to food gathering.
The correspondence between beak shape and its use in obtaining food suggested to Darwin that finch species had diversified and adapted to eat different foods.

Modern research has verified Darwin's selection hypothesis.
Natural selection acts on variation in beak morphology, favoring larger-beaked birds during extended droughts and smaller-beaked birds during long periods of heavy rains.

Because this variation is heritable, evolutionary change occurs in the frequencies of beak sizes in subsequent generations.

21.2 Peppered Moths and Industrial Melanism: More Evidence of Selection

Light-colored moths decreased in polluted areas.
In polluted areas where soot built up on tree trunks, the dark-colored form of the peppered moth became more common. In unpolluted areas, light-colored forms remained predominant.

Experiments suggested that predation by birds was the cause; light-colored moths stand out on dark trunks, and vice versa.

When environmental conditions reverse, so does selection pressure.
In the last 50 years, pollution has decreased in many areas and the frequency of light-colored moths has rebounded.

The agent of selection may be difficult to pin down.
Some have questioned whether bird predation is the agent of selection, but recent research supports this hypothesis. Regardless, the observation that the dark-colored form has increased during times of pollution and then declined as pollution abates indicates that natural selection has acted on moth coloration.

21.3 Artificial Selection: Human-Initiated Change
(figure 21.5)

Experimental selection produces changes in populations.
Laboratory experiments in directional selection have shown that substantial evolutionary change can occur in these controlled populations.

Agricultural selection has led to extensive modification of crops and livestock.

Domesticated breeds have arisen from artificial selection.
Crop plants and domesticated animal breeds are often substantially different from their wild ancestors.

If artificial selection can rapidly create substantial change over short periods of time, then it is reasonable to assume that natural selection could have created the Earth's diversity of life over millions of years.

21.4 Fossil Evidence of Evolution

The age of fossils can be estimated.
Specimens become fossilized in different ways. Fossils in rock can be dated by calculating the extent of radioactive decay based on half-lives of known isotopes.

Fossils present a history of evolutionary change.

Fossils document evolutionary transitions.
The history of life on Earth can be traced through the fossil record. In recent years, new fossil discoveries have provided more detailed understanding of major evolutionary transitions.

The evolution of horses is a prime example of evidence from fossils.
The fossil record indicates that horses have evolved from small, forest-dwelling animals to the large and fast plains-dwelling species alive today.

Over the course of 50 million years, evolution has not been constant and uniform. Rather, change has been rapid at some times, slow at others. Although a general trend toward increase in size is evident, some species evolved to smaller sizes.

21.5 Anatomical Evidence for Evolution

Homologous structures suggest common derivation.
Homologous structures may have different appearances and functions even though derived from the same common ancestral body part.

Early embryonic development shows similarities in some groups.
Embryonic development shows similarity in developmental patterns among species whose adult phenotypes are very different.

Species that have lost a feature that was present in an ancestral form often develop and then lose that feature during embryological development.

Some structures are imperfectly suited to their use.
Natural selection can influence only the variation present in a population; because of this, evolution often results in workable, but imperfect structures, such as the vertebrate eye.

Vestigial structures can be explained as holdovers from the past.
The existence of vestigial structures supports the concept of common ancestry among organisms that share them.

21.6 Convergent Evolution and the Biogeographical Record

Marsupials and placentals demonstrate convergence.
Convergent evolution may occur in species or populations exposed to similar selective pressures. Marsupial mammals in Australia have converged upon features of their placental counterparts elsewhere.

Convergent evolution is a widespread phenomenon.
Examples include hydrodynamic streamlining in marine species and the evolution of tree species on islands from ancestral forms that were not treelike.

Biogeographical studies provide further evidence of evolution.
Island species usually are closely related to species on nearby continents even if the environments are different. Early island colonizers often evolve into diverse species because other, competing species are scarce.

21.7 Darwin's Critics

Darwin's theory of evolution by natural selection is almost universally accepted by biologists. Many criticisms have been made both historically and recently, but most stem from a lack of understanding of scientific principles, the theory's actual content, or the time spans involved in evolution.

Review Questions

UNDERSTAND

1. Artificial selection is different from natural selection because
   a. artificial selection is not capable of producing large changes.
   b. artificial selection does not require genetic variation.
   c. natural selection cannot produce new species.
   d. breeders (people) choose which individuals reproduce based on desirability of traits.

2. Gaps in the fossil record
   a. demonstrate our inability to date geological sediments.
   b. are expected since the probability that any organism will fossilize is extremely low.
   c. have not been filled in as new fossils have been discovered.
   d. weaken the theory of evolution.

3. The evolution of modern horses *(Equus)* is best described as
   a. the constant change and replacement of one species by another over time.
   b. a complex history of lineages that changed over time, with many going extinct.
   c. a simple history of lineages that have always resembled extant horses.
   d. None of the choices is correct.

4. Homologous structures
   a. are structures in two or more species that originate as the same structure in a common ancestor.
   b. are structures that look the same in different species.
   c. cannot serve different functions in different species.
   d. must serve different functions in different species.

5. Convergent evolution
   a. is an example of stabilizing selection.
   b. depends on natural selection to independently produce similar phenotypic responses in different species or populations.
   c. occurs only on islands.
   d. is expected when different lineages are exposed to vastly different selective environments.

6. Darwin's finches are a noteworthy case study of evolution by natural selection because evidence suggests
   a. they are descendants of many different species that colonized the Galápagos.
   b. they radiated from a single species that colonized the Galápagos.
   c. they are more closely related to mainland species than to one another.
   d. None of the choices is correct.

7. The possession of fine fur in 5-month human embryos indicates
   a. that the womb is cold at that point in pregnancy.
   b. humans evolved from a hairy ancestor.
   c. hair is a defining feature of mammals.
   d. some parts of the embryo grow faster than others.

APPLY

1. In Darwin's finches,
   a. occurrence of wet and dry years preserves genetic variation for beak size.
   b. increasing beak size over time proves that beak size is inherited.
   c. large beak size is always favored.
   d. All of the choices are correct.

2. Artificial selection experiments in the laboratory such as in figure 21.5 are an example of
   a. stabilizing selection.
   b. negative frequency-dependent selection.
   c. directional selection.
   d. disruptive selection.

3. Convergent evolution is often seen among species on different islands because
   a. island populations are usually smaller and more affected by genetic drift.
   b. disruptive selection occurs commonly on islands.
   c. island species are usually most closely related to species in similar habitats elsewhere.
   d. when islands are first colonized, many ecological resources are unused, allowing descendants of a colonizing species to diversify and adapt to many different parts of the environment.

SYNTHESIZE

1. What conditions are necessary for evolution by natural selection?

2. Explain how data shown in figure 21.2*a* and *b* relate to the conditions identified by you in question 1.

3. On figure 21.2*b*, draw the relationship between offspring beak depth and parent beak depth, assuming that there is no genetic basis to beak depth in the medium ground finch.

4. Refer to figure 21.5, artificial selection in the laboratory. In this experiment, one population of *Drosophila* was selected for low numbers of bristles and the other for high numbers. Note that not only did the means of the populations change greatly in 35 generations, but also all individuals in both experimental populations lie outside the range of the initial population. What would happen if the direction of selection were reversed, such that a greater number of bristles was selected for in the low-bristle population, and vice versa? How would the rate of evolutionary change compare with that in the initial part of the experiment before selection was reversed?

5. The ancestor of horses was a small, many-toed animal that lived in forests, whereas today's horses are large animals with a single hoof that live on open plains. A series of intermediate fossils illustrate how this transition has occurred, and for this reason, many old treatments of horse evolution portrayed it as a steady increase through time in body size accompanied by a steady decrease in toe number. Why is this interpretation incorrect?

CHAPTER 22

The Origin of Species

Chapter Contents

22.1 The Nature of Species and the Biological Species Concept
22.2 Natural Selection and Reproductive Isolation
22.3 The Role of Genetic Drift and Natural Selection in Speciation
22.4 The Geography of Speciation
22.5 Adaptive Radiation and Biological Diversity
22.6 The Pace of Evolution
22.7 Speciation and Extinction Through Time

Introduction

Although Darwin titled his book *On the Origin of Species,* he never actually discussed what he referred to as that "mystery of mysteries"—how one species gives rise to another. Rather, his argument concerned evolution by natural selection—that is, how one species evolves through time to adapt to its changing environment. Although an important mechanism of evolutionary change, the process of adaptation does not explain how one species becomes another, a process we call *speciation*. As we shall see, adaptation may be involved in the speciation process, but it does not have to be.

Before we can discuss how one species gives rise to another, we need to understand exactly what a species is. Even though the definition of a species is of fundamental importance to evolutionary biology, this issue has still not been completely settled and is currently the subject of considerable research and debate.

22.1 The Nature of Species and the Biological Species Concept

Learning Outcomes

1. Understand the biological species concept and why it does not explain all observations.
2. Define the two kinds of reproductive isolating mechanisms.
3. Describe the relationship of reproductive isolating mechanisms to the biological species concept.

Any concept of a species must account for two phenomena: the distinctiveness of species that occur together at a single locality, and the connection that exists among different populations belonging to the same species.

Sympatric species inhabit the same locale but remain distinct

Put out birdfeeders on your balcony or in your back yard, and you will attract a wide variety of birds (especially if you include different kinds of foods). In the midwestern United States, for example, you might routinely see cardinals, blue jays, downy woodpeckers, house finches—even hummingbirds in the summer.

Although it might take a few days of careful observation, you would soon be able to readily distinguish the many different species. The reason is that species that occur together (termed **sympatric**) are distinctive entities that are phenotypically different, utilize different parts of the habitat, and behave differently. This observation is generally true not only for birds, but also for most other types of organisms.

Occasionally, two species occur together that appear to be nearly identical. In such cases, we need to go beyond visual similarities. When other aspects of the phenotype are examined, such as the mating calls or the chemicals exuded by each species, they usually reveal great differences. In other words, even though we might have trouble distinguishing them, the organisms themselves have no such difficulties.

Populations of a species exhibit geographic variation

Within a single species, individuals in populations that occur in different areas may be distinct from one another. In areas where these populations occur close to one another, individuals often exhibit combinations of features characteristic of both populations (figure 22.1). In other words, even though geographically distant populations may appear distinct, they are usually connected by intervening populations that are intermediate in their characteristics.

The biological species concept focuses on the ability to exchange genes

What can account for both the distinctiveness of sympatric species and the connectedness of geographically separate populations of the same species? One obvious possibility is that each species exchanges genetic material only with other members of its species. If sympatric species commonly exchanged genes, which they generally do not, we might expect such species to rapidly lose their distinctions, as the **gene pools** (that is, all of the alleles present in a species) of the different species became homogenized. Conversely, the ability of geographically distant populations of a single species to share genes through the process of gene flow may keep these populations integrated as members of the same species.

Based on these ideas, in 1942 the evolutionary biologist Ernst Mayr set forth the **biological species concept,** which defines *species* as ". . . groups of actually or potentially interbreeding natural populations which are reproductively isolated from other such groups."

In other words, the biological species concept says that a species is composed of populations whose members mate with each other and produce fertile offspring—or would do so if they came into contact. Conversely, populations whose members do not mate with each other or who cannot produce fertile offspring are said to be **reproductively isolated** and, therefore, are members of different species.

What causes reproductive isolation? If organisms cannot interbreed or cannot produce fertile offspring, they clearly belong to different species. However, some populations that are considered separate species can interbreed and produce fertile offspring, but they ordinarily do not do so under natural conditions. They are still considered reproductively isolated because genes from one species generally will not enter the gene pool of the other.

Table 22.1 summarizes the steps at which barriers to successful reproduction may occur. Such barriers are termed **reproductive isolating mechanisms** because they prevent genetic exchange between species. We will discuss examples of these next, beginning with those that prevent the formation of zygotes, which are called **prezygotic isolating mechanisms.** Mechanisms that

Figure 22.1 Geographic variation in the eastern rat snake, *Pantherophis alleghaniensis*. Although populations at the eastern, western and northern ends of the species' range are phenotypically quite distinctive from one another, they are connected by populations that are phenotypically intermediate.

TABLE 22.1 Reproductive Isolating Mechanisms

Mechanism	Description
PREZYGOTIC ISOLATING MECHANISMS	
Ecological isolation	Species occur in the same area, but they occupy different habitats and rarely encounter each other.
Behavioral isolation	Species differ in their mating rituals.
Temporal isolation	Species reproduce in different seasons or at different times of the day.
Mechanical isolation	Structural differences between species prevent mating.
Prevention of gamete fusion	Gametes of one species function poorly with the gametes of another species or within the reproductive tract of another species.
POSTZYGOTIC ISOLATING MECHANISMS	
Hybrid inviability or infertility	Hybrid embryos do not develop properly, hybrid adults do not survive in nature, or hybrid adults are sterile or have reduced fertility.

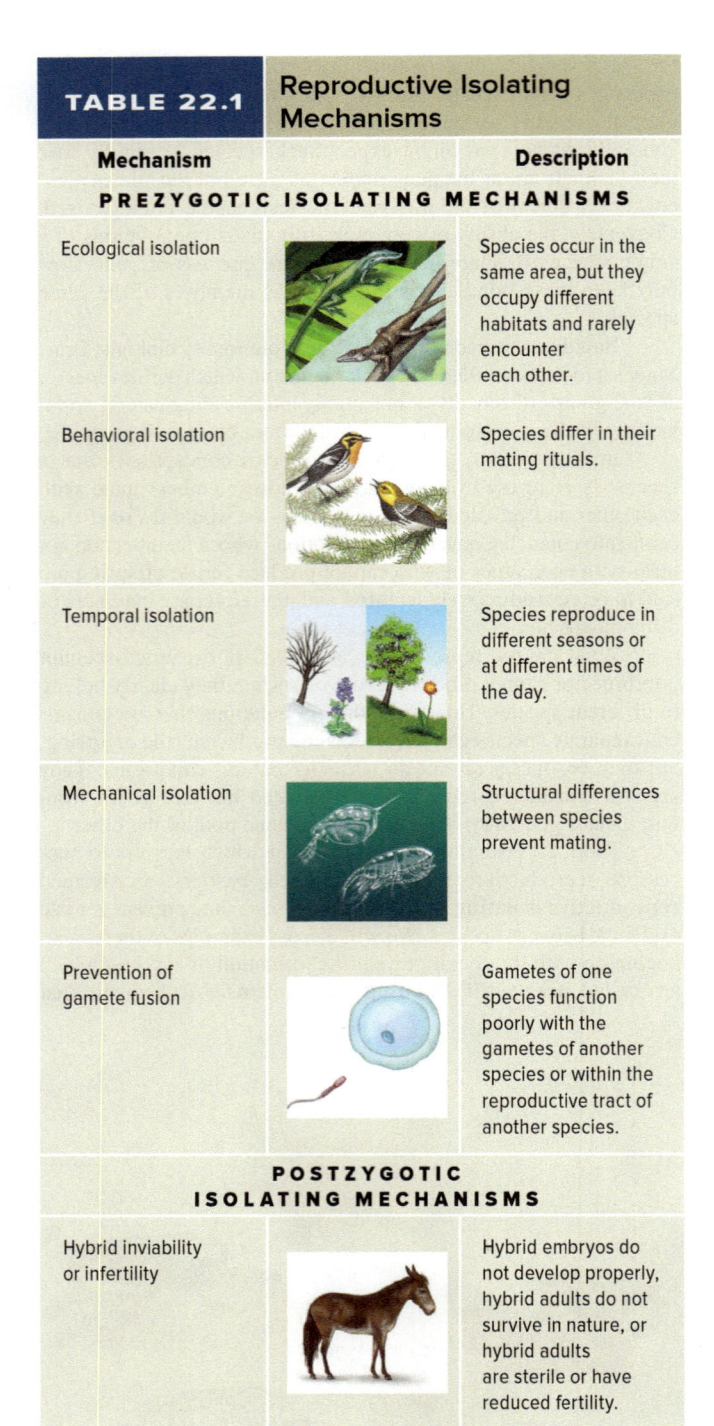

prevent the proper functioning of zygotes after they form are called **postzygotic isolating mechanisms**.

Prezygotic isolating mechanisms prevent the formation of a zygote

Mechanisms that prevent formation of a zygote include ecological or environmental isolation, behavioral isolation, temporal isolation, mechanical isolation, and prevention of gamete fusion.

Ecological isolation

Even if two species occur in the same area, they may utilize different portions of the environment and thus not hybridize because they do not encounter each other. For example, in India, the ranges of lions and tigers overlapped until about 150 years ago. Even so, there were no records of natural hybrids. Lions stayed mainly in the open grassland and hunted in groups called prides; tigers tended to be solitary creatures of the forest (figure 22.2). Because of their ecological and behavioral differences, lions and tigers rarely came into direct contact with each other, even though their ranges overlapped over thousands of square kilometers.

In another example, the ranges of two toads, *Bufo woodhousei* and *B. americanus*, overlap in some areas. Although these two species can produce viable hybrids, they usually do not interbreed because they utilize different portions of the habitat for breeding. *B. woodhousei* prefers to breed in streams, and *B. americanus* breeds in rainwater puddles.

Similar situations occur among plants. Two species of oaks occur widely in California: the valley oak, *Quercus lobata*, and the scrub oak, *Q. dumosa*. The valley oak, a graceful deciduous tree that can be as tall as 35 m, occurs in the fertile soils of open grassland on gentle slopes and valley floors. In contrast, the scrub oak is an evergreen shrub, usually only 1 to 3 m tall, which often forms the kind of dense scrub known as chaparral. The scrub oak is found on steep slopes in less fertile soils. Hybrids between these different oaks do occur and are fully fertile, but they are rare. The sharply distinct habitats of their parents limit their occurrence together, and there is little intermediate habitat where the hybrids might flourish.

Behavioral isolation

Chapter 54 describes the often elaborate courtship and mating rituals of some groups of animals. Related species of organisms such as birds often differ in their courtship rituals, which tends to keep these species distinct in nature even if they inhabit the

Figure 22.2 Lions and tigers are ecologically isolated.
The ranges of lions and tigers overlap in India. However, lions and tigers do not hybridize in the wild because they utilize different portions of the habitat. Hybrids, such as this tiglon, have been successfully produced in captivity, but hybridization does not occur in the wild.

Figure 22.3 Differences in courtship rituals can isolate related bird species. These Galápagos blue-footed boobies select their mates only after an elaborate courtship display. This male is lifting his feet in a ritualized high-step that shows off his bright blue feet. The display behavior of the two other species of boobies that occur in the Galápagos is very different, as is the color of their feet.

same places (figure 22.3). Sympatric species avoid mating with members of the wrong species in a variety of ways; every mode of communication imaginable appears to be used by some species. Differences in visual signals, as just discussed, are common; however, other types of animals rely more on other sensory modes for communication. Many species, such as frogs, birds, and a variety of insects, use sound to attract mates. Predictably, sympatric species of these animals produce different calls. Similarly, the "songs" of lacewings are produced when they vibrate their abdomens against the surface on which they are sitting, and sympatric species produce different vibration patterns (figure 22.4).

Other species rely on the detection of chemical signals, called **pheromones.** The use of pheromones in moths has been particularly well studied. When female moths are ready to mate, they emit a pheromone that males can detect at great distances. Sympatric species differ in the pheromone they produce: Either they use different chemical compounds, or, if using the same compounds, the proportions used are different. Laboratory studies indicate that males are remarkably adept at distinguishing the pheromones of their own species from those of other species or even from synthetic compounds similar, but not identical, to that of their own species.

Some species even use electroreception. African and South Asian electric fish independently have evolved specialized organs in their tails that produce electrical discharges and electroreceptors on their skins to detect them. These discharges are used to communicate in social interactions; field experiments indicate that males can distinguish between signals produced by their own and other, co-occurring species, probably on the basis of differences in the timing of the electrical pulses.

Temporal isolation

Two species of wild lettuce, *Lactuca graminifolia* and *L. canadensis*, grow together along roadsides throughout the southeastern United States. Hybrids between these two species are easily made experimentally and are completely fertile. But these hybrids are rare in nature because *L. graminifolia* flowers in early spring and *L. canadensis* flowers in summer. When their blooming periods overlap, as happens occasionally, the two species do form hybrids, which may become locally abundant.

Many species of closely related amphibians have different breeding seasons that prevent hybridization. For example, five species of frogs of the genus *Rana* occur together in most of the eastern United States, but hybrids are rare because the peak breeding time is different for each of them.

Mechanical isolation

Structural differences prevent mating between some related species of animals. Aside from such obvious features as size, the structure of the male and female copulatory organs may be incompatible. In many insect and other arthropod groups, the sexual organs, particularly those of the male, are so diverse that they are used as a primary basis for distinguishing species.

Similarly, flowers of related species of plants often differ significantly in their proportions and structures. Some of these differences limit the transfer of pollen from one plant species to another. For example, bees may carry the pollen of one species on a certain place on their bodies; if this area does not come into contact with the receptive structures of the flowers of another plant species, the pollen is not transferred.

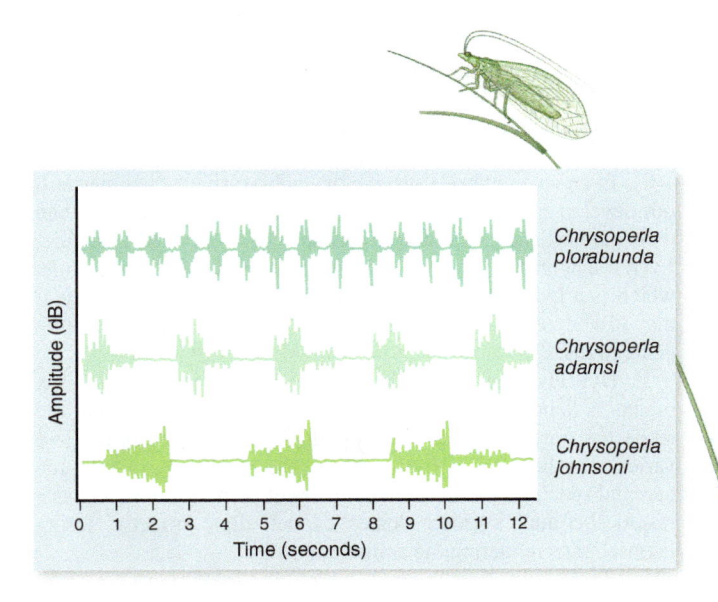

Figure 22.4 Differences in courtship song of sympatric species of lacewings. Lacewings are small insects that rely on signals produced by moving their abdomens to vibrate the surface on which they are sitting to attract mates. As these recordings indicate, the vibration patterns produced by sympatric species differ greatly. Females, which detect the calls as they are transmitted through solid surfaces such as branches, are able to distinguish calls of different species and only respond to individuals producing their own species' call.

Prevention of gamete fusion

In animals that shed gametes directly into water, the eggs and sperm derived from different species may not attract or fuse with one another. Many animals that have internal fertilization may not hybridize successfully because the sperm of one species functions so poorly within the reproductive tract of another that fertilization never takes place. In plants, the growth of pollen tubes may be impeded in hybrids between different species. In both plants and animals, isolating mechanisms such as these prevent the union of gametes, even following successful mating.

Postzygotic isolating mechanisms prevent normal development into reproducing adults

All of the factors we have discussed so far tend to prevent hybridization. If hybrid matings do occur and zygotes are produced, many factors may still prevent those zygotes from developing into normally functioning, fertile individuals.

As you saw in chapter 19, development is a complex process. In hybrids, the genetic complements of two species may be so different that they cannot function together normally in embryonic development. For example, hybridization between sheep and goats usually produces embryos that die in the earliest developmental stages.

The leopard frogs (*Rana pipiens* complex) of the eastern United States are a group of similar species, assumed for a long time to constitute a single species (figure 22.5). Careful examination, however, revealed that although the frogs appear similar, successful mating between them is rare because of problems that occur as the fertilized eggs develop. Many of the hybrid combinations cannot be produced even in the laboratory. Examples of this kind, in which the existence of multiple species has been recognized only as a result of hybridization experiments, are common in plants.

Even when hybrids survive the embryo stage, they may still not develop normally. If the hybrids are less physically fit than their parents, they will almost certainly be eliminated in nature. Even if a hybrid is vigorous and strong, as in the case of the mule, which is a hybrid between a female horse and a male donkey, it may still be sterile and thus incapable of contributing to succeeding generations.

Hybrids may be sterile because the development of sex organs is abnormal, because the chromosomes derived from the respective parents cannot pair properly during meiosis, or due to a variety of other causes. In the case of the mule, for example, the parental species have different numbers of chromosomes, and as a result, the mule's chromosomes cannot align properly during meiosis, thus rendering the animal infertile.

The biological species concept does not explain all observations

The biological species concept has proved to be an effective way of understanding the existence of species in nature. Nonetheless, it fails to take into account all observations, leading some biologists to propose alternative species concepts.

One criticism of the biological species concept concerns the extent to which all species truly are reproductively isolated. By

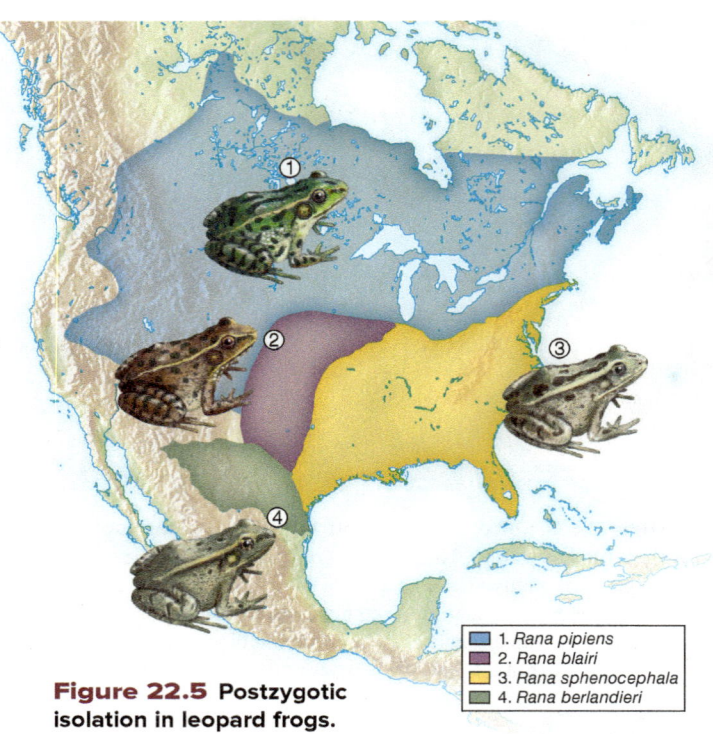

Figure 22.5 Postzygotic isolation in leopard frogs.
These four species resemble one another closely in their external features. Their status as separate species first was suspected when hybrids between some pairs of these species were found to produce defective embryos in the laboratory. Subsequent research revealed that the mating calls of the four species differ substantially, indicating that the species have both pre- and postzygotic isolating mechanisms.

1. *Rana pipiens*
2. *Rana blairi*
3. *Rana sphenocephala*
4. *Rana berlandieri*

definition, under the biological species concept, species should not interbreed and produce fertile offspring. But in recent years, biologists have detected much greater amounts of interspecies hybridization than was previously thought to occur between populations that seem to coexist as distinct biological entities.

Botanists have always been aware that plant species often undergo substantial amounts of hybridization. More than 50% of California plant species included in one study, for example, were not well defined by genetic isolation. This coexistence without genetic isolation can be long-lasting: Fossil data show that balsam poplars and cottonwoods have been phenotypically distinct for 12 million years, but they also have routinely produced hybrids throughout this time. Consequently, many botanists have long felt that the biological species concept is misnamed and only applies to animals.

New evidence, however, increasingly indicates that hybridization is not all that uncommon in animals, either. In recent years, many cases of substantial hybridization between animal species have been documented. One survey indicated that almost 10% of the world's more than 10,000 bird species are known to have hybridized in nature.

The Galápagos finches provide a particularly well-studied example. Three species on the island of Daphne Major—the medium ground finch, the cactus finch, and the small ground finch—are clearly distinct morphologically, and they occupy different ecological niches. Studies over the past 30 years by Peter and Rosemary Grant found that, on average, 2% of the medium ground finches and 1% of the cactus ground finches mated with other species every year. Furthermore, hybrid offspring appeared to be at no disadvantage in terms of survival or subsequent reproduction. This is not a trivial amount of genetic exchange, and one might expect to see the species coalesce into one genetically variable population—but the species are maintaining their distinctiveness.

Hybridization is not rampant throughout the animal world, however. Most bird species do not hybridize, and probably even fewer experience significant amounts of hybridization. Still, hybridization is common enough to cast doubt on whether reproductive isolation is the only force maintaining the integrity of species.

Natural selection and the maintenance of species' differences

An alternative hypothesis proposes that the distinctions among species are maintained by natural selection. The idea is that each species has adapted to its own specific part of the environment. Stabilizing selection, described in chapter 20, then maintains the species' different adaptations. Hybridization has little effect because alleles introduced into one species' gene pool from other species are quickly eliminated by natural selection.

You probably recall from chapter 20 that the interaction between gene flow and natural selection can have many outcomes. In some cases, strong selection can overwhelm any effects of gene flow—this is how natural selection can maintain species' differences in the face of gene flow. But, in other situations, gene flow can prevent natural selection from eliminating less successful alleles from a population; in this case, the biological species concept would be applicable.

Other weaknesses of the biological species concept

The biological species concept has been criticized for other reasons as well. For example, it can be difficult to apply the concept to populations that are geographically separated in nature. Because individuals of these populations do not encounter each other, it is not possible to observe whether they would interbreed naturally.

Although experiments can determine whether fertile hybrids can be produced, this information is not enough. Many species that coexist without interbreeding in nature will readily hybridize in the artificial settings of the laboratory or zoo. Consequently, evaluating whether such populations constitute different species is ultimately a judgment call. In addition, the concept is more limited than its name would imply. Many organisms are asexual and reproduce without mating. Reproductive isolation therefore has no meaning for such organisms.

For these reasons, a variety of other ideas have been put forward to establish criteria for defining species. Many of these are specific to a particular type of organism, and none has universal applicability. In reality, there may be no single explanation for what maintains the identity of species. Given the incredible variation evident in plants, animals, and microorganisms in all aspects of their biology, it would not be surprising to find that different processes are operating in different organisms.

In addition, some scientists have turned from emphasizing the processes that maintain species distinctions to examining the evolutionary history of populations. These genealogical species concepts are currently a topic of great debate and are discussed further in chapter 23.

Learning Outcomes Review 22.1

Species are populations of organisms that are distinct from other, co-occurring species, and are interconnected geographically. The biological species concept therefore defines species based on their ability to interbreed. Reproductive isolating mechanisms prevent successful interbreeding between species. An alternative approach emphasizes the role of adaptation and natural selection as a force for maintaining separation of species.

- How does the ability to exchange genes explain why sympatric species remain distinct and geographic populations of one species remain connected?
- How could natural selection explain this phenomenon?

22.2 Natural Selection and Reproductive Isolation

Learning Outcomes

1. Define reinforcement in the context of reproductive isolation.
2. Explain the possible outcomes when two populations that are partially reproductively isolated become sympatric.

One of the oldest questions in the field of evolution is: How does one ancestral species become divided into two descendant species? If species are defined by the existence of reproductive isolation, then the process of speciation is identical to the evolution of reproductive isolating mechanisms.

Selection may reinforce isolating mechanisms

The formation of species is a continuous process, and as a result, two populations may have gone only part of the way toward becoming different species and thus be only partially reproductively isolated. For example, because of behavioral or ecological differences, individuals of two populations may be more likely to mate with members of their own population, but between-population matings may still occur. If mating occurs and fertilization produces a zygote, postzygotic barriers may also be incomplete: Developmental problems may result in lower embryo survival or reduced fertility, but some individuals may survive and reproduce.

How reinforcement can complete the speciation process

What happens when two populations come into contact thus depends on the extent to which isolating mechanisms have already evolved. If isolating mechanisms have not evolved at all, then the two populations will interbreed freely, and whatever other differences have evolved between them should disappear over the course of time, as genetic exchange homogenizes the populations. Conversely, if the populations are completely reproductively isolated, then no genetic exchange will occur, and the two populations will remain different species.

An intermediate state, in which hybrids are produced but are only partly sterile, or not as well adapted to the existing habitats as their parents, is more interesting. Selection would favor any alleles in the parental populations that prevented hybridization, because individuals that did not engage in hybridization would produce more successful offspring.

The result would be that selection would improve prezygotic isolating mechanisms until the two populations were completely reproductively isolated. This process is termed **reinforcement** because initially incomplete isolating mechanisms are reinforced by natural selection until they are completely effective.

An example of reinforcement is provided by wildflowers called phlox (figure 22.6). Most phlox flowers are a pale blue color, sometimes with a pink tinge, and Drummond's phlox, a beautiful species that grows profusely along highways, is no exception. The one exception is in certain parts of Texas, where it is sympatric with another species, the pointed phlox. When the two species are found together, the pointed phlox maintains the pale blue coloration, but Drummond's phlox is a deep shade of red.

How has this color difference evolved, and why? The color of these flowers is a result of red and blue color pigments, produced by different genes and leading to the light blue, slightly pinkish color, and another gene that affects the amount of all pigments produced, affecting the intensity of the color. In sympatry, an allele that causes an increase in pigment production has replaced the normal allele. By itself, this change would lead to increased production of both red and blue pigments, leading to a much deeper shade of the same color. However, a new allele at a second gene has knocked out production of the blue pigments, leaving the red pigment to predominate. The result is the deep red color of the sympatric Drummond's phlox flowers.

But why has this changed occurred? The answer lies in the interaction between the two phlox species. The flowers are able to interbreed, but their offspring have very low fertility. Consequently, any factor that would diminish interbreeding would be favored by natural selection. And that's just what color does. Both species are pollinated by butterflies that go from one flower to another, and it turns out that individual butterflies have color preferences. As a result, some butterflies visit only the light blue flowers and others only the dark red ones, and thus pollen from one species does not fertilize flowers of the other species. This example illustrates how reproductive isolation initially resulting from postzygotic infertility was reinforced by the evolution of premating isolation.

How gene flow may counter speciation

Reinforcement is not inevitable, however. When incompletely isolated populations come together, gene flow immediately begins to occur between them. Although hybrids may be inferior, they are not completely inviable or infertile—if they were, the species would already be completely reproductively isolated. When these surviving hybrids reproduce with members of either population, they will serve as a conduit of genetic exchange from one population to the other, and the two populations will tend to lose their genetic distinctiveness. Thus, a race ensues: Can complete reproductive isolation evolve before gene flow erases the differences between the populations? The outcome depends on the initial conditions and the particular natural history of the species involved, but many experts consider reinforcement generally to be the less common outcome.

Learning Outcomes Review 22.2

Natural selection may favor the evolution of increased prezygotic reproductive isolation between sympatric populations when postzygotic isolation initially exists and prezygotic isolation is only partial. This phenomenon is termed reinforcement, and it may lead to populations becoming completely reproductively isolated. In contrast, however, genetic exchange between populations may decrease genetic differences among populations, thus preventing speciation from occurring.

- How might the initial degree of reproductive isolation affect the probability that reinforcement will occur when two populations come into sympatry?

22.3 The Role of Genetic Drift and Natural Selection in Speciation

Learning Outcomes

1. Describe the effects of genetic drift on a population.
2. Explain how genetic drift and natural selection can lead to speciation.

What role does natural selection play in the speciation process? Certainly, the process of reinforcement is driven by natural selection, favoring the evolution of complete reproductive isolation. But reinforcement may not be common. In situations other than reinforcement, does natural selection play a role in the evolution of reproductive isolating mechanisms?

Figure 22.6 Reinforcement in phlox. Drummond's phlox, *Phlox drummondii*, is light blue with a pink tinge throughout most of its range (bottom left). However, where it is sympatric with the pointed phlox, *D. cuspidata* (bottom right), it is deep red in color, making it easier for the butterflies that pollinate them to distinguish between the two species.

Random changes may cause reproductive isolation

As mentioned in chapter 20, populations may diverge as a result of genetic drift. Random change in small populations, founder effects, and population bottlenecks all may lead to changes in traits that cause reproductive isolation.

For example, in the Hawaiian Islands, closely related species of *Drosophila* often differ greatly in their courtship behavior. Colonization of new islands by these fruit flies probably involved a founder effect, in which one or a few flies—perhaps only a single pregnant female—were blown by strong winds to the new island. Changes in courtship behavior between ancestor and descendant populations may be the result of such founder events.

Given enough time, any two isolated populations will diverge because of genetic drift (remember that even large populations experience drift, but at a lower rate than in small populations). In some cases, this random divergence may affect traits responsible for reproductive isolation, and speciation may occur.

Adaptation can lead to speciation

Although random processes may sometimes be responsible, in many cases natural selection probably plays a role in the speciation process. As populations of a species adapt to different circumstances, they likely accumulate many differences that may lead to reproductive isolation. For example, if one population of flies adapts to wet habitats and another to dry areas, then natural selection will favor a variety of corresponding differences in physiological and sensory traits. These differences may produce ecological and behavioral isolation and may cause any hybrids the two populations produce to be poorly adapted to either habitat.

Selection might also act directly on mating behavior. Male *Anolis* lizards, for example, court females by extending a colorful flap of skin, called a *dewlap,* located under their throats (figure 22.7). The ability of one lizard to see the dewlap of another lizard depends not only on the color of the dewlap, but on the environment in which the lizards occur. A light-colored dewlap is most effective in reflecting light in a dim forest, whereas dark colors are more apparent in the bright glare of open habitats. As a result, when these lizards occupy new habitats, natural selection favors evolutionary change in dewlap color because males whose dewlaps cannot be seen will attract few mates. But the lizards also distinguish members of their own species from other species by the color of the dewlap. Adaptive change in mating signals in new environments to enhance visibility to conspecifics could therefore produce reproductive isolation from populations in the ancestral environment; if individuals from the two populations ever came into contact, because of differences in their dewlap colors, they might not recognize each other as members of the same species.

Laboratory scientists have conducted experiments on fruit flies and other fast-reproducing organisms in which they isolate populations in different laboratory chambers and measure how much reproductive isolation evolves. These experiments indicate that genetic drift by itself can lead to some degree of reproductive isolation, but in general, reproductive isolation evolves more rapidly when the populations are forced to adapt to different laboratory environments (such as differences in temperature or food type). Although natural selection in the experiment does not directly favor traits because they lead to reproductive isolation, the incidental effect of adaptive divergence is that populations in different environments become reproductively isolated. For this reason, some biologists believe that the term *isolating mechanisms* is misguided, because it implies that the traits evolved specifically for the purpose of genetically isolating a species, which in most cases—except reinforcement—is probably incorrect.

Learning Outcomes Review 22.3

Genetic drift refers to randomly generated changes in a population's genetic makeup. Isolated populations will eventually diverge because of genetic drift. Adaptation to different environments may also lead to populations becoming reproductively isolated from each other and, in general, is a more potent force producing reproduction isolation.

- *How is the evolution of reproductive isolation in populations adapting to different environments different from the process of reinforcement?*

22.4 The Geography of Speciation

Learning Outcomes

1. Compare and contrast sympatric and allopatric speciation.
2. Explain the conditions required for sympatric speciation to occur.

Figure 22.7 Dewlaps of different species of Caribbean *Anolis* lizards. Males use their dewlaps in both territorial and courtship displays. Coexisting species almost always differ in their dewlaps, which are used in species recognition. Darker-colored dewlaps, such as those of the two species on the left, are easier to see in open habitats, whereas lighter-colored dewlaps, like those of the two species on the right, are more visible in shaded environments.

Figure 22.8 Populations can become geographically isolated for a variety of reasons. *a.* Colonization of remote areas by one or a few individuals can establish populations in a distant place. *b.* Barriers to movement can split an ancestral population into two isolated populations. *c.* Extinction of intermediate populations can leave the remaining populations isolated from one another.

Speciation is a two-part process. First, initially identical populations must diverge, and second, reproductive isolation must evolve to maintain these differences. The difficulty with this process, as we have seen, is that the homogenizing effect of gene flow between populations is constantly acting to erase any differences that may arise by genetic drift or natural selection. Gene flow only occurs between populations that are in contact, however, and populations can become geographically isolated for a variety of reasons (figure 22.8). Consequently, evolutionary biologists have long recognized that speciation is much more likely in geographically isolated populations.

Allopatric speciation takes place when populations are geographically isolated

Ernst Mayr was the first biologist to demonstrate that geographically separated, or *allopatric,* populations appear much more likely to have evolved substantial differences leading to speciation. Marshalling data from a wide variety of organisms and localities, Mayr made a strong case for allopatric speciation as the primary means of speciation.

For example, the little paradise kingfisher varies little throughout its wide range in New Guinea, despite the great variation in the island's topography and climate. By contrast, isolated populations on nearby islands are strikingly different from one another and from the mainland population (figure 22.9). Thus, geographic isolation seems to have been an important prerequisite for the evolution of differences between populations.

Many other examples indicate that speciation can occur under allopatric conditions. Because we would expect isolated populations to diverge over time, by either drift or selection, this result is not surprising. Rather, the more intriguing question becomes: Is geographic isolation *required* for speciation to occur?

Sympatric speciation occurs without geographic separation

For decades, biologists have debated whether one species can split into two at a single locality, without the two new species ever

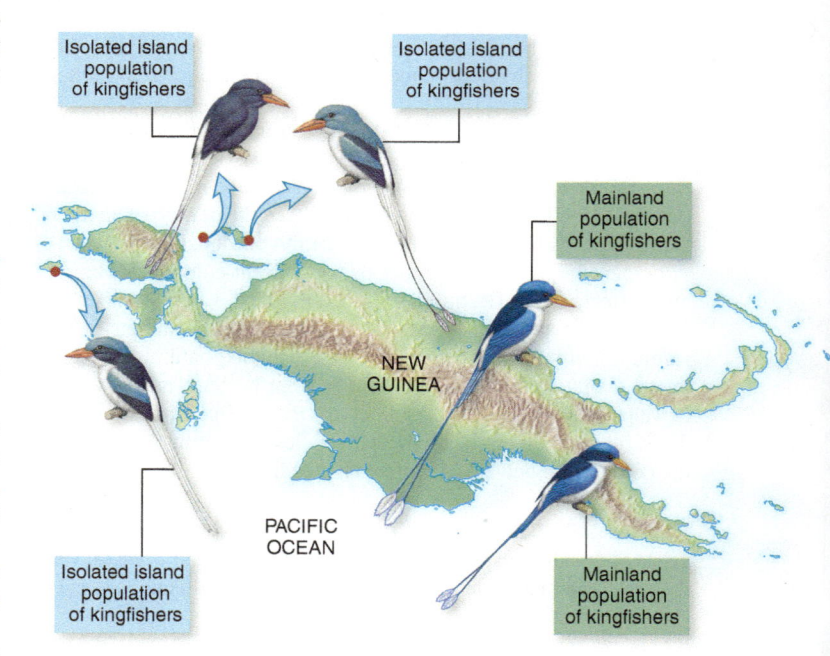

Figure 22.9 Phenotypic differentiation in the little paradise kingfisher, *Tanysiptera hydrocharis,* in New Guinea. Isolated island populations *(left)* are quite distinctive, showing variation in tail feather structure and length, plumage coloration, and bill size, whereas kingfishers on the mainland *(right)* show little variation.

having been geographically separated. Investigators have suggested that this sympatric speciation could occur either instantaneously or over the course of multiple generations. Although most of the hypotheses suggested so far are highly controversial, one type of instantaneous sympatric speciation is known to occur commonly, as the result of polyploidy.

Instantaneous speciation through polyploidy

Instantaneous sympatric speciation occurs when an individual is born that is reproductively isolated from all other members of its species. In most cases, a mutation that would cause an individual to be greatly different from others of its species would have many adverse side effects due to the many pleiotropic effects of most genes (see chapter 12), and the individual likely would not survive. One exception often seen in plants, however, occurs through the process of **polyploidy,** which produces individuals that have more than two sets of chromosomes.

Polyploid individuals can arise in two ways. In **autopolyploidy,** all of the chromosomes may arise from a single species. This might happen, for example, due to an error in cell division that causes a doubling of chromosomes. Such individuals, termed *tetraploids* because they have four sets of chromosomes, can self-fertilize or mate with other tetraploids, but cannot mate and produce fertile offspring with normal diploids. The reason is that the tetraploid species produce "diploid" gametes that produce triploid offspring (having three sets of chromosomes) when combined with haploid gametes from normal diploids. Triploids are sterile because the odd number of chromosomes prevent proper pairing during meiosis.

A more common type of polyploid speciation is **allopolyploidy,** which may happen when two species hybridize (figure 22.10). The resulting offspring, having one copy of the chromosomes of each species, is usually infertile because the chromosomes do not pair correctly in meiosis. However, such individuals are often otherwise healthy and can reproduce asexually. For example, if the chromosomes of such an individual were to spontaneously double, as just described, the resulting tetraploid would have two copies of each set of chromosomes. Consequently, pairing would no longer be a problem in meiosis. As a result, such tetraploids would be able to interbreed, and a new species would have been created.

It is estimated that about half of the approximately 260,000 species of plants have a polyploid episode in their history, including many of great commercial importance, such as bread wheat, cotton, tobacco, sugarcane, bananas, and potatoes. Speciation by polyploidy is also known to occur in a variety of animals, including insects, fish, and salamanders, although much more rarely than in plants.

Sympatric speciation by disruptive selection

Some investigators believe that sympatric speciation can occur over the course of multiple generations through the process of disruptive selection. As noted in chapter 20, disruptive selection can cause a population to contain individuals exhibiting two different phenotypes.

One might think that if selection is strong enough, these two phenotypes would evolve over a number of generations into different species. But before the two phenotypes could become different species, they would have to evolve reproductive isolating mechanisms. Initially, the two phenotypes would not be reproductively isolated at all, and genetic exchange between individuals of the two phenotypes would tend to prevent genetic divergence in mating preferences or other isolating mechanisms. As a result, the two phenotypes would be retained as polymorphisms within a single population. For this reason, most biologists consider sympatric speciation of this type to be a rare event.

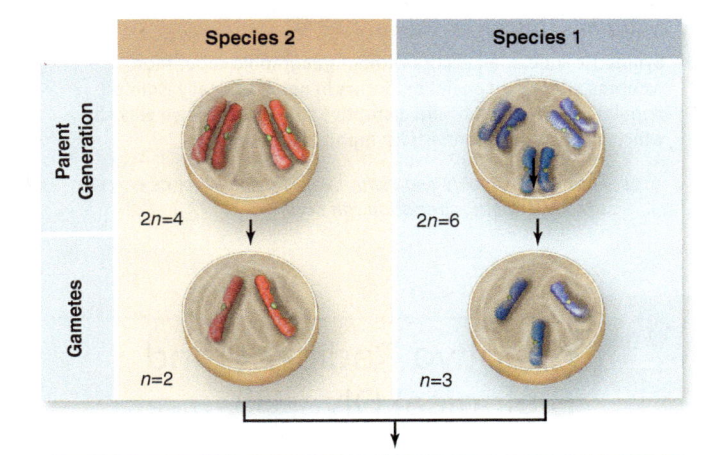

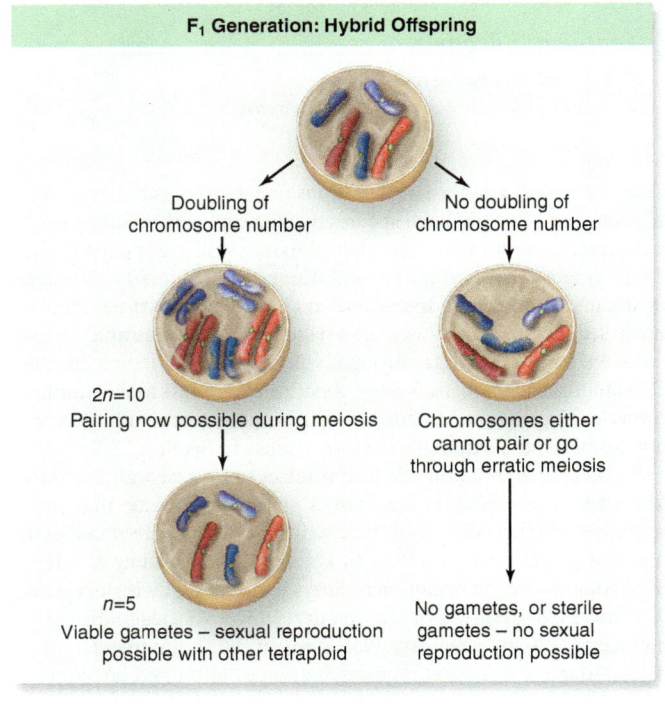

Figure 22.10 Allopolyploid speciation. Hybrid offspring from parents with different numbers of chromosomes often cannot reproduce sexually. Sometimes, the number of chromosomes in such hybrids doubles to produce a tetraploid individual that can undergo meiosis and reproduce with similar tetraploid individuals.

In recent years, however, a number of cases have appeared that are difficult to interpret in any way other than as sympatric speciation. For example, Lake Barombi Mbo in Cameroon is an extremely small and ecologically homogeneous lake, with no opportunity for within-lake isolation. Nonetheless, 11 species of closely related cichlid fish occur in the lake; all of the species are more closely related evolutionarily to one another than to any species outside of the lake. The most reasonable explanation is that an ancestral species colonized the lake and subsequently underwent sympatric speciation multiple times.

Learning Outcomes Review 22.4

Sympatric speciation occurs without geographic separation, whereas allopatric speciation occurs in geographically isolated populations. Polyploidy and disruptive selection are two ways by which a species may undergo sympatric speciation.

- *How do polyploidy and disruptive selection differ as ways in which sympatric speciation can occur?*

22.5 Adaptive Radiation and Biological Diversity

Learning Outcomes
1. Describe adaptive radiation.
2. List conditions that may lead to adaptive radiation.

One of the most visible manifestations of evolution is the existence of groups of closely related species that have recently evolved from a common ancestor and have adapted to many different parts of the environment. These **adaptive radiations** are particularly common in situations in which a species occurs in an environment with few other species and many available resources. One example is the creation of new islands through volcanic activity, such as the Hawaiian and Galápagos Islands. Another example is a catastrophic event leading to the extinction of most other species, a phenomenon termed mass extinction that we discuss in Section 22.7.

Adaptive radiation can also result when a new trait, called a **key innovation**, evolves within a species allowing it to use resources or other aspects of the environment that were previously inaccessible. Classic examples of key innovation leading to adaptive radiation are the evolution of lungs in fish and of wings in birds and insects, both of which allowed descendant species to diversify and adapt to many newly available parts of the environment.

Adaptive radiation requires both speciation and adaptation to different habitats. A classic model postulates that a species colonizes multiple islands in an archipelago. Speciation subsequently occurs allopatrically, and then the newly arisen species colonize other islands, producing multiple species per island (figure 22.11).

Adaptation to new habitats can occur either during the allopatric phase, as the species respond to different environments on the different islands, or after two species become sympatric. In the latter case, this adaptation may be driven by the need to minimize competition for available resources with other species. In this process, termed **character displacement,** two reproductively isolated but ecologically similar species come into contact. Because the two species use the same resources, natural selection in each species favors those individuals that use resources not used by the

1. An ancestral species flies from mainland to colonize one island.

2. The ancestral species spreads to different islands.

3. Populations on different islands evolve to become different species.

a. or *b.*

4. Species evolve different adaptations in allopatry.

4. Colonization of islands.

5. Colonization of islands.

5. Species evolve different adaptations to minimize competition with other species (character displacement).

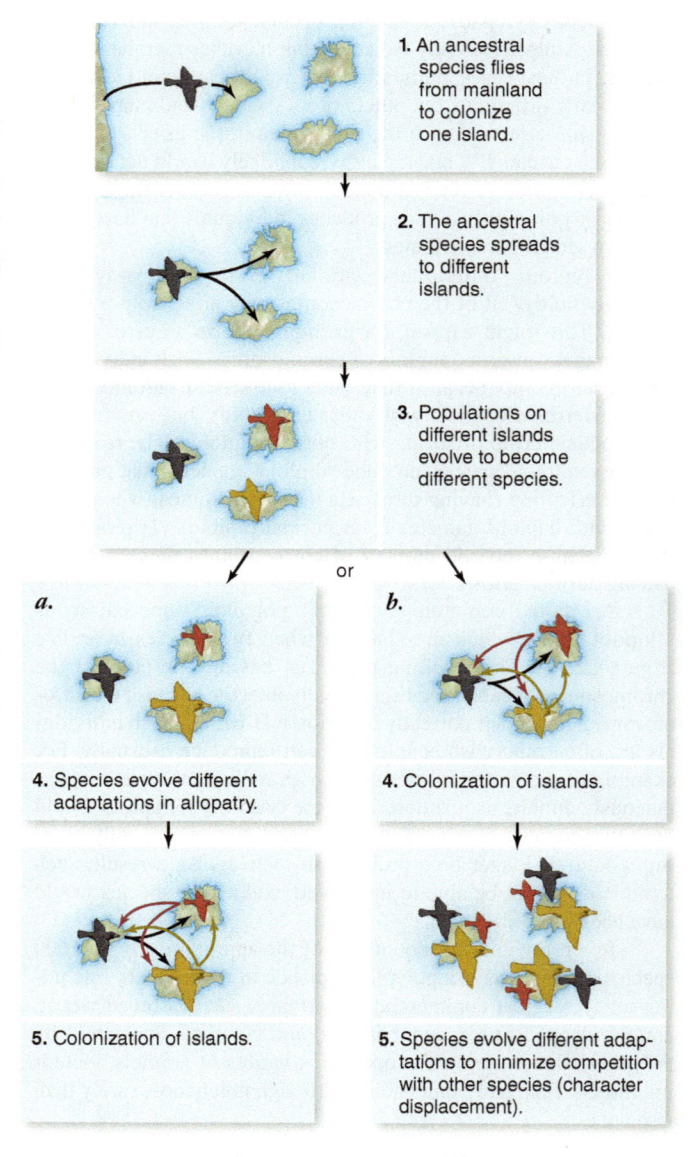

Figure 22.11 Classic model of adaptive radiation on island archipelagoes. *(1)* An ancestral species colonizes an island in an archipelago. Subsequently, the population colonizes other islands *(2),* after which the populations on the different islands speciate in allopatry *(3).* Then some of these new species colonize other islands, leading to local communities of two or more species. Adaptive differences can either evolve when species are in allopatry in response to different environmental conditions *(a)* or as the result of ecological interactions between species *(b)* by the process of character displacement.

other species. Because those individuals will have greater fitness, whatever traits cause the differences in resource use will increase in frequency (assuming that a genetic basis exists for these differences), and, over time, the species will diverge (figure 22.12).

One example of character displacement involves three-spined sticklebacks, small fish found in temperate lakes throughout the northern hemisphere. In British Columbia, many lakes were created in the past 12,000 years when the glaciers melted at the end of the last Ice Age. Some of these lakes near the ocean have been invaded by marine populations of the stickleback. These fish tend to occur in habitats throughout the lake and evolved to be different from their marine ancestors. So different, in fact, that in the few lakes in which a second colonization occurred, the two stickleback populations had evolved a high degree of reproductive isolation. As a result, natural selection led to the evolution of differences in the two populations to minimize competition for food. One stickleback adapted to using open water, and evolved a streamlined body form for greater speed and more effective gill structures for straining out the zooplankton that occur in the open, whereas the other population adapted to foraging near the margins and bottom of the lake, evolving a larger and stouter body and altered the shape of the mouth to better feed on large invertebrates from the lake floor (figure 22.13). Similar character displacement occurred in seven doubly colonized lakes.

An alternative possibility is that adaptive radiation occurs through repeated instances of sympatric speciation, producing a suite of species adapted to different habitats. As discussed in section 22.4, such scenarios are hotly debated.

In this section, we discuss four exemplary cases of adaptive radiation.

Inquiry question How would the scenario for adaptive radiation differ depending on whether speciation is allopatric or sympatric? What is the relationship between character displacement and sympatric speciation?

Hawaiian plants and animals exploited a rich, diverse habitat

More than 1000 species in the fly genus *Drosophila* occur on the Hawaiian Islands. New species of *Drosophila* are still being discovered in Hawaii, although the rapid destruction of the native vegetation is making the search more difficult.

Aside from their sheer number, Hawaiian *Drosophila* species are unusual because of their incredible diversity of morphological

---

SCIENTIFIC THINKING

Question: Does competition for resources cause character displacement?
Hypothesis: Competition with similar species will cause natural selection to promote evolutionary divergence.
Experiment: Place a species of fish in a pond with another, similar fish species and measure the form of selection. As a control, place a population of the same species in a pond without the second species. Note that the size of food that these fish eat is related to the size of the fish.
Result: In the pond with two species, directional selection favors those individuals which have phenotypes most dissimilar from the other species, and thus are most different in resource use. Directional selection does not occur in the control population.

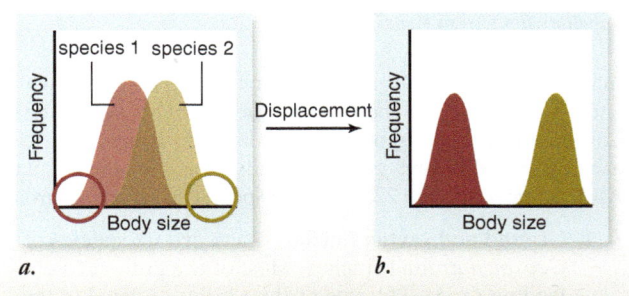

Interpretation: Would you expect character displacement to occur if resources were unlimited?

Figure 22.12 Character displacement. *a.* Two species are initially similar and thus overlap greatly in resource use, as might happen if the two species were similar in size (in many species, body size and food size are closely related). Individuals in each species that are most different from the other species *(circled)* will be favored by natural selection, because they will not have to compete with the other species. For example, the smallest individuals of one species and the largest of the other would not compete with the other species for food and thus would be favored. *b.* As a result, the species will diverge in resource use and minimize competition between the species.

Figure 22.13 Stickleback fish adapted to using different habitats. Fish living in open water (*a*) are more streamlined in body form, whereas those that occur near the substrate (*b*) are larger, bulkier, and have a mouth better equipped to pick invertebrates off the lake bottom.

Figure 22.14 Hawaiian *Drosophila*. The hundreds of species that have evolved on the Hawaiian Islands are extremely variable in appearance; note in particular the differences in head shape and body color between these two species. ***a.*** *Drosophila heteroneura.* ***b.*** *Drosophila silvestris.*

and behavioral traits (figure 22.14). Evidently, when their ancestors first reached these islands, they encountered many "empty" habitats that other kinds of insects and other animals occupied elsewhere. As a result, the species have adapted to all manners of fruit fly life and include predators, parasites, and herbivores, as well as species specialized for eating the detritus in leaf litter and the nectar of flowers. The larvae of various species live in rotting stems, fruits, bark, leaves, or roots, or feed on sap. No comparable diversity of *Drosophila* species is found anywhere else in the world.

The great diversity of Hawaiian species is a result of the geological history of these islands. New islands have continually arisen from the sea in this region. As they have done so, they appear to have been invaded successively by the various *Drosophila* groups present on the older islands. New species thus have evolved as new islands have been colonized.

In addition, the Hawaiian Islands are among the most volcanically active islands in the world. Periodic lava flows often have created patches of habitat within an island surrounded by a "sea" of barren rock. These land islands are termed *kipukas*. *Drosophila* populations isolated in these kipukas often undergo speciation. In these ways, rampant speciation combined with ecological opportunity has led to an unparalleled diversity of insect life.

Similar patterns are shown by many other animal and plant groups. For example, lobeliads are a group of 125 species that have radiated to produce plants ranging in growth form from vines and shrubs to tall trees and that live in habitats as diverse as cliffsides, high-elevation bogs, rainforests, deserts, and coastlines (figure 22.15).

Darwin's finch species adapted to use different food types

The diversity of Darwin's finches on the Galápagos Islands was first mentioned in chapter 21. Presumably, the ancestor of Darwin's finches reached these islands before other land birds, and as a result many of the types of habitats that other types of birds use on the mainland were unoccupied.

As the new arrivals moved into these vacant ecological niches and adopted new lifestyles, they were subjected to many different sets of selective pressures. Under these circumstances, and aided by the geographic isolation afforded by the many islands of the Galápagos archipelago, the ancestral finches rapidly

Figure 22.15 Hawaiian lobeliads. *(a)* A coastal cliff dwelling species, *Brighamia rockii*, on the cliffs of Molokaì, Hawaiì, *(b)* large tree form of *Cyanea hamatiflora* from Maui, *(c)* small tree of *Cyanea koolauensis* from Oahu, and *(d)* bog habitat species *Lobelia villosa* from Kauai.

split into a series of diverse populations, some of which evolved into separate species. These species now occupy many different habitats on the Galápagos Islands, which are comparable to the habitats several distinct groups of birds occupy on the mainland. As illustrated in figure 22.16, the 14 species fall into four groups:

1. **Ground and cactus finches.** There are six species of *Geospiza* ground finches. Most of the ground finches feed on seeds. The size of their bills is related to the size of the seeds they eat. Some of the ground finches feed primarily on cactus flowers and fruits, and they have a longer, larger, and more pointed bill than the others.
2. **Tree finches.** There are five species of insect-eating tree finches. Four species have bills suitable for feeding on insects. The woodpecker finch has a chisel-like beak. This unusual bird carries around a twig or a cactus spine, which it uses to probe for insects in deep crevices.
3. **Vegetarian tree finch.** The very heavy bill of this species is used to wrench buds from branches.
4. **Warbler finches.** These unusual birds play the same ecological role in the Galápagos woods that warblers

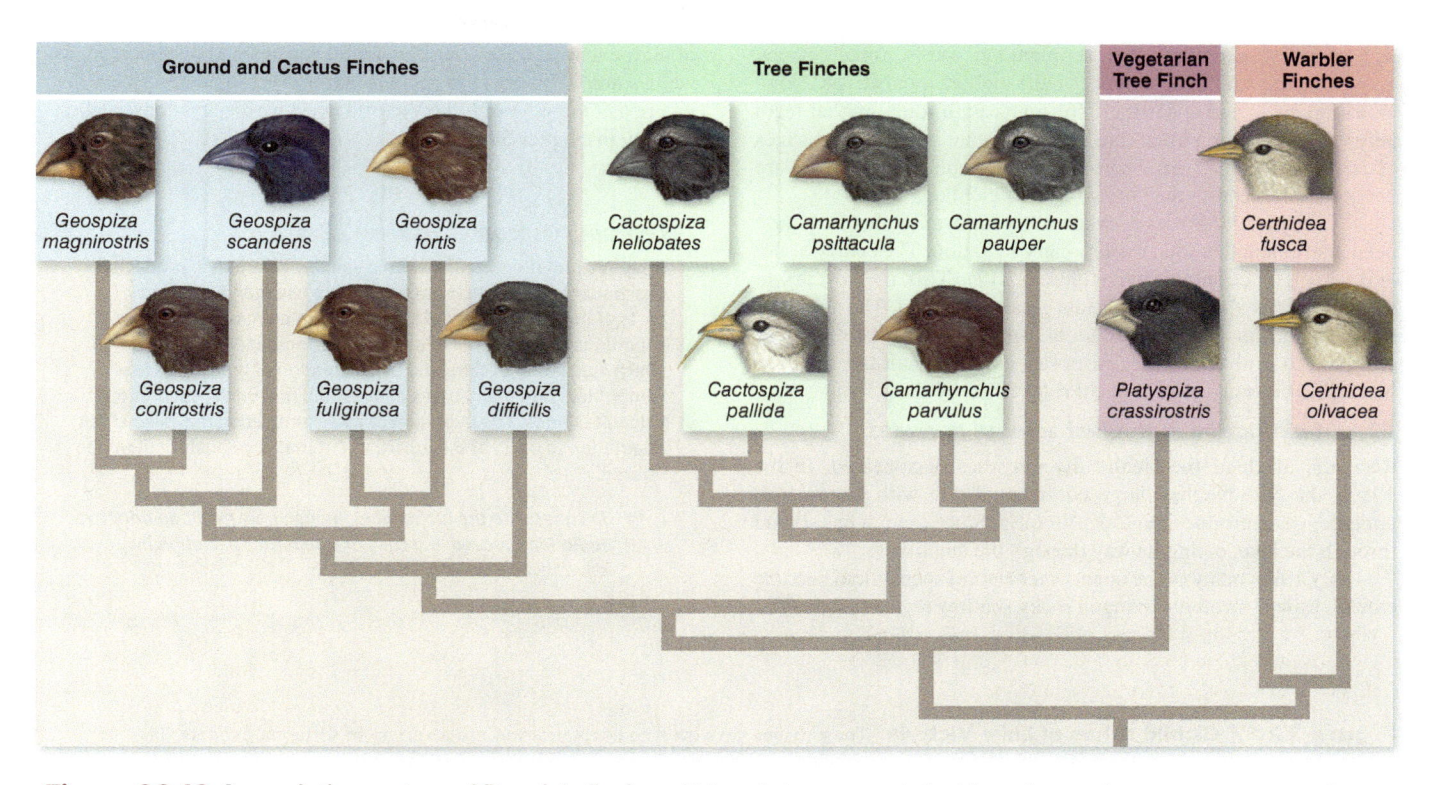

Figure 22.16 An evolutionary tree of Darwin's finches. This evolutionary tree, derived from the complete genome sequences of Darwin's finches, indicates that warbler finches are an early offshoot. Ground and tree finches subsequently diverged, and then species within each group specialized to use different resources. Analysis of these genomic data suggested that for several species, such as *G. difficilis*, what we recognize as a single species that occurs on different islands may instead represent several different species, and these species may not be closely related to each other. Further study is needed to examine this hypothesis.

play on the mainland, searching continuously over the leaves and branches for insects. They have slender, warblerlike beaks.

Recently, scientists have examined the complete genome sequences of Darwin's finches to study their evolutionary history. These studies suggest that the deepest branches in the finch evolutionary tree lead to warbler finches, which implies that warbler finches were among the first types to evolve after colonization of the islands. The ground species are closely related to one another, as are all of the tree finches. Nonetheless, within each group, species differ in beak size and other attributes, as well as in resource use.

Field studies, conducted in conjunction with those discussed in chapter 21, demonstrate that ground species compete for resources; the differences between species likely resulted from character displacement as initially similar species diverged to minimize competitive pressures.

Recent studies are beginning to uncover the underlying genetic and developmental basis for Darwin's finch diversification. Work to date has focused on the beaks of the ground finches. Beak variation among species is a result of differential rates of growth in beak height, width, and length. Researchers have identified five genes affecting these aspects of beak growth: *Bmp4*, *CaM*, *TGFβIIr*, *β-catenin* and *Dkk3*. Genetic changes in different combinations of these genes are responsible for the interspecific differences in adult beaks.

Lake Victoria cichlid fishes diversified very rapidly

Lake Victoria is an immense, shallow, freshwater sea about the size of Switzerland in the heart of equatorial East Africa. Until recently, the lake was home to an incredibly diverse collection of over 450 species of cichlid fishes.

Geologically recent radiation

The cluster of cichlid species appears to have evolved recently and quite rapidly. By sequencing the cytochrome *b* gene in many of the lake's fish, scientists have been able to estimate that the first cichlids entered Lake Victoria only 200,000 years ago.

Dramatic changes in water level encouraged species formation. As the lake rose, it flooded new areas and opened up new habitats. Many of the species may have originated after the lake dried down 14,000 years ago, isolating local populations in small lakes until the water level rose again.

Cichlid diversity

Cichlids are small, perchlike fishes ranging from 5 to 25 centimeters (cm) in length, and the males come in endless varieties of colors. The ecological and morphological diversity of these fish is remarkable, particularly given the short span of time over which they have evolved.

We can gain some sense of the vast range of types by looking at how different species eat. There are mud biters, algae scrapers, leaf chewers, snail crushers, zooplankton eaters, insect eaters, prawn eaters, and fish eaters. Snail shellers pounce on slow-crawling snails and spear their soft parts with long, curved teeth before the snail can retreat into its shell. Scale scrapers rasp slices of scales off other fish. There are even cichlid species that are "pedophages," eating the young of other cichlids.

Cichlid fish have a remarkable key innovation that may have been instrumental in their evolutionary radiation: They carry a second set of functioning jaws (figure 22.17). This trait occurs in many other fish, but in cichlids it is greatly enlarged. The ability of these second jaws to manipulate and process food has freed the oral jaws to evolve for other purposes, and the result has been the incredible diversity of ecological roles filled by these fish.

Abrupt extinction in the last several decades

Recently, much of the cichlid diversity has disappeared. In the 1950s, the Nile perch, a large commercial fish with a voracious appetite, was introduced to Lake Victoria. Since then, it has spread through the lake, eating its way through the cichlids.

By 1990, many of the open-water cichlid species had become extinct, as well as others living in rocky shallow regions. Over 70% of all the named Lake Victoria cichlid species had disappeared, as well as untold numbers of species that had yet to be described. We will revisit the story of Lake Victoria when we discuss conservation biology in chapter 59.

Learning Outcomes Review 22.5

Adaptive radiation occurs when a species diversifies, producing descendant species that are adapted to use many different parts of the environment. Adaptive radiation may occur under conditions of recurrent isolation, which increases the rate at which speciation occurs, and by occupation of areas with few competitors and many types of available resources, such as on volcanic islands. The evolution of a key innovation may also allow adaptation to parts of the environment that previously couldn't be utilized.

■ *In contrast to the archipelago model, how might an adaptive radiation proceed in a case of sympatric speciation by disruptive selection?*

Figure 22.17 Cichlid fishes of Lake Victoria. These fishes have evolved adaptations to use a variety of different habitats. The enlarged second set of jaws located in the throat of these fish has provided evolutionary flexibility, allowing oral jaws to be modified in many ways.

22.6 The Pace of Evolution

Learning Outcomes
1. Define stasis and compare it to gradual evolutionary change.
2. Explain the components of the punctuated equilibrium hypothesis.

We have discussed the manner in which speciation may occur, but we haven't yet considered the relationship between speciation and the evolutionary change that occurs within a species. Two hypotheses, *gradualism* and *punctuated equilibrium,* have been advanced to explain the relationship.

Gradualism is the accumulation of small changes

For more than a century after the publication of *On the Origin of Species,* the standard view was that evolution occurred very slowly. Such change would be nearly imperceptible from generation to generation, but would accumulate such that, over the course of thousands and millions of years, major changes could occur. This view is termed **gradualism** (figure 22.18*a*).

a. Gradualism   *b.* Punctuated equilibrium

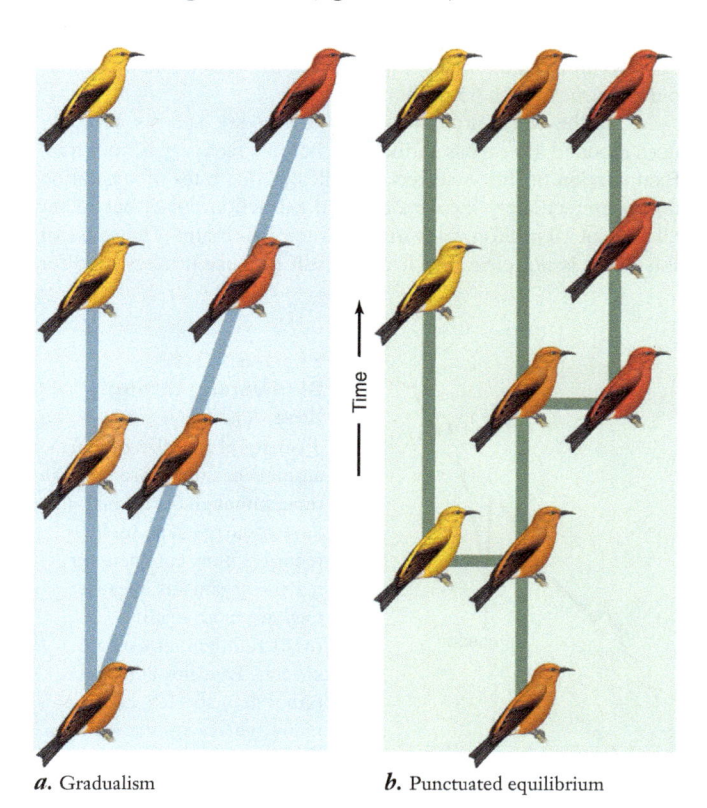

Figure 22.18 Two views of the pace of macroevolution.
a. Gradualism suggests that evolutionary change occurs slowly through time and is not linked to speciation, whereas (*b*) punctuated equilibrium requires that phenotypic change occurs in bursts associated with speciation, separated by long periods of little or no change.

Punctuated equilibrium is long periods of stasis followed by relatively rapid change

An alternative possibility is that species experience long periods of little or no evolutionary change (termed **stasis**), punctuated by bursts of evolutionary change occurring over geologically short time intervals. This phenomenon is termed **punctuated equilibrium** (figure 22.18*b*); some have argued that these periods of rapid change occurred only during the speciation process.

We have seen in chapters 20 and 21 that when natural selection is strong, evolutionary change can occur rapidly, so the "punctuated" part of the theory is not controversial. A more difficult question involves the long periods of stasis (the equilibrium): Why would species exist for thousands, or even millions, of years without changing?

Although a number of possible reasons have been suggested, most researchers now believe that a combination of stabilizing and oscillating selection is responsible for stasis. If the environment does not change over long periods of time, or if environmental changes oscillate back and forth, then stasis may occur for long periods. One factor that may enhance this stasis is the ability of species to shift their ranges; for example, during the ice ages, when the global climate cooled, the geographic ranges of many species shifted southward, so that the species continued to experience similar environmental conditions.

> **Inquiry question** Why would changes in geographic ranges promote evolutionary stasis?

Evolution may include both types of change

Some well-documented groups, such as African mammals, clearly have evolved gradually, not in spurts. Other groups, such as marine bryozoa, seem to show the irregular pattern of evolutionary change predicted by the punctuated equilibrium model. It appears, in fact, that gradualism and punctuated equilibrium are two ends of a continuum. Although some groups have evolved solely in a gradual manner and others only in a punctuated mode, many other groups show evidence of both gradual and punctuated episodes at different times in their evolutionary history.

The idea that speciation is necessarily linked to phenotypic change has not been supported, however. On one hand, it is now clear that speciation can occur without substantial phenotypic change. For example, many closely related salamander species are nearly indistinguishable. On the other, it is also clear that phenotypic change can occur within species in the absence of speciation.

Learning Outcomes Review 22.6
Gradualism is the accumulation of almost imperceptible changes that eventually results in major differences. Punctuated equilibrium proposes that long periods of stasis are interrupted (punctuated) by periods of rapid change. Evidence for both gradualism and punctuated equilibrium has been found in different groups. Stasis refers to a period in which little or no evolutionary change occurs. Stasis may result from stabilizing or oscillating selection.

■ *Could evolutionary change be punctuated in time (that is, rapid and episodic), but not linked to speciation?*

22.7 Speciation and Extinction Through Time

Learning Outcomes
1. Describe the pattern of species diversity through time.
2. Define mass extinction and identify when major mass extinctions have occurred.

Biological diversity has increased vastly over the last 600 million years, but the trend has been far from consistent. After a rapid rise, diversity reached a plateau for about 200 million years, but since then has risen steadily. Because changes in the number of species reflect the rate of origin of new species relative to the rate at which existing species disappear, this long-term trend reveals that speciation has, in general, surpassed extinction.

Nonetheless, speciation has not always outpaced extinction. In particular, interspersed in the long-term increase in species diversity have been a number of sharp declines, termed **mass extinctions.**

Five mass extinctions have occurred in the distant past

Five major mass extinctions have been identified, the most severe one occurring at the end of the Permian period, approximately 250 million years ago (figure 22.19). At that time, more than half of all plant and animal families and as much as 96% of all species may have perished.

The most famous and well-studied extinction, although not as drastic, occurred at the end of the Cretaceous period (65 million years ago), at which time the dinosaurs and a variety of other organisms went extinct. Recent findings have supported the hypothesis that this extinction event was triggered when a large asteroid slammed into Earth, perhaps causing global forest fires and obscuring the Sun for months by throwing particles into the air. The cause of other mass extinction events is less certain. Some scientists suggest that asteroids may have played a role in at least some of the other mass extinction events; other hypotheses implicate global climate change and other causes.

One important result of mass extinctions is that not all groups of organisms are affected equally. For example, in the extinction at the end of the Cretaceous, not only dinosaurs, but also marine and flying reptiles, and ammonites (a type of mollusk) went extinct. Marsupials, flowering plants, birds, and some forms of plankton were greatly reduced in diversity. In contrast, turtles, crocodilians, and amphibians seemed to have been unscathed. Why some groups were harder hit than others is not clear, but one hypothesis suggests that survivors were those animals that could shelter underground or in water, and that could either scavenge or required little food in the cool temperatures that resulted from the blockage of sunlight.

A consequence of mass extinctions is that previously dominant groups may perish, thus changing the course of evolution. This is certainly true of the Cretaceous extinction. During the Cretaceous period, placental mammals were a minor group composed of species that were mostly no larger than a house cat. When the dinosaurs, which had dominated the world for more than 100 million years, disappeared at the end of this period, the placental mammals underwent a significant adaptive radiation. It is humbling to think that humans might never have arisen had that asteroid not struck Earth 65 million years ago.

As the world around us illustrates today, species diversity does rebound after mass extinctions, but this recovery is not rapid. Examination of the fossil record indicates that rates of speciation do not immediately increase after an extinction pulse, but rather take about 10 million years to reach their maximum. The cause of this delay is not clear, but it may result because it takes time for

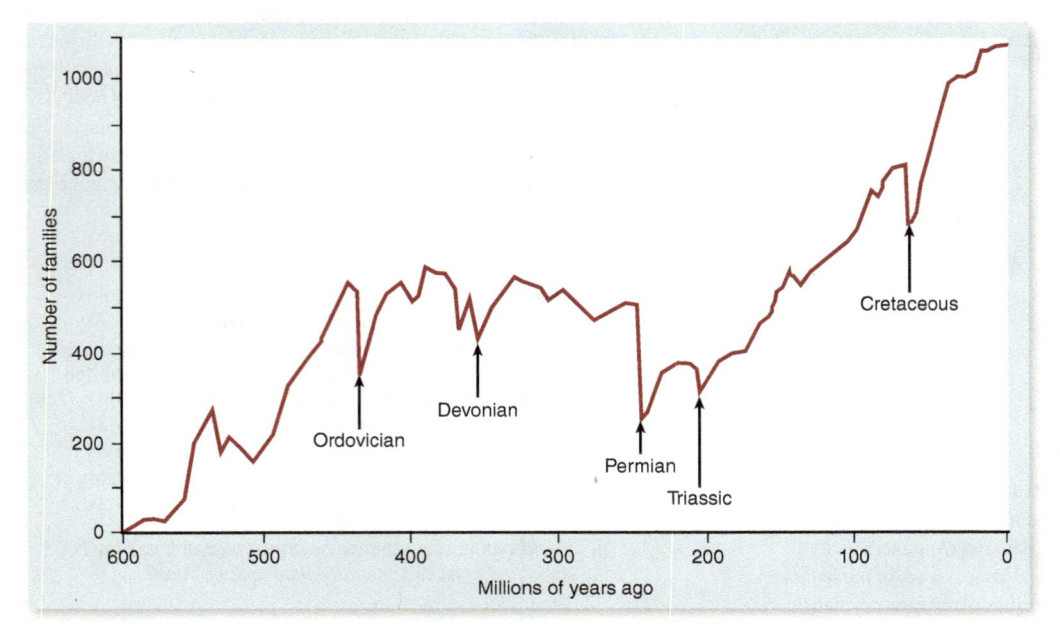

Figure 22.19
Biodiversity through time. The taxonomic diversity of families of marine animals has increased through time, although occasional dips have occurred. The fossil record is most complete for marine organisms because they are more readily fossilized than terrestrial species. Families are shown, rather than species, because many species are known from only one specimen, thus introducing error into counts of the number of species present at a particular point in time. Arrows indicate the five major mass extinction events.

ecosystems to recover and for the processes of speciation and adaptive diversification to begin. Consequently, species diversity may require 10 million years, or even much longer, to attain its previous level.

A sixth extinction is under way

The number of species in the world in recent times is greater than it has ever been. Unfortunately, that number is decreasing at an alarming rate due to human activities (see chapter 59).

Some estimate that as much as one-fourth of all species will become extinct in the near future, a rate of extinction not seen on Earth since the Cretaceous mass extinction. Moreover, the rebound in species diversity may be even slower than following previous mass extinction events because, instead of the ecologically impoverished, but energy-rich environment that existed after previous mass extinction events, a large proportion of the world's resources will be taken up already by human activities, leaving few resources available for adaptive radiation.

Learning Outcomes Review 22.7

The number of species has increased through time, although not at a constant rate. Five major extinction events have substantially, although briefly, reduced the number of species. Diversity rebounds, but the recovery is not rapid, and the groups making up that diversity are not the same as those that existed before the extinction event. Unfortunately, humans are currently causing a sixth mass extinction event.

■ *In what ways are the current mass extinction event different from those that have occurred in the past?*

Chapter Review

22.1 The Nature of Species and the Biological Species Concept

Sympatric species inhabit the same locale but remain distinct.
Most sympatric species are readily distinguishable phenotypically and ecologically. Others usually can be identified by careful study.

Populations of a species exhibit geographic variation.
Populations that differ greatly in phenotype or ecologically are usually connected by geographically intermediate populations.

The biological species concept focuses on the ability to exchange genes.
In the biological species concept, species are defined as those populations that interbreed, or have the potential to do so, and produce fertile offspring. Reproductive isolating mechanisms prevent genetic exchange between species.

Prezygotic isolating mechanisms prevent the formation of a zygote.
Prezygotic isolating mechanisms prevent a viable zygote from being created. These mechanisms include ecological, behavioral, temporal, and mechanical isolation, as well as failure of gametes to unite to form a zygote.

Postzygotic isolating mechanisms prevent normal development into reproducing adults.
Postzygotic isolating mechanisms prevent a zygote from developing into a viable and fertile individual.

The biological species concept does not explain all observations.
An alternative explanation for the existence of distinct species focuses on the role of natural selection and differences among species in their ecological requirements.

In reality, no one species concept can explain all the diversity of life.

22.2 Natural Selection and Reproductive Isolation

Selection may reinforce isolating mechanisms.
Populations may evolve complete reproductive isolation in allopatry. If populations that have evolved only partial reproductive isolation come into contact, natural selection can lead to increased reproductive isolation, a process termed "reinforcement."

Alternatively, genetic exchange between the populations can lead to homogenization and thus prevent speciation from occurring.

22.3 The Role of Genetic Drift and Natural Selection in Speciation

Random changes may cause reproductive isolation.
In small populations, genetic drift may cause populations to diverge; the differences that evolve may cause the populations to become reproductively isolated.

Adaptation can lead to speciation.
Adaptation to different situations or environments may incidentally lead to reproductive isolation. In contrast to reinforcement, these differences are not directly favored by natural selection because they prevent hybridization.

22.4 The Geography of Speciation (figure 22.9)

Allopatric speciation takes place when populations are geographically isolated.
Allopatric, or geographically isolated, populations may evolve into separate species because no gene flow occurs between them. Most speciation probably occurs in allopatry.

Sympatric speciation occurs without geographic separation.
Sympatric speciation can occur in two ways. One is polyploidy, which instantly creates a new species. Disruptive selection also can cause one species to divide into two, but scientists debate how commonly it occurs.

22.5 Adaptive Radiation and Biological Diversity

Adaptive radiation occurs when a species finds itself in a new or suddenly changed environment with many resources and few competing species (figure 22.11).

The evolution of a new trait that allows individuals to use previously inaccessible parts of the environment, termed a "key innovation," may also trigger an adaptive radiation.

Hawaiian plants and animals exploited a rich, diverse habitat.
At least 1000 species of Hawaiian *Drosophila* have been identified. Several groups of plants have also diversified to occupy many habitats.

Darwin's finch species adapted to use different food types.
Fourteen species in four genera have evolved to exploit four different habitats based on type of food.

Lake Victoria cichlid fishes diversified very rapidly.
Cichlids underwent rapid radiation to form more than 450 species, although today more than 70% of these species are now extinct.

22.6 The Pace of Evolution

Gradualism is the accumulation of small changes.
Historically, scientists took the view that speciation occurred gradually through very small cumulative changes.

Punctuated equilibrium is long periods of stasis followed by relatively rapid change.
The punctuated equilibrium hypothesis contends that not only is change rapid and episodic, but that it is only associated with the speciation process.

Evolution may include both types of change.
Scientists generally agree that evolutionary change occurs on a continuum, with gradualism and punctuated change being the extremes.

22.7 Speciation and Extinction Through Time

In general, the number of species have increased through time (figure 22.19).

Five mass extinctions have occurred in the distant past.
Mass extinctions have led to dramatic decreases in species diversity, which take millions of years to recover. Five such mass extinctions have occurred in the distant past due to asteroids hitting the Earth and global climate change, among other events.

A sixth mass extinction is under way.
Humans are currently causing a sixth mass extinction.

Review Questions

UNDERSTAND

1. Prezygotic isolating mechanisms include all of the following except
   a. hybrid sterility.
   b. courtship rituals.
   c. habitat separation.
   d. seasonal reproduction.

2. Reproductive isolation is
   a. a result of individuals not mating with each other.
   b. a specific type of postzygotic isolating mechanism.
   c. required by the biological species concept.
   d. None of the choices is correct.

3. Problems with the biological species concept include the fact that
   a. many species reproduce asexually.
   b. postzygotic isolating mechanisms decrease hybrid viability.
   c. prezygotic isolating mechanisms are extremely rare.
   d. All of the choices are correct.

4. Allopatric speciation
   a. is less common than sympatric speciation.
   b. involves geographic isolation of some kind.
   c. is the only kind of speciation that occurs in plants.
   d. requires polyploidy.

5. Gradualism and punctuated equilibrium are
   a. two ends of the continuum of the rate of evolutionary change over time.
   b. mutually exclusive views about how all evolutionary change takes place.
   c. mechanisms of reproductive isolation.
   d. None of the choices is correct.

6. Prezygotic isolation
   a. always involves mechanisms that prevent interbreeding of members of different species.
   b. includes the death of a zygote shortly after fertilization.
   c. only occurs in plants.
   d. becomes stronger as the result of reinforcement.

7. Speciation by allopolyploidy
   a. takes a long time.
   b. is common in birds.
   c. leads to reduced numbers of chromosomes.
   d. occurs after hybridization between two species.

8. Adaptive radiation
   a. is the result of enriched uranium used in power plants.
   b. is the evolution of closely related species adapted to use different parts of the environment.
   c. results from genetic drift.
   d. is the outcome of stabilizing selection favoring the maintenance of adaptive traits.

9. Leopard frogs from different geographic populations of the *Rana pipiens* complex
   a. are members of a single species because they look very similar to one another.
   b. are different species shown to have pre- and postzygotic isolating mechanisms.
   c. frequently interbreed to produce viable hybrids.
   d. are genetically identical due to effective reproductive isolation.

10. Character displacement
    a. arises through competition and natural selection, favoring divergence in resource use.
    b. arises through competition and natural selection, favoring convergence in resource use.
    c. does not promote speciation.
    d. reduced speciation rates in Galápagos finches.

11. During the history of life on Earth
    a. there have been major extinction events.
    b. species diversity has steadily increased.
    c. species diversity has stayed relatively constant.
    d. extinction rates have been completely offset by speciation rates.

APPLY

1. If reinforcement is weak and hybrids are not completely infertile,
   a. genetic divergence between populations may be overcome by gene flow.
   b. speciation will occur 100% of the time.
   c. gene flow between populations will be impossible.
   d. the speciation will be more likely than if hybrids were completely infertile.

2. Natural selection can
   a. enhance the probability of speciation.
   b. enhance reproductive isolation.
   c. act against hybrid survival and reproduction.
   d. All of the choices are correct.

3. Hybridization between incompletely isolated populations
   a. always leads to reinforcement due to the inferiority of hybrids.
   b. can serve as a mechanism for preserving gene flow between populations, thus preventing speciation.
   c. only occurs in plants.
   d. never affects rates of speciation.

4. Natural selection can lead to speciation
   a. by causing small populations to diverge more than large populations.
   b. because the evolutionary changes that two populations acquire while adapting to different habitats may have the effect of making them reproductively isolated.
   c. by favoring the same evolutionary change in multiple populations.
   d. by favoring intermediate phenotypes.

SYNTHESIZE

1. Natural selection can lead to the evolution of prezygotic isolating mechanisms, but not postzygotic isolating mechanisms. Explain.

2. If there is no universally accepted definition of a species, what good is the term? Will the idea of and need for a "species concept" be eliminated in the future?

3. Refer to figure 22.6. In Texas, Drummond's phlox is a dark shade of red where it occurs with the pale blue pointed phlox. Elsewhere, where it occurs by itself, it is a pale pinkish-blue. In this case, there is no competition for ecological resources as in other cases of character divergence discussed. Why has the color of Drummond's phlox evolved when it is sympatric with pointed phlox?

4. Refer to figure 22.16. *Geospiza fuliginosa* and *Geospiza fortis* are found in sympatry on at least one island in the Galápagos and in allopatry on several islands in the same archipelago. Compare your expectations about degree of morphological similarity of the two species in these two contexts, given the hypothesis that competition for food played a large role in the adaptive radiation of this group. Would your expectations be the same for a pair of finch species that are not as closely related? Explain.

CHAPTER 23

Systematics, Phylogenies, and Comparative Biology

Chapter Contents

23.1 Systematics
23.2 Cladistics
23.3 Systematics and Classification
23.4 Phylogenetics and Comparative Biology
23.5 Phylogenetics and Disease Evolution

Introduction

All organisms share many biological characteristics. They are composed of one or more cells, carry out metabolism and transfer energy with ATP, and encode hereditary information in DNA. Yet, there is also a tremendous diversity of life, ranging from bacteria and amoebas to blue whales and sequoia trees. For generations, biologists have tried to group organisms based on shared characteristics. The most meaningful groupings are based on the study of evolutionary relationships among organisms. New methods for constructing evolutionary trees and a sea of molecular sequence data are leading to improved evolutionary hypotheses to explain life's diversification.

23.1 Systematics

Learning Outcomes
1. Understand what a phylogeny represents.
2. Explain why phenotypic similarity does not necessarily indicate close evolutionary relationship.

One of the great challenges of modern science is to understand the history of ancestor–descendant relationships that unites all forms of life on Earth, from the earliest single-celled organisms to the complex organisms we see around us today. If the fossil record were perfect, we could trace the evolutionary history of species and examine how each arose and proliferated; however, as discussed in chapter 21, the fossil record is far from complete. Although it answers many questions about life's diversification, it leaves many others unsettled.

Consequently, scientists must rely on other types of evidence to establish the best hypothesis of evolutionary relationships. Bear in mind that the outcomes of such studies *are* hypotheses, and as such, they require further testing. All hypotheses may be disproved by new data, leading to the formation of better, more accurate scientific ideas.

The study of evolutionary relationships is called **systematics**. By looking at the similarities and differences between species, systematists can construct an evolutionary tree, or **phylogeny**, which represents a hypothesis about patterns of relationship among species.

Branching diagrams depict evolutionary relationships

Darwin envisioned that all species were descended from a single common ancestor, and that the history of life could be depicted as a branching tree (figure 23.1). In Darwin's view, the twigs of the tree represent existing species. As one works down the tree, the joining of twigs and branches reflects the pattern of common ancestry back in time to the single common ancestor of all life. The process of descent with modification from common ancestry results in all species being related in this branching, hierarchical fashion, and their evolutionary history can be depicted using branching diagrams or phylogenetic trees. Figure 23.1b shows how evolutionary relationships are depicted with a branching diagram. Humans and chimpanzees are descended from a common ancestor and are each other's closest living relative (the position of this common ancestor is indicated by the node labeled 1). Humans, chimps, and gorillas share an older common ancestor (node 2), and all apes share a more distant common ancestor (node 3).

One key to interpreting a phylogeny is to look at how recently species share a common ancestor, rather than looking at the arrangement of species across the top of the tree. If you compare the three versions of the phylogeny of figure 23.1b, you can see that the relationships are the same: Regardless of where they are positioned, chimpanzees and humans are still more closely related to each other than to any other species.

Moreover, even though humans are placed next to gibbons in version 1 of figure 23.1b, the pattern of relationships still indicates that humans are more closely related (that is, share a more recent common ancestor) with gorillas and orangutans than with gibbons. Phylogenies are also sometimes displayed on their side, rather than upright (figure 23.1b, version 3), but this arrangement also does not affect its interpretation.

Similarity may not accurately predict evolutionary relationships

We might expect that the greater the time since two species diverged from a common ancestor, the more different they would be. Early systematists relied on this reasoning and constructed phylogenies based on overall similarity. If, in fact, species evolved at a constant rate, then the amount of divergence between two species would be a function of how long they had been diverging, and thus phylogenies based on degree of similarity would be accurate. As a result, we might think that chimps and gorillas are more closely related to each other than either is to humans.

But as chapter 22 revealed, evolution can occur very rapidly at some times and very slowly at others. Species invading new habitats are likely to experience new selective pressures and may

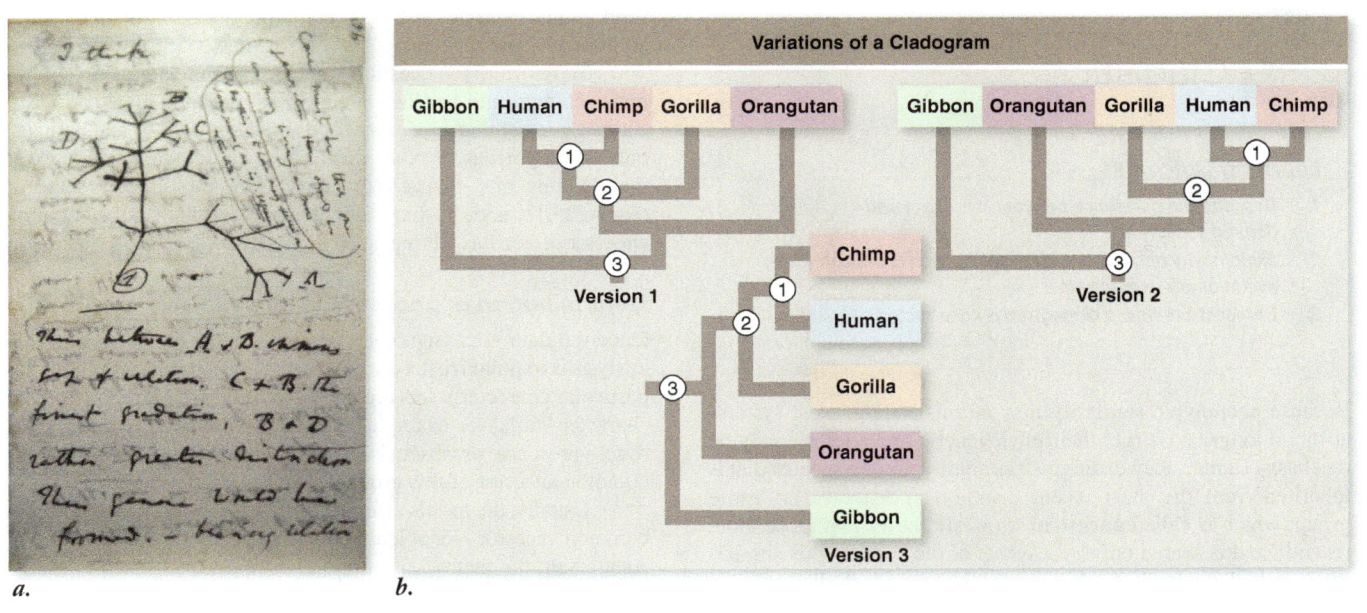

Figure 23.1 Phylogenies depict evolutionary relationships. *a.* A drawing from one of Darwin's notebooks, written in 1837 as he developed his ideas that led to *On the Origin of Species*. Darwin viewed life as a branching process akin to a tree, with species on the twigs, and evolutionary change represented by the branching pattern displayed by a tree as it grows. *b.* An example of a phylogeny. Note that these three versions convey the same information despite the differences in arrangement of species and orientation. Humans and chimpanzees are more closely related to each other than they are to any other living species. This is apparent because they share a common ancestor (the node labeled 1) that was not an ancestor of other species. Similarly, humans, chimpanzees, and gorillas are more closely related to one another than any of them is to orangutans because they share a common ancestor (node 2) that was not ancestral to orangutans. Node 3 represents the common ancestor of all apes. At each node, the two descendants can be rotated without changing the meaning. For example, one difference between versions 1 and 2 is that the descendants of node 2 have been rotated so that gorilla branches to the right in version 1 and to the left in version 2. However, this does not affect the interpretation that human and chimp are more closely related to each other than either is to gorilla.

change greatly; those staying in the same habitats as their ancestors may change only a little. For this reason, rates of change may vary markedly among species. Consequently, some will diverge more than others, and as a result, similarity is not necessarily a good predictor of how long it has been since two species shared a common ancestor.

A second fundamental problem exists as well: Evolution is not always divergent. In chapter 21, we discussed convergent evolution, in which two species independently evolve the same features. Often, species evolve convergently because they use similar habitats, in which similar adaptations are favored. As a result, two species that are not closely related may end up more similar to each other than they are to their close relatives. Evolutionary reversal, the process in which a species re-evolves the characteristics of an ancestral species, also has this effect.

Learning Outcomes Review 23.1
Systematics is the study of evolutionary relationships. Phylogenies, or phylogenetic trees, are graphic representations of relationships among species. Similarity of organisms alone does not necessarily correlate with their relatedness because evolutionary change is not constant in rate and direction.

- Why might a species be most phenotypically similar to a species that is not its closest evolutionary relative?

23.2 Cladistics

Learning Outcomes
1. Describe the difference between ancestral and derived similarities.
2. Explain why only shared, derived characters indicate close evolutionary relationship.
3. Demonstrate how a cladogram is constructed.

Because phenotypic similarity may be misleading, most systematists no longer construct their phylogenetic hypotheses solely on this basis. Rather, they distinguish similarity among species that is inherited from the most recent common ancestor of an entire group, which is called **ancestral,** from similarity that arose more recently and is shared only by a subset of the species; this similarity is called **derived.** In this approach, termed **cladistics,** only **shared derived characters** are considered informative in determining evolutionary relationships.

The cladistic method requires that character variation be identified as ancestral or derived

To employ the method of cladistics, systematists first gather data on a number of characters for all the species in the analysis. Characters can be any aspect of the phenotype, including morphology, physiology, behavior, and DNA. As chapters 18 and 24 show, the revolution in genomics is now providing a vast body of data revolutionizing our ability to identify and study character variation.

To be useful, the characters should exist in recognizable **character states.** For example, consider the character "tail" in vertebrates. This character has two states: presence in most vertebrates and absence in humans, apes, and some other groups such as frogs.

A cladistic analysis begins by determining the states for a number of characters for each **taxon** (taxa are species or higher level groups, such as genera or families) in the analysis. In the example in figure 23.2, six characters are used for seven taxa. The first character, presence of jaws, is scored as absent in lamprey (denoted by 0) and present in the other species (denoted with a 1). Another character, hair, is absent in all of the taxa except the three mammal species.

Examples of ancestral versus derived characters
The presence of hair is a shared derived feature of mammals (figure 23.2); in contrast, the presence of lungs in mammals is an ancestral feature because it is also present in amphibians and reptiles (represented by a salamander and a lizard) and therefore presumably evolved prior to the common ancestor of mammals (figure 23.2). The presence of lungs, therefore, does not tell us that mammal species are all more closely related to one another than to reptiles or amphibians, but the shared, derived feature of hair suggests that all mammal species share a common ancestor that existed more recently than the common ancestor of mammals, amphibians, and reptiles.

To return to the question concerning the relationships of humans, chimps, and gorillas, a number of morphological and DNA characters exist that are derived and shared by chimps and humans, but not by gorillas or other great apes. These characters suggest that chimps and humans diverged from a common ancestor (figure 23.1b, node 1) that existed more recently than the common ancestor of gorillas, chimps, and humans (node 2).

Determination of ancestral versus derived
Once the data are assembled, the first step in a manual cladistic analysis is to **polarize** the characters—that is, to determine whether particular character states are ancestral or derived. To polarize the character "tail," for example, systematists must determine which state—presence or absence—was exhibited by the most recent common ancestor of this group.

Usually, the fossils available do not represent the most recent common ancestor—or at least we cannot be confident that they do. As a result, the method of *outgroup comparison* is used to assign character polarity. To use this method, a species or group of species that is closely related to, but not a member of, the group under study is designated as the **outgroup.** When the group under study exhibits multiple character states, and one of those states is exhibited by the outgroup, then that state is considered to be ancestral and other states are considered to be derived. However, outgroup species also evolve from their ancestors, so the outgroup species will not always exhibit the ancestral condition.

Polarity assignments are most reliable when the same character state is exhibited by several different outgroups. In the preceding example, a tail is generally present in the nearest

Traits: Organism	Jaws	Lungs	Amniotic Membrane	Hair	No Tail	Bipedal
Lamprey	0	0	0	0	0	0
Shark	1	0	0	0	0	0
Salamander	1	1	0	0	0	0
Lizard	1	1	1	0	0	0
Tiger	1	1	1	1	0	0
Gorilla	1	1	1	1	1	0
Human	1	1	1	1	1	1

a.

b.

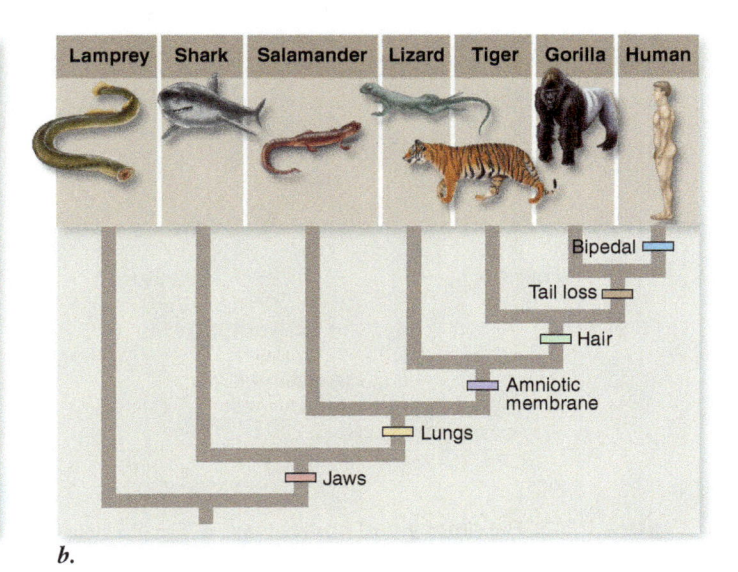

Figure 23.2 A cladogram. *a.* Morphological data for a group of seven vertebrates are tabulated. A "1" indicates possession of the derived character state, and a "0" indicates possession of the ancestral character state (note that the derived state for character "no tail" is the absence of a tail; for all other traits, absence of the trait is the ancestral character state). *b.* A tree, or cladogram, diagrams the relationships among the organisms based on the presence of derived characters. The derived characters between the cladogram branch points are shared by all organisms above the branch points and are not present in any below them. The outgroup (in this case, the lamprey) does not possess any of the derived characters.

outgroups of vertebrates, as well as in most species of vertebrates themselves. Consequently, the presence of a tail in vertebrates is considered ancestral, and its absence in some species is considered derived.

Construction of a cladogram

Once all characters have been polarized, systematists use this information to construct a **cladogram,** which depicts a hypothesis of evolutionary relationships. Species that share a common ancestor, as indicated by the possession of shared derived characters, are said to belong to a **clade.** Clades are thus evolutionary units and refer to a common ancestor and all of its descendants. A derived character shared by clade members is called a **synapomorphy** of that clade. Figure 23.2*b* illustrates that a simple cladogram is a nested set of clades, each characterized by its own synapomorphies. For example, amniotes are a clade for which the evolution of an amniotic membrane is a synapomorphy. Within that clade, mammals are a clade, with hair as a synapomorphy, and so on.

Ancestral states are called **plesiomorphies,** and shared ancestral states are called **symplesiomorphies.** In contrast to synapomorphies, symplesiomorphies are not informative about phylogenetic relationships.

Let's return to the character state "presence of a tail," which is exhibited by lampreys, sharks, salamanders, lizards, and tigers. Does this mean that tigers are more closely related to—and shared a more recent common ancestor with—lizards and sharks than to apes and humans, their fellow mammals? The answer, of course, is no: Because symplesiomorphies reflect character states inherited from a distant ancestor, they do not imply that species exhibiting that state are closely related.

Homoplasy complicates cladistic analysis

In real-world cases, phylogenetic studies are rarely as simple as the examples we have shown so far. The reason is that in some cases, the same character has evolved independently in several species. These characters would be categorized as shared derived characters, but they would be false signals of a close evolutionary relationship. In addition, derived characters may sometimes be lost as species within a clade re-evolve to the ancestral state.

Homoplasy refers to a shared derived character state that has not been inherited from a common ancestor exhibiting that character state. Homoplasy can result from convergent evolution or from evolutionary reversal. For example, adult frogs do not have a tail. Thus, absence of a tail is a synapomorphy that unites not only gorillas and humans, but also frogs. However, frogs are not closely related to gorillas and humans; they have neither an amniotic membrane nor hair, both of which are synapomorphies for clades that contain gorillas and humans.

In cases such as this, when there are conflicts among the characters, systematists rely on the **principle of parsimony,** which favors the hypothesis that requires the fewest assumptions. As a result, the phylogeny that requires the fewest evolutionary events is considered the best hypothesis of phylogenetic relationships (figure 23.3). In the example just stated, therefore, grouping frogs with salamanders is favored because it requires only one instance of homoplasy (the multiple origins of taillessness), whereas a phylogeny in which frogs were most closely related to humans and gorillas would require two homoplastic evolutionary events (the loss of both amniotic membranes and hair in frogs).

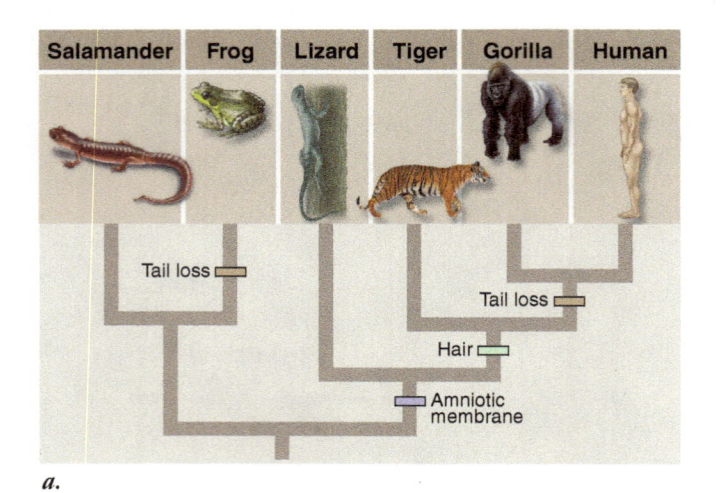

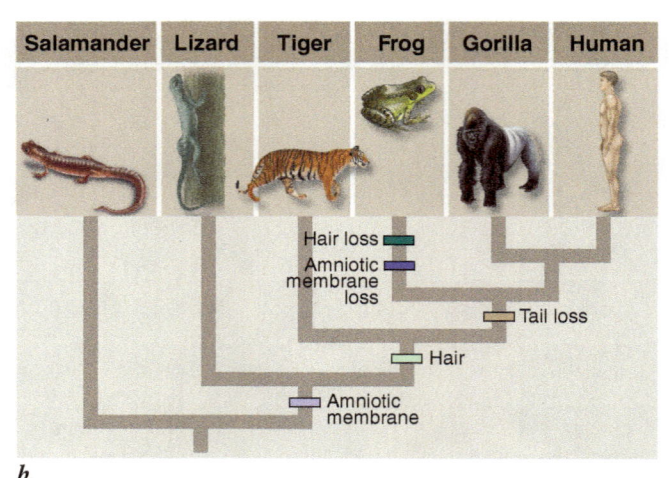

Figure 23.3 Parsimony and homoplasy. *a.* The placement of frogs as closely related to salamanders requires that tail loss evolved twice, an example of homoplasy. *b.* If frogs are closely related to gorillas and humans, then tail loss only had to evolve once. However, this arrangement would require two additional evolutionary changes: Frogs would have had to have lost the amniotic membrane and hair (alternatively, hair could have evolved independently in tigers and the clade of humans and gorillas; this interpretation would require two evolutionary changes in the hair character, just like the interpretation shown in the figure, in which hair evolved only once, but then was lost in frogs). Based on the principle of parsimony, the cladogram that requires the fewest number of evolutionary changes is favored; in this case the cladogram in *(a)* requires four changes, whereas that in *(b)* requires five, so *(a)* is considered the preferred hypothesis of evolutionary relationships.

Data analysis Construct a data matrix like in figure 23.2, showing the character states for all six species for the traits hair, amniotic membrane, and tail.

The examples presented so far have all involved morphological characters, but systematists increasingly use DNA sequence data to construct phylogenies because of the large number of characters that can be obtained through sequencing. Cladistics analyzes sequence data in the same manner as any other type of data: Character states are polarized by reference to the sequence of an outgroup, and a cladogram is constructed that minimizes the amount of character evolution required (figure 23.4).

Other phylogenetic methods work better than cladistics in some situations

If characters evolve from one state to another at a slow rate compared with the frequency of speciation events, then the principle of parsimony works well in reconstructing evolutionary relationships. In this situation, the principle's underlying assumption—that shared derived similarity is indicative of recent common ancestry—is

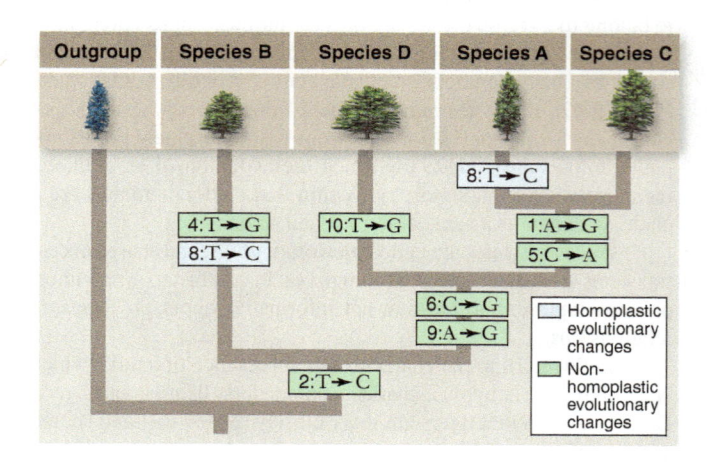

Site	1	2	3	4	5	6	7	8	9	10
Species A	G	C	A	T	A	G	G	C	G	T
Species B	A	C	A	G	C	C	G	C	A	T
Species C	G	C	A	T	A	G	G	T	G	T
Species D	A	C	A	T	C	G	G	T	G	G
Outgroup	A	T	A	T	C	C	G	T	A	T

Figure 23.4 Cladistic analysis of DNA sequence data. Sequence data are analyzed just like any other data. The most parsimonious interpretation of the DNA sequence data requires nine evolutionary changes. Each of these changes is indicated on the phylogeny. Change in site 8 is homoplastic: Species A and B independently evolved from thymine to cytosine at that site. Nonhomoplastic changes are those in which all species that possess the trait derived it from the same common ancestor. Such traits are referred to as homologous.

usually correct. In recent years, however, systematists have realized that some characters evolve so rapidly that the principle of parsimony may be misleading.

Rapid rates of evolutionary change and homoplasy

Of particular interest is the rate at which some parts of the genome evolve. As discussed in chapter 18, some stretches of DNA do not appear to have any function. As a result, mutations that occur in these parts of the DNA are not eliminated by natural selection, and thus the rate of evolution of new character states can be quite high in these regions as a result of genetic drift.

Moreover, because only four character states are possible for any nucleotide base (A, C, G, or T), there is a high probability that two species will independently evolve the same derived character state at any particular base position. If such homoplasy dominates the character data set, then the assumptions of the principle of parsimony are violated: similarity of species in this situation would more likely result from homoplasy than from inheritance from a common ancestor. As a result, under these circumstances, phylogenies inferred using the cladistic method are likely to be inaccurate.

> **Inquiry question** Why do high rates of evolutionary change and a limited number of character states cause problems for parsimony analyses?

Statistical approaches

Because evolution can sometimes proceed rapidly, systematists in recent years have been exploring other methods based on statistical approaches, such as maximum likelihood, to infer phylogenies. These methods start with an assumption about the rate at which characters evolve and then fit the data to these models to derive the phylogeny that best accords (that is, "maximally likely") with these assumptions.

One advantage of these methods is that different assumptions of rate of evolution can be used for different characters. If some DNA characters evolve more slowly than others—for example, because they are constrained by natural selection—then the methods can employ different models of evolution for the different characters. This approach is more effective than parsimony in dealing with homoplasy when rates of evolutionary changes are high.

The molecular clock

In general, cladograms such as the one in figure 23.2 only indicate the order of evolutionary branching events; they do not contain information about the timing of these events. In some cases, however, branching events can be timed, either by reference to fossils, or by making assumptions about the rate at which characters change. One widely used but controversial method is the **molecular clock**, which states that the rate of evolution of a molecule is constant through time. In this model, divergence in DNA can be used to calculate the times at which branching events have occurred. To make such estimates, the timing of one or more divergence events must be confidently estimated. For example, the fossil record may indicate that two clades diverged from a common ancestor at a particular time. Alternatively, the timing of separation of two clades may be estimated from geological events that likely led to their divergence, such as the rise of a mountain that now separates the two clades. With this information, the amount of DNA divergence separating two clades can be divided by the length of time separating the two clades, which produces an estimate of the rate of DNA divergence per unit of time (usually, per million years). Assuming a molecular clock, this rate can then be used to date other divergence events in a cladogram.

Although the molecular clock appears to hold true in some cases, in many others the data indicate that rates of evolution have not been constant through time across all branches in an evolutionary tree. For this reason, evolutionary dates derived from molecular data must be treated cautiously. Recently, methods have been developed to date evolutionary events without assuming that molecular evolution has been clocklike. These methods hold great promise for providing more reliable estimates of evolutionary timing.

Learning Outcomes Review 23.2

In cladistics, derived character states are distinguished from ancestral character states, and species are grouped based on shared derived character states. Derived characters are determined from comparison to a group known to be closely related, termed an outgroup. A clade contains all descendants of a common ancestor. A cladogram is a hypothetical representation of evolutionary relationships based on derived character states. Homoplasies may give a false picture of relationships.

- Why is cladistics more successful at inferring phylogenetic relationships in some cases than in others?
- Why are only shared derived, instead of all derived, characters useful in cladistics for reconstructing phylogenies?

23.3 Systematics and Classification

Learning Outcomes

1. Explain the taxonomic classification system.
2. Differentiate among monophyletic, paraphyletic, and polyphyletic groups.
3. Explain the meaning of the phylogenetic species concept and why it is controversial.

Whereas systematics is the reconstruction and study of evolutionary relationships, **classification** refers to how we place species and higher groups—genus, family, class, and so forth—into the taxonomic hierarchy.

The taxonomic classification system is hierarchical

Taxonomy is the science of classifying living things. A group of organisms at a particular level in a classification system is called a *taxon* (plural, *taxa*). By agreement among taxonomists throughout the world, no two organisms can have the same scientific name. The scientific name of an organism is the same anywhere in the world and avoids confusion caused by common names that may differ from one place to the next.

Also by agreement, the first word of the binomial name is the genus to which the organism belongs. This word is always capitalized.

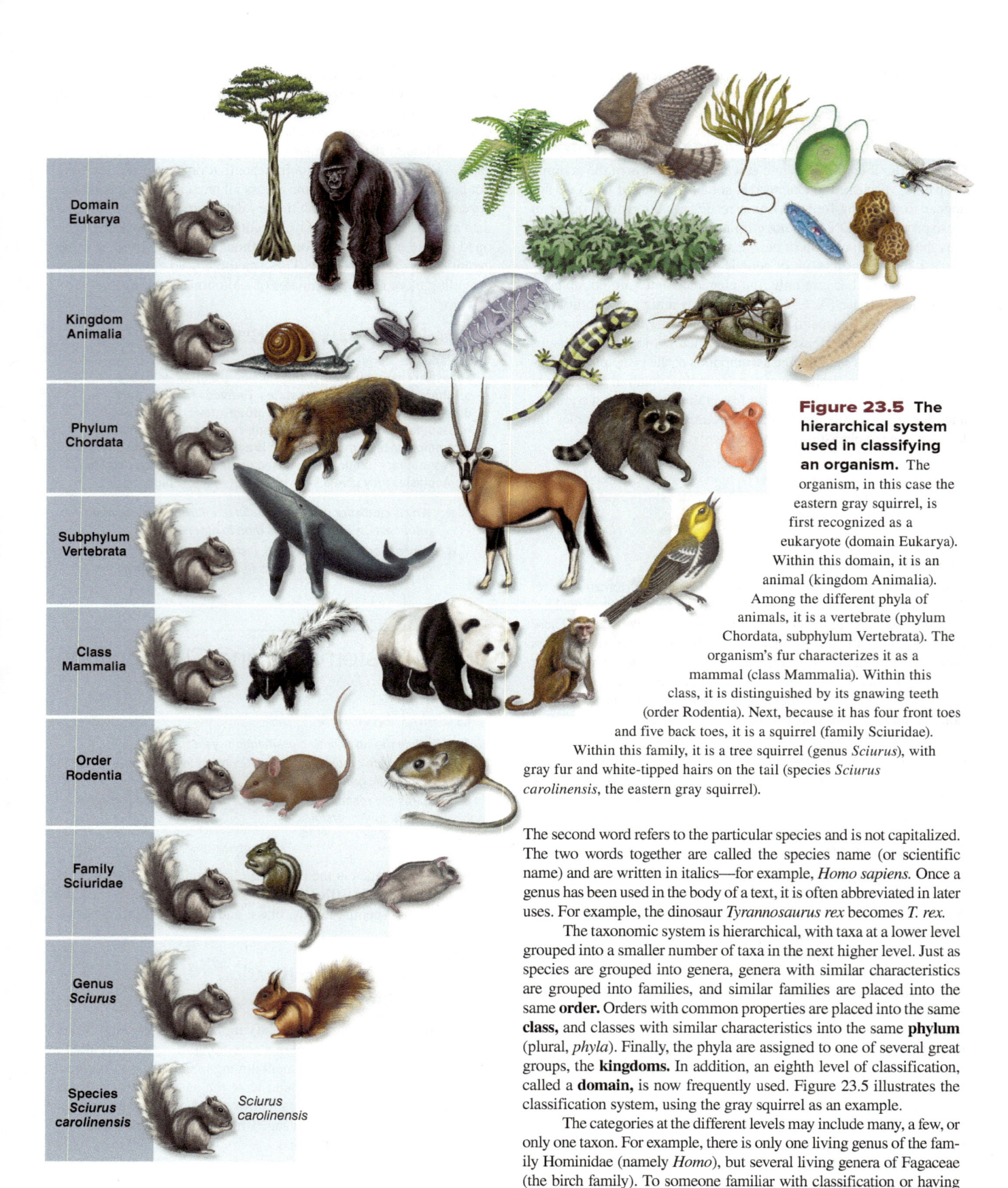

Figure 23.5 The hierarchical system used in classifying an organism. The organism, in this case the eastern gray squirrel, is first recognized as a eukaryote (domain Eukarya). Within this domain, it is an animal (kingdom Animalia). Among the different phyla of animals, it is a vertebrate (phylum Chordata, subphylum Vertebrata). The organism's fur characterizes it as a mammal (class Mammalia). Within this class, it is distinguished by its gnawing teeth (order Rodentia). Next, because it has four front toes and five back toes, it is a squirrel (family Sciuridae). Within this family, it is a tree squirrel (genus *Sciurus*), with gray fur and white-tipped hairs on the tail (species *Sciurus carolinensis*, the eastern gray squirrel).

The second word refers to the particular species and is not capitalized. The two words together are called the species name (or scientific name) and are written in italics—for example, *Homo sapiens*. Once a genus has been used in the body of a text, it is often abbreviated in later uses. For example, the dinosaur *Tyrannosaurus rex* becomes *T. rex*.

The taxonomic system is hierarchical, with taxa at a lower level grouped into a smaller number of taxa in the next higher level. Just as species are grouped into genera, genera with similar characteristics are grouped into families, and similar families are placed into the same **order.** Orders with common properties are placed into the same **class,** and classes with similar characteristics into the same **phylum** (plural, *phyla*). Finally, the phyla are assigned to one of several great groups, the **kingdoms.** In addition, an eighth level of classification, called a **domain,** is now frequently used. Figure 23.5 illustrates the classification system, using the gray squirrel as an example.

The categories at the different levels may include many, a few, or only one taxon. For example, there is only one living genus of the family Hominidae (namely *Homo*), but several living genera of Fagaceae (the birch family). To someone familiar with classification or having access to the appropriate reference books, each taxon implies both a set of characteristics and a group of organisms belonging to the taxon.

Current classification sometimes does not reflect evolutionary relationships

Systematics and traditional classification are not always congruent; to understand why, we need to consider how species may be grouped based on their phylogenetic relationships. A **monophyletic** group includes the most recent common ancestor of the group and all of its descendants. By definition, a clade is a monophyletic group. A **paraphyletic** group includes the most recent common ancestor of the group, but not all its descendants, and a **polyphyletic** group does not include the most recent common ancestor of all members of the group (figure 23.6).

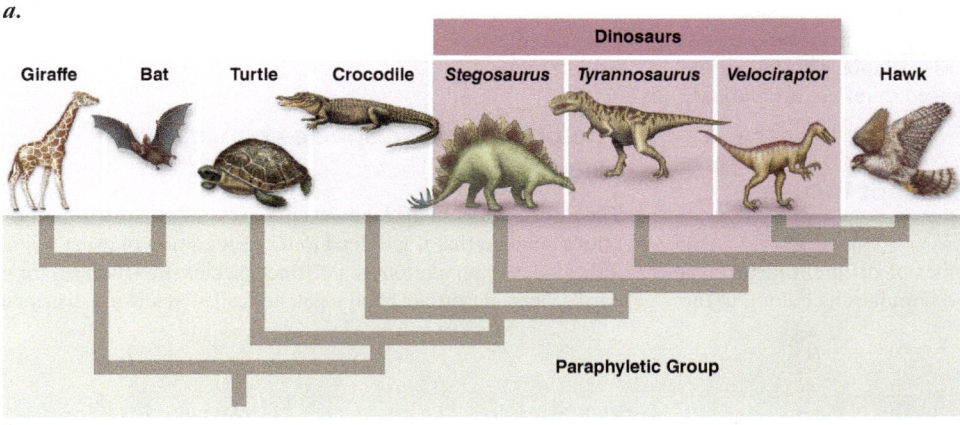

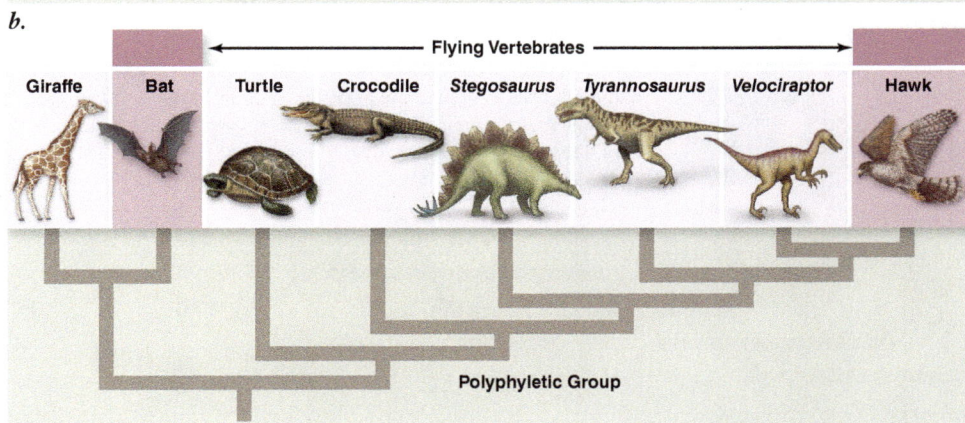

Figure 23.6 Monophyletic, paraphyletic, and polyphyletic groups.
a. A monophyletic group consists of the most recent common ancestor and all of its descendants. For example, the name "Archosaurs" is given to the monophyletic group that includes a crocodile, *Stegosaurus*, *Tyrannosaurus*, *Velociraptor*, and a hawk. *b.* A paraphyletic group consists of the most recent common ancestor and some of its descendants. For example, some, but not all, taxonomists traditionally give the name "dinosaurs" to the paraphyletic group that includes *Stegosaurus*, *Tyrannosaurus*, and *Velociraptor*. This group is paraphyletic because one descendant of the most recent ancestor of these species, birds, is not included in the group. Other taxonomists include birds within the Dinosauria because *Tyrannosaurus* and *Velociraptor* are more closely related to birds than to other dinosaurs. *c.* A polyphyletic group does not contain the most recent common ancestor of the group. For example, bats and birds could be classified in the same group, which we might call "flying vertebrates," because they have similar shapes, anatomical features, and habitats. However, their similarities reflect convergent evolution, not common ancestry.

? Inquiry question Based on this phylogeny, are there any alternatives to convergence to explain the presence of wings in birds and bats? What types of data might be used to test these hypotheses?

Taxonomic hierarchies are based on shared traits, and ideally they should reflect evolutionary relationships. Traditional taxonomic groups, however, do not always fit well with the new understanding of phylogenetic relationships. For example, birds have historically been placed in the class Aves, and dinosaurs have been considered part of the class Reptilia. But recent phylogenetic advances make clear that birds evolved from dinosaurs. The last common ancestor of all birds and a dinosaur was a meat-eating dinosaur (figure 23.6).

Therefore, having two separate monophyletic groups, one for birds and one for reptiles (including dinosaurs and crocodiles, as well as lizards, snakes, and turtles), is not possible based on phylogeny. And yet the terms Aves and Reptilia are so familiar and well established that suddenly referring to birds as a type of dinosaur, and thus a type of reptile, is difficult for some. Nonetheless, biologists increasingly refer to birds as a type of dinosaur and hence a type of reptile.

Situations like this are not uncommon. Another example concerns the classification of plants. Traditionally, three major groups were recognized: green algae, bryophytes, and vascular plants (figure 23.7). However, recent research reveals that neither the green algae nor the bryophytes constitute monophyletic groups. Rather, some bryophyte groups are more closely related to vascular plants than they are to other bryophytes, and some green algae are more closely related to bryophytes and vascular plants than they are to other green algae. As a result, systematists no longer recognize green algae or bryophytes as evolutionary groups, and the classification system has been changed to reflect evolutionary relationships.

The phylogenetic species concept focuses on shared derived characters

In chapter 22, you read about a number of different ideas concerning what determines whether two populations belong to the same species. The biological species concept (BSC) defines species as groups of interbreeding populations that are reproductively isolated from other groups. In recent years, a phylogenetic perspective has emerged and has been applied to the question of species concepts. Advocates of the **phylogenetic species concept (PSC)** propose that the term *species* should be applied to groups of populations that have been evolving independently of other groups of populations. Moreover, they suggest that phylogenetic analysis is the way to identify such species. In this view, a species is a population or set of populations characterized by one or more shared derived character—the rationale being that the possession of a derived character implies a period of separate evolution.

This approach solves two of the problems with the BSC that were discussed in chapter 22. First, the BSC cannot be applied to allopatric populations because scientists cannot determine whether individuals of the populations would interbreed and produce fertile offspring if they ever came together. The PSC solves this problem: Instead of trying to predict what will happen in the future if allopatric populations ever come into contact (that is, would the populations be reproductively isolated?), the PSC looks to the past to determine whether a population (or groups of populations) has evolved independently for a long enough time to develop its own derived characters.

Second, the PSC can be applied equally well to both sexual and asexual species, in contrast to the BSC, which deals only with sexual forms.

The phylogenetic species concept also has drawbacks

The PSC is controversial, however, for several reasons. First, some critics contend that it will lead to the recognition of every slightly differentiated population as a distinct species. In Missouri, for example, open, desertlike habitat patches called glades are distributed

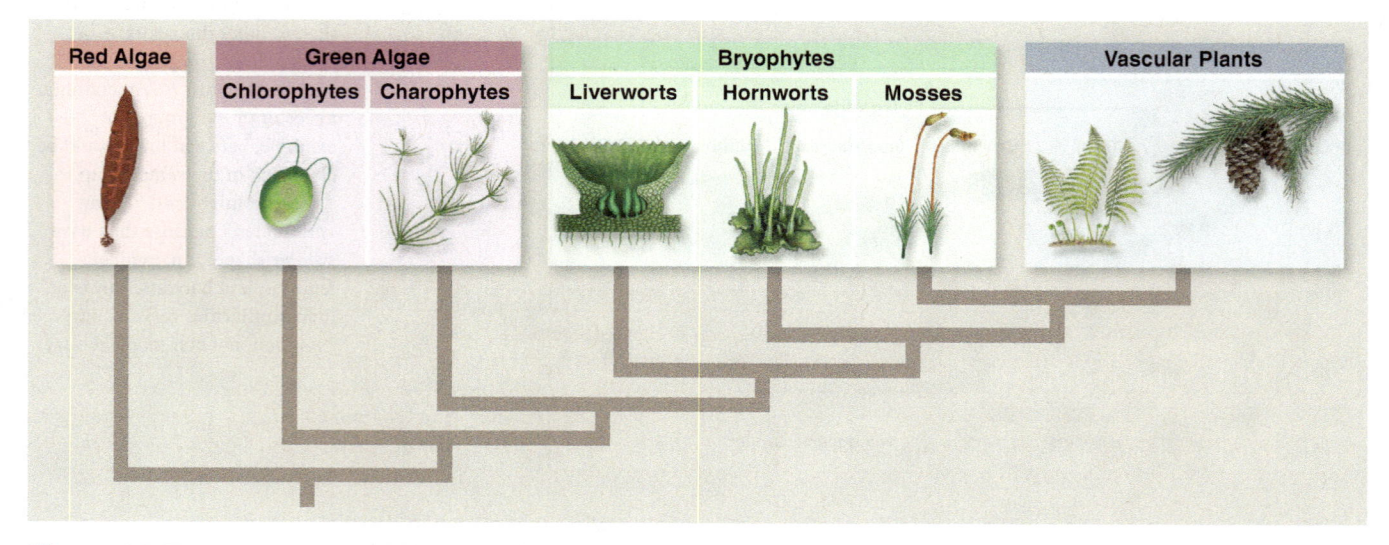

Figure 23.7 Phylogenetic information transforms plant classification. The traditional classification included two groups that we now realize are not monophyletic: the green algae and bryophytes. For this reason, plant systematists have developed a new classification of plants that does not include these groups (discussed in chapter 30).

throughout much of the state. These glades contain a variety of warmth-loving species of plants and animals that do not occur in the forests that separate the glades. Glades have been isolated from one another for a few thousand years, allowing enough time for populations on each glade to evolve differences in some rapidly evolving parts of the genome. Does that mean that each of the hundreds, if not thousands, of Missouri glades contains its own species of lizards, grasshoppers, and scorpions? Some scientists argue that if one takes the PSC to its logical extreme, that is exactly what would result.

A second problem is that species may not always be monophyletic, contrary to the definition of some versions of the phylogenetic species concept. Consider, for example, a species composed of five populations, with evolutionary relationships like those indicated in figure 23.8. Suppose that population C becomes isolated and evolves differences that make it qualify as a species by any concept (for example, reproductively isolated, ecologically differentiated). But this distinction would mean that the remaining populations, which might still be perfectly capable of exchanging genes, would be paraphyletic, rather than monophyletic. Such situations probably occur often in the natural world.

Phylogenetic species concepts, of which there are many different permutations, are increasingly used, but are also contentious for the reasons just discussed. Evolutionary biologists are trying to find ways to reconcile the historical perspective of the PSC with the process-oriented perspective of the BSC and other species concepts.

Learning Outcomes Review 23.3

The traditional classification is hierarchical, placing each species into a nested set of taxonomic categories. This classification is not always consistent with modern understanding of phylogenetic relationships. By definition, a clade is monophyletic. A paraphyletic group contains the most recent common ancestor, but not all its descendants; a polyphyletic group does not contain the most recent common ancestor of all members. The phylogenetic species concept focuses on the possession of shared derived characters, in contrast to the biological species concept, which emphasizes reproductive isolation. The PSC solves some problems of the BSC but has difficulties of its own.

- *Under the biological species concept, is it possible for a species to be polyphyletic?*

23.4 Phylogenetics and Comparative Biology

Learning Outcomes
1. Explain the importance of homoplasy for interpreting patterns of evolutionary change.
2. Describe how phylogenetic trees can reveal the existence of homoplasy.
3. Discuss how a phylogenetic tree can indicate the timing of species diversification.

Phylogenies not only provide information about evolutionary relationships among species, but they are also indispensable for understanding how evolution has occurred. By examining the distribution of traits among species in the context of their phylogenetic relationships, much can be learned about how and why evolution has proceeded. In this way, phylogenetics is the basis of all comparative biology.

Homologous features are derived from the same ancestral source; homoplastic features are not

In chapter 21, we pointed out that homologous structures are those that are derived from the same body part in a common ancestor. Thus, the forelegs of a dolphin (flipper) and of a horse (leg) are homologous because they are derived from the same bones in an ancestral vertebrate. By contrast, the wings of birds and those of dragonflies are homoplastic structures because they are not derived from the same structure in a common ancestor. Phylogenetic analysis can help determine whether structures are homologous or homoplastic.

Homologous parental care in dinosaurs, crocodiles, and birds

Recent fossil discoveries have revealed that many species of dinosaurs exhibited parental care. They incubated eggs laid in nests and

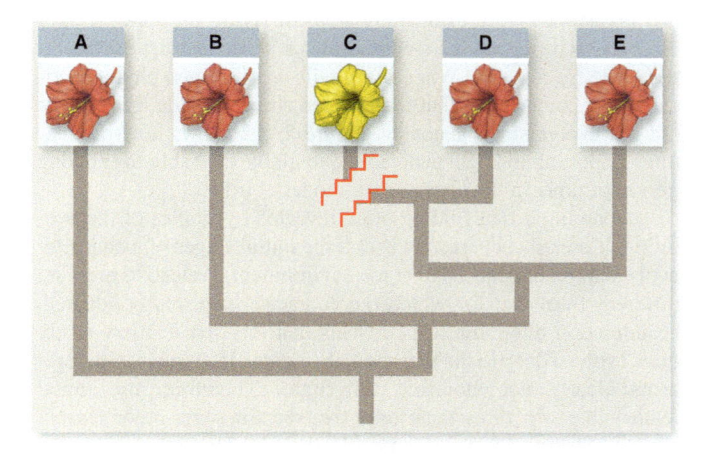

Figure 23.8 Paraphyly and the phylogenetic species concept. The five populations initially were all members of the same species, with their historical relationships indicated by the cladogram. Then, population C evolved in some ways to become greatly differentiated ecologically and reproductively from the other populations. By all species concepts, this population would qualify as a different species. However, the remaining four species do not form a clade; they are paraphyletic because population C has been removed and placed in a different species. This scenario may occur commonly in nature, but most versions of the phylogenetic species concept do not recognize paraphyletic species.

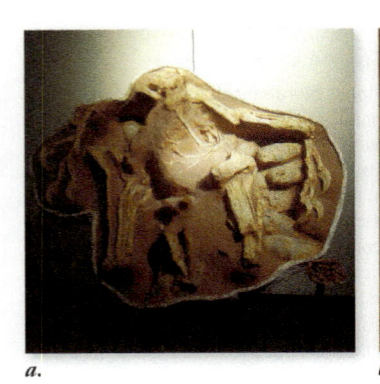

Figure 23.9 Parental care in dinosaurs and crocodiles *a.* Fossil dinosaur incubating its eggs. This remarkable fossil of *Oviraptor* shows the dinosaur sitting on its nest of eggs just as chickens do today. Not only is the dinosaur squatting on the nest, but its forelimbs are outstretched, perhaps to shade the eggs. *b.* Crocodile exhibiting parental care. Female crocodilians build nests and then remain nearby, guarding them, while the eggs incubate. When they are ready to hatch, the baby crocodiles vocalize; females respond by digging up the eggs and carrying the babies to the water.

took care of the growing baby dinosaurs, many of which could not have fended for themselves. Some recent fossils show dinosaurs sitting on a nest in exactly the same posture used by birds today (figure 23.9*a*)! Initially, these discoveries were treated as remarkable and unexpected—dinosaurs apparently had independently evolved behaviors similar to those of modern-day organisms. But examination of the phylogenetic position of dinosaurs (see figure 23.6) indicates that they are most closely related to two living groups of animals—crocodiles and birds—both of which exhibit parental care (figure 23.9*b*).

It appears likely, therefore, that the parental care exhibited by crocodiles, dinosaurs, and birds did not evolve convergently from different ancestors that did not exhibit parental care; rather, the behaviors are homologous, inherited by each of these groups from their common ancestor that cared for its young.

Homoplastic convergence: Saber teeth and plant conducting tubes

In other cases, by contrast, phylogenetic analysis can indicate that similar traits have evolved independently in different clades. This convergent evolution from different ancestral sources indicates that such traits represent homoplasies. As one example, the fossil record reveals that extremely elongated canine teeth (saber teeth) occurred in a number of different groups of extinct carnivorous mammals. Although how these teeth were actually used is still debated, all saber-toothed carnivores had body proportions similar to those of cats, which suggests that these different types of carnivores all evolved into a similar predatory lifestyle. Examination of the saber-toothed character state in a phylogenetic context reveals that it most likely evolved independently at least four times (figure 23.10).

Conducting tubes in plants provide a similar example. The tracheophytes, a large group of land plants discussed in chapter 30, transport photosynthetic products, hormones, and other molecules over long distances through elongated, tubular cells that have perforated walls at the end. These structures are stacked upon each other to create a conduit called a sieve tube. Sieve tubes facilitate long-distance transport that is essential for the survival of tall plants on land.

Most members of the brown algae, which includes kelp, also have sieve elements (see figure 23.11 for a comparison of the sieve plates in brown algae and angiosperms) that aid in the rapid transport of materials. The land plants and brown algae are distantly related (figure 23.11), and their last common ancestor was a single-celled organism that could not have had a multicellular transport system. This indicates that the strong structural and functional similarity of sieve elements in these plant groups is an example of convergent evolution.

Complex characters evolve through a sequence of evolutionary changes

Most complex characters do not evolve, fully formed, in one step. Rather, they are often built up, step-by-step, in a series of evolutionary transitions. Phylogenetic analysis can help discover these evolutionary sequences.

Modern-day birds—with their wings, feathers, light bones, and breastbone—are exquisitely adapted flying machines. Fossil discoveries in recent years now allow us to reconstruct the evolution of these features. When the fossils are arranged phylogenetically, it becomes clear that the features characterizing living birds did not evolve simultaneously. Figure 23.12 shows how the features important to flight evolved sequentially, probably over a long period of time, in the ancestors of modern birds.

One important finding often revealed by studies of the evolution of complex characters is that the initial stages of a character evolved as an adaptation to some environmental selective pressure different from that for which the character is currently adapted. Examination of figure 23.12 reveals that the first feathery structures evolved deep in the theropod phylogeny, in animals with forearms clearly not modified for flight. Therefore, the initial feather-like structures must have evolved for some other reason, perhaps serving as insulation or decoration. Through time, these structures have become modified to the extent that modern feathers produce excellent aerodynamic performance.

Phylogenetic methods can be used to distinguish between competing hypotheses

Understanding the causes of patterns of biological diversity observed today can be difficult because a single pattern often could have resulted from several different processes. In many cases, scientists can use phylogenies to distinguish between competing hypotheses.

SCIENTIFIC THINKING

Question: How many times have saber teeth evolved in mammals?

Hypothesis: Saber teeth are homologous and have only evolved once in mammals (or, conversely, saber teeth are convergent and have evolved multiple times in mammals).

Phylogenic Analysis: Examine the distribution of saber teeth on a phylogeny of mammals, and use parsimony to infer the history of saber tooth evolution (note that not all branches within marsupials and placentals are shown on the phylogeny).

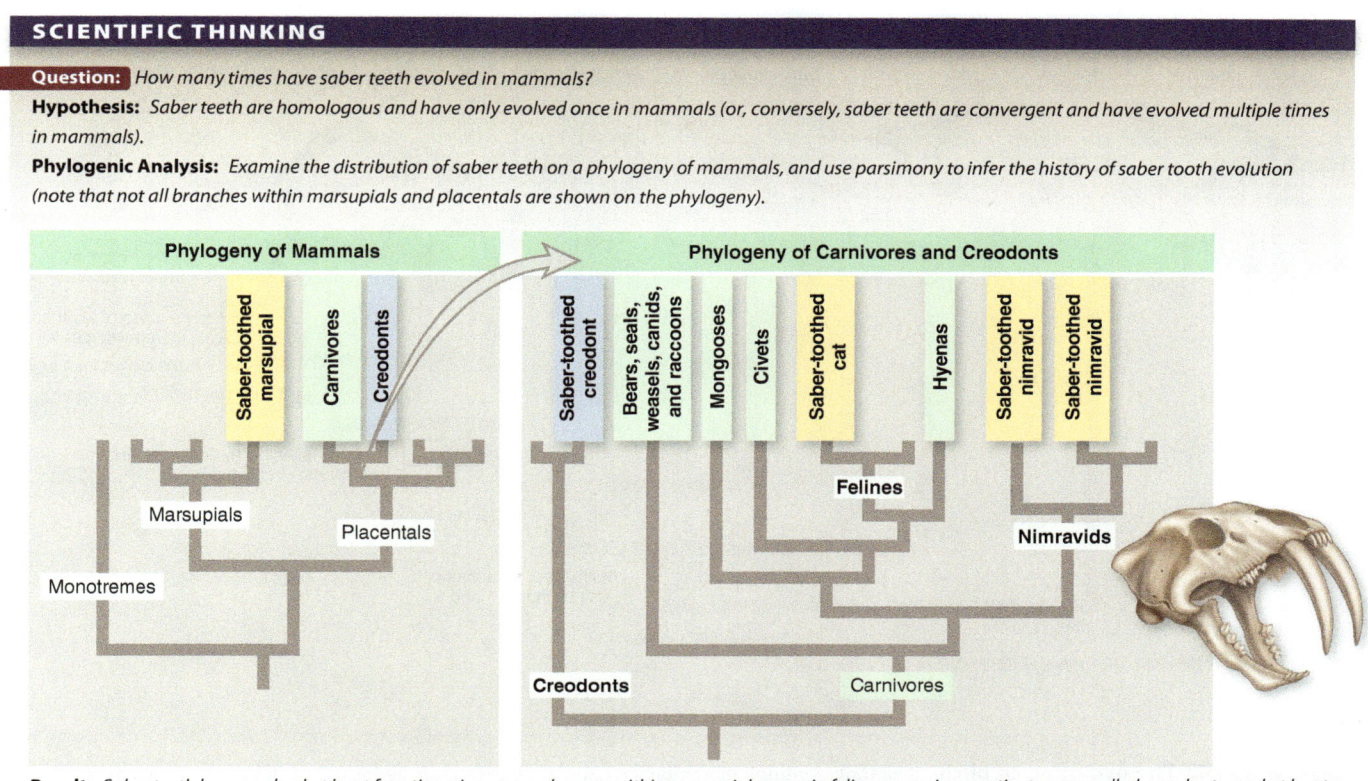

Result: Saber teeth have evolved at least four times in mammals: once within marsupials, once in felines, once in an extinct group called creodonts, and at least once in another group of now-extinct catlike carnivores called nimravids.

Interpretation: Note that it is possible that saber teeth evolved twice in nimravids, but another possibility that requires the same number of evolutionary changes (and thus is equally parsimonious) is that saber teeth evolved only once in the ancestor of nimravids and then were subsequently lost in one group of nimravids. (Note that for clarity, not all branches within marsupials and placentals are shown in this illustration.)

Figure 23.10 Distribution of saber-toothed mammals.

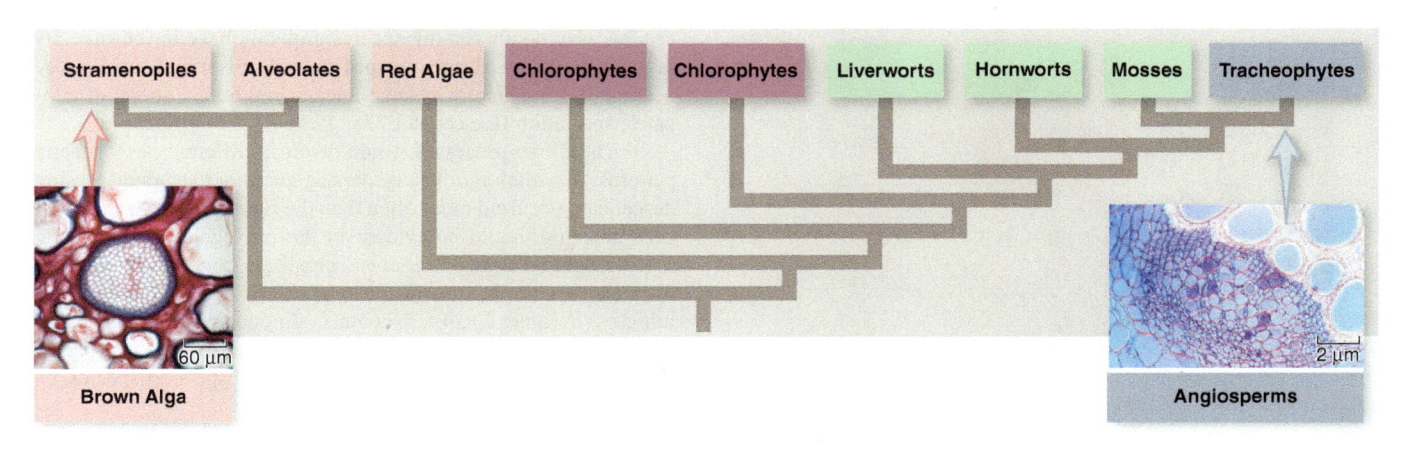

Figure 23.11 Convergent evolution of conducting tubes. Sieve tubes, which transport hormones and other substances throughout the plant, have evolved in two distantly related plant groups (brown algae are stramenopiles and angiosperms are tracheophytes).

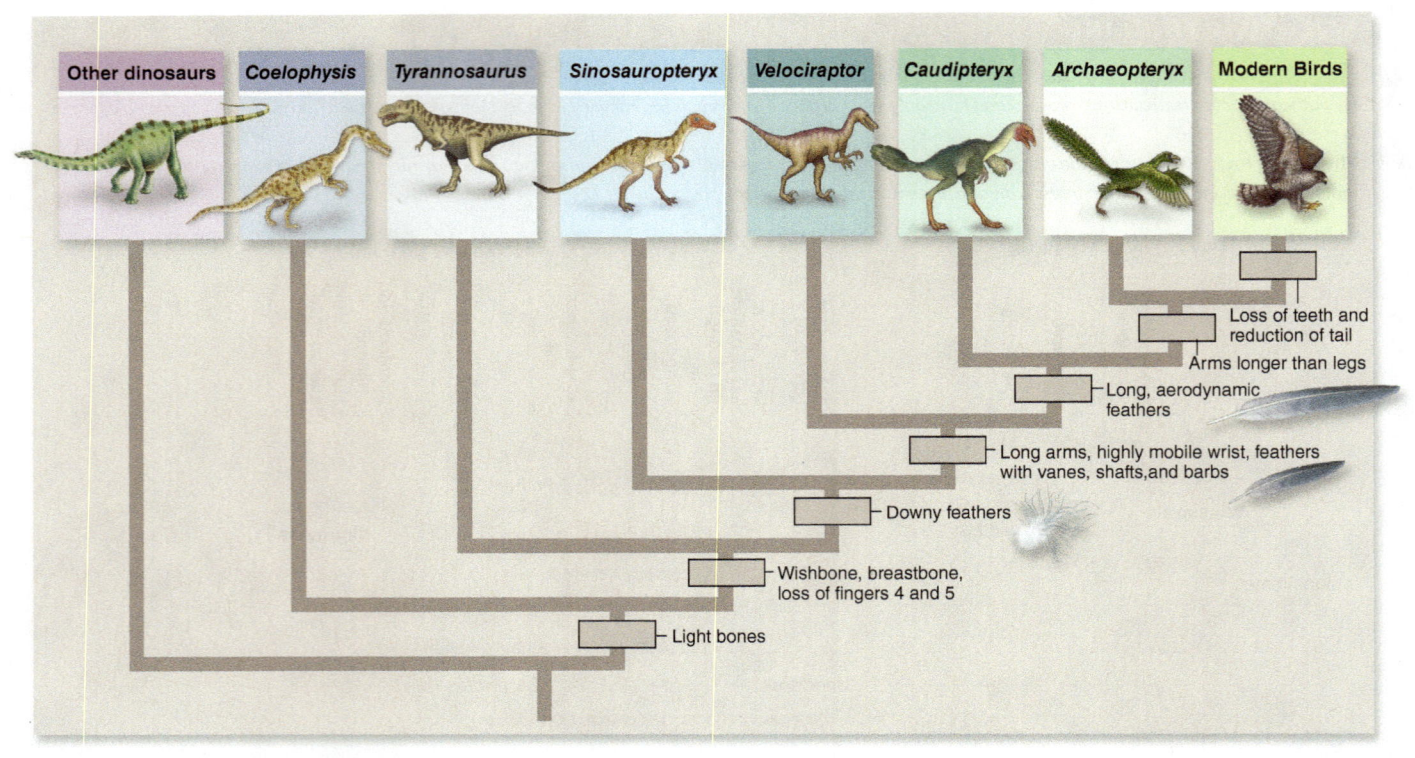

Figure 23.12 The evolution of birds. The traits we think of as characteristic of modern birds have evolved in stages over many millions of years.

Larval dispersal in marine snails

An example of this use of phylogenetic analysis concerns the evolution of larval forms in marine snails. Most species of snails produce microscopic larvae that drift in the ocean currents, sometimes traveling hundreds or thousands of miles before becoming established and transforming into adults. Some species, however, have evolved larvae that settle to the ocean bottom very quickly and thus don't disperse far from their place of origin. Studies of fossil snails indicate that the proportion of species that produce nondispersing larvae has increased through geological time (figure 23.13).

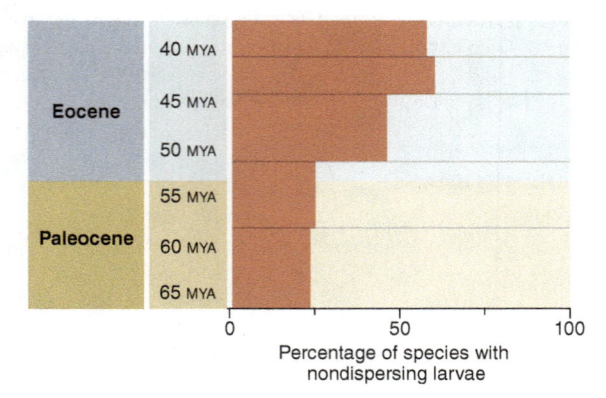

Figure 23.13 Larvae dispersal. Increase through time in the proportion of species whose larvae do not disperse far from their place of birth.

Two processes could produce an increase in nondispersing larvae through time. First, if evolutionary change from dispersing to nondispersing occurs more often than change in the opposite direction, then the proportion of species that are nondispersing would increase through time.

Alternatively, if species that are nondispersing speciate more frequently, or become extinct less frequently, than dispersing species, then through time the proportion of nondispersers would also increase (assuming that the descendants of nondispersing species also were nondispersing). This latter case is a reasonable hypothesis because nondispersing species probably have lower amounts of gene flow than dispersing species, and thus might more easily become geographically isolated, increasing the likelihood of allopatric speciation (see chapter 22).

These two processes would result in different phylogenetic patterns. If evolution from a dispersing ancestor to a nondispersing descendant occurred more often than the reverse, then an excess of such changes should be evident in the phylogeny, as shown by more transitions from dispersing to nondispersing larvae (indicated by pluses) than changes in the opposite direction (indicated by minuses) in figure 23.14a. In contrast, if nondispersing species underwent greater speciation, then clades of nondispersing species would contain more species than clades of dispersing species, as shown in figure 23.14b.

Evidence for both processes was revealed in an examination of the phylogeny of marine snails in the genus *Conus*, in which 30% of species are nondispersing (figure 23.14c). The phylogeny indicates that possession of dispersing larvae was the ancestral

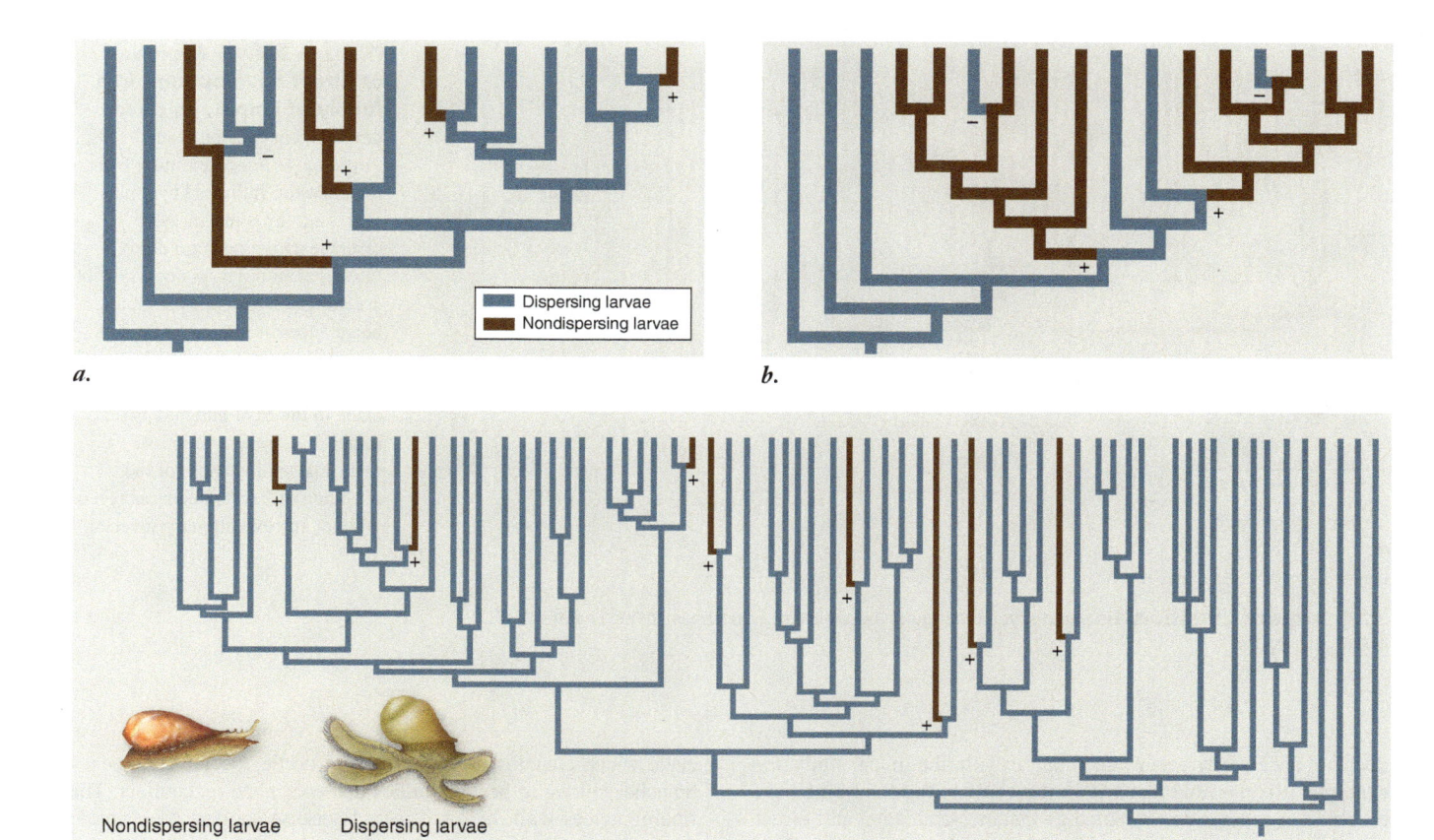

Figure 23.14 Phylogenetic investigation of the evolution of nondispersing larvae. *a.* In this hypothetical example, the evolutionary transition from dispersing to nondispersing larvae occurs more frequently (four times, indicated by a plus sign) than the converse (once, indicated by a minus sign). By contrast, in *(b)* the number of evolutionary changes in each direction is the same, but clades that have nondispersing larvae diversify to a greater extent due to higher rates of speciation or lower rates of extinction (assuming that extinct forms are not shown). *c.* Phylogeny for *Conus,* a genus of marine snails. Nondispersing larvae have evolved eight separate times from dispersing larvae, with no instances of evolution in the reverse direction. This phylogeny does not show all species, however; nondispersing clades contain on average 3.5 times as many species as dispersing clades.

state; nondispersing larvae are inferred to have evolved eight times, with no evidence for evolutionary reversal from nondispersing to dispersing larvae.

At the same time, clades of nondispersing larvae tend to have on average 3.5 times as many species as dispersing larvae, which suggests that in nondispersing species, rates of speciation are higher, rates of extinction are lower, or both.

This analysis therefore indicates that the evolutionary increase in nondispersing larvae through time may be a result both of a bias in the direction in which evolution proceeds plus an increase in rate of diversification (that is, speciation rate minus extinction rate) in nondispersing clades.

The lack of evolutionary reversal is not surprising because when larvae evolve to become nondispersing, they often lose a variety of structures used for feeding while drifting in the ocean current. In most cases, once a structure is lost, it rarely re-evolves, and thus the standard view is that the evolution of nondispersing larvae is a one-way street, with few examples of re-evolution of dispersing larvae.

Loss of the larval stage in marine invertebrates

A related phenomenon in many marine invertebrates is the loss of the larval stage entirely. Most marine invertebrates—in groups as diverse as snails, sea stars, and anemones—pass through a larval stage as they develop. But in a number of different types of organisms, the larval stage is omitted, and the eggs develop directly into adults.

The evolutionary loss of the larval stage has been suggested as another example of a nonreversible evolutionary change because once the larval stages are lost, it is difficult for them to re-evolve— or so the argument goes. A recent study on one group of marine limpets, shelled marine organisms related to snails, shows that this is not necessarily the case. Among these limpets, direct development has evolved many times; however, in three cases, the phylogeny strongly suggests that evolution reversed and a larval stage re-evolved (figure 23.15*a*).

It is important to remember that patterns of evolution suggested by phylogenetic analysis are not always correct—evolution

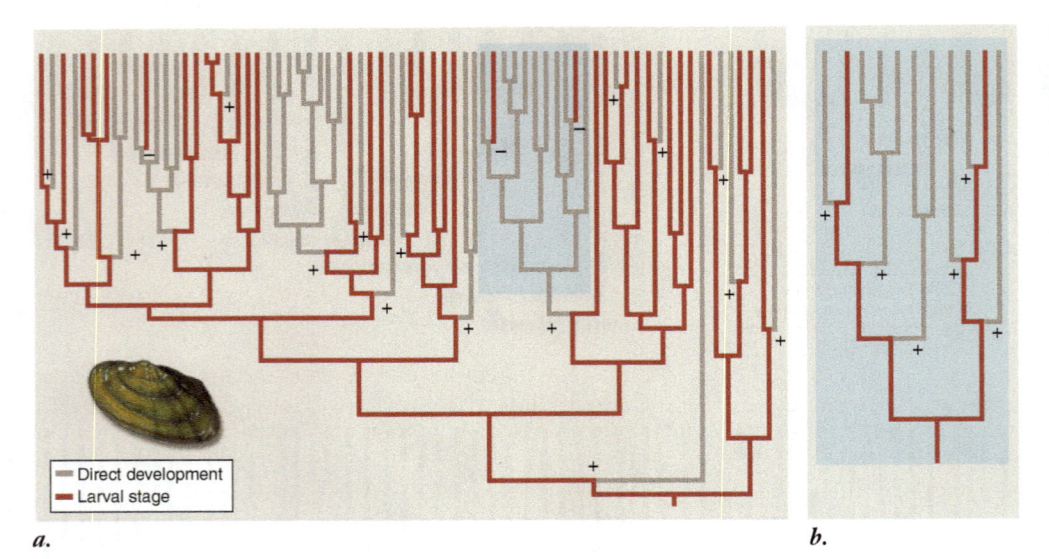

Figure 23.15 Evolution of direct development in a family of limpets. *a.* Direct development evolved many times (*beige lines* stemming from a red ancestor, indicated by a plus sign), and three instances of reversed evolution from direct development to larval development are indicated (*red lines* from a beige ancestor, indicated by a minus sign). *b.* A less parsimonious interpretation of evolution in the clade in the light blue box is that, rather than two evolutionary reversals, six instances of the evolution of development occurred without any evolutionary reversal.

Inquiry question How might you distinguish between the hypotheses in parts *a* and *b* of this figure?

does not necessarily occur parsimoniously. In the limpet study, for example, it is possible that within the clade in the light blue box, presence of a larva was retained as the ancestral state, and direct development evolved independently six times (figure 23.15*b*). Phylogenetic analysis cannot rule out this possibility, even if it is less phylogenetically parsimonious.

If the re-evolution of lost traits seems unlikely, then the alternative hypothesis that direct development evolved six times—rather than only once at the base of the clade, with two instances of evolutionary reversal—should be considered. For example, studies of the morphology or embryology of direct-developing species might shed light on whether such structures are homologous or convergent. In some cases, artificial selection experiments in the laboratory or genetic manipulations can test the hypothesis that it is difficult for lost structures to re-evolve. Conclusions from phylogenetic analyses are always stronger when supported by results of other types of studies.

Phylogenetics helps explain species diversification

One of the central goals of evolutionary biology is to explain patterns of species diversity: Why do some types of plants and animals exhibit more **species richness**—a greater number of species per clade—than others? Phylogenetic analysis can be used both to suggest and to test hypotheses about such differences.

Species richness in beetles

Beetles (order Coleoptera) are the most diverse group of animals. Approximately 60% of all animal species are insects, and approximately 80% of all insect species are beetles. Among beetles, families that are herbivorous are particularly species-rich.

Examination of the phylogeny provides insight into beetle evolutionary diversification (figure 23.16). Among the Phytophaga, the clade which contains most herbivorous beetle species, the deepest branches belong to beetle families that specialize on conifers. This finding agrees with the fossil record because conifers were among the earliest seed plant groups to evolve. By contrast, the flowering plants (angiosperms) evolved more recently, in the Cretaceous, and beetle families specializing on them have shorter evolutionary branches, indicating their more recent evolutionary appearance.

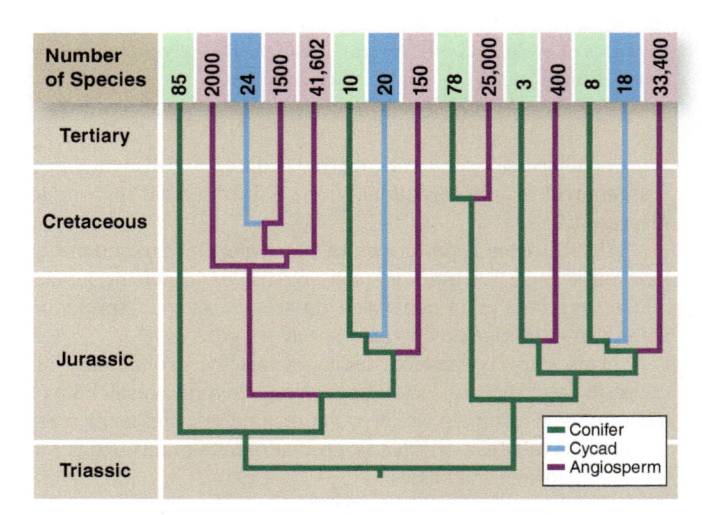

Figure 23.16 Evolutionary diversification of the Phytophaga, the largest clade of herbivorous beetles. Clades that originated deep in the phylogenetic tree feed on conifers; clades that feed on angiosperms, which evolved more recently, originated more recently. Age of clades is established by examination of fossil beetles.

This correspondence between phylogenetic position and timing of plant origins suggests that beetles have been remarkably conservative in their diet. The family Nemonychidae, for example, appears to have remained specialized on conifers since the beginning of the Jurassic, approximately 210 MYA.

Phylogenetic explanations for beetle diversification

The phylogenetic perspective suggests factors that may be responsible for the incredible diversity of beetles. The phylogeny for the Phytophaga indicates that it is not the evolution of herbivory itself that is linked to great species richness. Rather, specialization on angiosperms seems to have been a prerequisite for great species diversification. Specialization on angiosperms appears to have arisen five times independently within herbivorous beetles; in each case, the angiosperm-specializing clade is substantially more species-rich than the clade to which it is most closely related (termed a *sister clade*) and which specializes on some other type of plant.

Why specialization on angiosperms has led to great species diversity is not yet clear and is the focus of much current research. One possibility is that this diversity is linked to the great species-richness of angiosperms themselves. With more than 250,000 species of angiosperms, beetle clades specializing on them may have had a multitude of opportunities to adapt to feed on different species, thus promoting divergence and speciation.

Learning Outcomes Review 23.4

Homologous traits are derived from the same ancestral character states, whereas homoplastic traits are not, even though they may have similar function. Phylogenetic analysis can help determine whether homology or homoplasy has occurred. By correlating phylogenetic branching with known evolutionary events, the timing and cause of diversification can be inferred.

- *Does the possession of the same character state by all members of a clade mean that the ancestor of that clade necessarily possessed that character state?*

23.5 Phylogenetics and Disease Evolution

Learning Outcome

1. Discuss how phylogenetic analysis can help identify patterns of disease transmission.

The examples so far have illustrated the use of phylogenetic analysis to examine relationships among species. Such analyses can also be conducted on virtually any group of biological entities, as long as evolutionary divergence in these groups occurs by a branching process, with little or no genetic exchange between different groups. No example illustrates this better than recent attempts to understand the evolution of the virus that causes acquired immunodeficiency syndrome (AIDS).

HIV has evolved from a simian viral counterpart

AIDS was first recognized in the early 1980s, and it rapidly became epidemic in the human population. Current estimates are that more than 35 million people are infected with the human immunodeficiency virus (HIV), of whom more than 1.5 million die each year.

At first, scientists were perplexed about where HIV had originated and how it had infected humans. In the mid-1980s, however, scientists discovered a related virus in laboratory monkeys, termed simian immunodeficiency virus (SIV). In biochemical terms, the viruses are very similar, although genetic differences exist. At last count, SIV has been detected in 36 species of primates, but only in species found in sub-Saharan Africa. Interestingly, SIV—which appears to be transmitted sexually—does not appear to cause illness in some of these species.

Based on the degree of genetic differentiation among strains of SIV, scientists estimate that SIV may have been around for more than a million years in these primates, perhaps providing enough time for these species to adapt to the virus and thus prevent it from having adverse effects.

Phylogenetic analysis identifies the path of transmission

Phylogenetic analysis of strains of HIV and SIV reveals three clear findings. First, HIV obviously descended from SIV. All strains of HIV are phylogenetically nested within clades of SIV strains, indicating that HIV is derived from SIV (figure 23.17).

Second, a number of different strains of HIV exist, and they appear to represent independent transfers from different primate species. Each of the human strains is more closely related to a strain of SIV than it is to other HIV strains, indicating separate origins of the HIV strains.

Finally, humans have acquired HIV from different host species. HIV-1, which is the virus responsible for the global epidemic, has four subtypes. Two of these subtypes are most closely related to strains in chimpanzees, whereas a third is most closely related to a strain in gorillas, indicating transmission from both ape species. The origin of the fourth subtype is not clear; further sampling will likely reveal it to be the sister taxon to a currently undiscovered chimp or a gorilla strain.

By contrast, subtypes of HIV-2, which is much less widespread (in some cases known from only one individual), are related to SIV found in West African monkeys, primarily the sooty mangabey (*Cercocebus atys*). Moreover, the subtypes of HIV-2 also appear to represent multiple cross-species transmissions to humans.

Transmission from other primates to humans

Several hypotheses have been proposed to explain how SIV jumped from chimps and monkeys to humans. The most likely idea is that

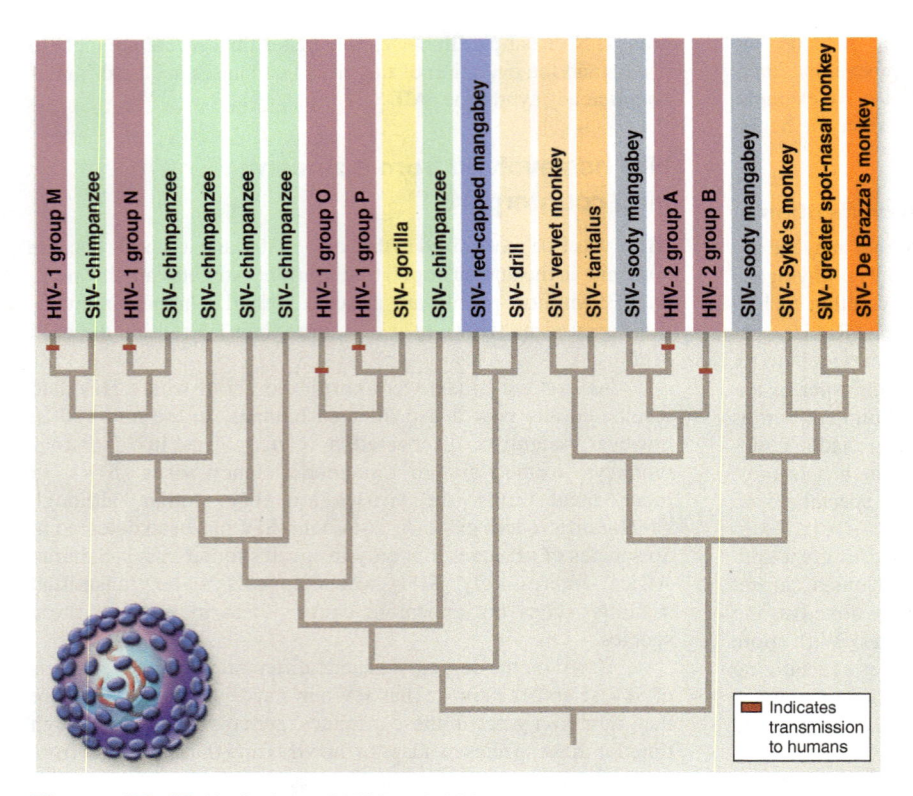

Figure 23.17 Evolution of HIV and SIV. HIV has evolved multiple times and from strains of SIV in different primate species (each primate species indicated by a different color; drill and tantalus are species of monkey). Boxes indicate transmission to humans from other primate species.

transmission occurred as the result of blood-to-blood contact that may occur when humans kill and butcher monkeys and apes. Recent years have seen a huge increase in the rate at which primates are hunted for the "bushmeat" market, particularly in central and western Africa. This increase has resulted from a combination of increased human populations desiring ever greater amounts of protein, combined with increased access to the habitats in which these primates live as the result of road building and economic development. The unfortunate result is that population sizes of many primate species, including our closest relatives, are plummeting toward extinction. A second consequence of this hunting is that humans are increasingly brought into contact with bodily fluids of these animals, and it is easy to imagine how during the butchering process, blood from a recently killed animal might enter the human bloodstream through cuts in the skin, perhaps obtained during the hunting process.

Establishing the crossover time line and location

Where and when did this cross-species transmission occur? HIV strains are most diverse in Africa, and the incidence of HIV is higher there than elsewhere in the world. Combined with the evidence that HIV is related to SIV in African primates, it seems certain that AIDS appeared first in Africa.

As for when the jump from other primates to humans occurred, the fact that AIDS was not recognized until the 1980s suggests that HIV probably arose recently. Descendants of slaves brought to North America from West Africa in the 19th century lacked the disease, indicating that it probably did not occur at the time of the slave trade.

Once the disease was recognized in the 1980s, scientists scoured repositories of old blood samples to see whether HIV could be detected in blood samples from the past. The earliest HIV-positive result was found in a sample from 1959, pushing the date of origin back at least two decades. Based on the amount of genetic difference between strains of HIV-1, including the 1959 sample, and assuming the operation of a molecular clock, scientists estimate that the deadly strain of AIDS probably crossed into humans some time before 1940.

Phylogenies can be used to track the evolution of AIDS among individuals

The AIDS virus evolves extremely rapidly, so much so that different strains can exist within a single individual in a single population. As a result, phylogenetic analysis can be applied to answer very specific questions; just as phylogeny proved useful in determining the source of HIV, it can also pinpoint the source of infection for particular individuals.

This ability became apparent in a court case in Louisiana in 1998, in which a dentist was accused of injecting his former girlfriend with blood drawn from an HIV-infected patient. The dentist's records revealed that he had drawn blood from the patient and had done so in a suspicious manner. Scientists sequenced the viral strains from the victim, the patient, and from a large number of HIV-infected people in the local community. The phylogenetic analysis clearly demonstrated that the victim's viral strain was most closely related to the patient's (figure 23.18). This analysis, which for the first time established phylogenetics as a legally admissible form of evidence in courts in the United States, helped convict the dentist, who is now serving a 50-year sentence for attempted murder.

Learning Outcome Review 23.5

Modern phylogenetic techniques and analysis can track the evolution of disease strains, uncovering sources and progression. The HIV virus provides a prime example: Analysis of viral strains has shown that the progression from simian immunodeficiency virus (SIV) into human hosts has occurred several times. Phylogenetic analysis is also used to track the transmission of human disease.

- *Could HIV have arisen in humans and then have been transmitted to other primate species?*

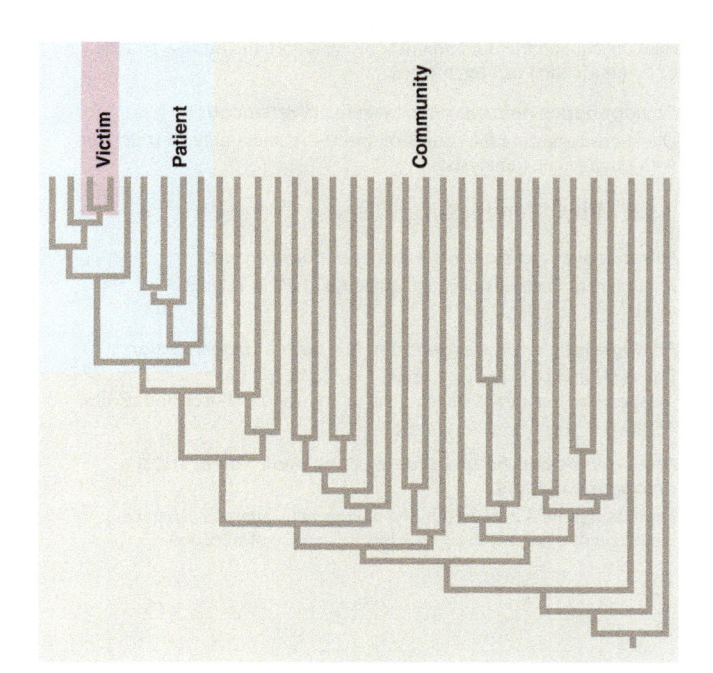

Figure 23.18 Evolution of HIV strains reveals the source of infection. HIV mutates so rapidly that a single HIV-infected individual often contains multiple genotypes in his or her body. As a result, it is possible to create a phylogeny of HIV strains and to identify the source of infection of a particular individual. In this case, the HIV strains of the victim (*pink*) clearly are derived from strains in the body of another individual, the patient (*blue*). Other HIV strains are from HIV-infected individuals in the local community.

? Inquiry question What would the phylogeny look like if the victim had not gotten HIV from the patient?

Chapter Review

23.1 Systematics

Branching diagrams depict evolutionary relationships.

Systematics is the study of evolutionary relationships, which are depicted on branching evolutionary trees, called phylogenies.

Similarity may not accurately predict evolutionary relationships.

The rate of evolution can vary among species and can even reverse direction. Closely related species can therefore be dissimilar in phenotypic characteristics.

Conversely, convergent evolution results in distantly related species being phenotypically similar.

23.2 Cladistics

The cladistic method requires that character variation be identified as ancestral or derived.

Derived character states are those that differ from the ancestral condition. Only shared derived characters are useful for inferring phylogenies.

Character polarity is established using an outgroup comparison in which the outgroup consists of one or a group of species, relative to the group under study.

Character states exhibited by the outgroup are assumed to be ancestral, and other character states are considered derived.

A cladogram is a graphically represented hypothesis of evolutionary relationships.

Homoplasy complicates cladistic analysis.

Homoplasy refers to a shared character state, such as wings of birds and wings of insects, that has not been inherited from a common ancestor.

Cladograms are constructed based on the principle of parsimony, which indicates that the phylogeny requiring the fewest evolutionary changes is accepted as the best working hypothesis.

Other phylogenetic methods work better than cladistics in some situations.

When evolutionary change is rapid, other methods, such as statistical approaches and the use of the molecular clock, are sometimes more useful.

23.3 Systematics and Classification

The taxonomic classification system is hierarchical.

Species are grouped into genera, genera into families, and so on through the ranks of order, class, phylum, kingdom, and domain.

Current classification sometimes does not reflect evolutionary relationships.

A monophyletic group consists of the most recent common ancestor and all of its descendants.

A paraphyletic group consists of the most recent common ancestor and some of its descendants.

A polyphyletic group does not contain the most recent ancestor of the group.

Some currently recognized taxa are not monophyletic, such as reptiles, which are paraphyletic with respect to birds.

The phylogenetic species concept focuses on shared derived characters.

The phylogenetic species concept (PSC) emphasizes the possession of shared derived characters, whereas the biological species concept focuses on reproductive isolation. Many versions of this concept recognize species that are monophyletic.

The phylogenetic species concept also has drawbacks.
Among criticisms of the PSC are that it subdivides groups too far via impractical distinctions, and that the PSC definition of a group may not always apply as selection proceeds.

23.4 Phylogenetics and Comparative Biology

Homologous features are derived from the same ancestral source; homoplastic features are not.
Homologous structures can be identified by phylogenetic analysis, establishing whether or not different structures have been built from the same ancestral structure.

Complex characters evolve through a sequence of evolutionary changes.
Most complex features do not evolve in a single step but include stages of transition. They may have begun as an adaptation to a selective pressure different from the one for which the feature is currently adapted.

Phylogenetic methods can be used to distinguish between competing hypotheses.
Different evolutionary scenarios can be distinguished by phylogenetic analysis. The minimum number of times a trait may have evolved can be established, and the direction of trait evolution, the timing, and the cause of diversification can be inferred.

Phylogenetics helps explain species diversification.
Questions regarding the causes of species richness may be addressed with phylogenetic analysis.

23.5 Phylogenetics and Disease Evolution

HIV has evolved from a simian viral counterpart.
Phylogenetic methods have indicated that HIV is related to SIV.

Phylogenetic analysis identifies the path of transmission.
It is clear that HIV has descended from SIV, and that independent transfers from simians to humans have occurred several times.

Phylogenies can be used to track the evolution of AIDS among individuals.
Even though HIV evolves rapidly, phylogenetic analysis can trace the origin of a current strain to a specific source of infection.

Review Questions

UNDERSTAND

1. Overall similarity of phenotypes may not always reflect evolutionary relationships
   a. due to convergent evolution.
   b. because of variation in rates of evolutionary change of different kinds of characters.
   c. due to homoplasy.
   d. All of the choices are correct.

2. Cladistics
   a. is based on overall similarity of phenotypes.
   b. requires distinguishing similarity due to inheritance from a common ancestor from other reasons for similarity.
   c. is not affected by homoplasy.
   d. None of the choices is correct.

3. The principle of parsimony
   a. helps evolutionary biologists distinguish among competing phylogenetic hypotheses.
   b. does not require that the polarity of traits be determined.
   c. is a way to avoid having to use outgroups in a phylogenetic analysis.
   d. cannot be applied to molecular traits.

4. Parsimony suggests that parental care in birds, crocodiles, and some dinosaurs
   a. evolved independently multiple times by convergent evolution.
   b. evolved once in an ancestor common to all three groups.
   c. is a homoplastic trait.
   d. is not a homologous trait.

5. The forelimb of a bird and the forelimb of a rhinoceros
   a. are homologous and symplesiomorphic.
   b. are not homologous but are symplesiomorphic.
   c. are homologous and synapomorphic.
   d. are not homologous but are synapomorphic.

6. In order to determine polarity for different states of a character
   a. there must be a fossil record of the groups in question.
   b. genetic sequence data must be available.
   c. an appropriate name for the taxonomic group must be selected.
   d. an outgroup must be identified.

7. In a paraphyletic group
   a. all species are more closely related to each other than they are to a species outside the group.
   b. evolutionary reversal is common.
   c. polyphyly also usually occurs.
   d. some species are more closely related to species outside the group than they are to some species within the group.

8. A paraphyletic group includes
   a. an ancestor and all of its descendants.
   b. an ancestor and some of its descendants.
   c. descendants of more than one common ancestor.
   d. All of the choices are correct.

9. Sieve tubes and sieve elements are
   a. homoplastic because they have different function.
   b. homologous because they have similar function.
   c. homoplastic because their common ancestor was single-celled.
   d. structures involved in transport within animals.

APPLY

1. A taxonomic group that contains species that have similar phenotypes due to convergent evolution is
   a. paraphyletic.
   b. monophyletic.
   c. polyphyletic.
   d. a good cladistic group.

2. Rapid rates of character change relative to the rate of speciation pose a problem for cladistics because
   a. the frequency with which distantly related species evolve the same derived character state may be high.
   b. evolutionary reversals may occur frequently.
   c. homoplasy will be common.
   d. All of the choices are correct.

3. Species recognized by the phylogenetic species concept
   a. sometimes also would be recognized as species by the biological species concept.
   b. are sometimes paraphyletic.
   c. are characterized by symplesiomorphies.
   d. are more frequent in plants than in animals.

SYNTHESIZE

1. List the synapomorphy and the taxa defined by that synapomorphy for the groups pictured in figure 23.2. Name each group defined by a set of synapomorphies in a way that might be construed as informative about what kind of characters define the group.

2. Identifying "outgroups" is a central component of cladistic analysis. As described on page 462, a group is chosen that is closely related to, but not a part of the group under study. If one does not know the relationships of members of the group under study, how can one be certain that an appropriate outgroup is chosen? Can you think of any approaches that would minimize the effect of a poor choice of outgroup?

3. As noted in your reading, cladistics is a widely utilized method of systematics, and our classification system (taxonomy) is increasingly becoming reflective of our knowledge of evolutionary relationships. Using birds as an example, discuss the advantages and disadvantages of recognizing them as reptiles versus as a group separate and equal to reptiles.

4. Across many species of limpets, loss of larval development and reversal from direct development appears to have occurred multiple times. Under the simple principle of parsimony, are changes in either direction merely counted equally in evaluating the most parsimonious hypothesis? If it is much more likely to lose a larval mode than to re-evolve it from direct development, should that be taken into account? If so, how?

5. Birds, pterosaurs (a type of flying reptile that lived in the Cretaceous period), and bats all have modified their forelimbs to serve as wings, but they have done so in different ways. Are these structures homologous? If so, how can they be both homologous and convergent? Do any organisms possess wings that are convergent, but not homologous?

6. In what sense does the biological species concept focus on evolutionary mechanisms and the phylogenetic species concept on evolutionary patterns? Which, if either, is correct?

CHAPTER 24

Genome Evolution

Chapter Contents

24.1 Comparative Genomics
24.2 Genome Size
24.3 Evolution Within Genomes
24.4 Gene Function and Expression Patterns
24.5 Applying Comparative Genomics

Introduction

Genomes contain the raw material for evolution, and many clues to evolution are hidden in the ever-changing nature of genomes. As more genomes have been sequenced, the new and exciting field of comparative genomics has emerged and has yielded some surprising results as well as many, many questions. Comparing whole genomes, not just individual genes, enhances our ability to understand the workings of evolution, to improve crops, and to identify the genetic basis of disease so that we might develop more effective treatments with minimal side effects. The focus of this chapter is the role of comparative genomics in enhancing our understanding of genome evolution and how this new knowledge can be applied to improve our lives.

24.1 Comparative Genomics

Learning Outcomes

1. *Describe the kinds of differences that can be found between genomes.*
2. *Relate timescale to genome evolution.*
3. *Explain why genomes may evolve at different rates.*

A key challenge of modern evolutionary biology is finding a way to link changes in DNA sequences, which we are now able to study in great detail, with the evolution of the complex morphological characters used to construct a traditional phylogeny. Many different genes contribute to complex characters. Making the connection between a specific change in a gene and a modification in a morphological character is particularly difficult.

Comparing genomes (entire DNA sequences) provides a powerful tool for exploring evolutionary divergence among organisms to connect DNA level changes with morphological differences. Genomes are more than instruction books for building and maintaining an organism; they also record the history of life. The growing number of fully sequenced genomes in all kingdoms is leading to a revolution in comparative evolutionary biology (figure 24.1). Genetic differences between species can be explored in a very direct way, examining the footprints on the evolutionary path between different species.

Evolutionary differences accumulate over long periods

Genomes of viruses and bacteria can evolve in a matter of days, whereas complex eukaryotic species evolve over millions of years. To illustrate this point, we compare three vertebrate genomes: human, tiger pufferfish *(Fugu rubripes)*, and mouse *(Mus musculus)*.

Comparison between human and pufferfish genomes

The draft (preliminary) sequence of the tiger pufferfish was completed in 2002; it was only the second vertebrate genome to be sequenced. For the first time, we were able to compare the genomes

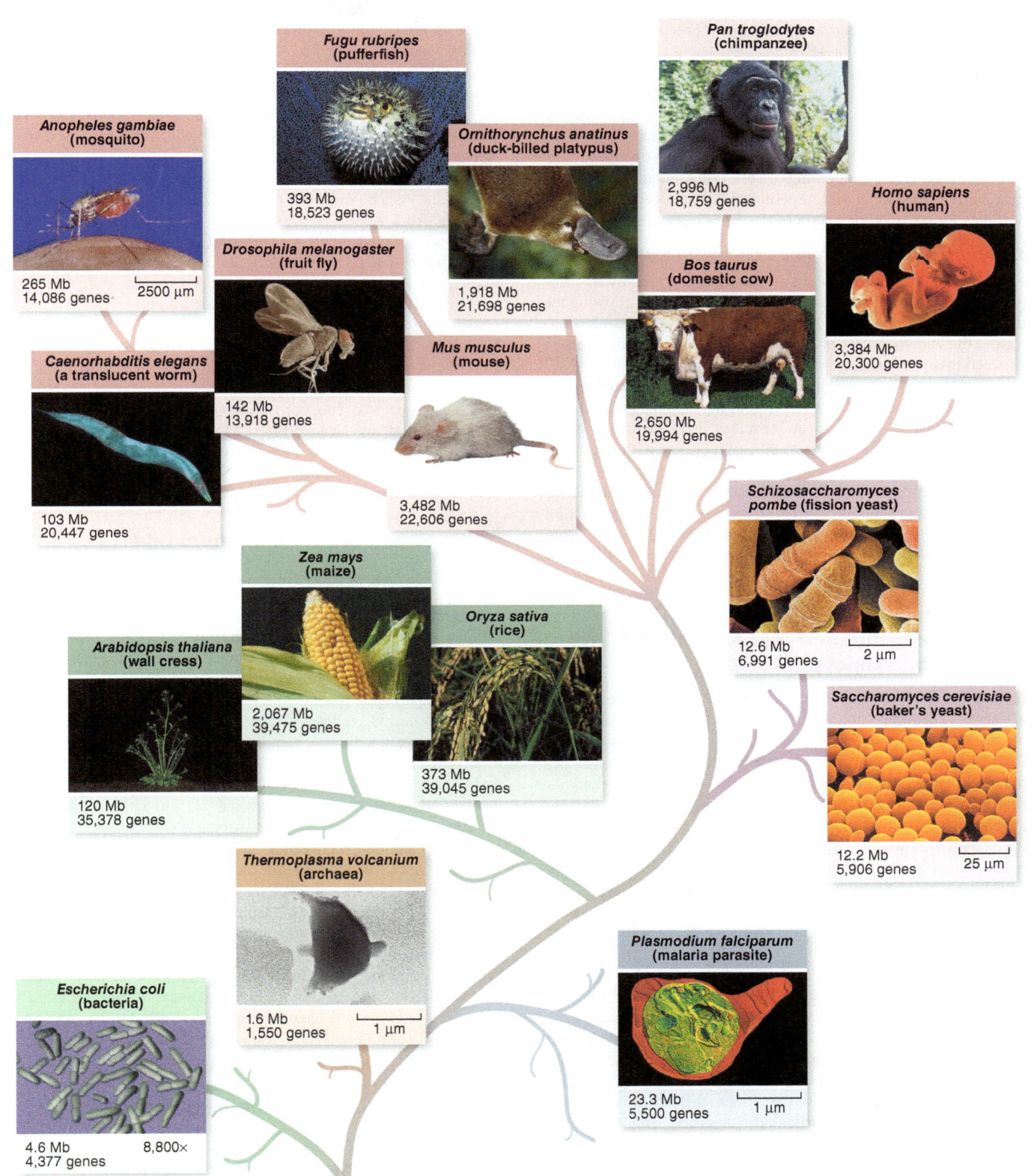

Figure 24.1 Milestones of comparative eukaryotic genomics. The sizes of genomes listed are the latest found in a variety of genome databases. The number of genes refers specifically to protein coding gene, or so-called coding sequences (CDS). This does not include genes for microRNAs and other forms of RNA.

of two vertebrates: humans and pufferfish. These two animals last shared a common ancestor 450 MYA.

Some human and pufferfish genes have been conserved during evolution, but others are unique to each species. About 25% of human genes have no counterparts in *Fugu*. Also, extensive genome rearrangements have occurred during the 450 million years since the mammal lineage and the teleost fish diverged, indicating a considerable scrambling of gene order. Finally, the human genome contains about 50% repetitive DNA, but repetitive DNA accounts for less than one-sixth of the *Fugu* sequence.

Comparison between human and mouse genomes

Later in 2002, a draft sequence of the mouse genome was completed by an international consortium of investigators, allowing for the first time a comparison of two mammalian genomes. In contrast to the human–pufferfish genome comparison, the differences between these two mammalian genomes are minuscule.

The human genome has about 400 million more nucleotides than that of the mouse. A comparison of the genomes reveals that both have about 20,000 genes, and that they share the bulk of them; in fact, the human genome shares 99% of its genes with mice. Humans and mice diverged about 75 MYA, approximately one-sixth of the amount of time that separates humans from pufferfish. There are only 300 genes unique to either human or mouse, constituting about 1% of the genome. Even 450 million years after last sharing a common ancestor, 75% of the genes found in humans have counterparts in pufferfish.

The human genome shares similarities with existing and extinct primates

Humans and chimpanzees, *Pan troglodytes*, diverged about 4.1 MYA. The chimp genome was sequenced in 2005, providing a comparative window between us and our closest living relative. Comparative studies indicate that the human and chimpanzee genomes differ by only 1.23% in terms of nucleotide substitutions. At first glance, the largest difference between our genomes actually appears to be in transposable elements (see chapter 18). In humans, the elements known as SINEs have been threefold more active than in the chimp, but the chimp has acquired two elements that are not found in the human genome. The differences due to insertion and deletion (indels) of bases are fewer than substitutions but account for about 1.5% of the euchromatic sequence being unique in each genome.

Comparing chimp and human indels to outgroups can identify whether the indel was present in a common ancestor and thus identify which species has the derived condition. Fifty-three of the potentially human-specific indels lead to loss-of-function changes that might correlate with some of the traits that distinguish us from chimps, including a larger cranium and lack of body hair. As discussed in section 24.4, mutations leading to differences in the patterns of gene expression are particularly important in understanding the phenotypes that differentiate humans and chimps.

Comparisons of single-nucleotide substitutions reveal that only 2.7% of the two genomes have consistent differences in single nucleotides. Mutations in coding DNA are classified into two groups: those that alter the amino acids coded for in the sequence (nonsynonymous changes) and those that do not (synonymous changes). For example, a synonymous mutation that changed UUU to UUC would still code for phenylalanine (use the genetic code in table 15.1 to see if you can identify examples of possible synonymous and nonsynonymous changes).

The genome of an even closer but extinct relative, *Homo neanderthalensis* (Neanderthal), is remarkably similar to our own. About half a million years ago, the two shared a common ancestor in Africa. Then, Neanderthals headed north to eventually settle in Europe and Asia before dying out about 28,000 years ago. Humans migrated from Africa later and, for about 100,000 years, overlapped with Neanderthals in Europe. The most compelling evidence that humans and Neanderthals interbred in Europe is the finding that Europeans and Asians, but not Africans, share 1 to 4% of their genome with Neanderthals (figure 24.2).

Using DNA isolated from 38,000- and 44,000-year-old Neanderthal bones found in Croatia, the Neanderthal genome could be sequenced and reconstructed. Of the nearly 10 million amino acids encoded in our genome, only 78 differences that could alter the shape and/or function of a protein were consistently found in humans and not Neanderthals. More than 200 other regions within the genome appear to have evolved since the human and Neanderthal branches split, but it is nothing that clearly points to our uniquely human traits.

Genomes evolve at different rates

As noted earlier in this section, the genomes of viruses and bacteria can evolve in a matter of days. Evolution of insect genomes occurs more rapidly than mammalian genomes, which evolve over millions of years. A short generation time may be responsible for these rate differences. However, even in some complex eukaryotic organisms, genome evolution can occur more rapidly.

Comparing 30 plant genomes that have been sequenced with sequenced animal genomes yielded a surprising finding. Plant genomes change much more rapidly than animal genomes. This is especially evident in noncoding DNA, which tends to change more rapidly than protein-coding DNA. Many conserved, noncoding regions of DNA can be identified when a fish and a primate genome are aligned, even though the two diverged 400 MYA. But, very few similarities are found between the noncoding DNA of *Arabidopsis* and rice, *Oryza sativa*, although the two diverged only 200 MYA. The sequencing of the corn genome has revealed that two varieties can differ by as much as 20%. Contrast this with the differences between human and chimp genomes.

Genomic change in plants is so fast that geneticist Detlef Weigel has concluded that plants practice "genomic anarchy." What accounts for all the variation in plant genomes? Transposable elements and other mobile elements frequently remodel the genome. Different bits of different genes come together in new combinations, and, as a result, new regulatory elements are created.

Plant, fungal, and animal genomes have unique and shared genes

We now step back farther and consider genomic differences among the eukaryotic kingdoms that diverged long before the examples just discussed. You have already seen that many genes are highly conserved in animals. Plant genes are also highly conserved, but new families of genes have emerged at each stage of plant evolution.

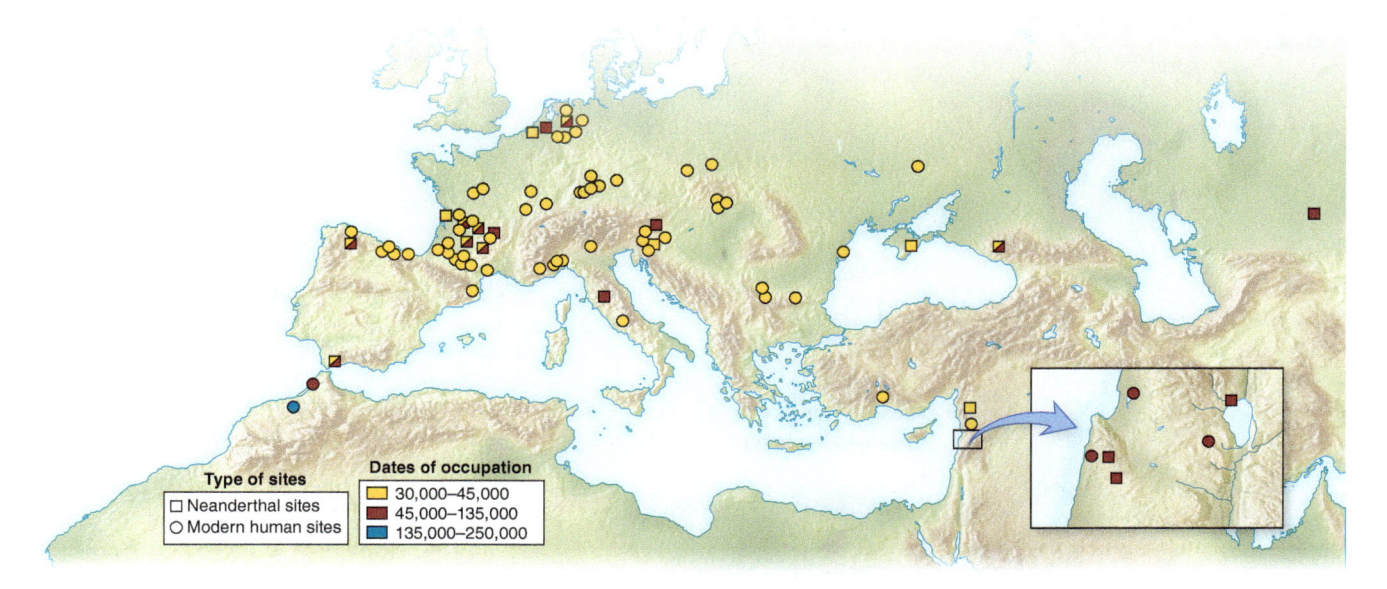

Figure 24.2 Europeans and Asians, but not Africans, share between 1% and 4% of their genomes with Neanderthals. From 30,000 to 45,000 years ago humans and Neanderthals coexisted in Europe, and possible in the Middle East as early as 80,000 years ago. Interbreeding likely occurred after Europeans and Asians ancestors left Africa.

Evolution of plant-specific genes

Sequenced plant genomes offer a window into a billion years of evolution. A total of 3814 gene families are shared by every plant, from the algae to grapes and corn. All the essentials for photosynthesis are found in the genome of the single-celled alga, *Chlamydomonas reinhardtii*, along with a gene that is now used to make flowers in the flowering plants. Although multicellularity was a huge evolutionary step, the multicellular alga *Volvox* has just about the same number of genes as *Chlamydomonas*. Lots of extra genes were not required for this major innovation.

The move onto land is represented by the genome of the moss *Physcomitrella patens* that arose 450 MYA. New genes that allowed the plant to survive in an arid environment arose, resulting in 3006 new gene families. Although moss are very different from flowering plants, making spores instead of seed, for example, they share 80% of the toolkit of developmental genes found in the flowering plant *Arabidoposis*. The shift from mosses to plants with internal transport systems required another 516 gene families. The transition to flowering plants, which now dominate land-based flora, required another 1350 gene families. It is perhaps not surprising that the biggest genomic change in the billion years of plant history was in the gene families required for colonizing land, a completely new environment.

Comparison of plants with animals and fungi

About one-third of the genes in *Arabidopsis* and rice appear to be in some sense "plant" genes—that is, genes not found in any animal or fungal genome sequenced so far. These include the many thousands of genes involved in photosynthesis and photosynthetic anatomy. Few plant genomes have been sequenced to date, however.

Among the remaining genes found in plants, many are very similar to those found in animal and fungal genomes, particularly the genes involved in basic intermediary metabolism, in genome replication and repair, and in transcription and protein synthesis. Prior to the availability of whole-genome sequences, assessment of the extent of genetic similarity and difference among diverse organisms had been difficult at best.

Learning Outcomes Review 24.1

Genomes vary in number of similar genes, arrangement of genes, total number of base-pairs, and base-pair differences. Closely related animals such as humans and chimps exhibit highly similar genomes, but greater variation is found in related plant genomes.

- Would you expect a high degree of similarity between genomes of a bony fish, such as a swordfish, and a cartilaginous fish, such as a shark? What genes might be different?

24.2 Genome Size

Learning Outcomes

1. Differentiate between autopolyploidy and allopolyploidy.
2. Explain why most crosses between two species do not result in a new polyploid species.
3. Explain why the genome of a polyploid is not identical to the sum of the two parental genomes.
4. Explain why genome size and genome number do not correlate.

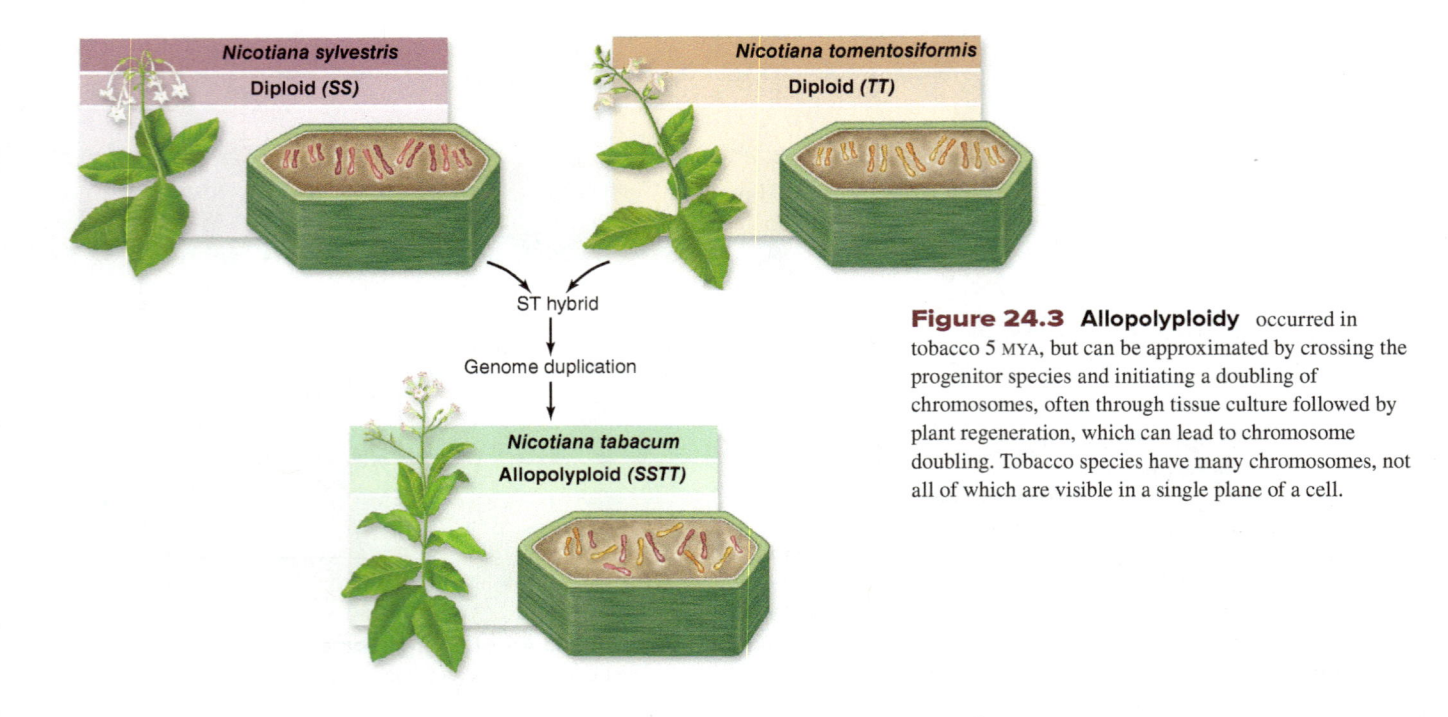

Figure 24.3 Allopolyploidy occurred in tobacco 5 MYA, but can be approximated by crossing the progenitor species and initiating a doubling of chromosomes, often through tissue culture followed by plant regeneration, which can lead to chromosome doubling. Tobacco species have many chromosomes, not all of which are visible in a single plane of a cell.

Both genome size and gene number vary greatly among the eukaryotes species (see figure 24.1). One contributing factor is whole-genome duplication, which results in polyploidy, the focus of this section. Genome duplication alone, however, cannot explain why rice has about 40,000 genes distributed among 370 Mb of DNA, but within the about 3400 Mb of the human genome, only about 20,000 genes are found. To more fully understand the relationship between gene number and genome size in eukaryotes, the noncoding DNA must also be considered.

Data analysis Estimate and compare the number of genes per mega base-pair of genomic DNA in humans and rice. Which genomes in figure 24.1 are most like humans in terms of the ratio of genes to base-pairs?

Ancient and newly created polyploids guide studies of genome evolution

Polyploidy (three or more chromosome sets) can give rise to new species, as you learned in chapter 22. **Autopolyploids** arise by genome duplication within a single lineage, and **allopolyploids** arise by hybridization of two lineages followed by genome duplication (figure 24.3). Genome duplication is more prevalent in plants than in animals, although animal polyploids are known, especially in fish and amphibians. In fact, the vertebrate lineage is thought to have undergone two ancient whole-genome duplication events.

Two avenues of research have lead to intriguing insights into genome alterations following polyploidization. The first approach studies ancient polyploids, called **paleopolyploids**. Sequence comparisons and phylogenetic tools establish the time and patterns of polyploidy events (figure 24.4). Sequence divergence between homologues, as well as the presence or absence of duplicated gene pairs from the hybridization, can be used for historical reconstructions of genome evolution. All copies of duplicate gene pairs arising through polyploidy are not necessarily found thousands or millions of years after polyploidization. The loss of duplicate genes is considered later in this section.

The second approach is to create **synthetic polyploids** by crossing plants most closely related to the ancestral species and then chemically inducing chromosome doubling. Unless the hybrid genome becomes doubled, the plant will be sterile because it will

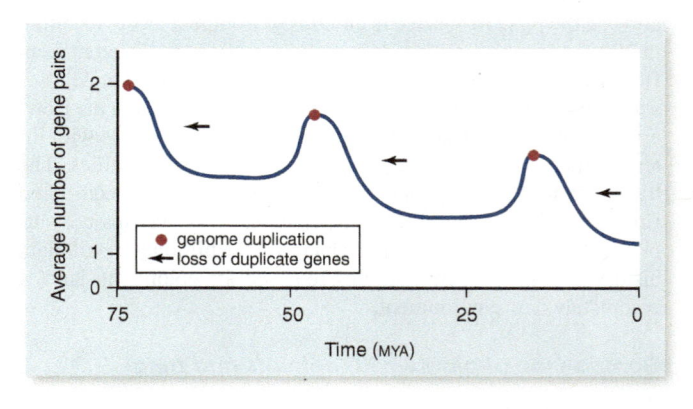

Figure 24.4 Sequence comparisons of numerous genes in a polyploid genome tell us how long ago allopolyploidy or autopolyploidy events occurred. Complex analyses of sequence divergence among duplicate gene pairs and presence or absence of duplicate gene pairs provide information about when both genome duplication and gene loss occurred. The graph reveals multiple polyploidy events over evolutionary time.

lack homologous chromosomes that need to pair during metaphase I of meiosis.

Because meiosis requires an even number of chromosome sets, species with ploidy levels that are multiples of two can reproduce sexually. However, meiosis is problematic in a 3n organism, such as asexually propagated commercial bananas, since three sets of chromosomes can't be evenly divided between two cells. Triploid bananas are seedless. The aborted ovules appear as the little brown dots in the center of any cross section of a banana.

> **Inquiry question** Sketch out what would happen in meiosis in a 3n banana cell, referring back to chapter 11 if necessary.

Later in this section, we examine further the effect of polyploidization on genomes. Plant examples have been selected to illustrate key points in this section because polyploidization occurs more frequently in plants. The somewhat surprising findings, however, are not limited to the plant kingdom. For example, triploid populations of whiptail lizards *(Cnemidophorus tesselatus)* are found in southeastern Colorado. These lizards are all female, and they reproduce parthenogenetically (see chapter 52). That is, the triploid eggs are able to produce identical triploid, female offspring through mitosis, with no meiosis or fertilization needed for reproduction.

Evidence of ancient polyploidy is found in plant genomes

Polyploidy has occurred numerous times in the evolution of the flowering plants (figure 24.5). The legume clade that includes the soybean *(Glycine max)*, the plant *Medicago truncatula* (a forage legume often used in research), and the garden pea *(Pisum sativum)* underwent a major polyploidization event 44 to 58 MYA and again 15 MYA.

A quick comparison of the genomes of soybean and *M. truncatula* reveal a huge difference in genome size (figure 24.6). In addition to increasing genome size through polyploidization,

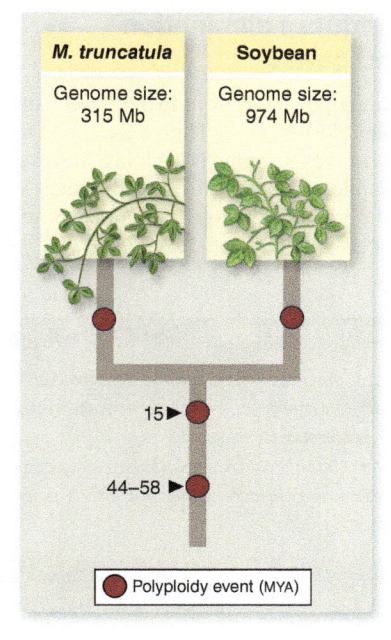

Figure 24.6 Genome downsizing. Genome downsizing must have occurred in *Medicago truncatula* (the numbers indicate millions of years ago—MYA).

genomes like *M. truncatula* definitely downsized over evolutionary time as well. The total size of a genome cannot be explained solely on the basis of polyploidization.

A number of independent whole-genome duplications in plants cluster around 65 MYA, coinciding with a mass extinction event caused by a catastrophic change due to an asteroid impact or increased volcanic activity. Polyploids may have had a better chance of surviving the extinction event with increased genes and alleles available for selection.

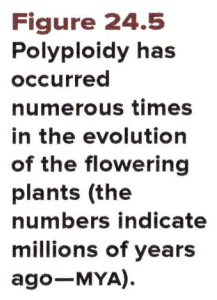

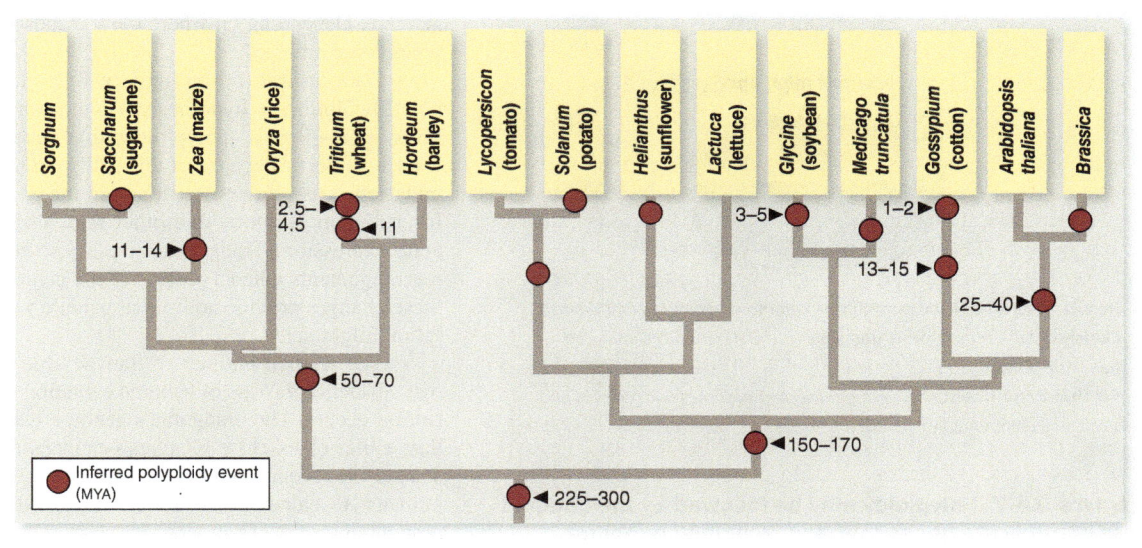

Figure 24.5 Polyploidy has occurred numerous times in the evolution of the flowering plants (the numbers indicate millions of years ago—MYA).

Polyploidy induces elimination of duplicated genes

Formation of an allopolyploid from two different species is often followed by a rapid loss of genes (figure 24.7) and rearrangement or loss of chromosomes. In some polyploids, however, loss of one copy of many duplicated genes arises over a much longer period. In some species a great deal of gene loss occurs in the first few generations after polyploidization.

SCIENTIFIC THINKING

Hypothesis: Some duplicated genes may be eliminated after polyploidy.
Prediction: More duplicate genes will be present when an allopolyploid forms than a few generations later.
Test: Make a synthetic polyploid from two Nicotiana species and look at the chromosomes under a microscope then and after 3 generations of self-fertilizing offspring.

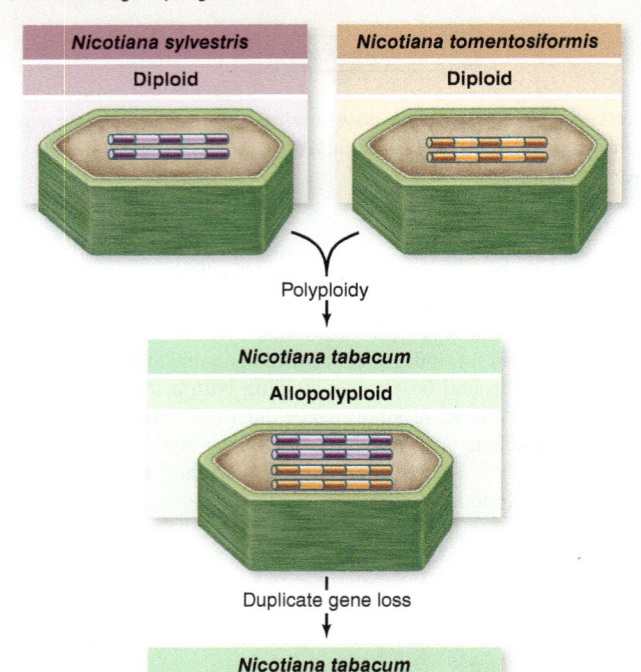

Nicotiana sylvestris Diploid

Nicotiana tomentosiformis Diploid

Polyploidy

Nicotiana tabacum Allopolyploid

Duplicate gene loss

Nicotiana tabacum

Result: Over time, N. tomentosiformis chromosomes are lost or shortened.
Conclusion: Chromosomes and genes are preferentially eliminated following polyploidy.
Further Experiments: Why might the chromosomes and genes of one species be preferentially eliminated? How could you test your explanation?

Figure 24.7 Polyploidy may be followed by the unequal loss of duplicate genes from the combined genomes.

Modern tobacco, *Nicotiana tabacum*, arose from the hybridization and genome duplication of a cross between *Nicotiana sylvestris* (female parent) and *N. tomentosiformis* (male parent). To recreate a cross that occurred over 5 MYA, researchers constructed synthetic *N. tabacum* and observed the chromosome loss that followed. Curiously, the loss of chromosomes is not even. More *N. tomentosiformis* chromosomes were jettisoned than those of *N. sylvestris*. Similar unequal chromosome loss has been observed in synthetic wheat hybrids, in which 13% of the genome of one parent is lost in contrast to 0.5% of the other parental genome. A comparison of closely related *Arabidopsis thaliana* and *Arabidopsis lyrata* found that *A. lyrata* has 500 more genes than *A. thaliana*. In the 10,000 years since the two species diverged, hundreds of thousands of small regions of the *A. thaliana* genome have been deleted. It is possible that different rates of genome replication could explain the differential loss, as is true for synthetic human–mouse hybrid cultured cells.

Inquiry question Why does the number of duplicate genes decrease after multiple rounds of polyploidization?

Polyploidy can alter gene expression

A striking discovery is the change in gene expression that occurs in the early generations after polyploidization. Some of this may be connected to an increase in the methylation of cytosines in the DNA. Methylated genes are usually not transcribed, as described in chapter 16. Simply put, polyploidization can lead to a short-term silencing of some genes. In subsequent generations, methylation decreases.

Transposons are mobilized by polyploidization

Barbara McClintock, in her Nobel Prize-winning work on transposable (mobile) genetic elements, referred to these jumping DNA regions as *controlling elements*. She hypothesized that transposons could respond to genome shock and jump into a new position in the genome. Depending on where the transposon moved, new phenotypes could emerge.

Data on transposon activity following hybridization support McClintock's hypothesis. Again, during the early generations following a polyploidization event, new transposon insertions occur because of unusually active transposition. Gene mutations arise if the insertion is in the coding region of a gene. Insertions in promoter and other regulatory regions can change gene expression. Transposition can also result in chromosomal rearrangements when a portion of the genome is relocated. All of these changes provide additional genetic variation on which evolution might act.

The relative number of transposable elements in a species may influence the rate of genome evolution whether or not hybridization occurs. The orangutan's genome has evolved more slowly than either of its close relatives—humans and chimps. Comparisons of the recently sequenced orangutan genome with chimp and humans revealed many fewer transposable elements in the orangutan.

Polyploidy alone cannot account for variation in genome size

It is difficult to account for all genome size variation with polyploidy. Even with gene or chromosome elimination following hybridization, polyploidy is unlikely to result in such a range of ratios of gene number to genome size. Humans have nine times the amount of DNA found in the 3.93×10^8-bp pufferfish genome, but about the same number of genes. Plants have an even greater range of genome sizes. As much as a 200-fold difference has been found, yet all these plants weigh in with about 24,000 to 60,000 genes. Tulips, for example, have 170 times more DNA than *Arabidopsis*.

Noncoding DNA inflates genome size

Why do humans have so much extra DNA compared with pufferfish? Much of it appears to be in the form of introns, noncoding segments (ncDNA) within a gene's sequence, that are substantially bigger than those in pufferfish. The *Fugu* genome has only a handful of "giant" genes containing long introns; studying them should provide insight into the evolutionary forces that have driven the change in genome size during vertebrate evolution.

Large expanses of retrotransposon DNA contribute to the differences in genome size from one species to another. Although part of the genome, ncDNA does not contain genes in the usual sense. As another example, *Drosophila* exhibits less ncDNA than *Anopheles*, although the evolutionary force driving this reduction in noncoding regions is unclear.

The 370-Mb rice genome and the 2-Gb maize genome are now fully sequenced. Both have about 40,000 protein-coding genes, roughly twice the number of genes in the human genome. Maize contains lots of repetitive DNA, which has increased its DNA content, but not gene content. Comparisons between the rice, maize, and wheat genomes should provide clues about the genome of their common ancestor and the dynamic evolutionary balance between opposing forces that increase genome size (polyploidy, transposable element proliferation, and gene duplication) and those that decrease genome size (mutational loss).

Learning Outcomes Review 24.2

Autopolyploids arise from duplication of a species' genome; allopolyploids occur from hybridization between two species. Unless the number of chromosomes doubles, an allopolyploid will likely be sterile because of lack of homologous pairs at meiosis. Polyploidization can lead to major changes in genome structure, including loss of genes, alteration of gene expression, increased transposon hopping, and chromosomal rearrangements. Polyploidy is considered important in the generation of biodiversity and adaptation, especially in plants, but it cannot explain why increases or decreases in genome size do not correlate with the number of genes. Evidently DNA content is not the same as gene content. Polyploidy in plants does not by itself explain differences in genome size. Often a greater amount of DNA is explained by the presence of introns and non-protein-coding sequences than by gene duplicates.

■ *How might a genome with a small number of genes and a small number of total base-pairs evolve into a genome with the same small number of genes and a thousand-fold larger genome?*

24.3 Evolution Within Genomes

Learning Outcomes

1. Define the terms segmental duplication, genome rearrangement, and pseudogene.
2. Explain why horizontal gene transfer can complicate evolutionary hypotheses.

Genome evolution can occur at the level of whole-genome duplication, but also occurs within genomes. Alterations are seen at all levels from individual genes to chromosomal rearrangements (see chapter 13). Duplication of a genomic region, either within a chromosome, or to another chromosome, allows genes with the same function to diverge because a "backup pair" of genes exist. A duplicated gene can lose function; it is not selected against because there is another functioning gene, not just an allele. The duplicated gene can also diverge and acquire new functions because the original persists.

Individual chromosomes may be duplicated

Aneuploidy refers to the duplication or loss of an individual chromosome rather than of an entire genome. Failure of a pair of homologous chromosomes or sister chromatids to separate during meiosis is the most common way that aneuploidy occurs. In general, plants are better able to tolerate aneuploidy than animals, but the explanation for this difference is elusive.

DNA segments may be duplicated

One of the greatest sources of novel traits in genomes is duplication of segments of DNA. Two genes within an organism that have arisen from the duplication of a single gene in an ancestor are called paralogues. In contrast, orthologues reflect the inheritance of a single gene from a common ancestor—a conserved gene.

When a gene is duplicated, the most likely fates of the duplicate gene are (1) losing function through subsequent mutation (becomes a pseudogene), (2) gaining a novel function through subsequent mutation (neofunctionalization), and (3) having the total function of the ancestral gene partitioned into the two duplicates (subfunctionalization). Gene families grow through gene duplication. In reality, however, most duplicate genes lose function.

Why then, do we think that gene duplication is a major evolutionary force for genes to gain new function? Part of the answer lies in where gene duplication is most likely to occur in the genome. In humans, the highest rates of duplication lie in the three most gene-rich chromosomes. Conversely, the seven chromosomes with the fewest genes show the least amount of duplication. Note that gene-poor does not necessarily mean less total DNA.

Even more compelling, certain types of genes appear more likely to be duplicated: growth and development genes, immune system genes, and cell-surface receptors. Finally, and importantly, gene duplication is thought to be a major evolutionary force for gene innovation because the duplicated genes are found to have different patterns of gene expression (see chapter 25 for examples).

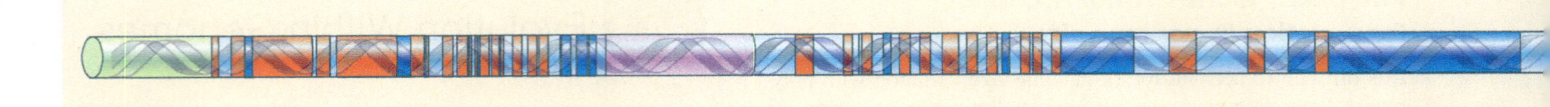

Figure 24.8 Segmental duplication on the human Y chromosome. Each orange region has 98% sequence similarity with a sequence on a different human chromosome. Each dark blue region has 98% sequence similarity with a sequence elsewhere on the Y chromosome.

For example, the two duplicated copies may be expressed in different or overlapping sets of tissues or organs during development.

About 5% of the human genome consists of segmental duplications, which are duplications of blocks of genes (figure 24.8). Segmental duplications may account for differences between human and chimp. Species-specific segmental duplications tend to contain genes that are differentially expressed between the species. A simple explanation is that there are more copies of the gene being expressed, but experimental evidence shows that it is more likely that the regulatory DNA near the duplicate is different.

Both rice and *Arabidopsis* have higher *copy numbers* for gene families (multiple slightly divergent copies of a gene) than are seen in animals or fungi, suggesting that these plants may have undergone numerous episodes of segmental duplication during the 150 to 200 million years since rice and *Arabidopsis* diverged from a common ancestor. As whole-genome duplications have also occurred since they diverged, additional analysis will be needed to determine the relative effects of polyploidy and segmental duplication in plant evolution.

Genomes may become rearranged

Humans have one fewer chromosome than chimpanzees, gorillas, and orangutans (figure 24.9). It's not that we have lost a chromosome. Rather, at some point in time, two midsized ape chromosomes fused to make what is now human chromosome 2, the second largest chromosome in our genome.

The fusion leading to human chromosome 2 is an example of the sort of genome reorganization that has occurred in many species. Rearrangements like this can provide evolutionary clues, but they are not always definitive proof of how closely related two species are.

Consider the organization of known orthologues shared by humans, chickens, and mice. One study estimated that 72 chromosome rearrangements had occurred since the chicken and human last shared a common ancestor. This number is substantially less than the estimated 128 rearrangements between chicken and mouse, or 171 between mouse and human.

This does not mean that chickens and humans are more closely related than mice and humans or mice and chickens. What these data actually show is that chromosome rearrangements have occurred at a much lower frequency in the lineages that led to humans and to chickens than in the lineages leading to mice. Chromosomal rearrangements in mouse ancestors seem to have occurred at twice the rate seen in the human line. These different rates of change help counter the notion that humans existed hundreds of millions of years ago.

Genomes that have undergone relatively slow chromosome change are the most helpful in reconstructing the hypothetical genomes of ancestral vertebrates. If regions of chromosomes have changed little in distantly related vertebrates over the last

Figure 24.9 Living great apes. All living great apes, with the exception of humans, have a haploid chromosome number of 24. Humans have not lost a chromosome; rather, two smaller chromosomes fused to make a single chromosome.

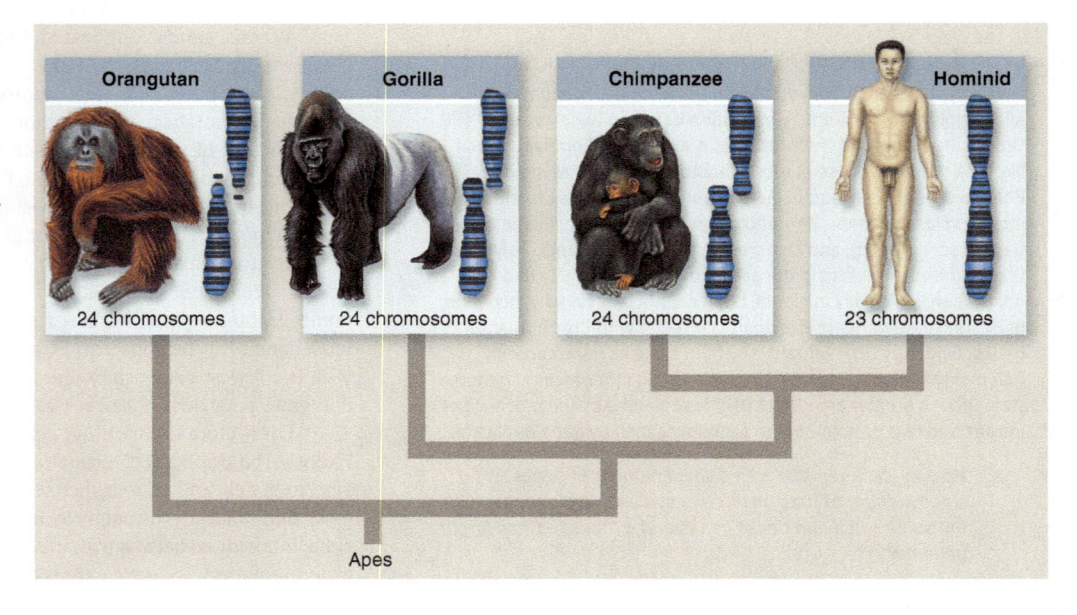

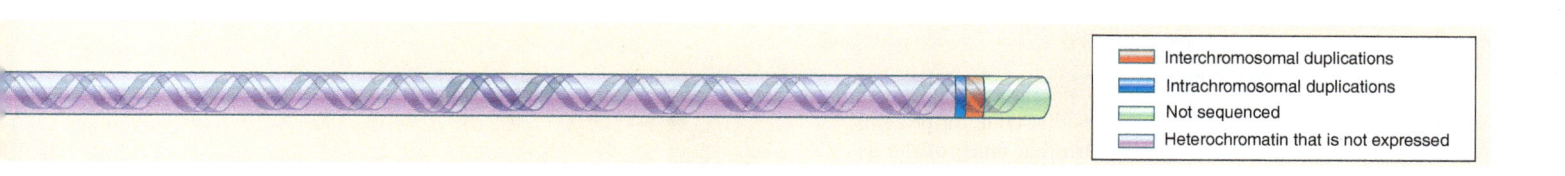

	Interchromosomal duplications
	Intrachromosomal duplications
	Not sequenced
	Heterochromatin that is not expressed

300 million years, then we can reasonably infer that the common ancestor of these vertebrates had genomic similarities.

Variation in the organization of genomes is as intriguing as gene sequence differences. Chromosome rearrangements are common, yet over long segments of chromosomes, the linear order of mouse and human genes is the same—the common ancestral sequence has been preserved in both species. This **conservation of synteny** (see chapter 18) was anticipated from earlier gene-mapping studies, and it provides strong evidence that evolution actively shapes the organization of the eukaryotic genome. As seen in figure 24.10, the conservation of synteny allows researchers to more readily locate a gene in a different species using information about synteny, thus underscoring the power of a comparative genomic approach.

Gene inactivation results in pseudogenes

The loss of gene function is another important way genomes evolve. Consider the olfactory receptor (OR) genes that are responsible for our sense of smell. These genes code for receptors that bind odorants, initiating a cascade of signaling events that eventually lead to our perception of scents.

Gene inactivation seems to be the best explanation for our reduced sense of smell relative to that of the great apes and other mammals. Mice have the largest mammalian, OR gene family, with about 1500 OR genes—about 40% more OR genes than humans. Only 20% of the mouse OR genes are pseudogenes, compared to 63% in humans, which are inactive **pseudogenes** (genes that do not produce a functional product due to premature stop codons, missense mutations, or deletions).

In contrast, half the chimpanzee and gorilla OR genes function effectively, and over 95% of New World monkey OR genes appear to be functional. The most likely explanation for these differences is that humans came to rely on other senses, reducing the selection pressure against loss of OR gene function by random mutation.

An older question about the possibility of positive selection for OR genes in chimps was resolved with the completion of the chimp genome. A careful analysis indicated that both humans and chimps are gradually losing OR genes to pseudogenes and that there is no evidence to support positive selection for any of the OR genes in the chimp.

Rearranged DNA can acquire new functions

Errors in meiosis that rearrange parts of genes most often create pseudogenes, but occasionally a broken piece of a gene can end up in a new spot in the genome where it acquires a new function. One of the most intriguing examples occurred within a family of fish in the suborder Notothenioidae found in the Antarctic Ocean. These fish are called icefish because they survive the frigid temperatures in the Antarctic, in part because of a protein in their blood that works like antifreeze. Reconstructing evolutionary history using comparative genomics reveals that 9-bp fragment of a gene coding for a digestive enzyme moved to a new location where it evolved to encode part of an antifreeze protein. The series of errors that gave rise to the new protein persisted only because the change coincided with a massive cooling of the Antarctic waters. Natural selection acted on this mutation over millions of years.

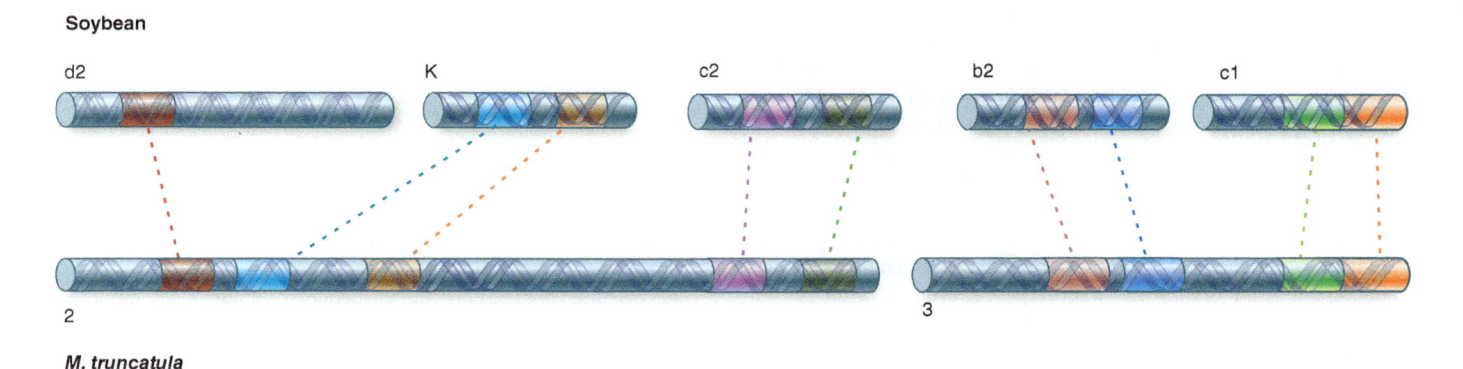

Figure 24.10 Synteny and gene identification. Genes sequenced in the model legume, *Medicago truncatula*, can be used to identify homologous genes in soybean, *Glycine max*, because large regions of the genomes are syntenic, as illustrated for some of the linkage groups (chromosomes) of the two species. Regions of the same color represent homologous genes.

Noncoding DNA can acquire regulatory functions

So far, we have primarily compared genes that code for proteins. As more genomes are sequenced, we learn that much of the genome is composed of ncDNA. The repetitive DNA is often retrotransposon DNA, contributing to as much as 30% of animal genomes and 40 to 80% of plant genomes. (Refer to chapter 18 for more information on repetitive DNA in genomes.) Yet there are conserved noncoding regions (CNCs) that evolve much more slowly than expected, assuming there was no associated function. An analysis of CNCs in 40 vertebrates led to the conclusion that as the rate of base-pair substitution slowed down over evolutionary time, these CNCs had begun to regulate the expression of transcription factors, genes that specifically influence development, and genes that encode receptor-binding proteins that are important in signaling.

Horizontal gene transfer complicates matters

Evolutionary biologists build phylogenies on the assumption that genes are passed from generation to generation, a process called **vertical gene transfer (VGT)**. Hitchhiking genes from other species, a process referred to as **horizontal gene transfer (HGT)** and sometimes called *lateral gene transfer,* can lead to phylogenetic complexity. HGT was likely most prevalent very early in the history of life, when the boundaries between individual cells and species seem to have been less firm than they are now and DNA more readily moved among different organisms. Earlier in the history of life, gene swapping between species was rampant. Today HGT continues in prokaryotes and eukaryotes, but less frequently. An intriguing example of more recent HGT between moss and a flowering plant is described in chapter 31.

Gene swapping in early lineages

The extensive gene swapping among early organisms has caused many researchers to reexamine the base of the tree of life. Early phylogenies based on ribosomal RNA (rRNA) sequences indicate that an early prokaryote gave rise to two major domains: the Bacteria and the Archaea. From one of these lineages, the domain Eukarya emerged; its organelles originated as unicellular organisms engulfed specialized prokaryotes.

This rRNA phylogeny is being revised as more microbial genomes are sequenced. In 2015, the National Center for Biotechnology Information (NCBI) databases contained 33,543 prokaryotic genomes. With new sequencing technology, microbial genomes can be sequenced in less than a day. Phylogenies built with rRNA sequences suggest that the domain Archaea is more closely related to the Eukarya than to the Bacteria. But as more microbial genomes are sequenced, investigators find bacterial and archaeal genes showing up in the same organism! The most likely conclusion is that organisms swapped genes, even absorbing DNA obtained from a food source. Perhaps the base of the tree of life is better viewed as a web than a branch (figure 24.11).

Evidence for gene swapping in the human genome

Let's move closer to home and look at the human genome, which is riddled with foreign DNA, often in the form of transposons. The many transposons of the human genome provide a paleontological record over several hundred million years.

Comparisons of versions of a transposon that has duplicated many times allow researchers to construct a "family tree" to identify the ancestral form of the transposon. The percent of sequence divergence found in duplicates allows an estimate of the time at which that particular transposon originally invaded the human genome. In humans, most of the DNA hitchhiking seems to have occurred millions of years ago in very distant ancestor genomes.

Our genome carries many more ancient transposons than do the genomes of *Drosophila, C. elegans,* and *Arabidopsis.* One explanation for the observed low level of transposons in *Drosophila* is that fruit flies somehow eliminate unnecessary DNA from their genome 75 times faster than humans do. Our genome has simply hung on to hitchhiking DNA more often.

The human genome has had minimal transposon activity in the past 50 million years; mice, by contrast, are continuing to acquire new transposable elements. This difference may explain in part the more rapid change in chromosome organization in mice than in humans.

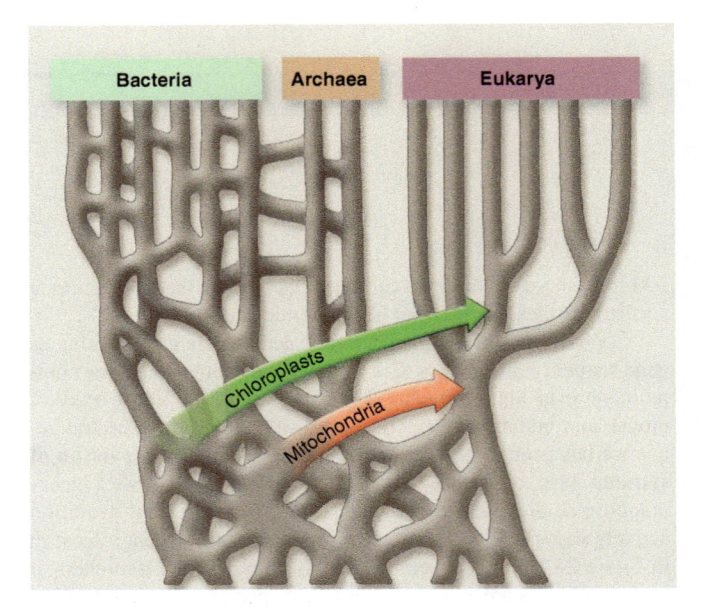

Figure 24.11 Horizontal gene transfer. Early in the history of life, organisms may have freely exchanged genes beyond multiple endosymbiotic events. To a lesser extent, this transfer continues today. The tree of life may be more like a web or a net.

Learning Outcomes Review 24.3

In segmental duplication, part of a chromosome and the genes it contains are duplicated. In genome rearrangement, segments of chromosomes may change places or chromosomes may fuse with one another. Pseudogenes have become inactivated in the course of evolution but still persist in the genome. All these changes have evolutionary consequences. Horizontal gene transfer has led to an unexpected mixing of genes among organisms, creating many phylogenetic questions.

- *How would you determine whether a gene was a pseudogene or an example of horizontal gene transfer?*

24.4 Gene Function and Expression Patterns

Learning Outcomes
1. Explain how species with nearly identical genes can look very different.
2. Describe the action of the **FOXP2** gene across species.

Gene function can be inferred by comparing genes in different species. When the human genome was first compared to another mammalian genome, the mouse, 1000 genes with unknown function had a function assigned. One of the major puzzles arising from comparative genomics is that organisms with very different forms can share so many conserved genes in their genomic toolkit.

The best explanation for why a mouse develops into a mouse and not a human is that the same or similar genes are expressed at different times, in different tissues, and in different amounts and combinations. For example, the cystic fibrosis gene (cystic fibrosis transmembrane conductance regulator, *CFTR*), which has been identified in both species and affects a chloride ion channel, illustrates this point. Defects in the human *CFTR* gene cause especially devastating effects in the lungs, but mice with the mutant *CFTR* gene do not have lung symptoms. Possibly variations in expression of *CFTR* between mouse and human explain the difference in lung symptoms when *CFTR* is defective.

Chimp and human gene transcription patterns differ

Humans and chimps diverged from a common ancestor only about 4.1 MYA—too little time for much genetic differentiation to evolve, but enough for significant morphological and behavioral differences to have developed. Sequence comparisons indicate that chimp DNA is 98.77% identical to human DNA. If just the gene sequences encoding proteins are considered, the similarity increases to 99.2%. How could two species differ so much in body and behavior, and yet have almost equivalent sets of genes?

One potential answer to this question is based on the observation that chimp and human genomes show very different patterns of gene transcription activity, at least in brain cells. Investigators used microarrays (see chapter 18) containing up to 18,000 human genes to analyze RNA isolated from cells in the fluid extracted from several regions of living brains of chimps and humans. The RNA was linked with a fluorescent tag and then incubated with the microarray under conditions that allow the formation of DNA–RNA hybrids if the sequences are complementary. If the transcript of a particular gene is present in the cells, then the microarray spot corresponding to that gene lights up under UV light. The more copies of the RNA, the more intense the signal.

Because the chimp genome is so similar to that of humans, the microarray detects the activity of chimp genes reasonably well. Although the same genes were transcribed in chimp and human brain cells, the patterns and levels of transcription varied. Much of the difference between human and chimp brains lies in which genes are transcribed and when and where that transcription occurs.

Inquiry question You are given a microarray of ape genes and RNA from both human and ape brain cells. Using the experimental technique described for comparison of humans and chimps, what would you expect to find in terms of genes being transcribed? What about levels of transcription?

Posttranscriptional differences may also play a role in building distinct organisms from similar genomes. As research continues to push the frontiers of proteomics and functional genomics, a more detailed picture of the subtle differences in the developmental and physiological processes of closely related species will be revealed. The integration of development and genome evolution is explored in depth in chapter 25.

Speech is uniquely human: An example of complex expression

Development of human culture is closely tied to the capacity to control the larynx and mouth to produce speech. Humans with a single point mutation in the transcription factor gene *FOXP2* have impaired speech and grammar but not impaired language comprehension.

The *FOXP2* gene is also found in chimpanzees, gorillas, orangutans, rhesus macaques, and even the mouse, yet none of these mammals speak (figure 24.12). The gene is expressed in areas of the brain that affect motor function, including the complex coordination needed to create words.

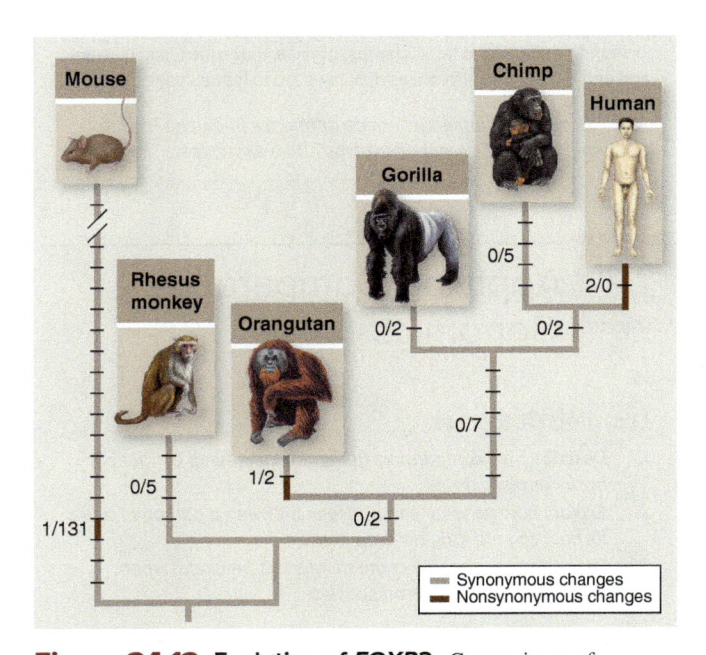

Figure 24.12 Evolution of *FOXP2*. Comparisons of synonymous and nonsynonymous changes in mouse and primate *FOXP2* genes indicate that changing two amino acids in the gene corresponds to the emergence of human language. Beige bars represent synonymous changes and brown bars represent nonsynonymous changes.

FOXP2 protein in mice and humans differs by only three amino acids. There is only a single amino acid difference between mouse and chimp, gorilla, and rhesus macaque, which all have identical amino acid sequences for FOXP2. Two more amino acid differences exist between humans and the sequence shared by chimp, gorilla, and macaque. The difference of only two amino acids between human and other primate FOXP2 may be linked to language acquisition in humans. Evidence points to strong selective pressure for the two *FOXP2* mutations that allow brain, larynx, and mouth to coordinate to produce speech.

Is it possible that two amino acid changes lead to speech, language, and ultimately human culture? This box of mysteries will take a long time to unpack, but hints indicate that the changes are linked to signaling and gene expression. The two altered amino acids may change the ability of FOXP2 transcription factor to be phosphorylated. One way signaling pathways operate is through activation or inactivation of an existing transcription factor by phosphorylation.

Comparative genomics efforts are now extending beyond primates. A role for *FOXP2* in songbird singing and vocal learning has been proposed. Mice communicate via squeaks, with lost young mice emitting high-pitched squeaks. *FOXP2* mutations leave mice squeakless and transgenic baby mice with the human *FOXP2* squeak differently. For both mice and songbirds, it is a stretch to claim that *FOXP2* is a language gene—but it likely is needed in the neuromuscular pathway to make sounds.

Learning Outcomes Review 24.4

To understand functional differences between genes shared by species, one must look beyond sequence similarity. Alterations in the time and place of gene expression can lead to marked differences in phenotype. As an example, the FOXP2 factor appears to be involved in sound production in mice, chimps, gorillas, macaques, and humans, and only very small differences may have led to human speech.

- How can a single-nucleotide difference in a gene lead to a noticeably different phenotype? Give examples.

24.5 Applying Comparative Genomics

Learning Outcomes

1. Describe how comparative genomics can reveal the genetic basis for disease.
2. Explain how genome comparisons between a pathogen and its host can aid drug development.
3. Describe how genome comparisons can be useful when working with endangered species.

Comparisons among individual human genomes continue to provide information on genetic disease detection and the best course of treatment. An even broader array of possibilities arises when comparisons are made among species. There are advantages to comparing both closely and distantly related pairs of species, as well as comparing the genomes of a pathogen and its host. Genome-level comparisons are also assisting conservation biologists in developing breeding programs. Examples of the benefits of each type of genome comparison follow.

Distantly related genomes offer clues for causes of disease

Sequences that are conserved between humans and pufferfish provide valuable clues for understanding the genetic basis of many human diseases. Amino acids critical to protein function tend to be preserved over the course of evolution, and changes at such sites within genes are more likely to cause disease.

It is difficult to distinguish functionally conserved sites when comparing human proteins with those of other mammals because not enough time has elapsed for sufficient changes to accumulate at nonconserved sites. A promising exception is the duck-billed platypus *(Ornithorhynchus anatinus)*, which diverged from other mammals about 166 MYA and whose genome provides clues to the evolution of the immune system. Because the pufferfish genome is only distantly related to humans, conserved sequences are far more easily distinguished than even in the platypus.

Closely related organisms enhance medical research

Because studies in humans must be strictly regulated for ethical reasons, it is much easier to design experiments to identify gene function in an experimental system like that of the mouse. Comparing mouse and human genomes quickly revealed the function of 1000 previously unidentified human genes. The effects of these genes can be studied in mice, and the results can be used in potential treatments for human diseases.

A draft of the rat genome has been completed, and even more exciting news about the evolution of mammalian genomes may emerge from comparisons of these species. One of the most exciting aspects of comparing rat and mice genomes is the potential to capitalize on the extensive research on rat physiology, especially heart disease, and the long history of genetics in mice. Linking genes to disease has become much easier.

As described in section 24.3, the human genome contains segmental duplications that are absent in the chimp genome. Some of these duplications contain genes with alleles that produce human disease. For example, some of the regions duplicated only in humans correspond to regions implicated in Prader–Willi syndrome and spinal muscular atrophy. The significance of these observations in not clear, but there is great interest in analyzing regions that differ between humans and closely related species.

Pathogen–host genome differences reveal drug targets

With genome sequences in hand, pharmaceutical researchers are more likely to find suitable drug targets to eliminate pathogens without harming the host. Diseases—including malaria and Chagas disease—in many developing countries have both human and insect hosts. Both these infections are caused by protists (see chapter 29), and the value of comparative genomics in drug discovery to treat them is illustrated later in this section.

Malaria

Anopheles gambiae, the malaria-carrying mosquito, along with *Plasmodium falciparum,* the protistan parasite it transmits, together have an enormous effect on human health, resulting in 1.7 to 2.5 million deaths each year from malaria. The genomes of both *Anopheles* and *Plasmodium* were sequenced in 2002.

P. falciparum, which causes malaria, has a relatively small genome of 2.33×10^7 bp that proved very difficult to sequence. It has an unusually high proportion of adenine and thymine, making it hard to distinguish one portion of the genome from the next. The project took five years to complete. *P. falciparum* appears to have about 5500 genes, with those of related function clustered together, suggesting that they might share the same regulatory DNA.

P. falciparum is a particularly crafty organism that hides from our immune system inside red blood cells, regularly changing the proteins it presents on the surface of the red blood cell. This "cloaking" has made developing a vaccine or other treatment for malaria particularly difficult.

Recently, a link to chloroplast-like structures in *P. falciparum* has raised other possibilities for treatment. An odd subcellular component called the *apicoplast,* found only in *Plasmodium* and its relatives, appears to be derived from a chloroplast appropriated from algae engulfed by the parasite's ancestor (figure 24.13).

Analysis of the *Plasmodium* genome reveals that about 12% of all the parasite's proteins, encoded by the nuclear genome, head for the apicoplast. These proteins act there to produce fatty acids. The apicoplast is the only location in which the parasite makes the fatty acids, suggesting that drugs targeted at this biochemical pathway might be very effective against malaria.

Another disease-prevention possibility is to look at chloroplast-specific herbicides, which might kill *Plasmodium* by targeting the chloroplast-derived apicoplast. Coupled with a newer vaccine, this treatment could substantially reduce the incidence of malaria.

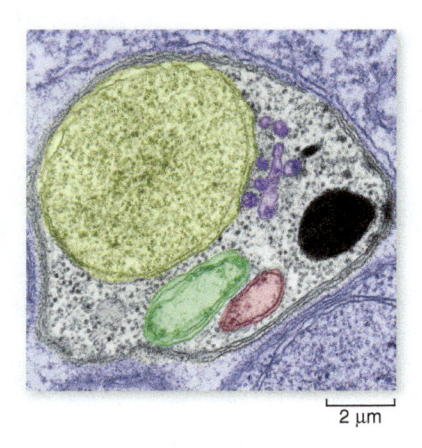

Figure 24.13 *Plasmodium* **apicoplast.** Drugs targeting enzymes used for fatty acid biosynthesis within *Plasmodium* apicoplasts (colored *dark green*) offer hope for treating malaria.

Chagas disease

Trypanosoma cruzi, an insect-borne protozoan, kills about 21,000 people in Central and South America each year. As many as 18 million suffer from this infection, called Chagas disease, the symptoms of which include damage to the heart and other internal organs. Genome sequencing of *T. cruzi* was completed in 2005.

A surprising and hopeful finding is that a common core of 6200 genes is shared among *T. cruzi* and two other insect-borne pathogens: *Trypanosoma brucei* and *Leishmania major*. *T. brucei* causes African sleeping sickness, and *L. major* infections result in lesions of the skin of the limbs and face. These core genes are being considered as possible targets for drug treatments.

Currently, no effective vaccines and only a few drugs with limited effectiveness are available to treat any of these diseases. The genomic similarities may aid not only in targeting drug development, but perhaps also result in a treatment or vaccine that is effective against all three devastating illnesses (figure 24.14).

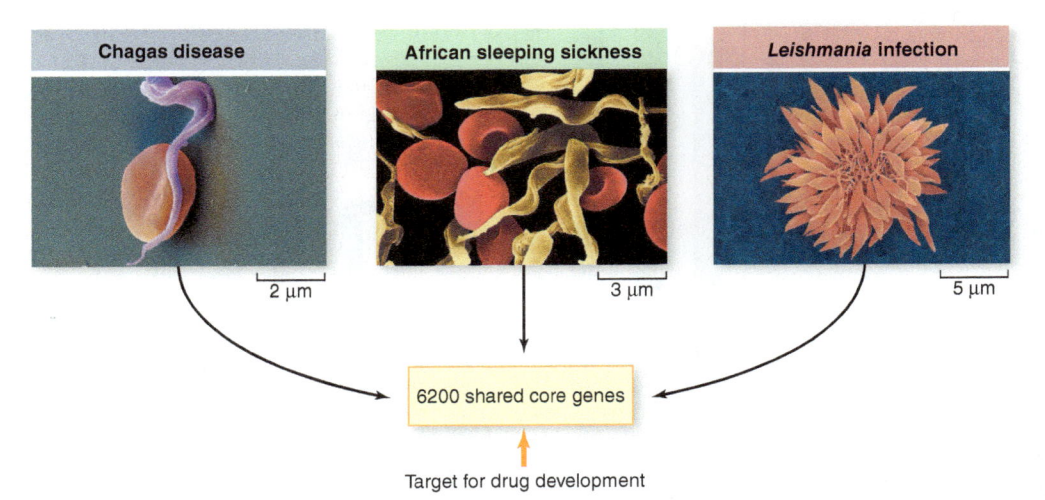

Figure 24.14 Comparative genomics may aid in drug development. The organisms that cause Chagas disease, African sleeping sickness, and leishmaniasis, which claim millions of lives in developing nations each year, share 6200 core genes. Drug development targeted at proteins encoded by the shared core genes could yield a single treatment for all three diseases.

Figure 24.15 DFTD is a fatal condition for Tasmanian devils. Tumors interfere with eating, and devils die within months of the first appearance of lesions.

Genome comparisons inform conservation biology

Genomes of species on the brink of extinction are being mined for information that could contribute to disease reduction and conservation efforts. Here we consider the application of genomics to three endangered species, the Tasmanian devil *(Sarcophilus harrisii)*, the giant panda *(Ailuropoda melanoleura)*, and the polar bear *(Ursus maritimus)*.

The Tasmanian devil, a marsupial on the Australian island of Tasmania, faces decimation from devil facial tumor disease (DFTD, figure 24.15). Close to 90% of the population is affected, and, since 1996, 60% of the devils have been wiped out due to DFTD. A comparison of genomes from two devils, named Cedric and Spirit, from distant corners of Tasmania revealed extremely low genetic diversity that can be traced back 100 years before the DFTD outbreak. Only the now extinct Tasmanian tiger had less genetic diversity (figure 24.16). Breeding programs can use the genomic information to preserve what diversity there is in the population.

Sequencing of the genome of the giant panda offered more promising news about population diversity (figure 24.16). Almost three times as many SNPs (2.7 million) were identified in the panda than for the devil. Bamboo is the mainstay of a panda diet, and habitat destruction is a factor in the decline of pandas. Curiously, the panda's relatives are carnivores, and pandas maintain the genes associated with carnivores. Furthermore, they lack the genes coding for enzymes needed to fully digest bamboo. Attention is now focused on the microbes in the panda's gut that digest bamboo. Collectively, this information may assist conservationists in keeping the population from dipping below its current level of 2500 to 3000 individuals.

An unexpected finding arose when the mitochondrial genomes of modern and fossil polar and brown bears (including DNA from the jaw of a 110,000- to 130,000-year-old polar bear) were sequenced. Polar bears evolved about 150,000 years ago. Geneticists were shocked to find that the entire maternal line of polar bears can be traced back to a brown bear living in Ireland between 20,000 and 50,000 years ago. Despite the close relation to brown bears and the ability to interbreed, this finding does not offer much hope for the polar bears facing extinction as warmer temperatures melt their sea ice homes. Average temperatures for the next 50 years are predicted to be much warmer than anything polar bears have experienced in their 150,000 years on Earth or the last 20,000 or so years since they mated with brown bears.

Learning Outcomes Review 24.5

DNA sequences conserved over evolutionary time tend to be those critical for protein function and survival. Variations in these conserved sequences may provide clues to diseases with a hereditary component. Knowledge of a pathogenic organism's genome and its differences from a host's genome may allow targeting of drugs and vaccines that affect the invader but leave the host unharmed. Comparative genomics within species can provide information of population diversity of endangered species. Conservation biologists can use this information to develop breeding strategies.

- *A pathogen makes a critical protein that differs from the human version by only seven amino acids. What approaches might lead to an effective drug against the pathogen? What drawbacks might be encountered?*

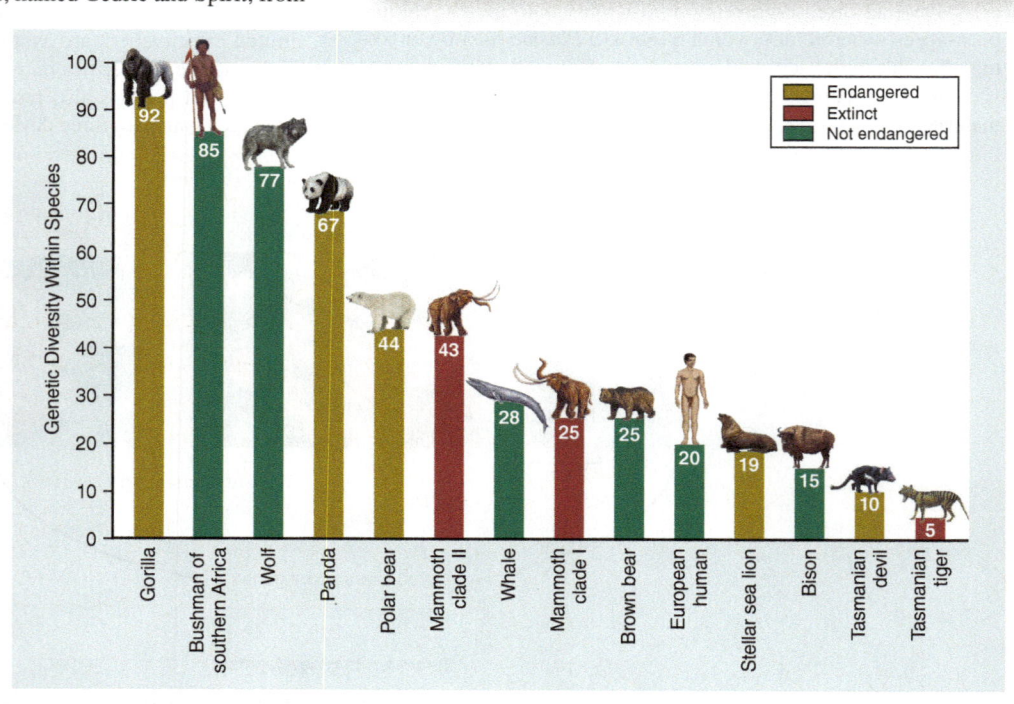

Figure 24.16 Genetic diversity of mitochondrial genomes based on average number of differences between pairs of individuals in the populations indicated.

Gorilla: 92 (Endangered)
Bushman of southern Africa: 85 (Not endangered)
Wolf: 77 (Not endangered)
Panda: 67 (Endangered)
Polar bear: 44 (Endangered)
Mammoth clade II: 43 (Extinct)
Whale: 28 (Not endangered)
Mammoth clade I: 25 (Extinct)
Brown bear: 25 (Not endangered)
European human: 20 (Not endangered)
Stellar sea lion: 19 (Not endangered)
Bison: 15 (Not endangered)
Tasmanian devil: 10 (Endangered)
Tasmanian tiger: 5 (Extinct)

Chapter Review

24.1 Comparative Genomics

Evolutionary differences accumulate over long periods.

Even distantly related species may often have many genes in common. Changes in DNA codons that do not alter the amino acid specified are termed synonymous changes.

The human genome shares similarities with existing and extinct primates.

A variety of comparisons of human and chimp genomes indicated differences in nucleotide substitutions, small insertions and deletions are on the order of about 1.5%. Comparison with the sequence of the extinct Neanderthal genome indicate that Europeans and Asians, but not Africans, have 1 to 4% of their genome from Neanderthals.

Genomes evolve at different rates.

Viral, bacterial, and even insect genomes evolve more rapidly than mammalian genomes. Plant genomes change more quickly than animal genomes, possibly as the result of massive genome remodeling from extensive transposition of mobile elements.

Plant, fungal, and animal genomes have unique and shared genes.

Plant genomes have changed much more rapidly than animal genomes. It appears about one-third of plant genes are unique to plants. Of the remaining plant genes, many are also found in animal and fungal genomes and are required for metabolism and gene expression.

24.2 Genome Size

Ancient and newly created polyploids guide studies of genome evolution.

Autopolyploidy results from an error in meiosis that leads to a duplicated genome; allopolyploidy is the result of hybridization between species (figure 24.3).

Evidence of ancient polyploidy is found in plant genomes.

Polyploidy has occurred numerous times in the evolution of flowering plants, and downsizing of genomes is common.

Polyploidy induces elimination of duplicated genes.

Downsizing of a polyploid genome can be caused by unequal loss of duplicated genes (figure 24.7).

Polyploidy can alter gene expression.

Polyploidization can lead to short-term silencing of genes via methylation of cytosines in the DNA.

Transposons are mobilized by polyploidization.

Transposons become highly active after polyploidization; their insertions into new positions may lead to new phenotypes.

Polyploidy alone cannot account for variation in genome size.

The vast range of ratios of gene number to genome size, even in closely related species, cannot be explained by polyploidy alone.

Noncoding DNA inflates genome size.

Genome size is most often inflated due to the presence of introns and non-protein-coding sequences. Genome size does not correlate with the number of genes. Some species have very little noncoding DNA, and others have extensive amounts of ncDNA, often retrotransposon DNA. Two species, such as rice and maize, can vary substantially in the amount of ncDNA and yet have similar numbers of genes.

24.3 Evolution Within Genomes

Individual chromosomes may be duplicated.

Aneuploidy, the duplication or loss of individual chromosomes, results from errors in meiosis. It is tolerated better in plants than in animals.

DNA segments may be duplicated (figure 24.8).

Paralogues are duplicated ancestral genes; orthologues are conserved ancestral genes. Duplicated DNA is common for genes associated with growth and development, immunity, and cell-surface receptors.

Genomes may become rearranged.

Genomes may be rearranged by moving gene locations within a chromosome or by the fusion of two chromosomes.

Conservation of synteny refers to preservation of long segments of ancestral chromosome sequences identifiable in related species (figure 24.10).

Gene inactivation results in pseudogenes.

Some ancestral genes become inactivated as they acquire mutations and are termed pseudogenes.

Rearranged DNA can acquire new functions.

Occasionally part of a gene can end up in a new spot in the genome where its function changes.

Noncoding DNA can acquire regulatory functions.

DNA that has no function in an organism can acquire mutations without any detrimental effect. However, conserved noncoding regions (CNCs) evolved more slowly than expected. These CNC regions appear to have gained functions that regulate the expression of neighboring protein-coding genes.

Horizontal gene transfer complicates matters.

Horizontal gene transfer creates many phylogenetic questions, such as the origins of the three major domains (figure 24.11).

24.4 Gene Function and Expression Patterns

Chimp and human gene transcription patterns differ.

Even when species have highly similar genes, expression of these genes may vary greatly. Posttranscriptional differences may also contribute to species differences.

Speech is uniquely human: An example of complex expression.

Small evolutionary changes in the FOXP2 protein and its expression may have led to human speech (figure 24.12).

24.5 Applying Comparative Genomics

Distantly related genomes offer clues for causes of disease.

Changes in amino acid sequences of critical proteins are a likely cause of diseases, and these differences can be identified by genome comparison.

Closely related organisms enhance medical research.

By comparing related organisms, researchers can focus on genes that cause diseases and devise possible treatments.

Pathogen–host genome differences reveal drug targets.

Analysis of the genomes of pathogenic organisms may provide new avenues of treatment and prevention (figure 24.14).

Genome comparisons inform conservation biology.

Comparative genomics within species can provide information of population diversity of endangered species. Conservation biologists can use this information to develop breeding strategies.

Review Questions

UNDERSTAND

1. Humans and pufferfish diverged from a common ancestor about 450 mya, and these two genomes have
   a. very few of the same genes in common.
   b. all the same genes.
   c. a large proportion of the genes in common.
   d. no nucleotide divergence.

2. Genome comparisons have suggested that mouse DNA has mutated about twice as fast as human DNA. What is a possible explanation for this discrepancy?
   a. Mice are much smaller than humans.
   b. Mice live in much less sanitary conditions than humans and are therefore exposed to a wider range of mutation-causing substances.
   c. Mice have a smaller genome size.
   d. Mice have a much shorter generation time.

3. Polyploidy in plants
   a. has only arisen once and therefore is very rare.
   b. only occurs naturally when there is a hybridization event between two species.
   c. is common, but never occurs in animals.
   d. is common, and does occur in some animals.

4. Homologous genes in distantly related organisms can often be easily located on chromosomes due to
   a. horizontal gene transfer.
   b. conservation of synteny.
   c. gene inactivation.
   d. pseudogenes.

5. All of the following are believed to contribute to genomic diversity among various species, *except*
   a. gene duplication.
   b. gene transcription.
   c. lateral gene transfer.
   d. chromosomal rearrangements.

6. What is the fate of *most* duplicated genes?
   a. Gene inactivation
   b. Gain of a novel function through subsequent mutation
   c. They are transferred to a new organism using lateral gene transfer.
   d. They become orthologues.

APPLY

1. Chimp and human DNA whole-genome sequences differ by about 1.23%. Determine which of the following explanations is most consistent with the substantial differences in morphology and behavior between the two species.
   a. It must be due largely to gene expression.
   b. It must be due exclusively to environmental differences.
   c. It cannot be explained with current genetic theory.
   d. The differences are caused by random effects during development.

2. You are offered a summer research opportunity to investigate a region of ncDNA in maize. A friend politely smiles and says that only graduate students get to work on the coding regions of DNA. How would you critique your friend's statement?
   a. The friend has a point; ncDNA is "junk" DNA and therefore not very important.
   b. The ncDNA produces protein through mechanisms other than transcription.
   c. Most ncDNA is usually translated.
   d. Often ncDNA produces RNA transcripts that themselves have regulatory function.

3. Analyze the conclusion that the *Medicago truncatula* genome has been downsized relative to its ancestral legume, and circle the evidence that is consistent with this conclusion.
   a. *Medicago* has a proportional decrease in the number of genes.
   b. *Medicago* has a proportional increase in the number of genes.
   c. *Medicago* has an increase in the amount of DNA.
   d. *Medicago* has a decrease in the amount of DNA.

4. Analyze why an herbicide that targets the chloroplast is effective against malaria.
   a. Because *Plasmodium* needs a functional apicoplast
   b. Because the main vector for malaria is a plant
   c. Because mosquitoes require plant leaves for food
   d. Because *Plasmodium* mitochondria are very similar to chloroplasts

SYNTHESIZE

1. The *FOXP2* gene is associated with speech in humans. It is also found in chimpanzees, gorillas, orangutans, rhesus macaques, and even the mouse, yet none of these mammals speak. Develop a hypothesis that explains why *FOXP2* supports speech in humans but not other mammals.

2. One of the common misconceptions about sequencing projects (especially the high-profile Human Genome Project) is that creating a complete road map of the DNA will lead directly to cures for genetically based diseases. Given the percentage similarity in DNA between humans and chimps, is this simplistic view justified? Explain.

3. How does horizontal gene transfer (HGT) complicate phylogenetic analysis?

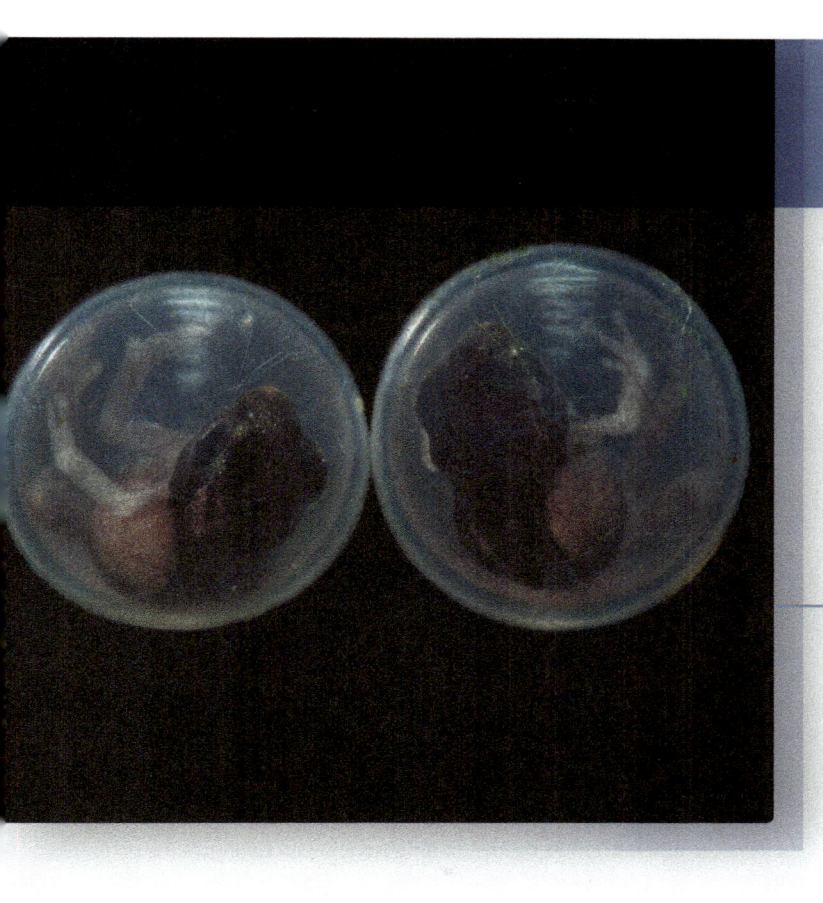

CHAPTER 25

Evolution of Development

Chapter Contents

25.1 Evolution of Developmental Patterns

25.2 Single-Gene Changes and the Alteration of Form and Function

25.3 Different Ways to Evolve the Same Structure

25.4 Diversity of Eyes in the Natural World: A Case Study

Introduction

How is it that closely related species of frogs can have completely different patterns of development? One frog goes from fertilized egg to adult frog with no intermediate tadpole stage; as shown above, the embryo develops directly into the same body form as the adult. The sister species has an extra developmental stage neatly slipped in between early development and the formation of limbs—the tadpole stage. The answer to this and other such evolutionary differences in development that yield novel phenotypes are now being investigated with modern genetic and genomic tools. Research findings are accentuating the biological puzzle that many developmental genes are highly conserved even among very different organisms. In this chapter, we explore the emerging field of evolution of development, a field that brings together previously distinct fields of biology.

25.1 Evolution of Developmental Patterns

Learning Outcomes

1. Explain how the same gene can produce different morphologies in different species.
2. Identify types of genes most likely to affect morphology.
3. Evaluate the limitations of comparative genomics in exploring the evolution of development.

Ultimately, evolutionary change occurs by changes in development—closely related species differ because of genetic changes that lead to differences in how they develop, producing different phenotypes.

Phenotypic diversity could either result from changes in protein-coding regions of many different genes or could be explained by changes in a much smaller set of genes that regulate the expression of protein-coding genes. The latter is true for two sea urchin species.

Closely related sea urchins have been discovered that have very distinctive developmental patterns (figure 25.1). One species exhibits an intermediate developmental stage, a free-swimming larva called a *pluteus*. The direct-developing urchin never makes a

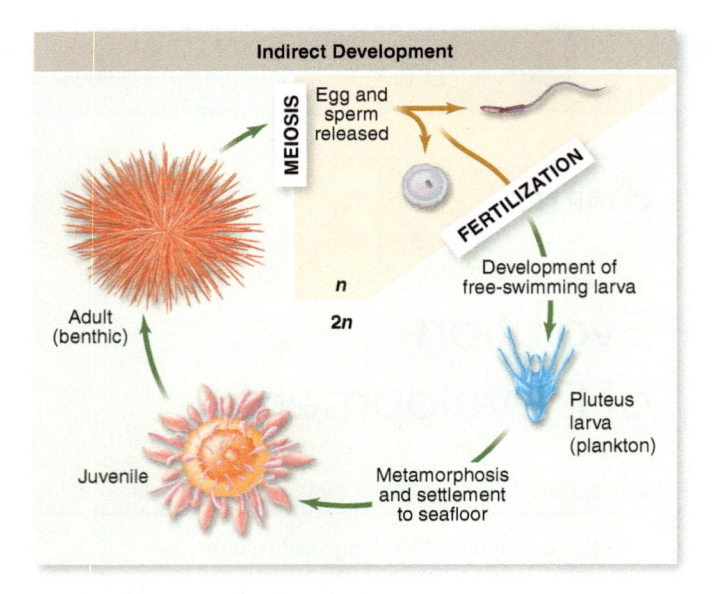

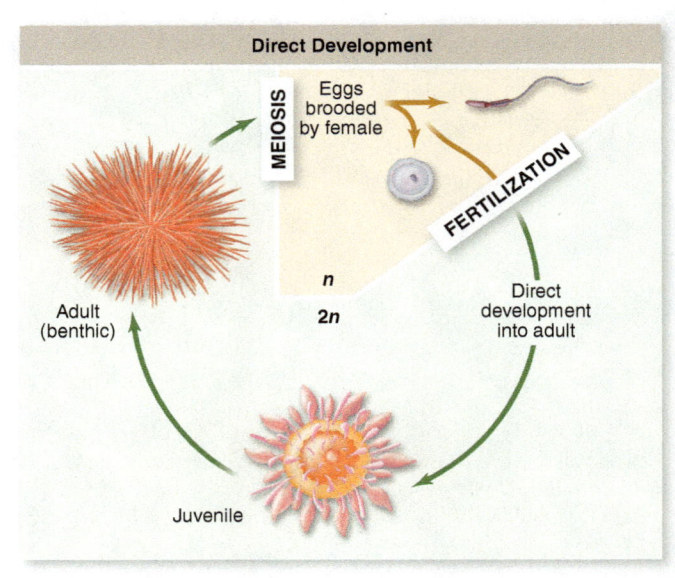

Figure 25.1 Direct and indirect sea urchin development. Phylogenetic analysis shows that a developmental sequence that includes a pluteus larval stage was the ancestral state. Direct-developing sea urchins have lost this intermediate stage of development.

pluteus larva—it just jumps ahead to its adult form. One possible explanation is that the two forms have different developmental genes—one species with genes that produce the larva, the other without—but it turns out that this is not the case. Instead, the two forms have the same sets of genes, but differ dramatically in patterns of developmental gene expression, even though their adult form is nearly the same. In this case, patterns of gene expression have changed.

Highly conserved genes produce diverse morphologies

Transcription factors and genes involved in signaling pathways are responsible for coordination of development. As you saw in chapter 9, key elements of kinase and G protein–signaling pathways are also highly conserved among organisms. Even subtle changes in a signaling pathway can alter the enzyme that is activated or repressed, the transcription factor that is activated or repressed, or the activation or repression of gene expression. Any of these changes can have dramatic effects on the development of an organism.

A relatively small number of gene families, about two dozen, regulate animal and plant development. The developmental roles of several of these families, including *Hox* gene transcription factors, are described in chapter 19.

Hox (homeobox) genes appeared before the divergence of plants and animals; in plants, they have a role in shoot growth and leaf development, and in animals they establish body plans. These genes code for proteins with a highly conserved homeodomain that binds to the regulatory region of other genes to activate or repress these genes' expression. *Hox* genes specify when and where genes are expressed.

Another family of transcription factors, box genes, are found throughout the eukaryotes. The *MADS* box also codes for a DNA-binding motif. Large numbers of *MADS* box genes establish the body plan of plants, especially the flowers. Although the *MADS* box region is highly conserved, variation exists in other regions of the coding sequence. Later in this chapter, we consider how there came to be so many *MADS* box genes in plants and how such similar genes can have very different functions.

Developmental mechanisms evolve

Understanding how development evolves requires integration of knowledge about genes, gene expression, development, and evolution. Either transcription factors or signaling molecules can be modified during evolution, changing the timing or position of gene expression and, as a result, gene function.

Heterochrony

Alterations in timing of developmental events due to a genetic change are called **heterochrony.** A heterochronic mutation could affect a gene that controls when a plant transitions from the juvenile to the adult stage, at which point it can produce reproductive organs. A mutation in a gene that controls flowering time in plants can result in a small plant that flowers quickly rather than requiring months or years of growth before it flowers.

Most mutations that affect developmental regulatory genes are lethal, but every so often a novel phenotype emerges that persists because of increased fitness. If a mutation leading to early flowering increased the fitness of a plant, the new phenotype will persist. For example, a tundra plant that flowers earlier, enabling it to be fertilized and set seed quickly, could have increased fitness over an individual of the same species that flowers later, just as the short summer comes to a close.

Homeosis

Alternations in the spatial pattern of gene expression can result in *homeosis*, which occurs when one body segment takes on the identity of another. A four-winged *Drosophila* fly is an example of a change in gene expression pattern with a dramatic effect on morphology—the production of a second set of large wings on a body segment that usually does not have large wings (homeotic mutations are discussed in chapter 19). Mutations in three genes in the *Bithorax* complex are required to produce this phenotype, which resembles more ancestral insects with four rather than two wings (see figure 19.17).

The *Drosophila Antennapedia* mutant, which has a leg on its head where an antenna should be, is another example of a homeotic mutation. Mutations in genes such as *Antennapedia* can arise spontaneously in the natural world or by mutagenesis in the laboratory, but their bizarre phenotypes would have little survival value in nature.

Although neither of these examples produces a phenotype that is viable, it is easy to see how occasionally such a mutation could produce a new phenotype that provides a fitness advantage and thus would evolve by natural selection.

Changes in transcription

The timing or location of gene expression can be modified in several ways, giving rise to heterochrony or homeosis. The coding sequence of a gene can contain multiple regions with different functions (figure 25.2). The DNA-binding motifs, exemplified by *MADS* box and *Hox* genes, could be altered so that they no longer bind to their target genes; as a result, that developmental pathway would cease to function. Alternatively, the modified transcription factor might bind to a different target and initiate a new sequence of developmental events.

The regulatory region of a gene encoding a transcription factor may also be altered. A sequence change in the promoter could prevent transcription of the transcription factor gene. Alternatively a changed regulatory region might bind a different transcription factor. In this case, the downstream targets would be the same, but the cells that express the target genes or the time at which these target genes are expressed could change.

Changes in signaling pathways

Signaling pathways help cells coordinate information about neighboring cells and the external environment that is essential for successful development. If the structure of a ligand changes, it may no longer bind to its target receptor; it could bind to a different receptor or no receptor at all. If, as a result of a genetic change, a receptor is produced in a different cell type, a homeotic phenotype may appear. And, as mentioned earlier, small changes in signaling molecules can alter their targets.

The remainder of this section provides specific examples of the evolution of diverse morphologies. For each example, consider how the mechanism of development has been altered and what the outcome is. Keep in mind that these are the successful examples; most morphological novelties that arise quickly, but do not improve fitness, are quickly eliminated from the gene pool.

Understanding evolution of development requires functional analysis

Comparative genomics is amazingly useful for understanding morphological diversity. Limitations exist, however, in the inferences

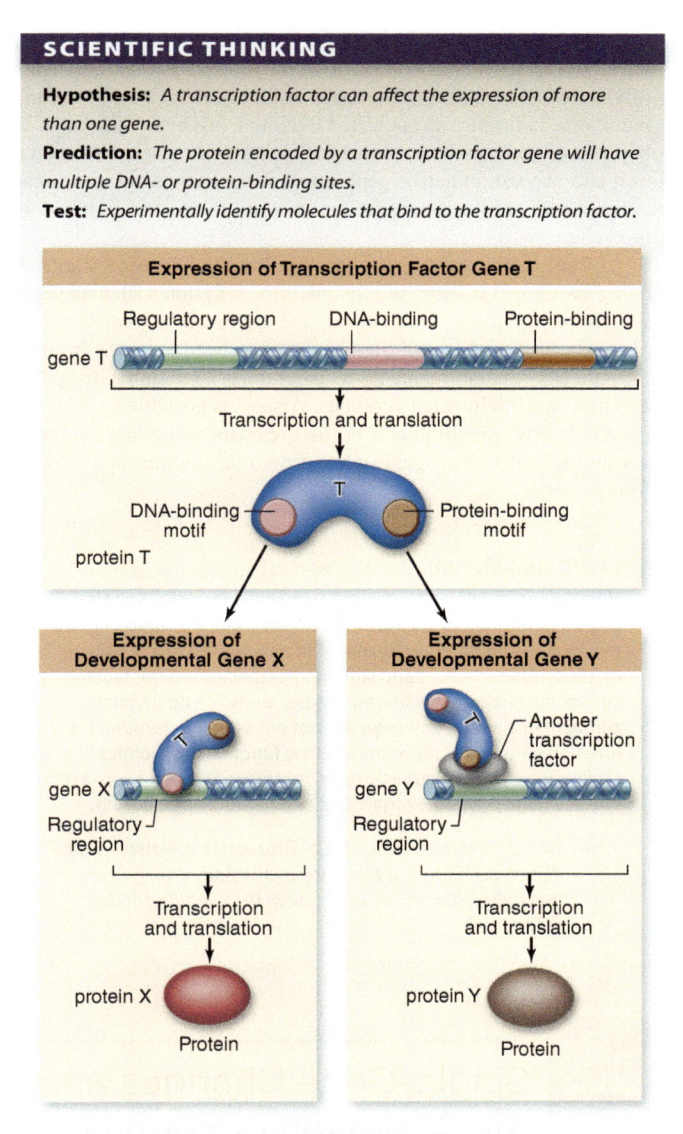

SCIENTIFIC THINKING

Hypothesis: *A transcription factor can affect the expression of more than one gene.*
Prediction: *The protein encoded by a transcription factor gene will have multiple DNA- or protein-binding sites.*
Test: *Experimentally identify molecules that bind to the transcription factor.*

Result: *This transcription factor has a site that binds to the regulatory region of a gene and a site that binds to a transcription factor that regulates expression of a second gene.*
Conclusion: *A single transcription factor can regulate the expression of more than one gene.*
Further Experiments: *Determine the specific developmental role of each binding domain by creating mutations in regions of the gene that code for specific binding sites in the protein.*

Figure 25.2 Transcription factors have a key role in the evolution of development.

about evolution of development that we can draw from sequence comparison alone. *Functional genomics* includes a range of experiments designed to test the actual function of a gene in different species as explained in chapter 18.

Sequence comparisons among organisms are essential for both phylogenetic and comparative developmental studies. Rapidly evolving research using bioinformatics, which utilizes computer programming to analyze DNA and protein data, leads to hypotheses that can be tested experimentally. A single base mutation can change an active gene into an inactive pseudogene, and experiments are necessary to demonstrate the actual function of the gene.

Tools for functional analysis exist in model systems and are being developed in other organisms. Model systems such as yeast, the flowering plant *Arabidopsis,* the nematode *Caenorhabditis elegans, Drosophila,* and the mouse have been selected because they are easy to manipulate in the laboratory, have short life cycles, and have well-delineated genomes. Also, it is possible to visualize gene expression within parts of the organism using labeled markers and to create transgenic organisms that contain and express foreign genes.

Learning Outcomes Review 25.1

Highly conserved genes can undergo small changes in their coding or regulatory regions that alter the place or time of gene expression and function, resulting in new body plans. Changes in transcription factors and signaling pathways are the most common source of new morphologies. Genetic and genomic comparisons alone, however, are not enough to determine the function of genes in different species; functional genomics studies whether conserved genes operate in the same way across species utilizing model organisms and genetic engineering.

- Two closely related species of Drosophila in Hawaii can be distinguished by the presence of one pair of wings versus two pairs. How would you explain the evolution of this difference?

25.2 Single-Gene Changes and the Alteration of Form and Function

Learning Outcomes

1. Explain how a small number of mutations can give rise to new morphologies and even new species.
2. Explain how a gene could acquire a new function.
3. Describe how duplicated genes could give rise to new functions in an organism.

In chapter 24, we discussed the similarity between human and mouse genomes. If all but 300 of the approximately 20,000 human genes are shared with mice, why are mice and humans so different? Part of the answer is that genes with similar sequences in two different species may work in slightly or even dramatically different ways. Here we explore several examples of single-gene mutations that altered the form and function of particular species

of plants and animals to better understand how such small changes can sometimes have such dramatic effects on the overall body plan of an organism.

Cauliflower and broccoli began with a stop codon

The species *Brassica oleracea* is particularly fascinating because individual members can have extraordinarily diverse phenotypes (figure 25.3). Wild cabbage, kale, tree kale, red cabbage, green cabbage, brussels sprouts, broccoli, and cauliflower are all members of the same species. Some flower early, some late. Some have long stems, others have short ones. Some form a few flowers, and others, like broccoli and cauliflower, initiate many flowers, but development of the flowers is arrested, forming the large head you're familiar with from your dinner plate. How can one species produce such a diversity of phenotypes?

One piece of the puzzle lies with the gene *CAL (Cauliflower),* which was first cloned in a close *Brassica* relative, *Arabidopsis.* In combination with another mutation, *Apetala1, Arabidopsis* plants can be turned from plants with a limited number of simple flowers into miniature broccoli or cauliflower plants with masses of arrested flower meristems or flower buds. These two genes are needed for the transition to making flowers and arose through duplication of a single ancestral gene within the brassica group. When they are absent, meristems continue to make branches, but are delayed in producing flowers.

The *CAL* gene was cloned from large numbers of *B. oleracea* subspecies, and a stop codon, TAG, was found in the middle of the

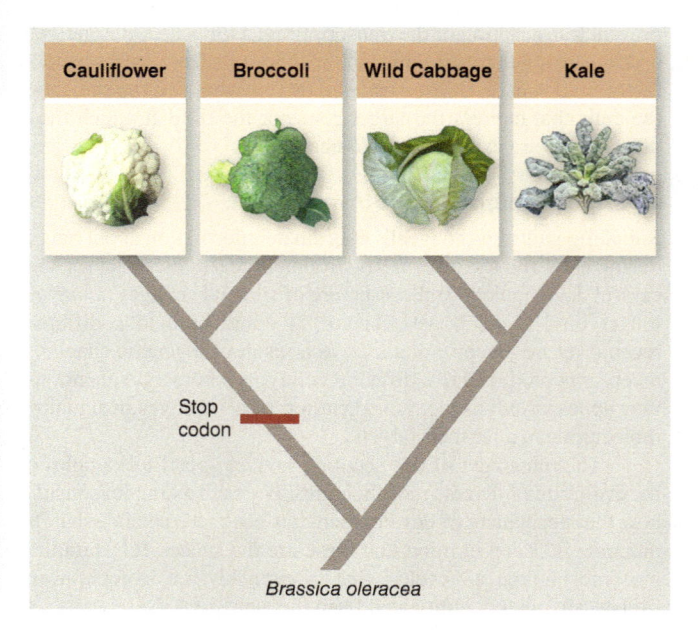

Figure 25.3 Evolution of cauliflower and broccoli.
A point mutation that converted an amino acid–coding region into a stop codon resulted in the extensive reproductive branching pattern that was artificially selected for in cauliflower and broccoli. All four vegetables in this figure are members of the same species, *Brassica oleracea.*

CAL coding sequences of broccoli and cauliflower. A phylogenetic analysis of *B. oleracea* coupled with the *CAL* sequence analysis leads to the conclusion that this stop codon appeared after the ancestors of broccoli and cauliflower diverged from other subspecies members, but before broccoli and cauliflower diverged from each other (figure 25.3).

> **Inquiry question** Knowing that cauliflower and broccoli have a stop codon in the middle of the *CAL* gene-coding sequence, predict the wild-type function of *CAL*. What additional evolutionary events may have occurred since broccoli and cauliflower diverged?

A second, somewhat unusual feature of this example is that the driving selective force for this diversity was artificial. Wild relatives are still found scattered along the rocky coasts of Spain and the Mediterranean region. The most likely scenario is that humans found a *cal* mutant and selected for that phenotype through cultivation. The large heads of broccoli and cauliflower offer a larger amount of a vegetable material than the wild kale plants and a tasty alternative to *Brassica* leaves.

Cichlid fish jaws demonstrate morphological diversity

Single-gene mutations can lead to rapid evolutionary change. Here we consider modifications in form and function from adaptive radiation of cichlid fish in Lake Malawi in East Africa. In less than a few million years, hundreds of species have evolved in the lake from a common ancestor. The great diversity that has been produced in this radiation is illustrated in figure 22.16.

One possible explanation for the evolutionary exuberance of these fish is that different species' feeding habits have led to different morphologies. There are bottom feeders, biters, and rammers. The rammers have particularly long snouts with which to ram their prey; biters have an intermediate snout; and the bottom feeders have short snouts adapted to scrounging for food at the base of the lake (figure 25.4).

How did these fish acquire such different snout forms? An extensive genetic analysis revealed that a small number of genes are likely responsible for the shape and size of the jaw. Two of these genes in particular play an important role in determining jaw length and height. Presumably, mutations in these genes in the ancestral cichlid allowed individuals to access food in different parts of the habitat, leading to evolutionary divergence and adaptive radiation.

Stickleback fish lose their armor with a single mutation

The freshwater three-spine stickleback fish, *Gasterousteus aculeatus,* originated after the last ice age from marine populations with bony plates that protect the fish from predators. Freshwater populations, subject to less predation, have lost their bony armor. The *Ectodysplasin (Eda)* gene is one of a few associated with reduced armor in freshwater three-spine sticklebacks. The *Eda* allele that causes reduced armor, however, is not the result of a new mutation that occurred in freshwater populations. Rather, the mutation originated about 2 MYA in marine sticklebacks and persists with a frequency of about 1% in marine environments. Why natural selection has not completely eliminated the allele in marine environments is unknown, but what clearly has happened is that in freshwater environments, this allele is beneficial and thus has evolved to high frequency as a result of natural selection.

To test the hypothesis that freshwater fish with the low armor allele have higher fitness, marine sticklebacks that were heterozygous for the *Eda* allele were moved to four freshwater environments and allowed to breed. Positive selection for the reduced armor allele was observed and was correlated with longer length in juvenile fish, likely because fewer resources were allocated to armor development, which allowed the fish to grow more rapidly.

Ancestral genes may be co-opted for new functions

The evolution of chordates can partially be explained by the co-option of an existing gene for a new function. Ascidians are basal chordates that have a notochord, but no vertebrae (see chapter 35). The *Brachyury* gene of ascidians encodes a

Figure 25.4 Diversity of cichlid fish jaws. A difference in two genes is responsible for a short snout in *Labeotropheus fuelleborni* and a long snout in *Metriaclima zebra*. Genes that affect jaw length can affect body shape as well because of the constraints the size of the jaw places on muscle development.

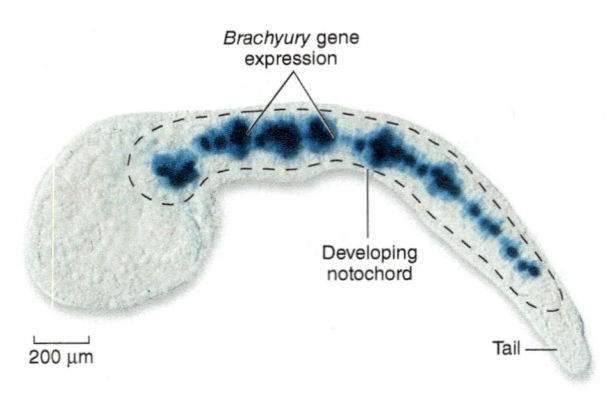

Figure 25.5 Co-opting a gene for a new function. *Brachyury* is a gene found in invertebrates that has been used for notochord development in this ascidian, a basal chordate. By attaching the *Brachyury* promoter to a gene with a protein product that stains blue, it is possible to see that *Brachyury* gene expression in ascidians is associated with the development of the notochord, a novel function compared with its function in organisms lacking a notochord. For example, the homologous gene in the nematode *Caenorhabditis elegans* is important for hindgut and male tail development.

transcription factor and is expressed in the developing notochord (figure 25.5).

Brachyury is not a novel gene that appeared as vertebrates evolved. It is also found in invertebrates. For example, a mollusk homologue of *Brachyury* is associated with anterior–posterior axis specification. Most likely, an ancestral *Brachyury* gene was co-opted for a new role in notochord development in chordates.

Brachyury is a member of a gene family with a specific sequence motif—that is, a conserved sequence of base-pairs within the gene. A region of *Brachyury* encodes a protein domain called the **T box,** which is a transcription factor. So, *Brachyury*-encoded protein turns on a gene or genes. The details of which genes are regulated by *Brachyury* are only now being discovered.

How a single genetic toolkit can be used to build an insect, a bird, a bat, a whale, or a human is an intriguing puzzle for evolutionary developmental biologists, exemplified with the *Brachyury* gene. One explanation is that *Brachyury* turns on different genes or combinations of genes in different animals. Although there are not yet enough data to sort out the details of *Brachyury,* we can look at limb formation for an explanation of how such a change could have evolved.

Limbs have developed through modification of transcriptional regulation

Most tetrapods have four limbs—two hindlimbs and two forelimbs—although two or more limbs have been lost many times in lizards, including snakes (which in evolutionary terms are lizards). The wing in a bird is actually a forelimb. Our arm is a forelimb. Clearly, these are two very different structures, but they have a common evolutionary origin. As you learned in chapter 23, these are termed *homologous structures*.

At the genetic level, humans and birds both express the *Tbx5* gene in developing forelimb buds and *Tbx4* in hindlimbs. Like *Brachyury*, *Tbx5* is a member of a transcription factor gene family with T box motif—that is, a conserved sequence of base-pairs within the gene. *Tbx5*-encoded protein turns on *Fibroblast growth factor-10* (Fgf10) that is needed to make a limb. Mutations in the human *Tbx5* gene cause Holt–Oram syndrome, resulting in forelimb and heart abnormalities.

The link between *Tbx5*, which initiates forelimb development, and human heart development can be traced back to limbless amphioxus, a chordate that lacks vertebrates. Amphioxus has a homologue, *AmphiTbx4/5*, expressed in the heart region (figure 25.6).

The evolutionary tale of *Tbx5* is one of both gene co-option and duplication. Two whole-genome duplications accompanied the emergence of the vertebrates, and the duplicated *AmphiTbx4/5* gave rise to both *Tbx5* and *Tbx4,* which were co-opted for forelimb and hindlimb development, respectively. Two scenarios for the evolution of *Tbx5* are possible. The coding region could be modified so the transcription factor interacted with other genes. Or, the regulatory region could be altered. To distinguish between the two possibilities, transgenic mice were made using the *AmphiTbx4/5* gene.

If the regulatory, not the coding sequence, evolved a new function, then swapping the mouse *Tbx5* coding region with the amphioxus *AmphiTbx4/5* coding region should not affect forelimb development in the mouse. Transgenic mice with a *Tbx5* regulatory region and an *AmphiTbx4/5* coding region form forelimbs (figure 25.6). A second experiment is needed to determine if the regulatory region changed in the 520 million years since amphioxus and mice last shared a common ancestor. When the amphioxus *AmphiTbx4/5* regulatory region is swapped with the mouse *Tbx5* regulatory region, no forelimbs develop although the *Tbx5* coding region is present (figure 25.6). This experiment demonstrates that the key vertebrate innovation was a new regulatory region, specific to the forelimb area in the case of *Tbx5*.

Changes in gene regulation also explain the evolution of digit formation of limbs. The two whole genome duplications in early vertebrate evolution gave rise to four clusters of *Hox* genes, with *Hoxc* and *Hoxd* arising from a common ancestor in the second duplication. Of the two, only *Hoxd* contributes to limb development, specifically digit formation. The HOXD12 protein is expressed in the forelimb, but not the HOXC12 protein. To sort out the role of regulatory elements, the entire upstream chromosomal regions of mouse *Hoxc* and *Hoxd* were swapped. In mice with the shuffled genes, the HOXC12 protein was able to replace the function of HOXD12 in digit formation and rescue forelimb development (figure 25.7).

In the case of both *Tbx5* and *Hoxd12*, the whole-genome duplications in early vertebrate development yielded duplicate genes with redundant function upon which natural selection could act. In both cases, regulatory regions rather than coding regions experienced change essential for limb formation.

Gene duplications provide opportunities for new gene functions

The analysis of the evolution of limb development emphasized how changes in gene regulatory sequences can lead to new gene

SCIENTIFIC THINKING

Background: AmphiTbx4/5 gave rise to the vertebrate Tbx4 and Tbx5 after whole-genome duplication. The three DNA sequences are very similar.
Hypothesis 1: Tbx4 and Tbx5 were co-opted for limb development in vertebrates.
Hypothesis 2: Tbx5 underwent changes in its regulatory rather than coding DNA.
Prediction 1: AmphiTbx4/5 will be expressed in the heart region of amphioxus and Tbx5 and Tbx4 in the forelimb and hindlimb, respectively, of a mouse.
Experiment 1: Transgenic amphioxi and mice are made with AmphiTbx4/5, Tbx5, and Tbx4 promoters placed in front of a β-galactosidase gene that will enzymatically convert an added stain blue in cells where it is expressed.

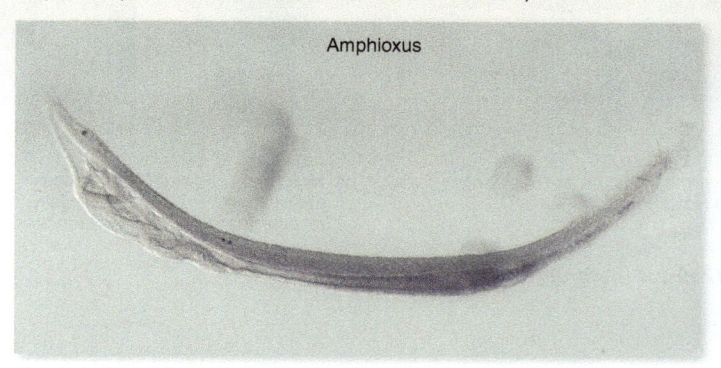

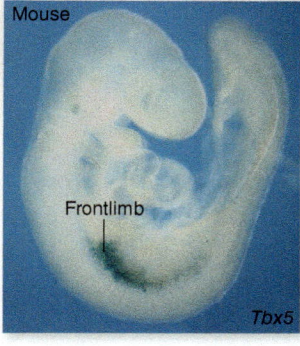

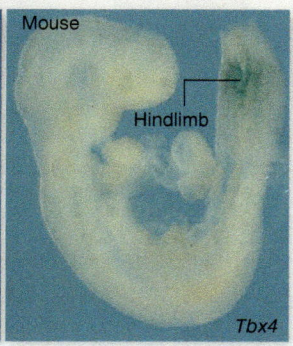

Result 1: Expression patterns are as predicted.

Prediction 2: AmphiTbx4/5 will support forelimb bud development in mice missing the Tbx5k coding region, if it is under regulatory control of Tbx5 but not AmphiTbx4/5.
Experiment 2: Transgenic mice containing AmphiTbx4/5 and the regulatory but not coding region of Tbx5 and transgenic mice with AmphiTbx4/5 and its 70-kb upstream and downstream regulatory regions, but no Tbx5 sequences were constructed and allowed to develop.

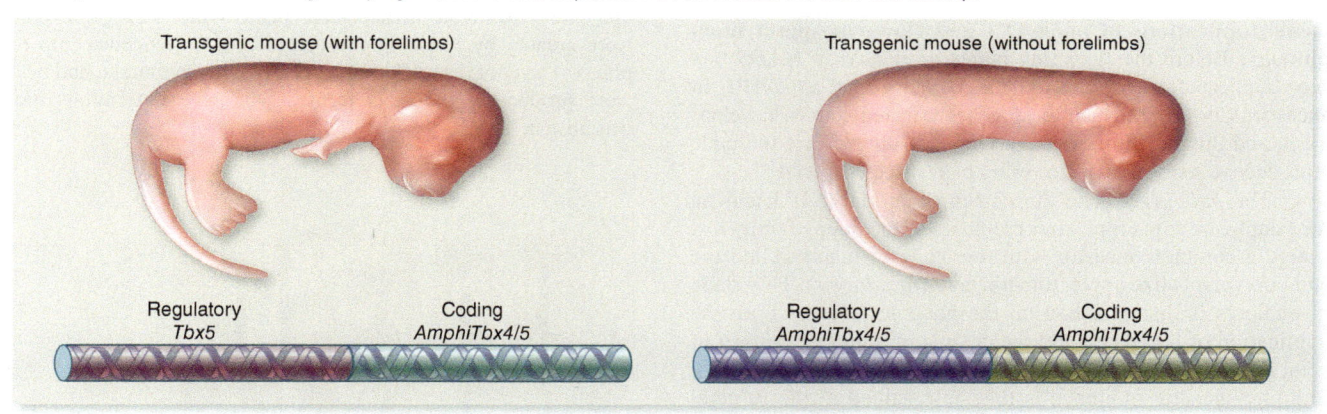

Result 2: AmphiTbx4/5 supports limb development under Tbx5, but not AmphiTbx4/5 regulatory control.
Conclusion: New regulatory regions for Tbx5 and Tbx4 lead to limb bud formation.

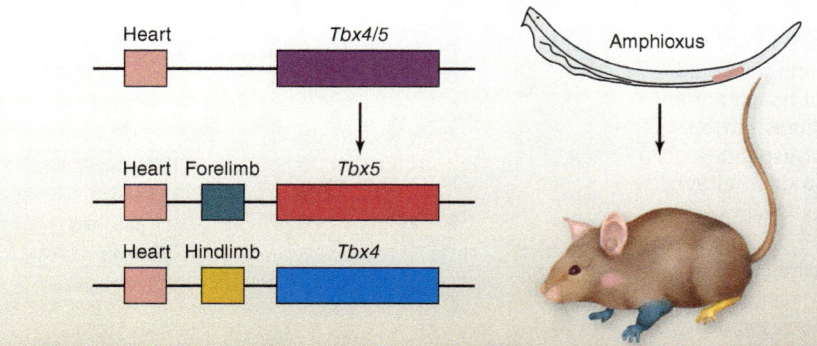

Further Experiments: Identify the regulatory region that results in Tbx5 expression in the forelimb. How would you design these experiments?

Figure 25.6 Amphioxus heart gene co-opted for limb development in vertebrates through changes in gene regulation.

a. **Wild-type mouse:** *Hoxd12* supports forelimb digit development and *Hoxc12* genes are not expressed.

b. Mouse forelimb digits are aberrant when *Hoxc* and *Hoxd* genes are absent.

c. *Hoxd* and *Hoxc* regulatory regions are swapped. *Hoxc12* can partially rescue *Hoxd12* activity in the forelimb.

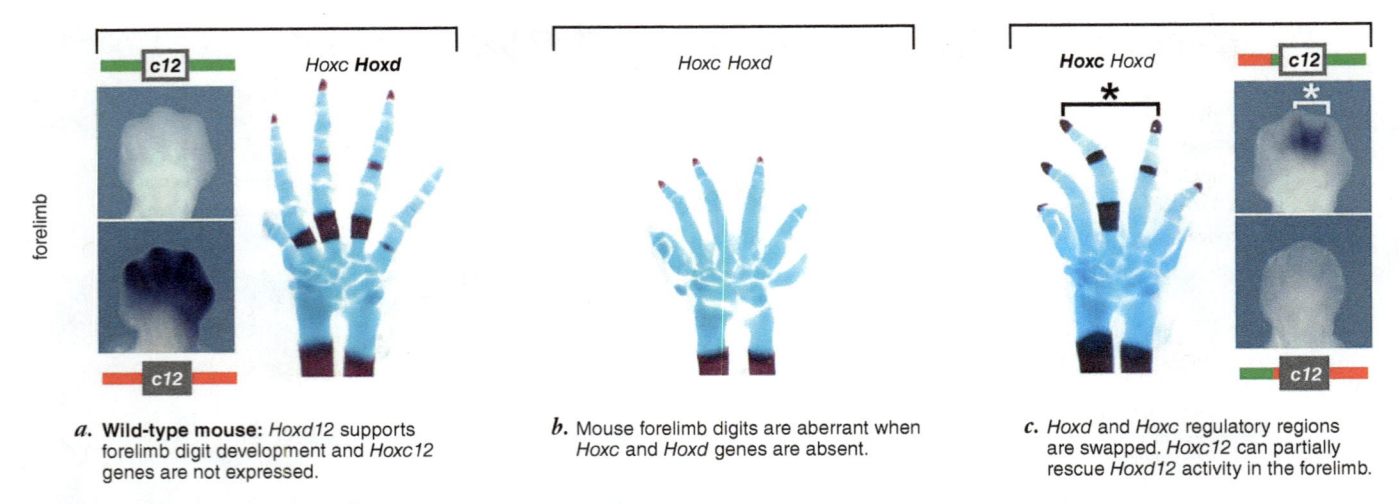

Figure 25.7 *Hoxc12* can partially substitute for *Hoxd12* in forelimb digit development.

functions. It is also an example of the importance of gene duplication in the evolution of new gene functions. A duplicate gene provides a backup gene that can mutate without being lethal to the organism. In this section, we explore a specific example of the evolution of development through gene duplication and divergence in flower form.

Gene duplications of *paleoAP3* and flowering-plant morphology. Before the flowering plants originated, a *MADS* box gene duplicated, giving rise to genes called *PI* and *paleoAP3*. In ancestral flowering plants, these genes affected stamen development, and this function has been retained (stamens are the male reproductive structures of flowering plants; see chapter 41).

The *paleoAP3* gene duplicated to produce *AP3* and an *AP3* duplicate some time after members of the poppy family last shared a common ancestor with the clade of plants called the eudicots (plants like apple, tomato, and *Arabidopsis*). This clade of eudicots is distinguished on the genome level by both the duplication of *paleoAP3* and the origins of a precise pattern of petal development in their last common ancestor (figure 25.8). The phylogenetic inference is that *AP3* gained a role in petal development.

Alteration in gene divergence of *AP3* function and controlling petal development. Although the occurrence of *AP3* through duplication corresponds with a uniform developmental process for specifying petal development, the correlation could simply be coincidental. Experiments that mix and match parts of the *AP3* and *PI* genes, and then introduce them into *ap3* mutant plants, confirm that the phylogenetic correspondence is not a coincidence. The *ap3* plants do not produce either petals or stamens. A summary of the experiments is shown in figure 25.9.

In section 25.1, the *MADS* box transcription factor gene family was introduced. One region of a *MADS* gene codes for a DNA-binding motif; other regions code for different functions, including protein–protein binding. The PI and AP3 proteins can bind to each other and, as a result, can regulate the transcription of genes needed for stamen and petal formation.

Both AP3 and PI have distinct sequences at the C (carboxy) protein terminus (coded for by the 3′ end of the genes). The C-terminus sequence of the AP3 protein is essential for specifying petal function, and it contains a conserved sequence shared among the eudicots. The *AP3* C-terminus DNA sequence was deleted from the wild-type gene, and the new construct was inserted into *ap3* plants to create a transgenic plant. Other transgenic plants were created by inserting the complete *AP3* sequence into *ap3* plants. The complete *AP3* sequence rescued the mutant, and petals were produced. No petals formed when the C-terminus motif was absent.

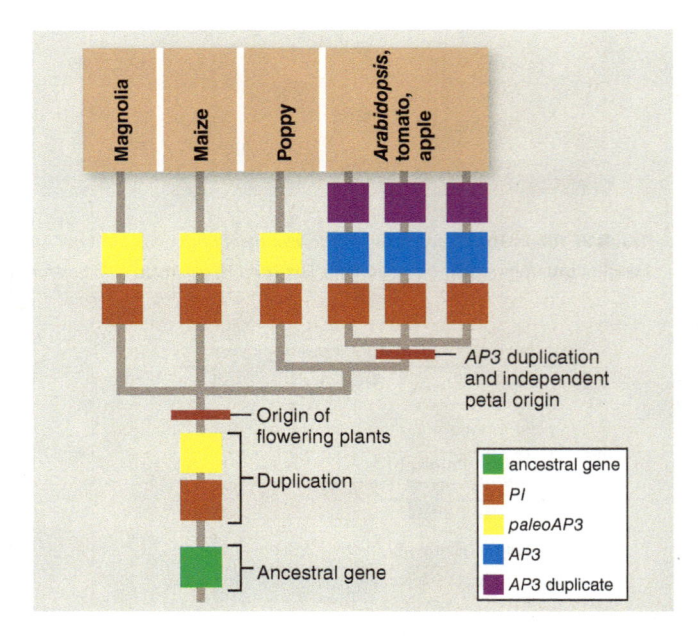

Figure 25.8 Petal evolution through gene duplication. Two gene duplications resulted in the *AP3* gene in the eudicots that has acquired a role in petal development.

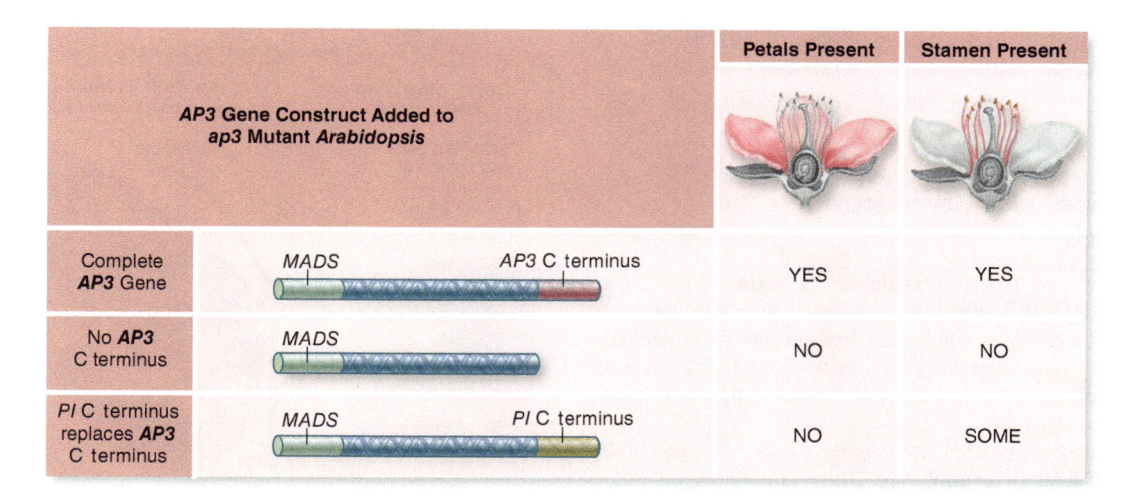

AP3 Gene Construct Added to ap3 Mutant *Arabidopsis*		Petals Present	Stamen Present
Complete *AP3* Gene	MADS — AP3 C terminus	YES	YES
No *AP3* C terminus	MADS	NO	NO
PI C terminus replaces *AP3* C terminus	MADS — PI C terminus	NO	SOME

Figure 25.9 *AP3* has acquired a domain necessary for petal development. The *AP3* gene includes *MADS* box encoding a DNA-binding domain and a highly specific sequence near the C terminus. Without the 3' region of the *AP3* gene, the *Arabidopsis* plant will not make petals.

AP3 is also needed for stamen development, an ancestral trait found in *paleoAP3*. Plants lacking *AP3* fail to produce either stamens or petals. Transgenic plants with the C-terminus deletion construct also failed to produce stamens.

The *pi* mutant phenotype also lacks stamens and petals. To test whether the *PI* C terminus could substitute for the *AP3* C terminus in specifying petal formation, the *PI* C terminus was added to the truncated *AP3* gene. No petals formed, but stamen development was partially rescued. These experiments demonstrate that *AP3* has acquired an essential role in petal development, encoded in a sequence at the 3' end of the gene.

? Inquiry question Explain how functional analysis was used to support the claim that petal development evolved through the acquisition of petal function in the *AP3* gene of *Arabidopsis*.

Learning Outcomes Review 25.2

Although most mutations are lethal, some confer a fitness advantage. These may consist of very small mutations, such as a change to a single codon, that have large effects on development and morphology. During the long course of evolution, genes have been co-opted for new functions. A change in the protein-coding region of a transcription factor can change the genes it can bind to and regulate. A change in the regulatory region of a gene can change where or when that gene is expressed, which can lead to altered morphology. Gene duplication allows for divergence that can lead to novel function.

■ *A marine three-spine stickleback fish is mated with a freshwater three-spine with reduced armor, and all the offspring have reduced armor. Both populations have identical Eda coding regions. Could this difference in the Eda gene be the cause of the difference in armor? How could you test this?*

25.3 Different Ways to Evolve the Same Structure

Learning Outcomes

1. Differentiate between homologous structures and homoplastic structures.
2. Explain how two very similar morphologies can arise from different developmental pathways.

Homoplastic structures (see chapter 23), also known as convergent structures, have the same or similar functions, but arose independently—unlike homologous structures that arose once from a common ancestor. Phylogenies reveal convergent events, but the origin of the convergence may not be easily understood. In many cases different developmental pathways have been modified, as is the case with the spots on butterfly wings. In other cases, such as flower shape, it is not always as clear whether the same or different genes are responsible for convergent evolution.

Insect wing patterns demonstrate homoplastic convergence

Insect wings, especially those of moths and butterflies, have beautiful patterns that can protect them from predation. The origins of these patterns are best explained by co-option, the recruitment of existing regulatory programs for new functions.

Distal-less is one of the genes co-opted for butterfly spot development. Limb development in insects and arthropods requires *Distal-less*, but the expression of this gene also predicts where spots will form on butterfly wings (figure 25.10). *Distal-less* determines the center of the spot, but several other genes have been co-opted to determine the overall size and pattern of different spots.

Flower shapes also demonstrate convergence

Flowers exhibit two types of symmetry. Looking down on a **radially symmetrical** flower, you see a circle. No matter how you cut that flower, as long as you have a straight line that intersects with the center, you end up with two identical parts. Examples of radially symmetrical flowers are daisies, roses, tulips, and many other flowers.

Bilaterally symmetrical flowers have mirror-image halves on each side of a single central axis. If they are cut in any other direction, two nonsimilar shapes result. Plants with bilaterally symmetrical flowers include snapdragons, mints, and peas. Bilaterally symmetrical flowers are attractive to their pollinators, and the shape may have been an important factor in their evolutionary success.

At the crossroads of evolution and development, two questions arise: First, what genes are involved in bilateral symmetry? And second, are the same genes involved in the numerous, independent origins of asymmetrical flowers?

Cycloidia (CYC) is a snapdragon gene responsible for the bilateral symmetry of the flower. Snapdragons with mutations in *CYC* have radially symmetrical flowers (see figure 41.17b). Beginning with robust phylogenies, researchers have selected flowers that evolved bilateral symmetry independently from snapdragons and cloned the *CYC* gene. The *CYC* gene in closely related symmetrical flowers has also been sequenced.

Comparisons of the *CYC* gene sequence among phylogenetically diverse flowers indicate that both radial symmetry and bilateral symmetry evolved in multiple ways in flowers. Although radial symmetry is the ancestral condition, some radially symmetrical flowers have a bilaterally symmetrical ancestor. Loss of *CYC* function accounts for the loss of bilateral symmetry in some of these plants.

Gain of bilateral symmetry arose independently among some species because of the *CYC* gene. This change is an example of convergent evolution through mutations of the same gene. In other cases, *CYC* is not clearly responsible for the bilateral symmetry. Other genes also played a role in the convergent evolution of bilaterally symmetrical flowers.

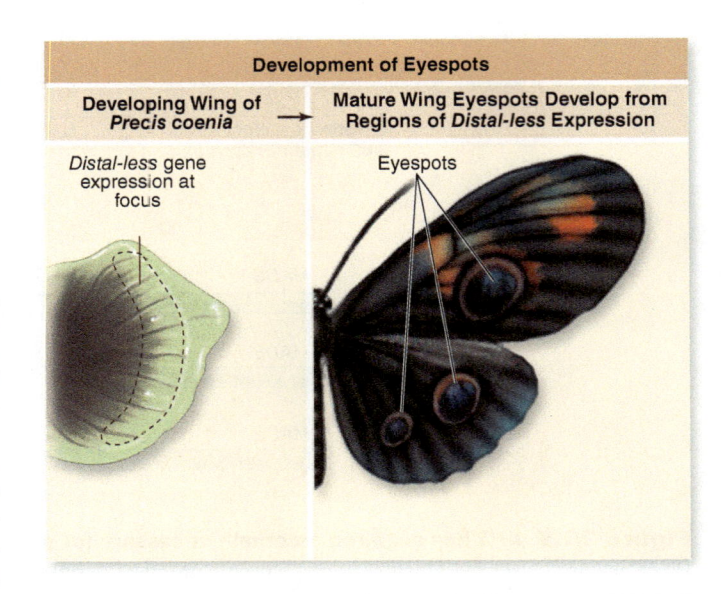

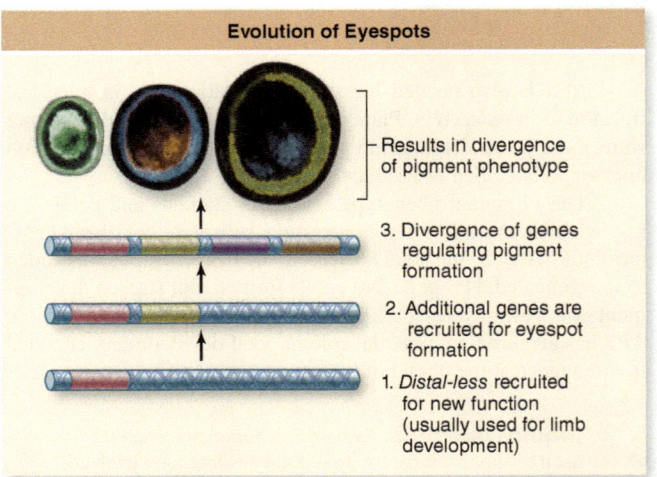

Figure 25.10 Butterfly eyespot evolution. The *Distal-less* gene, usually used for limb development, was recruited for eyespot development on the wings of the butterfly, *Precis coenia. Distal-less* initiates the development of different-colored spots in different butterfly species by regulating different pigment genes in different species. Eyespots can protect butterflies by startling predators. Although many butterflies have evolved spots, some have co-opted genes other than *distal-less* for their regulation.

Learning Outcomes Review 25.3

Homoplastic features have a similar function but different evolutionary origins and are examples of convergent evolution. Homologous features have the same evolutionary origins, but may have a different function. Knowledge of the underlying genes often reveals that a feature thought to be homoplastic has a more common origin than expected.

- Both sharks and whales have pectoral fins. Are these features homologous or homoplastic?

25.4 Diversity of Eyes in the Natural World: A Case Study

Learning Outcome

1. Explain how the compound eye of a fly, the human eye, and the eyespot of a ribbon worm could have a common evolutionary origin.

The eye is one of the most complex organs. Biologists have studied it for centuries. Indeed, explaining how such a complicated structure could evolve was one of the great challenges facing Darwin. If all parts of a structure such as an eye are required for proper functioning, how could natural selection build such a structure?

Darwin's response was that even intermediate structures—which provide, for example, the ability to distinguish light from dark—would be advantageous compared with the ancestral state of no visual capability whatsoever, and thus these structures would be favored by natural selection. In this way, by incremental improvements in function, natural selection could build a complicated structure.

Morphological evidence indicates eyes evolved at least thirty times

Comparative anatomists long have noted that the structures of the eyes of different types of animals are quite different. Consider, for example, the difference in the eyes of a vertebrate, an insect, a mollusk (octopus), and a jellyfish (figure 25.11; see also figure 44.15). The eyes of these organisms are extremely different in many ways, ranging from compound eyes, to simple eyes, to mere eyespots.

Consequently, morphologists concluded that eyes are examples of convergent evolution and are homoplastic. For this reason, evolutionary biologists traditionally viewed the eyes of different organisms as having independently evolved, perhaps as many as 30 times. Moreover, this view holds that the most recent common ancestor of all these forms was a primitive animal with no ability to detect light. Although this was the conclusion of morphologists, molecular studies are pointing to a different conclusion.

The same gene, *Pax6*, initiates fly and mouse eye development

In the early 1990s, biologists studied the development of the eye in both vertebrates and insects. In each case, a gene was discovered that codes for a transcription factor important in lens formation; the mouse gene was given the name **Pax6**, whereas the fly gene was called *eyeless*. A mutation in the *eyeless* gene led to a lack of production of the transcription factor, and thus the complete absence of eye development, giving the gene its name.

When these genes were sequenced, it became apparent that they were highly similar; in essence, the homologous *Pax6* gene was responsible for triggering lens formation in both insects and vertebrates. A stunning demonstration of this homology was conducted by the Swiss biologist Walter Gehring, who inserted the mouse version of the *Pax6* into the genome of a fruit fly, creating a transgenic fly. In this fly, the *Pax6* gene was turned on by regulatory factors in the fly's leg and an eye formed on the leg of the fly (figure 25.12)!

These results were truly shocking to the evolutionary biology community. Insects and vertebrates diverged from a common ancestor more than 500 MYA. Moreover, given the large differences in structure of the vertebrate eye and insect eye, the standard assumption was that the eyes evolved independently, and thus that their development would be controlled by completely different genes. That eye development was affected by the same homologous gene, and that these genes were so similar that the vertebrate gene seemed to function normally in the insect genome, was completely unexpected.

The *Pax6* story extends to eyeless fish found in caves (figure 25.13). Fish that live in dark caves need to rely on senses other than sight. In cavefish, *Pax6* gene expression is greatly reduced. Eyes start to develop, but then degenerate.

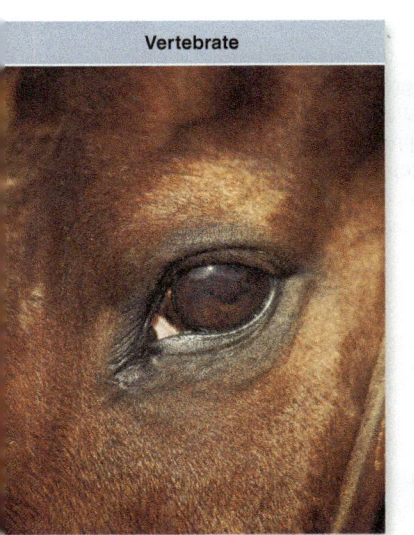

Figure 25.11 A diversity of eyes. Morphological and anatomical comparisons of eyes are consistent with the hypothesis of independent, convergent evolution of eyes in diverse species such as flies and humans.

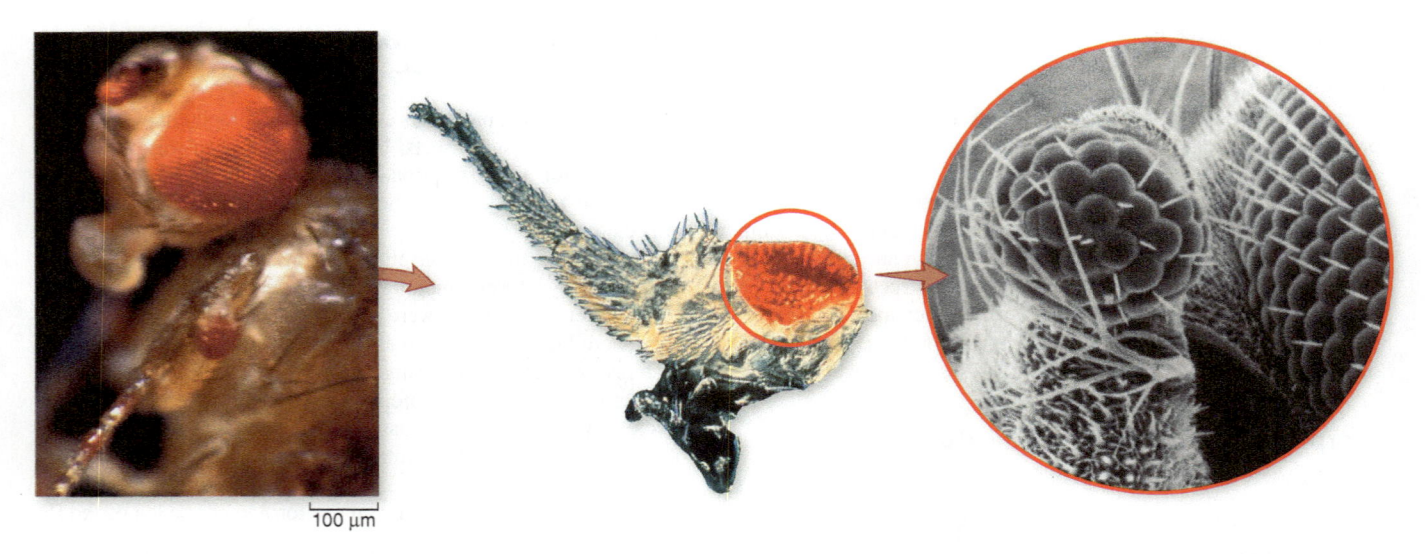

Figure 25.12 Mouse *Pax6* makes an eye on the leg of a fly. *Pax6* and *eyeless* are functional homologues. The *Pax6* master regulator gene can initiate compound eye development in a fruit fly or simple eye development in a mouse.

Ribbon worms, but not planaria, use *Pax6* for eye development

Research on other bilaterians (animals with mirror-image symmetry) yielded further surprises about the *Pax6* gene. Even the very simple ribbon worm, *Lineus sanguineus*, relies on *Pax6* for development of its eyespots. A *Pax6* homologue has been cloned and has been shown to express at the sites where eyespots develop.

Ribbon worms can regenerate their head region if it is removed. In an elegant experiment, the head of a ribbon worm was removed, and biologists followed the regeneration of eyespots. At the same time, the expression of the *Pax6* homologue was observed using in situ hybridization. To observe *Pax6* gene expression, an antisense RNA sequence of the *Pax6* was made and labeled with a color marker. When the regenerating ribbon worms were exposed to the antisense *Pax6* probe, the antisense RNA paired with expressed *Pax6* RNA transcripts and could be seen as colored spots under the microscope (figure 25.14).

Although the evidence supporting the role of *Pax6* for eye development in bilaterians continues to grow, there are exceptions. The flatworm planaria can also regenerate eyespots when it is cut in half lengthwise, but no *Pax6* gene expression is associated with regenerating the eyespots. These *Pax6*-related genes are, however, expressed in the central nervous system. A *Pax6*-responsive element, P3-enhancer, has also been identified and shown to be active in planaria.

Figure 25.13 Cavefish have lost their sight. Mexican tetras, *Astyanax mexicanus*, have (*a*) surface-dwelling members and (*b*) cave-dwelling members of the same species. The cavefish have very tiny eyes, partly because of reduced expression of *Pax6*.

Cnidarians use other *Pax* genes for eye development

Pax genes were involved in eye development much earlier in evolutionary history than was imagined. In the radially symmetrical cnidarians, which include the jellyfish, two different groups of jellyfish, the hydrozoans and cubozoans, rely on *PaxA* and *PaxB* genes, respectively, for eye development. Walter Gehring repeated his startling transgenic fly experiment (see figure 25.12) by misexpressing a jellyfish *PaxA* gene in a fly, which made a compound eye on its leg again!

SCIENTIFIC THINKING

Hypothesis: Pax6 is necessary for eyespot regeneration in ribbon worms.
Prediction: Eyespots will regenerate where Pax6 is expressed.
Test:

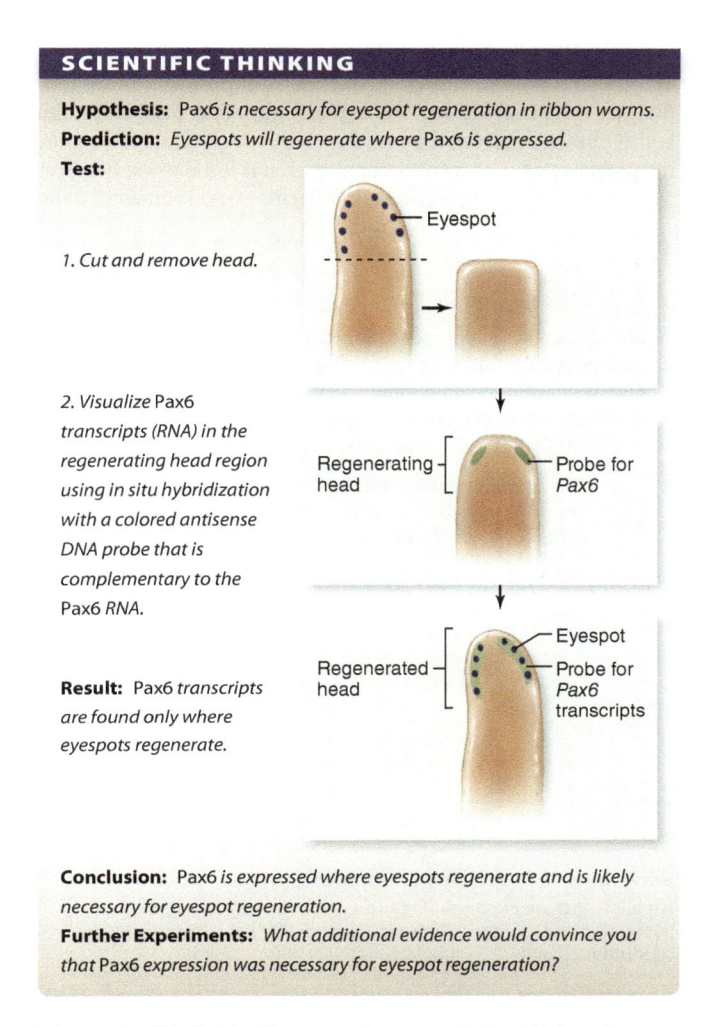

1. Cut and remove head.

2. Visualize Pax6 transcripts (RNA) in the regenerating head region using in situ hybridization with a colored antisense DNA probe that is complementary to the Pax6 RNA.

Result: Pax6 transcripts are found only where eyespots regenerate.

Conclusion: Pax6 is expressed where eyespots regenerate and is likely necessary for eyespot regeneration.
Further Experiments: What additional evidence would convince you that Pax6 expression was necessary for eyespot regeneration?

Figure 25.14 *Pax6* expression correlates with ribbon worm eyespot regeneration.

As more *Pax* family genes are analyzed, an intriguing phylogeny has emerged (figure 25.15). Different *Pax* genes members have been recruited by different animal lineages for eye development after early gene duplications. Learning more about the *Pax* genes that occur in the simplest animals—sponges—should help solve the mystery of the *Pax* genes and the origins of the eye.

The initiation of eye development may have evolved just once

Several explanations are possible for these findings. One is that eyes in different types of animals evolved truly independently, as originally believed. But if this is the case, why are the *Pax* genes so structurally similar and able to play a similar role in so many different groups? Opponents of single evolution of the eye point out that *Pax6* is involved not only in development of the eye, but also in development of the entire forehead region of many organisms. Consequently, it is possible that if *Pax6* had a regulatory role in the forehead of early bilaterians, perhaps it has been independently co-opted time and time again to serve a role in eye development. This role would be consistent with the data on planarians and the use of *PaxA* and *PaxB* in jellyfish eye development.

Many other biologists find this interpretation unlikely. The consistent use of *Pax6* in eye development in so many organisms, the fact that it functions in the same role in each case, and the great similarity in DNA sequence and even functional replaceability suggest to many that *Pax6* acquired its evolutionary role in eye development only a single time, in the common ancestor of all extant organisms that use *Pax6* in eye development. The replaceability of *Pax6* with jellyfish *PaxA* in fly eye development further strengthens the case.

The use of different *Pax* family member for eye development is consistent with single origins also. It can be argued that after the early duplications, all the *Pax* genes served redundant functions, including eyespot or light perception functions, and then individual

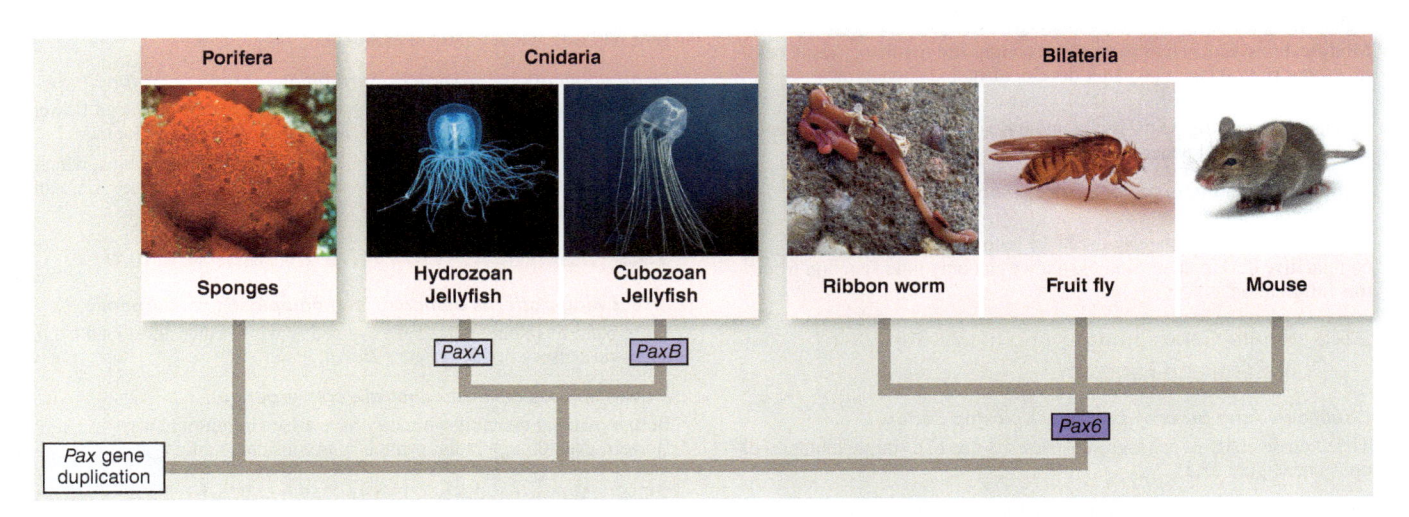

Figure 25.15 Three different *Pax* gene family members were recruited by different animal lineages for eye development.

genes became specialized for specific functions. In different lineages, this subfunctionalization evolved differently, but had a common origin. Support for this conclusion comes from findings that other genes are conserved in eye development.

Given the great dissimilarity among eyes of different groups, how can this be? One hypothesis is that the common ancestor of these groups was not completely blind, as traditionally assumed. Rather, that organism may have had some sort of rudimentary visual system—maybe no more than a pigmented photoreceptor cell, maybe a slightly more elaborate organ that could distinguish light from dark.

Whatever the exact phenotype, the important point is that some sort of basic visual system existed that used a *Pax* gene or genes in its development. Subsequently, the descendants of this ancestor diversified independently, evolving the sophisticated and complex image-forming eyes exhibited by different animal groups today.

Most evolutionary and developmental biologists today support some form of this hypothesis. Nonetheless, no independent evidence exists that the common ancestor of most of today's animal groups, a primitive form that lived probably more than 500 MYA, had any ability to detect light. The reason for this belief comes not from the fossil record, which could not record the presence of such a soft structure even if it did exist, but from a synthesis of phylogenetic and molecular developmental data.

Learning Outcome Review 25.4
Multidisciplinary approaches can clarify the evolutionary history of the world's biological diversity. The *Pax6* gene and its many homologues indicate that eye development, although highly diverse in outcome, may have a single evolutionary origin.

- Why would mutations leading to defective *Pax6* persist in cavefish? If these fish were introduced into a habitat with light, what would you expect to occur?

Chapter Review

25.1 Evolution of Developmental Patterns

Highly conserved genes produce diverse morphologies.
The *Hox* genes establish body form in animals; *MADS* box genes have a similar function in plants. Changes in these transcription factors and in genes involved in signaling pathways are responsible for new morphologies.

Developmental mechanisms evolve.
Heterochrony refers to alteration of timing of developmental events due to genetic changes; homeosis refers to alterations in the spatial pattern of gene expression.
Modifications of different parts of the coding and regulatory sequences of a transcription factor can alter development and phenotypic expression (figure 25.2).
Changes in signaling pathways, including mutations in receptors, can alter developmental patterns.

Understanding evolution of development requires functional analysis.
Sequence comparisons are essential for both phylogenetic and comparative development studies, but we can only infer function from this information.

25.2 Single-Gene Changes and the Alteration of Form and Function

Cauliflower and broccoli began with a stop codon.
The wide diversity of cabbage subspecies is due to a simple mutation of one gene (figure 25.3).

Cichlid fish jaws demonstrate morphological diversity.
Jaw morphology in fish has also been modified by mutations in one or a few genes.

Stickleback fish lose their armor with a single mutation.
An allele that alters morphology can persist at very low levels in a population and rapidly increase in frequency when environmental conditions change.

Ancestral genes may be co-opted for new functions (figure 25.6).
A single gene may act on different genes or combinations of genes in different species.

Limbs have developed through modification of transcriptional regulation.
Species differences in limbs have resulted from evolutionary changes in gene expression and timing of expression.

Gene duplications provide opportunities for new gene functions.
Duplication of the *AP3* gene and subsequent divergence produced flower petals, whereas the ancestral form affected only stamen development.
Studies have narrowed the active region in *AP3* to the C terminus, which acts differently from the C terminus of the related *PI* gene; only *AP3* can produce petals (figure 25.9).

25.3 Different Ways to Evolve the Same Structure

Insect wing patterns demonstrate homoplastic convergence.
Wing patterns in butterflies have evolved as wing scales developed from ancestral sensory bristles (figure 25.10).

Flower shapes also demonstrate convergence.
Both radial and bilateral symmetry have arisen in multiple ways in flowers, even though radial symmetry is considered ancestral.

25.4 Diversity of Eyes in the Natural World: A Case Study

Morphological evidence indicates eyes evolved at least thirty times.

Homoplasy and convergent evolution are supported as explaining the diversity of eyes found in the animal kingdom.

The same gene, Pax6, initiates fly and mouse eye development.
Transgenic experiments showed that the *Pax6* gene from a mouse, inserted into the genome of *Drosophila*, could cause development of an eye (figure 25.12).

Ribbon worms, but not planaria, use Pax6 for eye development.
Further evidence for the use of *Pax6* in eye development in the bilaterians comes from ribbon worms, but there are exceptions, including the planaria (figure 25.14).

Cnidarians use other Pax genes for eye development.
Jellyfish rely on *PaxA* and *PaxB* genes, rather than *Pax6* for eye development and also used other conserved genes for eye development (figure 25.15).

The initiation of eye development may have evolved just once.
It appears that at some distant point in evolutionary time, *Pax* genes were part of a visual system that later diverged many times, with specific *Pax* family members diverging in terms of their functions.

Review Questions

UNDERSTAND

1. Heterochrony is
   a. the alteration of the spatial pattern of gene expression.
   b. a change in the relative position of a body part.
   c. a change in the relative timing of developmental events.
   d. a change in a signaling pathway.

2. Vast differences in the phenotypes of organisms as different as fruit flies and humans
   a. must result from differences among many thousands of genes controlling development.
   b. have apparently arisen largely through manipulation of the timing and regulation of expression of probably less than 100 highly conserved genes.
   c. can be entirely explained by heterochrony.
   d. can be entirely explained by homeotic factors.

3. Homoplastic structures
   a. can involve convergence of completely unrelated developmental pathways.
   b. always are morphologically distinct.
   c. are produced by divergent evolution of homologous structures.
   d. are derived from the same structure in a shared common ancestor.

4. *Hox* genes are
   a. found in both plants and animals.
   b. found only in animals.
   c. found only in plants.
   d. only associated with genes in the *MADS* complex.

5. The *Brachyury* and *Tbx5* in vertebrates and the *Ap3* gene in flowering plants
   a. are examples of *Hox* genes.
   b. are examples of co-opting a gene for a new function.
   c. are homologues for determining the body plan of eukaryotes.
   d. help regulate the formation of appendages.

6. Which of the following statements about *Pax6* is false?
   a. *Pax6* has a similar function in mice and flies.
   b. *Pax6* is involved in eyespot formation in ribbon worms.
   c. *Pax6* is required for eye formation in *Drosophila*.
   d. *Pax6* is required for eyespot formation in planaria.

7. Which of the following statements about *Tbx5* is true?
   a. *Tbx5*, *Tbx4*, and *AmphiTbx4/5* have very similar coding regions.
   b. *Tbx5* is involved in tail development in vertebrates.
   c. *Tbx5*, *Tbx4*, and *AmphiTbx4/5* have very similar regulatory regions.
   d. *Tbx5* initiates hindlimb development.

8. Homeosis
   a. results from a maintained and unchanging genetic environment.
   b. results from a temporal change in gene expression.
   c. results from a spatial change in gene expression.
   d. is not an important genetic mechanism in development.

9. Transcription factors are
   a. genes.
   b. sequences of RNA.
   c. proteins that affect the expression of genes.
   d. None of the choices are correct.

10. Independently derived mutations of the *CYC* gene in plants
    a. suggests bilateral floral symmetry among all plants is homologous.
    b. establishes that radial floral symmetry is preferred by pollinators.
    c. establishes that radial floral symmetry is derived for all plants.
    d. None of the choices are correct.

APPLY

1. Choose the statement that best explains how the *Tbx5* protein can be responsible for the development of the heart in amphioxus but initiates forelimb development in mice.
   a. *Tbx5* is a key component of bone.
   b. *Tbx5* is a transcription factor that binds to Fgf-10.
   c. *Tbx5*'s regulatory region has evolved since the two vertebrate whole-genome duplications.
   d. *Tbx5* is a signaling molecule involved in the signaling pathway for limb development in both species.

2. Analyze why it was important to create transgenic plants to determine the role of *AP3* in petal formation and choose the most compelling reason.
   a. It provided a functional test of the role of *AP3* in petal development.
   b. Duplication of *AP3* could not be resolved on the phylogeny.
   c. To check if the phylogenetic position of *AP3* is really derived.
   d. Because tests already established the role of *paleoAP3* in stamen development.

3. The *Eda* allele that causes reduced armor originated about 2 MYA in marine stickleback fish and persists with a frequency of about 1% in marine environments. The frequency is much higher in freshwater populations. Apply your understanding of natural selection to determine the most likely reason for the difference in *Eda* allele frequency.
   a. The *Eda* allele causing reduced armor has evolved repeatedly in marine and freshwater environments.
   b. Fish have many predators in the marine environment and fish with reduced armor have reduced fitness.
   c. There is negative selection for armor in freshwater because building armor is energetically expensive.
   d. Both b and c are valid.

4. In *Drosophila* species, the yellow *(y)* gene is responsible for the patterning of black pigment on the body and the wings. A comparison of two species reveals that one has a black spot on each wing, and the other species lacks black pigmentation on the wing. The sequence of the coding region for *y* is identical in both species. Critique the following explanations and choose the most plausible one.
   a. The protein coded for by the *y* gene has a different structure in species with the black spots than the species without.
   b. The species without black pigmentation on the wing has a mutation in a regulatory region of the *y* gene.
   c. A deletion mutation is present in one of the exons of the *y* gene in the species without black pigment in the wings.
   d. Convergent evolution explains why both species develop black wing spots.

SYNTHESIZE

1. The *paired-like homeodomain transcription factor 1 (pitx1)* is expressed in the hindlimbs of developing mouse embryos, and its homologue is expressed in the pelvic region of the nine-spine stickleback fish *(Pungitius pungitius)*. You are beginning a research project on *pitx1*, and your supervisor is convinced that *pitx1* has been co-opted to make armor around the pelvic region of this species of stickleback. You begin by isolating RNA from a freshwater *Pungitius pungitius* that has very reduced armor. You convert the RNA to cDNA and amplify the *pitx1* cDNA using PCR. You have your *pitx1* PCR product sequenced and find it has exactly the same sequence as the marine *Pungitius pungitius* with armor. In light of this evidence, analyze your supervisor's hypothesis.

2. From the chapter on evolution of development it would seem that the generation of new developmental patterns would be fairly easy and fast, leading to the ability of organisms to adapt quickly to environmental changes. Construct an explanation for why it can take millions of years, typically, for many of the traits examined to evolve. (*Hint*: Consider the differences in *Eda* allele frequency in marine and freshwater three-spine stickleback fish.)

3. Phenotypic diversity among major groups of organisms can be explained in several ways. On one end of the spectrum, such differences could arise out of differences in many genes that control development. On the other end, small sets of genes might differ in how they regulate the expression of various parts of the genome. Evaluate which view represents our current understanding.

4. Critique the argument that eyes have multiple evolutionary origins.

5. Based on the information in figure 25.8, construct an explanation for the evolutionary differences in genes related to *AP3* in maize and tomato. Be sure to consider what happened before and after the two species diverged from a common ancestor.

6. Having read all of this chapter, return to the claim that the difference between direct and indirect development in sea urchins is caused by a change in gene expression, not differences in genes. Starting at the level of DNA sequence, formulate an argument in support of the claim.

Diversity of Life on Earth

Part V Diversity of Life on Earth

CHAPTER 26

The Origin and Diversity of Life

Chapter Contents

- 26.1 Deep Time
- 26.2 Origins of Life
- 26.3 Evidence for Early Life
- 26.4 Earth's Changing System
- 26.5 Ever-Changing Life on Earth

Introduction

All life descended from a common ancestor and have many things in common: They are composed of one or more cells, they carry out metabolism and transfer energy with ATP, and they encode hereditary information in DNA. But species are also highly diverse, ranging from bacteria and amoebas to blue whales and sequoia trees. Coral reefs, such as the one pictured here, are microcosms of diversity, comprising many life-forms and sheltering an enormous array of life. The origins and history of life on Earth are intertwined with Earth's ever-changing geology, climate, and atmosphere. To understand the diversity of life, questions must be asked about the history of Earth on a geological time scale and about the effects life itself has had on Earth's systems.

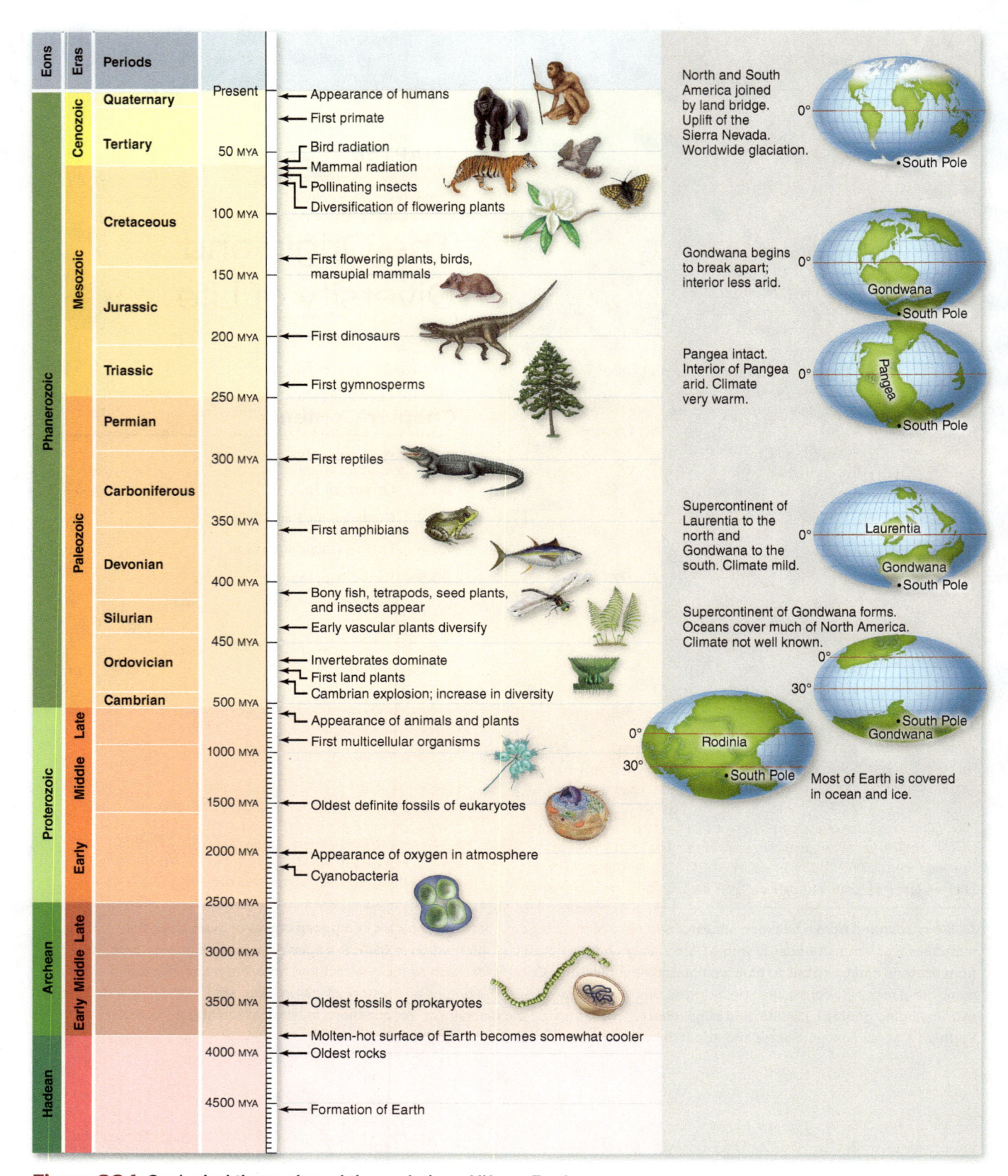

Figure 26.1 Geological timescale and the evolution of life on Earth.

26.1 Deep Time

Learning Outcomes
1. Describe the history of the Earth.
2. Understand the relationship between geological events and the evolution of life.

Earth changed over geological time

Many of the fundamental questions about the history of Earth are geological. To explore the origins and diversification of life over billions of years, it is important to understand geological, or deep, time. Geological time is divided into four eons, stretching over 4.6 billion years (figure 26.1). Eons are subdivided into eras, which are further subdivided into periods.

So much change has occurred since the Earth formed that no rocks exist from the first 500 to 700 million years of Earth's history (called the Hadean eon) that preceded the earliest fossils. Although it is impossible to be certain what early Earth was like, geological evidence is consistent with a meteor hitting the Earth almost 4.6 BYA with such force that that debris from the impact formed the Moon. The rocky mantle of the Earth literally melted as atmospheric temperatures exceeded 2000°C.

Hadean Earth was also pummeled by asteroids, which could potentially vaporize entire oceans. Shifting between a fiery and a sometimes frozen Earth, it was a wildly dynamic environment that was unlikely to have supported life.

CO_2 levels shifted, affecting temperature

The early atmosphere likely had high levels of CO_2, and water slowly vaporized from the molten rock. The Earth cooled over a 2-million-year period. As the temperature cooled, clouds made of silicate condensed in the atmosphere and rained down, forming a warm ocean under a CO_2 atmosphere. CO_2 levels dropped and the Earth cooled, and for a period of time the ocean froze.

Decreases in CO_2 contributed to the decrease in temperature because less radiant energy was absorbed in the atmosphere. What changed the CO_2 level in the atmosphere? The ocean and atmosphere were in equilibrium in terms of CO_2 levels. Volcanic eruptions added CO_2 to the atmosphere and ocean, while the weathering of rock decreased CO_2 levels.

Weathering occurred more rapidly under hot, wet conditions than cold, dry conditions. Weathering is the conversion of silicate rock to soil. It occurs when the CO_2 in the atmosphere combines with water (H_2O) to create a carbonic acid (H_2CO_3) rain. The carbonic acid interacted with the rock, releasing bicarbonate ions (HCO_3^-) and Ca^{2+}. The weathered solutes moved through rivers and oceans and formed calcium carbonate ($CaCO_3$), which precipitated and sequestered the CO_2 in the ocean (figure 26.2).

Continents moved over geological time

Earth's crust formed rigid slabs of rock called plates under both continents and oceans. These huge slabs shift a few centimeters each year, a process called plate tectonics. The term *tectonics* comes from the Greek word for build or builder, and plate movements built and continue to build the geological features of Earth. Most major earthquakes and volcanoes occur at the edges of the plates when they move.

Although the plates move slowly, over deep geological time the cumulative effects are astounding. Several times in Earth's history, all the continents have come together to form a single supercontinent (see figure 26.1). Two proposed supercontinents, Rodinia (all the continents) and Gondwana (composed of all the current Southern Hemisphere continents), existed at different times and both occupied the Southern Hemisphere. Gondwana contributed to the supercontinent Pangea, which was fully formed 225 MYA and began to separate 25 to 50 million years later. Geologists have the greatest confidence in plate tectonic evidence from the last 200 million years, beginning with the breakup of Pangea.

Life emerged in the Archean

At some point, life emerged. Some fossil evidence exists from the Archean eon that followed the Hadean. Two billion years into Earth's history, the Proterozoic (meaning "early life") eon occurred. It was characterized by the formation of the supercontinent Rodinia, which 650 MYA broke up into several continents before the start of the Phanerozoic ("visible life") eon. Collectively, the Hadean, Archean, and Proterozoic eons are referred to as the Precambrian. With the start of the Phanerozoic's Paleozoic era, marked first by the Cambrian period, a remarkable diversification of multicellular organisms took place. Beginning with the Phanerozoic both geologists and biologists begin to focus on chunks of time on the order of periods and shorter.

The Phanerozoic eon represents only 12% of Earth's history, yet it contains most of the biological history of the diversification of multicellular life (see figure 26.1). Birds and mammals have existed for 4% of Earth's existence, whereas humans have existed for 0.2% of the history of Earth.

Learning Outcomes Review 26.1

The history of Earth extends over 4.6 billion years, but the geological record of what occurred during the Hadean eon before life emerged is limited. Initially, the Earth was inhospitable to life, but as conditions changed, life emerged more than 3 billion years ago. Multicellular species appeared only within the last billion years.

- Given an opportunity to study the diversity of life in the fossil record, which eon would you choose? Support your choice.

26.2 Origins of Life

Learning Outcomes
1. Contrast today's atmosphere with the likely atmosphere at the end of the Hadean eon.
2. Describe key steps necessary for life to originate.

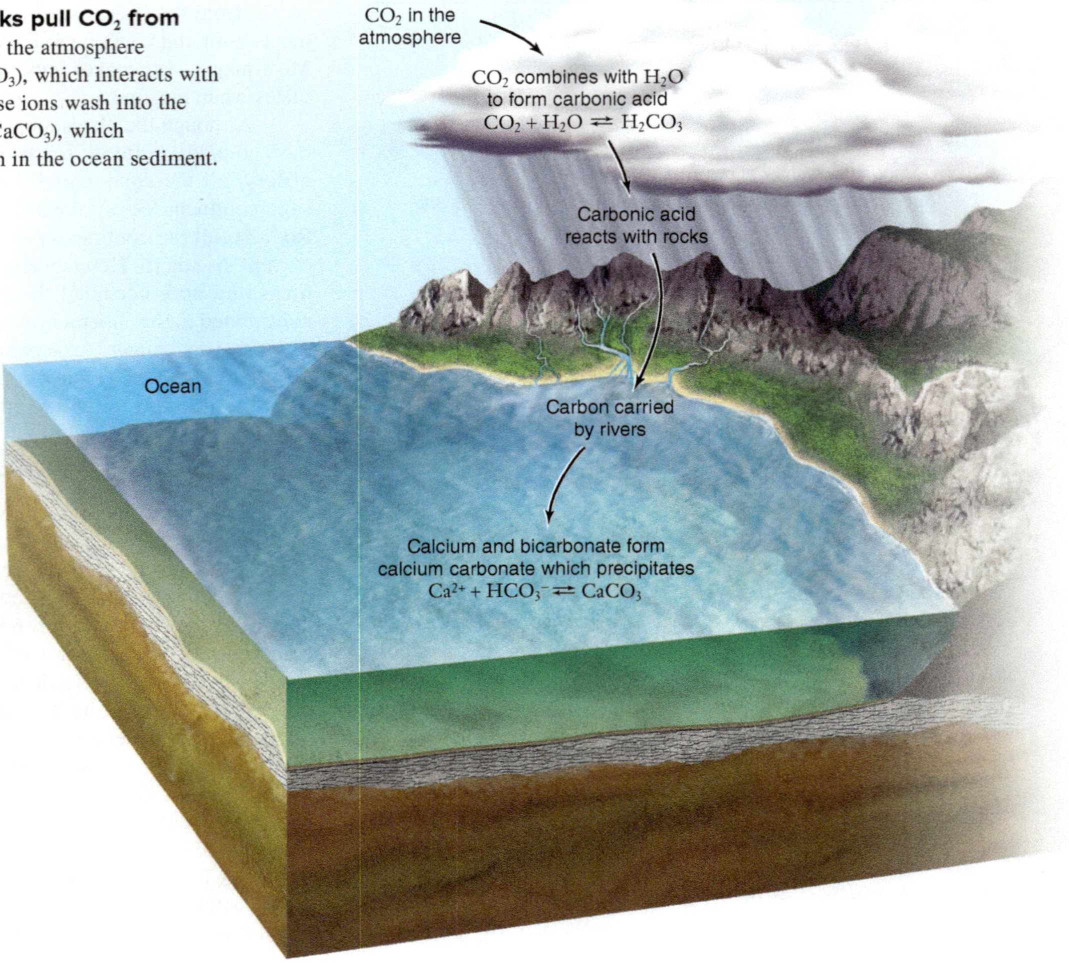

Figure 26.2 Weathering rocks pull CO_2 from the atmosphere. H_2O and CO_2 in the atmosphere combine to form carbonic acid (H_2CO_3), which interacts with rock to release HCO_3^- and Ca^{2+}. These ions wash into the ocean and form calcium carbonate ($CaCO_3$), which precipitates and sequesters the carbon in the ocean sediment.

CO_2 in the atmosphere

CO_2 combines with H_2O to form carbonic acid
$CO_2 + H_2O \rightleftharpoons H_2CO_3$

Carbonic acid reacts with rocks

Ocean

Carbon carried by rivers

Calcium and bicarbonate form calcium carbonate which precipitates
$Ca^{2+} + HCO_3^- \rightleftharpoons CaCO_3$

We really don't know how life began on Earth. Because we cannot recreate the process now, we have to use various lines of scientific exploration to piece together the puzzle of life's origins, beginning with the geology of early Earth. Hadean Earth was a hot mass of molten rock about 4.6 BYA. As it cooled, much of the water vapor present in Earth's atmosphere condensed into liquid water that accumulated on the surface in chemically rich oceans. One scenario for the origin of life is that it originated in this dilute, hot, smelly soup of ammonia, formaldehyde, formic acid, cyanide, methane, hydrogen sulfide, and organic hydrocarbons. Whether at the oceans' edges, in hydrothermal deep-sea vents, or elsewhere, life likely arose spontaneously from these early waters. Although the way in which this happened remains a puzzle, we cannot escape a certain curiosity about the earliest steps that eventually led to the origin of all living things on Earth, including ourselves. How did organisms evolve from the complex molecules that swirled in the early oceans?

Long before there were cells with the properties of life, organic (carbon-based) molecules formed from inorganic molecules. The formation of proteins, nucleic acids, carbohydrates, and lipids were essential, but not sufficient for life. The evolution of cells required early organic molecules to assemble into a functional, interdependent unit.

Early organic molecules may have originated in various ways

Organic molecules may have had extraterrestrial origins

Organic molecules are the basis of all living organisms. How the first organic molecules formed is not known, and some could have extraterrestrial origins. Hundreds of thousands of meteorites and comets are known to have slammed into the early Earth, and recent findings suggest that at least some may have carried organic materials. For example, chemical analysis of the Tagish Lake meteorite, a rocky, carbon-based meteorite that landed in British Columbia in 2000, found that nearly 3% of the weight was organic matter. Soluble organic compounds in the meteorite included carboxylic and sulfonic acids, along with trace levels of amino acids. Glycine is the most abundant amino acid in the meteorite and its carbon isotope ratios are inconsistent with rocks found on Earth, supporting the claim that some organic molecules may have had extraterrestrial origins.

Organic molecules may have originated on early Earth

Very few geochemists agree on the exact composition of the early atmosphere. One popular view is that it contained principally carbon dioxide (CO_2) and nitrogen gas (N_2), along with significant amounts of water vapor (H_2O). It is possible that the early atmosphere also contained hydrogen gas (H_2) and compounds in which hydrogen atoms were bonded to the other light elements (sulfur, nitrogen, and carbon), producing hydrogen sulfide (H_2S), ammonia (NH_3), and methane (CH_4).

We refer to such an atmosphere as a *reducing atmosphere* because of the ample availability of hydrogen atoms and their electrons. Because a reducing atmosphere would not have required as much energy to drive chemical reactions as it would today, it would have made it easier to form the carbon-rich molecules from which life evolved.

An early attempt to determine what kinds of organic molecules might have been produced on the early Earth was carried out in 1953 by American chemists Stanley L. Miller and Harold C. Urey. In what has become a classic experiment, they attempted to reproduce the conditions in the Earth's primitive oceans under a reducing atmosphere. Even if their hypothesis proves incorrect—the jury is still out on this—this experiment is critically important because it ushered in the field of prebiotic chemistry.

To carry out their experiment, Miller and Urey (1) assembled a reducing atmosphere rich in hydrogen and excluding gaseous oxygen; (2) placed this atmosphere over liquid water; (3) maintained this mixture at a temperature somewhat below 100°C; and (4) simulated lightning by bombarding it with energy in the form of sparks (figure 26.3).

They found that within a week, 15% of the carbon originally present as methane gas (CH_4) had converted into other simple carbon compounds. Among these compounds were formaldehyde (CH_2O) and hydrogen cyanide (HCN). These compounds then combined to form simple molecules, such as formic acid (HCOOH) and urea (NH_2CONH_2), and more complex molecules containing carbon–carbon bonds, including the amino acids glycine and alanine.

In similar experiments performed later by other scientists, more than 30 different carbon compounds were identified, including the amino acids glycine and alanine, but also glutamic acid, valine, proline, and aspartic acid. As we saw in chapter 3, amino acids are the basic building blocks of proteins, and proteins are one of the major kinds of molecules of which organisms are composed. Other biologically important molecules were also formed in these experiments. For example, hydrogen cyanide contributed to the production of a complex ring-shaped molecule called adenine—one of the bases found in DNA and RNA. Thus, the key molecules of life could have formed in the reducing atmosphere of the early Earth.

Metabolic pathways may have emerged in various ways

Many hypotheses for the emergence of metabolic pathways exist. One scenario assumes that primitive organisms were autotrophic, building all the complex organic molecules they required from simple inorganic compounds, rather than heterotrophic and acquiring all organic compounds from the surrounding environment. For example, glucose may have been synthesized from formaldehyde, CH_2O, in the alkaline conditions that could have existed on early Earth. Glycolysis and a version of the Krebs cycle (see chapter 7) that functioned without enzymes are proposed to be the core from which other metabolic pathways emerged. Early autotrophs could have made, stored, and later used glucose as an energy source.

Enzymes to catalyze metabolic pathways also emerged. Although most enzymes are proteins, RNA can catalyze reactions as well as store genetic information. According to the hypothesis of an RNA world, RNA, rather than DNA, was the first nucleic acid that permitted self-replication, an important step toward life.

Figure 26.3 The Miller–Urey experiment. The apparatus consisted of a closed tube connecting two chambers. The upper chamber contained a mixture of gases thought to resemble the primitive Earth's atmosphere. Electrodes discharged sparks through this mixture, simulating lightning. Condensers then cooled the gases, causing water droplets to form, which passed into the second heated chamber, the "ocean." Any complex molecules formed in the atmosphere chamber would be dissolved in these droplets and carried to the ocean chamber, from which samples were withdrawn for analysis.

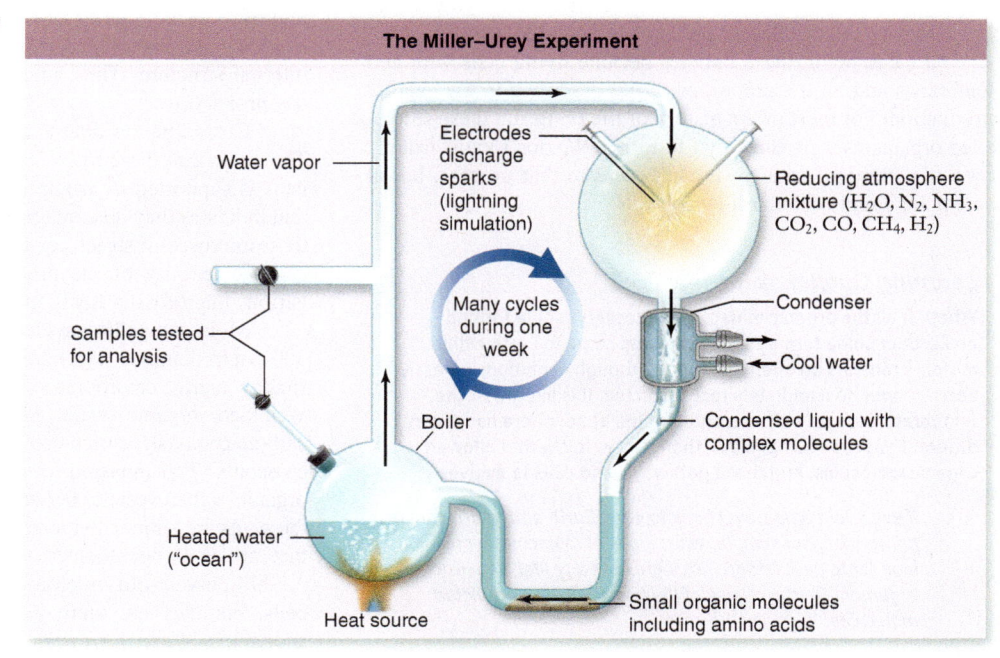

Later DNA, which is more stable than RNA, took over the information storage function. Proteins that have a greater variety of building blocks (amino acids) gained the enzymatic function.

Ribozymes are RNA sequences with an enzymatic function. Strong evidence supporting the RNA world hypothesis comes from the ribosome, which is used in cells for translation of RNA into proteins. Although the ribosome is composed of both protein and RNA, it is an RNA sequence that is involved in the central mechanism for translation. This is consistent with the hypothesis that early cells used RNA to catalyze the synthesis of peptides from an RNA sequence.

Although much evidence supports an early RNA world, there have been some questions about the hypothesis. With what is known about prebiotic Earth, it is unlikely that there was much ribose sugar then, which is essential for the sugar–phosphate backbone of RNA. However, research has shown that it is possible to synthesize RNA nucleotides without pure ribose under the conditions postulated to have existed on prebiotic Earth.

Another challenge to the hypothesis was the puzzle of how long chains of RNA nucleotides formed. Evidence suggests that nucleotides could have concentrated on clay surfaces and that bonds would have formed, linking the concentrated nucleotides. Increased concentrations of RNA nucleotides, fostering bond formation, may also have occurred in ice crystals in salty water. The findings support the RNA world hypothesis.

Single cells were the first life-forms

In addition to metabolism, cells require membranes. Constraining organic molecules to a physical space within a lipid or protein bubble could lead to an increased concentration of specific molecules. This in turn could increase the probability of metabolic reactions occurring.

Although modern membranes are made up of a bilayer of phospholipids (see chapter 5), early membranes may have been composed of fatty acids. These are simpler molecules than phospholipids and were more likely to form under prebiotic conditions. Just like phospholipids, they have hydrophilic heads and hydrophobic tails so they can form bilayers and enclosed cell-like structures.

At some point, these bubbles became living cells with cell membranes and all the properties of life described in the chapter introduction. For most of the history of life on Earth, these single-celled organisms were the only life-forms. We don't know exactly how cells formed because we can't recreate that process, but at some point simple cellular life evolved.

Learning Outcomes Review 26.2

Whether all the organic molecules necessary for life formed on Earth or some formed elsewhere and came to Earth within meteors remains an open question. Although conditions on early Earth cannot be completely reconstructed, it is likely that the temperatures were extreme and that the atmosphere had a very different gaseous composition than it does today that allowed organic molecules, metabolic pathways, and cells to evolve.

- If you could time travel back to early Earth and return with a primordial pool sample, what types of molecules would you look for to understand the origins of early life? Construct an argument for the origins of life based on your predicted molecules.

26.3 Evidence for Early Life

Learning Outcome

1. Evaluate the strength of fossil evidence dating the origins of life.

The more we learn about the Earth's early history, the more likely it seems that Earth's first organisms emerged and lived at very high temperatures. By about 3.8 BYA, ocean temperatures are thought to have dropped to a hot 49° to 88°C (120°–190°F). Around 3.8 BYA, life first appeared, promptly after the Earth was habitable. Thus, as intolerable as early Earth's infernal temperatures seem to us today, they gave birth to life. Here we explore the evidence for life dating back 3.2 to 3.8 billion years.

Fossil evidence indicates life may have originated 3.2 BYA

Early life may have arisen during the Archean, but evidence of life in the form of microfossils is difficult both to find and to interpret. Nonbiological processes can produce microfossil-like structures, and rocks older than 3 billion years are rarely unchanged by geological action over time. Two main formations of 3.5- to 3.8-billion-year-old rocks have been found that are mostly intact: the Kaapvaal craton in South Africa and the Pilbara craton in western Australia. (A *craton* is a rock layer of undisturbed continental crust.) Structures have been found in each of these formations and others that are interpreted to be biological in origin. Although this interpretation has been controversial, the accumulation of evidence over time favors these structures as being true fossil cells.

Microfossils are fossilized forms of microscopic life. Many microfossils are small (1 to 2 μm in diameter) and appear to be single-celled, lack external appendages, and have little evidence of internal structure. Thus, microfossils seem to resemble present-day prokaryotes.

Currently, the oldest microfossils are 3.5 billion years old. The claim that these microfossils are the remains of living organisms is supported by isotopic data and by spectroscopic analysis that indicates they do contain complex carbon molecules. Whether these microscopic structures are true fossil cells is still controversial, and the identity of the prokaryotic groups represented by the various microfossils is still unclear.

More compelling evidence comes from microfossils that are 300 μm in diameter and were part of what were microbial mats in shallow marine environments in South Africa 3.2 BYA (figure 26.4). Wrinkled organic walls (160 nm thick) surrounding these carbonaceous structures can be seen with scanning electron microscopy. Transmission electron microscopy revealed hollow organic-walled vesicles between compressed walls. The features of the walls are similar to those observed in Proterozoic microfossils that are well documented to be biotic in origin. The size of 3.2-billion-year-old microfossils is consistent with eukaryotic cells, but they are more likely to be cyanobacteria (discussed shortly).

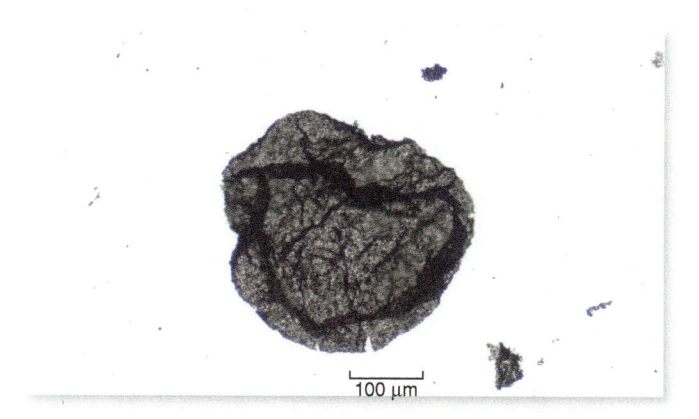

Figure 26.4 A 3.2-billion-year-old microfossil from South Africa.

In addition to these microfossils, indirect evidence for ancient life can be found in the form of sedimentary deposits called **stromatolites** (figure 26.5). These structures are commonly interpreted as a combination of sedimentary deposits and precipitated material that are held in place by mats of micro-organisms. The microorganisms that make up the mats are thought to be cyanobacteria. Formations of stromatolites are as old as 2.7 billion years. Because relatively modern stromatolites are also known, the formation and biological nature of these structures is less contentious.

Isotopic data indicate that carbon fixation is an ancient process

Another way to ask when life began is to look for the signature of living systems in the geological record. Living systems alter their environments, and sometimes this change can be detected. The most obvious change is that living systems are selective in the isotopes of carbon in compounds they use. Living organisms incorporate ^{12}C into their cells before any other carbon isotope, and thus they can alter the ratios of these isotopes in the atmosphere. They also have a higher level of ^{12}C in their fossilized bodies than does the nonorganic rock around them.

Much work has been done on dating and analyzing carbon compounds in the oldest rocks, looking for signatures of life. Analysis of carbon signatures indicates carbon fixation, the incorporation of inorganic carbon into organic form, was active as long as 3.8 BYA, consistent with the dating of the oldest microfossils.

The ancient fixation of carbon happened via two main pathways. The most common pathway for carbon fixation is the Calvin cycle (see chapter 8). This is the pathway used by cyanobacteria, algae, and modern land plants that perform oxygenic photosynthesis using two photosystems. The Calvin cycle is also active in green and purple sulfur bacteria that perform anoxygenic photosynthesis using a single photosystem. This anoxygenic form of photosynthesis could account for ancient carbon fixation.

To date, the entire Calvin cycle has not been demonstrated in the archaea, a group of prokaryotes, although the key enzyme for this pathway has been identified in a few archaeal isolates. Instead, some archaea use a reductive version of the Krebs cycle (see chapter 7). This pathway of carbon fixation is also used by some lithotrophic bacteria, which derive energy from the oxidation of inorganic compounds, and by the green sulfur bacteria. Two other pathways may also occur in the lithotrophs, archaea, and the green nonsulfur bacteria. Evidence suggests that the ability to fix carbon has evolved more than once over the course of evolution.

Some hydrocarbons found in ancient rocks may have biological origins

Another way to look for evidence of ancient life is to look for organic molecules, which are clearly of biological origin; such molecules are called *biomarkers*. Although the process sounds simple, it has proved difficult to find such markers. Hydrocarbons, which are derived from the fatty acid tails of lipids, are one type of biomarker. These can be analyzed for their carbon isotope ratios to indicate biological origin. The analysis of extractable hydrocarbons from the Pilbara formation in Australia found lipids that are indicative of cyanobacteria as long ago as 2.7 billion years. The search for definitive chemical markers for living systems in the oldest rocks and in meteorites is an area of intense interest.

Figure 26.5 Stromatolites. Mats of bacterial cells that trap mineral deposits and form the characteristic dome shapes seen here.

Learning Outcome Review 26.3

Evidence for the earliest cells exists in microfossils. The earliest microfossils are controversial, but they are at least 3.5 billion years old. Other evidence for early life includes isotopic ratios that are skewed by biological activity. The Calvin cycle and a reductive version of the Krebs cycle, as well as other pathways, appear to have led to carbon fixation in ancient life. Some hydrocarbons appear to be biomarkers and may therefore also indicate ancient life-forms.

- *You have discovered a fossil that may be an early bacterial cell. What evidence would be sufficient to convince you that this is indeed an early bacterial cell?*

26.4 Earth's Changing System

Learning Outcomes
1. Construct an explanation for relationship between CO_2 levels and glaciation.
2. Argue that plate tectonics has affected the evolution of life on Earth.

Climate (temperature and water availability) and atmosphere (including levels of CO_2 and O_2) are among the many factors that affect the ability of organisms to survive and reproduce. Over the course of Earth's history, repeated and dramatic shifts in all these factors have led to mass extinctions and otherwise influenced the course of evolution (the phenomenon of mass extinction is discussed in chapter 22). Shifting tectonic plates have led to volcanic eruptions that alter the atmosphere, including simply blocking sunlight. Oscillating CO_2 levels over geological time correlate with temperature changes as increased atmospheric concentrations of CO_2 trap the heat radiating from the Earth, creating a greenhouse effect (see chapter 58). Some changes affecting the climate and atmosphere are strictly geological, but living organisms account for other changes. For example, the evolution of photosynthesis increased O_2 in the atmosphere. In this section, we explore how shifts in climate and the atmosphere have affected Earth and life on Earth over geological time.

Earth's climate has been ever-changing

The range of temperatures and rainfall over Earth's history is astounding. Early Earth experienced temperatures ranging from over 2000°C at some times and global mean temperatures of –50°C at others. Temperature shifts immediately affect terrestrial, aquatic, and marine organisms. Over a slightly longer time frame, the level of oceans is affected, further disrupting life.

Earth has been cooling since its formation, but several sudden drops in temperature, including three exceptionally sharp drops that occurred both early and late in the Proterozoic, have decimated life. These extreme drops in temperature resulted in glacial ice covering Earth from pole to pole, a phenomenon called *Snowball Earth* (figure 26.6). Under such conditions, the ice covering the Earth's surface reflected back most of the radiant energy from the Sun, maintaining the cold temperatures. Even at the equator, temperatures would have climbed no higher than –20°C, equivalent to current Antarctic temperatures. With frozen oceans, temperatures were less moderated, and the shifts in temperature were more marked than on Earth today.

Glaciation results in massive extinctions of species. It is difficult to imagine life surviving pole-to-pole glaciation, yet only 80 million years after the last Snowball Earth event, bilaterally symmetrical animals are found in the fossil record.

Geological changes and living organisms explain shifts in the atmosphere

Geological changes can explain many, but not all, of the shifts in the composition of the atmosphere. Living organisms have also

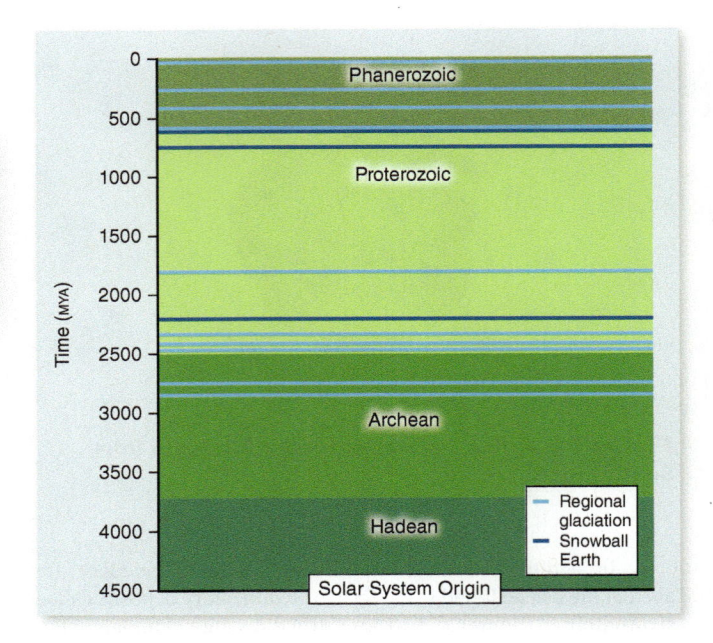

Figure 26.6 Three global glaciation events occurred during the Proterozoic.

substantially altered the atmosphere, as discussed in the section 26.5. Here we focus on CO_2 as one example, recognizing that changes in concentration of other atmospheric gases, including O_2, N_2, and CH_4, have affected the evolution of life on Earth.

At the time of the last two Snowball Earth events in the late Proterozoic, most of the continents were in the tropics (see figure 26.1). During normal times, the hot, wet climate of the tropics accelerated weathering, described earlier, which led to a significant decline in the concentration of atmospheric CO_2. The accompanying decrease in temperature slowed the weathering, stabilizing both temperature and CO_2 levels. Weathering is hypothesized to have led to the late Proterozoic glaciations.

In addition to changes in weathering associated with a hot, wet climate, plate tectonics can also affect weathering and thus atmospheric levels of CO_2. As a large continent breaks into smaller pieces, more area is exposed to the ocean as shoreline and becomes wetter. The increased moisture leads to increased weathering and decreased atmospheric CO_2 concentrations. Thus increased weathering from continental shifts during the late Proterozoic also contributed to glaciation.

Continental motion affected evolution

Although they contribute to a changing atmosphere and climate, shifting plates also affect evolution by reproductively isolating populations or allowing previously separate populations to interbreed (see chapter 22). The continents sit on submerged plates that are in motion. During the Proterozoic eon, all the landmass existed as the single supercontinent Rodinia that began to break up into smaller continents about 700 MYA. This and other shifts have continued throughout Earth's history (see figure 26.1).

The Paleozoic era began with the expansive diversification of life of the Cambrian period and a number of separate continents. But, as the Paleozoic came to a close, Earth once again had a single supercontinent—Pangea. During the Carboniferous and Permian periods of the Paleozoic, major collisions of continents completed the development of the Appalachian mountains of North America, when eastern North America and northwestern Africa smashed together. Pangea lasted for 100 million years and began to break up during the Late Triassic and Early Jurassic periods of the Mesozoic era.

The current Cenozoic era began 65 MYA. Australia and Antarctica separated, as did Greenland and North America. The Atlantic Ocean continued to grow as plates in the mid-Atlantic spread. Greenhouse conditions during the Cretaceous period led to a rise in sea level and extensive continental areas were submerged.

During the last 2 million years, cold periods led to glaciation and a drop in sea level. As a result, a land bridge formed between Asia and North America. Humans and other animals migrated between the once disconnected continents. A connection between Australia and southeast Asia also allowed for migration. Understanding when landmasses were connected and which species existed at the time is critical in explaining the evolution of the diversity of life on Earth.

Life changes Earth

Oxygenic photosynthesis produced atmospheric O_2

The early atmosphere contained CO_2, but, unlike the modern atmosphere, it lacked oxygen gas (O_2). The evolution of photosynthesis added O_2 to the ocean and atmosphere, providing an environment conducive to the evolution of cellular respiration (figure 26.7).

Geological evidence shows a 200-million-year lag between the origins of photosynthesis and substantial concentrations of O_2 in the environment. One explanation for the lag is the formation and precipitation of iron oxide in the ocean. Much of the O_2 released in the ocean during the first 200 million years reacted with elemental iron to form iron oxide, which prevented an increase in atmospheric O_2 concentrations.

As O_2 increased in the atmosphere, some of it interacted with ultraviolet (UV) radiation from the Sun and formed O_3 (ozone). The ozone layer protects terrestrial Earth from UV radiation, reducing the rate of mutations and making life on land possible.

Did plants contribute to glaciations?

There is growing evidence that plants contributed to two glaciations. The initial colonization of land by plants was also followed by gradual cooling and abrupt glaciation 488 to 444 MYA in the Ordovician period. The CO_2 levels just prior to the cooling were in the range of 20 times higher than present-day concentrations, and climate models predict the levels would need to be reduced by 50% to trigger glaciation. Geological weathering could explain some cooling through decreased CO_2, but the decrease would not be sufficient to trigger glaciation.

The early land plants lacked roots, but studies using extant relatives show they released organic acids that can increase rock weathering. Although additional weathering caused by plants could accelerate decreases in CO_2 and temperature, it would still be insufficient to trigger glaciation. The most convincing proposal is that phosphorous, an essential nutrient for plant growth, was released from the rocks through weathering and entered the ocean where it supported extensive algal growth. The rapid growth of the photosynthetic algae led to increased uptake of CO_2 for photosynthesis, decreasing atmospheric CO_2 levels and triggering glaciation. Increased phosphorous in sedimentary rock from that time supports this conclusion. The entire system equilibrated after the initial release of phosphorous and plants began recycling phosphorous in the soil, eliminating the more massive runoff into aquatic environments.

A second glaciation was concurrent with the diversification of vascular plants 400 to 360 MYA during the Devonian period. Vascular plants' extensive root systems increased weathering of rocks, which decreased atmospheric CO_2 levels. Plant roots release the same organic acids as the rootless early land plants that weather rocks, releasing essential nutrients including phosphorous. As the vascular plants colonized Earth, global cooling and glaciation at the poles followed.

Plant-induced glaciation not only changed Earth, but also affected other life. Glaciation often led to rapid drops in sea level. This in turn resulted in extinction of many marine species.

Learning Outcomes Review 26.4

Changes in climate, atmosphere, and location of continents has supported the diversification of life and led to extinctions, as a result of glaciation. In concert, life has changed the Earth's system including land, water, and atmosphere. Both abiotic and biotic factors can lead to decreased CO_2 levels, which can precipitate glaciation.

- Compare and contrast the events that triggered the glaciation at the end of the Proterozoic with the glaciation early in the Ordovician period.

26.5 Ever-Changing Life on Earth

Learning Outcomes

1. Compare and contrast the evolution of the endomembrane system and mitochondria.
2. Discuss the types of cell specialization necessary for multicellularity.

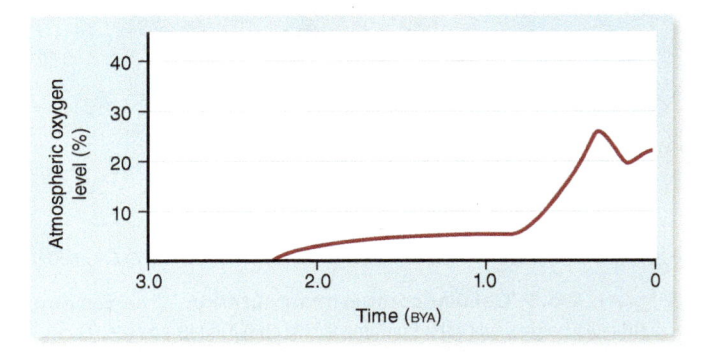

Figure 26.7 Atmospheric O_2 levels over time.

Beginning with a single cell in the Archean eon, life has evolved into three monophyletic clades called domains: Eubacteria, Archaea, and Eukaryotes. Six supergroups have been identified within the eukaryotes, based on their phylogenetic relationships: Excavata (organisms lacking typical mitochondria), Chromalveolata (organisms with chloroplasts obtained through secondary endosymbiosis), Archaeplastida (organisms with chloroplasts for photosynthesis), Rhizaria (organisms with slender pseudopods used for movement), Amoebozoa (organisms with blunt pseudopods used for movement), and Opisthokonta (fungi, animal ancestors, and animals) (figure 26.8). Here we will consider the key evolutionary events supporting this incredible diversity of life evolving in response to an ever-changing environment.

Compartmentalization of cells enabled the advent of eukaryotes

For at least 1 billion years, bacteria and archaea ruled the Earth. No other types of organisms existed to eat them or compete with them, and their tiny cells formed the world's oldest fossils. Archaea are more closely related to eukaryotes and may be found in environments that are extreme by today's standards, including environments characterized by high temperature, high pressure, or high salt. Both bacteria and archaea are distinct from eukaryotes in that they lack compartmentalization of their cells.

Members of the third great domain of life, the eukaryotes, appear in the fossil record much later, only about 1.5 BYA. But despite the metabolic similarity of eukaryotic cells to prokaryotic cells, their structure and function enabled these cells to be larger, and eventually, allowed multicellular life to evolve.

Evolution of the endomembrane system

The hallmark of eukaryotes is complex cellular organization, highlighted by an extensive endomembrane system that subdivides the eukaryotic cell into functional compartments, including the nucleus (figure 26.9; see also chapter 4). The evolution of a nuclear membrane, not found in bacteria and archaea, accounts for increased complexity in eukaryotes. In eukaryotes, RNA transcripts from nuclear DNA are processed and transported across the nuclear membrane into the cytosol, where translation occurs. The physical separation of transcription and translation in eukaryotes adds additional levels of control to the process of gene expression.

The Golgi apparatus and endoplasmic reticulum are key innovations that facilitate intracellular transport and the localization of proteins in specific regions of the cell (see chapter 4). These membrane systems, as well as the nuclear membrane, arose through the infolding of the cellular membrane. Not all cellular compartments, however, are derived from the endomembrane system.

Endosymbiosis and the origin of eukaryotes

With few exceptions, modern eukaryotic cells possess the energy-producing organelles termed *mitochondria*, and photosynthetic eukaryotic cells possess *chloroplasts*, the energy-harvesting organelles. Mitochondria and chloroplasts are both believed to have entered early eukaryotic cells by a process called **endosymbiosis,** which is discussed in more detail in chapter 29.

Mitochondria are the descendants of relatives of purple sulfur bacteria and the parasite *Rickettsia* that were incorporated into eukaryotic cells early in the history of the group. Chloroplasts are derived from cyanobacteria. The red and green algae acquired their chloroplasts by directly incorporating a cyanobacterium. The brown algae most likely incorporated red algae to obtain chloroplasts (figure 26.10).

Molecular phylogenetic data indicate that eukaryotes arose early in the Proterozoic and diversified later in that eon. Microfossil evidence from the Proterozoic supports the existence of eukaryotes as early as 1.5 BYA.

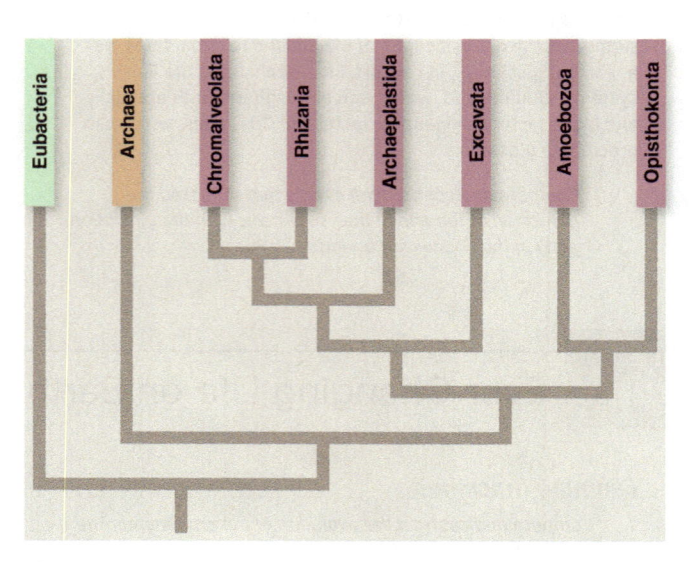

Figure 26.8 Six supergroups have been identified within the Eukaryote domain, one of three domains of life on Earth.

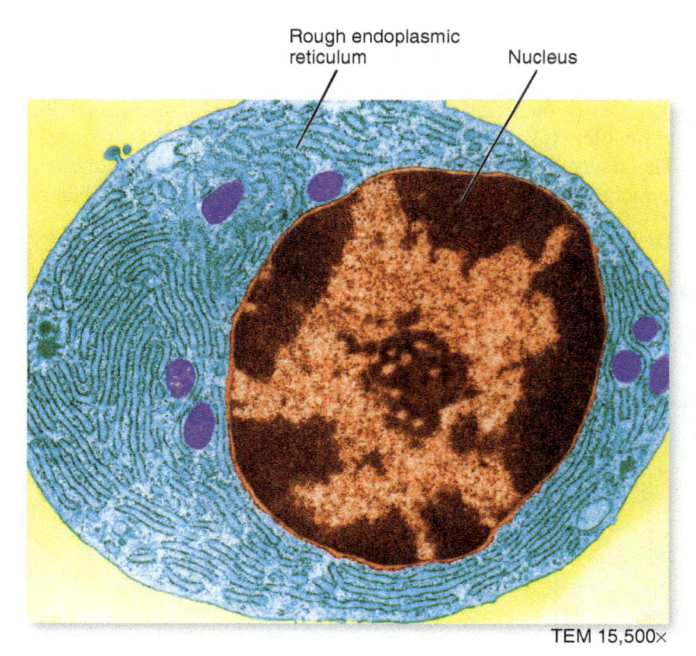

Figure 26.9 Cellular compartmentalization. These complex, single-celled organisms called paramecia are classified as protists. The yeasts, stained purple in this photograph, have been consumed and are enclosed within membrane-bounded sacs called digestive vacuoles.

Bacteria and many single-celled eukaryotes form colonial aggregates of many cells, but the cells in the aggregates have little differentiation or integration of function (figure 26.11).

Multicellularity has arisen independently in different eukaryotic supergroups. For example, multicellularity arose independently in the red, brown, and green algae. One lineage of multicellular green algae was the ancestor of the plants (see chapter 29). A different unicellular ancestor in the Opisthokonts gave rise to all multicellular animals.

Multicellularity required that cells connect to each other and communicate. Although each cell has identical genetic information, gene expression varies among cells to allow specialization. Mechanisms for coordinating gene expression and cell differentiation evolved.

Sexual reproduction increases genetic diversity

Another major characteristic of eukaryotic species as a group is sexual reproduction. Although some interchange of genetic material occurs in bacteria, it is certainly not a regular, predictable mechanism in the same sense that sex is in eukaryotes. Sexual reproduction allows greater genetic diversity through the processes of meiosis and crossing over, as you learned in chapter 13.

In some of the eukaryotic supergroups, sexual reproduction occurs only occasionally. The first eukaryotes were probably haploid; diploids seem to have arisen on a number of separate occasions by the fusion of haploid cells, which then eventually divided by mitosis.

Rapid diversification occurred during the Cambrian

The evolutionary innovations in cells occurred while life was primarily aquatic and established the foundations that led to the tremendous diversity of life on Earth today. As the Proterozoic

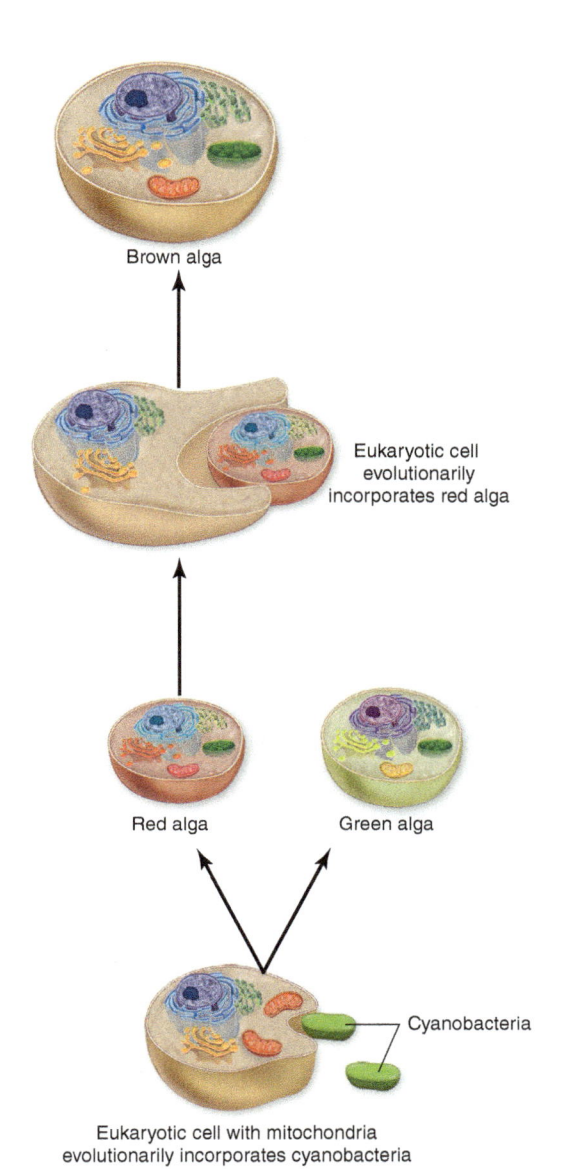

Figure 26.10 All chloroplasts are monophyletic.
The same cyanobacteria were evolutionarily incorporated by multiple hosts that were ancestral to the red and green algae. Brown algae share the same ancestral chloroplast DNA, but most likely gained it by engulfing red algae.

Multicellularity leads to cell specialization

The unicellular body plan has been tremendously successful, with unicellular prokaryotes and eukaryotes constituting about half of the biomass on Earth. But a single cell has limits, even with the within-cell specialization provided by compartmentalization in eukaryotes. The evolution of multicellularity allowed organisms to deal with their environments in novel ways through differentiation of cell types into tissues and organs.

True multicellularity, in which the activities of individual cells are coordinated and the cells themselves are in contact, occurs only in eukaryotes and is one of their major characteristics.

Figure 26.11 Aggregation of *Dictyostelium discoideum* to form a colonial organism.

chapter 26 *The Origin and Diversity of Life* 523

Figure 26.12 Fossil from the Cambrian Explosion. An unusually large number of fossils with soft body parts, such as this *Marrella splendens fossil* from the Burgess Shale in Yoho National Park in British Columbia, provide evidence for the rapid diversification of animal life during this period.

came to a close with the end of the third Snowball Earth event, the Paleozoic began with the Cambrian period lasting from 542 to 488 MYA. This was a period of extremely rapid expansion of life referred to as the Cambrian explosion. We know a great deal about this period because not only the hard parts of organisms, but also the soft parts were preserved in the fossil record in three sites in British Columbia, Greenland, and China (figure 26.12).

For 3 billion years, life, with the exception of a few groups of algae, had been unicellular. In the period leading up to the Cambrian radiation, the first multicellular animals appeared. During the 50 million years that followed, ancestors of almost every group of animals evolved.

Major innovations allowed for the move onto land

The Cambrian radiation was confined to the ocean. Shortly after, plants and then animals colonized terrestrial environments. The evolution of photosynthesis, which resulted in an O_2-rich atmosphere, also resulted in the ozone layer, which protects life on the surface from UV radiation.

Successfully moving from a watery to a terrestrial environment required innovations to prevent desiccation and to obtain water. Gas exchange also required new strategies. In animals, lungs were more effective than gills or gas exchange through moist skin. Plants evolved stomata, regulated openings in the surface of the plant, to facilitate gas exchange and prevent water loss. These and other major innovations are discussed in later chapters 27 through 35.

Learning Outcomes Review 26.5

The history of life on Earth is the history of the diversification of cells and organisms. Eukaryotic cells are highly compartmentalized with endomembrane systems, and they have acquired mitochondria and chloroplasts by endosymbiosis. Multicellularity led to cellular specialization, and meiosis and sexual reproduction increased genetic diversity. Bursts of rapid radiation of species, as well as periodic extinctions, have led to the diversity of marine and terrestrial life in existence today.

- Analyze ways evolutionary innovations prior to the Cambrian may have contributed to the rapid radiation.

Chapter Review

26.1 Deep Time

Earth changed over geological time (figure 26.1).
During the 4.6-billion-year history of life on Earth, Earth has experienced extreme shifts in temperature that correspond with shifting CO_2 levels. Continents formed, collided to make supercontinents, and separated again multiple times. Most of the biological history of life on Earth has occurred in the last 12% of Earth's history.

26.2 Origins of Life

Early organic molecules may have originated in various ways.
Some organic molecules may have had extraterrestrial origins, but most likely formed under the reducing environment of early Earth.

Metabolic pathways may have emerged in various ways.
We are not quite sure how metabolic pathways arose. Autotrophic organisms may have evolved ways to incorporate inorganic molecules in the environment into organic molecules. Early pathways for utilizing energy did not require enzymes.

Single cells were the first life-forms.
At some point, simple cellular life evolved as membranes formed and bounded metabolic and self-replicating activities.

26.3 Evidence for Early Life

Fossil evidence indicates life may have originated 3.2 BYA.
Microfossils with organic wells have been studied with scanning electron microscopy and date back at least 3.2 billion years. More substantial

evidence comes from 2.7-billion-year-old stromatolites, sedimentary deposits of microorganisms.

Isotopic data indicate that carbon fixation is an ancient process.
Carbon dating revealed that organisms were sequestering carbon using an ancient form of photosynthesis as early as 3.8 BYA.

Some hydrocarbons found in ancient rocks may have biological origins.
Organic molecules that clearly have biological origins, including lipids, have been identified in 2.7-billion-year-old rock formations in Australia.

26.4 Earth's Changing System

Earth's climate has been ever-changing.
Earth has been cooling since its formation, but extreme shifts in temperature, correlating with changes in CO_2 levels, have been associated with glaciation events, including three glaciations that covered the entire globe. Glaciations can lead to mass extinction and affect the course of evolution.

Geological changes and living organisms explain shifts in the atmosphere.
Weathering in hot, wet climates and on increased moist surface areas resulting from the breakup of supercontinents resulted in the sequestration of CO_2 in the oceans. In some cases, the drop in CO_2 concentrations can be sufficient to trigger glaciations.

Continental motion affected evolution.
Plate tectonics explains the gradual movement of continents that can change the climate and affect whether or not populations are able to mix and interbreed.

Life changes Earth.
Plant life has contributed to two glaciation events by pulling down CO_2 from the atmosphere through both photosynthesis and the weathering of rock to release phosphorous, a necessary plant nutrient.

26.5 Ever-Changing Life on Earth

Compartmentalization of cells enabled the advent of eukaryotes.
Through both the infolding of membranes to form endomembrane systems and endosymbiosis, eukaryotic cells have gained compartments that enhance cellular function (figure 26.10).

Multicellularity leads to cell specialization.
Multicellularity arose independently in eukaryotic supergroups.

Sexual reproduction increases genetic diversity.
The evolution of meiosis increased the genetic diversity available for natural selection.

Rapid diversification occurred during the Cambrian.
The evolution of compartmentalization, multicellularity, and sexual reproduction established the foundations for the rapid evolution of animal life during the Cambrian period (542 to 488 MYA).

Major innovations allowed for the move onto land.
Moving to a terrestrial environment required adaptations to prevent desiccation, obtain water, and facilitate gas exchange in a dry environment.

Review Questions

UNDERSTAND

1. The Miller–Urey experiment demonstrated that
   a. life originated on Earth.
   b. organic molecules could have originated in the early atmosphere.
   c. the early genetic material on the planet was DNA.
   d. the early atmosphere contained large amounts of oxygen.

2. Plate tectonics can contribute to
   a. volcanoes and earthquakes.
   b. formation of supercontinents.
   c. increased weathering and CO_2 sequestration.
   d. All of the choices are correct.

3. Identify which of the following statements is false and correct the statement.
   a. Brown and red algae are not closely related phylogenetically.
   b. Chloroplasts in brown and red algae are monophyletic.
   c. Brown algae gained chloroplasts by engulfing green algae (endosymbiosis).
   d. None of the statements are false.

4. Which of the following events occurred first in eukaryotic evolution?
   a. Endosymbiosis and mitochondria evolution
   b. Endosymbiosis and chloroplast evolution
   c. Compartmentalization and formation of the nucleus
   d. Formation of multicellular organisms

5. Fossil data indicate that
   a. life originated 4 BYA.
   b. life may have originated 3.5 BYA, but definitely by 3.2 BYA.
   c. the Cambrian explosion led to the origin of life.
   d. plants played an important role in causing glaciation.

APPLY

1. A global glaciation would be unlikely to occur if
   a. a supercontinent formed near the equator and there was extensive rainfall.
   b. millions of acres of forest were cleared.
   c. vast amounts of phosphorous found its way into aquatic and oceanic environments.
   d. there was a rapid expansion of algal populations in the ocean.

2. Although we do not know how early life arose, the following likely happened:
   a. All organic molecules were transported to Earth by meteors.
   b. High levels of oxygen were essential for glycolysis.
   c. Lipids organized to form cell membranes.
   d. Organic molecules formed once temperatures reached moderate levels, equivalent to today's environment.

3. Which bits of evidence would convince you that life originated as early as 3.2 BYA?
   a. You look at a microfossil under a scanning electron microscope and it is the same shape as a cell.
   b. A high-quality transmission electron micrograph of a fossil cell reveals cellular compartments, including a possible nucleus.
   c. Potassium dating of a fossil containing a possible cell indicates that the fossil is 3.2 billion years old.
   d. Using transmission and scanning electron microscopy, you find evidence of a carbon-based material in what appears to be a cell wall of a fossil isotopically dated as 3.2 billion years old.

4. During which times would you expect that geographic isolation would be particularly important in the evolution of life?
   a. Cambrian period
   b. End of the Paleozoic era
   c. The beginning of the Cenozoic era
   d. Both a and c are correct.

5. The chloroplasts of brown algae
   a. have a chromosome with a very different DNA sequence than the chromosome of a red alga.
   b. are surrounded by four membranes.
   c. are surrounded by two membranes.
   d. have a chromosome with a DNA sequence that is similar to red algae, but very different from green algae.

SYNTHESIZE

1. Vascular plants sequestered large amounts of CO_2 through photosynthesis when they first expanded across terrestrial environment, leading to a glaciation. Evaluate the claim that a similar glaciation event occurred when the first plants colonized land. Be sure to consider whether or not photosynthesis by these new species alone could cause a glaciation.

2. Synthesizing what you have learned in this chapter, analyze the multiple effects the collision of two plates could have on the evolution of life on Earth.

3. Analyze the factors that may have contributed to the Cambrian explosion.

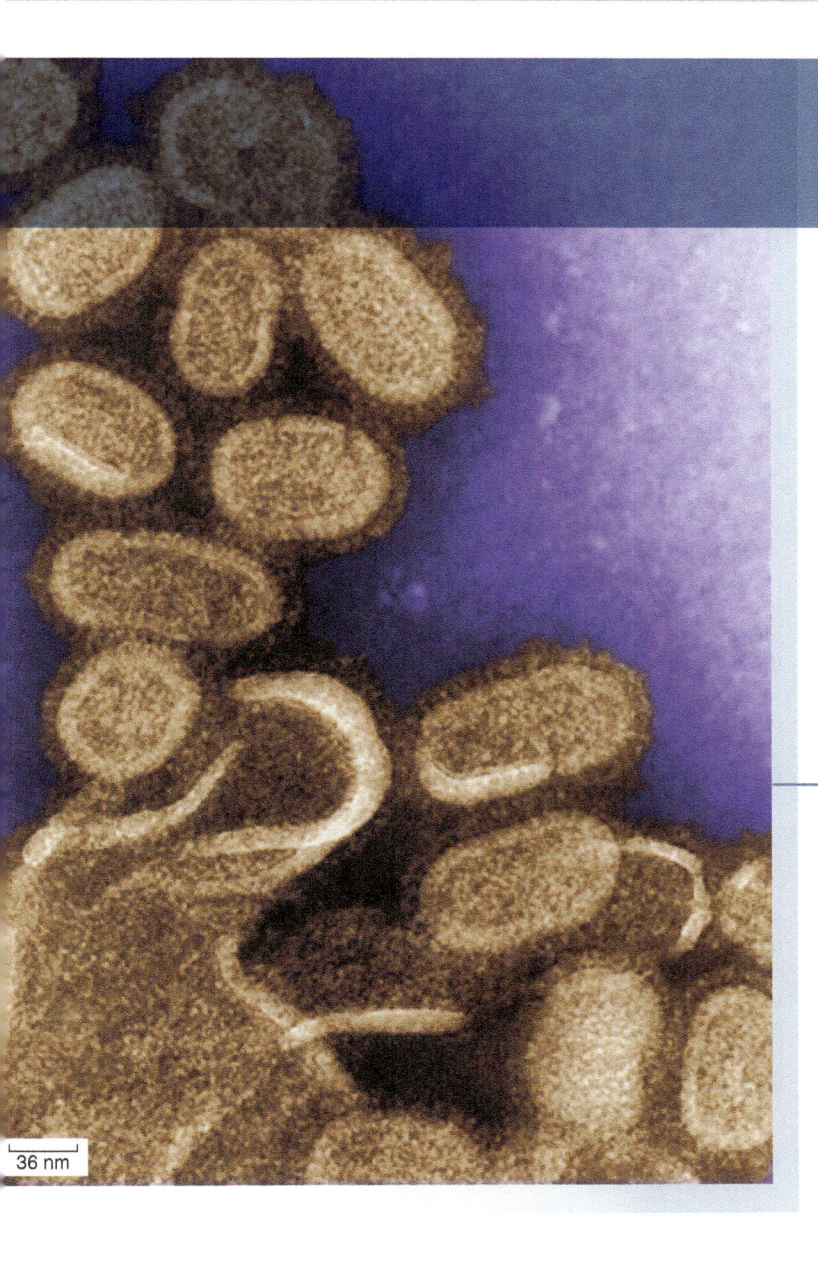

CHAPTER 27

Viruses

Chapter Contents

27.1 The Nature of Viruses
27.2 Bacteriophage: Bacterial Viruses
27.3 Human Immunodeficiency Virus (HIV)
27.4 Other Viral Diseases
27.5 Prions and Viroids: Subviral Particles

36 nm

Introduction

We begin our exploration of the diversity of life with viruses. Viruses are genetic elements enclosed in protein; they are not considered organisms because they lack many of the features associated with life, including cellular structure, and independent metabolism or replication. For this reason viral particles are not called viral cells, but *virions*, and they are generally not described as living or dead but as active or inactive. Because of their disease-producing potential, however, viruses are important biological entities. The virus particles pictured here are responsible for causing influenza—flu for short. In the flu season of 1918 to 1919, an influenza pandemic killed an estimated 50 to 100 million people worldwide, twice as many as were killed in combat during World War I. Other viruses cause such diseases as AIDS, SARS, and hemorrhagic fever, and some cause certain forms of cancer.

For more than four decades, viral studies have been thoroughly intertwined with those of genetics and molecular biology. Classic studies using viruses that infect bacteria (known as *bacteriophage*) have led to the discovery of restriction enzymes and the identification of nucleic acid, not protein, as the hereditary material. Currently, viruses are one of the principal tools used to experimentally carry genes from one organism to another. Applications of this technology could include treating genetic illnesses and fighting cancer.

27.1 The Nature of Viruses

Learning Outcomes
1. Describe the different structures found in viruses.
2. Understand the basic mechanism of viral replication.

All viruses have the same basic structure—a core of nucleic acid surrounded by protein. This structure lacks cytoplasm, and it is not a cell. Individual viruses contain only a single type of nucleic acid, either DNA or RNA. The DNA or RNA genome may be linear or circular; single-stranded or double-stranded.

RNA viruses may be segmented, with multiple RNA molecules within a virion, or nonsegmented, with a single RNA molecule. Viruses are classified, in part, by the nature of their genomes: RNA viruses, DNA viruses, or retroviruses.

Viruses are strands of nucleic acids encased in a protein coat

Nearly all viruses form a protein sheath, or **capsid,** around their nucleic acid core (figure 27.1). The capsid is composed of one to a few different protein molecules repeated many times. The repeating units are called capsomeres.

In several viruses, specialized enzymes are stored with the nucleic acid, inside the capsid. One example is reverse transcriptase, which is required for retroviruses to complete their cycle and is not found in the host. This enzyme is needed early in the infection process and is carried within each virion.

Many animal viruses have an *envelope* around the capsid that is rich in proteins, lipids, and glycoproteins. The lipids found in the envelope are derived from the host cell; however, the proteins in a viral envelope are generally virally encoded.

Viral hosts include virtually every kind of organism

Viruses occur as obligate intracellular parasites in every kind of organism that has been investigated for their presence. Viruses infect fungal cells, bacterial cells, and protists as well as cells of plants and animals; however, each type of virus can replicate in only a very limited number of cell types. A virus that infects bacteria would be ill-equipped to infect a human or plant cell.

The suitable cells for a particular virus are collectively referred to as its **host range.** Once inside a multicellular host, many viruses also exhibit **tissue tropism,** targeting only a specific set of cells. For example, rabies virus grows within neurons, and hepatitis virus replicates within liver cells. Once inside a host cell, some viruses, such as the highly dangerous Ebola virus, wreak havoc on the cells they infect; others produce little or no damage. Still other viruses remain dormant until a specific signal or event triggers their expression.

As one example, a person can get chicken pox as a child, recover, and develop the disease shingles decades later. Both chicken pox and shingles are caused by the same virus, varicella-zoster. This virus can remain dormant, or **latent,** for years. Stresses to the immune system may trigger an outbreak of shingles in people who have had chicken pox in the past. This is caused by the same virus, but the infection may be called herpes zoster because the virus is actually a herpesvirus.

Any given organism may often be susceptible to more than one kind of virus. This observation suggests that many more kinds of viruses may exist than there are kinds of organisms—perhaps trillions of different viruses. Only a few thousand viruses have been described at this point.

Structure	a.	b.	c.	d.
Virus	Plant virus (TMV)	Animal virus (adenovirus)	Bacterial virus	Animal virus (influenza)
Viral shape	Helical capsid	Icosahedral capsid	Icosahedral head: helical tail	Helical capsid within envelope

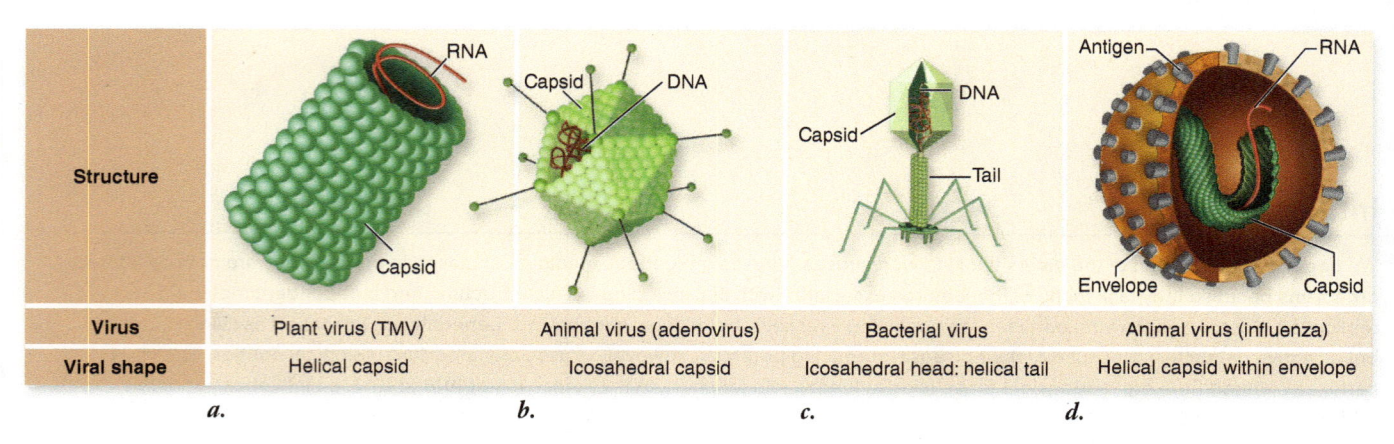

Figure 27.1 Structure of virions. Viruses are characterized as helical, icosahedral, binal, or polymorphic, depending on their symmetry. *a.* The capsid may have helical symmetry such as the tobacco mosaic virus (TMV). TMV infects plants and consists of 2130 identical protein molecules *(green)* that form a cylindrical coat around the single strand of RNA *(red)*. *b.* The capsid of icosahedral viruses has 20 facets made of equilateral triangles. These viruses can come in many different sizes all based on the same basic shape. *c.* Bacteriophage come in a variety of shapes, but binal symmetry is exclusively seen in phages such as the T4 phage of *E. coli*. This form of symmetry is characterized by an icosahedral head, which contains the viral genome, and a helical tail. *d.* Viruses can also have an envelope surrounding the capsid such as the influenza virus. This gives the virus a polymorphic shape. This virus has eight RNA segments, each within a helical capsid.

Viruses replicate by taking over host machinery

An infecting virus can be thought of as a set of instructions, not unlike a computer program. A cell is normally directed by chromosomal DNA-encoded instructions, just as a computer's operation is controlled by the instructions in its operating system. A virus is simply a set of instructions, the viral genome, that can trick the cell's replication and metabolic enzymes into making copies of the virus. Computer viruses get their name because they perform similar actions, taking over a computer and directing its activities. Like a computer with a virus, a cell with a virus is often damaged by infection.

Viruses can reproduce only when they enter cells. When they are outside of a cell, viral particles are called *virions* and are metabolically inert. Viruses lack ribosomes and the enzymes necessary for protein synthesis and most, if not all, of the enzymes for nucleic acid replication. Inside cells, the virus hijacks the transcription and translation systems to produce viral proteins from *early genes,* which are the genes in the viral genome expressed first. This is followed by the expression of *middle genes* and eventually *late genes.* This cascade of gene expression leads to replication of viral nucleic acid and production of viral capsid proteins. The late genes generally code for proteins important in assembly and release of viral particles from a host cell.

Most viruses come in two simple shapes

Most viruses have an overall structure that is either *helical* or *icosahedral*. Helical viruses, such as the tobacco mosaic virus in figure 27.1a, have a rodlike or threadlike appearance. Icosahedral viruses have a soccer ball shape, the geometry of which is revealed only under the highest magnification with an electron microscope.

The **icosahedron** is a structure with 20 equilateral triangular facets. Most animal viruses are icosahedral in basic structure (figure 27.1b). The icosahedron is the basic design of the geodesic dome, and it is the most efficient symmetrical arrangement that subunits can take to form a shell with maximum internal capacity (figure 27.2).

Some viruses, such as the T-even bacteriophage shown in figure 27.3, are complex. Complex viruses have a *binal,* or two-fold, symmetry that is not either purely icosahedral or helical. The

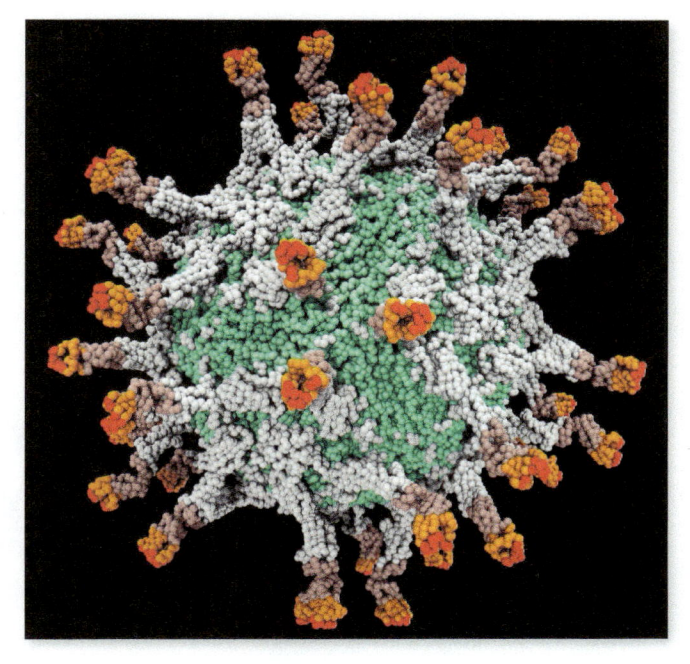

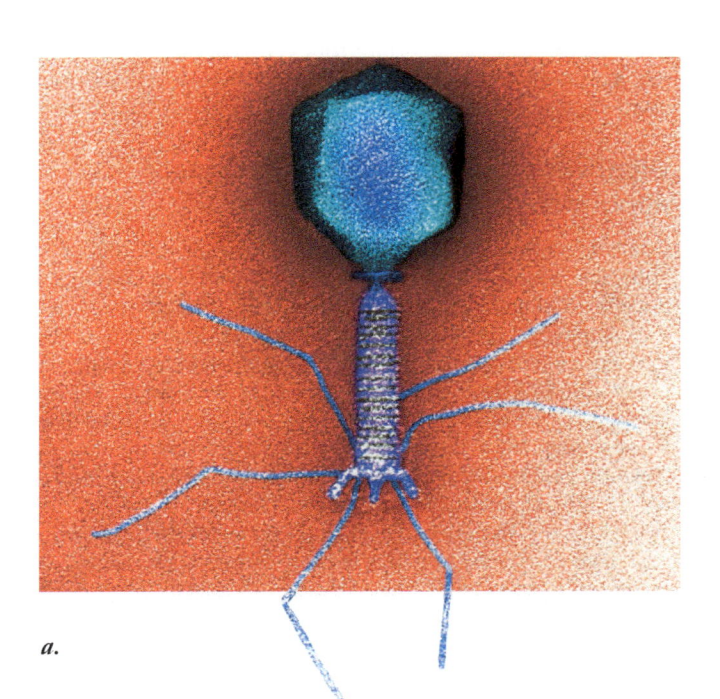

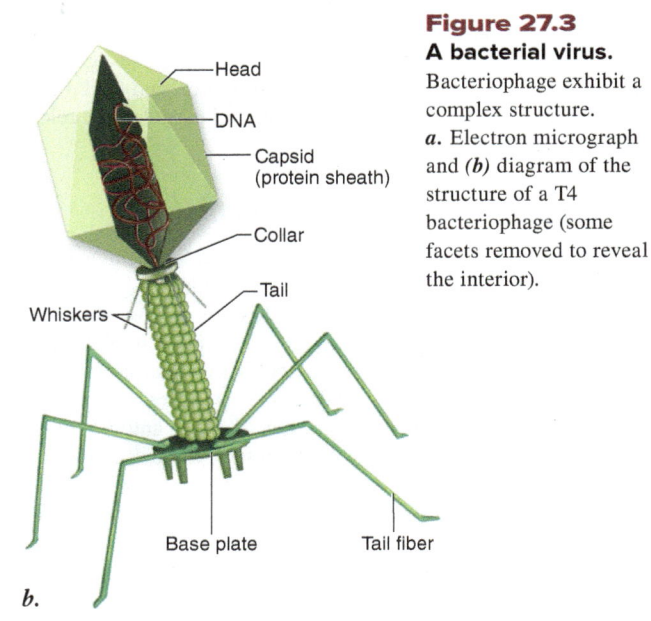

Figure 27.2 Icosahedral virion. The poliovirus has icosahedral symmetry. The capsid is formed from multiple copies of four different proteins. Image is a digitally generated 3D model.

Figure 27.3 A bacterial virus. Bacteriophage exhibit a complex structure. *a.* Electron micrograph and (*b*) diagram of the structure of a T4 bacteriophage (some facets removed to reveal the interior).

a.

b.

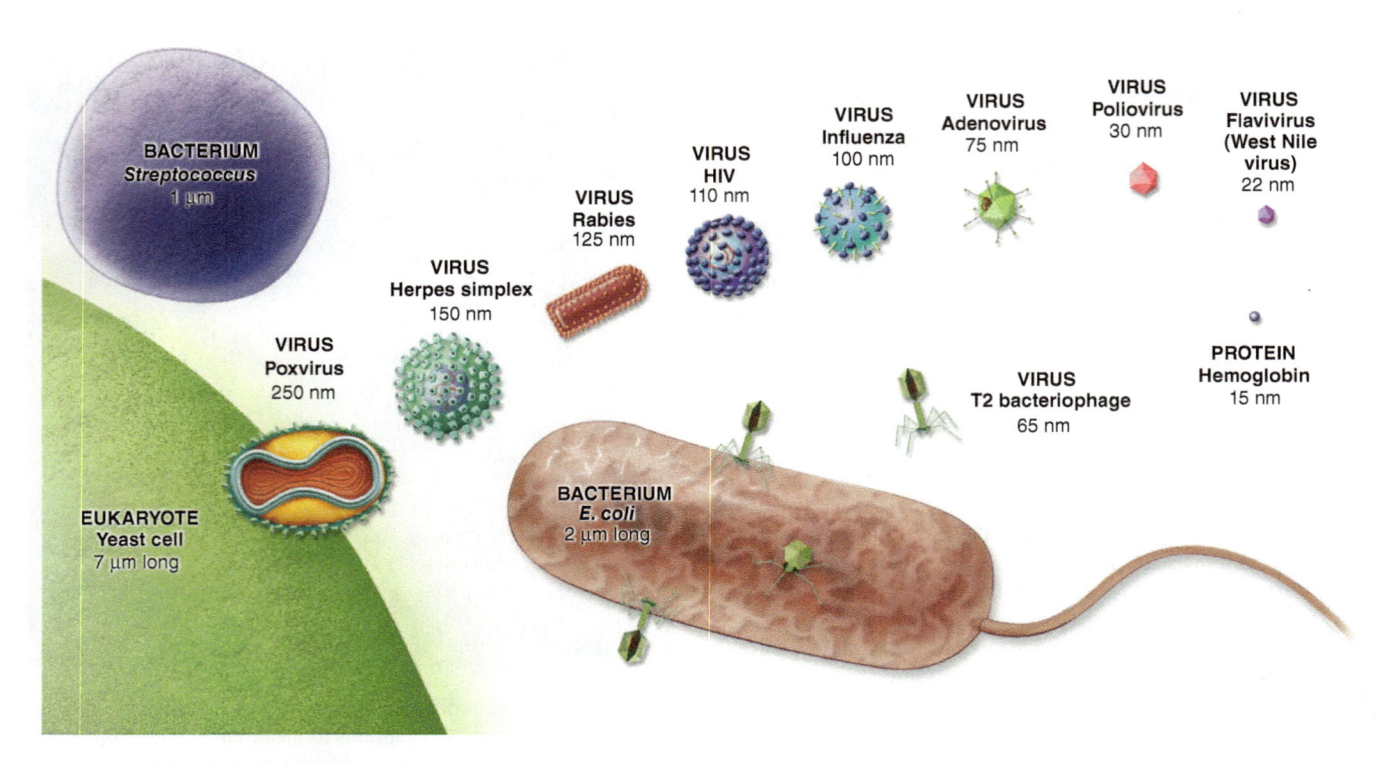

Figure 27.4 Viruses vary in size and shape. Note the dramatic differences in the size of a eukaryotic yeast cell, prokaryotic bacterial cells, and the many different viruses.

T-even phage shown has a head structure that is an elongated icosahedron. A collar connects the head to a hollow tube with helical symmetry that ends in a complex baseplate with tail fibers. Although animal viruses do not have this binal symmetry, some, such as the poxviruses, do have a complex multilayered capsid structure. Some enveloped viruses, such as influenza, are *polymorphic*, having no distinctive symmetry.

As shown in figure 27.4, the very smallest viruses, such as the poliovirus, have actually been synthesized in a lab from sequence data. The larger viruses, such as the poxviruses, generally carry more genes, have more complex structures, and tend to have a very short cycle time between entry of viral particles and release of newly formed virions.

Viral genomes exhibit great variation

Viral genomes vary greatly in both type of nucleic acid and number of strands (table 27.1). Some viruses, including those that cause flu, measles, and AIDS, possess RNA genomes. Most RNA viruses are single-stranded and are replicated and assembled in the cytosol of infected eukaryotic cells. RNA virus replication is error-prone, leading to high rates of mutation. This makes them difficult targets for the host immune system, vaccines, and antiviral drugs.

In single-stranded RNA viruses, if the genome has the same base sequence as the mRNA used to produce viral proteins, then the genomic RNA can serve as the mRNA. Such viruses are called *positive-strand viruses*. In contrast, if the genome is complementary to the viral mRNA, then the virus is called a *negative-strand virus*.

A special class of RNA viruses, called *retroviruses*, have an RNA genome that is reverse-transcribed into DNA by the enzyme *reverse transcriptase*. The DNA fragments produced by reverse transcription are often integrated into a host's chromosomal DNA. *Human immunodeficiency virus (HIV)*, the agent that causes *acquired immunodeficiency syndrome (AIDS)*, is a retrovirus. (We describe HIV in detail in section 27.3.)

Other viruses, such as the viruses causing smallpox and herpes, have DNA genomes. Most DNA viruses are double-stranded, and their DNA is replicated in the nucleus of eukaryotic host cells.

Giant viruses challenge assumptions about viruses

One common practical definition of viruses was that they were "filterable agents." That is, infectious agents that could pass through a 500-nm filter. The first virus to fail this centuries-old test was Mimivirus, originally misidentified as a bacterium that infected amoebas. When Mimivirus was analyzed, it proved to have a virion 750 nm in diameter, and a genome of more than 1 Mb (megabase, or 1 million bases). A number of other so-called giant viruses were isolated, and the current largest known virus is Pandoravirus, isolated in 2013, with micrometer-sized virions and a DNA genome of 2.5 Mb.

Although the size of these viruses is enough to challenge assumptions about the distinction between viruses and cells, the genomes of some of these giant viruses also contain features not previously observed in viral genomes. The Mimivirus genome contains nine genes central to the translation process, including

TABLE 27.1 Important Human Viral Diseases

Disease	Pathogen	Genome	Vector/Epidemiology
Chicken pox	Varicella-zoster virus	Double-stranded DNA	Spread through contact with infected individuals. No cure. Rarely fatal. Vaccine approved in U.S. in early 1995. May exhibit latency leading to shingles. A vaccine for shingles that consists of a higher dose of the '95 vaccine was approved in 2006.
Hepatitis B (viral)	Hepadnavirus	Double-stranded DNA	Highly infectious through contact with infected body fluids. Approximately 1% of U.S. population infected. Vaccine available. No cure. Can be fatal.
Herpes	Herpes simplex virus	Double-stranded DNA	Blisters; spread primarily through skin-to-skin contact with cold sores/blisters. Very prevalent worldwide. No cure. Exhibits latency—the disease can be dormant for several years.
Mononucleosis	Epstein–Barr virus	Double-stranded DNA	Spread through contact with infected saliva. May last several weeks; common in young adults. No cure. Rarely fatal.
Smallpox	Variola virus	Double-stranded DNA	Historically a major killer; the last recorded case of smallpox was in 1977. A worldwide vaccination campaign wiped out the disease completely.
AIDS	HIV	(+) Single-stranded RNA (two copies)	Destroys immune defenses, resulting in death by opportunistic infection or cancer. In 2013, WHO estimated that 35 million people are living with AIDS, with an estimated 2.1 million new HIV infections and an estimated 1.5 million deaths.
Polio	Enterovirus	(+) Single-stranded RNA	Acute viral infection of the CNS that can lead to paralysis and is often fatal. Prior to the development of Salk's vaccine in 1954, 60,000 people a year contracted the disease in the U.S. alone.
West Nile fever	Flavivirus	(+) Single-stranded RNA	Spread by mosquitoes and can be amplified in bird hosts. Can lead to neurological problems. Present in U.S. since 1999.
Ebola	Filoviruses	(−) Single-stranded RNA	Acute hemorrhagic fever; virus attacks connective tissue, leading to massive hemorrhaging and death. Peak mortality is 50–90% if untreated. Outbreaks confined to local regions of central Africa.
Influenza	Influenza viruses	(−) Single-stranded RNA (eight segments)	Historically a major killer (20–50 million died during 18 months in 1918–1919); wild Asian ducks, chickens, and pigs are major reservoirs. The ducks are not affected by the flu virus, which shuffles its antigen genes while multiplying within them, leading to new flu strains. Vaccines are available.
Measles	Paramyxoviruses	(−) Single-stranded RNA	Extremely contagious through contact with infected individuals. Vaccine available. Usually contracted in childhood, when it is not serious; more dangerous to adults.
SARS	Coronavirus	(−) Single-stranded RNA	Acute respiratory infection; an emerging disease, can be fatal, especially in the elderly. Commonly infected animals include bats, foxes, skunks, and raccoons. Domestic animals can be infected.
Rabies	Rhabdovirus	(−) Single-stranded RNA	An acute viral encephalomyelitis transmitted by the bite of an infected animal. Fatal if untreated. Commonly infected animals include bats, foxes, skunks, and raccoons. Domestic animals can be infected.

aminoacyl-tRNA synthetases. The lack of proteins involved in translation was considered a hallmark of viruses, and presence of any part of the translation machinery implies these viruses could have once been autonomous. These viruses also replicate in the cytoplasm and encode virtually all proteins necessary for DNA replication. Together, these observations tend to blur the line between a virus and an obligate intracellular parasitic bacterium.

Learning Outcomes Review 27.1

Viruses have a very simple structure that includes a nucleic acid genome encased in a protein coat. Viruses replicate by taking over a host's cell systems and are thus obligate intracellular parasites. Viruses show diverse genomes that are composed of DNA or RNA, which may be single- or double-stranded; most DNA viruses are double-stranded. Giant viruses have virions and genomes similar in size to a bacterium.

- *Why can't viruses replicate outside of a cell?*

27.2 Bacteriophage: Bacterial Viruses

Learning Outcomes
1. Distinguish between lytic and lysogenic cycles in bacteriophage.
2. Describe how viruses can contribute DNA to their hosts.

Bacteriophage (both singular and plural) are viruses that infect bacteria. They are diverse, both structurally and functionally, and are united solely by their occurrence in bacterial hosts. Many of these types of bacteriophage, called *phage* for short, are large and complex, with relatively large amounts of DNA and proteins.

Phage that infect *E. coli* were among the first to be discovered and are still some of the best studied. Some of these viruses have been named as members of a "T" series (T1, T2, etc.). To illustrate the diversity of these viruses, T3 and T7 phage are icosahedral and have short tails. In contrast, the so-called T-even phage are more complex with an icosahedral head, a neck with a collar and "whiskers," a long tail, and a complex base plate (see figure 27.3).

Archaeal viruses have diverse morphologies

Archaeal viruses were initially thought to be similar to bacterial viruses, but current evidence argues against this. Surveys of viruses in several extreme environments dominated by archaeal species have uncovered an unexpected diversity of viral forms. In addition to the viral types described in section 27.1, viruses with a two-tailed structure, with a bottle-shaped structure, and with a spindle-shaped structure have all been observed. All of these viruses have double-stranded DNA genomes, and most appear to be unrelated to any bacteriophage. The characterization of these viruses is in the early stages, so we will not discuss them further.

Bacterial viruses exhibit two reproductive cycles

The usual result of viral infection is production of new virus particles, which are released, generally killing the cell. Viral release allows a new cycle of infection, or horizontal transmission of the virus. The infection cycle is called a lytic cycle because the virus usually causes the cells to rupture, or lyse. Some bacterial viruses can also enter a latent phase, called the lysogenic cycle, after the initial infection. These latent viruses are then transmitted vertically by cell division.

The lytic cycle

The lytic cycle of virus reproduction is illustrated in figure 27.5. The basic steps of a phage lytic cycle are similar to those of a non-enveloped animal virus. We will use the lytic cycle as an example of a viral life cycle. Although the basic outline of this infective cycle is common to most viruses, the details vary as do the viruses themselves.

The first step is called **attachment** (or absorption), in which the virus contacts the cell and becomes specifically bound to the cell. This step limits the host range of the virus as they bind to specific proteins on the surface of the cell. Different phages may target different parts of the outer surface of a bacterial cell. The next step is called **penetration** and results in the release of the viral genome into the host. This has been studied in detail in the binal phage, such as T4. Once contact is established, the tail contracts, and the tail tube passes through an opening that appears in the base plate, piercing the bacterial cell wall. The viral genome is literally injected into the host cytoplasm.

Once inside the bacterial cell, in the **synthesis** phase, the virus takes over the cell's replication and protein synthesis machinery in order to synthesize viral components. These components are then assembled (**assembly** phase) to produce mature virus particles.

In the **release** phase, mature virus particles are released, either through the action of enzymes that lyse the host cell or by budding through the host cell wall. The time between adsorption and the formation of new viral particles is called an eclipse period because if a cell is lysed at this point, few if any active virions can be released.

The lysogenic cycle

Phage that are capable of latent infection usually do this by integrating their nucleic acid into the genome of the infected host cell. This integration allows a virus to be replicated along with the host cell's DNA. These viruses are called **temperate,** or **lysogenic,** phage. The integrated phage genome is called a **prophage,** and the cell containing a prophage is called a **lysogen.**

The best-studied lysogenic phage is lambda (λ), which infects *E. coli*. When phage λ infects a cell, early events determine whether the virus will replicate and destroy the cell or become a lysogen and be passively replicated with the cell's genome. This lysis/lysogeny "decision" depends on the expression of early genes. Two regulatory proteins are produced that compete for binding to sites on the phage's DNA. Depending on which protein "wins," either the genes necessary for replication of the genome are expressed starting the lytic cycle, or the enzymes necessary for integration of the viral genome are expressed producing a lysogen (figure 27.5, *right*).

Figure 27.5 Comparison of phage lytic and lysogenic cycles. In the lytic cycle the viral DNA directs the production of new viral particles by the host cell until the virus kills the cell by lysis. In the lysogenic cycle, the bacteriophage DNA is integrated into the host chromosome. This prophage is replicated along with the host DNA as the bacterium divides. It may persist as a prophage, or enter the lytic cycle and kill the cell. Bacteriophage not drawn to scale in this diagram.

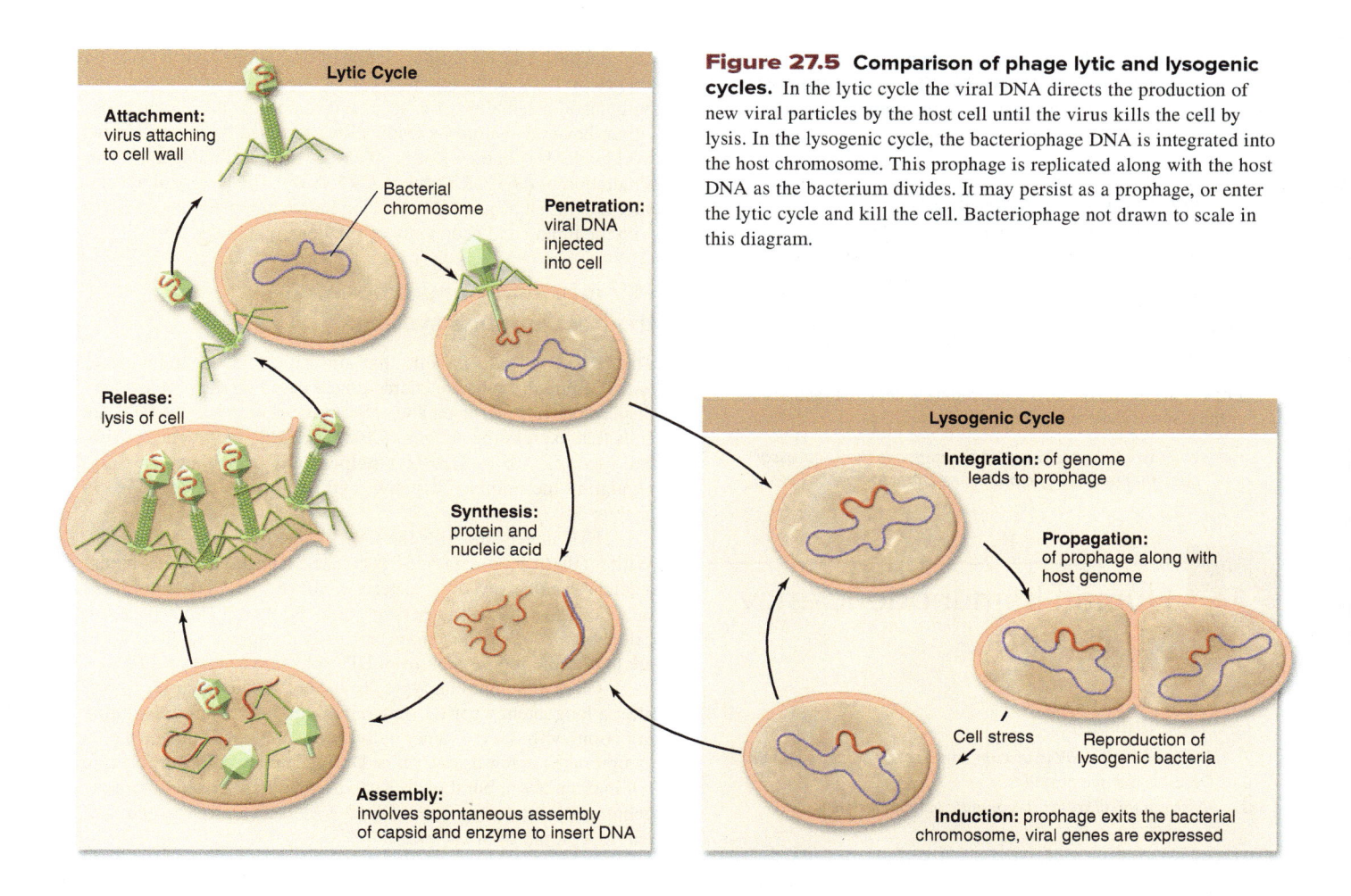

In a lysogen expression of the prophage genome is repressed (see chapter 16) by one of the two viral regulatory proteins involved in the initial decision. Although the lysogen is quite stable, it can be induced to enter the lytic cycle. If the cell is stressed, the prophage can be derepressed, expressing the enzymes necessary for excision of the genome, and the lytic cycle can commence.

The switch from a lysogenic prophage to a lytic cycle is called induction because it requires turning on the gene expression necessary for the lytic cycle (see chapter 16). It can be stimulated in the laboratory by stressors such as starvation or ultraviolet radiation. The molecular basis of induction is a stress-induced protease that can destroy the repressor protein that keeps the prophage silent. The normal function of this protease is to degrade a host repressor that controls DNA repair genes.

Bacteriophage can contribute genes to the host genome

During the integrated portion of a lysogenic reproductive cycle, a few viral genes may be expressed at the same time as host cell genes. Sometimes the expression of these genes has an important effect on the host cell, altering it in novel ways. When the pheno-type or characteristics of the lysogenic bacterium is altered by the prophage, the alteration is called **phage conversion**.

Phage conversion of the cholera-causing bacterium

The bacterium *Vibrio cholerae* usually exists in a harmless form, but a second, disease-causing form also occurs. In this latter form, the bacterium is responsible for the deadly disease cholera, but how the bacteria changed from harmless to deadly was not known until recently.

Research now shows that a lysogenic phage that infects *V. cholerae* introduces into the host bacterial cell a gene that codes for the cholera toxin. This gene, along with the rest of the phage genome, becomes incorporated into the bacterial chromosome. The toxin gene is expressed along with the other host genes, thereby converting the benign bacterium to a disease-causing agent.

The receptors used by this toxin-encoding phage are pili (hairlike projections) found on the outer surface of *V. cholerae* (see chapter 28); in recent experiments, it was determined that mutant bacteria that did not have pili were resistant to infection by the bacteriophage. This discovery has important implications in efforts to develop vaccines against cholera, which have been unsuccessful up to this point. Phage conversion could change any pili-expressing, nontoxigenic *V. cholerae* into a toxin-producing, potentially deadly form.

Another example involved in human disease is the toxin found in *Corynebacterium diphtheriae*. This toxin is the product of phage conversion, as are the changes to the outer surface of certain infectious *Salmonella* species.

> **Learning Outcomes Review 27.2**
>
> Bacterial viruses are called bacteriophage, or just phage. They have two possible life cycles: a lytic cycle that results in immediate death of the host, and a lysogenic cycle where the viral genome is integrated into the host genome, creating a lysogen carrying a prophage. The prophage is transmitted vertically by cell division. The lytic cycle by can be induced in a lysogen by environmental conditions. A prophage can contribute genes to the host. The toxin produced by *V. cholerae* responsible for the disease cholera is encoded by a prophage.
>
> ■ *What would be the result of a mutation in the λ repressor gene that resulted in a protein resistant to host protease?*

27.3 Human Immunodeficiency Virus (HIV)

> **Learning Outcomes**
> 1. Explain how the HIV virus compromises the immune system.
> 2. Describe the disease AIDS.
> 3. Illustrate the different therapeutic options for AIDS.

A diverse array of viruses occurs among animals. A good way to gain a general idea of the characteristics of these viruses is to look at one animal virus in detail. Here we examine the virus responsible for a comparatively new and fatal viral disease, *acquired immunodeficiency syndrome (AIDS)*.

AIDS is caused by HIV

The disease now known as AIDS was first reported in the United States in 1981, although a few dozen people in the United States had likely died of AIDS prior to that time and had not been diagnosed. Frozen plasma samples and estimates based on evolutionary speed and current diversity of HIV strains trace the origins of HIV in the human population to Africa in the 1950s. It was not long before the infectious agent, a retrovirus, was identified by laboratories in France. Study of HIV revealed it to be closely related to a chimpanzee virus (simian immunodeficiency virus, SIV), suggesting a recent host expansion to humans from chimpanzees in central Africa.

Infected humans have varying degrees of resistance to HIV. Some have little resistance to infection and rapidly progress from having HIV-positive status to developing AIDS and eventually die. Others, even after repeated exposure, fail to become HIV-positive or may become HIV-positive without developing AIDS.

One hypothesis to explain this great variability in susceptibility is genetic variation among these groups due to the selective pressure put on the human population by the smallpox virus (*Variola major*) over the centuries. Because of successful vaccination and immunization, smallpox has been eradicated from the human population; however, before its eradication, it caused billions of deaths worldwide. It is known that people resistant to HIV infection have a mutation in the CCR5 gene (CCR5-Δ32). The historical appearance and distribution of this allele in human populations correlates with the timing and geography of smallpox outbreaks.

HIV infection compromises the host immune system

The HIV virus targets cells that are critical to the immune response. Immune cells are characterized based on the proteins they display on their surface. The relevant cells targeted by HIV are cells that express the antigen CD4, thus **CD4$^+$ cells.** The specific cell type infected by HIV is **T-helper cells,** which are critical to regulating the immune response, and their action is described in chapter 51.

HIV infects and kills the CD4$^+$ cells until very few are left. Without these crucial immune system cells, the body cannot mount a defense against invading bacteria or viruses. AIDS patients die of infections that a healthy person could fight off. These diseases, called *opportunistic infections*, normally do not cause disease and are part of the progression from HIV infection to having AIDS.

Clinical symptoms typically do not begin to develop until after a long latency period, generally 8 to 10 years after the initial infection with HIV. Some individuals, however, may develop symptoms in as few as two years. During latency, HIV particles are not in circulation, but the virus can be found integrated within the genome of macrophages and CD4$^+$ T cells as a provirus (equivalent to a prophage in bacteria).

HIV testing

HIV tests do not test for the presence of circulating virus but rather for the presence of antibody against HIV. Because only those people exposed to HIV in their bloodstream at one time or another would have anti-HIV antibodies, this screening provides an effective way to determine whether further testing is needed to confirm HIV-positive status.

The spread of AIDS

Although carriers of HIV have no clinical symptoms during the long latency period, they are apparently fully infectious, which makes the spread of HIV very difficult to control. The reason HIV remains hidden for so long seems to be that its infection cycle continues throughout the 8- to 10-year latency period without doing serious harm to the infected person because of an effective immune response. Eventually, however, a random mutational event in the virus or a failure of the immune response allows the virus to quickly overcome the immune defense, beginning the course of AIDS.

HIV infects key immune-system cells

The way in which HIV infects humans provides a good example of how animal viruses replicate (figure 27.6). Most other viral infections follow a similar course, although the details of entry and replication differ in individual cases.

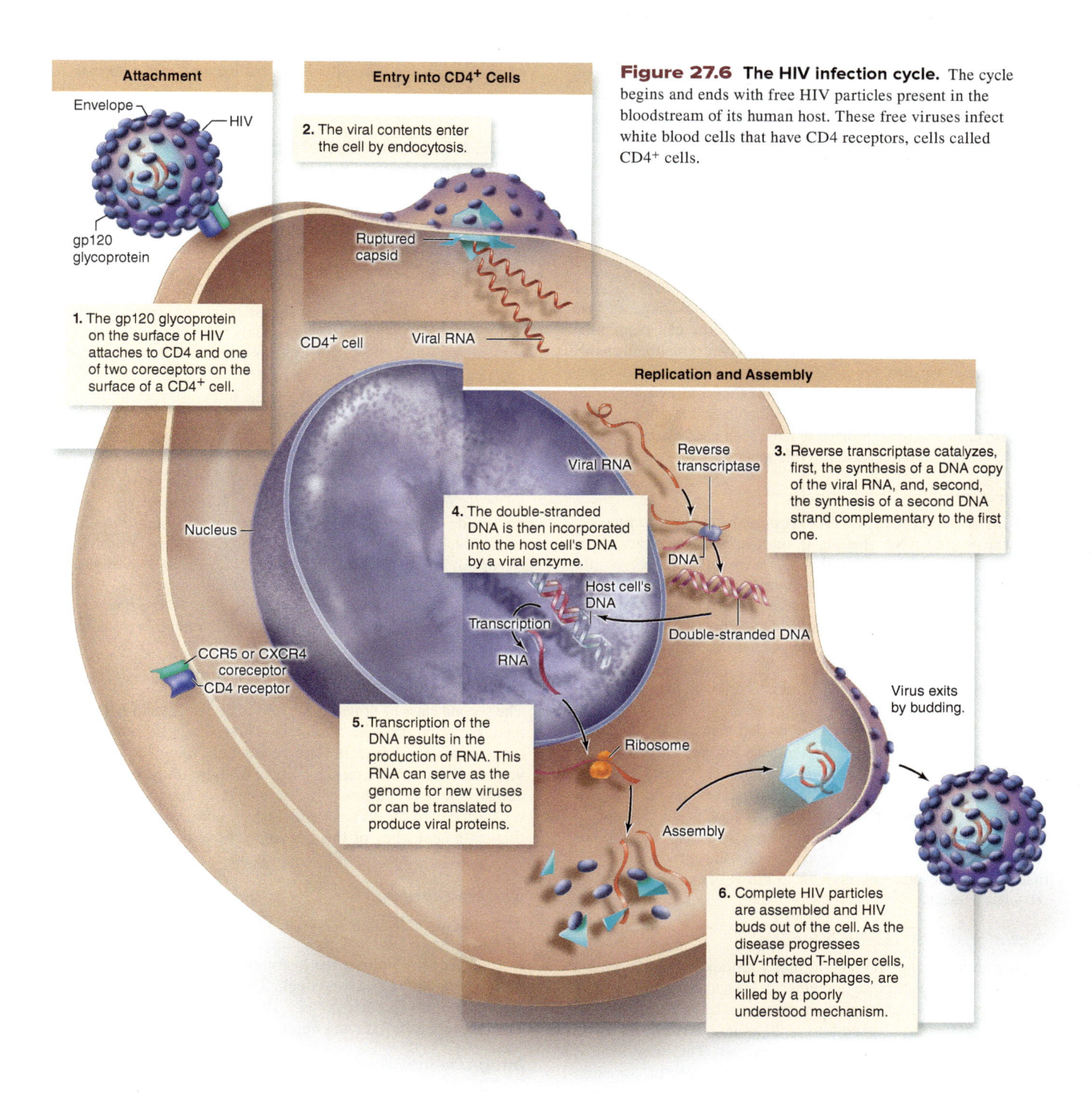

Figure 27.6 The HIV infection cycle. The cycle begins and ends with free HIV particles present in the bloodstream of its human host. These free viruses infect white blood cells that have CD4 receptors, cells called CD4+ cells.

Attachment

1. The gp120 glycoprotein on the surface of HIV attaches to CD4 and one of two coreceptors on the surface of a CD4+ cell.

Entry into CD4+ Cells

2. The viral contents enter the cell by endocytosis.

Replication and Assembly

3. Reverse transcriptase catalyzes, first, the synthesis of a DNA copy of the viral RNA, and, second, the synthesis of a second DNA strand complementary to the first one.

4. The double-stranded DNA is then incorporated into the host cell's DNA by a viral enzyme.

5. Transcription of the DNA results in the production of RNA. This RNA can serve as the genome for new viruses or can be translated to produce viral proteins.

6. Complete HIV particles are assembled and HIV buds out of the cell. As the disease progresses HIV-infected T-helper cells, but not macrophages, are killed by a poorly understood mechanism.

Attachment

When HIV is introduced into the human bloodstream, the virus particles circulate throughout the body but only infect CD4+ cells. Most other animal viruses are similarly narrow in their requirements; hepatitis goes only to the liver, and rabies to the brain. This tissue tropism is determined by the proteins found on a cell surface and on a viral surface.

For example, the common cold virus uses the ICAM-1 membrane protein as a receptor to enter cells. ICAM-1 is a protein that is up-regulated (increased) in times of immune activation and stress. So, the more inflammation and stress in an area, the more receptors exist for the virus to enter a cell and continue the disease process.

How does a virus such as HIV recognize a target cell? Recall from chapter 4 that every kind of cell in the human body has a specific array of cell-surface glycoprotein markers that serve to identify them to other, similar cells. Invading viruses take advantage of this to bind to specific cell types. Each HIV particle

chapter **27** *Viruses* 535

possesses a glycoprotein called gp120 on its surface that precisely fits the cell-surface marker protein CD4 on the surfaces of the immune system macrophages and T cells. Macrophages, another type of white blood cell, are infected first. Because macrophages commonly interact with CD4+ T cells, this may be one way that the T cells are infected. Several coreceptors also significantly affect the likelihood of viral entry into cells, including the CCR5 receptor, which is mutated in HIV-immune individuals.

Entry of virus

After docking onto the CD4 receptor of a cell, HIV requires a coreceptor such as CCR5, to pull itself across the cell membrane. After gp120 binds to CD4, it goes through a conformational change that allows it to then bind the coreceptor. Receptor binding is thought to ultimately result in fusion of the viral and target cell membranes and entry of the virus through a fusion pore. The coreceptor, CCR5, is hypothesized to have been used by the smallpox virus as was mentioned earlier in this section.

Replication

Once inside the host cell, the HIV particle sheds its protective coat. This leaves viral RNA floating in the cytoplasm, along with the reverse transcriptase enzyme that was also within the virion. Reverse transcriptase synthesizes a double strand of DNA complementary to the virus RNA, often making mistakes and introducing new mutations. This double-stranded DNA then enters the nucleus along with a viral enzyme that incorporates the viral DNA into the host cell's DNA. After a variable period of dormancy the HIV provirus directs the host cell's machinery to produce many copies of the virus.

As is the case with most enveloped viruses, HIV does not directly rupture and kill the cells it infects. Instead, the new viruses are released from the cell by *budding*, a process much like exocytosis. HIV synthesizes large numbers of viruses in this way, challenging the immune system over a period of years. In contrast, naked viruses, those lacking an envelope, generally lyse the host cell in order to exit. Some enveloped viruses may produce enzymes that damage the host cell enough to kill it or may produce lytic enzymes as well.

Evolution of HIV during infection

During an infection, HIV is constantly replicating and mutating. The reverse transcriptase enzyme is less accurate than DNA polymerases, leading to a high mutation rate. Eventually, by chance, variants in the gene for gp120 arise that cause the gp120 protein to alter its second-receptor partner. This new form of gp120 protein will bind to a different second receptor, for example CXCR4, instead of CCR5. During the early phase of an infection, HIV primarily targets immune cells with the CCR5 receptor. Eventually the virus mutates to infect a broader range of cells. Ultimately, infection results in the destruction and loss of critical T-helper cells.

This destruction of T cells blocks the body's immune response and leads directly to the onset of AIDS, with cancers and opportunistic infections free to invade the defenseless victim. Most deaths due to AIDS are not a direct result of HIV, but are from other diseases that normally do not harm a host with a normal immune system.

AIDS treatment targets different phases of the HIV life cycle

The federal Food and Drug Administration (FDA) currently lists 34 antiretroviral drugs that are used in AIDS therapy. These target four aspects of the HIV life cycle: viral entry, genome replication, integration of viral DNA, and maturation of HIV proteins (figure 27.7). Of these, the vast majority are inhibitors of the replication enzyme, reverse transcriptase, and the protease that is involved in maturation of proteins. Only two drugs block viral entry: one that blocks the fusion of virus with the cell membrane, and one that blocks the chemokine coreceptor CCR5. A single drug has also been approved that targets the integrase protein that integrates the viral genome into a chromosome.

Reverse transcriptase inhibitors

The first drug licensed for clinical use was AZT, a reverse transcriptase inhibitor. This class of drugs falls into two categories: nucleotide or nucleoside reverse transcriptase inhibitors (NRTI) such as AZT, and nonnucleoside reverse transcriptase inhibitors (NNRTI). These drugs are selective for HIV (or other retroviruses) because reverse transcriptase is not a cellular enzyme. So although the NRTI drugs affect cellular enzymes, they inhibit reverse transcriptase at much lower doses. Nineteen RT inhibitors are currently approved by the FDA.

Protease inhibitors

The second class of drugs discovered to be effective in AIDS treatment were the protease inhibitors. These drugs target a protease that cleaves a polyprotein into the smaller proteins necessary for viral replication and assembly. Some of these drugs are actually a good example of the concept of rational drug design. Drug designers started with the protease enzyme, then targeted drugs at transition state analogues of this enzyme. There are now 11 of these drugs that have received FDA approval.

Blocking viral entry

Two drugs have been approved by the FDA that block the entry of HIV into the cell. One of these, the fusion inhibitor, was approved in 2003. The drug blocks the fusion of the viral envelope with the plasma membrane of a target cell. This entry also requires recognizing the CD4 receptor protein, and a coreceptor such as CCR5. The coreceptor blocker was approved in mid-2007.

Integrase inhibitors

A number of companies have been working on drugs targeting the viral integrase protein. This protein catalyzes the integration reaction of the viral genome. Two drugs that target this protein have been approved.

Combination therapy

The most successful form of therapy has been to use combinations of the previously discussed drugs. The standard treatment regimen consists of a minimum of three active drugs: an NNRTI or protease

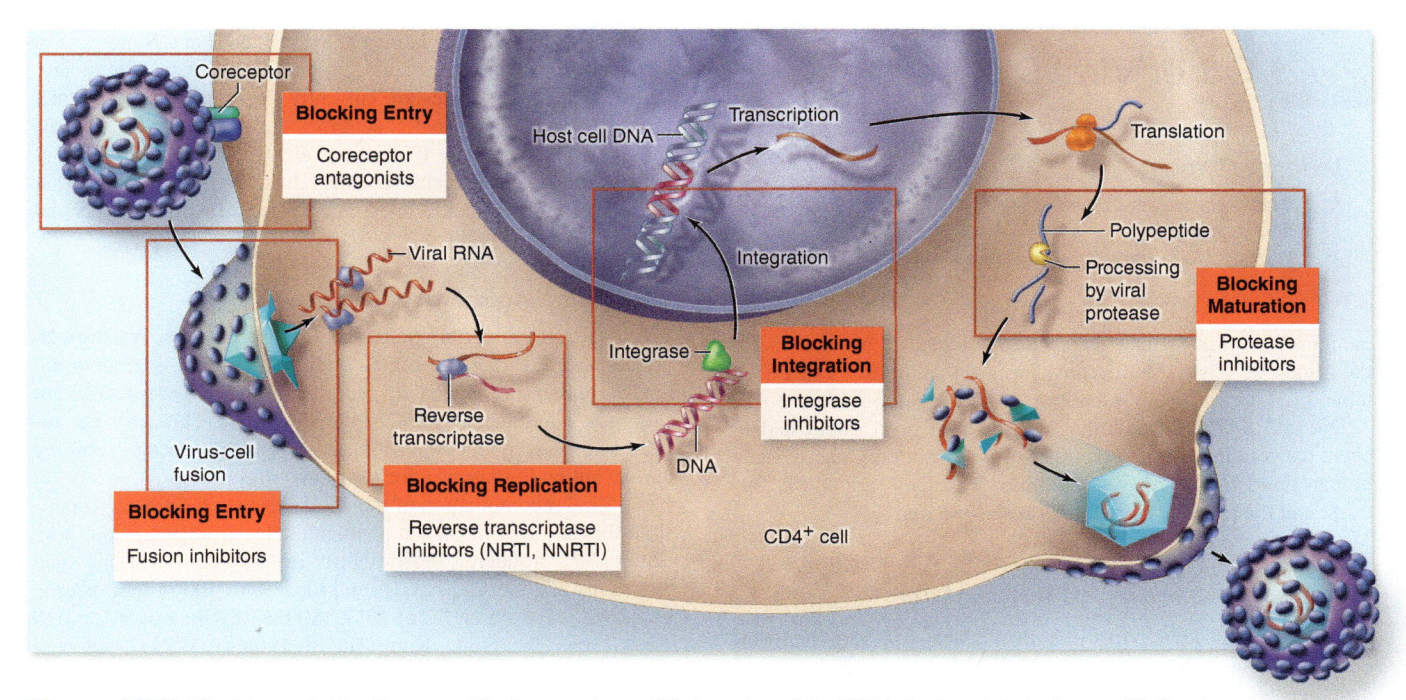

Figure 27.7 Viral targets for therapeutic drugs. A simplified version of the HIV infection cycle is shown with the steps targeted for drug intervention. Four parts of the life cycle have been targeted: viral entry, genome replication, integration of the genome, and maturation of viral proteins. NRTI, nucleoside reverse transcriptase inhibitor; NNRTI, nonnucleoside reverse transcriptase inhibitor.

inhibitor combined with two different NRTIs. This form of combination therapy has entirely eliminated the HIV virus from many patients' bloodstreams. All of these patients began to receive the combination drug therapy within three months of contracting the virus, before their bodies had an opportunity to develop tolerance to any single drug. Widespread use of this *combination therapy,* otherwise known as *highly active antiretroviral therapy (HAART),* has cut the U.S. AIDS death rate by three-fourths since its introduction in the mid-1990s.

Unfortunately, this sort of combination therapy does not appear to actually eliminate HIV from the body. Although the virus disappears from the bloodstream, traces of it can still be detected in the patient's lymph tissue. When combination therapy is discontinued, virus levels in the bloodstream once again rise.

In addition to the search for new drugs to add to the mix of HAART, much effort has been put into simplifying the drug regimens. The more complex the regimen of drugs, the less likely that patients will stick with it. Three drugs that combine NRTI and NNRTI compounds into a single pill to reduce this complexity have been approved by the FDA.

Vaccine development for HIV has been unsuccessful

A large amount of effort has been put toward developing an anti-HIV vaccine. Thus far, this has been totally unsuccessful. A recent large-scale international HIV vaccine trial was halted when an early examination of data indicated that the vaccine was essentially useless for preventing infection or lowering viral load. This has led to a retooling of clinical trials to have periodic examination of data instead of waiting for endpoints, but has not helped in terms of the vaccine itself. This vaccine was a subunit vaccine in which a specific HIV protein was engineered into an adenovirus vector.

Although the high mutation rate of HIV has always been seen as a problem for vaccine development, it appears that the reason for vaccine failures may be more basic, and harder to surmount. A vaccine needs to produce a strong cellular immune response, and thus far no trial HIV subunit vaccine has done this. The only type of vaccine in an animal system that has been shown to provide protection against infection was a vaccine made from attenuated SIV. Unfortunately over time, the attenuated virus was able to mutate into an infective virus, and the experimental animals eventually developed simian AIDS.

Learning Outcomes Review 27.3

HIV is a retrovirus that enters cells via membrane fusion. The virus primarily infects host CD4$^+$ T cells, ultimately resulting in massive death of these cells, which thereby compromises the host's immune system. Most deaths result from cancer and infections that typically do not harm hosts with normal immune systems. Combination drug therapy is a treatment modality in developed countries, and much research is being done to develop vaccines or to find agents that can prevent infection.

- *Does combination therapy such as HAART represent a cure for AIDS?*

27.4 Other Viral Diseases

Learning Outcomes
1. Explain why we need a new vaccine each year for influenza.
2. Illustrate where emerging viruses come from.

Humans have known and feared diseases caused by viruses for thousands of years. Among the diseases that viruses cause (see table 27.1) are influenza, smallpox, hepatitis, yellow fever, polio, AIDS, and SARS. In addition, viruses have been implicated in some cancers including leukemias. Viruses not only cause many human diseases, but also cause major losses in agriculture, forestry, and the productivity of natural ecosystems.

The flu is caused by influenza virus

Perhaps the most lethal virus in human history has been the influenza virus. The influenza pandemic of 1918 and 1919 is thought to have infected a third of the world's population.

Types and subtypes

Flu viruses are enveloped segmented RNA viruses that infect animals. An individual flu virus resembles a rod studded with spikes composed of two kinds of protein. The three general "types" of flu virus are distinguished by their capsid protein, which surrounds the viral RNA segments and is different for each type: **Type A flu virus** causes most of the serious flu epidemics in humans and also occurs in mammals and birds. Type B and type C viruses are restricted to humans and rarely cause serious health problems.

Different strains of flu virus, called subtypes, differ in their protein spikes. One of these proteins, hemagglutinin (H), aids the virus in gaining access to the cell interior. The other, neuraminidase (N), helps the daughter viruses break free of the host cell once virus replication has been completed.

Parts of the H molecule contain "hotspots" that display an unusual tendency to change as a result of mutation of the viral RNA during imprecise replication. Point mutations cause changes in these spike proteins in 1 of 100,000 viruses during the course of each generation. These highly variable segments of the H molecule are targets against which the body's antibodies are directed. These constantly changing H-molecule regions improve the reproductive capacity of the virus and hinder our ability to make effective vaccines.

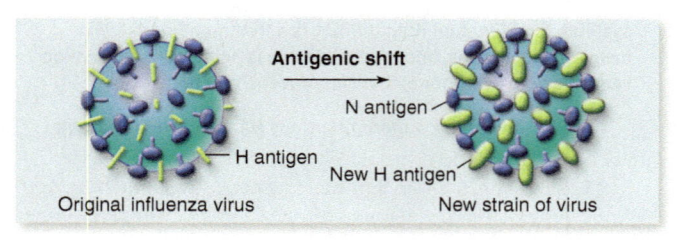

Because of accumulating changes in the H and N molecules, different flu vaccines are required to protect against different subtypes. Type A flu viruses are currently classified into 18 distinct H subtypes and 11 distinct N subtypes, for a total of 198 possible subtypes. There are only 13 subtypes identified in humans, each of which requires a different vaccine. Thus, the type A virus that caused the 1918 influenza pandemic has type 1H and type 1N and is designated A(H1N1).

The importance of recombination

The greatest problem in combating flu viruses arises not through mutation, but through recombination. Viral RNA segments are readily reassorted by genetic recombination when two different subtypes simultaneously infect the same cell. This may put together novel combinations of H and N spikes unrecognizable by human antibodies specific for the old configuration.

Viral recombination of this kind seems to have been responsible for the three major flu pandemics that occurred in the 20th century, by producing drastic shifts in H–N combinations. The "Spanish flu" of 1918, A(H1N1), killed 50 to 100 million people worldwide. The Asian flu of 1957, A(H2N2), killed over 100,000 Americans. The Hong Kong flu of 1968, A(H3N2), infected 50 million people in the United States alone, of whom 70,000 died.

Origin of new strains

It is no accident that new strains of flu usually originate in the Far East. The most common hosts of influenza virus are ducks, chickens, and pigs, which in Asia often live in close proximity to each other and to humans. Pigs are subject to infection by both bird and human strains of the virus, and individual animals are often simultaneously infected with multiple strains. This creates conditions favoring genetic recombination between strains, producing new combinations of H and N subtypes.

In 1997, a form of avian influenza, A(H5N1), was discovered that could infect humans. Avian influenza, or "bird flu," is highly contagious and deadly among domestic bird populations, and this H5N1 strain is transmitted between domestic birds, which live in close contact with humans, and wild birds, which migrate worldwide. This new influenza variant caused concern because, although the number of infections worldwide is low, the mortality rate in these rare infections is quite high. However, this strain does not appear to spread by human-to-human contact.

While H5N1 captured the interest of the press worldwide, another viral reassortment of avian, human, and swine viruses occurred. This led to the H1N1 pandemic of 2009. The virus first appeared in an outbreak of influenza in Veracruz, Mexico, and spread worldwide over 2009–10. The WHO raised the event to pandemic status in May 2009. This pandemic is interesting on many levels, both scientific and social. Events of this type are instructive in terms of how societies deal with a crisis, whether it ends up being as severe as expected or not. In the end, the infection rate was similar to previous pandemics, but this is not much different from seasonal forms. The mortality rate was not high (around 1.5%), but the age distribution was unusual affecting more younger people with older people exhibiting some immunity. In the end, the death toll was similar to seasonal influenza, but if measured in terms of life expectancy, the greater toll on the young raises this number.

Antigenic shift

Original influenza virus — H antigen, N antigen → New strain of virus — New H antigen

New viruses emerge by infecting new hosts

Sometimes viruses that originate in one organism pass to another, thus expanding their host range. Often, this expansion is deadly to the new host. HIV, for example, is thought to have arisen in chimpanzees and relatively recently passed to humans. Influenza is fundamentally a bird virus. Viruses that originate in one organism and then pass to another and cause disease are called **emerging viruses**. They represent a considerable threat in an age when airplane travel potentially allows infected individuals to move about the world quickly, spreading an infection.

Hantavirus

An emerging virus caused a sudden outbreak of a deadly pneumonia in the southwestern United States in 1993. This disease was traced to a species of **hantavirus** and was called the *sin nombre*, or *no-name*, virus. Hantavirus is a single-stranded RNA virus associated with rodents. This virus was eventually traced to deer mice. The deer mouse hantavirus is transmitted to humans through fecal and urine contamination in areas of human habitation. Controlling the deer mouse population has limited the disease.

Hemorrhagic fever: Ebola

Sometimes the origin of an emerging virus is unknown, making control of outbreaks difficult. Among the most lethal emerging viruses are a collection of filamentous viruses arising in central Africa that cause severe hemorrhagic fever. The lethality rate for these filoviruses can exceed 50%. One, Ebola virus (figure 27.8), has exhibited lethality rates in excess of 90% in isolated outbreaks in central Africa.

Until 2014, Ebola was responsible for a series of limited outbreaks characterized by high lethality, but few people were infected. This changed in 2014 with an outbreak in West Africa that captured worldwide attention as it ravaged the countries of Guinea, Liberia, and Sierra Leone. By mid-2015, the outbreak was finally running down, but left in its wake an estimated 25,000 infected, of which 15,000 were "laboratory confirmed" with more than 10,000 dead. The natural host of Ebola remains unknown.

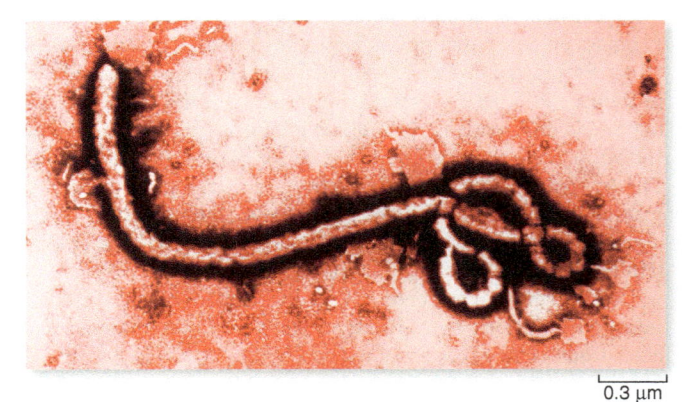

Figure 27.8 The Ebola virus. This virus, with a fatality rate that can exceed 90%, appears sporadically in central Africa. The natural host of the virus currently is unknown.

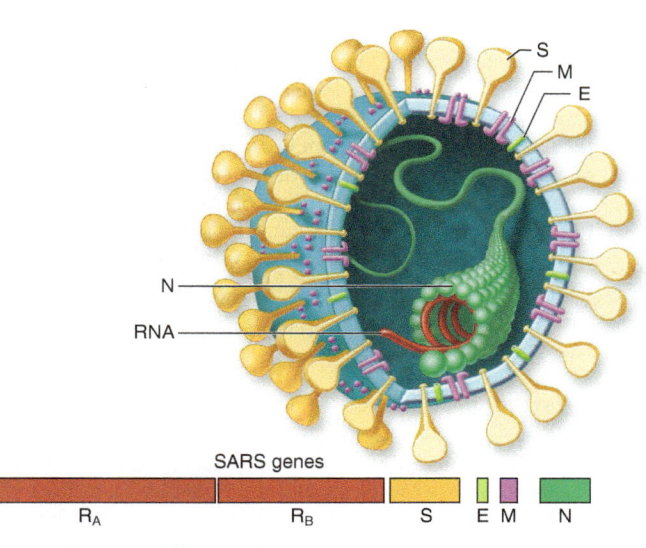

Figure 27.9 SARS coronavirus. The 29,751-nucleotide SARS genome is composed of RNA and contains six principal genes: R_A and R_B, replicases; S, spike proteins; E, envelope glycoproteins; M, membrane glycoprotein; N, nucleocapsid protein.

SARS

A recently emerged species of coronavirus (figure 27.9) was responsible for the 2003 worldwide outbreak of **severe acute respiratory syndrome (SARS)**, a respiratory infection with pneumonia-like symptoms that is fatal in over 8% of cases. When the 29,751-nucleotide RNA genome of the SARS coronavirus was sequenced, it proved to be a new form of coronavirus, not closely related to any of the three previously described forms.

Genome sequences have been analyzed from SARS patients at various stages of the outbreak, and these analyses indicate that the virus's mutation rate is low, in marked contrast to HIV, another RNA virus. The stable genome of the SARS virus should make development of a SARS vaccine practical. The lessons learned from developing antiviral agents against other RNA viruses, such as HIV and influenza, have helped develop drugs to treat SARS. Several anti-SARS agents and vaccines are being tested in laboratories across the world.

Viruses can cause cancer

Through epidemiological studies and research, scientists have established a link between some viral infections and the subsequent development of cancer. Examples include the association between chronic hepatitis B infections and the development of liver cancer, and the development of cervical carcinoma following infections with certain strains of human papillomaviruses (HPV).

Viruses may contribute to about 15% of all human cancer cases worldwide. They are capable of altering the growth properties of human cells they infect by triggering the expression of cancer-causing genes called oncogenes (see chapter 10). Changes in the normal function of these genes leads to cancer.

These changes can occur because viral proteins interfere with the regulation of oncogene expression. Alternatively, the integration of a viral genome into a host chromosome may disrupt a gene required to control the cell cycle. Viruses themselves may encode these oncogenes as well. Virus-induced cancer involves complex interactions with cellular genes and requires a series of events in order to develop. The association of viruses with some forms of cancer led to research on vaccine development for prevention of these cancers. In 2006, the FDA approved an HPV vaccine for women and young girls from the age of 11 to prevent cervical cancer. The CDC now recommends this for all 11- to 12-year-olds, both male and female.

> **Learning Outcomes Review 27.4**
>
> Many types of viruses have caused disease in humans for as long as we have recorded history. Some of these, such as influenza, have caused pandemics responsible for millions of deaths worldwide. Recombination is common in the influenza virus, making natural immunity and development of vaccines problematic. Emerging diseases can occur when viruses switch hosts—that is, jump from another species to humans. Hantavirus, Ebola, and SARS all fall into this category. Virus infection has also been linked to the development of certain cancers.
>
> ■ Why does the effectiveness of flu shots vary from year to year?

27.5 Prions and Viroids: Subviral Particles

Learning Outcomes
1. Explain why prion replication was a "heretical" concept.
2. Describe the mechanism of prion transmission.

For decades, scientists have been fascinated by a peculiar group of fatal brain diseases. These diseases have an unusual property: Years and often decades pass after infection before the disease is detected in infected individuals. The brains of infected individuals develop numerous small cavities as neurons die, producing a marked spongy appearance. Called *transmissible spongiform encephalopathies (TSEs)*, these diseases include scrapie in sheep; bovine spongiform encephalopathy (BSE), or "mad cow" disease in cattle; chronic wasting disease in deer and elk; and kuru, Creutzfeldt–Jakob disease (CJD), and variant Creutzfeldt–Jakob disease (vCJD) in humans.

TSEs can be transmitted experimentally by injecting infected brain tissue into a recipient animal's brain. TSEs can also spread via tissue transplants and, apparently, tainted food. The disease kuru was once common in the Fore people of Papua New Guinea, because they practiced ritual cannibalism, eating the brains of infected individuals. Mad cow disease spread widely among the cattle herds of England in the 1990s because cows were fed bonemeal prepared from sheep and cattle carcasses to increase the protein content of their diet. Like the Fore, the British cattle were eating the tissue of cattle that had died of the disease.

In the years following the outbreak of BSE, there has been a significant increase in CJD incidence in England. Some cases appear to be genetic. Mysteriously, patients with no family history of CJD were being diagnosed with the disease. This led to the discovery of a new form of CJD called variant CJD, or vCJD, that is acquired from eating meat of BSE-infected animals. Concern exists that vCJD may be transmitted from person to person through blood products, similar to the transmission of HIV through blood and blood products.

Prion replication was a heretical suggestion

In the 1960s, British researchers Tikvah Alper and John Stanley Griffith noted that infectious TSE preparations remained infectious even after exposure to radiation that would destroy DNA or RNA. They suggested that the infectious agent was a protein. They speculated that the protein could sometimes misfold, and then catalyze other proteins to do the same, the misfolding spreading like a chain reaction. This "heretical" suggestion was not accepted by the scientific community, because it violated a key tenet of molecular biology: Only DNA or RNA act as hereditary material, transmitting information from one generation to the next.

Evidence has accumulated that prions cause TSEs

In the early 1970s, American physician Stanley Prusiner began to study TSEs. Try as he might, Prusiner could find no evidence of nucleic acids or viruses in the infectious TSE preparations. He concluded, as Alper and Griffith had, that the infectious agent was a *protein*, which in a 1982 paper he named a **prion**, for "proteinaceous infectious particle."

Prusiner went on to isolate a distinctive prion protein, and to amass evidence that prions play a key role in triggering TSEs. Every host tested to date expresses a normal prion protein (PrP^c) in their cells. The disease-causing prions are the same protein, but folded differently (PrP^{sc}). These misfolded proteins have been shown in vitro to serve as a template for normal PrP to misfold. The misfolded PrP proteins are very resistant to degradation, making it possible for them to pass through the acidic digestive tract intact and therefore to be transmitted by ingesting tainted food.

Experimental evidence has accumulated to support this idea. Injection of prions with different abnormal conformations into hosts leads to the same abnormal conformations as the parent prions. Mice genetically engineered to lack PrP^c are immune to TSE infection (figure 27.10). If brain tissue with the prion protein is grafted into the mice, the grafted tissue—but not the rest of the

SCIENTIFIC THINKING

Hypothesis: Normal cellular PrP protein is required for scrapie infection.
Prediction: Infection of mice lacking PrP with scrapie agent should not lead to scrapie.
Test: Construct "knockout" mice with PrP deleted. Inject PrP knockout mice, and control mice with normal PrP, with brain homogenates from scrapie infected mice.

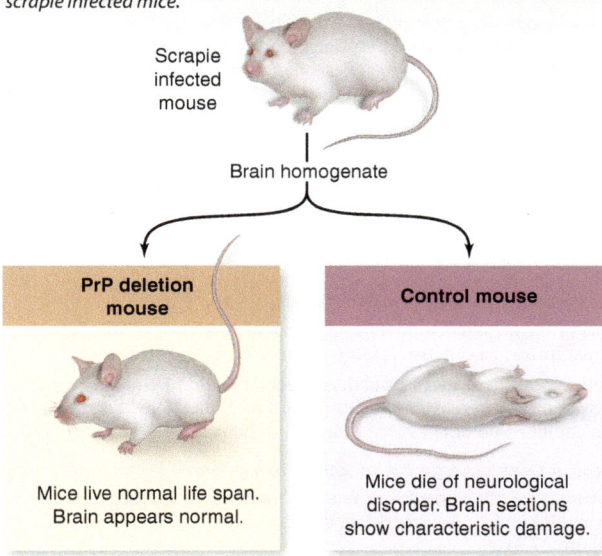

Scrapie infected mouse
↓
Brain homogenate

PrP deletion mouse: Mice live normal life span. Brain appears normal.

Control mouse: Mice die of neurological disorder. Brain sections show characteristic damage.

Result: All control mice died with scrapie like neurological damage. Periodic sacrifice and histological examination showed normal disease progress. None of the PrP knockout mice showed any sign of neurological disorder.
Conclusion: Normal PrP protein is necessary for productive infection by scrapie infectious particle.
Further Experiments: What kind of experiment would conclusively prove that the infectious agent of scrapie is a misfolded PrP protein?

Figure 27.10 Demonstration that normal prion protein is necessary for scrapie infectivity.

brain—can then be infected with TSE. However, infectious PrP has not been generated in vitro in quantities that would produce disease in nature. The mechanism of pathogenesis also remains controversial.

Prions have also been found in yeast and other fungi. Three different "genes" that behave in a non-Mendelian fashion have all been shown to be prions in yeast. There has been greater progress in the in vitro conversion of normal cellular proteins to the prion form in the yeast system.

Viroids are infectious RNA with no protein coat

Viroids are tiny, naked molecules of circular RNA, only a few hundred nucleotides long, that are important infectious disease agents in plants. A recent viroid outbreak killed over 10 million coconut palms in the Philippines. Despite their small size, viroids autonomously replicate without a helper virus, although they obviously use host proteins. Despite being nucleic acids, the important information they carry appears to be in their three-dimensional structure. There are also some intriguing hints that they might use the plant siRNA machinery to affect gene expression.

Learning Outcomes Review 27.5

Prions and viroids are smaller and simpler than viruses. Prions are infectious particles that do not seem to contain any nucleic acid. They appear to be misfolded proteins that cause related cellular proteins to also misfold. Prions are the causative agent of TSEs. Viroids are infectious RNAs that are implicated in some plant diseases.

- *If prions are infectious proteins, then what is the form of genetic information that they carry?*

Chapter Review

27.1 The Nature of Viruses

Viruses are strands of nucleic acids encased in a protein coat.
Viral genomes can consist of either DNA or RNA and can be classified as DNA viruses, RNA viruses, or retroviruses.

Most viruses have a protein sheath or capsid around their nucleic acid core. Many animal viruses have an envelope around the capsid composed of proteins that are virally encoded, lipids from the host cell, and glycoproteins.

Some viruses also have enzymes inside their capsid that are important in early infection.

Viral hosts include virtually every kind of organism.
Each virus has a limited host range, and many also exhibit tissue tropism.

Viruses replicate by taking over host machinery.
Viruses are obligate intracellular parasites lacking ribosomes and proteins needed for replication. Viruses take over host machinery and direct their own nucleic acid and protein synthesis.

Most viruses come in two simple shapes.
Viruses vary in size and come in two simple shapes: helical (rodlike) or icosahedral (spherical) (figures 27.1 & 27.4).

Viral genomes exhibit great variation.
The DNA or RNA viral genome may be linear or circular, single- or double-stranded. RNA viruses may have multiple RNA molecules (segmented), or only one RNA molecule (nonsegmented). Retroviruses contain RNA that is transcribed into DNA by reverse transcriptase.

Giant viruses challenge assumptions about viruses
Mimivirus is the size of a small bacterium, with a megabase-sized genome. The genome also encodes parts of the translation apparatus.

27.2 Bacteriophage: Bacterial Viruses

Archaeal viruses have diverse morphologies.

Bacterial viruses exhibit two reproductive cycles.
The lytic cycle kills the host cell, whereas the lysogenic cycle incorporates the virus into the host genome as a prophage (figure 27.5). A cell containing a prophage is called a lysogen.

The prophage can be induced by DNA damage and other environmental cues to reenter the lytic cycle.

For most phage, steps in infection include attachment, injection of DNA (penetration), macromolecular synthesis, assembly of new phage, and release of progeny phage.

Bacteriophage can contribute genes to the host genome.
Phage conversion occurs when foreign DNA is contributed to the host by a bacterial virus.

27.3 Human Immunodeficiency Virus (HIV)

AIDS is caused by HIV.
Human immunodeficiency virus (HIV) causes acquired immunodeficiency syndrome (AIDS) (figure 27.6).

HIV infection compromises the host immune system.
HIV specifically targets macrophages and CD4+ cells, a type of helper T-lymphocyte cell. With the loss of these cells the body cannot fight off opportunistic infections, which ultimately lead to death.

HIV infects key immune-system cells.
The viral glycoprotein gp120 precisely fits on the cell-surface marker protein CD4+ on macrophages and T cells. When HIV attaches to two receptors, CD4+ and CCR5, receptor-mediated endocytosis is activated, bringing the HIV particle into the cell.

Once inside the cell, the protective coat is shed, releasing viral RNA and reverse transcriptase into the cytoplasm. Reverse transcriptase makes double-stranded DNA complementary to the viral RNA. This DNA may be incorporated into the host DNA as a provirus.

Replicated viruses are budded off the host cell by exocytosis.

HIV has a high mutation rate because the reverse transcriptase enzyme is much less accurate than DNA polymerases. Mutations lead to an altered glycoprotein gp120, which now binds instead to the CXCR4 receptor found only on the surface of CD4+ cells. Incorporation of the altered HIV particle leads to a rapid decline in T cells and immune response.

AIDS treatment targets different phases of the HIV life cycle.
Drugs target reverse transcriptase, a protease involved in protein maturation, viral entry, and the integration of the genome. Most approved drugs are reverse transcriptase and protease inhibitors. Combination drug therapy using nucleoside analogues and protease inhibitors eliminates HIV from the bloodstream but not totally from the body.

Vaccine development for HIV has been unsuccessful.
Successful vaccines must elicit a strong immune response, and attempts to create immunity against a specific HIV subunit have failed. Attenuated viruses in animal tests have also proved capable of mutating into infectious forms.

27.4 Other Viral Diseases

The flu is caused by influenza virus.
One of the most lethal viruses in human history is type A influenza virus. Influenza can also infect other mammals and birds.

Genes in influenza viruses undergo recombination frequently, so they are not recognized by antibodies against past infections. Each year the composition of flu vaccines must be changed.

New viruses emerge by infecting new hosts.
Viruses can extend their host range by jumping to another species. Examples include hantavirus, hemorrhagic fever, and SARS (figure 27.9).

Viruses can cause cancer.
Viruses have been linked to formation of cancers, including liver cancer and cervical papillomas.

27.5 Prions and Viroids: Subviral Particles

Prion replication was a heretical suggestion.
Prions are proteinaceous infectious particles consisting of a misfolded form of a protein. This misfolding catalyzes a chain reaction of misfolding in normal proteins, causing disease.

Evidence has accumulated that prions cause TSEs (figure 27.10).
TSE disease-causing prions (PrP^{sc}) are the same protein as the normal version (PrP^c) but misfolded.

Viroids are infectious RNA with no protein coat.
Viroids are circular, naked molecules of RNA that infect plants. They use host protein to replicate.

Review Questions

UNDERSTAND

1. The reverse transcriptase enzyme is active in which class of viruses?
   a. Positive-strand RNA viruses
   b. Double-stranded DNA viruses
   c. Retroviruses
   d. Negative-strand RNA viruses

2. Which of the following is NOT part of a virus?
   a. Capsid
   b. Ribosomes
   c. Genetic material
   d. All of the choices are found in viruses.

3. Which of the following is common in animal viruses but NOT in bacteriophage?
   a. DNA
   b. Capsid
   c. Envelope
   d. Icosahedral shape

4. Which of the following would NOT be part of the life cycle of a lytic virus?
   a. Macromolecular synthesis
   b. Attachment to host cell
   c. Assembly of progeny virus
   d. Integration into the host genome

5. A process by which a virus may change a benign bacteria into a virulent strain is called
   a. induction.
   b. phage conversion.
   c. lysogeny.
   d. replication.

6. Prior to entry, the _____ glycoprotein of the HIV virus recognizes the _____ receptor on the surface of the macrophage.
   a. CCR5; gp120
   b. CXCR4; CCR5
   c. CD4; CCR5
   d. gp120; CD4

7. The use of multiple drugs in HAART to treat AIDS has
   a. completely removed the virus from infected individuals.
   b. reduced the viral level in the bloodstream to undetectable levels.
   c. been a complete failure.
   d. been supplanted by the new HIV vaccine.

APPLY

1. The varying degrees of resistance to HIV in populations has been suggested to be related to the patterns of smallpox outbreaks over human history. This explanation hinges on
   a. the similarity in the genomes of the two viruses.
   b. the fact that both viruses use reverse transcriptase.
   c. both viruses using the same receptor to bind to host cells.
   d. the fact that both viruses compromise the immune system.

2. The idea of a protein that was an infectious agent was heretical because
   a. proteins are not that important in cells.
   b. proteins are not the informational molecules in cells.
   c. the function of proteins does not depend on their structure.
   d. proteins require nucleic acids for their function.

3. Bacterial viruses and animal viruses are similar in that they both
   a. have only DNA as genetic material.
   b. have only RNA as genetic material.
   c. require host functions for some aspect of their life cycle.
   d. do not require any host proteins.

4. The drugs used against HIV in AIDS therapy are not effective against the flu because
   a. HIV is an RNA virus and influenza is a DNA virus.
   b. HIV is a DNA virus and influenza is an RNA virus.
   c. the two viruses have different-sized genomes.
   d. the proteins targeted by HIV drugs are not found in influenza.

5. Phage conversion in which viruses add genes to a bacterial cell can be considered to be a form of
   a. standard inheritance.
   b. horizontal gene transfer.
   c. vertical gene transfer.
   d. parasitism.

6. According to the prion hypothesis, the infectious agent for scrapie must have "genetic" information in
   a. the sequence of amino acids in the scrapie protein.
   b. the sequence of bases in the scrapie gene.
   c. the three-dimensional structure of the scrapie prion.
   d. All of the choices are correct.

7. The difficulty designing a single flu vaccine that will work forever is that influenza
   a. is an RNA virus.
   b. is a DNA virus.
   c. both mutates and can be recombined to form new viruses.
   d. infects only humans.

8. The SARS outbreak is an example of
   a. a virus jumping from one species to another.
   b. mutation of a virus that only infects humans.
   c. how viruses can disable the human immune system.
   d. two viruses combining to form a new virus.

SYNTHESIZE

1. *E. coli* lysogens derived from infection by phage λ can be induced to form progeny viruses by exposure to radiation. The inductive event is the destruction of a repressor protein that keeps the prophage genome unexpressed. What might be the normal role of the protein that recognizes and destroys the λ repressor?

2. Most biologists believe that viruses evolved following the origin of the first cells. Defend or critique this concept.

3. Much effort has been expended to produce a vaccine for HIV. To date, this has not been successful. Why has this been such a difficult task? Are any other viruses equally resistant to a vaccine, and are the reasons the same? Why do you think that we could make a vaccine against the smallpox virus that allowed us to completely eradicate this virus?

4. What do we mean by the term "emerging virus"? How is this a medical problem, and what is a recent example?

5. How might phage λ be used to transfer *E. coli* genes between different bacterial cells? Could this be used to transfer any gene?

CHAPTER 28

Prokaryotes

Chapter Contents

28.1 Prokaryotic Diversity
28.2 Prokaryotic Cell Structure
28.3 Prokaryotic Genetics
28.4 Prokaryotic Metabolism
28.5 Human Bacterial Disease
28.6 Beneficial Prokaryotes

0.6 μm

Introduction

One of the hallmarks of living organisms is their cellular organization. You learned in chapter 4 that living things come in two basic cell types: prokaryotes and eukaryotes. It might surprise you to learn that prokaryotes, despite being considerably smaller, are more numerous than their eukaryotic counterparts. Based on sheer numbers, prokaryotes are the dominant life-forms on the planet. If we examined a human closely, we would discover that for every single cell of the human body there are approximately 10 prokaryotic cells—and there are trillions of human cells.

Prokaryotic microbes play an important role in global ecology as well. We no longer think of geochemistry, but now of biogeochemistry. Most biologists think prokaryotes were the first organisms to evolve. The diversity of eukaryotic organisms that currently live on Earth could not exist without prokaryotes because they make possible many of the essential functions of ecosystems. Prokaryotic photosynthesis, for example, is thought to have been the source of the oxygen in ancient Earth's atmosphere, and it still contributes significantly to oxygen production today. An understanding of prokaryotes is essential to understanding all life on Earth, past and present.

28.1 Prokaryotic Diversity

Learning Outcomes

1. Differentiate among archaea, bacteria, and eukarya.
2. Describe the basic features of bacteria and archaea.
3. Explain classification methods for prokaryotes.

A brief history of microbiology

The size of prokaryotic cells led to their being undiscovered for most of human history. Despite this, as early as 1546, the Italian physician Girolamo Fracastoro suggested that disease was caused by unseen organisms. The history of microbiology entwines the twin strands of technology for visualizing the unseen (microscopy) and the investigation of the nature of infectious diseases. This history also includes the prolonged debate about the very nature of how living things arise: from other living things or by spontaneous generation from nonliving matter.

Although he did not invent the microscope, Dutch scientist Antony van Leeuwenhoek was the first to observe and accurately describe microbial life. From these earliest observations we have built more and more sophisticated microscopes to study this unseen realm (see chapter 4). The modern electron microscope allows us to see extensive substructure within cells.

The controversy over spontaneous generation continued into the 19th century, eventually being refuted by French microbiologist Louis Pasteur (see figure 1.4) and others. During this same period in the 19th century, evidence accumulated that microorganisms could cause disease: the so-called germ theory of disease. Ultimately, German physician Robert Koch studied the disease anthrax in detail and was able to culture the causative agent. He proposed four postulates to prove a causal relationship between a microorganism and a disease:

1. The microorganism must be present in every case of the disease and absent from healthy individuals.
2. The putative causative agent must be isolated and grown in pure culture.
3. The same disease must result when the cultured microorganism is used to infect a healthy host.
4. The same microorganism must be isolated again from the diseased host.

The study of microorganisms and disease also led to advances in knowledge about the immune system, and how challenge with a disease-causing agent can prevent future infection (see chapter 51). These many different historical strands produced a vibrant field—microbiology—which is still relevant today. In this chapter we will concentrate on the study of prokaryotic microorganisms, and later in chapter 29 we will consider their eukaryotic counterparts.

Given the rich history of microbiology, which was barely touched upon here, it is amazing that at the end of the 20th century, new techniques that allow us to survey microbial diversity without having to grow cells in culture led to the conclusion that we have barely begun to understand the diversity of prokaryotic life. Although thousands of different kinds of prokaryotes are currently recognized, it is estimated that only between 1 and 10% of all prokaryotic species are known and characterized, leaving between 90 and 99% unknown and undescribed. Every place microbiologists look, new species are being discovered, often altering the way we think about prokaryotes.

Archaea and bacteria are the oldest, structurally simplest, and most abundant forms of life. They are also the only organisms with prokaryotic cellular organization. Prokaryotes were abundant for over a billion years before eukaryotes appeared in the world. Early photosynthetic bacteria (cyanobacteria) altered the Earth's atmosphere by producing oxygen, which stimulated extreme bacterial and eukaryotic diversity.

Prokaryotes are ubiquitous and live everywhere eukaryotes do; they are also able to thrive in places no eukaryote could live. Bacteria and archaea have been found in deep-sea caves, volcanic rims, and deep within glaciers. Some of the extreme environments in which prokaryotes can be found would be lethal to any other life-form.

Many archaea are extremophiles. They live in hot springs that would cook other organisms (figure 28.1), in hypersaline environments that would dehydrate other cells, and in atmospheres rich in otherwise-toxic gases such as methane or hydrogen sulfide. They have even been recovered living beneath 435 m of ice in Antarctica!

Figure 28.1 Hot springs such as these in Yellowstone National Park contain thermophilic prokaryotes.

These harsh environments may be similar to the conditions present on the early Earth when life first began. It is likely that prokaryotes evolved to dwell in these harsh conditions early on and have retained the ability to exploit these areas as the rest of the atmosphere has changed.

Prokaryotes are fundamentally different from eukaryotes

Prokaryotes differ from eukaryotes in numerous important features. These differences represent some of the most fundamental distinctions that separate any groups of organisms:

Unicellularity. With a few exceptions, prokaryotes are fundamentally single-celled. In some types, individual cells adhere to one another within a matrix and form filaments; however, the cells retain their individuality. Even in this case, the characteristics of the organism are the characteristics of the individual cells, unlike in a multicellular organism.

In their natural environments, most bacteria appear to be capable of forming a complex community of different species called a **biofilm** (figure 28.2). These associations

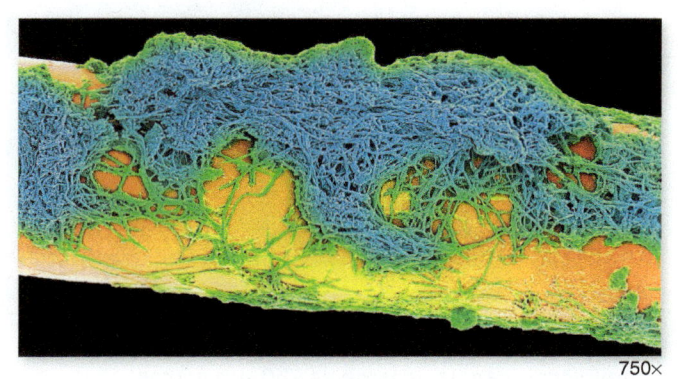

750×

Figure 28.2 Used toothbrush bristle shows a biofilm of the bacteria that cause dental plaque.

are more resistant to antibiotics, desiccation, and other environmental stressors than is a simple colony of a single type of microbe, such as laboratory culture.

Cell size. As new species of prokaryotes are discovered, investigators are finding that the size of prokaryotic cells varies tremendously, by as much as five orders of magnitude. The largest bacterial cells currently characterized are from *Thiomargarita namibiensis* (figure 28.3). A single cell from this species is up to 750 μm across, which is visible to the naked eye and is roughly the size of the eye of a bumblebee. Most prokaryotic cells, however, are only 1 μm or less in diameter, whereas most eukaryotic cells are well over 10 times bigger. This generality is misleading, however, because there are very small eukaryotes as well as very large prokaryotes.

Nucleoid. Prokaryotes lack a membrane-bounded nucleus; instead they usually have a single circular chromosome made up of DNA and histonelike proteins in a nucleoid region of the cell. The exceptions to this are *Vibrio cholerae*, which has two circular chromosomes, and *Borrelia* species, which have multiple linear chromosomes. Prokaryotic cells often have accessory DNA molecules called plasmids as well.

Cell division and genetic recombination. Cell division in eukaryotes takes place by mitosis and involves spindles made up of microtubules. Cell division in prokaryotes takes place mainly by binary fission (see chapter 10), which does not use a spindle. Prokaryotes also do not have a sexual cycle, although we now know that they exchange genetic material extensively. These mechanisms are collectively called horizontal gene transfer and are not related to reproduction.

Internal compartmentalization. The cytoplasm of prokaryotes, unlike that of eukaryotes, does not have extensive internal compartments and no membrane-bounded organelles (see chapter 4). Instead the plasma membrane of prokaryotes can be extensively infolded, and both respiration and photosynthesis take place on organized portions of the plasma membrane.

Flagella. Prokaryotic flagella do not show the 9 + 2 architecture found in eukaryotes. Instead, they are composed of a single fiber of the protein flagellin. Bacterial flagella also are rigid and spin like propellers, whereas eukaryotic flagella have a whiplike motion (described in more detail in section 28.2 and in figure 28.9).

Metabolic diversity. Photosynthetic bacteria have two basic patterns of photosynthesis: (1) oxygenic, producing oxygen; and (2) anoxygenic, nonoxygen producing. Anoxygenic photosynthesis involves the formation of products such as sulfur and sulfate instead of oxygen. Prokaryotic cells are also the only chemolithotrophic organisms, meaning that they use the energy stored in chemical bonds of inorganic molecules to synthesize carbohydrates (we will explore this metabolic diversity in section 28.4).

Despite similarities, bacteria and archaea differ fundamentally

Archaea and bacteria are similar in that both have a prokaryotic cellular structure, but they vary considerably at the biochemical and molecular levels. They differ in four key areas: plasma membranes, cell walls, DNA replication, and gene expression.

Plasma membranes. All prokaryotes have plasma membranes with a fluid mosaic architecture (see chapter 5). The plasma membranes of archaea differ from both bacteria and eukaryotes. Archaeal membrane lipids are composed of glycerol linked to hydrocarbon chains by ether linkages, not the ester linkages seen in bacteria and eukaryotes (figure 28.4a). These hydrocarbons may also be branched, and they may be organized as tetraethers that form a monolayer instead of a bilayer (figure 28.4b).

In the case of some hyperthermophiles, the majority of the membrane may be this tetraether monolayer. This structural feature is part of what allows these archaea to withstand high temperatures.

Cell wall. Both kinds of prokaryotes typically have cell walls covering the plasma membrane that strengthen the cell. The cell walls of bacteria are constructed, minimally, of **peptidoglycan,** which is formed from carbohydrate polymers linked together by peptide cross-bridges. The peptide cross-bridges also contain D-amino acids, which are never found in cellular protein. The cell walls of archaea lack peptidoglycan, although some have **pseudomurein,** which is similar to peptidoglycan in structure and function. This wall layer is also a carbohydrate polymer with peptide cross-bridges, but the carbohydrates are different, and the peptide cross-bridge structure also differs. Other archaeal cell walls have been found to be composed of a variety of proteins and carbohydrates, making generalizations difficult.

DNA replication. Although both archaea and bacteria have a single replication origin, the nature of this origin and the proteins that act there are quite different. Archaeal initiation of DNA replication is more similar to that of eukaryotes (see chapter 14).

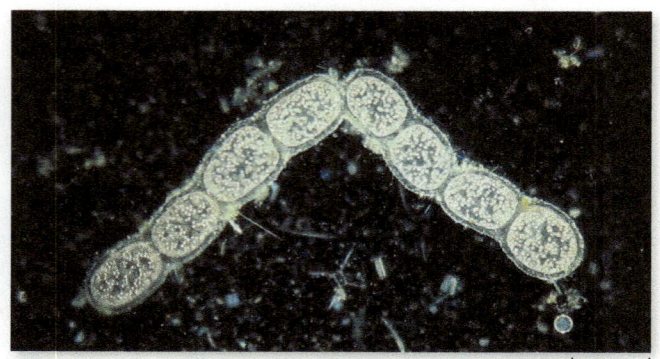

Figure 28.3 Cells of the bacterium *Thiomargarita namibiensis* can be as large as 750 μm across.

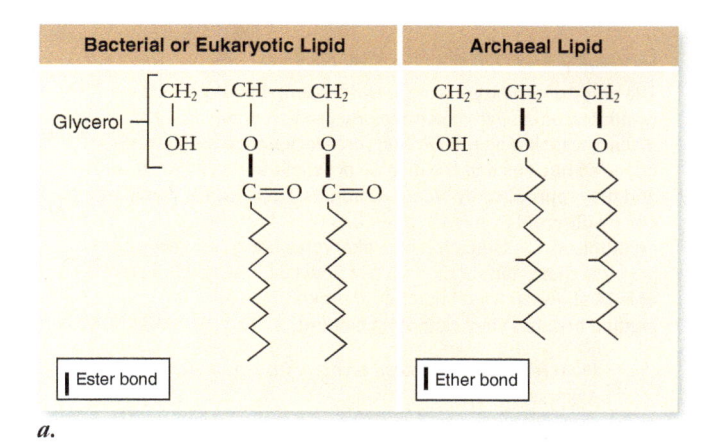

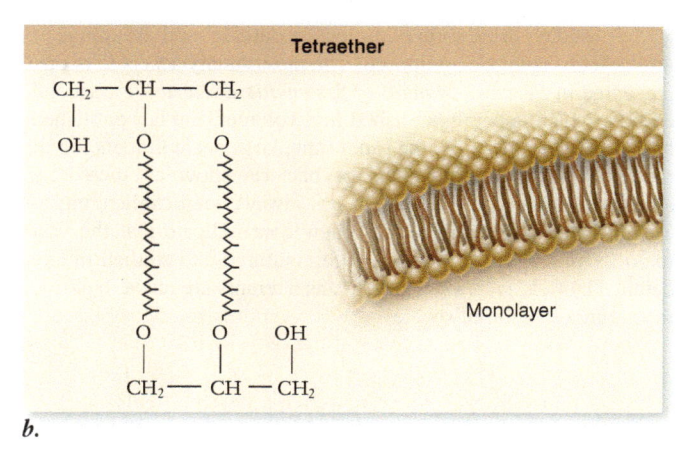

Figure 28.4 Archaeal membrane lipids. *a.* Archaeal membrane lipids are formed on a glycerol skeleton similar to bacterial and eukaryotic lipids, but the hydrocarbon chains are connected to the glycerol by ether linkages not ester linkages. The hydrocarbons can also be branched and even contain rings. *b.* These lipids can also form as tetraethers instead of diethers. The tetraether forms a monolayer as it includes two polar regions connected by hydrophobic hydrocarbons.

Gene expression. The machinery used for gene expression also differs between archaea and bacteria. The archaea may have more than one RNA polymerase, and these enzymes more closely resemble the eukaryotic RNA polymerases than they do the single bacterial RNA polymerase. Some of the translation machinery is also more similar to that of eukaryotes (see chapter 15).

Most prokaryotes have not been characterized

Prokaryotes are not easily classified according to their forms, and only recently has enough been learned about their biochemical and metabolic characteristics to develop a satisfactory overall classification scheme comparable to that used for other organisms.

Early classification characteristics

Early systems for classifying prokaryotes relied on differential stains such as the Gram stain and differences in the observable phenotype of the organism. Key characteristics once used in classifying prokaryotes were as follow:

1. photosynthetic or nonphotosynthetic,
2. motile or nonmotile,
3. unicellular or colony-forming or filamentous,
4. formation of spores or division by transverse binary fission, and
5. importance as human pathogens or not.

Molecular approaches to classification

With the development of genetic and molecular approaches, prokaryotic classifications may help reflect true evolutionary relatedness. Molecular approaches include the following:

1. the analysis of the amino acid sequences of key proteins,
2. the analysis of nucleic acid–base sequences by establishing the percent of guanine (G) and cytosine (C),
3. nucleic acid hybridization, which is essentially the mixing of single-stranded DNA from two species and determining the amount of base-pairing (closely related species will have more bases pairing),
4. gene and RNA sequencing, especially looking at ribosomal RNA, and
5. whole-genome sequencing.

The three-domain system of phylogeny, originated by Carl Woese (figure 28.5 and chapter 26) relies on all of these molecular methods, but emphasizes the comparison of rRNA sequences to establish the evolutionary relatedness of all organisms. The rRNA sequences were chosen for their high degree of evolutionary conservation to ask questions about these most ancient splits in the tree of life.

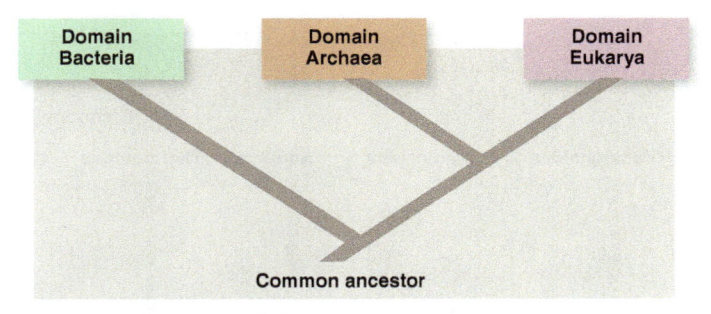

Figure 28.5 The three domains of life. The two prokaryotic domains, Archaea and Bacteria, are not closely related, although both are prokaryotes. In many ways (see text), archaea more closely resemble eukaryotes than bacteria. This tree is based on rRNA sequences.

Based on these sorts of molecular data, several groupings of prokaryotes have been proposed. The most widely accepted is that presented in *Bergey's Manual of Systematic Bacteriology,* second edition, which is being published in 5 volumes, the last published in 2012 (figure 28.6). At the same time, large-scale sequencing of randomly sampled collections of bacteria show an incredible amount of diversity. Although it has always been challenging to assign bacteria to species, these new data indicate that the vast majority of bacteria have never been cultured and studied in any detail. The field is in a state of flux as attempts are made to define the nature of bacterial species.

Learning Outcomes Review 28.1

The study of prokaryotic organisms was first made possible by microscopes. Early microbiology grew from the study of disease-causing agents. Prokaryotes lack both a membrane-bounded nucleus and the diverse organelles seen in eukaryotes, and they reproduce by binary fission. Bacteria and archaea are clearly different from each other based on both structure and metabolism. Classification of prokaryotes had been based on physical characteristics, and it has now been aided by the use of DNA analysis. A vast number of prokaryotes have not been studied because they cannot be cultured.

- What features distinguish archaea from both bacteria and eukaryotes?

Euryarchaeota

1.37 μm

Archaea differ greatly from bacteria. Although both are prokaryotes, archaeal cell walls lack peptidoglycan; plasma membranes are made of different kinds of lipids than bacterial plasma membranes; RNA and ribosomal proteins are more like eukaryotes than bacteria. Mostly anaerobic. Examples include *Methanococcus, Thermoproteus, Halobacterium.*

Aquificae

1 μm

The Aquificae represent the deepest or oldest branch of bacteria. *Aquifex pyrophilus* is a rod-shaped hyper-thermophile with a temperature optimum at 85°C; a chemoautotroph, it oxidizes hydrogen or sulfur. Several other related phyla are also thermophiles.

Bacilli

4 μm

Gram-positive bacteria. Largely solitary; many form endospores. Responsible for many significant human diseases, including anthrax (*Bacillus anthracis*); botulism (*Clostridium botulinum*); and other common diseases (*Staphylococcus, Streptococcus*).

Actinobacteria

22 μm

Some gram-positive bacteria form branching filaments and some produce spores; often mistaken for fungi. Produce many commonly used antibiotics, including streptomycin and tetracycline. One of the most common types of soil bacteria; also common in dental plaque. *Streptomyces, Actinomyces.*

Spirochetes

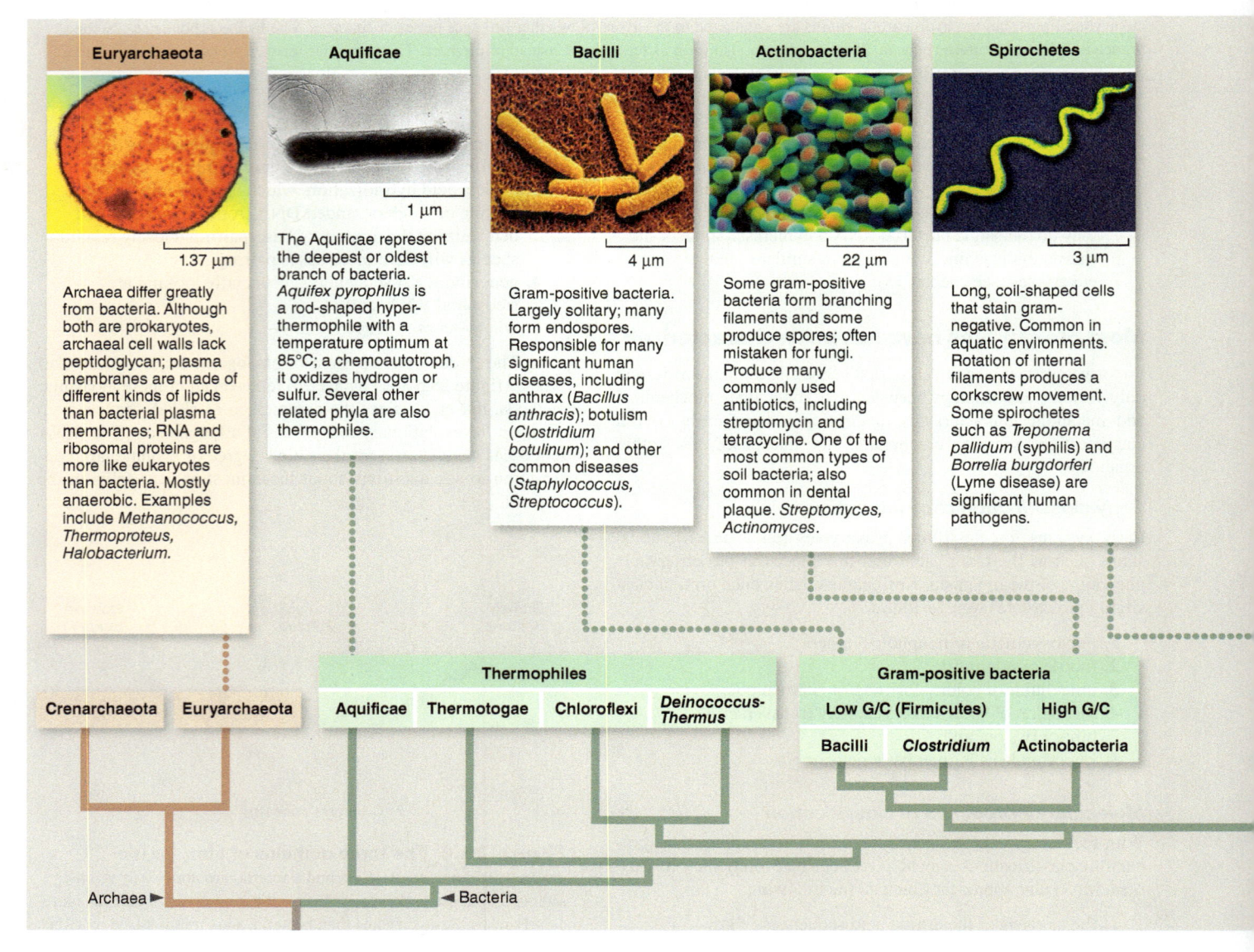

3 μm

Long, coil-shaped cells that stain gram-negative. Common in aquatic environments. Rotation of internal filaments produces a corkscrew movement. Some spirochetes such as *Treponema pallidum* (syphilis) and *Borrelia burgdorferi* (Lyme disease) are significant human pathogens.

Crenarchaeota	Euryarchaeota		Aquificae	Thermotogae	Chloroflexi	Deinococcus-Thermus	Low G/C (Firmicutes)		High G/C
				Thermophiles				**Gram-positive bacteria**	
							Bacilli	*Clostridium*	Actinobacteria

◄ Archaea   ◄ Bacteria

Figure 28.6 Some major clades of prokaryotes. The classification adopted here is that of *Bergey's Manual of Systematic Bacteriology,* second edition, 2001. G/C refers to %G/C in genome.

28.2 Prokaryotic Cell Structure

Learning Outcomes
1. Describe the general features common to all prokaryotic cells.
2. Explain the differences between gram-positive and gram-negative bacterial cells.
3. Describe the features that distinguish different kinds of prokaryotic cells.

Prokaryotic cells are relatively simple, but they can be categorized based on cell shape. They also have some variations in structure that give them different staining properties for certain dyes. Other features are found in some types of cells but not in others.

Prokaryotes have three basic forms: Rods, cocci, and spirals

Although it is an oversimplification, it is useful to divide bacteria based on easily definable morphologies. Most prokaryotes exhibit one of three basic shapes: rod-shaped, often called a *bacillus*

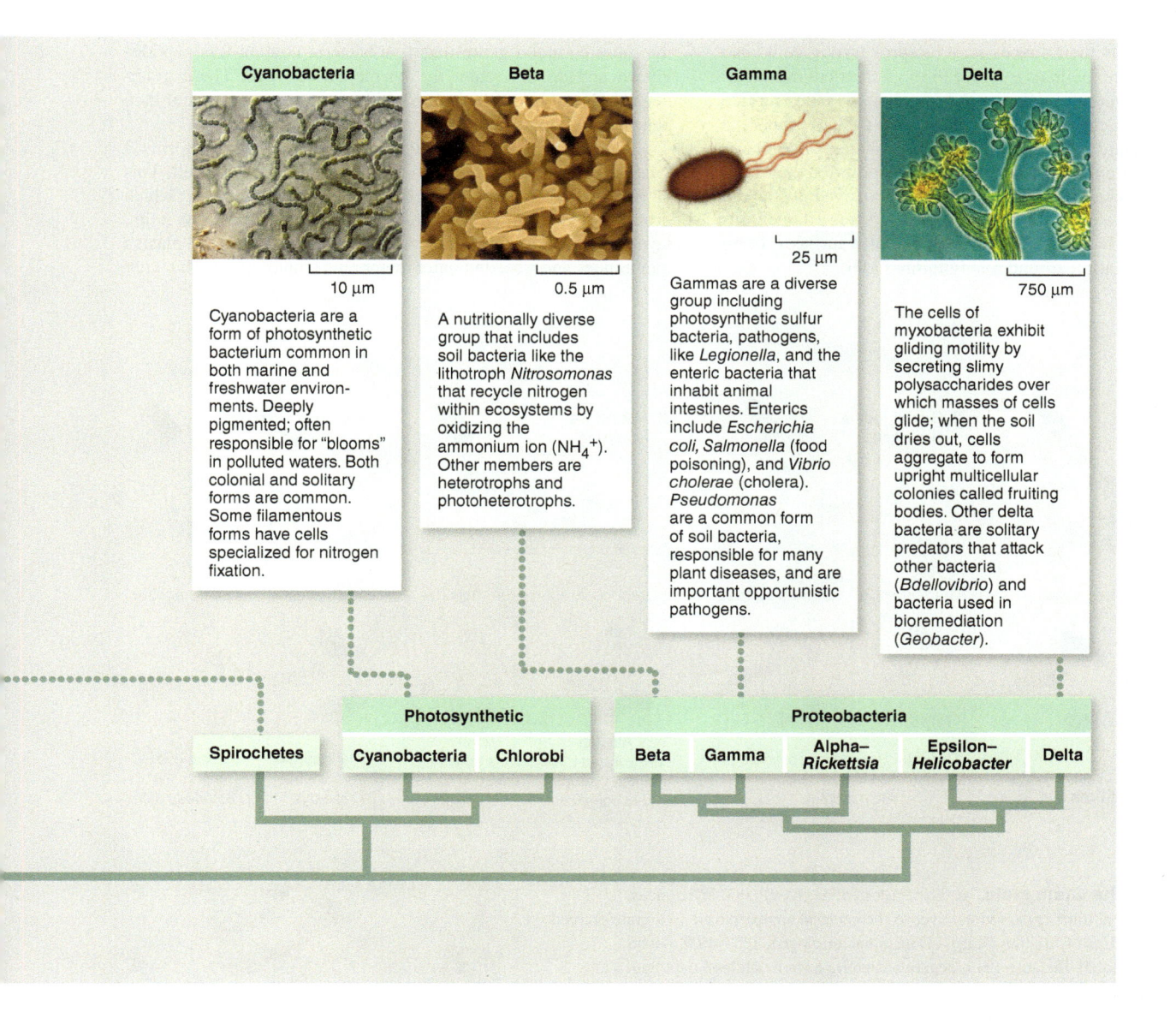

Cyanobacteria

10 μm

Cyanobacteria are a form of photosynthetic bacterium common in both marine and freshwater environments. Deeply pigmented; often responsible for "blooms" in polluted waters. Both colonial and solitary forms are common. Some filamentous forms have cells specialized for nitrogen fixation.

Beta

0.5 μm

A nutritionally diverse group that includes soil bacteria like the lithotroph *Nitrosomonas* that recycle nitrogen within ecosystems by oxidizing the ammonium ion (NH_4^+). Other members are heterotrophs and photoheterotrophs.

Gamma

25 μm

Gammas are a diverse group including photosynthetic sulfur bacteria, pathogens, like *Legionella*, and the enteric bacteria that inhabit animal intestines. Enterics include *Escherichia coli*, *Salmonella* (food poisoning), and *Vibrio cholerae* (cholera). *Pseudomonas* are a common form of soil bacteria, responsible for many plant diseases, and are important opportunistic pathogens.

Delta

750 μm

The cells of myxobacteria exhibit gliding motility by secreting slimy polysaccharides over which masses of cells glide; when the soil dries out, cells aggregate to form upright multicellular colonies called fruiting bodies. Other delta bacteria are solitary predators that attack other bacteria (*Bdellovibrio*) and bacteria used in bioremediation (*Geobacter*).

Spirochetes | Photosynthetic: Cyanobacteria, Chlorobi | Proteobacteria: Beta, Gamma, Alpha–Rickettsia, Epsilon–Helicobacter, Delta

(plural, *bacilli*); *coccus* (plural, *cocci*), spherical- or ovoid-shaped; and *spirillum* (plural, *spirilla*), long and helical-shaped; these bacteria are also called *spirochetes*.

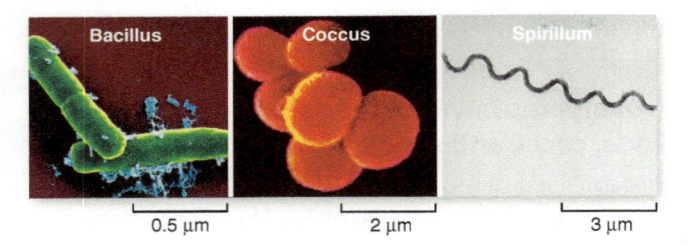

The bacterial cell wall is the single most important contributor to cell shape. Bacteria that normally lack cell walls, such as the mycoplasmas, do not have a set shape.

As diverse as their shapes may be, prokaryotic cells also have many different methods to move through their environment. A *flagellum* or several flagella may be found on the outer surface of many prokaryotic cells. These structures are used to propel the organisms in a fluid environment. Some rod-shaped and spherical bacteria form colonies, adhering end-to-end after they have divided, forming chains. Some bacterial cells change into stalked structures or grow long, branched filaments. Some filamentous bacteria are capable of a gliding motion on solid surfaces, often combined with rotation around a longitudinal axis.

Prokaryotes have a tough cell wall and other external structures

The prokaryotic cell wall is often complex, consisting of many layers. Minimally it consists of peptidoglycan, a polymer unique to bacteria. This polymer forms a rigid network of polysaccharide strands cross-linked by peptide side chains. It is an important structure because it maintains the shape of the cell and protects the cell from swelling and rupturing in hypotonic solutions, which are most commonly found in the environment. The archaea do not possess peptidoglycan, but some have a similar structure called pseudomurein, or pseudopeptidoglycan.

Gram-positive and gram-negative bacteria

Two types of bacteria can be identified using a staining process called the **Gram stain,** hence their names. **Gram-positive** bacteria have a thicker peptidoglycan wall and stain a purple color, whereas the more common **gram-negative** bacteria contain less peptidoglycan and do not retain the purple-colored dye. These gram-negative bacteria can be stained with a red counterstain and then appear dark pink (figure 28.7).

In the gram-positive bacteria, the peptidoglycan forms a thick, complex network around the outer surface of the cell. This network also contains lipoteichoic and teichoic acid, which protrudes from the cell wall. In the gram-negative bacteria, a thin layer of peptidoglycan is sandwiched between the plasma membranes and a second outer membrane (figure 28.8). The outer

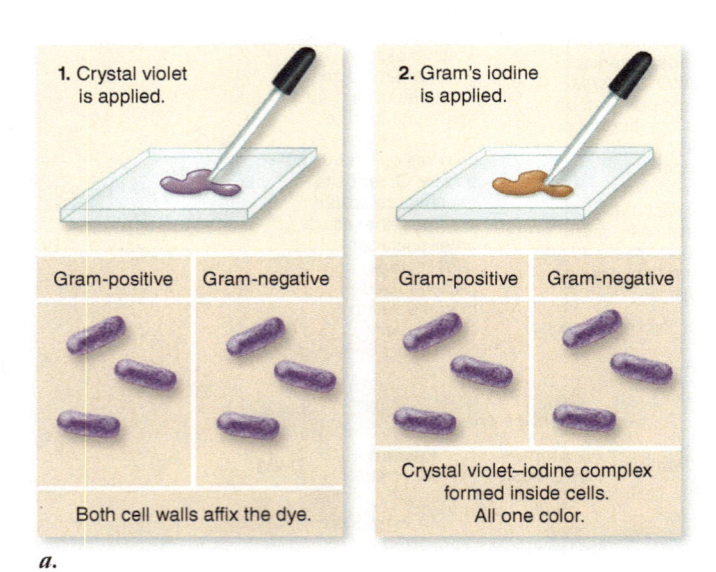

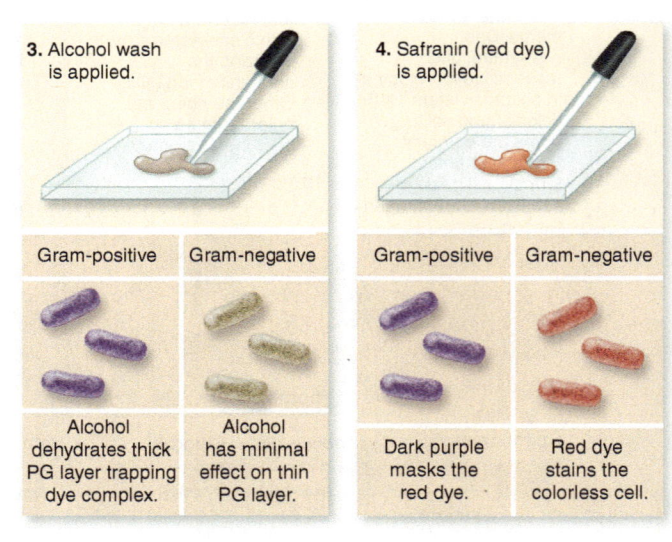

Figure 28.7 The Gram stain. *a.* The thick peptidoglycan (PG) layer encasing gram-positive bacteria traps crystal violet dye, so the bacteria appear purple in a gram-stained smear (named after Hans Christian Gram—Danish bacteriologist, 1853–1938—who developed the technique). Because gram-negative bacteria have much less peptidoglycan (located between the plasma membrane and an outer membrane), they do not retain the crystal violet dye and so exhibit the red counterstain (usually a safranin dye). *b.* A micrograph showing the results of a Gram stain with both gram-positive and gram-negative cells.

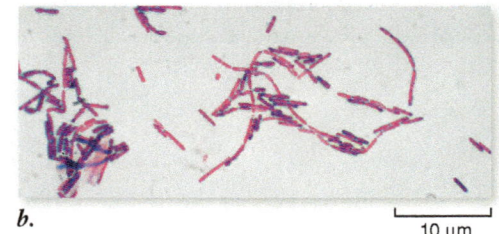

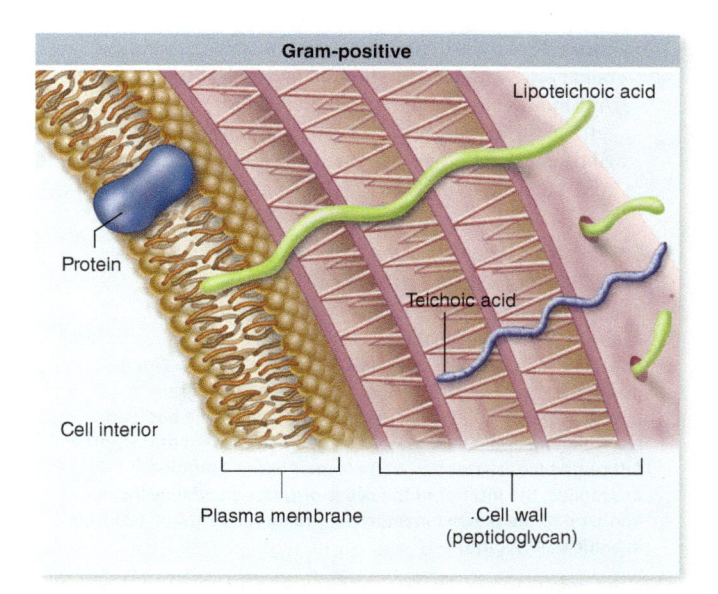

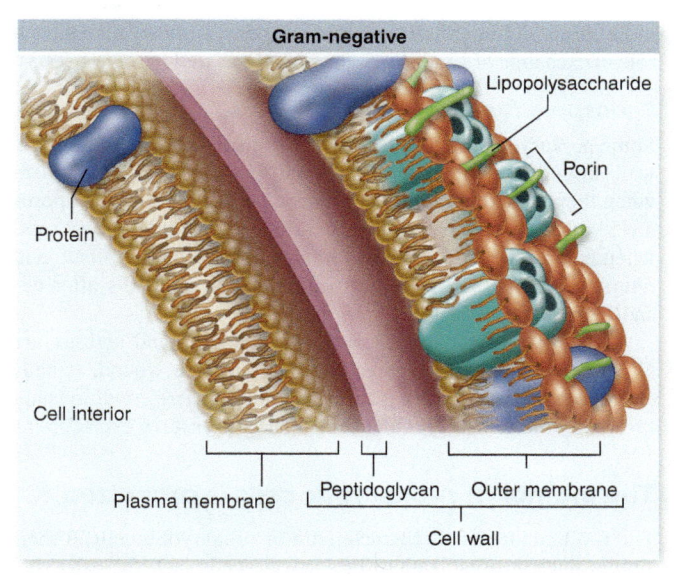

Figure 28.8 The structure of gram-positive and gram-negative cell walls. The gram-positive cell wall is much simpler, composed of a thick layer of cross-linked peptidoglycan chains. Molecules of lipoteichoic acid and teichoic acid are also embedded in the wall and exposed on the surface of the cell. The gram-negative cell wall is composed of multiple layers. The peptidoglycan layer is thinner than in gram-positive bacteria and is surrounded by an additional membrane composed of lipopolysaccharide. Porin proteins form aqueous pores in the outer membrane. The space between the outer membrane and peptidoglycan is called the periplasmic space.

membrane contains large molecules of **lipopolysaccharide,** lipids with polysaccharide chains attached. The outer membrane layer makes gram-negative bacteria resistant to many antibiotics that interfere with cell-wall synthesis in gram-positive bacteria. For example, penicillin acts to inhibit the cross-linking of peptidoglycan in a gram-positive cell wall, killing growing bacterial populations.

S-layer

In some bacteria and archaea, an additional protein or glycoprotein layer forms a rigid paracrystalline surface called an *S-layer* outside of the peptidoglycan or outer membrane layers of gram-positive and gram-negative bacteria, respectively. Among the archaea, the S-layer is almost universal and can be found outside of a pseudopeptidoglycan layer or, in contrast to the bacteria, may be the only rigid layer surrounding the cell. The functions of S-layers are diverse and variable but often involve adhesion to surfaces or protection.

The capsule

In some bacteria, an additional gelatinous layer, the **capsule,** surrounds the other wall layers. A capsule enables a prokaryotic cell to adhere to surfaces and to other cells, and, most important, to evade an immune response by interfering with recognition by phagocytic cells. Therefore, a capsule often contributes to the ability of bacteria to cause disease.

Bacterial flagella and pili

Many kinds of prokaryotes have slender, rigid, helical flagella composed of the protein **flagellin** (figure 28.9). These flagella range from 3 to 12 μm in length and are very thin—only 10 to 20 nm thick. They are anchored in the cell wall and spin like a propeller, moving the cell through a liquid environment. Bacterial cells that have lost the genes for flagellin are not able to swim.

Pili (singular, *pilus*) are other hairlike structures that occur on the cells of some gram-negative prokaryotes. They are shorter than prokaryotic flagella and about 7.5 to 10 nm thick. Pili are

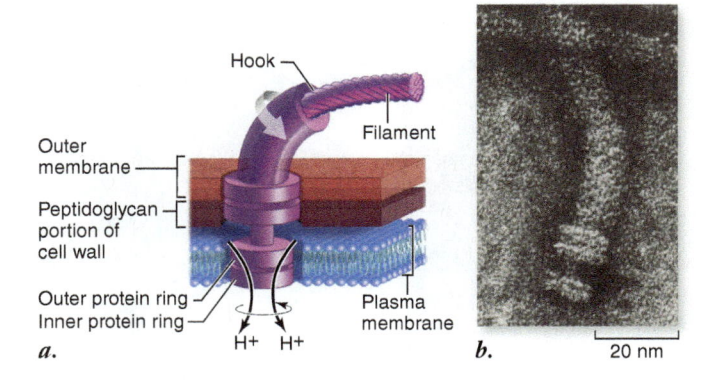

Figure 28.9 The flagellar motor of a gram-negative bacterium. *a.* A protein filament, composed of the protein flagellin, is attached to a protein rod that passes through a sleeve in the outer membrane and through a hole in the peptidoglycan layer to rings of protein anchored in the cell wall and plasma membrane, like rings of ball bearings. The rod rotates when the inner protein ring attached to the rod turns with respect to the outer ring fixed to the cell wall. The inner ring is an H^+ ion channel, a proton pump that uses the flow of protons into the cell to power the movement of the inner ring past the outer one. The membrane wall anchor of the flagellum is called the basal body. *b.* Electron micrograph of bacterial flagellum.

more important in adhesion than movement, and they also have a role in exchange of genetic information (discussed in section 28.3).

Endospore formation

Some prokaryotes are able to form **endospores,** developing a thick wall around their genome and a small portion of the cytoplasm when they are exposed to environmental stress. These endospores are highly resistant to environmental stress, especially heat, and when environmental conditions improve, they can germinate and return to normal cell division to form new individuals after decades or even centuries.

The bacteria that cause tetanus, botulism, and anthrax are all capable of forming spores. With a puncture wound, tetanus endospores may be driven deep into the skin where conditions are favorable for them to germinate and cause disease, or even death.

The interior of prokaryotic cells is organized

The most fundamental characteristic of prokaryotic cells is their simple interior organization. Prokaryotic cells lack the extensive functional compartmentalization seen within eukaryotic cells, but they do have the following structures:

Internal membranes. Many prokaryotes possess invaginated regions of the plasma membrane that function in respiration or photosynthesis (figure 28.10).

Nucleoid region. Prokaryotes lack nuclei and generally do not possess linear chromosomes. Instead, their genes are encoded within a single double-stranded ring of DNA that is highly condensed to form a visible region of the cell known as the **nucleoid region.** Many prokaryotic cells also possess plasmids, which are small, independently replicating circles of DNA. Plasmids contain only a few genes, and although these genes may confer a selective advantage, they are not essential for the cell's survival.

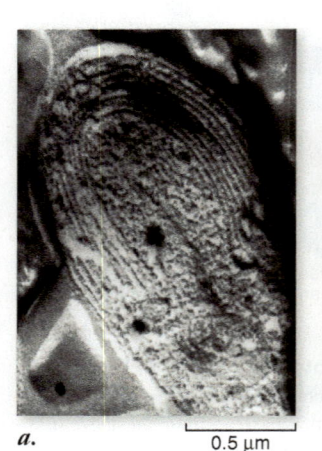

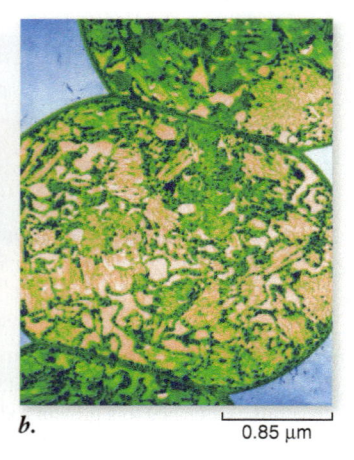

a. ⊢ 0.5 μm ⊣   *b.* ⊢ 0.85 μm ⊣

Figure 28.10 Prokaryotic cells often have complex internal membranes. *a.* This aerobic bacterium exhibits extensive respiratory membranes (long dark curves that hug the cell wall) within its cytoplasm, not unlike those seen in mitochondria. *b.* This cyanobacterium has thylakoid-like membranes (ripplelike shapes along the edges and in the center) that provide a site for photosynthesis.

Ribosomes. Prokaryotic ribosomes are smaller than those of eukaryotes and differ in protein and RNA content. Antibiotics such as tetracycline and chloramphenicol can tell the difference, however—they bind to prokaryotic ribosomes and block protein synthesis, but they do not bind to eukaryotic ribosomes.

Learning Outcomes Review 28.2

The three basic shapes of prokaryotes are rod-shaped, spherical, and spiral-shaped. Bacteria have a cell wall containing peptidoglycan, which is the basis for the Gram stain. Gram-positive bacteria have a thick cell wall, relative to gram-negative species. Many also have an external capsule. Some bacteria have flagella and pili. Some can form heat-resistant endospores. Although prokaryotes do not have membrane-bounded organelles, the interior of the cell is organized and may include infolding of the plasma membrane. Prokaryotic DNA is localized in a nucleoid region.

- What would be the simplest method to determine whether two bacteria belong to the same species?

28.3 Prokaryotic Genetics

Learning Outcomes

1. Contrast the mechanisms of DNA exchange in prokaryotes.
2. Explain genetic mapping in E. coli.
3. Describe how genetics explains the spread of antibiotic resistance.

In sexually reproducing populations, traits are transferred vertically from parent to child. Prokaryotes do not reproduce sexually, but they can exchange DNA between different cells. This horizontal gene transfer occurs when genes move from one cell to another by **conjugation,** requiring cell-to-cell contact, or by means of viruses *(transduction).* Some species of bacteria can also pick up genetic material directly from the environment *(transformation).*

All of these processes have been observed in archaea, but the study of archaeal genetics is hampered by the difficulty in culturing most species. We concentrate here on bacterial systems, primarily *E. coli,* which has been studied extensively.

Conjugation depends on the presence of a conjugative plasmid

Plasmids may encode functions that are not necessary to the organism but that provide a selective advantage in particular environments. For example, antibiotic resistance is not necessary, but is obviously advantageous in the presence of the antibiotic. Some plasmids can be transferred from one cell to another via conjugation. The best-known transmissible plasmid is the *E. coli* F plasmid (fertility factor). Cells containing **F plasmids** are termed **F⁺** cells, and cells that lack the F plasmid are **F⁻** cells. The F plasmid is an

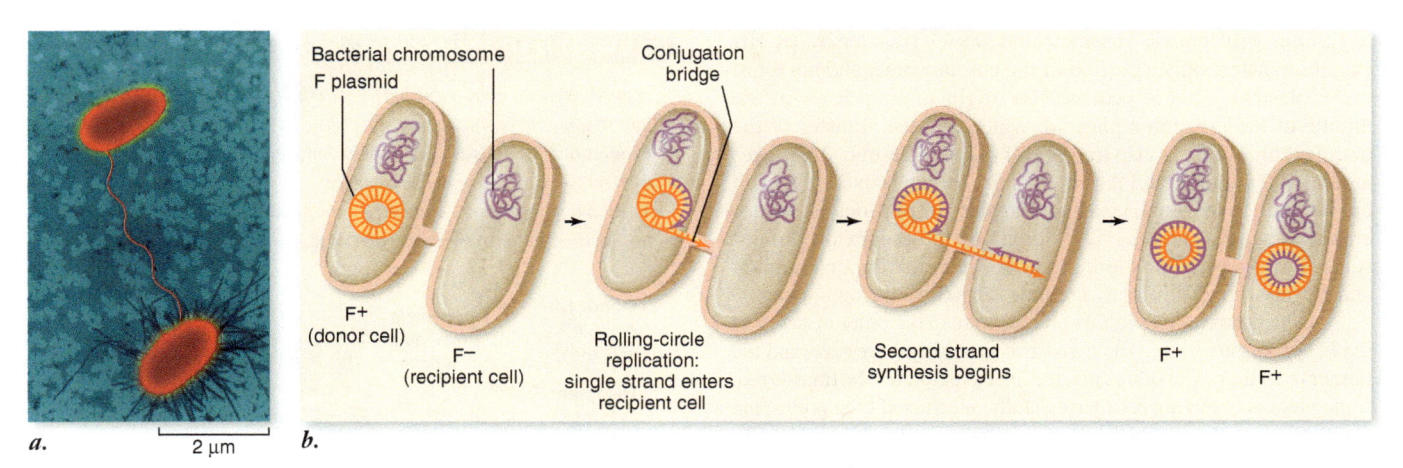

Figure 28.11 Conjugation bridge and transfer of F plasmid between F⁺ and F⁻ cell. *a.* The electron micrograph shows two *E. coli* cells caught in the act of conjugation. The connection between the cells is the extended F pilus. *b.* F⁻ cells are converted to F⁺ cells by the transfer of the F plasmid. The cells are joined by a conjugation bridge and the plasmid is replicated in the donor cell, displacing one parental strand. The displaced strand is transferred to the recipient cell, then replicated. After successful transfer, the recipient cell becomes an F⁺ cell capable of expressing genes for the F pilus and acting as a donor.

independent genetic entity that uses cellular machinery for replication. Studies involving the F plasmid were critical to unraveling the genetics and organization of the *E. coli* chromosome.

F plasmid transfer

The F plasmid contains a DNA replication origin and several genes that promote its transfer to other cells. These genes encode protein subunits that assemble on the surface of the bacterial cell, forming a hollow pilus that is necessary for the transfer process (figure 28.11*a*).

First, the F plasmid binds to a site on the interior of the F⁺ cell just beneath the pilus, now called a *conjugation bridge*. Then, by a process called *rolling-circle replication*, the F plasmid begins to copy its DNA at the binding point. As it is replicated, the displaced single strand of the plasmid passes into the other cell. There, a complementary strand is added, creating a new, stable F plasmid (figure 28.11*b*).

Recombination between the F plasmid and host chromosome

The F plasmid can integrate into the host chromosome by recombining with it. This process is similar to the crossing over (recombination) observed during meiosis in eukaryotes (see chapter 13). This process is also called homologous recombination, as it occurs between homologous regions of DNA. In the case of the F plasmid, there are regions of homology with the *E. coli* chromosome called insertion sequences (IS). Recombination between the F plasmid and the chromosome integrates the plasmid into the chromosome (figure 28.12). This process uses host proteins, and depends on the presence of IS elements. These IS elements are actually transposable elements that probably moved from the chromosome to the F plasmid.

We call a cell with an integrated F plasmid an **Hfr cell** for high frequency of recombination, because now transfer by the

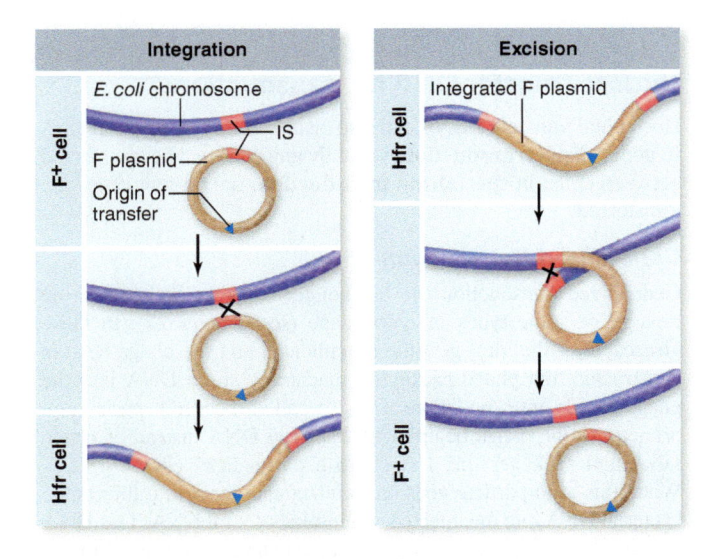

Figure 28.12 Integration and excision of F plasmid. The F plasmid contains short insertion sequences (IS) that also exist in the chromosome. This allows the plasmid to pair with the chromosome, and a single recombination event between two circles leads to a larger circle. This integrates the plasmid into the chromosome, creating an Hfr cell, as shown on the left. The process is reversible because the IS elements in the integrated plasmid can pair, and now a recombination event will return the two circles and convert the Hfr back to an F⁺ cell as shown on the right.

> **Data analysis** If the excision of an F plasmid is not precise, and some *E. coli* DNA is added to the F plasmid, what are the consequences when this plasmid is transferred to a new cell?

F plasmid will include chromosomal DNA. Transfer by an Hfr usually transfers only a portion of the chromosome, and not all of the F plasmid. This is because the origin of transfer is in the middle of the integrated plasmid, which requires transfer of the entire chromosome to also transfer all of the F plasmid. This takes around 100 minutes, and the conjugation bridge is usually broken before that time. This leads to transfer of portions of donor chromosome that can replace regions of the recipient chromosome by two recombination events between the donor fragment and chromosome, similar to a double crossover during meiosis.

Geneticists have used this process to map the order of genes in the *E. coli* chromosome. The transfer of genes is progressive, and the farther from the origin of transfer, the later a region will be transferred. If the process of mating is experimentally interrupted time points, the gene order can be mapped based on time of entry (figure 28.13). Genes can be followed by using a donor with wild type alleles that replace mutant alleles in the recipient. These experiments have shown that the genetic map of *E. coli* is circular, with minutes as the units of the map, and the entire chromosome is 100 minutes long.

The F plasmid can also excise itself by reversing the integration process (see figure 28.12). If excision is inaccurate, the F plasmid can pick up some chromosomal DNA in the process to create an F′ plasmid. Normal transfer of this F′ will produce a cell that is diploid for the material carried by the plasmid. Such partial diploids were used in classical bacterial genetics for a variety of purposes.

Viruses transfer DNA by transduction

Horizontal transfer of DNA can also be mediated by bacteriophage. In **generalized transduction,** virtually any gene can be transferred between cells; in **specialized transduction,** only a few genes are transferred.

Generalized transduction

Generalized transduction can be thought of as an accident of the biology of some types of lytic phage (see chapter 27). In these viruses, after the viral genome is replicated and the phage head is constructed, the phage packaging machinery stuffs DNA into the phage head until no more fits, so-called headfull packaging. Sometimes the phage begins with bacterial DNA instead of phage DNA and packages this DNA into a phage head (figure 28.14). When this viral particle goes on to infect another cell, it injects the bacterial DNA into the infected cell instead of viral DNA. This DNA can then be incorporated into the recipient chromosome by homologous recombination. Similar to transfer by Hfr cells described earlier in this section, two recombination events are necessary to integrate the linear piece of DNA into the circular chromosome (figure 28.14).

Generalized transduction has also been used for mapping purposes in *E. coli*, although the logic is different from that in conjugation. In transduction, the closer together two genes are, the more likely it is that they will be transferred in a single transduction event. This can be expressed mathematically as the *cotransduction frequency*. Correlation of maps from the two methods allows an empirical conversion between cotransduction frequency and minutes in the genetic map.

Specialized transduction

Specialized transduction is limited to phage that exhibit a lysogenic life cycle (see chapter 27). The prototype for this is phage λ from *E. coli*. When λ infects a cell and its genome integrates into

SCIENTIFIC THINKING

Hypothesis: Conjugation using Hfr strains involves the linear transfer of information from donor to recipient cell.

Prediction: If there is a linear transfer of information, then different markers should appear in a time sequence.

Test: Mating strains are agitated at time points to break the conjugation bridge, then plated to determine genotype.

Donor Hfr
azi^r ton^r
lac^+ gal^+

Recipient F⁻
azi^s ton^s
lac^- gal^-

Sample placed in blender. → Sample plated.

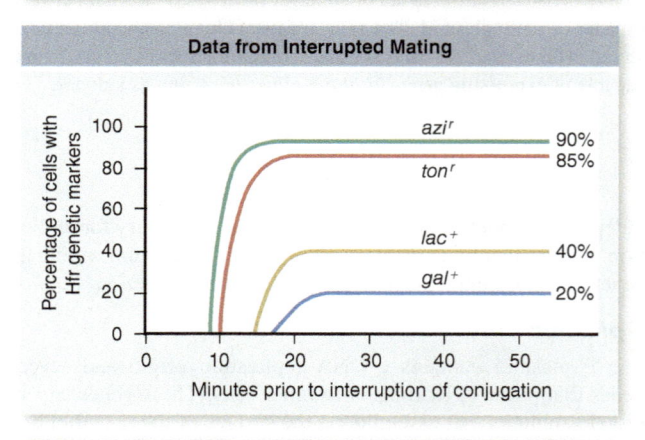

Mating interrupted by agitation in blender

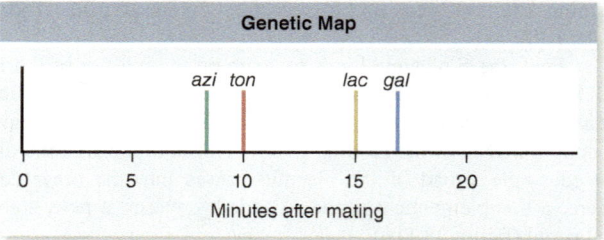

Data from Interrupted Mating

azi^r — 90%
ton^r — 85%
lac^+ — 40%
gal^+ — 20%

(Percentage of cells with Hfr genetic markers vs. Minutes prior to interruption of conjugation)

Genetic Map

azi ton   lac gal
0   5   10   15   20
Minutes after mating

Result: The different genes from the donor strain appear in a linear time sequence.

Conclusion: The transfer of genetic information is linear. This sequence can be used to construct a genetic map ordering the genes on the chromosomes.

Further Experiments: Can other methods of DNA exchange also be used for genetic mapping?

Figure 28.13 Interrupted mating experiment allows construction of genetic map.

the host chromosome, it does not destroy the cell but is passed on by cell division. This integration event is similar to the integration of the F plasmid, except that in the case of λ the recombination is a site-specific event mediated by phage-encoded proteins.

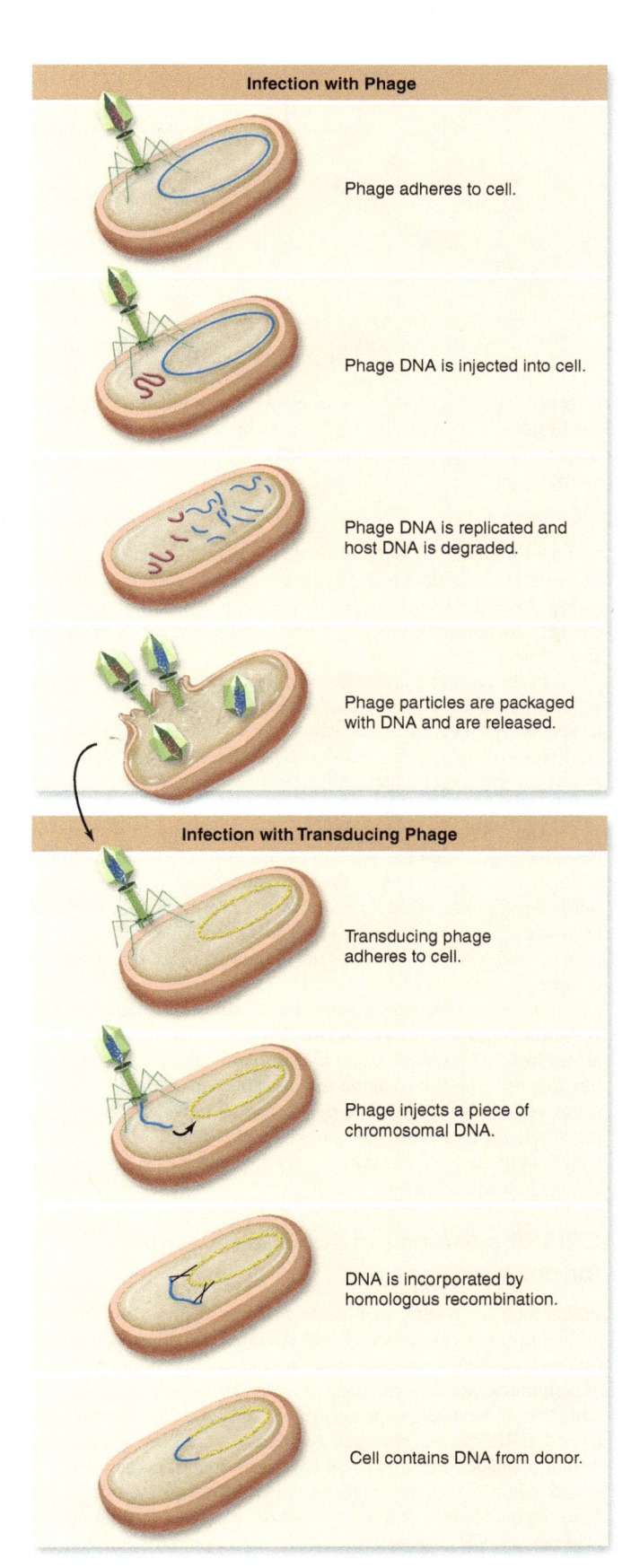

Figure 28.14 Transduction by generalized transducing phage. When some phage infect cells, they degrade the host DNA into pieces. When the phage package their DNA, they can package host DNA in place of phage DNA to produce a transducing phage as shown on the top. When a transducing phage infects a cell, it injects host DNA that can then be integrated into the host genome by homologous recombination. With a linear piece of DNA, it requires two recombination events, which replace the chromosomal DNA with the transducing DNA as shown on the bottom. If the new allele is different from the old, the cell's phenotype will change.

In this lysogenic state, the phage is called a prophage and it is dormant. The prophage encodes the functions necessary to eventually excise itself and undergo lytic growth, leading to the death of the cell. If this excision event is imprecise, it may take some chromosomal DNA with it, in the process making a specialized transducing phage. These phage carry both phage genes and chromosomal genes, unlike generalized transducing phage that carry only chromosomal DNA.

Because the phage head can carry only as much DNA as is found in the phage genome, imprecise excision results in deletion of phage genes. Thus specialized transducing phage may be defective if genes necessary for phage growth are lost in the process.

Specialized transducing phage particles can then integrate into the chromosome, just like wild-type phage, also making the cell diploid for the genes carried by the phage. Phage particles that can integrate as prophages may become trapped in the host genome if the genes necessary for excision become defective by mutation or are lost. The *E. coli* genome contains a number of such cryptic prophage, some of which encode functions important to the cell and must now be considered part of the host genome.

Transformation is the uptake of DNA directly from the environment

Transformation is a naturally occurring process in some species, such as the bacteria that were studied by Frederick Griffith (see chapter 14). Griffith discovered the process despite not knowing what chemical component was transferred. Transformation occurs when one bacterial cell has died and ruptured, spilling its fragmented DNA into the surrounding environment. This DNA can be taken up by another cell and incorporated into its genome, thereby transforming it (figure 28.15). When the uptake occurs under natural conditions, it is termed natural transformation. Some species of both gram-positive and gram-negative bacteria exhibit natural transformation, although the mechanisms seem to differ between the groups.

The proteins involved in the process of natural transformation are all encoded by the bacterial chromosome. The implication is that natural transformation may be the only one of the mechanisms of DNA exchange that evolved as part of normal cellular machinery. The transfer of chromosomal DNA by either conjugation or transduction can be thought of as accidents of plasmid or phage biology, respectively.

Transformation is also important in molecular cloning, but *E. coli* does not exhibit natural transformation. When transformation is accomplished in the laboratory it is called artificial transformation. Artificial transformation is useful for cloning and DNA manipulation (see chapter 17).

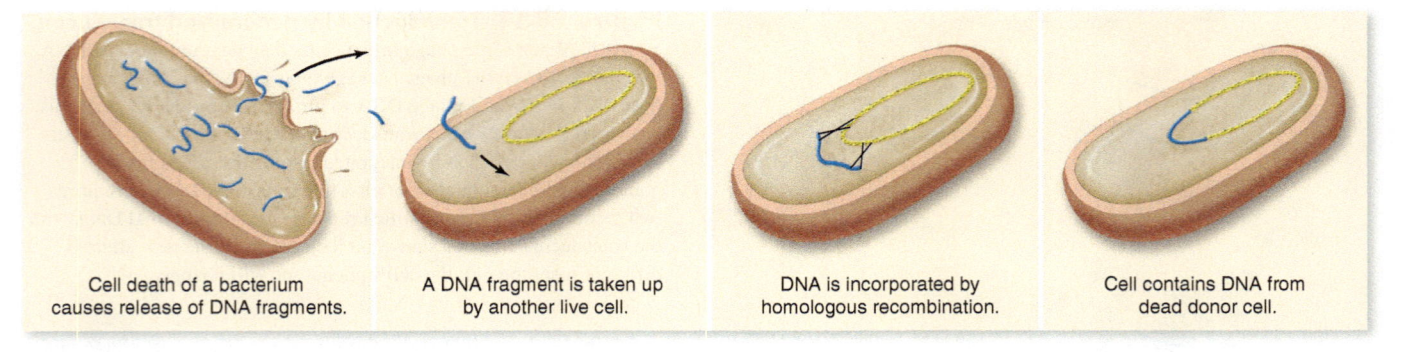

| Cell death of a bacterium causes release of DNA fragments. | A DNA fragment is taken up by another live cell. | DNA is incorporated by homologous recombination. | Cell contains DNA from dead donor cell. |

Figure 28.15 Natural transformation. Natural transformation occurs when one cell dies and releases its contents to the surrounding environment. The DNA is usually fragmented, and small pieces can be taken up by other, living cells. The DNA taken up can replace chromosomal DNA by homologous recombination as in conjugation and transduction. If the new DNA contains different alleles from the chromosome, the phenotype of the cell changes, possibly providing a selective advantage.

Antibiotic resistance can be transferred by resistance plasmids

Some conjugative plasmids pick up antibiotic resistance genes, becoming resistance plasmids, or **R plasmids.** The rapid transfer of newly acquired, antibiotic resistance genes by plasmids has been an important factor in the appearance of the resistant strains of the pathogen *Staphylococcus aureus* discussed later in this section.

Resistance plasmids often acquire antibiotic resistance genes through transposable elements (see chapter 18). These elements can move from chromosome to chromosome or from plasmid to chromosome and back again, and they can transfer antibiotic resistance genes in the process. If a conjugative plasmid picks up these genes, then the bacterium carrying it has a selective advantage in the presence of those antibiotics.

An important example in terms of human health involves the Enterobacteriaceae, the family of bacteria to which the common intestinal bacterium *E. coli* belongs. This family contains many pathogenic bacteria, including the organisms that cause dysentery, typhoid, and other major diseases. At times, some of the genetic material from these pathogenic species is exchanged with or transferred to *E. coli* by transmissible plasmids or bacteriophage. Because of its abundance in the human digestive tract, *E. coli* poses a special threat if it acquires harmful traits, as seen by the outbreaks of the food-borne O157:H7 strain of *E. coli*. Infection with this strain of *E. coli* can lead to serious illness. This is a new strain of *E. coli* that evolved by acquiring genes for pathogenic traits. Evidence suggests this occurred by both transduction and the acquisition of a large virulence plasmid by conjugation.

Variation can also arise by mutation

Just as with any organism, mutations can arise spontaneously in bacteria. Certain factors, especially those that damage DNA, such as radiation, ultraviolet light, and various chemicals, increase the likelihood of mutation.

A typical bacterium such as *E. coli* contains about 5000 genes. The probability of mutation occurring by chance is about in one out of every million copies of a gene. With 5000 genes in a bacterium, we can predict that approximately 1 out of every 200 bacteria will have a mutation. With adequate food and nutrients, a population of *E. coli* can double in 20 minutes. Because bacteria multiply so rapidly, mutations can spread rapidly in a population and can change the characteristics of that population in a relatively short time.

In the laboratory, bacteria are grown on different substrates, called *growth media,* that reflect their nutritional needs. For a particular species, the medium that contains only those nutrients required for wild-type growth is termed a *minimal medium.* A mutant that can no longer survive on minimal medium and needs particular nutritional supplements, such as an amino acid, is called an **auxotroph.**

The ability of prokaryotes to change rapidly in response to new challenges often has adverse effects on humans. A number of antibiotic-resistant strains of the bacterium *Staphylococcus aureus* (termed methicillin-resistant *Staphylococcus aureus,* or MRSA) had been known in hospital settings for some time. More recently these have been observed in infections out of the hospital setting, so-called community acquired MRSA.

Of most concern among these strains is **vancomycin-resistant *Staphylococcus aureus*** (VRSA). This appears to have arisen rapidly by mutation and is alarming because vancomycin is the drug of last resort, making these strains and the infections they cause very difficult to stop. *Staphylococcus* infections, or "staph" infections for short, provide an excellent example of the way in which mutation and intensive selection can bring about rapid change in bacterial populations.

CRISPR systems provide adaptive immunity for prokaryotes

Prokaryotes use a variety of innate defenses against viral infection including restriction-modification systems (the source of restriction endonucleases), and toxin-antitoxin systems. With the advent of genomics, screens of many prokaryotic genomes revealed a structure of repeated sequences with "spacer regions" that were named **CRISPR,** for clustered regularly interspaced short palindromic repeats. Recently, these have been identified as a form of adaptive immunity to viral infection. In response to viral challenge, bacteria and archaea will integrate short segments of viral nucleic acid in these CRISPR loci, then use them to produce an RNA that can be used to guide a complex that can degrade viral nucleic acid. Much like restriction endonucleases, the CRISPR system has proved useful for molecular biologists for genome editing (see chapter 17).

A 2012 survey found CRISPR loci in 48% of bacterial genomes, and 95% of archaeal genomes sequenced. Although few of these have been investigated in detail, those that have reveal three distinct types of system with the common function of incorporating genetic information from infections, then using these to direct a response involving destruction of invading genetic information. These can be directed against both plasmid and viral nucleic acids, and they represent a primitive form of adaptive immunity. If you consider how viruses outnumber prokaryotic cells in virtually every environmental context, this represents very strong selective pressure for a system that can recognize previous infection.

Learning Outcomes Review 28.3

Prokaryotic DNA exchange is horizontal, from donor cell to recipient cell. DNA can be exchanged by conjugation via plasmids, by transduction via viruses, and by transformation through the direct uptake of DNA from the environment. These forms of DNA exchange can be used experimentally to map genes. Variation in prokaryotes also arises by mutation. Extensive use of antibiotics has led to selection for resistant organisms. Resistance genes can be transferred, rapidly spreading resistance.

- How does transfer of genetic information in bacteria differ from eukaryotic sex?

28.4 Prokaryotic Metabolism

Learning Outcomes
1. Describe the different ways that prokaryotes acquire energy and carbon.
2. Explain how bacterial proteins can cause disease in humans.

The variation seen in prokaryotes manifests itself most noticeably in biochemical rather than morphological diversity. Wide variation has been found in the types of metabolism prokaryotes exhibit, especially in the means by which they acquire energy and carbon.

Prokaryotes acquire carbon and energy in four basic ways

Prokaryotes have evolved many mechanisms to acquire the energy and carbon they need for growth and reproduction. Many are *autotrophs* that obtain their carbon from inorganic CO_2. Other prokaryotes are *heterotrophs* that obtain at least some of their carbon from organic molecules, such as glucose. Depending on the method by which they acquire energy, autotrophs and heterotrophs are categorized as follows:

Photoautotrophs. Many bacteria carry out photosynthesis, using the energy of sunlight to build organic molecules from carbon dioxide. The **cyanobacteria** use chlorophyll *a* as the key light-capturing pigment and H_2O as an electron donor, releasing oxygen gas as a by-product. They are therefore oxygenic, and their method of photosynthesis is very similar to that found in algae and plants.

Other bacteria use bacteriochlorophyll as their light-capturing pigment and H_2S as an electron donor, leaving elemental sulfur as the by-product. These bacteria do not produce oxygen (anoxygenic) and have a simpler method of photosynthesis. These are the purple and green sulfur bacteria.

Archaeal species also carry out photosynthesis, the simplest form known. This involves a single protein, bacteriorhodopsin, that uses energy from light to translocate protons across a membrane. This then provides a proton motive force for ATP synthesis. Recent surveys of microbial diversity in marine ecosystems using DNA sequencing have found a new relative of the rhodopsin family called proteorhodopsin. First found in a bacterial species, these proteorhodopsins are quite widespread, found in bacterial, archaeal, and even algal species. This raises the possibility that photosynthesis in marine systems may be more widespread and complex than previously thought.

Chemolithoautotrophs. Some prokaryotes obtain energy by oxidizing inorganic substances. Nitrifiers, for example, oxidize ammonia or nitrite to obtain energy, producing the nitrate that is taken up by plants. This process is called **nitrification,** and it is essential in terrestrial ecosystems because plants primarily absorb nitrogen in the form of nitrate.

Other chemolithoautotrophs oxidize sulfur, hydrogen gas, and other inorganic molecules. On the dark ocean floor at depths of 2500 m, entire ecosystems subsist on prokaryotes that oxidize hydrogen sulfide as it escapes from thermal vents.

Photoheterotrophs. The so-called purple and green nonsulfur bacteria use light as their source of energy but obtain carbon from organic molecules, such as carbohydrates or alcohols that have been produced by other organisms.

Chemoheterotrophs. The majority of prokaryotes obtain both carbon atoms and energy from organic molecules. These include decomposers and most pathogens. Human beings and all nonphotosynthetic eukaryotes are chemoheterotrophs as well.

Some bacteria can attack other cells directly

Invading pathogens of the genera *Yersinia* can introduce proteins directly into host cells by a specialized form of secretion. (*Yersinia pestis* is the bacterial species responsible for bubonic plague.) Most proteins secreted by gram-negative bacteria have special signal sequences that allow them to pass through the bacterium's double membrane. The proteins secreted by *Yersinia* lacked a key signal sequence that two known secretion mechanisms require for transport. The proteins must therefore have been secreted by means of a third type of system, which researchers called the *type III system*. This kind of system acts like a kind of molecular syringe allowing the pathogen to inject proteins directly into the cytoplasm of host cells.

As more bacterial species are studied, the genes coding for the type III system are turning up in other gram-negative animal pathogens, and even in more distantly related plant pathogens. The genes seem more closely related to one another than are the bacteria. Furthermore, the genes are similar to those that code for bacterial flagella.

These proteins are used to transfer other virulence proteins, such as toxins, into nearby eukaryotic cells. Given the similarity of the type III genes to the genes that code for flagella, the transfer proteins may form a flagellum-like structure that shoots virulence proteins into the host cells. Once in the eukaryotic cells, the virulence proteins affect the host's response to the pathogen.

In *Yersinia,* proteins secreted by the type III system are injected into macrophages; the proteins disrupt signals that tell the macrophages to engulf bacteria. *Salmonella* and *Shigella* use their type III proteins to enter the cytoplasm of eukaryotic cells, and thus they are protected from the immune system of their host. The proteins secreted by certain strains of *E. coli* alter the cytoskeleton of nearby intestinal eukaryotic cells, resulting in a bulge onto which the bacterial cells can tightly bind.

Bacteria are costly plant pathogens

Although the majority of commercially relevant plant pathogens are fungi, many diseases of plants are associated with particular heterotrophic bacteria. Almost every kind of plant is susceptible to one or more kinds of bacterial disease, including blights, soft rots, and wilts. Fire blight, which destroys pear and apple trees and related plants, is a well-known example of bacterial disease.

The early symptoms of these plant diseases vary, but they are commonly manifested as spots of various sizes on the stems, leaves, flowers, or fruits. Most bacteria that cause plant diseases are members of the group of rod-shaped gram-negative bacteria known as pseudomonads.

Learning Outcomes Review 28.4

Prokaryotes exhibit amazing metabolic diversity with both autotrophic and heterotrophic species. Photoautotrophs use light as an energy source; chemolithoautotrophs oxidize inorganic compounds. Photoheterotrophs use light as an energy source and organic compounds as carbon sources. Chemoheterotrophs use organic compounds for both energy and carbon. Bacterial animal pathogens attack host cells with toxic proteins that disrupt the host's immune response, among other effects.

■ Why is metabolism a better way than morphology to characterize prokaryotes?

28.5 Human Bacterial Disease

Learning Outcomes

1. Describe common human bacterial pathogens.
2. Explain how bacteria can cause ulcers.
3. Identify sexually transmitted diseases caused by bacteria.

In the early 20th century, before the discovery and widespread use of antibiotics, infectious diseases killed nearly 20% of all U.S. children before they reached the age of five. Sanitation and antibiotics considerably improved the situation. In recent years, however, we have seen the appearance or reappearance of many bacterial diseases, including cholera, leprosy, tetanus, bacterial pneumonia, whooping cough, diphtheria, and Lyme disease (table 28.1). Members of the genus *Streptococcus* are associated with scarlet fever, rheumatic fever, pneumonia, "flesh-eating disease," and other infections. Tuberculosis, another bacterial disease, is still a leading cause of death in humans worldwide.

Bacteria have many different methods to spread through a susceptible population. Tuberculosis and many other bacterial diseases of the respiratory tract are mostly spread through the air in droplets of mucus or saliva. Diseases such as typhoid fever, paratyphoid fever, and bacillary dysentery are spread by fecal contamination of food or water. Lyme disease and Rocky Mountain spotted fever are spread to humans by tick vectors.

Tuberculosis has infected humans for all of recorded history

Tuberculosis (TB) is second only to HIV/AIDS as the number one killer worldwide from a single infectious agent. But, this is not a new problem—there is evidence that peoples from ancient Egypt and pre-Columbian South America died from TB; the TB bacillus (*Mycobacterium tuberculosis*) has been identified in prehistoric mummies. TB afflicts the respiratory system and is easily transmitted from person to person through the air.

The spread of tuberculosis

Currently, about one-third of all people worldwide are regularly exposed to *Mycobacterium tuberculosis*. This caused an estimated 9 million new infections and 1.5 million deaths in 2013. Both of these rates have been declining, although this is leveling off. In the period from 1990–2013, the death rate fell 45% and TB prevalence fell 41%.

Since the mid-1980s, the United States has experienced a resurgence of TB. This peaked in the mid-1990s and has been declining since, although the rate of decline is leveling off. There were 9412 new TB cases in 2014, down 2.2% from 2013.

Tuberculosis treatment

Most TB patients are placed on multiple, expensive antibiotics for six to twelve months. Alarming outbreaks of **multidrug-resistant (MDR) strains** of TB have occurred, however, in the United States and worldwide. These MDR strains are resistant to most of the best available anti-TB medications. MDR TB is of particular concern because it requires much more time and is more expensive to treat. Also, it is more likely to prove fatal.

This spread of MDR TB is likely due to the extremely long course of antibiotics required to treat the disease. Patients often quit taking the antibiotics before completing the course, setting up conditions in their bodies to allow drug-resistant bacteria to thrive.

The basic principles of TB treatment and control are to make sure all patients complete a full course of medication, so

TABLE 28.1 Important Human Bacterial Diseases

Disease	Pathogen	Vector/Reservoir	Epidemiology
Anthrax	*Bacillus anthracis*	Animals, including processed skins	Bacterial infection that can be transmitted through contact or ingestion. Rare except in sporadic outbreaks. May be fatal.
Botulism	*Clostridium botulinum*	Improperly prepared food	Contracted through ingestion or contact with wound. Produces acute toxic poison; can be fatal.
Chlamydia	*Chlamydia trachomatis*	Humans, sexually transmitted disease (STD)	Urogenital infections with possible spread to eyes and respiratory tract. Increasingly common over past 30 years.
Cholera	*Vibrio cholerae*	Human feces, plankton	Causes severe diarrhea that can lead to death by dehydration; 50% peak mortality rate if untreated. A major killer in times of crowding and poor sanitation; after the 2010 earthquake in Haiti, 180,000 were infected in just 70 days.
Dental caries	*Streptococcus mutans, Streptococcus sobrinus*	Humans	A dense collection of these bacteria on the surface of teeth leads to secretion of acids that destroy minerals in tooth enamel; sugar alone does not cause caries.
Diphtheria	*Corynebacterium diphtheriae*	Humans	Acute inflammation and lesions of respiratory mucous membranes. Spread through respiratory droplets. Vaccine available.
Gonorrhea	*Neisseria gonorrhoeae*	Humans only	STD, on the increase worldwide. Usually not fatal.
Hansen disease (leprosy)	*Mycobacterium leprae*	Humans, feral armadillos	Chronic infection of the skin; worldwide incidence about 10–12 million, especially in southeast Asia. Spread through contact with infected individuals.
Lyme disease	*Borrelia burgdorferi*	Ticks, deer, small rodents	Spread through bite of infected tick. Lesion followed by malaise, fever, fatigue, pain, stiff neck, and headache.
Peptic ulcers	*Helicobacter pylori*	Humans	Originally thought to be caused by stress or diet, most peptic ulcers now appear to be caused by this bacterium; good news for ulcer sufferers because it can be treated with antibiotics.
Plague	*Yersinia pestis*	Fleas of wild rodents: rats and squirrels	Killed one-fourth of the population of Europe in the 14th century; endemic in wild rodent populations of the western United States today.
Pneumonia	*Streptococcus, Mycoplasma, Chlamydia, Haemophilus*	Humans	Acute infection of the lungs; often fatal without treatment. Vaccine for streptococcal pneumonia available.
Tuberculosis	*Mycobacterium tuberculosis*	Humans	An acute bacterial infection of the lungs, lymph, and meninges. Incidence had been on the rise but is now falling. Infections are complicated by the existence of drug-resistant forms of the bacterium.
Typhoid fever	*Salmonella typhi*	Humans	A systemic bacterial disease of worldwide incidence. Fewer than 500 cases a year are reported in the United States. Spread through contaminated water or foods (such as improperly washed fruits and vegetables). Vaccines are available for travelers.
Typhus	*Rickettsia typhi*	Lice, rat fleas, humans	Historically a major killer in times of crowding and poor sanitation; transmitted from human to human through the bite of infected lice and fleas. Peak untreated mortality rate of 70%.

that all of the bacteria causing the infection are killed and drug-resistant strains do not develop. Great efforts are being made to ensure that high-risk individuals who are infected but not yet sick receive preventive therapy under observation. Such programs are approximately 90% effective in reducing the likelihood of developing active TB and spreading it to others. The efforts are having a significant effect. From 2000–2008 the TB rate decreased almost 4% annually, and in 2009 the case rate decreased 11%.

Bacterial biofilms are involved in tooth decay

Bacteria and other organisms may form mixed cultures on certain surfaces that are extremely difficult to treat. On teeth, this biofilm, or plaque, consists largely of bacterial cells surrounded by a polysaccharide matrix. Most of the bacteria in plaque are filaments of rod-shaped cells classified as various species of *Actinomyces,* which extend out perpendicular to the surface of the tooth. Many other bacterial species are also present in plaque.

Tooth decay, or dental caries, is caused by the bacteria present in the plaque, which persist especially in places that are difficult to reach with a toothbrush. Diets that are high in simple sugars are especially harmful to teeth because certain bacteria, notably *Streptococcus sobrinus* and *S. mutans,* ferment the sugars to lactic acid. This acid production reduces the pH in the area around the plaque, breaking down the structure of the hydroxyapatite that makes tooth enamel hard. As the enamel degenerates, the remaining soft matrix of the tooth becomes vulnerable to bacterial attack.

Bacteria can cause ulcers

Bacteria can also be the cause of disease states that on the surface appear to have no infectious basis. Peptic ulcer disease is due to craterlike lesions in the gastrointestinal tract that are exposed to peptic acid. Ulcers can be caused by drugs, such as nonsteroidal anti-inflammatory drugs, and also by some tumors of the pancreas that cause an oversecretion of peptic acid. In 1982, a bacterium named *Campylobacter pylori* (now named *Helicobacter pylori*) was isolated from gastric juices. Over the years evidence has accumulated that this bacterium is actually the causative agent in the majority of cases of peptic ulcer disease.

Antibiotic therapy can now eliminate *H. pylori,* treating the cause of the disease, and not just the symptoms. The discovery of the action of this bacterial species illustrates how even disease states that appear to be unrelated to infectious disease may actually be caused by cryptic (unknown) infection.

Many sexually transmitted diseases are bacterial

A number of bacteria cause sexually transmitted diseases (STDs), three particularly important examples of which are gonorrhea, syphilis, and chlamydia (figure 28.16).

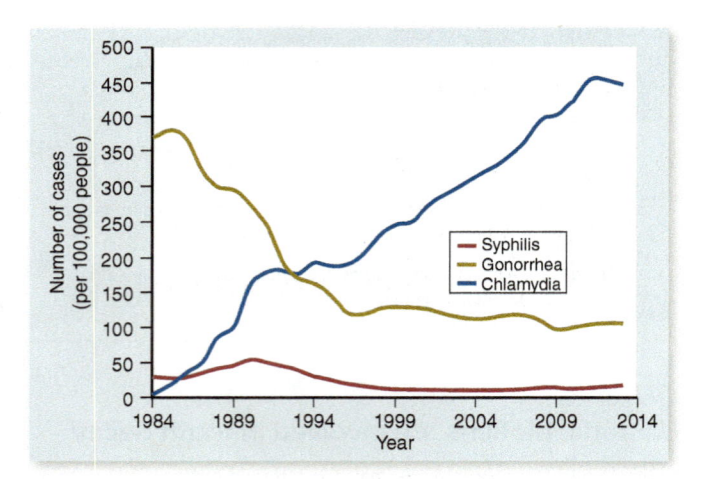

Figure 28.16 Trends in sexually transmitted diseases in the United States.

? Inquiry question How is it possible for the incidence of one STD (chlamydia) to rise as another (gonorrhea) falls?

Gonorrhea

Gonorrhea is one of the most prevalent communicable diseases in North America. Caused by the bacterium *Neisseria gonorrhoeae,* gonorrhea can be transmitted through sexual intercourse or any other sexual contact in which body fluids are exchanged, such as oral or anal intercourse. It can also pass from mother to baby during delivery through the birth canal.

The incidence of gonorrhea has been on the decline in the United States, although that decline appears to have leveled off. It remains a serious threat worldwide; especially of concern is the appearance of antibiotic-resistant strains of *N. gonorrhoeae.*

Syphilis

Syphilis, a very destructive STD, was once prevalent and deadly but is now less common due to the advent of blood-screening procedures and antibiotics. Syphilis is caused by a spirochete bacterium, *Treponema pallidum,* transmitted during sexual intercourse or through direct contact with an open syphilis chancre sore. The bacterium can also be transmitted from a mother to her fetus, often causing damage to the heart, eyes, and nervous system of the baby.

Once inside the body, the disease progresses in four distinct stages. The first, or primary stage, is characterized by the appearance of a small, painless, often unnoticed sore called a *chancre.* The chancre resembles a blister and occurs at the location where the bacterium entered the body about three weeks following exposure. This stage of the disease is highly infectious, and an infected person may unwittingly transmit the disease to others. This sore heals without treatment in approximately four weeks, deceptively indicating a "cure" of the disease, although the bacterium remains in the body.

The second stage of syphilis, or secondary syphilis, is marked by a rash, a sore throat, and sores in the mouth. The bacteria can be transmitted at this stage through kissing or contact with an open sore. Commonly at this point, the disease enters the third stage, a latent period. This latent stage of syphilis is symptomless and may last for several years. At this point, the person is no longer infectious, but the bacteria are still present in the body, attacking the internal organs.

The final stage of syphilis is the most debilitating, as the damage done by the bacteria in the third stage becomes evident. Sufferers at this stage of syphilis experience heart disease, mental deficiency, and nerve damage, which may include loss of motor functions or blindness.

Chlamydia

Chlamydia is caused by an unusual bacterium. *Chlamydia trachomatis* is genetically a bacterium but is an obligate intracellular parasite, much like a virus in this respect. It is susceptible to antibiotics but it depends on its host to replicate its genetic material. The bacterium is transmitted through vaginal, anal, or oral intercourse with an infected person.

Chlamydia is called the "silent STD" because women usually experience no symptoms until after the infection has become established. In part because of this symptomless nature, the incidence of chlamydia has skyrocketed, increasing from 6.5 cases per 100,000 population in 1984 to 447 cases per 100,000 population in 2013. Although 1.4 million cases were reported in 2013, CDC estimates almost 2.9 million infections occur annually. The rate is particularly high among young people with the prevalence among 14- to 24-year-olds, three times that among persons 25 to 39 years old.

The effects of an established chlamydia infection on the female body are extremely serious. Chlamydia can cause pelvic inflammatory disease (PID), which can lead to sterility and sometimes death.

Infection of the male or female reproductive tract by chlamydia can cause heart disease. Chlamydiae produce a peptide similar to one produced by cardiac muscle. As the body's immune system tries to fight off the infection, it recognizes and reacts to this peptide. The similarity between the bacterial and cardiac peptides confuses the immune system, and T cells attack cardiac muscle fibers, inadvertently causing inflammation of the heart and other problems.

The treatment for the disease is antibiotics, usually tetracycline, which can penetrate the eukaryotic plasma membrane to attack the bacterium. Any woman who experiences the symptoms associated with this STD or who is at risk of developing an STD should be tested for the presence of the chlamydia bacterium; otherwise, her fertility may be at risk.

Learning Outcomes Review 28.5

Many human diseases are due to bacterial infection, including tuberculosis, streptococcal and staphylococcal infection, and sexually transmitted diseases. The causative agent of most peptic ulcers is *Helicobacter pylori,* an inhabitant of the digestive tract. Bacteria are responsible for many STDs, including gonorrhea, syphilis, and chlamydia. In many cases symptoms of infection disappear although the disease is still present, and all can have serious consequences if untreated, especially for women.

■ *Why is infection by most pathogens not fatal?*

28.6 Beneficial Prokaryotes

Learning Outcomes

1. Recognize the role of prokaryotes in the global cycling of elements.
2. Describe examples of bacterial/eukaryotic symbiosis.
3. Explain how bacteria can be used for bioremediation.

Prokaryotes were largely responsible for creating the current properties of the atmosphere and the soil through billions of years of their activity. Today, they still affect the Earth and human life in many important ways.

Prokaryotes are involved in cycling important elements

Life on Earth is critically dependent on the cycling of chemical elements between organisms and the physical environments in which they live—that is, between the living and nonliving elements of ecosystems. Prokaryotes, algae, and fungi play many key roles in this chemical cycling, a process discussed in detail in chapter 57.

Decomposition

The carbon, nitrogen, phosphorus, sulfur, and other atoms of biological systems all have come from the physical environment, and when organisms die and decay, these elements all return to it. The prokaryotes and fungi that carry out the decomposition portion of chemical cycles, releasing a dead organism's atoms to the environment, are called *decomposers.*

Fixation

Other prokaryotes play important roles in fixation, the other half of chemical cycles, helping to return elements from inorganic forms to organic forms that heterotrophic organisms can use.

Carbon. The role of photosynthetic prokaryotes in fixing carbon is obvious. The organic compounds that plants, algae, and photosynthetic prokaryotes produce from CO_2 pass up through food chains to form the bodies of all the ecosystem's heterotrophs. Ancient cyanobacteria are thought to have added oxygen to the Earth's atmosphere as a by-product of their photosynthesis. Modern photosynthetic prokaryotes continue to contribute to the production of oxygen.

Nitrogen. Less obvious, but no less critical to life, is the role of prokaryotes in recycling nitrogen. The nitrogen in the Earth's atmosphere is in the form of N_2 gas. A triple covalent bond links the two nitrogen atoms and is not easy to break. Among the Earth's organisms, only a very few species of prokaryotes are able to accomplish this feat, reducing N_2 to ammonia (NH_3), which is used to build amino acids and other nitrogen-containing biological molecules. When the organisms that contain these molecules die, decomposers return nitrogen to the soil as ammonia. This is then converted to nitrate (NO_3^-) by nitrifying bacteria, making nitrogen available for plants. The nitrate can also be converted back into molecular nitrogen by *denitrifiers* that return the nitrogen to the atmosphere, completing the cycle.

To fix atmospheric nitrogen, prokaryotes employ an enzyme complex called nitrogenase, encoded by a set of genes called *nif* ("nitrogen fixation") genes. The nitrogenase complex is extremely sensitive to oxygen and is found in a wide range of free-living prokaryotes.

In aquatic environments, nitrogen fixation is carried out largely by cyanobacteria such as *Anabaena,* which forms long chains of cells. Because the nitrogen fixation process is strictly anaerobic, individual cyanobacteria cells may develop into *heterocysts,* specialized nitrogen-fixing cells impermeable to oxygen.

In soil, nitrogen fixation occurs in the roots of plants that harbor symbiotic colonies of nitrogen-fixing bacteria. These associations include *Rhizobium* (a genus of proteobacteria; see figure 28.6) with legumes, *Frankia* (an actinomycete) with many woody shrubs, and *Anabaena* with water ferns.

Prokaryotes may live in symbiotic associations with eukaryotes

Many prokaryotes live in symbiotic association with eukaryotes. **Symbiosis** refers to the ecological relationship between different species that live in direct contact with each other. The symbiotic association of nitrogen-fixing bacteria with plant roots is an example of *mutualism,* a form of symbiosis in which both parties benefit. The bacteria supply the plant with useful nitrogen, and the plant supplies the bacteria with sugars and other organic nutrients (see chapter 38).

Many bacteria live symbiotically within the digestive tracts of animals, providing nutrients to their hosts. Cattle and other grazing mammals are unable to digest cellulose in the grass and plants they eat because they lack the required cellulase enzyme. Colonies of cellulase-producing bacteria inhabiting the gut allow cattle to digest their food (see chapter 47 for a fuller account). Similarly, humans maintain large colonies of bacteria in the large intestine that produce vitamins—particularly B_{12} and K—that the body cannot make.

Many bacteria inhabit the outer surfaces of animals and plants without doing damage. These associations are examples of *commensalism,* in which one organism (the bacterium) receives benefits while the animal or plant is neither benefited nor harmed.

Parasitism is a form of symbiosis in which one member (in this case, the bacterium) benefits, and the other (the infected animal or plant) is harmed. Infection might be considered a form of parasitism.

Bacteria are used in genetic engineering

Because the genetic code is universal, a gene from a human can be inserted into a bacterial cell, and the bacterium produces a human protein. The use of bacteria in genetic engineering was discussed in chapter 17, and it is a large part of modern molecular biology.

In addition to the production of pharmaceutical agents such as insulin, applying genetic engineering methods to produce improved strains of bacteria for commercial use holds promise for the future. Bacteria are now widely used as "biofactories" in the commercial production of a variety of enzymes, vitamins, and antibiotics. Immense cultures of bacteria, often genetically modified to enhance performance, are used to produce commercial acetone and other industrially important compounds.

Bacteria can be used for bioremediation

The use of organisms to remove pollutants from water, air, and soil is called *bioremediation.* The normal functioning of sewage treatment plants depends on the activity of microorganisms. In sewage treatment plants, the solid matter from raw sewage is broken down by bacteria and archaea naturally present in the sewage. The end product, methane gas (CH_4) is often used as an energy source to heat the treatment plant.

Biostimulation—that is, the addition of nutrients such as nitrogen and phosphorus sources—has been used to encourage the growth of naturally occurring microbes that can degrade crude oil spills. This approach was used successfully to clean up the Alaskan shoreline after the crude oil spill of the *Exxon Valdez* in 1998. Similarly, biostimulation has been used to encourage the growth of naturally occurring microbial flora in contaminated groundwater. Current efforts include those concentrated on the use of endogenous microbes such as *Geobacter* (see figure 28.6) to eliminate radioactive uranium from groundwater contaminated during the cold war.

Chlorinated compounds released into the environment by a variety of sources are another serious pollutant. Some bacteria can actually use these compounds for energy by performing reductive dehalogenation that is linked to electron transport, a process termed *halorespiration.* Although still at the development stage, the use of such bacteria to remove halogenated compounds from toxic waste holds great promise.

Learning Outcomes Review 28.6

Prokaryotes are vital to ecosystems for both recycling elements and fixation, or making elements available in organic form. Bacteria are involved in fixation of both carbon and nitrogen and are the only organisms that can fix nitrogen. These nitrogen-fixing bacteria may live in symbiotic association with plants. Bacteria are a key component of waste treatment, and they are also being used in bioremediation to remove toxic compounds introduced into the environment.

■ *Does the information about nitrogen fixation shed any light on the practice of crop rotation?*

Chapter Review

28.1 Prokaryotic Diversity

A brief history of microbiology.
Microbiology grew out of the study of infectious disease, and was aided by technology to view the unseen world. Spontaneous generation was disproved by experimentation, and Koch's postulates provide guidelines to assign a causative agent to a disease.

Prokaryotes are fundamentally different from eukaryotes.
Prokaryotic features include unicellularity, small circular DNA, division by binary fission, lack of internal compartmentalization, a singular flagellum, and metabolic diversity.

Despite similarities, bacteria and archaea differ fundamentally.
Bacteria and archaea differ in four key areas: plasma membranes, cell walls, DNA replication, and gene expression.

Archaeal lipids have ether instead of ester linkages and can form tetraether monolayers. The cell walls of bacteria contain peptidoglycans, but those of archaea do not.

Both bacterial and archaeal DNA have a single replication origin, but the origin and the replication proteins are different. Archaeal initiation of DNA replication and RNA polymerases are more like those of eukaryotes.

Most prokaryotes have not been characterized.
Nine clades of prokaryotes have been found so far, but many bacteria have not been studied (figure 28.6).

28.2 Prokaryotic Cell Structure

Prokaryotes have three basic forms: Rods, cocci, and spirals.

Prokaryotes have a tough cell wall and other external structures.
Bacteria are classified as gram-positive or gram-negative based on the Gram stain (figure 28.7). Gram-positive bacteria have a thick peptidoglycan layer in the cell wall that contains teichoic acid (figure 28.8). Gram-negative bacteria have a thin peptidoglycan layer and an outer membrane containing lipopolysaccharides in their cell wall (figure 28.8).

Some bacteria have a gelatinous layer, the capsule, enabling the bacterium to adhere to surfaces and evade an immune response.

Many bacteria have a slender, rigid, helical flagellum composed of flagellin, which can rotate to drive movement (figure 28.9). Some bacteria have hairlike pili that have roles in adhesion and exchange of genetic information.

Some bacteria form highly resistant endospores in response to environmental stress.

The interior of prokaryotic cells is organized.
In prokaryotes, invaginated regions of the plasma membrane function in respiration and photosynthesis. The nucleoid region contains a compacted circular DNA with no bounding membrane.

Prokaryotic ribosomes are smaller than those of eukaryotes and some antibiotics work by binding to these ribosomes, blocking protein synthesis.

28.3 Prokaryotic Genetics

Conjugation depends on the presence of a conjugative plasmid.
DNA can be exchanged by conjugation (figure 28.11), which depends on the presence of conjugative plasmids like the F plasmid in *E. coli*. The F^+ donor cell transfers the F plasmid to the F^- recipient cell.

The F plasmid can also integrate into the bacterial genome. Excision may be imprecise, so that the F plasmid carries genetic information from the host.

Viruses transfer DNA by transduction (figure 28.14).
Generalized transduction occurs when viruses package host DNA and transfer it on subsequent infection. Specialized transduction is limited to lysogenic phage.

Transformation is the uptake of DNA directly from the environment (figure 28.15).
Transformation occurs when cells take up DNA from the surrounding medium. It can be induced artificially in the laboratory.

Antibiotic resistance can be transferred by resistance plasmids.
R plasmids have played a significant role in the appearance of strains resistant to antibiotics, such as *S. aureus* and *E. coli* O157:H7.

Variation can also arise by mutation.
Mutations can occur spontaneously in bacteria due to radiation, UV, and various chemicals.

CRISPR systems provide adaptive immunity for prokaryotes.
CRISPR systems incorporate DNA from infecting phage and then use this to degrade DNA from future infections.

28.4 Prokaryotic Metabolism

Prokaryotes acquire carbon and energy in four basic ways.
Photoautotrophs carry out photosynthesis and obtain carbon from carbon dioxide. Chemolithoautotrophs obtain energy by oxidizing inorganic substances. Photoheterotrophs use light for energy but obtain carbon from organic molecules. Chemoheterotrophs, the largest group, obtain carbon and energy from organic molecules.

Some bacteria can attack other cells directly.
Some bacteria release proteins through their cell walls, and these proteins may transfer other, virulent proteins into eukaryotic cells.

Bacteria are costly plant pathogens.
Gram-negative bacteria known as pseudomonads are responsible for most plant diseases.

28.5 Human Bacterial Disease (table 28.1)

Bacterial diseases are spread through mucus or saliva droplets, contaminated food and water, and insect vectors.

Tuberculosis has infected humans for all of recorded history.
Tuberculosis continues to be a major public health problem. Treatment requires a long course of antibiotics.

Bacterial biofilms are involved in tooth decay.

Bacteria can cause ulcers.
Most stomach ulcers are caused by infection with *Helicobacter pylori*.

Many sexually transmitted diseases are bacterial.
The potentially dangerous sexually transmitted diseases gonorrhea, syphilis, and chlamydia are caused by bacteria.

28.6 Beneficial Prokaryotes

Prokaryotes are involved in cycling important elements.
Prokaryotes are involved in the recycling of carbon and nitrogen; only bacteria can fix nitrogen.

Prokaryotes may live in symbiotic associations with eukaryotes.

Bacteria are used in genetic engineering.
Genetically engineered prokaryotes can be used to produce human pharmaceutical agents and other useful products.

Bacteria can be used for bioremediation.

Review Questions

UNDERSTAND

1. Which of the following would be an example of a biomarker?
   a. A microfossil found in a meteorite
   b. A hydrocarbon found in an ancient rock layer
   c. An area that is high in carbon-12 concentration in a rock layer
   d. A newly discovered formation of stromatolites

2. A cell that can use energy from the sun, and CO_2 as a carbon source is a
   a. photoautotroph.
   b. chemoautotroph.
   c. photoheterotroph.
   d. chemoheterotroph.

3. Gram-positive (+) and gram-negative (−) bacteria are characterized by differences in
   a. the cell wall: gram+ have peptidoglycan, gram− have pseudo-peptidoglycan.
   b. the plasma membrane: gram+ have ester-linked lipids, gram− have ether-linked lipids.
   c. the cell wall: gram+ have a thick layer of peptidoglycan and gram− have an outer membrane.
   d. chromosomal structure: gram+ have circular chromosomes, gram− have linear chromosomes.

4. Which of the following characteristics is unique to the archaea?
   a. A fluid mosaic model of plasma membrane structure
   b. The use of an RNA polymerase during gene expression
   c. Ether-linked phospholipids
   d. A single origin of DNA replication

5. The horizontal transfer of DNA using a plasmid is an example of
   a. generalized transduction.
   b. binary fission.
   c. transformation.
   d. conjugation.

6. The disease tuberculosis is
   a. caused by a bacterial pathogen.
   b. an emerging disease that is now worldwide.
   c. caused by a viral pathogen.
   d. not treatable with antibiotics.

7. Prokaryotes participate in the global cycling of
   a. proteins and nucleic acids.
   b. carbon and nitrogen.
   c. carbohydrates and lipids.
   d. All of the choices are correct.

APPLY

1. Which of the following is typically NOT associated with a prokaryote?
   a. Horizontal transfer of genetic information
   b. A lack of internal compartmentalization
   c. Multiple, linear chromosomes
   d. A cell size of 1 μm

2. The mechanisms of DNA exchange in prokaryotes share the feature of
   a. vertical transmission of information.
   b. horizontal transfer of information.
   c. requiring cell contact.
   d. the presence of a plasmid in one cell.

3. The cell wall in both gram-positive and gram-negative cells is
   a. composed of phospholipids.
   b. a target for antibiotics that affect peptidoglycan synthesis.
   c. composed of peptidoglycan.
   d. surrounded by a membrane.

4. The three domains of life
   a. represent variations of the same basic cell type.
   b. include two different basic cell types.
   c. consist of three different basic cell types.
   d. describe current cells but say nothing about their history.

5. Ulcers and tooth decay do not appear related, but in fact both
   a. are due to eating particular kinds of foods.
   b. are caused by viral infection.
   c. are caused by environmental factors.
   d. can be due to bacterial infection.

6. Bacteria lack independent internal membrane systems, but are able to perform photosynthesis and respiration, both of which use membranes. They are able to perform these functions because
   a. they actually have internal membranes, but only for these functions.
   b. invaginations of the plasma membrane can provide an internal membrane surface.
   c. they take place outside of the cell between the membrane and the cell wall.
   d. they use protein-based structures to take the place of internal membranes.

7. Plants cannot fix nitrogen, yet some plants do NOT need nitrogen from the soil. This is because
   a. of a symbiotic association with a bacterium that can fix nitrogen.
   b. these plants are the exceptions that can fix nitrogen.
   c. they have been infected by a parasitic virus that can fix nitrogen.
   d. they are able to obtain nitrogen from the air.

SYNTHESIZE

1. If a new form of carbon fixation was discovered that was not biased toward carbon-12, would this affect our analysis of the earliest evidence for life?

2. Frederick Griffith's experiments (see chapter 14) played an important role in showing that DNA is the genetic material. Griffith showed that dead virulent bacteria mixed with live nonvirulent bacteria could cause pneumonia in mice. Live virulent bacteria could also be cultured from the infected mice. The difference between the two strains is a polysaccharide capsule found in the virulent strain. Given what you have learned in this chapter, how would you explain these observations?

3. In the 1960s, it was common practice to prescribe multiple antibiotics to fight bacterial infections. It is also often the case that patients do not always take the entire "course" of their antibiotics. Antibiotic resistance genes are often found on conjugative plasmids. How do these factors affect the evolution of antibiotic resistance and of resistance to multiple antibiotics in particular?

4. Soil-based nitrogen-fixing bacteria appear to be highly vulnerable to exposure to UV radiation. Suppose that the ozone level continues to be depleted, what are the long-term effects on the planet?

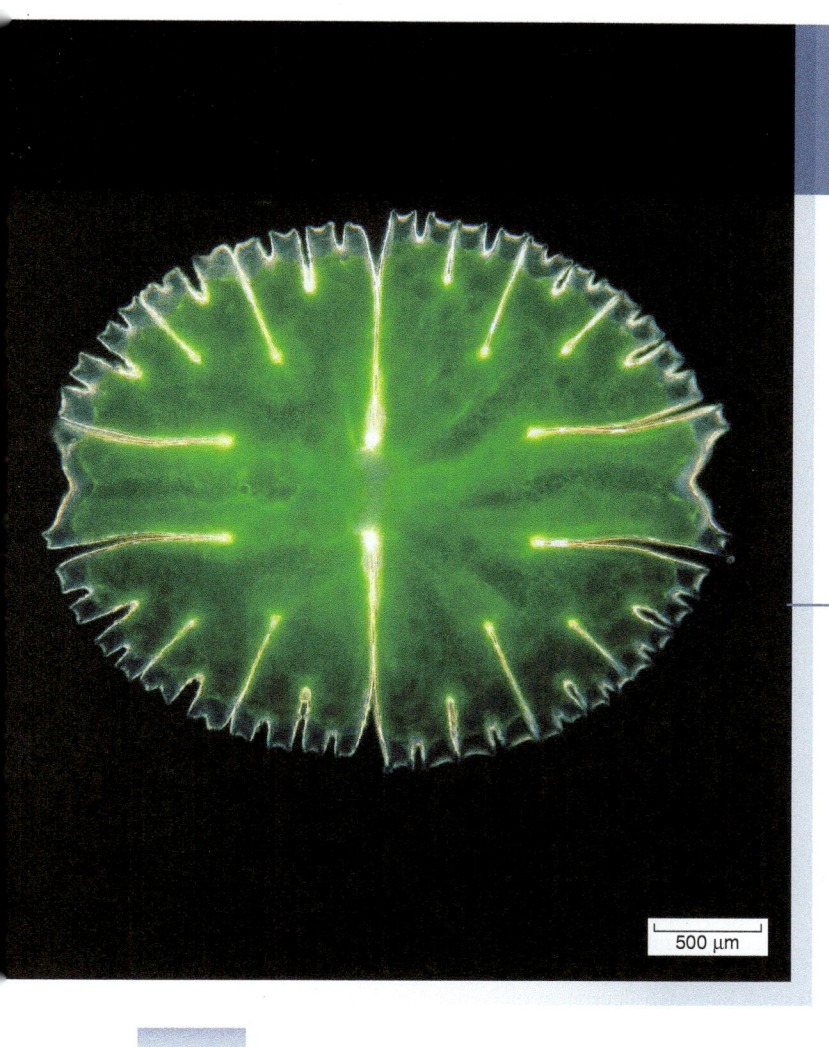

CHAPTER 29

Protists

Chapter Contents

29.1 Eukaryotic Origins and Endosymbiosis
29.2 Overview of Protists
29.3 Feeding Grove in Excavata
29.4 Secondary Endosymbiosis in Chromalveolata
29.5 Chloroplasts in Archaeplastida
29.6 Slender Pseudopods in Rhizaria
29.7 Blunt Pseudopods in Amoebozoa
29.8 Propulsion via a Single Posterior Flagellum in Opisthokonta

Introduction

For more than half of the long history of life on Earth, all life was microscopic. The biggest organisms that existed for over 2 billion years were single-celled bacteria. These prokaryotes lacked internal membranes, except for folds in surface membranes of photosynthetic bacteria.

The first evidence of a different type of organism is found in tiny fossils in rock 1.5 billion years old. These fossil cells are much larger than bacteria (up to 10 times larger) and contain internal membranes and what appear to be small, membrane-bounded structures. The complexity and diversity of form among these single cells is astonishing. The step from relatively simple to quite complex cells marks one of the most important events in the evolution of life—the appearance of a new kind of organism, the eukaryote. Eukaryotes that are clearly not animals, plants, or fungi are collectively referred to as protists.

29.1 Eukaryotic Origins and Endosymbiosis

Learning Outcomes

1. List the defining features of eukaryotes.
2. Define endosymbiosis and explain how it relates to the evolution of mitochondria and chloroplasts.
3. Describe how mitosis in fungi and some protists differs from that in other eukaryotes.

Protists were the first eukaryotes. Eukaryotic cells are distinguished from prokaryotes by the presence of a cytoskeleton and compartmentalization that includes a nuclear envelope and organelles. The exact sequence of events that led to large, complex eukaryotic cells is unknown, but several key events are agreed upon. Loss of a rigid cell wall allowed membranes to fold inward, increasing surface area. Membrane flexibility also made it possible for one cell to engulf another.

Fossil evidence dates the origins of eukaryotes

In rocks about 1.5 billion years old, scientists have found well-preserved microfossils that are noticeably different in appearance from earlier, simpler cells. An example of fossil algae is shown

Figure 29.1 Early eukaryotic fossil. Fossil algae that lived in Siberia 1 BYA.

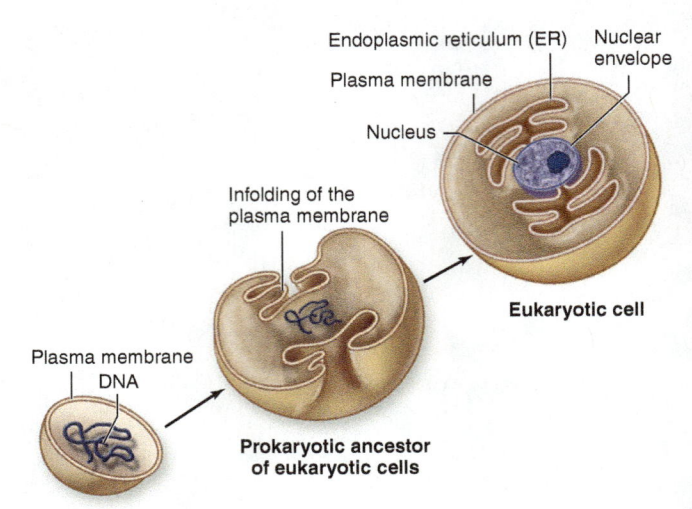

Figure 29.2 Origin of the nucleus and endoplasmic reticulum. Many prokaryotes today have infoldings of the plasma membrane (see also figure 28.10). The eukaryotic internal membrane system, called the endoplasmic reticulum (ER), and the nuclear envelope may have evolved from such infoldings of the plasma membrane, encasing the DNA of prokaryotic cells that gave rise to eukaryotic cells.

in figure 29.1. These cells are much larger than those of prokaryotes and have internal membranes and thicker walls. Recently, large microfossils more than twice that old (3.2 BYA) have been discovered. Although these cells are large, they are not complex. It is not yet known whether these ancient organisms were prokaryotes or members of a new type of life-form, the eukaryotes, described next.

Early fossils mark a major event in the evolution of life: A new type of organism had appeared. These new cells are called eukaryotes, from the Greek words meaning "true nucleus," because they possess an internal structure called a nucleus. All organisms are either prokaryotes (which lack a nucleus) or eukaryotes (which possess a nucleus).

During discussions of the origin of eukaryotic internal structure, keep in mind that, as discussed in chapter 24, horizontal gene transfer occurred frequently while eukaryotic cells were evolving. Eukaryotic cells evolved not only through horizontal gene transfer, but through infolding of membranes and by engulfing other cells. Today's eukaryotic cell is the result of cutting and pasting of DNA and organelles from different species.

The nucleus and endoplasmic reticulum arose from membrane infoldings

Many prokaryotes have infoldings of their outer membranes extending into the cytoplasm that serve as passageways to the surface. The network of internal membranes in eukaryotes is called the endoplasmic reticulum (ER), and the nuclear envelope, an extension of the ER network that isolates and protects the nucleus, is thought to have evolved from such infoldings (figure 29.2).

Mitochondria evolved from engulfed aerobic bacteria

Bacteria that live within other cells and perform specific functions for their host cells are called *endosymbiotic bacteria*. Their widespread presence in nature led biologist Lynn Margulis in the early 1970s to champion the theory of endosymbiosis, which was first proposed by Konstantin Mereschkowsky in 1905. Endosymbiosis means living together in close association.

The endosymbiotic theory supports the concept that a critical stage in the evolution of eukaryotic cells involved endosymbiotic relationships with prokaryotic organisms. According to this theory, energy-producing bacteria may have come to reside within larger bacteria, eventually evolving into what we now know as mitochondria (figure 29.3). Possibly the original host cell was anaerobic with hydrogen-dependent metabolic pathways. The symbiont had a form of respiration that produced H_2. The host depended on the symbiont for H_2 under anaerobic conditions and was able later to adapt to an O_2-rich atmosphere using the symbiont's respiratory pathways.

Chloroplasts evolved from engulfed photosynthetic bacteria

Photosynthetic bacteria were likely engulfed by other larger bacteria, leading to the evolution of chloroplasts, which are the photosynthetic organelles of plants and algae (figure 29.3). The history of chloroplast evolution illustrates the care that must be taken in phylogenetic studies. All chloroplasts are likely derived from a single line of cyanobacteria, but the organisms that host these chloroplasts are not monophyletic. This apparent paradox is resolved by considering the possibility of secondary, and even tertiary endosymbiosis. Figure 26.12 explains how red and green algae both obtained their chloroplasts by engulfing photosynthetic cyanobacteria. The brown algae most likely obtained their chloroplasts by engulfing one or more red algae, a process called **secondary endosymbiosis** (figure 29.4).

A phylogenetic tree based only on chloroplast gene sequences from brown, red, and green algae reveals a close evolutionary relationship. This tree is misleading, however, because it is not possible to tell just from these data how much the algal lines had diverged at the time they engulfed the same line of cyanobacteria.

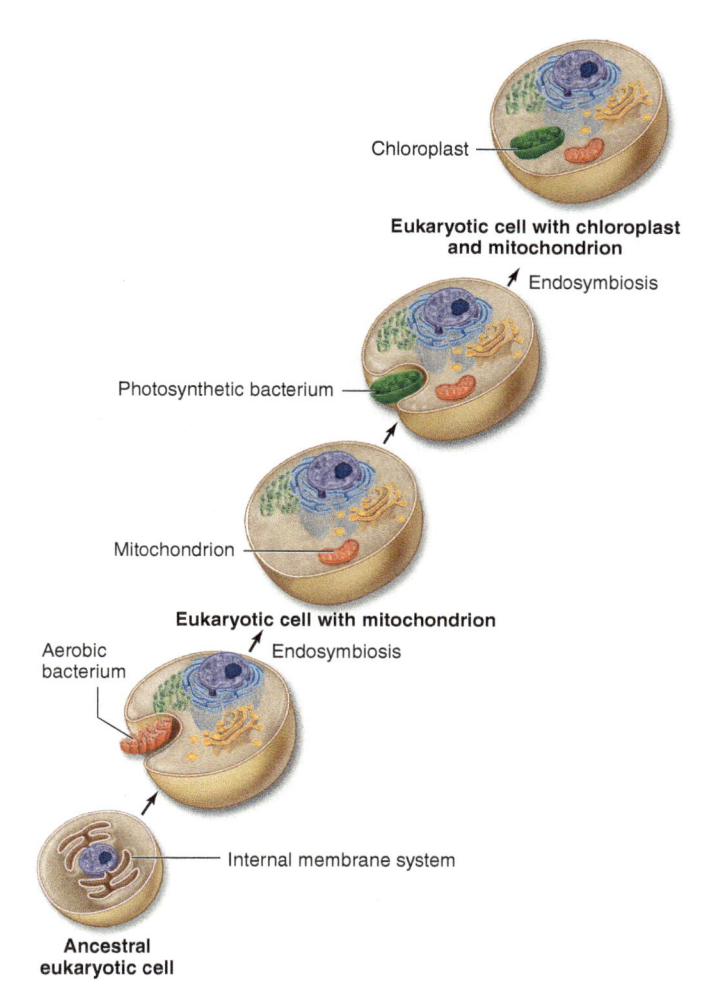

Figure 29.3 The theory of endosymbiosis. Scientists propose that ancestral eukaryotic cells, which already had an internal system of membranes, engulfed aerobic bacteria, which then became mitochondria within the eukaryotic cell. Chloroplasts also originated this way, with eukaryotic cells engulfing photosynthetic bacteria.

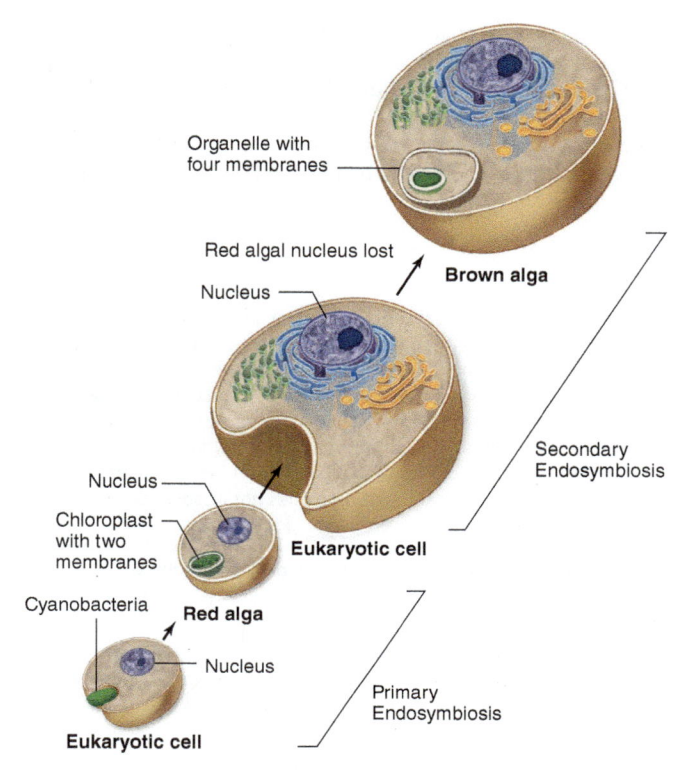

Figure 29.4 Endosymbiotic origins of chloroplasts in red and brown algae.

Nuclear gene sequences, as well as morphological and chemical traits, are more helpful than chloroplast gene sequences in sorting out algal relations.

Endosymbiosis is supported by a range of evidence

The fact that we now witness so many symbiotic relationships lends general support to the endosymbiotic theory. Even stronger support comes from the observation that present-day organelles such as mitochondria and chloroplasts contain their own DNA, which is remarkably similar to the DNA of bacteria in size and character. During the billion and a half years in which mitochondria have existed as endosymbionts within eukaryotic cells, most of their genes have been transferred to the chromosomes of the host cells—but not all. Each mitochondrion still has its own genome, a circular, closed molecule of DNA similar to that found in bacteria, on which are located genes encoding the essential proteins of oxidative metabolism. These genes are transcribed within the mitochondrion, using mitochondrial ribosomes that are smaller than those of eukaryotic cells, very much like bacterial ribosomes in size and structure. Many antibiotics that inhibit protein synthesis in bacteria also inhibit protein synthesis in mitochondria and chloroplasts, but not in the cytoplasm. Chloroplasts and mitochondria replicate via binary fission, not mitosis, further supporting bacterial origins.

Mitosis evolved in eukaryotes

A prokaryote carries its genes on a single circular DNA molecule. In contrast, a eukaryote carries its genes on multiple chromosomes, which are usually present in pairs. During the evolution of eukaryotes, a systematic process needed to be developed to divide the chromosomes and other cell contents during cell division. The separation of chromosomes is called mitosis, whereas the division of the cytoplasm is called cytokinesis. In fungi and in some groups of protists, the nuclear membrane does not dissolve during mitosis, as it does in plants, animals, and most other protists. Consequently, mitosis is confined to the nucleus. When mitosis is complete in these organisms, the nucleus divides into two daughter nuclei, and only then does the rest of the cell divide. In all other eukaryotes, the nuclear membrane breaks down prior to mitosis, so the separation of chromosomes occurs in the cytoplasm rather than the nucleus. We do not know whether mitosis without nuclear membrane dissolution represents an intermediate step on the evolutionary journey, or simply a different way of solving the same problem. We cannot see the interiors of dividing cells well enough in fossils to be able to trace the history of mitosis.

> **Learning Outcomes Review 29.1**
>
> Eukaryotes are organisms that contain a nucleus and other membrane-bounded organelles. Endoplasmic reticulum and the nuclear membrane are believed to have evolved from infoldings of the outer membranes. According to the endosymbiont theory, mitochondria and chloroplasts evolved from engulfed bacteria that remained intact. Mitochondria and chloroplasts have their own DNA, which is similar to that of prokaryotes. Unlike other eukaryotes, the nuclear membrane does not dissolve during cell division in fungi and some protists.
>
> ■ What evidence supports the endosymbiont theory?

Inquiry question How could you distinguish between primary and secondary endosymbiosis by looking at micrographs of cells with chloroplasts?

29.2 Overview of Protists

Learning Outcomes
1. Describe how an organism would be classified as a protist.
2. Recognize the six supergroups that contain the protists.
3. List the two main means of locomotion used by protists.
4. Distinguish between phototrophs, phagotrophs, and osmotrophs.

Unlike other groups of organisms, protists have no unifying features. A eukaryotic organism is considered to be a protist if it lacks features that would place it with the fungi, plants, or animals. Many are unicellular, but numerous colonial and multicellular groups also exist. Most are microscopic, but some are as large as trees. They represent all symmetries and exhibit all types of nutrition. The origin of eukaryotes, which began with ancestral protists, is among the most significant events in the evolution of life.

Eukaryotes are organized into six supergroups that all contain protists

Unlike plants, fungi, and animals, the over 200,000 different protists are not monophyletic. Eukaryotes diverged rapidly in a world that was shifting from anaerobic to aerobic conditions. We may never be able to completely sort out the relationships among different lineages during this major evolutionary transition. Applications of a variety of molecular methods are providing insights into the evolutionary relationships among protists. Molecular systematics is helping to sort out the protists, which are now contained in six supergroups: Excavata, Chromalveolata, Archaeplastida, Rhizaria, Amoebozoa, and Ophisthokonta. With the exception of Chromalveolata—the supergroups are monophyletic.

In this chapter, we explore the diverse and fascinating world of the protists based on our current understanding of phylogeny (figure 29.5). Understanding the evolution of protists is key to understanding the origins of plants, fungi, and animals. The supergroups encompass all of eukaryotic diversity and many protists are closely related to plants, animals, or fungi. The remaining chapters in this diversity unit (see chapters 30 to 35) focus on the monophyletic fungi, plants, and animals that trace back to common ancestors shared with different protist lineages.

Protist cell surfaces vary widely

Protists possess a varied array of cell surfaces. Some protists, such as amoebas, are surrounded only by their plasma membrane. All other protists have a plasma membrane with an extracellular matrix (ECM) deposited on the outside of the membrane. Some ECMs form strong cell walls; for instance, diatoms and foraminifera secrete glassy shells of silica.

Many protists with delicate surfaces are capable of surviving unfavorable environmental conditions. How do they manage to survive so well? They form cysts, which are dormant forms with resistant outer coverings in which cell metabolism is almost completely shut down. Not all cysts are so sturdy, however. Vertebrate parasitic amoebas, for example, form cysts that are quite resistant to gastric acidity, but will not tolerate desiccation or high temperature.

Protists have several means of locomotion

Protists move by diverse mechanisms. Protist movement is chiefly by either flagellar rotation or pseudopods. Many protists wave one or more flagella to propel themselves through the water, and others use banks of short, flagella-like structures called cilia to create water currents for their feeding or propulsion. Pseudopods (Greek, meaning "false feet") are the chief means of locomotion among amoebas. Other protists have long, thin pseudopods that can be extended or retracted. Because the tips can adhere to adjacent surfaces, the cell can move by a rolling motion, shortening the pseudopods in front and extending those in the rear.

Protists have a range of nutritional strategies

Protists can be heterotrophic or autotrophic. The heterotrophs obtain energy from organic molecules synthesized by other organisms. Some autotrophic protists are photosynthetic, whereas others are chemoautotrophic.

Among the heterotrophic protists are some called *phagotrophs*, which ingest visible particles of food by pulling them into intracellular vesicles called food vacuoles or phagosomes. Lysosomes fuse with the food vacuoles, introducing enzymes that

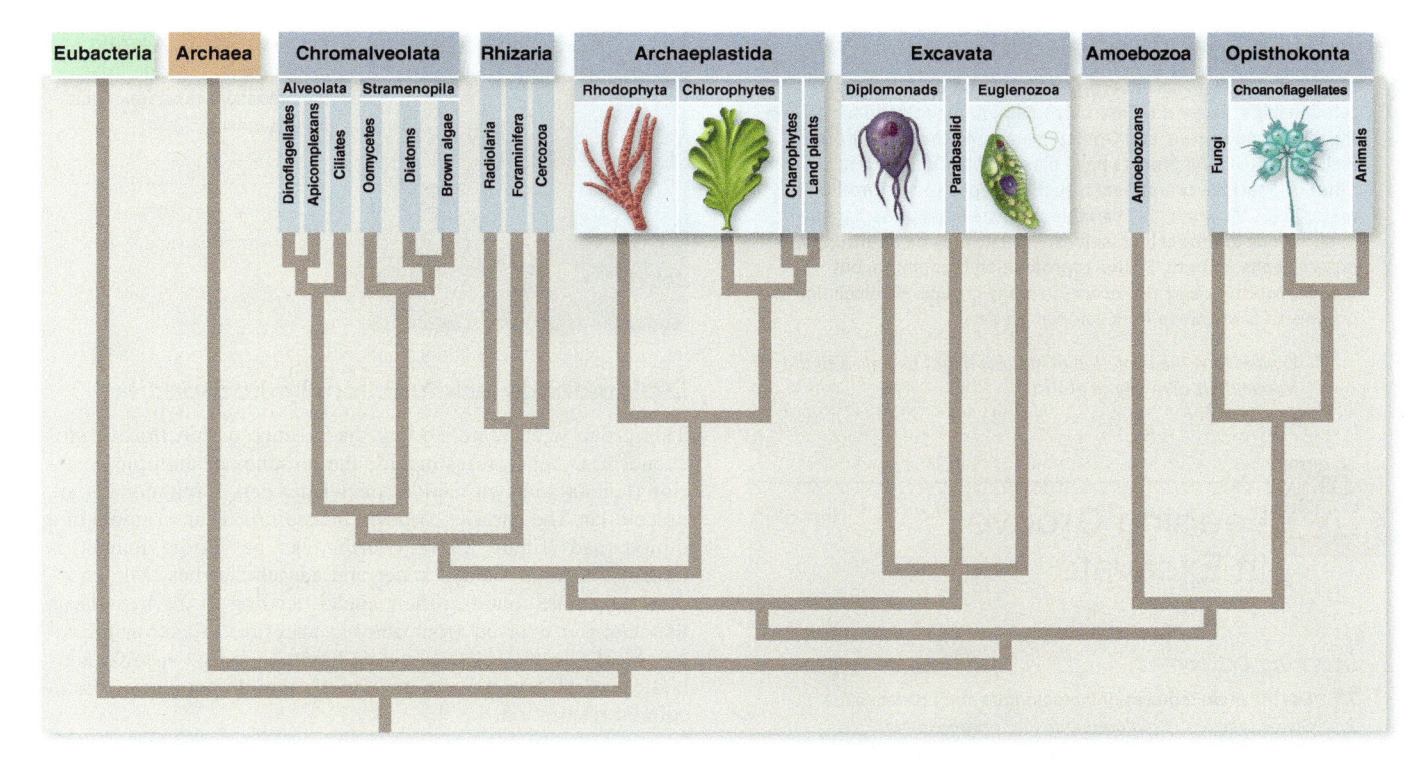

Figure 29.5 Eukaryotic evolutionary relationships. Analyses of current comparative molecular data group the eukaryotes into six supergroups shown here. Within the eukaryotes, plants, fungi, and animals are monophyletic clades, but protists are polyphyletic. Protist lineages are shaded in blue.

digest the food particles within. Digested molecules are absorbed across the vacuolar membrane.

Another example of the protists' tremendous nutritional flexibility is seen in *mixotrophs,* protists that are both phototrophic and heterotrophic.

Protists reproduce asexually and sexually

Protists typically reproduce asexually. In addition, some undergo sexual reproduction regularly, whereas others undergo sexual reproduction at times of stress, including food shortages.

Asexual reproduction

Asexual reproduction involves mitosis, but the process often differs from the mitosis in multicellular animals. For example, the nuclear membrane often persists throughout mitosis, with the microtubular spindle forming within it.

In some species, a cell simply splits into nearly equal halves after mitosis. Sometimes the daughter cell is considerably smaller than its parent and then grows to adult size—a type of cell division called **budding.** In *schizogony,* common among some protists, cell division is preceded by several nuclear divisions. This allows cytokinesis to produce several individuals almost simultaneously.

Sexual reproduction

Most eukaryotic cells also possess the ability to reproduce sexually, something prokaryotes cannot do at all. Meiosis (see chapter 11) is a major evolutionary innovation that arose in ancestral protists and allows for the production of haploid cells from diploid cells. Sexual reproduction is the process of producing offspring by fertilization, the union of two haploid cells, which were generated by meiosis. The great advantage of sexual reproduction is that it allows for frequent genetic recombination, which generates the variation required for evolution. The evolution of meiosis and sexual reproduction contributed to the tremendous explosion of diversity among the eukaryotes.

Protists are the bridge to multicellularity

Diversity in eukaryotes was also promoted by the development of *multicellularity.* Some single eukaryotic cells began living in association with others, in colonies. Eventually, individual members of the colony began to assume different duties, and the colony began to take on the characteristics of a single individual. Multicellularity has arisen many times among the eukaryotes. Practically every organism big enough to be seen with the unaided eye, including all animals and plants, is multicellular. The great advantage of multicellularity is that it fosters specialization; some cells devote all of their energies to one task, other cells to another. Few innovations have had as great an influence on the history of life as the specialization made possible by multicellularity.

Learning Outcomes Review 29.2

Eukaryotes fall into six supergroups based on evolutionary relationships: Excavata, Chromalveolata, Archaeplastida, Rhizaria, Amoebozoa, and Ophisthokonta. All the supergroups include protists. All protists have plasma membranes, but other cell-surface components, such as deposited extracellular material (ECM), are highly variable. Protists mainly use flagella or pseudopods to propel themselves. Protists may be autotrophs, heterotrophs, or both. Sexual reproduction is common, but asexual reproduction also occurs in many groups. Multicellular organisms likely arose from colonial protists.

- Explain how the evolution of meiosis might be linked to the tremendous diversity of protists.

29.3 Feeding Groove in Excavata

Learning Outcomes
1. List the main features of diplomonads and parabasalids.
2. Give examples of diplomonads and parabasalids.
3. Explain why euglenozoa cannot be classified as either plants or animals.
4. Describe the distinguishing feature of kinetoplastids.

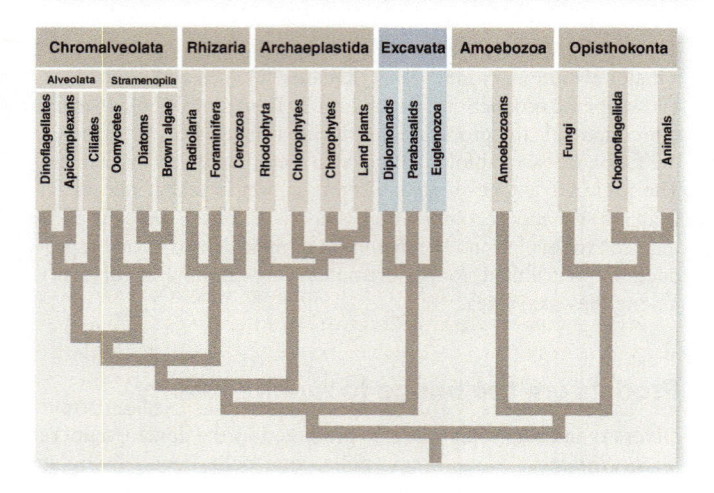

The *diplomonads*, *parabasalids*, and euglenozoans are unicellular protists grouped as Excavata (the excavates) based on cytoskeletal and DNA sequence similarities showing evolutionary relatedness. Many have a characteristic feeding groove that looks like it was excavated from the side of their bodies, hence the name excavate. Although these groups have similar features, differences separate them into three distinct groups. Parabasalids and diplomonads have two nuclei, multiple flagella, and modified mitochondria; euglenozoans have unique flagella and several have acquired chloroplasts through endosymbiosis. None of the algae are closely related to euglenozoa, a reminder that endosymbiosis is widespread.

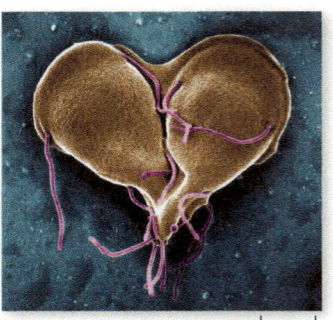

Figure 29.6 *Giardia intestinalis.* This parasitic diplomonad lacks a functional mitochondrion.

Diplomonads lack functional mitochondria

This group is characterized by the absence of functional mitochondria. Other features include the presence of multiple posterior flagella and two haploid nuclei per cell. Diplomonads are unicellular. The parasite, *Giardia intestinalis,* is an example of a diplomonad (figure 29.6). *Giardia* can pass from human to human via contaminated water and causes diarrhea. Mitochondrial genes are found in their nuclei, leading to the hypothesis that *Giardia* evolved from aerobic ancestors. Electron micrographs of *Giardia* cells stained with mitochondrial-specific antibodies reveal degenerate mitochondria that do not participate in cellular respiration.

Parabasalids have undulating membranes

The distinguishing feature of parabasalids is an undulating membrane that is used for locomotion (figure 29.7). They also use flagella to propel themselves. Unlike diplomonads, parabasalids have semifunctional mitochondria and have one nucleus per cell. Parabasalids contain an intriguing array of species. Some live in the gut of termites and digest cellulose, the main component of the termite's wood-based diet. The symbiotic relationship is one layer more complex because these parabasalids have a symbiotic relationship with bacteria that also aid in the digestion of cellulose. The persistent activity of these three symbiotic organisms from three different kingdoms can lead to the collapse of a home built of wood or recycle tons of fallen trees in a forest. Another parabasalid, *Trichomonas vaginalis,* causes trichomoniasis, a common sexually transmitted disease in humans.

Euglenozoans change shape while swimming

The most distinguishing feature of euglenozoans is apparent when they swim. Their bodies change shape, alternating between

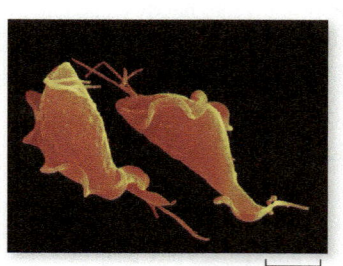

Figure 29.7 Undulating membrane characteristic of parabasalids. Trichomoniasis is caused by this parasite species, *Trichomonas vaginalis.*

being stretched out and rounded up. Euglenozoans can change shape because they lack a cell wall. Instead, strips of protein encircle the cell. These strips can slide over each other, providing flexibility. **Euglenozoa** diverged early and were among the earliest free-living eukaryotes to possess mitochondria. Both free-living euglenids and parasitic kinetoplastids are considered euglenozoans.

Euglenids

Euglenids clearly illustrate the impossibility of distinguishing plantlike protists from animal-like protists. About one-third of the approximately 170 genera of euglenids have chloroplasts and are fully autotrophic; the others lack chloroplasts, ingest their food, and are heterotrophic.

Some euglenids with chloroplasts may become heterotrophic in the dark; the chloroplasts become small and nonfunctional. If they are put back in the light, they may become green within a few hours. Photosynthetic euglenids may sometimes feed on dissolved or particulate food.

Reproduction is asexual and occurs by mitotic cell division. The nuclear envelope remains intact throughout mitosis. No sexual reproduction is known to occur in this group.

In *Euglena* (figure 29.8), the genus for which the euglenid phylum is named, two flagella are attached at the base of a flask-shaped opening called the *reservoir*, which is located at the anterior end of the cell. One of the flagella is long and has a row of very fine, short, hairlike projections along one side. A second, shorter flagellum is located within the reservoir but does not emerge from it. Contractile vacuoles collect excess water from all parts of the organism and empty it into the reservoir, which apparently helps regulate the osmotic pressure within the organism. A light-sensitive stigma, which also occurs in the green algae (Chlorophyta), helps these photosynthetic organisms move toward light.

Cells of *Euglena* contain numerous small chloroplasts. These chloroplasts, like those of the green algae and plants, contain chlorophylls *a* and *b,* together with carotenoids. Although the chloroplasts of euglenids differ somewhat in structure from those of green algae, they probably had a common origin. *Euglena's* photosynthetic pigments are light-sensitive (figure 29.9). It seems likely that euglenid chloroplasts ultimately evolved from a symbiotic relationship through ingestion of green algae.

Parasitic kinetoplastids

A second major group within the euglenozoa is the *kinetoplastids*. The name kinetoplastid refers to a unique, single mitochondrion in each cell. Each mitochondrion has minicircles and maxicircles of DNA. (Remember that prokaryotes have circular DNA, and mitochondria had prokaryotic origins.) This mitochondrial DNA is responsible for very rapid glycolysis and also for an unusual kind of editing of the RNA by guide RNAs encoded in the minicircles.

Parasitism has evolved multiple times within the kinetoplastids. Trypanosomes are a group of kinetoplastids that cause many serious human diseases, the most familiar being trypanosomiasis, also known as African sleeping sickness, which causes extreme lethargy and fatigue (figure 29.10).

Leishmaniasis, which is transmitted by sand flies infected with the protist parasite *Leishmania,* is a trypanosomic disease that causes skin sores and in some cases can affect internal organs, leading to death. About 1.3 million new cases are reported each year. The rise in leishmaniasis in South America correlates with the move of infected individuals from rural to urban environments, where there is a greater chance of spreading the parasite.

Chagas disease is caused by *Trypanosoma cruzi*. At least 90 million people, from the southern United States to Argentina, are at risk of contracting *T. cruzi* from small wild mammals that carry the parasite and can spread it to other mammals and humans through skin contact with urine and feces. Blood transfusions have also increased the spread of the infection. Chagas disease can lead to severe cardiac and digestive problems in humans and domestic animals, but it appears to be tolerated in the wild mammals.

Control is especially difficult because of the unique attributes of these parasites. For example, tsetse fly-transmitted trypanosomes have evolved an elaborate genetic mechanism for repeatedly changing the antigenic nature of their protective glycoprotein coat, thus dodging the antibodies their hosts produce against them (see chapter 51). Only a single gene out of some 1000 variable-surface glycoprotein (VSG) genes is expressed at a time. A VSG gene is usually duplicated and moved to 1 of about 20 expression sites near the telomere where it is transcribed.

In the guts of the flies that spread them, trypanosomes are noninfective. When they are ready to transfer to the skin or

Figure 29.8
Euglenids.
a. Micrograph of *Euglena gracilis.*
b. Diagram of *Euglena.* Paramylon granules are areas where food reserves are stored.

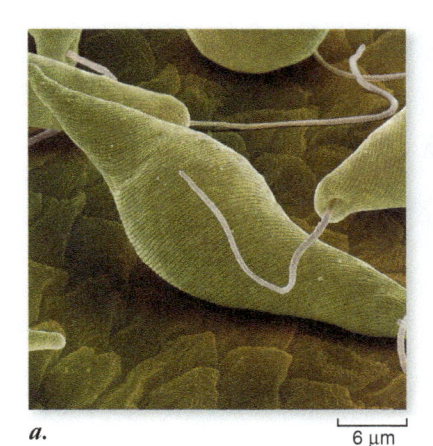

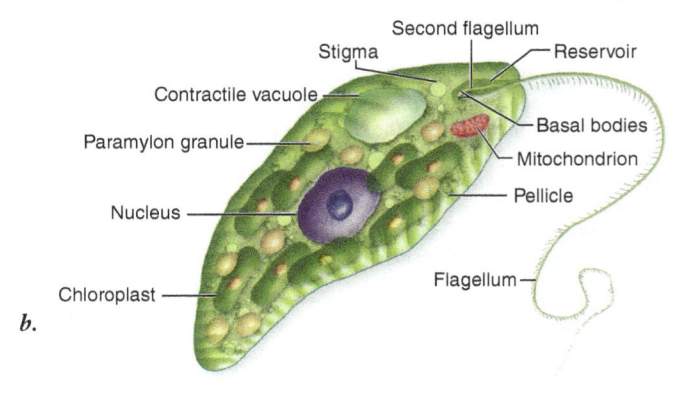

SCIENTIFIC THINKING

Hypothesis: Euglena *cells do not retain photosynthetic pigments in a dark environment.*
Prediction: Photosynthetic pigments will be degraded when light-grown Euglena *cells are transferred to the dark and new pigment will not be produced.*
Test: Grow Euglena *under normal light conditions. Transfer the culture to two flasks. Take a sample from each flask and measure the amount of photosynthetic pigments in each. Maintain one flask in the light and transfer the other to the dark. After several days, extract the photosynthetic pigments from each flask, and compare amounts with each other and with initial levels.*

Grow culture of *Euglena* under light. → Partition into two flasks. → Take a sample from each flask and quantify photosynthetic pigment. → Put one flask in dark, and expose the other to light. → Allow growth for several days. → Quantify photosynthetic pigment in each flask.

Result: *Photosynthetic pigment levels are lower in the dark-grown flask than in the light-grown one. Pigment levels in the dark-grown flask are lower than at the beginning of the experiment. Pigment levels in the light-grown flask are unchanged.*
Conclusion: *The hypothesis is supported. Maintenance of* Euglena *in the dark resulted in a loss of photosynthetic pigment. Pigments were degraded in the dark-grown flask.*
Further Experiments: *Transfer dark-grown flasks back to the light and measure changes in pigment levels over time. Are original pigment levels restored after growth in light?*

Figure 29.9 Effect of light on *Euglena* photosynthetic pigments.

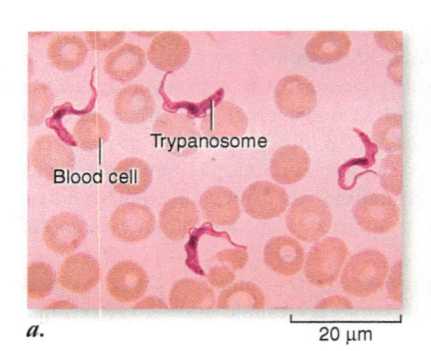

Figure 29.10 A kinetoplastid.
a. Trypanosoma among red blood cells. The nuclei (*dark-staining bodies*), anterior flagella, and undulating, changeable shape of the trypanosomes are visible in this photomicrograph. *b.* The tsetse fly, shown here sucking blood from a human arm, can carry trypanosomes.

bloodstream of their host, trypanosomes migrate to the salivary glands and acquire the thick coat of glycoprotein antigens that protect them from the host's antibodies. Later, when they are taken up by a tsetse fly, the trypanosomes again shed their coats.

The production of vaccines against such a system is complex, but tests of a protein kinase inhibitor hold promise. Releasing sterilized flies to impede the reproduction of populations is another technique being tried to control the fly population. Traps made of dark cloth and scented like cows, but poisoned with insecticides, have likewise proved effective.

The sequencing of the genomes of the three kinetoplastids described earlier revealed a core of common genes in all three, as described in chapter 24. The devastating toll all three take on human life could be alleviated by the development of a single drug targeted at one or more of the core proteins shared by the three parasites.

Learning Outcomes Review 29.3

Excavata are grouped based on cytoskeletal structure and DNA sequence similarities. Within the excavates, diplomonads lack functional mitochondria but may contain mitochondrial genes. They are unicellular, have two nuclei, and move with flagella; an example is *Giardia*. Parabasalids produce semifunctional mitochondria and use flagella and undulating membranes for locomotion; an example is *Trichomonas*. The euglenids includes phototrophs and heterotrophs. Some members have chloroplasts that remain nonfunctional unless light is present, and some phototrophs may feed if food particles are present. The kinetoplastids contain a single mitochondrion with two types of DNA and the ability to edit RNA. Trypanosomes are disease-causing kinetoplastids.

- ■ *In what type of habitat would it be useful to use undulating membranes for locomotion?*
- ■ *How does a contractile vacuole regulate osmotic pressure in a* Euglena *cell?*

29.4 Secondary Endosymbiosis in Chromalveolata

Learning Outcomes
1. Identify the distinguishing feature of the members of alveolates.
2. Explain the function of the apical complex in apicomplexans.
3. Describe the characteristic features of the stramenopiles.
4. Explain how the oomycetes are distinguished from other protists.

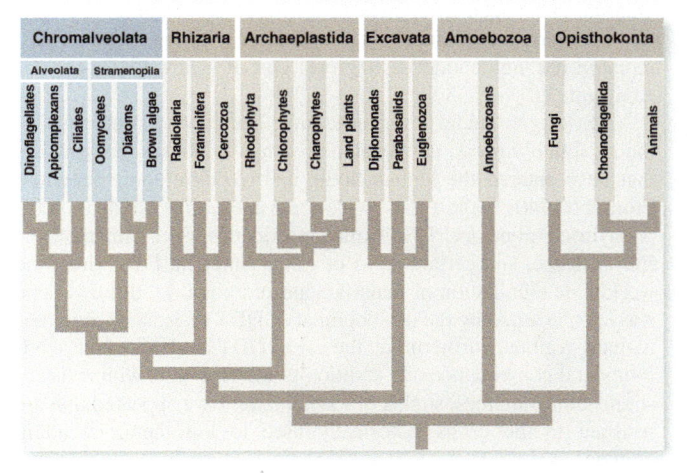

Chromalveolates constitute a supergroup that may have arisen by one or more secondary endosymbiotic events. Phylogenies based on sequence data do not support a monophyletic Chromalveolata supergroup, but the exact evolutionary history remains to be resolved. One branch is the **alveolates,** including the *dinoflagellates, apicomplexans,* and *ciliates,* all of which have a common lineage but diverse modes of locomotion. A common trait is the presence of flattened vesicles called alveoli (hence the name Alveolata) stacked in a continuous layer below their plasma membranes (figure 29.11). The alveoli may function in membrane transport, similar to Golgi bodies. A second branch of chromalveolates is the stramenopiles, including *brown algae, diatoms,* and the *oomycetes* (water molds). The name refers to unique, fine hairs found on the flagella of members of this group (figure 29.12), although a few species have lost their hairs during evolution.

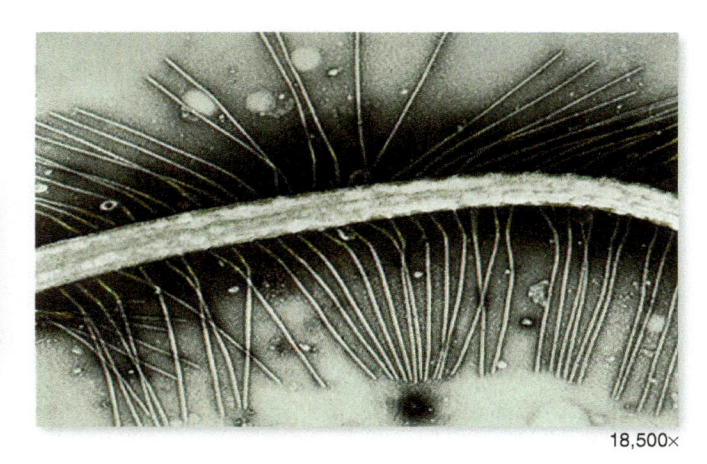

Figure 29.12 Stramenopiles have very fine hairs on their flagella.

Dinoflagellates are photosynthesizers with distinctive features

Most dinoflagellates are photosynthetic unicells with two flagella. Dinoflagellates live in both marine and freshwater environments. Some dinoflagellates are luminous and contribute to the twinkling or flashing effects seen in the sea at night, especially in the tropics.

The flagella, protective coats, and biochemistry of dinoflagellates are distinctive, and the dinoflagellates do not appear to be directly related to any other phylum. Plates made of a cellulose-like material, often encrusted with silica, encase the dinoflagellate cells (figure 29.13). Grooves at the junctures of these plates usually house the flagella, one encircling the cell like a belt, and the other perpendicular to it. By beating in their grooves, these flagella cause the dinoflagellate to spin as it moves.

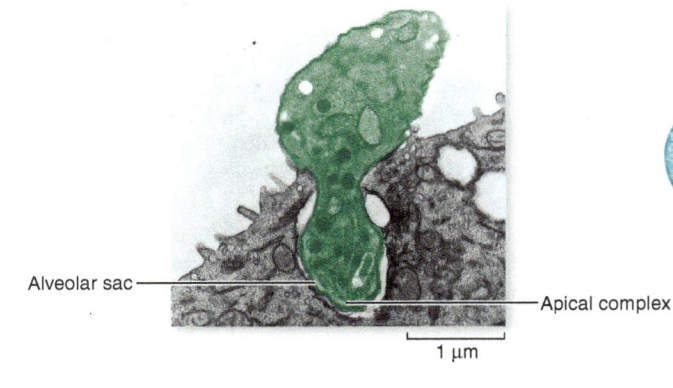

Figure 29.11 An alveolus is a continuum of vesicles just below the plasma membrane of dinoflagellates, apicomplexans, and ciliates. The apical complex of the alveolus forces the parasite into host cells.

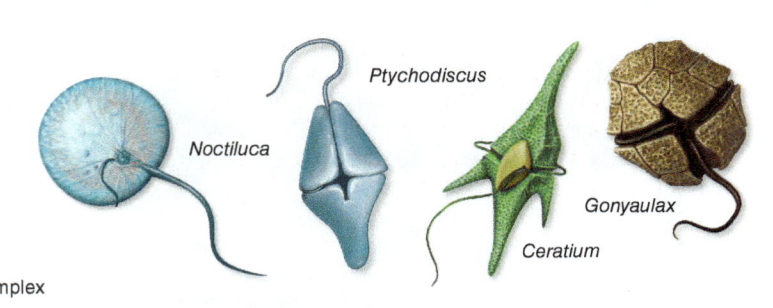

Figure 29.13 Some dinoflagellates. *Noctiluca,* which lacks the heavy cellulose armor characteristic of most dinoflagellates, is one of the bioluminescent organisms that cause the waves to sparkle in warm seas. In the other three genera, the shorter, encircling flagellum is seen in its groove, with the longer one projecting away from the body of the dinoflagellate. (Not drawn to scale.)

chapter 29 Protists 573

Most dinoflagellates have chlorophylls *a* and *c*, in addition to carotenoids, so that in the biochemistry of their chloroplasts, they resemble the diatoms and the brown algae. Possibly this lineage acquired such chloroplasts by forming endosymbiotic relationships with members of those groups.

Although sexual reproduction does occur under starvation conditions, dinoflagellates reproduce primarily by asexual cell division. Asexual cell division relies on a unique form of mitosis in which the permanently condensed chromosomes divide within a permanent nuclear envelope. After the numerous chromosomes duplicate, the nucleus divides into two daughter nuclei.

Also, the dinoflagellate chromosome is unique among eukaryotes in that the DNA is not generally complexed with histone proteins. In all other eukaryotes, the chromosomal DNA is complexed with histones to form nucleosomes, structures that represent the first order of DNA packaging in the nucleus (see chapter 10). How dinoflagellates maintain distinct chromosomes with a small amount of histones remains a mystery.

Red tide: Overgrowth of dinoflagellates

The poisonous and destructive "red tides" that occur frequently in coastal areas are often associated with great population explosions, or "blooms," of dinoflagellates, whose pigments color the water (figure 29.14). The blooms are most often triggered by excess nutrients from agricultural and other human activity. Red tides have a profound, detrimental effect on the fishing industry worldwide. Some 20 species of dinoflagellates produce powerful neurotoxins that inhibit the diaphragm, causing respiratory failure in many vertebrates. When the toxic dinoflagellates are abundant, many fishes, birds, and marine mammals may die.

Apicomplexans include the malaria parasite

Apicomplexans are spore-forming parasites of animals. They are called apicomplexans because of a unique arrangement of fibrils, microtubules, vacuoles, and other cell organelles at one end of the cell, termed an *apical complex* (see figure 29.11). The apical complex is a cytoskeletal and secretory complex that enables the apicomplexan to invade its host. The best-known apicomplexan is the malarial parasite *Plasmodium*. (The use of the genome sequence of the parasite and the mosquito that carries it is discussed in chapter 24.)

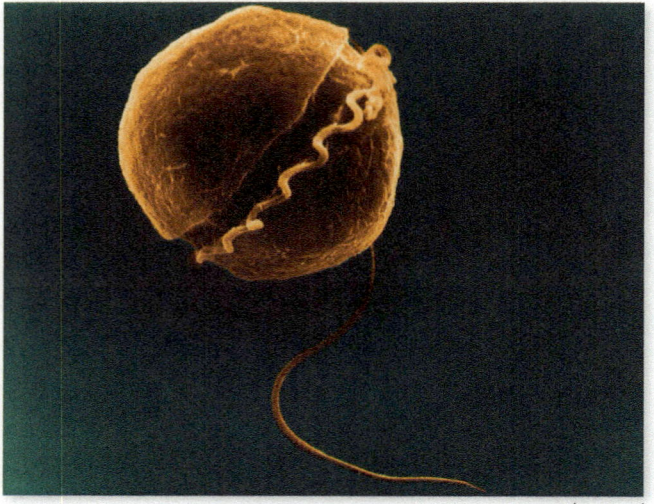

Plasmodium *and malaria*

Plasmodium glides inside the red blood cells of its host with amoeboid-like contractility. Like other apicomplexans, *Plasmodium* has a complex life cycle involving sexual and asexual phases and alternation between different hosts, in this case mosquitoes (*Anopheles gambiae*) and humans (figure 29.15). Even though *Plasmodium* has mitochondria, it grows best in a low-O_2, high-CO_2 environment.

Efforts to eradicate malaria have focused on (1) eliminating the mosquito vectors; (2) developing drugs to poison the parasites that have entered the human body; and (3) developing vaccines. From the 1940s to the 1960s, wide-scale applications of dichlorodiphenyltrichloroethane (DDT) killed mosquitoes in the United States, Italy, Greece, and certain areas of Latin America. For a time, the worldwide elimination of malaria appeared possible. But this hope was soon crushed by the development of DDT-resistant mosquitoes in many regions. Furthermore, the use of DDT has had serious environmental consequences. In addition to the problems with resistant strains of mosquitoes, strains of *Plasmodium* have appeared that are resistant to the drugs historically used to kill them, including quinine.

An experimental vaccine containing a surface protein of one malaria-causing parasite, *P. falciparum*, seems to induce the immune system to defend against future infections. In tests, six out of seven vaccinated people did not get malaria after being bitten by mosquitoes that carried *P. falciparum*. Many are hopeful that this new vaccine may be able to fight malaria. More than two dozen malaria vaccine candidates are at various stages of development. (Chapter 24 contains a discussion of the genome sequences of both *Plasmodium* and its mosquito vector.)

Gregarines

Gregarines are another group of apicomplexans that use their distinctive apical complex to attach themselves in the intestinal epithelium of arthropods, annelids, and mollusks. Most of the gregarine body, aside from the apical complex, is in the intestinal cavity, and nutrients appear to be obtained through the apicomplexan attachment to the cell (figure 29.16).

Toxoplasma

Using its apical complex, *Toxoplasma gondii* invades the epithelial cells of the human gut. Most individuals infected with the parasite mount an immune response, preventing any permanent damage. In the absence of a fully functional immune system, however, *Toxoplasma* can damage brain (figure 29.17), heart, and skeletal tissues, in addition to gut and lymph tissue, during extended infections. Individuals with AIDS are particularly susceptible to *Toxoplasma* infection. If a pregnant women touches a litter box of an infected cat, *Toxoplasma* parasites from the cat can, if ingested,

0.8 μm

Figure 29.14 Red tide. Although small in size, huge populations of dinoflagellates, including this *Gymnopodium* species, can color the sea red and release toxins into the water.

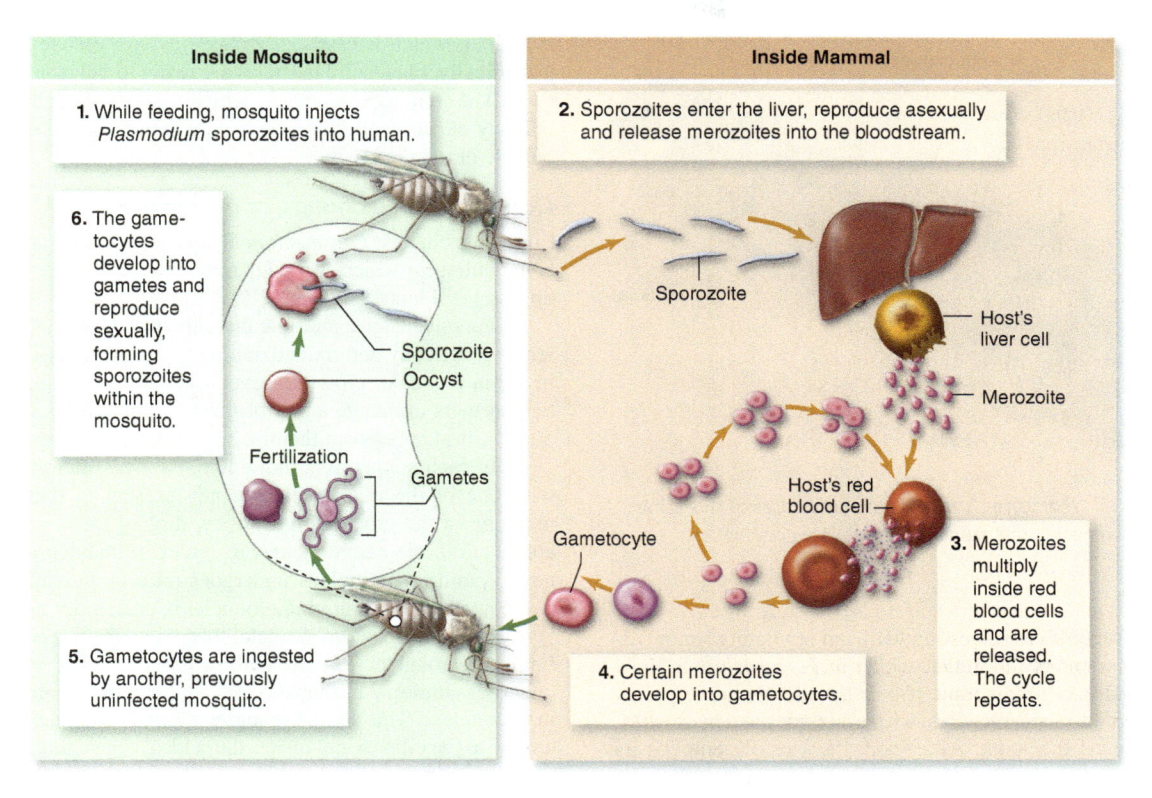

Figure 29.15 The life cycle of *Plasmodium*. *Plasmodium,* the apicomplexan that causes malaria, has a complex life cycle that alternates between mosquitoes and mammals.

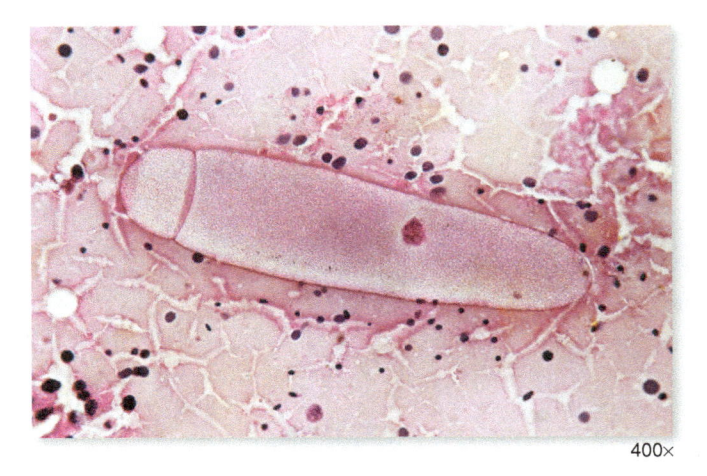

cross the placental barrier and harm the developing fetus with an immature immune system.

Ciliates are characterized by their mode of locomotion

As the name indicates, most ciliates feature large numbers of cilia (tiny beating hairs). Their cilia are usually arranged either in longitudinal rows or in spirals around the cell. Cilia are anchored to microtubules beneath the plasma membrane (see chapter 5), and they beat in a coordinated fashion. In some groups, the cilia have specialized functions, becoming fused into sheets, spikes, and rods that may then function as mouths, paddles, teeth, or feet.

The ciliates have a pellicle, a tough but flexible outer covering, that enables them to squeeze through or move around obstacles.

Micronucleus and macronucleus

All known ciliates have two different types of nuclei within their cells: a small **micronucleus** and a larger **macronucleus** (figure 29.18). DNA in the macronucleus is transcribed for the daily activities of the organism. The macronucleus is typically polyploid and may contain up to 1000 copies of each chromosome. The micronucleus is diploid and used only as the germ line for sexual reproduction. DNA in the micronucleus is not transcribed.

Figure 29.16 Gregarine entering a cell.

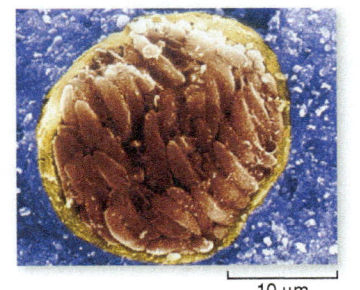

Figure 29.17 Micrograph of a cyst filled with *Toxoplasma*. *Toxoplasma* can enter the brain and form cysts filled with slowly replicating parasites.

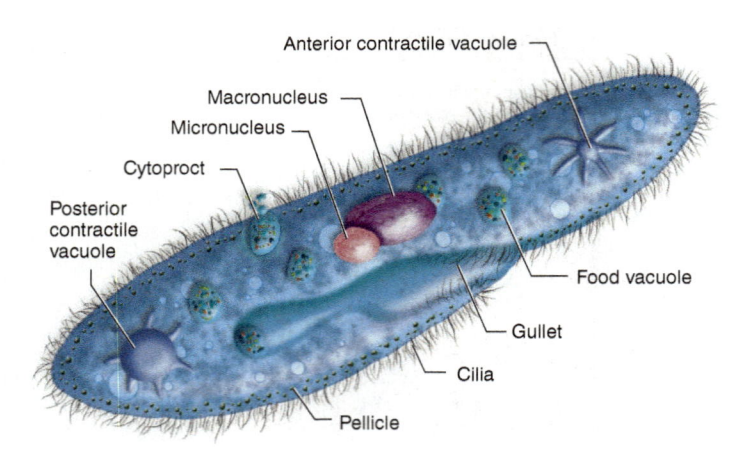

Figure 29.18 Paramecium. The main features of this ciliate include cilia, two nuclei, and numerous specialized organelles.

Vacuoles

Ciliates form vacuoles for ingesting food and regulating water balance. Food first enters the gullet, which in *Paramecium* is lined with cilia fused into a membrane (figure 29.18). From the gullet, the food passes into food vacuoles, where enzymes and hydrochloric acid aid in its digestion. Afterward, the vacuole empties its waste contents through a special pore in the pellicle called the *cytoproct,* which is essentially an exocytotic vesicle that appears periodically when solid particles are ready to be expelled.

The contractile vacuoles, which regulate water balance, periodically expand and contract as they empty their contents to the outside of the organism.

Conjugation: Exchange of micronuclei

Like most ciliates, *Paramecium* undergoes a sexual process called conjugation, in which two cells remain attached to each other for up to several hours (figure 29.19).

Paramecia have multiple mating types. Only cells of two different genetically determined mating types can conjugate. Meiosis in the micronuclei produces several haploid micronuclei, and the two partners exchange a pair of their micronuclei through a cytoplasmic bridge between them.

In each conjugating individual, the new micronucleus fuses with one of the micronuclei already present in that individual, resulting in the production of a new diploid micronucleus. After conjugation, the macronucleus in each cell disintegrates, and the new diploid micronucleus undergoes mitosis, thus giving rise to two new identical diploid micronuclei in each individual.

One of these micronuclei becomes the precursor of the future micronuclei of that cell, while the other micronucleus undergoes multiple rounds of DNA replication, becoming the new macronucleus. This complete segregation of the genetic material is unique to the ciliates and makes them ideal organisms for the study of certain aspects of genetics.

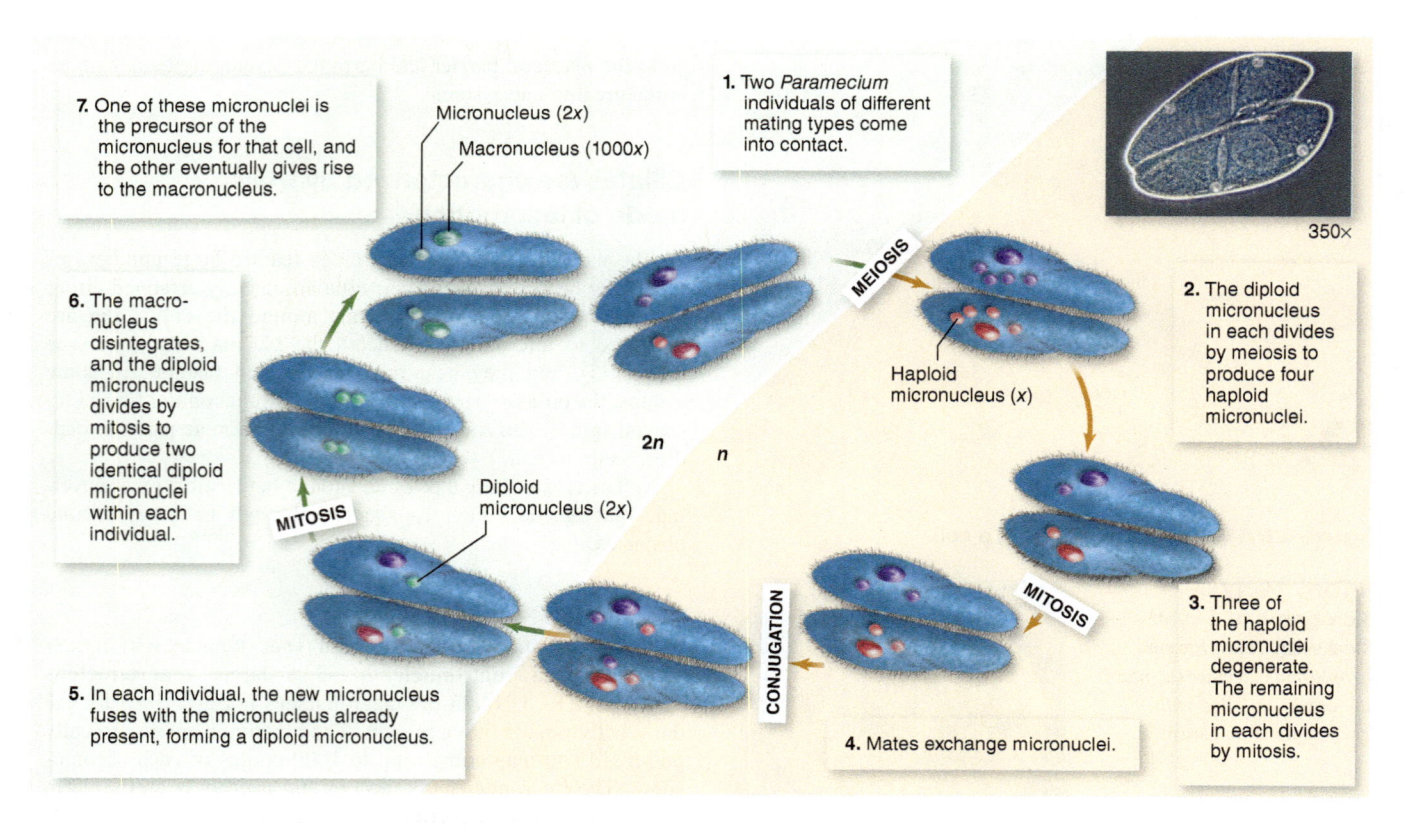

Figure 29.19 Life cycle of *Paramecium*. In sexual reproduction, two mature cells fuse in a process called conjugation.

"Killer" strains

Paramecium strains that kill other, sensitive strains of *Paramecium* long puzzled researchers. Initially, killer strains were believed to have genes coding for a substance toxic to sensitive strains. The true source of the toxin turned out to be an endosymbiotic bacterium in the "killer" strains. If this bacterium is engulfed by a "nonkiller" strain, the toxin is released, and the sensitive *Paramecium* dies.

Brown algae include large seaweeds

Brown algae, along with diatoms and oomycetes ("water molds") make up the stramenopiles within the chromalveolates. Brown algae are the most conspicuous seaweeds in many northern regions (figure 29.20). The **haplodiplontic** life cycle of the brown algae is marked by alternation of generations between a multicellular sporophyte (diploid) and a multicellular gametophyte (haploid) (figure 29.21). Some sporophyte cells go through meiosis and produce spores. These spores germinate and undergo mitosis to produce the large individuals we recognize, such as the kelps. The gametophytes are often much smaller, filamentous individuals, perhaps a few centimeters in width.

Even in an aquatic environment, transport can be a challenge for the very large brown algal species. Distinctive transport cells that stack one upon the other enhance transport within some species (see figure 23.10). However, even though the large kelp look like plants, it is important to realize that they

Figure 29.20 Brown alga. The giant kelp, *Macrocystis pyrifera*, grows in relatively shallow water along the coasts throughout the world and provides food and shelter for many different kinds of organisms.

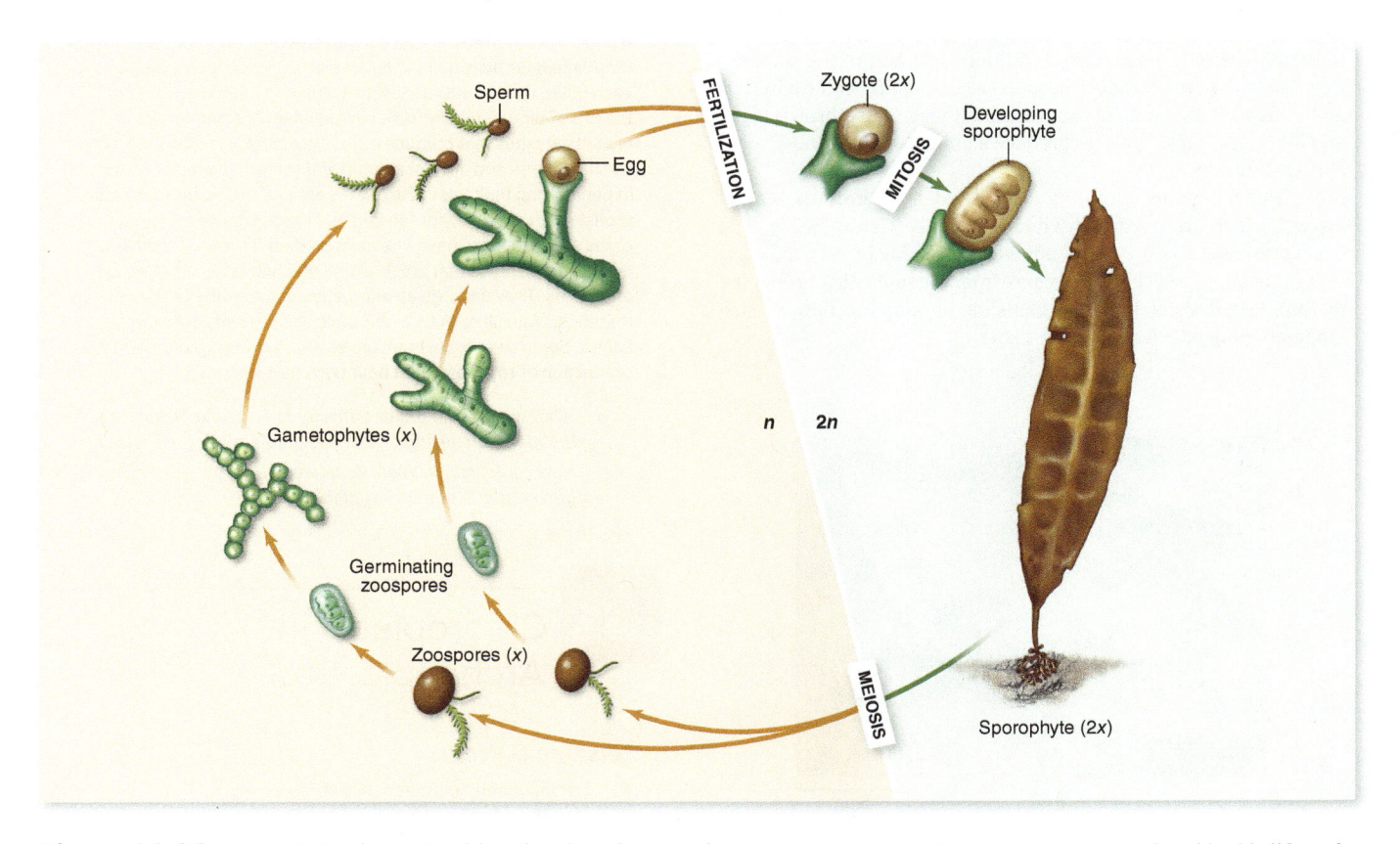

Figure 29.21 Haplodiplontic cycle of *Laminaria*, a brown alga. Multicellular haploid and diploid stages are found in this life cycle, although the male and female gametophytes are quite small.

Figure 29.22 Diatoms. These different radially symmetrical diatoms have unique silica, two-part shells.

do not contain the complex tissues such as xylem that are found in plants.

Diatoms are unicellular organisms with double shells

Diatoms, members of the phylum Chrysophyta, are photosynthetic, unicellular organisms with unique double shells made of opaline silica, which are often strikingly marked (figure 29.22). The shells of diatoms are like small boxes with lids, one half of the shell fitting inside the other. Their chloroplasts, containing chlorophylls *a* and *c*, as well as carotenoids, resemble those of the brown algae and dinoflagellates. Diatoms produce a unique carbohydrate called chrysolaminarin.

Some diatoms move by using two long grooves, called *raphes*, which are lined with vibrating fibrils (figure 29.23). The exact mechanism is still being unraveled and may involve the ejection of mucopolysaccharide streams from the raphe that propel the diatom. Pencil-shaped diatoms can slide back and forth over each other, creating an ever-changing shape.

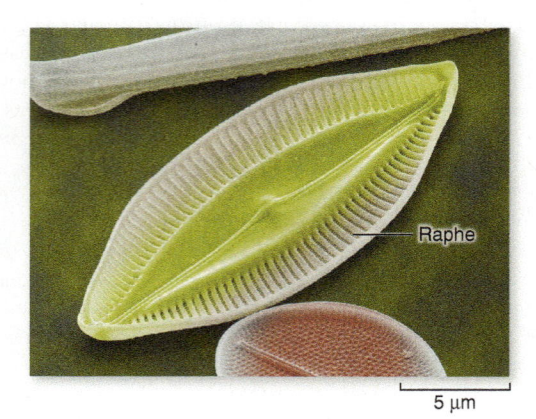

Figure 29.23 Diatom raphes are lined with fibrils that aid in locomotion.

Oomycetes, the "water molds," have some pathogenic members

All oomycetes are either parasites or saprobes (organisms that live by feeding on dead organic matter). At one time, these organisms were considered fungi, which is the origin of the term *water mold* and why their name contains the root word *-mycetes*.

They are distinguished from other protists by the structure of their motile spores, or zoospores, which bear two unequal flagella, one pointed forward and the other backward. Zoospores are produced asexually in a sporangium. Sexual reproduction involves the formation of male and female reproductive organs that produce gametes. Most oomycetes are found in water, but their terrestrial relatives are plant pathogens.

Phytophthora infestans, which causes late blight of potatoes, was one of multiple factors, including weather and oppressive social conditions that contributed to the Irish potato famine of 1845 and 1847. During the famine, about 400,000 people starved to death or died of diseases complicated by starvation, and about 2 million Irish immigrated to the United States and elsewhere.

Another oomycete, *Saprolegnia,* is a fish pathogen that can cause serious losses in fish hatcheries. When these fish are released into lakes, the pathogen can infect amphibians and kill millions of amphibian eggs at a time at certain locations. This pathogen is thought to contribute to the phenomenon of amphibian decline.

Learning Outcomes Review 29.4

All members of the Alveolata contain flattened vesicles called alveoli. Dinoflagellates have pairs of flagella arranged perpendicularly to each other, which causes them to swim with a spinning motion. Blooms of dinoflagellates result in red tides. Apicomplexans are animal parasites that produce a structure called an apical complex, which is composed of cytoskeleton and secretory structures and aids in penetrating their host. The ciliates are unicellular, heterotrophic protists with cilia used for feeding and propulsion. Most members of the stramenopiles have fine hairs on their flagella. Brown algae are large seaweeds that provide food and habitat for marine organisms. They undergo an alternation of generations. Diatoms are unicellular with silica in their cell walls, which forms a shell with two halves. Some can propel themselves. Oomycetes are unique in the production of zoospores that bear two unequal flagella.

- *What would be a major difficulty in finding a poison to fight the malaria-causing protist* Plasmodium?
- *How could you distinguish between the sporophyte and the gametophyte of a brown alga?*

29.5 Chloroplasts in Archaeplastida

Learning Outcomes

1. List the major characteristics of red algae.
2. Describe how humans use red algae.
3. Explain why charophytes are considered the closest relatives of land plants.

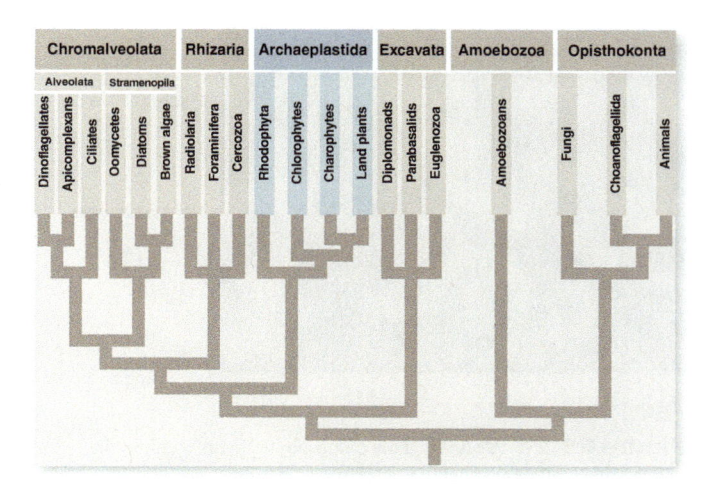

Archaeplastids acquired their chloroplasts through primary endosymbiosis, unlike the brown algae in which secondary endosymbiosis resulted in photosynthetic cells (see figures 29.3 and 29.4). As noted in section 29.1, the common origins of the chloroplast genome in all the algae complicates phylogenetic research, especially because some of the chloroplast genes moved into the nuclear genome. Extensive comparisons of brown, red, and green algal DNA sequences are consistent with a closer evolutionary relationship between the reds and the greens than the browns, which belong to the chromalveolates. A smaller group of algae, the glaucophytes, are also grouped with the Archaeplastida. Glaucophyte plastids were acquired through primary endosymbiosis and have a peptidoglycan (sugar and amino acid polymer) layer that harks back to the cyanobacterial endosymbiont.

The green algae have two groups—Chlorophyta and Charophyta—with the Charophyta sharing a common ancestor with the land plants. Land plants are the focus of chapters 30 and 31, with other Archaeplastida discussed here.

> **Data analysis** To construct a phylogeny with brown, red, and green algae, look for the amount of sequence variation in several genes from the chloroplast genomes of species in each of the groups and nuclear genomes. Few nucleotide differences are to be found among the chloroplast genes. More nucleotide differences occur between the brown algal and either the red or green algal nuclear genes than between the red and green nuclear genes. Draw a phylogeny and explain your reasoning.

Rhodophyta are red algae

The red algae, formally known as **Rhodophyta**, include more than 6000 species, ranging in size from microscopic organisms to *Schizymenia borealis* with blades as long as 2 m (figure 29.24). Many of the algae that build coral reefs are in this group. Sushi rolls are wrapped in nori, a red alga. Red algal polysaccharides are used commercially to thicken ice cream and cosmetics.

This lineage lacks flagella and centrioles and has the accessory photosynthetic pigments phycoerythrin, phycocyanin, and allophycocyanin, which are often red. As with the brown algae, red algae have both haploid and diploid phases in their life cycles.

Chlorophytes and charophytes are green algae

Green algae have two distinct lineages: the chlorophytes, discussed here, and another lineage, the charophytes, that gave rise to the land plants (see figure 29.5). The chlorophytes are of special interest here because of their unusual diversity and lines of specialization. The chlorophytes have an extensive fossil record dating back 900 million years. Modern chlorophytes closely resemble land plants, especially in the biochemical composition of their chloroplasts. They contain chlorophylls *a* and *b,* as well as carotenoids.

Unicellular chlorophytes

Early green algae probably resembled *Chlamydomonas reinhardtii,* diverging from land plants over 1 BYA (figure 29.25). Individuals are microscopic, green and rounded, and they have two flagella at the anterior end. They are soil dwellers that move rapidly in water by beating their flagella in opposite directions. Most individuals of *Chlamydomonas* are haploid. *Chlamydomonas* reproduces asexually as well as sexually, but because it is always unicellular, the life cycle is not haplodiplontic (figure 29.25).

Several lines of evolutionary specialization have been derived from organisms such as *Chlamydomonas,* including the evolution of nonmotile, unicellular green algae. *Chlamydomonas* is capable of retracting its flagella and settling down as an immobile unicellular organism if the pond in which it lives dries out. Some common algae found in soil and bark, such as *Chlorella,* are essentially like *Chlamydomonas* in this trait, but they do not have the ability to form flagella.

The genome of *Chlamydomonas reinhardtii* has been sequenced, and this organism has become a workhorse for comparative genetic studies and gene function analyses. In fact, it is a widely

Figure 29.24 Red algae come in many forms and sizes.

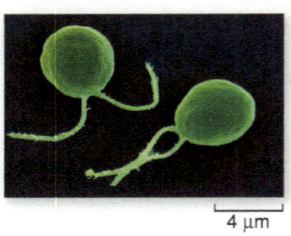

Figure 29.25 Chlamydomonas life cycle. This single-celled chlorophyte has both asexual and sexual reproduction. Unlike multicellular green plants, gamete fusion is not followed by mitosis.

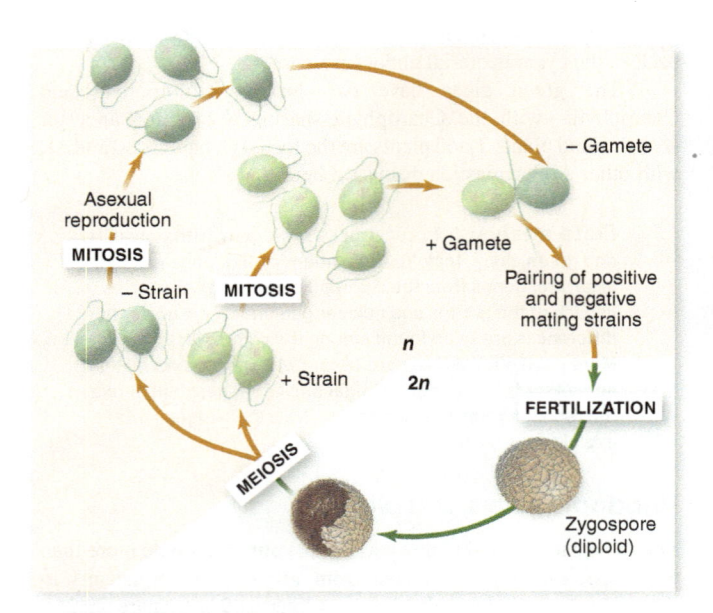

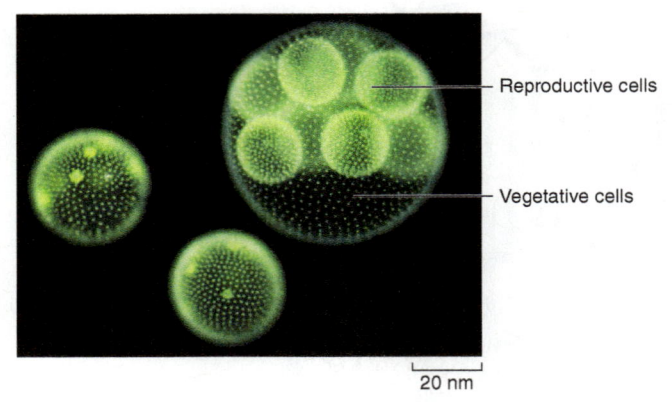

Figure 29.26 Volvox. This chlorophyte forms a colony in which some cells are specialized for reproduction.

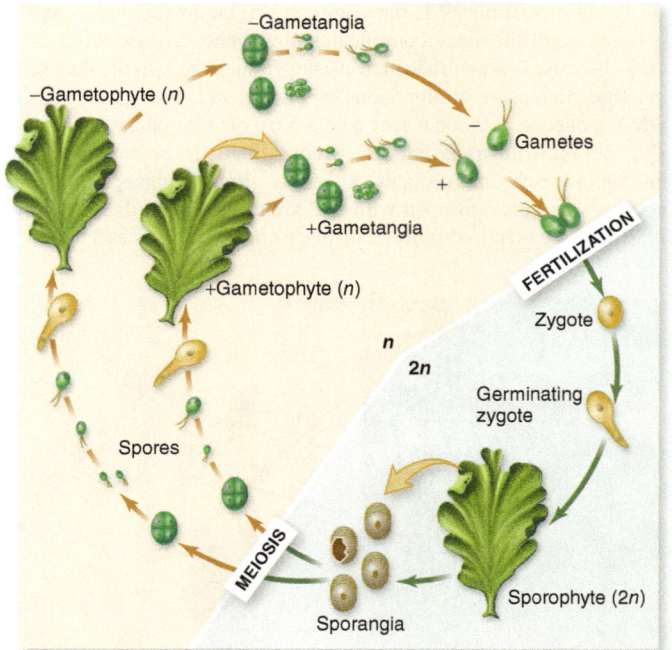

Figure 29.27 Life cycle of Ulva. This chlorophyte alga has a haplodiplontic life cycle. The gametophyte and sporophyte are multicellular and identical in appearance.

used host for recombinant protein expression and the biomanufacture of antibodies, immunotoxins, hormones, and industrial enzymes.

Cell specialization in colonial chlorophytes

Multicellularity arose many times in the eukaryotes. Colonial chlorophytes provide examples of cellular specialization, an aspect of multicellularity.

The most elaborate of these organisms is *Volvox* (figure 29.26), a hollow sphere made up of a single layer of 500 to 60,000 individual cells, each cell having two flagella. Only a small number of the cells are reproductive. Some reproductive cells may divide asexually, bulge inward, and give rise to new colonies that initially remain within the parent colony. Other reproductive cells produce gametes for sexual reproduction.

Haplodiplontic life cycles in multicellular chlorophytes

Haplodiplontic life cycles are found in some chlorophytes and the streptophytes, which include both charophytes and land plants. *Ulva*, a multicellular chlorophyte, has identical gametophyte and sporophyte generations that consist of flattened sheets two cells thick (figure 29.27). Unlike the charophytes, none of the ancestral chlorophytes gave rise to land plants.

> **Inquiry question** Are *Ulva* gametes formed by meiosis? Explain your response.

Charophytes: Closest Relatives to Land Plants

Charophytes are also green algae, and they are distinguished from chlorophytes by their close phylogenetic relationship to the land plants. Currently, the molecular evidence from rRNA and DNA sequences favors the charophytes as the green algal clade most closely related to land plants. Charophytes also have haplontic life cycles.

Identifying which of the charophyte clades is sister (most closely related) to the land plants puzzled biologists for a long time, as the charophyte algae fossil record is scarce. The two candidate charophyte clades have been the Charales and the Coleochaetales, (figure 29.28). Both lineages are primarily freshwater algae, but the Charales are huge, relative to the microscopic Coleochaetales. Both clades have similarities to land plants. *Coleochaete* and its relatives have cytoplasmic linkages between cells called *plasmodesmata*, which are found in land plants. The species *Chara* in the Charales undergoes mitosis and cytokinesis like land plant cells. Sexual reproduction in both relies on a large, nonmotile egg and flagellated sperm. These gametes are more similar to those of land plants than many charophyte relatives. Both charophyte clades form green mats around the edges of freshwater ponds and marshes. One species must have successfully inched its way onto land through adaptations to drying.

Learning Outcomes Review 29.5

Red algae vary greatly in size and produce accessory pigments that may give them a red color. They lack centrioles and flagella, and they reproduce using an alternation of generations. Humans use red algae as a food and a thickening agent. The chlorophytes have chloroplasts very similar to those of land plants. Specializations in this group include nonmotile, unicellular species that can tolerate drying and colonial organisms that exhibit some cell specialization. Charophytes are the green algae that are most likely sisters to the land plants based on molecular evidence, morphology, and reproduction

- Why would you expect to get different results from phylogenetic analysis of nuclear DNA versus plastid DNA?
- What major barrier must be overcome for sexual reproduction of land-based organisms?

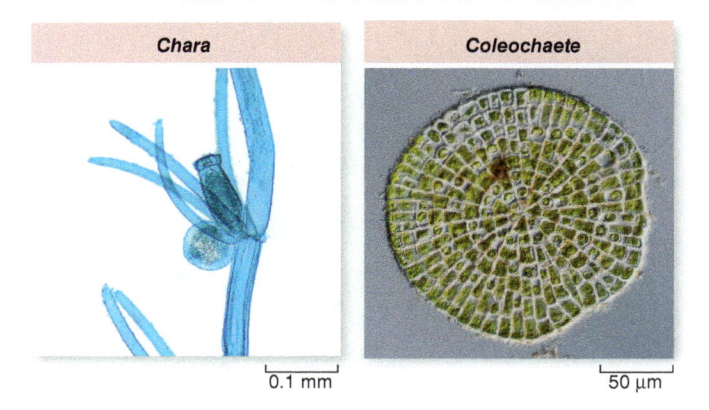

Figure 29.28 *Chara,* a member of the Charales, and *Coleochaete,* a member of the Coleochaetales, represent the two clades most closely related to land plants.

29.6 Slender Pseudopods in Rhizaria

Learning Outcomes

1. Distinguish between the shells of most foraminifera and those of diatoms.
2. Distinguish between cellular and plasmodial slime molds.

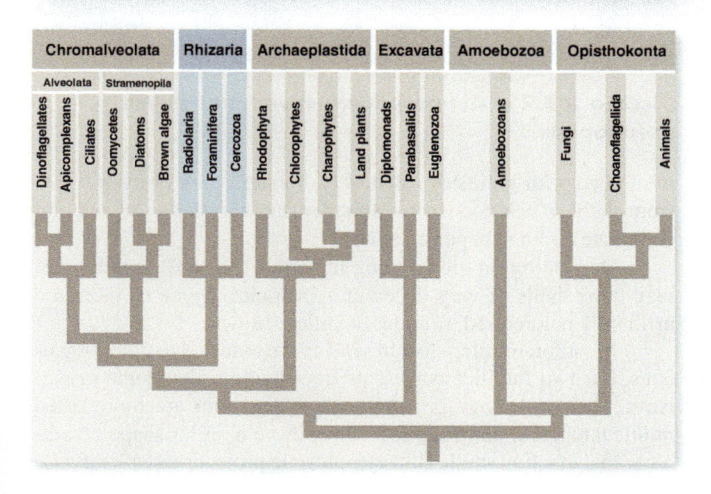

Both Rhizarians and Amoebozoans (described later in this section) use pseudopods for locomotion. Pseudopods are flowing projections of cytoplasm that extend to pull the organism forward. Amoeboid motion was once used as a trait to group protists, but the inclusion of molecular data in phylogenetic reconstructions led to the realization that locomotion alone was not a useful trait in evolutionary analyses. Within the Rhizaria, three distinct monophyletic groups have been identified. Radiolaria are common marine phytoplankton. Foraminifera are also marine phytoplankton, and their calcium carbonate shells contribute to chalky deposits over time. Cercozoa are found in the soil.

Radiolarians have silica exoskeletons

Radiolarians secrete glassy exoskeletons made of silica. These skeletons give the unicellular organisms a distinct shape, exhibiting either bilateral or radial symmetry. The shells of different species form many elaborate and beautiful shapes, with pseudopods extruding outward along spiky projections of the skeleton (figure 29.29). Microtubules support these cytoplasmic projections.

Foraminifera fossils created huge limestone deposits

Members of the phylum Foraminifera are heterotrophic marine protists. They range in size from microscopic to several centimeters. They resemble tiny snails and can form 3-m-deep layers in marine sediments. Characteristic of the group are pore-studded shells (called *tests*) composed of organic materials usually

Figure 29.29 *Actinosphaerium* with needlelike pseudopods.

reinforced with grains of calcium carbonate, sand, or even plates from shells of echinoderms or spicules (minute needles of calcium carbonate) from sponge skeletons.

Depending on the building materials they use, foraminifera may have shells of very different appearance. Some of them are brilliantly colored red, salmon, or yellow-brown.

Most foraminifera live in sand or are attached to other organisms, but two families consist of free-floating planktonic organisms. Their tests may be single-chambered, but are more often multichambered, and they sometimes have a spiral shape resembling that of a tiny snail. Thin cytoplasmic projections called *podia* emerge through openings in the tests (figure 29.30). Podia are used for swimming, gathering materials for the tests, and feeding. Foraminifera eat a wide variety of small organisms.

The life cycles of foraminifera are extremely complex, alternating between haploid and diploid generations. Foraminifera have contributed massive accumulations of their tests to the fossil record for more than 200 million years. Because of the excellent preservation of their tests and the striking differences among them, forams are very important as geological markers. The pattern of occurrence of different forams is often used as a guide in searching for oil-bearing strata. Limestones all over the world, including the famous White Cliffs of Dover in southern Great Britain, are often rich in forams (figure 29.31).

Figure 29.31 White Cliffs of Dover. The limestone that forms these cliffs is composed almost entirely of fossil shells of protists, including foraminifera.

Cercozoa locomote with flagella or pseudopods

Cercozoa are a morphologically diverse group of primarily soil protists. Some rely on flagella for locomotion, but others extend pseudopods for amoeboid movement. Some have silica-based shells made up of scales or plates. A cercozoan, *Paulinella chromatophora*, may have ingested a cyanobacterium as recently as 60 MYA. (figure 29.32). Except for plants, this is the only known case of primary endosymbiosis. Consequently, this species is being investigated for clues to the evolution of endosymbiosis.

Learning Outcomes Review 29.6

Rhizaria move with the aid of needlelike pseudopods. Many members of the foraminifera occupy marine habitats. Most of them have calcium carbonate shells and are responsible for large fossil deposits of limestone, whereas cercozoans and radiolarians have silica shells.

- *How would you determine if amoeboid locomotion using pseudopods was a good trait to use in reconstructing protist phylogenies?*

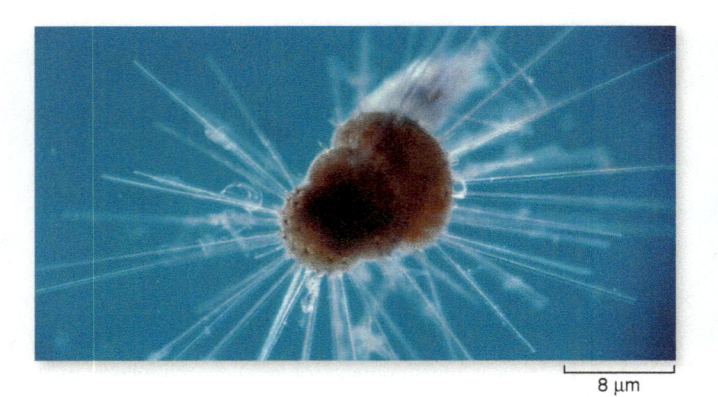

Figure 29.30 A representative of the foraminifera. *Podia*, thin cytoplasmic projections, extend through pores in the calcareous test, or shell, of this living foram.

Figure 29.32 The cercozoan *Paulinella chromatophora*. Cercozoans may be a more recent endosymbiont, offering insights into the evolution of endosymbiosis.

29.7 Blunt Pseudopods in Amoebozoa

Learning Outcome
1. Explain how amoebas move.

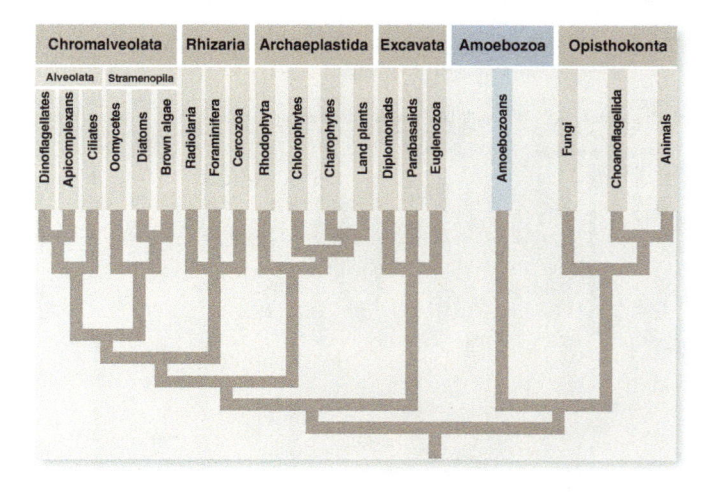

Amoebas move from place to place by means of their pseudopods. As mentioned in section 29.6, pseudopods are flowing projections of cytoplasm that extend and pull the amoeba forward or engulf food particles. An amoeba puts a pseudopod forward and then flows into it (figure 29.33). Microfilaments of actin and myosin similar to those found in muscles are associated with these movements. The pseudopods can form at any point on the cell body so that it can move in any direction. The amoebas in the Amoebozoa supergroup are most closely related to the Opisthokonta, which include the close relatives of fungi and animals. Members of these two supergroups have a single flagellum, in contrast with two or more in the other supergroups. Amoebozoans and Opisthokonts are collectively referred to as Unikonts, with "uni" referring to the single flagellum; however, they are distinct supergroups.

Figure 29.33 *Amoeba proteus*. The projections are pseudopods; an amoeba moves by flowing into them.

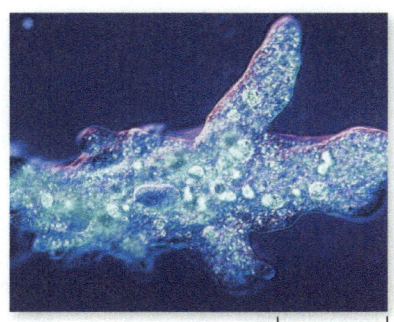

Figure 29.34 A plasmodial protist. This multinucleate pretzel slime mold, *Hemitrichia serpula*, moves about in search of the bacteria and other organic particles that it ingests.

Most amoeba are free-living, but some are parasitic

Found in the soil, as well as freshwater, most of these Amoebozoans are free-living and important to the soil ecosystem. A few rare cases of a species acting as a human pathogen have been reported. In immune-suppressed individuals, the *Acanthamoeba* enters the body through a wound and is able to cross the blood-brain barrier into the brain. Once in the brain, inflammation and death follow.

Plasmodial slime molds are multinucleate

Plasmodial slime molds stream along as a **plasmodium,** a non-walled, multinucleate mass of cytoplasm that resembles a moving mass of slime (figure 29.34). This form is called the *feeding phase,* and the plasmodium may be orange, yellow, or another color.

Plasmodia show a back-and-forth streaming of cytoplasm that is very conspicuous, especially under a microscope. They are able to pass through the mesh in cloth or simply to flow around or through other obstacles. As they move, they engulf and digest bacteria, yeasts, and other small particles of organic matter.

A multinucleated *Plasmodium* cell undergoes mitosis synchronously, with the nuclear envelope breaking down, but only at late anaphase or telophase. Centrioles are absent.

When either food or moisture is in short supply, the plasmodium migrates relatively rapidly to a new area. Here it stops moving and either forms a mass in which spores differentiate or divides into a large number of small mounds, each of which produces a single, mature **sporangium,** the structure in which spores are produced. These sporangia are often beautiful and extremely complex in form (figure 29.35). The spores are highly resistant to unfavorable environmental influences and may last for years if kept dry.

Cellular slime molds exhibit cell differentiation

The cellular slime molds have become an important group for the study of cell differentiation because of their relatively simple developmental systems (figure 29.36). The individual organisms behave as separate amoebas, moving through the soil and ingesting bacteria. When food becomes scarce, the individuals aggregate to

Figure 29.35 Sporangia of a plasmodial slime mold.
These *Arcyria* sporangia are found in the phylum Myxomycota.

form a moving "slug." Cyclic adenosine monophosphate (cAMP) is sent out in pulses by some of the cells, and other cells move in the direction of the cAMP to form the slug. In the cellular slime mold *Dictyostelium discoideum*, this slug goes through morphogenesis to make stalk and spore cells. The spores then go on to form a new amoeba if they land in a moist habitat.

Learning Outcome Review 29.7
At least three lineages of slime mold exist. Cellular slime molds are multicellular, and plasmodial slime molds consist of large multinucleate single cells.

- *Would you say that cellular slime molds are closely related to plasmodial slime molds?*

29.8 Propulsion via a Single Posterior Flagellum in Opisthokonta

Learning Outcome
1. Describe the evolutionary significance of the choanoflagellates.

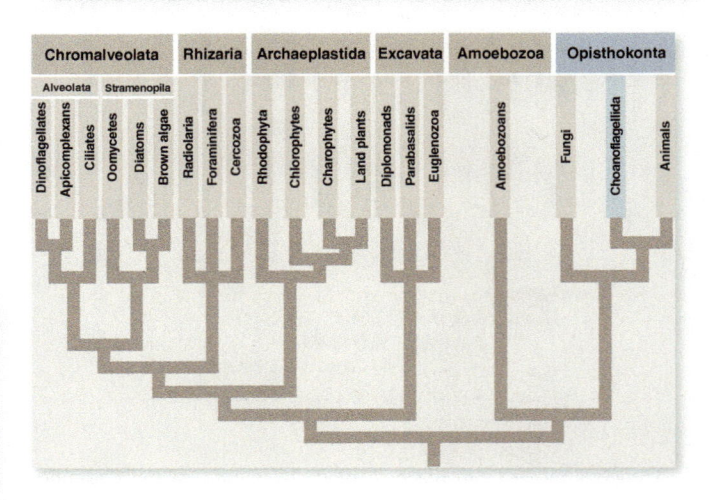

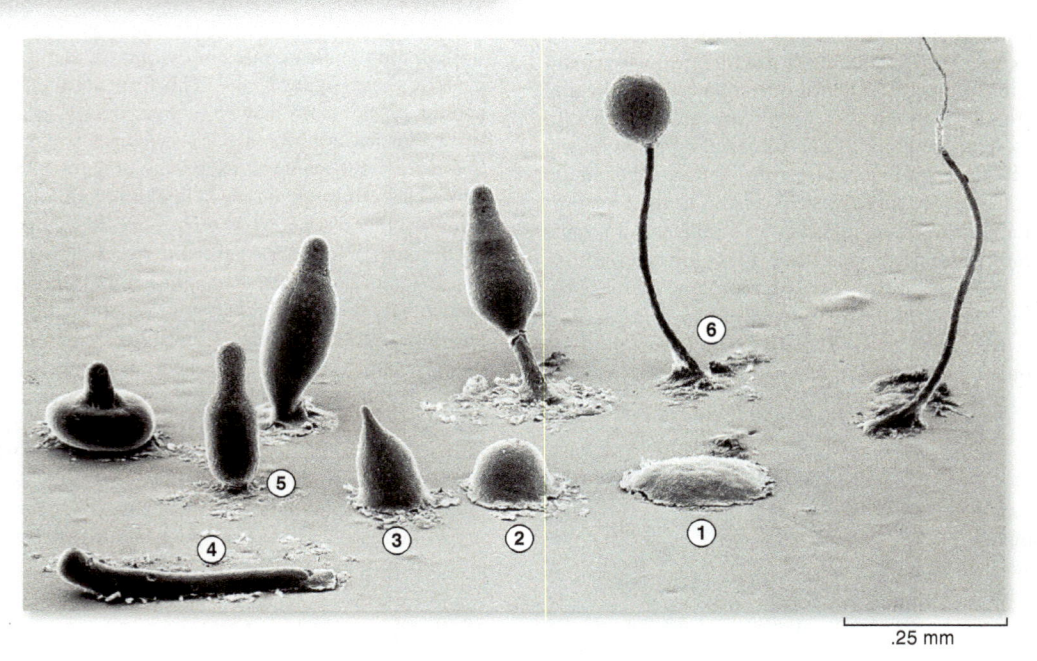

Figure 29.36 Development in *Dictyostelium discoideum*, a cellular slime mold. *(1)* First, a spore germinates, forming an amoeba that feeds and reproduces until the food runs out. At that point, amoebas aggregate and move toward a fixed center. *(2)* The aggregated amoebas begin to form a mound. *(3)* The mound produces a tip and begins to fall sideways. *(4)* Next, the aggregate forms a multicellular "slug," 2–3 mm long, that migrates toward light. *(5)* The slug stops moving, and a process called culmination begins. Cells differentiate into stalk and spore cells. *(6)* In the mature fruiting body, amoebas become encysted as spores.

Fungi and animals are more closely related to each other than to plants because they share a common ancestor, which leads them to be grouped as Opisthokonts. Of particular interest in the Opisthokonts are the *choanoflagellates,* unicellular organisms that are most like the common ancestor of the sponges and, indeed, all animals. Choanoflagellates have a single emergent flagellum surrounded by a funnel-shaped, contractile collar composed of closely placed filaments, a structure that is exactly matched in the sponges, which are animals. These protists feed on bacteria strained out of the water by their collar. Colonial forms resemble freshwater sponges (figure 29.37).

The close relationship of choanoflagellates to animals was further demonstrated by the strong homology between a surface receptor (a tyrosine kinase receptor) found in choanoflagellates and sponges. This surface receptor initiates a signaling pathway involving phosphorylation (see chapter 9).

Figure 29.37 Colonial choanoflagellates resemble their close animal relatives, the sponges.

30 μm

Learning Outcome Review 29.8

The choanoflagellates are believed to be the closest relatives of animals. Colonial forms are similar to freshwater sponges, and both organisms have a homologous cell-surface receptor.

- What other types of studies might connect choanoflagellates with sponges?

Chapter Review

29.1 Eukaryotic Origins and Endosymbiosis

Fossil evidence dates the origins of eukaryotes.
Although eukaryotes may have arisen earlier, the fossil evidence of their appearance dates back to 1.5 BYA.

The nucleus and endoplasmic reticulum arose from membrane infoldings (figure 29.2).

Mitochondria evolved from engulfed aerobic bacteria.
According to the theory of endosymbiosis, ancestral eukaryotic cells engulfed aerobic bacteria, which then became mitochondria (figure 29.3).

Chloroplasts evolved from engulfed photosynthetic bacteria.
Chloroplasts are believed to have arisen when ancestral eukaryotic cells engulfed photosynthetic bacteria (figure 29.4). Brown algae arose through a secondary endosymbiotic event when a red alga was engulfed by a single-celled organism.

Endosymbiosis is supported by a range of evidence.
In support of endosymbiosis, several organelles are found to contain their own DNA, which closely resembles that of prokaryotes. Over a billion and a half years, many of the chloroplast and mitochondrial genes have migrated to the nuclear genome.

Mitosis evolved in eukaryotes.
Mechanisms of mitosis vary among organisms, suggesting that the process did not evolve all at once.

29.2 Overview of Protists

Eukaryotes are organized into six supergroups that all contain protists.
Molecular systematics is helping to sort out the protists, which are now grouped into six supergroups: Excavata, Chromalveolata, Archaeplastida, Rhizaria, Amoebozoa, and Ophisthokonta (figure 29.5). With the exception of Chromalveolata, the supergroups are monophyletic.

Monophyletic clades have been identified among the protists.

Protist cell surfaces vary widely.
Extracellular material (ECM) may cover the plasma membrane.

Protists have several means of locomotion.
Protists mainly use flagella or pseudopods for locomotion, although many other means of propulsion are found.

Protists have a range of nutritional strategies.
Protists include phototrophs, heterotrophs, (phagotrophs or osmotrophs), and mixotrophs capable of both modes.

Protists reproduce asexually and sexually.
Protists can reproduce asexually by mitosis, budding, or schizogony. They may also carry out sexual reproduction.

Protists are the bridge to multicellularity.
Colonial protists may be the precursors of multicellular organisms.

29.3 Feeding Groove in Excavata

Diplomonads lack functional mitochondria.
Diplomonads are unicellular, move with flagella, and have two nuclei.

Parabasalids have undulating membranes.
Parabasalids use flagella and undulating membranes for locomotion.

Euglenozoans change shape while swimming.
Free-living euglenids, exemplified by *Euglena*, can produce chloroplasts to carry out photosynthesis in the light. They contain a

pellicle and move via anterior flagella. Kinetoplastids are parasitic and are distinctive in having a single, unique mitochondrion with two types of circular DNA.

29.4 Secondary Endosymbiosis in Chromalveolata

Dinoflagellates are photosynthesizers with distinctive features.
Dinoflagellates have pairs of flagella arranged so that they swim with a spinning motion. Blooms of dinoflagellates cause red tides (figure 29.13).

Apicomplexans include the malaria parasite.
Apicomplexans are spore-forming animal parasites (figure 29.15). They have a unique arrangement of organelles at one end of the cell, called the apical complex, which is used to invade the host.

Ciliates are characterized by their mode of locomotion.
Ciliates are unicellular, heterotrophic protists that use numerous cilia for feeding and propulsion. Each cell has a macronucleus and a micronucleus. Micronuclei are exchanged during conjugation. (figure 29.19)

Brown algae include large seaweeds.
Brown algae are typically large seaweeds that have a haplodiplontic life cycle, producing gametophyte and sporophyte stages (figure 29.21).

Diatoms are unicellular organisms with double shells.
Diatoms have silica in their cell walls. Each diatom produces two overlapping glassy shells that fit like a box and lid.

Oomycetes, the "water molds," have some pathogenic members.
Oomycetes are parasitic and are unique in the production of asexual spores (zoospores) that bear two unequal flagella.

29.5 Chloroplasts in Archaeplastida

Rhodophyta are red algae.
Red algae produce accessory pigments that may give them a red color. They lack centrioles and flagella, and they reproduce using an alternation of generations.

Chlorophytes and charophytes are green algae.
Unicellular chlorophytes include *Chlamydomonas*, which has two flagella, and *Chlorella*, which has no flagella and reproduces asexually.

Volvox is an example of a colonial green alga; some cells are specialized for producing gametes or for asexual reproduction. It may represent a step on the way to multicellularity. Multicellular chlorophytes can have haplodiplontic life cycles.

Ulva has sporophyte and gametophyte generations; however, the chlorophytes did not give rise to land plants although they are the closest relatives to land plants.

Both groups within the charophytes, Charales and Coleochaetales, have plasmodesmata, cytoplasmic links between cells. They also undergo mitosis and cytokinesis like terrestrial plants. Charophytes are most closely related to the land plants of all the algae.

29.6 Slender Pseudopods in Rhizaria

Radiolarians have silica exoskeletons.
Glassy exoskeletons made of silica give radiolarians a distinct shape. Microtubules support pseudopods that extrude toward the tips of the spiky exoskeleton.

Foraminifera fossils created huge limestone deposits.
The Foraminifera are heterotrophic marine protists with pore-studded shells primarily formed by deposit of calcium carbonate.

Cercozoa locomote with flagella or pseudopods.
Like the marine radiolarians, cercozoans have silica exoskeletons, but most are found in soils. Cercozoans ingest green algae and may provide clues to the evolution of endosymbiosis.

29.7 Blunt Pseudopods in Amoebozoa

Most amoeba are free-living, but some are parasitic.
Many protists are free-living and found in the soil and freshwater ecosystems, but a few species are pathogenic to humans.

Plasmodial slime molds are multinucleate.
Plasmodia in their feeding phase are macroscopically visible masses of oozing slime (figure 29.34). These colorful large cells undergo repeated rounds of mitosis without cell division.

Cellular slime molds exhibit cell differentiation.
Cellular slime molds such as *Dictyostelium discoideum* can signal and interact with neighboring cells, resulting in differentiated cell types in an aggregate organism.

29.8 Propulsion via a Single Posterior Flagellum in Opisthokonta

Colonial choanoflagellates are structurally similar to freshwater sponges, and molecular similarities have been found.

Review Questions

UNDERSTAND

1. Fossil evidence of eukaryotes dates back to
   a. 2.5 BYA.
   b. 1.5 BYA.
   c. 2.5 MYA.
   d. 1.5 MYA.

2. DNA is not found in this organelle.
   a. Endoplasmic reticulum
   b. Nucleus
   c. Chloroplast
   d. Mitochondrion

3. The products of budding are
   a. two cells of equal size.
   b. two cells, one of which is smaller than the other.
   c. many cells of equal size.
   d. many cells of variable size.

4. Both diplomonads and parabasalids
   a. contain chloroplasts.
   b. have multinucleate cells.
   c. lack mitochondria.
   d. have silica in their cell walls.

5. Trypanosomes are examples of
   a. euglenoids.
   b. diplomonads.
   c. parabasalids.
   d. kinetoplastids.

6. The function of the apical complex in apicomplexans is to
   a. propel the cell through water.
   b. penetrate host tissue.
   c. absorb food.
   d. detect light.

7. If a cell contains a pellicle, it
   a. can change shape readily.
   b. is shaped like a sphere.
   c. is shaped like a torpedo.
   d. must have a contractile vacuole.

8. Stramenopila are
   a. tiny flagella.
   b. large cilia.
   c. small hairs on flagella.
   d. pairs of large flagella.

9. Choose all of the following that exhibit an alternation of multicellular generations.
   a. Dinoflagellates
   b. Brown algae
   c. Red algae
   d. Diatoms

10. Choose all of the following that are photosynthetic.
    a. Diatoms
    b. Ciliates
    c. Apicomplexans
    d. Dinoflagellates

11. Which is most likely the ancestor of animals?
    a. Trypanosomes
    b. Diplomonads
    c. Ciliates
    d. Choanoflagellates

12. When food is scarce, cells of this organism communicate with each other to form a multicellular slug.
    a. Cellular slime molds
    b. True amoebas
    c. Foraminifera
    d. Diatoms

APPLY

1. Analyze the following statements and choose the one that most accurately supports the endosymbiotic theory.
   a. Mitochondria rely on mitosis for replication.
   b. Chloroplasts contain DNA but translation does not occur in chloroplasts.
   c. Vacuoles have double membranes.
   d. Antibiotics that inhibit protein synthesis in bacteria can have the same effect on mitochondria.

2. Determine which feature of the choanoflagellates was likely the most significant for the evolution of animals?
   a. Flagellum with a funnel-shaped, contractile collar also found in sponges
   b. A tyrosine kinase receptor on the surface of choanoflagellates that has strong homology to fungi
   c. A colonial form that resembles some fungi
   d. Eyespots that are similar to ribbon worms

3. Examine the life cycle of cellular slime molds, and determine which feature affords the greatest advantage for surviving food shortages.
   a. Cellular slime molds produce spores when starved.
   b. Cellular slime molds are saprobes.
   c. A diet of bacteria ensures there will never be a shortage of food.
   d. Cellular slime molds use cAMP to guide each other to food sources.

SYNTHESIZE

1. Modern taxonomic treatments rely heavily on phylogenetic data to classify organisms. In the past, taxonomists often used a morphological species concept, in which species were defined based on similarities in growth form. Give an example of how a morphological species concept would group a set of protists differently from how a phylogenetic species concept would.

2. Three methods have been used to try to eradicate malaria. One is to eliminate the mosquito vectors of the parasite, a second is to kill the parasites after they entered the human body, and the third is to develop a vaccine against the parasite, allowing the human immune system to provide protection from the disease. Which do you suppose is the most promising in the long run? Why? Think about both the biology of the disease and the efficacy of carrying out each of the methods on a large scale.

3. Design an experiment to demonstrate that cells of cellular slime molds are attracted to cyclic AMP. Then, design a follow-up experiment to determine whether they are always attracted to cAMP or only when resources are scarce.

CHAPTER 30

Seedless Plants

Chapter Contents

- 30.1 Origin of Land Plants
- 30.2 Bryophytes: Dominant Gametophyte Generation
- 30.3 Tracheophyte Plants: Dominant Sporophyte Generation
- 30.4 Lycophytes: Divergent from Main Lineage of Vascular Plants
- 30.5 Pterophytes: Ferns and Their Relatives

Introduction

Colonization of land by plants fundamentally altered the history of life on Earth. A terrestrial environment offers abundant CO_2 and solar radiation for photosynthesis. But for at least 500 million years, the lack of water and the higher ultraviolet (UV) radiation on land confined green algal ancestors of plants to an aquatic environment. Evolutionary innovations for reproduction, structural support, and prevention of water loss are key in the story of plant adaptation to land. The evolutionary shift on land to life cycles dominated by a diploid generation masks recessive mutations arising from higher UV exposure. As a result, larger numbers of alleles persist in the gene pool, creating greater genetic diversity. Long before seeds and flowers evolved, the seedless plants covered the Earth. In this chapter we consider the evolutionary innovations in seedless plants during the first 100 million years of terrestrial life.

30.1 Origin of Land Plants

Learning Outcomes

1. Explain the relationship between the different algae clades and plants.
2. Describe the haplodiplontic life cycle.
3. Distinguish between a sporophyte and a gametophyte.
4. Identify two major environmental challenges for land plants and associated adaptations.

Imagine Earth 500 million years ago. Most of the continents were in the southern hemisphere. A diverse array of animals swam through the seas, feeding on an abundance of algae. In stark contrast, the terrestrial environment was a barren landscape. Slowly, though, sea life began to explore the possibilities of living on land. Eventually, land plants evolved from green algae. This began the conversion of bare ground to the rich and diverse ecosystems found on our planet today.

Green algae and the land plants shared a common ancestor a little over 1 BYA and are collectively referred to as the green plants. DNA sequence data are consistent with the claim that a single individual gave rise to all green plants. The green plants are photoautotrophic, but not all photoautotrophs are plants.

Although the plant kingdom includes the green algae, it does not include the fungi, which are more closely related to metazoan animals (see chapter 33). Fungi, however, were essential to the colonization of land by plants, enhancing plants' nutrient uptake from the soil.

One of the most significant evolutionary events in the billion-year-old history of the green plants is the adaptation to terrestrial living. Thus this chapter and chapter 31 focus on the land plants. Land plants had to develop mechanisms to avoid water loss, and protection from the harmful effects of solar radiation. In addition, although all land plants alternate between haploid gametophyte and diploid sporophyte generations, there is a trend toward dominance of the sporophyte in the more advanced groups.

Land plants evolved from freshwater algae

Some saltwater algae evolved to thrive in a freshwater environment. Just a single species of freshwater green algae gave rise to the entire terrestrial plant lineage, from mosses through the flowering plants (angiosperms). Exactly what this ancestral alga was is still a mystery, but close relatives, members of the charophytes, exist in freshwater lakes today.

The green algae split into two major clades: the chlorophytes, which never made it to land, and the charophytes, which are a sister clade to all the land plants (figure 30.1). Together charophytes and land plants are referred to as streptophytes. Land plants have multicellular haploid and diploid stages, unlike the charophytes. Diploid embryos are also land plant innovations. As land plants have evolved, the trend has been toward more embryo protection and a smaller haploid stage in the life cycle.

Land plants have adapted to terrestrial life

Unlike their freshwater ancestors, most land plants are not immersed in water. As an adaptation to living on land, most plants are protected from desiccation—drying out—by a waxy surface material called the cuticle that is secreted onto parts that are exposed to the air. The cuticle is relatively impermeable, preventing water loss. This solution, however, limits the gas exchange essential for respiration and photosynthesis. Gas diffusion into and out of a plant occurs through tiny mouth-shaped openings called **stomata** (singular, *stoma*) on plant leaves and stems, which allow water to diffuse out at the same time. Chapter 37 describes how stomata can be closed at times to limit water loss.

Moving water within plants is a challenge that increases with plant size. Bryophytes cannot grow tall because they lack a system to efficiently transport water. Tracheophytes are more advanced, producing specialized vascular tissue for transport over long distances. The tissue that carries water is called xylem, whereas food is carried by phloem (see chapter 37).

Terrestrial plants are exposed to higher intensities of UV irradiation than aquatic algae, increasing the chance of mutation. These land plants minimize the effects of mutation by carrying two copies of every gene. That is, their bodies are diploid, meaning that deleterious recessive mutations are masked. All land plants have both haploid and diploid generations, but there is an evolutionary shift toward a dominant diploid generation.

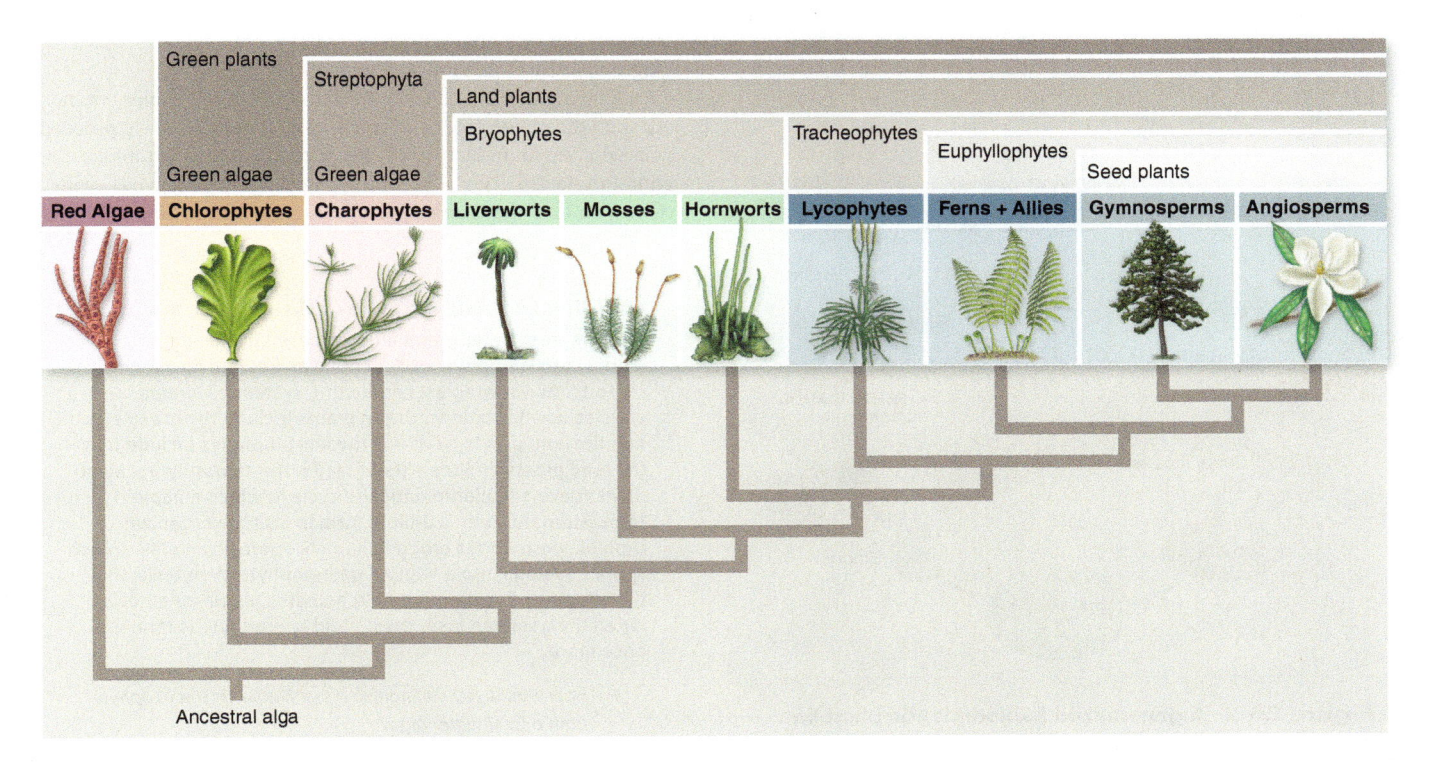

Figure 30.1 Green plant phylogeny.

The haplodiplontic cycle produces alternation of generations

Humans and other animals have a **diplontic** life cycle, meaning that only the diploid stage is multicellular. In animals, meiosis produces single-celled haploid gametes (egg and sperm cells). These gametes fuse at fertilization to produce a diploid embryo, which divides by mitosis to become the next generation. The products of meiosis never divide by mitosis, so there is no multicellular haploid generation in animals. By contrast, the land plant life cycle is **haplodiplontic,** having multicellular haploid and diploid stages. They undergo mitosis after both fertilization (the diploid stage) and meiosis (the haploid stage). The result is a multicellular diploid individual (called the sporophyte) and a multicellular haploid individual (called the gametophyte)—The haplodiplontic cycle is a fundamental feature of plants and is summarized in figure 30.2.

Many brown, red, and green algae are also haplodiplontic. Humans produce gametes via meiosis, but land plants actually produce gametes by *mitosis* in a multicellular, haploid individual. Meiosis produced a haploid spore that divided by mitosis to produce a haploid gametophyte. The multicellular diploid generation, or **sporophyte,** alternates with the multicellular haploid generation, or **gametophyte.** Sporophyte means "spore plant," and gametophyte means "gamete plant." These terms indicate the types of reproductive cells the respective generations produce.

The diploid sporophyte produces haploid spores (not gametes) by meiosis. Meiosis takes place in structures called **sporangia,** where diploid **spore mother cells (sporocytes)** undergo meiosis, each producing four haploid **spores.** Spores are the first cells of the gametophyte generation. Spores divide by mitosis, producing a multicellular, haploid gametophyte.

The haploid gametophyte produces gametes by mitosis. Consequently, the gametes (egg cells and sperm cells) are also haploid. When two gametes fuse, the zygote they form is diploid and is the first cell of the next sporophyte generation. The zygote grows into a diploid sporophyte by mitosis and produces sporangia in which meiosis ultimately occurs.

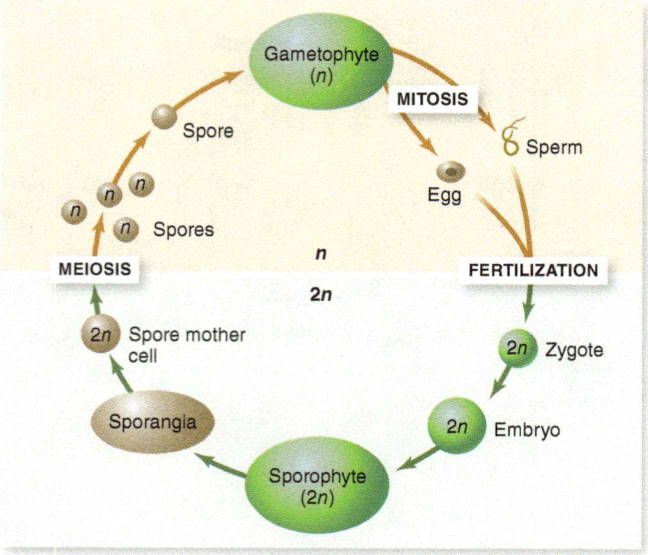

The relative sizes of haploid and diploid generations vary

All land plants are haplodiplontic; however, the haploid generation consumes a much larger portion of the life cycle in mosses and ferns than it does in the seed plants—the gymnosperms and angiosperms. In mosses, and ferns, the gametophyte is photosynthetic and free-living. When you see a moss, you are actually looking at the gametophyte (haploid) generation. The sporophyte is a small structure attached to the gametophyte. Although ferns produce photosynthetic gametophytes, they are barely visible to the naked eye. When you see a fern plant, you are looking at the sporophyte generation. The gametophyte is even smaller in the seed plants. It is typically microscopic. So, as with the ferns, the sporophyte is the dominant generation in seed plants.

Although the sporophyte generation can get very large, the size of the gametophyte is limited in all plants. When the gametophyte generation of mosses produces gametes at its tips, the egg remains stationary, whereas the sperm swims to the egg in a droplet of water. In the more advanced plants, the tracheophytes, the gametophyte is composed of only a few cells and the sperm no longer need to swim to reach the egg cell.

Having completed an overview of plant life cycles, we next consider the major groups of seedless land plants. As we proceed, you will see a reduction of the size of the gametophyte and **gametangia** (structures in which gametes are produced), with a corresponding increase in specialization for life on land.

Figure 30.2 A generalized haplodiplontic plant life cycle. Note that both haploid and diploid individuals are multicellular. Also, spores are produced by meiosis, but gametes are produced by mitosis.

Learning Outcomes Review 30.1

A single freshwater green alga successfully invaded land, establishing the land plants. Land plants developed reproductive strategies, conducting systems, stomata, and cuticles as adaptations. Green plants include the green algae and the land plants, whereas the streptophytes include only the land plants and their sister clade, the charophytes. Most plants have a haplodiplontic life cycle in which a haploid form alternates with a diploid form in a single organism. Diploid sporophytes produce haploid spores by meiosis. Each spore develops into a haploid gametophyte by mitosis; the gametophyte produces haploid gametes, again by mitosis. When the gametes fuse, the diploid sporophyte is formed once more.

- How would you distinguish a small aquatic tracheophyte from a freshwater alga?
- What distinguishes gamete formation in plants from gamete formation in humans?

30.2 Bryophytes: Dominant Gametophyte Generation

Learning Outcome
1. Describe adaptations of bryophytes for terrestrial environments.

Bryophytes are the closest living descendants of the first land plants. Plants in this group are also called nontracheophytes because they lack the transport cells called *tracheids*. Bryophytes are comprised of liverworts (phylum Hepaticophyta), mosses (phylum Bryophyta), and hornworts (phylum Anthocerotophyta). They are shown in the cladogram.

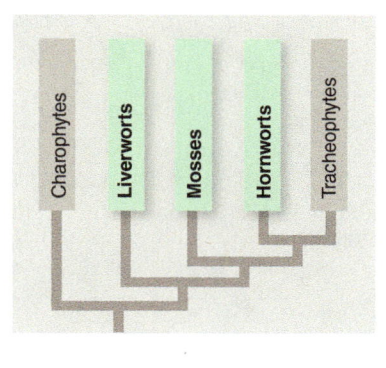

Water and gas availability were limiting factors for early terrestrial plants. These plants likely had little ability to regulate internal water levels and tolerated desiccation, traits found in most modern (extant) mosses.

Fungi and early land plants cohabited, and the fungi formed close associations with the plants, enhancing water uptake. The beneficial symbiotic relationship between fungi and plants, called **mycorrhizal associations,** are also found in many existing bryophytes. More information on mycorrhizal fungi is found in chapter 32.

Bryophytes have a photosynthetic gametophyte

The approximately 16,000 species of bryophytes are simple but are found in a diversity of terrestrial environments, even deserts. The gametophyte is small but specialized for light capture and photosynthesis. Water and nutrients are transported throughout the gametophyte via simple conducting cells. Egg and sperm cells produced by the gametophyte unite to form the sporophyte generation, which typically grows upright from the surface of the ground-dwelling gametophyte. Because the sporophyte is above the ground, it effectively disperses the spores it produces. These spores are carried by air currents and when they land, they germinate into new gametophyte plants.

Scientists now agree that bryophytes consist of three quite distinct clades of relatively unspecialized plants: liverworts, mosses, and hornworts. Like ferns and certain other vascular (tracheophyte) plants, bryophytes require water (such as rainwater) to reproduce sexually, tracing back to their aquatic origins. The sperm must swim to the egg. It is not surprising that bryophytes are especially common in moist places, both in the tropics and temperate regions. They are commonly seen on damp forest floors and by stream beds.

Liverworts are an ancient phylum

The Old English word *wyrt* means "plant" or "herb." Some common liverworts have flattened gametophytes with lobes resembling those of liver—hence the name "liverwort." Although the lobed liverworts are the best-known representatives of this phylum, they constitute only about 20% of the species (figure 30.3). The other 80% are leafy and superficially resemble mosses. The gametophytes are flat instead of upright, with single-celled rhizoids that aid in absorption. Although rhizoids function like roots, they are not organs because they are unicellular.

Some liverworts have air chambers containing rows of photosynthetic cells, each chamber having a pore at the top to facilitate gas exchange. Unlike stomata, the pores are fixed open and cannot close.

Sexual reproduction in liverworts is similar to that in mosses. Lobed liverworts may form gametangia in umbrella-like structures. In this case, the sporophyte is suspended above the ground by the gametophyte. This is in contrast to most bryophytes, in which the sporophyte itself forms a stalk above the ground to aid in spore dispersal. Asexual reproduction occurs when lens-shaped pieces of tissue that are released from the gametophyte grow to form new gametophytes.

Mosses have rhizoids and water-conducting tissue

Unlike other bryophytes, the gametophytes of mosses typically consist of small, leaflike photosynthetic structures (not true leaves, which contain vascular tissue) arranged around a stemlike axis (figure 30.4); the axis is anchored to its substrate by means of rhizoids. Each rhizoid consists of several cells that absorb water, but not nearly the volume of water that is absorbed by a vascular plant root.

Moss leaflike structures have little in common with leaves of vascular plants, except for the superficial appearance of the green, photosynthetic, flattened blade and slightly thickened midrib that runs lengthwise down the middle. Only one cell layer thick (except

Figure 30.3 A common liverwort, *Marchantia* (phylum Hepaticophyta). The microscopic sporophytes are formed by fertilization within the tissues of the umbrella-shaped structures that arise from the surface of the flat, green, creeping gametophyte.

Figure 30.4 A hair-cup moss, *Polytrichum* (phylum Bryophyta). The leaflike structures belong to the gametophyte. Each of the yellowish brown stalks with a capsule, or sporangium, at its summit is a sporophyte.

at the midrib), they lack vascular strands and stomata, and all the cells are haploid. However, mosses do have stomata on the capsule portion of the sporophyte and because of that are the basal land group with stomata.

Although mosses lack a vascular system, water may be passively carried up a strand of specialized cells in the center of a moss gametophyte. Some mosses also have specialized food-conducting cells surrounding those that conduct water.

Moss reproduction

Multicellular gametangia (gamete-producing structures) are formed at the tips of the leafy moss gametophytes (figure 30.5). Female gametangia (**archegonia**) may develop either on the same gametophyte as the male gametangia (**antheridia**) or on separate plants, depending on the species. A single egg is produced in the swollen lower part of an archegonium, whereas numerous flagellated sperm are produced in an antheridium. Because gametophytes are haploid, eggs and sperm are produced by mitosis (not meiosis) in the gametangia.

When sperm are released from an antheridium, they swim with the aid of flagella through a film of dew or rainwater to the archegonia. One sperm (which is haploid) unites with an egg (also haploid), forming a diploid zygote. The zygote divides by mitosis and develops into the sporophyte, a slender, stalk with a swollen capsule, the *sporangium*, at its tip. The base of the sporophyte is embedded in gametophyte tissue, its nutritional source.

The sporangium is often cylindrical or club-shaped. Spore mother cells within the sporangium undergo meiosis, each producing four genetically different haploid spores. In many mosses at maturity, the top of the sporangium pops off, and the spores are released. A spore that lands in a suitable damp location may germinate and grow via mitosis, into a threadlike structure, which branches to form rhizoids and "buds" that grow upright. Each bud develops into a new leafy gametophyte plant.

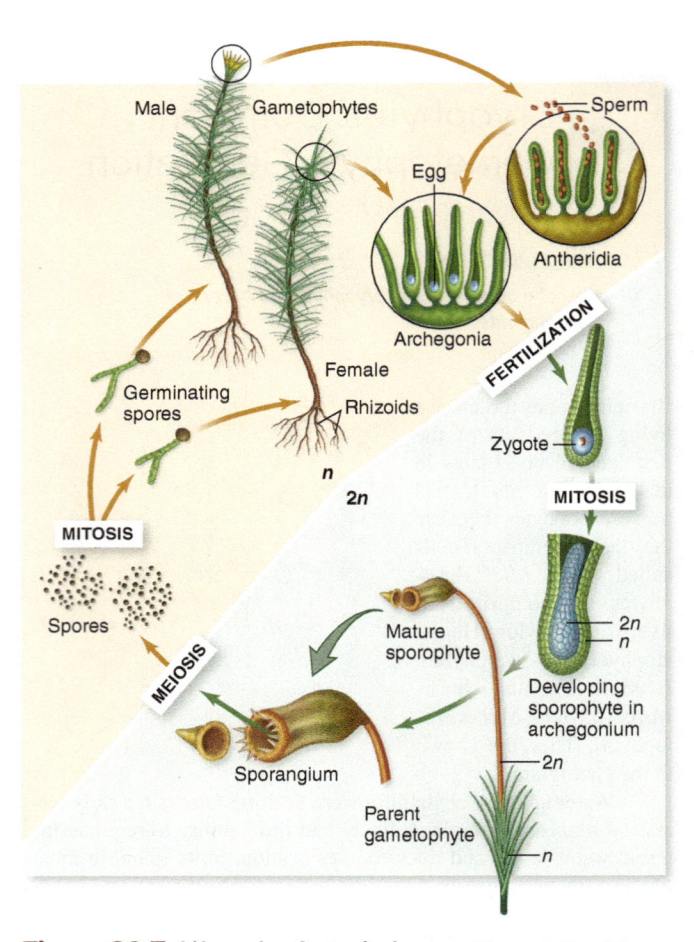

Figure 30.5 Life cycle of a typical moss. The majority of the life cycle of a moss is in the haploid state. The leafy gametophyte (*n*) is photosynthetic, but the smaller sporophyte (2*n*) is not and is nutritionally dependent on the gametophyte. Water is required to carry sperm to the egg.

Moss distribution

In the Arctic and the Antarctic, mosses are the most abundant plants. The greatest diversity of moss species, however, is found in the tropics. Many mosses are able to withstand prolonged periods of drought, although mosses are not common in deserts.

Most mosses are highly sensitive to air pollution and are rarely found in abundance in or near cities or other areas with poor air quality. Some mosses, such as the peat mosses *(Sphagnum),* can absorb up to 25 times their weight in water and are valuable commercially in potting soils or as a fuel when dry.

The moss genome

Moss plants can survive extreme water loss—an adaptive trait in the early colonization of land that has been lost from vegetative tissues of tracheophytes. Desiccation tolerance and phylogenetic position were among the traits that led researchers to choose the moss *Physcomitrella patens* as the first nontracheophyte plant to sequence. DNA sequence data reveal that the loss of genes associated with a watery life, including flagellar arms, were lost in the earliest common ancestor of the land plants. Genes associated with tolerance of terrestrial stresses, including temperature and water availability, are absent in the green alga *Chlamydomonas*, but present in moss. For example, the plant hormone abscisic acid (ABA) is important in stress responses in moss and other land plants and genes needed for ABA signaling are not found in algae.

Hornworts are the sister group to tracheophytes

The origin of hornworts (phylum Anthocerotophyta) is a puzzle. They are most likely among the earliest land plants, yet the earliest hornwort fossil spores date from the Cretaceous period (65 to 145 MYA), when angiosperms were emerging. Only about 200 hornwort species are found worldwide. However, the hornworts form an important link in the study of plant evolution because they are the sister group to the tracheophytes. This group is likely to provide insights into the transition from the dominant gametophyte generation in the bryophtyes to the dominant sporophyte generation in the tracheophytes.

Among the bryophytes sporophytes, those of the hornworts are most similar to vascular plant sporophytes. Hornwort sporophytes look like green horns, rising from gametophytes usually less than 2 cm in diameter (figure 30.6). The sporophyte base is embedded in gametophyte tissue, from which it derives some of its nutrition. It has stomata to regulate gas exchange, is photosynthetic, and provides much of the energy needed for growth and reproduction. Meiosis occurs within the "horn" and spores are released along its entire length. Hornwort cells usually have a single large chloroplast similar to that found in algae. Hornwort gametophytes typically live in symbiotic association with cyanobacteria, which provide them with nitrogen (see chapter 28).

Learning Outcome Review 30.2

The bryophytes exhibit adaptations to terrestrial life. Moss adaptations include rhizoids to anchor the moss body and to absorb water, and water-conducting tissues. Mosses are found in a variety of habitats, and some can survive droughts. Liverworts are the oldest bryophyte phylum, whereas hornworts are the closest relatives to the tracheophytes.

- *What might account for the abundance of mosses in the Arctic and Antarctic?*

Figure 30.6 Hornworts (phylum Anthocerotophyta). Hornwort sporophytes are seen in this photo. Unlike the sporophytes of other bryophytes, most hornwort sporophytes are photosynthetic.

30.3 Tracheophyte Plants: Dominant Sporophyte Generation

Learning Outcomes
1. Explain the evolutionary significance of tracheids.
2. Evaluate the significance of roots, stems, and leaves in the tracheophytes.

Tracheophytes, also known as vascular plants, first appeared about 410 MYA. The first tracheophytes with a relatively complete record belonged to the phylum Rhyniophyta. We are not certain what the earliest of these vascular plants looked like, but fossils of *Cooksonia* provide some insight into their characteristics (figure 30.7).

Cooksonia, the first known vascular land plant, appeared in the late Silurian period about 420 MYA, but is now extinct. It was successful partly because it encountered little competition as it spread out over vast tracts of barren land. The plants were only a few centimeters tall and had no roots or leaves. They consisted of little more than a branching axis, the branches forking evenly and expanding slightly toward the tips. They were **homosporous** (producing only one type of spore). Sporangia formed at branch tips. Other ancient vascular plants that followed evolved more complex arrangements of sporangia. A major feature of the tracheophytes is the dominant sporophyte generation. We do not yet know how the transition from the dominant gametophyte in the bryophytes to the dominant sporophyte in the tracheophytes occurred. However, the evolutionary advantage of having two sets of chromosomes

Figure 30.7 *Cooksonia*, the first known vascular land plant. This fossil represents a plant that lived some 420 MYA. *Cooksonia* belongs to phylum Rhyniophyta, consisting entirely of extinct plants. Its upright, branched stems, which were no more than a few centimeters tall, terminated in sporangia, as seen here. It probably lived in moist environments such as mudflats, had a resistant cuticle, and produced spores typical of vascular plants.

in the conspicuous generation is significant, as discussed in section 30.1.

Vascular tissue allows for distribution of nutrients

Cooksonia and the other early tracheophytes became successful colonizers of the land by developing efficient water- and food-conducting systems called *vascular tissues*. These tissues consist of strands of specialized cells that form a network throughout a plant, extending from the roots, through the stems, and into the leaves. One type of vascular tissue, **xylem,** conducts water and dissolved minerals upward from the roots; another type of tissue, **phloem,** conducts sucrose and hormones throughout the plant. In the early vascular plants, tracheids were the only cells that conducted water in xylem tissue. Later, plants developed even more effective water transport cells, called vessel elements. Vascular tissue allows tracheophytes to achieve great size, like the coastal redwood that can grow to over 300 feet high. It develops in the sporophyte, but (with a few exceptions) not in the gametophyte. It is important to note here that the sporophyte is the dominant generation in the tracheophytes. A green moss plant is a gametophyte, whereas a green tree, for example, is a sporophyte. (Vascular tissue structure is discussed more fully in chapter 37.) Other features found in all vascular plant sporophytes include a waxy cuticle on above-ground parts and pores called stomata. Both features reduce water loss by land plants.

> **Inquiry question** Explain why tracheophytes may have had a selective advantage over bryophytes during the evolution of land plants.

Tracheophytes are grouped in three clades

Three clades of vascular plants exist today: (1) lycophytes (club mosses), (2) pterophytes (ferns and their relatives), and (3) seed plants. The first two clades are explored in this chapter (table 30.1). Advances in molecular systematics have changed the way we view the evolutionary history of vascular plants. Whisk ferns and horsetails were long believed to be distinct phyla that were transitional between bryophytes and vascular plants. Phylogenetic evidence now shows they are the closest living relatives to ferns, and they are grouped as pterophytes.

Tracheophytes dominate terrestrial habitats everywhere, except for the highest mountains and the tundra. The haplodiplontic life cycle persists, but the gametophyte has been reduced in size

TABLE 30.1 The Phyla of Extant Seedless Vascular Plants

Phylum	Examples	Key Characteristics	Approximate Number of Living Species
Lycophyta	Club mosses	Homosporous or heterosporous. Sperm motile. External water necessary for fertilization. About 17 genera.	1,275
Pterophyta	Ferns	Primarily homosporous (a few heterosporous). Sperm motile. External water necessary for fertilization. Leaves uncoil as they mature. Sporophytes and virtually all gametophytes are photosynthetic. About 365 genera.	11,000
	Horsetails	Homosporous. Sperm motile. External water necessary for fertilization. Stems ribbed, jointed, either photosynthetic or nonphotosynthetic. Leaves scalelike, in whorls; nonphotosynthetic at maturity. One genus.	15
	Whisk ferns	Homosporous. Sperm motile. External water necessary for fertilization. No differentiation between root and shoot. No leaves; one of the two genera has scalelike extensions and the other leaflike appendages.	6

relative to the sporophyte during the evolution of tracheophytes. A similar reduction in multicellular gametangia has occurred as well. The reduction in the size of the gametophyte is a major theme in the evolution of land plants.

Stems evolved prior to roots

Fossils of early vascular plants reveal stems, but no roots or leaves. The earliest vascular plants, including *Cooksonia*, had transport cells in their stems, but the lack of roots limited the size of these plants in two ways. First, roots anchor plants to the ground, keeping them from falling over. Tall plants need strong roots. Second, a large plant has a large surface area for water loss. An extensive root system is needed to keep up with water demand.

Roots provide structural support and transport capability

True roots are found only in the tracheophytes. Other, somewhat similar structures enhance either transport or support in nontracheophytes, but only roots have a dual function: providing both transport and support. Lycophytes diverged from other tracheophytes before roots appeared, based on fossil evidence. It appears that roots evolved at least two separate times.

Leaves evolved more than once

Leaves increase surface area of the sporophyte, enhancing photosynthetic capacity. Lycophytes have single vascular strands supporting relatively small leaves called lycophylls. True leaves, called euphylls, are found only in ferns and seed plants, having distinct origins from lycophylls (figure 30.8). Lycophylls may have resulted from vascular tissue penetrating small, leafy protuberances on stems. Euphylls most likely arose from branching stems that became webbed with leaf tissue.

Data analysis Compare the amount of time that passed between the early diversification of land plants, the origins of tracheophytes, and the origins of leaves.

Seeds are another innovation in the advanced tracheophyte phyla

Seeds are highly resistant structures well suited to protecting the plant embryo from environmental stresses and to some extent from predators. In addition, almost all seeds contain a supply of food for the young plant. Lycophytes and pterophytes do not have seeds. They are dispersed via haploid spores.

Fruits in the flowering plants (angiosperms) add a layer of protection to seeds and have adaptations that assist in seed dispersal, expanding the potential range of the species. Flowers allow plants to secure the benefits of wide outcrossing in promoting genetic diversity, as discussed in chapter 31. Before moving on to the specifics of lycophytes and pterophytes, review the evolutionary history of terrestrial innovations in the land plants illustrated in figure 30.9.

Learning Outcomes Review 30.3

Most tracheophytes have well-developed vascular tissues, including tracheids, that enable efficient delivery of water and nutrients throughout the organism. They also exhibit specialized roots, stems, leaves, cuticles, and stomata. Many produce seeds, which protect and nourish embryos.

- Why would vascular tissue be prevalent in the sporophyte, but not the gametophyte, generation?

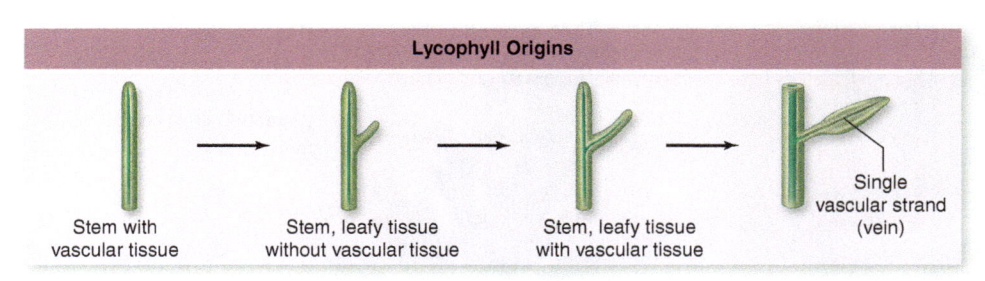

Figure 30.8 The evolution of leaves independently in two plant lineages. Lycophylls are small simple leaves, whereas those of euphylls may be large with a complex system of veins.

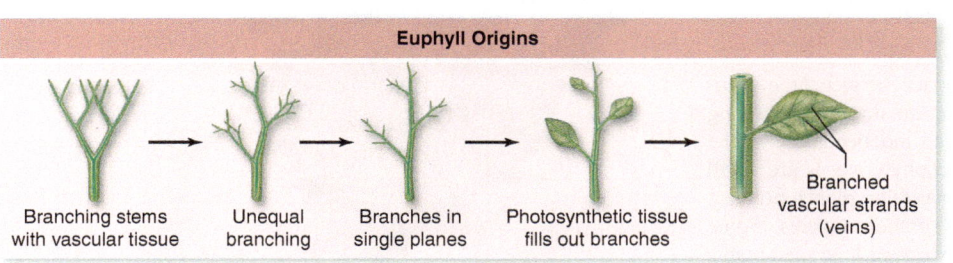

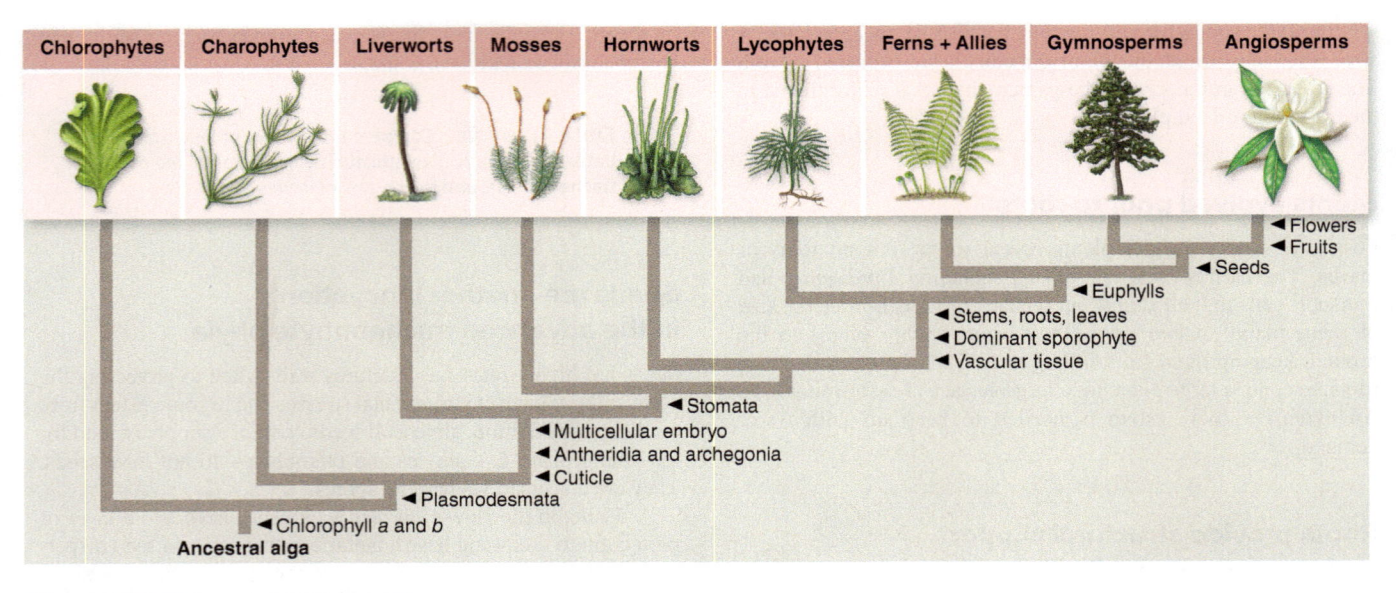

Figure 30.9 Land plant innovations.

30.4 Lycophytes: Divergent from Main Lineage of Vascular Plants

Learning Outcomes
1. Explain features that differentiate lycophytes from bryophytes.
2. Distinguish between lycophytes and other vascular plants.

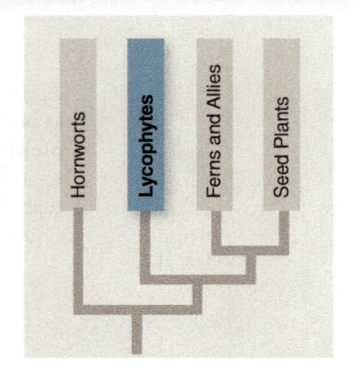

The earliest vascular plants lacked seeds. Members of four phyla of living vascular plants also lack seeds, as do at least three other phyla known only from fossils. As we explore the adaptations of the vascular plants, we focus on both reproductive strategies and the advantages of increasingly complex transport systems.

The lycophytes (club mosses) are the sister group to all other vascular plants (figure 30.10). Despite the early divergence between the lycophytes and all other vascular plants, there are similarities among the groups. For example, leaves developed independently in the lycophytes and other vascular plants. Lycophyte leaves are small unbranched veins, whereas the higher vascular plants produce larger leaves with a complex pattern of branched veins. Some ancient club mosses looked like the trees we see in the seed plants of today. The treelike growth form is another example of independent origins of similar features in the lycophytes and the other vascular plants. Today, club mosses are worldwide in distribution but are most abundant in the tropics and moist temperate regions. However, modern club mosses are small. None have the treelike growth form of their ancient ancestors.

Club mosses superficially resemble true mosses, but once their internal vascular structure and reproductive processes became known, it was clear that they are unrelated to mosses. The sporophyte stage is the dominant (obvious) stage. Specialized leaves bear spores, which are produced by meiosis. Each haploid spore germinates into a small free-living (in most lycophytes) gametophyte

Figure 30.10
A club moss. *Selaginella moellendorffii's* sporophyte generation grows on moist forest floors.

that produces both sperm and egg cells. The gametophyte of some lycophytes is photosynthetic, whereas that of other species grows below-ground and obtains its nourishment from a symbiotic association with soil fungi. After fertilization, the small sporophyte remains nutritionally dependent on the gametophyte until its green leaves develop and start to carry out photosynthesis.

The lycophyte *Selaginella moellendorffii* is the first seedless vascular plant with a fully sequenced genome. A few clues to the evolution of vascular plants, hidden in the genome, emerged in comparisons with genomes of flowering plants. They supported the independent origins of leaflike structures in different vascular plant lineages. Genome differences also reflect differences in developmental pathways leading to reproductive maturity in the sporophyte generation in lycophytes and flowering plants.

Comparing predicted proteins in the *Chlamydomonas* (green alga), *Physcomitrella* (moss), and *Selaginella* with 15 angiosperms revealed 3814 gene families that all the green plants share—the essential instructions for building a green plant. About 3000 new genes were needed in the transition from the single-celled green alga to the multicellular moss, but only 516 genes were needed for the transition from nonvascular to vascular plants. This is a first step in sorting out the evolutionary steps that led to the vascular plants.

Learning Outcomes Review 30.4

Lycophytes are basal to all other vascular plants. Although they superficially resemble bryophytes, they have a dominant sporophyte generation, they contain tracheid-based vascular tissues, and their reproductive cycle is like that of other vascular plants; however, their leaves are simple, with unbranched veins.

- *What factors may have led to the independent origins of similar growth forms in the lycophytes and the other vascular plants?*

30.5 Pterophytes: Ferns and Their Relatives

Learning Outcomes

1. List the features exhibited by pterophytes.
2. Contrast pterophyte and moss sporophytes.

The phylogenetic relationships among ferns and their near relations is intriguing. A common ancestor gave rise to two clades: One clade diverged to produce a line of ferns and horsetails; the other diverged to yield another line of ferns and whisk ferns—ancient-looking plants.

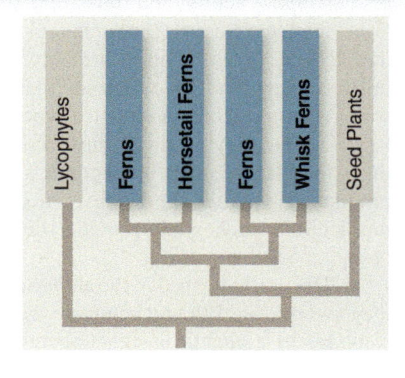

Whisk ferns and horsetails are close relatives of ferns. Like lycophytes and bryophytes, they all form antheridia (containing sperm) and archegonia (containing eggs). Free water is required for the process of fertilization, during which the sperm, which have flagella, swim to and unite with the eggs. In contrast, most seed plants have nonflagellated sperm.

Whisk ferns lack roots and leaves

In whisk ferns, which occur in the tropics and subtropics, the sporophytic generation consists merely of evenly forking green stems without roots (figure 30.11). The two or three species of the genus *Psilotum* do, however, have tiny, green, spirally arranged flaps of leaflike tissue lacking veins and stomata. Another genus, *Tmesipteris,* has appendages that look even more like leaves. Currently, systematists believe that whisk ferns lost their leaves and roots when they diverged from others in the fern lineage.

Given the simple structure of whisk ferns, it was particularly surprising to discover that they are monophyletic with ferns. The gametophytes of whisk ferns are essentially colorless and very small. As with some lycophytes, these gametophytes form symbiotic associations with fungi, which furnish their nutrients. Some develop elements of vascular tissue and have the distinction of being the only gametophytes known to do so.

Horsetails have jointed stems with brushlike leaves

The 15 living species of horsetails are members of a single genus, *Equisetum.* Fossil forms of *Equisetum* extend back 300 million years to an era when some of their relatives were treelike. Today, they are widely scattered around the world, mostly in damp places. Some that grow among the coastal redwoods of California may reach a height of 3 m, but most are less than a meter tall (figure 30.12).

Figure 30.11 A whisk fern. Whisk ferns have no roots or leaves. The green, photosynthetic stems have yellow sporangia attached. Meiosis in the sporangia produces haploid spores, which are released and germinate into gametophytes.

**Figure 30.12
A horsetail, *Equisetum telmateia*.** This species forms two kinds of erect stems; one is green and photosynthetic, and the other, which terminates in a spore-producing "cone," is mostly light brown.

**Figure 30.13
A tree fern (phylum Pterophyta) in the forests of Malaysia.** The ferns are by far the largest group of seedless vascular plants.

Horsetail sporophytes consist of ribbed, jointed, photosynthetic stems that arise from branching underground *rhizomes* with roots at their nodes. A whorl of nonphotosynthetic, scalelike leaves emerges at each node. The hollow stems have silica deposits in the epidermal cells. Horsetails are also called scouring rushes because pioneers of the American West used them to scrub pans. Conelike structures on horsetail sporophytes produce haploid spores via meiosis. Spores are carried away by air currents and, when they land on the ground, they germinate into photosynthetic gametophtyes. A horsetail gametophyte looks like a small lobed liverwort plant. Each gametophyte produces both eggs and flagellated (swimming) sperm, which unite at fertilization. The young sporophyte will remain dependent on the gametophyte until it produces photosynthetic tissue.

Ferns have fronds that bear sori

Ferns are the most abundant group of seedless vascular plants, with about 11,000 living species. Recent research indicates that they may be the closest relatives of the seed plants.

Rain forests and swamps of fern trees growing in the eastern United States and Europe over 300 MYA formed the coal currently being mined. Today, ferns flourish in a wide range of habitats throughout the world; however, about 75% of the species occur in the tropics.

The conspicuous sporophytes may be less than a centimeter in diameter (as in small aquatic ferns such as Azolla), or more than 24 m tall, with leaves up to 5 m or longer in the tree ferns (figure 30.13). The sporophytes and the much smaller gametophytes, which rarely reach 6 mm in diameter, are both photosynthetic.

The fern life cycle (figure 30.14) differs from that of a moss primarily in the much greater development, independence, and dominance of the fern's sporophyte. The fern sporophyte is structurally more complex than the moss sporophyte, having vascular tissue and well-differentiated roots, stems, and leaves. The gametophyte, however, lacks the vascular tissue found in the sporophyte.

Fern morphology

Fern sporophytes, like horsetails, have rhizomes (horizontal underground stems). Leaves, referred to as *fronds,* usually develop at the tip of the rhizome as tightly rolled-up coils ("fiddleheads") that unroll and expand (figure 30.15). The tight coil of the fiddlehead allows the delicate fern fronds to push out of the soil from their base on underground rhizomes. Once in the air, the fronds can unfurl without being damaged. Fiddleheads are considered a delicacy in several cuisines, but some species contain secondary compounds linked to stomach cancer.

Many fronds are highly dissected and feathery, making the ferns that produce them prized as ornamental garden plants. Some ferns, such as *Marsilea,* have fronds that resemble a four-leaf clover, but *Marsilea* fronds still begin as coiled fiddleheads. Other ferns produce a mixture of photosynthetic fronds and nonphotosynthetic reproductive fronds that tend to be brownish in color.

Fern reproduction

Ferns produce distinctive sporangia, usually in clusters called **sori** (singular, *sorus*), typically on the underside of the fronds. Sori are often protected during their development by a transparent, umbrella-like covering. (At first glance, one might mistake the sori for an infection on the plant.) Diploid spore mother cells in each sporangium undergo meiosis, producing haploid spores.

At maturity, the spores are catapulted from the sporangium by a snapping action, and those that land in suitable damp locations may germinate, producing photosynthetic gametophytes that are often heart-shaped, are only one cell layer thick (except in the center), and have rhizoids that anchor them to their substrate. These rhizoids are not true roots because they lack vascular tissue, but they do aid in transporting water and nutrients from the soil. Flask-shaped archegonia and globular antheridia are produced on either the same or a different gametophyte. The multicellular archegonia provide some protection for the developing embryo.

The sperm formed in the antheridia have flagella, with which they swim toward the archegonia when water is present, often in response to a chemical signal secreted by the archegonia. One

Figure 30.14
Life cycle of a typical fern. Both the gametophyte (*n*) and sporophyte (2*n*) are photosynthetic and can live independently. Water is necessary for fertilization. Sperm are released on the underside of the gametophyte and swim in moist soil to neighboring gametophytes. Spores are dispersed by wind.

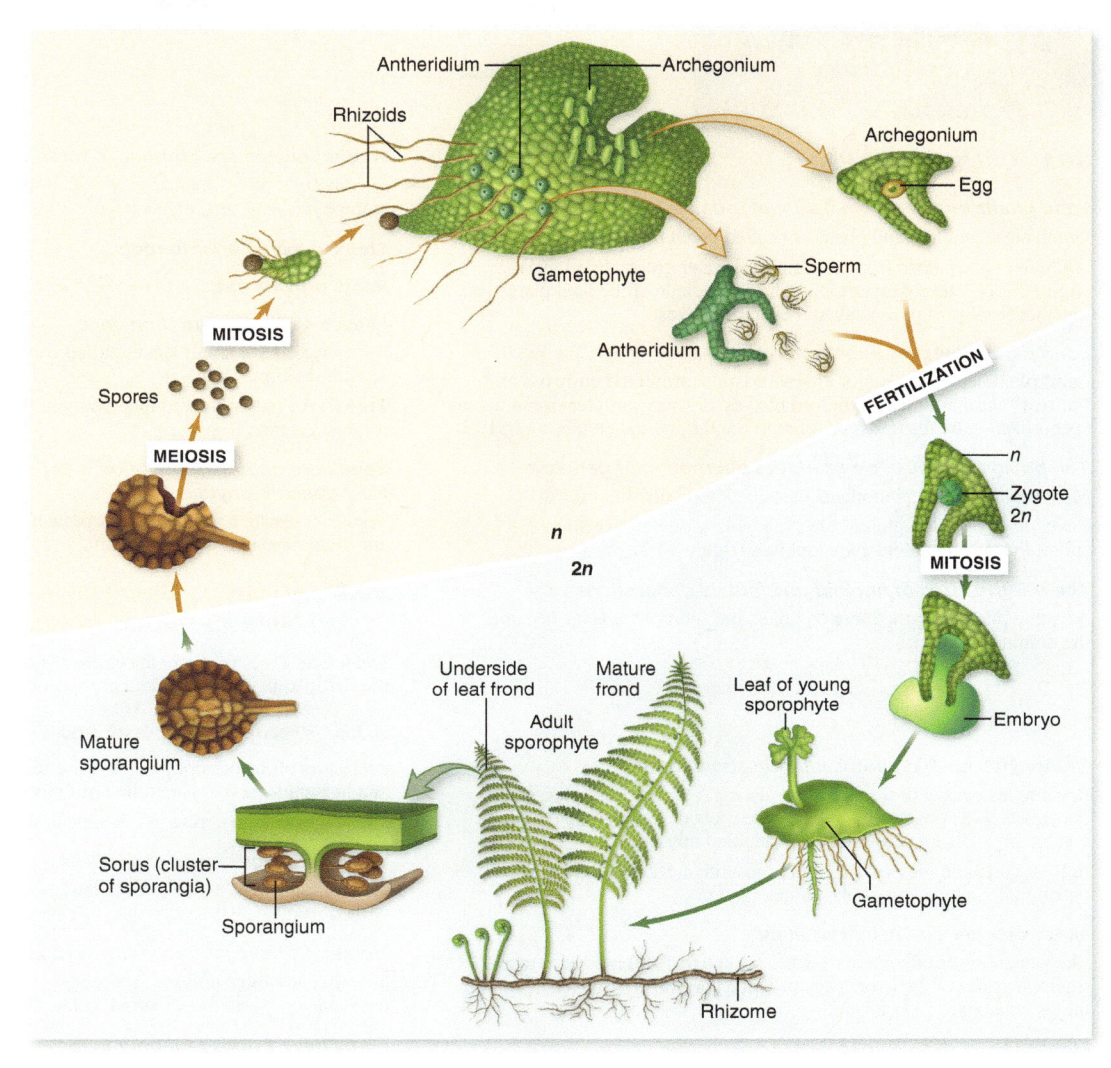

Figure 30.15 Fern "fiddlehead." Fronds develop in a coil and slowly unfold on ferns, including the tree fern fronds in these photos.

sperm unites with the single egg toward the base of an archegonium, forming a zygote. The zygote then develops into a new sporophyte, completing the life cycle (see figure 30.14).

The developing fern embryo has substantially more protection from the environment than a charophyte zygote, but it cannot enter a dormant phase to survive a harsh winter the way a seed plant embryo can.

Learning Outcomes Review 30.5

Ferns and their relatives have a large and conspicuous sporophyte with vascular tissue. Many have well-differentiated roots, stems, and leaves (fronds). The gametophyte generation is small and green, but lacks vascular tissue.

■ *Why are the fiddleheads of many fern species poisonous?*

chapter 30 *Seedless Plants* 599

Chapter Review

30.1 Origin of Land Plants

Land plants evolved from freshwater algae.
Green algae and the land plants are called green plants.
All green plants arose from a single freshwater green algal species (figure 30.1). The charophytes are the sister clade of the land plants and together these groups are called the streptophytes.

Land plants have adapted to terrestrial life.
Land plants have two major characteristics: protected embryos and multicellular haploid and diploid phases. A waxy cuticle, stomata, and specialized cells for transport of water and minerals enhance survival.

The haplodiplontic cycle produces alternation of generations.
Most multicellular green plants have haplodiplontic life cycles.
Plants have a haplodiplontic life cycle with multicellular diploid sporophytes and haploid gametophytes (figure 30.2).

The relative sizes of haploid and diploid generations vary.
As some plants became more complex, the sporophyte stage became the dominant phase.

30.2 Bryophytes: Dominant Gametophyte Generation

Bryophytes have a photosynthetic gametophyte.
Bryophytes consist of three distinct clades: liverworts, mosses, and hornworts. Bryophytes do not have true roots or tracheids, but do have conducting cells for movement of water and nutrients.
In liverworts and mosses, the nonphotosynthetic sporophyte is nutritionally dependent on the gametophyte.

Liverworts are an ancient phylum.
The gametophyte of some liverworts is flattened and has lobes that resemble those of the liver. They produce upright structures that contain the gametangia.

Mosses have rhizoids and water-conducting tissue.
Mosses exhibit alternation of generations and have widespread distribution. Many are able to withstand droughts.

Hornworts are the sister group to tracheophytes.
Stomata in the sporophyte can open and close to regulate gas exchange. The sporophyte is also photosynthetic.

30.3 Tracheophyte Plants: Dominant Sporophyte Generation (table 30.1)

Vascular tissue allows for distribution of nutrients.
The evolution of tracheids allowed more efficient vascular systems to develop. This vascular tissue develops in the sporophyte.
Vascular plants have a much reduced gametophyte.

Tracheophytes are grouped in three clades.
The vascular plants found today exist in three clades: lycophytes, pterophytes, and seed plants (figure 30.9).

Stems evolved prior to roots.

Roots provide structural support and transport capability.

Leaves evolved more than once.
Lycophytes have small leaves called lycophylls that lack vascularization.
True leaves (euphylls) are found only in ferns and seed plants and have origins different from lycophylls.

Seeds are another innovation in the advanced tracheophyte phyla.
Seeds are resistant structures that protect the embryo from desiccation and to some extent from predators.

30.4 Lycophytes: Divergent from Main Lineage of Vascular Plants

Lycophyte ancestors were the earliest vascular plants and were among the first plants to have a dominant sporophyte generation.

30.5 Pterophytes: Ferns and Their Relatives

Ancestors of the pterophytes gave rise to two clades: one line of ferns and horsetails, and a second line of ferns and whisk ferns.
Pterophytes require water for fertilization and are seedless.

Whisk ferns lack roots and leaves.
The sporophyte of a whisk fern consists of evenly forking green stems without roots.

Horsetails have jointed stems with brushlike leaves.
Scalelike leaves of horsetail sporophytes emerge in a whorl. The stems have silica deposits in epidermal cells of their ribs.

Ferns have fronds that bear sori.
The leaves of ferns, called fronds, develop as tightly rolled coils that unwind to expand. Sporangia called sori develop on the underside of the fronds. The gametophyte is often heart shaped and can live independently.

Review Questions

UNDERSTAND

1. Which of the following plant structures is NOT matched to its correct function?
   a. Stomata—allow gas transfer
   b. Tracheids—allow the movement of water and minerals
   c. Cuticle—prevents desiccation
   d. All of the choices are matched correctly.

2. Which of the following genera most likely directly gave rise to the land plants?
   a. *Volvox*
   b. *Chlamydomonas*
   c. *Ulva*
   d. *Chara*

3. Which of the following would NOT be found in a bryophyte?
   a. Mycorrhizal associations
   b. Rhizoids
   c. Tracheid cells
   d. Photosynthetic gametophytes

4. Which of the following statements is correct regarding the bryophytes?
   a. The bryophytes represent a monophyletic clade.
   b. The sporophyte stage of all bryophytes is photosynthetic.
   c. Archegonium and antheridium represent haploid structures that produce reproductive cells.
   d. Stomata are common to all bryophytes.

5. Evolutionary innovations that increase desiccation tolerance include
   a. waxy cuticles.
   b. abscisic acid–signaling pathways.
   c. rhizoids.
   d. All of the choices are correct.

6. Which of the following statements about the pterophytes is accurate?
   a. Horsetails and whisk ferns form a single clade.
   b. Ferns form a single clade.
   c. Whisk ferns have euphylls.
   d. All pterophytes have a dominant sporophyte generation.

APPLY

1. Compare what happens to a spore mother cell as it gives rise to a spore with what happens to a spore as it gives rise to a gametophyte.
   a. The spore mother cell and the spore both go through meiosis.
   b. The spore mother cell and the spore both go through mitosis.
   c. The spore mother cell goes through mitosis, and the spore goes through meiosis.
   d. The spore mother cell goes through meiosis, and the spore goes through mitosis.

2. How could a plant without roots obtain sufficient nutrients from the soil?
   a. It cannot; all land plants have roots.
   b. Mycorrhizal fungi associate with the plant and assist with the transfer of nutrients.
   c. Charophytes associate with the plant and assist with the transfer of nutrients.
   d. It relies on its xylem in the absence of a root.

3. A major innovation of land plants is embryo protection. How is a moss embryo protected from desiccation?
   a. By the seed
   b. By the antheridium
   c. By the archegonium
   d. By the lycophyll

4. In comparing the *Selaginella* and *Physcomitrella* genomes, you would expect to find that
   a. both have genes needed for flagellar arms.
   b. *Selaginella* has 3000 new genes not found in *Physcomitrella*.
   c. *Selaginella*, but not *Physcomitrella*, has abscisic acid genes and other abiotic stress genes.
   d. some of the novel genes in *Selaginella* encode proteins necessary for tracheid development.

5. The following evolutionary trends are seen in the seedless land plants.
   a. Gametophytes became photosynthetic.
   b. A key innovation accompanying the appearance of pterophytes was the haplodiplontic life cycle.
   c. The gametophyte generation became the dominant generation.
   d. There is increased protection of the sporophyte generation in its early developmental stages.

6. Identify which of the following statements is true.
   a. Moss and lycophytes have leaves.
   b. Leaves of lycophytes and pterophytes have different origins.
   c. Leaves were a bryophyte innovation.
   d. Leaf polarity genes in *Selaginella* support the hypothesis that leaves evolved one time in all the land plants.

SYNTHESIZE

1. You have access to the sequenced genomes for moss and the lycophyte *Selaginella*. Your goal in analyzing the data is to write a ground-breaking paper that answers an important question about the evolution of land plants. What question would you try to answer?

2. Would you expect the sporophyte generation of a moss or a fern to have more mitotic divisions? Why?

3. Imagine hypothetical moss and fern trees, each 10 m tall. Which would face the greater barriers to sexual reproduction? Why?

CHAPTER 31

Seed Plants

Chapter Contents

31.1 The Evolution of Seed Plants
31.2 Gymnosperms: Plants with "Naked Seeds"
31.3 Angiosperms: The Flowering Plants
31.4 Seeds
31.5 Fruits

Introduction

The history of the land plants is replete with evolutionary innovations allowing the ancestors of aquatic algae to colonize the harsh and varied terrestrial terrains. Early innovations made survival on land possible, later followed by an explosion of plant life that continues to change the land and atmosphere, and support terrestrial animal life. Seed-producing plants have come to dominate the terrestrial landscape over the last several hundred million years. Much of the remarkable success of seed plants can be attributed to the evolution of the seed, an innovation that protects and provides food for delicate embryos. Seeds allow embryos to "stop the clock" and germinate after a harsh winter or extremely dry season has passed. Fruits, a later innovation, enhanced the dispersal of embryos across a broader landscape. Within the seed plants, the flowering plants have coevolved intricate relationships with animals to enhance sexual reproduction, as well as dispersal of the next generation.

31.1 The Evolution of Seed Plants

Learning Outcomes

1. List the evolutionary advantages of seeds.
2. Distinguish between pollen and sperm in seed plants.

Numerous evolutionary solutions to terrestrial challenges have resulted in over 400,000 species of seed plants dominating all terrestrial communities today. Collectively these seed plants affect almost every aspect of our lives, from improving environmental quality to providing pharmaceuticals, food, fuels, building materials, and clothing.

Seed plants appear to have evolved from spore-bearing plants known as progymnosperms. Progymnosperms shared several features with modern gymnosperms, including secondary vascular tissues (which allow for an increase in girth). Some progymnosperms had leaves. Their reproduction was very simple, and it is not certain which group of progymnosperms gave rise to seed plants.

Biologists have long been intrigued by the rapid success and diversity of seed plants, particularly the diversification of the flowering plants (angiosperms). Likely the diversity can be attributed to a confluence of features, with the evolution of the seed being one important event. Phylogenetic analysis of seedless and seed plants with sequenced genomes revealed vast numbers of gene duplications around 319 MYA, when the early seed plants (ancestral to gymnosperms and angiosperms) began to diversify from their seedless plant ancestor, and again 192 MYA when the earliest angiosperms (flowering plants) were appearing. The gene duplications are at such a scale that the most logical explanation is that whole-genome duplications occurred at the origins of both seed plants and angiosperms. As discussed in chapter 24, whole-genome duplications can accelerate evolution through mutations in the extra copies of genes.

The seed protects the embryo

From an evolutionary and ecological perspective, the seed represents an important advance. The embryo is protected by an extra layer or two of sporophyte tissue called the **integument,** creating the **ovule** (figure 31.1). Within the ovule, meiosis occurs in the megasporangium, producing a haploid megaspore. The megaspore divides by mitosis to produce a female gametophyte carrying the female gamete, an egg. The egg combines with the male gamete, a sperm, resulting in the zygote. The single-celled zygote divides by mitosis to produce the young sporophyte, an embryo. Seeds also contain a food supply for the developing embryo.

During development, the integuments harden to produce the seed coat, which protects the embryo. The seed is easily dispersed. Perhaps even more significantly, the presence of seeds introduces into the life cycle a dormant phase that allows the embryo to survive until environmental conditions are favorable for further growth.

Water is not required to transport the male gametophyte to the female gametophyte

Seed plants produce two kinds of gametophytes—male and female—each of which consists of just a few cells. A consistent feature in the evolution of plants is a reduction in the size of the gametophyte and a corresponding increase in dominance of the sporophyte generation.

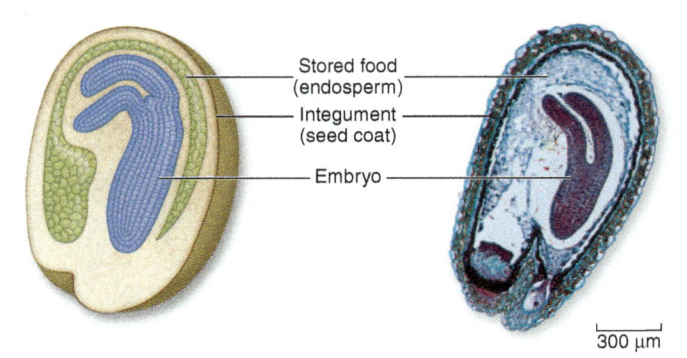

Figure 31.1 Cross-section of an ovule.

A pollen grain is a multicellular male gametophyte carrying the male gamete, a sperm cell. Pollen grains are carried to the female gametophyte by wind or by a pollinator. In some seed plants, the sperm moves toward the female gametophyte through a growing **pollen tube.** This eliminates the need for external water. In contrast to the seedless plants, the whole male gametophyte, rather than just the sperm, moves to the female gametophyte.

A female gametophyte forms within the protection of the integuments, collectively forming the ovule. In angiosperms, the ovules are completely enclosed within additional diploid sporophyte tissue. The ovule and the surrounding protective tissue are called the ovary. The ovary develops into the fruit.

Learning Outcomes Review 31.1

The gymnosperms and angiosperms evolved from a common seed-producing ancestor. Whole-genome duplication likely contributed to the rise and dominance of seed plants. Seeds protect the embryo, aid in dispersal, and can allow for an extended pause in the life cycle. Seed plants produce male and female gametophytes; the male gametophyte is a pollen grain, which is carried to the female gametophyte by wind or other means. The male gametophyte carries a sperm cell, whereas the female gametophyte carries an egg cell.

■ Why is water not essential for fertilization in seed plants?

31.2 Gymnosperms: Plants with "Naked Seeds"

Learning Outcomes

1. Describe the distinguishing features of a gymnosperm.
2. List the four groups of living gymnosperms.

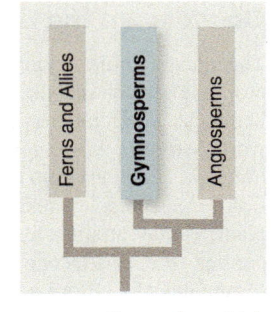

Seeds distinguish the gymnosperms from the pterophytes (ferns and allies). **Gymnosperms** encompass four of the five lineages of seed plants. The four groups of gymnosperms are the coniferophytes, cycadophytes, gnetophytes, and ginkgophytes. The fifth lineage of seed plants is the angiosperms. Although both gymnosperms and angiosperms produce seeds, only the angiosperms produce flowers and fruits (table 31.1). In gymnosperms, the ovule, which becomes a seed, rests exposed on a scale (a modified shoot or leaf) and is not completely enclosed by sporophyte tissues at the time of pollination. The name *gymnosperm* means "naked seed" because the seed develops on the surface of the scale.

Gymnosperms vary in their reproduction and growth form. For example, cycads and *Ginkgo* have motile sperm, whereas conifers and gnetophytes have sperm with no flagella. All sperm are carried within a pollen tube that forms when a pollen grain

TABLE 31.1	The Five Phyla of Extant Seed Plants		
Phylum	**Examples**	**Key Characteristics**	**Approximate Number of Living Species**
Coniferophyta	Conifers (including pines, spruces, firs, yews, redwoods, and others)	Heterosporous seed plants. Sperm not motile; conducted to egg by a pollen tube. Leaves mostly needlelike or scalelike. Trees, shrubs. About 68 genera. Many produce seeds in cones.	630
Cycadophyta	Cycads	Heterosporous. Sperm flagellated and motile but confined within a pollen tube that grows to the vicinity of the egg. Palmlike plants with pinnate leaves. Secondary growth slow compared with that of the conifers. Ten genera. Seeds in cones.	306
Gnetophyta	Gnetophytes	Heterosporous. Sperm not motile; conducted to egg by a pollen tube. The only gymnosperms with vessels. Trees, shrubs, vines. Three very diverse genera (*Ephedra, Gnetum, Welwitschia*).	65
Ginkgophyta	*Ginkgo*	Heterosporous. Sperm flagellated and motile but conducted to the vicinity of the egg by a pollen tube. Deciduous tree with fan-shaped leaves that have evenly forking veins. Seeds resemble a small plum with fleshy, foul-smelling outer covering. One genus.	1
Anthophyta	Flowering plants (angiosperms)	Heterosporous. Sperm not motile; conducted to egg by a pollen tube. Seeds enclosed within a fruit. Leaves greatly varied in size and form. Herbs, vines, shrubs, trees. About 14,000 genera.	>300,000

germinates. Most gymnosperms are terrestrial, although a few aquatic species exist.

Conifers are the largest gymnosperm phylum

The most familiar gymnosperms are **conifers** (phylum Coniferophyta), which include pines (figure 31.2), spruces, firs, cedars, hemlocks, yews, larches, cypresses, and others. Nearly 40% of the world's forests are composed of conifers. The coastal redwood *(Sequoia sempervirens)*, a conifer native to northwestern California and southwestern Oregon, is the tallest living vascular plant; it may attain a height of nearly 100 m (300 ft). Another conifer, the bristlecone pine *(Pinus longaeva)* of the White Mountains of California, is the oldest living tree; one specimen is 4900 years old.

Conifers are found in the colder temperate and sometimes drier regions of the world. Various species are sources of timber, biofuel feedstock, paper, resin, paclitaxel (Taxol, which is used to treat cancer), and other economically important products. Genomes of three conifers—loblolly pine, sugar pine, and Douglas fir—are currently being sequenced because of their economic importance. With genomes approximately 10-fold larger than the human genome this is a substantial challenge.

Pines are an exemplary conifer genus

More than 100 species of pines exist today, all native to the northern hemisphere. Pines and spruces, which belong to the same

Figure 31.2 Conifers.
Loblolly pine, *Pinus taeda*, is found on 58 million acres of land in the southeastern part of the United States. Worldwide 16% of the annual timber supply comes from this conifer.

family, are members of the vast coniferous forests that lie between the arctic tundra and the temperate deciduous forests and prairies to their south. During the past century, pines have been extensively planted in the southern hemisphere.

Pine morphology

Pines have tough, needle-like leaves produced mostly in clusters of two to five. Among the conifers, only pines have clustered leaves. The leaves, which have a thick cuticle and recessed stomata, represent an evolutionary adaptation for minimizing water loss. This strategy is important because many of the trees grow in areas where the topsoil is frozen for part of the year, making it difficult for the roots to obtain water.

The leaves and other parts of the sporophyte have canals into which surrounding cells secrete resin. The resin deters insect and fungal attacks. The resin of certain pines is harvested commercially for its volatile liquid portion, called *turpentine,* and for the solid *rosin,* which is used on bowed stringed instruments. The wood of pines lacks some of the more rigid cell types found in other trees, and it is considered a "soft" rather than a "hard" wood.

The thick bark of pines is an adaptation for surviving fires and subzero temperatures. Some cones actually depend on fire to open them, releasing seeds to reforest burned areas.

Reproductive structures

All seed plants produce two types of spores that give rise to two types of gametophytes (figure 31.3). The male gametophytes (pollen grains) of pines develop from microspores, which are produced by meiosis in male cones that develop in clusters, typically at the tips of the lower branches; there may be hundreds of such clusters on any single tree.

The male pine cones generally are 1 to 4 cm long and consist of small, papery scales arranged in a spiral or in whorls. A pair of microsporangia form as sacs within each scale. Numerous microspore mother cells in the microsporangia undergo meiosis, each becoming four microspores. The microspores divide by mitosis to produce four-celled pollen grains, each with a pair of air sacs that give them added buoyancy when released into the air. A single cluster of male pine cones may produce more than a million pollen grains.

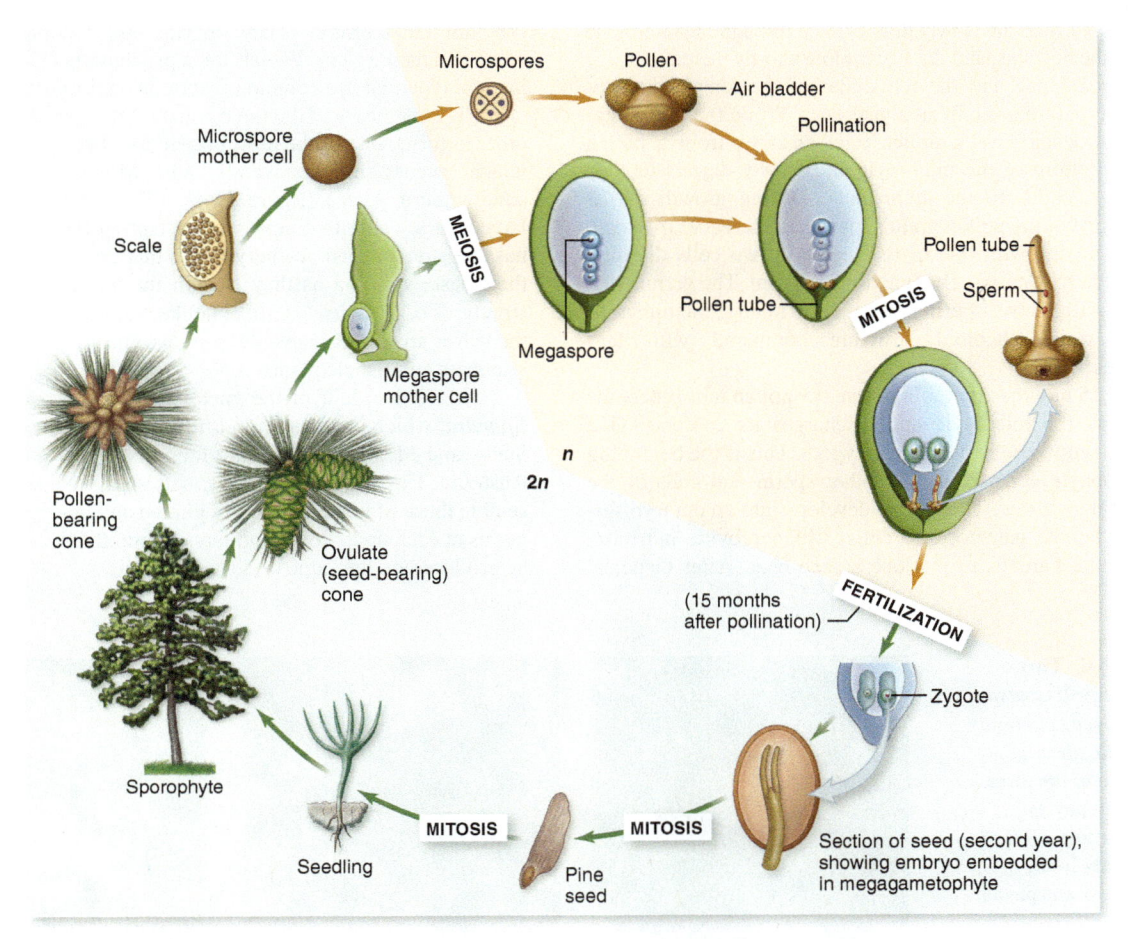

Figure 31.3 Life cycle of a typical pine. The $2n$ stage is the sporophyte, whereas the n stage is the gametophyte. Wind generally disperses the male gametophyte (pollen), which carries sperm. Pollen tube growth delivers the sperm to the egg on the female cone. Additional protection for the embryo is provided by the integument, which develops into the seed coat.

Female pine cones typically are produced on the upper branches of the same tree that produces male cones. Female cones are larger than male cones, and their scales become woody.

Two ovules develop toward the base of each scale. Each ovule contains a megasporangium called the **nucellus.** The nucellus itself is completely surrounded by a thick layer of cells called the integument that has a small opening (the **micropyle**) toward one end. One of the layers of the integument later becomes the seed coat. A single megaspore mother cell within each megasporangium undergoes meiosis, becoming a row of four megaspores. Three of the megaspores break down, but the remaining one, over the better part of a year, slowly develops into a female gametophyte through mitotic divisions. The female gametophyte at maturity may consist of thousands of cells, with two to six archegonia formed at the micropylar end. Each archegonium contains an egg so large it can be seen without a microscope.

Fertilization and seed formation

Female cones usually take two or more seasons to mature. At first they may be reddish or purplish in color, but they soon turn green, and during the first spring, the scales spread apart. While the scales are open, pollen grains carried by the wind drift down between them, some catching in sticky fluid oozing out of the micropyle. The pollen grains within the sticky fluid are slowly drawn down through the micropyle to the top of the nucellus, and the scales close shortly thereafter.

The archegonia and the remainder of the female gametophyte are not mature until about a year later. While the female gametophyte is developing, a pollen tube emerges from a pollen grain at the bottom of the micropyle and slowly digests its way through the nucellus to the archegonia. During growth of the pollen tube, one of the pollen grain's four cells, the *generative cell*, divides by mitosis, with one of the resulting two cells dividing once more. These last two cells function as sperm. The germinated pollen grain with its two sperm is the mature male gametophyte, a small and simple haploid structure compared with fern gametophytes.

About 15 months after pollination, the pollen tube enters the ovule through the micropyle and discharges its contents. One sperm unites with the egg, forming a zygote. This is the beginning of the sporophyte generation. The other sperm and cells of the pollen grain degenerate. The zygote develops into an embryo surrounded by female gametophyte cells, which provide nutrition. The integument hardens to produce a seed coat. After dispersal and germination of the seed, the young sporophyte of the next generation develops into a tree.

Cycads resemble palms, but are not flowering plants

Cycads (phylum Cycadophyta) are slow-growing gymnosperms of tropical and subtropical regions. The sporophytes of most of the 250 known species resemble palm trees (figure 31.4a) with trunks that can attain heights of 15 m or more. Unlike palm trees, which are flowering plants, cycads produce cones and have a life cycle similar to that of pines.

The female cones, which develop upright among the leaf bases, are huge in some species and can weigh up to 45 kg. The sperm of cycads, although formed within a pollen tube, are released within the ovule to swim to an archegonium. These sperm are the largest sperm cells among all living organisms. Several species of cycads are facing extinction in the wild and soon may exist only in botanical gardens.

Gnetophytes have xylem vessels

There are three genera and about 65 living species of gnetophytes (phylum Gnetophyta). They are the only gymnosperms with vessels in their xylem. **Vessels** are a particularly efficient conducting cell type that is a common feature in angiosperms.

The members of the three genera differ greatly from one another in form. One of the most bizarre of all plants is *Welwitschia*, which occurs in the Namib and Mossamedes deserts of southwestern Africa (figure 31.4b). The stem is shaped like a large, shallow cup that tapers into a taproot below the surface. It has two strap-shaped, leathery leaves that grow continuously from their base, splitting as they flap in the wind. The reproductive structures of *Welwitschia* are conelike, appear toward the bases of the leaves around the rims of the stems, and are produced on separate male and female plants.

More than half of the gnetophyte species are in the genus *Ephedra*, which is common in arid regions of the western United States and Mexico. Species are found on every continent except Australia. The plants are shrubby, with stems that superficially resemble those of horsetails, being jointed and having tiny, scalelike leaves at each node. Male and female reproductive structures may be produced on the same or different plants.

Figure 31.4 Three phyla of gymnosperms.
a. A cycad, *Cycas circinalis*.
b. Welwitschia mirabilis represents one of the three genera of gnetophytes.
c. Maidenhair tree, *Ginkgo biloba*, the only living representative of the phylum Ginkgophyta.

The drug ephedrine, widely used in the treatment of respiratory problems, was in the past extracted from the Chinese species of *Ephedra,* but it has now been largely replaced with synthetic preparations (pseudoephedrine). Because ephedrine found in herbal remedies for weight loss was linked to strokes and heart attacks, it was withdrawn from the market in April of 2004. Sales restrictions were placed on pseudoephedrine-containing products in 2006 because it can be used to manufacture the illegal drug methamphetamine.

The best-known species of the third genus, *Gnetum,* is a tropical tree, but most species are vinelike. All species have broad leaves similar to those of angiosperms. One *Gnetum* species is cultivated in Java for its tender shoots, which are cooked as a vegetable.

Only one species of the ginkgophytes remains extant

The fossil record indicates that members of the ginkgophytes (phylum Ginkgophyta) were once widely distributed, particularly in the northern hemisphere; today, only one living species, *Ginkgo biloba,* remains (figure 31.4c). This tree, which sheds its leaves in the fall, was first encountered by Europeans in cultivation in Japan and China; it apparently no longer exists in the wild.

Like the sperm of cycads, those of *Ginkgo* have flagella. The ginkgo is **dioecious**—that is, the male and female reproductive structures are produced on separate trees. The fleshy outer coverings of the seeds of female ginkgo plants exude the foul smell of rancid butter, caused by the presence of butyric and isobutyric acids, which are also found in butter and vomit. As a result, male plants vegetatively propagated from shoots are preferred for cultivation. Because of its beauty and resistance to air pollution, *Ginkgo* is commonly planted along city streets.

Learning Outcomes Review 31.2

Gymnosperms are mostly cone-bearing seed plants. In gymnosperms, the ovules are not completely enclosed by sporophyte tissue at pollination, and thus have "naked seeds." The four groups of gymnosperms are conifers, cycads, gnetophytes, and ginkgophytes.

■ *What adaptation do conifers exhibit to capture wind-borne pollen?*

31.3 Angiosperms: The Flowering Plants

Learning Outcomes

1. List the defining features of angiosperms.
2. Describe the roles of some animals in the angiosperm life cycle.
3. Explain double fertilization and its outcome.

Over 300,000 known species of flowering plants are called **angiosperms** because their ovules, unlike those of gymnosperms, are enclosed within diploid tissues at the time of pollination. The *carpel,* a modified leaf that encapsulates seeds, develops into the fruit, a unique angiosperm feature (figure 31.5). Although some gymnosperms, including the yew (*Taxus*

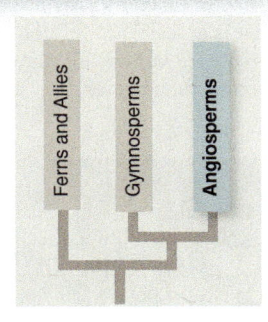

Figure 31.5 Evolution of fruit. Unlike gymnosperms, the ovule of angiosperms is surrounded by sporophyte tissue derived from leaves. This tissue is called the carpel and develops into the fruit. The leaf origins of carpels are still visible in the pea pod.

species), have fleshlike tissue around their seeds, it is of a different origin and not a true fruit.

Sorting out angiosperm origins is complicated

The origins of the angiosperms puzzled even Darwin who referred to their origin as an "abominable mystery." What particularly puzzled Darwin was the very rapid diversification and success of the angiosperms. Notice in table 31.1 that there are over 300,000 species of angiosperms, but only about 1000 species of all other seed plants combined. The rapid explosion of angiosperm diversity changed Earth's landscape from one dominated by ferns, cycads, and conifers to the more varied and familiar ecosystems we see today. There are many potential explanations for the takeover of the world's biomes by the angiosperms. An interesting new hypothesis is that the breakup of Pangea led to climate change and opportunities for the expansion of the angiosperms. The middle latitudes became more humid and temperate, and deserts between the tropic and temperate zones became smaller. The ability of angiosperms to move into these new environments was enhanced by their unique features, including flower production, the ability to attract insect pollinators, and the development of broad leaves with dense veins.

In Liaoning province of China, a complete angiosperm fossil that is at least 125 million years old has been found (figure 31.6). *Archaefructus* fossils have both male and female reproductive structures; however, they lack the sepals and petals that evolved in

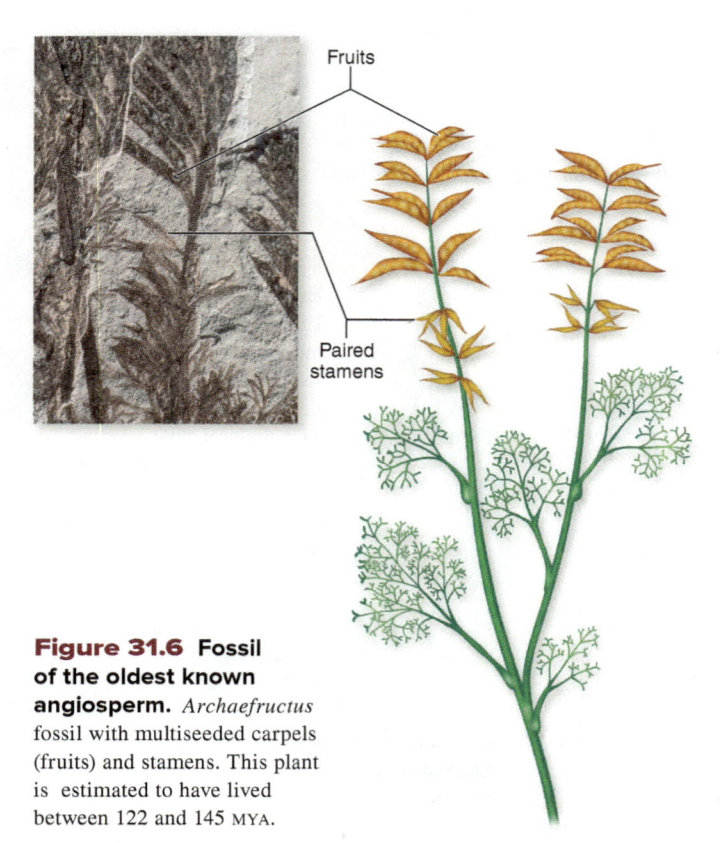

Figure 31.7 An ancient living angiosperm, *Amborella trichopoda*. This plant is believed to be the closest living relative to the original angiosperm.

later angiosperms to attract pollinators. The fossils were so well preserved that fossil pollen could be examined using scanning electron microscopy. Although *Archaefructus* is ancient, it is unlikely to be the very first angiosperm. Still, the incredibly well-preserved fossils provide valuable detail on angiosperms in the Upper Jurassic to Lower Cretaceous period, when dinosaurs roamed the Earth.

Consensus has also been growing on the most basal living angiosperm—*Amborella trichopoda* (figure 31.7). *Amborella*, with small, cream-colored flowers, is even more primitive than water lilies. This small shrub, found only on the island of New Caledonia in the South Pacific, is the last remaining species of the earliest extant lineage of the angiosperms that arose about 135 MYA.

Although *Amborella* is not the original angiosperm, it is sufficiently close that studying its reproductive biology may help us understand the early radiation of the angiosperms. The angiosperm phylogeny reflects an evolutionary hypothesis that is driving new research on angiosperm origins (figure 31.8).

Horizontal gene transfer occurred in land plants

Amborella trichopoda is the closest living relative to the earliest angiosperms, yet 20 out of 31 of its known mitochondrial protein genes hopped into the mitochondrial genome from other land plants through horizontal gene transfer (HGT). In addition, three different moss species contributed to the mix (figure 31.9).

? Inquiry question Explain why a phylogenetic tree based on comparisons of a single gene could result in an inaccurate evolutionary hypothesis.

Amborella is not typical of most extant flowering plants. It is the only existing member of its genus and is native only to the tropical rainforests of New Caledonia, an island group east of Australia that has been isolated for some 70 million years and contains many ancient endemic species. Here parasitic plants (plants that derive

Figure 31.6 Fossil of the oldest known angiosperm. *Archaefructus* fossil with multiseeded carpels (fruits) and stamens. This plant is estimated to have lived between 122 and 145 MYA.

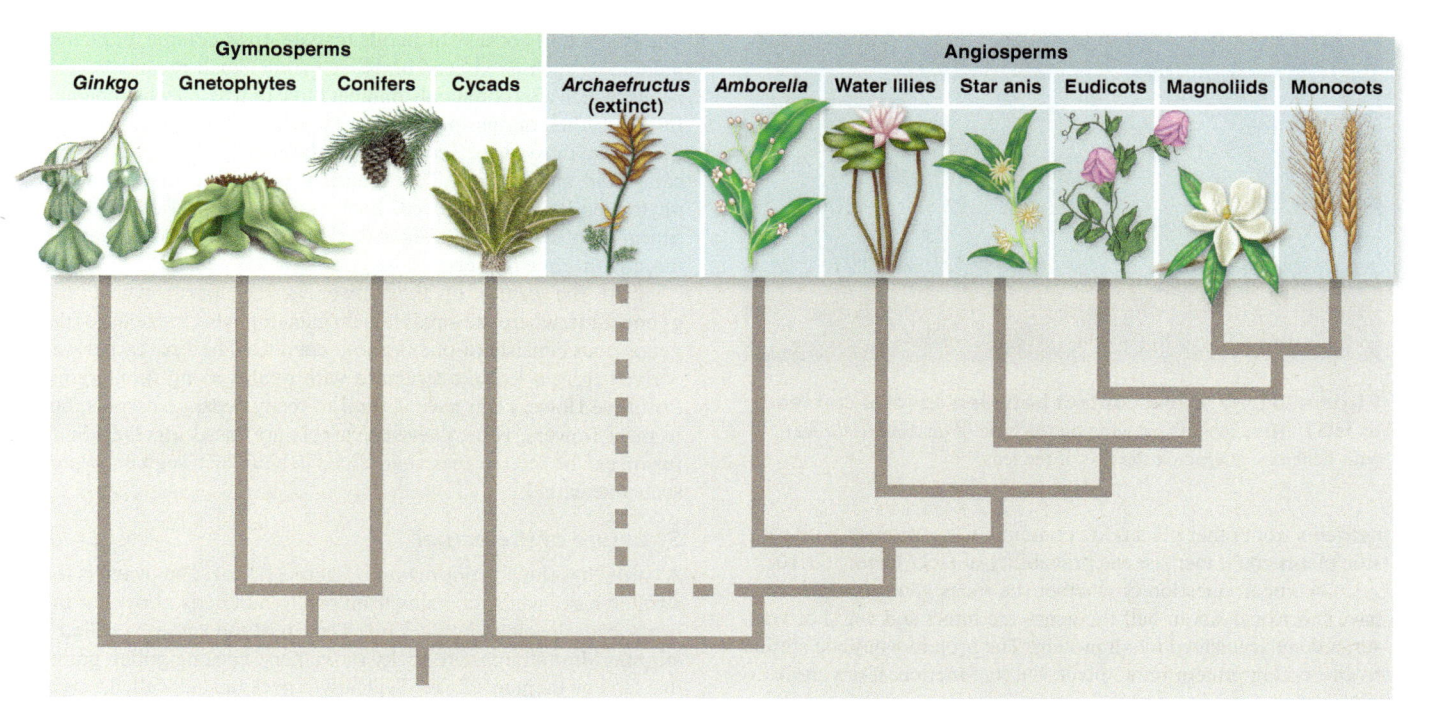

Figure 31.8 *Archaefructus* may be the sister clade to all other angiosperms. All members of the *Archaefructus* lineage are extinct, leaving *Amborella* as the most basal, living angiosperm. Gymnosperm species' labels are shaded green.

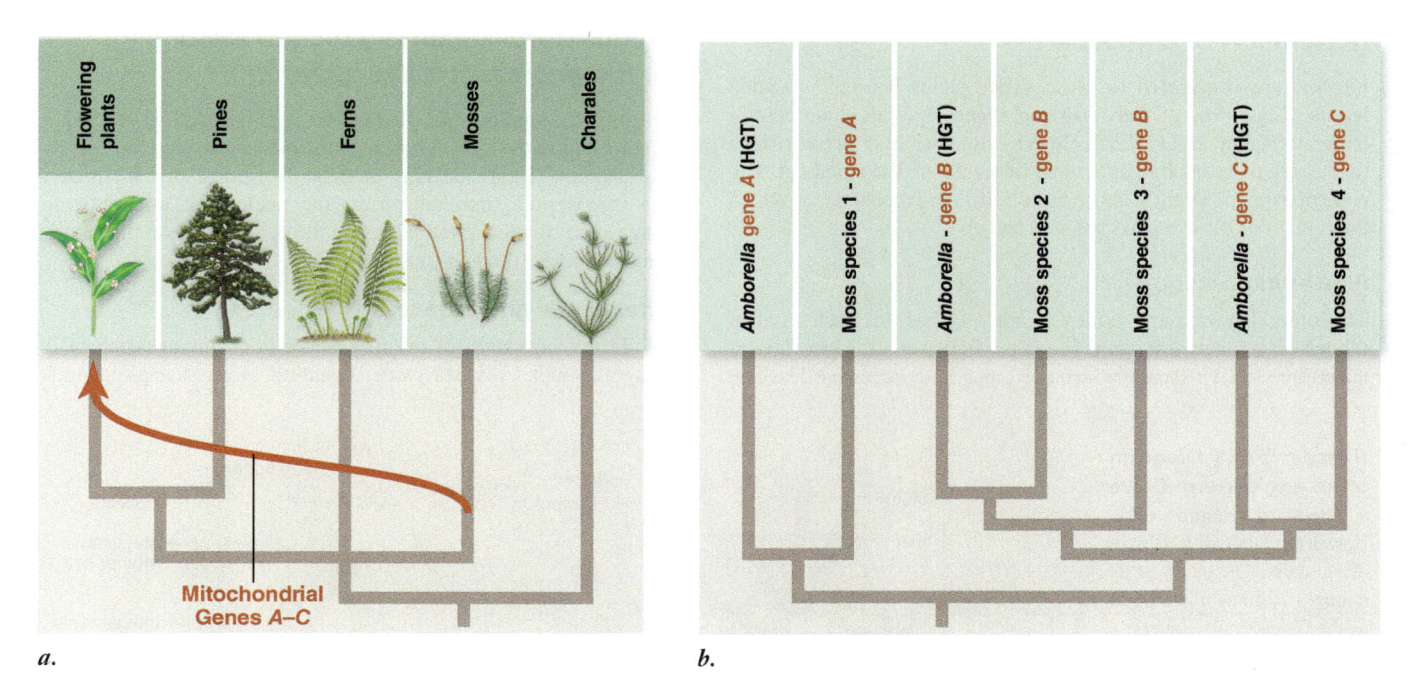

Figure 31.9 The flowering plant *Amborella* acquired three moss genes through horizontal gene transfer (HGT).
a. Phylogenetic relationship of *Amborella* to other land plants. As shown by the arrow connecting moss and the flowering plants, HGT is the only plausible explanation for the presence of moss mitochondrial genes in *Amborella*. ***b.*** Phylogenetic relationships among the horizontally transferred gene.

Figure 31.10 Close contact between species can lead to HGT. Here moss is growing on the base of an *Amborella* leaf with lichens scattered on the rest of the leaf.

nutrients from other plants) are common. Close contact with parasitic plants could increase the probability of HGT (figure 31.10).

An open question is whether the moss genes in *Amborella* have functions. About half the genes are intact and could be transcribed and translated into a protein. The protein would be similar to an existing protein in the plant, but its function, if any, remains to be determined.

Inquiry question How would you determine if a moss gene in *Amborella* had a function? (*Hint:* Refer to chapter 24.)

Flowers house the gametophyte generation of angiosperms

Flowers are considered to be modified stems bearing modified leaves. Regardless of their size and shape, they all share certain features (figure 31.11). Each flower originates as a **primordium** that develops into a bud at the end of a stalk called a pedicel. The pedicel expands slightly at the tip to form the receptacle, to which the remaining flower parts are attached.

Flower morphology

The other flower parts typically are attached in circles called **whorls.** The outermost whorl is composed of **sepals.** Most flowers have three to five sepals, which are green and somewhat leaflike.

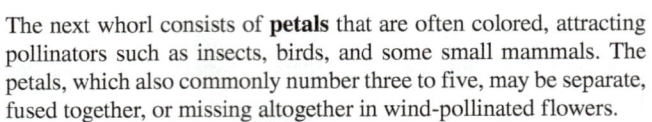

The next whorl consists of **petals** that are often colored, attracting pollinators such as insects, birds, and some small mammals. The petals, which also commonly number three to five, may be separate, fused together, or missing altogether in wind-pollinated flowers.

The third whorl consists of **stamens** and is collectively called the androecium. This whorl is where the male gametophytes, pollen, are produced. Each stamen consists of a pollen-bearing **anther** and a stalk called a **filament,** which may be missing in some flowers.

At the center of the flower is the fourth whorl, the **gynoecium,** where the small female gametophytes are housed; the gynoecium consists of one or more **carpels.** The first carpel was derived from a leaflike structure with ovules along its margins. Primitive flowers can have several to many separate carpels, but in most flowers, two to several carpels are fused together. Such fusion can be seen in an orange sliced in half; each segment represents one carpel.

Structure of the carpel

A carpel has three major regions (figure 31.11*a*). The **ovary** is the swollen base, which contains from one to hundreds of ovules; the ovary later develops into a **fruit.** The tip of the carpel is called a **stigma.** Most stigmas are sticky or feathery, causing pollen grains that land on them to adhere. Typically, a neck or stalk called a **style** connects the stigma and the ovary; in some flowers, the style may be very short or even missing.

Many flowers have nectar-secreting glands called *nectaries,* often located toward the base of the ovary. Nectar is a fluid containing sugars, amino acids, and other molecules that attracts insects, birds, and other animals to flowers.

Double fertilization is a unique feature of the angiosperm life cycle

During development of a flower bud, a single diploid megaspore mother cell in the ovule undergoes meiosis, producing four haploid megaspores (figure 31.12). In most flowering plants, three of the megaspores soon disappear; the nucleus of the remaining megaspore divides mitotically, and the cell slowly expands until it becomes many times its original size.

The female gametophyte

While the expansion of the megaspore is occurring, each of the daughter nuclei divides twice, resulting in eight haploid nuclei

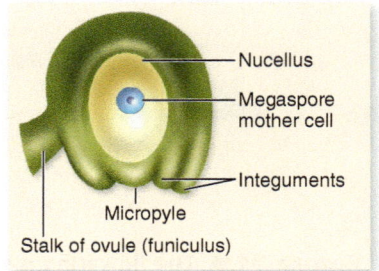

Figure 31.11 Diagram of an angiosperm flower. *a.* The main structures of the flower are labeled. *b.* Details of an ovule. The ovary as it matures will become a fruit; as the ovule's outer layers (integuments) mature, they will become a seed coat.

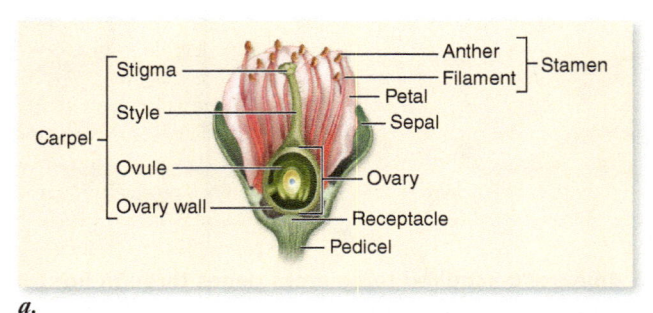

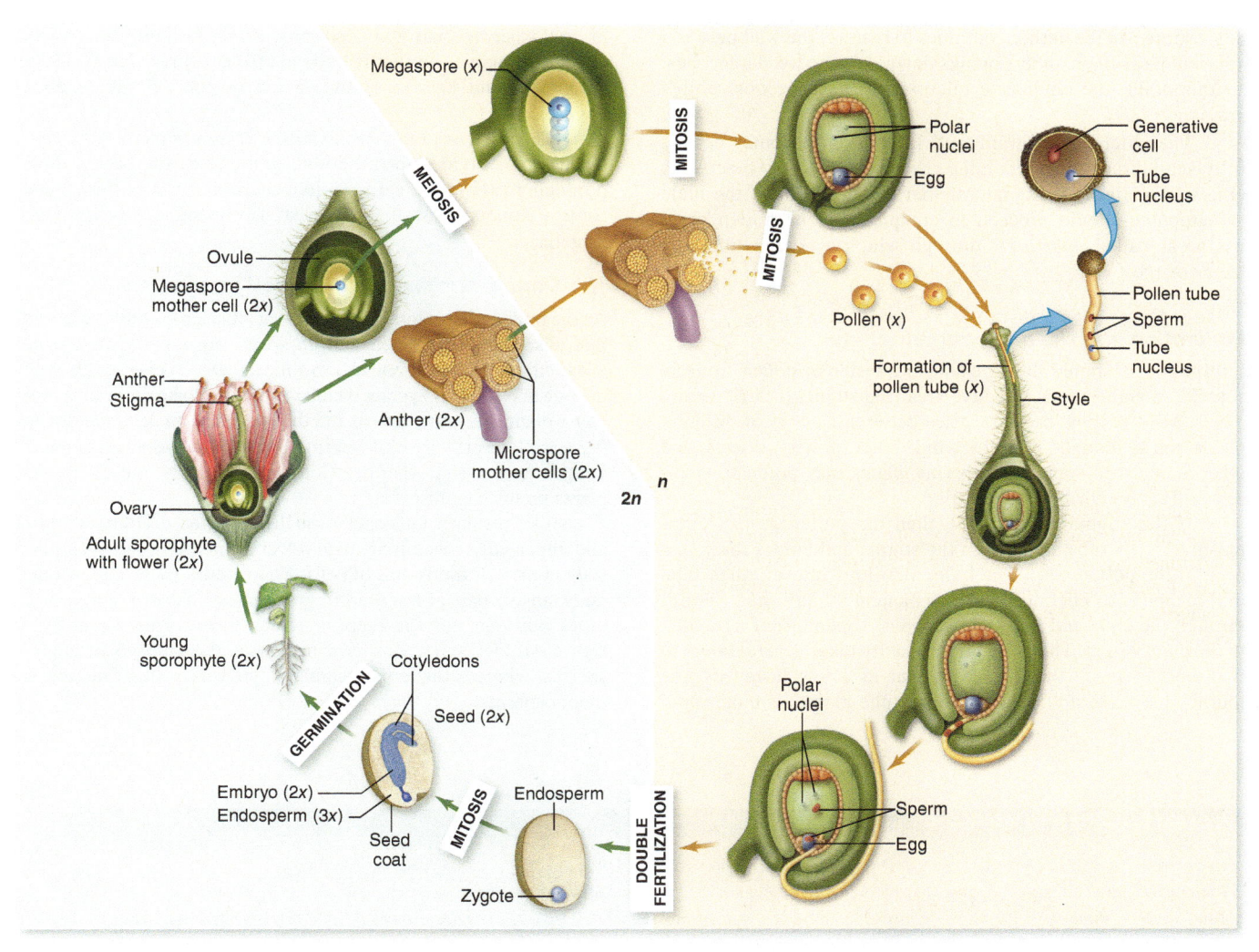

Figure 31.12 Life cycle of a typical angiosperm. As in pines, external water is no longer required for fertilization. In most species of angiosperms, animals carry pollen to the carpel. The outer wall of the carpel forms the fruit, which often entices animals to disperse the seed.

arranged in two groups of four. One nucleus from each group of four migrates toward the center, where they function as polar nuclei. A cell wall forms around the two polar nuclei to create the central cell. Cell walls also form around the remaining nuclei. In the group closest to the micropyle, one cell functions as the egg; the other two cells are called synergids. At the other end, the three cells are now called antipodals; they have no apparent function and eventually break down and disappear. At the same time, two layers of the ovule, the integuments, differentiate and become the *seed coat* of a seed. The integuments, as they develop, form the micropyle, a small gap or pore at one end that was described earlier in this section (see figure 31.11*b*).

The large sac with eight nuclei in seven cells is called an embryo sac; it constitutes the female gametophyte. Although it is completely dependent on the sporophyte for nutrition, it is a multicellular, haploid individual.

Pollen production

While the female gametophyte is developing, a similar but less complex process takes place in the anthers (figure 31.12). Most anthers have patches of tissue (usually four) that eventually become chambers lined with nutritive cells. The tissue in each patch is composed of many diploid microspore mother cells that undergo meiosis, each producing four microspores.

The four microspores at first remain together as a quartet, or tetrad, and the nucleus of each microspore divides once; in most species, the microspores of each quartet then separate. At the same time, a two-layered wall develops around each microspore. As the anther continues to mature, the wall between adjacent pairs of chambers breaks down, leaving two larger sacs. At this point, the binucleate microspores have become pollen grains.

The outer pollen grain wall layer often becomes beautifully sculptured, and it contains chemicals that may react with others in a stigma to signal whether development of the male gametophyte should proceed to completion. The pollen grain has areas called *apertures,* through which a pollen tube may later emerge.

Pollination and the male gametophyte

Pollination is simply the mechanical transfer of pollen from its source (an anther) to a receptive area (the stigma of a flowering plant). Most pollination takes place between flowers of different plants and is brought about by wind, water, gravity, insects, and other animals. In some angiosperms, plants may pollinate themselves. This is called self-pollination.

If the stigma is receptive, then the pollen grain's cytoplasm absorbs substances from the stigma and bulges through a pore in the pollen wall. The bulge develops into a pollen tube that responds to chemical and mechanical stimuli that guide it through the style and into the micropyle (opening) of an ovule. to the embryo sac. The pollen tube usually takes several hours to two days to reach the micropyle, but in a few instances, the journey may take up to a year. Pollen tube growth is more rapid in angiosperms than gymnosperms. Rapid pollen tube growth rate is an innovation that is believed to have preceded fruit development and been essential in the origins of angiosperms (figure 31.13)

The nucleus of one of the pollen grain's two cells, the *generative cell,* divides in the pollen grain or in the pollen tube, producing two sperm cells. Unlike sperm in mosses, ferns, and some gymnosperms, the sperm of flowering plants have no flagella.

Double fertilization and seed production

As the pollen tube enters the embryo sac, it destroys a synergid and then discharges its contents. Both sperm are functional, and an event called **double fertilization** follows. One sperm unites with the egg and forms a zygote, which develops into an embryo sporophyte plant. The other sperm and the two polar nuclei unite, forming a triploid primary endosperm nucleus. Double fertilization is unique to the angiosperms. Consequently, only the flowering plants produce endosperm.

The primary endosperm nucleus begins dividing rapidly and repeatedly, becoming triploid endosperm tissue that may soon consist of thousands of cells. Endosperm tissue can become an extensive part of the seed in grasses such as corn, and it provides nutrients for the embryo in most flowering plants (see figure 41.35). The endosperm nourishes the embryo in angiosperms, whereas the female gametophyte serves that function in gymnosperms.

SCIENTIFIC THINKING

Hypothesis: *An increase in pollen tube growth rate accompanied angiosperm origins.*

Prediction: *Pollen tube growth rates will be slower in gymnosperms than in close relatives of the first angiosperm.*

Test: *Pollinate three angiosperm species, including Amborella, that are most closely related to the first angiosperm. At different time intervals, remove the style and measure the lengths of pollen tubes using a fluorescence microscope. Calculate and compare the rate of pollen tube growth with results published for gymnosperms.*

Species	Pollen Tube Growth Rate (μm/hour)
Close relatives of ancestral angiosperm	
Amborella trichopoda	80.0
Nuphar polysepala	589.2
Austrobaileya scandens	271.7
Gymnosperms	
Gnetum gnemon	6.4
Ginkgo biloba	0.3

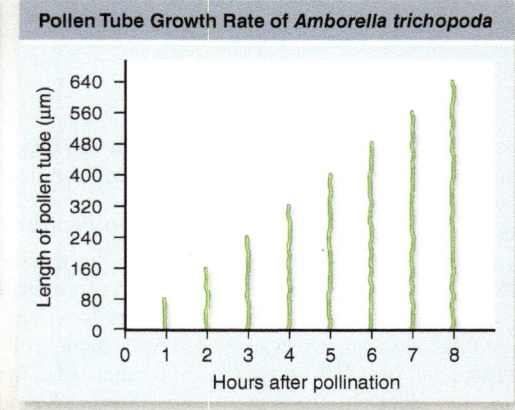

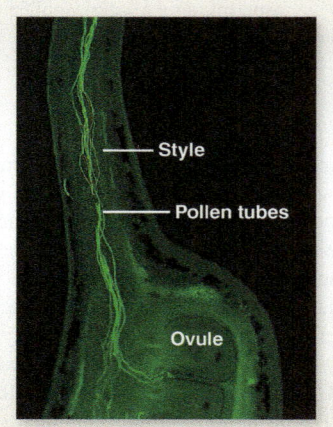

Conclusion: *Pollen tube growth rate is higher in relatives of ancestral angiosperms than in gymnosperms, supporting the hypothesis.*

Further Experiments: *Develop a hypothesis to explain the difference in pollen tube growth rates among the tested angiosperms. Is there a correlation between the distance a pollen tube must travel to fertilize an egg and the rate of growth? How could you test your hypothesis?*

Figure 31.13 Rapid pollen tube growth accompanied the origin of angiosperms.

> **Inquiry question** If the endosperm failed to develop in a seed, how do you think the fitness of that seed's embryo would be affected? Explain your answer.

> **Data analysis** Using the graph in figure 31.13, calculate the pollen tube growth rate, being sure to show your work. Compare your rate with the rate in the table in figure 31.13 and propose an explanation for any differences.

Learning Outcomes Review 31.3

Amborella, the closest living relative of the first angiosperms, provides clues to the origins of this very successful group. Although angiosperms are a clade, horizontal gene transfer has contributed genes from distantly related plant species. Angiosperms are characterized by ovules that at pollination are enclosed within an ovary at the base of a carpel, a structure unique to the phylum; the ovary develops into a fruit. Evolutionary innovations of angiosperms include flowers to attract pollinators, fruits to protect embryos and aid in their dispersal, rapid pollen tube growth, and double fertilization, which provides endosperm to help nourish the embryo.

- How could genes from moss be transported into a flowering plant, other than via parasitic plants?

31.4 Seeds

Learning Outcomes

1. Describe four ways in which seeds help to ensure the survival of a plant's offspring.
2. List environmental conditions that can lead to seed germination in some plants.

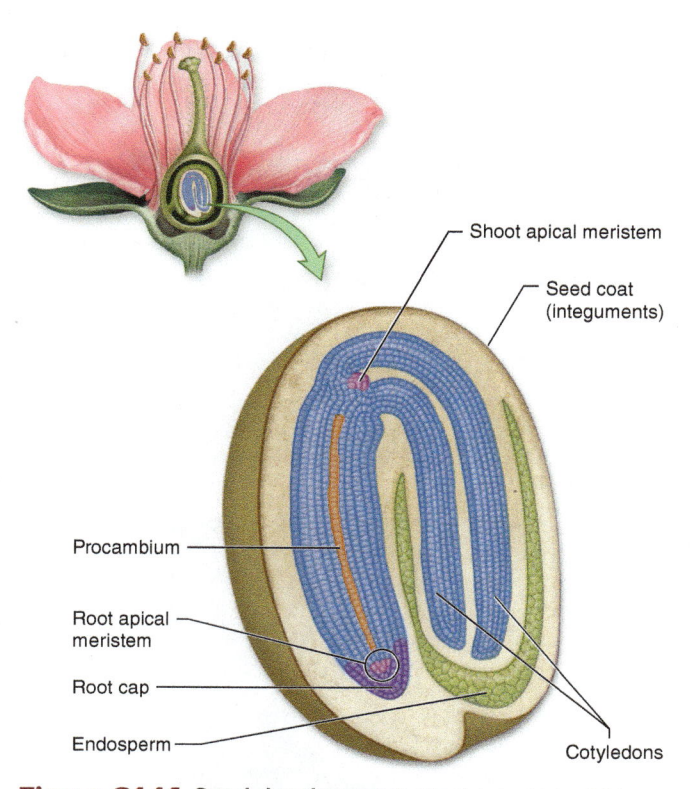

Figure 31.14 Seed development. The integuments of this mature angiosperm ovule are forming the seed coat. Note that the two cotyledons have grown into a bent shape to accommodate the tight confines of the seed.

Following double fertilization in angiosperms and single fertilization in gymnosperms leading to the production of an embryo, a profoundly important event occurs: The embryo stops developing. In this section, we focus on seed development in the angiosperms. In many plants, development of the embryo is arrested soon after the meristems and cotyledons (seed leaves) form. The integuments—the outer cell layers of the ovule—develop into a relatively impermeable **seed coat,** which encloses the seed, which contains a dormant embryo and stored food (figure 31.14).

Seeds protect the embryo

The seed is a vehicle for dispersing the embryo to distant sites. Being encased in the protective layers of a seed allows a plant embryo to survive in environments that might kill a mature plant.

Seeds are an important adaptation in at least four ways:

1. Seeds maintain dormancy under unfavorable conditions and postpone development until better conditions arise.
2. Seeds afford maximum protection to the young plant at its most vulnerable stage of development.
3. Seeds contain stored food that allows a young plant to grow and develop before photosynthetic activity begins.
4. Perhaps most important, seeds are adapted for dispersal, facilitating the migration of plant genotypes into new habitats.

A mature seed contains only about 5 to 20% water. Under these conditions, the seed and the young plant within it are very stable; its arrested growth is primarily due to the progressive and severe desiccation of the embryo and the associated reduction in metabolic activity. Germination cannot take place until water and oxygen reach the embryo. Seeds of some plants have been known to remain viable for hundreds and, in rare instances, thousands of years.

Specialized seed adaptations improve survival

Specific adaptations often help ensure that seeds will germinate only under appropriate conditions. Sometimes, seeds lie within tough cones that do not open until they are exposed to the heat of a fire (figure 31.15). This strategy causes the seed to germinate in an

Figure 31.15 Fire induces seed release in some pines.
Fire can destroy adult jack pines, but stimulate growth of the next generation. *a.* The cones of a jack pine are tightly sealed and cannot release the seeds protected by the scales. *b.* High temperatures lead to the release of the seeds.

open, fire-cleared habitat where nutrients are relatively abundant, having been released from plants burned in the fire.

Seeds of other plants germinate only when inhibitory chemicals leach from their seed coats, thus guaranteeing their germination when sufficient water is available. Still other seeds germinate only after they pass through the intestines of birds or mammals or are regurgitated by them, which both weakens the seed coats and ensures dispersal. Sometimes seeds of plants thought to be extinct in a particular area may germinate under unique or improved environmental circumstances, and the plants may then reestablish themselves.

Learning Outcomes Review 31.4

The seed coat originates from the integuments and encloses the embryo and stored nutrients. The four advantages conferred by seeds are dormancy, protection of the embryo, nourishment, and a method of dispersal. Fire, heavy rains, or passage through an animal's digestive tract may be required for germination in some species.

- *What type of seed dormancy would you expect to find in trees living in climates with cold winters?*

31.5 Fruits

Learning Outcomes
1. Identify the structures from which fruits develop.
2. Distinguish among berries, legumes, drupes, and samaras.

Survival of angiosperm embryos depends on fruit development as well as seed development. Fruits are most simply defined as mature ovaries (carpels). During seed formation, the ovary begins to develop into fruit (figure 31.16).

It is possible for fruits to develop without seeds. Commercial bananas, for example, are fruits that lack viable seeds. Bananas must be propagated asexually.

Fruits are adapted for dispersal

Fruits form in many ways and exhibit a wide array of adaptations for dispersal. Three layers of ovary wall, also called the *pericarp*, can have distinct fates, which account for the diversity of fruit

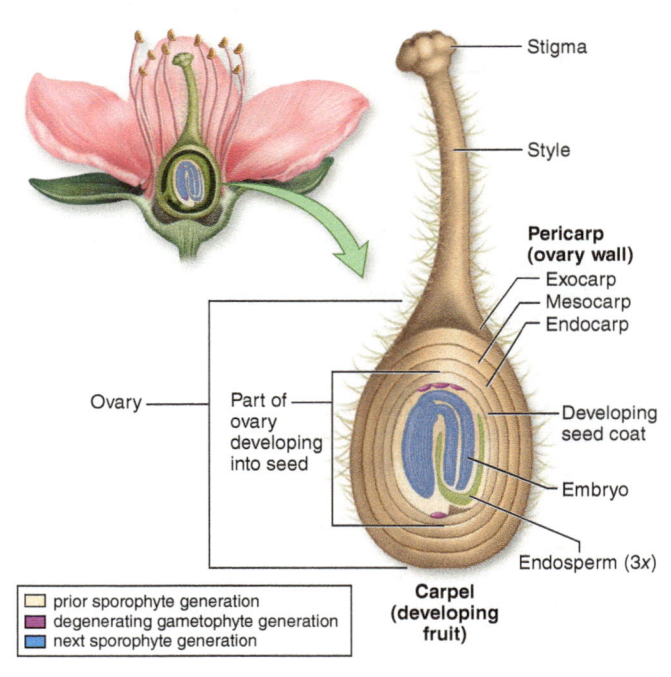

Figure 31.16 Fruit development. The carpel (ovary) wall is composed of three layers: the exocarp, mesocarp, and endocarp. One, some, or all of these layers develops to contribute to the recognized fruit in different species. The seed matures within this developing fruit.

? Inquiry question Three generations are represented in this diagram. Label the ploidy levels of the tissues of different generations shown here.

Figure 31.17 Examples of some kinds of fruits. Legumes and samaras are examples of dry fruits. Legumes open to release their seeds, but samaras do not. Drupes and true berries are simple fleshy fruits; they develop from a flower with a single pistil composed of one or more carpels. Aggregate and multiple fruits are compound fleshy fruits; they develop from flowers with more than one pistil or from more than one flower.

types from fleshy and soft to dry and hard. The differences among some of the fruit types are shown in figure 31.17.

Fruits contain three genotypes in one package. The fruit and seed coat are from the prior sporophyte generation. The embryo represents the next sporophyte generation (see figure 31.16). Finally, the endosperm is a transient, triploid product of fertilization.

Fruits allow angiosperms to colonize large areas

The diversity of fruit types contributes to an array of dispersal methods. Fruits with fleshy coverings normally are dispersed by birds or other vertebrates (figure 31.18a). Like red flowers, red fruits signal an abundant food supply. By feeding on these fruits, birds and other animals may carry seeds from place to place and thus transfer plants from one suitable habitat to another. Such seeds require a hard seed coat to resist stomach acids and digestive enzymes.

Fruits with hooked spines, such as those of burrs (figure 31.18b), are typical of several genera of plants that occur in the northern deciduous forests. Such fruits are often disseminated by mammals, including humans, when they hitch a ride on fur or clothing. Squirrels and similar mammals disperse and bury fruits such as acorns and other nuts. Some of these sprout when conditions become favorable, such as after the spring thaw.

Other fruits, including those of maples, elms, and ashes, have wings that aid in their distribution by the wind. Orchids have minute, dustlike seeds, which are likewise blown away by the wind. The dandelion provides another familiar example of a fruit

Figure 31.18 Animal-dispersed fruits. *a.* The bright red berries of this honeysuckle, *Lonicera hispidula,* are highly attractive to birds. After eating the fruits, birds may carry the seeds they contain for great distances either internally or, because of their sticky pulp, stuck to their feet or other body parts. *b.* You will know if you have ever stepped on the fruits of *Cenchrus incertus;* their spines adhere readily to any passing animal. *c.* False dandelion, *Pyrrhopappus carolinianus,* has "parachutes" that widely disperse the fruits in the wind, much to the gardener's despair. *d.* This fruit of the coconut palm, *Cocos nucifera,* is sprouting on a sandy beach. Coconuts, one of the most useful fruits for humans in the tropics, have become established on other islands by drifting there on waves.

type that is wind-dispersed (figure 31.18c), and the dispersal of seeds from plants such as milkweeds, willows, and cottonwoods is similar. Water dispersal adaptations include air-filled chambers surrounded by impermeable membranes to prevent waterlogging.

Coconuts and other plants that characteristically occur on or near beaches are regularly spread throughout a region by floating in water (figure 31.18d). This sort of dispersal is especially important in the colonization of distant island groups, such as the Hawaiian Islands.

It has been calculated that the seeds of about 175 angiosperms, nearly one-third from North America, must have reached Hawaii to evolve into the roughly 970 species found there today. Some of these seeds blew through the air, others were transported on the feathers or in the guts of birds, and still others floated across the Pacific. Although the distances are rarely as great as the distance between Hawaii and the mainland, dispersal is just as important for mainland plant species that have discontinuous habitats, such as mountaintops, marshes, or rivers.

Learning Outcomes Review 31.5

As a seed develops, the pericarp layers of the ovary wall develop into the fruit. A berry has a fleshy pericarp; a legume has a dry pericarp that opens to release seeds; the outer layers of a drupe pericarp are fleshy; and a samara is a dry structure with a wing. Animals often distribute the seeds of fleshy fruits and fruits with spines or hooks. Wind disperses lightweight seeds and samaras.

■ *What features of fruits might encourage animals to eat them?*

Chapter Review

31.1 The Evolution of Seed Plants

The seed protects the embryo.
An extra layer of sporophyte tissue surrounds the embryo, creating the ovule, which later hardens (figure 31.1). The seed protects the embryo, helps it resist drying out, and allows a dormant stage that pauses the life cycle until environmental conditions are favorable.

Water is not required to transport the male gametophyte to the female gametophyte.
The gametophytes of seed plants typically consist of only a few cells. Pollen grains are male gametophytes; each pollen grain contains sperm cells. Water is not required for fertilization. The female gametophyte develops within an ovule that forms the seed.

31.2 Gymnosperms: Plants with "Naked Seeds" (figure 31.3)

Gymnosperms have ovules that are not completely enclosed in sporophyte (diploid) tissue at the time of pollination.

The four groups of gymnosperms are coniferophytes, cycadophytes, gnetophytes, and ginkgophytes; all lack flowers and true fruits.

Conifers are the largest gymnosperm phylum.
Conifers include pines, spruces, firs, cedars, and many other groups. Both the tallest and the oldest vascular plants are conifers.

Pines are an exemplary conifer genus.
Pines have tough, needlelike leaves in clusters of two to five. They produce male and female cones.

In the male cones, microspore mother cells in the microsporangia give rise to microspores that then develop into four-celled pollen grains, the male gametophytes.

In the female cones, a megasporangium (nucellus) produces a single megaspore mother cell that becomes four megaspores; three of these break down, and the remaining one develops into a female gametophyte, which produces archegonia that carry eggs.

Upon fertilization, a pollen tube emerges from the pollen grain and grows through the nucellus. Eventually two sperm cells migrate through the tube, and one unites with the egg; the other disintegrates.

Cycads resemble palms, but are not flowering plants.

Most cycads can reach heights of 15 m or more. Cycads, unlike palms, produce cones; in some species the female cones are huge. Their life cycle is like that of conifers.

Gnetophytes have xylem vessels.

Xylem vessels, a common feature in angiosperms, are highly efficient conducting cells. Gnetophytes are the only gymnosperms that have them.

This group includes the strange *Welwitschia* genus of southwestern Africa and the numerous worldwide *Ephedra* species.

Only one species of the ginkgophytes remains extant.

Ginkgo biloba is a gymnosperm with broad leaves that it sheds in the fall. It is dioecious, and the fleshy seeds of the female tree have a foul odor.

31.3 Angiosperms: The Flowering Plants (figure 31.8)

Angiosperms are distinct from gymnosperms and other plants because their ovules are enclosed within diploid tissue called the ovary at the time of fertilization, and they form fruits.

Sorting out angiosperm origins is complicated.

No one is certain how the angiosperms arose, although the breakup of Pangea may have led to climate change that favored the expansion of the angiosperms. The earliest extant angiosperm appears to be *Amborella trichopoda,* found on the island of New Caledonia.

Horizontal gene transfer occurred in land plants.

Moss genes have been found in the *Amborella* genome, possibly transferred by parasitic plants, although it is unknown if they have a function in *Amborella.*

Flowers house the gametophyte generation of angiosperms.

Flowers are considered to be modified stems that bear modified leaves. Flower parts are organized into four whorls: sepals, petals, androecium, and gynoecium (figure 31.11*a*).

The male androecium consists of the stamens where haploid pollen, the male gametophyte, is produced.

The female gynoecium consists of one or more carpels that contain the female gametophyte. The carpel has three major regions: the ovary, which later becomes the fruit; the stigma, which is the tip of the carpel; and the style, a stalk that connects the stigma and ovary.

Double fertilization is a unique feature of the angiosperm life cycle (figure 31.12).

The megaspore produces eight haploid nuclei. The female gametophyte consists of a large embryo sac with the eight nuclei in seven cells. The egg and the two polar nuclei (in a single cell) are most important.

After landing on a receptive stigma, a pollen grain develops a pollen tube that grows toward the embryo sac. Eventually, two sperm pass through this tube. One fuses with the egg to form a zygote, and the other unites with the polar bodies to form a triploid endosperm nucleus that develops into endosperm to nourish the embryo.

31.4 Seeds (figure 31.14)

Seeds protect the embryo.

Seeds help to ensure the survival of the next generation by maintaining dormancy during unfavorable conditions, protecting the embryo, providing food for the embryo, and providing a means for dispersal.

Specialized seed adaptations improve survival.

Before a seed germinates, its seed coat must become permeable so that water and oxygen can reach the embryo. Adaptations have evolved to ensure germination under appropriate survival conditions. In certain gymnosperms, seeds may be released from cones after a fire. Alternatively, seeds may require passage through a digestive tract, freeze–thaw cycles, or abundant moisture.

31.5 Fruits (figure 31.16)

Fruits are adapted for dispersal.

In angiosperms, a fruit is a mature ovary. Fruit development is coordinated with embryo, endosperm, and seed coat development.

Angiosperms produce many types of fruit, which vary depending on the fate of the pericarp (carpel wall). Fruits can be dry or fleshy, and they can be simple (single carpel), aggregate (multiple carpels), or multiple (multiple flowers).

A fruit is genetically unique because it contains tissues from the parent sporophyte (the seed coat and fruit tissue), the gametophyte (remnants in the developing seed), and the offspring sporophyte (the embryo).

Fruits allow angiosperms to colonize large areas.

Fruits exhibit a wide array of dispersal mechanisms. They may be ingested and transported by animals, buried in caches by herbivores, carried away by birds and mammals, blown by the wind, or float away on water.

Review Questions

UNDERSTAND

1. The lack of seeds is a characteristic of all
   a. lycophytes.
   b. conifers.
   c. tracheophytes.
   d. gnetophytes.

2. Which of the following adaptations allows plants to pause their life cycle until environmental conditions are optimal?
   a. Stomata
   b. Phloem and xylem
   c. Seeds
   d. Flowers

3. Which of the following gymnosperms possesses a form of vascular tissue that is similar to that found in the angiosperms?
   a. Cycads
   b. Gnetophytes
   c. Ginkgophytes
   d. Conifers

4. In a pine tree, the microspores and megaspores are produced by the process of
   a. fertilization.
   b. mitosis.
   c. fusion.
   d. meiosis.

5. Which of the following terms is NOT associated with a male portion of a plant?
   a. Megaspore
   b. Antheridium
   c. Pollen grains
   d. Microspore

6. Which of the following potentially represents the oldest known living species of angiosperm?
   a. *Cooksonia*
   b. *Archaefructus*
   c. *Chlamydomonas*
   d. *Amborella*

7. The integuments of an ovule will develop into the
   a. embryo.
   b. endosperm.
   c. fruit.
   d. seed coat.

8. An example of a drupe is a
   a. strawberry.
   b. plum.
   c. bean.
   d. pineapple.

9. The pericarp is the
   a. ovary wall.
   b. developing seed coat.
   c. ovary.
   d. mature endosperm.

APPLY

1. Reproduction in angiosperms can occur more quickly than in gymnosperms because
   a. gymnosperm sperm requires water to swim to the egg.
   b. flowers always increase the rate of reproduction.
   c. angiosperm pollen tubes grow more quickly than gymnosperm pollen tubes.
   d. angiosperms have nectaries.

2. In double fertilization, one sperm produces a diploid ___, and the other produces a triploid ___.
   a. zygote; primary endosperm
   b. primary endosperm; microspore
   c. antipodal; zygote
   d. polar nuclei; zygote

3. Apply your understanding of angiosperms to identify which innovations likely contributed to the tremendous success of angiosperms.
   a. Homospory in angiosperms
   b. Fruits that attract animal dispersers
   c. Cones that protect the seed
   d. Dominant gametophyte generation

4. Comparing stems of two plant specimens under the microscope, you identify vessels in one sample and conclude the specimen
   a. with vessels must be an angiosperm or a gnetophyte.
   b. with vessels is either *Ephedra* or a cycad.
   c. without vessels is a pterophyte.
   d. without vessels must be a tracheophyte.

5. In a flower after fertilization, the following tissues are diploid:
   a. carpel, integuments, and megaspore mother cell.
   b. carpel, integuments, and megaspore.
   c. carpel, megaspore, and zygote.
   d. carpel, megaspore mother cell, and endosperm.

6. Fruits are complex organs that are specialized for dispersal of seeds. Which of the following plant tissues does NOT contribute to mature fruit?
   a. Sporophytic tissue from the previous generation
   b. Gametophytic tissue from the previous generation
   c. Sporophytic tissue from the next generation
   d. Gametophytic tissue from the next generation

SYNTHESIZE

1. You have been hired as a research assistant to investigate the origins of the angiosperms, specifically the boundary between a gymnosperm and an angiosperm. Which characteristics would you use to clearly define a new fossil as a gymnosperm? An angiosperm?

2. Assess the benefits and drawbacks of self-pollination for a flowering plant? Explain your answer.

3. The relationship between flowering plants and pollinators is often used as an example of coevolution. Many flowering plant species have flower structures that are adaptive to a single species of pollinator. Evaluate the benefits and drawbacks of using such a specialized relationship.

4. As you are eating an apple one day, you decide that you'd like to save the seeds and plant them. You do so, but they fail to germinate. Discuss all possible reasons that the seeds did not germinate and strategies you could try to improve your chances of success.

CHAPTER 32

Fungi

Chapter Contents

- 32.1 Defining Fungi
- 32.2 Microsporidia: Unicellular Parasites
- 32.3 Chytridiomycota and Relatives: Fungi with Zoospores
- 32.4 Zygomycota: Fungi That Produce Zygotes
- 32.5 Glomeromycota: Asexual Plant Symbionts
- 32.6 Basidiomycota: The Club (Basidium) Fungi
- 32.7 Ascomycota: The Sac (Ascus) Fungi
- 32.8 Ecology of Fungi
- 32.9 Fungal Parasites and Pathogens

Introduction

The fungi, an often overlooked group of unicellular and multicellular organisms, have a profound influence on ecology and human health. Along with bacteria, they are important decomposers and disease-causing organisms. Fungi are found everywhere—from the tropics to the tundra and in both terrestrial and aquatic environments. Fungi made it possible for plants to colonize land by associating with rootless stems and aiding in the uptake of nutrients and water. Mushrooms and toadstools grow rapidly under proper conditions and produce large numbers of fungal spores. A single *Armillaria* fungus can cover 15 hectares underground and weigh 100 tons, making it the largest organism in the world based on area. Some puffball fungi are almost a meter in diameter and may contain 7 trillion spores—enough to circle the Earth's equator!

Yeasts are fungi that are used to make bread and beer and some fungi are prized as food, but other fungi cause disease in plants and animals. These eukaryotic microbes are particularly problematic because fungi are animals' closest relatives. Drugs that can kill fungi often have toxic effects on animals, including humans. In this chapter we present the major groups of this intriguing life-form.

32.1 Defining Fungi

Learning Outcomes

1. Identify characteristics that distinguish fungi from other eukaryotes.
2. Compare mitosis in fungi and animals.
3. Explain why fungi are useful for bioremediation.

Mycologists, scientists who study fungi and funguslike protists, believe there may be as many as 1.5 million fungal species. Fungi exist either as single-celled yeasts or in multicellular form. Their reproduction may be either sexual or asexual, and they exhibit an unusual form of mitosis. They are specialized to extract and absorb nutrients from their surroundings by secreting enzymes. Other major features of fungi include a cell wall composed of chitin and reproduction by spores. Recent phylogenetic analysis of DNA sequences and protein sequences indicates that fungi are more closely related to animals than to plants. Fossils and molecular data indicate that animals and fungi last shared a common ancestor

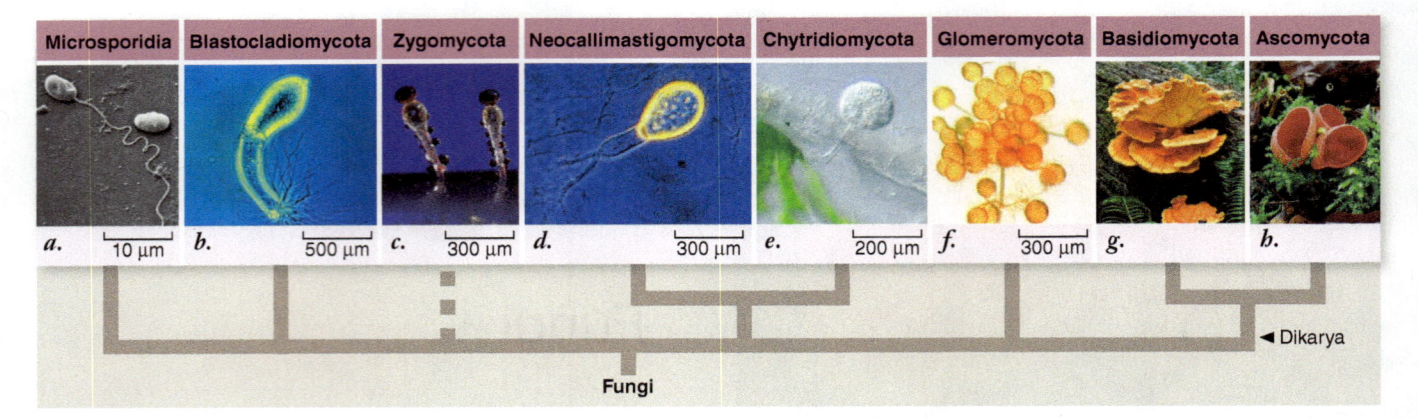

Figure 32.1 Seven major phyla of fungi. All phyla, except Zygomycota, are monophyletic. *a.* Microsporidia, including *Encephalitozoon cuniculi,* are animal parasites. *b. Allomyces arbuscula,* a water mold, is a blastocladiomycete. *c. Pilobolus,* a zygomycete, grows on animal dung and also on culture medium. Stalks about 10 mm long contain dark, spore-bearing sacs. *d.* Neocallimastigales, including *Piromyces communis,* decompose cellulose in the rumens of herbivores. *e.* Some chytrids, including members of the genus *Rhizophydium,* parasitize green algae. *f.* Spores of *Glomus intraradices,* a glomeromycete associated with roots. *g. Laetiporus sulphureus,* a shelf fungus. *h.* The cup fungus *Cookeina tricholoma* is an ascomycete from the rainforest of Costa Rica. In the cup fungi, the spore-producing structures line the cup; in basidiomycetes that form mushrooms such as *Amanita,* they line the gills beneath the cap of the mushroom. All visible structures of fungi, including the ones shown here, arise from an extensive network of filamentous hyphae that penetrates and is interwoven with the substrate on which they grow.

at least 460 MYA, but inconsistencies remain. The last common ancestor with animals was a single cell, and unique solutions to multicellularity evolved both in fungi and in animals.

The phylogenetic relationships among fungi have been the cause of much debate. Traditionally, four fungal phyla were recognized, based primarily on characteristics of the cells undergoing meiosis: Chytridiomycota ("chytrids"), Zygomycota ("zygomycetes"), Ascomycota ("ascomycetes"), and Basidiomycota ("basidiomycetes"). The chytrids and zygomycetes are not monophyletic. The understanding of fungal phylogeny is going through rapid changes, aided by increasing molecular sequence data (figure 32.1 and table 32.1). In 2007, mycologists agreed on seven monophyletic phyla: Microsporidia, Blastocladiomycota, Neocallimastigomycota, Chytridiomycota, Glomeromycota, Basidiomycota, and Ascomycota (figure 32.1). Blastomycetes and neocallimastigomycetes were formerly grouped with the chytrids. Confounding our understanding of the complex relationships among fungi is a growing body of evidence that horizontal gene transfer has occurred in many fungal species.

Fungi may be unicellular or filamentous

The Chytridiomycota and Microsporidia are ancient fungi and, as such, are unicellular organisms. During the evolution of the higher

TABLE 32.1	Fungi		
Group	**Typical Examples**	**Key Characteristics**	**Approximate Number of Living Species**
Chytridiomycota	*Allomyces*	Aquatic, flagellated fungi that produce haploid gametes by sexual reproduction or diploid zoospores by asexual reproduction.	750
Zygomycota	*Rhizopus, Pilobolus*	Multinucleate hyphae lack septa, except for reproductive structures; fusion of hyphae leads directly to formation of a zygote in zygosporangium, in which meiosis occurs just before it germinates; asexual reproduction is most common.	1050
Glomeromycota	*Glomus*	Form arbuscular mycorrhizae. Multinucleate hyphae lack septa. Reproduce asexually.	230
Ascomycota	Truffles, morels	In sexual reproduction, ascospores are formed inside a sac called an ascus; asexual reproduction is also common.	64,000
Basidiomycota	Mushrooms, toadstools, rusts	In sexual reproduction, basidiospores are borne on club-shaped structures called basidia; asexual reproduction occurs occasionally.	33,000

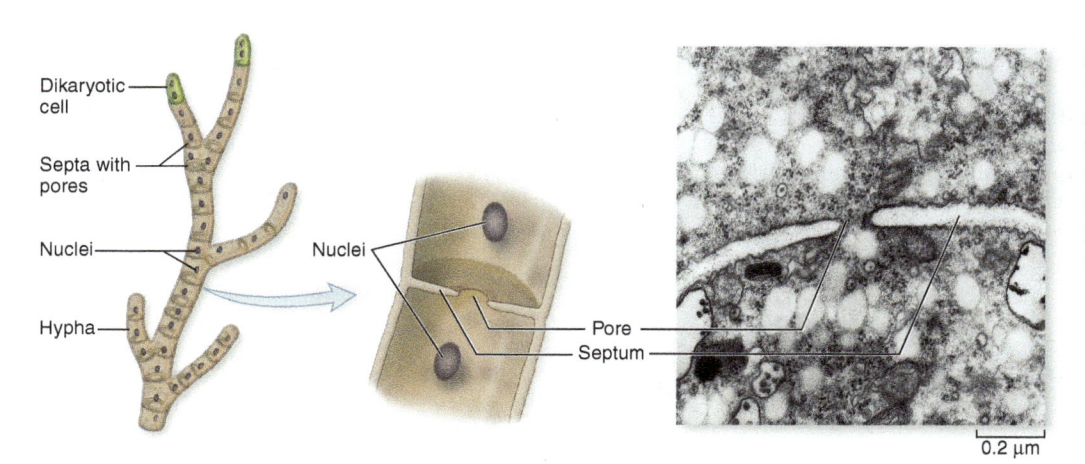

Figure 32.2 A septum. This transmission electron micrograph (and the accompanying schematic drawing) of a section through a hypha of an ascomycete shows a pore through which the cytoplasm streams.

fungi, filamentous growth forms developed. The filaments are strings of cells called hyphae. Some hyphae are continuous or branching tubes filled with cytoplasm and multiple nuclei. The hyphae of the more advanced fungal groups (Ascomycota and Basidiomycota) are typically made up of long chains of cells joined end-to-end and divided by cross-walls called septa (singular, *septum*). The septa rarely form a complete barrier, except when they separate the reproductive cells. Even fungi with septa can be considered one long cell.

Cytoplasm characteristically flows or streams freely throughout the hyphae, passing through major pores in the septa (figure 32.2). Because of this streaming, nutrients synthesized or obtained by any part of the hyphae may be carried to the actively growing tips. As a result, fungal hyphae may grow very rapidly when food and water are abundant and the temperature is optimum. For example, you may have seen mushrooms suddenly appear in a lawn overnight after a rain in summer.

The mycelium

A mass of connected hyphae is called a **mycelium** (plural, *mycelia*). (This word and the term **mycology,** the study of fungi, are both derived from the Greek word *mykes,* meaning fungi.) The mycelium of a fungus (figure 32.3) constitutes a system that may be many meters long. The mycelium grows into the soil, wood, or other material and digestion of the material begins quickly. In two of the four major groups of fungi, reproductive structures formed of interwoven hyphae, such as mushrooms, puffballs, and morels, are produced at certain stages of the life cycle. These structures expand rapidly because of rapid growth of the hyphae.

Cell walls with chitin

The cell walls of most fungi are formed of polysaccharides, including chitin. In contrast, cell walls of plants and many protists contain cellulose, not chitin. Chitin is a modified cellulose consisting of linked glucose units to which nitrogen groups have been added; this polymer is then cross-linked with proteins. Chitin is the same material that makes up the major portion of the hard shells, or exoskeletons, of arthropods, a group of animals that includes insects and crustaceans (see chapter 34). Chitin is one of the shared traits that has led scientists to believe that fungi are more closely related to animals than they are to plants.

Fungal cells may have more than one nucleus

Fungi are different from most animals and plants in that each cell (or hypha) can house one, two, or more nuclei. A hypha that has only one nucleus is called **monokaryotic;** a cell with two nuclei is **dikaryotic.** In a dikaryotic cell, the two haploid nuclei exist independently. Dikaryotic hyphae have some genetic properties of diploids, because both genomes are transcribed.

Sometimes, many nuclei intermingle in the common cytoplasm of a fungal mycelium, which can lack distinct cells. If a dikaryotic or multinucleate hypha has nuclei that are derived from two genetically distinct individuals, the hypha is called **heterokaryotic.** Hyphae whose nuclei are genetically similar to one another are called **homokaryotic.**

Mitosis is not followed by cell division

Mitosis in multicellular fungi differs from that in most other organisms. Because of the linked nature of the cells, the cell itself is not

Figure 32.3 Fungal mycelium. This mycelium, composed of hyphae, is growing through leaves on the forest floor.

the relevant unit of reproduction; instead, the nucleus is. The nuclear envelope does not break down and reform; instead, the spindle apparatus is formed *within* it.

Centrioles are absent in all fungi except chytrids; instead, fungi regulate the formation of microtubules during mitosis with small, relatively amorphous structures called *spindle plaques*. This unique combination of features strongly suggests that fungi originated from some unknown group of single-celled eukaryotes with these characteristics.

**Figure 32.4
Fungal spores.**
Scanning electron micrograph of fungal spores from *Aspergillus*.

10 µm

Fungi can reproduce both sexually and asexually

Many fungi are capable of producing both sexual and asexual spores. When a fungus reproduces sexually, two haploid hyphae of compatible mating types may come together and fuse.

The dikaryon stage

In animals, plants, and some fungi, the fusion of two haploid cells during reproduction immediately results in a diploid cell ($2x$). But in other fungi, namely basidiomycetes and ascomycetes, an intervening dikaryotic stage ($1x + 1x$) occurs before the parental nuclei fuse and form a diploid nucleus. In ascomycetes, this dikaryon stage is brief, occurring in only a few cells of the sexual reproductive structure. In basidiomycetes, however, it can last for most of the life of the fungus, including both the feeding and sexual spore-producing structures.

Reproductive structures

Some fungal species produce specialized mycelial structures to house the production of spores. Examples are the mushrooms we see above ground, the "shelf" fungus that appears on the trunks of dead trees, and puffballs, which can house billions of spores. Because the mycelium is typically underground, the production of above-ground structures aids in spore dispersal.

As noted earlier in this section, the cytoplasm in fungal hyphae normally flows through perforated septa or moves freely in their absence. Reproductive structures are an important exception to this general pattern. When reproductive structures form, they are cut off by complete septa that lack perforations or that have perforations that soon become blocked.

Spores

Spores are the most common means of reproduction among fungi. They may form as a result of either asexual or sexual processes, and they are often dispersed by the wind. When spores land in a suitable place, they germinate, giving rise to a new fungal mycelium.

Because the spores are very small (figure 32.4), they can remain suspended in the air for a long time. Many of the fungi that cause diseases in plants and animals are spread rapidly by such means. The spores of other fungi are routinely dispersed by insects or other small animals. A few fungal phyla retain the ancestral flagella and produce zoospores (motile spores).

Fungi are heterotrophs that absorb nutrients

All fungi obtain their food by secreting digestive enzymes into their surroundings and then absorbing the organic molecules produced by this external digestion. The fungal body plan reflects this approach. Unicellular fungi have the greatest surface area-to-volume ratio of any fungus, maximizing the surface area for absorption. Extensive networks of hyphae also provide an enormous surface area for absorptive nutrition in a fungal mycelium.

Many fungi are able to break down the cellulose in wood, cleaving the linkages between glucose subunits and then absorbing the glucose molecules as food. Most fungi also digest lignin, an insoluble organic compound that strengthens plant cell walls. The specialized metabolic pathways of fungi allow them to obtain nutrients from dead trees and from an extraordinary range of organic compounds, including tiny roundworms called nematodes (figure 32.5a).

The mycelium of the edible oyster mushroom *Pleurotus ostreatus* (figure 32.5b) excretes a substance that paralyzes nematodes that feed on the fungus. When the worms become sluggish and inactive, the fungal hyphae envelop and penetrate their bodies. Then the fungus secretes digestive juices and absorbs the nematode's nutritious contents, just like it would from a plant source.

This fungus usually grows within living trees or on old stumps, obtaining the bulk of its glucose through the enzymatic digestion of cellulose and lignin from plant cell walls. The nematodes it consumes apparently serve mainly as a source of nitrogen—a substance almost always in short supply in biological

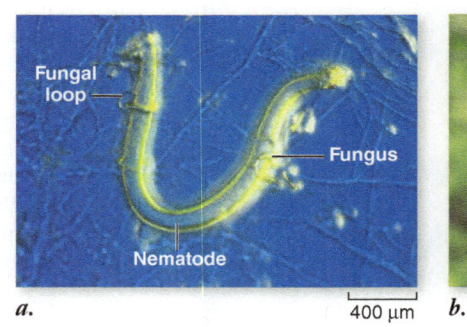

a. Fungal loop — Fungus — Nematode — 400 µm   *b.*

Figure 32.5 Carnivorous fungi. *a.* Fungus obtaining nutrients from a nematode. *b.* The oyster mushroom *Pleurotus ostreatus* not only decomposes wood but also immobilizes nematodes, which the fungus uses as a source of nitrogen.

systems. Other fungi are even more active predators than *Pleurotus,* snaring, trapping, or firing projectiles into nematodes, rotifers, and other small animals on which they prey (figure 32.5a).

Because of their ability to break down almost any carbon-containing compound—even jet fuel—fungi are of interest for use in bioremediation, using organisms to clean up soil or water that is environmentally contaminated. As one example, some fungal species can remove selenium, an element that is toxic in high accumulations, from soils by combining it with other harmless volatile compounds.

Learning Outcomes Review 32.1

Fungi are more closely related to animals than to plants. Many, but not all, fungi form seven monophyletic phyla. A fungus consists of a mass of hyphae (cells) termed a mycelium; cell walls contain the polysaccharide chitin. Mitosis in fungi divides the nucleus but not the hypha itself. Sexual reproduction may occur when hyphae of two different mating types fuse. Haploid nuclei from each type may persist separately in some groups, termed a dikaryon stage. Spores are produced sexually or asexually and are spread by wind or animals. Fungi secrete digestive enzymes externally and then absorb the products of the digestion. Fungi can break down almost any organic compound.

- What differentiates fungi from animals, since both are heterotrophs?
- What protects hyphae from being digested by the fungus's own secreted enzymes?

32.2 Microsporidia: Unicellular Parasites

Learning Outcomes

1. Explain characteristics that led microsporidians to be classified as protists.
2. Describe evidence for placing microsporidians with fungi.

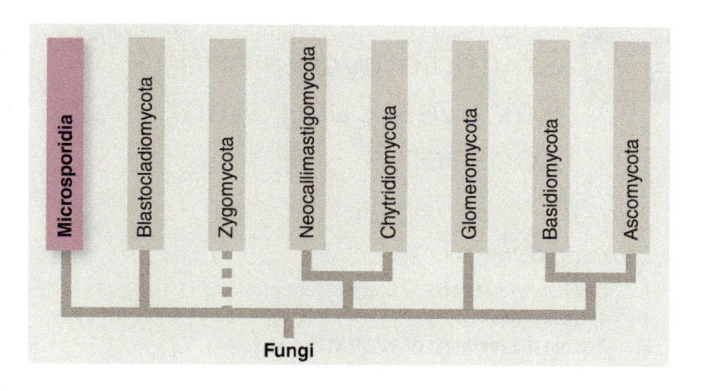

Microsporidia are obligate, unicellular parasites of animals. They are characterized by a specialized organelle, called a polar tube, that allows them to invade host cells. Most microsporidians infect insects, crustaceans, and fish. About 10% of microsporidian species parasitize vertebrates, including humans. Microsporidians do not possess mitochondria and, consequently, cannot carry out aerobic respiration. They are obligate anaerobes. The spores produced by microsporidians are often highly resistant to environmental stresses and can survive for several years.

Encephalitozoon cuniculi and other microsporidians commonly cause disease in immunosuppressed patients, such as those with AIDS and those who have received organ transplants. Microsporidians infect hosts with their spores, which contain a polar tube (figure 32.6). The polar tube extrudes the contents of the spore into the cell and the parasite sets up housekeeping in a vacuole. *E. cuniculi* infects intestinal and neuronal cells, leading to diarrhea and neurodegenerative disease.

Learning Outcomes Review 32.2

Microsporidia lack mitochondria, so their energy demands must be met by anaerobic respiration. As obligate parasites, microsporidians cause diseases in animals, including humans.

- How would you distinguish a microsporidian from the parasitic protistan *Plasmodium*?

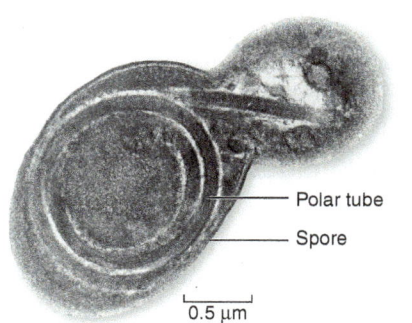

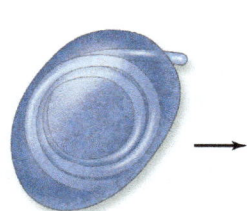

Figure 32.6 The polar tube of the microsporidian *Encephalitozoon cuniculi* spore infects cells. Through an explosive mechanism, a germinating spore discharges the polar tube, which penetrates into the host cell. The contents of the spore, including its cytoplasm and nucleus, travel through the tube into the host cell.

32.3 Chytridiomycota and Relatives: Fungi with Zoospores

Learning Outcomes
1. Distinguish between blastocladiomycetes and microsporidians.
2. Explain the meaning of "chytrid."
3. Discuss possible uses of neocallimastigomycetes.

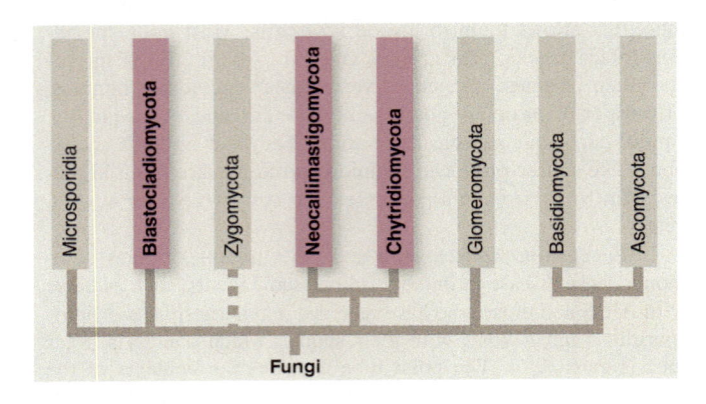

Members of phylum Chytridiomycota, the **chitridiomycetes** or **chytrids**, are aquatic, flagellated fungi that are closely related to ancestral fungi. A distinguishing character of this fungal group is the production of zoospores, which are flagellated asexual spores. Chytrid has its origins in the Greek word *chytridion*, meaning "little pot," referring to the structure that releases the zoospores (figure 32.7).

Chytrids include *Batrachochytrium dendrobatidis*, which has been implicated in the die-off of amphibians. Other chytrids have been identified as plant pathogens (figure 32.8). Traditionally

Figure 32.7 "Little pot." The potlike structure (*chytridion* in Greek) containing the zoospores gives chytrids their name.

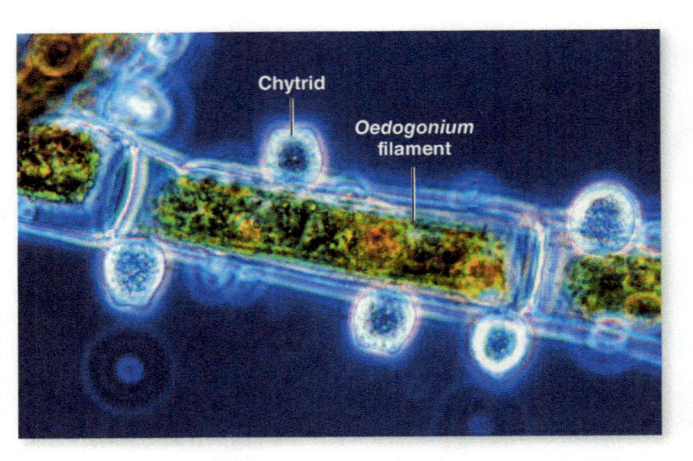

Figure 32.8 Chytridiomycota can be pathogens of plants and algae. The chytrid *Rhizophydium granulosporum* sporangia on a filament of the green alga *Oedogonium*.

the blastocladiomycetes and the neocallimastigomycetes have been grouped with the chytrids. Recent phylogenetic inferences have accorded the two groups their own phyla. Both proposed phyla are discussed here because of their many similarities with the chytrids.

Blastocladiomycota exhibit an alternation of generations

Blastocladiomycetes, also known as blastoclads, have uniflagellated zoospores. Blastocladiomycetes, neocallimastigomycetes, and chytridiomycetes were originally grouped as a single phylum because they all have flagella that have since been lost in other groups, except the microsporidians. Inclusion of multiple genes in phylogenetic analyses has established that the three groups form three separate, monophyletic phyla. Another characteristic feature of blastoclads is the production of ornate thick-walled resting sporangia, which produce zoospores when they germinate.

The blastoclads are the only fungal group that have been found to exhibit a true haploidiplontic life cycle, with an alternation between haploid and diploid generations. *Allomyces* is a water mold with a typical blastoclad life cycle (figure 32.9). Reproduction in *Allomyces* species is enhanced by the secretion of a pheromone, a long-distance chemical signal, from the female gametes that attracts male gametes. Pheromones are similar to hormones, but work between organisms, rather than within organisms. The *Allomyces* pheromone is called sirenin (after the Sirens in Greek mythology) and was the first fungal sex hormone to be identified chemically.

Neocallimastigomycota digest cellulose in ruminant herbivores

Within the rumen (first digestive chamber) of mammalian herbivores, **neocallimastigomycetes** enzymatically digest the cellulose and lignin of the plant biomass in their grassy diet. Sheep, cows,

Figure 32.9
***Allomyces*, a blastocladiomycete that grows in the soil.** *a.* The spherical sporangia can produce either diploid zoospores via mitosis or haploid spores via meiosis. *b.* Life cycle of *Allomyces*, which has both haploid and diploid multicellular stages (alternation of generations).

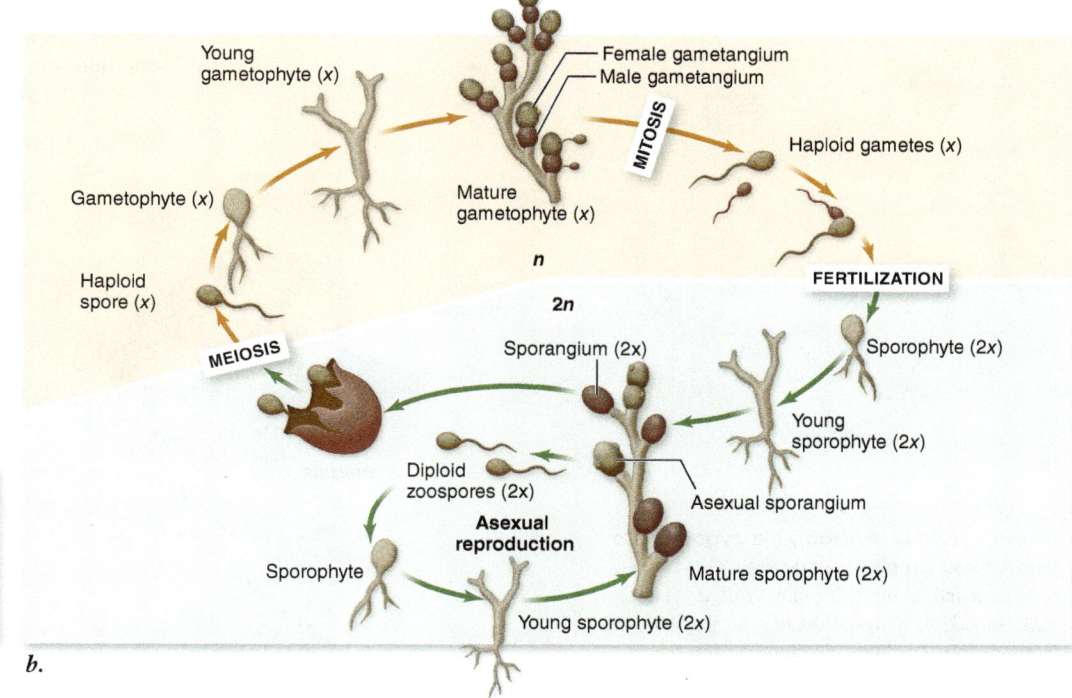

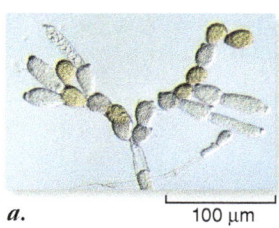

---

kangaroos, and elephants all depend on these fungi to obtain sufficient calories. These anaerobic fungi have greatly reduced mitochondria that lack cristae. Their zoospores have multiple flagella. "Mastig" in Neocalli*mastigo*mycota is Latin for "whips," referencing the multiple flagella.

The genus *Neocallimastix* can survive on cellulose alone. Genes encoding digestive enzymes such as cellulase made their way into the *Neocallimastix* genomes via horizontal gene transfer from bacteria. Horizontal gene transfer is covered in chapter 24. Anaerobic fungi are increasingly being recognized as major contributors to rumen metabolism. As such, they make important contributions to the nutrition of ruminant animals, such as cows, sheep, and deer.

The many enzymes that neocallimastigomycetes use to digest cellulose and lignin in plant cell walls may be useful in biofuel production from cellulose. Although it is possible to obtain ethanol from cellulose, breaking down the cellulose is a major technical hurdle. The use of neocallimastigomycetes fungi to produce cellulosic ethanol is a promising, cost-effective approach.

Learning Outcomes Review 32.3

Three closely related phyla, Chytridiomycota, Blastocladiomycota, and Neocallimastigomycota produce zoospores. Chytrid refers to the potlike shape of the structure releasing the zoospores. *Allomyces*, a representative blastocladiomycete, has uniflagellated zoospores with giant mitochondria, and a haplodiplontic life cycle. Neocallimastigomycetes acquired cellulases from bacteria. They aid ruminant animals in digesting cellulose from plants and may be useful in production of biofuels.

■ *What two features differentiate blastocladiomycetes from microsporidians?*

32.4 Zygomycota: Fungi That Produce Zygotes

Learning Outcomes
1. Describe the defining feature of the zygomycetes.
2. Explain the advantage of zygospore formation.

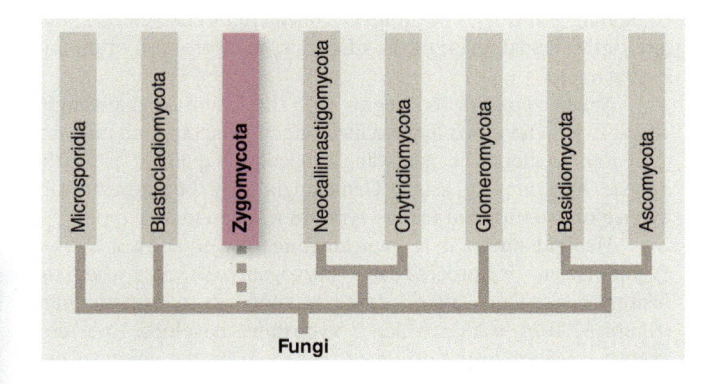

Zygomycetes (phylum Zygomycota) include only about 1050 named species, but they are incredibly diverse. The Zygomycota are not monophyletic, but are included in this chapter as a group while research on their evolutionary history continues. Among them are some of the more common bread molds, which have been assigned to the monophyletic subphylum Mucoromycotina (figure 32.10), as well as a variety of others found on decaying

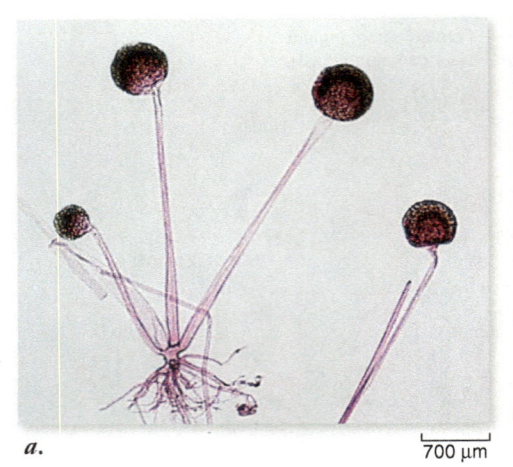

Figure 32.10 Rhizopus, a zygomycete that grows on simple sugars. This fungus is often found on moist bread or fruit. *a.* The dark, spherical, spore-producing sporangia are on hyphae about 1 cm tall. The rootlike hyphae (rhizoids) anchor the sporangia. *b.* Life cycle of *Rhizopus*. The Zygomycota group is named for the zygosporangia characteristic of *Rhizopus*. The (+) and (−) denote mating types.

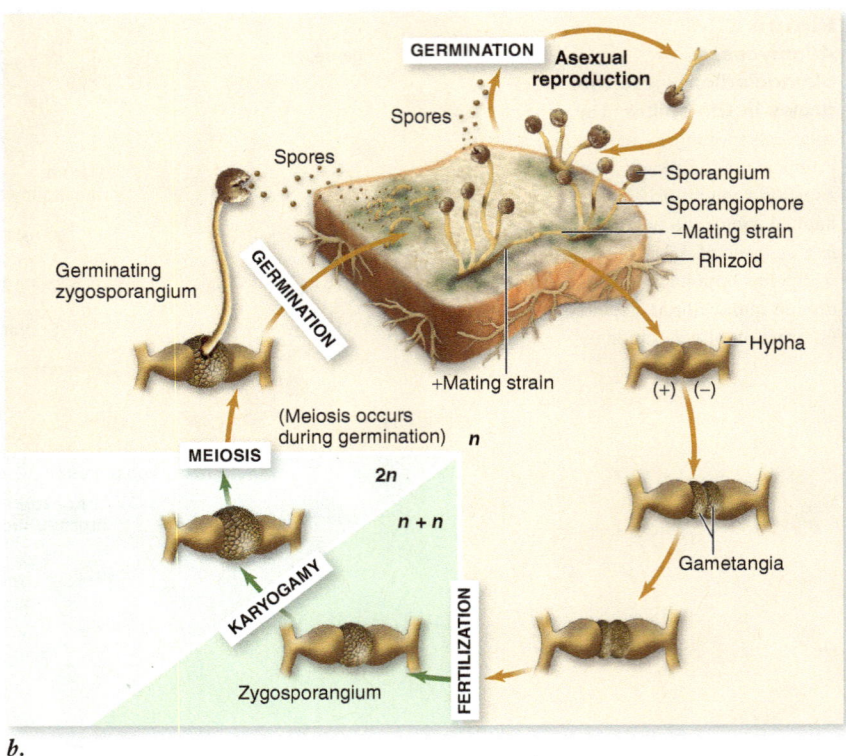

organic material, including strawberries and other fruits. A few human pathogens are in this group.

Sexual reproduction produces a diploid zygote

The zygomycetes lack septa in their hyphae except when they form sporangia (structures containing spores) or gametangia (structures containing gametes). The group is named after a characteristic feature of the sexual phase of the life cycle, the formation of diploid zygote nuclei.

Sexual reproduction begins with the fusion of gametangia, which form where two hyphae meet. Each gametangium contains numerous nuclei. The gametangia become separated from the hyphae by complete septa. Gametangia may be formed from hyphae of two different mating types or a single mating type.

Haploid nuclei in a gametangium fuse to form a diploid zygote nucleus in a process called karyogamy. The area where the fusion has taken place develops into a zygosporangium (figure 32.10*b*), within which a **zygospore** develops. The term zygospore refers to the origin of this spore by the union of two haploid nuclei, analogous to the production of a zygote by fertilization in plants and animals. The zygospore, which may contain one or more diploid nuclei, acquires a thick coat that helps the fungus survive conditions not favorable for growth. The zygomycetes are unique in producing a diploid zygote that does not immediately undergo meiosis to produce haploid spores.

Meiosis, followed by mitosis, occurs during the germination of the zygospore, which releases haploid spores. Haploid hyphae grow when these haploid spores germinate. Except for the zygote nuclei, all nuclei of the zygomycetes are haploid.

Asexual reproduction is common

Asexual reproduction occurs much more frequently than sexual reproduction in the zygomycetes. During asexual reproduction, hyphae produce clumps of erect stalks, called **sporangiophores.** The tips of the sporangiophores form sporangia, which are separated by septa. Thin-walled haploid spores are produced within the sporangia. These spores are shed above the food substrate, in a position where they may be picked up by the wind and dispersed to a new food source. These airborne spores are widely distributed. Bread becomes moldy when spores in the air land and find it to be a suitable substrate on which to germinate and form a mycelium.

Learning Outcomes Review 32.4

Zygomycetes are named for the production of diploid zygote nuclei by fusion of haploid nuclei, a process called karyogamy. The hyphae of zygomycetes are multinucleate, with septa only where gametangia or sporangia are separated. Many zygomycetes form characteristic resting structures called zygosporangia, which contain zygospores that are able to withstand harsh conditions.

■ *Under what conditions would you expect a zygomycete to produce zygospores rather than haploid spores?*

32.5 Glomeromycota: Asexual Plant Symbionts

Learning Outcomes
1. Explain why Glomeromycota is now considered separate from Zygomycota.
2. Provide evidence for the role of the Glomeromycota in the colonization of land by plants.

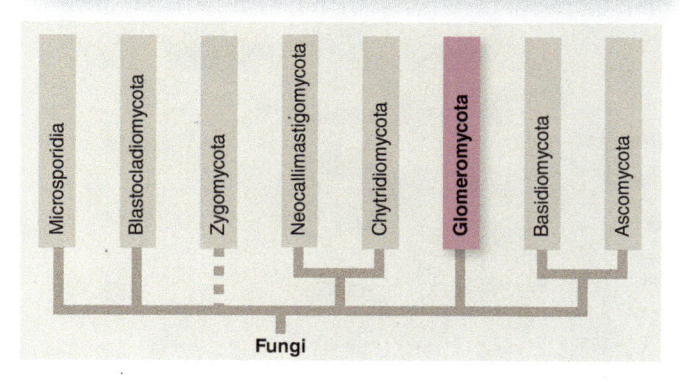

The glomeromycetes are composed of about 230 species. Glomeromycete hyphae are nonseptate and sexual reproduction has not been documented. Despite the small number of species in this group, it was likely critical in the transition of early plants from an aquatic environment to a terrestrial one. Tips of hyphae grow within the root cells of most trees and herbaceous plants, carrying nutrients from the soil into the plant. Plant roots with beneficial fungi growing in their cells are called **arbuscular mycorrhizae.** The specifics of arbuscular mycorrhizal associations and other forms of mycorrhizal interactions are detailed in section 32.8.

Glomeromycetes cannot survive without a host plant, although the host can survive without the fungus. The symbiotic relationship is mutualistic, with the glomeromycetes providing essential minerals, especially phosphorus, and the plants providing carbohydrates. These fungi are being studied for their contributions to sustainable agricultural systems.

The lack of evidence for sexual reproduction makes characterizing glomeromycetes difficult. Glomeromycetes were once grouped with zygomycetes because both lack septa in their hyphae, but comparisons of DNA and small-subunit rRNA sequences indicate that glomeromycetes are a monophyletic clade distinct from zygomycetes. Glomeromycota arose well before the split of the Ascomycota and Basidiomycota, which we will consider in sections 32.6 and 32.7.

Learning Outcomes Review 32.5
Glomeromycetes are a monophyletic fungal lineage based on analysis of small-subunit rRNAs. Their obligate symbiotic relationship with the roots of many extant plants appears to be ancient and may have made it possible for terrestrial plants to evolve.

■ *Why do glomeromycetes require a host plant?*

32.6 Basidiomycota: The Club (Basidium) Fungi

Learning Outcomes
1. Explain which cells in the life cycle of a basidiomycete are diploid.
2. Distinguish between primary and secondary mycelium in basidiomycetes.

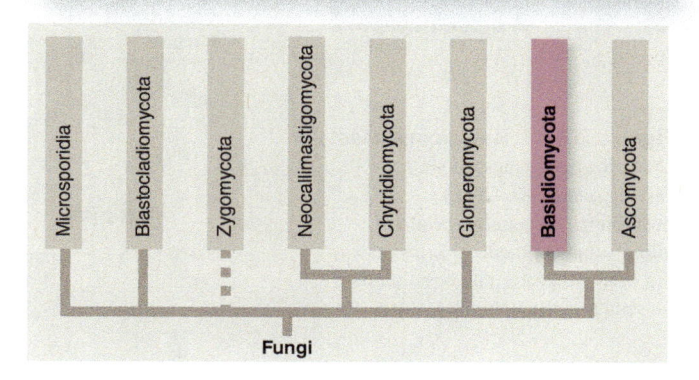

The basidiomycetes (phylum Basidiomycota) include some of the most familiar fungi. Among the basidiomycetes are not only the mushrooms, toadstools, puffballs, jelly fungi, and shelf fungi, but also many important plant pathogens, including rusts and smuts (figure 32.11a). Rust infections resemble rusting metals, and smut infections appear black and powdery due to the spores. Many mushrooms are used as food, but others are hallucinogenic or deadly poisonous.

Basidiomycetes produce sexual spores on basidia

Basidiomycetes are named for their characteristic sexual reproductive structure, the club-shaped **basidium** (plural, *basidia*). Karyogamy (fusion of two nuclei) occurs within the basidium, giving rise to the only diploid cell of the life cycle (figure 32.11b). Meiosis occurs immediately after karyogamy. In the basidiomycetes, the four haploid products of meiosis are incorporated into **basidiospores.** In most members of this phylum, the basidiospores are borne at the end of the basidia on slender projections (sterigmata).

The secondary mycelium of basidiomycetes is heterokaryotic

The life cycle of a basidiomycete continues with the production of monokaryotic hyphae after spore germination. These hyphae lack septa early in development. Eventually, septa form between the nuclei of the monokaryotic hyphae. A basidiomycete mycelium made up of monokaryotic hyphae is called a *primary mycelium.*

Different mating types of monokaryotic hyphae may fuse, forming a dikaryotic mycelium, or *secondary mycelium.* Such a mycelium is heterokaryotic, with two nuclei representing the two

Figure 32.11 Basidiomycetes.
a. The death cap mushroom, *Amanita phalloides*. These mushrooms are usually lethal to animals that eat them. *b.* Life cycle of a basidiomycete. The basidium is the sexual reproductive structure.

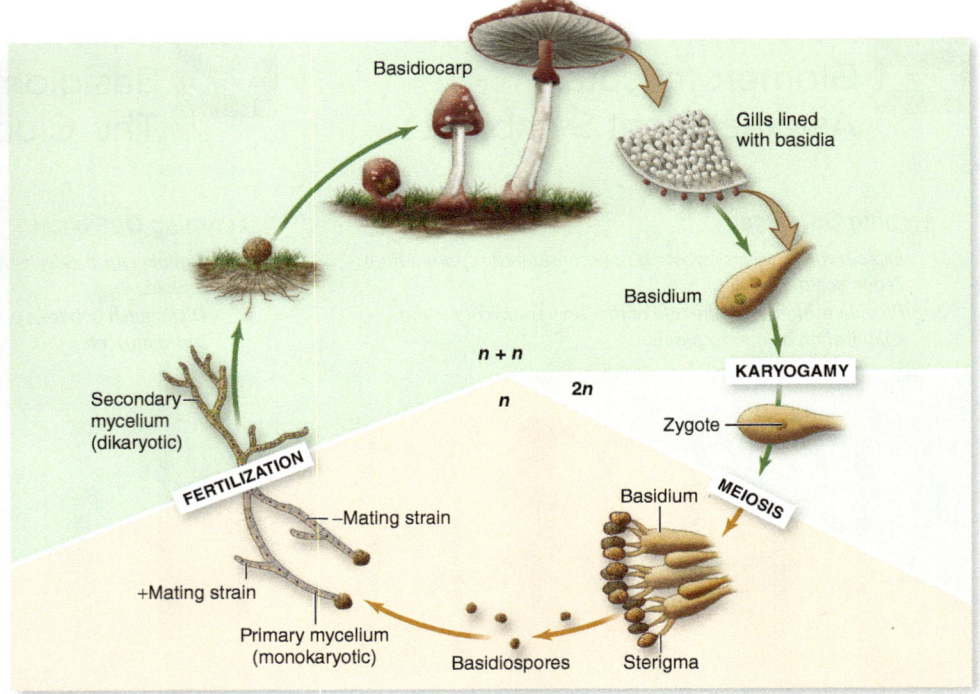

---

different mating types between each pair of septa. This stage is the dikaryon stage described earlier as being a distinguishing feature of fungi. It is found in both the ascomycetes and the basidiomycetes. The two phyla are grouped as the subkingdom Dikarya (two nuclei) because of this commonality.

The **basidiocarps,** or mushrooms, are formed entirely of secondary (dikaryotic) mycelium. Gills, sheets of tissue on the undersurface of the cap of a mushroom, produce vast numbers of minute spores. Because the fungal mycelium is typically underground, above-ground structures are produced to aid in spore dispersal. It has been estimated that a mushroom can produce as many as 40 million spores per hour!

Data analysis You compared spore production of the basidiocarp *Cryptoporus volvatus* growing on fir trees by placing a stainless steel funnel under the fungus to collect and count spores produced between June 1 and September 1. One of your samples had a surface area of 1.10 cm² and produced a total of 1.31×10^8 spores contrasted with a second with an 8.37 cm² surface area that produced 6.87×10^9 spores. Make a quantitative statement about the relative amount of spore production per square centimeter in the two samples.

Learning Outcomes Review 32.6

Basidiomycetes undergo karyogamy, after which meiosis occurs within club-shaped basidia. The primary mycelium consists of monokaryotic hyphae resulting from spore germination. The secondary mycelium of basidiomycetes is the dikaryon stage, in which two nuclei exist within a single hyphal segment.

■ *What distinguishes a dikaryotic cell from a diploid cell?*

32.7 Ascomycota: The Sac (Ascus) Fungi

Learning Outcomes

1. Compare the ascomycetes and the basidiomycetes.
2. List the ways ascomycetes affect humans.

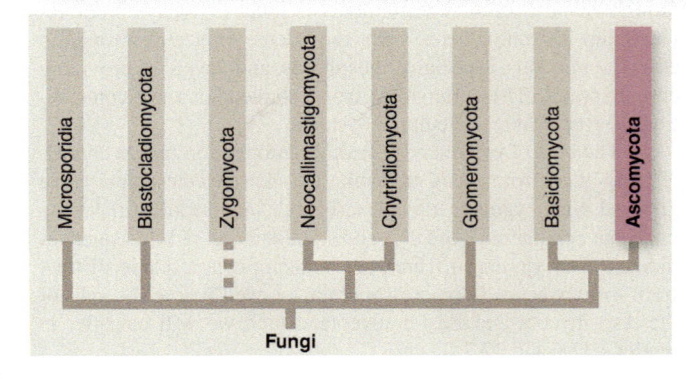

The phylum **Ascomycota** contains about 75% of the known species of fungi. Among the ascomycetes are such familiar and economically important fungi as bread yeasts, common molds, morels (figure 32.12*a*), cup fungi (figure 32.12*b*), and truffles. Also included in this phylum are many serious plant pathogens, including those that produce chestnut blight, *Cryphonectria parasitica,* and Dutch elm disease, *Ophiostoma ulmi.* Penicillin-producing ascomycetes are in the genus *Penicillium*.

Sexual reproduction occurs within the ascus

The ascomycetes are named for their characteristic reproductive structure, the microscopic, saclike **ascus** (plural, *asci*). Karyogamy, the fusion of two haploid nuclei, produces the only diploid nucleus of the ascomycete life cycle (figure 32.12c), occurs within the ascus. The structure of an ascus differs from that of a basidium, although functionally the two are identical. If you compare the life cycles, you will see that basidiospores are produced on stalks, whereas ascospores are produced in a sac. Both types of spores, though, are products of meiosis, so they are haploid and genetically variable.

Asci are differentiated within a structure made up of densely interwoven hyphae, corresponding to the visible portions of a morel or cup fungus, called the **ascocarp.** Meiosis immediately follows karyogamy, forming four haploid daughter nuclei. These usually divide again by mitosis, producing eight haploid nuclei that become walled **ascospores.**

In many ascomycetes, the ascus becomes highly turgid at maturity and ultimately bursts. Ascospores may be thrown as far as 32 cm, an amazing distance considering that most ascospores are only about 10 μm long. This would be equivalent to throwing a baseball 1.25 km—about 10 times the length of a home run!

Asexual reproduction occurs within conidiophores

Asexual reproduction is very common in the ascomycetes. It takes place by means of **conidia** (singular, *conidium*), asexual spores cut off by septa at the ends of modified hyphae called conidiophores. Conidia allow for the rapid colonization of a new food source. The hyphae of ascomycetes are divided by septa, but the septa are perforated, and the cytoplasm flows along the length of each hypha. The septa that cut off the asci and conidia are initially perforated, but later become blocked.

Some ascomycetes have yeast morphology

Most yeasts are ascomycetes with a single-celled lifestyle. Most reproduction in yeasts is asexual and takes place by cell fission or budding, when a smaller cell forms from a larger one (figure 32.13). Sometimes two yeast cells fuse, forming one cell containing two nuclei. This cell may then function as an ascus, with karyogamy followed immediately by meiosis. The resulting ascospores function directly as new yeast cells.

The ability of yeasts to ferment carbohydrates, breaking down glucose to produce ethanol and carbon dioxide, is fundamental to the production of bread, beer, wine, and distilled spirits such

Figure 32.12 Ascomycetes. *a.* This morel, *Morchella esculenta,* is a delicious edible ascomycete that appears in early spring. *b.* A cup fungus. *c.* Life cycle of an ascomycete. Haploid ascospores form within the ascus.

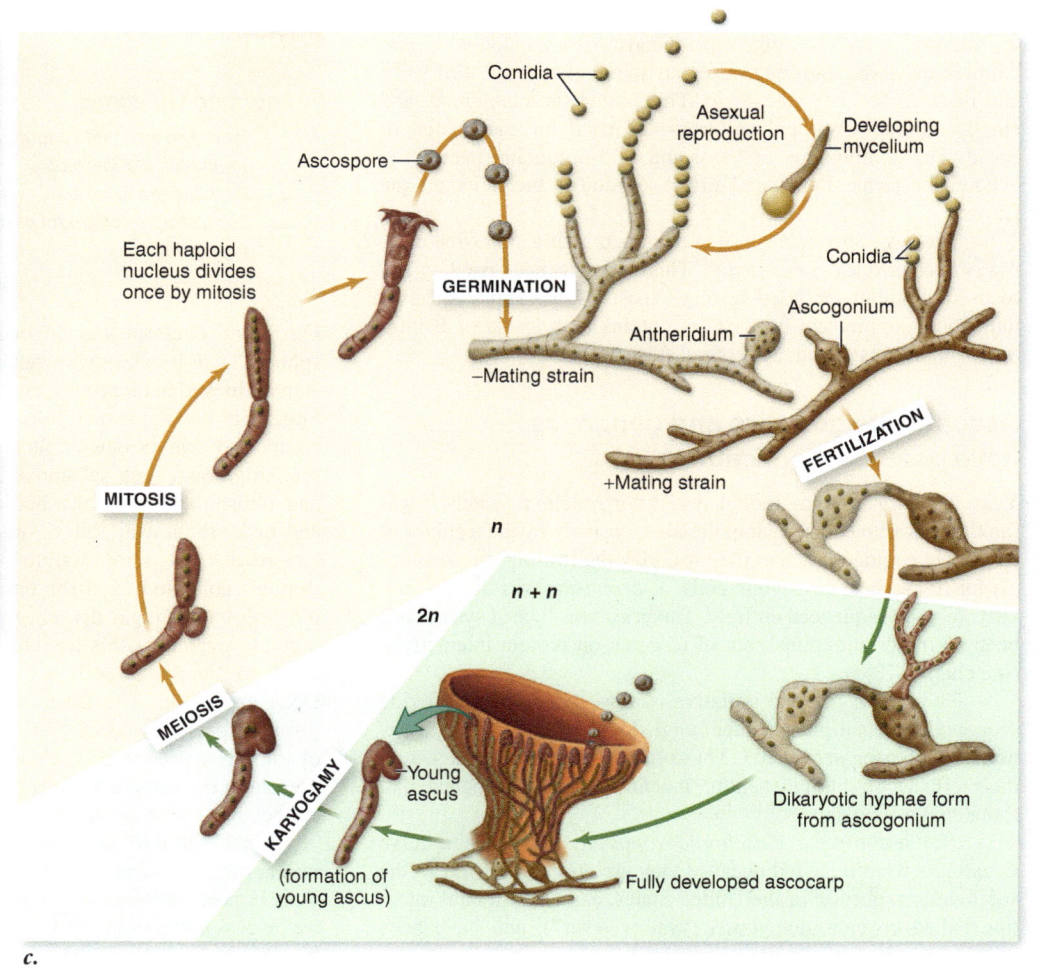

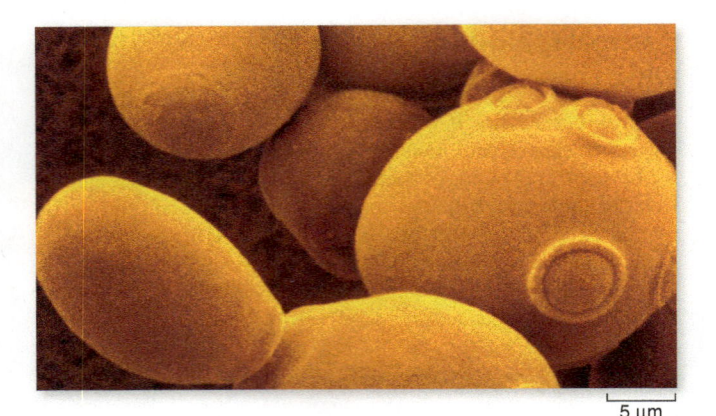

Figure 32.13 Budding in *Saccharomyces*. The smaller cell in the lower left corner of the image is budding off of a larger cell.

as whiskey. About 4 billion of these tiny organisms can fit in a teaspoon. Many different strains of yeast have been domesticated and selected for fermentation, using the sugars in rice, barley, wheat, and corn. Wild yeasts—ones that occur naturally in the areas where wine is made—were important in wine making historically, but domesticated cultured yeasts are normally used now.

Wild yeast, often *Candida milleri,* is still important in making sourdough bread. Unlike most breads that are made with pure cultures of yeast, sourdough uses an active culture of wild yeast and bacteria that generates acid. This culture is maintained, and small amounts—"starter" cultures—are used for each batch of bread. The combination of yeast and acid-producing bacteria is needed for fermentation and gives sourdough bread its unique flavor.

The most important yeast in baking, brewing, and wine making is *Saccharomyces cerevisiae*. This yeast has been used by humans throughout recorded history. Yeast is also employed as a nutritional supplement because it contains high levels of B vitamins and because about 50% of yeast is protein.

Ascomycete genetics and genomics have practical applications

Yeast is a long-standing model system for genetic research. It was the first eukaryote to be manipulated extensively by the techniques of genetic engineering, and they still play the leading role as models for research in eukaryotic cells. *S. cerevisiae* was the first eukaryote to be sequenced entirely. The yeast two-hybrid system has been an important component of research on protein interactions (see chapter 18).

The fungal genome initiative is now under way to provide sequence information on other fungi. More than 300 fungal genomes have been sequenced. These fungi were selected based on their effects on human health, including plant pathogens that threaten our food supply.

The ascomycete *Coccidioides posadasii* and its relative *C. immitis* were included because they are endemic in soil in the southwestern portion of the United States, can cause a fatal infection called coccidioidomycosis ("valley fever"), and have been considered a possible bioterrorism threat.

A second important criterion in selecting which fungi to sequence was the potential to provide information on fungal evolution. This new information will complement and expand our understanding of the diverse fungal kingdom.

Learning Outcomes Review 32.7

Ascomycetes undergo karyogamy (nuclear fusion) within a saclike structure, the ascus. The function of the ascus is therefore identical to that of the basidium, although they are structurally different. Meiosis follows, resulting in the production of haploid ascospores. Asexual reproduction is common in the ascomycetes. Yeasts are unicellular ascomycetes. Ascomycetes include both beneficial forms used as foods, and in the production of foods, and harmful forms responsible for diseases and spoilage. Some ascomycetes are employed in scientific research.

■ *Coccidioidomycosis is caused by inhaling spores; it often occurs in farmworkers in the southwest United States. What would help prevent this disease?*

32.8 Ecology of Fungi

Learning Outcomes

1. Identify a trait that contributes to the value of fungi in symbiotic relationships.
2. Describe the living components of a lichen.
3. List examples of fungal associations with different organisms.

Fungi and bacteria are the principal decomposers in the biosphere. They break down organic materials and return the substances locked in those molecules to circulation in the ecosystem. Fungi can break down cellulose and lignin, an insoluble organic compound that is one of the major constituents of wood. By breaking down such substances, fungi release carbon, nitrogen, and phosphorus from the bodies of living or dead organisms and make them available to other organisms. Many fungal species are difficult or impossible to culture in the lab. Consequently, samples collected from the natural environment represent only the tip of the fungal diversity iceberg. A promising new technique to overcome this limitation is the study of environmental DNA. A sample of soil or water, for example, is subjected to DNA extraction. The DNA is then amplified using PCR with primers specific to DNA that encodes ribosomal RNA. Analysis of the DNA produced by PCR reveals the organisms that were present in the sample, even if the organisms themselves were not collected.

In addition to their role as decomposers, fungi have entered into fascinating relationships with a variety of life-forms. Interactions between different species are described in chapter 57 where we discuss community ecology, but we cover fungal relationships briefly here because of their unique character.

Fungi have a range of symbioses

The interactions, or symbioses, between fungi and other living organisms fall into a broad range of categories. In some cases, the symbiosis is an **obligate symbiosis** (essential for survival), and in other cases it is a **facultative symbiosis** (the fungus can survive without the host). Within a group of closely related fungi, several different types of symbiosis can be found.

First, here is a summary of ways in which living things can interact. **Pathogens** and **parasites** gain resources from their host, but they have a negative effect on the host that can even lead to death. The difference between pathogens and parasites is that pathogens cause disease, but parasites do not, except in extreme cases.

Commensal relationships benefit one partner but do not harm the other. Fungi that are in a **mutualistic** relationship benefit both themselves and their hosts. Many of these relationships are described in this section.

Endophytes live inside plants and may protect plants from parasites

Endophytic fungi live inside plants, actually in the intercellular spaces. Found throughout the plant kingdom, many of these relationships may be examples of parasitism or commensalism.

There is growing evidence that some of these fungi protect their hosts from herbivores by producing chemical toxins or deterrents. Most often, the fungus synthesizes alkaloids that protect the plant. As you will learn in chapter 39, plants also synthesize a wide range of alkaloids, many of which serve to defend the plant. Some endophytes are being studied for their production of anticancer substances, including taxol.

One way to assess whether an endophyte is enhancing the health of its host plant is to grow plots of plants with and without the same endophyte. An experiment with perennial ryegrass, *Lolium perenne*, demonstrated that it is more resistant to aphid reproduction when an endophytic fungus, *Neotyphodium*, is present (figure 32.14).

Lichens are an example of symbiosis between different kingdoms

Lichens (figure 32.15) are symbiotic associations between a fungus and a photosynthetic partner. Although many lichens are excellent examples of mutualism, some fungi are parasitic on their photosynthetic host.

Composition of a lichen

Ascomycetes are the fungal partners in all but about 20 of the approximately 20,000 species of lichens estimated to exist. Most of the visible body of a lichen consists of its fungus, but between the filaments of that fungus are cyanobacteria, green algae, or sometimes both (figure 32.16).

Specialized fungal hyphae penetrate or envelop the photosynthetic cell walls within them and transfer nutrients directly to the fungal partner. Note that although fungi penetrate the cell wall, they do not penetrate the plasma membrane. Biochemical

SCIENTIFIC THINKING

Hypothesis: An endophytic fungus can reduce aphid reproduction on perennial ryegrass.

Prediction: There will be fewer aphids on perennial ryegrass (Lolium perenne) infected with the endophytic fungus than on the uninfected ryegrass plant.

Test: Place five adult aphids on each pot of 2-week-old grass plants with and without endophytic fungi. Place pots in perforated bags and grow for 36 days. Count the number of aphids in each pot. Repeat the experiment three times.

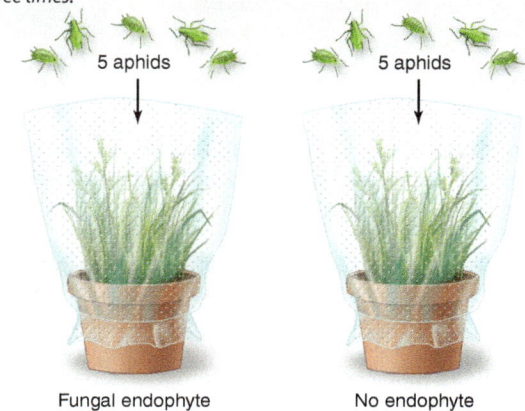

Result: Significantly more aphids were found on the uninfected grass plants. Error bars indicate variability among the three trials.

Conclusion: Aphid reproduction was lower on perennial ryegrass infected with an endophytic fungus.

Further Experiments: How do you think the fungus protects the plants from herbivory? If it secretes chemical toxins, could you use this basic experimental design to test specific fungal compounds?

Figure 32.14 Effect of the fungal endophyte *Neotyphodium* on the aphid population living on perennial ryegrass, *Lolium perenne*.

Figure 32.15 Lichens are found in a variety of habitats. *a.* A fruticose lichen, growing on soil. *b.* A foliose ("leafy") lichen, growing on the bark of a tree. *c.* A crustose lichen, growing on rocks, leading to the breakdown of rock into soil.

signals sent out by the fungus apparently direct its cyanobacterial or green algal component to produce metabolic substances that it does not produce when growing independently of the fungus.

The fungi in lichens are unable to grow normally without their photosynthetic partners, and the fungi protect their partners from strong light and desiccation. When fungal components of lichens have been experimentally isolated from their photosynthetic partner, they survive, but grow very slowly.

Ecology of lichens

The durable construction of the fungus combined with the photosynthetic properties of its partner have enabled lichens to invade the harshest habitats—the tops of mountains, the farthest northern and southern latitudes, and dry, bare rock faces in the desert. In harsh, exposed areas, lichens are often the first colonists, breaking down the rocks and setting the stage for the invasion of other organisms.

Lichens are often strikingly colored because of the presence of pigments that probably play a role in protecting the photosynthetic partner from the destructive action of the Sun's rays. These same pigments may be extracted from the lichens and used as natural dyes. The traditional method of manufacturing Scotland's famous Harris tweed used fungal dyes.

Lichens vary in sensitivity to pollutants in the atmosphere, and some species are used as bioindicators of air quality. Their sensitivity results from their ability to absorb substances dissolved in rain and dew. Lichens are generally absent in and around cities because of automobile traffic and industrial activity, but some are adapted to these conditions. As pollution decreases, lichen populations tend to increase.

Mycorrhizae are fungi associated with roots of plants

The roots of about 90% of all plant families have species that are involved in mutualistic symbiotic relationships with certain fungi. It has been estimated that these fungi probably amount to 15% of the total weight of the world's plant roots. Associations of this kind are termed **mycorrhizae,** from the Greek words for fungus and root.

The fungi in mycorrhizal associations function as extensions of the root system. The fungal hyphae dramatically increase the amount of soil contact and total surface area for absorption. When mycorrhizae are present, they aid in the direct transfer of phosphorus, zinc, copper, and other mineral nutrients from the soil into the roots. The plant, on the other hand, supplies organic carbon to the fungus, so the system is an example of mutualism.

There are two principal types of mycorrhizae (figure 32.17). In arbuscular mycorrhizae, the fungal hyphae penetrate the outer cells of the plant root, forming coils, swellings, and minute branches; they also extend out into the surrounding soil. In **ectomycorrhizae,** the hyphae surround but do not penetrate the

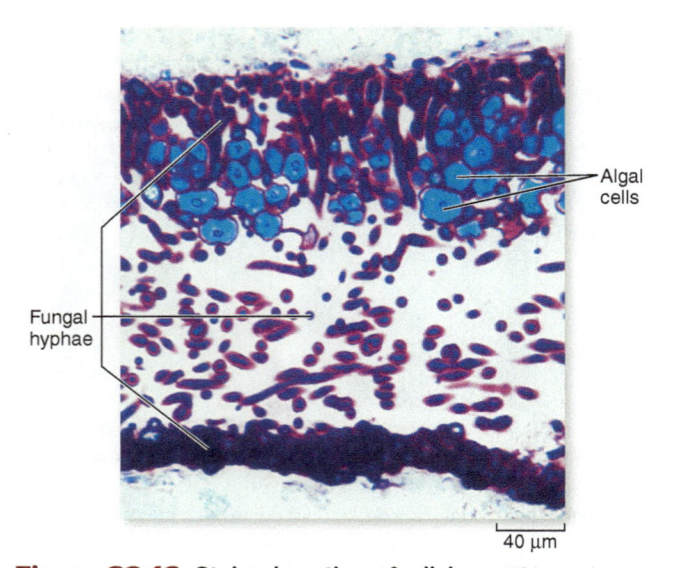

Figure 32.16 Stained section of a lichen. This section shows fungal hyphae *(purple)* more densely packed into a protective layer on the top and, especially, the bottom layer of the lichen. The blue cells near the upper surface of the lichen are those of a green alga. These cells supply carbohydrate to the fungus.

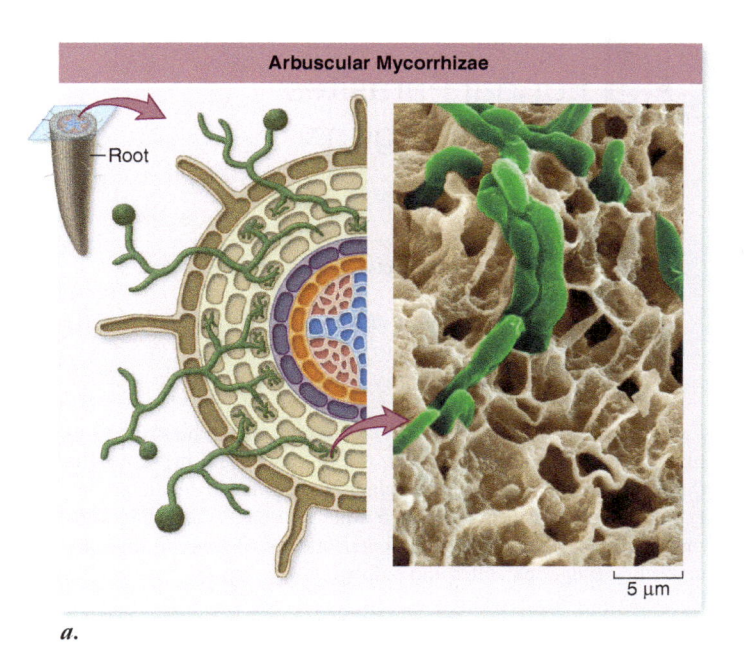

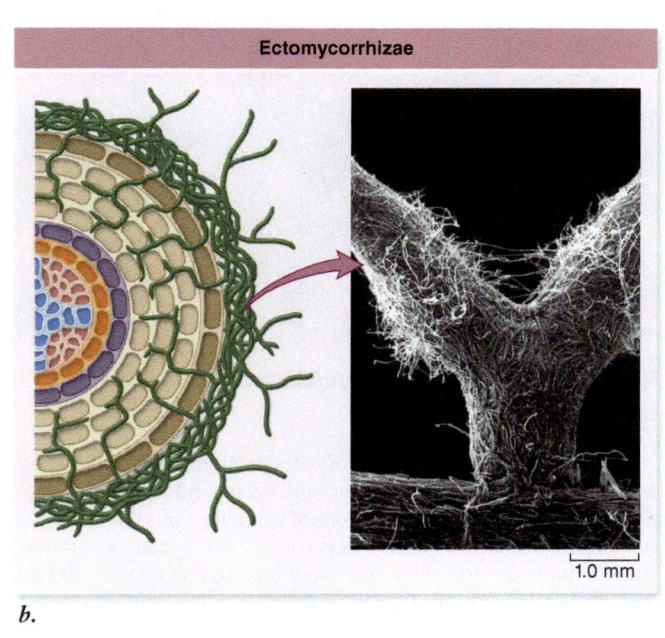

Figure 32.17 Arbuscular mycorrhizae and ectomycorrhizae. *a.* In arbuscular mycorrhizae, fungal hyphae penetrate the root cell wall of plants but not the plant membranes. *b.* Ectomycorrhizae on the roots of a Eucalyptus tree do not penetrate root cells, but grow around and extend between the cells.

cell walls of the roots. In both kinds of mycorrhizae, the mycelium extends far out into the soil. A single root may associate with many fungal species.

Arbuscular mycorrhizae

Arbuscular mycorrhizae are by far the more common of the two types, involving roughly 75% of all plant species (figure 32.17*a*). The fungal component is the glomeromycetes.

Unlike mushrooms, none of the glomeromycetes produce above-ground fruiting structures, and as a result, it is difficult to arrive at an accurate count of the number of extant species. Arbuscular mycorrhizal fungi are being studied intensively because they are potentially capable of increasing crop yields with lower phosphate and energy inputs.

The earliest fossil plants often show arbuscular mycorrhizal roots. Such associations may have played an important role in allowing plants to colonize land. The soils available at such times would have been sterile and lacking in organic matter. Plants that form mycorrhizal associations are particularly successful in infertile soils; considering the fossil evidence, it seems reasonable that mycorrhizal associations helped the earliest plants succeed on such soils. In addition, the closest living relatives of early vascular plants surviving today continue to depend strongly on mycorrhizae.

Some nonphotosynthetic plants also have mycorrhizal associations, but the symbiosis is one-way because the plant has no photosynthetic resources to offer. Instead of a two-partner symbiosis, a tripartite symbiosis is established. The fungal mycelium extends between a photosynthetic plant and a nonphotosynthetic, parasitic plant. This third, nonphotosynthetic member of the symbiosis is called an *epiparasite*. Not only does it obtain phosphate from the fungus, but it uses the fungus to channel carbohydrates from the photosynthetic plant to itself. Epiparasitism also occurs in ectomycorrhizal symbiosis.

Ectomycorrhizae

Ectomycorrhizae (figure 32.17*b*) involve far fewer kinds of plants than do arbuscular mycorrhizae—perhaps a few thousand. Most ectomycorrhizal hosts are forest trees, such as pines, oaks, birches, willows, eucalyptus, and many others. The fungal components in most ectomycorrhizae are basidiomycetes, but some are ascomycetes.

Most ectomycorrhizal fungi are not restricted to a single species of plant, and most ectomycorrhizal plants form associations with many ectomycorrhizal fungi. Different combinations have different effects on the physiological characteristics of the plant and its ability to survive under different environmental conditions. At least 5000 species of fungi are involved in ectomycorrhizal relationships.

Fungi also form mutual symbioses with animals

A range of mutualistic fungi–animal symbioses has been identified. Ruminant animals host neocallimastigomycete fungi in their gut. The fungus gains a nutrient-rich environment in exchange for releasing nutrients from grasses with high cellulose and lignin content.

One tripartite symbiosis involves ants, plants, and fungi. Leaf-cutter ants are the dominant herbivore in the New World tropics. These ants have an obligate symbiosis with specific fungi that they have domesticated and maintain in an underground garden. The ants provide fungi with leaves to eat and protection from pathogens and other predators (figure 32.18). The fungi are the ants' food source.

Figure 32.18 Ant–fungi symbiosis. Ants farming their fungal garden.

Depending on the species of ant, the ant nest can be as small as a golf ball or as large as 50 cm in diameter and many feet deep. Some nests are inhabited by millions of leaf-cutter ants that maintain fungal gardens. These social insects have a caste system, and different ants have specific roles. Traveling on trails as long as 200 m, leaf-cutter ants search for foliage for their fungi. A colony of ants can defoliate an entire tree in a day. This ant farmer–fungi symbiosis has evolved multiple times and may have occurred as early as 50 MYA.

Learning Outcomes Review 32.8

Fungi and bacteria are the primary decomposers in ecosystems. A range of symbiotic relationships have evolved between fungi and plants. Endophytes live inside tissues of a plant and may protect it from parasites. Lichens are a complex symbiosis between fungal species and cyanobacteria or green algae. Mycorrhizal associations between fungi and plant roots are mutually beneficial and in some cases are obligate symbioses. Fungi have also coevolved with animals in mutualistic relationships.

- How might the symbiosis between fungi and ants have evolved?

32.9 Fungal Parasites and Pathogens

Learning Outcomes

1. Review the pathogenic effects of fungi and the targets they affect.
2. Explain why treating fungal disease in animals is particularly difficult.

Fungi can destroy plants and create significant problems for human health. A major problem in the treatment and prevention is that fungi are eukaryotes, as are plants and animals. Understanding how fungi are distinct from these other two eukaryotic kingdoms may lead to safer and more efficient means of treating diseases caused by fungal parasites and pathogens.

Fungal infestation can harm plants and those who eat them

Fungal species cause many diseases in plants (figure 32.19), and they are responsible for billions of dollars in agricultural losses every year. Not only are fungi among the most harmful pests of living plants, but they also spoil food products that have been harvested and stored. In addition, fungi can secrete substances into the foods they are infesting that make these foods unpalatable, carcinogenic, or poisonous.

Pathogenic fungi–plant symbioses are numerous, and fungal pathogens of plants can also harm the animals that consume the plants. *Fusarium* species growing on spoiled food produce highly toxic substances, including vomitoxin, which has been implicated in brain damage in humans and animals in the southwestern United States.

Figure 32.19 World's largest organism?
a. Armillaria, a pathogenic fungus shown here afflicting three discrete regions of coniferous forest in Montana, grows out from a central focus as a single, circular clone. The large patch at the bottom of the picture is almost 8 hectares. *b.* Close-up of tree destroyed by *Armillaria. c. Armillaria* growing on a tree.

Figure 32.20 Maize (corn) fungal infections. *a. Ustilago maydis* infections of maize are a delicacy in Hispanic cuisine. *b.* A photomicrograph of *Aspergillus flavus* conidia. *Aspergillus flavus* infects maize and can produce aflatoxins that are harmful to animals.

Figure 32.21 Frog killed by chytridiomycosis. Lesions formed by the chytrid can be seen on the abdomen of this frog.

Aflatoxins, which are among the most carcinogenic compounds known, are produced by some *Aspergillus flavus* strains growing on corn, peanuts, and cotton seed (figure 32.20). Aflatoxins can also damage the kidneys and the nervous system of animals, including humans. Most developed countries have legal limits on the concentration of aflatoxin permitted in foods. More recently, aflatoxins have been considered as possible bioterrorism agents.

In contrast, corn smut is a maize fungal disease that is harmful to the plant, but not the animals that consume it (figure 32.20). Corn smut is caused by the basidiomycete *Ustilago maydis* and is edible.

Fungi can harm humans and other animals

Human and animal diseases can also be fungal in origin. Some common diseases, such as ringworm (which is not a worm but a fungus), athlete's foot, and nail fungus, can be treated with topical antifungal ointments and in some cases with oral medication.

Fungi can create devastating human diseases that are often difficult to treat because of the close phylogenetic relationship between fungi and animals. Yeast ascomycetes are important pathogens that cause diseases such as thrush, an infection of the mouth; the yeast *Candida* causes common oral or vaginal infections. *Pneumocystis jiroveci* (formerly *P. carinii*) invades the lungs, disrupting breathing, and can spread to other organs. In immunosuppressed AIDS patients, this infection can lead to death. The main sterol on fungal cell membranes is ergosterol, whereas that in animals is cholesterol. Consequently, antifungal drugs often target ergosterol.

Mold allergies are common, and mold-infested "sick" buildings pose concerns for inhabitants. Individuals with suppressed immune systems and people undergoing steroid treatments for inflammatory disorders are particularly at risk for fungal disease.

An example of a parasitic fungi–animal symbiosis is **chytridiomycosis,** first identified in 1998 as an emerging infectious disease of amphibians. Amphibian populations have been declining worldwide for over three decades. The many possible causes for the decline in amphibian numbers are covered in chapter 59. In this chapter, a primary causative agent, fungal infection, is explored. The decline correlates with the presence of the chytrid *Batrachochytrium dendrobatidis* encased in the skin (figure 32.21), identified after extensive studies of frog carcasses. Sick and dead frogs were more likely than healthy frogs to have flasklike structures encased in their skin, which proved to be associated with chytrid spore production.

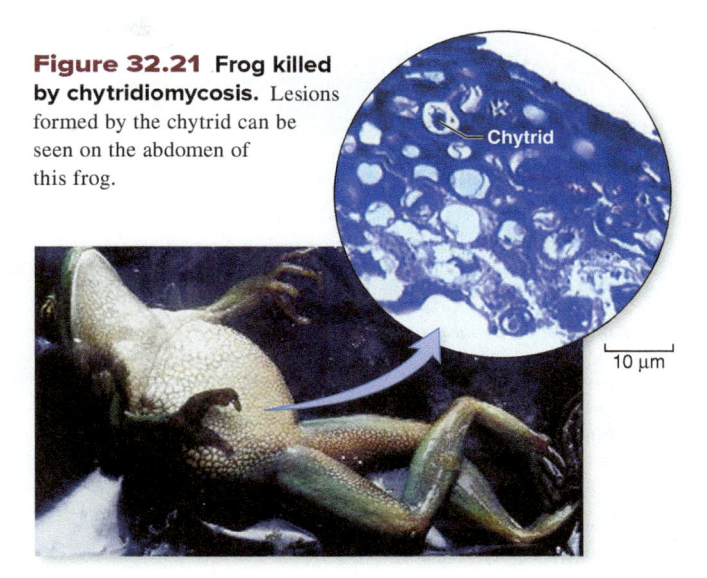

The connection with *B. dendrobatidis* has been supported by DNA sequence data, by isolating and culturing the chytrid, and by infecting healthy frogs with the organism and replicating disease symptoms. The *B. dendrobatidis* genome has now been sequenced. The infection reduces sodium and potassium transport across the skin, altering electrolyte balance, which leads to cardiac arrest. Bathing frogs in antifungal drugs can halt the disease or even eliminate the chytrids.

The pathogenic nature of some fungi can be important for the control of insect pests. For example, microsporidians help to keep populations of European corn borer, spruce budworm, and red fire ants in check. The microsporidian *Vavraia culicis* infects mosquitoes and shortens their lives. This parasite is being tested for the control of malaria.

Learning Outcomes Review 32.9

Fungi can severely harm or kill both plants and animals, either by direct infection or by secretion of toxins and carcinogens. Treatment of fungal disease and parasitism in animals is made difficult by the close relationship between fungi and animals; what is damaging to the fungus may also have ill effects on the host.

- *What is likely to be the most common mechanism for the spread of fungal disease?*

Chapter Review

32.1 Defining Fungi

Fungi are more closely related to animals than to plants and form seven monophyletic phyla (figure 32.1).

Fungi are heterotrophic and have hyphal cells; their cell walls contain chitin. They may have a dikaryon stage and undergo nuclear mitosis.

Fungi may be unicellular or filamentous.
A mass of connected hyphae is termed a mycelium. Hyphae can be continuous and multinucleate, or they may be divided into long chains of cells separated by cross-walls called septa.

The chitin found in fungi cell walls is the same material found in the exoskeletons of arthropods.

Fungal cells may have more than one nucleus.
A hypha with only one nucleus is monokaryotic; a hypha with two nuclei is dikaryotic. The two haploid nuclei exist independently, but both genomes are transcribed so that some properties of diploids may be observed.

Mitosis is not followed by cell division.
Because cells are linked, the cell is not the relevant unit of reproduction, but rather the nucleus is. The spindle forms inside the nuclear envelope, which does not break down and re-form.

Fungi can reproduce both sexually and asexually.
Fungi can reproduce sexually by fusion of hyphae from two compatible mating types or hyphae from the same fungus. Spores can form either by asexual or sexual reproduction and are usually dispersed by the wind.

Fungi are heterotrophs that absorb nutrients.
Fungi obtain their nutrients through excreting enzymes for external digestion and then absorbing the products.

32.2 Microsporidia: Unicellular Parasites

Microsporidia are obligate cellular parasites that lack mitochondria but may have had them at one time; they were previously classed with the protists.

32.3 Chytridiomycota and Relatives: Fungi with Zoospores

Chytrids form symbiotic relationships; they have been implicated in the decline of amphibian species.

Blastocladiomycota exhibit an alternation of generations.
Allomyces, a blastocladiomycete, is an example of a fungus with a haplodiplontic life cycle (figure 32.9).

Neocallimastigomycota digest cellulose in ruminant herbivores.
Neocallimastigomycetes have enzymes that can digest cellulose and lignin; they may have uses in production of biofuels.

32.4 Zygomycota: Fungi That Produce Zygotes

Sexual reproduction produces a diploid zygote (figure 32.10).
Zygomycetes all produce a diploid zygote. In sexual reproduction, fusion (karyogamy) of the haploid nuclei of gametangia produces diploid zygote nuclei. These become zygospores.

Asexual reproduction is common.
Sporangia produce airborne haploid spores; bread mold is a common example of a zygomycete.

32.5 Glomeromycota: Asexual Plant Symbionts

Glomeromycete hyphae form intracellular associations with plant roots and are called arbuscular mycorrhizae.

The glomeromycetes show no evidence of sexual reproduction.

32.6 Basidiomycota: The Club (Basidium) Fungi

The basidiocarp is the visible reproductive structure of this group, which includes mushrooms, toadstools, puffballs, and others.

Basidiomycetes produce sexual spores on basidia (figure 32.11).
Karyogamy occurs within the basidia, giving rise to a diploid cell. Meiosis then ultimately results in four haploid basidiospores.

The secondary mycelium of basidiomycetes is heterokaryotic.
Primary mycelium is monokaryotic, but different mating types may fuse to form the secondary mycelium. Maintenance of two haploid genomes allows greater genetic plasticity.

32.7 Ascomycota: The Sac (Ascus) Fungi

Sexual reproduction occurs within the ascus (figure 32.12).
Karyogamy occurs only in the ascus and results in a diploid nucleus. Meiosis and mitosis then result in eight haploid nuclei in walled ascospores.

Asexual reproduction occurs within conidiophores.
Asexual reproduction is very common and occurs by means of conidia formed at the end of modified hyphae called conidiophores.

Some ascomycetes have yeast morphology.
Yeasts usually reproduce by cell fission or budding.

Ascomycete genetics and genomics have practical applications.

32.8 Ecology of Fungi

Fungi are organisms capable of breaking down cellulose and lignin.

Fungi have a range of symbioses.
Fungi can be pathogenic or parasitic, commensal, or mutualistic.

Endophytes live inside plants and may protect plants from parasites.

Lichens are an example of symbiosis between different kingdoms.
A lichen is composed of a fungus, usually an ascomycete, along with cyanobacteria, green algae, or both.

Mycorrhizae are fungi associated with roots of plants.
Arbuscular mycorrhizae are common and involve glomeromycetes; ectomycorrhizae are primarily found in forest trees and involve basidiomycetes and a few ascomycetes.

Fungi also form mutual symbioses with animals.
Some ants grow "farms" of fungi by providing plant material.

32.9 Fungal Parasites and Pathogens

Fungal infestation can harm plants and those who eat them.
Fungi spread via spores and can secrete chemicals that make food unpalatable, carcinogenic, or poisonous.

Fungi can harm humans and other animals.
Treatment of fungal diseases in animals is difficult because of the similarities between the two kingdoms.

Review Questions

UNDERSTAND

1. Which of the following is NOT a characteristic of a fungus?
   a. Cell walls made of chitin
   b. A form of mitosis different from plants and animals
   c. Ability to conduct photosynthesis
   d. Filamentous structure

2. A fungal cell that contains two genetically different nuclei would be classified as
   a. monokaryotic.
   b. bikaryotic.
   c. homokaryotic.
   d. heterokaryotic.

3. Which of the following groups of fungi is NOT monophyletic?
   a. Zygomycota
   b. Basidiomycota
   c. Glomeromycota
   d. Ascomycota

4. Based on physical characteristics, the _____ represent the most ancient phylum of fungi.
   a. Basidiomycota
   b. Zygomycota
   c. Ascomycota
   d. Chytridiomycota

5. The early evolution of terrestrial plants was made possible by mycorrhizal relationships with the
   a. Zygomycetes.
   b. Glomeromycota.
   c. Ascomycota.
   d. Basidiomycota.

6. Symbiotic relationships occur between the fungi and
   a. plants.
   b. bacteria.
   c. animals.
   d. All of the choices are correct.

7. Which of the following species of fungi is NOT associated with diseases in humans?
   a. *Pneumocystis jiroveci*
   b. *Aspergillus flavus*
   c. *Candida albicans*
   d. *Batrachochytrium dendrobatidis*

APPLY

1. In a culture of hyphae of unknown origin you notice that the hyphae lack septa and that the fungi reproduce asexually by using clumps of erect stalks. However, at times sexual reproduction can be observed. To what group of fungi would you assign it?
   a. Chytridiomycota
   b. Basidiomycota
   c. Ascomycota
   d. Zygomycota

2. Examine the life cycle of a typical basidiomycetes and determine where you would expect to find a dikaryotic cell.
   a. Primary mycelium
   b. Secondary mycelium
   c. In the basidiospores
   d. In the zygote

3. Determine which of the following is correct regarding the yeast *Saccharomyces cerevisiae*.
   a. It reproduces asexually by a process called budding.
   b. It produces an ascocarp during reproduction.
   c. It belongs in the group Zygomycota.
   d. All of the choices are correct.

4. Appraise the fungal relationship between a forest tree and basidiomycete and determine the most suitable classification for the symbiosis.
   a. Parasitism only
   b. An arbuscular mycorrhizae
   c. Ectomycorrhizae
   d. A lichen

5. Choose which of the following best reflects the symbiotic relationships between animals and fungi.
   a. Protection from bacteria
   b. Colonization of land
   c. Protection from desiccation
   d. Exchange of nutrients

SYNTHESIZE

1. Historically fungi have been classified as being more plantlike despite their lack of photosynthetic ability. Although we now know that fungi are more closely related to the animals than the plants, review characteristics that initially led scientists to place them closer to the plants.

2. The importance of fungi in the evolution of terrestrial life is typically understated. Evaluate the importance of fungi in the colonization of land.

3. Based on your understanding of fungi, hypothesize why antibiotics won't work in the treatment of a fungal infection.

4. You are looking through a seed catalog and discover you have a choice of pea seed that is treated or not treated with a fungicide. Identify one reason why you might choose the treated seed and one reason you might prefer the untreated seed.

CHAPTER 33

Animal Diversity and the Evolution of Body Plans

Chapter Contents

- 33.1 Some General Features of Animals
- 33.2 Evolution of the Animal Body Plan
- 33.3 Animal Phylogeny
- 33.4 Parazoa: Animals That Lack Specialized Tissues
- 33.5 Eumetazoa: Animals with True Tissues
- 33.6 The Bilateria

Introduction

Animals are among the most abundant living organisms. Found in almost every habitat, they bewilder us with their diversity of form, habitat, behavior, and lifestyle. About a million and a half species have been described, and several million more are thought to await discovery. Despite their great diversity, animals have much in common. For example, locomotion is a distinctive characteristic, although not all animals can move about—early naturalists thought that sponges and corals were plants because the adults are attached to the surface on which they live. In this chapter and chapters 34 and 35, we explore the great diversity of modern animals, the result of a long evolutionary history. We start in this chapter with a discussion of the features that characterize animals, and review the phyla at the base of the animal phylogeny. In chapters 34 and 35, we discuss the two major groups of animals: the protostomes and deuterostomes.

33.1 Some General Features of Animals

Learning Outcome
1. Identify three features that characterize all animals and three that characterize only some types of animals.

Animals are so diverse that few criteria fit them all. But some, such as animals being eaters, or consumers, apply to all. Others, such as animals being mobile (they can move about), have exceptions. Taken together, the universal characteristics and other features of major importance exhibited by most species are convincing evidence that animals are a monophyletic group—all members more closely related to each other than any members are to another type of organism.

Animals share some general characteristics

Heterotrophy
All animals are heterotrophs—that is, they obtain energy and organic molecules by ingesting other organisms. Unlike autotrophic plants and algae, animals cannot construct organic molecules from inorganic chemicals. Some animals (herbivores) consume autotrophs; other animals (carnivores) consume heterotrophs; some animals (omnivores) consume both autotrophs and heterotrophs; and still others (detritivores) consume decomposing organisms.

Multicellularity
All animals are multicellular; many have complex bodies. Unicellular heterotrophic organisms, which were at one time regarded as simple animals, are now considered members of several different clades within the large and diverse group of protists, discussed in chapter 29.

No cell walls
Animal cells differ from those of other multicellular organisms: They lack rigid cell walls and are usually quite flexible. The many cells of animal bodies are held together by extracellular frames of structural proteins such as collagen. Other proteins form unique intercellular junctions between animal cells.

Active movement
Although other single-celled organisms are able to travel from place to place, animals move more rapidly and in more complex ways—this ability is perhaps their most striking characteristic, one directly related to the flexibility of their cells and the evolution of nerve and muscle tissues. A remarkable form of movement unique to animals is flying, a capability that is well developed among vertebrates and insects such as the butterfly (phylum Arthropoda) shown in figure 33.1*a*. Many animals cannot move from place to place (they are sessile) or do so rarely or slowly (they are sedentary), although they have muscles or muscle fibers that allow parts of their bodies to move. Sponges, however, have little capacity for movement.

Figure 33.1 Some characteristics shared by animals. *a.* Many, but not all, animals are able to move from place to place. Other organisms are also able to move, but flying is unique to animals. *b.* Animals exhibit a wide array of different sizes and forms, but the vast majority of animals are invertebrates, lacking a backbone, such as this millipede. *c.* Animals undergo a process of development beginning with series of cell divisions called cleavage, producing a multicelled structure called a blastula.

Diversity of form

Animals vary greatly in form, ranging in size from organisms too small to see with the unaided eye to enormous whales and giant squids. Almost all animals lack a backbone—they are therefore called invertebrates, like the millipede (phylum Arthropoda) in figure 33.1b. Of the million known living animal species, fewer than 60,000 are vertebrates (that is, those species that possess a backbone). Probably most of the many millions of animal species awaiting discovery are invertebrates.

Diversity of habitat

Animals are grouped into 35 to 40 phyla, most of which have members that occur only in the sea. Members of fewer phyla occur in freshwater, and members of fewer still occur on land. Members of three phyla that are successful in the marine environment—Arthropoda, Mollusca, and Chordata—also dominate animal life on land. Only one animal phylum, Onychophora (velvet worms), is entirely terrestrial.

Sexual reproduction

Most animals reproduce sexually. Animal eggs, which are non-mobile, are much larger than the small, usually flagellated sperm. In animals, cells formed in meiosis function as gametes. These haploid cells do not divide by mitosis first, as they do in plants and fungi, but rather fuse directly with each other to form the zygote. Consequently, there is no counterpart among animals to the alternation of haploid (gametophyte) and diploid (sporophyte) generations characteristic of plants. Some individuals of some species and all individuals of a very few others are incapable of sexual reproduction.

Embryonic development

Shortly after fertilization, an animal zygote first undergoes a series of mitotic divisions, called cleavage, and like the dividing frog's egg in figure 33.1c, cleavage produces a ball of cells, the blastula. In most animals, the blastula folds inward at one point to form a hollow sac with an opening at one end called the blastopore. An embryo at this stage is called a gastrula. The subsequent growth and movement of the cells of the gastrula differ from one group of animals to another, reflecting the evolutionary history of the group (see chapter 25). Embryos of most kinds of animals develop into a larva, which looks unlike the adult of the species, lives in a different habitat, and eats different sorts of food; in most groups, it is very small. A larva undergoes metamorphosis, a radical reorganization, to transform into the adult body form.

Tissues

The cells of all animals except sponges are organized into structural and functional units called tissues—collections of cells that together are specialized to perform specific tasks.

Learning Outcome Review 33.1

All animals are multicellular and heterotrophic, and their cells lack cell walls. Most animals can move from place to place, can reproduce sexually, and possess unique tissues. Animals can be found in almost all habitats.

- Why could animals not have been the first type of life to have evolved?

33.2 Evolution of the Animal Body Plan

Learning Outcomes

1. Differentiate between a pseudocoelom and a coelom.
2. Explain the difference between protostomes and deuterostomes.
3. Describe the advantages of segmentation.

The features described in section 33.1 evolved over the course of millions of years. We can understand how the history of life has proceeded by examining the types of animal bodies and body plans present in fossils and in existence today. Five key innovations can be noted in animal evolution:

1. symmetry;
2. tissues, allowing specialized structures and functions,
3. a body cavity;
4. various patterns of embryonic development; and
5. segmentation, or repeated body units.

These innovations are explained in this section. Some innovations appear to have evolved only once, some twice or more. Innovations that have only evolved a single time can provide evidence of close evolutionary relationship, thus serving as a synapomorphy (shared derived character) for an evolutionary group composed of an ancestral species and all of its descendants, termed a clade (see chapter 23). On the other hand, some innovations evolve more than once in different clades. This is the phenomenon of convergent evolution (see chapter 23). Although not indicative of close evolutionary relationship, convergently evolved innovations may be evidence that distantly related species have adapted in similar ways to similar environmental conditions.

Most animals exhibit radial or bilateral symmetry

A typical sponge lacks definite symmetry, growing as an irregular mass. Virtually all other animals have a definite shape and symmetry that can be defined along an imaginary axis drawn through the animal's body. The two main types of symmetry are radial and bilateral (note that many animals are not perfectly symmetrical, but close enough that they are considered symmetrical).

Radial symmetry

The body of a member of phylum Cnidaria (jellyfish, sea anemones, and corals: The C of Cnidaria is silent; see chapter 34) exhibits **radial symmetry.** Its parts are arranged in such a way that any longitudinal plane passing through the central axis divides the organism into halves that are approximate mirror images. A pie, for example, is radially symmetrical; you can cut it with a knife from above and as long as you cut through the center, the two halves will be identical. The same is true for a sea anemone (figure 33.2a).

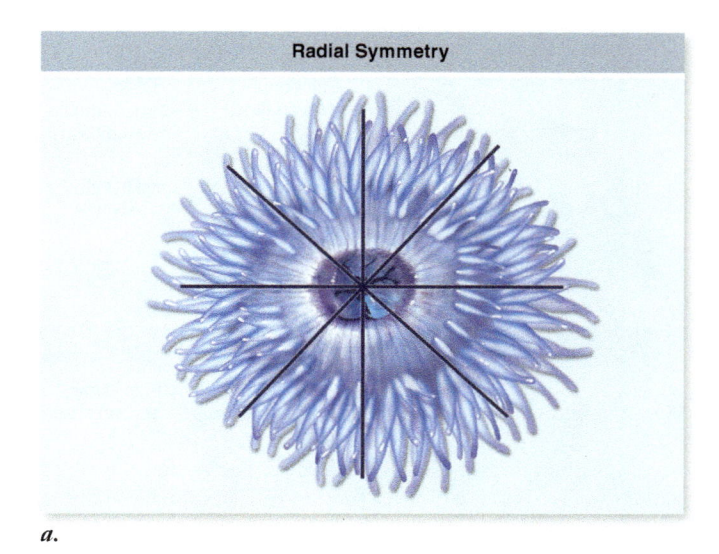

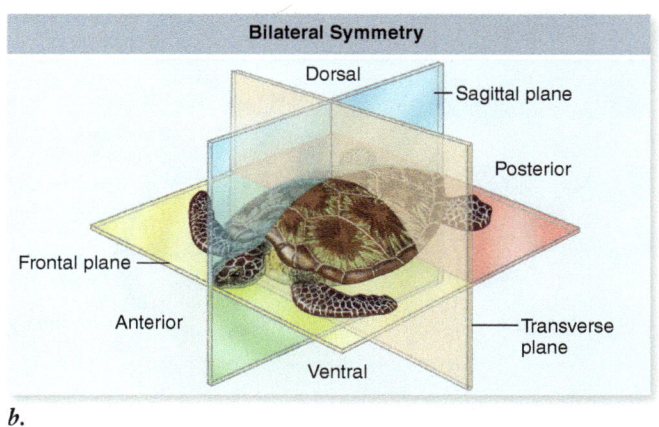

Figure 33.2 A comparison of radial and bilateral symmetry. *a.* Radially symmetrical animals, such as this sea anemone (phylum Cnidaria), can be bisected into equal halves by any longitudinal plane that passes through the central axis. *b.* Bilaterally symmetrical animals, such as this turtle (phylum Chordata), can only be bisected into equal halves in one plane (the sagittal plane).

Bilateral symmetry

The bodies of most animals other than sponges and cnidarians exhibit **bilateral symmetry,** in which the body has right and left halves that are mirror images of each other. Animals with this body plan are collectively termed the Bilateria. The sagittal plane defines these halves (figure 33.2b). A bilaterally symmetrical body has, in addition to left and right halves, dorsal and ventral portions, which are divided by the frontal plane, and anterior (front) and posterior (rear) ends, which are divided by the transverse plane (in an animal that walks on all fours, dorsal is the top side).

Sometimes, deciding whether an organism should be considered bilaterally or radially symmetrical is not easy. For example, in echinoderms (sea stars and their relatives), adults are radially symmetrical (actually pentaradially symmetrical, because the body has five clear sections), but the larvae are bilaterally symmetrical. In this case, examination of animal phylogeny indicates that the radial symmetry of the adult is an evolutionarily derived condition from a bilaterally symmetrical ancestral condition; combined with the form of the larvae, scientists consider echinoderms to be bilaterally symmetrical.

Bilateral symmetry constitutes a major evolutionary advance in the animal body plan. Bilaterally symmetrical animals have the ability to move through the environment in a consistent direction (typically with the anterior end leading)—a feat that is difficult for radially symmetrical animals. Associated with directional movement is the grouping of nerve cells into a brain, and sensory structures, such as eyes and ears, at the anterior end of the body. This concentration of nervous tissue at the anterior end, which appears to have occurred early in evolution, is called **cephalization.** Much of the layout of the nervous system in bilaterally symmetrical animals is centered on one or more major longitudinal nerve cords that transmit information from the anterior sense organs and brain to the rest of the body. Cephalization is often considered a consequence of the development of bilateral symmetry.

The evolution of tissues allowed for specialized structures and functions

The zygote (a fertilized egg), has the capability of giving rise to all the kinds of cells in an animal's body. That is, it is totipotent (all powerful). During embryonic development, cells specialize in carrying out particular functions. In all animals except sponges, the process is irreversible: Once a cell differentiates to serve a function, it and its descendants can never serve any other.

A sponge cell that had specialized to serve one function (such as lining the cavity where feeding occurs) can lose the special attributes that serve that function and change to serve another function (such as being a gamete). Thus a sponge cell can dedifferentiate and redifferentiate. Cells of all other animals are organized into tissues, each of which is characterized by cells of particular morphology and capability. But their ability to dedifferentiate prevents sponge cells from becoming specialized to form clearly defined tissues (and therefore, of course, organs, which are composed of tissues).

Sponges are unique in being able to dedifferentiate and then redifferentiate. The fact that all other animals differentiate irreversibly suggests that organisms with bodies containing cells specialized to serve particular functions may have an advantage over those with cells that potentially have multiple functions. The evolution of specialized tissues may be a key innovation—a trait that fosters evolutionary diversification (see chapter 22).

A body cavity made the development of advanced organ systems possible

In the process of embryonic development, the cells of animals of most groups organize into three layers (called germ layers): an outer **ectoderm,** an inner **endoderm,** and an intermediate

mesoderm. Animals with three embryonic cell layers are said to be triploblastic. During maturation from the embryo, certain organs and organ systems develop from each germ layer. The ectoderm gives rise to the outer covering of the body and the nervous system; the endoderm gives rise to the digestive system, including the intestine; and the skeleton and muscles develop from the mesoderm. Cnidarians have only two layers (thus they are diploblastic)—the endoderm and the ectoderm—and lack organs. Sponges lack germ layers altogether; they, of course, have no tissues or organs. All triploblastic animals are members of the Bilateria.

A key innovation in the body plan of some bilaterians was a body cavity isolated from the exterior of the animal. This is different from the digestive cavity, which is open to the exterior at least through the mouth, and in most animals at the opposite end as well, via the anus. The evolution of efficient organ systems within the animal body was not possible until a body cavity evolved for accommodating and supporting organs (such as our heart and lungs), distributing materials, and fostering complex developmental interactions. The cavity is filled with fluid: In most animals, the fluid is liquid, but in vertebrates, it is gas—the body cavity of humans filling with liquid represents a life-threatening condition. A very few types of bilaterians have no body cavity, the space between tissues that develop from the mesoderm and those that develop from the endoderm being filled with cells and connective tissue. These are the so-called acoelomate animals (figure 33.3).

Body cavities

Body cavities appear to have evolved multiple times in the Bilateria. In some animals, a body cavity called the **pseudocoelom** develops embryologically between the mesoderm and endoderm and thus occurs in the adult between tissues derived from the mesoderm and those derived from endoderm; animals with this type of body cavity are termed pseudocoelomates (figure 33.3). Although the word *pseudocoelom* means "false coelom," this is a true body space and characterizes many groups of animals. A **coelom** is a cavity that develops entirely within the mesoderm. The coelom is surrounded by a layer of epithelial cells derived from the mesoderm and termed the peritoneum.

The circulatory system

In many small animals, nutrients and oxygen are distributed and wastes are removed by fluid in the body cavity. Most larger animals, in contrast, have a **circulatory system,** a network of vessels that carry fluids to and from the parts of the body distant from the sites of digestion (gut) and gas exchange (gills or lungs). The circulating fluid carries nutrients and oxygen to the tissues and removes wastes, including carbon dioxide, by diffusion between the circulatory fluid and the other cells of the body.

In an **open circulatory system,** the blood passes from vessels into sinuses (open areas within the body), mixes with body fluid that bathes the cells of tissues, then reenters vessels in another location. In a **closed circulatory system,** the blood flows entirely within blood vessels, so it is physically separated from other body fluids. Blood moves through a closed circulatory system faster and more efficiently than it does through an open

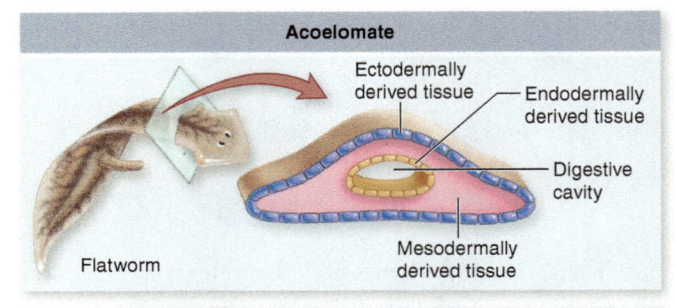

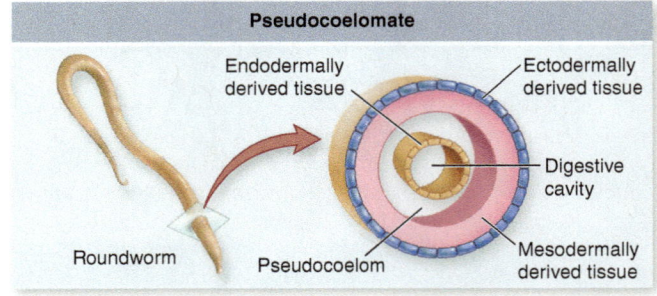

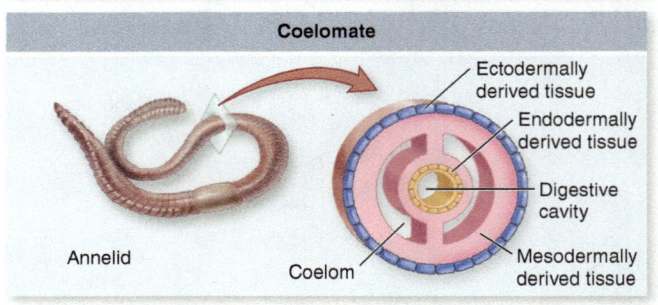

Figure 33.3 Three body plans for bilaterally symmetrical animals. Acoelomates, such as flatworms, have no body cavity between the digestive tract (derived from the endoderm) and the musculature layer (derived from the mesoderm). Pseudocoelomates have a body cavity, the pseudocoelom, between tissues derived from the endoderm and those derived from the mesoderm. Coelomates have a body cavity, the coelom, that develops entirely within tissues derived from the mesoderm, and so is lined on both sides by tissue derived from the mesoderm.

system; open systems are typical of animals that are relatively inactive and so do not have a high demand for oxygen. In small animals, blood can be pushed through a closed circulatory system by the animal's movement. In larger animals, the body musculature does not provide enough force, so the blood must be propelled by contraction of one or more hearts, which are specialized, muscular parts of the blood vessels.

Bilaterians have two main types of development

The processes of embryonic development in animals is discussed fully in chapter 53. Briefly, development of a bilaterally symmetrical animal begins with mitotic cell divisions (called cleavages) of the egg that lead to the formation of a hollow ball of cells, which subsequently indents to form a two-layered ball. The internal space that is created through such indentation (see figure 53.11) is the **archenteron** (literally the "primitive gut"); it communicates with the outside by a **blastopore**.

In a protostome, the mouth of the adult animal develops from the blastopore or from an opening near the blastopore (protostome means "first mouth"—the first opening becomes the mouth). **Protostomes** include most bilaterians, including flatworms, nematodes, mollusks, annelids, and arthropods. In some protostomes, both mouth and anus form from the embryonic blastopore; in other protostomes, the anus forms later in another region of the embryo. Two outwardly dissimilar groups—the echinoderms and the chordates—together with a few other small phyla, constitute the **deuterostomes,** in which the mouth of the adult animal does not develop from the blastopore. The deuterostome blastopore gives rise to the organism's anus, and the mouth develops from a second pore that arises later in development (deuterostome means "second mouth"). Protostomes and deuterostomes differ in several other aspects of embryology too, as discussed later in this section.

Cleavage patterns

The cleavage pattern relative to the embryo's polar axis (which runs from top to bottom through the center of the organism) determines how the resulting cells lie with respect to one another. In some protostomes, each new cell cleaves off to the right or left, producing a layer of cells shifted to one side or the other, relative to those from which they cleaved. This pattern is called **spiral cleavage** because a line drawn through a sequence of dividing cells spirals outward from the polar axis (figure 33.4, *top*). Spiral cleavage is characteristic of annelids, mollusks, nemerteans, and related phyla; the clade of animals with this cleavage pattern is therefore known as the Spiralia.

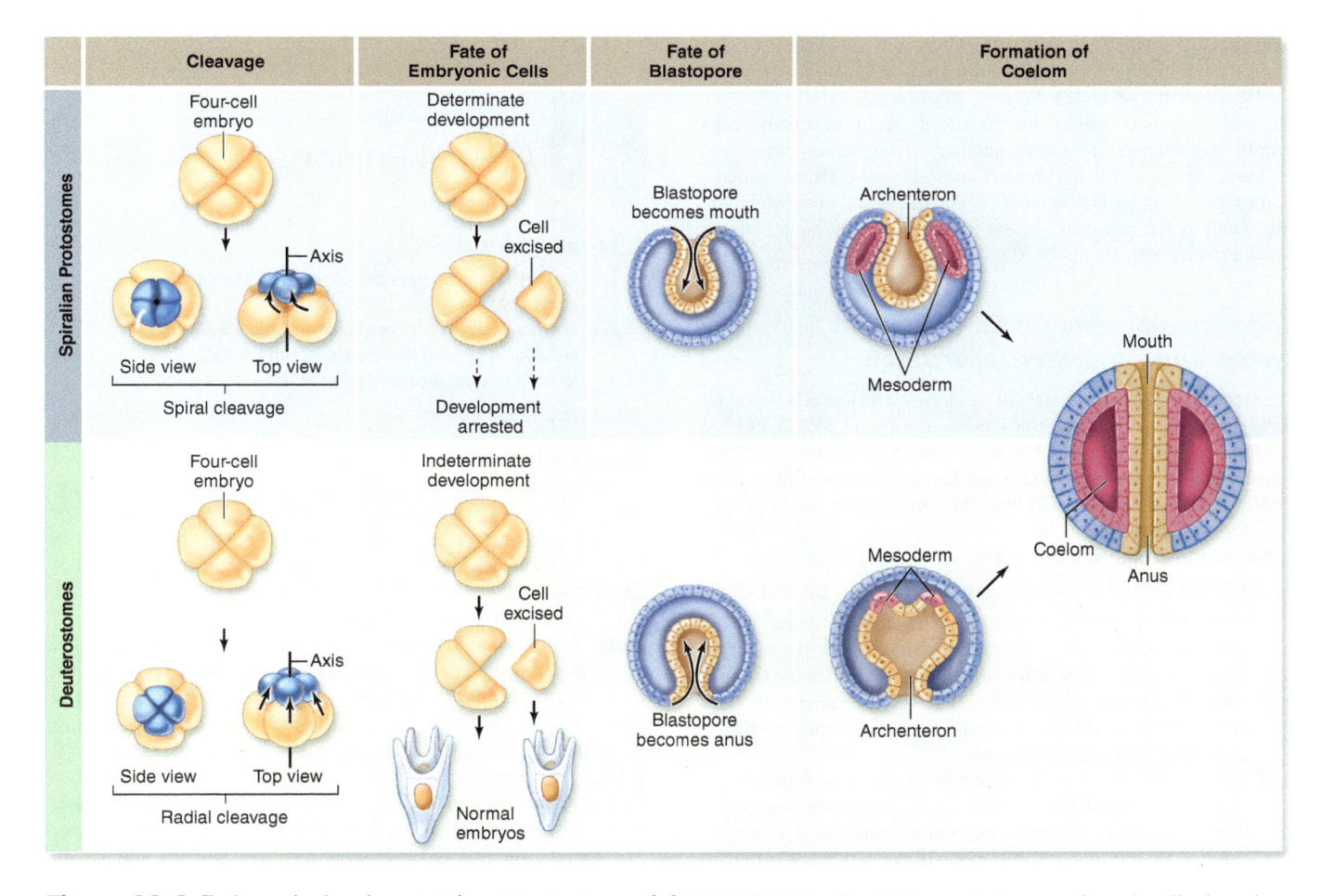

Figure 33.4 Embryonic development in protostomes and deuterostomes. In spiralian protostomes, embryonic cells cleave in a spiral pattern and exhibit determinate development; the blastopore becomes the animal's mouth, and the coelom originates from a split among endodermal cells. In deuterostomes, embryonic cells cleave radially and exhibit indeterminate development; the blastopore becomes the animal's anus, and the coelom originates from an invagination of the archenteron.

In all deuterostomes, by contrast, the cells divide parallel to and at right angles to the polar axis. As a result, the pairs of cells from each division are positioned directly above and below one another, a process that gives rise to a loosely packed ball. This pattern is called **radial cleavage** because a line drawn through a sequence of dividing cells describes a radius outward from the polar axis (figure 33.4, *bottom*).

Determinate versus indeterminate development

Many protostomes exhibit **determinate development,** in which the type of tissue each embryonic cell will form in the adult is determined early, in many lineages even before cleavage begins, when the molecules that act as developmental signals are localized in different regions of the egg. Consequently, each embryonic cell is destined to occur only in particular parts of the adult body, so if the cells are separated, development cannot proceed.

Deuterostomes, conversely, display **indeterminate development.** The first few cell divisions of the zygote produce identical daughter cells. If the cells are separated, any one can develop into a complete organism (this is how identical twins are formed). Thus, each cell remains totipotent and its fate is not determined for several cleavages.

Formation of the coelom

The coelom arises within the mesoderm. In protostomes, cells simply move apart from one another to create an expanding coelomic cavity within the mass of mesodermal cells. In deuterostomes, groups of cells pouch off the end of the archenteron, which you recall is the primitive gut—the hollow in the center of the developing embryo that is lined with endoderm.

Segmentation allowed for redundant systems and improved locomotion

Segmented animals consist of a series of linearly arrayed compartments that typically look alike (see figure 34.21), at least early in development, but that may have specialized functions. Development of segmentation is mediated at the molecular level by *Hox* genes (see chapters 19 and 25). During early development, segments first are obvious in the mesoderm but later are reflected in the ectoderm and endoderm. Two advantages result from early embryonic segmentation:

1. In highly segmental animals, such as earthworms (phylum Annelida), each segment may develop a more or less complete set of adult organ systems. Because these are redundant systems, damage to any one segment need not be fatal because other segments duplicate the damaged segment's functions.
2. Locomotion is more efficient when individual segments can move semi-independently. Because partitions isolate the segments, each can contract or expand independently. Therefore, a long body can move in ways that are often quite complex.

Segmentation underlies the organization of body plans of the most morphologically complex animals. In some adult arthropods, the segments are fused, but segmentation is usually apparent in embryological development. In vertebrates, the backbone and muscle blocks are segmented, although segmentation is often disguised in the adult form.

Previously, zoologists considered that true segmentation was found only in annelids, arthropods, and chordates, but segmentation is now recognized to be more widespread. Animals such as onychophorans (velvet worms), tardigrades (water bears), and kinorhynchs (mud dragons) are also segmented.

Learning Outcomes Review 33.2

Animals are distinguished on the basis of symmetry, tissues, type of body cavity, sequence of embryonic development, and segmentation. Bilateral animals are those that have bodies with a left and right side, which are mirror images. Within bilaterians, protostomes develop the mouth prior to the anus; deuterostomes develop the mouth after the anus has formed. Segmentation allows redundant systems and more efficient locomotion. A pseudocoelom is a space that develops between the mesoderm and endoderm; a coelom develops entirely within mesoderm.

- How is cephalization related to body symmetry?

33.3 Animal Phylogeny

Learning Outcomes

1. Identify the characters that distinguish the major animal phyla.
2. Identify patterns of convergent evolution in significant morphological and developmental characters.
3. Identify the placement of humans among the animal phyla.

There is little disagreement among biologists about the placement of most animals into phyla, although zoologists disagree on the status of some phyla, particularly those with few members or recently discovered ones. The diversity of animals is obvious in tables 33.1 and 33.2, which describe key characteristics of 20 of the animal phyla.

Traditionally, the phylogeny of animals has been inferred using features of anatomy and aspects of embryological development, from which a broad consensus emerged over the last century concerning the main branches of the animal tree of life. However, in the past 30 years, gene sequence data have accumulated at an accelerating pace for all animal groups. Phylogenies developed from different molecules sometimes suggest quite different evolutionary relationships among the same groups of animals. However, combining data from multiple genes has resolved the relationships of most phyla. Current studies are using sequences from hundreds of genes to try to fully resolve the animal tree of life.

Molecular data are helping to resolve some problems with the traditional phylogeny, such as puzzling groups that did not fit well into the widely accepted phylogeny. These data may be

TABLE 33.1 Animal Phyla with the Most Species

Phylum	Typical Examples	Key Characteristics	Approximate Number of Named Species
Arthropoda (arthropods)	Beetles, other insects, crabs, spiders, krill, scorpions, centipedes, millipedes	Chitinous exoskeleton covers segmented, coelomate body. With paired, jointed appendages; many types of insects have wings. Occupy marine, terrestrial, and freshwater habitats. Most arthropods are insects (as are most animals!).	1,200,000
Mollusca (mollusks)	Snails, oysters, clams, octopuses, slugs	Coelomate body of many mollusks is covered by one or more shells secreted by a part of the body termed the mantle. Many kinds possess a unique rasping tongue, a radula. Members occupy marine, terrestrial, and freshwater habitats.	150,000
Chordata (chordates)	Mammals, fish, reptiles, amphibians	Each coelomate individual possesses a notochord, a dorsal nerve cord, pharyngeal slits, and a postanal tail at some stage of life. In vertebrates, the notochord is replaced during development by the spinal column. Members occupy marine, terrestrial, and freshwater habitats.	56,000
Platyhelminthes (flatworms)	Planarians, tapeworms, liver and blood flukes	Unsegmented, acoelomate, bilaterally symmetrical worms. Digestive cavity has only one opening; tapeworms lack a gut. Many species are parasites of medical and veterinary importance. Members occupy marine, terrestrial, and freshwater habitats (as well as the bodies of other animals).	55,000
Nematoda (roundworms)	*Ascaris*, pinworms, hookworms, filarial worms	Pseudocoelomate, unsegmented, bilaterally symmetrical worms; Tubular digestive tract has mouth and anus. Members occupy marine, terrestrial, and freshwater habitats; some are important parasites of plants and animals, including humans.	61,000 (but it is thought by some that the number of nematode species may be much greater)
Annelida (segmented worms)	Earthworms, polychaetes, tube worms, leeches	Segmented, bilaterally symmetrical, coelomate worms with a complete digestive tract; most have bristles (chaetae) on each segment that anchor them in tubes or aid in crawling. Occupy marine, terrestrial, and freshwater habitats.	32,000
Porifera (sponges)	Barrel sponges, boring sponges, basket sponges, bath sponges	Bodies of most asymmetrical: defining "an individual" is difficult. Body lacks tissues or organs, being a meshwork of cells surrounding channels that open to the outside through pores, and that expand into internal cavities lined with food-filtering flagellated cells (choanocytes). Most species are marine (150 species live in freshwater).	26,000
Echinodermata (echinoderms)	Sea stars, sea urchins, sand dollars, sea cucumbers	Adult body pentaradial (fivefold) in symmetry. Water-vascular system is a coelomic space; endoskeleton of calcium carbonate plates. Many can regenerate lost body parts. Fossils are more diverse in body plan than extant species. Exclusively marine.	12,000
Cnidaria (cnidarians)	Jellyfish, *Hydra*, corals, sea anemones, sea fans	Radially symmetrical, acoelomate body has tissues but no organs. Mouth opens into a simple digestive sac and is surrounded by tentacles armed with stinging capsules (nematocysts). In some groups, individuals are joined into colonies; some can secrete a hard exoskeleton. The very few nonmarine species live in freshwater.	10,000
Bryozoa (moss animals) (also called Polyzoa and Ectoprocta)	Sea mats, sea moss	The only exclusively colonial phylum; each colony comprises numerous small, coelomate individuals (zooids) connected by an exoskeleton (calcareous in marine species, organic in most freshwater ones). A ring of ciliated tentacles (lophophore) surrounds the mouth of each zooid; the anus lies beyond the lophophore.	10,000

TABLE 33.2 Some Important Animal Phyla with Fewer Species—and Three Recently Discovered Ones

Phylum	Typical Examples	Key Characteristics	Approximate Number of Named Species
Rotifera (wheel animals)	Rotifers	Small pseudocoelomates with a complete digestive tract including a set of complex jaws. Cilia at the anterior end beat so they resemble a revolving wheel. Some are very important in marine and freshwater habitats as food for predators such as fishes.	2500
Nemertea (ribbon worms) (also called Rhynchocoela)	Lineus	Protostome worms notable for their fragility—when disturbed, they fragment in pieces. Long, extensible proboscis occupies their coelom; that of some tipped by a spearlike stylet. Most marine, but some live in freshwater, and a few are terrestrial.	2400
Tardigrada (water bears)	Hypsibius	Microscopic protostomes with five body segments and four pairs of clawed legs. An individual lives a week or less but can enter a state of suspended animation ("cryptobiosis") in which it can survive for many decades. Occupy marine, freshwater, and terrestrial habitats.	1300
Brachiopoda (lamp shells)	Lingula	Protostomous animals encased in two shells that are oriented with respect to the body differently than in bivalved mollusks. A ring of ciliated tentacles (lophophore) surrounds the mouth. More than 30,000 fossil species are known.	335
Ctenophora (sea walnuts)	Comb jellies, sea walnuts	Gelatinous, almost transparent, often bioluminescent marine animals; eight bands of cilia; largest animals that use cilia for locomotion; complete digestive tract with anal pore.	150
Chaetognatha (arrow worms)	Sagitta	Small, bilaterally symmetrical, transparent marine worms with a fin along each side, powerful bristly jaws, and lateral nerve cords. Some inject toxin into prey and some have large eyes. It is uncertain if they are coelomates, and, if so, whether protostomes or deuterostomes.	130
Onychophora (velvet worms)	Peripatus	Segmented protostomous worms with a chitinous soft exoskeleton and unsegmented appendages. Related to arthropods. The only exclusively terrestrial phylum, but what are interpreted as their Cambrian ancestors were marine.	110
Loricifera (loriciferans)	Nanaloricus mysticus	Tiny marine pseudocoelomates that live in spaces between grains of sand. The mouth is borne on the tip of a flexible tube. Discovered in 1983.	35
Cycliophora (cycliophorans)	Symbion	Microscopic animals that live on mouthparts of claw lobsters. Discovered in 1995.	3
Micrognathozoa (micrognathozoans)	Limnognathia	Microscopic animals with complicated jaws. Discovered in 2000 in Greenland.	1

especially helpful in clarifying relationships that conventional data cannot, as, for example, in animals such as parasites. Through dependence on their host, the anatomy, physiology, and behavior of parasites tends to be greatly altered, so features that may reveal the phylogenetic affinities of free-living animals can be highly modified or lost.

Current understanding of phylogeny differs from traditional views

Although they differ from one another in some respects, phylogenies utilizing molecular data share some deep structure with the traditional animal tree of life. Figure 33.5 is a summary of animal phylogeny developed from morphological, molecular, life history, and other types of relevant data. Some parts of this phylogeny are not firmly established, and new studies are constantly appearing, often with somewhat different conclusions. It is an exciting time to be a systematist, but shifts in understanding of relationships among groups of animals can be frustrating to some! Like any scientific idea, a phylogeny is a hypothesis, open to challenge and to being revised in light of additional data.

One consistent result is that Porifera (sponges) constitutes a monophyletic group that shares a common ancestor with other animals. Some systematists had considered sponges to comprise two (or three) groups that are not particularly closely related, but molecular data support what had been the majority view, that phylum Porifera is monophyletic.

Among remaining animals, termed Eumetazoa, molecular data are in accord with the traditional view that cnidarians (hydras, sea jellies, and corals) branch off the tree before the origin of animals with bilateral symmetry, the Bilateria. Our understanding of the phylogeny of the deuterostome branch of Bilateria (discussed in chapter 35) has not changed much, but our understanding of the phylogeny of protostomes has been altered by molecular data.

The most revolutionary development is that annelids and arthropods, which had been considered closely related because both exhibit segmentation, are in fact not close relatives, but rather belong to separate clades. Now arthropods are grouped with protostomes that molt their cuticles at least once during their life. These are termed ecdysozoans, which means "molting animals" (see chapter 34). Molecular sequence data can help test our ideas of which morphological features reveal evolutionary relationships best; in this case, molecular data allowed us to see that, contrary to our hypothesis, segmentation seems to have evolved convergently, but molting did not.

But not all features are easy to diagnose, and molecular data do not resolve all uncertainties. The enigmatic phylum Ctenophora (comb jellies)—pronounced with a silent C—has been considered both diploblastic and triploblastic and has been thought to have both a complete gut and a blind gut. Likewise the enigmatic phylum Chaetognatha (arrow worms) has been considered both coelomate and pseudocoelomate, and if coelomate, both protostome and deuterostome. Their placement in phylogenies varies, seeming to depend on the features and methods used to construct the tree. Further research is needed to resolve these uncertainties.

Morphology-based phylogeny focused on the state of the coelom

Current understanding of animal phylogeny requires a rethinking of how some characters have evolved. Zoologists previously inferred that the first animals were acoelomate, that some of their descendants evolved a pseudocoelom, and that some pseudocoelomate descendants evolved the coelom. This view was so widespread that classification systems were based on the state of the coelom. However, as you saw in chapter 21, evolution rarely occurs in such a linear and directional way. Our understanding of animal phylogeny now makes clear that the state of the internal cavity has evolved many more times than previously realized, and consequently that it is not a reliable character to infer phylogenetic relationships.

In particular, a coelom appears to have evolved just once, in the ancestor of the clade comprising protostomes and deuterostomes (figure 33.5). Subsequently, one clade reverted secondarily to the acoelomate condition and the pseudocoelomate condition arose several times from coelomate or secondarily acoelomate ancestors within the protostome clade. As a result, all deuterostomes have coeloms, but the condition in protostomes is mixed. In addition, although the acoelomate condition is the ancestral state for animals, some animals have lost the body space, becoming acoelomate secondarily.

Protostomes consist of spiralians and ecdysozoans

Early in their history, protostomes divided into two clades: the spiralians and the ecdysozoans (figure 33.5). Spiralian animals grow by gradual addition of mass to the body and undergo spiral cleavage (see figure 33.4). There are two main groups of spiralians: Lophotrochozoa and Platyzoa. Lophotrochozoans move by muscular contractions and include most protostomes with a coelom. Most platyzoans are acoelomates and move by ciliary action.

Ecdysozoans are animals that molt (figure 33.6), a phenomenon that seems to have evolved only once in the animal kingdom. Of the numerous phyla of protostomes assigned to the Ecdysozoa, Arthropoda contains the largest number of described species of any phylum.

Deuterostomes include chordates and echinoderms

Deuterostomes consist of fewer phyla and species than protostomes, and are more uniform in many ways, despite great differences in appearance. Echinoderms such as sea stars, and chordates such as humans, share a mode of development that is evidence of their evolution from a common ancestor, and separates them clearly from other animals.

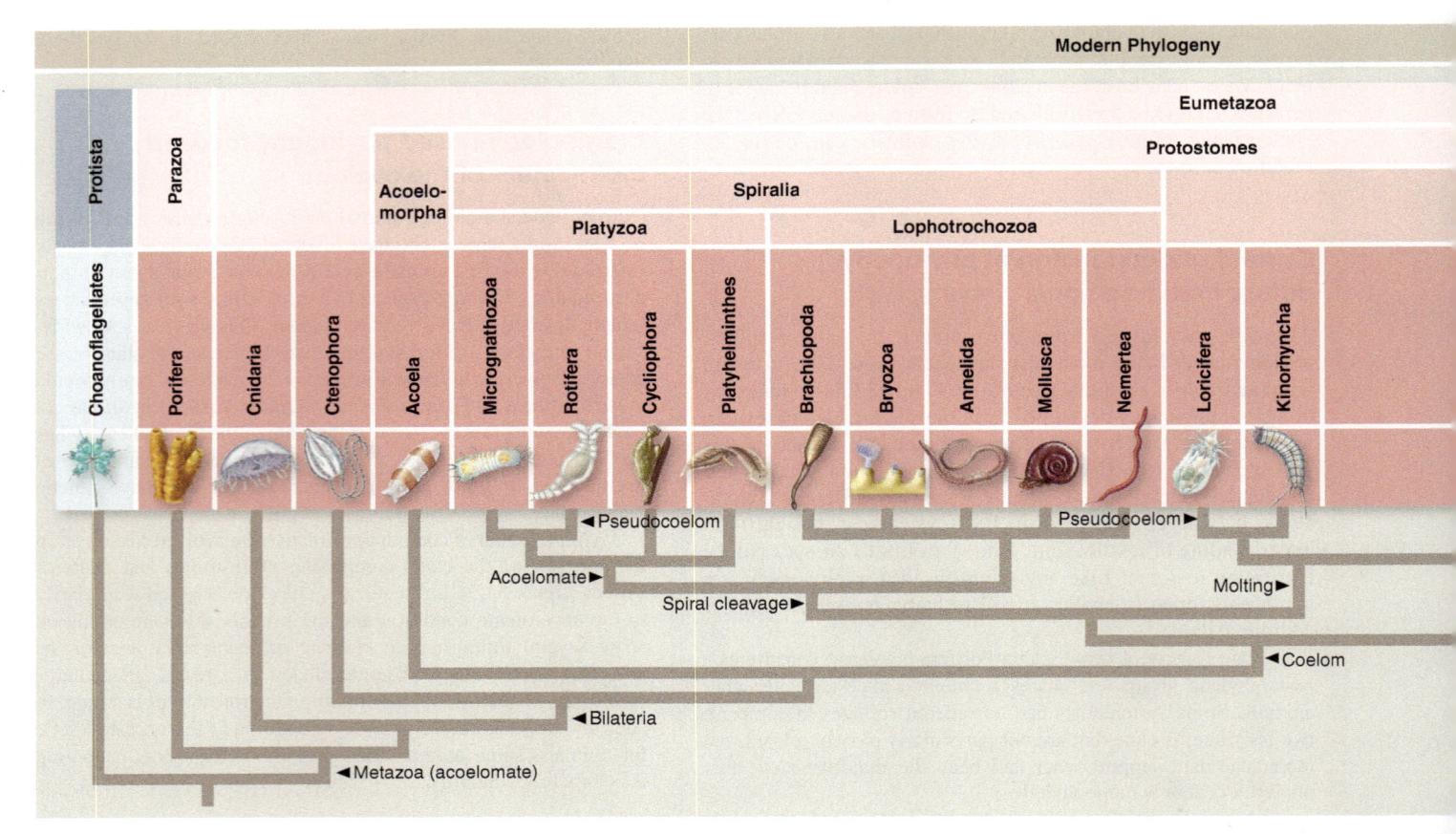

Figure 33.5 Proposed revision of the animal tree of life. A phylogeny of many of the 35–40 phyla reflects a consensus based on interpretation of anatomical and developmental data as well as results derived from molecular phylogenetic studies. Whether Chaetognatha is a protostome or a deuterostome is unclear and the position of ctenophores is currently being debated.

Inquiry question Which condition, acoelomate, pseudocoelomate, or coelomate, is a reliable indicator of phylogenetic relationships?

Learning Outcomes Review 33.3

Scientists have defined phyla based on tissues, symmetry, characteristics such as presence or absence of a coelom or pseudocoelom, protostome versus deuterostome development, growth pattern and larval stages, and molecular data. Among protostomes, spiralian organisms have a growth pattern in which their body size simply increases; ecdysozoans must molt in order to grow larger. The condition of the coelom and the occurrence of segmentation have evolved convergently among animals, whereas other traits, such as molting, have arisen only once. Among the deuterostome phyla are echinoderms and chordates, the latter which include humans.

- *Why do systematists attempt to characterize each group of animals by one or more features that have evolved only once?*

Figure 33.6 Blue crab undergoing ecdysis (molting). Members of Ecdysozoa grow step-wise because their external skeleton is rigid.

648 part V Diversity of Life on Earth

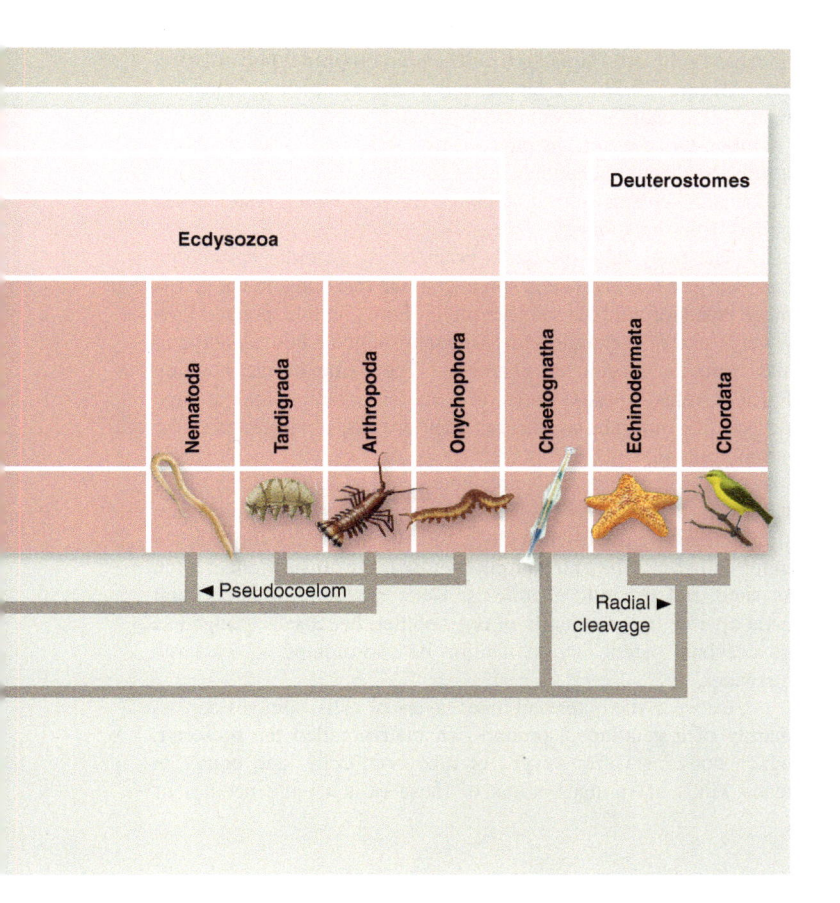

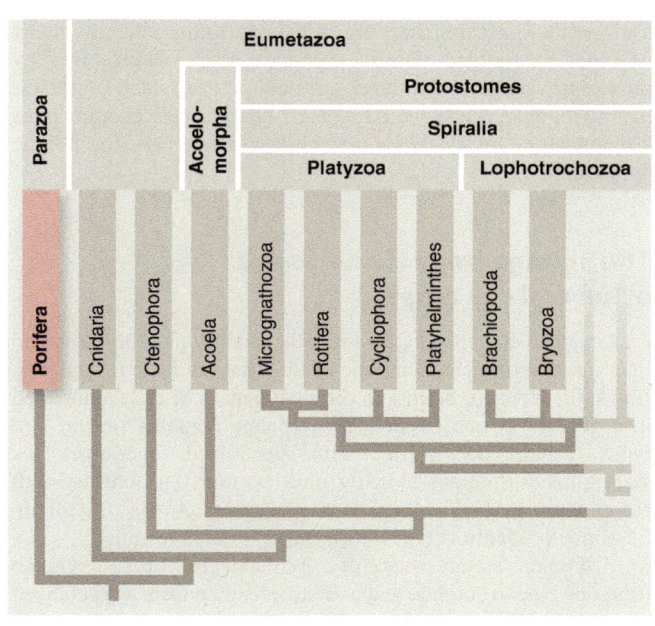

33.4 Parazoa: Animals That Lack Specialized Tissues

Learning Outcomes
1. Describe the different types of cells in the sponge body.
2. Explain the function of choanocytes.

In the remainder of this chapter and chapters 34 and 35, we will explore the great diversity of animals. We start with the morphologically simplest members of the animal kingdom—sponges, jellyfish, and some types of worms (fully a third of animal phyla are based on a "wormy" body plan!). Despite their simplicity, these animals can carry out all the essential functions of life, just as do more morphologically complex animals—eating, respiring, reproducing, and protecting themselves. The major organization of the animal body first evolved in these animals, the basic body plan from which all the rest of animals evolved.

Systematists traditionally divided the kingdom Animalia (also termed Metazoa) into two main branches. Parazoa ("near animals") comprises animals that, for the most part, lack definite symmetry and do not possess tissues. These are the sponges, phylum Porifera. Because they are so different in so many ways from other animals, some scientists inferred that sponges were not closely related to other animals, which would mean that what we consider animals had two separate origins. Eumetazoa ("true animals") are animals that have a definite shape and symmetry. All have tissues, and most have organs and organ systems. Now most systematists agree that Parazoa and Eumetazoa are descended from a common ancestor, so animal life had a single origin. And although most phylogenies constructed with molecular data consider Parazoa to be at the base of the animal tree of life, some do not.

Sponges have a loose body organization

The major group of parazoans is phylum **Porifera,** the sponges. Despite their morphological simplicity, like all animals, sponges are truly multicellular.

More than 26,000 species of sponges live in the sea, and perhaps 150 species live in freshwater. Marine sponges occur at all depths, and may be among the most abundant animals in the deepest part of the oceans. Although some sponges are small (no more than a few millimeters across), some may reach 2 m or more in diameter.

A few small sponges are radially symmetrical, but most members of this phylum lack symmetry. Some have a low and encrusting form and grow covering various sorts of surfaces; others are erect and lobed, some in complex patterns (figure 33.7a).

As is true of many marine invertebrate animals, larval sponges are free-swimming. After a sponge larva attaches to an appropriate surface, it metamorphoses into an adult and remains attached to that surface for the rest of its life. Thus adult sponges are sessile—that is, they are anchored, immobile, on rocks or other submerged objects. Sponges defend themselves by producing chemicals that repel potential predators and organisms that might overgrow them. As a result, pharmaceutical companies are interested in possibly using these chemicals for human applications.

The sponge body is composed of several cell types

Lacking a head, appendages, a mouth, an anus, and the organized internal structure characteristic of all other animals, a sponge at first sight seems to be little more than a mass of cells embedded in a gelatinous matrix. In fact, a sponge contains several cell types (figure 33.7b), each with specialized functions. This distinguishes sponges as truly multicellular, by contrast with colonial protists, which may form aggregates of cells, but all are functionally identical (except for the reproductive cells).

A unique feature of sponge cells is their ability to differentiate from one type to another, and to dedifferentiate from a specialized state to an unspecialized one. If a sponge is put through a fine sieve or coarse cloth so that the cells are separated, they will seek one another out and reassemble the entire sponge—a phenomenon that does not occur in any other animal.

A small, anatomically simple sponge has a vaselike shape. The walls of the "vase" have three functional layers. Facing the internal cavity are flagellated cells called **choanocytes,** or collar cells (figure 33.7b). A larger and more complex sponge has many small chambers connected by channels rather than a single chamber. Once water has passed through a flagellated chamber, it travels through channels that converge at a large opening called an **osculum** (plural, *oscula*), through which water is expelled from the sponge.

The body of a sponge is bounded by an outer epithelium consisting of flattened cells somewhat like those that make up the outer layers of animals in other phyla. Pores on the sponge allow water to enter the channels that course through its body, leading to and from the flagellated chambers. The name of the phylum, Porifera, refers to these pores, or ostia (singular, *ostium*); a large sponge has multiple oscula, but they are far, far fewer than the number of ostia.

Some epithelial cells are specialized to surround the ostia; they can contract when touched or exposed to appropriate stimuli, causing the ostia to close, thereby protecting the delicate inner cells from the entry of potentially harmful substances such as sand and noxious chemicals. Cells surrounding individual ostia operate independently of one another; because a sponge has no nervous system, actions cannot be coordinated across large distances.

Between the outer and inner layers of cells, sponges consist mainly of a gelatinous, protein-rich matrix called the **mesohyl,** which contains various types of amoeboid cells (and eggs). In many kinds of sponges, some of these cells secrete needles of

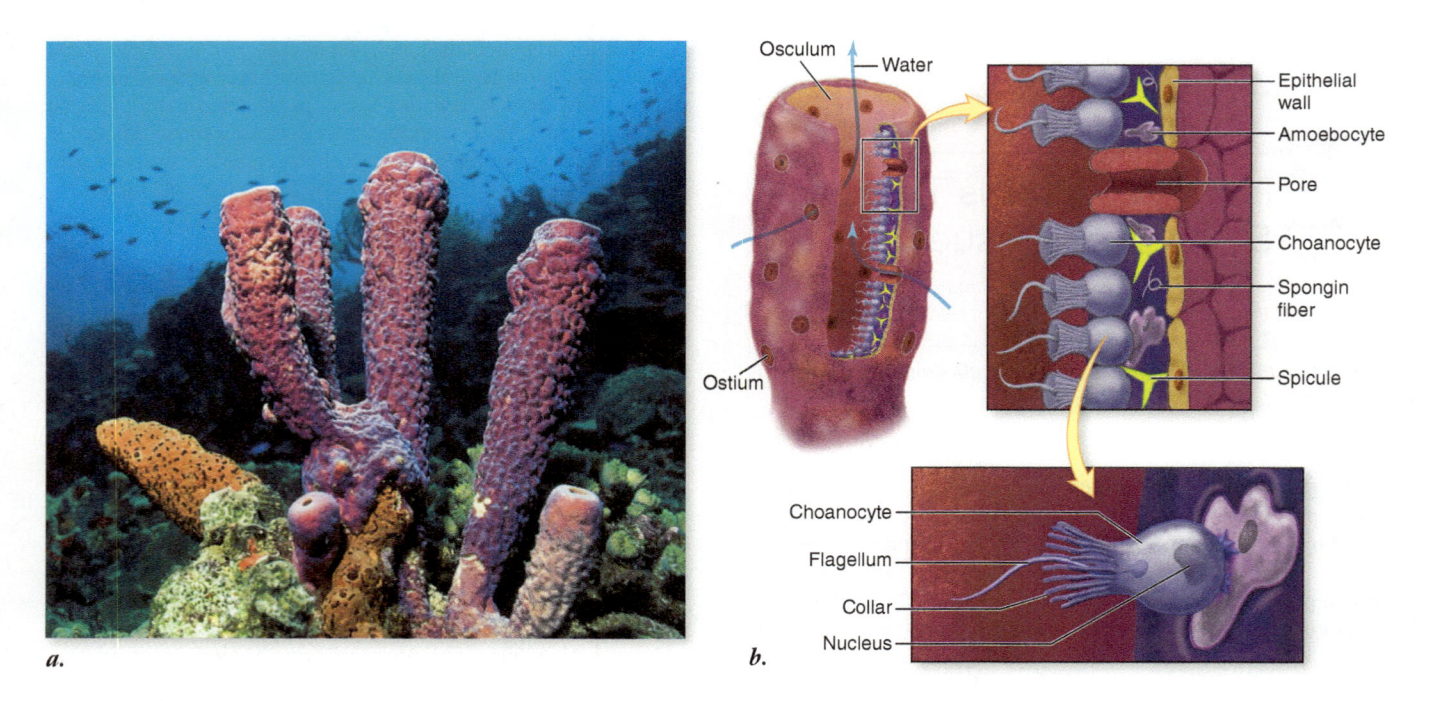

Figure 33.7 Phylum Porifera: Sponges. *a. Aplysina longissima.* This beautiful, bright orange and purple elongated sponge is found on deep coral reefs. *b.* Diagrammatic drawing of the simplest type of sponge. Sponges are composed of several distinct cell types, the activities of which are coordinated. The sponge body has no organized tissues, and most are not symmetrical.

calcium carbonate or silica known as **spicules,** or fibers of a tough protein called **spongin.** Spicules and spongin form the skeleton of the sponge, strengthening its body. The siliceous spicules of some deep-sea sponges can reach a meter in length! A genuine bath sponge is the spongin skeleton of a marine sponge; artificial sponges made of cellulose or plastic are modeled on the body design of this animal, with a porous body adapted to contain large amounts of water.

Choanocytes help circulate water through the sponge

Each choanocyte resembles a protist with a single flagellum (plural, *flagella*) (figure 33.7b). The pressure created by the beating flagella in the cavity contributes to circulating the water that brings in food and oxygen and carries out wastes. In large sponges, the inner wall of the body interior is convoluted, increasing the surface area and, therefore, the number of flagella. In such a sponge, 1 cm^3 of sponge can propel more than 20 L of water per day. Choanocytes also capture food particles from the passing water, engulfing and digesting them. Obviously, this arrangement restricts a sponge to feeding on particles considerably smaller than choanocytes—largely bacteria.

Sponges reproduce both asexually and sexually

Some sponges can reproduce asexually simply by breaking into fragments. Each fragment is able to continue growing as a new individual.

Sponge sperm are created by the transformation of choanocytes, which are then released into the water where they may be carried into another sponge of the same species. When a sperm is captured by a choanocyte, it is carried to an egg cell, which is in the mesohyl (and in some sponges is also a transformed choanocyte). In many sponges, development of the externally ciliated larva occurs within the mother. In sponges of other species, the fertilized egg is released into the water, where development occurs. Whether a fertilized egg or a ciliated larva is released, after a short planktonic (drifting) stage, the larva settles on a suitable substrate where it transforms into an adult.

> ### Learning Outcomes Review 33.4
> Sponges possess multicellularity but have neither tissue-level development nor body symmetry. Cells that compose a sponge include a layer of choanocytes, a layer of epithelial cells, and amoeboid cells in the mesohyl between the two layers. Choanocytes have flagella that beat to circulate water through the sponge body, allowing food particles to be trapped.
>
> ■ *In what ways is the ability of sponge cells to dedifferentiate an advantage and a disadvantage compared to other animals?*

33.5 Eumetazoa: Animals with True Tissues

Learning Outcomes
1. Explain the defining feature of cnidarians.
2. Differentiate between cnidarians and ctenophores.
3. Discuss the question of symmetry of ctenophores.

The Eumetazoa contains animals that evolved the first key transition in the animal body plan: distinct tissues. The embryonic cell layers differentiate into the tissues of the adult body, giving rise to the body plan characteristic of each group of animals.

Recall that the outer covering of the body (the epidermis) and the nervous system develop from the embryonic ectoderm, and the digestive tissue (called the **gastrodermis**) develops from the embryonic endoderm. In the Bilateria, the embryonic mesoderm, which lies between endoderm and ectoderm, forms the muscles. Eumetazoans also evolved body symmetry, the two main types of which are radial and bilateral symmetry.

All cnidarians are carnivores

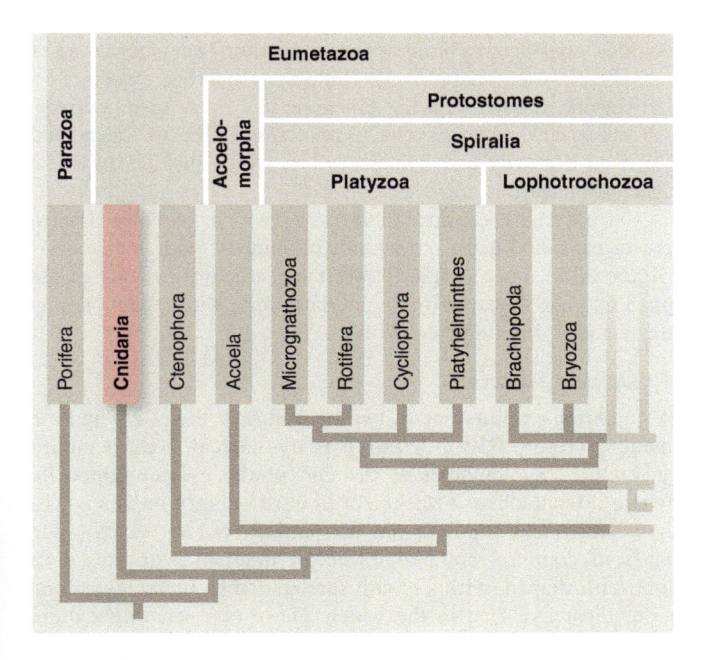

Most of the 10,000 species of cnidarians are marine; a very few live in freshwater. The bodies of these fascinating and simply constructed diploblastic animals are radially symmetrical and made of distinct tissues, although they do not have organs. Despite having no reproductive, circulatory, digestive, or excretory systems, cnidarians reproduce, exchange gas, capture and digest

Figure 33.8 Phylum Cnidaria: Cnidarians. The cells of a cnidarian such as this *Hydra* are organized into specialized tissues. The interior gut cavity is specialized for extracellular digestion—that is, digestion begins within the gut cavity rather than within a cell. The epidermis contains nematocysts for defense and for capturing prey. This *Hydra* is undergoing asexual reproduction—budding off a new individual.

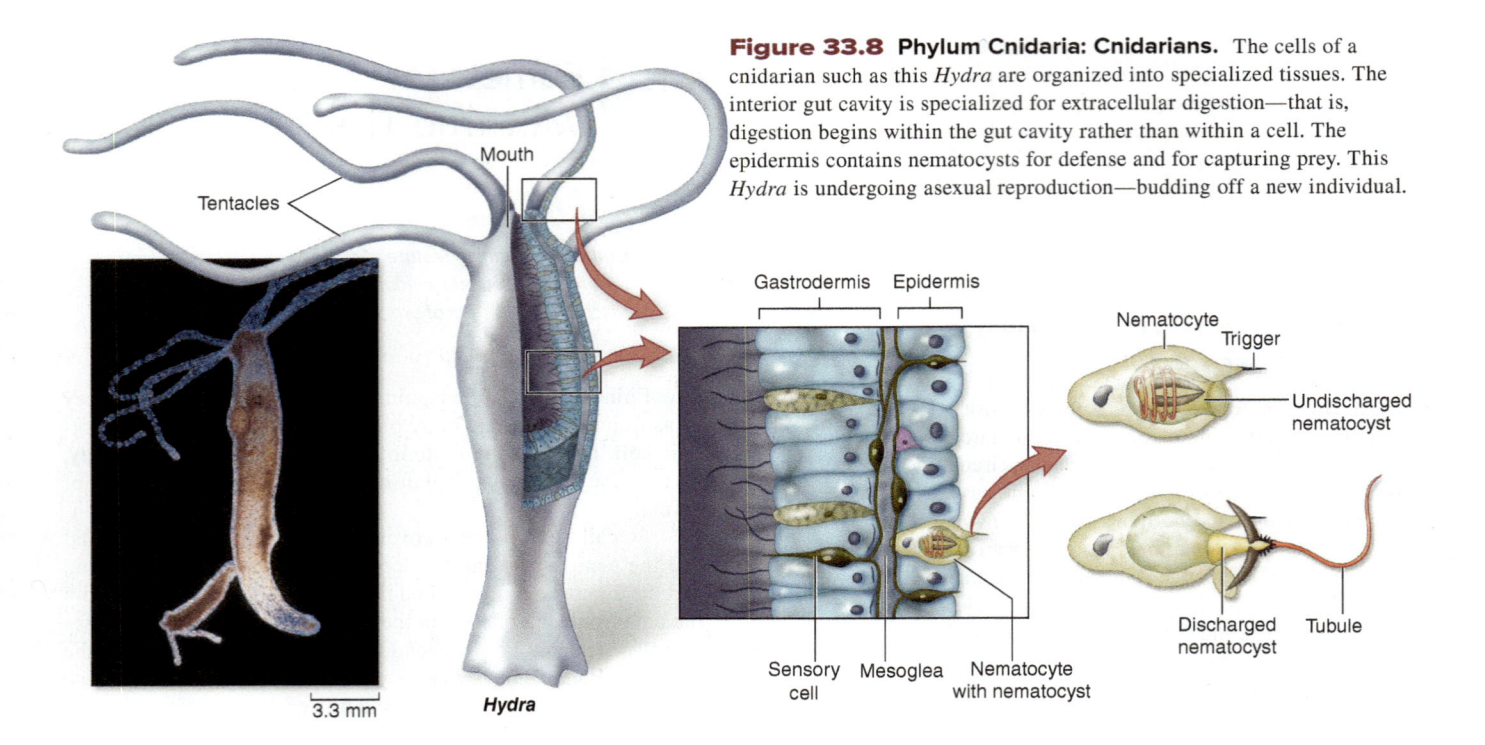

prey, and distribute the resulting organic molecules to all their cells. A cnidarian has no concentration of nervous tissue that could be considered a brain or even a ganglion (a smaller group of nerve cells—see chapter 43). Rather, its nervous system is a latticework, the cells having junctions like those of bilaterians. All cnidarians have nervous receptors sensitive to touch, and some have gravity and light receptors, which include, in a few species, image-forming eyes.

Cnidarians capture their prey (which includes fishes, crustaceans, and many other kinds of animals) with nematocysts (figure 33.8), microscopic intracellular structures unique to the phylum. Food captured by a cnidarian is brought to the mouth by the tentacles that ring the mouth.

Basic body plans

A cnidarian exhibits one of two body forms: the polyp and the medusa (figure 33.9). A **polyp** is cylindrical, with a mouth surrounded by tentacles at the end of the cylinder opposite where it is attached. Attachment in most solitary polyps is to a firm substrate, but polyps that are part of a colony attach to the mass of common colonial tissue. A **medusa** is discoidal or umbrella-shaped, with a mouth surrounded by tentacles on one side; most live free in the water. These two seemingly quite different body forms share the same morphology—the mouth opens into a saclike gastrovascular cavity and is surrounded by tentacles.

The cnidarian body plan has a single opening leading to the gastrovascular space, which is the site of digestion, most gas exchange, waste discharge, and, in many cnidarians, formation of gametes. The body wall is composed of two layers: the epidermis, which covers the surfaces in contact with the outside environment, and the gastrodermis, which covers the gastrovascular cavity. Between these two layers is the **mesoglea,** which varies from acellular, being no more than a glue holding gastrodermis to epidermis in *Hydra*, to thick and rubbery with many cells in large medusae called jellyfish (see figures 33.8 and 33.9).

The gastrovascular space also serves as a hydrostatic skeleton (see chapter 46). A hydrostatic skeleton serves two of the roles that a skeleton made of bone or shells does: It provides a rigid structure against which muscles can operate, and it gives the animal shape. For muscles to operate against it, the fluid-filled space must be closed sufficiently tightly so the fluid is under pressure and does not escape when muscles contract around the space. Think of the gastrovascular space of a cnidarian

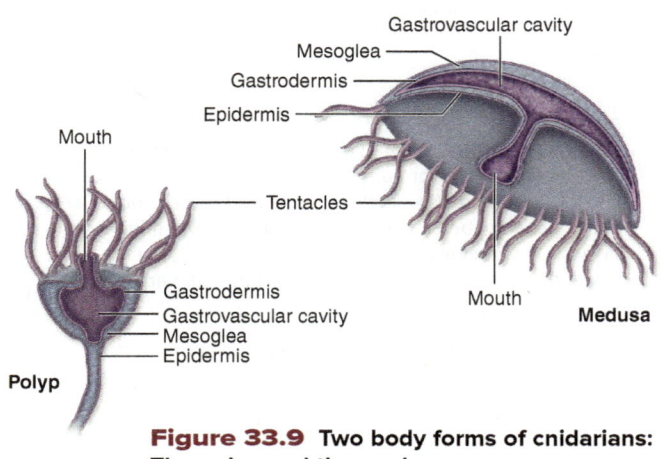

Figure 33.9 Two body forms of cnidarians: The polyp and the medusa.

full of water like an inflated, elongated, air-filled balloon. Because it is firm, muscles that shorten this space cause it to broaden, and muscles that decrease its diameter lengthen it. However, if you inflate the balloon with air, but then do not hold the opening closed, the air will escape and the balloon will become flaccid. Similarly, when a cnidarian contracts greatly, it must open its mouth so water can escape. However, the mouth can close sufficiently tightly to retain water in the gastrovascular cavity, so the animal is firm and can bend and extend its tentacles.

In addition to the hydrostatic skeleton of all polyps, those of many species build an exoskeleton of chitin or calcium carbonate around themselves; in many colonial species, the skeleton links the members of the colony, forming what we call coral. The polyps of a smaller number of species secrete internal skeletal elements. Polyps of some groups of cnidarians, such as sea anemones, have no skeleton at all. All medusae are solitary (that is, they do not form colonies) and form no skeleton.

The cnidarian life cycle

Cnidarians exhibit considerable variation in life history. Some cnidarians occur only as polyps, and others exist only as medusae, but many alternate between these two phases (figure 33.9); both phases consist of diploid individuals. In general, in species having both polyp and medusa in the life cycle, the medusa forms gametes. The sexes are separate; an individual is either male or female. An egg and a sperm unite to form a zygote, which develops into a planktonic ciliated **planula** larva that metamorphoses into a polyp. The polyp produces medusae asexually. Dispersal occurs in both medusa and larvae. In most species with such a life cycle, a polyp may also produce other polyps asexually; if they remain attached to one other, the resulting group is referred to as a colony.

In a small number of species, the planula produced by medusae may develop directly or indirectly into another medusa without passing through the polyp stage. More commonly, a polyp occurs but no medusa in the life cycle. In such cnidarians, the polyp can form gametes, and the resulting planula develops into another polyp. In some, but by no means all, of these species, the polyp can also produce other polyps asexually, by dividing, budding, or breaking off bits of itself that regenerate (grow the missing parts).

Digestion

A major evolutionary innovation in cnidarians is extracellular digestion of food inside the animal. Recall that in a sponge, digestion occurs within choanocytes, so food particles must be small enough to be engulfed by a choanocyte. In a cnidarian, digestion takes place partly in the gastrovascular cavity. Digestive enzymes, released from cells lining the cavity, partially break down food. Other cells lining the space engulf those food fragments by phagocytosis. This allows a cnidarian to feed on prey larger than a sponge can handle.

Nematocysts

Cnidarians capture food with the aid of **nematocysts,** microscopic stinging capsules unique to the phylum (see figure 33.8). Although often referred to as "stinging cells," they are capsules, each secreted within a cell called a nematocyte. Cnidarians may be morphologically simple, but nematocysts are the most complex structures secreted by single animal cells. When appropriately stimulated, the closure at the end of the capsule springs open, and a tubule is emitted. Nematocyst discharge is one of the fastest cellular processes in nature. The mechanism of discharge of a nematocyst is unknown—several hypotheses for it have been proposed. Although the action of a nematocyst is often described as harpoon-like, the tubule actually everts (turns inside-out). It may penetrate or wrap around an object; the tubules of some nematocysts are barbed and some carry venom. In a very few species, the venoms are strong enough to kill a human.

Many thousand nematocytes typically are part of the epidermis of each tentacle. Each one can be used only once. In some types of cnidarians, nematocysts occur in parts of the body other than the tentacles, including in the gastrovascular cavity, where they may aid in digestion. In addition to being used offensively, nematocysts are the only defense of cnidarians—they are the reason jellyfish and fire coral sting. Some cnidarians have stinging capsules of types other than nematocysts, which are termed cnidae—hence the name of the phylum, Cnidaria.

Cnidarians are grouped into five classes

Traditionally, three classes of Cnidaria were recognized: Anthozoa (sea anemones, corals, sea fans), Hydrozoa (hydroids, *Hydra,* Portuguese man-of-war), and Scyphozoa (jellyfish). Now nearly all biologists accept that some scyphozoans are sufficiently distinct in life cycle and morphology that they constitute their own class, Cubozoa (box jellies), some of which have a sting sufficiently toxic to kill humans. Less widely accepted is the separation of some other scyphozoans into their own class, Staurozoa (star jellies).

Class Anthozoa: Sea anemones and corals

The largest class of Cnidaria is **Anthozoa,** consisting of approximately 6200 species of solitary and colonial polyps. They include soft-bodied sea anemones (figure 33.10), stony corals (figure 33.11), and other groups known by such fanciful names as sea pens, sea pansies, sea fans, and sea whips. Many of these names reflect the plantlike appearance of the individual polyps or colonies.

Figure 33.10
Class Anthozoa. Crimson anemone, *Cribrinopsis fernaldi.*

Figure 33.11 Cup coral. *Tubastraea coccinea*, a non–reef-building species from Malaysia.

An anthozoan polyp differs from a polyp of the other classes in that its gastrovascular cavity is compartmentalized by radially arrayed, longitudinal sheets of tissue called mesenteries. The gametes of an anthozoan develop in the mesenteries. The tentacles of an anthozoan are hollow, whereas those of many other cnidarians are solid.

Sea anemones (see figure 33.10) constitute a group of just over 1000 species of highly muscular and relatively complex soft-bodied anthozoans. Living in waters of all depths throughout the world, they range from a few millimeters to over a meter in diameter, and many are equally long.

Most corals are anthozoans, and most hard corals (about 1400 species; see figure 33.11) are closely related to sea anemones. The polyp of a coral secretes an exoskeleton of calcium carbonate around and under itself (this is the same material of which chalk is composed); as the individual or colony grows upward, dead skeleton accumulates below it. In shallow water of the tropics, these accumulations form coral reefs. Most waters in which coral reefs develop are nutrient-poor, but the corals are able to grow well because they contain within their cells symbiotic dinoflagellates (zooxanthellae) that photosynthesize, thereby providing energy for the animals. Obviously coral reefs are restricted to water sufficiently shallow for sunlight to reach the living animals (typically no more than 100 m), but corals that do not form reefs can grow deeper—to about 5000 m. Only about half the species of hard corals participate in forming reefs.

Coral reefs are economically important, serving as refuges for the young of many species of crustaceans and fishes that are eaten by humans, as well as protecting coasts of many tropical islands. Unfortunately, they are threatened by global climate change. Water warmer than usual can cause the symbiosis with zooxanthellae to break down—a phenomenon known as "coral bleaching" because the underlying white skeleton of the animal becomes visible through its body in the absence of the symbionts. Bleaching does not always kill a coral, but it is stressful. As carbon dioxide in the atmosphere rises, it dissolves in water, forming carbonic acid, which lowers the pH of the water. This makes calcium carbonate less available; some species of calcifying marine plants and animals, including corals, have been shown to form less robust skeletons, and to form them more slowly, as pH drops.

Some "soft corals" secrete needles of calcium carbonate much like sponge spicules; these sclerites form a case around the polyp or are embedded in their tissues, presumably providing protection against predation. Some of these animals also secrete a horny rod that provides flexible support to the members of the colony growing around it (sea fans are such colonies).

Class Cubozoa: The box jellies

As their name implies, medusae of class **Cubozoa** are box-shaped, with a tentacle or group of tentacles hanging from each corner of the box (figure 33.12). Most of the 90 or so species are only a few centimeters in height, although some reach 25 cm. Box jellies are strong swimmers and voracious predators of fish in tropical and subtropical waters; both of these facts are related to some cubozoans having image-forming eyes! The stings of some species can be fatal to humans. The polyp stage is inconspicuous and in many cases unknown.

Class Hydrozoa: The hydroids

Most of the approximately 6000 species of class **Hydrozoa** have both polyp and medusa stages in their life cycle (see figure 33.9). The polyp stage of most species is colonial. The polyps of a colony may not be identical in structure or function: Some may be specialized to feed but be unable to reproduce, whereas the opposite may be true of others (they are nourished by material transported from feeding polyps through the gastrovascular space connecting members of the colony). The Portuguese man-of-war *(Physalia physalis)* (figure 33.13) is not a jellyfish, but is a floating colony of highly integrated polypoid and medusoid individuals! Some marine hydroids and medusae are bioluminescent.

Figure 33.12 Class Cubozoa. *Chironex fleckeri*, a box jelly.

Figure 33.13
Portuguese man-of-war, *Physalia physalis*. This species has a very painful sting.

Figure 33.15
Star jelly. Stalked jellyfish, *Haliclystus auricula*.

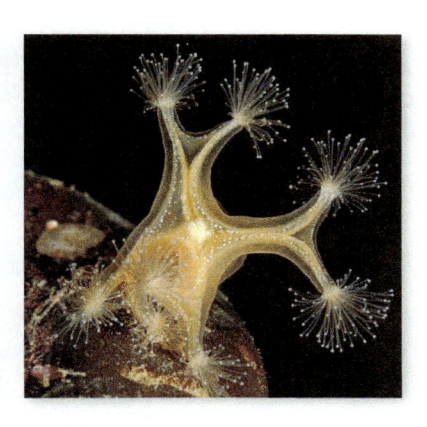

Hydrozoa is the only class of Cnidaria with freshwater members. A well-known hydroid is the freshwater *Hydra*, which is exceptional not only in its habitat, but in having no medusa stage and being solitary (see figure 33.8). Each polyp attaches by a basal disk, on which it can glide, aided by mucus secretions. It can also somersault—by bending over and attaching to the substrate by its tentacles, then looping over to a new location. If the polyp detaches from the substrate, it can float to the surface.

Class Scyphozoa: The jellyfish

In the approximately 350 species of class **Scyphozoa**, the medusa is much more conspicuous and complex than the polyp. Although many scyphomedusae are essentially colorless (figure 33.14), some are a striking orange, blue, or pink. The polyp is small, inconspicuous, simple in structure, and typically white in color, but is entirely lacking in a few scyphozoans that live in the open ocean.

The epithelium around the margin of a jellyfish contains a ring of muscle cells that can contract rhythmically to propel the animal through the water by jetting water from the gastrovascular cavity or space beneath it; by contracting those on just one side, the animal can steer. A scyphomedusa *(Cyanea capilata)* was the killer in the Sherlock Holmes tale, *The Adventure of the Lion's Mane*, but, although some can inflict pain on humans, they are unlikely to be capable of killing.

Figure 33.14
Class Scyphozoa. *Aurelia aurita*, a jellyfish.

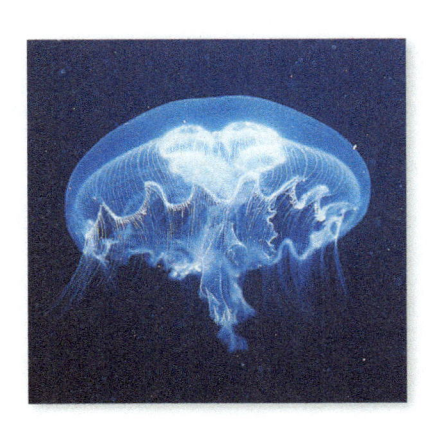

Class Staurozoa: The star jellies

Among the reasons the 100 species of this group were separated from the Scyphozoa and placed in their own class (Staurozoa) is that the animal resembles a medusa in most ways but is attached to the substratum by a sort of stalk that emerges from the side opposite the mouth (figure 33.15). In addition, in all staurozoans known, the planula larva creeps rather than swims or drifts.

The comb jellies use cilia for movement

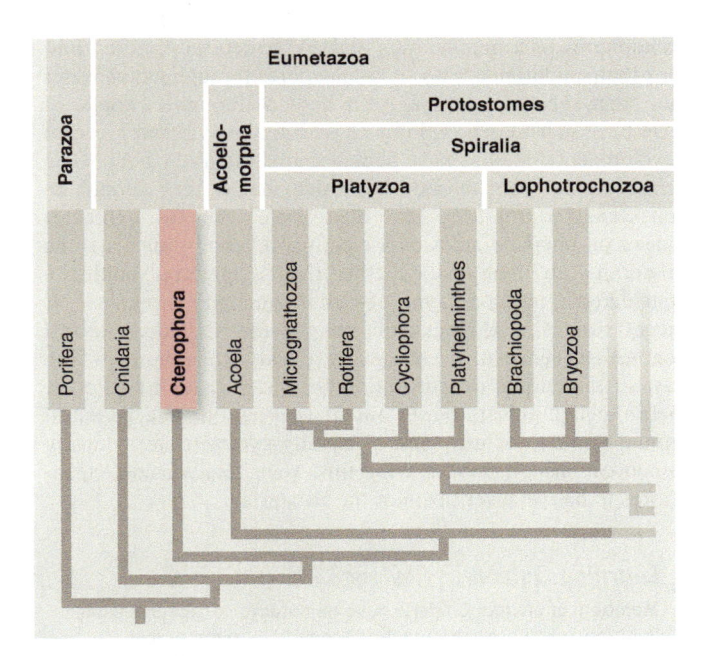

Pelagic members of the small marine phylum **Ctenophora** range from spherical to ribbon-like and are known as comb jellies, sea walnuts, or sea gooseberries. Abundant in the open ocean, most of these animals are transparent and only a few centimeters long, but rare ones are deeply pigmented and reach 1 m in length. They propel themselves through the water with eight rows of comblike plates of fused cilia that beat in a coordinated fashion, scattering light so they appear like rainbows (figure 33.16). Ctenophores

Figure 33.16 A comb jelly (phylum Ctenophora). Note the iridescent rows of comblike plates of fused cilia.

are the largest animals that use cilia for locomotion. Many ctenophores are bioluminescent, giving off bright flashes of light that are particularly evident in the deep sea or on the surface of the ocean at night. Most ctenophores have two long, retractable tentacles used in prey capture. The epithelium of the tentacles contains **colloblasts,** a type of cell that bursts to discharge a strong adhesive material on contact with animal prey.

The phylogenetic position of Ctenophora is unclear. It was formerly considered closely related to Cnidaria because of the gelatinous, medusa-like form of most species. However, ctenophores lack nematocysts and are structurally more complex than cnidarians. They have anal pores through which water and other substances can exit the body. Although ctenophores have been considered diploblasts with radial symmetry, recent developmental studies have demonstrated them to have muscle cells derived from mesoderm, which means they should be considered triploblasts, like bilaterians. The two tentacles placed on opposite sides of a ctenophore's body impart some bilaterality to them, but whether their symmetry should be considered a modified type of radial symmetry is unclear. A recent molecular phylogenetic analysis that included the full genome sequence of a ctenophore places ctenophores at the base of the animal tree of life, as the sister group to all other metazoans. If this is borne out, it implies that the ancestral animal was triploblastic and bilaterally symmetrical, contrary to conventional wisdom. For the time being, we consider ctenophores to be the sister taxon of the Bilateria.

Learning Outcomes Review 33.5

Members of phylum Cnidaria have nematocysts, capsules used in defense and prey capture, and are radially symmetrical. Members of phylum Ctenophora lack nematocysts; they propel themselves by means of eight rows of comblike plates of fused cilia. Ctenophores have cells called colloblasts that assist in prey capture. The symmetry of ctenophores is a subject of debate.

■ *Why is the phylogenetic position of ctenophores uncertain?*

33.6 The Bilateria

Learning Outcomes
1. List the distinguishing features of the Bilateria.
2. Understand the phylogeny of the major groups of Bilateria.

The Bilateria is characterized by a key transition in the animal body plan—bilateral symmetry—which allowed animals to achieve high levels of specialization within parts of their bodies, such as the concentration of sensory structures at the anterior. (Even if ctenophores are truly bilateral as just mentioned, they lack these other specializations.) The Bilateria is divided into two clades. One comprises the protostomes and deuterostomes, which are discussed in chapters 34 and 35. Here we briefly mention the other clade, the acoel flatworms.

Acoel flatworms appear to be distinct from other flatworms

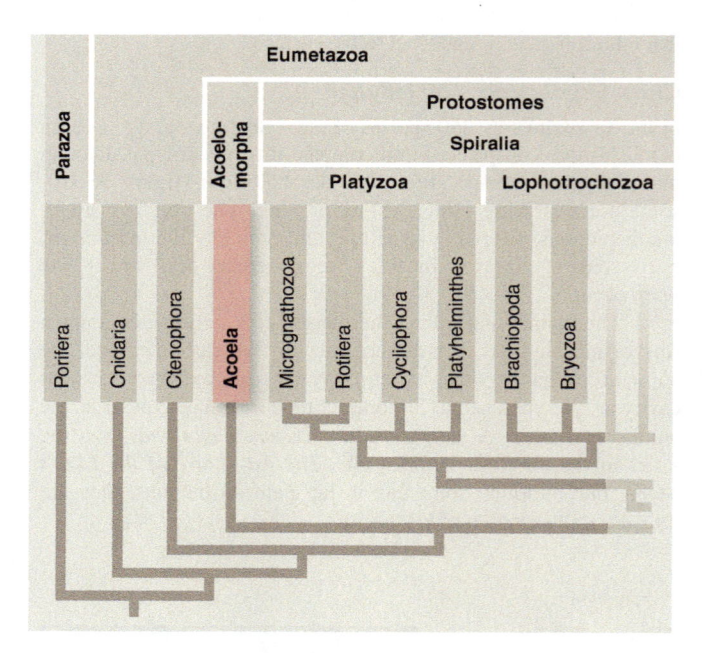

As its name implies, an acoel flatworm (figure 33.17) lacks a coelom. In addition, it has a nervous system that consists of a simple network of nerves with a minor concentration of neurons in the anterior end of the body. It lacks a permanent digestive cavity, so the mouth leads to a solid digestive syncytium (a mass of cells that have no cell membranes separating them).

These characteristics traditionally have been interpreted to classify acoel flatworms with flatworms in the phylum Platyhelminthes, which are discussed in chapter 34. However, based on molecular evidence, scientists have concluded that acoels are not closely related to members of phylum Platyhelminthes—

Figure 33.17 Phylum Acoela. An acoel flatworm of the genus *Waminoa*. These flatworms, long thought to be relatives of the Platyhelminthes, have a primitive nervous system, and lack a permanent digestive cavity.

similarities between the two groups are convergent. They belong in their own phylum, Acoela. Their precise position in the phylogenetic tree differs depending on the features used to construct it; one hypothesis, followed here, is that they are an early offshoot of the Bilateria, thus constituting the sister taxon to the clade comprising the protostomes and deuterostomes.

Learning Outcomes Review 33.6

The Bilateria is a clade characterized by bilateral symmetry and composed of two clades: the acoel flatworms and the clade comprising protostomes and deuterostomes. Traditionally, acoel flatworms were considered to be related to other types of flatworms, but recent molecular phylogenetic studies indicate that they are not closely related, and consequently that their morphological similarity is a result of convergent evolution.

■ *The acoel flatworms lack a coelom. Do any members of their sister taxon exhibit a similar condition?*

Chapter Review

33.1 Some General Features of Animals

Animals share some general characteristics.
Features common to all animals are multicellularity, heterotrophic lifestyle, and lack of a cell wall. Other features include specialized tissues, ability to move, and sexual reproduction.

33.2 Evolution of the Animal Body Plan

Most animals exhibit radial or bilateral symmetry.
Most sponges are asymmetrical, but other animals are bilaterally or radially symmetrical at some time during their life. The body parts of radially symmetrical animals are arranged around a central axis. The body of a bilaterally symmetrical animal has left and right halves. Most bilaterally symmetrical organisms are cephalized and can move directionally.

The evolution of tissues allowed for specialized structures and functions.
Each tissue consists of differentiated cells that have characteristic forms and functions.

A body cavity made the development of advanced organ systems possible.
Most bilaterian animals possess a body cavity other than the gut. A coelom is a cavity that lies within tissues derived from mesoderm. A pseudocoelom lies between tissues derived from mesoderm and the gut (which develops from endoderm). The acoelomate condition and the pseudocoelom appear to have evolved more than once, but the coelom evolved only once.

A circulatory system is an example of a specialized organ system that assists with distribution of nutrients and removal of wastes.

Bilaterians have two main types of development.
In a protostome, the mouth develops from or near the blastopore. A protostome has determinate development, and many have spiral cleavage.

In a deuterostome, the anus develops from the blastopore. A deuterostome has indeterminate development and radial cleavage.

Segmentation allowed for redundant systems and improved locomotion.
Segmentation, which evolved multiple times, allows for efficient and flexible movement because each segment can move somewhat independently. Another advantage to segmentation is redundant organ systems.

33.3 Animal Phylogeny

Animals are classified into 35 to 40 phyla based on shared characteristics.

Current understanding of phylogeny differs from traditional views.
By incorporating molecular data into phylogenetic analyses, some alterations have been made to the traditional view of how members of these phyla are related (figure 33.5).

Porifera (sponges) constitutes a monophyletic group that shares a common ancestor with other animals. Cnidarians (hydras, sea jellies, and corals) evolved before the origin of bilaterally symmetrical animals.

Annelids and arthropods had been considered closely related based on segmentation, but now arthropods are grouped with protostomes that molt their cuticles at least once during their life.

Morphology-based phylogeny focused on the state of the coelom.
The two groups of bilaterally symmetrical animals (the Bilateria)—protostomes and deuterostomes—differ in embryology. Possession of a coelom is an ancestral character of the clade comprising the protostome and deuterostomes. All deuterostomes have a coelom, but some protostomes have evolved a pseudocoelom and others have lost a coelom entirely.

Protostomes consist of spiralians and ecdysozoans.
Spiralia comprises the clades Lophotrochozoa and Platyzoa. Spiralian animals grow by gradual addition of mass to the body and undergo spiral cleavage.

Ecdysozoans grow by molting the external skeleton; Ecdysozoa includes many varied species, ranging from the pseudocoelomate, unsegmented Nematoda to the coelomate, segmented Arthropoda.

Deuterostomes include chordates and echinoderms.
The major groups of deuterostomes are the echinoderms, which include animals such as sea stars and sea urchins, and the chordates, which include vertebrates.

Deuterostome development indicates that echinoderms and chordates evolved from a common ancestor, distinguishing them clearly from other animals.

33.4 Parazoa: Animals That Lack Specialized Tissues

Parazoa ("near animals") comprises animals that, for the most part, lack definite symmetry, and that do not possess tissues.

Sponges have a loose body organization.
Sponges lack tissues and organs and a definite symmetry, but they do have a complex multicellularity. Larval sponges are free-swimming, and the adults are anchored onto submerged objects.

The sponge body is composed of several cell types.
Sponges have three layers: an external protective epithelial layer; a central protein-rich matrix called mesohyl with amoeboid cells; and an inner layer of choanocytes that circulate water and capture food particles (figure 33.7b).

Choanocytes help circulate water through the sponge.
The mesohyl may contain spicules and/or fibers of a tough protein called spongin that strengthen the body of the sponge.

Sponges reproduce both asexually and sexually.
Fragments of a sponge are able to grow into complete individuals. Sperm and eggs may be produced by mature individuals; these undergo fertilization to form zygotes that develop into free-swimming larvae that eventually become sessile adults.

33.5 Eumetazoa: Animals with True Tissues

The Eumetazoa contains animals with distinct tissues. The embryonic cell layers differentiate into the tissues of the adult body, giving rise to the body plan characteristic of each group of animals (figure 33.8).

All cnidarians are carnivores.
Members of the carnivorous and radially symmetrical Cnidaria have distinct tissues but no organs. They are diploblastic and have two body forms: a sessile, cylindrical polyp and a free-floating medusa.

Cnidarians are distinguished by capsules called nematocysts that are used in offense and defense.

Cnidarians have no circulatory, excretory, or respiratory systems. They have a latticework of nerve cells and are sensitive to touch; some have gravity receptors and light receptors.

Cnidarians are grouped into five classes.
The five classes of Cnidaria are the Anthozoa (sea anemones, corals, seafans); Cubozoa (box jellies); Hydrozoa (hydroids, Hydra), Scyphozoa (jellyfish); and Staurozoa (star jellies). The Staurozoa class is not accepted by all scientists.

The comb jellies use cilia for movement.
Ctenophora (comb jellies) is a small phylum of medusa-like animals that propel themselves with bands of fused cilia. They may be triploblastic. They capture prey with colloblasts, cells that release an adhesive.

33.6 The Bilateria

The Bilateria are characterized by bilateral symmetry, which allows for functional specialization such as having nerve receptors at the anterior end of the body.

Acoel flatworms appear to be distinct from other flatworms.
Acoel flatworms, phylum Acoela, were once considered basal to the phylum Platyhelminthes, but they appear to have evolved early in the history of the Bilateria and are the sister taxon to the clade comprised of protostomes and deuterostomes.

Review Questions

UNDERSTAND

1. Which of the following characteristics is unique to all animals?
   a. Sexual reproduction
   b. Multicellularity
   c. Lack of cell walls
   d. Heterotrophy

2. Animals are unique in the fact that they possess _____ for movement and _____ for conducting signals between cells.
   a. brains; muscles
   b. muscle tissue; nervous tissue
   c. limbs; spinal cords
   d. flagella; nerves

3. In animal sexual reproduction the gametes are formed by the process of
   a. meiosis.
   b. mitosis.
   c. fusion.
   d. binary fission.

4. The evolution of bilateral symmetry was a necessary precursor for the evolution of
   a. tissues.
   b. segmentation.
   c. a body cavity.
   d. cephalization.

5. A fluid-filled cavity that develops completely within mesodermal tissue is a characteristic of a
   a. coelomate.
   b. pseudocoelomate.
   c. acoelomate.
   d. All of the choices are correct.

6. Which of the following statements is NOT true regarding segmentation?
   a. Segmentation allows the evolution of redundant systems.
   b. Segmentation is a requirement for a closed circulatory system.
   c. Segmentation enhances locomotion.
   d. Segmentation represents an example of convergent evolution.

7. Which of the following characteristics is used to distinguish between a parazoan and a eumetazoan?
   a. Presence of a true coelom
   b. Segmentation
   c. Cephalization
   d. Tissues

8. With regard to classification of animals, the study of which of the following is changing our understanding of how animals are organized?
   a. Molecular systematics
   b. Origin of tissues
   c. Patterns of segmentation
   d. Evolution of morphological characteristics

9. The coelom
   a. is a synapomorphy for the clade comprising protostomes and deuterostomes.
   b. evolved multiple times convergently.
   c. is always an indicator of close evolutionary relationship.
   d. forms within the endoderm.

10. A coelomate organism may have which of the following characteristics?
    a. Circulatory system
    b. Internal skeleton
    c. Larger size than a pseudocoelomate
    d. All of the choices are correct.

11. Which of the following cell types of a sponge possesses a flagellum?
    a. Choanocyte   c. Epithelial
    b. Amoebocyte   d. Spicules

12. The larval stage of a cnidarian is known as a
    a. medusa.   c. polyp.
    b. planula.   d. cnidocyte.

13. Acoel flatworms
    a. are closely related to other flatworms.
    b. have evolved an acoelom.
    c. are a member of the Parazoa.
    d. are the sister taxon to the clade comprised of protostomes and deuterostomes.

APPLY

1. The following diagram is of the blastopore stage of embryonic development. Based on the information in the diagram, which of the following statements is correct?

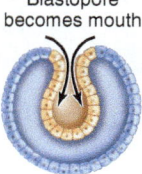

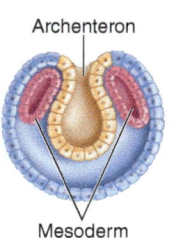

   a. Most animals exhibit this pattern of development.
   b. It would have formed by spiralian cleavage.
   c. It would exhibit determinate development.
   d. All of the choices are correct.

2. Which of the following characteristics would not apply to a species in the Ecdysozoa?
   a. Bilateral
   b. Indeterminate cleavage
   c. Molt at least once in their life cycle
   d. Metazoan

3. In the rainforest you discover a new species that is terrestrial, has determinate development, molts during its lifetime, and possesses jointed appendages. To which phylum of animals should it be assigned?

4. Which of the following cell layers is not necessary to be considered a eumetazoan?
   a. Ectoderm
   b. Endoderm
   c. Mesoderm
   d. All of the above are found in all eumetazoans.

SYNTHESIZE

1. Worm evolution represents an excellent means of understanding the evolution of a body cavity. Using the phyla of worms Nematoda, Annelida, Platyhelminthes, and Nemertea, construct a phylogenetic tree based only on the form of body cavity (refer to figure 33.3 and table 33.1 for assistance). How does this differ from the phylogeny in figure 33.5? Should body cavity be used as the sole characteristic for classifying a worm?

2. Most students find it hard to believe that Echinodermata and Chordata are closely related phyla. If it were not for how they share a similar pattern of development, where would you place Echinodermata in the animal kingdom? Defend your answer.

CHAPTER 34

Protostomes

Chapter Contents

34.1 The Clades of Protostomes
34.2 Platyzoans: Flatworms (Platyhelminthes)
34.3 Platyzoans: Rotifers (Rotifera)
34.4 Lophotrochozoans: Mollusks (Mollusca)
34.5 Lophotrochozoans: Ribbon Worms (Nemertea)
34.6 Lophotrochozoans: Annelids (Annelida)
34.7 Lophophorates: Bryozoans (Bryozoa) and Brachiopods (Brachiopoda)
34.8 Ecdysozoans: Roundworms (Nematoda)
34.9 Ecdysozoans: Arthropods (Arthropoda)

Introduction

Almost all bilaterian animals belong to one of two clades, the protostomes and the deuterostomes. These clades differ in how they develop embryologically—in protostomes, the mouth of the adult animal develops from the blastopore or from an opening near it, whereas in deuterostomes, the blastopore gives rise to the anus and the mouth develops in other ways. The great majority of all animal species are protostomes, including insects, mollusks, worms, and many others.

34.1 The Clades of Protostomes

Learning Outcome
1. List the major clades of protostomes and identify their distinguishing characteristics.

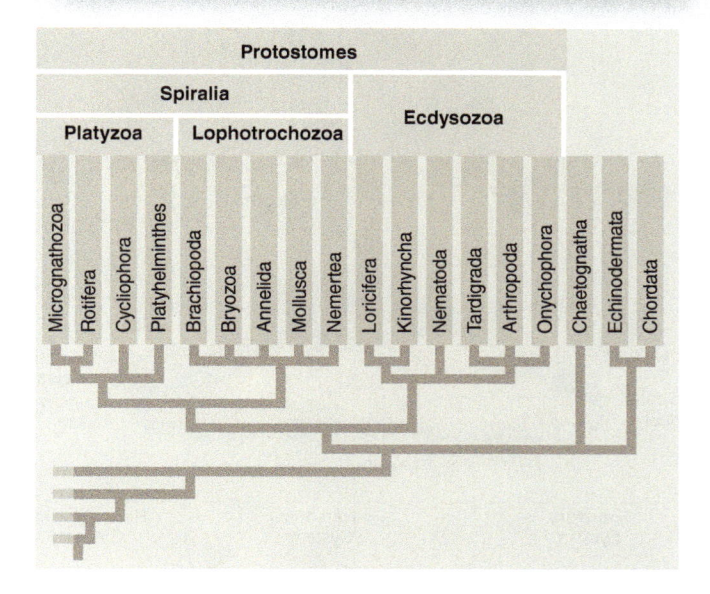

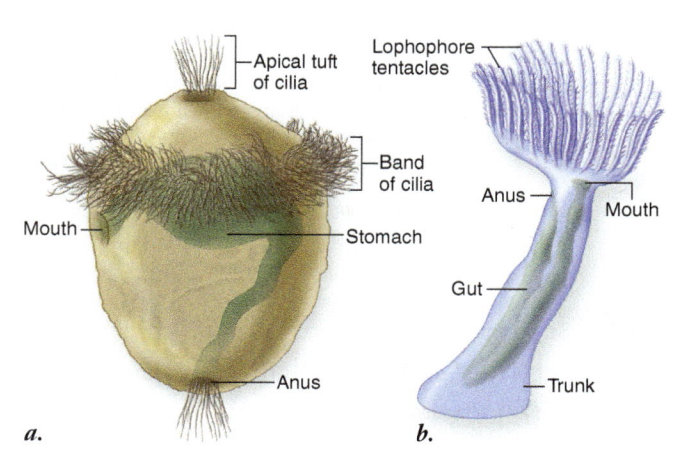

Figure 34.1 Trochophore *(a)* and lophophore *(b)*.

All protostomes belong to either the Spiralia or the Ecdysozoa (see figure 33.5). Before discussing the diversity of protostomes, we will briefly review the characteristics of these groups.

Spiralia

Spiralians develop as embryos using spiral cleavage (see figure 33.4). Most live in water and move through it using cilia or contractions of the body musculature. The two clades of spiralians are platyzoans and lophotrochozoans. The most prominent group of platyzoans is the flatworms (phylum Platyhelminthes), animals that have a simple body with no circulatory or respiratory system but a complex reproductive system. This group includes marine and freshwater planarians as well as the parasitic flukes and tapeworms.

Lophotrochozoa consists of two major phyla and several smaller ones. Many of the animals have a type of free-living larva known as a **trochophore** (figure 34.1*a*), and some have a feeding structure termed a **lophophore** (figure 34.1*b*), a horseshoe-shaped crown of ciliated tentacles around the mouth used in filter feeding. The phyla characterized by a lophophore are Bryozoa and Brachiopoda. Lophophorate animals are sessile (anchored in place).

Among the lophotrochozoans with a trochophore are phyla Mollusca and Annelida. Mollusks are unsegmented, and their coelom is reduced to a hemocoel (open circulatory space) and some other small body spaces. This phylum includes animals as diverse as octopuses, snails, and clams. Annelids are segmented coelomate worms, the most familiar of which is the earthworm, but also includes leeches and the largely marine polychaetes.

Ecdysozoa

The other major clade of protostomes is the Ecdysozoa, which contains animals that molt. When an animal grows large enough that it completely fills its hard external skeleton, it must lose that skeleton by molting (a process also called ecdysis, hence the name Ecdysozoa). While the animal grows, it forms a new exoskeleton underneath the existing one. The first step in molting is for the body to swell until the existing exoskeleton cracks open and is shed (figure 34.2). Upon molting that skeleton, the animal inflates the soft, new one, expanding it using body fluids (and, in many insects and spiders, air as well). When the new one hardens, it is larger than the molted one had been and has room for growth. Thus, rather than being continuous, as in other animals, the growth of ecdysozoans is step-wise.

The Ecdysozoa contains the arthropods, which contains more species than any other phylum, and nematodes (roundworms), which are so numerous that, it is said, if everything in the world except nematodes were to disappear, an outline of what had been there would be visible in the remaining nematodes! These two phyla each contain one of the model organisms used in laboratory studies that have informed much of our current understanding of genetics and development: the fruit fly *Drosophila melanogaster* and the roundworm *Caenorhabditis elegans*.

In this chapter, we review the diversity of the major phyla of protostomes, starting with the platyzoans, followed by the lophotrochozoans and then the ecdysozoans.

Figure 34.2 A damselfly, a type of insect, molting.

Learning Outcome Review 34.1

Protostomes are those animals in which the mouth forms from the blastopore during development. Protostomes are divided into spiralians, characterized by spiral cleavage, and ecdysozoans, which molt.

- What is the phylogenetic significance of molting and cleavage type in protostome phylogeny?

34.2 Platyzoans: Flatworms (Platyhelminthes)

Learning Outcomes

1. List the distinguishing features of bilaterian flatworms.
2. Explain why the scolex of a tapeworm is not a head.

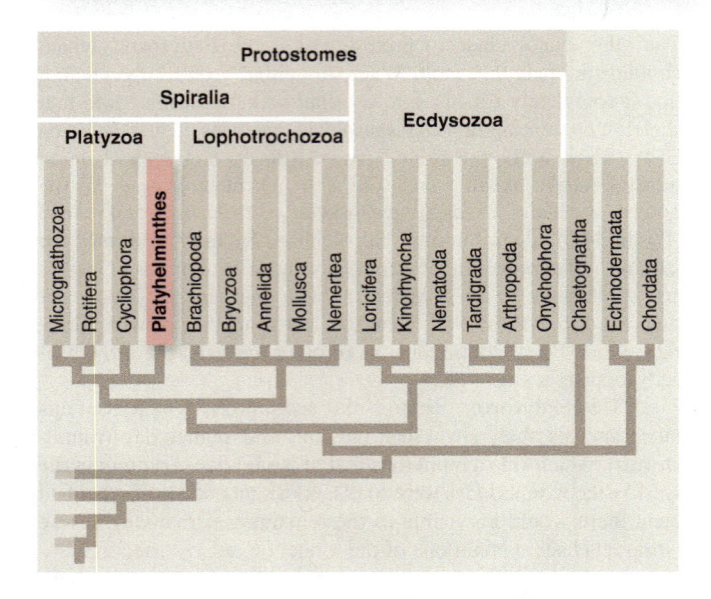

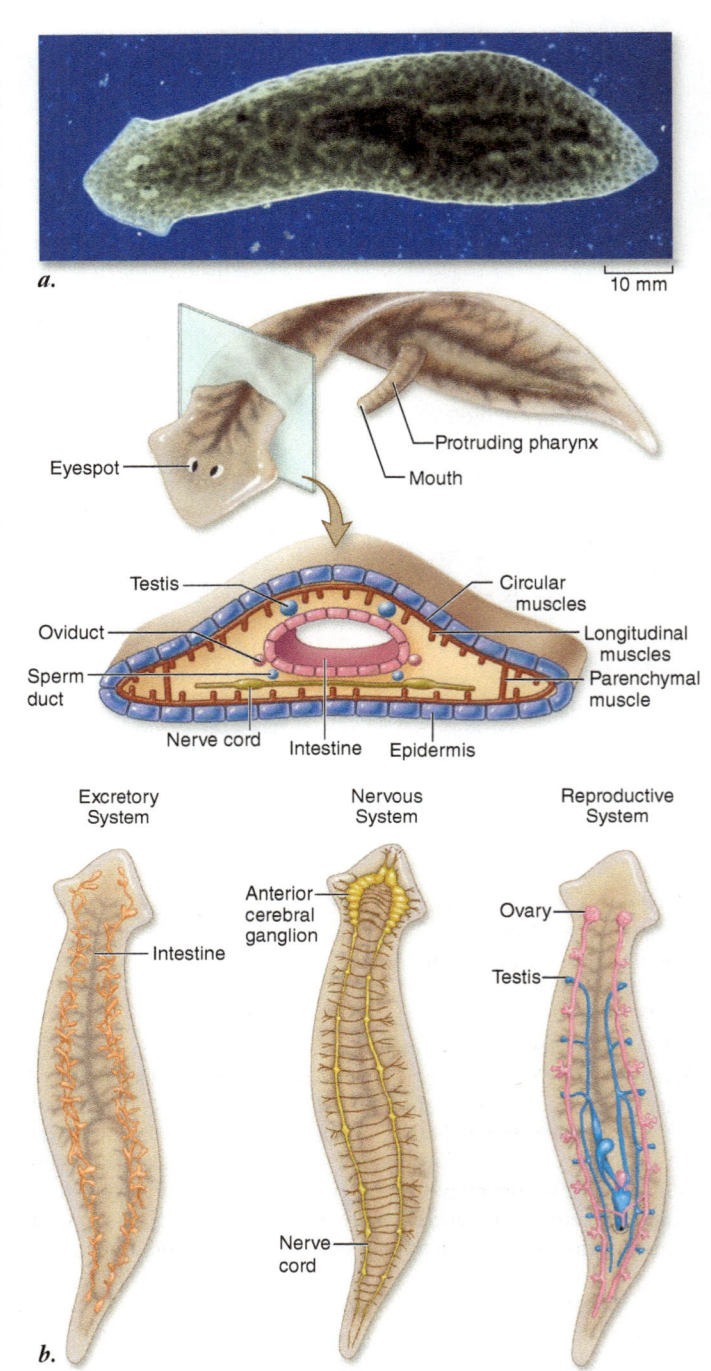

Figure 34.3 Architecture of a flatworm. A photo (*a*) and idealized diagrams (*b*) of the genus *Dugesia*, the familiar freshwater planarian of ponds and rivers. Upper schematic shows a whole animal and a transverse section through the anterior part of the body. Schematics below show the digestive, central nervous, and reproductive systems.

Flatworms have an incomplete gut

The phylum Platyhelminthes consists of some 55,000 species. These ciliated, soft-bodied animals are flattened dorsoventrally, the anatomy that gives them the name flatworms. Flatworm bodies are solid aside from an incomplete digestive cavity (figure 34.3). Although among the morphologically simplest of bilaterally symmetrical animals, they have some complex structures, like their reproductive apparatus, and they have the most complex life cycles among animals.

Free-living flatworms occur in a wide variety of marine, freshwater, and even moist terrestrial habitats. They are carnivores and scavengers, eating small animals and bits of organic debris. They move around by means of ciliated epithelial cells, which are particularly concentrated on their ventral surfaces, but they also have well-developed musculature. Not all flatworms are free-living, however; many species of flatworms are parasitic, living inside the bodies of other animals. Flatworms range in length from 1 mm or less to many meters (some tapeworms).

Digestion in flatworms

Most flatworms have only a single opening for their digestive cavity, a mouth located on the bottom side of the animal at midbody. A flatworm ingests its food and tears it into small bits using muscular contractions in the upper end of the gut, the pharynx.

Like sponges, cnidarians, and ctenophores, a flatworm lacks a circulatory system for the transport of oxygen and food molecules. The thin body of a flatworm allows gas to diffuse between its cells and the surrounding environment (oxygen diffuses in and carbon dioxide diffuses out). Branches of the gut extend throughout the body, so the gut functions in both digestion and distribution of food. Cells that line the gut engulf most of the food particles by phagocytosis and digest them; but, as in cnidarians and most bilaterians, some particles are partly digested extracellularly. Tapeworms, which are parasitic, have a mouth at the front of their body and no digestive cavity at all; they absorb food directly through the body wall.

Excretion and osmoregulation

Unlike cnidarians and ctenophores, flatworms have an excretory system, which consists of a network of fine tubules that run throughout the body. Bulblike **flame cells,** so named because of the flickering movements of the flagella beating inside them, are located on the side branches of the tubules. Flagella in the flame cells move water and excretory substances into the tubules and then to pores located between the epidermal cells through which the liquid is expelled. Flame cells primarily regulate water balance of the organism; the excretory function appears to be secondary, and much of the metabolic waste of a flatworm diffuses into the gut and is eliminated through the mouth.

Nervous system and sensory organs

The flatworm nervous system is composed of an anterior cerebral ganglion and nerve cords that run down the body, with cross-connections that give it a ladderlike shape. Free-living flatworms have eyespots on their heads (figure 34.3). These inverted, pigmented cups, which contain light-sensitive cells connected to the nervous system, enable a worm to distinguish light from dark: Most flatworms tend to move away from strong light.

Flatworm reproduction

The reproductive systems of flatworms are complex. Most are **hermaphroditic,** each individual containing both male and female sexual structures (figure 34.*b*). In many members of Platyhelminthes, copulation is required between two individuals, and fertilization is internal, each partner depositing sperm in the copulatory sac of the other. The sperm travel along special tubes to reach the eggs.

In most freshwater flatworms, fertilized eggs are laid in cocoons strung in ribbons and hatch into miniature adults. In contrast, some marine species develop indirectly, the fertilized egg undergoing spiral cleavage, and the embryo giving rise to a larva that swims or drifts until metamorphosing, when it settles in an appropriate habitat.

Flatworms are known for their regenerative capacity: When a single individual of some species is divided into two or more parts, an entirely new flatworm can regrow what is missing from each bit.

Flatworms consist of two major groups

Most flatworms are parasitic; those that are not are referred to as free-living. Phylogenetic evidence indicates that the parasitic lifestyle evolved only once in platyhelminths, from free-living ancestors.

Turbellaria: Free-living flatworms

Flatworm phylogeny is in a state of flux. The group of free-living flatworms, Turbellaria, has been considered a class, but recent studies show it is not monophyletic so it is likely to be divided into several classes. One of the most familiar members of this group are freshwater members of the genus *Dugesia,* the common planarian used in biology laboratories.

Neodermata: Parasitic flatworms

All parasitic flatworms are placed in the subphylum Neodermata. That name means "new skin" and refers to the animal's outer surface, termed the neodermis. All neodermatans live as ectoparasites on or endoparasites in the bodies of other animals for some period of their lives. The neodermis is resistant to the digestive enzymes and immune defenses produced by the animals parasitized by these flatworms. These animals also lack other features of free-living flatworms such as eyespots, which are of no adaptive value to an organism living inside the body of another animal. Neodermata contains two subgroups: Trematoda, the flukes, and Cercomeromorpha, the tapeworms and their relatives.

Trematoda: The flukes

The more than 10,000 named species of flukes range in length from less than 1 mm to more than 8 cm. Flukes attach themselves within the bodies of their hosts by means of suckers, anchors, or hooks. A fluke takes in food (cells or fluids of the host) through its mouth, like its free-living relatives. The life cycle of some species involves only one host, usually a fish, but the life cycle of most flukes involves two or more hosts. The first intermediate host is almost always a snail, and the final host (in which the adult fluke lives and reproduces sexually) is almost always a vertebrate; in between there may be other intermediate hosts. Although the life of a parasite is secure within a host, which provides food and shelter, getting from one host to another is extremely risky, and most individuals die in the transition.

Flukes that cause disease in humans

The oriental liver fluke, *Clonorchis sinensis,* is an example of a flatworm that parasitizes humans (as well as cats, dogs, and pigs), living in the bile duct of the liver (figure 34.4). It is especially common in Asia. Each worm is 1 to 2 cm long and, like all flukes, has a complex life cycle. A fertilized egg containing a ciliated first-stage larva, the **miracidium,** is passed in the feces. If the larva reaches water, it may be ingested by an aquatic snail, but most do not reach water and most that do are not ingested. The prodigious number of eggs a parasitic flatworm produces is an adaptation to this life cycle full of risks. Within the snail, the ciliated larva transforms into a **sporocyst,** a baglike structure containing embryonic germ cells, each of which develops into a **redia** (plural, *rediae*), an elongated, nonciliated larva. Each of these larvae grows within the snail, then gives rise to several individuals of the next larval stage, the tadpole-like **cercaria** (plural, *cercariae*).

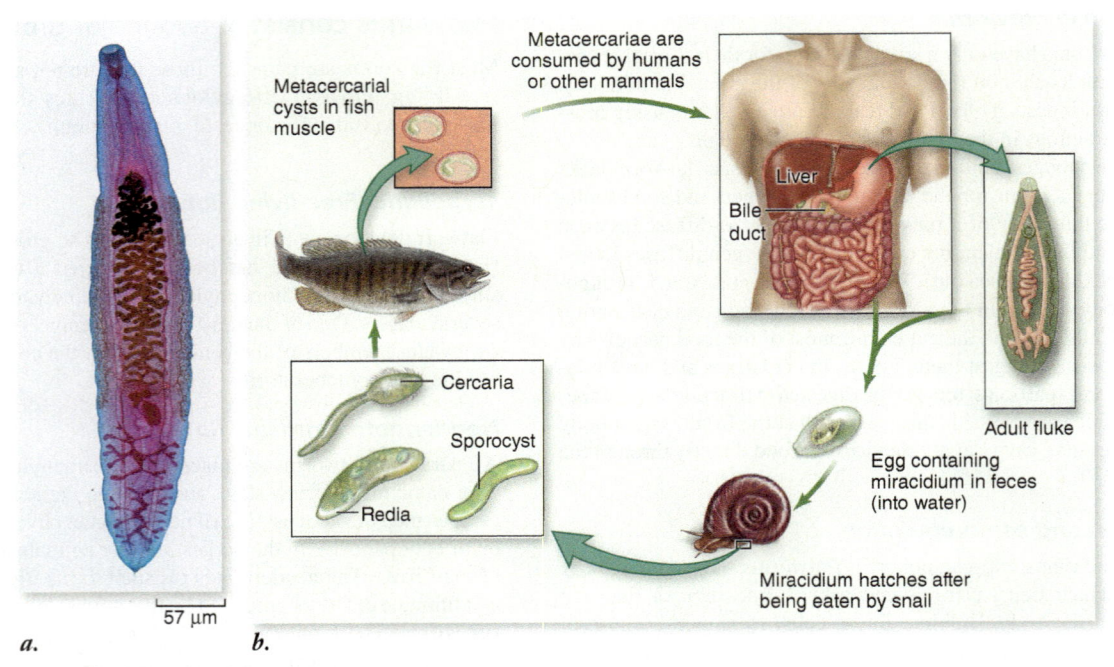

Figure 34.4 Life cycle of the oriental liver fluke, *Clonorchis sinensis*. *a.* Micrograph. *b.* Schematic of the life cycle.

Cercariae escape into the water, where they swim about freely. When one encounters a fish of the family Cyprinidae—the family that includes carp and goldfish—it bores into the muscles, loses its tail, and encysts, transforming into a metacercaria. If a human or other mammal eats raw fish containing metacercariae, the cyst dissolves in the intestine, and the young fluke migrates to the bile duct, where it matures, thereby completing the cycle. Even if infected fish is cooked, the parasite can be transmitted if metacercariae stick to cutting boards or the hands of a person handling the raw fish flesh are then ingested. An individual fluke may live for 15 to 30 years in the liver; a heavy infection of liver flukes may cause cirrhosis of the liver and death in humans.

Perhaps the most important trematodes to human health are blood flukes of the genus *Schistosoma*. They afflict about 200 million people in tropical Asia, Africa, Latin America, and the Middle East. About 200,000 people die each year from the disease called **schistosomiasis,** or bilharzia.

Schistosomes live in blood vessels associated with the intestine or the urinary bladder, depending on the species (figure 34.5). As a result, its fertilized eggs must first break through the wall of the blood vessel to get into the intestine or urinary bladder, from which it can exit the body. The damage done to the blood vessels and gut or bladder wall is considerable, especially considering that an individual worm may release 300 to 3000 eggs per day and live for many years.

Great effort is being made to control schistosomiasis. The worms protect themselves from the body's immune system in part by coating themselves with some of the host's own antigens that effectively render the worm immunologically invisible (see chapter 51). The search is on for a vaccine that would cause the host to develop antibodies to one of the antigens of young worms before they can protect themselves.

Figure 34.5 Schistosomes. The male lies within the groove on the female's ventral side, with its front and hind ends protruding.

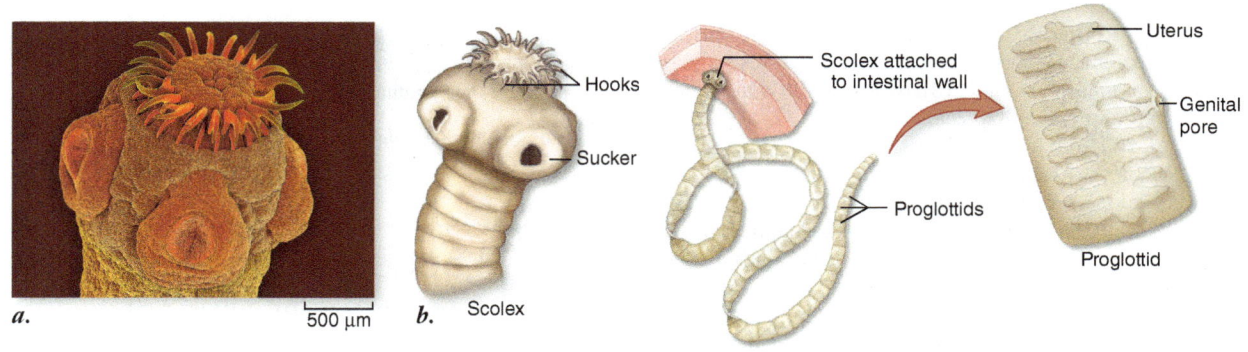

Figure 34.6 Tapeworms. *a.* Flatworms such as this beef tapeworm, *Taenia saginata,* live parasitically in the intestine of mammals. *b.* Schematic.

Cercomeromorpha: Tapeworms and their relatives

An adult tapeworm hangs onto the inner wall of its host's intestine by means of a terminal attachment structure. It lacks a digestive cavity as well as digestive enzymes, absorbing food from the host's gut through its outer surface. Most species of tapeworm occur in the intestines of vertebrates; about a dozen of them regularly occur in humans.

The long, flat body of a tapeworm is divided into three zones: the **scolex,** or attachment structure; the neck; and a series of repetitive sections, the **proglottids** (figure 34.6). The scolex of many species bears four suckers and may also have hooks. The scolex is not a head: It has neither concentrated nervous tissue nor a mouth. Each proglottid is a complete hermaphroditic unit, containing both male and female reproductive organs. Proglottids are formed continuously in a growth zone at the base of the neck, with maturing ones being pushed posteriorly as new ones are formed. The gonads in the proglottids progressively mature away from the neck, fertilization occurs, and the embryos form, the proglottids nearer the end of the body being more mature. The most terminal proglottid, filled with embryonated eggs, breaks off; either it ruptures, and the embryos, each surrounded by a shell, are carried out with the host's feces, or the entire proglottid is carried out, whereupon the embryos emerge from it through a pore or the ruptured body wall. Embryos are scattered in the environment, on leaves, in water, or in other places where they may be picked up by another animal.

The beef tapeworm, *Taenia saginata,* occurs as a juvenile in the intermuscular tissue of cattle; as an adult, it inhabits the intestines of humans, where a mature worm may reach a length of 10 m or more. A worm attaches itself to the intestinal wall of the host by a scolex with four suckers. Shed proglottids pass from the human in the feces and may crawl onto vegetation, a favorable place to be picked up by cattle; they may remain viable for up to five months. If one is ingested by cattle, the larva burrows through the wall of the intestine and ultimately reaches muscle tissues through the blood or lymph vessels. When humans eat infected beef that is cooked "rare," infection by these tapeworms can occur. As a result, the beef tapeworm is a frequent parasite of humans, particularly in countries with poor sanitation.

Learning Outcomes Review 34.2

Flatworms are compact, bilaterally symmetrical animals. Most have a blind digestive cavity with only one opening; they also have an excretory system consisting of flame cells within tubules that run throughout the body. Many are free-living, but some cause human diseases. The scolex of a tapeworm is not a head, but simply an anchoring device to hold the animal in the intestine where it acts as a parasite.

■ How does the anatomy of a tapeworm relate to its way of life?

34.3 Platyzoans: Rotifers (Rotifera)

Learning Outcome

1. Explain why rotifers are referred to as "wheel animals."

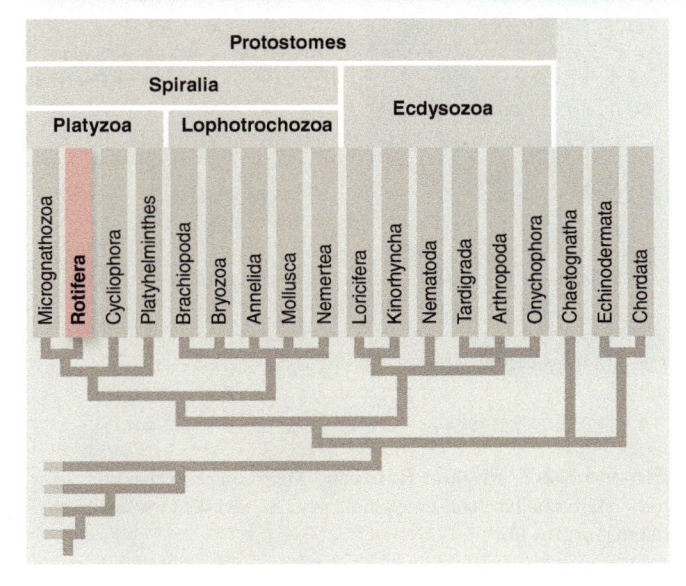

The rotifers, phylum Rotifera, are tiny

Rotifers (phylum **Rotifera**) are bilaterally symmetrical, unsegmented pseudocoelomates (figure 34.7a). Several features suggest their ancestors may have resembled flatworms, with which they are classified in the spiralian Platyzoa.

At 50 to 500 µm long, rotifers are smaller than some ciliate protists. But they have complex bodies with three cell layers, highly developed internal organs, and a complete gut (figure 34.7b). An extensive pseudocoelom acts as a hydrostatic skeleton; the cytoskeleton provides rigidity. A rotifer has a rigid external covering, but its body can lengthen and shorten greatly because the posterior part is tapered so it can fold up like a telescope. Many have adhesive toes used for clinging to vegetation and other such objects.

Diversity and distribution

About 2500 species are known, most of which occur in freshwater; a few rotifers live in soil, mosses, and the ocean. The life span of a rotifer is typically no longer than one or two weeks, but some species can survive in a desiccated, inactive state on the leaves of plants; when rain falls, the rotifers become active and feed in the film of water that temporarily covers the leaf.

Food gathering

The corona, a conspicuous ring of cilia at the anterior end (see figure 34.7b), is the source of the common name "wheel animals" for rotifers because its beating cilia make it appear that a wheel is rotating around the head of the animal. The corona is used for locomotion, but its cilia also sweep food into the rotifer's mouth. Once food is swallowed, it is crushed with a complex jaw in the pharynx.

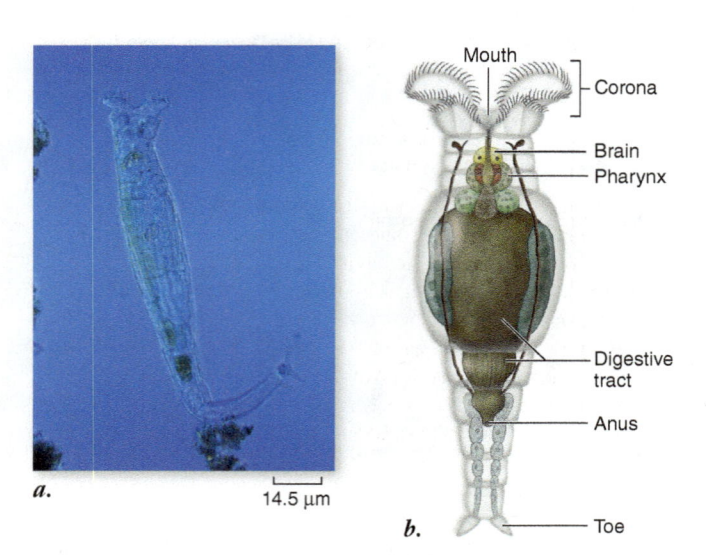

Figure 34.7 Phylum Rotifera. Microscopic in size (*a*), rotifers are smaller than some ciliate protists, and yet have complex internal organs (*b*).

Learning Outcome Review 34.3

Rotifers are extremely small but with a highly complex body structure; the name "wheel animals" comes from the apparent motion of their beating cilia.

- What is the role of the cilia in rotifers?

34.4 Lophotrochozoans: Mollusks (Mollusca)

Learning Outcomes

1. List the defining features of phylum Mollusca.
2. Describe representatives of the four best-known groups of mollusks.
3. Explain the distinguishing features of cephalopods.

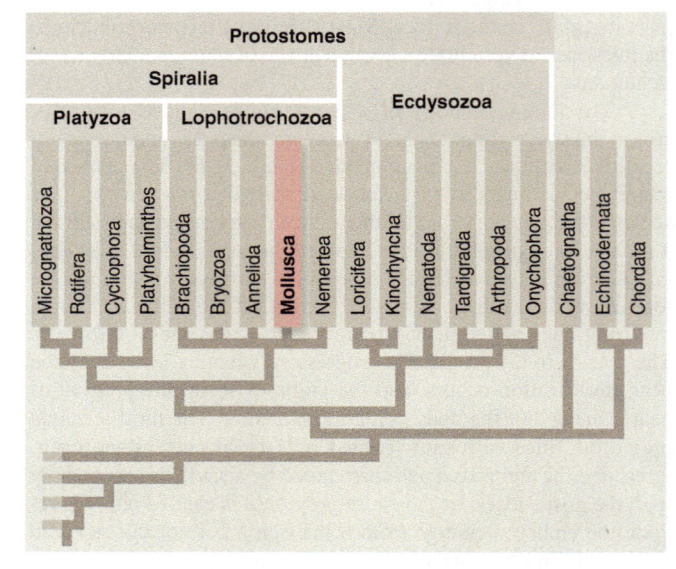

Mollusks (phylum Mollusca) are diverse morphologically and live in many types of environments. With more than 150,000 described species, the phylum is second in size only to arthropods. Mollusks include snails, slugs, clams, scallops, oysters, cuttlefish, octopuses, and many other familiar animals (figures 34.8 and 34.9). The shells of many mollusks are beautiful and elegant; they have long been collected, preserved, and studied by professional scientists and amateurs. Some mollusks, however, lack a shell.

Mollusks are extremely diverse— and important to humans

Mollusks range in size from almost microscopic to huge. Although most measure a few millimeters to centimeters in their largest

a. *b.* *c.* *d.*

Figure 34.8 Mollusk diversity. Mollusks exhibit a broad range of variation. *a.* The flame scallop, *Lima scabra,* is a filter feeder. *b.* The common octopus, *Octopus vulgaris,* is one of the few mollusks dangerous to humans. Strikingly beautiful, it is equipped with a sharp beak that can deliver a poisonous bite! *c.* Nautiluses, such as this chambered nautilus, *Nautilus pompilius,* have been around since before the age of the dinosaurs. *d.* The banana slug, *Ariolimax columbianus,* native to the Pacific Northwest, is the second largest slug in the world, attaining a length of 25 cm.

dimension, the giant squid may grow to more than 15 m long and weigh as much as 250 kg. It is therefore one of the heaviest invertebrates (although nemerteans can be longer, as discussed in section 34.5). Other large mollusks are the giant clams of the genus *Tridacna,* which may be as long as 1.5 m and may weigh as much as 270 kg (figure 34.9).

Like all major animal groups, mollusks evolved in the oceans, and most groups have remained there. Marine mollusks are widespread and many are abundant. Snails and slugs have invaded freshwater and terrestrial habitats, and freshwater mussels live in lakes and streams (the flat foot of a snail or slug allows it to crawl, but the foot of clams, mussels, and other bivalved mollusks is adapted to digging, so they cannot move about on land). Some places where terrestrial mollusks live, such as crevices of desert rocks, may appear dry, but if mollusks live there, the habitat has at least a temporary supply of water.

Mollusks—including oysters, clams, scallops, mussels, octopuses, and squids—are an important source of food for humans. They are also economically significant in other ways. For example, the material called mother-of-pearl (nacre), which is used for jewelry and other decorative objects, and formerly for buttons, comes from mollusk shells, most notably that of the abalone. Mollusks can also be pests. Bivalves called ship-worms burrow through wood exposed to the sea, damaging boats, docks, and pilings. The zebra mussel (*Dreissena polymorpha*) (see figure 59.16) has recently invaded many North American freshwater ecosystems. Many slugs and snails damage flowers, vegetable gardens, and crops. Other mollusks serve as hosts to the larval stages of many serious parasites, as discussed in section 34.2.

The mollusk body plan is complex and varied

Some mollusk body plans are illustrated in figure 34.10. The **mantle,** a thick epidermal sheet, covers the dorsal side of the body, and bounds the mantle cavity. The mantle secretes the calcium carbonate of the shell in those mollusks with a shell. The muscular foot is the primary means of locomotion in mollusks other than cephalopods (octopuses, squid, and chambered nautilus). The head may be well developed or not. Mollusks are bilaterally symmetrical, but this symmetry is modified during development of many gastropods (snails and their relatives) which undergo torsion (a twisting of the body discussed in detail later in this section).

The muscular foot is variously adapted for locomotion, attachment, food capture, digging, or combinations of these functions. The foot of slugs and snails secretes mucus, forming a path that it glides along. In cephalopods, the foot is divided into tentacles and, in some cephalopods, arms. Clams burrow into mud and

Figure 34.9 Giant clam. Second only to the arthropods in number of described species, members of the phylum Mollusca occupy almost every habitat on Earth. This giant clam, *Tridacna maxima,* is one of the heaviest invertebrates in the world.

Internal organs

In all mollusks, the coelom is highly reduced, being limited to small spaces around the excretory organs, heart, and part of the intestine. The coelom plays an important role in some other invertebrates by forming the hydrostatic skeleton, but in mollusks, the shell takes on this function. The digestive, excretory, and reproductive organs are concentrated in a visceral mass.

In aquatic mollusks, the gills (respiratory structures technically termed "ctenidia") project into the mantle cavity (figure 34.10). They consist of filaments rich in blood vessels that greatly increase the surface area and capacity for gas exchange. The continuous stream of water that passes through the mantle cavity, propelled by cilia on the gills of all mollusks except cephalopods, carries oxygen in and carbon dioxide away. Mollusk ctenidia are so efficient that they can extract 50% or more of the dissolved oxygen from the water that passes through the mantle cavity. In addition to extracting oxygen from incoming water, the gills of most bivalves filter out food. Because the outlets from the excretory, reproductive, and digestive organs open into the mantle cavity, wastes and gametes are carried away from a mollusk's body with the exiting water stream.

Shells

One of the best-known characters of the phylum is the shell. A mollusk shell, which is secreted by the outer surface of the mantle, protects against predators and adverse environmental conditions. However, a shell is clearly not essential: Reduction, internalization, and loss of the shell have evolved repeatedly. Examples of shell-less mollusks are cuttlefish, squids, and octopuses (cephalopods) as well as slugs (gastropods).

A typical mollusk shell consists of two layers of calcium carbonate, which is precipitated extracellularly. The outer layer consists of densely packed crystals. In some species, the inner layer is pearly in appearance, and is called mother-of-pearl or nacre. Pearls are formed when a foreign object, such as a grain of sand, becomes lodged between the mantle and the inner shell layer. The mantle coats the object with layer upon layer of nacre to reduce the irritation caused by the object. The most beautiful and highly valued pearls are produced by oysters depositing nacre.

Feeding and prey capture: Radula

A characteristic feature of most mollusks is the **radula** (plural, *radulae*), a rasping, tonguelike structure used in feeding. It consists of dozens to hundreds of microscopic, chitinous teeth arranged in rows on an underlying membrane and lies in a chamber at the anterior end of the gut (figure 34.11). The membrane wraps around a muscular support structure so the radula can be protruded through the mouth and move something like a sanding belt over what is being rasped. Benthic mollusks use their radulae to scrape up algae and other food materials.

In some predatory gastropods (such as moon snails), the radula is modified to drill through clam shells so the snail can eat the clams. In snails of the genus *Conus,* the radula has been transformed into a harpoon that is associated with a venom gland; cone snails use this harpoon to capture prey such as fish, and some cone snails can harm—or even kill—humans.

Bivalves are the only mollusks that have no radula. The gills of most bivalves are adapted to filter food particles from the water,

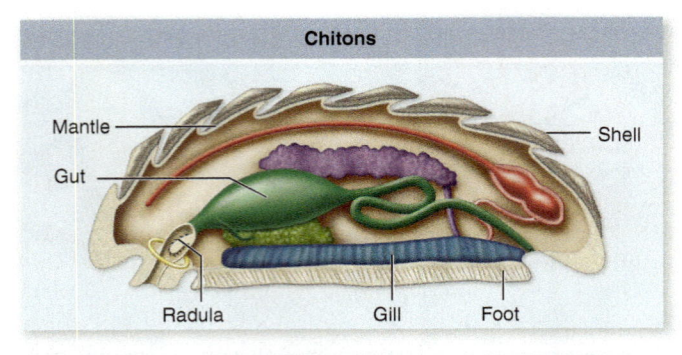

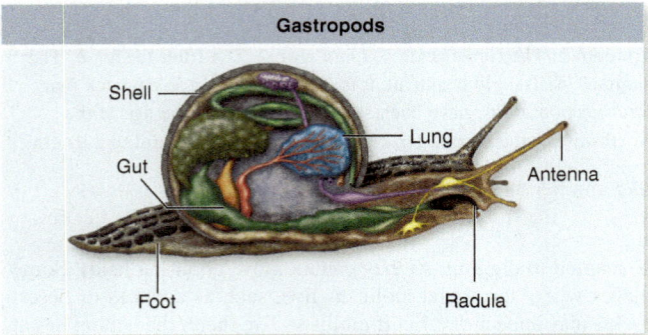

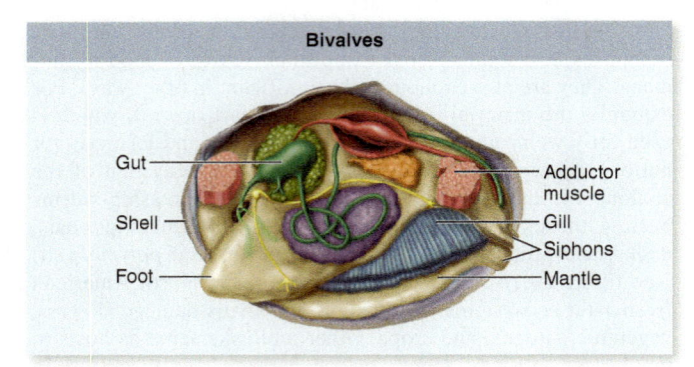

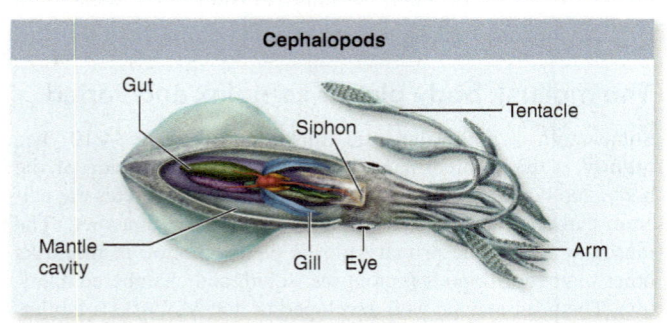

Figure 34.10 Body plans of some mollusks.

sand using their hatchet-shaped foot. In some mollusks that live in the open ocean, the foot is modified into winglike projections that can beat like wings and that have a large surface area that slows the rate of sinking.

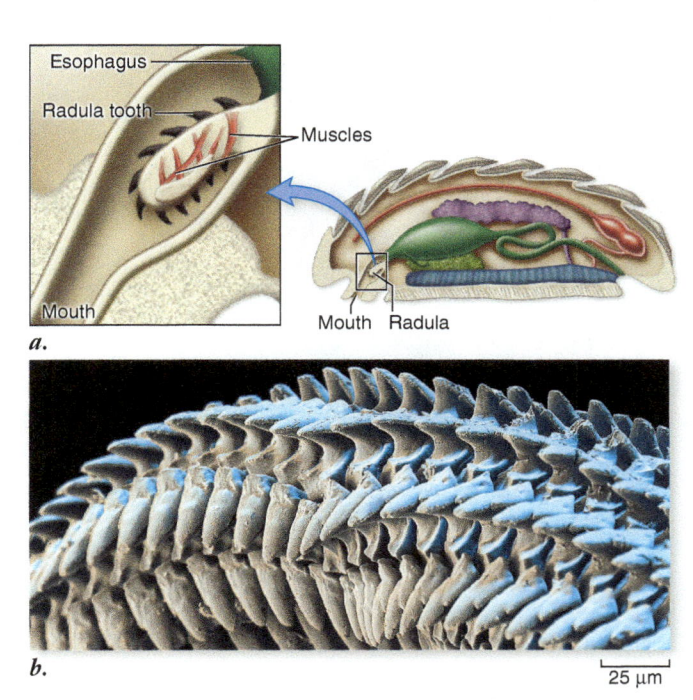

Figure 34.11 Structure of the radular apparatus in a chiton. The radula consists of rows of teeth made of chitin. As the animal feeds, its mouth opens, and the radula is thrust out to scrape food off surfaces. *a.* Schematic. *b.* Micrograph.

although primitive bivalves pick up bits of food from soft sediments (this is, they are deposit feeders) using appendages around the mouth.

Removal of wastes

Nitrogenous wastes are removed from the mollusk body by the **nephridium** (plural, *nephridia*), a sort of kidney. A typical nephridium has an open funnel, the nephrostome, which is lined with cilia. A coiled tubule runs from the **nephrostome** into a bladder, which in turn connects to an excretory pore. Wastes are gathered by the nephridia from the coelomic cavity and discharged into the mantle cavity. Sugars, salts, water, and other materials are reabsorbed by the walls of the nephridia and returned to the animal's body as needed to maintain osmotic balance.

The circulatory system

The main coelomic cavity of a mollusk is a hemocoel, which includes several sinuses and a network of vessels in the gills, where gas exchange takes place. Except for cephalopods, all mollusks have such an open circulatory system; blood (technically termed "hemolymph") is propelled by a heart through the aorta vessel that empties into the hemocoel. The blood moves through the hemocoel being recaptured by other venous vessels before re-entering the heart (see chapter 49). The heart of most, but not all, mollusks has three chambers: two that collect aerated blood from the gills, and a third that pumps it into the hemocoel. The blood of the closed circulatory system of cephalopods is contained in a continuous system of vessels, so it does not contact other tissues directly.

Reproduction

Most mollusks have separate sexes although a few bivalves and many freshwater and terrestrial gastropods are hermaphroditic. Most hermaphroditic mollusks engage in cross-fertilization. Some oysters are able to change sex.

As is typical of animals living in the sea, many marine mollusks have external fertilization: Gametes are released by males and females into the water where fertilization occurs. Most gastropods, however, have internal fertilization, with the male inserting sperm directly into the female's body (not, that is, into a preformed channel). Internal fertilization, a foot, and an efficient excretory system that prevents desiccation are some of the key adaptations that allowed gastropods to colonize the land.

A mollusk zygote undergoes spiral cleavage (thus Mollusca is part of the Spiralia; see chapter 33). The embryo develops into a free-swimming larva called a **trochophore** (figure 34.12*a*) that closely resembles the larval stage of many marine annelids and other lophotrochozoans. A trochophore swims by means of cilia that encircle the middle of its body. In most marine snails and in bivalves, the trochophore develops into a second free-swimming stage, the **veliger.** The veliger forms the beginnings of a foot, shell, and mantle (figure 34.12*b*). Trochophores and veligers drift widely in the ocean, dispersing these otherwise sedentary mollusks.

Four classes of mollusks show the diversity of the phylum

We examine four of the seven or eight recognized classes of mollusks: (1) Polyplacophora—chitons; (2) Gastropoda—limpets, snails, slugs, and their relatives; (3) Bivalvia—clams, oysters, scallops, and their relatives; and (4) Cephalopoda—squids, octopuses, cuttlefishes, and the chambered nautilus.

Chitons (Polyplacophora)

Chitons are exclusively marine and comprise perhaps 1000 species. The oval body is covered dorsally with eight overlapping dorsal

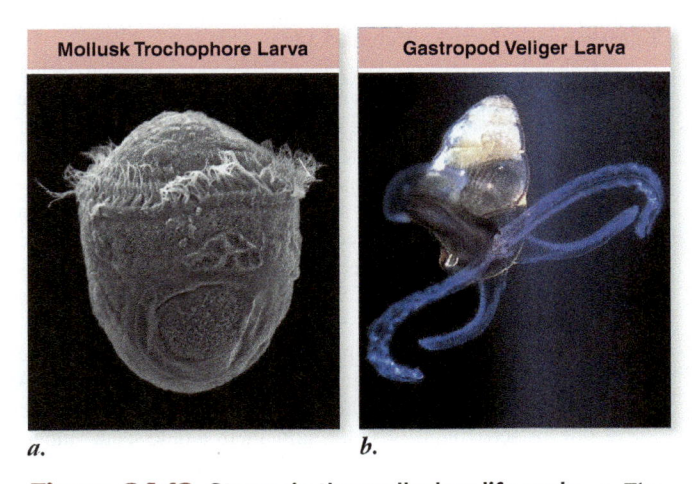

Figure 34.12 Stages in the molluskan life cycle. *a.* The trochophore larva. Similar larvae are characteristic of some annelid worms as well as a few other phyla. *b.* The veliger stage of a gastropod.

Figure 34.13 The noble chiton, *Eudoxochiton nobilis*, from New Zealand. The foot of the chiton can grip the substrate very strongly, making chitons hard to dislodge by waves or predators.

Figure 34.14 A Gastropod mollusk. The terrestrial Oregon forest snail, *Allogona townsendiana*.

calcareous plates (figure 34.13). The body is not segmented, but chitons do have eight sets of dorsoventral pedal retractor muscles and serially repeated gills. The broad, flat ventral foot, on which the animal creeps, is surrounded by a groove, the mantle cavity, in which the gills are suspended. Most chitons are grazing herbivores and live in shallow marine habitats, but chitons occur to depths of more than 7000 m.

Snails and slugs (Gastropoda)

The 60,000 or so species of limpets, snails, and slugs belong to this class, which is primarily marine. However, this group contains many freshwater species and the only terrestrial mollusks (figure 34.14). Most gastropods have a single shell, but some, such as slugs and nudibranchs (or sea slugs), have lost the shell through the course of evolution. Most gastropods creep on a foot, but in some it is modified for swimming.

The head of most gastropods has a pair of tentacles, which serve as chemo- or mechanoreceptors, with eyes at the base. A typical garden snail may have two sets of tentacles, one of them bearing eyes at the ends (figure 34.14).

Uniquely among animals, gastropods undergo torsion during larval life. **Torsion** involves twisting of the body so the mantle cavity and anus are moved from a posterior location to the front of the body. Torsion may lead to the reduction or disappearance of the nephridium, gonad, or other internal organs on one side. Most adult gastropods are therefore not bilaterally symmetrical. Torsion should not be confused with coiling, the spiral winding of the shell. Coiling also occurs in cephalopods. The fossil record suggests that the first gastropods were coiled but did not undergo torsion.

Like many gastropods, nudibranchs (sea slugs) are active predators. Nudibranchs get their name from their gills, which, instead of being enclosed within the mantle cavity, are exposed along the dorsal surface (figure 34.15). Nudibranchs would seem vulnerable to predation, but many secrete distasteful chemicals. They prey on animals that are avoided by other predators. Some are specialist feeders on sponges, most species of which form spicules and noxious chemicals. Other nudibranchs specialize in feeding on cnidarians, which are protected from most predators by their nematocysts. Some of these nudibranchs have the extraordinary ability to extract the nematocysts undischarged, transfer them through their digestive tract to the surface of their bodies, and use them for their own protection.

In terrestrial gastropods, the mantle cavity, which is occupied by gills in aquatic snails, is extremely rich in blood vessels and serves as a lung. This lung absorbs oxygen from the air much more effectively than a gill could; however, a snail will drown if its lung fills with water; for this reason, terrestrial gastropods can close the opening of the lung to the outside.

Clams, mussels, and cockles (Bivalvia)

Most of the 10,000 species of bivalves are marine, but some live in freshwater. More than 500 species of pearly freshwater mussels, or naiads, occur in the rivers and lakes of North America.

Unlike other mollusks, a bivalve does not have a radula or distinct head (figure 34.16). The foot of most is wedge-shaped,

Figure 34.15 A nudibranch (or sea slug). The bright colors of many nudibranchs such as this alert predators to their repulsive taste.

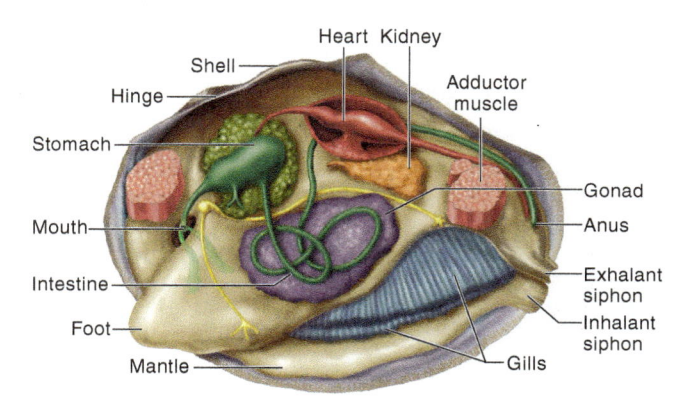

Figure 34.16 Diagram of a clam. The left shell and mantle are removed to show the internal organs and the foot. Bivalves such as this clam circulate water across their gills and filter out food particles.

Figure 34.17 Problem-solving by an octopus. This two-month-old common octopus, *Octopus vulgaris,* was presented with a crab in a jar. It tried to unscrew the lid to get to the crab. Although it failed in this attempt, in some cases it was successful.

adapted for burrowing or anchoring the animal in its burrow. Some species of clams can dig into sand or mud very rapidly by means of muscular contractions of their foot. Some species of scallops and file clams can move swiftly by clapping their shells rapidly together (although they cannot control the direction of their movement); the adductor muscle that allows this clapping is the part of a scallop eaten by humans. Projecting from the edge of a scallop's mantle are tentacle-like projections having complex eyes between them.

Bivalves, as their name implies, have two shells (valves) that are hinged dorsally so the shells are oriented laterally (left and right). A ligament lying along the hinge is structured so it causes the shells to gape open. One or two large adductor muscles link the shells internally (figure 34.16), and when they contract, they counteract the hinge ligament to draw the shells together. The mantle covers the internal surface of the shells, enveloping the visceral mass on its inner side and secreting the shells on its outer side. As is typical of mollusks, the respiratory structures, a set of complexly folded gills on each side of the visceral mass, lie in the mantle cavity.

The edges of the mantle may be partly fused. In bivalves, in which this is the case, typically two areas are not fused and may be drawn out to form tubes called siphons (figure 34.16). Water enters the mantle chamber through the **inhalant siphon** bringing oxygen and food, and water exits through the **exhalant siphon,** taking wastes and gametes with it. In bivalves that live buried deeply in mud or burrow into rock, the siphons allow the animals to eat and breathe, functioning essentially as snorkels.

Octopuses, squids, and nautiluses (Cephalopoda)

The more than 700 species of cephalopods are strictly marine. They are active predators that swim, often swiftly, and they are the only mollusks with a closed circulatory system. The foot has evolved into a series of arms equipped with suction cups, adhesive structures, or hooks that seize prey. Octopuses, as their name suggests, have eight arms; squids have eight arms and two tentacles; and the chambered nautilus has 80 to 90 tentacles (which lack suckers). After snaring prey with its arms, a cephalopod bites the prey with its strong, beaklike jaws, then pulls it into its mouth by the action of the radula. The salivary gland of many cephalopods secretes a toxin that can be injected into prey; the tiny common octopus of Australia (see figure 34.8*b*) can kill a human with its deadly bite.

Cephalopods have the largest relative brain sizes among invertebrates and highly developed nervous systems. Many exhibit complex patterns of behavior and are highly intelligent (figure 34.17); octopuses can be easily trained to distinguish among classes of objects and have been known to climb out of their aquaria, move across a laboratory floor, enter another aquarium to capture and devour a crab, and then return to their own tank, leaving lab personnel to wonder why their crabs were disappearing. Cephalopod eyes are much like those of vertebrates, although they evolved independently (see chapter 44).

Aside from the chambered nautiluses, living cephalopods lack an external shell. Shelled cephalopods were formerly far more diverse, as evidenced by the many fossil cephalopods such as ammonites and belemnites. These cephalopods were extraordinarily successful because they could move in open water instead of on the sea bottom like other mollusks. However, once more maneuverable fishes evolved, shelled mollusks declined, some dying out, others experiencing evolutionary reduction and eventual loss of their heavy shells. The cuttlebone of cuttlefish and the pen of squids are internal shells that support these animals and give them some buoyancy. Even the internal shell has disappeared in the lineage that gave rise to octopuses.

As in other mollusks, water passes through the mantle cavity. In a cephalopod, it is pumped in by muscles and exits through a siphon, which allows the animal to move by jet propulsion, and which can be directed to steer. The ink sac of cephalopods, which typically contains a purplish fluid, can eject its contents through the siphon as a cloud that may hide the cephalopod and confuse predators (figure 34.18).

Figure 34.18 Ink defense by a giant Pacific octopus, *Octopus dofleini*. When threatened, octopuses and squids expel a dark cloudy liquid.

Most octopuses and squids are capable of changing skin color and texture to match their background or to communicate with one another. They do so using chromatophores, epithelial cells that contain pigments. Some deep-sea squids harbor symbiotic luminescent bacteria. These may be emitted with the ink to produce a glimmering cloud (ink would not be seen at depths where sunlight cannot penetrate) or they may inhabit cells like chromatophores so they can light up the surface of the animal.

Another difference with many other mollusks is that cephalopods have direct development—that is, they lack a larval stage, hatching as miniature adults.

Learning Outcomes Review 34.4

Mollusks have a reduced coelom. Most have efficient excretory systems, gills for respiration, and a rasping structure—the radula—for gathering food. The mantle of mollusks not only secretes their protective shell but also forms structures essential to body functions. Chitons, gastropods, bivalves, and cephalopods are the four best-known groups.

- Why might the coelom of mollusks be reduced?

34.5 Lophotrochozoans: Ribbon Worms (Nemertea)

Learning Outcome

1. Describe the ways in which ribbon worms are similar and different from flatworms.

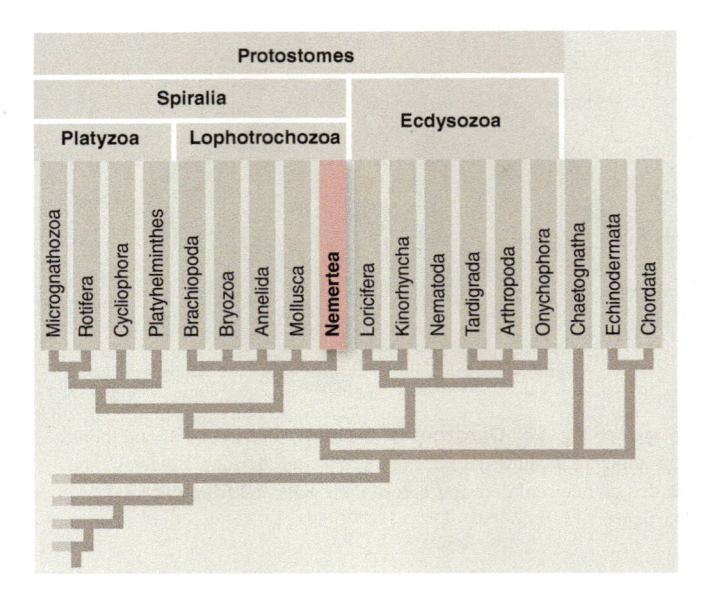

Nemerteans (phylum Nemertea) consist of about 2400 species of cylindrical to flattened very long worms (figure 34.19). Most nemerteans are marine; a few species live in freshwater and humid terrestrial habitats. An individual may reach 10 to 20 cm in length, although the animals are difficult to measure because they can stretch, and many species break into pieces when disturbed or handled. The species *Lineus longissimus* has been reported to measure 60 m in length—the longest animal known!

The nemertean body plan resembles that of a flatworm, with networks of fine tubules constituting the excretory system, and with internal organs not lying in a body cavity. A bit of cephalization is present, with two lateral nerve cords extending posteriorly from an anterior ganglion; some animals have eyespots on the head. But, by contrast with a platyhelminth, a nemertean has a complete gut, with both mouth and anus, connected by a straight tube. Nemerteans also possess a fluid-filled cavity called a rhynchocoel. This sac serves as a hydraulic power source for the proboscis, a long muscular tube that can be thrust out quickly from a sheath to capture animal prey.

Figure 34.19 Phylum Nemertea: A ribbon worm of the genus *Lineus*. Some nemerteans can stretch to several meters in length.

Nemerteans all reproduce sexually. Some are capable of asexual reproduction by fragmentation. However, in most species, most fragments resulting from disturbance die, so nemertean regenerative powers may not be as great as is sometimes stated.

Blood of nemerteans flows entirely in vessels that are derived from the coelom. That and the rhynchocoel are good evidence that nemerteans are not related to flatworms, which they resemble superficially, but belong to the Lophotrochozoa, along with mollusks.

Learning Outcome Review 34.5

Nemerteans are very long worms that have a coelomic cavity and blood vessels derived from the coelom. They capture prey with a muscular proboscis. Unlike the acoelomate flatworms, nemerteans have a complete gut with mouth and anus.

- *What would be some advantages of a flow-through digestive tract?*

34.6 Lophotrochozoans: Annelids (Annelida)

Learning Outcomes

1. *Explain how circular and longitudinal muscles in a segmented body facilitate movement.*
2. *Distinguish between the classes Polychaeta and Clitellata.*
3. *Describe adaptations in leeches for feeding on the blood of animals.*

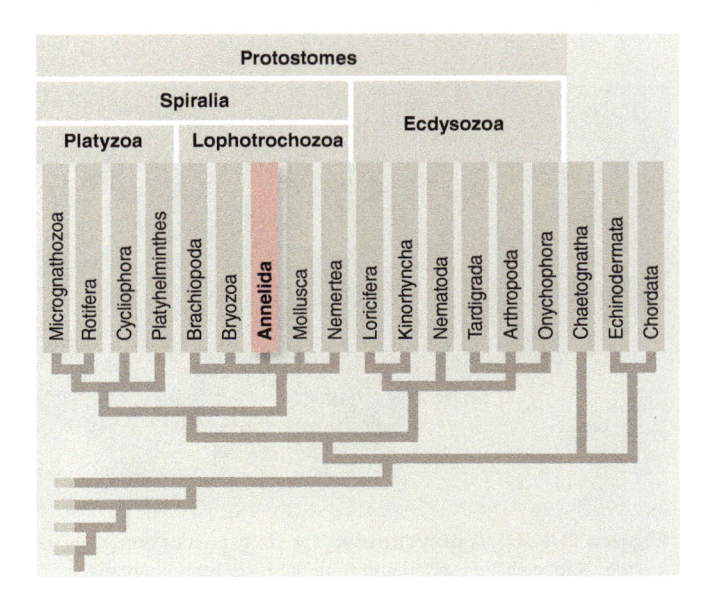

Figure 34.20 A polychaete annelid. *Nereis virens* is a wide-ranging, predatory, marine polychaete worm equipped with feathery parapodia for movement and respiration, as well as jaws for hunting. You may have purchased *Nereis* as fishing bait!

An important innovation in the animal body plan was segmentation, the building of a body from a series of repeated units (see chapter 33), which has evolved multiple times. One advantage of a segmented body is that the development and function of individual segments or groups of segments can differ. For example, some segments may be specialized for reproduction, whereas others are adapted for locomotion or excretion.

All animals that have been regarded as annelids are segmented (figure 34.20), so the animals were considered to constitute a natural group, but the monophyly of Annelida is being reconsidered because some unsegmented worms may belong to this clade.

The annelid body is composed of ringlike segments

The head, which contains a well-developed cerebral ganglion, or brain, and sensory organs occurs at the anterior end (front) of a series of ringlike segments that resemble a stack of coins (figure 34.21). Many species have eyes, which in some species

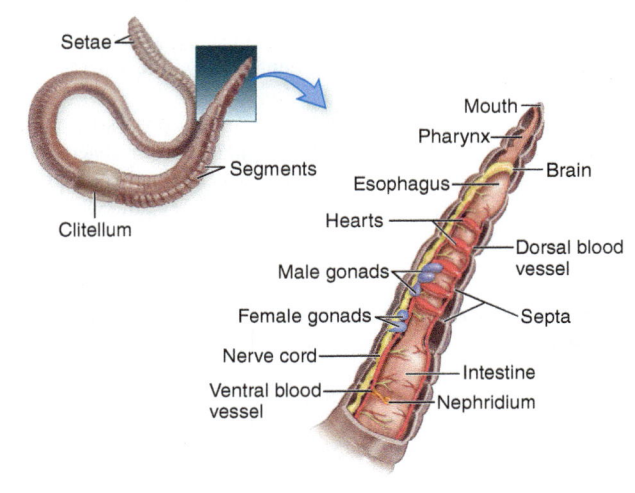

Figure 34.21 Phylum Annelida: An oligochaete. The earthworm body plan is based on repeated body segments. Segments are separated internally from each other by septa.

have lenses and retinas. Technically the head is not a segment, nor is the posterior end of the worm, the pygidium. In embryonic development, the head and tail form first, and then segments form between them; if a worm is cut in pieces, generally only those parts containing either head or tail can regenerate the missing parts, and those parts coming from the middle die.

Internally, the segments are divided from one another by partitions called septa, just as bulkheads separate the compartments of a submarine. Each segment has a pair of excretory organs, a ganglion, and locomotory structure; in most marine annelids, each also has a set of reproductive organs.

Although septa separate the segments, materials and biological signals do pass between segments. A closed circulatory system carries blood the length of the animal, anteriorly in the dorsal vessel and posteriorly in the ventral one. Connections from ventral to dorsal vessel in each segment bring the blood near enough to each cell so oxygen and food molecules diffuse from the blood into the cells of the body wall, and carbon dioxide and other wastes diffuse from the cells into the blood. A ventral nerve cord connects the ganglia in each segment with one another and with the brain. These neural connections allow the worm to function as a unified and coordinated organism.

Annelids move by contracting their segments

The basic annelid body plan is a tube within a tube, the digestive tract—extending from mouth to anus—passing through the septa, and suspended within the spacious coelom, which is surrounded by the body wall. Each portion of the digestive tract—pharynx, esophagus, crop, gizzard, and intestine—is specialized for a different function.

The coelomic fluid creates a hydrostatic skeleton that gives each segment rigidity, like an inflated balloon (see chapter 46). Annelids move by contraction of the circular and longitudinal muscles against the hydrostatic skeleton. When circular muscles are contracted around a segment, the segment decreases in diameter, so the coelomic fluid causes the segment to elongate. When longitudinal muscles contract, the segment shortens, so the coelomic fluid causes the segment to increase in diameter. Alternating these contractions and confining them to only some segments allows the worms to move in complex ways.

In most annelid groups, each segment possesses bristles of chitin called **chaetae** (or setae—singular, seta or chaeta). By extending the chaetae in some segments so that they protrude into the substrate and retracting them in other segments, the worm can extend its body, but not slip (see figure 46.1).

Annelids have a closed circulatory system but a segmented excretory system

Unlike mollusks, except for cephalopods, annelids have a closed circulatory system. Annelids exchange oxygen and carbon dioxide with the environment through their body surfaces, although some aquatic species have gills along the sides of the body or at the anterior end. Gases (and food molecules) are distributed throughout the body in blood vessels. Some of the vessels at the anterior end of the body are enlarged and heavily muscular,

serving as hearts that pump the blood. An earthworm has five pulsating blood vessels on each side that help to move blood from the main dorsal vessel, the major pumping structure, to the ventral vessel.

The excretory system of annelids consists of ciliated, funnel-shaped nephridia like those of mollusks. Each segment has a pair of nephridia that collect wastes and transport them out of the body by way of excretory tubes. Some polychaete worms have protonephridia like the flame cells of planarians.

Annelida is composed of two— or three—classes

The roughly 32,000 described species of annelids occur in many habitats. They range in length from as little as 0.5 mm to giant Australian earthworms more than 3 m long. Although traditionally annelids have been classified into classes Polychaeta (mostly marine worms), Oligochaeta (mostly terrestrial worms, including earthworms), and Hirudinea (leeches), the monophyly of polychaetes is not well established. The classification of annelids may change in the near future, but we adopt the current two-class system of Polychaeta and Clitellata (which combines Oligochaeta and Hirudinea).

Polychaetes (Polychaeta)

Polychaetes include clamworms, scaleworms, lugworms, sea mice, tubeworms, and many others. Polychaetes are a crucial part of many marine food chains and are extremely abundant in particular habitats. Some of these worms are beautiful, with unusual forms and iridescent colors (figure 34.22).

On most segments, a polychaete has paired, fleshy, paddle-like lateral projections called **parapodia** (see figures 34.20 and 34.22). The parapodia bear chaetae; the word *polychaeta* means "many chaetae." Parapodia are used in swimming, burrowing, or crawling, and those of polychaetes that live in burrows or

Figure 34.22 A polychaete. The shiny bristleworm, *Oenone fulgida*. Notice chaetae extending from the iridescent parapodia.

tubes may have chaetae with hooks that help anchor the worm. Parapodia can also play an important role in gas exchange because they greatly increase the surface area of the body, and in some species they bear or are even transformed into gill-like structures.

Polychaetes can swim or crawl, and some are active predators with powerful jaws. Other polychaetes live in tubes or burrows of hardened mud, sand, mucus-like secretions, or calcium carbonate. Sedentary polychaetes may project feathery tentacles that sweep the water for food, filter feeding; the tentacles may also serve as gills, exchanging gas. Some deep-sea tubeworms such as *Riftia* (figure 34.23) are gutless as adults. Projections from the body of these worms house sulfur-oxidizing bacteria that synthesize organic compounds used by the worm. These worms aggregate near hydrothermal vents where sulfur is plentiful and can grow to more than a meter in length.

Polychaete gametes typically are released into the water, where fertilization occurs externally. Many polychaetes have gonads in most segments, but in some groups, gonads are confined to certain segments. In palolo worms *(Palola viridis)* and their relatives, these segments are at the end of the body; spawning involves the end of the worm breaking off and swimming to the surface of the sea, where it ruptures, releasing the gametes. The gamete-filled terminal parts of palolo worms are considered delicacies by some people in the South Pacific.

Fertilization results in spiral cleavage followed by the production of ciliated, mobile trochophore larvae similar to that of mollusks. Metamorphosis of a trochophore involves differentiation of a head and tail end, with development of segments in between, from a posterior growth zone.

Earthworms and leeches (Clitellata)

Some authorities still consider earthworms to belong to class Oligochaeta and leeches to class Hirudinea, but most now put earthworms and leeches into a single class, although, confusingly, it sometimes may be called Oligochaeta. More often it is called Clitellata, because of the feature that unites these seemingly quite different animals, the clitellum—a thickened band on the body, which is the familiar "saddle" of an earthworm (see figure 34.21).

Earthworms. The body of a typical earthworm consists of 100 to 175 similar segments. The head is not well differentiated. An earthworm has no parapodia, and its chaetae (which are fewer than in polychaetes—the word *oligochaeta* means "few chaetae") project directly from the body wall.

Earthworms eat their way through the soil, ingesting it by muscular action of their strong pharynx; organic material is ground in the gizzard. What passes through an earthworm is deposited outside the opening of its burrow as castings that form irregular mounds. In this way, earthworms contribute to loosening, aeration, and enrichment of the soil. In view of their underground lifestyle, it is not surprising that earthworms have no eyes. But they do have light-, chemo-, and touch-sensitive cells, most concentrated in segments near each end of the body—those regions most likely to encounter light or other stimuli.

Earthworms are hermaphroditic, another way in which they differ from most polychaetes, which have separate sexes, but they cross-fertilize through mating (figure 34.24). The clitellum secretes mucus that holds the worms together during copulation, their anterior ends pointing in opposite directions, their ventral surfaces touching. Sperm cells are released from pores in specialized segments of one partner into the sperm receptacles of the other, the process going in both directions simultaneously. Several days after the worms separate, the clitellum of each worm secretes a mucus cocoon, surrounded by a protective layer of chitin. As the worm moves, this sheath passes over the female pores of the body, receiving eggs and incorporating the deposited sperm so that fertilization takes place within the cocoon. When the cocoon passes over the end of the worm, its edges pinch together and the sealed cocoon is left behind. Within the cocoon, the fertilized eggs develop directly into young worms similar to adults.

Leeches. Most leeches live in freshwater, although a few are marine and some tropical leeches occupy terrestrial habitats. Dorsoventrally flattened, most leeches are 2 to 6 cm long, but one

Figure 34.23 Giant tubeworms, *Riftia pachyptila*, living in deep-sea hydrothermal vents near the Galápagos. These tubeworms and associated animals (notice the small crab near the *top right*) are an example of a community dependent on hydrogen sulfide, rather than the Sun, as an energy source (see chapter 58).

Figure 34.24 Earthworms mating. The anterior ends are pointing in opposite directions.

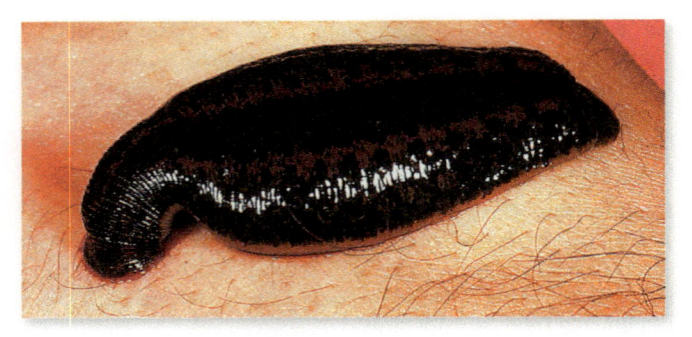

Figure 34.25 A leech. *Hirudo medicinalis*, the medicinal leech, feeding on a human arm. Leeches use chitinous, bladelike jaws to make an incision to access blood; they secrete an anticoagulant to keep the blood from clotting. Both the anticoagulant and the leech itself have made important contributions to modern medicine.

tropical species reaches 30 cm. Like earthworms, leeches are hermaphroditic, but the clitellum develops only during the breeding season. Leeches also cross-fertilize.

Unlike that of earthworms and polychaetes, the coelom of a leech is reduced and not divided into segments. The suckers at one or both ends of a leech's body are used for locomotion and to attach to prey. A leech with suckers at both ends moves by attaching first one and then the other end to the substrate, looping along. Many species are also capable of swimming. Except for one species, leeches have no chaetae.

About half the known species of leeches eat detritus or devour small animals. The others suck blood or other fluids from their hosts (figure 34.25). Blood-sucking leeches secrete an anticoagulant into the wound to prevent the blood from clotting, and vasodilators to keep the blood flowing; the leech's powerful pharynx pumps the blood out quickly once a hole has been opened. Anesthetics injected into the prey prevent the leech from being noticed while piercing the skin; they are usually detected only after detaching, when blood starts to flow from the wound. Freshwater parasitic leeches may remain on their hosts for long periods, sucking the host's blood from time to time. Leeches detect prey by sensing gradients of carbon dioxide in the environment. In some tropical forests, a few seconds after you stop on a trail, you can see dozens of leeches approaching you from all directions!

One of the best-known species, the medicinal leech *Hirudo medicinalis* (figure 34.25), reaches 10 to 12 cm long, and has bladelike, chitinous jaws that can rasp through an animal's skin. Leeches were used in medicine for hundreds of years to treat patients whose diseases were mistakenly believed to be caused by an excess of blood. Today, however, leeches are used to remove excess blood after surgery or to keep blood from coagulating in severed appendages that have been reattached. Accumulations of blood can cause the tissue to die; when leeches remove such blood, new capillaries form in about a week, and the tissues remain healthy. The anticoagulant and anesthetic properties of the leech are also being investigated by pharmaceutical companies.

Learning Outcomes Review 34.6

Annelids generally exhibit segmentation. Each segment has its own excretory and locomotor elements; circular and longitudinal muscles in segments cause the body to extend and contract, respectively. Worms of class Polychaeta, which are mostly marine, have parapodia on their segments. Leeches and earthworms were formerly in separate classes, but a major morphological similarity, the clitellum, has been used to group them into a single class, Clitellata.

- What would be the advantages of having nervous and circulatory systems that serve the entire body instead of segmented systems?

34.7 Lophophorates: Bryozoans (Bryozoa) and Brachiopods (Brachiopoda)

Learning Outcomes

1. Describe the lophophore and its function.
2. Distinguish between bryozoans and brachiopods.

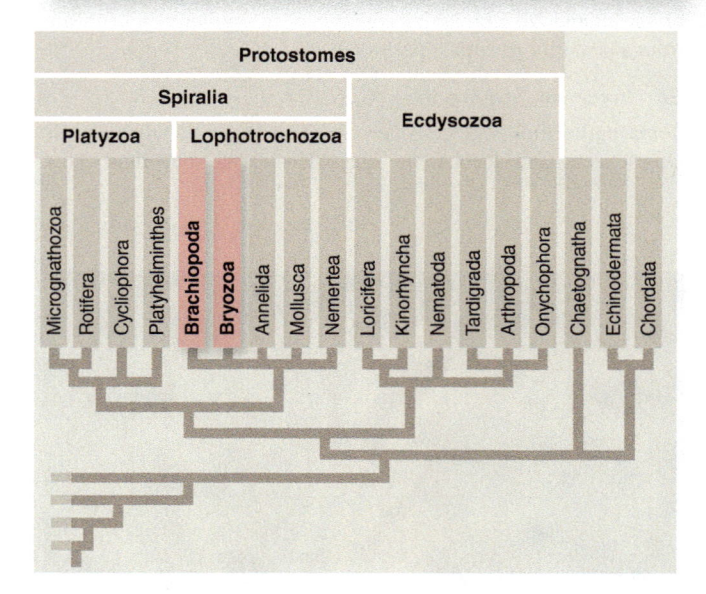

Two phyla of mostly marine animals—Bryozoa and Brachiopoda—are characterized by a lophophore, a circular or U-shaped ridge around the mouth bearing one or two rows of ciliated tentacles into which the coelom extends. The lophophore functions as a surface for gas exchange, and the cilia of the lophophore serve to guide the organic detritus and plankton on which the animal feeds to the mouth. Because of the lophophore, bryozoans and brachiopods have been considered related to one another, but some recent data indicate that the structures may have evolved convergently.

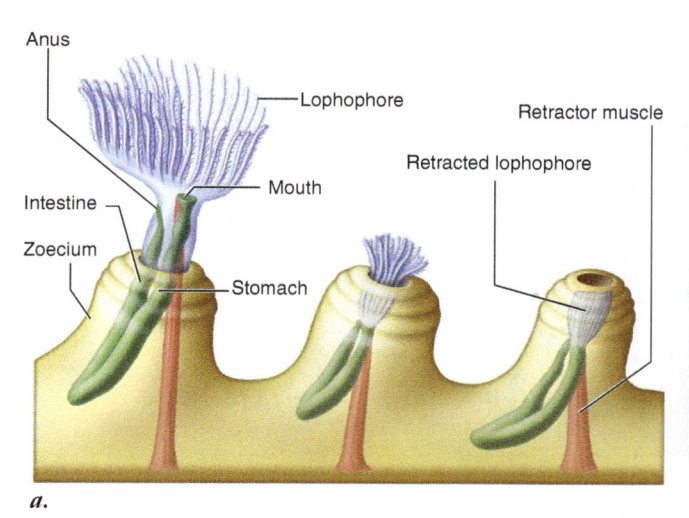

Figure 34.26 Bryozoans (phylum Bryozoa). *a.* This drawing depicts a small portion of a colony of the freshwater bryozoan genus *Plumatella*, which grows on rocks. The individual at the left has a fully extended lophophore. The tiny individuals disappear into the zoecium when disturbed. *b. Plumatella repens*, another freshwater bryozoan.

Brachiopods share some features with protostomes and others with deuterostomes. Cleavage in both brachiopods and bryozoans is mostly radial, as in deuterostomes. The formation of the coelom varies. In phoronids (once considered a phylum on their own but now considered part of Brachiopoda), the mouth forms from the blastopore (a feature of protostomes), whereas in the rest of brachiopods and in the bryozoans, it forms from the end of the embryo opposite the blastopore (a feature of deuterostomes). Molecular evidence allies lophophorates with protostomes. Because of the discrepancies between anatomical and developmental characters and molecular characters, the phylogeny of these animals continues to be a fascinating puzzle.

Bryozoans are the only exclusively colonial animals

Bryozoans are small—usually less than 0.5 mm long—and live in colonies that look like patches of moss on the surfaces of submerged objects (figure 34.26). Their common name, "moss-animals," is a direct translation of the Latin word *bryozoa*. The digestive system is U-shaped, with the anus opening near the mouth, as in many sessile animals.

The 10,000 species of bryozoans include both marine and freshwater forms. Each individual bryozoan—a zooid—secretes a tiny chitinous chamber called a zoecium (plural, *zoecia*) that is attached to rocks or other substrates such as the leaves of marine plants and algae. Calcium carbonate is deposited in the wall of a zoecium in many marine bryozoans, and in early geological times, bryzoans formed reefs just as corals do today. A zooid can divide or bud to create asexually another zooid beside the existing one so one wall of the new zooid's zoecium is shared with that of the existing one; this expanding group of zoecia constitutes a colony. Individuals in the colony communicate chemically through pores between the zoecia. Not all zoecia of a colony may be identical; some are specialized for functions such as feeding, reproduction, or defense.

Brachiopods and phoronids are solitary lophophorates

Brachiopods, or lamp shells, superficially resemble clams because they have two calcified valves (figure 34.27). Recall that the shells of bivalves are lateral, but in brachiopods, the valves are

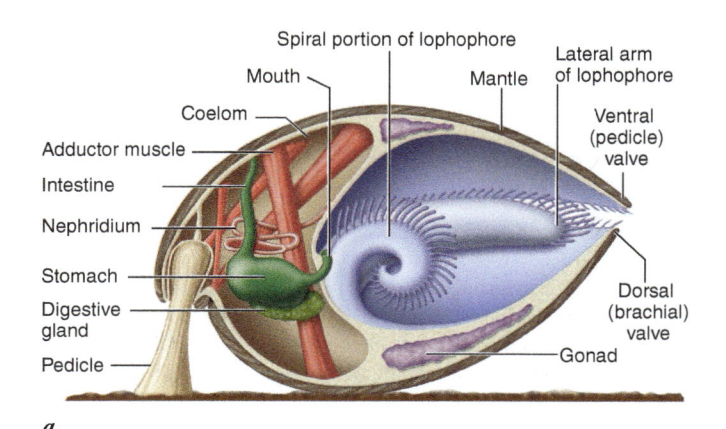

Figure 34.27 Brachiopods (phylum Brachiopoda). *a.* All the body structures except the pedicle lay within two calcified shells, or valves. *b.* The brachiopod, *Terebratulina septentrionalis*, is slightly opened so the lophophore is visible.

dorsal and ventral. Many species attach to rocks or sand by the pedicle (a stalk) that protrudes through an opening in one shell, whereas in others one valve is cemented to the substrate and the animal lacks a pedicle. The lophophore lies on the body, between the shells. The gut in some brachiopods is U-shaped, as in bryozoans, whereas in others there is no anus at all.

Three hundred thirty-five species of brachiopods exist today, but more than 30,000 fossil species are known. Because brachiopods were common in the Earth's oceans for millions of years and because their shells fossilize readily, many are used as index fossils, defining a particular geological period or sediment type.

Each **phoronid** (figure 34.28) secretes a chitinous tube around itself from which it can extend its lophophore to feed. The animal quickly withdraws into the tube when disturbed, as some polychaete worms do. Only 18 phoronid species are known, ranging in length from a few millimeters to 30 cm. Individuals of some species lie buried in sand; others are attached to rocks, either singly or in groups, forming loose colonies.

Learning Outcomes Review 34.7

The two phyla of lophophorates probably share a common ancestor, and they show a mixture of protostome and deuterostome characteristics. The lophophore is a characteristic feeding and gas exchange structure with ciliated tentacles. Bryozoans are exclusively colonial, whereas brachiopods and phoronids are solitary. Brachiopods have two shells with dorsal and ventral valves, not lateral.

- How is a U-shaped digestive tract an advantage to bryozoans and brachiopods?

34.8 Ecdysozoans: Roundworms (Nematoda)

Learning Outcomes

1. Describe the musculature of a nematode that allows it to wriggle in a highly characteristic manner.
2. Explain the life cycle of a nematode and how it produces disease in humans.

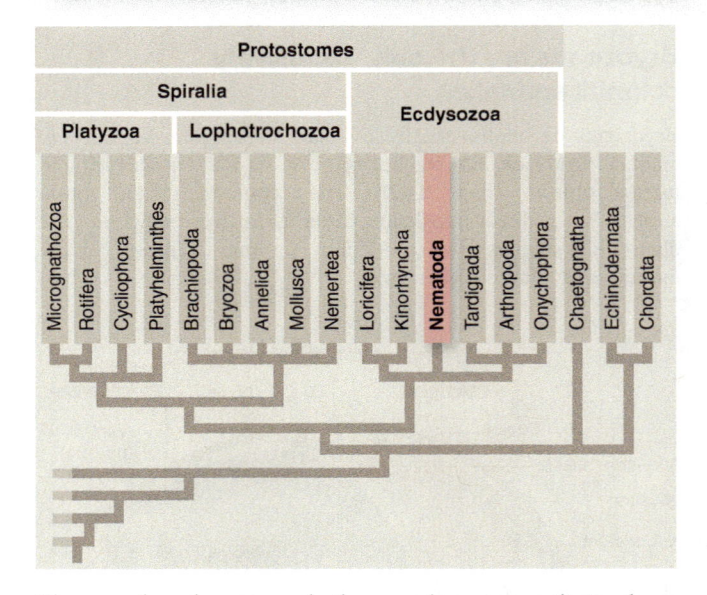

Vinegar eels, eelworms, and other roundworms constitute a large phylum, **Nematoda,** with some 61,000 recognized species; scientists estimate that the actual number might approach 100 times that many. Nematodes are abundant and diverse in marine and freshwater habitats, and many members of this phylum are parasites of plants and animals (figure 34.29). Many nematodes are microscopic and live in soil. A spadeful of fertile soil may contain, on the average, a million nematodes.

Figure 34.28 Phoronids. A phoronid lives in a chitinous tube that the animal secretes. The lophophore consists of two horseshoe-shaped ridges of tentacles and can be withdrawn into the tube when the animal is disturbed.

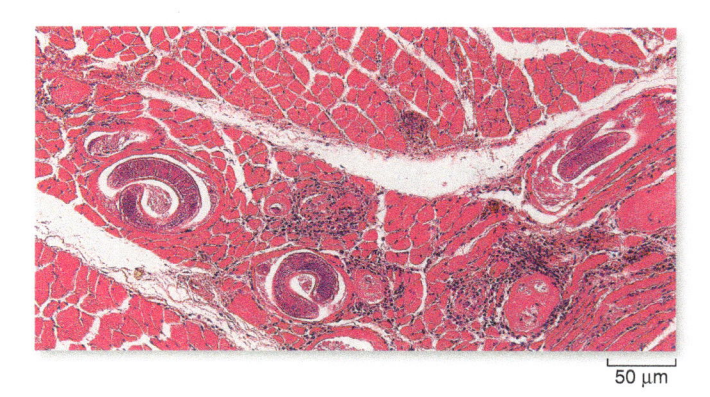

Figure 34.29 Trichinella *nematode encysted in pork*. The serious disease trichinosis can result from eating undercooked pork or bear meat containing such cysts.

Nematode structure

Nematodes are bilaterally symmetrical, unsegmented worms covered by a flexible, thick cuticle that is molted as they grow—parasitic nematodes molt four times. Lacking specialized respiratory organs, nematodes exchange oxygen and carbon dioxide through their cuticles. Muscles beneath the epidermis, which underlies the cuticle, extend longitudinally, from anterior to posterior. Nematodes are unusual among worms in that they lack circular musculature, so they can shorten but not change diameter. The pulling of these longitudinal muscles against both the cuticle and the pseudocoelom produces the characteristic wriggling motion of nematodes.

Nematodes possess a well-developed digestive system and feed on a diversity of food sources. Near the mouth, at the anterior end, are hairlike sensory structures. The mouth may be equipped with piercing organs called **stylets.** Food passes into the mouth as a result of the sucking action produced by the rhythmic contraction of a muscular pharynx and continues through the intestine; waste is eliminated through the anus (figure 34.30).

Reproduction and development

Reproduction in nematodes is sexual. Nematode males and females differ in form, a state known as **sexual dimorphism** (meaning "two bodies"): The tail end of the smaller male is hooked (figure 34.30), whereas that of a female is straight. Fertilization is internal, the male using its hooked end and associated structure to help inseminate the female. Development is indirect, meaning that an egg hatches into a larva, which does not grow directly into an adult, but must pass through several molts and, in parasitic species, transfer from one host to another.

The adults of some species consist of a fixed number of cells. For this reason, nematodes have become extremely important subjects for genetic and developmental studies (see chapter 19). The 1-mm-long *Caenorhabditis elegans* matures in three days; its body is transparent, and it has precisely 959 cells. It is the only animal whose complete developmental cellular anatomy is known.

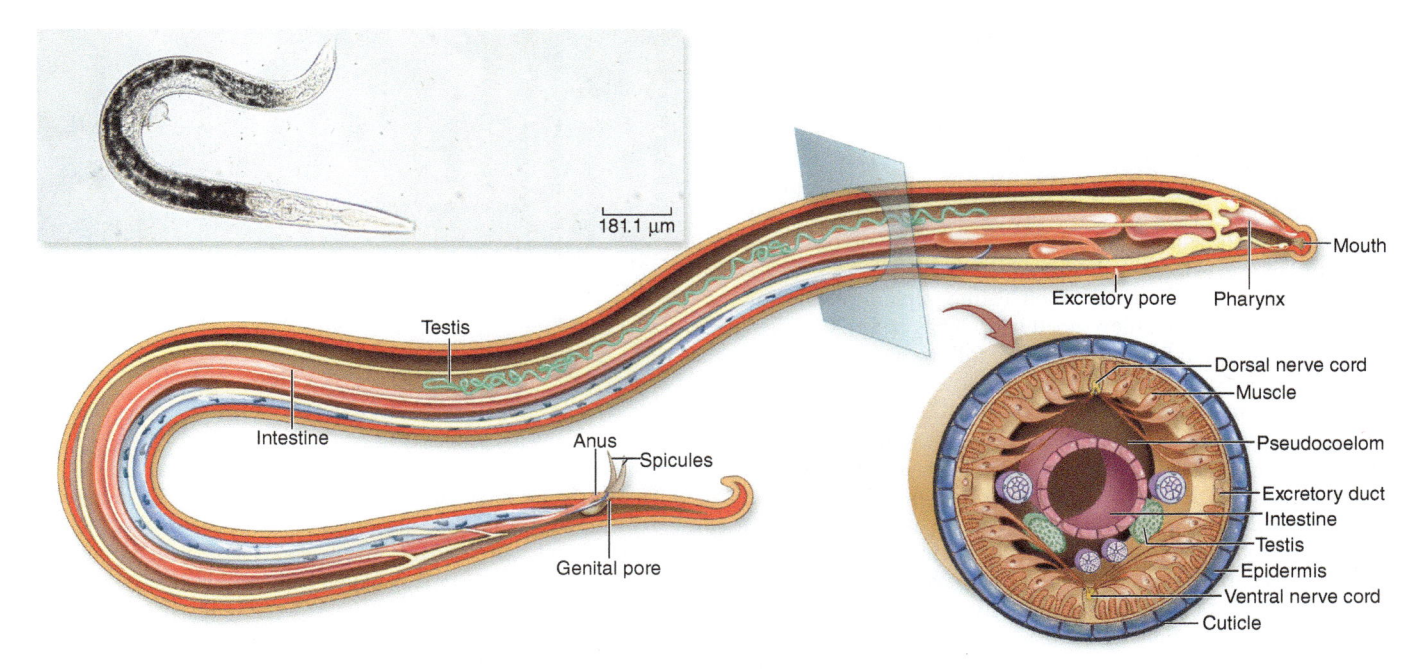

Figure 34.30 Phylum Nematoda: Roundworms. Roundworms such as this male nematode possess a body cavity between the gut and the body wall called the pseudocoelom. It allows nutrients to circulate throughout the body and prevents organs from being deformed by muscle movements.

Nematode lifestyles

Many nematodes are active hunters, preying on protists and other small animals. Many are parasites of plants or live within the bodies of larger animals. Almost every species of plant and animal that has been studied has been found to have at least one parasitic species of nematode. The largest known nematode, which can attain a length of 9 m, parasitizes the placenta of sperm whales.

Nematode-caused human diseases

About 50 species of nematodes, including several that are rather common in the United States, regularly parasitize humans. Hookworms, most of the genus *Necator*, can be common in southern states. By sucking blood through the intestinal wall, they can produce anemia.

The most serious and common nematode-caused disease in temperate regions is trichinosis. Worms of the genus *Trichinella* (see figure 34.29) live in the small intestine of some mammals, especially pigs and bears, where fertilized females burrow through the intestinal wall and release live young (as many as 1500 per female). The young enter the lymph channels, which transport them to muscles throughout the body. There they mature and form highly resistant, calcified cysts. Eating undercooked or raw pork or bear in which cysts are present transmits the worm. Fatal infections, which can occur if the worms are abundant, are rare: In the United States, only 11 cases are reported, on average, every year. In other countries, however, it can be much more common; in China, for example, there are 10,000 cases every year.

It is estimated that pinworms, *Enterobius vermicularis*, infect 11% of the people in the United States and is particularly prevalent in children. Adult pinworms live in the human rectum where they usually cause nothing more serious than itching of the anus; large numbers, however, can lead to prolapse of the rectum. The worms can easily be killed by drugs.

The intestinal roundworm *Ascaris lumbricoides* infects approximately one of six people worldwide, but is rare in areas with modern plumbing. An adult female, which can be as much as 30 cm long, can release as many as 20,000 fertilized eggs each day into the gut of its host. The eggs are carried from the body in the host's feces and can remain viable for years in the soil; dust may carry them onto food, eating implements, or lips. Because the embryo has developed in each egg before it is shed, once it is ingested, it hatches. The larva follows a circuitous path through the body, and metamorphoses into an adult, which lives in the human intestine.

Some nematode-caused diseases are extremely serious in the tropics. Filariasis is caused by several species of nematodes that infect at least 250 million people worldwide. Filarial worms of some species live in the circulatory system. Infection by *Wuchereria bancrofti* may produce the condition known as elephantiasis, in which the lower extremities may swell to disfiguring proportions. This occurs because worms clog the lymph nodes, causing severe inflammation and resulting in swelling by preventing the lymph from circulating. The larval filarial worms are transmitted by an intermediate host, typically a blood-sucking insect such as a mosquito.

Learning Outcomes Review 34.8

Nematodes are bilaterally symmetrical, unsegmented worms that have longitudinal muscle, but not muscle in a circle around the body. Aside from many free-living species, some nematodes are parasites of animals and plants. In some parasitic species, females burrow through the intestinal walls of some mammals and release their eggs, which become lodged in the muscles. Eating undercooked meat of such animals can transmit the worms to humans, which can cause an occasionally fatal infection.

■ *In what ways can nematode-borne diseases be avoided?*

34.9 Ecdysozoans: Arthropods (Arthropoda)

Learning Outcomes

1. Name the key features of arthropods.
2. List the four extant classes of arthropods and a characteristic that distinguishes them from one another.
3. Describe the advantages and drawbacks of an exoskeleton.

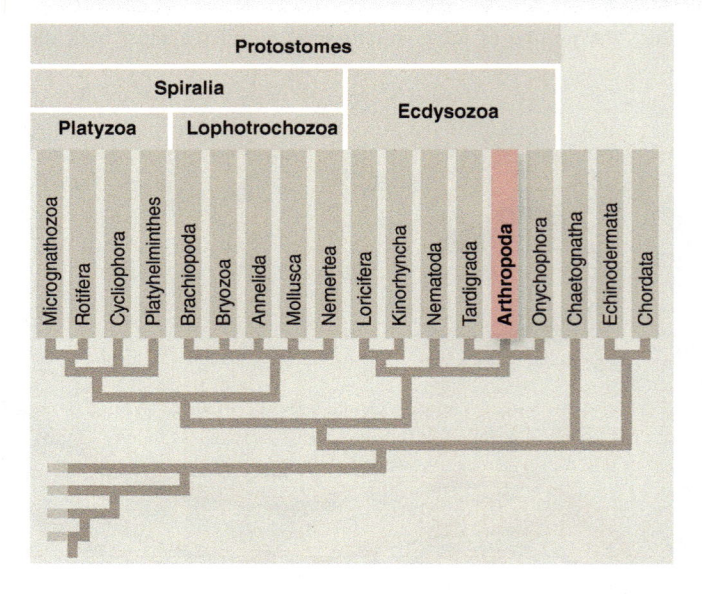

Arthropods are by far the most successful of all animals (table 34.1); 1,200,000 species—about two-thirds of all the named species on Earth—are members of the phylum Arthropoda (figure 34.31). One scientist recently estimated that insects alone may include as many as 30 million species. About 200 million individual insects are alive at any time for each human! Insects (figure 34.31) and other arthropods abound in every habitat on the planet, but there are few marine insects. Members of the phylum are small, generally a few millimeters in length, but adults range in size from about 80 μm long (some parasitic mites) to 3 m across (Japanese spider crabs).

TABLE 34.1	Major Groups of the Phylum Arthropoda	
Class	**Characteristics**	**Members**
Chelicerata	Mouthparts are chelicerae (pincers or fangs).	Spiders, mites, ticks, scorpions, daddy long-legs, horseshoe crabs
Crustacea	Mouthparts are mandibles; appendages are biramous ("two-branched"); the head has two pairs of antennae.	Lobsters, crabs, shrimps, isopods, barnacles
Hexapoda	Mouthparts are mandibles; the body consists of three regions: a head with one pair of antennae, a thorax, and an abdomen; appendages are uniramous ("single-branched").	Insects (beetles, bees, flies, fleas, true bugs, grasshoppers, butterflies, termites), springtails
Myriapoda	Mouthparts are mandibles; the body consists of a head with one pair of antennae, and numerous segments, each bearing paired uniramous appendages.	Centipedes, millipedes

Arthropods are of enormous economic importance, affecting all aspects of human life. They pollinate crops and are valuable as food for humans and other animals, but they also compete with humans for food and damage crops. Diseases spread by insects and ticks strike every kind of plant and animal, including humans. Insects are by far the most important herbivores in terrestrial ecosystems: virtually every kind of plant is eaten by one or more species.

Although our understanding of the phylogenetic relationships among the groups of arthropods and their relationships to other animals may shift with new findings, taxonomists currently recognize four extant classes (a fifth, the trilobites, is extinct): chelicerates, crustaceans, hexapods, and myriapods. Mouthparts of chelicerates are chelicerae (pincers), whereas those of the other three classes are **mandibles** (biting jaws). Mandibles are inferred to have arisen (probably from a pair of limbs) in the common ancestor of crustaceans, hexapods, and myriapods, which means that these groups are more closely related to one another than any of them is to chelicerates.

Arthropods exhibit key features and organ systems

Part of arthropod success is explained by the modularity of the segmented body, the **exoskeleton,** and the jointed appendages. The advantages of segmentation were discussed in section 34.6. A hard exoskeleton confers protection against predators, but it acts something like a straight-jacket, restricting motion. Joints in the appendages maintain protection while providing some flexibility. With this system, arthropods have developed many efficient modes of locomotion, both in the oceans, where they originated, and on land, which they colonized early in the Devonian period, more than 400 MYA.

Segmentation

In members of some classes of arthropods, many body segments look alike. In others, the segments are specialized into functional groups, or **tagmata** (singular, *tagma*), such as the head, thorax, and abdomen of an insect (figure 34.32). The fusion of segments,

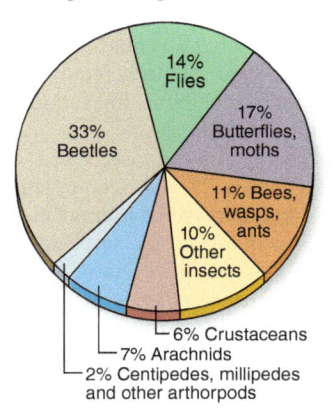

Figure 34.31 Arthropods are a successful group. About two-thirds of all named species are arthropods. About 85% of all arthropods are insects, and more than a third of the named species of insects are beetles.

Figure 34.32 Phylum Arthropoda. This bee, like all insects and other arthropods, has a segmented body and jointed appendages. An insect body is composed of three tagmata: head, thorax, and abdomen. All arthropods have an exoskeleton made of chitin. Some insects evolved wings that permit them to fly.

known as tagmatization, is of central importance in the evolution of arthropods. Typically, the segments can be distinguished during larval development, but fusion in development obliterates them. All arthropods have a distinct head; in many crustaceans and chelicerates, head and thorax fuse to form the cephalothorax, or **prosoma.**

An exoskeleton

The rigid external skeleton, or exoskeleton, is made of chitin and protein. In any animal, the skeleton provides antagonism for muscles (and in many animals, a surface for muscle attachment), support for the body, and protection against physical forces. The arthropod exoskeleton protects against water loss, which was a powerful advantage in insects colonizing land. As you learned in chapter 3, chitin is chemically similar to cellulose, the dominant structural component of plants, and shares with it properties of toughness and flexibility. The chitin and protein of an arthropod exoskeleton provide a covering that is very strong while being capable of flexing in response to the contraction of muscles attached to it.

An exoskeleton has inherent limitations. As arthropods increase in size, their exoskeletons must get disproportionately thick to bear the pull of the muscles. If beetles were as large as eagles, or crabs the size of cows, the exoskeleton would be so thick that the animal would be unable to move its great weight. Few terrestrial arthropods weigh more than a few grams, but aquatic ones can be heavier because water, being denser than air, provides more support. Another limitation is that, because the body is encased in a rigid skeleton, arthropods periodically must undergo **ecdysis,** or molting. Controlled by ecdysteroid hormones (see chapter 45), molting was explained in chapter 33. The anterior and posterior regions of the digestive tract as well as the **compound eyes** are covered with cuticle and therefore are also shed at ecdysis. The animal is especially vulnerable during molting while the exoskeleton is soft, and it may hide from predators until the new exoskeleton hardens.

Jointed appendages

The name *arthropod* means "jointed feet"; all arthropods have jointed appendages. Appendages may be modified into antennae, mouthparts of various kinds, or legs.

One advantage of jointed appendages is that they can be extended and retracted by bending. Imagine how difficult life would be if your arms and legs could not bend! In addition, joints serve as a fulcrum, or stable point, for appendage movement, so leverage is possible. A small muscle force on a lever can produce a large movement; for example, extending your lower arm takes advantage of the fulcrum of the elbow. A small contraction distance in your muscles moves your hand through a large arc.

Circulatory system

The circulatory system of arthropods is open. The principal component of an insect's circulatory system is a longitudinal muscular vessel called the heart, which is near the dorsal surface of the thorax and abdomen (figure 34.33). When the heart contracts, blood is pumped anteriorly. From there it gradually flows through the spaces between the tissues toward the posterior end. When the heart relaxes, blood returns to it from those

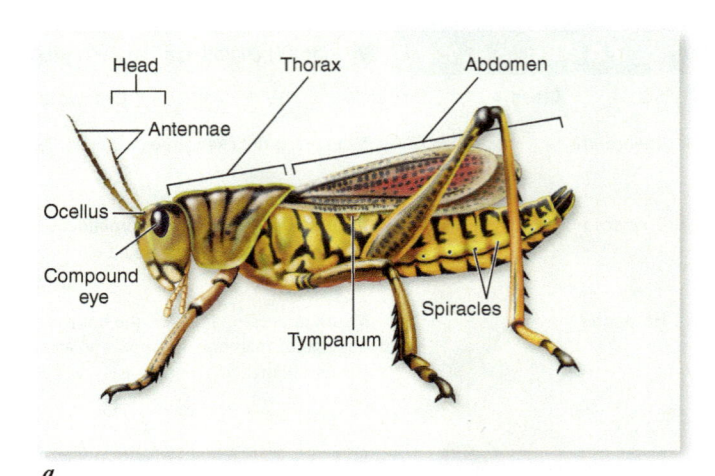

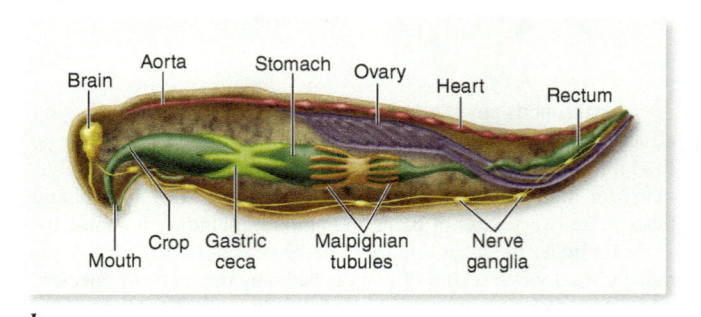

Figure 34.33 A Grasshopper (order Orthoptera). This grasshopper illustrates the major structural features of the insects, the arthropod group with the greatest number of species. *a.* External anatomy. *b.* Internal anatomy.

spaces through one-way valves in the posterior region of the heart.

Nervous system

The central feature of the arthropod nervous system is a double chain of segmented ganglia along the animal's ventral surface (figure 34.33). At the anterior end of the animal are three fused pairs of dorsal ganglia, which constitute the brain; however, ventral ganglia (generally a pair per segment) control much of the animal's activities. Therefore, an arthropod can carry out functions such as eating, moving, and copulating even if the brain has been removed. The brain seems to be a control point, or inhibitor, for various actions, rather than a stimulator, as it is in vertebrates.

Compound eyes (figure 34.34) occur in many insects, crustaceans, centipedes, and the extinct trilobites. They are composed of hundreds or more independent visual units called **ommatidia** (singular, *ommatidium*), each covered with a lens and including a complex of eight retinular cells and a light-sensitive central core, the rhabdom. Simple eyes, or **ocelli** (singular, *ocellus*) with single lenses, occur in some arthropods including those with compound eyes. Ocelli distinguish light from darkness. The ocelli of locusts and dragonflies function as horizon detectors to help the insect visually stabilize flight.

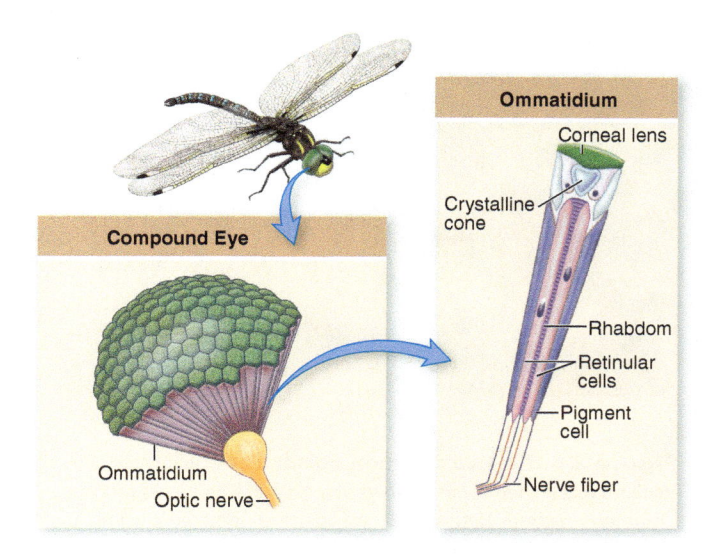

Figure 34.34 The compound eye. The compound eyes in insects are complex structures composed of many independent visual units called ommatidia.

Respiratory system

Marine arthropods such as crustaceans have gills, and marine chelicerates (such as horseshoe crabs) have book gills, flaps under the prosoma that appear to have evolved from legs. Some tiny arthropods lack any structures for exchanging oxygen, and their outer epithelium or gut have a respiratory function.

The respiratory system of most terrestrial arthropods consists of small, branched, cuticle-lined ducts called **tracheae** (singular, *trachea*) (figure 34.35) (the lining of which is shed at ecdysis). Tracheae ultimately branch into very small **tracheoles,** which are in direct contact with individual cells, allowing oxygen and carbon dioxide to diffuse across the plasma membranes. Because insects depend on the respiratory system rather than the circulatory system to carry oxygen to their tissues, all parts of the body must be near a respiratory passage. Along with the weight of the exoskeleton, this places severe limitations on arthropod size.

Air passes into the tracheae through openings in the exoskeleton called **spiracles** (see figures 34.32, 34.33, and 34.35), which, in most insects, can be opened and closed by valves. The ability to prevent water loss by closing the spiracles was a key adaptation that facilitated the arthropod invasion of land.

Many spiders have book lungs instead of or in addition to tracheae. A **book lung** is a series of leaflike plates within a chamber into which air is drawn and from which it is expelled by muscular contraction.

Excretory system

Various kinds of excretory systems occur in arthropods. In aquatic arthropods, much of the waste may diffuse from the blood in the gills.

Malpighian tubules, which occur in terrestrial insects, myriapods, and chelicerates, are slender projections from the digestive tract attached at the junction of the midgut and hindgut (see figures 34.32 and 34.33). Fluid passes through the walls of the Malpighian tubules to and from the blood in which the tubules are bathed. Nitrogenous wastes are precipitated as concentrated uric acid or guanine, emptied into the hindgut, and eliminated. Most of the water and salts in the fluid are reabsorbed by the hindgut and rectum to be returned to the arthropod's body. This efficient conservation of water by Malpighian tubules was another key adaptation facilitating invasion of the land by arthropods.

Spiders, mites, ticks, and horseshoe crabs (Chelicerata) have specialized anterior appendages

In the class Chelicerata, with some 70,000 species, the most anterior appendages, called **chelicerae** (singular, *chelicera*), may function as fangs or pincers. The body of a chelicerate is divided into two tagmata: the anterior prosoma, which bears all the appendages, and the posterior **opisthosoma,** which contains the reproductive organs. Chelicerates include familiar largely terrestrial arthropods such as spiders, ticks, mites, scorpions, and daddy long-legs. However, 4000 known species of mites and one species of spider live in freshwater habitats, and a few mites live in the sea. Exclusively marine groups of chelicerates are horseshoe crabs and sea spiders.

In addition to a pair of chelicerae, a chelicerate has a pair of **pedipalps,** and four pairs of walking legs on its prosoma. The pedipalps (often simply called palps) resemble legs but have one fewer segment and are not used for locomotion. In male spiders, the pedipalps are copulatory organs; in scorpions, they are large pincers; and in most other chelicerates, they are sensorial, acting like the antennae of other arthropods.

Most chelicerates are carnivorous, but mites are largely herbivorous. Aside from the daddy long-legs, which can ingest small particles, most cannot consume solid food. They subsist on liquids, including solid food that they liquefy by injecting with digestive enzymes and then suck up with the muscular pharynx.

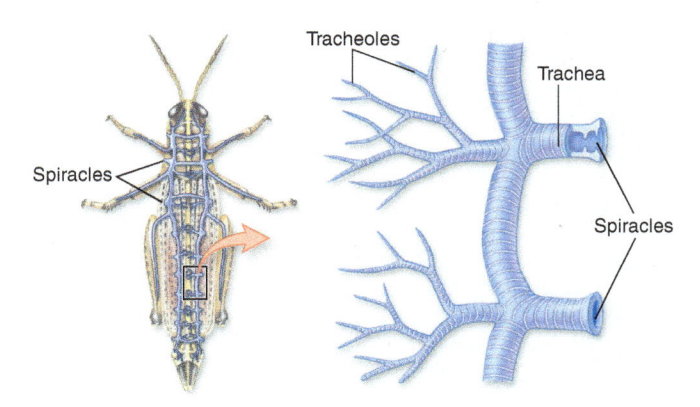

Figure 34.35 Tracheae and tracheoles. Tracheae and tracheoles are connected to the exterior by openings called spiracles and carry oxygen to all parts of a terrestrial insect's body.

Figure 34.36 *Horseshoe crabs.* Horseshoe crabs are aquatic chelicerates. This horseshoe crab, *Limulus polyphemus*, is found along the east coast of North America.

Figure 34.37 Two common poisonous spiders. *a.* The southern black widow, *Latrodectus mactans*. *b.* The brown recluse, *Loxosceles reclusa*. Both species are common throughout temperate and subtropical North America.

The four species of horseshoe crabs (figure 34.36) live off the North American Atlantic coast and in Southeast Asia. Another type of chelicerate, the sea spiders are such strange marine animals that some authorities exclude them from the Chelicerata. Most are small, but some can reach 150 mm or so across; many live in association with other marine animals, such as hydroids.

The 35,000 named species of spiders (order Araneae) play a major role in almost all terrestrial ecosystems. They are particularly important as predators of insects and other small animals. Spiders hunt their prey or catch it in silk webs of remarkable diversity. Silk is formed from a fluid protein that is forced out of **spinnerets** on the posterior portion of the spider's abdomen. Trap-door spiders construct silk-lined burrows with lids, seizing their prey as it passes by. Spiders such as the familiar wolf spiders and tarantulas hunt rather than spin webs.

All spiders have poison glands with channels through their chelicerae, which are pointed and are used to bite and paralyze prey. The bites of some, such as the western black widow, *Latrodectus hesperus*, and brown recluse, *Loxosceles reclusa*, (figure 34.37), can be fatal to humans and other large mammals.

The order Acari is the most diverse of the chelicerates. About 50,000 species of mites and ticks have been named, but scientists estimate that more than a million members of this order may exist. Acarines are found in nearly every habitat; they feed on a variety of organisms as predators and parasites.

Most mites are less than 1 mm long, but adults range from 100 nm to 2 cm. In most mites, the cephalothorax and abdomen are fused into an unsegmented, ovoid body. Respiration is by means of tracheae or directly through the body surface. Many mites pass through several stages during their life cycle. In most, an inactive eight-legged prelarva gives rise to an active six-legged larva, which in turn produces a succession of three eight-legged stages, and finally, the adult.

Ticks are parasites that attach to the surface of humans and other animals, causing discomfort by sucking blood. Ticks, which are larger than most other members of the order, can carry disease-causing agents. For example, Rocky Mountain spotted fever is caused by bacteria; Lyme disease is caused by spirochetes; and red-water fever, or Texas fever, is an important tick-borne protozoan disease of cattle, horses, sheep, and dogs.

Crabs, shrimps, lobsters, and pill bugs (Crustacea) are largely marine organisms

Crustaceans (class Crustacea) include 65,000 species of largely marine organisms, such as crabs, shrimps, lobsters, and barnacles. However, some groups, such as crayfish, occur in freshwater, and some crabs and copepods (figure 34.38) are among the most abundant multicellular organisms on Earth. A small number are terrestrial, including about half the estimated 10,000 species of order Isopoda, the pill bugs; and some sand fleas or beach fleas (order Amphipoda) are semiterrestrial.

Figure 34.38 Freshwater crustacean. A copepod with attached eggs. Members of the marine and freshwater order Copepoda are important components of the plankton. Most are a few millimeters long.

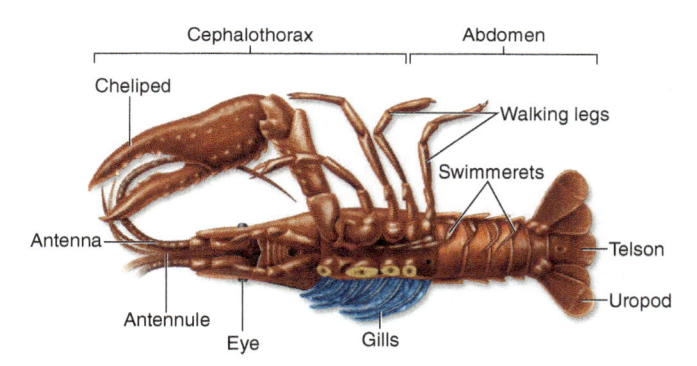

Figure 34.39 Decapod crustacean. Ventral view of a lobster, *Homarus americanus,* with some of its principal features labeled.

Some crustaceans (such as lobsters and crayfish) are valued as food for humans; planktonic crustaceans (such as krill) and larval crustaceans, which are abundant in the plankton, are the primary food of baleen whales and many smaller marine animals.

Crustacean body plans

A typical crustacean has three tagmata; the anteriormost two—the cephalon and thorax—may fuse to form the cephalothorax (figure 34.39). Most crustaceans have two pairs of antennae, three pairs of appendages for chewing and manipulating food, and various pairs of legs. Crustacean appendages, with the possible exception of the first pair of antennae, are biramous ("two-branched"). Crustaceans differ from hexapods, but resemble myriapods, in having appendages on their abdomen as well as their thorax. They are the only arthropods with two pairs of antennae.

Large crustaceans have feathery gills for respiration near the bases of their legs (figure 34.39). Oxygen extracted from the gills is distributed through the circulatory system. In smaller crustaceans, gas exchange takes place directly through the thinner areas of the cuticle or the entire body.

Crustacean reproduction

In crustaceans, many kinds of copulation occur, and the members of some groups carry their eggs, either singly or in egg pouches, until they hatch. Crustaceans develop through a **nauplius** (plural, *nauplii*) stage (figure 34.40). The nauplius hatches with three pairs of appendages and undergoes metamorphosis through several stages before reaching maturity. In many groups, this nauplius stage is passed in the egg, and the hatchling resembles a miniature adult.

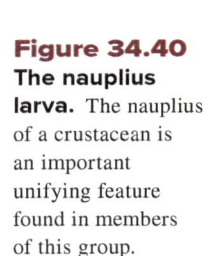

The nauplius larva is characteristic of Crustacea, providing evidence that all members of this diverse group descended from a common ancestor that had a nauplius in its life cycle. The sessile barnacles, with their shell-like exoskeleton, had been thought to be related to mollusks until they were discovered to have a nauplius larva. More recently, the wormy pentastomids, which parasitize the respiratory tracts of vertebrates, were determined to be crustaceans in part through their nauplii (recall that parasites may become morphologically simplified and so can be difficult to place phylogenetically).

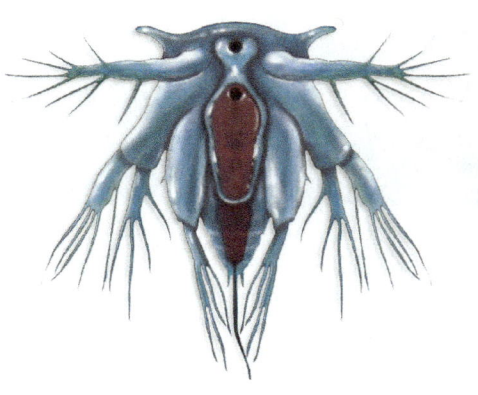

Shrimps, lobsters, crabs, and crayfish (Decapoda)

Large, primarily marine crustaceans such as shrimps, lobsters, and crabs, along with their freshwater relatives, the crayfish, belong to order Decapoda (see figure 34.39), which means "ten-footed" because of their five pairs of thoracic appendages. The exoskeleton of most is reinforced with calcium carbonate. The cephalothorax is covered by a dorsal shield, or carapace, which arises from the head. The pincers of many decapod crustaceans are used in obtaining food—for example, by crushing mollusk shells.

In lobsters and crayfish, appendages called **swimmerets** that occur along the ventral surface of the abdomen are used in reproduction and swimming. At the posterior end of the abdomen, paired flattened appendages known as **uropods** form a kind of paddle between which is a **telson,** a tail spine (see figure 34.39). When a lobster or crayfish contracts its abdominal muscle, the uropods and telson push water anteriorly, propelling the animal rapidly and forcefully backward through the water. It is this very large muscle that constitutes the "lobster tail" so valued by human diners!

One difference between crabs and lobsters is that the crab carapace is relatively much broader and the abdomen is just a small vestige tucked under the cephalothorax (figure 34.41). The abdomen of a male crab is much narrower than that of a female

Figure 34.40 The nauplius larva. The nauplius of a crustacean is an important unifying feature found in members of this group.

Figure 34.41 Cephalothorax. The cephalothorax of a Chesapeake Bay blue crab, *Callinectes sapidus.*

chapter **34** *Protostomes* 685

Figure 34.42 Gooseneck barnacles, *Lepas anatifera*. These barnacles are filter feeding, which they do by sweeping their jointed legs through the water, gathering small particles of food. These are stalked barnacles; others lack a stalk.

of the same species and size: the female carries her eggs attached to appendages of the abdomen between it and the thorax.

Sessile crustaceans: Barnacles (Cirripedia)

Barnacles (order Cirripedia; figure 34.42) are sessile as adults. At the end of its larval life, a barnacle nauplius attaches by its head to a piling, rock, or other submerged object, metamorphoses, grows calcareous plates around it, and spends the rest of its life capturing food with its jointed, feathery legs. The hermaphroditic state of barnacles is thought to be related to their sessility. Barnacles have the longest penis in the animal kingdom, relative to their size, which, considering that they can't move, is useful for mating with other barnacles located some distance away.

Insects (Hexapoda) are the most abundant animals on Earth

The insects, members of class Hexapoda, are by far the largest group of animals on Earth, in terms of number of species and number of individuals. Insects live in every habitat on land and in freshwater, but very few have invaded the sea. More than half of

Figure 34.43 Insect diversity. *a.* Luna moth, *Actias luna*. Luna moths and their relatives are among the most spectacular insects (order Lepidoptera). *b.* An oak treehopper, *Platycotis vittata*. *c.* Boll weevil, *Anthonomus grandis*. Weevils are one of the largest groups of beetles. *d.* Soldier fly, *Ptecticus trivittatus*. *e.* Lubber grasshopper, *Romalea guttata*. *f.* Like ants, termites, such as these big-headed subterranean termites (*Macrotermes* sp.), have several castes with individuals specialized for different tasks (see chapter 54). The individual on the left is a soldier; its large jaws aid in defending the colony.

named animal species are insects, and the actual proportion may be higher because millions of forms await detection, classification, and naming.

Approximately 90,000 described species occur in the United States and Canada; the actual number probably approaches 125,000. A hectare of lowland tropical forest is estimated to be inhabited by as many as 41,000 species of insects, and many suburban gardens may have 1500 or more species. It has been estimated that approximately a quintillion (10^{18}) individual insects are alive at any one time. A glimpse into the enormous diversity of insects is presented in figure 34.43 and table 34.2.

External features

Insects are primarily terrestrial, and aquatic insects probably had terrestrial ancestors. Most are small, ranging from 0.1 mm to about 30 cm in length or wingspan. Insect mouthparts all have the same basic structure; modifications reflect feeding habits (figure 34.44). Most insects have compound eyes, and many also have ocelli.

An insect body has three regions: the head, thorax, and abdomen (see figures 34.32 and 34.33). The thorax consists of three segments, each with a pair of legs, which accounts for the name of the group, hexa (six) and poda (legs). Legs are absent in the larvae of certain groups—for example, most flies (order Diptera) and mosquitoes (figure 34.45). In addition, an insect may have one or two pairs of wings, which are not homologous to the other appendages, and which attach to the middle and posterior segments of the thorax (see figure 34.32). The wings, which consist of chitin and protein, arise as saclike outgrowths of the body wall; wings of moths and butterflies are covered with detachable scales that provide most of their bright colors (figure 34.46). Veins strengthen wings. An insect's thorax is almost entirely filled with

TABLE 34.2 Major Orders of Insects

Order	Typical Examples	Key Characteristics	Approximate Number of Named Species
Coleoptera	Beetles	Two pairs of wings, the front one hard, protecting the rear one; heavily armored exoskeleton; biting and chewing mouthparts. Complete metamorphosis. The most diverse animal order.	350,000
Lepidoptera	Butterflies, moths	Two pairs of broad, scaly, flying wings, often brightly colored. Hairy body; tubelike, sucking mouthparts. Complete metamorphosis.	180,000
Diptera	Flies	Front flying wings transparent; hindwings reduced to knobby balancing organs called halteres. Sucking, piercing, or lapping mouthparts; some bite people and other mammals. Complete metamorphosis.	150,000
Hymenoptera	Bees, wasps, ants	Two pairs of transparent flying wings; mobile head and well-developed compound eyes; often possess stingers; chewing and sucking mouthparts. Many social. Complete metamorphosis.	115,000
Hemiptera and Homoptera	True bugs, bedbugs, leafhoppers, aphids, cicadas	Wingless or with two pairs of wings; piercing, sucking mouthparts, with which some draw blood, some feed on plants. Simple metamorphosis.	60,000
Orthoptera	Grasshoppers, crickets	Wingless or with two pairs of wings; among the largest insects; biting and chewing mouthparts in adults. Third pair of legs modified for jumping. Simple metamorphosis.	20,000
Odonata	Dragonflies	Two pairs of transparent flying wings that cannot fold back; large, long, and slender body; chewing mouthparts. Simple metamorphosis.	5,000
Isoptera	Termites	Two pairs of wings, but some stages wingless; chewing mouthparts; simple metamorphosis. Social organization; labor divided among several body types. Some are among the few types of animals able to digest wood. Complete metamorphosis.	2,600
Siphonaptera	Fleas	Wingless; flattened body with jumping legs; piercing and sucking mouthparts. Small; known for irritating bites. Complete metamorphosis.	2,000

**Figure 34.44
Mouthparts in three kinds of insects.** Mouthparts are modified for *(a)* piercing in this mosquito of the genus *Culex;* *(b)* sucking nectar from flowers in the alfalfa butterfly of the genus *Colias;* and *(c)* sopping up liquids in the housefly, *Musca domestica*.

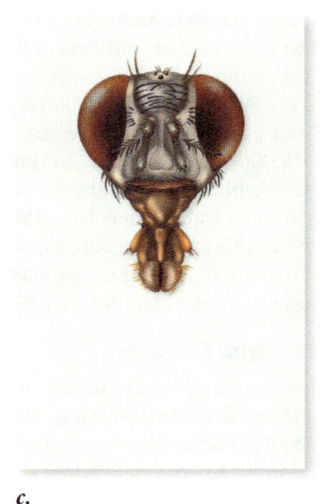

a. *b.* *c.*

muscles that operate the legs and wings. Fleas and lice are considered secondarily wingless, having descended from ancestors with wings. However, the ancestors of springtails and silverfish evolved before wings did, so those hexapods are considered primarily wingless.

Internal organization

The internal features of insects resemble those of other arthropods in many ways. The digestive tract is a tube about the same length as the body in some groups. However, in insects that feed on juices and so have sucking mouthparts, such as leafhoppers, cicadas, and many flies, the greatly coiled digestive tube may be several times longer than the body. Digestion takes place primarily in the stomach, or midgut, and excretion takes place through Malpighian tubules. Digestive enzymes are mainly secreted from the cells that line the midgut, although some are contributed by the salivary glands near the mouth.

In many winged insects, tracheae are dilated in various parts of the body, forming air sacs, which are surrounded by muscles to form a kind of bellows system that forces air deep into the body. The spiracles through which air enters the tracheal system are located on or between the segments along the sides of the thorax and abdomen. In most insects, the spiracles can be opened by muscular action. In some parasitic and aquatic groups of insects, the spiracles are permanently closed. In these groups, the tracheae are just below the surface of the insect, and gas exchange takes place by diffusion.

Sensory receptors

In addition to eyes, insects have several characteristic kinds of sense receptors. **Sensory setae** are hairlike structures that are widely distributed over the body. They are sensitive to mechanical and chemical stimulation and are linked to nerve cells. They are particularly abundant on the antennae and legs, the parts of the insect most likely to come into contact with other objects.

Figure 34.45 Larva of a mosquito, *Culex pipiens*. The aquatic larvae of mosquitoes are quite active. They breathe through tubes at the surface of the water, as shown here. Covering the water with a thin film of oil suffocates them.

Figure 34.46 Scales on the wing of *Parnassius imperator*, a butterfly from China. Scales of this sort account for most of the colored patterns on the wings of butterflies and moths.

Sound, which may be of vital importance to insects such as grasshoppers, crickets, cicadas, and some moths, is detected by a thin membrane, each called a **tympanum** (see figure 34.33), associated with the tracheal air sacs. In other groups of insects, sound waves are detected by sensory hairs. Male mosquitoes use thousands of sensory hairs on their antennae to detect sounds made by the vibrating wings of female mosquitoes.

In addition to sound, nearly all insects communicate by means of chemicals known as pheromones. These extremely diverse compounds are sent forth into the environment, where they convey a variety of messages, including mating signals and trail markers.

Insect life histories

During the course of their development, many insects undergo metamorphosis. For those such as grasshoppers, in which immature individuals are quite similar to adults, a series of molts results in an individual gradually getting bigger and more developed; this is termed simple metamorphosis. Those such as moths and butterflies have a life history involving a wormlike larval stage, a resting stage called a **pupa** or **chrysalis,** during which metamorphosis occurs, and then a final molt into the adult form or imago; this is termed complete metamorphosis.

Centipedes and millipedes (Myriapoda) have numerous legs

The body of a centipede and millipede consists of a head region posterior to which are numerous, more or less similar segments. Nearly all segments of a centipede have one pair of appendages (figure 34.47a), and nearly all segments of a millipede have two pairs of appendages (figure 34.47b). Each segment of a millipede is a simple tagma derived evolutionarily from two ancestral segments, which explains why millipedes have twice as many legs per segment as centipedes. Although the name *centipede* would imply an animal with 100 legs and the name *millipede* one with 1000, an adult centipede usually has fewer than 100 legs (most have 15, 21, or 23 pairs of legs); an adult millipede never reaches 1000 legs, most having 100 or fewer.

In both centipedes and millipedes, fertilization is internal, and all lay eggs. Young millipedes usually hatch with three pairs of legs; they add segments and legs as they pass through growth stages, but they do not change in general appearance. Centipedes have several types of development, young of some species hatching with their final number of legs and others adding legs after hatching. Centipedes that do not add legs as they grow tend to take care of their young, a behavior rather uncommon among invertebrates.

Centipedes, of which more than 5000 species are known, are carnivorous, feeding mainly on insects. The appendages of the first trunk segment are modified into a pair of poison fangs. The poison may be toxic to humans, and although extremely painful, centipede bites are never fatal. In contrast, most millipedes are herbivores, feeding mainly on decaying vegetation such as leaf litter and rotting logs (which are typical habitats for the animals). Many millipedes can roll their bodies into a flat coil or sphere to defend themselves. More than 12,000 species of millipedes have been named, but this is estimated to be no more than one-sixth of the number of species that exists.

In each segment of their body, many millipedes have a pair of complex glands that produce a bad-smelling fluid, which they exude for defense through openings along the sides of the body. The chemistry of this material interests biologists because of the diversity of the compounds involved and their effectiveness in protecting millipedes from attack. Some species produce cyanide gas from segments near their head.

Learning Outcomes Review 34.9

Arthropods are segmented animals with exoskeletons and jointed appendages. The four living classes of arthropods are Chelicerata, with mouthparts (chelicerae) that function as fangs or pincers; Crustacea, in which all members have a nauplius developmental stage; Hexapoda, the insects, having three pairs of legs attached to fused segments of the thorax as adults; and Myriopoda, with one or two pairs of appendages on every body segment. An exoskeleton provides protection and muscle attachment, but growth requires molting. In some cases, larvae undergo metamorphosis into an adult with very different appearance.

- *What would explain why the largest arthropods are found in marine environments?*

Figure 34.47 Myriapods. *a.* Centipedes, such as this member of the genus *Scolopendra,* are active predators. *b.* Millipedes, such as this member of the genus *Sigmoria,* are important herbivores and detritivores. Centipedes have one pair of legs per body segment, millipedes have two pairs per segment.

Chapter Review

34.1 The Clades of Protosomes

Spiralians develop as embryos using spiral cleavage. Most live in water and move through it using cilia or contractions of the body musculature. The two clades of spiralians are platyzoans and lophotrochozoans.

Ecdysozoans are animals that can molt.

34.2 Platyzoans: Flatworms (Platyhelminthes)

Flatworms have an incomplete gut.

Free-living flatworms move by muscles and ciliated epithelial cells. They also exhibit a head and an incomplete gut.

Flatworms have an excretory system containing a fine network of tubules with flame cells. The primary function of this system is water balance.

Flatworms reproduce sexually and are hermaphroditic. They also have the capacity for asexual regeneration.

Flatworms consist of two major groups.

Free-living flatworms belong to the groups Turbellaria, which is likely not to be monophyletic.

Parasitic flatworms belong to the group Neodermata, of which there are two groups: the flukes (Trematoda), and the tapeworms and their relatives (Cercomeromorpha). Flukes and tapeworms can cause disease in humans.

34.3 Platyzoans: Rotifers (Rotifera)

The rotifers, phylum Rotifera, are tiny.

Rotifers propel themselves and gather food with cilia and break down food with a complex jaw located in the pharynx. They are either free-swimming or sessile.

34.4 Lophotrochozoans: Mollusks (Mollusca)

Mollusks are extremely diverse—and important to humans.

Mollusks range from microscopic to huge and exhibit many different forms. They all have a coelom surrounding the heart.

The mollusk body plan is complex and varied.

Mollusks are generally bilaterally symmetrical, at least at some point in their lives (figure 34.10). They use a muscular foot (podium) for locomotion, attachment, food capture, or a combination.

The mantle is a thick epidermal sheet that forms the mantle cavity. It houses the respiratory structures (ctenidia or gills), and digestive, excretory, and reproductive products are discharged into it.

The outer mantle secretes a protective calcium carbonate shell. Some mollusks have internal or reduced shells, or none at all.

All mollusks except bivalves have a radula, a rasplike structure used in feeding (figure 34.10). Most have an open circulatory system, but cephalopods have a closed circulatory system.

In many mollusks, an embryo develops into a free-swimming trochophore larva (figure 34.12a). In some bivalves and gastropods, the embryo becomes a free-swimming veliger larva (figure 34.12b).

Four classes of mollusks show the diversity of the phylum.

The four best-known classes are Polyplacophora (chitons), Gastropoda (snails and slugs), Bivalvia (clams, mussels, and cockles), and Cephalopoda (octopuses, squids, and nautiluses).

34.5 Lophotrochozoans: Ribbon Worms (Nemertea)

Nemerteans superficially resemble acoelomate flatworms, but they have a closed circulatory system and a complete digestive tract.

34.6 Lophotrochozoans: Annelids (Annelida)

The annelid body is composed of ringlike segments.

The segments of the annelid body are separated by septa. Each segment contains a pair of excretory organs, a ganglion, and, in most marine annelids, a set of reproductive organs. Anterior and posterior segments contain light-, chemo-, and touch receptors.

Segments are connected by a ventral nerve cord that includes an anterior brain region, and by a closed circulatory system.

Annelids move by contracting their segments.

The fluid-filled coelom acts as a hydrostatic skeleton. Each segment typically possesses chaetae, chitin bristles that help anchor the worm.

Annelids have a closed circulatory system but a segmented excretory system.

Annelids have a closed circulatory system; the dorsal vessel is connected to the ventral vessel by smaller vessels in the body wall.

Each segment contains a pair of nephridia that excrete wastes out of the body via the coelom and excretory tubes.

Annelida is composed of two—or three—classes.

Annelids have been grouped into three classes, but the currently accepted classification recognizes two: the Polychaeta, which exhibit parapodia and are mostly marine; and the Clitellata, which includes earthworms and leeches.

34.7 Lophophorates: Bryozoans (Bryozoa) and Brachiopods (Brachiopoda)

Bryozoa and Brachiopoda are characterized by a lophophore, a U-shaped ridge around the mouth bearing ciliated tentacles.

Bryozoans are the only exclusively colonial animals.

Each individual zooid produces a chitinous chamber called a zoecium that attaches to substrates and other colony members. They have deuterostome development.

Brachiopods and phoronids are solitary lophophorates.

The body of brachiopods is enclosed between two calcified shells that are dorsal and ventral, not lateral as in bivalves (figure 34.27). They exhibit some deuterostome characteristics and some protostome characteristics.

The phoronids, tube worms, are included with bryozoans because they are solitary lophophorates; however, they are protostomes.

34.8 Ecdysozoans: Roundworms (Nematoda)

Nematodes reproduce sexually and exhibit sexual dimorphism. More species of nematodes may exist than species of arthropods.

Some important human, veterinary, and plant diseases are caused by nematodes, including hookworm, pinworm, trichinosis, intestinal roundworm, and filariasis.

34.9 Ecdysozoans: Arthropods (Arthropoda)

Arthropods exhibit key features and organ systems.

Arthropods are segmented and exhibit an exoskeleton with muscles attached to the inside. The exoskeleton is molted during ecdysis, allowing the arthropod to grow. In some species, segments are fused into units called tagmata.

Arthropods' jointed appendages may be modified into mouthparts, antennae, or legs.

Many arthropods have compound eyes composed of ommatidia (figure 34.34); others have simple eyes (ocelli).

An open circulatory system includes a muscular heart. The respiratory system in terrestrial arthropods comprises spiracles, tracheae, and tracheoles (figure 34.35). The main components of the nervous system are an inhibitory brain and a ventral nerve cord.

The excretory system of terrestrial hexapods, myriapods, and chelicerates consists of Malpighian tubules that eliminate uric acid or guanine.

Spiders, mites, ticks, and horseshoe crabs (Chelicerata) have specialized anterior appendages.

Chelicerates have specialized anterior appendages, the chelicerae, that function as fangs or pincers.

Crabs, shrimps, lobsters, and pillbugs (Crustacea) are largely marine organisms.

Crustacea is characterized by a nauplius larva (figure 34.40). Crustaceans have three tagmata, and the two anterior ones fused to form a cephalothorax (figure 34.41).

Insects (Hexapoda) are the most abundant animals on Earth.

The hexapods are extraordinarily diverse. Insects have three pairs of legs attached to the thorax and may have wings. They have highly developed sensory systems.

Centipedes and millipedes (Myriapoda) have numerous legs.

Centipedes have one pair of appendages per segment, whereas millipedes have two pairs.

Review Questions

UNDERSTAND

1. In the flatworm, flame cells are involved in what metabolic process?
   a. Reproduction
   b. Digestion
   c. Locomotion
   d. Osmoregulation

2. The _____ of a mollusk is a highly efficient respiratory structure.
   a. nephridium
   b. radula
   c. ctenidium
   d. eliger

3. Torsion is a unique characteristic of the
   a. bivalves.
   b. gastropods.
   c. chitons.
   d. cephalopods.

4. Intelligence and complex behaviors are characteristics of the
   a. cephalopods.
   b. polychaetes.
   c. brachiopods.
   d. echinoderms.

5. Serial segmentation is a key characteristic of which of the following phyla?
   a. Mollusca
   b. Brachiopoda
   c. Bryozoa
   d. Annelida

6. The distinguishing feature of the Bryozoa and Brachiopoda is
   a. a coelom.
   b. segmentation.
   c. chaetae.
   d. a lophophore.

7. Which of the following nematodes cause disease?
   a. Filarial worms
   b. Pinworms
   c. Trichina worms
   d. All of the choices are correct.

8. In terms of numbers of species, the most successful phylum on the planet is the
   a. Mollusca.
   b. Arthropoda.
   c. Echinodermata.
   d. Annelida.

9. Which of the following characteristics is NOT found in the arthropods?
   a. Jointed appendages
   b. Segmentation
   c. Closed circulatory system
   d. Segmented ganglia

10. Which of the following classes of arthropod possess chelicerae?
    a. Chilopoda
    b. Crustacea
    c. Hexapoda
    d. Chelicerata

11. Examples of decapods are
    a. centipedes and millipedes.
    b. barnacles.
    c. ticks and mites.
    d. lobsters and crayfish.

APPLY

1. To which of the following groups would a species that does not molt, possesses a coelom, and has a trochophore larva belong?
   a. Ecdysozoa
   b. Parazoa
   c. Platyzoa
   d. Lophotrochozoa

2. Which of the following is most closely related to lobsters and what characteristics are significant in determining this relationship?
   a. Bivalves
   b. Centipedes
   c. Earthworms
   d. Nematodes

3. Nematodes once were thought to be closely related to rotifers due to the presence of a pseudocoelom, but are now considered closer to the arthropods due to
   a. molting.
   b. jointed appendages.
   c. wings.
   d. a coelom.

SYNTHESIZE

1. Scientists studying the Chesapeake Bay have discovered that the decline in populations of clams and scallops has led to a drastic increase in the levels of pollution. What characteristic of this group could be attributed to this observation?

2. Chitin is present in a number of invertebrates, as well as the fungi. What does this tell you about the origin and importance of this substance?

3. What benefit would being a hermaphrodite confer on a parasitic species?

4. Does the lack of a digestive system in tapeworms indicate that it is a primitive, ancestral form of Platyhelminthes? Explain your answer.

CHAPTER 35

Deuterostomes

Chapter Contents

35.1 Echinoderms
35.2 Chordates
35.3 Nonvertebrate Chordates
35.4 Vertebrate Chordates
35.5 Fishes
35.6 Amphibians
35.7 Reptiles
35.8 Birds
35.9 Mammals
35.10 Evolution of the Primates

Introduction

At first glance, echinoderms such as sea stars and sea cucumbers and chordates such as hawks and elephants would seem to have little in common. Yet, both exhibit the same pattern of development, in which the anus is derived from the blastopore and the mouth originates from another part of the embryo. This synapomorphy unites the two clades as members of the Deuterostomia.

35.1 Echinoderms

Learning Outcomes
1. List the specific characteristics of echinoderms.
2. Explain what is meant by pentaradial symmetry.
3. Describe the five extant classes of echinoderms.

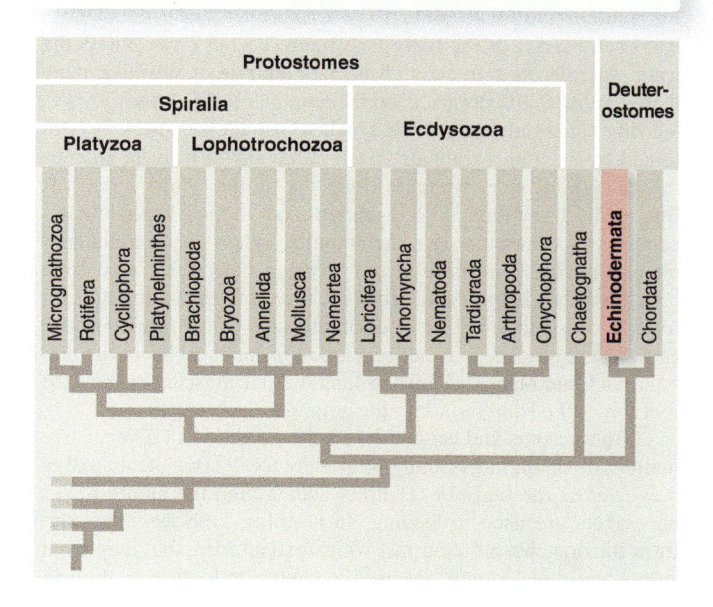

Members of the exclusively marine phylum Echinodermata are characterized by deuterostome development (see chapter 33) and possession of an internal skeleton. This endoskeleton is composed of hard, calcium carbonate plates that lie just beneath the delicate skin. The term *echinoderm,* which means "spiny skin," refers to the spines or bumps that occur on these plates in many species. Echinoderms, including sea stars (figure 35.1), brittle stars, sea urchins, sand dollars, and sea cucumbers, are some of the most familiar animals on the seashore. All have five axes of symmetry (that is, you could draw lines through their body in five different places that produced mirror images on each side) and thus are said to be **pentaradially symmetrical.**

Echinoderms are ancient and unmistakable

Echinodermata is an ancient group of marine animals that appeared nearly 600 MYA and that contains about 12,000 living species. Echinoderms have an endoskeleton covered by living tissue. A unique feature of echinoderms is the hydraulic system that aids in movement or feeding. This fluid-filled system, called a **water-vascular system,** is a modification of one of several coelomic spaces and is composed of a central ring canal from which **radial canals** extend.

Although an excellent fossil record extends back into the Cambrian period, the origin of echinoderms remains unclear. Their phylogenetic placement (figure 33.5) suggests that they evolved from bilaterally symmetrical ancestors; this hypothesis is

Figure 35.1 Phylum Echinodermata. *a.* Echinoderms, such as sea stars (class Asteroidea), are coelomates with a deuterostomate pattern of development and an endoskeleton made of calcium carbonate plates. The water-vascular system of an echinoderm is shown in detail. Radial canals transport liquid to the tube feet. As the ampulla in each tube foot contracts, the tube foot extends and can attach to the substrate. When the muscles in the tube feet contract, the tube foot bends, pulling the animal forward. *b.* Extended tube feet of the sea star, *Luidia magnifica.*

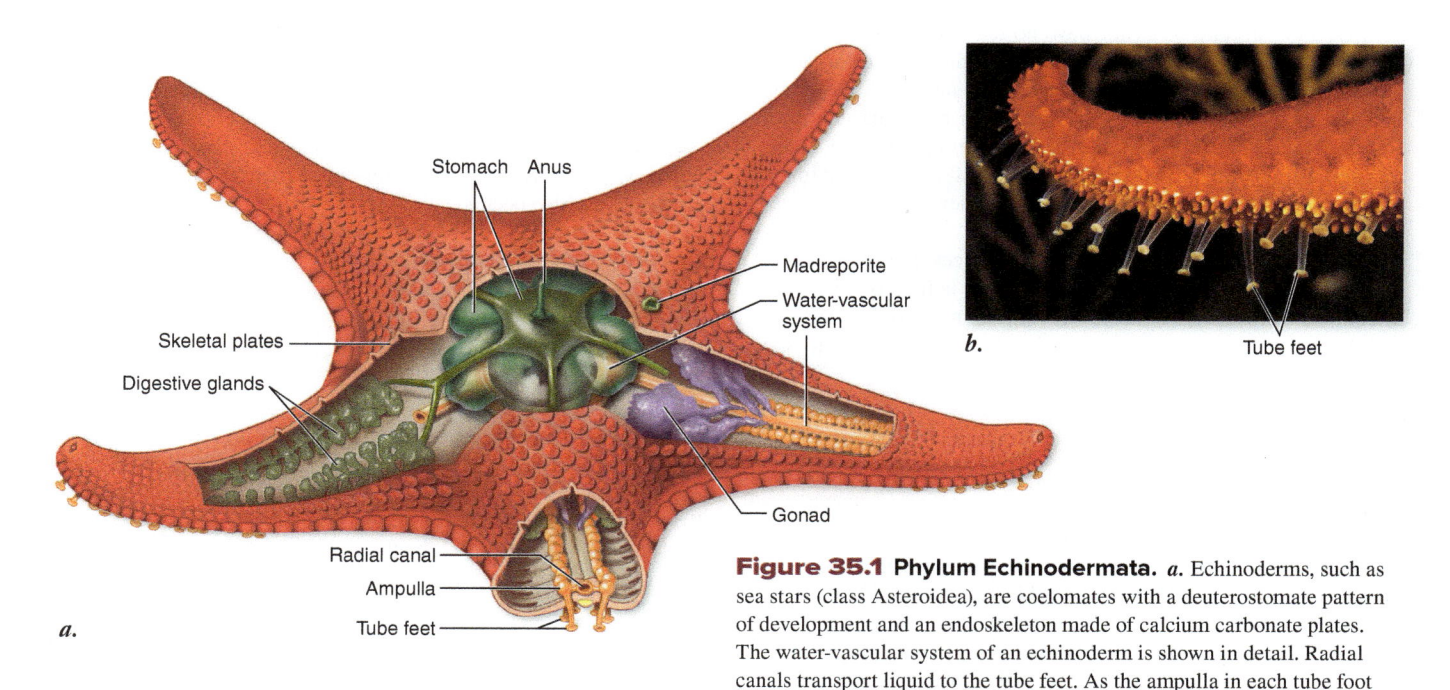

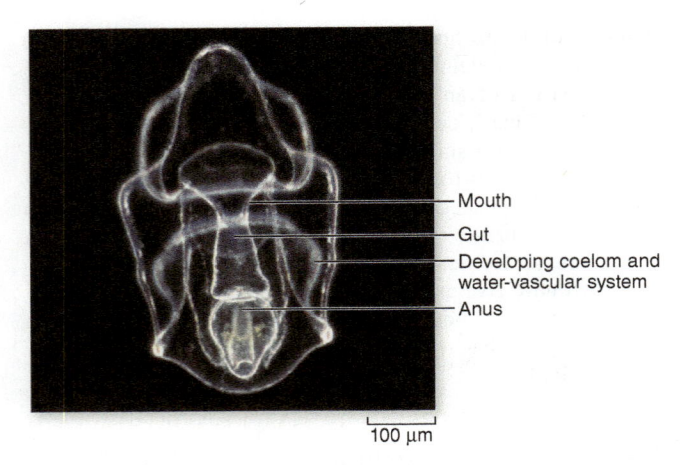

Figure 35.2 The free-swimming larva of the common sea star, *Asterias rubens*. Such bilaterally symmetrical larvae suggest that the ancestors of the echinoderms may not have been radially symmetrical.

supported by the observation that echinoderm larvae are bilaterally symmetrical (figure 35.2).

Echinoderm symmetry is bilateral in larvae but pentaradial in adults

The body of echinoderms undergoes a fundamental shift during development, from bilaterally symmetrical larvae (figure 35.2) to pentaradially symmetrical adults (see figure 35.1). Dorsal, ventral, anterior, and posterior have no meaning without a head or tail for orientation. Thus, orientation of the body of an adult echinoderm is described in reference to the mouth. All systems of an adult echinoderm are organized with branches radiating from a center. For example, the nervous system consists of a central nerve ring from which branches arise; although the animals are capable of complex behavior patterns, there is no centralization of function.

In many echinoderms, the mouth faces the substratum, although in sea cucumbers, the animal's axis is horizontal, so the animal crawls with its mouth foremost, and in crinoids (sea lilies and feather stars), the oral surface is located opposite to the substrate.

The endoskeleton

Echinoderms have a delicate epidermis that stretches over an endoskeleton composed of calcium carbonate (calcite) plates called **ossicles.** In echinoderms such as asteroids (sea stars), the individual skeletal elements are loosely joined to one another. In others, especially echinoids (sea urchins and sand dollars), the ossicles abut one another tightly, forming a rigid shell (called a test). In sea cucumbers, by contrast, the ossicles are widely scattered, so the body wall is flexible. The ossicles in certain portions of the body of some echinoderms are perforated by pores. Tube feet, part of the water-vascular system discussed later, extend through these pores.

Echinoderms have collagenous tissue, which can change in texture from tough and rubbery to weak and fluid. This amazing tissue accounts for attributes of echinoderms such as the ability to autotomize (cast off) parts. This tissue is also responsible for a sea cucumber's being able to change from almost rigid to flaccid in a matter of seconds.

The water-vascular system

The water-vascular system is radially organized. From the ring canal, which encircles the animal's esophagus, a radial canal extends into each branch of the body (see figure 35.1). Water enters the water-vascular system through a **madreporite,** a sievelike plate that, in most echinoderms, is on the animal's surface, and flows to the ring canal through a stone canal, so named because it is reinforced by calcium carbonate. Each radial canal, in turn, extends through short side branches into the hollow tube feet (see figure 35.1*b*). At the base of each tube foot in most types of echinoderms is a muscular sac, the ampulla. When the ampulla contracts, the fluid, prevented from entering the radial canal by a one-way valve, is forced into the tube foot, thus extending it. Contraction of longitudinal muscles on one side of the tube foot wall causes the tube foot to bend; relaxation of the muscles in the ampulla and contraction of all the longitudinal muscles in the tube foot forces the fluid back into the ampulla.

In asteroids and echinoids, concerted action of a very large number of small, individually weak tube feet causes the animal to move across the seafloor. The tube feet around the mouth of sea cucumbers are used in feeding. In crinoids, tube feet that arise from the branches of the arms, which extend from the margins of an upward-directed cup, are used in capturing food from the surrounding water. The tube feet of brittle stars are pointed and specialized for feeding.

Gas exchange in most echinoderms is through the body surface and tube feet. In addition, a sea cucumber has paired respiratory trees, which branch off the hindgut. Water is drawn into them and exits from them through the anus. In an asteroid, one of the coelomic spaces other than the water-vascular system has branches into protrusions from the epidermis called papulae, through which gas exchange also occurs.

Regeneration and reproduction

Many echinoderms are able to regenerate lost parts. Some echinoderms can lose and then regrow an arm. When some sea cucumbers are disturbed, they cast out their entire digestive system and subsequently regrow it over a period of several months.

Some echinoderms can reproduce asexually by splitting. However, most reproduction in echinoderms is sexual. Gametes are generally released into the water, where fertilization occurs, and free-swimming bilaterally symmetrical larvae develop (figure 35.2). Each class of echinoderms has a characteristic type of larva. These larvae develop while floating in the ocean as plankton until they metamorphose into the sedentary adults.

Echinodermata contains five extant classes

There are five classes of living echinoderms, but more than 20 others have gone extinct. The living ones are (figure 35.3)

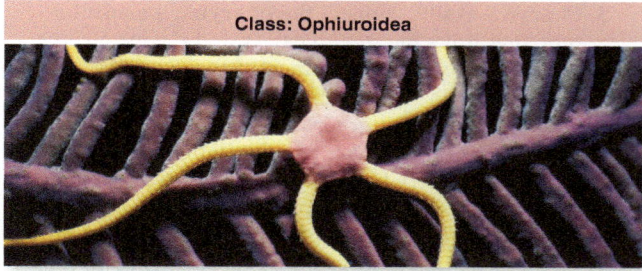

a. Class: Asteroidea
b. Class: Holothuroidea
c. Class: Echinoidea
d. Class: Crinoidea
e. Class: Ophiuroidea

Figure 35.3 Diversity in echinoderms. *a.* Sea star, *Oreaster occidentalis*, in the Gulf of California, Mexico. *b.* Warty sea cucumber, *Parastichopus parvimensis*, Philippines. *c.* Sea urchin of the genus *Echinometra*, Carmel Bay, California. *d.* Feather star of the genus *Comatheria*, from Indonesia. *e.* Gaudy brittle star, *Ophioderma ensiferum*, Grand Turk Island, Caribbean Sea.

(1) sea stars, or starfish, and sea daisies (Asteroidea); (2) sea cucumbers (Holothuroidea); (3) sea urchins and sand dollars (Echinoidea); (4) sea lilies and feather stars (Crinoidea); and (5) brittle stars (Ophiuroidea).

Adults of all classes exhibit a five-part body plan. Even sea cucumbers, which are shaped like their fruit namesake, have five longitudinal grooves along their bodies. Here we discuss three of the living classes that are most characteristic of the phylum.

Sea stars

Sea stars are perhaps the most familiar echinoderms. Important predators in many marine ecosystems, they range in size from a centimeter to a meter across. They are abundant in the intertidal zone, but also occur at the greatest depths of the ocean—10,000 m. Around 1500 species of sea stars are known.

A sea star consists of tapering arms that gradually merge into a central disk (see figure 35.1). Most sea stars have five arms, but others have many more, typically in multiples of five. The digestive space and gonads extend into the arms. The body is somewhat flattened, flexible, and covered with a pigmented epidermis. Asteroidea includes sea daisies, which were discovered in 1986 and were once considered to constitute their own class.

Brittle stars

Brittle stars not only constitute the largest class of echinoderms, with about 2000 species, but are probably the most abundant as well. They resemble sea stars, but differ morphologically in several ways. Their arms are of nearly equal diameter across their entire length; they taper only slightly from base to tip, merge abruptly into the central disk. They are nearly solid and can easily be broken off (the source of the name "brittle"). The tube feet lack ampullae and suckers, being used for feeding, not locomotion. The animal has no anus; waste is eliminated through its mouth.

The most mobile of echinoderms, brittle stars move by pulling themselves along by "rowing" over the substrate by moving their slender arms from side to side. Some brittle stars use their arms to swim, an unusual habit among echinoderms. A brittle star has five arms, but in some larger ones (the basket stars), each arm may bifurcate several times. Brittle stars avoid light and are more active at night.

Sea urchins and sand dollars

Sand dollars and sea urchins lack arms. Five double rows of tube feet protrude through the plates of the calcareous skeleton. The protective, moveable spines that are attached to the skeleton by muscles and connective tissue are also arrayed pentamerally.

About 950 living species constitute the class Echinoidea. Their calcareous plates preserve well, so sea urchins and sand dollars are well represented in the fossil record, with more than 5000 extinct species described. Sand dollars are essentially flattened sea urchins; heart urchins are intermediate between the two.

Like holothurians, sea urchins are eaten by humans. In particular, the gonads, known as *uni* in Japan, are considered a delicacy.

Learning Outcomes Review 35.1

Echinoderms are marine deuterostomes with endoskeletons. They are characterized by pentaradial symmetry in the adult, in which a line drawn in five directions produces mirror images. The water-vascular system and tube feet that act as suction cups aid in movement and feeding. The five living classes of echinoderms are Asteroidea (sea stars), Crinoidea (sea lilies), Echinoidea (sea urchins and sand dollars), Holothuroidea (sea cucumbers), and Ophiuroidea (brittle stars).

- *How does phylogenetic information lead to the conclusion that echinoderms were bilaterally symmetrical?*

35.2 Chordates

Learning Outcomes
1. List the defining characteristics of chordates.
2. Describe the evolutionary relationships of chordates to other taxa.

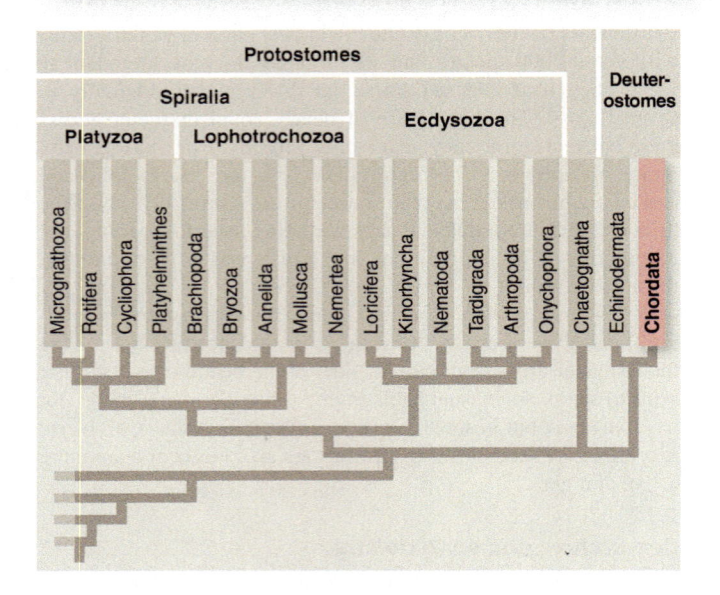

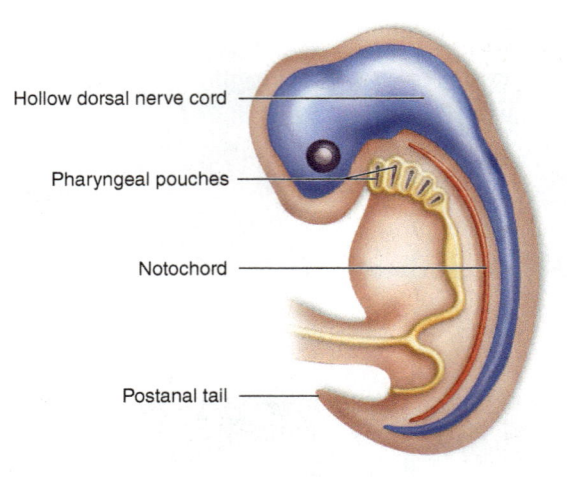

Figure 35.4 The four principal features of the chordates, as shown in a generalized embryo.

Members of the phylum Chordata exhibit great changes in the endoskeleton from that seen in echinoderms. The endoskeleton of echinoderms is functionally similar to the exoskeleton of arthropods—a hard shell with muscles attached to its inner surface. Chordates employ a very different kind of endoskeleton—one that is truly internal. Members of the phylum Chordata have a flexible rod that develops down the length of the back of the embryo. Muscles attached to this rod allowed early chordates to swing their bodies from side to side, swimming through the water. This key evolutionary advance, attaching muscles to an internal element, started chordates along an evolutionary path that led to the vertebrates—and, for the first time, to truly large animals.

Four features characterize the chordates and have played an important role in the evolution of the phylum, which includes fish, amphibians, reptiles, birds, and mammals (figure 35.4):

1. A single, hollow **nerve cord** runs just beneath the dorsal surface of the animal. In vertebrates, the dorsal nerve cord differentiates into the brain and spinal cord.
2. A flexible rod, the **notochord,** forms on the dorsal side of the primitive gut in the early embryo and is present at some developmental stage in all chordates. The notochord is located just below the nerve cord. The notochord may persist in some chordates; in others it is replaced during embryonic development by the vertebral column that forms around the nerve cord.
3. **Pharyngeal slits** connect the **pharynx,** a muscular tube that links the mouth cavity and the esophagus, with the external environment. In terrestrial vertebrates, the slits do not actually connect to the outside and are better termed **pharyngeal pouches.** Pharyngeal pouches are present in the embryos of all vertebrates. They become slits, open to the outside in animals with gills, but leave no external trace in those lacking gills. The presence of these structures in all vertebrate embryos provides evidence of their aquatic ancestry.
4. Chordates have a **postanal tail** that extends beyond the anus, at least during their embryonic development. Nearly all other animals have a terminal anus.

All chordates have all four of these characteristics at some time in their lives. For example, humans as embryos have pharyngeal pouches, a dorsal nerve cord, a postanal tail, and a notochord. As adults, the nerve cord remains, and the notochord is replaced by the vertebral column. All but one pair of pharyngeal pouches is lost; this remaining pair forms the Eustachian tubes that connect the throat to the middle ear. The postanal tail regresses, forming the tail bone (coccyx). The presence of these traits in human embryos is an example of how patterns of development reflect evolutionary change, as discussed in chapter 21.

A number of other characteristics also distinguish the chordates from other animals. Chordate muscles are arranged in segmented blocks, an arrangement which affects the basic organization of the chordate body and can often be clearly seen in embryos of this phylum (figure 35.5). Most chordates have an internal skeleton against which the muscles work. Either this internal skeleton or the notochord (figure 35.6) makes possible the extraordinary powers of locomotion characteristic of this group.

Figure 35.5 A mouse embryo. At 11.5 days of development, the mesoderm is already divided into segments called somites (*dark stain* in this photo), reflecting the segmented nature of all chordates.

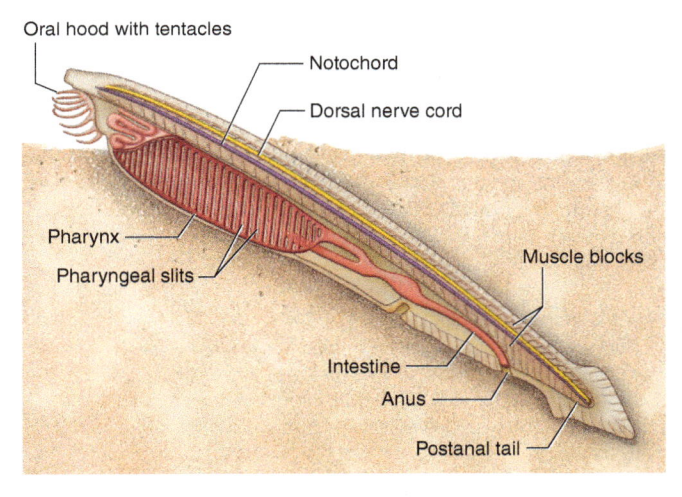

Figure 35.6 Characteristics of chordates. Vertebrates, tunicates, and lancelets are chordates (phylum Chordata), coelomate animals with a flexible rod, the notochord, that provides resistance to muscle contraction and permits rapid lateral body movements. Chordates also possess pharyngeal pouches or slits (reflecting their aquatic ancestry and present habitat in some) and a hollow dorsal nerve cord. In nearly all vertebrates, the notochord is replaced during embryonic development by the vertebral column.

Learning Outcomes Review 35.2

Chordates are characterized by a hollow dorsal nerve cord, a notochord, pharyngeal pouches, and a postanal tail at some point in their development. The flexible notochord anchors internal muscles and allows rapid, versatile movement. Chordates are deuterostomes; their nearest relatives are the echinoderms.

- *What distinguishes a chordate from an echinoderm?*

35.3 Nonvertebrate Chordates

Learning Outcome

1. Describe the nonvertebrate chordates and their characteristics.

Phylum Chordata can be divided into three subphyla. Two of these, Urochordata and Cephalochordata, are nonvertebrate; the third subphylum is Vertebrata. The nonvertebrate chordates do not form vertebrae or other bones, and in the case of the urochordates, their adult form is greatly different from what we expect chordates to look like.

Tunicates have chordate larval forms

The tunicates and salps (subphylum Urochordata) are a group of about 2000 species of marine animals. Most of them are immobile as adults, with only the larvae having a notochord and nerve cord. As adults, they exhibit neither a major body cavity nor visible signs of segmentation (figure 35.7*a, b*). Most species occur in shallow waters, but some are found at great depths. In some tunicates, adults are colonial, living in masses on the ocean floor. The pharynx is lined with numerous cilia; beating of the cilia draws a stream of water into the pharynx, where microscopic food particles become trapped in a mucous sheet secreted from a structure called an *endostyle*.

The tadpole-like larvae of tunicates exhibit all of the basic characteristics of chordates (figure 35.7*c*). The larvae do not feed and have a poorly developed gut. They remain free-swimming for only a few days before settling to the bottom and attaching themselves to a suitable substrate by means of a sucker.

Tunicates change so much as they mature and adjust developmentally to an immobile, filter-feeding existence that it is difficult to discern their evolutionary relationships solely by examining an adult. Many adult tunicates secrete a tunic, a tough sac composed mainly of cellulose, a substance frequently found in the cell walls of plants and algae but rarely found in animals. The tunic surrounds the animal and gives the subphylum its name. Colonial tunicates may have a common sac and a common opening to the outside.

One group of Urochordates, the Larvacea, retains the tail and notochord into adulthood. One theory of vertebrate origins involves a larval form, perhaps that of a tunicate, which acquired the ability to reproduce.

Lancelets are small marine chordates

Lancelets (subphylum Cephalochordata) were given their English name because they resemble a lancet—a small, two-edged surgical knife. These scaleless chordates, a few centimeters long, occur widely in shallow water throughout the oceans of the world. There are about 30 species of this subphylum. Most of them belong to the genus *Branchiostoma*, formerly called *Amphioxus*, a

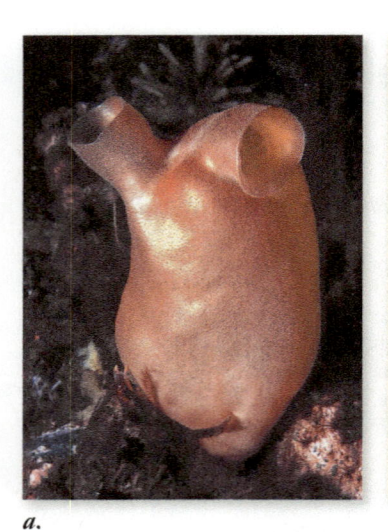

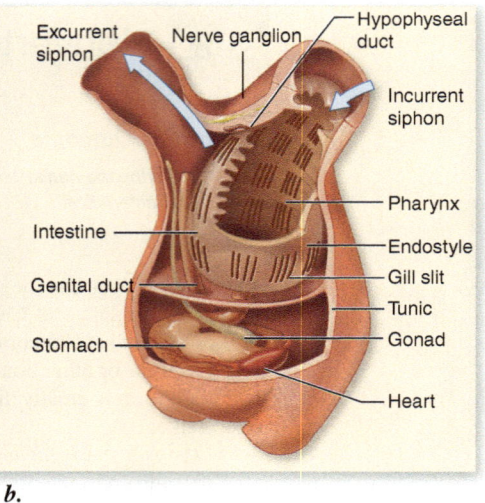

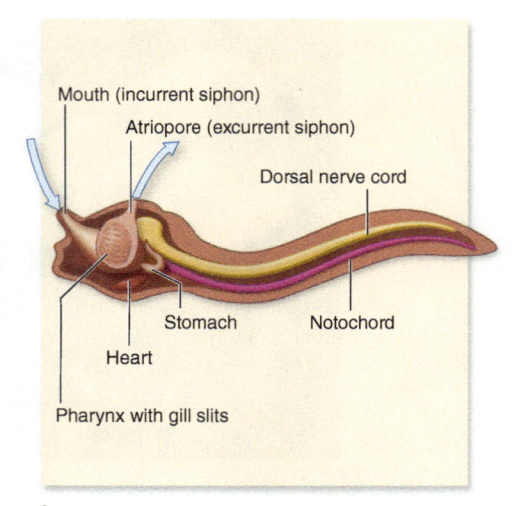

Figure 35.7 Tunicates. *a.* The sea peach, *Halocynthia auranthium*, like other tunicates, does not move as an adult, but rather is firmly attached to the seafloor. *b.* Diagram of the structure of an adult tunicate. *c.* Diagram of the structure of a larval tunicate, showing the characteristic tadpole-like form. Larval tunicates resemble the postulated common ancestor of the chordates.

name still used widely. In lancelets, the notochord runs the entire length of the dorsal nerve cord and persists throughout the animal's life.

Lancelets spend most of their time partly buried in sandy or muddy substrates, with only their anterior ends protruding (figure 35.8). They can swim, although they rarely do so. Their muscles can easily be seen through their thin, transparent skin as a series of discrete blocks, called myomeres. Lancelets have many more pharyngeal gill slits than fishes do. Their skin lacks pigment and has only a single layer of cells, unlike the multi-layered skin of vertebrates. The lancelet body is pointed at both ends. There is no distinguishable head or sensory structure other than pigmented light receptors.

Lancelets feed on microscopic plankton, using a current created by beating cilia that line the oral hood, pharynx, and gill slits. The gill slits provide an exit for the water and are an adaptation for filter feeding. The oral hood projects beyond the mouth and bears sensory tentacles, which also ring the mouth.

The recent discovery of fossil forms similar to living lancelets in rocks 550 million years old argues for the antiquity of this group. Recent studies by molecular systematists further support the hypothesis that lancelets are the closest relatives of vertebrates.

Learning Outcome Review 35.3

Nonvertebrate chordates have notochords but no vertebrae or bones. Urochordates, such as tunicates, have obviously chordate larval forms but drastically different adult forms. Cephalochordates, such as lancelets, do not change body form as adults.

- *How do lancelets and tunicates differ from each other, and from vertebrates?*

Figure 35.8 Lancelets. Two lancelets, *Branchiostoma lanceolatum*, partly buried in shell gravel, with their anterior ends protruding. The muscle segments are clearly visible.

35.4 Vertebrate Chordates

Learning Outcomes

1. Distinguish vertebrates from other chordates.
2. Explain how cartilage and bone contributed to increased size in vertebrates.

Vertebrates (subphylum Vertebrata) are chordates with a spinal column. The name *vertebrate* comes from the individual bony or cartilaginous segments, called vertebrae, that make up the spine.

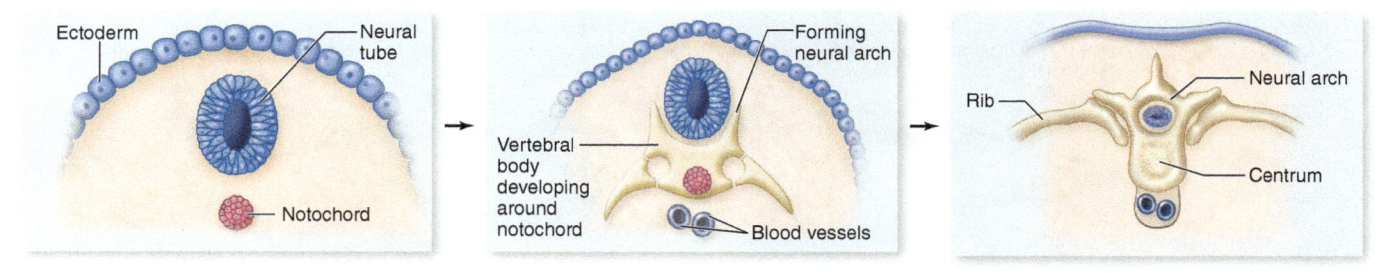

Figure 35.9 Embryonic development of a vertebra. During development, the flexible notochord is surrounded and eventually replaced by a cartilaginous or bony covering, the centrum. The neural tube is protected by an arch above the centrum. The vertebral column functions as a strong, flexible rod that the muscles pull against when the animal swims or moves.

Vertebrates have vertebrae, a distinct head, and other features

Vertebrates differ from the tunicates and lancelets in two important respects:

Vertebral column. In all vertebrates except the earliest diverging fishes, the notochord is replaced during embryonic development by a vertebral column (figure 35.9). The column is a series of bony or cartilaginous vertebrae that enclose and protect the dorsal nerve cord like a sleeve.

Head. Vertebrates have a distinct and well-differentiated head with three pairs of well-developed sensory organs, and the brain is encased within a protective box, the skull, made of bone or cartilage.

In addition to these two key characteristics, vertebrates differ from other chordates in other important respects (figure 35.10):

Neural crest. A unique group of embryonic cells called the **neural crest** contributes to the development of many vertebrate structures. These cells develop on the crest of the neural tube as it forms by invagination and pinching together of the neural plate (see chapter 53 for a detailed account). Neural crest cells then migrate to various locations in the developing embryo, where they participate in the development of many different structures.

Internal organs. Internal organs characteristic of vertebrates include a liver, kidneys, and endocrine glands. The ductless endocrine glands secrete hormones that help regulate many of the body's functions. All vertebrates have a heart and a closed circulatory system. In both their circulatory and their excretory functions, vertebrates differ markedly from other animals.

Endoskeleton. The endoskeleton of most vertebrates is made of cartilage or bone. Cartilage and bone are specialized tissues containing fibers of the protein collagen compacted together (see chapter 46). Bone also contains crystals of a calcium phosphate salt.

Vertebrates evolved half a billion years ago: An overview

The first vertebrates evolved in the oceans about 545 MYA, during the Cambrian period. Many of them looked like a flattened hot dog, with a mouth at one end and a fin at the other. The appearance of a hinged jaw was a major advance, opening up new food-gathering options, and jawed fishes became the dominant creatures in the sea. Their descendants, the amphibians, invaded the land. Amphibians, in turn, gave rise to the first reptiles about 300 MYA. Within 50 million years, reptiles, better suited to living out of water, replaced amphibians as the dominant land vertebrates.

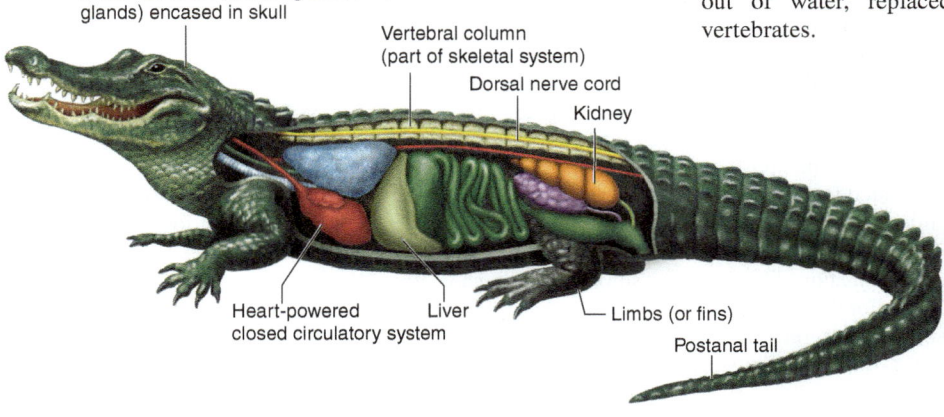

Figure 35.10 Major characteristics of vertebrates. Adult vertebrates are characterized by an internal skeleton of cartilage or bone, including a vertebral column and a skull. Several other internal and external features are characteristic of vertebrates.

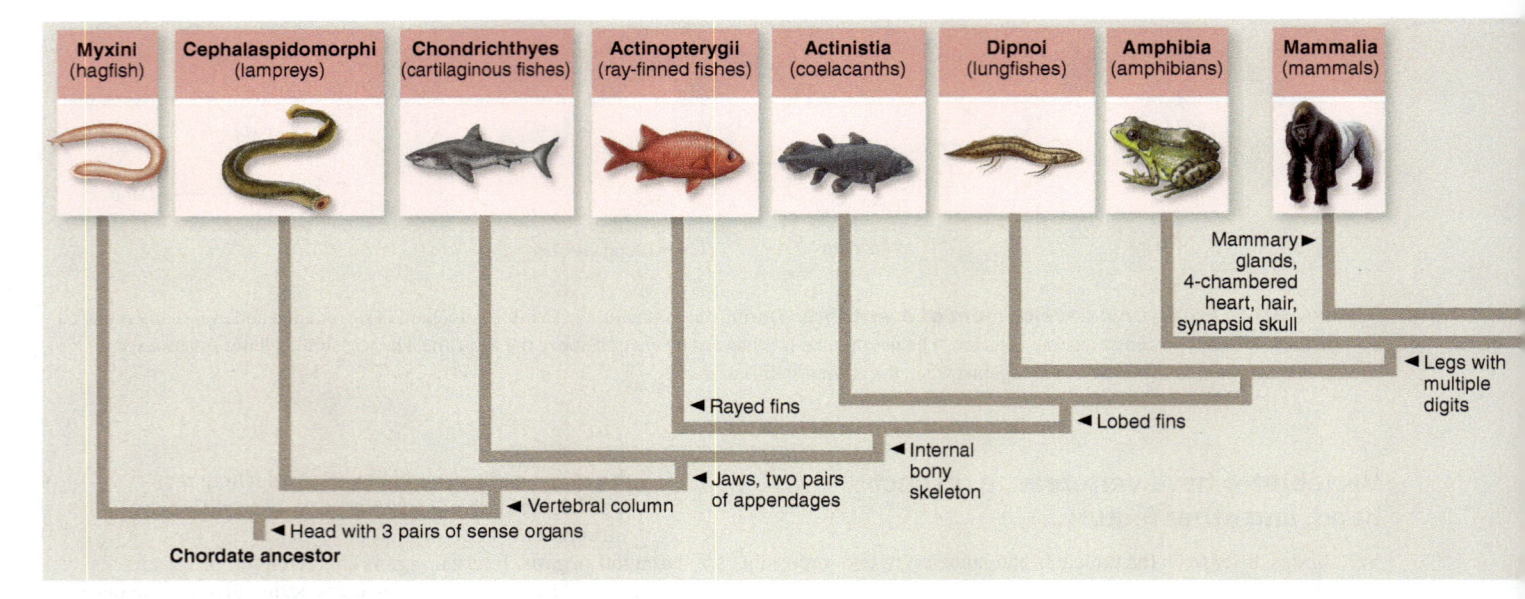

Figure 35.11 Phylogeny of the living vertebrates. Some of the key characteristics that evolved among the vertebrate groups are shown in this phylogeny.

Inquiry question Which vertebrate classes are paraphyletic and why?

With the success of reptiles, vertebrates truly came to dominate the surface of the Earth. Many kinds of reptiles evolved, ranging in size from smaller than a chicken to bigger than a bus, and including some that flew and others that swam. Among them evolved reptiles that gave rise to the two remaining great lines of terrestrial vertebrates: birds and mammals.

Dinosaurs and mammals appear at about the same time in the fossil record, 220 MYA. For over 150 million years, dinosaurs dominated the face of the Earth. Over all these million-and-a-half centuries, the largest mammal was no bigger than a medium-sized dog. Then, in the Cretaceous mass extinction, about 65 MYA, the dinosaurs and other types of reptiles abruptly disappeared. In their absence, mammals and birds quickly took their place, becoming abundant and diverse.

The history of vertebrates has been a series of evolutionary advances that have allowed vertebrates to first invade the land and then the air. In this chapter, we examine the key evolutionary advances that permitted vertebrates to invade the land successfully. As you will see, this invasion was a staggering evolutionary achievement, involving fundamental changes in many body systems.

While considering vertebrate evolution, keep in mind that two vertebrate groups—fish and reptiles—are paraphyletic (figure 35.11). Some fish are more closely related to other vertebrates than they are to some other fish. Similarly, some reptiles are more closely related to birds than they are to other reptiles. Biologists are currently divided about how to handle situations like this, as we discussed in chapter 23 (see figure 23.6). For the time being, we continue to use the traditional classification of vertebrates, bearing in mind the evolutionary implications of the paraphyly of fish and reptiles.

Learning Outcomes Review 35.4

Vertebrates are characterized by a vertebral column and a distinct head. Other distinguishing features are the development of a neural crest, a closed circulatory system, specialized organs, and a bony or cartilaginous endoskeleton that has the strength to support larger body size and powerful movements.

- *In what ways would an exoskeleton limit the size of an organism?*

35.5 Fishes

Learning Outcomes

1. Describe the major groups of fishes.
2. Discuss the significance of the evolutionary innovations of fishes.

Over half of all vertebrates are fishes. The most diverse vertebrate group, fishes provided the evolutionary base for invasion of land by amphibians. In many ways, amphibians can be viewed as transitional—"fish out of water."

The story of vertebrate evolution started in the ancient seas of the Cambrian period (545 to 490 MYA). Figure 35.11 shows the key vertebrate characteristics that evolved subsequently. Wriggling through the water, jawless and toothless, the first fishes sucked up small food particles from the ocean floor like miniature vacuum cleaners. Most were less than a foot long, respired with gills, and had no paired fins or vertebrae (although some had rudimentary vertebrae); they did have a head and a primitive tail to push them through the water.

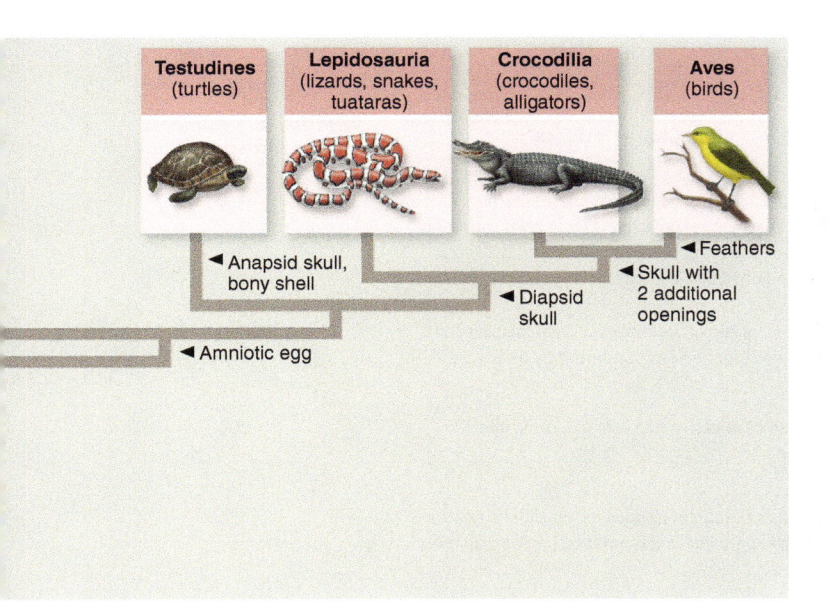

For 50 million years, during the Ordovician period (490 to 438 MYA), these simple fishes were the only vertebrates. By the end of this period, fish had developed primitive fins to help them swim and massive shields of bone for protection. Jawed fishes first appeared during the Silurian period (438 to 408 MYA), and along with them came a new mode of feeding.

Fishes exhibit five key characteristics

From 18-m-long whale sharks to tiny gobies no larger than your fingernail, fishes vary considerably in size, shape, color, and appearance (figure 35.12). Some live in freezing arctic seas, others in warm, freshwater lakes, and still others spend a lot of time entirely out of water. However varied, all fishes have important characteristics in common:

1. **Vertebral column.** Fish have an internal skeleton with a bony or cartilaginous spine surrounding the dorsal nerve cord, and a bony or cartilaginous skull encasing the brain. Exceptions are the jawless hagfish and lampreys. In hagfish, a cartilaginous skull is present, but vertebrae are not; the notochord persists and provides support. In lampreys, a cartilaginous skeleton and notochord are present, but rudimentary cartilaginous vertebrae also surround the notochord in places.
2. **Jaws and paired appendages.** Fishes other than lampreys and hagfish all have jaws and paired appendages, features that are also seen in tetrapods (see figure 35.11). Jaws allowed these fish to capture larger and more active prey. Most fishes have two pairs of fins: a pair of pectoral fins at the shoulder, and a pair of pelvic fins at the hip. In the lobe-finned fish, these pairs of fins became jointed.

Figure 35.12 Fish. Fish are the most diverse vertebrates and include more species than all other kinds of vertebrates combined. *a.* Ribbon eel, *Rhinomuraena quaesita;* (*b*) leafy sea-dragon, *Phycodurus eques;* (*c*) yellowfin tuna, *Thunnus albacares.*

3. **Internal gills.** Fishes are water-dwelling creatures and must extract oxygen dissolved in the water around them. They do this by directing a flow of water through their mouths and across their gills (see chapter 48). The gills are composed of fine filaments of tissue that are rich in blood vessels.
4. **Single-loop blood circulation.** Blood is pumped from the heart to the gills. From the gills, the oxygenated blood passes to the rest of the body, and then returns to the heart. The heart is a muscular tube-pump made of two chambers that contract in sequence.
5. **Nutritional deficiencies.** Fishes are unable to synthesize the aromatic amino acids (phenylalanine, tryptophan, and tyrosine; see chapter 3), so they must consume them in their foods. This inability has been inherited by all of their vertebrate descendants.

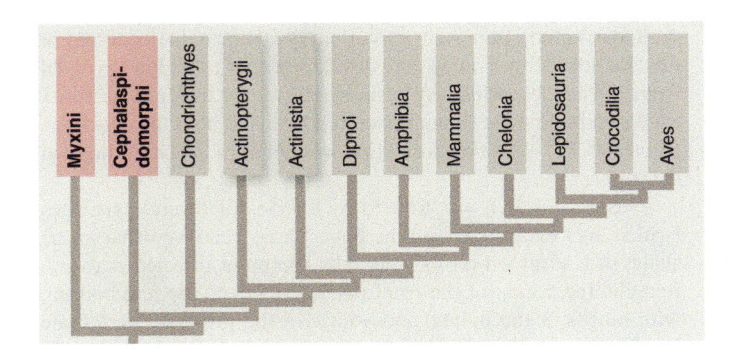

The first fishes

The first fishes did not have jaws, and instead had only a mouth at the front end of the body that could be opened to take in food.

TABLE 35.1 Major Classes of Fishes

Class	Typical Examples	Key Characteristics	Approximate Number of Living Species
Sarcopterygii	Lobe-finned fishes	Largely extinct group of bony fishes possessing paired lobed fins; living species are coelacanths (Actinistia) and lungfish (Dipnoi); because tetrapods (legged vertebrates) arose from a lobe-finned fish, the Sarcopterygii is paraphyletic	8
Actinopterygii	Ray-finned fishes	Most diverse group of vertebrates; swim bladders and bony skeletons; paired fins supported by bony rays	30,000
Chondrichthyes	Sharks, skates, rays	Cartilaginous skeletons; no swim bladders; internal fertilization	1,000
Cephalaspidomorphi	Lampreys	Largely extinct group of jawless fishes with no paired appendages; parasitic and nonparasitic types; all breed in freshwater	38
Myxini	Hagfishes	Jawless fishes with no paired appendages; scavengers; mostly blind, but having a well-developed sense of smell	60
Placodermi	Armored fishes	Jawed fishes with heavily armored heads; many were quite large	Extinct
Acanthodii and Ostracoderms	Spiny fishes	Fishes with (acanthodians) or without (placoderms) jaws; paired fins supported by sharp spines; head shields made of bone; rest of skeleton cartilaginous	Extinct

Two groups survive today as hagfish (class Myxini; table 35.1) and lampreys (class Cephalaspidomorphi).

Another group were the *ostracoderms* (a word meaning "shell-skinned"). Only their head-shields were made of bone; their elaborate internal skeletons were constructed of cartilage. Many ostracoderms were bottom-dwellers, with a jawless mouth underneath a flat head, and eyes on the upper surface. Ostracoderms thrived in the Ordovician and Silurian periods (490 to 408 MYA), only to become almost extinct at the close of the Devonian period (408 to 360 MYA).

Evolution of the jaw

A fundamentally important evolutionary advance that occurred in the late Silurian period was the development of jaws. Jaws evolved from gill arches, which in jawless fish were cartilaginous struts used to reinforce the tissue between gill slits to hold the slits open (figure 35.13). This transformation was not as radical as it might at first appear.

Each gill arch was formed by a series of several cartilages (which later evolved to become bones) arranged somewhat in the shape of a V turned on its side, with the point directed outward. Imagine the fusion of the front pair of arches at top and bottom, with hinges at the points, and you have the primitive vertebrate jaw. The top half of the jaw is not attached to the skull directly except at the rear. Teeth developed on the jaws from modified scales on the skin that lined the mouth (figure 35.13).

Armored fishes called placoderms and spiny fishes called acanthodians both had jaws. Spiny fishes were very common during the early Devonian period, largely replacing ostracoderms, but they became extinct themselves at the close of the Permian. Like ostracoderms, they had internal skeletons made of cartilage, but their scales contained small plates of bone, foreshadowing the much larger role bone would play in the future of vertebrates. Spiny fishes were jawed predators and far better swimmers than ostracoderms, with as many as seven fins to aid their swimming. All of these fins were reinforced with strong spines, giving these fishes their name.

By the mid-Devonian period, the heavily armored placoderms became common. A very diverse and successful group, seven orders of placoderms dominated the seas of the late Devonian, only to become extinct at the end of that period. The placoderm jaw was much improved over the primitive jaw of spiny fishes,

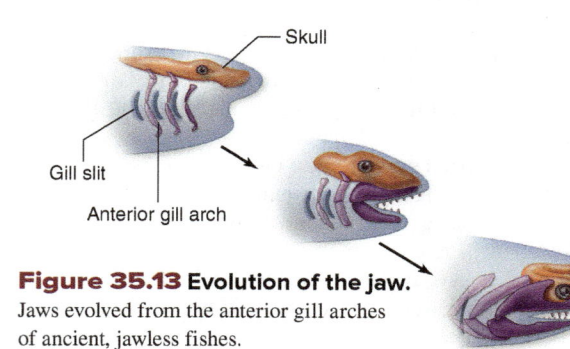

Figure 35.13 Evolution of the jaw.
Jaws evolved from the anterior gill arches of ancient, jawless fishes.

with the upper jaw fused to the skull and the skull hinged on the shoulder. Many of the placoderms grew to enormous sizes, some over 30 feet long, with 2-foot skulls that had an enormous bite.

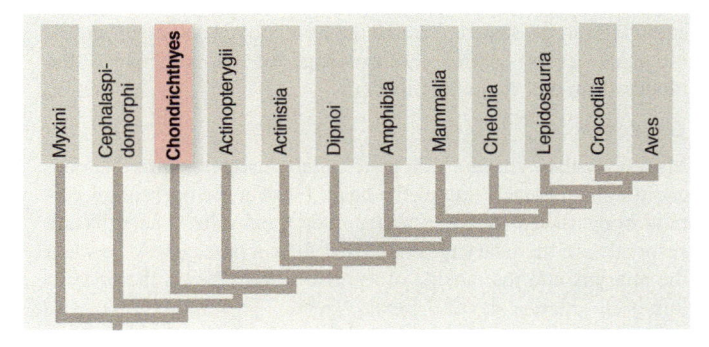

Sharks, with cartilaginous skeletons, became top predators

At the end of the Devonian period, essentially all of these pioneer vertebrates disappeared, replaced by sharks and bony fishes in one of several mass extinctions that occurred during Earth's history (see chapter 22). Sharks and bony fishes first evolved in the early Devonian, 400 MYA. In these fishes, the jaw was improved even further, with the upper part of the first gill arch behind the jaws being transformed into a supporting strut or prop, joining the rear of the lower jaw to the back of the skull. This allowed the mouth to open much wider than was previously possible.

During the Carboniferous period (360 to 280 MYA), sharks became the dominant predators in the sea. Sharks, as well as skates and rays (all in the class Chondrichthyes), have a skeleton made of cartilage, like primitive fishes, but it is "calcified," strengthened by granules of calcium carbonate deposited in the outer layers of cartilage. The result is a very light and strong skeleton.

Streamlined, with paired fins and a light, flexible skeleton, sharks are superior swimmers (figure 35.14). Their pectoral fins are particularly large, jutting out stiffly like airplane wings—and that is how they function, adding lift to compensate for the downward thrust of the tail fin. Sharks are very aggressive predators, and some early sharks reached enormous size.

The evolution of teeth

Sharks were among the first vertebrates to develop teeth. These teeth evolved from rough scales on the skin and are not set into the jaw as human teeth are, but rather sit atop it. They are not firmly anchored and are easily lost. In a shark's mouth, the teeth are arrayed in up to 20 rows; the teeth in front do the biting and cutting, and behind them other teeth grow and wait their turn. When a tooth breaks or is worn down, a replacement from the next row moves forward. A single shark may eventually use more than 20,000 teeth in its lifetime.

A shark's skin is covered with tiny, toothlike scales, giving it a rough "sandpaper" texture. Like the teeth, these scales are constantly replaced throughout the shark's life.

The lateral line system

Sharks, as well as bony fishes, possess a fully developed lateral line system. The **lateral line system** consists of a series of sensory organs that project into a canal beneath the surface of the skin. The canal runs the length of the fish's body and is open to the exterior through a series of sunken pits. Movement of water past the fish forces water through the canal. The pits are oriented so that some are stimulated no matter what direction the water moves. Details of the lateral line system's function are described in chapter 44. In a very real sense, the lateral line system is a fish's equivalent of hearing and therefore is a means of mechanoreception.

Reproduction in cartilaginous fishes

Reproduction in the class Chondrichthyes differs from that of most other fishes. Shark eggs are fertilized internally. During mating, the male grasps the female with modified fins called claspers, and sperm run from the male into the female through grooves in the claspers. Although a few species lay fertilized eggs, the eggs of most species develop within the female's body, and the pups are born alive (see figure 52.4).

This reproductive system can work to the detriment of sharks. Because of their long gestation periods and relatively few offspring, shark populations are not able to recover quickly from population declines. Unfortunately, in recent times sharks have been fished very heavily because shark fin soup has become popular in Asia and elsewhere. As a result, shark populations have declined greatly, and there is concern that many species may soon face extinction.

Shark evolution

Many of the early evolutionary lines of sharks died out during the great extinction at the end of the Permian period (248 MYA). The survivors thrived and underwent a burst of diversification during the Mesozoic era (248 to 65 MYA), when most of the modern groups of sharks appeared. Skates and rays, which are dorsoventrally flattened relatives of sharks, evolved at this time, some 200 million years after the sharks first appeared.

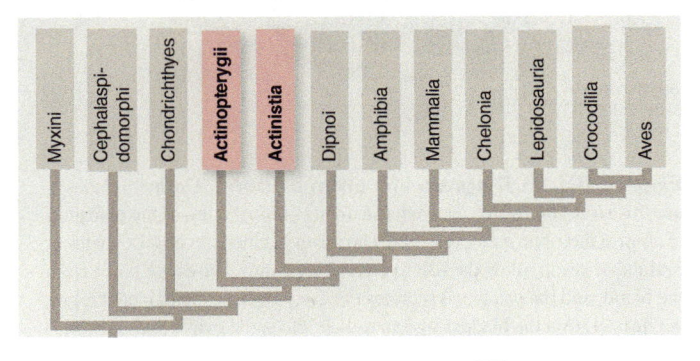

Figure 35.14 Chondrichthyes. Members of the class Chondrichthyes, such as this blue shark, *Prionace glauca*, are mainly predators or scavengers.

Bony fishes dominate the waters

Bony fishes evolved at the same time as sharks, some 400 MYA, but took quite a different evolutionary road. Instead of gaining speed through lightness, as sharks did, bony fishes adopted a heavy internal skeleton made completely of bone.

Bone is very strong, providing a base against which very strong muscles can pull. Not only is the internal skeleton made of bone, but so is the outer covering of plates and scales. Most bony fishes have highly mobile fins, very thin scales, and completely symmetrical tails (which keep the fish on a straight course as it swims through the water). Bony fishes are the most species-rich group of fishes, indeed of all vertebrates. There are several dozen orders containing more than 30,000 living species.

The remarkable success of the bony fishes has resulted from a series of significant adaptations that have enabled them to dominate life in the water. These include the swim bladder and the gill cover (figure 35.15).

Swim bladder

Although bones are heavier than cartilaginous skeletons, most bony fishes are still buoyant because they possess a **swim bladder**, a gas-filled sac that allows them to regulate their buoyant density and so remain suspended at any depth in the water effortlessly. Sharks, by contrast, must move through the water or sink because, lacking a swim bladder, their bodies are denser than water.

In primitive bony fishes, the swim bladder is a dorsal outpocketing of the pharynx behind the throat, and these species fill the swim bladder by simply gulping air at the surface of the water. In most of today's bony fishes, the swim bladder is an independent organ that is filled and drained of gases, mostly nitrogen and oxygen, internally.

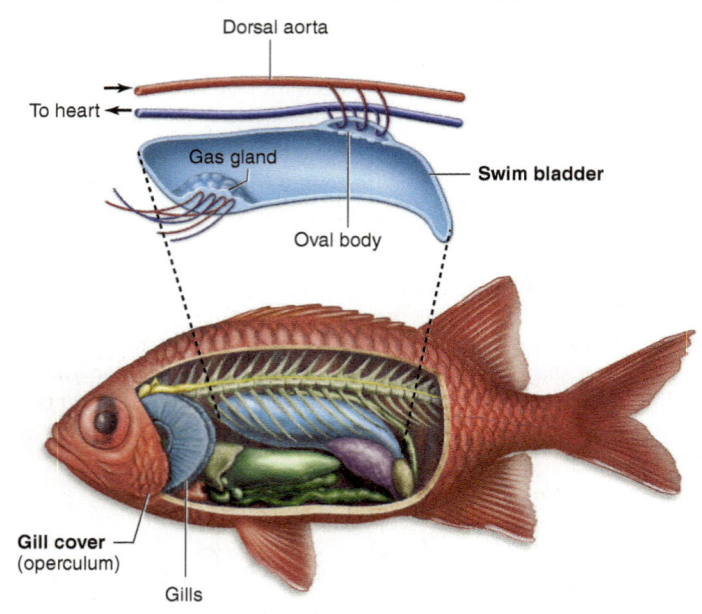

Figure 35.15 Diagram of a swim bladder. The bony fishes use this structure, which evolved as a dorsal outpocketing of the pharynx, to control their buoyancy in water. The swim bladder can be filled with or drained of gas to allow the fish to control buoyancy. Gases are taken from the blood, and the gas gland secretes the gases into the swim bladder; gas is released from the bladder by a muscular valve, the oval body.

How do bony fishes manage this remarkable trick? It turns out that the gases are harvested from the blood by a unique gland that discharges the gases into the bladder when more buoyancy is required. To reduce buoyancy, gas is reabsorbed into the bloodstream through a structure called the oval body. A variety of physiological factors control the exchange of gases between the bloodstream and the swim bladder.

Gill cover

Most bony fishes have a hard plate called the **operculum** that covers the gills on each side of the head. Flexing the operculum permits bony fishes to pump water over their gills. The gills are suspended in the pharyngeal slits that form a passageway between the pharynx and the outside of the fish's body. When the operculum is closed, it seals off the exit.

When the mouth is open, closing the operculum increases the volume of the mouth cavity, so that water is drawn into the mouth. When the mouth is closed, opening the operculum decreases the volume of the mouth cavity, forcing water past the gills to the outside. Using this very efficient bellows, bony fishes can pass water over the gills while remaining stationary in the water. That is what a goldfish is doing when it seems to be gulping in a fish tank.

The evolutionary path to land ran through the coelacanths

Two major groups of bony fish are the ray-finned fishes (class Actinopterygii; figure 35.16a) and coelacanths (class Sarcopterygii). The groups differ in the structure of their fins (figure 35.16b). In ray-finned fishes, the internal skeleton of the fin is composed of parallel bony rays that support and stiffen each fin. There are no muscles within the fins; rather, the fins are moved by muscles within the body.

By contrast, coelacanths have paired fins that consist of a long fleshy muscular lobe (hence their name), supported by a central core of bones that form fully articulated joints with one another. There are bony rays only at the tips of each lobed fin. Muscles within each lobe can move the fin rays independently of one another, a feat no ray-finned fish could match.

Coelacanths evolved 390 MYA, shortly after the first bony fishes appeared. Only eight species survive today, two species of coelacanth (subclass Actinistia; figure 35.16b) and six species of lungfish (subclass Dipnoi). Although rare today, lobe-finned fishes played an important part in the evolutionary story of vertebrates because it was a coelacanths that crawled out of water and colonized land, giving rise to the first amphibians. Lungfish are more closely related to tetrapods (legged vertebrates) than they are to coelacanths, thus making the Sarcopterygii a paraphyletic group (see figure 35.11).

Learning Outcomes Review 35.5

Fishes are generally characterized by the possession of a vertebral column, jaws, paired appendages, a lateral line system, internal gills, and single-loop circulation. Cartilaginous fishes have lightweight skeletons and were among the first vertebrates to develop teeth. The very successful bony fishes have unique characteristics such as swim bladders and gill covers, as well as ossified skeletons. One type of bony fish, the lobe-finned fish, gave rise to the ancestors of amphibians.

- *What advantages do lobed fins have over ray fins?*

Biology, 11th Edition

35.6 Amphibians

Learning Outcomes
1. Describe the characteristics and major groups of amphibians.
2. Explain the challenges of moving from an aquatic to a terrestrial environment.

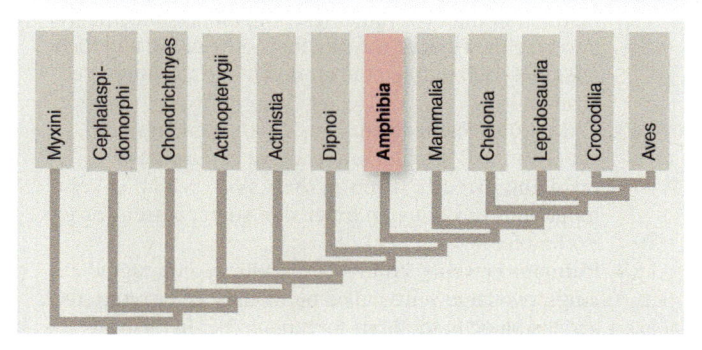

Frogs, salamanders, and caecilians, the damp-skinned vertebrates, are direct descendants of fishes. They are the sole survivors of a very successful group, the amphibians (class Amphibia)—the first vertebrates to walk on land. Most present-day amphibians are small and live largely unnoticed by humans, but they are among the most numerous of terrestrial vertebrates. Throughout the world, amphibians play key roles in terrestrial food chains.

Living amphibians have five distinguishing features

Biologists have classified living species of amphibians into three orders (table 35.2): Frogs and salamanders are well-known; less familiar are the wormlike, nearly blind organisms called caecilians

Figure 35.16 Ray-finned and lobe-finned fishes. *a.* The ray-finned fishes, such as this Koi carp *(Cyprinus carpio)*, are characterized by fins of only parallel bony rays. *b.* By contrast, the fins of lobe-finned fish have a central core of bones as well as rays. The coelacanth, *Latimeria chalumnae*, a lobe-finned fish (class Sarcopterygii, subclass Actinistia), was discovered in the western Indian Ocean in 1938. This coelacanth represents a group of fishes thought to have been extinct for about 70 million years. Scientists who studied living individuals in their natural habitat at depths of 100–200 m observed them drifting in the current and hunting other fishes at night. Some individuals are nearly 3 m long; they have a slender, fat-filled swim bladder.

TABLE 35.2		Orders of Amphibians	
Order	Typical Examples	Key Characteristics	Approximate Number of Living Species
Anura	Frogs, toads	Compact, tailless body; large head fused to the trunk; rear limbs specialized for jumping	6500
Caudata	Salamanders, newts	Slender body; long tail and limbs set out at right angles to the body	665
Apoda	Caecilians	Tropical group with a snakelike body; no limbs; little or no tail	200

chapter 35 *Deuterostomes*

that live in the tropics and make up the order Apoda ("without legs"). These amphibians have several key characteristics in common:

1. **Legs.** Frogs and most salamanders have four legs and can move about on land quite well. Legs were one of the key adaptations to life on land. Caecilians have lost their legs during the course of adapting to a burrowing existence.
2. **Lungs.** Most amphibians possess a pair of lungs, although the internal surfaces have much less surface area than do reptilian or mammalian lungs. Amphibians breathe by lowering the floor of the mouth to suck in air, and then raising it back to force the air down into the lungs (see chapter 48).
3. **Cutaneous respiration.** Frogs, salamanders, and caecilians all supplement the use of lungs by respiring directly through their skin, which is kept moist and provides an extensive surface area for gas exchange.
4. **Pulmonary veins.** After blood is pumped through the lungs, two large veins called pulmonary veins return the aerated blood to the heart for repumping. In this way aerated blood is pumped to the tissues at a much higher pressure.
5. **Partially divided heart.** A dividing wall helps prevent aerated blood from the lungs from mixing with nonaerated blood being returned to the heart from the rest of the body. The blood circulation is thus divided into two separate paths: pulmonary and systemic. The separation is imperfect, however, because no dividing wall exists in one chamber of the heart, the ventricle (see chapter 49).

Amphibians overcame terrestrial challenges

The word *amphibia* means "double life," and it nicely describes the essential quality of modern-day amphibians, reflecting their ability to live in two worlds—the aquatic world of their fish ancestors and the terrestrial world they first invaded. Here, we review the checkered history of this group, almost all of whose members have been extinct for the last 200 million years. Then we examine in more detail what the few kinds of surviving amphibians are like.

The successful invasion of land by vertebrates posed a number of major challenges:

- Because amphibian ancestors had relatively large bodies, supporting the body's weight on land as well as enabling movement from place to place was a challenge (figure 35.17). Legs evolved to meet this need.
- Far more oxygen is available to gills in air than in water, but the delicate structure of fish gills requires the buoyancy of water to support them, and they cannot function in air. Therefore, other methods of obtaining oxygen were required.
- Delivering great amounts of oxygen to the larger muscles needed for movement on land required modifications to the heart and circulatory system.

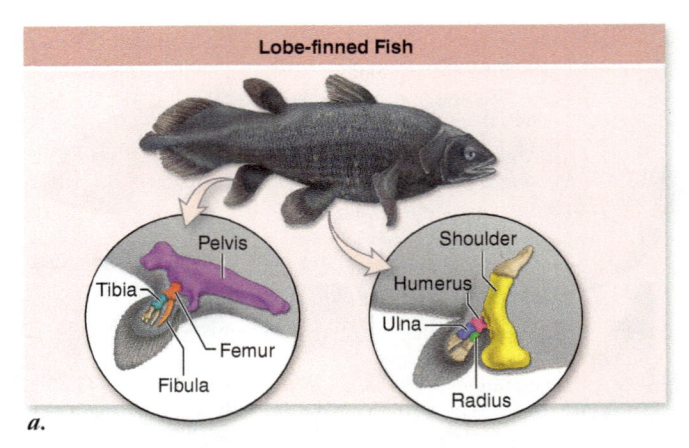

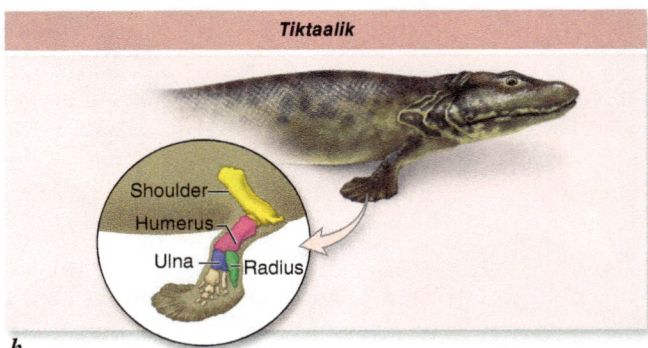

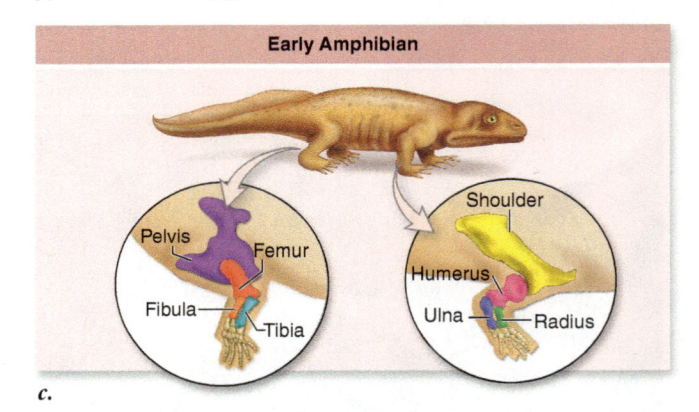

Figure 35.17 A comparison between the limbs of a lobe-finned fish, *Tiktaalik*, and a primitive amphibian.
a. A lobe-finned fish. Some of these animals could probably move on land. *b. Tiktaalik.* The shoulder and limb bones are like those of an amphibian, but the fins are like those of a lobe-finned fish. The fossil of *Tiktaalik* did not contain the hindlimbs. *c.* A primitive amphibian. As illustrated by their skeletal structure, the legs of such an animal clearly could function better than those of its ancestors for movement on land.

- Reproduction still had to be carried out in water so that eggs would not dry out.
- Most importantly, the body itself had to be prevented from drying out.

The first amphibian

Amphibians solved these problems only partially, but their solutions worked well enough that amphibians have survived for 340 million years. Paleontologists agree that amphibians evolved from lobe-finned fish. *Ichthyostega,* one of the earliest amphibian fossils (figure 35.18), was found in a 370-million-year-old rock in Greenland. At that time, Greenland was part of what is now the North American continent and lay near the equator. All amphibian fossils from the next 100 million years are found in North America. Only when Asia and the southern continents merged with North America to form the supercontinent Pangea did amphibians spread throughout the world.

Ichthyostega was a strongly built animal, with sturdy forelegs well supported by shoulder bones. Unlike the skeleton of fish, the shoulder bones were no longer attached to the skull, so the limbs could support the animal's weight and the legs could move independently of the head. Because the hindlimbs were flipper-shaped, *Ichthyostega* probably moved like a seal, with the forelimbs providing the propulsive force for locomotion and the hindlimbs being dragged along with the rest of the body. To strengthen the backbone further, long, broad ribs evolved that overlap each other and formed a solid cage for the lungs and heart. The rib cage was so solid that it probably couldn't expand and contract for breathing. Instead, *Ichthyostega* probably obtained oxygen as many amphibians do today, by lowering the floor of the mouth to draw air in, and then raising it to push air down the windpipe into the lungs.

In 2006, an important transitional fossil between fish and *Ichthyostega* was discovered in northern Canada. *Tiktaalik,* which lived 375 MYA, had gills and scales like a fish, but a neck like an amphibian. Particularly significant, however, was the form of its forelimbs (see figure 35.17b): its shoulder, forearm, and wrist bones were like those of amphibians, but at the end of the limb was a lobed fin, rather than the toes of an amphibian. Ecologically, the 3-m-long *Tiktaalik* was probably also intermediate between fish and amphibians, spending most of its time in the water, but capable of hauling itself out onto land to capture food or escape predators.

The rise and fall of amphibians

By moving onto land, amphibians were able to utilize many more resources and to access many habitats. Amphibians first became common during the Carboniferous period (360 to 280 MYA). Fourteen families of amphibians are known from the early Carboniferous, nearly all of them aquatic or semiaquatic, like *Ichthyostega* (see figure 35.18). By the late Carboniferous, much of North America was covered by low-lying tropical swamplands, and 34 families of amphibians thrived in this wet terrestrial environment, sharing it with pelycosaurs and other early reptiles.

In the early Permian period that followed (280 to 248 MYA), a remarkable change occurred among amphibians—they began to leave the marshes for dry uplands. Many of these terrestrial amphibians had bony plates and armor covering their bodies and grew to be very large, some as big as a pony. Both their large size and the complete covering of their bodies indicate that these amphibians did not use the cutaneous respiratory system of present-day amphibians, but rather had an impermeable leathery skin to prevent water loss. Consequently, they must have relied entirely on their lungs for respiration. By the mid-Permian period, there were 40 families of amphibians. Only 25% of them were still semiaquatic like *Ichthyostega;* 60% of the amphibians were fully terrestrial, and 15% were semiterrestrial. This was the peak of amphibian success, sometimes called the Age of Amphibians.

By the end of the Permian period, reptiles had evolved from amphibians. One group, therapsids, had become common and ousted the amphibians from their newly acquired niche on land. Following the mass extinction event at the end of the Permian, therapsids were the dominant land vertebrate, and most amphibians were aquatic. This trend continued in the following Triassic period (248 to 213 MYA), which saw the virtual extinction of amphibians from land.

Modern amphibians belong to three groups

All of today's amphibians descended from the three families of amphibians that survived the Age of the Dinosaurs. During the Tertiary period (65 to 2 MYA), these moist-skinned amphibians accomplished a highly successful invasion of wet habitats all over the world, and today there are over 7000 species of amphibians in the orders Anura, Caudata, and Apoda.

Order Anura: Frogs and toads

Frogs and toads, amphibians without tails, live in a variety of environments, from deserts and mountains to ponds and puddles (figure 35.19a). Frogs have smooth, moist skin, a broad body, and long hind legs that make them excellent jumpers. Most frogs live in or near water and go through an aquatic tadpole stage before metamorphosing into frogs; however, some tropical species that don't live near water bypass this stage and hatch out as little froglets.

Figure 35.18 Amphibians were the first vertebrates to walk on land. *Ichthyostega,* one of the first amphibians, had efficient limbs for crawling on land, an improved olfactory sense associated with a lengthened snout, and a relatively advanced ear structure for picking up airborne sounds. Despite these features, *Ichthyostega,* which lived about 350 MYA, was still quite fishlike in overall appearance and may have spent much of its life in water.

Figure 35.19 Amphibians *a.* Red-eyed tree frog, *Agalychnis callidryas*. *b.* An adult tiger salamander, *Ambystoma tigrinum*. *c.* A caecilian, *Caecilia tentaculata*.

Unlike frogs, toads have a dry, bumpy skin and short legs, and are well adapted to dry environments. Toads do not form a monophyletic group; that is, all toads are not more closely related to each other than they are to some other frogs. Rather, the term *toad* is applied to those anurans that have adapted to dry environments by evolving a suite of adaptive characteristics; this convergent evolution has occurred many times among distantly related anurans.

Most frogs and toads return to water to reproduce, laying their eggs directly in water. Their eggs lack watertight external membranes and would dry out quickly on land. Eggs are fertilized externally and hatch into swimming larval forms called tadpoles. Tadpoles live in the water, where they generally feed on algae. After considerable growth, the body of the tadpole gradually undergoes metamorphosis into that of an adult frog.

Order Caudata: Salamanders

Salamanders have elongated bodies, long tails, and smooth, moist skin (figure 35.19*b*). They typically range in length from a few inches to a foot, although giant Asiatic salamanders of the genus *Andrias* are as much as 1.5 m long and weigh up to 33 kg. Most salamanders live in moist places, such as under stones or logs, or among the leaves of tropical plants. Some salamanders live entirely in water.

Salamanders lay their eggs in water or in moist places. Most species practice a type of internal fertilization in which the male deposits packets containing sperm on the bottom of ponds or streams and the female then straddles the packet, taking it into her cloaca, where the sperm can fertilize her eggs. Like anurans, many salamanders go through a larval stage before metamorphosing into adults. However, unlike anurans, in which the tadpole is strikingly different from the adult frog, larval salamanders are quite similar to adults, although most live in water and have external gills and gill slits that disappear at metamorphosis.

Order Apoda: Caecilians

Caecilians, members of the order Apoda (also called Gymnophiona), are a highly specialized group of tropical burrowing amphibians (figure 35.19*c*). These legless, wormlike creatures average about 30 cm long, but can be up to 1.3 m long. They have very small eyes and are often blind. They resemble worms but have jaws with teeth. They eat worms and other soil invertebrates. Fertilization is internal.

Learning Outcomes Review 35.6

Amphibians, which includes frogs and toads, salamanders, and caecilians, are generally characterized by legs, lungs, cutaneous respiration, and a more complex and a partially divided heart. All of these developed as adaptations to life on land. Most species rely on a water habitat for reproduction. Although some early forms reached the size of a pony, modern amphibians are generally quite small.

- What challenges did amphibians overcome to make the transition to living on land?
- If reptiles evolved from amphibians, why aren't amphibians portrayed as being paraphyletic in figure 35.11?

35.7 Reptiles

Learning Outcomes

1. Describe the characteristics and major groups of reptiles.
2. Distinguish between synapsids and diapsids.
3. Explain the significance of the evolution of the amniotic egg.

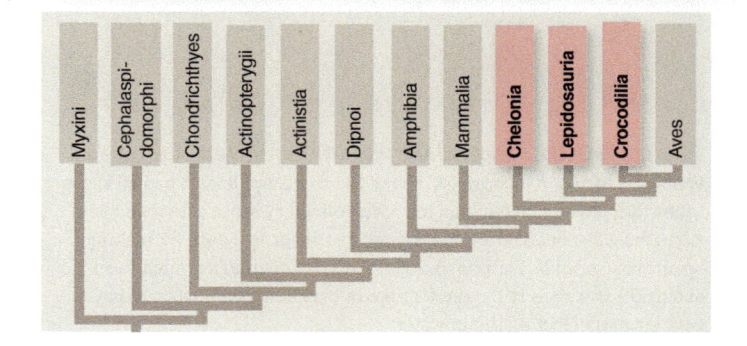

If we think of amphibians as a first draft of a manuscript about survival on land, then reptiles are the finished book. For each of the five key challenges of living on land, reptiles improved on the innovations of amphibians. The arrangement of legs evolved to support the body's weight more effectively, allowing reptile bodies to be bigger and to run. Lungs and heart became more efficient. The skin was covered with dry plates or scales to minimize water loss, and watertight coverings evolved for eggs.

Over 10,000 species of reptiles (class Reptilia) now live on Earth (table 35.3). They are a highly successful group in today's world; there are more living species of snakes and lizards than there are of mammals.

TABLE 35.3 Major Orders of Reptiles

Order	Typical Examples	Key Characteristics	Approximate Number of Living Species
Squamata, suborder Sauria	Lizards	Limbs set at right angles to body; anus is in transverse (sideways) slit; most are terrestrial	6200
Squamata, suborder Serpentes	Snakes	No legs; move by slithering; scaly skin is shed periodically in a single piece; most are terrestrial	3500
Rhynchocephalia	Tuataras	Sole survivors of a once successful group that largely disappeared before dinosaurs; fused, wedgelike, socketless teeth; primitive third eye under skin of forehead	1
Chelonia	Turtles, tortoises, sea turtles	Armored reptiles with shell of bony plates to which vertebrae and ribs are fused; sharp, horny beak without teeth	340
Crocodylia	Crocodiles, alligators	Large reptiles with four-chambered heart and socketed teeth; anus is a longitudinal (lengthwise) slit; closest living relatives to birds	25
Ornithischia	Stegosaur	Dinosaurs with two pelvic bones facing backward, like a bird's pelvis; herbivores; legs under body	Extinct
Saurischia	Tyrannosaur	Dinosaurs with one pelvic bone facing forward, the other back, like a lizard's pelvis; both plant- and flesh-eaters; legs under body; birds evolved from Saurischian dinosaurs.	Extinct (except for bird descendants)
Pterosauria	Pterosaur	Flying reptiles; wings were made of skin stretched between fourth fingers and body; wingspans of early forms typically 60 cm, later forms nearly 8 m	Extinct
Plesiosaura	Plesiosaur	Barrel-shaped marine reptiles with sharp teeth and large, paddle-shaped fins; some had snakelike necks twice as long as their bodies	Extinct
Ichthyosauria	Ichthyosaur	Streamlined marine reptiles with many body similarities to sharks and modern fishes	Extinct

Reptiles exhibit three key characteristics

All living reptiles share certain fundamental characteristics, features they retain from the time when they replaced amphibians as the dominant terrestrial vertebrates. Among the most important are:

1. **Amniotic eggs.** Amphibians' eggs must be laid in water or a moist setting to avoid drying out. Most reptiles lay watertight eggs that contain a food source (the yolk) and a series of four membranes: the yolk sac, the amnion, the allantois, and the chorion (figure 35.20). Each membrane plays a role in making the egg an independent life-support system. All modern reptiles, as well as birds and mammals, show exactly this same pattern of membranes within the egg. These three classes are called **amniotes.**

    The outermost membrane of the egg is the **chorion,** which lies just beneath the porous shell. It allows exchange of respiratory gases but retains water. The **amnion** encases the developing embryo within a fluid-filled cavity. The **yolk sac** provides food from the yolk for the embryo via blood vessels connecting to the embryo's gut. The **allantois** surrounds a cavity into which waste products from the embryo are excreted.
2. **Dry skin.** Most living amphibians have moist skin and must remain in moist places to avoid drying out. Reptiles have dry, watertight skin. A layer of scales covers their bodies, greatly reducing water loss. These scales develop as surface cells fill with keratin, the same protein that forms claws, fingernails, hair, and bird feathers.
3. **Thoracic breathing.** Amphibians breathe by squeezing their throat to pump air into their lungs; this limits their breathing capacity to the volume of their mouths. Reptiles developed pulmonary breathing, expanding and contracting the rib cage and diaphragm to suck air into the lungs and then force it out. The capacity of this system is limited only by the volume of the lungs.

Reptiles dominated the Earth for 250 million years

During the 250 million years that reptiles were the dominant large terrestrial vertebrates, a series of different reptile groups appeared and then disappeared.

Synapsids

An important feature of reptile classification is the presence and number of openings behind the eyes (figure 35.21). Reptiles' jaw muscles were anchored to these holes, which allowed them to bite more powerfully. The first group to rise to dominance were the **synapsids,** whose skulls had a hole behind the openings for the eyes.

Pelycosaurs, an important group of early synapsids, were dominant for 50 million years and made up 70% of all land vertebrates; some species weighed as much as 200 kg. With long, sharp, "steak knife" teeth, these pelycosaurs were the first land vertebrates to kill beasts their own size (figure 35.22).

About 250 MYA, pelycosaurs were replaced by another type of synapsid, the therapsids (figure 35.23). Some evidence indicates that they may have been endotherms, able to produce heat internally, and perhaps even possessed hair. This would have permitted therapsids to be far more active than other vertebrates of that time, when winters were cold and long.

For 20 million years, therapsids (also called "mammal-like reptiles") were the dominant land vertebrate, until they were largely replaced 230 MYA by another group of reptiles, the diapsids. Most therapsids became extinct 170 MYA, but one group survived and has living descendants today—the mammals.

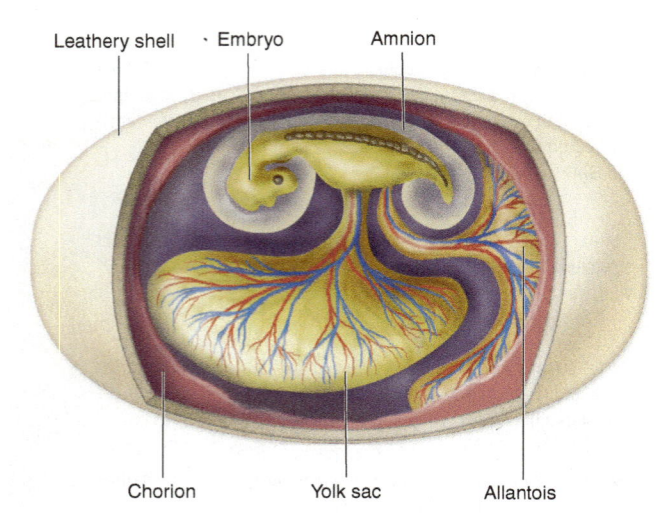

Figure 35.20 The watertight egg. The amniotic egg is perhaps the most important feature that allows reptiles to live in a wide variety of terrestrial habitats.

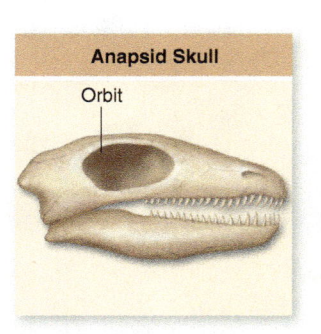

Figure 35.21 Skulls of reptile groups. Reptile groups are distinguished by the number of holes on the side of the skull behind the eye orbit: 0 (anapsids), 1 (synapsids), or 2 (diapsids). Turtles are the only living anapsids, although several extinct groups also had this condition.

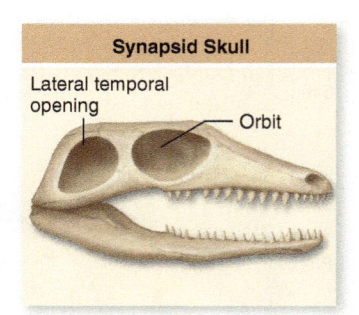

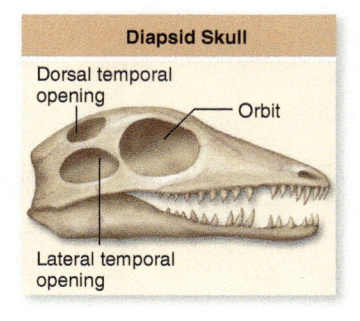

Figure 35.22 A pelycosaur. *Dimetrodon,* a 4-m-long carnivorous pelycosaur, had a dorsal sail that is thought to have been used to regulate body temperature by dissipating body heat or gaining it by basking.

Archosaurs

Diapsids have skulls with two pairs of holes on each side of the head, and like amphibians and early reptiles, they were ectotherms. A variety of different diapsids occurred in the Triassic period (213 to 248 MYA), but one group, the archosaurs, were of particular evolutionary significance because they gave rise to crocodiles, pterosaurs, dinosaurs, and birds (figure 35.24).

Among the early archosaurs were the largest animals the world had seen, up to that point, and the first land vertebrates to be bipedal—to stand and walk on two feet. By the end of the Triassic period, however, one archosaur group rose to prominence: the dinosaurs.

Dinosaurs evolved about 220 MYA. Unlike previous bipedal diapsids, their legs were positioned directly underneath their bodies (figure 35.25). This design placed the weight of the body directly over the legs, which allowed dinosaurs to run with great speed and agility. Subsequently, a number of types of dinosaur evolved to enormous size and reverted to a four-legged posture to support their massive weight. Other types—the theropods—became the most fearsome land predators the Earth has ever seen, and one theropod line evolved to become birds. Dinosaurs went on to become the most successful of all land vertebrates, dominating for more than 150 million years. All dinosaurs, except their bird descendants, became extinct rather abruptly 65 MYA, apparently as a result of an asteroid's impact.

Important characteristics of modern reptiles

As you might imagine from the structure of the amniotic egg, reptiles and other amniotes do not practice external fertilization as most amphibians do. Sperm would be unable to penetrate the membrane barriers protecting the egg. Instead, the male places sperm inside the female, where sperm fertilizes the egg before the protective membranes are formed. This is called internal fertilization.

Figure 35.24 An early archosaur. *Euparkeria* had rows of bony plates along the sides of the backbone, as seen in modern crocodiles and alligators.

Figure 35.23 A therapsid. This small, weasel-like cynodont therapsid, *Megazostrodon,* may have had fur. Living in the late Triassic period, this therapsid is so similar to modern mammals that some paleontologists consider it the first mammal.

Figure 35.25 Mounted skeleton of *Afrovenator*. This bipedal carnivore was about 30 feet long and lived in Africa about 130 MYA.

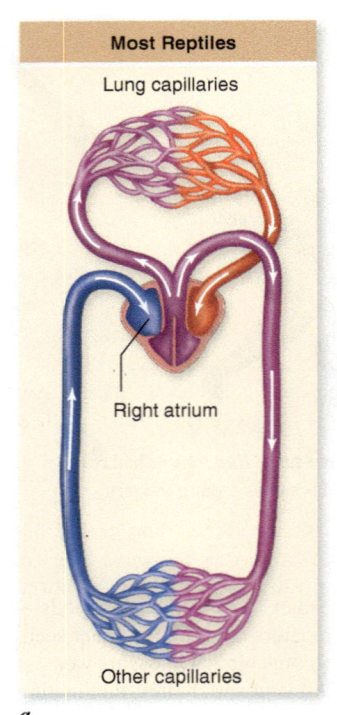

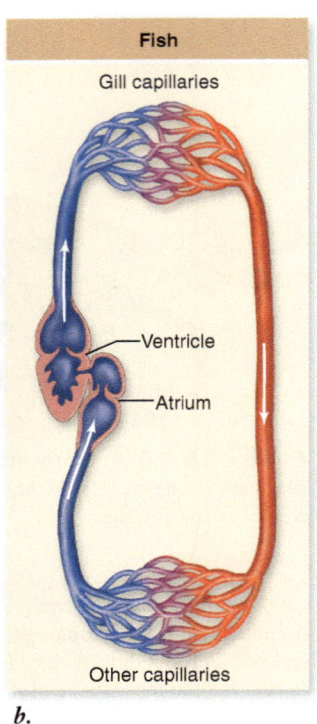

Figure 35.26 A comparison of reptile and fish circulation. *a.* In most reptiles, oxygenated blood *(red)* is repumped after leaving the lungs, and circulation to the rest of the body remains vigorous. *b.* The blood in fishes flows from the gills directly to the rest of the body, resulting in slower circulation.

The circulatory system of reptiles is an improvement over that of fish and amphibians, providing oxygen to the body more efficiently (figure 35.26; see chapter 49). The improvement is achieved by extending the septum within the heart from the atrium partway across the ventricle. This septum creates a partial wall that tends to lessen mixing of oxygen-poor blood with oxygen-rich blood within the ventricle. In crocodiles, the septum completely divides the ventricle, creating a four-chambered heart, just as it does in birds and mammals (and probably did in dinosaurs).

All living reptiles are **ectothermic,** obtaining their heat from external sources. In contrast, **endothermic** animals are able to generate their heat internally (see chapter 42). Although they are ectothermic, that does not mean that reptiles cannot control their body temperature. Many species are able to precisely regulate their temperature by moving in and out of sunlight. In this way, some desert lizards can keep their bodies at a constant temperature throughout the course of an entire day. Of course, on cloudy days, or for species that live in shaded habitats, such thermoregulation is not possible, and in these cases, body temperature is the same as the temperature of the surrounding environment (see chapter 55).

Modern reptiles belong to four orders

The four surviving orders of reptiles contain more than 10,000 species. Reptiles occur worldwide except in the coldest regions, where it is impossible for ectotherms to survive. Reptiles are among the most numerous and diverse of terrestrial vertebrates.

Order Chelonia: Turtles and tortoises

The order Chelonia (figure 35.27*a*) consists of about 340 species of turtles (most of which are aquatic) and tortoises (which are

Figure 35.27 Living orders of reptiles. *a.* **Chelonia.** The Indian tent turtle, *Pangshura tentoria (left)* is shown basking, an effective means by which many ectotherms can precisely regulate their body temperature. The domed shells of tortoises, such as this Sri Lankan star tortoise, *Geochelone elegans,* provide protection against predators for these entirely terrestrial chelonians. *b.* **Tuatara (*Sphenodon punctatus*).** The sole living member of the ancient group Rhynchocephalia. Although they look like lizards, the common ancestor of rhynchocephalians and lizards diverged more than 250 MYA. *c.* **Squamata.** A collared lizard, *Crotaphytus collaris,* is shown left, and a smooth green snake, *Liochlorophis vernalis,* on the right. *d.* **Crocodylia.** Most crocodilians, such as the crocodile, *Crocodylus acutus,* and the gharial, *Gavialis gangeticus (right),* resemble birds and mammals in having four-chambered hearts; all other living reptiles have three-chambered hearts. Like birds, crocodiles are more closely related to dinosaurs than to any of the other living reptiles.

terrestrial). Turtles and tortoises lack teeth but have sharp beaks. They differ from all other reptiles because their bodies are encased within a protective shell. Many of them can pull their head and legs into the shell as well, for total protection from predators.

The shell consists of two basic parts. The carapace is the dorsal covering, and the plastron is the ventral portion. The vertebrae and ribs of most turtle and tortoise species are fused to the inside of the carapace, and all of the support for muscle attachment comes from the shell.

Whereas most tortoises have a dome-shaped shell into which they can retract their head and limbs, water-dwelling turtles have a streamlined, disc-shaped shell that permits rapid turning in water. Freshwater turtles have webbed toes; in marine turtles, the forelimbs have evolved into flippers. Just as toads are not a monophyletic group, but rather a number of frog groups that have independently evolved to adapt to living in dry conditions, tortoises also are not monophyletic; rather, the high domed shells and robust legs have evolved a number of times in turtles adapted to living entirely on land.

Although marine turtles spend their lives at sea, they must return to land to lay their eggs. Many species migrate long distances to do this. Atlantic green turtles *(Chelonia mydas)* migrate from their feeding grounds off the coast of Brazil to Ascension Island in the middle of the South Atlantic—a distance of more than 2000 km—to lay their eggs on the same beaches where they themselves hatched.

Order Rhynchocephalia: Tuataras

Today, the order Rhynchocephalia contains a single species, the tuatara, a large, lizard-like animal about half a meter long (figure 35.27*b*). The only place in the world where this endangered species is found is a cluster of small islands off the coast of New Zealand. The limited diversity of modern rhynchocephalians belies their rich evolutionary past: In the Triassic period, rhynchocephalians experienced a great adaptive radiation, producing many species that differed greatly in size and habitat.

An unusual feature of the tuatara (and some lizards) is the inconspicuous "third eye" on the top of its head, called a parietal eye. Concealed under a thin layer of scales, the eye has a lens and a retina and is connected by nerves to the brain. Why have an eye, if it is covered up? The parietal eye may function to alert the tuatara when it has been exposed to too much sunlight, protecting it against overheating. Unlike most reptiles, tuataras are most active at low temperatures. They burrow during the day and feed at night on insects, worms, and other small animals.

Rhynchocephalians are the closest relatives of snakes and lizards, with whom they form the group **Lepidosauria.**

Order Squamata: Lizards and snakes

The order Squamata (figure 35.27*c*) includes 6200 species of lizards and about 3500 species of snakes. A distinguishing characteristic of this order is the presence of two copulatory organs in the male. In addition, changes to the morphology of the head and jaws allow greater strength and mobility. Most lizards and snakes are carnivores, preying on insects and small animals, and these improvements in jaw design have made a major contribution to their evolutionary success.

Snakes, which evolved from a lizard ancestor, are characterized by the lack of limbs, movable eyelids, and external ears, as well as a great number of vertebrae (sometimes more than 300). Limblessness has actually evolved more than a dozen times in lizards; snakes are simply the most extreme case of this evolutionary trend.

Common lizards include iguanas, chameleons, geckos, and anoles. Most are small, measuring less than 0.3 m in length. The largest lizards belong to the monitor family. The largest of all monitor lizards is the Komodo dragon of Indonesia, which reaches 3 m in length and can weigh more than 100 kg. Snakes also vary in length from only a few centimeters to more than 10 m.

Many lizards and snakes rely on agility and speed to catch prey and elude predators. Only two species of lizard are venomous, the Gila monster of the southwestern United States and the beaded lizard of western Mexico. Similarly, most species of snakes are nonvenomous, although some, like vipers and cobras, can be deadly.

Many lizards, including anoles, skinks, and geckos, have the ability to lose their tails and then regenerate a new one. This ability allows these lizards to escape from predators.

Order Crocodylia: Crocodiles and alligators

The order Crocodylia is composed of 25 species of large, primarily aquatic reptiles (figure 35.27*d*). In addition to crocodiles and

Order Squamata

Order Crocodylia

c. *d.*

alligators, the order includes two less familiar animals: the caimans and the gavials, or gharials. Although all crocodilians are fairly similar in appearance today, much greater diversity existed in the past, including species that were entirely terrestrial and others that achieved a total length in excess of 50 feet.

Crocodiles are largely nocturnal animals that live in or near water in tropical or subtropical regions of Africa, Asia, and the Americas. The American crocodile (*Crocodylus acutus*) is found in southern Florida, Cuba, and throughout tropical Central America. Nile crocodiles (*Crocodylus niloticus*) and estuarine crocodiles (*Crocodylus porosus*) can grow to enormous size and are responsible for many human fatalities each year.

There are only two species of alligators: one living in the southern United States (*Alligator mississippiensis*) and the other a rare endangered species living in China (*Alligator sinensis*). Caimans, which resemble alligators, are native to Central and South America. Gharials are a group of fish-eating crocodilians with long, slender snouts that live only in India and Nepal.

All crocodilians are carnivores. They generally hunt by stealth, waiting in ambush for prey, and then attacking with a quick lunge. Their bodies are well adapted for this form of hunting with eyes on top of their heads and their nostrils on top of their snouts, so they can see and breathe while lying quietly submerged in water. They have enormous mouths, studded with sharp teeth, and very strong necks. A valve in the back of the mouth prevents water from entering the air passage when a crocodilian feeds underwater.

In many ways, crocodiles resemble birds far more than they do other living reptiles. For example, crocodiles build nests and care for their young (traits they share with at least some dinosaurs), and they have a four-chambered heart, as birds do. Why are crocodiles more similar to birds than to other living reptiles? Most biologists agree that birds are in fact the direct descendants of dinosaurs. Both crocodiles and birds are more closely related to dinosaurs, and to each other, than they are to lizards and snakes.

Learning Outcomes Review 35.7

Reptiles have a hard, scaly skin that minimizes water loss, thoracic breathing, and an enclosed, amniotic egg that does not need to be laid in water. Synapsids had a single hole in the skull behind the eyes, and included ancestors of mammals; diapsids had two holes and are ancestors of modern reptiles and birds. The four living orders of reptiles include the turtles and tortoises, tuataras, lizards and snakes, and crocodiles.

■ *How do reptile eggs differ from those of amphibians?*

35.8 Birds

Learning Outcomes

1. Name the key characteristics of birds.
2. Explain why some consider birds to be one type of reptile.

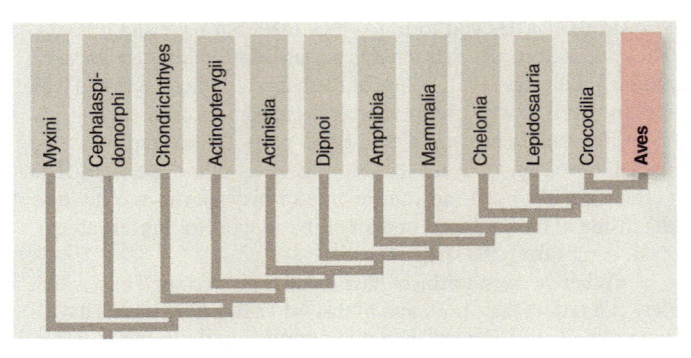

The success of birds lies in the development of a structure unique in the animal world—the feather. Developed from reptilian scales, feathers are the ideal adaptation for flight, serving as lightweight airfoils that are easily replaced if damaged (unlike the vulnerable skin wings of bats and the extinct pterosaurs). Today, birds (class **Aves**) are as diverse as reptiles, with more than 10,000 species, more than half of which occur in the order Passeriformes (see figure 35.32; see also table 35.4).

Key characteristics of birds are feathers and a lightweight skeleton

Modern birds lack teeth and have only vestigial tails, but they still retain many reptilian characteristics. For instance, birds lay amniotic eggs. Also, reptilian scales are present on the feet and lower legs of birds. Two primary characteristics distinguish birds from living reptiles:

1. **Feathers.** Feathers are modified reptilian scales made of keratin, just like hair and scales. Feathers serve two functions: providing lift for flight and conserving heat. The structure of feathers combines maximum flexibility and strength with minimum weight (figure 35.28).

   Feathers develop from tiny pits in the skin called follicles. In a typical flight feather, a shaft emerges from the follicle, and pairs of vanes develop from its opposite sides.

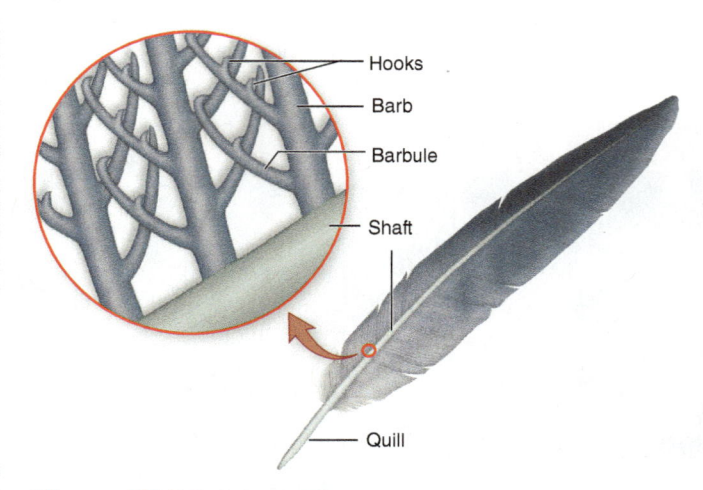

Figure 35.28 A feather. The enlargement shows how the secondary branches and barbs of the vanes are linked together by microscopic barbules.

TABLE 35.4 Major Orders of Birds

Order	Typical Examples	Key Characteristics	Approximate Number of Living Species
Passeriformes	Crows, mockingbirds, robins, sparrows, starlings, warblers	Well-developed vocal organs; perching feet; dependent young	5300
Apodiformes	Hummingbirds, swifts	Fast fliers; short legs; small bodies; rapid wing beat	450
Piciformes	Honeyguides, toucans, woodpeckers	Grasping feet; chisel-like, sharp bills can break down wood	400
Psittaciformes	Cockatoos, parrots	Large, powerful bills for crushing seeds; well-developed vocal organs	375
Charadriiformes	Auks, gulls, plovers, sandpipers, terns	Shorebirds; long, stiltlike legs; slender, probing bills	350
Columbiformes	Doves, pigeons	Perching feet; rounded, stout bodies	310
Falconiformes	Eagles, falcons, hawks	Carnivorous; keen vision; sharp, pointed beaks for tearing flesh; active during the day	290
Galliformes	Chickens, grouse, pheasants, quail	Often limited flying ability; rounded bodies	290
Gruiformes	Bitterns, coots, cranes, rails	Marsh birds; long, stiltlike legs; diverse body shapes; marsh-dwellers	209
Anseriformes	Ducks, geese, swans	Waterfowl; webbed toes; broad bill with filtering ridges	150
Strigiformes	Barn owls, screech owls	Nocturnal birds of prey; strong beaks; powerful feet	146
Ciconiiformes	Herons, ibises, storks	Waders; long-legged; large bodies	114
Procellariformes	Albatrosses, petrels	Seabirds; tube-shaped bills; capable of flying for long periods of time	104
Sphenisciformes	Emperor penguins, crested penguins	Marine; modified wings for swimming; flightless; found only in southern hemisphere; thick coats of insulating feathers	18
Dinornithiformes	Kiwis	Flightless; small; confined to New Zealand	5
Struthioniformes	Ostriches	Powerful running legs; flightless; only two toes; very large	1

At maturity, each vane has many branches called barbs. The barbs, in turn, have many projections called barbules that are equipped with microscopic hooks. These hooks link the barbs to one another, giving the feather a continuous surface and a sturdy but flexible shape.

Like scales, feathers can be replaced. Among living animals, feathers are unique to birds. Recent fossil finds suggest that some dinosaurs may have had feathers.

2. **Flight skeleton.** The bones of birds are thin and hollow. Many of the bones are fused, making the bird skeleton more rigid than a reptilian skeleton. The fused sections of backbone and of the shoulder and hip girdles form a sturdy frame that anchors muscles during flight. The power for active flight comes from large breast muscles that can make up 30% of a bird's total body weight. They stretch down from the wing and attach to the breastbone, which is greatly enlarged and bears a prominent keel for muscle attachment. Breast muscles also attach to the fused collarbones that form the so-called wishbone. No other living vertebrates have a fused collarbone or a keeled breastbone.

Birds arose about 150 MYA

A 150-million-year-old fossil of the first known bird, *Archaeopteryx* (figure 35.29; see also figure 21.10) was found in 1862 in a limestone quarry in Bavaria, the impression of its feathers stamped clearly into the rocks. The skeleton of *Archaeopteryx* shares many features with small theropod dinosaurs. About the size of a crow, *Archeopteryx* had a skull with teeth, and very few of its bones were fused to one another. Its bones are thought to have been solid, not hollow like a bird's. Also, it had a long, reptilian tail and no enlarged breastbone such as modern birds use to anchor flight muscles. Finally, the skeletal structure of the forelimbs was nearly identical to those of theropods.

Figure 35.29 *Archaeopteryx.* Closely related to its ancestors among the bipedal dinosaurs, the crow-sized *Archaeopteryx* lived in the forests of central Europe 150 MYA. The true feather colors of *Archaeopteryx* are not known.

Because of its many dinosaur features, several *Archaeopteryx* fossils in which the feathers had not been preserved were originally misclassified as *Compsognathus*, a small theropod dinosaur of similar size. What makes *Archaeopteryx* distinctly avian is the presence of feathers on its wings and tail.

The remarkable similarity of *Archaeopteryx* to *Compsognathus* has led almost all paleontologists to conclude that *Archaeopteryx* is the direct descendant of dinosaurs—indeed, that today's birds are "feathered dinosaurs." Some even speak flippantly of "carving the dinosaur" at Thanksgiving dinner. The recent discovery of fossils of a feathered dinosaur in China lends strong support to this inference. The dinosaur *Caudipteryx*, for example, is clearly intermediate between *Archaeopteryx* and dinosaurs, having large feathers on its tail and arms but also many features of dinosaurs like *Velociraptor* (figure 35.30; see also figure 23.12). Because the arms of *Caudipteryx* were

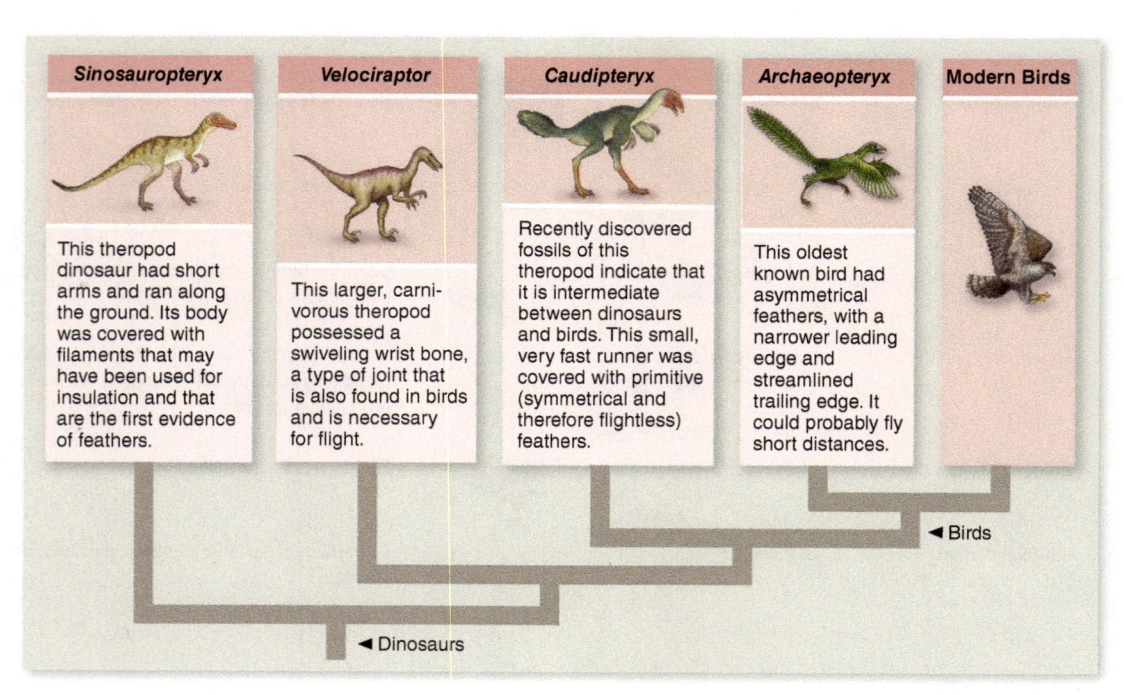

Figure 35.30 The evolutionary path to the birds. Almost all paleontologists now accept the theory that birds are the direct descendants of theropod dinosaurs.

Sinosauropteryx: This theropod dinosaur had short arms and ran along the ground. Its body was covered with filaments that may have been used for insulation and that are the first evidence of feathers.

Velociraptor: This larger, carnivorous theropod possessed a swiveling wrist bone, a type of joint that is also found in birds and is necessary for flight.

Caudipteryx: Recently discovered fossils of this theropod indicate that it is intermediate between dinosaurs and birds. This small, very fast runner was covered with primitive (symmetrical and therefore flightless) feathers.

Archaeopteryx: This oldest known bird had asymmetrical feathers, with a narrower leading edge and streamlined trailing edge. It could probably fly short distances.

Modern Birds

Figure 35.31 A fossil bird from the early Cretaceous. *Confuciornis* had long tail feathers. Some fossil specimens of this species lack the long tail feathers, suggesting that this trait was present in only one sex, as in some modern birds.

too short to use as wings, feathers probably didn't evolve for flight, but instead served as insulation, much as fur does for mammals, or perhaps as decorations used in sexual displays. This is an example of the early stages of a trait evolving for one reason, only subsequently to be further elaborated evolutionarily by natural selection for some other purpose, as described in chapter 23.

Flight is an ability certain kinds of dinosaurs achieved as they evolved longer arms. We call these dinosaurs birds. Despite their close affinity to dinosaurs, birds exhibit three evolutionary novelties: feathers, hollow bones, and physiological mechanisms such as superefficient lungs that permit sustained, powered flight.

By the early Cretaceous period, only a few million years after *Archaeopteryx* lived, a diverse array of birds had evolved, with many of the features of modern birds. Fossils in Mongolia, Spain, and China discovered within the last few years reveal a diverse collection of toothed birds with the hollow bones and breastbones necessary for sustained flight (figure 35.31). Other fossils reveal highly specialized, flightless diving birds. These diverse birds shared the skies with pterosaurs for 70 million years until the flying reptiles went extinct at the end of the Cretaceous.

Because the impression of feathers is rarely fossilized and modern birds have hollow, delicate bones, the fossil record of birds is incomplete. Relationships among the many families of modern birds are mostly inferred from studies of the anatomy and degree of DNA similarity among living birds.

Modern birds are diverse but share several characteristics

You can tell a great deal about the habits and food of a bird by examining its beak and feet. For instance, carnivorous birds such as owls have curved talons for seizing prey and sharp beaks for tearing apart the prey's flesh. The beaks of ducks are flat for shoveling through mud, whereas the beaks of finches are short, thick seed-crushers.

Many adaptations enabled birds to cope with the heavy energy demands of flight, including respiratory and circulatory adaptations and endothermy.

Efficient respiration

Flight muscles consume an enormous amount of oxygen during active flight. The reptilian lung has a limited internal surface area,

Figure 35.32 Diversity of *Passeriformes*, the largest order of birds. *a.* Variable Sunbird, *Cinnyris venustus*; (*b*) Indigo Bunting, *Passerina cyanea*; (*c*) Blue jay, *Cyanocitta cristata*; (*d*) Blue-winged Pitta, *Pitta moluccensis*.

not nearly enough to absorb all the oxygen needed. Mammalian lungs have a greater surface area, but bird lungs satisfy this challenge with a radical redesign.

When a bird inhales, the air goes past the lungs to a series of air sacs located near and within the hollow bones of the back; from there, the air travels to the lungs and then to a set of anterior air sacs before being exhaled. Because air passes all the way through the lungs in a single direction, gas exchange is highly efficient. Respiration in birds is described in more detail in chapter 48.

Efficient circulation

The revved-up metabolism needed to power active flight also requires very efficient blood circulation, so that the oxygen captured by the lungs can be delivered to the flight muscles quickly. In the heart of most living reptiles, oxygen-rich blood coming from the lungs mixes with oxygen-poor blood returning from the body because the wall dividing the ventricle into two chambers is not complete. In birds, the wall dividing the ventricle is complete, and the two blood circulations do not mix—so flight muscles receive fully oxygenated blood (see chapter 49).

In comparison with reptiles and most other vertebrates, birds have a rapid heartbeat. A hummingbird's heart beats about 600 times a minute, and an active chickadee's heart beats 1000 times a minute. In contrast, the heart of the large, flightless ostrich averages 70 beats per minute—the same rate as the human heart.

Endothermy

Birds, like mammals, are endothermic (an example of convergent evolution). Many paleontologists believe the dinosaurs from which birds evolved were endothermic as well. Birds maintain body temperatures significantly higher than those of most mammals, ranging from 40° to 42°C (human body temperature is 37°C). Feathers provide excellent insulation, helping to conserve body heat.

The high temperatures maintained by endothermy permit metabolism in the bird's flight muscles to proceed at a rapid pace, providing the ATP necessary to drive rapid muscle contraction.

Learning Outcomes Review 35.8

Birds have the greatest diversity of species of all terrestrial vertebrates. Archaeopteryx, the oldest fossil bird, exhibited many traits shared with theropod dinosaurs. Key features of birds are feathers and a lightweight, hollow skeleton; additional features include auxiliary air sacs and a four-chambered heart.

■ What traits do birds share with reptiles?

35.9 Mammals

Learning Outcomes

1. Describe the characteristics of mammals.
2. Compare the three groups of living mammals.

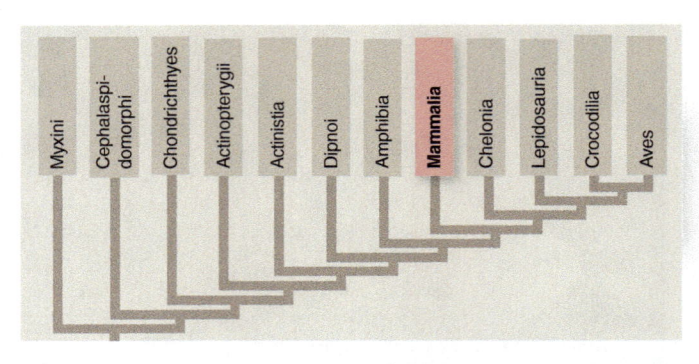

There are about 5000 living species of mammals (class Mammalia), fewer than the number of fishes, amphibians, reptiles, or birds. Most large, land-dwelling vertebrates are mammals. When we look out over an African plain, we see the big mammals—the lions, zebras, gazelles, and antelope. But the typical mammal is not that large. Of the 5000 species of mammals, almost 4000 are rodents, bats, shrews, or moles.

Mammals have hair, mammary glands, and other characteristics

Mammals are distinguished from all other classes of vertebrates by two fundamental characteristics—hair and mammary glands—and are marked by several other notable features:

1. **Hair.** All mammals have hair. Even apparently hairless whales and dolphins grow sensitive bristles on their snouts. The evolution of fur and the ability to regulate body temperature enabled mammals to invade colder climates that ectothermic reptiles do not inhabit. Mammals are endothermic animals and typically maintain body temperatures higher than the temperature of their surroundings. The dense undercoat of many mammals reduces the amount of body heat that escapes.

    Another function of hair is camouflage. The coloration and pattern of a mammal's coat usually matches its background. A little brown mouse is practically invisible against the brown leaf litter of a forest floor, and the orange and black stripes of a Bengal tiger disappear against the orange-brown color of the tall grass in which it hunts. Hairs also function as sensory structures. The whiskers of cats and dogs are stiff hairs that are very sensitive to touch. Mammals that are active at night or live underground often rely on their whiskers to locate prey or to avoid colliding with objects. Finally, hair can serve as a defensive weapon. Porcupines and hedgehogs protect themselves with long, sharp, stiff hairs called quills.

    Unlike feathers, which evolved from modified reptilian scales, mammalian hair is a completely different form of skin structure. An individual mammalian hair is a long, protein-rich filament that extends like a stiff thread from a bulblike foundation beneath the skin known as a hair follicle. The filament is composed mainly of dead cells filled with the fibrous protein keratin.

2. **Mammary glands.** All female mammals possess mammary glands that can secrete milk. Newborn mammals, born without teeth, suckle this milk as their primary food. Even baby whales are nursed by their mother's milk. Milk is a very high-calorie food, important because of the high energy needs of a rapidly growing newborn mammal. About 50% of the energy in the milk comes from fat.
3. **Endothermy.** As stated previously, mammals are endothermic, a crucial adaptation that has allowed them to be active at any time of the day or night and to colonize extreme environments, from deserts to ice fields. Also, more efficient blood circulation provided by the four-chambered heart (see chapter 49) and more efficient respiration provided by the *diaphragm* (a special sheet of muscles below the rib cage that aids breathing; see chapter 48) make possible the higher metabolic rate on which endothermy depends.
4. **Placenta.** In most mammal species, females carry their developing young internally in a uterus, nourishing them through the placenta, and give birth to live young. The **placenta** is a specialized organ that brings the bloodstream of the fetus into close contact with the bloodstream of the mother (figure 35.33). Food, water, and oxygen can pass across from mother to child, and wastes can pass over to the mother's blood and be carried away.

In addition to these main characteristics, the mammalian lineage gave rise to several other adaptations in certain groups. These include specialized teeth, the ability of grazing animals to digest plants, hooves and horns made of keratin, and adaptations for flight in the bats.

Specialized teeth

Mammals have different types of teeth that are highly specialized to match particular eating habits (figure 35.34). It is usually possible to determine a mammal's diet simply by examining its teeth. A dog's long canine teeth, for example, are well suited for biting and holding prey, and some of its premolar and molar teeth are triangular and sharp for ripping off chunks of flesh.

In contrast, large herbivores such as deer lack canine teeth; instead, deer clip off mouthfuls of plants with flat, chisel-like incisors on its lower jaw. The deer's molars are large and covered with ridges to effectively shred and grind tough plant tissues.

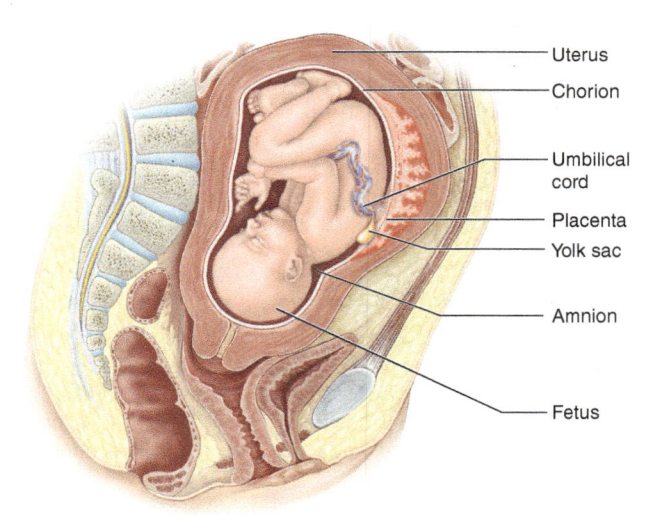

Figure 35.33 The placenta. The placenta is characteristic of the largest group of mammals, the placental mammals. It evolved from membranes in the amniotic egg. The umbilical cord evolved from the allantois. The chorion, or outermost part of the amniotic egg, forms most of the placenta itself. The placenta serves as the provisional lungs, intestine, and kidneys of the fetus, without ever mixing maternal and fetal blood.

? Inquiry question How similar is the placenta to the egg of a reptile or bird? What can explain that similarity?

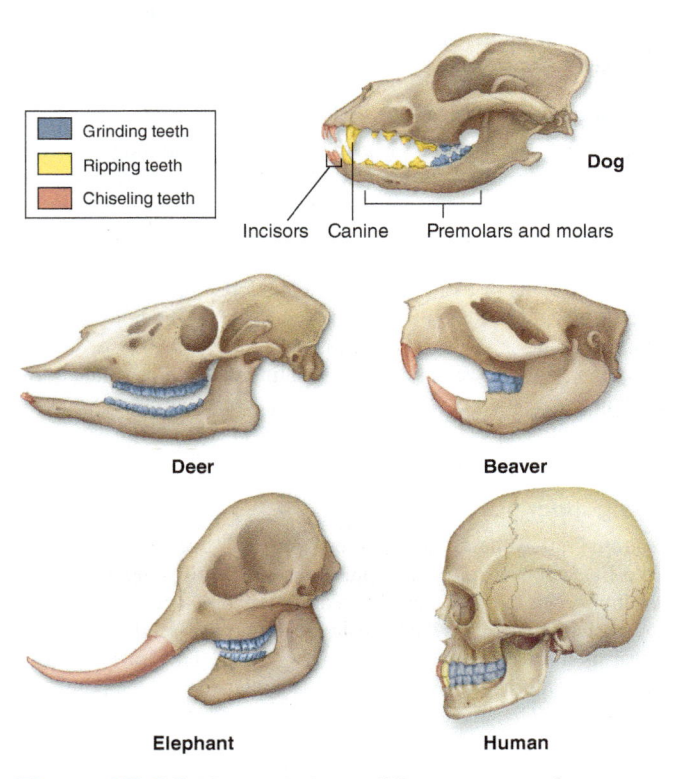

Figure 35.34 Mammals have different types of specialized teeth. Carnivores, such as dogs, have canine teeth that are able to rip food; some of the premolars and molars in dogs are also ripping teeth. Herbivores, such as deer, have incisors to chisel off vegetation and molars designed to grind up the plant material. In the beaver, the chiseling incisors dominate. In the elephant, the incisors have become specialized weapons, and molars grind up vegetation. Humans are omnivores; we have all three types: grinding, ripping, and chiseling teeth.

Digestion of plants

Most mammals are herbivores, eating mostly or only plants. Cellulose forms the bulk of a plant's body and is a major source of food for mammalian herbivores. Mammals do not have the necessary enzymes, however, for breaking the links between glucose molecules in cellulose. Herbivorous mammals rely on a mutualistic partnership with bacteria in their digestive tracts that have the necessary cellulose-splitting enzymes (see chapter 47).

Mammals such as cows, buffalo, antelopes, goats, deer, and giraffes have huge, four-chambered fermentation vats derived from the esophagus and stomach. The first chamber is the largest and holds a dense population of cellulose-digesting bacteria. Chewed plant material passes into this chamber, where the bacteria attack the cellulose. The material is then digested further in the other three chambers.

Rodents, horses, rabbits, and elephants, by contrast, have relatively small stomachs, and instead digest plant material in their large intestine, like a termite. The bacteria that actually carry out the digestion of the cellulose live in a pouch called the cecum that branches from the end of the small intestine.

Even with these complex adaptations for digesting cellulose, a mouthful of plant is less nutritious than a mouthful of meat. Herbivores must consume large amounts of plant material to gain sufficient nutrition. An elephant eats 135 to 150 kg of plant foods each day.

Development of hooves and horns

Keratin, the protein of hair, is also the structural building material in claws, fingernails, and hooves. Hooves are specialized keratin pads on the toes of horses, cows, sheep, and antelopes. The pads are hard and horny, protecting the toe and cushioning it from impact.

The horns of cattle, sheep, and antelope are composed of a core of bone surrounded by a sheath of keratin. The bony core is attached to the skull, and the horn is not shed.

Deer antlers are made not of keratin, but of bone. Male deer grow and shed a set of antlers each year. As they grow during the summer, antlers are covered by a thin layer of skin known as velvet that provides a blood supply for the growing bone.

Flying mammals: Bats

Bats are the only mammals capable of powered flight (figure 35.35). Like the wings of birds and pterosaurs, bat wings are modified forelimbs, but even though these three types of vertebrates have convergently evolved wings, they have done so by modifying their forearms in different ways (see figure 46.22). The bat wing is a leathery membrane of skin and muscle stretched over the bones of four fingers. The edges of the membrane attach to the side of the body and to the hind leg. When resting, most bats prefer to hang upside down by their toe claws.

After rodents, bats are the second largest order of mammals. They have been a particularly successful group because many species have been able to utilize a food resource that most birds do not use—night-flying insects.

How do bats navigate in the dark? Late in the 18th century, the Italian biologist Lazzaro Spallanzani showed that a blinded bat could fly without crashing into things and still capture insects. Clearly another sense other than vision was being used by bats to navigate in the dark. When Spallanzani plugged the ears of a bat, however, it was unable to navigate and collided with objects. Spallanzani concluded that bats "hear" their way through the night world.

Figure 35.35 Greater horseshoe bat, *Rhinolophus ferrumequinum*. Bats are the only mammal capable of true flight.

Mammals diverged about 220 MYA

Mammals have been around since the time of the dinosaurs, about 220 MYA. Tiny, shrewlike creatures that lived in trees eating insects, the earliest mammals were only a minor element in a land that quickly came to be dominated by dinosaurs. Fossils reveal that these early mammals had large eye sockets, evidence that they may have been active at night. Early mammals also had a single lower jawbone. The fossil record shows a change in therapsids (the ancestors of mammals) from the reptile lower jaw, having several bones, to a jaw closer to the mammalian type. Two of the bones forming the therapsid jaw joint moved into the middle ear of mammals, linking with a bone already there to produce a three-bone structure that amplifies sound better than the reptilian ear (see figure 44.7).

The Age of Mammals

At the end of the Cretaceous period, 65 MYA, the dinosaurs and numerous other land and marine animals became extinct, but mammals survived, possibly because of the insulation their fur provided. In the Tertiary period (lasting from 65 MYA to 2 MYA), mammals rapidly diversified, taking over many of the ecological roles once dominated by dinosaurs.

Mammals reached their maximum diversity late in the Tertiary period, about 15 MYA. At that time, tropical conditions existed over much of the world. During the last 15 million years, world climates have changed, and the area covered by tropical habitats has decreased, causing a decline in the total number of mammalian species.

Modern mammals are placed into three groups

For 155 million years, while the dinosaurs flourished, mammals were a minor group of small insectivores and herbivores. The most primitive mammals were members of the subclass Prototheria.

Most prototherians were small and resembled modern shrews. The only prototherians surviving today are the monotremes, the only egg-laying mammals.

The other major mammalian group is the subclass **Theria**. Therians are viviparous (that is, their young are born alive). The two living therian groups are marsupials, or pouched mammals (including kangaroos, opossums, and koalas), and the placental mammals (dogs, cats, humans, horses, and most other mammals).

Monotremes: Egg-laying mammals

The duck-billed platypus and two species of echidna are the only living **monotremes** (figure 35.36a). Among living mammals, only monotremes lay shelled eggs. The structure of their shoulder and pelvis is more similar to that of the early reptiles than to any other living mammal. Also like reptiles, monotremes have a cloaca, a single opening through which feces, urine, and reproductive products leave the body.

Despite the retention of some reptilian features, monotremes have the diagnostic mammalian characters: a single bone on each side of the lower jaw, fur, and mammary glands. Young monotremes drink their mother's milk after they hatch from eggs. Females lack well-developed nipples; instead, the milk oozes onto the mother's fur, and the babies lap it off with their tongues.

The platypus, found only in Australia, lives much of its life in the water and is a good swimmer. It uses its bill much as a duck does, rooting in the mud for worms and other soft-bodied animals. However, unlike ducks, the platypus has electroreceptors in its bill that can detect the electrical discharges produced by muscle contractions in their prey, helping them to locate their next meal. Echidnas of Australia (*Tachyglossus aculeatus,* the short-nosed echidna) and New Guinea (*Zaglossus bruijni,* the long-nosed echidna) have very strong, sharp claws, which they use for burrowing and digging. The echidna probes with its snout for insects, especially ants and termites.

Marsupials: Pouched mammals

The major difference between **marsupials** (figure 35.36b) and other mammals is their pattern of embryonic development. In marsupials, a fertilized egg is surrounded by chorion and amniotic membranes, but no shell forms around the egg as it does in monotremes. During most of its early development, the marsupial embryo is nourished by an abundant yolk within the egg. Shortly before birth, a short-lived placenta forms from the chorion membrane. Soon after, sometimes within eight days of fertilization, the embryonic marsupial is born. It emerges tiny and hairless, and crawls into the marsupial pouch, where it latches onto a mammary-gland nipple and continues its development.

Marsupials evolved shortly before placental mammals, about 125 MYA. Today, most species of marsupials live in Australia and South America, areas that have undergone long periods of geographic isolation. Marsupials in Australia and New Guinea have diversified to fill ecological positions occupied by placental mammals elsewhere in the world (see figure 21.18). The placental mammals in Australia and New Guinea today arrived relatively recently and include some introduced by humans. The only marsupial found in North America is the Virginia opossum, which has migrated north from Central America within the last 3 million years.

Figure 35.36 Today's mammals. *a.* Monotremes, the short-nosed echidna, *Tachyglossus aculeatus (left),* and the duck-billed platypus, *Ornithorhynchus anatinus (right);* **(b)** marsupials, the red kangaroo, *Macropus rufus (left)* and the opossum, *Didelphis virginiana, (right);* **(c)** placental mammals, the lion, *Panthera leo (left)* and the bottle-nosed dolphin, *Tursiops truncatus (right).*

Placental mammals

A placenta that nourishes the embryo throughout its entire development forms in the uterus of placental mammals (figure 35.36c). Most species of mammals living today, including humans, are in this group, which includes all but two orders of living mammals (although some scientists recognize four orders of marsupials, rather than one). Table 35.5 shows some of these orders. They are a very diverse group, ranging in size from 1.5-g pygmy shrews to 100,000-kg whales.

Early in the course of embryonic development, the placenta forms. Both fetal and maternal blood vessels are abundant in the placenta, and substances can be exchanged efficiently between the bloodstreams of mother and offspring (see figure 35.33). The fetal placenta is formed from the membranes of the chorion and allantois. In placental mammals, unlike in marsupials, the young undergo a considerable period of development before they are born.

TABLE 35.5 Major Orders of Placental Mammals

Order	Typical Examples	Key Characteristics	Approximate Number of Living Species
Rodentia	Beavers, mice, porcupines, rats	Small plant-eaters; chisel-like incisor teeth	2277
Chiroptera	Bats	Flying mammals; primarily fruit- or insect-eaters; elongated fingers; thin wing membrane; mostly nocturnal; navigate by sonar	1240
Soricomorpha	Moles, shrews	Small, burrowing mammals; insect-eaters; spend most of their time underground	420
Carnivora	Bears, cats, raccoons, weasels, dogs	Carnivorous predators; teeth adapted for shearing flesh; no native families in Australia	280
Primates	Apes, humans, lemurs, monkeys	Primarily tree-dwellers; large brain size; binocular vision; opposable thumb	270
Artiodactyla	Cattle, deer, giraffes, pigs	Hoofed mammals with two or four toes most species herbivorous ruminants	220
Cetacea	Dolphins, porpoises, whales	Fully marine mammals; streamlined bodies; front limbs modified into flippers; no hindlimbs; blowholes on top of head; no hair except on muzzle	90
Lagomorpha	Rabbits, hares, pika	Four upper incisors (rather than the two seen in rodents); hind legs often longer than forelegs, an adaptation for jumping	80
Edentata	Anteaters, armadillos, sloths	Insectivorous; many are toothless, but some have small, peglike teeth	30
Perissodactyla	Horses, rhinoceroses, tapirs	Hoofed mammals with odd number of toes; herbivorous teeth adapted for chewing	17
Proboscidea	Elephants	Long-trunked herbivores; two upper incisors elongated as tusks; largest living land animal	2

Learning Outcomes Review 35.9

Mammals are the only animals with hair and mammary glands. Other mammalian specializations include endothermy, the placenta, a tooth design suited to diet, and specialized sensory systems. Today three subgroups of mammals are recognized: monotremes, which lay eggs; marsupials, which feed embryonic young in a marsupial pouch; and placental mammals, in which the placenta nourishes the embryo throughout its development.

- What features found in both mammals and birds are examples of convergent evolution?

35.10 Evolution of the Primates

Learning Outcomes

1. Describe the characteristics and major groups of primates.
2. List the distinguishing characteristics of hominids.
3. Explain the variations that form the basis for human races and why races do not represent evolutionarily distinct entities.

Primates are the mammalian group that gave rise to our own species. Primates evolved two distinct features that allowed them to succeed as arboreal (tree-dwelling) insectivores:

1. **Grasping fingers and toes.** Unlike the clawed feet of tree shrews and squirrels, primates have grasping hands and feet that enable them to grip tree limbs, hang from branches, seize food, and in some primates, use tools. The first digit in many primates, the thumb, is opposable, and at least some, if not all, of the digits have nails.
2. **Binocular vision.** Unlike the eyes of shrews and squirrels, which sit on each side of the head, the eyes of primates are shifted forward to the front of the face. This produces overlapping binocular vision that lets the brain judge distance precisely—important to an animal moving through the trees and trying to grab or pick up food items.

Other mammals—for example, carnivorous predators—have binocular vision, but only primates have both binocular vision and grasping hands, making them particularly well adapted to their arboreal environment.

Figure 35.37 A tarsier.
This Philippine Tarsier, *Tarsius syrichta*, is native to tropical Asia. It shows the characteristic features of primates: grasping fingers and toes and binocular vision.

The anthropoid lineage led to the earliest humans

Living primates are divided into three groups. Tarsiers, lemurs, and lorises used to be considered prosimians, but it is now realized that this group is paraphyletic, with tarsiers more closely related to monkeys and apes than they are to lemurs and lorises (figure 35.37). Tarsiers, lorises, and some lemurs are small and nocturnal, but other lemur species are larger and diurnal. Lemurs occur only on the island of Madagascar, where they experienced an adaptive radiation in the absence of competition from monkeys, which do not occur there.

Anthropoids

Anthropoids include monkeys, apes, and humans. Anthropoids are almost all diurnal—that is, active during the day—feeding mainly on fruits and leaves. Natural selection favored many changes in eye design, including color vision, that were adaptations to daytime foraging. An expanded brain governs the improved senses, with the braincase forming a larger portion of the head.

Anthropoids, like the diurnal lemurs, live in groups with complex social interactions. They tend to care for their young for prolonged periods, allowing for a long childhood of learning and brain development.

About 30 MYA, some anthropoids migrated to South America. Their descendants, known as the New World monkeys (figure 35.38a), are easy to identify: All are arboreal; they have flat, spreading noses; and many of them grasp objects with long, prehensile tails.

Figure 35.38
Anthropoids. *a.* New World Monkey, the squirrel monkey, *Saimiri oerstedii;* (*b*) Old World Monkey, the mandrill, *Mandrillus sphinx;* (*c*) hominoids, human, *Homo sapiens* (left) and gorilla, *Gorilla gorilla* (right).

a. New World Monkeys   *b.* Old World Monkeys   *c.* Hominoids

Anthropoids that remained in Africa gave rise to two lineages: the Old World monkeys (figure 35.38b) and the hominoids (apes and humans; figure 35.38c). Old World monkeys include ground-dwelling as well as arboreal species. None of them have prehensile tails, their nostrils are close together, their noses point downward, and some have toughened pads of skin on their rumps for prolonged sitting.

Hominoids

The **hominoids** include the apes and the **hominids** (humans and their direct ancestors). The living apes consist of the gibbon (genus *Hylobates*), orangutan *(Pongo),* gorilla *(Gorilla),* and chimpanzee *(Pan).* Apes have larger brains than monkeys, and they lack tails. With the exception of the gibbon, which is small, all living apes are larger than any monkey. Apes exhibit the most adaptable behavior of any mammal except humans. Once widespread in Africa and Asia, apes are rare today, living in relatively small areas. No apes ever occurred in North or South America.

Studies of ape DNA have explained a great deal about how the living apes evolved. The Asian apes evolved first. The line of apes leading to gibbons diverged from other apes about 15 MYA, whereas orangutans split off about 10 MYA (figure 35.39). Neither group is closely related to humans.

The African apes evolved more recently, between 6 and 10 MYA. These apes are the closest living relatives to humans. The taxonomic group "apes" is a paraphyletic group; some apes are more closely related to hominids than they are to other apes. For this reason, some taxonomists have advocated placing humans and the African apes in the same zoological family, the Hominidae.

Fossils of the earliest hominids (humans and their direct ancestors), described later in this section, suggest that the common ancestor of the hominids was more like a chimpanzee than a gorilla. Based on genetic differences, scientists estimate that gorillas diverged from the line leading to chimpanzees and humans some 8 MYA.

Soon after the gorilla lineage diverged, the common ancestor of all hominids split off from the chimpanzee line to begin the evolutionary journey leading to humans. Because this split was so recent, few genetic differences between humans and chimpanzees have had time to evolve. For example, a human hemoglobin molecule differs from its chimpanzee counterpart in only a single amino acid. In general, humans and chimpanzees exhibit a level of genetic similarity normally found between closely related species of the same genus! In fact, when geneticists sequenced the genome of the chimpanzee, they found that it was 99% identical to the human genome (see chapter 24).

Comparing apes with hominids

The common ancestor of apes and hominids is thought to have been an arboreal climber. Much of the subsequent evolution of the hominoids reflected different approaches to locomotion. Hominids became bipedal, walking upright; in contrast, the apes evolved knuckle-walking, supporting their weight on the dorsal sides of their fingers. (Monkeys, by contrast, walk using the palms of their hands.)

Humans depart from apes in several areas of anatomy related to bipedal locomotion. Because humans walk on two legs, their

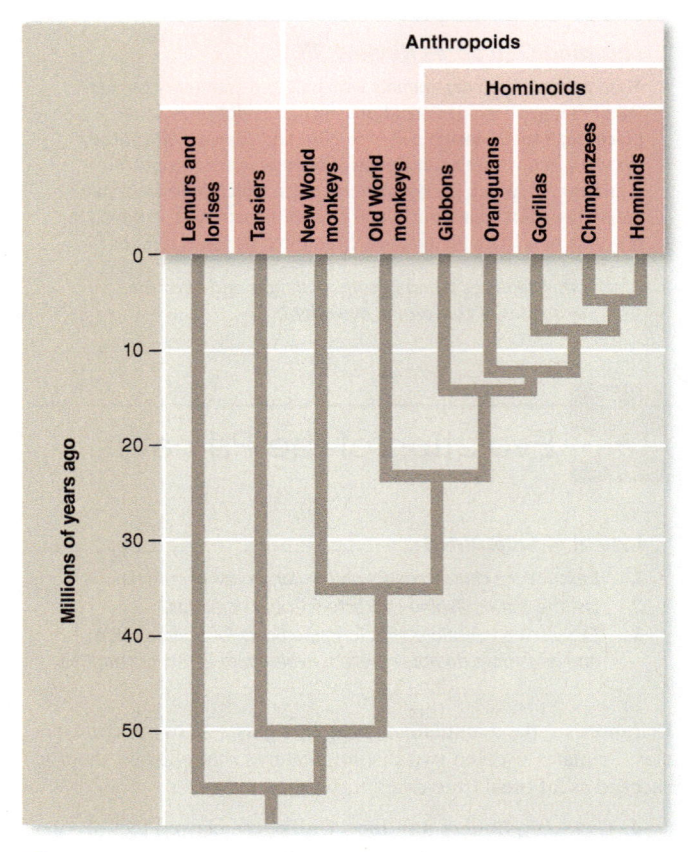

Figure 35.39 A primate evolutionary tree. Lemurs, lorises, and tarsiers used to be considered a group, the prosimians, which we now realize to be paraphyletic. Similarly, apes constitute a paraphyletic group because some apes are more closely related to nonape species (hominids) than they are to other apes.

vertebral column is more curved than an ape's, and the human spinal cord exits from the bottom rather than the back of the skull. The human pelvis has become broader and more bowl-shaped, with the bones curving forward to center the weight of the body over the legs. The hip, knee, and foot have all changed proportions.

Being bipedal, humans carry much of the body's weight on the lower limbs, which make up 32 to 38% of the body's weight and are longer than the upper limbs; human upper limbs do not bear the body's weight and make up only 7 to 9% of human body weight. African apes walk on all fours, with the upper and lower limbs both bearing the body's weight; in gorillas, the longer upper limbs account for 14 to 16% of body weight, the somewhat shorter lower limbs for about 18%.

Australopithecines were early hominids

Five to 10 MYA, the world's climate began to get cooler, and the great forests of Africa were largely replaced with savannas and open woodland. In response to these changes, a new kind of hominoid was evolving, one that was bipedal. These new hominoids are classified as hominids—that is, of the human line.

The major groups of hominids include three to seven species of the genus *Homo* (depending how you count them), seven species of the older, smaller-brained genus *Australopithecus,* and several even older lineages (figure 35.40). In every case where the fossils allow a determination to be made, the hominids are bipedal, the hallmark of hominid evolution.

In recent years, anthropologists have found a remarkable series of early hominid fossils extending as far back as 6 to 7 million years. Often displaying a mixture of primitive and modern traits, these fossils have thrown the study of early hominids into turmoil. Although the inclusion of these fossils among the hominids seems warranted, only a few specimens have been discovered, and they do not provide enough information to determine with certainty their relationships to australopithecines and humans. The search for additional early hominid fossils continues.

Early australopithecines

Our knowledge of australopithecines is based on hundreds of fossils, all found in South and East Africa (except for one specimen from Chad in West Africa). Australopithecines may have lived over a much broader area of Africa, but rocks of the proper age that might contain fossils are not exposed elsewhere. The evolution of hominids seems to have begun with an initial radiation of numerous species. The seven species identified so far provide ample evidence that australopithecines were a diverse group.

These early hominids weighed about 18 kg and were about 1 m tall. Their dentition was distinctly hominid, but their brains were no larger than those of apes, generally 500 cubic centimeters (cm^3) or less. *Homo* brains, by comparison, are usually larger than 600 cm^3; modern *H. sapiens* brains average 1350 cm^3.

The structure of australopithecine fossils clearly indicates that they walked upright. Evidence of bipedalism includes a set of some 69 hominid footprints found at Laetoli, East Africa. Two individuals, one larger than the other, walked upright side-by-side for 27 m, their footprints preserved in a layer of 3.7-million-year-old volcanic ash. Importantly, the big toe is not splayed out to the side as in a monkey or ape, indicating that these footprints were clearly made by hominids.

Bipedalism

The evolution of bipedalism marks the beginning of hominids. Bipedalism seems to have evolved as australopithecines left dense forests for grasslands and open woodland.

Whether larger brains or bipedalism evolved first was a matter of debate for some time. One school of thought proposed that hominid brains enlarged first, and then hominids became bipedal. Another school of thought saw bipedalism as a precursor to larger brains, arguing that bipedalism freed the forelimbs to manufacture and use tools, leading to the evolution of bigger brains. Recently, fossils unearthed in Africa have settled the debate. These fossils demonstrate that bipedalism extended back 4 million years; knee joint, pelvis, and leg

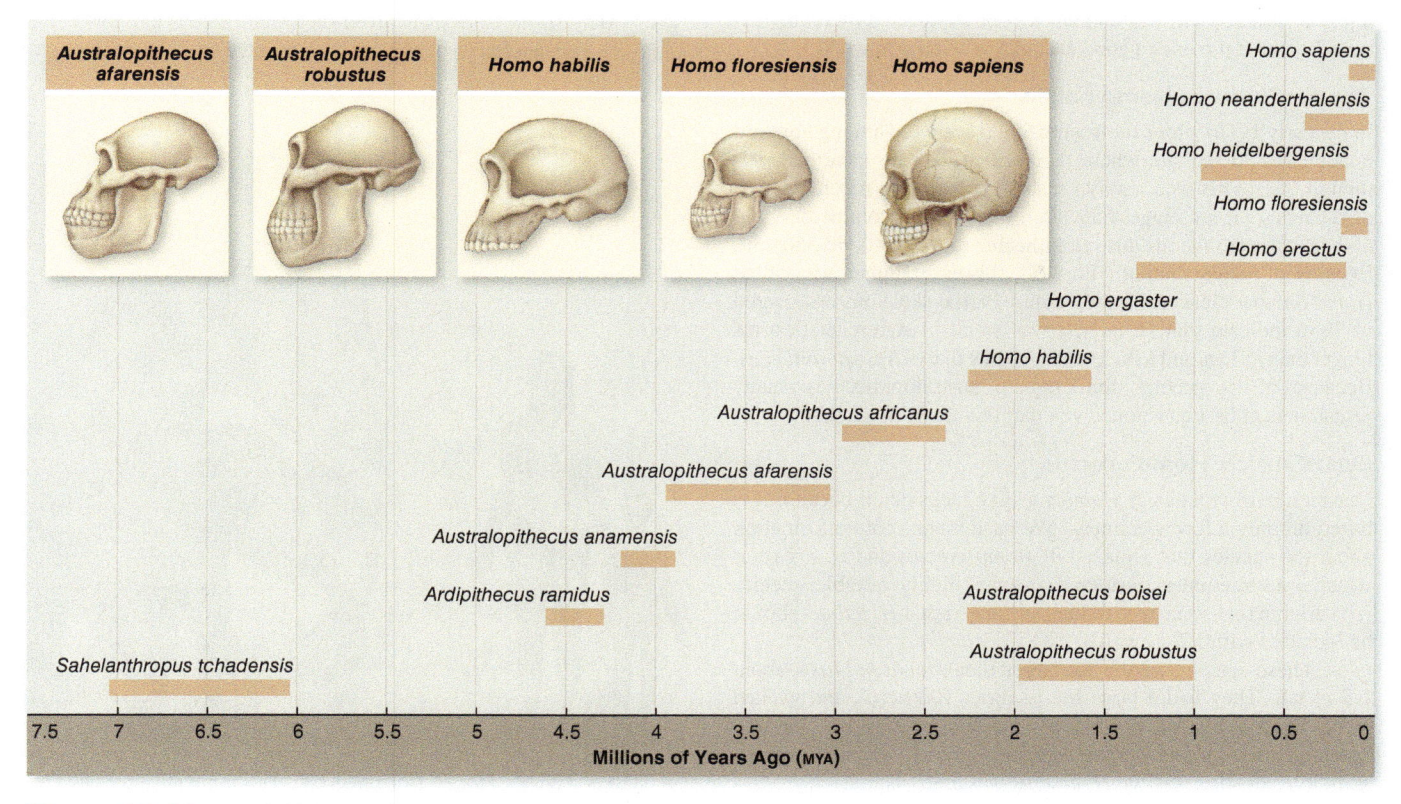

Figure 35.40 Hominid fossil history. Most, but not all, fossil hominoids are indicated here. New species are regularly being discovered, though great debate sometimes exists about whether a particular specimen represents a new species; conversely, in recent years, some scientists have suggested that too many species have been recognized and that some named species are really not distinct. For example, it was recently proposed that *H. ergaster* and *H. habilis* should be considered populations of *H. erectus*.

bones all exhibit the hallmarks of an upright stance. Substantial brain expansion, on the other hand, did not appear until roughly 2 MYA. In hominid evolution, upright walking clearly preceded large brains.

The reason bipedalism evolved in hominids remains a matter of controversy. No tools appeared until 2.5 MYA, so tool making seems an unlikely cause. Alternative ideas suggest that walking upright is faster and uses less energy than walking on four legs; that an upright posture permits hominids to pick fruit from trees and see over tall grass; that being upright reduces the body surface exposed to the Sun's rays; that an upright stance aided the wading of semiaquatic hominids; and that bipedalism frees the forelimbs of males to carry food back to females, encouraging pair-bonding. All of these suggestions have their proponents, and none is universally accepted. The origin of bipedalism, the key event in the evolution of hominids, remains a mystery.

The genus *Homo* arose roughly 2 MYA

The first humans (genus *Homo*) evolved from australopithecine ancestors about 2 MYA. The exact ancestor has not been clearly identified, but is commonly thought to be *Australopithecus afarensis* (the famous fossil "Lucy" was a member of this species). Only within the last 30 years have a significant number of fossils of early *Homo* been uncovered. An explosion of interest has fueled intensive field exploration, and new finds are announced regularly; every year, our picture of the base of the human evolutionary tree grows clearer. The following historical account will undoubtedly be supplanted by future discoveries, but it provides a good example of science at work.

The first human: Homo habilis

In the early 1960s, stone tools were found scattered among hominid bones close to the site where *A. boisei* had been unearthed. Although the fossils were badly crushed, painstaking reconstruction of the many pieces suggested a skull with a brain volume of about 680 cm^3, larger than the australopithecine range of 400 to 550 cm^3. Because of its association with tools, this early human was called *Homo habilis,* meaning "handy man." Partial skeletons discovered in 1986 indicate that *H. habilis* was small in stature, with arms longer than its legs and a skeleton much like that of *Australopithecus.* Because of its general similarity to australopithecines, many researchers at first questioned whether this fossil was human.

Out of Africa: Homo erectus

Our picture of what early *Homo* was like lacks detail because it is based on only a few specimens. We have much more information about the species that replaced it, *Homo erectus* and *H. ergaster,* which are sometimes considered a single, highly variable, species (so variable that some researchers suggested that *H. habilis* should be included within it as well!).

These species were a lot larger than *Homo habilis*—about 1.5 m tall. They had a large brain, about 1000 cm^3, and walked erect. Their skull had prominent brow ridges and, like modern humans, a rounded jaw. Most interesting of all, the shape of the skull interior suggests that *H. erectus* was able to talk.

Far more successful than *H. habilis, H. erectus* quickly became widespread and abundant in Africa, and within 1 million years had migrated into Asia and Europe. A social species, *H. erectus* lived in tribes of 20 to 50 people, often dwelling in caves. They successfully hunted large animals, butchered them using flint and bone tools, and cooked them over fires—a site in China contains the remains of horses, bears, elephants, and rhinoceroses.

Homo erectus survived for over a million years, longer than any other species of human. These very adaptable humans only disappeared in Africa about 500,000 years ago, as modern humans were emerging. Interestingly, they survived even longer in Asia, until 250,000 years ago.

A new addition to the human family: Homo floresiensis

The world was stunned in 2004 with the announcement of the discovery of fossils of a new human species from the tiny Indonesian island of Flores (figure 35.41). *Homo floresiensis* was notable for its diminutive stature; standing only a meter tall, and with a brain size of just 380 cm^3, the species was quickly nicknamed "the Hobbit" after the characters in J. R. R. Tolkien's books. Just as surprising was the age of the fossils, the youngest of which was only 15,000 years old.

Despite its recency, a number of skeletal features suggest to most scientists that *H. floresiensis* is more closely related to *H. erectus* than to *H. sapiens.* If correct (and not all scientists agree), this result would indicate that the *H. erectus* lineage persisted much longer than previously thought—almost to the present day. It also would mean that until very recently, *H. sapiens* was not the only species of human on the planet. We can only speculate about how *H. sapiens* and *H. floresiensis* may have interacted, and how these interactions may have been affected by the great difference in body size.

Why *H. floresiensis* evolved such small size is unknown, although a number of experts have pointed to the phenomenon

Figure 35.41 Homo floresiensis. This diminutive species (compare the modern human female on the right with the female *Homo floresiensis* on the *left*) occurred on the small island of Flores in what is now Indonesia. *H. floresiensis* preyed upon a dwarf species of elephant, *Stegodon sondaari,* which also occurred on Flores (compare with the larger African elephant, *Loxodonta africana,* in *gray*).

of "island dwarfism," in which mammal species evolve to be much smaller on islands. Indeed, *H. floresiensis* coexisted with and preyed on a miniature species of elephant that also lived on Flores, but which also has gone extinct. These findings have rekindled interest in explaining why island dwarfism occurs.

Modern humans

The evolutionary journey entered its final phase when modern humans first appeared in Africa about 600,000 years ago. Investigators who focus on human diversity denote three species of modern humans: *Homo heidelbergensis, H. neanderthalensis,* and *H. sapiens* ("wise man").

The oldest modern human, *Homo heidelbergensis,* is known from a 600,000-year-old fossil from Ethiopia. Although it coexisted with *H. erectus* in Africa, *H. heidelbergensis* has more advanced anatomical features, including a bony keel running along the midline of the skull, a thick ridge over the eye sockets, and a large brain. Also, its forehead and nasal bones are very much like those of *H. sapiens.*

As *H. erectus* was becoming rarer, about 130,000 years ago, a new species of human arrived in Europe from Africa. *Homo neanderthalensis* likely branched off the ancestral line leading to modern humans as long as 500,000 years ago. Compared with modern humans, Neanderthals were short, stocky, and powerfully built; their skulls were massive, with protruding faces, heavy, bony ridges over the brows, and larger braincases.

Cro-Magnons and Neanderthals

The Neanderthals (classified by many paleontologists as a separate species, *Homo neanderthalensis*) were named after the Neander Valley of Germany where their fossils were first discovered in 1856. Rare at first outside of Africa, they became progressively more abundant in Europe and Asia, and by 70,000 years ago had become common.

The Neanderthals made diverse tools, including scrapers, spearheads, and hand axes. They lived in huts or caves. Neanderthals took care of their injured and sick and commonly buried their dead, often placing food, weapons, and even flowers with the bodies. Such attention to the dead strongly suggests that they believed in a life after death. This is the first evidence of the symbolic thinking characteristic of modern humans.

Fossils of *H. neanderthalensis* abruptly disappear from the fossil record about 34,000 years ago and are replaced by fossils of *H. sapiens* called the Cro-Magnons (named after the valley in France where their fossils were first discovered). Scientists have long debated whether Cro-Magnons outcompeted Neanderthals, or whether the two species interbred, merging their gene pools. In recent years, this debate has been settled by the sequencing of the entire genome of several Neanderthals. Using precise laboratory methods to prevent contamination with modern human DNA, researchers have been able to extract DNA from fossil Neanderthal bones preserved in cool caves. Analysis of the genome sequences derived from these fossils revealed that the Neanderthal genome is quite distinct from that of modern humans, indicating that Neanderthals had a gene pool distinct from our Cro-Magnon ancestors. Nonetheless, detailed comparisons indicate that in some modern human populations, as much as 4% of the DNA is Neanderthal in origin, evidence of at least some degree of interbreeding in the past.

Genome sequences yield a surprise: The Denisovans

The Neanderthal genome project took a surprising turn in 2009 when scientists sequenced the genome from DNA extracted from a 41,000-year-old finger bone found in a cave in Siberia. This cave contains remnants of both Neanderthals and modern humans, so the researchers expected the bone to belong to one or the other. However, to everyone's surprise, the DNA did not belong to either. It was more similar to Neanderthal DNA than to that of modern humans DNA; the degree of difference suggests that this individual belonged to a previously unknown species of hominid that is estimated to have diverged from Neanderthals more than 600,000 years ago.

This species—now known as the Denisovans after the name of the cave in which the fossils were found—is a mystery. All scientists have to study are the finger bone and two molar teeth. However, another unexpected twist occurred when scientists compared the Denisovan genome to that of modern humans. Although there are no traces of Denisovan DNA in most modern populations, as much as 4% of the DNA of some populations, such as those occurring in parts of the South Pacific, appears to derive from the Denisovans, suggesting that at some time in the past, these mysterious hominids interbred with the ancestors of some modern human populations, although how a group living in a cave in Siberia could get their genes to the South Pacific is one of many unanswered questions. Obviously, much—almost everything!—remains to be learned about the Denisovans, but these findings have opened up an entirely new field of study, *paleogenomics,* which holds the prospect of discovering ancient species unknown from the fossil record.

Our own species: Homo sapiens

A variety of evidence indicates that Cro-Magnons came from Africa—fossils of essentially modern aspect but as much as 100,000 years old have been found there. Cro-Magnons seem to have replaced the Neanderthals completely in the Middle East by 40,000 years ago, and then spread across Europe, coexisting with the Neanderthals for several thousand years. The Cro-Magnons that replaced the Neanderthals had a complex social organization and are thought to have had full language capabilities. Elaborate and often beautiful cave paintings made by Cro-Magnons can be seen throughout Europe (figure 35.42).

Humans of modern appearance eventually spread across Siberia to North America, where they arrived at least 13,000 years ago, after the ice had begun to retreat and a land bridge still connected Siberia and Alaska. By 10,000 years ago, about 5 million people inhabited the entire world (compared with more than 6 billion today).

Homo sapiens is the only surviving species of the genus *Homo* and indeed the only surviving hominid. Some of the best fossils of *H. sapiens* are 20 well-preserved skeletons with skulls found in a cave near Nazareth in Israel. Modern dating techniques estimate these humans to be between 90,000 and 100,000 years old. The skulls are modern in appearance and size, with high, short braincases, vertical foreheads with only slight brow ridges, and a cranial capacity of roughly 1550 cm^3. Our evolution has been marked by a progressive increase in brain size, distinguishing us from other animals in several ways. First, humans are able to make and use tools more effectively than any other animal—a capability that, more than any other factor, has been responsible for our dominant position in the world. Second, although not the only animal capable of conceptual thought, humans have refined and extended this ability until it has become the hallmark

Figure 35.42 Cro-Magnon art. Rhinoceroses are among the animals depicted in this remarkable cave painting found in 1995 near Vallon-Pont d'Arc, France.

of our species. Finally, we use symbolic language and can, with words, shape concepts out of experience and transmit that accumulated experience from one generation to another.

Humans have undergone what no other animal ever has: extensive cultural evolution. Through culture, we have found ways to change and mold our environment, rather than changing evolutionarily in response to the environment's demands. We control our biological future in a way never before possible—an exciting potential and a frightening responsibility.

Human races

Humans, like all other species, have differentiated in their characteristics as they have spread throughout the world. Local populations in one area often appear significantly different from those that live elsewhere. For example, northern Europeans often have blond hair, fair skin, and blue eyes, whereas Africans often have black hair, dark skin, and brown eyes. These traits may play a role in adapting the particular populations to their environments. Blood groups may be associated with immunity to diseases more common in certain geographical areas, and dark skin shields the body from the damaging effects of ultraviolet radiation, which is much stronger in the tropics than in temperate regions.

All humans are capable of mating with one another and producing fertile offspring. The reasons that they do or do not choose to associate with one another are purely psychological and behavioral (cultural).

The number of groups into which the human species might logically be divided has long been a point of contention. Some contemporary anthropologists divide people into as many as 30 "races," others as few as three: Caucasoid, Negroid, and Oriental. American Indians, Bushmen, and Aborigines are examples of particularly distinctive subunits that are sometimes regarded as distinct groups.

The problem with classifying people or other organisms into races in this fashion is that the characteristics used to define the races are usually not well correlated with one another, and so the determination of race is always somewhat arbitrary. Humans are visually oriented; consequently, we have relied on visual cues—primarily skin color—to define races. However, when other types of characteristics, such as blood groups, are examined, patterns of variation correspond very poorly with visually determined racial classes. Indeed, if one were to break the human species into subunits based on overall genetic similarity, the groupings would be very different from those based on skin color or other visual features (figure 35.43).

a.

b.

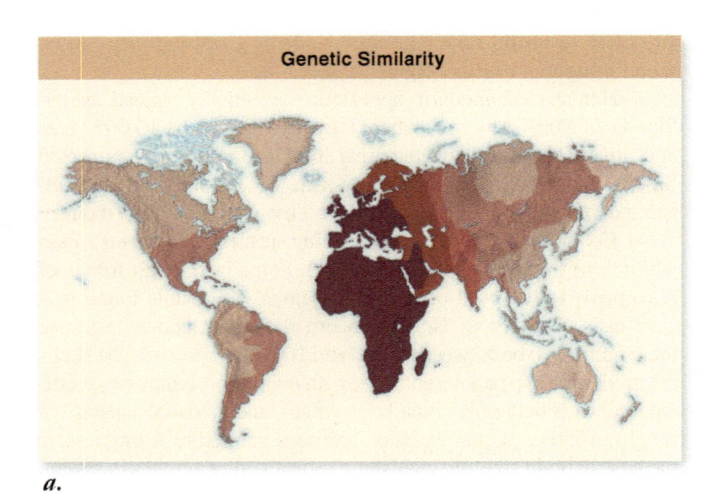

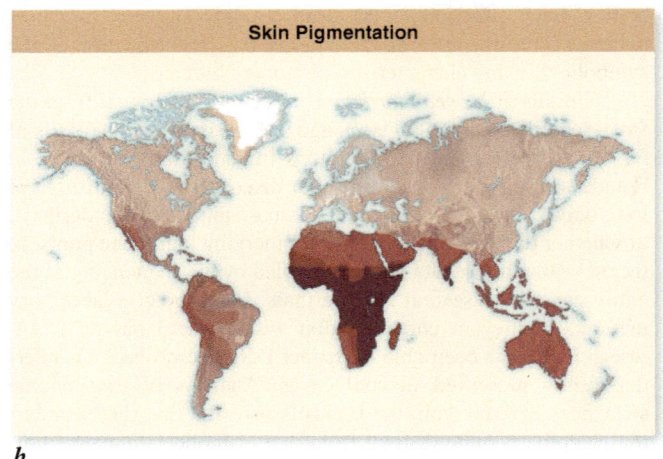

Figure 35.43 Patterns of genetic variation in human populations differ from patterns of skin color variation. *a.* Genetic variation among *Homo sapiens*. The more similar areas are in color, the more similar they are genetically based on many enzyme and blood group genetic loci. *b.* Similarity among *Homo sapiens* based on skin color. The color of an area represents the skin pigmentation of the people native to that region.

In humans, it is simply not possible to delimit clearly defined races that reflect biologically differentiated and well-defined groupings. The reason is simple: Different groups of people have constantly intermingled and interbred with one another during the entire course of history. This constant gene flow has prevented the human species from fragmenting into highly differentiated subspecies. Those characteristics that are differentiated among populations, such as skin color, represent classic examples of the antagonism between gene flow and natural selection. As you saw in chapter 20, when selection is strong enough, as it is for dark coloration in tropical regions, populations can differentiate even in the presence of gene flow. However, even in cases such as this, gene flow will still ensure that populations are relatively homogeneous for genetic variation at other loci.

For this reason, relatively little of the variation in the human species represents differences between the described races. Indeed, one study calculated that only 8% of all genetic variation among humans could be accounted for as differences that exist among racial groups; in other words, the human racial categories do a very poor job in describing the vast majority of genetic variation that exists in humans. For this reason, most modern biologists reject human racial classifications as a means of reflecting patterns of biological differentiation in the human species. This is a sound biological basis for dealing with each human being on his or her own merits and not as a member of a particular "race."

Learning Outcomes Review 35.10

Primates include prosimians, monkeys, apes, and humans (hominids). Primates have grasping fingers and toes and binocular vision. Hominids diverged from other primates by developing bipedal locomotion that led to a number of additional adaptations, including modification of the spine, pelvis, and limbs. Several species of *Homo* evolved in Africa, and some migrated from there to Europe and Asia. Our own species, *Homo sapiens*, is proficient at conceptual thought and tool use and is the only animal that uses symbolic language. Considerable variation occurs among human populations, but the recognized races, which are based primarily on skin color, do not reflect significant biological differences.

■ *Which of these groups is monophyletic: prosimians, monkeys, apes, hominids?*

Chapter Review

35.1 Echinoderms

Echinoderms are ancient and unmistakable.
Echinoderms are characterized by deuterostome development, pentameral symmetry, an endoskeleton covered by a delicate epidermis, and a water-vascular system.

Echinoderm symmetry is bilateral in larvae but pentaradial in adults.
The larvae of echinoderms are bilateral, but the adult is pentaradial.
The water-vascular system, which is one of several coelomic compartments, aids in movement, feeding, circulation, respiration, and excretion.

Echinodermata contains five extant classes.
The five classes of echinoderms are Asteroidea (sea stars), Crinoidea (sea lilies), Echinoidea (sea urchins and sand dollars), Holothuroidea (sea cucumbers), and Ophiuroidea (brittle stars).

35.2 Chordates

Chordates share four features at some time during development: a single, hollow nerve cord; a flexible rod, the notochord; pharyngeal slits or pouches; and a postanal tail (figure 35.4).

35.3 Nonvertebrate Chordates

Tunicates have chordate larval forms.
Tunicates have a swimming larval form exhibiting all the features of a chordate, but their adult form is sessile and baglike.

Lancelets are small marine chordates.
Lancelets have chordate features throughout life, but as adults they lack bones and have no distinct head (figure 35.6).

35.4 Vertebrate Chordates

Vertebrates have vertebrae, a distinct head, and other features.
In vertebrates, a vertebral column encloses and protects the dorsal nerve cord. The distinct and well-differentiated head carries sensory organs. Vertebrates also have specialized internal organs and a bony or cartilaginous endoskeleton.

Vertebrates evolved half a billion years ago: An overview.
In fishes, evolution of a hinged jaw was a major advance. Other changes allowed a move into the terrestrial environment, giving rise to amphibians, reptiles, birds, and mammals (figure 35.11).

35.5 Fishes

Fishes exhibit five key characteristics.
The key characteristics of fishes are a vertebral column of bone or cartilage, jaws, paired appendages, internal gills, and a closed circulatory system.

Sharks, with cartilaginous skeletons, became top predators.
Sharks, rays, and skates are cartilaginous fishes. Sharks were streamlined for fast swimming, and they evolved teeth that enabled them to readily grab, kill, and devour prey.
The lateral line system of sharks and bony fishes is a sensory system that detects changes in pressure waves.

Bony fishes dominate the waters.
Bony fishes belong either to the ray-finned fishes (Actinopterygii), or the lobe-finned fishes (Sarcopterygii). Ray-finned fishes have fins stiffened with bony parallel rays.

The evolutionary path to land ran through the lobe-finned fishes.
The lobe-finned fishes have muscular lobes with bones connected by joints (figure 35.16). These structures could evolve into limbs capable of movement on land.

35.6 Amphibians

Living amphibians have five distinguishing features.
Amphibian adaptations include legs, lungs, cutaneous respiration, pulmonary veins, and a partially divided heart.

Amphibians overcame terrestrial challenges.
Terrestrial adaptations included being able to support large bodies against gravity, to breathe out of water, and to avoid desiccation.

Modern amphibians belong to three groups.
The Anura (frogs and toads) lack tails as adults; many have a larval tadpole stage. The Caudata (salamanders) have tails as adults and larvae similar to the adult form. The Apoda (caecilians) are legless.

35.7 Reptiles

Reptiles exhibit three key characteristics.
Reptiles posses a watertight amniotic egg; dry, watertight skin; and thoracic breathing (figure 35.20).
Modern reptiles practice internal fertilization and are ectothermic.

Reptiles dominated the Earth for 250 million years.
Synapsids gave rise to the therapsids that became the mammalian line. Diapsids gave rise to modern reptiles and the birds.

Modern reptiles belong to four orders.
The four orders of reptiles are Chelonia (turtles and tortoises); Rhynchocephalia (tuataras); Squamata (lizards and snakes); and Crocodylia (crocodiles and alligators).

35.8 Birds

Key characteristics of birds are feathers and a lightweight skeleton.
The feather is a modified reptilian scale. Feathers provide lift in gliding or flight and conserve heat (figure 35.28).
The lightweight skeleton of birds is an adaptation to flight.

Birds arose about 150 MYA.
Birds evolved from theropod dinosaurs. Feathers probably first arose to provide insulation, only later being modified for flight.

Modern birds are diverse but share several characteristics.
In addition to the key characteristics, birds have efficient respiration and circulation and are endothermic.

35.9 Mammals

Mammals have hair, mammary glands, and other characteristics.
Mammals are distinguished by fur and by mammary glands, which provide milk to feed the young. Mammals are also endothermic.

Mammals diverged about 220 MYA.
Mammals evolved from therapsids (synapsids) and reached maximum diversity about 15 MYA.

Modern mammals are placed into three groups.
The monotremes lay shelled eggs. In marsupials, an embryo completes development in a pouch. Placental mammals produce a placenta in the uterus to nourish the embryo.

35.10 Evolution of the Primates (figure 35.39)

Primates share two innovations: grasping fingers and toes, and binocular vision.

The anthropoid lineage led to the earliest humans.
The earliest primates were the prosimians; anthropoids, which include monkeys, apes, and humans, evolved later. Hominoids include the apes and the hominids, or humans.

Austrolopithecines were early hominids.
The distinguishing characteristics of hominids are upright posture and bipedal locomotion.

The genus Homo arose roughly 2 MYA.
Common features of early *Homo* species include a larger body and brain size. *Homo sapiens* is the only extant species. Humans exhibit conceptual thought, tool use, and symbolic language.

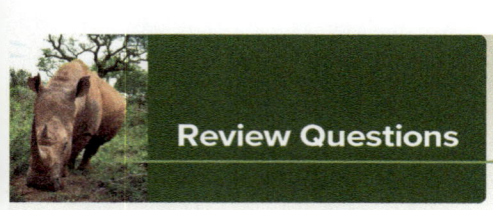

Review Questions

UNDERSTAND

1. Which of the following structures is not a component of the water-vascular system of an echinoderm?
   a. Ossicles
   b. Ampullae
   c. Radial canals
   d. Madreporites

2. Which of the following statements regarding all species of chordates is false?
   a. Chordates are deuterostomes.
   b. A notochord is present in the embryo.
   c. The notochord is surrounded by bone or cartilage.
   d. All possess a postanal tail during embryonic development.

3. In the figure, item A is the _____ and item B is the _____.
   a. complete digestive system; notochord
   b. spinal cord; nerve cord
   c. notochord; nerve cord
   d. pharyngeal slits; notochord

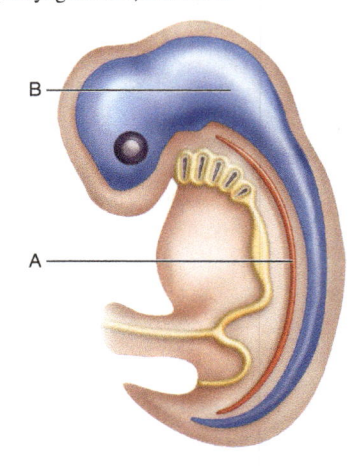

4. During embryonic development, a neural crest would be found in all of the following chordates, except
   a. cephalochordates.   c. birds.
   b. reptiles.   d. mammals.

5. The ___ of the bony fish evolved to counter the effects of increased bone density.
   a. gills   c. swim bladder
   b. jaws   d. teeth

6. Why was the evolution of the pulmonary veins important for amphibians?
   a. To move oxygen to and from the lungs
   b. To increase the metabolic rate
   c. For increased blood circulation to the brain
   d. None of the choices are correct.

7. Which of the following groups lacks a four-chambered heart?
   a. Birds   c. Mammals
   b. Crocodilians   d. Amphibians

8. All of the following are characteristics of reptiles, except
   a. cutaneous respiration.   c. thoracic breathing.
   b. an amniotic egg.   d. dry, watertight skin.

9. Which of the following evolutionary adaptations allows the birds to become efficient at flying?
   a. Structure of the feather
   b. High metabolic temperatures
   c. Increased respiratory efficiency
   d. All of the choices are correct.

APPLY

1. The reason that birds and crocodilians both build nests might be because they
   a. are both warm-blooded.
   b. both eat fish.
   c. both inherited the trait from a common ancestor.
   d. both are ectothermic.

2. Which of the following is the closest relative of lungfish?
   a. Hagfish   c. Ray-finned fish
   b. Sharks   d. Mammals

3. The fact that monotremes lay eggs
   a. indicates that they are more closely related to some reptiles than they are to some mammals.
   b. is a plesiomorphic trait.
   c. demonstrates that the amniotic egg evolved multiple times.
   d. is a result of ectothermy.

SYNTHESIZE

1. Some scientists believe the feathers did not evolve initially for flight, but rather for insulation. What benefits would this have had for early flightless birds?

2. Some people state that the dinosaurs have not "gone extinct," they are with us today. What evidence can be used to support this statement?

3. In what respect is the evolutionary diversification of hominids similar to that of horses discussed in chapter 21?

Ecology and Behavior

Part VIII Ecology and Behavior

CHAPTER 54

Behavioral Biology

Chapter Contents

54.1 The Natural History of Behavior
54.2 Nerve Cells, Neurotransmitters, Hormones, and Behavior
54.3 Behavioral Genetics
54.4 Learning
54.5 The Development of Behavior
54.6 Animal Cognition
54.7 Orientation and Migratory Behavior
54.8 Animal Communication
54.9 Behavioral Ecology
54.10 Reproductive Strategies and Sexual Selection
54.11 Altruism
54.12 The Evolution of Group Living and Animal Societies

Introduction

The study of behavior is at the center of many disciplines of biology. Observing behavior provides important insights into the workings of the brain and nervous system, the influences of genes and the environment, when and how animals reproduce, and how they adapt to their environment. Behavior is shaped by natural selection and is controlled by internal mechanisms involving genes, hormones, neurotransmitters, and neural circuits. In this chapter, we explore how behavioral biology integrates approaches from several branches of biological science to provide a detailed understanding of the mechanisms that underscore behavior and its evolution.

54.1 The Natural History of Behavior

Learning Outcomes
1. Contrast the proximate and ultimate causation of behavior.
2. Explain the relationship between a key stimulus and a fixed action pattern.
3. Describe the physiological factors that might be the basis for innate behaviors.

Observing animal behavior and making inferences about what one sees is at once simple and profound. Behavior is what an animal does. It is the most immediate way an animal responds to its environment by tracking environmental cues and signals such as odors, sounds, or visual signals associated with food, predators, or mates. Behavior also concerns thinking and cognition, monitoring one's social environment, and making decisions as to whether or not to cooperate or act altruistically. Behavior allows animals to survive and reproduce and is thus critical to the evolutionary process. The work of behavioral biologists has provided important insights into animal behavior, including the very meaning of human behavior.

Behavior can be analyzed in terms of mechanisms and evolutionary origin

Why does an animal behave in a particular way? Consider hearing a bird sing. We could ask how it vocalizes or determine the time of the year it sings most frequently. We could also ask about the function of the song, that is, ask why it sings. Answers to questions about how birds sing consider the role of internal factors such as hormones and nerve cells and other physiological processes. Such questions concern proximate causation: the mechanisms that produce the behavior. To analyze the proximate cause of birdsong, we could measure hormone levels or study the development of brain regions and neural circuits associated with singing. For example, a male songbird may sing during the breeding season because of an increased level of the steroid sex hormone testosterone, which binds to receptors in the brain and triggers the production of song. Additionally, neural connections between the brain and the syrinx (the bird's vocal organ) must develop to allow songs to be produced. These explanations describe the proximate cause of birdsong.

Asking about the function of a behavior (once again, birdsong) is to ask why it evolved. To answer this question, we would determine how it influenced survival or reproductive success. A male bird sings to defend a territory from other males and to attract a female with which to reproduce. This is the ultimate, or evolutionary, explanation for the male's vocalization. Now we can understand its ultimate causation, or adaptive value. Researchers often study behavior from both perspectives to fully appreciate its mechanisms and ecological function, and thus its role in evolution. Behavior can be analyzed at four levels: (1) physiology (how it is influenced by hormones, nerve cells, and other internal factors); (2) ontogeny (how it develops in an individual); (3) phylogeny (its origin in groups of related species); and (4) adaptive significance (its role in survival and fitness). We'll begin by tracing the history of the study of mechanisms of behavior by focusing on the work of ethologists—biologists who first began to study behavior at the turn of the 20th century.

Ethology emphasizes the study of instinct and its origins

Ethology is the study of the natural history of behavior, with an emphasis on behaviors that form an animal's instincts, or programmed behaviors. Ethologists observed that individuals of a given species behaved in stereotyped ways, showing the same pattern of behavior in response to a particular stimulus. Because their behavior seemed reflexive, they considered it to be instinctive, or *innate*. Behaviors were thought to be programmed by the nervous system, which in turn was designed by genes, and responses would occur without experience. Ethologists based their instinct model on observations and experiments of simple behaviors such as egg retrieval by geese. Geese incubate their eggs in a nest. If an egg falls out of the nest, the goose will roll the egg back into the nest with a side-to-side motion of its neck while the egg is tucked beneath its bill (figure 54.1). Even if the egg is removed during retrieval, the goose will still complete the egg-retrieval sequence, as if driven by a program activated by the initial sight of the egg outside the nest.

This example is one paradigm of instinct theory and illustrates the way ethologists conceptualized the mechanisms of behavior. Egg retrieval behavior is triggered by a *key stimulus* (sometimes called a *sign stimulus*); this is the egg out of the nest. Early ethologists thought the nervous system regulated behavior via the *innate releasing mechanism,* a neural circuit involved in the perception of the key stimulus and the triggering of a motor program, the *fixed action pattern,* in this case the act of guiding the egg back to the nest. Ethologists generalized that the key stimulus

Figure 54.1 Innate egg-rolling response in geese. The series of movements used by a goose to retrieve an egg is a fixed action pattern. Once it detects the sign stimulus (in this case, an egg outside the nest), the goose goes through the entire set of movements: It will extend its neck toward the egg, get up, and roll the egg back into the nest by moving its neck from side-to-side with the egg positioned under its bill.

is a cue or signal in the environment that initiates neural events that cause behavior. The innate releasing mechanism involves the sensory apparatus that detects the signal and the neural circuit controlling muscles to generate the fixed action pattern.

> ### Learning Outcomes Review 54.1
> Proximate causation of behavior involves the immediate mechanisms that bring about an action; ultimate causation refers to the adaptive value of a behavior. Ethology is the study of the nature of behavior, emphasizing instinct and the regulation of behavior by internal factors such as genes, nerve cells, and hormones. Instinctual behavior is thought to occur when key stimuli affect innate releasing mechanisms, which trigger fixed motor programs.
>
> ■ Why is it important to understand the phylogeny (evolutionary origins) of behavior?

54.2 Nerve Cells, Neurotransmitters, Hormones, and Behavior

> ### Learning Outcomes
> 1. Relate the structure of neural circuits to their function.
> 2. Describe the role of hormones and neurotransmitters in behavior.

Although early ethologists had little understanding of neurobiology, they hypothesized that elements of the nervous system (the innate releasing mechanism) controlled behavior. Today, neuroethologists—researchers who examine the neurobiology of behavior—can describe in detail how information in the environment is processed by sensory cells and how nerve impulses are transmitted to other neurons and muscles to form neural circuits that regulate behaviors important to survival. Behavior reflects the organization of the peripheral and central nervous system, and studying behavior can help us understand how neurons function individually and in combination with other neurons in circuits (see chapters 43 and 44).

Behaviors that must occur rapidly, like those used to capture prey or flee predators, involve neural mechanisms that enable such functions. Some moths have an earlike sensory organ equipped with sensory neurons designed to detect the ultrasonic cries of bats; the neuron fires so rapidly that the moth can initiate evasive action before the bat is even aware of the moth's presence. Specialized cells in a frog's retina detect moving objects like insects and release the tongue in fractions of a second once suitable prey is sighted. Likewise, the jaws of a predatory ant snap shut when prey trigger sensory hairs between the mandibles. Rapid responses to predators or prey often involve large nerve cell axons that can quickly transmit impulses to muscles. In the example of "trap jaw" ants, large axons of the mandibular motor neuron—the fastest neuron yet identified—fire nerve impulses that close the jaws in only 33 msec. Neural circuits that enable quick responses often are made up of few sensory and motor neurons, and their connecting nerve cells.

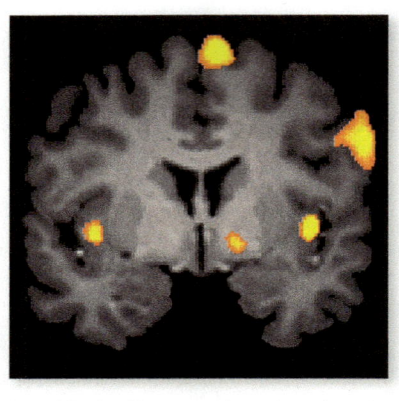

Figure 54.2 Functional magnetic resonance imaging (fMRI). MRIs reveal neural activity in specific regions of the brain. In this case, activity in part of the brain called the nucleus accumbens is associated with viewing images of food.

Behavioral biologists examine the relationship of hormones to behavior to understand the endocrine mechanisms that are the foundation of reproduction, parental care, aggression, and stress (see chapter 45). In this way, the effects of the steroid sex hormones estrogen and testosterone on the behavior of vertebrates have been determined. Testosterone in the male, for example, regulates territorial behavior and courtship, whereas estrogen in the female controls her mating behavior. Glucocorticoid hormones are involved in stress.

Neuroscientists may measure levels of neurotransmitters such as serotonin and dopamine in the nervous system or blood and associate these chemicals with behavior (see chapter 43). These chemicals are released by nerve cells and can affect activity in different brain regions. Serotonin has been shown to influence aggression in an incredibly wide range of animals including lobsters, mice, and humans. Researchers may inject a neurotransmitter or pharmacologically change its level in the brain to examine how it affects behavior.

The techniques of neuroethology include identifying and mapping individual neurons, their dendrites and connections to other neurons, and studying how their impulses and neurochemicals regulate behavior. Today, techniques such as functional magnetic resonance imaging (fMRI) are generating exciting data on the specialized functions of different regions of the human brain. One striking example concerns how the brain responds to images of food (figure 54.2). In contrast to expectation, the brain's response does not occur in the visual cortex, the region associated with object recognition, but in a circuit in the nucleus accumbens in the forebrain, normally involved in reward and pleasure.

> ### Learning Outcomes Review 54.2
> Instinctive behaviors appear to involve programmed circuits in the nervous system that are likely to be genetically controlled. Research in neuroethology supports the instinct concept of behavior by describing the organization of neural circuits governing behavior. Chemical signals provided by hormones and by neurotransmitters such as serotonin and dopamine cause behaviors to occur.
>
> ■ If a male songbird is injected with testosterone two weeks earlier than when these birds normally start to sing in the spring, what would you expect to happen?

54.3 Behavioral Genetics

Learning Outcomes
1. Discuss the types of studies that have provided evidence to link genes and behavior.
2. Explain how single genes can influence behavior.
3. Describe the role of genes in complex behaviors such as aggression, parental care, and pair bonding.

Instinct theory assumed that genes play a role in behavior, but ethologists did not conclusively demonstrate the role genes can play. The study of genes and behavior has often been highly controversial, as ethologists and social scientists engaged in a seemingly endless debate over whether behavior is determined more by an individual's genes (nature) or by its learning and experience (nurture). One problem with this nature/nurture controversy is that the question is framed as an "either/or" proposition, which fails to consider that both instinct and experience can have significant roles, often interacting in complex ways to shape behavior.

Behavioral genetics deals with the contribution that heredity makes to behavior. It is obvious that genes, the units of heredity, are passed from one generation to the next and guide the development of the nervous system and potentially the behavioral responses it regulates. But animals may also develop in a rich social environment and have experiences that guide behavior. The importance of "nature" and "nurture" to behavior can be seen by first reviewing the history of studies in behavioral genetics and next examining the importance of experience and development. We'll then consider their interaction.

Artificial selection, hybrid, and twin studies link genes and behavior

Pioneering research indicated that behavioral differences among individuals result from genetic differences. Research on a variety of animals demonstrated that hybrids showed behaviors involved in nest building and courtship that were intermediate between those of parents, as would be expected given that hybrids have genes from both parents. These early efforts to define the role of genes in behavior demonstrated that behavior can have a heritable component, but fell short of identifying the genes involved. With the development of molecular biology, far greater precision was added to the analysis of the genetics of behavior.

Learning itself can be influenced by genes. In one classic study, rats had to find their way through a maze of blind alleys and only one exit, where a reward of food awaited them. Some rats quickly learned to zip through the maze to the food, making few mistakes, but other rats made more errors in learning the correct path. Researchers bred rats that made few errors with one another to establish a "maze-bright" group, and error-prone rats were interbred, forming a "maze-dull" group. Offspring in each group were then tested for their maze-learning ability. The offspring of maze-bright rats learned to negotiate the maze with fewer errors than their parents, whereas the offspring of maze-dull parents performed more poorly. Repeating this artificial selection method for several generations led to two behaviorally distinct types of rat with very different maze-learning abilities (figure 54.3). This type of study suggests the ways in which natural selection could shape behavior over time, making genes for certain abilities more prevalent.

The interaction of genes and the environment can be seen in humans by comparing the behavior of identical twins (who are genetically the same), raised in the same environment or separated at birth and raised apart in different environments. Data on human twins raised together or raised apart allows researchers to determine whether similarities in behavior result from their genetic similarity or from shared environmental experiences. Twins studies indicate many similarities in a wide range of personality traits even though twins were raised in very different environments. On the other hand, these studies show that some traits, such as antisocial behavior, result from a combination of genes and experience during childhood. These studies indicate that genetics plays a role in behavior even in humans, although the relative importance of genetics versus environment is still debated.

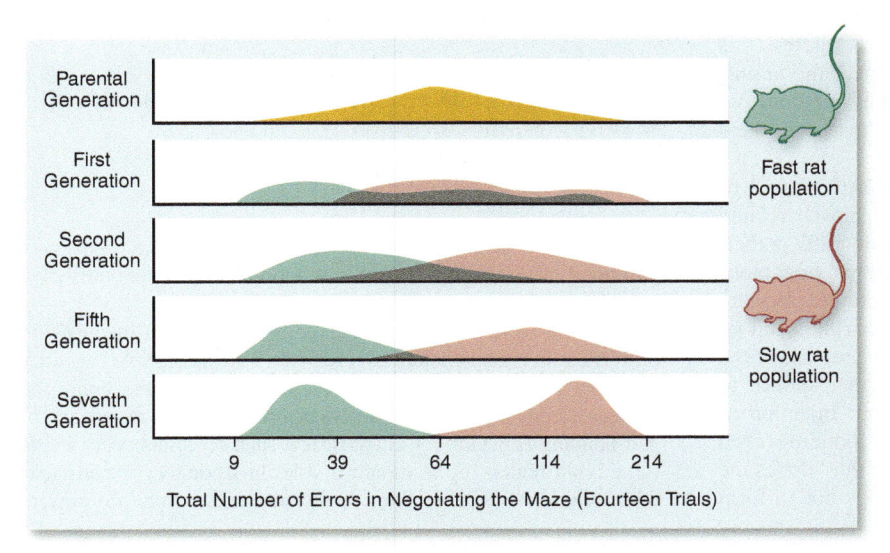

Figure 54.3 The genetics of learning. Rats that made the fewest errors in the parental population were interbred to select for rats that had improved maze-learning ability *(green)*, and rats that made the most errors were interbred to select for rats that were error-prone *(red)*.

Inquiry question What would happen if, after the seventh generation, rats were randomly assigned mates regardless of their ability to learn the maze?

Data analysis By approximately how much has the mean number of errors changed from the parental population to the seventh generation of the fast rat population? What type of selection does this illustrate?

Some behaviors appear to be controlled by a single gene

Artificial selection, hybrid, and twin studies indicated a role for genes in behavior, but more recent work has taken advantage of advances in molecular biology and identified the actual genes involved. Fruit flies, *Drosophila*, have traditionally provided a useful model system in which the effect of single genes have been identified. Single genes have also been shown to influence behavior in animals ranging from mice to humans.

In fruit flies, individuals that possess alternative alleles for a particular gene differ greatly in their feeding behavior as larvae: Larvae with one allele move around a great deal as they eat, whereas individuals with the alternative allele move hardly at all. A wide variety of experimentally induced mutations at other genes affect courtship behavior in males and females. For example, *fru* is a regulatory gene whose transcription products govern the design of the courtship center of the fruit fly brain. This gene turns on other genes involved in the neural circuitry of courtship.

Single genes in mice are associated with spatial memory and parenting. For example, some mice with a particular mutation have trouble remembering recently learned information about where objects are located. This is apparently because they lack the ability to produce the enzyme α-calcium-calmodulin-dependent kinase II, which plays an important role in the functioning of the hippocampus, a part of the brain important for spatial learning.

Single genes can play a role even in behaviors as complex as maternal care. For example, the *fosB* gene determines whether female mice nurture their young effectively. Females with both *fosB* alleles disabled initially investigate their newborn babies, but then ignore them, in stark contrast to the caring and protective maternal behavior displayed by normal females (figure 54.4). The cause of this inattentiveness appears to result from a chain reaction. When mothers of new babies initially inspect them, information from their auditory, olfactory, and tactile senses is transmitted to the hypothalamus, where *fosB* alleles are activated. The *fosB* alleles produce a protein, which in turn activates other enzymes and genes that affect the neural circuitry of the hypothalamus. These modifications in the brain cause the female to behave maternally. If mothers lack the *fosB* alleles, this process is stopped midway. No protein is activated, the brain's neural circuitry is not rewired, and maternal behavior does not result. The "maternal instincts" of mice can thus be defined genetically!

Another fascinating example of the genetic basis of behavior concerns prairie and montane voles, two closely related species of North American rodents that differ profoundly in their social behavior. Male and female prairie voles form monogamous pair bonds and share parental care, whereas montane voles are promiscuous (meaning they mate with multiple partners and go their separate ways). The act of mating leads to the release of the neuropeptides vasopressin and oxytocin, and the response to these peptides differs dramatically in the species. Injection of either peptide into prairie voles leads to pair bonding even without mating. Conversely, injecting a chemical that blocks the action of these neuropeptides causes prairie voles not to form

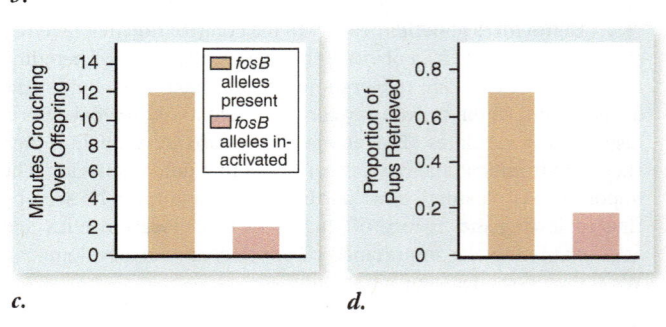

Figure 54.4 Genetically caused defect in maternal care.
a. In mice, normal mothers take very good care of their offspring, retrieving them if they move away and crouching over them.
b. Mothers with the mutant *fosB* allele perform neither of these behaviors, leaving their pups exposed. *c.* Amount of time female mice were observed crouching in a nursing posture over offspring.
d. Proportion of pups retrieved when they were experimentally moved.

Inquiry question Why does the lack of *fosB* alleles lead to maternal inattentiveness?

pair bonds after mating. By contrast, montane voles are unaffected by either of these manipulations.

These different responses have been traced to interspecific differences in brain structure (figure 54.5). The prairie vole has many receptors for these peptides in a particular part of the brain, the nucleus accumbens, which seems to be involved in the expression of pair-bonding behavior. By contrast, few such receptors occur in the same brain region in the montane vole. In laboratory experiments with prairie voles, blocking these receptors tends to prevent

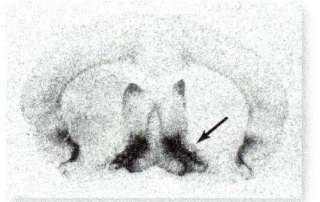

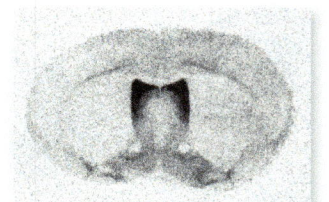

a. Prairie vole   *b.* Montane vole

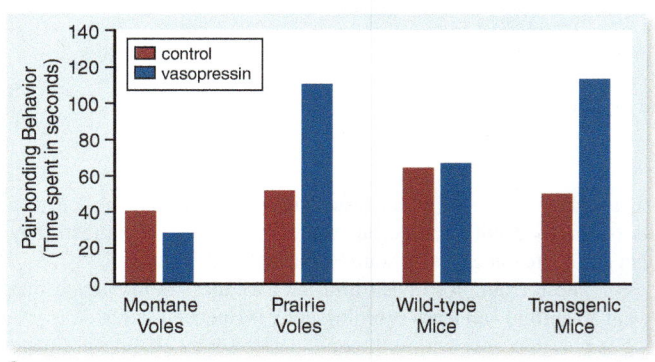

c.

Figure 54.5 Genetic basis of differences in pair-bonding behavior in two rodent species. *a.* and *b.* The prairie vole, *Microtus ochrogaster,* and montane vole, *M. montanus,* differ in the distribution of one type of vasopressin receptor in the brain, as indicated by the dark coloring in the part of the brain indicated by the arrow, called the nucleus accumbens. *c.* Prairie voles respond to the injection of vasopressin by exhibiting heightened levels of pair-bonding behavior in 5-min trials compared to their response to a control injection. By contrast, montaine voles do not show such an increase, nor do wild-type mice. However, transgenic laboratory mice created with the prairie-vole version of the receptor genes respond to the injection of vasopressin in a manner almost identical to the response of prairie voles.

> **Data analysis** Which group of mice shows the highest level of pair-bonding behavior under normal conditions? Which shows the least?

pair-bonding, whereas stimulating them leads to pair-bonding behavior. The gene that codes for the peptide receptors has also been identified, and a difference in the DNA structure between the species has been discovered. To test the hypothesis that this genetic difference was responsible for the differences in behavior, scientists created transgenic laboratory mice with the prairie vole version of the gene, and sure enough, when injected with vasopressin, the transgenic mice exhibited pair-bonding behavior very similar to that of prairie voles, whereas normal mice showed no response (figure 54.5).

The vasopressin receptor gene also varies in structure among primate species that vary in degree of pair bonding. In human males, the gene has recently been found to be associated with the strength of marital bonds and satisfaction in marriage.

Another example of single-gene effects involves the production of monoamine oxidases (MAOs), enzymes that degrade neurotransmitters such as serotonin and dopamine. Transgenic mice that lack MAOA (monoamine oxidase-A) are highly aggressive. In humans, a single point mutation results in the lack of the ability to produce MAOA, resulting in antisocial behavior and violence. MAO abnormalities are also associated with mood disorders in humans.

Learning Outcomes Review 54.3

A relationship between genes and behavior has been demonstrated in many ways, including artificial selection experiments and studies on the effects of single genes. Genes can regulate behavior by producing molecular factors that influence the function of the nervous system; mutations altering these factors have been found to affect behavior.

- *What would you infer about the role of genes in pair-bonding in prairie voles if you learned that males sometimes seek to copulate with females other than their own mate?*

54.4 Learning

Learning Outcomes

1. Describe the mechanisms of learning.
2. Define learning preparedness.
3. Explain how instinct influences learning preparedness.

Instincts can guide an animal's actions, but behavior can also develop from previous experiences, a process termed learning. Traditionally, psychologists studied the mechanisms of learning using laboratory rodents, but today both proximate and ultimate causes of learning are understood by integrating learning into an ecological and evolutionary framework.

Learning mechanisms include habituation and association

Habituation is a simple form of learning defined as a decrease in response to a repeated stimulus that has no positive or negative consequences. Initially, the stimulus may evoke a strong response, but the response declines with repeated exposure. For example, young birds see many types of objects moving overhead. At first, they may respond by crouching and remaining still. But frequently seen objects, such as falling leaves or members of their own species flying overhead, have no positive or negative consequence to the nestlings. Over time, the young birds may habituate to such stimuli and stop responding. Thus, habituation can be thought of as learning not to respond to a stimulus.

One ecological context in which habituation has adaptive value is prey defense. Birds that feed on insects search for suitable prey in a visually complex environment. Insects that have camouflaged bodies appear to be twigs or leaves. Because birds see such objects very frequently, they habituate to their appearance. Insects that look like twigs or leaves are therefore protected because they do not trigger an attack, and they survive to reproduce.

Figure 54.6 Learning what is edible. Associative learning is involved in predator–prey interactions. *a.* A naive toad is offered a bumblebee as food. *b.* The toad is stung, and *(c)* subsequently avoids feeding on bumblebees or any other insects having black-and-yellow coloration. The toad has associated the appearance of the insect with pain and modifies its behavior.

More complex forms of learning concern changes in behavior through an association between two stimuli or between a stimulus and a response. In *associative learning,* for example, (figure 54.6) a behavior is modified, or conditioned, through the association. The two major types of associative learning—classical conditioning and operant conditioning—differ in the way the associations are established. In **classical conditioning,** the paired presentation of two different kinds of stimuli causes the animal to form an association between the stimuli. Classical conditioning is also called **Pavlovian conditioning,** after the Russian psychologist Ivan Pavlov, who first described it.

Pavlov presented meat powder, an unconditioned stimulus, to a dog and noted that the dog responded by salivating, an unconditioned response. If an unrelated stimulus, such as the ringing of a bell, was repeatedly presented at the same time as the meat powder, the dog would soon salivate in response to the sound of the bell alone. The dog had learned to associate the unrelated sound stimulus with the meat powder stimulus. Its response to the sound stimulus was, therefore, conditioned, and the sound of the bell is referred to as a conditioned stimulus.

In **operant conditioning,** an animal learns to associate its behavior with a reward or punishment. American psychologist B. F. Skinner studied operant conditioning in rats by placing them in an apparatus that came to be called a "Skinner box." As the rat explored the box, it would occasionally press a lever by accident, causing a pellet of food to appear. Soon it learned to associate pressing the lever (the behavioral response) with obtaining food (the reward). This sort of trial-and-error learning is of major importance to most vertebrates. Learning provides flexibility that allows behavior to be fine-tuned to an environment.

Instinct governs learning preparedness

Psychologists once believed that any two stimuli could be linked through learning and that animals could be conditioned to perform any learnable behavior. This view has changed. Today, researchers believe that instinct guides learning by determining what type of information can be learned. Animals may have innate predispositions toward forming certain associations and not others. For example, if a rat is offered a food pellet at the same time it is exposed to X-rays (which later produce nausea), the rat remembers the taste of the food pellet but not its size, and in the future will avoid food with that taste, but will readily eat pellets of the same size if they have a different taste. Similarly, pigeons can learn to associate food with colors, but not with sounds. In contrast, they can associate danger with sounds, but not with colors.

These examples of learning preparedness demonstrate that what an animal can learn is biologically influenced—that is, learning is possible only within the boundaries set by evolution. Innate programs for learning have evolved because they lead to adaptive responses. In nature, food that is toxic to a rat is likely to have a particular taste; thus, it is adaptive to be able to associate a taste with a feeling of sickness that may develop hours later. Similarly, the seed a pigeon eats may have a distinctive color that the pigeon can see, but it makes no sound the pigeon can hear, so it makes sense that the ability to associate illness with color and not sound has evolved.

Learning Outcomes Review 54.4

Habituation is a diminishing response to a repeated stimulus that is neither positive nor negative. Association may occur as either classical conditioning or operant conditioning. Animals can change their behavior through learning in a variety of ways. Although learning mechanisms may be similar across species, animals also differ in their learning abilities according to their ecology.

- In some rodents, males travel far, whereas females remain close to the nest. Do males or females have greater spatial memory? What experiment could you conduct to test your hypothesis?

54.5 The Development of Behavior

Learning Outcomes

1. Discuss the role of the critical period in imprinting.
2. Explain how social contact can influence growth and development.
3. Explain how the study of song learning in white-crowned sparrows illustrates the interaction of instinct and learning.

Behavioral biologists recognize that behavior has both genetic and learned components. Thus far in this chapter, we have discussed the influence of genes and learning separately. But as you will see, these factors interact during development to shape behavior.

Parent–offspring interactions influence how behavior develops

As an animal matures, it may form social attachments to other individuals or develop preferences that will influence behavior later in life. This process of behavioral development is called **imprinting.** The success of imprinting is highest during a critical period (roughly 13 to 16 hours after hatching in geese). During this time, information required for normal development must be acquired. In **filial imprinting,** social attachments form between parents and offspring. For example, young birds like ducks and geese begin to follow their mother within a few hours after hatching, and their following response results in a social bond between mother and young. The young birds' initial experience, through imprinting, can determine how social behavior develops later in life. The ethologist Konrad Lorenz showed that geese will follow the first object they see after hatching and direct their social behavior toward that object, even if it is not their mother! Lorenz raised geese from eggs, and when he offered himself as a model for imprinting, the goslings treated him as if he were their parent, following him dutifully (figure 54.7).

Interactions between parents and offspring are key to the normal development of social behavior. The psychologist Harry Harlow gave orphaned rhesus monkey infants the opportunity to

Figure 54.7 An unlikely parent. The eager goslings follow Konrad Lorenz as if he were their mother. He is the first object they saw when they hatched, and they have used him as a model for imprinting. Lorenz won the 1973 Nobel Prize in Physiology or Medicine for this work.

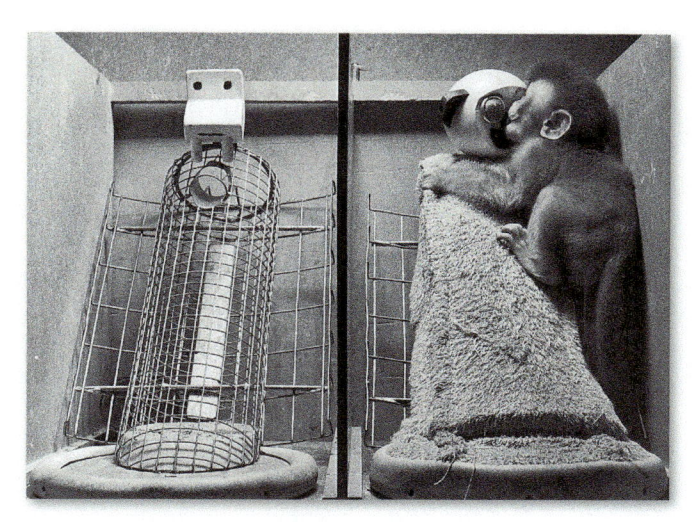

Figure 54.8 Choice trial on infant monkeys. Given a choice between a wire frame that provided food and a similar frame covered with cloth and given a monkey like head (but that did not provide food), orphaned rhesus monkeys, *Macaca mulatta,* chose the monkeylike figure over the food.

form social attachments with two surrogate "mothers," one made of soft cloth covering a wire frame and the other made only of wire (figure 54.8). The infants chose to spend time with the cloth mother, even if only the wire mother provided food, indicating that texture and tactile contact, rather than provision of food, may be among the key qualities in a mother that promote infant social attachment. If infant monkeys are deprived of normal social contact, their development is abnormal. Greater degrees of deprivation lead to greater abnormalities in social behavior during childhood and adulthood. Studies of orphaned human infants similarly suggest that a constant "mother figure" is required for normal growth and psychological development.

A biological need exists for the stimulation that occurs during parent–offspring interactions early in life. Female rats lick their pups after birth, and this stimulation inhibits the release of a brain peptide that can block normal growth. Pups that receive normal tactile stimulation also have more brain receptors for glucocorticoid hormones, thus a greater tolerance for stress, and longer-lived brain cells. Premature human infants who are massaged gain weight rapidly. These studies indicate that the need for normal social interaction is based in the brain, and that touch and other aspects of contact between parents and offspring are important for physical as well as behavioral development.

Instinct and learning may interact as behavior develops

We began this chapter by considering the proximate and ultimate causation of birdsong. Let's continue with the classic studies by neurobiologist Peter Marler on song learning in white-crowned sparrows to examine how innate programs and experience each contribute to the development of behavior.

Mature male white-crowned sparrows sing a species-specific courtship song during the mating season. Through a series

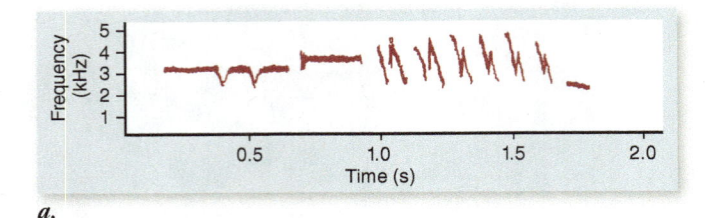

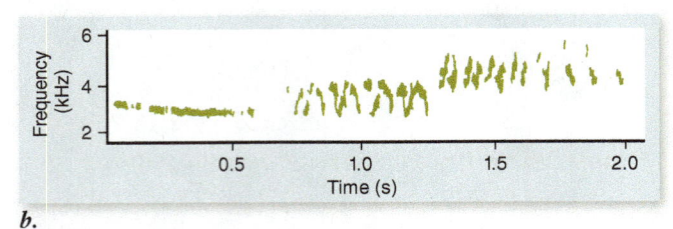

Figure 54.9 Song development in birds. *a.* The sonograms of songs produced by male white-crowned sparrows, *Zonotrichia leucophrys*, that had been exposed to their own species' song during development are different from (*b*) those of male sparrows that heard no song during rearing. This difference indicates that the genetic program itself is insufficient to produce a normal song.

of elegant experiments, Marler asked if the song was the result of an instinctive program, learning, or both. Marler reared male birds in soundproof incubators equipped with speakers and microphones to control what a bird heard as it matured, and then recorded the song it produced as an adult. Males that heard no song at all during development sang a poorly developed song as adults (figure 54.9), indicating that instinct alone did not guide song production. In a second study, males were played only the song of a different species, the song sparrow. These males sang a poorly structured song as well. This experiment showed that males would not imitate any song they heard to learn to sing. But birds that heard the song of their own species, or that heard the songs of both the white-crowned sparrow and the song sparrow, sang a fully developed, white-crowned sparrow song as adults.

These results suggest males have a selective **genetic template**, or innate program, that guides them to learn the appropriate song. During a critical period in development, the template will accept the white-crowned sparrow song as a model. Thus, song acquisition depends on learning, but only the song of the correct species can be learned; the genetic template limits what can be learned.

But learning plays a prominent role as well. If a young male becomes deaf after it hears its species' song during the critical period, it will sing a poorly developed song as an adult. Therefore, the bird must hear the correct song at the right time, and then "practice" listening to himself sing, matching what he hears to the model his genetic template has accepted.

Although this explanation of song development stood unchallenged for many years, white-crowned sparrow males can learn another species' song under certain conditions. If a live male strawberry finch *(Amandava amandava)* is placed in a cage next to a young male sparrow, the young sparrow will learn to sing the strawberry finch's song. This finding indicates that social stimuli—in this case, being able to see, hear, and interact with another bird—is more effective than a tape-recorded song in altering the innate template that guides song development.

Figure 54.10 Brood parasite. Cuckoos lay their eggs in the nests of other species of birds. Because the young cuckoos (large bird to the *right*) are raised by a different species (such as this meadow pipit, smaller bird to the *left*), they have no opportunity to learn the cuckoo song; the cuckoo song they later sing is innate.

The males of some bird species may have no opportunity to hear the song of their own species. In such cases, it appears that the males instinctively "know" their own species' song. For example, cuckoos are **brood parasites;** females lay their eggs in the nest of another species of bird, and the young that hatch are reared by the foster parents (figure 54.10). When the cuckoos become adults, they sing the song of their own species rather than that of their foster parents. Because male brood parasites would most likely hear the song of their host species during development, it is adaptive for them to ignore such "incorrect" stimuli. They hear no adult males of their own species singing, so no correct song models are available. In these species, natural selection has produced a completely genetically guided song. Other birds can also sing a correct species-typical song, even if reared in isolation.

? Inquiry question Imagine there is only one bird species on an island. Do you think instinct, learning, or both will guide song development?

Learning Outcomes Review 54.5

During the critical period, offspring must engage in certain social interactions for normal behavioral development. Parent–offspring contact stimulates the release of physiological factors, such as hormones and brain receptors, crucial to growth and brain development. In white-crowned sparrows, young males must hear their species' song to sing it correctly, indicating that both instinct and learning affect song development.

- *Some researchers have tried to link IQ and genes in humans. Why would this research be seen as controversial?*

54.6 Animal Cognition

Learning Outcome
1. Explain why behavioral biologists today are more open to considering that animals can think.

For many decades, students of animal behavior flatly rejected the notion that nonhuman animals can think. Now serious attention is given to animal awareness. The central question of whether animals show **cognitive behavior**—that is, they process information and respond in a manner that suggests thinking—is widely supported (figure 54.11).

In a series of classic experiments conducted in the 1920s, a chimpanzee was left in a room with bananas hanging from the ceiling out of reach. Also in the room were several boxes lying on the floor. After some unsuccessful attempts to jump up and grab the bananas, the chimp stacked the boxes beneath the suspended bananas, and climbed up to claim its prize (figure 54.12). Field researchers have observed that Japanese macaques learned to wash sand off potatoes and to float grain to separate it from sand. Chimpanzees pull leaves off a tree branch and then stick the branch into the entrance of a termite nest to "fish" for food. Chimps also crack open nuts using pieces of wood in a "hammer and anvil" technique. Even more remarkable is that parents appear to teach nut cracking to their offspring!

Recent studies have found that chimpanzees and other primates show behaviors that provide strong evidence of cognition. For example, chimps cooperate with other chimps in ways that suggest an understanding of past success. Cognitive ability is not limited to primates: Ravens and other corvid birds also show extraordinary insight and problem-solving ability (figure 54.13) as do octopuses (see figure 34.17).

Figure 54.12 Problem solving by a chimpanzee. Unable to get the bananas by jumping, the chimpanzee devises a solution.

Learning Outcome Review 54.6
Research has provided compelling evidence that some nonhuman animals are able to solve problems and use reasoning, cognitive abilities once thought uniquely human.

- How could you determine whether a chimpanzee has the ability to count objects?

Figure 54.11 Animal thinking? *a.* This chimpanzee is stripping the leaves from a twig, which it will then use to probe a termite nest. This behavior strongly suggests that the chimpanzee is consciously planning ahead, with full knowledge of what it intends to do. *b.* This sea otter is using a rock as an "anvil," against which it bashes a clam to break it open. A sea otter will often keep a favorite rock for a long time, as though it has a clear idea of its future use of the rock. Behaviors such as these suggest that animals have cognitive abilities.

Figure 54.13 Problem solving by a raven. Confronted with a problem it has never previously faced, the raven figures out how to get the meat at the end of the string by repeatedly pulling up a bit of string and stepping on it.

54.7 Orientation and Migratory Behavior

Learning Outcomes
1. Define migration.
2. Distinguish between orientation and navigation.
3. Describe different systems for navigation.

Monarch butterflies and many birds travel thousands of miles over continents to overwintering sites in the tropics. Many animals travel away from a nest and then return. To do so, they track cues in the environment, often showing exceptional skill at orientation. Animals with a homing instinct, such as pigeons, recognize complex features of the environment to return to their home. Despite decades of study, our understanding of animal orientation is far from complete.

Migration often involves populations moving large distances

Long-range, two-way movements are known as migrations. Each fall, ducks, geese, and many other birds migrate south along flyways from Canada across the United States, heading as far as South America, and then returning each spring.

Monarch butterflies also migrate each fall from central and eastern North America to their overwintering sites in several small, geographically isolated areas of coniferous forest in the mountains of central Mexico (figure 54.14). Each August, the butterflies begin a flight southward and at the end of winter, the monarchs begin the return flight to their summer breeding ranges. Two to five generations may be produced as the butterflies fly north: Butterflies that migrate in the autumn to the precise locations in Mexico have never been there before!

Recent geographic range expansions by some migrating birds have revealed how migratory patterns change. When colonies of bobolinks became established in the western United States, far from their normal range in the Midwest and East, they did not migrate directly to their winter range in South America. Instead, they migrated east to their ancestral range, and then south along the original flyway (figure 54.15). Rather than changing the original migration pattern, they simply added a new segment. Scientists continue to study the western bobolinks to learn whether, in time, a more efficient migration path will evolve or whether the birds will always follow their ancestral course. The behavior of butterflies and birds accentuate the mysteries of the mechanism employed during migration.

Migrating animals must be capable of orientation and navigation

Birds and other animals navigate by looking at the sun during the day and the stars at night. The indigo bunting is a short-distance nocturnal migrant bird. It flies during the day using the Sun as a guide, and compensates for the movement of the Sun in the sky as the day progresses. These birds use the positions of constellations around the North Star in the night sky as a compass.

Many migrating birds also have the ability to detect Earth's magnetic field and to orient themselves with respect to it when cues from the Sun or stars are not available. In an indoor cage, they will attempt to move in the correct geographic direction, even though there are no visible external cues. However, the placement of a magnet near the cage can alter the direction in which the birds attempt to move. Researchers have found magnetite, a magnetized iron ore, in the eyes and upper beaks of some birds, but how these sensory organs function is not known.

The first migration of a bird appears to be innately guided by both celestial cues (the birds fly mainly at night) and Earth's

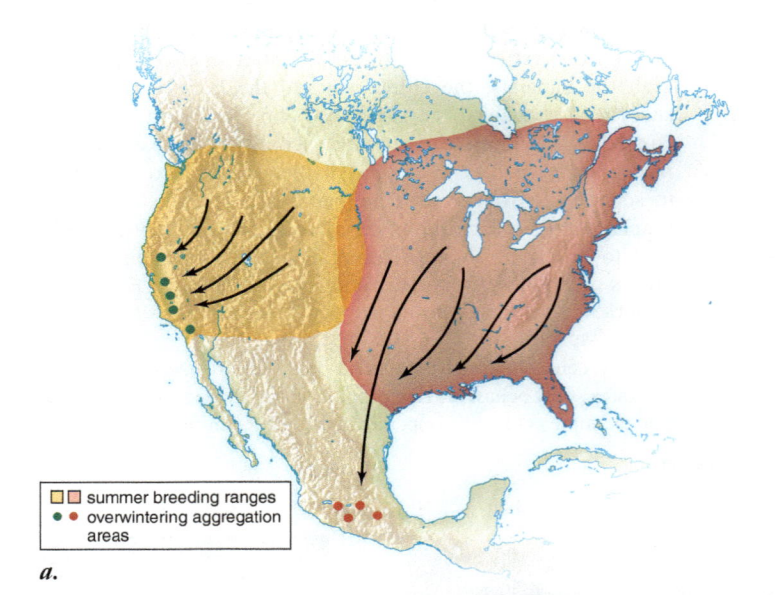

Figure 54.14 Migration of monarch butterflies, *Danaus plexippus*. *a.* Monarchs from western North America overwinter in areas of mild climate along the Pacific coast. Those from eastern North America migrate over 3000 km to Mexico. *b.* Monarch butterflies arrive at the remote forests of the overwintering grounds in Mexico, where they *(c)* form aggregations on the tree trunks.

magnetic field. When the two cues are experimentally manipulated to give conflicting directions, the information provided by the stars seems to override the magnetic information. Recent studies, however, indicate that celestial cues indicate the general direction for migration, whereas magnetic cues indicate the specific migratory path (perhaps a turn the bird must make mid-route). Experiments on starlings indicate that inexperienced birds migrate by following a set direction to get to their destination, but older birds that have migrated previously use true navigation (figure 54.16).

We know relatively little about how other migrating animals navigate. For instance, green sea turtles migrate from Brazil halfway across the Atlantic Ocean to Ascension Island, where the females lay their eggs. How do they find this tiny island in the middle of the ocean, which they haven't seen for perhaps 30 years? How do the young that hatch on the island know how to find their way to Brazil? Newly hatched turtles use

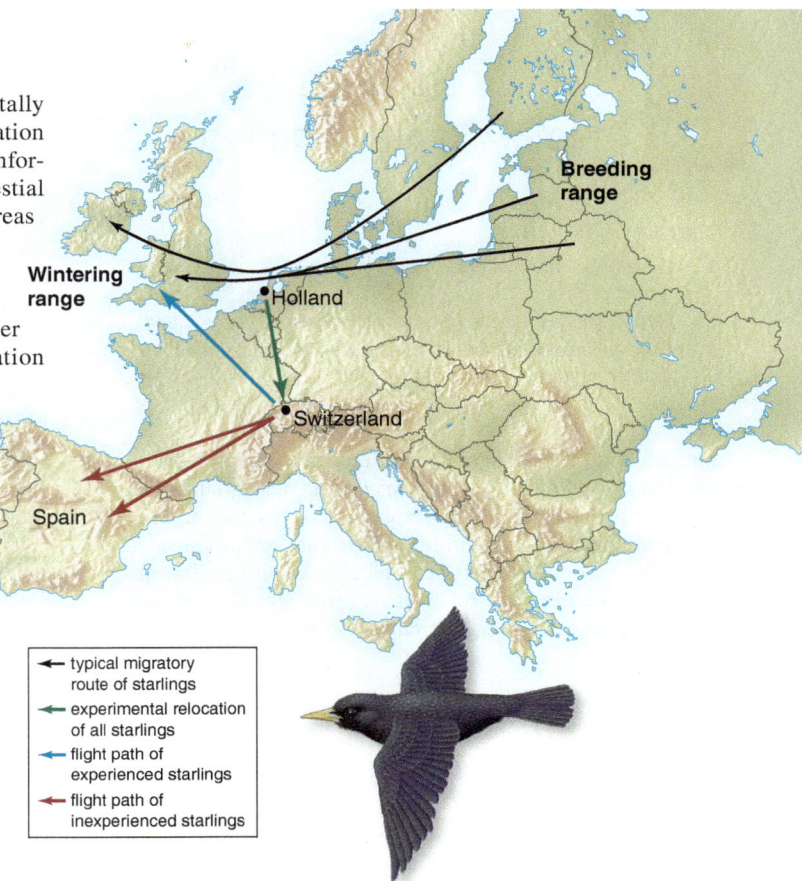

Figure 54.16 Migratory behavior of starlings, *Sturnus vulgaris*. The navigational abilities of inexperienced birds differ from those of adults that have made the migratory journey before. Starlings were captured in Holland, halfway along their full migratory route from Baltic breeding grounds to wintering grounds in the British Isles; these birds were transported to Switzerland and released. Experienced older birds compensated for the displacement and flew toward the normal wintering grounds *(blue arrow)*. Inexperienced young birds kept flying in the same direction, on a course that took them toward Spain *(red arrows)*.

wave action as a cue to head to sea. Some sea turtles use the Earth's magnetic field to maintain position in the North Atlantic, but turtle migration is still largely a mystery.

Learning Outcomes Review 54.7

Migration is the long-distance movement of a population, often in a cyclic way. Orientation refers to following a bearing or a direction; navigation involves setting a bearing or direction based on some sort of map or memory. Many species use celestial navigation; they may also be able to detect magnetic fields when those cues are absent. The precision of animal migration remains a mystery in many species.

- Animals as diverse as butterflies and birds migrate over long distances. Would you expect them to use different navigation systems? Why or why not?

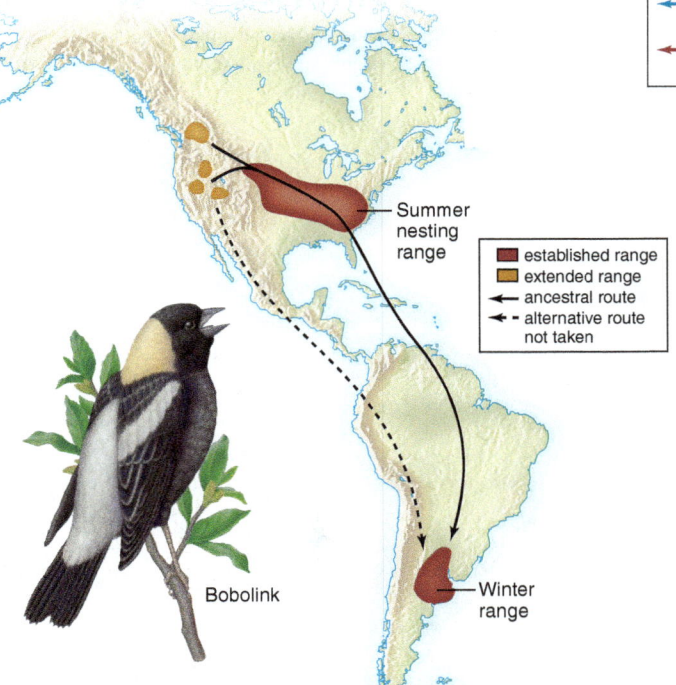

Figure 54.15 Birds on the move. The summer range of bobolinks, *Dolichonyx oryzivorus*, recently extended to the far western United States from their established range in the Midwest. When birds in these newly established populations migrate to South America in the winter, they do not fly directly to the winter range; instead, they first fly to the Midwest and then use the ancestral flyway, going much farther than if they flew directly to their winter range.

54.8 Animal Communication

Learning Outcomes
1. Explain the nature of signals used in mate attraction.
2. Explain the role of courtship signals in reproductive isolation.
3. Describe how honeybees communicate information about the location of new food sources.

Communication is central to species recognition and reproductive isolation, and to the interactions that are essential to social behavior. Much research in behavior analyzes the nature of communication signals, determining how they are produced and received, and identifies their ecological roles and evolutionary origins. Communication involves several signal modalities, including visual, acoustic, chemical, electric, and vibrational signals.

Successful reproduction depends on appropriate signals and responses

During courtship, animals produce signals to communicate with potential mates. A stimulus–response chain sometimes occurs, in which the behavior of the male in turn releases a behavior in the female, resulting in mating (figure 54.17). These signals are usually highly species-specific. Many studies on communication involve designing experiments to determine which key stimuli associated with an animal's visual appearance, sounds, or odors convey information about the nature of the signals produced by the sender. One classical study analyzed territorial defense and courtship communication in stickleback fish (figure 54.18).

Finding a mate: Communicating information about species identity

Courtship signals often restrict communication to members of the same species and in doing so serve a key function in reproductive isolation (see chapter 22). The flashes of fireflies (which are actually

Figure 54.17 A stimulus–response chain. Stickleback courtship involves a sequence of behaviors leading to the fertilization of eggs.

1. Female gives head-up display to male
2. Male swims zigzag to female and then leads her to nest
3. Male shows female entrance to nest
4. Female enters nest and spawns while male stimulates tail
5. Male enters nest and fertilizes eggs

SCIENTIFIC THINKING

Hypothesis: The red underside of male stickleback is the key stimulus that releases an aggressive response by a territory-holding male.

Prediction: Models with red coloration will trigger an attack by a resident male.

Test: Construct plastic models, some of which accurately resemble a stickleback male, but lack the red underside. Construct other models that vary in their fishlike appearance, but have red-colored undersides. Expose a territorial male to the models one at a time, and record the number of attacks.

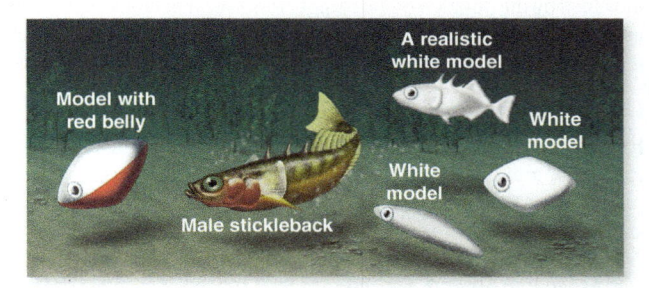

Result: Realistic models lacking red elicit no response. Odd-shape models trigger an attack if they have red undersides, even if they poorly resemble fish.

Conclusion: The red underside of a male stickleback is the key stimulus that triggers aggressive behavior.

Further Experiments: How would you determine if the color of a male stickleback was a releaser of aggressive behavior? How would you know if sound was important? Could you determine if stimuli have additive effects? How? (See also figure 54.17.)

Figure 54.18 Key stimulus in stickleback fish.

Figure 54.19 Firefly fireworks. The bioluminescent displays of these lampyrid beetles are species-specific and serve as behavioral mechanisms of reproductive isolation. Each number represents the flash pattern of a male of a different species.

beetles) are species-specific signals: females recognize conspecific males by their flash pattern (figure 54.19), and males recognize conspecific females by their flash response. This series of reciprocal responses provides a continuous "check" on the species identity of potential mates.

Pheromones, chemical messengers used for communication between individuals of the same species, serve as sex attractants in many animals. Female silk moths *(Bombyx mori)* produce a sex pheromone called bombykol in a gland associated with the reproductive system. The male's antennae contain numerous highly sensitive sensory receptors, and neurophysiological studies show they specifically detect bombykol. In some moth species, males can detect extremely low concentrations of sex pheromone and locate females from as far as 7 km away!

Many insects, amphibians, and birds produce species-specific acoustic signals to attract mates. Bullfrog males call by inflating and discharging air from their vocal sacs, located beneath the lower jaw. Females can distinguish a conspecific male's call from those of other frogs that may be in the same habitat and calling at the same time. As mentioned in section 54.1 male birds sing to advertise their presence and to attract females. In many species, variations in the males' songs identify individual males in a population. In these species, the song is individually specific as well as species-specific.

Vibrations, like sound signals, are a form of mechanical communication used by insects, amphibians, and other animals. For example, lacewings distinguish one species from another by the pattern of vibrations they make on branches with their abdomens (see figure 22.4).

Courtship behaviors play a major role in sexual selection, which we discuss in section 54.10.

Communication enables information exchange among group members

Many insects, fish, birds, and mammals live in social groups in which information is communicated between group members. For example, some individuals in mammalian societies serve as sentinels, vigilantly on the lookout for danger. When a predator appears, they give an alarm call, and group members respond by seeking shelter (figure 54.20). Social insects such as ants and honeybees produce alarm pheromones that trigger attack behavior. Ants also deposit trail pheromones between the nest and a food source to lead other colony members to food. Honeybees have an extremely complex dance language that directs hivemates to nectar sources.

The dance language of the honeybee

The European honeybee lives in colonies of tens of thousands of individuals whose behaviors are integrated into a complex,

Figure 54.20 Alarm calling by a prairie dog, *Cynomys ludovicianus*. When a prairie dog sees a predator, it stands on its hind legs and gives an alarm call, which causes other prairie dogs to rapidly return to their burrows.

cooperative society. Worker bees may forage miles from the hive, collecting nectar and pollen from a variety of plants and switching between plant species depending on their energetic rewards. Food sources used by bees tend to occur in patches, and each patch offers much more food than a single bee can transport to the hive. A colony is able to exploit the resources of a patch because of the behavior of scout bees, which locate patches and communicate their location to hivemates through a dance language. Over many years, Nobel laureate Karl von Frisch (who shared the 1973 prize with Tinbergen and Lorenz), together with generations of students and colleagues, was able to unravel the details of dance language communication.

When a scout bee returns after finding a distant food source, she performs a remarkable behavior pattern called a waggle dance on a vertical comb in the darkness of the hive. The path of the bee during the dance resembles a figure-eight. On the straight part of the path (indicated with dashes in figure 54.21), the bee vibrates ("waggles") her abdomen while producing bursts of sound. The bee may stop periodically to give her hivemates a sample of the nectar carried in her crop. As she dances, she is followed closely by other bees, which soon appear at the new food source to assist in collecting food.

Von Frisch and his colleagues performed experiments to show that hivemates use information in the waggle dance to locate new food sources. The scout bee indicates the direction of the food source by representing the angle between the food source, the hive, and the Sun as the deviation from vertical of the straight run of the dance performed on the hive comb. Thus if the bee danced with the straight run pointing directly up, then the food source would be in the direction of the Sun. If the food is at a 30° angle to the right of the Sun's position, then the straight run would be oriented upward at a 30° angle to the right of vertical) (figure 54.21*a*). The distance to the food source is indicated by the duration of the straight run. One ingenious experiment used computer-controlled robot bees to give hivemates incorrect information, demonstrating that bees use the directions coded in the dance!

Language in nonhuman primates and humans

Evolutionary biologists have sought the origins of human language in the communication systems of monkeys and apes. Some nonhuman primates have a "vocabulary" that allows individuals to signal the identity of specific predators. Different vocalizations of African vervet monkeys, for example, indicate eagles, leopards, or snakes, among other threats (figure 54.22).

The complexity of human language would at first appear to defy biological explanation, but closer examination suggests that the differences are in fact superficial—all languages share many basic similarities. All of the roughly 3000 languages draw from the same set of 40 consonant and vowel sounds (English uses two dozen of them), and humans of all cultures can acquire and learn them. Researchers believe these similarities reflect the way our brains handle abstract information.

Learning Outcomes Review 54.8

Animal communication involves production and reception of signals, in the form of sounds, chemicals, or movements, that primarily have an ecological function. Courtship signals are highly species-specific and serve as a mechanism of reproductive isolation. Animals living in social groups, such as honeybees, may use complex systems of communication to exchange information about food and predators.

- *Two species of moth use the same sex pheromone to locate mates. Explain how these species could nevertheless be reproductively isolated.*

Figure 54.21 The waggle dance of honeybees, *Apis mellifera*. *a.* The angle between the food source, the nest, and the Sun is represented by a dancing bee as the angle between the straight part of the dance and vertical. The food is 30° to the right of the Sun, and the straight part of the bee's dance on the hive is 30° to the right of vertical. *b.* A scout bee dances on a comb in the hive.

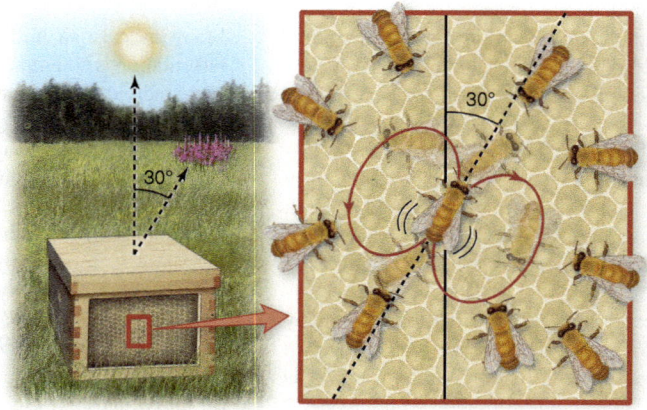

a.   *b.*

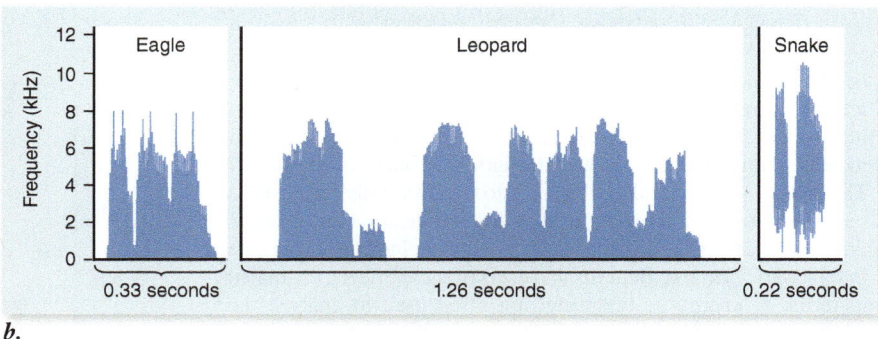

Figure 54.22 Primate semantics. Vervet monkeys, *Cercopithecus aethiops,* give different alarm calls (*a*) when troop members sight an eagle, leopard, or snake. *b.* Each distinctive call elicits a different and adaptive escape behavior.

54.9 Behavioral Ecology

Learning Outcomes
1. Describe behavioral ecology.
2. Discuss the economic analysis of behaviors.

Niko Tinbergen pioneered the study of the adaptive function of behavior. Stated simply, this is how behavior allows an animal to stay alive and keep its offspring alive. For example, Tinbergen observed that after gull nestlings hatch, the parents remove the eggshells from the nest. To understand why (ultimate causation), he painted chicken eggs to resemble gull eggs (figure 54.23), which had camouflage coloration to allow them to be inconspicuous against the natural background. He distributed them throughout the area in which the gulls were nesting, placing broken eggshells with their prominent white interiors next to some of the eggs. As a control, he left other camouflaged eggs alone without eggshells. He then noted which eggs were found more easily by crows. Because the crows could use the white interior of a broken eggshell as a cue, they ate more of the camouflaged eggs that were near eggshells. Tinbergen concluded that eggshell removal behavior is adaptive: It reduces predation and thus increases the offspring's chances of survival.

Tinbergen is credited with being one of the founders of **behavioral ecology,** the study of how natural selection shapes behavior. This branch of ecology examines the adaptive significance of behavior, or how behavior may increase survival and reproduction. Current research in behavioral ecology focuses on how behavior contributes to an animal's reproductive success, or fitness. As we saw in section 54.3, differences in behavior among individuals often result from genetic differences. Therefore, natural selection operating on behavior has the potential to produce evolutionary change.

Consequently, the field of behavioral ecology is concerned with two questions. First, is behavior adaptive? Although it is tempting to assume that behavior must in some way represent an adaptive response to the environment, this need not be the case. As you saw in chapter 20, traits can appear for many reasons other than natural selection, such as genetic drift, gene flow, or the correlated consequences of selection on other traits. Moreover, traits may be present in a population because they evolved as adaptations in the past, but are no longer useful. These possibilities hold true for behavioral traits as much as for any other kind of trait.

Figure 54.23 The adaptive value of egg coloration.
Niko Tinbergen, a winner of the 1973 Nobel Prize in Physiology or Medicine, painted chicken eggs to resemble the mottled brown camouflage of gull eggs. The eggs were used to test the hypothesis that camouflaged eggs are more difficult for predators to find and thus increase the young's chances of survival.

If behavior is adaptive, the next question is: "How is it adaptive?" Although the ultimate criterion is reproductive success, behavioral ecologists are interested in how behavior can lead to greater reproductive success. Does a behavior enhance energy intake, thus increasing the number of offspring produced? Does it increase mating success? Does it decrease the chance of predation? The job of a behavioral ecologist is to determine the effect of a behavioral trait—for example, foraging efficiency—on each of these activities and then to discover whether increases translate into increased fitness. Benefits and costs of these effects, estimated in terms of energy or offspring, are often used to analyze the adaptive nature of behavior.

Foraging behavior can directly influence energy intake and individual fitness

A useful way to understand the approach of behavioral ecology is by focusing on foraging behavior. For many animals, food comes in a variety of sizes. Larger foods may contain more energy but may be harder to capture and less abundant. In addition, animals may forage for some types of food that are farther away than other types. For these animals foraging involves a trade-off between a food's energy content and the cost of obtaining it. The net energy (estimated in calories or joules) gained by feeding on prey of each size is simply the energy content of the prey minus the energy costs of pursuing and handling it. According to **optimal foraging theory,** natural selection favors individuals whose foraging behavior is as energetically efficient as possible. In other words, animals tend to feed on prey that maximize their net energy intake per unit of foraging time.

A number of studies have demonstrated that foragers do prefer prey that maximize energy return. Shore crabs, for example, tend to feed primarily on intermediate-sized mussels, which provide the greatest energy return; larger mussels yield more energy, but also take considerably more energy to crack open (figure 54.24).

This optimal foraging approach assumes natural selection will favor behavior that maximizes energy acquisition if the increased energy reserves lead to increases in reproductive success. In both Columbian ground squirrels (*Spermophilus columbianus*) and captive zebra finches, a direct relationship exists between net energy intake and the number of offspring raised; similarly, the reproductive success of orb-weaving spiders is related to how much food they can capture.

Animals have other needs beside energy, however, and sometimes these needs conflict. One obvious need is the avoidance of predators: Often, the behavior that maximizes energy intake is not the one that minimizes predation risk. In this case, the behavior that maximizes fitness often may reflect a trade-off between obtaining the most energy at the least risk of being eaten. Not surprisingly, many studies have shown that a wide variety of animal species alter their foraging behavior—becoming less active, spending more time watching for predators, or staying nearer to cover—when predators are present. Compromises, in this case a trade-off between vigilance and feeding, may thus be made during foraging.

Optimal foraging theory assumes that energy-maximizing behavior has evolved by natural selection. Therefore, it must have a genetic basis. For example, female zebra finches particularly successful in maximizing net energy intake tend to have similarly successful offspring. In this study, young birds were removed from their mothers before they were able to leave the nest, so this similarity indicates that foraging behavior probably has a genetic component. Studies on other animals show that age, experience, and learning are also important to the development of efficient foraging.

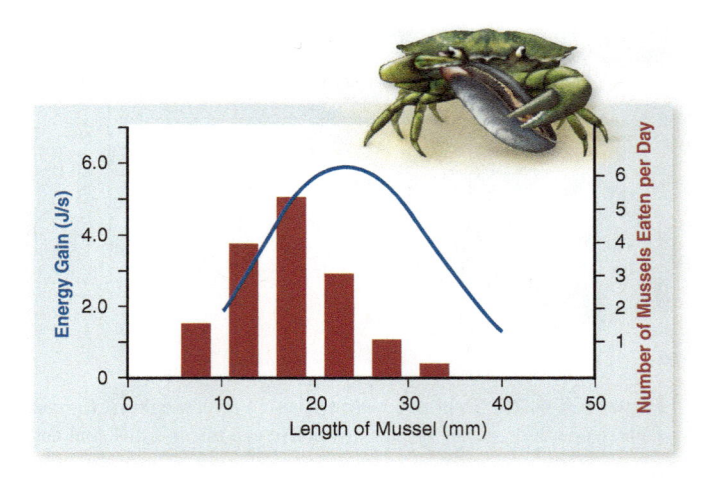

Figure 54.24 Optimal diet. The shore crab selects a diet of energetically profitable prey. The curve describes the net energy gain (equal to energy gained minus energy expended) derived from feeding on different sizes of mussels. The bar graph shows the numbers of mussels of each size in the diet. Shore crabs tend to feed on those mussels that provide the most energy.

Inquiry question What factors might be responsible for the slight difference in peak prey length relative to the length optimal for maximum energy gain?

Data analysis Suppose a raccoon forages for 10 hours a day. It can either look for frogs in a stream or insects in the dirt. It takes on average 30 minutes to find a frog, but only 12 minutes to find a cricket. The raccoon obtains 10 joules of energy eating a frog and 2 joules for each cricket. Where should the raccoon forage, the stream or the dirt?
  In reality, we must also include the time that it takes to capture and eat prey (called "handling time" as well as the time it takes to find them). Suppose it takes 15 minutes to subdue and ingest every frog, as opposed to 3 minutes per cricket. Does that change your answer?

Territorial behavior evolves if the benefits of holding a territory exceed the costs

Animals often move over a large area, or home range, during their course of activity. In many species, the home range of several individuals overlaps in time or in space, but each individual defends a portion of its home range and uses it and its resources exclusively. This behavior is called **territoriality** (figure 54.25).

The defining characteristic of territorial behavior is defense against intrusion and resource use by other individuals. Territories

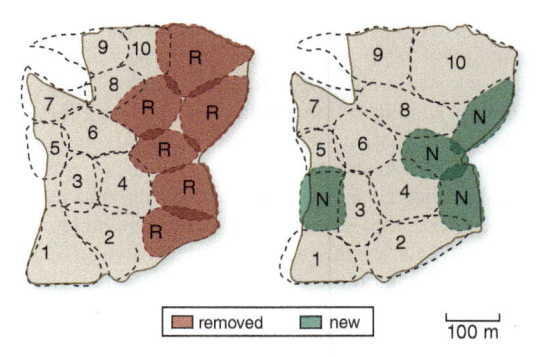

Figure 54.25 Competition for space. Territory size in birds is adjusted according to the number of competitors. When six pairs of great tits, *Parus major,* were removed from their territories (indicated by R in the *left* figure), their territories were taken over by other birds in the area and by four new pairs (indicated by N in the *right* figure). Numbers correspond to the birds present before and after.

are defended by displays advertising that territories are occupied, and by overt aggression. A bird sings from its perch within a territory to prevent take-over by a neighboring bird. If a potential usurper is not deterred by the song, the territory owner may attack and try to drive it away. But territorial defense has its costs. Singing is energetically expensive, and attacks can lead to injury. Using a signal (a song or visual display) to advertise occupancy can reveal a bird's position to a predator.

Why does an animal bear the costs of territorial defense? Energetic benefits of territoriality may take the form of increased food intake due to exclusive use of resources, access to mates, or access to refuges from predators. Studies of nectar-feeding birds such as hummingbirds and sunbirds provide an example (figure 54.26). A bird benefits from having the exclusive use of a patch of flowers because it can efficiently harvest the nectar the flowers produce. To maintain exclusive use, the bird must actively defend the patch. The benefits of exclusive use outweigh the costs of defense only under certain conditions.

Figure 54.26 The benefit of territoriality. Sunbirds (on the *left*), found in Africa and ecologically similar to New World hummingbirds (on the *right*), protect their food source by attacking other sunbirds that approach flowers in their territory.

Sunbirds, for example, expend 3000 calories per hour chasing intruders from a territory. Whether the benefit of defending a territory will exceed this cost depends on the amount of nectar in the flowers and how efficiently the bird can collect it. When flowers are very scarce or nectar levels are very low, a nectar-feeding bird may not gain enough energy to balance the energy used in defense. Under these conditions, it is not energetically advantageous to be territorial. Similarly, when flowers are very abundant, a bird can efficiently meet its daily energy requirements without behaving territorially and adding the costs of defense. Again, from an energetic standpoint, defending abundant resources isn't worth the cost. Territoriality therefore only occurs at intermediate levels of flower availability and nectar production, when the benefits of defense outweigh the costs.

In many species, access to females is a more important determinant of territory size for males than is food availability. In some lizards, for example, males maintain enormous territories during the breeding season. These territories, which encompass the territories of several females, are much larger than would be required to supply enough food, and they are defended vigorously. In the nonbreeding season, by contrast, male territory size decreases dramatically, as does aggressive territorial behavior.

Learning Outcomes Review 54.9

Behavioral ecology is the study of the adaptive significance of behavior—that is, how it affects survival and reproductive success. An economic approach estimates the energy benefits and costs of a behavior and assumes that animals gain more from a behavior than they expend, obtaining a fitness advantage. Foraging behavior and defense of a territory can be analyzed in this way. Apart from energy gains, considerations such as avoiding predators are also important to fitness.

- *The Hawaiian honeycreeper, a nectar-feeding bird, fails to defend flowers that are either infrequently encountered or very abundant. Why?*

54.10 Reproductive Strategies and Sexual Selection

Learning Outcomes

1. Explain parental investment and the prediction it makes about mate choice.
2. Describe how sexual selection leads to the evolution of secondary sexual characteristics.
3. Explain why some species are generally monogamous and others are polygynous.

During the breeding season, animals make several important life-history "decisions" concerning their choice of mates, how many mates to have, and how much time and energy to devote to rearing offspring. These decisions are all aspects of an animal's **reproductive strategy,** a set of behaviors that presumably have evolved to maximize reproductive success. Energetic costs of reproduction appear to have been critically important to behavioral differences between females and males. Ecological factors such as the way food resources, nest sites, and members of the opposite sex are spatially distributed in the environment, as well as disease, are important in the evolution of reproductive decisions.

The sexes often have different reproductive strategies

Males and females have the common goal of improving the quantity and quality of offspring they produce, but usually differ in the way they attempt to maximize fitness. Such a difference in reproductive behavior is clearly seen in mate choice. Darwin was the first to observe that females often do not mate with the first male they encounter, but instead seem to evaluate a male's quality and then decide whether to mate. Peahens prefer to mate with peacocks that have more eyespots on their elaborate tail feathers (figure 54.27b, c). Similarly, female frogs prefer to mate with males having more acoustically complex, and thus attractive, calls. This behavior, called mate choice, is well known in many invertebrate and vertebrate species.

Males are selective in choosing a mate much less frequently than females. Why should this be? Many of the differences in reproductive strategies between the sexes can be understood by comparing the parental investment made by males and females. **Parental investment** refers to the energy and time each sex "invests" in producing and rearing offspring; it is, in effect, an estimate of the energy expended by males and females in each reproductive event.

Numerous studies have shown that females generally have a higher parental investment. One reason is that eggs are much larger than sperm—195,000 times larger in humans! Eggs contain proteins and lipids in the yolk and other nutrients for the developing embryo, but sperm are little more than mobile DNA packages. In some groups of animals (mammals, for example), females are responsible for gestation and lactation, costly reproductive functions only they can carry out.

The consequence of such inequalities in reproductive investment is that the sexes face very different selective pressures. Because any single reproductive event is relatively inexpensive for males, they can best increase their fitness by mating with as many females as possible. This is because male fitness is likely not limited by the amount of sperm they can produce. By contrast, each reproductive event for females is much more costly, and the number of eggs that can be produced often limits reproductive success. For this reason, a female should be choosy, trying to pick the male that can provide the greatest benefit to her offspring and thus improve her fitness.

These conclusions hold only when female reproductive investment is much greater than that of males. In species with biparental care, males may contribute equally to the cost of raising young; in this case, the degree of mate choice should be more equal between the sexes.

In some cases, male investment exceeds that of females. For example, male Mormon crickets transfer a protein-containing packet (a *spermatophore*) to females during mating. Almost 30% of a male's body weight is made up by the spermatophore, which provides nutrition for the female and helps her develop her eggs. As we might expect from our model of mate choice, in this case it is the females that compete with one another for access to males, which are the choosy sex. Indeed, males are quite selective, favoring heavier females. Heavier females have more eggs; thus, males that choose larger females leave more offspring (figure 54.28).

a. *b.*

c.

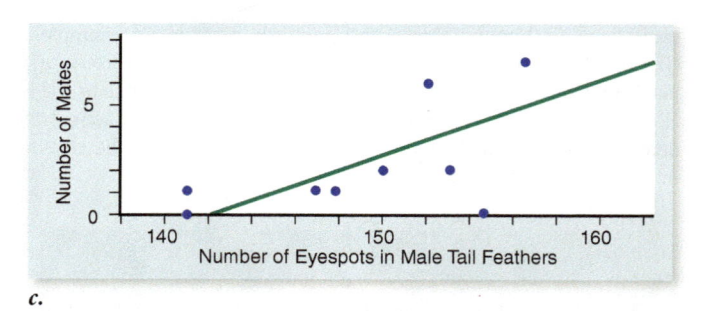

Figure 54.27 Products of sexual selection. In bird species such as (*a*) the raggiana bird of paradise, *Paradisaea raggiana*, and (*b*) the peacock, *Pavo cristatus*, males use their much longer tail feathers in courtship displays. *c.* Female peahens prefer to mate with males having greater numbers of eyespots in their tail feathers.

? Inquiry question Why do females prefer males with more spots?

Data analysis Suppose a male had 155 eyespots. How many females would you expect him to mate with? If you had multiple males with 155 eyespots, do you think they all would mate with the same number of females?

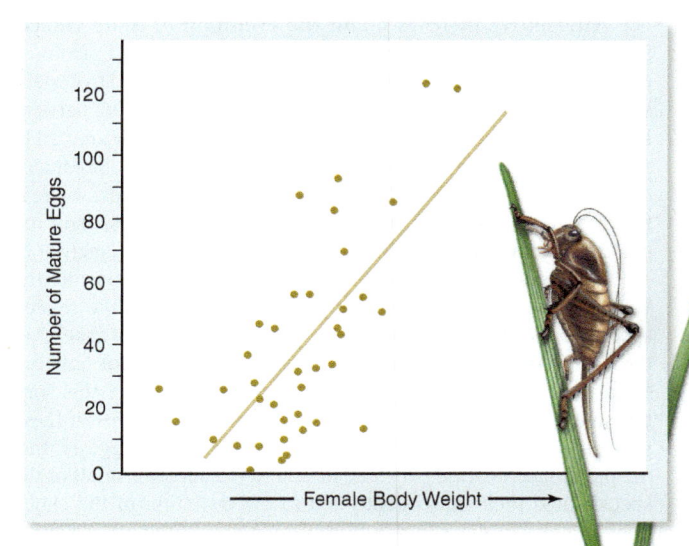

Figure 54.28 The advantage of male mate choice. Male Mormon crickets, *Anabrus simplex*, choose heavier females as mates, and larger females have more eggs. Thus, male mate selection increases fitness.

> **? Inquiry question** Is there a benefit to females for mating with large males?

Males care for eggs and developing young in many species, including seahorses and a number of birds and insects. As with Mormon crickets, these males are often choosy, and females compete for mates.

Sexual selection occurs through mate competition and mate choice

As discussed in chapter 20, the reproductive success of an individual is determined by how long the individual lives, how frequently it mates, and how many offspring it produces per mating. The second of these factors, competition for mates, is termed **sexual selection.** Some people consider sexual selection to be distinctive from natural selection, but others see it as a subset of natural selection, just one of the many factors affecting an organism's fitness.

Sexual selection involves both **intrasexual selection,** or competitive interactions between members of one sex ("the power to conquer other males in battle," as Darwin put it), and **intersexual selection,** which is another name for mate choice ("the power to charm"). Sexual selection leads to the evolution of structures used in combat with other males, such as a deer's antlers and a ram's horns, as well as ornaments used to "persuade" members of the opposite sex to mate, such as long tail feathers and bright plumage (see figure 54.27a, b). These traits are called **secondary sexual characteristics.**

Selection strongly favors any trait that confers greater ability in mate competition. Larger body size is a great advantage if dominance is important, as it is in territorial species. Males may thus be considerably larger than females. Such differences between the sexes are referred to as **sexual dimorphism.** In other species, structures used for fighting, such as horns, antlers, and large canine teeth, have evolved to be larger in males because of the advantage they give in intrasexual competition.

Sometimes **sperm competition** occurs between the sperm of different males if females mate with multiple males. This type of competition, which occurs after mating, has selected for features that increase the probability that a male's sperm will fertilize the eggs. Such features include sperm-transfer organs designed to remove the sperm of a prior mating, large testes to produce more sperm per mating, and sperm that hook themselves together to swim more rapidly.

Intrasexual selection

In many species, individuals of one sex—usually males—compete with each other for the opportunity to mate. Competition among males can occur for a territory in which females feed or bear young. Males may also directly compete for the females themselves. A few successful males may engage in an inordinate number of matings, whereas most males do not mate at all. For example, elephant seal males control territories on breeding beaches and a few dominant males do most of the breeding (figure 54.29). On one beach, for example, eight males impregnated 348 females, whereas the remaining males mated rarely, if at all.

Intersexual selection

Intersexual selection concerns the active choice of a mate. Mate choice has both direct and indirect benefits.

Direct benefits of mate choice. In some cases, the female benefits directly from her choice of mates. If males help raise offspring, females benefit by choosing the male that can provide the best care—the better the parent, the more offspring she is likely to rear. In other species, males provide no care, but maintain territories that provide food, nesting sites, and predator refuges. In red deer, males that hold territories with the highest quality grasses

Figure 54.29 Male elephant seals, *Mirounga angustirostris*, fight with one another for possession of territories. Only the largest males can hold territories, which contain many females.

mate with the most females. In this case, there is a direct benefit of a female mating with such a territory owner: She feeds with little disturbance on quality food.

Indirect benefits of mate choice. In many species, however, males provide no direct benefits of any kind to females. In such cases, it is not intuitively obvious what females have to gain by being "choosy." Moreover, what could be the possible benefit of choosing a male with an extremely long tail or a complex song? One possible indirect benefit of mate choice is that it could lead to higher quality offspring.

A number of theories have been proposed as to how this might work. One idea is that females choose the male that is the healthiest or oldest. Large males, for example, have probably been successful at living long, acquiring a lot of food, and resisting parasites and disease. In other species, features other than size may indicate a male's condition. In guppies and some birds, the brightness of a male's color reflects the quality of his diet and overall health. Females may gain two benefits from mating with the healthiest males. First, healthy males are less likely to be carrying diseases, which might be transmitted to the female during mating; this would be a direct benefit. Second, to the extent that the males' success in living long and prospering is the result of his genetic makeup, the female will be ensuring that her offspring receive good genes from their father, an indirect benefit.

Several experimental studies in fish and moths have examined whether female mate choice leads to greater reproductive success. In these experiments, females in one group were allowed to choose the males with which they would mate, whereas males were randomly mated to a different group of females. Offspring of females that chose their mates were more vigorous and survived better than offspring from females given no choice, which suggests that females preferred males with a better genetic makeup.

A variant of this theory goes one step further. In some cases, females prefer mates with traits that appear to be detrimental to survival (see figure 54.27c). The long tail of the peacock is a hindrance in flying and makes males more vulnerable to predators. Why should females prefer males with such traits? The **handicap hypothesis** states that only genetically superior mates can survive with such a handicap. By choosing a male with the largest handicap, the female is ensuring that her offspring will receive these quality genes. Of course, the male offspring will also inherit the genes for the handicap. For this reason, evolutionary biologists are still debating the merit of this hypothesis.

Alternative theories about the evolution of mate choice. Some courtship displays appear to have evolved from a predisposition in the female's sensory system to respond to certain stimuli. For example, females may be better able to detect particular colors or sounds at a certain frequency, and thus be attracted to such signals. **Sensory exploitation** involves the evolution in males of a signal that "exploits" these preexisting biases. For example, if females are particularly adept at detecting red objects, then red coloration may evolve in males as part of a courtship display.

To understand the evolution of courtship calls, consider the vocalizations of the Túngara frog (figure 54.30). Unlike related species, males include a short burst of sound, termed a "chuck," at the end of their calls. Recent research suggests that not only are females of this species particularly attracted to calls of this sort, but so are females of related species, even though males of these species do not produce "chucks." This pattern suggests that the preference evolved for some reason in the ancestor of all of the species, and then subsequently led to the evolution of the chuck only in males of the Túngara frog. Why males of the other species haven't evolved to produce chucks is a good question.

A great variety of other hypotheses have been proposed to explain the evolution of mating preferences. Many of these hypotheses may be correct in some circumstances, but none seems capable of explaining all of the variation in mating behavior in the animal world. This is an area of vibrant research, with new discoveries appearing regularly.

Mating systems reflect the ability of parents to care for offspring and are influenced by ecology

The number of individuals with which an animal mates during the breeding season varies among species. Mating systems include monogamy (one male mates with one female), polygyny (one male mates with more than one female; see figure 54.29), and polyandry (one female mates with more than one male). Only monogamous mating includes a pair bond (like prairie voles). Like mate choice, mating systems have evolved to allow females and males to maximize fitness.

The option of having more than one mate may be constrained by the need for offspring care. If females and males are able to care for young, then the presence of both parents may be necessary for young to be reared successfully. Monogamy may thus be favored. Generally this is the case for birds, in which over 90% of all species appear to be monogamous. A male may

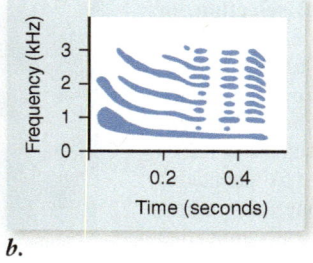

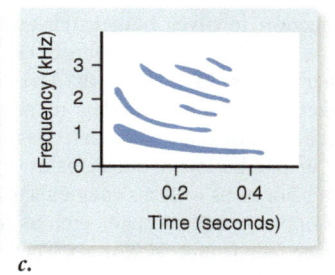

Figure 54.30 Male Túngara frog, *Physalaemus pustulosus*, calling. Female frogs of several species in the genus *Physalaemus* prefer males that include a "chuck" in their call. However, only males of the Túngaru frog (*a*) produce such calls (*b*); males of other species do not (*c*).

either remain with his mate and provide care for the offspring or desert that mate to search for others; both strategies may increase his fitness. The strategy that natural selection will favor depends on the requirement for male assistance in feeding or defending the offspring. In some species (like humans), offspring are **altricial**—they require prolonged and extensive care. In these species, the need for care by two parents reduces the tendency for the male to desert his mate and seek other matings. In species in which the young are **precocial** (requiring little parental care), males may be more likely to be polygynous because the need for their parenting is lower. In mammals, only females lactate, freeing males from feeding offspring. It follows that most mammals are polygynous.

Mating systems are strongly influenced by ecology. A male may defend a territory that holds nest sites or food sources sufficient for more than one female. If territories vary in quality or quantity of resources, a female's fitness is maximized if she mates with a male holding a high-quality territory. Although a male may already have a mate, it is still more advantageous for the female to breed with a mated male holding a high-quality territory than with an unmated male holding a low-quality territory. This favors the evolution of polygyny.

Polyandry is relatively rare, but the evolution of multiple mating by females is becoming better understood. It is best known in birds like spotted sandpipers and jacanas living in highly productive environments such as marshes and wetlands. Here, females take advantage of the increased resources available to rear offspring by laying clutches of eggs with more than one male. Males provide all incubation and parenting, and females mate and leave eggs with two or more males.

Females may also mate with several males to genetically diversify their offspring, which in turn increases disease resistance. This appears to be the case in honey bees, for example, in which a queen may mate with many males.

Extra-pair copulations

The apparent "monogamy" of many bird species has been reevaluated as DNA fingerprinting (see chapter 17) has become commonly used to determine paternity and precisely quantify the reproductive success of individual males (figure 54.31a). In red-winged blackbirds (figure 54.31b), researchers established that half of all nests contained at least one hatchling fertilized by a male other than the territory owner; overall, 20% of the offspring were the result of such **extra-pair copulations (EPCs).**

What is the evolutionary advantage of EPCs? For males, the answer is obvious: increased reproductive success. Females, on the other hand, may mate with genetically superior individuals even if already paired with a male, thus enhancing the genes passed on to their offspring. The female doesn't produce more offspring, but offspring of better genetic quality. In some birds and other animals, EPCs may help females increase the amount of care they get from males to raise their offspring. This is exactly what happens in a common English bird. Females mate not only with the territory owner, but also with subordinate males that hang around the edge of the territory. If these subordinates mate a sufficient number of times with a female, they will help raise her young, presumably because they may have fathered some of these young.

Alternative mating strategies

Natural selection has led to the evolution of many ways of increasing reproductive success. For example, in many species of fish,

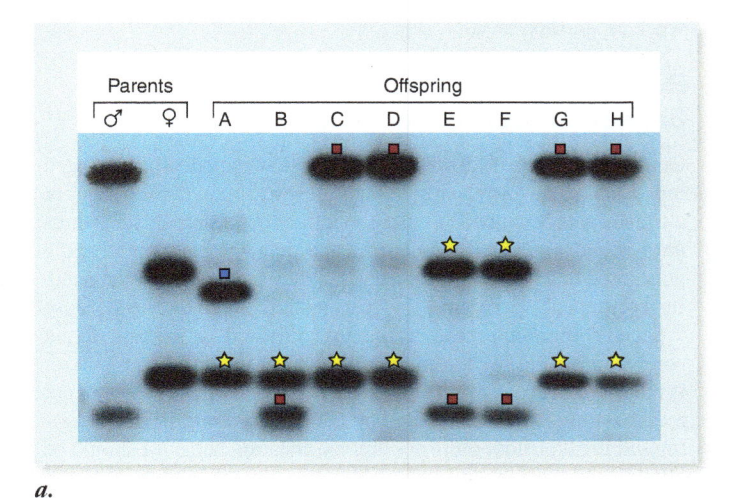

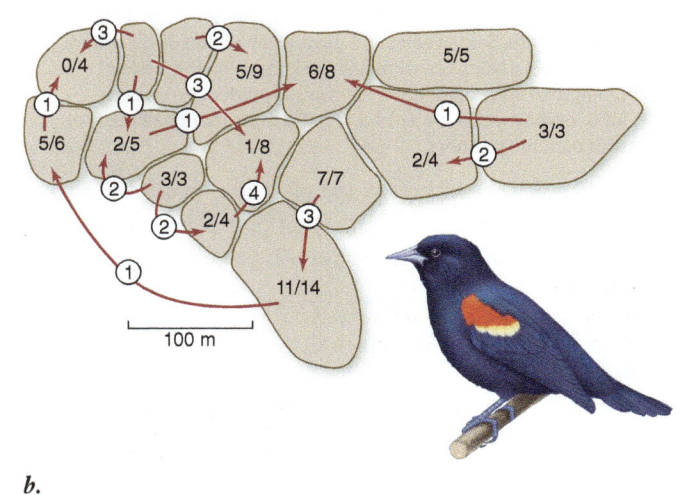

a.

b.

Source: a: (unpubl, Douglas Ross)

Figure 54.31 The study of paternity. *a.* DNA fingerprinting gel from parents and offspring in a nest of pied flycatchers, *Ficedula hypoleuca*. The bands represent fragments of different DNA lengths. The female who lays the egg and her mate are both heterozygous and do not share bands, allowing assignment of parentage. All offspring have one of the female's bands, indicated by a yellow star, as would be expected given that the eggs are in her nest. Seven of the offspring have an allele from the father, indicated by a red symbol, but the eighth, A, has an allele not present in either parent, indicated by a blue symbol, demonstrating that this bird was the result of an extra-pair copulation with another male in the population. *b.* Results of a DNA fingerprinting study in red-winged blackbirds, *Agelaius phoeniceus*. Fractions indicate the proportion of offspring fathered by the male in whose territory the nest occurred. Arrows indicate how many offspring were fathered by particular males outside of each territory. Nests on some territories were not sampled.

there are two genetic classes of males. One group is large and defends territories to obtain matings. The other group is small and adopts a completely different strategy. These males do not maintain territories, but loiter at the edge of the territories of large males. Just at the end of a male's courtship, when the female is laying her eggs and the territorial male is depositing sperm, the smaller male darts in and releases its own sperm into the water, thus fertilizing some of the eggs. If this strategy is successful, natural selection will favor the evolution of these two different male reproductive strategies.

Similar patterns are seen in other organisms. In some dung beetles, territorial males have large horns that they use to guard the chambers in which females reside, whereas genetically small males don't have horns. The smaller males can't possibly hope to outcompete the larger ones in one-on-one battles. However, they have a different strategy. Because of the size of their horns, the large males cannot enter the females' chambers. Taking advantage of this, the smaller males dig side tunnels and attempt to intercept and mate with the female inside her chamber. Isopods take this one step further by having three genetic size classes. The medium-sized males pass for females and enter a large male's territory in this way; males in the smallest class are so tiny, they are able to sneak in completely undetected.

This is just a glimpse of the rich diversity in mating systems and mating tactics that have evolved. The bottom line is: If there is a way of increasing reproductive success, natural selection will favor its evolution.

Learning Outcomes Review 54.10

The sex that invests more in reproduction (parental investment) tends to exhibit mate choice. Females or males can be selective, depending on the energy and time they devote to parental care. Sexual selection governs evolution of secondary sex characteristics in that mates are chosen on the basis of phenotype and competitive success. Reproductive success influences whether males and females mate monogamously or with multiple partners.

- *Pipefish males incubate young in a brood pouch. Which sex would you expect to show mate choice? Why?*

54.11 Altruism

Learning Outcomes

1. Explain altruism and its benefits.
2. Explain kin selection and inclusive fitness.
3. Discuss how haplodiploidy influences kin selection in eusocial insects.

Understanding the evolution of altruism has been a particular challenge to evolutionary biologists, including Darwin himself. Why should an individual decrease his or her own fitness to help another? How could genes for altruism be favored by natural selection, given that the frequency of such genes should decrease in populations through time?

In fact, there can be great benefits to being an altruist, even if the altruism leads an individual to forego reproduction or even sacrifice its own life. Let's examine how this can work.

Altruism is behavior that benefits another individual at a cost to the actor. Humans sacrificing themselves in times of war or placing themselves in jeopardy to help their children are examples, but altruism also has been described in an extraordinary variety of organisms. In many bird species, for example, there are "helpers at the nest"—birds other than parents who assist in raising their young. In both mammals and birds, individuals that spy a predator may give an alarm call, alerting other members of their group to allow them to escape, even though such an act might call the predator's attention to the caller. And in social insects like ants, workers are sterile offspring that help their mother, the colony's queen, to reproduce.

A number of explanations have been put forward to explain the evolution of altruism. Once it was thought that altruism evolved for the "good of the species." Individuals that fail to mate, for example, have been called "altruists" because their lack of success in competition has been misinterpreted as a willingness to forego reproduction so that the population or species does not increase in size, exhaust its resources, and go extinct. This group selection explanation (selection acting on a population or species) is simply incorrect because individuals that fail to secure mates and not breed will not leave any offspring. Therefore, their "altruism" would not be favored by selection.

Current studies of altruism note that seemingly altruistic acts are in fact selfish. For example, helpers at the nest are often young birds that gain valuable parenting experience by assisting established breeders; this may give them an advantage when they breed. Moreover, they may have limited opportunities to reproduce on their own, and by hanging around breeding pairs, may inherit the territory when established breeders die.

Reciprocity theory explains altruism between unrelated individuals

One explanation of altruism proposes that genetically unrelated individuals may form "partnerships" in which mutual exchanges of altruistic acts occur because they benefit both participants. Partners are willing to give aid at one time and delay "repayment" for the good deed to a time in the future when they themselves are in need. In **reciprocal altruism,** the partnerships are stable because "cheaters" (nonreciprocators) are discriminated against and do not receive future aid. According to this hypothesis, if the altruistic act is relatively inexpensive, the small benefit a cheater receives by not reciprocating is far outweighed by the potential cost of not receiving future aid. Under these conditions, cheating behavior should be eliminated by selection.

Vampire bats roost in hollow trees, caves, and mines in groups of 8 to 12 individuals (figure 54.32). Because bats have a high metabolic rate, individuals that have not fed recently may die. Bats that have found a host imbibe a great deal of blood, so giving up a small amount to keep a roostmate from starvation presents no great energy cost to the donor. Vampire bats tend to share blood with past reciprocators that are not necessarily relatives. If an individual fails to give blood to a bat from which it received blood in the past, it will be excluded from future bloodsharing. Reciprocity routinely occurs in many primates, including humans.

Figure 54.32 Truth is stranger than fiction: Reciprocal altruism in vampire bats, *Desmodus rotundus*. Vampire bats do feed on the blood of large mammals, but they don't transform into people and sleep in coffins. Vampires live in groups and share blood meals. They remember which bats have provided them with blood in the past and are more likely to share with those bats that have shared with them previously. The bats here are feeding on cattle in Brazil.

Kin selection theory proposes a direct genetic advantage to altruism

The great population geneticist J. B. S. Haldane once passionately said in a pub that he would willingly lay down his life for two brothers or eight first cousins.

Evolutionarily speaking, this sacrifice makes sense, because for each allele Haldane received from his parents, his brothers each had a 50% chance of receiving the same allele (figure 54.33). Statistically, it is expected that two of his brothers would pass on as many of Haldane's particular combination of alleles to the next generation as Haldane himself would. Similarly, Haldane and a first cousin would share an eighth of their alleles (figure 54.33). Their parents, who are siblings, would each share half their alleles, and each of their children would receive half of these, of which half on the average would be in common: 1/2 × 1/2 × 1/2 = 1/8. Eight first cousins would therefore pass on as many of those alleles to the next generation as Haldane himself would.

The most compelling explanation for the kin-related origin of altruism was presented by one of the most influential evolutionary biologists of our time, William D. Hamilton, in 1964. Hamilton understood Haldane's point: Natural selection will favor any behavior, including the sacrifice of life, that increases the propagation of an individual's alleles.

Hamilton mathematically showed that by directing aid toward close genetic relatives, an altruist may increase the reproductive success of its relatives enough to not only compensate for the reduction in its own fitness, but even increase its fitness beyond what would be possible without assisting relatives. Because the altruist's behavior increases the propagation of alleles in relatives, it will be favored by natural selection. Selection that favors altruism directed toward relatives is called kin selection. Although the behaviors are altruistic, the genes are actually "behaving selfishly," because they encourage the organism to favor the success of copies of themselves

in relatives. In other words, if an individual has a dominant allele that causes altruism, any action that increases the frequency of this allele in future generations will be favored, even if that action is detrimental to the actor.

Hamilton then defined reproductive success with a new concept—**inclusive fitness.** Inclusive fitness considers gene propagation through both direct (personal fitness) and indirect (the

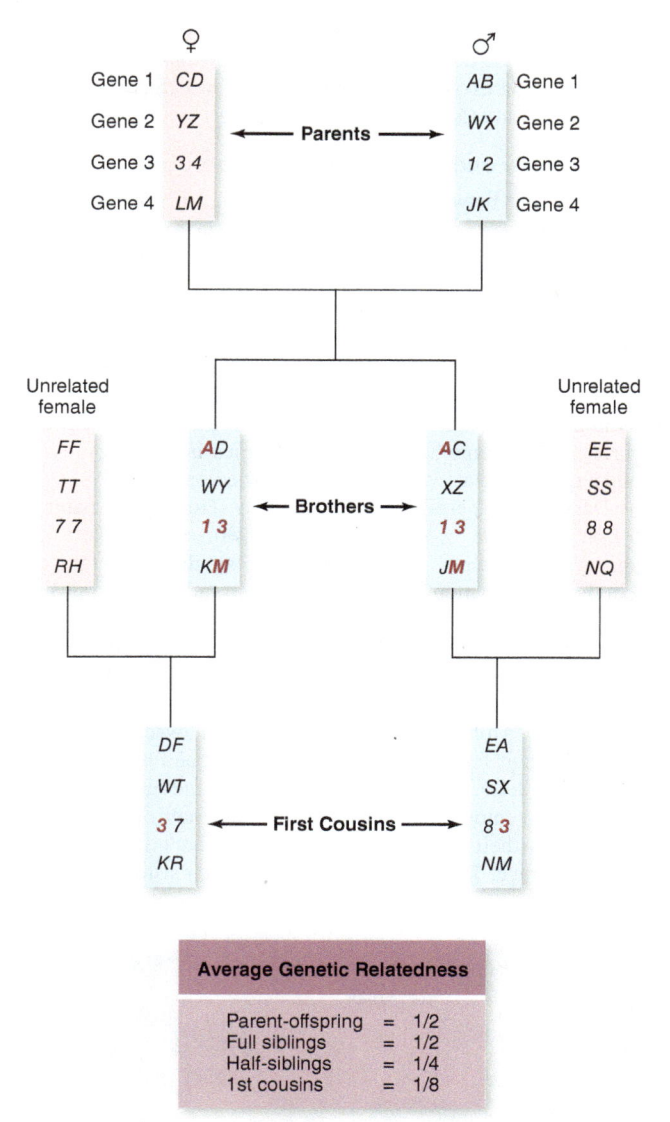

Average Genetic Relatedness	
Parent-offspring	= 1/2
Full siblings	= 1/2
Half-siblings	= 1/4
1st cousins	= 1/8

Figure 54.33 Hypothetical example of genetic relationships. On average, full siblings share half of their alleles (for illustrative purposes, the parents are shown as each having two different alleles for each gene; for example, the father has alleles *A* and *B* for Gene 1, whereas the mother has alleles *C* and *D*). By contrast, cousins only share one-eighth of their alleles on average. Each letter and number represents a different allele.

Data analysis On average, how genetically similar are you to your aunt?

fitness of relatives) reproduction. Hamilton's kin selection model predicts that altruism is likely to be directed toward close relatives. The more closely related two individuals are, the greater the potential genetic payoff, and the greater the inclusive fitness. This is described by Hamilton's rule, which states that altruistic acts are favored when $rb > c$. In this expression, b and c are the benefits and costs of the altruistic act, respectively, and r is the coefficient of relatedness, the proportion of alleles shared by two individuals through common descent. For example, an individual should be willing to have one less child ($c = 1$) if such actions allow a half-sibling, which shares one-quarter of its genes ($r = 0.25$), to have five or more additional offspring ($b = 5$).

Haplodiploidy and altruism in ants, bees, and wasps

The relationship between genetic relatedness, kin selection, and altruism can be best understood using social insects as an example. A hive of honeybees consists of a single queen, who is the sole egg-layer, and tens of thousands of her offspring, female workers with nonfunctional ovaries (figure 54.34). Honeybees are eusocial ("truly" social): Their societies are defined by reproductive division of labor (only the queen reproduces), cooperative care of the brood (workers nurse, clean, and forage), and overlap of generations (the queen lives with several generations of her offspring). Males only have one role: They leave the hive and try to find a queen to mate with and start a new colony.

Darwin was perplexed by eusociality. How could natural selection favor the evolution of sterile workers that left no offspring? It remained for Hamilton to explain the origin of eusociality in hymenopterans (bees, wasps, and ants) using his kin selection model. In these insects, males are haploid (produced from unfertilized eggs) and females are diploid. This system of sex determination, called haplodiploidy, leads to unusual genetic relatedness among colony members. If the queen is fertilized by a single male, then all female offspring will inherit exactly the same alleles from their father (because he is haploid and has only one copy of each allele). Female offspring (workers and future queens) will also share among themselves, on average, half of the alleles they get from their mother, the queen. Consequently, they will share, on average, 75% of their alleles with each sister (to verify this, rework figure 54.33, but allow the father to only have one allele for each gene).

Now recall Haldane's statement of commitment to family while you read this section. If a worker should have offspring of her own, she would share only half of her alleles with her young (the other half would come from their father). Thus, because of the close genetic relatedness of sisters due to haplodiploidy, workers would propagate more of their own alleles by giving up their own reproduction to assist their mother in rearing their sisters, some of whom will be new queens, start new colonies, and reproduce.

In this way, the unusual haplodiploid system may have set the "genetic stage" for the evolution of eusociality. Indeed, eusociality has evolved at least 12 separate times in the Hymenoptera. One wrinkle in this theory, however, is that eusocial systems have evolved in other insects (thrips, some weevils, and termites), and mammals (naked mole rats). Although the thrips and weevils are also haplodiploid, termites and naked mole rats are not. Thus, although haplodiploidy may have facilitated the evolution of eusociality, other factors can influence social evolution.

> **Data analysis** Suppose that males are haploid (only have one chromosome, and thus only one allele for each gene), as in hymenopterans. How closely related would female first cousins be if their moms were sisters? How closely related would they be if, instead, their fathers were brothers (remember that first cousins are the offspring of siblings (either two brothers, two sisters, or a brother and a sister)? You may want to refer to figure 54.33 to help you organize your approach to this question.

Other examples of kin selection

Kin selection may explain altruism in other animals. Belding's ground squirrels (*Spermophilus beldingi*) give alarm calls when they spot a predator such as a coyote or a badger. Such predators may attack a calling squirrel, so giving the signal places the caller at risk. A ground squirrel colony consists of a female and her daughters, sisters, aunts, and nieces. When males mature, they disperse long distances from where they are born, so adult males in the colony are not genetically related to the females. By marking all squirrels in a colony with an individual dye pattern on their fur and by recording which individuals gave calls and the social circumstances of their calling, researchers found that females who have relatives living nearby are more likely to give alarm calls than females with no kin nearby. Males tend to call much less frequently, as would be expected because they are not related to most colony members.

Another example of kin selection is provided by the white-fronted bee-eater, a bird that lives along river banks in Africa in colonies of 100 to 200 individuals (figure 54.35). In contrast to ground squirrels, the male bee-eaters usually remain in the colony in which they were born, and the females disperse to join new colonies. Many bee-eaters do not raise their own offspring, but instead help others. Most helpers are young birds, but older birds whose nesting attempts have failed may also be helpers. The presence of a single helper, on average, doubles the number of offspring that survive. Two lines of evidence support the idea that kin selection is important in determining helping behavior in this species. First, helpers are normally males, which are usually related to other birds in the colony, and not females, which are not related. Second, when birds

Figure 54.34 Reproductive division of labor in honeybees. The queen (*center*) is the sole egg-layer. Her daughters are sterile workers.

Figure 54.35 Kin selection in the white-fronted bee-eater, *Merops bullockoides*. Bee-eaters are small insectivorous birds that live in Africa in large colonies. Bee-eaters often help others raise their young; helpers usually choose to help close relatives.

have the choice of helping different parents, they almost invariably choose the parents to which they are most closely related.

Learning Outcomes Review 54.11

Genetic and ecological factors have contributed to evolution of altruism, a behavior that benefits another individual at a cost to the actor. Individuals may benefit directly if cooperative acts are reciprocated among unrelated interactants. Kin selection explains how altruistic acts directed toward relatives, which share alleles, increase an individual's inclusive fitness. Haplodiploidy has resulted in eusociality among some insects by increasing genetic relatedness; it is not found in vertebrates.

- *Imagine that you witness older group members rescuing infants in a troupe of monkeys when a predator appears. How would you test whether the altruistic act you see is reciprocity or kin selection?*

54.12 The Evolution of Group Living and Animal Societies

Learning Outcomes

1. Explain the possible advantages of group living.
2. Contrast the nature of insect and vertebrate societies.
3. Discuss social organization in African weaver birds and how it is influenced by ecology.

Organisms from cnidarians and insects to fish, whales, chimpanzees, and humans live in social groups. To encompass the wide variety of social phenomena, we can broadly define a society as a group of organisms of the same species that are organized in a cooperative manner.

Why have individuals in some species given up a solitary existence to become members of a group? One hypothesis is that individuals in groups benefit directly from social living. For example, a bird in a flock may be better protected from predators. As flock size increases, the risk of predation decreases because there are more individuals to scan the environment for predators.

A member of a flock may also increase its feeding success if it can acquire information from other flock members about the location of new, rich food sources. In some predators, hunting in groups can increase success and allow the group to tackle prey too large for any one individual.

Insect societies form efficient colonies containing specialized castes

In section 54.11 we discussed the origin of eusociality in the insect order Hymenoptera (ants, bees, and wasps). Additionally, all termites (order Isoptera) are also eusocial, and a few other insect and arthropod species are eusocial. Social insect colonies are composed of different *castes,* groups of individuals that differ in reproductive ability (queens versus workers), size, and morphology and perform different tasks. Workers nurse, maintain the nest, and forage; soldiers are large and have powerful jaws specialized for defense.

The structure of an insect society is illustrated by leaf-cutters, which form colonies of as many as several million individuals. These ants cut leaves and use it to grow crops of fungi beneath the ground. Workers divide the tasks of leaf cutting, defense, mulching the fungus garden, and implanting fungal hyphae according to their body size (figure 54.36).

The structure of a vertebrate society is related to ecology

In contrast to the highly structured and integrated insect societies and their remarkable forms of altruism, vertebrate social groups are usually less rigidly organized and less cohesive. It seems paradoxical that vertebrates, which have larger brains and are capable of more complex behaviors, are generally less altruistic than insects (the exception, of course, is humans). Reciprocity and kin-selected altruism are common in vertebrate societies, although there is often more conflict and aggression among group members. Conflicts generally center on access to food and mates and occur because a vertebrate society is made up of individuals striving to improve their own fitness.

Social groups of vertebrates have a size, stability of members, number of breeding males and females, and type of mating

Figure 54.36 Castes of ants. These leaf-cutter ants are members of different castes. The large ant is a worker carrying leaves to the nest, whereas the smaller ant is protecting the worker from attack by small parasitic flies.

system characteristic of a given species. Diet and predation are important factors in shaping social groups. For example, meerkats take turns watching for predators, while other group members forage for food (figure 54.37).

African weaver birds, which construct nests from vegetation, provide an excellent example of the relationship between ecology and social organization. Their roughly 90 species can be divided according to the type of social group they form. One group of species lives in the forest and builds camouflaged, solitary nests. Males and females are monogamous; they forage for insects to feed their young. The second group of species nests in colonies in trees on the savanna. They are polygynous and feed in flocks on seeds.

The feeding and nesting habits of these two groups of species are correlated with their mating systems. In the forest, insects are hard to find, and both parents must cooperate in feeding the young. The camouflaged nests do not call the attention of predators to their brood. On the open savanna, building a hidden nest is not an option. Rather, savanna-dwelling weaver birds protect their young from predators by nesting in trees, which are not very abundant. This shortage of safe nest sites means that birds must nest together in colonies. Because seeds occur abundantly, a female can acquire all the food needed to rear young without a male's help. The male, free from the duties of parenting, spends his time courting many females—a polygynous mating system.

One exception to the general rule that vertebrate societies are not organized like those of insects is the naked mole rat (*Heterocephalus glaber*), a small, hairless rodent that lives in and near East Africa. Unlike other kinds of mole rats, which live alone or in small family groups, naked mole rats form large underground colonies with a far-ranging system of tunnels and a central nesting area. It is not unusual for a colony to contain 80 individuals.

Naked mole rats feed on bulbs, roots, and tubers, which they locate by constant tunneling. As in insect societies, there is a division of labor among the colony members, with some individuals working as

Figure 54.37 Foraging and predator avoidance. A meerkat sentinel on duty. Meerkats, *Suricata suricata,* are a species of highly social mongoose living in the semiarid sands of the Kalahari Desert in southern Africa. This meerkat is taking its turn to act as a lookout for predators. Under the security of its vigilance, the other group members can focus their attention on foraging.

tunnelers, while others perform different tasks, depending on the size of their bodies. Large mole rats defend the colony and dig tunnels.

Naked mole rat colonies have a reproductive division of labor similar to the one normally associated with the eusocial insects. All of the breeding is done by a single female, or "queen," who has one or two male consorts. The workers, consisting of both sexes, keep the tunnels clear and forage for food.

Learning Outcomes Review 54.12

Advantages of group living include protection from predators and increased feeding success. Eusocial insects form complex, highly altruistic societies that increase the fitness of the colony. The members of vertebrate societies exhibit more conflict and competition, but also cooperate and behave altruistically, especially toward kin. African weaver birds have developed different types of societies depending on the ecology of their habitat, particularly the safety of nesting sites.

- *What are the benefits and costs associated with living in social groups?*
- *Why is altruism directed toward kin considered to be selfish behavior?*
- *Is a human army more like an insect society or a vertebrate society? Explain your answer.*

Chapter Review

54.1 The Natural History of Behavior

Behavior can be analyzed in terms of mechanisms and evolutionary origin.
Proximate causation refers to the mechanisms of behavior. Ultimate causation examines a behavior's evolutionary significance.

Ethology emphasizes the study of instinct and its origins.
Innate, or instinctive, behavior is a response to an environmental stimulus or trigger that does not require learning (figure 54.1).

54.2 Nerve Cells, Neurotransmitters, Hormones, and Behavior

Instinctive behaviors are accomplished by neural circuits, which develop under genetic control. Hormones and neurotransmitters can act to regulate behavior.

54.3 Behavioral Genetics

Artificial selection, hybrid, and twin studies link genes and behavior.
Breeding fast-learning and slow-learning rats for several generations produced two distinct behavioral populations (figure 54.3).

Some behaviors appear to be controlled by a single gene.

54.4 Learning

Learning mechanisms include habituation and association.
Habituation, a form of nonassociative learning, is a decrease in response to repeated nonessential stimuli. Associative learning is a change in behavior by association of two stimuli or of a behavior and a response (conditioning).
Classical (Pavlovian) conditioning occurs when two stimuli are associated with each other. Operant conditioning occurs when an animal associates a behavior with reward or punishment.

Instinct governs learning preparedness.
What an animal can learn is biologically influenced—that is, learning is possible only within the boundaries set by evolution.

54.5 The Development of Behavior

Parent–offspring interactions influence how behavior develops.
In imprinting, a young animal forms an attachment to other individuals or develops preferences that influence later behavior.

Instinct and learning may interact as behavior develops.
Animals may have an innate genetic template that guides their learning as behavior develops, such as song development in birds. Studies on twins reveal a role for both genes and environment in human behavior.

54.6 Animal Cognition

Some animals exhibit cognitive behavior and can devise solutions to novel situations (figures 54.11 & 54.12).

54.7 Orientation and Migratory Behavior

Migration often involves populations moving large distances.
Migrating animals must be capable of orientation and navigation (figure 54.15).
Orientation is the mechanism by which animals move by tracking environmental stimuli such as celestial clues or Earth's magnetic field. Navigation is following a route based on orientation and some sort of "map." The nature of the map in animals is not known.

54.8 Animal Communication

Successful reproduction depends on appropriate signals and responses.
Courtship signals are usually species-specific and help to ensure reproductive isolation (figure 54.18).

Communication enables information exchange among group members (figures 54.19 & 54.20).

54.9 Behavioral Ecology

Foraging behavior can directly influence energy intake and individual fitness.
Natural selection favors optimal foraging strategies in which energy acquisition (cost) is minimized and reproductive success (benefit) is maximized.

Territorial behavior evolves if the benefits of holding a territory exceed the costs.

54.10 Reproductive Strategies and Sexual Selection

The sexes often have different reproductive strategies.
One sex may be choosier than the other, and which one often depends on the degree of parental investment.

Sexual selection occurs through mate competition and mate choice.
Intrasexual selection involves competition among members of the same sex for the chance to mate. Intersexual selection is one sex choosing a mate.
Mate choice may provide direct benefits (increased resource availability or parental care) or indirect benefits (genetic quality of the mate).

Mating systems reflect the ability of parents to care for offspring and are influenced by ecology.
Mating systems include monogamy, polygyny, and polyandry; they are influenced by ecology and constrained by needs of offspring.

54.11 Altruism

Reciprocity theory explains altruism between unrelated individuals.
Mutual exchanges benefit both participants; a participant that does not reciprocate would not receive future aid.

Kin selection theory proposes a direct genetic advantage to altruism.
Kin selection increases the reproductive success of relatives and increased frequency of alleles shared by kin, and thus increases an individual's inclusive fitness.
Ants, bees, and wasps have haplodiploid reproduction, and therefore a high degree of gene sharing.

54.12 The Evolution of Group Living and Animal Societies

A social system is a group organized in a cooperative manner.

Insect societies form efficient colonies containing specialized castes (figure 54.36).
Social insect societies are composed of different castes that are specialized to reproduce or to perform certain colony maintenance tasks.

The structure of a vertebrate society is related to ecology.
Vertebrate social systems are less rigidly organized and cohesive and are influenced by food availability and predation.

Review Questions

UNDERSTAND

1. A key stimulus, innate releasing mechanism, and fixed action pattern
   a. are mechanisms associated with behaviors that are learned.
   b. are components of behaviors that are innate.
   c. involve behaviors that cannot be explained in terms of ultimate causation.
   d. involve behaviors that are not subject to natural selection.

2. In operant conditioning
   a. an animal learns that a particular behavior leads to a reward or punishment.
   b. an animal associates an unconditioned stimulus with a conditioned response.
   c. learning is unnecessary.
   d. habituation is required for an appropriate response.

3. The study of song development in sparrows showed that
   a. the acquisition of a species-specific song is innate.
   b. there are two components to this behavior: a genetic template and learning.
   c. song acquisition is an example of associative learning.
   d. All of the choices are correct.

4. The difference between following a set of driving directions given to you by somebody on the street (for example ". . . take a right at the next light, go four blocks and turn left . . .") and using a map to find your destination is
   a. the difference between navigation and orientation, respectively.
   b. the difference between learning and migration, respectively.
   c. the difference between orientation and navigation, respectively.
   d. why birds are not capable of orientation.

5. In courtship communication
   a. the signal itself is always species-specific.
   b. the sign communicates species identity.
   c. a stimulus–response chain is involved.
   d. courtship signals are produced only by males.

6. Behavioral ecology assumes
   a. that all behavioral traits are innate.
   b. learning is the dominant determinant of behavior.
   c. behavioral traits are subject to natural selection.
   d. behavioral traits do not affect fitness.

7. According to optimal foraging theory
   a. individuals minimize energy intake per unit of time.
   b. energy content of a food item is the only determinant of a forager's food choice.
   c. time taken to capture a food item is the only determinant of a forager's food choice.
   d. a higher energy item might be less valuable than a lower energy item if it takes too much time to capture the larger item.

8. The elaborate tail feathers of a male peacock evolved because they
   a. improve reproductive success of males and females.
   b. improve male survival.
   c. reduce survival.
   d. None of the choices is correct.

9. From the perspective of females, extra-pair copulations (EPCs)
   a. are always disadvantageous to females.
   b. can be associated with receiving male aid.
   c. are unimportant to fitness because they rarely lead to fertilization of eggs.
   d. can only be of benefit if the EPC male has elaborate secondary sexual traits.

10. In the haplodiploidy system of sex determination, males are
    a. haploid.
    b. diploid.
    c. sterile.
    d. not present because bees exist as single-sex populations.

11. According to kin selection, saving the life of your _____ would do the least for increasing your inclusive fitness.
    a. mother
    b. brother
    c. sister-in-law
    d. niece

12. Altruism
    a. is only possible through reciprocity.
    b. is only possible through kin selection.
    c. can only be explained by group selection.
    d. will only occur when the fitness benefit of a given act is greater than the fitness cost.

APPLY

1. Refer to figure 54.24. Data on size of mussels eaten by shore crabs suggest they eat sizes smaller than expected by an optimal foraging model. Suggest a hypothesis for why and describe an experiment to test your hypothesis.

2. Refer to figure 54.25. Six pairs of birds were removed but only four pairs moved in. Where did the new pairs come from? Additionally, it appears that many of the birds that were not removed expanded their territories and that the new residents ended up with smaller territories than the pairs they replaced. Explain.

3. Refer to figure 54.27. Peahens prefer to mate with peacocks that have more eyespots in their tail feathers (that is, longer tail feathers). It has also been suggested that the longer the tail feathers, the more impaired the flight of the males. One possible hypothesis to explain such a preference by females is that the males with the longest tail feathers experience the most severe handicap, and if they can nevertheless survive, it reflects their "vigor." Suggest some studies that would allow you to test this idea. Your description should include the kinds of traits that you would measure and why.

4. An altruistic act is defined as one that benefits another individual at a cost to the actor. There are two theories to explain how such behavior evolves: reciprocity and kin selection. How would you distinguish between the two in a field study? In the context of natural selection, is an altruistic act "costly" to an individual who performs it?

SYNTHESIZE

1. Insects that sting or contain toxic chemicals often have black and yellow coloration and consequentially are not eaten by predators. How could you determine if a predator has an innate avoidance of insects that are colored this way, or if the avoidance is learned? If avoidance is learned, how would you determine the learning mechanism involved? How would you measure the adaptive significance of the black and yellow coloration to the prey insect?

2. Behavioral genetics has made great advances from detailed studies of a single animal such as the fruit fly as a model system to develop general principles of how genes regulate behavior. What are advantages and disadvantages of this "model system" approach? How would you determine how broadly applicable the results of such studies are to other animals?

3. If a female bird chooses to live in the territory of a particular male, why might she mate with a male other than the territory owner?

CHAPTER 55

Ecology of Individuals and Populations

Chapter Contents

- 55.1 The Environmental Challenges
- 55.2 Populations: Groups of a Single Species in One Place
- 55.3 Population Demography and Dynamics
- 55.4 Life History and the Cost of Reproduction
- 55.5 Environmental Limits to Population Growth
- 55.6 Factors That Regulate Populations
- 55.7 Human Population Growth

Introduction

Ecology, the study of how organisms relate to one another and to their environments, is a complex and fascinating area of biology that has important implications for each of us. In our exploration of ecological principles, we first consider how organisms respond to the abiotic environment in which they exist and how these responses affect the properties of populations, emphasizing population dynamics. In chapter 56, we discuss communities of coexisting species and the interactions that occur among them. In chapters 57 through 59, we discuss the functioning of entire ecosystems and of the biosphere, concluding with a consideration of the problems facing our planet and our fellow species.

55.1 The Environmental Challenges

Learning Outcomes

1. List some challenges that organisms face in their environments.
2. Describe ways in which individuals respond to environmental changes.
3. Explain how species adapt to environmental conditions.

The nature of the physical environment in large measure determines which organisms live in a particular climate or region. Key elements of the environment include:

Temperature. Most organisms are adapted to live within a relatively narrow range of temperatures and will not thrive if temperatures are colder or warmer. The growing season of plants, for example, is importantly influenced by temperature.

Water. All organisms require water. On land, water is often scarce, so patterns of rainfall have a major influence on life.

Figure 55.1 Meeting the challenge of obtaining moisture. On the dry sand dunes of the Namib Desert in southwestern Africa, the fog-basking beetle, *Onymacris unguicularis*, collects moisture from the fog by holding its abdomen up at the crest of a dune to gather condensed water; water condenses as droplets and trickles down to the beetle's mouth.

Sunlight. Almost all ecosystems rely on energy captured by photosynthesis; the availability of sunlight influences the amount of life an ecosystem can support, particularly below the surface in marine communities.

Soil. The physical consistency, pH, and mineral composition of the soil often severely limit terrestrial plant growth, particularly the availability of nitrogen and phosphorus.

An individual encountering environmental variation may maintain a "steady-state" internal environment, a condition known as *homeostasis*. Many animals and plants actively employ physiological, morphological, or behavioral mechanisms to maintain homeostasis. The beetle in figure 55.1 is using a behavioral mechanism to cope with limited water availability. Other animals and plants are known as conformers because they conform to the environment in which they find themselves, their bodies adopting the temperature, salinity, and other physical aspects of their surroundings.

Responses to environmental variation can be seen over both the short and the long term. In the short term, spanning periods of a few minutes to an individual's lifetime, organisms have a variety of ways of coping with environmental change. Over longer periods, natural selection can operate to make a population better adapted to its environment.

Organisms are capable of responding to environmental changes that occur during their lifetime

During the course of a day, a season, or a lifetime, an individual organism must cope with a range of living conditions. They do so through the physiological, morphological, and behavioral abilities they possess. These abilities are a product of natural selection acting in a particular environmental setting over time, which explains why an individual organism that is moved to a different environment may not survive.

TABLE 55.1 Physiological Changes at High Elevation

Increased rate of breathing
Increased erythrocyte production, raising the amount of hemoglobin in the blood
Decreased binding capacity of hemoglobin, increasing the rate at which oxygen is unloaded in body tissues
Increased density of mitochondria, capillaries, and muscle myoglobin

Physiological responses

Many organisms are able to respond to environmental change by making physiological adjustments. For example, you sweat when it is hot, increasing evaporative heat loss and thus preventing overheating. Similarly, people who visit high altitudes may initially experience altitude sickness—the symptoms of which include heart palpitations, nausea, fatigue, headache, mental impairment, and in serious cases, pulmonary edema—because of the lower atmospheric pressure and consequent lower oxygen availability in the air. After several days, however, the same people usually feel fine, because a number of physiological changes have increased the delivery of oxygen to their body tissues (table 55.1).

Some insects avoid freezing in the winter by adding glycerol "antifreeze" to their blood; others tolerate freezing by converting much of their glycogen reserves into alcohols that protect their cell membranes from freeze damage.

Morphological capabilities

Animals that maintain a constant internal temperature (endotherms) in a cold environment have adaptations that tend to minimize energy expenditure. For example, many mammals grow thicker coats during the winter, their fur acting as insulation to retain body heat. In general, the thicker the fur, the greater the insulation (figure 55.2). Thus, a wolf's fur is about three times thicker in winter than in summer and insulates more than twice as well.

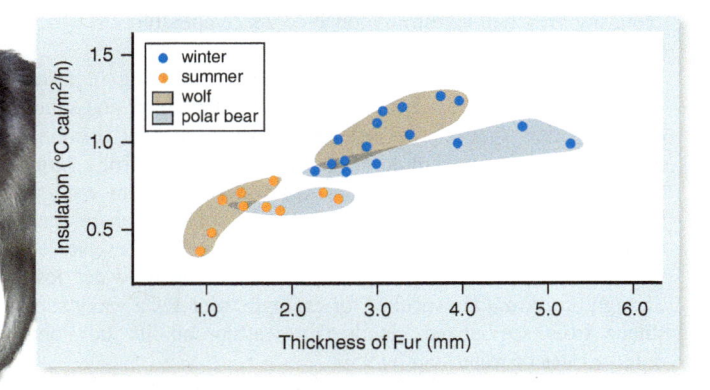

Figure 55.2 Morphological adaptation. Fur thickness in North American mammals has a major effect on the degree of insulation the fur provides.

Behavioral responses

Many animals deal with variation in the environment by moving from one patch of habitat to another, avoiding areas that are unsuitable. The tropical lizard in figure 55.3 manages to maintain a fairly uniform body temperature in an open habitat by basking in patches of sunlight and then retreating to the shade when it becomes too hot. By contrast, in shaded forests, the same lizard does not have the opportunity to regulate its body temperature through behavioral means. Thus, it becomes a conformer and adopts the temperature of its surroundings.

Behavioral adaptations can be extreme. Spadefoot toads (genus *Scaphiophus*), which are widely distributed in North America, can burrow nearly a meter below the surface and remain there for as long as nine months of each year, their metabolic rates greatly reduced as they live on fat reserves. When moist, cool conditions return, the toads emerge and breed. The young toads mature rapidly and then burrow back underground to await their turn to breed in the next year.

Natural selection leads to evolutionary adaptation to environmental conditions

The ability of an individual to alter its physiology, morphology, or behavior is itself an evolutionary adaptation, the result of natural selection. The results of natural selection can also be detected by comparing closely related species that live in different environments. In such cases, species often have evolved striking adaptations to the particular environment in which they live.

For example, animals that live in different climates show many differences. Mammals from colder climates tend to have shorter ears and limbs—a phenomenon termed *Allen's rule*—which reduces the surface area across which animals lose heat. Lizards that live in different climates exhibit physiological adaptations for coping with life at different temperatures. Desert lizards are unaffected by high temperatures that would kill a lizard from northern Europe, but the northern lizards are capable of running, capturing prey, and digesting food at cooler temperatures at which desert lizards would be completely immobilized.

Many species also exhibit adaptations to living in areas where water is scarce. Everyone knows of the camel and other desert animals that can go extended periods without drinking water. Another example of desert adaptation is seen in frogs. Most frogs have moist skins through which water permeates readily. Such animals could not survive in arid climates because they would rapidly dehydrate and die. However, some frogs have solved this problem by evolving a greatly reduced rate of water loss through the skin. One species, for example, secretes a waxy substance from specialized glands that waterproofs its skin and reduces rates of water loss by 95%.

Adaptation to different environments can also be studied experimentally. For example, when strains of *E. coli* were grown at high temperatures (42°C), the speed at which the bacteria utilized resources improved through time. After 2000 generations, this ability increased 30% over what it had been when the experiment started. The mechanism by which efficiency of resource use increased is unknown and is the focus of current research.

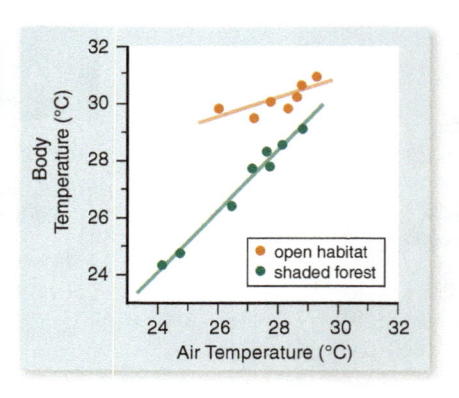

Figure 55.3 Behavioral adaptation. In open habitats, the Puerto Rican crested lizard, *Anolis cristatellus,* maintains a relatively constant temperature by seeking out and basking in patches of sunlight; as a result, it can maintain a relatively high temperature even when the air is cool. In contrast, in shaded forests, this behavior is not possible, and the lizard's body temperature conforms to that of its surroundings.

Inquiry question When given the opportunity, lizards regulate their body temperature to maintain a temperature optimal for physiological functioning. Would lizards in open habitats exhibit different escape behaviors from lizards in shaded forest?

Data analysis Can the slope of the line tell us something about the behavior of the lizard?

Learning Outcomes Review 55.1

Environmental conditions include temperature, water and light availability, and soil characteristics. When the environment changes, individual organisms use a variety of physiological, morphological, and behavioral mechanisms to adjust. Over time, adaptations to different environments may evolve in populations.

■ *How might a species respond if its environment grew steadily warmer over time?*

55.2 Populations: Groups of a Single Species in One Place

Learning Outcomes

1. *Distinguish between a population and a metapopulation.*
2. *Understand what causes a species' geographic range to change through time.*

Organisms live as members of a **population,** groups of individuals that occur together at one place and time. In the rest of this chapter, we consider the properties of populations, focusing on factors that influence whether a population grows or shrinks, and at what rate. The explosive growth of the world's human population in the last few centuries provides a focus for our inquiry.

The term population can be defined narrowly or broadly. This flexibility allows us to speak in similar terms of the world's human population, the population of protists in the gut of a termite, or the population of deer that inhabit a forest. Sometimes the boundaries defining a population are sharp, such as the edge of an isolated mountain lake for trout, and sometimes they are fuzzier, as when deer readily move back and forth between two forests separated by a cornfield.

Three characteristics of population ecology are particularly important: (1) population range, the area throughout which a population occurs; (2) the pattern of spacing of individuals within that range; and (3) how the population changes in size through time.

A population's geographic distribution is termed its range

No population, not even one composed of humans, occurs in all habitats throughout the world. Most species, in fact, have relatively limited geographic ranges, and the range of some species is minuscule. For example, the Devil's Hole pupfish lives in a single spring in southern Nevada (figure 55.4), and the Socorro isopod *(Thermosphaeroma thermophilus)* is known from a single spring system in New Mexico. At the other extreme, some species are widely distributed. The common dolphin *(Delphinus delphis)*, for example, is found throughout all the world's oceans.

As discussed in section 55.1, organisms must be adapted to the environment in which they occur. Polar bears are exquisitely adapted to survive the cold of the Arctic, but you won't find them in the tropical rainforest. Certain prokaryotes can live in the near-boiling waters of Yellowstone's geysers, but they do not occur in cooler streams nearby. Each population has its own requirements—temperature, humidity, certain types of food, and a host of other factors—that determine where it can live and reproduce and where it can't. In addition, in places that are otherwise suitable, the presence of predators, competitors, or parasites may prevent a population from occupying an area, a topic we will take up in chapter 56.

Ranges undergo expansion and contraction

Population ranges are not static but change through time. These changes occur for two reasons. In some cases, the environment changes. As the glaciers retreated at the end of the last ice age, approximately 10,000 years ago, many North American plant and animal populations expanded northward. At the same time, as climates warmed, species experienced shifts in the elevation at which they could live (figure 55.5).

In addition, populations can expand their ranges when they are able to circumvent inhospitable habitat to colonize suitable, previously unoccupied areas. For example, the cattle egret is native to Africa. Some time in the late 1800s, these birds appeared in northern South America, having made the nearly 3500-km

Figure 55.4 The Devil's Hole pupfish, *Cyprinodon diabolis*. This fish has the smallest range of any vertebrate species in the world.

transatlantic crossing, perhaps aided by strong winds. Since then, they have steadily expanded their range and now can be found throughout most of the United States (figure 55.6).

The human effect

By altering the environment, humans have allowed some species, such as coyotes, to expand their ranges and move into areas they previously did not occupy. Moreover, humans have served

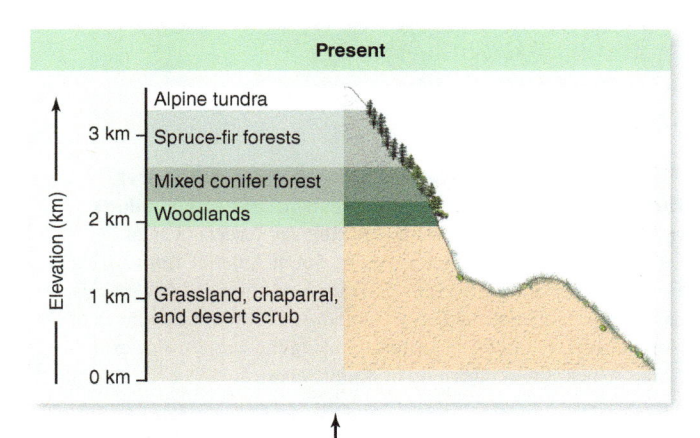

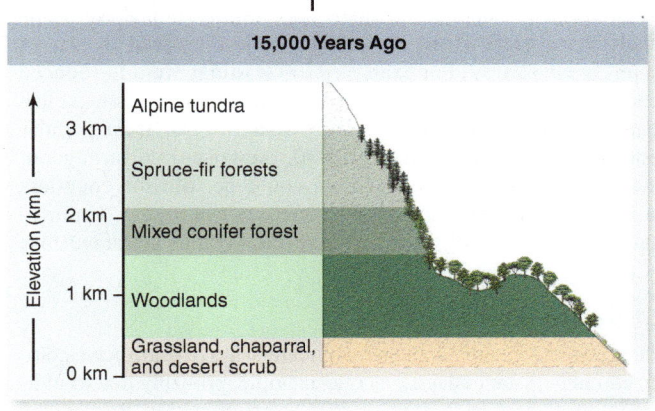

Figure 55.5 Shifts in altitudinal distributions of trees in the mountains of southwestern North America. During the glacial period 15,000 years ago, conditions were cooler than they are now. As the climate warmed, tree species that require colder temperatures shifted their range upward in altitude so that they continue to live in the climatic conditions to which they are adapted.

Figure 55.6 Range expansion of the cattle egret, Bubulcus ibis. The cattle egret—so named because it follows cattle and other hoofed animals, catching any insects or small vertebrates it disturbs—first arrived in South America from Africa in the late 1800s. Since the 1930s, the range expansion of this species has been well documented, as it has moved northward into much of North America, as well as southward along the western side of the Andes to near the southern tip of South America.

distances before falling to the ground. Still others are enclosed in fleshy fruits. These seeds can pass through the digestive systems of mammals or birds and then germinate where they are defecated. Finally, seeds of mistletoes (*Arceuthobium*) are violently propelled from the base of the fruit in an explosive discharge. Although the probability of long-distance dispersal events leading to successful establishment of new populations is low, over millions of years, many such dispersals have occurred.

Individuals in populations exhibit different spacing patterns

Another key characteristic of population structure is the way in which individuals of a population are distributed. They may be randomly spaced, uniformly spaced, or clumped.

Random spacing

Random spacing of individuals within populations occurs when they do not interact strongly with one another and when they are not affected by nonuniform aspects of their environment. Random distributions are not common in nature. Some species of trees, however, appear to exhibit random distributions in Panamanian rainforests.

Uniform spacing

Uniform spacing within a population may often, but not always, result from competition for resources. This spacing is accomplished, however, in many different ways.

as an agent of dispersal for many species. Some of these transplants have been widely successful, as is discussed in greater detail in chapter 59. For example, 100 starlings were introduced into New York City in 1896 in a misguided attempt to establish every species of bird mentioned by Shakespeare. Their population steadily spread so that by 1980, they occurred throughout the United States. Similar stories could be told for countless plants and animals, and the list increases every year. Unfortunately, the success of these invaders often comes at the expense of native species.

Dispersal mechanisms

Dispersal to new areas can occur in many ways. Lizards have colonized many distant islands, as one example, probably due to individuals or their eggs floating or drifting on vegetation. Bats are the only mammals on many distant islands because they can fly to them.

Seeds of plants are designed to disperse in many ways (figure 55.7). Some seeds are aerodynamically designed to be blown long distances by the wind. Others have structures that stick to the fur or feathers of animals, so that they are carried long

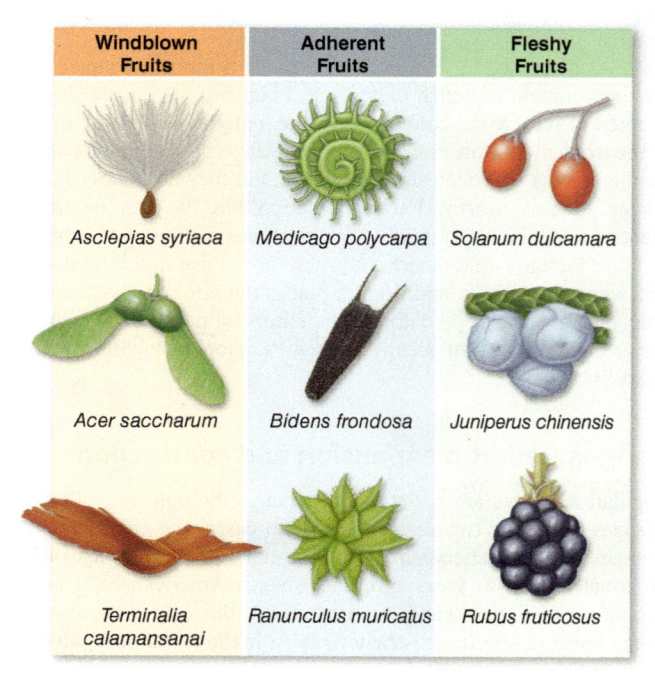

Windblown Fruits	Adherent Fruits	Fleshy Fruits
Asclepias syriaca	Medicago polycarpa	Solanum dulcamara
Acer saccharum	Bidens frondosa	Juniperus chinensis
Terminalia calamansanai	Ranunculus muricatus	Rubus fruticosus

Figure 55.7 Some of the many adaptations for dispersal of seeds. Seeds have evolved a number of different means of facilitating dispersal from their maternal plant. Some seeds can be transported great distances by the wind, whereas seeds enclosed in adherent or fleshy fruits can be transported by animals.

In animals, uniform spacing often results from behavioral interactions, as described in chapter 54. In many species, individuals of one or both sexes defend a territory from which other individuals are excluded. These territories provide the owner with exclusive access to resources, such as food, water, hiding refuges, or mates, and tend to space individuals evenly across the habitat. Even in nonterritorial species, individuals often maintain a defended space into which other animals are not allowed to intrude.

Among plants, uniform spacing is also a common result of competition for resources. Closely spaced individual plants compete for available sunlight, nutrients, or water. These contests can be direct, as when one plant casts a shadow over another, or indirect, as when two plants compete by extracting nutrients or water from a shared area. In addition, some plants, such as the creosote bush, produce chemicals in the surrounding soil that are toxic to other members of their species. In all of these cases, only plants that are spaced an adequate distance from each other will be able to coexist, leading to uniform spacing.

Clumped spacing

Individuals clump into groups or clusters in response to uneven distribution of resources in their immediate environments. Clumped distributions are common in nature because individual animals, plants, and microorganisms tend to occur in habitats defined by soil type, moisture, or other aspects of the environment to which they are best adapted.

Social interactions also can lead to clumped distributions. Many species live and move around in large groups, which go by a variety of names (for example, flock, herd, pride). These groupings can provide many advantages, including increased awareness of and defense against predators, decreased energy cost of moving through air and water, and access to the knowledge of all group members.

On a broader scale, populations are often most densely populated in the interior of their range and less densely distributed toward the edges. Such patterns usually result from the manner in which the environment changes in different areas. Populations are often best adapted to the conditions in the interior of their distribution. As environmental conditions change, individuals are less well adapted, and thus densities decrease. Ultimately, the point is reached at which individuals cannot persist at all; this marks the edge of a population's range.

A metapopulation comprises distinct populations that may exchange members

Species often exist as a network of distinct populations that exchange individuals. Such networks, termed **metapopulations,** usually occur in areas in which suitable habitat is patchily distributed and is separated by intervening stretches of unsuitable habitat.

Dispersal and habitat occupancy

The degree to which populations within a metapopulation interact depends on the amount of dispersal; this interaction is often not symmetrical: Populations increasing in size tend to send out many dispersers, whereas populations at low levels tend to receive more immigrants than they send off. In addition, relatively isolated populations tend to receive relatively few arrivals.

Not all suitable habitats within a metapopulation's area may be occupied at any one time. For a number of reasons, some individual populations may become extinct, perhaps as a result of an epidemic disease, a catastrophic fire, or the loss of genetic variation following a population bottleneck (see chapter 20). Dispersal from other populations, however, may eventually recolonize such areas. In some cases, the number of habitats occupied in a metapopulation may represent an equilibrium in which the rate of extinction of existing populations is balanced by the rate of colonization of empty habitats.

Source–sink metapopulations

A species may also exhibit a metapopulation structure in areas in which some habitats are suitable for long-term population maintenance, but others are not. In these situations, termed **source–sink metapopulations,** the populations in the better areas (the sources) continually send out dispersers that bolster the populations in the poorer habitats (the sinks). In the absence of such continual replenishment, sink populations would have a negative growth rate and would eventually become extinct.

Metapopulations of butterflies have been studied particularly intensively. In one study, researchers sampled populations of the Glanville fritillary butterfly at 1600 meadows in southwestern Finland (figure 55.8). On average, every year, 200 populations

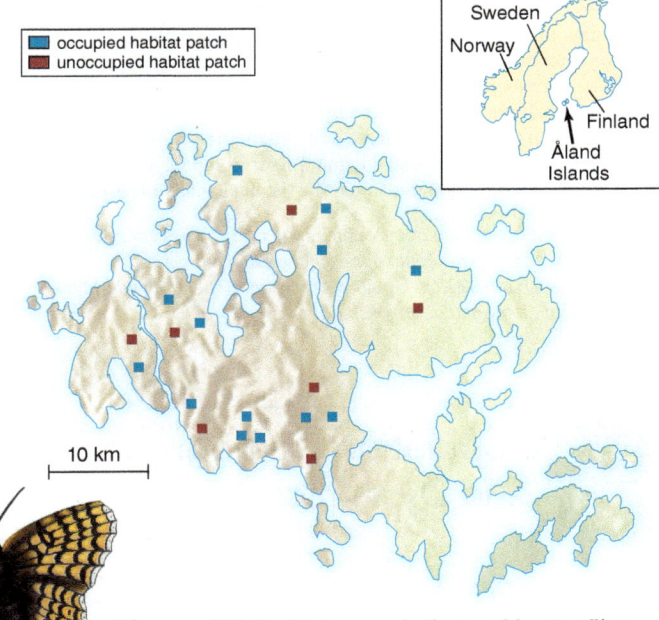

Figure 55.8 Metapopulations of butterflies. The Glanville fritillary butterfly, *Melitaea cinxia*, occurs in metapopulations in southwestern Finland on the Åland Islands. None of the populations is large enough to survive for long on its own, but continual immigration of individuals from other populations allows some populations to survive. In addition, continual establishment of new populations tends to offset extinction of established populations, although in recent years, extinctions have outnumbered colonizations.

became extinct, but 114 empty meadows were colonized. A variety of factors seemed to increase the likelihood of a population's extinction, including small population size, isolation from sources of immigrants, low resource availability (as indicated by the number of flowers on a meadow), and lack of genetic variation within the population.

The researchers attribute the greater number of extinctions than colonizations to a string of very dry summers. Because none of the populations is large enough to survive on its own, continued survival of the species in southwestern Finland would appear to require the continued existence of a metapopulation network in which new populations are continually created and existing populations are supplemented by immigrants. Continued bad weather thus may doom the species, at least in this part of its range.

Metapopulations, where they occur, can have two important implications for the range of a species. First, through continuous colonization of empty patches, metapopulations prevent long-term extinction. If no such dispersal existed, then each population might eventually perish, leading to disappearance of the species from the entire area. Moreover, in source–sink metapopulations, the species occupies a larger area than it otherwise might, including marginal areas that could not support a population without a continual influx of immigrants. For these reasons, the study of metapopulations has become very important in conservation biology as natural habitats become increasingly fragmented.

> **Learning Outcomes Review 55.2**
>
> A population is a group of individuals of a single species existing together in an area. A population's range, the area it occupies, changes over time. Populations, in turn, may form a network, or metapopulation, connected by individuals that move from one group to another. Within a population, the distribution of individuals can be random, uniform, or clumped, and the distribution is determined in part by the availability of resources.
>
> ■ How might the geographic range of a species change if populations could not exchange individuals with each other? How would your answer depend on what type of metapopulation existed?

55.3 Population Demography and Dynamics

Learning Outcomes
1. Define demography.
2. Describe the factors that influence a species' demography.
3. Explain the significance of survivorship curves.

The dynamics of a population—how it changes through time—are affected by many factors. One important factor is the age distribution of individuals—that is, what proportion of individuals are adults, juveniles, and young.

Demography is the quantitative study of populations. How the size of a population changes through time can be studied at two levels: as a whole or broken down into parts. At the most inclusive level, we can study the whole population to determine whether it is increasing, decreasing, or remaining constant. Put simply, populations grow if births outnumber deaths and shrink if deaths outnumber births (ignoring, for the moment, immigration and extinction). Understanding these trends is often easier, however, if we break the population into smaller units composed of individuals of the same age (for example, 1-year-olds) and study the factors affecting birth and death rates for each unit separately.

Sex ratio and generation time affect population growth rates

Population growth can be influenced by the population's sex ratio. The number of births in a population is usually directly related to the number of females; births may not be as closely related to the number of males in species in which a single male can mate with several females. In many species, males compete for the opportunity to mate with females, as you learned in chapter 54; consequently, a few males have many matings, and many males do not mate at all. In such species, the sex ratio of reproducing animals is female-biased and does not affect population growth rates; reduction in the number of males simply changes the identities of the reproductive males without reducing the number of births. By contrast, among monogamous species, pairs may form long-lasting reproductive relationships, and a reduction in the number of males can then directly reduce the number of births.

Generation time is the average interval between the birth of an individual and the birth of its offspring. This factor can also affect population growth rates. Species differ greatly in generation time. Differences in body size can explain much of this variation—mice go through approximately 100 generations during the course of one elephant generation (figure 55.9). But small size does not always mean short generation time. Newts, for example, are smaller than mice, but have considerably longer generation times.

In general, populations with short generations can increase in size more quickly than populations with long generations. Conversely, because generation time and life span are usually closely correlated, populations with short generation times may also diminish in size more rapidly if birth rates suddenly decrease.

Age structure is determined by the numbers of individuals in different age groups

A group of individuals of the same age is referred to as a cohort. In most species, the probability that an individual will reproduce or die varies through its life span. As a result, within a population, every cohort has a characteristic birth rate, or **fecundity**, defined as the number of offspring produced in a standard time (for example, per year), and death rate, or **mortality**, the number of individuals that die in that period.

The relative number of individuals in each cohort defines a population's **age structure.** Because different cohorts have different fecundity and death rates, age structure has a critical influence

Figure 55.9 The relationship between body size and generation time. In general, larger organisms have longer generation times, although there are exceptions.

? Inquiry question
If resources became more abundant, would you expect smaller or larger species to increase in population size more quickly?

Data analysis
How well could you predict generation time from body size?

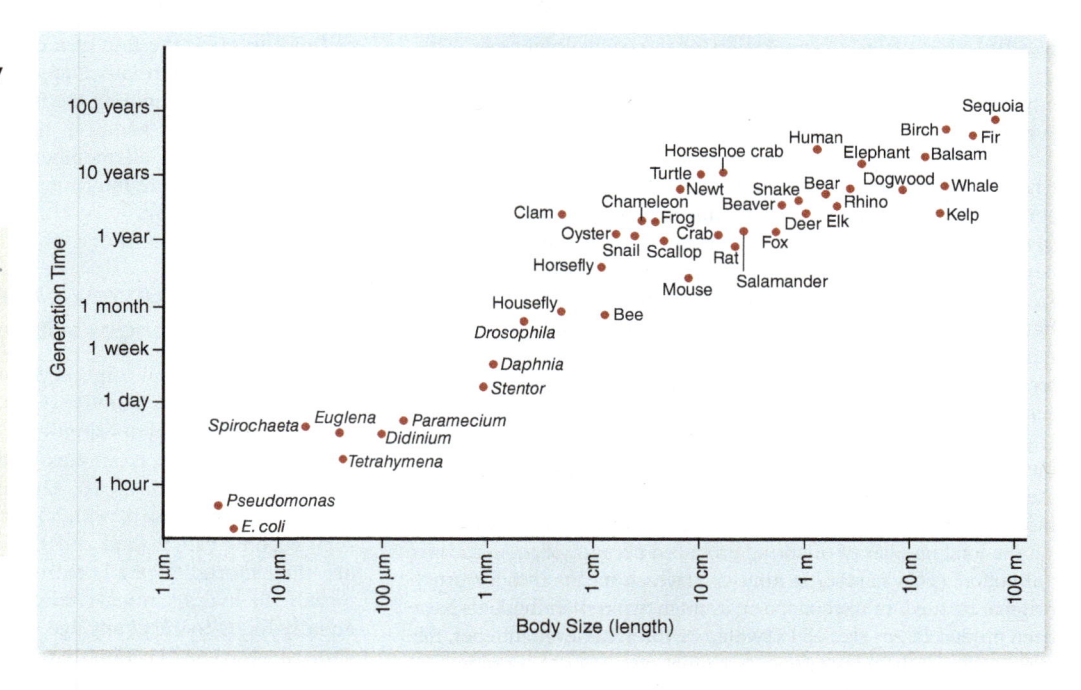

on a population's growth rate. Populations with a large proportion of young individuals, for example, tend to grow rapidly because an increasing proportion of their individuals are reproductive. Human populations in many developing countries are an example, as will be discussed later in section 55.7. Conversely, if a large proportion of a population is relatively old, populations may decline. This phenomenon now characterizes Japan and some countries in Europe.

Life tables show probability of survival and reproduction through a cohort's life span

To assess how populations in nature are changing, ecologists use a **life table,** which tabulates the fate of a cohort from birth until death, showing the number of offspring produced and the number of individuals that die each time period. Table 55.2 is an example

TABLE 55.2 Life Table of the Annual Bluegrass (*Poa annua*) for a Cohort Containing 843 Seedlings

Age (in 3-month intervals)	Number Alive at Beginning of Time Interval	Proportion of Cohort Alive at Beginning of Time Interval (survivorship)	Deaths During Time Interval	Mortality Rate During Time Interval	Seeds Produced During Time Interval	Seeds Produced per Surviving Individual (fecundity)	Seeds Produced per Member of Cohort (fecundity × survivorship)
0	843	1.000	121	0.144	0	0.00	0.00
1	722	0.856	195	0.270	303	0.42	0.36
2	527	0.625	211	0.400	622	1.18	0.74
3	316	0.375	172	0.544	430	1.36	0.51
4	144	0.171	90	0.626	210	1.46	0.25
5	54	0.064	39	0.722	60	1.11	0.07
6	15	0.018	12	0.800	30	2.00	0.04
7	3	0.004	3	1.000	10	3.33	0.01
8	0	0.000	—		Total = 1665		Total = 1.98

of a life table analysis from a study of the annual bluegrass. This study follows the fate of 843 individuals through time, charting how many survive in each interval and how many offspring each survivor produces.

In table 55.2, the first column indicates the age of the cohort (that is, the number of 3-month intervals from the start of the study). The second and third columns indicate the number of survivors and the proportion of the original cohort still alive at the beginning of that interval. The fifth column presents the **mortality rate,** the proportion of individuals that started that interval alive but died by the end of it. The seventh column indicates the average number of seeds produced by each surviving individual in that interval, and the last column shows the number of seeds produced relative to the size of the original cohort.

Much can be learned by examining life tables. In the case of the annual bluegrass, we see that both the probability of dying and the number of offspring produced per surviving individual steadily increases with age. By adding up the numbers in the last column, we get the total number of offspring produced per individual in the initial cohort. This number is almost 2, which means that for every original member of the cohort, on average two new individuals have been produced. A value of 1.0 would be the break-even number, the point at which the population was neither growing nor shrinking. In this case, the population appears to be growing rapidly.

> **Data analysis** If annual bluegrass were an undesirable weed and you wanted to target one age group to remove, would you focus on the age class that produces the most seeds per individual? Can you imagine any circumstance in which your answer would be different?

In most cases, life table analysis is more complicated than this. First, except for organisms with short life spans, it is difficult to track the fate of a cohort until the death of the last individual. An alternative approach is to construct a cross-sectional study, examining the fate of cohorts of different ages in a single period. In addition, many factors—such as offspring reproducing before all members of their parents' cohort have died—complicate the interpretation of whether populations are growing or shrinking.

Survivorship curves demonstrate how survival probability changes with age

The percentage of an original population that survives to a given age is called its **survivorship.** One way to express some aspects of the age distribution of populations is through a *survivorship curve.* Examples of different survivorship curves are shown in figure 55.10. Oysters produce vast numbers of offspring, only a few of which live to reproduce. However, once they become established and grow into reproductive individuals, their mortality rate is extremely low (type III survivorship curve). In hydra, animals related to jellyfish, individuals are equally likely to die at any age. The result is a straight survivorship curve (type II). Finally, mortality rates in humans, as in many other animals and in protists, rise steeply later in life (type I survivorship curve).

Of course, these descriptions are just generalizations, and many organisms show more complicated patterns. Examination of the data for annual bluegrass, for example, reveals that it is most similar to a type II survivorship curve (figure 55.11).

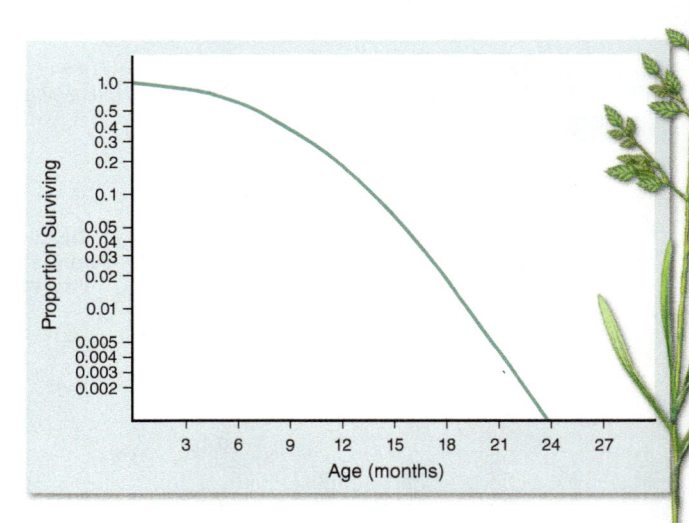

Figure 55.11 Survivorship curve for a cohort of annual bluegrass. After several months of age, mortality occurs at a constant rate through time.

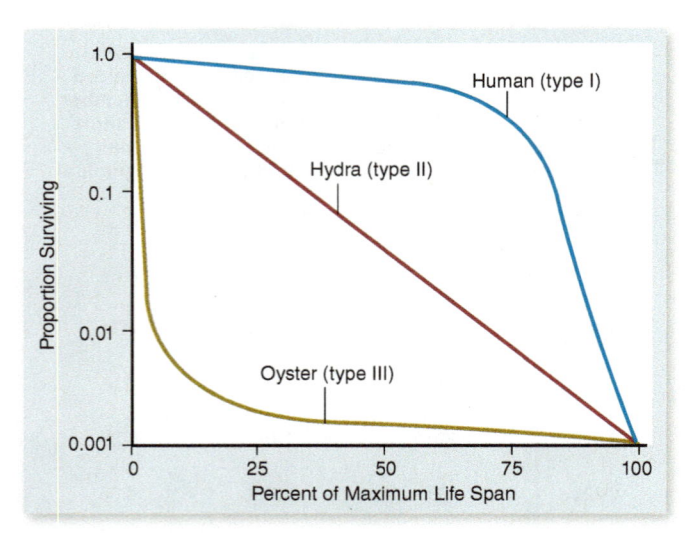

Figure 55.10 Survivorship curves. By convention, survival (the vertical axis) is plotted on a log scale, which produces a straight line if the rate of mortality is constant through time. Humans have a type I life cycle, hydra (an animal related to jellyfish) type II, and oysters type III.

> **Inquiry question** Suppose you wanted to keep annual bluegrass in your room as a houseplant. Suppose, too, that you wanted to buy an individual plant that was likely to live as long as possible. What age plant would you buy? How might the shape of the survivorship curve affect your answer?

Learning Outcomes Review 55.3

Demography is the quantitative study of populations. Demographic characteristics include age structure, life span, sex ratio, generation time, and birth and mortality rates. The age structure of a population and the manner in which mortality and birth rates vary among different age cohorts determine whether a population will increase or decrease in size.

- *Will populations with higher survivorship rates always have higher population growth rates than populations with lower survivorship rates? Explain.*

55.4 Life History and the Cost of Reproduction

Learning Outcomes

1. Describe reproductive trade-offs in an organism's life history.
2. Compare the costs and benefits of allocating resources to reproduction.

Natural selection favors traits that maximize the number of surviving offspring left in the next generation by an individual organism. Two factors affect this quantity: how long an individual lives, and how many young it produces each year.

Why doesn't every organism reproduce immediately after its own birth, produce large families of offspring, care for them intensively, and perform these functions repeatedly throughout a long life? The answer is that usually no one organism can do all of this, simply because not enough resources are available. Consequently, organisms allocate resources either to current reproduction or to increasing their prospects of surviving and reproducing at later life stages.

The complete life cycle of an organism constitutes its life history. All life histories involve significant trade-offs. Because resources are limited, a change that increases reproduction may decrease survival and reduce future reproduction. As one example, a Douglas fir tree that produces more cones increases its current reproductive success—but it also grows more slowly. Because the number of cones produced is a function of how large a tree is, this diminished growth will decrease the number of cones it can produce in the future. Similarly, birds that have more offspring each year have a higher probability of dying during that year or of producing smaller clutches the following year (figure 55.12). Conversely, individuals that delay reproduction may grow faster and larger, enhancing future reproduction.

In one elegant experiment, researchers changed the number of eggs in the nests of a bird, the collared flycatcher (figure 55.13). Birds whose clutch size (the number of eggs produced in one breeding event) was decreased expended less energy raising their young and thus were able to lay more eggs the next year, whereas those given more eggs worked harder and consequently produced fewer eggs the following year. Ecologists refer to the reduction in future reproductive potential resulting from current reproductive efforts as the **cost of reproduction.**

Natural selection favors the life history that maximizes lifetime reproductive success. When the cost of reproduction is low, individuals should produce as many offspring as possible because there is little cost. Low costs of reproduction may occur when resources are abundant and may also be relatively low when overall mortality rates are high. In the latter case, individuals may be unlikely to survive to the next breeding season anyway, so the

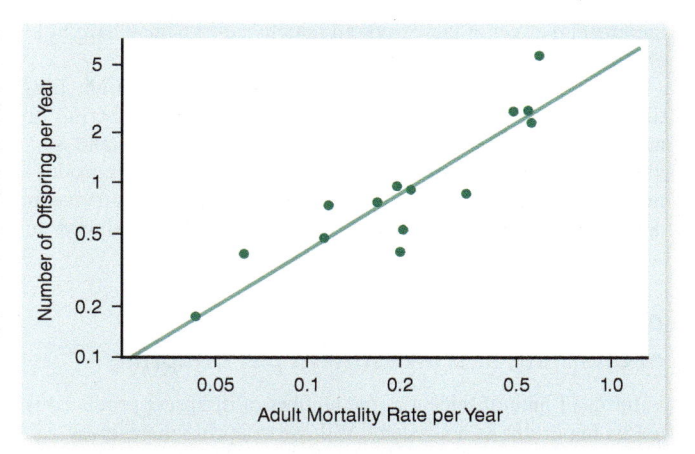

Figure 55.12 Reproduction has a price. Data from many bird species indicate that increased fecundity in birds correlates with higher mortality, ranging from the albatross (lowest) to the sparrow (highest). Species that raise more offspring per year have a higher probability of dying during that year.

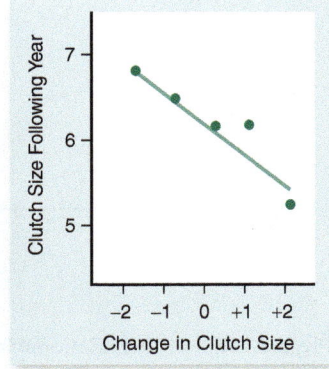

Figure 55.13 Reproductive events per lifetime. Adding eggs to nests of collared flycatchers, *Ficedula albicollis,* which increases the reproductive efforts of the female rearing the young, decreases clutch size the following year; removing eggs from the nest increases the next year's clutch size. This experiment demonstrates the trade-off between current reproductive effort and future reproductive success.

Data analysis Which has a greater effect, removing or adding eggs to a nest?

incremental effect of increased reproductive efforts may have little effect on future survival.

Alternatively, when costs of reproduction are high, lifetime reproductive success may be maximized by deferring or minimizing current reproduction to enhance growth and survival rates. This situation may occur when costs of reproduction significantly affect the ability of an individual to survive or decrease the number of offspring that can be produced in the future.

A trade-off exists between number of offspring and investment per offspring

In terms of natural selection, the number of offspring produced is not as important as how many of those offspring themselves survive to reproduce. Assuming that the amount of energy to be invested in offspring is limited, a balance must be reached between the number of offspring produced and the size of each offspring (figure 55.14). This trade-off has been experimentally demonstrated in the side-blotched lizard, which normally lays between four and five eggs at a time. When some of the egg follicles are removed surgically early in the reproductive cycle, the female lizard produces only one to three eggs, but supplies each of these eggs with greater amounts of yolk, producing eggs and, subsequently, hatchlings that are much larger than normal (figure 55.15).

In the side-blotched lizard and many other species, the size of offspring is critical—larger offspring have a greater chance of survival. Producing many offspring with little chance of survival might not be the best strategy, but producing only a single, extraordinarily robust offspring also would not maximize the number of surviving offspring. Rather, an intermediate situation, in which several fairly large offspring are produced, should lead to the maximum number of surviving offspring.

Figure 55.15 Variation in the size of baby side-blotched lizards, *Uta stansburiana*, produced by experimental manipulations. In clutches in which some developing eggs were surgically removed, the remaining offspring were larger *(center)* than lizards produced in control clutches in which all the eggs were allowed to develop *(right)*. The causal relationship between yolk provided by the mother and size of the offspring was demonstrated in experiments in which some of the yolk was removed from the eggs and smaller lizards resulted *(left)*.

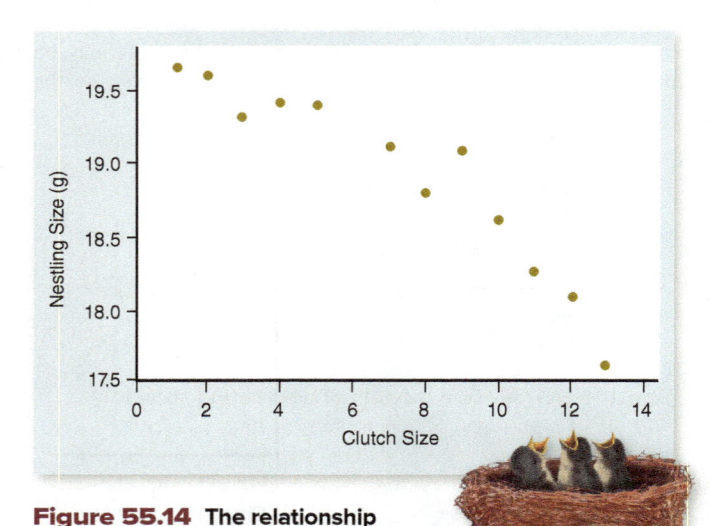

Figure 55.14 The relationship between clutch size and offspring size. In great tits, *Parus major*, the size of the nestlings is inversely related to the number of eggs laid. The more mouths they have to feed, the less the parents can provide to any one nestling.

Data analysis Overall, what clutch size probably requires the greatest energy expenditure by the parents?

Inquiry question Would natural selection favor producing many small young or a few large ones?

Reproductive events per lifetime represent an additional trade-off

The trade-off between age and fecundity plays a key role in many life histories. Annual plants and most insects focus all their reproductive resources on a single large event and then die. This life history adaptation is called **semelparity**. Organisms that produce offspring several times over many seasons exhibit a life history adaptation called **iteroparity**.

Species that reproduce yearly must avoid overtaxing themselves in any one reproductive episode so that they will be able to survive and reproduce in the future. Semelparity, or "big bang" reproduction, is usually found in short-lived species that have a low probability of staying alive to reproduce again, such as plants growing in harsh climates. Semelparity is also favored when fecundity entails large reproductive cost, exemplified by Pacific salmon migrating upriver to their spawning grounds. In these species, rather than investing some resources in an unlikely bid to survive until the next breeding season, individuals put all their resources into one reproductive event.

Age at first reproduction correlates with life span

Among mammals and many other animals, longer-lived species put off reproduction longer than short-lived species, relative to expected life span. The advantage of delayed reproduction is that juveniles gain experience and can grow more rapidly before expending the high costs of reproduction. In long-lived animals, these advantages outweigh the costs in lost reproduction of not breeding earlier.

In shorter-lived animals, on the other hand, time is of the essence; thus, quick reproduction is more critical than juvenile training and faster growth, and reproduction tends to occur earlier.

Learning Outcomes Review 55.4

Life history adaptations involve many trade-offs between reproductive cost and investment in survival. These trade-offs take a variety of forms, from laying fewer than the maximum possible number of eggs to putting all energy into a single bout of reproduction. Natural selection favors maximizing lifetime reproductive success.

- How might the life histories of two species differ if one was subject to high levels of predation and the other had few predators?

55.5 Environmental Limits to Population Growth

Learning Outcomes

1. Explain exponential growth.
2. Discuss why populations cannot grow exponentially forever.
3. Define carrying capacity and explain what might cause it to change.

Populations often remain at a relatively constant size, regardless of how many offspring are born. As you saw in chapter 1, Darwin based his theory of natural selection partly on this seeming contradiction. Natural selection occurs because of checks on reproduction, with some individuals producing fewer surviving offspring than others. To understand populations, we must consider how they grow and what factors in nature limit population growth.

The exponential growth model applies to populations with no growth limits

The rate of population increase, r, is defined as the difference between the birth rate, b, and the death rate, d, corrected for movement of individuals in or out of the population (e, rate of movement out of the area; i, rate of movement into the area). Thus,

$$r = (b - d) + (i - e)$$

Movements of individuals can have a major influence on population growth rates. For example, the increase in human population in the United States during the closing decades of the 20th century was mostly due to immigration.

The simplest model of population growth assumes that a population grows without limits at its maximal rate and also that rates of immigration and emigration are equal. This rate, called the **biotic potential**, is the rate at which a population of a given species increases when no limits are placed on its rate of growth. In mathematical terms, this is defined by the following formula:

$$\frac{dN}{dt} = r_i N$$

where N is the number of individuals in the population, dN/dt is the rate of change in its numbers over time, and r_i is the intrinsic rate of natural increase for that population—its innate capacity for growth.

The biotic potential of any population is exponential (*red line* in figure 55.16). Even when the *rate* of increase remains constant, the actual *number* of individuals accelerates rapidly as the size of the population grows. The result of unchecked exponential growth is a population explosion.

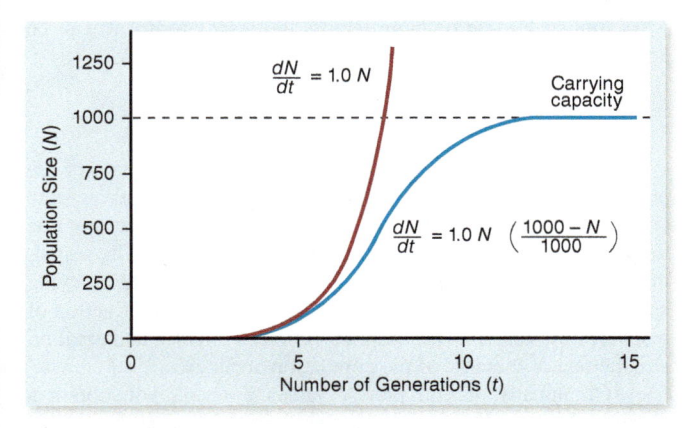

Figure 55.16 Two models of population growth. The red line illustrates the exponential growth model for a population with an r of 1.0. The blue line illustrates the logistic growth model in a population with $r = 1.0$ and $K = 1000$ individuals. At first, logistic growth accelerates exponentially; then, as resources become limited, the death rate increases and growth slows. Growth ceases when the death rate equals the birth rate. The carrying capacity (K) ultimately depends on the resources available in the environment.

Data analysis Suppose in one species of mouse, each pair produced four surviving offspring, whereas in another population, each pair produced five offspring. If this rate of population increase remained unchanged, after 10 generations, how much larger would the population of the second species be?

A single pair of houseflies, laying 120 eggs per generation, could produce more than 5 trillion descendants in a year. In 10 years, their descendants would form a swarm more than 2 m thick over the entire surface of the Earth! In practice, such patterns of unrestrained growth prevail only for short periods, usually when an organism reaches a new habitat with abundant resources. Natural examples of such a short period of unrestrained growth include dandelions arriving in the fields, lawns, and meadows of North America from Europe for the first time; algae colonizing a newly formed pond; or cats introduced to an island with many birds, but previously lacking predators.

Carrying capacity

No matter how rapidly populations grow, they eventually reach a limit imposed by shortages of important environmental factors, such as space, light, water, or nutrients. A population ultimately may stabilize at a certain size, called the **carrying capacity** of the particular place where it lives. The carrying capacity, symbolized by K, is the maximum number of individuals that the environment can support.

The logistic growth model applies to populations that approach their carrying capacity

As a population approaches its carrying capacity, its rate of growth slows greatly, because fewer resources remain for each new individual to use. The growth curve of such a population, which is always limited by one or more factors in the environment, can be approximated by the following logistic growth equation:

$$\frac{dN}{dt} = rN\left(\frac{K-N}{K}\right)$$

In this model of population growth, the growth rate of the population (dN/dt) is equal to its intrinsic rate of natural increase (r multiplied by N, the number of individuals present at any one time), adjusted for the amount of resources available. The adjustment is made by multiplying rN by the fraction of K, the carrying capacity, still unused $[(K − N)/K]$. As N increases, the fraction of resources by which r is multiplied becomes smaller and smaller, and the rate of increase of the population declines.

Graphically, if you plot N versus t (time), you obtain a **sigmoidal growth curve** characteristic of many biological

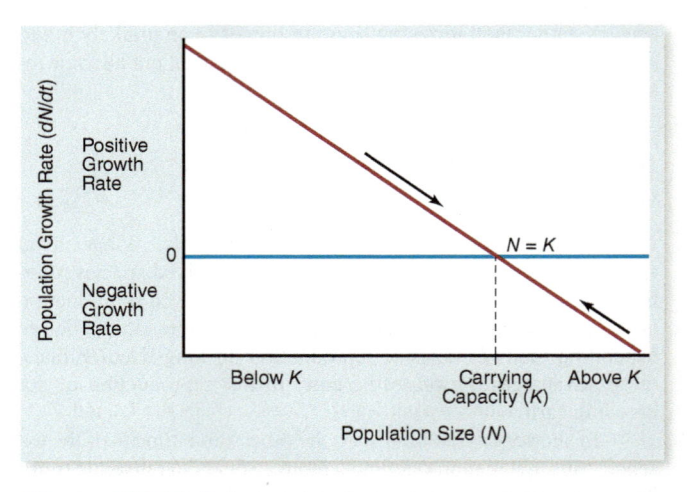

Figure 55.17 Relationship between population growth rate and population size. Populations far from the carrying capacity (K) have high growth rates—positive if the population is below K, and negative if it is above K. As the population approaches K, growth rates approach zero.

? Inquiry question Why does the growth rate converge on zero?

populations. The curve is called "sigmoidal" because its shape has a double curve like the letter S. As the size of a population stabilizes at the carrying capacity, its rate of growth slows, eventually coming to a halt (*blue line* in figure 55.16).

In mathematical terms, as N approaches K, the *rate* of population growth (dN/dt) begins to slow, reaching 0 when $N = K$ (figure 55.17). Conversely, if the population size exceeds the carrying capacity, then $K − N$ will be negative, and the population will experience a negative growth rate. As the population size then declines toward the carrying capacity, the magnitude of this negative growth rate will decrease until it reaches 0 when $N = K$.

Notice that the population tends to move toward the carrying capacity regardless of whether it is initially above or below it. For this reason, logistic growth tends to return a population to the same size. In this sense, such populations are considered to be in

Figure 55.18 Many populations exhibit logistic growth. *a.* A fur seal, *Callorhinus ursinus*, population on St. Paul Island, Alaska. *b.* Two laboratory populations of the cladoceran, *Bosmina longirostris*. Note that the populations first exceeded the carrying capacity, before decreasing to what appears to be a size that was then maintained.

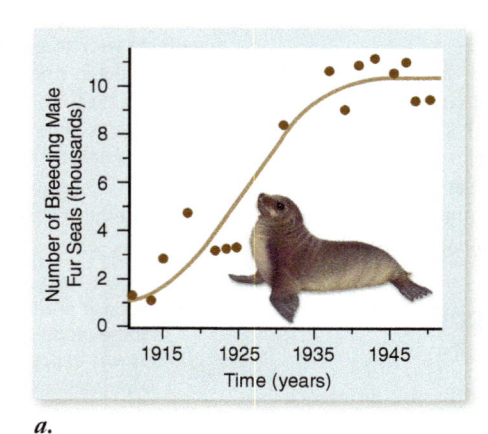

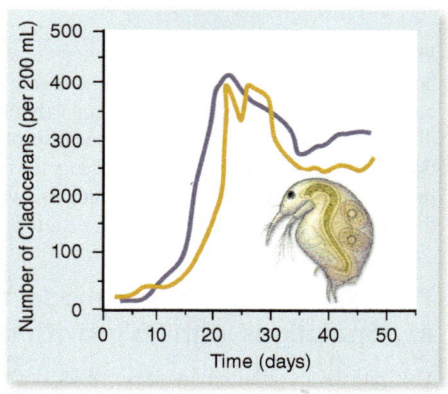

equilibrium because they would be expected to be at or near the carrying capacity at most times.

In many cases, real populations display trends corresponding to a logistic growth curve. This is true not only in the laboratory, but also in natural populations (figure 55.18a). In some cases, however, the fit is not perfect (figure 55.18b), and as we shall see shortly, many populations exhibit other patterns.

Learning Outcomes Review 55.5

Exponential growth refers to population growth in which the number of individuals accelerates even when the rate of increase remains constant; it results in a population explosion. Exponential growth is eventually limited by resource availability. The size at which a population in a particular location stabilizes is defined as the carrying capacity of that location for that species. Populations often grow to the carrying capacity of their environment.

- *What might cause a population's carrying capacity to change, and how would the population respond?*

55.6 Factors That Regulate Populations

Learning Outcomes

1. Compare density-dependent and density-independent factors.
2. Evaluate why the size of some populations cycles.
3. Consider how the life history adaptations of species may differ depending on how often populations are at their carrying capacity.

A number of factors may affect population size through time. Some of these factors depend on population size and are therefore termed *density-dependent*. Other factors, such as natural disasters, affect populations regardless of size; these factors are termed *density-independent*. Many populations exhibit cyclic fluctuations in size that may result from complex interactions of factors.

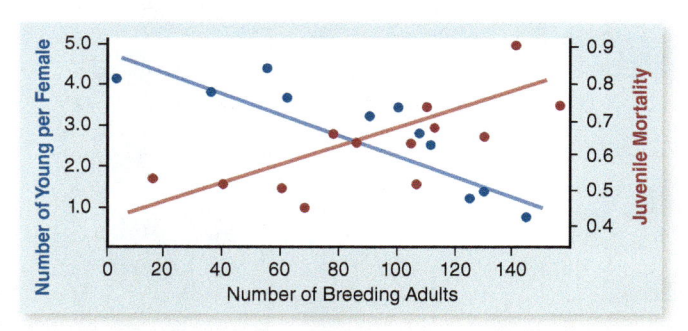

Figure 55.20 Density dependence in the song sparrow, *Melospiza melodia,* on Mandarte Island. Reproductive success decreases and mortality rates increase as population size increases.

Inquiry question What would happen if researchers supplemented the food available to the birds?

Density-dependent effects occur when reproduction and survival are affected by population size

The reason population growth rates are affected by population size is that many important processes have **density-dependent effects.** That is, as population size increases, either reproductive rates decline or mortality rates increase, or both, a phenomenon termed *negative feedback* (figure 55.19).

Populations can be regulated in many different ways. When populations approach their carrying capacity, competition for resources can be severe, leading both to a decreased birth rate and an increased risk of death (figure 55.20). In addition, predators often focus their attention on a particularly common prey species, which also results in increasing rates of mortality as populations increase. High population densities can also lead to an accumulation of toxic wastes in the environment.

Behavioral changes may also affect population growth rates. Some species of rodents, for example, become antisocial, fighting more, breeding less, and generally acting stressed-out. These behavioral changes result from hormonal actions, but their ultimate cause is not yet clear; most likely, they

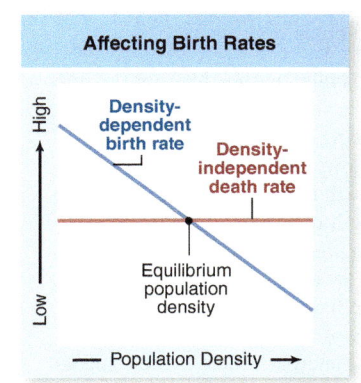

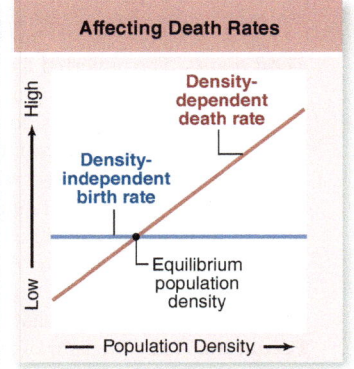

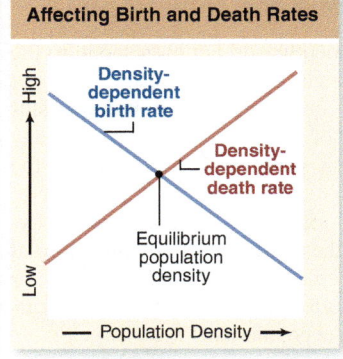

Figure 55.19 Density-dependent population regulation. Density-dependent factors can affect birth rates, death rates, or both.

Inquiry question Why might birth rates be density-dependent?

Figure 55.21 Density-dependent effects. Migratory locusts, *Locusta migratoria*, are a legendary plague of large areas of Africa and Eurasia. At high population densities, the locusts have different hormonal and physical characteristics and take off as a swarm.

have evolved as adaptive responses to situations in which resources are scarce. In addition, in crowded populations, the population growth rate may decrease because of an increased rate of emigration of individuals attempting to find better conditions elsewhere (figure 55.21).

However, not all density-dependent factors are negatively related to population size. In some cases, growth rates increase with population size. This phenomenon is referred to as the **Allee effect** (after American zoologist Warder Allee, who first described it), and is an example of *positive feedback*. The Allee effect can take several forms. Most obviously, in populations that are too sparsely distributed, individuals may have difficulty finding mates. Moreover, some species may rely on large groups to deter predators or to provide the necessary stimulation for breeding activities. The Allee effect is a major threat for many endangered species, which may never recover from decreased population sizes caused by habitat destruction, overexploitation, or other causes (see chapter 59).

Density-independent effects include environmental disruptions and catastrophes

Growth rates in populations sometimes do not correspond to the logistic growth equation. In many cases, such patterns result because growth is under the control of **density-independent effects.** In other words, the rate of growth of a population at any instant is limited by something unrelated to the size of the population.

A variety of factors may affect populations in a density-independent manner. Most of these are aspects of the external environment, such as extremely cold winters, droughts, storms, or volcanic eruptions. Individuals often are affected by these occurrences regardless of the size of the population.

Populations in areas where such events occur relatively frequently display erratic growth patterns in which the populations increase rapidly when conditions are benign, but exhibit large reductions whenever the environment turns hostile (figure 55.22). Needless to say, such populations do not produce the sigmoidal growth curves characteristic of the logistic equation.

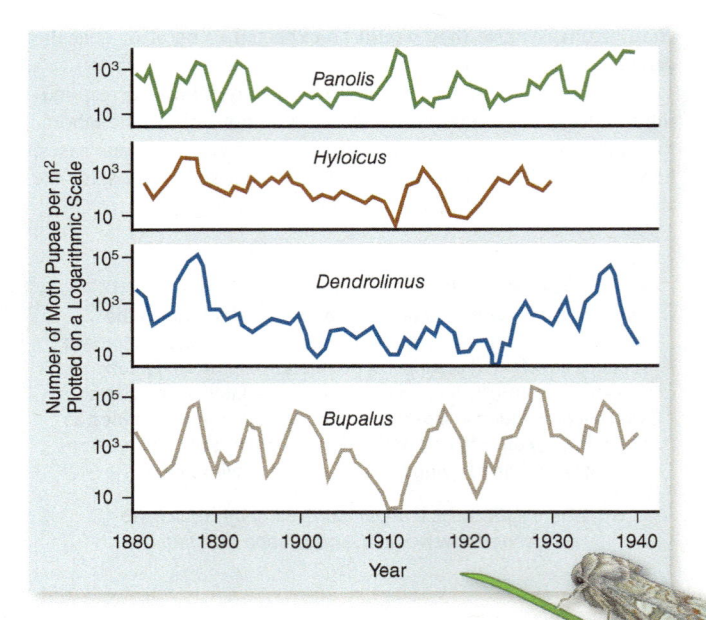

Figure 55.22 Fluctuations in the number of pupae of four moth species in Germany. The population fluctuations suggest that density-independent factors are regulating population size. The concordance in trends through time suggests that the same factors are regulating population size in all four species.

Population cycles may reflect complex interactions

In some populations, density-dependent effects lead not to an equilibrium population size but to cyclic patterns of increase and decrease. For example, ecologists have studied cycles in hare populations since the 1820s. They have found that the North American snowshoe hare *(Lepus americanus)* follows a "10-year cycle" (in reality, the cycle varies from 8 to 11 years). Hare population numbers fall 10-fold to 30-fold in a typical cycle, and 100-fold changes can occur (figure 55.23). Two factors appear to be generating the cycle—food plants and predators:

Food plants. The preferred foods of snowshoe hares are willow and birch twigs. As hare density increases, the quantity of these twigs decreases, forcing the hares to feed on high-fiber (low-quality) food. Lower birth rates, low juvenile survivorship, and low growth rates follow. The hares also spend more time searching for food, an activity that increases their exposure to predation. The result is a precipitous decline in willow and birch twig abundance, and a corresponding fall in hare abundance. It takes two to three years for the quantity of mature twigs to recover.

Predators. A key predator of the snowshoe hare is the Canada lynx. The Canada lynx shows a "10-year" cycle of abundance that seems remarkably entrained to the hare abundance cycle (figure 55.23). As hare numbers increase, lynx numbers do too, rising in response to the increased availability of the lynx's food. When hare numbers fall, so do lynx numbers, their food supply depleted.

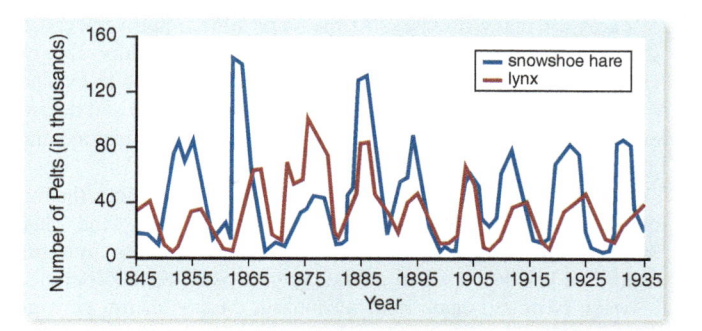

Figure 55.23 Linked population cycles of the snowshoe hare, *Lepus americanus*, and the northern lynx, *Lynx canadensis*. These data are based on records of fur returns from trappers in the Hudson Bay region of Canada. The lynx population carefully tracks that of the snowshoe hare, but lags behind it slightly.

> **Inquiry question** Suppose experimenters artificially kept the hare population at a high and constant level; what would happen to the lynx population? Conversely, if experimenters artificially kept the lynx population at a high and constant level, what would happen to the hare population?

Which factor is responsible for the predator–prey oscillations? Do increasing numbers of hares lead to overharvesting of plants (a hare–plant cycle), or do increasing numbers of lynx lead to overharvesting of hares (a hare–lynx cycle)? Field experiments carried out by Charles Krebs and coworkers in 1992 provide an answer.

In Canada's Yukon, Krebs set up experimental plots that contained hare populations. If food is added (no food shortage effect) and predators are excluded (no predator effect) in an experimental area, hare numbers increase 10-fold and stay there—the cycle is lost. However, the cycle is retained if either of the factors is allowed to operate alone: exclude predators but don't add food (food shortage effect alone), or add food in the presence of predators (predator effect alone). Thus, both factors can affect the cycle, which in practice seems to be generated by the interaction between the two.

Population cycles are not as rare as traditionally thought; one review of nearly 700 long-term (25 years or more) studies of trends within populations found that cycles were not uncommon; nearly 30% of the studies—including birds, mammals, fish, and crustaceans—provided evidence of some cyclic pattern in population size through time, although most of these cycles are nowhere near as dramatic in amplitude as the hare–lynx cycles. In some cases, such as that of the snowshoe hare and lynx, density-dependent factors may be involved, whereas in other cases, density-independent factors, such as cyclic climatic patterns, may be responsible.

Resource availability affects life history adaptations

As you have seen, some species usually maintain stable population sizes near the carrying capacity, whereas in other species population sizes fluctuate markedly and are often far below carrying capacity. The selective factors affecting such species differ markedly. Individuals in populations near their carrying capacity may face stiff competition for limited resources; by contrast, individuals in populations far below carrying capacity have access to abundant resources.

When resources are limited, the cost of reproduction often will be very high. Consequently, selection will favor individuals that can compete effectively and utilize resources efficiently. Such adaptations often come at the cost of lowered reproductive rates, as organisms wait longer to reproduce so that they can grow larger and stronger, and produce fewer, larger offspring. Such populations are termed ***K*-selected** because they are adapted to thrive when the population is near its carrying capacity (*K*). Table 55.3 lists some of the typical features of *K*-selected populations. Examples of *K*-selected species include coconut palms, whooping cranes, whales, and humans.

By contrast, in populations far below the carrying capacity, resources may be abundant. Costs of reproduction are low, and selection favors those individuals that can produce the maximum number of offspring. Selection here favors individuals with the highest reproductive rates; such populations are termed ***r*-selected**. Examples of organisms displaying *r*-selected life history adaptations include dandelions, aphids, mice, and cockroaches.

Most natural populations show life history adaptations that exist along a continuum ranging from completely *r*-selected traits to completely *K*-selected traits. Although these

TABLE 55.3 *r*-Selected and *K*-Selected Life History Adaptations

Adaptation	*r*-Selected Populations	*K*-Selected Populations
Age at first reproduction	Early	Late
Life span	Short	Long
Maturation time	Short	Long
Mortality rate	Often high	Usually low
Number of offspring produced per reproductive episode	Many	Few
Number of reproductions per lifetime	Few	Many
Parental care	None	Often extensive
Size of offspring or eggs	Small	Large

tendencies hold true as generalities, few populations are purely *r*- or *K*-selected and show all of the traits listed in table 55.3. These attributes should be treated as generalities, with the recognition that many exceptions exist.

> **Learning Outcomes Review 55.6**
>
> Density-dependent factors such as resource availability come into play particularly when population size is larger; density-independent factors such as natural disasters operate regardless of population size. Population density may be cyclic due to complex interactions such as resource cycles and predator effects. Populations with density-dependent regulation often are near their carrying capacity; in species with populations well below carrying capacity, natural selection may favor high rates of reproduction when resources are abundant.
>
> ■ *Can a population experience both positive and negative density-dependent effects?*

55.7 Human Population Growth

> **Learning Outcomes**
>
> 1. Explain how the rate of human population growth has changed through time.
> 2. Describe the effects of age distribution on future growth.
> 3. Evaluate the relative importance of rapid population growth and resource consumption as threats to the biosphere and human welfare.

Humans exhibit many *K*-selected life history traits, including small brood size, late reproduction, and a high degree of parental care. These life history traits evolved during the early history of hominids, when the limited resources available from the environment controlled population size. Throughout most of human history, our populations have been regulated by food availability, disease, and predators. Although unusual disturbances, including floods, plagues, and droughts, no doubt affected the pattern of human population growth, the overall size of the human population grew slowly during our early history.

Two thousand years ago, perhaps 130 million people populated the Earth. It took a thousand years for that number to double, and it was 1650 before it had doubled again, to about 500 million. In other words, for most of its existence, the human population was characterized by very slow growth. In this respect, human populations resembled many other species with predominantly *K*-selected life history adaptations.

Human populations have grown exponentially

Starting in the early 1700s, changes in technology gave humans more control over their food supply, enabled them to develop superior weapons to ward off predators, and led to the development of cures for many diseases. At the same time, improvements in shelter and storage capabilities made humans less vulnerable to climatic uncertainties. These changes allowed humans to expand the carrying capacity of the habitats in which they lived and thus to escape the confines of logistic growth and re-enter the exponential phase of the sigmoidal growth curve.

Responding to the lack of environmental constraints, the human population has grown explosively over the last 300 years. Although the birth rate has decreased over time, the death rate has decreased to a much greater extent. The difference between birth and death rates meant that the population grew as much as 2% per year, although the rate has now declined to 1.1% per year.

A 1.1% annual growth rate may not seem large, but it has produced a current human population that has surpassed 7 billion people (figure 55.24). At this growth rate, 75 million people would be added to the world population in the next year, and the human population would double in 63 years. Both the current human population level and the projected growth rate have potentially grave consequences for our future.

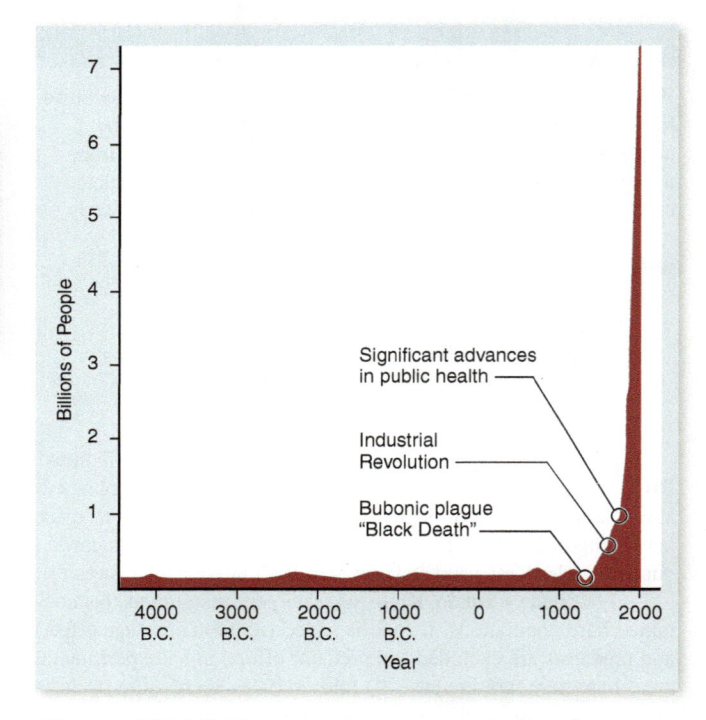

Figure 55.24 History of human population size. Temporary increases in death rate, even a severe one such as that occurring during the Black Death of the 1300s, have little lasting effect. Explosive growth began with the Industrial Revolution in the 1800s, which produced a significant, long-term lowering of the death rate. The current world population has recently exceeded 7 billion, and if it continued to grow at the present rate, would double in 63 years.

? Inquiry question Based on what we have learned about population growth, what do you predict will happen to human population size?

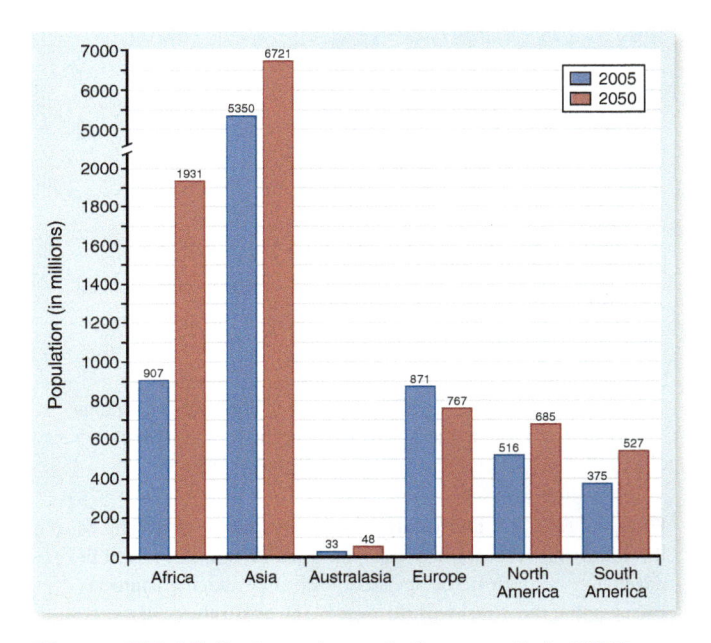

Figure 55.25 Projected population growth in 2050. Developed countries are predicted to grow little; almost all of the population increase will occur in less-developed countries.

Population pyramids show birth and death trends

Although the human population as a whole continues to grow rapidly at the beginning of the 21st century, this growth is not occurring uniformly over the planet. Rather, most of the population growth is occurring in Africa and Asia (figure 55.25). By contrast, populations are actually decreasing in some countries in Europe.

The rate at which a population can be expected to grow in the future can be assessed graphically by means of a **population pyramid,** a bar graph displaying the numbers of people in each age category (figure 55.26). Males are conventionally shown to the left of the vertical age axis, females to the right. A human population pyramid thus displays the age composition of a population by sex. In most human population pyramids, the number of older females is disproportionately large compared with the number of older males, because females in most regions have a longer life expectancy than males.

Viewing such a pyramid, we can predict demographic trends in births and deaths. In general, a rectangular pyramid is characteristic of countries whose populations are stable, neither growing nor shrinking. A triangular pyramid is characteristic of a country that will exhibit rapid future growth because most of its population has not yet entered the childbearing years. Inverted triangles are characteristic of populations that are shrinking, usually as a result of sharply declining birth rates.

Examples of population pyramids for Sweden and Kenya in 2014 are shown in figure 55.26. The two countries exhibit very different age distributions. The nearly rectangular population pyramid for Sweden indicates that its population is not expanding because birth rates have decreased and average life span has increased. The very triangular pyramid of Kenya, by contrast, results from relatively high birth rates and shorter average life spans, which can lead to explosive future growth. The difference is most apparent when we consider that only 17% of Sweden's population is less

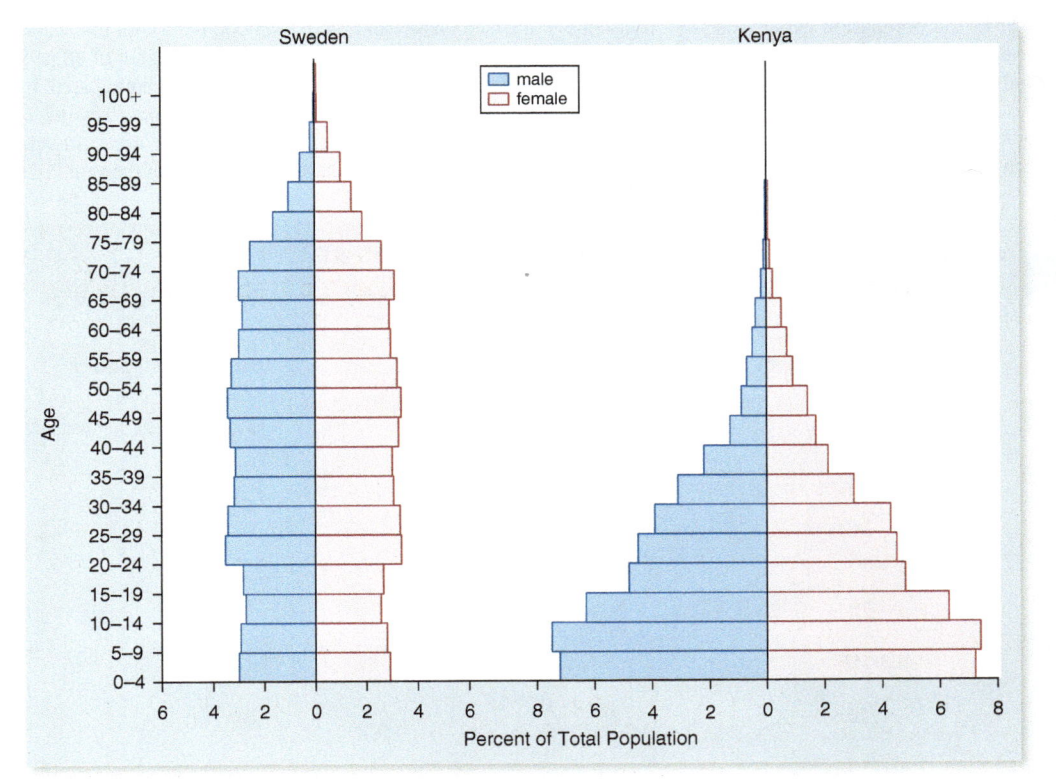

Figure 55.26
Comparison of Swedish and Kenyan population pyramids. Population pyramids from 2014 are graphed according to a population's age distribution. Kenya's pyramid has a broad base because of the great number of individuals below childbearing age. When the young people begin to bear children, the population will experience rapid growth. The Swedish pyramid exhibits a slight bulge among middle-aged Swedes, the result of the "baby boom" that occurred in the middle of the 20th century, and many postreproductive individuals resulting from Sweden's long average life span.

Inquiry question
What will the population distributions look like in 20 years?

than 15 years old, compared with more than 40% of all Kenyans. Moreover, the fertility rate (offspring per woman) in Sweden is 1.9; in Kenya, it is 4.5. As a result, Kenya's population could double in less than 30 years, whereas Sweden's will remain stable.

Humanity's future growth is uncertain

Earth's rapidly growing human population constitutes perhaps the greatest challenge to the future of the biosphere, the world's interacting community of living things. Humanity is adding 75 million people a year to its population—over a million every 5 days, 150 every minute! In more rapidly growing countries, the resulting population increase is staggering (table 55.4). India, for example, had a population of 1.05 billion in 2002; by 2050, its population likely will exceed 1.6 billion.

A key element in the world's population growth is its uneven distribution among countries. Of the billion people added to the world's population in the 1990s, 90% live in developing countries (figure 55.27). The fraction of the world's population that lives in industrialized countries is therefore diminishing. In 1950, fully one-third of the world's population lived in industrialized countries; by 1996, that proportion had fallen to one-quarter; and in 2020, the proportion will have fallen to one-sixth. In the future, the world's population growth will be centered in the parts of the world least equipped to deal with the pressures of rapid growth.

Rapid population growth in developing countries has had the harsh consequence of increasing the gap between rich and poor. Today, the 19% of the world's population that lives in the industrialized world have a per capita income of $22,060, but 81% of the world's population lives in developing countries and has a per capita income of only $3,580. Furthermore, of the people in the developing world, about one-quarter of the population gets by on $1 per day. Eighty percent of all the energy used today is consumed by the industrialized world, but only 20% is used by developing countries.

No one knows whether the world can sustain today's population of over 7 billion people, much less the far greater numbers

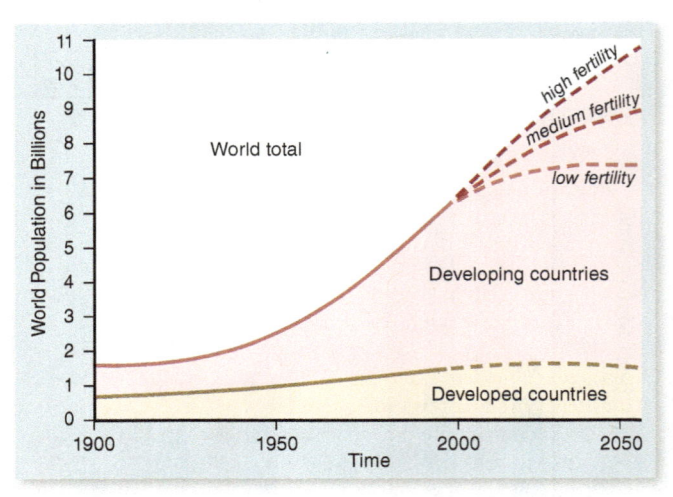

Figure 55.27 Distribution of population growth. Most of the worldwide increase in population since 1950 has occurred in developing countries. The age structures of developing countries indicate that this trend will increase in the near future. World population in 2050 likely will be between 7.3 and 10.7 billion, according to a recent study. Depending on fertility rates, the population at that time will either be increasing rapidly or slightly, or in the best case, declining slightly.

expected in the future. As chapter 57 outlines, the world ecosystem is already under considerable stress. We cannot reasonably expect to expand its carrying capacity indefinitely, and indeed we already seem to be stretching the limits.

Despite using an estimated 45% of the total biological productivity of Earth's landmasses and more than one-half of all renewable sources of freshwater, between one-fourth and one-eighth of all people in the world are malnourished. Moreover, as anticipated by Thomas Malthus in his famous 1798 work, *Essay on the Principle of Population,* death rates are beginning to rise in some areas. In sub-Saharan Africa, for example, population projections for the year 2025

TABLE 55.4	A Comparison of 2010 Population Data in Developed and Developing Countries		
	United States (highly developed)	**Brazil (moderately developed)**	**Ethiopia (poorly developed)**
Fertility rate	1.9	1.8	4.6
Doubling time at current rate (years)	99	77	27
Infant mortality rate (per 1000 births)	6	19	56
Life expectancy at birth (years)	79	74	63
Per capita GDP (U.S. $)*	$53,042	$11,208	$505
Population < 15 years old (%)	20	24	41

*GDP, gross domestic product.

have been scaled back from 1.33 billion to 1.05 billion (a 21% decrease) because of the effect of AIDS. Similar decreases are projected for Russia as a result of higher death rates due to disease.

If we are to avoid catastrophic increases in the death rate, birth rates must fall dramatically. Faced with this grim dichotomy, significant efforts are under way worldwide to lower birth rates.

The population growth rate has declined

The world population growth rate is declining, from a high of 2.0% in the period 1965–1970 to 1.1% in 2015. Nonetheless, because of the larger population, this amounts to an increase of 75 million people per year to the world population, compared with 53 million per year in the 1960s.

Most countries are devoting considerable attention to slowing the growth rate of their populations, and there are genuine signs of progress. For example, from 1984 to 2008, family planning programs in Kenya succeeded in reducing the fertility rate from 8.0 to 4.5 children per couple, thus lowering the population growth rate from 4.0% per year to 2.4% per year. Because of these efforts, the global population may stabilize at about 8.9 billion people by the middle of the current century. How many people the planet can support sustainably depends on the quality of life that we want to achieve; there are already more people than can be sustainably supported with current technologies.

Consumption in the developed world further depletes resources

Population size is not the only factor that determines resource use; per capita consumption is also important. In this respect, we in the industrialized world need to pay more attention to lessening the impact each of us makes because, even though the vast majority of the world's population is in developing countries, the overwhelming percentage of consumption of resources occurs in the industrialized countries. Indeed, the wealthiest 20% of the world's population accounts for 86% of the world's consumption of resources and produces 53% of the world's carbon dioxide emissions, whereas the poorest 20% of the world is responsible for only 1.3% of consumption and 3% of carbon dioxide emissions. Looked at another way, in terms of resource use, a child born today in the industrialized world will consume many more resources over the course of his or her life than a child born in the developing world.

One way of quantifying this disparity is by calculating what has been termed the **ecological footprint,** which is the amount of productive land required to support an individual at the standard of living of a particular population through the course of his or her life. This figure estimates the acreage used for the production of food (both plant and animal), forest products, and housing, as well as the area of forest required to absorb carbon dioxide produced by the combustion of fossil fuels. As figure 55.28 illustrates, the ecological footprint of an individual in the United States is more than 10 times greater than that of someone in India.

Based on these measurements, researchers have calculated that resource use by humans is now one-third greater than the amount that nature can sustainably replace. Moreover, consumption is increasing rapidly in parts of the developing world; if all humans lived at the standard of living in the industrialized world, two additional planet Earths would be needed.

Building a sustainable world is the most important task facing humanity's future. The quality of life available to our children will depend to a large extent on our success in limiting both population growth and the amount of per capita resource consumption.

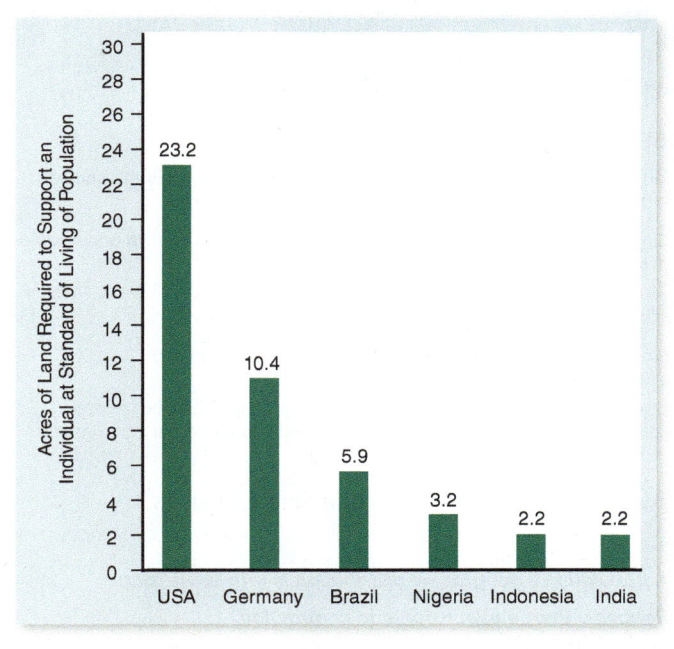

Figure 55.28 Ecological footprints of individuals in different countries. An ecological footprint calculates how much land is required to support a person through his or her life, including the acreage used for production of food, forest products, and housing, in addition to the forest required to absorb the carbon dioxide produced by the combustion of fossil fuels.

> **Inquiry question** Which is a more important cause of resource depletion, overpopulation or overconsumption? Explain.

Learning Outcomes Review 55.7

For most of its history, the *K*-selected human population increased gradually. In the last 400 years, with resource control, the human population has grown exponentially; at the current rate, it would double in 63 years. A population pyramid shows the number of individuals in different age categories. Pyramids with a wide base are undergoing faster growth than those that are uniform from top to bottom. Growth rates overall are declining, but consumption per capita in the developed world is still a significant drain on resources.

- *Which is more important, reducing global population growth or reducing resource consumption levels in developed countries?*

Chapter Review

55.1 The Environmental Challenges

Key environmental factors include temperature, water, sunlight, and soil type. Individuals seek to maintain internal homeostasis.

Organisms are capable of responding to environmental changes that occur during their lifetime.
Most individuals can cope with variations in their natural habitat, such as short-term changes in temperature and water availability.

Natural selection leads to evolutionary adaptation to environmental conditions.
Over evolutionary time, physiological, morphological, or behavioral adaptations evolve that make organisms better suited to the environment in which they live.

55.2 Populations: Groups of a Single Species in One Place

A population's geographic distribution is termed its range.

Ranges undergo expansion and contraction.
Most populations have limited geographic ranges that can expand or contract through time as the environment changes.

Dispersal mechanisms may allow some species to cross a barrier and expand their range. Human actions have led to range expansion of some species, often with detrimental effects.

Individuals in populations exhibit different spacing patterns.
Within a population, individuals are distributed randomly, uniformly, or are clumped. Nonrandom distributions may reflect resource distributions or competition for resources.

A metapopulation comprises distinct populations that may exchange members.
The degree of exchange between populations in a metapopulation is highest when populations are large and more connected.

Metapopulations may act as a buffer against extinction by permitting recolonization of vacant areas or marginal areas.

55.3 Population Demography and Dynamics

Sex ratio and generation time affect population growth rates.
Abundant females, a short generation time, or both can be responsible for more rapid population growth.

Age structure is determined by the numbers of individuals in different age groups.
Every age cohort has a characteristic fecundity and death rate, and so the age structure of a population affects growth.

Life tables show probability of survival and reproduction through a cohort's life span.

Survivorship curves demonstrate how survival probability changes with age (figures 55.10 & 55.11).
In some populations, survivorship is high until old age, whereas in others, survivorship is lowest among the youngest individuals.

55.4 Life History and the Cost of Reproduction

Because resources are limited, reproduction has a cost. Resources allocated toward current reproduction cannot be used to enhance survival and future reproduction (figure 55.13).

A trade-off exists between number of offspring and investment per offspring.
When reproductive cost is high, fitness can be maximized by deferring reproduction, or by producing a few large-sized young that have a greater chance of survival.

Reproductive events per lifetime represent an additional trade-off.
Semelparity is reproduction once in a single large event. Iteroparity is production of offspring several times over many seasons.

Age at first reproduction correlates with life span.
Longer-lived species delay first reproduction longer compared with short-lived species, in which time is of the essence.

55.5 Environmental Limits to Population Growth

The exponential growth model applies to populations with no growth limits.
The rate of population increase, r, is defined as the difference between birth rate, b, and death rate, d.

Exponential growth occurs when a population is not limited by resources or by other species (figure 55.16).

The logistic growth model applies to populations that approach their carrying capacity.
Logistic growth is observed as a population reaches its carrying capacity. Usually, a population's growth rate slows to a plateau. In some cases the population overshoots and then drops back to the carrying capacity.

55.6 Factors That Regulate Populations

Density-dependent effects occur when reproduction and survival are affected by population size.
Density-dependent factors include increased competition and disease. To stabilize a population size, birth rates must decline, death rates must increase, or both.

Density-independent effects include environmental disruptions and catastrophes.
Density-independent factors are not related to population size and include environmental events that result in mortality.

Population cycles may reflect complex interactions.
In some cases, population size is cyclic because of the interaction of factors such as food supply and predation (figure 55.23).

Resource availability affects life history adaptations.
Populations at carrying capacity have adaptations to compete for limited resources; populations well below carrying capacity exhibit a high reproductive rate to use abundant resources.

55.7 Human Population Growth

Human populations have grown exponentially.
Technology and other innovations have simultaneously increased the carrying capacity and decreased mortality in the past 300 years.

Population pyramids show birth and death trends.
Populations with many young individuals are likely to experience high growth rates as these individuals reach reproductive age.

Humanity's future growth is uncertain.
The human population is unevenly distributed. Rapid growth in developing countries has resulted in poverty, whereas most resources are utilized by the industrialized world.

The population growth rate has declined.
Even at lower growth rates, the number of individuals on the planet is likely to plateau at 7 to 10 billion.

Consumption in the developed world further depletes resources.
Resource consumption rates in the developed world are very high; a sustainable future requires limits both to population growth and to per capita resource consumption.

Review Questions

UNDERSTAND

1. Source–sink metapopulations are distinct from other types of metapopulations because
   a. exchange of individuals only occurs in the former.
   b. populations with negative growth rates are a part of the former.
   c. populations never go extinct in the former.
   d. all populations eventually go extinct in the former.

2. The potential for social interactions among individuals should be maximized when individuals
   a. are randomly distributed in their environment.
   b. are uniformly distributed in their environment.
   c. have a clumped distribution in their environment.
   d. None of the choices is correct.

3. When ecologists talk about the cost of reproduction they mean
   a. the reduction in future reproductive output as a consequence of current reproduction.
   b. the amount of calories it takes for all the activity used in successful reproduction.
   c. the amount of calories contained in eggs or offspring.
   d. None of the choices is correct.

4. A life history trade-off between clutch size and offspring size
   a. means that as clutch size increases, offspring size increases.
   b. means that as clutch size increases, offspring size decreases.
   c. means that as clutch size increases, adult size increases.
   d. means that as clutch size increases, adult size decreases.

5. The difference between exponential and logistic growth rates is
   a. exponential growth depends on birth and death rates and logistic does not.
   b. in logistic growth, emigration and immigration are unimportant.
   c. that both are affected by density, but logistic growth is slower.
   d. that only logistic growth reflects density-dependent effects on births or deaths.

6. The logistic population growth model, $dN/dt = rN[(K - N)/K]$, describes a population's growth when an upper limit to growth is assumed. As N approaches (numerically) the value of K
   a. dN/dt increases rapidly.
   b. dN/dt approaches 0.
   c. dN/dt increases slowly.
   d. the population becomes threatened by extinction.

7. Which of the following is an example of a density-dependent effect on population growth?
   a. An extremely cold winter
   b. A tornado
   c. An extremely hot summer in which cool burrow retreats are fewer than number of individuals in the population
   d. A drought

APPLY

1. If the size of a population is reduced due to a natural disaster such as a flood
   a. population growth rates may increase because the population is no longer near its carrying capacity.
   b. population growth rates may decrease because individuals have trouble finding mates.
   c. population rates may remain unchanged if the population was already well below the carrying capacity.
   d. All of the choices are correct.

2. In populations subjected to high levels of predation
   a. individuals should invest little in reproduction so as to maximize their survival.
   b. individuals should produce few offspring and invest little in any of them.
   c. individuals should invest greatly in reproduction because their chance of surviving to another breeding season is low.
   d. individuals should stop reproducing altogether.

3. In a population in which individuals are uniformly distributed
   a. the population is probably well below its carrying capacity.
   b. natural selection should favor traits that maximize the ability to compete for resources.
   c. immigration from other populations is probably keeping the population from going extinct.
   d. None of the choices is correct.

4. The elimination of predators by humans
   a. will cause its prey to experience exponential growth until new predators arrive or evolve.
   b. will lead to an increase in the carrying capacity of the environment.
   c. may increase the population size of a prey species if that prey's population was being regulated by predation from the predator.
   d. will lead to an Allee effect.

SYNTHESIZE

1. Refer to figure 55.8. What are the implications for evolutionary divergence among populations that are part of a metapopulation versus populations that are independent of other populations?

2. Refer to figure 55.13. Given a trade-off between current reproductive effort and future reproductive success (the so-called cost of reproduction), would you expect old individuals to have the same "optimal" reproductive effort as young individuals?

3. Refer to figure 55.14. Because the number of offspring that a parent can produce is often a trade-off with the size of individual offspring, many circumstances lead to an intermediate number and size of offspring being favored. If the size of an offspring was completely unrelated to the quality of that offspring (its chances of surviving until it reaches reproductive age), would you expect parents to fall on the left or right side of the x-axis (clutch size)? Explain.

4. Refer to figure 55.26. Would increasing the mean generation time have the same kind of effect on population growth rate as reducing the number of children that an individual female has over her lifetime? Which effect would have a bigger influence on population growth rate? Explain.

CHAPTER 56

Community Ecology

Chapter Contents

56.1 Biological Communities: Species Living Together

56.2 The Ecological Niche Concept

56.3 Predator–Prey Relationships

56.4 The Many Types of Species Interactions

56.5 Ecological Succession, Disturbance, and Species Richness

Introduction

All the organisms that live together in a place are members of a community. The myriad of species that inhabit a tropical rainforest are a community as are the species that live in a desert oasis. Indeed, every inhabited place on Earth supports its own particular array of organisms. Over time, the different species that live together have made many complex adjustments to community living, evolving together and forging relationships that give the community its character and stability. Both competition and cooperation have played key roles; in this chapter, we look at these and other factors in community ecology.

56.1 Biological Communities: Species Living Together

Learning Outcomes
1. Define community.
2. Describe how community composition may change across a geographic landscape.

Almost any place on Earth is occupied by species, sometimes by many of them, as in the rainforests of the Amazon, and sometimes by only a few, as in the near-boiling waters of Yellowstone's geysers (where a number of microbial species live). The term **community** refers to the species that occur at any particular locality (figure 56.1). Communities can be characterized either by their constituent species or by their properties, such as **species richness** (the number of species present) or **primary productivity** (the amount of energy produced).

Interactions among community members govern many ecological and evolutionary processes. These interactions, such as predation and mutualism, affect the population biology of particular species—whether a population increases or decreases in abundance, for example—as well as the ways in which energy and nutrients cycle through the ecosystem. Moreover, the community context affects the patterns of natural selection faced by a species, and thus the evolutionary course it takes.

Scientists study biological communities in many ways, ranging from detailed observations to elaborate, large-scale experiments. In some cases, studies focus on the entire community, whereas in other cases only a subset of species that are likely to interact with one another are studied.

Communities change over space and time

For the most part, species seem to respond independently to changing environmental conditions. As a result, community composition changes gradually across landscapes as some species appear and become more abundant, while others decrease in abundance and eventually disappear.

A famous example of this pattern is the abundance of tree species in the Santa Catalina Mountains of Arizona along a geographic gradient running from very dry to very moist. Figure 56.2 shows that species can change abundance in patterns that are for the most part independent of one another. As a result, tree communities at different localities in these mountains fall on a continuum, one merging into the next, rather than representing discretely different sets of species.

Similar patterns through time are seen in paleontological studies. For example, a very good fossil record exists for the trees and small mammals that occurred in North America over the past 20,000 years. Examination of prehistoric communities shows little similarity to those that occur today. Many species that occur together today were never found together in the past. Conversely, species that used to occur in the same communities often do not overlap in their geographic ranges today. These findings suggest that as climate has changed during the waxing and waning of the Ice Ages, species have responded independently, rather than shifting their distributions together.

Nonetheless, in some cases the abundance of species in a community does change geographically in a synchronous pattern. Often, this occurs at **ecotones**, places where the environment changes abruptly. For example, in the western United States, certain patches of habitat have serpentine soils. This soil differs from normal soil in many ways—for example, high concentrations of nickel, chromium, and iron; low concentrations of copper and calcium. Comparison of the plant species that occur on different soils shows that distinct communities exist on each type, with an abrupt transition from one to the other over a short distance (figure 56.3). Similar transitions are seen wherever greatly different habitats come into contact, such as at the interface between terrestrial and aquatic habitats or where grassland and forest meet.

Figure 56.1 An African savanna community. A community consists of all the species—plants, animals, fungi, protists, and prokaryotes—that occur at a locality, in this case Etosha National Park in Namibia.

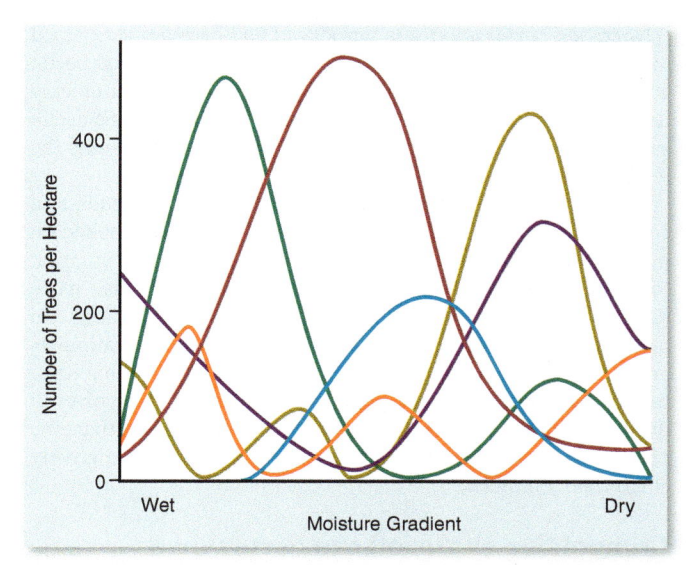

Figure 56.2 Abundance of tree species along a moisture gradient in the Santa Catalina Mountains of southeastern Arizona. Each line represents the abundance of a different tree species. The species' patterns of abundance are independent of one another. Thus, community composition changes continually along the gradient.

> **? Inquiry question** Why do species exhibit different patterns of response to change in moisture?

Learning Outcomes Review 56.1

A community comprises all species that occur at one site. In most cases, the abundance of community members appears to vary independently across space and through time. Community composition also changes gradually depending on environmental factors when moving from one location to another, such as from a very dry area to a very moist area.

■ *In a community, would you expect greater variation over time in abundance of animal life or plant life? Why?*

56.2 The Ecological Niche Concept

Learning Outcomes

1. Define niche and resource partitioning.
2. Differentiate between fundamental and realized niches.
3. Explain how the presence of other species can affect a species' realized niche.

Each organism in a community confronts the challenge of survival in a different way. The **niche** an organism occupies is the total of

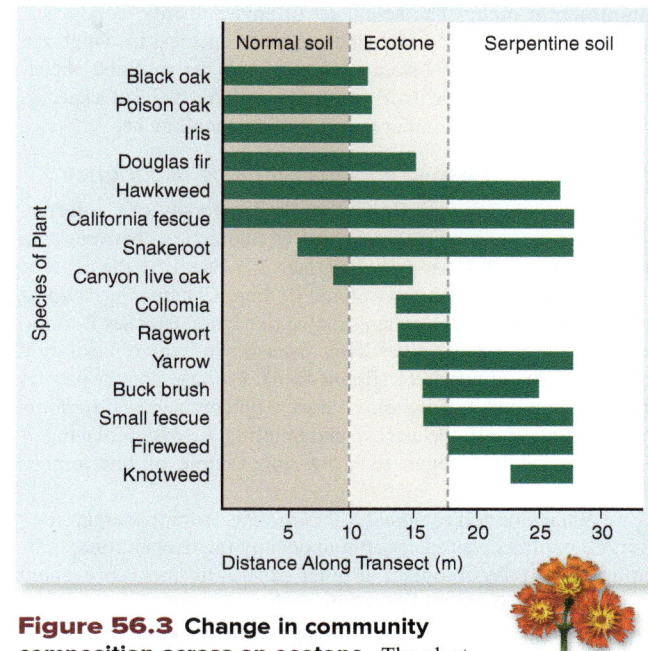

Figure 56.3 Change in community composition across an ecotone. The plant communities on normal and serpentine soils are greatly different, and the transition from one community to another occurs over a short distance.

> **🔍 Data analysis** In this field study, how close is the serpentine soil to the normal soil?

> **? Inquiry question** Why is there a sharp transition between the two community types?

all the ways it uses the resources of its environment. A niche may be described in terms of space utilization, food consumption, temperature range, appropriate conditions for mating, requirements for moisture, and other factors.

Sometimes species are not able to occupy their entire niche because of the presence or absence of other species. Species can interact with one another in a number of ways, and these interactions can either have positive or negative effects. One type of interaction, **interspecific competition,** occurs when two species use the same resource and there is not enough to satisfy both. Physical interactions over access to resources—such as fighting to defend a territory or displacing an individual from a particular location—are referred to as **interference competition;** consuming the same resources is called **exploitative competition.**

Fundamental niches are potential; realized niches are actual

The entire niche that a species is capable of using, based on its physiological tolerance limits and resource needs, is called the

chapter **56** *Community Ecology* **1189**

fundamental niche. The actual set of environmental conditions, including the presence or absence of other species, in which the species can establish a stable population is its **realized niche.** Because of interspecific interactions, the realized niche of a species may be considerably smaller than its fundamental niche.

Competition between species for niche occupancy

In a classic study, Joseph Connell of the University of California, Santa Barbara, investigated competitive interactions between two species of barnacles that grow together on rocks along the coast of Scotland. Of the two species Connell studied, *Chthamalus stellatus* lives in shallower water, where tidal action often exposes it to air, and *Semibalanus balanoides* lives lower down, where it is rarely exposed to the atmosphere (figure 56.4). In these areas, space is at a premium and *S. balanoides* always outcompeted *C. stellatus* by crowding it off the rocks, undercutting it, and replacing it even where it had begun to grow, an example of interference competition.

When Connell removed *S. balanoides* from this area, however, *C. stellatus* was easily able to occupy the deeper zone, indicating that no physiological or other general obstacles prevented it from becoming established there. In contrast, *S. balanoides* could not survive in the shallow-water habitats where *C. stellatus* normally occurs; it does not have the physiological adaptations to warmer temperatures that allow *C. stellatus* to occupy this zone. Thus, the fundamental niche of *C. stellatus* includes both shallow and deeper zones, but its realized niche is much narrower because *C. stellatus* can be outcompeted by *S. balanoides* in parts of its fundamental niche. By contrast, the realized and fundamental niches of *S. balanoides* appear to be identical.

Other causes of niche restriction

Processes other than competition can also restrict the realized niche of a species. For example, the plant St. John's wort (*Hypericum perforatum*) was introduced and became widespread in open rangeland habitats in California until a specialized beetle was introduced to control it. Population size of the plant quickly decreased, and it is now only found in shady sites where the beetle cannot thrive. In this case, the presence of a predator limits the realized niche of a plant.

In some cases, the absence of another species leads to a smaller realized niche. Many North American plants depend on insects for pollination; indeed, the value of insect pollination for American agriculture has been estimated as more than $9 billion per year. However, pollinator populations are currently declining for several reasons. Conservationists are concerned that if these insects disappear from some habitats, the realized niche of many plant species will decrease or even disappear entirely. In this case, the absence—rather than the presence—of another species will be the cause of a relatively small realized niche.

Competitive exclusion can occur when species compete for limited resources

In classic experiments carried out in 1934 and 1935, Russian ecologist Georgii Gause studied competition among three species of *Paramecium*, a tiny protist. Each of the three species grew well in culture tubes by themselves, preying on bacteria and yeasts that fed on oatmeal suspended in the culture fluid (figure 56.5*a*). However, when Gause grew *P. aurelia* together with *P. caudatum* in the same culture tube, the numbers of *P. caudatum* always declined to extinction, leaving *P. aurelia* the only survivor (figure 56.5*b*). Why did this happen? Gause found that *P. aurelia* could grow six times faster than its competitor *P. caudatum* because it was able to better utilize the limited available resources, an example of exploitative competition.

From experiments such as this, Gause formulated what is now called the principle of **competitive exclusion.** This principle

Figure 56.4
Competition among two species of barnacles.
The fundamental niche of *Chthamalus stellatus* includes both deep and shallow zones. *C. stellatus* is forced out of the part of its fundamental niche that overlaps the realized niche of *Semibalanus balanoides*.

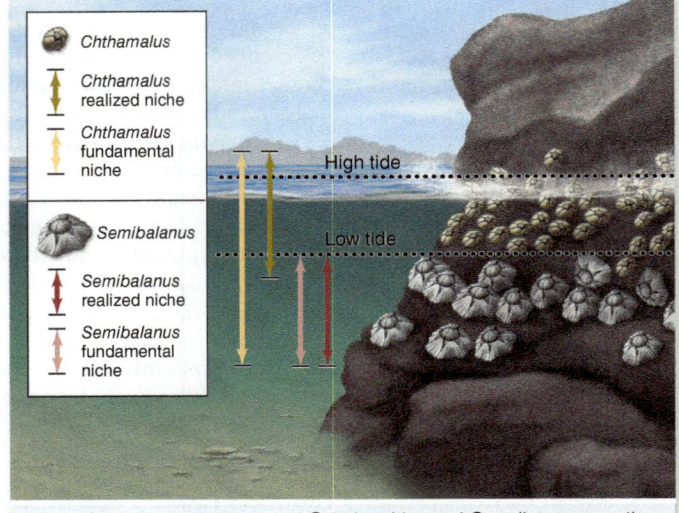

S. balanoides and *C. stellatus* competing

C. stellatus fundamental and realized niches are identical when *S. balanoides* is removed.

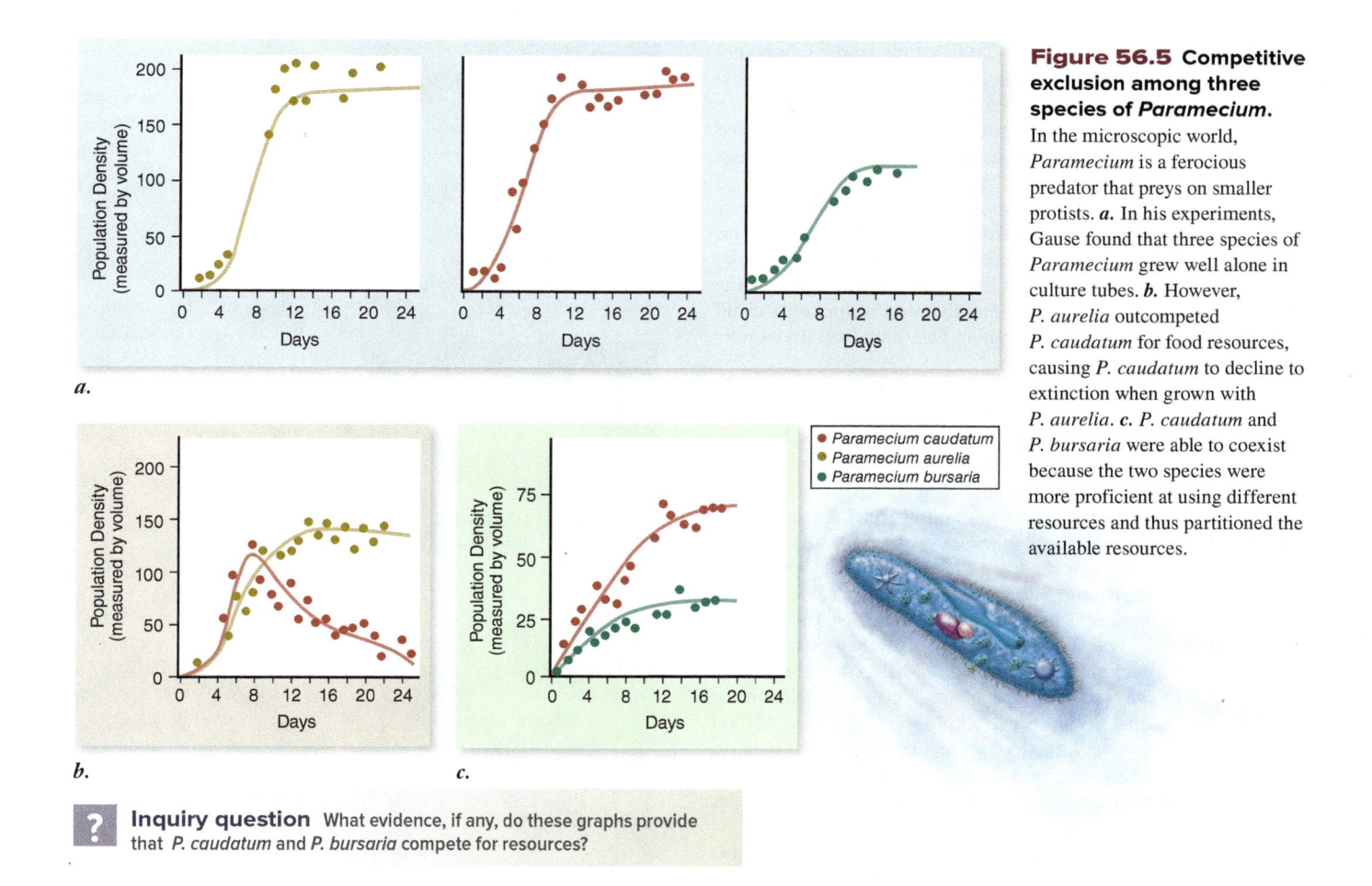

Figure 56.5 Competitive exclusion among three species of *Paramecium*. In the microscopic world, *Paramecium* is a ferocious predator that preys on smaller protists. ***a.*** In his experiments, Gause found that three species of *Paramecium* grew well alone in culture tubes. ***b.*** However, *P. aurelia* outcompeted *P. caudatum* for food resources, causing *P. caudatum* to decline to extinction when grown with *P. aurelia*. ***c.*** *P. caudatum* and *P. bursaria* were able to coexist because the two species were more proficient at using different resources and thus partitioned the available resources.

Inquiry question What evidence, if any, do these graphs provide that *P. caudatum* and *P. bursaria* compete for resources?

states that if two species are competing for a limited resource such as food or water, the species that uses the resource more efficiently will eventually eliminate the other locally. In other words, no two species with the same niche can coexist when resources are limiting.

Niche overlap and coexistence

In a revealing experiment, Gause challenged *Paramecium caudatum*—the defeated species in his earlier experiments—with a third species, *P. bursaria*. Because he expected these two species to also compete for the limited bacterial food supply, Gause thought one would win out, as had happened in his previous experiments. But that's not what happened. Instead, both species survived in the culture tubes, dividing the food resources.

The explanation for the species' coexistence is simple. In the upper part of the culture tubes, where the oxygen concentration and bacterial density were high, *P. caudatum* dominated because it was better able to feed on bacteria. In the lower part of the tubes, however, the lower oxygen concentration favored the growth of a different potential food, yeast, and *P. bursaria* was better able to eat this food. The fundamental niche of each species was the whole culture tube, but the realized niches of the species were different portions of the tube, allowing them to coexist. However,

competition did have a negative effect on the participants (figure 56.5*c*). When grown without a competitor, both species reached densities three times greater than when they were grown with a competitor.

Competitive exclusion refined

Gause's principle of competitive exclusion can be restated as: No two species can occupy identical niches *indefinitely* when resources are limiting. Certainly species can and do coexist while competing for some of the same resources, but Gause's hypothesis predicts that when two species coexist on a long-term basis, either resources must not be limited or their niches will always differ in one or more features; otherwise, one species will outcompete the other, and the extinction of the second species will inevitably result.

Competition may lead to resource partitioning

Gause's competitive exclusion principle has a very important consequence: If competition for a limited resource is intense, then either one species will drive the other to extinction, or the species will evolve differences that reduce the competition between them.

When the ecologist Robert MacArthur studied five species of warblers, small insect-eating forest songbirds, he discovered that they appeared to be competing for the same resources. But when he investigated in greater detail he found that each species actually fed in a different part of spruce trees and so ate different subsets of insects. One species fed on insects near the tips of branches, a second within the dense foliage, a third on the lower branches, a fourth high on the trees, and a fifth at the very apex of the trees. Thus, each species of warbler had evolved so as to utilize a different portion of the spruce tree resource. They had *subdivided the niche* to avoid direct competition with one another. This niche subdivision is termed **resource partitioning.**

Resource partitioning is often seen in similar species that occupy the same geographic area. Such sympatric species (that is, species that occur together) often avoid competition by living in different portions of the habitat or by using different food or other resources (figure 56.6) This pattern of resource partitioning is thought to result from the process of natural selection causing initially similar species to diverge in resource use to reduce competitive pressures.

Whether such evolutionary divergence has occurred can be investigated by comparing species whose ranges only partially overlap. Where the two species occur together, they often tend to exhibit greater differences in morphology (the form and structure of an organism) and resource use than do allopatric populations of the same species that do not occur with the other species. Called *character displacement,* the differences evident between sympatric species are thought to have been favored by natural selection as a means of partitioning resources and thus reducing competition (see chapter 22).

As an example, the two Darwin's finches in figure 56.7 have bills of similar size where the finches are allopatric (that is, each living on an island where the other does not occur). On islands where they are sympatric, the two species have evolved beaks of different sizes, one adapted to larger seeds and the other to smaller ones. Character displacement such as this may play an important role in adaptive radiation, leading new species to adapt to different parts of the environment, as discussed in chapter 22.

Detecting interspecific competition can be difficult

It is not simple to determine when two species are competing. The fact that two species use the same resources need not imply competition if that resource is not in limited supply. Even if the population sizes of two species are negatively correlated, such that where one species has a large population, the other species has a small population and vice versa, the two species may not be competing for the same limiting resource. Instead, the two species might be independently responding to the same feature of the environment—perhaps one species thrives best in warm conditions and the other where it's cool.

Experimental studies of competition

Some of the best evidence for the existence of competition comes from experimental field studies. By setting up experiments in

Figure 56.6 Resource partitioning among sympatric lizard species. Species of *Anolis* lizards on Caribbean islands partition their habitats in a variety of ways. Some species (*a*) occupy leaves and branches in the canopy of trees, (*b*) others use twigs on the periphery, and (*c*) still others are found at the base of the trunk. In addition, (*d*) some use grassy areas in the open. When two species occupy the same part of the tree, they either utilize different-sized insects as food or partition the thermal microhabitat; for example, one might only be found in the shade, whereas the other would only be found in the open, basking in the Sun.

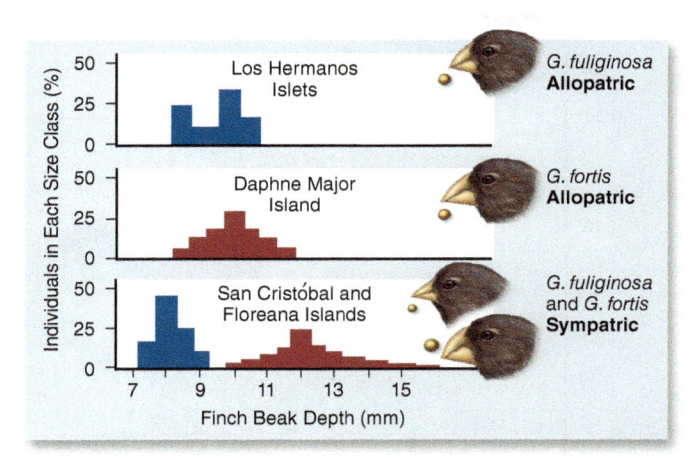

Figure 56.7 Character displacement in Darwin's finches. These two species of finches (genus *Geospiza*) have beaks of similar size when allopatric, but different size when sympatric.

> **Inquiry question** Why do the two species have bills of the same size when they are allopatric?

which two species occur either alone or together, scientists can determine whether the presence of one species has a negative effect on a population of the second species.

For example, a variety of seed-eating rodents occur in North American deserts. In one experiment, researchers set up a series of 50-m × 50-m enclosures to investigate the effect of kangaroo rats on smaller, seed-eating rodents. Kangaroo rats were removed from half of the enclosures, but not from the others. Over the course of the next three years, the researchers monitored the number of the smaller rodents present in the plots. As figure 56.8 illustrates, the number of other rodents was substantially higher in the absence of kangaroo rats, indicating that kangaroo rats compete with the other rodents and limit their population sizes.

A great number of similar experiments have indicated that interspecific competition occurs between many species of plants and animals. The effects of competition can be seen in aspects of population biology other than population size, such as behavior and individual growth rates. For example, two species of *Anolis* lizards occur on the Caribbean island of St. Maarten. When one of the species, *A. gingivinus*, is placed in 12-m × 12-m enclosures without the other species, individual lizards grow faster and perch lower than do lizards of the same species when placed in enclosures in which *A. pogus*, a species normally found near the ground, is also present.

Limitations of experimental studies

Experimental studies are a powerful means of understanding interactions between coexisting species and are now commonly conducted by ecologists. Nonetheless, they have their limitations.

First, care is necessary in interpreting the results of field experiments. Negative effects of one species on another do not automatically indicate the existence of competition. For example, many fish have a negative effect on the population numbers of other, similar-sized fish species, but it results not from competition, but from the fact that adults of each species prey on juveniles of the other species.

In addition, the presence of one species may attract predators or parasites, which then also prey on the second species. In this case, even if the two species are not competing, the second species may have a lower population size in the presence of the first species due to predators or parasites attracted by the first species. Indeed, we can't rule out this possibility with the results of

---

SCIENTIFIC THINKING

Question: Does interspecific interaction occur between rodent species?
Hypothesis: The larger kangaroo rat will have a negative effect on other species.
Experiment: Build large cages in desert areas. Remove kangaroo rats from some cages, leaving them present in others.
Result: In the absence of kangaroo rats, the number of other rodents increases quickly and remains higher than in the control cages throughout the course of the experiment.

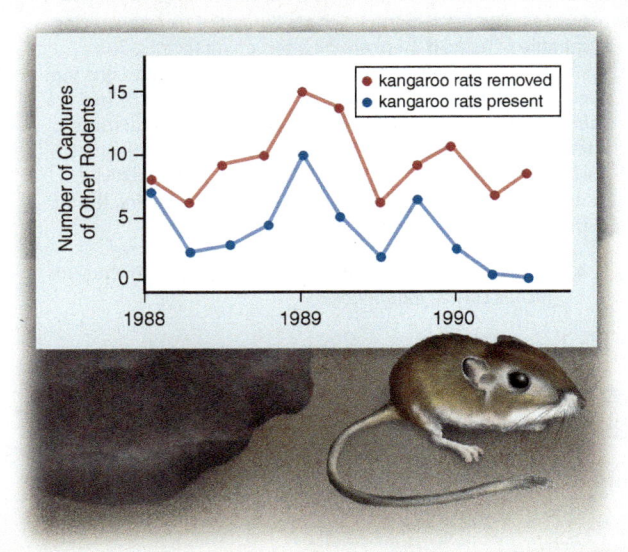

Interpretation: Why do you think population sizes rise and fall in synchrony in the two cages?

Figure 56.8 Detecting interspecific competition.
This experiment tested how removal of kangaroo rats affected the population size of other rodents. Immediately after kangaroo rats were removed, the number of other rodents increased relative to the enclosures that still contained kangaroo rats. Notice that population sizes (as estimated by number of captures) changed in synchrony in the two treatments, probably reflecting changes in the weather.

> **Inquiry question** Why are there more individuals of other rodent species when kangaroo rats are excluded?

the kangaroo rat exclusion study just mentioned, although the close proximity of the enclosures (they were adjacent) would suggest that the same predators and parasites were present in all of them. Thus, experimental studies are most effective when combined with detailed examination of the ecological mechanisms causing the observed effect of one species on another.

Second, experimental studies are not always feasible. For example, the coyote population has increased in the United States over the past three centuries concurrently with the decline of the grey wolf. Is this trend an indication that the species compete? Because of the size of the animals and the large geographic areas occupied by each individual, manipulative experiments involving fenced areas with only one or both species—with each experimental treatment replicated several times for statistical analysis—are not practical. Similarly, studies of slow-growing trees might require many centuries to detect competition between adult trees. In such cases, detailed studies of the ecological requirements of each species are our best bet for understanding interspecific interactions.

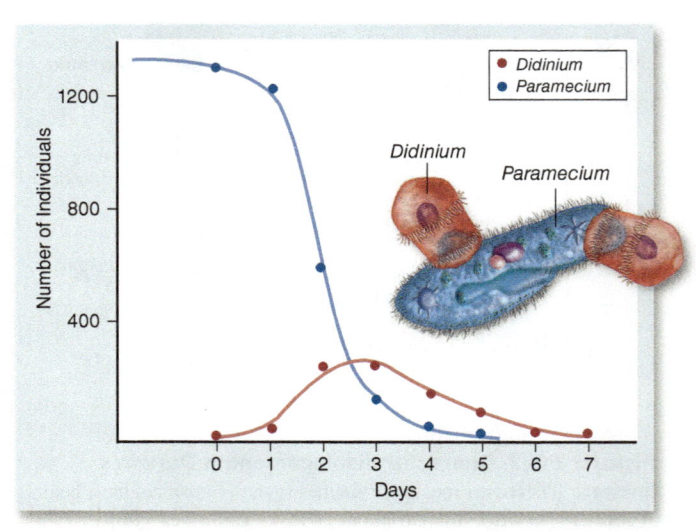

Figure 56.9 Predator–prey in the microscopic world. When the predatory *Didinium* is added to a *Paramecium* population, the numbers of *Didinium* initially rise, and the numbers of *Paramecium* steadily fall. When the *Paramecium* population is depleted, however, the *Didinium* individuals also die.

? **Inquiry question** Can you think of any ways this experiment could be changed so that *Paramecium* might not go extinct?

Learning Outcomes Review 56.2

A niche is the total of all the ways a species uses environmental resources. The fundamental niche is the entire niche a species is capable of using if there are no intervening factors. The realized niche is the set of actual environmental conditions that allow establishment of a stable population. Realized niches are usually smaller than fundamental niches because interspecific interactions limit a species' use of some resources. Resource partitioning allows two sympatric species to occupy a niche, reducing competition between them and also lessening the size of the realized niche.

- Under what circumstances can two species with identical niches coexist indefinitely?

Predation strongly influences prey populations

In nature, predators often have large effects on prey populations. As the previous example indicates, however, the interaction is a two-way street: Prey can also affect the dynamics of predator populations. The outcomes of such interactions are complex and depend on a variety of factors.

Prey population explosions and crashes

Some of the most dramatic examples of the interconnection between predators and their prey involve situations in which humans have either added or eliminated predators from an area. For example, the elimination of large carnivores from much of the eastern United States has led to population explosions of white-tailed deer, which strip the habitat of all edible plant life within their reach. Similarly, when sea otters were hunted to near extinction on the western coast of the United States, populations of sea urchins, a principal prey item of the otters, exploded.

56.3 Predator–Prey Relationships

Learning Outcomes

1. Define predation.
2. Describe the effects predation can have on a population.

Predation is the consuming of one organism by another. In this sense, predation includes everything from a leopard capturing and eating an antelope, to a deer grazing on spring grass.

When experimental populations are set up under simple laboratory conditions, as illustrated in figure 56.9 with the predatory protist *Didinium* and its prey *Paramecium,* the predator often exterminates its prey and then becomes extinct itself, having nothing left to eat. If refuges are provided for the *Paramecium,* however, its population drops to low levels but not to extinction. Low prey population levels then provide inadequate food for the *Didinium,* causing the predator population to decrease. When this occurs, the prey population can recover.

Conversely, the introduction of rats, dogs, and cats to many islands around the world has led to the decimation of native fauna. Populations of Galápagos tortoises on several islands are endangered by introduced rats, pigs, dogs, and cats, which eat the eggs and the young tortoises. Similarly, in New Zealand, several species of birds and reptiles have been eradicated by rat predation and now only occur on a few offshore islands that the rats have not reached. On one such island, every individual of the now-extinct Stephens Island wren was killed by a single lighthouse keeper's cat.

A classic example of the role predation can play in a community involves the introduction of prickly pear cactus to Australia in the 19th century. In the absence of predators, the cactus spread rapidly, so that by 1925 it occupied 12 million hectares of rangeland in an impenetrable morass of spines that made cattle ranching difficult. To control the cactus, a predator from its natural habitat in Argentina, the moth *Cactoblastis cactorum,* was introduced, beginning in 1926. By 1940, populations had been greatly reduced and the cactus now usually occurs in small populations.

Predation and coevolution

Predation provides strong selective pressures on prey populations. Any feature that would decrease the probability of capture should be strongly favored. In turn, the evolution of such features causes natural selection to favor counteradaptations in predator populations. The process by which these adaptations are selected in lockstep fashion in two or more interacting species is termed **coevolution.** A coevolutionary "arms race" may ensue in which predators and prey are constantly evolving better defenses and better means of circumventing these defenses. In the sections that follow, you'll learn more about these defenses and responses.

Plant adaptations defend against herbivores

Plants have evolved many mechanisms to defend themselves from herbivores. The most obvious are morphological defenses: Thorns, spines, and prickles play an important role in discouraging large plant eaters, and plant hairs, especially those that have a glandular, sticky tip, deter insect herbivores. Some plants, such as grasses, deposit silica in their leaves, both strengthening and protecting themselves. If enough silica is present, these plants are simply too tough to eat.

Chemical defenses

As significant as morphological adaptations are, the chemical defenses that occur so widely in plants are even more widespread. Plants exhibit some amazing chemical adaptations to combat herbivores. For example, recent work demonstrates that when attacked by caterpillars, wild tobacco plants emit a chemical into the air that attracts a species of bug that feeds on that caterpillar (discussed in greater detail in chapter 39).

The best-known and perhaps most important of the chemical defenses of plants against herbivores are *secondary chemical compounds.* These chemicals are distinguished from primary compounds, which are the components of a major metabolic pathway, such as respiration. Many plants, and apparently many algae as well, contain structurally diverse secondary compounds that are either toxic to most herbivores or disturb their metabolism greatly, preventing, for example, the normal development of larval insects. Consequently, most herbivores tend to avoid the plants that possess these compounds.

The mustard family (Brassicaceae) produces a group of chemicals known as mustard oils. These substances give the pungent aromas and tastes to plants such as mustard, cabbage, watercress, radish, and horseradish. Although we enjoy these flavors, the chemicals are toxic to many groups of insects.

Similarly, plants of the milkweed family (Asclepiadaceae) and the related dogbane family (Apocynaceae) produce a milky sap that deters herbivores from eating them. In addition, these plants usually contain cardiac glycosides, molecules that can produce drastic deleterious effects on the heart function of vertebrates.

The coevolutionary response of herbivores

Some herbivores have evolved ways of circumventing plant defenses, allowing them to feed on these plants without harm, often as their exclusive food source.

For example, cabbage butterfly caterpillars feed almost exclusively on plants of the mustard and caper families, as well as on a few other small families of plants that also contain mustard oils (figure 56.10). Similarly, caterpillars of monarch butterflies and their relatives feed on plants of the milkweed and dogbane families. How do these animals manage to avoid the chemical defenses of the plants, and what are the evolutionary precursors and ecological consequences of such patterns of specialization?

We can offer a potential explanation for the evolution of these particular patterns. Once the ability to manufacture mustard oils evolved in the ancestors of the caper and mustard families, the plants were protected for a time against most or all herbivores that were feeding on other plants in their area. At some point, certain groups of insects—for example, the cabbage butterflies—evolved the ability to break down mustard oils and thus feed on these plants without harming themselves. Having developed this new capability, the butterflies were able to use a new resource without competing with other herbivores for it. As we saw in chapter 22, exposure to an underutilized resource often leads to evolutionary diversification and adaptive radiation.

**Figure 56.10
Insect herbivores well suited to their plant hosts.** *a.* The green caterpillars of the cabbage white butterfly, *Pieris rapae,* are camouflaged on the leaves of cabbage and other plants on which they feed. Although mustard oils protect these plants against most herbivores, the cabbage white butterfly caterpillars are able to break down the mustard oil compounds. *b.* An adult cabbage white butterfly.

Animal adaptations defend against predators

Some animals that feed on plants rich in secondary compounds receive an extra benefit. For example, when the caterpillars of monarch butterflies feed on plants of the milkweed family, they do not break down the cardiac glycosides that protect these plants from herbivores. Instead, the caterpillars concentrate and store these compounds in fat bodies; they then retain them through the chrysalis stage to the adult and even to the eggs of the next generation.

The incorporation of cardiac glycosides protects all stages of the monarch life cycle from predators. A bird that eats a monarch butterfly quickly regurgitates it (figure 56.11) and in the future avoids the conspicuous orange-and-black pattern that characterizes the adult monarch. Some bird species have evolved the ability to tolerate the protective chemicals; these birds eat the monarchs.

Chemical defenses

Animals also manufacture and use a startling array of defensive substances. Bees, wasps, predatory bugs, scorpions, spiders, and many other arthropods use chemicals to defend themselves and to kill their own prey. In addition, various chemical defenses have evolved among many marine invertebrates, as well as a variety of vertebrates, including frogs, snakes, lizards, fishes, and some birds.

The poison-dart frogs of the family Dendrobatidae produce toxic alkaloids in the mucus that covers their brightly colored skin; these alkaloids are distasteful and sometimes deadly to animals that try to eat the frogs (figure 56.12). Some of these toxins are so powerful that a few micrograms will kill a person if injected into the bloodstream. More than 200 different alkaloids have been isolated from these frogs, and some are playing important roles in neuromuscular research. Similarly intensive investigations of marine animals, venomous reptiles, algae, and flowering plants are under way in search of new drugs to fight cancer and other diseases, or to use as sources of antibiotics.

Figure 56.12 Vertebrate chemical defenses. Frogs of the family Dendrobatidae, abundant in the forests of Central and South America, are extremely poisonous to vertebrates; 80 different toxic alkaloids have been identified from different species in this genus. Dendrobatids advertise their toxicity with bright coloration. As a result of either instinct or learning, predators avoid such brightly colored species that might otherwise be suitable prey.

Defensive coloration

Many insects that feed on milkweed plants are brightly colored; they advertise their poisonous nature using an ecological strategy known as warning coloration.

Showy coloration is characteristic of animals that use poisons and stings to repel predators; organisms that lack specific chemical defenses are seldom brightly colored. In fact, many have cryptic coloration—color that blends with the surroundings and thus hides the individual from predators (figures 56.13; see also figure 56.10). Camouflaged animals usually do not live together in groups because a predator that discovers one individual gains a valuable clue to the presence of others.

Mimicry allows one species to capitalize on defensive strategies of another

During the course of their evolution, many species have come to resemble distasteful ones that exhibit warning coloration. The mimic gains an advantage by looking like the distasteful model. Two types of mimicry have been identified: Batesian mimicry and Müllerian mimicry.

Batesian mimicry

Batesian mimicry is named for Henry Bates, the British naturalist who first brought this type of mimicry to general attention in 1857. In his journeys to the Amazon region of South America, Bates discovered many instances of palatable insects that resembled brightly colored, distasteful species. He reasoned that the mimics would be avoided by predators, who would be fooled by the disguise into thinking the mimic was the distasteful species.

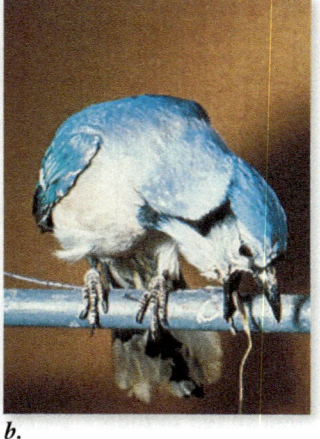

Figure 56.11 A blue jay learns not to eat monarch butterflies. *a.* This cage-reared jay had never seen a monarch butterfly before it tried eating one. *b.* The same jay regurgitated the butterfly a few minutes later. After having such a bad experience, birds usually do not try to capture orange-and-black insects of any kind.

Figure 56.13 Cryptic coloration and form. A waved umber, *Menophra abruptaria,* camouflaged caterpillar closely resembles the twig on which it is hanging.

Many of the best-known examples of Batesian mimicry occur among butterflies and moths. Predators of these insects must use visual cues to hunt for their prey; otherwise, similar color patterns would not matter to potential predators. Increasing evidence indicates that Batesian mimicry can involve nonvisual cues, such as olfaction, although such examples are less obvious to humans.

The kinds of butterflies that provide the models in Batesian mimicry are, not surprisingly, members of groups whose caterpillars feed on only one or a few closely related plant families. The plant families on which they feed are strongly protected by toxic chemicals. The model butterflies incorporate the poisonous molecules from these plants into their bodies. The mimic butterflies, in contrast, belong to groups in which the feeding habits of the caterpillars are not so restricted. As caterpillars, these butterflies feed on a number of different plant families that are unprotected by toxic chemicals.

One often-studied mimic among North American butterflies is the tiger swallowtail, whose range occurs throughout the eastern United States and into Canada (figure 56.14*a*). In areas in which the poisonous pipevine swallowtail occurs, female tiger swallowtails are polymorphic and one color form is extremely similar in appearance to the pipevine swallowtail.

The caterpillars of the tiger swallowtail feed on a variety of trees, including tulip, aspen, and cherry, and neither caterpillars nor adults are distasteful to birds. Interestingly, the Batesian mimicry seen in the adult tiger swallowtail butterfly does not extend to the caterpillars: Tiger swallowtail caterpillars are camouflaged on leaves, resembling bird droppings, but the pipevine swallowtail's distasteful caterpillars are very conspicuous.

Müllerian mimicry

Another kind of mimicry, **Müllerian mimicry,** was named for the German biologist Fritz Müller, who first described it in 1878. In Müllerian mimicry, several unrelated but protected animal species come to resemble one another (figure 56.14*b*). If animals that resemble one another are all poisonous or dangerous, they gain an advantage because a predator will learn more quickly to avoid them. In some cases, predator populations even evolve an innate avoidance of species; such evolution may occur more quickly when multiple dangerous prey look alike.

In both Batesian and Müllerian mimicry, mimic and model must not only look alike but also act alike. For example, the members of several families of insects that closely resemble wasps behave surprisingly like the wasps they mimic, flying often and actively from place to place.

Learning Outcomes Review 56.3

Predation is the consuming of one organism by another. High predation can drive prey populations to extinction; conversely, in the absence of predators, prey populations often explode and exhaust their resources. Defensive adaptations may evolve in prey species, such as becoming distasteful or poisonous, or having defensive structures, appearance, or capabilities.

- *A nonpoisonous scarlet king snake has red, black, and yellow bands of color similar to that of the poisonous eastern coral snake. What type of mimicry is being exhibited?*

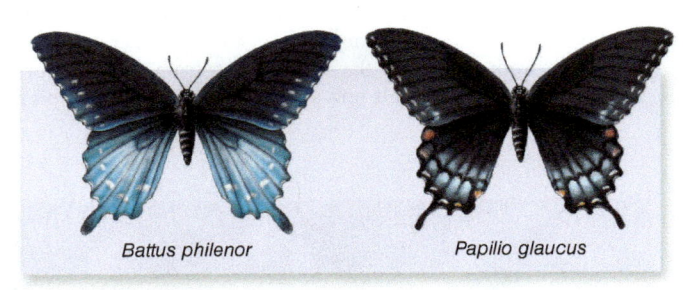

a. Batesian mimicry

b. Müllerian mimicry

Figure 56.14 Mimicry. *a.* Batesian mimicry. Pipevine swallowtail butterflies, *Battus philenor,* are protected from birds and other predators by the poisonous compounds they derive from the food they eat as caterpillars and store in their bodies. Adult pipevine swallowtails advertise their poisonous nature with warning coloration. Tiger swallowtails, *Papilio glaucus,* are Batesian mimics of the poisonous pipevine swallowtail and are not chemically protected. *b.* Pairs of Müllerian mimics. *Heliconius erato* and *H. melpomene* are sympatric, and *H. sapho* and *H. cydno* are sympatric. All of these butterflies are distasteful. They have evolved similar coloration patterns in sympatry to minimize predation; predators need only learn one pattern to avoid.

56.4 The Many Types of Species Interactions

Learning Outcomes
1. Explain the different forms of symbiosis.
2. Describe how coevolution occurs between mutualistic partners.
3. Explain how the occurrence of one ecological process may affect the outcome of another occurring at the same time.

The plants, animals, protists, fungi, and prokaryotes that live together in communities have changed and adjusted to one another continually over millions of years. We have already discussed competition and predation, but other types of ecological interactions commonly occur. For example, many features of flowering plants have evolved to attract beneficial animals (figure 56.15). These animals, in turn, have evolved a number of special traits that enable them to obtain food or other resources efficiently from the plants they visit, often from their flowers. While doing so, the animals provide a service to the plants, picking up pollen, which they may deposit on the next plant they visit, or eating fruits, which contain seeds that they transport elsewhere in the environment, sometimes a great distance from the parent plant.

Symbiosis involves long-term interactions

In symbiosis, two or more kinds of organisms interact in often elaborate and more-or-less permanent relationships. All symbiotic relationships carry the potential for coevolution between the organisms involved, and in many instances the results of this coevolution are fascinatingly complex.

Lichens, which are associations of certain fungi with green algae or cyanobacteria, are one example of symbiosis. Another important example are mycorrhizae, associations between fungi and the roots of most kinds of plants. The fungi expedite the plant's absorption of certain nutrients, and the plants in turn provide the fungi with carbohydrates (both mycorrhizae and lichens are discussed in greater detail in chapter 32). Similarly, root nodules that occur in legumes and certain other kinds of plants contain bacteria that fix atmospheric nitrogen and make it available to their host plants.

In the tropics, leaf-cutter ants are often so abundant that they can remove a quarter or more of the total leaf surface of the plants in a given area in a single year (see figure 32.18). They do not eat these leaves directly; rather, they take them to underground nests, where they chew them up and inoculate them with the spores of particular fungi. These fungi are cultivated by the ants and brought from one specially prepared bed to another, where they grow and reproduce. In turn, the fungi constitute the primary food of the ants and their larvae. The relationship between leaf-cutter ants and these fungi is an excellent example of symbiosis. Recent phylogenetic studies using DNA and assuming a molecular clock (see chapter 23) suggest that these symbioses are ancient, perhaps originating more than 50 MYA.

The major kinds of symbiotic relationships include (1) **mutualism,** in which both participating species benefit; (2) **parasitism,** in which one species benefits but the other is harmed; and (3) **commensalism,** in which one species benefits and the other neither benefits nor is harmed. Parasitism can also be viewed as a form of predation, although the organism that is preyed on does not necessarily die.

Mutualism benefits both species

Mutualism is a symbiotic relationship between organisms in which both species benefit. Mutualistic relationships are of fundamental importance in determining the structure of biological communities.

Mutualism and coevolution

Some of the most spectacular examples of mutualism occur among flowering plants and their animal visitors, including insects, birds, and bats. During the course of flowering-plant evolution, the characteristics of flowers evolved in relation to the characteristics of the animals that visit them for food and, in the process, spread their pollen from individual to individual. At the same time, characteristics of the animals have changed, increasing their

Figure 56.15 Pollination by a bat. Many flowers have coevolved with other species to facilitate pollen transfer. Insects are widely known as pollinators, but they're not the only ones: Birds, bats, and even small marsupials and lizards serve as pollinators for some species. Notice the cargo of pollen on the bat's snout.

specialization for obtaining food or other substances from particular kinds of flowers.

Another example of mutualism involves ants and aphids. Aphids are small insects that suck fluids from the phloem of living plants with their piercing mouthparts. They extract a certain amount of the sucrose and other nutrients from this fluid, but they excrete much of it in an altered form through their anus. Certain ants have taken advantage of this—in effect, domesticating the aphids. Like ranchers taking cattle to fresh fields to graze, the ants carry the aphids to new plants and then consume as food the "honeydew" that the aphids excrete.

Ants and acacias: A prime example of mutualism

A particularly striking example of mutualism involves ants and certain Latin American tree species of the genus *Acacia*. In these species, certain leaf parts, called stipules, are modified as paired, hollow thorns. The thorns are inhabited by stinging ants of the genus *Pseudomyrmex,* which do not nest anywhere else (figure 56.16). Like all thorns that occur on plants, the acacia thorns serve to deter herbivores.

At the tip of the leaflets of these acacias are unique, protein-rich bodies called Beltian bodies, named after the 19th-century British naturalist Thomas Belt. Beltian bodies do not occur in species of *Acacia* that are not inhabited by ants, and their role is clear: they serve as a primary food for the ants. In addition, the plants secrete nectar from glands near the bases of their leaves. The ants consume this nectar as well, feeding it and the Beltian bodies to their larvae.

Obviously, this association is beneficial to the ants, and one can readily see why they inhabit acacias of this group. The ants and their larvae are protected within the swollen thorns, and the trees provide a balanced diet, including the sugar-rich nectar and the protein-rich Beltian bodies. What, if anything, do the ants do for the plants?

Whenever any herbivorous insect lands on the branches or leaves of an acacia inhabited by ants, the ants, which continually patrol the acacia's branches, immediately attack and devour the herbivore. The ants that live in the acacias also help their hosts compete with other plants by cutting away any encroaching branches that touch the acacia in which they are living. They create, in effect, a tunnel of light through which the acacia can grow, even in the lush tropical rainforests of lowland Central America. In fact, when an ant colony is experimentally removed from a tree, the acacia is unable to compete successfully. Finally, the ants bring organic material into their nests. The parts they do not consume, together with their excretions, provide the acacias with an abundant source of nitrogen.

When mutualism may not be mutualism

Things are not always as they seem. Ant–acacia associations also occur in Africa; in Kenya, several species of acacia ants occur, but only a single species is found on any one tree. One species, *Crematogaster nigriceps,* is competitively inferior to two of the other species. To prevent invasion by these other ant species, *C. nigriceps* prunes the branches of the acacia, preventing it from coming into contact with branches of other trees, which would serve as a bridge for invaders. Although this behavior is beneficial to the ant, it is detrimental to the tree because it destroys the tissue from which flowers are produced, essentially sterilizing the tree. In this case, what initially evolved as a mutualistic interaction has instead become a parasitic one.

Parasitism benefits one species at the expense of another

Parasitism is harmful to the prey organism and beneficial to the parasite. In many cases, the parasite kills its host, and thus the ecological effects of parasitism can be similar to those of predation.

External parasites

Parasites that feed on the exterior surface of an organism are external parasites, or ectoparasites (figure 56.17). Many instances of external parasitism are known in both plants and animals. **Parasitoids** are insects that lay eggs in or on living hosts. This behavior is common among wasps, whose larvae feed on the body of the unfortunate host, often killing it.

Internal parasites

Parasites that live within the body of their hosts, termed **endoparasites,** occur in many different phyla of animals and protists. Internal parasitism is generally marked by much more extreme specialization than external parasitism, as shown by the many protist and invertebrate parasites that infect humans.

The more closely the life of the parasite is linked with that of its host, the more its morphology and behavior are likely to have been modified during the course of its evolution (the same is true of symbiotic relationships of all sorts). Conditions within the body of an organism are different from those encountered outside and are apt to be much more constant. Consequently, the structure of an

Figure 56.16 Mutualism: Ants and acacias. Ants of the genus *Pseudomyrmex* live within the hollow thorns of certain species of acacia trees in Latin America. The nectaries at the bases of the leaves and the Beltian bodies at the ends of the leaflets provide food for the ants. The ants, in turn, supply the acacias with organic nutrients and protect the acacias from herbivores and shading from other plants.

Figure 56.17 An external parasite. The yellow vines are the flowering plant dodder, *Cuscuta*, a parasite that has lost its chlorophyll and its leaves in the course of its evolution. Because it is heterotrophic (unable to manufacture its own food), dodder obtains its food from the host plants it grows on.

Figure 56.18 Parasitic manipulation of host behavior. Due to a parasite in its brain, an ant climbs to the top of a grass blade, where it may be eaten by a grazing herbivore, thus passing the parasite from insect to mammal.

internal parasite is often simplified, and unnecessary armaments and structures are lost as it evolves (for example, see descriptions of tapeworms in chapter 34).

Parasites and host behaviors

Many parasites have complex life cycles that require several different hosts for growth to adulthood and reproduction. Recent research has revealed the remarkable adaptations of certain parasites that alter the behavior of the host and thus facilitate transmission from one host to the next. For example, many parasites cause their hosts to behave in ways that make them more vulnerable to their predators; when the host is ingested, the parasite is able to infect the predator.

One of the most famous examples involves a parasitic flatworm, *Dicrocoelium dendriticum*, which lives in ants as an intermediate host, but reaches adulthood in large herbivorous mammals such as cattle and deer. Transmission from an ant to a deer might seem difficult because deer do not normally eat insects. The flatworm, however, has evolved a remarkable adaptation. When an ant is infected, one of the flatworms migrates to the brain and causes the ant to climb to the top of vegetation and lock its mandibles onto a grass blade at the end of the day, just when herbivores are grazing (figure 56.18). The result is that the ant is eaten along with the grass, leading to infection of the grazer.

Commensalism benefits one species and is neutral to the other

In commensalism, one species benefits and the other is neither hurt nor helped by the interaction. In nature, individuals of one species are often physically attached to members of another. For example, epiphytes are plants that grow on the branches of other plants. In general, the host plant is unharmed, and the epiphyte that grows on it benefits. An example is Spanish moss, which hangs on trees in the southern United States. This plant and other members of its genus, which is in the pineapple family, grow on trees to gain access to sunlight; they generally do not harm the trees (figure 56.19).

Similarly, various marine animals, such as barnacles, grow on other, often actively moving sea animals, such as whales, and thus are carried passively from place to place. These "passengers" presumably gain more protection from predation than they would if they were fixed in one place, and they also reach new sources of food. The increased water circulation that these animals receive as their host moves around may also be of great importance, particularly if the passengers are filter feeders. Unless the number of these passengers gets too large, the host species is usually unaffected.

When commensalism may not be commensalism

One of the best-known examples of symbiosis involves the relationships between certain small tropical fishes (clownfish) and sea anemones, shown in the opening figure of this chapter. The fish have evolved the ability to live among the stinging tentacles of

Figure 56.19 An example of commensalism. Spanish moss, *Tillandsia usneoides*, benefits from using trees as a substrate, but the trees generally are not affected positively or negatively.

Figure 56.20 Commensalism, mutualism, or parasitism? In this symbiotic relationship, oxpeckers definitely receive a benefit in the form of nutrition from the ticks and other parasites they pick off their host in this case, an impala, *Aepyceros melampus*. But the effect on the host is not always clear. If the ticks are harmful, their removal benefits the host, and the relationship is mutually beneficial. If the oxpeckers also pick at scabs, causing blood loss and possible infection, the relationship may be parasitic. If the hosts are unharmed by either the ticks or the oxpeckers, the relationship may be an example of commensalism.

sea anemones, even though these tentacles would quickly paralyze other fishes that touched them. The clownfish feed on food particles left from the meals of the host anemone, remaining uninjured under remarkable circumstances.

On land, an analogous relationship exists between birds called oxpeckers and grazing animals such as cattle or antelopes (figure 56.20). The birds spend most of their time clinging to the animals, picking off parasites and other insects, carrying out their entire life cycles in close association with the host animals.

No clear-cut boundary exists between commensalism and mutualism; in each of these cases, it is difficult to be certain whether the second partner receives a benefit or not. A sea anemone may benefit by having particles of food removed from its tentacles because it may then be better able to catch other prey. Similarly, although often thought of as commensalism, the association of grazing mammals and gleaning birds is actually an example of mutualism. The mammal benefits by having parasites and other insects removed from its body, but the birds also benefit by gaining a dependable source of food.

On the other hand, commensalism can easily transform itself into parasitism. Oxpeckers are also known to pick not only parasites, but also scabs off their grazing hosts. Once the scab is picked, the birds drink the blood that flows from the wound. Occasionally, the cumulative effect of persistent attacks can greatly weaken the herbivore, particularly when conditions are not favorable, such as during droughts.

Ecological processes have interactive effects

We have seen the different ways in which species can interact with one another. In nature, however, more than one type of interaction often occurs at the same time. In many cases, the outcome of one type of interaction is modified or even reversed when another type of interaction is also occurring.

Predation reduces competition

When resources are limiting, a superior competitor can eliminate other species from a community through competitive exclusion. However, predators can prevent or greatly reduce exclusion by lowering the numbers of individuals of competing species.

A given predator may often feed on two, three, or more kinds of plants or animals in a given community. The predator's choice depends partly on the relative abundance of the prey options. In other words, a predator may feed on species A when it is abundant and then switch to species B when A is rare. Similarly, a given prey species may become a primary source of food for increasing numbers of species as it becomes more abundant. In this way, superior competitors may be prevented from competitively excluding other species.

Such patterns are often characteristic of communities in marine intertidal habitats. For example, by preying selectively on bivalves, sea stars prevent bivalves from monopolizing a habitat, opening up space for many other organisms (figure 56.21). When sea stars are removed from a habitat, species diversity falls

SCIENTIFIC THINKING

Question: Does predation affect the outcome of interspecific competitive interactions?

Hypothesis: In the absence of predators, prey populations will increase until resources are limiting, and some species will be competitively excluded.

Experiment: Remove predatory sea stars (Pisaster ochraceus) from some areas of rocky intertidal shoreline and monitor populations of species the sea stars prey upon. In control areas, pick up sea stars, but replace them where they were found to control for the effects of people walking through the study area.

a. *b.*

Result: In the absence of sea stars, the population of the mussel Mytilus californianus exploded, occupying all available space and eliminating many other species from the community.

Interpretation: What would happen if sea stars were returned to the experimental plots?

Figure 56.21 Predation reduces competition. *a.* In a controlled experiment in a coastal ecosystem, Robert Paine of the University of Washington removed a key predator, sea stars, *Pisaster*. *b.* In response, fiercely competitive mussels, a type of bivalve mollusk, exploded in population growth, effectively crowding out seven other indigenous species.

precipitously, and the seafloor community comes to be dominated by a few species of bivalves.

Predation tends to reduce competition in natural communities, so it is usually a mistake to attempt to eliminate a major predator, such as wolves or mountain lions, from a community. The result may be a decrease in biological diversity.

Parasitism may counter competition

Parasites may affect sympatric species differently and thus influence the outcome of interspecific interactions. One classic experiment investigated interactions between two sympatric flour beetles, *Tribolium castaneum* and *T. confusum,* with and without an intracellular parasite. In the absence of the parasite, *T. castaneum* is dominant, and *T. confusum* normally becomes extinct. When the parasite is present, however, the outcome is reversed, and *T. castaneum* perishes.

Similar effects of parasites in natural systems have been observed in many species. For example, in the *Anolis* lizards of St. Maarten, mentioned in section 56.2, the competitively inferior species is resistant to lizard malaria (a disease related to human malaria), whereas the other species is highly susceptible. In places where the parasite occurs, the competitively inferior species can hold its own and the two species coexist; elsewhere, the competitively dominant species outcompetes and eliminates it.

Indirect effects

In some cases, species may not directly interact, yet the presence of one species may affect a second by way of interactions with a third. Such effects are termed indirect effects.

Many desert rodents eat seeds, and so do the ants in their community; thus, we might expect them to compete with each other. But when all rodents were removed from experimental enclosures and not allowed back in, ant populations first increased, but then declined (figure 56.22).

The initial increase was the expected result of removing a competitor. Why did it then reverse? The answer reveals the intricacies of natural ecosystems. Rodents prefer large seeds, whereas ants prefer smaller ones. Furthermore, in this system, plants with large seeds are competitively superior to plants with small seeds. The removal of rodents therefore led to an increase in the number of plants with large seeds, which reduced the number of small seeds available to ants, which in turn led to a decline in ant populations. In summary, the effect of rodents on ants is complicated: a direct, negative effect of resource competition and an indirect, positive effect mediated by plant competition.

Biology, 11th Edition

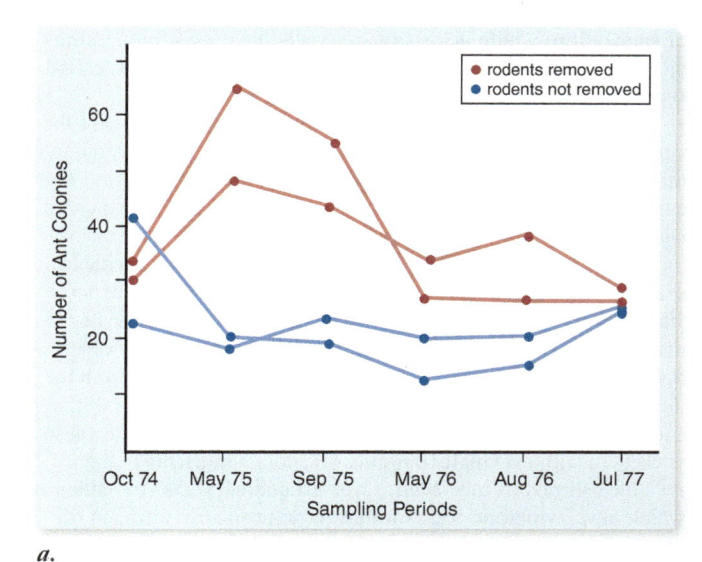

Figure 56.23 Example of a keystone species. Beavers, by constructing dams and transforming flowing streams into ponds, create new habitats for many plant and animal species.

Keystone species have major effects on communities

Species whose effects on the composition of communities are greater than one might expect based on their abundance are termed **keystone species.** Predators, such as the sea star described earlier in this section, can often serve as keystone species by preventing one species from outcompeting others, thus maintaining high levels of species richness in a community.

A wide variety of other types of keystone species also exist. Some species manipulate the environment in ways that create new habitats for others. Beavers, for example, change running streams into small impoundments, altering the flow of water and flooding areas (figure 56.23). Similarly, alligators excavate deep holes at the bottoms of lakes. In times of drought, these holes are the only areas where water remains, thus allowing aquatic species that otherwise would perish to persist until the drought ends and the lake refills.

Figure 56.22 Direct and indirect effects in an ecological community. *a.* In the enclosures in which rodents had been removed, ants initially increased in population size relative to the ants in the control enclosures, but then these ant populations declined. *b.* Rodents and ants both eat seeds, so the presence of rodents has a direct negative effect on ants, and vice versa. However, the presence of rodents has a negative effect on large seeds. In turn, the number of plants with large seeds has a negative effect on plants that produce small seeds, which the ants eat. Hence, the presence of rodents should increase the number of small seeds. In turn, the number of small seeds has a positive effect on ant populations. Thus, indirectly, the presence of rodents has a positive effect on ant population size.

Data analysis Based on the flow chart in *(b),* explain why the number of ant colonies first increases and then declines when rodents are removed.

Inquiry question How would you test the hypothesis that plant competition mediates the positive effect of rodents on ants?

Learning Outcomes Review 56.4

The types of symbiosis include mutualism, in which both participants benefit; commensalism, in which one benefits and the other is neutrally affected; and parasitism, in which one benefits at the expense of the other. Mutualistic species often undergo coevolution, such as the shape of flowers and the features of animals that feed on and pollinate them. Ecological interactions can affect many processes in a community; for example, predation and parasitism may lessen resource competition.

■ *How could the presence of a predator positively affect populations of a species on which it preys?*

chapter 56 *Community Ecology* 1203

56.5 Ecological Succession, Disturbance, and Species Richness

Learning Outcomes
1. Define succession and distinguish primary versus secondary.
2. Describe how early colonizers may affect subsequent occurrence of other species.
3. Explain how disturbance can either positively or negatively affect species richness.

Even when the climate of an area remains stable year after year, communities have a tendency to change from simple to complex in a process known as **succession.** This process is familiar to anyone who has seen a vacant lot or cleared woods slowly become occupied by an increasing number of species.

Succession produces a change in species composition

If a wooded area is cleared or burned and left alone, plants will slowly reclaim the area. Eventually, all traces of the clearing will disappear, and the area will again be woods. This kind of succession, which occurs in areas where an existing community has been disturbed but organisms still remain, is called **secondary succession.**

In contrast, **primary succession** occurs on bare, lifeless substrate, such as rocks, or in open water, where organisms gradually move into an area and change its nature. Primary succession occurs on land exposed after the retreat of glaciers, and on volcanic islands that rise from the sea (figure 56.24).

Primary succession in formerly glaciated areas provides an example (figure 56.24). On the bare, mineral-poor ground exposed when glaciers recede, soil pH is basic as a result of carbonates in the rocks, and nitrogen levels are low. Lichens are the first vegetation able to grow under such conditions. Acidic secretions from the lichens help break down the substrate and reduce the pH, as well as adding to the accumulation of soil. Mosses then colonize these pockets of soil, eventually building up enough nutrients in the soil for alder shrubs to take hold. Over a hundred years, the alders, which have symbiotic bacteria that fix atmospheric nitrogen (described in chapter 28), increase soil nitrogen levels, and their acidic leaves further lower soil pH. Eventually, spruce trees grow above the alders and shade them, crowding them out entirely and forming a dense spruce forest.

In a similar example, an *oligotrophic* lake—one poor in nutrients—may gradually, by the accumulation of organic matter, become *eutrophic*—rich in nutrients. As this occurs, the composition of communities will change, first increasing in species richness and then declining.

Why succession happens

Succession happens because species alter the habitat and the resources available in it in ways that favor other species. Three dynamic concepts are of critical importance in the process: establishment, facilitation, and inhibition:

1. **Establishment.** Early successional stages are characterized by weedy, *r*-selected species that are tolerant of the harsh, abiotic conditions in barren areas (chapter 55 discussed *r*-selected and *K*-selected species).

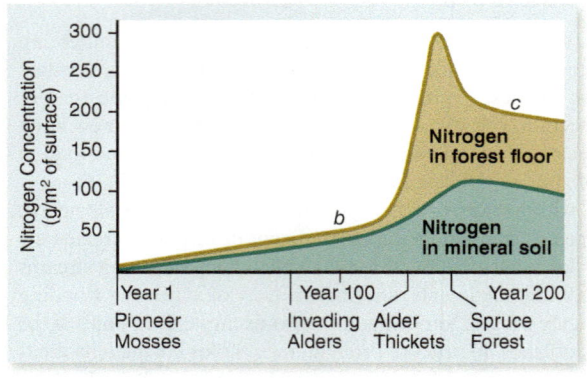

Figure 56.24 Primary succession at Alaska's Glacier Bay.
a. Initially, the recently deglaciated area at Glacier Bay, Alaska, had little soil nitrogen *b.* The first invaders of these exposed sites are pioneer moss species with nitrogen-fixing, mutualistic microbes. *c.* Within 20 years, young alder shrubs take hold. Rapidly fixing nitrogen, they soon form dense thickets. *d.* Eventually spruce overgrow the mature alders, forming a forest.

2. **Facilitation.** The weedy early successional stages introduce local changes in the habitat that favor other, less weedy species. Thus, the mosses in the Glacier Bay succession convert nitrogen to a form that allows alders to invade (figure 56.24). Similarly, the nitrogen build-up produced by the alders, though not necessary for spruce establishment, leads to more robust forests of spruce better able to resist attack by insects.
3. **Inhibition.** Sometimes the changes in the habitat caused by one species, while favoring other species, also inhibit the growth of the original species that caused the changes. Alders, for example, do not grow as well in acidic soil as the spruce and hemlock that replace them.

Over the course of succession, the number of species typically increases as the environment becomes more hospitable. In some cases, however, as ecosystems mature, more K-selected species replace r-selected ones, and superior competitors force out other species, leading ultimately to a decline in species richness.

Succession in animal communities

The species of animals present in a community also change through time in a successional pattern. As the vegetation changes during succession, habitat disappears for some species and appears for others.

A particularly striking example occurred on the Krakatau islands, which were devastated by an enormous volcanic eruption in 1883. Initially composed of nothing but barren ash-fields, the three islands of the group experienced rapid successional change as vegetation became reestablished. A few blades of grass appeared the next year, and within 15 years the coastal vegetation was well established and the interior was covered with dense grasslands. By 1930, the islands were almost entirely forested (figure 56.25).

The fauna of Krakatau changed in synchrony with the vegetation. Nine months after the eruption, the only animal found was a single spider, but by 1908, 200 animal species were found in a three-day exploration. For the most part, the first animals were grassland inhabitants, but as trees became established, some of the early colonists, such as the zebra dove and the long-tailed shrike (a type of predatory bird), disappeared and were replaced by forest-inhabiting species, such as fruit bats and fruit-eating birds.

Although patterns of succession of animal species have typically been caused by vegetational succession, changes in the composition of the animal community in turn have affected plant occurrences. In particular, many plant species that are animal-dispersed or pollinated could not colonize Krakatau until their dispersers or pollinators had become established. For example, fruit bats were slow to colonize Krakatau, and until they appeared, few bat-dispersed plant species were present.

Disturbances can play an important role in structuring communities

Traditionally, many ecologists considered biological communities to be in a state of equilibrium, a stable condition that resisted change and fairly quickly returned to its original state if disturbed by humans or natural events. Such stability was usually attributed to the process of interspecific competition.

In recent years, this viewpoint has been reevaluated. Increasingly, scientists are recognizing that communities are constantly changing as a result of climatic changes, species invasions, and disturbance events. As a result, many ecologists now invoke nonequilibrium models that emphasize change, rather than stability. A particular focus of ecological research concerns the role that disturbances play in determining the structure of communities.

Disturbances can be widespread or local. Severe disturbances, such as forest fires, drought, and floods, may affect large areas. Animals may also cause severe disruptions. Gypsy moths can devastate a forest by consuming all of the leaves on its trees. Unregulated deer populations may grow explosively, the deer

Figure 56.25 Succession after a volcanic eruption. A major volcanic explosion in 1883 on the island of Krakatau destroyed all life on the island. *a.* This photo shows a later, much less destructive eruption of the volcano. *b.* Krakatau, forested and populated by animals.

a. *b.*

overgrazing and so destroying the forest in which they live. On the other hand, local disturbances may affect only a small area, as when a tree falls in a forest or an animal digs a hole and uproots vegetation.

Intermediate disturbance hypothesis

In some cases, disturbance may act to increase the species richness of an area. According to the *intermediate disturbance hypothesis,* communities experiencing moderate amounts of disturbance will have higher levels of species richness than communities experiencing either little or great amounts of disturbance.

Two factors could account for this pattern. First, in communities where moderate amounts of disturbance occur, patches of habitat exist at different successional stages. Within the area as a whole, then, species diversity is greatest because the full range of species—those characteristic of all stages of succession—are present. For example, a pattern of intermittent episodic disturbance that produces gaps in the rainforest (as when a tree falls) allows invasion of the gap by other species (figure 56.26). Eventually, the species inhabiting the gap will go through a successional sequence, one tree replacing another, until a canopy tree species comes again to occupy the gap. But if there are many gaps of different ages in the forest, many different species will be coexisting, some in young gaps and others in older ones.

Second, moderate levels of disturbance may prevent communities from reaching the final stages of succession in which a few dominant competitors eliminate most of the other species. In contrast, too much disturbance might leave the community continually in the earliest stages of succession, when species richness is relatively low.

Ecologists are increasingly realizing that disturbance is common, rather than exceptional, in many communities. As a result, the idea that communities inexorably move along a successional trajectory culminating in the development of a predictable end-state, or "climax," community is no longer widely accepted. Rather, predicting the state of a community in the future may be difficult because the unpredictable occurrence of disturbances will often counter successional changes. Understanding the role that disturbances play in structuring communities is currently an important area of investigation in ecology.

Figure 56.26 Intermediate disturbance. A single fallen tree created a small light gap in the tropical rainforest of Panama. Such gaps play a key role in maintaining the high species diversity of the rainforest. In this case, a sunlight-loving plant is able to sprout up among the dense foliage of trees in the forest.

Learning Outcomes Review 56.5

Communities change through time by a process termed succession. Primary succession occurs on bare, lifeless substrate; secondary succession occurs where an existing community has been disturbed. Early arriving species alter the environment in ways that allow other species to colonize, and new colonizers may have negative effects on species already present. Sometimes, moderate levels of disturbance can lead to increased species richness because species characteristic of all levels of succession may be present.

- *From a community point of view, would clear-cutting a forest be better than selective harvest of individual trees? Why or why not?*

Chapter Review

56.1 Biological Communities: Species Living Together

A community is a group of different species that occupy a given location.

Communities change over space and time.
In accordance with the individualistic view, species generally respond independently to environmental conditions, and community composition gradually changes over space and time. However, in locations where conditions rapidly change, species composition may change greatly over short distances.

56.2 The Ecological Niche Concept

Fundamental niches are potential; realized niches are actual.
A niche is the total of all the ways a species uses environmental resources. The fundamental niche is the entire niche a species is capable of using if there are no intervening factors. The realized niche is the set of actual environmental conditions that allow establishment of a stable population.

Realized niches are usually smaller than fundamental niches because interspecific interactions limit a species' use of some resources.

Competitive exclusion can occur when species compete for limited resources.

The principle of competitive exclusion states that if resources are limiting, two species cannot simultaneously occupy the same niche; rather, one species will be eliminated.

Competition may lead to resource partitioning.

By using different resources (partitioning), sympatric species can avoid competing with each other and can coexist with reduced realized niches.

Detecting interspecific competition can be difficult.

Although experimentation is a powerful means of testing the hypothesis that species compete, practical limitations exist. Detailed knowledge of the ecology of species is important to evaluate the results of experiments and possible interactions.

56.3 Predator–Prey Relationships

Predation strongly influences prey populations.

Predation is the consuming of one organism by another, and includes not only one animal eating another, but also an animal eating a plant.

Natural selection strongly favors adaptations of prey species to prevent predation. In turn, sometimes predators evolve counter-adaptations, leading to an evolutionary "arms race."

Plant adaptations defend against herbivores.

Plants produce secondary chemical compounds that deter herbivores. Sometimes the herbivores evolve an ability to ingest the compounds and use them for their own defense.

Animal adaptations defend against predators.

Animal adaptations include chemical defenses and defensive coloration such as warning coloration or camouflage.

Mimicry allows one species to capitalize on defensive strategies of another.

In Batesian mimicry, a species that is edible or nontoxic evolves warning coloration similar to that of an inedible or poisonous species. In Müllerian mimicry, two species that are both toxic evolve similar warning coloration.

56.4 The Many Types of Species Interactions

Symbiosis involves long-term interactions.

Many symbiotic species have coevolved and have permanent relationships.

Mutualism benefits both species.

One example is the case of ants and acacias, in which *Acacia* plants provide a home and food for a species of stinging ants that protect them from herbivores.

Parasitism benefits one species at the expense of another.

Many organisms have parasitic lifestyles, living on or inside one or more host species and causing damage or disease as a result.

Commensalism benefits one species and is neutral to the other.

Examples of commensal relationships include epiphytes growing on large plants and barnacles growing on sea animals.

Ecological processes have interactive effects.

Because many processes may occur simultaneously, species may affect one another not only through direct interactions but also through their effects on other species in the community.

Keystone species have major effects on communities.

Keystone species are those that maintain a more diverse community by reducing competition between species or by altering the environment to create new habitats.

56.5 Ecological Succession, Disturbance, and Species Richness

Succession produces a change in species composition.

Primary succession begins with a barren, lifeless substrate, whereas secondary succession occurs after an existing community is disrupted by fire, clearing, or other events.

Disturbances can play an important role in structuring communities.

Community composition changes as a result of local and global disturbances that "reset" succession.

Intermediate levels of such disturbance may maximize species richness in two ways: by creating a patchwork of different habitats harboring different species, and by preventing communities from reaching the final stage of succession, which may be dominated by only a few, competitively superior species.

Review Questions

UNDERSTAND

1. Studies that demonstrate that species living in an ecological community change independently of one another in space and time
   a. indicate that the species arrived in the community at different times.
   b. indicate that the species' realized niches are regulated by different aspects of the environment.
   c. suggest species interactions are the sole determinant of which species coexist in a community.
   d. None of the choices is correct.

2. If two species have very similar realized niches and are forced to coexist and share a limiting resource indefinitely,
   a. both species would be expected to coexist.
   b. both species would be expected to go extinct.
   c. the species that uses the limiting resource most efficiently should drive the other species extinct.
   d. both species would be expected to become more similar to one another.

3. According to the idea of coevolution between predator and prey, when a prey species evolves a novel defense against a predator
   a. the predator is expected to always go extinct.
   b. the prey population should increase irreversibly out of control of the predator.
   c. the predator population should increase.
   d. evolution of a predator response should be favored by natural selection.

4. In order for mimicry to be effective in protecting a species from predation, it must
   a. occur in a palatable species that looks like a distasteful species.
   b. have cryptic coloration.
   c. occur such that mimics look and act like models.
   d. occur in only poisonous or dangerous species.

5. Which of the following is an example of commensalism?
   a. A tapeworm living in the gut of its host
   b. A clownfish living among the tentacles of a sea anemone
   c. An acacia tree and acacia ants
   d. Bees feeding on nectar from a flower

6. A species whose effect on the composition of a community is greater than expected based on its abundance can be called a
   a. predator.
   b. primary succession species.
   c. secondary succession species.
   d. keystone species.

7. When a predator preferentially eats the superior competitor in a pair of competing species
   a. the inferior competitor is more likely to go extinct.
   b. the superior competitor is more likely to persist.
   c. coexistence of the competing species is more likely.
   d. None of the choices is correct.

8. Species that are the first colonists in a habitat undergoing primary succession
   a. are usually the fiercest competitors.
   b. help maintain their habitat constant so their persistence is ensured.
   c. may change their habitat in a way that favors the invasion of other species.
   d. must first be successful secondary succession specialists.

APPLY

1. Which of the following can cause the realized niche of a species to be smaller than its fundamental niche?
   a. Predation
   b. Competition
   c. Parasitism
   d. All of the choices are correct.

2. The presence of a predatory species
   a. always drives a prey species to extinction.
   b. can positively affect a prey species by having a detrimental effect on competing species.
   c. indicates that the climax stage of succession has been reached.
   d. None of the choices is correct.

3. Resource partitioning by sympatric species
   a. always occurs when species have identical niches.
   b. may not occur in the presence of a predator, which reduces prey population sizes.
   c. results in the fundamental and realized niches being the same.
   d. is more common in herbivores than carnivores.

4. Parasitism differs from predation because
   a. the presence of parasitism doesn't lead to selection for defensive adaptations in parasitized species.
   b. parasites and the species they parasitize never engage in an evolutionary "arms race."
   c. parasites don't have strong effects on the populations of the species they parasitize.
   d. None of the choices is correct.

5. The presence of one species (A) in a community may benefit another species (B) if
   a. a commensalistic relationship exists between the two.
   b. The first species (A) preys on a predator of the second species (B).
   c. The first species (A) preys on a species that competes with a species that is eaten by the second species (B).
   d. All of the choices are correct.

SYNTHESIZE

1. Competition is traditionally indicated by documenting the effect of one species on the population of another. Are there alternative ways to study the potential effects of competition on organisms that are impractical to study with experimental manipulations because they are too big or live too long?

2. Refer to figure 56.9. If the single prey species of *Paramecium* was replaced by several different potential prey species that varied in their palatability or ease of subduing by the predator (leading to different levels of preference by the predator) what would you expect the dynamics of the system to look like; that is, would the system be more or less likely to go to extinction?

3. Refer to figure 56.22. Are there alternative hypotheses that might explain the increase followed by the decrease in ant colony numbers subsequent to rodent removal in the experiment described in figure 56.22? If so, how would you test the mechanism hypothesized in the figure?

4. Refer to figure 56.7. Examine the pattern of beak size distributions of two species of finches on the Galápagos Islands. One hypothesis that can be drawn from this pattern is that character displacement has taken place. Are there other hypotheses? If so, how would you test them?

5. Is it possible that some species function together as an integrated, holistic community, whereas other species at the same locality behave more individualistically? If so, what factors might determine which species function in which way?

CHAPTER 57

Dynamics of Ecosystems

Chapter Contents

57.1 Biogeochemical Cycles
57.2 The Flow of Energy in Ecosystems
57.3 Trophic-Level Interactions
57.4 Biodiversity and Ecosystem Stability
57.5 Island Biogeography

Introduction

The Earth is a relatively closed system with respect to chemicals. It is an open system in terms of energy, however, because it receives energy at visible and near-visible wavelengths from the Sun and steadily emits thermal energy to outer space in the form of infrared radiation. The organisms in ecosystems interact in complex ways as they participate in the cycling of chemicals and as they capture and expend energy. All organisms, including humans, depend on the specialized abilities of other organisms—plants, algae, animals, fungi, and prokaryotes—to acquire the essentials of life. Interactions between species determine both how chemicals and energy move through ecosystems and how communities are structured.

57.1 Biogeochemical Cycles

Learning Outcomes
1. Define ecosystem.
2. List four chemicals whose cyclic interactions are critical to organisms.
3. Describe how human activities disrupt these cycles.

An ecosystem includes all the organisms that live in a particular place, plus the abiotic (nonliving) environment at that location. Ecosystems are intrinsically dynamic in a number of ways, including their processing of matter and energy. We start with matter.

The atomic constituents of matter cycle within ecosystems

During the biological processing of matter, the atoms of which it is composed, such as the atoms of carbon or oxygen, maintain their integrity even as they are assembled into new compounds and the compounds are later broken down. The Earth has an essentially fixed number of each of the types of atoms of biological importance, and the atoms are recycled.

Each organism assembles its body from atoms that previously were in the soil, the atmosphere, other parts of the abiotic environment, or other organisms. When the organism dies, its atoms are released unaltered to be used by other organisms or returned to the abiotic environment. Because of the cycling of the atomic constituents of matter, your body is likely during your life to contain a carbon or oxygen atom that once was part of Julius Caesar's body or Cleopatra's.

The atoms of the various chemical elements are said to move through ecosystems in **biogeochemical cycles,** a term emphasizing that the cycles of chemical elements involve not only biological organisms and processes, but also geological (abiotic) systems and processes. Biogeochemical cycles include processes that occur on many spatial scales, from cellular to planetary, and they also include processes that occur on multiple time-scales, from seconds (biochemical reactions) to millennia (weathering of rocks).

Biogeochemical cycles usually cross the boundaries of ecosystems to some extent, rather than being self-contained within individual ecosystems. For example, one ecosystem might import or export carbon to others.

In this section, we consider the cycles of some major elements along with the compound water. We also present an example of biogeochemical cycles in a forest ecosystem.

Carbon, the basis of organic compounds, cycles through most ecosystems

Carbon is a major constituent of the bodies of organisms because carbon atoms help form the framework of all organic compounds (see chapter 3); almost 20% of the weight of the human body is carbon. From the viewpoint of the day-to-day dynamics of ecosystems, carbon dioxide (CO_2) is the most significant carbon-containing compound in the abiotic environments of organisms. It makes up 0.03% of the volume of the atmosphere, meaning the atmosphere contains about 750 billion metric tons of carbon. In aquatic ecosystems, CO_2 reacts spontaneously with the water to form bicarbonate ions (HCO_3^-).

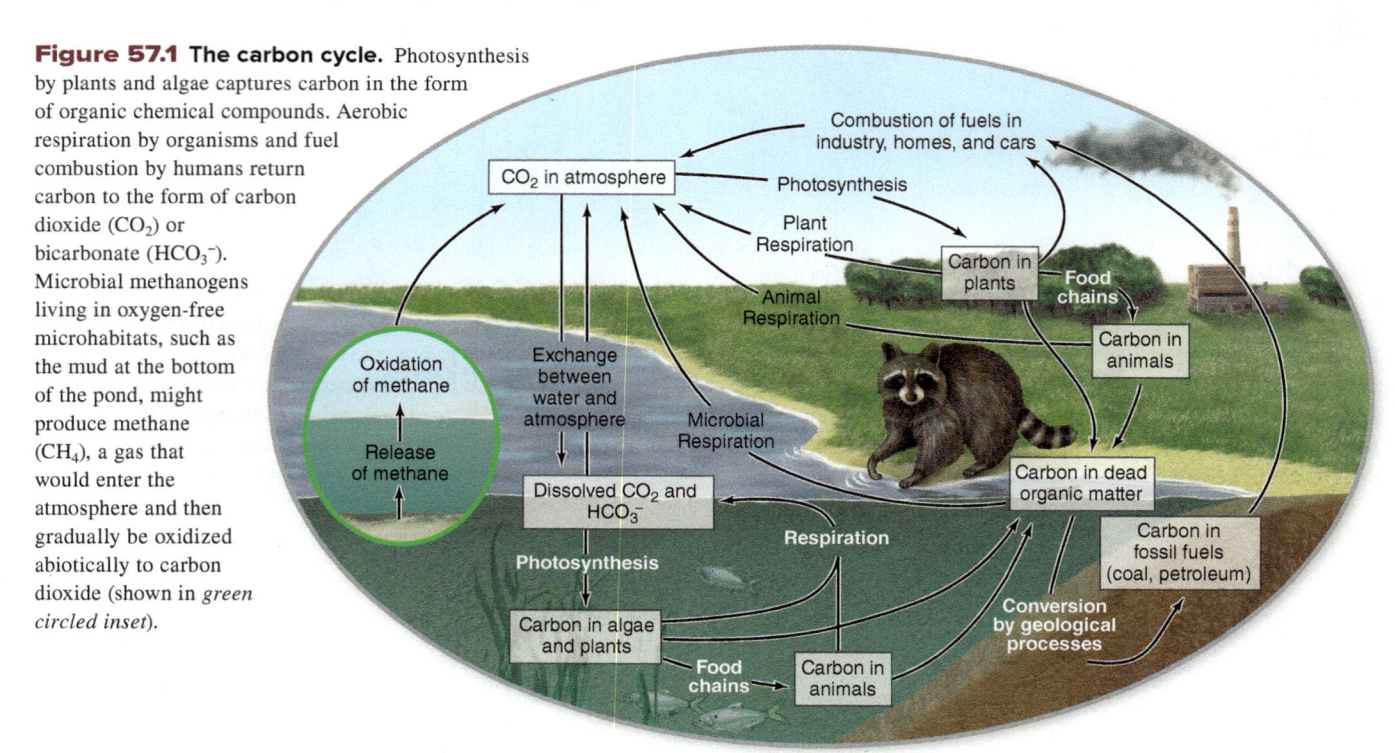

Figure 57.1 The carbon cycle. Photosynthesis by plants and algae captures carbon in the form of organic chemical compounds. Aerobic respiration by organisms and fuel combustion by humans return carbon to the form of carbon dioxide (CO_2) or bicarbonate (HCO_3^-). Microbial methanogens living in oxygen-free microhabitats, such as the mud at the bottom of the pond, might produce methane (CH_4), a gas that would enter the atmosphere and then gradually be oxidized abiotically to carbon dioxide (shown in *green circled inset*).

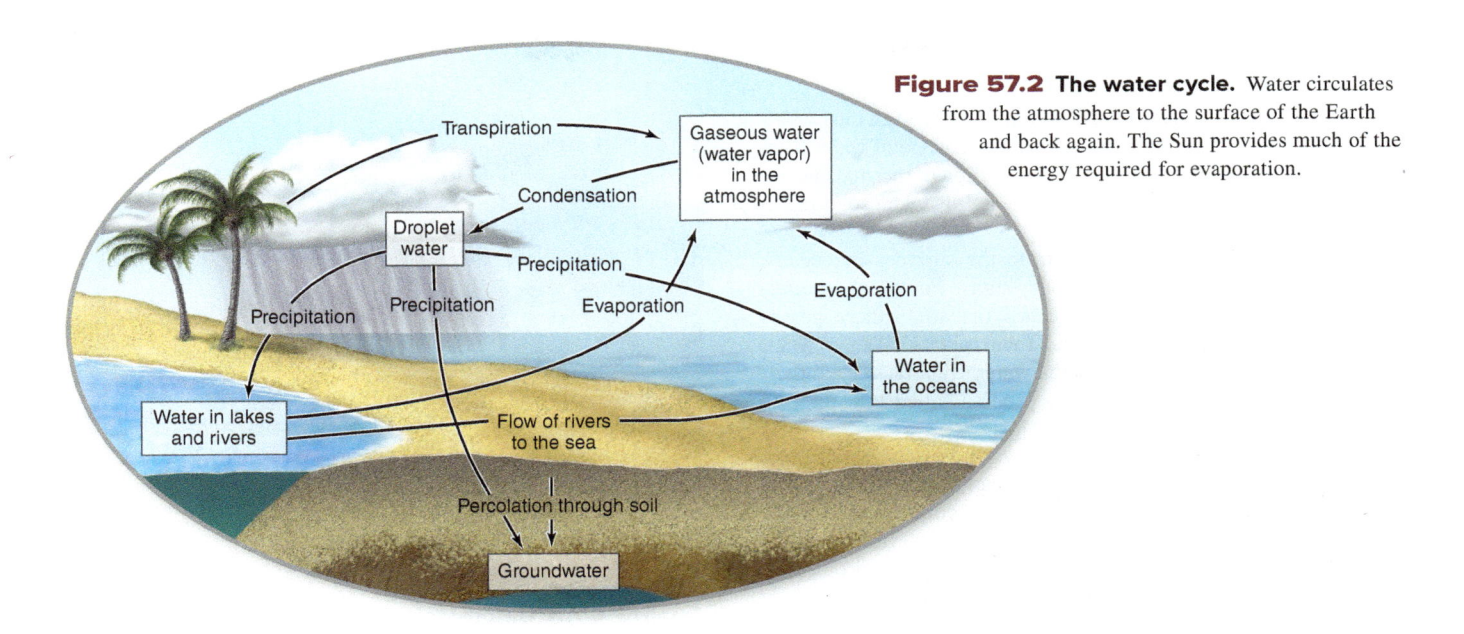

Figure 57.2 The water cycle. Water circulates from the atmosphere to the surface of the Earth and back again. The Sun provides much of the energy required for evaporation.

The basic carbon cycle

The carbon cycle is straightforward, as shown in figure 57.1. In terrestrial ecosystems, plants and other photosynthetic organisms take in CO_2 from the atmosphere and use it in photosynthesis to synthesize the carbon-containing organic compounds of which they are composed (see chapter 8). The process is sometimes called carbon fixation; *fixation* refers to metabolic reactions that make nongaseous compounds from gaseous ones.

Animals eat the photosynthetic organisms and build their own tissues by making use of the carbon atoms in the organic compounds they ingest. Both the photosynthetic organisms and the animals obtain energy during their lives by breaking down some of the organic compounds available to them, through aerobic cellular respiration (see chapter 7). When they do this, they produce CO_2. Decaying organisms also produce CO_2. Carbon atoms returned to the form of CO_2 are available once more to be used in photosynthesis to synthesize new organic compounds (carbon molecules occurring within an organism are referred to as "organic," whereas those outside of an organism are called "inorganic").

In aquatic ecosystems, the carbon cycle is fundamentally similar, except that inorganic carbon is present in the water not only as dissolved CO_2, but also as HCO_3^- ions, both of which act as sources of carbon for photosynthesis by algae and aquatic plants.

Methane producers

Microbes that break down organic compounds by anaerobic cellular respiration (see chapter 7) provide an additional dimension to the global carbon cycle. Methanogens, for example, are microbes that produce methane (CH_4) instead of CO_2. One major source of CH_4 is wetland ecosystems, where methanogens live in the oxygen-free sediments. Methane that enters the atmosphere is oxidized abiotically to CO_2, but CH_4 that remains isolated from oxygen can persist for great lengths of time.

The rise of atmospheric carbon dioxide

Another dimension of the global carbon cycle is that over long stretches of time, some parts of the cycle may proceed more rapidly than others. These differences in rate have ordinarily been relatively minor on a year-to-year basis; in any one year, the amount of CO_2 made by breakdown of organic compounds almost matches the amount of CO_2 used to synthesize new organic compounds.

Small mismatches, however, can have large consequences if continued for many years. The Earth's present reserves of coal were built up over geologic time. Organic compounds such as cellulose accumulated by being synthesized faster than they were broken down, and then they were transformed by geological processes into the fossil fuels. Most scientists believe that the world's petroleum and natural gas reserves were created in the same way.

Human burning of fossil fuels today is creating large contemporary imbalances in the carbon cycle. Carbon that took millions of years to accumulate in the reserves of fossil fuels is being rapidly returned to the atmosphere, driving the concentration of CO_2 in the atmosphere upward year by year and leading to global warming (see chapter 58).

The availability of water is fundamental to terrestrial ecosystems

The water cycle, seen in figure 57.2, is probably the most familiar of all biogeochemical cycles. All life depends on the presence of water. Indeed, the bodies of most organisms consist mainly of water—for example, humans are about 60% water by weight. The amount of water available in an ecosystem often determines the nature and abundance of the organisms present, as illustrated by the difference between forests and deserts (see chapter 58).

Each type of biogeochemical cycle has distinctive features. A distinctive feature of the water cycle is that water is a compound, not an element, and thus it can be synthesized and broken down. It is synthesized during aerobic cellular respiration (see chapter 7) and chemically split during photosynthesis (see chapter 8). The rates of these processes are ordinarily about equal, and therefore a relatively constant amount of water cycles through the biosphere.

The basic water cycle

One key part of the water cycle is that liquid water from the Earth's surface evaporates into the atmosphere. The change of water from a liquid to a gas requires a considerable addition of thermal energy, explaining why evaporation occurs more rapidly when solar radiation beats down on a surface.

Evaporation occurs directly from the surfaces of oceans, lakes, and rivers. In terrestrial ecosystems, however, approximately 90% of the water that reaches the atmosphere passes through plants. Trees, grasses, and other plants take up water from soil via their roots, and then the water evaporates from their leaves and other surfaces through a process called transpiration (see chapter 37).

Evaporated water exists in the atmosphere as a gas, just like any other atmospheric gas. The water can condense back into liquid form, however, mostly because of cooling of the air. Condensation of gaseous water (water vapor) into droplets or crystals causes the formation of clouds, and if the droplets or crystals are large enough, they fall to the surface of the Earth as precipitation (rain or snow).

Groundwater

Less obvious than surface water, which we see in rivers and lakes, is water under ground—termed groundwater. Groundwater occurs in **aquifers,** which are permeable, underground layers of rock, sand, and gravel that are often saturated with water. Groundwater is the most important reservoir of water on land in many parts of the world, representing over 95% of all freshwater in the United States, for example.

Groundwater consists of two subparts. The upper layers of the groundwater constitute the water table, which is unconfined in the sense that it flows into streams and is partly accessible to the roots of plants. The lower, confined layers of the groundwater are generally out of reach to streams and plants, but can be tapped by wells. Groundwater is recharged by water that percolates downward from above, such as from precipitation. Water in an aquifer flows much more slowly than surface water, anywhere from a few millimeters to a meter or so per day.

In the United States, groundwater provides about 25% of the water used by humans for all purposes, and it supplies about 50% of the population with drinking water. In the Great Plains states, the deep Ogallala Aquifer is tapped extensively as a water source for agricultural and domestic needs. The aquifer is being depleted faster than it is recharged—a local imbalance in the water cycle—posing an ominous threat to the agricultural production of the area. Similar threats exist in many of the drier portions of the globe.

Changes in ecosystems brought about by changes in the water cycle

Water is so crucial for life that changes in its supply in an ecosystem can radically alter the nature of the ecosystem. Such changes have occurred often during the Earth's geological history.

Consider, for example, the ecosystem of the Serengeti Plain in Tanzania, famous for its seemingly endless grasslands occupied by vast herds of antelopes and other grazing animals. The semiarid grasslands of today's Serengeti were rainforests 25 MYA.

Figure 57.3 Deforestation disrupts the local water cycle. Tropical deforestation can have severe consequences, such as the extensive erosion in this area in the Amazon region of Brazil.

Starting at about that time, mountains such as Mount Kilimanjaro rose up between the rainforests and the Indian Ocean, their source of moisture. The presence of the mountains forced winds from the Indian Ocean upward, cooling the air and causing much of its moisture to precipitate before the air reached the rainforests. The land became much drier, and the forests turned to grasslands.

Today, human activities can alter the water cycle so profoundly that major changes occur in ecosystems. Changes in rainforests caused by deforestation provide an example. In healthy tropical rainforests, more than 90% of the moisture that falls as rain is taken up by plants and returned to the air by transpiration. Plants, in a very real sense, create their own rain: The moisture returned to the atmosphere falls back on the forests.

When human populations cut down or burn the rainforests in an area, the local water cycle is broken. Water that falls as rain thereafter drains away in rivers instead of rising to form clouds and fall again on the forests. Just such a transformation is occurring today in many tropical rainforests (figure 57.3). Large areas in Brazil, for example, were transformed in the 20th century from lush tropical forest to semiarid desert, depriving many unique plant and animal species of their native habitat.

The nitrogen cycle depends on nitrogen fixation by microbes

Nitrogen is a component of all proteins and nucleic acids and is required in substantial amounts by all organisms; proteins are 16% nitrogen by weight. In many ecosystems, nitrogen is the chemical element in shortest supply relative to the needs of organisms. A paradox is that the atmosphere is 78% nitrogen by volume.

Nitrogen availability

How can nitrogen be in short supply if the atmosphere is so rich with it? The answer is that the nitrogen in the atmosphere is in its elemental form—molecules of nitrogen gas (N_2)—and the vast majority of organisms, including all plants and animals, have no way to use nitrogen in this chemical form.

For animals, the ultimate source of nitrogen is nitrogen-containing organic compounds synthesized by plants or by algae or other microbes. Herbivorous animals, for example, eat plant or algal proteins and use the nitrogen-containing amino acids in them to synthesize their own proteins.

Plants and algae use a number of simple nitrogen-containing compounds as their sources of nitrogen to synthesize proteins and other nitrogen-containing organic compounds in their tissues. Two commonly used nitrogen sources are ammonia (NH_3) and nitrate ions (NO_3^-). As described in chapter 38, certain prokaryotic microbes can synthesize ammonia and nitrate from N_2 in the atmosphere, thereby constituting a part of the nitrogen cycle that makes atmospheric nitrogen accessible to plants and algae (figure 57.4). Other prokaryotes turn NH_3 and NO_3^- into N_2, making the nitrogen inaccessible. The balance of the activities of these two sets of microbes determines the accessibility of nitrogen to plants and algae.

Microbial nitrogen fixation, nitrification, and denitrification

The synthesis of nitrogen-containing compounds from N_2 is known as **nitrogen fixation.** The first step in this process is the synthesis of NH_3 from N_2, and biochemists sometimes use the term *nitrogen fixation* to refer specifically to this step. After NH_3 has been synthesized, other prokaryotic microbes oxidize part of it to form NO_3^-, a process called **nitrification.**

Certain genera of prokaryotes have the ability to accomplish nitrogen fixation using a system of enzymes known as the nitrogenase complex (the *nif* gene complex; see chapter 28). Most of the microbes are free-living, but on land some are found in symbiotic relationships with the roots of legumes (plants of the pea family, Fabaceae), alders, myrtles, and other plants.

Additional prokaryotic microbes (including both bacteria and archaea) are able to convert the nitrogen in NO_3^- into N_2 (or other nitrogen gases such as N_2O), a process termed **denitrification.** Ammonia can be subjected to denitrification indirectly by being converted first to NO_3^- and then to N_2.

Nitrogenous wastes and fertilizer use

Most animals, when they break down proteins in their metabolism, excrete the nitrogen from the proteins as NH_3. Humans and other mammals excrete nitrogen as urea in their urine (see chapter 50); a number of types of microbes convert the urea to NH_3. The NH_3 from animal excretion can be picked up by plants and algae as a source of nitrogen.

Human populations are radically altering the global nitrogen cycle by the use of fertilizers on lawns and agricultural fields. The fertilizers contain forms of fixed nitrogen that crops can use, such as ammonium (NH_4) salts manufactured industrially from atmospheric N_2. Partly because of the production of fertilizers, humans have already doubled the rate of transfer of N_2 in usable forms into soils and waters.

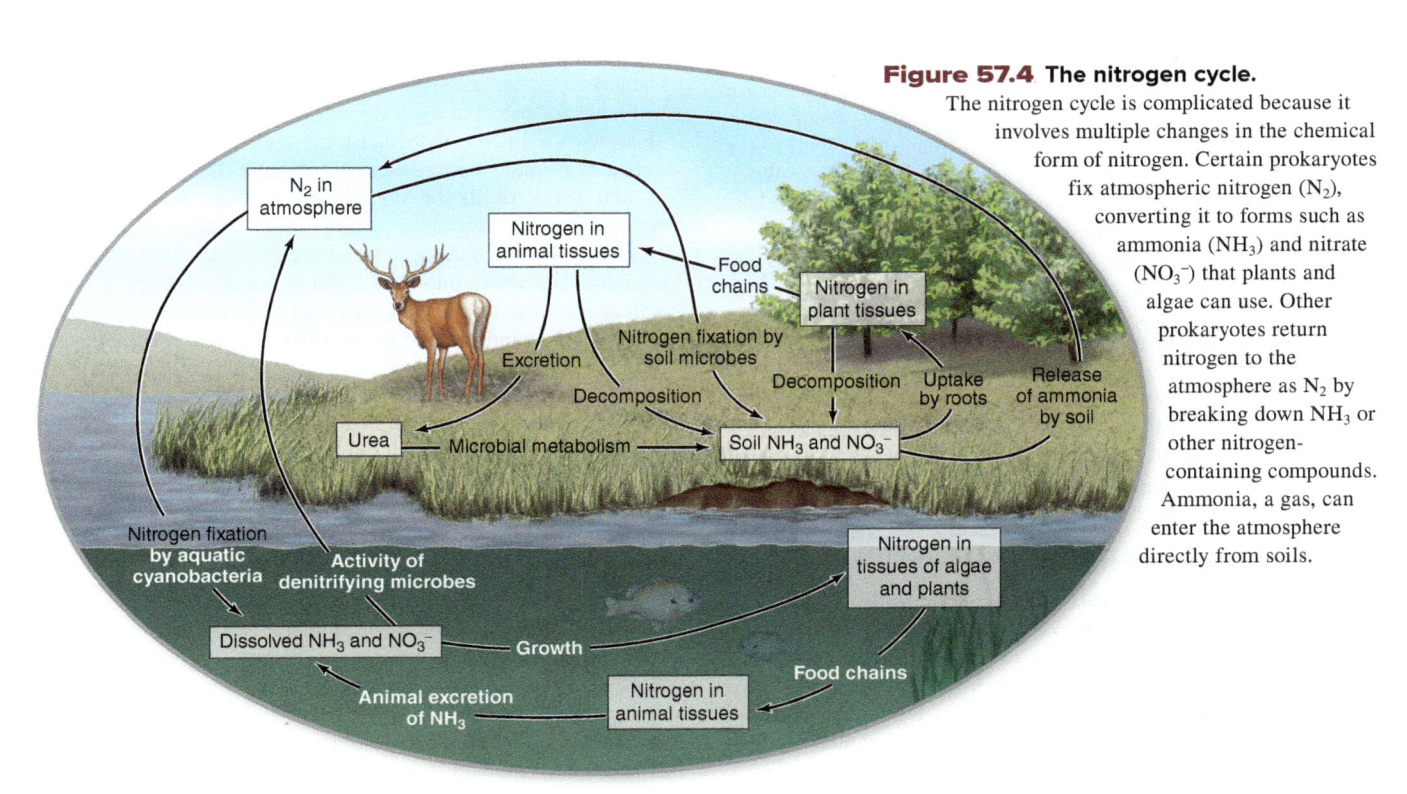

Figure 57.4 The nitrogen cycle. The nitrogen cycle is complicated because it involves multiple changes in the chemical form of nitrogen. Certain prokaryotes fix atmospheric nitrogen (N_2), converting it to forms such as ammonia (NH_3) and nitrate (NO_3^-) that plants and algae can use. Other prokaryotes return nitrogen to the atmosphere as N_2 by breaking down NH_3 or other nitrogen-containing compounds. Ammonia, a gas, can enter the atmosphere directly from soils.

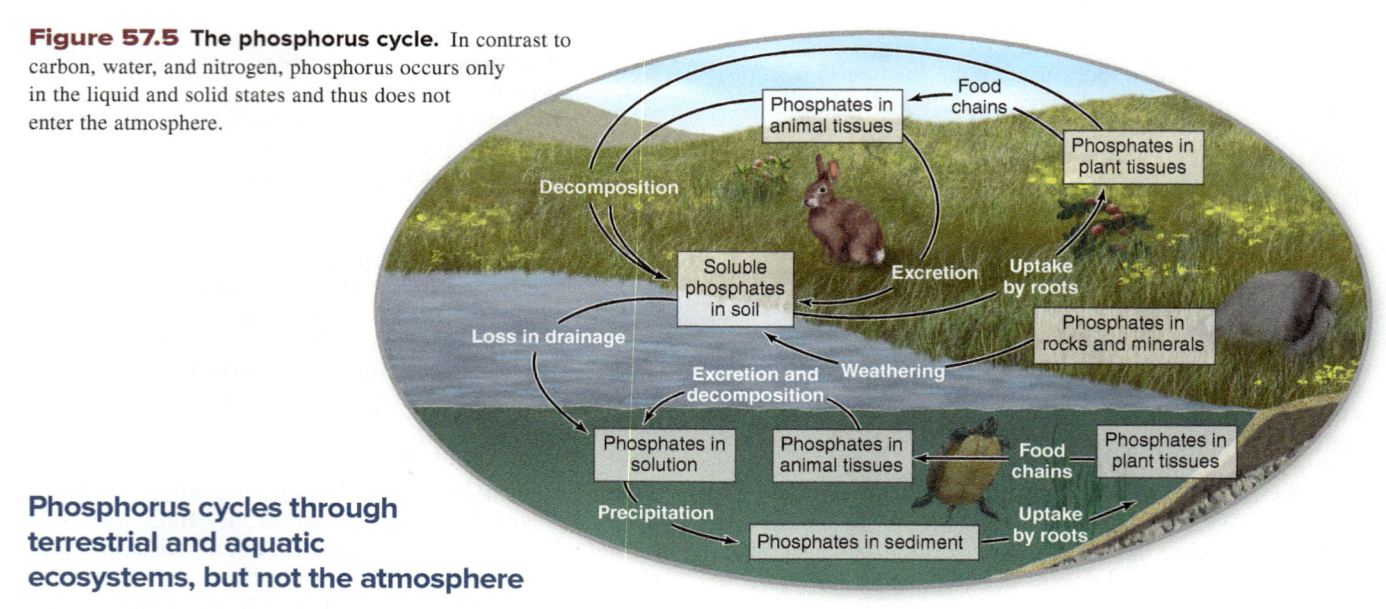

Figure 57.5 The phosphorus cycle. In contrast to carbon, water, and nitrogen, phosphorus occurs only in the liquid and solid states and thus does not enter the atmosphere.

Phosphorus cycles through terrestrial and aquatic ecosystems, but not the atmosphere

Phosphorus is required in substantial quantities by all organisms; it occurs in nucleic acids, membrane phospholipids, and other essential compounds, such as adenosine triphosphate (ATP).

Unlike carbon, water, and nitrogen, phosphorus (P) has no significant gaseous form and does not cycle through the atmosphere (figure 57.5). In this respect, the phosphorus cycle exemplifies the sorts of cycles also exhibited by calcium, silicon, and many other mineral elements. Another feature that greatly simplifies the phosphorus cycle compared with the nitrogen cycle is that phosphorus exists in ecosystems in just a single oxidation state, phosphate (PO_4^{3-}).

Phosphate availability

Plants and algae use free inorganic PO_4^{3-} in the soil or water for synthesizing their phosphorus-containing organic compounds. Animals then tap the phosphorus in plant or algal tissue compounds to build their own phosphorus compounds. When organisms die, decay microbes—in a process called phosphate remineralization—break up the organic compounds in their bodies, releasing phosphorus as inorganic PO_4^{3-} that plants and algae again can use.

The phosphorus cycle includes critical abiotic chemical and physical processes. Free PO_4^{3-} exists in soil in only low concentrations both because it combines with other soil constituents to form insoluble compounds and because it tends to be washed away by streams and rivers. Weathering of many sorts of rocks releases new PO_4^{3-} into terrestrial systems, but then rivers carry the PO_4^{3-} into the ocean basins. There is a large one-way flux of PO_4^{3-} from terrestrial rocks to deep-sea sediments.

Phosphates as fertilizers

Human activities have greatly modified the global phosphorus cycle since the advent of crop fertilization. Fertilizers are typically designed to provide PO_4^{3-} because crops might otherwise be short of it; the PO_4^{3-} in fertilizers is typically derived from crushed phosphate-rich rocks and bones. Detergents are another potential culprit in adding PO_4^{3-} to ecosystems, but laws now mandate low-phosphate detergents in much of the world.

Limiting nutrients in ecosystems are those in short supply relative to need

A chain is only as strong as its weakest link. For the plants and algae in an ecosystem to grow—and to thereby provide food for animals—they need many different chemical elements. The simplest theory is that in any particular ecosystem, one element will be in shortest supply relative to the needs for it by the plants and algae. That element is the limiting nutrient—the weak link—in the ecosystem.

The cycle of a limiting nutrient is particularly important because it determines the rate at which the nutrient is made available for use. We gave the nitrogen and phosphorus cycles close attention precisely because those elements are the limiting nutrients in many ecosystems. Nitrogen is the limiting nutrient in about two-thirds of the oceans and in many terrestrial ecosystems.

Iron is the limiting nutrient for algal populations (phytoplankton) in about one-third of the world's oceans. In these waters, wind-borne soil dust seems often to be the chief source of iron. When wind brings in iron-rich dust, algal populations proliferate, provided the iron is in a usable chemical form. In this way, sand storms in the Sahara Desert, by increasing the dust in global winds, can increase algal productivity in Pacific waters (figure 57.6).

Biogeochemical cycling in a forest ecosystem has been studied experimentally

An ongoing series of studies at the Hubbard Brook Experimental Forest in New Hampshire has yielded much of the available information about the cycling of nutrients in forest ecosystems. Hubbard Brook is the central stream of a large watershed that drains the hillsides of a mountain range covered with temperate deciduous forest. Multiple tributary streams carry water off the hillsides into Hubbard Brook.

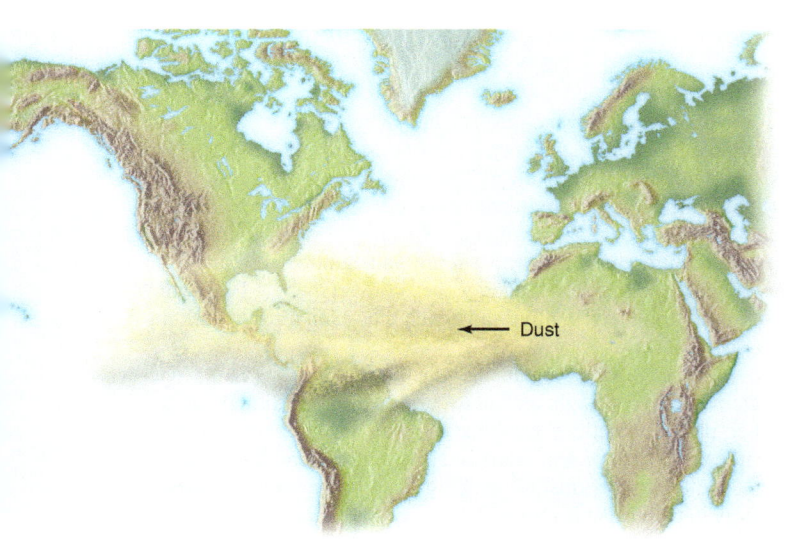

Figure 57.6 One world. Every year, millions of metric tons of iron-rich dust is carried westward by the Trade Winds from the Sahara Desert and neighboring Sahel area. A working hypothesis of many oceanographers is that this dust fertilizes parts of the ocean, including parts of the Pacific Ocean, where iron is the limiting nutrient. Land use practices in Africa, which are increasing the size of the north African desert, can thus affect ecosystems on the other side of the globe.

Six tributary streams, each draining a particular valley, were equipped with measurement devices when the study was started. All of the water that flowed out of each valley had to pass through the measurement system, where the flow of water and concentrations of nutrients was quantified.

The undisturbed forests around Hubbard Brook are efficient at retaining nutrients. In a year, only small quantities of nutrients enter a valley from outside, doing so mostly as a result of precipitation. The quantities carried out in stream waters are small also. When we say "small," we mean the influxes and outfluxes represent just minor fractions of the total amounts of nutrients in the system—about 1% in the case of calcium, for example.

In 1965 and 1966, the investigators felled all the trees and cleared all shrubs in one of the six valleys and prevented regrowth (figure 57.7a). The effects were dramatic. The amount of water running out of that valley increased by 40%, indicating that water previously taken up by vegetation and evaporated into the atmosphere was now running off. The amounts of a number of nutrients running out of the system also greatly increased. For example, the rate of loss of calcium increased ninefold. Phosphorus, on the other hand, did not increase in the stream water; it apparently was locked up in insoluble compounds in the soil.

The change in the status of nitrogen in the disturbed valley was especially striking (figure 57.7b). The undisturbed forest in this valley had been accumulating NO_3^- at a rate of about 5 kg per hectare per year, but the deforested ecosystem lost NO_3^- at a rate of about 53 kg per hectare per year. The NO_3^- concentration in the stream water rapidly increased. The fertility of the valley decreased dramatically, while the runoff of nitrate generated massive algal blooms downstream.

This experiment is particularly instructive in the 21st century because forested land continues to be cleared worldwide (see chapter 58).

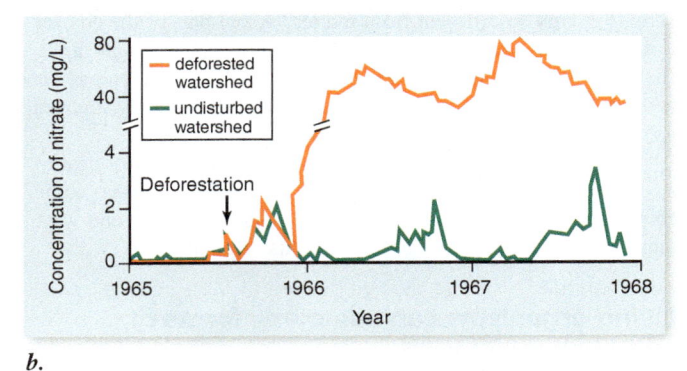

Figure 57.7 The Hubbard Brook experiment. *a.* A 38-acre watershed was completely deforested, and the runoff monitored for several years. *b.* Deforestation greatly increased the loss of nutrients in runoff water from the ecosystem. (Note: The values on the y-axis scale are interrupted [double line] to save space.) The orange line shows the nitrate concentration in the runoff water from the deforested watershed; the green line shows the nitrate concentration in runoff water from an undisturbed neighboring watershed.

Data analysis About how many times greater was the nitrate runoff in the deforested area than in the neighboring undisturbed area?

> **Learning Outcomes Review 57.1**
>
> An ecosystem consists of the living and nonliving components of a particular place. Biogeochemical cycles describe how elements move between these components. Carbon, nitrogen, and phosphorus cycle in known ways, as does water, which is critical to ecosystems. Human populations disrupt these cycles with artificial fertilization, deforestation, diversion of water, and burning of fossil fuels.
>
> ■ *Would fertilization with animal manure be less disruptive than fertilization with purified chemicals? Why or why not?*

57.2 The Flow of Energy in Ecosystems

> **Learning Outcomes**
> 1. Describe the different trophic levels.
> 2. Distinguish between energy and heat.
> 3. Explain how energy moves through trophic levels.

The dynamic nature of ecosystems includes the processing of energy as well as that of matter. Energy, however, follows very different principles than does matter. Energy is never recycled. Instead, radiant energy from the Sun that reaches the Earth makes a one-way pass through our planet's ecosystems before being converted to heat and radiated back into space, signifying that the Earth is an open system for energy.

Energy can neither be created nor destroyed, but changes form

Why is energy so different from matter? A key part of the answer is that energy exists in several different forms, such as light, chemical-bond energy, motion, and heat. Although energy is neither created nor destroyed in the biosphere, it frequently changes form (the First Law of Thermodynamics).

A second key point is that organisms cannot convert heat to any of the other forms of energy. Thus, if organisms convert some chemical-bond or light energy to heat, the conversion is one-way; they cannot cycle that energy back into its original form.

Living organisms can use many forms of energy, but not heat

To understand why the Earth must function as an open system with regard to energy, two additional principles need to be recognized. The first is that organisms can use only certain forms of energy. For animals to live, they must have energy specifically as chemical-bond energy, which they acquire from their foods. Plants must have energy as light. Neither animals nor plants (nor any other organisms) can use heat as a source of energy.

The second principle is that whenever organisms use chemical-bond or light energy, some of it is converted to heat. Consistent with the Second Law of Thermodynamics (see section 6.2), which states that entropy tends to increase through time, a partial conversion to heat is inevitable. Put another way, animals and plants require chemical-bond energy and light to stay alive, but as they use these forms of energy, they convert them to heat, which they cannot use to stay alive and which they cannot cycle back into the original forms.

Fortunately for organisms, the Earth functions as an open system for energy. Light arrives every day from the Sun. Plants and other photosynthetic organisms use the newly arrived light to synthesize organic compounds and stay alive. Animals then eat the photosynthetic organisms, making use of the chemical-bond energy in their organic molecules to stay alive. Light and chemical-bond energy are partially converted to heat at every step. In fact, the light and chemical-bond energy are ultimately converted completely to heat. The heat leaves the Earth by being radiated into outer space at invisible, infrared wavelengths of the electromagnetic spectrum. For life to continue, new light energy is always required.

The Earth's incoming and outgoing flows of radiant energy must be equal for global temperature to stay constant. One concern is that human activities are changing the composition of the atmosphere in ways that impede the outgoing flow—the so-called *greenhouse effect*—leading to an increase in global temperatures (see chapter 58).

Energy flows through trophic levels of ecosystems

In chapter 7, we introduced the concepts of autotrophs ("self-feeders") and heterotrophs ("fed by others"). **Autotrophs** synthesize the organic compounds of their bodies from inorganic precursors such as CO_2, water, and NO_3^- using energy from an abiotic source. Some autotrophs use light as their source of energy and therefore are **photoautotrophs;** they are the photosynthetic organisms, including plants, algae, and cyanobacteria. Other autotrophs are **chemoautotrophs** and obtain energy by means of inorganic oxidation reactions, such as the microbes that use hydrogen sulfide available at deep water vents (see chapter 58). All chemoautotrophs are prokaryotic. The photoautotrophs are of greatest importance in most ecosystems, and we focus on them in the remainder of this chapter.

Heterotrophs are organisms that cannot synthesize organic compounds from inorganic precursors, but instead live by taking in organic compounds that other organisms have made. They obtain the energy they need to live by breaking up some of the organic compounds available to them, thereby liberating chemical-bond energy for metabolic use (see chapter 7). Animals, fungi, and many microbes are heterotrophs.

When living in their native environments, species are often organized into chains that eat each other sequentially. For example, a species of insect might eat plants, and then a species of shrew might eat the insect, and a species of hawk might eat the shrew. Food passes through the four species in the sequence: plants → insect → shrew → hawk. A sequence of species like this is termed a food chain.

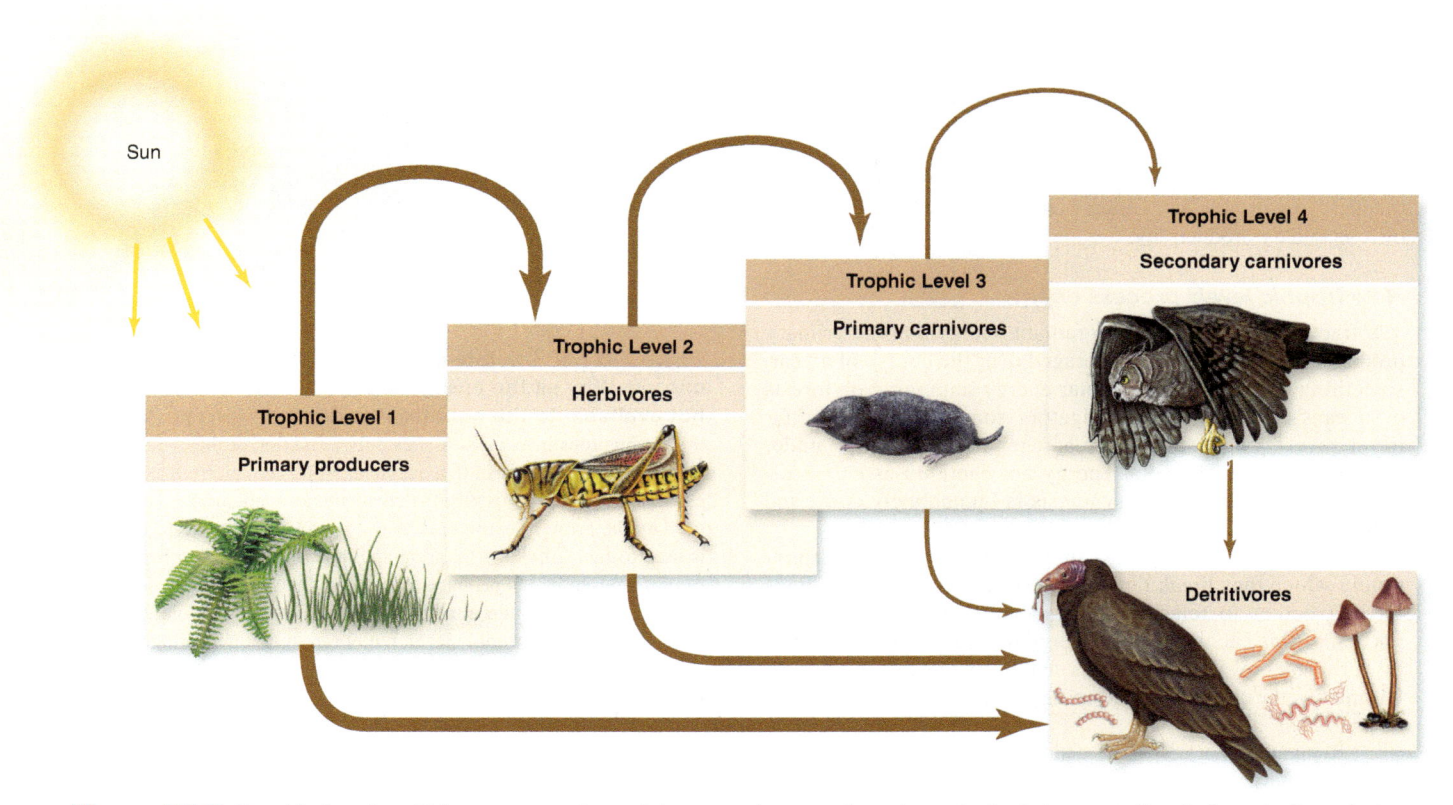

Figure 57.8 Trophic levels within an ecosystem. Primary producers such as plants obtain their energy directly from the Sun, placing them in trophic level 1. Animals that eat plants, such as plant-eating insects, are herbivores and are in trophic level 2. Animals that eat the herbivores, such as shrews, are primary carnivores and are in trophic level 3. Animals that eat the primary carnivores, such as owls, are secondary carnivores in trophic level 4. Each trophic level, although illustrated here by a particular species, consists of all the species in the ecosystem that function in a similar way in terms of what they eat. The organisms in the detritivore trophic level consume dead organic matter they obtain from all the other trophic levels.

In a whole ecosystem, many species play similar roles; there is typically not just a single species in each role. For example, the animals that eat plants might include not just a single insect species, but perhaps 30 species of insects, plus perhaps 10 species of mammals. To organize this complexity, ecologists recognize a limited number of feeding, or **trophic, levels** (figure 57.8).

Definitions of trophic levels

The first trophic level in an ecosystem, called the **primary producers,** consists of all the autotrophs in the system. The other trophic levels consist of the heterotrophs—the **consumers.** All the heterotrophs that feed directly on the primary producers are placed together in a trophic level called the **herbivores.** In turn, the heterotrophs that feed on the herbivores (eating them or being parasitic on them) are collectively termed **primary carnivores,** and those that feed on the primary carnivores are called **secondary carnivores.**

Advanced studies of ecosystems need to take into account that organisms often do not line up in simple linear sequences in terms of what they eat; some animals, for example, eat both primary producers and other animals; we call such animals *omnivores*. Similarly, some carnivores eat both herbivores and lower level carnivores. Nonetheless, a linear sequence of trophic levels is a useful organizing principle for many purposes. We will follow this approach here, but keep in mind that it is a simplification. Because animal species often eat prey at multiple trophic levels, ecosystems are more like food webs than food chains.

An additional consumer level is the **detritivore** trophic level. Detritivores differ from the organisms in the other trophic levels in that they feed on the remains of already dead organisms; detritus is dead organic matter. A subcategory of detritivores is the **decomposers,** which are mostly microbes and other minute organisms that live on and break up dead organic matter.

Concepts to describe trophic levels

Trophic levels consist of whole populations of organisms. For example, the primary-producer trophic level consists of the whole populations of all the autotrophic species in an ecosystem. Ecologists have developed a special set of terms to refer to the properties of populations and trophic levels.

The **productivity** of a trophic level is the rate at which the organisms in the trophic level collectively synthesize new organic matter (new tissue substance). **Primary productivity** is the productivity of the primary producers. An important complexity in analyzing the primary producers is that not only do they synthesize new organic matter by photosynthesis, but they also break down some of the organic matter to release energy by means of aerobic cellular respiration (see chapter 7). The **respiration** of the primary producers, in this context, is the rate at which they break down organic compounds. **Gross primary productivity (GPP)** is simply the raw rate at which the primary producers synthesize new organic matter; **net primary productivity (NPP)** is the GPP minus the respiration of the primary producers. The NPP represents the organic matter available for herbivores to use as food.

The productivity of a heterotroph trophic level is termed **secondary productivity.** For instance, the rate that new organic matter is made by means of individual growth and reproduction in all the herbivores in an ecosystem is the secondary productivity of the herbivore trophic level. Each heterotroph trophic level has its own secondary productivity.

How trophic levels process energy

The fraction of incoming solar radiant energy that the primary producers capture is small. Averaged over the course of a year, something around 1% of the solar energy impinging on forests or oceans is captured. Investigators sometimes observe far lower levels, but also see percentages as high as 5% under some conditions. The solar energy not captured as chemical-bond energy through photosynthesis is immediately converted to heat.

The primary producers, as noted before, carry out respiration in which they break down some of the organic compounds in their bodies to release chemical-bond energy. They use a portion of this chemical-bond energy to make ATP, which they in turn use to power various energy-requiring processes. Ultimately, the chemical-bond energy they release by respiration turns to heat.

Remember that organisms cannot use heat to stay alive. As a result, whenever energy changes form to become heat, it loses much or all of its usefulness for organisms as a fuel source. What we have seen so far is that about 99% of the solar energy impinging on an ecosystem turns to heat because it fails to be used by photosynthesis. Then some of the energy captured by photosynthesis also becomes heat because of respiration by the primary producers. All the heterotrophs in an ecosystem must live on the chemical-bond energy that is left.

An example of energy loss between trophic levels

As chemical-bond energy is passed from one heterotroph trophic level to the next, a great deal of the energy is diverted all along the way. This principle has dramatic consequences. It means that, over any particular period of time, the amount of chemical-bond energy available to primary carnivores is far less than that available to herbivores, and the amount available to secondary carnivores is far less than that available to primary carnivores.

Why does the amount of chemical-bond energy decrease as energy is passed from one trophic level to the next? Consider the use of energy by the herbivore trophic level as an example (figure 57.9). After an herbivore such as a leaf-eating insect ingests some food, it produces feces. The chemical-bond energy in the compounds in the feces is not passed along to the primary carnivore trophic level. The remaining chemical-bond energy of the food is assimilated by the herbivore and is used for a number of functions. Part of the assimilated energy is liberated by cellular respiration to be used for tissue repair, body movements, and other such functions. The energy used in these ways turns to heat and is not passed along to the carnivore trophic level. The remainder of the chemical-bond energy is built into the tissues of the herbivore and can serve as food for a carnivore. However, some herbivore individuals die of disease or accident rather than being eaten by predators. As a result, only a fraction of the initial chemical-bond energy acquired from the leaf is built into the

17% growth
33% cellular respiration
50% feces

Figure 57.9 The fate of ingested chemical-bond energy: Why all the energy ingested by a heterotroph is not available to the next trophic level. A heterotroph such as this herbivorous insect assimilates only a fraction of the chemical-bond energy it ingests. In this example, 50% is not assimilated and is eliminated in feces; this eliminated chemical-bond energy cannot be used by the primary carnivores. A third (33%) of the ingested energy is used to fuel cellular respiration and thus is converted to heat, which cannot be used by the primary carnivores. Only 17% of the ingested energy is converted into insect biomass through growth and can serve as food for the next trophic level, but not even that percentage is certain to be used in that way because some of the insects die before they are eaten.

tissues of herbivore individuals that are eaten by primary carnivores. The same scenario is repeated at each step in a series of trophic levels (figure 57.10).

Ecologists figure as a rule of thumb that the amount of chemical-bond energy available to a trophic level over time is about 10% of that available to the preceding level over the same period of time. In some instances the percentage is higher, even as high as 30%.

Heat as the final energy product

Essentially all of the chemical-bond energy captured by photosynthesis in an ecosystem eventually becomes heat as the chemical-bond energy is used by various trophic levels. To see this important point, recognize that when the detritivores in the ecosystem metabolize all the dead bodies, feces, and other materials made available to them, they produce heat just like the other trophic levels do.

Productive ecosystems

Ecosystems vary considerably in their NPP. Wetlands and tropical rainforests are examples of particularly productive ecosystems (figure 57.11); in them, the NPP, measured as dry weight of new organic matter produced, is often around 2000g/m^2/year. By contrast, the corresponding figures for some other types of ecosystems are 1200 to 1300 for temperate forests, 900 for savanna, and 90 for deserts. (These general ecosystem types, termed *biomes*, are described in chapter 58.)

The number of trophic levels is limited by energy availability

The rate at which chemical-bond energy is made available to organisms in different trophic levels decreases exponentially as energy makes its way from primary producers to herbivores and

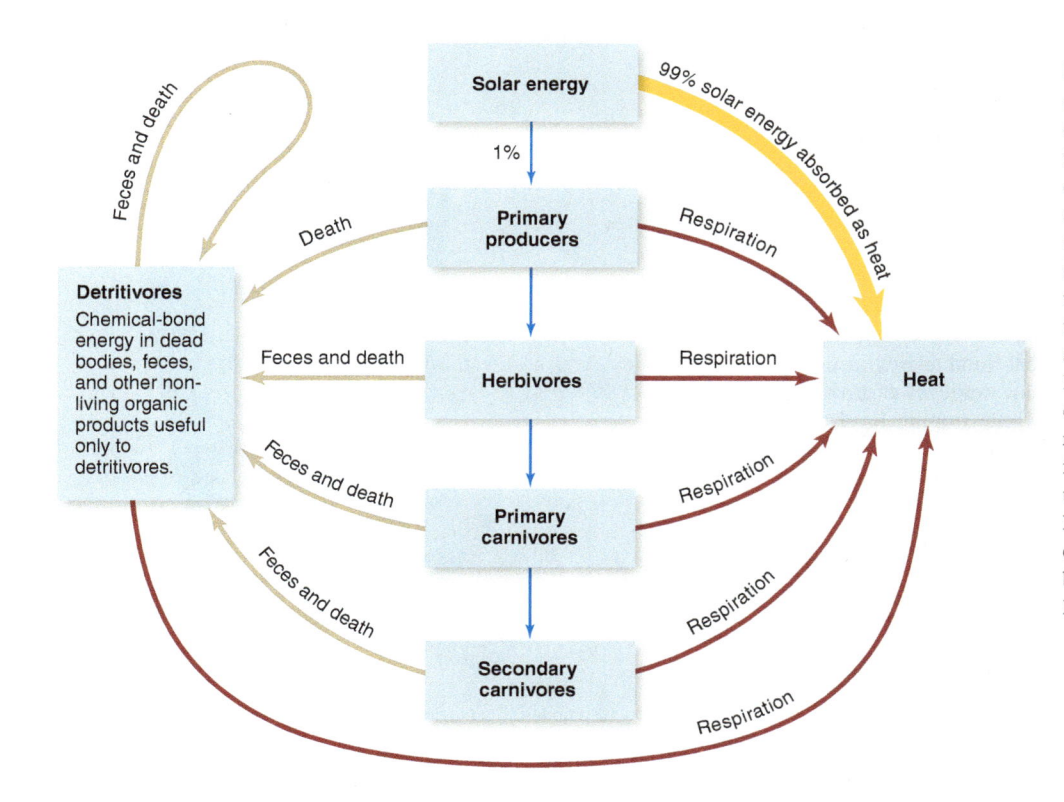

Figure 57.10 The flow of energy through an ecosystem. Blue arrows represent the flow of energy that enters the ecosystem as light and is then passed along as chemical-bond energy to successive trophic levels. At each step energy is diverted, meaning that the chemical-bond energy available to each trophic level is less than that available to the preceding trophic level. Red arrows represent diversions of energy into heat. Tan arrows represent diversions of energy into feces and other organic materials useful only to the detritivores. Detritivores may be eaten by carnivores, so some of the chemical-bond energy returns to higher trophic levels.

then to various levels of carnivores. To envision this critical point, assume for simplicity that the primary producers in an ecosystem gain 1000 units of chemical-bond energy over a period of time. If the energy input to each trophic level is 10% of the input to the preceding level, then the input of chemical-bond energy to the herbivore trophic level is 100 units, to the primary carnivores, 10 units, and to the secondary carnivores, 1 unit over the same period of time.

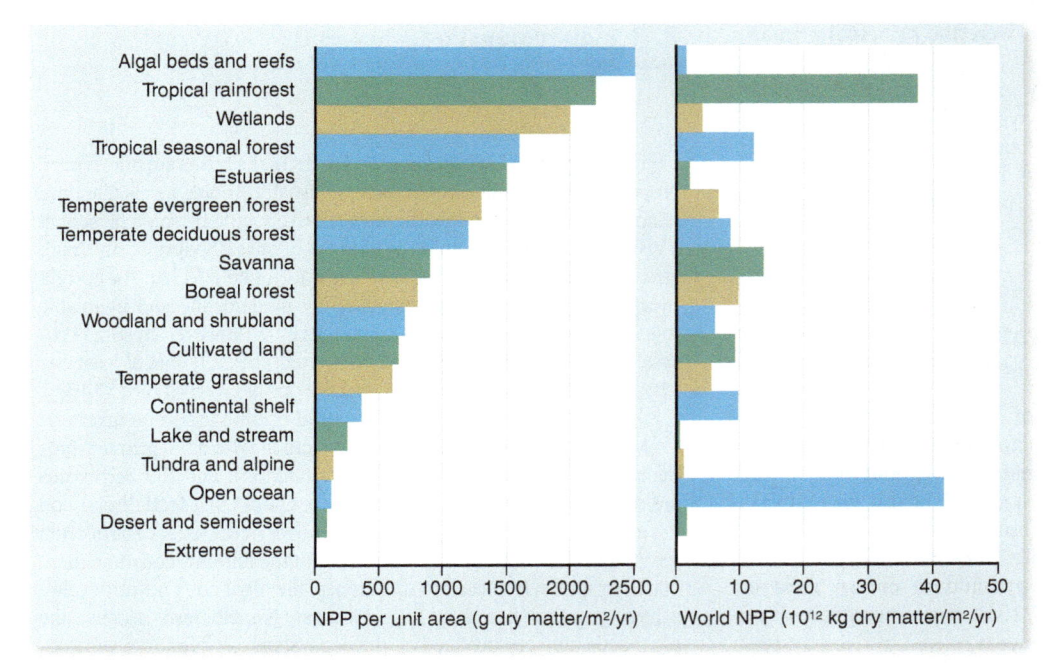

Figure 57.11 Ecosystem productivity per year. The first column of data shows the average net primary productivity (NPP) per square meter per year. The second column of data factors in the area covered by the ecosystem type; it is the product of the productivity per square meter per year multiplied by the number of square meters occupied by the ecosystem type worldwide. Note that an ecosystem type that is very productive on a square-meter basis may not contribute much to global productivity if it is an uncommon type, such as wetlands. On the other hand, a very widespread ecosystem type, such as the open ocean, can contribute greatly to global productivity even if its productivity per square meter is low.

Data analysis Is there a relationship among the habitat types between the NPP per unit area and the World NPP?

Limits on top carnivores

The exponential decline of chemical-bond energy in a trophic chain limits the lengths of trophic chains and the numbers of top carnivores an ecosystem can support. According to our model calculations, if an ecosystem includes secondary carnivores, only about one-thousandth of the energy captured by photosynthesis passes all the way through the series of trophic levels to reach these animals as usable chemical-bond energy. Tertiary carnivores would receive only one ten-thousandth. This helps explain why no predators subsist solely on eagles or lions.

The decline of available chemical-bond energy also helps explain why the numbers of individual top-level carnivores in an ecosystem tend to be low. The whole trophic level of top carnivores receives relatively little energy, and yet such carnivores tend to be big: They have relatively large individual body sizes and great individual energy needs. Because of these two factors, the population numbers of top predators tend to be small.

The longest trophic chains probably occur in the oceans. Some tunas and other top-level ocean predators probably function as third- and fourth-level carnivores at times. The challenge of explaining such long trophic chains is obvious, but the solutions are not well understood presently.

Humans as consumers: A case study

The flow of energy in Cayuga Lake in upstate New York (figure 57.12) helps illustrate how the energetics of trophic levels can affect the human food supply. Researchers calculated from the actual properties of this ecosystem that about 150 of each 1000 calories of chemical-bond energy captured by primary producers in the lake were transferred into the bodies of herbivores. Of these calories, about 30 were transferred into the bodies of smelt, small fish that were the principal primary carnivores in the system.

If humans ate the smelt, they gained about 6 of the 1000 calories that originally entered the system. If trout ate the smelt and humans ate the trout, the humans gained only about 1.2 calories. For human populations in general, more energy is available if plants or other primary producers are eaten than if animals are eaten—and more energy is available if herbivores rather than carnivores are consumed.

Ecological pyramids illustrate the relationship of trophic levels

Imagine that the trophic levels of an ecosystem are represented as boxes stacked on top of each other. Imagine also that the width of each box is proportional to the productivity of the trophic level it represents. The stack of boxes will always have the shape of a pyramid; each box is narrower than the one under it because of the inviolable rules of energy flow. A diagram of this sort is called a pyramid of energy flow or pyramid of productivity (figure 57.13a). It is an example of an ecological pyramid.

There are several types of ecological pyramids. Pyramid diagrams can be used to represent standing crop biomass (that is, the biomass of all individuals alive at the same time) or numbers of individuals, as well as productivity.

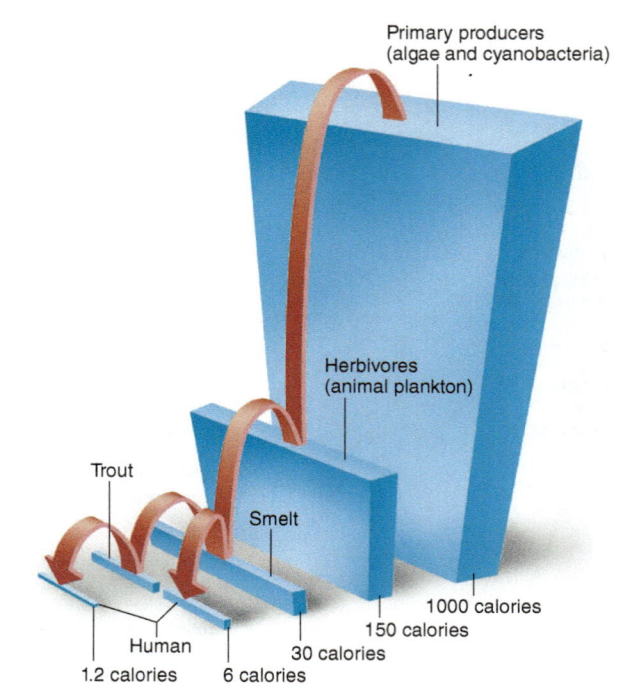

Figure 57.12 Flow of energy through the trophic levels of Cayuga Lake. Autotrophic plankton (algae and cyanobacteria) fix the energy of the Sun, the herbivores (animal plankton) feed on them, and both are consumed by smelt. The smelt are eaten by trout. The amount of fish flesh produced per unit time for human consumption is at least five times greater if people eat smelt rather than trout, but people typically prefer to eat trout.

? Inquiry question Why does it take so many calories of algae to support so few calories of humans?

In a **pyramid of biomass,** the widths of the boxes are drawn to be proportional to standing crop biomass. Usually, trophic levels that have relatively low productivity also have relatively little biomass present at a given time. Thus, pyramids of biomass are usually upright, meaning each box is narrower than the one below it (figure 57.13b). An upright pyramid of biomass is not mandated by fundamental and inviolable rules like an upright pyramid of productivity is, however. In some ecosystems, the pyramid of biomass is inverted, meaning that at least one trophic level has greater biomass than the one below it (figure 57.13c).

How is it possible for the pyramid of biomass to be inverted? Consider a common sort of aquatic system in which the primary producers are single-celled algae (phytoplankton), and the herbivores are rice grain-sized animals (such as copepods) that feed directly on the algal cells. In such a system, the turnover of the algal cells is often very rapid: The cells multiply rapidly, but the animals consume them equally rapidly. In these circumstances, the algal cells never develop a large population size or large biomass. Nonetheless, because the algal cells are very productive, the ecosystem can support a substantial biomass of the animals, a biomass larger than that ever observed in the algal population. In other words, even though the productivity of the algae is much higher than that of the copepods, the biomass at any point in time of the copepods is greater than that of the algae.

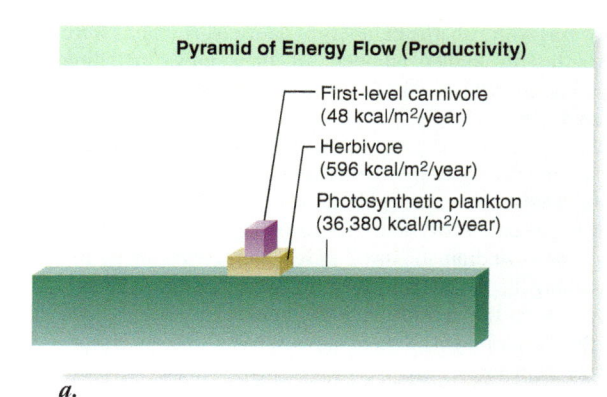

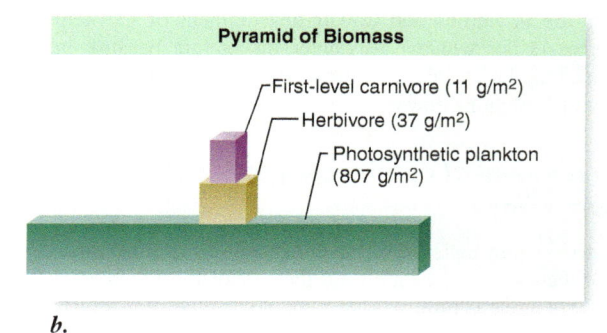

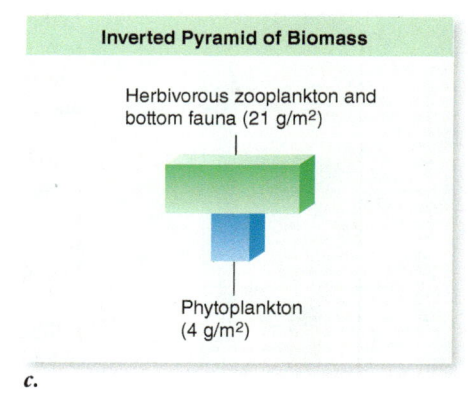

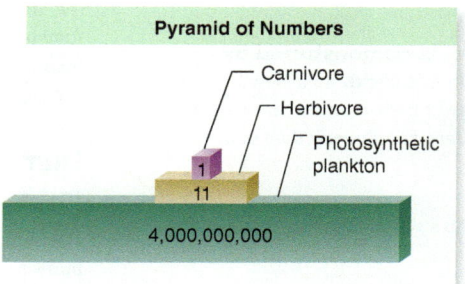

Figure 57.13 Ecological pyramids. In an ecological pyramid, successive trophic levels in an ecosystem are represented as stacked boxes, and the widths of the boxes represent the magnitude of an ecological property in the various trophic levels. Ecological pyramids can represent several different properties. *a.* Pyramid of energy flow (productivity). *b.* Pyramid of biomass of the ordinary type. *c.* Inverted pyramid of biomass. *d.* Pyramid of numbers.

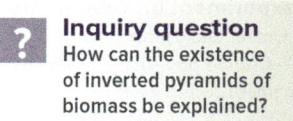

? Inquiry question
How can the existence of inverted pyramids of biomass be explained?

In a pyramid of numbers, the widths of the boxes are proportional to the numbers of individuals present in the various trophic levels (figure 57.13d). Such pyramids are usually, but not always, upright.

Learning Outcomes Review 57.2

Trophic levels in an ecosystem include primary producers, herbivores, primary carnivores, and secondary carnivores. Detritivores consume dead or waste matter from all levels. As energy passes from one level to another, some is inevitably lost as heat, which cannot be reclaimed. Photosynthetic primary producers capture about 1% of solar energy as chemical-bond energy. As this energy is passed through the other trophic levels, some is diverted at each step into heat, feces, and dead matter; only about 10% is available to the next level.

- Describe the different ways that matter, such as carbon atoms, and energy move through ecosystems?

57.3 Trophic-Level Interactions

Learning Outcomes

1. Explain the meaning of trophic cascade.
2. Distinguish between top-down and bottom-up effects.

The existence of food chains creates the possibility that species in any one trophic level may have effects on more than one trophic level.

Primary carnivores, for example, may have effects not only on the animals they eat, but also, indirectly, on the plants or algae eaten by their prey. The process by which effects exerted at an upper trophic level flow down to influence two or more lower levels is termed a **trophic cascade**. The effects themselves are called **top-down effects**.

Conversely, increases in primary productivity may provide more food not just to herbivores, but also, indirectly, to carnivores. When an effect flows up through a trophic chain, such as from primary producers to higher trophic levels, it is termed a **bottom-up effect**.

Top-down effects occur when changes in the top trophic level affect primary producers

The existence of top-down effects has been confirmed by controlled experiments in some types of ecosystems, particularly freshwater ones. For example, in one study, sections of a stream were enclosed with a mesh that prevented fish from entering. Brown trout—predators on invertebrates—were added to some enclosures but not others. After 10 days, the numbers of invertebrates in the enclosures with trout were only two-thirds as great as the numbers in the no-fish enclosures (figure 57.14). In turn, the biomass of algae, which the invertebrates ate, was five times greater in the trout enclosures than the no-fish ones.

The logic of the trophic cascade just described leads to the expectation that if secondary carnivores are added to enclosures, they would also cause cascading effects. The secondary carnivores would be predicted to keep populations of primary carnivores in check, which would lead to a profusion of herbivores and a scarcity of primary producers.

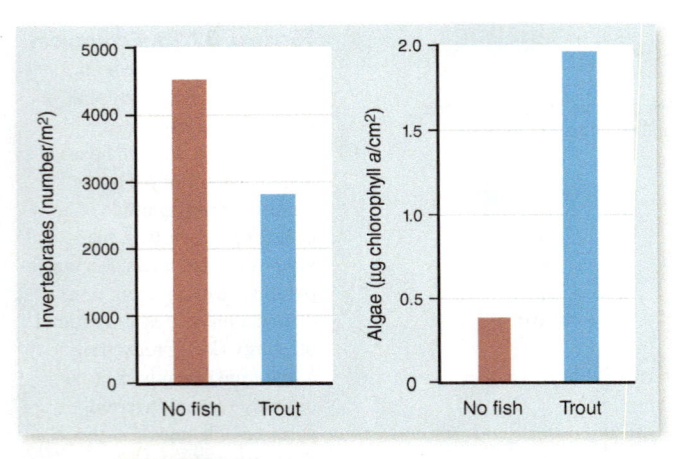

Figure 57.14 Top-down effects demonstrated by experiment in a simple trophic cascade. In a New Zealand stream, enclosures with trout had fewer herbivorous invertebrates *(left-hand panel)* and more algae *(right-hand panel)* than ones without trout.

? Inquiry question Why do streams with trout have more algae?

In an experiment similar to the one just described, enclosures were created in free-flowing streams in northern California. In these streams, the principal primary carnivores were damselfly larvae (termed nymphs). Fish that preyed on the nymphs and on other primary carnivores were added to some enclosures but not others.

In the enclosures with fish, the numbers of damselfly nymphs were reduced, leading to higher numbers of their prey, including herbivorous insects, which led in turn to a decreased biomass of algae (figure 57.15).

Trophic cascades in large-scale ecosystems are not as easy to verify by experiment as ones in stream enclosures, and the workings of such cascades are not thoroughly known. Nonetheless, certain cascades in large-scale ecosystems are recognized by most ecologists. One of the most dramatic involves sea otters, sea urchins, and kelp forests along the West Coast of North America (figure 57.16).

The otters eat the urchins, and the urchins eat young kelps, inhibiting the development of kelp forests. When the otters are abundant, the kelp forests are well developed because there are relatively few urchins in the system. But when the otters are sparse, the urchins are numerous and impair development of the kelp forests. Orcas (killer whales) also enter the picture because in recent years they have started to prey intensively on the otters, driving otter populations down.

Human removal of carnivores produces top-down effects

Human activities are believed to have had top-down effects in a number of ecosystems, usually by the removal of top-level carnivores. The great naturalist Aldo Leopold posited such effects long before the trophic cascade hypothesis had been scientifically articulated when he wrote in *Sand County Almanac*:

"I have lived to see state after state extirpate its wolves. I have watched the face of many a new wolfless mountain, and seen the south-facing slopes wrinkle with a maze of new deer trails.

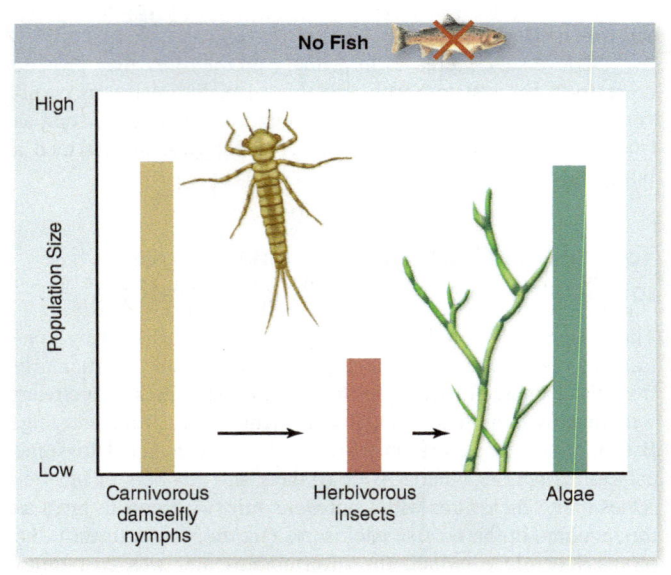

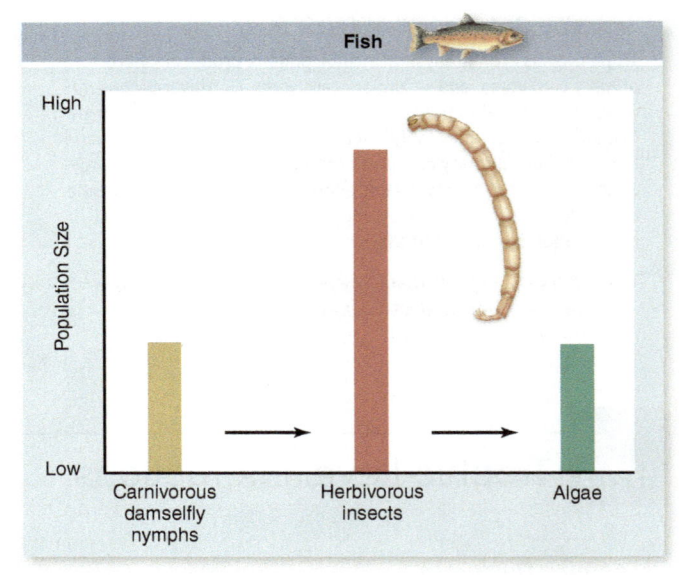

Figure 57.15 Top-down effects demonstrated by an experiment in a four-level trophic cascade. Stream enclosures with large, carnivorous fish *(on right)* have fewer primary carnivores, such as damselfly nymphs, more herbivorous insects (exemplified here by the number of chironomids, a type of aquatic insect), and lower levels of algae.

? Inquiry question Explain why the presence of fish leads to less algae in this experiment, and more algae in the experiment in figure 57.14. What might be the effect if snakes that prey on fish were added to the enclosures?

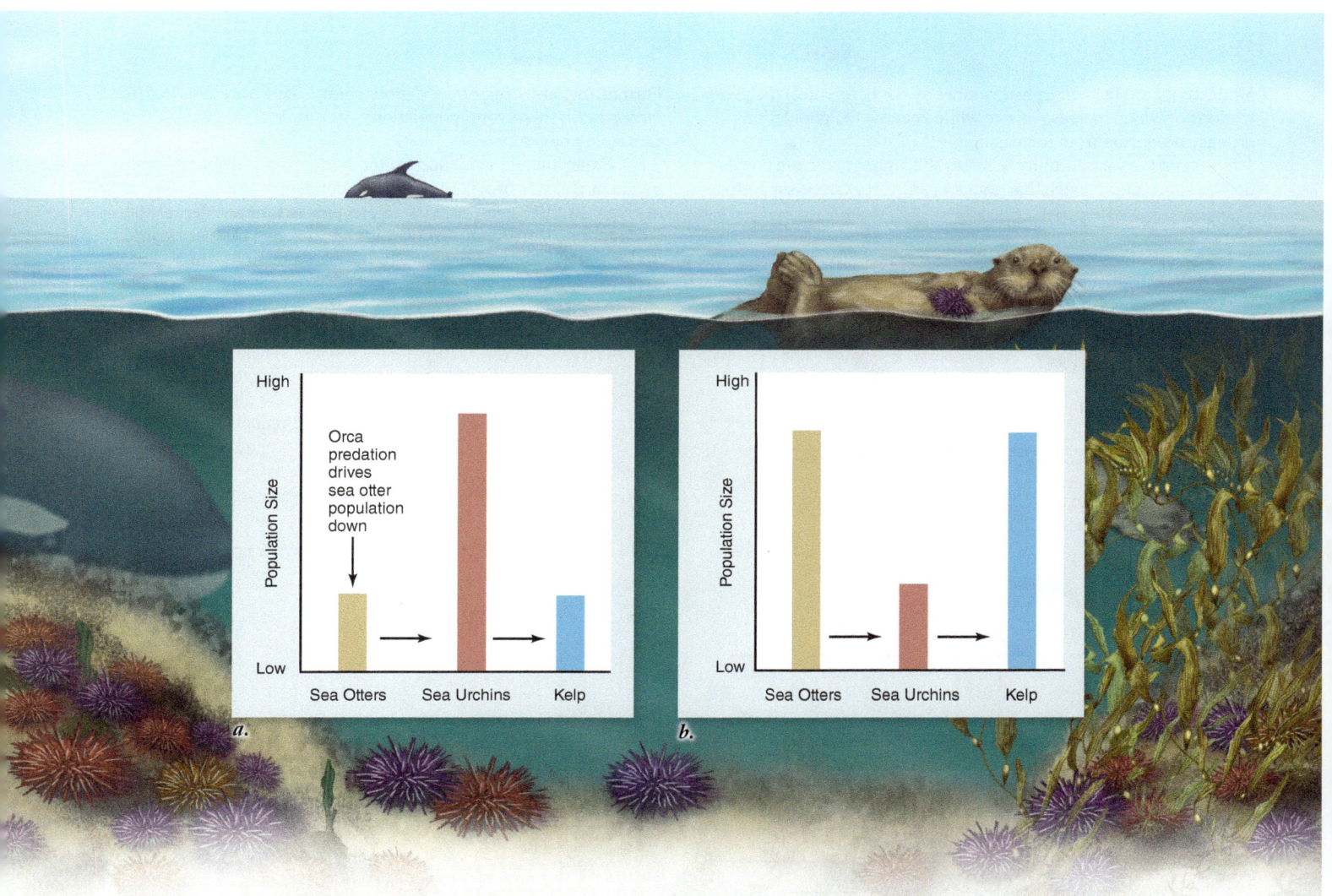

Figure 57.16 A trophic cascade in a large-scale ecosystem. Along the West Coast of North America, the sea otter/sea urchin/kelp system exists in two states: In the state shown in panel (*a*) low populations of sea otters permit high populations of urchins, which suppress kelp populations; in the state shown in panel (*b*) high populations of otters keep urchins in check, permitting profuse kelp growth. According to a recent hypothesis, a switch of orcas to preying on otters rather than other mammals is leading the ecosystem today to be mostly in the state represented on the left.

I have seen every edible bush and seedling browsed, first to anemic desuetude, and then to death. I have seen every edible tree defoliated to the height of a saddle horn."

Many similar examples exist in which the removal of predators has led to cascading effects on lower trophic levels. Large predators such as jaguars and mountain lions are absent on Barro Colorado Island, a hilltop turned into an island by the construction of the Panama Canal at the beginning of the last century. As a result, smaller predators whose populations are normally held in check—including monkeys, peccaries (a relative of the pig), coatimundis, and armadillos—have become extraordinarily abundant. These animals eat almost anything they find. Ground-nesting birds are particularly vulnerable, and many species have declined; at least 15 bird species have vanished from the island entirely.

Similarly, in the world's oceans, large predatory fish such as billfish and cod have been reduced by overfishing to an average of 10% of their previous numbers in virtually all parts of the world's oceans. In some regions, the prey of cod—such as certain shrimp and crabs—have become many times more abundant than they were before, and further cascading effects are evident at still lower trophic levels.

Bottom-up effects occur when changes to primary producers affect higher trophic levels

Bottom-up effects occur when enhanced primary productivity flows upward through the food chain, increasing food for not only herbivores, but also for the carnivores that eat them. In predicting bottom-up effects, ecologists must take account of the life histories of the organisms present. A model of bottom-up effects thought to apply to a number of types of ecosystems is diagrammed in figure 57.17.

According to the model, when primary productivity is low, producer populations cannot support significant herbivore populations. As primary productivity increases, herbivore populations become a feature of the ecosystem. Increases in primary productivity are then entirely devoured by the herbivores, the populations of which increase in size while keeping the populations of primary producers from increasing.

As primary productivity becomes still higher, herbivore populations become large enough that primary carnivores can be supported.

Further increases in primary productivity then does not lead to increases in herbivore populations, but rather to increases in carnivore populations.

Experimental evidence for the bottom-up effects predicted by the model was provided by a study conducted in enclosures on a river (figure 57.18). The enclosures excluded large fish (secondary carnivores). A roof was placed above each enclosure. Some roofs were clear, whereas others were tinted to various degrees, so that the enclosures differed in the amount of sunlight entering them.

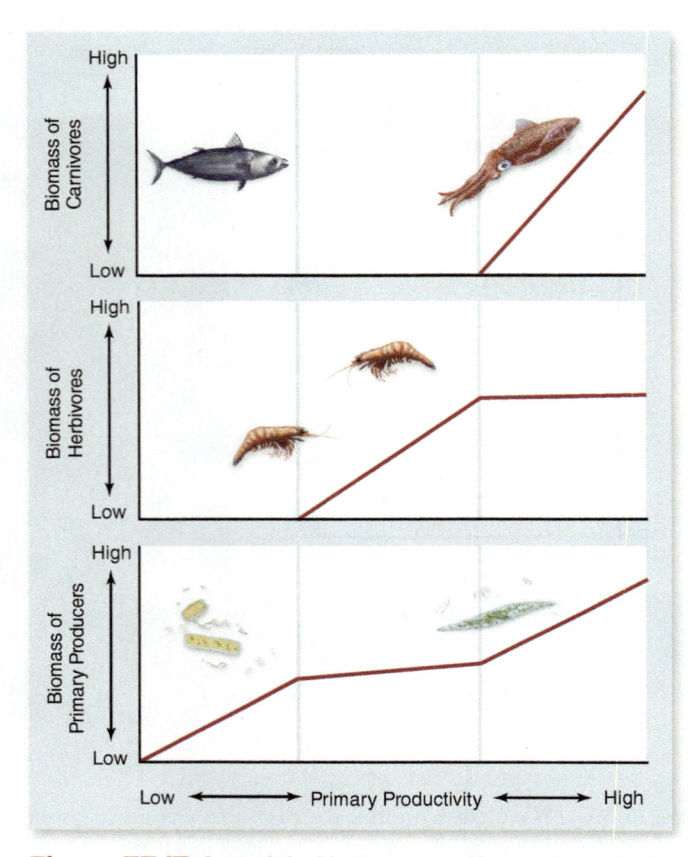

Figure 57.17 A model of bottom-up effects. At low levels of primary productivity, herbivore populations cannot obtain enough food to be maintained; without herbivory, the standing crop biomass of the primary producers such as these diatoms increases as their productivity increases. Above some threshold, increases in primary productivity lead to increases in herbivore populations and herbivore biomass; the biomass of the primary producers then does not increase as primary productivity increases because the increasing productivity is cropped by the herbivores. Above another threshold, populations of primary carnivores can be sustained. As primary productivity increases above this threshold, the carnivores consume the increasing productivity of the herbivores, so the biomass of the herbivore populations remains relatively constant while the biomass of the carnivore populations increases. The biomass of the primary producers is no longer constrained by increases in the herbivore populations and thus also increases with increasing primary productivity. A key to understanding the model is to maintain a distinction between the concepts of productivity and standing crop biomass.

Inquiry question How is it possible for the biomass of the primary producers to stay relatively constant as the primary productivity increases?

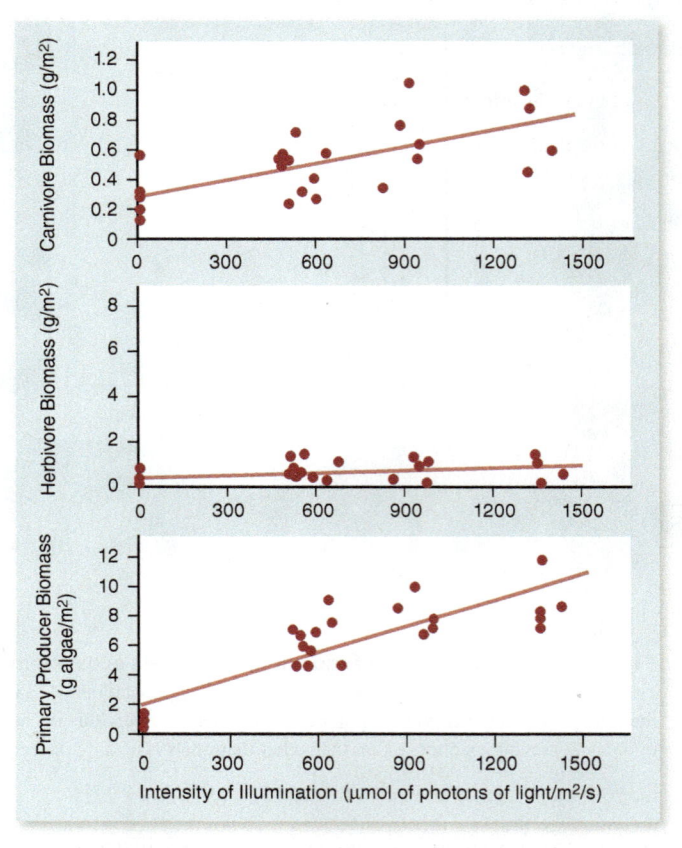

Figure 57.18 An experimental study of bottom-up effects in a river ecosystem. This system, studied on the Eel River in northern California, exhibited the patterns modeled by the red graphs of figure 57.17. Increases in the intensity of illumination led to increases in primary productivity and in the biomass of the primary producers. The biomass of the carnivore populations also increased. However, herbivore biomass did not increase much with increasing primary productivity because increases in herbivore productivity were consumed by the carnivores.

Data analysis What is the significance of the angle of the regression line in the three graphs?

Inquiry question Suppose secondary carnivores existed at all light levels in this experimental ecosystem. How would you expect the biomass of primary producers, herbivores, and primary carnivores to change with increasing light levels?

The primary productivity was highest in the enclosures with clear roofs and lowest in the ones with darkly tinted roofs. As primary productivity increased in parallel with illumination, the biomass of the primary producers increased, as did the biomass of the carnivores. However, the biomass of the trophic level sandwiched in between, the herbivores, did not increase much, as predicted by the model in figure 57.17 (see *red graph lines*).

Learning Outcomes Review 57.3

Populations of species at different trophic levels affect one another, and these effects can propagate through the levels. Top-down effects, termed trophic cascades, are observed when changes in carnivore populations affect lower trophic levels. Bottom-up effects are observed when changes in primary productivity affect the higher trophic levels.

- Could top-down and bottom-up effects occur simultaneously?

57.4 Biodiversity and Ecosystem Stability

Learning Outcomes

1. Define ecosystem stability.
2. Describe the effects of species richness on ecosystem function.
3. Name possible factors that contribute to species richness in the tropics.

In chapter 56, we discussed *species richness*—the number of species present in a community. Ecologists have long debated the consequences of differences in species richness between communities. One theory is that species-rich communities are more stable—that is, more constant in composition and better able to resist disturbance. This hypothesis has been elegantly studied by David Tilman and colleagues at the University of Minnesota's Cedar Creek Natural History Area.

Species richness may increase stability: The Cedar Creek studies

Workers monitored 207 small rectangular plots of land (8 to 16 m²) for 11 years (figure 57.19*a*). In each plot, they counted the number of prairie plant species and measured the total amount of plant biomass (that is, the mass of all plants on the plot). Over the course of the study, plant species richness was related to community stability—plots with more species showed less year-to-year variation in biomass. Moreover, in two drought years, the decline in biomass was negatively related to species richness—that is, plots with more species were less affected by drought.

These findings were subsequently confirmed by an experiment in which plots were seeded with different numbers

SCIENTIFIC THINKING

Question: *Does species richness affect the invasibility of a community?*

Hypothesis: *The rate of successful invasion will be lower in communities with greater richness.*

Experiment: *Add seeds from the same number of non-native plants to experimental plots that differ in the number of plant species.*

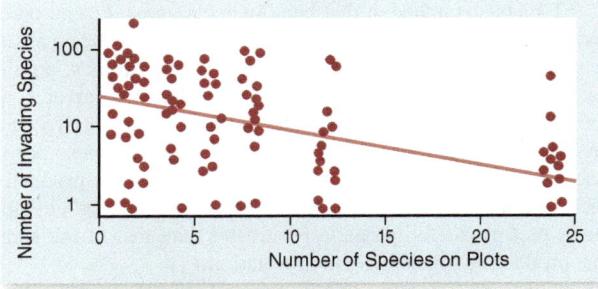

a.

b.

Result: *Although the number of successful invasive species is highly variable, more species-rich plots on average are invaded by fewer species.*

Interpretation: *What might explain why so much variation exists in the number of successful invading species in communities with the same species richness?*

Figure 57.19 Effect of species richness on ecosystem stability. *a.* One of the Cedar Creek experimental plots. *b.* Community stability can be assessed by looking at the effect of species richness on community invasibility. Each dot represents data from one experimental plot in the Cedar Creek experimental fields. Plots with more species are harder to invade by nonnative species.

Data analysis How well can you predict the number of invading species from knowing the number of species on a plot?

of species. Again, more species-rich plots had greater year-to-year stability in biomass over a 10-year period.

In a related experiment, when seeds of other plant species were added to different plots, the ability of these species to become established was negatively related to species richness (figure 57.19*b*). More diverse communities, in other words, are more resistant to invasion by new species, which is another measure of community stability.

Species richness may also affect other ecosystem processes. Tilman and colleagues monitored 147 experimental plots that varied in number of species to estimate how much growth was occurring and how much nitrogen the growing plants were taking up from the soil. They found that the more species a plot had, the greater the nitrogen uptake and total amount of biomass produced. In his study, increased biodiversity clearly appeared to lead to greater productivity.

Laboratory studies on artificial ecosystems have provided similar results. In one elaborate study, ecosystems covering 1 m² were constructed in laboratory growth chambers that controlled temperature, light levels, air currents, and atmospheric gas concentrations. A variety of plants, insects, and other animals were introduced to construct ecosystems composed of 9, 15, or 31 species, with the lower diversity treatments containing a subset of the species in the higher diversity enclosures. As with Tilman's experiments, the amount of biomass produced was related to species richness, as was the amount of carbon dioxide consumed, another measure of the productivity of the ecosystem.

Tilman's conclusion that healthy ecosystems depend on diversity is not accepted by all ecologists, however. Critics question the validity and relevance of these biodiversity studies, arguing that the more species are added to a plot, the greater the probability that one species will be highly productive. To show that high productivity results from high species richness per se, rather than from the presence of particular highly productive species, experimental plots have to exhibit "overyielding"; in other words, plot productivity has to be greater than that of the single most productive species grown in isolation.

Although this point is still debated, recent work at Cedar Creek and elsewhere has provided evidence of overyielding, supporting the claim that species richness of communities enhances community productivity and stability.

Species richness is influenced by ecosystem characteristics

A number of factors are known or hypothesized to affect species richness in a community. We discussed some in chapter 56, such as loss of keystone species and moderate physical disturbance. Here we discuss three more: primary productivity, habitat heterogeneity, and climatic factors.

Primary productivity

Ecosystems differ substantially in primary productivity (see figure 57.11). Some evidence indicates that species richness is related to primary productivity, but the relationship between them is not linear. In a number of cases, for example, ecosystems with intermediate levels of productivity tend to have the greatest number of species (figure 57.20a).

Why this is so is debated. One possibility is that levels of productivity are linked with numbers of consumers. Applying this concept to plant species richness, the argument is that at low productivity, there are few herbivores, and superior competitors among the plants are able to eliminate most other plant species. In contrast, at high productivity so many herbivores are present that only the plant species most resistant to grazing survive, reducing species diversity. As a result, the greatest numbers of plant species coexist at intermediate levels of productivity and herbivory.

Habitat heterogeneity

Spatially heterogeneous abiotic environments are those that consist of many habitat types—such as soil types, for example. These heterogeneous environments can be expected to accommodate more species of plants than spatially homogeneous environments. What's more, the species richness of animals can be expected to reflect the species richness of plants present. An example of this latter effect is seen in figure 57.20b: The number of lizard species

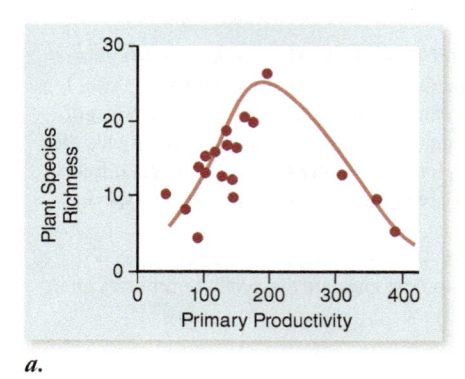

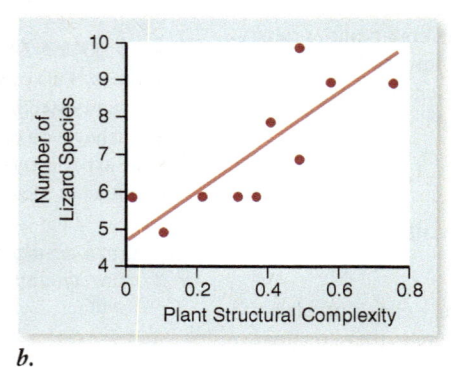

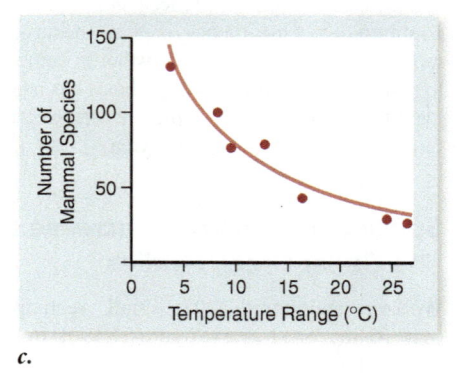

Figure 57.20 Factors that affect species richness. *a. Productivity:* In plant communities of mountainous areas of South Africa, species richness of plants peaks at intermediate levels of productivity (biomass). *b. Spatial heterogeneity:* The species richness of desert lizards is positively correlated with the structural complexity of the plant cover in desert sites in the American Southwest. *c. Climate:* The species richness of mammals is inversely correlated with monthly mean temperature range along the West Coast of North America.

Inquiry question (a) Why is species richness greatest at intermediate levels of productivity? (b) Why do more structurally complex areas have more species? (c) Why do areas with less variation in temperature have more species?

at various sites in the American Southwest mirrors the local structural diversity of the plants.

Climatic factors

The role of climatic factors is more difficult to predict. On the one hand, more species might be expected to coexist in a seasonal environment than in a constant one because a changing climate may favor different species at different times of the year. On the other hand, stable environments are able to support specialized species that would be unable to survive where conditions fluctuate. The number of mammal species at locations along the West Coast of North America is inversely correlated with the amount of local temperature variation—the wider the variation, the fewer mammalian species—supporting the latter line of argument (figure 57.20c).

Tropical regions have the highest diversity, although the reasons are unclear

Biologists have long recognized that more different kinds of animals and plants inhabit the tropics than the temperate regions. For many types of organisms, there is a steady increase in species richness from the arctic to the tropics. This species diversity gradient in numbers of species correlated with latitude has been reported for plants and animals, including birds (figure 57.21), mammals, and reptiles.

For the better part of a century, ecologists have puzzled over the species diversity gradient from the arctic to the tropics. The difficulty has not been in forming a reasonable hypothesis of why more species exist in the tropics, but rather in sorting through these many reasonable hypotheses. Here, we consider five of the most commonly discussed suggestions.

Evolutionary age of tropical regions

Scientists have frequently proposed that the tropics have more species than temperate regions because the tropics have existed over long, uninterrupted periods of evolutionary time, whereas temperate regions have been subject to repeated glaciations. The greater age of tropical communities would have allowed complex population interactions to coevolve within them, fostering a greater variety of plants and animals.

Recent work suggests that the long-term stability of tropical communities has been greatly exaggerated, however. An examination of pollen within soil cores reveals that during glaciations, the tropical forests contracted to a few small refuges surrounded by grassland. This suggests that tropical communities have not been stable over long time periods, negating the premise of this hypothesis.

On the other hand, recent phylogenetic studies (see chapter 23) have suggested that many plants and animals evolved in the tropics and then colonized temperate areas. As a result, these groups have had a longer presence in the tropics, and thus their greater species richness there may just reflect the longer period over which diversification has occurred.

Increased productivity

A second often-advanced hypothesis is that the tropics contain more species because this part of the Earth receives more solar radiation than do temperate regions. The argument is that more solar energy, coupled to a year-round growing season, greatly increases the overall photosynthetic activity of tropical plants.

If we visualize the tropical forest's total resources as a pie, and its species niches as slices of the pie, we can see that a larger pie accommodates more slices. But as noted earlier in this section, many field studies have indicated that species richness is highest at intermediate levels of productivity. Accordingly, increasing productivity would be expected to lead to lower, not higher, species richness.

Stability/constancy of conditions

Seasonal variation, though it does exist in the tropics, is generally substantially less than in temperate areas. This reduced seasonality might encourage specialization, with niches subdivided to partition resources and so avoid competition. The expected result would be a larger number of more specialized species in the tropics, which is what we see. Many field tests of this hypothesis have been carried out, and almost all support it, reporting larger numbers of narrower niches in tropical communities than in temperate areas.

Predation

Many reports indicate that predation may be more intense in the tropics. In theory, more intense predation could reduce the importance of competition, permitting greater niche overlap and thus promoting greater species richness.

Spatial heterogeneity

As noted earlier in this section, spatial heterogeneity promotes species richness. Tropical forests, by virtue of their complexity, create a variety of microhabitats and so may foster larger numbers

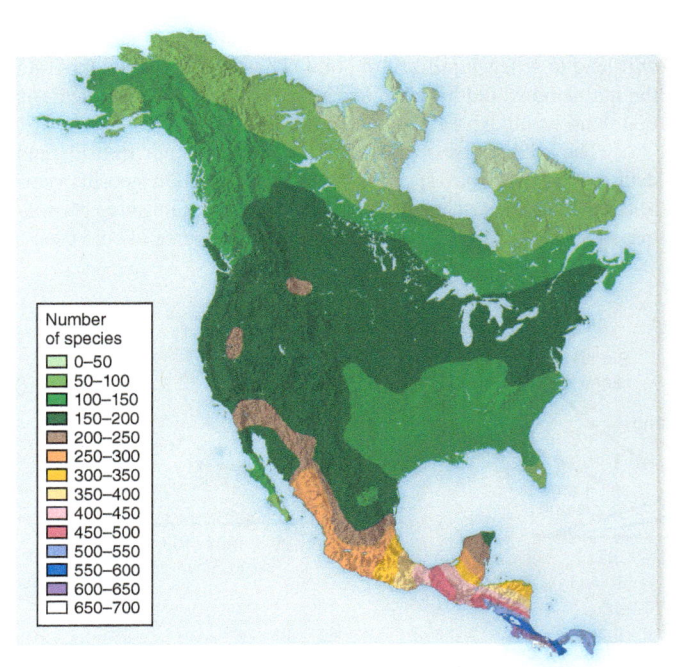

Figure 57.21 A latitudinal gradient in species richness. Among North and Central American birds, a marked increase in the number of species occurs moving toward the tropics. Fewer than 100 species are found at arctic latitudes, but more than 600 species live in southern Central America.

of species. Perhaps the long vertical column of vegetation through which light passes in a tropical forest produces a wide range of light frequencies and intensities, creating a greater variety of light environments and so promoting species diversity.

Learning Outcomes Review 57.4

An ecosystem is stable if it remains relatively constant in composition and is able to resist disturbance. Experimental field studies support the conclusion that species-rich communities are better able to resist invasion by new species, as well as have increased biomass production at the primary level, although not all ecologists agree with these conclusions. Species richness is greatest in the tropics, and the reasons may include habitat variation, increased sunlight, and long-term climate and seasonal stability.

- What might be the effects on primary productivity if air pollution decreased the amount of sunlight reaching Earth's surface?

57.5 Island Biogeography

Learning Outcomes

1. Describe the species–area relationship.
2. Explain how area and isolation affect rates of colonization and extinction.

One of the most reliable patterns in ecology is the observation that larger islands contain more species than do smaller islands. In 1967, Robert MacArthur of Princeton University and Edward O. Wilson of Harvard University proposed that this species–area relationship was a result of the effect of geographic area and isolation on the likelihood of species extinction and colonization.

The equilibrium model proposes that extinction and colonization reach a balance point

MacArthur and Wilson reasoned that species are constantly being dispersed to islands, so islands have a tendency to accumulate more and more species. At the same time that new species are added, however, other species are lost by extinction. As the number of species on an initially empty island increases, the rate of colonization must decrease as the pool of potential colonizing species not already present on the island becomes depleted. At the same time, the rate of extinction should increase—the more species on an island, the greater the likelihood that any given species will perish.

As a result, at some point the number of extinctions and colonizations should be equal, and the number of species should then remain constant. Every island of a given size, then, has a characteristic equilibrium number of species that tends to persist through time (the intersection point in figure 57.22a)—though the species composition will change as some species become extinct and new species colonize.

MacArthur and Wilson's equilibrium model proposes that island species richness is a dynamic equilibrium between colonization and extinction. Both island size and distance from the mainland would affect colonization and extinction. We would expect smaller islands to have higher rates of extinction because their population sizes would, on average, be smaller. This is reflected in a higher curve charting the relationship between number of species and extinction rate for smaller islands (figure 57.22b). Also, we would expect fewer colonizers to reach islands that lie farther from the mainland, leading to a lower curve for the relationship between number of species and colonization rate for more distant islands (figure 57.22b).

The equilibrium species richness occurs when the two colonization and extinction curves intersect—changing either curve will lead to a new equilibrium point. Thus, small islands far from the mainland would have the fewest species; large islands near the mainland would have the most (figure 57.22b).

The predictions of this simple model bear out well in field data. Asian Pacific bird species (figure 57.22c) exhibit a positive correlation of species richness with island size, but a negative correlation of species richness with distance from the source of colonists.

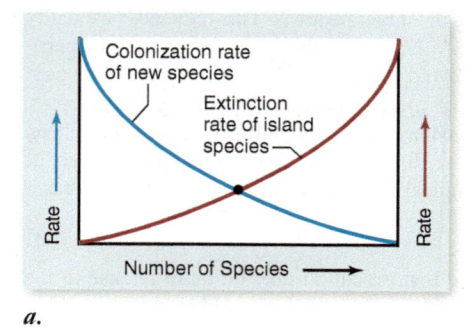

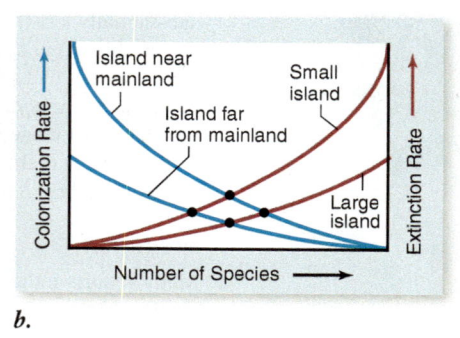

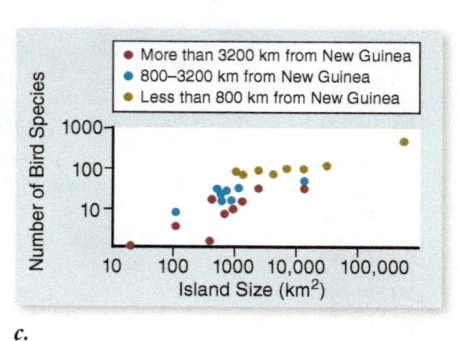

Figure 57.22 The equilibrium model of island biogeography. *a.* Island species richness reaches an equilibrium *(black dot)* when the colonization rate of new species equals the extinction rate of species on the island. *b.* The equilibrium shifts depending on the rate of colonization, the size of an island, and its distance to sources of colonists. Species richness is positively correlated with island size and inversely correlated with distance from the mainland. Smaller islands have higher extinction rates, shifting the equilibrium point to the left. Similarly, more distant islands have lower colonization rates, again shifting the equilibrium point leftward. *c.* The effect of distance from a larger island, which can be the source of colonizing species, is readily apparent. More distant islands have fewer bird species than do nearer islands of the same size.

The equilibrium model is still being tested

Wilson and Dan Simberloff, then a graduate student, performed initial studies in the mid-1960s on small mangrove islands in the Florida Keys. These islands were censused, cleared of animal life by fumigation, and then allowed to recolonize, with censuses being performed at regular intervals. These and other such field studies have tended to support the equilibrium model.

Long-term experimental field studies, however, are suggesting that the situation is more complicated than MacArthur and Wilson envisioned. Their model predicts a high level of species turnover as some species perish and others arrive. But studies of island birds and spiders indicate that very little turnover occurs from year to year. Those species that do come and go, moreover, comprise a subset of species that never attain high populations. A substantial proportion of the species appear to maintain high populations and rarely go extinct.

These studies have been going on for a relatively short period of time. It is possible that over periods of centuries, the equilibrium model is a good description of what determines island species richness.

Learning Outcomes Review 57.5

The species–area relationship is an observation that an island of larger area contains more species. Species richness on islands appears to be a dynamic equilibrium between colonization and extinction. Distance from a mainland also affects the rates of colonization and extinction, and therefore fewer species would be found on small, isolated islands far from a mainland.

- Under what circumstances would a smaller island be expected to have more species than a larger island?

Chapter Review

57.1 Biogeochemical Cycles

The atomic constituents of matter cycle within ecosystems.
The atoms of chemical elements move through ecosystems in biogeochemical cycles.

Carbon, the basis of organic compounds, cycles through most ecosystems.
The carbon cycle usually involves carbon dioxide, which is fixed through photosynthesis and released by respiration. Carbon is also present as bicarbonate ions and as methane. Burning of fossil fuels has created an imbalance in the carbon cycle (figure 57.1).

The availability of water is fundamental to terrestrial ecosystems (figure 57.2).
Water enters the atmosphere via evaporation and transpiration and returns to the Earth's surface as precipitation. It is broken down during photosynthesis and also produced during cellular respiration. Much of the Earth's water, including the groundwater in aquifers, is polluted, and human activities alter the water supply of ecosystems (figure 57.3).

The nitrogen cycle depends on nitrogen fixation by microbes.
Nitrogen is usually the element in shortest supply even though N_2 makes up 78% of the atmosphere. Nitrogen must be converted into usable forms by nitrogen-fixing microorganisms. Human use of nitrates in fertilizers has doubled the available nitrogen (figure 57.4).

Phosphorus cycles through terrestrial and aquatic ecosystems, but not the atmosphere.
Phosphorus, another limiting nutrient, is released by weathering of rocks; it flows into the oceans where it is deposited in deep-sea sediments. Humans also use phosphates as fertilizers (figure 57.5).

Limiting nutrients in ecosystems are those in short supply relative to need.
The cycle of a limiting nutrient, such as nitrogen, determines the rate at which the nutrient is made available for use.

Biogeochemical cycling in a forest ecosystem has been studied experimentally.
Ongoing experiments indicate that severe disturbance of an ecosystem results in mineral depletion and runoff of water.

57.2 The Flow of Energy in Ecosystems

Energy can neither be created nor destroyed, but changes form.
Energy exists in forms such as light, stored chemical-bond energy, motion, and heat. In any conversion, some energy is lost.

Living organisms can use many forms of energy, but not heat.
The Second Law of Thermodynamics states that whenever organisms use chemical-bond or light energy, some of it is inevitably converted to heat and cannot be retrieved.

Energy flows through trophic levels of ecosystems.
Organic compounds are synthesized by autotrophs and are utilized by both autotrophs and heterotrophs. As energy passes from organism to organism, each level is termed a trophic level, and the sequence through progressive trophic levels is called a food chain (figure 57.8).

The base trophic level includes the primary producers; herbivores that consume primary producers are the next level. They in turn are eaten by primary carnivores, which may be consumed by secondary carnivores. Detritivores feed on waste and the remains of dead organisms.

Only about 1% of the solar energy that impinges on the Earth is captured by photosynthesis. As energy moves through each trophic level, very little (approximately 10%) remains from the preceding trophic level (figure 57.10).

The number of trophic levels is limited by energy availability.
The exponential decline of energy between trophic levels limits the length of food chains and the numbers of top carnivores that can be supported.

Ecological pyramids illustrate the relationship of trophic levels.
Ecological pyramids based on energy flow, biomass, or numbers of organisms are usually upright. Inverted pyramids of biomass or numbers are possible if at least one trophic level has a greater biomass or more organisms than the level below it (figure 57.13).

57.3 Trophic-Level Interactions

Top-down effects occur when changes in the top trophic level affect primary producers.
A trophic cascade, or top-down effect, occurs when a change exerted at an upper trophic level affects a lower level (figure 57.15).

Human removal of carnivores produces top-down effects.
Removal of carnivores causes an increase in the abundance of species in lower trophic levels, such as an increase in deer populations when wolves or other predators are destroyed.

Bottom-up effects occur when changes to primary producers affect higher trophic levels.
An increase of producers may lead to the appearance or increase of herbivores; however, further increase in producers may then lead to increase in carnivores, without a comparable increase in herbivores (figure 57.17).

57.4 Biodiversity and Ecosystem Stability

Species richness may increase stability: The Cedar Creek studies.
The Cedar Creek studies indicate that higher species richness results in less year-to-year variation in biomass and in greater resistance to drought and invasion by nonnative species.

Species richness is influenced by ecosystem characteristics.
Primary production, habitat heterogeneity, and climatic factors all affect the number of species in an ecosystem (figure 57.20).

Tropical regions have the highest diversity, although the reasons are unclear.
The higher diversity of tropical regions may reflect long evolutionary time, higher productivity from increased sunlight, less seasonal variation, greater predation that reduces competition, or spatial heterogeneity (figure 57.21).

57.5 Island Biogeography

The species–area relationship reflects that larger islands contain more species than do smaller ones.

The equilibrium model proposes that extinction and colonization reach a balance point (figure 57.22).
Smaller islands have fewer species because of higher rates of extinction. Islands near a mainland have more species than distant islands because of higher rates of colonization. An equilibrium is reached when the extinction rate balances the colonization rate.

The equilibrium model is still being tested.
Long-term studies are needed to clarify all the factors involved.

Review Questions

UNDERSTAND

1. Which of the statements about groundwater is NOT accurate?
   a. In the United States, groundwater provides 50% of the population with drinking water.
   b. Groundwaters are being depleted faster than they can be recharged.
   c. Groundwaters are becoming increasingly polluted.
   d. Removal of pollutants from groundwaters is easily achieved.

2. Photosynthetic organisms
   a. fix carbon dioxide.
   b. release carbon dioxide.
   c. fix oxygen.
   d. Both a and b are correct.
   e. Both a and c are correct.

3. Some bacteria have the ability to "fix" nitrogen. This means
   a. they convert ammonia into nitrites and nitrates.
   b. they convert atmospheric nitrogen gas into biologically useful forms of nitrogen.
   c. they break down nitrogen-rich compounds and release ammonium ions.
   d. they convert nitrate into nitrogen gas.

4. Nitrogen is often a limiting nutrient in many ecosystems because
   a. there is much less nitrogen in the atmosphere than carbon.
   b. elemental nitrogen is very rapidly used by most organisms.
   c. nitrogen availability is being reduced by pollution due to fertilizer use.
   d. most organisms cannot use nitrogen in its elemental form.

5. Which of the following statements about the phosphorus cycle is correct?
   a. Phosphorus is fixed by plants and algae.
   b. Most phosphorus released from rocks is carried to the oceans by rivers.
   c. Animals cannot get their phosphorus from eating plants and algae.
   d. Fertilizer use has not affected the global phosphorus budget.

6. As a general rule, how much energy is lost in the transmission of energy from one trophic level to the one immediately above it?
   a. 1%
   b. 10%
   c. 90%
   d. 50%

7. Inverted ecological pyramids of real systems usually involve
   a. energy flow.
   b. biomass.
   c. energy flow and biomass.
   d. None of the choices is correct.

8. Bottom-up effects on trophic structure result from
   a. a limitation of energy flowing to the next higher trophic level.
   b. actions of top predators on lower trophic levels.
   c. climatic disruptions on top consumers.
   d. stability of detritivores in ecosystems.

9. Species diversity
   a. increases with latitude as you move away from the equator to the arctic.
   b. decreases with latitude as you move away from the equator to the arctic.
   c. stays the same as you move away from the equator to the arctic.
   d. increases with latitude as you move north of the equator and decreases with latitude as you move south of the equator.

10. The equilibrium model of island biogeography suggests all of the following *except*
    a. larger islands have more species than smaller islands.
    b. the species richness of an island is determined by colonization and extinction.
    c. smaller islands have lower rates of extinction.
    d. islands closer to the mainland will have higher colonization rates.

APPLY

1. Based on results from studies at Hubbard Brook Experimental Forest, what would be the predicted effect of clearing trees from a watershed?
   a. Increased loss of water and nutrients from a watershed
   b. Decreased loss of water and nutrients from a watershed
   c. Increased availability of phosphorus
   d. Increased availability of nitrate

2. According to the trophic cascade hypothesis, the removal of carnivores from an ecosystem may result in
   a. a decline in the number of herbivores and a decline in the amount of vegetation.
   b. a decline in the number of herbivores and an increase in the amount of vegetation.
   c. an increase in the number of herbivores and an increase in the amount of vegetation.
   d. an increase in the number of herbivores and a decrease in the amount of vegetation.

3. At Cedar Creek Natural History Area, experimental plots showed reduced numbers of invaders as species diversity of plots increased
   a. suggesting that low species diversity increases stability of ecosystems.
   b. suggesting that ecosystem stability is a function of primary productivity only.
   c. consistent with the theory that intermediate disturbance results in the highest stability.
   d. None of the choices is correct.

SYNTHESIZE

1. Given that ectotherms do not utilize a large fraction of ingested food energy to maintain a high and constant body temperature (generate heat), how would you expect the food chains of systems dominated by ectothermic herbivores and carnivores to compare with systems dominated by endothermic herbivores and carnivores?

2. Given that, in general, energy input is greatest at the bottom trophic level (primary producers) and decreases with increasing transfers across trophic levels, how is it possible for many lakes to show much greater standing biomass in herbivorous zooplankton than in the phytoplankton they consume?

3. Ecologists often worry about the potential effects of the loss of species (e.g., due to pollution, habitat degradation, or other human-induced factors) on an ecosystem for reasons other than just the direct loss of the species. Using figure 57.17 explain why.

4. Explain several detailed ways in which increasing plant structural complexity could lead to greater species richness of lizards (figure 57.20b). Could any of these ideas be tested? How?

Principles of Biology I and II

CHAPTER 58

The Biosphere

Chapter Contents

58.1 Ecosystem Effects of Sun, Wind, and Water
58.2 Earth's Biomes
58.3 Freshwater Habitats
58.4 Marine Habitats
58.5 Human Impacts on the Biosphere: Pollution and Resource Depletion
58.6 Human Impacts on the Biosphere: Climate Change

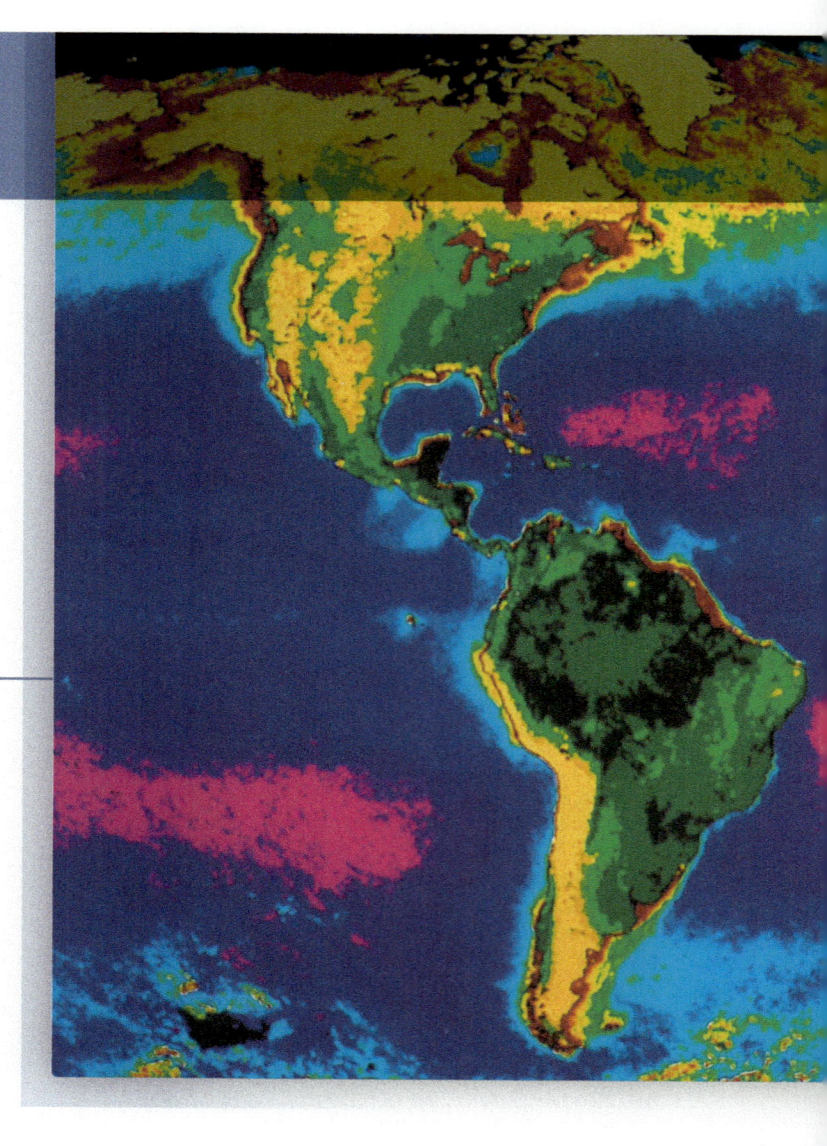

Introduction

The biosphere includes all living communities on Earth, from the profusion of life in the tropical rainforests to the planktonic communities in the world's oceans. In a very general sense, the distribution of life on Earth reflects variations in the world's abiotic environments, such as the variations in temperature and availability of water from one terrestrial environment to another. The figure on this page is a satellite image of the Americas. The colors are keyed to the relative abundance of chlorophyll, an indicator of the richness of biological communities. Green and dark green areas on land are areas with high primary productivity (such as thriving forests), whereas yellow areas include the deserts of the Americas and the tundra of the far north, which have lower productivity.

58.1 Ecosystem Effects of Sun, Wind, and Water

Learning Outcomes

1. Describe changes in wind and current direction with latitude.
2. Explain the Coriolis effect.
3. Describe how temperature changes with altitude and latitude.

The great global patterns of life on Earth are heavily influenced by (1) the amount of solar radiation that reaches different parts of the Earth and seasonal variations in that radiation; and (2) the patterns of global atmospheric circulation and the resulting patterns of oceanic circulation. Local characteristics, such as soil types and the altitude of the land, interact with the global patterns in sunlight, winds, and water currents to determine the conditions under which life exists and thus the distributions of ecosystems.

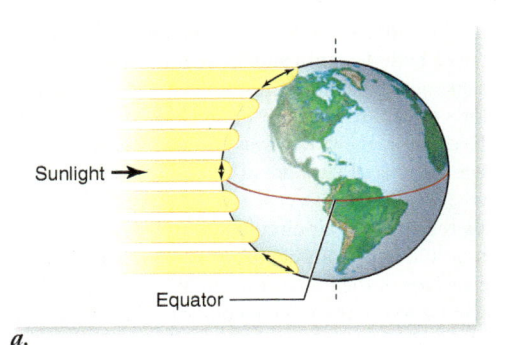

Figure 58.1 Relationships between the Earth and the Sun are critical in determining the nature and distribution of life on Earth.
a. A beam of solar energy striking the Earth in the middle latitudes of the northern hemisphere (or the southern) spreads over a wider area of the Earth's surface than an equivalent beam striking the Earth at the equator.
b. The fact that the Earth orbits the Sun each year has a profound effect on climate. In the northern and southern hemispheres, temperature changes in an annual cycle because the Earth's axis is not perpendicular to its orbital plane and, consequently, each hemisphere tilts toward the Sun in some months but away from the Sun in others.

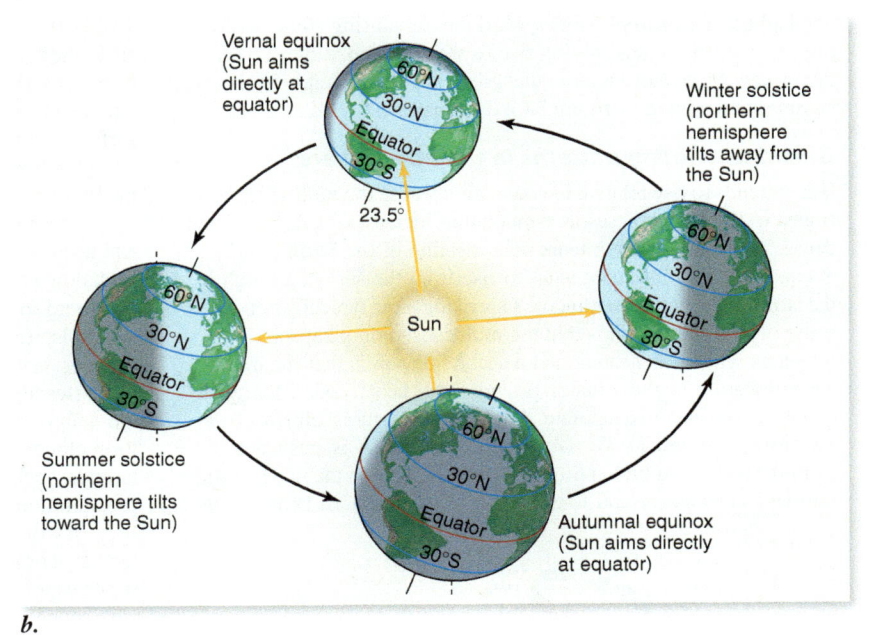

Solar energy and the Earth's rotation affect atmospheric circulation

The Earth receives energy from the Sun at a high rate in the form of electromagnetic radiation at visible and near-visible wavelengths. Each square meter of the upper atmosphere receives about 1400 joules per second (J/sec), which is equivalent to the output of fourteen 100-watt (W) lightbulbs.

As the solar radiant energy passes through the atmosphere, its intensity and wavelength composition are modified. About half of the energy is absorbed within the atmosphere, and half reaches the Earth's surface. The gases in the atmosphere absorb some wavelengths strongly while allowing other wavelengths to pass freely through. As a result, the wavelength composition of the solar energy that reaches the Earth's surface is different from that emitted by the Sun. For example, the band of ultraviolet (UV) wavelengths known as UV-B is strongly absorbed by ozone (O_3) in the atmosphere, and thus this wavelength is greatly reduced in the solar energy that reaches the Earth's surface.

How solar radiation affects climate

Some parts of the Earth's surface receive more energy from the Sun than others. These differences have a great effect on climate.

A major reason for differences in solar radiation from place to place is the fact that Earth is a sphere, or nearly so (figure 58.1*a*). The tropics are particularly warm because the Sun's rays arrive almost perpendicular to the surface of the Earth in regions near the equator. Closer to the poles, the angle at which the Sun's rays strike, called the *angle of incidence,* spreads the solar energy out over more of the Earth's surface, providing less energy per unit of surface area. As figure 58.2 shows, the highest annual mean temperatures occur near the equator (0° latitude).

The Earth's annual orbit around the Sun and its daily rotation on its own axis are also important in determining patterns of solar radiation and their effects on climate (figure 58.1*b*). The axis of rotation of the Earth is not perpendicular to the plane in which the earth orbits the Sun. Because the axis is tilted by approximately 23.5°, a progression of seasons occurs on all parts of the Earth, especially at latitudes far from the equator. The northern

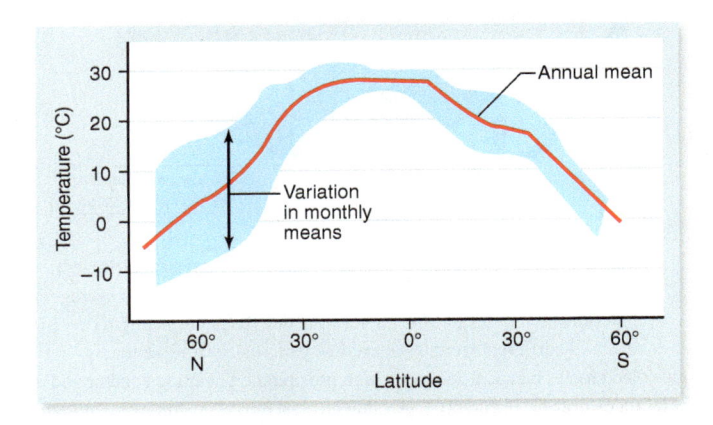

Figure 58.2 Annual mean temperature varies with latitude. The red line represents the annual mean temperature at various latitudes, ranging from near the North Pole at the left to near Antarctica at the right; the equator is at 0° latitude. At each latitude, the upper edge of the blue zone is the highest mean monthly temperature observed in all the months of the year, and the lower edge is the lowest mean monthly temperature.

Data analysis How is seasonal variation in temperature related to latitude and why?

hemisphere, for example, tilts toward the Sun during some months but away during others, giving rise to summer and winter; the farther away from the equator, the greater the difference between summer and winter, as figure 58.2 demonstrates.

Global circulation patterns in the atmosphere

Hot air tends to rise relative to cooler air because the motion of molecules in the air increases as temperature increases, making it less dense. Accordingly, the intense solar heating of the Earth's surface at equatorial latitudes causes air to rise from the surface to high in the atmosphere at these latitudes. This rising air is typically rich with water vapor; one reason is that the moisture-holding capacity of air increases when it is heated, and a second reason is that the intense solar radiation at the equator provides the heat needed for great quantities of water to evaporate. After the warm, moist air rises from the surface (figure 58.3), rising air underneath it is pushed away from the equator at high altitudes (above 10 km), to the north in the northern hemisphere and to the south in the southern hemisphere.

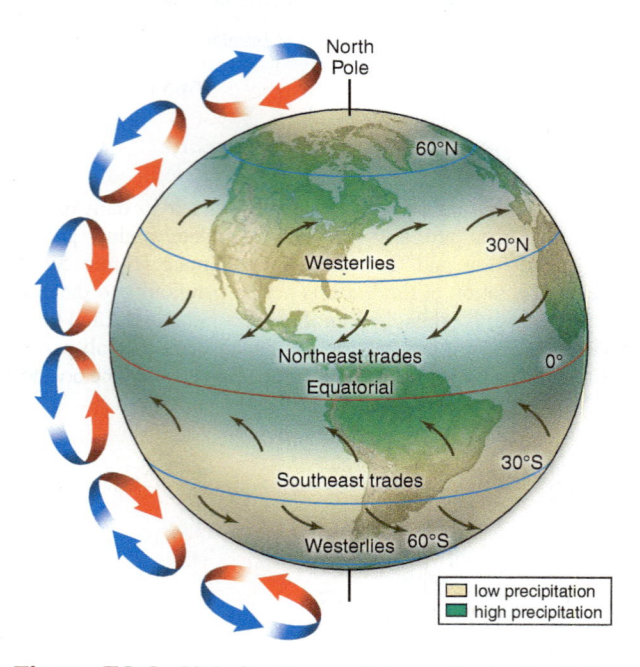

Figure 58.3 Global patterns of atmospheric circulation. The diagram shows the patterns of air circulation that prevail on average over weeks and months of time (on any one day the patterns might be dramatically different from these average patterns). Rising air that is cooled creates bands of relatively high precipitation near the equator and at latitudes near 60°N and 60°S. Air that has lost most of its moisture at high altitudes tends to descend to the surface of the Earth at latitudes near 30°N and 30°S, creating bands of relatively low precipitation. The red arrows show the winds blowing at the surface of the Earth; the blue arrows show the direction the winds blow at high altitude. The winds travel in curved paths relative to the Earth's surface because the Earth is rotating on its axis under them (the Coriolis effect). A terminological problem to recognize is that the formal names given to winds refer to the directions from which they come, rather than the directions toward which they go; thus, the winds between 30° and 60° are called Westerlies because they come out of the west. Unfortunately, oceanographers use the opposite approach, naming water currents for the directions in which they go.

To take the place of the rising air, cooler air flows toward the equator along the surface from both the north and the south. These air movements give rise to one of the major features of the global atmospheric circulation: air flows toward the equator in both hemispheres at the surface, rises at the equator, and flows away from the equator at high altitudes. The exact patterns of flow are affected by the spinning of the Earth on its axis; we discuss this effect shortly.

For complex reasons, the air circulating up from the equator and away at high altitudes in both hemispheres tends to circulate back down to the surface of the Earth at about 30° of latitude, both north and south (figure 58.3). During the course of this movement, the moisture content of the air changes radically because of the changes in temperature the air undergoes. Cooling dramatically decreases air's ability to hold water vapor. Consequently, much of the water vapor in the air rising from the equator condenses to form clouds and rain as the air moves upward. This rain falls in the latitudes near the equator, latitudes that experience the greatest precipitation on Earth.

By the time the air starts to descend back to the Earth's surface at latitudes near 30°, it is cold and thus has lost most of its water vapor. Although the air rewarms as it descends, it does not gain much water vapor on the way down. Many of the greatest deserts occur at latitudes near 30° because of the steady descent of dry air to the surface at those latitudes. The Sahara Desert is the most dramatic example.

The air that descends at latitudes near 30° flows only partly toward the equator after reaching the surface of the Earth. Some of it flows toward the poles, helping to give rise in each hemisphere to winds that blow over the Earth's surface from 30° toward 60° latitude. At latitudes near 60° air tends to rise from the surface toward high altitudes.

> **? Inquiry question** Why is it hotter at latitudes near 0°?

The Coriolis effect

If Earth did not rotate on its axis, global air movements would follow the simple patterns already described. Air currents—the winds—move across a rotating surface, however. Because the solid Earth rotates under the winds, the winds move in curved paths across the surface, rather than straight paths. The curvature of the paths of the winds due to Earth's rotation is termed the **Coriolis effect,** after the 19th-century French mathematician, Gaspard-Gustave Coriolis, who described it.

If you were standing on the North Pole, the Earth would appear to be rotating counterclockwise on its axis, but if you were at the South Pole, the Earth would appear to be rotating clockwise. This property of a rotating sphere, that its direction of rotation is opposite when viewed from its two poles, explains why the direction of the Coriolis effect is opposite in the two hemispheres. In the northern hemisphere, winds always curve to the right of their direction of motion; in the southern hemisphere, they always curve to the left.

The reason for these wind patterns is that the circumference of a sphere, the Earth, changes with latitude. It is zero at the poles and 38,000 km at the equator. Thus, land surface speed changes from about 0 to 1500 km per hour going from the poles to the equator. Air descending at 30° north latitude may be going roughly the same speed as the land surface below it. As it moves toward the equator, however, it is moving more slowly than the surface below it, so it is deflected to its right in the northern hemisphere and to its left in the southern hemisphere. In other words, in both the northern and southern hemispheres, the winds blow westward as well as toward the equator. The result (figure 58.3) is that winds on both sides of the equator—called the Trade Winds—blow out of the east and toward the west.

Conversely, air masses moving north from 30° are moving more rapidly than underlying land surfaces and thus are deflected again to their right, which in this case is eastward. Similarly, in the southern hemisphere, air masses between 30° and 60° are deflected eastward, to the left. In both hemispheres, therefore, winds between 30° and 60° blow out of the west and toward the east; these winds are called Westerlies.

Global currents are largely driven by winds

The major ocean currents are driven by the winds at the surface of the Earth, which means that indirectly the currents are driven by solar energy. The radiant input of heat from the Sun sets the atmosphere in motion as already described, and then the winds set the ocean in motion.

In the north Atlantic Ocean (figure 58.4), the global winds follow this pattern: Surface winds tend to blow out of the east and toward the west near the equator, but out of the west and toward the east at mid-latitudes (between 30° and 60°). Consequently, surface waters of the north Atlantic Ocean tend to move in a giant closed curve—called a **gyre**—flowing from North America toward Europe at mid-latitudes, then returning from Europe and Africa to North America at latitudes near the equator.

Water currents are affected by the Coriolis effect. Thus, the Coriolis effect contributes to this clockwise closed-curve motion. Water flowing across the Atlantic toward Europe at mid-latitudes tends to curve to the right and enters the flow from east to west near the equator. This latter flow also tends to curve to its right and enters the flow from west to east at mid-latitudes. In the south Atlantic Ocean, the same processes occur in a sort of mirror image, and similar clockwise and counterclockwise gyres occur in the north and south Pacific Ocean as well.

Regional and local differences affect terrestrial ecosystems

The environmental conditions at a particular place are affected by regional and local effects of solar radiation, air circulation, and water circulation, not just the global patterns of these processes. In this section we look at just a few examples of regional and local effects, focusing on terrestrial systems. These include rain shadows, monsoon winds, elevation, and presence of microclimate factors.

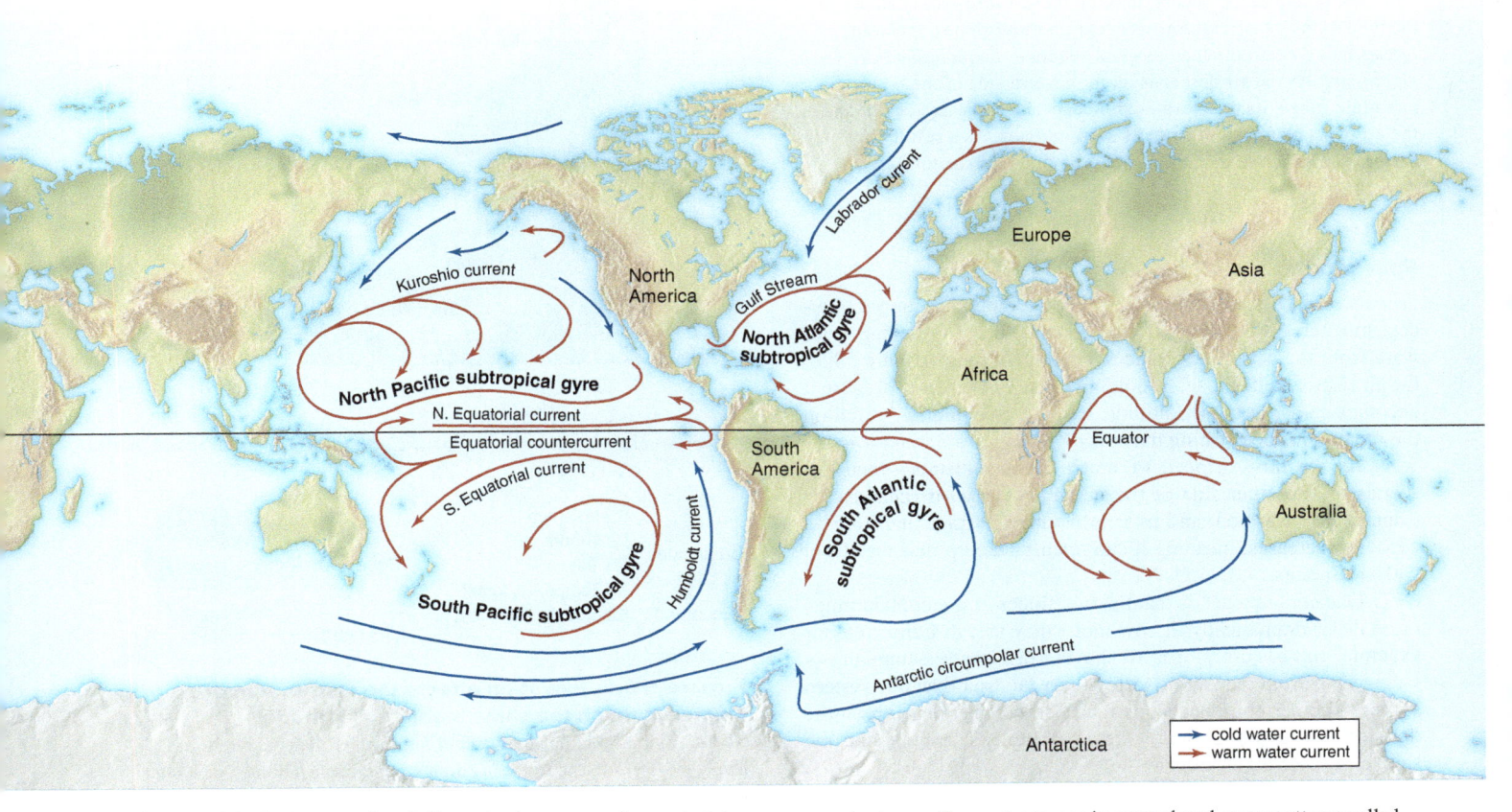

Figure 58.4 Ocean circulation. In the centers of several of the great ocean basins, surface water moves in great closed-curve patterns called gyres. These water movements affect biological productivity in the oceans and sometimes profoundly affect the climate on adjacent landmasses, as when the Gulf Stream brings warm water to the region of the British Isles.

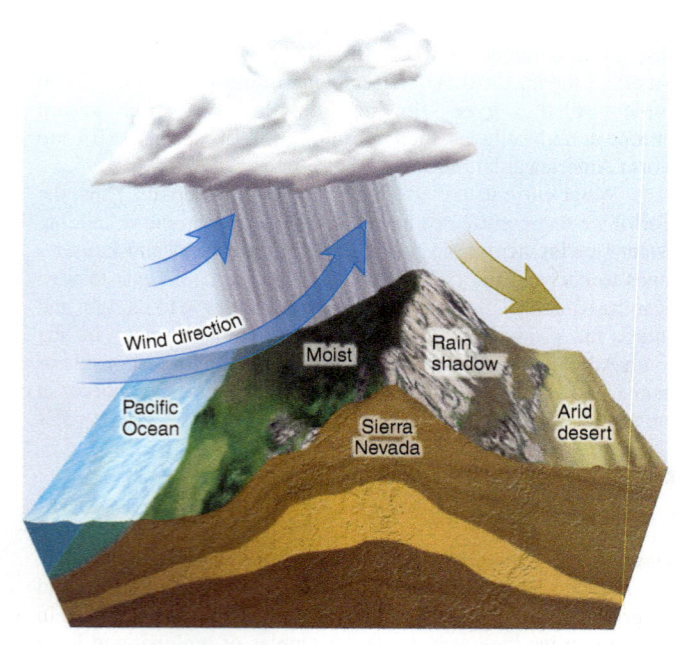

Figure 58.5 The rain shadow effect exemplified in California. Moisture-laden winds from the Pacific Ocean rise and are cooled when they encounter the Sierra Nevada Mountains. As the moisture-holding capacity of the air decreases at colder, higher altitudes, precipitation occurs, making the seaward-facing slopes of the mountains moist; tall forests occur on those slopes, including forests that contain the famous giant sequoias, *Sequoiadendron giganteum*. As the air descends on the eastern side of the mountain range, its moisture-holding capacity increases again, and the air picks up moisture from its surroundings. As a result, the eastern slopes of the mountains are arid, and rain shadow deserts sometimes occur.

Rain shadows

Deserts on land sometimes occur because mountain ranges intercept moisture-laden winds from the sea. When air flowing landward from the oceans encounters a mountain range (figure 58.5), the air rises, and its moisture-holding capacity decreases because it becomes cooler at higher altitude, causing precipitation to fall on the mountain slopes facing the sea.

As the air—stripped of much of its moisture—then descends on the other side of the mountain range, it remains dry even as it is warmed, and as it is warmed its moisture-holding capacity increases, meaning it can readily take up moisture from soils and plants.

One consequence is that the two slopes of a mountain range often differ dramatically in how moist they are; in California, for example, the eastern slopes of the Sierra Nevada Mountains—facing away from the Pacific Ocean—are far drier than the western slopes. Another consequence is that a desert may develop on the dry side, the Mojave Desert being an example. The mountains are said to produce a rain shadow.

Monsoons

The continent of Asia is so huge that heating and cooling of its surface during the passage of the seasons causes massive regional shifts in wind patterns. During summer, the surface of the Asian landmass heats up more than the surrounding oceans, but during winter the landmass cools more than the oceans. The consequence is that winds tend to blow off the water into the interior of the Asian continent in summer, particularly in the region of the Indian Ocean and western tropical Pacific Ocean. These winds reverse to flow off the continent out over the oceans in winter. These seasonally shifting winds are called the monsoons. They affect rainfall patterns, and their duration and strength can spell the difference between food sufficiency and starvation for hundreds of millions of people in the region each year.

Elevation

Another significant regional pattern is that in mountainous regions, temperature and other conditions change with elevation. At any given latitude, air temperature falls about 6°C for every 1000-m increase in elevation. The ecological consequences of the change of temperature with elevation are similar to those of the change of temperature with latitude (figure 58.6).

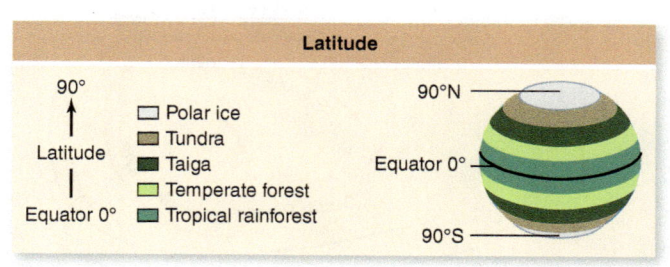

Figure 58.6 Elevation affects the distribution of biomes in much the same manner as latitude does. Biomes that normally occur far north of the equator at sea level also occur in the tropics at high mountain elevations. Thus, on a tall mountain in the tropics, one might see a sequence of biomes like the one illustrated above. In North America, a 1000-m increase in elevation results in a temperature drop equal to that of an 880-km increase in latitude.

Microclimates

Conditions also vary in significant ways on very small spatial scales. For example, in a forest, a bird sitting in an open patch may experience intense solar radiation, a high air temperature, and a low humidity, even while a mouse hiding under a log 10 feet away may experience shade, a cool temperature, and air saturated with water vapor. Such highly localized sets of climatic conditions are called microclimates.

In some cases, species avoid competing by adapting to use different microclimates. Sympatric salamanders, for example, may be specialized for the different levels of moisture found in different parts of the habitat.

Learning Outcomes Review 58.1

More intense solar heating of some global regions relative to others sets up global patterns of atmospheric circulation, which in turn cause global patterns of water circulation in the oceans. The Coriolis effect is caused by the Earth spinning beneath the moving air masses of the atmosphere. These patterns—plus seasonal changes—strongly affect the conditions that exist for living organisms in different parts of the world. In general, temperature declines as altitude or latitude increases.

- How would global air movement patterns be different if the Earth turned in the opposite direction?

58.2 Earth's Biomes

Learning Outcomes

1. Define biome.
2. Explain the primary factors that determine which type of biome is found in a particular place.

Biomes are major types of ecosystems on land. Each biome has a characteristic appearance and is distributed over wide areas of land defined largely by sets of regional climatic conditions. Biomes are named according to their vegetational structures, but they also include the animals that are present.

As you might imagine from the broad definition given for biomes, there are a number of ways to classify terrestrial ecosystems into biomes. Here we recognize eight principal biomes: (1) tropical rainforest, (2) savanna, (3) desert, (4) temperate grassland, (5) temperate deciduous forest, (6) temperate evergreen forest, (7) taiga, and (8) tundra.

Six additional biomes recognized by some ecologists are: polar ice, mountain zone, chaparral, warm moist evergreen forest, tropical monsoon forest, and semidesert. Other ecologists lump these six with the eight major ones. Figure 58.7 shows the distributions of all 14 biomes.

- polar ice
- tundra
- taiga
- mountain zone
- temperate deciduous forest
- temperate evergreen forest
- warm, moist evergreen forest
- tropical monsoon forest
- tropical rainforest
- chaparral
- temperate grassland
- savanna
- semidesert
- desert

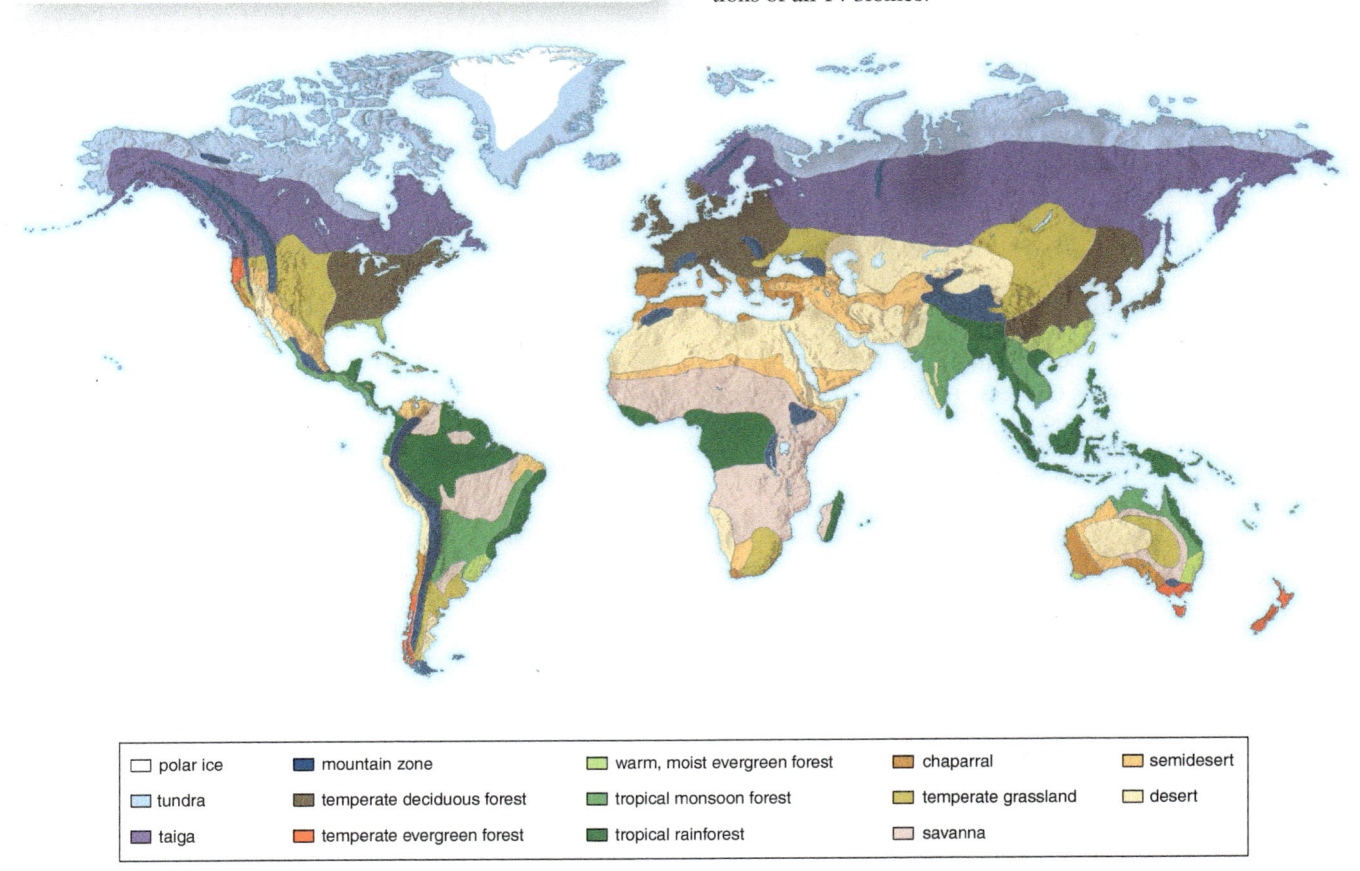

Figure 58.7 The distributions of biomes. Each biome is similar in vegetational structure and appearance wherever it occurs.

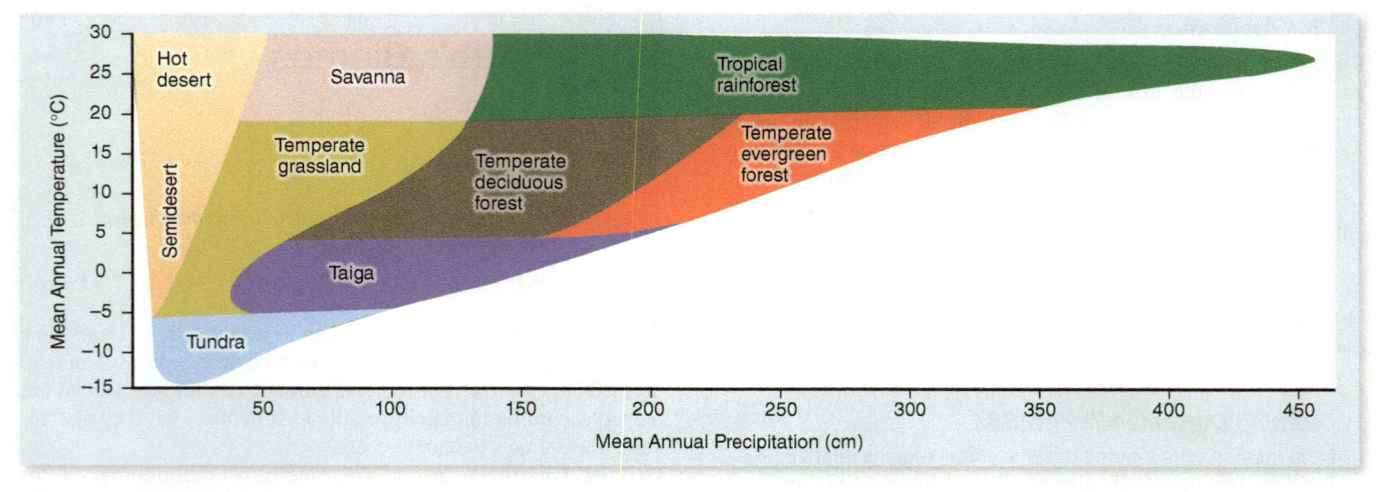

Figure 58.8 Predictors of biome distribution. Temperature and precipitation are good predictors of biome distribution, although other factors sometimes also play critical roles.

Biomes are defined by their characteristic vegetational structures and associated climatic conditions, rather than by the presence of particular plant species. Two regions assigned to the same biome thus may differ in the species that dominate the landscape. Tropical rainforests around the world, for example, are all composed of tall, lushly vegetated trees, but the tree species that dominate a South American tropical rainforest are different from those in an Indonesian one. The similarity between such forests results from convergent evolution (see chapter 21).

Temperature and moisture often determine biomes

In determining which biomes are found where, two key environmental factors are temperature and moisture. As seen in figure 58.8, if you know the mean annual temperature and mean annual precipitation in a terrestrial region, you often can predict the biome that dominates. Temperature and moisture affect ecosystems in a number of ways. One reason they are so influential is that primary productivity is strongly correlated with them, as described in chapter 57 (figure 58.9).

Different places that are similar in mean annual temperature and precipitation sometimes support different biomes, indicating that temperature and moisture are not the only factors that can be important. Soil structure and mineral composition (see chapter 38) are among the other factors that can be influential. The biome that is present may also depend on whether the conditions of temperature and precipitation are strongly seasonal or relatively constant.

Tropical rainforests are highly productive equatorial systems

Tropical rainforests, which typically require 140 to 450 cm of rain per year, are the richest ecosystems on land (figure 58.10). They are very productive because they enjoy the advantages of

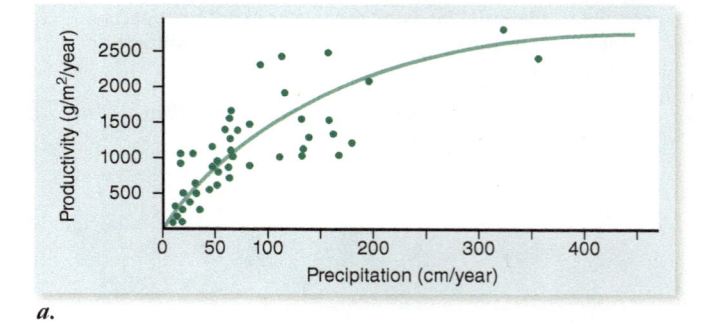

Figure 58.9 The correlations of primary productivity with precipitation and temperature. The net primary productivity of ecosystems at 52 locations around the globe correlates significantly with (*a*) mean annual precipitation and (*b*) mean annual temperature.

Inquiry question Why might you expect primary productivity to increase with increasing precipitation and temperature?

Data analysis Figure 58.9a indicates great variation in the productivity of ecosystems with intermediate amounts of precipitation (~125 cm/year). How can information in figures 58.9b and 58.8 explain this?

Figure 58.10 Tropical rainforest.

both high temperature and high precipitation (see figure 58.9). They also exhibit very high biodiversity, being home to at least half of all the species of terrestrial plants and animals in the world—over 2 million species! In a single square mile of Brazilian rainforest, there can be 1200 species of butterflies—twice the number found in all of North America. Tropical rainforests recycle nutrients rapidly, so their soils often lack great reservoirs of nutrients.

Savannas are tropical grasslands with seasonal rainfall

Savannas are tropical or subtropical grasslands, often dotted with widely spaced trees or shrubs. On a global scale, savannas often occur as transitional ecosystems between tropical rainforests and deserts; they are characteristic of warm places where annual rainfall (50 to 125 cm) is too little to support rainforest, but not so little as to produce desert conditions.

Rainfall is often highly seasonal in savannas. The Serengeti ecosystem in East Africa is probably the world's most famous example of the savanna biome. In most of the Serengeti, no rain falls for many months of the year, but during other months rain is abundant. The huge herds of grazing animals in the ecosystem respond to the seasonality of the rain; a number of species migrate away from permanently flowing rivers only during the months when rain falls.

Deserts are regions with little rainfall

Deserts are dry places where rain is both sparse (annual rainfall often less than 25 to 40 cm) and unpredictable. The unpredictability means that plants and animals cannot depend on experiencing rain even once each year. As mentioned in section 58.1, many of the largest deserts occur at latitudes near 30°N and 30°S because of global air circulation patterns (see figure 58.3). Other deserts result from rain shadows (see figure 58.5).

Vegetation is sparse in deserts, and survival of both plants and animals depends on water conservation. Many desert organisms enter inactive stages during rainless periods. To avoid extreme temperatures, small desert vertebrates often live in deep, cool, and sometimes even somewhat moist burrows. Some emerge only at night. Among large desert animals, camels drink large quantities of water when it is available and then conserve it so well that they can survive for weeks without drinking. Oryxes (large, desert-dwelling antelopes) survive opportunistically on moisture in leaves or roots that they dig up, as well as drinking water when possible.

Temperate grasslands have rich soils

Halfway between the equator and the poles are temperate regions where rich temperate grasslands grow. These grasslands, also called **prairies,** once covered much of the interior of North America, and they were widespread in Eurasia and South America as well.

The roots of perennial grasses characteristically penetrate far into the soil, and grassland soils tend to be deep and fertile. Temperate grasslands are often highly productive when converted to agricultural use, and vast areas have been transformed in this way. In North America prior to this change in land use, huge herds of bison and pronghorn antelope inhabited the temperate grasslands, migrating seasonally as resources changed over the course of the year. Natural temperate grasslands are one of the biomes adapted to periodic fire and therefore need fires to prosper.

Temperate deciduous forests are adapted to seasonal change

Mild but seasonal climates (warm summers and cold winters), plus plentiful rains, promote the growth of temperate deciduous forests in the eastern United States, eastern Canada, and Eurasia (figure 58.11). A deciduous tree is one that drops its leaves in the winter. Deer, bears, beavers, and raccoons are familiar animals of these forests.

Temperate evergreen forests are coastal

Temperate evergreen forests occur along coastlines with temperate climates, such as in the northwest of the United States. The dominant vegetation includes trees, such as spruces, pines, and redwoods, that do not drop their leaves (thus, they are *ever green*).

Taiga is the northern forest where winters are harsh

Taiga and tundra (described next) differ from other biomes in that both stretch in great unbroken circles around the entire globe

Figure 58.11 Temperate deciduous forest.

(see figure 58.7). The taiga consists of a great band of northern forest dominated by coniferous trees (spruce, hemlock, and fir) that retain their needle-like leaves all year long.

The taiga is one of the largest biomes on Earth. Winters in taiga regions are severely long and cold, and most of the limited precipitation falls in the summer. Many large herbivores, including elk, moose, and deer, plus carnivores such as wolves, bears, lynx, and wolverines, are characteristic of the taiga.

Tundra is a largely frozen treeless area with a short growing season

In the far north, at latitudes above the taiga but south of the polar ice, few trees grow. The landscape that occurs in this band, called tundra, is open, windswept, and often boggy. This enormous biome covers one-fifth of the Earth's land surface. Little rain or snow falls. *Permafrost*—soil ice that persists throughout all seasons—usually exists within a meter of the ground surface.

What trees can be found are small and mostly confined to the margins of streams and lakes. Large grazing mammals, including musk-oxen and reindeer (caribou), and carnivores such as wolves, foxes, and lynx, live in the tundra. Populations of lemmings (a small rodent native to the Arctic) rise and fall dramatically, with important consequences for the animals that prey on them.

Learning Outcomes Review 58.2

Major types of ecosystems called biomes can be distinguished in different climatic regions on land. These biomes are much the same wherever they are found on the Earth. Annual mean temperature and precipitation are effective predictors of biome type; however, the range of seasonal variation and the soil characteristics of a region also come into play.

■ *Why do different biomes occur at different latitudes?*

58.3 Freshwater Habitats

Learning Outcomes

1. Define photic zone.
2. Explain what causes spring and fall overturns in lakes.
3. Distinguish between eutrophic and oligotrophic lakes.

Of the major habitats, freshwater covers by far the smallest percentage of the Earth's surface: Only 2%, compared with 27% for land and 71% for ocean. The formation of freshwater starts with the evaporation of water into the atmosphere, which removes most dissolved constituents, much like distillation does. When water falls back to the Earth's surface as rain or snow, it arrives in an almost pure state, although it may have picked up biologically significant dissolved or particulate matter from the atmosphere.

Freshwater wetlands—marshes, swamps, and bogs—represent intermediate habitats between the freshwater and terrestrial realms. Wetlands are highly productive (see figure 57.11). They also play key additional roles, such as acting as water storage basins that moderate flooding.

Primary production in freshwater bodies is carried out by single-celled algae (phytoplankton) floating in the water, by algae growing as films on the bottom, and by rooted plants such as water lilies. In addition, a considerable amount of organic matter—such as dead leaves—enters some bodies of freshwater from plant communities growing on the land nearby.

Life in freshwater habitats depends on oxygen availability

The concentration of dissolved oxygen (O_2) is a major determinant of the properties of freshwater communities. Oxygen dissolves in water just like sugar or salt does. Fish and other aquatic organisms obtain the oxygen they need by taking it up from solution. The solubility of oxygen is therefore critically important.

In reality, oxygen is not very soluble in water. Consequently, even when freshwater is fully aerated and at equilibrium with the atmosphere, the amount of oxygen it contains per liter is only 5%, or less, of that in air. This means that in terms of acquiring the oxygen they need, freshwater organisms have a far smaller margin of safety than air-breathing ones.

Oxygen is constantly added to and removed from any body of freshwater. Oxygen is added by photosynthesis and by aeration from the atmosphere, and it is removed by animals and other heterotrophs. If a lot of decaying organic matter is present in a body of water, the oxygen demand of the decay microbes can be high, possibly depleting the amount of oxygen available for other species. Under conditions in which the rate of oxygen removal from water exceeds the rate of addition, the concentration of dissolved oxygen can fall so low that many aquatic animals cannot survive in it.

Lake and pond habitats change with water depth

Bodies of relatively still freshwater are called lakes if large and ponds if small. Water absorbs light passing through it, and the intensity of sunlight available for photosynthesis decreases sharply with increasing depth. In deep lakes, only water relatively near the surface receives enough light for phytoplankton to exhibit a positive net primary productivity (figure 58.12). Those waters are described as the **photic zone.**

The photic zone

The thickness of the photic zone depends on how much particulate matter is in the water. Water that is relatively clear and relatively free of particulate matter allows light to penetrate to a depth of 10 m at sufficient intensity to support phytoplankton. Water that is thick with surface algal cells or soil from erosion may not allow light to penetrate very far before its intensity becomes too diminished for algal growth.

The supply of dissolved oxygen to the deep waters of a lake can be a problem because oxygen enters aquatic systems near their surface, where gas exchange occurs with the air and photosynthesis occurs in aquatic organisms. In the still waters of a lake, mixing between the surface and deeper layers may not occur except occasionally as a result of thermal stratification; consequently, oxygen only rarely enters the deep waters from the surface waters.

Thermal stratification

Thermal stratification is characteristic of many lakes and large ponds. In summer, as shown at the bottom of figure 58.13, water warmed by the Sun forms a layer known as the *epilimnion* at the surface—because warm water is less dense than cold water and

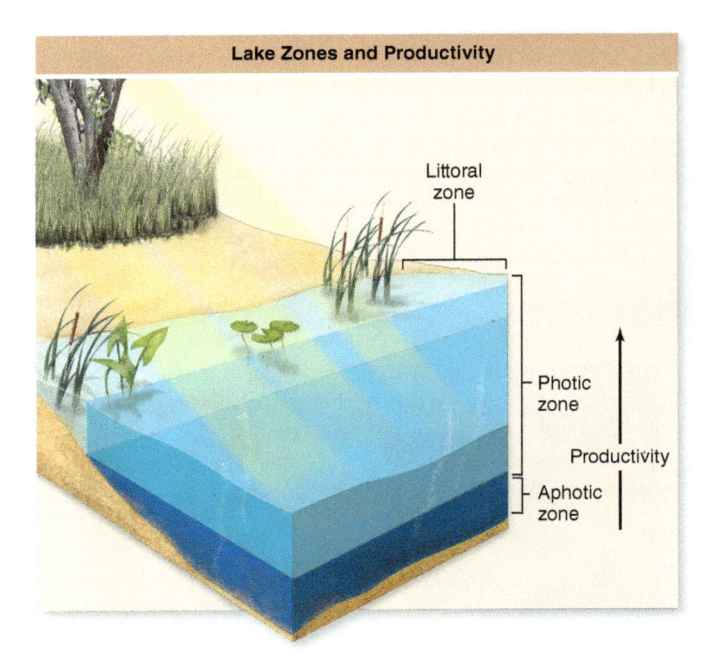

Figure 58.12 Light in a lake. The intensity of the sunlight available for photosynthesis decreases with depth in a lake. Consequently, only some of the upper waters—termed the photic zone—receive sufficient light for the net primary productivity of phytoplankton to be positive. The depth of the photic zone depends on how cloudy the water is. The shallows at the edge of a lake are called the littoral zone. They are well-illuminated to the bottom, so rooted plants and bottom algae can thrive there.

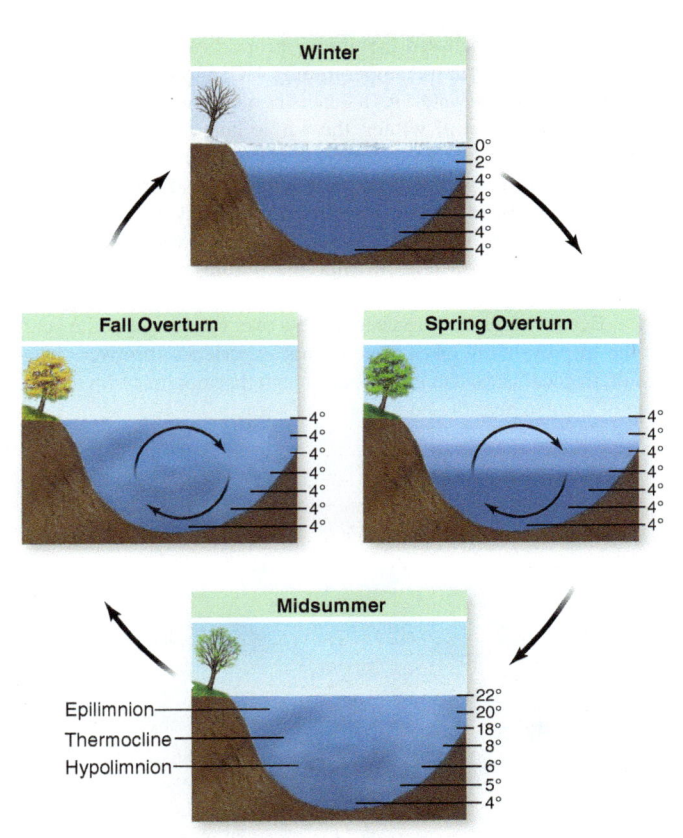

Figure 58.13 The annual cycle of thermal stratification in a temperate-zone lake. During the summer *(lower diagram)*, water warmed by the Sun (the epilimnion) floats on top of colder, denser water (the hypolimnion). The lake is also thermally stratified in winter *(upper diagram)* when water that is near freezing or frozen floats on top of water that is at 4°C (the temperature of greatest density for freshwater). Stratification is disrupted in the spring and fall overturns, when the lake is at an approximately uniform temperature and winds mix it from top to bottom.

tends to float on top. Colder, denser water, called the *hypolimnion*, lies below. Between the warm and cold layers is a transitional layer, the thermocline. Although here we are focusing on freshwater, a similar thermal structuring of the water column occurs also in many parts of the ocean.

In a lake, thermal stratification tends to cut off the oxygen supply to the bottom waters; a consequence of the stratification is that the upper waters that receive oxygen do not mix with the bottom waters. The concentration of oxygen at the bottom may then gradually decline over time as the organisms living there use oxygen faster than it is replaced. If the rate of oxygen use is high, the bottom waters may run out of oxygen and become oxygen-free before summer is over. Oxygen-free conditions, if they occur, kill most animals in a lake.

In autumn, the temperature of the upper waters in a stratified lake drops until it is about the same as the temperature of the deep waters. The densities of the two water layers become similar, and the tendency for them to stay apart is weakened. Winds can then force the layers to mix, a phenomenon called the fall overturn (figure 58.13). High oxygen concentrations are then restored in the bottom waters.

Chapter 2 discussed the unique properties of water. Freshwater is densest when its temperature is 4°C, and ice, at 0°, floats on top of this dense water. As a lake is cooled toward the freezing point with the onset of winter, the whole lake first reaches 4°C. Then, some water cools to an even lower temperature, and when it does, it becomes less dense and rises to the top. Further cooling of this surface water causes it to freeze into a layer of ice covering the lake. In spring, the ice melts, the surface water warms up, and again winds are able to mix the whole lake—the spring overturn.

Because temperature changes less over the course of the year in the tropics, many lakes there do not experience turnover. As a result, tropical lakes can have a permanent thermocline with depletion of oxygen near the bottom.

Lakes differ in oxygen and nutrient content

Bodies of freshwater that are low in algal nutrients (such as nitrate or phosphate) and low in the amount of algal material per unit of volume are termed *oligotrophic*. Such waters are often crystal clear. Oligotrophic streams and rivers tend to be high in dissolved oxygen because the movement of the flowing water aerates them; the small amount of organic matter in the water means that oxygen is used at a relatively low rate. Similarly, oligotrophic lakes and ponds tend to be high in dissolved oxygen at all depths all year because they also have a low rate of oxygen use. Because the water is relatively clear, light can penetrate the waters readily, allowing photosynthesis to occur through much of the water column, from top to bottom (figure 58.14).

Eutrophic bodies of water are high in algal nutrients and often populated densely with algae. They are more likely to be low in dissolved oxygen, especially in summer. In a eutrophic body of water, decay microbes often place high demands on the oxygen available because when thick populations of algae die, large amounts of organic matter are made available for decomposition. Moreover, light does not penetrate eutrophic waters well because of all the organic matter in the water; photosynthetic oxygen addition is therefore limited to just a relatively thin layer of water at the top.

Human activities have often transformed oligotrophic lakes into eutrophic ones. For example, when people overfertilize their lawns or fields, nitrate and phosphate from the fertilizers wash off into local water systems. Lakes that receive these nutrients become more eutrophic. A consequence is that the bottom waters are more likely to become oxygen-free during the summer. Many species of fish that are characteristic of oligotrophic lakes, such as trout, are very sensitive to oxygen deprivation. When lakes become eutrophic, these species of fish disappear and are replaced with species like carp that can better tolerate low oxygen concentrations. Lakes can return toward an oligotrophic state over time if steps are taken

Oligotrophic Lake

a.

Eutrophic Lake

b.

Figure 58.14 Oligotrophic and eutrophic lakes. *a.* Oligotrophic lakes are low in algal nutrients, have high levels of dissolved oxygen, and are clear. *b.* Eutrophic lakes have high levels of algal nutrients and low levels of dissolved oxygen. Light does not penetrate deeply in such lakes.

to eliminate the addition of excess nitrates, phosphates, and foreign organic matter such as sewage.

Learning Outcomes Review 58.3

The photic zone is the layer near the surface into which light penetrates. Photosynthesis can occur only in the photic zone. Thermal stratification is a major determinant of oxygen levels. In temperate lakes, mixing of different layers occurs when the layers reach the same temperature in spring and fall, and winds can cause the layers to mix. This overturn prevents oxygen depletion near the lake bottom. Eutrophic lakes are high in nutrients for algae but are low in dissolved oxygen; oligotrophic lakes are low in nutrients but high in dissolved oxygen at all depths.

- Why do tropical lakes often not experience seasonal turnover, and what effect is this likely to have on the ecosystems of these lakes?

58.4 Marine Habitats

Learning Outcomes
1. Know the different marine habitats.
2. Explain why El Niño events occur.

About 71% of the Earth's surface is covered by ocean. The continental shelves occur near the coastlines, where the water is not especially deep (figure 58.15); the shelves, in essence, represent the submerged edges of the continents. Worldwide, the shelves average about 80 km wide, and the depth of the water over them increases from 1 m to about 130 m as one travels from the coast toward the open ocean.

Beyond the continental shelves, the depth suddenly becomes much greater. The average depth of the open ocean is 4000 to 5000 m, and some parts—called trenches—are far deeper, reaching 11,000 m in the Marianas Trench in the western Pacific Ocean.

In most of the ocean, the principal primary producers are phytoplankton floating in the well-lit surface waters. A revolution is currently under way in scientific understanding of the limiting nutrients for ocean phytoplankton (see chapter 57). Primary production by the phytoplankton is presently understood to be nitrogen-limited in about two-thirds of the world's ocean, but iron-limited in about one-third. The principal known iron-limited areas are the great Southern Ocean surrounding Antarctica, parts of the equatorial Pacific Ocean, and parts of the subarctic, northeast Pacific Ocean. Where the water is shallow along coastlines, primary production is carried out not just by phytoplankton, but also by rooted plants such as seagrasses and by bottom-dwelling algae, including seaweeds.

The world's oceans are so vast that they include many different types of ecosystems. Some, such as coral reefs and estuaries, are high in their net primary productivity per unit of area

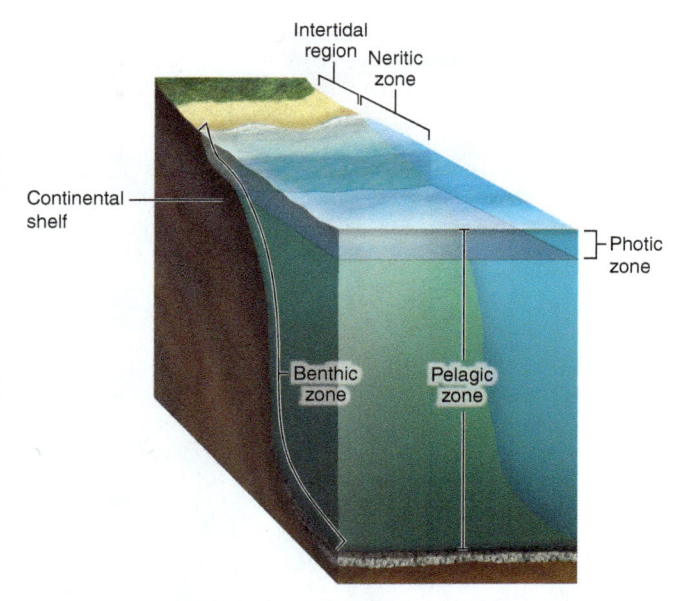

Figure 58.15 Basic concepts and terminology used in describing marine ecosystems. The continental shelf is the submerged edge of the continent. The waters over it are termed neritic and, on a worldwide average basis, are only 130 m deep at their deepest. The region where the tides rise and fall along the shoreline is called the intertidal region. The bottom is called the benthic zone, whereas the water column in the open ocean is called the pelagic zone. The photic zone is the part of the pelagic zone in which enough light penetrates for the phytoplankton to have a positive net primary productivity. The vertical scale of this drawing is highly compressed; whereas the outer edge of the continental shelf is 130 m deep, the open ocean in fact averages 35 times deeper (4000–5000 m deep).

(see figure 57.11), but others are low in productivity per unit area. Ocean ecosystems are of four major types: open oceans, continental shelf ecosystems, upwelling regions, and deep sea.

Open oceans have low primary productivity

In speaking of the open oceans, we mean the waters far from land (beyond the continental shelves) that are near enough to the surface to receive sunlight or to interact on a daily or weekly basis with those waters. We will discuss the deep sea separately later in this section.

The intensity of solar illumination in the open oceans drops from being high at the surface to being essentially zero at 200 m of depth; photosynthesis is limited to this level of the ocean. However, nutrients for phytoplankton, such as nitrate, tend to be present at low concentrations in the photic zone because over eons of time in the past, ecological processes have exported nitrate and other nutrients from the upper waters to the deep waters, and no vigorous forces exist in the open ocean to return the nutrients to the sunlit waters.

Because of the low concentrations of nutrients in the photic zone, large parts of the open oceans are low in primary productivity per unit area (see figure 57.11) and aptly called a

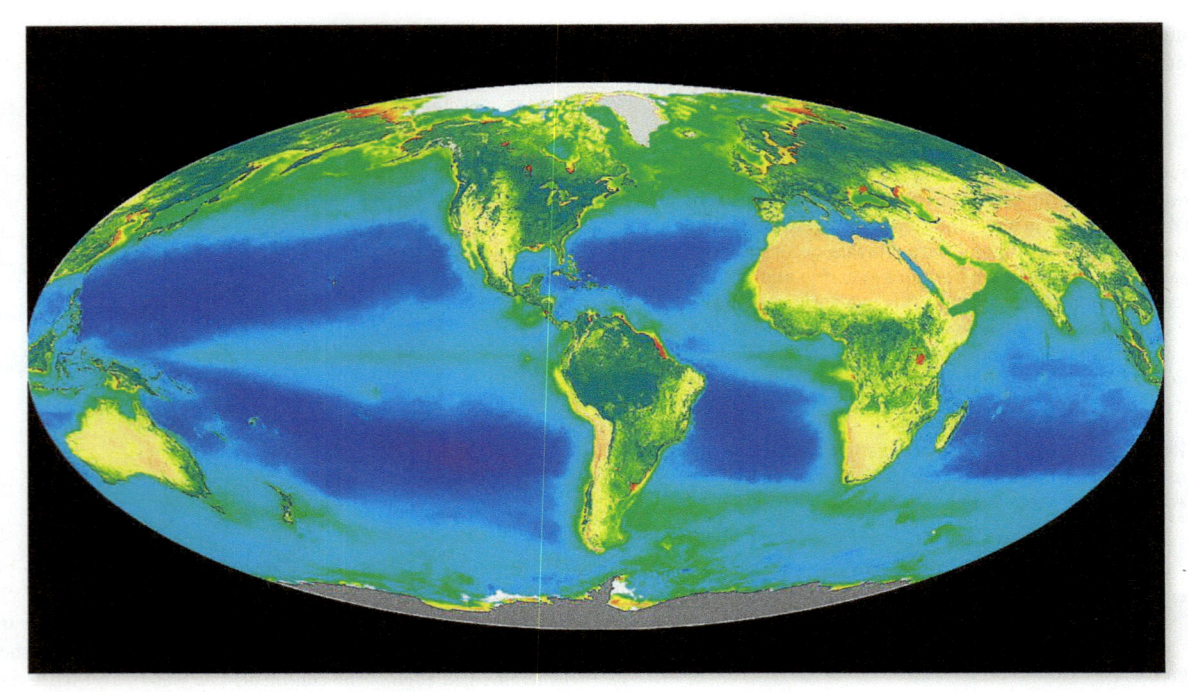

Figure 58.16 Major functional regions of the ocean. The regions classed as oligotrophic ocean (*colored dark blue*) are "biological deserts" with low productivity per unit area. Continental shelf ecosystems (*green* at the edge of continents) are typically medium to high in productivity. Upwelling regions (*yellow* at the edge of continents) such as along the western coasts of North and South America and southern Africa are the highest in productivity per unit area and rank with the most productive of all ecosystems on Earth.

"biological desert." These parts—which correspond to the centers of the great midocean gyres (see figure 58.4)—are often collectively termed the *oligotrophic ocean* (figure 58.16) in reference to their low nutrient levels and low productivity.

People fish the open oceans today for only a few species, such as tunas and some species of squids and whales. Fishing in the open oceans is limited to relatively few species for two reasons. First, because of the low primary productivity per unit of area, animals tend to be thinly distributed in the open oceans. The only ones that are commercially profitable to catch are those that are individually large or that tend to gather together in tight schools. Second, costs for traveling far from land are high. All authorities agree that as we turn to the sea to help feed the burgeoning human population, we cannot expect the open ocean regions to supply great quantities of food.

Continental shelf ecosystems provide abundant resources

Many of the ecosystems on the continental shelves are relatively high in productivity per unit area. An important reason is that the waters over the shelves—termed the *neritic waters* (see figure 58.15)—tend to have relatively high concentrations of nitrate and other nutrients, averaged over the year.

Because the waters over the shelves are shallow, they have not been subject, over the eons of time, to the loss of nutrients into the deep sea, as the open oceans have. Over the shelves, nutrient-rich materials that sink hit the shallow bottom, and the nutrients they contain are stirred back into the water column by stormy weather. In addition, nutrients are continually replenished by runoff from nearby land.

Around 99% of the food people harvest from the ocean comes from continental shelf ecosystems or nearby upwelling regions. The shelf ecosystems are also particularly important to humankind in other ways. Mineral resources taken from the ocean, such as petroleum, come almost exclusively from the shelves. In addition, almost all recreational uses of the ocean, from sailing to scuba diving, take place on the shelves. The shelves feature prominently in these ways because they are close to coastlines and relatively shallow.

Estuaries

Estuaries are one of the types of shelf ecosystems. An estuary is a place along a coastline, such as a bay, that is partially surrounded by land and in which freshwater from streams or rivers mixes with ocean water, creating intermediate (brackish) salinities.

Estuaries, besides being bodies of water, include intertidal marshes or swamps. An *intertidal* habitat is an area that is exposed to air at low tide, but under water at high tide. The marshes of the intertidal zone are called *salt marshes*. Intertidal swamps called *mangrove swamps* (dominated by trees and bushes) occur in tropical and subtropical parts of the world.

Estuaries are a vital and highly productive ecosystem—they provide shelter and food for many aquatic animals, especially the larvae and young, that people harvest for food. Estuaries are also important to a very large number of other animal species, such as migrating birds.

Banks and coral reefs

Other types of shelf ecosystems include banks and coral reefs. **Banks** are local shallow areas on the shelves, often extremely important as fishing grounds. Georges Bank, 100 km off the shore of Massachusetts, was formerly one of the most productive and famous; much of this area has been closed to fishing since the mid-1990s because of overexploitation.

Coral reef ecosystems occur in subtropical and tropical latitudes. Their defining feature is that in them, stony corals—corals that secrete a solid, calcified type of skeleton—build three-dimensional frameworks that form a unique habitat in which many other distinctive organisms live, including reef fish and soft corals (figure 58.17).

All the 700 or so species of reef-building corals are animal–algal symbioses; the animals are cnidarians, and dinoflagellate symbionts live within the cells of their inner cell layer (the gastrodermis); see chapter 33. These corals depend on photosynthesis by the algal symbionts, and thus require clear waters through which sunlight can readily penetrate. Reef-building corals are threatened worldwide, as described in section 58.5.

Upwelling regions experience mixing of nutrients and oxygen

The upwelling regions of the ocean are localized places where deep water is drawn consistently to the surface because of the action of local forces such as local winds. The deep water is often rich in nitrate and other nutrients. Upwelling therefore steadily brings nutrients into the well-lit surface layers. Phytoplankton respond to the abundance of nutrients and light with prolific growth and reproduction. Upwelling regions have the highest primary productivity per unit area in the world's ocean.

The most famous upwelling region (see figure 58.16) is found along the coast of Peru and Ecuador, where upwelling occurs year-round. Another important upwelling region is the coastline of California, along which upwelling occurs during about half the year in the summer, explaining why swimmers find cold water at the beaches even in July and August.

Upwelling regions support prolific but vulnerable fisheries. Sardine fishing in the California upwelling region crashed a few decades ago, but previously was enormously important to the region, as Nobel Prize–winning author John Steinbeck chronicled in a number of his books, most notably *Cannery Row*.

El Niño Southern Oscillation (ENSO)

The phenomenon named El Niño first came to the attention of science in studies of the Peru–Ecuador upwelling region. In that region, every two to seven years on an irregular and relatively unpredictable basis, the water along the coastline becomes profoundly warm, and simultaneously the primary productivity becomes unusually low.

Because of the low primary productivity, the ordinarily prolific fish populations weaken, and populations of seabirds and sea mammals that depend on the fish are stressed and plummet. The local people had named a mild annual warming event, which occurred around Christmas each year, "El Niño" (literally, "the child," after the Christ Child). Scientists adopted the term El Niño Southern Oscillation (ENSO) to refer to those dramatic warming events.

The immediate cause of El Niño took several decades to figure out, but research ultimately showed that the cause is a weakening of the east-to-west Trade Winds in the region. The Trade Winds ordinarily blow warm surface water to the west, away from the Peru–Ecuador coast. This thins the warm surface layer of water along the coast, so that deep water—cold but highly rich in nutrients—is drawn to the surface, leading to high primary production.

Weakening of the Trade Winds allows the warm surface layer to become thicker. Upwelling continues, but under such circumstances it merely recirculates the thick warm surface layer, which is nutrient-depleted.

After these fundamentals had been discovered, researchers in the 1980s realized that the weakening of the Trade Winds is actually part of a change in wind circulation patterns that recurs irregularly. One reason the Trade Winds blow east-to-west in ordinary times is that the surface waters in the western equatorial Pacific are warmer than those in the eastern equatorial Pacific; air rises from the warm western areas, creating low pressure at the surface there, and air blows out of the east into the low pressure. During an El Niño, the warmer the eastern ocean gets, the more similar it becomes to the western ocean, reducing the difference in pressure across the ocean. Thus, once the Trade Winds weaken a bit, the pressure difference that makes them blow is lessened, weakening the Trade Winds further. Warm water ordinarily kept in the west by the Trade Winds creeps progressively eastward at equatorial latitudes because of this self-reinforcing series of events. Ultimately, effects of El Niño occur across large parts of the world's weather systems, affecting sea temperatures in California,

Figure 58.17 A coral reef ecosystem. Reef-building corals, which consist of symbioses between cnidarians and algae, construct the three-dimensional structure of the reef and carry out considerable primary production. Fish and many other kinds of animals find food and shelter, making these ecosystems among the most diverse. About 20% of all fish species occur specifically in coral reef ecosystems.

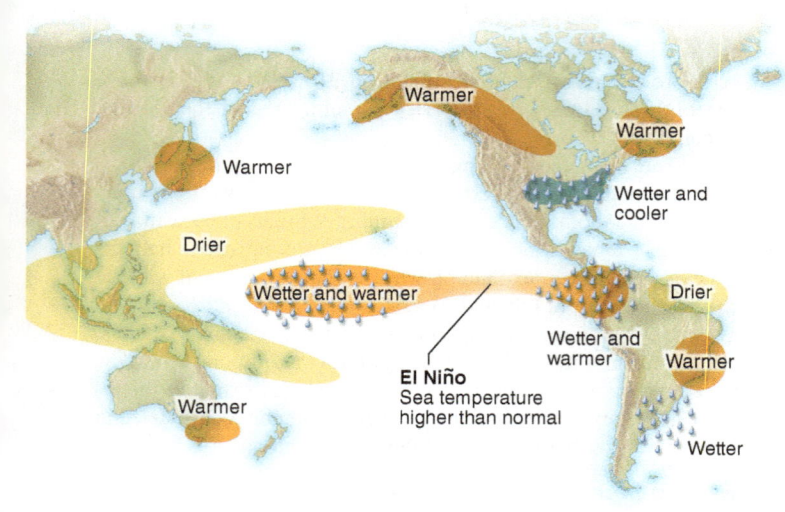

Figure 58.18 An El Niño winter. This diagram shows just some of the worldwide alterations of weather that are often associated with the El Niño phenomenon.

rainfall in the southwestern United States, and even systems as far distant as Africa.

One specific result is to shift the weather systems of the western Pacific Ocean 6000 km eastward. The tropical rainstorms that usually drench Indonesia and the Philippines occur when warm seawater abutting these islands causes the air above it to rise, cool, and condense its moisture into clouds. When the warm water moves east, so do the clouds, leaving the previously rainy areas in drought. Conversely, the western edge of Peru and Ecuador, which usually receives little precipitation, gets a soaking.

El Niño can wreak havoc on ecosystems. During an El Niño event, plankton can drop to one-twentieth of their normal abundance in the waters of Peru and Ecuador, and because of the drop in plankton productivity, commercial fish stocks virtually disappear (figure 58.18). In the Galápagos Islands, for example, seabird and sea lion populations crash as animals starve due to the lack of fish. By contrast, on land, the heavy rains produce a bumper crop of seeds, and land birds flourish. In Chile, similar effects on seed abundance propagate up the food chain, leading first to increased rodent populations and then to increased predator populations, a nice example of a bottom-up trophic cascade, as was discussed in chapter 57.

The deep sea is a cold, dark place with some fascinating communities

The deep sea is by far the single largest habitat on Earth, in the sense that it is a huge region characterized by relatively uniform conditions throughout the globe. The deep sea is seasonless, cold (2°–5°C), totally dark, and under high pressure (400 to 500 atm where the bottom is 4000 to 5000 m deep).

In most regions of the deep sea, food originates from photosynthesis in the sunlit waters far above. Such food—in the form of carcasses, fecal pellets, and mucus—can take as much as a month to drift down from the surface to the bottom, and along the way about 99% of it is eaten by animals living in the water column. Thus, the bottom communities receive only about 1% of the primary production and are food-poor. Nonetheless, a great many species of animals—most of them small-bodied and thinly distributed—are now known to live in the deep sea. Some of the animals are bioluminescent (figure 58.19a) and thereby able to communicate or attract prey by use of light.

Hydrothermal vent communities

The most astounding communities in the deep sea are the hydrothermal vent communities. Unlike most parts of the deep sea, these communities are thick with life (figure 58.19b), including large-bodied animals such as worms the size of baseball bats. The reason such a profusion of life can be supported is that these communities live on vigorous, local primary production rather than depending on the photic zone far above.

The hydrothermal vent communities occur at places where tectonic plates are moving apart, and seawater—circulating through porous rock—is able to come into contact with very hot rock under the seafloor. This water is heated to temperatures in excess of 350°C and, in the process, becomes rich in hydrogen sulfide.

As the water rises up out of the porous rock, free-living and symbiotic bacteria oxidize the sulfide, and from this reaction they obtain energy, which, in a manner analogous to photosynthesis, they use to synthesize their own cellular substance, grow, and reproduce. These sulfur-oxidizing bacteria are chemoautotrophs (see chapter 57). Animals in the communities either survive on the

Figure 58.19 Life in the deep sea. *a.* The luminous spot below the eye of this deep-sea fish results from the presence of a symbiotic colony of bioluminescent bacteria. Bioluminescence is a fairly common feature of mobile animals in the parts of the ocean that are so deep as to be dark. It is more common among species living partway down to the bottom than in ones living at the bottom. *b.* These large worms (shown in the larger photograph) live along vents where hot water containing hydrogen sulfide rises through cracks in the seafloor crust (see *inset*). Much of the body of each worm is devoted to a colony of symbiotic sulfur-oxidizing bacteria. The worms transport sulfide and oxygen to the bacteria, which oxidize the sulfur and use the energy thereby obtained for primary production of new organic compounds, which they share with their worm hosts.

a.   *b.*

bacteria or eat other animals that do. The hydrothermal vent communities are among the few communities on Earth that do not depend on the Sun's energy for primary production.

Learning Outcomes Review 58.4

The oligotrophic ocean includes the open ocean and the deep sea, where little primary productivity occurs. Continental shelf ecosystems tend to be moderate to high in productivity; they include estuaries, salt marshes, fishing banks, and coral reefs. The highest levels of productivity are found in upwelling regions, such as those along the west coasts of North and South America, where prolific but vulnerable fisheries can be found. Periodic weakening of the Trade Winds in this region can prevent the upwelling of cold water and subsequently cause weather changes in an event termed El Niño.

- *What sort of population cycles would you expect to see in regions that are affected by the ENSO?*

58.5 Human Impacts on the Biosphere: Pollution and Resource Depletion

Learning Outcomes

1. Name the major human threats to ecosystems.
2. Differentiate between point-source pollution and diffuse pollution.
3. Explain the effect of deforestation.
4. Describe how habitat fragmentation can lead to increased incidence of disease in humans.

We all know that human activities can cause adverse changes in ecosystems. In discussing these, it is important to recognize that creative people can often come up with rational solutions to such problems.

An outstanding example is provided by the history of DDT in the United States. DDT is a highly effective insecticide that was sprayed widely in the decades following World War II, often on wetlands to control mosquitoes. During the years of heavy DDT use, populations of ospreys, bald eagles, and brown pelicans—all birds that catch large fish—plummeted. Ultimately, the use of DDT was connected with the demise of these birds.

Scientists established that DDT and its metabolic products became more and more concentrated in the tissues of animals as the compounds were passed along food chains (figure 58.20). Animals at the bottom of food chains accumulated relatively low concentrations in their fatty tissues. But the primary carnivores that preyed on them accumulated higher concentrations from eating great numbers, and the secondary carnivores accumulated higher concentrations yet. Top-level carnivores, such as the birds that eat large fish, were dramatically affected by the DDT. In these birds, scientists found that metabolic products of DDT disrupted the formation of eggshells. The birds laid eggs with such thin shells that they often cracked before the young could hatch.

Researchers concluded that the demise of the fish-eating birds could be reversed by a rational plan to clean ecosystems of DDT, and laws were passed banning its use. Now, over four decades later, populations of ospreys, eagles, and pelicans are rebounding dramatically. For some people, a major reason to study science is the opportunity to be part of success stories of this sort.

Freshwater habitats are threatened by pollution and resource depletion

Freshwater is not just the smallest of the major habitats, but also the most threatened. One of the simplest yet most ominous threats to freshwater is that burgeoning human populations often extract excessive amounts of water from rivers, lakes, or streams. The Colorado River, for example, is one of the greatest rivers in North America, originating with snow melt in the Rocky Mountains and flowing through Utah, Arizona, Nevada, California, and northern Mexico before emptying into the ocean. Today, water is pumped out of the river all along its way to meet the water needs of cities (even ones as distant as Los Angeles) and to irrigate crops. The river now frequently runs out of water and dries up in the desert, never reaching the sea. Worldwide, many crises in the supply of freshwater loom on the horizon.

Pollution: Point source versus diffuse

Pollution of freshwater is a global problem. Point-source pollution comes from an identifiable location—such as easily identified

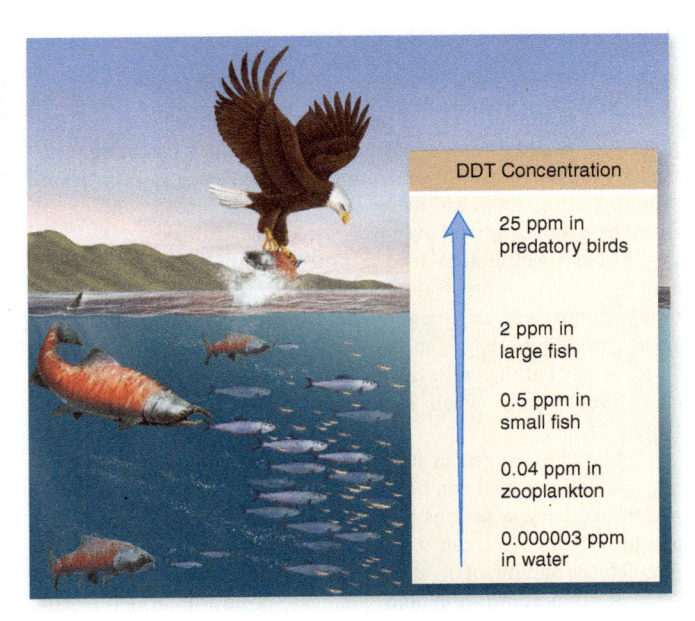

Figure 58.20 Biological magnification of DDT concentration. Because all the DDT an animal eats in its food tends to accumulate in its fatty tissues, DDT becomes increasingly concentrated in animals at higher levels of the food chain. The concentrations at the right are in parts per million (ppm). Before DDT was banned in the United States, bird species that eat large fish underwent drastic population declines because metabolic products of DDT made their eggshells so thin that the shells broke during incubation.

DDT Concentration:
- 25 ppm in predatory birds
- 2 ppm in large fish
- 0.5 ppm in small fish
- 0.04 ppm in zooplankton
- 0.000003 ppm in water

factories or other facilities that add pollutants at defined locations, such as an outfall pipe. Examples include sewage-treatment plants, which discharge treated effluents at specific spots on rivers, and factories that sometimes discharge water contaminated with heavy metals or chemicals. Laws and technologies can readily be brought to bear to moderate point-source pollution because the exact locations and types of pollution are well defined. In many countries, great progress has been made, but in other countries, often in the developing world, water pollution is still a major problem.

Diffuse pollution is exemplified by eutrophication caused by excessive runoff of nitrates and phosphates from lawn and agricultural field fertilization. When excessive nitrates and phosphates enter rivers and lakes, the character of the bodies of water is changed for the worse; the concentration of dissolved oxygen declines, and fish species such as carp take the place of more desirable species. The problem is exacerbated when rivers empty into the ocean. The eutrophication caused by the accumulation of chemicals can lead to enormous areas of water with no oxygen, causing massive die-offs of fish and other animals. The most famous such area, covering approximately 17,500 km^2 in 2011, occurs where the Mississippi River empties into the Gulf of Mexico, but other "dead zones" occur in places around the world.

The nitrates and phosphates that cause these problems originate on thousands of farms and lawns spread over whole watersheds, and they often enter freshwaters at virtually countless locations. The diffuseness of this sort of pollution renders it difficult to modify by simple technical fixes. Instead, solutions often depend on public education and political action.

Pollution from coal burning: Acid precipitation

A type of pollution that has properties intermediate between the point-source and diffuse types is the pollution that can arise from burning of coal for power generation. Although each smokestack is a point source, there are many stacks, and the smoke and gases from these stacks spread over wide areas.

Acid precipitation is one aspect of this problem. When coal is burned, sulfur in the coal is oxidized. The sulfur oxides, unless controlled, are spewed into the atmosphere in the stack smoke, and there they combine with water vapor to produce sulfuric acid. Falling rain or snow picks up the acid and is excessively acidic when it reaches the surface of the Earth (figure 58.21).

Mercury emitted in stack smoke is a second potential problem. Burning of coal can be one of the major sources of environmental mercury, a serious public health issue because just small amounts of mercury can interfere with brain development in human fetuses and infants.

Acid precipitation and mercury pollution affect freshwater ecosystems. At pH levels below 5.0, many fish species and other aquatic animals die, unable to reproduce. Thousands of lakes and ponds around the world no longer support fish because of pH shifts induced by acid precipitation. Mercury that falls from atmospheric emissions into lakes and ponds accumulates in the tissues of food fish. In the Great Lakes region of the United States, people—especially pregnant women—are advised to eat little or no locally caught fish because of its mercury content.

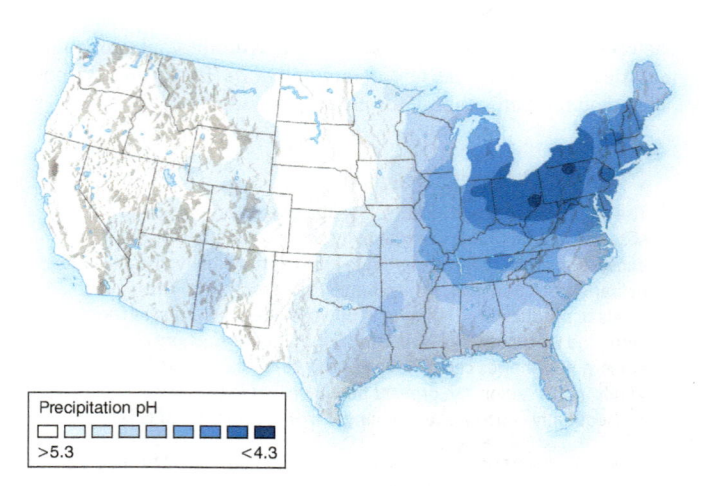

Figure 58.21 pH values of rainwater in the United States. pH values of less than 7 represent acid conditions; the lower the values, the greater the acidity. Precipitation in parts of the United States, especially in the Northeast, is commonly more acidic than natural rainwater, which has a pH of 5.6 or higher.

Forest ecosystems are threatened in tropical and temperate regions

Probably the single greatest problem for terrestrial habitats worldwide is deforestation by cutting or burning. There are many reasons for deforestation. In poverty-stricken countries, deforestation is often carried out diffusely by the general population; people burn wood to cook or stay warm, and they collect it from the local forests.

At the other extreme, corporations still cut large tracts of virgin forests in an industrialized fashion, often shipping the wood halfway around the world to buyers. Tropical hardwoods, such as mahogany, from Southeast Asian rainforests are shipped to the United States for use in furniture, and softwood logs are shipped from Alaska to East Asia for pulping and paper production. Forests are sometimes simply burned to open up land for farming or ranching (figure 58.22a).

Loss of habitat

The loss of forest habitat can have dire consequences. Particularly diverse sets of species depend on tropical rainforests for their habitat, for example. Thus, when rainforests are cleared, the loss of biodiversity can be extreme. Many tropical forest regions have been severely degraded, and recent estimates suggest that less than half of the world's tropical rainforests remain in pristine condition. All of the world's tropical rainforests will be degraded or gone in about 30 years at present rates of destruction.

Besides loss of habitat, deforestation can have numerous secondary consequences, depending on local contexts. In the Sahel region, south of the Sahara Desert in Africa, deforestation has been a major contributing factor in increased desertification. In the forests of the northeastern United States, as the Hubbard Brook experiment shows (see figure 57.7), deforestation can lead to both a loss of nutrients from forest soils and a simultaneous nutrient enrichment of bodies of water downstream.

Figure 58.22 Destroying the tropical rainforests.
a. These fires are destroying a tropical rainforest in Brazil to clear it for cattle pasture. *b.* The consequences of deforestation can be seen on these middle-elevation slopes in Madagascar, which once supported tropical rainforest, but now support only low-grade pastures and permit topsoil to erode into the rivers—note the color of the water, stained brown by high levels of soil erosion. This sort of picture is seen in a number of places around the world, including Ecuador and Haiti as well as Madagascar.

Disruption of the water cycle

As discussed in chapter 57, cutting of a tropical rainforest often interrupts the local water cycle in ways that permanently alter the landscape. After an area of tropical rainforest is cleared, rainwater often runs off the land to distant places, rather than being returned to the atmosphere immediately above by transpiration. This change may render conditions unsuitable for the rainforest trees that originally lived there. Then the poorly vegetated land—exposed and no longer stabilized by thick root systems—may be ravaged by erosion (figure 58.22*b*).

Acid rain

Deforestation can be a problem in temperate regions, as well as in the tropics. In addition, acid rain affects forests as well as lakes and streams; large tracts of trees in temperate regions have been adversely affected by acid rain. By changing the acidity of the soil, acid rain can lead to widespread tree mortality (figure 58.23).

Disease transmission from animals to people

Human interactions with the forest and its inhabitants are leading to an increase in new human health problems. Zoonotic diseases are contagious diseases that spread from animals to humans (of course, it works in the other direction as well). Throughout our species' history, we have been the unhappy recipient of many such maladies. The plague, the flu, cholera, salmonella, and many more—60% of all human pathogens are zoonotic in origin. But in recent years, a great number of new diseases have made the jump to humans, including West Nile virus, hantavirus, Middle Eastern respiratory syndrome (MERS), and severe acute respiratory syndrome (SARS). Ebola is another zoonotic disease that has long been known, but erupted in 2014, killing more than 11,000 people in West Africa.

There are a number of reasons for the recent occurrence of so many new zoonotic diseases, but many have resulted from humans increasingly using forests and coming into contact with their inhabitants that harbor the disease organisms. Human immunodeficiency virus (HIV) was transmitted to humans from apes and monkeys in West Africa, very likely occurring when the animal's blood entered a human during hunting or butchering (see section 23.5).

Another example is the Nipah virus, which is normally found in fruit bats, but that led to an outbreak that killed more than 100 people in Malaysia in 1999. The immediate cause of the transmission likely resulted from the establishment of large pig farms in the vicinity of fruit trees that the bats frequented. Bats are messy eaters and drop lots of fruit; bat saliva contains the virus and can be found in half-eaten fruit. It is easy to imagine how a pig, picking up such fruit, could catch the disease and then pass it on to pig farmers.

Zoonotic diseases can also result when humans disrupt the forest ecosystem. A clear example is the outbreak of Lyme disease (see table 28.1), which is diagnosed in 300,000 people in the United States every year, most of them in the northeastern states. The disease is caused by the bacterium *Borrelia burgdorferi*, which is commonly found in the white-footed mouse *(Peromyscus leucopus)*. The bacterium is transmitted from one mouse to another by the blacklegged tick *(Ixodes scapularis)*, which also can transmit the disease to humans.

One reason why the disease has become so prevalent in the northeastern United States is that the forest there has been broken into many small patches by roads, fields, and cities. Small forest patches tend to contain relatively few predators, and as a result, the population density of the mice increases. This, in turn, leads not only to a much higher incidence of ticks (three times higher in patches less than 1.2 hectares in size compared to larger patches), but also to a greater incidence of infection in the ticks (seven times more likely to be infected in the small patches). As a result, people

Figure 58.23 Damage to trees by acid precipitation at Clingman's Dome, Tennessee. Acid precipitation weakens trees and makes them more susceptible to pests and predators.

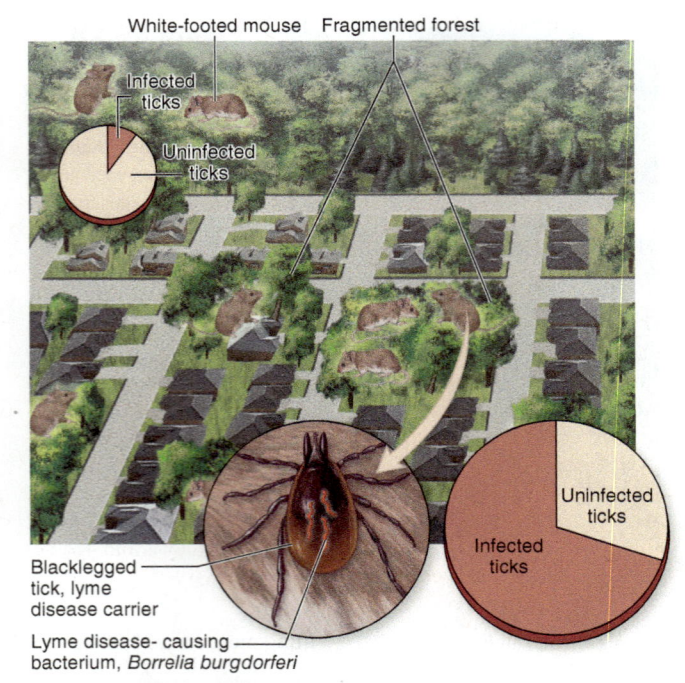

Figure 58.24 How forest fragmentation leads to Lyme disease outbreaks. In a fragmented forest, there are fewer predators, so mouse populations increase. This in turn leads to a higher number of ticks, and of that larger population of ticks, more are infected with Lyme disease–causing bacterium.

Source: National Science Foundation

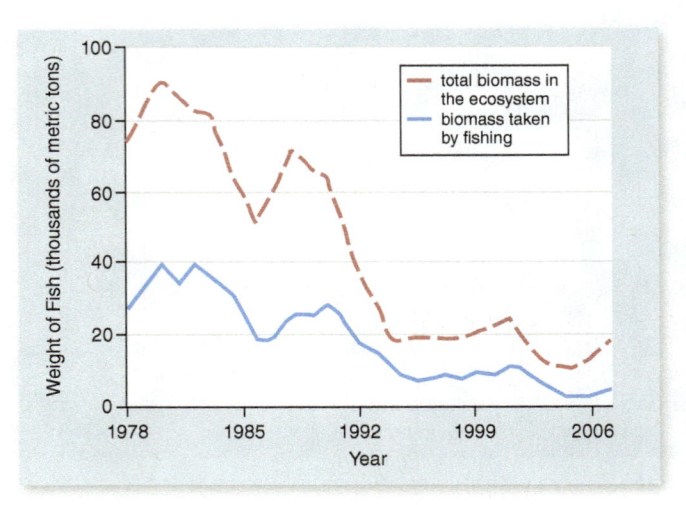

Figure 58.25 The collapse of a fishery. The red line shows the biomass of cod, *Gadus morhua,* in the Georges Bank ecosystem as estimated by the U.S. National Marine Fisheries Service based on data collected by scientific sampling. The biomass declined steeply between the 1970s and 1990s because of fishing pressure. As the years passed, commercial landings of cod *(blue line)* remained fairly constant, in part because ships worked harder and harder to catch cod, until catches fell precipitously toward zero and the fishery collapsed in the mid-1990s. Regulatory agencies closed the fishery in the mid-1990s to permit the cod to recover, but even in 2013 recovery of cod was weak, and production from the fishery was far below historical norms.

living near such patches are much more likely to be bitten by a tick and get Lyme disease (figure 58.24).

Unfortunately, the conditions for the development of new zoonoses are becoming increasingly common, and new diseases keep emerging. The real concern is that a disease that is easily transmissible between humans—for example, by contact or coughing—could cause a major epidemic. So far, we have been lucky, but public health officials fear that without constant vigilance, a major outbreak may be inevitable.

Marine habitats are being depleted of fish and other species

Overfishing of the ocean has risen to crisis proportions in recent decades and probably represents the single greatest current problem in the ocean realm. The ocean is so huge that it has tended to be more immune than freshwater or terrestrial ecosystems to global human alteration. Nonetheless, the total world fish catch has been pushed to its maximum for over two decades, even as demand for fish has continued to rise. Fishing pressure is so excessive that 25 to 30% of the world's ocean fish stocks are presently officially rated as being overexploited, depleted, or in recovery; another 40 to 50% are rated as being maximally exploited.

Major cod fisheries in waters off of Nova Scotia, Massachusetts, and Great Britain have been closed to fishing in the past 20 years because of collapse (figure 58.25). Overfishing can have disturbing indirect effects. In impoverished parts of Africa, poaching on primates and other wild mammals in national parks increases when fish catches decline.

Aquaculture: At present only a quick fix

Production of fish by aquaculture has grown steadily in the last two decades, and it is often viewed as a straightforward solution to the fisheries problem. But the dietary protein needs of many aquacultured fish, such as salmon, are met largely with wild-caught fish. In this case, exploitation has simply shifted to different species.

In addition, current aquaculture practices often damage natural ocean ecosystems. One example is the clearing of mangrove swamps along coasts to create shrimp and fish ponds, which are abandoned when their productivity declines. Research is needed to ameliorate these problems.

Pollution effects

As large as the ocean is, enough pollutants are being added that at the start of the 21st century, polluting materials are easily detectable on a global basis. An expedition to some of the most remote, uninhabited islands in the vast Pacific Ocean recently reported, for example, that considerable amounts of plastic could be found washed up on the beaches. Similarly, even the waters of the Arctic are laced with toxic chemicals; biopsy samples of tissue from Arctic killer whales *(Orcinus orca)* revealed extremely high levels of many chemicals, including pesticides and a flame-retardant chemical often used in carpets. Nonetheless, because of the ocean's vastness, concentrations of pollutants are not at crisis levels in the ocean at large.

Just as in freshwater habitats, pollution from coal burning and other sources is causing increased acidity of the ocean, but in this case, the culprit is carbon dioxide. Scientists have only recently realized that one consequence of the massive amounts of carbon dioxide entering the atmosphere (see section 58.6) is that much of it ends up being dissolved into the ocean. And when it does, it combines with water to form hydrogen (H^+) and bicarbonate (HCO_3^-) ions. By some measurements, the result has been an increase in the ocean's acidity by 30% since the 1950s, making the ocean's pH lower than it has been any time in the last 20 million years, with a projected increase of 150% by the end of this century.

The biological consequences of this acidification are not clear, but a primary concern is that many marine organisms—including corals, echinoderms, oyster larvae, and many others—use calcium carbonate to build their skeletons and other structures. As the ocean becomes more acidic, calcium carbonate ($CaCO_3$) in the water combines with hydrogen ions to form bicarbonate, making it more difficult for animals to build the structures they need to survive and grow. In fact, the opposite can occur, and calcium carbonate already incorporated into skeletons or other structures may actually dissolve. The magnitude of these effects is not yet clear, but many scientists fear that coral reefs and many other elements of ocean ecosystems may be in peril.

Destruction of coastal ecosystems

Deterioration of coastal ecosystems is also a grave problem. Estuaries along coastlines are often subject to severe eutrophication; since about 1970, for example, the bottom waters of the Chesapeake Bay near Washington, DC, have become oxygen-free each summer because of the decay of excessive amounts of organic matter.

Another coastal problem is destruction of salt marshes, which (like freshwater wetlands) are often perceived as disposable.

Most authorities believe that the loss of salt marshes in the 20th century was a major contributing factor to the destruction of New Orleans by Hurricane Katrina in 2005; had the salt marshes and cypress swamps been present at their full extent, they would have absorbed a great deal of the flooding water and buffered the city from some of the storm's violence.

Stratospheric ozone depletion has led to an ozone "hole"

The colors of the satellite photo in figure 58.26a represent different concentrations of ozone (O_3) located 20 to 25 km above the Earth's surface in the stratosphere. Stratospheric ozone is depleted over Antarctica (purple region in the figure) to between one-half and one-third of its historically normal concentration, a phenomenon called the ozone hole.

Although depletion of stratospheric ozone is most dramatic over Antarctica, it is a worldwide phenomenon. Over the United States, the ozone concentration has been reduced by about 4%, according to the U.S. Environmental Protection Agency.

Stratospheric ozone and UV-B

Stratospheric ozone is important because it absorbs ultraviolet (UV) radiation—specifically the wavelengths called **UV-B**—from incoming solar radiation. UV-B is damaging to living organisms in a number of ways; for instance, it increases risks of cataracts and skin cancer in people. Depletion of stratospheric ozone permits more UV-B to reach the Earth's surface and therefore increases the risks of UV-B damage. Every 1% drop in stratospheric ozone is estimated to lead to a 6% increase in the incidence of skin cancer, for example. In the southern hemisphere, where ozone depletion is

a.

b.

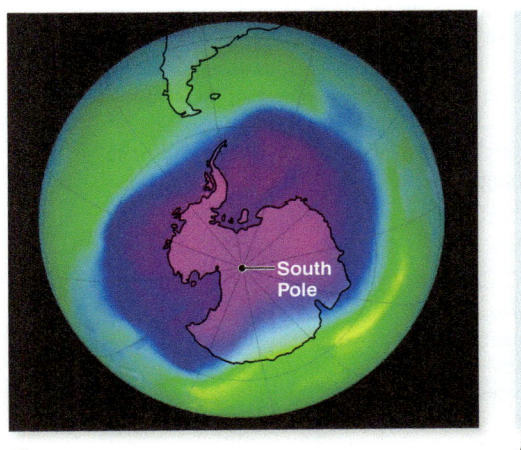

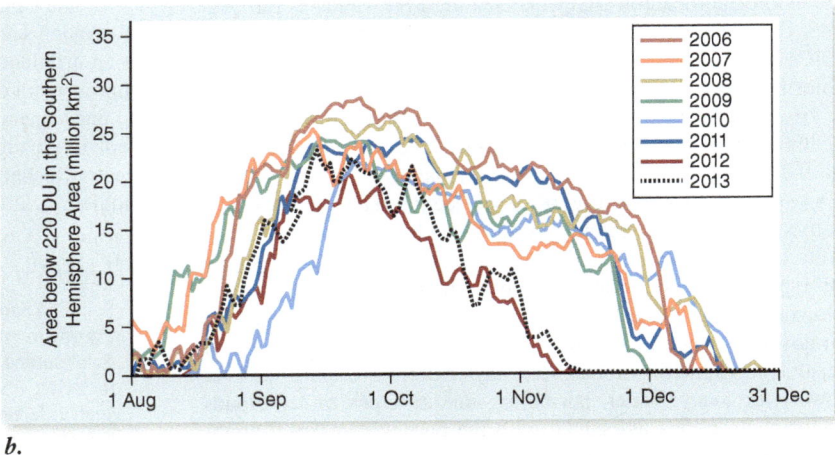

Figure 58.26 The ozone hole over Antarctica. NASA satellites currently track the extent of ozone depletion in the stratosphere over Antarctica each year. Every year since about 1980, an area of profound ozone depletion, called the ozone hole, has appeared in August (early spring in the southern hemisphere) when sunlight triggers chemical reactions in cold air trapped over the South Pole during the Antarctic winter. The hole intensifies during September before tailing off as temperature rises in November–December. *a.* In September 2006, the 11.4-million-square-mile hole (*purple* in the satellite image shown) covered an area larger than the United States, Canada, and Mexico combined, the largest hole ever recorded. *b.* Concentrations of ozone-depleting chemical compounds in the atmosphere have probably peaked in the last few years and are expected to decline slowly over the decades ahead.

SCIENTIFIC THINKING

Question: Does exposure to UV radiation affect the survival of amphibian eggs?
Hypothesis: Direct UV exposure is detrimental to eggs.
Experiment: Fertilized eggs from several frog species are placed into enclosures in full sunlight. All enclosures have screens, some of which filter out UV radiation, whereas others do not affect UV transmission. Eggs are monitored to see whether they survive to hatching or whether they die.

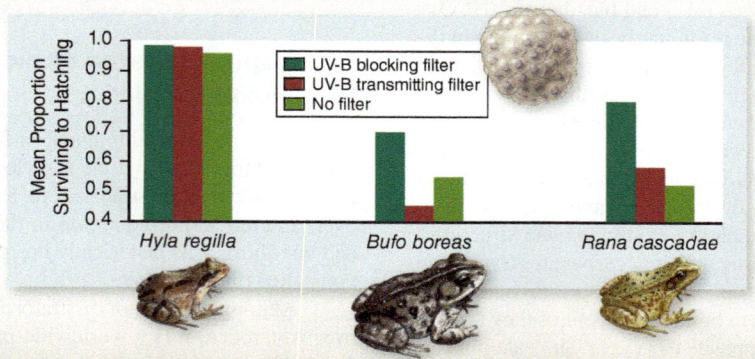

Result: Egg survival was greatly decreased in two of three species in the enclosures where UV radiation was not filtered out, as compared to survival in the filtered enclosures. Therefore, the hypothesis is confirmed: UV exposure is detrimental to amphibian eggs.
Further Questions: What factors might explain why some species are affected by UV exposure and others are not? How could your hypotheses be tested?

Figure 58.27 The effect of UV radiation on amphibian eggs.

the greatest, rates of skin cancer are much higher than elsewhere in the world; in Australia, for example, two out of every three people will develop skin cancer by age 70. UV exposure also may be detrimental to many types of animals, such as amphibians (figure 58.27).

Ozone depletion and CFCs

The major cause of the depletion of stratospheric ozone is the addition of industrially produced chlorine- and bromine-containing compounds to the atmosphere. Of particular concern are chlorofluorocarbons (CFCs), used until the 1990s as refrigerants in air conditioners and refrigerators, and in manufacturing. CFCs released into the atmosphere can ultimately liberate free chlorine atoms, which in the stratosphere catalyze the breakdown of ozone molecules (O_3) to form ordinary oxygen (O_2). Ozone is continually being made and broken down, and free chlorine atoms tilt the balance toward a faster rate of breakdown.

The extreme depletion of ozone seen in the ozone hole is a consequence of the unique weather conditions that exist over Antarctica. During the continuous dark of the Antarctic winter, a strong stratospheric wind, the polar-night jet, develops and, blowing around the full circumference of the Earth, isolates the stratosphere over Antarctica from the rest of the atmosphere. As a result, the Antarctic stratosphere stays extremely cold (–80°C or lower) for many weeks, permitting unique types of ice clouds to form. Reactions associated with the particles in these clouds lead to accumulation of diatomic chlorine, Cl_2. When sunlight returns in the early Antarctic spring, the diatomic chlorine is photochemically broken up to form free chlorine atoms in great abundance, and the ozone-depleting reactions ensue.

Phase-out of CFCs

After research revealed the causes of stratospheric ozone depletion, worldwide agreements were reached to phase out the production of CFCs and other compounds that lead to ozone depletion. Manufacture of such compounds ceased in the United States in 1996, and there is now general awareness about the importance of using "ozone-safe" alternative chemicals. The atmosphere will cleanse itself of ozone-depleting compounds only slowly because the substances are chemically stable. Nonetheless, the problem of ozone depletion is diminishing and is expected to be substantially corrected by the second half of the 21st century (figure 58.26b).

The CFC story is an excellent example of how environmental problems arise and can be solved. Initially, CFCs were heralded as an efficient and cost-effective way to provide cooling, a clear improvement over previous technologies. At that time, their harmful consequences were unknown. Once the problems were identified, international agreements led to an effective solution, and creative technological advances led to replacements that solved the problem at little cost.

Learning Outcomes Review 58.5

Pollution and resource depletion are the major human effects on the environment, with freshwater habitats being most threatened. Point-source pollution comes from identifiable locations, such as factories, whereas diffuse pollution comes from numerous sources, such as fertilized lawns. Deforestation is a major problem in that it destroys habitat, disrupts communities, depletes resources, and changes the local water cycle and weather patterns. Habitat fragmentation and other environmental disturbances are playing a role in the transmission of new diseases from animals to humans. Overfishing is the greatest problem in the oceans.

- Were CFCs an example of point-source or diffuse pollution? In general, how do efforts to combat pollution depend on their source?

58.6 Human Impacts on the Biosphere: Climate Change

Learning Outcomes
1. Explain the link between atmospheric carbon dioxide and global warming.
2. Describe the consequences of global warming on ecosystems and human health.

By studying the Earth's history and making comparisons with other planets, scientists have determined that concentrations of gases in our atmosphere, particularly CO_2, maintain the average temperature on Earth about 25°C higher than it would be if these gases were absent. This fact emphasizes that the composition of our atmosphere is a key consideration for life on Earth. Unfortunately, human activities are now changing the composition of the atmosphere in ways that most authorities conclude will be damaging or, in the long run, disastrous.

Because of changes in atmospheric composition, the average temperature of the Earth's surface is increasing, a phenomenon called global warming. As you might imagine from what we said at the beginning of this chapter, changes in temperature alter global wind and water-current patterns in complicated ways. This means that as the average global temperature increases, some particular regions of the world warm to a lesser extent, whereas other regions heat up to a greater extent (figure 58.28). It also means that rainfall patterns are altered because global precipitation patterns depend on global wind patterns. Enormous computer models are used to calculate the effects predicted in all parts of the world.

Independent computer models predict global changes

The Intergovernmental Panel on Climate Change, which shared the 2007 Nobel Peace Prize with Al Gore for their work on global climate change, recently released its fifth assessment report. Based on a variety of different scenarios, computer models predicted that global temperatures would increase 2.5°C to 7.8°C (4.5°–14.0°F) by the end of this century.

More ominous perhaps than temperature are some of the predictions for precipitation. For example, although northern Europe is expected to receive more precipitation than today, another recent study predicted that parts of southern Europe will receive about 20% less, disrupting natural ecosystems, agriculture, and human water supplies. Some European countries may come out ahead economically, but others will come out behind, and political relationships among countries will likely change as some shift from being food exporters to the more tenuous role of requiring food imports.

Carbon dioxide is a major greenhouse gas

Carbon dioxide is the gas usually emphasized in discussing the cause of global warming (figure 58.29), although other atmospheric gases are also involved. A monitoring station on the top of the Mauna Loa volcano on the island of Hawaii has monitored the concentration of atmospheric CO_2 since the 1950s. This station is particularly important because it is in the middle of the Pacific Ocean, far from the great continental landmasses where most people live, and it is therefore able to monitor the state of the global atmosphere without confounding influences of local events.

In 1958, the atmosphere was 0.031% CO_2. By 2014, the concentration had risen to 0.040%. All authorities agree that the cause of this steady rise in atmospheric CO_2 is the burning of coal and

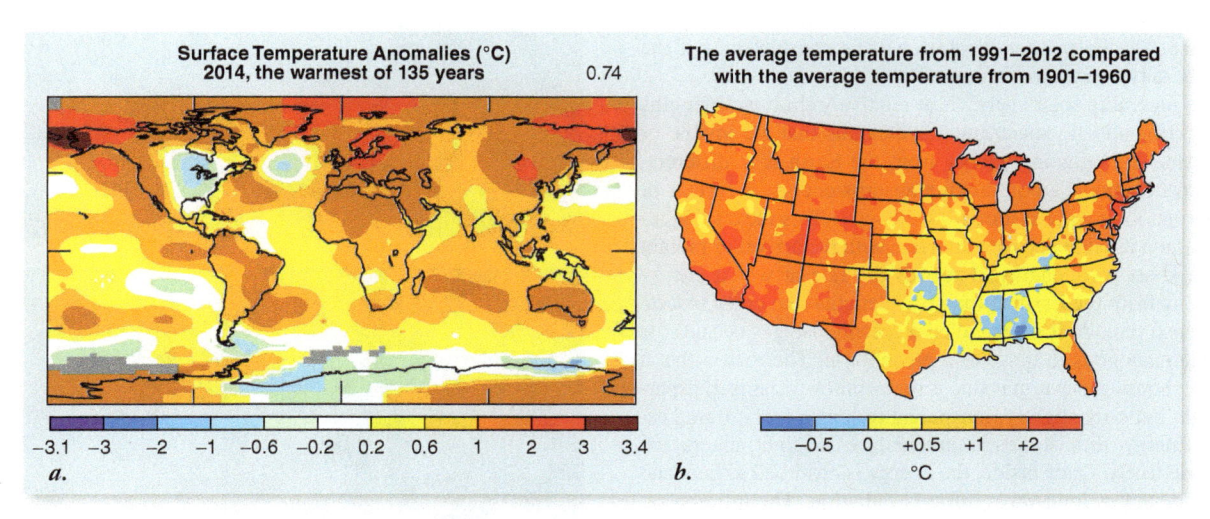

Figure 58.28 Geographic variation in global warming. The year 2014 was the warmest year since record-keeping began in 1880, and the 15 warmest years have all occurred since 1998. (*a*) Some areas of the globe heated up more than others. Colors indicate how much warming occurred in 2014 relative to the mean temperature during a reference period (1901–1960) prior to full onset of the modern greenhouse effect. (*b*) Similarly, some areas of the United States have gotten warmer more than others. This figure illustrates average temperatures from 1991–2012 compared with average temperatures from 1901–1960.

Source: National Climatic Data Center

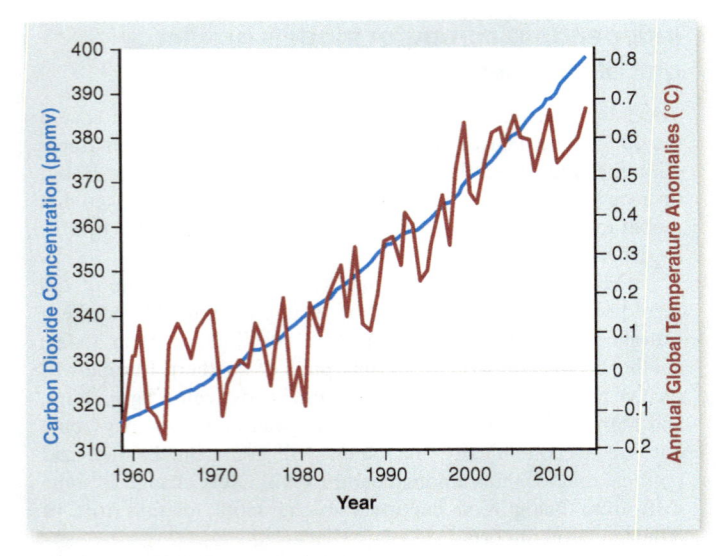

Figure 58.29 The greenhouse effect. The concentration of carbon dioxide in the atmosphere has increased steadily since the 1950s, as shown by the blue line. The red line shows the change in average global temperatures for the same period. The temperatures are recorded as anomalies relative to the mean temperature during a reference period (1951–1980).

petroleum products by the increasing (and increasingly energy-demanding) human population.

How carbon dioxide affects temperature

The atmospheric concentration of CO_2 affects global temperature because carbon dioxide strongly absorbs electromagnetic radiant energy at some of the wavelengths that are critical for the global heat budget. As stressed in chapter 57, the Earth not only receives radiant energy from the Sun, but also emits radiant energy into outer space. The Earth's temperature will be constant only if the rates of these two processes are equal.

The incoming solar energy is at relatively short wavelengths of the electromagnetic spectrum: wavelengths that are visible or near-visible. The outgoing energy from the Earth is at different, longer wavelengths. Carbon dioxide absorbs energy at certain of the important long-wave infrared wavelengths. This means that although carbon dioxide does not interfere with the arrival of radiant energy at short wavelengths, it retards the rate at which energy travels away from the Earth at long wavelengths into outer space.

Carbon dioxide is often called a greenhouse gas because its effects are analogous to those of a greenhouse. The reason that a glass greenhouse gets warm inside is that window glass is transparent to light but only slightly transparent to long-wave infrared radiation. Energy that strikes a greenhouse as light enters the greenhouse freely. Once inside, the energy is absorbed as heat and then re-radiated as long-wave infrared radiation. The infrared radiation cannot easily get out through the glass, and therefore energy accumulates inside.

Other greenhouse gases

Carbon dioxide is not the only greenhouse gas. Others include methane and nitrous oxide. The effect of any particular greenhouse gas depends on its molecular properties and concentration. For example, molecule-for-molecule, methane has about 20 times the heat-trapping effect of carbon dioxide; on the other hand, methane is less concentrated and less long-lived in the atmosphere than carbon dioxide.

Methane is produced in globally significant quantities in anaerobic soils and in the fermentation reactions of ruminant mammals, such as cows. Huge amounts of methane are presently locked up in Arctic permafrost. Melting of the permafrost could cause a sudden and large perturbation in global temperature by releasing methane rapidly.

Agricultural use of fertilizers is the largest source of nitrous oxide emissions, with energy consumption second and industrial use third.

Global temperature change has affected ecosystems in the past and is doing so now

Evidence for warming can be seen in many ways. For example, on a worldwide statistical basis, ice on lakes and rivers forms later and melts sooner than it used to; on average, ice-free seasons are now 2.5 weeks longer than they were a century ago. Also, the extent of ice at the North Pole has decreased substantially, and glaciers are retreating around the world (figure 58.30).

Global warming—and cooling—have occurred in the past, most recently during the ice ages and intervening warm periods. Species often responded by shifting their geographic ranges, tracking their environments. For example, a number of cold-adapted North American tree species that are now found only in the far north, or at high elevations, lived much farther south or at substantially lower elevations 10,000 to 20,000 years ago, when

Figure 58.30 Disappearing glaciers. Mount Kilimanjaro in Tanzania in 1970 *(top)* and 2014 *(bottom)*. Note the decrease in glacier coverage over three decades.

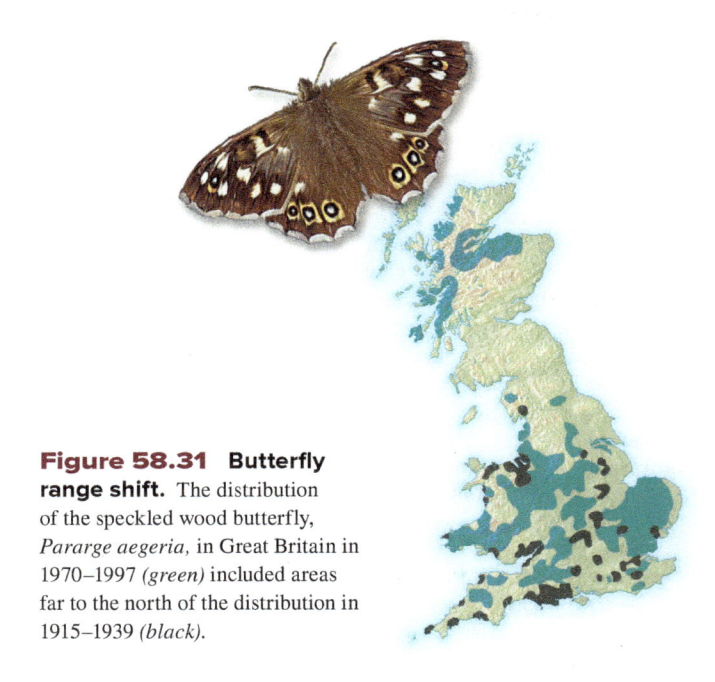

Figure 58.31 Butterfly range shift. The distribution of the speckled wood butterfly, *Pararge aegeria*, in Great Britain in 1970–1997 *(green)* included areas far to the north of the distribution in 1915–1939 *(black)*.

conditions were colder. Present-day global warming is having similar effects. For example, many butterfly and bird species have shifted northward in recent decades (figure 58.31).

Many migratory birds arrive earlier at their summer breeding grounds than they did decades ago. Many insects and amphibians breed earlier in the year, and many plants flower earlier. In Australia, recent research shows that wild fruit fly populations have undergone changes in gene frequencies in the past 20 years, such that populations in cool parts of the country now genetically resemble ones in warm parts.

Reef-building corals seem to have narrow margins of safety between the sea temperatures to which they are accustomed and the maximum temperatures they can survive. Global warming seems already to be threatening some corals by inducing mass "bleaching," a disruption of the normal and necessary symbiosis between the cnidarians and algal cells.

There are reasons to think that the effects of global warming on natural ecosystems today may, overall, be more severe than those of warming events in the distant past. One concern is that the rate of warming today is rapid, and therefore evolutionary adaptations would need to occur over relatively few generations to aid the survival of species. Another concern is that natural areas no longer cover the whole landscape, but often take the form of parks that are completely surrounded by cities or farms. The parks are at fixed geographic locations and in general cannot be moved. If climatic conditions in a park become unsuitable for its inhabitants, the species cannot shift their range because the park is surrounded by inhospitable habitats; as a result, the species may disappear and the park's biodiversity will plummet.

Similarly, as temperatures increase, many montane species have shifted to higher altitudes to find their preferred habitat. However, eventually they can shift no higher because they reach the mountain's peak. As the temperature continues to increase, the species' habitat disappears entirely. A number of Costa Rican frog species are thought to have become extinct for this reason. The same fate may befall many Arctic species as their habitat melts away.

Global warming affects human populations as well

Global warming could affect human health and welfare in a variety of ways. Some of these changes may be beneficial, but even if they are detrimental, some countries—particularly the wealthier ones—will be able to adjust. Poorer countries, however, may not be able to transform as quickly, and some changes will require extremely costly countermeasures that even wealthy countries will be hard-pressed to afford.

Rising sea levels

Since the late 20th century, sea level has risen more than 3 mm per year. The U.S. Environmental Protection Agency predicts that sea level is likely to rise two or three times faster in the 21st century because of two effects of global warming: (1) the melting of polar ice and glaciers, adding water to the ocean; and (2) the increase of average ocean temperature, causing an increase in volume because water expands as it warms. Such an increase would cause increased erosion and inundation of low-lying land and coastal marshes, and other habitats would also be imperiled. As many as 200 million people would be affected by increased flooding. Should sea levels continue to rise, coastal cities and some entire islands, such as the Maldives in the Indian Ocean, would be in danger of becoming submerged.

Other climatic effects

Global warming is predicted to have a variety of effects besides increased temperatures. In particular, the frequency or severity of extreme meteorological events—such as heat waves, droughts, severe storms, and hurricanes—is expected to increase, and El Niño events, with their attendant climatic effects, may become more common.

In addition, rainfall patterns are likely to shift, and those geographic areas that are already water-stressed, which are currently home to nearly 2 billion people, will likely face even graver water shortage problems in the years to come. Some evidence suggests that these effects are already evident in the increase in powerful storms, hurricanes, and the frequency of El Niño events over the past few years.

Effects on agriculture

Global warming may have both positive and negative effects on agriculture. On the positive side, warmer temperatures and increased atmospheric carbon dioxide tend to increase growth of some crops and thus may increase agricultural yields. Other crops, however, may be negatively affected. Furthermore, most crops will be affected by increased frequencies of droughts. Moreover, although crops in north temperate regions may flourish with higher temperatures, many tropical crops are already growing at their maximal temperatures, so increased temperatures may lead to reduced crop yields.

Also on the negative side, changes in rainfall patterns, temperature, pest distributions, and various other factors will require many adjustments. Such changes may come relatively easily for farmers in the developed world, but the associated costs may be devastating for those in the developing countries.

Effects on human health

Increasingly frequent storms, flooding, and drought will have adverse consequences on human health. Aside from their direct effect, such events often disrupt the fragile infrastructure of developing countries, leading to the loss of safe drinking water and other problems. As a result, epidemics of cholera and other diseases may be expected to occur more often.

In addition, as temperatures rise, areas suitable for tropical organisms will expand northward. Of particular concern are those organisms that cause human diseases. Many diseases that used to be limited to tropical areas are expanding their ranges and becoming problematic in nontropical countries. Diseases transmitted by mosquitoes, such as malaria (see chapter 29), dengue fever, and several types of encephalitis, are examples. The distribution of mosquitoes is limited by cold; winter freezes kill many mosquitoes and their eggs. As a result, malaria only occurs in areas where temperatures are usually above 16°C, and yellow fever and dengue fever, transmitted by a different mosquito species from malaria, occur in areas where temperatures are normally above 10°C. Moreover, at higher temperatures, the malaria pathogen matures more rapidly.

Malaria already kills about 500,000 people every year; some projections suggest that the percentage of the human population at risk for malaria may increase by 33% by the end of the 21st century. Moreover, as predicted, malaria already appears to be on the move. By 1980, malaria had been eradicated from all of the United States except California, but in recent years it has appeared in a variety of southern, and even a few northern, states.

Dengue fever (sometimes called "breakbone fever" because of the pain it causes) is also spreading. Previously a disease restricted to the tropics and subtropics, where it infects 50 to 100 million people a year, it now occurs in the United States, southern South America, and northern Australia.

One of the most alarming aspects of these diseases is that no vaccines are available. Drug treatment is available (for malaria), but the parasites are rapidly evolving resistance and rendering the drugs ineffective. There is no drug treatment for dengue fever.

Solving the problem

The release of the IPCC's fifth assessment in 2013 may come to be seen as a turning point in humanity's response to climate change. Global warming is now recognized, even by former skeptics, as an ongoing phenomenon caused in large part by human actions. Even formerly recalcitrant governments now seem poised to take action, and corporations are recognizing the opportunities provided by the need to reverse human impacts. The resulting "green" technologies and practices are becoming increasingly common. With concerted efforts from citizens, corporations, and governments, the more serious consequences of global climate change hopefully can be averted, just as ozone depletion was reversed in the last century.

Learning Outcomes Review 58.6

Carbon dioxide is a significant greenhouse gas, meaning that it prevents heat from escaping the Earth so that temperatures rise. Global warming caused by changes in atmospheric composition—most notably CO_2 accumulation—may increase desertification and cause some habitats and species to disappear. Global warming may also melt ice caps and glaciers, altering coastlines as water levels rise. Violent weather events, disruption of water availability, and flooding of low-lying areas, as well as increased incidence of tropical diseases, may also occur.

- In what ways does global climate change pose different questions from those posed by ozone depletion?

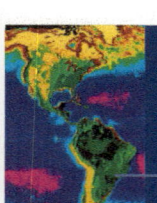

Chapter Review

58.1 Ecosystem Effects of Sun, Wind, and Water

Solar energy and the Earth's rotation affect atmospheric circulation.
The amount of solar radiation reaching the Earth's surface has a great effect on climate. The seasons result from changes in the Earth's position relative to the Sun (figure 58.1). Hot air with its increased water content rises at the equator, then cools and loses its moisture, creating the equatorial rainforests (figure 58.3).

As the drier cool air of the upper atmosphere moves away from the equator and then descends to Earth, it removes moisture from the Earth's surface and creates deserts on its way back to the equator.

Winds travel in curved paths relative to the Earth's surface because the Earth rotates on its axis (the Coriolis effect; see figure 58.3).

Global currents are largely driven by winds (figure 58.4).
Four large circular gyres in ocean currents can be found, driven by wind direction. These also are influenced by the Coriolis effect.

Regional and local differences affect terrestrial ecosystems.
A rain shadow occurs when a range of mountains removes moisture from air moving over it from the windward side, creating a drier environment on the opposite side (figure 58.5).

For every 1000-m increase in elevation, temperature drops approximately 6°C (figure 58.6).

Microclimates are small-scale differences in conditions.

58.2 Earth's Biomes

Temperature and moisture often determine biomes.
Average annual temperature and rainfall, as well as the range of seasonal variation, determine different biomes. Eight major types of biomes are recognized.

Tropical rainforests are highly productive equatorial systems.

Savannas are tropical grasslands with seasonal rainfall.

Deserts are regions with little rainfall.

Temperate grasslands have rich soils.

Temperate deciduous forests are adapted to seasonal change.

Temperate evergreen forests are coastal.

Taiga is the northern forest where winters are harsh.

Tundra is a largely frozen treeless area with a short growing season.

58.3 Freshwater Habitats

Life in freshwater habitats depends on oxygen availability.

Oxygen is not very soluble in water. Oxygen is constantly added by photosynthesis of aquatic plants and removed by heterotrophs.

Lake and pond habitats change with water depth.

The photic zone, near the surface, is the zone of primary productivity; its depth varies with water clarity (figure 58.12).

In the summer, the warmer water (epilimnion) floats on top of the colder water (hypolimnion). Freshwater lakes turn over twice a year as the temperature at the surface and at depth become the same, and the layers are set in motion by wind (figure 58.13).

Lakes differ in oxygen and nutrient content.

Oligotrophic lakes have high oxygen and low nutrients, whereas eutrophic lakes are the opposite.

58.4 Marine Habitats

The ocean is divided into several zones: intertidal, neritic, photic, benthic, and pelagic zones (figure 58.15).

Open oceans have low primary productivity.

Phytoplankton is the primary producer in open waters, and primary production is low due to low nutrient levels.

Continental shelf ecosystems provide abundant resources.

Neritic waters are found over continental shelves and have higher nutrient levels (figure 58.15). Estuaries frequently contain rich intertidal zones. Other ecosystems include productive banks on continental shelves and symbiotic coral reef ecosystems.

Upwelling regions experience mixing of nutrients and oxygen.

In upwelling regions, local winds bring up nutrient-rich deep waters, creating the highest rates of primary production. El Niño events occur when Trade Winds weaken, restricting upwelling to surface waters rather than to the deeper nutrient-rich waters.

The deep sea is a cold, dark place with some fascinating communities.

The deep sea is the single largest habitat. Hydrothermal vent communities occur where tectonic plates are moving apart; chemoautotrophs living there obtain energy from oxidation of sulfur.

58.5 Human Impacts on the Biosphere: Pollution and Resource Depletion

Dangerous chemicals like DDT are biomagnified as energy moves up the food chain (figure 58.20).

Freshwater habitats are threatened by pollution and resource depletion.

Point-source and diffuse pollution, acid precipitation, and overuse threaten freshwater habitats (figure 58.21).

Forest ecosystems are threatened in tropical and temperate regions.

Deforestation leads to loss of habitat, disruption of the water cycle, and loss of nutrients. Acid rain has a major detrimental effect on forests as well as on lakes and streams (figure 58.23).

Disease transmission from animals to people

Marine habitats are being depleted of fish and other species.

Many fisheries, such as the Georges Bank ecosystem, have collapsed and have not recovered.

Stratospheric ozone depletion has led to an ozone "hole."

Increased transmission of UV-B radiation is harmful to life. Global regulation of CFCs seems to be reversing ozone depletion.

58.6 Human Impacts on the Biosphere: Climate Change

Independent computer models predict global changes.

Carbon dioxide is a major greenhouse gas.

Carbon dioxide allows solar radiation to pass through the atmosphere but prevents heat from leaving the Earth, creating warmer conditions.

Global temperature change has affected ecosystems in the past and is doing so now.

If temperatures change rapidly, natural selection cannot occur rapidly enough to prevent many species from becoming extinct.

Global warming affects human populations as well.

Changing sea levels, increased frequency of extreme climatic events, direct and indirect effects on agriculture, and the expansion of tropical diseases can all affect human life.

Human environmental disturbance is leading to the transmission of new diseases from animals to humans.

Review Questions

UNDERSTAND

1. The Coriolis effect
   a. drives the rotation of the Earth.
   b. is responsible for the relative lack of seasonality at the equator.
   c. drives global wind circulation patterns.
   d. drives global wind and ocean circulation patterns.

2. What two factors are most important in biome distribution?
   a. Temperature and latitude
   b. Rainfall and temperature
   c. Latitude and rainfall
   d. Temperature and soil type

3. In a rain shadow, air is cooled as it rises and heated as it descends, often producing a wet and dry side because the water-holding capacity of the air
   a. is directly related to air temperature.
   b. is inversely related to air temperature.
   c. is unaffected by air temperature.
   d. produces changes in air temperature.

4. Thermal stratification in a lake
   a. is not modified by fall and spring overturn.
   b. leads to higher oxygen in deep versus surface waters.
   c. leads to higher oxygen in surface versus deep waters.
   d. is reduced when ice forms on the surface of the lake.

5. Oligotrophic lakes have
   a. low oxygen, and high nutrient availability.
   b. high oxygen, and high nutrient availability.
   c. high oxygen, and low nutrient availability.
   d. low oxygen, and low nutrient availability.

6. Deep-sea hydrothermal vent communities
   a. get their energy from photosynthesis in the photic zone near the surface.
   b. use bioluminescence to generate food.
   c. are built on the energy produced by the activity of chemoautotrophs that oxidize sulfur.
   d. contain only bacteria and other microorganisms.

7. Biological magnification occurs when
   a. pollutants increase in concentration in tissues at higher trophic levels.
   b. the effect of a pollutant is magnified by chemical interactions within organisms.
   c. an organism is placed under a dissecting scope.
   d. a pollutant has a greater than expected effect once ingested by an organism.

8. Which of the following is a point source of pollution?
   a. Lawns
   b. Smokestacks of coal-fired power plants
   c. Factory effluent pipe draining into a river
   d. Acid rain

APPLY

1. If the Earth were not tilted on its axis of rotation, the annual cycle of seasons in the northern and southern hemispheres
   a. would be reversed.
   b. would stay the same.
   c. would be reduced.
   d. would not exist.

2. Oligotrophic lakes can be turned into eutrophic lakes as a result of human activities such as
   a. overfishing of sensitive species, which disrupts fish communities.
   b. introducing nutrients into the water, which stimulates plant and algal growth.
   c. disrupting terrestrial vegetation near the shore, which causes soil to run into the lake.
   d. spraying pesticides into the water to control aquatic insect populations.

3. If a pesticide is harmless at low concentrations (such as DDT) and used properly, how can it become a threat to nontarget organisms?
   a. Because after exposure to DDT, some species develop allergic reactions even at low levels of exposure
   b. Because DDT molecules can combine so that their concentration increases through time
   c. Because the concentration of chemicals such as DDT is increasingly concentrated at higher trophic levels
   d. Because global warming and exposure to UV-B radiation renders molecules such as DDT increasingly potent

4. If there are many greenhouse gases, why is only carbon dioxide considered a cause of global warming?
   a. The other gases do not cause global warming.
   b. Scientists are concerned about other causes; for example, release of methane from melting permafrost could have significant effects on global warming.
   c. Other gases occur in such low quantities that they have little effect on the climate.
   d. Carbon dioxide is the only gas that absorbs long-wavelength infrared radiation.

SYNTHESIZE

1. Discuss how figure 58.1 explains the pattern observed in figure 58.2.

2. Why are most of the Earth's deserts found at approximately 30° latitude?

3. If the world has experienced global warming many times in the past, why should we be concerned about it happening again now?

4. What can be done to prevent new diseases from moving from animals to humans?

CHAPTER 59

Conservation Biology

Chapter Contents

59.1 Overview of the Biodiversity Crisis
59.2 The Value of Biodiversity
59.3 Factors Responsible for Extinction
59.4 Approaches for Preserving Endangered Species and Ecosystems

Introduction

Among the greatest challenges facing the biosphere is the accelerating pace of species extinctions. Not since the end of the Cretaceous period (65 MYA) have so many species become extinct in so short a time span. This challenge has led to the emergence of the discipline of conservation biology. Conservation biology is an applied science that seeks to learn how to preserve species, communities, and ecosystems. It studies the causes of declines in species richness and attempts to develop methods for preventing such declines. In this chapter, we first examine the biodiversity crisis and its importance. Then, using case histories, we identify and study factors that have played key roles in many extinctions. We finish with a review of recovery efforts at the species and community levels.

59.1 Overview of the Biodiversity Crisis

Learning Outcomes

1. Describe the history of extinction through time.
2. Explain the importance of hotspots to biodiversity conservation.

Extinction is a fact of life. Most species—probably all—become extinct eventually. More than 99% of species known to science (most from the fossil record) are now extinct. Current rates of extinction are alarmingly high, however. Taking into account the current rapid and accelerating loss of habitat, especially in the tropics, it has been calculated that as much as 20% of the world's biodiversity may disappear by the middle of this century. In addition, many of these species may disappear before we are even aware of their existence. Scientists estimate that no more than 15% of the world's eukaryotic organisms have been discovered and given scientific names, and this proportion is probably much lower for tropical species.

These losses will affect more than poorly known groups. As many as 50,000 species of the world's total of 250,000 species of plants, 4000 of the world's 20,000 species of butterflies, and nearly 2000 of the world's 10,000 species of birds could be lost during this

Figure 59.1 North America before human inhabitants. Animals found in North America prior to the arrival of humans included birds and large mammals, such as the saber-toothed cat, giant ground sloth, and teratorn vulture. This painting by Charles R. Knight is titled, *Rancho La Brea Tar Pit*. It's displayed in the George C. Page Museum in Los Angeles.

period. Considering that the human species has been in existence for less than 200,000 years of the world's 4.5-billion-year history, and that our ancestors developed agriculture only about 10,000 years ago, this is an astonishing—and dubious—accomplishment.

Prehistoric humans were responsible for local extinctions

A great deal can be learned about current rates of extinction by studying the past. In prehistoric times, members of *Homo sapiens* wreaked havoc whenever they entered a new area. For example, at the end of the last Ice Age, approximately 12,000 years ago, the fauna of North America was composed of a diversity of large mammals similar to those living in Africa today: mammoths and mastodons, horses, camels, giant ground sloths, saber-toothed cats, and lions, among others (figure 59.1).

Shortly after humans arrived, 74 to 86% of the megafauna (that is, animals weighing more than 100 lb) became extinct. These extinctions are thought to have been caused by hunting and, indirectly, by burning and clearing of forests. (Some scientists attribute these extinctions to climate change, but that hypothesis doesn't explain why the ends of earlier Ice Ages were not associated with mass extinctions, nor does it explain why extinctions occurred primarily among larger animals, with smaller species being relatively unaffected.)

Around the globe, similar results have followed the arrival of humans. Forty thousand years ago, Australia was occupied by a wide variety of large animals, including marsupials similar in size and ecology to hippos and leopards, a kangaroo 9 ft tall, and a 20-ft-long monitor lizard. These all disappeared at approximately the same time as humans arrived.

Smaller islands have also been devastated. Madagascar has seen the extinction of at least 15 species of lemurs, including one the size of a gorilla; a pygmy hippopotamus; and the flightless elephant bird, *Aepyornis*, the largest bird to ever live (more than 3 m tall and weighing 450 kg). On New Zealand, 30 species of birds went extinct, including all 13 species of moas, another group of large, flightless birds. Interestingly, one continent that seems to have been spared these megafaunal extinctions is Africa. Scientists speculate that this lack of extinction in prehistoric Africa may have resulted because much of human evolution occurred in Africa. Consequently, African species had been coevolving with humans for several million years and thus had evolved counteradaptations to human predation.

Extinctions have continued in historical time

Historical extinction rates are best known for birds and mammals because these species are conspicuous—that is, relatively large and well studied. Estimates of extinction rates for other species are much rougher. The data presented in table 59.1, based on the best available evidence, show recorded extinctions from 1600 to the present. These estimates indicate that about 85 species of mammals and 113 species of birds have become extinct since the year 1600. That is about 1.7% of known mammal species and 1.1% of known birds.

The majority of extinctions have occurred since 1900: five species of plants and animals per year during the 20th century. This increase in the rate of extinction is the heart of the biodiversity crisis.

Unfortunately, the situation is worsening. For example, the number of bird species recognized as "endangered" increased 13% from 1999 to 2009, and a recent report suggested that as many as half of Earth's plant species may be threatened with extinction. Some researchers predict that two-thirds of all vertebrate species could perish by the end of this century.

The majority of historic extinctions—though by no means all of them—have occurred on islands. For example, of the 85 species of mammals that have gone extinct in the last 400 years, 60% lived on islands. The particular vulnerability of island species probably results from a number of factors: Such species

TABLE 59.1 Recorded Extinctions Since 1600

Taxon	Mainland	Island	Ocean	Total	Approximate Number of Species	Percent of Taxon Extinct
Mammals	30	51	4	85	5,000	1.7
Birds	21	92	0	113	10,000	1.1
Reptiles	1	20	0	21	10,000	0.2
Fish	22	1	0	23	31,000	0.1
Invertebrates*	49	48	1	98	1,000,000+	0.01
Flowering plants	245	139	0	384	250,000	0.2

*Number of extinct invertebrates is probably greatly underestimated due to lack of knowledge for many species (other groups are probably underestimated to a lesser extent for the same reason).

have often evolved in the absence of predators, and so have lost their ability to escape both humans and introduced predators such as rats and cats. In addition, humans have introduced competitors and diseases; malaria, for example, has devastated the bird fauna of the Hawaiian Islands. Finally, island populations are often relatively small, and thus particularly vulnerable to extinction, as we shall see in section 59.3.

In recent years, the extinction crisis has moved from islands to continents. Most species now threatened with extinction occur on continents, and these areas will bear the brunt of the extinction crisis in this century.

Some people have argued that we should not be concerned, because extinctions are a natural event and mass extinctions have occurred in the past. Indeed, mass extinctions have taken place several times over the past half-billion years (see section 22.7). However, the current mass extinction event is notable in several respects. First, it is the only such event triggered by a single species (us!). Moreover, although species diversity usually recovers after a few million years, this is a long time to deny our descendants the benefits and joys of biodiversity.

In addition, it is not clear that biodiversity will rebound this time. After previous mass extinctions, new species have evolved to utilize resources newly available due to extinctions of the species that previously used them. Today, however, such resources are unlikely to be available, because humans are destroying the habitats and taking the resources for their own use.

Endemic species hotspots are especially threatened

A species found naturally in only one geographic area and no place else is said to be endemic to that area. The area over which an endemic species is found may be very large. For example, the black cherry tree *(Prunus serotina)* is endemic to all of temperate North America. More typically, however, endemic species occupy restricted ranges. The Komodo dragon *(Varanus komodoensis)* lives only on a few small islands in the Indonesian archipelago, and the Mauna Kea and Haleakala silverswords *(Argyroxiphium sandwicense sandwicense* and *A. s. macrocephalum)* each lives in a single volcano crater on the island of Hawaii (figure 59.2). Isolated geographic areas, such as oceanic islands, lakes, and mountain peaks, often have high percentages of endemic species, many in significant danger of extinction.

The number of endemic plant species can vary greatly from one place to another. In the United States, for example, 379 plant species are found in Texas and nowhere else, whereas New York has only one endemic plant species. California, with its varied array of habitats, including deserts, mountains, seacoast, old-growth forests, and grasslands, is home to more endemic plant species than any other state.

Species hotspots

Worldwide, notable concentrations of endemic species occur in particular regions. Conservationists have identified areas, termed hotspots, that have high endemism and are disappearing at a rapid rate. Such hotspots include Madagascar, a variety of tropical

Figure 59.2 Haleakala silversword, *Argyroxiphium sandwicense*. Many species of silverswords are endemic to very small areas. This photo illustrates two stages in the plant's life cycle.

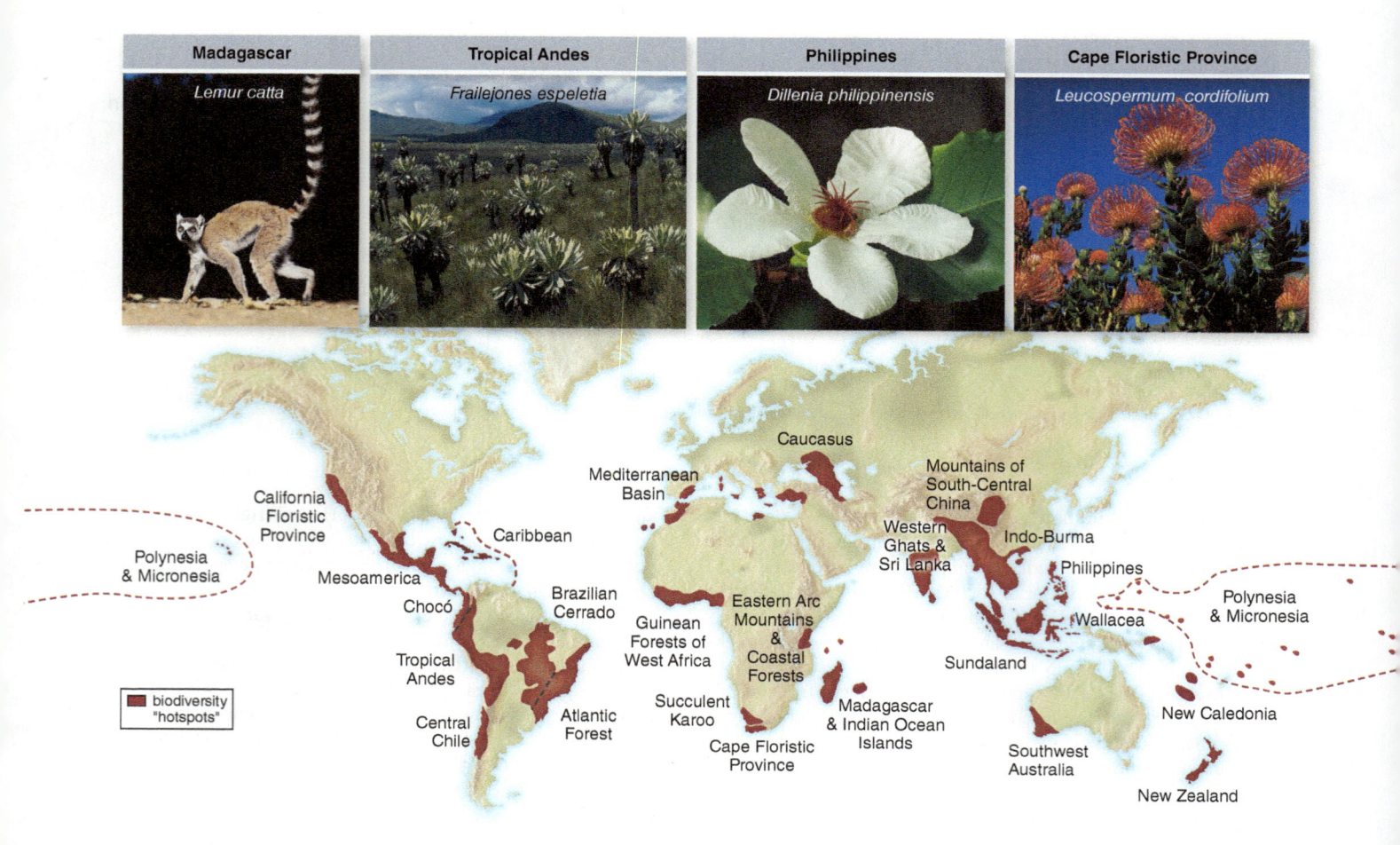

Figure 59.3 Hotspots of high endemism. These areas are rich in endemic species under threat of imminent extinction.

rainforests, the eastern Himalayas, areas with Mediterranean climates such as California, South Africa, and Australia, and several other areas (figure 59.3 and table 59.2). Overall, 25 such hotspots have been identified, which in total contain nearly half of all the terrestrial species in the world.

Why these areas contain so many endemic species is a topic of active scientific research. Some of these hotspots occur in areas of high species diversity; for these hotspots, the explanations for high species diversity in general, such as high productivity, probably apply (see chapter 57). In addition, some hotspots occur on isolated islands, such as New Zealand, New Caledonia, and the Hawaiian Islands, where evolutionary diversification over long periods has resulted in rich biotas composed of plant and animal species found nowhere else in the world.

Human population growth in hotspots

Because of the great number of endemic species that hotspots contain, conserving their biological diversity must be an important component of efforts to safeguard the world's biological heritage. Or, to look at it another way, by protecting just 1.4% of the world's land surface, 44% of the world's vascular plants and 35% of its terrestrial vertebrates can be preserved.

Unfortunately, hotspots contain not only many endemic species, but also growing human populations. In 1995, these areas contained 1.1 billion people—20% of the world's population—sometimes at high densities (figure 59.4a). More important, human populations were growing in all but one of these hotspots both because birth rates are much higher than death rates and because rates of immigration into these areas are often high. Overall, the rate of growth exceeded the global average in 23 hotspots (figure 59.4b). In some hotspots, the rate of growth is nearly twice that of the rest of the world.

Not surprisingly, many of these areas are experiencing high rates of habitat destruction as land is cleared for agriculture, housing, and economic development. More than 70% of the original area of each hotspot has already disappeared, and in 14 hotspots, 15% or less of the original habitat remains. In Madagascar, it is estimated that 90% of the original forest has already been lost—this on an island where 85% of the species are found nowhere else in the world. In the forests of the Atlantic

TABLE 59.2	Numbers of Endemic Species in Some Hotspot Areas			
Region	Mammals	Reptiles	Amphibians	Plants
Atlantic coastal forest (Brazil)	160	60	253	6,000
South American Chocó	60	63	210	2,250
Philippines	115	159	65	5,832
Tropical Andes	68	218	604	20,000
Southwestern Australia	7	50	24	4,331
Madagascar	84	301	187	9,704
Cape region (South Africa)	9	19	19	5,682
California Floristic Province	30	16	17	2,125
New Caledonia	6	56	0	2,551
South-Central China	75	16	51	3,500

coast of Brazil, the extent of deforestation is even higher: 95% of the original forest is gone.

Population pressure is not the only cause of habitat destruction in hotspots. Commercial exploitation to meet the demands of more affluent people in the developed world also plays an important role. For example, large-scale logging of tropical rainforests occurs in countries around the world to provide lumber, most of which ends up in the United States, western Europe, and Japan.

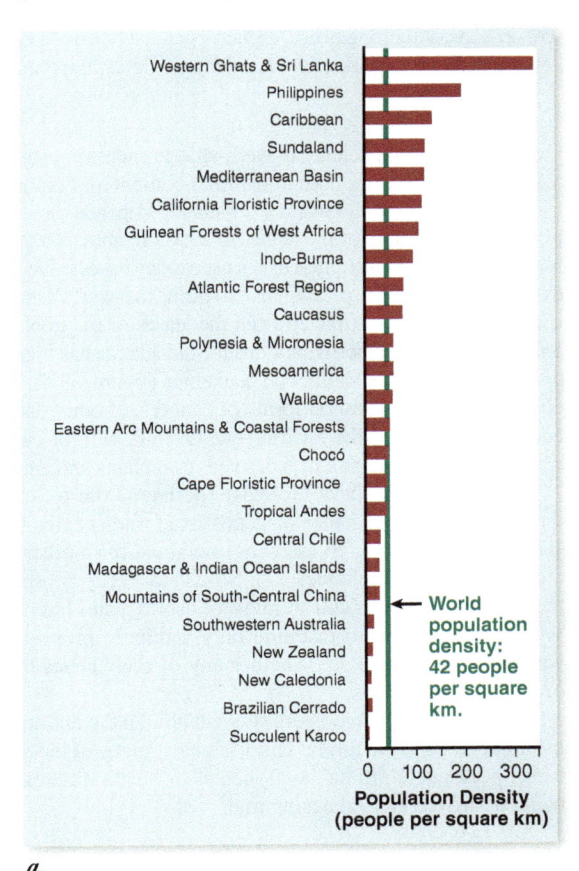

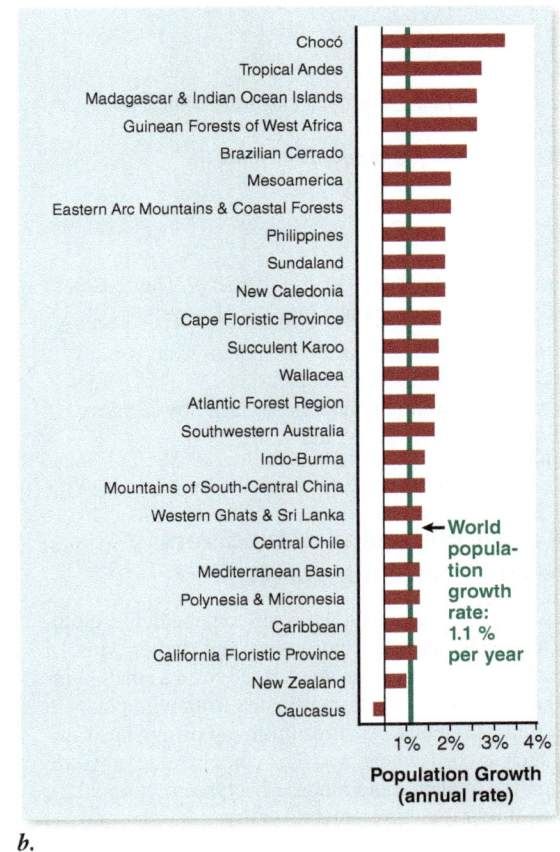

Figure 59.4 Human populations in hotspots. *a.* Human population density and *(b)* population growth rate in biodiversity hotspots.

Inquiry question Why do population density and growth rates differ among hotspots?

Data analysis If we assume that population density is a good indicator of threat to the environment, which hotspots are not in particularly bad shape now, but are likely to become problematic in the future because their populations are currently growing quickly?

Similarly, many forests in Central and South America are cleared to make way for cattle ranches that produce cheap meat for fast-food restaurants. Hotspots in more affluent countries are often at risk because they occur in areas where land has great value for real estate and commercial purposes, such as in Florida and California in the United States.

Learning Outcomes Review 59.1

The most recent losses of biodiversity have resulted from human activities, in both prehistoric and historical times. Endemic species are found in only a single region on Earth; regions with a high number of endemic species, known as hotspots, are particularly threatened by human encroachment and their preservation is critical.

- Why does so much resource exploitation occur in areas that are biodiversity hotspots?

59.2 The Value of Biodiversity

Learning Outcomes

1. Distinguish between the direct and indirect economic value of biodiversity.
2. Explain what is meant by the aesthetic value of biodiversity.

Why should we worry about loss of biodiversity? The reason is that biodiversity is valuable to us in a number of ways:

- *direct economic value* of products we obtain from species of plants, animals, and other groups;
- *indirect economic value* of benefits produced by species without our consuming them; and
- *ethical and aesthetic values.*

The direct economic value of biodiversity includes resources for our survival

Many species have direct value as sources of food, medicine, clothing, biomass (for energy and other purposes), and shelter. Most of the world's food crops, for example, are derived from a small number of plants that were originally domesticated from wild plants in tropical and semiarid regions. As a result, many of our most important crops, such as corn, wheat, and rice, contain relatively little genetic variation (equivalent to a founder effect; see section 20.3), whereas their wild relatives have great diversity.

In the future, genetic variation from wild strains of these species may be needed if we are to improve yields or find a way to breed resistance to new pests. In fact, recent agricultural breeding experiments have illustrated the value of conserving wild relatives of common crops. For example, by breeding commercial varieties of tomato with a small, oddly colored wild tomato species from the mountains of Peru, scientists were able to increase crop yields by 50%, while increasing both nutritional content and color.

About 70% of the world's population depends directly on wild plants as their source of medicine. In addition, about 40% of the prescription and nonprescription drugs used today have active ingredients extracted from plants or animals. Aspirin, the world's most widely used drug, was first extracted from the leaves of the tropical willow, *Salix alba*. The rosy periwinkle from Madagascar has yielded potent drugs for combating childhood leukemia (figure 59.5), and drugs effective in treating several forms of cancer and other diseases have been produced from the Pacific yew. Overall, 62% of cancer drugs were developed from products derived from plants and animals.

Only recently have biologists perfected the techniques that make possible the transfer of genes from one species to another. We are just beginning to use genes obtained from other species to our advantage (see chapter 17). So-called "gene prospecting" of the genomes of plants and animals for useful genes has only begun. We have been able to examine only a minute proportion of the world's organisms to see whether any of their genes have useful properties for humans.

By conserving biodiversity, we maintain the option of finding useful benefits in the future. Unfortunately, many of the most promising species occur in habitats, such as tropical rainforests, that are being destroyed at an alarming rate.

a. *b.*

Figure 59.5 Plants of pharmaceutical importance.
a. Two drugs extracted from the rosy periwinkle, *Catharanthus roseus*, vinblastine and vincristine, effectively treat common forms of childhood leukemia, increasing chances of survival from 20 to over 95%. *b.* Cancer-fighting drugs, such as paclitaxel (Taxol), have been developed from the bark of the Pacific yew, *Taxus brevifolia*.

Indirect economic value is derived from ecosystem services

Diverse biological communities are of vital importance to healthy ecosystems. They help maintain the chemical quality of natural water, buffer ecosystems against storms and drought, preserve soils and prevent loss of minerals and nutrients, moderate local and regional climate, absorb pollution, and promote the breakdown of organic wastes and the cycling of minerals.

In chapter 57, we discussed the evidence that the stability and productivity of ecosystems is related to species richness. By destroying biodiversity, we are creating conditions of instability and lessened productivity and promoting desertification, waterlogging, mineralization, and many other undesirable outcomes throughout the world.

The value of intact habitats

Economists have recently been able to compare the societal value, in monetary terms, of intact habitats compared with the value of destroying those habitats. In many studies, intact ecosystems are more valuable than the products derived by destroying them. In Thailand, as one example, coastal mangrove habitats are commonly cleared so that shrimp farms can be established. Although the shrimp produced are valuable, their value is vastly outweighed by the benefits in timber, charcoal production, offshore fisheries, and storm protection provided by the mangroves (figure 59.6a).

Similarly, intact tropical rainforest in Cameroon, West Africa, provides fruit and other forest materials. Clearing the forest for agriculture or palm plantations leads to stream-polluting erosion as well as increased flooding. Combining all the costs and benefits of the three options, maintaining intact forests has the highest economic value (figure 59.6b).

Case study: New York City watersheds

Probably the most famous example of the value of intact ecosystems is provided by the watersheds of New York City. Ninety percent of the water for the New York area's 8.5 million residents comes from the Catskill Mountains and the nearby headwaters of the Delaware River (figure 59.7). Water that runs off from over 4000 km² of rural, mountainous areas is collected into reservoirs and then transported by aqueduct more than 130 km to New York City at a rate of 4.9 billion liters per day.

In the 1990s, New York City faced a dilemma. New federal water regulations were requiring ever cleaner water, even as development and pollution in the source areas of the water were threatening to compromise water quality. The city had two choices: either work to protect the functioning ecosystem so that it could produce clean water, or construct filtration plants to clean it on arrival. Economic analysis made the choice clear: Building the plants would cost $6 billion, with annual operating costs of $300 million, whereas spending a billion dollars over 10 years could preserve the ecosystem and maintain water purity. The decision was easy.

Economic trade-offs

These examples provide some idea of the value of the services that ecosystems provide. But maintaining ecosystems is not always more valuable than converting them to other uses. Certainly, when the United States was being settled and land was plentiful, ecosystem conversion was beneficial. Even today, habitat destruction sometimes makes economic sense. Nonetheless, we

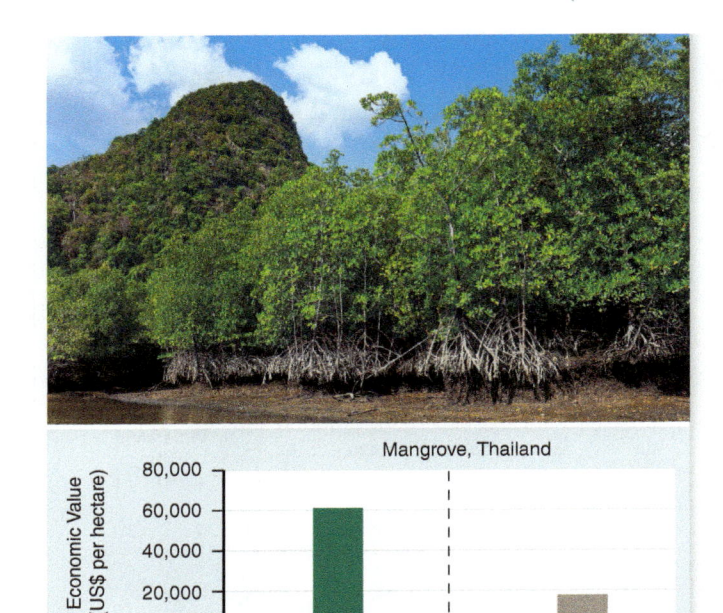

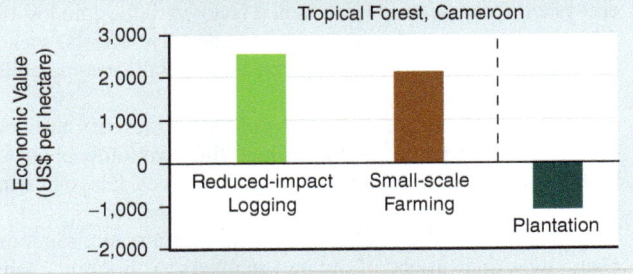

Figure 59.6 The economic value of maintaining habitats. *a.* Mangroves in Thailand are more valuable than shrimp farms. *b.* Rainforests in Cameroon provide more economic benefits if they are left standing than if they are destroyed and the land used for other purposes.

Data analysis If shrimp farms established on cleared mangrove habitats make money, how can clearing mangroves not be an economic plus?

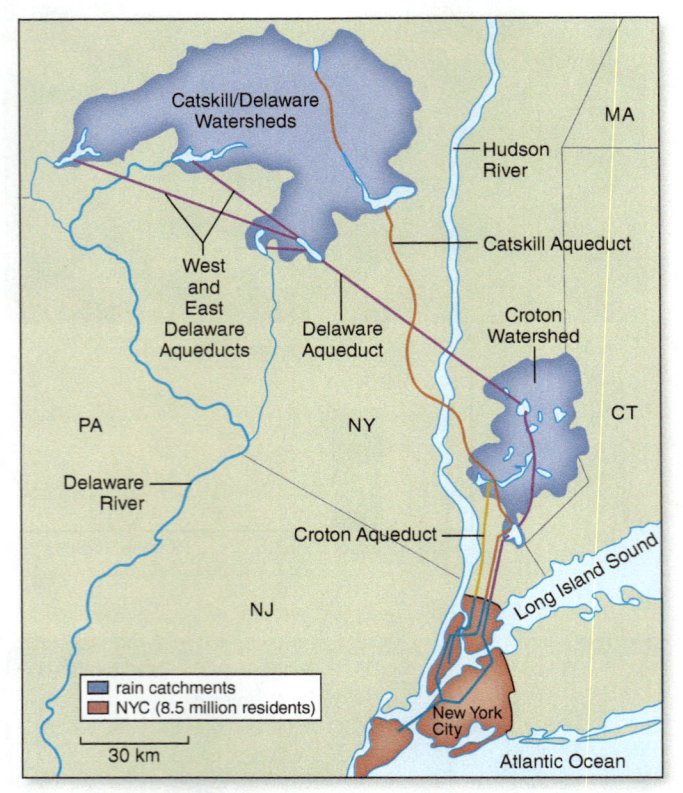

Figure 59.7 New York City's water source. New York gets its water from distant rain catchments. Preserving the ecological integrity of these areas is cheaper than building new water treatment plants.

the people who pay the costs. For instance, in the Thai mangrove example, the shrimp farmers reap the financial rewards, while the local people bear the costs. The same is true of factories that produce air or water pollution. Environmental economists are devising ways to appropriately value and regulate the use of the environment in ways that maximize the benefits relative to the costs to society as a whole.

Ethical and aesthetic values are based on our conscience and our consciousness

Many people believe that preserving biodiversity is an ethical issue because every species is of value in its own right, even if humans are not able to exploit or benefit from it. These people feel that along with the power to exploit and destroy other species comes responsibility: As the only organisms capable of eliminating large numbers of species and entire ecosystems, and as the only organisms capable of reflecting on what we are doing, humans should act as guardians or stewards for the diversity of life around us.

Almost no one would deny the aesthetic value of biodiversity—a wild mountain range, a beautiful flower, or a noble elephant—but how do we place a value on beauty or on the renewal many of us feel when we are in natural surroundings? Perhaps the best we can do is to consider the deep sense of loss we would feel if it no longer existed.

still have only a rudimentary knowledge of the many ways intact ecosystems provide services. Often, it is not until they are lost that the value becomes clear, as unexpected negative effects, such as increased flooding and pollution, decreased rainfall, or vulnerability to hurricanes become apparent.

The same argument can be made for preserving particular species within ecosystems. Given how little we know about the biology of most species, particularly in the tropics, it is impossible to predict all the consequences of removing a species.

Imagine taking a parts list for an airplane and randomly changing a digit in one of the part numbers. You might change a seat cushion into a roll of toilet paper—or you might just as easily change a key bolt holding up a wing into a pencil. By removing biodiversity, we are gambling with the future of the ecosystems on which we depend and whose functioning we understand very little.

In recent years, the field of ecological economics has developed to study how the societal benefits provided by species and ecosystems can be appropriately valued. The problem is twofold. First, until recently, we have not had a good estimate of the monetary value of services provided by ecosystems, a situation which, as you've just seen, is now changing.

The second problem, however, is that the people who gain the benefits of environmental degradation are often not the same as

Learning Outcomes Review 59.2

The direct value of biodiversity includes resources for our survival, such as natural products and medicines that enhance our lives and can be used in a sustainable way. Indirect value includes economic benefits provided by healthy ecosystems, such as availability of clean water and recreational benefits. The aesthetic value of biodiversity refers to our sense of beauty and peace when experiencing a natural environment.

■ *What arguments could you use to convince shrimp farmers to stop operations and remediate the area they are using?*

59.3 Factors Responsible for Extinction

Learning Outcomes

1. List the major causes of species extinction.
2. Explain how these causes can interact to bring about extinction.

A variety of causes, independently or in concert, are responsible for extinctions (table 59.3). Historically, overexploitation was the major cause of extinction; although it is still a factor, habitat loss is the major problem for most groups today, and introduced

TABLE 59.3	Causes of Extinctions				
	PERCENTAGE OF SPECIES INFLUENCED BY A GIVEN FACTOR*				
Group	Habitat Loss	Overexploitation	Species Introduction	Other	Unknown
EXTINCTIONS						
Mammals	19	23	20	2	36
Birds	20	11	22	2	45
Reptiles	5	32	42	0	21
Fish	35	4	30	4	48
THREATENED EXTINCTIONS						
Mammals	68	54	6	20	—
Birds	58	30	28	2	—
Reptiles	53	63	17	9	—
Fish	78	12	28	2	—

*Some species may be influenced by more than one factor; thus, some rows may exceed 100%.

species rank second. Many other factors can contribute to species extinctions as well, including disruption of ecosystem interactions, pollution, loss of genetic variation, and catastrophic disturbances, either natural or human-caused.

More than one of these factors may affect a species. In fact, a chain reaction is possible in which the action of one factor predisposes a species to be more severely affected by another factor. For example, habitat destruction may lead to decreased birth rates and increased mortality rates. As a result, populations become smaller and more fragmented, making them more vulnerable to disasters such as floods or forest fires, which may eliminate populations. Also, as the habitat becomes more fragmented, populations become isolated, so that genetic interchange ceases and areas devastated by disasters are not recolonized. Finally, as populations become very small, inbreeding increases, and genetic variation is lost through genetic drift, further decreasing population fitness. Which factor acts as the final coup de grace may be irrelevant; many factors, and the interactions between them, may have contributed to a species' eventual extinction.

Amphibians are on the decline: A case study

In 1963, herpetologist Jay Savage was hiking through pristine cloud forest in Costa Rica. Reaching a windswept ridge, he couldn't believe his eyes. Before him was a huge aggregation of breeding toads. What was so amazing was the color of the toads: bright, eye-dazzling orange, unlike anything he had ever seen before (figure 59.8).

The color of the toads was so amazing and unexpected that Savage briefly considered the possibility that his colleagues had played a practical joke, getting to the clearing before him and somehow coloring normal toads orange. Realizing that this could not be, he went on to study the toads, eventually describing a species new to science, the golden toad, *Bufo periglenes*.

For the next 24 years, large numbers of toads were seen during the breeding season each spring. Their home was legally recognized as the Monteverde Cloud Forest Reserve, a well-protected, intact, and functioning ecosystem, seemingly a successful model of conservation. Then, in 1988, few toads were seen, and in 1989, only a single male was observed. Since then, despite exhaustive efforts, no more golden toads have been found.

Despite living in a well-protected ecosystem, with no obvious threats from pollution, introduced species, overexploitation, or any other factor, the species appears to have gone extinct, right under the eyes of watchful scientists and conservationists. How could this happen?

Frogs in trouble

At the first World Herpetological Congress in 1989 in Canterbury, England, frog experts from around the world met to discuss conservation issues relating to frogs and toads. At this meeting, it became clear that the golden toad story was not unique. Experts reported case after case of similar losses: Frog populations that had once been abundant were now decreasing or entirely gone.

Figure 59.8 An extinct species. A golden toad, *Bufo periglenes*, which was last seen in the wild in 1989.

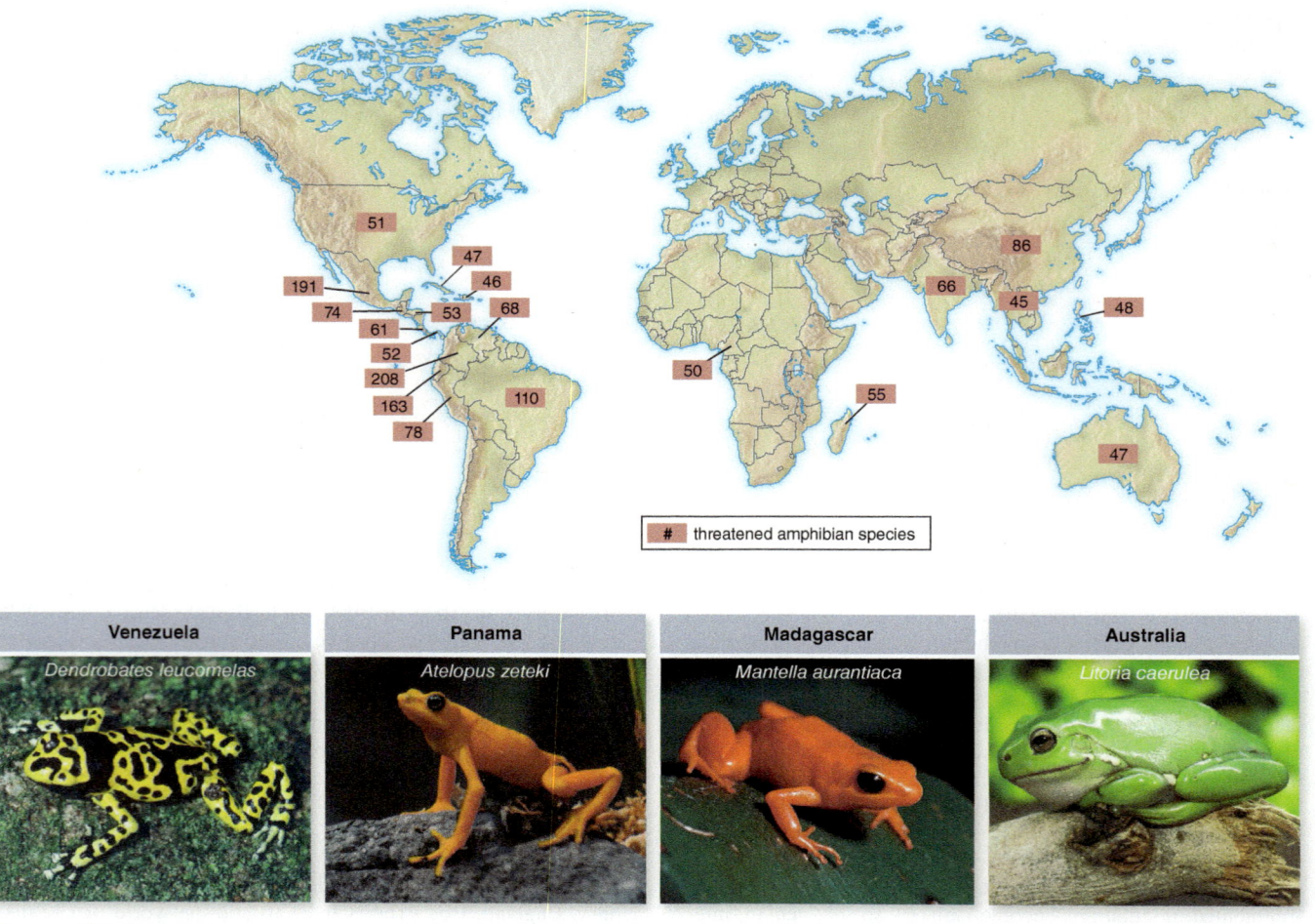

Figure 59.9 Amphibian extinction crisis. Boxes indicate the number of threatened species around the world in 2004. These numbers will undoubtedly be revised upward as scientists focus their attention on little-known species, many of which turn out to be in grave danger.

Since then, scientists have devoted a great deal of time and effort to determining whether frogs and other amphibian species truly are in trouble and, if so, why. Unfortunately, the situation appears to be even worse than originally suspected. Amphibian experts recently reported that 43% of all amphibian species have experienced decreases in population size, and one-third of all amphibian species are threatened with extinction in countries as different as Ecuador, Venezuela, Australia, and the United States (figure 59.9).

Moreover, these numbers are probably underestimates; little information exists from many areas of the world, such as Southeast Asia and central Africa. Indeed researchers think that as many as 100 species from the island nation of Sri Lanka have recently gone extinct, perhaps not surprisingly because 95% of that nation's rainforests have also disappeared in recent times.

Cause for concern

Amphibian declines are worrisome for several reasons. First, many of the species—including the golden toad—have declined in pristine, well-protected habitats. If species are becoming extinct in such areas, it brings into question our ability to preserve global biological diversity.

Second, many amphibian species are particularly sensitive to the state of the environment because of their moist skin, which allows chemicals from the environment to pass into the body, and their use of aquatic habitats for larval stages, which requires unpolluted water (see section 35.6). In other words, amphibians may be analogous to the canaries formerly used in coal mines to detect problems with air quality: If the canaries keeled over, the miners knew they had to get out.

Third, no single cause for amphibian declines is apparent. Although a single cause would be of concern, it would also suggest that a coordinated global effort could reverse the trend, as happened with chlorofluorocarbons and decreasing ozone levels (see chapter 58). However, different species are afflicted by different problems, including habitat destruction, the effects of global warming, pollution, decreased stratospheric ozone levels, disease epidemics, and introduced species.

The implication is that the global environment is deteriorating in many different ways. Could amphibians be global "canaries," serving as indicators that the world's environment is in serious trouble?

Habitat loss devastates species richness

As table 59.3 indicates, habitat loss is the most important cause of modern-day extinction. Given the tremendous amounts of ongoing destruction of all types of habitat, from rainforest to ocean floor, this should come as no surprise. Natural habitats may be adversely affected by humans in four ways:

1. **destruction,**
2. **pollution,**
3. **disruption,** and
4. **habitat fragmentation.**

In addition to these causes, global climate change, discussed in chapter 58, is an insidious threat that combines many of these factors. As climate changes, habitats will change—or disappear entirely, as is the case for polar bears *(Thalarctos maritimus),* which require ice floes on which to hunt their seal prey. Some studies estimate that as many as 30% of all species may be imperiled by global warming.

Destruction of habitat

A proportion of the habitat available to a particular species may simply be destroyed. This destruction is a common occurrence in the "clear-cut" harvesting of timber, in the burning of tropical forest to produce grazing land, and in urban and industrial development. Deforestation has been, and continues to be, by far the most pervasive form of habitat disruption (figure 59.10). Many tropical forests are being cut or burned at a rate of 1% or more per year.

To estimate the effect of reductions in habitat available to a species, biologists often use the well-established observation that larger areas support more species (see figure 57.22). Although this relationship varies according to geographic area, type of organism, and type of area, in general a 10-fold increase in area leads to approximately a doubling in the number of species. This relationship suggests, conversely, that if the area of a habitat is reduced by 90%, so that only 10% remains, then half of all species will be lost. Evidence for this hypothesis comes from a study in Finland of extinction rates of birds on habitat islands (that is, islands of a particular type of habitat surrounded by unsuitable habitat) where the population extinction rate was found to be inversely proportional to island size (figure 59.11).

Pollution

Habitat may be degraded by pollution to the extent that some species can no longer survive there. Degradation occurs as a result of many forms of pollution, from acid rain to pesticides. Aquatic environments are particularly vulnerable; for example, many northern lakes in both Europe and North America have been essentially sterilized by acid rain (see chapter 58).

Disruption

Human activities may disrupt a habitat enough to make it untenable for some species. For example, visitors to caves in Alabama

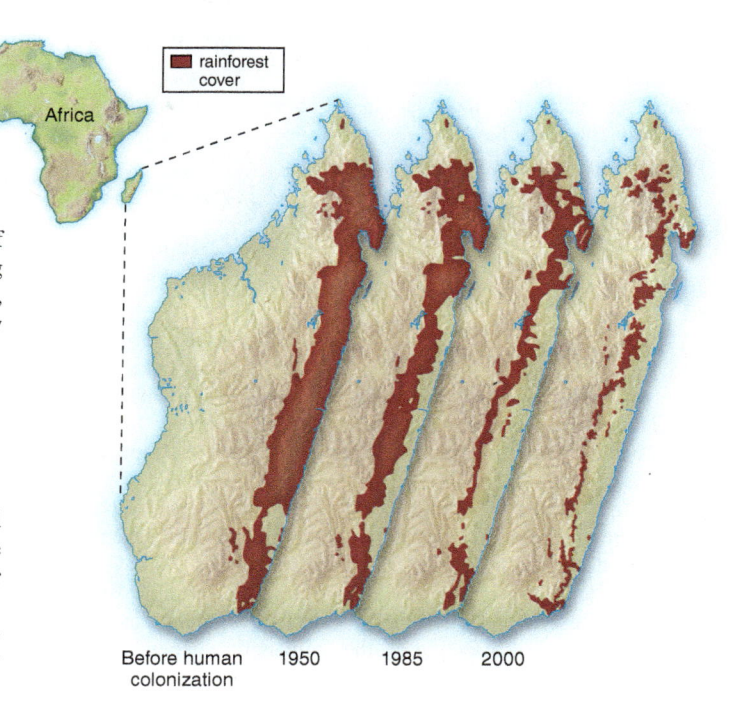

Figure 59.10 Extinction and habitat destruction. The rainforest covering the eastern coast of Madagascar, an island off the coast of East Africa, has been progressively destroyed and fragmented as the island's human population has grown. Ninety percent of the eastern coast's original forest cover is now gone. Many species have become extinct, and many others are threatened, including 16 of Madagascar's 31 primate species.

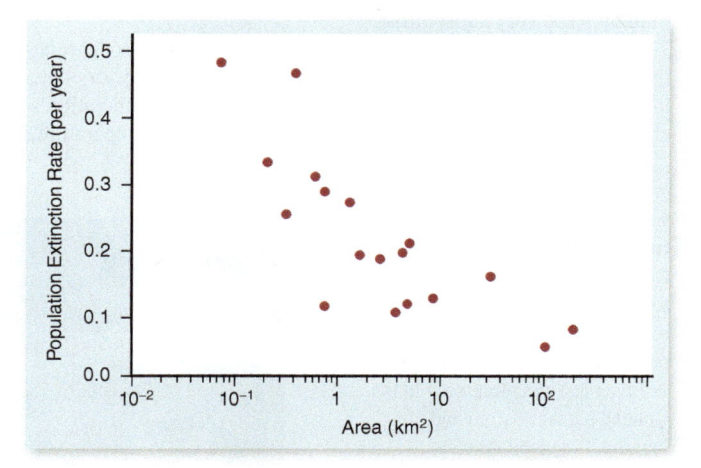

Figure 59.11 Extinction and island area. The data present percent extinction rates for populations as a function of habitat area for birds on a series of Finnish habitat islands. Smaller islands experience far greater extinction rates.

Data analysis Would the extinction rate increase if an area were decreased in size by 90%? If so, by how much?

Figure 59.12 Fragmentation of woodland habitat. From the time of settlement of Cadiz Township, Wisconsin, the forest has been progressively reduced from a nearly continuous cover to isolated woodlots covering less than 1% of the original area.

1831　　1882　　1902　　1950

and Tennessee caused significant population declines in bats over an eight-year period, some as great as 100%. When visits were fewer than one per month, less than 20% of bats were lost, but caves having more than four visits per month suffered population declines of 86 to 95%.

More generally, humans often alter the interactions that occur among species, such as the predator–prey or symbiotic relationships discussed in chapter 56. These disruptions can have far-ranging effects throughout an ecosystem. For example, when pollinating insects are killed off by insecticides, many plants do not reproduce, thus affecting all the animals that depend on the plants and their seeds for food.

Habitat fragmentation

Loss of habitat by a species frequently results not only in lowered population numbers, but also in fragmentation of the population into unconnected patches (figure 59.12). A habitat also may become fragmented in nonobvious ways, as when roads and habitation intrude into forest. The effect is to carve the populations living in the habitat into a series of smaller populations, often with disastrous consequences because of the relationship between range size and extinction rate. Although detailed data are not available, fragmentation of wildlife habitat in developed temperate areas is thought to be very substantial.

As habitats become fragmented and shrink in size, the relative proportion of the habitat that occurs on the boundary, or edge, increases. **Edge effects** can significantly degrade a population's chances of survival. Changes in microclimate (such as temperature, wind, humidity) near the edge may reduce appropriate habitat for many species more than the physical fragmentation suggests. In isolated fragments of rainforest, for example, trees on the edge are exposed to direct sunlight. As a result, these trees experience hotter and drier conditions than those normally encountered in the cool, moist forest interior, leading to negative effects on their survival and growth. In one study, the biomass of trees within 100 m of the forest edge decreased by 36% in the first 17 years after fragment isolation.

Also, increasing habitat edges opens up opportunities for some parasite and predator species that are more effective at edges. As fragments decrease in size, the proportion of habitat that is distant from any edge decreases, and consequently, more and more of the habitat is within the range of these species. Habitat fragmentation is blamed for local extinctions in a wide range of species.

The impact of habitat fragmentation can be seen clearly in a study conducted in Manaus, Brazil, where the rainforest was commercially logged. Landowners agreed to preserve patches of rainforest of various sizes, and censuses of these patches were taken before the logging started, while they were still part of a continuous forest. After logging, species began to disappear from the now-isolated patches (figure 59.13). First to go were the monkeys, which have large home ranges. Birds that prey on the insects flushed out by marching army ants followed, disappearing from patches too small to maintain enough army ant colonies to support them. As expected, the extinction rate was negatively related to patch size, but even the largest patches (100 hectares) lost half of their bird species in less than 15 years.

Figure 59.13 A study of habitat fragmentation. Landowners in Manaus, Brazil, agreed to preserve patches of rainforest of different sizes to examine the effect of patch size on species extinction. Biodiversity was monitored in the isolated patches before and after logging. Fragmentation led to significant species loss within patches. Army ants were one of the species that disappeared from smaller patches.

Because some species, such as monkeys, require large patches, large fragments are indispensable if we wish to preserve high levels of biodiversity. The take-home lesson is that preservation programs will need to provide suitably large habitat fragments to avoid this effect.

Case study: Songbird declines

Every year since 1966, the U.S. Fish and Wildlife Service has organized thousands of amateur ornithologists and birdwatchers in an annual bird count called the Breeding Bird Survey. In recent years, a shocking trend has emerged. Although year-round residents that prosper around humans, such as robins, starlings, and blackbirds, have increased their numbers and distribution over the last 50 years, forest songbirds have declined severely. The decline has been greatest among long-distance migrants such as thrushes, orioles, tanagers, vireos, buntings, and warblers. These birds nest in northern forests in the summer, but spend their winters in South or Central America or the Caribbean Islands.

In many areas of the eastern United States, more than three-quarters of the tropical migrant bird species have declined significantly. Rock Creek Park in Washington, D.C., for example, has lost 90% of its long-distance migrants in the past 20 years. Nationwide, American redstarts declined about 50% in the single decade of the 1970s. Studies of radar images from National Weather Service stations in Texas and Louisiana indicate that only about half as many birds fly over the Gulf of Mexico each spring as did in the 1960s. This suggests a total loss of about half a billion birds.

The culprit responsible for this widespread decline appears to be habitat fragmentation and loss. Fragmentation of breeding habitat and nesting failures in the summer nesting grounds of the United States and Canada have had a major negative effect on the breeding of woodland songbirds. Many of the most threatened species are adapted to deep woods and need an area of 25 acres or more per pair to breed and raise their young. As woodlands are broken up by roads and developments, it is becoming increasingly difficult for them to find enough contiguous woods to nest successfully.

A second and perhaps even more important factor is the availability of critical winter habitat in Central and South America. Studies of the American redstart clearly indicate that birds with better winter habitat have a superior chance of successfully migrating back to their breeding grounds in the spring. In one study, scientists were able to determine the quality of the habitat that particular birds used during the winter by examining levels of the stable carbon isotope ^{13}C in their blood. Plants growing in the best habitats in Jamaica and Honduras (mangroves and wetland forests) have low levels of ^{13}C, and so do the redstarts that feed on the insects that live in them. Of these wet-forest birds, 65% maintained or gained weight over the winter.

By contrast, plants growing in substandard dry scrub have high levels of ^{13}C, and so do the redstarts that feed in those habitats. Scrub-dwelling birds lost up to 11% of their body mass over the winter. Now here's the key: Birds that winter in the substandard scrub leave later in the spring on the long flight to northern breeding grounds, arrive later at their summer homes, and have fewer young (figure 59.14).

The proportion of ^{13}C in birds arriving in New Hampshire breeding grounds increases as spring wears on and scrub-overwintering stragglers belatedly arrive. Thus, loss of mangrove habitat in the neotropics is having a quantifiable negative influence. As the best habitat disappears, overwintering birds fare poorly, and this leads to decreased reproduction and population declines.

Unfortunately, the Caribbean lost about 10% of its mangroves in the 1980s, and continues to lose about 1% per year. This loss of key habitat appears to be a driving force in the looming extinction of some songbirds.

Overexploitation wipes out species quickly

Species that are hunted or harvested by humans have historically been at grave risk of extinction, even when the species is initially very abundant. A century ago, the skies of North America were darkened by huge flocks of passenger pigeons, but after being hunted as free and tasty food, they were driven to extinction. The bison that used to migrate in enormous herds across the central plains of North America only narrowly escaped the same fate.

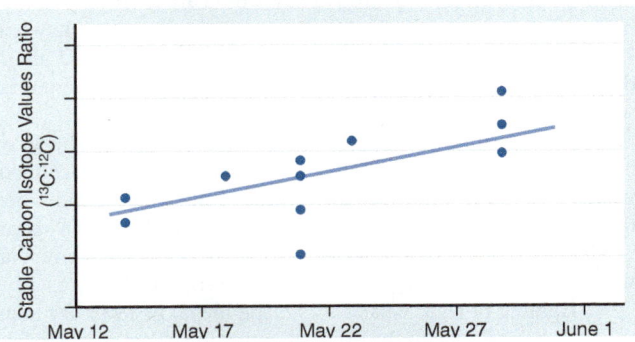

Figure 59.14 The American redstart, *Setophaga ruticilla*, a migratory songbird. The numbers of this species are in serious decline. The graph presents data on the ratio of ^{13}C to ^{12}C in male redstarts arriving at summer breeding grounds. Early arrivals, which have higher reproductive success, have lower proportions of ^{13}C to ^{12}C, indicating they wintered in more favorable mangrove–wetland forest habitats.

Data analysis Based on the information in the text, what would the relationship between arrival date and reproductive success look like?

Commercial motivation for exploitation

The existence of a commercial market often leads to overexploitation of a species. The international trade in furs, for example, has severely reduced the numbers of chinchilla, vicuña, otter, and many cat species. The harvesting of commercially valuable trees provides another example: Almost all West Indies mahogany trees *(Swietenia mahogani)* have been logged, and the extensive cedar forests of Lebanon, once widespread at high elevations, now survive in only a few isolated groves.

A particularly telling example of overexploitation is the commercial harvesting of fish in the North Atlantic. During the 1980s, fishing fleets continued to harvest large amounts of cod off the coast of Newfoundland, even as the population numbers declined precipitously. By 1992, the cod population had dropped to less than 1% of its original numbers. The American and Canadian governments have closed the fishery, but thus far the populations have not recovered. The Atlantic bluefin tuna has experienced a 90% population decline in recent years and swordfish has declined even further. In both cases, the drop has led to even more intense fishing of the remaining populations.

Case study: Whales

Whales, the largest living animals that ever evolved, are rare in the world's oceans today, their numbers driven down by commercial whaling. Before the advent of cheap, high-grade oils manufactured from petroleum in the early 20th century, oil made from whale blubber was an important commercial product in the worldwide marketplace. In addition, the fine, lattice-like structure termed "baleen" used by baleen whales to filter-feed plankton from seawater was used in women's undergarments. Because a whale is such a large animal, each individual captured is of significant commercial value.

In the 18th century, right whales were the first to bear the brunt of commercial whaling. They were called "right" whales because they were slow, easy to capture, did not sink when killed and provided up to 150 barrels of blubber oil and abundant baleen, making them the right whale for a commercial whaler to hunt.

As the right whale declined, whalers turned to the gray, humpback, and bowhead whales. As their numbers declined, whalers turned to the blue, the largest of all whales, and when those were decimated, to the fin, then the Sei, and then the sperm whales. As each species of whale became the focus of commercial whaling, its numbers began a steep decline (figure 59.15).

Hunting of right whales was made illegal in 1935. By then, they had been driven to the brink of extinction, their numbers less than 5% of what they had been. Although protected ever since, their numbers have not recovered in either the North Atlantic or the North Pacific. By 1946, several other whale species faced imminent extinction, and whaling nations formed the International Whaling Commission (IWC) to regulate commercial whale hunting. Like a fox guarding the henhouse, the IWC for decades did little to limit whale harvests, and whale numbers continued to decline steeply.

Finally, in 1974, when the numbers of all but the small minke whales had been driven down, the IWC banned hunting of blue, gray, and humpback whales, and instituted partial bans on other species. The rule was violated so often, however, that the IWC in 1986 instituted a worldwide moratorium on all commercial killing of whales. Although some commercial whaling continues, often under the guise of harvesting for scientific studies, annual whale harvests have dropped dramatically in the last 30 years.

Some species appear to be recovering, but others are not. Humpback numbers have increased substantially since the early 1960s, and Pacific gray whales have fully recovered to their previous numbers of about 20,000 animals, after having been hunted to fewer than 1000. Sei, fin, and North Atlantic right whales have not recovered, and no one knows whether they ever will.

Introduced species threaten native species and habitats

Colonization, a natural process by which a species expands its geographic range, occurs in many ways: A flock of birds gets blown off course, a bird eats a fruit and defecates its seed miles away, or lowered sea levels connect two previously isolated landmasses, allowing species to freely move back and forth. Such events—particularly those leading to successful establishment of a new population—probably occur rarely, but when they do, the resulting change to natural communities can be large. The reason is that colonization brings together species with no history of interaction. Consequently, ecological interactions may be particularly strong because the species have not evolved ways of adjusting to the presence of one

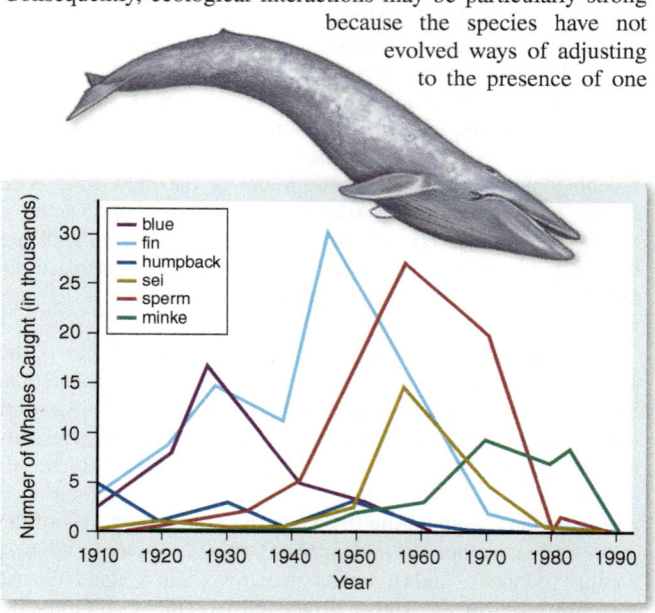

Figure 59.15 World catch of some whale species in the 20th century. Each species was hunted in turn until its numbers fell so low that hunting it became commercially unprofitable.

Inquiry question Why might whale populations fail to recover once hunting is stopped?

Data analysis Only two species of whales were heavily hunted at the start of the 20th century. Does this graph provide an explanation for the timing of increased hunting of the other species?

another, such as adaptations to avoid predation or to minimize competitive effects.

The paleontological record documents many cases in which geologic changes brought previously isolated species together, such as when the Isthmus of Panama emerged above the sea approximately 3 MYA, connecting the previously isolated fauna and flora of North and South America. In some cases, the result has been an increase in species diversity, but in other cases, invading species have led to the extinction of natives.

Human influence on colonization

Unfortunately, what was naturally a rare process has become all too common in recent years, thanks to the actions of humans. Species introductions due to human activities occur in many ways, sometimes intentionally, but usually not. Plants and animals can be transported in the ballast of large ocean vessels; in nursery plants; as stowaways in boats, cars, and planes; as beetle larvae within wood products—even as seeds or spores in the mud stuck to the bottom of a shoe. Overall, some researchers estimate that as many as 50,000 species have been introduced into the United States.

The effects of introductions on humans have been enormous. In the United States alone, nonnative species cost the economy an estimated $140 billion per year. For example, dozens of foreign weeds in Colorado have covered more than a million acres. Just three of these species cost wheat farmers tens of millions of dollars. At the same time, leafy spurge, a plant from Europe, outcompetes native grasses, ruining rangeland for cattle at a price tag of $144 million per year.

The zebra mussel, a mollusk native to the Black Sea region, is a huge problem throughout much of the eastern and central United States. Attaining densities as high as 700,000/m^2, it clogs pipes, including those for water and power plants, and causes an estimated $3 to $5 billion damage a year (figure 59.16).

Introduced species can also affect human health. For example, West Nile fever was probably introduced from Africa or the Middle East to the United States in the late 1990s. By the mid-2010s, thousands of people were infected with the disease each year; in 2014 alone, infections were reported in 47 states and the District of Columbia.

The effect of species introductions on native ecosystems is equally dramatic. Islands have been particularly affected. For example, as mentioned in chapter 56, a single lighthouse keeper's cat wiped out an entire species, the Stephens Island wren. Rats had a devastating effect throughout the South Pacific where bird species nested on the ground and had no defense against the voracious predators to which they were evolutionarily naive. More recently, the brown tree snake, introduced to the island of Guam, essentially eliminated almost all species of forest birds.

In Hawaii, the problem has been slightly different: Introduced mosquitoes brought with them bird malaria, to which the native species had evolved no resistance. The result is that more than 100 species (over 70% of the native fauna) either became extinct or are now restricted to higher and cooler elevations where the mosquitoes don't occur (figure 59.17).

The effects of introduced species may reverberate throughout an ecosystem. For example, the Argentine ant has spread through much of the southern United States, greatly reducing populations of most native ant species with which it comes in contact. The extinction of these ant species has had a dramatic negative effect on the coast horned lizard *(Phrynosoma coronatum)*, which feeds on the larger native species. In their absence, the lizards have shifted to less-preferred prey species. In addition, the native ant species consume seeds, and

Figure 59.16 Zebra mussels, *Dreissena polymorpha*, clogging a pipe. These mussels were introduced from Europe, and are now a major problem in North American rivers.

Figure 59.17 The akiapolaau, *Hemignathus munroi,* and the palila, *Loxioides bailleui,* endangered Hawaiian birds. More than two-thirds of Hawaii's native bird species are now extinct or have been greatly reduced in population size. Bird faunas on islands around the world have experienced similar declines after human arrival.

in the process, play an important role in seed dispersal. Argentine ants, by contrast, do not eat seeds. In South Africa, where the Argentine ant has also appeared, at least one plant species has experienced decreased reproductive success due to the loss of its dispersal agent.

The most dramatic effects of introduced species, however, occur when entire ecosystems are transformed. Some plant species can completely overrun a habitat, displacing all native species and turning the area into a monoculture (that is, an area occupied by a single species). In California, the yellow star thistle (*Centaurea solstitialis*) now covers 4 million hectares of what was once highly productive grassland. In Hawaii, a small tree native to the Canary Islands, *Myrica faya,* has spread widely. Because it is able to fix nitrogen at high rates, it has caused a 90-fold increase in the nitrogen content of the soil, thus allowing other, nitrogen-requiring species to invade.

Efforts to combat introduced species

Once an introduced species becomes established, eradicating it is often extremely difficult, expensive, and time-consuming. Some efforts—such as the removal of goats and rabbits from certain small islands—have been successful, but many other efforts have failed. The best hope for stopping the ravages of introduced species is to prevent them from being introduced in the first place. Although easier said than done, government agencies are now working strenuously to put into place procedures that can intercept species in transit, before they have the opportunity to become established.

Case study: Lake Victoria cichlids

Lake Victoria, an immense, shallow, freshwater sea about the size of Switzerland in the heart of equatorial East Africa, used to be home to an incredibly diverse collection of over 450 species of cichlid fishes (see figure 22.17). These small, perchlike fish range from 5 to 13 cm in length, with males having endless varieties of color. Today, most of these cichlid species are threatened, endangered, or extinct.

What happened to bring about the abrupt loss of so many endemic cichlid species? In 1954, the Nile perch, a commercial fish with a voracious appetite, was purposely introduced on the Ugandan shore of Lake Victoria. Nile perch, which grow to over a meter in length, were to form the basis of a new fishing industry (figure 59.18). For decades, these perch did not seem to have a significant effect; over 30 years later, in 1978, Nile perch still made up less than 2% of the fish harvested from the lake.

Then something happened to cause the Nile perch population to explode and to spread rapidly through the lake, eating their way through the cichlids. By 1986, Nile perch constituted nearly 80% of the total catch of fish from the lake and the endemic cichlid species were virtually gone. Over 70% of cichlid species disappeared, including all open-water species.

So what happened to kick-start the mass extinction of the cichlids? The trigger seems to have been eutrophication. Before 1978, Lake Victoria had high oxygen levels at all depths, down to the bottom layers more than 60 m deep. However, by 1989 high inputs of nutrients from agricultural runoff and sewage from towns and villages had led to algal blooms that severely depleted oxygen levels in deeper parts of the lake. Cichlids feed on algae, and

Figure 59.18 Nile perch, *Lates niloticus*. This predatory fish, which can reach a length of 2 m and a mass of 200 kg, was introduced into Lake Victoria as a potential food source. It is responsible for the virtual extinction of hundreds of species of cichlid fishes.

initially their population numbers are thought to have risen in response to this increase in their food supply, but unlike the conditions during similar algal blooms of the past, the Nile perch was present to take advantage of the situation. With a sudden increase in its food supply (cichlids), the numbers of Nile perch exploded, and they simply ate all available cichlids.

Since 1990, the situation has been compounded by the introduction into Lake Victoria of a floating water weed from South America, the water hyacinth *Eichhornia crassipes*. Reproducing quickly under eutrophic conditions, thick mats of water hyacinth soon covered entire bays and inlets, choking off the coastal habitats of non-open-water cichlids.

Disruption of ecosystems can cause an extinction cascade

Species often become vulnerable to extinction when their web of ecological interactions becomes seriously disrupted. Because of the many relationships linking species in an ecosystem (see chapter 57), human activities that affect one species can have ramifications throughout an ecosystem, ultimately affecting many other species.

A recent case in point involves the sea otters that live in the cold waters off Alaska and the Aleutian Islands. Otter populations have declined sharply in recent years. In a 500-mile stretch of coastline, otter numbers have dropped from 53,000 in the 1970s to an estimated 6000, a plunge of nearly 90%. Investigating this catastrophic decline, marine ecologists uncovered a chain of interactions among the species of the ocean and kelp forest ecosystems, a falling-domino series of lethal effects that illustrates the concepts of both top-down and bottom-up trophic cascades discussed in chapter 57.

Case study: Alaskan near-shore habitat

The first in a series of events leading to the sea otter's decline seems to have been the heavy commercial harvesting of whales, described earlier in this section. Without whales to keep their numbers in check, ocean zooplankton thrived, leading in turn to proliferation of a species of fish called pollock that feeds on the abundant zooplankton. Given this ample food supply, the pollock outcompeted other northern Pacific fish, such as herring and ocean perch, so that levels of these other fish fell steeply in the 1970s.

Then the falling chain of dominoes began to accelerate. The decline in the nutritious forage fish led to an ensuing crash in Alaskan populations of sea lions and harbor seals, for which pollock did not provide sufficient nourishment. This decline may also have been hastened by orcas (also called killer whales) switching from feeding on the less-available whales to feeding on seals and sea lions; the numbers of these pinniped species have fallen precipitously since the 1970s.

When pinniped numbers crashed, some orcas, faced with a food shortage, turned to the next best thing: sea otters. In one bay where the entrance from the sea was too narrow and shallow for orcas to enter, only 12% of the sea otters have disappeared, while in a similar bay that orcas could enter easily, two-thirds of the otters disappeared in a year's time.

Without otters to eat them, the population of sea urchins exploded, eating the kelp and thus "deforesting" the kelp forests and denuding the ecosystem (figure 59.19). As a result, fish species that live in the kelp forest, such as sculpins and greenlings, are declining.

Loss of keystone species

As discussed in chapter 56, a keystone species is a species that exerts a greater influence on the structure and functioning of an ecosystem than might be expected solely on the basis of its abundance. The sea otters of figure 59.19 are a keystone species of the kelp forest ecosystem, and their removal can have disastrous consequences.

No hard-and-fast line allows us to clearly identify keystone species. Rather, it is a qualitative concept, a statement that indicates a species plays a particularly important role in its community. Keystone species are usually characterized by the strength of their effect on their community.

Case study: Flying foxes

The severe decline of many species of "flying foxes," a type of bat (figure 59.20), in the Old World tropics is an example of how the loss of a keystone species can dramatically affect the other species living within an ecosystem, sometimes even leading to a cascade of further extinctions.

These bats have very close relationships with important plant species on the islands of the Pacific and Indian Oceans. The family Pteropodidae contains nearly 200 species, approximately one-quarter of them in the genus *Pteropus,* and is widespread on the islands of the South Pacific, where they are the most important— and often the only—pollinators and seed dispersers.

Figure 59.19 Disruption of the kelp forest ecosystem. Overharvesting by commercial whalers altered the balance of fish in the ocean ecosystem, inducing killer whales to feed on sea otters, a keystone species of the kelp forest ecosystem.

1. **Whales** Overharvesting of plankton-eating whales may have caused an increase in plankton-eating pollock populations.

2. **Nutritious fish** Populations of nutritious fish like ocean perch and herring declined, likely due to competition with pollock.

3. **Sea lions and harbor seals** Sea lion and harbor seal populations drastically declined in Alaska, probably because the less-nutritious pollock could not sustain them.

4. **Killer whales** With the decline in their prey populations of sea lions and seals, killer whales turned to a new source of food: sea otters.

5. **Sea otters** Sea otter populations declined so dramatically that they disappeared in some areas.

6. **Sea urchins** Usually the preferred food of sea otters, sea urchin populations now exploded and fed on kelp.

7. **Kelp forests** Severely thinned by the sea urchins, the kelp beds no longer support a diversity of fish species, which may lead to a decline in populations of eagles that feed on the fish.

Figure 59.20 The importance of keystone species.
Flying foxes, a type of fruit-eating bat, are keystone species on many Old World tropical islands. It pollinates many plants and is a key disperser of seeds. Its elimination due to hunting and habitat loss is having a devastating effect on the ecosystems of many South Pacific Islands.

A study in Samoa found that 80 to 100% of the seeds landing on the ground during the dry season were deposited by flying foxes, which eat the fruits and defecate the seeds, often moving them great distances in the process. Many species are entirely dependent on these bats for pollination. Some have evolved features such as night-blooming flowers that prevent any other potential pollinators from taking over the role of the fruit bats.

In Guam, the two local species of flying fox have recently been driven extinct or nearly so, with a substantial impact on the ecosystem. Botanists have found that some plant species are not fruiting or are doing so only marginally, producing fewer fruits than normal. Fruits are not being dispersed away from parent plants, so seedlings are forced to compete, usually unsuccessfully, with adult trees.

Flying foxes are being driven to extinction by human hunters who kill them for food and for sport, and by orchard farmers who consider them pests. Flying foxes are particularly vulnerable because they live in large and obvious groups of up to a million individuals. Because they move in regular and predictable patterns and can be easily tracked to their home roost, hunters can easily kill thousands at a time.

Programs aimed at preserving particular species of flying foxes are only just beginning. One particularly successful example is the program to save the Rodrigues fruit bat, *Pteropus rodricensis,* which occurs only on Rodrigues Island in the Indian Ocean near Madagascar. The population dropped from about 1000 individuals in 1955 to fewer than 100 by 1974, largely due to the loss of the fruit bat's forest habitat to farming. Since 1974, the species has been legally protected, and the forest area of the island is being increased through a tree-planting program. Eleven captive-breeding colonies have been established, and the bat population in the wild has increased to 4000 bats. The combination of legal protection, habitat restoration, and captive breeding has in this instance produced a very effective preservation program.

Small populations are particularly vulnerable

Because of the factors just discussed, populations of many species are fragmented and reduced in size. Such populations are particularly prone to extinction.

Demographic factors

Small populations are vulnerable to events that decrease survival or reproduction. For example, by nature of their size, small populations are ill-equipped to withstand catastrophes, such as a flood, forest fire, or disease epidemic. One example is provided by the history of the heath hen. Although the species was once common throughout the eastern United States, hunting pressure in the 18th and 19th centuries eventually eliminated all but one population, on the island of Martha's Vineyard near Cape Cod, Massachusetts. Protected in a nature preserve, the population was increasing in number until a fire destroyed most of the preserve's habitat. The small surviving population was then ravaged the next year by an unusual congregation of predatory birds, followed shortly thereafter by a disease epidemic. The last sighting of a heath hen, a male, was in 1932 (figure 59.21*a*).

When populations become extremely small, bad luck can spell the end. For example, the dusky seaside sparrow (figure 59.21*b*), a now-extinct subspecies that was found on the east coast of Florida, dwindled to a population of five individuals, all of which happened to be males. In a large population, the probability that all individuals will be of one sex is infinitesimal. But in small populations, just by the luck of the draw, it is possible that 5 or 10 or even 20 consecutive births will all be individuals of one sex, and that can be enough to send a species to extinction. In addition, when populations are small, individuals may have trouble finding each other (the Allee effect discussed in chapter 55), thus leading the population into a downward spiral toward extinction.

Figure 59.21 Alive no more. *a.* A museum specimen of the heath hen, *Tympanuchus cupido cupido,* which went extinct in 1932. *b.* This male was one of the last dusky seaside sparrows, *Ammodramus maritimus nigrescens.*

Lack of genetic variability

Small populations face a second dilemma. Because of their low numbers, such populations are prone to the loss of genetic variation as a result of genetic drift (figure 59.22). Indeed, many small populations contain little or no genetic variability. The result of such genetic homogeneity can be catastrophic. Genetic variation is beneficial to a population both because of heterozygote advantage (see chapter 20) and because genetically variable individuals tend not to have two copies of deleterious recessive alleles. Populations lacking variation are often composed of sickly, unfit, or sterile individuals. Laboratory groups of rodents and fruit flies that are maintained at small population sizes often perish after a few generations as each generation becomes less robust and fertile than the preceding one.

Although it is difficult to demonstrate that a species has gone extinct because of lack of genetic variation, studies of both zoo and natural populations clearly reveal that more genetically variable individuals have greater fitness. Furthermore, in the longer term, populations with limited genetic variation have diminished ability to adapt to changing environments, a particular concern given the way humans are changing the environment in so many ways (see chapter 58).

Interaction of demographic and genetic factors

As populations decrease in size, demographic and genetic factors combine to cause what has been termed an "extinction vortex." That is, as a population gets smaller, it becomes more vulnerable to demographic catastrophes. In turn, genetic variation starts to be lost, causing reproductive rates to decline and population numbers to decline even further, and so on. Eventually, the population disappears entirely, but attributing its demise to one particular factor would be misleading.

Case study: Prairie chickens

The greater prairie chicken, a close relative of the now-extinct heath hen, is a showy, 2-lb bird renowned for its flamboyant mating rituals (figure 59.23). Abundant in many midwestern states, the prairie chickens in Illinois have in the past seven decades undergone a population collapse.

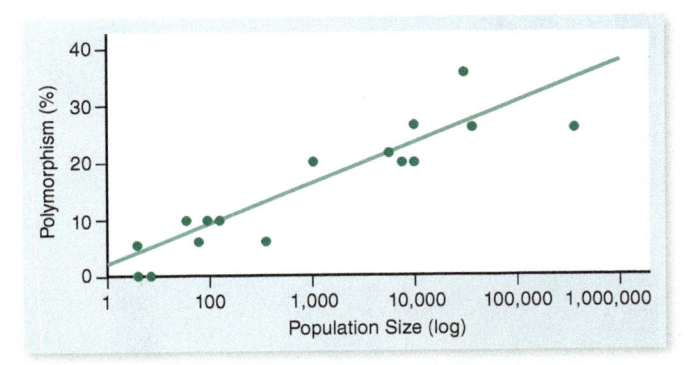

Figure 59.22 Loss of genetic variability in small populations. The percentage of genes that are polymorphic in isolated populations of the tree *Halocarpus bidwillii* in the mountains of New Zealand is a sensitive function of population size.

? Inquiry question Why do small populations lose genetic variation?

Figure 59.23 A mating ritual. The male greater prairie chicken, *Tympanuchus cupido pinnatus*, inflates bright orange air sacs, part of his esophagus, into balloons on each side of his head. As air is drawn into the sacs, it creates a three-syllable low-frequency "boom-boom-boom" that can be heard for kilometers.

Once, enormous numbers of birds occurred throughout the state, but with the 1837 introduction of the steel plow, the first that could slice through the deep, dense root systems of prairie grasses, the Illinois prairie began to be replaced by farmland. By the turn of the 20th century, the prairie had all but vanished, and by 1931, the heath hen had become locally extinct in Illinois. The greater prairie chicken fared little better, its numbers falling to 25,000 statewide in 1933 and then to 2000 by 1962. In surrounding states with less intensive agriculture, it continued to prosper.

In 1962 and 1967, sanctuaries were established in Illinois to attempt to preserve the greater prairie chicken. But privately owned grasslands kept disappearing, along with their prairie chickens, and by the 1980s the birds were extinct in Illinois except for two preserves, and even there, their numbers kept falling. By 1990, the egg hatching rate, which at one time had averaged between 91 and 100%, had dropped to an extremely low 38%. By the mid-1990s, the count of males had dropped to as low as six in each sanctuary.

What was wrong with the sanctuary populations? One suggestion was that because of very small population sizes and a mating ritual whereby one male may dominate a flock, the Illinois prairie chickens had lost so much genetic variability as to create serious genetic problems. To test this idea, biologists at the University of Illinois compared DNA from frozen tissue samples of birds that had died in Illinois between 1974 and 1993, and found that Illinois birds had indeed become genetically less diverse.

The researchers then extracted DNA from tissue in the roots of feathers from stuffed birds collected in the 1930s from the same population. They found that Illinois birds had lost fully one-third of the genetic diversity of birds living in the same place before the population collapse of the 1970s. By contrast, prairie chicken populations in other states still contained much of the genetic variation that had disappeared from Illinois populations.

Now the stage was set to halt the Illinois prairie chicken's race toward extinction in Illinois. Wildlife managers began to transplant birds from genetically diverse populations of Minnesota, Kansas, and Nebraska to Illinois. Between 1992 and 1996, a total of 518 out-of-state prairie chickens were brought in to interbreed with the Illinois birds, and hatching rates were back up to 94% by 1998. As the population rebounded, it looked as though the prairie chickens had been saved from extinction in Illinois. However, a 2011 hailstorm and a 2013 drought were major setbacks, and whether the population can bounce back yet again remains to be seen.

The key lesson here is the importance of not allowing things to go too far—not to drop down to a single isolated population. Without the outlying genetically different populations, the prairie chickens in Illinois could not have been saved. When the last population of the dusky seaside sparrow lost its last female, there was no other source of females, and the subspecies went extinct.

Learning Outcomes Review 59.3

Factors responsible for extinction include habitat destruction, pollution, disruption, and fragmentation. Overexploitation can reduce populations to low levels or eliminate them entirely. Introduced species can wreak havoc on native communities. Finally, small populations have less ability to rebound from catastrophes and are vulnerable to loss of genetic variation. Interaction of all these factors can hasten species' decline into extinction.

- *Does it make sense to take endangered species out of the wild to preserve them if their habitat is allowed to disappear? Explain.*

59.4 Approaches for Preserving Endangered Species and Ecosystems

Learning Outcomes

1. Distinguish between restoration of species and restoration of ecosystem functioning.
2. List the strategies for habitat restoration.
3. Explain the rationale for captive breeding programs.

Once the cause of a species' endangerment is known, it becomes possible to design a recovery plan. If the cause is commercial over-harvesting, regulations can be issued to restrict harvesting and protect the threatened species. If the cause is habitat loss, plans can be instituted to restore the habitat. Loss of genetic variability in isolated subpopulations can be countered by transplanting individuals from genetically different populations. Populations in immediate danger of extinction can be captured, introduced into a captive-breeding program, and later reintroduced to other suitable habitat.

All of these solutions are extremely expensive. But it is much more economical to prevent "environmental trainwrecks" from occurring than to clean them up afterward. Preserving ecosystems and monitoring species before they are threatened is the most effective means of protecting the environment and preventing extinctions.

Destroyed habitats can sometimes be restored

Conservation biology typically concerns itself with preserving populations and species in danger of decline or extinction. Conservation, however, requires that there be something left to preserve; in many situations, conservation is no longer an option. Species, and in some cases whole communities, have disappeared or been irretrievably modified. The clear-cutting of the temperate forests of Washington State leaves little behind to conserve, as does converting a piece of land into a wheat field or an asphalt parking lot. Redeeming these situations requires restoration rather than conservation.

Three quite different sorts of habitat restoration programs might be undertaken, depending on the cause of the habitat loss.

Pristine restoration

In ecosystems where all species have been effectively removed, conservationists might attempt to restore the plants and animals that are the natural inhabitants of the area, if these species can be identified. When abandoned farmland is to be restored to prairie, as in figure 59.24, how would conservationists know what to plant?

Figure 59.24 Habitat restoration. The University of Wisconsin–Madison Arboretum has pioneered restoration ecology. *a.* The restoration of the prairie was at an early stage in November 1935. *b.* The prairie as it looks today. This picture was taken at approximately the same location as the 1935 photograph.

Although it is in principle possible to reestablish each of the original species in their original proportions, rebuilding a community requires knowing the identities of all the original inhabitants and the ecologies of each of the species. We rarely have this much information, so no restoration is ever truly pristine.

Increasingly, restoration biologists are working on restoring the functioning of an ecosystem, rather than trying to recreate the same community composition. This approach shifts the focus from restoring species to reconstructing the processes that operated in the natural habitat.

Removing introduced species

Sometimes the habitat has been destroyed by a single introduced species. In such a case, habitat restoration involves removing the introduced species. Restoration of the once-diverse cichlid fishes to Lake Victoria will require more than breeding and restocking the endangered species. The introduced water hyacinth and Nile perch populations will have to be brought under control or removed, and eutrophication will have to be reversed.

It is important to act quickly if an introduced species is to be removed. When aggressive African bees (the so-called "killer bees") were inadvertently released in Brazil, they remained confined to the local area for only one season. Now they occupy much of the western hemisphere.

Cleanup and rehabilitation

Habitats seriously degraded by chemical pollution cannot be restored until the pollution is cleaned up. The successful restoration of the Nashua River in New England is one example of how a concerted effort can succeed in restoring a heavily polluted habitat to a relatively pristine condition.

Once so heavily polluted by chemicals from dye manufacturing plants that it was different colors in different places, the river is now clean and used for many recreational activities.

Captive breeding programs have saved some species

Recovery programs, particularly those focused on one or a few species, must sometimes involve direct intervention in natural populations to avoid an immediate threat of extinction.

Case study: The peregrine falcon

American populations of birds of prey, such as the peregrine falcon, began an abrupt decline shortly after World War II. Of the approximately 350 breeding pairs east of the Mississippi River in 1942, all had disappeared by 1960. The culprit proved to be the chemical pesticide DDT (see chapter 58).

The use of DDT was banned by federal law in 1972, causing levels in the eastern United States to fall quickly. However, no peregrine falcons were left in the eastern United States to reestablish a natural population. Falcons from other parts of the country were used to establish a captive-breeding program at Cornell University in 1970, with the intent of reestablishing the peregrine falcon in the eastern United States by releasing offspring of these birds. Since 1974, more than 6000 birds were released in North America, producing an astonishingly strong recovery that has restored the species to its entire historic range (figure 59.25).

Case study: The California condor

The number of California condors *(Gymnogyps californianus)*, a large, vulture-like bird with a wingspan of nearly 3 m, has been declining gradually for more than 200 years. By 1985, condor numbers had dropped so low that the bird was on the verge of extinction. Six of the remaining 15 wild birds disappeared in that year alone. The entire breeding population of the species consisted of the birds remaining in the wild and an additional 21 birds in captivity.

In a last-ditch attempt to save the condor from extinction, the remaining birds were captured and placed in a captive-breeding population. The breeding program was set up in zoos, with the goal of releasing offspring on a large, 5300-hectare ranch in prime condor habitat. Birds were isolated from human contact as much as possible, and closely related individuals were prevented from breeding.

By early 2009, the captive population of California condors had reached over 160 individuals. After extensive pre-release training to avoid power poles and people, captive-reared condors have been released successfully in California at two sites in the mountains north of Los Angeles, as well as at the Grand Canyon. Many of the released birds are doing well, and the wild population now numbers over 200 birds. Biologists are particularly excited by breeding activities that

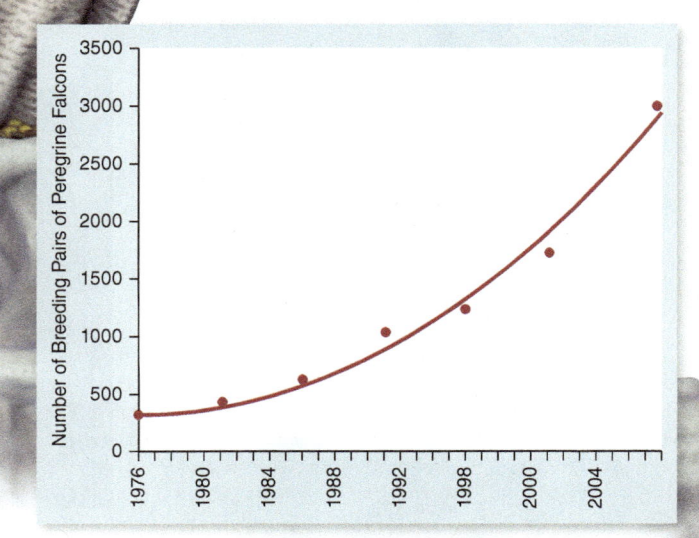

Figure 59.25 Success of captive breeding. The peregrine falcon, *Falco peregrinus,* has been reestablished in the eastern United States by releasing captive-bred birds over a period of 25 years.

resulted in the first-ever offspring produced in the wild by captive-reared parents in both California and Arizona.

Case study: Yellowstone wolves

The ultimate goal of captive-breeding programs is not simply to preserve interesting species, but rather to restore ecosystems to a balanced, functional state. Yellowstone Park has been an ecosystem out of balance, due in large part to the systematic extermination of the gray wolf (*Canis lupus*) in the park early in the 20th century. Without these predators to keep their numbers in check, herds of elk and deer expanded rapidly, damaging vegetation so that the elk themselves starve in time of scarcity.

In an attempt to restore the park's natural balance, two complete wolf packs from Canada were released into the park in 1995 and 1996. The wolves adapted well, breeding so successfully that by 2002 the park contained 16 free-ranging packs and by 2013 the greater Yellowstone area contained over 400 wolves.

Although ranchers near the park have been unhappy about the return of the wolves, little damage to livestock has been noted, and the ecological equilibrium of Yellowstone Park seems well on the way to being regained. Elk are congregating in larger herds and are avoiding areas near rivers where they are vulnerable. As a result, in some areas riverside trees such as willows are increasing in number, in turn providing food for beavers, whose dams lead to the creation of ponds, a habitat type that had become rare in Yellowstone. This newly restored habitat, in turn, has led to increases in some species of birds such as the redstart that had been in decline for decades or disappeared entirely. Why this recovery has occurred in some parts of Yellowstone and not others is currently being debated.

Current conservation approaches are multidimensional

Historically, conservationists strived to solve the problems of habitat fragmentation by focusing solely on preserving as much land as possible in a pristine state in national parks and reserves. Increasingly, however, it has become apparent that the amount of land that can be preserved in such a state is limited; moreover, many areas that are not completely protected nonetheless provide suitable habitat for many species.

As a result, conservation plans are becoming multidimensional, including not only pristine areas, but also surrounding areas in which some level of human disturbance is permitted. As discussed in section 59.3, isolated patches of habitat lose species far more rapidly than large preserves do. By including these other, less pristine areas, the total amount of area available for many species is increased.

The key to managing such large tracts of land successfully over a long time is to operate them in a way compatible with local land use. For example, although no economic activity is allowed in the core pristine area, the remainder of the land may be used for nondestructive harvesting of resources. Even areas in which hunting of some species is allowed provide protection for many other species.

Corridors of dispersal are also being provided that link the pristine areas, thus effectively increasing population sizes and allowing recolonization if a population disappears in one area due to a catastrophe. Corridors can also provide protection to species that move over great distances during the course of a year. Corridors in East Africa have protected the migration routes of ungulates. In Costa Rica, a corridor linking the lowland rainforest at the La Selva Biological Station to the montane rainforest in Braulio Carrillo National Park permits the altitudinal migration of many species of birds, mammals, and butterflies (figure 59.26).

Figure 59.26 Corridor connecting two reserves. *a.* The Organization of Tropical Studies' La Selva Biological Station in Costa Rica is connected to Braulio Carrillo National Park. *b.* The corridor allows migration of birds, mammals, butterflies, and other animals from La Selva at 35 m above sea level to mountainous habitats up to 2900 m elevation.

In addition to this focus on maintaining large enough reserves, conservation biologists also have recognized that the best way to preserve biodiversity is to focus on preserving intact ecosystems, rather than particular species. For this reason, attention in many cases is turning to identifying those ecosystems most in need of preservation and devising the means to protect not only the species within the ecosystem, but the functioning of the ecosystem itself. This entails making sure that reserves are not only large enough, but also that they protect elements such as watersheds so that activities outside the reserve won't threaten the ecosystem within it.

> ### Learning Outcomes Review 59.4
> Restoration of species may prevent extinction, but only if restoration of habitat or an entire ecosystem is also undertaken. Removal of introduced species and cleanup of pollutants are primary strategies for habitat restoration. In cases where extinction appears imminent, removal of individuals from the wild and preservation in captive breeding programs may be necessary while habitat is restored.
>
> ■ Can habitat restoration ever approach a pristine state? Why or why not?

Chapter Review

59.1 Overview of the Biodiversity Crisis

Prehistoric humans were responsible for local extinctions.
Shortly after humans arrived in North America after the last Ice Age, at least 75% of large mammals became extinct. The same pattern has been observed in other parts of the world.

Extinctions have continued in historical time.
The majority of historical extinctions have occurred since 1900. The current mass extinction is the only such event triggered by one species, *Homo sapiens,* and the only one in which resources will not be widely available for evolutionary recovery afterward.

Endemic species hotspots are especially threatened.
Endemic species are found in one restricted range and are thus vulnerable to extinction. Hotspots are areas with many endemic species; many hotspots are the site of large human population growth and high rates of extinction.

59.2 The Value of Biodiversity

The direct economic value of biodiversity includes resources for our survival.
Many products are obtained from different species and ecosystems, including food, materials for clothing and shelter, and medicines.

Indirect economic value is derived from ecosystem services.
Intact ecosystems provide services such as maintaining water quality, preserving soils and nutrients, moderating local climates, and recycling nutrients. The value of intact ecosystems is often not apparent until they are lost.

Ethical and aesthetic values are based on our conscience and our consciousness.
Humans can and should make ethical decisions to protect the esthetic, ecological, and economic values of ecosystems.

59.3 Factors Responsible for Extinction

Amphibians are on the decline: A case study.
Almost half of all amphibian species have experienced decreases in population size. No single cause has been identified, which implies that global environmental changes may be responsible.

Habitat loss devastates species richness.
Habitat may be destroyed, polluted, disrupted, or fragmented. As habitats become more fragmented, the relative proportion of the remaining habitat that occurs on the boundary or edge increases rapidly, exposing species to parasites, nonnative invasive species, and predators (figure 59.12).

Overexploitation wipes out species quickly.
Hunting and harvesting of wild species pose a risk of extinction. The collapse of the cod fisheries of the North Atlantic and the decline of whale species are only two of many examples.

Introduced species threaten native species and habitats.
Natural or accidental introductions of new species results in large and often negative changes to a community because of lack of checks and balances on introduced species' growth in the form of species interactions.

Disruption of ecosystems can cause an extinction cascade.
Extinction cascades may occur either top-down or bottom-up through the trophic levels. Loss of a keystone species may increase competition and greatly alter ecosystem structure and function.

Small populations are particularly vulnerable.
Catastrophes, lack of mates, and loss of genetic variability all make reduced populations more likely to become extinct.

59.4 Approaches for Preserving Endangered Species and Ecosystems

Destroyed habitats can sometimes be restored.
Restoration by removal of introduced species is very difficult and is most successful if done very soon after a new species is introduced. Severely polluted or damaged habitats sometimes cannot be restored to original conditions, but they may be restored to provide different environmental services.

Captive breeding programs have saved some species.
Species may be bred in captivity and returned to the wild when the factors that caused their endangerment are no longer a threat. Preservation of habitat may be a key in successful reintroduction.

Current conservation approaches are multidimensional.
The best way to preserve biodiversity is to preserve intact ecosystems rather than individual species. The key to management of large tracts of land is to operate them in a way compatible with local human needs.

Corridors of dispersal can link habitat fragments with one another and with larger habitats, allowing for increased population size, genetic exchange, and recolonization.

Review Questions

UNDERSTAND

1. Conservation hotspots are best described as
   a. areas with large numbers of endemic species, in many of which species are disappearing rapidly.
   b. areas where people are particularly active supporters of biological diversity.
   c. islands that are experiencing high rates of extinction.
   d. areas where native species are being replaced with introduced species.

2. The economic value of indirect ecosystem services
   a. is unlikely to exceed the economic value derived from uses after ecosystem conversion.
   b. has never been carefully determined.
   c. can greatly exceed the value derived after ecosystem conversion.
   d. is entirely aesthetic.

3. The amphibian decline is best described as
   a. global disappearance of amphibian populations due to the pervasiveness of local habitat destruction.
   b. global shrinkage of amphibian populations due to global climate change.
   c. the unexplained disappearance of golden toads in Costa Rica.
   d. None of the choices is correct.

4. Habitat fragmentation can negatively affect populations by
   a. restricting gene flow among areas that were previously continuous.
   b. increasing the relative amount of edge in suitable habitat patches.
   c. creating patches that are too small to support a breeding population.
   d. All of the choices are correct.

5. When populations are drastically reduced in size, genetic diversity and heterozygosity
   a. are likely to increase, enhancing the probability of extinction.
   b. are likely to decrease, enhancing the probability of extinction.
   c. are usually not factors that influence the probability of extinction.
   d. automatically respond in a way that protects populations from future changes.

6. A captive-breeding program followed by release to the wild
   a. is very likely, all by itself, to save a species threatened by extinction.
   b. is only likely to succeed when genetic variation of wild populations is very low.
   c. may be successful when combined with proper regulations and habitat restoration.
   d. None of the choices is correct.

APPLY

1. Historically, island species have tended to become extinct faster than species living on a mainland. Which of the following reasons can be used to explain this phenomenon?
   a. Island species have often evolved in the absence of predators and have no natural avoidance strategies.
   b. Humans have introduced diseases and competitors to islands, which negatively affect island populations.
   c. Island populations are usually smaller than mainland populations.
   d. All of the choices are correct.

2. Ninety-nine percent of all the species that ever existed have gone extinct,
   a. serving as evidence that current extinction rates are not higher than normal.
   b. but most of these losses have occurred in the last 400 years.
   c. which argues that the world just had too many species.
   d. None of the choices is correct.

3. To effectively address the biodiversity crisis, the protection of individual species
   a. must be used in concert with a principle of ecosystem management and restoration.
   b. is a sufficient management approach that merely needs to be expanded to more species.
   c. has no role to play in addressing the biodiversity crisis.
   d. usually conflicts with the principle of ecosystem management.

4. The introduction of a nonnative predator to an ecosystem could cause extinction by
   a. causing a top-down trophic cascade (see chapter 57).
   b. outcompeting a native carnivore (see chapter 56).
   c. transmitting parasites to which the native species are not adapted.
   d. All of the choices are correct.

SYNTHESIZE

1. If 99% of the species that ever existed are now extinct, why is there such concern over the extinction rates over the last several centuries?

2. Ecosystem conversion always has a cost and a benefit. Usually the benefit flows to a segment of society (a business or one group of people, for instance), but the costs are borne by all of society. That is what makes decisions about how and when to convert ecosystems difficult. However, is that a problem unique to conversion of ecosystems in the way we understand it today (for example, the conversion of the mangrove to a shrimp farm)? Are there other examples we can look to for guidance in how to make these decisions?

3. There is concern and evidence that amphibian populations are declining worldwide as a consequence of factors acting globally. Given that we know that species extinction is a natural process, how do we determine if there is a global decline that is different from normal species extinction?

4. Given what you learned in chapter 56 about interactions between species and in chapter 57 about interactions among trophic levels, how can the extinction of one species have far-ranging effects on an ecosystem? Is it possible to predict which species would be particularly likely to affect many other species if they were to go extinct?

5. All populations become small before going extinct. Is small population size really a cause of extinction, or just something that happens as a result of other factors that cause extinction?

Appendix A

Answer Key

CHAPTER 1

LEARNING OUTCOME QUESTIONS

1.1 No. The study of biology encompasses information/tools from chemistry, physics, and geology—in fact all of the "natural sciences."

1.2 A scientific theory has been tested by experimentation. A hypothesis is a starting point for explaining a body of observations. When predictions generated using the hypothesis have been tested, it gains the confidence associated with a theory. A theory still cannot be "proved," however, as new data can always force us to reevaluate a theory.

1.3 No. Natural selection explains the patterns of living organisms we see at present and allows us to work back in time, but it is not intended to explain how life arose. This does not mean that we can never explain this, but merely that natural selection does not do this.

1.4 Viruses do not fit well into our definition of living systems. It is a matter of controversy whether viruses should be considered "alive." They lack the basic cellular machinery, but they do have genetic information. Some theories for the origin of cells view viruses as being a step from organic molecules to cell, but looking at current organisms, they do not fulfill our definition of life.

INQUIRY AND DATA ANALYSIS QUESTIONS

Page 10 Data analysis: The curve would be shifted to the right, but the basic shape would not be changed. It would delay the time it takes for population size to outstrip resources.

Page 10 Inquiry question: It can be achieved by lowering family size or delaying childbearing.

Page 11 Inquiry question: A snake would fall somewhere near the bird, as birds and snakes are closely related.

UNDERSTAND

1. b   2. c   3. a   4. b   5. d   6. b   7. c   8. c

APPLY

1. d   2. d   3. c   4. d   5. d   6. d   7. a

SYNTHESIZE

1. For something to be considered living it would demonstrate organization, possibly including a cellular structure. The organism would gain and use energy to maintain homeostasis, respond to its environment, and grow and reproduce. These latter properties would be difficult to determine if the evidence of life from other planets comes from fossils. Similarly, the ability of an alien organism to evolve could be difficult to establish.
2. a. The variables that were held the same between the two experiments include the broth, the flask, and the sterilization step.
   b. The shape of the flask influences the experiment because any cells present in the air can enter the flask with the broken neck, but they are trapped in the neck of the other flask.
   c. If cells can arise spontaneously, then cell growth will occur in both flasks. If cells can only arise from preexisting cells (cells in the air), then only the flask with the broken neck will grow cells. Breaking the neck exposes the broth to a source of cells.
   d. If the sterilization step did not actually remove all cells, then growth would have occurred in both flasks. This result would seem to support the hypothesis that life can arise spontaneously.

CHAPTER 2

LEARNING OUTCOME QUESTIONS

2.1 If the number of protons exceeds neutrons, there is no effect on charge; if the number of protons exceeds electrons, then the charge is (+).

2.2 Atoms are reactive when their outer electron shell is not filled with electrons. The noble gases have filled outer electron shells and are thus unreactive.

2.3 An ionic bond results when there is a transfer of electrons, resulting in positive and negative ions that are attracted to each other. A covalent bond is the result of two atoms sharing electrons. Polar covalent bonds involve unequal sharing of electrons. This produces regions of partial charge, but not ions.

2.4 C and H have about the same electronegativity, and thus form nonpolar covalent bonds. This would not result in a cohesive or adhesive fluid.

2.5 Since ice floats, a lake will freeze from the top down, not the bottom up. This means that water remains fluid on the bottom of the lake, allowing living things to overwinter.

2.6 Since pH is a log scale, this would be a change of 100-fold in $[H^+]$.

INQUIRY AND DATA ANALYSIS QUESTION

Page 30 Data analysis: From the graph, about four volumes of base must be added to change the pH from 4 to 6.

UNDERSTAND

1. b   2. d   3. b   4. a   5. c   6. d   7. b

APPLY

1. c   2. b   3. a   4. c   5. d   6. Chemical reactions involve changes in the electronic configuration of atoms. Radioactive decay involves the actual decay of the nucleus itself, producing another atom and emitting radiation.

SYNTHESIZE

1. A cation is an element that tends to lose an electron from its outer energy level, leaving behind a net positive charge due to the presence of the protons in the atomic nucleus. Electrons are only lost from the outer energy level if that loss is energetically favorable; that is, if it makes the atom more stable by virtue of obtaining a filled outer energy level (the octet rule). The atoms that fall into this category will be those in the first two columns. Examples include lithium (Li), magnesium (Mg), and beryllium (Be).
2. Silicon has an atomic number of 14. This means that there are four unpaired electrons in its outer energy level (comparable to carbon). Based on this fact, you can conclude that silicon, like carbon, could form four covalent bonds. Silicon also falls within the group of elements with atomic masses less than 21, a property of the elements known to participate in the formation of biologically important molecules. Interestingly, silicon is much more prevalent than carbon on Earth. Although silicon dioxide is found in the cell walls of plants and single-celled organisms called diatoms, silicon-based life has not been identified on this planet. Given the abundance of silicon on Earth you can conclude that some other aspect of the chemistry of this atom makes it incompatible with the formation of molecules that make up living organisms.
3. Water is considered to be a critical molecule for the evolution of life on Earth. It is reasonable to assume that water on other planets could play a similar role. The key properties of water that would support its role in the evolution of life are:
   - The ability of water to act as a solvent. Molecules dissolved in water could move and interact in ways that would allow for the formation of larger, more complex molecules such as those found in living organisms.
   - The high specific heat of water. Water can modulate and maintain its temperature, thereby protecting the molecules or organisms within it from temperature extremes—an important feature on other planets.

- The difference in density between ice and liquid water. The fact that ice floats is a simple, but important feature of water environments since it allows living organisms to remain in a liquid environment protected under a surface of ice. This possibility is especially intriguing, given recent evidence of ice-covered oceans on Europa, a moon of the planet Jupiter.

Chapter 3
LEARNING OUTCOME QUESTIONS

3.1 Hydrolysis is the reverse reaction of dehydration. Dehydration is a synthetic reaction involving the loss of water and hydrolysis is cleavage by addition of water.

3.2 Starch and glycogen are both energy-storage molecules. Their highly branched nature allows the formation of droplets, and the similarity in the bonds holding adjacent glucose molecules together mean that the enzymes we have to break down glycogen allow us to break down starch. The same enzymes do not allow us to break down cellulose. The structure of cellulose leads to the formation of tough fibers.

3.3 The sequence of bases in an RNA would be identical to one strand and complementary to the other strand of the DNA with the exception that U would be in place of T (complementary to A).

3.4 If an unknown protein has sequence similarity to a known protein, we can infer its function is also similar. If an unknown protein has known functional domains or motifs, we can also use these to help predict function.

3.5 Phospholipids have a charged group replacing one of the fatty acids in a triglyceride. This leads to an amphipathic molecule that has both hydrophobic and hydrophilic regions. This will spontaneously form bilayer membranes in water.

UNDERSTAND
1. b 2. a 3. d 4. c 5. b 6. b 7. c 8. b

APPLY
1. c 2. d 3. b 4. d 5. b 6. b 7. d

SYNTHESIZE

1. The four biological macromolecules all have different structure and function. In comparing carbohydrates, nucleic acids, and proteins, we can think of these as being polymers with different monomers. In the case of carbohydrates, the polymers are all polymers of the simple sugar glucose. These are energy-storage molecules (with many C—H bonds) and structural molecules such as cellulose that make tough fibers.
   Nucleic acids are formed of nucleotide monomers, each of which consists of ribose, phosphate, and a nitrogenous base. These molecules are informational molecules that encode information in the sequence of bases. The bases interact in specific ways: A base-pairs with T, and G base-pairs with C. This is the basis for their informational storage.
   Proteins are formed of amino acid polymers. There are 20 different amino acids, and thus an incredible number of different proteins. These can have an almost unlimited number of functions. These functions arise from the amazing flexibility in structure of protein chains.

2. *Nucleic Acids*—Hydrogen bonds are the basis for complementary base-pairing between the two strands of the double helix. Complementary base-pairing can also occur within the single nucleic acid strand of an RNA molecule. Hydrophobic interactions between stacked bases also stabilize the double helix.
   Proteins—The α helices and β-pleated sheets of secondary structure are stabilized by hydrogen bond formation between the amino and carboxyl groups of the amino acid backbone. Hydrogen bond formation between R groups helps stabilize the three-dimensional folding of the protein at the tertiary level of structure. Hydrophobic interactions also stabilize tertiary structure.
   Carbohydrates—Hydrogen bonds are less important for carbohydrates; however, these bonds are responsible for the formation of the fibers of cellulose that make up the cell walls of plants.
   Lipids—Lipids do not form hydrogen bonds, but hydrophobic interactions are important. The hydrophobic nature of the fatty acids gives membranes structures.

3. The enzymes that bind to starch in their active site would not be expected to bind to cellulose, as the shape of the bonds is different. Thus, these enzymes are unlikely to be involved in cellulose synthesis.

Chapter 4
LEARNING OUTCOME QUESTIONS

4.1 The statement about all cells coming from preexisting cells might need to be modified. It would really depend on whether these Martian life-forms had a similar molecular/cellular basis as that of terrestrial life.

4.2 Bacteria and archaea both tend to be single cells that lack a membrane-bounded nucleus, and lack extensive internal endomembrane systems. They both have a cell wall, although the composition is different. They do not undergo mitosis, although the proteins involved in DNA replication and cell division are not similar.

4.3 Part of what gives different organs their unique identities are the specialized cell types found in each. That does not mean that there will not be some cell types common to all (epidermal cells for example) but organs tend to have specialized cell types.

4.4 They don't!

4.5 The nuclear genes that encode organellar proteins moved from the organelle to the nucleus. There is evidence for a lot of "horizontal gene transfer" across domains; this is an example of how that can occur.

4.6 It provides structure and support for larger cells, especially in animal cells that lack a cell wall.

4.7 Microtubules and microfilaments are both involved in cell motility and in movement of substance around cells. Intermediate filaments do not have this dynamic role, but are more structural.

4.8 Cell junctions help to put together cells into higher level structures that are organized and joined in different ways. Different kinds of junctions can be used for different functional purposes.

INQUIRY AND DATA ANALYSIS QUESTIONS

Page 63 Inquiry question: Stretch, dent, convolute, fold, add more than one nucleus, anything which would increase the amount of diffusion between the cytoplasm and the external environment.

Page 75 Inquiry question: Both mitochondria and chloroplasts have highly infolded inner membranes, where many of the reactions take place leading to the production of ATP. They also have an inner chamber with soluble material containing enzymes.

Page 80 Inquiry question: Ciliated cells in the trachea help to remove particulate matter from the respiratory tract so that it can be expelled or swallowed and processed in the digestive tract.

UNDERSTAND
1. d 2. d 3. c 4. a 5. c 6. d 7. b

APPLY
1. c 2. b 3. c 4. b 5. c 6. b 7. a

SYNTHESIZE

1. Your diagram should start at the SER and then move to the RER, Golgi apparatus, and finally to the plasma membrane. Small transport vesicles are the mechanism that would carry a phospholipid molecule between two membrane compartments. Transport vesicles are small "membrane bubbles" composed of a phospholipid bilayer.

2. If these organelles were free-living bacteria, they would have the features found in bacteria. Mitochondria and chloroplasts both have DNA but no nucleus, and they lack the complex organelles found in eukaryotes. At first glance, the cristae may seem to be an internal membrane system, but they are actually infoldings of the inner membrane. If endosymbiosis occurred, this would be the plasma membrane of the endosymbiont, and the outer membrane would be the plasma membrane of the engulfing cell. Another test would be to compare DNA in these organelles with current bacteria. This has actually shown similarities that make us confident of the identity of the endosymbionts.

3. The prokaryotic and eukaryotic flagella are examples of an analogous trait. Both flagella function to propel the cell through its environment by converting chemical energy into mechanical force. The key difference is in the structure of the flagella. The bacterial flagellum is composed of a single protein emerging from a basal body anchored within the cell's plasma membrane and using the potential energy of a proton gradient to cause a rotary movement. In contrast, the flagellum of the eukaryote is composed of many proteins assembled into a complex axoneme structure that uses ATP energy to cause an undulating motion.

4. The origins of mitochondria and chloroplasts are hypothesized to be the result of a bit of cellular "indigestion" in which aerobic or photosynthetic prokaryotes were engulfed but not digested by the larger ancestor eukaryote. Given this information, there are two possible scenarios for the origin of *Giardia*. In the first scenario, the ancestor of *Giardia* split off from the eukaryotic lineage after the evolution of the nucleus but before the acquisition of mitochondria. In the second scenario, the ancestor of *Giardia* split off after the acquisition of mitochondria and subsequently lost the mitochondria. At present, neither of these two scenarios can be rejected. The first case was long thought to be the best explanation, but recently it has been challenged by evidence for the second case.

CHAPTER 5

LEARNING OUTCOME QUESTIONS

5.1 Cells would not be able to control their contents. Nonpolar molecules would be able to cross the membrane by diffusion, as would small polar molecules, but without proteins to control the passage of specific molecules, it would not function as a semipermeable membrane.

5.2 No. The nonpolar interior of the bilayer would not be soluble in the solvent. The molecules would organize with their nonpolar tails in the solvent, but the negative charge on the phosphates would repel other phosphates.

5.3 Transmembrane domains anchor protein in the membrane. They associate with the hydrophobic interior, thus they must be hydrophobic as well. If they slide out of the interior, they are repelled by water.

5.4 The concentration of the IV will be isotonic with your blood cells. If it were hypotonic, your blood cells would take on water and burst; if it were hypertonic, your blood cells would lose water and shrink.

5.5 Channel proteins are aqueous pores that allow facilitated diffusion. They cannot actively transport ions. Carrier proteins bind to their substrates and can couple transport to some form of energy for active transport.

5.6 In all cases, there is recognition and specific binding of a molecule by a protein. In each case this binding is necessary for biological function.

INQUIRY AND DATA ANALYSIS QUESTIONS

Page 95 Inquiry question: As the name suggests for the fluid mosaic model, cell membranes have some degree of fluidity. The degree of fluidity varies with the composition of the membrane, but in all membranes, phospholipids are able to move about within the membrane. Also, due to the hydrophobic and hydrophilic opposite ends of phospholipid molecules, phospholipid bilayers form spontaneously. Therefore, if stressing forces happen to damage a membrane, adjacent phospholipids automatically move to fill in the opening.

Page 96 Inquiry question: Integral membrane proteins are those that are embedded within the membrane structure and provide passageways across the membrane. Because integral membrane proteins must pass through both polar and nonpolar regions of the phospholipid bilayer, the protein portion held within the nonpolar fatty acid interior of the membrane must also be nonpolar. The amino acid sequence of an integral protein would have polar amino acids at both ends, with nonpolar amino acids making up the middle portion of the protein.

UNDERSTAND

1. d 2. a 3. d 4. d 5. b 6. d 7. a

APPLY

1. c 2. b 3. d 4. c 5. d

SYNTHESIZE

1. Since the membrane proteins become intermixed in the absence of the energy molecule, ATP, one can conclude that chemical energy is not required for their movement. Since the proteins do not move and intermix when the temperature is cold, one can also conclude that the movement is temperature-sensitive. The passive diffusion of molecules also depends on temperature and does not require chemical energy; therefore, it is possible to conclude that membrane fluidity occurs as a consequence of passive diffusion.

2. The inner half of the bilayer of the various endomembranes becomes the outer half of the bilayer of the plasma membrane.

3. Lipids can be inserted into one leaflet to produce asymmetry. When lipids are synthesized in the SER, they can be assembled into asymmetric membranes. There are also enzymes that can flip lipids from one leaflet to the other.

CHAPTER 6

LEARNING OUTCOME QUESTIONS

6.1 At the bottom of the ocean, light is not an option as it does not penetrate that deep. However, there is a large source of energy in the form of reduced minerals, such as sulfur compounds, that can be oxidized. These are abundant at hydrothermal vents found at the junctions of tectonic plates. This supports whole ecosystems dependent on bacteria that oxidize reduced minerals.

6.2 In a word, no. Enzymes only alter the rate of a reaction. The action of an enzyme does not change the ΔG for the reaction.

6.3 In the text, it stated that the average person turns over approximately their body weight in ATP per day. This gives us enough information to determine approximately the amount of energy released:

$$100 \text{ kg} = 1.0 \times 10^5 \text{ g}$$
$$(1.0 \times 10^5 \text{ g})/(507.18 \text{ g/mol}) = 197.2 \text{ mol}$$
$$(197.2 \text{ mol})(7.3 \text{ kcal/mol}) = 1439 \text{ kcal}$$

6.4 This is a question that cannot be definitely answered, but we can give some reasonable conjectures. First, DNA's location is in the nucleus and not the cytoplasm, where most enzymes are found. Second, the double-stranded structure of DNA works well for information storage, but would not necessarily function well as an enzyme. Each base interacts with a base on the opposite strand, which makes for a very stable linear molecule, but does not encourage folding into the kind of complex 3-D shape found in enzymes.

6.5 Feedback inhibition is common in pathways that synthesize metabolites. In these anabolic pathways, when the end-product builds up, it feeds back to inhibit its own production. Catabolic pathways are involved in the degradation of compounds. Feedback inhibition makes less biochemical sense in a pathway that degrades compounds as the end-product is destroyed or removed and cannot feed back.

INQUIRY AND DATA ANALYSIS QUESTION

Page 113 Data analysis: The overall ΔG would be the sum of the individual ΔGs, or −3.9 kcal/mol. This makes the overall process exergonic.

UNDERSTAND

1. b 2. a 3. b 4. a 5. d 6. b 7. d

APPLY

1. b 2. c 3. a 4. c 5. c 6. c

SYNTHESIZE

1. a. At 40°C the enzyme is at its optimum. The rate of the reaction is at its highest level.
   b. Temperature is a factor that influences enzyme function. This enzyme does not appear to function at either very cold or very hot temperatures. The shape of the enzyme is affected by temperature, and the enzyme's structure is altered enough at extreme temperatures that it no longer binds substrate. Alternatively, the enzyme may be denatured—that is a complete loss of normal three-dimensional shape at extreme temperatures. Think about frying an egg: What happens to the proteins in the egg?
   c. Everyone's body is slightly different. If the temperature optimum was very narrow, then the cells that make up a body would be vulnerable. Having a broad range of temperature optimums keeps the enzyme functioning.

2. a. The reaction rate would be slow because of the low concentration of the substrate ATP. The rate of reaction depends on substrate concentration.
   b. ATP acts like a noncompetitive, allosteric inhibitor when ATP levels are very high. If ATP binds to the allosteric site, then the reaction should slow down.
   c. When ATP levels are high, the excess ATP molecules bind to the allosteric site and inhibit the enzyme. The allosteric inhibitor functions by causing a change in the shape of the active site in the enzyme. This reaction is an example of feedback regulation because ATP is a final

product of the overall series of reactions associated with glycolysis. The cell regulates glycolysis by regulating this early step catalyzed by phosphofructokinase; the allosteric inhibitor is the "product" of glycolysis (and later stages) which is ATP.

CHAPTER 7

LEARNING OUTCOME QUESTIONS

7.1 Cells require energy for a wide variety of functions. The reactions involved in the oxidation of glucose are complex, and linking these to the different metabolic functions that require energy would be inefficient. Thus cells make and use ATP as a reusable source of energy.

7.2 The location of glycolysis does not argue for or against the endosymbiotic origin of mitochondria. It could have been located in the mitochondria previously and moved to the cytoplasm or could have always been located in the cytoplasm in eukaryotes.

7.3 For an enzyme like pyruvate decarboxylase the complex reduces the distance for the diffusion of substrates for the different stages of the reaction. Any possible unwanted side reactions are prevented. Finally the reactions occur within a single unit and thus can be controlled in a coordinated fashion. The main disadvantage is that since the enzymes are all part of a complex their evolution is more constrained than if they were independent.

7.4 At the end of the Krebs cycle, the electrons removed from glucose are all carried by soluble electron carriers. Most of these are in NADH, and a few are in $FADH_2$. All of these are fed into the electron transport chain under aerobic conditions where they are used to produce a proton gradient.

7.5 A hole in the outer membrane would allow protons in the intermembrane space to leak out. This would destroy the proton gradient across the inner membrane, stopping the phosphorylation of ADP by ATP synthase.

7.6 Chemiosmosis means that the ATP/NADH ratio is not dependent on the number of "pumping" stations. Instead, it is dependent on the number of protons needed for one turn of the enzyme, and on the number of binding sites on the enzyme for ADP/ATP.

7.7 Glycolysis, which is the starting point for respiration from sugars, is regulated at the enzyme phosphofructokinase. This enzyme is just before the 6-C skeleton is split into two 3-C molecules. The allosteric effectors for this enzyme include ATP and citrate. Thus the "end-product" ATP, and an intermediate from the Krebs cycle, both feed back to inhibit the first part of this process.

7.8 The first obvious point is that the most likely type of ecosystem would be one where oxygen is nonexistent or limiting. This includes marine, aquatic, and soil environments. Any place where oxygen is in short supply is expected to be dominated by anaerobic organisms, and respiration produces more energy than fermentation.

7.9 The short answer is no. The reason is twofold. First the oxidation of fatty acids feeds acetyl units into the Krebs cycle. The primary output of the Krebs cycle is electrons that are fed into the electron transport chain to eventually produce ATP by chemiosmosis. The second reason is that the process of β oxidation that produces the acetyl units is oxygen dependent as well. This is because β oxidation uses FAD as a cofactor for oxidation, and the $FADH_2$ is oxidized by the electron transport chain.

7.10 The evidence for the origins of metabolism is indirect. The presence of O_2 in the atmosphere is the result of photosynthesis, so the record of when we went from a reducing to an oxidizing atmosphere chronicles the rise of oxygenic photosynthesis. Glycolysis is a universal pathway that is found in virtually all types of cells. This indicates that it is an ancient pathway that likely evolved prior to other types of energy metabolism. Nitrogen fixation probably evolved in the reducing atmosphere that preceded oxygenic photosynthesis as it is poisoned by oxygen, and aided by the reducing atmosphere.

INQUIRY AND DATA ANALYSIS QUESTION

Page 142 Data analysis: During the catabolism of fats, each round of β oxidation uses one molecule of ATP and generates one molecule each of $FADH_2$ and NADH. For a 16-carbon fatty acid, seven rounds of β oxidation would convert the fatty acid into eight molecules of acetyl-CoA. The oxidation of each acetyl-CoA in the Krebs cycle produces 10 molecules of ATP. The overall ATP yield from a 16-carbon fatty acid would be a net gain of 21 ATP from seven rounds of β oxidation [gain of 4 ATP per round – 1 per round to prime reactions] + 80 ATP from the oxidation of 8 acetyl-CoAs = 101 molecules of ATP.

UNDERSTAND

1. d 2. d 3. c 4. c 5. a 6. d 7. c

APPLY

1. b 2. b 3. c 4. c 5. c 6. b 7. b

SYNTHESIZE

1.

Molecules	Glycolysis	Cellular Respiration
Glucose	Is the starting material for the reaction	Does not directly use glucose; however, does use pyruvate derived from glucose
Pyruvate	The end product of glycolysis	The starting material for cellular respiration
Oxygen	Not required	Required for aerobic respiration, but not for anaerobic respiration
ATP	Produced through substrate-level phosphorylation	Produced through oxidative phosphorylation. More produced than in glycolysis
CO_2	Not produced	Produced during pyruvate oxidation and Krebs cycle

2. The electron transport chain of the inner membrane of the mitochondria functions to create a hydrogen ion concentration gradient by pumping protons into the intermembrane space. In a typical mitochondrion, the protons can only diffuse back down their concentration gradient by moving through the ATP synthase and generating ATP. If protons can move through another transport protein, then the potential energy of the hydrogen ion concentration gradient would be "lost" as heat.

3. If brown fat persists in adults, then the uncoupling mechanism to generate heat described above could result in weight loss under cold conditions. There is now some evidence to indicate that this may be the case.

CHAPTER 8

LEARNING OUTCOME QUESTIONS

8.1 Both chloroplasts and mitochondria have an outer membrane and an elaborate inner membrane. These inner membrane systems have electron transport chains that move protons across the membrane to allow for the synthesis of ATP by chemiosmosis. They also both have a soluble compartment in which a variety of enzymes carry out reactions.

8.2 All of the carbon in your body comes from carbon fixation by autotrophs. Thus, all of the carbon in your body was once CO_2 in the atmosphere, before it was fixed by plants.

8.3 The action spectrum for photosynthesis refers to the most effective wavelengths. The absorption spectrum for an individual pigment shows how much light is absorbed at different wavelengths.

8.4 Before the discovery of photosystems, we assumed that each chlorophyll molecule absorbed photons resulting in excited electrons.

8.5 Without a proton gradient, synthesis of ATP by chemiosmosis would be impossible. However, NADPH could still be synthesized because electron transport would still occur as long as photons were still being absorbed to begin the process.

8.6 A portion of the Calvin cycle is the reverse of glycolysis (the reduction of 3-phosphoglycerate to glyceraldehyde -3-phosphate).

8.7 Both C_4 plants and CAM plants fix carbon by incorporating CO_2 into the 4-carbon malate and then use this to produce high local levels of CO_2 for the Calvin cycle. The main difference is that in C_4 plants, this occurs in different cells, and in CAM plants this occurs at different times.

INQUIRY AND DATA ANALYSIS QUESTIONS

Page 150 Data analysis: Light energy is used to produce NADPH and ATP that can then be used to reduce CO_2. Molecules of chlorophyll absorb photons of light energy, so increasing light intensity increases photosynthesis, but will saturate when all chlorophyll molecules are in use.

Page 154 Data analysis: The curves would be similar to the observed curve shown, but would plateau at a higher and lower level, respectively. Even the higher level would still be below the expected curve for all chlorophyll molecules being used.

Page 157 Data analysis: You could conclude that the two photosystems function together and not in series.

UNDERSTAND
1. c   2. a   3. a   4. b   5. c   6. c   7. a   8. b

APPLY
1. d   2. b   3. d   4. c   5. d   6. d   7. a   8. a

SYNTHESIZE

1. In C_3 plants CO_2 reacts with ribulose 1,5-bisphosphate (RuBP) to yield 2 molecules of PGA. This reaction is catalyzed by the enzyme rubisco, which also catalyzes the oxidation of RuBP. Which reaction predominates depends on the relative concentrations of reactants. The reactions of the Calvin cycle reduce the PGA to G3P, which can be used to make a variety of sugars including RuBP. In C_4 and CAM plants, an initial fixation reaction incorporates CO_2 into malate. The malate then can be decarboxylated to pyruvate and CO_2 to produce locally high levels of CO_2. The high levels of CO_2 get around the oxidation of RuBP by rubisco. In C_4 plants malate is produced in one cell, then shunted into an adjacent cell that lacks stomata to produce high levels of CO_2. CAM plants fix carbon into malate at night when their stomata are open, then use this during the day to fuel the Calvin cycle. Both are evolutionary innovations that have arisen in hot dry climates to more efficiently fix carbon and prevent desiccation.

2. Figure 8.19 diagrams this relationship. The oxygen produced by photosynthesis is used as a final electron acceptor for electron transport in respiration. The CO_2 that results from the oxidation of glucose (or fatty acids) is incorporated into organic compounds via the Calvin cycle. Respiration also produces water, while photosynthesis consumes water.

3. Yes. Plants use their chloroplasts to convert light energy into chemical energy. During light reactions ATP and NADPH are created, but these molecules are consumed by the Calvin cycle. The G3P produced by the Calvin cycle stores the chemical energy from the light reactions. Ultimately, this energy is stored in glucose and utilized by the cell via glycolysis and cellular respiration.

CHAPTER 9
LEARNING OUTCOME QUESTIONS

9.1 Ligands bind to receptors based on complementary shapes. This interaction based on molecular recognition is similar to the way enzymes interact with their ligands.

9.2 Hydrophobic molecules can cross the membrane and are thus more likely to have an internal receptor.

9.3 Intracellular receptors have direct effects on gene expression. This generally leads to effects with longer duration.

9.4 Ras protein occupies a central role in signaling pathways involving growth factors. A number of different kinds of growth factors act through Ras. So it is not surprising mutated Ras protein is a factor in a number of different cancers.

9.5 GPCRs are a very ancient and flexible receptor/signaling pathway. The genes encoding these receptors have been duplicated and then have diversified over evolutionary time, so now there are many members of this gene family.

UNDERSTAND
1. b   2. b   3. c   4. d   5. b   6. d   7. c   8. a

APPLY
1. b   2. c   3. c   4. d   5. d   6. c

SYNTHESIZE

1. All signaling events start with a ligand binding to a receptor. The receptor initiates a chain of events that ultimately leads to a change in cellular behavior. In some cases the change is immediate—for example, the opening of an ion channel. In other cases the change requires more time before it occurs, such as when the MAP kinase pathway becomes activated multiple different kinases become activated and deactivated. Some signals only affect a cell for a short time (the channel example), but other signals can permanently change the cell by changing gene expression, and therefore the number and kind of proteins found in the cell.

2. a. This system involves *both* autocrine and paracrine signaling because Netrin-1 can influence the cells within the crypt that are responsible for its production and the neighboring cells.
   b. The binding of Netrin-1 to its receptor produces the signal for cell growth. This signal would be strongest in the regions of the tissue with the greatest amount of Netrin-1—that is, in the crypts. A concentration gradient of Netrin-1 exists such that the levels of this ligand are lowest at the tips of the villi. Consequently, the greatest amount of cell death would occur at the villi tips.
   c. Tumors occur when cell growth goes on unregulated. In the absence of Netrin-1, the Netrin-1 receptor can trigger cell death—controlling the number of cells that make up the epithelial tissue. Without this mechanism for controlling cell number, tumor formation is more likely.

CHAPTER 10
LEARNING OUTCOME QUESTIONS

10.1 The concerted replication and segregation of chromosomes works well with one small chromosome, but would likely not work as well with many chromosomes.

10.2 No.

10.3 The first irreversible step is the commitment to DNA replication.

10.4 Loss of cohesins would mean that the products of DNA replication would not be kept together. This would make normal mitosis impossible, and thus lead to aneuploid cells and probably be lethal.

10.5 The segregation of chromatids that lose cohesin would be random because they could no longer be held at metaphase attached to opposite poles. This would likely lead to gain and loss of this chromosome in daughter cells.

10.6 Tumor suppressor genes are genetically recessive, but proto-oncogenes are dominant. Loss of function for a tumor suppressor gene leads to cancer, whereas inappropriate expression or gain of function lead to cancer with proto-oncogenes.

UNDERSTAND
1. d   2. b   3. b   4. b   5. a   6. c   7. b   8. d

APPLY
1. a   2. c   3. b   4. d   5. c   6. d

SYNTHESIZE

1. With no Wee-1, the cell will not inhibit Cdk by phosphorylation. If Cdk remains active, then it will continue to signal the cell to move through the G_2/M checkpoint, but now in an unregulated manner. Cells undergo multiple rounds of cell division without the growth associated with G_2. Thus, daughter cells will become smaller and smaller with each division—hence the name of the protein!

2. Growth factor = ligand
   1. Ligand binds to receptor (the growth factor will bind to a growth factor receptor).
   2. A signal is transduced (carried) into the cytoplasm.
   3. A signal cascade is triggered. Multiple intermediate proteins or second messengers will be affected.
   4. A transcription factor will be activated to bind to a specific site on the DNA.
   5. Transcription occurs and the mRNA enters the cytoplasm.
   6. The mRNA is translated and a protein is formed.
   7. The protein functions within the cytoplasm—possibly triggering S phase.

   If you study figure 10.22 you will see a similar pathway for the formation of S phase proteins following receptor–ligand binding by a growth factor. Ultimately, the Rb protein that regulated the transcription factor E2F becomes phosphorylated, which allows E2F to bind to the genes for S phase proteins and cyclins.

3. Proto-oncogenes tend to encode proteins that function in signal transduction pathways that control cell division, When the regulation of these proteins is aberrant, or they are stuck in the "on" state by mutation, it can lead to cancer. Tumor suppressor genes, on the other hand, tend to be in genes that

encode proteins that suppress instead of activate cell division. Thus loss of function for a tumor suppressor gene leads to cancer.

CHAPTER 11
LEARNING OUTCOME QUESTIONS

11.1 Stem cells divide by mitosis to produce one cell that can undergo meiosis and another stem cell.

11.2 No. Keeping sister chromatids together at the first division is key to this reductive division. Homologues segregate at the first division, reducing the number of chromosomes by half.

11.3 An improper disjunction at anaphase I would result in 4 aneuploid gametes: 2 with an extra chromosome and 2 that are missing a chromosome. Nondisjunction at anaphase II would result in 2 normal gametes and 2 aneuploid gametes: 1 with an extra chromosome and 1 missing a chromosome.

11.4 The independent alignment of homologous pairs at metaphase I and the process of crossing over. The first shuffles the genome at the level of entire chromosomes, and the second shuffles the genome at the level of individual chromosomes.

INQUIRY AND DATA ANALYSIS QUESTION

Page 217 Inquiry question: No. At the conclusion of meiosis I each cell has a single copy of each homologue. So, even if the attachment of sister chromatids were lost after a meiosis I division, the results would not be the same as mitosis.

UNDERSTAND
1. c   2. d   3. a   4. b   5. b   6. a   7. b

APPLY
1. c   2. b   3. b   4. d   5. b   6. a

SYNTHESIZE

1. Compare your figure with figure 11.8.
   a. There would be three homologous pairs of chromosomes for an organism with a diploid number of six.
   b. For each pair of homologues, you should now have a maternal and paternal pair.
   c. Many possible arrangements are possible. The key to your image is that it must show the homologues aligned pairwise—not single-file along the metaphase plate. The maternal and paternal homologues *do not* have to align on the same side of the cell. Independent assortment means that the pairs can be mixed.
   d. A diagram of metaphase II would not include the homologous pairs. The pairs have separated during anaphase of meiosis I. Your picture should diagram the haploid number of chromosomes, in this case three, aligned single-file along the metaphase plate. Remember that meiosis II is similar to mitosis.

2. The diploid chromosome number of a mule is 63. The mule receives 32 chromosomes from its horse parent (diploid 64: haploid 32) and another 31 chromosomes from its donkey parent (diploid 61: haploid 31). 32 + 31 = 63. The haploid number for the mule would be one half the diploid number 63 ÷ 2 = 31.5. Can there be a 0.5 chromosome? Even if the horse and donkey chromosomes can pair (no guarantee of that), there will be one chromosome without a partner. This will lead to aneuploid gametes that are not viable.

3. Independent assortment involves the random distribution of maternal versus paternal homologues into the daughter cells produced during meiosis I. The number of possible gametes is equal to 2^n, where *n* is the haploid number of chromosomes. Crossing over involves the physical exchange of genetic material between homologous chromosomes, creating new combinations of genes on a single chromosome. Crossing over is a relatively rare event that affects large blocks of genetic material, so independent assortment likely has the greatest influence on genetic diversity.

4. a. Nondisjunction occurs at the point when the chromosomes are being pulled to opposite poles. This occurs during anaphase.
   b. Use an image like figure 11.8 and illustrate nondisjunction at anaphase I versus anaphase II

Anaphase I nondisjunction:

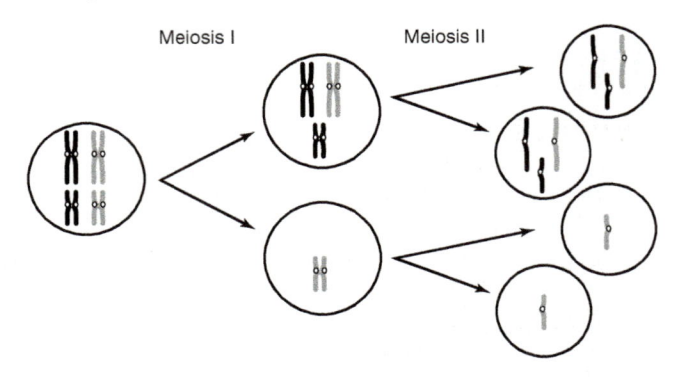

CHAPTER 12
LEARNING OUTCOME QUESTIONS

12.1 Both had an effect, but the approach is probably the most important. In theory, his approach would have worked for any plant, or even animal, he chose. In practice, the ease of both cross- and self-fertilization was helpful.

12.2 1/3 of tall F_2 plants are true-breeding.

12.3 The events of meiosis I are much more important in explaining Mendel's laws. During anaphase I homologues separate and are thus segregated, and the alignment of different homologous pairs at metaphase I is independent.

12.4 The cross is *Aa Bb* × *aa Bb* and the probability for (*A_ B_*) = (3/4) (1/2) =3/8.

12.5 1:1:1:1 dom-dom:dom-rec:rec-dom:rec-rec

12.6 7/16.

INQUIRY AND DATA ANALYSIS QUESTIONS

Page 223 Inquiry question: The ability to control whether the plants self-fertilized or cross-fertilized was of paramount importance in Mendel's studies. Results due to cross-fertilization would have had confounding influences on the predicted number of offspring with a particular phenotype.

Page 225 Data analysis: If the purple F_1 were backcrossed to the white parent, the phenotypic ratio would be 1 purple:1 white, and the genotypic ratio would be 1 non-true-breeding dominant (heterozygous):1 true-breeding recessive (homozygous recessive).

Page 227 Data analysis: Yes. The affected females each had one unaffected parent and thus are heterozygous. This means that the probability of affected offspring would be 50%.

Page 228 Inquiry question: Genetic defects that remain hidden or dormant as heterozygotes in the recessive state are more likely to be revealed in homozygous state among closely related individuals.

Page 231 Data analysis: The probability of being purple-flowered, round, and yellow is the same as the probability of being dominant for all three traits. The probability for the dominant phenotype from a cross of two heterozygotes is 3/4, so the probably of being dominant for all three traits in this cross would be (3/4)(3/4)(3/4) = 27/64.

Page 232 Data analysis:

TABLE 12.2	Dihybrid Testcross	
Actual Genotype	**Results of Testcross**	
	Trait A	Trait B
AABB	Trait A breeds true	Trait B breeds true
AaBB	——	Trait B breeds true
AABb	Trait A breeds true	——
AaBb	——	——

Page 233 Data analysis: Cuenot would have observed 2 yellow mice:1 wild type. This is because the homozygous yellow mice die, leaving 2 heterozygotes to 1 homozygous recessive.

Page 235 Inquiry question: Almost certainly, differences in major phenotypic traits of twins would be due to environmental factors such as diet.

Page 236 Data analysis: This is a case of epistasis, in which the albino gene obscures the effects of the black/brown locus. The offspring that are B_ aa (where the albino locus is designated by a) will appear the same as bb aa offspring. Thus the overall ratio is 9 black:3 brown:3+1 albino.

UNDERSTAND
1. b 2. c 3. c 4. c 5. b 6. d

APPLY
1. d 2. d 3. b 4. a 5. c 6. d

SYNTHESIZE

1. The approach to solving this type of problem is to identify the possible gametes. Separate the possible gamete combinations into the boxes along the top and side. Fill in the Punnett square by combining alleles from each parent.
   a. A monohybrid cross between individuals with the genotype Aa and A

	A	a
A	AA	Aa
a	aA	aa

   Phenotypic ratio: 3 dominant to 1 recessive

   b. A dihybrid cross between two individuals with the genotype AaBb

	AB	Ab	aB	ab
AB	AABB	AABb	AaBB	AaBb
Ab	AAbB	AAbb	AabB	Aabb
aB	aABB	aABb	aaBB	aaBb
ab	aAbB	aAbb	aabB	aabb

   Phenotypic ratio: 9 dominant dominant to 3 dominant recessive to 3 recessive dominant to 1 recessive recessive

   Using the product rule: Prob(A_ B_) = (¾)(¾) = 9/16
                           Prob(A_ bb) = (¾)(¼) = 3/16
                           Prob(aa B_) = (¼)(¾) = 3/16
                           Prob(aa bb) = (¼)(¼) = 1/16

   c. A dihybrid cross between individuals with the genotype AaBb and aabb

	AB	Ab	aB	ab
ab	aAbB	aAbb	aabB	aabb

   Using the product rule: Prob(A_ B_) = (¼)(1) = 1/16
                           Prob(A_ bb) = (¼)(1) = 1/16
                           Prob(aa B_) = (¼)(1) = 1/16
                           Prob(aa bb) = (¼)(1) = 1/16

2. The segregation of different alleles for any gene occurs due to the pairing of homologous chromosomes, and the subsequent separation of these homologues during anaphase I. The independent assortment of traits, more accurately the independent segregation of different allele pairs, is due to the independent alignment of chromosomes during metaphase I of meiosis.

3. There seems to be the loss of a genotype since there are only 3 possible outcomes (2 yellow and 1 black). A yellow gene that has a dominant effect on coat color, but also causes lethality when homozygous, could explain the observations. Therefore, a yellow mouse is heterozygous, and crossing two yellow mice yields 1 homozygous yellow (dead):2 heterozygous (appears yellow):1 black. You could test this by crossing the yellow to homozygous black. You should get 1 yellow:1 black, and all black offspring should be true-breeding, and all yellow should behave as above.

4. Two genes are involved, one of which is epistatic to the other. At one gene, there are two alleles: black and brown; at the other gene, there are two alleles: albino and colored. The albino gene is epistatic to the brown gene so that when the animal is homozygous recessive for albino, it is albino regardless of whether it is black or brown at the other locus. This leads to the 4 albino in a Mendelian kind of crossing scheme.

CHAPTER 13

LEARNING OUTCOME QUESTIONS

13.1 Females would be all wild type; males would be all white-eyed.

13.2 Yes, should be viable and appear female.

13.3 The mt⁻ DNA could be degraded by a nuclease similar to how bacteria deal with invading viruses. Alternatively, the mt⁻-containing mitochondria could be excluded from the zygote.

13.4 No, not by genetic crosses.

13.5 Yes. First-division nondisjunction yields four aneuploid gametes, but second-division yields only two aneuploid gametes.

INQUIRY AND DATA ANALYSIS QUESTIONS

Page 244 Inquiry question: There would probably be very little if any recombination, so the expected assortment ratios would have been skewed from the expected 9:3:3:1.

Page 247 Data analysis: Instead of equal proportions of four types of gametes, you would get 45% of each of the two parental types and only 5% of each of the recombinant types. How this would affect a dihybrid cross would depend on the original parents, but it would clearly skew the results such that the 9:3:3:1 phenotypic ratio would be completely obscured. It would have been impossible to conclude that the two loci were behaving independently, as in fact, they are not.

Page 251 Data analysis: What has changed is the mother's age. The older the woman, the higher the risk she has of nondisjunction during meiosis. Thus, she also has a much greater risk of producing a child with Down syndrome.

Page 251 Inquiry question: Nondisjunction produces an XX egg that is fertilized by a Y sperm. A normal X egg is fertilized by an XY sperm produced by nondisjunction. Note that the nondisjunction event that produces the XX egg could be either MI or MII, but the event that produces the XY sperm would have to be MI.

Page 253 Inquiry question: Advanced maternal age, a previous child with birth defects, or a family history of birth defects.

UNDERSTAND
1. c 2. d 3. d 4. a 5. c 6. c 7. c

APPLY
1. c 2. b 3. c 4. b 5. d 6. b

SYNTHESIZE

1. Theoretically, 25% of the children from this cross will be color blind. All of the color blind children will be male, and 50% of the males will be color blind.

2. Parents of heterozygous plant were green wrinkled × yellow round
   Frequency of recombinants is 36+29/1300 = 0.05
   Map distance = 5 cM

3. The coloration that is associated with calico cats is the product of X inactivation. X inactivation only occurs in females as a response to dosage levels of the X-linked genes. The only way to get a male calico is to be heterozygous for the color gene and to be the equivalent of a Klinefelter male (XXY).

CHAPTER 14

LEARNING OUTCOME QUESTIONS

14.1 The 20 different amino acid building blocks offers chemical complexity. This appears to offer informational complexity as well.

14.2 The proper tautomeric forms are necessary for proper base-pairing, which is critical to DNA structure.

14.3 Prior to replication in light N (i.e., ^{14}N) isotope there would be only one band. After one round of replication, there would be two bands with denatured DNA: one heavy and one light.

14.4 The 5′ to 3′ activity is used to remove RNA primers. The 3′ to 5′ activity is used to remove mispaired bases (proofreading).

14.5 A shortening of chromosome ends would eventually affect DNA that encodes important functions.

14.6 No. The number of DNA-damaging agents, in addition to replication errors, would cause lethal damage (this has been tested in yeast).

INQUIRY AND DATA ANALYSIS QUESTIONS

Page 262 Data analysis: The Watson–Crick helical structure fits the measurements made from Franklin's X-ray diffraction pictures. The base-pairing explains why Chargaff observed the regularities in quantities of G = C and A = T, and the base-pairing depends on the proper tautomeric forms of the bases. Taken together, the model rationalizes all of these data.

Page 264 Data analysis: If the Meselson–Stahl experiment was allowed to run for another round, the pattern of bands would be the same as the second round (light molecules and hybrid molecules), but the light, ^{14}N, band would be darker. This is because the products from the ^{14}N alone would remain ^{14}N, and the products from the hybrid molecules ($^{15}N/^{14}N$) would produce one hybrid molecule and one all-^{14}N molecule.

Page 265 Inquiry question: The covalent bonds create a strong backbone for the molecule, making it difficult to disrupt. Individual hydrogen bonds are more easily broken, allowing enzymes to separate the two strands without disrupting the inherent structure of the molecule.

Page 269 Inquiry question: DNA ligase is important in connecting Okazaki fragments during DNA replication. Without it, the lagging strand would not be complete.

Page 273 Inquiry question: The linear structure of chromosomes creates the end problem discussed in the text. It is impossible to finish the ends of linear chromosomes using unidirectional polymerases that require RNA primers. The size of eukaryotic genomes also means that the time necessary to replicate the genome is much greater than in prokaryotes with smaller genomes. Thus the use of multiple origins of replication.

Page 275 Inquiry question: Cells have a variety of DNA repair pathways that allow them to restore damaged DNA to its normal constitution. If DNA repair pathways are compromised, the cell will have a higher mutation rate. This can lead to higher rates of cancer in a multicellular organism such as humans.

UNDERSTAND
1. d   2. a   3. c   4. a   5. c   6. b   7. b

APPLY
1. c   2. b   3. c   4. c   5. a   6. b   7. d   8. c

SYNTHESIZE

1. a. If both bacteria are heat-killed, then the transfer of DNA will have no effect since pathogenicity requires the production of proteins encoded by the DNA. Protein synthesis will not occur in a dead cell.
    b. The nonpathogenic cells will be transformed to pathogenic cells. Loss of proteins will not alter DNA.
    c. The nonpathogenic cells remain nonpathogenic. If the DNA is digested, it will not be transferred and no transformation will occur.

2. To ask questions about the two strands of a DNA molecule, it is necessary to separate them. Denaturing the DNA prior to separation by ultracentrifugation allows the density of single strands to be assessed. If DNA from the first round of replication is denatured, it should produce two bands. Using the same procedure with all heavy, and all light, DNA will reveal the position of heavy and light single strands.

3. a. *DNA gyrase* functions to relieve torsional strain on the DNA. If DNA gyrase were not functioning, the DNA molecule would undergo supercoiling, causing the DNA to wind up on itself, preventing the continued binding of the polymerases necessary for replication.
    b. *DNA polymerase III* is the primary polymerase involved in the addition of new nucleotides to the growing polymer and in the formation of the phosphodiester bonds that make up the sugar–phosphate backbone. If this enzyme were not functioning, then no new DNA strand would be synthesized and there would be no replication.
    c. *DNA ligase* is involved in the formation of phosphodiester bonds between Okazaki fragments. If this enzyme were not functioning, then the fragments would remain disconnected and would be more susceptible to digestion by nucleases.
    d. *DNA polymerase I* functions to remove and replace the RNA primers that are required for DNA polymerase III function. If DNA polymerase I were not available, then the RNA primers would remain and the replicated DNA would become a mix of DNA and RNA.

CHAPTER 15
LEARNING OUTCOME QUESTIONS

15.1 There is no molecular basis for recognition between amino acids and nucleotides. The tRNA is able to interact with nucleic acid by base-pairing, and an enzyme can covalently attach amino acids to it.

15.2 There would be no specificity to the genetic code. Each codon must specify a single amino acid, although amino acids can have more than one codon.

15.3 Transcription/translation coupling cannot exist in eukaryotes where the two processes are separated in both space and time.

15.4 No. This is a result of the evolutionary history of eukaryotes but is not necessitated by genome complexity.

15.5 Alternative splicing offers flexibility in coding information. One gene can encode multiple proteins.

15.6 This tRNA would be able to "read" STOP codons. This could allow nonsense mutations to be viable, but would cause problems making longer than normal proteins. Most bacterial genes actually have more than one STOP at the end of the gene.

15.7 Attaching amino acids to tRNAs, bringing charged tRNAs to the ribosome, and ribosome translocation all require energy.

15.9 No. It depends on where the breakpoints are that created the inversion, or duplication. For duplications it also depends on the genes that are duplicated.

INQUIRY AND DATA ANALYSIS QUESTIONS

Page 280 Data analysis: This double-mutant strain would only grow on media supplemented with all of the intermediates. This is because ArgE mutants are blocked at the first step. In general, these kinds of double mutants allow you to determine which gene occurs earlier in a pathway: the double mutant will look like an organism with a mutation at the earliest point in the pathway.

Page 281 Inquiry question: One would expect higher amounts of error in transcription over DNA replication. Proofreading is important in DNA replication because errors in DNA replication will be passed on to offspring as mutations. However, RNAs have very short life spans in the cytoplasm, and therefore mistakes are not permanent.

Page 284 Inquiry question: The universality of the genetic code argues that all life on Earth is descended from the same origin event.

Page 285 Inquiry question: The promoter acts as a binding site for RNA polymerase. The structure of the promoter provides information both about where to bind and the direction of transcription. If the two sites were identical, the polymerase would need some other cue for the direction of transcription.

Page 289 Data analysis: Splicing can produce multiple transcripts from the same gene. This is shown below with the exons shown in different colors.

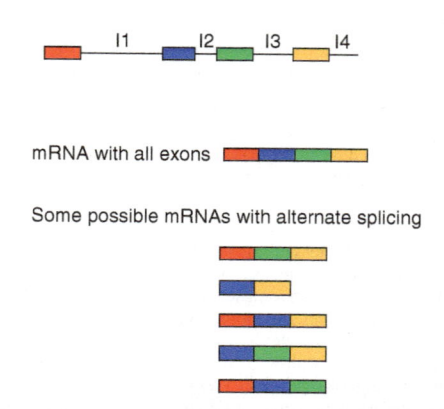

Page 297 Inquiry question: Wobble not only explains the number of tRNAs that are observed due to the increased flexibility in the 5′ position, it also accounts for the degeneracy that is observed in the genetic code. The degenerate base is the one in the wobble position.

UNDERSTAND
1. d   2. c   3. d   4. b   5. c   6. b   7. c

APPLY
1. b 2. c 3. b 4. d 5. a 6. b 7. b

SYNTHESIZE

1. The predicted sequence of the mRNA for this gene is
   5′–GCAAUGGGCUCGGCAUGCUAAUCC–3′
   The predicted amino acid sequence of the protein is
   Met-Gly-Ser-Ala-Cys-STOP

2. A frameshift essentially turns the sequence of bases into a "random" sequence. If you consider the genetic code, 3 of the 64 codons are STOP, so the probability of hitting a STOP in a random sequence is 3/64, or about 1 in every 20 codons.

3. a. mRNA = 5′–GCA AUG GGC UCG GCA UUG CUA AUC C–3′
   The amino acid sequence would then be:
   Met-Gly-Ser-Ala-Leu-Leu-Iso-.
   There is no STOP codon. This is an example of a frameshift mutation. The addition of a nucleotide alters the "reading frame," resulting in a change in the type and number of amino acids in this protein.
   b. mRNA = 5′–GCA AUG GGC UAG GCA UGC UAA UCC–3′
   The amino acid sequence would then be: Met-Gly-STOP.
   This is an example of a nonsense mutation. A single nucleotide change has resulted in the early termination of protein synthesis by altering the codon for Ser into a STOP codon.
   c. mRNA = 5′–GCA AUG GGC UCG GCA AGC UAA UCC –3′
   The amino acid sequence would then be: Met-Gly-Ser-Ala-Ser-STOP.
   This base substitution has affected the codon that would normally encode Cys (UGC) and resulted in the addition of Ser (AGC).

4. The split genes of eukaryotes offers the opportunity to control the splicing process, which does not exist in prokaryotes. This is also true for polyadenylation in eukaryotes. In prokaryotes, transcription/translation coupling offers the opportunity for the process of translation to have an effect on transcription.

CHAPTER 16
LEARNING OUTCOME QUESTIONS

16.1 The control of gene expression would be more like that in humans (fellow eukaryotes) than in *E. coli*.

16.2 The two helices both interact with DNA, so the spacing between the helices is important for both to be able to bind to DNA.

16.3 The operon would be on all of the time (constitutive expression).

16.4 The loss of a general transcription factor would likely be lethal as it would affect all transcription. The loss of a specific factor would affect only those genes controlled by the factor.

16.5 These genes are necessary for the ordinary functions of the cell. That is, the role of these genes is in ordinary housekeeping and not in any special functions.

16.6 RNA interference offers a way to specifically affect gene expression using drugs made of siRNAs.

16.7 Because there are many proteins in a cell doing a variety of functions, uncontrolled degradation of proteins would be devastating to the cell.

INQUIRY AND DATA ANALYSIS QUESTIONS

Page 308 Inquiry question: The presence of more than one gene in the operon allows for increased control over the elements of the pathway and therefore the product. A single regulatory system can regulate several adjacent genes.

Page 309 Data analysis: A mutation that prevents the repressor from binding to DNA would express the genes for lactose metabolism all the time (same phenotype as operator mutants that fail to bind repressor). A mutation that prevents the inducer from binding to the repressor would bind to DNA all the time, and thus never express the genes for lactose metabolism (uninducible).

Page 315 Inquiry question: Regulation occurs when various genes have the same regulatory sequences, which bind the same proteins.

Page 324 Inquiry question: Ubiquitin is added to proteins that need to be removed because they are nonfunctional or those that are degraded as part of a normal cellular cycle.

UNDERSTAND
1. c 2. d 3. a 4. c 5. b 6. c 7. b

APPLY
1. c 2. c 3. b 4. d 5. c 6. a 7. c

SYNTHESIZE

1. Mutations that affect binding sites for proteins on DNA will control the expression of genes covalently linked to them. Introducing a wild-type binding site on a plasmid will not affect this. We call this being *cis*-dominant. Mutations in proteins that bind to DNA would be recessive to a wild-type gene introduced on a plasmid.

2. Negative control of transcription occurs when the ability to initiate transcription is reduced. Positive control occurs when the ability to initiate transcription is enhanced. The *lac* operon is regulated by the presence or absence of lactose. The proteins encoded within the operon are specific to the catabolism (breakdown) of lactose. For this reason, operon expression is only required when there is lactose in the environment. Allolactose is formed when lactose is present in the cell. The allolactose binds to a repressor protein, altering its conformation and allowing RNA polymerase to bind. In addition to the role of lactose, there is also a role for the activator protein CAP in regulation of *lac*. When cAMP levels are high then CAP can bind to DNA and make it easier for RNA polymerase to bind to the promoter. The *lac* operon is an example of both positive and negative control.
   The *trp* operon encodes protein manufacture of tryptophan in a cell. This operon must be expressed when cellular levels of tryptophan are low. Conversely, when tryptophan is available in the cell, there is no need to transcribe the operon. The tryptophan repressor must bind tryptophan before it can take on the right shape to bind to the operator. This is an example of negative control.

3. Forms that control gene expression that are unique to eukaryotes include alternative splicing, control of chromatin structure, control of transport of mRNA from the nucleus to the cytoplasm, control of translation by small RNAs, and control of protein levels by ubiquitin-directed destruction. Of these, most are obviously part of the unique features of eukaryotic cells. The only mechanisms that could work in prokaryotes would be translational control by small RNAs and controlled destruction of proteins.

4. Mutation is a permanent change in the DNA. Regulation is a short-term change controlled by the cell. Like mutations, regulation can alter the number of proteins in a cell, change the size of a protein, or eliminate the protein altogether. The key difference is that gene regulation can be reversed in response to changes in the cell's environment. Mutations do not allow for this kind of rapid response.

CHAPTER 17
LEARNING OUTCOME QUESTIONS

17.1 It is important to be able to convert mRNA to cDNA so that the only the protein coding sequences of genes, and the specific genes being expressed in a particular tissue or at a particular developmental time point, can be cloned and studied. mRNA cannot be cloned or easily manipulated so must be converted to cDNA.

17.2 Both PCR and DNA replication result in new copies of DNA being made. Both processes involve separating two strands of DNA, however, in DNA replication in cells this is achieved by the action of an enzyme, and in PCR by heating the DNA to break hydrogen bonds. In both PCR and DNA replication a DNA polymerase reads information in a template strand and joins nucleotides together to make a complementary strand of DNA following the Watson-Crick base-pairing rules. In both DNA replication and PCR a primer is needed to initiate DNA replication; however, in cells that primer is made by an enzyme and is made of RNA, whereas in PCR it is a synthetically synthesized piece of DNA. In PCR a heat-stable polymerase is used, whereas in DNA replication in most cells, the DNA polymerase is sensitive to heat. In PCR billions of copies of DNA are made, but in DNA replication, only one copy of each DNA molecule is made per cell division cycle.

17.3 #1 Genome editing acts directly on the genomic gene sequence inside the cell using the TALEN or CRISPR/Cas9 system. In vitro mutagenesis is done outside the cell, often on cDNA sequences, which then need to be introduced back into the cells. Even if the in vitro mutated gene or cDNA can be integrated into the chromosomal DNA, it is not easy, or always possible, to replace the normal alleles of the gene.

#2 The main difference between the TALE and the CRISPR/Cas9 systems is in how the gene of interest is targeted. In the TALE system specific amino acids in the TALE-repeat domain of TALE proteins are altered to target a specific nucleotide. Combining different TALE-repeat domains allows for a specific sequence of DNA to be targeted. In the CRISPR/Cas9 system the targeting is done using a guide RNA that matches a short region of the gene of interest. The CRISPR/Cas9 system is more easily manipulated and used by scientists than is the TALEN system.

17.4 It would be wise to make a conditional knockout mouse if a gene was known to be essential during early development, but effects later in development want to be studied. It would also be appropriate to make a conditional knockout mouse if the targeted gene affected function in many tissues and the effects in specific tissues wanted to be investigated. A knockin mouse would be good for situations where it is desirable to introduce a particular genetic alteration in a gene so the effects of that can be studied in vivo.

17.5 Bacteria are easily genetically manipulated, so it is possible to envision how changes to metabolic pathways could increase the aerobic or anaerobic breakdown of organic waste. Of concern would be the ease by which such modified organisms might end up accidentally released into the natural environment.

17.6 A gene or cDNA of interest can be isolated by PCR, cloned into a suitable vector such as a plasmid, and introduced into an appropriate host organism. Expression of the gene or cDNA product could be induced from a promoter in the cloning vector, or if the whole-gene sequence was used, from its own promoter sequences. Mutation of the cDNA or gene sequence may be needed to ensure optimal expression of the protein.

17.7 Many plants, grasses included, do not use insects to move pollen from plant to plant for pollination. Instead, wind carries the pollen. Genetically altered plants producing pollen distributed by wind, and thus by water, could rapidly spread glyphosate-resistance genes to other grass populations.

INQUIRY AND DATA ANALYSIS QUESTIONS

Page 332 Data analysis: 3×10^9 base pairs divided by 500 bp per clone fragment equates to 600,000 different fragments to obtain complete coverage with all the clones.

Page 335 Inquiry question: No. cDNA is made from mRNA, and the mRNA used has been processed to remove intron sequences via splicing.

Page 342 Inquiry question: Algae can be genetically altered in some of the same ways as plant cells. Generally, it would be useful to introduce additional copies of genes involved in acetyl-CoA synthesis or other aspects of fatty acid biosynthesis. Alternatively, it is interesting to speculate on using knockdown technologies to remove gene products that negatively regulate fatty acid biosynthesis, thereby stimulating the process. It may also be possible to introduce or remove genes to increase the efficiency of light capture or carbon dioxide fixation to increase photosynthetic rates.

Page 345 Data analysis: 184. In a normal diploid human cell, there are 23 pairs of chromosomes—the autosomes 1 through 22 and a pair of sex chromosomes (two Xs or an X and a Y). Each chromosome has two ends and so each chromosome would have two dots—one on each end. Twenty-three pairs of chromosomes would be 92 dots. However, once a cell replicates its DNA and is in G_2, there are sister chromatids making up each chromosome so the number of ends doubles from 92 to 184.

UNDERSTAND
1. c 2. b 3. c 4. d 5. a 6. b

APPLY
1. c 2. c 3. b

SYNTHESIZE
1. If the production was to be conducted in a bacterium such as *E. coli*, the lack of organelles would mean that the posttranslational associations of the subunits into functional hemoglobin may be significantly impaired. It is possible that the mature hemoglobin molecules would not even form. Also, given that hemoglobin has four bound iron atoms, it is not guaranteed that the appropriate amounts of iron would be available if the proteins were overexpressed. An additional issue could be the formation of disulfide bonds in the folding of the subunits. In a non-animal eukaryotic cell, such as yeast, there is no guarantee that folding would be correct.

2. It may be possible to construct a knockin mouse where different mutations have been introduced into the APP gene. The effects of specific mutations could be seen in mice phenotypes. Alternatively, if normal and mutant APP can be expressed in *E. coli*, yeasts, or cultured human cells, it may be possible to create mutations in cDNA that are then introduced into suitable cells so that aberrant or normal folding could be measured.

CHAPTER 18
LEARNING OUTCOME QUESTIONS

18.1 A centimorgan is a distance that corresponds to recombination frequency. Recombination frequencies can vary based on DNA sequence and local chromatin structure. The recombination frequency in one piece of DNA 100 kbp long can be different than the recombination frequency in another piece of DNA of the same length. Thus, two markers the same physical distance apart could be closer or more distant, based on the frequency with which they can be separated by recombination.

18.2 Whole-genome assembly via shotgun sequencing is quick and relatively cheap but increases in difficulty to the point of being impossible as genome size and complexity increases. In highly redundant genomes with large amounts of repetitive sequence, it can be impossible to find enough overlap between clones to perform a reliable assembly. Clone-contig assembly is more labor intensive but works well on larger, more complex genomes with large amounts of redundant sequence.

18.3 Genomes such as that of wheat are often complex, large, and/or highly repetitive due to hybridization events during evolution. The repetitive nature of these genomes makes genome assembly difficult; it is hard to find sequences of DNA that have enough differences to allow contigs to be created. The redundant nature of the genome also makes annotation difficult because it is hard to know to which version of the genome—A, B, or D in the case of the wheat genome—a particular sequence belongs.

18.4 One possibility is that transposable elements can move within the genome and by inserting into genes, or gene regulatory sequences can introduce genetic variation upon which natural selection can act. It is also possible that the movement of transposons affects chromatin structure in ways that affect gene expression. Again, such variation would be subject to natural selection.

18.5 Even if the proteome could not be determined, the transcriptome would allow the determination of relative levels of each transcript in the cell. It is possible to determine how those levels change in response to particular stimuli. The proteome can be inferred from the transcriptome, although certain caveats to this exist. For example, a given transcript may be stable, but the protein product may be inherently unstable and present at very low levels in the cell. Also, alternative splicing can produce a variety of proteins.

18.6 Life is defined by the presence of certain characteristics and the absence of others. The synthetic cell described is clearly living, and if it has many or all of the characteristics of the cell from where its genome came, it is probably best not characterized as a new form of life, or even as a new species.

INQUIRY AND DATA ANALYSIS QUESTIONS

Page 355 Inquiry question: There is one copy of *bcr* (seen with the green probe) and one copy of *abl* (seen with the red probe). The other *bcr* and *abl* genes are fused together and they are seen as yellow because the green and red probes overlap to produce yellow.

Page 360 Data analysis: 1500 bp. If 1% of a 3 billion base-pair genome codes for genes, then approximately 30 million base-pairs code for genes. If there are 20,000 genes, then each gene is about 1500 base-pairs long. Of course, this is a hypothetical mathematical average, and actual average gene lengths are much higher than this due to skewed distribution of gene lengths.

Page 362 Inquiry question: You could use computer algorithms to look for open-reading frames that have characteristics of introns and exons. You could also look for consensus promoter sequences that signal the binding site for RNA polymerase. Your estimate would be inaccurate because not all genes produce proteins while other genes produce mRNAs that can make several proteins due to alternative splicing.

Page 364 Inquiry question: Your first challenge would be to clarify your definition of "functional." If ESTs corresponding to the sequence could be identified, then it is likely that a functional product (an RNA or protein) is produced from the sequence. If you could clone the sequence, you could run biochemical tests to see if certain proteins, such as those involved in chromatin structure regulation or RNA polymerase, bound to the DNA. You could also look to see what kinds of epigenetic modifications (methylation, acetylation, phosphorylation) were associated with the chromatin at that sequence.

Page 369 Inquiry question: It is sometimes possible to reverse-engineer the gene sequence from a protein's amino acid sequence. Based on the sequence of amino acids, it might be possible to predict the presence of protein domains with particular functions. For example, many protein and lipid kinase domains are highly conserved, and this information would be important in assigning function to a gene. There may also be consensus amino acid sequences for posttranslation modifications that affect protein structure, localization, or function.

Page 370 Inquiry question: The genome contains the sequences that will be transcribed to produce the transcriptome, but also contains sequences that will not be transcribed. Not all transcripts are, or produce, functional products, however.

The proteome is the collection of proteins, determined by the proteome, at a given time, under given conditions, in a cell or organism.

Page 372 Inquiry question: If the gene is synthetic, then it is not natural and thus not protected against patent. This would be strengthened if the sequence of the gene was significantly different from any existing gene; for example, perhaps the sequence is modified so that it is expressed at higher than normal levels.

UNDERSTAND

1. b 2. a 3. c 4. d 5. b 6. b 7. c 8. d

APPLY

1. a 2. a 3. b 4. a 5. c 6. d 7. a

SYNTHESIZE

1. The STSs represent unique sequences in the genome. They can be used to align the clones into one contiguous sequence of the genome based on the presence or absence of an STS in a clone. The contig, with aligned clones, would look like this:

```
            STS 1  STS 2
Clone E   ————+——————+——
                 STS 2        STS 3
Clone B        ——+——————————————+——
                          STS 3      STS 4 STS 5
Clone A              ——————+——————————+———+——
                          STS 3    STS 4
Clone D              ——————+————————+——
                                        STS 5       STS 6
Clone C                            ————+———————————+——

         STS 1  STS 2    STS 3    STS 4 STS 5    STS 6
Contig ———+——————+————————+————————+———+—————————+——
```

2. The anthrax genome has been sequenced. Investigators would look for differences in the genome between existing natural strains, those collected from a suspected outbreak, and those known to be used in research laboratories. The genome of an infectious agent can be modified, or "weaponized," to make it more deadly. Also, single-nucleotide polymorphisms could be used to identify the source of the anthrax.

Chapter 19

LEARNING OUTCOME QUESTIONS

19.2 The early cell divisions are very rapid and do not involve an increase in size between divisions. Interphase is greatly reduced allowing very fast cell divisions.

19.3 This requires experimentation to isolate cell from contact, transplantation, or to follow a particular cell's lineage.

19.4 The nucleus must be reprogrammed. What this means exactly on the molecular level is not clear, but probably involves changes in chromatin structure and methylation patterns.

19.5 Homeotic genes seem to have arisen very early in the evolutionary history of bilaterians. These have been duplicated, and they have diversified with increasing morphological complexity.

19.6 Cell death can be a patterning mechanism. Your fingers were sculpted from a paddle-like structure by cell death.

INQUIRY AND DATA ANALYSIS QUESTIONS

Page 381 Data analysis: If you use a reagent to block FGF signaling, you would get only muscle precursor cells and nerve cord precursor cells. If you had a reagent that causes constitutive signaling, then you would get only mesenchyme precursor cells and notochord precursor cells.

Page 384 Data analysis: The difference in pigmentation pattern acts as a marker for the nuclear and cytoplasmic donors. Since the difference is genetic, the phenotype of the resulting sheep tells you which nucleus gave rise to this individual. Thus Dolly resembled the nuclear donor and not the cytoplasmic donor or the surrogate mother.

UNDERSTAND

1. b 2. d 3. c 4. a 5. b 6. c 7. b

APPLY

1. d 2. a 3. b 4. a 5. c 6. c 7. c

SYNTHESIZE

1. The horizontal lines of the fate map represent cell divisions. Starting with the egg, four cell divisions are required to establish a population of cells that will become nervous tissue. It takes another eight to nine divisions to produce the final number of cells that will make up the nervous system of the worm. It takes seven to eight rounds of cell division to generate the population of cells that will become the gonads. Once established, another seven to eight cell divisions are required to produce the actual gonad cells.

2. Not every cell in a developing embryo will survive. The process of apoptosis is responsible for eliminating cells from the embryo. In *C. elegans*, the process of apoptosis is regulated by three genes: *ced-3*, *ced-4*, and *ced-9*. Both *ced-3* and *ced-4* encode proteases, which can destroy the cell from the inside-out. The *ced-9* gene functions to repress the activity of the protease-encoding genes, thereby preventing apoptosis.

3. a. N-cadherin plays a specific role in differentiating cells of the nervous system from ectodermal cells. Ectodermal cells express E-cadherin, but neural cells express N-cadherin. The difference in cell-surface cadherins means that the neural cells lose their contact with the surrounding ectodermal cells and establish new contacts with other neural cells. In the absence of N-cadherin, the nervous system would not form. If you assume that E-cadherin expression is also lost (as would occur normally in development), then these cells would lose all cell–cell contacts and would probably undergo apoptosis.
   b. Integrins mediate the connection between a cell and its surrounding environment, the extracellular matrix (ECM). The loss of integrins would result in the loss of cell adhesion to the ECM. These cells would not be able to move, and, therefore, gastrulation and other developmental processes would be disrupted.
   c. Integrins function by linking the cell's cytoskeleton to the ECM. This connection is critical for cell movement. The deletion of the cytoplasmic domain of the integrin would not affect the ability of integrin to attach to the ECM, but it would prevent the cytoskeleton from getting a "grip." This deletion would likely result in a disruption of development similar to the complete loss of integrin.

4. Adult cells from the patient would be cultured with factors that reprogram the nucleus into pluripotent cells. These cells would then be grown in culture with factors necessary to induce differentiation into a specific cell type that could be transplanted into the patient. This would be easiest for tissue like a liver that regenerates, but could in theory be used for a variety of cell types.

Chapter 20

LEARNING OUTCOME QUESTIONS

20.1 Natural selection occurs when some individuals are better suited to their environment than others. These individuals live longer and reproduce more, leaving more offspring with the traits that enabled their parents to thrive. In essence, genetic variation within a population provides the raw material on which natural selection can act thereby leading to evolution.

20.2 #1 To determine if a population is in Hardy–Weinberg equilibrium, it is first necessary to determine the actual allele frequencies, which can be calculated based on the genotype frequencies. After assigning variables p and q to the allele frequencies, we then use the Hardy–Weinberg equation, $p^2 + 2pq + q^2 = 1$ to determine the expected genotype frequencies. If the actual and expected genotype frequencies are the same (or, at least not significantly different), it is safe to say that the population is in Hardy–Weinberg equilibrium.
#2 You would conclude that one or more of the five evolutionary agents were acting to cause the lack of equilibrium. The next step would be to design studies to test hypotheses about which assumption is not being met.

20.3 There are five mechanisms of evolution: natural selection, mutation, gene flow (migration), genetic drift, and nonrandom mating. Any of these mechanisms can alter allele frequencies within a population, although usually a change in allele frequency results from more than one mechanism working in concert (e.g., mutation can introduce a beneficial new allele into the population, and natural selection will select for that allele such that its frequency increases over the course of two or more generations). Natural selection, the first mechanism and probably the most influential in bringing about evolutionary change, is also the only one to produce adaptive change, that is, change that results in the population being better adapted to its environment. Mutation is the only way in which new alleles can be introduced—it is the ultimate source of all variation. Because it is a relatively rare event, mutation by itself is not a strong agent of allele frequency change; however, in concert with other mechanisms, especially natural selection, it can drastically change the allele frequencies in a population. Gene flow can introduce new alleles into a population from another population of the same species, thus changing the allele frequency within both the recipient and donor populations. Genetic drift is the random, chance factor of evolution—although the results of genetic drift can be negligible in a large population, small populations can undergo drastic changes in allele frequency due to this agent. Finally, nonrandom mating results in populations that vary from Hardy–Weinberg equilibrium not by changing allele frequencies but by changing genotype frequencies—nonrandom mating reduces the proportion of heterozygotes in a population.

20.4 Reproductive success relative to other individuals within an organism's population is referred to as that organism's fitness. Its fitness is determined by its longevity, mating frequency, and the number of offspring it produces for each mating. None of these factors is always the most important in determining reproductive success—instead it is the cumulative effects of all three factors that determine an individual's reproductive success. For example, an individual that has a very long life span but mates only infrequently might have lower fitness than a conspecific that lives only half as long but mates more frequently and with greater success. As seen with the water strider example in this section, traits that are favored for one component of fitness, say, for example, longevity, may be disadvantageous for other components of fitness, say, lifetime fecundity.

20.5 #1 In a population wherein heterozygotes had the lowest fitness, natural selection should favor both homozygous forms. This would result in disruptive selection and a bimodal distribution of traits within the population. Over enough time, it could lead to a speciation event.

#2 Negative frequency-dependent selection occurs when rare alleles have a fitness advantage over common alleles. As a result, selection is maintained because once an allele becomes rare, selection operates to increase its frequency. Oscillating selection refers to the situation in which the environment changes, first favoring one allele, then another. Over the long term, negative frequency-dependent selection is more likely to preserve variation, because rare alleles are always favored. By contrast, if the environment does not change (oscillate) for an unusually long period of time, the disadvantaged allele may disappear from the population.

20.6 Directional selection occurs when one phenotype has an adaptive advantage over other phenotypes in the population, regardless of its relative frequency within the population. Frequency-dependent selection, on the other hand, results when either a common (positive frequency-dependent selection) or rare (negative frequency-dependent selection) has a selective advantage simply by virtue of its commonality or rarity. In other words, if a mutation introduces a novel allele into a population, directional selection may result in evolution because the allele is advantageous, not because it is rare.

20.7 Background color matching is a form of camouflage used by many species to avoid predation; however, this example of natural selection runs counter to sexual selection—males want to be inconspicuous to predators but attractive to potential mates. For example, to test the effects of predation on background color matching in a species of butterfly, one might raise captive populations of butterflies with a normal variation in coloration. After a few generations, add natural predators to half of the enclosures. After several generations, one would expect the butterflies in the predatory environment to have a high degree of background color matching in order to avoid predation, while the nonpredatory environment would have promoted brightly colored individuals where color would correlate with mating success.

20.8 The dynamics among the different evolutionary mechanisms are very intricate, and it is often difficult, if not impossible, to discern in which direction each process is operating within a population—it is much easier to simply see the final cumulative effects of the various agents of evolutionary change. However, in some cases more than one evolutionary process will operate in the same direction, with the resulting population changing, or evolving, more rapidly than it would have under only one evolutionary mechanism. For example, mutation may introduce a beneficial allele into a population; gene flow could then spread the new allele to other populations. Natural selection will favor this allele within each population, resulting in relatively rapid evolutionary adaptation of a novel phenotype.

20.9 Pleiotropic effects occur with many genes; in other words, a single gene has multiple effects on the phenotype of the individual. Whereas natural selection might favor a particular aspect of the pleiotropic gene, it might select against another aspect of the same gene; thus, pleiotropy often limits the degree to which a phenotype can be altered by natural selection. Epistasis occurs when the expression of one gene is controlled or altered by the existence or expression of another gene. Thus, the outcome of natural selection will depend not just on the genotype of one gene, but the other genotype as well.

INQUIRY AND DATA ANALYSIS QUESTIONS

Page 402 Data analysis: In the example of figure 20.3, the frequency of the recessive white genotype is 0.16. The remaining 84 cats (out of 100) in the population are homozygous or heterozygous black. If the 16 white cats died, they will not contribute recessive white genes to the next generation. Only heterozygous black cats will produce white kittens in a 3:1 ratio of black to white. Homozygous × homozygous black and homozygous × heterozygous black cats will have all black kittens. Since there are 36 homozygous black cats and 48 heterozygous black cats, with a new total of 84 cats, the new frequency of homozygous black cats is 36/84, or 43%, with the heterozygous black cats now comprising 57% of the population. If $p^2 = 0.43$, then $p = 0.65$ (approximately), then $1 - p = q$, and $q = 0.35$. The frequency of white kittens in the next generation, q^2, is 0.12, or 12%.

Page 405 Inquiry question: Rare alleles are more likely to be lost in a population bottleneck specifically because they are rare; the chance that a small population will not include an individual that has that allele is relatively high. For this reason, rare alleles are at risk of being lost. However, if by chance an individual with a rare allele is in the small population, the allele's frequency becomes much greater because of the small number of individuals in the population.

Page 408 Inquiry question: Differential predation might favor brown toads over green toads, green toads might be more susceptible to disease, or green toads might be less able to tolerate variations in climate, among other possibilities.

Page 409 Inquiry question: Since the intermediate-sized water strider has the highest level of fitness, we would expect that the intermediate size would become more prevalent in the population. If the number of eggs laid per day was not affected by body size, the small water striders would be favored because of their tendency to live longer than their larger counterparts.

Page 409 Data analysis: 12-mm striders lay about 2 eggs per day and live about 32 days, for an expected total egg number of 64; the corresponding numbers for 15-mm water striders are 6 and 18, respectively, for a total of 108.

Page 410 Data analysis: In frequency-dependent selection, the fitness of a phenotype is related to its frequency. Thus, for each color type, fitness should be related to frequency, and the color types should not exhibit consistent differences in fitness. By contrast, directional selection is not frequency-dependent; color types should differ in fitness, and those differences should not be affected by frequency.

Page 413 (figure 20.15) Inquiry question: The proportion of flies moving toward light (positive phototropism) would again begin to increase in successive generations.

Page 413 (figure 20.16) Inquiry question: The distribution of birth weights in the human population would expand somewhat to include more babies of higher and lower birth weights.

Page 413 Data analysis: Rates of infant mortality are highest at very low and very high birth weights, and lowest at intermediate birth weights. However, many more babies are born with intermediate birth weights than with extreme births. For example, there are 40 times more individuals born with a birth weight of 6.5 pounds than 2 pounds. As a result, even though the mortality rate is 8 times higher for 2-pound babies than 6.5-pound babies, the absolute number of deaths would be higher for the intermediate-weight size. Overall, the relationship between number of deaths and birth weight would show a peak at intermediate birth weights.

Page 415 Inquiry question: Guppy predators evidently locate their prey using visual cues. The more colorful the guppy, the more likely it is to be seen and thus the more likely it will become prey.

Page 416 Data analysis: On the right-hand side, the index declines with distance from the mine, which suggests that whatever process is operating, its effect is negatively related to distance from the mine. For a given distance from the mine in both directions, index values are higher on the right side, suggesting that the mechanism is stronger on that side. Dispersal of tolerance alleles driven by wind patterns can explain both results: why the allele frequency tails off with distance from the mine, and why the levels are higher on the downwind side of the mine.

Page 417 Inquiry question: Thoroughbred horse breeders have been using selective breeding for certain traits over many decades, effectively removing variation from the population of thoroughbred horses. Unless mutation produces a faster horse, it remains unlikely that winning speeds will improve.

UNDERSTAND

1. a   2. b   3. d   4. a   5. d   6. a   7. d

APPLY

1. d   2. d   3. d

SYNTHESIZE

1. The results depend on coloration of guppies increasing their conspicuousness to predators such that an individual's probability of survival is lower than if it was a drab morph. In the laboratory it may be possible to conduct trials in simulated environments; we would predict, based on the hypothesis of predation, that the predator would capture more of the colorful morph than the drab morph when given access to both. Design of the simulated environment would obviously be critical, but results from such an experiment, if successful, would be a powerful addition to the work already accomplished.

2. On the large lava flows, where the background is almost entirely black, those individuals with black coloration within a population will have a selective advantage because they will be more cryptic to predators. On the other hand, on small flows, which are disrupted by light sand and green plants, dark individuals would be at an adaptive disadvantage for the same reason. You can read more about this in chapter 21; the black peppered moths had an advantage on the trees lacking lichen, but a disadvantage on lichen-covered trees.

3. Ultimately, genetic variation is produced by the process of mutation. However, compared with the speed at which natural selection can reduce variation in traits that are closely related to fitness, mutation alone cannot account for the persistence of genetic variation in traits that are under strong selection. Other processes can account for the observation that genetic variation can persist under strong selection. They include gene flow. Populations are often distributed along environmental gradients of some type. To the extent that different environments favor slightly different variants of phenotypes that have a genetic basis, gene flow among areas in the habitat gradient can introduce new genetic variation or help maintain existing variation. Similarly, just as populations frequently encounter different selective environments across their range (think of the guppies living above and below the waterfalls in Trinidad), a single population also encounters variation in selective environments across time (oscillating selection). Traits favored this year may not be the same as those favored next year, leading to a switching of natural selection and the maintenance of genetic variation.

CHAPTER 21

LEARNING OUTCOME QUESTIONS

21.1 No. If eating hard seeds caused individuals to develop bigger beaks, then the phenotype is a result of the environment, not the genotype. Natural selection can only act upon those traits with a genetic component. Just as a body builder develops large muscles in his or her lifetime but does not have well-muscled offspring, birds that develop large beaks in their lifetime will not necessarily have offspring with larger beaks.

21.2 An experimental design to test this hypothesis could be as simple as producing enclosures for the moths and placing equal numbers of both morphs into each enclosure and then presenting predatory birds to each enclosure. One enclosure could be used as a control. One enclosure would have a dark background, and the other would have a light background. After several generations, measuring the phenotype frequency of the moths should reveal very clear trends—the enclosure with the dark background should consist of mostly dark moths, the enclosure with the light background mostly light moths, and the neutral enclosure should have an approximately equal ratio of light to dark moths.

21.3 If the trait that is being artificially selected for is due to the environment rather than underlying genotype, then the individuals selected that have that trait will not necessarily pass it on to their offspring.

21.4 The major selective agent in most cases of natural selection is the environment; thus, climatic changes, major continental shifts, and other major geological changes would result in dramatic changes in selective pressure; during these times the rate and direction of evolutionary change would likely be affected in many, if not most, species. On the other hand, during periods of relative environmental stability, the selective pressure does not change and we would not expect to see many major evolutionary events.

21.5 The only other explanation that could be used to explain homologous characteristics and vestigial structures could be mutation. Especially in the case of vestigial structures, if one resulted from a mutation that had pleiotropic effects, and the other effects of the genetic anomaly were selected for, then the vestigial structure would also be selected for, much like a rider on a Congressional bill.

21.6 Convergence occurs when distantly related species experience similar environmental pressures and respond, through natural selection, in similar ways. For example, penguins (birds), sharks (fish), sea lions (mammals), and even the extinct ichthyosaur (reptile) all exhibit the fusiform shape. Each of these animals has similar environmental pressures in that they are all aquatic predators and need to be able to move swiftly and agilely through the water. Clearly their most recent common ancestor does not have the fusiform body shape; thus the similarities are due to convergence (environment) rather than homology (ancestry). However, similar environmental pressures will not always result in convergent evolution. Most importantly, in order for a trait to appear for the first time in a lineage, there must have been a mutation; however, mutations are rare events, and even rarer is a beneficial mutation. There may also be other species that already occupy a particular niche; in these cases it would be unlikely that natural selection would favor traits that would increase the competition between two species.

21.7 It is really neither a hypothesis nor a theory. Theories are the building blocks of scientific knowledge; they have withstood the most rigorous testing and review. Hypotheses, on the other hand, are tentative answers to a question. Unfortunately, a good hypothesis must be testable and falsifiable, and stating that humans came from Mars is not realistically testable or falsifiable; thus, it is, in the realm of biological science, a nonsense statement.

INQUIRY AND DATA ANALYSIS QUESTIONS

Page 423 Inquiry question: Assuming that the size of seeds that parents fed to offspring was unrelated to parent adult size, then no relationship would exist between parent and offspring beak depth—some offspring, by chance, would be fed more small seeds and would develop shallower beaks, whereas others would be fed larger seeds and develop deeper beaks. This result would indicate that beak depth was not a genetically determined trait. On the other hand, if parents with deeper beaks fed their offspring larger seeds, and if seed size determined beak depth, then the same relationship pictured in figure 21.2*b* would result. In this case, genes and environment would be correlated, and researchers wouldn't know which factor was responsible for beak depth. To distinguish between the two, researchers bring nestling birds into the laboratory and feed them different-sized seeds to determine if diet determined beak depth (this would be difficult in the Galápagos, though, where Darwin's finches are strictly protected).

Page 423 Data analysis: The parents' mean beak depth would be 9 mm, so we would expect the offspring to have a beak depth of approximately 9 mm. The figure does not allow us to tell whether the sex of the parent matters; however, the fact that there is relatively little scatter around the line (i.e., most of the points lie close to the regression line) suggests that the sex of the parents (i.e., which parent is larger) doesn't have a substantial effect.

Page 425 Inquiry question: Such a parallel trend would suggest that similar processes are operating in both localities. Thus, we would conduct a study to identify similarities. In this case, both areas have experienced coincident reductions in air pollution, which most likely is the cause of the parallel evolutionary trends.

Page 426 Inquiry question: Assuming that small and large individuals would breed with each other, then middle-sized offspring would still be born (the result of matings between small and large flies). Nonetheless, there would also be many small and large individuals (the result of small × small and large × large matings). Thus, the frequency distribution of body sizes would be much broader than the distributions in the figures.

Page 430 Inquiry question: This evolutionary decrease could occur for many reasons. For example, maybe *Nannippus* adapted to forested habitats and thus selection favored smaller size, as it had in the ancestral horses before horses moved into open, grassland habitats. Another possibility is that many species of horses were present at that time, and different-sized horses ate different types of food. By evolving small size, *Nannippus* may have been able to eat a type of food not eaten by the others.

UNDERSTAND

1. d   2. b   3. b   4. a   5. b   6. b   7. b

APPLY

1. a   2. d   3. d

SYNTHESIZE

1. Briefly, they are:
   a. There must be variation among individuals within a population.
   b. Variation among individuals must be related to differences among individuals in their success in producing offspring over their lifetime.
   c. Variation related to lifetime reproductive success must have a genetic (heritable) basis.

2. Figure 21.2*a* shows in an indirect way that beak depth varies from year to year. Presumably this is a function of variation among individuals in beak size. However, the most important point of figure 21.2*a* is that it shows the result of selection. That is, if the three conditions hold, we might expect to

see average beak depth change accordingly as precipitation varies from year to year. Figure 21.2b is more directly relevant to the conditions noted for natural selection to occur. The figure shows that beak size varies among individuals, *and* that it tends to be inherited.

3. The relationship would be given by a cloud of points with no obvious linear trend in any direction different from a zero slope. In other words, it would be a horizontal line through an approximately circular cloud of points. Such data would suggest that whether a parent(s) has a large or small beak has no bearing on the beak size of its offspring.

4. The direction of evolution would reverse, and the two populations would evolve to be more similar in bristle number, until the differences disappeared entirely. However, the rate at which evolution occurred probably would be slower than in the first part of the experiment. Because selection had initially favored low and high bristle numbers in the two populations at the expense of intermediate bristle numbers, some of the genetic variation for intermediate numbers probably would have been lost from the population. As a result, the rate at which intermediate numbers re-evolved would probably be slower because there was less appropriate genetic variation available.

5. The evolution of horses was not a linear event; instead it occurred over 55 million years and included descendants of 34 different genera. By examining the fossil record, one can see that horse evolution did not occur gradually and steadily; instead several major evolutionary events occurred in response to drastic changes in environmental pressures. The fossil record of horse evolution is remarkably detailed, and shows that although there have been trends toward certain characteristics, change has not been fluid and constant over time, nor has it been entirely consistent across all of the horse lineages. For example, some lineages experienced rapid increases in body size over relatively short periods of geological time, but other lineages actually saw decreases in body size.

CHAPTER 22

LEARNING OUTCOME QUESTIONS

22.1 **#1** The biological species concept states that different species are capable of mating and producing viable, fertile offspring. If sympatric species are unable to do so, they will remain reproductively isolated and thus distinct species. Along the same lines, gene flow between populations of the same species allow for homogenization of the two populations such that they remain the same species.

#2 The ecological species concept states that sympatric species are adapted to use different parts of the environment, and thus hybrids between them would not be well adapted to either habitat and thus would not survive. Even if they did survive and reproduce, genes from one species that made their way into the other species' gene pool would likely be eliminated by natural selection. The concept would explain the connection of geographic populations of a species. As a result, these populations occupy similar parts of the environment and thus experience similar selective pressures.

22.2 In order for reinforcement to occur and complete the process of speciation, two populations must have some reproductive barriers in place prior to sympatry. In the absence of this initial reproductive isolation we would expect rapid exchange of genes and thus homogenization resulting from gene flow. On the other hand, if two populations are already somewhat reproductively isolated (due to hybrid infertility or a prezygotic barrier such as behavioral isolation), then we would expect natural selection to continue improving the fitness of the nonhybrid offspring, eventually resulting in speciation.

22.3 Reproductive isolation that occurs due to different environments is a factor of natural selection; the environmental pressure favors individuals best suited for that environment. As isolated populations continue to develop, they accumulate differences due to natural selection that eventually result in two populations so different that they are reproductively isolated. Reinforcement, on the other hand, is a process that specifically relates to reproductive isolation. It occurs when natural selection favors nonhybrids because of hybrid infertility or are simply less fit than their parents. In this way, populations that may have been only partly reproductively isolated become completely reproductively isolated.

22.4 Polyploidy occurs instantaneously; in a single generation, the offspring of two different parental species may be reproductively isolated; however, if it is capable of self-fertilization, then it is, according to the biological species concept, a new species. Disruptive selection, on the other hand, requires many generations as reproductive barriers between the two populations must evolve and be reinforced before the two would be considered separate species.

22.5 In the archipelago model, adaptive radiation occurs as each individual island population adapts to its different environmental pressures. On the other hand, in sympatric speciation resulting from disruptive selection, the traits selected are not necessarily best suited for a novel environment but are best able to reduce competition with other individuals. It is in the latter scenario wherein adaptive radiation due to a key innovation is most likely to occur.

22.6 It depends on what species concept you are using to define a given species. Certainly evolutionary change can be punctuated, but in times of changing environmental pressures we would expect adaptation to occur. The adaptations, however, do not necessarily have to lead to the splitting of a species—instead one species could simply adapt in accordance with the environmental changes to which it is subjected. This would be an example of nonbranching, as opposed to branching, evolution; but again, whether the end-result organism is a different species from its ancestral organism that preceded the punctuated event is subject to interpretation.

INQUIRY AND DATA ANALYSIS QUESTIONS

Page 451 Inquiry question: Speciation can occur under allopatric conditions because isolated populations are more likely to diverge over time due to drift or selection. Adaptive radiation tends to occur in places inhabited by only a few other species or where many resources in a habitat are unused. Different environmental conditions typical of adaptive radiation tend to favor certain traits within a population. Allopatric conditions would then generally favor adaptive radiation.

In character displacement, natural selection in each species favors individuals able to use resources not used by the other species. Two species might have evolved from two populations of the same species located in the same environment (sympatric species). Individuals at the extremes of each population are able to use resources not used by the other group. Competition for a resource would be reduced for these individuals, possibly favoring their survival and leading to selection for the tendency to use the new resource. Character displacement tends to compliment sympatric speciation.

Page 455 Inquiry question: If one area experiences an unfavorable change in climate, a mobile species can move to another area where the climate was like it was before the change. With little environmental change to drive natural selection within that species, stasis would be favored.

UNDERSTAND

1. a 2. c 3. a 4. b 5. a 6. d 7. d 8. b 9. b 10. a

APPLY

1. a 2. d 3. b 4. b

SYNTHESIZE

1. If hybrids between two species have reduced viability or fertility, then natural selection will favor any trait that prevents hybrid matings. This way individuals don't waste time, energy, or resources on such matings and will have greater fitness if they instead spend the time, energy, and resources on mating with members of their own species. For this reason, natural selection will favor any trait that decreases the probability of hybridization. By contrast, once hybridization has occurred, the time, energy, and resources have already been expended. Thus, there is no reason that less fit hybrids would be favored over more fit ones. The only exception is for species that invest considerable time and energy in incubating eggs and rearing the young; for those species, selection may favor reduced viability of hybrids because parents of such individuals will not waste further time and energy on them.

2. The biological species concept, despite its limitations, reveals the continuum of biological processes and the complexity and dynamics of organic evolution. At the very least, the biological species concept provides a mechanism for biologists to communicate about taxa and know that they are talking about the same thing! Perhaps even more significantly, discussion and debate about the meaning of "species" fuels a deeper understanding about biology and evolution in general. It is unlikely that we will ever have a single unifying concept of species, given the vast diversity of life, both extinct and extant.

3. The principle is the same as in character displacement. In sympatry, individuals of the two species that look alike may mate with each other. If the species are not completely interfertile, then individuals hybridizing will be at a selective disadvantage. If a trait appears in one species that allows that species to more easily recognize members of its own species and thus avoid hybridization, then individuals bearing that trait will have higher fitness and that trait will spread through the population.

4. The two species would be expected to have more similar morphology when they are found alone (allopatry) than when they are found together (sympatry), assuming that food resources were the same from one island to the next. This would be the result of character displacement expected under

a hypothesis of competition for food when the two species occur in sympatry. A species pair that is more distantly related might not be expected to show the pattern of character displacement since they show greater differences in morphology (and presumably in ecology and behavior as well), which should reduce the potential for competition to drive character divergence.

CHAPTER 23

LEARNING OUTCOME QUESTIONS

23.1 Because of convergent evolution; two distantly related species subjected to the same environmental pressures may be more phenotypically similar than two species with different environmental pressures but a more recent common ancestor. Other reasons for the possible dissimilarity between closely related species include oscillating selection and rapid adaptive radiations in which species rapidly adapt to a new available niche.

23.2 #1 In some cases wherein characters diverge rapidly relative to the frequency of speciation, it can be difficult to construct a phylogeny using cladistics because the most parsimonious phylogeny may not be the most accurate. In most cases, however, cladistics is a very useful tool for inferring phylogenetic relationships among groups of organisms.

#2 Only shared derived characters indicate that two or more taxa are descended from an ancestor that was not an ancestor of taxa that do not possess the character state in question. Thus, taxa possessing the state are more closely related to each other than they are to taxa without the trait. As the text discusses, this assumption is usually correct, but not when rates of homoplasy are high.

23.3 Yes, in some instances this is possible. For example, assume two populations of a species become geographically isolated from one another in similar environments, and each population diverges and speciation occurs, with one group retaining its ancestral traits and the other deriving new traits. The ancestral group in each population may be part of the same biological species but would be considered polyphyletic because to include their common ancestor would also necessitate including the other, more-derived species (which may have diverged enough to be reproductively isolated).

23.4 Not necessarily; it is possible that the character changed since the common ancestor and is present in each group due to convergence. Although the most recent common ancestor possessing the character is the most parsimonious, and thus the most likely, explanation, it is possible, especially for small clades, that similar environmental pressures resulted in the emergence of the same character state repeatedly during the course of the clade's evolution.

23.5 Hypothetically it is possible; however, the viral analyses and phylogenetic analyses have provided strong evidence that HIV emergence was the other way around; it began as a simian disease and mutated to a human form, and this has occurred several times.

INQUIRY AND DATA ANALYSIS QUESTIONS

Page 464 Data analysis: The data matrix would be similar to this one:

	Traits		
Organisms	Hair	Amniotic membrane	Tail
Salamander	0	0	1
Frog	0	0	0
Lizard	0	1	1
Tiger	1	1	1
Gorilla	1	1	0
Human	1	1	0

Page 465 Inquiry question: In parsimony analyses of phylogenies, the least complex explanation is favored. High rates of evolutionary change and few character states complicate matters. High rates of evolutionary change, such as occur when mutations arise in noncoding portions of DNA, can be misleading when constructing phylogenies. Mutations arising in noncoding DNA are not eliminated by natural selection in the same manner as mutations in coding (functional) DNA. Also, evolution of new character states can be very high in nonfunctional DNA, which can lead to genetic drift. Since DNA has only four nucleotides (four character states), it is highly likely that two species could evolve the same derived character at a particular base position. This leads to a violation of the assumptions of parsimony—that the fewest evolutionary events lead to the best hypothesis of phylogenetic relationships—and resulting phylogenies are inaccurate.

Page 467 Inquiry question: The only other hypothesis is that the most recent common ancestor of birds and bats was also winged. Of course, this scenario is much less parsimonious (and thus much more unlikely) than the convergence hypothesis, especially given the vast number of reptiles and mammals without wings. Most phylogenies are constructed based on the rule of parsimony; in the absence of fossil evidence of other winged animals and molecular data supporting a closer relationship between birds and bats than previously thought, there is no way to test the hypothesis that bird and bat wings are homologous rather than analogous.

Page 474 Inquiry question: One possibility would be to closely examine the patterns of development in different species. In part *a*, all of the species in the blue box with direct development would have inherited that character state from a common ancestor. Consequently, we might expect that the developmental patterns would be similar. By contrast, in part *b* direct development evolved many different times in those species. In the latter case, it is possible that direct development could have evolved in many different ways. Thus, if you compared how development occurs in these species and found that there were different patterns of development in different species, then the hypothesis in part *b* might seem more likely. However, even if all the species inherited direct development from their common ancestor (part *a*), species could diverge evolutionarily, so it is always possible that differences you found might have arisen after divergence from a common ancestor. For this reason, definitive conclusions are always difficult in situations such as this.

Page 477 Inquiry question: If the victim had contracted HIV from a source other than the patient, the most recent common ancestor of the two strains would be much more distant. As it is, the phylogeny shows that the victim and patient strains share a relatively recent ancestor, and that the victim's strain is derived from the patient's strain.

UNDERSTAND

1. d   2. b   3. a   4. b   5. a   6. d   7. d   8. b   9. c

APPLY

1. c   2. d   3. a

SYNTHESIZE

1. Naming of groups can vary; names provided here are just examples. Jaws—shark, salamander, lizard, tiger, gorilla, human (jawed vertebrates); lungs—salamander, lizard, tiger, gorilla, human (terrestrial tetrapods); amniotic membrane—lizard, tiger, gorilla, human (amniote tetrapods); hair—tiger, gorilla, human (mammals); no tail—gorilla, human (humanoid primate); bipedal—human (human).

2. It would seem to be somewhat of a conundrum, or potentially circular chase; choosing a closely related species as an outgroup when we do not even know the relationships of the species of interest. One way of guarding against a poor choice for an outgroup is to choose several species as outgroups and examine how the phylogenetic hypothesis for the group of interest changes as a consequence of using different outgroups. If the choice of outgroup makes little difference, then that might increase one's confidence in the phylogenetic hypotheses for the species of interest. On the other hand, if the choice makes a big difference (different phylogenetic hypotheses result when choosing different outgroups), that might at least lead to the conclusion that we cannot be confident in inferring a robust phylogenetic hypothesis for the group of interest without collecting more data.

3. Recognizing that birds are reptiles potentially provides insight to the biology of both birds and reptiles. For example, some characteristics of birds are clearly of reptilian origin, such as feathers (modified scales), nasal salt-secreting glands, and strategies of osmoregulation/excretion (excreting nitrogenous waste products as uric acid) representing ancestral traits that continue to serve birds well in their environments. On the other hand, some differences from other reptiles (again, feathers) seem to have such profound significance biologically, that they overwhelm similarities visible in shared ancestral characteristics. For example, no extant nonavian reptiles can fly or are endothermic, and these two traits have created a fundamental distinction in the minds of many biologists. Indeed, many vertebrate biologists prefer to continue to distinguish birds from reptiles rather than emphasize their similarities even though they recognize the power of cladistic analysis in helping to shape classification. Ultimately, it may be nothing much more substantial than habit which drives the preference of some biologists to traditional classification schemes.

4. In fact, such evolutionary transitions (the loss of the larval mode, and the re-evolution of a larval mode from direct development) are treated with equal weight under the simplest form of parsimony. However, if it is known

from independent methods (e.g., developmental biology) that one kind of change is less likely than another (loss vs. a reversal), these should and can be taken into account in various ways. The simplest way might be to assign weights based on likelihoods; two transitions from larval development to direct development is equal to one reversal from direct development back to a larval mode. In fact, such methods exist, and they are similar in spirit to the statistical approaches used to build specific models of evolutionary change rather than rely on simple parsimony.

5. The structures are both homologous, as forelimbs, and convergent, as wings. In other words, the most recent common ancestor of birds, pterosaurs, and bats had a forelimb similar in morphology to that which these organisms possess—it has similar bones and articulations. Thus, the forelimb itself among these organisms is homologous. The wing, however, is clearly convergent; the most recent common ancestor surely did not have wings (or all other mammals and reptiles would have had to have lost the wing, which violates the rule of parsimony). The wing of flying insects is purely convergent with the vertebrate wing, as the forelimb of the insect is not homologous with the vertebrate forelimb.

6. The biological species concept focuses on processes, in particular those that result in the evolution of a population to the degree that it becomes reproductively isolated from its ancestral population. The process of speciation as utilized by the biological species concept occurs through the interrelatedness of evolutionary mechanisms such as natural selection, mutation, and genetic drift. On the other hand, the phylogenetic species concept focuses not on process but on history, on the evolutionary patterns that led to the divergence between populations. Neither species concept is more right or more wrong; species concepts are, by their very nature, subjective and potentially controversial.

CHAPTER 24

LEARNING OUTCOME QUESTIONS

24.1 There should be a high degree of similarity between the two genomes because they are relatively closely related. There could be differences in the relative amounts of noncoding DNA. Genes that are necessary for bony skeletal development might be found in the bony fish and not cartilaginous fish.

24.2 The additional DNA would likely be noncoding DNA that might represent large expanses of retrotransposon DNA that duplicated and transposed multiple times.

24.3 Compare the sequence of the pseudogene with other species. If, for example, it is a pseudogene of an olfactory gene that is found in mice or chimps, the sequences will be much more similar than in a more distantly related species. If horizontal gene transfer explains the origins of the gene, there may not be a very similar gene in closely related species.

24.4 A SNP can change a single amino acid in the coded peptide. If the new R group is very different, the protein may fold in a different way and not function effectively. SNPs in the *FOXP2* gene may, in part, explain why humans have speech and chimps do not. Other examples from earlier in the text include cystic fibrosis and sickle cell anemia.

24.5 An effective drug would bind to a region of the pathogen protein that is distinct from the human protein. This makes the drug selective for the pathogen protein. If the seven amino acids that differ are scattered throughout the protein, they might have a minimal effect on structure, and it would be difficult to develop an effective drug.

INQUIRY AND DATA ANALYSIS QUESTIONS

Page 484 Data analysis: Rice has 43,000 genes in 430 Mb of DNA (100 genes/Mb), whereas humans have 22,698 genes in 2500 Mb (9 genes/Mb). Rice has over 10 times as many genes as humans per Mb. The organisms closest to humans in terms of number of genes/Mb would be the duck-billed platypus with 8.4 genes/Mb and next would be the domestic cow with 7.6 genes/Mb.

Page 485 Inquiry question: Meiosis in a 3*n* cell would be impossible because three sets of chromosomes cannot be divided equally between two cells. In a 3*n* cell, all three homologous chromosomes would pair in prophase I, then align during anaphase I. As the homologous chromosomes separate, two of a triplet might go to one cell and the third chromosome would go to the other cell. The same would be true for each set of homologues. Daughter cells would have an unpredictable number of chromosomes.

Page 486 Inquiry question: Polyploidization seems to induce the elimination of duplicated genes. Duplicate genes code for the same gene product. It is reasonable that duplicate genes would be eliminated to decrease the redundancy arising from the translation of several copies of the same gene.

Page 491 Inquiry question: Ape and human genomes show very different patterns of gene transcription activity, even though genes encoding proteins are over 99% similar between chimps and humans. Different genes would be transcribed when comparing apes with humans, and the levels of transcription would vary widely.

UNDERSTAND

1. c 2. d 3. d 4. b 5. b 6. a

APPLY

1. a 2. d 3. d 4. a

SYNTHESIZE

1. The two amino acid difference between the FOXP2 protein in humans and closely related primates must alter the way the protein functions in the brain. The protein affects motor function in the brain, allowing coordination of larynx, mouth, and brain for speech in humans. For example, if the protein affects transcription, there could be differences in the genes that are regulated by FOXP2 in humans and chimps.

2. Human and chimp DNA is close to 99% similar, yet our phenotypes are conspicuously different in many ways. This suggests that a catalog of genes is just the first step to identifying the mechanisms underlying genetically influenced diseases like cancer or cystic fibrosis. Clearly, gene expression, which might involve the actions of multiple noncoding segments of the DNA and other potentially complex regulatory mechanics, are important sources of how phenotypes are formed, and it is likely that many genetically determined diseases result from such complex underlying mechanisms, making the gene identification of genomics just the first step; a necessary but not nearly sufficient strategy. What complete genomes do offer is a starting point to correlating sequence differences among humans with genetic disease, as well as the opportunity to examine how multiple genes and regulatory sequences interact to cause disease.

3. Phylogenetic analysis usually assumes that most genetic and phenotypic variation arises from descent with modification (vertical inheritance). If genetic and phenotypic characteristics can be passed horizontally (i.e., not vertically through genetic lineages), then using patterns of shared character variation to infer genealogical relationships will be subject to potentially significant error. We might expect that organisms with higher rates of HGT will have phylogenetic hypotheses that are less reliable or at least are not resolved as a neatly branching tree.

CHAPTER 25

LEARNING OUTCOME QUESTIONS

25.1 A change in the promoter of a gene necessary for wing development might lead to the repression of wing development in a second segment of a fly in a species that has double wings.

25.2 Yes, although this is not the only explanation. The coding regions could be identical, but the promoter or other regulatory regions could have been altered by mutation, leading to altered patterns of gene expression. To test this hypothesis, the *Eda* gene should be sequenced in both fish and compared.

25.3 The pectoral fins are homoplastic because sharks and whales are only distantly related and pectoral fins are not found in whales' more recent ancestors.

25.4 There is no need for eyes in the dark. Perhaps the fish expend less energy when eyes are not produced and that offered a selective advantage in cavefish. In a habitat with light, a mutation that resulted in a functional *Pax6* would likely be selected for, and over time more of the fish would have eyes. Keep in mind that the probability of a mutation restoring *Pax6* function is very low, but real.

INQUIRY AND DATA ANALYSIS QUESTIONS

Page 501 Inquiry question: Because there is a stop codon located in the middle of the *CAL* (cauliflower) gene-coding sequence, the wild-type function of *CAL* must be concerned with producing flowers. Unlike the ancestral *Brassica oleracea*, cauliflower and broccoli keep branching instead of producing a flower. One interpretation is that a meristem lacking CAL protein is delayed in producing flower parts and instead keeps producing branches, resulting in the broccoli and cauliflower heads. Cauliflower and broccoli heads differ in terms of how advanced floral development is (only meristems in cauliflower and early floral buds in broccoli). This difference could be the result of a later evolutionary event. Additional evolutionary events possibly include large flower heads, unusual head coloration, protective leaves covering flower heads, or head size variants, among other possibilities.

Page 505 Inquiry question: Functional analysis involves the use of a variety of experiments designed to test the function of a specific gene in different species. By mixing and matching parts of the *AP3* and *PI* genes and introducing them into *ap3* mutant plants, it was found that the C terminus sequence of the *AP3* protein is essential for specifying petal function. Without the 3 region of the *AP3* gene, the *Arabidopsis* plant cannot make petals.

UNDERSTAND
1. c 2. b 3. a 4. a 5. b 6. d 7. a 8. c 9. c 10. d

APPLY
1. c 2. a 3. d 4. b

SYNTHESIZE
1. Your supervisor might still be correct. The mutation could be in the regulatory region and the *pitx1* gene could be transcribed in cells in different tissues in the two species. You could test this by testing whether or not you are able to obtain the RNA from hindlimbs and the pelvis of both species at different times in development.

2. Development is a highly conserved and constrained process; small perturbations can have drastic consequences, and most of these are negative. Given the thousands or hundreds of thousands of variables that can change in even a simple developmental pathway, most perturbations lead to negative outcomes. Over millions of years, some of these changes will arise under the right circumstances to produce a benefit. In this way, developmental perturbations are not different from what we know about mutations in general. Beneficial mutations are rare, but with enough time they will emerge and spread under specific circumstances.

    Not all mutations provide a selective advantage. For example, reduced body armor increases the fitness of fish in fresh water, but it was not selected for in a marine environment where the armor was important for protection from predators. The new trait can persist at low levels for a very long time until a change in environmental conditions results in an increase in fitness for individuals exhibiting the trait.

3. The latter view represents our current understanding. There are many examples of small gene families (e.g., *Hox*, *MADS*) whose apparent role in generating phenotypic diversity among major groupings of organisms is in altering the expression of other genes. Alterations in timing (heterochrony) or spatial pattern of expression (homeosis) can lead to shifts in developmental events, giving rise to new phenotypes. Many examples are presented in the chapter, such as the developmental variants of two species of sea urchins, one with a normal larval phase, and another with direct development. In this case the two species do not have different sets of developmental genes; rather the expression of those genes differ. Another example that makes the same point is the evolution of an image-forming eye. Recent studies suggest, in contrast to the view that eyes across the animal kingdom evolved independently multiple times, that image-forming eyes from very distantly related taxa (e.g., insects and vertebrates) may trace back to the common origin of the *Pax6* gene. If that view is correct, then genes controlling major developmental patterns would seem to be highly conserved across long periods of time, with expression being the major form of variation.

4. Unless the *Pax6* gene was derived multiple times, it is difficult to hypothesize multiple origins of eyes. Pax6 initiates eye development in many species. The variation in eyes among animals is a result of which genes are expressed and when after Pax6 initiates eye development.

5. Maize relies on *paleoAP3* and *PI* for flower development, whereas tomato has three genes because of a duplication of *paleoAP3*. This duplication event in the ancestor of tomato, but not maize, is correlated with independent petal origin.

6. The direct developing sea urchin has an ancestor that had one or more mutations in genes that were needed to regulate the expression of other genes needed for larval stage development. When those genes were not expressed, there was no larval development and the genes necessary for adult development were expressed.

CHAPTER 26
LEARNING OUTCOME QUESTIONS

26.1 The Phanerozoic eon includes the greatest diversity of life and would be the most informative in terms of fossil record reflecting diverse life-forms. Earlier eons have a more limited fossil record, with primarily unicellular organisms.

26.2 Amino acids, including glycine and alanine, nitrogenous bases like adenine that are found in DNA and RNA, lipids, and simple carbon compounds like CH_2O might be found. If the early life was based on RNA, then ribonucleotides rather than deoxyribonucleotides would be found. Early RNA sequences would have had both enzymatic and information functions. Once the machinery was in place for RNA to encode proteins, amino acid sequences would become less random and gain functions in catalyzing metabolism as well as structural roles and regulating nucleic acid replication, transcription, and translation. Lipids would form early cell membranes, and metabolic pathways would continue to evolve within the confines of the cell membrane. Nucleic acid replication and cell division would evolve early on also.

26.3 Electron microscopy revealing organic walls and vesicles would provide good evidence, combined with spectroscopic analysis indicating the presence of complex carbon molecules.

26.4 Weathering is proposed as a key event associated with both glaciations. Weathering in the late Proterozoic was caused by warm, moist air and by shifts in plate tectonics that increased surface area for weathering. In the Ordovician period, land was being colonized for the first time. Weathering of rocks because of atmospheric changes is not sufficient to explain the CO_2 shift leading to glaciation. Rather, early land plants secreted an organic acid that weathered rocks, releasing phosphorous that ran into the oceans. In the oceans, phosphorous triggered algal blooms that pulled down CO_2 from the atmosphere, triggering a rapid drop in temperature.

26.5 Major pre-Cambrian evolutionary innovations that set the stage for rapid radiation included the evolution of eukaryotic cells with chloroplasts and mitochondria, and meiosis that allowed for sexual reproduction. The new gene combinations created by meiosis provided a set of phenotypes for natural selection to act upon.

UNDERSTAND
1. b 2. d 3. c The correct statement would read, "Brown algae gained chloroplasts by engulfing red algae (endosymbiosis)." 4. c 5. b

APPLY
1. b 2. c 3. d 4. c 5. c

SYNTHESIZE
1. Early land plants lacked roots and had substantially less biomass than the later vascular plants. Alone they could not have sequestered sufficient CO_2 to trigger glaciation. However, just like the vascular plants, they secreted organic acid that enhanced weathering of rocks, leading to the release of phosphorous. The excess phosphorous washed into the oceans. As an essential plant nutrient, the excess phosphorous triggered extensive algal blooms, which did sequester sufficient CO_2, coupled with the land plants, to trigger a glaciation.

2. The collision of tectonic plates could have multiple effects on evolution of life. When the plates collide, less surface area is exposed to water, which could reduce weathering. Reduced weathering would pull less CO_2 out of the atmosphere, which could affect climate. When plates collide, organisms that were previously separated come together. This potentially affects interbreeding and also changes selective pressures affecting evolution.

3. The evolution of meiosis provided a much greater range of genetic diversity for natural selection to act upon. The evolution of multicellularity immediately preceded the Cambrian explosion and was likely a contributing factor to the rapid diversification when ancestors of almost every group of animals evolved. The climate must also have been more hospitable to life during the Cambrian, a period following a global ice age.

CHAPTER 27
LEARNING OUTCOME QUESTIONS

27.1 Viruses use cellular machinery for replication. They do not make all of the proteins necessary for complete replication.

27.2 A prophage carrying such a mutation could not be induced to undergo the lytic cycle.

27.3 This therapy, at present, does not remove all detectable viruses, so it cannot be considered a true cure.

27.4 In addition to a high mutation rate, the influenza genome consists of multiple RNA segments that can recombine during infection. This causes the main antigens for the immune system to shift rapidly.

27.5 Prions carry information in their three-dimensional structure. This 3-D information is different from the essentially one-dimensional genetic information in DNA.

UNDERSTAND
1. c 2. b 3. c 4. d 5. b 6. d 7. b

APPLY
1. c 2. b 3. c 4. d 5. b 6. c 7. c 8. a

SYNTHESIZE
1. A set of genes that are involved in the response to DNA damage are normally induced by the same system. The protein involved destroys a repressor that keeps DNA repair genes unexpressed. Lambda has evolved to use this system to its advantage.
2. Since viruses require the replication machinery of a host cell to replicate, it is unlikely that they existed before the origin of the first cells.
3. This is a complex situation. The relevant factors include the high mutation rate of the virus and the fact that the virus targets the very cells that mount an immune response. The influenza virus also requires a new vaccine every year due to rapid changes in the virus. The smallpox virus was a DNA virus that had antigenic determinants that did not change rapidly, thus making a vaccine possible.
4. Emerging viruses are those that jump species and thus are new to humans. Recent examples include SARS and Ebola.
5. If excision of the λ prophage is imprecise, then the phage produced will carry *E. coli* genes adjacent to the integration site.

CHAPTER 28

LEARNING OUTCOME QUESTIONS

28.1 Archaea have ether-linked instead of ester-linked phospholipids; their cell wall is made of unique material.

28.2 Compare their DNA. The many metabolic tests we have used for years have been supplanted by DNA analysis.

28.3 Transfer of genetic information in bacteria is directional: from donor to recipient and does not involve fusion of gametes.

28.4 Prokaryotes do not have a lot of morphological features, but do have diverse metabolic functions.

28.5 Pathogens tend to evolve to be less virulent. If they are too good at killing, their lifestyle becomes an evolutionary dead end.

28.6 Rotating a crop that has a symbiotic association with nitrogen-fixing bacteria will return nitrogen to the soil depleted by other plants.

INQUIRY AND DATA ANALYSIS QUESTIONS

Page 553 Data analysis: An imprecise excision of the F plasmid produces a so-called F′ plasmid that carries *E. coli* DNA in addition to the normal plasmid DNA. If this is introduced into a new F⁻ cell, it will then produce a new cell that is "diploid" for the region carried on the F′ plasmid. These so-called partial diploids have many genetic uses.

Page 560 Inquiry question: The simplest explanation is that the two STDs are occurring in different populations, and one population has rising levels of sexual activity, while the other has falling levels. However, the rise in incidence of an STD can reflect many parameters other than level of sexual activity. The virulence or infectivity of one or both disease agents may be changing, for example, or some aspect of exposed people may be changing in such a way as to alter susceptibility. Only a thorough public health study can sort this out.

UNDERSTAND
1. b 2. a 3. c 4. c 5. d 6. a 7. b

APPLY
1. c 2. b 3. b 4. c 5. d 6. b 7. a

SYNTHESIZE
1. The study of carbon signatures in rocks using isotopic data assumes that ancient carbon fixation involves one of two pathways that each show a bias toward incorporation of carbon-12. If this bias were not present, it is not possible to infer early carbon fixation by this pathway. This pathway could have arisen even earlier and we would have no way to detect it.
2. The heat killing of the virulent S strain of *Streptococcus* released the genome of the virulent smooth strain into the environment. These strains of *Streptococcus* bacteria are capable of natural transformation. At least some of the rough-strain cells took up smooth-strain genes that encoded the polysaccharide coat from the environment. These genes entered into the rough-strain genome by recombination, and then were expressed. These transformed cells were now smooth bacteria.
3. Use of multiple antibiotics is not a bad idea if all of the bacteria are killed. In the case of some persistent infections, this is an effective strategy. However, it does provide very strong selective pressure for rare genetic events that produce multiple resistances in a single bacterial species. For this reason, it is not a good idea for it to be the normal practice. The more bacteria that undergo this selection for multiple resistance, the more likely it will arise. This is helped by patients not taking their entire course of antibiotic because bacteria may survive by chance and proliferate, with each generation providing the opportunity for new mutations. This is also complicated by the horizontal transfer of resistance via resistance plasmids, and by the existence of transposable genetic elements that can move genes from one piece of DNA to another.
4. Most species on the planet are incapable of fixing nitrogen without the assistance of bacteria. Without nitrogen, amino acids and other compounds cannot be synthesized. Thus a loss of the nitrogen-fixing bacteria due to increased UV radiation levels would reduce the ability of plants to grow, severely limiting the food sources of the animals.

CHAPTER 29

LEARNING OUTCOME QUESTIONS

29.1 Mitochondria and chloroplasts contain their own DNA. Mitochondrial genes are transcribed within the mitochondrion, using mitochondrial ribosomes that are smaller than those of eukaryotic cells and quite similar to bacterial ribosomes. Antibiotics that inhibit protein translation in bacteria also inhibit protein translation in mitochondria. Also, both chloroplasts and mitochondria divide using binary fission like bacteria.

29.2 Meiosis makes sexual reproduction possible. Each generation, new combinations of alleles provide much greater genetic diversity for evolution to act on than is available with asexual reproduction. Different gene combinations were advantageous in different settings, and species began to diverge as natural selection acted on the genetic diversity.

29.3 #1 Undulating membranes would be effective on surfaces with curvature that may not always be smooth, such as intestinal walls.
#2 Contractile vacuoles collect and remove excess water from within the *Euglena*.

29.4 #1 The *Plasmodium* often becomes resistant to new poisons and drugs. And, because the *Plasmodium* has multiple hosts, a drug for humans wouldn't eradicate the *Plasmodium* in other stages of its life cycle in the mosquito.
#2 Just based on outward appearances, the sporophyte of brown algae forms a larger bladelike structure, whereas the gametophyte is a small filamentous structure.

29.5 #1 Both the red and green algae obtained their chloroplasts through endosymbiosis, possibly of the same lineage of photosynthetic bacteria. The red and green algae had diverged before the endosymbiotic events, and the history recorded in their nuclear DNA is a different evolutionary history than that recorded in the plastids derived through endosymbiosis.
#2 The lack of water is the major barrier for sperm that move through water to reach the egg. It is more difficult for sperm to reach the egg on land.

29.6 Construct phylogenies with other traits, including DNA sequences. Then map the presence of amoeboid locomotion with pseudopods on to the other phylogenies. If the pseudopod trait is not clustered, it is likely that it evolved independently multiple times and is thus not a good trait to use in reconstructing protist phylogenies.

29.7 It is unlikely that cellular and plasmodial slime molds are closely related. They both appear in the last section of this chapter because they have yet to be assigned to clades. The substantial differences in their cell biology are inconsistent with a close phylogenetic relationship.

29.8 Comparative genomic studies of choanoflagellates and sponges would be helpful. Considering the similarities among a broader range of genes than just the conserved tyrosine kinase receptor would provide additional evidence.

INQUIRY AND DATA ANALYSIS QUESTIONS

Page 568 Inquiry question: Red and green algae obtained chloroplasts by engulfing photosynthetic bacteria by primary endosymbiosis; chloroplasts in these cells have two membranes. Brown algae obtained chloroplasts by engulfing cells of red algae through secondary endosymbiosis; chloroplasts in cells of brown algae have four membranes. Counting the number of cell membranes of chloroplasts indicates primary or secondary endosymbiosis.

Page 579 Data analysis: Your phylogeny should show a common ancestor for all three algae. The brown alga should branch from a common ancestor of the red and green algae. (See the phylogeny on p. 575 of your text and trace the three algae.)

Page 581 Inquiry question: No, they are formed by mitosis. Meiosis produces haploid spores, which divide mitotically to produce multicellular gametophytes. Gametes are produced from haploid gametophyte cells by mitosis.

UNDERSTAND

1. b 2. a 3. b 4. c 5. d 6. b 7. a 8. c 9. b, c 10. a, d 11. d 12. a

APPLY

1. d 2. a 3. a

SYNTHESIZE

1. Cellular and plasmodial slime molds both exhibit group behavior and can produce mobile slime mold masses. However, these two groups are very distantly related phylogenetically.

2. The development of a vaccine, though challenging, will be the most promising in the long run. It is difficult to eradicate all the mosquito vectors, and many eradication methods can be harmful to the environment. Treatments to kill the parasites are also difficult because the parasite is likely to become resistant to each new poison or drug. A vaccine would provide long-term protection without the need to use harmful pesticides or drugs for which drug resistance is a real possibility.

3. For the first experiment, plate the cellular slime molds on a plate that has no bacteria. Spot cyclic AMP and designated places on the plate and determine if the bacteria aggregate around the cAMP.
   For the second experiment, repeat the first experiment using plates that have a uniform coating of bacteria as well as plates with no bacteria. If the cellular slime molds aggregate on both plates, resource scarcity is not an issue. If the cells aggregate only in the absence of bacteria, you can conclude that the attraction to cAMP occurs only under starvation conditions.

CHAPTER 30

LEARNING OUTCOME QUESTIONS

30.1 #1 Make sections and examine them under the microscope to look for tracheids. Only the tracheophytes will have tracheids.
#2 Gametes in plants are produced by mitosis. Human gametes are produced directly by meiosis.

30.2 Mosses are extremely desiccation-tolerant and can withstand the lack of water. Also, freezing temperatures at the poles are less damaging when mosses have a lower water content.

30.3 The sporophyte generation has evolved to be the larger generation, and therefore an effective means of transporting water and nutrients over greater distances would be advantageous.

30.4 Comparing the genomes of lycophytes and other vascular plants supports the independent origin of leaflike structures. Leaflike structures in lycophytes and other vascular plants would have provided these plants with enhanced photosynthetic capabilities. The trend of reproductive maturity in the sporophyte generation gave lycophytes and other vascular plants an evolutionary advantage related to genetic variation in the populations.

30.5 The fiddleheads are the immature fronds, emerging from the soil. The production of secondary compounds that can cause stomach cancer is most likely a defense mechanism to protect the emerging plants.

INQUIRY AND DATA ANALYSIS QUESTIONS

Page 594 Inquiry question: Tracheophytes developed vascular tissue, enabling them to have efficient water- and food-conducting systems. Vascular tissue allowed tracheophytes to grow larger, possibly then enabling them to outcompete smaller, nonvascular land plants. A protective cuticle and stomata that can close during dry conditions also conferred a selective advantage.

Page 595 Data analysis: Tracheophytes first appeared 40 million years after land plants first began diverging. Another 40 million years passed before true leaves appeared.

UNDERSTAND

1. d 2. d 3. c 4. c 5. a 6. d

APPLY

1. d 2. b 3. c 4. d 5. d 6. b

SYNTHESIZE

1. Moss has a dominant gametophyte generation, whereas lycophytes have a dominant sporophyte generation. Perhaps a comparison of the two genomes would provide insight into the genomic differences associated with the evolutionary shift from dominant gametophyte to dominant sporophyte.

2. Because the sporophyte generation of the fern is much larger than the sporophyte generation of the moss, it is expected that the fern sporophyte would have more mitotic divisions in order for the plant to be larger.

3. Moss would face the greater challenge. Without vascular tissue, it would be difficult to move water and nutrients efficiently over 10 m of height. Although the fern gametophyte would be small and on the ground where water would be available for sperm to travel a short distance to fertilize eggs, getting egg and sperm together at the tops of two different moss gametophytes would be challenging at a height of 10 m.

CHAPTER 31

LEARNING OUTCOME QUESTIONS

31.1 The pollen tube grows toward the egg, carrying the sperm within the pollen tube; therefore water is not needed for fertilization.

31.2 The ovule rests, exposed on the scale (a modified leaf), which allows easier access to the ovule through wind pollination.

31.3 Any number of vectors that could acquire the moss gene and also affect the flowering plant could account for horizontal gene transfer. Pathogens including viruses and bacteria, as well as herbivores could potentially facilitate horizontal gene transfer.

31.4 We'd expect a dormancy that required a period of chilling to be broken and/or time in the ground for the seed coat to be weakened by microbes in the soil over the winter so the seedling can force its way out of the seed.

31.5 Fleshy fruits are more likely to encourage animals to eat them. Different colors will attract different animals. For example, birds are attracted to red berries.

INQUIRY AND DATA ANALYSIS QUESTIONS

Page 608 Inquiry question: Comparisons of a single gene could result in an inaccurate phylogenetic tree because it fails to take into account the effects of horizontal gene transfer. For example, the clade of *Amborella trichopoda* is a sister clade to all other flowering plants, but roughly 2/3 of its mitochondrial genes are present due to horizontal gene transfer from other land plants, including more distantly related mosses.

Page 610 Inquiry question: To determine if a moss gene had a function you would employ functional analysis, using a variety of experiments, to test for possible functions of the moss gene in *Amborella*. For example, you could create a mutation in the moss gene and see if there were any phenotypic differences in *Amborella* plants homozygous for the mutant gene. You could place the moss gene promoter in front of a reporter gene and visualize where, if any place, in the plant the gene was expressed.

Page 613 Inquiry question: Endosperm provides nutrients for the developing embryo in most flowering plants. The embryo cannot derive nutrition from soil prior to root development; therefore, without endosperm, the embryo is unlikely to survive, and fitness would be zero.

Page 613 Data analysis: $(640\ \mu m - 80\ \mu m)/(8\ hr - 1\ hr) = 80\ \mu m/hr$. This matches the value in the data table. Differences could be the result of reading the graph incorrectly or of rounding in doing the calculation.

Page 614 Inquiry question: The prior sporophyte generation (tan) should be labeled $2n$. The degenerating gametophyte generation (purple) should be labeled $1n$. The next sporophyte generation (blue) should be labeled $2n$.

UNDERSTAND

1. a 2. c 3. b 4. d 5. a 6. d 7. d 8. b 9. a

APPLY

1. c 2. a 3. b 4. a 5. a 6. d

SYNTHESIZE

1. Answers to this question may vary. However, gymnosperms are defined as "naked" seed plants. Therefore, an ovule that is not completely protected by sporophyte tissue would be characteristic of a gymnosperm. To be classified as an angiosperm, evidence of flower structures and double fertilization are key characteristics, although double fertilization has been observed in some gnetophytes.

2. The purpose of pollination is to bring together the male and female gametes for sexual reproduction. Sexual reproduction is designed to increase the genetic variability of a species. If a plant allows self-pollination, then the amount of genetic diversity will be reduced, but this is a better alternative than not reproducing at all. This would be especially useful in species in which the individuals are widely dispersed.

3. The benefit is that by developing a relationship with a specific pollinator, the plant species increases the chance that its pollen will be brought to another member of its species for pollination. If the pollinator is a generalist, then the pollinator might not travel to another member of the same species, and pollination would not occur. The drawback is that if something happens to the pollinator (extinction or drop in population size), then the plant species would be left with either a reduced or nonexistent means of pollination.

4. The seeds may need to be chilled before they can germinate. You can store them in the refrigerator for several weeks or months and try again. The surface of the seed may need to be scarified (damaged) before it can germinate. Usually this would happen from the effects of weather or if the seed goes through the digestive track of an animal where the seed coat is weakened by acid in the gut of the animal. You could substitute for natural scarification by rubbing your seeds on sandpaper before germinating them. It is possible that your seed needs to be exposed to light or received insufficient water when you first planted it. You may need to soak your seed in water for a bit to imbibe it. Exposing the imbibed seed to sunlight might also increase the chances of germination.

CHAPTER 32

LEARNING OUTCOME QUESTIONS

32.1 #1 In fungi, mitosis results in duplicated nuclei, but the nuclei remain within a single cell. This lack of cell division following mitosis is very unusual in animals.
 #2 Hyphae are protected by chitin, which is not digested by fungal enzymes.

32.2 Microsporidians lack mitochondria, which are found in *Plasmodium*.

32.3 Blastocladiomycetes are free-living and have mitochondria. Microsporidians are obligate parasites and lack mitochondria.

32.4 Zygospores are more likely to be produced when environmental conditions are not favorable. Sexual reproduction increases the chances of offspring with new combinations of genes that will have an advantage in a changing environment. Also, the zygospore can stay dormant until conditions improve.

32.5 Glomeromycetes depend on the host plant for carbohydrates.

32.6 A dikaryotic cell has two nuclei, each with a single set of chromosomes. A diploid cell has a single nucleus with two sets of chromosomes.

32.7 Preventing the spread of the fungal infection using fungicides and good cultivation practices could help. If farmworkers must tend to infected fields, masks that filter out the spores could protect the workers.

32.8 The fungi that ants consumed may have originally been growing on leaves. Over evolutionary time, mutations that altered ant behavior so the ants would bring leaves to a stash of fungi would have been favored, and the tripartite symbiosis evolved.

32.9 Wind can spread spores over large distances, resulting in the spread of fungal disease.

INQUIRY AND DATA ANALYSIS QUESTION

Page 628 Data analysis: The first fungus had 1.31×10^8 spores/1.1 cm^2 = 1.2×10^8 spores/cm^2 versus the second with 6.87×10^9 spores/8.37 cm^2 = 8.2×10^8 spores/cm^2. The second fungus produced a little over 6 times as many spores per cm^2 as the first fungus.

UNDERSTAND

1. c 2. d 3. a 4. d 5. b 6. d 7. d

APPLY

1. d 2. b 3. a 4. c 5. d

SYNTHESIZE

1. Fungi possess cell walls. Although the composition of these cell walls differs from that of the plants, cell walls are completely absent in animals. Fungi are also immobile (except for chytrids), and mobility is a key characteristic of the animals.

2. The mycorrhizal relationships between the fungi and plants allow plants to make use of nutrient-poor soil. Without the colonization of land by plants, it is unlikely that animals would have diversified to the level they have achieved today. Lichens are important organisms in the colonization of land. Early landmasses would have been composed primarily of barren rock, with little or no soil for plant colonization. As lichens colonize an area they begin the process of soil formation, which allows other plants to grow.

3. Antibiotics are designed to combat prokaryotic organisms, and fungi are eukaryotic. In addition, fungi possess a cell wall that has a different chemical constitution (chitin) from that of prokaryotes.

4. The fungicide will increase the germination success of your seeds by preventing fungi from harming the seed and emerging seedling. However, because of the close phylogenetic relationship between fungi and animals, a compound that is lethal to fungi may also be harmful to humans. You would not want to get the fungicide on your hands.

CHAPTER 33

LEARNING OUTCOME QUESTIONS

33.1 The rules of parsimony state that the simplest phylogeny is most likely the true phylogeny. As there are living organisms that are both multicellular and unicellular, it stands to reason that the first organisms were unicellular, and multicellularity followed. Animals are also all heterotrophs; if they were the first type of life to have evolved, there would not have been any autotrophs on which they could feed.

33.2 Cephalization, the concentration of nervous tissue in a distinct head region, is intrinsically connected to the onset of bilateral symmetry. Bilateral symmetry promotes the development of a central nerve center, which in turn favors the nervous tissue concentration in the head. In addition, the onset of both cephalization and bilateral symmetry allows for the marriage of directional movement (bilateral symmetry) and the presence of sensory organs facing the direction in which the animal is moving (cephalization).

33.3 This allows systematists to classify animals based solely on derived characteristics. Using features that have only evolved once implies that the species that have that characteristic are more closely related to each other than they are to species that do not have the characteristic.

33.4 The cells of a truly colonial organism, such as a colonial protist, are all structurally and functionally identical; however, sponge cells are differentiated and these cells coordinate to perform functions required by the whole organism. Unlike all other animals, however, sponges do appear much like colonial organisms in that they are not composed of true tissues, and the cells are capable of differentiating from one type to another.

33.5 Two key morphological characters are the number of body layers, making an animal diploblastic or triploblastic, and the type of symmetry. Ctenophores were traditionally thought to be diploblastic, which is the ancestral condition. However, recent studies have suggested that they are, in fact, triploblastic, which is a synapomorphy of bilaterians. In addition, ctenophores exhibit a modified type of radial symmetry, which suggests that this character may not link them with Cnidaria.

33.6 Yes, the coelom has been lost in several clades of protostomes.

INQUIRY AND DATA ANALYSIS QUESTION

Page 648 Inquiry question: None of these characters is a completely reliable indicator of phylogenetic relationships. The coelom only evolved once and is a synapomorphy for the clade comprising protostomes and deuterostomes. Thus, all species with a coelom belong to this clade, but, a number of members of this clade have evolved a pseudocoelom or have become acoelomic. As a result, some species with a coelom are more closely related to species with a pseudocoelom or no coelom than they are to other species with a coelom. A pseudocoelom has evolved convergently in several clades. Thus, the possession of a pseudocoelom is not an indicator of phylogenetic relationships among major animal clades. The acoelomic condition is ancestral for animals, but also re-evolved within Bilateria, so this character state is also not a reliable phylogenetic trait.

UNDERSTAND

1. c 2. b 3. a 4. d 5. a 6. b 7. d 8. a 9. a 10. d 11. a 12. b 13. d

APPLY

1. d 2. b 3. Determinate development indicates that it is a protostome and the fact that it molts places it within the Ecdysozoa. The presence of jointed appendages makes it an arthropod. 4. d

SYNTHESIZE

1. The tree should contain Platyhelminthes and Nemetera on one branch, a second branch should contain nematodes, and a third branch should contain the annelids and the hemichordates. This does not coincide with the information in figure 33.5. Therefore, some of the different types of body cavities have evolved multiple times, and the body cavities are not good characteristics from which to infer phylogenetic relationships.
2. Answers may vary depending on the classification used. Many students will place the Echinoderms near the Cnidaria due to radial symmetry; others will place them closer to the annelids.

Chapter 34

LEARNING OUTCOME QUESTIONS

34.1 Molting is a synapomorphy of ecdysozoans, and spiral cleavage of spiralians.

34.2 Tapeworms are parasitic platyhelminthes that live in the digestive system of their host. Tapeworms have a scolex, or head, with hooks for attaching to the wall of their host's digestive system. Another way in which the anatomy of a tapeworm relates to its way of life is its dorsoventrally flattened body and corresponding lack of a digestive system. Tapeworms live in their food; as such they absorb their nutrients directly through the body wall, and their flat bodies facilitate this form of nutrient delivery.

34.3 The cilia both sweep food into the rotifer's mouth and provide the propulsion for locomotion.

34.4 In most invertebrates with a coelom, an important function is internal support by hydrostatic pressure; however, this is unnecessary in most mollusks because the shell serves this purpose.

34.5 With a flow-through digestive tract, food moves in only one direction. This allows for specialization within the tract; sections may be specialized for mechanical and chemical digestion, some for storage, and yet others for absorption. Overall, the specialization yields greater efficiency than does a gastrovascular cavity.

34.6 The main advantage is coordination. A nervous system that serves the entire body allows for coordinated movement and coordinated physiological activities such as reproduction and excretion, even if those systems themselves are segmented. Likewise, a body-wide circulatory system enables efficient oxygen delivery to all of the body cells regardless of the nature of the organism's individual segments.

34.7 Lophophorates are sessile suspension-feeding animals. Much of their body also remains submerged in the ocean floor. Thus, a traditional tubular digestive system would require either the mouth or the anus to be inaccessible to the water column—meaning the animal either could not feed or would have to excrete waste into a closed environment. The U-shaped gut allows them to both acquire nutrients from and excrete waste into their environment.

34.8 *Ascaris lumbricoides*, the intestinal roundworm, infects humans when the human swallows food or water contaminated with roundworm eggs. The most effective ways of preventing the spread of intestinal roundworms is to increase sanitation (especially in food handling), to promote education about the infection process, and to cease using human feces as fertilizer. Not surprisingly, infection by these parasites is most common in areas without modern plumbing.

34.9 One of the defining features of the arthropods is the presence of a chitinous exoskeleton. As arthropods increase in size, the exoskeleton must increase in thickness disproportionately, in order to bear the pull of the animal's muscles. This puts a limit on the size a terrestrial arthropod can reach because the increased bulk of the exoskeleton would prohibit the animal's ability to move. Water is denser than air and thus provides more support; for this reason aquatic arthropods are able to be larger than terrestrial arthropods.

UNDERSTAND

1. d 2. c 3. b 4. a 5. d 6. d 7. d 8. b 9. c 10. d 11. d

APPLY

1. d 2. b Lobsters, centipedes, and nematods are all ecdysozoans, but centipedes and lobsters have segmented bodies with appendages, which makes them more closely related. 3. a

SYNTHESIZE

1. Clams and scallops are bivalves, which are filter feeders that siphon large amounts of water through their bodies to obtain food. They act as natural pollution-control systems for bays and estuaries. A loss of bivalves (from overfishing, predation, or toxic chemicals) would upset the aquatic ecosystem and allow pollution levels to rise.
2. Chitin is an example of convergent evolution since these organisms do not share a common chitin-equipped ancestor. Chitin is often used in structures that need to withstand the rigors of stress (chaetae, exoskeletons, zoecium, etc.).
3. Since the population size of a parasitic species may be very small (just a few individuals), possessing both male and female reproductive structures would allow the benefits of sexual reproduction.
4. Answers may vary. However, it is known that the tapeworm is not the ancestral form of platyhelminthes; instead it has lost its digestive tract due to its role as an intestinal parasite. As an intestinal parasite, the tapeworm relies on the digestive system of its host to break down nutrients into their building blocks for absorption.

Chapter 35

LEARNING OUTCOME QUESTIONS

35.1 Echinoderms are members of the Bilateria, a clade that also includes chordates and protostomes (see figure 33.4). A synapomorphy of the Bilateria is the evolution of bilateral symmetry. Because all bilaterians are bilaterally symmetrical, the evolution of pentaradial symmetry must have occurred at the base of the Echinodermata from a bilaterally symmetrical ancestor.

35.2 Chordates have a truly internal skeleton (an endoskeleton), as opposed to the endoskeleton on echinoderms, which is functionally similar to the exoskeleton of arthropods. Whereas an echinoderm uses tube feet attached to an internal water-vascular system for locomotion, a chordate has muscular attachments to its endoskeleton. Finally, chordates have a suite of four characteristics that are unique to the phylum: a nerve chord, a notochord, pharyngeal slits, and a postanal tail.

35.3 Although mature and immature lancelets are similar in form, the tadpole-like tunicate larvae are markedly different from the sessile, vaselike adult form. Both tunicates and lancelets are chordates, but they differ from vertebrates in that they do not have vertebrae or internal bony skeletons.

35.4 The functions of an exoskeleton include protection and locomotion—arthropod exoskeletons, for example, provide a fulcrum to which the animals' muscles attach. In order to resist the pull of increasingly large muscles, the exoskeleton must dramatically increase in thickness as the animal grows larger. There is thus a limit on the size of an organism with an exoskeleton—if it gets too large it will be unable to move due to the weight and heft of its exoskeleton.

35.5 Lobe-finned fish are able to move their fins independently, whereas ray-finned fish must move their fins simultaneously. This ability to "walk" with their fins indicates that lobe-finned fish are most certainly the ancestors of amphibians.

35.6 #1 The challenges of moving onto land were plentiful for the amphibians. First, amphibians needed to be able to support their body weight and locomote on land; this challenge was overcome by the evolution of legs. Second, amphibians needed to be able to exchange oxygen with the atmosphere; this was accomplished by the evolution of more efficient lungs than their lungfish ancestors as well as cutaneous respiration. Third, since movement on land requires more energy than movement in the water, amphibians needed a more efficient oxygen delivery system to supply their larger muscles; this was accomplished by the evolution of double-loop circulation and a partially divided heart. Finally, the first amphibians needed to develop a way of staying hydrated in a nonaquatic environment, and these early amphibians developed leathery skin that helped prevent desiccation.

#2 Extant amphibian orders are monophyletic. However, reptiles evolved from a type of amphibian, now extinct. So, if one includes all amphibians, extinct and extant, then amphibians are paraphyletic. But because those amphibian groups more closely related to reptiles have gone extinct, extant amphibians are monophyletic.

35.7 Amphibians remain tied to the water for their reproduction; their eggs are jelly-like and if laid on the land will quickly desiccate. Reptile eggs, on the other hand, are amniotic eggs—they are watertight and contain a yolk, which nourishes the developing embryo, and a series of four protective and nutritive membranes.

35.8 Two primary traits are shared between birds and reptiles. First, both lay amniotic eggs. Second, they both possess scales (which cover the entire reptile body but solely the legs and feet of birds). Birds also share characteristics (e.g., a four-chambered heart) only with one group of reptiles—the crocodilians.

35.9 The most striking convergence between birds and mammals is endothermy, the ability to regulate body temperature internally. Less striking is flight; found in most birds and only one mammal, the ability to fly is another example of convergent evolution.

35.10 Only the hominids constitute a monophyletic group. Prosimians, monkeys, and apes are all paraphyletic—they include the common ancestor but not all descendants: the clade that prosimians share with the common prosimian ancestor excludes all anthropoids, the clade that monkeys share with the common monkey ancestor excludes hominoids, and the clade that apes share with the common ape ancestor excludes hominids.

INQUIRY AND DATA ANALYSIS QUESTIONS

Page 700 Inquiry question: Fish are paraphyletic because the clade comprising all other vertebrates arose from within fishes and is the sister taxon to Sarcopterygii. Reptiles are paraphyletic because birds arose from within and are sister taxon to—among living reptiles—crocodilians. If one considers extinct amphibians, then Amphibia also is paraphyletic, because some extinct amphibian taxa are more closely related to amniotes (the clade comprising reptiles, birds, and mammals) than they are to other amphibians.

Page 719 Inquiry question: The placenta is very similar to reptile and bird eggs, sharing a number of features: chorion, amnion, and yolk sac, and the umbilical cord is derived from the allantois. The reason is that the placenta is the structure that evolved in placental mammals from the egg of their egg-laying ancestors.

UNDERSTAND

1. a 2. c 3. c 4. a 5. c 6. a 7. d 8. a 9. d

APPLY

1. c 2. c 3. b

SYNTHESIZE

1. Increased insulation would have allowed birds to become endothermic and thus to be active at times that ectothermic species could not be active. High body temperature may also allow flight muscles to function more efficiently.
2. Birds evolved from one type of dinosaurs. Thus, in phylogenetic terms, birds are a type of dinosaur.
3. Like the evolution of modern-day horses, the evolution of hominids was not a straight and steady progression to today's *Homo sapiens*. Hominid evolution started with an initial radiation of numerous species. From this group, the evolutionary trend was toward increasing size, similar to what is seen in the evolution of horses. However, like in horse evolution, there are examples of evolutionary decreases in body size as seen in *Homo floresiensis*. Hominid evolution also reveals the coexistence of related species, as seen with *Homo neaderthalensis* and *Homo sapiens*. Hominid evolution, like horse evolution, was not a straight and steady progression to the animal that exists today.

CHAPTER 36

LEARNING OUTCOME QUESTIONS

36.1 Primary growth contributes to the increase in plant height, as well as branching, with shoot and root apical meristems as the source of the growth. Secondary growth makes substantial contributions to the increase in girth of the plant, allowing for a much larger sporophyte generation, with lateral meristems contributing to secondary growth.

36.2 Vessels transport water and are part of the xylem. The cells are dead and only the walls remain. Cylinders of stacked vessels move water from the roots to the leaves of plants. Sieve-tube members are part of the phloem and transport nutrients. Sieve-tube members are living cells, but they lack a nucleus. They rely on neighboring companion cells to carry out some metabolic functions. Like vessels, sieve-tube members are stacked to form a cylinder.

36.3 If abundant water were available, a mutant plant with decreased numbers of root hairs could survive.

36.4 The axillary buds in the axils of the leaves would grow in the absence of the shoot tip. These shoots would eventually flower and reproduce.

36.5 Unlike eudicot leaves, the mesophyll of monocot leaves is not divided into palisade and spongy layers. Instead, cells surrounding the vascular tissues are specialized for carbon fixation. This anatomical variation allows carbon fixation to occur in a part of the leaf where the oxygen concentration is lower, increasing the efficiency of photosynthesis.

INQUIRY AND DATA ANALYSIS QUESTIONS

Page 738 Inquiry question: Three dermal tissue traits that are adaptive for a terrestrial lifestyle include guard cells, trichomes, and root hairs. Guard cells flank an epidermal opening called a stoma and regulate its opening and closing. Stomata are closed when water is scarce, thus conserving water. Trichomes are hairlike outgrowths of the epidermis of stems, leaves, and reproductive organs. Trichomes help to cool leaf surfaces and reduce evaporation from stomata. Root hairs are epidermal extensions of certain cells in young roots and greatly increase the surface area for absorption.

Page 743 Data analysis: Each time a meristem cell divides, one daughter cell contributes to new tissue and the other continues as a meristem cell. Each cell would have to divide 10 times for every 100 root cap cells that need to be replaced.

UNDERSTAND

1. d 2. d 3. c 4. b 5. b 6. c 7. a 8. b 9. b 10. b 11. d

APPLY

1. b 2. c 3. d 4. c 5. d 6. d 7. c 8. a 9. b 10. a

SYNTHESIZE

1. Roots lack leaves with axillary buds at nodes, although there may be lateral roots that originated from deep within the root. The vascular tissue would have a different pattern in roots and stems. If there is a vascular stele at the core with a pericycle surrounded by a Casparian strip, you are looking at a root.
2. Lenticels increase gas exchange. In wet soil, the opportunity for gas exchange decreases. Lenticels could compensate for decreased gas exchange, which would be adaptive.
3. There are many ways to answer this question. Root modifications could result in increased surface area for absorption under dry conditions, leaf modifications could better balance the loss of water through stomata by optimizing the level of CO_2 relative to O_2 in the leaf. A shoot modification could lead to increased nutrient storage.
4. Whenever water is scarce, the corn plant will wilt. Without the large cross-section vessels, there are insufficient conduits for water in the plant.
5. The plant would grow quickly, use minimal water, have leaf anatomy and physiology with optimized photosynthesis, and have a modified root or shoot that can be used as a food source for humans.

CHAPTER 37

LEARNING OUTCOME QUESTIONS

37.1 Physical pressures include gravity and transpiration, as well as turgor pressure as an expanding cell presses against its cell wall. Increased turgor pressure and other physical pressures are associated with increases in water potential. Solute concentration determines whether water enters or leaves a cell via osmosis. The smallest amount of pressure on the side of the cell membrane with the greater solute concentration that is necessary to stop osmosis is the solute potential. Water potential is the sum of the pressure from physical forces and from the solute potential.

37.2 The apoplastic route. The Casparian strip blocks water movement through the cell walls of the endodermal cells.

37.3 The driving force for transpiration is the gradient between 100% humidity inside the leaf and the external humidity. When the external humidity is low, the rate of transpiration is high, limited primarily by the amount of water available for uptake through the root system. Guttation occurs when transpiration is low, but ions continue to move into the root because of water potential differences. In turn, more water enters the root and can push the liquid up. This push is in contrast to the pull of transpiration.

37.4 The flowchart would start with blue light triggering proton transport, creating a proton gradient that drives the opening of the K+ channels. This leads to an influx of water, an increase in turgor in the guard cells, and the opening of the stomata. Increasing temperatures could inhibit the opening of the stomata as well as dry conditions.

37.5 Drought tolerance: Deeply embedded stomata decrease water loss through transpiration because water tension is altered in the crypts.

Flooding tolerance: Pneumatophores are modified roots that grow above flood waters, allowing for oxygen exchange, which is limited for roots under water.

Salt tolerance: Production of high levels of organic molecules in the roots results in uptake of water, even in saline soil, by changing the water potential inside the root.

37.6 Phloem transport can be bidirectional, whereas xylem transport is always from the base to the tip of the plant. Xylem transport depends on transpiration to move the water and dissolved minerals upwards. Phloem transport relies on active transport for loading and unloading at the source and sink, respectively.

INQUIRY AND DATA ANALYSIS QUESTIONS

Page 761 Data analysis: Before equilibrium, the solute potential of the solution is −0.5 MPa, and that of the cell is −0.2 MPa. Since the solution contains more solute than does the cell, water will leave the cell to the point that the cell is plasmolyzed. Initial turgor pressure (Ψ_p) of the cell = 0.05 MPa, and that of the solution is 0 MPa. At equilibrium, both the solution and the cell will have the same Ψ_w. Ψ_{cell} = −0.2 MPa + 0.5 MPa = 0.3 MPa before equilibrium is reached. At equilibrium, $\Psi_{w\,cell}$ = $\Psi_{solution}$ = −0.5 MPa, thus $\Psi_{w\,cell}$ = −0.5 MPa. At equilibrium, the plasmolyzed cell Ψ_p = 0 MPa. Finally, using the relationship $\Psi_{w(cell)} = \Psi_p + \Psi_s$ and $\Psi_{w(cell)}$ = −0.5 MPa, $\Psi_{P(cell)}$ = 0 MPa, then $\Psi_{s(cell)}$ = −0.5 MPa.

Page 763 Inquiry question: The fastest route for water movement through cells has the least hindrance, and thus is the symplast route. The route that exerts the most control over what substances enter and leave the cell is the transmembrane route, which is then the best route for moving nutrients into the plant.

Page 765 Data analysis: #1 The volume that can move through a vessel or tracheid is proportional to r^4. If a mutation increases the radius, r, of a xylem vessel threefold, then the movement of water through the vessel would increase 81-fold ($r^4 = 3^4 = 81$). A plant with larger-diameter vessels can move much more water up its stems.

#2 The volume that can move through a vessel or tracheid is proportional to r^4. $500^4/80^4$ = 1525 times faster. In the context of a plant, temperature, humidity, and whether or not the guard cells are open affects the rate of transpiration. Structural considerations, including cavitation, can also affect the rate of transpiration.

Page 770 Data analysis: Substances in phloem can move at a rate of 50 to 100 cm/hr. Sucrose could move 25 cm in 15 to 30 min. 25 cm/(50 cm/hr) = 0.5 hr = 30 min. 25 cm/(100 cm/hr) = 0.25 hr = 15 min.

UNDERSTAND

1. a 2. c 3. d 4. a 5. c 6. d 7. b 8. a 9. d 10. b

APPLY

1. b 2. d 3. b 4. a 5. b

SYNTHESIZE

1. The solute concentration outside the root cells is greater than inside the cells. Thus the solute potential is more negative outside the cell, and water moves out of the root cells and into the soil. Without access to water, your plant wilts.

2. Look for wilty plants since the rate of water movement across the membrane would decrease in the aquaporin mutants.

3. At the level of membrane transport, plants and animals are very similar. Plant cell walls allow plant cells to take up more water than most animal cells, which rupture without the supportive walls. At the level of epidermal cells there is substantial variation among animals. Amphibians exchange water across the skin. Plants have waterproof epidermal tissue but lose water through stomata. Humans sweat, but dogs do not. Some animals have adaptations for living in aquatic or high-saline environment, as do plants. Vascular plants move vast amounts of water through the plant body via the xylem, using evaporation to fuel the transport. Animals with closed circulatory systems can move water throughout the organism and also excrete excess water through the urinary system, which is responsible for osmoregulation.

4. The rate of transpiration is greater during the day than the night. Since water loss first occurs in the upper part of the tree where more leaves with stomata are located, the decrease in water volume in the xylem would first be observed in the upper portion, followed by the lower portion of the tree.

5. Spring year 1—The new carrot seedling undergoes photosynthesis in developing leaves and the sucrose moves toward the growing tip.

Summer year 1—The developing leaves are sources of carbohydrate, which now moves to the developing root and also the growing young tip.

Fall year 1—The carrot root is now the sink for all carbohydrates produced by the shoot.

Spring year 2—Stored carbohydrate in the root begins to move upward into the shoot.

Summer year 2—The shoot is flowering, and the developing flowers are the primary sink for carbohydrates from the root and also from photosynthesis in the leaves.

Fall year 2—Seeds are developing and they are the primary sink. The root reserves have been utilized, and any remaining carbohydrates from photosynthesis are transported to developing seeds.

CHAPTER 38

LEARNING OUTCOME QUESTIONS

38.1 The water potential around the root decreases, and water no longer moves into the root. Transpiration stops, and the plant has insufficient water to maintain life functions.

38.2 Magnesium is found in the center of the chlorophyll molecule. Without sufficient magnesium, chlorophyll deficiencies will result in decreased photosynthesis and decreased yield per acre.

38.3 *Rhizobium* and legume roots have a symbiotic relationship in which the bacteria enter the root through an infection thread, differentiate, and produce NH_3 for the plant in exchange for carbohydrates. Although nitrogen-fixing bacteria are generally limited to the legumes, more than 90% of vascular plants have a symbiotic relationship with mycorrhizal fungi that extend the surface area of the roots and enhance phosphate transfer. Mycorrhizal symbiosis existed first, and some of the signaling pathways in this symbiosis are also found in *Rhizobium* and legume symbioses.

38.4 #1 As the amount of nitrogen, relative to carbon, decreased in an organism, less protein is made. For example, a plant may be the same size under high- and low-CO_2 atmospheres, but the relative amount of protein is lower under the high-CO_2 conditions. For herbivores, including humans, more plant material would need to be consumed to obtain the same amount of protein under higher versus lower concentrations of CO_2.

#2 Increasing the amount of available nitrogen in the soil is one strategy. This can be accomplished by using chemically produced ammonia for fertilizing, intercropping with nitrogen-fixing legumes, or using organic matter rich in nitrogen for enriching the soil. Efforts to reduce the relative amounts of atmospheric CO_2 would also be helpful.

38.5 Large poplar trees that are not palatable to animals offer a partial solution. Fencing in areas that are undergoing phytoremediation is another possibility, but it would be difficult to isolate all animals, especially birds. Plants that naturally deter herbivores with secondary compounds, including some mustard species (*Brassica* species) could be effective for phytoremediation.

INQUIRY AND DATA ANALYSIS QUESTIONS

Page 780 Inquiry question: CO_2 from the atmosphere and H_2O from the soil are incorporated into carbohydrates during photosynthesis to create much of the mass of the tree. The largest percentage of the mass comes from atmospheric CO_2.

Page 786 Inquiry question: At low and high temperature extremes, enzymes involved in plant respiration are denatured. Plants tend to acclimate to slower long-term changes in temperature, and rates of respiration are able to adjust. Short-term, more dramatic, changes might slow or halt respiration, especially if a temperature change is large enough to cause enzymes to denature.

UNDERSTAND

1. b 2. a 3. c 4. d 5. a 6. b 7. c 8. a

APPLY

1. c 2. a 3. a 4. d

appendix A **A-23**

5. a. For the micronutrient problems you would also use the estimate of 1000 kg of potatoes. The conversion you need to use is that 1 ppm is the same as 1 mg/kg. 4 ppm of copper is the same as 4 mg/kg. Multiply this by 1000 kg of potato, and you have 4000 mg or 4 grams of copper up to 30,000 mg or 33 grams.
  b. 15 ppm of zinc is the same as 15 mg/kg. Multiply this by 1000 kg of potato = 15,000 mg or 15 grams up to 100,000 mg or 100 grams.
  c. For potassium, you calculate 0.5% of 1000, which is 5 kg. You would do the same type of calculation for 6%, which is 60 kg.
  d. 25 ppm of iron is the same as 25 mg/kg. Multiply by 1000 kg of potato = 25,000 mg or 25 grams up to 300 grams.

SYNTHESIZE

1. Bacteria that are important for nitrogen fixation could be destroyed. Other microorganisms that make nutrients available to plants could also be destroyed.
2. Grow the tomatoes hydroponically in a complete nutrient solution minus boron and complete nutrient solution with varying concentrations of boron. Compare the coloration of the leaves, the rate of growth (number of new leaves per unit time), and number and size of fruits produced on plants in each treatment group. It would also be helpful to compare the dry weights of plants from each treatment group at the end of the study.
3. Other inputs include both the macronutrients and micronutrients. Nitrogen, potassium, and phosphorous are common macronutrients in fertilizers, and their levels in the circulating water, along with micronutrients, need to be monitored. CO_2 levels in the air also could be monitored.

CHAPTER 39

LEARNING OUTCOME QUESTIONS

39.1 The lipid-based compounds help to create a water-impermeable layer on the leaves.

39.2 A drug prepared from a whole plant or plant tissue would contain a number of different compounds, in addition to the active ingredient. Chemically synthesized or purified substances contain one or more known substances in known quantities.

39.3 It is unlikely that wasps will kill all the caterpillars. When attacked by a caterpillar, the plant releases a volatile substance that attracts the wasp. But, the wasp has to be within the vicinity of the signal when the plant releases the signal. As a result, some caterpillars will escape detection by wasps.

39.4 The local death of cells creates a barrier between the pathogen and the rest of the plant.

INQUIRY AND DATA ANALYSIS QUESTION

Page 797 Inquiry question: Ricin functions as a ribosome-binding protein that limits translation. A very small quantity of ricin was injected into Markov's thigh from the modified tip of his assassin's umbrella. Without translation of proteins in cells, enzymes and other gene products are no longer produced, causing the victim's metabolism to shut down and leading to death.

UNDERSTAND

1. d 2. b 3. b 4. d 5. c 6. a 7. c 8. a 9. d

APPLY

1. c 2. d 3. d 4. b 5. c 6. a

SYNTHESIZE

1. Humans learn quickly, so plants with toxins that made people ill would not become a dietary mainstay. If there was variation in the levels of toxin in the same species in different areas, humans would likely have continued to harvest plants from the area where plants had reduced toxin levels. As domestication continued, seeds would be collected from the plants with reduced toxin levels and grown the following year.
2. For parasitoid wasps to effectively control caterpillars, sufficiently large populations of wasps would need to be maintained in the area where the infestation occurred. As wasps migrate away from the area, new wasps would need to be introduced. The density of wasps is critical because the wasp has to be in the vicinity of the plant being attacked by the caterpillar when the plant releases its volatile chemical signal. Maintaining sufficient density is a major barrier to success.
3. If a plant is flowering or has fruits developing, the systemin will move toward the fruit or flowers, providing protection for the developing seed. If the plant is a biennial, such as a carrot plant, in its first year of growth, systemin will likely be diverted to the root or other storage organ that will reserve food stores for the plant for the following year.

CHAPTER 40

LEARNING OUTCOME QUESTIONS

40.1 Chlorophyll is essential for photosynthesis. Phytochromes regulate plant growth and development using light as a signal. Phytochrome-mediated responses align the plant with the light environment so photosynthesis is maximized, which is advantageous for the plant.

40.2 The plant would not have normal gravitropic responses. Other environmental signals, including light, would determine the direction of plant growth.

40.3 Folding leaves can startle an herbivore that lands on the plant. The herbivore departs, and the plant is protected.

40.4 During the winter months the leaves would cease photosynthesis except on a few warm days. If the weather warmed briefly, water would move into the leaves and photosynthesis would begin. Unfortunately, the minute the temperature dropped, the leaves would freeze and be permanently damaged. Come the spring, the leaves would not be able to function, and the tree would die. It is to the tree's advantage to shed its leaves and grow new, viable leaves in the spring when the danger of freezing is past.

40.5 Gibberellins and brassinosteroids are likely candidates. Applying brassinosteroids or gibberellins to mutant plants grown under low light intensity would be a good initial experiment. If the plants became taller than the wild-type plants, then you would have a candidate hormone. No response does not indicate that neither hormone pathway is altered. It could be that the mutation is in the receptor for the hormone. One way to determine which hormone receptor could be mutated would be to make the mutant transgenic with a wild-type receptor for either gibberellins or brassinosteroids.

INQUIRY AND DATA ANALYSIS QUESTIONS

Page 806 Inquiry question: A number of red-light–mediated responses are linked to phytochrome action alone, including seed germination, shoot elongation, and plant spacing. Only some of the red-light–mediated responses leading to gene expression depend on the action of protein kinases. When phytochrome converts to the Pfr form, a protein kinase triggers phosphorylation that, in turn, initiates a signaling cascade that triggers the translation of certain light-regulated genes. Not all red-light–mediated responses are disrupted in a plant with a mutation in the protein kinase domain of phytochrome.

Page 807 Data analysis: Red light is absorbed by the leaves at a greater rate than far-red light as the sunlight passes through the canopy. Plant-signaling pathways integrate information about the relative amounts of red and far-red light via phytochrome to regulate plant growth and development. For example, seed germination is inhibited as the relative amount of far-red light increases. Thus, below the shady canopy, seed germination would be less likely to occur. With reduced sunlight, the seedlings would have had less chance of survival because of reduced photosynthesis. Alternatively, the information about reduced sunlight could signal rapid stem elongation to get the plant closer to sufficient sunlight for photosynthesis.

Page 809 Inquiry question: Auxin is involved in the phototropic growth responses of plants, including the bending of stems and leaves toward light. Auxin increases the plasticity of plant cells and signals their elongation. The highest concentration of auxin would most likely occur at the tips of stems where sun exposure is maximal.

Page 817 Inquiry question: A chemical substance, such as the hormone auxin, could trigger the elongation of cells on the shaded side of a stem, causing the stem to bend toward the light.

Page 818 Data analysis: A protractor would be a simple tool for determining the angle of the bend relative to either a vertical or horizontal axis.

UNDERSTAND

1. c 2. a 3. d 4. d 5. b 6. d

APPLY

1. b 2. a 3. b 4. c 5. c 6. b 7. d

SYNTHESIZE

1. You are observing etiolation. Etiolation is an energy conservation strategy to help plants growing in the dark to reach the light before they die. They don't green up until light becomes available, and they divert energy to internode elongation. This strategy is useful for potato shoots. The sprouts will be long so they can get to the surface more quickly. They will remain white until exposed to sunlight which will signal the production of chlorophyll.

2. Tropism refers to the growth of an organism in response to an environmental signal such as light. Taxis refers to the movement of an organism in response to an environmental signal. Since plants cannot move, they will not exhibit taxis, but they do exhibit tropisms.

3. Auxin accumulates on the lower side of a stem in a gravitropism, resulting in elongation of cells on the lower side. If auxin or vesicles containing auxin responded to a gravitational field, it would be possible to have a gravitropic response without amyloplasts.

4. Farmers are causing a thigmotropic response. In response to touch, the internodes of the seedlings will increase in diameter. The larger stems will be more resistant to wind and rain once they are moved to the field. The seedlings will be less likely to snap once they are moved to the more challenging environment.

CHAPTER 41

LEARNING OUTCOME QUESTIONS

41.1 Without flowering in angiosperms, sexual reproduction is impossible and the fitness of a plant drops to zero.

41.2 Set up an experiment in a controlled growth chamber with a day–night light regimen that promotes flowering. Then interrupt the night length with a brief exposure to light. If day length is the determining factor, the brief flash of light will not affect flowering. If night length is the determining factor, the light flash may affect the outcome. For example, if the plant requires a long night, interrupting the night will prevent flowering. If the plant flowers whether or not you interrupt the night with light, it may be a short-night plant. In that case, you would want to set up a second experiment where you lengthen the night length. That should prevent the plant from flowering.

41.3 Flowers can attract pollinators, enhancing the probability of reproduction.

41.4 No, because the gametes are formed by meiosis, which allows for new combinations of alleles to combine. You may want to review Mendel's law of independent assortment.

41.5 In gymnosperms, the megagametophyte is the nutritional source for the embryo, in contrast with the double-fertilization event in angiosperms that produces the endosperm and the embryo.

41.6 Retaining the seed in the ground might provide greater stability for the seedling until its root system is established.

41.7 When conditions are uniform and the plant is well adapted to those constant conditions, genetic variation would not be advantageous. Rather, vegetative reproduction would ensure that the genotypes that are well adapted to the current conditions are maintained.

41.8 A biennial life cycle allows an organism to store up substantial reserves to be used to support reproduction during the second season. The downside to this strategy is that the plant might not survive the winter between the two growing seasons, and its fitness would therefore be reduced to zero.

INQUIRY AND DATA ANALYSIS QUESTIONS

Page 835 Inquiry question: Strict levels of CONSTANS (CO) gene protein are maintained according to the circadian clock. Phytochrome, the pigment that perceives photoperiod, regulates the transcription of CO. By examining posttranslational regulation of CO, it might be possible to determine whether protein levels are modulated by means other than transcription. An additional level of control might be needed to ensure that the activation of floral meristem genes coincides with the activation of genes that code for individual flower organs.

Page 836 Inquiry question: Flower production employs up to four genetically regulated pathways. These pathways ensure that the plant flowers when it has reached adult size, when temperature and light levels are optimal, and when nutrition is sufficient to support flowering. All of these factors combine to ensure the success of flowering and the subsequent survival of the plant species.

Page 837 Inquiry question: Once vernalization occurred and nutrition was optimal, flowering could occur in the absence of flower-repressing genes, even if the plant had not achieved adult size. Thus flowering might occur earlier than normal.

Page 850 Data analysis: Seven cell divisions. One cell divides to produce 2 cells, which divide to produce 4, which divide to produce 8, which proceed through four more divisions to produce a total of 128.

Page 851 Inquiry question: #1 Because it is mutant for monopteros, it does not make a functional MONOPTEROS protein, which is a transcription factor needed for transcribing genes required for root development.
#2 If MONOPTEROS could not be repressed, root development would occur whenever there was auxin available. Roots might develop in unexpected places on the plant.

UNDERSTAND

1. a 2. c 3. a 4. d 5. d 6. c 7. d 8. c 9. a 10. d 11. a 12. c 13. c 14. a 15. b 16. c 17. a

APPLY

1. b 2. c 3. b 4. a 5. b 6. c 7. c 8. b 9. c

SYNTHESIZE

1. Poinsettias are short-day plants. The lights from the cars on the new highway interrupt the long night and prevent flowering.

2. Spinach is a long-day plant, and you want to harvest the vegetative, not the reproductive, parts of the plant. Spinach will flower during the summer as the days get longer, but only leaves will be produced during the spring. If you grow and harvest your spinach in the early spring, you will be able to harvest the leaves before the plant flowers and begins to senesce.

3. Cross-pollination increases the genetic diversity of the next generation. However, self-pollination is better than no pollination. The floral morphology of columbine favors cross-pollination, but self-pollination is a backup option. Should this backup option be utilized, there is still one more opportunity for cross-pollination to override self-pollination because the pollen tube from the other plant can still grow through the style more rapidly than the pollen tube from the same plant.

4. Potatoes grown from true seed take longer to produce new potatoes than potatoes grown from tubers. Seeds are easier to store between growing seasons and require much less storage space than whole potatoes. The seed-grown potatoes will have greater genetic diversity than the asexually propagated potato tubers. If environmental conditions vary from year to year, the seed-grown potatoes may have a better yield because different plants will have an advantage under different environmental conditions. The tuber-grown potatoes will be identical. If conditions are optimal for that genotype, the tuber-grown potatoes will outperform the more-variable seed-grown potatoes. But, if conditions are not optimal for the asexually propagated potatoes, the seed-grown potatoes may have the higher yield.

5. Place *Fucus* zygotes on a screen and shine a light from the bottom. If light is more important, the rhizoid will form toward the light, even though that is the opposite direction gravity would dictate. If gravity is more important, the rhizoid will form away from the direction of the light.

CHAPTER 42

LEARNING OUTCOME QUESTIONS

42.1 Organs may be made of multiple tissue types. For example, the heart contains muscle, connective tissue, and epithelial tissue.

42.2 The epithelium in glandular tissue produces secretions, the epithelium has microvilli on the apical surface that increase surface area for absorption.

42.3 Blood is a form of connective tissue because it contains abundant extracellular material: the plasma.

42.4 The function of heart cell requires their being electrically connected. The gap junctions allow the flow of ions between cells.

42.5 Neurons may be a meter long, but this is a very thin projection that still can allow diffusion of materials along its length. They do require specialized transport along microtubules to move proteins from the cell body to the synapse.

42.6 The organ systems may overlap. Consider the respiratory and circulatory systems. These systems are interdependent.

42.7 Yes.

42.8 The distinction should be between the ability to generate metabolic heat to modulate temperature, and the lack of that ability. Thus ectotherms and endotherms have replaced cold-blooded and warm-blooded.

INQUIRY AND DATA ANALYSIS QUESTIONS

Page 881 Inquiry question: After 2 min of shivering, the thoracic muscles have warmed up enough to engage in full contractions. The muscle contractions that allow the full range of motion of the wings utilize kinetic energy in the movement of the wings, rather than releasing the energy as heat, which occurred in the shivering response.

Page 883 Data analysis: Small mammals, with a proportionately larger surface area, dissipate heat readily, which is helpful in a warm environment, but detrimental in a cold environment. In cold conditions, small mammals must seek shelter or have adaptations, such as insulating hair, to maintain body temperature. Because of a greater volume and proportionately less surface area, large mammals are better adapted to cold environments since it takes much longer for them to lose body heat. Hot environments pose a greater challenge for them for the same reason.

Page 884 Data analysis: Pyrogens target the integrating center in the hypothalamus. They act by increasing the "set point," thus leading to a consistently higher body temperature.

UNDERSTAND
1. a 2. c 3. c 4. d 5. a 6. d 7. b 8. b 9. b

APPLY
1. b 2. c 3. c 4. d 5. a 6. c

SYNTHESIZE
1. Yes, both the gut and the skin include epithelial tissue. A disease that affects epithelial cells could affect both the digestive system and the skin. For example, cystic fibrosis affects the ion transport system in epithelial membranes. It is manifested in the lungs, gut, and sweat glands.
2. The nervous system and endocrine system are involved in regulation and maintenance. They function in sensing the internal and external environments and cause changes in the body to respond to changes to maintain homeostasis. The digestive, circulatory, and respiratory systems are also involved in regulation and maintenance. They are grouped together because they all provide necessary nutrients for the body. The digestive system is responsible for the acquisition of nutrients from food; the respiratory system provides oxygen and removes waste (carbon monoxide). The circulatory system transports nutrients to the cells of the body and removes metabolic wastes.
3. Hunger is a negative feedback stimulus. Hunger stimulates an individual to eat, which in turn causes a feeling of fullness that removes hunger. Hunger is the stimulus; eating is the response that removes the stimulus.
4. The internal environment is constantly changing. As you move through your day, muscle activity raises your body temperature, but when you sit down to eat or rest, your temperature cools. The body must constantly adjust to changes in activity or the environment.

CHAPTER 43

LEARNING OUTCOME QUESTIONS

43.1 The somatic nervous system is under conscious control.

43.2 A positive current inward (influx of Na^+) depolarizes the membrane, and a positive current outward (efflux of K^+) repolarizes the membrane.

43.3 Tobacco contains the compound nicotine, which can bind some acetylcholine receptors. This leads to the classic symptoms of addiction due to underlying habituation involving changes to receptor numbers and responses.

43.4 Reflex arcs allow you to respond to a stimulus that is damaging before the information actually arrives at your brain.

43.5 These two systems work in opposition. This may seem counterintuitive, but it is the basis for much of homeostasis.

INQUIRY AND DATA ANALYSIS QUESTIONS

Page 894 Data analysis: The sum of all three is shown in the picture. The other possible pairwise sums of potentials are shown in the following graphs.

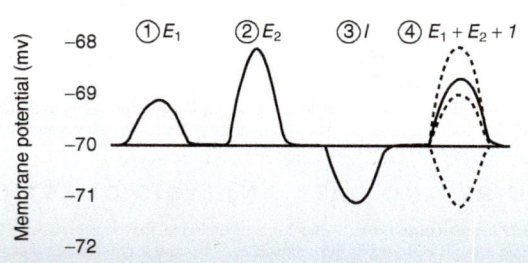

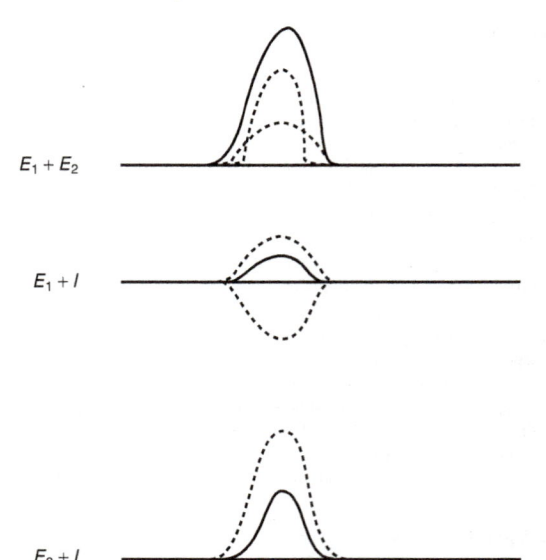

Page 899 Data analysis: The excitatory and inhibitory potentials shown in the figure would sum to a membrane potential close to, or at, the resting potential.

UNDERSTAND
1. a 2. c 3. a 4. a 5. d 6. d 7. a

APPLY
1. d 2. b 3. c 4. a 5. a 6. c 7. d

SYNTHESIZE
1. TEA blocks K^+ channels so that they will not permit the passage of K^+ out of the cell, thereby preventing the cell from returning to the resting potential. Voltage-gated Na^+ channels would still be functional and Na^+ would still flow into the cell, but there would be no repolarization. Na^+ would continue to flow into the cell until an electrochemical equilibrium was reached for Na^+, which is +60 mV. After the membrane potential reached +60 mV, there would be no net movement of Na^+, but the membrane would also not be able to repolarize back to the resting membrane potential. The neuron would no longer be able to function.

   The effects on the postsynaptic cell would be somewhat similar if TEA were applied to the presynaptic cell. The presynaptic cell would depolarize and would continue to release neurotransmitter until it had exhausted its store of synaptic vesicles. As a result, the postsynaptic cell would be bombarded with neurotransmitters and would be stimulated continuously until the stores of presynaptic neurotransmitter were depleted. The postsynaptic cell would, however, recover, being able to repolarize its membrane, and would return to the resting membrane potential.

2. Rising: Na^+ gates open, K^+ closed

   Falling: Na^+ inactivation gate closed, K^+ open

   Undershoot: Na^+ activation gate closed, inactivation gate open, K^+ gate closing.

3. Action potential arrives at the end of the axon.

   Ca^{2+} channels open.

   Ca^{2+} causes synaptic vesicles to fuse with the axon membrane at the synapse.

   Synaptic vesicles release their neurotransmitter.

   Neurotransmitter molecules diffuse across synaptic cleft.

   Postsynaptic receptor proteins bind neurotransmitter.

   Postsynaptic membrane depolarizes.

   If this were an inhibitory synapse, the binding of receptor protein and neurotransmitter would cause the postsynaptic membrane to hyperpolarize.

4. Cells exposed to a stimulus repeatedly may lose their ability to respond. This is known as habituation. Karen's postsynaptic cells may have decreased the number of receptor proteins they produce because the stimulatory signal is so abundant. The result is that it now takes more stimuli to achieve the same result.

CHAPTER 44

LEARNING OUTCOME QUESTIONS

44.1 When the log values of the intensity of the stimulus and the frequency of the resulting action potentials are plotted against each other, a straight line results; this is referred to as a logarithmic relationship.

44.2 Proprioceptors detect the stretching of muscles and subsequently relay information about the relative position and movement of different parts of the organism's body to the central nervous system. This knowledge is critical for the central nervous system; it must be able to respond to these data by signaling the appropriate muscular responses, allowing for balance, coordinated locomotion, and reflexive responses.

44.3 The lateral line system supplements the sense of hearing in fish and amphibian larvae by allowing the organism to detect minute changes in the pressure and vibrations of its environment. This is facilitated by the density of water; without an aquatic environment, the adult, terrestrial amphibian will no longer be able to make use of this system. On land, sound waves are more easily detectable by the sense of hearing than are vibratory or pressure waves by the similar structures of a lateral line.

44.4 Many insects, such as the housefly, have chemoreceptors on their feet with which they can detect the presence of edible materials as they move through their environment. These insects can thus "taste" what they are walking on, and when they encounter an edible substrate they can then descend their proboscis and consume the food.

44.5 Individuals with complete red-green color blindness (those who have no red cones or no green cones, as opposed to those who lack only some red or green cones) would be highly unlikely to be able to learn to distinguish these colors. In order for an individual to perceive the colors in the red-green area of the spectrum, both red and green cones are required; without both cones there is no reference point by which individuals could compare the signals between the retina and the brain. If there are other cues available, such as color saturation or object shape and size, individuals with a less severe form of color blindness may be able to learn to distinguish these colors. In the absence of other references, however, it would be very difficult for individuals with even partial red-green color blindness to distinguish these colors.

44.6 The body temperature of an ectothermic organism is not necessarily the same as the ambient temperature; for example, other reptiles may bask in the sun, wherein on a chilly day the sun may warm the animal's body temperature above the ambient temperature. In this situation, the heat-sensing organs of, for example, a pit viper would still be effective in hunting because it would be able to distinguish the differences between the environment and its ectothermic prey.

INQUIRY AND DATA ANALYSIS QUESTIONS

Page 921 Inquiry question: As the injured fish thrashed around, it would produce vibrations—rapid changes in the pressure of the water. The lateral line system in fish consists of canals filled with sensory cells that send a signal to the brain in response to the changes in water pressure.

Page 928 Inquiry question: Both taste and smell utilize chemoreceptors as sensory receptors, wherein the binding of specific proteins to the receptor induces an action potential, which is sent as a sensory signal to the brain. The chemicals detected by both systems must first be dissolved in extracellular fluid before they can be detected. One major difference between the two systems is that the olfactory system does not route a signal through the thalamus; instead, action potentials are routed directly to the olfactory cortex. Another difference is that olfactory receptors occur in larger numbers—tens of millions, as opposed to tens or hundreds of thousands for taste receptors.

Page 932 Inquiry question: Additional kinds of cones can enhance color vision in two ways. First, they may be sensitive to regions of the color spectrum not covered by other cones, such as those that detect ultraviolet light in some animals. In addition, they may allow enhanced discrimination of color even in regions to which other cones are sensitive. Consider figure 45.19 and the color yellow. Humans have three cones, each of which is sensitive to a varying extent to spectra with a wavelength of approximately 520 nm. The human brain integrates the response of the three cones and interprets the color as yellow. Imagine, however, if we had a fourth cone whose peak sensitivity was in the region of 520 nm. Output from that cone would provide additional discrimination not possible based on the other three cones, and thus would allow an individual to identify shades of color not distinguishable by individuals lacking that cone.

Page 933 Data analysis: Ultraviolet light has shorter wavelengths than the blue light in figure 44.19, and so the peak absorption would fall to the left of the blue cone's peak absorption of 420 nm. According to figure 8.4, the UV portion of the electromagnetic spectrum is between 10 nm and 400 nm, which is the end of the visual spectrum. The peak would fall in that range, probably closer to the 400 nm to be considered "near-ultraviolet."

UNDERSTAND

1. d 2. b 3. d 4. b 5. a 6. c 7. b 8. d 9. a

APPLY

1. c 2. a 3. c 4. b

SYNTHESIZE

1. When blood pH becomes acidic, chemoreceptors in the circulatory and the nervous systems notify the brain and the body responds by increasing the breathing rate. This causes an increase in the release of carbon dioxide through the lungs. Decreased carbon dioxide levels in the blood cause a decrease in carbonic acid, which, in turn, causes the pH to rise.

2. In order to reach the retina and generate action potentials on the optic nerve, light must first pass through the ganglion and bipolar cells to reach the rods and cones that synapse with the bipolar cells. The bipolar cells then synapse with ganglion cells. These in turn send action potentials to the brain. Because the retina comprises three layers, with the rods and cones located farthest from the pupil, light must travel to the deepest level to set off reactions that move up through the more superficial levels and result in optic signals.

3. Without gravity to force the otoliths down toward the hair cells, the otolith organ will not function properly. The otolith membrane would not rest on the hair cells and would not move in response to movement of the body parallel or perpendicular to the pull of gravity. Consequently, the hair cell would not bend and so would not produce receptor potentials. Because the astronauts can see, they would have an impression of motion—they can see themselves move in relation to objects around them—but with their eyes closed, they would not know if they were moving in relation to their surroundings. Because their proprioceptors would still function, they would be able to sense when they moved their arms or legs, but they would not have the sensation of their enter body moving through space.

   The semicircular canals would not function equally well in zero-gravity conditions. Although the fluid in the semicircular canals is still able to move around, some sensation of angular movement would most likely occur, but the full function of the semicircular canals requires the force of gravity to aid in the directional movement of the fluid in the canals.

CHAPTER 45

LEARNING OUTCOME QUESTIONS

45.1 Neurotransmitters are released at a synapse and act on the postsynaptic membrane. Hormones enter the circulatory system and are thus delivered to the entire body.

45.2 The response of a particular tissue depends first on the receptors on its surface, and second on the response pathways active in a cell. Different receptor subtypes can bind the same hormone, and the same receptor can stimulate different response pathways.

45.3 This might lower the amount of GH in circulation. As a treatment, it may have unwanted side effects.

45.4 With two hormones that have antagonistic effects, the body can maintain a fine-tuned level of blood sugar.

45.5 Reducing blood volume should also reduce blood pressure.

INQUIRY AND DATA ANALYSIS QUESTIONS

Page 954 Data analysis: If hypophysectomy is performed on a very early tadpole, the entire signaling axis is disrupted and metamorphosis is blocked. If the surgery is performed later, TSH signaling has already begun, so metamorphosis is not prevented because the pituitary hormones are no longer needed.

Page 958 Data analysis: The use of juvenile hormone as an insecticide is not lethal, but it does prevent metamorphosis and thus prevents reaching sexual maturity, effectively sterilizing the insect. For an insect whose larval form does damage, preventing the next generation will not stop damage from the present generation and, in fact, may increase it.

UNDERSTAND
1. b 2. d 3. c 4. b 5. d 6. b 7. a

APPLY
1. a 2. c 3. c 4. b 5. b 6. b 7. b

SYNTHESIZE
1. If the target cell for a common hormone or paracrine regulator becomes cancerous, it may become hypersensitive to the messenger. This may in turn cause overproduction of cells, which would result in tumor formation. By blocking the production of the hormone specific to that tissue (for example, sex steroids in breast or prostate tissue), it would be possible to slow the growth rate and decrease the size of the tumor.
2. The same hormone can affect two different organs in different ways because the second messengers triggered by the hormone have different targets inside the cell, owing to the cell's different functions. Epinephrine affects the cells of the heart by increasing metabolism so that their contractions are faster and stronger. However, liver cells do not contract and so the second messenger in liver cells triggers the conversion of glycogen into glucose. That is why hormones are so valuable but also economical to the body. One hormone can be produced, one receptor can be made, and one second-messenger system can be used, but there can be two different targets inside the cell.
3. With hormones such as thyroxine, whose effects are slower and have a broader range of activity, a negative feedback system using one hormone adequately controls the system. However, for certain parameters that have a very narrow range and change constantly within that range, a regulatory system that uses up-and-down regulation is desirable. Too much or too little Ca^{2+} or glucose in the blood can have devastating effects on the body, and so those levels must be controlled within a very narrow range. Relying on negative feedback loops would restrict the quick "on" and "off" responses needed to keep the parameters within a very narrow range.

CHAPTER 46
LEARNING OUTCOME QUESTIONS

46.1 Terrestrial invertebrates experience three limitations due to an exoskeleton. First, animals with an exoskeleton can only grow by shedding, or molting, the exoskeleton, leaving them vulnerable to predation. Second, muscles that act on the exoskeleton cannot strengthen and grow because they are confined within a defined space. Finally, the exoskeleton, in concert with the respiratory system of many terrestrial invertebrates, limits the size to which these animals can grow. In order for the exoskeleton of a terrestrial animal to be strong, it has to have a sufficient surface area, and thus it has to increase in thickness as the animal gets larger. The weight of a thicker exoskeleton would impose debilitating constraints on the animal's ability to move.

46.2 Vitamin D is important for the absorption of dietary calcium as well as the deposition of calcium phosphate in bone. Children undergo a great deal of skeletal growth and development; without sufficient calcium deposition, their bones can become soft and pliable, leading to a condition known as rickets, which causes a bending or bowing of the lower limbs. In the elderly, bone remodeling without adequate mineralization of the bony tissue can lead to brittle bones, a condition known as osteoporosis.

46.3 First, unlike the chitinous exoskeleton, a bony endoskeleton is made of living tissue; thus, the endoskeleton can grow along with the organism. Second, because the muscles that act on the bony endoskeleton are not confined within a rigid structure, they are able to strengthen and grow with increased use. Finally, the size limitations imposed by a heavy exoskeleton that covers the entire organism are overcome by the internal bony skeleton, which can support a greater size and weight without itself becoming too cumbersome.

46.4 Slow-twitch fibers are found primarily in muscles adapted for endurance rather than strength and power. Myoglobin provides oxygen to the muscles for the aerobic respiration of glucose, thus providing a higher ATP yield than anaerobic respiration. Increased mitochondria also increases the ATP productivity of the muscle by increasing availability of cellular respiration and thus allows for sustained aerobic activity.

46.5 Locomotion via alternation of legs requires a greater degree of nervous system coordination and balance; the animal needs to constantly monitor its center of gravity in order to maintain stability. In addition, a series of leaps will cover more ground per unit time and energy expenditure than will movement by alternation of legs.

INQUIRY AND DATA ANALYSIS QUESTIONS

Page 971 Inquiry question: The idea is very similar; both quadrupeds and insects such as grasshoppers have flexors and extensors that exert antagonistic control over many of their muscles. The main difference is in structure rather than function—in a grasshopper, the muscles are covered by the skeletal elements, whereas in organisms with an endoskeleton, the muscles overlie the bony skeleton.

Page 976 (figure 46.18) Inquiry question: Increasing the frequency of stimulation to a maximum rate will yield the maximum amplitude of a summated muscle contraction. The strength of a contraction increases because little or no relaxation time occurs between successive twitches.

Page 976 (figure 46.19) Inquiry question: A rough estimate of the composition of the calf muscle could be obtained by measuring the amount of time the calf muscle takes to reach maximum tension and compare that amount with the contraction speed of muscles of known fiber composition. Alternatively, a small sample of muscle could be extracted and examined for histological differences in fiber composition.

UNDERSTAND
1. d 2. b 3. d 4. a 5. c 6. a 7. b 8. a

APPLY
1. c 2. b 3. d 4. b 5. b 6. b

SYNTHESIZE
1. Although a hydrostatic skeleton might have advantages in terms of ease of transport and flexibility of movement, the exoskeleton would probably do a better job at protecting the delicate instruments within. This agrees with our observations of these support systems on Earth. Worms and marine invertebrates use hydrostatic skeletons, although arthropods ("hard bodies") use an exoskeleton. Worms are very flexible, but easily crushed.
2. The first 90 seconds of muscle activity are anaerobic in which the cells utilize quick sources of energy (creatine phosphate, lactic acid fermentation) to generate ATP. After that, the respiratory and circulatory systems will catch up and begin delivering more oxygen to the muscles, which allows them to use aerobic respiration, a much more efficient method of generating ATP from glucose.
3. If acetylcholinesterase is inhibited, acetylcholine will continue to stimulate muscles to contract. As a result, muscle twitching, and eventually paralysis, will occur. In March 1995, canisters of Sarin were released into a subway system in Tokyo. Twelve people were killed and hundreds injured.
4. Natural selection is not goal-oriented. In other words, evolution does not anticipate environmental pressures, and the structures that result from evolution by natural selection are those most well suited to the previous generations' environment. Since vertebrate wing development occurred several times during evolution, it is probable that the animals in question—birds, pterosaurs, and bats—all encountered different evolutionary pressures during wing evolution.

CHAPTER 47
LEARNING OUTCOME QUESTIONS

47.1 The cells and tissues of a one-way digestive system are specialized such that ingestion, digestion, and elimination can happen concurrently, making food processing and energy utilization more efficient. With a gastrovascular cavity, however, all of the cells are exposed to all aspects of digestion.

47.2 Herbivores and omnivores chew their food, using the flat surfaces of their molars to break it into small components and to introduce saliva, which begins the digestive process. Carnivores tear off their food with their sharp teeth (canines and carnasial premolars) and swallow whole chunks of flesh, providing little time or opportunity for digestion to occur in the mouth.

47.3 A chicken sandwich includes carbohydrate (bread), protein (chicken), and fat (mayonnaise). The breakdown of carbohydrates begins with salivary amylase in the mouth. The breakdown of proteins begins in the stomach with pepsinogen, and the emulsification of fats begins in the duodenum with the introduction of bile. Therefore, the chicken begins its breakdown in the stomach.

47.4 Fats are broken down, by emulsification, into fatty acids and monoglycerides, both of which are nonpolar molecules. Nonpolar molecules are able to enter the epithelial cells by simple diffusion.

47.5 The success of any mutation depends on the selective pressures to which that species is subjected. Thus, if two different species are subjected to similar environmental conditions and undergo the same mutation, then, yes, the mutation should be similarly successful. If two species undergo the same mutation but are under different selective pressures, then the mutation may not be successful in both species.

47.6 The sight, taste, and, yes, smell of food are the triggers the digestive system needs to release digestive enzymes and hormones. The saliva and gastric secretions that are required for proper digestion and are triggered by the sense of smell would be affected by anosmia.

47.7 Any ingested compounds that might be dangerous are metabolized first by the liver, thus reducing the risk to the rest of the body.

47.8 Even with normal leptin levels, individuals with reduced sensitivity in the brain to the signaling molecule may still become obese.

INQUIRY AND DATA ANALYSIS QUESTIONS

Page 986 Inquiry question: Recall the old expression: a jack of all trades is master of none. Omnivores are proficient at eating many different types of food, but they're not specialized for any particular type, and thus are less efficient. However, herbivores, with teeth specialized for cutting and grinding, are more effective at eating plant material, and carnivores, with teeth specialized for slicing and tearing, are more efficient at eating muscle tissue. Thus, if a species' diet is solely one type, it is better off having teeth specialized for that purpose.

Page 987 Inquiry question: If the epiglottis does not properly seal off the larynx and trachea, food can accidentally become lodged in the airway, causing choking.

Page 988 Inquiry question: The digestive system secretes a mucus layer that helps to protect the delicate tissues of the alimentary canal from the acidic secretions of the stomach.

Page 989 Data analysis: One meter equals 3.28 feet. The area of a square is length squared. Consider a square with sides 1 meter long. The area is $1 \times 1 = 1$ square meter (1 m^2). But because a meter is 3.28 feet, the area of this square in feet is $3.28 \times 3.28 = 10.76$. Hence, $1 \text{ m}^2 = 10.76 \text{ feet}^2$. Therefore the exact area of the small intestine that is $300 \text{ m}^2 = 3228 \text{ feet}^2$.

Page 994 Data analysis: The amino acid sequences for lysozyme evolved convergently among ruminants and langur monkeys. Thus, if a phylogeny was constructed using solely the lysozyme molecular data, these species—ruminants and langur monkeys—would be adjacent to each other on the phylogenetic tree.

Page 999 Inquiry question: GIP and CCK send inhibitory signals to the hypothalamus when food is ingested. If the hypothalamus sensors did not work properly, leptin levels would increase; increased leptin levels would result in a loss of appetite.

UNDERSTAND
1. b 2. c 3. c 4. b 5. d 6. a 7. c

APPLY
1. d 2. b 3. c 4. c

SYNTHESIZE

1. Birds feed their young with food they acquire from the environment. The adult bird consumes the food but stores it in her crop. When she returns to the nest, she regurgitates the food into the mouths of the fledglings. Mammals on the other hand, feed their young with milk that is produced in the mother's mammary glands. Their young feed by latching onto the mother's nipples and sucking the milk. Mammals have no need for a crop in their digestive system because they feed their young a liquid, which does not require the grinding provided by the crop in birds.

2. Leptin is produced by the adipose cells and serves as a signal for feeding behavior. Since low blood leptin levels signal the brain to initiate feeding, a treatment for obesity would need to raise leptin levels, thereby decreasing appetite.

3. The liver plays many important roles in maintaining homeostasis. Two of those roles are detoxifying drugs and chemicals and producing plasma proteins. A drop in plasma protein levels is indicative of liver disease, which in turn could be caused by abuse of alcohol or other drugs.

4. The selective pressures that guide the adaptation of mutated alleles within a population were the same in these two groups of organisms. Both ruminants and langur monkeys eat tough, fibrous plant materials, which are broken down by intestinal bacteria. The ruminants and langurs then absorb the nutrients from the cellulose by digesting those bacteria; this is accomplished through the use of these adapted lysozymes. Normal lysozymes, found in saliva and other secretions, work in a relatively neutral pH environment. These intestinal lysozymes, however, needed to adapt to an acidic environment, which explains the level of convergence.

5. Whereas mammalian dentition is adapted to processing different food types, birds are able to process different types of food by breaking up food particles in the gizzard. Bird diets are comparably diverse to mammalian diets; some birds are carnivores, others are insectivores or frugivores, still others omnivores.

CHAPTER 48

LEARNING OUTCOME QUESTIONS

48.1 Fick's Law states that the rate of diffusion (R) can be increased by increasing the surface area of a respiratory surface, increasing the concentration difference between respiratory gases, and decreasing the distance the gases must diffuse: $R = \dfrac{DA\Delta p}{d}$. Continually beating cilia increase the concentration difference (Δp).

48.2 Countercurrent flow systems maximize the oxygenation of the blood by increasing Δp; therefore, maintaining a higher oxygen concentration in the water than in the blood throughout the entire diffusion pathway is required. This process is enhanced by allowing water to flow in only one direction, counter to the blood flow. The lamellae, found within a fish's gill filaments, achieves this.

48.3 Birds have a more efficient respiratory system than other terrestrial vertebrates. Birds that live or fly at high altitudes are subjected to lower oxygen partial pressure and thus have evolved a respiratory system that is capable of maximizing the diffusion and retention of oxygen in the lungs. In addition, efficient oxygen exchange is crucial during flight; flying is more energetically taxing than most forms of locomotion, and without efficient oxygen exchange birds would be unable to fly even short distances safely.

48.4 There are both structural and functional differences in bird and mammalian respiration. Both mammals and birds have lungs, but only birds also have air sacs, which they use to move air in and out of the respiratory system, whereas only mammals have a muscular diaphragm used to move air in and out of the lungs. Mammalian lungs are pliable, and gas exchange occurs within small closed-ended sacs called alveoli in the mammalian lung. In contrast, bird lungs are rigid, and gas exchange occurs in the unidirectional parabronchi. In addition, because air flow in mammals is bidirectional, oxygenated and deoxygenated air are mixed, but the unidirectional air flow in birds increases the purity of the oxygen entering the capillaries. Mammalian respiration is less efficient than avian respiration; birds transfer more oxygen with each breath than do mammals. Finally, mammals only have one respiratory cycle, whereas birds have two complete cycles.

48.5 Most oxygen is transported in the blood bound to hemoglobin (forming oxyhemoglobin), whereas only a small percentage is dissolved in the plasma. Carbon dioxide, on the other hand, is predominantly transported as bicarbonate (having first been combined with water to form carbonic acid and then dissociated into bicarbonate and hydrogen ions). Carbon dioxide is also transported dissolved in the plasma and bound to hemoglobin.

INQUIRY AND DATA ANALYSIS QUESTIONS

Page 1004 Data analysis: The rate is a function of area (A) multiplied by differences in pressure (Δp). Thus, increases in either directly increase the rate, so whichever increases to a greater extent would have a greater effect on increasing the rate of diffusion.

Page 1007 Data analysis: Exchange is greatest when the difference in O_2 concentration is greatest. So, for the countercurrent exchange example, the difference is about the same throughout, so exchange would be about the same along the length of the capillaries, although the greatest difference occurs at the top, where the difference is 15% (100% − 85%). In concurrent exchange, the greatest difference is at the bottom, where it is 80% (90% − 10%).

Page 1014 Inquiry question: Fick's Law of Diffusion states that for a dissolved gas, the rate of diffusion is directly proportional to the pressure difference between the two sides of the membrane and to the area over which the diffusion occurs. In emphysema, alveolar walls break down and alveoli increase in size, effectively reducing the surface area for gas exchange. Emphysema thus reduces the diffusion of gases.

Page 1015 Data analysis: The difference in oxygen content between arteries and veins during rest and exercise shows how much oxygen was unloaded to the tissues.

Page 1016 Data analysis: Not really. A healthy individual still has a substantial oxygen reserve in the blood even after intense exercise.

Page 1016 (figure 48.16) Data analysis: Oxyhemoglobin saturation decreases with increasing temperature, and so for any given P_{O_2}, oxygen will unload at higher temperatures, resulting in lower percent saturation and will remain bound at lower temperatures, resulting in higher percent saturation.

Page 1016 Inquiry question: It increases it. At any pH or temperature, the percentage of O_2 saturation falls (e.g., more O_2 is delivered to tissues) as pressure increases.

UNDERSTAND

1. c 2. d 3. d 4. d 5. b 6. a 7. d 8. c

APPLY

1. d 2. c 3. a 4. c 5. a 6. c

SYNTHESIZE

1. Fish gills have not only a large respiratory surface area but also a counter-current flow system, which maintains an oxygen concentration gradient throughout the entire exchange pathway, thus providing the most efficient system for the oxygenation of blood. Amphibian respiratory systems are not very efficient. They practice positive-pressure breathing. Bird lungs are quite effective, in that they have a large surface area and one-way air flow; mammals, on the other hand, have only a large surface area but no mechanism to ensure the maintenance of a strong concentration gradient.

2. During exercise, cellular respiration increases the amount of carbon dioxide released, thus decreasing the pH of the blood. In addition, the increased cellular respiration increases temperature, as heat is released during glucose metabolism. Decreased pH and increased temperature both facilitate an approximately 20% increase in oxygen unloading in the peripheral tissues.

3. Unicellular prokaryotic organisms, protists, and many invertebrates are small enough that gas exchange can occur over the body surface directly from the environment. Only larger organisms, for which most cells are not in direct contact with the environment with which gases must be exchanged, require specialized structures for gas exchange.

CHAPTER 49

LEARNING OUTCOME QUESTIONS

49.1 Following an injury to a vessel, vasoconstriction is followed by the accumulation of platelets at the site of injury and the subsequent formation of a platelet plug. This triggers a positive feedback enzyme cascade, attracting more platelets, clotting factors, and other chemicals, each of which continually attracts additional clotting molecules until the clot is formed. The enzyme cascade also causes fibrinogen to come out of solution as fibrin, forming a fibrin clot that will eventually replace the platelet plug.

49.2 When the insect heart contracts, it forces hemolymph out through the vessels and into the body cavities. When it relaxes, the resulting negative pressure gradient, combined with muscular contractions in the body, draws the blood back to the heart.

49.3 The primary advantage of having two ventricles rather than one is that oxygenated blood is separated from deoxygenated blood. In fish and amphibians, oxygenated and deoxygenated blood mix, leading to less oxygen being delivered to the body's cells.

49.4 The delay following auricular contraction allows the atrioventricular valves to close prior to ventricular contraction. Without that delay, the contraction of the ventricles would force blood back up through the valves into the atria.

49.5 During systemic gas exchange, only about 90% of the fluid that diffuses out of the capillaries returns to the blood vessels; the rest moves into the lymphatic vessels, which then return the fluid to the circulatory system via the left and right subclavian veins.

49.6 Breathing rate is regulated to ensure ample oxygen is available to the body. However, the heart rate must be regulated to ensure efficient delivery of the available oxygen to the body cells and tissues. For example, during exertion, respiratory rates will increase in order to increase oxygenation and allow for increased aerobic cellular respiration. But simply increasing the oxygen availability is not enough—the heart rate must also increase so that the additional oxygen can be quickly delivered to the muscles undergoing cellular respiration.

INQUIRY DATA ANALYSIS QUESTIONS

Page 1022 Inquiry question: Erythropoietin is a hormone that stimulates the production of erythrocytes from the myeloid stem cells. If more erythrocytes are produced, the oxygen-carrying capacity of the blood is increased. This could potentially enhance athletic performance and is why erythropoietin is banned from use during the Olympics and other sporting events.

Page 1023 Inquiry question: It depends what organisms you are studying. In mammals, the nucleus is ejected during maturation of the erythrocyte, and thus nuclear DNA cannot be obtained. In other vertebrates, this is not true, and thus nuclear DNA is present in mature red blood cells.

Page 1032 Data analysis: The difference between the two is 15-fold (60/4). Because resistance increases to the fourth power in inverse proportion, the difference in resistance would be 15 to the fourth power, or 15^4, which equals 50,625.

Page 1036 Data analysis: Output equals the number of beats per minute multiplied by volume per beat (output = beats × volume). At rest, output is 5 L/min, so to increase it to 25 L/min requires a fivefold increase. In theory, this could be accomplished by increasing the number of heartbeats fivefold, to 360/min, while keeping volume the same, but this is not physically possible. Similarly, one could keep the number of beats constant but increase the volume fivefold, to 350 mL/beat, but this is also impossible. In reality, both heartbeat number and volume must increase by amounts that, when multiplied together, equal 5. So, for example, if heartbeat doubled to 144 beats/min and volume increased by 2.5-fold to 175 mL/beat, then the total increase (beat × volume) would be a fivefold increase.

UNDERSTAND

1. a 2. b 3. c 4. a 5. c 6. d 7. c 8. a

APPLY

1. a 2. b 3. c 4. a

SYNTHESIZE

1. Antidiuretic hormone (ADH) is secreted by the posterior pituitary but its target cells are in the kidney. In response to the presence of ADH, the kidneys increase the amount of water reabsorbed. This water eventually returns to the plasma where it causes an increase in volume and subsequent increase in blood pressure. Another hormone, aldosterone, also causes an increase in blood pressure by causing the kidney to retain Na^+, which sets up a concentration gradient that also pulls water back into the blood.

2. Blood includes plasma (composed primarily of water with dissolved proteins) and formed elements (red blood cells, white blood cells, and platelets). Lymph is composed of interstitial fluid and is found only within the lymphatic vessels and organs. Both blood and lymph are found in organisms with closed circulatory systems. Hemolymph is both the circulating fluid and the interstitial fluid found in organisms with open circulatory systems.

3. Many argue that the evolution of endothermy was less an adaptation to maintain a constant internal temperature and more an adaptation to function in environments low in oxygen. If this is the case, then yes, it makes sense that the evolution of the four-chambered heart, an adaptation that increases the availability of oxygen in the body tissues and which would be highly beneficial in an oxygen-poor environment, and the evolution of endothermy were related. These two adaptations can also be looked at as related in that the more efficient heart would be able to provide the oxygen necessary for the increased metabolic activity that accompanies endothermy.

4. The SA node acts as a natural pacemaker. If it is malfunctioning, one would expect a slow or irregular heartbeat or irregular electrical activity between the atria and the ventricles.

CHAPTER 50

LEARNING OUTCOME QUESTIONS

50.1 Water moves toward regions of higher osmolarity.

50.2 Nitrogenous waste is a problem because it is toxic.

50.3 They are both involved in water conservation.

50.4 This may have arisen independently in both the mammalian and avian lineages or have been lost from the reptilian lineage.

50.5 This would increase the osmolarity within the tubule system and thus should decrease the reabsorption of water, leading to water loss.

50.6 Blocking aquaporin channels would prevent reabsorption of water from the collecting duct.

INQUIRY AND DATA ANALYSIS QUESTION

Page 1045 Data analysis: An increase of any solute in the blood will reduce the amount of water that is reabsorbed.

UNDERSTAND
1. d 2. a 3. c 4. c 5. d 6. d 7. d

APPLY
1. d 2. c 3. b 4. b 5. c 6. b

SYNTHESIZE

1. a. Antidiuretic hormone (ADH) is produced in the hypothalamus and is secreted by the posterior pituitary. ADH targets the collecting duct of the nephron and stimulates the reabsorption of water from the urine by increasing the permeability of water in the walls of the duct. The primary stimulus for ADH secretion is an increase in the osmolarity of blood.
   b. Aldosterone is produced and secreted by the adrenal cortex in response to a drop in blood Na^+ concentration. Aldosterone stimulates the distal convoluted tubules to reabsorb Na^+, thus decreasing the excretion of Na^+ in the urine. The reabsorption of Na^+ is followed by the reabsorption of both Cl^- and water, and so aldosterone has the net effect of retaining both salt and water. Aldosterone secretion, however, is not stimulated by a decrease in blood osmolarity, but rather by a decrease in blood volume. A group of cells located at the base of the glomerulus, called the juxtaglomerular apparatus, detects drops in blood volume that then stimulates the renin–angiotensin–aldosterone system.
   c. Atrial natriuretic hormone (ANH) is produced and secreted by the right atrium of the heart, in response to an increase in blood volume. The secretion of ANH results in the reduction of aldosterone secretion. With the secretion of ANH, the distal convoluted tubules reduce the amount of Na^+ that is reabsorbed, and likewise reduce the amount of Cl^- and water that is reabsorbed. The final result is a reduction in blood volume.
   John's normal renal blood flow rate would be 21% of cardiac output, or 7.2 L/min × 0.21 = 1.5 L/min. If John's kidneys are not affected by his circulatory condition, his renal blood flow rate should be about 1.5 L/min.

2. Weight = 90 kg and blood volume = 80 mL/kg
   Blood volume for 90 kg individual
   80 mL/kg × 90 kg = 7200 mL = 7.2 l
   Total blood volume is pumped in 1 min
   Therefore cardiac flow rate 7.2 L/min
   Renal flow is 21% of cardiac
   7.2 L/min × 0.21 = 1/5 L/min

CHAPTER 51

LEARNING OUTCOME QUESTIONS

51.1 No, innate immunity shows some specificity for classes of molecules common to pathogens.

51.2 Hematopoietic stem cells.

51.3 T-cell receptors are rearranged to generate a large number of different receptors with specific binding abilities. Toll-like receptors are not rearranged and recognize specific classes of molecules, not specific molecules.

51.4 Ig receptors are rearranged to generate many different specificities. TLR innate receptors are not rearranged and bind to specific classes of molecules.

51.5 Allergies are a case of the immune system overreacting, whereas autoimmune disorders involve the immune system being compromised.

51.6 Diagnostic kits use monoclonal antibodies because they are developed against a single specific epitope of an antigen. They are also more efficient to produce because they use cells that can be grown in culture and do not require a more complex process of immunizing animals, bleeding them, and then isolating the antibodies from their sera.

51.7 The main difference between polio and influenza is the rate at which the viruses can change. The poliovirus is a RNA virus with a genome that consists of a single RNA. The viral surface proteins do not change rapidly, allowing immunity via a vaccine. Influenza is an RNA virus with a high mutation rate, which means that surface proteins change rapidly. Influenza has a genome that consists of multiple RNAs, which allows recombination of the different viral RNAs during infection with different strains.

INQUIRY AND DATA ANALYSIS QUESTIONS

Page 1061 Inquiry question: The viruses would be liberated into the body where they could infect numerous additional cells.

Page 1065 Inquiry question: The antigenic properties of the two viruses must be similar enough that immunity to cowpox also enables protection against smallpox.

Page 1076 Inquiry question: The common structure and mechanism of formation of B-cell immunoglobulins (Igs) and T-cell receptors (TCRs) suggests a common ancestral form of adaptive immunity gave rise to the two cell lines existing today.

Page 1080 Data analysis: A high level of hCG in a urine sample will block the binding of the antibody to hCG-coated particles, thus preventing any agglutination.

Page 1081 Inquiry question: Influenza frequently alters its surface antigens, making it impossible to produce a vaccine with a long-term effect. Smallpox virus has a considerably more stable structure.

UNDERSTAND
1. b 2. a 3. b 4. c 5. c 6. b 7. c

APPLY
1. d 2. c 3. c 4. b, a, d, c 5. c 6. b 7. a

SYNTHESIZE

1. It would be difficult to advertise this lotion as immune enhancing. The skin serves as a barrier to infection because it is oily and acidic. Applying a lotion that is watery and alkaline will dilute the protective effects of the skin secretions, thereby inhibiting the immune functions. Perhaps it is time to look for another job.

2. The scratch has caused an inflammatory response. Although it is very likely that some pathogens entered her body through the broken skin, the response is actually generated by the injury to her tissue. The redness is a result of the increased dilation of blood vessels caused by the release of histamine. This also increases the temperature of the skin by bringing warm blood closer to the surface. Leakage of fluid from the vessels causes swelling in the area of the injury, which can cause pressure on the pain sensors in the skin. All of these serve to draw defensive cells and molecules to the injury site, thereby helping to defend her against infection.

3. You would need to first determine what receptor the virus is binding to. Then compare the structure of this receptor in susceptible vs. nonsusceptible individuals.

4. These data imply that innate immunity is a very ancient defense mechanism. The presence of these proteins in cnidarians indicates that they arose soon after multicellularity.

CHAPTER 52

LEARNING OUTCOME QUESTIONS

52.1 Genetic sex determination essentially guarantees equal sex ratios; when sex ratios are not equal, the predominant sex is selected against because those individuals have more competition for mates. Temperature-dependent sex determination can result in skewed sex ratios in which one sex or the other is selected against. Genetic sex determination, on the other hand, can provide much greater stability within the population, and consequently the genetic characteristics that provide that stability are selected for.

52.2 Estrous cycles occur in most mammals, and most mammalian species have relatively complex social organizations and mating behaviors. The cycling of sexual receptivity allows for these complex mating systems. Specifically, in social groups where male infanticide is a danger, synchronized estrous among females may be selected for as it would eliminate the ability of the male to quickly impregnate the group females. Physiologically, estrous cycles result in the maturation of the egg accompanying the hormones that promote sexual receptivity.

52.3 In mating systems where males compete for mates, sperm competition, a form of sexual selection, is very common. In these social groups, multiple males may mate with a given female, and thus those individuals who produced the highest number of sperm would have a reproductive advantage—a higher likelihood of siring the offspring.

52.4 The answer varies depending on the circumstances. In a species that is very r-selected, in other words, one that reproduces early in life and often but does not invest much in the form of parental care, multiple offspring per pregnancy would definitely be favored by natural selection. In K-selected species where parental care is very high, on the other hand, single births might be favored because the likelihood of offspring survival is greater if the parental resources are not divided among the offspring.

52.5 The birth control pill works by hormonally controlling the ovulation cycle in women. By releasing progesterone continuously the pill prevents ovulation. Ovulation is a cyclical event and under hormonal control, thus it is relatively easy for the process to be controlled artificially. In addition, the female birth control pill only has to halt the release of a single ovum. An analogous male birth control pill, on the other hand, would have to completely cease sperm production (and men produce millions of sperm each day), and such hormonal upheaval in the male could lead to infertility or other intolerable side effects.

INQUIRY AND DATA ANALYSIS QUESTIONS

Page 1087 Inquiry question: The ultimate goal of any organism is to maximize its relative fitness. Small females are able to reproduce but once they become very large they would be better able to maximize their reproductive success by becoming male, especially in groups where only a few males mate with all of the females. Protandry might evolve in species where the supply of mates is limited and there is relatively little space; a male of such a species in close proximity to another male would have higher reproductive success by becoming female and mating with the available male than by waiting for a female (and then having to compete for her with the other male).

Page 1090 Inquiry question: The evolutionary progression from oviparity to viviparity is a complex process; requiring the development of a placenta or comparable structure. Once a complex structure evolves, it is rare for an evolutionary reversal to occur. Perhaps more importantly, viviparity has several advantages over oviparity, especially in cold environments where eggs are vulnerable to mortality due to cold weather (and predation). In aquatic reptiles such as sea snakes, viviparity allows the female to remain at sea and avoid coming ashore, where both she and her eggs would be exposed to predators.

Page 1096 Inquiry question: Under normal circumstances, the testes produce hormones, testosterone and inhibin, which exert negative feedback inhibition on the hormones produced and secreted by the anterior pituitary (luteinizing hormone and follicle-stimulating hormone). Following castration, testosterone and inhibin are no longer produced, and thus the brain will overproduce LH and FSH. For this reason, hormone therapy is usually prescribed following castration.

Page 1097 Data analysis: In the menstrual cycle FSH exceeds the levels of LH for only a short period at the end of one cycle and continuing to the beginning of the next (approx. days 27–28 and 1–2, respectively).

UNDERSTAND

1. c   2. d   3. d   4. d   5. c   6. c   7. a   8. c   9. a

APPLY

1. a   2. b   3. d   4. a

SYNTHESIZE

1. A mutation that makes *SRY* nonfunctional would mean that the embryo would lack the signal to form male structures during development. Therefore, the embryo would have female genitalia at birth.
2. Amphibians and fish that rely on external fertilization also have access to water. Lizards, birds, and mammals have adaptations that allow them to reproduce away from a watery environment. These adaptations include eggs that have protective shells or internal development, or both.
3. FSH and LH are produced by the anterior pituitary in both males and females. In both cases, they play roles in the production of sex hormones and gametogenesis. However, FSH stimulates spermatogenesis in males and oogenesis in females, whereas LH promotes the production of testosterone in males and estradiol in females.
4. It could indeed work. The hormone hCG is produced by the zygote to prevent menstruation, which would in turn prevent implantation in the uterine lining. Blocking the hormone receptors would prevent implantation and therefore pregnancy.
5. Parthenogenic species reproduce from gametes that remain diploid. Sperm are haploid, whereas eggs do not complete meiosis (becoming haploid) until after fertilization. Therefore, only eggs could develop without DNA from an outside source. In addition, only eggs have the cellular structures needed for development. Therefore only females can undergo parthenogenesis.

CHAPTER 53

LEARNING OUTCOME QUESTIONS

53.1 Ca^{2+} ions act as second messengers and bring about changes in protein activity that result in blocking polyspermy and increasing the rate of protein synthesis within the egg.

53.2 In a mammal, the cells at the four-cell stage are still uncommitted, so separating them will still allow for normal development. In frogs, on the other hand, yolk distribution results in displaced cleavage; thus, at the four-cell stage the cells do not each contain a nucleus, which contains the genetic information required for normal development.

53.3 The cellular behaviors necessary for gastrulation differ across organisms; however, some processes are necessary for any gastrulation to occur. Specifically, cells must rearrange and migrate throughout the developing embryo.

53.4 No—neural crest cell fate is determined by its migratory pathway.

53.5 Marginal zone cells in both the ventral and dorsal regions express bone morphogenetic protein 4 (BMP4). The fate of these cells is determined by the number of receptors on the cell membrane to bind to BMP4; greater BMP4 binding will induce a ventral mesodermal fate. The organizer cells, which previously were thought to activate dorsal development, have been found to actually inhibit ventral development by secreting one of many proteins that block the BMP4 receptors on the dorsal cells.

53.6 Most of the differentiation of the embryo, in which the initial structure formation occurs, happens during the first trimester; the second and third trimesters are primarily times of growth and organ maturation, rather than the actual development and differentiation of structures. Thus, teratogens are most potent during this time of rapid organogenesis.

INQUIRY AND DATA ANALYSIS QUESTIONS

Page 1122 Data analysis: Since both mutants cause a loss of pigment cells, and one is a signaling molecule, the other is likely to be the receptor for the signaling molecule. In the mouse system, the signaling molecule is called Steel and the receptor is called c-Kit (an RTK, see chapter 9 for details).

Page 1129 Inquiry question: High levels of estradiol and progesterone in the absence of pregnancy would still affect the body in the same way. High levels of both hormones would inhibit the release of FSH and LH, thereby preventing ovulation. This is how birth control pills work. The pills contain synthetic forms of either both estradiol and progesterone or just progesterone. The high levels of these hormones in the pill trick the body into thinking that it is pregnant and so the body does not ovulate.

UNDERSTAND

1. d   2. b   3. d   4. c   5. d   6. b   7. a

APPLY

1. c   2. b   3. a   4. d   5. b   6. c

SYNTHESIZE

1. By starting with a series of embryos at various stages, you could try removing cells at each stage. Embryos that failed to compensate for the removal (evidenced by missing structures at maturity) would be those that lost cells after they had become committed; that is, when their fate has been determined.
2. Homeoboxes are sequences of conserved genes that play crucial roles in development of both mammals (Fifi) and *Drosophila* (the fruit fly). In fact, we know that they are more similar than dissimilar; research has demonstrated that both groups use the same transcription factors during organogenesis. The major difference between them is in the genes that are transcribed. Homeoboxes in mammals turn on genes that cause the development of mammalian structures, and those in insects would generate insect structures.
3. After fertilization, the zygote produces hCG, which inhibits menstruation and maintains the corpus luteum. At 10 weeks' gestation, the placenta stops releasing hCG, but it continues to release estradiol and progesterone, which maintain the uterine lining and inhibit the pituitary production of FSH and LH. Without FSH and LH, no ovulation and no menstruation occur.
4. Spemann and Mangold removed cells from the dorsal lip of one amphibian embryo and transplanted them to a different location on a second embryo. The transplanted cells caused cells that would normally form skin and belly to instead form somites and the structures associated with the dorsal area. Because of this and because the secondary dorsal structures contained both

host and transplanted cells, Spemann and Mangold concluded that the transplanted cells acted as organizers for dorsal development.

Chapter 54

LEARNING OUTCOME QUESTIONS

54.1 Just as with morphological characteristics that enhance an individual's fitness, behavioral characteristics can also affect an individual's survivability and reproductive success. Understanding the evolutionary origins of many behaviors allows biologists insights into animal behavior, including that of humans.

54.2 A male songbird injected with testosterone prior to the usual mating season would likely begin singing prior to the usual mating season. However, since female mating behavior is largely controlled by hormones (estrogen) as well, most likely that male will not have increased fitness (and may actually have decreased fitness, if the singing stops before the females are ready to mate, or if the energetic expenditure from singing for two additional weeks is compensated by reduced sperm production).

54.3 The genetic control over pair-bonding in prairie voles has been fairly well established. The fact that males sometimes seek extra-pair copulations indicates that the formation of pair-bonds is under not only genetic control but also behavioral control.

54.4 In species in which males travel farther from the nest (and thus have larger range sizes), there should be significant sex differences in spatial memory. However, in species without sexual differences in range sizes, males and females should not express sex differences in spatial memory. To test the hypothesis you could perform maze tests on males and females of species with sex differences in range size as well as those species without range size differences between the sexes. (NOTE: Such experiments have been performed and do support the hypothesis that there is a significant correlation between range size and spatial memory, so in species with sex differences in range size there are indeed sex differences in spatial memory. See C.M. Jones, V.A. Braithwaite, S.D. Healy. The Evolution of Sex Differences in Spatial Ability. *Behav Neurosci* 2003;117(3): 403–411.

54.5 Although there may be a link between IQ and genes in humans, there is most certainly also an environmental component to IQ. The danger of assigning a genetic correlation to IQ lies in the prospect of selective "breeding" and the emergence of "designer babies."

54.6 One experiment that has been implemented in testing counting ability among different primate and bird species is to present the animal with a number and have him match the target number to one of several arrays containing that number of objects. In another experiment, the animal may be asked to select the appropriate number of individual items within an array of items that equals the target number.

54.7 Butterflies and birds have extremely different anatomy and physiology and thus most likely use very different navigation systems. Birds generally migrate bidirectionally; moving south during the cold months and back north during the warmer months in the northern hemisphere. Usually, then, migrations are multi-generational events and it could be argued that younger birds can learn migratory routes from older generations. Butterflies, on the other hand, fly south to breed and die. Their offspring must then fly north having never been there before.

54.8 In addition to chemical reproductive barriers, many species also employ behavioral and morphological reproductive barriers, such that even if a female moth is attracted by the pheromones of a male of another species, the two may be behaviorally or anatomically incompatible.

54.9 The benefits of territorial behavior must outweigh the potential costs, which may include physical danger due to conflict, energy expenditure, and the loss of foraging or mating time. In a flower that is infrequently encountered, the honeycreeper would lose more energy defending the resource than it could gain by utilizing it. On the other hand, there is usually low competitive pressure for highly abundant resources, thus the bird would expend unnecessary energy defending a resource to which its access is not limited.

54.10 The males should exhibit mate choice because they are the sex with the greater parental investment and energy expenditure; thus, like females of most species, they should be the "choosier" sex.

54.11 Generally, reciprocal behaviors are low-cost, whereas behaviors due to kin selection may be low- or high-cost. Protecting infants from a predator is definitely a high-risk/potentially high-cost behavior; thus it would seem that the behavior is due to kin selection. The only way to truly test this hypothesis, however, is to conduct genetic tests or, in a particularly well-studied population, consult a pedigree.

54.12 #1 Living in a group is associated with both costs and benefits. The primary cost is increased competition for resources, and the primary benefit is protection from predation.

#2 Altruism toward kin is considered selfish because helping individuals closely related to you will directly affect your inclusive fitness.

#3 Most armies more closely resemble insect societies than vertebrate societies. Insect societies consist of multitudes of individuals congregated for the purpose of supporting and defending a select few individuals. One could think of these few protected and revered individuals as the society the army is charged with protecting. These insect societies, like human armies, are composed of individuals each "assigned" to a particular task. Most vertebrate societies, on the other hand, are less altruistic and express increased competition and aggression between group members. In short, vertebrate societies are composed of individuals whose primary concern is usually their own fitness, whereas insect societies are composed of individuals whose primary concern is the colony itself.

INQUIRY AND DATA ANALYSIS QUESTIONS

Page 1137 Inquiry question: Selection for learning ability would cease, and thus change from one generation to the next in maze-learning ability; would only result from random genetic drift.

Page 1137 Data analysis: In populations with a symmetrical distribution with a peak in the middle (a "bell curve" also called a "normal" curve is an example), the mean value usually corresponds to the peak of the curve. In somewhat skewed distributions, such as those in the seventh generation, the mean usually lies a little bit away from the peak, toward the side of the long tail. In this case, the mean of the parental population is about 64, and that of the fast rat population in generation 7 is around 39, so the difference would be approximately 25. This is an example of directional selection (see chapter 21).

Page 1138 Inquiry question: Normal *fosB* alleles produce a protein that in turn affects enzymes that affect the brain. Ultimately, these enzymes trigger maternal behavior. In the absence of the enzymes, normal maternal behavior does not occur.

Page 1139 Data analysis: This experiment tests the effect on rodent pair-bonding behavior of injecting them with the neuropeptide vasopressin. In the control (normal) treatment, vasopressin is not injected (sometimes scientists inject saline (= salt water) solutions, which have no physiological effect but control for the effect of giving an injection. In the control treatment (red bars), wild-type mice exhibit the highest level of pair-bonding behavior, and montane voles exhibit the lowest.

Page 1142 Inquiry question: Peter Marler's experiments addressed this question and determined that both instinct and learning are instrumental in song development in birds.

Page 1150 Inquiry question: Many factors affect the behavior of an animal other than its attempts to maximize energy intake. For example, avoiding predation is also important. Thus, it may be that larger prey require more time to subdue and ingest, thus making the crabs more vulnerable to predators. Hence, the crabs may trade off decreased energy gain for decreased vulnerability for predators. Many other similar explanations are possible.

Page 1150 Data analysis: Assuming that it takes no time to subdue or eat the prey, the raccoon can get 20 J/hr eating frogs (one every 30 min) and 10 J/hr eating crickets (5 crickets/hr—one every 12 minutes—multiplied by 2 J/cricket). The raccoon should search in the stream for frogs.

Follow-up Question: In this scenario, it takes 45 min to find and eat a frog, which means that, on average, a raccoon can eat 1 1/3 frogs/hr. Multiplied by 10 J/frog, results in an average of 13.33 J/hr. By contrast, it takes 15 min to find and devour a cricket, so four can be eaten per hour, leading to a rate of 8 J/hr. Frogs are still favored, but less so.

Page 1152 Inquiry question: This question is the subject of much current research. Ideas include the possibility that males with longer tails (and therefore more spots) are in better condition (because males in poor condition couldn't survive the disadvantage imposed by the tail). The advantage to a female mating with a male in better condition might be either that the male is less likely to be parasitized and thus less likely to pass that parasite on to the female, or the male may have better genes, which in turn would be passed on to the offspring. Another possibility is the visual system for some reason is better able to detect males with long tails, and thus long-tailed males are preferred by females simply because the longer tails are more easily detected and responded to.

Page 1152 Data analysis: To get the predicted number of mates, find 155 on the *x*-axis and then see the corresponding *y*-axis value where the regression line hits 155 on the *x*-axis. You can visualize this by drawing a vertical line from 155 on the *x*-axis, and then drawing a horizontal line from the intersection of the vertical line and the regression line. Where the horizontal line crosses the *y*-axis is the predicted value, which is approximately 4.

Regression lines are calculated by determining the line that best fits the data. The "best fit" line is the one that minimizes the squared vertical distance of each point from the line. To test this, measure the vertical distance from each point to the regression line, square it, and then add them all together. Then draw a different line and do the same thing. The sum of the squared distances in the new line will be greater.

Note, however, that most points don't fall on the line. That means there is variation in the population, and all of that variation is not accounted for by the regression line. As a result, the regression line can only give a predicted value for a male with a particular number of spots, but in reality, most males will deviate from that line by some unpredictable amount. As a result, you wouldn't expect multiple males with the same number of eyespots to have the same number of mates.

Page 1153 Inquiry question: Yes, the larger the male, the larger the prenuptial gift, which provides energy that the female converts into egg production.

Page 1157 Data analysis: Your aunt is the sister of one of your parents. Siblings share 1/2 of their genes, so your aunt is 1/2 genetically related to your mother or father. You also are 1/2 related to your mother or father. Thus, of the 1/2 of your parent's genes that your aunt has, you are likely to have 1/2 of them. 1/2 of 1/2 = 1/4.

Page 1158 Data analysis: Males get all of their genes from their mother. Because the mother has two copies of each gene, on average, two brothers would be related by 1/2. Because males are haploid, however, their offspring get all of their genes. As a result, cousins that are daughters of brothers would also be related by 1/2 in the genes they get from their father. Because their mothers are unrelated, they would share no genes from their mother, and thus would be related by 1/4. Alternatively, if the cousins were daughters of sisters, the two sisters would share 3/4 of their genes, so the two cousins would share 3/8 of their genes. Because the males were unrelated, the cousins would be 3/16 related.

UNDERSTAND

1. b 2. a 3. a 4. c 5. a 6. c 7. d 8. a 9. b 10. a 11. c 12. d

APPLY

1. Presumably, the model is basic, taking into account only size and energetic value of mussels. However, it may be that larger mussels are in places where shore crabs would be exposed to higher levels of predation or greater physiological stress. Similarly, it could be that the model underestimated time costs or energy returns as a function of mussel size. In the case of large mussels being in a place where shore crabs are exposed to costs not considered by the model, one could test the hypothesis in several ways. First, how are the sizes of mussels distributed in space? If they are completely interspersed that would tend to reject the hypothesis. Alternatively, if the mussels were differentially distributed such that the hypothesis was reasonable, mussels could be experimentally relocated (change their distribution in space) and the diets would be expected to shift to match more closely the situation predicted by the model.

2. The four new pairs may have been living in surrounding habitat that was of lower quality, or they may have been individuals that could not compete for a limited number of suitable territories for breeding. Often, the best territories are won by the most aggressive or largest or otherwise best competitors, meaning that the new territory holders would likely have been less fierce competitors. If new residents were weaker competitors (due to aggression or body size), then the birds not removed would have been able to expand their territories to acquire even more critical resources.

3. The key here is that if the tail feathers are a handicap, then by reducing the handicap in these males should enhance their survival compared with males with naturally shorter tail feathers. The logic is simple. If the male with long tail feathers is superior such that it can survive the negative effect of the long tail feathers, then that superior phenotype should be "exposed" with the removal or reduction of the tail feathers. Various aspects of performance could be measured since it is thought that the tail feathers hinder flying. Can males with shorter tails fly faster? Can males with shorter tails turn better? Ultimately, whether males with shorter tails survive better than males with unmanipulated tails can be measured.

4. Both reciprocity and kin selection explain the evolution of altruistic acts by examining the hidden benefits of the behavior. In both cases, altruism actually *benefits* the individual performing the act in terms of its fitness effects. If it didn't, it would be very hard to explain how such behavior could be maintained because actions that reduce the fitness of an individual should be selected against. Definition of the behavior reflects the apparent paradox of the behavior because it focuses on the cost and not the benefit that also accrues to the actor.

SYNTHESIZE

1. The best experiment for determining whether predatory avoidance of certain coloration patterns would involve rearing a predator without an opportunity to learn avoidance and subsequently presenting the predator with prey with different patterns. If the predator avoids the black and yellow coloration more frequently than expected, the avoidance is most likely innate. If the predator does not express any preference but after injury from a prey with the specific coloration does begin to express a preference, then the avoidance is most likely learned. In this case, the learning is operant conditioning; the predator has learned to associate the coloration with pain and thus subsequently avoids prey with that coloration. To measure the adaptive significance of black and yellow coloration, both poisonous (or stinging) and harmless prey species with the coloration and without the coloration pattern could be presented to predators; if predators avoid both the harmful and the harmless prey, the coloration is evolutionary significant.

2. In many cases the organisms in question are unavailable for or unrealistic to study in a laboratory setting. Model organisms allow behavioral geneticists to overcome this obstacle by determining general patterns and then applying these patterns and findings to other, similar organisms. The primary disadvantage of the model system is, of course, the vast differences that are usually found between groups of taxa; however, when applying general principles, in particular those of genetic behavioral regulation, the benefits of using a model outweigh the costs. Phylogenetic analysis is the best way to determine the scale of applicability when using model organisms.

3. Extra-pair copulations and mating with males that are outside a female's territory are, by and large, more beneficial than costly to the female. By mating with males outside her territory, she reduces the likelihood that a male challenging the owner of her territory would target her offspring; males of many species are infanticidal but would not likely attack infants that could be their own. Historical data have actually shown that in many cases, females are more attracted to infanticidal males if those males win territory prior to their infanticidal behavior.

CHAPTER 55

LEARNING OUTCOME QUESTIONS

55.1 It depends on the type of species in question. Conformers are able to adapt to their environment by adjusting their body temperature and making other physiological adjustments. Over a longer period of time, individuals within a nonconforming species might not adjust to the changing environment, but we would expect the population as a whole to adapt due to natural selection.

55.2 If the populations in question comprised source-sink metapopulations, then the lack of immigration into the sink populations would, most likely, eventually result in the extinction of those populations. The source populations would likely then increase their geographic ranges.

55.3 It depends on the initial sizes of the populations in question; a small population with a high survivorship rate will not necessarily grow faster than a large population with a lower survivorship rate.

55.4 A species with high levels of predation would likely exhibit an earlier age at first reproduction and shorter interbirth intervals in order to maximize its fitness under the selective pressure of the predation. On the other hand, species with few predators have the luxury of waiting until they are more mature before reproducing and can increase the interbirth interval (and thus invest more in each offspring) because their risk of early mortality is decreased.

55.5 Many different factors might affect the carrying capacity of a population. For example, climate changes, even on a relatively small scale, could have large effects on carrying capacity by altering the available water and vegetation, as well as the phenology and distribution of the vegetation. Regardless of the type of change in the environment, however, most populations will move toward carrying capacity; thus, if the carrying capacity is lowered, the population should decrease, and if the carrying capacity is raised, the population should increase.

55.6 A given population can experience both positive and negative density-dependent effects, but not at the same time. Negative density-dependent effects, such as low food availability or high predation pressure, would decrease the population size. On the other hand, positive density-dependent effects, such as is seen with the Allee effect, results in a rapid increase in population size. Since a population cannot both increase and decrease at the same time, the two cannot occur concurrently. However, the selective pressures on a population are on a positive–negative continuum, and the forces shaping population size can not only vary in intensity but can also change direction from negative to positive or positive to negative.

55.7 The two are closely tied together, and both are extremely important if the human population is not to exceed the Earth's carrying capacity. As population growth increases, the human population approaches the planet's carrying capacity; as consumption increases, the carrying capacity is lowered—thus, both trends must be reversed.

INQUIRY AND DATA ANALYSIS QUESTIONS

Page 1166 Inquiry question: Very possibly. How fast a lizard runs is a function of its body temperature. Researchers have shown that lizards in shaded habitats have lower temperatures and thus lower maximal running speeds. In such circumstances, lizards often adopt alternative escape tactics that rely less on rapidly running away from potential predators.

Page 1166 Data analysis: Yes. A large value for the slope (at the extreme, 1.0) indicates that as air temperature increases, body temperature also increases. This indicates that the lizard is conforming to environmental conditions; as the air gets hotter (or colder), so does the lizard. In contrast, a low slope (at the extreme, 0.0) indicates that the lizard's body temperature is not affected by air temperature. This could result if the lizard is extremely effective at regulating its body temperature by moving into and out of the shade.

Page 1171 Inquiry question: Because of their shorter generation times, smaller species tend to reproduce more quickly, and thus would be able to respond more quickly to increased resources in the environment.

Page 1171 Data analysis: In general terms, you could make a reasonable prediction: the bigger the animal, the longer the life span. But it's important to realize that a lot of variation still exists, independent of body size. Notice that the axes of this graph are not arithmetic; each interval covers a larger span than the one before it (e.g., the time in the intervals jumps from one day to one week to one month to one year). If you look carefully at the data, you will see that some species are very similar in body size, but differ greatly in life span. Compare, for example, the mouse and the newt, which are about the same size, but the mouse lives for months and the newt for years.

Page 1172 Data analysis: No. The age classes that produce the most seeds are the two oldest, but so few individuals survive to that age that they make a very minor contribution to the overall reproduction of the population. Instead, you might focus on the Age 2 cohort, as their combination of survivors multiplied by seeds per survivor made the greatest contribution to reproduction. One circumstance in which your answer might be different is if the oldest individuals produced a huge number of seeds per individual. In that case, even though there would only be a few old individuals, they still might make the most important contribution to the population. Such a phenomenon occurs in some fish and other organisms, in which the oldest individuals are also the largest, and the number of eggs produced increases exponentially with body size.

Page 1172 Inquiry question: Based on the survivorship curve of meadow grass, the older the plant, the less likely it is to survive. It would be best to choose a plant that is very young to ensure the longest survival as a house plant. A survivorship curve that is shaped like a type I curve, in which most individuals survive to an old age and then die would also lead you to select a younger plant. A type III survivorship curve, in which only a few individuals manage to survive to an older age, would suggest the selection of a middle-aged plant that had survived the early stages of life since it would also be more likely to survive to old age.

Page 1173 Data analysis: The effect seems to be about equal. Adding two eggs reduces clutch size the following year approximately 0.8; taking away two eggs increases clutch size by almost exactly the same amount! Note that the one outlying point in the figure is what happens when one egg is added. Most likely, this is a random fluctuation, the sort of noise that one encounters in any study of natural populations. Perhaps, for example, just by chance those nests that had an egg added were situated in trees that were especially near food sources, so that the effect of adding an egg was not as great as it might have been otherwise. This sort of natural variation can complicate research on natural populations.

Page 1174 Data analysis: Many factors could affect the amount of energy required, but one obvious one is the total amount of offspring mass produced; two chicks 3/4 the size of one larger chick probably require more food than the singleton. In this example, you can see that although the average nestling size decreases with increasing clutch size, the amount of the increase is relatively slight. Each nestling in the clutch of 13 still weights about 90% of the weight of the singleton. Hence, parents raising 13 such birds must have put in a lot more effort than the parents of the lone chick in the smallest clutch.

Page 1174 Inquiry question: It depends on the situation. If only large individuals are likely to reproduce (as is the case in some territorial species, in which only large males can hold a territory), then a few large offspring would be favored; alternatively, if body size does not affect survival or reproduction, then producing as many offspring as possible would maximize the representation of an individual's genes in subsequent generations. In many cases, intermediate values are favored by natural selection.

Page 1175 Data analysis: The first population would double in population each generation (2 adults producing 4 offspring), whereas the second population would increase by 2.5 each generation. After one generation, the second species would be 25% greater in population size (2.5/2). After two generations, the second species would have a population of 12.5 (don't ask what half a mouse looks like) and the first species, 8, a difference of 1.56. After 10 generations, the second species would have a population of 19,073, the first species, 1024, a difference of 1863%! In exponential growth, small differences in rate, compounded yearly, can add up to major differences in total numbers.

Page 1176 Inquiry question: Because when the population is below carrying capacity, the population increases in size. As it approaches the carrying capacity, growth rate slows down either from increased death rates, or decreased birth rates, or both, becoming zero as the population hits the carrying capacity. Similarly, populations well above the carrying capacity will experience large decreases in growth rate, resulting either from low birth rates or high death rates, that also approach zero as the population hits the carrying capacity.

Page 1177 (figure 55.19) Inquiry question: There are many possible reasons. Perhaps resources become limited, so that females are not able to produce as many offspring. Another possibility is that space is limited so that, at higher populations, individuals spend more time in interactions with other individuals and squander energy that otherwise could be invested in producing and raising more young.

Page 1177 (figure 55.20) Inquiry question: The answer depends on whether food is the factor regulating population size. If it is, then the number of young produced at a given population size would increase and the juvenile mortality rate would decrease. However, if other factors, such as the availability of water or predators, regulated population size, then food supplementation might have no effect.

Page 1179 Inquiry question: If hare population levels were kept high, then we would expect lynx populations to stay high as well because lynx populations respond to food availability. If lynx populations were maintained at a high level, we would expect hare populations to remain low because increased reproduction of hares would lead to increased food for the lynxes.

Page 1180 Inquiry question: If human populations are regulated by density-dependent factors, then as the population approaches the carrying capacity, either birth rates will decrease or death rates will increase, or both. If populations are regulated by density-independent factors, and if environmental conditions change, then either both rates will decline, death rates will increase, or both.

Page 1181 Inquiry question: The answer depends on whether age-specific birth and death rates stay unchanged. If they do, then the Swedish distribution would remain about the same. By contrast, because birth rates are far outstripping death rates, the Kenyan distribution will become increasingly unbalanced as the bulge of young individuals enter their reproductive years and start producing even more offspring.

Page 1183 Inquiry question: Both are important causes, and the relative importance of the two depends on which resource we are discussing. One thing is clear: The world cannot support its current population size if everyone lived at the level of resource consumption of people in the United States.

UNDERSTAND

1. b 2. c 3. a 4. b 5. d 6. b 7. c

APPLY

1. d 2. c 3. b 4. c

SYNTHESIZE

1. The genetic makeup of isolated populations will change over time based on the basic mechanisms of evolutionary change; that is, natural selection, mutation, assortative mating, and drift. These same processes affect the genetic makeup of populations in a metapopulation, but the outcomes are likely to be much more complicated. For example, if immigration between a source and a sink population is very high, then local selection in a sink population may be swamped by the regular flow of individuals carrying alleles of lower fitness from a source population where natural selection may not be acting against those alleles; divergence might be slowed or even stopped under some circumstances. On the other hand, if sinks go through repeated population declines such that they often are made up of a very small number of individuals, then they may lose considerable genetic diversity due to drift. If immigration from source populations is greater than zero but not large, these small populations might begin to diverge substantially from other populations in the metapopulation due to drift. The difference is that in the metapopulation, such populations might actually be able to persist and diverge, rather than just going extinct due to small numbers of individuals and no ability to be rescued by neighboring sources.

2. The probability that an animal lives to the next year should decline with age (note that in figure 55.11, all the curves decrease with age), so the cost of reproduction for an old animal would, all else being equal, be lower than for a young animal. The reason is that the cost of reproduction is measured by

changes in fitness. Imagine a very old animal that has almost no chance in surviving to another reproductive event; it should spend all its effort on a current reproductive effort since its future success is likely to be zero anyway.

3. If offspring size does not affect offspring quality, then it is in the parent's interest to produce absolutely as many small offspring as possible. In doing so, it would be maximizing its fitness by increasing the number of related individuals in the next generation. This would shift the curve to the right side of the *x*-axis with larger clutch sizes.

4. By increasing the mean generation time (increasing the age at which an individual can begin reproducing; age at first reproduction), keeping all else equal, one would expect that the population growth rate would be reduced. That comes simply from the fact of reducing the number of individuals that are producing offspring in the adult age classes; lower population birth rates would lead to a reduced population growth rate. As to which would have a larger influence, that is hard to say. If the change in generation time (increased age at first reproduction) had an overall larger effect on the total number of offspring an individual female had than a reduced fecundity at any age, then population growth rate would probably be more sensitive to the change in generation time. Under different scenarios, the comparison of these two effects could become more complicated, however. Suffice it to say that population growth control can come from more than one source: fecundity and age at first reproduction.

CHAPTER 56

LEARNING OUTCOME QUESTIONS

56.1 The answer depends on the habitat of the community in question. Some habitats are more hospitable to animals, and others to plants. The abundance of plants and animals in most habitats is also closely tied; thus the variation in abundance of one would affect the variation in abundance of the other.

56.2 It depends on whether we are talking about fundamental niche or realized niche. Two species can certainly have identical fundamental niches and coexist indefinitely, because they could develop different realized niches within the fundamental niche. In order for two species with identical realized niches to coexist indefinitely, the resources within the niche must not be limited.

56.3 This is an example of Batesian mimicry, in which a nonpoisonous species evolves coloration similar to that of a poisonous species.

56.4 In an ecosystem with limited resources and multiple prey species, one prey species could outcompete another to extinction in the absence of a predator. In the presence of the predator, however, the prey species that would have otherwise been driven to extinction by competitive exclusion is able to persist in the community. The predators that lower the likelihood of competitive exclusion are known as keystone predators.

56.5 Selective harvesting of individual trees would be preferable from a community point of view. According to the intermediate disturbance hypothesis, moderate degrees of disturbance, as occur in selective harvesting, increase species richness and biodiversity more than severe disturbances, such as from clear-cutting.

INQUIRY AND DATA ANALYSIS QUESTIONS

Page 1189 (figure 56.2) Inquiry question: The different soil types require very different adaptations, and thus different species are adapted to each soil type.

Page 1189 Data analysis: The ecotone, the area that separates the normal soil from the serpentine soil, is approximately 7–8 m, so the normal and serpentine soils are approximately 7–8 m apart.

Page 1189 (figure 56.3) Inquiry question: The sharp transition between the communities occurs because two different habitats are in close contact with each other.

Page 1191 Inquiry question: Yes. Both species reach higher population densities when they are by themselves than when they are in the presence of each other. The most likely explanation—although there are others—is that they compete for resources, limiting the carrying capacity each species experiences in the presence of the others.

Page 1193 (figure 56.7) Inquiry question: Presumably the resource distribution is the same on the two islands, thus favoring the same intermediate beak size when there is only one species on an island

Page 1193 (figure 56.8) Inquiry question: The kangaroo rats competed with all the other rodent species for resources, keeping the size of other rodent populations smaller. In the absence of competition when the kangaroo rats were removed, more resources were available, which allowed the other rodent populations to increase in size.

Page 1194 Inquiry question: This could be accomplished in a variety of ways. One option would be to provide refuges to give some *Paramecium* a way of escaping the predators. Another option would be to include predators of the *Didinium*, which would limit their populations (see chapter 57).

Page 1203 Data analysis: When rodents are removed, more seeds are available to the ants, so the number of ant colonies initially increases. However, through time, the greater number of large seeds leads to an increase in plants that produce large seeds. The increase in these plants leads to a decrease in the number of plants that produce small seeds, which are more useful to ants. As the number of small seeds declines, the number of ant colonies declines.

Page 1203 Inquiry question: By removing the kangaroo rats from the experimental enclosures and measuring the effects on both plants and ants. At first, the number of small seeds available to ants increases due to the absence of rodents. However, over time, plants that produce large seeds outcompete plants that produce small seeds, and thus fewer small seeds are produced and available to ants; hence, ant populations decline.

UNDERSTAND

1. b 2. a 3. d 4. a 5. b 6. d 7. c 8. c

APPLY

1. d 2. b 3. d 4. d 5. d

SYNTHESIZE

1. Experiments are useful means to test hypotheses about ecological limitations, but they are generally limited to rapidly reproducing species that occur in relatively small areas. Alternative means of studying species' interactions include detailed studies of the mechanisms by which species might interact; sometimes, for long-lived species, instead of monitoring changes in population size, which may take a very long time, other indices can be measured, such as growth or reproductive rate. Another means of assessing interspecific interactions is to study one species in different areas, in only some of which a second species occurs. Such studies must be interpreted cautiously, however, because there may be many important differences between the areas in addition to the difference in the presence or absence of the second species.

2. Adding differentially preferred prey species might have the same effect as putting in a refuge for prey in the single-species system. One way to think about it is that if a highly preferred species becomes rare due to removal by the predator, then a predator might switch to a less desirable species, even if it doesn't taste as good or is harder to catch, simply because it still provides a better return than chasing after a very rare preferred species. Although the predator has switched, there might be enough time for the preferred species to rebound. All of these dynamics depend on the time it takes a predator to reduce the population size of its prey relative to the time it takes for those prey populations to rebound once the predator pressure is removed.

3. Although the mechanism might be known in this system, hidden interactions might affect interpretations in many ways because ecological systems are complex. For example, what if some other activity of the rodents besides their reduction of large seeds leading to an increase in the number of small seeds was responsible for the positive effect of rodents on ants? One way to test the specific mechanism would be to increase the abundance of small seeds experimentally independent of any manipulation of rodents. Under the current hypothesis, an increase in ant population size would be expected and should be sustained, unlike the initial increase followed by a decrease seen when rodents are removed.

4. By itself, the pattern shown in figure 56.7 suggests character displacement, but alternative hypotheses are possible. For example, what if the distribution of seeds available on the two islands where the species are found alone is different from that seen where they are found in sympatry? If there were no large and small seeds seen on Los Hermanos or Daphne, just medium-sized ones, then it would be hard to conclude that the bill size on San Cristóbal has diverged relative to that on the other islands just due to competition. This is a general criticism of inferring the process of character displacement with just comparing the size distributions in allopatry and sympatry. In this case, however, the Galápagos system has been very well studied. It has been established that the size distribution of seeds available is not measurably different. Furthermore, natural selection–induced changes seen in the bill size of birds on a single island, in response to drought-induced changes in seed size lend further support to the role of competition in establishing and maintaining these patterns.

5. It is possible, because the definition of an ecosystem depends on scale. In some ecosystems, there may be other, smaller ecosystems operating within it. For example, within a rainforest ecosystem, there are small aquatic ecosystems, ecosystems within the soil, ecosystems on an individual tree. Research seems to indicate that most species behave individualistically, but in some instances groups of species do depend on one another and do function holistically. We would expect this kind of dual-community structure especially in areas of overlap between distinct ecosystems, where ecotones exist.

CHAPTER 57

LEARNING OUTCOME QUESTIONS

57.1 Yes, fertilization with natural materials such as manure is less disruptive to the ecosystem than is chemical fertilization. Many chemical fertilizers, for example, contain higher levels of phosphates than does manure, and thus chemical fertilization has disrupted the natural global phosphorus cycle.

57.2 Both matter and energy flow through ecosystems by changing form, but neither can be created or destroyed. Both matter and energy also flow through the trophic levels within an ecosystem. The flow of matter such as carbon atoms is more complex and multileveled than is energy flow, largely because it is truly a cycle. The atoms in the carbon cycle truly cycle through the ecosystem, with no clear beginning or end. The carbon is changed during the process of cycling from a solid to a gaseous state and back again. On the other hand, energy flow is unidirectional. The ultimate source of the energy in an ecosystem is the Sun. The solar energy is captured by the primary producers at the first trophic level and is changed in form from solar to chemical energy. The chemical energy is transferred from one trophic level to another, until only heat, low-quality energy, remains.

57.3 Yes, there are certainly situations in ecosystems in which the top predators in one trophic chain affect the lower trophic levels, and within the same ecosystem the primary producers affect the higher trophic levels within another trophic chain.

57.4 It depends on whether the amount of sunlight captured by the primary producers was affected. Currently, only approximately 1% of the solar energy in Earth's atmosphere is captured by primary producers for photosynthesis. If less sunlight reached Earth's surface, but a correlating increase in energy capture accompanied the decrease in sunlight, then the primary productivity should not be affected.

57.5 The equilibrium model of island biogeography describes the relationship between species richness and not only island size but also distance from the mainland. A small island closer to the mainland would be expected to have more species than would a larger island that is farther from the mainland.

INQUIRY AND DATA ANALYSIS QUESTIONS

Page 1215 Data analysis: The nitrogen concentration in the runoff from the undisturbed area was between 1 and 2 mg/L, whereas the nitrogen concentration in the runoff in the deforested area varied between about 35 and 80 mg/L. This was at least a 17-fold increase (from 2 to 35) and even more if you consider the maximal values after deforestation. Note that the figure doesn't seem to reflect such a large change, but notice the break in the y-axis from 4 to 40 mg/L. This reflects a change in the scale of the y-axis and is sometimes done in scientific figures to save space when two sets of data vary by a large amount. It is understood that the data within that area continued the progression trend shown.

Page 1219 Data analysis: No. The two highest habitats for world NPP are habitats that differ greatly in NPP per unit area (tropical rainforest, high; open ocean, low). Overall, you can see that the habitats are arranged by NPP per unit area, with highest at the top, but when you look at world NPP, there is no general trend from top to bottom.

Page 1220 Inquiry question: At each link in the food chain, only a small fraction of the energy at one level is converted into mass of organisms at the next level. Much energy is dissipated as heat or excreted.

Page 1221 Inquiry question: In the inverted pyramid, the primary producers reproduce quickly and are eaten quickly, so that at any given time, a small population of primary producers exist relative to the heterotroph population.

Page 1222 (figure 57.14) Inquiry question: Because the trout eat the invertebrates, which graze the algae. With fewer grazers, there is more algae.

Page 1222 (figure 57.15) Inquiry question: The food chain has four levels in this experiment and only three in the previous one. The snakes might reduce the number of fish, which would allow an increase in damselflies, which would reduce the number of chironomids and increase the algae. In other words, lower levels of the food chain would be identical for the "snake and fish" and "no fish and no snake" treatments. Both would differ from the enclosures with only fish.

Page 1224 (figure 57.17) Inquiry question: Herbivores consume much of the algal biomass even as primary productivity increases. Increases in primary productivity can lead to increased herbivore populations. The additional herbivores crop the biomass of the algae even while primary productivity increases.

Page 1224 Data analysis: The angle of the line is called the slope. It indicates how the two variables are positively related. A large slope (greater angle) indicates that as the x-variable increases, so does the y-variable, as in the top and bottom diagrams. A low angle (as in the middle diagram) indicates no relationship: as the x-variable increases, the y-variable changes little. Of course, a negative slope, not indicated here, would indicate a negative relationship, as x increases, y decreases.

Page 1224 (figure 57.18) Inquiry question: Secondary carnivores would keep the primary carnivore biomass at a relatively constant, low level, which would allow herbivore biomass to increase with increasing light level, which would keep primary producer biomass at a low and constant level. In other words, the increased productivity at higher light levels would flow up the food chain. This situation is an extension of figure 57.17 to the addition of a fourth food chain level.

Page 1225 Data analysis: As figure 57.19b illustrates, there is a general declining relationship between number of species in a plot and number of invading species (what scientists called a "statistically significant" relationship). Nonetheless, as the figure illustrates, there is quite a lot of scatter in the data. For example, when there are 24 species in a plot, the number of invaders in different plots ranged from 1 to a very large number. So, in answer to the question, there is an overall relationship, but the ability to make an accurate prediction for any given plot is relatively low because there is so much variation. One would be correct to expect that in general, plots with more species would have invaders, but it is very possible that if one randomly chose one plot with many species and one with few, the opposite outcome might occur.

Page 1226 Inquiry question: (a) Perhaps because an intermediate number of predators is enough to keep numbers of superior competitors down. (b) Perhaps because more habitats are available and thus more different ways of surviving in the environment. (c) Hard to say. Possibly more-stable environments permit greater specialization, thus permitting coexistence of more species.

UNDERSTAND

1. d   2. d   3. b   4. d   5. a   6. c   7. b   8. a   9. a   10. c

APPLY

1. a   2. d   3. d

SYNTHESIZE

1. Because the length of food chains appears to be ultimately limited by the amount of energy entering a system, and the characteristic loss of usable energy (about 90%) as energy is transferred to each higher level, it would be reasonable to expect that the ectotherm-dominated food chains would be longer than the endotherm-dominated chains. In fact, there is some indirect evidence for this from real food chains, and it is also predicted by some advanced ecological models. However, it is difficult to determine whether, in reality, this is the case because all of the complex factors that determine food chain length and structure. Moreover, many practical difficulties are associated with measuring actual food chain length in natural systems.

2. It is critical to distinguish, as this chapter points out, between energy and mass transfer in trophic dynamics of ecosystems. The standing biomass of phytoplankton is not necessarily a reliable measure of the energy contained in the trophic level. If phytoplankton are eaten as quickly as they are produced, they may contribute a tremendous amount of energy, which can never be directly measured by a static biomass sample. The standing crop, therefore, is an incomplete measure of the productivity of the trophic level.

3. As figure 57.17 suggests, trophic structure and dynamics are interrelated and are primary determinants of ecosystem characteristics and behavior. For example, if a particularly abundant herbivore is threatened, energy that is abundant at the level of primary productivity in an ecosystem may be relatively unavailable to higher trophic levels (e.g., carnivores). That is, the herbivores are an important link in transducing energy through an ecosystem. Cascading effects, whether they are driven from the bottom up or from the top down are a characteristic of energy transfer in ecosystems, and that translates into the reality that effects on any particular species are unlikely to be limited to that species itself.

4. There are many ways to answer this question, but the obvious place to start is to think about the many ways plant structural diversity potentially affects animals that are not eating the plants directly. For example, plants may provide shelter, refuges, food for prey, substrate for nesting, among other things. Therefore, increasing complexity might increase the ability of lizards to partition the habitat in more ways, allow more species to escape their

predators or seek refuge from harsh physical factors (e.g., cold or hot temperatures), provide a greater substrate for potential prey in terms of food resources, for instance. If we want to know the exact mechanisms for the relationship, we would need to conduct experiments to test specific hypotheses. For example, if we hypothesize that some species require greater structural complexity in order to persist in a particular habitat, we could modify the habitat (reduce plant structure) and test whether species originally present were reduced in numbers or became unable to persist.

CHAPTER 58

LEARNING OUTCOME QUESTIONS

58.1 If the Earth rotated in the opposite direction, the Coriolis effect would be reversed. In other words, winds descending between 30° north or 30° south and the equator would still be moving more slowly than the underlying surface so it would be deflected; however, they would be deflected to the left in the northern hemisphere and to the right in the southern hemisphere. The pattern would be reversed between 30° and 60° because the winds would be moving more rapidly than the underlying surface and would thus be deflected again in the opposite directions from normal—to the left in the northern hemisphere and to the right in the southern hemisphere. All of this would result in Trade Winds that blew from west to east and "Westerlies" that were actually "Easterlies," blowing east to west.

58.2 As with elevation, latitude is a primary determinant of climate and precipitation, which together largely determine the vegetational structure of a particular area, which in turn defines biomes.

58.3 The spring and fall turnovers that occur in freshwater lakes found in temperate climates result in the oxygen-poor water near the bottom of the lake getting remixed with the oxygen-rich water near the top of the lake, essentially eliminating, at least temporarily, the thermocline layer. In the tropics, there is less temperature fluctuation; thus the thermocline layer is more permanent and the oxygen depletion (and resulting paucity of animal life) is sustained.

58.4 Regions affected by the ENSO, or El Niño Southern Oscillation events, experience cyclical warming events in the waters around the coastline. The warmed water lowers the primary productivity, which stresses and subsequently decreases the populations of fish, seabirds, and sea mammals.

58.5 CFCs, or chlorofluorocarbons, are an example of point-source pollution. CFCs and other types of point-source pollutants are, in general, easier to combat because their sources are more easily identified and thus the pollutants more easily eliminated.

58.6 Global climate change and ozone depletion may be interconnected. However, although climate change and ozone depletion are both global environmental concerns due to the impact each has on human health, the environment, economics, and politics, there are some different approaches to combating and understanding each dilemma. Ozone depletion results in an increase in the ultraviolet radiation reaching the Earth's surface. Global climate change, on the other hand, results in long-term changes in sea level, ice flow, and storm activity.

INQUIRY AND DATA ANALYSIS QUESTIONS

Page 1233 Data analysis: Seasonal variation is positively related to the distance from the equator because the farther north (or south), the stronger the Sun's rays are in the winter and the weaker they are in the summer. The effect is much stronger in the northern hemisphere for complicated reasons related to the differing amounts of landmass in the two hemispheres and other factors.

Page 1234 Inquiry question: Because of the tilt of the Earth's axis and the spherical shape of the planet, the light (and heat) from the Sun hits the equator and nearby latitudes more directly than it does at the poles.

Page 1238 Inquiry question: Increased precipitation and temperature allows for the sustainability of a larger variety and biomass of vegetation, and primary productivity is a measure of the rate at which plants convert solar energy into chemical energy.

Page 1238 Data analysis: Figure 58.8 indicates that biomes with approximately 125 cm of precipitation/year vary tremendously in mean annual temperature, from hot savannas to cold taigas. Figure 58.9b indicates that productivity is a function of temperature. Hence the variation in figure 58.9a results from the different temperatures of ecosystems with intermediate amounts of precipitation.

UNDERSTAND

1. d   2. b   3. a   4. c   5. c   6. c   7. a   8. c

APPLY

1. d   2. b   3. c   4. b

SYNTHESIZE

1. The Earth is tilted on its axis such that regions away from the equator receive less incident solar radiation per unit surface area (because the angle of incidence is oblique). The northern and southern hemispheres alternate between angling toward versus away from the Sun on the Earth's annual orbit. These two facts mean that the annual mean temperature declines as you move away from the equator, and that variation in the mean temperatures of the northern and southern hemispheres is complementary to each other; when one is hot, the other is cold.

2. Energy absorbed by the Earth is maximized at the equator because of the angle of incidence. Because there are large expanses of ocean at the equator, warmed air picks up moisture and rises. As it rises, equatorial air, now saturated with moisture, cools and releases rain, the air falling back to Earth's surface displaced north and south to approximately 30°. The air, warming as it descends, absorbs moisture from the land and vegetation below, resulting in desiccation in the latitudes around 30°.

3. Even though global climate changes have occurred in the past, conservation biologists are concerned about the current warming trend for two reasons. First, the warming rate is rapid, thus the selective pressures on the most vulnerable organisms may be too strong for the species to adapt. Second, the natural areas that covered most of the globe during past climatic changes are now in much more limited, restricted areas, thus greatly impeding the ability of organisms to migrate to more suitable habitats.

CHAPTER 59

LEARNING OUTCOME QUESTIONS

59.1 Unfortunately, most of the Earth's biodiversity hotspots are also areas of the greatest human population growth; human population growth is accompanied by increased resource utilization and exploitation.

59.2 I would tell the shrimp farmers that if they were to shut down the shrimp farm and remediate the natural mangrove swamp on which their property sits, other, more economically lucrative businesses could be developed, such as timber, charcoal production, and offshore fishing.

59.3 Absolutely. The hope of conservation biologists is that even if a species is endangered to the brink of extinction due to habitat degradation, the habitat may someday be restored. The endangered species can be bred in captivity (which also allows for the maintenance of genetic diversity within the species) and either reintroduced to a restored habitat or even introduced to another suitable habitat.

59.4 It depends on the reason for the degradation of the habitat in the first place, but yes, in some cases, habitat restoration can approach a pristine state. For example, the Nashau River in New England was heavily polluted, but habitat restoration efforts returned it to a relatively pristine state. However, because habitat degradation affects so many species within the ecosystem, and the depth and complexity of the trophic relationships within the ecosystem are difficult if not impossible to fully understand, restoration is rarely if ever truly pristine.

INQUIRY AND DATA ANALYSIS QUESTIONS

Page 1263 Inquiry question: Many factors affect human population trends, including resource availability, governmental support for settlement in new areas or for protecting natural areas, and the extent to which governments attempt to manage population growth.

Page 1263 Data analysis: Some hotspots that don't have particularly high population densities today, but are growing quickly are the Chocó, tropical Andes, Madagascar, and Indian Ocean Islands. The Brazilian Cerrado has a very high population growth rate, but because densities there are currently very low, this area is of less immediate concern.

Page 1265 Data analysis: The mangroves provide many economic services. For example, without them, fisheries become less productive and storm damage increases. However, because the people who benefit from these services do not own the mangroves, governmental action is needed to ensure that the value of what economists call "common goods" is protected.

Page 1269 Data analysis: Examination of the figure indicates that for every 90% reduction in area, extinction rate approximately doubles. For example, at an area of 100 (10^2) km², extinction rate is approximately 0.08, but at 10 km², it increases to approximately 0.15, and at 1 km², it drops to about 0.23.

Page 1271 Data analysis: Reproductive success would decline through time, perhaps with the same slope, but negative instead of positive, as shown in the figure.

Page 1272 Inquiry question: As discussed in this chapter, populations that are small face many problems that can reinforce one another and eventually cause extinction.

Pate 1272 Data analysis: Yes. Additional species were hunted when the number of the "catch" in the number of whales already being hunted started to decline. Thus, the number of sperm and sei whales increases markedly at the same time that the number of fin whales decreases (and note that the number of fin whales caught soared when blue whales declined). Then, when sei and sperm whale numbers begin to decline, minke whaling accelerated.

Page 1277 Inquiry question: As we discussed in chapter 21, allele frequencies change randomly in a process called genetic drift. The smaller the population size, the greater these random fluctuations will be. Thus, small populations are particularly prone to one allele being lost from a population due to these random changes.

UNDERSTAND

1. a   2. c   3. d   4. d   5. b   6. c

APPLY

1. d   2. d   3. a   4. d

SYNTHESIZE

1. Although it is true that extinction is a natural part of the existence of a species, several pieces of evidence suggest that current rates of extinction are considerably elevated over the natural background level and the disappearance is associated with human activities (which many of the most pronounced extinction events in the history of the Earth were not). It is important to appreciate the length of time over which the estimate of 99% is made. The history of life on Earth extends back billions of years. Certainly, clear patterns of the emergence and extinction of species in the fossil record extend back many hundreds of millions of years. Since the average time of species' existence is short relative to the great expanse of time over which we can estimate the percentage of species that have disappeared, the perception might be that extinction rates have always been high, when in fact the high number is driven by the great expanse of time of measurement. We have very good evidence that modern extinction rates (over human history) are considerably elevated above background levels. Furthermore, the circumstances of the extinctions may be very different because they are also associated with habitat and resource removal; thus potentially limiting the natural processes that replace extinct species.

2. The problem is not unique and not new. It represents a classic conflict that is the basic source of societal laws and regulations, especially in the management of resources. For example, whether or not to place air pollution scrubbers on the smoke stacks of coal-fired power plants is precisely the same issue. In this case, it is not ecosystem conversion, per se, but the fact that the businesses that run the power plants benefit from their operation, but the public "owns" and relies on the atmosphere is a conflict between public and private interests. Some of the ways to navigate the dilemma is for society to create regulations to protect the public interest. The problem is difficult and clearly does not depend solely on economic valuation of the costs and benefits because there can be considerable debate about those estimates. One only has to look at the global climate change problem to suggest how hard it will be to make progress in an expedient manner.

3. This is not a trivial undertaking, which is why, since the first concerns were raised in the late 1980s, it has taken nearly 15 years to collect evidence showing a decline is likely. Although progress has been made on identifying potential causes, much work remains to be done. Many amphibians are secretive, relatively long-lived, and subject to extreme population fluctuations. Given those facts about their biology, documenting population fluctuations (conducting censuses of the number of individuals in populations) for long periods of time is the only way to ultimately establish the likely fate of populations, and that process is time-consuming and costly.

4. Within an ecosystem, every species is dependent on and depended on by any number of other species. Even the smallest organisms, bacteria, are often specific about the species they feed on, live within, parasitize, and so on. So, the extinction of a single species anywhere in the ecosystem will affect not only the organisms it directly feeds on and that directly feed on it, but also those related more distantly. In the simplest terms, if, for example, a species of rodent goes extinct, the insects and vegetation on which it feeds would no longer be under the same predation pressure and thus could grow out of control, outcompeting other species and leading to their demise. In addition, the predators of the rodent would have to find other prey, which would result in competition with those species' predators. And so on, and so on. The effects could be catastrophic to the entire ecosystem. By looking at the trophic chains in which a particular organism is involved, you could predict the effects its extinction would have on other species.

5. Population size is not necessarily a direct cause of extinction, but it certainly is an indirect cause. Smaller populations have a number of problems that themselves can lead directly to extinction, such as loss of diversity (and thus increased susceptibility to pathogens) and greater vulnerability to natural catastrophes.

Glossary

A

ABO blood group A set of four phenotypes produced by different combinations of three alleles at a single locus; blood types are A, B, AB, and O, depending on which antigens are on the red blood cell surface.

abscission In vascular plants, the dropping of leaves, flowers, fruits, or stems at the end of the growing season, as the result of the formation of a layer of specialized cells (the abscission zone) and the action of a hormone (ethylene).

absorption spectrum The relationship of absorbance vs. wavelength for a pigment molecule. This indicates which wavelengths are absorbed maximally by a pigment. For example, chlorophyll *a* absorbs most strongly in the violet-blue and red regions.

acceptor stem The 3′ end of a tRNA molecule; the portion that amino acids become attached to during the tRNA charging reaction.

accessory pigment A secondary light-absorbing pigment used in photosynthesis, including chlorophyll *b* and the carotenoids, that complement the absorption spectrum of chlorophyll *a*.

acoelomate An animal, such as a flatworm, having a body plan that has no body cavity; the space between mesoderm and endoderm is filled with cells and organic materials.

acetyl-CoA Metabolic intermediate consisting of an acetyl group linked to coenzyme A. Produced by the transition reaction in cellular respiration and β oxidation of fatty acids.

achiasmate segregation The accurate segregation of homologues during meiosis I without the formation of chiasmata between homologues.

acid Any substance that dissociates in water to increase the hydrogen ion (H^+) concentration and thus lower the pH.

actin One of the two major proteins that make up vertebrate muscle; the other is myosin.

action potential A transient, all-or-none reversal of the electric potential across a membrane; in neurons, an action potential initiates transmission of a nerve impulse.

action spectrum A measure of the efficiency of different wavelengths of light for photosynthesis. In plants it corresponds to the absorption spectrum of chlorophylls.

activation energy The energy that must be processed by a molecule in order for it to undergo a specific chemical reaction.

active site The region of an enzyme surface to which a specific set of substrates binds, lowering the activation energy required for a particular chemical reaction and so facilitating it.

active transport The pumping of ions or other molecules across a cellular membrane from a region of lower concentration to one of higher concentration (i.e., against a concentration gradient); this transport process requires energy.

adaptation A peculiarity of structure, physiology, or behavior that promotes the likelihood of an organism's survival and reproduction in a particular environment.

adapter protein Any of a class of proteins that acts as a link between a receptor and other proteins to initiate signal transduction.

adaptive radiation The evolution of several divergent forms from a primitive and unspecialized ancestor.

adenosine triphosphate (ATP) A nucleotide consisting of adenine, ribose sugar, and three phosphate groups; ATP is the energy currency of cellular metabolism in all organisms.

adherens junction An anchoring junction that connects the actin filaments of one cell with those of adjacent cells or with the extracellular matrix.

ATP synthase The enzyme responsible for producing ATP in oxidative phosphorylation; it uses the energy from a proton gradient to catalyze the reaction $ADP + P_i \rightarrow ATP$.

adenylyl cyclase An enzyme that produces large amounts of cAMP from ATP; the cAMP acts as a second messenger in a target cell.

adhesion The tendency of water to cling to other polar compounds due to hydrogen bonding.

adipose cells Fat cells, found in loose connective tissue, usually in large groups that form adipose tissue. Each adipose cell can store a droplet of fat (triacylglyceride).

adventitious Referring to a structure arising from an unusual place, such as stems from roots or roots from stems.

aerenchyma In plants, loose parenchymal tissue with large air spaces in it; often found in plants that grow in water.

aerobic Requiring free oxygen; any biological process that can occur in the presence of gaseous oxygen.

aerobic respiration The process that results in the complete oxidation of glucose using oxygen as the final electron acceptor. Oxygen acts as the final electron acceptor for an electron transport chain that produces a proton gradient for the chemiosmotic synthesis of ATP.

aleurone In plants, the outer layer of the endosperm in a seed; on germination, the aleurone produces α-amylase that breaks down the carbohydrates of the endosperm to nourish the embryo.

alga, pl. algae A unicellular or simple multicellular photosynthetic organism lacking multicellular sex organs.

allantois A membrane of the amniotic egg that functions in respiration and excretion in birds and reptiles and plays an important role in the development of the placenta in most mammals.

allele One of two or more alternative states of a gene.

allele frequency A measure of the occurrence of an allele in a population, expressed as proportion of the entire population, for example, an occurrence of 0.84 (84%).

allometric growth A pattern of growth in which different components grow at different rates.

allelopathy The release of a substance from the roots of one plant that block the germination of nearby seeds or inhibits the growth of a neighboring plant.

allopatric speciation The differentiation of geographically isolated populations into distinct species.

allopolyploid A polyploid organism that contains the genomes of two or more different species.

allosteric activator A substance that binds to an enzyme's allosteric site and keeps the enzyme in its active configuration.

allosteric inhibitor A noncompetitive inhibitor that binds to an enzyme's allosteric site and prevents the enzyme from changing to its active configuration.

allosteric site A part of an enzyme, away from its active site, that serves as an on/off switch for the function of the enzyme.

alpha (α) helix A form of secondary structure in proteins where the polypeptide chain is wound into a spiral due to interactions between amino and carboxyl groups in the peptide backbone.

alternation of generations A reproductive cycle in which a haploid (*n*) phase (the gametophyte), gives rise to gametes, which, after fusion to form a zygote, germinate to produce a diploid (2*n*) phase (the sporophyte). Spores produced by meiotic division from the sporophyte give rise to new gametophytes, completing the cycle.

alternative splicing In eukaryotes, the production of different mRNAs from a single primary transcript by including different sets of exons.

altruism Self-sacrifice for the benefit of others; in formal terms, the behavior that increases the fitness of the recipient, while reducing the fitness of the altruistic individual.

alveolus, pl. alveoli One of many small, thin-walled air sacs within the lungs in which the bronchioles terminate.

amino acid The subunit structure from which proteins are produced, consisting of a central carbon atom with a carboxyl group (–COOH), an amino group (–NH_2), a hydrogen, and a side group (*R* group); only the side group differs from one amino acid to another.

aminoacyl-tRNA synthetase Any of a group of enzymes that attach specific amino acids to the

amniocentesis Indirect examination of a fetus by tests on cell cultures grown from fetal cells obtained from a sample of the amniotic fluid or tests on the fluid itself.

amnion The innermost of the extraembryonic membranes; the amnion forms a fluid-filled sac around the embryo in amniotic eggs.

amniote A vertebrate that produces an egg surrounded by four membranes, one of which is the amnion; amniote groups are the reptiles, birds, and mammals.

amniotic egg An egg that is isolated and protected from the environment by a more or less impervious shell during the period of its development and that is completely self-sufficient, requiring only oxygen.

ampulla In echinoderms, a muscular sac at the base of a tube foot that contracts to extend the tube foot.

amyloplast A plant organelle called a plastid that specializes in storing starch.

anabolism The biosynthetic or constructive part of metabolism; those chemical reactions involved in biosynthesis.

anaerobic Any process that can occur without oxygen, such as anaerobic fermentation or H_2S photosynthesis.

anaerobic respiration The use of electron transport to generate a proton gradient for chemiosmotic synthesis of ATP using a final electron acceptor other than oxygen.

analogous Structures that are similar in function but different in evolutionary origin, such as the wing of a bat and the wing of a butterfly.

anaphase In mitosis and meiosis II, the stage initiated by the separation of sister chromatids, during which the daughter chromosomes move to opposite poles of the cell; in meiosis I, marked by separation of replicated homologous chromosomes.

anaphase-promoting complex (APC) A protein complex that triggers anaphase; it initiates a series of reactions that ultimately degrades cohesin, the protein complex that holds the sister chromatids together. The sister chromatids are then released and move toward opposite poles in the cell.

anchoring junction A type of cell junction that mechanically attaches the cytoskeleton of a cell to the cytoskeletons of adjacent cells or to the extracellular matrix.

androecium The floral whorl that comprises the stamens.

aneuploidy The condition in an organism whose cells have lost or gained a chromosome; Down syndrome, which results from an extra copy of human chromosome 21, is an example of aneuploidy in humans.

angiosperms The flowering plants, one of five phyla of seed plants. In angiosperms, the ovules at the time of pollination are completely enclosed by tissues.

animal pole In fish and other aquatic vertebrates with asymmetrical yolk distribution in their eggs, the hemisphere of the blastula comprising cells relatively poor in yolk.

anion A negatively charged ion.

correct tRNA during the tRNA-charging reaction. Each of the 20 amino acids has a corresponding enzyme.

annotation In genomics, the process of identifying and making note of "landmarks" in a DNA sequence to assist with recognition of coding and transcribed regions.

anonymous markers Genetic markers in a genome that do not cause a detectable phenotype, but that can be detected using molecular techniques.

antenna complex A complex of hundreds of pigment molecules in a photosystem that collects photons and feeds the light energy to a reaction center.

anther In angiosperm flowers, the pollen-bearing portion of a stamen.

antheridium, pl. antheridia A sperm-producing organ.

anthropoid Any member of the mammalian group consisting of monkeys, apes, and humans.

antibody A protein called immunoglobulin that is produced by lymphocytes in response to a foreign substance (antigen) and released into the bloodstream.

anticodon The three-nucleotide sequence at the end of a transfer RNA molecule that is complementary to, and base-pairs with, an amino-acid–specifying codon in messenger RNA.

antigen A foreign substance, usually a protein or polysaccharide, that stimulates an immune response.

antiporter A carrier protein in a cell's membrane that transports two molecules in opposite directions across the membrane.

anus The terminal opening of the gut; the solid residues of digestion are eliminated through the anus.

aorta (Gr. *aeirein*, to lift) The major artery of vertebrate systemic blood circulation; in mammals, carries oxygenated blood away from the heart to all regions of the body except the lungs.

apical meristem In vascular plants, the growing point at the tip of the root or stem.

apoplast route In plant roots, the pathway for movement of water and minerals that leads through cell walls and between cells.

apoptosis A process of programmed cell death, in which dying cells shrivel and shrink; used in all animal cell development to produce planned and orderly elimination of cells not destined to be present in the final tissue.

aposematic coloration An ecological strategy of some organisms that "advertise" their poisonous nature by the use of bright colors.

aquaporin A membrane channel that allows water to cross the membrane more easily than by diffusion through the membrane.

aquifers Permeable, saturated, underground layers of rock, sand, and gravel, which serve as reservoirs for groundwater.

archegonium, pl. archegonia The multicellular egg-producing organ in bryophytes and some vascular plants.

archenteron The principal cavity of a vertebrate embryo in the gastrula stage; lined with endoderm, it opens up to the outside and represents the future digestive cavity.

arteriole A smaller artery, leading from the arteries to the capillaries.

artificial selection Change in the genetic structure of populations due to selective breeding by

humans. Many domestic animal breeds and crop varieties have been produced through artificial selection.

ascomycetes A large group comprising part of the "true fungi." They are characterized by separate hyphae, asexually produced conidiospores, and sexually produced ascospores within asci.

ascus, pl. asci A specialized cell, characteristic of the ascomycetes, in which two haploid nuclei fuse to produce a zygote that divides immediately by meiosis; at maturity, an ascus contains ascospores.

asexual reproduction The process by which an individual inherits all of its chromosomes from a single parent, thus being genetically identical to that parent; cell division is by mitosis only.

A site In a ribosome, the aminoacyl site, which binds to the tRNA carrying the next amino acid to be added to a polypeptide chain.

assortative mating A type of nonrandom mating in which phenotypically similar individuals mate more frequently.

aster In animal cell mitosis, a radial array of microtubules extending from the centrioles toward the plasma membrane, possibly serving to brace the centrioles for retraction of the spindle.

atom The smallest unit of an element that contains all the characteristics of that element. Atoms are the building blocks of matter.

atrial peptide Any of a group of small polypeptide hormones that may be useful in treatment of high blood pressure and kidney failure; produced by cells in the atria of the heart.

atrioventricular (AV) node A slender connection of cardiac muscle cells that receives the heartbeat impulses from the sinoatrial node and conducts them by way of the bundle of His.

atrium An antechamber; in the heart, a thin-walled chamber that receives venous blood and passes it on to the thick-walled ventricle; in the ear, the tympanic cavity.

autonomic nervous system The involuntary neurons and ganglia of the peripheral nervous system of vertebrates; regulates the heart, glands, visceral organs, and smooth muscle.

autopolyploid A polyploid organism that contains a duplicated genome of the same species; may result from a meiotic error.

autosome Any eukaryotic chromosome that is not a sex chromosome; autosomes are present in the same number and kind in both males and females of the species.

autotroph An organism able to build all the complex organic molecules that it requires as its own food source, using only simple inorganic compounds.

auxin (Gr. *auxein*, to increase) A plant hormone that controls cell elongation, among other effects.

auxotroph A mutation, or the organism that carries it, that affects a biochemical pathway causing a nutritional requirement.

avirulent pathogen Any type of normally pathogenic organism or virus that utilizes host resources but does not cause extensive damage or death.

axil In plants, the angle between a leaf's petiole and the stem to which it is attached.

axillary bud In plants, a bud found in the axil of a stem and leaf; an axillary bud may develop into a new shoot or may become a flower.

axon A process extending out from a neuron that conducts impulses away from the cell body.

B

b₆–f complex See cytochrome b₆–f complex.

bacteriophage A virus that infects bacterial cells; also called a *phage*.

Barr body A deeply staining structure, seen in the interphase nucleus of a cell of an individual with more than one X chromosome, that is a condensed and inactivated X. Only one X remains active in each cell after early embryogenesis.

basal body A self-reproducing, cylindrical, cytoplasmic organelle composed of nine triplets of microtubules from which the flagella or cilia arise.

base Any substance that dissociates in water to absorb and therefore decrease the hydrogen ion (H^+) concentration and thus raise the pH.

base-pair A complementary pair of nucleotide bases, consisting of a purine and a pyrimidine.

basidium, pl. basidia A specialized reproductive cell of the basidiomycetes, often club-shaped, in which nuclear fusion and meiosis occur.

basophil A leukocyte containing granules that rupture and release chemicals that enhance the inflammatory response. Important in causing allergic responses.

Batesian mimicry A survival strategy in which a palatable or nontoxic organism resembles another kind of organism that is distasteful or toxic. Both species exhibit warning coloration.

B cell A type of lymphocyte that, when confronted with a suitable antigen, is capable of secreting a specific antibody protein.

behavioral ecology The study of how natural selection shapes behavior.

biennial A plant that normally requires two growing seasons to complete its life cycle. Biennials flower in the second year of their lives.

bilateral symmetry A single plane divides an organism into two structural halves that are mirror images of each other.

bile salts A solution of organic salts that is secreted by the vertebrate liver and temporarily stored in the gallbladder; emulsifies fats in the small intestine.

binary fission Asexual reproduction by division of one cell or body into two equal or nearly equal parts.

binomial name The scientific name of a species that consists of two parts, the genus name and the specific species name, for example, *Apis mellifera*.

biochemical pathway A sequence of chemical reactions in which the product of one reaction becomes the substrate of the next reaction. The Krebs cycle is a biochemical pathway.

biodiversity The number of species and their range of behavioral, ecological, physiological, and other adaptations, in an area.

biofilm A complex bacterial community comprising different species; plaque on teeth is a biofilm.

biofuel Fuel produced via biological rather than geological processes. Usually made from plant material, or agricultural, commercial or industrial wastes.

biogeography The study of the geographic distribution of species.

biological community All the populations of different species living together in one place; for example, all populations that inhabit a mountain meadow.

biological species concept (BSC) The concept that defines species as groups of populations that have the potential to interbreed and that are reproductively isolated from other groups.

biomass The total mass of all the living organisms in a given population, area, or other unit being measured.

biome One of the major terrestrial ecosystems, characterized by climatic and soil conditions; the largest ecological unit.

bipolar cell A specialized type of neuron connecting cone cells to ganglion cells in the visual system. Bipolar cells receive a hyperpolarized stimulus from the cone cell and then transmit a depolarization stimulus to the ganglion cell.

biramous Two-branched; describes the appendages of crustaceans.

blade The broad, expanded part of a leaf; also called the lamina.

blastocoel The central cavity of the blastula stage of vertebrate embryos.

blastodisc In the development of birds, a disclike area on the surface of a large, yolky egg that undergoes cleavage and gives rise to the embryo.

blastomere One of the cells of a blastula.

blastopore In vertebrate development, the opening that connects the archenteron cavity of a gastrula stage embryo with the outside.

blastula In vertebrates, an early embryonic stage consisting of a hollow, fluid-filled ball of cells one layer thick; a vertebrate embryo after cleavage and before gastrulation.

Bohr effect The release of oxygen by hemoglobin molecules in response to elevated ambient levels of CO_2.

bottleneck effect A loss of genetic variability that occurs when a population is reduced drastically in size.

Bowman's capsule In the vertebrate kidney, the bulbous unit of the nephron, which surrounds the glomerulus.

β oxidation The oxygen-dependent reactions where 2-carbon units of fatty acids are cleaved and combined with CoA to produce acetyl-CoA, which then enters the Krebs cycle. This occurs cyclically until the entire fatty acid is oxidized.

β sheet A form of secondary structure in proteins where the polypeptide folds back on itself one or more times to form a planar structure stabilized by hydrogen bonding between amino and carboxyl groups in the peptide backbone. Also known as a β-pleated sheet.

book lung In some spiders, a unique respiratory system consisting of leaflike plates within a chamber over which gas exchange occurs.

bronchus, pl. bronchi One of a pair of respiratory tubes branching from the lower end of the trachea (windpipe) into either lung.

bud An asexually produced outgrowth that develops into a new individual. In plants, an embryonic shoot, often protected by young leaves; buds may give rise to branch shoots.

buffer A substance that resists changes in pH. It releases hydrogen ions (H^+) when a base is added and absorbs H^+ when an acid is added.

C

C_3 photosynthesis The main cycle of the dark reactions of photosynthesis, in which CO2 binds to ribulose 1,5-bisphosphate (RuBP) to form two 3-carbon phosphoglycerate (PGA) molecules.

C_4 photosynthesis A process of CO_2 fixation in photosynthesis by which the first product is the 4-carbon oxaloacetate molecule.

cadherin One of a large group of transmembrane proteins that contain a Ca^{2+}-mediated binding between cells; these proteins are responsible for cell-to-cell adhesion between cells of the same type.

callus Undifferentiated tissue; a term used in tissue culture, grafting, and wound healing.

Calvin cycle The dark reactions of C_3 photosynthesis; also called the Calvin–Benson cycle.

calyx The sepals collectively; the outermost flower whorl.

CAM plant Plants that use C_4 carbon fixation at night, then use the stored malate to generate CO_2 during the day to minimize dessication.

Cambrian explosion The huge increase in animal diversity that occurred at the beginning of the Cambrian period.

cAMP response protein (CRP) See catabolite activator protein (CAP)

cancer The unrestrained growth and division of cells; it results from a failure of cell division control.

capillary The smallest of the blood vessels; the very thin walls of capillaries are permeable to many molecules, and exchanges between blood and the tissues occur across them; the vessels that connect arteries with veins.

capsid The outermost protein covering of a virus.

capsule In bacteria, a gelatinous layer surrounding the cell wall.

carapace (Fr. from Sp. *carapacho*, shell) Shieldlike plate covering the cephalothorax of decapod crustaceans; the dorsal part of the shell of a turtle.

carbohydrate An organic compound consisting of a chain or ring of carbon atoms to which hydrogen and oxygen atoms are attached in a ratio of approximately 2:1; having the generalized formula $(CH_2O)_n$; carbohydrates include sugars, starch, glycogen, and cellulose.

carbon fixation The conversion of CO_2 into organic compounds during photosynthesis; the first stage of the dark reactions of photosynthesis, in which carbon dioxide from the air is combined with ribulose 1,5-bisphosphate.

carotenoid Any of a group of accessory pigments found in plants; in addition to absorbing light energy, these pigments act as antioxidants, scavenging potentially damaging free radicals.

carpel A leaflike organ in angiosperms that encloses one or more ovules.

carrier protein A membrane protein that binds to a specific molecule that cannot cross the membrane and allows passage through the membrane.

carrying capacity The maximum population size that a habitat can support.

cartilage A connective tissue in skeletons of vertebrates. Cartilage forms much of the skeleton of embryos, very young vertebrates, and some adult vertebrates, such as sharks and their relatives.

Casparian strip In plants, a band that encircles the cell wall of root endodermal cells. Adjacent cells' strips connect, forming a layer through which water cannot pass; therefore, all water entering roots must pass through cell membranes and cytoplasm.

catabolism In a cell, those metabolic reactions that result in the breakdown of complex molecules into simpler compounds, often with the release of energy.

catabolite activator protein (CAP) A protein that, when bound to cAMP, can bind to DNA and activate transcription. The level of cAMP is inversely related to the level of glucose, and CAP/cAMP in *E. coli* activates the *lac* (lactose) operon. Also called *cAMP response protein* (*CRP*).

catalysis The process by which chemical subunits of larger organic molecules are held and positioned by enzymes that stress their chemical bonds, leading to the disassembly of the larger molecule into its subunits, often with the release of energy.

cation A positively charged ion.

cavitation In plants and animals, the blockage of a vessel by an air bubble that breaks the cohesion of the solution in the vessel; in animals more often called embolism.

CD4+ cell A subtype of helper T cell that is identified by the presence of the CD4 protein on its surface. This cell type is targeted by the HIV virus that causes AIDS.

cecum In vertebrates, a blind pouch at the beginning of the large intestine.

cell cycle The repeating sequence of growth and division through which cells pass each generation.

cell determination The molecular "decision" process by which a cell becomes destined for a particular developmental pathway. This occurs before overt differentiation and can be a stepwise process.

cell-mediated immunity Arm of the adaptive immune system mediated by T cells, which includes cytotoxic cells and cells that assist the rest of the immune system.

cell plate The structure that forms at the equator of the spindle during early telophase in the dividing cells of plants and a few green algae.

cell-surface marker A glycoprotein or glycolipid on the outer surface of a cell's membrane that acts as an identifier; different cell types carry different markers.

cell-surface receptor A cell surface protein that binds a signal molecule and converts the extracellular signal into an intracellular one.

cellular blastoderm In insect embryonic development, the stage during which the nuclei of the syncytial blastoderm become separate cells through membrane formation.

cellular respiration The metabolic harvesting of energy by oxidation, ultimately dependent on molecular oxygen; carried out by the Krebs cycle and oxidative phosphorylation.

cellulose The chief constituent of the cell wall in all green plants, some algae, and a few other organisms; an insoluble complex carbohydrate formed of microfibrils of glucose molecules.

cell wall The rigid, outermost layer of the cells of plants, some protists, and most bacteria; the cell wall surrounds the plasma membrane.

central nervous system (CNS) That portion of the nervous system where most association occurs; in vertebrates, it is composed of the brain and spinal cord; in invertebrates, it usually consists of one or more cords of nervous tissue, together with their associated ganglia.

central vacuole A large, membrane-bounded sac found in plant cells that stores proteins, pigments, and waste materials, and is involved in water balance.

centriole A cytoplasmic organelle located outside the nuclear membrane, identical in structure to a basal body; found in animal cells and in the flagellated cells of other groups; divides and organizes spindle fibers during mitosis and meiosis.

centromere A visible point of constriction on a chromosome that contains repeated DNA sequences that bind specific proteins. These proteins make up the kinetochore to which microtubules attach during cell division.

cephalization The evolution of a head and brain area in the anterior end of animals; thought to be a consequence of bilateral symmetry.

cerebellum The hindbrain region of the vertebrate brain that lies above the medulla (brainstem) and behind the forebrain; it integrates information about body position and motion, coordinates muscular activities, and maintains equilibrium.

cerebral cortex The thin surface layer of neurons and glial cells covering the cerebrum; well developed only in mammals, and particularly prominent in humans. The cerebral cortex is the seat of conscious sensations and voluntary muscular activity.

cerebrum The portion of the vertebrate brain (the forebrain) that occupies the upper part of the skull, consisting of two cerebral hemispheres united by the corpus callosum. It is the primary association center of the brain. It coordinates and processes sensory input and coordinates motor responses.

chaetae Bristles of chitin on each body segment that help anchor annelid worms during locomotion.

channel protein (ion channel) A transmembrane protein with a hydrophilic interior that provides an aqueous channel allowing diffusion of species that cannot cross the membrane. Usually allows passage of specific ions such as K^+, Na^+, or Ca^{2+} across the membrane.

chaperone protein A class of enzymes that help proteins fold into the correct configuration and can refold proteins that have been misfolded or denatured.

character displacement A process in which natural selection favors individuals in a species that use resources not used by other species. This results in evolutionary change leading to species dissimilar in resource use.

character state In cladistics, one of two or more distinguishable forms of a character, such as the presence or absence of teeth in amniote vertebrates.

charging reaction The reaction by which an aminoacyl-tRNA synthetase attaches a specific amino acid to the correct tRNA using energy from ATP.

chelicera, pl. chelicerae The first pair of appendages in horseshoe crabs, sea spiders, and arachnids—the chelicerates, a group of arthropods. Chelicerae usually take the form of pincers or fangs.

chemical synapse A close association that allows chemical communication between neurons. A chemical signal (neurotransmitter) released by the first neuron binds to receptors in the membrane of the second neurons.

chemiosmosis The mechanism by which ATP is generated in mitochondria and chloroplasts; energetic electrons excited by light (in chloroplasts) or extracted by oxidation in the Krebs cycle (in mitochondria) are used to drive proton pumps, creating a proton concentration gradient; when protons subsequently flow back across the membrane, they pass through channels that couple their movement to the synthesis of ATP.

chiasma An X-shaped figure that can be seen in the light microscope during meiosis; evidence of crossing over, where two chromatids have exchanged parts; chiasmata move to the ends of the chromosome arms as the homologues separate.

chitin A tough, resistant, nitrogen-containing polysaccharide that forms the cell walls of certain fungi, the exoskeleton of arthropods, and the epidermal cuticle of other surface structures of certain other invertebrates.

chlorophyll The primary type of light-absorbing pigment in photosynthesis. Chlorophyll *a* absorbs light in the violet-blue and the red ranges of the visible light spectrum; chlorophyll *b* is an accessory pigment to chlorophyll *a*, absorbing light in the blue and red-orange ranges. Neither pigment absorbs light in the green range, 500–600 nm.

chloroplast A cell-like organelle present in algae and plants that contains chlorophyll (and usually other pigments) and carries out photosynthesis.

choanocyte A specialized flagellated cell found in sponges; choanocytes line the body interior.

chorion The outer member of the double membrane that surrounds the embryo of reptiles, birds, and mammals; in placental mammals, it contributes to the structure of the placenta.

chorionic villus sampling A technique in which fetal cells are sampled from the chorion of the placenta rather than from the amniotic fluid; this less invasive technique can be used earlier in pregnancy than amniocentesis.

chromatid One of the two daughter strands of a duplicated chromosome that is joined by a single centromere.

chromatin The complex of DNA and proteins of which eukaryotic chromosomes are composed; chromatin is highly uncoiled and diffuse in interphase nuclei, condensing to form the visible chromosomes in prophase.

chromatin-remodeling complex A large protein complex that has been found to modify histones and DNA and that can change the structure of chromatin, moving or transferring nucleosomes.

chromosomal mutation Any mutation that affects chromosome structure.

chromosome The vehicle by which hereditary information is physically transmitted from one generation to the next; in a bacterium, the chromosome consists of a single naked circle of DNA; in eukaryotes, each chromosome consists of a single linear DNA molecule and associated proteins.

chromosomal theory of inheritance The theory stating that hereditary traits are carried on chromosomes.

cilium A short cellular projection from the surface of a eukaryotic cell, having the same internal structure of microtubules in a 9 + 2 arrangement as seen in a flagellum.

circadian rhythm An endogenous cyclical rhythm that oscillates on a daily (24-hour) basis.

circulatory system A network of vessels in coelomate animals that carries fluids to and from different areas of the body.

cisterna A small collecting vessel that pinches off from the end of a Golgi body to form a transport vesicle that moves materials through the cytoplasm.

cisternal space The inner region of a membrane-bounded structure. Usually used to describe the interior of the endoplasmic reticulum; also called the *lumen*.

clade A taxonomic group composed of an ancestor and all its descendents.

cladistics A taxonomic technique used for creating hierarchies of organisms that represent true phylogenetic relationship and descent.

class A taxonomic category between phyla and orders. A class contains one or more orders, and belongs to a particular phylum.

classical conditioning The repeated presentation of a stimulus in association with a response that causes the brain to form an association between the stimulus and the response, even if they have never been associated before.

clathrin A protein located just inside the plasma membrane in eukaryotic cells, in indentations called clathrin-coated pits.

cleavage In vertebrates, a rapid series of successive cell divisions of a fertilized egg, forming a hollow sphere of cells, the blastula.

cleavage furrow The constriction that forms during cytokinesis in animal cells that is responsible for dividing the cell into two daughter cells.

climax vegetation Vegetation encountered in a self-perpetuating community of plants that has proceeded through all the stages of succession and stabilized.

cloaca In some animals, the common exit chamber from the digestive, reproductive, and urinary system; in others, the cloaca may also serve as a respiratory duct.

clone-by-clone sequencing A method of genome sequencing in which a physical map is constructed first, followed by sequencing of fragments and identifying overlap regions.

clonal selection Amplification of a clone of immune cells initiated by antigen recognition.

cloning Producing a cell line or culture all of whose members contain identical copies of a particular nucleotide sequence; an essential element in genetic engineering.

closed circulatory system A circulatory system in which the blood is physically separated from other body fluids.

coacervate A spherical aggregation of lipid molecules in water, held together by hydrophobic forces.

coactivator A protein that functions to link transcriptional activators to the transcription complex consisting of RNA polymerase II and general transcription factors.

cochlea In terrestrial vertebrates, a tubular cavity of the inner ear containing the essential organs for hearing.

coding strand The strand of a DNA duplex that is the same as the RNA encoded by a gene. This strand is not used as a template in transcription, it is complementary to the template.

codominance Describes a case in which two or more alleles of a gene are each dominant to other alleles but not to each other. The phenotype of a heterozygote for codominant alleles exhibit characteristics of each of the homozygous forms. For example, in human blood types, a cross between an AA individual and a BB individual yields AB individuals.

codon The basic unit of the genetic code; a sequence of three adjacent nucleotides in DNA or mRNA that codes for one amino acid.

coelom In animals, a fluid-filled body cavity that develops entirely within the mesoderm.

coenzyme A nonprotein organic molecule such as NAD that plays an accessory role in enzyme-catalyzed processes, often by acting as a donor or acceptor of electrons.

coevolution The simultaneous development of adaptations in two or more populations, species, or other categories that interact so closely that each is a strong selective force on the other.

cofactor One or more nonprotein components required by enzymes in order to function; many cofactors are metal ions, others are organic coenzymes.

cohesin A protein complex that holds sister chromatids together during cell division. The loss of cohesins at the centromere allows the anaphase movement of chromosomes.

collenchyma cell In plants, the cells that form a supporting tissue called collenchyma; often found in regions of primary growth in stems and in some leaves.

colloblast A specialized type of cell found in members of the animal phylum Ctenophora (comb jellies) that bursts on contact with zooplankton, releasing an adhesive substance to help capture this prey.

colonial flagellate hypothesis The proposal first put forth by Haeckel that metazoans descended from colonial protists; supported by the similarity of sponges to choanoflagellate protists.

commensalism A relationship in which one individual lives close to or on another and benefits, and the host is unaffected; a kind of symbiosis.

community All of the species inhabiting a common environment and interacting with one another.

companion cell A specialized parenchyma cell that is associated with each sieve-tube member in the phloem of a plant.

competitive exclusion The hypothesis that two species with identical ecological requirements cannot exist in the same locality indefinitely, and that the more efficient of the two in utilizing the available scarce resources will exclude the other; also known as Gause's principle.

competitive inhibitor An inhibitor that binds to the same active site as an enzyme's substrate, thereby competing with the substrate.

complementary Describes genetic information in which each nucleotide base has a complementary partner with which it forms a base-pair.

complementary DNA (cDNA) A DNA copy of an mRNA transcript; produced by the action of the enzyme reverse transcriptase.

complement system The chemical defense of a vertebrate body that consists of a battery of proteins that become activated by the walls of bacteria and fungi.

complete digestive system A digestive system that has both a mouth and an anus, allowing unidirectional flow of ingested food.

compound eye An organ of sight in many arthropods composed of many independent visual units called ommatidia.

concentration gradient A difference in concentration of a substance from one location to another, often across a membrane.

condensin A protein complex involved in condensation of chromosomes during mitosis and meiosis.

cone (1) In plants, the reproductive structure of a conifer. (2) In vertebrates, a type of light-sensitive neuron in the retina concerned with the perception of color and with the most acute discrimination of detail.

conidia An asexually produced fungal spore.

conjugation Temporary union of two unicellular organisms, during which genetic material is transferred from one cell to the other; occurs in bacteria, protists, and certain algae and fungi.

consensus sequence In genome sequencing, the overall sequence that is consistent with the sequences of individual fragments; computer programs are used to compare sequences and generate a consensus sequence.

conservation of synteny The preservation over evolutionary time of arrangements of DNA segments in related species.

contig A contiguous segment of DNA assembled by analyzing sequence overlaps from smaller fragments.

continuous variation Variation in a trait that occurs along a continuum, such as the trait of height in human beings; often occurs when a trait is determined by more than one gene.

contractile vacuole In protists and some animals, a clear fluid-filled vacuole that takes up water from within the cell and then contracts, releasing it to the outside through a pore in a cyclical manner; functions primarily in osmoregulation and excretion.

conus arteriosus The anteriormost chamber of the embryonic heart in vertebrate animals.

convergent evolution The independent development of similar structures in organisms that are not directly related; often found in organisms living in similar environments.

cork cambium The lateral meristem that forms the periderm, producing cork (phellem) toward the surface (outside) of the plant and phelloderm toward the inside.

cornea The transparent outer layer of the vertebrate eye.

corolla The petals, collectively; usually the conspicuously colored flower whorl.

corpus callosum The band of nerve fibers that connects the two hemispheres of the cerebrum in humans and other primates.

corpus luteum A structure that develops from a ruptured follicle in the ovary after ovulation.

cortex The outer layer of a structure; in animals, the outer, as opposed to the inner, part of an organ; in vascular plants, the primary ground tissue of a stem or root.

cotyledon A seed leaf that generally stores food in dicots or absorbs it in monocots, providing nourishment used during seed germination.

crassulacean acid metabolism (CAM) A mode of carbon dioxide fixation by which CO_2 enters open leaf stomata at night and is used in photosynthesis during the day, when stomata are closed to prevent water loss.

crista A folded extension of the inner membrane of a mitochondrion. Mitochondria contain numerous cristae.

CRISPR (Clustered Regular Interspersed Short Palindromic Repeats) Repeat elements found in prokaryotic DNA that with cas enzymes form the basis for adaptive immunity to viruses. Also the source of molecular tools now used to edit genomes.

cross-current flow In bird lungs, the latticework of capillaries arranged across the air flow, at a 90° angle.

crossing over In meiosis, the exchange of corresponding chromatid segments between homologous chromosomes; responsible for genetic recombination between homologous chromosomes.

ctenidia Respiratory gills of mollusks; they consist of a system of filamentous projections of the mantle that are rich in blood vessels.

cuticle A waxy or fatty, noncellular layer (formed of a substance called cutin) on the outer wall of epidermal cells.

cutin In plants, a fatty layer produced by the epidermis that forms the cuticle on the outside surface.

cyanobacteria A group of photosynthetic bacteria, sometimes called the "blue-green algae," that contain the chlorophyll pigments most abundant in plants and algae, as well as other pigments.

cyclic AMP (cAMP) A form of adenosine monophosphate (AMP) in which the atoms of the phosphate group form a ring; found in almost all organisms, cAMP functions as an intracellular second messenger that regulates a diverse array of metabolic activities.

cyclic photophosphorylation Reactions that begin with the absorption of light by reaction center chlorophyll that excites an electron. The excited electron returns to the photosystem, generating ATP by chemiosmosis in the process. This is found in the single bacterial photosystem, and can occur in plants in photosystem I.

cyclin Any of a number of proteins that are produced in synchrony with the cell cycle and combine with certain protein kinases, the cyclin-dependent kinases, at certain points during cell division.

cyclin-dependent kinase (Cdk) Any of a group of protein kinase enzymes that control progress through the cell cycle. These enzymes are only active when complexed with cyclin. The cdc2 protein, produced by the *cdc2* gene, was the first Cdk enzyme discovered.

cytochrome Any of several iron-containing protein pigments that serve as electron carriers in transport chains of photosynthesis and cellular respiration.

cytochrome b$_6$–f complex A proton pump found in the thylakoid membrane. This complex uses energy from excited electrons to pump protons from the stroma into the thylakoid compartment.

cytokinesis Division of the cytoplasm of a cell after nuclear division.

cytokine Signaling molecules secreted by immune cells that affect other immune cells.

cytoplasm The material within a cell, excluding the nucleus; the protoplasm.

cytoskeleton A network of protein microfilaments and microtubules within the cytoplasm of a eukaryotic cell that maintains the shape of the cell, anchors its organelles, and is involved in animal cell motility.

cytosol The fluid portion of the cytoplasm; it contains dissolved organic molecules and ions.

cytotoxic T cell A special T cell activated during cell-mediated immune response that recognizes and destroys infected body cells.

D

deamination The removal of an amino group; part of the degradation of proteins into compounds that can enter the Krebs cycle.

deductive reasoning The logical application of general principles to predict a specific result. In science, deductive reasoning is used to test the validity of general ideas.

dehydration reaction A type of chemical reaction in which two molecules join to form one larger molecule, simultaneously splitting out a molecule of water; one molecule is stripped of a hydrogen atom, and another is stripped of a hydroxyl group (–OH), resulting in the joining of the two molecules, while the H and –OH released may combine to form a water molecule.

dehydrogenation Chemical reaction involving the loss of a hydrogen atom. This is an oxidation that combines loss of an electron with loss of a proton.

deletion A mutation in which a portion of a chromosome is lost; if too much information is lost, the deletion can be fatal.

demography The properties of the rate of growth and the age structure of populations.

denaturation The loss of the native configuration of a protein or nucleic acid as a result of excessive heat, extremes of pH, chemical modification, or changes in solvent ionic strength or polarity that disrupt hydrophobic interactions; usually accompanied by loss of biological activity.

dendrite A process extending from the cell body of a neuron, typically branched, that conducts impulses toward the cell body.

deoxyribonucleic acid (DNA) The genetic material of all organisms; composed of two complementary chains of nucleotides wound in a double helix.

dephosphorylation The removal of a phosphate group, usually by a phosphatase enzyme. Many proteins can be activated or inactivated by dephosphorylation.

depolarization The movement of ions across a plasma membrane that locally wipes out an electrical potential difference.

derived character A characteristic used in taxonomic analysis representing a departure from the primitive form.

dermal tissue In multicellular organisms, a type of tissue that forms the outer layer of the body and is in contact with the environment; it has a protective function.

desmosome A type of anchoring junction that links adjacent cells by connecting their cytoskeletons with cadherin proteins.

derepression Seen in anabolic operons where the operon that encodes the enzymes for a biochemical pathway is repressed in the presence of the end product of the pathway and derepressed in the absence of the end product. This allows production of the enzymes only when they are necessary.

determinate development A type of development in animals in which each embryonic cell has a predetermined fate in terms of what kind of tissue it will form in the adult.

deuterostome Any member of a grouping of bilaterally symmetrical animals in which the anus develops first and the mouth second; echinoderms and vertebrates are deuterostome animals.

diacylglycerol (DAG) A second messenger that is released, along with inositol-1,4,5-trisphosphate (IP$_3$), when phospholipase C cleaves PIP$_2$. DAG can have a variety of cellular effects through activation of protein kinases.

diaphragm (1) In mammals, a sheet of muscle tissue that separates the abdominal and thoracic cavities and functions in breathing. (2) A contraceptive device used to block the entrance to the uterus temporarily and thus prevent sperm from entering during sexual intercourse.

diapsid Any of a group of reptiles that have two pairs of temporal openings in the skull, one lateral and one more dorsal; one lineage of this group gave rise to dinosaurs, modern reptiles, and birds.

diastolic pressure In the measurement of human blood pressure, the minimum pressure between heartbeats (repolarization of the ventricles). *Compare with* systolic pressure.

dicer An enzyme that generates small RNA molecules in a cell by chopping up double-stranded RNAs; dicer produces miRNAs and siRNAs.

dicot Short for dicotyledon; a class of flowering plants generally characterized as having two cotyledons, net-veined leaves, and flower parts usually in fours or fives.

dideoxynucleotide A nucleotide lacking –OH groups at both the 2′ and 3′ positions; used as a chain terminator in the enzymatic sequencing of DNA.

differentiation A developmental process by which a relatively unspecialized cell undergoes a progressive change to a more specialized form or function.

diffusion The net movement of dissolved molecules or other particles from a region where they are more concentrated to a region where they are less concentrated.

dihybrid An individual heterozygous at two different loci; for example, *A/a B/b*.

dihybrid cross A single genetic cross involving two different traits, such as flower color and plant height.

dikaryotic In fungi, having pairs of nuclei within each cell.

dioecious Having the male and female elements on different individuals.

diploid Having two sets of chromosomes (2*n*); in animals, twice the number characteristic of gametes; in plants, the chromosome number characteristic of the sporophyte generation; in contrast to haploid (*n*).

directional selection A form of selection in which selection acts to eliminate one extreme from an array of phenotypes.

disaccharide A carbohydrate formed of two simple sugar molecules bonded covalently.

disruptive selection A form of selection in which selection acts to eliminate rather than favor the intermediate type.

dissociation In proteins, the reversible separation of protein subunits from a quaternary structure without altering their tertiary structure. Also refers to the dissolving of ionic compounds in water.

disassortative mating A type of nonrandom mating in which phenotypically different individuals mate more frequently.

diurnal Active during the day.

DNA-binding motif A region found in a regulatory protein that is capable of binding to a specific base sequence in DNA; a critical part of the protein's DNA-binding domain.

DNA fingerprinting An identification technique that makes use of a variety of molecular techniques to identify differences in the DNA of individuals.

DNA gyrase A topoisomerase involved in DNA replication; it relieves the torsional strain caused by unwinding the DNA strands.

DNA library A collection of DNAs in a vector (a plasmid, phage, or artificial chromosome) that taken together represent a complex mixture of DNAs, such as the entire genome, or the cDNAs made from all of the mRNA in a specific cell type.

DNA ligase The enzyme responsible for formation of phosphodiester bonds between adjacent nucleotides in DNA.

DNA microarray An array of DNA fragments on a microscope slide or silicon chip, used in hybridization experiments with labeled mRNA or DNA to identify active and inactive genes, or the presence or absence of particular sequences.

DNA polymerase A class of enzymes that all synthesize DNA from a preexisting template. All synthesize only in the 5′-to-3′ direction, and require a primer to extend.

domain (1) A distinct modular region of a protein that serves a particular function in the action of the protein, such as a regulatory domain or a DNA-binding domain. (2) In taxonomy, the level higher than kingdom. The three domains currently recognized are Bacteria, Archaea, and Eukarya.

Domain Archaea In the three-domain system of taxonomy, the group that contains only the Archaea, a highly diverse group of unicellular prokaryotes.

Domain Bacteria In the three-domain system of taxonomy, the group that contains only the Bacteria, a vast group of unicellular prokaryotes.

Domain Eukarya In the three-domain system of taxonomy, the group that contains eukaryotic organisms including protists, fungi, plants, and animals.

dominant An allele that is expressed when present in either the heterozygous or the homozygous condition.

dosage compensation A phenomenon by which the expression of genes carried on sex chromosomes is kept the same in males and females, despite a different number of sex chromosomes. In mammals, inactivation of one of the X chromosomes in female cells accomplishes dosage compensation.

double fertilization The fusion of the egg and sperm (resulting in a 2n fertilized egg, the zygote) and the simultaneous fusion of the second male gamete with the polar nuclei (resulting in a primary endosperm nucleus, which is often triploid, 3n); a unique characteristic of all angiosperms.

double helix The structure of DNA, in which two complementary polynucleotide strands coil around a common helical axis.

duodenum In vertebrates, the upper portion of the small intestine.

duplication A mutation in which a portion of a chromosome is duplicated; if the duplicated region does not lie within a gene, the duplication may have no effect.

E

ecdysis Shedding of outer, cuticular layer; molting, as in insects or crustaceans.

ecdysone Molting hormone of arthropods, which triggers when ecdysis occurs.

ecology The study of interactions of organisms with one another and with their physical environment.

ecosystem A major interacting system that includes organisms and their nonliving environment.

ecotype A locally adapted variant of an organism; differing genetically from other ecotypes.

ectoderm One of the three embryonic germ layers of early vertebrate embryos; ectoderm gives rise to the outer epithelium of the body (skin, hair, nails) and to the nerve tissue, including the sense organs, brain, and spinal cord.

ectomycorrhizae Externally developing mycorrhizae that do not penetrate the cells they surround.

ectotherms Animals such as reptiles, fish, or amphibians, whose body temperature is regulated by their behavior or by their surroundings.

electronegativity A property of atomic nuclei that refers to the affinity of the nuclei for valence electrons; a nucleus that is more electronegative has a greater pull on electrons than one that is less electronegative.

electron transport chain The passage of energetic electrons through a series of membrane-associated electron-carrier molecules to proton pumps embedded within mitochondrial or chloroplast membranes. *See* chemiosmosis.

elongation factor (Ef-Tu) In protein synthesis in *E. coli*, a factor that binds to GTP and to a charged tRNA to accomplish binding of the charged tRNA to the A site of the ribosome, so that elongation of the polypeptide chain can occur.

embryo A multicellular developmental stage that follows cell division of the zygote.

embryonic stem cell (ES cell) A stem cell derived from an early embryo that can develop into different adult tissues and give rise to an adult organism when injected into a blastocyst.

emergent properties Novel properties arising from the way in which components interact. Emergent properties often cannot be deduced solely from knowledge of the individual components.

emerging virus Any virus that originates in one organism but then passes to another; usually refers to transmission to humans.

endergonic Describes a chemical reaction in which the products contain more energy than the reactants, so that free energy must be put into the reaction from an outside source to allow it to proceed.

endocrine gland Ductless gland that secretes hormones into the extracellular spaces, from which they diffuse into the circulatory system.

endocytosis The uptake of material into cells by inclusion within an invagination of the plasma membrane; the uptake of solid material is phagocytosis, and that of dissolved material is pinocytosis.

endoderm One of the three embryonic germ layers of early vertebrate embryos, destined to give rise to the epithelium that lines internal structures and most of the digestive and respiratory tracts.

endodermis In vascular plants, a layer of cells forming the innermost layer of the cortex in roots and some stems.

endomembrane system A system of connected membranous compartments found in eukaryotic cells.

endometrium The lining of the uterus in mammals; thickens in response to secretion of estrogens and progesterone and is sloughed off in menstruation.

endomycorrhizae Mycorrhizae that develop within cells.

endonuclease An enzyme capable of cleaving phosphodiester bonds between nucleotides located internally in a DNA strand.

endoplasmic reticulum (ER) Internal membrane system that forms a netlike array of channels and interconnections within the cytoplasm of eukaryotic cells. The ER is divided into rough (RER) and smooth (SER) compartments.

endorphin One of a group of small neuropeptides produced by the vertebrate brain; like morphine, endorphins modulate pain perception.

endosperm A storage tissue characteristic of the seeds of angiosperms, which develops from the union of a male nucleus and the polar nuclei of the embryo sac. The endosperm is digested by the growing sporophyte either before maturation of the seed or during its germination.

endospore A highly resistant, thick-walled bacterial spore that can survive harsh environmental stress, such as heat or dessication, and then germinate when conditions become favorable.

endosymbiosis Theory that proposes that eukaryotic cells evolved from a symbiosis between different species of prokaryotes.

endotherm An animal capable of maintaining a constant body temperature. *See* homeotherm.

energy level A discrete level, or quantum, of energy that an electron in an atom possesses. To change energy levels, an electron must absorb or release energy.

enhancer A site of regulatory protein binding on the DNA molecule distant from the promoter and start site for a gene's transcription.

enthalpy In a chemical reaction, the energy contained in the chemical bonds of the molecule, symbolized as *H*; in a cellular reaction, the free energy is equal to the enthalpy of the reactant molecules in the reaction.

entropy A measure of the randomness or disorder of a system; a measure of how much energy in a system has become so dispersed (usually as evenly distributed heat) that it is no longer available to do work.

enzyme A protein that is capable of speeding up specific chemical reactions by lowering the required activation energy.

enzyme–substrate complex The complex formed when an enzyme binds with its substrate. This complex often has an altered configuration compared with the nonbound enzyme.

epicotyl The region just above where the cotyledons are attached.

epidermal cell In plants, a cell that collectively forms the outermost layer of the primary plant body; includes specialized cells such as trichomes and guard cells.

epidermis The outermost layers of cells; in plants, the exterior primary tissue of leaves, young stems, and roots; in vertebrates, the nonvascular external layer of skin, of ectodermal origin; in invertebrates, a single layer of ectodermal epithelium.

epididymis A sperm storage vessel; a coiled part of the sperm duct that lies near the testis.

epistasis Interaction between two nonallelic genes in which one of them modifies the phenotypic expression of the other.

epithelium In animals, a type of tissue that covers an exposed surface or lines a tube or cavity.

equilibrium A stable condition; the point at which a chemical reaction proceeds as rapidly in the reverse direction as it does in the forward direction, so that there is no further net change in the concentrations of products or reactants. In ecology, a stable condition that resists change and fairly quickly returns to its original state if disturbed by humans or natural events.

erythrocyte Red blood cell, the carrier of hemoglobin.

erythropoiesis The manufacture of blood cells in the bone marrow.

E site In a ribosome, the exit site that binds to the tRNA that carried the previous amino acid added to the polypeptide chain.

estrus The period of maximum female sexual receptivity, associated with ovulation of the egg.

ethology The study of patterns of animal behavior in nature.

euchromatin That portion of a eukaryotic chromosome that is transcribed into mRNA; contains active genes that are not tightly condensed during interphase.

eukaryote A cell characterized by membrane-bounded organelles, most notably the nucleus, and one that possesses chromosomes whose DNA is associated with proteins; an organism composed of such cells.

eutherian A placental mammal.

eutrophic Refers to a lake in which an abundant supply of minerals and organic matter exists.

evolution Genetic change in a population of organisms; in general, evolution leads to progressive change from simple to complex.

excision repair A nonspecific mechanism to repair damage to DNA during synthesis. The damaged or mismatched region is excised, and DNA polymerase replaces the region removed.

exergonic Describes a chemical reaction in which the products contain less free energy than the reactants, so that free energy is released in the reaction.

exhalant siphon In bivalve mollusks, the siphon through which outgoing water leaves the body.

exocrine gland A type of gland that releases its secretion through a duct, such as a digestive gland or a sweat gland.

exocytosis A type of bulk transport out of cells in which a vacuole fuses with the plasma membrane, discharging the vacuole's contents to the outside.

exon A segment of DNA that is both transcribed into RNA and translated into protein. See intron.

exonuclease An enzyme capable of cutting phosphodiester bonds between nucleotides located at an end of a DNA strand. This allows sequential removal of nucleotides from the end of DNA.

exoskeleton An external skeleton, as in arthropods.

experiment A test of one or more hypotheses. Hypotheses make contrasting predictions that can be tested experimentally in control and test experiments where a single variable is altered.

expressed sequence tag (EST) A short sequence of a cDNA that unambiguously identifies the cDNA.

expression vector A type of vector (plasmid or phage) that contains the sequences necessary to drive expression of inserted DNA in a specific cell type.

exteroceptor A receptor that is excited by stimuli from the external world.

extremophile An archaeal organism that lives in extreme environments; different archaeal species may live in hot springs (thermophiles), highly saline environments (halophiles), highly acidic or basic environments, or under high pressure at the bottom of oceans.

F

5′ cap In eukaryotes, a structure added to the 5′ end of an mRNA consisting of methylated GTP attached by a 5′ to 5′ bond. The cap protects this end from degradation and is involved in the initiation of translation.

facilitated diffusion The diffusion of molecules or ions through carrier proteins or ion channels. Requires a concentration gradient, but not any input of energy.

family A taxonomic grouping of similar species above the level of genus.

fat A molecule composed of glycerol and three fatty acid molecules.

feedback inhibition Control mechanism whereby an increase in the concentration of some molecules inhibits the synthesis of that molecule.

fermentation The enzyme-catalyzed extraction of energy from organic compounds without the involvement of oxygen.

fertilization The fusion of two haploid gamete nuclei to form a diploid zygote nucleus.

fibroblast A flat, irregularly branching cell of connective tissue that secretes structurally strong proteins into the matrix between the cells.

first filial (F_1) generation The offspring resulting from a cross between a parental generation (P); in experimental crosses, these parents usually have different phenotypes.

First Law of Thermodynamics Energy cannot be created or destroyed, but can only undergo conversion from one form to another; thus, the amount of energy in the universe is unchangeable.

fitness The genetic contribution of an individual to succeeding generations. Relative fitness refers to the fitness of an individual relative to other individuals in a population.

fixed action pattern A stereotyped animal behavior response, thought by ethologists to be based on programmed neural circuits.

flagellin The protein composing bacterial flagella, which allow a cell to move through an aqueous environment.

flagellum A long, threadlike structure protruding from the surface of a cell and used in locomotion.

flame cell A specialized cell found in the network of tubules inside flatworms that assists in water regulation and some waste excretion.

flavin adenine dinucleotide (FAD, $FADH_2$) A cofactor that acts as a soluble (not membrane-bound) electron carrier (can be reversibly oxidized and reduced).

fluorescent in situ hybridization (FISH) A cytological method used to find specific DNA sequences on chromosomes with a specific fluorescently labeled probe.

food security Having access to sufficient, safe food to avoid malnutrition and starvation; a global human issue.

foraging behavior A collective term for the many complex, evolved behaviors that influence what an animal eats and how the food is obtained.

founder effect The effect by which rare alleles and combinations of alleles may be enhanced in new populations.

fovea A small depression in the center of the retina with a high concentration of cones; the area of sharpest vision.

frameshift mutation A mutation in which a base is added or deleted from the DNA sequence. These changes alter the reading frame downstream of the mutation.

free energy Energy available to do work.

free radical An ionized atom with one or more unpaired electrons, resulting from electrons that have been energized by ionizing radiation being ejected from the atom; free radicals react violently with other molecules, such as DNA, causing damage by mutation.

frequency-dependent selection A type of selection that depends on how frequently or infrequently a phenotype occurs in a population.

fruit In angiosperms, a mature, ripened ovary (or group of ovaries), containing the seeds.

functional genomics The study of the function of genes and their products, beyond simply ascertaining gene sequences.

functional group A molecular group attached to a hydrocarbon that confers chemical properties or reactivities. Examples include hydroxyl (–OH), carboxylic acid (–COOH) and amino groups ($-NH_2$).

fundamental niche Also referred to as the hypothetical niche, this is the entire niche an organism could fill if there were no other interacting factors (such as competition or predation).

G

G_0 phase The stage of the cell cycle occupied by cells that are not actively dividing.

G_1 phase The phase of the cell cycle after cytokinesis and before DNA replication called the first "gap" phase. This phase is the primary growth phase of a cell.

G₁/S checkpoint The primary control point at which a cell "decides" whether or not to divide. Also called START and the restriction point.

G₂ phase The phase of the cell cycle between DNA replication and mitosis called the second "gap" phase. During this phase, the cell prepares for mitosis.

G₂/M checkpoint The second cell-division control point, at which division can be delayed if DNA has not been properly replicated or is damaged.

gametangium, pl. gametangia A cell or organ in which gametes are formed.

gamete A haploid reproductive cell.

gametocytes Cells in the malarial sporozoite life cycle capable of giving rise to gametes when in the correct host.

gametophyte In plants, the haploid (n), gamete-producing generation, which alternates with the diploid ($2n$) sporophyte.

ganglion, pl. ganglia An aggregation of nerve cell bodies; in invertebrates, ganglia are the integrative centers; in vertebrates, the term is restricted to aggregations of nerve cell bodies located outside the central nervous system.

gap gene Any of certain genes in *Drosophila* development that divide the embryo into large blocks in the process of segmentation; *hunchback* is a gap gene.

gap junction A junction between adjacent animal cells that allows the passage of materials between the cells.

gastrodermis In eumetazoan animals, the layer of digestive tissue that develops from the endoderm.

gastrula In vertebrates, the embryonic stage in which the blastula with its single layer of cells turns into a three-layered embryo made up of ectoderm, mesoderm, and endoderm.

gastrulation Developmental process that converts blastula into embryo with three embryonic germ layers: endoderm, mesoderm, and ectoderm. Involves massive cell migration to convert the hollow structure into a three-layered structure.

gene The basic unit of heredity; a sequence of DNA nucleotides on a chromosome that encodes a protein, tRNA, or rRNA molecule, or regulates the transcription of such a sequence.

gene conversion Alteration of one homologous chromosome by the cell's error-detection and repair system to make it resemble the sequence on the other homologue.

gene expression The conversion of the genotype into the phenotype; the process by which DNA is transcribed into RNA, which is then translated into a protein product.

gene pool All the alleles present in a species.

gene-for-gene hypothesis A plant defense mechanism in which a specific protein encoded by a viral, bacterial, or fungal pathogen binds to a protein encoded by a plant gene and triggers a defense response in the plant.

general transcription factor Any of a group of transcription factors that are required for formation of an initiation complex by RNA polymerase II at a promoter. This allows a basal level that can be increased by the action of specific factors.

generalized transduction A form of gene transfer in prokaryotes in which any gene can be transferred between cells. This uses a lytic bacteriophage as a carrier where the virion is accidentally packaged with host DNA.

genetic counseling The process of evaluating the risk of genetic defects occurring in offspring, testing for these defects in unborn children, and providing the parents with information about these risks and conditions.

genetic drift Random fluctuation in allele frequencies over time by chance.

genetic map An abstract map that places the relative location of genes on a chromosome based on recombination frequency.

Genetic marker Any genetic difference that can be detected and used to discriminate between individuals.

genome The entire DNA sequence of an organism.

genomic imprinting Describes an exception to Mendelian genetics in some mammals in which the phenotype caused by an allele is exhibited when the allele comes from one parent, but not from the other.

genomic library A DNA library that contains a representation of the entire genome of an organism.

genomics The study of genomes as opposed to individual genes.

genotype The genetic constitution underlying a single trait or set of traits.

genotype frequency A measure of the occurrence of a genotype in a population, expressed as a proportion of the entire population, for example, an occurrence of 0.25 (25%) for a homozygous recessive genotype.

genus, pl. genera A taxonomic group that ranks below a family and above a species.

germination The resumption of growth and development by a spore or seed.

germ layers The three cell layers formed at gastrulation of the embryo that foreshadow the future organization of tissues; the layers, from the outside inward, are the ectoderm, the mesoderm, and the endoderm.

germ-line cells During zygote development, cells that are set aside from the somatic cells and that will eventually undergo meiosis to produce gametes.

gill (1) In aquatic animals, a respiratory organ, usually a thin-walled projection from some part of the external body surface, endowed with a rich capillary bed and having a large surface area. (2) In basidiomycete fungi, the plates on the underside of the cap.

globular protein Proteins with a compact tertiary structure with hydrophobic amino acids mainly in the interior.

glomerular filtrate The fluid that passes out of the capillaries of each glomerulus.

glomerulus A cluster of capillaries enclosed by Bowman's capsule.

glucagon A vertebrate hormone produced in the pancreas that acts to initiate the breakdown of glycogen to glucose subunits.

gluconeogenesis The synthesis of glucose from noncarbohydrates (such as proteins or fats).

glucose A common six-carbon sugar ($C_6H_{12}O_6$); the most common monosaccharide in most organisms.

glucose repression In *E. coli*, the preferential use of glucose even when other sugars are present; transcription of mRNA encoding the enzymes for utilizing the other sugars does not occur.

glycocalyx A "sugar coating" on the surface of a cell resulting from the presence of polysaccharides on glycolipids and glycoproteins embedded in the outer layer of the plasma membrane.

glycogen Animal starch; a complex branched polysaccharide that serves as a food reserve in animals, bacteria, and fungi.

glycolipid Lipid molecule modified within the Golgi complex by having a short sugar chain (polysaccharide) attached.

glycolysis The anaerobic breakdown of glucose; this enzyme-catalyzed process yields two molecules of pyruvate with a net of two molecules of ATP.

glycoprotein Protein molecule modified within the Golgi complex by having a short sugar chain (polysaccharide) attached.

glyoxysome A small cellular organelle or microbody containing enzymes necessary for conversion of fats into carbohydrates.

glyphosate A biodegradable herbicide that works by inhibiting EPSP synthetase, a plant enzyme that makes aromatic amino acids; genetic engineering has allowed crop species to be created that are resistant to glyphosate.

Golgi apparatus (Golgi body) A collection of flattened stacks of membranes in the cytoplasm of eukaryotic cells; functions in collection, packaging, and distribution of molecules synthesized in the cell.

G protein A protein that binds guanosine triphosphate (GTP) and assists in the function of cell-surface receptors. When the receptor binds its signal molecule, the G protein binds GTP and is activated to start a chain of events within the cell.

G protein-coupled receptor (GPCR) A receptor that acts through a heterotrimeric (three component) G protein to activate effector proteins. The effector proteins then function as enzymes to produce second messengers such as cAMP or IP_3.

gradualism The view that species change very slowly in ways that may be imperceptible from one generation to the next but that accumulate and lead to major changes over thousands or millions of years.

Gram stain Staining technique that divides bacteria into gram-negative or gram-positive based on retention of a violet dye. Differences in staining are due to cell wall construction.

granum (pl. grana) A stacked column of flattened, interconnected disks (thylakoids) that are part of the thylakoid membrane system in chloroplasts.

gravitropism Growth response to gravity in plants; formerly called geotropism.

ground meristem The primary meristem, or meristematic tissue, that gives rise to the plant body (except for the epidermis and vascular tissues).

ground tissue In plants, a type of tissue that performs many functions, including support, storage, secretion, and photosynthesis; may consist of many cell types.

growth factor Any of a number of proteins that bind to membrane receptors and initiate intracellular signaling systems that result in cell growth and division.

guard cell In plants, one of a pair of sausage-shaped cells flanking a stoma; the guard cells open and close the stomata.

guttation The exudation of liquid water from leaves due to root pressure.

gymnosperm A seed plant with seeds not enclosed in an ovary; conifers are gymnosperms.

gynoecium The aggregate of carpels in the flower of a seed plant.

H

habitat The environment of an organism; the place where it is usually found.

habituation A form of learning; a diminishing response to a repeated stimulus.

halophyte A plant that is salt-tolerant.

haplodiploidy A phenomenon occurring in certain organisms such as wasps, wherein both haploid (male) and diploid (female) individuals are encountered.

haploid Having only one set of chromosomes (n), in contrast to diploid ($2n$).

haplotype A region of a chromosome that is usually inherited intact, that is, it does not undergo recombination. These are identified based on analysis of SNPs.

Hardy-Weinberg equilibrium A mathematical description of the fact that allele and genotype frequencies remain constant in a random-mating population in the absence of inbreeding, selection, or other evolutionary forces; usually stated: if the frequency of allele a is p and the frequency of allele b is q, then the genotype frequencies after one generation of random mating will always be $p_2 + 2pq + q_2 = 1$.

Haversian canal Narrow channels that run parallel to the length of a bone and contain blood vessels and nerve cells.

heat A measure of the random motion of molecules; the greater the heat, the greater the motion. Heat is one form of kinetic energy.

heat of vaporization The amount of energy required to change 1 g of a substance from a liquid to a gas.

heavy metal Any of the metallic elements with high atomic numbers, such as arsenic, cadmium, lead, etc. Many heavy metals are toxic to animals even in small amounts.

helicase Any of a group of enzymes that unwind the two DNA strands in the double helix to facilitate DNA replication.

helix-turn-helix motif A common DNA-binding motif found in regulatory proteins; it consists of two α-helices linked by a nonhelical segment (the "turn").

helper T cell A class of white blood cells that initiates both the cell-mediated immune response and the humoral immune response; helper T cells are the targets of the AIDS virus (HIV).

hemoglobin A globular protein in vertebrate red blood cells and in the plasma of many invertebrates that carries oxygen and carbon dioxide.

hemopoietic stem cell The cells in bone marrow where blood cells are formed.

hermaphroditism Condition in which an organism has both male and female functional reproductive organs.

heterochromatin The portion of a eukaryotic chromosome that is not transcribed into RNA; remains condensed in interphase and stains intensely in histological preparations.

heterochrony An alteration in the timing of developmental events due to a genetic change; for example, a mutation that delays flowering in plants.

heterokaryotic In fungi, having two or more genetically distinct types of nuclei within the same mycelium.

heterosporous In vascular plants, having spores of two kinds, namely, microspores and megaspores.

heterotroph An organism that cannot derive energy from photosynthesis or inorganic chemicals, and so must feed on other plants and animals, obtaining chemical energy by degrading their organic molecules.

heterozygote advantage The situation in which individuals heterozygous for a trait have a selective advantage over those who are homozygous; an example is sickle cell anemia.

heterozygous Having two different alleles of the same gene; the term is usually applied to one or more specific loci, as in "heterozygous with respect to the *W* locus" (that is, the genotype is *W/w*).

Hfr cell An *E. coli* cell that has a high frequency of recombination due to integration of an F plasmid into its genome.

histone protein Any of eight proteins with an overall positive charge that associate in a complex. The DNA duplex coils around a core of eight histone proteins, held by its negatively charged phosphate groups, forming a nucleosome.

holoblastic cleavage Process in vertebrate embryos in which the cleavage divisions all occur at the same rate, yielding a uniform cell size in the blastula.

homeobox A sequence of 180 nucleotides located in homeotic genes that produces a 60-amino-acid peptide sequence (the homeodomain) active in transcription factors.

homeosis A change in the normal spatial pattern of gene expression that can result in homeotic mutants where a wild-type structure develops in the wrong place in or on the organism.

homeostasis The maintenance of a relatively stable internal physiological environment in an organism; usually involves some form of feedback self-regulation.

homeotherm An organism, such as a bird or mammal, capable of maintaining a stable body temperature independent of the environmental temperature. See endotherm.

homeotic gene One of a series of "master switch" genes that determine the form of segments developing in the embryo.

hominid Any primate in the human family, Hominidae. *Homo sapiens* is the only living representative.

hominoid Collectively, hominids and apes; the monkeys and hominoids constitute the anthropoid primates.

homokaryotic In fungi, having nuclei with the same genetic makeup within a mycelium.

homologue One of a pair of chromosomes of the same kind located in a diploid cell; one copy of each pair of homologues comes from each gamete that formed the zygote.

homologous (1) Refers to similar structures that have the same evolutionary origin. (2) Refers to a pair of the same kind of chromosome in a diploid cell.

homoplasy In cladistics, a shared character state that has not been inherited from a common ancestor exhibiting that state; may result from convergent evolution or evolutionary reversal. The wings of birds and of bats, which are convergent structures, are examples.

homosporous In some plants, production of only one type of spore rather than differentiated types. *Compare with* heterosporous.

homozygous Being a homozygote, having two identical alleles of the same gene; the term is usually applied to one or more specific loci, as in "homozygous with respect to the *W* locus" (i.e., the genotype is *W/W* or *w/w*).

horizontal gene transfer (HGT) The passing of genes laterally between species; more prevalent very early in the history of life.

hormone A molecule, usually a peptide or steroid, that is produced in one part of an organism and triggers a specific cellular reaction in target tissues and organs some distance away.

host range The range of organisms that can be infected by a particular virus.

***Hox* gene** A group of homeobox-containing genes that control developmental events, usually found organized into clusters of genes. These genes have been conserved in many different multicellular animals, both invertebrates and vertebrates, although the number of clusters changes in lineages, leading to four clusters in vertebrates.

humoral immunity Arm of the adaptive immune system involving B cells that produce soluble antibodies specific for foreign antigens.

humus Partly decayed organic material found in topsoil.

hybridization The mating of unlike parents.

hydration shell A "cloud" of water molecules surrounding a dissolved substance, such as sucrose or Na^+ and Cl^- ions.

hydrogen bond A weak association formed with hydrogen in polar covalent bonds. The partially positive hydrogen is attracted to partially negative atoms in polar covalent bonds. In water, oxygen and hydrogen in different water molecules form hydrogen bonds.

hydrolysis reaction A reaction that breaks a bond by the addition of water. This is the reverse of dehydration, a reaction that joins molecules with the loss of water.

hydrophilic Literally translates as "water-loving" and describes substances that are soluble in water. These must be either polar or charged (ions).

hydrophobic Literally translates as "water-fearing" and describes nonpolar substances that are not soluble in water. Nonpolar molecules in water associate with each other and form droplets.

hydrophobic exclusion The tendency of nonpolar molecules to aggregate together when placed in water. Exclusion refers to the action of water in forcing these molecules together.

hydrostatic skeleton The skeleton of most soft-bodied invertebrates that have neither an internal nor an external skeleton. They use the relative incompressibility of the water within their bodies as a kind of skeleton.

hyperpolarization Above-normal negativity of a cell membrane during its resting potential.

hypersensitive response Plants respond to pathogens by selectively killing plant cells to block the spread of the pathogen.

hypertonic A solution with a higher concentration of solutes than the cell. A cell in a hypertonic solution tends to lose water by osmosis.

hypha, pl. **hyphae** A filament of a fungus or oomycete; collectively, the hyphae constitute the mycelium.

hypocotyl The region immediately below where the cotyledons are attached.

hypothalamus A region of the vertebrate brain just below the cerebral hemispheres, under the thalamus; a center of the autonomic nervous system, responsible for the integration and correlation of many neural and endocrine functions.

hypotonic A solution with a lower concentration of solutes than the cell. A cell in a hypotonic solution tends to take in water by osmosis.

I

icosahedron A structure consisting of 20 equilateral triangular facets; this is commonly seen in viruses and forms one kind of viral capsid.

imaginal disk One of about a dozen groups of cells set aside in the abdomen of a larval insect and committed to forming key parts of the adult insect's body.

immune response In vertebrates, a defensive reaction of the body to invasion by a foreign substance or organism. *See* antibody and B cell.

immunoglobulin An antibody molecule.

immunological tolerance Process where immune system learns to not react to self-antigens.

in vitro mutagenesis The ability to create mutations at any site in a cloned gene to examine the mutations' effects on function.

inbreeding The breeding of genetically related plants or animals; inbreeding tends to increase homozygosity.

inclusive fitness Describes the sum of the number of genes directly passed on in an individual's offspring and those genes passed on indirectly by kin (other than offspring) whose existence results from the benefit of the individual's altruism.

incomplete dominance Describes a case in which two or more alleles of a gene do not display clear dominance. The phenotype of a heterozygote is intermediate between the homozygous forms. For example, crossing red-flowered with white-flowered four o'clocks yields pink heterozygotes.

independent assortment In a dihybrid cross, describes the random assortment of alleles for each of the genes. For genes on different chromosomes this results from the random orientations of different homologous pairs during metaphase I of meiosis. For genes on the same chromosome, this occurs when the two loci are far enough apart for roughly equal numbers of odd- and even-numbered multiple crossover events.

indeterminate development A type of development in animals in which the first few embryonic cells are identical daughter cells, any one of which could develop separately into a complete organism; their fate is indeterminate.

inducer exclusion Part of the mechanism of glucose repression in *E. coli* in which the presence of glucose prevents the entry of lactose such that the *lac* operon cannot be induced.

induction (1) Production of enzymes in response to a substrate; a mechanism by which binding of an inducer to a repressor allows transcription of an operon. This is seen in catabolic operons and results in production of enzymes to degrade a compound only when it is available. (2) In embryonic development, the process by which the development of a cell is influenced by interaction with an adjacent cell.

inductive reasoning The logical application of specific observations to make a generalization. In science, inductive reasoning is used to formulate testable hypotheses.

industrial melanism Phrase used to describe the evolutionary process in which initially light-colored organisms become dark as a result of natural selection.

inflammatory response A generalized nonspecific response to infection that acts to clear an infected area of infecting microbes and dead tissue cells so that tissue repair can begin.

inhalant siphon In bivalve mollusks, the siphon through which incoming water enters the body.

inheritance of acquired characteristics Also known as Lamarckism; the theory, now discounted, that individuals genetically pass on to their offspring physical and behavioral changes developed during the individuals' own lifetime.

inhibitor A substance that binds to an enzyme and decreases its activity.

initiation factor One of several proteins involved in the formation of an initiation complex in prokaryote polypeptide synthesis.

initiator tRNA A tRNA molecule involved in the beginning of translation. In prokaryotes, the initiator tRNA is charged with *N*-formylmethionine (tRNA^{fMet}); in eukaryotes, the tRNA is charged simply with methionine.

inorganic phosphate A phosphate molecule that is not a part of an organic molecule; inorganic phosphate groups are added and removed in the formation and breakdown of ATP and in many other cellular reactions.

inositol-1,4,5-trisphosphate (IP$_3$) Second messenger produced by the cleavage of phosphatidylinositol-4,5-bisphosphate.

insertional inactivation Destruction of a gene's function by the insertion of a transposon.

instar A larval developmental stage in insects.

integrin Any of a group of cell-surface proteins involved in adhesion of cells to substrates. Critical to migrating cells moving through the cell matrix in tissues such as connective tissue.

intercalary meristem A type of meristem that arises in stem internodes in some plants, such as corn and horsetails; responsible for elongation of the internodes.

interferon In vertebrates, a protein produced in virus-infected cells that inhibits viral multiplication.

intermembrane space The outer compartment of a mitochondrion that lies between the two membranes.

interneuron (association neuron) A nerve cell found only in the middle of the spinal cord that acts as a functional link between sensory neurons and motor neurons.

internode In plants, the region of a stem between two successive nodes.

interoceptor A receptor that senses information related to the body itself, its internal condition, and its position.

interphase The period between two mitotic or meiotic divisions in which a cell grows and its DNA replicates; includes G$_1$, S, and G$_2$ phases.

intron Portion of mRNA as transcribed from eukaryotic DNA that is removed by enzymes before the mature mRNA is translated into protein. *See* exon.

inversion A reversal in order of a segment of a chromosome; also, to turn inside out, as in embryogenesis of sponges or discharge of a nematocyst.

ionizing radiation High-energy radiation that is highly mutagenic, producing free radicals that react with DNA; includes X-rays and γ-rays.

isomer One of a group of molecules identical in atomic composition but differing in structural arrangement; for example, glucose and fructose.

isotonic A solution having the same concentration of solutes as the cell. A cell in an isotonic solution takes in and loses the same amount of water.

isotope Different forms of the same element with the same number of protons but different numbers of neutrons.

J

jasmonic acid An organic molecule that is part of a plant's wound response; it signals the production of a proteinase inhibitor.

K

karyotype The morphology of the chromosomes of an organism as viewed with a light microscope.

keratin A tough, fibrous protein formed in epidermal tissues and modified into skin, feathers, hair, and hard structures such as horns and nails.

key innovation A newly evolved trait in a species that allows members to use resources or other aspects of the environment that were previously inaccessible.

kidney In vertebrates, the organ that filters the blood to remove nitrogenous wastes and regulates the balance of water and solutes in blood plasma.

kilocalorie Unit describing the amount of heat required to raise the temperature of a kilogram of water by 1°C; sometimes called a Calorie, equivalent to 1000 calories.

kinase cascade A series of protein kinases that phosphorylate each other in succession; a kinase cascade can amplify signals during the signal transduction process.

kinesis Changes in activity level in an animal that are dependent on stimulus intensity. *See* kinetic energy.

kinetic energy The energy of motion.

kinetochore Disk-shaped protein structure within the centromere to which the spindle fibers attach during mitosis or meiosis. *See* centromere.

kingdom The second highest commonly used taxonomic category.

kin selection Selection favoring relatives; an increase in the frequency of related individuals (kin) in a population, leading to an increase in the relative frequency in the population of those alleles shared by members of the kin group.

knockout mice Mice in which a known gene is inactivated ("knocked out") using recombinant DNA and ES cells.

Krebs cycle Another name for the citric acid cycle; also called the tricarboxylic acid (TCA) cycle.

L

labrum The upper lip of insects and crustaceans situated above or in front of the mandibles.

lac **operon** In *E. coli*, the operon containing genes that encode the enzymes to metabolize lactose.

lagging strand The DNA strand that must be synthesized discontinuously because of the 5′-to-3′ directionality of DNA polymerase during replication, and the antiparallel nature of DNA. Compare *leading strand*.

larva A developmental stage that is unlike the adult found in organisms that undergo metamorphosis. Embryos develop into larvae that produce the adult form by metamorphosis.

larynx The voice box; a cartilaginous organ that lies between the pharynx and trachea and is responsible for sound production in vertebrates.

lateral line system A sensory system encountered in fish, through which mechanoreceptors in a line down the side of the fish are sensitive to motion.

lateral meristems In vascular plants, the meristems that give rise to secondary tissue; the vascular cambium and cork cambium.

Law of Independent Assortment Mendel's second law of heredity, stating that genes located on nonhomologous chromosomes assort independently of one another.

Law of Segregation Mendel's first law of heredity, stating that alternative alleles for the same gene segregate from each other in production of gametes.

leading strand The DNA strand that can be synthesized continuously from the origin of replication. Compare *lagging strand*.

leaf primordium, pl. primordia A lateral outgrowth from the apical meristem that will eventually become a leaf.

lenticels Spongy areas in the cork surfaces of stem, roots, and other plant parts that allow interchange of gases between internal tissues and the atmosphere through the periderm.

leucine zipper motif A motif in regulatory proteins in which two different protein subunits associate to form a single DNA-binding site; the proteins are connected by an association between hydrophobic regions containing leucines (the "zipper").

leucoplast In plant cells, a colorless plastid in which starch grains are stored; usually found in cells not exposed to light.

leukocyte A white blood cell; a diverse array of nonhemoglobin-containing blood cells, including phagocytic macrophages and antibody-producing lymphocytes.

lichen Symbiotic association between a fungus and a photosynthetic organism such as a green alga or cyanobacterium.

ligand A signaling molecule that binds to a specific receptor protein, initiating signal transduction in cells.

light-dependent reactions In photosynthesis, the reactions in which light energy is captured and used in production of ATP and NADPH. In plants this involves the action of two linked photosystems.

light-independent reactions In photosynthesis, the reactions of the Calvin cycle in which ATP and NADPH from the light-dependent reactions are used to reduce CO_2 and produce organic compounds such as glucose. This involves the process of carbon fixation, or the conversion of inorganic carbon (CO_2) to organic carbon (ultimately carbohydrates).

lignin A highly branched polymer that makes plant cell walls more rigid; an important component of wood.

limbic system The hypothalamus, together with the network of neurons that link the hypothalamus to some areas of the cerebral cortex. Responsible for many of the most deep-seated drives and emotions of vertebrates, including pain, anger, sex, hunger, thirst, and pleasure.

linked genes Genes that are physically close together and therefore tend to segregate together; recombination occurring between linked genes can be used to produce a map of genetic distance for a chromosome.

linkage disequilibrium Association of alleles for 2 or more loci in a population that is higher than expected by chance.

lipase An enzyme that catalyzes the hydrolysis of fats.

lipid A nonpolar hydrophobic organic molecule that is insoluble in water (which is polar) but dissolves readily in nonpolar organic solvents; includes fats, oils, waxes, steroids, phospholipids, and carotenoids.

lipid bilayer The structure of a cellular membrane, in which two layers of phospholipids spontaneously align so that the hydrophilic head groups are exposed to water, while the hydrophobic fatty acid tails are pointed toward the center of the membrane.

lipopolysaccharide A lipid with a polysaccharide molecule attached; found in the outer membrane layer of gram-negative bacteria; the outer membrane layer protects the cell wall from antibiotic attack.

locus The position on a chromosome where a gene is located.

long interspersed element (LINE) Any of a type of large transposable element found in humans and other primates that contains all the biochemical machinery needed for transposition.

long terminal repeat (LTR) A particular type of retrotransposon that has repeated elements at its ends.

loop of Henle In the kidney of birds and mammals, a hairpin-shaped portion of the renal tubule in which water and salt are reabsorbed from the glomerular filtrate by diffusion.

lophophore A horseshoe-shaped crown of ciliated tentacles that surrounds the mouth of certain spiralian animals; seen in the phyla Brachiopoda and Bryozoa.

lumen A term for any bounded opening; for example, the cisternal space of the endoplasmic reticulum of eukaryotic cells, the passage through which blood flows inside a blood vessel, and the passage through which material moves inside the intestine during digestion.

luteal phase The second phase of the female reproductive cycle, during which the mature eggs are released into the Fallopian tubes, a process called ovulation.

lymph In animals, a colorless fluid derived from blood by filtration through capillary walls in the tissues.

lymphatic system In animals, an open vascular system that reclaims water that has entered interstitial regions from the bloodstream (lymph); includes the lymph nodes, spleen, thymus, and tonsils.

lymphocyte A type of white blood cell. Lymphocytes are responsible for the immune response; there are two principal classes: B cells and T cells.

lymphokine A regulatory molecule that is secreted by lymphocytes. In the immune response, lymphokines secreted by helper T cells unleash the cell-mediated immune response.

lysis Disintegration of a cell by rupture of its plasma membrane.

lysogenic cycle A viral cycle in which the viral DNA becomes integrated into the host chromosome and is replicated during cell reproduction. Results in vertical rather than horizontal transmission.

lysosome A membrane-bounded vesicle containing digestive enzymes that is produced by the Golgi apparatus in eukaryotic cells.

lytic cycle A viral cycle in which the host cell is killed (lysed) by the virus after viral duplication to release viral particles.

M

macroevolution The creation of new species and the extinction of old ones.

macromolecule An extremely large biological molecule; refers specifically to proteins, nucleic acids, polysaccharides, lipids, and complexes of these.

macronutrients Inorganic chemical elements required in large amounts for plant growth, such as nitrogen, potassium, calcium, phosphorus, magnesium, and sulfur.

macrophage A large phagocytic cell that is able to engulf and digest cellular debris and invading bacteria.

madreporite A sievelike plate on the surface of echinoderms through which water enters the water–vascular system.

***MADS* box gene** Any of a family of genes identified by possessing shared motifs that are the predominant homeotic genes of plants; a small number of *MADS* box genes are also found in animals.

major groove The larger of the two grooves in a DNA helix, where the paired nucleotides' hydrogen bonds are accessible; regulatory proteins can recognize and bind to regions in the major groove.

major histocompatibility complex (MHC) A set of protein cell-surface markers anchored in the plasma membrane, which the immune system uses to identify "self." All the cells of a given individual have the same "self" marker, called an MHC protein.

Malpighian tubules Blind tubules opening into the hindgut of terrestrial arthropods; they function as excretory organs.

mandibles In crustaceans, insects, and myriapods, the appendages immediately posterior to the antennae; used to seize, hold, bite, or chew food.

mantle The soft, outermost layer of the body wall in mollusks; the mantle secretes the shell.

map unit Each 1% of recombination frequency between two genetic loci; the unit is termed a centimorgan (cM) or simply a map unit (m.u.).

marsupial A mammal in which the young are born early in their development, sometimes as soon as eight days after fertilization, and are retained in a pouch.

mass extinction A relatively sudden, sharp decline in the number of species; for example, the extinction at the end of the Cretaceous period in which the dinosaurs and a variety of other organisms disappeared.

mass flow hypothesis The overall process by which materials move in the phloem of plants.

mast cells Leukocytes with granules containing molecules that initiate inflammation.

maternal inheritance A mode of uniparental inheritance from the female parent; for example, in humans mitochondria and their genomes are inherited from the mother.

matrix In mitochondria, the solution in the interior space surrounded by the cristae that contains the enzymes and other molecules involved in oxidative respiration; more generally, that part of a tissue within which an organ or process is embedded.

medusa A free-floating, often umbrella-shaped body form found in cnidarian animals, such as jellyfish.

megapascal (MPa) A unit of measure used for pressure in water potential.

megaphyll In plants, a leaf that has several to many veins connecting it to the vascular cylinder of the stem; most plants have megaphylls.

mesenchymal stem cell A type of stem cell that gives rise to connective tissues and used in the majority of approved stem cell therapies.

mesoglea A layer of gelatinous material found between the epidermis and gastrodermis of eumetazoans; it contains the muscles in most of these animals.

mesohyl A gelatinous, protein-rich matrix found between the choanocyte layer and the epithelial layer of the body of a sponge; various types of amoeboid cells may occur in the mesohyl.

metacercaria An encysted form of a larval liver fluke, found in muscle tissue of an infected animal; if the muscle is eaten, cysts dissolve in the digestive tract, releasing the flukes into the body of the new host.

methylation The addition of a methyl group to bases (primarily cytosine) in DNA. Cytosine methylation is correlated with DNA that is not expressed.

meiosis I The first round of cell division in meiosis; it is referred to as a "reduction division" because homologous chromosomes separate, and the daughter cells have only the haploid number of chromosomes.

meiosis II The second round of division in meiosis, during which the two haploid cells from meiosis I undergo a mitosis-like division without DNA replication to produce four haploid daughter cells.

membrane receptor A signal receptor present as an integral protein in the cell membrane, such as GPCRs, chemically gated ion channels in neurons, and RTKs.

Mendelian ratio The characteristic dominant-to-recessive phenotypic ratios that Mendel observed in his genetics experiments. For example, the F_2 generation in a monohybrid cross shows a ratio of 3:1; the F_2 generation in a dihybrid cross shows a ratio of 9:3:3:1.

menstruation Periodic sloughing off of the blood-enriched lining of the uterus when pregnancy does not occur.

meristem Undifferentiated plant tissue from which new cells arise.

meroblastic cleavage A type of cleavage in the eggs of reptiles, birds, and some fish. Occurs only on the blastodisc.

mesoderm One of the three embryonic germ layers that form in the gastrula; gives rise to muscle, bone and other connective tissue, the peritoneum, the circulatory system, and most of the excretory and reproductive systems.

mesophyll The photosynthetic parenchyma of a leaf, located within the epidermis.

messenger RNA (mRNA) The RNA transcribed from structural genes; RNA molecules complementary to a portion of one strand of DNA, which are translated by the ribosomes to form protein.

metabolism The sum of all chemical processes occurring within a living cell or organism.

metamorphosis Process in which a marked change in form takes place during postembryonic development as, for example, from tadpole to frog.

metaphase The stage of mitosis or meiosis during which microtubules become organized into a spindle and the chromosomes come to lie in the spindle's equatorial plane.

metastasis The process by which cancer cells move from their point of origin to other locations in the body; also, a population of cancer cells in a secondary location, the result of movement from the primary tumor.

methanogens Obligate, anaerobic archaebacteria that produce methane.

microbody A cellular organelle bounded by a single membrane and containing a variety of enzymes; generally derived from endoplasmic reticulum; includes peroxisomes and glyoxysomes.

microevolution Refers to the evolutionary process itself. Evolution within a species. Also called adaptation.

micronutrient A mineral required in only minute amounts for plant growth, such as iron, chlorine, copper, manganese, zinc, molybdenum, and boron.

microphyll In plants, a leaf that has only one vein connecting it to the vascular cylinder of the stem; the club mosses in particular have microphylls.

micropyle In the ovules of seed plants, an opening in the integuments through which the pollen tube usually enters.

micro-RNA (miRNA) A class of very short RNAs that are involved in controlling gene expression. These act as part of an RNA-induced silencing complex (RISC). *See also* small interfering RNAs (siRNAs).

microtubule In eukaryotic cells, a long, hollow protein cylinder, composed of the protein tubulin; these influence cell shape, move the chromosomes in cell division, and provide the functional internal structure of cilia and flagella.

microvillus Cytoplasmic projection from epithelial cells; microvilli greatly increase the surface area of the small intestine.

middle lamella The layer of intercellular material, rich in pectic compounds, that cements together the primary walls of adjacent plant cells.

mimicry The resemblance in form, color, or behavior of certain organisms (mimics) to other more powerful or more protected ones (models).

miracidium The ciliated first-stage larva inside the egg of the liver fluke; eggs are passed in feces, and if they reach water they may be eaten by a host snail in which they continue their life cycle.

missense mutation A base substitution mutation that results in the alteration of a single amino acid.

mitochondrion The organelle called the powerhouse of the cell. Consists of an outer membrane, an elaborate inner membrane that supports electron transport and chemiosmotic synthesis of ATP, and a soluble matrix containing Krebs cycle enzymes.

mitogen-activated protein (MAP) kinase Any of a class of protein kinases that activate transcription factors to alter gene expression. A mitogen is any molecule that stimulates cell division. MAP kinases are activated by kinase cascades.

mitosis Somatic cell division; nuclear division in which the duplicated chromosomes separate to form two genetically identical daughter nuclei.

molar concentration Concentration expressed as moles of a substance in 1 L of pure water.

mole The weight of a substance in grams that corresponds to the atomic masses of all the component atoms in a molecule of that substance. One mole of a compound always contains 6.023×10^{23} molecules.

molecular clock method In evolutionary theory, the method in which the rate of evolution of a molecule is constant through time.

molecular cloning The isolation and amplification of a specific sequence of DNA.

monocot Short for monocotyledon; flowering plant in which the embryos have only one cotyledon, the floral parts are generally in threes, and the leaves typically are parallel-veined.

monocyte A type of leukocyte that becomes a phagocytic cell (macrophage) after moving into tissues.

monoecious A plant in which the staminate and pistillate flowers are separate, but borne on the same individual.

monomer The smallest chemical subunit of a polymer. The monosaccharide α-glucose is the monomer found in plant starch, a polysaccharide.

monophyletic In phylogenetic classification, a group that includes the most recent common ancestor of the group and all its descendants. A clade is a monophyletic group.

monosaccharide A simple sugar that cannot be decomposed into smaller sugar molecules.

monosomic Describes the condition in which a chromosome has been lost due to nondisjunction during meiosis, producing a diploid embryo with only one of these autosomes.

monotreme An egg-laying mammal.

morphogen A signal molecule produced by an embryonic organizer region that informs surrounding cells of their distance from the organizer, thus determining relative positions of cells during development.

morphogenesis The development of an organism's body form, namely its organs and anatomical features; it may involve apoptosis as well as cell division, differentiation, and changes in cell shape.

morphology The form and structure of an organism.

morula Solid ball of cells in the early stage of embryonic development.

mosaic development A pattern of embryonic development in which initial cells produced by cleavage divisions contain different developmental signals (determinants) from the egg, setting the individual cells on different developmental paths.

motif A substructure in proteins that confers function and can be found in multiple proteins. One example is the helix-turn-helix motif found in a number of proteins that is used to bind to DNA.

motor (efferent) neuron Neuron that transmits nerve impulses from the central nervous system to an effector, which is typically a muscle or gland.

M phase The phase of cell division during which chromosomes are separated. The spindle assembles, binds to the chromosomes, and moves the sister chromatids apart.

M phase-promoting factor (MPF) A Cdk enzyme active at the G_2/M checkpoint.

Müllerian mimicry A phenomenon in which two or more unrelated but protected species resemble one another, thus achieving a kind of group defense.

multidrug-resistant (MDR) strain Any bacterial strain that has become resistant to more than one antibiotic drug; MDR *Staphylococcus* strains, for example, are responsible for many infection deaths.

multienzyme complex An assembly consisting of several enzymes catalyzing different steps in a sequence of reactions. Close proximity of these related enzymes speeds the overall process, making it more efficient.

multigene family A collection of related genes on a single chromosome or on different chromosomes.

muscle fiber A long, cylindrical, multinucleated cell containing numerous myofibrils, which is capable of contraction when stimulated.

mutagen An agent that induces changes in DNA (mutations); includes physical agents that damage DNA and chemicals that alter DNA bases.

mutation A permanent change in a cell's DNA; includes changes in nucleotide sequence, alteration of gene position, gene loss or duplication, and insertion of foreign sequences.

mutualism A symbiotic association in which two (or more) organisms live together, and both members benefit.

mycelium, pl. **mycelia** In fungi, a mass of hyphae.

mycorrhiza, pl. **mycorrhizae** A symbiotic association between fungi and the roots of a plant.

myelin sheath A fatty layer surrounding the long axons of motor neurons in the peripheral nervous system of vertebrates.

myofilament A contractile microfilament, composed largely of actin and myosin, within muscle.

myosin One of the two protein components of myofilaments (the other is actin); a principal component of vertebrate muscle. Functions as a motor protein that can use energy from ATP hydrolysis to generate force.

N

natural killer cell A cell that does not kill invading microbes, but rather, the cells infected by them.

natural selection The differential reproduction of genotypes; caused by factors in the environment; leads to evolutionary change.

nauplius A larval form characteristic of crustaceans.

negative control A type of control at the level of DNA transcription initiation in which the frequency of initiation is decreased; repressor proteins mediate negative control.

negative feedback A homeostatic control mechanism whereby an increase in some substance or activity inhibits the process leading to the increase; also known as feedback inhibition.

nematocyst A harpoonlike structure found in the cnidocytes of animals in the phylum Cnidaria, which includes the jellyfish among other groups; the nematocyst, when released, stings and helps capture prey.

nephridium, pl. **nephridia** In invertebrates, a tubular excretory structure.

nephrid organ A filtration system of many freshwater invertebrates in which water and waste pass from the body across the membrane into a collecting organ, from which they are expelled to the outside through a pore.

nephron Functional unit of the vertebrate kidney; one of numerous tubules involved in filtration and selective reabsorption of blood; each nephron consists of a Bowman's capsule, an enclosed glomerulus, and a long attached tubule; in humans, called a renal tubule.

nephrostome The funnel-shaped opening that leads to the nephridium, which is the excretory organ of mollusks.

nerve A group or bundle of nerve fibers (axons) with accompanying neurological cells, held together by connective tissue; located in the peripheral nervous system.

nerve cord One of the distinguishing features of chordates, running lengthwise just beneath the embryo's dorsal surface; in vertebrates, differentiates into the brain and spinal cord.

neural crest A migratory cell population that arises from the neural tube and gives rise to the peripheral nervous system and other structures.

neural groove The long groove formed along the long axis of the embryo by a layer of ectodermal cells.

neural tube The dorsal tube, formed from the neural plate, that differentiates into the brain and spinal cord.

neuroglia Nonconducting nerve cells that are intimately associated with neurons and appear to provide nutritional support.

neuromuscular junction The structure formed when the tips of axons contact (innervate) a muscle fiber.

neuron A nerve cell specialized for signal transmission; includes cell body, dendrites, and axon.

neurotransmitter A chemical released at the axon terminal of a neuron that travels across the synaptic cleft, binds a specific receptor on the far side, and depending on the nature of the receptor, depolarizes or hyperpolarizes a second neuron or a muscle or gland cell.

neurulation A process in early embryonic development by which a dorsal band of ectoderm thickens and rolls into the neural tube.

neutrophil An abundant type of granulocyte capable of engulfing microorganisms and other foreign particles; neutrophils comprise about 50–70% of the total number of white blood cells.

niche The role played by a particular species in its environment.

nicotinamide adenine dinucleotide (NAD) A molecule that becomes reduced (to NADH) as it carries high-energy electrons from oxidized molecules and delivers them to ATP-producing pathways in the cell.

NADH dehydrogenase An enzyme located on the inner mitochondrial membrane that catalyzes the oxidation by NAD^+ of pyruvate to acetyl-CoA. This reaction links glycolysis and the Krebs cycle.

nitrification The oxidization of ammonia or nitrite to produce nitrate, the form of nitrogen taken up by plants; some bacteria are capable of nitrification.

nociceptor A naked dendrite that acts as a receptor in response to a pain stimulus.

nocturnal Active primarily at night.

node The part of a plant stem where one or more leaves are attached. See internode.

node of Ranvier A gap formed at the point where two Schwann cells meet and where the axon is in direct contact with the surrounding intercellular fluid.

nodule In plants, a specialized tissue that surrounds and houses beneficial bacteria, such as root nodules of legumes that contain nitrogen-fixing bacteria.

nonassociative learning A learned behavior that does not require an animal to form an association between two stimuli, or between a stimulus and a response.

noncompetitive inhibitor An inhibitor that binds to a location other than the active site of an enzyme, changing the enzyme's shape so that it cannot bind the substrate.

noncyclic photophosphorylation The set of light-dependent reactions of the two plant photosystems, in which excited electrons are shuttled between the two photosystems, producing a proton gradient that is used for the chemiosmotic synthesis of ATP. The electrons are used to reduce NADP to NADPH. Lost electrons are replaced by the oxidation of water producing O_2.

nondisjunction The failure of homologues or sister chromatids to separate during mitosis or meiosis, resulting in an aneuploid cell or gamete.

nonextreme archaea Archaeal groups that are not extremophiles, living in more moderate environments on Earth today.

nonpolar Said of a covalent bond that involves equal sharing of electrons. Can also refer to a compound held together by nonpolar covalent bonds.

nonsense codon One of three codons (UAA, UAG, and UGA) that are not recognized by tRNAs, thus serving as "stop" signals in the mRNA message and terminating translation.

nonsense mutation A base substitution in which a codon is changed into a stop codon. The protein is truncated because of premature termination.

Northern blot A blotting technique used to identify a specific mRNA sequence in a complex mixture. See Southern blot.

notochord In chordates, a dorsal rod of cartilage that runs the length of the body and forms the primitive axial skeleton in the embryos of all chordates.

nucellus Tissue composing the chief pair of young ovules, in which the embryo sac develops; equivalent to a megasporangium.

nuclear envelope The bounding structure of the eukaryotic nucleus. Composed of two phospholipid bilayers with the outer one connected to the endoplasmic reticulum.

nuclear pore One of a multitude of tiny but complex openings in the nuclear envelope that allow selective passage of proteins and nucleic acids into and out of the nucleus.

nuclear receptor Intracellular receptors, primarily for steroid hormones, that are found in both the cytoplasm and the nucleus. The site of action of the hormone–receptor complex is in the nucleus where they modify gene expression.

nucleic acid A nucleotide polymer; chief types are deoxyribonucleic acid (DNA), which is double-stranded, and ribonucleic acid (RNA), which is typically single-stranded.

nucleoid The area of a prokaryotic cell, usually near the center, that contains the genome in the form of DNA compacted with protein.

nucleolus In eukaryotes, the site of rRNA synthesis; a spherical body composed chiefly of rRNA in the process of being transcribed from multiple copies of rRNA genes.

nucleosome A complex consisting of a DNA duplex wound around a core of eight histone proteins.

nucleotide A single unit of nucleic acid, composed of a phosphate, a 5-carbon sugar (either ribose or deoxyribose), and a purine or a pyrimidine.

nucleus In atoms, the central core, containing positively charged protons and (in all but hydrogen) electrically neutral neutrons; in eukaryotic cells, the membranous organelle that houses the chromosomal DNA; in the central nervous system, a cluster of nerve cell bodies.

nutritional mutation A mutation affecting a synthetic pathway for a vital compound, such as an amino acid or vitamin; microorganisms with a nutritional mutation must be grown on medium that supplies the missing nutrient.

O

ocellus, pl. ocelli A simple light receptor common among invertebrates.

octet rule Rule to describe patterns of chemical bonding in main group elements that require a total of eight electrons to complete their outer electron shell.

Okazaki fragment A short segment of DNA produced by discontinuous replication elongating in the 5′-to-3′ direction away from the replication.

olfaction The function of smelling.

ommatidium, pl. ommatidia The visual unit in the compound eye of arthropods; contains light-sensitive cells and a lens able to form an image.

oncogene A mutant form of a growth-regulating gene that is inappropriately "on," causing unrestrained cell growth and division.

oocyst The zygote in a sporozoan life cycle. It is surrounded by a tough cyst to prevent dehydration or other damage.

open circulatory system A circulatory system in which the blood flows into sinuses in which it mixes with body fluid and then reenters the vessels in another location.

open reading frame (ORF) A region of DNA that encodes a sequence of amino acids with no stop codons in the reading frame.

operant conditioning A learning mechanism in which the reward follows only after the correct behavioral response.

operator A regulatory site on DNA to which a repressor can bind to prevent or decrease initiation of transcription.

operculum A flat, bony, external protective covering over the gill chamber in fish.

operon A cluster of adjacent structural genes transcribed as a unit into a single mRNA molecule.

opisthosoma The posterior portion of the body of an arachnid.

oral surface The surface on which the mouth is found; used as a reference when describing the body structure of echinoderms because of their adult radial symmetry.

orbital A region around the nucleus of an atom with a high probability of containing an electron. The position of electrons can only be described by these probability distributions.

order A category of classification above the level of family and below that of class.

organ A body structure composed of several different tissues grouped in a structural and functional unit.

organelle Specialized part of a cell; literally, a small cytoplasmic organ.

orthologues Genes that reflect the conservation of a single gene found in an ancestor.

oscillating selection The situation in which selection alternately favors one phenotype at one time, and a different phenotype at a another time, for example, during drought conditions versus during wet conditions.

osculum A specialized, larger pore in sponges through which filtered water is forced to the outside of the body.

osmoconformer An animal that maintains the osmotic concentration of its body fluids at about the same level as that of the medium in which it is living.

osmosis The diffusion of water across a selectively permeable membrane (a membrane that permits the free passage of water but prevents or retards the passage of a solute); in the absence of differences in pressure or volume, the net movement of water is from the side containing a lower concentration of solute to the side containing a higher concentration.

osmotic concentration The property of a solution that takes into account all dissolved solutes in the solution; if two solutions with different osmotic concentrations are separated by a water-permeable membrane, water will move from the solution with lower osmotic concentration to the solution with higher osmotic concentration.

osmotic pressure The potential pressure developed by a solution separated from pure water by a differentially permeable membrane. The higher the solute concentration, the greater the osmotic potential of the solution; also called *osmotic potential*.

ossicle Any of a number of movable or fixed calcium-rich plates that collectively make up the endoskeleton of echinoderms.

osteoblast A bone-forming cell.

osteocyte A mature osteoblast.

outcrossing Breeding with individuals other than oneself or one's close relatives.

ovary (1) In animals, the organ in which eggs are produced. (2) In flowering plants, the enlarged basal portion of a carpel that contains the ovule(s); the ovary matures to become the fruit.

oviduct In vertebrates, the passageway through which ova (eggs) travel from the ovary to the uterus.

oviparity Refers to a type of reproduction in which the eggs are developed after leaving the body of the mother, as in reptiles.

ovoviviparity Refers to a type of reproduction in which young hatch from eggs that are retained in the mother's uterus.

ovulation In animals, the release of an egg or eggs from the ovary.

ovum, pl. ova The egg cell; female gamete.

oxidation Loss of an electron by an atom or molecule; in metabolism, often associated with a gain of oxygen or a loss of hydrogen.

oxidation–reduction reaction A type of paired reaction in living systems in which electrons lost from one atom (oxidation) are gained by another atom (reduction). Termed a *redox reaction* for short.

oxidative phosphorylation Synthesis of ATP by ATP synthase using energy from a proton gradient. The proton gradient is generated by electron transport, which requires oxygen.

oxygen debt The amount of oxygen required to convert the lactic acid generated in the muscles during exercise back into glucose.

oxytocin A hormone of the posterior-pituitary gland that affects uterine contractions during childbirth and stimulates lactation.

ozone O_3, a stratospheric layer of the Earth's atmosphere responsible for filtering out ultraviolet radiation supplied by the Sun.

P

***p53* gene** The gene that produces the p53 protein that monitors DNA integrity and halts cell division if DNA damage is detected. Many types of cancer are associated with a damaged or absent *p53* gene.

pacemaker A patch of excitatory tissue in the vertebrate heart that initiates the heartbeat.

pair-rule gene Any of certain genes in *Drosophila* development controlled by the gap genes that are expressed in stripes that subdivide the embryo in the process of segmentation.

paleopolyploid An ancient polyploid organism used in analysis of polyploidy events in the study of a species' genome evolution.

palisade parenchyma In plant leaves, the columnar, chloroplast-containing parenchyma cells of the mesophyll. Also called *palisade cells*.

panspermia The hypothesis that meteors or cosmic dust may have brought significant amounts of complex organic molecules to Earth, kicking off the evolution of life.

papilla A small projection of tissue.

paracrine A type of chemical signaling between cells in which the effects are local and short-lived.

paralogues Two genes within an organism that arose from the duplication of one gene in an ancestor.

paraphyletic In phylogenetic classification, a group that includes the most recent common ancestor of the group, but not all its descendants.

parapodia One of the paired lateral processes on each side of most segments in polychaete annelids.

parasexuality In certain fungi, the fusion and segregation of heterokaryotic haploid nuclei to produce recombinant nuclei.

parasitism A living arrangement in which an organism lives on or in an organism of a different species and derives nutrients from it.

parenchyma cell The most common type of plant cell; characterized by large vacuoles, thin walls, and functional nuclei.

parthenogenesis The development of an egg without fertilization, as in aphids, bees, ants, and some lizards.

partial diploid (merodiploid) Describes an *E. coli* cell that carries an F′ plasmid with host genes. This makes the cell diploid for the genes carried by the F′ plasmid.

partial pressure The components of each individual gas—such as nitrogen, oxygen, and carbon dioxide—that together constitute the total air pressure.

passive transport The movement of substances across a cell's membrane without the expenditure of energy.

pedigree A consistent graphic representation of matings and offspring over multiple generations for a particular genetic trait, such as albinism or hemophilia.

pedipalps A pair of specialized appendages found in arachnids; in male spiders, these are specialized as copulatory organs, whereas in scorpions they are large pincers.

pelagic Free-swimming, usually in open water.

pellicle A tough, flexible covering in ciliates and euglenoids.

pentaradial symmetry The five-part radial symmetry characteristic of adult echinoderms.

peptide bond The type of bond that links amino acids together in proteins through a dehydration reaction.

peptidoglycan A component of the cell wall of bacteria, consisting of carbohydrate polymers linked by protein cross-bridges.

peptidyl transferase In translation, the enzyme responsible for catalyzing the formation of a peptide bond between each new amino acid and the previous amino acid in a growing polypeptide chain.

perianth In flowering plants, the petals and sepals taken together.

pericycle In vascular plants, one or more cell layers surrounding the vascular tissues of the root, bounded externally by the endodermis and internally by the phloem.

periderm Outer protective tissue in vascular plants that is produced by the cork cambium and functionally replaces epidermis when it is destroyed during secondary growth; the periderm includes the cork, cork cambium, and phelloderm.

peristalsis In animals, a series of alternating contracting and relaxing muscle movements along the length of a tube such as the oviduct or alimentary canal that tend to force material such as an egg cell or food through the tube.

peroxisome A microbody that plays an important role in the breakdown of highly oxidative hydrogen peroxide by catalase.

petal A flower part, usually conspicuously colored; one of the units of the corolla.

petiole The stalk of a leaf.

phage conversion The phenomenon by which DNA from a virus, incorporated into a host cell's genome, alters the host cell's function in a significant way; for example, the conversion of *Vibrio cholerae* bacteria into a pathogenic form that releases cholera toxin.

phage lambda (λ) A well-known bacteriophage that has been widely used in genetic studies and is often a vector for DNA libraries.

phagocyte Any cell that engulfs and devours microorganisms or other particles.

phagocytosis Endocytosis of a solid particle; the plasma membrane folds inward around the particle (which may be another cell) and engulfs it to form a vacuole.

pharyngeal pouches In chordates, embryonic regions that become pharyngeal slits in aquatic and marine chordates and vertebrates, but do not develop openings to the outside in terrestrial vertebrates.

pharyngeal slits One of the distinguishing features of chordates; a group of openings on each side of the anterior region that form a passageway from the pharynx and esophagus to the external environment.

pharynx A muscular structure lying posterior to the mouth in many animals; aids in propelling food into the digestive tract.

phenotype The realized expression of the genotype; the physical appearance or functional expression of a trait.

pheromone Chemical substance released by one organism that influences the behavior or physiological processes of another organism of the same species. Pheromones serve as sex attractants, as trail markers, and as alarm signals.

phloem In vascular plants, a food-conducting tissue basically composed of sieve elements, various kinds of parenchyma cells, fibers, and sclereids.

phoronid Any of a group of lophophorate invertebrates, now classified in the phylum Brachiopoda, that burrows into soft underwater substrates and secretes a chitinous tube in which it lives out its life; it extends its lophophore tentacles to feed on drifting food particles.

phosphatase Any of a number of enzymes that removes a phosphate group from a protein, reversing the action of a kinase.

phosphodiester bond The linkage between two sugars in the backbone of a nucleic acid molecule; the phosphate group connects the pentose sugars through a pair of ester bonds.

phospholipid Similar in structure to a fat, but having only two fatty acids attached to the glycerol backbone, with the third space linked to a phosphorylated molecule; contains a polar hydrophilic "head" end (phosphate group) and a nonpolar hydrophobic "tail" end (fatty acids).

phospholipid bilayer The main component of cell membranes; phospholipids naturally associate in a bilayer with hydrophobic fatty acids oriented to the inside and hydrophilic phosphate groups facing outward on both sides.

phosphorylation Chemical reaction resulting in the addition of a phosphate group to an organic molecule. Phosphorylation of ADP yields ATP. Many proteins are also activated or inactivated by phosphorylation.

photic zone The area in an aquatic habitat that receives sufficient light for photosynthesis to occur and net primary productivity to be positive.

photoelectric effect The ability of a beam of light to excite electrons, creating an electrical current.

photon A particle of light having a discrete amount of energy. The wave concept of light explains the different colors of the spectrum, whereas the particle concept of light explains the energy transfers during photosynthesis.

photoperiodism The tendency of biological reactions to respond to the duration and timing of day and night; a mechanism for measuring seasonal time.

photoreceptor A light-sensitive sensory cell.

photorespiration Action of the enzyme rubisco, which catalyzes the oxidization of RuBP, releasing CO_2; this reverses carbon fixation and can reduce the yield of photosynthesis.

photosystem An organized complex of chlorophyll, other pigments, and proteins that traps light energy as excited electrons. Plants have two linked photosystems in the thylakoid membrane of chloroplasts. Photosystem II passes an excited electron through an electron transport chain to photosystem I to replace an excited electron passed to NADPH. The electron lost from photosystem II is replaced by the oxidation of water.

phototropism In plants, a growth response to a light stimulus.

pH scale A scale used to measure acidity and basicity. Defined as the negative log of H^+ concentration. Ranges from 0 to 14. A value of 7 is neutral; below 7 is acidic and above 7 is basic.

phycobiliprotein A type of accessory pigment found in cyanobacteria and some algae. Complexes of phycobiliprotein are able to absorb light energy in the green range.

phycologist A scientist who studies algae.

phyllotaxy In plants, a spiral pattern of leaf arrangement on a stem in which sequential leaves are at a 137.5° angle to one another, an angle related to the golden mean.

phylogenetic species concept (PSC) The concept that defines species on the basis of their phylogenetic relationships.

phylogenetic tree A pattern of descent generated by analysis of similarities and differences among organisms. Modern gene-sequencing techniques have produced phylogenetic trees showing the evolutionary history of individual genes.

phylogeny The evolutionary history of an organism, including which species are closely related and in what order related species evolved; often represented in the form of an evolutionary tree.

phylum, pl. phyla A major category, between kingdom and class, of taxonomic classifications.

physical map A map of the DNA sequence of a chromosome or genome based on actual landmarks within the DNA.

phytochrome A plant pigment that is associated with the absorption of light; photoreceptor for red to far-red light.

phytoestrogen One of a number of secondary metabolites in some plants that are structurally and functionally similar to the animal hormone estrogen.

phytoremediation The process that uses plants to remove contamination from soil or water.

pigment A molecule that absorbs light.

pilus, pl. pili Extensions of a bacterial cell enabling it to transfer genetic materials from one individual to another or to adhere to substrates.

pinocytosis The process of fluid uptake by endocytosis in a cell.

pistil Central organ of flowers, typically consisting of ovary, style, and stigma; a pistil may consist of one or more fused carpels and is more technically and better known as the gynoecium.

pith The ground tissue occupying the center of the stem or root within the vascular cylinder.

pituitary gland Endocrine gland at the base of the hypothalamus composed of anterior and posterior

placenta, pl. **placentae** (1) In flowering plants, the part of the ovary wall to which the ovules or seeds are attached. (2) In mammals, a tissue formed in part from the inner lining of the uterus and in part from other membranes, through which the embryo (later the fetus) is nourished while in the uterus and through which wastes are carried away.

plankton Free-floating, mostly microscopic, aquatic organisms.

plant receptor kinase Any of a group of plant membrane receptors that, when activated by binding ligand, have kinase enzymatic activity. These receptors phosphorylate serine or threonine, unlike RTKs in animals that phosphorylate tyrosine.

planula A ciliated, free-swimming larva produced by the medusae of cnidarian animals.

plasma The fluid of vertebrate blood; contains dissolved salts, metabolic wastes, hormones, and a variety of proteins, including antibodies and albumin; blood minus the blood cells.

plasma cell An antibody-producing cell resulting from the multiplication and differentiation of a B lymphocyte that has interacted with an antigen.

plasma membrane The membrane surrounding the cytoplasm of a cell; consists of a single phospholipid bilayer with embedded proteins.

plasmid A small fragment of extrachromosomal DNA, usually circular, that replicates independently of the main chromosome, although it may have been derived from it.

plasmodesmata In plants, cytoplasmic connections between adjacent cells.

plasmodium Stage in the life cycle of myxomycetes (plasmodial slime molds); a multinucleate mass of protoplasm surrounded by a membrane.

plasmolysis The shrinking of a plant cell in a hypertonic solution such that it pulls away from the cell wall.

plastid An organelle in the cells of photosynthetic eukaryotes that is the site of photosynthesis and, in plants and green algae, of starch storage.

platelet In mammals, a fragment of a white blood cell that circulates in the blood and functions in the formation of blood clots at sites of injury.

pleiotropy Condition in which an individual allele has more than one effect on production of the phenotype.

plesiomorphy In cladistics, another term for an ancestral character state.

plumule The epicotyl of a plant with its two young leaves.

point mutation An alteration of one nucleotide in a chromosomal DNA molecule.

polar body Minute, nonfunctioning cell produced during the meiotic divisions leading to gamete formation in vertebrates.

polar covalent bond A covalent bond in which electrons are shared unequally due to differences in electronegativity of the atoms involved. One atom has a partial negative charge and the other a partial positive charge, even though the molecule is electrically neutral overall.

polarity (1) Refers to unequal charge distribution in a molecule such as water, which has a positive region and a negative region although it is neutral overall. (2) Refers to axial differences in a developing embryo that result in anterior–posterior and dorsal–ventral axes in a bilaterally symmetrical animal.

polarize In cladistics, to determine whether character states are ancestral or derived.

pollen tube A tube formed after germination of the pollen grain; carries the male gametes into the ovule.

pollination The transfer of pollen from an anther to a stigma.

polyandry The condition in which a female mates with more than one male.

polyclonal antibody An antibody response in which an antigen elicits many different antibodies, each fitting a different portion of the antigen surface.

polygenic inheritance Describes a mode of inheritance in which more than one gene affects a trait, such as height in human beings; polygenic inheritance may produce a continuous range of phenotypic values, rather than discrete either–or values.

polygyny A mating choice in which a male mates with more than one female.

polymer A molecule composed of many similar or identical molecular subunits; starch is a polymer of glucose.

polymerase chain reaction (PCR) A process by which DNA polymerase is used to copy a sequence of interest repeatedly, making millions of copies of the same DNA.

polymorphism The presence in a population of more than one allele of a gene at a frequency greater than that of newly arising mutations.

polyp A typically sessile, cylindrical body form found in cnidarian animals, such as hydras.

polypeptide A molecule consisting of many joined amino acids; not usually as complex as a protein.

polyphyletic In phylogenetic classification, a group that does not include the most recent common ancestor of all members of the group.

polyploidy Condition in which one or more entire sets of chromosomes is added to the diploid genome.

polysaccharide A carbohydrate composed of many monosaccharide sugar subunits linked together in a long chain; examples are glycogen, starch, and cellulose.

polyunsaturated fat A fat molecule having at least two double bonds between adjacent carbons in one or more of the fatty acid chains.

population Any group of individuals, usually of a single species, occupying a given area at the same time.

population genetics The study of the properties of genes in populations.

positive control A type of control at the level of DNA transcription initiation in which the frequency of initiation is increased; activator proteins mediate positive control.

posttranscriptional control A mechanism of control over gene expression that operates after the transcription of mRNA is complete.

postzygotic isolating mechanism A type of reproductive isolation in which zygotes are produced but are unable to develop into reproducing adults; these mechanisms may range from inviability of zygotes or embryos to adults that are sterile.

potential energy Energy that is not being used, but could be; energy in a potentially usable form; often called "energy of position."

precapillary sphincter A ring of muscle that guards each capillary loop and that, when closed, blocks flow through the capillary.

pre-mRNA splicing In eukaryotes, the process by which introns are removed from the primary transcript to produce mature mRNA; pre-mRNA splicing occurs in the nucleus.

pressure potential In plants, the turgor pressure resulting from pressure against the cell wall.

prezygotic isolating mechanism A type of reproductive isolation in which the formation of a zygote is prevented; these mechanisms may range from physical separation in different habitats to gametic in which gametes are incapable of fusing.

primary endosperm nucleus In flowering plants, the result of the fusion of a sperm nucleus and the (usually) two polar nuclei.

primary growth In vascular plants, growth originating in the apical meristems of shoots and roots; results in an increase in length.

primary immune response The first response of an immune system to a foreign antigen. If the system is challenged again with the same antigen, the memory cells created during the primary response will respond more quickly.

primary induction Inductions between the three primary tissue types: mesoderm and endoderm.

primary meristem Any of the three meristems produced by the apical meristem; primary meristems give rise to the dermal, vascular, and ground tissues.

primary nondisjunction Failure of chromosomes to separate properly at meiosis I.

primary phloem The cells involved in food conduction in plants.

primary plant body The part of a plant consisting of young, soft shoots and roots derived from apical meristem tissues.

primary productivity The amount of energy produced by photosynthetic organisms in a community.

primary structure The specific amino acid sequence of a protein.

primary tissues Tissues that make up the primary plant body.

primary transcript The initial mRNA molecule copied from a gene by RNA polymerase, containing a faithful copy of the entire gene, including introns as well as exons.

primary wall In plants, the wall layer deposited during the period of cell expansion.

primase The enzyme that synthesizes the RNA primers required by DNA polymerases.

primate Monkeys and apes (including humans).

primitive streak In the early embryos of birds, reptiles, and mammals, a dorsal, longitudinal strip of ectoderm and mesoderm that is equivalent to the blastopore in other forms.

primordium In plants, a bulge on the young shoot produced by the apical meristem; primordia can differentiate into leaves, other shoots, or flowers.

principle of parsimony Principle stating that scientists should favor the hypothesis that requires the fewest assumptions.

prions Infectious proteinaceous particles.

procambium In vascular plants, a primary meristematic tissue that gives rise to primary vascular tissues.

product rule *See* rule of multiplication.

proglottid A repeated body segment in tapeworms that contains both male and female reproductive organs; proglottids eventually form eggs and embryos, which leave the host's body in feces.

prokaryote A cell lacking a membrane-bounded nucleus or membrane-bounded organelles.

prometaphase The transitional phase between prophase and metaphase during which the spindle attaches to the kinetochores of sister chromatids.

promoter A DNA sequence that provides a recognition and attachment site for RNA polymerase to begin the process of gene transcription; it is located upstream from the transcription start site.

prophase The phase of cell division that begins when the condensed chromosomes become visible and ends when the nuclear envelope breaks down. The assembly of the spindle takes place during prophase.

proprioceptor In vertebrates, a sensory receptor that senses the body's position and movements.

prosimian Any member of the mammalian group that is a sister group to the anthropoids; prosimian means "before monkeys." Members include the lemurs, lorises, and tarsiers.

prosoma The anterior portion of the body of an arachnid, which bears all the appendages.

prostaglandins A group of modified fatty acids that function as chemical messengers.

prostate gland In male mammals, a mass of glandular tissue at the base of the urethra that secretes an alkaline fluid that has a stimulating effect on the sperm as they are released.

protease An enzyme that degrades proteins by breaking peptide bonds; in cells, proteases are often compartmentalized into vesicles such as lysosomes.

proteasome A large, cylindrical cellular organelle that degrades proteins marked with ubiquitin.

protein A chain of amino acids joined by peptide bonds.

protein kinase An enzyme that adds phosphate groups to proteins, changing their activity.

protein microarray An array of proteins on a microscope slide or silicon chip. The array may be used with a variety of probes, including antibodies, to analyze the presence or absence of specific proteins in a complex mixture.

proteome All the proteins coded for by a particular genome.

proteomics The study of the proteomes of organisms. This is related to functional genomics as the proteome is responsible for much of the function encoded by a genome.

protoderm The primary meristem that gives rise to the dermal tissue.

proton pump A protein channel in a membrane of the cell that expends energy to transport protons against a concentration gradient; involved in the chemiosmotic generation of ATP.

proto-oncogene A normal cellular gene that can act as an oncogene when mutated.

protostome Any member of a grouping of bilaterally symmetrical animals in which the mouth develops first and the anus second; flatworms, nematodes, mollusks, annelids, and arthropods are protostomes.

pseudocoel A body cavity located between the endoderm and mesoderm.

pseudogene A copy of a gene that is not transcribed.

pseudomurien A component of the cell wall of archaea; it is similar to peptidoglycan in structure and function but contains different components.

pseudopod A nonpermanent cytoplasmic extension of the cell body.

P site In a ribosome, the peptidyl site that binds to the tRNA attached to the growing polypeptide chain.

punctuated equilibrium A hypothesis about the mechanism of evolutionary change proposing that long periods of little or no change are punctuated by periods of rapid evolution.

Punnett square A diagrammatic way of showing the possible genotypes and phenotypes of genetic crosses.

pupa A developmental stage of some insects in which the organism is nonfeeding, immotile, and sometimes encapsulated or in a cocoon; the pupal stage occurs between the larval and adult phases.

purine The larger of the two general kinds of nucleotide base found in DNA and RNA; a nitrogenous base with a double-ring structure, such as adenine or guanine.

pyrimidine The smaller of two general kinds of nucleotide base found in DNA and RNA; a nitrogenous base with a single-ring structure, such as cytosine, thymine, or uracil.

pyruvate A 3-carbon molecule that is the end product of glycolysis; each glucose molecule yields two pyruvate molecules.

Q

quantitative trait A trait that is determined by the effects of more than one gene; such a trait usually exhibits continuous variation rather than discrete either–or values.

quaternary structure The structural level of a protein composed of more than one polypeptide chain, each of which has its own tertiary structure; the individual chains are called subunits.

R

radial canal Any of five canals that connect to the ring canal of an echinoderm's water–vascular system.

radial cleavage The embryonic cleavage pattern of deuterostome animals in which cells divide parallel to and at right angles to the polar axis of the embryo.

radial symmetry A type of structural symmetry with a circular plan, such that dividing the body or structure through the midpoint in any direction yields two identical sections.

radicle The part of the plant embryo that develops into the root.

radioactive isotope An isotope that is unstable and undergoes radioactive decay, releasing energy.

radioactivity The emission of nuclear particles and rays by unstable atoms as they decay into more stable forms.

radula Rasping tongue found in most mollusks.

reaction center A transmembrane protein complex in a photosystem that receives energy from the antenna complex exciting an electron that is passed to an acceptor molecule.

reading frame The correct succession of nucleotides in triplet codons that specify amino acids on translation. The reading frame is established by the first codon in the sequence as there are no spaces in the genetic code.

realized niche The actual niche occupied by an organism when all biotic and abiotic interactions are taken into account.

receptor-mediated endocytosis Process by which specific macromolecules are transported into eukaryotic cells at clathrin-coated pits, after binding to specific cell-surface receptors.

receptor protein A cellular protein that can bind to a ligand (signaling molecule) and initiate a signal transduction pathway. Can be either a membrane protein or intracellular depending on the nature of the ligand.

receptor tyrosine kinase (RTK) A diverse group of membrane receptors that when activated have kinase enzymatic activity. Specifically, they phosphorylate proteins on tyrosine. Their activation can lead to diverse cellular responses.

recessive An allele that is only expressed when present in the homozygous condition, but being "hidden" by the expression of a dominant allele in the heterozygous condition.

redia A secondary, nonciliated larva produced in the sporocysts of liver flukes.

regulatory protein Any of a group of proteins that modulates the ability of RNA polymerase to bind to a promoter and begin DNA transcription.

replicon An origin of DNA replication and the DNA whose replication is controlled by this origin. In prokaryotic replication, the chromosome plus the origin consist of a single replicon; eukaryotic chromosomes consist of multiple replicons.

replisome The macromolecular assembly of enzymes involved in DNA replication; analogous to the ribosome in protein synthesis.

reciprocal altruism Performance of an altruistic act with the expectation that the favor will be returned. A key and very controversial assumption of many theories dealing with the evolution of social behavior. *See* altruism.

reciprocal cross A genetic cross involving a single trait in which the sex of the parents is reversed; for example, if pollen from a white-flowered plant is used to fertilize a purple-flowered plant, the reciprocal cross would be pollen from a purple-flowered plant used to fertilize a white-flowered plant.

reciprocal recombination A mechanism of genetic recombination that occurs only in eukaryotic organisms, in which two chromosomes trade segments; can occur between nonhomologous chromosomes as well as the more usual exchange between homologous chromosomes in meiosis.

recombinant DNA Fragments of DNA from two different species, such as a bacterium and a mammal, joined together in the laboratory into a single molecule.

recombination frequency The value obtained by dividing the number of recombinant progeny by the total progeny in a test cross. This value is converted into a percentage, and each 1% is termed a map unit.

reduction The gain of an electron by an atom, often with an associated proton.

reflex In the nervous system, a motor response subject to little associative modification; a reflex is among the simplest neural pathways, involving only a sensory neuron, sometimes (but not always) an interneuron, and one or more motor neurons.

reflex arc The nerve path in the body that leads from stimulus to reflex action.

refractory period The recovery period after membrane depolarization during which the membrane is unable to respond to additional stimulation.

reinforcement In speciation, the process by which partial reproductive isolation between populations is increased by selection against mating between members of the two populations, eventually resulting in complete reproductive isolation.

replica plating A method of transferring bacterial colonies from one plate to another to make a copy of the original plate; an impression of colonies growing on a Petri plate is made on a velvet surface, which is then used to transfer the colonies to plates containing different media, such that auxotrophs can be identified.

replication fork The Y-shaped end of a growing replication bubble in a DNA molecule undergoing replication.

repolarization Return of the ions in a nerve to their resting potential distribution following depolarization.

repression In general, control of gene expression by preventing transcription. Specifically, in bacteria such as *E. coli* this is mediated by repressor proteins. In anabolic operons, repressors bind DNA in the absence of corepressors to repress an operon.

repressor A protein that regulates DNA transcription by preventing RNA polymerase from attaching to the promoter and transcribing the structural gene. *See* operator.

reproductive isolating mechanism Any barrier that prevents genetic exchange between species.

residual volume The amount of air remaining in the lungs after the maximum amount of air has been exhaled.

resting membrane potential The charge difference (difference in electric potential) that exists across a neuron at rest (about 70 mV).

restriction endonuclease An enzyme that cleaves a DNA duplex molecule at a particular base sequence, usually within or near a palindromic sequence; also called a restriction enzyme.

restriction fragment length polymorphism (RFLP) Restriction enzymes recognize very specific DNA sequences. Alleles of the same gene or surrounding sequences may have base-pair differences, so that DNA near one allele is cut into a different-length fragment than DNA near the other allele. These different fragments separate based on size on electrophoresis gels.

retina The photosensitive layer of the vertebrate eye; contains several layers of neurons and light receptors (rods and cones); receives the image formed by the lens and transmits it to the brain via the optic nerve.

retinoblastoma susceptibility gene (*Rb*) A gene that, when mutated, predisposes individuals to a rare form of cancer of the retina; one of the first tumor-suppressor genes discovered.

retrovirus An RNA virus. When a retrovirus enters a cell, a viral enzyme (reverse transcriptase) transcribes viral RNA into duplex DNA, which the cell's machinery then replicates and transcribes as if it were its own.

reverse genetics An approach by which a researcher uses a cloned gene of unknown function, creates a mutation, and introduces the mutant gene back into the organism to assess the effect of the mutation.

reverse transcriptase A viral enzyme found in retroviruses that is capable of converting their RNA genome into a DNA copy.

Rh blood group A set of cell-surface markers (antigens) on the surface of red blood cells in humans and rhesus monkeys (for which it is named); although there are several alleles, they are grouped into two main types: Rh-positive and Rh-negative.

rhizome In vascular plants, a more or less horizontal underground stem; may be enlarged for storage or may function in vegetative reproduction.

rhynchocoel A true coelomic cavity in ribbonworms that serves as a hydraulic power source for extending the proboscis.

ribonucleic acid (RNA) A class of nucleic acids characterized by the presence of the sugar ribose and the pyrimidine uracil; includes mRNA, tRNA, and rRNA.

ribosomal RNA (rRNA) A class of RNA molecules found, together with characteristic proteins, in ribosomes; transcribed from the DNA of the nucleolus.

ribosome The molecular machine that carries out protein synthesis; the most complicated aggregation of proteins in a cell, also containing three different rRNA molecules.

ribosome-binding sequence (RBS) In prokaryotes, a conserved sequence at the 5′ end of mRNA that is complementary to the 3′ end of a small subunit rRNA and helps to position the ribosome during initiation.

ribozyme An RNA molecule that can behave as an enzyme, sometimes catalyzing its own assembly; rRNA also acts as a ribozyme in the polymerization of amino acids to form protein.

ribulose 1,5-bisphosphate (RuBP) In the Calvin cycle, the 5-carbon sugar to which CO_2 is attached, accomplishing carbon fixation. This reaction is catalyzed by the enzyme rubisco.

ribulose bisphosphate carboxylase/oxygenase (rubisco) The four-subunit enzyme in the chloroplast that catalyzes the carbon fixation reaction joining CO_2 to RuBP.

RNA interference A type of gene silencing in which the mRNA transcript is prevented from being translated; small interfering RNAs (siRNAs) have been found to bind to mRNA and target its degradation prior to its translation.

RNA polymerase An enzyme that catalyzes the assembly of an mRNA molecule, the sequence of which is complementary to a DNA molecule used as a template. *See* transcription.

RNA primer In DNA replication, a sequence of about 10 RNA nucleotides complementary to unwound DNA that attaches at a replication fork; the DNA polymerase uses the RNA primer as a starting point for addition of DNA nucleotides to form the new DNA strand; the RNA primer is later removed and replaced by DNA nucleotides.

RNA splicing A nuclear process by which intron sequences of a primary mRNA transcript are cut out and the exon sequences spliced together to give the correct linkages of genetic information that will be used in protein construction.

rod Light-sensitive nerve cell found in the vertebrate retina; sensitive to very dim light; responsible for "night vision."

root The usually descending axis of a plant, normally below ground, which anchors the plant and serves as the major point of entry for water and minerals.

root cap In plants, a tissue structure at the growing tips of roots that protects the root apical meristem as the root pushes through the soil; cells of the root cap are continually lost and replaced.

root hair In plants, a tubular extension from an epidermal cell located just behind the root tip; root hairs greatly increase the surface area for absorption.

root pressure In plants, pressure exerted by water in the roots in response to a solute potential in the absence of transpiration; often occurs at night. Root pressure can result in guttation, excretion of water from cells of leaves as dew.

root system In plants, the portion of the plant body that anchors the plant and absorbs ions and water.

R plasmid A resistance plasmid; a conjugative plasmid that picks up antibiotic resistance genes and can therefore transfer resistance from one bacterium to another.

rule of addition The rule stating that for two independent events, the probability of either event occurring is the sum of the individual probabilities.

rule of multiplication The rule stating that for two independent events, the probability of both events occurring is the product of the individual probabilities.

rumen An "extra stomach" in cows and related mammals wherein digestion of cellulose occurs and from which partially digested material can be ejected back into the mouth.

S

salicylic acid In plants, an organic molecule that is a long-distance signal in systemic acquired resistance.

saltatory conduction A very fast form of nerve impulse conduction in which the impulses leap from node to node over insulated portions.

saprobes Heterotrophic organisms that digest their food externally (e.g., most fungi).

sarcolemma The specialized cell membrane in a muscle cell.

sarcomere Fundamental unit of contraction in skeletal muscle; repeating bands of actin and myosin that appear between two Z lines.

sarcoplasmic reticulum The endoplasmic reticulum of a muscle cell. A sleeve of membrane that wraps around each myofilament.

satellite DNA A nontranscribed region of the chromosome with a distinctive base composition; a short nucleotide sequence repeated tandemly many thousands of times.

saturated fat A fat composed of fatty acids in which all the internal carbon atoms contain the maximum possible number of hydrogen atoms.

Schwann cells The supporting cells associated with projecting axons, along with all the other nerve cells that make up the peripheral nervous system.

sclereid In vascular plants, a sclerenchyma cell with a thick, lignified, secondary wall having many pits; not elongate like a fiber.

sclerenchyma cell Tough, thick-walled cells that strengthen plant tissues.

scolex The attachment organ at the anterior end of a tapeworm.

scrotum The pouch that contains the testes in most mammals.

scuttellum The modified cotyledon in cereal grains.

second filial (F₂) generation The offspring resulting from a cross between members of the first filial (F₁) generation.

secondary cell wall In plants, the innermost layer of the cell wall. Secondary walls have a highly organized microfibrillar structure and are often impregnated with lignin.

secondary growth In vascular plants, an increase in stem and root diameter made possible by cell division of the lateral meristems.

secondary immune response The swifter response of the body the second time it is invaded by the same pathogen because of the presence of memory cells, which quickly become antibody-producing plasma cells.

secondary induction An induction between tissues that have already differentiated.

secondary metabolite A molecule not directly involved in growth, development, or reproduction of an organism; in plants these molecules, which include nicotine, caffeine, tannins, and menthols, can discourage herbivores.

secondary plant body The part of a plant consisting of secondary tissues from lateral meristem tissues; the older trunk, branches, and roots of woody plants.

secondary structure In a protein, hydrogen-bonding interactions between —CO and —NH groups of the primary structure.

secondary tissue Any tissue formed from lateral meristems in trees and shrubs.

Second Law of Thermodynamics A statement concerning the transformation of potential energy into heat; it says that disorder (entropy) is continually increasing in the universe as energy changes occur, so disorder is more likely than order.

second messenger A small molecule or ion that carries the message from a receptor on the target cell surface into the cytoplasm.

seed bank Ungerminated seeds in the soil of an area. Regeneration of plants after events such as fire often depends on the presence of a seed bank.

seed coat In plants, the outer layers of the ovule, which become a relatively impermeable barrier to protect the dormant embryo and stored food.

segment polarity gene Any of certain genes in *Drosophila* development that are expressed in stripes that subdivide the stripes created by the pair-rule genes in the process of segmentation.

segmentation The division of the developing animal body into repeated units; segmentation allows for redundant systems and more efficient locomotion.

segmentation gene Any of the three classes of genes that control development of the segmented body plan of insects; includes the gap genes, pair-rule genes, and segment polarity genes.

segregation The process by which alternative forms of traits are expressed in offspring rather than blending each trait of the parents in the offspring.

selection The process by which some organisms leave more offspring than competing ones, and their genetic traits tend to appear in greater proportions among members of succeeding generations than the traits of those individuals that leave fewer offspring.

selectively permeable Condition in which a membrane is permeable to some substances but not to others.

self-fertilization The union of egg and sperm produced by a single hermaphroditic organism.

semen In reptiles and mammals, sperm-bearing fluid expelled from the penis during male orgasm.

semicircular canal Any of three fluid-filled canals in the inner ear that help to maintain balance.

semiconservative replication DNA replication in which each strand of the original duplex serves as the template for construction of a totally new complementary strand, so the original duplex is partially conserved in each of the two new DNA molecules.

senescent Aged, or in the process of aging.

sensory (afferent) neuron A neuron that transmits nerve impulses from a sensory receptor to the central nervous system or central ganglion.

sensory setae In insects, bristles attached to the nervous system that are sensitive mechanical and chemical stimulation; most abundant on antennae and legs.

sepal A member of the outermost floral whorl of a flowering plant.

septation In prokaryotic cell division, the formation of a septum where new cell membrane and cell wall is formed to separate the two daughter cells.

septum, pl. septa A wall between two cavities.

sequence-tagged site (STS) A small stretch of DNA that is unique in a genome, that is, it occurs only once; useful as a physical marker on genomic maps.

seta, pl. setae (L., bristle) In an annelid, bristles of chitin that help anchor the worm during locomotion or when it is in its burrow.

severe acute respiratory syndrome (SARS) A respiratory infection with an 8% mortality rate that is caused by a coronavirus.

sex chromosome A chromosome that is related to sex; in humans, the sex chromosomes are the X and Y chromosomes.

sex-linked A trait determined by a gene carried on the X chromosome and absent on the Y chromosome.

Sexual dimorphism Morphological differences between the sexes of a species.

sexual reproduction The process of producing offspring through an alternation of fertilization (producing diploid cells) and meiotic reduction in chromosome number (producing haploid cells).

sexual selection A type of differential reproduction that results from variable success in obtaining mates.

shared derived character In cladistics, character states that are shared by species and that are different from the ancestral character state.

shoot In vascular plants, the aboveground portions, such as the stem and leaves.

short interspersed element (SINE) Any of a type of retrotransposon found in humans and other primates that does not contain the biochemical machinery needed for transposition.

shotgun sequencing The method of DNA sequencing in which the DNA is randomly cut into small fragments, and the fragments cloned and sequenced. A computer is then used to assemble a final sequence.

sieve cell In the phloem of vascular plants, a long, slender element with relatively unspecialized sieve areas and with tapering end walls that lack sieve plates.

signal recognition particle (SRP) In eukaryotes, a cytoplasmic complex of proteins that recognizes and binds to the signal sequence of a polypeptide, and then docks with a receptor that forms a channel in the ER membrane. In this way the polypeptide is released into the lumen of the ER.

signal transduction The events that occur within a cell on receipt of a signal, ligand binding to a receptor protein. Signal transduction pathways produce the cellular response to a signaling molecule.

simple sequence repeat (SSR) A one- to three-nucleotide sequence such as CA or CCG that is repeated thousands of times.

single-nucleotide polymorphism (SNP) A site present in at least 1% of the population at which individuals differ by a single nucleotide. These can be used as genetic markers to map unknown genes or traits.

sinus A cavity or space in tissues or in bone.

sister chromatid One of two identical copies of each chromosome, still linked at the centromere, produced as the chromosomes duplicate for mitotic division; similarly, one of two identical copies of each homologous chromosome present in a tetrad at meiosis.

small interfering RNAs (siRNAs) A class of micro-RNAs that appear to be involved in control of gene transcription and that play a role in protecting cells from viral attack.

small nuclear ribonucleoprotein particles (snRNP) In eukaryotes, a complex composed of snRNA and protein that clusters together with other snRNPs to form the spliceosome, which removes introns from the primary transcript.

small nuclear RNA (snRNA) In eukaryotes, a small RNA sequence that, as part of a small nuclear ribonucleoprotein complex, facilitates recognition and excision of introns by base-pairing with the 5′ end of an intron or at a branch site of the same intron.

sodium–potassium pump Transmembrane channels engaged in the active (ATP-driven) transport of Na^+, exchanging them for K^+, where both ions are being moved against their respective concentration gradients; maintains the resting membrane potential of neurons and other cells.

solute A molecule dissolved in some solution; as a general rule, solutes dissolve only in solutions of similar polarity; for example, glucose (polar) dissolves in (forms hydrogen bonds with) water (also polar), but not in vegetable oil (nonpolar).

solute potential The amount of osmotic pressure arising from the presence of a solute or solutes in water; measure by counterbalancing the pressure until osmosis stops.

solvent The medium in which one or more solutes is dissolved.

somatic cell Any of the cells of a multicellular organism except those that are destined to form gametes (germ-line cells).

somatic cell nuclear transfer (SCNT) The transfer of the nucleus of a somatic cell into an enucleated egg cell that then undergoes development. Can be used to make ES cells and to create cloned animals.

somatic mutation A change in genetic information (mutation) occurring in one of the somatic cells of a multicellular organism, not passed from one generation to the next.

somatic nervous system In vertebrates, the neurons of the peripheral nervous system that control skeletal muscle.

somite One of the blocks, or segments, of tissue into which the mesoderm is divided during differentiation of the vertebrate embryo.

Southern blot A technique in which DNA fragments are separated by gel electrophoresis, denatured into single-stranded DNA, and then "blotted" onto a sheet of filter paper; the filter is then incubated with a labeled probe to locate DNA sequences of interest.

S phase The phase of the cell cycle during which DNA replication occurs.

specialized transduction The transfer of only a few specific genes into a bacterium, using a lysogenic bacteriophage as a carrier.

speciation The process by which new species arise, either by transformation of one species into another, or by the splitting of one ancestral species into two descendant species.

species, pl. **species** A kind of organism; species are designated by binomial names written in italics.

specific heat The amount of heat that must be absorbed or lost by 1 g of a substance to raise or lower its temperature 1°C.

specific transcription factor Any of a great number of transcription factors that act in a time- or tissue-dependent manner to increase DNA transcription above the basal level.

spectrin A scaffold of proteins that links plasma membrane proteins to actin filaments in the cytoplasm of red blood cells, producing their characteristic biconcave shape.

spermatid In animals, each of four haploid (n) cells that result from the meiotic divisions of a spermatocyte; each spermatid differentiates into a sperm cell.

spermatozoa The male gamete, usually smaller than the female gamete, and usually motile.

sphincter In vertebrate animals, a ring-shaped muscle capable of closing a tubular opening by constriction (e.g., between stomach and small intestine or between anus and exterior).

sphingolipid A phospholipid built on the carbon skeleton of the hydrocarbon ceramide. Usually contains two chains of saturated hydrocarbons with a polar head group.

spicule Any of a number of minute needles of silica or calcium carbonate made in the mesohyl by some kinds of sponges as a structural component.

spindle apparatus The assembly that carries out the separation of chromosomes during cell division; composed of microtubules (spindle fibers) and assembled during prophase at the equator of the dividing cell.

spindle checkpoint The third cell-division checkpoint, at which all chromosomes must be attached to the spindle. Passage through this checkpoint commits the cell to anaphase.

spinnerets Organs at the posterior end of a spider's abdomen that secrete a fluid protein that becomes silk.

spiracle External opening of a trachea in arthropods.

spiral cleavage The embryonic cleavage pattern of some protostome animals in which cells divide at an angle oblique to the polar axis of the embryo; a line drawn through the sequence of dividing cells forms a spiral.

spiralian A member of a group of invertebrate animals; many groups exhibit spiral cleavage. Mollusks, annelids, and flatworms are examples of spiralians.

spliceosome In eukaryotes, a complex composed of multiple snRNPs and other associated proteins that is responsible for excision of introns and joining of exons to convert the primary transcript into the mature mRNA.

spongin A tough protein made by many kinds of sponges as a structural component within the mesohyl.

spongy parenchyma A leaf tissue composed of loosely arranged, chloroplast-bearing cells. *See* palisade parenchyma.

sporangium, pl. **sporangia** A structure in which spores are produced.

spore A haploid reproductive cell, usually unicellular, capable of developing into an adult without fusion with another cell.

sporophyte The spore-producing, diploid (2n) phase in the life cycle of a plant having alternation of generations.

stabilizing selection A form of selection in which selection acts to eliminate both extremes from a range of phenotypes.

stamen The organ of a flower that produces the pollen; usually consists of anther and filament; collectively, the stamens make up the androecium.

starch An insoluble polymer of glucose; the chief food storage substance of plants.

start codon The AUG triplet, which indicates the site of the beginning of mRNA translation; this codon also codes for the amino acid methionine.

stasis A period of time during which little evolutionary change occurs.

statocyst Sensory receptor sensitive to gravity and motion.

stele The central vascular cylinder of stems and roots.

stem cell A relatively undifferentiated cell in animal tissue that can divide to produce more differentiated tissue cells.

stereoscopic vision Ability to perceive a single, three-dimensional image from the simultaneous but slightly divergent two-dimensional images delivered to the brain by each eye.

stigma (1) In angiosperm flowers, the region of a carpel that serves as a receptive surface for pollen grains. (2) Light-sensitive eyespot of some algae.

stipules Leaflike appendages that occur at the base of some flowering plant leaves or stems.

stolon A stem that grows horizontally along the ground surface and may form adventitious roots, such as runners of the strawberry plant.

stoma, pl. **stomata** In plants, a minute opening bordered by guard cells in the epidermis of leaves and stems; water passes out of a plant mainly through the stomata.

stop codon Any of the three codons UAA, UAG, and UGA, that indicate the point at which mRNA translation is to be terminated.

stratify To hold plant seeds at a cold temperature for a certain period of time; seeds of many plants will not germinate without exposure to cold and subsequent warming.

stratum corneum The outer layer of the epidermis of the skin of the vertebrate body.

striated muscle Skeletal voluntary muscle and cardiac muscle.

stroma In chloroplasts, the semiliquid substance that surrounds the thylakoid system and that contains the enzymes needed to assemble organic molecules from CO_2.

stromatolite A fossilized mat of ancient bacteria formed as long as 2 BYA, in which the bacterial remains individually resemble some modern-day bacteria.

style In flowers, the slender column of tissue that arises from the top of the ovary and through which the pollen tube grows.

stylet A piercing organ, usually a mouthpart, in some species of invertebrates.

suberin In plants, a fatty acid chain that forms the impermeable barrier in the Casparian strip of root endoderm.

subspecies A geographically defined population or group of populations within a single species that has distinctive characteristics.

substrate (1) The foundation to which an organism is attached. (2) A molecule on which an enzyme acts.

succession In ecology, the slow, orderly progression of changes in community composition that takes place through time.

summation Repetitive activation of the motor neuron resulting in maximum sustained contraction of a muscle.

supercoiling The coiling in space of double-stranded DNA molecules due to torsional strain, such as occurs when the helix is unwound.

surface tension A tautness of the surface of a liquid, caused by the cohesion of the molecules of liquid. Water has an extremely high surface tension.

surface area-to-volume ratio Relationship of the surface area of a structure, such as a cell, to the volume it contains.

suspensor In gymnosperms and angiosperms, the suspensor develops from one of the first two cells of a dividing zygote; the suspensor of an angiosperm is a nutrient conduit from maternal tissue to the embryo. In gymnosperms the suspensor positions the embryo closer to stored food reserves.

swim bladder An organ encountered only in the bony fish that helps the fish regulate its buoyancy by increasing or decreasing the amount of gas in the bladder via the esophagus or a specialized network of capillaries.

swimmerets In lobsters and crayfish, appendages that occur in lines along the ventral surface of the abdomen and are used in swimming and reproduction.

symbiosis The condition in which two or more dissimilar organisms live together in close association; includes parasitism (harmful to one of the organisms), commensalism (beneficial to one, of no significance to the other), and mutualism (advantageous to both).

sympatric speciation The differentiation of populations within a common geographic area into species.

symplast route In plant roots, the pathway for movement of water and minerals within the cell cytoplasm that leads through plasmodesmata that connect cells.

symplesiomorphy In cladistics, another term for a shared ancestral character state.

symporter A carrier protein in a cell's membrane that transports two molecules or ions in the same direction across the membrane.

synapomorphy In systematics, a derived character that is shared by clade members.

synapse A junction between a neuron and another neuron or muscle cell; the two cells do not touch, the gap being bridged by neurotransmitter molecules.

synapsid Any of an early group of reptiles that had a pair of temporal openings in the skull behind the eye sockets; jaw muscles attached to these openings. Early ancestors of mammals belonged to this group.

synapsis The point-by-point alignment (pairing) of homologous chromosomes that occurs before the first meiotic division; crossing over takes place during synapsis.

synaptic cleft The space between two adjacent neurons.

synaptic vesicle A vesicle of a neurotransmitter produced by the axon terminal of a nerve. The filled vesicle migrates to the presynaptic membrane, fuses with it, and releases the neurotransmitter into the synaptic cleft.

synaptonemal complex A protein lattice that forms between two homologous chromosomes in prophase I of meiosis, holding the replicated chromosomes in precise register with each other so that base-pairs can form between nonsister chromatids for crossing over that is usually exact within a gene sequence.

syncytial blastoderm A structure composed of a single large cytoplasm containing about 4000 nuclei in embryonic development of insects such as *Drosophila*.

syngamy The process by which two haploid cells (gametes) fuse to form a diploid zygote; fertilization.

synthetic polyploidy A polyploidy organism created by crossing organisms most closely related to an ancestral species and then manipulating the offspring.

systematics The reconstruction and study of evolutionary relationships.

systemic acquired resistance (SAR) In plants, a longer-term response to a pathogen or pest attack that can last days to weeks and allow the plant to respond quickly to later attacks by a range of pathogens.

systemin In plants, an 18-amino-acid peptide that is produced by damaged or injured leaves that leads to the wound response.

systolic pressure A measurement of how hard the heart is contracting. When measured during a blood pressure reading, ventricular systole (contraction) is what is being monitored.

T

3′ poly-A tail In eukaryotes, a series of 1–200 adenine residues added to the 3′ end of an mRNA; the tail appears to enhance the stability of the mRNA by protecting it from degradation.

tagma, pl. **tagmata** A compound body section of an arthropod resulting from embryonic fusion of two or more segments; for example, head, thorax, abdomen.

Taq polymerase A DNA polymerase isolated from the thermophilic bacterium *Thermus aquaticus* (Taq); this polymerase is functional at higher temperatures, and is used in PCR amplification of DNA.

TATA box In eukaryotes, a sequence located upstream of the transcription start site. The TATA box is one element of eukaryotic core promoters for RNA polymerase II.

taxis, pl. **taxes** An orientation movement by a (usually) simple organism in response to an environmental stimulus.

taxonomy The science of classifying living things. By agreement among taxonomists, no two organisms can have the same name, and all names are expressed in Latin.

T cell A type of lymphocyte involved in cell-mediated immunity and interactions with B cells; the "T" refers to the fact that T cells are produced in the thymus.

telencephalon The most anterior portion of the brain, including the cerebrum and associated structures.

telomerase An enzyme that synthesizes telomeres on eukaryotic chromosomes using an internal RNA template.

telomere A specialized nontranscribed structure that caps each end of a chromosome.

telophase The phase of cell division during which the spindle breaks down, the nuclear envelope of each daughter cell forms, and the chromosomes uncoil and become diffuse.

telson The tail spine of lobsters and crayfish.

temperate (lysogenic) phage A virus that is capable of incorporating its DNA into the host cell's DNA, where it remains for an indeterminate length of time and is replicated as the cell's DNA replicates.

template strand The DNA strand that is used as a template in transcription. This strand is copied to produce a complementary mRNA transcript.

tendon (Gr. *tendon*, stretch) A strap of cartilage that attaches muscle to bone.

tensile strength A measure of the cohesiveness of a substance; its resistance to being broken apart. Water in narrow plant vessels has tensile strength that helps keep the water column continuous.

tertiary structure The folded shape of a protein, produced by hydrophobic interactions with water, ionic and covalent bonding between side chains of different amino acids, and van der Waal's forces; may be changed by denaturation so that the protein becomes inactive.

testcross A mating between a phenotypically dominant individual of unknown genotype and a homozygous "tester," done to determine whether the phenotypically dominant individual is homozygous or heterozygous for the relevant gene.

testis, pl. **testes** In mammals, the sperm-producing organ.

tetanus Sustained forceful muscle contraction with no relaxation.

thalamus That part of the vertebrate forebrain just posterior to the cerebrum; governs the flow of information from all other parts of the nervous system to the cerebrum.

therapeutic cloning The use of somatic cell nuclear transfer to create stem cells from a single individual that may be reimplanted in that individual to replace damaged cells, such as in a skin graft.

thermodynamics The study of transformations of energy, using heat as the most convenient form of measurement of energy.

thermogenesis Generation of internal heat by endothermic animals to modulate temperature.

thigmotropism In plants, unequal growth in some structure that comes about as a result of physical contact with an object.

threshold The minimum amount of stimulus required for a nerve to fire (depolarize).

thylakoid In chloroplasts, a complex, organized internal membrane composed of flattened disks, which contain the photosystems involved in the light-dependent reactions of photosynthesis.

Ti (tumor-inducing) plasmid A plasmid found in the plant bacterium *Agrobacterium tumefaciens* that has been extensively used to introduce recombinant DNA into broadleaf plants. Recent modifications have also allowed it to be used with cereal crop plants.

tight junction Region of actual fusion of plasma membranes between two adjacent animal cells that prevents materials from leaking through the tissue.

tissue A group of similar cells organized into a structural and functional unit.

tissue plasminogen activator (TPA) A human protein that causes blood clots to dissolve; if used within 3 hours of an ischemic stroke, TPA may prevent disability.

tissue-specific stem cell A stem cell that is capable of developing into the cells of a certain tissue, such as muscle or epithelium; these cells persist even in adults.

tissue system In plants, any of the three types of tissue; called a system because the tissue extends throughout the roots and shoots.

tissue tropism The affinity of a virus for certain cells within a multicellular host; for example, hepatitis B virus targets liver cells.

tonoplast The membrane surrounding the central vacuole in plant cells that contains water channels; helps maintain the cell's osmotic balance.

topoisomerase Any of a class of enzymes that can change the topological state of DNA to relieve torsion caused by unwinding.

torsion The process in embryonic development of gastropods by which the mantle cavity and anus move from a posterior location to the front of the body, closer to the location of the mouth.

totipotent A cell that possesses the full genetic potential of the organism.

trachea, pl. **tracheae** A tube for breathing; in terrestrial vertebrates, the windpipe that carries air between the larynx and bronchi (which leads to the lungs); in insects and some other terrestrial arthropods, a system of chitin-lined air ducts.

tracheids In plant xylem, dead cells that taper at the ends and overlap one another.

tracheole The smallest branches of the respiratory system of terrestrial arthropods; tracheoles convey air from the tracheae, which connect to the outside of the body at spiracles.

trait In genetics, a characteristic that has alternative forms, such as purple or white flower color in pea plants or different blood type in humans.

transcription The enzyme-catalyzed assembly of an RNA molecule complementary to a strand of DNA.

transcription complex The complex of RNA polymerase II plus necessary activators, coactivators, transcription factors, and other factors that are engaged in actively transcribing DNA.

transcription factor One of a set of proteins required for RNA polymerase to bind to a eukaryotic promoter region, become stabilized, and begin the transcription process.

transcription bubble The region containing the RNA polymerase, the DNA template, and the RNA transcript, so called because of the locally unwound "bubble" of DNA.

transcription unit The region of DNA between a promoter and a terminator.

transcriptome All the RNA present in a cell or tissue at a given time.

transfection The transformation of eukaryotic cells in culture.

transfer RNA (tRNA) A class of small RNAs (about 80 nucleotides) with two functional sites; at one site, an "activating enzyme" adds a specific amino acid, while the other site carries the nucleotide triplet (anticodon) specific for that amino acid.

transformation The uptake of DNA directly from the environment; a natural process in some bacterial species.

transgenic organism An organism into which a gene has been introduced without conventional breeding, that is, through genetic engineering techniques.

translation The assembly of a protein on the ribosomes, using mRNA to specify the order of amino acids.

translation repressor protein One of a number of proteins that prevents translation of mRNA by binding to the beginning of the transcript and preventing its attachment to a ribosome.

translocation (1) In plants, the long-distance transport of soluble food molecules (mostly sucrose), which occurs primarily in the sieve tubes of phloem tissue. (2) In genetics, the interchange of chromosome segments between nonhomologous chromosomes.

transmembrane domain Hydrophobic region of a transmembrane protein that anchors it in the membrane. Often composed of α-helices, but sometimes utilizing β-pleated sheets to form a barrel-shaped pore.

transmembrane route In plant roots, the pathway for movement of water and minerals that crosses the cell membrane and also the membrane of vacuoles inside the cell.

transpiration The loss of water vapor by plant parts; most transpiration occurs through the stomata.

transposable elements Segments of DNA that are able to move from one location on a chromosome to another. Also termed *transposons* or *mobile genetic elements*.

transposition Type of genetic recombination in which transposable elements (transposons) move from one site in the DNA sequence to another, apparently randomly.

trichome In plants, a hairlike outgrowth from an epidermal cell; glandular trichomes secrete oils or other substances that deter insects.

triglyceride (triacylglycerol) An individual fat molecule, composed of a glycerol and three fatty acids.

triploid Possessing three sets of chromosomes.

trisomic Describes the condition in which an additional chromosome has been gained due to nondisjunction during meiosis, and the diploid embryo therefore has three of these autosomes. In humans, trisomic individuals may survive if the autosome is small; Down syndrome individuals are trisomic for chromosome 21.

trochophore A specialized type of free-living larva found in lophotrochozoans.

trophic level A step in the movement of energy through an ecosystem.

trophoblast In vertebrate embryos, the outer ectodermal layer of the blastodermic vesicle; in mammals, it is part of the chorion and attaches to the uterine wall.

tropism Response to an external stimulus.

tropomyosin Low-molecular-weight protein surrounding the actin filaments of striated muscle.

troponin Complex of globular proteins positioned at intervals along the actin filament of skeletal muscle; thought to serve as a calcium-dependent "switch" in muscle contraction.

***trp* operon** In *E. coli*, the operon containing genes that code for enzymes that synthesize tryptophan.

true-breeding Said of a breed or variety of organism in which offspring are uniform and consistent from one generation to the next. This is due to the genotypes that determine relevant traits being homozygous.

tube foot In echinoderms, a flexible, external extension of the water–vascular system that is capable of attaching to a surface through suction.

tubulin Globular protein subunit forming the hollow cylinder of microtubules.

tumor-suppressor gene A gene that normally functions to inhibit cell division; mutated forms can lead to the unrestrained cell division of cancer, but only when both copies of the gene are mutant.

turgor pressure The internal pressure inside a plant cell, resulting from osmotic intake of water, that presses its cell membrane tightly against the cell wall, making the cell rigid. Also known as *hydrostatic pressure*.

tympanum In some groups of insects, a thin membrane associated with the tracheal air sacs that functions as a sound receptor; paired on each side of the abdomen.

U

ubiquitin A 76-amino-acid protein that virtually all eukaryotic cells attach as a marker to proteins that are to be degraded.

unequal crossing over A process by which a crossover in a small region of misalignment at synapsis causes two homologous chromosomes to exchange segments of unequal length.

uniporter A carrier protein in a cell's membrane that transports only a single type of molecule or ion.

uniramous Single-branched; describes the appendages of insects.

unsaturated fat A fat molecule in which one or more of the fatty acids contain fewer than the maximum number of hydrogens attached to their carbons.

urea An organic molecule formed in the vertebrate liver; the principal form of disposal of nitrogenous wastes by mammals.

urethra The tube carrying urine from the bladder to the exterior of mammals.

uric acid Insoluble nitrogenous waste products produced largely by reptiles, birds, and insects.

urine The liquid waste filtered from the blood by the kidney and stored in the bladder pending elimination through the urethra.

uropod One of a group of flattened appendages at the end of the abdomen of lobsters and crayfish that collectively act as a tail for a rapid burst of speed.

uterus In mammals, a chamber in which the developing embryo is contained and nurtured during pregnancy.

V

vacuole A membrane-bounded sac in the cytoplasm of some cells, used for storage or digestion purposes in different kinds of cells; plant cells often contain a large central vacuole that stores water, proteins, and waste materials.

valence electron An electron in the outermost energy level of an atom.

variable A factor that influences a process, outcome, or observation. In experiments, scientists attempt to isolate variables to test hypotheses.

vascular cambium In vascular plants, a cylindrical sheath of meristematic cells, the division of which produces secondary phloem outwardly and secondary xylem inwardly; the activity of the vascular cambium increases stem or root diameter.

vascular tissue Containing or concerning vessels that conduct fluid.

vas deferens In mammals, the tube carrying sperm from the testes to the urethra.

vasopressin A posterior-pituitary hormone that regulates the kidney's retention of water.

vector In molecular biology, a plasmid, phage, or artificial chromosome that allows propagation of recombinant DNA in a host cell into which it is introduced.

vegetal pole The hemisphere of the zygote comprising cells rich in yolk.

vein (1) In plants, a vascular bundle forming a part of the framework of the conducting and supporting tissue of a stem or leaf. (2) In animals, a blood vessel carrying blood from the tissues to the heart.

veliger The second larval stage of mollusks following the trochophore stage, during which the beginning of a foot, shell, and mantle can be seen.

ventricle A muscular chamber of the heart that receives blood from an atrium and pumps blood out to either the lungs or the body tissues.

vertebrate A chordate with a spinal column; in vertebrates, the notochord develops into the vertebral column composed of a series of vertebrae that enclose and protect the dorsal nerve cord.

vertical gene transfer (VGT) The passing of genes from one generation to the next within a species.

vesicle A small intracellular, membrane-bounded sac in which various substances are transported or stored.

vessel element In vascular plants, a typically elongated cell, dead at maturity, which conducts water and solutes in the xylem.

vestibular apparatus The complicated sensory apparatus of the inner ear that provides for balance and orientation of the head in vertebrates.

vestigial structure A morphological feature that has no apparent current function and is thought to be an evolutionary relic; for example, the vestigial hip bones of boa constrictors.

villus, pl. villi In vertebrates, one of the minute, fingerlike projections lining the small intestine that serve to increase the absorptive surface area of the intestine.

virion A single virus particle.

viroid Any of a group of small, naked RNA molecules that are capable of causing plant diseases, presumably by disrupting chromosome integrity.

virus Any of a group of complex biochemical entities consisting of genetic material wrapped in protein; viruses can reproduce only within living host cells.

visceral mass Internal organs in the body cavity of an animal.

vitamin An organic substance that cannot be synthesized by a particular organism but is required in small amounts for normal metabolic function.

viviparity Refers to reproduction in which eggs develop within the mother's body and young are born free-living.

voltage-gated ion channel A transmembrane pathway for an ion that is opened or closed by a change in the voltage, or charge difference, across the plasma membrane.

W

water potential The potential energy of water molecules. Regardless of the reason (e.g., gravity, pressure, concentration of solute particles) for the water potential, water moves from a region where water potential is greater to a region where water potential is lower.

water–vascular system A fluid-filled hydraulic system found only in echinoderms that provides body support and a unique type of locomotion via extensions called tube feet.

Western blot A blotting technique used to identify specific protein sequences in a complex mixture. *See* Southern blot.

wild type In genetics, the phenotype or genotype that is characteristic of the majority of individuals of a species in a natural environment.

wobble pairing Refers to flexibility in the pairing between the base at the 5′ end of a tRNA anticodon and the base at the 3′ end of an mRNA codon. This flexibility allows a single tRNA to read more than one mRNA codon.

wound response In plants, a signaling pathway initiated by leaf damage, such as being chewed by a herbivore, and lead to the production of proteinase inhibitors that give herbivores indigestion.

X

X chromosome One of two sex chromosomes; in mammals and in *Drosophila*, female individuals have two X chromosomes.

xylem In vascular plants, a specialized tissue, composed primarily of elongate, thick-walled conducting cells, which transports water and solutes through the plant body.

Y

Y chromosome One of two sex chromosomes; in mammals and in *Drosophila*, male individuals have a Y chromosome and an X chromosome; the Y determines maleness.

yolk plug A plug occurring in the blastopore of amphibians during formation of the archenteron in embryological development.

yolk sac The membrane that surrounds the yolk of an egg and connects the yolk, a rich food supply, to the embryo via blood vessels.

Z

zinc finger motif A type of DNA-binding motif in regulatory proteins that incorporates zinc atoms in its structure.

zona pellucida An outer membrane that encases a mammalian egg.

zone of cell division In plants, the part of the young root that includes the root apical meristem and the cells just posterior to it; cells in this zone divide every 12–36 hr.

zone of elongation In plants, the part of the young root that lies just posterior to the zone of cell division; cells in this zone elongate, causing the root to lengthen.

zone of maturation In plants, the part of the root that lies posterior to the zone of elongation; cells in this zone differentiate into specific cell types.

zoospore A motile spore.

zooxanthellae Symbiotic photosynthetic protists in the tissues of corals.

zygomycetes A type of fungus whose chief characteristic is the production of sexual structures called zygosporangia, which result from the fusion of two of its simple reproductive organs.

zygote The diploid (2n) cell resulting from the fusion of male and female gametes (fertilization).

Credits

Biology, 11th Edition 967

Photo Credits

Front Matter:
Page iv (right): © Ian J. Quitadamo, Ph.D.

Contents:
I: © Soames Summerhays/Natural Visions; II: © Dr. Gopal Murti/Science Source; III: © Stephen P. Lynch; IV: © Cathy & Gordon ILLG; V: © Mint Images/Frans Lanting/Getty Images; VI: © Susan Singer; VII: © Dr. Roger C. Wagner, Professor Emeritus of Biological Sciences, University of Delaware; VIII: © K. Ammann/Bruce Coleman Inc./Photoshot.

Chapter 1
Opener: © Soames Summerhays/Natural Visions; 1.1 (organelle): © Keith R. Porter/Science Source; 1.1 (cell): © Steve Gschmeissner/Getty Images; 1.1 (tissue): © Ed Reschke; 1.1 (organism): © Russell Illig/Getty Images RF; 1.1 (population): © George Ostertag/agefotostock; 1.1 (species): © PhotoLink/Getty Images RF; 1.1 (community): © Ryan McGinnis/Alamy; 1.1 (ecosystem): © Robert and Jean Pollock; 1.1 (biosphere): NASA Goddard Space Flight Center Image by Reto Stöckli (land surface, shallow water, clouds). Enhancements by Robert Simmon (ocean color, compositing, 3D globes, animation). Data and technical support: MODIS Land Group. MODIS Science Data Support Team. MODIS Atmosphere Group. MODIS Ocean Group Additional data: USGS EROS Data Center (topography). USGS Terrestrial Remote Sensing Flagstaff Field Center (Antarctica). Defense Meteorological Satellite Program (city lights); 1.5: © Huntington Library/SuperStock; 1.11a: © Dennis Kunkel Microscopy, Inc./Phototake; 1.11b: © Karl E. Deckart/Phototake; 1.12 (plantae left): © Alan L. Detrick/Science Source; 1.12 (plantae middle): © Steven P. Lynch; 1.12 (plantae right): © Iconotec/Glow Images RF; 1.12 (fungi left): © Royalty-Free/Corbis; 1.12 (fungi middle): © Mediscan/Corbis; 1.12 (fungi right): © Russell Illig/Getty Images RF; 1.12 (animalia left): © Corbis Super RF/Alamy; 1.12 (animalia middle): © Naturepix/Alamy RF; 1.12 (animalia right): © Alan and Sandy Carey/Getty Images RF; 1.12 (protista left): © Elmer Frederick Fischer/Corbis RF; 1.12 (protista middle): © Comstock Images/Getty Images RF; 1.12 (protista right): © Douglas P. Wilson/Frank Lane Picture Agency/Corbis; 1.12 (archaea left): © Power and Syred/Science Source; 1.12 (archaea right): © Kari Lounatmaa/Science Source; 1.12 (bacteria left): © Dwight R. Kuhn; 1.12 (bacteria right): © Alfred Pasieka/SPL/Science Source; pp. 15, 16: © Soames Summerhays/Natural Visions.

Chapter 2
Opener: © Virtualphoto/Getty Images RF; 2.2: Image Courtesy of Bruker Corporation; 2.10a: © Glen Allison/Getty Images RF; 2.10b: © PhotoLink/Getty Images RF; 2.10c: © Jeff Vanuga/Corbis; 2.13: © Hermann Eisenbeiss/National Audubon Society Collection/Science Source; pp. 31, 32: © Virtualphoto/Getty Images RF.

Chapter 3
Opener: © Deco/Alamy; 3.10b: © Asa Thoresen/Science Source; 3.10c: © J.L. Carson/CustomMedical; 3.11b: © Jim Zuckerman/agefotostock; 3.12: OAR/National Undersea Research Program (NURP); 3.13a: © Driscoll, Youngquist & Baldeschwieler, Caltech/SPL/Science Source; 3.13b: © M. Freeman/PhotoLink/Getty Images RF; pp. 56, 57: © Deco/Alamy.

Chapter 4
Opener: © Dr. Gopal Murti/Science Source; Table 4.1 (1): © Nancy Nehring/Getty Images RF; Table 4.1 (2): © Laguna Design/Science Source; Table 4.1 (3): De Agostini Picture Library/agefotostock; Table 4.1 (4): © Imago Stock&people/Newscom; Table 4.1 (5): © Dr. Torsten Wittmann/Science Source; Table 4.1 (6): © Med. Mic. Sciences, Cardiff Uni./Wellcome Images; Table 4.1 (7): © Microworks/Phototake; Table 4.1 (8): © Steve Gschmeissner/Science Source; p. 62 (bottom middle): © Don W. Fawcett/Science Source; 4.3: © PTP/Phototake; 4.4: Image by John Waterbury © Woods Hole Oceanographic Institution; 4.5a: © Eye of Science/Science Source; 4.8b: © Don W. Fawcett/Science Source; 4.8c: © John T. Hansen, Ph.D./Phototake; 4.8d: © Dr. Ueli Aebi; 4.10: © Don W. Fawcett/Science Source; 4.11: © Dennis Kunkel Microscopy, Inc./Phototake; 4.14: © Don W. Fawcett/Science Source; 4.15: © Biophoto Associates/Science Source; 4.16: © Keith R. Porter/Science Source; 4.17: © Dr. Jeremy Burgess/Science Source; 4.23 (top, bottom): © William Dentler, University of Kansas; 4.24a-b: © SPL/Science Source; 4.25: © Biophoto Associates/Science Source; 4.28a: Courtesy of Daniel Goodenough; 4.28b-c: © Don W. Fawcett/Science Source; pp. 85, 86: © Dr. Gopal Murti/Science Source.

Chapter 5
Opener: © Dr. Gopal Murti/Science Source; 5.2b: © Whitney L. Stutts, University of Florida; p. 91 & 5.4 (4): © Don W. Fawcett/Science Source; 5.12 (left, middle, right): © David M. Phillips/Science Source; 5.15a: Dr. Edwin P. Ewing, Jr./CDC; 5.15b-c: © Don W. Fawcett/Science Source; 5.16b: © Dr. Birgit Satir; pp. 105, 106: © Dr. Gopal Murti/Science Source.

Chapter 6
Opener: © Robert Caputo/Aurora Photos; 6.3 (left, right): © Jill Braaten; 6.11b: © Professor Emeritus Lester J. Reed, University of Texas at Austin; pp. 119, 120: © Robert Caputo/Aurora Photos.

Chapter 7
Opener: © Creatas/PunchStock RF; 7.18a: © Wolfgang Baumeister/Science Source; 7.18b: National Park Service, Photo by Jim Peaco; pp. 144, 145: © Creatas/PunchStock RF.

Chapter 8
Opener: © Royalty-Free/Corbis; 8.1 (middle right): Courtesy Dr. Kenneth Miller, Brown University; 8.8 (top, bottom): © Eric Soder/pixsource.com; 8.20: © Dr. Jeremy Burgess/Science Source; 8.22a: © Steven P. Lynch; 8.22b: © Joseph Nettis/National Audubon Society Collection/Science Source; 8.24: © Jessica Solomatenko/Getty Images RF; pp. 166, 167: © Royalty-Free/Corbis.

Chapter 9
Opener & pp. 183, 184: © Science Photo Library RF/Getty Images.

Chapter 10
Opener: © Stem Jems/Science Source; 10.2 (left, right): Courtesy of William Margolin; 10.4: © Biophoto Associates/Science Source; 10.6: © CNRI/Science Source; 10.10: Courtesy of Péter Lénárt and Jan-Michael Peters; 10.11 (1-7), 10.12: © Andrew S. Bajer, University of Oregon; 10.13 (top, bottom): © Dr. Jeremy Pickett-Heaps; 10.14a: © Don W. Fawcett/Science Source; 10.14b: © Guenter Albrecht-Buehler, Northwestern University, Chicago; 10.15 (top): © Biophoto Associates/Science Source; pp. 205, 206: © Stem Jems/Science Source.

Chapter 11
Opener: © Stephen P. Lynch; 11.6 (1-8): © Ed Reschke/Getty Images; pp. 218, 219: © Stephen P. Lynch.

Chapter 12
Opener: © Royalty-Free/Corbis; 12.1: © Norbert Schaefer/Corbis; 12.2: © Peter Fakler/Alamy RF; 12.3: © Leslie Holzer/Science Source; 12.11: © McGraw-Hill Education/Photo by David Hyde and Wayne Falda; 12.14: © DK Limited/Corbis; pp. 237, 238: © Royalty-Free/Corbis.

Chapter 13
Opener: © Adrian T. Sumner/Science Source; 13.1 (left, right): © Cabisco/Phototake; p. 241: © Biophoto Associates/Science Source; 13.3: © Bettmann/Corbis; p. 243 (left): From Brian P. Chadwick and Huntington F. Willard, "Multiple spatially distinct types of facultative heterochromatin on the human inactive X chromosome," PNAS, 101 (50): 17450-17455, Fig. 3 © 2004 National Academy of Sciences, U.S.A.; 13.4: © Kenneth Mason; 13.11: © Jackie Lewin, Royal Free Hospital/Science Source; 13.12: © Colorado Genetics Laboratory, University of Colorado, Anschutz Medical Campus; pp. 253, 254: © Adrian T. Sumner/Science Source.

Chapter 14
Opener: © PhotoDisc/Volume 29 RF; 14.5a: © Heritage Image Partnership Ltd/Alamy; 14.5b: © Omikron/Science Source; 14.6: © Barrington Brown/Science Source; 14.11: From M. Meselson and F.W. Stahl, PNAS, 44 (1958): 671; 14.16a-b: © Ramon Andrade 3Dciencia/SPL/Science Source; 14.20: © Don W. Fawcett/Science Source; pp. 275, 276: © PhotoDisc/Volume 29 RF.

Chapter 15
Opener: © Dr. Gopal Murti/Science Source; 15.3: © Science Source; 15.5: Image courtesy of the

University of Missouri-Columbia, Agricultural Information; 15.9: © Professor Oscar Miller/Science Source; 15.12b: Courtesy of Dr. Bert O'Malley, Baylor College of Medicine; 15.14 (3D Space-filled Model): Created by John Beaver using ProteinWorkshop, a product of the RCSB PDB, and built using the Molecular Biology Toolkit developed by John Moreland and Apostol Gramada (mbt.sdsc. edu). The MBT is financed by grant GM63208; 15.17: © Science Photo Library RF/agefotostock; pp. 302, 303: © Dr. Gopal Murti/Science Source.

Chapter 16
Opener: © Science Source; 16.9 (top, bottom): Courtesy of Dr. Harrison Echols and Dr. Sydney Kustu; 16.21: Image of PDB ID 4B4T (Unverdorben, P., et al., "Deep classification of a large cryo-EM dataset defines the conformational landscape of the 26S proteasome," *PNAS* (2014) 111: 5444-5549). © 2015 David Goodsell & RCSB Protein Data Bank (www.rcsb.org). Molecule of the Month DOI 10.2210/rcsb_pdb/mom_2013_10; pp. 324, 325: © Science Source.

Chapter 17
Opener: © Prof. Stanley Cohen/Science Source; 17.8: © SSPL/The Image Works; 17.18 (right): © Bill Barksdale/agefotostock; pp. 350, 351: © Prof. Stanley Cohen/Science Source.

Chapter 18
Opener: © William C. Ray, Director, Bioinformatics and Computational Biology Division, Biophysics Program, The Ohio State University; 18.2b (left, right): W. Achkar, A. Wafa, H. Mkrtchyan, F. Moassass and T. Liehr, "Novel complex translocation involving 5 different chromosomes in a chronic myeloid leukemia with Philadelphia chromosome: a case report," *Molecular Cytogenetics*, 2: 21 (2009). doi: 10.1186/1755-8166-2-21; 18.14: Image of PDB ID 1AZ2 (Harrison, D.H., Bohren, K.M., Petsko, G.A., Ringe, D., Gabbay, K.H., (1997), "The alrestatin double-decker: binding of two inhibitor molecules to human aldose reductase reveals a new specificity determinant," *Biochemistry*, 36 (51): 16134-16140) created using ProteinWorkshop, a product of the RCSB PDB, and built using the Molecular Biology Toolkit developed by John Moreland and Apostol Gramada (mbt.sdsc.edu). The MBT is financed by grant GM63208; 18.15: © Thomas Deerinck and Mark Ellisman, NCMIR, and John Glass, JCVI; pp. 372, 373: © William C. Ray, Director, Bioinformatics and Computational Biology Division, Biophysics Program, The Ohio State University.

Chapter 19
Opener: © Andrew Paul Leonard/Science Source; 19.1a-c: © Carolina Biological Supply Company/Phototake; 19.5b: © J. Richard Whittaker, used by permission; 19.8b: © Deco Images II/Alamy; 19.9 (right): © APTV/AP Photo; 19.13 (top left, top right): © Steve Paddock and Sean Carroll; 19.13 (bottom left, bottom right): © Jim Langeland, Steve Paddock and Sean Carroll; 19.16a: © Dr. Daniel St. Johnston/Wellcome Images; 19.16b: © Schupbach, T. and van Buskirk, C.; 19.16c: From Roth et al., 1989, courtesy of Siegfried Roth; 19.17: © David Scharf/Corbis; 19.20a-b: From Boucaut et al., 1984, courtesy of J-C Boucaut; pp. 396, 397: © Andrew Paul Leonard/Science Source.

Chapter 20
Opener: © Cathy & Gordon ILLG; 20.2: © Rosita So Image/Getty Images RF; pp. 418, 419: © Cathy & Gordon ILLG.

Chapter 21
Opener: Photo courtesy of the Georgia Southern University Museum; 21.3 (left): © Stephen Dalton/NHPA/Photoshot/Newscom; 21.3 (right): © The Natural History Museum/Alamy; 21.8 (left, right): Courtesy of Lyudmila N. Trut, Institute of Cytology & Genetics, Siberian Dept. of the Russian Academy of Sciences; 21.10: © Jason Edwards/Getty Images RF; 21.14: © James Hanken, Museum of Comparative Zoology, Harvard University, Cambridge; 21.17: © Prof. Dr. Uwe Kils; pp. 438, 439: Photo courtesy of the Georgia Southern University Museum.

Chapter 22
Opener: © Chris Johns/National Geographic/Getty Images; 22.2: © Porterfield/Chickering/Science Source; 22.3: © Mark Jones Roving Tortoise Photos/Getty Images; 22.6(1-3): © Robin Hopkins; 22.7 (1–4): © Jonathan Losos; 22.13a-b: © Gerrit J. Velema; 22.14a-b: © Karl Magnacca; 22.15a-d: © K. R. Wood/NTBG; pp. 457, 458: © Chris Johns/National Geographic/Getty Images.

Chapter 23
Opener: © Imagemore Co, Ltd./Getty Images RF; 23.1a: © Letz/SIPA/Newscom; 23.9a: © David Clark/Dinosaurs Alive!/Giant Screen Films; 23.9b: © Nature Picture Library/Alamy; 23.11 (left): © Lee W. Wilcox; 23.11 (right): © Clouds Hill Imaging Ltd./Corbis; pp. 477, 478: © Imagemore Co, Ltd./Getty Images RF.

Chapter 24
Opener: © Daily Mail/Rex/Alamy; 24.1 (mosquito): James Gathany/CDC; 24.1 (pufferfish): © Digital Vision/Getty Images RF; 24.1 (duck-billed platypus): © Nicole Duplaix/National Geographic/Getty Images; 24.1 (chimpanzee): © Digital Vision/Getty Images RF; 24.1 (*C. elegans*): © Heiti Paves/Alamy; 24.1 (fruit fly): © Thomas Deerinck, NCMIR/Getty Images; 24.1 (mouse): © G.K & Vikki Hart/Getty Images RF; 24.1 (domestic cow): © Ian Murray/Getty Images; 24.1 (human): © James Stevenson/Science Source; 24.1 (wall cress): © Nigel Cattlin/Alamy; 24.1 (maize): © Royalty-Free/Corbis; 24.1 (rice): Photo by Gary Kramer, USDA Natural Resources Conservation Service; 24.1 (fission yeast, baker's yeast): © Steve Gschmeissner/Science Source; 24.1 (bacteria): © BSIP/agefotostock RF; 24.1 (archea): © Dr. Harald Huber and Prof. Dr. Reinhard Rachel, University of Regensburg, Regensburg, Germany; 24.1 (malaria parasite): © T. Brain/Science Source; 24.13: Courtesy of Dr. Lewis G. Tilney and Dr. David S. Roos, University of Pennsylvania; 24.14 (left): © Eye of Science/Science Source; 24.14 (middle): © LSHTM/Stone/Getty Images; 24.14 (right): © Phototake/AP Photo; 24.15: © Tasmanian Department of Primary Industries/AP Photo; pp. 495, 496: © Daily Mail/Rex/Alamy.

Chapter 25
Opener: © Danté Fenolio/Science Source; 25.4 (left): © Michael Persson; 25.4 (right): © Christian Alfredsson; 25.5: © Dr. Anna Di Gregorio, Weill Cornell Medical College; 25.6: From Carolina Minguillon, Jeremy J. Gibson-Brown, and Malcolm P. Logan, "*Tbx4/5* gene duplication and the origin of vertebrate paired appendages," *PNAS*, 106 (51): 21726-21730, Fig. 4 © 2009 National Academy of Sciences, U.S.A.; 25.7: From Patrick Tschopp, Nadine Fraudeau, Frédérique Béna, and Denis Duboule, "Reshuffling genomic landscapes to study the regulatory evolution of *Hox* gene clusters," *PNAS*, 108 (26): 10632-10637, Fig. 3 © 2011 National Academy of Sciences, U.S.A.; 25.11 (left): © Chuck Pefley/Getty Images; 25.11 (middle left): © Darwin Dale/Science Source; 25.11 (middle right): © Dirscherl Reinhard/agefotostock; 25.11 (right): Hiroshi Suga, Patrick Tschopp, Daria F. Graziussi, Michael Stierwald, Volker Schmid, and Walter J. Gehring, (2010) "Flexibly deployed *Pax* genes in eye development at the early evolution of animals demonstrated by studies on a hydrozoan jellyfish," *PNAS*, 107: L4263-L4268; 25.12 (left, middle, right): Courtesy of Walter J. Gehring; 25.13a-b: Courtesy of Dr. William Jeffrey; 25.15 (sponges): © Datacraft/agefotostock RF; 25.15 (hydrozoa): © Mark Conlin/Getty Images; 25.15 (cubozoa): © Karen Gowlett-Holmes/Getty Images; 25.15 (ribbonworm): © Acteon Marine Biology Consultancy www.acteon.nl; 25.15 (fruit fly): © Herman Eisenbeiss/Science Source; 25.15 (mouse): © Digital Zoo/Getty Images RF; pp. 510, 511: © Danté Fenolio/Science Source.

Chapter 26
Opener: © Jeff Hunter/The Image Bank/Getty Images; 26.4: Picture courtesy of E. Javaux; 26.5: © Roger Garwood & Trish Ainslie/Corbis; 26.9: © Don W. Fawcett/Science Source; 26.11: © Rupert Mutzel; 26.12: © O. Louis Mazzatenta/National Geographic/Getty Images; pp. 524, 525: © Jeff Hunter/The Image Bank/Getty Images.

Chapter 27
Opener: Cynthia Goldsmith/CDC; 27.2: © Calysta Images/Getty Images RF; 27.3a: © Dept. of Biology, Biozentrum/SPL/Science Source; 27.8: © Royalty-Free/Corbis; pp. 541, 542: Cynthia Goldsmith/CDC.

Chapter 28
Opener: © David M. Phillips/Science Source; 28.1: © Robert Glusic/Corbis RF; 28.2: © Steve Gschmeissner/Science Source; 28.3: © Dr. Heribert Cypionka www.microbial-world.com; 28.6 (euryarchaeota): © SPL/Science Source; 28.6 (aquificae): © Prof. Dr. R. Rachel and Prof. Dr. K. O. Stetter, University of Regensburg, Regensburg, Germany; 28.6 (bacilli, spirochetes): © Andrew Syred/SPL/Science Source; 28.6 (actinobacteria): © Microfield Scientific Ltd./SPL/Science Source; 28.6 (cyanobacteria): © McGraw-Hill Education/Don Rubbelke, photographer; 28.6 (beta, gamma): © Dennis Kunkel Microscopy, Inc./Phototake; 28.6 (delta): © Prof. Dr. Hans Reichenbach, Helmholtz Centre for Infection Research, Braunschweig; p. 550 (Bacillus): © Dr. Gary Gaugler/SPL/Science Source; p. 550 (Coccus): © CNRI/Science Source; p. 550 (Spirillum): CDC; 28.7b: Dr. William A. Clark/CDC; 28.9b: © Julius Adler; 28.10a: © Dr. W.J. Ingledew/Science Source; 28.10b: © Biophoto Associates/SPL/Science Source; 28.11a: © Phototake/AP Photo; pp. 563, 564: © David M. Phillips/Science Source.

Chapter 29
Opener: © Stephen Durr; 29.1: © Andrew H. Knoll/Harvard University; 29.6: Dr. Stan Erlandsen/CDC; 29.7: © David M. Phillips/Science Source; 29.8a: © Andrew Syred/Science Source; 29.10a: Dr. Myron G. Schultz/CDC; 29.10b: © Edward S. Ross; 29.11: © Vern Carruthers, David Elliott; 29.12: © David J. Patterson; 29.14: © David M. Phillips/Science Source;

29.16: © De Agostini Picture Library/Getty Images; 29.17: © Prof. David J.P. Ferguson, Oxford University; 29.19 (top right): © M. I. Walker/Science Source; 29.20: © Elmer Frederick Fischer/Corbis RF; 29.22: © Dennis Kunkel Microscopy, Inc./Phototake; 29.23: © Andrew Syred/Science Source; 29.24 (top left, top right): © Steven P. Lynch; 29.24 (bottom): © Premaphotos/Alamy; 29.25: © Aaron J. Bell/Science Source; 29.26: © Stephen Durr; 29.27: © Dr. Diane S. Littler; 29.28 (left) © Steven P. Lynch; 29.28 (right): © Lee W. Wilcox; 29.29: © NaturimBild/A Wellma/agefotostock; 29.30: © Peter Parks/Oceans-Image/Photoshot/Newscom; 29.31: © Markus Keller/agefotostock; 29.32: © David J. Patterson; 29.33-29.35: © Eye of Science/Science Source; 29.36: © Mark J. Grimson and Richard L. Blanton, Biological Sciences Electron Microscopy Laboratory, Texas Tech University; 29.37: © William Bourland; pp. 585, 586: © Stephen Durr.

Chapter 30
Opener: © Dr. Parvinder Sethi; 30.3: © Steven P. Lynch; 30.4: © Edward S. Ross; 30.6: © Lee W. Wilcox; 30.7: Courtesy of Hans Steur, The Netherlands; 30.10: © Dr. Jody Banks, Purdue University; 30.11: © Kingsley Stern; 30.12: © Stephen P. Parker/Science Source; 30.13: © NHPA/Photoshot; 30.15 (left): © Mike Zens/Corbis; 30.15 (right): © Ed Reschke; pp. 600, 601: © Dr. Parvinder Sethi.

Chapter 31
Opener: © Zoonar/O Popova/agefotostock RF; 31.1: © Biology Media/Science Source; 31.2: © Scott Nodine/Alamy RF; 31.4a: © Luca Invernizzi Tetto/agefotostock; 31.4b: © Juan Carlos Muñoz/agefotostock RF; 31.4c: © Nancy Nehring/Getty Images RF; 31.5: © Goodshoot/Alamy RF; 31.6: © David Dilcher and Ge Sun; 31.7: Courtesy of Sandra Floyd; 31.10: © Sean W. Graham, UBC Botanical Garden & Centre for Plant Research, University of British Columbia; 31.13: © Dr. Joseph Williams; 31.15a: © Ed Reschke; 31.15b: NPS Photo by Don Despain; 31.17 (top left, top middle, bottom middle): © Kingsley Stern; 31.17 (top right): © Steven P. Lynch; 31.17 (bottom left): © Royalty-Free/Corbis; 31.17 (bottom right): © Charles D. Winters/Science Source; 31.18a: © Edward S. Ross; 31.18b: © Steven P. Lynch; 31.18c: © Phil Ashley/Getty Images; 31.18d: © John Kaprielian/Science Source; pp. 616, 617: © Zoonar/O Popova/agefotostock RF.

Chapter 32
Opener: © David Clapp/Getty Images; 32.1a: © Dr. Ronny Larsson; 32.1b, d: Contributed by Don Barr, Mycological Society of America; 32.1c: © Carolina Biological Supply Company/Phototake; 32.1e: © Dr. Yuuji Tsukii; 32.1f: © Yolande Dalpe, Agriculture and Agri-Food Canada; 32.1g: © Inga Spence/Alamy; 32.1h: © Felix Labhardt/Getty Images RF; 32.2: © Biophoto Associates/Science Source; 32.3 (inset): © Micro Discovery/Corbis; 32.3 (right): © Darlyne A. Murawski/Science Source; 32.4: © Eye of Science/Science Source; 32.5a: © Carolina Biological Supply Company/Phototake; 32.5b: © L. West/Science Source; 32.6 (left): © Daniel P. Fedorko; 32.7: Contributed by Daniel Wubah, Mycological Society of America; 32.8: Contributed by Don Barr, Mycological Society of America; 32.9a, 32.10a: © Carolina Biological Supply Company/Phototake; 32.11a: © Alexandra Lowry/National Audubon Society Collection/Science Source; 32.12a: © Premium/UIG/Getty Images; 32.12b: © Ed Reschke/Getty Images; 32.13: © David Scharf/Science Source; 32.14 (left): © Nigel Cattlin/Alamy; 32.14 (right): © B. Borrell Casal/Frank Lane Picture Agency/Corbis; 32.15a: © Ken Wagner/Phototake; 32.15b: © Yogesh More/Alamy RF; 32.15c: © Robert Lee/Science Source; 32.16: © Ed Reschke; 32.17a: © Eye of Science/Science Source; 32.17b: R.L. Peterson, H.B. Massicotte, L.H. Melville and F. Phillips, 2006. *Mycorrhizas: anatomy and cell biology images*, CD-ROM, NRC Research Press, ISBN 0-660-19404-X. Permission granted by authors and NRC Research Press; 32.18: © Scott Camazine/Science Source; 32.19a: Courtesy of Ralph Williams/USDA Forest Service; 32.19b: © USDA Forest Service, Northern and Intermountain Region, USDA Forest Service, Bugwood.org; 32.19c: © agefotostock/SuperStock; 32.20a: © Scott Camazine/Alamy; 32.20b: © Kage Mikrofotografie/Phototake; 32.21 (left): © School of Biological Sciences, University of Canterbury, New Zealand; 32.21 (inset): Courtesy of Dr. Peter Daszak; pp. 636, 637: © David Clapp/Getty Images.

Chapter 33
Opener: © Corbis/Volume 53 RF; 33.1a: © Lisa Thornberg/Getty Images RF; 33.1b: © Edward S. Ross; 33.1c: © Carolina Biological Supply Company/Phototake; 33.6: © Steven C. Zinski; 33.7a: © Andrew J. Martinez/Science Source; 33.8 (inset): © Roland Birke/Phototake; 33.10: © Neil G. McDaniel/Science Source; 33.11: © Royalty-Free/Corbis; 33.12: © Visual&Written SL/Alamy; 33.13: © Stephen Frink/Getty Images; 33.14: © Amos Nachoum/Corbis; 33.15: © Wolfgang Poelzer/WaterFrame/agefotostock; 33.16: © Purestock/SuperStock RF; 33.17: © Darlyne A. Murawski/National Geographic/Getty Images; pp. 657, 658: © Corbis/Volume 53 RF.

Chapter 34
Opener: © C. Fredrickson Photography/Getty Images RF; 34.2: © TommyIX/Getty Images RF; 34.3a: © NHPA/M. I. Walker/Photoshot RF; 34.4a: © Dwight R. Kuhn; 34.5: © The Natural History Museum/Alamy; 34.6a: © Dennis Kunkel Microscopy, Inc./Phototake; 34.7a: © Melba Photo Agency/PunchStock RF; 34.8a: © Royalty-Free/Corbis; 34.8b: © Juniors Bildarchiv GmbH/Alamy RF; 34.8c: © Douglas Faulkner/Science Source; 34.8d: © agefotostock/SuperStock; 34.9: © A. Flowers & L. Newman/Science Source; 34.11b: © Eye of Science/Science Source; 34.12a: © Demian Koop, Kathryn Green, Daniel J. Jackson; 34.12b: © Kjell Sandved/Butterfly Alphabet; 34.13: © NHPA/Photoshot; 34.14: © Rosemary Calvert/Getty Images; 34.15: © PhotoDisc Green/Getty Images RF; 34.17: © Stringer/New Zealand/X01244/Reuters/Corbis; 34.18: © Jeff Rotman/Science Source; 34.19: © Kjell Sandved/Butterfly Alphabet; 34.20: © Derrick Alderman/Alamy; 34.22: © Ronald L. Shimek; 34.23: Photo by Fred Grassle © Woods Hole Oceanographic Institution; 34.24: © MikeLane45/Getty Images RF; 34.25: © Pascal Goetgheluck/Science Source; 34.26b: © Hecker/Sauer/agefotostock; 34.27b: © Gordon MacSkimming/agefotostock; 34.29: © Gary D. Gaugler/Science Source; 34.30: © Educational Images Ltd., Elmira, NY, USA. Used by Permission; 34.36: © Andrew J. Martinez/Science Source; 34.37a: © National Geographic/Getty Images; 34.37b: © S. Camazine/K. Visscher/Science Source; 34.38: © Roland Birke/Getty Images; 34.41: © David Liebman/Pink Guppy; 34.42: © Kjell Sandved/Butterfly Alphabet; 34.43a: © Cleveland P. Hickman; 34.43b: © NHPA/James Carmichael, Jr./Photoshot RF; 34.43c: © Gyorgy Csoka, Hungary Forest Research Institute, Bugwood.org; 34.43d: © Kjell Sandved/Butterfly Alphabet; 34.43e: © Greg Johnston/Lonely Planet Images/Getty Images; 34.43f: © Nature's Images/Science Source; 34.45: © Dwight R. Kuhn; 34.46: © Kjell Sandved/Butterfly Alphabet; 34.47a: © Matthijs Kuijpers/Alamy RF; 34.47b: © Edward S. Ross; pp. 690, 691: © C. Fredrickson Photography/Getty Images RF.

Chapter 35
Opener: © Ingram Publishing/SuperStock RF; 35.1b: © Frederic Pacorel/Getty Images; 35.2: © D.P. Wilson/FLPA/Science Source; 35.3a: NOAA National Estuarine Research Reserve Collection; 35.3b: © Randy Morse/GoldenStateImages.com; 35.3c: NOAA's Sanctuaries Collection; 35.3d: NOAA Okeanos Explorer Program, INDEX-SATAL 2010; 35.3e: © Jeff Rotman/Science Source; 35.5: © Eric N. Olson, Ph.D./The University of Texas MD Anderson Cancer Center; 35.7a: © Rick Harbo; 35.8: © Heather Angel/Natural Visions; 35.12a: © agefotostock/SuperStock; 35.12b: © Royalty-Free/Corbis; 35.12c: © Jeff Rotman/Getty Images; 35.14: © Royalty-Free/Corbis; 35.16a: © Federico Cabello/SuperStock; 35.16b: © Raymond Tercafs/Bruce Coleman Inc./Photoshot; 35.19a: © Digital Vision/Getty Images RF; 35.19b: © Suzanne L. Collins & Joseph T. Collins/Science Source; 35.19c: © Jany Sauvanet/Science Source; 35.25: © Didier Dutheil/Sygma/Corbis; 35.27a (left): © Getty Images/iStockphoto RF; 35.27a (right): © Frans Lemmens/Getty Images; 35.27b-c (left): © Jonathan Losos; 35.27c (right): © Rod Planck; 35.27d (left): © Corbis/Volume 6 RF; 35.27d (right): © AFP/Stringer/Getty Images; 35.31: © Layne Kennedy/Corbis; 35.32a: © Aubrey Stoll/Getty Images RF; 35.32b: USFWS; 35.32c: Frank Miles/USFWS; 35.32d: © Myron Tay/Getty Images; 35.35: © Stephen Dalton/National Audubon Society Collection/Science Source; 35.36a (left): © Sir Francis Canker Photography/Getty Images RF; 35.36a (right): © Dave Watts/Alamy; 35.36b (left): © Corbis/Volume 6 RF; 35.36b (right): © W. Perry Conway/Corbis; 35.36c (left): © Image Source/Getty Images RF; 35.36c (right): © Corbis RF/Alamy; 35.37: © Magdalena Biskup Travel Photography/Getty Images RF; 35.38a: © J & C Sohns/agefotostock; 35.38b: © PhotoDisc/Getty Images RF; 35.38c (left): © Tom Merton/OJO Images/Getty Images RF; 35.38c (right): Richard Ruggiero/USFWS; 35.42: © AP Photo; pp. 729, 730: © Ingram Publishing/SuperStock RF.

Chapter 36
Opener: © Susan Singer; 36.4 (top): © Steven P. Lynch; 36.4 (bottom): © Garry DeLong/Getty Images; 36.6a: © Jeremy Burgess/Getty Images; 36.6b: © Dr. Jeremy Burgess/SPL/Science Source; 36.6c: © EM Unit, Royal Holloway, University of London, Egham, Surrey; 36.7: © Jessica Lucas & Fred Sack; 36.8: © Andrew Syred/SPL/Science Source; 36.9a-b: Courtesy of Allan Lloyd; 36.10: © Ed Reschke/Getty Images; 36.11a, c: © Lee W. Wilcox; 36.11b: © Steven P. Lynch; 36.12 (top): © NC Brown Center for Ultrastructure Studies, SUNY, College of Environmental Science and Forestry,

Syracuse, NY; 36.12 (bottom): USDA Forest Service, Forest Products Laboratory, Madison, WI; 36.13b: © Kage Mikrofotografie/Phototake; 36.14: © Garry DeLong/Getty Images; 36.15b: © John Schiefelbein; 36.16b: Courtesy of Dr. Philip Benfey; 36.17 (top left): © Carolina Biological Supply Company/Phototake; 36.17 (top right, bottom right): Photos by George S. Ellmore; 36.17 (bottom left): © Lee W. Wilcox; 36.19a: © Nokhooknoi/Getty Images RF; 36.19b: © F. Scholz/picture alliance/Arco Images/Newscom; 36.19c: © FLPA/Mark Newman/agefotostock; 36.19d: © Dennis Albert; 36.19e: © Kingsley Stern; 36.20: Courtesy of J.H. Troughton and L. Donaldson/Industrial Research Ltd.; 36.23a-b, 36.26, 36.27a: © Ed Reschke; 36.27b: © Steven P. Lynch; 36.28a: © Author's Image/Glow Images RF; 36.28b, f: © Lee W. Wilcox; 36.28c: © Andrew McRobb/Dorling Kindersley/Getty Images; 36.28d: © Chase Studio Inc./Science Source; 36.28e: © Charles D. Winters/Science Source; 36.29 (left, right): © Scott Poethig, University of Pennsylvania; 36.30a: © Kjell Sandved/Butterfly Alphabet; 36.30b: © Steven P. Lynch; 36.31a: © Gusto/Science Source; 36.31b: © Peter Chadwick/Dorling Kindersley/Getty Images; 36.32 (all): © Dr. Julie Hofer; 36.34: © Ed Reschke; pp. 754, 755: © Susan Singer.

Chapter 37
Opener: © Richard Rowan's Collection Inc./Science Source; 37.5a-b: © David Cook/blueshiftstudios, photographer/Alamy RF; 37.7: © Ken Wagner/Phototake; 37.11 (top, bottom): © Dr. Ryder/Jason Borns/Phototake; 37.14: © Foteini Kakulas; 37.15a: © Ed Reschke; 37.15b: © Dr. Keith Wheeler/Science Source; 37.16: © Mark Boulton/Science Source; 37.18a: © Andrew Syred/Science Source; 37.18b: © Bruce Iverson Photomicrography; pp. 772, 773: © Richard Rowan's Collection Inc./Science Source.

Chapter 38
Opener: © lani barbitta/Getty Images RF; 38.4a: © Hulton Archive/Getty Images; 38.4b: © Karim Sahib/AFP/Getty Images; 38.6a: © McPhoto/SHU/agefotostock; 38.6b: © Nigel Cattlin/Alamy; 38.6c: © Dave Bevan/agefotostock; 38.6d: © Allen Barker; 38.8: © Phil Clarke Hill/In Pictures/Corbis; 38.9: © Ken Wagner/Phototake; 38.9 (inset): © Bruce Iverson Photomicrography; 38.11a: © Kjell Sandved/Butterfly Alphabet; 38.11b: © Steven P. Lynch; 38.11c: © Perennov Nuridsany/Science Source; 38.11d: © Barry Rice, 38.13: © Getty Images/iStockphoto RF; 38.15a: © Ed Reschke/Getty Images; 38.15b: © Herve Conge/ISM/Phototake; 38.16a-b: Courtesy of Nicholas School of the Environment and Earth Sciences, Duke University. Photos by Will Owens; 38.18: Greg Harvey USAF; 38.19a: © Daniel Beltra/epa/Corbis; 38.19b: © Marcelo del Pozo/Reuters; 38.19c: © AP Photo; pp. 790, 791: © lani barbitta/Getty Images RF.

Chapter 39
Opener: © Emily Keegin/fstop/Getty Images RF; 39.1: © David Cappaert/agefotostock; 39.2a: Photo by William Wergin and Richard Sayre/USDA/ARS; 39.2b: Photo by Scott Bauer/USDA/ARS; 39.5: © Clarence Styron/agefotostock; 39.6: © Adam Jones/Science Source; Table 39.1 (1): © Inga Spence/Science Source; Table 39.1 (2): © Aaron Roeth Photography; Table 39.1 (3): © Steven P. Lynch; Table 39.1 (4): © Heather Angel/Natural Visions; Table 39.1 (5): © Pallava Bagla/Corbis; 39.7: © Gilbert S. Grant/Science Source; 39.7 (inset): © Lee W. Wilcox; 39.8: © Martin Shields/Science Source; 39.12: Courtesy R. X. Latin. Reproduced, by permission, from Zitter, T. A., Hopkins, D. L., and Thomas, C. E. 1996, *Compendium of Cucurbit Diseases*, American Phytopathological Society, St. Paul, MN; pp. 802, 803: © Emily Keegin/fstop/Getty Images RF.

Chapter 40
Opener: © Educational Images/Getty Images; 40.5a-d: © Niko Geldner, UNIL; 40.8: © Ray F. Evert; 40.10a (all): © Jee Jung and Philip Benfey; 40.11: © Lee W. Wilcox; 40.13: © Frank Krahmer/Corbis; 40.16: © Don Johnston/agefotostock; 40.26a-c: © Prof. Malcolm B. Wilkins, Botany Dept., Glasgow University; 40.28: © Custom Life Science Images/Alamy; 40.30: © Science Source; 40.31: © Amnon Lichter, The Volcani Center; 40.36a: © McGraw-Hill Education/Evelyn Jo Johnson, photographer; 40.36b: From D. R. McCarty, C. B. Carson, P. S. Stinard, and D. S. Robertson, "Molecular analysis of *viviparous-1*: an abscisic acid-insensitive mutant of maize," *The Plant Cell*, 1 (5): 523-532 © 1989 American Society of Plant Biologists; 40.36c: © ISM/Phototake; pp. 829, 830: © Educational Images/Getty Images.

Chapter 41
Opener: © Heather Angel/Natural Visions; 41.3a: © Imagebrokers/Photoshot; 41.3b: © Pat Breen, Oregon State University; 41.4: Courtesy of Lingjing Chen & Renee Sung; 41.5a: © Brand X Pictures/PunchStock RF; 41.5b (top): © Blickwinkel/Alamy; 41.5b (bottom): © Ove Nilsson/Umeå Plant Science Centre and Detlef Weigel/Max Planck Institute for Developmental Biology; 41.7: © Design Pics/Don Hammond RF; 41.9 (1-4): Courtesy of John L. Bowman; 41.12: Dr. Thomas Barnes/University of Kentucky/USFWS; 41.13: © Traci Tatman; 41.14a-b: Courtesy of Enrico Coen; 41.16a-b: © L. DeVos/Free University of Brussels; 41.17: © Kingsley Stern; 41.18: © David Cappaert, Bugwood.org; 41.20: © Rolf Nussbaumer/Getty Images RF; 41.21a-b: © Thomas Eisner, Cornell University; 41.22: © Imagebroker/Alamy; 41.23: © Edward S. Ross; 41.24a: © Heiti Paves/Alamy; 41.24b: © Rafael Campillo/agefotostock; 41.29: © Edward Yeung, University of Calgary and David Meinke, Oklahoma State University; 41.30a-d: Kindly provided by Prof. Chun-Ming Liu, Institute of Botany, Chinese Academy of Sciences; 41.31: © Jan Lohmann, Max Planck Institute for Developmental Biology; 41.32c: © Ben Scheres, University of Utrecht; 41.32d-e: Courtesy of George Stamatiou and Thomas Berleth; 41.34 (left, right): © A. P. Mähönen; 41.35 (top right): © S. Kirchner/photocuisine/Corbis; 41.35 (bottom right): © Metta image/Alamy RF; 41.37a: © Nigel Cattlin/Alamy; 41.37b: © Martin Shields/Alamy; 41.39: © Jerome Wexler/Science Source; 41.40a-g: © Hans-Ulrich Koop; 41.41a: © Anthony Arendt/Alamy; 41.41b: © SuperStock/Alamy RF; pp. 860, 861: © Heather Angel/Natural Visions.

Chapter 42
Opener: © Dr. Roger C. Wagner, Professor Emeritus of Biological Sciences, University of Delaware; Table 42.1 (1, 3): © Ed Reschke; Table 42.1 (2, 4): © Victor P. Eroshenko; Table 42.1 (5): © McGraw-Hill Education/Al Telser, photographer; 42.3: © J. Gross, Biozentrum/Science Source; 42.4: © Biophoto Associates/Science Source; Table 42.2 (1, 4): © Ed Reschke; Table 42.2 (2): © McGraw-Hill Education; Table 42.2 (3): © Chuck Brown/Science Source; Table 42.2 (5): © Kenneth Eward/Science Source; Table 42.3 (1-3): © Ed Reschke; pp. 885, 886: © Dr. Roger C. Wagner, Professor Emeritus of Biological Sciences, University of Delaware.

Chapter 43
Opener: Courtesy of David I. Vaney, University of Queensland Australia; 43.3: © Don W. Fawcett/Science Source; 43.13: © John Heuser, Washington University School of Medicine, St. Louis, MO; 43.15: © Ed Reschke; 43.17b: © Kage Mikrofotografie/agefotostock; 43.25: © Dr. Marcus E. Raichle, Washington University, McDonnell Center for High Brain Function; 43.27: © Photo Lennart Nilsson/TT Nyhetsbyrån; 43.30: © Steve Gschmeissner/Science Source; pp. 913, 914: Courtesy of David I. Vaney, University of Queensland Australia.

Chapter 44
Opener: © Omikron/Science Source; 44.11d: © Image Source/Getty Images RF; 44.21: © Volker Steger/Science Source; 44.24: © Digital Vision/PunchStock RF; pp. 936, 937: © Omikron/Science Source.

Chapter 45
Opener: © Nature's Images/Science Source; 45.10: © Mike Goldwater/Alamy; 45.11: © Bettmann/Corbis; pp. 960, 961: © Nature's Images/Science Source.

Chapter 46
Opener: © Stockbyte RF; 46.3 (1, 7): © Ed Reschke; 46.3 (2): © David M. Phillips/Science Source; 46.3 (3): © Biophoto Associates/Science Source; 46.3 (4): © CNRI/Science Source; 46.3 (5): © Steve Gschmeissner/SPL/Getty Images RF; 46.3 (6, 8): © Ed Reschke/Getty Images; 46.10 (top, bottom): © Dr. H.E. Huxley; 46.21: © Treat Davidson/Science Source; pp. 980, 981: © Stockbyte RF.

Chapter 47
Opener: © Datacraft/UIG/agefotostock; 47.10: © Ron Boardman/Stone/Getty Images; 47.19: © The Rockefeller University/AP Photo; pp. 1001, 1002: © Datacraft/UIG/agefotostock.

Chapter 48
Opener: © Pixtal/agefotostock RF; 48.1: © Taketomo Shiratori/Oceans-Image/Photoshot/Newscom; 48.3: © Juniors Bildarchiv/Alamy; 48.13 (left): © Clark Overton/Phototake; 48.13 (right): © Martin Rotker/Phototake; 48.14: © Kenneth Eward/BioGrafx/Science Source; pp. 1018, 1019: © Pixtal/agefotostock RF.

Chapter 49
Opener: © Biophoto Associates/Science Source; 49.16a-b: © Ed Reschke; 49.16c: © McGraw-Hill Education/Al Telser, photographer; pp. 1037, 1038: © Biophoto Associates/Science Source.

Chapter 50
Opener & pp. 1055, 1056: © Rick & Nora Bowers/Alamy.

Chapter 51
Opener: © National Museum of Health and Medicine, Armed Forces Institute of Pathology/AP Photo; 51.3: © D. Phillips/SPL/Science Source; 51.6: © Wellcome

Library, London; 51.12a-b: © Dr. Andrejs Liepins/Science Source; 51.22: © CDC/Science Source; pp. 1083, 1084: © National Museum of Health and Medicine, Armed Forces Institute of Pathology/AP Photo.

Chapter 52
Opener: © Geordie Torr/Alamy; 52.1: © Dennis Kunkel Microscopy, Inc./Phototake; 52.2: © Fred McConnaughey/National Audubon Society Collection/Science Source; 52.4: © David Doubilet/Getty Images; 52.6: © George Grall/Getty Images; 52.7a: © Jonathan Losos; 52.7b: © Photoshot/SuperStock; 52.7c: © Andrea & Antonella Ferrari/NHPA/Photoshot/Newscom; 52.7d: © Danté Fenolio/Science Source; 52.8: © Michael Nolan/agefotostock; 52.9a: © Jean Phillippe Varin/Jacana/Science Source; 52.9b: © Tom McHugh/National Audubon Society Collection/Science Source; 52.9c: © Royalty-Free/Corbis; 52.12a: © David M. Phillips/Science Source; 52.17: © Ed Reschke; 52.21a: © Jonathan A. Meyers/Science Source; 52.21b: © McGraw-Hill Education/Jill Braaten, photographer; 52.21c: © McGraw-Hill Education/Bob Coyle, photographer; 52.21d: © Kumar Sriskandan/Alamy; pp. 1104, 1105: © Geordie Torr/Alamy.

Chapter 53
Opener: © Neil Bromhall/Science Source; 53.1c-d: © David M. Phillips/Science Source; 53.3a-d: Dr. Mathias Hafner, Mannheim University of Applied Sciences, Institute for Molecular Biology, Mannheim, Germany, and Dr. Gerald Schatten, Pittsburgh Development Centre Deputy Director, Magee-Woman's Research Institute, Professor and Vice-Chair of Obstetrics, Gynecology & Reproductive Sciences and Professor of Cell Biology & Physiology, Director, Division of Developmental and Regenerative Medicine, University of Pittsburgh School of Medicine Pittsburgh, PA 15213; 53.7: © David M. Phillips/Science Source; 53.8a: © Carolina Biological Supply Company/Phototake; 53.9: © David M. Phillips/Science Source; 53.11a-c: From, *An Atlas of the Development of the Sea Urchin Lytechinus variegatus*. Provided by Dr. John B. Morrill (left to right) Plate 20 p. 62 #I, Plate 33 p. 93 #C, Plate 38 p. 105 #G; 53.17 (top, bottom): Courtesy of Manfred Frasch; 53.20c (left, right): © Roger Fleischman, University of Kentucky; 53.26b: © Scott Camazine/Phototake; 53.27a: © Omikron/Science Source; 53.27b: © Kage Mikrofotografie/agefotostock; 53.27c: © Brand X Pictures/PunchStock RF; 53.27d: © Nestle/Petit Format/Science Source; pp. 1131, 1132: © Neil Bromhall/Science Source.

Chapter 54
Opener: © K. Ammann/Bruce Coleman Inc./Photoshot; 54.2: © Dr. Nicolette Siep; 54.4a: © Juniors Bildarchiv GmbH/Alamy; 54.4b: © Haruhiro Higashida, M.D., Ph.D.; 54.5a-b: From Hemanth P. Nair, Larry J. Young, "Vasopressin and pair-bond formation: genes to brain to behavior," *Physiology*, 21 (2): 146-152, Fig. 2 © 2006. Reprinted by permission from The American Physiological Society; 54.6a-c: © Boltin Picture Library/The Bridgeman Art Library; 54.7: © Thomas McAvoy, Life Magazine/Time, Inc./Getty Images; 54.8: © Harlow Primate Laboratory, University of Wisconsin-Madison USA; 54.10: © Roger Wilmhurst/National Audubon Society Collection/Science Source; 54.11a: © Linda Koebner/Bruce Coleman Inc./Photoshot; 54.11b: © Photoshot/SuperStock; 54.12a-c: © SuperStock; 54.13: Courtesy of Bernd Heinrich; 54.14b: © Roy Toft/Getty Images; 54.14c: © George Lepp/Getty Images; 54.17 (inset): © Dwight R. Kuhn; 54.20: © W. Perry Conway/Corbis RF; 54.21b: © Scott Camazine/Science Source; 54.22a: © Gerald Cubitt; 54.23: © Nina Leen, Life Magazine/Time, Inc./Getty Images; 54.26 (left): © Peter Steyn/Getty Images; 54.26 (right): © Gerald C. Kelley/Science Source; 54.27a: © Bruce Beehler/Science Source; 54.27b: © B. Chudleigh/VIREO; 54.29: © Cathy & Gordon ILLG; 54.30a: © W. Perry Conway/Corbis RF; 54.31a: © Dr. Douglas Ross; 54.32: © Nick Gordon/ardea.com; 54.34: © Steve Hopkin/Getty Images; 54.35: © Heinrich van den Berg/Getty Images; 54.36: © Amazon-Images/Alamy; 54.37: © Nigel Dennis/National Audubon Society Collection/Science Source; pp. 1161, 1162: © K. Ammann/Bruce Coleman Inc./Photoshot.

Chapter 55
Opener: © Ian Cartwright/Getty Images RF; 55.1: © Thorsten Milse/agefotostock; 55.3 (inset): © Melissa Losos; 55.4: © Stone Nature Photography/Alamy; 55.6 (inset): © Duncan Usher/ardea.com; 55.13 (left): © Christian Kerihuel; 55.15: © Barry Sinervo; 55.21 (left): © Juan Medina/Reuters/Corbis; 55.21 (right): © Juniors Tierbildarchiv/Photoshot; 55.23: © Alan & Sandy Carey/Science Source; pp. 1184, 1185: © Ian Cartwright/Getty Images RF.

Chapter 56
Opener: © Fuse/Getty Images RF; 56.1: © Daryl & Sharna Balfour/Okopia/Science Source; 56.6a-d: © Jonathan Losos; 56.10a: © Edward S. Ross; 56.10b: © Kevin D. Arvin, Bugwood.org; 56.11a-b: © Lincoln P. Brower; 56.12: © MedioImages/SuperStock RF; 56.13: © DEA Picture Library/Getty Images; 56.15: © Dr. Merlin D. Tuttle/Science Source; 56.16: © Dreedphotography/Getty Images RF; 56.17: © Charles T. Bryson, USDA Agricultural Research Service, Bugwood.org; 56.19: © JLFCapture/Getty Images RF; 56.20: © P. de Graaf/Getty Images RF; 56.21a: NOAA Central Library; 56.21b: © Ben Grib Design/Getty Images RF; 56.23: © David Hosking/National Audubon Society Collection/Science Source; 56.24b: © Cliff LeSergent/Alamy RF; 56.24c: © Allover images/Alamy; 56.24d: © Borut Trdina/Getty Images RF; 56.25a: © De Agostini/C. Dani & I. Jeske/Getty Images; 56.25b: © Prisma Bildagentur AG/Alamy; 56.26: © Jacques Jangoux/Alamy; pp. 1206, 1207: © Fuse/Getty Images RF.

Chapter 57
Opener: © 1001slilde/Getty Images RF; 57.3: © Worldwide Picture Library/Alamy; 57.6 (right): NASA Image courtesy Jeff Schmaltz, MODIS Land Rapid Response Team, NASA; 57.7a: Northern Research Center/US Forest Service; 57.19a: © Layne Kennedy/Corbis; pp. 1229, 1230: © 1001slilde/Getty Images RF.

Chapter 58
Opener: GSFC/NASA; 58.10: © Andoni Canela/agefotostock; 58.11: © Fuse/Getty Images RF; 58.14a: © Art Wolfe/Science Source; 58.14b: © McGraw-Hill Education/Pat Watson, photographer; 58.16: Provided by the SeaWiFS Project, NASA/Goddard Space Flight Center, and ORBIMAGE; 58.17: © Digital Vision/Getty Images RF; 58.19a: © Danté Fenolio/Science Source; 58.19b: © Ralph White/Corbis; 58.19b (inset): NOAA Pacific Marine Environmental Laboratory's Vents Program; 58.22a: © Environmental Images/agefotostock; 58.22b: © Frans Lanting/Corbis; 58.23: © Gilbert S. Grant/National Audubon Society Collection/Science Source; 58.26a: NASA/Goddard Space Flight Center Scientific Visualization Studio; 58.28a: Hansen, J., et al., "Global Temperature in 2014 and 2015" http://www.columbia.edu/~jeh1/mailings/ (January 16, 2015); 58.30 (top): © Dr. Bruno Messerli; 58.30 (bottom): © George Robertson/Alamy; pp. 1256, 1257: GSFC/NASA.

Chapter 59
Opener: © Norbert Rosing/National Geographic Creative; 59.1: Courtesy of Rhoda Knight Kalt & DN-Images; 59.2: © Robert Francis/Robert Harding/Getty Images; 59.3 (1): © Frank Krahmer/Masterfile; 59.3 (2): © Kevin Schafer/Corbis; 59.3 (3): © Heather Angel/Natural Visions; 59.3 (4): © NHPA/Photoshot; 59.5a: © Edward S. Ross; 59.5b: © Inga Spence/Alamy; 59.6a: © Juan Carlos Muñoz/agefotostock; 59.6b: © Andoni Canela/agefotostock; 59.8: © Bryan L Sage/ardea.com; 59.9 (1): © Brian Rogers/Natural Visions; 59.9 (2): © Elmer Martinez/AFP/Getty Images; 59.9 (3): © Craig K. Lorenz/Science Source; 59.9 (4): © Wattsy01/Getty Images RF; 59.13 (left): © Rob Bierregaard; 59.13 (right): © Dr. Morley Read/Science Source; 59.14: © FLPA/Neil Bowman/agefotostock; 59.16: © Peter Yates/SPL/Science Source; 59.17 (left): © Jack Jeffrey; 59.17 (right): © Photo Resource Hawaii/Alamy; 59.18: © Tom McHugh/Science Source; 59.20: © Dr. Merlin D. Tuttle/Science Source; 59.21a: ANSP © Steven Holt/stockpix.com; 59.21b: U.S. Fish and Wildlife Service; 59.23: © Keith Crowley/Alamy; 59.24a-b: © University of Wisconsin-Madison Arboretum; 59.26: © Bill Coster LA/Alamy; pp. 1281, 1282: © Norbert Rosing/National Geographic Creative.

Index

Boldface page numbers correspond with **boldface terms** in the text. Page numbers followed by an "f" indicate figures; page numbers followed by a "t" indicate tabular material.

A

ABA. *See* Abscisic acid
A bands, 971, 972f
ABC model, of floral organ specification, 836–839f
Abdominopelvic cavity, 865
ABO blood groups, 91t, 233t, 234–**235**, 235f, 400–401, 1079
Abortion, spontaneous, 214, 1128
Abscisic acid (ABA)
  for dormancy and suppression of growth, 814, 814f, 827–828, 828f
  functions of, 817t
  in stress response, 592–593
  transpiration rate and, **767**, 767f
Abscission, **813**–814, 814f
Abscission zone, 813, 814f
Absolute dating, 428
Absorption
  in digestive tract, 984, 991–992, 991f
  of virus to host, 532
  of water and minerals in plants, 761–764f
Absorption spectrum, of photosynthetic pigments, **152**, 152–153f
Abstinence, 1102
Acacia trees, mutualism with ants, 798, 798f, 1199, 1199f
Acanthamoeba, 583
Acanthodians, 702, 702t
Acari, 684
Acceptor stems, **291**–292, 291f
Accessory digestive organs, 985, 985f, 990–991, 990f, 996–997, 997f
Accessory pigments, **152**, 154
Accessory sex organs
  female, 1099–1100, 1100f
  male, 1094–1095, 1095f
Acellular bones, 968
Acetaldehyde, 35f, 129, 140
Acetic acid, 35f
Acetylcholine, 898–899, 898f, 910t, 911, 913f, 975, 1035
Acetylcholine (ACh) receptors, 172, 894f, 911, 913f
Acetylcholinesterase (AChE), 899
Acetyl coenzyme A (acetyl-CoA)
  in ß-oxidation, 141, 142f
  in Krebs cycle, 131, 131f, 132, 133f, 138, 138f
  from pyruvate, 129–130
  roles of, 142
ACh (acetylcholine) receptors, 172, 894f, 911, 913f
AChE (acetylcholinesterase), 899
Achiasmate segregation, **214**
Acid, stomach, 988–989
Acid growth hypothesis, **820**–821, 821f
Acidic soil, 777–778, 778f
Acid precipitation, 1248, 1248f
Acid rain, 1248, 1248f, 1249

Acids, **29**, 30
Acini, **990**
Acoela, 648f, 657, 657f
Acoel flatworms, 656–657, 657f
Acoelomates, 642, 642f, 647, 648f
Acoelomorpha, 648f
Aconitase, 321
Acquired thermotolerance, 815
Acromegaly, **952**
Acrosomal process, 1108
Acrosomes, **1108**
ACTH (adrenocorticotropic hormone), **950**
Actin, 44, 45t, **76**, 972f
Actin filaments, **76**, 76f, 77, 79, 83
Actinistia, 700f
Actinobacteria, 548f
Actinomyces, 548f, 559
Actinopterygii, 700f, 702t, 704, 705f
Actinosphaerium, 582f
Action potentials, **893**, 894–897
Action spectrum, of chlorophyll, **153**, 153f
Activation energy, **111**, 111f, 114
Activators, 117, 308, 313–315f, 316
Active immunity, 1065, 1076f
Active sites, **114**, 114f
Active transport, 99–102, 104t, 758f
Acute-phase proteins, 1061
Adaptation, speciation and, 447, 447f
Adapter proteins, **176**, 178
Adaptive immunity, 556–557, 1058, 1063–1067f, 1063–1068, 1064t
Adaptive radiation, 436, **450**–454
Adaptive significance, of behavior, 1149
Adaptive value, of egg coloration, 1149, 1149f
Adder's tongue fern, 189t
Addiction, to drugs, 901–902, 901f
Adenine, 42, 42f, 43, 259f, 260, 262, 822f
Adenohypophysis, **948**
Adenosine monophosphate (AMP), 112f, 113, 124
Adenosine triphosphate. *See* ATP
Adenovirus, 528f, 530f
Adenylyl cyclase, **179**–180, 180f, 182, 948
ADH. *See* Antidiuretic hormone
Adherens junctions, 82t, **83**
Adhesive junctions, 82t, **83**
Adhesive nature of water, 27, 27f
Adipose cells, 36f, **869**
Adipose tissue, **869**, 869f
ADP (adenosine diphosphate), 112f, 113, 125, 127, 128f, 158
Adrenal cortex, 955, 956, 956f
Adrenal gland, 955–956, 956f
Adrenal medulla, 955–956, 956f
Adrenocorticotropic hormone (ACTH), **950**
Adventitious plantlets, 857
Adventitious roots, **745**, 750, 750f, 856f
Aerenchyma, **769**, 769f
Aerial roots, 746, 746f
Aerobic capacity, 976–977
Aerobic metabolism, 129

Aerobic respiration
  ATP production from, 129, 137–138, 137f
  defined, **124**
  evolution of, 143
  glucose in, 132
  in mitochondria, 136f
  process overview, 126, 126f
  regulation of, 138, 138f
Aerobic treatment of wastewater, 342, 343f
Aesthetic value of biodiversity, 1266
Afferent arterioles, 1049
Aflatoxins, 635, 635f
African savanna, 1188f
African sleeping sickness (trypanosomiasis), 493f, 571
African violet, 751f
Afrovenator, 711f
Afterbirth, 1130
Age
  of Earth, 11
  at first reproduction, 1175
Agent Orange, 821
Age structure, of population, **1170**–1171
Aging, telomerase and, 273
Agricultural selection, 426, 427f
Agriculture
  biotechnology applications to, 347–349
  cultivation techniques, 347, 777
  global warming on, 1255
  pollution due to, 1253
  sustainability practices, 777
Agrobacterium tumefaciens, 340–341, 340f, 822, 823f
AIDS. *See* HIV/AIDS
Air, in soil, 776, 777f
Air pollution, lichens as bioindicators of, 632
Akiapolaau, 1273f
Alanine, 35f, 45, 47f, 141
Alaskan near-shore habitat, 1274–1275, 1275f
Albinism, 227t, 228, 228f, 249
Albumin, 99, **1021**
Albuminous cells, 742
Aldosterone, 956, **1037**, 1052, 1053, 1054f
Aleurone, **855**, 855f
Algae. *See also* Brown algae; Green algae; Red algae
  biofuels produced from, 341–342, 341f
  fossil algae, 565–566, 566f
  land plants, evolution from, 589, 589f
Alimentary canal, 984
Alkaloids, 795, 796t
Alkaptonuria, 227t, 279
Allantoin, **1042**
Allantois, **710**, 710f, 1091, 1118
Allee, Warder, 1178
Allee effect, **1178**
Alleles
  defined, **225**
  fitness and gene interactions, 417–418

  frequency of, 399–403, 402f, 405–406
  gene flow and, 404–405, 404f
  independent assortment of, 229
  multiple alleles for one gene, 233–235, 233t, 235f
  mutations and, 404
  recombination in crossing over, 244–245, 245f
  temperature-sensitive, 235, 235f
Allelopathy, **795**, 795f
Allen's rule, 1166
Allergies, 168, 635, 1077, 1078f
Alligators, 701f, 709t, 714
Allogona townsendiana, 670f
Allohexaploids, 361
Allolactose, 310
Allometric growth, **1130**
Allomyces, 620t, 624, 625f
Allopatric speciation, 448, 448f
Allopolyploidy, **449**, 449f, **484**, 484f, 486
Allosteric activators, **117**
Allosteric enzymes, 117
Allosteric inhibitors, **117**, 117f, 138
Allosteric sites, **117**, 118–119
Alper, Tikvah, 540
Alpha waves, 906
α helix, **48**, 49, 50, 95
α-ketoglutarate, 132, 133f, 141, 141f
Alternate leaf pattern, 747, 747f
Alternation of generations, 841
Alternative splicing, **290**, 320, 321–322f, 362, 368, 368f
Altman, Sidney, 116
Altricial young, **1155**
Altruism, **1156**–1159
Aluminum toxicity, in soil, 777–778
Alveolates, **573**
Alveoli, 573, 573f, **1009**–1010, 1010f
Alzheimer's disease, 323, 907
Amanita phalloides, 628f
Amborella trichopoda, 608, 608–610f, 610
American basswood, 828f
American woodcock (*Scolopax minor*), 934
Amino acid derivatives, 941
Amino acids
  catabolism of, 141, 141f
  chemical classes of, 45, 47f
  defined, **44**
  essential, 46, 1000
  function of, 37t
  as neurotransmitters, 899–900
  peptide bonds and, 46, 46f
  structure of, 36f, 44–46, 47f
  in TALE proteins, 336
Aminoacyl-tRNA synthetases, **291**, 291–292f, 292
Amino group, 35, 35f, 44
Amish populations, 406
Ammonia, 561, 1042, 1043f
Amniocentesis, **252**, 253f
Amnions, **710**, 710f, 1091, 1118
Amniotes, **710**
Amniotic eggs, 710, 710f, 714, 1091
Amniotic fluid, 1118
Amniotic membrane, 463, 463f, 1118

I-1

Amoebas
  cell surface of, 568
  movement of, 568, 583, 583f
  parasitic, 583
  slime molds, 583–584, 584f
Amoebozoa, 522, 522f, 568, 569f, 583–584
AMP (adenosine monophosphate), 112f, 113, 124
Amphibia, 700f, 705–708, 705t, 706–708f
Amphibians. *See also specific amphibians*
  brain of, 904, 904f
  circulation in, 706, 1025, 1026f
  classification of, 645t, 705t, 707–708, 708f
  development in, 378, 954, 954f
  eggs of, 710, 1252f
  evolution of, 699, 706–707
  features of, 705–706, 705t
  fertilization in, 1089–1090f, 1090
  fungal infections in, 635, 635f
  gastrulation in, 1116, 1116f
  heart in, 706, 1025, 1026f
  kidneys of, 1047
  legs of, 706, 706f
  lungs of, 706, 1009, 1009f
  nitrogenous wastes of, 1042, 1043f
  nuclear transplant in, 383
  orders of, 705t, 707–708, 708f
  phylogeny of, 700f
  population declines in, 1261t, 1267–1268f, 1267–1269
  prolactin in, 953
  reproduction in, 708
  respiration in, 706, 707, 1005–1006f, 1006
  temporal isolation of, 443
  terrestrial, 706–707, 707f
Amphioxus. See Branchiostoma
Amplification, 332
Ampullae of Lorenzini, **936**
Amygdala, **906**
Amylopectin, 40, 40f
Amyloplasts, **75**, 742, 810, 810f
Amylose, 39–40, 40f
Anabaena, 561
Anabolism, **117**
Anaerobic respiration, **124**, 139
Anaerobic treatment of wastewater, 342, 343f
Analogous structures, **11**
Anaphase
  meiosis I, 210, 211–212f, 214, 215, 216f
  meiosis II, 213f, 214, 215, 217f
  mitotic, 192, 192f, 195f, **196**, 197f, 211f, 216f
Anaphase A, 195f, 196
Anaphase B, 195f, 196, 197, 197f
Anaphase-promoting complex (APC), **201**, 202f
Anaphylactic shock, 1077
Anapsids, 710f
Anatomical dead space, 1012
Ancestral character variations, **462**–463, 463f
Anchitherium, 430f
Androecium, 610, 610f, 839
Androgens, 958
Anemia. *See also* Sickle cell anemia
  bone marrow transplants for, 347
  pernicious, 988
Aneuploid gametes, **214**
Aneuploidy, **250**, 344, **487**
Angelman syndrome (AS), 251–252
Angina pectoris, **1035**
Angiosperms, **607**. *See also* Flowering plants
Angle of incidence, 1233

Animal breeding
  artificial selection in, 427, 427f
  cattle, 383
  thoroughbred horses, 417, 417f
Animal cells
  cytokinesis in, 197, 197f
  division of, 188f, 376, 376f
  extracellular matrix of, 80–81, 81f
  general features of, 639
  mitosis in, 195
  sexual life cycle of, 208, 208f
  structure of, 66f, 81t
Animalia (kingdom), 13, 13f, 466f, 649
Animal pole, 380, **1112**, 1112f
Animals, 638–649. *See also specific animals*
  artificial selection of, 427, 427f
  body plan of, 640–644
  classification of, 644–649f
  coevolution of plants and, 795, 798
  cognition in, 1143, 1143f
  communication of, 1146–1149f
  defenses against predators, 1196, 1196f
  development in, 640, 642–644, 643f
  evolution of, 640–644, 647, 648–649f
  fertilization in, 1108–1111, 1108t, 1109–1111f
  fruit dispersal by, 615, 616f
  fungal diseases of, 635, 635f
  gap junctions in, 82t, **83–84**, 84f
  general features of, 639–640, 639f, 645–646t
  habitats of, 640
  modes of locomotion, 639, 639f, 977–979
  movement of, 639, 639f
  multicellularity in, 639
  in mutualistic relationships with fungi, 633–634, 634f
  nutrients for, 639
  phylogeny of, 644–649f
  plant defense by, 798–799
  pollination by, 843–844f, 843–845
  reproductive strategies of, 1086–1088
  sexual life cycle in, 208, 208f
  sexual reproduction in, 523, 640
  succession in communities of, 1205
  transgenic, 284f, 338–339f, 338–340
Anions, 19, 96
Annealing of primers, 332, 333f
Annelida (phylum), 645t, 648f, 673–676
Annelids
  body plan of, 642f, 673–674, 673f
  circulatory system of, 674, 1024
  classification of, 645t, 647, 674–676
  excretory system of, 674
  eyes of, 673–674, 929
  movement of, 674
  segmentation of, 644, 673–674, 674f
  trochophores of, 661
Annotation, genome, 361–362
Annual plants, 859, 859f
Anoles, 713
Anolis lizards, 447, 447f
Anonymous markers, **248**
Anopheles mosquitos, 360f, 481f, 493, 574, 798, 1081
Anoxygenic photosynthesis, 143, 148, 156, 546, 557
Anseriformes, 715t
Antagonistic effectors, 878–879, 878f
Anteaters, 435f
Antenna complex, **154**–155, 155f, 157, 158
Antennal glands, 1044
Antennapedia complex, **391**, 392, 392f
Antennapedia gene, 391, 499
Anterior pituitary gland, **948**, 949–953

Anterior–posterior axis formation, 387–390f
Antheridia, **592**, 592f, 597, 598
Anthers, **610**, 610f, 611–612, **839**, 839f
Anthocerotophyta, 593, 593f
Anthocyanins, 236, 814
Anthophytes, 604t
Anthozoa, **653**–654, 653f
Anthrax, 371, 371t, 545, 548f, 552, 559t
Anthropoids, 723–724, 723f
Antibiotics
  bacteria, susceptibility to, 64
  resistance to, 345, 407–408, 546, 551, 556
  for sexually transmitted diseases, 560, 561
  for tuberculosis, 558–559
  for ulcers, 560
Antibodies
  antigen-binding site on, 1072, 1073f
  functions of, 45t
  humoral immunity and, 1070–1076
  in immunoassays, 346, 346f
  in medical treatment and diagnosis, 180f, 1078–1080
  monoclonal, 1079–1080, 1080f
  polyclonal, 1079
  secretion of, **1065**
  specificity of, 1071–1072, 1072f
  in staining, 61–62
Anticodon loops, **291**–292, 291f
Antidiuretic hormone (ADH), **949**, 949f, **1037**, 1051, 1052–1053, 1053f
Antigen-binding sites, 1072, 1073f
Antigen drift, **1081**
Antigenic determinants, 1064
Antigen-presenting cells, **1068**
Antigens, **1063**–1064, 1064f, 1075–1076, 1080f
Antigen shift, **1081**
Antihistamines, 168
Antiparallel strands, in DNA, 262–263
Antiporters, **100**
Antiretroviral drugs, 536–537, 537f
Antlers, 720
Ants
  castes of, 1160f
  characteristics of, 687t
  leaf-cutter, 633, 634f
  mutualism with acacia trees, 798, 798f, 1199, 1199f
  nerve cells in, 1136
  nests of, 634
Anura, 705t, 707–708, 708f
Aorta, **1029**
Aortic bodies, **1013**, 1013f
Aortic valves, **1028**
APC (anaphase-promoting complex), **201**, 202f
Apertures, 612
Apes, 488, 488f, 724
Apex, **733**, 747, 747f
Aphasias, 906
Aphids, 687t, 770, 771f
Apical complex, 574
Apical meristems, **735**, 735–736f, 747, 747f, 820
Apical surface, 867, 989
Apicomplexans, 573, 573f, 574–575, 575f
Apicoplasts, 493, 493f
Aplysina longissima, 650f
Apoda, 705t, 706, 708, 708f
Apodiformes, 715t
Apomixis, 857
Apoplast route, **763**, 763f
Apoproteins, 805, 805f
Apoptosis, **394**–395, 394f, 1069, 1069f

Appendicular locomotion, 977
Appendix, 433, **992**, 992f
Applied research, 7–8
AP3 gene, in plants, 504–505
Aquaculture, 1250
Aquaporins, **98**, 104t, 758, 760f, 1052
Aqueous solutions, 97
Aquifers, **1212**
Aquifex pyrophilus, 548f
Aquificae, 548f
Arabidopsis
  aquaporins of, 758
  auxin transport in, 820
  branching mutants in, 822–823
  CAL gene in, 500
  cell division in, 378
  chilling of, 815
  columella cells in, 742
  CONSTANS gene in, 835–836
  copy numbers for gene families in, 488
  det2 mutant in, 807, 807f
  embryonic flower mutant, 833, 833f
  genome of, 360f, 361, 367f, 393, 481f, 482, 483, 485f, 486
  gibberellins in, 836
  GLABROUS3 mutant in, 738f
  HOBBIT gene in, 851, 852f
  hot mutants in, 815
  KANADI gene in, 751f
  LEAFY COTYLEDON gene in, 852
  LEAFY gene in, 833, 833f
  MADS-box genes in, 393
  MONOPTEROS gene in, 851, 852f
  overexpression of flowering gene in, 833f
  PHABULOSA gene in, 751f
  PHAVOLUTA gene in, 751f
  phytochrome genes in, 805–806, 805f
  SCARECROW gene in, 743–744, 743f, 810, 810f
  SHOOTMERISTEMLESS gene in, 851, 851f
  short root mutant in, 810, 810f
  suspensor mutant in, 850, 850f
  too many mouths mutant in, 737, 737f
  touch responses in, 811
  transposons in, 490
  trichome mutations in, 738f
  WEREWOLF gene in, 743, 743f
  WOODEN LEG gene in, 853, 853f
  YABBY gene in, 751f
Arachidonic acid, 945
Araneae, 684
Arbor, Werner, 328
Arbuscular mycorrhizae, **627**, 632, 633, 633f
Archaea (domain), 13, 13f, 481f, 490, 522, 522f, 547f
Archaeal viruses, 532
Archaebacteria. *See also* Prokaryotes
  bacteria vs., 546–547, 547f
  cell walls of, 64, 546
  gene expression in, 547
  lipid membrane of, 546, 547f
  photosynthetic, 557
  plasma membrane of, 64–65, 546
  replication in, 271, 546
  in wastewater treatment, 342
Archaefructus, 608, 608–609f
Archaeopteryx, 428–429, 429f, 472f, 716, 716f, 717
Archaeplastida, 522, 522f, 568, 569f, 578–581
Archegonia, **592**, 592f, 597, 598, 606
Archenteron, 1116
Archenterons, **643**, 643f, 644
Archosaurs, 467f, 711, 711f
Arenaviruses, 371t

ARFs (auxin response factors), 820, 820f
Argentine hemorrhagic fever, 371t
Arginine, 47f, 190, 279
Arginine vasotocin, 949
Arithmetic progressions, 10, 10f
Armillaria, 619, 634f
Armored fish, 702, 702t
Arousal (state of consciousness), 906
ART (assisted reproductive technology), **1104**
Arteries, 1029–1030, **1032**
Arterioles, **1032**
Arteriosclerosis, **1035**
Arthropoda (phylum), 645t, 649f, 680–689
Arthropods, 680–689. *See also* Insects
  body plan of, 681–683, 681t
  chelicerates, 681, 681t, 683–684, 684f
  circulatory system of, 682, 682f, 1023
  classification of, 645t, 647, 661
  Crustacea, 681t, 684–686
  diversity of, 680, 681f
  economic importance of, 681
  excretory system of, 683
  exoskeleton of, 681, 682
  eyes of, 929
  habitat of, 640
  jointed appendages of, 681, 682
  locomotion in, 978
  molting of, 682
  nervous system of, 682, 682–683f, 902f, 903
  respiratory system of, 683, 683f, 1008
  segmentation in, 644, 681–682, 681f
  taste in, 928, 928f
Artificial selection, **10**, 406, **426**–427, 1137
Artificial transformation, 555
Artiodactyla, 722t
AS (Angelman syndrome), 251–252
Ascaris, 208, 645t, 680
Ascidians, 501–502, 502f
Ascocarps, **629**, 629f
Ascomycetes, 622, 628–629, 629f, 631, 633
Ascomycota, 620, 620f, 620t, **628**–630
Ascospores, **629**, 629f
Ascus, **629**, 629f
Asexual reproduction
  cnidarians, 653
  defined, **569**
  echinoderms, 694
  forms of, 1086–1087, 1087f
  fungi, 622, 626, 629
  land plants, 591
  nemerteans, 673
  plants, 857–858, 857f
  protists, 569, 571, 574
  sponges, 651
A sites, **292**–293, 293f, 295, 296
Asparagine, 47f
Aspartate, 141
Aspartic acid, 47f
Aspen trees, 859
Aspergillus, 622f, 635, 635f
Assembly, of virus particles, **532**, 533f
Assisted reproductive technology (ART), **1104**
Association cortex, **906**
Assortative mating, **405**
Aster, **195**, 196f
Asteroidea, 694, 695, 695f
Asthma, **1014**
Atherosclerosis, 55, **1034**–1035, 1035f
Athlete's foot, 635

Atmosphere
  circulation of, 1233–1235
  of early Earth, 515, 516–517, 516f, 520, 521, 521f, 545, 561
  reducing, 517
Atomic mass, 18–19
Atomic number, 18
Atoms, 2f, **3, 18**–21
ATP (adenosine triphosphate)
  in active transport, 100–101f, 100–102
  from catabolism of fatty acids, 141
  in cellular respiration, 125–126
  components of, 43–44, 44f
  in coupled transport, 101–102, 101f
  cycle of, 113, 113f
  defined, **43**
  in electron transport chain, 124, 124f
  in endergonic reactions, 113, 113f, 125, 160
  in energy storage, 112–113
  in glycolysis, 126–127f, 127, 129
  in Krebs cycle, 131, 131f, 132, 133f
  in metabolism, 125
  in muscle contraction, 973, 973f
  in photosynthesis, 148, 149f, 151, 156–160, 156f, 158–159f
  in protein folding, 51, 51f
  in regulation of aerobic respiration, 138, 138f
  in sodium–potassium pump, 100–101, 100f
  structure of, 44f, 112, 112f
  synthesis of, 124–126, 134–136f, 135–136
  yield from aerobic respiration, 129, 137–138, 137f
ATP-dependent remodeling factor, 316, 317f
ATP synthase
  chemiosmotic mechanism for, 135, 135f
  in oxidative phosphorylation, **126**, 126f
  in photosynthesis, 156–159, 159f
  rotary motor of, 136, 136f
  structure of, 136, 136f
Atrial natriuretic hormone, 958, **1037**, 1053
Atrioventricular (AV) node, 1028, 1035
Atrioventricular (AV) valves, **1028**, 1029f
Atrium, **1025**, 1025f
Attachment, in HIV/AIDS infection cycle, 535–536, 535f
Attachment, of virus to host, **532**, 533f
Auditory tubes, 922
Australopithecines, 724–726, 725f, 726
Australopithecus, 725, 725f, 726
Autocrine signaling, 169, 940
Autoimmune diseases, 345–346, 1077
Autologous blood donation, 1079
Autonomic nervous system, **889**, 890f, 910t, 911, 911–913f, 912t
Autophosphorylation, 175–176, 175f
Autopolyploidy, 449, **484**, 484f
Autosomes, **241**, 250, 250f
Autotrophs, **123**, 557, 568, **1216**
Aux/IAA proteins, 820, 820f
Auxin-binding proteins, 819
Auxin receptors, 820
Auxin response factors (ARFs), 820, 820f
Auxins
  cytokinins and, 822, 822f
  discovery of, 817–818
  effects of, 818–819

  functions of, **815**, 816t
  in gene expression, 820, 820f
  gravitropism and, 809, 810
  mechanism of action of, 819–821
  phototropism and, 808, 819f
  synthetic, 819f, 821
  thigmotropism and, 811
Auxotrophs, **556**
Avascular bone, **968**
Avery, Oswald, 257, 258
Aves, 468, 701f, **714**–718, 714f, 715t, 716–717f
Avian cholera, 1063
Avian influenza, 538, 1081
Avirulence gene, 800, 801f
Axial locomotion, 977
Axillary buds, **747**, 748f
Axils, **747**
Axis formation, in vertebrates, 1124–1127
Axons
  action potentials propagated along, 895–896, 896f
  characteristics of, 873, 874t, **890**, 890f
  conduction velocities of, 896–897, 896t
  development of, 393
  diameter of, 896, 896t
  myelinated, 896, 896t, 897f
  unmyelinated, 896, 896f, 896t
Aznalcóllar mine spill (Spain), 789, 789f
Azolla ferns, 598
Azospirillum, 793

B

Bacillary dysentery, 558
Bacilli, 548f, 549–550, 550f
Bacillus anthracis. See Anthrax
Bacillus thuringiensis (Bt toxin), 348
Bacon, Francis, 5
BACs (bacterial artificial chromosomes), 331
Bacteria
  ancient, 519, 519f
  archaebacteria vs., 546–547, 547f
  benefits for plants, 793
  for bioremediation, 562
  capsule of, 63f, 64, 551
  cell division in, 187–188
  cell wall of, 64, 64f, 546
  conjugation in, 552–554
  diseases caused by, 556, 557–561, 559t, 560f
  endosymbiotic, 566
  flagella of, 64f, 65, 546, 551, 551f
  gene expression in, 547
  in genetic engineering, 562
  genome of, 481f
  Gram staining of, 550–551
  intestinal, 562
  nitrogen for plants provided by, 781–782
  photosynthetic, 64, 64f, 148, 150, 156, 156f, 545, 546, 557
  plasma membrane of, 64, 93, 546
  replication in, 546
  rhizobacteria, 793
  sulfur bacteria, 139, 139f
  transformation in, 257, 555–556, 556f
  in wastewater treatment, 342, 343f
Bacteria (domain), 13, 13f, 490, 547f
Bacterial artificial chromosomes (BACs), 331
Bacteriochlorophyll, 557
Bacteriophage
  conversion in, 533–534
  defined, **258**, 527

  DNA transfer by transduction, 554–555, 555f
  Hershey–Chase experiment with, 258–259, 258f
  lysogenic cycle of, 532–533, 533f
  lytic cycle of, 532, 533f
  structure of, 529, 529f
  T2, 530f
  T4, 529f, 532
  T-even, 529–530, 532
Bacteriorhodopsin, 95, 557
Ball-and-socket joints, **969**, 970f
Banks, of continental shelves, **1243**
BAP (6-benzylamino purine), 822f
Barley
  comparative genomics and, 366
  genome of, 485f
  transgenic, 347
Barnacles, 681t, 684–686, 686f, 1190, 1190f
Barometers, 1008
Baroreceptors, 920, **1036**–1037, 1036f
Barr bodies, **243**, 243f, 251
Barro Colorado Island, 1223
Basal bodies, **79**, 79f
Basal metabolic rate (BMR), **997**
Basal nuclei, 903t, 906
Basal surface, 867
Base-pairs, 43, 43f, **262**, 262f, 265, 265f
Bases, **30**
Base substitution mutations, **299**, 300f
Basic Local Alignment Search Tool (BLAST) algorithm, 362
Basic research, 7
Basidiocarps, **628**, 628f
Basidiomycetes, 622, **627**–628, 628f, 633
Basidiomycota, 620, 620f, 620t, 627–628, 628f
Basidiospores, **627**, 628f
Basidium, **627**, 628f
Basophils, 1021, 1021f, **1064**, 1064f
Bates, Henry, 1196
Batesian mimicry, **1196**–1197, 1197f
Batrachochytrium dendrobatidis, 624, 635
Bats
  classification of, 467f
  flight of, 720, 720f
  forelimbs of, 11f, 432f
  pollination by, 844, 1198f
  vampire, 1156, 1157f
B cells, 1064t, **1065**, 1065f, 1068t
Bdellovibrio, 549f
Beach fleas, 684
Beadle, George, 6, 279–280, 279f
Beans, 812, 813f, 854f, 856f
Beavers, 719f
Bedbugs, 687t
Bees, 189f, 687t, 843, 843f. *See also* Honeybees (*Apis mellifera*)
Beetles, 474–475, 474f, 645t, 687t
Behavior, 1134–1160. *See also specific types of behavior*
  adaptation to environmental changes, 1166, 1166f
  adaptive significance of, 1149
  altruistic, 1156–1159
  changes in, 1166, 1166f
  cognitive, 1143, 1143f
  communication and, 1146–1149f
  development of, 1140–1142
  foraging, 1150, 1150f
  hormones and neurotransmitters in, 1136
  innate, 1135, 1135f
  learning and, 907, 1137, 1137f, 1139–1140, 1140f, 1141–1142

migratory, 1144–1145
natural history of, 1135–1136
reproductive strategies, 1151–1156
study of, 1135–1136
territorial, 1150–1151, 1151f
Behavioral ecology, **1149–1151**
Behavioral genetics, **1137–1139**
Behavioral isolation, 442–443, 442t
Bent grass *(Agrostis tenuis)*, 416, 416f
6-Benzylamino Purine (BAP), 822f
Berg, Paul, 328
Bergey's Manual of Systematic Bacteriology, 548
Betacyanins, 814
β barrel, 95
ß-carotene, 153, 348, 349f
ß-oxidation, 141, 142f
ß subunit, 268, 268f, 269–270, 285
Beta waves, 906
β sheets, **48,** 49, 50, 95
β-α-β motif, 50, 50f
Bicarbonate ions, 30, 515, 1016–1017, 1017f
Bicoid proteins, **387**–388, 389f, 391
Bicuspid valves, **1028**
Biennial plants, **859**–860
Bilateral symmetry, 506, **641,** 641f, 694, 840, 840f
Bilateria
body cavity of, 642, 642f
classification of, 656–657, 657f
embryonic development in, 642–644, 643f
eye development in, 508, 509, 509f
muscle formation in, 651
phylogeny of, 647, 648f
symmetry of, 641
Bile, 985, 990f, 991
Bile pigments, 991
Bile salts, 991
Bilirubin, 1079
Binary fission, **187,** 187f
Binocular vision, 723, **934**
Binomial expansion, 403
Binomial name, 465
Biochemical pathways, **118**–119
Biochemical signatures, 364, 364f
Biodiesel, 341–342
Biodiversity. *See also* Species richness
crisis of, 1259–1264, 1261t, 1263t
descriptive nature of, 4
economic value of, 1264–1266
ecosystem stability and, 1225–1227f, 1225–1228
ethical and aesthetic values of, 1266
evolution and, 13
Bioenergetics, 107
Biofilms, 342, **545**–546, 545f, 559–560
Biofuels, **341**–342, 341f, 625
Biogenic amines, **900**
Biogeochemical cycles, 1210–1215
Biogeography, **435,** 436, 1228–1229, 1228f
Bioinformatics, **362,** 368–369, 500
Biological communities, 3f, **4, 1188,** 1188–1189f
Biological disturbances, 1205–1206, 1205f
Biological function, 364–365
Biological species concept (BSC), **441**–442, 442t, 444–445, 468
Biological weapons, 371
Biology, 1, 2
Bioluminescent species, 654, 656
Biomarkers, 345, 368, 519
Biomes, **1237**–1240
Bioremediation, 342, 562, 623
Biosphere, 3f, 4, 1247–1252
Biostimulation, 562

Biotechnology, 327–349
agricultural applications of, 347–349
defined, 327
environmental applications of, 341–342, 341f, 343f
genetic variation analysis, 335–337, 335f, 337f
medical applications of, 343–347
polymerase chain reactions in, 332–336
recombinant DNA in, 327–332
transgenic organisms and, 337–341
Bioterrorism, 371, 371t, 630, 635
Biotic potential, **1175**
Biotin, 1000t
Bipedalism, 463f, 724, 725–726
Bipolar cells, **932,** 932f
Birch *(Betula),* 845f
Bird flu, 538, 1081
Birds. *See also specific birds*
altruism in, 1158–1159, 1159f
beak sizes and selection, 8, 9f, 410, 422–423
bones of, 716
brain of, 904, 904f
circulation in, 718, 1026, 1027f
classification of, 468
cost of reproduction in, 1173, 1173f
courtship rituals of, 442–443, 443f
digestive tract of, 986, 986f
eggs of, 714
evolution of, 428–429, 429f, 470, 472f, 716–717
eyes of, 932, 933f, 934
feathers of, 470, 472f, 714, 714f, 716
features of, 714, 716
fertilization in, 1090–1091f, 1091
gastrulation in, 1116–1117, 1117f
habituation in, 1139
heart of, 1026, 1027f
hybridization between species of, 444
kidneys of, 1047, 1047f
locomotion in, 978–979, 979f
lungs of, 718, 1010, 1011f
magnetic field detection by, 936
migratory behavior of, 1144–1145, 1145f, 1255
modern, 717–718
nitrogenous wastes of, 1042, 1043f
orders of, 715t
parental care in, 470
phylogeny of, 701f
pollination by, 843–844, 844f
respiration in, 717–718, 1010, 1011f
sex chromosomes in, 241, 241t
songbirds, 1141–1142, 1142f, 1147, 1151, 1271
territorial behavior in, 1151, 1151f
thermoregulation in, 718
Birth control, 1101–1103, 1101f, 1101t, 1103f
Birth control pills, 1101f, 1101t, 1102
Bison, 494f
1,3 Bisphosphoglycerate, 127, 128f
Bithorax complex, **391,** 391f, 392, 499
Bivalent structures, 209
Bivalves, 667, 668–669, 668f
Bivalvia (class), 670–671, 671f
Blackman, F. F., 150
Black mangrove *(Avicennia germinans),* 769, 769f
Black walnut trees *(Juglans nigra),* 752f, 795, 795f
Black widow spiders, 684, 684f
Bladder
swim, 704, 704f
urinary, 1047, 1048f
Bladderworts *(Utricularia),* 784

Blades, of leaves, **751**
BLAST (Basic Local Alignment Search Tool) algorithm, 362
Blastocladiomycota, 620, 620f, 624, 625f
Blastocoel, **1112**
Blastocysts, 1112, 1114
Blastoderms, 387, 388f, 391, **1116,** 1117f
Blastodisc, **1113**
Blastomeres, **376,** 376f, 380, 1112, 1114
Blastopores, 395f, 640, **643,** 643f, 1116
Blastulas, 639f, 640, 1112
Bleaching, 1255
Blending inheritance, 402
Blinking, 909
Blood
classification of, 871
components of, 1020–1023f
functions of, 868t, 1020–1021
regulation of, 1035–1037, 1036f
Blood acidosis, 30
Blood alkalosis, 30
Blood cells, 1021–1022, 1021f
Blood clotting, 437, 1022, 1022f
Blood donation, 1079
Blood flow, 1035–1037
Blood groups, 91t, 233t, 234–235, 235f, 400–401, 1079
Blood plasma, **1020,** 1021, 1021f
Blood pressure, 920
measurement of, 1030–1031, 1031f
regulation of, 1035–1037, 1036f
Blood transfusions, 1079
Blood typing, 1079
Blood vessels
characteristics of, 1031–1035
four-chambered heart and, 1026, 1028–1031
paracrine regulation of, 944–945
tissue layers of, 1032, 1032f
Blue crabs, 648f, 685f
Blue-footed boobies, 443f
Bluehead wrasse *(Thalassoma bifasciatium),* 1087f
Blue-light receptors, in plants, 808, 808f
B lymphocytes. *See* B cells
BMP4 (bone morphogenetic protein 4), **1126,** 1126f
BMR (basal metabolic rate), **997**
Bobolinks *(Dolichonyx oryzivorus),* 1144, 1145f
Body cavities, 641–642, 642f, 865–866, 866f
Body plans
animals, 640–644
annelids, 642f, 673–674, 673f
arthropods, 681–683, 681f
cnidarians, 652–653, 652f
crustaceans, 685, 685f
echinoderms, 694, 695
fruit flies, 388–389f, 391
mollusks, 667–669
vertebrates, 865–866
Body position, sensing of, 925–926
Body size, metabolic heat and, 882–883, 883f
Body temperature regulation. *See* Thermoregulation
Bohr, Niels, 18
Bohr effect, **1016**
Bohr model of atoms, 19f, 20
Boll weevils *(Anthonomus grandis),* 686f
Bolus, **987**
Bone
avascular, 968
compact, 967f, 968
endochondral, 966f, 967–968

intramembranous, 965, 966f, 967
medullary, 967f, 968
remodeling of, 968–969
spongy, 967f, 968
structure of, 967f, 968
vascular, 968
Bone marrow, **1066**
Bone marrow transplants, 347
Bone morphogenetic protein 4 (BMP4), **1126,** 1126f
Bones
of birds, 716
cells, 871
development of, 965–968, 966f
functions of, 870t
homologous, 11, 11f
Bony fish, 703, 704, 704f, 1006–1008, 1006f, 1046
Book lungs, **683**
Borrelia, 546
Borrelia burgdorferi. See Lyme disease
Bottleneck effect, 405–406f, **406**
Bottom-up effects, **1221,** 1223–1225, 1224f
Botulism, 371t, 548f, 552, 559t
Bovine spongiform encephalopathy (BSE), 540
Bowman's capsule, 1045, **1049,** 1049f
Box jellyfish, 654, 654f
Boyer, Herbert, 328
Boysen-Jensen, Peter, 817
Brachiopoda, 646t, 648f, 661, 676–678
Brachydactyly, 227t
Brachyury gene, 501–502, 502f
Bracts, 753
Bradykinin, 944
Brain
amphibians, 904, 904f
birds, 904, 904f
divisions of, 903, 903t
evolution of, 904, 904f
fish, 903–904
forebrain, 903f, 903t, 904–906, 905f
hindbrain, 903, 903f, 903t
mammals, 904, 904f
midbrain, 903–904, 903f, 903t
primitive, 903f
reptiles, 904, 904f
size of, 904, 904f
vertebrates, 904, 904f
virus-induced diseases of, 540–541
Brainstem, 906
Branchial chambers, 1006
Branching diagrams, 461, 461f
Branching morphogenesis, **1120**
Branchiostoma (Amphioxus), 392, 502, 503f, 697–698, 698f
Branchless gene, 1120
Branch point, 290, 290f
Brassica
evolution of, 500–501, 500f
genome of, 485f
hyperaccumulation of toxic metals by, 789
Brassinolide, 825f
Brassinosteroids, 816t, 825, 825f–826f
BRCA1/BRCA2 genes, 372
Bread molds, 279, 625, 626
Breakbone fever, 1256
Breast cancer, 344, 345, 372, 797–798
Breathing
mechanics of, 1004, 1011–1014
negative pressure, 1009
positive pressure, 1009
rate of, 1012–1013
regulation of, 1013–1014, 1013f
Brenner, Sydney, 282, 283, 299
Briggs, Robert, 383
Briggs, Winslow, 818–819, 819f

Bright-field microscopes, 62t
Bristleworms, 674f
Brittle stars, 693–695, 695f
Broca's area, 906, 907
Bronchi, 1009, 1010f
Brood parasites, **1142**, 1142f
Brown, Robert, 65
Brown algae
 chloroplasts of, 522, 523f, 566, 567f
 conducting tubes in, 470, 471f
 fucus (zygote), 849, 850f
 haplodiplontic life cycle of, 577–578, 577f
 as stramenopile, 573
Brown bears, 494f
Brown recluse spiders, 684, 684f
Brush border, 989–990
Bryophytes, 468, 468f, 589, 589f, 591–593
Bryozoa, 455, 645t, 648f, 661, 676–677, 677f
BSC (biological species concept), **441**–442, 442t, 444–445, 468
BSE (bovine spongiform encephalopathy), 540
Bubonic plague, 371t, 557, 559t
Buchner, Eduard, 34f
Budding, 536, **569**, 629, 630f
Buffers, **30**, 30f
Bulbourethral gland, 1093f, 1094
Bulbs (plant), 749, 750f
Bulk transport, **102**–103
Bumblebees (*Bombus*), 843f
Bushmeat, 476
Buttercups (*Ranunculus*), 744f, 840
Butterflies
 characteristics of, 687t
 defenses against plant toxins, 795
 global warming and, 1255, 1255f
 life history of, 689
 metapopulations of, 1169–1170, 1169f
 migratory behavior in, 1144, 1144f
 mimicry in, 1197, 1197f
 mouthparts of, 688f
 movement of, 639, 639f
 thermoregulation by, 881
 wings of, 505–506, 506f, 687, 688f
Buttress roots, 746, 747f

C

Cabbage palmetto, 751f
Cactus finches (*Geospiza scandens*), 9f, 422f, 444, 445, 452, 453f
Cadherin domain, 395
Cadherins, **83**, 83–84f, 395
Caecilians, 705–706, 705t, 708, 708f
Caenorhabditis elegans
 apoptosis in, 394, 394f
 Brachyury gene in, 502f
 cell division in, 376, 377f
 development in, 679
 genome of, 360, 481f
 small RNAs in, 317
 transposons in, 490
Caimans, 714
Calcitonin, 320, 321f, 953, **954**–955
Calcitonin-gene-related peptide (CGRP), 320, 321f
Calcium
 in fertilization, 1110, 1110f
 homeostasis of, 954–955, 955f
 in muscle contraction, 974–975, 974f
 as second messenger, 181, 182f
Calcium carbonate, 515, 516f, 653, 654
CAL gene, 500–501, 500f
California condors (*Gymnogyps californianus*), 1279–1280
Callus (plant), 858

Calmodulin, 45t, 181, 181–182f
Calories, 108, 997
Calvin, Melvin, 161
Calvin cycle
 in carbon fixation, 149f, 160–163, 519
 defined, **161**
 phases of, 162
Calyx, **839**, 839f
Cambrian period, 515, 520, 523–524, 524f, 699, 700
CAM (crassulacean acid metabolism), 767
 plants, **164**, 165, 165f
cAMP. *See* Cyclic adenosine monophosphate
Camptodactyly, 227t
Campylobacter pylori, 560
Canaliculi, 871, 966f, **967**
Cancer. *See also* specific types of cancer
 cell cycle control in, 202–204
 chromosomal abnormalities and, 344
 defined, **202**
 genome projects, 361
 hormonal responses to, 959
 microarray analysis and, 368
 oncogenes and, 175, 204, 337, 361, 539–540
 telomerase and, 273
 virus-induced, 539–540
Candida, 630, 635
5′ cap, **288**, 288f
CAP (catabolite activator protein), 308f, 310–311, 310f
Capillaries, **1032**–1033
Capsids, viral, **528**, 528f
Capsule, of bacteria, 63f, 64, **551**
Captive breeding, 1279, 1279f
Carbohydrates
 catabolism of, 124, 138, 138f
 defined, **38**
 disaccharides, 38–39, 39f
 as macromolecules, 33, 37, 37t
 monosaccharides, 36f, 38, 38f
 polysaccharides, 39–41, 40f
 structure of, 36f, 37
 transport in plants, 759f, 770
Carbon
 covalent bonds of, 24, 34
 electronegativity of, 25t, 34
 functional groups and, 35, 35f
 isomers and, 35, 35f
 isotopes of, 19, 19f, 20
 in plants, 779, 779f, 784–787
 prokaryote acquisition of, 557
Carbon cycle, 1210–1211, 1210f
Carbon dioxide
 atmospheric, 515, 516f, 521, 1210, 1253
 as electron acceptor, 139
 from ethanol fermentation, 140, 140f
 from Krebs cycle, 131, 131f, 132, 133f
 in photosynthesis, 148, 149f, 150–151
 from pyruvate oxidation, 130, 130f
 in respiration, 1004
 transport by blood, 1016–1017, 1017f
Carbon fixation
 in C₄ plants, 165f
 defined, 1211
 isotopic data on, 519
 pathways of, 164f
 in photosynthesis, **148**, 149f, 151, 160–163
 prokaryotes and, 561
Carbonic acid, 30, 114, 515, 516f
Carbonic anhydrase, 114
Carbon-12 isotope, 19, 19f

Carbon-13 isotope, 19, 19f
Carbon-14 isotope, 19, 19f, 20
Carbon monoxide poisoning, 1017
Carbonyl group, 35, 35f
Carboxyl group, 35, 35f, 44, 46, 46f, 55
Carboxypeptidase, 117
Carboxy terminal domain (CTD), 288, 290
Cardiac cycle, **1028**, 1029f
Cardiac muscle, **872**–873, 872t, 873f
Cardiac output, **1035**–1036
Cardioacceleratory center, **1035**
Cardioinhibitory center, **1035**
Cardiovascular disease, 1034–1035, 1035f
Carnivores
 birds, 717
 characteristics of, 722t
 chelicerates, 683
 cnidarians, 651
 crocodilians, 714
 defined, **984**
 digestive system of, 993f
 fungi, 622, 622f
 human removal of, 1222–1223
 plants, 782–784
 primary, 1217, 1217f
 saber-toothed, 471f
 secondary, 1217, 1217f
 teeth of, 986, 986f
 top, 1220
 in trophic-level ecosystems, 1217, 1217f, 1220
Carotenoids, 152f, **153**, 153f, 155, 823f, 843
Carotid bodies, **1013**, 1013f
Carpels, 607, 607f, **610**, 610f, 839f, 840
Carrier proteins, 91t, **96**, 97, 97f, 104t
Carroll, Sean, 389f
Carrying capacity, **1176**
Cartilage, 870t, **871**, 965
Cartilaginous fish
 characteristics of, 702
 fertilization in, 1089
 immunity in, 1067
 kidneys in, 1046–1047
 lateral line system of, 703
 phylogeny of, 700f
 reproduction in, 703
 sharks, 463, 463f, 702t, 703, 703f
Casein, 45t
Casparian strips, 744f, **745**
Cassava (*Manihot esculenta*), 795, 796f
Castor beans (*Ricinus communis*), 797, 797f
Catabolism
 of carbohydrates, 124, 138, 138f
 defined, **117**
 of proteins and fats, 141–142
Catabolite activator protein (CAP), 308f, 310–311, 310f
Catalase, 73
Catalysis, 37, 44, **111**, 116
Catalysts, 25, 111–112, 111f
Catecholamines, 900, **941**, 955–956, 956f
Caterpillars, 122, 122f, 799
Cations, **19**, 96
Cats
 coat color in, 233t, 235, 235f, 243, 243f
 forelimbs of, 11f, 432f
 genetic variations in, 402–403, 402f
 ovary in, 1098f
 saber-toothed, 471f
Cattle, 383, 481f, 562
Caudal proteins, **388**, 390f
Caudata, 705t, 708, 708f

Caudipteryx, 472f, 716–717, 716f
Cavefish, 433, 507, 508f
Cavitation, **765**, 765f
Cayuga Lake, 1220, 1220f
CCK. *See* Cholecystokinin
CCR5 gene, 534, 536
CDC (Centers for Disease Control and Prevention), 371, 540
CD4+ cells, **534**, 535, 535f, 536
cdc2 gene, 199
Cdc2 kinase, 200–201
Cdk1, 200
Cdks. *See* Cyclin-dependent protein kinases
cDNA. *See* Complementary DNA
Cech, Thomas R., 116
Cecum, 433, 720, **992**, 992f
Cedar Creek experimental fields, 1225–1226, 1225f
Celera Genomics, 360
Cell adhesion, 82–85, 94f, 95
Cell body, of neurons, 873, 874f, **890**, 890f
Cell communication. *See* Cell signaling
Cell cycle, 192–204
 cancer cells and, 202–204
 checkpoints in, 198, 200, 200f
 control of, 198–204
 defined, **192**
 duration of, 192–193
 genetic analysis of, 199
 growth factors and, 202, 203f
 history of investigation into, 198–199
 interphase, 192, 192–194f, 193–194
 in multicellular eukaryotes, 201–202, 202f
 phases of, 192, 192f
Cell death, 394, 394f
Cell determination, 378–379f, **378**–380, 1119
Cell division. *See also* Cell cycle
 in animals, 188f
 bacterial, 187–188
 binary fission and, 187, 187f
 in development, 375, 376–377f, 376–378, 393
 in eukaryotes, 187, 393, 546
 in meristems, 734–735, 734f
 in prokaryotes, 187–188, 546
 in protists, 188f
 in yeasts, 188f
 zone of, 743–744, 743f
Cell-free fermentation, 33, 34f
Cell identity, 82, 82t
Cell junctions, 82–85, 82t, 83–85f
Cell-mediated immunity, **1065**, 1068–1070, 1068t, 1069f, 1071f
Cell migration, in development, 395–396, 395f
Cell plates, 195f, **197**, 198f
Cells
 blood, 1021–1022, 1021f
 earliest, 518–519, 519f
 as information-processing systems, 14
 in living organisms, 2, 2f, **3**, 12, 12f
 membranes, 81
 nucleus of, 62, 65–68, 78t, 81t
 origins of, 518
 of protists, 568
 size and shape of, 60, 61f, 393–394, 546, 546f
 of sponges, 650–651, 650f
 structural features of, 62–63, 62f
 synthetic, 370, 371f
 visualization of, 60–62, 62t
Cell signaling, 168–183
 cell-surface receptors in, 94f, 95, 171–173, 172f, 172t
 G protein–coupled receptors in, 172f, 172t, 173, 178–183, 179f

intracellular receptors in, 171, 172t, 173–174, 173f
mechanisms of, 169–170, 169f
overview, 168–170, 169f
receptor tyrosine kinases in, 174–178, 182–183
Cell-surface markers, 63, 82, 82t, 90, 91t, 94f, 95
Cell-surface receptors, 94f, 95, 171–173, 172f, 172t
Cell theory, **12**, 59–63
Cell-to-cell adhesion proteins, 94f, 95
Cell-to-cell interactions, 82–85, 82t
Cellular blastoderm, **387**, 388f, 1112
Cellular bones, 968
Cellular organization, as characteristic of life, 2, 2f, 3
Cellular respiration. *See also* Aerobic respiration
anaerobic, 124, 139
ATP in, 125–126
defined, **123**
evolution of, 142–143
harvesting efficiency of, 124–125
as oxidation of glucose, 123–124
Cellular slime molds, 583–584, 584f
Cellulose
breakdown of, 41, 622, 624–625, 630
function of, 37t
structure of, **39**, 40–41, 40f
Cell walls
of archaebacteria, 64, 546
of bacteria, 64, 64f, 546
of eukaryotes, 67f, 78t, 80, 80f, 81t
of fungi, 619, 621
of plants, 41, 67f, 80, 80f, 81t, 733–734, 734f
primary, 80, 80f
of prokaryotes, 63, 63–64, 64, 81t, 546, 550–551f, 550–552
secondary, 80, 80f
Celsius scale, 108
Cenchrus incertus, 616f
Centers for Disease Control and Prevention (CDC), 371, 540
Centimorgans (cM), 354
Centipedes, 645t, 681t, 689, 689f
Central chemoreceptors, **929**
Central dogma of molecular biology, **280**, 280f, 330
Central nervous system (CNS), 902–910
brain, 903–905f, 903–908
defined, **873**
diversity of, 902–903, 902f
divisions and subdivisions of, 903t
neurons of, 889, 890f
spinal cord, 908–910
Central vacuoles, **65**, 73, 73f
Centrioles, 66f, **76**–77, 77f, 81t, 194, 195
Centromeres, 189f, 191, 191f, **193**, 193f, 211, 215
Centrosomes, 76–77
Cephalaspidomorphi, 700f, 702t
Cephalization, **641**
Cephalochordata, 697
Cephalopoda, 671–672
Cephalopods, 667–668, 668f, 669, 671
Cephalothorax, 682, 684, 685, 685f
Ceratium, 573f
Cercariae, **663**–664, 664f
Cercomeromorpha, 663, 665
Cercozoa, 582, 582f
Cerebellum, **903**, 903t
Cerebral cortex, 903t, **905**–906, 905f, 933–935
Cerebral hemispheres, **904**, 905f
Cerebrum, 903t, **904**, 905f
Cervical cancer, 539

Cervical caps (birth control), 1101t, 1102
Cetacea, 722t
CFCs (chlorofluorocarbons), 1252
cGMP (cyclic guanosine monophosphate), 174, 933, 934f
CGRP (calcitonin-gene-related peptide), 320, 321f
Chaetae, 645t, **674**, 674f
Chaetognatha, 646t, 647, 649f
Chagas disease, 492, 493, 493f, 571
Chain terminators, 357
Chambered nautilus (*Nautilus pompilius*), 667, 667f, 671
Chancres, 560
Channel-linked receptors, 171–172, 172f, 172t
Channel proteins, 91t, **96**, 97f, 104t
Chaperone proteins, **51**, 51f
Chara, 581, 581f
Character displacement, **450**–451, 451f, 1192
Character states, **462**–463
Charadriiformes, 715t
Charales, 581, 581f
Chargaff, Erwin, 260
Chargaff's rules, 260, 262
Charging reactions, **292**, 292f
Charophyta (phylum), 579
Charophytes, 468f, **581**, 581f, 589f, 596f
Chase–Hershey experiment, 258–259, 258f
Checkpoints, in cell cycle, 198, 200, 200f
Chelicerae, 681, 681f, **683**
Chelicerates, 681, 681t, 683–684, 684f
Chelonia, 709f, 712–713, 712f
Chemical bonds, 23–25, 23f, 23t, 25t
Chemical defenses
of animals, 1196, 1196f
of plants, 794–798, 795f, 796f, 797f, 1195
Chemical digestion, 984
Chemically gated ion channels, 171–172, 172f, 172t
Chemical messengers, 940–941, 941f
Chemical reactions
activation energy for, 111, 111f
energy changes in, 110–111, 111f
factors influencing, 25
Chemical synapses, **170**, 897–898, 897f
Chemiosmosis, 134–136, 156, 158–159
Chemoheterotrophs, **557**, **1216**
Chemolithoautotrophs, **557**
Chemolithotrophs, 546
Chemoreceptors, 917, 926–929
Chernobyl nuclear disaster, 788
Chewing, 970, 984, 986
"Chewing the cud," 993
Chiasmata, **210**, 211, 211f, 215
Chicken pox, 528, 531f, 1063
Chickens
chromosomes in, 189t
clutch size in, 417
development in, 379, 1112f
genome of, 488
Chief cells, 988, 988f
Chihuahuas, 427, 427f
Childbirth
hormonal changes in, 1130, 1130f
oxytocin secretion in, 949
positive feedback during, 879f
Chilling, of plants, 814–815
Chimpanzees (*Pan*)
chromosomes in, 189t
cognitive behavior in, 1143, 1143f
evolutionary relationships of, 461, 461f, 462, 724
FOXP2 gene in, 491, 491f, 492

gene transcription patterns in, 491
genome of, 481f, 482, 488, 488f
HIV/AIDS in, 475, 534
olfactory receptor genes in, 489
Chiral molecules, 35, 35f
Chiroptera, 722t
Chitin, 37t, 41, 621, 653, 682, 964
Chitons, 668–669f, 669–670, 670f
Chitridiomycetes, **624**
Chlamydia, 559f, 560–561, 560f
Chlamydomonas, 244, 483, 579, 580f, 592, 597
Chloramphenicol, 552
Chlorella, 154, 579
Chlorenchyma, 739, 752
Chloroflexi, 548f
Chlorofluorocarbons (CFCs), 1252
Chlorophyll a, **152**, 152f, 153, 154, 557
Chlorophyll b, **152**, 152f
Chlorophylls
absorption spectra of, 152, 152f
action spectrum of, 153, 153f
in photosynthesis, **148**
structure of, 152–153, 152f
Chlorophyta (phylum), 579
Chlorophytes, 468f, 579–580, 580f, 589, 589f, 596f
Chloroplasts
of archaeplastids, 579
defined, **74**
DNA of, 74–75, 74f
energy cycle of, 162f, 163
of euglenozoans, 570, 571
functions of, 73, 74, 78t
genetic codes in, 284
light-harvesting complexes of, 155
maternal inheritance and, 244
origins of, 75, 75f, 522, 523f, 566–567, 567f
photosynthesis in, 148–149, 148f
photosystems in, 151, 156–160
in plant cells, 67f
structure of, 73, 74, 74f, 81t, 148, 148f
Choanocytes, 645t, **650**, 650f, 651
Choanoflagellates, 585, 585f, 648f
Cholecystokinin (CCK), **995**, 995f, 996t, 998, 999, 999f
Cholera, 64f, 180, 533, 546, 549f, 558, 559f
Cholesterol, 54, 54f, 92, 103, 1034
Chondrichthyes, 700f, 702t, 703, 703f, 1041, 1046
Chondroblasts, 966f
Chondrocytes, 871, 966f
Chondroitin, 871
Chordata (phylum), 466f, 645t, 649f, 696–700
Chordates
classification of, 645t, 647
endoskeleton of, 696
evolution of, 501–502, 502f
eyes of, 929, 930f
features of, 696, 696–697f
habitat of, 640
nonvertebrate, 697–698, 698f
segmentation in, 644
vertebrate, 698–700
Chorion, **710**, 710f, 1091, 1108
Chorionic villi sampling (CVS), **252**–253, 253f
Chromalveolata, 522, 522f, 568, 569f, 573–578
Chromatids. *See* Sister chromatids
Chromatin, **68**, 68f, **190**–191, 190f, 315–317f
Chromatin-remodeling complexes, 316
Chromatophores, 672
Chromophores, 805, 805f

Chromosomal mutations, **300**–301, 301f
Chromosomal theory of inheritance, **240**–241, 240f, 244
Chromosome maps, 354–355, 355f
Chromosomes. *See also* Cell cycle; Cell division; DNA (deoxyribonucleic acid)
artificial, 331
bacterial artificial, 331
banding patterns of, 354
defined, **65**
discovery of, 189, 207
duplication of, 487, 488–489f
of eukaryotes, 65, 78t, 81t, 189–191, 189t, 546
formation of, 43
fusion of, 488
homologous, 191, 191f, 209–210, 209f
linear, 271–272, 272f
in meiosis, 207–208, 208f
number of, 189, 189t, 207–208, 250, 250–251f
in polyploidy, 449
of prokaryotes, 546
replication of, 191, 193
sexual reproduction and, 208, 208f
structure of, 36f, 189–191
yeast artificial, 331
Chronic obstructive pulmonary disease (COPD), **1014**
Chronic wasting disease, 540
Chrysalis, **689**
Chrysolaminarin, 578
Chrysophyta (phylum), 578
Chylomicrons, **992**
Chyme, **989**, 995
Chymotrypsin, **990**
Chytridiomycosis, **635**, 635f
Chytridiomycota, 620, 620f, 620t, 624–625
Chytrids, 620, 622, **624**, 624f
Cicadas, 661t, 687t
Cichlid fish
in Lake Barombi Mbo, 450
in Lake Malawi, 501, 501f
in Lake Victoria, 453–454, 454f, 1274
pike cichlid, 414–415f, 415
Ciconiiformes, 715t
Cigarette smoking. *See* Smoking
Cilia, 66f, 78t, **79**–80, 81t, 568, 575, 655–656
Ciliates, 573, 573f, 575–577, 576f
Circadian rhythms, in plants, **808**, 812, 812–813f
Circulatory system, 1020–1037
amphibians, 706, 1025, 1026f
annelids, 674, 1024
arthropods, 682, 682f, 1023
birds, 718, 1026, 1027f
blood, components of, 1020–1023f
closed, 642, 674, 1023–1024, 1024f
defined, **642**
fish, 701, 712, 712f, 1025, 1025f
heart and blood vessels, 1028–1035
invertebrates, 1023–1024, 1024f
mammals, 719, 1026, 1027f
mollusks, 669, 1023
open, 642, 1023, 1024f
regulation of blood flow and blood pressure, 1035–1037, 1036f
reptiles, 712, 712f, 1026
vertebrates, 699
in vertebrates, 699, 875f, **877**, 1025–1027f
Cirrhosis, 664
Cirripedia, 686
Cisternae, of Golgi body, **70**, 71, 71f
Cisternal space, **70**

Citrate, 131, 132, 133f, 138
Citrate synthetase, 138, 138f
Citric acid, 131
Citric acid cycle. *See* Krebs cycle
CJD (Creutzfeldt–Jakob disease), 540
Clades, **463**, 465, 475, 640, 661
Cladistics, **462**–465
Cladograms, **463**, 463–464f, 464
Cladophylls, 750, 750f
Clams, 645t, 666, 667–668, 667f, 671, 671f
Claritin (loratadine), 168
Classical conditioning, **1140**
Classification
 of animals, 644–649f
 defined, **465**
 evolutionary relationships and, 467–468
 of plants, 468, 468f
 of prokaryotes, 547–548
 of protists, 568, 569f
 systematics and, 465–469
 taxonomic hierarchy of, 465–467, 466f, 468
Class (taxonomic), **466**, 466f
Clathrins, 91t, 103
Clean Air Acts, 425
Cleavage
 in cell division, **376**, 376–377f
 in fertilization, 1108t
 of glucose, 127, 128f
 holoblastic, 1112–1113, 1113f, 1113t
 in insects, 1112
 in mammals, 1114, 1114f
 meroblastic, 1113t, 1114, 1114f
 patterns of, 1112, 1113t
 radial, 643f, 644, 649f
 spiral, 643, 643f, 648f, 661
Cleavage furrows, 195f, **197**, 197f
Climate. *See also* Global climate change
 biomes and, 1238, 1238f
 of Earth, 520, 520f
 ecosystems, impact on, 1232–1237
 elevation and, 1236, 1236f
 El Niño Southern Oscillation, 1245–1246, 1246f
 extinction events and, 456
 human impact on, 1253–1256
 microclimates, 1237
 selection to match, 407
 solar energy and, 1233–1234f, 1233–1235
 species richness and, 1226f, 1227
Clitellata, 675–676
Clitellum, 673f, 675
Clitoris, **1096**, 1096f
Cloaca, 985
Clonal selection, **1065**
Clone-contig assembly, **359**, 359f
Cloning
 DNA libraries and, 330–332, 331f
 of mammals, 383–385f
 molecular, 330, 330f, 555
 of plants, 857–858, 858f
 reproductive, 384
 reversal of determination and, 383–385f
 therapeutic, 385–387f
Cloning vectors, 330–331
Clonorchis sinensis, 663, 664f
Closed circulatory system, **642**, 674, 1023–1024, 1024f
Clostridium botulinum. *See* Botulism
Clover, 834f
Club mosses, 594f, 596–597, 596f
Clumped spacing, 1169
Clustered regularly interspaced short palindromic repeats (CRISPR), 336–337, 337f, 556–557

Clutch size, in birds, 417, 1173, 1173–1174f
CNCs (conserved noncoding regions), 490
Cnidarians
 body plan of, 652–653, 652f
 circulatory system of, 1023, 1024f
 classification of, 645t, 647, 653–655
 digestion in, 653, 984, 984f
 eye development in, 508–509, 509f
 germ layers in, 642
 life cycle of, 653
 nervous system of, 902–903, 902f
 radial symmetry of, 640, 641f
 reproduction in, 1086–1087, 1087f
 structure of, 651–652, 652f
Cnidaria (phylum), 645t, 648f, 651–655
CNS. *See* Central nervous system
CNVs (copy number variations), 250, **301**
Coactivators, **174**, 313–314, 315f
Coal burning, 1248
Coastal ecosystems, destruction of, 1251
Cocaine, 901, 901f
Cocci, 550, 550f
Coccidioides, 630
Coccidioidomycosis, 630
Cochlea, **922**, 922f, 924, 924f
Cocklebur, 834f
Cockles, 670
Coconuts, 616, 616f
Coding sequences, 362
Coding strands, **281**, 285–286f
CODIS (Combined DNA Index System), 335
Codominance, 233t, **234**
Codons, **282**, 283–284, 283t, 296f, 297, 362
Coelacanths, 700f, 704, 705f
Coelomates, 642f
Coeloms, **642**, 642–643f, 644, 647, 648f, 865–866
Coelophysis, 472f
Coenzymes, **117**
Coevolution
 defined, **1195**
 mutualism and, 1198–1199, 1199f
 of plants and animals, 795, 798, 1195, 1195f
 predation and, 1195
Cofactors, **117**
Cognitive behavior, **1143**, 1143f
Cohen, Stanley, 328
Cohesins, **191**, 191f, 193, 193f, 201
Cohesiveness, of water, 26, 27f, 27t
Cold virus, 535
Coleochaetales, 581, 581f
Coleochaete, 581, 581f
Coleoptera, 686f, 687t
Coleoptiles, 807f, 855, 856f
Coleorhiza, 855, 856f
Collagen
 in bone development, 966f
 in cell migration, 395
 in extracellular matrix, 80, 81f
 fibers, 869, 869f, 871
 functions of, 44, 45t
Collar cells. *See* Choanocytes
Collecting ducts, 1049f, **1050**
Collenchyma cells, **739**–740, 739f
Colloblasts, **656**
Colon, **992**. *See also* Large intestine
Colon cancer, 992
Colonization, human influence on, 1273–1274
Coloration, defensive, 1196, 1197f
Color blindness, 227t, 242, **933**–934
Color vision, 931–932, 932f

Columbiformes, 715t
Columella cells, 742, 742f
Columnar cells, 867, 868t
Combination joints, **969**–970, 970f
Combination therapy, for HIV/AIDS, 536–537, 537f
Combined DNA Index System (CODIS), 335
Comb jellies, 646t, 647, 655–656, 656f
Commensalism, 562, **631, 1198**, 1200–1201, 1201f
Common cold virus, 535
Communicating junctions, 82t, **83**–84, 84f
Communication, of animals, 1146–1149f
Communities. *See* Biological communities
Communities, biological, 3f, 4, 1188, 1188–1189f
Compact bone, 967f, **968**
Compaction, 1114
Companion cells, 741, 741f
Comparative anatomy, 11, 11f
Comparative biology, 469–475
Comparative genomics
 applications of, 492–494
 evolution and, 480–483, 481f, 483f, 488, 488f
 limitations of, 499
 synteny and, 365–366, 366f, 489, 489f
Comparators, 878f
Compartmentalization
 in eukaryotes, 65, 522, 522f, 546
 in prokaryotes, 522f, 546
Competition
 in barnacle species, 1190, 1190f
 direct and indirect effects of, 1202, 1203f
 experimental studies of, 1192–1194, 1193f
 exploitative, 1189
 interference, 1189
 interspecific, 1189, 1192–1193, 1193f
 parasitism and, 1202
 reduction by predation, 1201–1202, 1202f
 for resources, 1191–1192
 sperm, 1153
Competitive exclusion, **1190**–1191, 1191f
Competitive inhibitors, **117**, 117f
Complementary base-pairing, **43**, 262, 262f, 265, 265f
Complementary DNA (cDNA), **330**, 331–333, 331f, 343–344, 362
Complement system, **1061**–1062
Complete flowers, 839, 839f
Complexity, as characteristic of life, 3
Compound eyes, 417, 417f, **682**, 682–683f, 687
Compound leaves, **752**, 752f
Compound microscopes, 61
Compounds
 ionic, 24
 molecular, 23
 organic, 22–23
Compsognathus, 716
Concentration
 in chemical reactions, 25
 osmotic, 98, 98f
Concentration gradient, 96, 100
Concurrent flow, 1007, 1007f
Condensation, 37, 132, 133f, 194
Condensin, 193, 201
Conditional inactivation, 339, 339f
Conditioning, of behavior, 1140
Condoms, 1101f, 1101t, 1102
Conducting tubes, in plants, 470, 471f
Conduction (heat transfer), 880f, 881
Cones (eye), **931**–932, 931f
Confocal microscopes, 62t

Confuciornis, 717f
Conidia, **629**, 629f
Conifers, 603, 604–605f, **604**–606, 604t, 609f, 813
Conjugation
 in bacteria, 552–554
 in ciliates, 576, 576f
 defined, **552**
 gene transfer by, 553–554, 553f
Conjugation bridge, 553, 553f
Connective tissue, **865**, 869–871, 869f, 870t
Connell, Joseph, 1190
Connexons, 84
Conservation biology, 494
Conservation of synteny, **489**, 489f
Conservation tillage, 777
Conservative replication, 263, 263f, 264
Conserved noncoding regions (CNCs), 490
CONSTANS gene, in *Arabidopsis*, 835–836
Constant region, 1072, 1072f
Constitutive heterochromatin, 363
Consumers, **1217**, 1217f, 1220
Consumption, of resources, 1183
Contact dermatitis, 1077
Contig, **354**, 359
Continental movement, 515, 520–521
Continental shelves, 1243f, 1244–1245
Continuous variation of traits, **232**, 233f
Contraception, 1100–1103, 1101f, 1101t, 1103f
Contractile roots, 746
Contractile vacuoles, 73, 99, 103, 571
Contraction, of muscles, 971–976f, 971–977
Control experiments, **6**
Controlling elements, 486
Conus, 668
Conus arteriosus, **1025**, 1025f
Convection (heat transfer), 880f, 881
Convergent evolution
 biogeographical record and, 434–436
 defined, **435**, 462
 of digestive system, 994, 994f
 flower shape and, 506
 homoplasies and, 470, 471f
 innovations and, 640
 of insect wing patterns, 505–506, 506f
Cooksonia, 593, 593f, 594, 595
COPD (chronic obstructive pulmonary disease), **1014**
Copepods, 684f
Coprophagy, 994
Copy numbers, 488
Copy number variations (CNVs), 250, **301**
Coral bleaching, 654
Coral reefs, 654, **1245**, 1245f, 1255
Corals, 640, 645t, 653, 654, 654f
Core polymerase, 284f, 285
Corepressors, 311–312, 316
Coriolis effect, **1234**–1235, 1234f
Cork cambium, **735**, 736f, 748, 749f
Cork cells, 748
Corms, 750
Corneas, **930**, 930f
Corn *(Zea mays)*
 abscisic acid and, 828f
 artificial selection of, 426, 427f
 chromosomes in, 189t
 comparative genomics and, 366, 366f
 cross-pollination of, 349
 endosperm of, 612, 853, 854f
 epistasis in, 236, 236f
 fungal infections in, 635, 635f
 genome of, 481f, 482, 485f, 487
 germination of, 856f

grain color in, 233t, 236, 236f, 245–246, 245f
oil content of kernels, 426
photorespiration and, 164
recombination in, 245, 245f
roots of, 742f, 745, 746f
stem of, 748f
stomata of, 737f
transgenic, 347–349
Corolla, **839,** 839f
Coronary arteries, **1029**
Coronary artery disease, 54
Coronavirus, 531t, 539, 539f
Corpora allata, 959
Corpus callosum, 903t, **904,** 905f
Corpus luteum, **1099**
Correns, Carl, 240, 244
Cortex (plant), 744f, **745,** 748
Cortical granules, **1110**
Cortical nephrons, **1049**
Corticosteroids, 956
Corticotropin, 950
Corticotropin-releasing hormone (CRH), 951
Cortisol, 173, 174, 825f, 945f, 956
Corynebacterium diphtheriae. See Diphtheria
Costa Rica, biosphere reserves in, 1280, 1280f
Cost of reproduction, 1173–1174f, **1173**–1175
Cotransduction frequency, 554
Cotton
genome of, 485f
transgenic, 347, 348
Cotyledons, **852**
Countercurrent flow, **1007,** 1007f
Countercurrent heat exchange, 882, 882f
Countertransport, 102
Coupled transport, 101–102, 101f, 104t
Courtship behavior. See also Mating and mating behavior
of *Anolis* lizards, 447, 447f
behavioral isolation and, 442–443, 443f
of blue-footed boobies, 443f
in frogs, 1154, 1154f
of fruit flies, 447
of lacewings, 443, 443f
of stickleback fish, 1146–1147
Covalent bonds, 23t, 24–25, 24f, 34
Cowper's gland, 1093f
Cowpox, 1063
Cows. See Cattle
COX (cyclooxygenase), 945
C₃ photosynthesis, **161,** 164, 164f
C₄ photosynthesis, 164–165, 752
C₃ plants, **164,** 164–165f
C₄ plants, **164**–165
Crabs, 645t, 648f, 681t, 684, 685–686, 685f
Cranial neural crest cells, 1122
Crassulacean acid metabolism (CAM), 767
Crassulacean acid pathway, 165
Cratons, 513
Crawling, cellular, 79
Crayfish, 684, 685
Creighton, Harriet, 245, 245f
Crenarchaeota, 548f
Creodonts, 471f
Creosote bushes, 859
Cre recombinase, 339–340, 339f
Cretinism, 954
Creutzfeldt–Jakob disease (CJD), 540
CRH (corticotropin-releasing hormone), 951

Crick, Francis, 259, 260–263, 261f, 280, 282, 283, 299
Crickets, 687t
Cri-du-chat syndrome, 300
Crinoidea, 694, 695, 695f
CRISPR (clustered regularly interspaced short palindromic repeats), 336–337, 337f, 556–557
Cristae, **74,** 74f
Crocodiles
characteristics of, 699f, 709t, 713–714
circulation in, 712
classification of, 467t, 713f
heart of, 1026
parental care in, 470, 470f
phylogeny of, 701f
Crocodylia, 701f, 709t, 713–714, 713f
Cro-Magnons, 727, 728f
Crops
artificial selection of, 426, 427f
global climate change and, 784–787
transgenic, 347–349
Cross-bridge cycle, 973, 973f
Cross-fertilization, 223, 223f
Crossing over
in meiosis, **209**–211, 209f, 211–212f, 215, 216f
multiple crossovers, 247, 247f
recombination of alleles in, 244–245, 245f
Cross-pollination, 223, 223f, 349, 842
Crown gall, 822, 823f
CRP (cyclic AMP response protein), 310–311, 310f
Crustacea, 681t, 684–686
Crustaceans
body plan of, 685, 685f
characteristics of, 681, 681t
decapods, 685–686, 685f
exoskeleton of, 41, 41f
habitats of, 684
reproduction in, 685, 685f
sessile, 686, 686f
Crustose lichens, 632f
CTD (carboxy terminal domain), 288, 290
Ctenidia, 668
Ctenophora, 646t, 647, 648f, **655**–656, 656f
Cuboidal cells, 867, 868t
Cubozoa, **654,** 654f
Cubuzoans, 508, 509f
Cuenot, Lucien, 233
Cultivation, 347, 776–777, 777f
Culture, differentiation in, 382
Cup fungi, 629, 629f
Cutaneous respiration, 706, **1008**
Cutaneous spinal reflex, 909f
Cuticle, of plants, 589, **737**
Cutin, **737**
Cuttlefish, 666, 668, 671
CVS (chorionic villi sampling), 252–253, 253f
Cyanide, 794
Cyanobacteria
chloroplasts from, 522, 523f, 566
classification of, 549f
electron micrograph of, 64, 64f
internal membranes of, 552f
in lichens, 631, 632
in nitrogen fixation, 561
photosynthetic, 143, 148, 545, 557
pigments in, 152, 153–154
in stromatolites, 519
Cyanogenic glycosides, 794, 795, 796t
Cycads, 603, 604t, **606,** 606f, 609f
Cycas circinalis, 606f
Cyclic adenosine monophosphate (cAMP)
hydrophilic hormones, 947–948
as second messenger, **173,** 179–181f, 182
in slime molds, 584

Cyclic AMP response protein (CRP), 310–311, 310f
Cyclic guanosine monophosphate (cGMP), 174, 933, 934f
Cyclic photophosphorylation, 156, 160
Cyclin B, 216
Cyclin dependent protein kinases (Cdks), 199–202, 376, 377f, 378
Cyclins, **199,** 199f, 323, 376, 377f
Cyliophora, 646t, 648f
Cycloidia gene, 506
CYCLOIDIA gene, of snapdragons, 840, 840f
Cyclooxygenase (COX), 945
Cyclosome, 201
Cyprinidae, 664
Cysteine, 35f, 45, 47f
Cystic fibrosis
characteristics of, 249t
gene for, 491
pleiotropic effects and, 233, 233t
protein folding mutations in, 51–52
as recessive trait, 227t
Cytochrome bc_1 complex, 134, 134f
Cytochrome, b_6-f complex, **157, 158,** 158–159f, 160
Cytochrome *c,* 134, 134f
Cytochromes, 45t, 124
Cytokines, **944, 1069**–1070, 1071f
Cytokinesis
in animal cells, 197, 197f
in cell cycle, **192,** 192f, 195f, **197**
in fungi, 197–198
in meiosis, 212–213f, 214
in plant cells, 197, 198f
in protists, 197–198
unequal, 393
Cytokinins, 816t, 821–823f
Cytological maps, 354
Cytoplasm, **62,** 621, 893t
Cytoplasmic determinants, 379–380, 379f
Cytoplasmic receptors, 1059
Cytoprocts, 576
Cytosine, 42, 42f, 43, 259f, 260, 262
Cytoskeleton
attachments to, 94f, 95
defined, **65**
functions of, 65, 77, 78t
structure of, 67f, 75–76f, 75–77
Cytosol, **62,** 70, 522
Cytotoxic T cells, 1064t, **1068**–1069, 1069f

D

Dachshund, 427, 427f
Daddy long-legs, 681t, 683
DAG (diacylglycerol), 180–181f, **181,** 948
Daltons (unit of mass), 19
Dance language, of honeybees, 1147–1148, 1148f
Dandelions, 615–616, 616f
Dark-field microscopes, 62t
Dark reactions, in photosynthesis, 150
Darwin, Charles
on branching diagrams, 461, 461f
critics of, 436–437
on eusociality, 1158
Malthus and, 10
on natural selection, 8–11, 406, 422–423
On the Origin of Species by, 8, 10, 399
on pace of evolution, 414
plant studies of, 608, 817
portrait of, 8f
on theory of evolution, 8–12, 400, 400f
voyage on *Beagle,* 1, 1f, 8, 9f, 10, 422

Darwin, Francis, 817
Dating, of fossils, 428, 428f, 519
Day-neutral plants, 835
DDT (dichlorodiphenyltrichloroethane), 574, 1247, 1247f
Deamination, **141,** 141f
Decapentaplegic protein, 1119–1120
Decapods, 685–686, 685f
Deciduous forests, 1239, 1240f
Deciduous plants, 859
Decomposers, 561, **1217**
Decomposition, 561
Deductive reasoning, **4**–5, 5f
Deep sea, 1246–1247, 1246f
Deep time, 514f, 515
Deer, 719, 719f, 720, 1092f
Deer mouse hantavirus, 539
Defensins, **795,** 1059, 1059f
Deforestation, 1212, 1212f, 1249
Degeneracy, of genetic code, 283–**284**
Dehydration reactions, **37,** 37f, 39, 42
Dehydrogenations, **123,** 130
Deinococcus-Thermus, 548f
Delamination, **1115**
Delayed hypersensitivity, 1077–1078
Deletions, 282, 282f, **300,** 301f
Delta waves, 906
de Mairan, Jean, 808
Demography, **1170**
Denaturation, of proteins, 48, **52**–53, 52f, 332, 333f
Dendrites, 873, 874t, **890,** 890f
Dendritic cells, 1064f, **1065**
Dendritic spines, 890
Dengue fever, 1256
Denisovans, 727
Denitrification, **1213**
Denitrifiers, 561
Dense connective tissue, **869,** 870t
Density, of water, 27t, 28
Density-dependent effect, 1177–1178
Density-independent effects, **1178,** 1178f
Dental caries, 545f, 559–560, 559t. See also Teeth
Deoxyhemoglobin, **1015**
Deoxyribonucleic acid. See DNA
Dephosphorylation, of proteins, **170,** 171f, 178
Depolarization, 893–894, 894f
Depression, 900
Derepression, 312
Derived character variations, 462–463, 463f, 468
Dermal tissue, of plants, **734,** 737–739f, 793, 851
Dermatitis, 1077
DES (diethystilbestrol), 959
Deserts, 1239
Desmosomes, 82t, **83,** 84f
Determinate development, 643f, **644**
Determination
cytoplasmic, 379–380, 379f
molecular basis of, 379
reversal of, 383–385f
standard test for, 378–379, 378f
Detritivore trophic level, **1217,** 1217f
Deuterostomes
cleavage patterns in, 643f, 644
defined, **643**
development in, 644, 1116
embryonic development in, 643, 643f
phylogeny of, 647, 649f
Development, 375–396
in animals, 640, 642–644, 643f
apoptosis in, 394–395, 394f
of behavior, 1140–1142
of bone, 965–968, 966f
cell division in, 375, 376–377f, 376–378, 393

cell migration in, 395–396, 395f
as characteristic of life, 3
defined, 375
determination in, 378–379f, 378–380
differentiation in, 375–376, 378–382
of echinoderms, 694, 694f
embryonic (See Embryonic development)
evolution of, 497–500
of eyes, 506–510
in fruit flies, 386–392, 499
gene expression in, 304
in humans, 1127–1130f, 1127–1131
induction and, 380–381
of limbs, 502, 503–504f
morphogenesis in, 376, 386, 393–396, 852–853, 853f
nuclear reprogramming in, 383–387f
pattern formation in, 376, 386–393
in plants, 377–378
in sea urchins, 497–498, 498f
transcription factors in, 379, 498, 499, 499f, 502, 504
Developmental knockouts, 340
Devil facial tumor disease (DFTD), 494, 494f
Dewlaps, of *Anolis* lizards, 447, 447f
Dextrorotatory chiral forms, 35
DFTD (devil facial tumor disease), 494, 494f
Diabetes insipidus, 98, 1052
Diabetes mellitus
therapeutic cloning and, 386, 386–387f
treatment of, 957
type II (non-insulin-dependent), 957
type I (insulin-dependent), 343, 957
Diacylglycerol (DAG), 180–181f, 181, 948
Diagnostics, 180f, 371, 1079–1080
Diaphragm (birth control), 1101f, 1101t, 1102
Diaphragm (muscle), **1011**, 1012f
Diapsids, 710, 710f, **711**
Diastole, **1028**, 1029f
Diastolic pressure, **1031**, 1031f
Diatoms, 568, 573, 578, 578f
Diazepam (Valium), 900
Dicer, 318, 319, 319f, 320
Dichlorodiphenyltrichloroethane (DDT), 574, 1247, 1247f
Dichlorophenoxyacetic acid (2,4-D), 819f, 821
Dichogamous plants, **846**
Dictyostelium discoideum, 523f, 584, 584f
Dideoxynucleotides, **357**, 357f
Dideoxy sequencing, 357, 357–358f
Didinium, 1194, 1194f
Diencephalon, 903t, 904
Diethylstilbestrol (DES), 959
Differential-interference-contrast microscopes, 62t
Differentiation, 14, 375–376, 378–382
Diffuse pollution, 1248
Diffusion, **96**–97, 104t, 1004
Digestion
chemical, 984
defined, 123
of plant material by mammals, 720
in small intestine, 989–992
in stomach, 988–989, 988f
Digestive system, 983–1000
accessory organs, 985, 985f, 990–991, 990f, 996–997, 997f
of birds, 986, 986f
of carnivores, 993f
in cnidarians, 653
esophagus and stomach, 987–989, 988f

in flatworms, 662, 662f, 663
in food energy and nutrition, 997–999, 1000t
of herbivores, 992, 993f
of insectivores, 993f
of insects, 688
intestines, 989–992
of invertebrates, 984, 984f
mouth and teeth, 986–987
of nematodes, 679, 984, 984f
neural and hormonal regulation of, 995, 995f, 996t
of ruminants, 993–994f
types of, 984–985
variations in, 993–994f, 993–995
in vertebrates, 875f, **877**, 984–985, 985f, 993–994f, 993–995
Digit formation, 502, 504f, 723
Dihybrid crosses, **228**–229, 229f, 230–231
Dihydroxyacetone phosphate, 128f
Dikaryon stage, 622, 628
Dikaryotic hyphae, **621**
Dimerization, 175
Dinoflagellates, 573–574, 573f, 654
Dinornithiformes, 715t
Dinosaurs
appearance of, 700
characteristics of, 709t
classification of, 467f, 468
evolution of, 710–711
extinction of, 456, 700, 711
feathered, 716
parental care in, 469–470, 470f
Dioecious plants, **607**, 846
Dioxin, 821
Diphtheria, 534, 558, 559t
Diploblastic animals, 642, 647
Diploids (*2n*), **191**, 208–209, 208f, 225, 449, 554, 1111
Diplomonads, 570, 570f
Diplontic life cycle, **590**
Dipnoi, 700f
Diptera, 686f, 687t
Direct contact signaling, 169, 169f, 170
Directional selection, **412**–413
Disaccharides, **38**–39, 39f
Disassortative mating, **405**
Disease. See also Human diseases; Plant diseases; *specific diseases*
bacterial, 556, 557–561, 559t, 560f
causes of, 492
evolution of pathogens, 475–477f
fungal, 634–635
genetic markers for, 345–346
genomics in treatment of, 370–371, 371f
germ theory of, 545
microorganisms and, 545
pathogen–host genome differences in, 492–493, 493f
of respiratory system, 1014, 1014f
sexually transmitted, 559t, 560–561, 560f, 570
skin as barrier to, 1058
stem cell therapy in treatment of, 346–347
transmission from animals to humans, 1249–1250
viral, 531t, 538–540
zoonotic, 1249–1250
Dispersive replication, 263f, 264, 265
Disruption, habitat loss and, 1269–1270
Disruptive selection, **411**–412, 412f, 449–450
Dissociation, of proteins, **53**
Distal convoluted tubules, 1049f, **1050**, 1051–1052
Distal-less gene, 505–506, 506f
Disturbances, biological, 1205–1206, 1205f

Disulfide bonds, 48, 48f, 344, 344f
Diurnal species, 723
DNA-binding motifs, 306–307, 307f
DNA-binding proteins, 48
DNA (deoxyribonucleic acid), 256–275. See also Genes; Genomes
antiparallel strands in, 262–263
in central dogma of molecular biology, 280, 280f, 330
of chloroplasts, 74–75, 74f
chromatin in, 68
in chromosomes (See Chromosomes)
cladistic analysis of sequence data, 464, 464f
in cloning (See Cloning)
coding strands, 281, 285–286f
complementary, 330, 331–333, 331f, 343–344, 362
defined, **12**, 41
double helix of, 41f, 43, 43f, 261–263, 262f
in eukaryotes, 62, 68
functions of, 37t, 41, 42–43
gel electrophoresis of, 329–330, 329f, 355f, 357
in genetic engineering (See Genetic engineering)
in heredity, 13
libraries of, 330–332, 331f, 336
major groove of, 262f, 306, 306f
methylation of, 252, 316, 316f
minor groove of, 262f, 306, 306f
of mitochondria, 74, 74f
nature of, 256–259
noncoding, 362–364, 363t, 487, 490
in prokaryotes, 62
protein-coding, 362
rearrangement of, 489
recombinant (See Recombinant DNA)
replication of (See Replication)
RNA vs., 43, 43f
segmental duplications of, 487–488
sequencing of, 50, 328, 328f, 334–335, 357, 358f
with sticky ends, 328, 328f, 330
structural, 363, 363t
structure of, 12, 41–43f, 42–43, 259–262f, 259–263
supercoiling of, 267, 267f
template strands of, 265, 265f, 280, 285–286f
three-dimensional structure of, 259–260f, 259–261
topological state of, 267
in transformation, 331, 331f
viruses and, 528, 528f, 531t
Watson–Crick model of, 261–262f, 262, 263
X-ray diffraction pattern of, 260, 261f
DNA fingerprinting, 335, 335f
DNA gyrase, **267**, 267f, 268t, 269, 269–270f
DNA helicase, **267**, 268t, 269f
DNA ligase, 268t, **269**, 269–270f, 328f, 330, 330f
DNA microarrays, 345–346, **366**–368, 367f
DNA polymerase delta, 271
DNA polymerase epsilon, 271
DNA polymerase I, **266**, 268–269, 268t, 269–270f
DNA polymerase II, **266**
DNA polymerase III, 265f, **266**, 268, 268t, 269–270
DNA polymerases
in *E. coli*, 266
proofreading function of, 273
in replication, **265**–266, 265f

DNA primase, 268–269, 268t, 269f, 271
DNA rearrangement, 1074–1075, 1075f
DNA repair, 273–275, 274f, 273–275, 274f
Docking, of proteins, 296f, 297
Dodder (*Cuscuta*), 746, 784
Doering, William, 798
Dogs
artificial selection of, 427, 427f
chromosomes in, 189t
teeth of, 719, 719f
"Dolly" (cloned sheep), 383–384, 385f
Dolphins, 429, 721f
Domains, of proteins, 50–51, 50f
Domains (taxonomic), 13, 13f, **466**, 466f, 522
Domestication, 427, 427f
Dominant hemisphere of brain, 906
Dominant traits
ABO blood groups, 233t, 234–235, 235f
codominance, 233t, 234
in humans, 227t
incomplete dominance, 233t, 234, 234f
in monohybrid crosses, 224–225f, **224**–228
Dopamine, **900**, 901, 1136
Dormancy
in plants, 813–814, 814f
in seeds, 814, 814f, 827–828, 828f
Dorsal body cavity, 865, 866f
Dorsal nerve cord, 379, 380f, 696, 696–697f, **1120**
Dorsal proteins, 388, 390f
Dorsal root, **910**
Dorsal root ganglia, **910**, 911f
Dorsal–ventral axis formation, 388, 390, 390f
Dosage compensation, **243**
Double circulation, 1025
Double covalent bonds, 24, 24f
Double fertilization, **612**, 847, 848f
Double helix, 41f, **43**, 43f, 261–263, 262f
Double-stranded RNA, 336
Douching, 1102
Down, J. Langdon, 250
Down syndrome, **250**, 250–251f, 252
Dragonflies, 687t
Drosophila. See Fruit flies
Drought
plant tolerance to, 768, 768f
soil degradation and, 777, 777f
Drugs. See also Antibiotics; *specific drugs*
addiction to, 901–902, 901f
for allergies, 1077
for depression, 900
for fungal infections, 635
for HIV/AIDS, 536–537, 537f
illegal manufacture of, 607
NSAIDS, 560, 945
from plants, 1264, 1264f
Duchenne muscular dystrophy, 227t, 249t
Duck-billed platypus (*Ornithorhynchus anatinus*), 481f, 492, 721, 721f, 936, 1092f
Dugesia, 662f, 663
Duke Experimental Forest, 785–786, 786f
Duodenum, **989**
Duplication (mutation), 486, 486f, 487–489f, 502, 504–505
Duplications, **300**, 301f
Dust Bowl, 776–777
Dwarfism, 727, 952
Dynactin complex, 77f, 79
Dynein, 77f, 79

E

Ears, 922–923f, 922–926
Earth
  age of, 11
  atmosphere of, 515, 516–517, 516f, 520, 521, 521f, 545, 561
  circumference of, 4, 5f
  climatic changes on, 520, 520f
  continental movement on, 515, 520–521
  evidence of early life on, 518–519, 519f
  formation of, 17
  geologic changes on, 514f, 515, 516f, 520
  orbit around sun, 1233, 1233f
  organic molecules on, 516–517
  origins of life on, 515–518
  plant-induced glaciation on, 521
  rotation of, 1233–1234f, 1233–1235
  water on, 25
Earthworms
  body plan of, 675
  circulatory system of, 674, 1024f
  digestive system of, 984, 984f
  locomotion in, 964, 964f
  nephridia of, 1044, 1044f
  nervous system of, 902f, 903
  reproduction in, 675, 675f
  segmentation of, 644, 645t, 673f, 675
Easter lilies *(Lilium candidum)*, 842f
Eastern rat snake *(Pantherophis alleghaniensis)*, 441f
Ebola virus, 371t, 528, 531t, 539, 539f, 1249
Ecdysis, 648f, **682**
Ecdysone, 959, 959f
Ecdysozoa, 678–689. *See also* Arthropods; Nematodes
  phylogeny of, 647, 649f, 661
  roundworms, 642f, 645t, 678–680
ECF (extracellular fluid), 893t
ECGs (electrocardiograms), 1029, 1030f
Echidna, 721, 721f
Echinodermata (phylum), 645t, 649f, 693–695
Echinoderms
  body plan of, 694, 695
  classification of, 645t, 647, 694–695, 695f
  development of, 694, 694f
  endoskeleton of, 693, 694
  nervous system of, 694, 902f
  regeneration in, 694
  reproduction in, 694
  respiration in, 1005f
  symmetry in, 641, 694
  water-vascular system of, 693, 693f, 694
Echinoidea, 694, 695, 695f
Echolocation, **925**
ECM (extracellular matrix), 80–81, 81f, 395, 568, 869
Ecological footprints, **1183**, 1183f
Ecological isolation, 442, 442f, 442t
Ecological niche, 1189–1194
Ecological pyramids, 1220–1221, 1221f
Ecological succession, 1204–1206
Ecology
  behavioral, 1149–1151
  of fungi, 630–634
  of lichens, 632
Economic value, of biodiversity, 1264–1266
Ecosystem level of organization, 3f, 4
Ecosystems
  biogeochemical cycles in, 1210–1215
  climate, impact on, 1232–1237

defined, **4**
  disruption of, 1274–1276
  energy flow through, 1216–1221
  global warming and, 1254–1255
  human activity, impact on, 1247–1252
  stability of, 1225–1227f, 1225–1228
  trophic levels in, 1217–1224f, 1217–1225
Ecotones, **1188**, 1189f
Ectoderm, 380, **641**–642, 642f, 865, 1115
Ectodysplasin gene, 501
Ectomycorrhizae, **632**–633, 633f
Ectoprocta, 645t
Ectotherms, 711, **712**, 881–882
Edema, 996, **1033**–1034
Edentata, 722t
Edge effects, **1270**
EEGs (electroencephalograms), 906
Eels, 977, 977f
Effector proteins, 179–182, 179f
Effectors, 877, 878–879, 878f
Efferent arterioles, 1049
EF-Tu, **295**, 295f
EGF (epidermal growth factor), 202, 204, 944
Egg-laying mammals, 471f, 721, 721f
Eggs
  activation of, 1110
  amniotic, 710, 710f, 714
  of amphibians, 710, 1252f
  of birds, 714
  coloration, adaptive value of, 1149, 1149f
  fertilization of, 1108–1111, 1108t, 1109–1111f
  of frogs, 1111, 1111f
  of reptiles, 710, 710f
Ejaculation, 1094–1095
Elasmobranchs, 936, 1042, 1046–1047
Elastin, 80, 81f, 395
Electrical currents, detection of, 936
Electrical synapses, **897**
Electrocardiograms (ECGs), 1029, 1030f
Electroencephalograms (EEGs), 906
Electromagnetic receptors, 917
Electromagnetic spectrum, 151, 151f
Electron acceptors, 124, 124–125f, 125, 139–140, 139f
Electron carriers, 124, 125f, 134, 134f
Electronegativity, **24**–25, 25t, 34
Electron microscopes, 61, 62t, 91–92, 518, 545
Electrons
  defined, 18, 19f
  energy levels of, **21**, 21f, 22, 23f
  orbitals of, 19, 20, 20f, 21
  valence, 22, 24
Electron transport chain
  ATP production in, 134–136f, 135–136
  chemiosmosis and, 134–136
  in Krebs cycle, 131, 132, 133f
  photosynthetic, 149, 156, 159f
  processes involved in, 124, 124f, **125**
Electrophoresis, 401
Electroreception, 443, 936
Elements
  defined, 18
  inert, 22
  in living systems, 22–23
  periodic table of, 22–23, 22f
  trace, 23, 1000
Elephantiasis, 680
Elephants, 719f, 720, 1004f
Elephant seals, 406, 406f
Elevation, climate and, 1236–1237, 1236f

El Niño Southern Oscillation, 1245–1246, 1246f
Elongation, zone of, **744**
Elongation factor, **295**, 295f
Embryogenesis, 376, 387, 850
Embryo implantation, prevention of, 1102
Embryonic development
  in animals, 640, 642–644, 643f
  evolutionary evidence from, 432, 433f
  in plants, 848–854
Embryonic flower mutant, in *Arabidopsis*, 833, 833f
Embryonic stem cells, **338**–339, 338f, **346**–347, 381–382, 382f
Embryo sacs, 611, **841**–842
Embryos (plant), 612, 613, **848**, 848f
Embryo transfer (ET), 1104
Emerald ash borer *(Agrilus planipennus)*, 793, 793f
Emergent properties, **4**, 14
Emerging viruses, **539**
Emerson, R. A., 236
Emphysema, **1014**
Enantiomers, 35, 35f
Encephalitozoon cuniculi, 360f, 623, 623f
Encyclopedia of DNA Elements (ENCODE) project, 364–365, 364f
Endangered species, 494, 1278–1281
Endemic species, 1261–1263f, 1261–1264, 1263t
Endergonic reactions, **110**, 111, 111f, 113, 113f, 125, 160
Endochondral development of bone, 966f, 967–968
Endocrine glands, 867, 940, 942–943t, 951–952. *See also specific glands*
Endocrine signaling, 169, 169f, **170**
Endocrine system, 939–959. *See also* Hormones
  defined, **875**, 875f, **940**
  hormones in, 945–947f, 945–948, 957–959f
  peripheral glands of, 953–957
  pituitary and hypothalamus, 948–953
  regulation of body processes, 940–941f, 940–945, 944f
Endocytosis, **102**–103, 102f, 104t
Endoderm, 380, **641**–642, 642f, 865, 1115
Endodermis, 744f, **745**, 763
Endogenous opiates, 900
Endomembrane system, **65**, 69–73, 522
Endometriosis, 1103
Endonucleases, **266**, 328, 328f, 330, 330f
Endoparasites, **1199**
Endophytes, 631, 631f
Endoplasmic reticulum (ER)
  cytoskeleton and, 77
  in eukaryotes, 65, **69**–70, 69f, 78t, 81f
  origin of, 566, 566f
  plasma membrane vs., 93–94, 93f
  proteins targeted to, 296f, 297
  rough, 69f, 70
  smooth, 69f, 70
Endorphins, **900**
Endoskeleton
  characteristics of, 964, **965**, 965f
  of chordates, 696
  of echinoderms, 693, 694
  of vertebrates, 699
Endosperm, 612, **848**, 848f, 853, 854f
Endospores, **552**
Endosteum, **968**
Endostyles, 697

Endosymbiosis, **75**, 75f, **522**, 566, 567, 567f
Endothelin, 944
Endothelium, 1032, 1032f
Endotherms
  birds as, 718, 877
  defined, 710, **712**, 881
  mammals as, 718, 719, 877
  metabolic heat and, 882–883, 883f
  reptiles as, 710, 712
Endurance training, 977
Energy
  in atoms, 21, 21f
  as characteristic of life, 3
  in chemical reactions, 110–111, 111f
  defined, 108
  expenditure of, 997, 999
  feeding behavior and, 999, 999f
  flow-through ecosystems, 1216–1221
  forms of, 108
  kinetic, 108, 108f
  laws of thermodynamics on, 109–110, 437
  light as, 151–152
  in living systems, 108–109
  in photons, 151–152
  potential, 21, 108, 108f, 110
  prokaryote acquisition of, 557
  solar, 1233–1234f, 1233–1235
Energy levels, **21**, 21f, 22, 23f
engrailed gene, 389f, 391
Enhancement effect, 157, 157f
Enhancers, **313**–314, 314f
Enkephalins, **900**
5-enolpyruvylshikimate-3-phosphate synthase, 347, 348f
Enteric bacteria, 549f
Enterobacteriaceae, 556
Enterobius vermicularis, 680
Enterogastrones, **995**, 998
Enterovirus, 531t
Enthalpy, **110**
Entropy, **110**, 110f
Environment
  biotechnology applications for, 341–342, 341f, 343f
  challenges in, 1164–1165
  enzyme function and, 116–117, 116f
  in gene expression, 233t, 235, 235f, 308–309
  individual responses to changes in, 1165–1166
  limitations on population growth, 1175–1176f, 1175–1177
  in sex determination, 242, 1088, 1088f
  transpiration rates and, 767, 767f
Environmental Protection Agency (EPA), 788, 1255
Enzymatic DNA sequencing, 328, 328f
Enzymatic receptors, 172–173, 172f, 172t
Enzyme catalysis, 44, 45t
Enzymes
  activation energy and, 114
  active sites of, 114, 114f
  catalytic cycle of, 114, 115f
  cofactors, 117
  computer-generated model of, 369–370, 370f
  digestive, 688, 996t
  in genetic disorders, 278–279
  in genetic variation in, 401
  inhibitors and activators, 117, 117f
  intracellular receptors as, 174
  in membranes, 94f, 95
  multienzyme complexes, 115–116, 115f
  nonprotein, 116
  pH and, 53, 116–117, 116f

Enzymes (continued)
 for replication, 268t, 269–270, 269f
 restriction, 328, 328–330f, 354, 355f
 RNA, 116
  temperature and, 53, 116, 116f
Enzyme–substrate complex, **114,** 115f
Eons, 514f, 515
Eosinophils, 1021, 1021f, **1064,** 1064f
EPA (Environmental Protection Agency), 788, 1255
EPCs (extra-pair copulations), **1155**
Ephedra, 604t, 606–607, 853
Ephedrine, 607
Ephemerals, 835
Epidermal cells, of plants, **737,** 737f, 739f, 744
Epidermal growth factor (EGF), 202, 204, 944
Epidermis
 of cnidarians, 652, 652f
 of plants, 734, **737,** 744f
Epididymis, **1094**
Epigenetic inheritance, **252**
Epigenetics, 316, **383**
Epiglottis, **987,** 987f
Epilimnion, 1241
Epinephrine, 95, 170, 182, **900**
Epiparasites, 1091
Epiphyseal growth plates, 967–968
Epiphyses, **967,** 967f
Epiphytic orchids, 746, 746f
Epistasis, **235–236,** 236f, 417
Epithelial tissue, **865,** 866–867, 868t
Epithelium, **866**
Epitopes, 1064, 1064f
EPSP (excitatory postsynaptic potential), 898, 900–901, 900f, 918
Epstein–Barr virus, 531t
Equilibrium constant, 111
Equilibrium model, of island biogeography, 1228–1229, 1228f
Equilibrium potential, **892,** 893t
Equisetum, 597, 598f
Equus, 430, 431, 431f
ER. See Endoplasmic reticulum
Eras, 514f, 515
Eratosthenes, 4, 5f
Erection, 1094
Erosion, 776–777, 777f
Erythrocytes
 characteristics of, **1021,** 1021f
 facilitated diffusion in, 97
 membrane of, 89–90, 91t
 production of, 871
Erythropoiesis, **1022**
Erythropoietin, 202, 958, **1022**
Escherichia coli (E. coli)
 adaptations of, 1166
 biochemical pathways in, 417–418
 cell division in, 187–188, 188f
 chaperonins in, 51
 classification of, 549f
 conjugation map of, 553–554
 DNA repair in, 274–275
 F plasmids of, 552–553, 553f
 genome of, 481f
 harmful traits of, 556
 insulin expression in, 343–344, 344f
 introduction of foreign DNA into, 330
 lac operon of, 308f
 lysogenic phage of, 532
 mutations in, 555, 556
 renaturation of, 53
 replication in, 266–270, 268t
 size of, 530f
E sites, **292–**293, 293f, 296
Esophagus, 985, 985f, 987–988, 988f
An Essay on the Principle of Population (Malthus), 10
Essential amino acids, 46, 1000

Essential nutrients
 for plants, 779t
 for vertebrates, 999–1000, 1000t
Estradiol, 1095t
Estrogen, 54, 173, 174, 958, 1095t, 1136
Estrous cycles, 1099
Estrus, **1092**
ESTs (expressed sequence tags), **362,** 362f
Estuaries, 1244–1245
ET (embryo transfer), 1104
Ethanol
 as biofuel, 341
 fermentation of, 140, 140f
 structure of, 35f
Ethical considerations
 in biodiversity, 1266
 in genomics, 371–372
 in stem cell research, 382
Ethology, 1135–1136
Ethylene, 811, 816t, 825–827, 827f
Etiolation, 807, 807f
Eubacteria (domain), 522, 522f
Eucalyptus, 752
Euchromatin, **190**
Eudicots
 clade of, 504, 609f
 leaves of, 751, 751f, 752
 roots of, 744f, 745
 stems of, 748, 748f
Eudoxochiton nobilis, 670f
Euglena gracilis, 571, 571–572f
Euglenids, 571, 571f
Euglenozoans, 570–572
Eukarya (domain), 13, 13f, 466f, 490, 522, 522f, 547f
Eukaryotes. *See also* Fungi; Protists
 ATP yield from aerobic respiration in, 137, 137f, 138
 cell cycle of, 192–204
 cell division in, 187, 393, 546
 cell transformation in, 330, 331f
 cell wall of, 67f, 78t, 80, 80f, 81t
 chloroplasts of, 74–75, 78t, 81t
 chromosomes of, 65, 78t, 81t, 189–191, 189t, 546
 compartmentalization in, 65, 522, 522f, 546
 cytoskeleton of, 65, 67f, 75–77, 76f, 78t, 79
 DNA in, 62, 68
 endomembrane system of, 65, 69–73, 522
 evolution of, 75, 75f, 188, 566–567
 fermentation of, 140
 flagella of, 66f, 78t, 79–80, 81t, 546
 gene expression in, 304, 305, 312–315f, 317–321, 322f
 gene organization in, 363t
 genome of, 360f, 481f
 initiation in, 295
 mitochondria of, 73–75f, 75, 78t, 81t
 multicellular, 201–202, 202f, 523, 523f
 noncoding DNA in, 362–363
 origin of, 522, 523f, 565–568
 plasma membrane of, 66–67f, 78t
 posttranscriptional control in, 317–322f
 pre-mRNA splicing in, 289–290f, 289–291
 prokaryotes vs., 545–546
 promoters of, 287–288, 288f, 313, 314f
 recombination in, 546
 replication in, 271–272f, 271–273
 ribosomes of, 68–69, 69f, 78t, 81t
 sequencing DNA in, 359

 sexual reproduction in, 523
 structure of, 13, 65–69, 78t, 81t
 transcriptional control in, 305, 312–315f, 322f
 transcription factors in, 287–288, 288f, 312–314
 transcription in, 287–289, 288f
 translation in, 295
Eumetazoa, 647, 648f, 649, 651–656
Euphyllophytes, 589f
Euphylls, 595, 595f
Euryarchaeota, 548f
Eusociality, 1158
Eutrophic lakes, 1204, 1204f, 1242, 1242f
Evaporation, 752, 880f, 881
EVCs (externally visible characteristics), 232
Evening primrose (*Oenothera biennis*), 843, 860
Evergreen forests, 1239
Evolution
 of aerobic respiration, 143
 agents of, 404–408
 age of Earth and, 11
 of amphibians, 699, 706–707
 anatomical record of, 432–434
 of animals, 640–644, 647, 648–649f
 of apes, 724
 of biochemical pathways, 118
 biodiversity and, 13
 biogeographical studies of, 435, 436
 of birds, 428–429, 429f, 470, 472f, 716–717
 of brain, 904, 904f
 of *Brassica,* 500–501, 500f
 of chordates, 501–502, 502f
 coevolution (*See* Coevolution)
 comparative anatomy and, 11, 11f
 comparative genomics and, 480–483, 481f, 483f, 488, 488f
 of complex characters, 470, 472f
 continental motion and, 520–521
 convergent (*See* Convergent evolution)
 critics of, 436–437
 Darwin on, 8–12, 400, 400f
 defined, 399
 of developmental patterns, 497–500
 of disease, 475–477f
 of dolphins, 429
 embryonic development and, 432, 433f
 of eukaryotes, 75, 75f, 188, 566–567
 experimental tests of, 413–415, 426, 426f
 of eyes, 417, 417f, 432–433, 433f, 506–510
 of eyespot on butterfly wings, 505–506, 506f
 of fish, 699, 700–701, 701f, 702–703, 703f, 704
 of flight, 470, 472f, 717
 of flowers, 474, 506, 839–842
 fossil record of, 10–11, 428–431, 436
 of gas exchange, 1004–1005, 1005f
 gene flow and, 404–405, 404f, 416, 416f
 genetic drift and, 404–405f, 405–406, 415–416
 genetic variation and, 399–401, 417, 417f
 of genomes, 482, 487–490
 of glycolysis, 129, 143
 of heart, 1026, 1026f
 heredity and, 11
 of homeobox genes, 392
 of homologous structures, 432, 432f
 of horses, 417, 417f, 429–431
 of humans, 406, 436, 461, 461f, 462, 723–729

 imperfect structures in, 432–433, 433f
 interactions among, 415–416, 416f
 on islands, 435, 448, 448f
 of land plants, 589, 589f
 of leaves, 595, 595f, 605
 of mammals, 700, 700f, 720
 of marsupials and placentals, 435, 435f
 of metabolism, 142–143
 of mitosis, 567–568
 molecular evidence for, 11–12, 11f
 of mollusks, 667
 mutations and, 301, 404, 404f, 415–416
 natural selection (*See* Natural selection)
 nitrogen fixation in, 143
 of oysters, 429
 of photosynthesis, 139, 143, 156, 520, 521
 of plants, 392
 of primates, 723–726f, 723–729, 728f
 of prosimians, 723
 rate of, 455, 455f, 465
 of reproductive isolation, 445, 446
 of reproductive system, 1089–1090, 1090f
 of reptiles, 699–700, 710–712
 reversal of, 462, 473
 of seed plants, 602–603
 selection and, 404f, 406–408
 of sharks, 703
 of small RNAs, 320
 of snakes, 429
 of social systems, 1159–1160
 speciation and, 455, 455f
 of stems, 595
 theory of, 8–12, 400, 400f, 436
 of vascular plants, 593–597
 of vertebrates, 429, 432, 699–701f, 1123, 1123f
 vestigial structures and, 433–434, 434f
 of whales, 429, 429f
 of wings, 505–506, 506f, 978–979, 979f
Evolutionary adaptation, 3
Evolutionary age, species richness and, 1227
Evolutionary biology, 414
Evolutionary conservation, 13–14
Excavata, 522, 522f, 568, 569f, 570–572
Excision repair, **274–**275, 274f
Excitation–contraction coupling, **974,** 974f
Excitatory postsynaptic potential (EPSP), 898, 900–901, 900f, 918
Excretion, by kidneys, 1050
Excretory system
 of annelids, 674
 of arthropods, 683
 of flatworms, 662f, 663
 of mollusks, 669
Exercise
 bone remodeling and, 969, 969f
 cardiac output and, 1035–1036
 metabolic rate and, 997
 muscle metabolism during, 976–977
Exergonic reactions, **110–**111, 111f, 112
Exhalant siphons, **671,** 671f
Exocrine glands, **867**
Exocytosis, **103,** 103f, 104t
Exons, 51, **289,** 289f, 290
Exon shuffling, 290
Exonucleases, **266,** 268, 269f
Exoskeleton
 of arthropods, **681,** 682
 characteristics of, **964–**965, 965f
 chitin and, 41, 41f, 653
Experimental selection, 426, 426f

Experiments, 5–6f, 6
Expiration. *See* Breathing
Exploitative competition, **1189**
Expressed sequence tags (ESTs), **362**, 362f
Extensor muscles, 971f
External fertilization, 1090, 1091f
External intercostal muscles, 1011
Externally visible characteristics (EVCs), 232
Exteroceptors, **917**
Extinction
   genetic variation and, 1277
   loss of keystone species and, 1275
   population size and, 1276–1278
Extinctions
   colonization and, 1228
   of dinosaurs, 456, 700, 711
   ecosystem disruptions and, 1274–1276
   factors responsible for, 1266–1277f, 1267f
   glaciation and, 520
   habitat loss and, 1269–1271, 1269f
   human role in, 1260
   introduced species and, 1272–1274
   of Lake Victoria cichlid fish, 454, 1274
   mass, 456–457, 456f, 520, 700
   overexploitation and, 1271–1272
   prehistoric, 1260, 1260t
   speciation and, 456–457
Extracellular fluid (ECF), 893t
Extracellular matrix (ECM), 80–81, 81f, 395, 568, 869
Extraembryonic coelom, 1118
Extraembryonic membrane, **1118**, 1118f
Extra-pair copulations (EPCs), **1155**
Extraterrestrial life, 516
Extremophiles, 545
Extrusion, 99
Exxon Valdez oil spill (1998), 562
Eyeless gene, 507, 508f
Eyes
   of annelids, 673–674, 929
   of arthropods, 929
   of birds, 932, 933f, 934
   of chordates, 929, 930f
   of cnidarians, 508–509, 509f
   color of, 232, 233t, 240–241, 240f
   compound eyes, 417, 417f, 682, 682–683f, 687
   development of, 506–510
   evolution of, 417, 417f, 432–433, 433f, 506–510
   of fish, 507, 508f
   focusing of, 931f
   of insects, 417, 417f, 507, 507f, 682, 682–683f, 687, 930f
   of invertebrates, 929–930, 929f
   of mollusks, 433, 433f, 507, 507f, 929, 930f
   parietal, 713
   photoreceptors, 432, 433f, 931–933
   of planaria, 508, 509
   potatoes, 750, 857
   structure of, 930, 930–931f
   of vertebrates, 432–433, 433f, 507, 507f, 930–933f, 1127f
Eyespots
   of butterflies, 505–506, 506f
   of flatworms, 508, 662f, 663, 929, 929f

F

FACE (Free Air CO$_2$ Enrichment) studies, 785, 786f
Facilitated diffusion, **96–97**, 97f, 104t
Facilitation, 1203
Facultative long-day plants, **835**
Facultative short-day plants, **835**
Facultative symbiosis, **631**
FAD (flavin adenine dinucleotide), **44**, 131, 131f, 132, 133f
FADH
   in ATP yield, 137, 137f, 141
   in electron transport chain, 134, 134f, 135
   from Krebs cycle, 131, 131f, 132, 133f
Falconiformes, 715t
Falling phase, of action potential, 894, 895f
Fallopian tubes, 1099, 1099f
Family (taxonomic), 466, 466f
Farming. *See* Agriculture
Farsightedness, 931f
Fast-twitch muscle fibers, 975, **976**, 976f
Fats. *See* Fatty acids; Lipids
Fatty acid desaturases, 93
Fatty acids
   absorption in small intestine, 990f, 991
   catabolism of, 141–142
   monounsaturated, 53, 54f
   in phospholipid molecules, 55, 55f
   polyunsaturated, 53, 54f
   saturated, 53, 54, 54f, 55
   structure of, 36f, 53, 54f
   trans-fatty acids, 54
   unsaturated, 53–54, 54f, 55
FDA (Food and Drug Administration), 54, 536, 540
Feathers, 470, 472f, 714, 714f, 716
Feather stars, 694, 695, 695f
Feces, 992
Fecundity, **1170**
Feedback inhibition, 117, **118**–119, 119f, 138
Feeding phase, of slime molds, 583
Felines. *See* Cats
Female infertility, 1103
Female reproductive system, 876f, 877, 1092, 1095t, 1096–1100
Fermentation
   cell-free, 33, 34f
   defined, **124**
   electron acceptors in, 139–140, 140f
   of ethanol, 140, 140f
   of lactic acid, 140, 140f
   in metabolism, 129
   yeasts in, 629–630
Ferns, 589f, 590, 594, 594t, 596–599f, 597–599
Ferredoxin, 158, 159f
Ferritin, 45t, 321
Fertilization
   in amphibian, 1089–1090f, 1090
   in animals, 1108–1111, 1108t, 1109–1111f
   in birds, 1090–1091f, 1091
   defined, **208**, 208f
   double, 612, 847, 848f
   external, 1090, 1091f
   in fish, 1089, 1089–1090f, 1090
   internal, 1089–1090f, 1089–1091
   in pine trees, 606
   in plants, 610–612, 611f, 847, 848f
   in reptiles, 711, 1090
   in vertebrates, 1089–1092
Fertilization envelope, 1110
Fertilizers
   nitrogen, 1213
   overuse of, 777
   phosphorus, 1214
   pollution from, 1254
   variable-rate application of, 777
Fever, **884**
F$_1$ generation. *See* First filial generation

F$_2$ generation. *See* Second filial generation
F$_3$ generation. *See* Third filial generation
FGF. *See* Fibroblast growth factor
Fiber, dietary, 992
Fibrin, 45t, **1022**
Fibrinogen, **1021**
Fibroblast cells, 59, 59f, 385, 966f
Fibroblast growth factor (FGF), 380, 380–381f, 381, 502, **1120**
Fibronectin, 80, 81f, 395–396, 395f
Fick's Law of Diffusion, **1004**, 1007
Fiddleheads, 598, 599f
Filaments (flower), **610**, 610f, 839, 839f
Filariasis, 680
Filial imprinting, **1141**
Filopodia, 1115
Filoviruses, 371t, 531t, 539
Filtrate, 1049–1050
Filtration, 1044, 1048–1049f, 1048–1050
Finches. *See also* Ground finches
   beaks of, 8, 9f, 410, 422–423, 452–453, 453f, 1192, 1193f
   evolution of, 452–453, 453f
   seedcracker, 412, 412f
Fingerprinting, 335, 335f
Fingers, grasping, 723
Fins, of fish, 701, 704
Fireflies, 1147, 1147f
First filial generation, **224**, 225f, 228–229, 229f
First Law of Thermodynamics, **109**
Fish. *See also specific fish*
   aquaculture, 1250
   armored, 702, 702t
   bony, 703, 704, 704f, 1006–1008, 1006f, 1046
   brain in, 903–904
   cartilaginous (*See* Cartilaginous fish)
   characteristics of, 701–703
   circulation in, 701, 712, 712f, 1025, 1025f
   classification of, 645t, 702t
   countercurrent heat exchange in, 882, 882f
   depletion of, 1250, 1250f
   evolution of, 699, 700–701, 701f, 702–703, 702f
   eye development in, 507, 508f
   fertilization in, 1089, 1089–1090f, 1090
   freshwater fish, 1046, 1046f
   gills of, 701, 704, 1006–1007f, 1006–1008
   hearing in, 921–922, 921f
   heart in, 1025, 1025f
   jawed, 501, 501f, 701, 702–703, 702f
   jawless, 702, 1067–1068
   kidneys in, 1046–1047, 1046f
   lobe-finned, 700f, 701, 702t, 704, 705–706f
   marine bony fish, 1046, 1046f
   nitrogenous wastes of, 1043f
   nutritional deficiencies, 701
   predation by, 409, 410f, 414–415
   prostaglandins in, 945
   ray-finned, 700f, 702t, 704, 705f
   reproduction in, 703
   respiration in, 1005–1007f, 1006–1008
   selection to match climatic conditions, 407
   sequential hermaphroditism in, 1087, 1087f
   sex determination in, 242
   spiny, 702, 702t
   swimming by, 977–978, 977f
   taste in, 927
   vestigial structures in, 433–434, 434f
   viviparous, 1089, 1089f
FISH (fluorescent in situ hybridization), 344–345, 345f, 355, 355f

Fission, 1086
Fitness, **408**–409, 409f, 417–418
Fixed action patterns, 1135
Flagella
   of bacteria, 64f, 65, 546, 551, 551f
   defined, **65**
   of eukaryotes, 66f, 78t, 79–80, 81t, 546
   of fungi, 624
   of prokaryotes, 63–64f, 65, 81t, 546, 550, 551, 551f
   of protists, 568, 570, 571, 573, 582
Flagellin, **551**, 551f
Flame cells, **663**
Flatworms
   acoel, 656–657, 657f
   body plan of, 642f, 661
   characteristics of, 645t
   classification of, 663–665
   digestion in, 662, 662f, 663
   excretion and osmoregulation in, 662f, 663
   eyespots of, 508, 662f, 663, 929, 929f
   free-living, 662, 663
   nervous system of, 662f, 663, 902f, 903
   osmoregulation in, 1043, 1044f
   parasitic, 662, 663–664, 1200
   reproduction in, 662f, 663
Flavin adenine dinucleotide (FAD), **44**, 131, 131f, 132, 133f
Flavivirus, 530f, 531t
Fleas, 684, 687t, 688
Flemming, Walther, 189, 207
Flesh-eating diseases, 558
Flexor muscles, 971f
Flies. *See also* Fruit flies (*Drosophila*)
   characteristics of, 687t
   eye development in, 507, 508–509f
   mouthparts of, 687, 688f
   selection for pesticide resistance in, 407
   taste in, 928, 928f
Flight, evolution of, 470, 472f, 717
Flight skeleton, 716
Flippers, 713
Floc aggregates, 342
Flooding, plant responses to, 768–769, 769f
Flor, H. H., 800, 801f
Floral leaves, 753
Floral meristem identity genes, 836–839f
Floral organ identity genes, 836–839f
Florigen, 836
Flowering hormones, 836
Flowering plants, 607–613. *See also* Flowers; Fruits
   autonomous pathway of, 836, 837f
   characteristics of, 604t
   dichogamous, 846
   dioecious, 846
   diversification of, 603
   embryonic development in, 848–854
   evolution of, 474
   fertilization in, 847, 848f
   flowering hormones in, 836
   gamete formation in, 841–842
   gene duplication in, 504, 504f
   genome of, 483
   gibberellin-dependent pathway in, 836, 837f
   innovations in, 595, 596f, 607
   life cycle of, 610–612, 611f, 832f
   light-dependent pathway in, 834–835f, 834–836
   monoecious, 846
   morphology of, 504, 504f
   origins of, 608, 608–609f
   phase change in, 832–833, 833f
   phylogeny, 589f

Flowering plants (continued)
　pollination in, 611–612, 612f
　polyploidy in, 485, 485f
　seeds in, 595
　temperature-dependent pathway in, 836, 837f
　trends in, 840, 840f
Flowers
　color of, 843–844, 844f
　complete, 839, 839f
　formation of floral meristems and floral organs, 836–839f
　incomplete, 839
　initiation of flowering, 832, 832f
　male and female structures of, 839–840, 846–847
　morphology of, 610, 610f, 639–640
　petal development, 504–505
　production of, 834–835f, 834–839f
　shape of, 506
　structure and evolution of, 474, 506, 839–842
　symmetry of, 506, 840, 840f
Fluidity, of membranes, 93, 93f
Fluid mosaic model, 89, 90f, 546
Flukes, 645t, 661, 663–664, 664f
Fluorescence microscopes, 62t
Fluorescent in situ hybridization (FISH), 344–345, 345f, 355, 355f
Fluoxetine (Prozac), 900
Flu virus. *See* Influenza virus
Flying foxes, 1275–1276, 1276f
Flying mammals, 720, 720f
Flying phalangers, 435f
Flying squirrels, 435f
fMRI (functional magnetic resonance imaging), 1136, 1136f
Focal adhesions, 82t, **83**
Folic acid, 1000t
Foliose lichens, 632f
Follicle-stimulating hormone (FSH), **950**, 951, 1092, 1095, 1095t, 1096f
Follicular phase of menstrual cycle, 1097f, 1098
Food and Drug Administration (FDA), 54, 536, 540
Food energy, 997–999
Food poisoning, 1081
Food preservation, 52–53
Foods. *See also* Nutrients
　caloric content of, 997
　genetically modified, 348–349, 349f
　intake, regulation of, 997–999
Food security, 781
Food storage, in plants, 746, 853–854, 854f
Food supply, population cycles and, 1178
Foraging behavior, 1150, 1150f
Foraminifera, 568, 581–582, 582f
Forebrain, 903f, 903t, 904–906, 905f
Forensics, 335, 335f, 371
Forest ecosystems
　biogeochemical cycles in, 1214–1215, 1215f
　deforestation and, 1212, 1212f, 1249
Fork head gene, 1119
Formed elements of blood cells, 1021
fosB gene, 1138
Fossil fuels, 1211
Fossil genes, 434
Fossil record
　of Cambrian period, 524, 524f
　community, 1188
　documentation of evolutionary change through, 428–429, 429f
　Earth's history from, 515, 518–519
　eukaryotic, 565–566, 566f
　of evolution, 10–11, 428–431, 436
　of flowering plants, 608, 608f
　of foraminifera tests, 582

gaps in, 428, 429, 436
　of ginkgophytes, 607
　microfossils, 518–519, 519f
Founder effect, **405**–406
Four o'clocks (*Mirabilis jalapa*), 233t, 234, 234f, 244
Fovea, 930f, **931**
Foxes, 427, 427f
FOXP2 gene, 491–492, 491f
F plasmid, **552**–554
F plasmid transfer, 553, 553f
Fracastoro, Girolamo, 544
Frameshift mutations, **283**, 299
Franklin, Rosalind, 260, 261f
Free Air CO$_2$ Enrichment (FACE) studies, 785, 786f
Free energy, **110**, 138
Free radicals, 153
Free water molecules, 98, 98f
Freeze-fracture microscopy, 91–92, 92f
Frequency-dependent selection, **409**–410, 410f
Freshwater habitat, 1240–1243, 1248
Frogs (*Rana*)
　characteristics of, 705t
　chromosomes in, 189t
　cladistic analysis of, 463, 464f
　classification of, 705–706, 705t
　courtship behavior in, 1154, 1154f
　development in, 376, 376–377f, 380, 432, 433f, 1112–1113f
　eggs of, 1111, 1111f
　fertilization in, 1090, 1091f
　fungal infections in, 635, 635f
　gastrulation in, 1116, 1116f
　habitat of, 707–708
　legs of, 706
　lungs of, 1009f
　nuclear reprogramming in, 383
　population declines in, 1267–1268f, 1267–1269
　postzygotic isolating mechanisms in, 444, 444f
　respiration in, 706
　temporal isolation of, 443
Fronds, 598, 599f
Frontal lobe, 904, 905f
Fructose, 38, 39, 39f
Fructose 1,6-bisphosphate, 128f, 138, 138f
Fructose 6-phosphate, 128f, 138, 138f, 161, 162
Fruit flies (*Drosophila*)
　alternative splicing in, 320
　axis formation in, 387–390
　behavioral genetics in, 1138
　body color in, 246–247, 246f
　body plan production in, 388–389f, 391
　branchless gene in, 1120
　bristle number in, 426, 426f
　cell cycle of, 192
　cell interactions in, 83
　chromosomes in, 189t
　courtship behavior of, 447
　development in, 386–392, 499, 1119–1120, 1119f
　eye color in, 240–241, 240f
　eyeless gene from, 507, 508f
　gene expression of, 304f
　genetic mapping of, 246–247, 246f, 350
　genome of, 360f, 365, 401, 481f, 487
　of Hawaiian Islands, 447, 451–452, 452f
　heart development in, 1120, 1120f
　hedgehog signaling molecule in, 1126
　homeobox genes in, 5
　homeotic genes in, 391–392

homeotic mutations in, 307, 391, 499
　meiosis in, 214
　Morgan's experiments with, 240–241, 240f
　pattern formation in, 386–392
　phototropism of, 412–413, 413f
　proteasomes of, 323f
　segmentation in, 387, 388–389f
　sex chromosomes of, 241, 243, 245
　toll-like receptors in, 1058
　transposons in, 490
　wing traits of, 246–247, 246f, 499
　X chromosomes of, 241, 243, 245
　Y chromosomes of, 241
Fruits
　adaptations of, 614–615
　defined, 614
　development of, **610**, 614, 614f
　dispersal of, 615–616, 616f
　evolution of, 607
　ripening of, 825–827, 827f
　seed protection function of, 595
　types of, 615, 615f
Fruticose lichens, 632f
FSH. *See* Follicle-stimulating hormone
FtsZ proteins, 63, 188, 188f
Fucus (zygote), 849, 850f
Fuel sources, renewable, 341–342
Fumarate, 132, 133f
Function, relationship with structure, 13
Functional genomics, 365, 366–368, 367f, 491, 499–500
Functional groups, **35**, 35f
Functional magnetic resonance imaging (fMRI), 1136, 1136f
Fundamental niche, 1189–**1190**, 1190f
Fungi, 619–635
　Ascomycota, 620, 620f, 620t, 628–630
　Basidiomycota, 620, 620f, 620t, 627–628, 628f
　benefits for plants, 793
　carnivorous, 622, 622f
　cell types in, 619
　cell walls of, 619, 621
　characteristics of, 620, 620t
　chromosomes in, 189t
　Chytridiomycota, 620, 620f, 620t, 624–625
　cytokinesis in, 197–198
　as decomposers, 630
　diseases caused by, 634–635, 793, 794f
　ecology of, 630–634
　endophytic, 631, 631f
　filamentous, 621
　genome of, 483
　genomes of, 630
　Glomeromycota, 620, 620f, 620t, 627
　lichens, 631–632, 632f
　mating behavior in, 177
　Microsporidia, 620, 620f, 623, 623f
　mitosis in, 621–622
　mycelium and, 621, 621f, 627–628
　mycorrhizae, 627, 632–633, 633f
　nucleus of, 621
　nutrients for, 619, 621, 622–623, 622f
　phylogeny of, 620, 620f, 620t
　reproduction in, 619, 622, 622f, 626, 629
　in rumen, 624, 625
　in symbiotic relationships, 591, 627, 631–634
　unicellularity of, 620, 623
　Zygomycota, 620, 620f, 620t, 625–626, 626f
Fungi (kingdom), 13, 13f
Fusarium, 634
Fusion inhibitors, 536, 537f

G

GABA (gamma-aminobutyric acid), 899, 900
Galactose, 38, 39, 39f
Galápagos Islands. *See also* Finches
　blue-footed boobies of, 443f
　creation of, 450
　founder effects on, 405
　hydrothermal vents near, 675f
Gallbladder, 990f, 991, 995
Galliformes, 715t
Gallstones, 991
Gametangia, **590**, 591, 592, 595, 626
Gametes
　aneuploid, 214
　chromosomes of, **208**, 208f
　fusion, prevention of, 442t, 444
　plant, 841–842
　uniting of, 1086
Gametic intrafallopian transfer (GIFT), 1104
Gametophytes
　in flowering plants, 610–611, 611f
　of flowers, 841–842
　in haplodiplontic life cycle, 577, 590, 590f, 594–595
　hornwort, 593
　photosynthetic, 591, 598
　in pine trees, 605–606
　pollination and, 612
　transport of, 603
Gametophytic self-incompatibility, 847, 847f
Gamma-aminobutyric acid (GABA), 899, 900
Ganglia, 682, 682f, 873, **910**, 911, 912f
Ganglion cells, **932**, 932f
Gap genes, 389f, **391**
Gap junctions, 82t, **83**–84, 84f
Gap phases, of cell cycle, **192**, 192f
Garden peas (*Pisum sativum*)
　branching mutants in, 822–823
　chromosomes in, 189t
　flower color in, 224–226f, 226–227, 231–232, 231f
　genome of, 485
　Knight's experiments with, 222
　Mendel's experiments with, 222–228
　practical considerations in study of, 222–223
　seed traits in, 224–225, 224f, 228–229, 229f
　stomata of, 737f
Garrod, Archibald, 278–279
Gases, noble, 22
Gas exchange
　in animals, 1005f
　evolution of, 1004–1005, 1005f
　leaves and, 752
　in lungs, 1011, 1012f
　in respiratory system, 1004–1005
　restriction of, 1014
　in single-celled organisms, 1005f
　in tissues, 1012f
Gasoline, 34, 124
Gastric inhibitory peptide (GIP), **995**, 995f, 996f, 998, 999, 999f
Gastric juice, **988**, 988f
Gastrin, **995**, 995f, 996t
Gastrodermis, **651**, 652, 652f
Gastropoda, 670, 670f
Gastropods, 667, 668f, 668–669, 670f
Gastrovascular cavity, 652–653, 984, **1023**, 1024f
Gastrulas, 378, 640
Gastrulation, **395**, 395f, 1108t, 1114–1118, 1115t, 1127
Gated ion channels, 97, 893–895f, 918
GA-TRXN protein, 824f

Gause, Georgii, 1190–1191
Gavials, 714
GDP (guanosine diphosphate), 132, 133f, 178, 178f
Geckos, 713
Geese, 432, 1135f
GEFs (guanine nucleotide exchange factors), 178, 178f
Gehring, Walter, 507, 508
Gel electrophoresis of DNA, 329–330, 329f, 355f, 357
Gene chips, 345–346
Gene duplication, 486, 486f, 487–489f, 502, 504–505
Gene expression, 304–324
   analysis of variations in, 335–337, 335f, 337f
   central dogma of molecular biology on, 280, 280f, 330
   chromatin structure and, 315–317f
   control of, 14, 42, 304–305
   in development, 1119
   environmental effects on, 233t, 235, 235f, 308–309
   in eukaryotes, 304, 305, 312–315f, 317–321, 322f
   modulation of, 337
   overview, 297, 298f, 298t
   in plants, 805–806, 820, 820f
   in polyploids, 486
   posttranscriptional regulation and, 317–322f
   in prokaryotes, 304, 305, 308–312, 547
   protein degradation and, 323–324
   regulatory proteins in, 305–307
   RNA in, 281
   serial-analysis of, 368
   steroid hormone receptors and, 173–174, 173f
   transcriptional control and, 305, 308, 312–315
   translational control in, 321
   universality of, 284
Gene flow
   alleles and, 404–405, 404f
   defined, **404**
   hybridization and, 349
   interactions among evolutionary forces and, 416, 416f
   speciation and, 441, 446
Gene-for-gene hypothesis, 801, 801f
Gene frequencies, **402**
Gene pools, **441**
Generalized transduction, **554,** 555f
Generally regarded as safe (GRAS) compounds, 54
General transcription factors, **313,** 313f
Generation time, **1170,** 1171f
Generative cells, 606, 612
Genes
   copy numbers, 488
   defined, **12**
   discovery of, 225
   functional analysis of, 499–500
   functions of, 502, 504–505
   inactivation of, 489
   nature of, 278–281
   one-gene/one-polypeptide hypothesis, 280
   ownership and patenting, 372
   pleiotropic effect of, 417
   pseudogenes, 363, 363t, 434
   segmental duplication of, 487–489f
   transgenes, 337, 340
Genetic code
   deciphering, 283, 283t
   degeneracy of, 283–284
   DNA and, 42–43

reading of, 282–283, 282f
universality of, 284, 337–338
Genetic counseling, **252**
Genetic disorders
   counseling for, 252
   detection of, 345
   enzyme deficiencies in, 278–279
   lysosomal storage disorders, 71
   overview, 249–253, 249t
   peroxisome biogenesis disorders, 72
   prenatal diagnosis of, 252–253, 253f
   sex linkage in, 242–243
Genetic drift, 404–405f, **405**–406, 415–416, 447, 465
Genetic engineering. *See also* Biotechnology
   bacteria in, 562
   bioremediation and, 342
   of insulin, 344, 344f
   social issues raised regarding, 349
   of transcription factors, 347
   transgenic organisms and, 337–341, 347–349
Genetic maps, 244–248, 353–357
Genetic markers, **354**–356
Genetic modifications. *See* Biotechnology; Genetic engineering
Genetic mosaics, **243**
Genetic recombination. *See* Recombination
Genetic relationships, 1157f
Genetics
   behavioral, 1137–1139
   population, 400
   of prokaryotes, 552–557
Genetic sex determination, 1088, 1088f
Genetic template, **1142**
Genetic variation
   allele frequency, changes in, 399–403, 402f
   in blood groups, 400–401
   defined, 399
   evolution and, 399–401, 417, 417f
   in human race, 728–729, 728f
   maintenance of, 409–411, 445
   natural selection, role in maintaining, 409–411, 410f, 445
   in nature, 400, 401f
Genistein, 796t
Genomes, 353–365. *See also* Genomics
   annotation and databases, 361–365, 362f, 364f
   coding and noncoding sequences of, 362–364
   of conifers, 604
   conserved regions in, 365
   defined, 13
   downsizing of, 485, 485f
   editing, 336–337, 337f
   eukaryotic, 360f, 481f
   evolution within, 482, 487–490
   of fungi, 630
   gene swapping evidence in, 490, 490f
   human (*See* Human genome)
   mapping, 353–357
   of mitochondria, 370, 567
   of moss, 592–593
   plants, 482–483, 485, 485f
   projects involving, 360–361
   rearrangement of, 488–489, 488f
   sequencing, 357–360, 364–365, 401
   size and complexity of, 360, 360f, 483–487
   of vascular plants, 597
   of viruses, 370, 529, 530
Genome-wide association studies (GWAS), 371
Genomic anarchy, 482
Genomic imprinting, **251**–252, 384

Genomics. *See also* Genomes
   analytic approaches in, 354
   comparative (*See* Comparative genomics)
   functional, 365, 366–368, 367f, 491, 499–500
   medical applications of, 370–372, 371t
   paleogenomics, 727
   social and ethical issues regarding, 371–372
   yeasts and, 630
Genotype, **226**
Genotype frequencies, 402f, **403**, 405
Genus, 465, 466f
Geobacter, 549f, 562
Geographic isolation, 448, 448f, 452
Geographic variation within species, 441, 441f
Geography of speciation, 447–450
Geological time, 514f, 515
Geometric progressions, 10, 10f
Germ cells, 384, 385f, 1093
Germ hypothesis, 6–7, 6f
Germinal centers, 1034
Germination
   of pollen grains, 603–604
   of seeds, 613–614, 807, **854**–856f
Germ layers, 641–642, **865**, 1115
Germ-line cells, **208**–209, 208f
Germ theory of disease, 545
Gharials, 713f, 714
GH (growth hormone), **950,** 952–953, 952f
GHIH (growth hormone-inhibiting hormone), 951
Ghrelin, **952,** 998
GHRH (growth hormone-releasing hormone), 951
Giant clams (*Tridacna maxima*), 667, 667f
Giant kelp, 577f
Giant pandas (*Ailuropoda melanoleura*), 494, 494f
Giant redwoods (*Sequoiadendron giganteum*), 859f
Giant squid, 667
Giardia intestinalis, 570, 570f
Gibberellins, 793, 816t, 823–824, 824f, 836, 837f
Gibbons, 461, 461f, 724
Gibbs free energy, 110
GIFT (gametic intrafallopian transfer), 1104
Gigantism, 952
Gill covers, 704
Gill filaments, 1007
Gills
   of fish, 701, 704, 1006–1007f, **1006**–1008
   internal, 701
   of mollusks, 668, 670
   of mushrooms, 628, 628f
Ginkgo biloba, 606f, 607, 609f
Ginkgophytes, 603, 604t, 607
GIP. *See* Gastric inhibitory peptide
Giraffes (*Giraffa camelopardalis*), 107, 107f, 400, 400f, 432, 467f
Girdling of plants, 741
Gizzard, 986
GLABROUS3 mutant, 738, 738f
Glaciers, 520, 520f, 521, 1254, 1254f
Glades, 468–469
Glaucoma, juvenile, 227–228, 227f
Glaucophytes, 579
Gliding joints, **969,** 970f
Global climate change
   biological communities and, 1188
   carbon dioxide and, 1253–1254, 1254f
   computer models of, 1253

crop production and, 784–787
ecosystems and, 1254–1255
geographic variation in, 1253, 1253f
human health and, 1256
species richness and, 1227
Globulins, **1021**
Glomeromycetes, 627, 633
Glomeromycota, 620, 620f, 620t, 627
Glomerulus, 1045–1046, 1045f, **1049,** 1049f
Glomus, 620t
Glottis, 1009, 1010f
Glucagon, 170, 182, **957,** 957f, 996–997, 997f
Glucocorticoids, 956, 956f
Gluconeogenesis, 142, 956, **997**
Glucose
   in aerobic respiration, 132
   catabolism of, 124, 138, 138f
   cleavage of, 127, 128f
   function of, 37t
   mobilization of, 182
   oxidation of, 25, 123–127
   from photosynthesis, 151
   polymers of, 40f
   priming of, 127, 128f
   reabsorption in kidneys, 1050
   regulation of, 996–997, 997f
   structure of, 38, 39f
   transport forms of, 38–39
Glucose 1-phosphate, 162
Glucose 6-phosphate, 128f, 997
Glucose repression, **310**–311, 310f
Glucose transporter, 45t, 97, 101, 101f
Glutamate, 141, 141f, 179, 899
Glutamic acid, 45, 47f, 49, 116
Glutamine, 47f
Glyceraldehyde 3-phosphate (G3P), **127,** 128f, 129, 161, 161f, 162
Glycerol, 53, 55, 55f, 89, 92, 93
Glycerol phosphate, 35f
Glycine, 45, 47f, 516, 899
Glycogen, 37f, **40,** 40f
Glycogenolysis, **997**
Glycolipids, 71, 82, **90,** 91t
Glycolysis
   defined, **125**
   electron transfer in, 132
   evolution of, 143
   history and evolution of, 129
   process overview, 126, 126–128f, 127–129
   recycling of NADH in, 129–130, 129f
   regulation of, 138, 138f
Glycolytic pathway, 128f, 142
Glycoprotein hormones, 949, 950
Glycoproteins, **70,** 71, 80, **90,** 91t
Glycosidases, 45t
Glyphosate, **347,** 348f
G_2/M checkpoint, **200,** 200–201f, 201
Gnetophytes, 603, 604t, 606–607, 606f, 609f
Gnetum, 604t, 607
GnRH (gonadotropin-releasing hormone), 951, 1095, 1096f
Goblet cells, 867
Goiter, 23, 951, 951f
Golden Rice, 348–349, 349f
Golgi, Camillo, 70
Golgi apparatus, **70**–71, 78f, 81t, 522
Golgi bodies, **70,** 742
Golgi tendon organs, **920**
Gonadotropin-releasing hormone (GnRH), 951, 1095, 1096f
Gonadotropins, 950
Gonorrhea, 559, 560, 560f, 1081
Gonyaulax, 573f
Gooseneck barnacles (*Lepas anatifera*), 686f
Gore, Al, 1253

Gorillas
  evolutionary relationships of, 461, 461f, 462, 724
  FOXP2 gene in, 491, 491f, 492
  genome of, 488, 488f, 494f
  HIV/AIDS in, 475
  homoplasy and, 464f
  morphological data for, 463f
  olfactory receptor genes in, 489
Gosling, Ray, 260
Gould, John, 422
Gout, 1042
GPCRs. *See* G protein–coupled receptors
G₀ phase, **192**–193, 202
G₁ phase, **192**, 192f, 193, 376, 377f
G₁/S checkpoint, **200**, 200–201f, 201
G₂ phase, **192**, 192f, 193–194, 194f, 376, 377f
GPP (gross primary productivity), **1217**
G protein–coupled receptors (GPCRs)
  in cell signaling, 172f, 172t, 173, **178**–183, 179f
  in hydrophilic hormones, 947–948, 947f
  membranes of, 95
G-proteins, **173**, 178–183, 178f, 911–912
G proteins, 948
Graafian follicles, **1098**, 1098f
Graded potentials, **893**–894, 894f
Gradualism, **455**, 455f
Gram-negative bacteria, 550–551f, **550**–552
Gram-positive bacteria, 548f, **550**–551
Gram stains, **550**–551
Grana, 74, 74f, 148, 148f, 160, 160f
Grant, Peter, 423, 444–445
Grant, Rosemary, 423, 444–445
Granular leukocytes, **1021**
Granulosa cells, **1096**
Grape flowers, 824, 824f
GRAS (generally regarded as safe) compounds, 54
Grasshoppers, 241, 241f, 682f, 686f, 687t, 689, 1024f
Grasslands, 1239
Gravitropism, **809**–810
Gravity sensing, in plants, 809, 809f
Gray matter, 891
Green algae
  chloroplasts of, 522, 523f, 566
  in classification system, 468, 468f
  evolutionary relationships of, 579–581
  flagella of, 80f
  in lichens, 631, 632
  multicellular, 523
  as photoautotrophs, 588
  phylogeny, 589, 589f
Greenbriar (*Smilax*), 744f
Greenhouse gases, 1253–1254, 1254f
Gregarines, 574, 575f
Greyhound dogs, 427, 427f
Griffith, Frederick, 257, 257f, 555
Griffith, John Stanley, 540
Gross primary productivity (GPP), **1217**
Ground finches
  adaptation of, 452, 453, 453f
  large, 9f, 422f
  medium, 410, 423, 423f, 444–445
  small, 444
Ground meristems, **735**, 736f, 745, 852
Ground substance, **869**
Ground tissue, **734**, 739–740, 739f, 852
Groundwater, 1212
Growth, as characteristic of life, 3
Growth factor receptors, 202
Growth factors, 202, 203f, **944**
Growth hormone (GH), **950**, 952–953, 952f
Growth hormone-inhibiting hormone (GHIH), 951

Growth hormone-releasing hormone (GHRH), 951
Growth media, 556
Gruiformes, 715t
GTP. *See* Guanosine triphosphate
Guanine, 42, 42f, 259f, 260, 262
Guanine nucleotide exchange factors (GEFs), 178, 178f
Guano, 1042
Guanosine diphosphate (GDP), 132, 133f, 178, 178f
Guanosine triphosphate (GTP), 132, 133f, 173, 178, 178f
Guard cells, **737**, 737f, 752, 752f, 766–767
Guppies, color variation in, 414–415
Gurdon, John, 383
Gurken proteins, 388, 390f
Gustation (taste), 927–928
Gut, 984, 998
Guttation, **764**
GWAS (genome-wide association studies), 371
Gymnopodium, 574f
Gymnosperms, 589f, 596f, **603**–607, 609f
Gynoecium, **610**, 610f, 839
Gyre, **1235**
Gyrus, 905

H

HAART (highly active antiretroviral therapy), 537
Haberlandt, Gottlieb, 821
Habitat fragmentation, 1270–1271, 1270f
Habitat occupancy, population dispersion and, 1169
Habitats
  of animals, 640, 667, 684, 707–708
  destruction of, 1269, 1269f
  economic value of, 1264–1265, 1265f
  freshwater, 1240–1243
  heterogeneity of, 1226–1227, 1226f
  lakes, 1241–1242f, 1241–1243
  loss of, 1247–1249, 1269–1271, 1269f
  marine, 1243–1246f, 1243–1247, 1250–1251
  ponds, 1241–1242f, 1241–1243
Habituation, **901**, 1139
Haemophilus influenza, 328, 353, 353f, 359, 360
Hagfish, 700f, 701, 702, 702t, 1041
Hair
  as ancestral vs. derived character, 462, 463f
  of mammals, 718
Hair cells, 921–922, 921f
Hair-cup moss (*Polytrichum*), 592f
Hairpins, 286, 286f
hairy gene, 389f, 391
Haldane, J. B. S., 301, 1157
Haleakala silversword (*Argyroxiphium sandwicense*), 1261, 1261f
Half-life, 20, 428, 428f
Halobacterium, 548f
Halophytes, **769**
Halorespiration, 562
Hamilton, William D., 1157–1158
Hamstrings, 970, 971f
Handicap hypothesis, **1154**
Hansen disease (leprosy), 558, 559t
Hantavirus, **539**
Haplodiploidy, **1158**
Haplodiplontic life cycle, **577**–578, 577f, 580, 580f, 590, 590f
Haploids (*n*), **191**, 208–209, 208f, 225
Haplopappus gracilis, 189t
Hardy, Godfrey H., 402
Hardy–Weinberg equation, 402–403

Hardy–Weinberg equilibrium, **402**–403, 402f, 405
Hardy–Weinberg principle, 402–404, 411
Harlow, Harry, 1141
Hashimoto thyroiditis, 1077
Haustoria, 746, 793
Haversian canals, 968
Haversian lamellae, 968
Haversian system, 967f, **968**
Hawaiian Islands
  birds of, 1273, 1273f
  creation of, 450
  founder effects on, 405
  fruit dispersal methods on, 616
  fruit flies of, 447, 451–452, 452f
  lobeliads of, 452, 452f
Hawks, 467f
H bands, 971, 972f
hCG (human chorionic gonadotropin), 1099, 1127
HDL (high-density lipoprotein), 54, 1034
HDN (hemolytic disease of newborns), 1079
Head, of vertebrates, 699, 699f
Hearing, 921–926
Heart
  in amphibians, 706, 1025, 1026f
  in arthropods, 682, 682f
  of birds, 1026, 1027f
  cardiac cycle, 1028, 1029f
  contraction of, 1028–1029
  development of, 502, 503f, 1120, 1120f
  evolution of, 1026, 1026f
  in fish, 1025, 1025f
  four-chambered, 1026, 1028–1031
  in insects, **1023**
  of mammals, 1026, 1027f
  in reptiles, 712, 712f, 1026
Heart attacks, **1035**
Heart disease, 54, 55, 561
Heat, 108, **109**, 110, 1218
Heat-losing center, 883
Heat of vaporization, 27t, **28**
Heat-promoting center, 883
Heat shock proteins (HSPs), 51, 815
Heat transfer, 880–881, 880f
Heavy chains (polypeptide), **1071**, 1072f
Heavy metals
  phytoremediation for, **788**–789, 789f
  plant tolerance to, 416, 416f
Hedgehog signaling molecule, 1126
Height, as continuously varying trait, 232, 233f
Helical viruses, 528f, 529
Helicase, **267**, 268t, 269f
Helicobacter, 549f
Helicobacter pylori, 559t, 560, 989
Helium, 22, 23f
Helix-turn-helix motif, 50, 50f, **306**–307, 307f, 311
Helper T cells, 1064t, 1068, 1069–1070, 1071f
Hematocrit, 1021
Hematopoiesis, **1022**, 1064–1065
Heme groups, 1015
Hemidesmosomes, 82t, **83**
Hemiptera, 687t
Hemocoel, 669
Hemocyanin, **1015**
Hemoglobin
  affinity for oxygen, 1016, 1016f
  defined, **1015**
  dissociation of, 53
  in erythrocytes, 1021
  evolution of, 11–12, 11f
  functions of, 44, 45t
  pH and temperature, effect on, 1016, 1016f

  in sickle cell anemia, 249–250
  size of, 530f
  structure of, 46, 48, 49, 1015, 1015f
  variations of, 49
Hemolymph, 669, **1023**
Hemolytic disease of newborns (HDN), 1079
Hemophilia, 227t, 242–243, 242f, 249t
Hemorrhagic fever, 539
Hensen's node, 1126
Hepadnavirus, 531t
Hepatitis virus, 528, 531t, 535, 539
Herbicide-resistant crops, 347, 348f, 349
Herbivores
  defined, **984**
  digestive system of, 992, 993f
  plant defenses against, 798–800, 1195, 1195f
  teeth of, 986, 986f
  in trophic level ecosystems, **1217**, 1217f
Heredity. *See also* Inheritance
  DNA and, 13
  evolution and, 11
  family resemblance and, 221, 222f
  historical assumptions regarding, 221–222
Hermaphrodites, **663**, 669, 675, 676, 1087, 1087f
Herpes simplex virus, 530f, 531t
Herpes zoster virus, 528
Hershey–Chase experiment, 258–259, 258f
HER2 gene, 344–345, 345f
Heterochromatin, **190**, 320
Heterochrony, **498**
Heterocysts, 561
Heterokaryotic hyphae, **621**, 627–628
Heterotherms, 881
Heterotrimeric G proteins, 179, 179f
Heterotrophs, **123**, 139, 557, 568–569, 622, 639, **1216**
Heterozygote advantages, **410**–411, 411f
Heterozygotes, **225**–227, 233t, 402–403
Hexapoda, 681, 681t, 686–689, 686f, 687t
Hfr cells, **553**–554
HGP. *See* Human Genome Project
HGT. *See* Horizontal gene transfer
Hibernation, 884
High-density lipoprotein (HDL), 54, 1034
Highly active antiretroviral therapy (HAART), 537
Hill, Robin, 151
Hindbrain, 903, 903f, 903t
Hinge joints, 969, 970f
Hippocampus, 903t, 904, 906
Hirudo medicinalis, 676, 676f
Histamine, 168
Histidine, 47
Histograms, 232, 233f
Histone, **190**, 190f, 316, 317f
HIV/AIDS
  evolution of, 475–477f
  immune system, impact on, 534, 1081–1082, 1082f
  immunoassays, 346, 346f
  immunosuppression in, 1081–1082
  inactivation of, 337
  infection cycle, 534–536, 535f
  latency period of, 534
  monitoring of, 1080
  progression of, 1082
  reverse transcriptase in, 330, 530
  size of, 530f
  spread of, 534
  structure of, 531t
  testing for, 534

I-15 Index

transmission of, 475–476, 534, 1249
treatment of, 536–537, 537f
HLAs (human leukocyte antigens), **1068**
H.M.S. *Beagle* (ship), 1, 1f, 8, 9f, 10, 422
H1N1 virus, 538, 1081
H5N1 virus, 538, 1081
HOBBIT gene, in *Arabidopsis*, 851, 852f
Holoblastic cleavage, **1112**–1113, 1113f, 1113t
Holoenzyme, 284f, 285
Holothuroidea, 695, 695f
Holt–Oram syndrome, 502
Homeobox genes, 5, 391–393, 498
Homeodomain, 391
Homeodomain motif, **307**
Homeodomain proteins, 14, 14f
Homeosis, 498–499
Homeostasis
  antagonistic effectors and, 878–879, 878f
  of calcium, 954–955, 955f
  as characteristic of life, 3
  defined, **3**, 877, 1165
  gene expression and, 305
  negative feedback loops and, 877, 878f
  positive feedback loops and, 879, 879f
  signaling networks and, 14
Homeotherms, 881
Homeotic genes, **391**–392
Homeotic mutations, 307, 391, 499
Hominids, 488f, 724–726, 725f
Hominoids, 723f, **724**
Homo erectus, 726, 727
Homo floresiensis, 725–726f, 726–727
Homo (genus), 725, 726–729, 726f, 728f
Homo habilis, 725f, 726
Homo heidelbergensis, 727
Homokaryotic hyphae, **621**
Homologous behaviors, 469–470, 470f
Homologous chromosomes, **191**, 191f, 209–210, 209f
Homologous pairing, 215, 216f
Homologous recombination, 553
Homologous structures, **11**, 11f, 432, 432f, 502
Homologues, **191**, 210, 211, 215
Homo neanderthalensis, 482, 483f, 727
Homoplasies, **463**–464, 464f, 465, 470
Homoplastic features, 470, 471f, 505–506
Homoptera, 686f, 687t
Homo sapiens, 725, 725f, 726, 727–728, 1260
Homosporous plants, **593**
Homozygotes, **225**–227, 233t
Honeybees *(Apis mellifera)*
  altruism in, 1158, 1158f
  chromosomes in, 189t, 241t
  color vision in, 932
  dance language of, 1147–1148, 1148f
  reproduction in, 1087
Honeysuckle, 616f
Hooke, Robert, 12, 59
Hookworms, 680
Hooves, 720
Horizontal gene transfer (HGT), **490**, 490f, 546, 608, 609–610f, 610
Hormone-activated transcription factor, 946
Hormone response elements, **946**
Hormones. *See also specific hormones*
  in behavior, 1136
  blood volume regulation by, 1037
  cancer and, 959
  in cell signaling, 169f, **170**
  chemical classes of, 941
  circulating, 940–941

defined, 940
digestive, 995, 995f, 996t
endocrine glands associated with, 942–943t
of female reproductive hormones, 1095f
functions of, 44, 45t
hydrophilic, 941, 944, 944f, 947–948, 947f
in infertility treatments, 1104
in insects, 958f, 959
lipophilic, 941, 944, 944f, 945–946f, 945–947
of male reproductive system, 1095, 1095t, 1096f
osmoregulatory function control by, 1052–1053, 1053f
plant *(See* Plant hormones*)*
protein, 44, 45t
steroid receptors, 173–174, 173f
Horns, of animals, 720
Hornworts, 468f, 589f, 593, 593f, 596f
Horses
  body structure of, 866f
  chromosomes in, 189t
  digestion in, 720
  evolution of, 417, 417f, 429–431
  forelimbs of, 11f, 432f
  teeth of, 986f
  thoroughbred, 417, 417f
Horseshoe crabs, 681t, 684, 684f
Horsetails, 189t, 594, 594f, 597–598, 598f
Host range, of viruses, **528**
Host restriction, 328
Hot mutants, in *Arabidopsis*, 815
Hotspots, 1261–1263, 1263t
Hox genes, **391**–392, 392f, 498, 499, 502, 644, 851
HPV (human papillomavirus), 539, 540
HR (hypersensitive response), **800**–801, 801f
HSPs (heat shock proteins), 51, 815
Hubbard Brook Experimental Forest, 1214–1215, 1215f
Human chorionic gonadotropin (hCG), 1099, 1127
Human chromosomes
  DNA of, 271f
  images of, 189f
  karyotypes of, 191, 191f
  number of, 189, 189t, 250, 250–251f
  segmental duplication on, 487, 488–489f
  sex chromosomes, 241–243, 241t, 248f
  structure of, 190
Human diseases
  bacterial, 556, 557–561, 559t, 560f
  from flukes, 663–664, 664f
  fungal, 635
  global warming and, 1256
  from nematodes, 679f, 680
  viral, 538–540
Human evolution, 406, 436, 461, 461f, 462, 723–729
Human Gene Mutation Database, 250
Human genome
  classes of DNA sequences found in, 362–363, 363f
  comparative genomics and, 365, 480, 481f, 482, 483f, 488, 488f
  functional elements in, 364–365, 364f
  gene swapping evidence in, 490
  genetic mapping of, 247–248, 248f
  segmental duplication in, 487–489f
  sequencing of, 334
  single nucleotide, polymorphisms in, 248

size of, 360, 360f
transposons in, 490
Human Genome Project (HGP), 1, 4, 50, 239, 252–253, 334, 360
Human immunodeficiency virus. *See* HIV/AIDS
Human leukocyte antigens (HLAs), **1068**
Human papillomavirus (HPV), 539, 540
Human population
  decline in growth rate, 1183
  in developing and developed countries, 1182–1183, 1182t, 1183f
  exponential, 1180, 1180f
  future outlook, 1182–1183
  growth of, 1180–1183, 1182t
  in hotspots, 1262–1264, 1263f
  population pyramids, 1181–1182, 1181f
Human race, 728–729, 728f
Humans
  biosphere, impact on, 1247–1252
  birth weight in, 413, 413f
  cleavage in, 1114
  development in, 1127–1130f, 1127–1131
  dominant and recessive traits in, 227–228, 227t
  embryonic stem cells in, 382
  evolution of, 406, 436, 461, 461f, 462, 723–729
  extinctions due to, 1260
  eye color in, 232, 233f
  forebrain of, 903f, 903t, 904–906, 905f
  forelimbs of, 11f, 432f
  FOXP2 gene in, 491, 491f, 492
  gastrulation in, 1117
  genetic diversity of, 494f
  gene transcription patterns in, 491
  genome *(See* Human genome*)*
  global warming and health of, 1256
  homoplasy and, 464f
  influence on flower morphology, 840
  language of, 1148
  morphological data for, 463f
  plant toxins, susceptibility to, 797, 797f
  in prehistoric times, 1260, 1260f
  sexual differentiation in, 1088
  skin of, 919–920, 919f
  survivorship curve for, 1172, 1172f
  teeth of, 719, 719f, 986, 986f
  vestigial structures in, 433
Hummingbirds, 718, 843, 844f, 884
Humoral immunity, **1065**, 1070–1076, 1073t, 1075–1076f
Humus, **775**, **776**
Hunchback proteins, **388**, 390f, 391
Huntington disease, 227t, 249, 249t, 300, 899
Hurricane Katrina, 1251
Hyalin, 1110
Hybrid inviability, 442t
Hybridization
  fluorescent in situ hybridization, 344–345, 345f, 355, 355f
  gene flow and, 349
  between species, **222**, 442, 442f, 443, 444–445
Hybridoma cells, 1079
Hydra, 392, 652, 652f, 655, 984f, 1023, 1024f
Hydration shells, 28, 29f, 98
Hydrocarbons, 34, 53, 342, 519
Hydrochloric acid, gastric, 988–989
Hydrocortisone, 945f, 956
Hydrogen
  atomic structure of, 19f
  covalent bonds of, 24, 24f
  electronegativity of, 25t, 34

Hydrogenated fats, 54
Hydrogen bonds
  basis and strength of interaction, 23t
  defined, **26**
  in proteins, 46, 48, 48f
  in water, 26, 26–27f
Hydrogen ion concentration, measurement of, 29–30
Hydrogen peroxide, 72–73
Hydrolysis reactions, 37, 37f
Hydrophilic hormones, 941, 944, 944f, 947–948, 947f
Hydrophilic molecules, **28**
Hydrophobic exclusion, **28**–29, 46, 48, 48f
Hydrophobic interactions, 23t
Hydrophobic molecules, **28**
Hydroponics, 780–781, 782f
Hydrostatic skeletons, 652, 674, **964**, 964f
Hydrothermal vents, 1246
Hydroxyapatite, 965
Hydroxyl group, 35, 35f, 42, 43
Hydrozoa, 508, 509f, **654**–655, 655f
Hymen, 1100
Hymenoptera, 687t
Hypercholesterolemia, 227t, 249t
Hyperosmotic solutions, 98
Hyperpolarization, **893**–894, 894f
Hypersensitive response (HR), **800**–801, 801f
Hypersensitivity, delayed, 1077–1078
Hypertension, 1031
Hyperthermophiles, 546
Hyperthyroidism, 953
Hypertonic solutions, **98**, 99f, 1041
Hyperventilation, 1012–1014
Hyphae, 621, 621f, 622, 627–628
Hypolimnion, 1242
Hypoosmotic solutions, 98
Hypophysectomy, 952
Hypophysis, **948**. *See also* Pituitary gland
Hypothalamohypophyseal portal system, 950
Hypothalamus
  anterior pituitary regulation by, 950–951, 950f
  functions of, 903t, 906
  neurohormones produced by, 949, 950–951
  temperature detection in, **877**
  thermoregulation by, 883–884, 884f
Hypothesis-driven science, 5–6f, 5–7
Hypothyroidism, 953
Hypotonic solutions, **98**, 99f, 1041
Hyracotherium, 430, 430f, 431

I

IAA (indoleacetic acid), **819**–820, 819f
IBA (indolebutyric acid), **821**
I bands, 971, 972f
ICAM-1 protein, 535
Icefish, 433–434, 434f, 489
Ichthyosauria, 709t
Ichthyosaurs, 435, 709t
Ichthyostega, 707, 707f
ICM (inner cell mass), **1114**
Icosahedrons, 528–529f, **529**, 530
ICSI (intracytoplasmic sperm injection), 1104
Ileum, **989**
Illumina next-generation sequencing, 358f
Immune system, 1057–1082. *See also* Immunity
  antibodies in medical treatment and diagnosis, 1078–1080, 1080f
  cell recognition in, 82
  cells of, 1064t
  HIV/AIDS, impact on, 534, 1081–1082, 1082f

Immune system (*continued*)
  organs of, 1066–1067
  pathogens that invade, 1081–1082, 1082f
  in vertebrates, 876f, **877**
Immunity
  active, 1065, 1076f
  adaptive, 556–557, 1058, 1063–1067f, 1063–1068, 1064t
  autoimmunity and hypersensitivity, 1076–1078, 1078f
  cell-mediated, 1065, 1068–1070, 1068t, 1069f, 1071f
  humoral, 1065, 1070–1076, 1073t, 1075–1076f
  innate, 1057–1062, 1070
  passive, 1065
Immunoassays, 346, 346f
Immunoglobulin A (IgA), 1073t, **1074**, 1076, 1081
Immunoglobulin D (IgD), 1073t, **1074**
Immunoglobulin E (IgE), 1073t, **1074**, 1076, 1077
Immunoglobulin G (IgG), 1073t, **1074**, 1076
Immunoglobulin M (IgM), **1073**, 1073t, 1076, 1079
Immunoglobulins (Ig), **1065**, 1065f, 1071–1074, 1072f, 1073t
Immunohistochemistry, 61
Immunological tolerance, **1077**
Immunosuppression, 1081–1082
Implantable contraceptives, 1101f, 1101t, 1102
Implantation, **1127**
Imprinting, 251–252, 384, **1141**
Inbreeding, 405
Inclusive fitness, **1157**–1158
Incomplete dominance, 233t, **234**, 234f
Incomplete flowers, 839
Incus, **922**
Independent assortment, **214**, 229
Independent events, 230
Indeterminate development, 643f, **644**
Indian pipes (*Hypopitys uniflora*), 784, 784f
Indoleacetic acid (IAA), **819**–820, 819f
Indolebutyric acid (IBA), **821**
Induced fit, 114, 114f
Inducer exclusion, **310**–311
Induction
  cell differentiation and, **380**–381
  in development, 1119
  of phage, 533, 533f
  primary, 1126
  of proteins, **308**, 310, 312
  secondary, 1126
Inductive reasoning, **5**
Industrial melanism, 424–**425**, 424–425f
Inert elements, 22
Infections. *See* Disease
Inferior vena cava, **1030**
Infertility
  female, 1103
  male, 1103
  as postzygotic isolating mechanism, 442t
  treatment of, 1104
Inflammatory diseases, 345–346
Inflammatory response, 1060–1061, 1062f
Influenza virus
  of 1918, 371, 538
  H1N1 strain, 538, 1081
  H5N1 strain, 538, 1081
  H subtypes, 538
  N subtypes, 538
  origin of new strains, 538, 538f
  recombination in, 538
  size of, 530f
  structure of, 528f, 530, 531t
  types and subtypes of, 538

Infrared radiation, sensing of, 935–936, 935f
Ingenhousz, Jan, 150
Ingression, **1115**
Inhalant siphons, **671**, 671f
Inheritance, 221–236. *See also* Dominant traits; Heredity
  chromosomal theory of, 240–241, 240f, 244
  dihybrid crosses and, 228–229, 229f, 230–231
  epigenetic, 252
  epistasis and, 235–236, 236f
  historical assumptions regarding, 221–222
  Mendel's experiments on, 11, 222–228
  molecular basis of, 12–13
  monohybrid crosses and, 224–228, 230
  polygenic, 232, 233f, 233t
  probability and, 230–231
  recessive traits, 224–225f, 224–228, 227t
  testcrossing and, 231–232, 231f
Inheritance of acquired characteristics, **400**
Inhibiting hormones, **950**, 951
Inhibition, 1205
Inhibitors, 117, 117f
Inhibitory molecules, 1126
Inhibitory postsynaptic potential (IPSP), 899, 899–900f, 900–901
Initiation complex, 288, 288f, **294**, 294f, 313, 313f
Initiation factors, 294, 294f
Initiator tRNA, **294**, 294f
Injectable contraceptives, 1101t
Ink sac, of cephalopods, 671, 672, 672f
Innate behavior, 1135, 1135f
Innate immunity, 1057–1062, 1070
Innate releasing mechanism, 1135
Inner cell mass (ICM), **1114**
Inner ear, **922**, 922f
Inorganic phosphate, **113**
Inositol phosphates, 180–181f, 181
Inositol triphosphate (IP$_3$), **948**
Inositol-1,4,5- triphosphate (IP$_3$), 180–181f, 181
Insecticides, 347–348
Insectivores, digestive system of, 993f
Insectivorous leaves, 753–754
Insects. *See also* specific insects
  Bt crops resistant to, 348
  characteristics of, 681t
  chromosomes in, 189t
  cleavage in, 1112
  courtship behavior of, 443, 443f
  digestive system of, 688
  diversity of, 686–687, 686f, 687t
  exoskeleton of, 41
  external features of, 687–688, 688f
  eyes of, 417, 417f, 507, 507f, 682, 682–683f, 687, 930f
  fish predation on, 409, 410f
  heart in, 1023
  hormones in, 958f, 959
  internal organization of, 688
  locomotion in, 978
  nitrogenous wastes of, 1042, 1043f
  orders of, 686f, 687t
  osmoregulation in, 1044–1045, 1044f
  pheromones of, 689
  pollination by, 843, 843–844f
  respiration in, 1005f
  selection for pesticide resistance in, 407, 408f
  sensory receptors of, 688–689
  sex chromosomes of, 241, 241t
  social systems of, 1159, 1160f
  taste in, 928, 928f

thermoregulation in, 881, 881f
wings of, 505–506, 506f, 687–688, 688f
Insertion sequences, 553, 553f
Instantaneous reactions, 111
Instincts, learning and, 1139, 1140, 1140f, 1141–1142
Institute for Genomic Research, 360
Insulin
  antagonistic actions of, 957f
  blood glucose regulation and, 996–997, 997f
  for diabetes, 957
  discovery of, **956**–957
  function of, 45f
  obesity and, 998
  production using recombinant DNA, 343–344, 344f
Insulin-like growth factors, 944, **952**
Insulin receptors, 176, 176f
Insulin response proteins, 176
Integral membrane proteins, 89, 90f, 95f
Integrase inhibitors, 536, 537f
Integrins, **80**–81, 81f, 83, 395–396
Integumentary system, 876f, **877**
Integuments, **603**, 603f, 606, 611, 613
Intelligent design theory, 436–437
Intercalary meristems, **735**
Intercalated disks, **872**, 1028
Intercropping, 777
Interference competition, **1189**
Interferons, 1059–1060
Intergovernmental Panel on Climate Change (IPCC), 784, 1253, 1256
Interior protein network, 89–90, 91t
Interleukin-1 (IL-1), **1061**
Intermediate disturbance hypothesis, 1206, 1206f
Intermediate filaments, 36f, **76**, 76f, 79, 83
Intermembrane space, of mitochondria, **74**, 74f, 135
Intermolecular catalysis, 116
Internal chemoreceptors, 929
Internal fertilization, 1089–1090f, 1089–1091
Internal membranes, of prokaryotes, 552, 552f
Internal organs
  of mollusks, 668
  of vertebrates, 699, 699f
Interneurons, 874f, **889**, 889f
Internodes, **747**, 748f
Interoceptors, **917**
Interoparity, **1174**
Interphase, of cell cycle, **192**, 192–194f, 193–194
Intersexual selection, **1153**–1154
Interspecific competition, **1189**, 1192–1193, 1193f
Intertidal region, 1243f, 1244
Intestines, 989–992. *See also* Large intestine; Small intestine
Intracellular receptors, 171, 172t, 173–174, 173f
Intracytoplasmic sperm injection (ICSI), 1104
Intramembranous development of bone, 965, 966f, 967
Intramolecular catalysis, 116
Intrasexual selection, **1153**
Intrauterine devices (IUDs), 1101t, 1102
Intrinsic factor, **988**
Introduced species, 1272–1274
Introns
  defined, **289**, 289f
  distribution of, 290

in gene expression, 298t
in human genome, 363, 363t, 487
splicing of, 321f
Invaginate, **1115**
Inversions, **300**, 301f
Invertebrates. *See also* specific invertebrates
  circulatory system of, 1023–1024, 1024f
  development of, 379
  digestive system of, 984, 984f
  eyes of, 929–930, 929f
  loss of larval stage in marine invertebrates, 473–474, 474f
  osmoregulatory organs of, 1043–1044, 1044f
  prevalence of, 640
In vitro fertilization (IVF), 1104
In vitro mutagenesis, 336
Involute, **1115**
Iodine, 951
Iodine deficiency, 23
Ion channels
  chemically gated, **171**–172, 172f, 172t, 893, 894f
  defined, **892**
  gated, 97, 893–895f, 918
  ligand-gated, 893
  membranes of, 91t, **97**, 97f
  stimulus-gated, 918, 918f
  transient receptor potential, 919
  voltage-gated, 893, 894, 895f
Ionic bonds, 23–24, 23f, 23t, 48, 48f
Ionic compounds, 24
Ionization of water, 29
Ions, 19, 24, 28, 96–97
IPCC (Intergovernmental Panel on Climate Change), 784, 1253, 1256
IPSP (inhibitory postsynaptic potential), 899, 899–900f, 900–901
Iris, **930**, 930f
Irish potato famine, 578
Irreducible complexity argument, 437
Island biogeography, 1228–1229, 1228f
Island dwarfism, 727
Islands
  adaptive radiation on, 450, 450f
  evolution on, 435, 448, 448f
  extinctions on, 1261, 1269, 1269f
Islets of Langerhans, 956, 957, 990f, **991**, 997
Isocitrate, 132, 133f
Isoleucine, 47f
Isomerization, 132, 133f
Isomers, **35**, 35f, 38, 39f
Isopods, 681t, 684
Isoptera, 686f, 687t
Isosmotic regulation, 99
Isosmotic solutions, 98
Isotonic solutions, **98**, 99f, 1041
Isotopes, **19**–20, 19f, 428, 428f
Isotopic dating, 428, 428f, 519
IUDs (intrauterine devices), 1101t, 1102
IVF (in vitro fertilization), 1104
Ivy, 747f

J

Jacob, François, 309
Jacob syndrome, 251
Jagendorf acid bath, 159f
Jasmonic acid, **799**, 800f
Jaundice, 991
Jaws
  as ancestral vs. derived character, 462, 463f
  evolution of, 702–703, 702f
  of fish, 501, 501f, 701, 702–703, 702f, 1067–1068
  joints in, 970

Jejunum, **989**
Jellyfish
　classification of, 645t, 655, 655f
　eye development in, 507, 507f, 508, 509, 509f
　medusae of, 652, 652f
　radial symmetry of, 640
Jenner, Edward, 1063, 1063f
Jimsonweed (*Datura stramonium*), 843
Jointed appendages, of arthropods, 681, 682
Joints, **969**–970, 970f
Joules, 108
Junk DNA, 363, 364
Juvenile glaucoma, 227–228, 227f
Juvenile hormone, **959**, 959f
Juxtamedullary nephrons, **1048**–1049

K

KANADI gene, 751f
Kangaroos, 721f, 1092f
Karyogamy, 626–629f, 629
Karyotypes, **191**, 191f, 355f
Kaufmann, Thomas, 391
Kelps, 577–578, 577f
Keratin, 45t, 76, 710, 718, 720, 867
Keratinized epithelium, 867
Kettlewell, Bernard, 424, 425
Key innovations, **450**
Key stimulus, 1135
Keystone species, **1203**, 1203f, 1275
Khorana, H. Gobind, 283
Kidneys
　of amphibians, 1047
　of birds, 1047, 1047f
　excretion in, 1050
　filtration in, 1048–1049f, 1048–1050
　of fish, 1046–1047, 1046f
　hormonal regulation of, 1052–1053, 1053f
　of mammals, 1047–1049f, 1047–1052
　reabsorption in, 1048, 1048f, 1050, 1051f, 1052–1053, 1053f
　of reptiles, 1047
　secretion of hormones by, 958, 1048
　transport function of, 1050–1052, 1052f
　of vertebrates, **1045**–1047
Killer strains, of *Paramecium*, 577
Killifish (*Rivulus hartii*), 414, 414f, 415
Kilocalories, 108, **997**
Kinase cascade, **176**–177, 177f, 178
Kinases, 45t, 171f
Kinesin, 77, 77f
Kinetic energy, **108**, 108f
Kinetin, 822f
Kinetochores, **193**, 193f, 196, 211, 211f, 216
Kinetoplastids, 571–572, 572f
King, Thomas, 383
Kingdoms (taxonomic), 13, 13f
Kingdom (taxonomic), **466**, 466f
Kingfishers, 448, 448f
Kinocilium, **921**, 921f
Kinorhyncha, 648f
Kinorhynchs, 644
Kin selection, 1158–1159, 1159f
Kipukas, 452
Kirll, 645t
Klinefelter syndrome, 251, 251f
Knee-jerk reflex, 908, 909f
Knight, T. A., 222
Knockin mice, 339, 340
Knockout mice, 338–339f, 338–340
Knottedlike homeobox (*knox*) genes, 393
Koch, Robert, 545

Koi carp (*Cyprinus carpio*), 705f
Kölreuter, Josef, 222
Komodo dragon (*Varanus komodoensis*), 713, 1261
Krebs, Charles, 1179
Krebs cycle
　in aerobic respiration, 126f
　in amino acid catabolism, 141
　ATP production in, 131, 131f, 132, 133f
　electron transfer in, 132
　entry to, 116
　in glucose oxidation, 125
　in pyruvate oxidation, 130
　reductive, 519
　regulation of, 138, 138f
　steps of, 118, 131, 131f
Krill, 685
K-selected populations, **1179**–1180, 1179t
Kurosawa, Eiichi, 824
Kuru, 540

L

Labyrinths, **922**, 925–926
Lacewings (*Chrysoperia*), 443, 443f
lac operon, **308**, 308–310f, 309–311
lac repressor, 45t, 309–310, 309f
Lactase, 39, 436
Lactate dehydrogenase, 140
Lactation, 1130
Lactic acid fermentation, 129, 140, 140f
Lactose, 39, 308–310
Lactose intolerance, 39, 990
Lacunae, **871**
Lagging strands, 267f, **268**–270f, 272f
Lagomorpha, 722t
Lake habitats, 1241–1242f, 1241–1243
Lake Malawi cichlid fish, 501, 501f
Lake Victoria cichlid fish, 453–454, 454f, 1274
Lamarck, Jean-Baptiste, 400, 400f
Lambda, 532
Lamellae, 1007
Lamellipodia, 1115
Laminaria, 577f
Lampreys
　characteristics of, 702, 702t
　morphological data for, 462, 463, 463f
　phylogeny of, 700f
　vertebrae in, 701
Lamp shells. See Brachiopoda
Lancelets, 697–698, 698f
Land plants, 588–599
　adaptations to terrestrial life, 589
　bryophytes, 468, 468f, 589, 589f, 591–593
　evolution of, 589, 589f
　haplodiplontic life cycle of, 590, 590f
　horizontal gene transfer in, 608, 609–610f, 610
　innovations in, 589, 595, 596f
　origin of, 588–590
　reproduction in, 591, 592, 592f
　tracheophytes, 470, 589, 589f, 593–599, 593f, 595–599f
Langerhans, Paul, 956
Language function, 906, 907f
Large ground finches (*Geospiza magnirostris*), 9f, 422f
Large intestine, 985, 985f, 992, 992f
Large offspring syndrome (LOS), 384
Lariats, 290, 290f
Larva
　of crustaceans, 685, 685f
　dispersal in marine snails, 472–473
　of echinoderms, 694, 694f
　of fruit flies, **387**, 388f
　of liver flukes, 663, 664f

marine invertebrates loss of larval stage, 473–474, 474f
　of sea urchins, 497–498, 498f
　of sponges, 649, 651
　of tunicates, 697, 698f
Larvacea, 697
Larynx, **987**, 987f
Lassa fever, 371t
Latent viruses, **528**
Lateral gene transfer. See Horizontal gene transfer (HGT)
Lateral geniculate nuclei, **933**, 935f
Lateral line system, **703**, 921–922, 921f
Lateral meristems, **735**, 736f
Lateral root cap cells, 742
Late tertiary follicles, **1098**
LDL (low-density lipoprotein), 54, 103, 1034
Leading strands, 267f, **268**, 269–270f, 272f
Leaf-cutter ants, 633, 634f
Leafhoppers, 687t
Leaflets, 752
LEAFY COTYLEDON gene, in *Arabidopsis*, 852
LEAFY gene, in *Arabidopsis*, 833, 833f
Leafy sea-dragons, 701f
Learning, 907, 1137, 1137f, 1139–1140, 1140f, 1141–1142
Leaves. See also Stomata
　abscission of, 813–814, 814f
　alternate, 747, 747f
　arrangement of, 747, 747f
　of carnivorous plants, 782–784, 783f
　compound, 752, 752f
　drought tolerance of, 768, 768f
　establishing top and bottom of, 751, 751f
　evolution of, 595, 595f, 605
　genetic regulation of, 752f
　herbivory and, 799, 799f
　modified, 753–754
　opposite, 747, 747f
　of pine trees, 605
　simple, 751–752, 752f
　structure of, 148, 148f, 751–753f
　whorled, 747, 747f
Leber's hereditary optic neuropathy (LHON), 244
Leeches, 645t, 675–676, 676f
Leeuwenhoek, Antony van, 12, 59, 545
Legionella, 549f
Legs, of amphibians, 706, 706f
Leishmaniasis, 493f, 571
Lemurs, 435f, 723
Lens, **930**, 930f
Lenticels, 748, 749f
Leopard frogs (*Rana pipiens*), 444, 444f
Leopold, Aldo, 1222
Lepidoptera, 686f, 687t
Lepidosauria, 701f, **713**
Leprosy, 558, 559t
Leptin, 998, 998f
let-7 gene, 318
Lettuce (*Lactuca*), 443, 485f
Leucine, 45, 46, 47f
Leucine zipper motif, **307**, 307f
Leucoplasts, 75
Leukemia, 347, 354
Leukocytes, 871, **1021**, 1060
Levorotatory chiral forms, 35
Lewis, Edward, 391
LH. See Luteinizing hormone
LHON (Leber's hereditary optic neuropathy), 244
Lice, 688
Lichens, **631**–632, 632f
Life
　characteristics of, 2–3, 3f
　evidence of early life on Earth, 518–519, 519f

　extraterrestrial, 516
　hierarchical organization of, 3–4, 3f
　origins of, 515–518
　science of, 1–4
Life cycle
　Allomyces, 625f
　ascomycetes, 629, 629f
　basidiomycetes, 627–628, 628f
　brown algae, 577–578, 577f
　Chlamydomonas, 580f
　cnidarian, 653
　ferns, 598–599, 599f
　flowering plants, 610–612, 611f, 832f
　haplodiplontic, 577–578, 577f, 580, 580f, 590, 590f
　of liver flukes, 663, 664f
　mollusks, 669, 669f
　mosses, 592
　Paramecium, 576f
　pine trees, 605–606, 605f
　plants, 590, 590f
　Plasmodium, 575f
　sexual, 208, 208f
　Ulva, 579f
Life history, 1173–1174f, 1173–1175
Life tables, **1171**–1172, 1171t
Ligand-gated channels, **893**
Ligands, **168**, 169f, 171
Light
　as energy form, 151–152
　pathway in flowering plants, 834–835f, 834–836
　plant responses to, 804–808
Light chains (polypeptide), 1071, 1072f
Light-dependent reactions, of photosynthesis, **148**, 149f, 150, 151, 156–160
Light-harvesting complexes, 155, 155f, 157
Light-independent reactions, of photosynthesis, **148**, 150, 150f, 151
Light microscopes, 61, 62t
Light-response genes, 805–806
Lignin, 622, 624, 630, **740**
Limb development, 502, 503–504f
Limbic system, 901, 903t, **906**
Limpets, 473, 474, 474f, 670
Linear chromosomes, 271–272, 272f
LINEs (long interspersed elements), **363**–364, 364f
Lineus, 508, 646t, 672, 672f
lin-4 gene, 317–318, 318f
lin-14 gene, 317, 318, 318f
Linkage maps. See Genetic maps
Lions (*Panthera leo*), 107, 107f, 442, 442f, 721f, 986f, 1134f
Lipase, **990**
Lipid membranes, 88, 546, 547f
Lipidomics, 92
Lipid rafts, 90
Lipids. See also Fatty acids; Phospholipids
　bilayers of, 56, 56f
　caloric content of, 55
　catabolism of, 141–142
　defined, **53**
　as energy-storage molecules, 54–55
　as macromolecules, 33, 37, 37t
　structure of, 36f, 37, 53–55f
　transport within cells, 71
Lipophilic hormones, 941, 944, 944f, 945–946f, 945–947
Lipopolysaccharide (LPS), **551**, 551f, 1058
Little paradise kingfishers (*Tanysiptera hydrocharis*), 448, 448f
Liver
　cancer of, 539
　cirrhosis of, 664
　homeostasis and, 996

Liver (*continued*)
    secretions of, 990f, **991**
    smooth endoplasmic reticulum in, 70
Liver flukes, 663–664, 664f
Liverworts, 468f, 589f, 591, 591f, 596f
Lizards
    behavioral adaptations of, 1166, 1166f
    characteristics of, 709t, 713
    collared, 713f
    homoplasy and, 464f
    mating behavior of, 447, 447f
    morphological data for, 463, 463f
    phylogeny of, 701f
    regeneration in, 713
    reproduction in, 1087
    resource partitioning among, 1193, 1193f
    triploid populations of, 485
    variation in size of, 1174, 1174f
Lobe-finned fish, 700f, 701, 702t, 704, 705–706f
Lobeliads, 452, 452f
Lobsters, 41, 41f, 681t, 684, 685, 685f
Local anaphylaxis, 1077
Locomotion
    in air, 978–979, 979f
    animals, 639, 639f, 977–979
    annelids, 674
    appendicular, 977
    arthropods, 681
    axial, 977
    bipedalism, 463f, 724, 725–726
    on land, 978, 978f
    protists, 568
    segmentation and, 644
    in water, 977–978, 977f
Logarithmic scales, 29
Lolium perenne, 631, 631f
Long, noncoding RNA, 363, 363t
Long-day plants, **834**–835, 834f
Long interspersed elements (LINEs), **363**–364, 364f
Long-term depression (LTD), 907, 908f
Long terminal repeats (LTRs), **364**, 364f
Long-term memory, 907
Long-term potentiation (LTP), 907, 908f
Loop of Henle, 1047, **1049**, 1049f, 1051
Loose connective tissue, **869**, 869f, 870t
Lophophores, 646f, **661**, 661f, 676, 677–678f, 678
Lophotrochozoa, 666–678. *See also* Annelids; Mollusks
    Brachiopoda, 646f, 648f, 661, 676–678
    Bryozoa, 455, 645t, 648f, 661, 676–677, 677f
    classification of, 647, 648f, 661
    ribbon worms, 508, 509f, 646t, 672–673, 672f
Loratadine (Claritin), 168
Lorenz, Konrad, 1141, 1141f, 1148
Loricifera, 646t, 648f
Lorises, 723
LOS (large offspring syndrome), 384
Low-density lipoprotein (LDL), 54, 103, 1034
Lox sequences, 339–340, 339f
LTD (long-term depression), 907, 908f
LTP (long-term potentiation), 907, 908f
LTRs (long terminal repeats), **364**, 364f
Lubber grasshoppers (*Romalea guttata*), 686f
Lumen, of endoplasmic reticulum, **70**
Luna moths (*Actias luna*), 686f
Lung cancer, 1014, 1014f
Lungfish, 700f, 704
Lungs
    of amphibians, 706, 1009, 1009f
    as ancestral vs. derived character, 462, 463f

of birds, 718, 1010, 1011f
evolution of, 1008
of mammals, 1009–1010, 1010f
of reptiles, 1009
structure and function of, 1011–1012, 1012f
Lupines, 401f
Luteal phase of menstrual cycle, 1097f, **1099**
Luteinizing hormone (LH), **950**, 952, 1092, 1095, 1095t, 1096f
Lycophylls, 595, 595f
Lycophytes, 589f, 594, 594t, 595–597
Lyme disease, 548f, 558, 559t, 684, 1249–1250
Lymph, **1034**
Lymphatic system, 876f, 1033–**1034**, 1034f
Lymph hearts, **1034**
Lymphocytes, 1021, 1021f, **1065**, 1065f
Lyon, Mary, 316
Lysenko, T. D., 836
Lysine, 47f, 49, 116, 190, 316
Lysogenic cycle, of bacteriophage, **532**–533, 533f
Lysogenic phage, **532**, 533
Lysogens, **532**, 533
Lysogeny, 532
Lysosomal storage disorders, 71
Lysosomes, **71**–72, 72f, 78t, 81t, 323, 568
Lysozyme, 114f, 1058
Lytic cycle, of bacteriophage, 532, 533f

M

MacArthur, Robert, 1192, 1228
macho-1 gene, 380–381, 381f
MacLeod, Colin, 257
MAC (membrane attack complex), **1062**, 1062f
Macrocystis pyrifera, 577f
Macromolecules, 2f, 33, 36–37f, 37, 37t
Macronucleus, **575**, 576f
Macronutrients, for plants, **779**–780, 779t
Macrophages, 536, **1060**, 1060f, 1064t
Macular degeneration, 386
Mad cow disease, 323, 540
Madreporites, **694**
MADS-box genes, **393**, 498, 499, 504, 505f
Magnetic fields, detection of, 936
Magnoliids, 609f
Maidenhair trees (*Ginkgo biloba*), 606f
Maize. *See* Corn (*Zea mays*)
Major groove, in DNA, 262f, 306, 306f
Major histocompatibility complex (MHC), 82, **1066**, 1067f
Malaria
    drug developments for, 492, 493, 493f, 798
    eradication efforts, 574
    genome of, 360f, 481f
    prevalence of, 1256
    sickle cell anemia and, 250, 411, 411f
    vaccine for, 574, 1081
Malate, 132, 133f
Male infertility, 1103
Male reproductive system, 876f, 877, 1092–1096f, 1095t
Malleus, **922**
Malpighian tubules, **683**, 688, 1044, 1044f
Malthus, Thomas, 10
MALT (mucosa-associated lymphoid tissue), 1066, 1066f, **1067**
Maltose, 39, 39f
Mammalia, 466f, 700f, 718–722, 722t

Mammals, 718–722. *See also specific mammals*
    ancestral vs. derived characters of, 462, 463f
    apoptosis in, 394, 394f
    brain of, 904, 904f
    circulatory system of, 719, 1026, 1027f
    classification of, 645t, 720–722, 721f, 722t
    cleavage in, 1114, 1114f
    egg-laying, 471f, 721, 721f
    estrous cycle of, 1099
    evolution of, 700, 700f, 720
    features of, 718–720
    flying, 720, 720f
    gastrulation in, 1117, 1117f
    heart of, 1026, 1027f
    kidneys of, 1047–1049f, 1047–1052
    lungs of, 1009–1010, 1010f
    marine, 722t
    nitrogenous wastes of, 1042, 1043f
    nuclear reprogramming in, 383–385f
    orders of, 722t
    phylogeny of, 700f
    placental (*See* Placental mammals)
    plant digestion by, 720
    pouched (*See* Marsupials)
    reproduction in, 1092, 1092f
    respiratory system of, 719, 1005f, 1011–1014
    saber-toothed, 470, 471f
    teeth of, 719, 719f
    thermoregulation in, 718, 719
Mammary glands, 719
Mammoths, 494f
Manatees, 433
Mandibles, of arthropods, **681**, 681t
Manganese, 117, 157
Mangold, Hilde, 1124
Mangroves, 769, 769f, 1244
Manihotoxin, 796c
Mannose-binding lectin (MBL) proteins, 1059, 1062
Mantle, **667**, 668f
Mantle cavity, 1006
MAOA (monoamine oxidase-A), 1139
MAOs (monoamine oxidases), 1139
MAP kinases. *See* Mitogen-activated protein kinases
Maps
    genetic, 244–248, 353–357
    physical, 354–357
Map units, 246
Marburg hemorrhagic fever, 371t
Marchantia, 591f
Margulis, Lynn, 566
Marine habitats, 1243–1246f, 1243–1247, 1250–1251
Markers, cell. *See* Cell-surface markers
Markov, Georgi, 797
Marler, Peter, 1141
Marrow cavity, **968**
Marsilea, 598
Marsupials
    accessory sex organs in, 1100
    characteristics of, **721**, 721f
    evolution of, 435, 435f
    reproduction in, 1092, 1092f
    saber-toothed, 471f
Mass, atomic, 18–19
Mass extinctions, **456**–457, 456f, 520, 700. *See also* Extinctions
Mass spectroscopy, 368, 369f
Mast cells, **1064**, 1064f
Mastication, 970, 984, 986
Mastiffs, 427, 427f
Maternal inheritance, **244**
Maternity plants, 857

Mating and mating behavior. *See also* Courtship behavior
    assortative, 405
    disassortative, 405
    fitness and, 408–409, 409f
    in fungi, 177
    nonrandom, 404f, 405
    pheromones in, 443
    selection acting on, 447, 447f
    sexual selection and, 1153–1154
Mating systems, 1154–1156, 1155f
Matrix
    of connective tissue, **869**
    of mitochondria, 74, 74f, 134
Matter, 18
Maturation, zone of, **744**–745, 744f
Mature mRNA, 289f
Mayr, Ernst, 441, 448
MBL (mannose-binding lectin) proteins, 1059, 1062
Mccarty, Maclyn, 257
McClintock, Barbara, 245, 245f, 486
MDRs (multidrug-resistant strains), **558**
Measles, 531t
Mechanical isolation, 442t, 443
Mechanoreceptors, 917–920
Mediators, 313–314, 315f
Medicago truncatula, 485, 485f, 489f
Medications. *See* Drugs; Pharmaceuticals
Medicine
    antibodies in diagnosis and treatment, 180f, 1078–1080
    biotechnology applications to, 343–347
    genomic applications to, 370–371, 371t, 492–494
    regenerative, 346, 382
Medium ground finches (*Geospiza fortis*), 410, 423, 423f, 444–445
Medulla oblongata, **903**, 903t
Medullary bone, 967f, **968**
Medullary cavity, 967f, 968
Medusa, **652**, 652f, 653, 654–655
Meerkats (*Suricata suricata*), 1160, 1160f
Megakaryocytes, 1022
Megapascals (MPa), **759**
Megaphylls, **751**
Megaspores, 606, 610
Meiosis, 207–218
    in chromosomes, 207–208, 208f
    comparison with mitosis, 210, 211f, 215–218
    defined, **208**
    errors in, 214
    features of, 209–210, 209f, 215
    stages of, 210–214
Meiosis I, **209**, 210–212f, 214, 215–216, 216f
Meiosis II, **209**, 213f, 214, 215, 217f
Meissner corpuscles, 919f, 920
Melanin, **953**
Melanism, industrial melanism, 424–**425**, 424–425f
Melanocyte-stimulating hormone (MSH), **950**, 953, 999
Melanotropin-inhibiting hormone (MIH), 951
Melatonin, **941**, 958
Membrane attack complex (MAC), **1062**, 1062f
Membrane potential, 97, 891–894
Membrane receptors, 171–173
Membranes. *See* Plasma membranes
Memory, 907
Menarche, 1096
Mendel, Gregor
    experiments performed by, 11, 222–229, 239, 402
    extensions to work of, 232–236

portrait of, 223f
testcrossing by, 231–232, 231f
Mendeleev, Dmitri, 22
Mendelian modified ratio, 236, 236f
Mendelian ratio, **225**
Meninges, 908, 1067
Meningitis, 1067
Menstrual cycle
  egg production in, 1097
  follicular (proliferative) phase, 1097f, 1098
  luteal phase of, 1097f, 1099
  menstrual phase of, 1099
  onset of, 1096
  ovulation phase of, 952, 1097f, 1098–1099, 1099f
  secretory phase of, 1099
Menstruation, **1092**
Mercury pollution, 1248
Mereschkowsky, Konstantin, 566
Meristems
  apical, **735**, 735–736f, 747, 747f, 822f
  cell division in, 734–735, 734f
  defined, **734**
  in development, 378, 392
  floral, 836–839f
  ground, 735, 736f, 745, 852
  intercalary, 735
  lateral, 735, 736f
  primary, 735, 851
Merkel cells, 919f, 920
Meroblastic cleavage, 1113t, **1114**, 1114f
MERS (Middle Eastern respiratory syndrome), 1249
Merychippus, 431f
Meselson–Stahl experiment, 264–265, 264f
Mesenchymal stem cells, **347**
Mesenchyme, 380, **965**, 966f, **1115**
Mesoderm, 380, **642**, 642f, 865, 1115
Mesoglea, **652**, 652f
Mesohippus, 430f
Mesohyl, **650**
Mesophylls, **752**, 753f
Messenger RNA (mRNA)
  as biomarkers, 345
  5′ cap, 288, 288f
  cDNA created from, 330, 331–333, 331f
  defined, **42**
  degradation of, 321
  in eukaryotes, 318, 318–319f, 319, 320–321, 321f
  functions of, 42, 43, 69
  in gene expression, 281
  mature, 289f
  poly-A tail of, 288f, 289
  posttranscriptional control in, 288–289, 288f
  pre-mRNA splicing, 289–290f, 289–291
  quantification of, 333–334, 334f
  in translation (See Translation)
  transport from nucleus, 321, 322f
Metabolic pathways, 517–518
Metabolism. *See also* Cellular respiration
  aerobic, 129
  ATP in, 125
  basal metabolic rate, 997
  biochemical pathways of, 118–119
  defined, **117**
  electron carriers in, 124
  evolution of, 142–143
  fermentation in, 129
  of hydrocarbons, 342
  in prokaryotes, 546, 557–558
Metamorphosis, **387**, 388f, 640, 675, 689, 959

Metaphase
  meiosis I, 210–212f, 216f
  meiosis II, 213f, 214, 217f
  mitotic, 192, 192f, 195–197f, **196**, 211f, 216f
Metaphase plate, 195–196f, 196, 211, 211f
Metapopulations, **1169**–1170, 1169f
Metazoa, 648f, 649
Methamphetamine, 607
Methane, 139, 342, 342f, 1211, 1254
Methanococcus, 548f
Methanogens, 139
Methicillin-resistant *Staphylococcus aureus* (MRSA), 556
Methionine, 45, 47f, 283, 295
Methylation, 252, 316, 316f
Methyl-G cap, **288**
Methyl group, 35f
MHC (major histocompatibility complex), 82, **1066**, 1067f
MHC proteins, 45t, 82, **1068**–1069
Micelles, 56, 56f
Mice (*Mus musculus*)
  behavioral genetics in, 1138, 1138f
  chromosomes in, 189t
  coat color in, 407, 407f
  embryonic stem cells in, 382
  embryo of, 697f
  eye development in, 507, 508–509f
  FOXP2 gene in, 491, 491f, 492
  fur color in, 233, 233f
  genome of, 360f, 365, 481f, 482, 488, 491, 491f
  homeodomain proteins in, 14f
  homeotic genes in, 392f
  knockin, 339, 340
  knockout, 338–339f, 338–340
  limb development in, 502, 503f
  marsupial, 435f
  ob gene in, 998, 998f
  olfactory receptor genes in, 489
  transposable elements in, 490
Microalgae, 341–342, 341f
Microarrays
  DNA, 345–346, **366**–368, 367f
  protein, 368
Microbial resistance, selection for, 407–408, 408f
Microbiology, 544–545
Microbodies, **72**–73, 78t
Microclimates, 1237
Microfilaments, 76, 76f, 77
Microfossils, 517, 519f, 565–566
Micrognathozoa, 646t, 648f
Micronucleus, **575**, 576f
Micronutrients, for plants, **779**–780, 779t
Microorganisms. *See also* Eukaryotes; Prokaryotes
  bioremediation of hydrocarbons by, 342
  disease and, 545
  pathogenic, 1058
  in soil, 776
  in wastewater treatment, 342, 343f
Microphylls, **751**
Micropyles, **606**, 611
Micro-ribonucleic acid (miRNA), **281**, 318–320, 363, 363t
Microscopes
  in discovery of cells, 12
  invention of, 59, 545
  resolution of, 60
  types of, 61, 62t
Microsporangia, 605
Microspores, 605, 611–612
Microsporidia, 620, 620f, 623, 623f
Microtiter plates, 346, 346f
Microtubule-organizing centers, 76

Microtubules, **76**–77, 76f, 79–80, 81t, 188f, 193, 196
Microvilli, 867, 927, **989**, 990f
Midbrain, 903–904, 903f, 903t
Middigital hair, 227t
Middle ear, **922**, 922f
Middle Eastern respiratory syndrome (MERS), 1249
Middle lamella, **80**, 80f, 197
Miescher, Friedrich, 259
Migratory behavior, 1144–1145, 1255
MIH (melanotropin-inhibiting hormone), 951
Milk (lactation), 1130
Milk let-down reflex, 1130
Milk sugar, 39
Miller, Stanley L., 517, 517f
Miller–Urey experiment, 517, 517f
Millipedes, 389f, 639f, 640, 645t, 681t, 689
Mimicry, 1196–1198f
Mimivirus, 530
Minerals
  absorption by plants, 761–764f
  for plant nutrition, 779t, 780, 780f
  in soil, 775–776, 776f
  transport in plants, 758, 758–759f, 765
Minimal medium, 279, 556
Minor groove, in DNA, 262f, 306, 306f
Miracidium, **663**, 664f
miRNA. *See* Micro-ribonucleic acid
Missense mutations, **299**, 300f
Mites, 681t, 683, 684
Mitochondria
  aerobic respiration in, 136f
  defined, **74**
  division of, 74
  DNA of, 74, 74f
  energy cycle of, 162f, 163
  functions of, 73, 78t
  genetic code in, 284
  genome of, 370, 567
  intermembrane space of, 74, 74f, 135
  maternal inheritance and, 244
  matrix of, 74, 74f, 134
  origins of, 75, 75f, 522, 523f, 566, 567f
  of protists, 570, 571
  ribosomes of, 74f
  structure of, 73, 74, 74f, 81t
Mitogen, 176
Mitogen-activated protein (MAP) kinases, 176–177, 177f, 178, 182–183, 202, 203f
Mitosis
  in asexual reproduction, 569
  in cell cycle, **192**, 192f
  comparison with meiosis, 210, 211f, 215–218
  defined, **189**
  evolution of, 567–568
  in fungi, 621–622
  phases of, 192–198, 192f, 194–195f, 216f
Mitral valves, **1028**
Mixotrophs, 569
Mobile genetic elements, 363
Models, in explanation of living systems, 7
Molar concentrations, **29**
Molds
  bread, 279, 625, 626
  water, 573, 578, 624
Mole (animal), 435f
Molecular biology, 4, 13, 280, 280f
Molecular clock, **465**
Molecular cloning, 330, 330f, 555. *See also* Cloning

Molecular evidence, for evolution, 11–12, 11f
Molecular formulas, 24
Molecular motors, 77, 77f, 79
Molecules, 2f, **3**, 23, 24
Mole (measurement), **29**
Mollicutes, 64
Mollusca (phylum), 645t, 648f, 666–672
Mollusks, 666–672
  body plan of, 667–669
  Brachyury gene in, 502
  characteristics of, 645t
  circulatory system of, 669, 1023
  classes of, 669–672
  diversity of, 666–667, 667f
  economic significance of, 667
  evolution of, 667
  excretion in, 669
  eyes of, 433, 433f, 507, 507f, 929, 930f
  feeding and prey capture in, 668–669, 669f
  habitat of, 640
  internal organs of, 668
  locomotion in, 978
  nervous system of, 902f
  reproduction in, 669, 669f
  shells of, 666, 668, 671
  size of, 666–667
  trochophores of, 661
Molting animals, 647, 648f, 661, 661f, 682
Molting hormone, 958f, **959**
Molybdenum, 117
Monarch butterflies (*Danaus plexippus*), 1144, 1144f
Monkeys, 489, 723–724, 723f
Monoamine oxidase-A (MAOA), 1139
Monoamine oxidases (MAOs), 1139
Monoclonal antibodies, 1079–1080, 1080f
Monocots
  clade of, 609f
  leaves of, 751, 751f, 752
  roots of, 744f, 745
  stems of, 748, 748f
Monocytes, 1021, 1021f, **1064**, 1064t
Monod, Jaques, 309
Monoecious plants, **846**
Monogamy, 1154–1155
Monohybrid crosses, 224–228, 230
Monokaryotic hyphae, **621**, 627
Monomers, 36f, **37**
Mononucleosis, 531t
Monophyletic groups, **467**, 467f, 468, 568
Monosaccharides, 36f, **38**, 38f
Monosomy, 189, **250**
Monotremes, 471f, **721**, 721f, **1092**, 1092f
Monounsaturated fatty acids, 53, 54f
Monsoons, 1236
Moon snails, 668
Morels, 620t, 629, 629f
Morgan, Thomas Hunt, 240–241, 245, 246
Morning after pill (birth control), 1102
Morphine, 795, 796t
Morphogenesis, 376, 386, **393**–396, 852–853, 853f
Morphogens, **387**–388, 390, 1112, 1124
Morphological adaptations to environmental change, 1165, 1165f
Mortality, **1170**
Mortality rate, **1172**
Mosquitos. *See also* Anopheles mosquitos; Malaria
  chromosomes in, 189t
  DDT-resistant, 574
  genome of, 360f, 481f
  larva of, 687, 688f

Mosquitos (*continued*)
  mouthparts of, 687, 688f
  sound and, 689
Mosses, 468f, 483, 589f, 590, 591–593, 592f, 596f
Moths
  characteristics of, 687t
  life history of, 689
  peppered, 424–425
  pheromones in, 443
  as pollinators, 844f
  thermoregulation by, 881, 881f
  wings of, 505, 687
Motifs, of proteins, **50,** 50f, 95, 370f
Motor effectors, 889
Motor neurons, 874t, **889,** 889f, 975
Motor proteins, 77, 77f, 194, 196, 973
Motor units, **975,** 975f
Mouthparts
  of arthropods, 681t
  in digestion, 986–987
  of insects, 687, 688f
MPF (M phase-promoting factor), 198–200f, 200, 201
M phase, 192, 194–198
M phase-promoting factor (MPF), 198–200f, 200, 201
mRNA. *See* Messenger RNA
MRSA (methicillin-resistant *Staphylococcus aureus*), 556
MSH (melanocyte-stimulating hormone), **950,** 953, 999
Mucigel, **742**
Mucoromycotina, 625
Mucosa, of gastrointestinal tract, **985,** 985f
Mucosa-associated lymphoid tissue (MALT), 1066, 1066f, **1067**
Mucus, 1058
Mucus-secreting cells, 988
Mules, 444
Mullein (*Verbascum thapsis*), 860
Müller, Fritz, 1197
Müllerian mimicry, **1197,** 1197f
Multicellularity
  in animals, 639
  cell cycle control and, 201–202, 202f
  development and, 376
  in eukaryotes, 201–202, 202f, 523, 523f
  in plants, 590, 590f
  in protists, 569
Multidrug-resistant strains (MDRs), **558**
Multienzyme complexes, **115**–116, 115f, 130
Multiple sclerosis, 346, 347
Multipotent cells, **381**
Muscle contraction, 971–976f, 971–977
Muscle fatigue, 140, **977**
Muscle fibers
  development of, 393
  in filaments, 972
  nuclei of, **872**
  types of, 975–976, 976f
Muscles
  determinants in tunicates, 379–380, 379f
  lactic acid accumulation in, 140
  metabolism during rest and exercise, 976–977
  organization of, 971f
  skeletal, 871–872, 872f, 970
Muscle spindles, **920,** 920f
Muscle tissue, **865,** 865f, 871–873, 872f, 873f
Muscular dystrophy (Duchenne), 227t, 249t
Muscularis, of gastrointestinal tract, **985,** 985f
Muscular system, 875–876, 875f

Musculoskeletal system, 963–979
  bone, 965–969
  joints, 969–970, 970f
  modes of locomotion and, 977–979
  muscle contraction, 971–976f, 971–977
  overview, **875**–876, 875f
  types of, 964–965
Mushrooms
  characteristics of, 620t
  growth of, 619, 619f
  harmful properties of, 627, 628f
  nutrients for, 621, 622
  reproductive structures of, 622
  spores produced by, 628, 628f
Mussels, 667, 670
Mutagen, **273**
Mutagenesis, 336
Mutations
  cancer and, 203–204f, 204, 361
  chromosomal, 300–301, 301f
  evolution and, 301, 404, 404f, 415–416
  frameshift, 283
  homeotic, 307, 391
  interactions among evolutionary forces and, 415–416, 416f, 501, 502
  point, 299–300, 335–336
  in prokaryotes, 556
  types of, 299–301
Mutualism
  animals and fungi in, 633–634, 634f
  coevolution and, 1198–1199, 1199f
  defined, **1198**
  plants and animals in, 798–799
  plants and fungi in, 627, **631,** 632
  prokaryotes and, 562
Mutually exclusive events, 230
Mycelium, **621,** 621f, 627–628
Mycobacterium leprae. See Leprosy
Mycobacterium tuberculosis. See Tuberculosis
Mycology, **621**
Mycoplasma, 64, 370, 371f, 559t
Mycorrhizae, 627, **632**–633, 633f, 782
Mycorrhizal associations, **591,** 632–633
Myelin sheath, 873, **891,** 891f
Myoblasts, 393
Myocardial infarctions, 1035
MyoD1 gene, 394
Myofibrils, **872, 971,** 971f
Myofilaments, **971,** 971–973f, 972–973
Myoglobin, 45f, **976,** 1015–1016
Myomeres, 698
Myosin, 44, 45t, **79,** 972, 972f, 973
Myriad Genetics, 372
Myriapoda, 681, 681t, 689, 689f
Myxini, 700f, 702t
Myxobacteria, 549f

N

NADH
  in ATP yield, 137, 137f, 141
  in electron transport chain, 132, 133f, 134, 135
  from glycolysis, 126–128f, 127
  from Krebs cycle, 131, 131f, 132, 133f
  in oxidation–reduction reactions, 123, 123f
  from pyruvate oxidation, 130, 130f
  recycling into NAD^+, 129–130, 129f
  structure of, 125f
NADH dehydrogenase, 134, 134f
NAD^+ (nicotinamide adenine dinucleotide)
  as electron acceptor, 125
  as electron carrier, 124, 125f

  functions of, **44**
  in oxidation–reduction reactions, 123–124, 123f
  regeneration of, 129–130, 129f, 140
  structure of, 125f
NADPH, in photosynthesis, 148, 149f, 151, 156–159f, 160
NADP reductase, 158, 159f
Nail fungus, 635
Naive lymphocytes, 1065
Naked mole rats (*Heterocephalus glaber*), 1160
Nannippus, 430–431f
Nanog gene, 385
Nanos proteins, **387**–388, 389f
Naphthalene acetic acid (NAA), 821
Nathans, Daniel, 328
National Center for Biotechnology Information (NCBI), 401, 490
Natural killer (NK) cells, **1060,** 1061f, 1064t
Natural selection
  adaptation to environmental conditions, 1166
  defined, **10**
  evidence of, 422–425
  as evolutionary mechanism, 8–10, 400, 406–407, 436–437
  experimental studies of, 413–415
  quantifying, 408–409
  reproductive isolation and, 445–446, 446f
  role in maintaining variation, 409–411, 410f, 445
  in speciation, 447
  testing predictions of, 10–12
Natural transformation, 555, 556f
Nauplius larva, **685,** 685f
Nautilus, 667f, 671
Navigation by sound, 924–925
NCBI (National Center for Biotechnology Information), 401, 490
NDI (nephrogenic diabetes insipidus), 98
Neanderthals, 482, 483f, 727
Nearsightedness, 931f
Necrosis, **394**
Nectar, 610
Nectaries, 610
Negative feedback loops, **877,** 878f, 951, 951f, 1177
Negative frequency-dependent selection, 409–411
Negative gravitropic response, 809f, 810
Negative pressure breathing, **1009**
Negative-strand viruses, 530
Negative transcriptional control, **308**–310
Neisseria gonorrhoeae. See Gonorrhea
Neisseria meningitidis, 1067, 1081
Nematocysts, 645, 652, 652f, **653**
Nematoda (phylum), 645f, 649f, **678**–680
Nematodes
  circulatory system of, 1024f
  classification of, 645t, 661
  digestive system of, 679, 984, 984f
  diseases caused by, 679f, 680
  as fungi nutrients, 622, 622f
  lifestyle of, 680
  parasitic, 678, 679f, 680
  as plant parasites, 793, 793–794f
  reproduction in, 679
  structure of, 679, 679f
Nemertea, 646f, 648f, 672–673, 672f
Neocallimastigomycetes, **624**–625, 633
Neocallismastigomycota, 620, 620f, 624–625
Neodermata, 663

Neohipparion, 431f
Neonates, 1130
Neotiella rutilans, 209f
Neotyphodium, 631, 631f
Nephridia, **669,** 674, 1044, 1044f
Nephrogenic diabetes insipidus (NDI), 98
Nephrons, **1045,** 1045f, 1048–1050
Nephrostomes, **669,** 1044
Nereis virens, 673f
Neritic waters, 1243f, 1244
Nerve cord, dorsal, 379, 380f, **696,** 696–697f
Nerve growth factor (NGF), 202, 394f, 944
Nerve nets, **903**
Nerve stimulation of muscle contraction, 974–975
Nerve tissue, **865,** 873, 874t
Nervous system, 888–913
  acoel flatworms, 656
  arthropods, 682, 682–683f, 902f, 903
  in bilaterally symmetrical animals, 641
  central (*See* Central nervous system (CNS))
  cnidarians, 902–903, 902f
  earthworms, 902f, 903
  echinoderms, 694, 902f
  flatworms, 662f, 663, 902f, 903
  heart rate and, 1035
  mollusks, 902f
  nerve impulse transmission, 891–897, 893t, 896t
  neurons and supporting cells of, 890–891
  organization of, 889–891
  peripheral, 873, 889, 890f, 910–913
  synapses of, 897–901f, 897–902
  in vertebrates, **874,** 875f
Net primary productivity (NPP), **1217**
Neural crest, **699,** 1121–1123, 1122f
Neural groove, **1120,** 1120f
Neural plate, 1120, 1120f
Neural tubes, **1120**–1121, 1120f
Neuroendocrine reflex, **949**
Neurofilaments, 76
Neuroglia, **873,** 891
Neurohormones, **940**–941, 949, 950–951
Neurohypophysis, **948**
Neuromodulators, **900**
Neuromuscular junction, **898,** 899f, 975
Neurons, **873,** 889, 889–890f, 890. *See also specific types of neurons*
Neuropeptides, **900,** 999
Neurospora
  Beadle and Tatum's experiment with, 279–280, 279f
  chromosomes in, 189t
  nutritional mutants in, 279–280
Neurotransmitters
  in behavior, 1136
  in cell signaling, 169f, **170**
  chemical compounds as, 898–899f, 898–900
  in drug addiction and, 901–902, 901f
  in synaptic vesicles, 898
Neurotropins, **944**
Neurulation, **395,** 1120–1121, 1121f
Neutrons, 18, 19, 19f
Neutrophils, 1021, 1021f, **1060, 1064,** 1064t
Newton, Isaac, 5, 7
Newts, 705t
New World monkeys, 489, 723, 723f
New York City, watersheds of, 1265, 1266f
Next-generation sequencing (NGS), 334, 357–359, 358f

N-formylmethionine, 294
NGF (nerve growth factor), 202, 394f, 944
NGS (next-generation sequencing), 334, 357–359, 358f
Niche, ecological, 1189–1194
Nicolson, Garth J., 89
Nicotinamide adenine dinucleotide. See NAD+
Nicotine, 901–902
Nicotine receptors, 901, 902
Nieuwkoop, Pieter, 1125
Nif genes, 561
Nile perch, 454, 1274, 1274f
Nimravids, 471f
9 + 2 structure, 79, 79f
Nipah virus, 1249
Nirenberg, Marshall, 283
Nitric oxide
   in blood pressure regulation, **1037**
   intracellular receptor for, 174
   as neurotransmitter, 900
Nitrification, **557**, 561, **1213**
Nitrogen
   covalent bonds of, 24f
   electronegativity of, 25t
   electron energy levels for, 23f
   in plants, 781–782, 781f, 784–786
Nitrogenase, 561
Nitrogen cycle, 1212–1213, 1213f
Nitrogen fixation
   in metabolic evolution, 143
   in nitrogen cycle, **1213**, 1213f
   in plants, 781–782, 781f
   prokaryotes and, 561–562
   in soil, 561–562
Nitrogenous bases
   in nucleotides, 42, 42f, 259–260, 259f
   tautomers of, 260–261
Nitrogenous wastes, 1042, 1043f, 1213
Nitroglycerin, 174
Nitrosomonas, 549f
Nitrous oxide, 1254
NK (natural killer) cells, **1060**, 1061f, 1064t
NMDA (*N*-methyl-D-aspartic acid), 907
N-methyl-D-aspartic acid (NMDA), 907
NMP (nicotinamide monophosphate), 124
NNRTIs (nonnucleoside reverse transcriptase inhibitors), 536, 537
Noble gases, 22
Nociceptors, **919**
Noctiluca, 573f
Nodes of Ranvier, 873, 890f, **891**
Nodes (plant stem), **747**, 748f
Nodules (plant), **781**, 781–782f
No-name virus, 539
Noncoding sequences, 362–364, 363t, 487, 490
Noncompetitive inhibitors, **117**, 117f
Noncyclic photophosphorylation, **157**–158, 158f, 160
Nondisjunction
   of autosomes, 250, 250f
   defined, **250**
   in meiosis, 214
   of sex chromosomes, 250–251, 251f
Nonequilibrium state, of living systems, 14
Nongranular leukocytes, **1021**
Nonnucleoside reverse transcriptase inhibitors (NNRTIs), 536, 537
Nonpolar covalent bonds, **25**
Nonpolar molecules, 28–29
Nonprotein enzymes, 116
Nonrandom mating, 404f, 405
Nonsense mutations, **299**, 300f
Nonspecific repair mechanisms, 274–275, 274f
Nonsteroidal anti-inflammatory drugs (NSAIDs), 560, 945, 1077

Nonvertebrate chordates, 697–698, 698f
Norepinephrine, **900**, 940
Normal distributions, 232
Northern elephant seals, 406, 406f
No-till planting, 347, 777
Notochords
   development of, 380, 501–502, 502f, 1120
   features of, **696**, 696–697f
   replacement by spinal column, 645t
   in tunicates, 379, 379–380f
NPP (net primary productivity), **1217**
NRTI (nucleoside reverse transcriptase inhibitors), 536, 537
NSAIDs (nonsteroidal anti-inflammatory drugs), 560, 945, 1077
Nucellus, **606**
Nuclear envelope
   breakdown of, 195
   defined, **62**
   in eukaryotes, **65**, 68, 68f, 188f, 566
Nuclear lamins, 65, 68f
Nuclear pores, **65**, 68f
Nuclear receptors, **174**
Nuclear receptor superfamily, 174
Nuclear reprogramming, 383–387f
Nucleic acids. *See also* DNA (deoxyribonucleic acid); RNA (ribonucleic acid)
   defined, **42**
   discovery of, 259
   function of, 41
   as macromolecules, 33, 37, 37t
   structure of, 36f, 37, 42, 42f
   viruses and, 528, 528f
Nuclein, 259
Nucleoids, **62**, 187, 546, **552**
Nucleolus, **65**, 68, 68f, 78t
Nucleoside reverse transcriptase inhibitors (NRTI), 536, 537
Nucleosomes, **190**, 190f, 316, 574
Nucleotides
   defined, **42**
   in DNA, 12, 13, 259–260, 259f
   function of, 37t, 43–44
   numbering carbon atoms in, 259–260, 259f
   origins of, 518
   in RNA, 259–260, 259f
   structure of, 36f, 42, 42f
Nucleus
   of atoms, 18, 19, 19f
   cellular, 62, 65–68, 78t, 81t
   cloning of animals and, 384, 384–385f
   defined, **65**
   of fungi, 621
   origin of, 566, 566f
   reprogramming of, 383–387f
   transport of RNA out of, 321, 322f
Nudibranchs, 670, 670f
Numbats, 435f
Nurse cells, 387, 388–389f
Nüsslein-Volhard, Christiane, 387, 389f
Nutrients
   for animals, 639
   essential, 999–1000, 1000t
   for fungi, 619, 621, 622–623, 622f
   limiting, 1214
   for mollusks, 668–669
   for plants, 779–781, 779t, 780–781f
   for protists, 568–569
Nutritional deficiencies, in fish, 701
Nutritional mutants, **279**–280, 279f

O

Oak trees (*Quercus*), 442, 741, 752f, 833f
Obesity, 997, 998

ob gene, in mice, 998, 998f
Obligate long-day plants, 835
Obligate short-day plants, 835
Obligate symbiosis, **631**, 633
Occipital lobe, 904, 905f
Ocean circulation, 1235, 1235f
Oceans, 1243–1244, 1244f
Ocelli, **682**, 683f
Ocelots, 435f
Octet rule, **22**, 24
Octopuses, 645t, 666, 667, 667f, 668, 671–672
Odonata, 687t
Offspring
   parent–offspring interactions, 1141, 1141f
   trade-off between number and investment, 1174, 1174f
Oil
   in corn kernels, 426
   in plants, 53, 55, 795
Oil-degrading bacteria, 562
Oil glands, 867, 1058
Oil spills, 562
Okazaki fragments, **268**–270f
Old World monkeys, 723f, 724
Olfaction (smell), 489, 928–929, 928f
Olfactory receptor (OR) genes, 489
Oligochaetes, 673f
Oligodendrocytes, **891**
Oligosaccharins, 816f, 825
Oligotrophic lakes, 1204, 1204f, 1242, 1242f
Oligotrophic oceans, 1244, 1244f
Ommatidia, 417, 417f, **682**, 683f
Omnivores, **984**, 986f
Oncogenes, 175, **204**, 337, 361, 539–540
One-gene/one-enzyme hypothesis, 6, 280
One-gene/one-polypeptide hypothesis, 6
1000 Genomes Project, 401
On the Origin of Species by Means of Natural Selection (Darwin), 8, 10, 399, 440
Onychophora, 649f
Onychophorans, 640, 644, 646t
Oocytes
   primary, 1097, 1098f
   secondary, 1098, 1098f
Oogenesis, 387, 1096f, 1098f
Oomycetes, 573, 578
Open circulatory system, **642**, 1023, 1024f
Open reading frames (ORFs), 362
Operant conditioning, **1140**
Operators, 308
Opercular cavities, 1006, 1006f
Operculum, **704**
Operons, **286**
Ophiuroidea, 695, 695f
Opiates, 900
Opisthokonta, 522, 522f, 523, 568, 569f, 584–585, 585f
Opisthosoma, **683**
Opium poppies (*Papaver somniferum*), 796t
Opossums, 189t, 721, 721f
Opportunistic infections, 534
Opposite leaf pattern, 747, 747f
Optic tectum, 903–904
Optimal foraging theory, **1150**
Optimum pH, 116, 116f
Optimum temperature, 116, 116f
Oral contraceptives, 1101f, 1101t, 1102
Orangutans
   evolutionary relationships of, 461, 461f, 724
   FOXP2 gene in, 491, 491f
   genome of, 486, 488, 488f

Orbitals, of electrons, **19**, 20, 20f, 21
Orchids, 746, 746f, 813, 840, 840f
Ordered complexity, as characteristic of life, 2–3
Order (taxonomic), **466**, 466f
ORFs (open reading frames), 362
Organelles, 2f, **3**, 62, 65
Organic compounds, 22–23, 139–140, 140f
Organic matter
   origins of, 516–517
   in soil, 776, 776f
   transport in plants, 770, 770f
Organismal level of organization, 3, 3f
Organizers, 1124–1126
Organogenesis, 1108f, 1118–1123
Organs and organ systems
   evolution of, 641–642
   in hierarchical organization, 3, 3f
   of immune system, 1066–1067
   in vertebrates, 865, 865f, 874–877
oriC site, 266, 269, 271
Orientation, migratory behavior and, 1144–1145
Origin of replication, 187–188, 187f, 266, 266f, 271
Origins of life, 515–518
Ornithischia, 709t
OR (olfactory receptor) genes, 489
Orthologues, 487, 488
Orthoptera, 682f, 686f, 687t
Oscillating selection, 410
Osculum, **650**, 650f
Osmoconformers, **1041**
Osmolarity, 1040–1042, 1041f
Osmoregulation, 663, 1043–1045, 1052–1053, 1053f
Osmoregulators, **1042**
Osmosis, 98–99, 104t, **758**
Osmotic balance, 99, 1040–1042, 1041f
Osmotic concentration, **98**, 98f
Osmotic pressure, **98**, 99f, 1041
Osmotic proteins, 45t
Ossicles, **694**
Osteoblasts, **965**, 966f
Osteoclasts, 966f
Osteocytes, **871**, 966f
Osteoporosis, 969
Ostracoderms, 702, 702t
Ostriches, 718
Otoliths, **922**
Outcrossing, 842, 846–847, 846f
Outer bark, 748
Outer ear, **922**, 922f
Outgroups, 462–463
Ovaries (plant), 603, **610**, 610f, 839f, **840**
Overexploitation, 1271–1272
Oviparity, 1089
Ovoviviparity, 1089
Ovulation, **952**, 1097f, 1098–1099, 1099f
Ovules, **603**, 603f, 839f, **840**
Oxaloacetate, 131, 132, 133f, 141
Oxidation
   β-oxidation, 141, 142f
   in cellular respiration, 123–126
   electrons and, **21**, 109, 109f
   of glucose, 25, 123–127
   in Krebs cycle, 132, 133f
   without oxygen, 139–140, 139f
   of pyruvate, 130, 130f, 138
Oxidation–reduction (redox) reactions, 21, **109**, 109f, 123–124, 123f
Oxidative phosphorylation, **126**, 126f
Oxygen
   atmospheric, 521, 521f
   atomic structure of, 19f
   covalent bonds of, 24f
   electronegativity of, 25t

Oxygen (continued)
  in freshwater ecosystems, 1240–1241
  oxidation without, 139–140, 139f
  partial pressure, 1008–1009
  in respiration, 1004
Oxygenic photosynthesis, 143, 148, 157, 521, 546, 557
Oxyhemoglobin, 1015, 1015–1016f, 1016
Oxytocin, 45t, **949**, 1095t
Oyster mushrooms (*Pleurotus ostreatus*), 622–623, 622f
Oysters
  classification of, 645t, 666
  evolution of, 429
  as food source, 667
Ozone, 273, 521, 1251–1252, 1251f

P

Paal, Arpad, 817
Pace, Norman, 116
Pacific giant octopus (*Octopus dofleini*), 672f
Pacific yew (*Taxus brevifolia*), 796t, 797
Pacilitaxel, 796t, **797**–798
Pacinian corpuscles, 919f, 920
Pain receptors, 919
Paired appendages, of fish, 701
Pair-rule genes, 389f, **391**
paleoAP3 gene, in plants, 504, 504f, 505
Paleogenomics, 727
Paleopolyploids, **484**
Palila, 1273f
Palindromes, 328
Palisade mesophyll, 752, 753f
Palolo worms, 675
PAMPs (pathogen-associated molecular patterns), 1058
Pancreas, 990–991, 990f
Pancreatic amylase, **990**
Pancreatic duct, 990, 990f
Pancreatic hormones, 956–957, 957f
Pancreatic juice, 985
Pandoravirus, 530
Pangea, 515, 521, 608
Papermaking, 741
Parabasalids, 570, 570f
Parabronchi, 1010
Paracrine regulators, **940,** 944–945
Paracrine signaling, 169, 169f, **170**
Paralogues, 487
Paramecium
  cellular compartmentalization of, 522f, 546
  competitive exclusion among, 1190–1191, 1191f
  conjugation of, 576, 576f
  features of, 576, 576f
  flagella of, 80f
  killer strains of, 577
  life cycle of, 576f
  osmotic balance, strategies for maintaining, 99
  predation by *Didinium*, 1194, 1194f
Paramyxoviruses, 531t
Paraphyletic groups, **467,** 467f, 700, 704
Paraphyly, 469f
Parapodia, **674**–675, 674f
Parasites
  brood, 1142, 1142f
  effect on competition, 1202
  epiparasites, 633
  external, 1199, 1200f
  flatworms, 662, 663–664, 1200
  fungi, 634–635
  internal, 1199–1200
  manipulation of host behavior, 1200f

nematodes, 678, 679f, 680
pathogens vs., **631**
plants as, 784, 784f
protists, 570, 571–572, 574, 578, 583
roots, 746
Parasitism, 562, **1198,** 1199–1200, 1200f
Parasitoids, **1199**
Parasitoid wasps, 799, 799f
Parasympathetic nervous system, 889, 890f, 911, 912–913f, 912t
Parathyroid gland, 953
Parathyroid hormone (PTH), **955,** 955f
Paratyphoid fever, 558
Parazoa, 648f, 649–651, 650f
Parenchyma cells, **739,** 739f, 741, 742, 745
Parental care, homologous, 469–470, 470f
Parental investment, **1152**
Parent–offspring interactions, 1141, 1141f
Parietal cells, 988, 988f
Parietal eyes, 713
Parietal lobe, 904, 905f
Parietal pleural membrane, **1011**
Parkinson disease, 323, 386, 900
Parnassius imperator, 688f
Parsimony, principle of, **463,** 464–465, 464f
Parthenogenesis, **1087**
Partial diploids, 554
Partial pressure, 1008–1009
Particle bombardment, 341
Passeriformes, 715t, 717f
Passive immunity, 1065
Passive transport, **96**–99, 104t, 758f
Pasteur, Louis, 6, 545, 1063
Pathogen-associated molecular patterns (PAMPs), 1058
Pathogen recognition, 800
Pathogens. *See also* Disease
  bacteria as, 558
  genomic research on, 371, 371t, 492–493, 493f
  of immune system, 1081–1082, 1082f
  parasites vs., **631**
  of plants, 624, 624f, 628
Pattern formation, 376, 386–393
Pattern recognition receptors (PRRs), 800
Paulinella chromatophora, 582, 582f
Pauling, Linus, 48
Pavlov, Ivan, 1140
Pavlovian conditioning, **1140**
Pax6 gene, 14f, **507**–509f, 509
PBDs (peroxisome biogenesis disorders), 72
PCNAs (proliferating cell nuclear antigens), 271
PCRs (polymerase chain reactions), 332–336
PDGF (platelet-derived growth factor), **202,** 204
Pearls, 668
Peas. *See* Garden peas (*Pisum sativum*)
Peat mosses, 592
Pectins, 40
Pedicels, 610, 610f
Pedigree analysis, **227,** 227–228f, 242–243, 242f, 252
Pedipalps, **683**
Peer review, 8
Pellicles, 575, 576, 576f
Pelvic inflammatory disease (PID), 561, 1103
Pelycosaurs, 710, 711f
Penetration, of virus particles, **532,** 533f
Penguins, 1091f
Penicillin, 64, 70, 407, 551
Penicillium, 628
Penis, 1093, 1094, 1095f

Pentaradial symmetry, **693,** 694
Peppered moths (*Biston betularia*), 424–425
PEP (phosphoenolpyruvate), 125f, 128f, 164, 165
Pepsin, 116f, 117, **988,** 989
Pepsinogen, **988**
Peptic ulcer disease, 559t, 560
Peptide bonds, **46,** 46f, 295, 296, 296f, 298f
Peptide hormones, **949,** 950
Peptides, 941
Peptidoglycan, 64, **546,** 550, 550–551f
Peptidyl transferase, **293**
Peregrine falcons (*Falco peregrinus*), 1279, 1279f
Perennial plants, **859,** 859f
Perforin, 1060
Pericardial cavity, **866,** 866f
Pericarps, 614, 615f
Pericentriolar material, **76**
Pericycle cells, 744f, **745**
Periderms, 748, 749f
Periodic table, 22–23, 22f
Periods, of time, 514f, 515
Peripheral chemoreceptors, **929**
Peripheral membrane proteins, 89, 90f
Peripheral nervous system (PNS), **873,** 889, 890f, 910–913
Perissodactyla, 722t
Peristalsis, **987,** 988f
Peritoneal cavity, 866, 866f
Peritubular capillaries, **1049,** 1049f
Periwinkle, 747f
Permafrost, 1240
Pernicious anemia, 988
Peroxisome biogenesis disorders (PBDs), 72
Peroxisomes, **72**–73, 72f
Pesticide resistance, selection for, 407, 408f
Pest-resistant crops, 347–348
Petals
  development of, 504–505
  morphology of, **610,** 610f
Petioles, 751
petunia (*Petunia hybrida*), 822–823
PGA (3-phosphoglycerate), 128f, 161, 161f, 162
6-PGD gene, 418
p53 gene, **202**–203, 203f, 204
PGPR (plant growth-promoting rhizobacteria), 793
pH
  of blood, 1016, 1016f
  enzymes and, 53, 116–117, 116f
  of rainwater, 1248, 1248f
  scale for, 29–30, 30f
  of soil, 777–778, 778f
  of urine, 1050
PHABULOSA gene, 751f
Phage, **258**–259, 258f. *See also* Bacteriophage
Phage conversions, 533–534
Phagocytosis, 71, 72, 102–**103,** 102f, 104t
Phagotrophs, 568–569
Pharmaceuticals. *See also* Drugs
  development of, 492–493, 493f
  genomic applications to, 371, 492–493, 493f
  from plants, 1264, 1264f
  plant secondary metabolites and, 797–798
Pharyngeal pouches, **696,** 696f
Pharyngeal slits, **696,** 697f
Pharynx, 679, 679f, **696,** 697f, 985, 985f
Phase change, in plants, **832**–833, 833f
Phase-contrast microscopes, 62t
PHAVOLUTA gene, 751f
Phenotype frequencies, 402, 402f

Phenotypes, **226,** 365, 406, 408, 409–410
Phenylalanine, 45, 46, 47f
Phenylketonuria (PKU), 249t, 252
Phenylthiocarbamide (PTC) sensitivity, 227t
Pheromones, **443,** 624, 689, 940, 1147
Phloem
  defined, 589
  plant transport in, 770–772f
  primary, 744f, 745
  vascular, **594,** 741–742, 741f
Phloem loading, **771**
Phlox, 446, 446f, 843
Phoronids, 677, **678,** 678f
Phosphatases, **170,** 171f
Phosphate group, 35, 35f, 42, 42f, 55, 259–260
Phosphate-to-oxygen ratio (P/O ratio), 137–138
Phosphatidylinositol-4,5-bisphosphate (PIP_2), 180f, 181
Phosphodiester backbone, of DNA, 261, 261–262f
Phosphodiester bonds, 42, 42–43f, **260,** 260f
Phosphoenolpyruvate (PEP), 125f, 128f, 164, 165
Phosphofructokinase, 128f, 138, 138f
2-Phosphoglycerate, 128f
3-Phosphoglycerate (PGA), 128f, 161, 161f, 162
Phospholipase C, 179, 180–181f, 181, 183
Phospholipids
  bilayers of, 89, 91t, 92–93
  functions of, 37t
  in membranes, 55–56, 89, 89f, 91t, 92–94
  structure of, 89, 89f, 92
Phosphorus cycle, 1214, 1214f
Phosphorylase kinase, 182, 182f
Phosphorylation
  in cell cycle control, 200–201
  defined, **170**
  oxidative, 126, 126f
  of proteins, 170–171, 171f, 198–199, 204
  substrate-level, 125–126, 125f, 127, 132, 133f
Phosphorylation cascade, 176–177, 177f
Phosphotyrosine, 176
Photic zone, **1241,** 1241f
Photoautotrophs, 557, 588, **1216**
Photobioreactors, 341–342, 341f
Photoefficiency, 153
Photoelectric effect, **152**
Photoheterotrophs, 557
Photolyase, 274, 274f
Photomorphogenesis, **805**
Photons, 149, **151**–152
Photoperiods, 834–835f, **834**–836
Photopigments, 931
Photopsins, **931**
Photoreceptors
  defined, 929
  in mollusks, 433, 433f
  sensory transduction in, 932–933, 934f
  in vertebrates, 432, 433f, 931–933
Photorepair, 274, 274f
Photorespiration, **163**–165, 752, 785, 785f, 786
Photosynthesis, 147–165
  anoxygenic, 143, 148, 156, 546, 557
  in bacteria, 64, 64f, 148, 150, 156, 156f, 545, 546, 557
  Calvin cycle in, 160–163
  carbon fixation in, 148, 149f, 151, 160–163
  carbon levels in plants and, 785–786
  chemical reaction of, 25

in chloroplasts, 148–149, 148f
C_3 photosynthesis, 161, 164, 164f
C_4 photosynthesis, 164–165, 752
defined, 147
discovery of, 149–151, 150f
electron transport system in, 149, 156, 159f
evolution of, 139, 143, 156, 520, 521
forms of, 148
light-dependent reactions of, 148, 149f, 150, 151, 156–160
oxygenic, 143, 148, 157, 521, 546, 557
photorespiration and, 163–164
pigments in, 151–154, 571, 572f
in plants, 148–149, 148f
saturation of, 154, 154f
stages of, 148, 149f
sunlight in, 108
Photosystem I, **157**, 158, 158–160f, 160
Photosystem II, **157**, 158–160f, 160
Photosystems
of bacteria, 156, 156f
in chloroplasts, 151, 156–160
evolution of, 156
organization of, 154–155
pigment molecules in, **148**–149, 149f
of plants, 156–160
Phototrophs, 569
Phototropin 1 (PHOT1), **808**, 808f
Phototropism
in fruit flies, 412–413, 413f
in plants, 807–808, 807f, 819f
pH scale, 29–30, 30f
Phycobiliproteins, **153**–154
Phyllotaxy, **747**
Phylogenetics
comparative biology and, 469–475
disease evolution and, 475–477f
species richness and, 474–475, 474f
Phylogenetic species concept (PSC), **468**–469, 469f
Phylogenetic trees, **12**, 461, 461f
Phylogenies
of animals, 644–649f
defined, **460**
evolutionary relationships and, 460–461, 461f, 469
of fungi, 620, 620f, 620t
of vertebrates, 700f
Phylum (taxonomic), **466**, 466f
Physcomitrella patens, 483, 592, 597
Physical defenses, of plants, 792–794
Physical maps, **354**–357
Physiological adaptations to environmental change, 1165, 1165t
Phytoaccumulation, 787f
Phytoalexins, **801**
Phytochrome, 805–806f, **805**–807
Phytodegradation, 787f
Phytoestrogens, 796t, **797**
Phytophthora infestans, 578
Phytoplankton, 1241
Phytoremediation, **787**–789
Phytovolatilization, 787f
Piciformes, 715t
PID (pelvic inflammatory disease), 561, 1103
PIF (prolactin-inhibiting factor), 951
PI gene, in plants, 504, 504f, 505
Pigments, photosynthetic, **151**–154, 571, 572f
Pike cichlids (*Crenicichla alta*), 414–415f, 415
Pili, **551**–552, 551f
Pill bugs, 684
Pilobolus, 620t
Pineal gland, 958
Pine cones, 605, 606

Pine needle, 605
Pine trees, 604–605f, 604–606
Pinnately compound leaves, 752f
Pinocytosis, 102–**103**, 102f, 104t
Pinworms, 645t, 680
Pitcher plants *(Nepenthes)*, 754, 782, 783f
Pith, 744f, 745, 748
Pit organs, **935**–936, 935f
Pits, of tracheids, 740f, 741
Pituitary dwarfism, **952**
Pituitary gland
anterior, 948–953
defined, **948**
posterior, 948–949, 949f
PKA (protein kinase A), 180, 180f, 182, 182f
PKC (protein kinase C), 181, 181f
Placenta
defined, **719**
formation of, 1114, 1127, 1128f
functions of, 1127
hormonal secretion by, 1128, 1129f
structure of, 719, 719f, 1128f
Placental mammals
characteristics of, 721–722, 721f
evolution of, 435, 435f
orders of, 722t
reproduction in, 1092, 1092f
saber-toothed, 471f
Placoderms, 702–703, 702t
Plague, 371t, 557, 559t
Planarians, 508, 509, 645t, 661, 984
Planet formation, 17
Plantae (kingdom), 13, 13f
Plant cells
cell walls of, 41, 67f, 80, 80f, 81t, 733–734, 734f
cytokinesis in, 197, 198f
mitosis in, 195
structure of, 67t, 81t
Plant defenses, 792–802
by animals, 798–799
chemical defenses, 794–798, 795f, 796t, 797f
against herbivores, 798–800, 1195, 1195f
pathogen-specific, 800–802, 801f
physical, 792–794
systemic responses, 799–802
toxins, 795–797, 796t, 797f
Plant diseases
bacterial, 558
fungal, 634–635, 793, 794f
nematodes and, 793, 793–794f
Plant growth-promoting rhizobacteria (PGPR), 793
Plant hormones, 815–828
for bud growth, 822–823, 823f
in cell division and differentiation, 821–822
in defense signaling, 825
for dormancy and suppression of growth, 827–828, 828f
for elongation and organization of body plan, 815, 817–821
flowering, 836
as fruit ripening aids and defenses, 825–827, 827f
functions of, 816–817t
for growth and nutrient utilization, 815, 823–824, 824f
similarity with animal hormones, 825
transport in phloem, 770, 771
Plantlets, 857
Plant receptor kinases, **175**
Plant reproduction, 831–860
asexual, 857–858
embryo development, 848–854
flowers and (*See* Flowering plants; Flowers)

germination, 854–856f
life spans, 858–860, 859f
pollination and fertilization, 842–848f
reproductive development, 832–833
Plants. *See also specific plants*
annual, 859, 859f
asexual reproduction in, 857–858, 857f
biennial, 859–860
biofuels from, 341
CAM plants, 164, 165, 165f
carnivorous, 782–784
cells (*See* Plant cells)
chilling of, 814–815
chromosomes in, 189t
circadian rhythms in, 808, 812, 812–813f
classification of, 468, 468f
cloning, 857–858, 858f
coevolution of animals and, 795, 798
conducting tubes in, 470, 471f
C_3 plants, 164, 164–165f
C_4 plants, 164–165
defenses of (*See* Plant defenses)
development in, 377–378
digestion by mammals, 720
diseases of (*See* Plant diseases)
dormancy in, 813–814, 814f
drought tolerance of, 768, 768f
ecological isolation of, 442
embryonic development in, 848–854
evolution of, 392
flooding, responses to, 768–769, 769f
flowering (*See* Flowering plants)
food storage in, 746, 853–854, 854f
fruit formation in, 610, 614, 614f
genome of, 482–486f, 485
glaciations and, 521
gravitropism in, 809–810
heavy metal tolerance by, 416, 416f
hormones (*See* Plant hormones)
hybridization between species of, 444
land, 588–599 (*See also* Land plants)
leaves, 751–754 (*See also* Leaves)
life cycles of, 590, 590f
life span of, 858–860, 859f
light, responses to, 804–808
morphogenesis in, 852–853, 853f
nutritional adaptations of, 781–784
nutritional requirements of, 779–781, 779t, 780–781f
organization of plant body, 733–736
parasitic, 784, 784f
pathogens of, 624, 624f, 628
pattern formation in, 392–393
perennial, 859, 859f
photomorphogenesis in, 805
photosynthesis in, 148–149, 148f
photosystems of, 156–160
phototropism in, 807–808, 807f, 819f
phytoremediation of, 787–789
plasmodesmata in, 67f, 82t, 84–85, 85f
polyploidy in, 449, 485, 485f
primary plant body, 735
primary tissues of, 735
reproduction in (*See* Plant reproduction)
roots, 742–746 (*See also* Roots)
in saline conditions, 769, 769f
secondary growth in, 735, 748, 749f
secondary metabolites of, 795, 796t, 797–798
secondary plant body, 735
secondary tissues of, 735
seed, 602–613 (*See also* Seed plants)
spacing of, 807
stems, 746–750 (*See also* Stems)

symbiotic relationship with fungi, 591, 627, 631–633
temporal isolation of, 443
thermotolerance in, 815
thigmotropism in, 811, 811f
tissue cultures, 858
tissues of, 594, 733–734f, 733–742, 745f, 851–852
transformation of, 340–341, 340f
transgenic, 340–341, 340f, 347–349
transport mechanisms (*See* Plant transport)
turgor movement in, 811–812, 812f
vacuoles of, 65, 73, 73f, 81t
vascular (*See* Vascular plants)
vegetative propagation of, 748–750
woody, 745, 748, 748f
wound response in, 799–800, 800f
Plant transport, 757–772
absorption in, 761–764f
in drought conditions, 768, 768f
in flooding conditions, 768–769, 769f
mechanisms of, 758–761
of minerals, 758, 758–759f, 765
in phloem, 770–772f
in saline conditions, 769, 769f
transpiration and, 758, 766–767
of water, 758–761
in xylem, 764–765
Planula, **653**, 655
Plaques. *See* Biofilms
Plasma, **1020**, 1021, 1021f
Plasma cells, 1064t
Plasma membranes, 88–104
active transport across, 99–102, 104t
of archaebacteria, 64–65, 546
of bacteria, 64, 93, 546
bulk transport across, 102–103
component groups of, 89–90, 91t
defined, **62**, 88
electron microscopy of, 91–92
endoplasmic reticulum vs., 93–94, 93f
of eukaryotes, 66–67f, 78t
fluid mosaic model of, 89, 90f, 546
lipids in, 88, 546, 547f
passive transport across, 96–99, 104t
phospholipids in, **55**–56, 89, 89f, 91t, 92–94
of prokaryotes, 63–64f, 64, 546
proteins in, 88, 94–95
of protists, 568
structure of, 62–63, 62f, 81t, 88–92
Plasmids
antibiotic resistant genes on, 556
cloning vector, 330–331
conjugative, 552–554, 553f
in prokaryotes, 546
tumor-inducing, 340, 347
Plasmodesmata, 67f, 82t, **84**–85, 85f, 581, 741
Plasmodial slime molds, 583, 584f
Plasmodium, **583**
Plasmodium, 574, 575f, 798, 1081. *See also* Malaria
Plasmolysis, **759**
Plasticity, synaptic, 907
Plastids, **75**
Plastocyanin, 158
Platelet-derived growth factor (PDGF), **202**, 204
Platelets, 202, **1020**, 1021–1022, 1021f
Plate tectonics, 515, 520
Platyhelminthes, 645t, 648f, 662–665, 662f, 664–665f, 903
Platyzoa, 647, 648f, 661, 662–666, 662f, 664–666f
Pleiotropic effect, 232–**233**, 233t, 417
Plesiomorphies, **463**
Plesiosaura, 709t
Plesiosaurs, 432, 709t

Pleural cavity, **866,** 866f, **1011**
Pleurotus ostreatus, 622–623, 622f
Plexuses, 985, 985f
Plumatella, 677f
Pluripotent stem cells, 381–382, 384–386, 385f, **1022**
Pluteus, 497–498, 498f
Pneumatophores, 746, 747f, 769
Pneumocystis jiroveci, 635, 1082
Pneumonia, 257, 539, 558, 559t
PNS. *See* Peripheral nervous system
Poa annua, 1171t
Podia, 582, 582f
POF (premature ovarian failure), 1103
Poikilotherms, 881
Poinsettias, 835f
Point mutations, **299**–300, 335–336
Point-source pollution, 1247
Poison glands, 684
Polar bears (*Ursus maritimus*), 494, 494f
Polar body, **1098**
Polar covalent bonds, **25**
Polarity, in development, **386,** 387, 388, 389f, 390
Polarized character states, **462**–463
Polar molecules, 25, 27, 28, 34
Polar nuclei, 842, 842f
Polar tubes, 623, 623f
Poliovirus, 529–530f, 530, 531t
Pollen
  allergies and, 168
  dispersal of, 840, 843, 843f, 845, 845f
  production of, 611–612
Pollen grains
  in fertilization, 847
  formation of, 612, **841,** 841–842f
  germination of, 603–604
  in pine trees, 605, 605f, 606
  transport of, 603
Pollen tubes, **603,** 606, 612, 612f, 847, 848f
Pollination, 842–847
  by animals, 843–844f, 843–845
  by bats, 1198f
  by birds, 843–844, 844f
  cross-pollination, 223, 223f, 349, 842
  defined, **612**
  in flowering plants, 611–612, 612f
  by insects, 843, 843–844f
  in pine trees, 605f, 606
  self-pollination, 612, 842, 845–846
  by wind, 842–843, 845, 845f
Pollinators, 603, **842**
Pollution
  biosphere, impact on, 1247–1252
  from coal burning, 1248
  diffuse, 1248
  of freshwater habitats, 1248
  habitat loss and, 1269
  of marine habitats, 1250–1251
  peppered moths and, 424, 425
  phytoremediation for, 787–789
  point-source, 1247
Polyandry, 1155
3′ poly-A tail, 288f, **289,** 321
Polychaeta, 674–675, 674f
Polychaetes, 645t, 673–675f, 674–675
Polyclonal antibodies, 1079
Polydactyly, 227t, 406
Polygenic inheritance, **232,** 233f, 233t
Polymerase chain reactions (PCRs), 332–336
Polymerases, 45t. *See also* DNA polymerases; RNA polymerases
Polymers, 36f, **37,** 39, 40f
Polymorphic variation, 401f
Polymorphic viruses, 530
Polymorphisms
  in DNA sequence, 401
  in enzymes, 401

restriction fragment length, 401
single nucleotide, 248, 367, 368, 371, 401
Polynucleotides, 42, 43f
Polypeptides, 36f, **46**
Polyphyletic groups, **467,** 467f
Polyplacophora, 669–670, 670f
Polyploidy
  defined, **449**
  elimination of duplicated genes and, 486, 486f
  fertilization and, 1110
  gene expression and, 486
  genome evolution and, 484–486
  in plants, 485, 485f
  speciation through, 449, 449f
  synthetic, 484–485
  transposon mobilization and, 486
Polyps, of cnidarians, **652,** 652f, 653, 654–655
Polysaccharides, **39**–41, 40f
Polyspermy, 1110
Polytrichum, 592f
Polyubiquitination, 323
Polyunsaturated fatty acids, **53,** 54, 54f
Polyzoa, 645t
Pond habitats, 1241–1242f, 1241–1243
Pons, **903,** 903f
Popper, Karl, 7
Populational level of organization, 4, 4f
Population cycles, 1178–1179, 1179f
Population dispersal, 1167–1169
Population genetics, **400**
Population growth
  factors affecting, 1170, 1171f
  in hotspots, 1262–1264, 1263f
  limitations by environment, 1175–1176f, 1175–1177
Population pyramids, **1181**–1182, 1181f
Populations
  age structure of, 1170–1171
  change through time, 1172
  defined, **4, 1166**
  demography and dynamics of, 1170–1172, 1171f
  geographic variation within, 441, 441f
  metapopulations, 1169–1170, 1169f
  range of, 1167–1168
  spacing patterns of, 1168–1169
  survivorship curves for, 1172, 1172f
Population size
  density-dependent effects, 1177–1178
  density-independent effects, 1178, 1178f
  extinction and, 1276–1278
  human, 1180, 1180f
Porifera (phylum), 509f, 645t, 647, 648f, **649**–651, 650f
Porin proteins, 95
Porphyrin rings, 152–153, 152f
Porpoises, forelimbs of, 11f, 432f
Portuguese man-of-war, 654, 655f
Positive feedback loops, 879, 879f, 952, 1178
Positive frequency-dependent selection, 410, 410f
Positive gravitropic response, 810, 810f
Positive pressure breathing, **1009**
Positive-strand viruses, 530
Positive transcriptional control, **308**
Postanal tail, **696,** 696–697f
Posterior pituitary gland, 948–949, 949f
Postganglionic neurons, 911
Postsynaptic cells, 897, 900–901, 900f
Posttranscriptional control
  alternative splicing in, 320, 321–322f
  in eukaryotes, 317–322f
  RNA editing in, 320–321
  small RNAs in, 317–320
  transcripts in, 320–322f

Postzygotic isolating mechanisms, **442,** 442t, 444, 444f
Potassium channels, voltage-gated, 894, 895f
Potatoes
  eyes of, 750, 857
  genome of, 485f
  Irish potato famine, 578
  as tubers, 750
Potential energy, 21, **108,** 108f, 110
Pouched mammals. *See* Marsupials
The Power of Movement of Plants (Darwin), 817
Power stroke, 973
Poxviruses, 530, 530f
p53 protein, 202, 203–204f
Prader–Willi syndrome (PWS), 251–252, 492
Prairie chickens (*Tympanuchus cupido pinnatus*), 1277–1278, 1277f
Prairie dogs (*Cynomys ludovivianus*), 1148f
Prairies, **1239**
Prebiotic chemistry, 517
Precocial young, **1155**
Predation
  defined, **1194**
Predators and predation
  animal defenses against, 1196, 1196f
  avoidance of, 407, 407f
  coevolution and, 1195
  competition reduction by, 1201–1202, 1202f
  evolution of prey populations, 414–415
  fish, 409, 410f, 414–415
  population cycles and, 1178–1179
  prey populations and, 1194–1197
  search images for prey, 409
  selection to avoid, 407, 407f
  sharks, 703, 703f
  species richness and, 1227
Predictions, in experiments, 5–6f, 6–7
Preganglionic neurons, 911
Pregnancies, high-risk, 252
Premature ovarian failure (POF), 1103
Pre-mRNA splicing, 289–290f, 289–291
Pressure–flow hypothesis, 770–772f
Pressure potential, **760,** 760f
Presynaptic cells, 897
Prey defense, 1139
Prezygotic isolating mechanisms, **441,** 442–443f, 442–444, 442t, 446
Priestly, Joseph, 149–150
Primary carnivores, **1217,** 1217f
Primary cell walls, **80,** 80f
Primary induction, **1126**
Primary lymphoid organs, **1066,** 1066–1067f
Primary meristems, **735,** 851
Primary mesenchyme cells, **1115**
Primary motor cortex, **905,** 905f
Primary mycelium, 627
Primary oocytes, **1097,** 1098f
Primary phloem, 744f, 745
Primary plant body, **735**
Primary producers, **1217,** 1217f
Primary productivity, **1188, 1217,** 1223–1225, 1224f, 1226, 1226f, 1238, 1238f
Primary somatosensory cortex, **905**–906, 905f
Primary structure, of proteins, **46,** 49f, 52f, 53
Primary succession, **1204,** 1204f
Primary tissue, **735,** 865
Primary transcript, 289, 289f
Primary xylem, 740, 744f, 745
Primates
  bushmeat market and, 476
  characteristics of, 722t

comparative genomics and, 482, 488, 488f
defined, **723**
evolution of, 723–726f, 723–729, 728f
language of, 1148, 1149f
transmission of HIV/AIDS to humans, 475–476
Primers, 268–269, 332, 335
Priming reactions, 127, 128f
Primitive streak, **1117**
Primordium, **610,** 610f
Primosome, 269
Principle of Independent Assortment, **229**
Principle of Segregation, **226,** 228
Prions, **540**–541, 541f
PRL (prolactin), **950,** 953, 1095t
Probability, 230–231, 437
Proboscidea, 722t
Procambium, **735,** 736f, 853
Procellariformes, 715t
Processivity, of DNA polymerases, 268
Prochloron cells, 64, 64f
Productivity
  defined, **1217**
  primary, 1188, 1217, 1223–1225, 1224f, 1226, 1226f, 1238, 1238f
  secondary, 1218
  species richness and, 1227
Product rule, **230**
Products, in chemical reactions, 25
Progesterone, 817, 958, 1095t
Proglottids, **665,** 665f
Progymnosperms, 602–603
Prokaryotes, 544–562. *See also* Archaebacteria; Bacteria
  adaptive immunity for, 556–557
  anaerobic respiration by, 139
  beneficial, 561–562
  cell division in, 187–188, 546
  cell walls of, 63, 63–64f, 64, 81t, 546, 550–551f, 550–552
  chromosomes of, 546
  classification of, 547–548
  defined, **62**
  disease-causing, 556, 557–561, 559t, 560f
  diversity of, 544–548
  DNA of, 62
  earliest, 519
  eukaryotes vs., 545–546
  flagella of, 63–64f, 65, 81t, 546, 550, 551, 551f
  gene expression in, 304, 305, 308–312, 547
  genetics of, 552–557
  genome of, 360f
  internal organization of, 552, 552f
  metabolic diversity of, 546, 557–558
  mutations in, 556
  plasma membranes in, 63–64f, 64, 546
  recombination in, 546
  replication in, 187–188, 187f, 266–270, 268t, 546
  ribosomes of, 63, 81t, 552
  shape of, 549–550, 550f
  size of, 546, 546f
  structure of, 13, 63–64, 81t, 549–552
  in symbiosis, 562
  transcriptional control in, 305, 308–312
  transcription in, 284–287f
  translation in, 294
  unicellularity of, 545–546
Prolactin-inhibiting factor (PIF), 951
Prolactin (PRL), **950,** 953, 1095t
Proliferating cell nuclear antigens (PCNAs), 271
Proliferative phase, **1098**
Proline, 45, 47f

Prometaphase, 192, 192f, 194f, **196**
Promoters
  of eukaryotes, 287–288, 288f, 313, 314f
  in prokaryotes, **285**, 285f
Proofreading function, of DNA polymerase, 273
Propagation, vegetative, 748–750
Propane, 34
Prophage, **532**, 533, 555
Prophase
  meiosis I, 209–212f, 215, 216f
  meiosis II, 213f, 214, 217f
  mitotic, 192, 192f, 194f, **195**, 216f
Proprioceptors, **920**
Prop roots, 745, 746f
Prosimians, 723
Prosoma, **682**, 683
Prosopagnosia, 907
Prostaglandins, 37t, 54, **945**
Prostate gland, **1094**
Protandry, 1087
Protease inhibitors, 536, 537f
Proteases, 45t, **323, 805**
Proteasomes, **323**–324
Protective coloring, in guppies, 414–415
Protective layer, of leaves, 813
Proteinase inhibitors, 799
Protein-encoding genes, 362, 363t
Protein hormones, **949**, 950
Protein kinase A (PKA), 180, 180f, 182, 182f
Protein kinase C (PKC), 181, 181f
Protein kinases, 170, 171f, **172**–173, 176–177, 177f, 947
Proteins, 44–53. *See also* Amino acids; Enzymes; Receptor proteins
  anchoring, 95, 95f
  carrier, 91t, 96, 97, 97f, 104t
  catabolism of, 141–142
  central dogma of molecular biology on, 280, 280f, 330
  channel, 91t, 96, 97f, 104t
  chaperone, 51, 51f
  degradation of, 322–324
  denaturation of, 48, 52–53, 52f
  dissociation of, 53
  domains of, **50**–51, 50f
  effector, 179–182, 179f
  in evolutionary theory, 437
  folding of, 51–52, 51f
  functions of, 44, 45t, 94–95, 94f
  homeodomain, 14, 14f
  as hormones, 941
  as macromolecules, 33, 37, 37t
  in membranes, 88, 94–95
  microarrays, 368
  motifs of, 50, 50f, 95, 370f
  movement of, 93, 93f
  nonpolar regions of, 46, 47–48f, 48–49
  nutritional deficiency of, 786
  one-gene/one-polypeptide hypothesis, 280
  phosphorylation of, 170–171, 171f, 198–199, 204
  polar regions of, 46, 47–48f
  predicting structure and function of, 369–370
  primary structure of, 46, 49f, 52f, 53
  production using recombinant DNA, 343–344, 344f
  quaternary structure of, 49, 49f, 53
  regulatory, 305–307
  renaturation of, 52f, 53
  secondary structure of, 46, **48**, 49f
  structure of, 36f, 37, 46, 48–50f, 48–51, 95
  synthesis of (*See* Translation)
  tertiary structure of, **48**–49, 49f, 52f, 53
  transmembrane, 84, 89, 91t, 95, 95f
  transporter, 44, 45t, 63, 94, 94f, 104t
  transport within cells, 71, 71f, 296f, 297
  ubiquitination of, 323–324, 323f
Proteobacteria, 549f
Proteoglycans, 80, 81f, 395
Proteomes, 368
Proteomics, 368–371
Proteorhodopsin, 557
Prothoracicotropic hormone (PTTH), **959**
Prothrombin, 1022
Protista (kingdom), 13, 13f, 648f
Protists, 568–585
  Amoebozoa, 522, 522f, 568, 569f, 583–584
  Archaeplastida, 522, 522f, 568, 569f, 578–581
  cell division in, 188f
  cell surface of, 568
  Chromalveolata, 522, 522f, 568, 569f, 573–578
  classification of, 568, 569f
  cysts of, 568
  cytokinesis in, 197–198
  Excavata, 522, 522f, 568, 569f, 570–572
  flagella of, 568, 570, 571, 573, 582
  mitochondria of, 570, 571
  movement of, 568
  multicellularity of, 569
  nutritional strategies of, 568–569
  Opisthokonta, 522, 522f, 523, 568, 569f, 584–585, 585f
  reproduction in, 569, 571, 574
  Rhizaria, 522, 522f, 568, 569f, 581–582, 582f
Protoderm, **735, 851**
Protogyny, 1087, 1087f
Protonephridia, **1043**, 1044f
Proton gradients, 131, 134, 134f, 135
Proton pumps, 45t, 71, 134–135, 134f
Protons, 18, 19, 19f
Proto-oncogenes, 203–**204**, 204f
Protoplasm, 33
Protoplasts, plant, 858, 858f
Protostomes
  clades of, 661
  cleavage patterns in, 643, 643f
  defined, **643**
  development in, 644
  embryonic development in, 643, 643f
  phylogeny of, 647, 648f
Prototheria, 720–721
Proximal convoluted tubules, **1049**, 1049f, 1050
Prozac (fluoxetine), 900
PRRs (pattern recognition receptors), 800
Prusiner, Stanley, 540
PSC (phylogenetic species concept), **468**–469, 469f
Pseudocoelomates, 642, 642f, 647, 666
Pseudocoeloms, **642**, 642f, 648–649f
Pseudoephedrine, 607
Pseudogenes, 363, 363t, **434, 489**
Pseudomonads, 558
Pseudomonas, 549f
Pseudomurein, **546**, 550
Pseudopods, 568, 581–583f, 583
Pseudostratified columnar cells, 868t
Psilotum, 597
P sites, **292**–293, 293f, 295, 296
Psittaciformes, 715t
PTC (phenylthiocarbamide) sensitivity, 227t
Pterophytes, 594, 594f, 595, 597–599

Pterosauria, 709t
Pterosaurs, 709t
PTH (parathyroid hormone), **955**, 955f
PTTH (prothoracicotropic hormone), **959**
Ptychodiscus, 573f
Pufferfish (*Fugu rubripes*), 360f, 480, 481f, 482, 487, 492
Pulmocutaneous circuit, 1025
Pulmonary arteries, 1025, **1029**
Pulmonary circulation, 1025
Pulmonary valves, **1028**, 1029f
Pulmonary veins, 706, 1025, 1026f, **1029**
Pulvini, 812, 812f
Pumpkins, 746
Punctuated equilibrium, **455**, 455f
Punnett, R. C., 227
Punnett squares, 226–**227**, 226f, 229, 229f, 403
Pupa, **689**
Purines, 42, 42f, 259f, **260**
PWS (Prader–Willi syndrome), 251–252, 492
Pyloric sphincter, 988f, 989
Pyramid of energy flow, 1220–1221, 1221f
Pyrimidines, 42, 42f, 259f, **260**
Pyrogens, **884**
Pyruvate
  conversion to acetyl-CoA, 129–130
  in ethanol fermentation, 140, 140f
  from glycolysis, 126, 126f, 127, 128f, 129–130, 129f
  in lactic acid fermentation, 140, 140f
  oxidation of, 130, 130f, 138
Pyruvate dehydrogenase, 115, 115f, 130, 138, 138f
Pyruvate kinase, 128f

Q

Q_{10}, 879–880
Quantitative traits, **232**
Quaternary structure, of proteins, **49**, 49f, 53
Quencher molecules, 334
Quiescent center, 743
Quinine, 574, 796t, **798**

R

Rabies virus, 528, 530f, 531t, 535
Race, human, 728–729, 728f
Radial canals, **693**, 694
Radial cleavage, 643f, **644**, 649f
Radial symmetry, **506, 640**, 641, 641f, 840
Radiation
  adaptive, 436, 450–454
  heat transfer by, 880, 880f
  ultraviolet, 521, 589, 1251–1252, 1252f
Radicles, **854**
Radioactive decay, 19–20, 428, 428f
Radioactive isotopes, **19**–20
Radiocarbon dating, 428
Radiolarians, 581
Radulae, 645t, **668**–669, 669f
Rain forests, tropical, **1238**–1239, 1239f, 1248–1249, 1249f
Rain shadows, 1236, 1236f
Ram ventilation, 1007
Random spacing, 1168
Raphes, 578, 578f
Rapid eye movement (REM) sleep, 906
Ras protein, 176, **178**, 178f, 203–204f
Rats
  genome of, 492
  selection for pesticide resistance in, 407
  tooth decay in, 426

Ravens, cognitive behavior in, 1143, 1143f
Ray-finned fish, 700f, 702t, 704, 705f
Ray initials, 741
Rays (fish), 702t, 703
Rays (parenchyma cells), 741
RBS (ribosome-binding sequence), **294**
Reabsorption, **1044**, 1048, 1048f, 1050, 1051f, 1052–1053, 1053f
Reactants, 25
Reaction centers, **155**, 155f, 157
Reading frame, **283**
Read length, in DNA sequencing, 359
Realized niche, **1190**, 1190f
Receptor-mediated endocytosis, 102, 102f, **103**, 104t
Receptor potential, 918
Receptor proteins
  cell-surface, 94f, 95, 171–173, 172f, 172t
  defined, **168**, 169f
  functions of, 63
  G protein–coupled, 172f, 172t, 173, 178–183, 179f
  intracellular, 171, 172t, 173–174, 173f
  in membranes, 91t, 95f, 95
Receptor tyrosine kinases (RTKs), 174–178, 182–183, 202, 947, 947f
Recessive traits, 224–225f, **224**–228, 227t
Reciprocal altruism, **1156**, 1157f
Reciprocal crosses, **223**
Recognition helix, 306
Recombinant DNA
  bacteria as host organism for, 330
  in biotechnology, 327–332
  construction of, 327–328, 330
  defined, **327**–328
  gel electrophoresis and, 329–330, 329f
  in insulin production, 343–344, 344f
  libraries of, 330–332, 331f
  restriction endonucleases and, 328, 328f
Recombination
  of alleles on homologues, 245–246, 245f
  in DNA repair, 275
  in eukaryotes, 546
  in genetic mapping, 245–246f, 246–247
  homologous, 553
  in meiosis, **209**–210, 215
  in prokaryotes, 546
  in viruses, 538
Recombination frequency, 246, 247, 247f
Recombination nodules, 210
Recruitment, **975**
Rectum, 992
Red algae
  chloroplasts of, 522, 523f, 566, 567f
  in classification system, 468f
  phylogeny, 589f
  sizes and forms of, 579, 579f
Red-bellied turtles (*Pseudemys rubriventris*), 712f
Red blood cells. *See* Erythrocytes
Red-eyed tree frog (*Agalychnis callidryas*), 708f
Red fibers, 976
Rediae, **663**, 664f
Red maple (*Acer rubrum*), 740f
Redox reactions, 21, **109**, 109f, 123–124, 123f
Red tide, 574, 574f
Reducing atmosphere, 517
Reduction, **21**, 109, 109f, 123, 123f
Reduction division, 210
Reductionism, **7**
Reduction potential, 160
Red-water fever, 684

Reflexes, **908**–909, 909f
Regeneration
 in echinoderms, 694
 of eyespots, 508, 509f
 of flatworms, 663
 in lizards, 713
 of NAD⁺, 129–130, 129f, 140
 of oxaloacetate, 132, 133f
 of spinal cord, 909–910
Regenerative medicine, 346, 382
Regulative development, 1114
Regulatory proteins, **305**–307
Reinforcement, **446**, 446f
Relative dating, 428
Release, of virus particles, **532**, 533f
Releasing hormones, **950**–951
Remodeling of bone, 968–969
REM (rapid eye movement) sleep, 906
Renal cortex, **1048**, 1048f
Renal medulla, **1048**, 1048f
Renal pelvis, 1048
Renaturation, of proteins, 52f, 53
Renewable fuel sources, 341–342
Replication
 of chromosomes, 191, 193
 conservative, 263, 263f, 264
 direction of, 265, 267–268, 267f, 272f
 dispersive, 263f, 264, 265
 elongation stage of, 265
 enzymes needed for, 268t, 269–270, 269f
 errors in, 273
 in eukaryotes, 271–272f, 271–273
 in HIV/AIDS infection cycle, 536
 initiation stage of, 265, 271
 lagging strands in, 267f, 268–270f, 272f
 leading strands in, 267f, 268, 269–270f, 272
 in meiosis, 210, 216
 Meselson–Stahl experiment on, 264–265, 264f
 Okazaki fragments in, 268–270f
 origin of, 187–188, 187f, 266, 266f
 in prokaryotes, 187–188, 187f, 266–270, 268t, 546
 rolling-circle, 553, 553f
 semiconservative, 263f, **264**, 265
 semidiscontinuous, 267–268, 267f
 termination stage of, 265, 269
 of viruses, 529
Replication fork, **268**–269, 269f
Replicons, **266**, 271
Replisomes, 266–267f, **269**–270
Repression of proteins, **308**, 312
Repressors, **308**, 309–312, 312f
Reproduction. *See also* Asexual reproduction; Sexual reproduction
 amphibians, 708
 bacterial viruses, 532–533, 533f
 bryophytes, 591, 592, 592f
 as characteristic of life, 3
 cnidarians, 1086–1087, 1087f
 cost of, 1173–1174f, 1173–1175
 crustaceans, 685, 685f
 earthworms, 675, 675f
 echinoderms, 694
 ferns, 598–599
 in fish, 703
 flatworms, 662f, 663
 fungi, 619, 622, 622f, 626, 629
 gymnosperms, 603–604
 in mammals, 1092, 1092f
 mollusks, 669, 669f
 nematodes, 679
 nemerteans, 673
 pine trees, 605–606, 605f
 in plants (*See* Plant reproduction)
 protists, 569, 571, 574

reproductive events per lifetime, 1174–1175
 yeasts, 629, 630f
Reproductive cloning, **384**
Reproductive isolating mechanisms, **441**–444, 442t, 445–446
Reproductive isolation, **441**, 444, 445–446, 446f, 447
Reproductive leaves, 753
Reproductive strategies, **1151**–1156
Reproductive system, 1086–1104. *See also* Reproduction
 of animals, 1086–1088
 evolution of, 1089–1090, 1090f
 female, 876f, **877**, 1092, 1095t, 1096–1100
 male, 876f, **877**, 1092–1096f, 1095t
 in vertebrates, 876f, **877**
Reptiles, 708–714. *See also specific reptiles*
 brain of, 904, 904f
 characteristics of, 710, 710f
 circulation in, 712, 712f, 1026
 classification of, 468, 645t, 709t
 eggs of, 710, 710f
 evolution of, 699–700, 710–712
 fertilization in, 711, 1090
 heart in, 712, 712f, 1026
 kidneys of, 1047
 lungs of, 1009
 modern, 712–714
 nitrogenous wastes of, 1042, 1043f
 orders of, 709t, 712–713f, 712–714
 respiration in, 710
 sex determination in, 242
 skin of, 710
 skull of, 710f
 thermoregulation in, 710, 712
RER (rough endoplasmic reticulum), 69f, **70**, 297
Research, basic vs. applied, 7–8
Resin, 605
Resolution, of microscopes, 60
Resource depletion, 1247–1252
Resource partitioning, **1192**, 1192f
Resources
 competition for, 1191–1192
 consumption of, 1183
Respiration. *See also* Aerobic respiration; Cellular respiration
 anaerobic, 124, 139
 cutaneous, 706, 1008
 in plants, 786–787
 of primary producers, 1217
 in single-celled organisms, 1005f
Respiratory control center, 1013
Respiratory system, 1003–1017
 amphibians, 706, 707, 1005–1006f, 1006
 arthropods, 683, 683f, 1008
 birds, 717–718, 1010, 1011f
 diseases of, 1014, 1014f
 echinoderms, 1005f
 fish, 1005–1007f, 1006–1008
 gas exchange in, 1004–1005
 gas transport in, 1014–1017
 gills, 1006–1007f, 1006–1008
 insects, 1005f
 lungs, 1008–1010
 mammals, 719, 1005f, 1011–1014
 reptiles, 710
 tracheal, 1008
 in vertebrates, 876f, **877**
Resting potential, 891, 892–893, 893f
Restoration ecology, 1278–1279, 1278f
Restriction endonucleases, 328, 328f, 330, 330f
Restriction fragment length polymorphisms (RFLPs), 401

Restriction maps, 354, 355f
Restriction point, 200
Restriction sites, 328f
Reticular-activating system, 906
Reticular formation, 906
Retinal, 153
Retinas, 930f, **932**, 932f
Retinoblastoma susceptibility gene *(Rb)*, 203f, **204**
Retrotransposons, 363, 487
Retroviruses, **280**, 330, 528, 530
Reverse transcriptase, **280**, 528, 530, 536
Reverse transcriptase inhibitors, 536, 537f
Reverse transcription, 330, 331–333, 331f
Reverse transcription PCR (RT-PCR), 333, 334f
Reverse transcription quantitative PCR (RT-qPCR), 333–334, 334f
Reversible reactions, 25, 111
RFLPs (restriction fragment length polymorphisms), 401
Rhabdovirus, 531t
Rhesus monkeys, 491, 491f, 492
Rheumatic fever, 558
Rh factor, 1079
Rhizaria, 522, 522f, 568, 569f, 581–582, 582f
Rhizobacteria, 793
Rhizobium, 561–562, 781–782f, 782, 793
Rhizoids, 591, 592, 592f, 598
Rhizomes, 598, **750**, 750f, 857
Rhizopus, 620t, 626f
Rh-negative individuals, 1079
Rhodophyta, **579**
Rhodopsin, 179, **931**
Rh-positive individuals, 1079
Rhynchocephalia, 709t, 712f, 713
Rhynchocoel, 672, 673
Rhyniophyta (phylum), 593
Ribbon eels, 701f
Ribbon worms, 508, 509f, 646t, 672–673, 672f
Ribonuclease (RNase), 52f, 53
Ribonucleic acid. *See* RNA
Ribosomal RNA (rRNA), **43**, **69**, **281**, 490
Ribosome-binding sequence (RBS), **294**
Ribosomes
 of eukaryotes, 68–69, 69f, 78f, 81t
 free, 69
 functions of, 293
 membrane-associated, 69
 of mitochondria, 74f
 of prokaryotes, 63, 81t, 552
 structure of, 293, 293f
 in translation, 293–297
 translocation of, 296
 tRNA-binding sites, 292–293, 293f
Ribozymes, 116, 518
Ribulose bisphosphate carboxylase/oxygenase (rubisco), **161**, 161f, 162–164
Ribulose 1,5-bisphosphate (RuBP), **161**, 161f, 162
Rice *(Oryza sativa)*
 comparative genomics and, 366, 366f
 copy numbers for gene families in, 488
 genome of, 360, 360f, 481f, 482, 483, 485f, 487
 golden, 348–349, 349f
 transgenic, 347
Ricin, 797, 797f
Rickettsia, 522, 549f, 559t
Riftia, 675, 675f
Rigor mortis, 973
Ringworm, 635
RISC (RNA-induced silencing complex), **318**–319, 319f, 320
Rising phase, of action potential, 894, 895f
RNA editing, 320–321

RNA-induced silencing complex (RISC), **318**–319, 319f, 320
RNA interference (RNAi), 319–320, 319f, 322f, 336
RNA polymerase I, 287
RNA polymerase II, 287, 288f, 313–315f
RNA polymerase III, 287
RNA polymerases
 core polymerase, 284f, 285
 in eukaryotes, 287–288
 holoenzyme, 284f, 285
 in priming, 268
 in prokaryotes, 284f, **285**
 in replication, 266, 271
 in transcription, 280, 280f, 305
RNA (ribonucleic acid)
 catalytic activity of, 116, 517
 in central dogma of molecular biology, 280, 280f, 330
 defined, **41**
 DNA vs., 43, 43f
 functions of, 37t, 41–43, 517–518
 in gene expression, 281
 sequencing of, 368
 structure of, 42, 42–43f, 43
 transcription and, 43, 280–281
 viruses and, 528, 528f, 530, 531t
RNase (ribonuclease), 52f, 53
Rocky Mountain spotted fever, 558, 684
Rodentia, 466f, 722t
Rods (eye), **931**–932, 931f
Rolling-circle replication, 553, 553f
Root caps, **742**–743, 742f
Root hairs, **738**, 739f, 745, 763, 763f
Root pressure, **764**
Roots
 adventitious, 745, 750, 750f, 856f
 functions of, 595
 gravitropic response in, 809, 809–810f, 810
 modified, 745–746, 746f
 mycorrhizae and, 632–633, 633f
 structure of, 742–744f, 742–745
 tissues of, 733–734, 734f
 of vascular plants, 733–734
Root systems, **733**
Rosin, 605
Rossmann fold, 50
Rotifera, 646t, 648f, 665–666, 666f
Rough endoplasmic reticulum (RER), 69f, **70**, 297
Roundworms, 642f, 645t, 678–680, 1023
R plasmids, **556**
rRNA (ribosomal RNA), **43**, **69**, **281**, 490
R-selected populations, **1179**–1180, 1179t
RTKs. *See* Receptor tyrosine kinases
RT-PCR (reverse transcription PCR), 333, 334f
RT-qPCR (reverse transcription quantitative PCR), 333–334, 334f
Rubisco, **161**, 161f, 162–164
Ruffini corpuscles, 919f, 920
Rule of addition, **230**
Rule of eight, 22, 23f
Rule of multiplication, **230**
Rumen, 624, 625, 993
Ruminants, 993, 993–994f
Rumination, 993
Runners (plant), 750, 750f, 857
Rusts, 620t, 627
Rutherford scattering experiment, 18f
Ryegrass, 631, 631f

S

Saber-toothed mammals, 470, 471f
Saccharomyces cerevisiae, 343, 360f, 481f, 630, 630f
Saccules, **925**–926, 925f
Sager, Ruth, 244

SAGE (serial-analysis of gene expression), 368
St. John's wort *(Hypericum perforatum)*, 1190
Salamanders, 463, 463–464f, 705, 705t, 706, 708, 984f
Salicylic acid, **800**
Salinity
 plant adaptations to, 769, 769f
 of soil, 769, 778–779
Saliva, 986
Salivary glands, 986–987, 1119–1120, 1119f
Salmonella
 classification of, 549f
 evasion of immune system, 1081
 phage conversion in, 534
 type III system in, 558
Salmonella typhi. See Typhoid fever
Salps, 697
Saltatory conduction, **896**–897, 897f
Salt bridges, 48
Salt marshes, 1244
Sand dollars, 645t, 693–695
Sand fleas, 684
Sanger, Frederick, 46, 357
Sap, **770**
Saprobes, 578
Saprolegnia, 578
Sarcomeres, **971**, 972–973f
Sarcoplasmic reticulum (SR), **974**, 974f
Sarcopterygii, 702t, 704, 705t
SARS (severe acute respiratory syndrome), 531t, **539**, 539f, 1249
SAR (systemic acquired resistance), **801**–802
Satiety factor, 998
Saturated fatty acids, **53**, 54, 54f, 55
Saturation, 97
Saurischia, 709t
Savannas, 1188f, **1239**
Scaffold proteins, 177, 177f
Scallops, 666, 667, 667f, 671
Scanning electron microscopes (SEMs), 61, 62t, 91, 518
Scanning tunneling microscopy, 18, 18f
SCARECROW gene, 743–744, 743f, 810, 810f
Scarlet fever, 558
Scars (leaf), 748, 748f
SCF complex, 820
Schistosomes, 664, 664f
Schistosomiasis, **664**
Schizogony, 569
Schizymenia borealis, 579
Schleiden, Matthias, 12, 60
Schwann, Theodor, 12, 60
Schwann cells, **891**, 891f
Science
 deductive and inductive reasoning in, 4–5, 5f
 defined, 4
 descriptive nature of, 4
 hypothesis-driven, 5–6f, 5–7
 of life, 1–4
Scientific method, 4, 7
Sciuridae (family), 466f
Sciurus carolinensis, 466f
Sciurus (genus), 466f
Sclera, **930**, 930f
Sclerenchyma cells, 739f, **740**
SCN (suprachiasmatic nucleus), 958
SCNT (somatic cell nuclear transfer), **383**, 384, 386
Scolex, **665**, 665f
Scorpions, 645t, 681t, 683
Scouring rushes. See Horsetails
Scrapie, 540, 541f
Scrotum, 1093
Scutellum, **854**, 855f

Scyphozoa, **655**, 655f
Sea anemones, 640, 641f, 645t, 653, 653f, 654
Sea cucumbers, 645t, 693–695, 695f
Sea daisies, 695
Sea fans, 645t
Sea level, effect of global warming on, 1255
Sea lilies, 694, 695
Sea mats, 645t
Sea moss, 645t
Sea peach *(Halocynthia auranthium)*, 698f
Sea slugs, 645t, 666, 667, 667f, 668, 670, 670f
Sea spiders, 684
Sea stars, 645t, 693, 693f, 694, 695, 695f
Sea turtles, 709t, 713
Sea urchins
 characteristics of, 695
 classification of, 645t, 693, 695, 695f
 development in, 497–498, 498f, 1112f
 fertilization in, 1108, 1109–1110f, 1110
 gastrulation in, 1115–1116, 1115f
 ossicles of, 694
Sea walnuts. See Comb jellies
Seaweeds, 577, 577f
Sebaceous (oil) glands, 867, 1058
Secondary carnivores, **1217**, 1217f
Secondary cell walls, **80**, 80f
Secondary chemical compounds, 1195
Secondary endosymbiosis, **566**, 567f
Secondary growth, in plants, **735**, 748, 749f
Secondary induction, **1126**
Secondary lymphoid organs, **1066**–1067, 1067f
Secondary metabolites, **795**, 796t, 797–798
Secondary mycelium, 627–628
Secondary oocytes, 1098, 1098f
Secondary plant body, **735**
Secondary productivity, **1218**
Secondary sexual characteristics, **1153**
Secondary structure, of proteins, 46, 48, 49f
Secondary succession, **1204**
Secondary tissues, of plants, **735**
Secondary xylem, 740
Second filial generation, **224**–225, 228–229, 229f
Second Law of Thermodynamics, **110**, 110f, 437, 880, 1216
Second messengers
 calcium as, 181, 182f
 cAMP and, 173, 179–181f, 182
 cGMP as, 174
 generation of, **173**
 for hydrophilic hormones, 947–948, 947f
 inositol phosphates as, 180–181f, 181
Secretin, 179, **995**, 996t
Secretion, **1045**
Secretions
 in kidneys, 958, 1048
 in liver, 990f, 991
 in stomach, 988–989
Secretory phase of menstrual cycle, **1099**
Securin, 201
Seed banks, 854
Seed coats, 603, 606, 611, **613**, 613f, 853
Seedcracker finches *(Pyrenestes ostrinus)*, 412, 412f
Seedlings, growth of, 854–855, 855f
Seed plants, 602–613. See also Seeds
 evolution of, 602–603
 extant phyla of, 603, 604t
 flowering (See Flowering plants)
 gymnosperms, 589f, 596f, 603–607

 haploid and diploid generations of, 590
 phylogeny, 589f
Seeds
 adaptations of, 613–614, 614f
 dispersal of, 1168, 1168f
 dormancy in, 814, 814f, 827–828, 828f
 embryo protection from, 603, 603f, 613
 in flowering plants, 595
 formation of, 605f, 606, 854–855
 germination of, 613–614, 807, 854–856f
Segmental duplications, 363, 363t, 487–489f
Segmentation
 in annelids, 644, 673–674, 674f
 in arthropods, 644, 681–682, 681f
 in earthworms, 644, 645t, 673f, 675
 evolution of, **644**
 in fruit flies, 387, 388–389f
Segmentation genes, **391**
Segment polarity genes, 389f, **391**
Segregation of traits, **222**, **226**
Selaginella moellendorffii, 596f, 597
Selectable markers, 331
Selected-effect function, 365
Selection. See also Natural selection
 agricultural, 426, 427f
 artificial, 10, 406, 426–427
 to avoid predators, 407, 407f
 on color in guppies, 414–415
 defined, **406**
 directional, 412–413
 disruptive, 411–412, 412f, 449–450
 evolution and, 404f, 406–408
 experimental, 426, 426f
 frequency-dependent, 409–410, 410f
 interactions among evolutionary forces and, 415–416, 416f
 limits of, 417–418, 417f
 to match climatic conditions, 407
 oscillating, 410
 for pesticide and microbial resistance, 407–408, 408f
 sexual, 408
 stabilizing, 413, 413f
Selectively permeable membranes, 96
Selective serotonin reuptake inhibitors (SSRIs), 900
Self-fertilization, **223**, 223f
Self-incompatibility, in plants, 847f, 847
Self-pollination, 612, 842, 845–846
Self vs. nonself recognition, 1068
Semelparity, **1174**
Semen, 1094
Semicircular canals, 925–**926**, 925–926f
Semiconservative replication, 263f, **264**, 265
Semidiscontinuous replication, 267–268, 267f
Semilunar valves, **1028**
SEMs (scanning electron microscopes), 61, 62t, 91, 518
Senescence, in plants, **859**
Sensitive plant *(Mimosa pudica)*, 808, 812, 812f
Sensors, 877, 878f
Sensory exploitation, **1154**
Sensory neurons, 874t, **889**, 889f
Sensory organs, of flatworms, 662f, 663
Sensory receptors, 688–689, 889, 917–918
Sensory setae, **688**
Sensory systems, 916–936
 chemoreceptors in, 917, 926–929
 defined, **874**
 diversity of, 935–936
 hearing, 921–926

 information path of, 917, 917f
 mechanoreceptors in, 917–920
 sensory receptors in, 688–689, 889, 917–918
 vision, 723, 929–935f
Sensory transduction, 918, 918f, 932–933, 934f
Sepals, **610**, 610f
Separase, 201
Separation layer, of leaves, 813
SEP genes, 837, 839f
Septate junctions, 82t, **83**
Septation (cell division), **188**, 188f
September 11, 2001, attacks, 335, 371
Septum
 in binary fusion, 187f, **188**
 in fungal hyphae, 621, 621f
 in segmentation, 674
Sequence-tagged site maps, 355–356, 356f
Sequence-tagged sites (STSs), 359, 362
Sequential hermaphroditism, 1087, 1087f
Serial-analysis of gene expression (SAGE), 368
Serine, 45, 47f, 170, 171f
Serosa, of gastrointestinal tract, **985**, 985f
Serotonin, **900**, 1136
Serotonin receptors, 320–321
SER (smooth endoplasmic reticulum), 69f, **70**
Serum, 202, **1021**
Serum albumin, 45t
Sessile crustaceans, 686, 686f
Set points, 878f
Severe acute respiratory syndrome (SARS), 531t, **539**, 539f, 1249
Sex chromosomes
 in birds, 241, 241t
 of fruit flies, **241**
 in humans, 241–243, 241t, 248f
 inheritance and, 240
 in insects, 241, 241t
 nondisjunction of, 250–251, 251f
Sex combs reduced gene, 1119
Sex determination
 chromosomes in, 241–242
 environmental factors in, 233t, 1088, 1088f
 genetic, 1088
 temperature-sensitive, 1088
Sex linkage, 240–**241**, 240f, 242–243
Sex steroids, 958
Sexual dimorphism, **679**, **1153**
Sexual life cycle, 208, 208f
Sexually transmitted disease (STDs), 559t, 560–561, 560f, 570, 1102
Sexual reproduction. See also Meiosis
 animals, 523, 640
 chromosomes and, 208, 208f
 defined, **208**
 echinoderms, 694
 essence of, 207
 fungi, 622, 626, 629
 genetic diversity and, 523
 land plants, 591
 nematodes, 679
 overview, 1086
 protists, 569, 574
 sponges, 651
Sexual selection, 408, **1153**–1154
Shade leaves, 753
Shared derived character variations, **462**, 468
Sharks
 characteristics of, 702t
 evolution of, 703
 morphological data for, 463, 463f
 as predators, 703, 703f

Sharks (*continued*)
  reproduction in, 703
  teeth of, 703
Sheep, cloning of, 383–385f
Shells
  of mollusks, 666, 668, 671
  of turtles and tortoises, 713
Shigella, 558
Shingles, 528
Shipworms, 667
SHOOTMERISTEMLESS gene, in *Arabidopsis*, 851, 851f
Shoots
  apex of, 747, 747f
  elongation of, 807, 807f
  gravitropic response in, 809, 809–810f, 810
  tissues of, 733–734, 734f
  in vascular plants, 733–734
Shoot systems, **733**
Shore crabs, 1150, 1150f
Short-day plants, **834**–835, 834f
Short interspersed elements (SINEs), **363**–364, 364f, 482
Short root mutant, in *Arabidopsis*, 810, 810f
Short tandem repeats (STRs), 335, 335f, 371
Short-term memory, 907
Shotgun assembly, **359**, 359f
Shrimp, 681t, 684, 685
Shugoshin, 215
Sickle cell anemia
  characteristics of, 249t
  as heterozygote advantage, 410–411, 411f
  malaria and, 250, 411, 411f
  mutations in, 299, 299f
  pathology of, 249–250, 249f
  pleiotropic effects and, 233, 233t
  as recessive trait, 227t
  structural causes of, 49
Sieve areas, 741
Sieve cells, 741, 742
Sieve plates, 741, 741f
Sieve-tube members, 741, 741f
Sieve tubes, 470, 471f
Sigmoidal growth curve, **1176**
Signaling molecule sonic hedgehog (Shh), 1126
Signal recognition particles (SRPs), **281**, 296f, **297**
Signal sequence, 296f, **297**
Signal transduction pathways
  cellular responses of, **170**
  components of, 168, 169f
  in development, 499
  models of, 7
  transmission of information across, 14
Sign stimulus, 1135, 1135f
Sildenafil (Viagra), 174, 900
Silent mutations, 299, 300f
Silkworm moth (*Bombyx mori*), 958f, 1147
Silkworms, 189t
Simberloff, Dan, 1229
Simian immunodeficiency virus (SIV), 475–476, 476f, 534
Simple epithelial membrane, 867
Simple epithelium, 867, 868t
Simple leaves, **751**–752, 752f
Simple sequence repeats (SSRs), **363**, 363t
SINEs (short interspersed elements), **363**–364, 364f, 482
Singer, S. Jonathan, 89
Single covalent bonds, 24, 24f
Single nucleotide polymorphisms (SNPs), **248**, 367, 368, 371, 401
Single nucleotide variations (SNVs), **299**–300

Single-strand-binding (SSB) proteins, 267, 268t, 269–270f
Single-strand RNA viruses, 530, 531t
Sinks (food storage tissues), 771, 772f
Sin nombre virus, 539
Sinoatrial (SA) node, **1025**, 1028
Sinosauropteryx, 472f, 716f
Sinus venosus, **1025**, 1025f
Siphonaptera, 687t
Siphons, **671**, 671f
Sirenin, 624
siRNA (small interfering ribonucleic acid), **281**, 318, 319–320, 319f
Sister chromatid cohesion, 210, 214, 215
Sister chromatids, **191**, 191f, 193, 193f, 196, 209, 209f
Sister clades, 475
Site-specific farming, 777
SIV (simian immunodeficiency virus), 475–476, 476f, 534
Skates, 702t, 703
Skeletal muscle, **871**–872, 872f, 970
Skeletal tissue, 965
Skeletons. *See also* Musculoskeletal system
  for flight, 716
  hydrostatic, 652, 674
  overview, 875–876, 875f
  types of, 964–965
Skin
  as barrier to infection, 1058
  of reptiles, 710
  sensory receptors in, 919–920, 919f
Skin grafts, 386
Skinks, 713
Skinner, B. F., 1140
Skinner boxes, 1140
Skulls, of reptiles, 710f
S-layers, 551
Sleep, 906
Sleep movement, in plants, 812, 813f
Sliding clamp structure, 268, 268–269f, 269
Slime molds, 180, 181f, 360f, 583–584, 584f
Slow-twitch muscle fibers, 975–**976**, 976f
Sludge digesters, 342
Slugs. *See* Sea slugs
Small ground finches (*Geospiza fuliginosa*), 444
Small interfering ribonucleic acid (siRNA), **281**, 318, 319–320, 319f
Small intestine, 985, 985f, 989–992
Small nuclear ribonucleic acid (snRNA), **281**
Small nuclear ribonucleoprotein particles (snRNP), **290**, 290f
Smallpox, 371t, 530, 531t, 534, 1063
SMC (structural maintenance of chromosome) proteins, 187
Smell (olfaction), 489, 928–929, 928f
Smith, Hamilton, 328
Smoking, 902, 1014, 1035, 1058
Smooth endoplasmic reticulum (SER), 69f, **70**
Smooth muscle, **871**, 872t
Snails
  classification of, 645t, 666
  habitat of, 667
  larval disposal in marine snails, 472–473
  types of, 670, 670f
Snakes
  characteristics of, 709t, 713
  classification of, 713f
  evolution of, 429
  geographic variation in species of, 441f

phylogeny of, 701f
sensing infrared radiation, 935–936, 935f
venom of, 45t, 437
vestigial structures in, 433
Snapdragons, 506, 840, 840f
Snowball Earth, 520, 520f, 524
SNPs (single nucleotide polymorphisms), **248**, 367, 368, 371, 401
snRNA (small nuclear ribonucleic acid), **281**
snRNP (small nuclear ribonucleoprotein particles), **290**, 290f
SNVs (single nucleotide variations), **299**–300
Social systems
  communication in, 1147–1148, 1148f
  evolution of, 1159–1160
Socrates, 797
Sodium, reabsorption in kidneys, 1051, 1051–1052f, 1052
Sodium channels, voltage-gated, 894, 895f
Sodium chloride, 23–24, 23f, 29f
Sodium–potassium pump
  in active transport, 100–101, 100f, 104t
  defined, **100**
  functions of, 45t
  in membranes, 91t, 891–892, 892f
Soil
  acidic, 777–778, 778f
  air in, 776, 777f
  charges on soil particles, 776, 776f
  composition of, 775–776, 776f
  defined, **775**
  degradation of, 776–777, 777f
  influence on life, 1165
  minerals in, 775–776, 776f
  nitrogen fixation in, 561–562
  organic matter in, 776, 776f
  salinity of, 769, 778–779
  water potential of, 764–765, 776, 777f
Solar energy, 1233–1234f, 1233–1235
Soldier flies (*Ptecticus trivittatus*), 686f
Solenoids, 190, 190f
Solitary bees, 843
Soluble receptors, 1059
Solute potential, **760**, 760f
Solutes, 28, **98**, 99f
Solvents, 28, **98**
Somatic cell nuclear transfer (SCNT), **383**, 384, 386
Somatic cells, **208**, 208f, 209, 384, 385f
Somatic motor neurons, 975
Somatic nervous system, **889**, 890f, 910t, 911
Somatostatin, 951
Somatotropin, 950
Somites, **1121**
Somitomeres, **1121**
Songbirds, 1141–1142, 1142f, 1147, 1151, 1271
Sonic hedgehog gene, 305
Sorghum, 164, 485f
Sori, **598**
Soricomorpha, 722t
SOS response, 275
Sounds, navigation by, 924–925
Source–sink metapopulations, **1169**–1170
Soybeans (*Glycine max*)
  phytoestrogens in, 796t, 797
  pollination of, 349
  transgenic, 347, 349
Spacing patterns, 1168–1169
Spallanzani, Lazzaro, 720
Spatial heterogeneity, species richness and, 1226–1228, 1226f
Spatial recognition, 906–907

Spatial summation, **901**
Special connective tissues, **869**, 871
Specialized transduction, **554**–555
Speciation
  adaptation and, 447, 447f
  allopatric, 448, 448f
  defined, **440**
  evolution and, 455, 455f
  extinction and, 456–457
  gene flow and, 441, 446
  genetic drift and, 447
  geography of, 447–450
  natural selection in, 447
  polyploidy and, 449, 449f
  reinforcement and, 446, 446f
  sympatric, 441, 443, 448–450, 449f
Species. *See also* Speciation; Species richness; *specific species*
  biological species concept, 441–442, 442t, 444–445, 468
  defined, 441, 468
  ecosystem characteristics and, 1226–1227, 1226f
  endemic, 1261–1263, 1261–1264, 1263t
  geographic variation within, 441, 441f
  in hotspots, 1261–1263, 1263t
  hybridization between, 442, 442f, 443
  keystone, 1203, 1203f, 1275
  nature of, 437–445
  in organization of living organisms, 3f, 4
  phylogenetic species concept, 468–469, 469f
  in taxonomic classification, 466
Species name, 466, 466f
Species richness. *See also* Biodiversity
  in beetles, 474–475, 474f
  climate and, 1226f, 1227
  defined, **474**, **1188**
  effects of, 1225–1226, 1225f
  evolutionary age and, 1227
  phylogenetics and, 474–475, 474f
  predation and, 1227
  productivity and, 1226, 1226f, 1227
  spatial heterogeneity and, 1226–1228, 1226f
  in tropical regions, 1227–1228, 1227f
Specific heat, 27t, **28**
Specific repair mechanisms, 274, 274f
Specific transcription factors, 313
Spectrin, 90, 91t, 95
Speech, genetic basis of, 491–492, 491f
Spemann, Hans, 1124
Spemann organizer, 1124–1126
Sperm
  blockage of, 1102
  destruction of, 1102
  fertilization of, 1108–1111, 1108t, 1109–1111f
  penetration of egg by, 1108, 1109f, 1111
  production of, 1093–1094, 1094f
  structure of, 1094, 1094f
Spermatids, **1094**
Spermatogenesis, 1093f
Spermatozoa, **1094**
Sperm competition, 1153
Spermicides, 1101f, 1101t, 1102
Sphagnum, 592
Sphenisciformes, 715t
Sphincter, 988
Sphingolipids, 89, 90, 90f, 92, 93
Sphingomyelin, 89, 90f
Sphygmomanometers, 1031
Spicules, 650f, **651**
Spiders, 645t, 681t, 683, 684, 684f
Spinal cord, 903f, 908–910
Spinal muscular atrophy, 385, 492
Spindle apparatus, 188f, **195**, 196

Spindle checkpoint, **200**, 200–201f
Spindle plaques, 622
Spines (plant), 753
Spinnerets, **684**
Spiny fish, 702, 702t
Spiracles, **683**, 683f, 688, 1008
Spiral cleavage, **643**, 643f, 648f, 661
Spiralia, 643, 647, 648f, 661, 669
Spirilla, 550, 550f
Spirochetes, 548–549f, 550, 684
Spliceosomes, **289**–290, 290f
Sponges, 509f, 641, 642, 649–651, 650f, 1023, 1024f
Spongin, 650f, **651**
Spongy bone, 967f, **968**
Spongy mesophyll, 752, 753f
Spontaneous abortion, 214, 1128
Spontaneous generation hypothesis, 6, 6f, 7
Spontaneous reactions, 111
Sporangia, **590**, 593, 598, 624
Sporangiophores, **626**, 626f
Sporangium, **583**, 584f, 592, 592f
Spore mother cells, **590**, 590f, 592, 598
Spores
  of ferns, 598, 599f
  of fungi, 622, 622f, 627–629
  megaspores, 606
  microspores, 605
  of moss, 592, 592f
  of pine trees, 605, 605f
  of plants, **590**, 590f
  zoospores, 578, 622, 624, 624–625f
Sporocysts, **663**, 664f
Sporocytes. *See* Spore mother cells
Sporophytes, 590–593, 598, 611
Sporophytic self-incompatibility, 847, 847f
Spotted cuscus, 435f
Springtails, 681t, 688
Squamata, 709t, 713, 713f
Squamous cells, 867, 868t
Squid, 667, 668, 671, 672
SRPs (signal recognition particles), **281**, 296f, **297**
SR (sarcoplasmic reticulum), **974**, 974f
SRY gene, 242
SSB (single-strand-binding) proteins, 267, 268t, 269–270f
SSRIs (selective serotonin reuptake inhibitors), 900
SSRs (simple sequence repeats), **363**, 363f
S (synthesis) phase, **192**, 192f, 193
Stabilizing selection, 412–413f, **413**
Stacked GM crops, 348
Stahl–Meselson experiment, 264–265, 264f
Stains, in visualization of cell structure, 61–62, 550–551
Stalked jellyfish, 655, 655f
Stamens, **610**, 610f, **839**, 839f
Stapes, **922**
Staphylococcus, 548f
Staphylococcus aureus, antibiotic resistance of, 407–408, 556
Star anis, 609f
Starches, 36f, 37t, **39**–40, 40f
Starfish. *See* Sea stars
Star formation, 17
Star jellies, 655, 655f
Starlings (*Sturnus vulgaris*), 1145, 1145f
START, in DNA synthesis, 199, 200
Start codons, **283**, 362
Starter cultures, 630
Start sites, **285**
Stasis, **455**
Statocysts, **925**
Staurozoa, 655, 655f

STDs. *See* Sexually transmitted disease
Stegosaurs, 467f, 709t
Steles, **745**
Stellar sea lions, 494f
Stem cells
  blood cells arising from, 1022, 1023f
  defined, 346, **381**
  differentiation and, 381
  embryonic, 338–339, 338f, 346–347, 381–382, 382f
  ethical considerations in research using, 382
  mesenchymal, 347
Stem cell therapy, 346–347
Stems
  evolution of, 595
  gravitropic response in, 809–910f
  modified, 748–750, 750f
  positive phototropism in, 807–808, 807f
  structure of, 747–749f
Ste5 protein, 177
Stereoisomers, 35, 35f, 38, 39f
Sterilization (birth control), 1102–1103, 1103f
Stern, Curt, 245
Steroid hormone receptors, 173–174, 173f
Steroids, 37t, 54, 55f, **941**, 955–956
Sterols, 92
Stickleback fish
  adaptation in, 451, 451f
  courtship signaling in, 1146–1147
  gene mutations in, 501
Stigma, of flowers, **610**, 610f, 612
Stimuli, 3, 917
Stimulus-gated ion channels, **918**, 918f
Stimulus–response chain, 1146, 1146f
Stipules, 748, **751**
Stolons, 750, 750f, 857
Stomach, **988**–989, 988f, 993, 994f
Stomach cancer, 598
Stomata
  in CAM plants, 165
  fungal penetration of, 793, 794f
  gas diffusion through, 589
  guard cells of, 737, 737f, 752, 752f, 766–767
  opening and closing of, 766–767
  in photorespiration, 163, 163f
  of tobacco plant, 163f
Stony corals, 653, 654f
Stop codons, **283**, 362, 500–501, 500f
Stramenopiles, 573, 573f
Stratified epithelial membrane, **867**
Stratified epithelium, 867, 868t
Stratified seeds, **854**
Stratospheric ozone depletion, 1251–1252
Streptococcus, 257–258, 257f, 530f, 548f, 558, 559t, 560
Streptomyces, 548f
Streptophytes, 589, 589f
Striated muscle, 871
Strigiformes, 715t
Strigolactones, 816f, 822–823, 823f
Strokes, **1035**
Stroke volume, 1035
Stroma, 74, 74f, **148**, 148–149f, 160, 160f
Stromatolites, **519**, 519f
STRs (short tandem repeats), 335, 335f, 371
Structural DNA, 363, 363f
Structural formulas, 24
Structural isomers, 35
Structural maintenance of chromosome (SMC) proteins, 187
Structure, relationship with function, 13
Struthioniformes, 715t
STSs (sequence-tagged sites), 359, 362

Stumps, 748, 749f
Sturtevant, Alfred, 246
Styles, of flowers, **610**, 610f
Stylets, **679**
Suberin, 745, **793**
Submucosa, of gastrointestinal tract, **985**, 985f
Substance P, **900**
Substrate-level phosphorylation, 125–126, 125f, 127, 132, 133f
Substrates, **114**, 114f
Subunit vaccines, 537
Succession, ecological, 1204–1206
Succinate, 131, 132, 133f
Succinyl-CoA, 132, 133f
Suckers, plant, 857
Sucrose, 28, 39, 39f, 770, 770f, 772f
Sugarcane, 164, 189t, 366f, 485f
Sugars. *See also* Carbohydrates
  catabolism of, 141f
  isomers of, 38, 39f
  in nucleotides, 42f
  transport in plants, 38, 39, 770, 770f, 772f
Sulfate respiration, 139
Sulfhydryl group, 35f
Sulfur bacteria, 139, 139f
Summation, **894**, 894f, 901, 975, 976f
Sundews (*Drosera*), 754, 783, 783–784f
Sunflowers (*Helianthus annuus*), 147f, 349, 485f, 748f, 788, 812f
Sunlight
  energy from, 108–109
  influence on life, 1165
  in photosynthesis, 108, 148, 150
  in stomatal opening and closing, 767
Supercoiling, of DNA, **267**, 267f
Superior vena cava, **1030**
Suprachiasmatic nucleus (SCN), 958
Surface area-to-volume ratio, **60**, 60f
Surface markers. *See* Cell surface markers
Surface tension, **26**, 27f
Survivorship, **1172**
Survivorship curves, **1172**, 1172f
Suspensors, **848**–850, 850f
Sustainable agriculture, 777
Sutherland, Earl, 947–948
Sutton, Walter, 240, 241
Swallowing, 987, 987f
Sweet woodruff, 747f
Swim bladders, **704**, 704f
Swimmerets, **685**, 685f
Swimming, 977–978, 977f
Symbiosis
  coevolution and, 1198
  defined, 75, **562**
  endosymbiosis, 75, 75f, 522, 566, 567, 567f
  facultative, 631
  fungi in, 591, 627, 631–634
  obligate, 631, 633
  prokaryotes in, 562
  tripartite, 633
Symmetry
  bilateral symmetry, 506, 641, 641f, 694, 840, 840f
  floral, 506, 840, 840f
  pentaradial, 693, 694
  radial symmetry, 506, 640, 641, 641f, 840
Sympathetic chain of ganglia, 911, 912f
Sympathetic nervous system, 889, 890f, 911, 912f, 912t
Sympatric speciation, **441**, 443, 448–450, 449f
Symplast route, **763**, 763f
Symplesiomorphies, **463**
Symporters, **100**
Synapomorphies, **463**, 640
Synapses, 170, 897–901f, **897**–902

Synapsids, **710**, 710f
Synapsis, **209**, 209f, 210, 212f
Synaptic clefts, **898**, 898f
Synaptic integration, 901
Synaptic plasticity, 907
Synaptic signaling, 169, 169f, **170**, 897f
Synaptic vesicles, **898**
Synaptonemal complex, **209**, 209f, 215
Syncytial blastoderm, **387**, 388f, 391, 1112
Syngamy, **208**
Synteny, 365–366, 366f, 489, 489f
Synthesis
  in cell cycle, **192**, 192f, 193
  of DNA, 332, 333f
  of virus particles, **532**, 533f
Synthetic cells, 370, 371f
Synthetic polyploids, **484**–485
Syphilis, 548f, 560, 560f
Systematics, 460–469
  cladistics and, 462–465
  classification and, 465–469
  defined, **460**
  phylogenies in, 460–461, 461f
Systemic acquired resistance (SAR), **801**–802
Systemic anaphylaxis, 1077
Systemic circulation, **1025**
Systemin, **799**
Systems biology, 14
Systole, **1028**, 1029f
Systolic pressure, **1031**, 1031f

T

Table salt. *See* Sodium chloride
Taenia saginata, 665, 665f
TAFs (transcription-associated factors), 313, 313f
Tagmata, **681**, 683, 685
Taiga, 1239–1240
Tails, as ancestral vs. derived character, 462–463, 463f
TALE (transcription activator-like effector) proteins, 336
Tandem duplications, 300
Tannins, 795
Tapeworms, 645f, 661, 663, 665, 665f, 1087
Taq polymerase, **332**, 333f, 334, 336
Tarantulas, 684
Tardigrada, 644, 646t, 649f
Tarsiers, 723, 723f
Tasmanian devils (*Sarcophilus harrisii*), 494, 494f
Tasmanian quolls, 435f
Tasmanian tigers, 494f
Taste buds, 927, 986
Taste (gustation), 927–928
Taste pores, 927
TATA sequence, 287
Tatum, Edward, 6, 279–280, 279f
Tautomers, of nitrogenous bases, 260–261
Taxol, 796t, 797–798
Taxon, **462**, 465, 466–467
Taxonomic hierarchy, 465–467, 466f, 468
Taxonomy, **465**
Tay–Sachs disease, 71, 249t, 252
T box, **502**
Tbx5 gene, 502, 503f
T-cell receptors (TCRs), **1066**, 1067f, 1075, 1076f
T cells
  in adaptive immunity, 1065, 1065f
  antigen recognition by, 1064f, 1068t
  cytotoxic, 1064f, 1068–1069, 1069f
  helper, 1064t, 1068, 1069–1070, 1071f
  HIV/AIDS infection of, 536
  HIV infection of, 1081–1082, 1082f
  transplant rejection and, 1070

TCEs (trichloroethylenes), 787–788f, 788
TCRs (T-cell receptors), **1066,** 1067f, 1075, 1076f
Tectonics, 515, 520
Teeth
   dental caries, 545f, 559–560, 559t
   evolution of, 703
   of horses, 430, 986f
   of humans, 719, 719f, 986, 986f
   of mammals, 719, 719f
   saber-toothed mammals, 470, 471f
   of sharks, 703
   of vertebrates, 986, 986f
Telencephalon, 903t, **904**
Telomerase, **272**–273, 272f
Telomeres, **271**–272, 272f
Telophase
   meiosis I, 212f, 214, 216f
   meiosis II, 213f, 214, 217f
   mitotic, 192, 192f, 195f, **197,** 216f
Telsons, **685,** 685f
Temperate deciduous forests, 1239, 1240f
Temperate evergreen forests, **1239**
Temperate grasslands, 1239
Temperate phage, **532**
Temperature. *See also* Thermoregulation
   adaptation to specific ranges of, 1164
   altitude and, 1236, 1236f
   annual mean, 1233, 1233f
   biomes and, 1238, 1238f
   carbon dioxide and, 1254
   chemical reactions and, 25
   dissociation curve, 1016, 1016f
   enzymes and, 53, 116, 116f
   flower production and, 836, 837f
   membrane fluidity and, 93
   oxyhemoglobin, effect on, 1016, 1016f
   plant respiration, impact on, 786–787
   plant responses to, 813–815, 814f
   in sex determination, 1088
   specific heat, 27t, **28**
   transpiration, impact on, 767
Template strands, 265, 265f, **280,** 285–286f
Temporal isolation, 442t, 443
Temporal lobe, 904, 905f
Temporal summation, **901**
TEMs (transmission electron microscopes), 61, 62f, 91, 518
Tendons, 970
Tendrils, 750, 750f, 811
Tensile strength, **765**
Terebratulina septentrionalis, 677f
Terminal buds, 748, 748f
Terminators, **285,** 285f, 286f
Termites, 570, 686f, 687t, 720
Terpenes, 37t, 54, 55f
Terpenoids, 796t
Territorial behavior, 1150–1151, 1151f
Territoriality, **1150**
Tertiary endosymbiosis, 566
Tertiary follicles, 1098
Tertiary structure, of proteins, 48–49, 49f, 52f, 53
Testcrossing, **231**–232, 231f, 241
Testes, 1093, 1093f, 1096f
Test experiments, **6**
Testosterone, 54, 173, 825f, 945f, 958, 1093, 1095t, 1136
Tests, of foraminifera, 581–582
Test tube babies, 1104
Testudines, 701f
Tetanus, 552, 558, **975**
Tetracycline, 552, 561
Tetrad structures, 209
Tetraethers, 546, 547f
Tetrahedrons, 26
Tetraploids, 384, 449, 449f

Tetrapods, 502
T-even bacteriophage, 529–530, 532
Thalamus, 903t, **904,** 906
Thalassemia, 49
T-helper cells, **534,** 536
Theories, **7,** 436
Theory of evolution, 8–12, 400, 400f, 436
Therapeutic cloning, 385–**386,** 386–387f
Therapsids, 710, 711f
Theria, **721**
Thermal stratification, **1241**–1242, 1241f
Thermodynamics
   defined, **108**
   First Law of, 109
   Second Law of, 110, 110f, 437, 880, 1216
Thermogenesis, **883**
Thermophiles, 139, 545f, 546, 548f
Thermoproteus, 548f
Thermoproteus tenax, 139f
Thermoreceptors, **919**
Thermoregulation
   in birds, 718
   by hypothalamus, 883–884, 884f
   in insects, 881, 881f
   in mammals, 718, 719
   negative feedback loops and, 877, 878f
   in reptiles, 710, 712
   in vertebrates, 879–884
Thermotogae, 548f
Thermotolerance, in plants, 815
Theropods, 711, 716
Theta waves, 906
Thick myofilaments, 972–973
Thigmomorphogenesis, **811**
Thigmonastic responses, 811
Thigmotropism, **811,** 811f
Thin myofilaments, 972–973
Thiomargarita namibia, 546, 546f
Third filial generation, 225, 225f
Thoracic breathing, 710
Thoracic cavity, 865
Thorns, 753
Thorn-shaped treehoppers (*Embonia crassiornis*), 686f
Three-point crosses, 247, 247f
Threonine, 45, 47f, 170, 171f
Threshold potential, **894**
Thrombocytes, 871
Thrush, 635
Thylacines, 435f
Thylakoids, **74,** 74f, 148, 148–149f, 160, 160f
Thymine, 42, 42f, 43, 259f, 260, 262
Thymine dimer, 274, 274f
Thymus, 958, **1066**
Thyroid gland, 951, 951f, 953–955
Thyroid hormones, **941,** 953, 954, 954f
Thyroid-stimulating hormone (TSH), 883, **950,** 951
Thyrotropin, 950
Thyrotropin-releasing hormone (TRH), 951
Thyroxine, 945f, **951,** 953, 954f
Ticks, 558, 681t, 683, 684
Tigers, 442, 442f, 463, 463–464f, 494f
Tiger salamanders (*Ambystoma tigrinum*), 708f
Tight junctions, 82t, **83,** 84f
Tiktaalik, 429, 706f, 707
Tinbergen, Niko, 1148, 1149, 1149f
Tissue cultures, plant, 858
Tissues
   of animals, 640
   connective tissue, 865, 869–871, 869f, 870f
   in digestive tract, 985, 985f
   epithelial, 865, 866–867, 868t
   evolution of, 641

   formation of, 82
   gas exchange in, 1012f
   of living organisms, 2f, **3**
   muscle tissue, 865, 865f, 871–873, 872f, 873f
   nerve tissue, 865, 873, 874t
   of plants, 594, 733–734f, 733–742, 745f, 851–852
   primary, 735, 865
   secondary, 735
   skeletal, 965
   in vertebrates, 865, 865f
Tissue-specific gene knockouts, 340
Tissue systems (plant), **734,** 851–852
Tissue tropism, **528,** 535
TLRs (toll-like receptors), 1058–1059
T lymphocytes. *See* T cells
Tmespiteris, 597
TMV (tobacco mosaic virus), 528f
TNT (trinitrotoluene), 788
Toads (*Bufo*)
   adaptations to behavior, 1166
   characteristics of, 705t
   ecological isolation in, 442
   extinction of, 1268
   habitat of, 707–708
Toadstools, 620t
Tobacco hornworms (*Manduca sexta*), 795, 795f
Tobacco mosaic virus (TMV), 528f
Tobacco (*Nicotiana tabacum*)
   genome of, 484f, 486, 486f
   hybridizations of, 222
   stomata of, 163f
Toes, grasping, 723
Toll-like receptors (TLRs), 1058–1059
Tomatoes (*Lycopersicon esculentum*)
   genome of, 485f
   hydroponically grown, 781f
   ripening of, 826, 827, 827f
   trichomes of, 738f
   wound response in, 800f
Tonicity, **1041**
Tonoplasts, **73,** 73f
Too many mouths mutation, in *Arabidopsis,* 737, 737f
Tooth. *See* Teeth
Top-down effects, **1221**–1223, 1222f
Topoisomerases, **267**
Topological state, of DNA, 267
Topsoil, **776,** 776f
Torpor, 884
Torsion, **670**
Tortoises, 709t, 712–713, 712f
Totipotent cells, **381,** 383, 384, 641
Touch, receptors in human skin for, 919–920, 919f
Toxins, plant, 795–797, 796f, 797f
Toxoplasma gondii, 574–575, 575f
Trace elements, 23, 1000
Trachea, 1009, 1010f
Tracheae, **683,** 683f, 688, 1008, 1120
Tracheids, 591, 740–741, 740f, 765
Tracheoles, **683,** 683f
Tracheophytes, 470, 589, 589f, 593–599, 593f, 595–599f
Trade Winds, 1235
Trailing arbutus (*Epigaea repens*), 835
Traits
   continuous variation of, 232, 233f
   segregation of, 222, 226
Transcription
   coupled to translation, 286, 287f
   defined, **280**
   DNA rearrangements and, 1074–1075, 1075f
   elongation phase of, 285–286, 288
   in eukaryotes, 287–289, 288f
   initiation of, 284f, 285, 287, 288f, 305, 322f

   posttranscriptional modifications, 288–289, 288f
   in prokaryotes, 284–287f
   in RNA production, 43, 280–281
   termination of, 286, 286f, 288–289
Transcription activator-like effector (TALE) proteins, 336
Transcriptional control
   in eukaryotes, 305, 312–315f, 322f
   negative, 308–310
   positive, 308
   in prokaryotes, 305, 308–312
Transcription-associated factors (TAFs), 313, 313f
Transcription bubble, 285–286f, **286**
Transcription complex, **314,** 315f
Transcription factors
   activation of, 202, 203f
   cytoplasmic determinants, 380, 381f
   in development, 379, 498, 499, 499f, 502, 504
   domains and, 50
   E2F, 203f
   in embryonic stem cells, 384–385
   in eukaryotes, 287–288, 288f, 312–314
   FOXP2, 492
   general, 313, 313f
   genetic engineering of, 347
   hormone-activated, 946
   regulation of, 490
   specific, 313
   TFIID, 313, 313f
Transcription units, **285**
Transcriptome, 366, 368
Transduction
   defined, 552
   generalized, 554, 555f
   sensory, 918, 918f
   specialized, 554–555
Trans-fatty acids, 54
Transfer RNA (tRNA)
   binding to ribosomes, 292–293, 293f
   charged, 292, 292f
   functions of, **43,** 69
   in gene expression, **281**
   initiator, 294, 294f
   structure of, 291–292, 291f
   in translation, 293–297
Transformation
   artificial, 555
   in bacteria, **257,** 552, 555–556, 556f
   DNA in, 331, 331f
   natural, 555, 556f
   in plants, 340–341, 340f
Transforming growth factor beta, 1124
Transforming principle, 257–258, 257f
Transfusions, blood, 1079
Transgenes, 337, 340, 349
Transgenic animals, 284f, 338–339f, 338–340
Transgenic organisms, 337–341
Transgenic plants, 340–341, 340f, 347–349
Transient receptor potential (TRP) ion channels, 919
Transition mutations, 299
Translation
   coupled to transcription, 286, 287f
   defined, **280**
   DNA rearrangements and, 1074–1075, 1075f
   elongation stage of, 294–296f, 295–297, 298f
   initiation of, 293–295, 294f, 298f, 298t, 321
   in prokaryotes, 294
   protein synthesis in, 281
   "start" and "stop" signals in, 283
   termination of, 296f, 297, 298f

Translational control, 321
Translation factors, 321
Translation repressor proteins, **321**
Translocation
　of chromosomes, 250, **301**, 301f
　in plants, **770**
　in translation, 295f, 296
Translocation Down syndrome, 250
Transmembrane proteins, 84, 89, 91t, **95**, 95f
Transmembrane route, **763**, 763f
Transmissible spongiform encephalopathies (TSEs), 540–541
Transmission electron microscopes (TEMs), 61, 62t, 91, 518
Transpiration
　defined, **740**, **758**
　environmental factors affecting, 767, 767f
　rate of, 766–767
Transport inhibitor response protein 1 (TIR1), 820
Transport proteins, 44, 45t, 63, 94, 94f, 104t
Transposable elements, **363**–364, 363t, 556
Transposons
　dead, 364, 364f
　defined, 363
　in fruit flies, 490
　in human genome, 490
　mobilization by polyploidization, 486
Transverse tubules (T tubules), **974**, 974f
Transversion mutations, 299
Trap-door spiders, 684
Tree ferns, 598, 598f
Tree finches (*Camarhynchus*), 452, 453f
Trees. *See specific trees*
Trematoda, 663–664
Treponema pallidum. See Syphilis
TRH (thyrotropin-releasing hormone), 951
Tricarboxylic acid cycle. *See* Krebs cycle
Trichinella, 679f, 680
Trichinosis, 679f, 680
Trichloroethylenes (TCEs), 787–788f, 788
Trichlorophenoxyacetic acid, 821
Trichomes, **738**, 738f, 789, 856f
Trichomonas vaginalis, 570, 570f
Tricuspid valves, **1028**
Triglycerides, 36f, 37t, **53**, 54
Triiodothyronine, 953
Trimesters, 1127–1129
Trinitrotoluene (TNT), 788
Tripartite symbiosis, 633
Triple covalent bonds, 24, 24f
Triple repeat expansion mutations, 299–300
Triplet-binding assay, 283
Triploblastic animals, 642, 647
Triploids, 449, 485
Trisomy, 189, **250**, 250f
tRNA. *See* Transfer RNA
Trochophores, **661**, 661f, **669**, 669f, 675
Trophectoderm, 382, 393
Trophic cascade, **1221**, 1222f–1223f
Trophic levels
　concepts for describing, 1217–1218
　defined, **1217**
　detritivore, 1217, 1217f
　energy loss between, 1218, 1218f
　energy processing in, 1218
　interactions among, 1221–1225
　number of, 1218–1220
Trophoblasts, **1114**, 1114f
Tropical ecosystems, species richness in, 1227–1228, 1227f

Tropical rain forests, **1238**–1239, 1239f, 1248–1249, 1249f
Tropic hormones, 949
Tropins, 949
Tropisms, 807
Tropomyosin, **974**, 974f
Troponin, **974**, 974f
trp operon, **308**, 311–312, 311f
trp promoter, 311, 311f
trp repressor, 311–312, 311f
TRP (transient receptor potential) channels, 919
True-breeding plants, **222**, 225f
Truffles, 620t
Trunk neural crest cells, 1122–1123, 1122f
Trypanosoma cruzi. See Chagas disease
Trypanosomes, 571–572, 572f
Trypanosomiasis (African sleeping sickness), 493f, 571
Trypsin, 116f, 37t, **990**–991
Tryptophan, 47f, 311–312, 312f, 819f
TSEs (transmissible spongiform encephalopathies), 540–541
Tsetse flies, 571, 572, 572f
TSH (thyroid-stimulating hormone), 883, **950**, 951
Tuataras, 701f, 709t, 712f, 713
Tubal ligation, 1103f
Tube feet, 693f, 694
Tuberculosis, 558–559, 559t, 1081
Tubers, 750
Tubeworms, 645t, 674, 675f
Tubulin, 76, 188, 188f, **194**
Tularemia, 371t
Tumor-inducing (Ti) plasmid, **340**, 347
Tumor-suppressor genes, **203**, 204, 204f, 361
Tuna, 701f
Tundra, 1239–1240
Tunicates
　chordate larval forms of, 697, 698f
　development in, 379–380
Tunneling microscopy, 18, 18f
Turbellaria, 663
Turgor movement, 811–812, 812f
Turgor pressure
　defined, **99**
　in osmotic balance, 99
　phloem transport and, 770–771
　in stomatal opening and closing, 766–767, 767f
　touch responses and, 811–812, 812f
　water potential and, 759, 760, 760f
Turner syndrome, 251, 251f
Turpentine, 605
Turtles, 467f, 641f, 701f, 709t, 712–713, 712f
Tutt, J. W., 424, 425
Twin studies, 1137
Twitches, muscle, 975–976, 976f
2,4-D (dichlorophenoxyacetic acid), 819f, **821**
Tympanum, **689**
Type A flu virus, **538**
Type III secretion system, 557–558
Typhoid fever, 558, 559t
Typhus, 559t
Tyrannosaurs, 467f, 472f, 709t
Tyrosine, 47f, 170, 171f

U

Ubiquinone, 134, 134f
Ubiquitin, 201, **323**–324, 323f
Ubiquitin ligase, 323, 323f
Ubiquitin–proteasome pathway, 324, 324f
Ulcers, 559t, 560, 989
Ultrabithorax gene, 391

Ultrasounds, 252
Ultraviolet radiation, 521, 589, 1251–1252, 1252f
Ulva, 580, 580f
Undershoot phase, of action potential, 894, 895f
Unicellularity
　of fungi, 620, 623
　of prokaryotes, 545–546
Uniform spacing, 1168–1169
Uniporters, **100**
Unipotent cells, 381
Unsaturated fatty acids, **53**–54, 54f, 55
Upwelling regions, 1245–1246
Uracil, 42, 42f, 43, 259f, 260
Urea, **1042**, 1043f
Ureter, **1047**, 1048f
Urethra, **1048**, 1048f
Urey, Harold C., 517, 517f
Uric acid, **1042**, 1043f
Uricase, 1042
Urinary bladder, **1047**, 1048f
Urinary system, 876f, **877**
Urine, 1042, 1044, 1050, 1058
Urochordata, 697
Uropods, **685**, 685f
Ustilago maydis, 635, 635f
Uterine contractions, 879, 879f, 949, 1130, 1130f
Uterus, **1099**, 1100f
Utricles, **925**–926, 925f
uvr genes, 274–275, 274f

V

Vaccines
　for HIV/AIDS, 537
　for human papillomavirus, 540
　for malaria, 574, 1081
　for smallpox, 1063
　for trypanosomes, 572
Vaccinia virus, 1063
Vacuoles
　in ciliates, 576, 576f
　contractile, 73, 99, 103, 571
　for food, 568
　of plant cells, 65, 73, 73f, 81t
Vaginal infections, 635
Vaginal secretions, 1058
Valence electrons, **22**, 24
Valine, 46, 47f, 49
Valium (diazepam), 900
Vampire bats (*Desmodus rotundus*), 1156, 1157f
van Beneden, Edouard, 207–208
Vancomycin, 64
Vancomycin-resistant *Staphylococcus aureus* (VRSA), **556**
van der Waals attractions, 23t, 48, 48f
Van Helmont, Jan Baptista, 149
Vanilla orchid, 746
van Niel, C. B., 150, 151
Variable lymphocyte receptors (VLRs), 1067–1068
Variable-rate fertilizer applicators, 777
Variable regions, of immunoglobulins, 1071–1072, 1072f
Variables, in experiments, **6**
Variable-surface glycoprotein (VSG) genes, 571
Varicella-zoster virus, 528, 531t, 1063
Variola virus, 531t, 1063
Vasa recta, **1049**, 1049f
Vascular bone, **968**
Vascular cambium, **735**, 736f, 748, 749f
Vascular plants
　classification of, 468, 468f
　diversification of, 521
　evolution of, 593–597
　extant phyla of, 594–595, 594f

genome of, 597
roots and shoots of, 733–734
tissue of, 594, 733–734, 740–741f, 740–742
Vascular tissue, **734**, 740–741f, 740–742, 852
Vas deferens, **1094**
Vasectomy, 1103f
Vasoconstriction, **1032**, 1033f
Vasodilation, **1032**, 1033f
Vasopressin, 45t, **1037**
Vectors, cloning, 330–331
Vegetal plate, **1115**
Vegetal pole, **1112**, 1112f
Vegetarian tree finches (*Platyspiza crassirostris*), 422f, 452, 453f
Vegetative propagation, 748–750
Vegetative reproduction, in plants, **857**, 857f
Veins
　in circulation, 706, 1029–1030, **1032**–1033
　in leaves, **751**
Veligers, **669**, 669f
Velociraptor, 467f, 472f, 716, 716f
Venous pump, **1033**
Venous valves, **1033**
Venter, Craig, 360, 370
Ventral body cavity, 865, 866f
Ventral root, **910**
Ventricles, **1025**, 1025f
Venules, 133, **1032**, 1032–1033f
Venus flytraps (*Dionaea muscipula*), 754, 783, 783–784f, 811, 821
Vernalization, **834**, 836
Vertebral column, 699, 699f, 701
Vertebrata, 466f, 697
Vertebrates, 698–729. *See also specific Vertebrates*
　axis formation in, 1124–1127
　brain of, 904, 904f
　chromosomes in, 189t
　circulatory system of, 699, 875f, 877, 1025–1027f
　comparative anatomy of, 11, 11f
　development in, 1120–1121, 1121f
　digestive system in, 875f, 877, 984–985, 985f, 993–994f, 993–995
　electrical currents detection by, 936
　embryonic development in, 432
　essential nutrients for, 999–1000, 1000t
　evolution of, 429, 432, 699–701f, 1123, 1123f
　eyes of, 432–433, 433f, 507, 507f, 930–933f, 1127f
　features of, 698–699
　fertilization and development in, 1089–1092
　hearing in, 922, 922–923f, 924
　invasion of land by, 706
　kidneys of, 1045–1047
　locomotion in, 978, 978f
　organization of body of, 865–866
　organ systems in, 865, 865f, 874–877
　photoreceptors in, 931–933
　photoreceptors of, 432, 433f
　segmentation in, 644
　smell in, 928–929, 928f
　social systems of, 1159–1160, 1160f
　taste in, 927–928, 927f
　teeth of, 986, 986f
　variations in, 993–994f, 993–995
Vertical gene transfer (VGT), **490**
Vervet monkey (*Ceropithecus aethiops*), 1148, 1149f
Vesicles, **65**, 71
Vessel elements, 594
Vessels (xylem), **606**, 740–741, 740f, 765
Vestibular apparatus, **926**

Vestigial structures, **433**–434, 434f
VGT (vertical gene transfer), **490**
Viagra (sildenafil), 174, 900
Vibrio cholerae. See Cholera
Victoria (Queen of England), 242–243, 242f
Villi, **989,** 990f
Vimentin, 76
Virions, 527, 528, 528f, 529
Viroids, **541**
Viruses, 527–541. *See also specific viruses*
 archaeal, 532
 bacteriophage (*See* Bacteriophage)
 brain diseases and, 540–541
 cancer and, 539–540
 disease-causing, 531t, 538–540
 DNA and, 528, 528f, 531t
 emerging, 539
 genome of, 370, 529, 530
 hosts of, 528
 immunoassays in diagnosis of, 346, 346f
 latent, 528
 nature of, 528–530f, 528–532, 531t
 recombination in, 538
 replication of, 529
 RNA and, 528, 528f, 530, 531t
 shape of, 528–530f, 529–530
 size of, 530, 530f
 structure of, 528–530, 528f–530f, 531t
 subviral particles, 540–541, 541f
 tissue tropism of, 528, 535
Viscera, 871
Visceral muscle, 871
Visceral pleural membrane, **1011**
Vision, 723, 929–935f. *See also* Eyes
Visual acuity, 933–934
Vitamin A, 348, 1000t
Vitamin B-complex vitamins, 117, 1000t
Vitamin C, 1000t
Vitamin D, 955, 955f, 1000t
Vitamin E, 1000t
Vitamin K, 994–995, 1000t
Vitamins, **1000,** 1000t
Vitelline envelope, 1108, 1110
Viviparity, 721, 1089, 1089f
VLRs (variable lymphocyte receptors), 1067–1068
Voltage-gated channels, **893, 894,** 895f
Volvox, 483, 580, 580f
Vomitoxin, 634
Von Frisch, Karl, 1148
VRSA (vancomycin-resistant *Staphylococcus aureus*), **556**
VSG (variable-surface glycoprotein) genes, 571

W

Wadlow, Robert, 952, 952f
Wallace, Alfred Russel, 10
Wall cress. *See* Arabidopsis
Warbler finches (*Certhidea olivacea*), 422, 422f, 452–453, 453f
Wasps, 687t, 799, 799f

Wastewater treatment, 342, 343f
Water. *See also* Water potential
 absorption by plants, 761–764f
 adhesive properties of, 27, 27f
 cohesive nature of, 26, 27f, 27t
 density of, 27f, 28
 on Earth, 25
 forms of, 26f
 heat of vaporization of, 27t, 28
 hydrogen bonds in, 26, 26–27f
 influence on life, 1164
 ionization of, 29
 lipids in, 56, 56f
 locomotion in, 977–978, 977f
 in osmosis, 98–99
 plant responses to, 813–815, 814f
 properties of, 27f, 28–29
 reabsorption in kidneys, 1050, 1051f, 1052–1053, 1053f
 as solvent, 27t, 28, 29f, 97–98
 specific heat of, 27t, 28
 structure of, 26, 26f
 transport in plants, 758–761
Water cycle, 1211–1212
Water-dispersed fruits, 616, 616f
Water lilies, 609f, 769f
Water molds, 573, 578, 624
Water potential
 calculation of, 760–761, 760f
 defined, **758**–759
 at equilibrium, 761, 761f
 gradient from roots to shoots, 762f, 764–765
 of soil, 764–765, 776, 777f
 turgor pressure and, 759, 760, 760f
Watersheds, of New York City, 1265, 1266f
Water storage roots, 746, 747f
Water striders, 27f, 408–409, 409f
Water-vascular system, **693,** 693f, 694
Waterwheels (*Aldrovanda*), 783–784
Watson, James, 259, 260–263, 261f
Weapons, biological, 371
Weathering, 515, 516f, 520, 521
Weigel, Detlef, 482
Weight, 18
Weinberg, Wilhelm, 402
Welwitschia, 604t, 606, 606f
Went, Frits, 818, 818f
WEREWOLF gene, 743, 743f, 744
Wernicke's area, 906
West Nile virus, 530f, 531f, 1249
Whales
 evolution of, 429, 429f
 genome of, 494f
 overexploitation of, 1272, 1272f
 vestigial structures in, 433, 434f
Whaling industry, 1272, 1272f
Wheat Genome Project, 361
Wheat (*Triticum*)
 apical meristem of, 747f
 chromosomes in, 189t
 comparative genomics and, 366, 366f
 genome of, 361, 485f, 486
 transgenic, 347

Whiptail lizards (*Cnemidophorus tesselatus*), 485
Whisk ferns, 594, 594t, 597, 597f
White blood cells. *See* Leukocytes
White Cliffs of Dover, 582, 582f
White fibers, 976
White-fronted bee-eaters (*Merops bullockoides*), 1158, 1159f
White matter, 891
White willow (*Salix alba*), 800
Whooping cough, 558
Whorled leaf pattern, 747, 747f
Whorls (flower parts), **610**
WHO (World Health Organization), 348, 538, 957, 1082
Wieschaus, Eric, 387
Wild geraniums (*Geranium maculatum*), 840f
Wilkins, Maurice, 260
Wilson, Edward O., 1228, 1229
Wind
 ocean circulation and, 1235, 1235f
 pollination by, 842–843, 845, 845f
Wind erosion, 776–777, 777f
Window leaves, 753
Wings
 of bats, 720
 evolution of, 505–506, 506f, 978–979, 979f
 of fruit flies, 246–247, 246f, 499
 of insects, 505–506, 506f, 687–688, 688f
Wnt pathway, 1125
Wobble pairing, 296–**297**
Woese, Carl, 547
Wolf spiders, 684
Wolves, 1165, 1165f
 captive breeding of, 1280
 domestication of, 427, 427f
 evolution of, 435f
 genome of, 494f
Woman River iron formation, 139
Woodpecker finches (*Cactospiza pallida*), 9f, 422, 422f, 452
Woodward, Robert, 798
Woody plants, 745, 748, 748f
World Health Organization (WHO), 348, 538, 957, 1082
Wound response, in plants, **799**–800, 800f
Wuchereria bancrofti, 680

X

X chromosomes
 of fruit flies, **241,** 243, 245
 gene map of, 248f
 in humans, 241–242, 248f
 inactivation of, 243, 243f, 316
 nondisjunction involving, 250–251, 251f
 in sex determination, 241t
Xenopus, 328, 1124, 1125
X-inactivation-specific transcript (Xist), 316

X-ray diffraction, 260, 261f
Xylem
 defined, 589
 plant transport in, 764–765
 primary, 740, 744f, 745
 secondary, 740
 in vascular plants, **594, 740**–741, 740f
 vessels in, 606

Y

YABBY gene, 751f
YACs (yeast artificial chromosomes), 331
Y chromosomes
 of fruit flies, **241**
 in humans, 241–242
 nondisjunction involving, 251, 251f
 segmental duplication on, 488–489f
 in sex determination, 241t
Yeast artificial chromosomes (YACs), 331
Yeasts
 cell cycle of, 200, 201f
 cell division in, 188f
 chromosomes in, 189t
 in ethanol fermentation, 140, 140f
 in fermentation, 629–630
 genome of, 360f, 481f
 mating behavior in, 177
 reproduction in, 629, 630f
Yellow fever, 1256
Yellowfin tuna, 701f
Yellowstone National Park
 hot springs in, 26f, 139f, 545f
 wolves of, 1280
Yersinia, 371t, 557, 558, 559t
Yolk plug, **1116,** 1116f
Yolk sacs, **710,** 710f, 1091, 1118

Z

Z diagrams, 157, 158f
Zebra mussels (*Dreissena polymorpha*), 667, 1273, 1273f
Zinc, 23, 117
Zinc fingers motif, **307**
Z lines, 971, 972f
Zoecium, 677
Zone cell of division, **743**–744, 743f
Zone of elongation, **744**
Zone of maturation, **744**–745, 744f
Zooids, 677
Zoonotic diseases, 1249–1250
Zoospores, 578, 622, 624, 624–625f
Zygomycetes, 620, **625**–626
Zygomycota, 620, 620f, 620t, 625–626, 626f
Zygosporangium, 626, 626f
Zygospores, **626,** 626f
Zygotes
 cell division in, 376
 chromosomes of, **208,** 208f
 defined, 877
 fungi, 626, 626f
 plant, 849, 850f
 prevention of formation of, 441, 442–443f, 442–444, 442t

Online Supplements

Principles of Biology I and II

McGraw Hill Education connect®

Connect Online Access for Biology, 11th Edition

McGraw-Hill Connect is a digital teaching and learning environment that improves performance over a variety of critical outcomes. With Connect, instructors can deliver assignments, quizzes and tests easily online. Students can practice important skills at their own pace and on their own schedule.

HOW TO REGISTER

Using a Print Book?
To register and activate your Connect account, simply follow these easy steps:
1. Go to the Connect course web address provided by your instructor or visit the Connect link set up on your instructor's course within your campus learning management system.
2. Click on the link to register.
3. When prompted, enter the Connect code found on the inside back cover of your book and click Submit. Complete the brief registration form that follows to begin using Connect.

Using an eBook?
To register and activate your Connect account, simply follow these easy steps:
1. Upon purchase of your eBook, you will be granted automatic access to Connect.
2. Go to the Connect course web address provided by your instructor or visit the Connect link set up on your instructor's course within your campus learning management system.
3. Sign in using the same email address and password you used to register on the eBookstore. Complete your registration and begin using Connect.

*Note: Access Code is for one use only. If you did not purchase this book new, the access code included in this book is no longer valid.

Need help? Visit mhhe.com/support